The Technology of Nuclear Reactor Safety

Volume 1
Reactor Physics and Control

The Technology of Nuclear Reactor Safety

VOLUME 1

Reactor Physics and Control

VOLUME 2

Reactor Materials and Engineering

The Technology of Nuclear Reactor Safety

VOLUME 1

Reactor Physics and Control

EDITORS
T. J. Thompson
J. G. Beckerley

Prepared under the auspices of the
Division of Technical Information
U. S. Atomic Energy Commission

Library of Congress Catalog Card Number 64-24957

PRINTED IN THE UNITED STATES OF AMERICA

Foreword

At the end of 1963 there were twenty civilian power reactors (including prototypes) in operation in the United States, with a combined electrical output of more than one million kilowatts and representing an operating experience of about seventy-five reactor-years. The safety record of this substantial nuclear power program has been excellent. All of this operating experience has shown that reactors can be operated safely.

The Atomic Energy Commission is confident that this gratifying record can be maintained. In order to do so, past experience must be continuously analyzed and applied to present and future problems, and present and future work must include activities designed to enlarge the knowledge of nuclear reactor safety.

The present volumes are directed primarily towards fulfilling the first of these needs — to sift past experience in reactor safety, to study it from many points of view and to learn its lessons. Guidance on the second need is also provided by the present volumes, since gaps in knowledge and limits of understanding, revealed in any careful study of past experience, are reliable indicators for future work.

While these books mainly reflect experience and attitudes in this country, the authors have made deliberate efforts to consider and to present experience and attitudes in other countries. Nuclear reactors are destined to become increasingly important sources of energy throughout the world; their safe operation is clearly of international importance.

Glenn T. Seaborg

Chairman

U.S. Atomic Energy Commission

Preface

The "SIFTOR" project evolved from a letter written by the Atomic Energy Commission Advisory Committee on Reactor Safeguards which suggested "a twenty-five man-year effort" to bring reactor safety information together in one place and to evaluate, organize, and generalize it. Subsequent informal discussions within the AEC staff and the Advisory Committee concluded that a full-time group of qualified authorities could not be drawn away from their regular duties long enough to complete such an effort. However, it was believed that these experts might be willing to participate on a part-time basis.

It was therefore proposed that each expert consider a limited technical area, discuss and coordinate his views with the other individuals, and then write a chapter to analyze the progress to date in that area, summarize the current situation, and identify the more pressing problems that remain. It was recognized that the product would not be as extensive and probably would not lead to as many broad conclusions as the full-time coordinated effort originally proposed; but it appeared to present a reasonable chance for success.

Accordingly, informal discussions were held in Washington during the spring and summer of 1961 to develop a tentative outline of what such a book might contain. In January of 1962, the Massachusetts Institute of Technology agreed to serve as coordinator for the effort, and subsequently a contract was concluded with the Atomic Energy Commission. Early in 1962, prospective authors were approached and asked to prepare detailed outlines of their chapters; and in June 1962 the authors first met as a group and discussed the complete draft outline. Subsequently, the outline was modified as a result of the "back-and-forth" among authors and between authors and Commission staff. A revised outline was approved by the Atomic Energy Commission in December 1962.

Because one of the foremost purposes of the Project was to produce a unified manuscript (not a "handbook"), considerable effort was spent at this and all other stages in author-to-author correspondence and discussion. A twelve-day conference of all authors was held in June 1963. At this conference all available draft material was considered in detail by groups of authors. It was concluded that a chapter on Doppler coefficients should be added and an author was recommended. (We are grateful to the author of this chapter for preparing it in such short time—a few months—and without the benefit of conferences with the other authors.) The authors' conference also recommended adding a chapter on accident experience and one of us volunteered to write it.

By early autumn a number of chapters, now in "semifinal" form, had been sent to outside reviewers for comment and to the AEC staff for comment and approval. Final revisions were then made, and the first manuscript went to press in December 1963.

A determined attempt has been made to report progress up to the end of 1963—with publication as soon as possible thereafter—since one of the hazards in producing large multi-authored books is that the technical content may be outdated by a long interval between the time the manuscript is prepared and the time the book is in the reader's hands. In reducing this time interval to the minimum, the clarity, consistency, and cross referencing may have suffered and for this we ask the reader's forbearance; we have felt it more important to get the work out promptly. The time from receipt of the last manuscript to publication has been reduced to well under six months.

The authors were selected by the Massachusetts Institute of Technology with the approval of the Atomic Energy Commission. We consider it most fortunate that such well-qualified men understood the need, and that they agreed to devote valuable time to this effort. Each author has the highest technical competence and is active in the field he writes about. The authors were purposely chosen to represent varied points of view and to reflect experience with all the principal reactor types. It turns out somewhat fortuitously that the authors are, in about equal proportions, from industrial concerns, Atomic Energy Commission laboratories, and universities.

The experience of the authors is sufficiently diverse to make it worth while to read several sections on identical or related subjects. For instance, the methods discussed for handling calculations in the various reactor kinetics chapters differ rather strikingly. This results partly from differences in the type of reactor under consideration and the particular problem being discussed, and partly from the differences in viewpoints of the various authors—reflecting their individual experiences and those of the groups in which they work. In these same chapters there are minor and sometimes major differences on such subjects as reactor control, reactivity coefficients, and many others. It is evident that some authors are more conservative, and some are more willing to state definite conclusions. This seeming disparity of opinions is in part a true indication of the state of the art: formal conclusions regarding safety can be reached for some types of reactors more readily than for others—and their behavior or performance is more easily predictable.

Each author has labored to present a clear and balanced perspective on his subject as it relates to the problems of reactor safety. Each author writes as an individual, independent of any organization. At the same time, it is clear that an author cannot help but reflect to some extent a measure of personal bias formed in experience.

Nuclear Reactor Safety is an enormous subject. This book does not pretend to encompass the entire field nor does it pretend to be the "last word" on the subject. The specific scope of the book—what it is and what it is not—is discussed in some detail in the Introduction.

We earnestly hope that the reader will find the books to be informative and a stimulant to greater effective activity in nuclear reactor safety.

Cambridge, Massachusetts
June 1964

T. J. Thompson
J. G. Beckerley

Acknowledgments

During the preparation of the books, every one of the authors gave advice and assistance to every other author. In some cases the help was well "above and beyond the call of duty". We would be remiss not to acknowledge this extraordinary spirit of cooperation.

Two individuals should be recognized as co-authors of certain parts of the books: William K. Ergen and David B. Hall. Both helped prepare the initial outline of the project. Both participated in conferences of authors and both read and commented on many draft manuscripts. We are particularly indebted to them.

Similarly, we are indebted to the Instrumentation Division of the Oak Ridge National Laboratory and, in particular, to S. H. Hanauer and E. P. Epler for helping in the initial stages of the project and for reviewing several chapters.

Drafts of chapters were sent for comment to many individuals and many organizations; however, not all of the chapters were reviewed by each of the individuals and organizations listed below. The responses invariably reflected genuine interest in the project and conscientious concern that the text be accurate and up to date. The specific comments and suggestions were extremely useful to all authors.

Of particular value were the extensive commentaries received from the United Kingdom Atomic Energy Authority (we are especially grateful to F. R. Farmer and H. M. Nicholson) and from Atomic Energy of Canada, Ltd. (we are indebted to W. B. Lewis).

Many industrial firms offered comments and suggestions and supplied useful information. We wish to express our gratitude to: Atomics International, Babcock and Wilcox Company, E. I. du Pont de Nemours and Company, General Electric Company, Phillips Petroleum Company, and Westinghouse Electric Corporation.

Besides providing the time and talents of some of our authors, the staffs of the National Laboratories have been very helpful in contributing comments, suggestions, and pertinent information. We particularly wish to acknowledge the assistance given by Argonne National Laboratory, Brookhaven National Laboratory, Los Alamos Scientific Laboratories, and Oak Ridge National Laboratory.

We wish also to express our thanks to the following members of the Atomic Energy Commission staff who encouraged us and assisted us in many ways and at many times during the course of the project:

M. Balicki, C. Beck, W. G. Belter, M. Biles, M. Booth, J. J. DiNunno, H. C. Field, M. Fox, R. Fraley, J. G. Gratton, A. B. Holt, R. J. Impara, D. C. Layman, R. R. Maccary, W. J. McCool, T. G. Schleiter, F. J. Shon, E. E. Sinclair, W. L. Smalley, S. Szawlewicz, A. Van Echo, W. S. Wilgus.

At various stages during the preparation of the manuscript we requested comments from many persons, and in the responses we received many valuable suggestions. Accordingly, we would like to express our sincere thanks to all those listed below (and to any other persons we may have inadvertently missed). Furthermore, we would like to make clear that any shortcomings of the books cannot (repeat, not) be ascribed to any of them.

W. B. Abbott, J. F. Ablitt, R. J. Allio, J. R. Beattie, E. Bevitt, G. Bright, G. L. Brooks, R. O. Brugge, S. H. Bush, B. G. Childs, H. K. Clark, E. D. Clayton, J. G. Collier, J. L. Crandall, R. J. Creagan, E. Critoph, G. Dessauer, F. E. Driggers, M. F. Duret, W. Francis, R. J. French, P. F. Gast, H. W. Graves, Jr., A. E. Green, V. Griffiths, C. Haire, M. J. Halsall, J. H. Horton,

D. L. Hunt, R. J. Huntoon, E. P. Irish, W. H. Irvine, R. G. Jarvis, R. L. Jenkins, A. A. Johnson, W. Joseph, B. F. Langer, W. D. Leggett, F. M. Leslie, J. Maloney, A. M. Marko, L. H. McEwen, C. H. Millar, P. L. Miller, S. Mirshak, T. G. Mitlo, A. J. Mooradian, J. G. Moore, J. S. Nelles, R. O'Neil, W. P. Overbeck, H. Pearlman, E. C. W. Perryman, A. Quinton, H. K. Rae, M. Remley, D. S. St.John, M. F. Sankovich, F. M. Sayers, F. Schroeder, J. A. Smith, A. Spano, J. D. Stewart, L. E. Strawbridge, W. R. Stratton, J. C. Tobin, L. S. Tong, J. Veeder, W. A. J. Wall, A. G. Ward, W. A. Wolfe, P. M. Wood, H. Worthington.

An appendix in each volume gives a tabulation of data relevant to reactor safety for a number of power reactors. Most of these data were obtained from responses to questionnaires sent to many reactor groups. For this help we are indebted to those who painstakingly responded to our queries.

Many of the tables and figures have been taken from other publications. We have indicated the source of the material by including the reference number after the first mention of each table and figure. Some of the figures have been redrawn or new labels added; in all cases we have exercised great care to see that no errors have been introduced. We are grateful to these authors and publishers for granting us permission to include this material.

The preparation of a multi-authored technical manuscript of this size requires not only authors' and editors' efforts but considerable additional careful and often tedious work. Chasing references, corralling figures, typing numerous drafts, corresponding with authors and reviewers, obtaining permissions to use other published material — these are some of the tasks performed at the "home base" by skilled individuals. Miss Dorothy W. Bishop has been the patient and capable leader of these activities. We are especially appreciative of her accomplishments. She has been ably assisted by Mrs. Susan M. Martin. Mr. Jake Wing Lung has drafted most of the drawings which were prepared here, with the exception of certain electrical drawings which were prepared by the M. I. T. Instrumentation Laboratory. We would also like to acknowledge the fine cooperation of the M. I. T. Graphic Arts Services.

T. J. T., J. G. B.

Table of Contents

VOLUME I

CHAPTER 1

Introduction

T. J. THOMPSON, J. G. BECKERLEY
Massachusetts Institute of Technology
Cambridge, Massachusetts

CHAPTER CONTENTS

1 GENERAL HISTORICAL BACKGROUND

Safety has been an important consideration from the very beginning of the development of nuclear reactors. On December 2, 1942 when the first atomic reactor was brought to criticality, Enrico Fermi had already made safety an important part of the experiment. In addition to a shutoff rod, other emergency procedures for shutting down the pile were prepared in advance. Fermi also considered the safety aspects of reactor operation. Shortly before the reactor was expected to reach criticality, Fermi noted the mounting tension of the crew. To make sure that the operation was carried out in a calm and considered manner, he directed that the experiment be shut down and that all adjourn for lunch. With such leadership in safety at the very beginning, it is no wonder that the operation of reactors to date has been singularly free of mishaps.

In the earliest large reactors, the plutonium production reactors at Hanford, the role of geographic isolation in protecting the safety of the general public was emphasized. After 1945, nuclear reactors began to be considered as potential sources of commercial power. To compensate for the lack of knowledge during the initial stages of power reactor development, large factors of safety were applied. In a sense, these were "ignorance factors."

At its first meeting in 1947 the Reactor Safeguards Committee of the Atomic Energy Commission considered the first proposal for a contained reactor, the SIR which was to be enclosed in a large spherical shell at West Milton, New York. From that time on, containment for protection of the general public has played an important role in reactor safety in the United States.

As knowledge of nuclear reactor technology has increased and as experience in the actual operation of nuclear power plants has accumulated, emphasis has slowly shifted from reliance on containment structures to the achievement of safety by proper design of the reactor plant itself. While providing for containment of fission products after an accident assures the safety of the general public, it is obviously better to design so that there can be no accident at all.

The goal that nuclear power plants compete economically with conventional (fossil fuel) plants has continued to create strong pressures to reduce capital costs, to increase reactor power levels, to lengthen core life, to achieve more efficient performance, and to bring reactors closer to metropolitan areas. The direction of many of these forces is towards reduced reactor safety. Happily, at the same time these pressures have developed, steady advances in the technology have made it possible to replace some of the very conservative safety measures, taken originally because of ignorance, by more realistic and at the same time more economic measures.

At the risk of oversimplification, it might be said that the first ten years of nuclear power development were devoted to demonstrating that power reactors could be designed, built, and operated; the second ten years were devoted to showing that power reactors might be operated economically; and the next ten years will, for the most part, be devoted to operating power reactors economically, devising improvements, and developing other different, and perhaps better, reactor types. Reactor safety has played a significant role in these developments and will continue to do so.

A review of the experience on reactor safety to date shows that the record of nuclear reactors has been very good indeed. There has been no accident that has been harmful to the public health and safety. Only seven fatalities have been directly caused by nuclear accidents. Four of these were caused by accidental critical assemblies and three were caused by a single reactor accident.

It is significant that the accident involving fatalities (the SL-1 accident) occurred at a time when the reactor was not in routine operation but was shut down and depressurized. In fact, there have been no accidents to reactors operating routinely at their normal power level.

Reactors have behaved well and have been "docile" even when purposely or accidentally mistreated. Experience has shown that reactors are more stable and reliable than anyone had a right to hope or believe in the earlier days.

Thus, the purpose of these volumes is not to show that nuclear reactors are unsafe (because, when properly designed, they are safe) but rather to point out problem areas and to review results and experience so that the present excellent record of nuclear reactor safety can be maintained or perhaps even improved.

2 BASIC SAFETY PHILOSOPHY

2.1 Goals of Reactor Safety

Basic criteria can be established to provide guidance in determining the emphasis to be placed on various aspects of reactor safety. One method is to establish the criteria according to the consequences of reactor accidents. Using this approach we defined, somewhat arbitrarily, four classes of accidents or mishaps, and on that basis we suggest a scale of relative reactor safety emphasis.

Briefly, reactor safety should be aimed towards achieving the following goals, in the listed order of importance:

1. There must be no release of radioactive material in dangerous quantities from a nuclear facility to the general public.—There must be no "Public Safety Accidents."
2. The likelihood of a serious accident which would result in severe damage to the nuclear facility should be kept as small as possible.—The "Economic Accident" should be prevented.
3. Every reasonable effort should be made to eliminate accidents involving plant employees.—The frequency of the "Industrial Personnel Accident" should be reduced to the lowest possible level, certainly lower than that of other comparable industries.
4. System malfunctions and deviations from normal behavior should be reduced to a minimum, especially since a system with minor faults is more likely to develop major ones.—The number of "Operational Problems" should be kept at a minimum.

Those agencies concerned with the interests of the general public devote their attention primarily to the achievement of the first of these four goals. Where the Public Safety Accident and the Economic Accident cannot be differentiated, these same agencies also have to consider the second goal. Usually the third goal and almost always the fourth concern primarily the operating groups. Only in cases where operational problems are abnormally frequent should they become a source of concern to public safety groups. Designers, builders, and operators of reactor facilities, however, must be concerned with achieving all four goals.

In the following subsections these concepts are developed in more detail in order to bring out their interrelation with basic safety philosophy.

2.1.1 Public Safety Accidents

An accident of this type involves the release of dangerous quantities of radioactive material (usually fission products) to the public environment. Such an accident is also likely to result in serious economic losses to the management responsible for operating the reactor. It may or may not involve personnel at the facility.

Prevention of Public Safety Accidents is the primary goal of reactor safety. The health and safety of the general public can be ensured by preventing accidents involving the reactor core and primary coolant circuit, or by providing structures to contain or confine the radioactivity released by any accident, or by geographically isolating reactor facilities. Up to the present time it has been customary in the United States to use a combination of all three of these methods.

Since this accident is the one directly involving the general public, it falls properly within the purview of safety groups representing the general public and should constitute the basis for safety reveiw by such bodies.

Very often in the United States, analysis of possible Public Safety Accidents has been approached by the selection of a so-called "maximum credible accident." This is the accident which could result from a series of unrelated mishaps or failures hypothecated as being "credible" by the reactor designer. It should be emphasized that the maximum credible accident is an imaginary accident and that it may or may not represent the maximum conceivable accident. It simply constitutes a framework within which to review the response of the system to accident conditions.

Frequently, in order to carry the analysis further, a discontinuity in the accident conditions is postulated. It is assumed that, without regard to how such an accident might happen, all or some selected fraction of the available fission products is released to the containment. In a sense, this procedure sets an upper limit to the fission product release that could constitute a hazard to the public, and therefore represents a maximum conceivable accident. An investigation is then carried out of the leakage of fission products through the containment, assuming the maximum specified leakage rate. Such a procedure assumes that the accident does not breach the containment, emphasizing the importance placed on the integrity of the containment. This general procedure is used to determine that the consequences are not too severe even in the event that a design weakness of the plant within the containment has escaped detection.

2.1.2 Economic Accidents

An Economic Accident involves major damage to the reactor facility and results in prolonged shutdown, extensive repairs, and so on. Usually safety measures which ensure the protection of the public also prevent economic loss. Except for containment features, the objectives of any reactor design chosen to protect the public are identical with those chosen to protect the plant. Because of this, such diverse groups as owners and operators, safety review and regulating agencies, and the general public have difficulty in differentiating between Economic Accidents and Public Safety Accidents. The only valid test of the difference is whether the real or hypothetical accident under consideration could affect the health and safety of the general public.

It should be recognized by all reactor designers, constructors, operators, and owners that the Economic Accident is far more likely to occur than a reactor accident involving the health and safety of the general public or the reactor operators. The serious reactor accidents to date have been primarily Economic Accidents. In a number of instances, reactor cores have melted down, or fission products have been released, in such a manner as to require extensive shutdowns and cleanup of sections of the reactor facility. Such shutdowns, if properly taken into account in the bookkeeping, can increase the costs of electrical power from a given facility sufficiently to make the plant uneconomic.

The reactor designer should emphasize those practical design features that ensure reliability, provide for easy maintenance, and result in a fail-safe system. The reactor fabricator should make certain that the quality of his workmanship is the highest possible. Installation, testing, operation and maintenance of equipment should be made with similar care.

The reactor owner must recognize that in the long run the safest reactor may well turn out to be the reactor most economical to operate. He should, by example and by direction, make certain that maintenance is effective and that problems are not allowed to go unattended because of the pressures to achieve a high-use factor; it is possible that failure to solve such problems may eventually result in even more hazardous situations. It is short-sighted and, sooner or later, expensive to operate a reactor in a sloppy, poorly maintained, or potentially hazardous manner.

2.1.3 Industrial Personnel Accidents

This category includes accidents which involve personnel working at the reactor facility and which differ little in their essential causes or consequences from accidents in other industries. Employees can fall down stairs, inhale noxious fumes, suffer chemical burns or electrical shocks, receive excessive radiation from radiographic equipment, and so on. They can misuse hazardous equipment or fail to be careful in operating machinery. Equipment itself can fail in a dangerous way.

Reactor plant personnel accept normal occupational risks of the type common throughout industry. Measures taken to prevent accidents involving only plant personnel are quite different from those taken to prevent Public Safety or Economic Accidents. Many accidents to plant personnel result from "conventional" hazards that are the subject of "Industrial Safety" activities. In these volumes we will consider "Reactor Safety" and "Industrial Safety" to be separate subjects; we will be concerned with "Reactor Safety," the unique hazards and safeguards associated with nuclear reactor facilities.

In every industry there are some hazards unique to that industry. Perhaps the best example of a hazard unique to the nuclear industry is that associated with a critical assembly. It is possible for an employee to enter an area in which a system is critical or where it may become critical while he is present. Every facility has safeguards, warning systems, and procedural controls to prevent such an occurrence. Still, it is not impossible for the safeguards to be circumvented. The likelihood is that there will occasionally be criticality accidents involving employees either in facilities designed for criticality tests or, even more likely, in an unplanned assembly. It is almost inconceivable that a critical assembly accident could endanger the health and safety of the general public since few, if any, fission products would be released and normally the fissile material itself would not be a serious hazard. Moreover, there is usually little or no economic loss as a result of a critical assembly accident. Consequently, such an accident is properly an Industrial Personnel Accident.

Accidents resulting from other hazards peculiar to nuclear facilities (e.g., presence of nuclear radiation fields, possibility of airborne radioactivity) are prevented primarily by onsite shielding, filtered air supplies, and specific operating procedures.

A single accident may combine an Economic Accident with an Industrial Personnel Accident. It seems fair to state that the SL-1 accident, which resulted in the loss of life of three men but no involvement of the public health and safety, was in this category. The loss of the reactor as well as the cleanup following the accident were expensive. It was thus both an Economic Accident and an Industrial Personnel Accident.

2.1.4 Operational Problems

Operational Problems are mishaps or accidents which do not affect the general public, the plant economics, or working personnel, but are simply aggravations or minor perturbations of the normal reactor operations. The operational Problem might be viewed as a very minor Economic Accident.

Some malfunctions which result in short shutdowns or perturbations in the normal operation of the reactor are to be expected in a reactor facility. An operational problem might be a change in pH of the water in a water-moderated reactor. It might even include such an accident as a sodium fire resulting from leakage of nonradioactive sodium from a reactor's intermediate heat exchanger system, assuming the fire were in an area where no conceivable fission product release might accompany any subsequent accident caused by the fire. If the fire can be readily extinguished and the consequences do not shut down the plant for an appreciable period of time, such an accident could properly be classified as an Operational Problem.* On the other hand, if the sodium fire were to destroy a major part of the plant, it would clearly be an Economic Accident; and if the fire were to release to the public environment large quantities of fission products, then it would be a Public Safety Accident.

2.2 Safety at Each Stage of a Reactor Project

It is revealing to consider reactor safety at each stage of a given reactor project. The stages can be designated as the design, construction, and operations stages. Frequently, these three stages are also clearly delineated by a change in contract responsibility at the end of each stage. In many existing reactors one group has been responsible for the reactor design, another group for the reactor construction, and a third group for the reactor operation.

2.2.1 The Design Stage

The reactor design stage includes both the conceptual design and the final detailed design. It is during the design stage that the ultimate power levels and capacity of the plant are set and conservative limits are established for its performance. Limits are set for the parameters which are amenable to measurement. All decisions relating to the safety of the design are carried out during this period.

The largest part of the material in the ensuing chapters is devoted to considerations which can properly be called design considerations. This is because the ultimate safety of the reactor itself, the adequacy of its containment, and other structures must be decided during the design stage. The important economic and safety decisions are made early in the reactor design stage. There may be no retreat from poorly made initial decisions. It is therefore essential that considerations of safety are given a vital role at this stage of a reactor project.

It is also reasonable to expect that the most rewarding safety reviews and evaluations can be carried out during this period. Normally, subsequent reviews of the design, perhaps after some operating period, are only for correcting or updating the information originally available or for considering the acceptability of changes in the limits of the system. Such changes are usually only a small fraction of the magnitudes chosen for the original values of the parameters.

Thus, the important design decisions and the important safety reviews should all be made as early as possible in a reactor project. For this reason, these books are primarily directed towards providing information on reactor technology needed for the design phase.

2.2.2 The Construction Stage

During the construction stage, the plans developed during the design period are implemented. In the ensuing chapters some mention is made of good construction practices. These include welding practice, inspection methods, testing of materials and components, and so on. However, in any reactor project it should be recognized that most of the basic decisions have been made before construction. During this phase of the project, it is simply necessary to see that the best possible construction practices are adopted.

Since many reactor projects have experienced difficulty due to inadequate workmanship, faulty materials, and other construction problems, the importance of this phase cannot be overemphasized. The execution of a reactor design, if not properly carried out, can nullify the safety features. Very little can be said in the way of guidance except that it is essential to maintain the highest standards of construction and installation for all parts of the reactor which could conceivably affect the neutron chain reaction primary system. The most rigorous inspection program and work surveillance should be instituted and maintained at all times. It is at this stage that agencies having responsibility for inspection should assure themselves that an adequate inspection program is being carried out and that the quality of workmanship in the installation is high. In a sense, this is a subjective judgment based upon experience and should be viewed as such. It need not normally involve a re-review of reactor design.

2.2.3 The Operation Stage

During initial operation, certain testing must be carried out including pressure vessel testing, containment testing, startup tests, etc. During normal operation, observation of such parameters as radiation levels, corrosion, temperatures, etc., must be made. Normal operations embrace by far the longest period of the entire project. During this time, safety depends very strongly upon the quality of the personnel involved. Their knowledge, their judgment, and their sense of responsibility are the primary safeguards for the reactor aside from those features which have already been incorporated during design and construction.

The primary responsibility and authority for the reactor operation and safety must be vested in those directly responsible for its operation. It is totally impractical and impossible for any regulatory group to exercise detailed surveillance and responsibility for the operation of a reactor facility. A facility can only be regulated in any real sense by inspections which exercise a subjective judgment of the adequacy of the personnel directly involved and the facility performance. The details of the operation of any given facility depends upon its manner of use, its function, its schedule, its location, and the type of personnel involved. Since these parameters cover the whole gamut of variability, it is essential to allow as much flexibility as possible to the reactor operating group in order that it may properly assume its responsibility with appropriate authority within limits set by the initial design and by agreements made in the initial safeguard review (during the design stage) or modifications thereto.

Other than the comments which appear here and a few others in the chapter on Accidents and Destructive Tests the subject of reactor operating

---

*In fact, it is understood that such an accident is so classified by the United Kingdom Atomic Energy Authority.

procedures is not considered within the scope of this project.

3 CREDIBILITY OF INDEPENDENT UNRELATED FAILURES

A reactor facility should be designed so that the failure of any single component will not immediately cause the failure of other components—that is, an arrangement like a "house of cards" should be avoided. For instance, if only a single electrical power source is available, its failure could cause loss of flow, loss of instrumentation power, loss of communications, etc. Clearly, all such failures must be designed against and prevented at all costs.

But what about independent or unrelated failures? Is it possible to experience accidents due to two or more indpendent events? Obviously, the probability of essentially simultaneous failures of two or more unrelated components is extremely small (see Sec. 1.4.3 of the chapter on Sensing and Control Instrumentation). However, the probability of two or more independent failures during any appreciable interval of time (i.e., seconds or longer) is not negligible.

The number of independent failures during a period of time deemed credible is a subject of controversy. Although everyone agrees that one failure can occur and can cause an accident, the credibility of two or more independent (and undetected) failures occurring in such a way and during such a time interval as to cause an accident has been debated. Some think it incredible that more than two such independent failures can occur. Others draw the line at a higher number.

A review of reactor accidents to date as carried out in the chapter on Accidents and Destructive Tests indicates rather clearly that often three or more independent causes for a reactor accident may exist. In our opinion three basic causes are involved in almost every reactor accident to date.

First, there is usually a design flaw—a hidden booby trap—which exists in the original reactor design and plays an important role in the subsequent accident. Elimination of design flaws is a most important function of a design team and of the safety review group. For example, the design of the SL-1 reactor required that each control rod be lifted by hand in order to attach it to the upper mechanism. At the same time, the design permitted sufficient reactivity control in a single rod so that the reactor could be made critical by withdrawal of this single rod.

Second, almost every accident involves a supervisor or human error of some sort. In the case of the SL-1 accident, the evidence is almost conclusive that the operators involved withdrew the central control rod far beyond the point called for by their orders. During the accident at Windscale, the supervisor, without aid of a properly annotated manual, made a command decision to reheat the reactor core, and this reheating basically was the trigger for the reactor fire.

The third parameter which is usually involved is some instrumentation problem. Included in the instrumentation problem is also lack of instrumentation. In both the SL-1 accident (no instrumentation operating except for one floor monitor) and in the Windscale accident, instrumentation played an important role.

Thus, at least three causes exist in most accidents experienced to date. In some cases more than three causes have been involved and, in fact, in such a way that the elimination of any one of several causes would have changed the course of the accident materially or would have prevented it.

It is clear that the credibility of a number of independent failures must be taken serious. Anyone who has read the story of the accident to the ship Andrea Doria or accounts of many other accidents will recognize that this is not a problem unique to nuclear reactors.

It must be emphasized that more than one failure can take place at essentially the same time if the failures are dependent—and the dependence may be very subtle. In fact, reactor designers and operators should beware of the label "independent." A structure as complex as a reactor and involving as many phenomena is likely to have relatively few completely independent components. Designers and operators must be continually alert to uncover relationships between potential or real malfunctions, and they must include consideration of reactor performance as a function of its life, since such an effect as misalignment or weakening due to wear (or corrosion or radiation distortion) can become an unrecognized common cause of a dangerous sequence of events.

In reactor facilities one of the chief booby traps exists because there are so many safeties involved. Some accidents have occurred because a relaxed or sloppy crew unknowingly has successively allowed various interlocks and safety measures to be breached one at a time; the logic of the operators always is that there are several and, therefore, the breaching of one is not important. Indeed, the results may be totally inconsequential until that time when the last in a long series is breached, and then the results may be very serious. This is one of the prime reasons why maintenance must ensure that all safety devices are operational at all times.*

4 NUCLEAR REACTOR SAFETY AND SAFETY IN OTHER INDUSTRIES

Except for the nuclear fission chain reaction and the associated nuclear radiations, the physical phenomena involved in reactor safety do not differ in any essential way from those normally associated with industrial plants. Industrial plants have hazards involving high pressures, high temperatures, potential chemical energy releases, corrosive and poisonous chemicals, fast moving mechanical parts, and so on.

Many industrial plants and storage and shipping facilities can, and sometimes do, experience energy releases far greater than even the worst ones which have been postulated as credible for nuclear reactors. Even the largest energy releases that have been estimated for hypothetical nuclear reactor accidents (which are believed to be well beyond the

*We exclude the built-in spare component from this discussion.

realm of credibility) are relatively small compared to energy releases possible in certain industrial operations.

Because a nuclear fission chain reaction is the energy source, the reactor power level can increase under some conditions with extreme rapidity. Several chapters in these volumes are concerned with the kinetics of the chain reaction. As these chapters point out, there are ways to design a reactor so that the chain reaction is self-limiting. Even though basically complex, the chain reaction is controllable and proper reactor design can greatly reduce or eliminate the potential hazards associated with excessive energy generation in a "nuclear run-away."

The presence of nuclear radiations in reactors complicates the measures required to ensure safety. Remote operation of some components is necessary; this often involves mechanical complexity and a strong dependence on instrumentation. Means for maintaining those parts of the reactor that are, or may become, radioactive must be anticipated in the reactor design. Because a nuclear radiation field may preclude replacement of components, ruggedness and reliability are very important.

In spite of these considerations, for the most part the safety problems of nuclear reactor facilities differ only in degree from those of other industrial plants. The safe handling of radioactive materials requires the same kind of techniques and precautions as observed in handling dangerous chemicals or toxic biological substances. The design, construction, and operation of nuclear reactors require great care at all stages just as similar activities in other complex and potentially hazardous industries.

5 EFFECT OF ECONOMIC FACTORS ON SAFETY

As nuclear reactors develop into commercial sources of power, it has become evident that competition with fossil fuel plants is very keen indeed. In some geographic areas nuclear plants are now economically competitive with conventional plants. This has resulted in pressures on nuclear plant designers, constructors, and operators to reduce capital costs, to increase core life, to increase performance, to increase core size, and to reduce distances from metropolitan areas.

It is argued that the larger the plant the lower will be the capital costs per kilowatt of generated power. Experience to date indicates this.* There is therefore a substantial incentive to increase as rapidly as possible the size of plants and to extrapolate existing experience to much larger plants. This, of course, involves serious safety problems that are not easily answered. In particular, increasing the size of components, such as the reactor vessel, valves, turbines, and so on, requires extrapolation of existing techniques and, consequently, some uncertainties result. Reactor physics may be somewhat different in larger, more loosely coupled cores with many more critical masses.

In an effort to reduce the capital costs, in particular those of transmission of electrical energy, considerable pressure has been exerted to move reactors closer to centers of use. The costs of rural transmission lines themselves range between \$50,000 and \$150,000 per mile, exclusive of right of way. Near urban centers where large quantities of electrical power are needed, the right of way is even more expensive and, in fact, in metropolitan areas, high voltage cables must be located underground. Such procedures are very expensive indeed and rapidly raise the costs of electricity. It is therefore clear that there are economic incentives to move reactors towards major centers of electrical energy consumption.

By increasing the performance of a given core, it is possible to improve the economics considerably. Almost every power reactor to date has been able to safely exceed its original design values and its power has been gradually raised. Increased core performance can be achieved by flattening the neutron flux distribution and thereby reducing the hot-channel factors. (The considerations involved in improving core performance are discussed in the chapter on The Reactor Core.) Pressures to increase core performance tend to force reactor designers to move closer to burnout conditions and to operate on narrower margins of safety as far as fuel is concerned. Thus, the burnout correlation and the validity of these correlations become very important indeed; the chapter on Heat Transfer emphasizes these aspects. As the flux is flattened, a larger and larger percentage of the core is being driven at or near the limiting thermal conditions. Thus, if there is a transient, fuel melting is likely to be more widespread than it would be in a reactor with a less flat flux.

As the safety margins are narrowed, it may be necessary to devise improved in-core instrumentation or to develop new means of burnout indication. To date efforts in this area have been limited; clearly, as performance has increased, such efforts become more and more necessary.

Another means of prolonging core life is by increasing the available reactivity. Since the total shutdown reactivity available in control rods in a power reactor is limited, it is necessary to limit the excess reactivity in such a way that the control rods can shut down the reactor in any conceivable situation. The only flexibility left is that which exists in core burnup or shutdown margin. Thus, the pressures to reduce the shutdown margin have been very great and in some cases the shutdown margin is now as low as 0.5% or less. This means that measurements of reactivity, reactivity effects, control rod worths, core lifetime effects, and so on, must be made much more carefully than formerly. At the moment, the state of the art in this area is not as good as it should be.

It should be noted that the incentives to prolong core life have also led to the design of new types of reactors which can be continuously fueled or refueled during operation on a partial basis. For

---

*For instance, the Yankee Atomic Electric Plant, operating at ~ 150 megawatts electrical, was built for approximately forty million dollars. The Connecticut Yankee Plant, operating at approximately three times this power level will only cost about twice as much. Experience is similar with boiling water reactors, such as those being built at Oyster Creek (New Jersey) and at Nine Mile Point (Oswego, New York).

example, the Canadian CANDU Reactor is a pressure-tube reactor which can be refueled while in operation. The British gas-cooled reactors have this general feature. Homogeneous reactors can be so designed that very little excess reactivity is required to carry the reactor along indefinitely without more than slow partial refueling.

It must be understood that, although economic incentives tend to push power reactor designers to the limits of existing knowledge, this does not necessarily mean a decrease in reactor safety. Rather it means that the frontiers of knowledge of reactor safety must be steadily expanded to provide adequate backup for newer and economically more attractive nuclear power plants.

6 ABOUT THE SIFTOR BOOKS

6.1 Purposes

To maintain or improve the already remarkably good safety record of nuclear reactors it will be necessary to adopt a two-pronged approach. First, those entering the field of nuclear technology or already in it must be made fully aware of the accumulated knowledge and experience concerning reactor safety so that they may apply this information to their work on future reactors. Because of this, the primary purpose of these volumes is to collect in one place in an organized fashion the essence of the safety information concerning reactor technology which has been built up over the past twenty years.

Second, the knowledge of reactor safety must be extended in certain areas. There must be continued alertness for the new and more subtle problems that may develop as reactor technology advances. There must be a continued drive towards more realism in reactor safety by proving better information to permit resolution of present "ignorance factors" or conservatism. Therefore, a secondary purpose of these volumes is to provide insight into the safety problems that need further investigation.

Since safety information cannot be divorced from the technology of reactors, a considerable fraction of the volumes is devoted to the technical problems of reactor design. The project title, "Safety Information For the Technology Of Reactors," describes both aim and content. (The abbreviation "SIFTOR" is intended to imply that the information has been "sifted," that is, not just assembled directly from the literature.) The project might also have been titled "Information For Safe Reactor Design." Information is supplied, trouble areas are pointed out, references are given. Few general conclusions are drawn, few over-all guide lines are stated, and almost no rules are proposed. These volumes are intended to lay the groundwork for such general principles as may some day emerge. This should be recognized as only the first step; it may be some time before other steps can, or should, be taken.

6.2 Intended Audience

As a whole, the two volumes are written for the technically trained man who has a basic knowledge of science and engineering. There has been little or no attempt to provide a discourse on fundamental concepts, but rather to start at a level which might be readily understood by a person with at least one degree in some branch of the physical sciences or engineering. Much of the first volume presupposes basic knowledge of nuclear reactor theory. Most of the second volume presupposes a knowledge of the fundamentals of engineering, such as mechanical design, materials and metallurgy, fluid flow, heat transfer, chemical engineering and processing, etc.

It is intended that the level of most of the chapters be such that a technically trained man with a strong background in one area—reactor physics, for instance—can read with understanding and appreciation the safety problems in another area—metallurgy or fluid flow, for instance. Of course, he may, and probably will, need additional background reading and study if he desires to understand the fine points.

This encouragement of "cross fertilization" between fields is one of the most important reasons for the assembly in one place of chapters stemming from diverse disciplines. It is hoped that this will broaden the interests and improve the appreciation of each reactor scientist or engineer for the problems which exist in technical areas other than his own.

For the expert in a given area, it is hoped that the chapters covering his area or closely related ones will serve as up-to-date reviews and will also include some information or viewpoints new to him. The expert in a given area needs only to be reminded of the problems that he already knows exist. It is hoped that the introductory chapters will serve as reminders and the more detailed chapters will present a more adequate coverage of the advanced facets of the state of the art.

The newcomer, having no backlog of experience, has no knowledge of what has come before and is greatly handicapped in acquiring that knowledge because of the diffuse nature of much of the relevant literature. It is hoped that these volumes will provide an artificial backlog of information and experience for the nonexpert, the safety evaluator, and for those starting out in reactor work.

The intended audience also includes reactor operators, where by "operator" is meant the group responsible for all reactor operations and not necessarily the man sitting at the control desk. Although designers can provide a variety of means for building safety into a reactor, these features may be negated by the operators. A safety device may go out of order, or a gauge yield an incorrect signal, or an operator ignore instrument responses that disclose malfunctions. Unless the operating organization is alert to all signs of abnormality, right down to the most subtle indications, the safety of the reactor is in jeopardy. The technical men involved in reactor operation, whether in management or in advisory positions, play an extremely important part in the safe operation of nuclear reactors. The present work is intended to reach these individuals.

6.3 Contents*

In the following paragraphs we outline what is and what is not in these books. Some indication of motives is also given.

---

*The reader will find it most useful to study the contents of each chapter as given in the lists of

It was the original intent to combine in one volume all of the chapters, showing thus the close interrelationships and the need for close cooperation between the diverse disciplines needed to achieve a safe and well balanced reactor design. However, as more subjects were added to the scope, the chapters proved to be longer than originally planned. Finally, the decision was reluctantly made to split the chapters between two volumes.

Division of the chapters into one group primarily concerned with reactor physics and a second group concerned with reactor engineering seemed the most logical procedure.

Accordingly, we decided to include in the first volume all the chapters on reactor kinetics. Instrumentation was also included since the safety aspects of reactor control are so closely related to the physics of the chain reaction. Similar considerations led us to put the chapter on accidents and destructive tests in the first group.

In the second volume we included all the chapters on engineering subjects—those on materials, fluid flow, heat transfer and chemical reactions. The chapters concerned with the release of fission products and structures to confine or contain them were also put in the second volume.

Several chapters are of such general interest that we would like to have included them in both volumes. However, for practical reasons (i.e., size and cost of each volume) we have minimized duplication. An exception is the Appendix in each volume summarizing data on reactor safety parameters; these data were obtained from responses to questionnaires sent to various reactor sites.

The principal disadvantage of splitting the material between two volumes is that some of the sense of the close interplay between reactor physics and reactor engineering is lost. But this is inherent in any split, no matter how the chapters are divided. It is our sincere hope that the reactor physicists will read and make use of the reactor engineering volume and vice versa.

An advantage of dividing the material between two volumes is that it may make one or the other of these volumes a suitable text or reference for certain academic Reactor Engineering or Reactor Physics courses. Of course, no arrangement will suit the personal choices of all readers and it seems likely that the present arrangement is best suited to the majority.

6.3.1 Volume 1 "Reactor Physics and Control"

Following this Introduction, Chapter 2, "The Reactor Core," provides an over-all view of the problems entering into the design of a safe reactor core. It is generally accepted as a requirement in nuclear plant design that all foreseeable hazards be guarded against, either by eliminating the causes or by providing safeguards to limit the consequences. Generally, when specific hazards are recognized, it is possible to satisfy this requirement. It follows, therefore, that the recognition of hazards and the proper assessment of their seriousness are of crucial importance. The goal of Chapter 2 is to introduce the reader to the fundamentals of reactor core design and to show the complex nature of the coordinated breadth of understanding and effort necessary to achieve a safe and still practical reactor core. Both the reactor physics and the reactor engineering problems are outlined. This chapter is intended to treat these problems and interrelations at an introductory level. It should be the first chapter read by virtually every reader.

Chapter 3, "General Reactor Dynamics," deals with the fundamentals of general reactor kinetics and is written primarily for the reactor physicist with considerable background. Such basic problems as nonlinearity in reactor kinetics and its relevance to reactor stability are considered. An appendix to the chapter reviews briefly certain digital computer codes available for studies of space-independent kinetics.

Chapter 4, "The Doppler Coefficient," reviews the theory and current state of understanding of Doppler coefficients. The fact that an entire chapter is devoted to a single reactivity coefficient is a reflection of the importance of the topic to reactor safety. The chapter is intended for reactor physicists with considerable background and sophistication. The information presented includes consideration of both the thermal and fast reactors.*

Chapter 5, "Criticality," includes a discussion of those aspects of startup that are important to the control of the neutron chain reaction. An improved set of delayed neutron constants is presented. Low source-level startups and problems of fuel storage are also discussed.

Chapter 6, "Sensing and Control Instrumentation," considers possible control and instrumentation philosophies.** It discusses control and instrumentation safety problems not only for the neutron chain reaction, but also for such other parameters as coolant flow, radiation levels, temperatures, and so on. The chapter is placed in this position since it is of general applicability to the material in the following chapters. This chapter, along with Chapters 2 and 5 should be of particular relevance to those whose interests lie in reactor instrumentation.

Chapter 7, "Mathematical Models of Fast Transients," discusses nuclear excursions from a semiempirical or phenomenological viewpoint. At the present time enough information is available on this important topic so that it seemed desirable to present it as a separate chapter. Specific instances of reactor transient behavior which are generalized in this chapter are included in subsequent chapters.

The next three chapters (Chapter 8, "Water Reactor Kinetics"; Chapter 9, "Kinetics of Solid Moderator Reactors" Chapter 10, "Fast Reactor Kinetics") deal with the kinetics of three classes

---

topics at the beginning of each chapter. The Subject Index at the end of each volume is also recommended to the reader.

*A discussion of the Doppler effect on an introductory level is also included in Chapter 2 since it is believed that the importance of this effect in reactor safety merits at least a basic understanding of it by all those involved in reactor design.

**Also discussed in Chapter 2.

of reactors. The classification is somewhat arbitary although it does tend to differentiate the reactor types according to their kinetic behavior. Because of this choice, the reader will find, for instance, some sodium-cooled reactors considered in Chapter 9 and others considered in Chapter 10—some are thermal reactors and some are fast reactors. In general, the three chapters on reactor kinetics complement each other. In the chapter on water reactors the use of transfer functions in kinetics analysis is discussed. In the chapter on solid-moderator reactors the digital computer code approach to kinetics analysis is emphasized. In part this difference in emphasis is due to the difference in the type reactor being considered, in part it reflects a difference in experience and viewpoint of the authors. It should be noted that these chapters deal with much more than just the physics of reactor kinetics; many problems of interest to metallurgists, engineers, and designers also are discussed in these chapters. Each of these three chapters considers the course and consequences of reactor accidents of various types and of varying severity.

The final chapter of the first volume, Chapter 11, "Accidents and Destructive Tests," reviews the negative experience to date with nuclear reactors. The nuclear reactor accidents on which reports are available are reviewed. Accidents to the end of 1963 in both critical assemblies as well as reactors are summarized in a series of tables. In general, the intent is to provide a record of experience and a series of object lessons, and to set forth conclusions for the discussions of the various safety problems contained in the other chapters.

In an appendix (in each volume) we have compiled most of the important characteristics and safety parameters of the power reactors in the United States. This information has been collected by contacts with the various responsible reactor organizations. Therefore, it represents the best calculated and measured values available for these parameters as of the end of 1963. A study of these data will indicate the range of variation of the parameters for the various reactor types and for various reactors of each type. It is believed that such information will be of considerable value to reactor safety evaluation groups.

6.3.2 Volume 2 "Reactor Materials and Engineering"

The first six chapters of the second volume deal with the fundamental safety problems of nuclear reactor engineering. These chapters plus the first ten chapters in Volume 1 are dedicated mainly to trying to prevent serious accidents to the reactor core. The remaining chapters of the second volume assume there has been a serious accident and provide information that should be of assistance in preventing the fission products from reaching the general public.

Chapter 12, "Materials and Metallurgy," provides an introduction to those materials problems that have serious safety implications in nuclear reactors. The choice and proper application of materials for use in a reactor are vital factors in its safety. The material in this chapter is closely related to that in the next two chapters which follow, as well as to Chapters 2, 8, 9, and 10 of the first volume.

Chapter 13, "Fuel Elements," has close interrelations with many other chapters in the second volume, especially Chapters 12, 15, 16, and 17. In addition, as mentioned above, it is closely related to the contents of Chapters 2, 8, 9, and 10 of the first volume. Since fission product release is a function of the form and metallurgy of the fuel, it is also related to the content of Chapter 18. The emphasis in the chapter is placed on safety problems which have been or may be encountered in fuel elements. In addition to an extensive discussion of fuel element design, there is a substantial section on specific fuel elements.

The principal safety problems of mechanical components are treated in Chapter 14, "Mechanical Design of Components." The reactor primary system is emphasized. Pressure vessel design is discussed in some detail. A discussion of the new ASME Unfired Pressure Vessel Code is included as well as the older Section VIII. Primary system components and control rod mechanisms are considered in some detail as they remain one of the principal safety problems of nuclear reactors. As mentioned above, this chapter is closely related to the others of the first six chapters of the second volume as well as Chapters 2, 6, 8, 9, and 10 of the first volume. It is also closely related to the chapters on containment and confinement structures.

While Chapter 15, "Fluid Flow," emphasizes the safety aspects of fluid flow, it reviews the equations of fluid flow (single-phase and dual-phase) and approaches the subject in such a way that engineers and scientists not familiar with the subject can readily use the text for other fluid flow problems. This chapter and the next are closely related. Transient as well as steady-state flow problems are discussed.

Chapter 16, "Heat Transfer," discusses certain aspects of heat transfer important to reactor safety. It discusses the thermal conduction problems connected with reactor safety analysis including thermal conductivity coefficients, thermal boundary resistances, and fuel element performance. Burnout or critical heat flux problems are considered in detail. Transient-boiling heat transfer, heat transfer during accidents, hot spot factors, and liquid metal heat transfer are discussed. Both this and the immediately preceding chapters are closely related to Chapters 2, 8, 9, and 10 of the first volume and to Chapters 13 and 14 of the second volume.

Chapter 17, "Chemical Reactions," treats those chemical reactions which may be important to reactor safety. Of most direct interest to reactor safety are the extensive discussions of metal-water, metal-air, graphite-gas, and hydrogen-oxygen reactions. In addition, however, the authors devote attention to slow (corrosion) reactions. While these reactions do not play a central part in accidents, they can be an initiating cause. Fuel meltdown, formation of eutectics which may lead to lowered fuel melting temperatures, thermite-like reactions, and so on are discussed. A short account of the cleanup and analysis of liquid metal coolants is also included.

Chapter 18, "Fission Product Release," deals with the loss of fission products from the fuel in event of a serious accident. In reactor safety analysis there is a serious problem in estimating the extent of release of the various fission products from the fuel to the containment. It is the principal purpose of this chapter to report the present status of the knowledge in this important area. If it is possible to show that certain fission products are not released or only released in small quantities it may be possible to relax requirements on containment. This chapter is closely related to the chapters which follow it and to some extent to the one which precedes it.

Chapter 19, "Fission Product Leakage," is concerned primarily with estimating the amount of leakage to the containment that may be expected under a variety of accident conditions. In particular, this chapter discusses possible methods for reducing the fraction of fission products that might be available for leakage from the containment in event of a very serious accident. Filtration, spray systems, foams, iodine collectors, and other means to reduce the release are discussed. In essence, Chapter 18 is concerned with what fraction of the fission products may be released to the containment under various circumstances while Chapter 19 is concerned with what special measures can be taken to reduce this fraction.

Chapter 20, "Radioactive Waste Management," discusses a problem which is both a part of regular reactor operations and an important consideration in serious reactor accidents. The initial sections of the chapter deal with how radioactive wastes occur in the various reactor types and how their quantity can be estimated. The middle part of the chapter discusses various ways of disposing of liquid, solid, and gaseous waste materials from nuclear reactor plants, under accident conditions as well as during normal operation. In the final sections a number of accidents are reviewed principally from the standpoint of how much radioactivity was released, what changes might have made the release less severe, and how the cleanup was effected. There is some minor overlap in reviews of accidents in this chapter and those in the last chapter of the first volume, i.e., Chapter 11. This is done for two reasons. First, Chapter 11 emphasizes the causes of the accidents and is primarily concerned with accident prevention, while the accident section of Chapter 20 emphasizes the radioactivity released. Second, the existence of accounts of accident experience in both volumes provides a ready reference in each concerning the general circumstances of most of the worst accidents.

In Chapter 21, "The Concept of Reactor Containment," and Chapter 22, "Containment and Confinement Structures," the various possible containment and confinement systems are described and the details of construction, testing, and maintenance discussed. These chapters relate very closely to the mechanical design problems considered in Chapter 14. Since large evolutions of heat are likely to occur, Chapters 15 and 16 are pertinent. In addition, Chapter 17 as well as the chapters on nuclear excursions of the first volume all relate to the important problem of estimating what stresses the containment must withstand and what materials it must hole.

6.3.3 Concluding Comments on Scope of Volumes

In concluding this description of the scope of the SIFTOR volumes we should point out that certain topics have been excluded.

First, our concern has been with the nuclear reactor plant and its containment, not with reactor location. The development of criteria for determining whether a site is or is not appropriate for nuclear reactor location is partly a technical matter and partly a matter of determining how the public interests are best represented. In the United States this problem has been carefully considered by the Atomic Energy Commission which has promulgated 10CFR, Part 100 as an initial step towards answering this question. The answers will differ in other countries. Therefore, the question is not considered here. In the SIFTOR books it is assumed that the reactor location has been set and the designer, builder, or operator must now design or build a safe reactor or operate a reactor safely at the given location.

Second, the text does not include discussion of the specific safety problems associated with "military reactors," that is, nuclear reactors incorporated in submarines, military aircraft, or other devices associated with military defense. Design of such reactors involves unique problems because of the conditions under which they are used. Geometric or weight limitations, insensitivity of various accelerations, capability for rapid startup, etc., are important. Such considerations mean that reactors designed for military use may have features neither useful nor appropriate in nonmilitary reactors. Moreover, public discussion of some aspects of reactors incorporated in military devices could compromise the security of these devices. It should be understood, however, that the text will be valuable to the military reactor designer or fabricator since many safety problems of military reactors are identical to those of reactors used in nonmilitary applications.

Third, the routine operational problems are mentioned in the text but not emphasized. Thus, there is no detailed discussion of shielding, health physics, refueling operations, etc., as long as these do not lead to serious accidents. These are considered for the most part to be maintenance problems.

Finally, even in considering the nuclear reactor itself, the text emphasizes only those aspects of reactors which are important from the point of view of plant safety. The primary interest is in those components or structures which require particularly careful design, fabrication, testing, and inspection because their failure or malfunction might result in a hazard to neighboring communities or to operators or might result in serious economic losses. The volumes reflect less interest in components which on failure might impair reactor performance but not reactor safety.

CHAPTER 2

The Reactor Core

J. R. DIETRICH
General Nuclear Engineering Corporation
Dunedin, Florida

CHAPTER CONTENTS*

1 INTRODUCTION

The design of a nuclear plant often requires so much specialized effort on details that it becomes difficult to achieve the coordinated breadth of understanding necessary for recognizing all hazardous relationships. This chapter is directed toward the alleviation of this difficulty, at least within the areas of core design and analysis. Its purpose is not to describe means of analyzing specific hazards but to remind the reader of the relationships which may require analysis.

Much of the chapter is devoted to considerations involving reactor physics: first, because reactor physics relationships so often may be the connecting links between apparently minor faults and serious hazards, and secondly, because intuitive understanding is usually least developed in the area of reactor physics. The chapter is not written for the reactor physicist. The relationships are discussed in as elementary a way as is considered compatible with a useful exposition, and it is hoped that much of the physics discussion will be understandable in

*Except for a few changes and additions made in proof, this chapter is based on information in the literature or known to the author prior to December 1963.

**Section 7 and subsection 6.3 were written by T. J. Thompson and J. R. Dietrich.

terms of the simple concepts outlined in the section immediately following; yet there has been no hesitancy to go beyond these concepts when it has been considered necessary to the purpose of the chapter.

So far, no fixed logical procedure has been developed for ferreting out all the possible hazardous relationships in a nuclear reactor design, and it seems improbable that one will be developed. Consequently, in this chapter, an organization of the material has been adopted which might be regarded as psychological rather than logical. A three-pronged approach has been taken, in an effort to stimulate the recognition of significant relationships by approaching them along as many different paths as possible. First, the various possible types of accidents are outlined in terms of the major reactor core relationships involved in each (Sec. 3). In a series of succeeding sections (4 through 9) the behavior of the reactor core is discussed, primarily in terms of the neutron chain reaction. Finally, (Sec. 10) the major characteristics related to safety are discussed for several specific reactor types, as illustrated by the designs of specific reactors. No apology is offered for the redundancy which is inherent in this approach: it is believed that redundancy in safety analysis may perform as useful a function as redundancy in the safety instrumentation of a reactor.

2 ELEMENTARY REACTOR PHYSICS CONCEPTS

The purpose of this section is to present briefly a number of reactor physics concepts which enter the discussions in later sections. The reader is assumed to be familiar with the elementary concepts of neutron flux, current density, and cross sections, and to be aware of the products of neutron-induced fission of the heavy elements. For those who need to refresh their memories, the first eight pages of reference [1] may prove helpful, and of course the more extensive treatments of these and other relationships which are given in standard reactor-physics texts will be even more useful.

Attention here is limited to those aspects of the subject which are needed in later discussions: these are confined mainly to questions of criticality and reactivity. Extensive treatments of the general principles of reactor physics may be found in references [2-6], while the methods used for specific reactor types are covered in references [7-39]. References [7-16] treat H_2O reactors, references [17-25] D_2O-moderated reactors, references [26-31] graphite-moderated reactors, and references [32-39] fast and intermediate spectrum reactors.

2.1 The Neutron Balance: Reactivity

The neutron-fission chain reaction is sustained by the actions of neutrons, and when such a reaction is going on in a medium all of its characteristics can be specified by specifying the neutron population in terms of its rate of growth and its distribution in space and energy. Every neutron in such a population must eventually come to an end in one of three ways: it may be absorbed by a fissile nucleus and cause that nucleus to fission, it may be captured without fission by a fissile or non-fissile nucleus, or it may wander to one of the boundaries of the medium and leak out. Those neutrons which do cause fissions thereby cause the production of fresh neutrons; the average number produced per fission is designated by the symbol ν. This quantity is a characteristic constant for a given species of fissile isotope at a given neutron energy. By the use of ν one can express the status of growth or decline of the neutron population in terms of the three processes cited above:

$$\frac{\text{Rate of birth of neutrons}}{\text{Rate of death of neutrons}} = \bar{\nu}\,\frac{F}{F + C}\,(1 - \mathcal{L}) = k_{eff}. \tag{2-1}$$

Here the bar is added above ν to indicate that an appropriate average is to be taken over the fissile species present and over the range of neutron energies. F is the fission rate in the medium and C is the neutron capture rate. The symbol $\mathcal{L}$ represents the fraction of neutrons produced which leak from the medium without causing fission or being captured.

When the ratio above, designated by k_{eff}, has the value unity, the neutron population remains constant and the fission chain reaction proceeds at a constant rate: the system is said to be critical. If the ratio is greater or less than one the neutron population will increase or decrease with time, and the rate of reaction—i.e., the power output of the chain-reacting assembly—will increase or decrease in direct proportion to the neutron population.

If it is assumed that a value, ℓ, can be assigned which will represent adequately the average lifetime per neutron, then the dynamics of the chain-reacting system can be expressed in terms of k_{eff}. If there are n neutrons present in the system at time t, the rate of death of neutrons will be n/ℓ, and the rate of increase of the neutron population, dn/dt will be:

$$\frac{dn}{dt} = \text{rate of birth of neutrons} - \text{rate of death of neutrons}$$

$$= \frac{\text{rate of birth} - \text{rate of death}}{\text{rate of death}} \times \text{rate of death}$$

$$= (k_{eff} - 1)\,\frac{n}{l}$$

or

$$\frac{1}{n}\frac{dn}{dt} = \frac{k_{eff} - 1}{l}$$

$$n = n_o\, e^{(k_{eff}-1)(t-t_o)/l}, \tag{2-2}$$

where n_o is the population at the initial time, t_o.

Since the death of each neutron, on the average, generates k_{eff} new neutrons, one may consider, instead of the average neutron lifetime, ℓ, the average neutron generation time, ℓ^*, where $\ell^* = \ell/k_{eff}$. In this form the kinetic equations are:

$$\frac{1}{n}\frac{dn}{dt} = \frac{k_{eff} - 1}{l^*\, k_{eff}} = \frac{\rho}{l^*}$$

$$n = n_o\, e^{\frac{\rho}{l^*}(t-t_o)}, \tag{2-3}$$

where $\rho = (k_{eff} - 1)/k_{eff}$.

In the actual case it is not possible to define an average lifetime or generation time which describes the neutron behavior adequately, for the reason that the generation times of the delayed neutrons are so vastly different from those of the prompt neutrons; but the kinetic behavior of the system can be described by more complex equations, analogous to Eq. (2-2) or (2-3), involving separate average generation times for the prompt neutrons and each group of delayed neutrons. In these equations the quantities k_{eff} and ρ play the same roles as in Eqs. (2-2) and (2-3). Conceptually, the important point is that the quantities k_{eff} or ρ form connecting links between the observable dynamic behavior of the reactor and the fundamental processes of the chain reaction, which are not directly observable, via Eq. (2-1) or its equivalent.

The quantity ρ is called the reactivity. Its meaning is (see Eq. 2-1).

$$\rho = \frac{k_{eff} - 1}{k_{eff}}$$

$$= \frac{\text{Rate of birth of neutrons} - \text{rate of death of neutrons}}{\text{Rate of birth of neutrons}}$$

= Fraction of neutrons born which are in excess of those required to hold the population constant.

Note that the reactivity is zero for a critical reactor, positive for a supercritical reactor, and negative for a subcritical reactor.

2.2 Components of the Neutron Balance

Useful as Eq. (2-1) may be conceptually, it does not go far toward evaluating the neutron balance in an actual system. In examining its components, it is again informative to assume that the probability of each of the events which may happen to a neutron in the system may be expressed by an average number. Thus the ratio F/(F + C) may be assumed to be given by $\bar{\Sigma}_F/(\bar{\Sigma}_F + \bar{\Sigma}_C)$ where $\bar{\Sigma}_F$ and $\bar{\Sigma}_C$ are appropriate effective macroscopic fission and capture cross sections. The neutron leakage is determined by the probability that the average neutron will diffuse to the boundary of the system before being absorbed. The appropriate average characteristic for expressing this probability is the mean square distance $\overline{r^2}$ that a neutron would travel, if the system were infinite in extent, from the time of its birth to the time of its absorption. To approximate the leakage, this characteristic must be related to the shape and size of the chain-reacting assembly by means of a partial differential equation which describes the neutron flow in terms of the flux, ϕ. The relations are:

$$\mathcal{L} = \frac{M^2 B_g^2}{1 + M^2 B_g^2}, \qquad (2\text{-}4)$$

where $M^2 = \overline{r^2}/6$, and B_g^2 is the solution of the partial differential equation:

$$\nabla^2 \phi + B_g^2 \phi = 0 \qquad (2\text{-}5)$$

for the boundary conditions appropriate to the system. This concept of the chain reaction is quite simple when it is recognized that Eq. (2-5) involves only the geometry of the system,* and that solutions have been worked out for the simple shapes which are likely to occur in practice. Thus, for example, if the system is a circular cylinder of extrapolated** radius R and extrapolated length L the value of B_g^2 is:

$$B_g^2 = \frac{2.405^2}{R^2} + \frac{\pi^2}{L^2}. \qquad (2\text{-}6)$$

B_g^2 is called the geometric buckling of the system. If attention is confined to those systems for which the evaluation of B_g presents no problems, it is easy to relate all of the characteristics determining the neutron balance to physical quantities which are conceptually measurable. Thus it has already been pointed out that F and C [Eq. (2-1)] may be expressed in terms of fission and capture cross sections. The quantity M^2 is a straightforward concept when defined in terms of $\overline{r^2}$, and is susceptible to direct measurement. It can also be expressed in terms of the neutron scattering cross section and absorption cross section of the medium, with due regard to the changes in neutron energy which accompany scattering. Hence the equation resulting from the combination of (2-1) and (2-4), and utilizing cross-section ratios:

$$k_{eff} = \left(\bar{\nu}\,\frac{\bar{\Sigma}_F}{\bar{\Sigma}_F + \bar{\Sigma}_C}\right)\left(\frac{1}{1 + M^2 B_g^2}\right) \qquad (2\text{-}7)$$

may be considered a valid representation of the processes going on in the system to the extent that it is valid to take averages as indicated. The limitations on the usefulness of the formulation arise because of the difficulty of finding appropriate averages over the large range of neutron energies which exist in most reactors.

---

*Obviously, the attempt here is to make available some relationships which can be used for practical approximations without going into the details of reactor theory. Texts on reactor physics should be consulted for further explanation. Unless that is done, the following qualification may not be very meaningful, but it is necessary for the sake of accuracy: the statement that Eq. (2-5) involves only the geometry of the system applies only if the system may be considered to have uniform composition and a boundary condition which can be stated, at all points, in terms of the ratio $\nabla \cdot \phi/\phi$; the same formalism may be used in complex systems, but the meaning of B_g^2 is then less apparent.

**R and L would be the actual radius and length, respectively, if the boundary condition were $\phi = 0$ at the boundary of the cylinder. In real systems, even those surrounded by vacuum, the neutron flux is not zero at the boundary, but extrapolates to zero a short distance beyond the actual boundary. The extrapolated radius is the radius at which ϕ extrapolates to zero; a similar definition holds for the extrapolated length.

2.3 Neutron Energy Distributions

The neutrons emitted in the fission process have energies in the Mev range. They diffuse through a medium (the reactor core) composed of atoms whose energies of thermal agitation lie mostly below 0.1 ev. As the neutrons collide with nuclei of these atoms they exchange energy with them, elastically at all energies and possibly inelastically when the neutron energies are in the Mev range.* Those neutrons which escape absorption and leakage long enough will eventually come into thermal equilibrium with the atoms. In "moderated" reactors this may be most of the neutrons, while in "unmoderated", or "fast", reactors it may be a negligible fraction; but in any case the neutron population will cover an energy range of some decades, and over this range the cross sections for the important neutron-nucleus interactions will vary importantly.

Figure 2-1 shows the variation of several cross sections for important reactor materials over the energy range from below 0.01 ev to 10 Mev, the range which is significant in "moderated" reactors. Also plotted in the figure is the fission spectrum of U^{235}—the energy distribution of the neutrons produced in fission.

If neutrons are slowing down by elastic collision in an infinite nonabsorbing medium, the neutron flux due to a source of Q neutrons per cm^3 sec at energy E_0 is, at any energy which is less than E_0 and which is well above thermal energy**:

$$\phi(E) = \frac{Q}{E} \cdot \frac{1}{\xi \Sigma_s(E)}, \qquad (2\text{-}8)$$

where $\Sigma_s(E)$ is the elastic scattering cross section. ξ is the average logarithmic change in energy per collision, and for elastic collisions is a function only of the mass of the nucleus involved in the collision, ranging from 1.0 for H to 0.008 for U. The flux ϕ (E) is the total flux per unit energy interval. Thus if Σ_s is constant with E, the flux per unit energy interval is inversely proportional to the energy. This behavior is approximated in many "moderated" reactors over an energy range of several decades above the thermal region.

For purposes of assessing qualitatively the effects of the variations of the important cross sections with energy (Fig. 2-1) it is convenient to divide the neutron energy spectrum into a number of regions. If the energy span of each region is made proportional to the energy at which the region occurs, then a flux distribution like that of Eq. (2-8) will put about the same total flux in each region. Such a breakdown is given in Fig. 2-2. The energy range from 0.001 ev to 10^7 ev is divided into 20 regions of equal logarithmic span—i.e., each region covers an energy range which is larger than the next lower region by the factor $\sqrt{10}$. The several parts of Fig. 2-2 illustrate, in an idealized way, some of the more important effects which determine the complete neutron energy spectrum in a reactor.

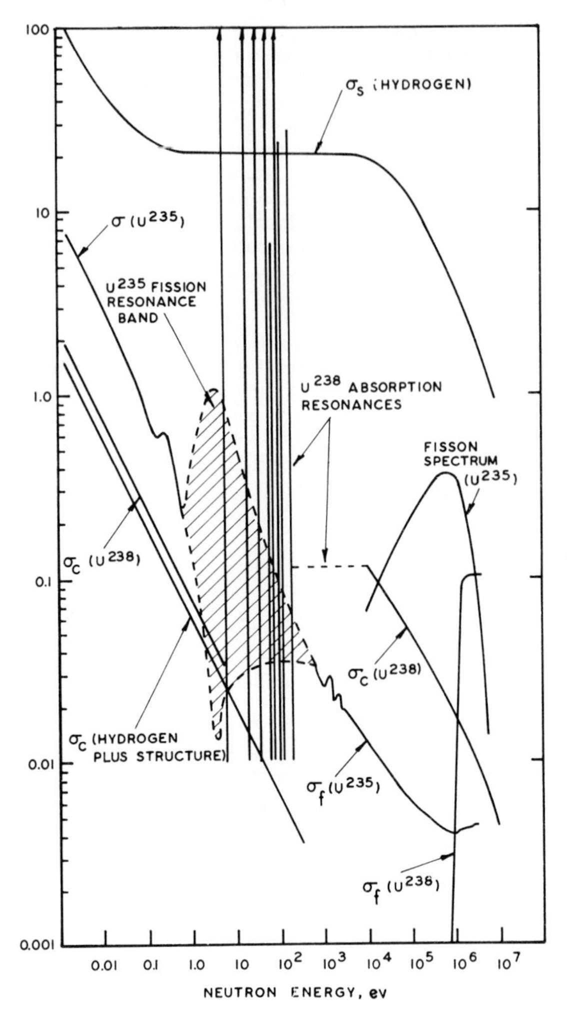

FIG. 2-1 Approximate behavior of important cross sections with energy in a typical water-moderated reactor. The reactor is assumed to be composed of H_2O, U^{235}, U^{238}, and structural material, in such ratios that the important cross sections (σ_f = fission cross section, σ_c = capture cross section), in barns per hydrogen atom, at 0.025 ev are as follows: σ_f (U^{235}) = 2.0; σ_c (U^{238}) = 0.5; σ_c (H_2O + structure) = 0.4. This would correspond roughly to a UO_2 lattice, of 2% enrichment, in cold water with a volume ratio of H_2O/UO_2 = 1.8. All cross sections in the figure are given in barns per hydrogen atom. The curves are intended only to define the important energy ranges; see cross section tabulations for precise values. The fission spectrum gives the fraction of neutrons produced in fission at energy E per unit energy interval, where the energy unit is 1 Mev.

Figure 2-2a shows an approximation to the spectrum which would exist in an ideal moderator (continous slowing down) infinite in extent, with a small 1/v absorption cross section, a constant elastic scattering cross section and no inelastic scattering, containing a uniformly distributed source of neutrons of energy 10^7 ev. The total flux in each logarithmic energy interval is constant over the range from source energy to about 1 ev. At lower energies the neutrons begin to come into thermal equilibrium with the moderator atoms, and Eq. (2-8) is no longer valid. At these energies the neutron absorption becomes important also, and the mag-

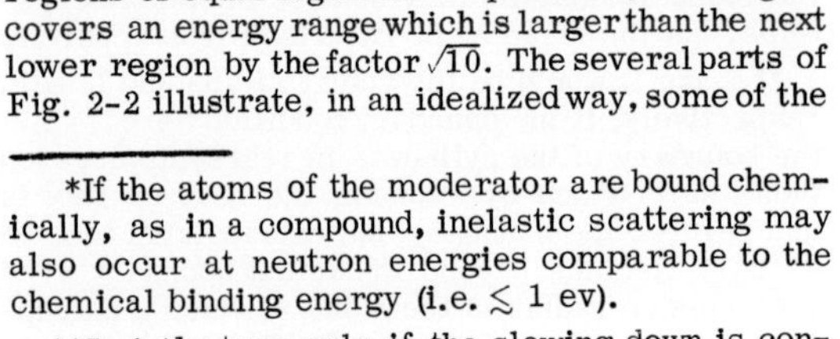
*If the atoms of the moderator are bound chemically, as in a compound, inelastic scattering may also occur at neutron energies comparable to the chemical binding energy (i.e. $\lesssim$ 1 ev).

**Strictly true only if the slowing down is continuous, i.e., if ξ is small.

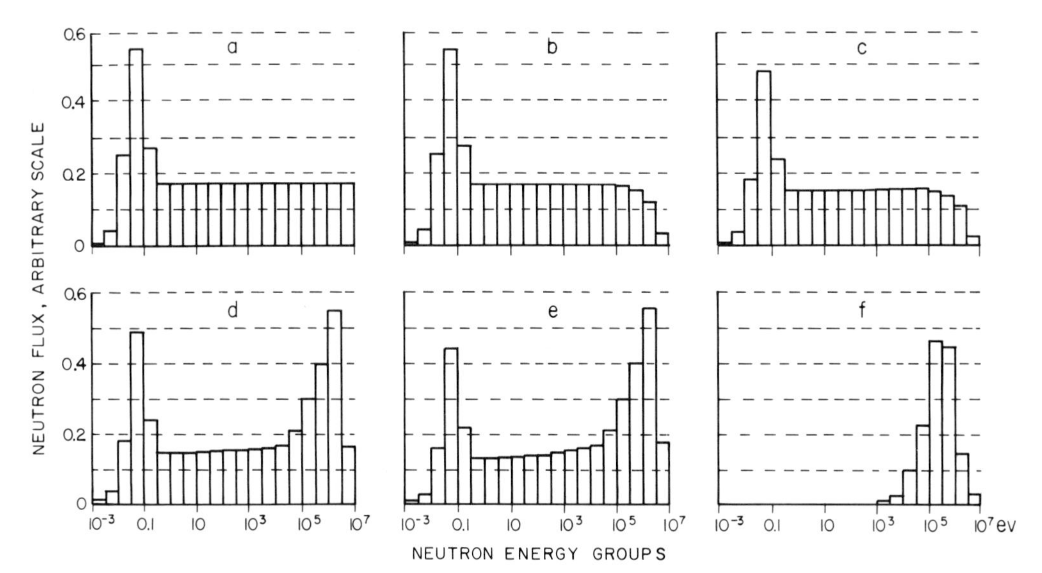

FIG. 2-2 Distribution of neutron energies within equal logarithmic energy groups for illustrative systems. The energy range from 10^{-3} to 10^7 ev has been divided into 20 energy groups, each group covering an energy range $\sqrt{10}$ larger than the range of the next lower group. Curves a, b, c, d, and e apply to thermal systems, while f is for a fast chain-reacting system. The neutron flux scales are arbitrary, but are constant and apply to the same fast neutron source strength, for all curves except f.

Curve a is for an idealized neutron source of 10 Mev neutrons, and is for a moderator like H_2O, except that the scattering cross section is held constant at 20 barns per hydrogen atom at energies up to 10 Mev. A 1/v-absorption cross section of 2.9 barns (at 2200 m/sec) per hydrogen atom is assumed to be present. In Curve b the source has been distributed in energy according to the fission spectrum of U^{235}. In Curve c, a leakage of 12% has been assumed. Most of the leakage occurs at high energies, but the fluxes in the thermal groups are reduced because of the resulting decrease of the slowing-down density. In Curve d, the effect of the dropoff in hydrogen scattering cross section at high energies is included (see Fig. 2-1). The leakage is still held at 12%. In Curve e, a 10% absorption by U^{238} resonances has been included.

Curve e is representative of the spectrum that might characterize an H_2O-moderated reactor having an absorption cross section of about 2.9 barns per hydrogen atom. In such a reactor, the fluxes in the highest energy groups might be somewhat lower because of the inelastic scattering by heavy materials, which has been neglected here.

Curve f is the calculated spectrum for a relatively large fast reactor fueled with plutonium, see reference [47]. It is the reactor described in the first column of Table 6-2.

nitude of the flux in each region varies approximately inversely with the amount of absorber present. Thus the neutron fluxes in the regions below about 0.1 volt may be much higher than those at higher energies or much lower, depending upon the general magnitude of the "thermal" absorption cross section.

In Fig. 2-2b the situation has been modified only by the assumption that the source neutrons have the fission energy spectrum. In Fig. 2-2c the additional assumption has been made that the system is not infinite, but is finite, and that leakage occurs both during slowing down and at thermal energies: both of these leakages reduce the magnitude of the flux in the regions below 1 ev.

Figure 2-2d illustrates the effect of the decrease of scattering cross section which occurs in most moderators at the higher energies, as indicated for hydrogen in Fig. 2-1. Figure 2-2e illustrates the effect of neutron absorption in resonances, such as those shown in Fig. 2-1 for U^{238}, in the energy range from about 10 to 1000 ev. This spectrum is roughly like that which would occur in a slightly-enriched, H_2O-moderated reactor.

In "unmoderated" or "fast" reactors no moderator is intentionally provided, but considerable slowing down of the neutrons nevertheless occurs by inelastic and elastic scattering in the materials present. Ordinarily no significant fraction of neutrons slows down to thermal energy in these reactors. The calculated spectrum for a large "unmoderated" reactor is given in Fig. 2-2f. "Intermediate" reactors exhibit spectra which lie between those of the "unmoderated" and the "thermal" reactors.

The breakdown of the neutron spectrum into a number of regions, as illustrated in Fig. 2-2 is just what is done in "multigroup" reactor calculations. The neutrons in each energy region are considered to constitute a "group." The energy regions (which need not be of equal logarithmic size) may be chosen of such size that properties can be averaged over the region with confidence. If many groups are used, the equations which express the reactivity become too complex for the intuitive understanding which is sought here. It is more useful to consider the implications of the energy variations illustrated in Figs. 2-1 and 2-2 for the very simple formulations such as Eq. (2-7).

Basically the major inadequacy of an equation like (2-7) is that the two sets of terms, in the two parentheses, are not, in general, independent. Thus, since both Σ_F and Σ_C vary strongly with energy, the ratio $\Sigma_F/(\Sigma_F + \Sigma_C)$ will depend upon how many neutrons are absorbed at each point in energy. But this energy distribution of absorptions will be affected by how much the flux has been depleted by leakage at each energy level; similarly the average leakage will depend upon the distribution of absorptions in energy. Hence, although rational values may exist in any given case for each quantity in the equation, it may be impossible to determine what these values are without going through a complex calulation which is itself tantamount to a calculation of k_{eff}. Nevertheless, the terms in (2-7) are usually independent to first order of accuracy, and inadequate as the equation may be in some cases for predicting k_{eff}, it can still be very useful for the recognition and qualitative evaluation of situations which produce important reactivity changes.

2.4 Thermal Reactors

In the special case of thermal reactors Eq. (2-7) may be modified to a quite useful form. Because of the rapid decrease of fission and (most) absorption cross sections with increasing neutron energy (Fig. 2-1), it is quite reasonable to assume, in thermal reactors (i.e., reactors in which the flux per unit logarithmic energy interval is substantially higher in the thermal region than in higher energy regions, as in Fig. 2-2e) that for the evaluation of neutron leakage all neutron absorptions may be considered to occur at thermal or near-thermal energies. Since the neutron leakage probability is nearly independent of neutron energy in the thermal region, the thermal energy spectrum is, conversely, nearly independent of leakage. Hence for these reactors the term $\bar{\nu}\bar{\Sigma}_F/(\bar{\Sigma}_F+\bar{\Sigma}_C)$ can be evaluated for an infinite system of the same composition as the finite reactor. This is a great simplification. For a truly "thermal" reactor—one in which all absorptions and fissions, except those involving thermal neutrons, were negligible—the expression for k_{eff} would be:

$$k_{eff} = \left(\frac{\bar{\nu}\,\bar{\Sigma}_F}{\bar{\Sigma}_F + \bar{\Sigma}_C}\right)_{thermal} (1 - \mathcal{L}) = k_\infty\,(1 - \mathcal{L})\,, \qquad (2\text{-}9)$$

where the symbol k_∞, called the infinite multiplication constant, is used to denote the quantity $\bar{\nu}\bar{\Sigma}_F/(\bar{\Sigma}_F+\bar{\Sigma}_C)$, evaluated for the infinite medium. The subscript "thermal" denotes that the averages are taken over only the thermal energy region.

In practice, few reactors are found in which all important absorption and fission events occur at thermal energies—only dilute reactors fueled with highly enriched fissile isotope approximate this condition. In the important case of a reactor containing a substantial quantity of fertile* isotope (U^{238} or thorium) a small, but usually significant, number of neutrons will be produced by the fission of fertile isotope by fast (Mev range) neutrons. Further, an important number of neutrons may be absorbed at epithermal energies by the fertile isotopes or by any other resonance absorbers which may be present. Usually, however, the importance of these processes is not strongly modified by neutron leakage effects, and hence it is still feasible to utilize the concept of the infinite multiplication constant (k_∞), even though the formulation of k_∞ must be modified from that indicated in Eq. (2-9).

In the usual formulation of k_∞ the "thermal" average indicated in Eq. (2-9) is retained, and the effects of fast fission and resonance absorption are included in the form of multipliers, thus:

*The term "fertile" is used to signify that the isotope can be converted to a fissile isotope by nuclear reactions involving the absorption of a neutron. It happens that the fertile isotopes are also fissionable by high energy neutrons.

$$k_\infty = \left(\frac{\bar{\nu}\,\bar{\Sigma}_F}{\bar{\Sigma}_F + \bar{\Sigma}_C}\right)_{\text{"thermal"}} \epsilon\, p\,. \qquad (2\text{-}10)$$

Here the quotation marks on the subscript, "thermal", signify that the averages are not confined strictly to the energy range of thermal equilibrium, but may be extended up to energies of the order 5 kT. The definitions of ϵ and p are:

ϵ = fast fission factor

$$= \frac{\text{"Total" number of fission-neutrons produced}}{\text{Number of neutrons produced by thermal fission}}$$

$$= 1 + \frac{\text{Net number of neutrons produced by fast fission}}{\text{Gross number of neutrons produced by thermal fission}}$$

p = resonance escape probability

= probability that a fast neutron, in the infinite system, will slow down into the thermal energy range without being captured by a resonance absorber. (2-11)

Note that these definitions give the following significance for k_∞:

$$k_\infty = \left[\frac{\text{Number of fission neutrons produced}}{\text{Number of neutrons absorbed at low energies}}\right]_{\text{in infinite system}}$$

or, more precisely:

$$k_\infty = \left[\frac{\text{Number of fission neutrons slowing down past the fast-fission threshold of the fertile isotope}}{\text{Number of neutrons absorbed at energies below the fast-fission threshold of the fertile isotope}}\right]_{\text{in infinite system}}$$

The equation:

$$k_{eff} = k_\infty\,(1 - \mathcal{L}) \qquad (2\text{-}12)$$

will then be consistent with Eq. (2-1) provided, in Eq. (2-12), the non-leakage fraction $(1 - \mathcal{L})$ is defined appropriately: as the fraction of those fission neutrons slowing down below the fission threshold of the fertile isotope which do not leak—i.e., which are absorbed in the finite reactor either at thermal or resonance energies.

In practice it is more convenient to identify neutron absorptions with materials (e.g., fuel, moderator) rather than with the processes of fission and capture; hence the first term of Eq. (2-12) is usually modified by the introduction of $\bar{\eta}$ defined as:

$$\bar{\eta} = \frac{\text{Number of fast neutrons produced from thermal fission}}{\text{Number of thermal neutrons captured in fuel}}$$

$$\bar{\eta} = \bar{\nu}\left[\frac{\bar{\Sigma}_F}{\bar{\Sigma}_F + \bar{\Sigma}_C}\right]_{\text{evaluated in fuel only}} \qquad (2\text{-}13)$$

and

$$\frac{\text{Number of fast neutrons produced from thermal fission}}{\text{Number of thermal neutrons captured in all materials}} =$$

$$\frac{\bar{\nu}\,\bar{\Sigma}_F}{\bar{\Sigma}_F + \bar{\Sigma}_C} = \frac{\bar{\nu}\,\bar{\Sigma}_F}{\bar{\Sigma}_F + \bar{\Sigma}_{Cf}} \cdot \frac{\bar{\Sigma}_F + \bar{\Sigma}_{Cf}}{\bar{\Sigma}_F + \bar{\Sigma}_{Cf} + \bar{\Sigma}_{Cr}} = \bar{\eta} \cdot f\,, \tag{2-14}$$

f = thermal utilization factor .

where $\bar{\Sigma}_{Cf}$ refers to the capture cross section of the fuel only and $\bar{\Sigma}_{Cr}$ refers to the capture cross section of all remaining materials, other than the fuel. Note that a variety of ways may be used to divide the reactor composition between "fuel" and other materials, as may be convenient. The fuel may be considered to be only the fissile isotope present, in which case $\bar{\eta}$ is a function of only the isotopic species and the neutron energy spectrum. Or the fuel may be considered to comprise the fissile isotope plus any fertile isotope present, plus perhaps other materials such as fuel cladding or fuel alloy components. In such a case $\bar{\eta}$ is a function of the composition of the "fuel", but may still be evaluated by Eq. (2-13).

Combining these concepts, it is customary to write k_∞ in terms of the four-factor formula:

$$k_\infty = \bar{\eta}\, f\, p\, \epsilon\,, \tag{2-15}$$

and the effective multiplication factor, in terms of the "one-group" picture is:

$$k_{eff} = k_\infty/(1 + M^2 B_g^2)\,. \tag{2-16}$$

In this discussion it has been possible only to outline conceptually the "four-factor" formulation. Many points of important detail emerge in more precise treatments. For example, there are hardly any practical reactors so well "thermalized" that the non-resonance absorptions and fissions occurring above the thermal energy range are truly negligible. What is done in practice is to extend the range of consideration up to as high an energy level as necessary to give an accurate value for $\bar{\eta}f$; this may be done by more or less "exact" calculations of the thermal and near-thermal energy spectrum (e.g., by the Wilkins or Wigner-Wilkins methods [40-44]) or by approximate treatments [45] which fit a "1/E tail" [cf Eq. (2-8)] onto a Maxwellian distribution of thermal neutron energies. It is sometimes necessary also to take into account fissions which occur in the resonance energy region. The fundamental assumptions which characterize the typical "thermal" reactor calculations, however, are that leakage can be neglected in determining the important features of the neutron energy spectrum (i.e., that the concept of k_∞ is a valid one) and that the leakage can be calculated as though all neutron absorptions occurred at thermal energies.

Considerable variations are possible in the details of definitions of the "four factors," within the framework of a consistent system. The various sets of definitions which have been used are discussed in reference [25].

The processes which enter into the determination of the four factors, and the effects of variations of these processes on the factors, and on reactivity, are discussed in Sec. 6.4.

2.5 One-Group Equation for Non-Thermal Reactors

Often it is useful to represent the neutron balance in a non-thermal reactor by a "one energy group" equation analagous* to Eq. (2-16):

$$k_{eff} = k/(1 + M^2 B_g^2)\,. \tag{2-17}$$

In this equation

k = number of neutrons produced per neutron absorbed in the finite reactor

and M^2 must be a value appropriate to the neutron spectrum in the finite reactor. Although the complexities of evaluating k and M^2 under these conditions may be so great as to render Eq. (2-17) relatively useless for the absolute calculation of k_{eff}, the equation is, nevertheless, often useful for approximating reactivity changes.

2.6 Spatial Averaging

In all the previous discussion average values have been used for the cross sections and for ν and η. This implies that appropriate averages are to be taken over both the neutron energy spectrum and the volume of the reactor. The spectrum averaging has been discussed above. Spatial averaging is necessary whenever the reactor composition is heterogeneous. If the heterogeneity is on a dimensional scale which is comparable to the reactor dimensions, only a very sophisticated type of averaging is appropriate (see discussion of neutron importance in Sec. 4.2.5); frequently no type of averaging is appropriate and the reactivity can be computed only by direct solution of the heterogeneous system. If the scale of the heterogeneity is, however, small relative to the reactor dimensions, and if the heterogeneity occurs in a uniform pattern over the reactor volume, it is necessary only to average cross sections in such a way as to yield the correct relative rates of nuclear interaction: i.e., one weights the cross section at a given point in the reactor by the neutron flux at that point.

The "regular" heterogeneity mentioned above occurs typically in solid-fuel reactors, where the fuel, moderator, and certain structural materials are usually arranged in a regularly recurring pattern, as illustrated for a simple case in Fig. 2-3. In such a case one may usually divide the reactor volume into a number of identical cells, compute the important flux distributions for a typical cell, on the assumption that the recurrent cell structure is infinite in extent, and compute the flux-weighted average cross sections. One may

---

*Note that, although the equations are the same in form their implications are quite different: there is no implication in Eq. (2-7) that either k or M^2 can be evaluated in an infinite system, although such an approximation may be usable in some specific cases.

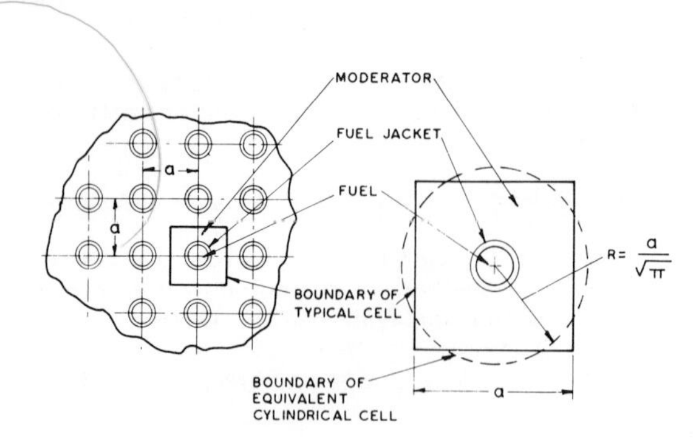

FIG. 2-3 Principle of representing a recurrent pattern of heterogeneity by a typical cell.

then use these average cross sections in computations of the actual reactor, as though the reactor were homogeneous in composition. Frequently the labor of computing the flux distributions in the cell can be greatly reduced by the approximation of replacing the actual cell by an approximately equivalent one of higher symmetry. The straightforward way of making such symmetry transformations is to preserve the volumes of all materials in the cell, but sometimes other transformations are more appropriate. These concepts are illustrated in Fig. 2-3.

The relative rates of nuclear interactions of a particular kind can be stated for any two materials, A and B, in terms of the volumes of the materials (V), the macroscopic cross sections for the particular interaction (Σ) in the two materials, and the average neutron fluxes (in the important energy range) in the two materials:

$$\frac{\text{Interaction rate in material A}}{\text{Interaction rate in material B}} = \frac{\bar{\phi}_A}{\bar{\phi}_B}\,\frac{V_A}{V_B}\,\frac{\Sigma_A}{\Sigma_B}. \qquad (2\text{-}18)$$

The flux ratio $\bar{\phi}_A / \bar{\phi}_B$ is usually called the disadvantage factor of material B relative to material A. However, terminology varies, and the ratio intended should be specifically identified when it is important.

2.7 Leakage Characteristics

Insofar as Eq. (2-4) specifies the fractional neutron leakage adequately, the effect of the reactor material properties on leakage can be specified by relating those properties to M^2. In a thermal reactor, the distance a neutron travels between birth and death is made up of two parts: the distance traveled as an epithermal neutron and the distance traveled as a thermal neutron. Since the directions of these two journeys are uncoordinated for any given neutron, the average distances covered do not add linearly to determine the total average distance travelled between birth and death. Rather, the mean square distances are additive. That is to say the migration area M^2 may be considered to be made up of two additive components:

$$M^2 = L^2 + L_f^2 . \qquad (2\text{-}19)$$

L^2, the thermal diffusion area, is $1/6\ \overline{r_{th}^2}$, where $\overline{r_{th}^2}$ is the mean square (crow-flight) distance traveled by a neutron between its point of thermalization and its point of absorption. L_f^2, the slowing down area, is $1/6\ \overline{r_f^2}$, where $\overline{r_f^2}$ is the mean square (crow-flight) distance traveled by a neutron between its point of birth and the point at which it becomes thermalized.

The thermal diffusion area is given by:

$$L^2 = \frac{D}{\bar{\Sigma}_a} \doteq \frac{1}{3\bar{\Sigma}_a\,\bar{\Sigma}_s\left(1-\frac{2}{3A}\right)}, \qquad (2\text{-}20)$$

where D_s is the thermal diffusion coefficient, A is the effective mass number of the scattering nucleus, and the absorption and scattering cross sections, $\bar{\Sigma}_a$ and $\bar{\Sigma}_s$, are averaged over the thermal-neutron spectrum.

The expression for L_f^2 differs somewhat among classes of moderators, but the Fermi age, τ, is a reasonable approximation in most cases. The value of τ for slowing down from energy E_0 to thermal (E_{th}) is given by:

$$\tau\,(E_o, E_{th}) = \int_{E_{th}}^{E_o} \frac{D\,(E)}{\xi\,\Sigma_s\,(E)}\,\frac{dE}{E}, \qquad (2\text{-}21)$$

where ξ is the average logarithmic energy change per scattering collision, Σ_s (E) is the macroscopic scattering cross section, and D (E) is the diffusion coefficient at energy E.

$$D\,(E) \doteq \frac{1}{3\cdot\left(1-\frac{2}{3A}\right)\Sigma_s\,(E)}. \qquad (2\text{-}22)$$

For the slowing down of fission neutrons, τ $(E_{fission}, E_{th})$ is approximately the weighted average of τ (E, E_{th}) over the fission spectrum. The value of M^2 is then given by:

$$M^2 = L^2 + \tau(E_{fission}, E_{th}). \qquad (2\text{-}23)$$

When the leakage is high Eqs. (2-4) and (2-16) may be rather poor approximations, for the leakage is not strictly expressible in terms of the first power of the buckling (B^2) alone. More accurate expressions for k_{eff} are the Fermi equation:

$$k_{eff} = k_\infty \Big/ (1 + L^2 B_g^2)\, e^{\tau(E_{fission}, E_{th}) B_g^2} \qquad (2\text{-}24)$$

or diffusion-group equations employing more than one energy group. The two-group equation is:

$$k_{eff} = k_\infty \Big/ (1 + L^2 B_g^2)(1 + L_f^2 B_g^2). \qquad (2\text{-}25)$$

2.8 Complex Reactors

It has been impractical here to speak of anything but the simplest reactors. These have been reactors which consist of a single, bare, multiplying medium, which is either homogeneous in composition or can be approximated by a medium of homogeneous composition with suitably averaged characteristics.

In critical reactors of this kind the spatial distributions of neutron fluxes in any energy range will be given by the solutions of Eq. (2-5), with simple zero-flux boundary conditions at the extrapolated boundaries of the medium. For example, the flux distribution in a critical cylindrical medium of extrapolated radius R and extrapolated length L is

$$\phi\ (r, z) = \phi_0 J_0(2.405 \frac{r}{R}) \cos (\frac{\pi z}{2L}), \quad (2\text{-}26)$$

where ϕ_0 is the flux at the origin of coordinates, which is taken at the center of the cylinder and J_0 (2.405 r/R) is the zero-order Bessel function of argument 2.405 r/R.

If a reflector is added, the reactor is no longer simple, but in many cases the neutron fluxes will follow the fundamental mode distributions [e.g., Eq. (2-26)] in the core, except near the core-reflector boundaries. In such cases the reactor will behave, in many important respects, much like a bare reactor of such a size as to have the same fundamental mode distribution, even though the energy spectrum of the neutrons may be highly distorted near the core-reflector boundary. When this is true, the difference in size, between the real core and a fictitious bare core with the same fundamental mode flux distributions, is called the reflector saving. This concept is illustrated in Fig. 2-4 [46].

Other complexities are usually harder to handle than the reflector effects. Large-scale spatial variations in reactor core properties are quite common in practical reactors. They may be caused by such things as the presence of strong control rods, or intentional nonuniform fuel loading in the

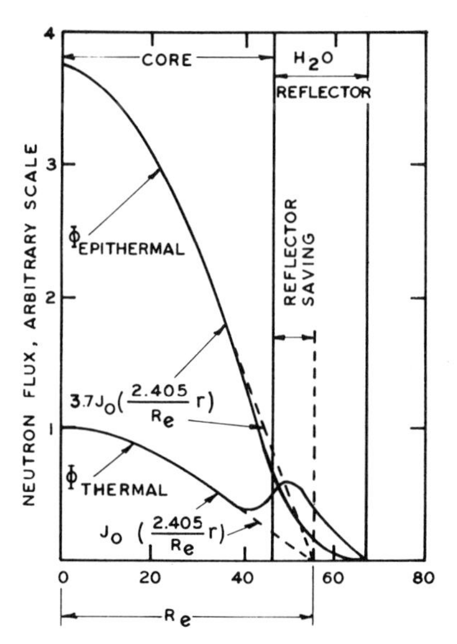

FIG. 2-4 Concept of reflector saving. This figure shows the radial distributions of epithermal and thermal neutron fluxes in a cylindrical reactor as calculated by the "two-group" approximation. The highly-enriched reactor utilizes zirconium as structural material, and is moderated and reflected by H_2O. In this particular case the ratio of epithermal to thermal neutron flux at the center of the core happens to be 3.7. Note that both epithermal and thermal fluxes are proportional to $J_0(2.405r/R_e)$ in the core, except near the core-reflector boundary. R_e is the "equivalent bare radius" or "extrapolated" radius of the core. The radial reflector saving is R_e - R, where R is the actual core radius.

core to flatten the power distribution, or nonuniform changes in fuel composition with reactor operation. Whenever such strong inhomogeneities exist, the simple concepts and equations outlined here will be inadequate for quantitative purposes. Yet they will seldom be wrong in their qualitative indications if they are applied with understanding. So applied, the very simple concepts of reactor physics can go a long way toward providing the intuitive understanding of reactor behavior which is essential to safe reactor design.

3 TYPES OF REACTOR ACCIDENTS AND RELATION TO CORE DESIGN

It is necessary first to define the term "reactor accident" as used here. Attention is restricted to those hazards which are uniquely associated with nuclear plants, and which do not come within the scope of routine health-physics control. Essentially this means hazards related to large releases of radioactivity. It is customary to recognize the existence in any reactor installation of a series of barriers against the possibility of any such release. For many installations this series of barriers consists of the fuel element jacket, the reactor vessel, the reactor containment shell or building, and, if a large exclusion area is used, the isolation provided by the exclusion area. In this chapter attention will be confined to those barriers which are integral parts of the reactor, and a reactor accident will be defined as any unintentional event which reduces the integrity of one or more of those barriers below the level allowed for in the design and the normal operating plans.

As so defined, a reactor accident does not necessarily involve any serious increase in the immediate hazard; however, an accident is likely to have resulted from or in a malfunction which needs correction before any further extensive operation of the reactor, to assure its integrity and to restore the original margin of safety.

The question of fuel element failures deserves some comment in relation to this definition. Any failure of a fuel element represents a failure of one of the barriers against fission product release; yet it is not reasonable in most cases to consider the failure of a single fuel element, or even the failure of several elements at intervals of time, as a reactor accident. Here it is necessary to recognize that the sum total of the fuel element jackets often represents a barrier which is slightly leaky, and that limited leakage is sometimes accepted as permissible. In many reactors, some fission product leakage into the coolant can be tolerated without a serious increase in hazard, and such minimal leakage is provided for in the design of the reactor system, particularly in the waste handling and disposal systems. When this is the case, one considers as an accident only fuel element failures grossly in excess of the normally expected failures, or substantial numbers of failures from causes other than those normally foreseen.

Although some fuel element failures may be tolerated by design, the importance of monitoring fuel element failures and fission product content of the coolant should not be underestimated. Fuel

Table 3-1

Types of Primary Accidents

General Type	Example	Possible Effects	Representative Causes
Reactivity Excursions	Uncontrolled control rod withdrawal	General core damage Nuclear energy burst	"Sticking" relays or switching circuits Failure of rod position indicators Operator error
	Too-fast control rod withdrawal (usually from a very low power level)	General core damage	Accidental reversal of scram motors Inadequate instrumentation at very low power Operator error
	Mechanical failure of control rod	Local to general core damage	Dropping of part of failed rod out of core Ejection of rod by unbalanced reactor pressure
	"Cold water" accident	Local to general core damage	Sudden initiation of flow in one of several parallel coolant loops
	Other coolant-borne reactivity	Local to general core damage	Nonuniform distribution of dissolved nuclear poisons "Hide-out" of fuel in circulating fuel reactors
	Loss of coolant (in reactors which have positive coolant void coefficients of reactivity)	Local to general core damage	Failure of coolant piping Nonscheduled boiling of coolant
Loss of Control	Control system malfunction or instability	Local to general core damage	Failure of components of automatic control system
	Operator error	Local to general core damage	Personal errors Inadequate instrumentation Inadequate knowledge of control positions Complexity of control job
	Loss of shutdown capability	Inability to shut reactor down if necessary for other reasons such as coolant failure Possibility of unexpected restart of shutdown reactor	Stuck rods Reactivity gains from Pu buildup or from burnout of burnable poison Core deterioration
	Stuck rods	Probably limited to local core damage unless aggravated by secondary effects or unless stuck rod has unusually high reactivity worth	Deformation, corrosion, or crudding of control rods or control rod channels Failure of control rod drive

to Operating Reactors

Primary Prevention By:	General Protection By:	Built-in Limitation	Relation to Core Design
Fail-safe control rod actuation circuits and position indicators "Dead-man" switches on control rod actuators	Period scram (if malfunction begins at low power level) Overpower scram (if malfunction begins at operating power)	Limits on rod speed and rod ganging	Independent means of scram Inherent limit on speed of rod withdrawal Inherent limitation on number of rods withdrawn simultaneously Prediction of control rod worths
Inherently nonreversible scram Adequate multiple instrumentation Adequate neutron source	Period scram Overpower scram	Limits on rod speed and rod ganging	Adequate source strength Adequate low level instrumentation Proper locations of source and instruments Limitations on rate of reactivity addition Independent scram capability Subdivision of control units
Fail-safe rod design Avoidance of large individual rod worths	Overpower scram Period scram	Fail-safe features Limitation on maximum reactivity that can be added by a single mechanical failure	Adequate materials for rods and linkages Choice of general type and orientation or rod installation Choice of number of control rods Determine requirements on safety factors and scram capability
Interlocks on pumps and/or valves of parallel coolant loops	Overpower scram Period scram	Limitations on coolant pumping rate	Adequate prediction of reactivity coefficients Adequate safety factor on thermal design Adequate scram capability
Appropriate design of the fluid system	Overpower scram Period scram	Limitations on possible quantities of coolant-borne materials	Search for possibilities of coolant-borne reactivity changes Adequate prediction of reactivity effects of dissolved poisons Adequate specifications for poison injection and removal systems Adequate scram capability
Safety factors in coolant system design	Overpower scram Coolant pressure scram Period scram	Limit on reactivity worth of coolant Pipe-size limits on rate of loss of coolant	Prediction of reactivity worth of coolant Avoidance of unwanted boiling of coolant Adequate scram capability
Fail-safe design of control system	Overpower scram Coolant temperature scram	Limitations on speed and/or worth of control rods in automatic system Limitation of total reactivity available	Prediction of reactivity coefficients Prediction of control rod worths Adequate design of control rod installation and drives
Adequate instrumentation Avoidance of inherent instability in reactor design	Overpower scram Coolant temperature scram	Limitations on speed of control rods Limitation of total reactivity available (e.g., Fermi)	Prediction of reactivity coefficients Prediction of control rod worths Accuracy of control rod indication Accuracy of reactivity-adjusting mechanism (back-lash, etc.)
Adequate design of reactivity control system	Independent emergency shutdown system Continuous monitoring of neutron levels Continuous cognizance of shut-down capability	Limitation on possible reactivity gain through reactor lifetime	Provide adequate shutdown capability Design to avoid rod sticking Design to prevent deterioration of reactivity-compensating elements Adequate prediction of long-term reactivity gains Adequate monitoring of shutdown margin Provide auxiliary shutdown system when required
Provision of spare shutdown margin	Overpower scram Independent emergency shut-down system	Limitations on worth of individual rods	Design to prevent sticking Provide adequate "stuck-rod" margin Base core design on a definite "stuck-rod" policy Provide information to assess significance of stuck rods

Table 3-1

Types of Primary Accidents

General Type	Example	Possible Effects	Representative Causes
Over-Power Operation	Control system malfunction	Local damage to or burnout of fuel elements	Component failures in automatic systems Control rod sticking
	Operator error	Local damage to or burnout of fuel elements	Personal errors Complexity of control job
	Instrument error	Local to general burnout of fuel elements and core damage	Instrument saturation Change of instrument calibration
Power-Coolant Mismatch - General	Control system malfunction	Local fuel element damage or burnout Unscheduled boiling of coolant	Component failure or drifts in automatic systems
	Operator error	Local fuel element damage or burnout Unscheduled boiling of coolant	Personal errors Complexity of control job
	Instrument error	Local fuel element damage or burnout Unscheduled boiling of coolant	Change of instrument calibration Operation with inadequate instrumentation as a result of past instrument failures
Power-Coolant Mismatch - Local	Hot spots; hot channels (including orificing)	Local fuel element damage or burnout Unscheduled boiling of coolant	Inadequate prediction of hot-spot and hot-channel factors Changes in power distribution as operation proceeds Changes in material properties with exposure
	Spatial oscillations of power	Local fuel element damage or burnout Unscheduled boiling of coolant	Effect of xenon in large, high-flux reactors
Coolant Failure - General	Circulation failure	Local to general burnout or core damage	Pump failure
	Loss of coolant	Local to general burnout or core damage Release of coolant radioactivity to reactor building	Large leak or rupture in primary coolant system
	Reactor vessel failure	General core damage Release of stored coolant energy	Flaws in vessel Thermal shock Radiation damage

to Operating Reactors (continued)

Primary Prevention By:	General Protection By:	Built-in Limitation	Relation to Core Design
Adequate design of control system; fail-safe features	Over-power scram Coolant-temperature scram	Varies with specific control systems	Design to prevent control rod sticking Provide adequate combination of over-power safety factor and scram capability
Adequate design of control rod system; simplification of operator's job	Over-power scram Coolant-temperature scram	Only those provided by scram systems	Provide adequate combination of over-power safety factor and scram capability Specify adequately control rod worths Simplify control operations
Multiple instruments, having staggered saturation ranges Automatic recalibration of nuclear power level instruments against thermal quantities	Over-power scram on separate instruments Coolant-temperature scram	Only those provided by scram systems	Specification of adequate instrument locations
Adequate design of control system Fail-safe features	Coolant-temperature scram or alarm Independent backup surveillance systems	Ordinarily none	Determination of safety factors for thermal design over range of coolant flows Possible installation of in-core instrumentation
Adequate design of control rod system Simplification of operator's job	Coolant-temperature scram or alarm Independent backup surveillance systems	Ordinarily none	Determination of safety factors for thermal design over range of coolant flows Possible installation of in-core instrumentation
Automatic recalibration of instruments Adequate initial installation	Coolant-temperature scram or alarm Independent backup surveillance systems	Ordinarily none	Determination of safety factors for thermal design over range of coolant flows Possible installation of in-core instrumentation
Adequate thermal design of core, taking into account changes with time (e.g., shim removal)	In-core instrumentation	Ordinarily none	Determination of power distribution and hot-spot and hot-channel factors Determination of need for in-core instrumentation, and design of installation
In-core instrumentation and appropriate control rod installation for those reactors in which oscillations can arise	In-core instrumentation	Maximum possible amplitude of oscillation, as determined by core geometry and reactivity coefficients	Predict occurrence of oscillations Design appropriate control system Design appropriate in-core instrumentation installation
Multiple pumps and/or coolant loops Emergency circulating equipment Low-flow and coolant temperature scrams		Normal coast-down characteristic of pumps and coolant loops Natural circulation capabilities	Factor loss-of-flow possibility into core thermal design Predict magnitude, distribution, and rate of decay of shutdown heat
Emergency cooling system (e.g., sprays; use of containment building as in GCR) Multiple coolant loops with isolation valves Loss-of-pressure scram Containment building; vapor quenching		Cooling provided by natural circulation after system has depressurized itself	Design core for cooling by emergency system and, if possible, for natural-circulation cooling Predict magnitude, distribution, and rate of decay of shutdown heat
Extreme care in design and fabrication of reactor vessel	Containment Building Vapor quenching Cooling sprays	Limited stored energy in coolant system	Thermal shielding of reactor vessel Shielding of vessel against radiation damage Provision for vessel surveillance

Table 3-1

Types of Primary Accidents

General Type	Example	Possible Effects	Representative Causes
Coolant Failure - Local	Starving of parallel-flow channels	Local fuel element damage or burnout	"Boiling disease"
	Deterioration of heat transfer effectiveness	Local fuel element damage or burnout	Crud deposition on heat transfer surfaces
	Local coolant obstructions	Local fuel element damage or burnout Unscheduled boiling of coolant	Plugging of coolant passages by crud deposits
Non-Nuclear Energy Release in Core	Release of stored energy by irradiated core materials	General overheating of core, with possible secondary fires	Wigner energy release in graphite-moderated reactors
	Chemical energy release	Rapid energy release Ignition	Buring of core or coolant in air (Na, organic) Direct chemical reaction of core materials Explosion of hydrogen-oxygen mixtures formed by radiolysis or electrolysis

element failures or fission product leaks may be important indicators of conditions in the core. The situation should be kept under constant surveillance and any sudden increase in the failure rate or in the fission product content of the coolant requires immediate and serious attention, even if permissible levels are not exceeded.

Reactor safety involves two considerations: the characteristics of the (as-built) plant, and the operating procedures. These considerations are complementary and interrelated. To some extent deficiences in one can be compensated by appropriate attention to the other, but careful and coordinated attention to both is of the highest importance. The reactor designer is responsible for producing a safe design and for developing the information needed to establish safe operating procedures. Since this chapter is a discussion of reactor core design, both of these activities of the designer are involved. The subject of operating procedures per se will not be considered, but the intent is to give a sufficiently basic understanding of the sources of hazards to indicate the safety relationships between design and operating procedures.

The designer's efforts are most effective in assuring the safety of the reactor when it is in its normal operating condition. In this condition, all the restrictions which the designer has built in to prevent hazardous conditions are presumably in effect, and most of the accidents which can be hypothesized under such conditions involve failures of components or equipment, or design errors. When the reactor is shut down for maintenance, refueling, or any operation which requires the partial disassembly of the system or the disconnecting of circuits, some of the restrictions which the designer has built in must necessarily be bypassed. Under these conditions, human error can play a much larger part in the range of to Operating Reactors (concluded)

Primary Prevention By:	General Protection By:	Built-in Limitation	Relation to Core Design
Adequate hydraulic design Local orificing	Possibly by in-core instrumentation	Limited number of coolant channels usually affected	Adequate thermal and hydraulic design Reactivity effects of localized boiling and/or fuel element distortion
Design testing Safety factor for surface fouling	-----	Usually occurs gradually and can therefore be recognized before becoming dangerous	Appropriate choice of materials, coolant quality, and hydraulic design
Design testing Surveillance by inspection and observation of pressure drop	In-core flow instrumentation	Fuel failures will usually occur individually, so investigation can be made before failure is general	Appropriate choice of materials, coolant quality, and hydraulic design
Minimization of energy storage by appropriate operating temperature Carefully controlled periodic annealing	-----	Limited amount of stored energy	Special consideration of these effects in core design
Avoidance of incompatible Materials Fire-control equipment Safe disposal of dissociation products	-----	Limited amount of material available for chemical reaction	Proper choice of core materials and coolants Prediction of radiolysis rates

possible accidents. Further, the very fact that the reactor is in the shut down condition may change appreciably its response to accident producing phenomena. Consequently, in the following discussion of possible accidents, these two conditions of the reactor are treated separately.

Finally, regardless of initial cause, the course of an accident will be very strongly affected by the characteristics of the reactor and its core. Because of this, Sec. 3.3 includes a discussion of the major factors which may come into play after the initiation of an accident and which may determine the course of the accident and its degree of seriousness from the point of view of safety.

3.1 Accidents to Operating Reactors

In this section the reactor is assumed to be in its "operating condition", that is, assembled with all circuits connected as specified in the design. The "operating condition" may include zero power, subcriticality, or even temporary shutdown.

In solid fuel reactors, most of the core-related accidents which may be visualized involve the overheating of the fuel and/or the fuel element. Overheating may result from excessive power generation in the fuel element, or from coolant failure, or from a mismatch between power generation and coolant flow. In addition to such accidents, there are other possibilities which involve failures of the reactor structure because of conditions imposed by the core design. Conceivable examples are the brittle failure of the reactor vessel due to radiation damage, and particularly in aqueous homogeneous reactors, vessel failures brought about by the chemical composition and the flow pattern of the fuel solution. The classes of reactor accidents related to core design are tabulated in Table 3-1 and typical examples are given. In each case, an attempt is made to specify the relationship to core design. Actual accident experiences are discussed in the chapter on Accidents. Further details and discussion of most of the entries in Table 3-1 are given in other chapters. (See Subject Index at end of this volume).

An explanation is in order relative to the treatment of control elements in the following. In almost all cases the discussion is in terms of neutron-absorbing control rods. The disadvantage of this lack of generality is compensated for by the advantages of more specific treatment. The reader should have little difficulty in extending the implications to any other type of control element which may interest him or to other movable parts, such as neutron-absorbing test pieces in a test reactor, which may be capable of changing reactivity even though not intended as control elements.

3.1.1 Reactivity Excursions

Since, in general, the power level of the reactor is determined by the amount of reactivity added and the schedule of reactivity addition, all over-power accidents are, in essence, reactivity accidents. Nevertheless, it has seemed convenient to subdivide such accidents, and the first subdivision in this rather arbitrary classification has been called Reactivity Excursions. This is meant to imply rather gross additions of reactivity resulting in rapid, usually nonoscillatory, increases of reactor power.

The conceivable effects of reactivity excursion accidents can vary widely from one reactor type to another, and from one reactor design to another with a given type. Some reactors can be shown to be relatively immune to them. However, as a class, they include some of the most severe accidents that can be visualized. If the normal protection circuits on a reactor fail to operate or are ineffectual, a reactivity excursion can result in local to general core damage, and may even release sufficient nuclear energy to produce a low level explosive effect. Consequently the characteristics given in Table 3-1 under Possible Effects should be considered only as approximate relative ratings within the class: The effects in any of the specific examples cited may vary over a wide range from reactor to reactor.

3.1.1.1 Uncontrolled Control Rod Withdrawal. This accident is frequently analyzed in hazard reports, not because it is considered a probable one, but because it is easily defined and, if the initial conditions are properly chosen, defines the upper limit of the possible consequences of control rod malfunction or maloperation. It may be visualized as the continuous withdrawal of control rods at their maximum attainable speed without regard for any warnings or stop signals which may be given by the reactor instrumentation. Conceptually, one can think of some malfunction of the circuitry to the control rod drives which would prevent them from being turned off once they were set in motion. Or one might visualize a continuous manual withdrawal by the operator, either for irrational reasons or because of failure of the instrumentation provided to indicate control rod position and reactor power level. The obvious means of preventing such an accident are suggested by the postulated causes. The presence in the control rod circuitry of relays which could conceivably stick in the position to produce control rod withdrawal should be avoided and, in general, the principles of fail-safe design should be applied to ensure that any conceivable failure in the control rod mechanism or circuitry will result either in insertion of the rods, or, at worst, in cessation of their motion. Similarly, the rod position indicators should provide that failure, if it occurs, will be obvious, and false readings will not be given. Manual switches governing control rod actuation should be of the momentary-contact, spring-return type, which must be held in the closed position to cause continuous withdrawal of control rods.

If the reactor in question has available any substantial amount of excess reactivity, it is unlikely that an uncontrolled rod withdrawal, to the total capability of the rods, could be tolerated without ultimate damage to the reactor, even though favorable (negative) power coefficients of reactivity might postpone the onset of damaging power levels. Hence, in most cases, the ultimate protection principle must be to provide a scram system, either on the control rods themselves, or on some other agency, or both, of such a nature that it is inherently independent of any malfunction which could conceivably have caused the uncontrolled rod withdrawal. Ordinarily, the action of such a scram system would be triggered either by the period circuits or by the over-power circuits, and the instruments and circuits providing these signals must of course be suitably reliable. The scram system must also be capable of stopping the power rise before the reactor is damaged. Usually this consideration is factored into the reactor design in the form of an upper limit which is placed on the the speed of rod withdrawal. To be meaningful, such a limit on rod speed must include provisions for prohibiting the possibility of multiple rod withdrawal which could increase the rate of reactivity addition above that nominally set as the limit.

In relation to the hypothesized accident of uncontrolled rod withdrawal, the considerations within the realm of core design (as distinguished from mechanism design, control system design, and instrumentation) are then the following: (1) provide a means of reactor scram independent of the normal control rod positioning system; (2) provide an inherent limitation on the speed of rod withdrawal to keep the possible rate of withdrawal within the capability of the scram system; (3) provide estimates of control rod worth which will make possible the design objectives of (1) and (2) above; (4) provide inherent protection against the simultaneous withdrawal of control rods in groups larger than those provided for in items (1) and (2) above. When it is possible, by suitable design, to provide inherent shutdown mechanisms or strong negative power coefficients of reactivity, these may be utilized to reduce the requirements of the scram system and to reduce any uncertainties as to its adequacy which may exist.

3.1.1.2 Too Fast Control Rod Withdrawal. A classic (conceptual) example of control rod withdrawal which is performed at too rapid a rate is the "startup accident". The hypothesis for this accident is usually that in bringing the reactor to criticality the reactor operator (or an automatic control system) is not fully informed of the multiplication level, through inadequacies in the neutron source and instrumentation; the operator then withdraws control rods at a rate which renders the reactor substantially supercritical before recognizing the true condition of the reactor. If the error were sufficiently great, the period of exponential power increase might be sufficiently short to cause a dangerous power rise before normal corrective action could be taken. The method of avoiding this type of accident is to provide a suitable combination of source strength, source location, and instrumentation to yield an adequate indication of the multiplication level before criticality is reached. In meeting this requirement, attention must be given to the problem of detecting low neutron fluxes against the strong background of gamma radiation produced by the decay of fission products. For ultimate protection against the accident, one usually depends upon a scram system whose action is designed to be independent of any conceivable startup accident. Quite obviously, if this scram system is adequate for protection against an uncontrolled rod withdrawal, it will also provide adequate protection against the too rapid withdrawal of rods by their normal operating mechanisms.

Another cause of too-rapid rod withdrawal may be malfunction of the rod operating mechanism. If, for example, the drive is by electric motors, then the possibility of the motor speed being increased, either through some kind of electrical failure or through improper connection of motor leads, must be examined. Particularly, if the normal rod actuating device is used also to provide rapid control rod insertion, either for scram or for rapid power cutback, then special attention must be given to the design of the reversing circuitry to ensure that no internal failures or connection errors could result in rapid withdrawal rather than rapid insertion of control rods. The only adequate protection against an accident of this kind is to eliminate the possibilities, for it is usually impossible to provide an independent scram system which can protect the reactor adequately against another scram system which, through malfunction, scrams in reverse.

The core design principles most strongly involved in the prevention of too rapid control rod withdrawal are: (1) to provide adequate neutron source strength for detection at subcritical levels; (2) to provide adequate instrumentation in the source range; (3) to locate neutron source and instrumentation so as to give an adequate indication of subcritical multiplication; (4) to ensure that limitations on control rod withdrawal speed cannot be violated through equipment failure or maloperation; (5) to provide adequate independent scram capability; (6) to consider the possibility of reducing the chances of over-speed rod withdrawal by subdivision of the units of rod withdrawal.

Although not a part of core design, adequate operating and maintenance procedures are also essential.

3.1.1.3 Mechanical Failure of Control Rod or Control Rod System. Fail-safe mechanical design is an important consideration in limiting the amount of reactivity that can be added and the speed with which it can be added when a control rod fails mechanically or when the linkage connecting rod to drive system faults. The guiding principle of course is to design so that the natural unbalanced forces acting on the control rod will result in a motion—if any—which will cause a decrease of reactivity. If such a design is not possible, appropriate design principles can be used at least to limit the amount of reactivity which can be added. For example, in a common type of control rod installation, the rods are supported from the top and hang in the reactor core. If a portion of the rod should break off and fall downward, the result might be either an increase or a decrease in reactivity, depending on where the break occurred and on the degree of withdrawal of the rod at the time of the failure. The amount of reactivity added cannot be large, however, if the total worth of the control rod is not large (i.e., if a large number of rods is used to control the total excess reactivity), and if a stop is provided to prevent the disengaged portion of the rod from falling through the core and out the bottom. The possibility of a reactivity increase by a failure of this kind can be eliminated if the control rods are driven from the bottom of the core and are pushed upward out of the core to increase reactivity. On the other hand, bottom-driven rods which must be pushed upward into the core to decrease reactivity may be sources of substantial reactivity additions if they are vulnerable to failures of any kind which allow them to drop downward out of the core.

In addition to the force of gravity, the forces which may be exerted by internal reactor pressure must be considered in control rod installations in reactors which are normally pressurized or which might be pressurized as the result of an accident. If, for example, top-mounted rods have extensions passing through bearings or seals in the pressure vessel head, then the designer must ensure that no failure of the linkage can allow the unbalanced reactor pressure to push the control rod out of the core through piston action on the rod and/or extension. Particular attention should be paid to the problem of assuring that multiple failures or actions of any sort could not drive out several rods at once. A similar consideration holds for bottom-mounted drives which pass through the pressure vessel if downward motion increases reactivity.

For those possibilities of reactivity addition by control rod failure which cannot be eliminated through proper design of the rod and its linkages, general protection must be provided either by the general scram system or by safety factors which will allow the reactor to tolerate the reactivity increase. Needless to say, the latter requirement must be taken into account in designing the scram system.

Finally, attention must be given to the possibility of accidents caused by errors in the indicated positions of control rods. A weakness of most present designs is that the indicating devices show the position of the control rod drive rather than the position of the control rod itself.

The most important core design considerations relevant to the possibility of control rod failures are: (1) to minimize the probabilities of failure by proper choice of materials for the control rod and control rod linkages; (2) to eliminate or limit the possible reactivity addition by appropriate selection and design of the control rod installation; (3) to consider reducing the difficulties by the use of an adequately large number of independently operated control rods; (4) to make an adequate determination of the possible consequences of control rod failures so as to establish proper specifications of over-power safety factors and scram capabilities; (5) to provide adequate and reliable indications of the positions of all control rods.

3.1.1.4 The Cold Water Accident. The most obvious possibility for the cold water accident occurs in the pressurized-water reactor, but it should be recognized as a member of a general class of "cold coolant accidents" which can be visualized in any reactor having a strong negative value of the coolant temperature coefficient of reactivity. Most pressurized water reactors do have a quite strong negative coefficient at operating temperature. Usually the accident is postulated for such reactors when they are provided with more than one coolant loop. If one of the loops were inoperative during reactor operation and consequently contained water which was cooler than normal for the operating condition, the subsequent startup of the loop would inject cold water into the reactor vessel at a rate governed by the pump capacity and cause a rapid increase of reactivity. This would cause a power increase unless counteracted by appropriate motion of the control rods; usually the required speed of control rod motion, for compensation of the reactivity increase, is greater than the design capability of the control system in normal (non-scram) operation. If the control system took no action at all, the cold water injection would result in a temporary power increase, which in many cases would exceed the safe limits of the core. Eventually, as the cold water heated up to normal operating temperature, the power level would drop back to a value equal to the dissipation capability of all the coolant loops then in operation. The possibilities of less severe cold water accidents include the temporary interruption of pumping in one of the primary loops and the reflection into the primary coolant system of temperature changes in the secondary system.

The sudden introduction of cold coolant can also produce large and rapid reactivity changes in boiling water reactors. In this case, the main component of the increase is due to the reduction of the steam void content of the core rather than to the change in moderator temperature, although the latter may also come into play if the cold water addition is large and sustained. Natural circulation boiling reactors which incorporate the circulation loop within the reactor vessel are usually susceptible only to relatively slow cold water additions, but forced circulation systems or natural circulation systems with multiple external loops may be subject to cold water injections analogous to those mentioned for the pressurized-water case.

The potential severity of a cold water accident varies with the amount of water involved, its temperature relative to the operating temperature of the reactor, the rate at which it may flow into the reactor, the temperature coefficient of reactivity, and the degree to which other inherent negative components of the power coefficient of reactivity may be expected to come into play. In many water-cooled reactors the potential consequences of a cold water accident are serious; and the possibility of a "cold coolant accident" should be investigated for any reactor which has an appreciable coolant temperature coefficient of reactivity. The direct means for preventing such an accident are through interlocks and procedures which prevent the sudden introduction of cold water, or through design features such as slow opening valves, limiting the rate at which loop circulation can be built up. General protection can be provided only through the scram system, coupled with an adequate safety factor on the power-removal capabilities of the reactor.

The main core design considerations relevant to the cold coolant accident are: (1) to predict adequately the reactivity effect of cold coolant introduction and the resulting increase in power level; (2) to give adequate attention to the over-power safety factor of the reactor and to integrate this information with the scram design to provide adequate general protection; (3) possibly to introduce instrumentation in the core or in the primary coolant system which will give early warning of cold coolant and/or initiate scram.

3.1.1.5 Other Coolant-Borne Reactivity Variations. In some cases it is conceivable that other sources of reactivity increase may be transported into the reactor by the flow of coolant. In circulating-fuel reactors, where the fuel circulates with the coolant, this possibility is obvious and is indeed recognized as an important consideration in the development and design of such reactors. In solid fuel reactors, the possibility is not usually serious but should always be examined. The most obvious sources of reactivity change are in those reactors employing soluble poisons in the coolant. Since the direct effect of the dissolved poison is to decrease reactivity, the opportunities for dangerous reactivity increases due to "hideout" of dissolved material are very much less than in circulating fuel reactors. Perhaps the most dangerous possibility is one analogous to the cold water accident. If a multi-loop reactor were in operation with dissolved poison in the coolant but poison had failed to dissolve in one loop because the loop was temporarily out of operation, then subsequent operation of the loop could give a substantial reactivity increase. For reactor systems of this kind, the direct means of protection against the accident are similar to those used for protection against the cold water accident.

Many reactors which do not use dissolved poison as a means of operating control are provided with soluble poison injection systems for backup emergency reactor shutdown. In designing such systems, proper care must be taken to ensure that, in addition to the reliable performance of its intended function, the system does not itself introduce a hazard. In particular, it is important to ensure that, after injection, the soluble poison is adequately mixed throughout the entire coolant system. Usually, when such systems are used, a scram of the normal control system is called for, either prior to, or simultaneously with the initiation of poison injection. This is a sound procedure, for serious troubles can be envisioned if, for example, the normal automatic control system were left in operation and it attempted to control the reactivity variations caused by the addition of the poison.

Another possibility in reactors employing soluble poisons in the coolant is that the absorbing material may gradually deposit on the fuel elements or core structure, and later be removed rapidly as a result of changes in the pH or in other characteristics of the coolant. The limited amount of experience gained with soluble poisons to date is not sufficient to rule out this possibility.

It is an elementary principle of safe reactor operation that a continuous accounting should be kept of the reactivity inventory and that the rationality of the apparent inventory should be under continuous examination. For reactors operating with dissolved poisons in the coolant, this implies that means for measuring the poison concentration should be provided and that the core designer should also provide the operator with the necessary information for assessing the significance of the record of poison concentration.

In many reactor systems the possibility of coolant-borne reactivity increases requires nothing more than a check and elimination. In reactors which employ soluble poisons during operation precautionary action is required. The core designer should attempt to: (1) predict adequately the reactivity effect of the soluble poison; (2) design the poison injection and removal systems, with due regard to the time constant represented by the circulation time of the coolant, to avoid significant coolant-borne fluctuations in reactivity; (3) provide for surveillance of the soluble poison content and its effect on reactivity, and (4) investigate possible unscheduled changes in poison content by chemical and solubility effects. For these and other systems containing dissolved poisons in which unpoisoned coolant can conceivably hide out in unused loops, the designs should analyze adequately the possible results of interlock and procedure failures and provide for general protection in the scram system.

3.1.1.6 Loss of Coolant; Change of Coolant Pressure. Although loss of coolant is in itself a serious occurrence, it is considered here only as a possible source of reactivity increase in those reactors which have positive coolant void coefficients of reactivity. The positive coolant void coefficient is considered in Sec. 3.3 as one of the factors which can affect the course of an accident started by other means. However, it must also be considered as a primary source of accident, since relatively trivial malfunctions, such as mismatches of power level and coolant flow, could conceivably initiate important reactivity changes through coolant boiling.

The occurrence of a positive coolant void coefficient of reactivity is often not an inherent characteristic of the coolant, but results both from the characteristics of the coolant and of the general core lattice. Thus, for example, the coolant (or moderator) void coefficient of reactivity in a water-moderated, slightly enriched reactor may be either positive or negative, depending on the ratio of fuel to coolant and on other factors; in some reactors it may be positive at low temperatures and negative at high temperatures. Consequently, one of the first considerations is to determine that the coefficient does indeed have the value intended in the reactor design. In some cases performance and economics require that the coolant void coefficient of reactivity be positive if the reactor type in question is to have any attraction. In these cases, the safety consideration must be factored into the entire reactor design and design compromises must be arrived at to allow adequate protection against the consequences of such reactivity increases, through suitable over-power safety factors, scrams, and operating levels. Some of the reactor types in which the coolant void coefficient of reactivity must be carefully examined are: boiling H_2O reactors; D_2O-moderated and -cooled reactors (when they employ internally-cooled fuel regions in a heterogeneous lattice); H_2O-cooled and organic-cooled D_2O-moderated reactors; H_2O-cooled, graphite-moderated reactors; and large sodium-cooled fast reactors. The pressurized-water reactor can be designed to have a positive coolant void coefficient of reactivity, but other considerations usually direct the design away from this, at least for those reactors which use slightly enriched fuel and which do not use soluble poison as a normal means of controlling reactivity.

As indicated above, if the coolant void coefficient of reactivity can be positive, then this possibility must be considered in terms of the entire core design and it is not possible to specify general means of solving the problem. The seriousness and difficulty of the problem may vary over a wide range. In some cases a slight change in the core lattice may eliminate the difficulty; in other cases the difficulty may be found to be so fundamental that it may render impractical the type of reactor under consideration.

Because the coolant void coefficient of reactivity usually becomes increasingly more negative in water-moderated reactors as the temperature is increased, a design which yields a negative void coefficient over the entire temperature range, from room temperature to operating temperature, may result in undesirably high values of the temperature defect, and, in boiling reactors, the power defect, of reactivity. In such cases it has sometimes been judged that a positive coefficient at low temperature represents a lesser hazard than the large temperature and power defects. This may be true if the design can provide adequate safeguards against the possibility of an accidental power excursion at low temperature.

A special case of the coolant void coefficient of reactivity can occur in integral boiling-superheating reactors. In these reactors it may be impossible to guarantee that the superheat coolant channels will never fill with water, and in some cases the flooding of the channels can be anticipated as a normal stage in the operation. In such instances, particular attention must be given to the reactivity effects of flooding and voiding the superheater, and special design provisions may be necessary to ensure that this possibility cannot lead to an accident.

The boiling water reactor, when properly designed with a negative coolant void coefficient of reactivity, may conceivably be subject to a reactivity increase which is the inverse of the loss-of-coolant effect. Thus any circumstances which might be visualized to cause a rapid gross increase in coolant pressure would lead to an increase of reactivity.

3.1.2 Loss of Control

A second rather arbitrarily defined group of reactivity accidents includes those involving loss of reactivity control by the automatic control system or the operator. It is distinguished from the class discussed above in that the loss of reactivity control is not conceived to result from any sudden reactivity addition.

One of the more obvious manifestations of loss of control is an instability in the system which comprises the reactor, the associated coolant and power conversion equipment, the control system, and, possibly, the operator. The probability of this kind of difficulty of course depends strongly on the dynamic characteristics of the reactor-power disposal system, and on the degree of inherent stability of that system. These characteristics vary over a wide range for different reactor types. Some pressurized-water reactors, for example, possess such strong power demand matching tendencies, by virtue of their large temperature coefficient of reactivity, that the entire plant may be inherently stable, provided no unstabilizing tendencies are built into the control system. Some other reactor types have such small power coefficients of reactivity that the stability of the system is largely a matter of the stability of the control system-operator complex. Still other reactor types are inherently unstable and depend entirely upon the provision of a stabilizing control system.

Most reactors are inherently sluggish in responding to small reactivity changes, because of the characteristics of the delayed neutron emission and because most power-dependent reactivity feedbacks work through reasonably long thermal time constants. This sluggishness makes the reactor power response to reactivity fluctuations and to inherent feedbacks slow, relative to the speed of control action that can be provided by electrical control circuits operating suitable reactivity adjustment devices. It is therefore possible, in principle, to provide automatic control systems which would stabilize the operation of almost any reactor system, despite any inherent unstable tendencies that it might have, and to provide automatic scram systems with adequate speed to protect against malfunctions of the control system. Nevertheless, there has been an understandable reluctance to build reactors which must depend upon fast automatic control systems for stability, although slow instabilities—those slow enough to be kept under control by considered action of the operator—are tolerated. The occasional instability of the EBR-I reactor (with Core II) was generally considered too "fast" to be acceptable for anything other than an experimental reactor, even though it was characterized by a sinusoidal period of 30 sec or more and was easily controllable manually in routine operation. On the other hand, it is known that the natural uranium, gas-cooled, graphite-moderated reactors show an instability due to the positive graphite temperature coefficient of reactivity after plutonium has built up in the fuel elements. This instability, which has a considerably longer time constant, is tolerated.

Fortunately, in most of the reactor types which appear attractive, suitable design can circumvent any serious inherent instabilities. In any case, it is apparent that the core designer must provide adequate predictions of the power and other coefficients of reactivity, along with their time constants, to make possible an accurate assessment of the inherent stability of the system. Beyond this, the provision of adequate independent protection facilities in the form of scram systems, etc., is necessary.

The following sections describe more specific sources of loss of control accidents, including those which are, in essence, instabilities.

3.1.2.1 Control System Malfunction or Inadequacy. If the reactor is controlled by an automatic system, it is always possible that the initial design or construction of the system is inadequate and that instability or loss of control can occur under certain operating conditions. Quite obviously, the system should be checked out carefully during the early operation of the reactor to find any problem areas of this sort. During this interval of check-out, the reactor is somewhat more susceptible to accident, but to a large extent this problem can be alleviated by procedures which entail greater-than-normal care by proceeding cautiously with operation in untried areas and perhaps by setting scram and protective circuits at more conservative levels than normal. In connection with this check-out procedure, the core designer should assess the uncertainties which may have gone into the core-connected aspects of the control system design and define the operating areas which may involve uncertainties.

Once the stability and accuracy of the control system have been checked out over all conceivable operating conditions, it will presumably function adequately and stably so long as components do not fail or deteriorate. The design of the system should be such that failures or deteriorations cannot lead to dangerous loss of control. Although this design consideration is primarily the concern of the control system designer, his decisions involve a number of questions of core design. Components of the system which are part of the core—usually the control rods—are possible sources of malfunction. If motion of the control rods is impeded by sticking, for example, important time constants of the system may be affected. If there is appreciable play or backlash in the linkage between the control rod and the drive system, the precision of the system will obviously be impaired and a troublesome source of neutronic noise may be introduced. Similar difficulties may arise from uncertainties in the sensing of control rod positions. To restrict the possible consequences of control system malfunction, limits are placed on the speed with which the automatic system can move control rods, and in some instances limits are placed on the amount of reactivity swing available to the automatic system. The physics of core design is strongly involved in relating mechanical limitations of this kind to limitations on reactivity. In addition to these considerations the core designer must provide an overpower safety factor and a scram capability adequate for protection of the reactor against any instability or malfunction of the automatic control system.

3.1.2.2 Operator Error. When the reactor is under direct manual control by the operator, loss of control is possible as a result of human error, or conceivably as a result of an instability-like runaway due to excessive demands on the skill, speed, or judgment of the operator. For prevention of human errors, the operator must have complete, accurate, and unambiguous information on those variables, such as control rod position, power level, and coolant flow, which are involved in the control operation. Items of information originating in the core constitute an important part of core design. To avoid instability of human operation, it is necessary that the rate of change of the controlled variables be geared to human reaction times, that the number of variables requiring adjustment be compatible with human capabilities, and that complex decisions involving technical judgment be minimized. A number of these requirements react upon core design. Particularly, the core designer should avoid the need for complex patterns of control rod adjustment, at least with respect to those adjustments which must be done fairly rapidly. Finally, the core designer must also consider the over-power safety factor and the scram capability in relation to the possibility of human error.

3.1.2.3 Loss of Shutdown Capability. The loss of shutdown capability is a quite different type of loss control. Ordinarily, adequate shutdown margin is built into the reactor initially and once this margin has been checked during the initial loading of the reactor, it is likely to remain adequate during the early life of the reactor. There are possible conditions, however, which can lead to a reduction of the shutdown margin during the life of the reactor, and it is conceivable that in extreme cases the normal control system may prove incapable of shutting the reactor down to the cold zero-power equilibrium condition. The most obvious possible source of this trouble is the sticking of control rods in withdrawn positions. This malfunction is usually readily apparent, and it is generally assumed that appropriate action will be taken to correct the condition before a serious stage is reached. Some of the less apparent means by which shutdown capability may be lost are the possible reactivity gains from plutonium buildup and/or from the burnout of burnable poison, and the gradual deterioration of the reactivity hold-down capacity from such causes as the corroding away of control rods or burnable poisons. These losses of shutdown capability are usually apparent only in terms of the critical positions of control rods.

In monitoring the shutdown capability it is important to take into consideration any short-term reactivity increments characteristic of the condition of the reactor when the shutdown margin is checked. Thus, a monitoring program involving periodic checks of critical rod configurations in the cold zero-power reactor is more straightforward and more suitable than a monitoring program involving only the surveillance of operating rod positions, since the latter are affected by reactivity effects like the temperature defect and the power coefficient of reactivity, themselves subject to change. Even when the shutdown margin is monitored by cold zero-power measurements, it is necessary to account for the effect of xenon. For example, a reactor which is adequately shut down shortly after extended power operation could, because of xenon decay, gain sufficient reactivity over a period of hours to become critical and restart itself. In cases where soluble poison is relied upon for cold shutdown, it is important to assure that no chemical or other means of reducing the poison content may come into play over long periods of shutdown.

Usually reactors are provided with an independent means of emergency shutdown which will have a shutdown capability even if the normal control system loses this capability. In such cases, the loss of capability in the normal system need not lead to serious consequences provided the deficiency is recognized and the emergency system is brought into action at the appropriate time. Frequently the emergency system acts more slowly than the normal system and, when this is true, the consequences must be taken into account. For example, one of the considerations determining the scram speed of the normal system is that of reducing power rapidly enough to avoid fuel element burnout in case of coolant failure. In some cases this requirement might not be met by the emergency system. On the other hand, in most reactors the normal control rod installation will maintain a substantial shutdown capability under power operating conditions even though it may have lost the capability of cold zero-power shutdown. This is because of the usual existence of a temperature defect and a negative power coefficient of reactivity.

The loss of shutdown capability is essentially a phenomenon of the core and consequently involves a number of considerations in core design. The principal ones are to: (1) provide adequate shutdown margin in the reactor as built; (2) design for the prevention of rod sticking and for the reliable indication of sticking if it should occur; (3) avoid the possibility of loss of shutdown margin through the deterioration of control elements or burnable poison elements; (4) assess accurately the reactivity gains from plutonium buildup, burnup of poison, and other possible sources; (5) provide for a method of monitoring shutdown capability: in particular, provide the means for adequate control rod calibrations and their interpretation which will permit the operator to make allowance for short-term effects like xenon decay when shutdown capability is checked; (6) consider the desirability of holding some shutdown capability in reserve whenever there is a possibility that reactivity may increase with time, as through the decay of xenon in a reactor which is shut down. This principle of "cocked control rods" is one which is well established in operating practice; (7) assess the need for an independent shutdown system and provide when necessary.

3.1.2.4 Stuck Control Rods. Although the sticking of control rods in withdrawn positions has already been mentioned in connection with the loss of shutdown capability, the sticking of control rods in itself represents a temporary loss of control, in the sense that the operator or the automatic control system can no longer exert control by means of the stuck rod. If the situation is recognized quickly and the control function is shifted to another rod, no serious consequences need arise from the sticking of an individual control rod. Indeed, in reactors which have large temperature defects and/or large negative power coefficients of reactivity, the sticking of a large fraction of the control rods during power operation could conceivably occur without any serious consequences to the reactor. The continued operation of the reactor with stuck control rods, over a substantial period of time, may however be a dangerous procedure, for the entire complex of safeguard engineering is usually based on the expectation that the control rods will be operative. It is possible that the capability of indefinite operation with stuck rods, up to a specified limit, may be quite desirable in some reactor applications (e.g. propulsion reactors) and in such cases this operation need entail no additional hazard to the reactor, but only provided that such operation has been factored into the complete design of the reactor and its safety system. In many reactors the control rods perform other functions besides the control of the general reactivity level. When the rods are important, for example, in determining the power distribution or in controlling flux tilts, these functions must also be taken into account in assessing the seriousness of rod sticking.

Another point of view that might be taken with respect to stuck rods is that the occurrence of sticking in one or more rods proves the susceptibility of all rods to this malfunction, and that such a condition, if it exists, should be corrected. This argument appears to be reasonable in some cases but not in all, and perhaps should be evaluated for particular cases rather than in general. If, for instance, it can be shown that a rod has stuck as a result of a slight overpressure in the reactor, then it is likely that a design flaw exists which may cause general rod sticking. On the other hand, if the sticking is due to a worn mechanism or a slightly bent shaft, then it is unlikely that all rods will behave similarly at the same time, and the situation is less serious. This consideration illustrates the close relationship between the design and operating activities and the need for careful coordination with the designer when operating procedures are drawn up.

In any event, the core design considerations involved in stuck-rod troubles are: (1) to eliminate the known sources of rod sticking so far as possible within the limits of technology; (2) to consider the shutdown margin in terms of a stuck-rod margin, and make adequate provision for shutdown in case of rod sticking; (3) to determine the policy with regard to stuck-rod operation during the core design period, and to base the core design on this decision; (4) to provide adequate information on control rod worths to make possible an assessment by the reactor operator of the significance of individual rod sticking. In some cases this may require considerations of power distribution and hot spot factors as well as control questions. It may also require assessment of the future, as well as the current, reactivity inventory.

3.1.2.5 Malfunction of Coolant System. Often the characteristics of the coolant system play an important part in determining the stability of the reactor system. In some cases the reactor may become unstable as a consequence of malfunction or failure of the coolant system. A boiling reactor for example may become unstable if the coolant pressure is reduced rapidly or by a large amount. Instabilities may also occur in a boiling reactor designed for forced circulation if the circulation is reduced drastically by pump failure. Similarly, a pressurized water reactor might go into unstable boiling operation if coolant system failures resulted in large decreases in the pressure or circulation rate of the coolant. Possibilities of this kind, as well as the question of adequate fuel element cooling, must be taken into account in determining the speed with which difficulties must be detected and shutdown must be initiated in the case of coolant system malfunction.

3.1.3 Overpower Operation

Certain conditions of overpower operation can arise which do not result from either reactivity excursions or loss of control of the reactor. These are usually ascribable to human errors, instrument errors, or errors in the control system due to malfunction or failure of minor parts. Protection against such errors, if they are not corrected by the operator, must be provided by the overpower protection system, which involves overpower scrams, coolant temperature scrams, etc. These overpower possibilities are considered in the following sections.

3.1.3.1 Control System Malfunction. Although fail-safe design can be applied where possible in the control system, there always are possible difficulties which may produce temporary or permanent overpower conditions. For instance, the sticking of a control rod under automatic control may temporarily perturb the control system. Other possibilities are the failure of the components of the circuitry and actuating mechanisms. These possibilities emphasize to the core designer the importance of an adequate combination of overpower safety factor and overpower scram. As in a number of other cases, the establishment of proper limitations on maximum rate of control rod withdrawal must also be considered. Finally, the designer should provide for the capability of checking functionally, and with safety, the performance of any movable component which can affect reactivity.

3.1.3.2 Operator Error. The possibilities in this category are analogous to those for the automatic control system. The core design considerations are also the same,and additional consideration should be given to simplifying the operations required of the operator so that he may concentrate effectively on the important variables under his control.

3.1.3.3 Instrument Error. Needless to say, instrument errors, if uncorrected, can lead to gross reactor overpower. The primary protection against this possibility is the use of multiple instrumentation. Although multiple instrumentation channels of the same type can provide an important safety factor, a still better system is to provide instruments which determine the power level by several different and independent techniques of observation and which monitor those quantities, such as coolant temperature, which usually give direct indications of overpower trouble. In core design, four main considerations affect the accuracy of instrument information: saturation, calibration, signal-to-noise ratio, and sampling.

The nuclear instruments, although they cover extremely wide ranges, do reach saturation when the radiation levels become too high. Ordinarily one expects the instrumentation engineer to avoid saturation levels, but the core designer often must specify the instrument locations which will provide coverage of the reactor power range without saturation.

Although nuclear instruments have short time constants when they are operating in reasonably high neutron fluxes and consequently are attractive as primary indicators of reactor power level, it is well known that the relationship between ion chamber output and reactor power is not fixed but may vary with control pattern, coolant temperature, and other reactor variables. It is therefore necessary to recalibrate this relationship either continuously or at frequent intervals. The core designer can minimize this problem by choosing instrument locations which will show the minimum sensitivity to such variables.

Finally, in installing sensing devices such as thermocouples, which sample only local conditions, the core designer must choose locations and specify a suitable multiplicity of measuring points to produce a representative sampling of the quantity being measured.

3.1.4 Power-Coolant Mismatch — General

For reactors in which coolant flow velocity or other characteristics of the coolant system are varied as the reactor power level is varied, it is possible that cooling may be inadequate through mismatch of the power and coolant flow velocity, even though the reactor power level does not exceed the design level. The sources of mismatch error are essentially the same as those discussed under Overpower Operation, in addition to possible errors in determining and adjusting the coolant characteristics. It is usually easy to incorporate rather large safety factors against such errors at reduced power, inasmuch as the economic and performance penalties for deviation (in the conservative direction) from the optimum power-coolant match are not usually large. For protection against such errors, one cannot ordinarily depend on the overpower protection systems, but must depend on scram and alarm systems which report some quantity, such as the coolant outlet temperature, directly related to the undesirable consequence of the error. From the point of view of core design, the new consideration, in addition to those discussed under Overpower Operation, is simply that the safety factors on power removal capability must be considered over the entire range of operating power, rather than simply at maximum power. A particular oversight which has led to some accidents in the past is the failure to reduce scram power levels when neutron flux tests were being carried out at very low power, with the coolant flow turned off.

A special problem of power-coolant matching, which occurs only in integral boiling-superheating reactors, is worth mentioning. In these reactors the coolant for the superheater part of the reactor core is steam supplied by the boiling of water in the boiling portion of the core, and the reactor operator ordinarily has no independent control over the supply of coolant to the superheater fuel elements. In this case, power-coolant matching in the superheater reduces to a problem of obtaining the proper ratio of power between the boiling and superheating sections of the core. The problem is complicated by the fact that the correct ratio is a function of the subcooling of the feedwater to the boiler section of the core. The provision and control of the appropriate boiler-superheater power split is a central problem in the design of boiling-superheating reactors; to a large extent the entire core design hinges upon the solution of this problem.

3.1.5 Power-Coolant Mismatch — Local

Matching the local coolant capability to the local power density is a problem which occupies a large fraction of the time of the core designer. It embraces all those questions of power distribution, and hot channel and hot spot factors, which occupy much of the time of the core physicist and heat transfer analyst during the detailed design of the reactor core. Fortunately, errors in this local matching of power density and cooling capacity are often not very serious from the safety point of view, since by their very nature they involve only limited regions of the core. Indeed, when one considers accidents of the very serious kind, one should perhaps consider the rather large maximum-average power density ratios and the rather high hot spot factors which tend to characterize nuclear reactors as credits rather than debits in the safety (but not the economic) ledger, since they force the designer to incorporate large safety factors with respect to the capability of the average portion of the core. Of course, if local failures can lead to general failures through a progressive failure mechanism, or if they can lead to unnatural power increases through some sort of positive reactivity effect, or if in some other logical way they can be compounded into more extensive failures, then the situation is quite different. Possibilities of this sort are treated more fully in Sec. 3.3. In any case, whether the consequences of local failures be large or small, the designer will continue to attempt to prevent them. The characteristics of two different types of local problems are considered briefly in the following paragraphs.

3.1.5.1 Static Problems. "Static problems" are all of those power-coolant distribution problems which are constant in time except for the effects of the over-all reactivity changes in the core, and the control rod motions which may be utilized to compensate them. These are the problems of heat removal design familiar to all core designers. They are discussed in Secs. 2 through 5 in the chapter on Heat Transfer. The one comment which may be offered here is that in-core instrumentation may help in solving these problems and, particularly, in keeping them under control throughout the life of the reactor. In some types of reactors in-core instrumentation may be of great assistance in determining directly what the local situation is; in other reactor types the currently available in-core instruments may yield little information that is useful. In any case, the reactor design must take into account the local effects that may occur throughout the life of the reactor. It is customary to give great attention to these questions at the time of initial reactor startup. Frequently it can be shown that the situation can do nothing but improve as time goes on, but this must not always be assumed to be true. In some cases, for example when partial core reloading is used or when poison shims are removed or when spikes are inserted, the possibility exists that a situation worse than the original one may be encountered.

3.1.5.2 Dynamic Problems: Spatial Oscillations of Power. It is generally known that in large, high flux reactors a sustained spatial oscillation of power distribution may develop as a result of the buildup and burnout of Xe^{135} in local core regions. This oscillation is quite slow, since its period is determined by the Xe^{135} half-life (9.13 hours), and experience has shown that it can be controlled in a straightforward manner, provided instrumentation is available to detect incipient spatial oscillations, and provided local control agencies are available for counteracting the xenon effects on a local scale. In those cores which may support xenon oscillation the use of local instrumentation is a necessity; usually the instrumentation must be located in-core, or must measure temperatures of coolant samples from localized core regions. In relation to this problem, the core designer must provide an adequate system of in-core instrumentation when it is needed, and control rods, or other control agencies, sufficiently well dispersed over the core to provide the means for countering any possible spatial oscillation of power.

Other possibilities of spatial oscillations of power exist, although the conditions for their occurrence are not so well defined as those for the xenon oscillation. Fortunately, most other local variations of reactivity will be coupled to the reactor power by negative power coefficients of reactivity, and hence will be unlikely to give rise to sustained oscillations. Spatial oscillations of coolant flow, however, are apt to be coupled only very loosely to reactor power, and the possibility of their occurrence should be given consideration, particularly in two-phase systems.

3.1.6 Coolant Failure — General

The loss of coolant or coolant flow may result in thermal damage to the fuel elements and possibly to other parts of the core structure. This primary effect is considered here. There may be secondary effects, conceivably equally, or even more serious; these effects are discussed in Sec. 3.3.

In any case of coolant failure, the first consideration is to shut the reactor down. If for any reason the shutdown circuits and mechanisms fail, the result can only be general thermal failure of the core in most reactors, although it is possible that certain reactors might shut themselves down, or at least reduce their power to a tolerable level through a strong negative coolant-void or temperature coefficient of reactivity. Once the reactor is successfully shut down after a coolant failure, the safety question resolves itself into one of cooling the fuel elements adequately to prevent wholesale thermal failure. Several degrees of coolant system failure are discussed in the following paragraphs.

3.1.6.1 Circulation Failure. Loss of coolant circulation can be caused by the failure of pumps or pump drives or by the blockage of coolant lines. The normal precaution against such failure is the provision of multiple pumps and/or multiple coolant loops. This must be supplemented by providing emergency power supplies for some or all of the pumps in event of a general power failure. Since several seconds are usually required for actuation of the scram system and decay of the reactor power to a low level, a means of prolonging the "coast-down" of the coolant flow, such as the installation of massive flywheels on the pumps, is sometimes found necessary or desirable. By such measures, the possibility of dangerous core damage through circulation failure may be reduced to quite a low level. In dealing with this type of accident, the core designer must specify the cooling requirements for the core, as a function of time after the incidence of the failure, and must investigate all possible secondary effects which might lead to complication of the total accident (see Sec. 3.3). In some cases, natural circulation of the coolant may be depended upon for cooling at some stage. In such cases, the requirements for effective natural circulation may play a considable part in setting the core design.

3.1.6.2 Loss of Coolant. The loss of all coolant or a substantial fraction of the coolant, as from a large leak or rupture in the primary coolant system, presents a much more difficult problem. The methods of coping with this problem vary from one reactor type to another. In such instances, the situation may be similar to loss of coolant flow, although it may be necessary to rely more heavily on natural circulation.

In water-cooled reactors, it is customary to rely upon emergency spray systems for cooling after loss of the normal coolant. In such cases, the reactor designer, in addition to specifying the cooling requirements, must provide a core structure which allows adequate contact of the sprayed coolant with all sections of the fuel elements, and must guard against any possibility of criticality resulting from the moderating effect of the emergency coolant. In reactors not normally cooled by water, the use of water sprays as emergency cooling systems may be impossible because of the incompatibility of the reactor materials with water, or may be undesirable because of deleterious effects if the emergency system should accidentally go into operation when not needed.

In gas-cooled reactors, the problem may be particularly severe because the use of air as an emergency coolant may be prohibited by considerations of materials compatibility, and because it may be impossible to achieve a sufficiently high gas density for adequate cooling, even of the shut down reactor, without pressurization. In such cases, the reactor containment building may be called upon to act as a secondary pressure vessel in the event of rupture of the reactor coolant system, and the size of the building may have to be made sufficiently small to ensure that the ambient gas pressure is well above atmospheric, even if the primary coolant gas is released the building. In cases such as this, the reactor core designer will be called upon for accurate estimates of the shutdown cooling requirements of the reactor.

3.1.6.3 Reactor Vessel Failure. In highly pressurized systems, the possible effects of a gross reactor vessel failure are so severe and so varied that it is probably impossible to arrive at principles of core design which will have any alleviating effect on the course of the accident, once it has been initiated. If the mechanical forces associated with such a severe accident can be conceived to leave the core in any way intact, the need for shutdown cooling will remain, and to the extent possible, the emergency cooling system should be designed so as to be, as nearly feasible, immune from the effects of reactor vessel failure. However, in core design the primary consideration relevant to such an accident is to avoid imposing on the pressure vessel any deteriorating effects. In particular, this means providing an adequate thermal shield in the vessel for the prevention of thermal stresses (as a result of absorbed radiation), providing any necessary coolant baffles to prevent thermal shock, and providing an adequate fast-neutron shield for the prevention of irradiation embrittlement. A further requirement is to provide means of surveillance of the effects of radiation on the pressure vessel through the incorporation of appropriate samples which may be examined periodically throughout the vessel life. Although the provision and installation of such samples is usually considered as a part of the reactor vessel design, the core designer will be called upon to assess such questions as the effects of the incident neutron spectrum and to find appropriate locations for the samples which will give meaningful results.

3.1.7 Coolant Failure — Local

Local failures or deteriorations of coolant flow or heat transfer efficiency may lead to local failures of fuel elements. If the reactor is so designed that failures of individual elements do not lead to serious contamination problems, such failures would not be expected to lead to serious accidents unless some kind of secondary effect were present to compound their effects. Ordinarily, the development of local difficulties shows up in the failure of one or a few fuel elements; the attentive operator can shut down and investigate the trouble before a serious condition develops. Needless to say, warnings of this kind should not be ignored by the operator.

Several possible causes of local coolant failure or deterioration of heat removal capability are discussed in the following sections.

3.1.7.1 Starving of Parallel Flow Channels. In practically all reactor cores, coolant supplied by a common circulating system must find its way through numerous parallel channels to cool the individual fuel elements. Two effects usually result. First, the available pressure drop from inlet to outlet of any particular channel is determined by the pressure drop through all channels. Thus, any particular coolant channel can be considered to have a fixed circulating pressure drop and, if any abnormal restriction to flow should occur in the individual channel, there will be no additional pressure available to compensate for it. This situation was recognized in the first reactor designs. In particular, it was postulated that if boiling should occur in one of many parallel water coolant channels, the increased specific volume of the two-phase mixture would result in a decrease of the mass flow to the boiling channel, so that still greater quantities of steam would be formed to carry away the heat. Thus a progressive increase in boiling was visualized which would eventually choke off the flow to the point where the fuel element would burn out. This possibility was often referred to as "boiling disease". Precautions were taken in the form of inlet orifices to the coolant channels. These introduced substantial pressure drops which were independent of the boiling in the fuel element channels and tended to make each coolant channel behave more as a constant-flow channel than a constant-pressure-drop channel. Subsequent tests and experience have shown that boiling disease is not as serious a problem as was at one time feared. This question is discussed in Secs. 3 and 4 in the chapter on Fluid Flow.

More recent reactor designs involve a second effect, associated with the use of coolant channels which are not only connected in parallel to the coolant circuit at their inlets, but which are interconnected over their entire lengths. This is typical in fuel rod bundles. Here the situation is quite complex, in that cross flow from channel to channel may occur at any point where there is a local obstruction: and when two phases are present, the two phases do not necessarily behave similarly with respect to interchannel flow. In water-cooled reactors, where burnout is the typical limitation to heat-removal capacity, the problem of interconnected parallel coolant channels has not been solved in any general way. The core designer relies on burnout correlations determined experimentally for configurations more or less similar to the one he is designing. These problems of parallel-flow heat-removal design are well known to the core designer, and it is unlikely that they will be slighted in any design project. Consequently any failures which do occur because of parallel-flow effects should probably be placed in the general category of design error rather than in a special category of reactor accidents; nevertheless, no discussion of core design in relation to reactor safety can ignore these very difficult questions.

In some instances the problems of parallel flow in interconnected channels are complicated by distortion of the channels which may occur during reactor operation as a result of such phenomena as the bowing of fuel rods or buckling of fuel-bearing plates under conditions of nonuniform heat generation or nonuniform heat removal. If such distortions can, in turn, produce significant reactivity changes, the problem may be complicated still further.

3.1.7.2 Deterioration of Effectiveness of Heat Transfer. In any reactor in which corrosion products or other materials may be deposited on fuel-element surfaces, the effects of such deposits on the effectiveness of heat transfer must be assessed. In water-cooled reactors such deposits probably do not affect the burnout limit unless they become large enough to constitute a significant obstruction to coolant flow; however they will cause an increase in the surface temperature of the fuel jacket, and may thereby increase the rate of corrosion. In other reactor types, such as those employing organic coolant, the effects may be more important. These possibilities and problems must always be taken into account in core design.

3.1.7.3 Local Coolant Obstructions. If the deposition of corrosion products or other materials is heavy enough to produce a significant restriction in the coolant flow to individual coolant channels, thermal failure of the fuel elements may result. Ordinarily this type of problem is characteristic of a particular coolant, and the appropriate measures for controlling it must be fitted to the particular coolant. In liquid sodium, for example, the necessity for preventing the deposition of oxide is well known; experience has also shown the importance of avoiding contamination of the sodium by organic materials. The gross fouling of fuel element surfaces and coolant channels is recognized as one of the basic problems with organic coolants. As a general comment it may be said that when the possibility of coolant obstruction is anticipated, in-core instrumentation for the measurement of local flow velocities can often be employed to advantage.

3.1.8 Non-nuclear Energy Release in the Core

Although considerations of non-nuclear energy release are usually most important as secondary effects in accidents initiated by other means (Sec. 3.3), they may, in some cases, be visualized as primary causes of accidents. Two possible sources of non-nuclear energy are discussed in the following paragraphs.

3.1.8.1 Stored Energy in Irradiated Materials. When the atoms of solid materials are displaced from their normal positions in the crystal lattice as a result of irradiation, they exist in a state of increased potential energy. This energy is released if the atoms return to their normal locations. The rate at which atoms return to their normal positions increases with temperature (annealing). Consequently, if a material containing many displaced atoms is raised to a sufficiently high temperature, the release of energy by atomic rearrangement may reach a sufficiently high rate that the gross temperature of the material is increased further, with the result that an autocatalytic energy release is produced. The only practical case of this phenomenon that has so far been observed is the Wigner energy release by graphite. In graphite-moderated reactors this is a serious possibility which must be taken into account in the design and operation of the reactor. The usual method of handling the problem is to design so that the normal operating temperature of the graphite is high enough to cause most of the displacements to anneal out at about the same rate as new displacements are produced. If a residual energy storage is anticipated, provision must be made for periodic annealing under carefully controlled procedures.

3.1.8.2 Chemical Energy. Needless to say, the use of materials which could conceivably react with one another chemically in the reactor at a rate sufficiently high to give a significant energy release must be avoided. Aside from this obvious precaution, further precautions should be taken against the possible accumulations of unwanted materials which might conceivably react to give significant chemical energy releases. An example of this kind is the hydrogen-oxygen mixture which is generated by radiolysis (and possibly by some electrolytic corrosion processes) in water-cooled reactors. This problem is particularly important in the aqueous homogeneous reactor, but significant amounts of oxygen and hydrogen may accumulate in almost any water-cooled reactor if steps are not taken to suppress the formation or to dispose of the product. Finally, the possibility of chemical reactions of core or coolant materials under conditions of equipment failure must be taken into account. Coolant fires may, for example, result from leaks in sodium or organic coolant systems; sodium-water reactions may result from failures in sodium-water heat exchangers, and graphite may burn if air is inadvertently introduced into a very hot graphite core. Core design questions are involved in the choice of core materials, in the prediction of chemical reaction and radiolysis rates in the core, and possibly in the selection of in-core instrumentation for the detection of troublesome chemical situations.

3.2 Accidents to Non-Operating Reactors

In many cases the reactor is more vulnerable

to accident when shut down and in the non-operating condition than when operating at a substantial power level. In particular, this is true when modifications are being made to the reactor, when it is being refueled, or when maintenance is being performed which requires operations on the reactor core or the control rod system. At such times some of the design features which normally prevent the injection of human error into manipulations of the reactor must, perforce, be bypassed or rendered inoperative. At the same time, in the shutdown condition the low power level and the low level of neutron flux minimize the amount of information that can be conveyed by the reactor instrumentation system and the low temperature and low power level delay the action of negative temperature and power coefficients in the event of accidental reactivity additions. Fortunately, the cold condition of the reactor and the absence of stored thermal or mechanical energy in the coolant may also reduce the severity of many of the possible accidents. But the extra vulnerability of the system to human errors requires that particular attention be given to the safety of the reactor during these crucial periods. In any case, the nature and causes of typical accidents which may occur during shutdown are rather different from those which might occur during operation at power. Consequently the two states of the reactor should be considered separately.

During periods of modification or repair, the reactor is inherently susceptible to the results of casual human errors of operation, as distinguished from errors of design. The objective of the designer should be to minimize the opportunity for such errors. In doing this, it is well to recognize the possibility of errors of two types: rational errors which result from inadequate knowledge or faulty—but considered—judgment, and irrational errors, which may result from temporary lapses of attention, carelessness, or, conceivably, irrational impulses. To prevent errors of the rational type, the designer must accept the responsibility for making available to the operators a complete and detailed understanding of the reactor and its particular characteristics. To minimize the possibility of irrational accidents, the designer must attempt to eliminate the possibilities of dangerous operations which can be carried out quickly and on the spur of the moment; he should also avoid the possibility of dangerous operations which can be carried out by a single person. Finally, no operation should be included in the servicing or modification of the reactor which requires unusual skill or manual dexterity for its safe execution. With these principles in mind, it is possible to discuss some of the classes of accidents to which the shutdown reactor may be vulnerable.

It goes without saying that the shutdown reactor must be provided with adequate cooling for the irradiated fuel elements, and that conventional sources of accidents, such as fires, must be avoided. Once these have been taken care of, most of the conceivable accidents to a shutdown reactor involve unexpected reactivity increases. The objectives of core design should be, first, to minimize the probability of occurrence of such accidents through operational errors, and, secondly, to provide the complete understanding of the reactor core which the operation must have in order to avoid mistakes. A number of possible sources of unexpected reactivity increases are discussed below. The information is summarized in Table 3-2.

3.2.1 Handling of Control Rods

From time to time it is necessary, in almost any reactor, to disconnect the control rods from their control circuits and their operating mechisms. Whenever this is done, certain restraints on their motion, which are normally built into the operating reactor, are abrogated and, unless the designer builds in further restraints, the reactor is vulnerable to reactivity increases which may be brought about by manipulation of the rods. The first design consideration must be that the control elements should not be subject to withdrawal under natural forces or inadvertently as a result of other operations on the reactor. For example, there should be no possibility for the control rods to fall out of the reactor under the action of gravity after they are disconnected from their drive mechanisms. If the control rod extensions pass through linear seals in the pressure vessel, the possibility of control rod ejection by internal reactor pressure, (should the rods be disconnected from their drives while the reactor is pressurized), must be eliminated. Again, if the control rod extensions pass through seals or bearings in the reactor vessel head, the possibility of inadvertently withdrawing the rods when the head is lifted off the vessel must be avoided. In compact cores, where control rods are closely adjacent to fuel assemblies, there should be no possibility of accidentally withdrawing a control rod along with a fuel element during refueling operations, by the catching of the control rod on the element or on the handling tool. Finally, the possibility of control rod ejection as a result of accidents which unexpectedly release nuclear or other energy must be avoided. Thus, for example, in pressurized-water reactors in which control rods could conceivably be ejected from the core by internal reactor pressure, steps should be taken to assure that the rods remain in the core even after the reactor has been depressurized, in order to avoid the possibility of ejection if a power excursion, caused by some other reactivity source, should occur.

Usually, at some time in the life of a reactor, it will be necessary to remove control rods from the reactor for inspection or replacement. Safety in this operation must depend, to some extent, on procedural controls, but in some cases the designer can build in restrictions which prevent or at least discourage deviation from safe procedures. If, for example, the safe and accepted procedure is to remove part or all of the fuel elements from the core before a control rod is removed, the designer can, in some cases, build in restrictions preventing removal of the rod before fuel elements have been removed. Usually the provision of a comfortable excess shutdown capacity, in the form of more than one excess rod, is of considerable help in assuring safe control rod handling. In any event, it should be made impossible to have some control

TABLE 3-2

Types of Primary Accidents to Reactors During Shutdown, Maintenance, and Refueling

Source of Accident	Type of Error	Examples	Core Design Principles Involved in Prevention
Handling of control rods	Inadvertent removal of control rods	Dropping after removal of drive Ejection by internal reactor pressure Accidental removal along with fuel element, or with reactor vessel head	Proper mechanical design of core and control rod installation
	Improper procedure in necessary control rod removal	Removal of control rod while fuel is in core, and without provision of alternate shutdown agency	Procedural controls Mechanical restrictions or interlocks on control rod removal
	Mistakes in rod-removal procedure due to inadequate information	Incorrect judgement of shutdown capability of remaining rods when one or more rods is removed	Procedural controls Adequate normal shutdown margin Complete information on shutdown margin and rod worths Reliability of control rod position indicators, even when reactor vessel is open
Fuel Handling	Loading without adequate shutdown capability	Loading fuel without a full complement of control rods Loading fuel in increments which are too large	Procedural controls Provide for installation of all control rods before fuel loading begins, or provide alternate shutdown agency
	Mistakes in loading pattern	Mix-up of enrichments where multiple enrichments are used Omission of shims when they are required for shutdown	Adequate marking of fuel elements and other reactivity-controlling components Mechanical interlocks to prevent improper installation of fuel, or mix-up of fuel
	Mistakes due to inadequate information	Shuffling errors Addition of fuel increments producing unexpectedly large reactivity increase Increase of reactivity as a result of fuel removal	Adequate information on reactivity worths of all possible fuel increments and on long-term reactivity changes Provision for frequent check of shutdown margin
Changes in Effective Core Geometry	Mechanical distortion of core geometry	Change of effective core diameter by toppling of insecurely stacked elements during core loading	Provide for positive retention of fuel elements in partially loaded core; restrain against motions due to falling objects, etc.
	Mistakes in assessing reactivity effects of geometric changes	Rapid increase of reactivity due to raising moderator level in a short core	Adequate information on all possible reactivity effects due to geometry changes, moderator level, etc.
Unexpected Slow Reactivity Gains	Misjudgement of shutdown margin	Supercriticality during shutdown activities due to undetected reduction of shutdown margin	Adequate information on predicted reactivity behavior Provision for frequent accurate check of shutdown margin

rods inadvertently partially withdrawn when control rod handling is begun. Some rods, however, may be intentionally held partially withdrawn (cocked) under a carefully considered procedure, to provide some reserve shutdown capability in case of an accidental reactivity increase.

The information on the shutdown margin, the worth of control rods, and the peculiarities of control rods should be complete. It is particularly important to point out any unusual characterisitics, such as a highly asymmetrical differential worth curve, or a particularly large maximum/average ratio in the differential worth curve, for a rod or a group of rods. Similarly, the worth of each individual rod in the completely rodded core, and the worths of individual rods in possible sequences of withdrawal, up to the withdrawal which would produce criticality, should be known. Particular attention should be given to any unusual shadowing effects, which make any particular rod have an unusually large worth when withdrawn alone or in some possible sequence. So far as safety is concerned, it is not necessary to know these quantities with a high degree of accuracy; the important consideration is to give the operator a sound and comprehensive knowledge of the approximate absolute and relative worths of control rods and control rod groups.

3.2.2 Fuel Handling

The question of fuel handling is closely related to questions of criticality, which are discussed in the chapter on Criticality. This section will treat those principles which are important in core design, and the information which the designer should supply for safety considerations.

The addition of fuel, either on initial core loading or upon reloading, is an operation which must have careful consideration from safety standpoints. The core designer should produce a design which makes the control of reactivity during fuel additions as safe as possible. For example, unless some auxiliary means of providing a large obvious excess shutdown margin is employed (such as gross poisoning of the moderator by soluble poison), it should be possible to install all control rods before fuel is loaded. In some cases it is possible to provide also for the actuation and scram of at least some of the control rods during the loading operation. In many cases the safety of the loading operation may be affected by the magnitude of the reactivity represented by the minimum increment of fuel addition (usually the worth of a single fuel assembly). In power reactors this consideration is not usually allowed to determine the size of the fuel assembly, but if other considerations, such as economic ones, dictate the addition of fuel in large increments, then the designer may find it wise to provide for an auxiliary shutdown mechanism during loading. The designer should minimize the possibility of mistakes in loading. When fuel assemblies of different enrichments are to be loaded into different core regions, very obvious means of identification should be provided, and the elements should be made incapable of installation in the wrong location if such an arrangement is compatible with the design objectives. If removable shims are used in the assemblies, means of determining the presence of the shims should be provided and, if feasible, the possibility of omission of the shims should be eliminated.

If fuel is to be shuffled in the reactor after exposure, all possibilities of reactivity increases upon shuffling should be investigated. These possibilities include the buildup of plutonium, the burnup of burnable poison, and the decay of xenon. Any unusual effects, or effects that might be unexpected by the operator should be investigated by the designer and pointed out. In particular, the possibility exists in some cores (e.g. in a highly-undermoderated D_2O core) that reactivity can actually be increased by the removal of fuel. Effects of this kind, because they are contrary to intuitive expectation, may be particularly dangerous.

3.2.3 Changes of Geometry

Considerations of changes in core geometry during shutdown manipulations are more important in critical experiments and in research and test reactors than in power reactors, where the possible variations in geometry are usually severely restricted by the permanent structure of the core. Nevertheless, in some cases the operator may fail to recognize how small a unit of a core (especially in water-moderated cores) may go critical, and how sensitive reactivity is to size in such small critical assemblies. It is the responsibility of the designer to point out these possibilities. For example, if a water-moderated reactor of high excess reactivity is brought to criticality with the control rods withdrawn, by raising the water level, it may go critical when the water level has been raised only to a small fraction of the total core height; under these conditions, a small additional increase in water level will result in a large reactivity increase.

The core designer should, of course, also design to prevent unexpected reactivity changes caused by changes in the geometry of the core. The core design should be arranged in such a way that, as additional fuel assemblies are added to the core, they are held firmly in their intended places. The interchange of absorbers and followers should be rendered impossible by component design in those reactors which employ control rods with fuel element followers. There should be no possibility of a gross increase of reactivity by shifting of the fuel elements (thus increasing the effective size of the core) either as a result of small disturbances of the reactor structure or the surrounding coolant, or as a result of gross disturbances such as the dropping of an additional fuel element from a crane.

3.2.4 Unexpected Slow Reactivity Gains

The reactor is subject to slow changes in reactivity over its lifetime and some of these, such as the xenon decay, can go on while the reactor is shut down. Although the net long-term effect of changes due to fuel exposure is ordinarily to decrease reactivity, and hence to increase the shutdown margin, reactivity increase can occur in the intermediate term, because of such effects as the buildup of plutonium and the burnup of burnable poisons. These effects will have been taken into account in the core design and adequate shutdown margin will usually have been provided. However, the possibility of inaccuracies in calculation always exists, and in addition certain unforeseen possibilities, such as the corroding away of burnable poison or control elements, may conceivably occur. For this reason it is important that the reactor operator be provided with adequate means for assessing the true shutdown capability of the reactor at any given time. Otherwise, it is conceivable that the shutdown margin might disappear during periods of shutdown, because of the xenon decay, or at least might be reduced to an uncomfortably low level.

For proper surveillance of the shutdown margin and for interpreting its significance, the reactor operator should be provided with the means for periodic accurate measurements of criticality, and with a knowledge of rod calibration in the configurations which are convenient for use in these criticality determinations. Further information is necessary for interpreting the criticality data in terms of shutdown margin and stuck-rod margin, and for predicting the situation after the decay of xenon. Most of these requirements are met by proper calibrations during the initial startup period, but the reactor designer must provide the means of accurate control rod position indication which are needed for precise measurements and usually must determine and specify what criticality determinations will be convenient and meaningful, and how they may be interpreted. The reactor designer and the operating reactor physicist must be sure that management understands the importance of these measurements and sees that they are carried out, even in spite of any economic pressures to keep the reactor running which may exist.

3.3 Secondary Factors Which May Affect the Course of Accidents

The foregoing discussions have to do with the initiation of reactor accidents; they point out that the usual effects of accidents are to overheat the reactor core, either as a result of excessive power generation, or inadequate cooling, or both. The temperature changes associated with the primary accident and the forces which they may give rise to will often change the conditions in the core in such a way as to modify the course of the accident, either for the better or for the worse. In the case of severe accidents, these secondary factors are usually the most significant ones in determining just what the total consequences will be. Usually their important effects will consist of changes which they may make to the energy release. Quite evidently, once it is assumed that an accident has progressed to the point where the integrity of the core structure is effected, the further consequences cannot be predicted in any detail in a general discussion. Each plausible accident must be analyzed in the reactor in which it is postulated and, even when such a specific analysis is made, the results may still contain large uncertainties. The following general discussion can only point out the types of effects that must be taken into account by the reactor designer and by the safety analyst.

3.3.1 Progressive Failures

As used here, the term "progressive failure" is intended to designate failures which begin as relatively minor and localized occurrences and grow to serious proportions through processes inherent in the malfunction itself. Such possibilities do not occur frequently, and it is not possible to give any specific rules for avoiding them. The only reasonable precaution that the core designer can take is to attempt to follow each recognizable possibility of malfunction through all of its conceivable consequences.

A rather obvious hypothetical case of a progressive failure is the self-propagating failure of fuel elements, which can be reasonably postulated if the fuel material in an element, in the event of a jacket failure, can react chemically, and with some degree of violence, with the reactor coolant. One can imagine that, should such a failure occur, the violence of the reaction might distort the remaining pieces of the failed fuel element sufficiently to block or restrict the coolant access to adjacent fuel elements. Failures of these, in turn, might produce similar effects in elements adjacent to them, and the failure thus might propagate throughout an extensive region of the core. The chances of this possibility were considered serious enough to play a large part, some years ago, in the rejection of metallic uranium in favor of the oxide as a fuel for slightly enriched water-cooled reactors.

Another example of progressive failure might be visualized in pressure tube reactors. Here the failure of a single pressure tube might conceivably cause mechanical dislocations which would cause failures in adjacent pressure tubes. This possibility however appears not to be an inherent one, but one which can be guarded against by adequate support and restraint of the individual tubes.

Progressive failures outside the reactor core can also be visualized. For example, leaks or minor breaks in coolant or auxiliary water systems might lead to serious loss-of-coolant-flow accidents if they could flood the main coolant pump motors to the point of causing them to fail.

3.3.2 Reactivity Coefficients

Almost any change in the temperature, density, or location of any of the materials in a reactor core will produce some change in reactivity. In this section those changes are considered which do not involve in themselves any permanent distortion of the core structure. The effects of larger distortions are considered in Sec. 3.3.3 below. The reactivity coefficients which characterize these changes determine the inherent stability of the reactor, and this aspect has already been mentioned in Sec. 3.1.2 above. At this point, the interest is in how they may affect the course of an accident which is already under way. The effect may be to limit or to increase the severity of the accident. In general secondary processes will increase the severity of an accident if their effect is to increase reactivity under the influence of the primary disturbance. The converse statement may not be categorically true. Some reactivity accidents may be visualized in which the speed of reactivity addition, rather than the total reactivity available, determines the final excess reactivity. In such cases it is conceivable that secondary effects which reduce reactivity might actually increase the total nuclear energy release of the accident. However, considerations of this type only emphasize the necessity for specific analyses of postulated accidents on the reactor under consideration; as a general design principle it is reasonable to state that secondary processes which decrease reactivity are favorable to reactor safety.

The features which determine the signs and magnitudes of the important reactivity coefficients are discussed in detail in Secs. 6 and 7. The present discussion considers only in a general way how they enter into possible accidents. These considerations are treated in much more detail for specific reactor types in the chapters on Fast Reactor Kinetics, Water Reactor Kinetics, and Solid-Moderator Reactor Kinetics.

As pointed out in Sec. 6, the manner in which the net reactivity behavior of a reactor is broken down into "coefficients" may be rather arbitrary, to fit the particular purpose at hand, and the terminology is not always consistent from one breakdown to another: that section should be consulted for the more precise definitions of various coefficients. The following discussion is general enough to avoid the need for precise definitions; the classification of the various reactivity effects is that which fits best the objectives of relating the dynamic behavior of the reactor to the characteristics of core design.

3.3.2.1 Temperature of Thermally Isolated Moderator. In many reactors the moderator, or an important fraction of the moderator, is isolated thermally (more or less) from both the fuel and the main coolant. It therefore has a thermal time constant which is substantially longer than those of the fuel and coolant, and from the point of view of dynamics it is convenient to associate a reactivity coefficient with the temperature (and/or density) of this moderator. This would be the case, for example, in heterogeneous graphite-moderated reactors and in D_2O-moderated reactors of the pressure-tube type. Although the temperature and density changes of such moderators may be rather slow, they and their effects on reactivity must be taken into account. All of the possible sources of temperature and density changes must be considered, including heat transfer, direct deposition of neutron and gamma energy, the release of stored Wigner energy, possible chemical reactions, and the generation of radiolytic gases. It is important to recognize that, since the moderator temperature determines the effective neutron temperature, the moderator temperature coefficient of reactivity may vary as the core composition varies (with fuel burnup): the coefficient may, for example, change from negative to positive as plutonium builds up in slightly enriched uranium fuel.

3.3.2.2 Coolant Temperature and Density. The reactivity effects associated with coolant temperature and density may be quite complex, because the coolant may not only act as a neutron absorber and a moderator in the usual sense, but it may also have important effects on disadvantage factors and upon resonance absorption. Consequently the designer should not rely on intuitive assessments of the reactivity effects associated with the coolant, but should seek the best possible calculations or experimental determinations.

In those cases where the coolant has significant effectiveness as a moderator, its temperature, per se, may have an effect on reactivity. Usually, however, the conceivable reactivity changes due to coolant density changes, either as the result of thermal expansion, boiling, or the actual leakage or expulsion of coolant from the core, will be larger and more important. Ordinarily the coolant characteristics will respond rather rapidly to changes in reactor power and, if the coolant temperature and coolant boiling coefficients of reactivity are negative, they may play an important part in limiting the severity of reactivity accidents. Conversely, temperature and boiling coefficients which are positive can produce autocatalytic effects and could conceivably amplify a relatively minor reactivity excursion into a serious accident.

When a positive coolant voiding coefficient of reactivity exists (i.e. when the coolant density coefficient of reactivity is negative) reactivity excursions may be initiated by events which in themselves involve no increase in reactor power. For example, a break in the primary coolant system, or even a pump failure, may result in a decrease in the average coolant density in the core, through leakage or vaporization; this possibility is particularly prominent when the system is pressurized. In such cases, particular care must be taken to ensure that the scram system is capable of shutting the reactor down before the reactivity increase can become serious.

3.3.2.3 Fuel Temperature. The "fuel temperature" coefficient of reactivity is an extremely important one, because it is "prompt" relative to the fuel temperature, which is to say that it is "prompt" relative to the most immediate undesirable consequence of the accident. In natural or slightly enriched fuels which are not mixed with moderating material, the fuel temperature coefficient of reactivity is negative for all cases that have been considered to date. The important component of this coefficient is due to the Doppler broadening of the U^{238} or thorium absorption resonances. In fully enriched, "unmoderated" fuels the "nuclear" component of the fuel temperature coefficient may be positive because of the Doppler broadening of the fission resonances. However, it is usually quite small and may be overridden by the effects of thermal expansion of the fuel or may be negligible relative to other coefficients. In the special case of fast neutron reactors, however, the possibility and the consequences of a positive fuel temperature coefficient of reactivity must be carefully examined (see the chapter on Fast Reactor Kinetics).

When the fuel is mixed with an appreciable quantity of moderating material, the effects of the thermal neutron temperature are important, and the net fuel temperature coefficient of reactivity may depend strongly on the fuel composition. In particular, plutonium fuels, when mixed with moderator, usually exhibit a positive temperature coefficient of reactivity due to the large fission resonance of Pu^{239} at 0.3 ev.

3.3.2.4 Small Movements of Core Parts. In some cases, the movements of core parts due to thermal expansion or other forces may produce significant reactivity effects, even though the motions do not exceed elastic limits or design limits of the materials or design tolerances. In compact cores, the thermal expansion of fuel elements may be enough to cause significant reactivity changes through the resulting change in core geometry and through the expulsion of coolant from the core. More often however the significant reactivity effects result from distortion of the core members as a result of nonuniform heating or nonuniform cooling. These distortions may occur in the fuel elements or in the core structure. The possible reactor physics effects involved are numerous. They include such effects as leakage changes due to changes in the gross geometry of the core, changes in disadvantage factors and resonance absorption because of motions of fuel elements toward or away from each other, and changes in control rod effectiveness due to motions relative to the fuel or moderator. The importance of reactivity changes due to core distortions is particularly great for fast neutron reactors, but the possibility should be considered for any reactors in which the reactivity effects will not obviously be overridden by other, much larger, reactivity coefficients.

3.3.3 Reactivity Effects of Gross Core Failure

If an accident has progressed to the point where the normal restraints on the core and the core components have been rendered ineffective, either because of melting or because of dynamic forces which have built up within the core, the possibilities for large reactivity changes become more numerous. The prediction of the reactivity behavior after core failure becomes a complex problem for any particular reactor; it is treated in some detail for several classes of reactors in the chapters on Fast Reactor Kinetics, Water Reactor Kinetics and Solid-Moderator Reactor Kinetics. In general, three classes of effects may be recognized; change of core geometry; concentration or dilution of the fuel due to melting; and changes in the heterogeneity of the core. Although these three merge into one another, they do, to a degree, involve different considerations.

3.3.3.1 Core Geometry. Gross changes in core geometry can be visualized as the results of forces which may build up within the core due to such things as the rapid vaporization of coolant. Ordinarily, one would expect these forces to be directed outward, and to tend to increase the diameter of the core. Often such an expansion of the core will be inherently connected with the formation of voids in the core, and the net effect of the two processes may be shown to decrease reactivity. It is conceivable, however, that in some cases a reactivity increase might result. Specific analyses are necessary for specific reactors. The subject is discussed in more detail in Sec. 6.1.

3.3.3.2 Concentration or Dilution of the Fuel upon Melting. If one visualizes extensive fuel melting, then the degree to which the fuel is mixed with other core materials may be changed substantially. The reactivity effect of such an occurrence again can be evaluated only for a specific reactor. In some cases the concentration of the fuel may increase reactivity, while in others it may decrease it.

In the special case of the fast-neutron reactor, concentration of the fuel, per se, will usually cause a reactivity increase. This effect may, however, be overridden by concurrent changes in geometry. One of the fundamental safety problems in the fast reactor is to insure that in the event of fuel melting, supercritical assemblies do not result from the new fuel configuration. This consideration of course, is not connected with the fact that the original reactor operates on fast neutrons but with the fact that the fast reactor normally contains large quantities of rather highly enriched fuel, which, if allowed to gather in a compact mass, are highly supercritical. The same consideration would apply for any reactor containing the same type of fuel inventory.

3.3.3.3 Changes of Heterogeneity. In many reactors the reactivity is determined strongly by the particular configuration in which the core materials are arranged. For example, many large reactors contain several critical masses, which are held under control by the local effects of control rods. Gross movements of portions of the core relative to the control rods can, of course, lead to substantial reactivity increases. Often the collection of fuel into more compact masses, as might occur if the fuel rods in a bundle melted or sagged together, will increase reactivity. Again, these considerations are particularly important in the case of fast neutron reactors. The effects are discussed more fully in Sec. 6.

3.3.4 Non-Nuclear Consequences of Gross Core Failure

The reactor portion of accident analysis is not complete until the final consequences have been predicted, in terms of the release of radiation and radioactive materials from the reactor and in terms of the materials and energy released which can exert forces on the reactor building or containment structure. The following paragraphs do no more than mention those processes which have also a relation to core design.

3.3.4.1 Release of Fission Products from Core. In accidents which involve the overheating or failure of fuel elements, one of the central questions concerns the degree to which the normal retention of fission products breaks down. For jacketed ceramic fuel elements, for example, the necessary analysis may involve the prediction of fission product release from the fuel body itself, the assessment of the probability of failure of the fuel element jacket, from overheating (Sec. 8 in the chapter on Heat Transfer) or fission-gas pressure (Sec. 3 in the chapter on Fuel Elements), and the retention of fission products in the core coolant. Beyond the latter stage, the question of fission product behavior may be considered to be outside the boundaries of core design.

3.3.4.2 Conversion of Nuclear Energy to Mechanical Energy. Although it has become customary to characterize reactivity accidents in terms of the nuclear energy release, this quantity does not have a precise general relation to the severity of the accident. To predict the mechanical consequences of the accident it is necessary to estimate the fraction of this energy which is converted to mechanical energy and the rate at which the conversion takes place. In most cases the conversion will depend upon the vaporization of some material: either the fuel itself, or the coolant, or both. Of these, the vaporization of the coolant probably involves the largest uncertainties.

If the accident leads to extensive melting of the fuel elements, the surface area of hot fuel available for heat transfer to the reactor coolant, or to any other liquid which may be present, may be increased by a very large factor. In such a case, the transfer of heat to the liquid may become very rapid, and explosive pressures may be generated. This effect apparently was observed in the terminal excursion of the BORAX-I reactor, when pressures far above the critical pressure of water were indicated. The characteristics of heat transfer, steam generation, and pressure buildup under such conditions are poorly understood, but the possibility of effects of this kind must be anticipated in severe accidents involving the extensive melting of fuel.

3.3.4.3 Chemical and Wigner Energy Releases. If an accident has progressed so far as to cause extensive fuel melting, the chemical relationships within the core may be quite different than those which ordinarily apply. If the melting of the fuel results in the production of quantities of finely divided materials, the conditions for chemical reaction may be made still more favorable, and any excess pressure which may have been generated in the course of the accident may enhance the probability still further. Consequently, the possibility of chemical energy releases must be considered whenever the thermodynamic situation is such as to make them possible (Secs. 3 and 4 of the Chemical Reactions chapter). The possibility of chemical reactions between metals and water has received considerable attention in connection with accidents to water-cooled reactors, but the results are not yet definitive. Needless to say, if in the course of the accident, obviously incompatible materials are brought together (e.g. if sodium leaks into water or vice-versa), the consequences must be taken into account.

In the case of accidents such as the loss-of-coolant accident, in which fuel may melt or become severely overheated without the occurrence of a reactivity excursion, the possibilities of chemical energy release must also be taken into account.

In the special case of graphite reactors, where there may be a significant quantity of stored (Wigner) energy, the possibility exists that it may be released in the course of any accident which causes moderator overheating. If in addition the moderator temperature coefficient of reactivity is positive, this may lead to still more serious consequences.

3.3.4.4 Dynamic Forces. It is important to recognize that in severe accidents forces may be generated and loads may be imposed which are quite different from those existing during normal operation of the reactor. A well known example occurred in the case of the SL-1 accident, in which the entire reactor vessel jumped several feet upward as a result of a "water hammer" initiated by a power excursion. Such vital components as the control rods should be so designed and installed that forces which may be generated in an accident cannot cause them to move in such a way as to contribute to the severity of the accident. Dynamic forces, such as those which were generated in the SL-1 accident, should be given special attention. In pressure tube reactors, for example, the "rocket-like" reaction of escaping steam from a ruptured pressure tube should be taken into account in the design of the tube supports, so that individual tube failures do not lead to the destruction of surrounding pressure tubes or core components. Naturally, in the design of pressure vessel reactors, the vessel must be provided with restraints that will be effective under all conceivable accident conditions.

4 REACTIVITY CONTROL

It should be clear from Section 3 that the reactivity control elements play important roles in most conceivable reactor accidents. In some cases they may be involved in the initiation of the accident, and in almost all cases they are called upon to shut the reactor down as part of the process of countering the effects of the accident. This section discusses the nuclear characteristics of the reactivity control elements and the relationship of those characteristics to safe design. Useful presentations of calculation methods for control elements may be found in references [48-51].

4.1 Reactivity Inventory and Shutdown Margin

One of the first considerations in nuclear design is to make available sufficient reactivity for operation under all proposed conditions over the design lifetime of the core. At the same time the designer must provide means for assuring the capability of shutdown at all times. Thus in a typical operating condition the reactor will be just critical and its reactivity, by definition, will be zero; but the reactivity can be increased or decreased by appropriate movements of the control elements. The amount that the reactivity can be increased or decreased depends upon the condition of the reactor, i.e., its temperature, pressure, time, in-core life, etc., and upon the control elements. The reactivity which would result if all control elements were adjusted to their maximum-reactivity states is known as the excess reactivity* of the reactor in that particular condition, ρ_{ex}. The reactivity which results when all control elements are adjusted to their minimum-reactivity states is designated here as ρ_{sd}. The difference between ρ_{sd} and the critical condition is generally called the shutdown margin. The difference $\rho_{ex} - \rho_{sd}$ may be considered to be the total reactivity worth of the control elements. Excess reactivities and shutdown margins may be specified (and will have characteristic values) at any time in the core lifetime and for any ambient conditions, but they are often specified at the beginning of core life for the cold, zero-power condition. Usually the difference, $\rho_{ex} - \rho_{sd}$ will also vary with core exposure and with ambient conditions.

---

*Sometimes also called the available excess reactivity, the available reactivity, or the built-in reactivity. The term excess reactivity is also used loosely in other senses. Usually the modifier excess is redundant, as indeed it is in the application above. The term available reactivity is a more logical one, but the terminology defined above is in more general use, at least in the United States.

4.1.1 Problems of Specifying the Reactivity Inventory

The above definitions of excess reactivity and shutdown margin are unambiguous and can even be described in terms of conceptual experiments. However, they define only the end-points of the reactivity scale for a reactor, whereas the designer and the safety analyst are usually interested in assigning components of the excess reactivity to those individual phenomena which make it necessary to provide excess reactivity. Useful as such a breakdown may be in concept, it leads to much confusion, for neither are the phenomena in question independent nor do the control elements affect the reactivity independently of each other. It is standard practice in hazard studies and reports to give a breakdown of the reactivity inventory, yet the precise meaning of the breakdown is usually not understandable by anyone except (possibly) the author of the study. Although much of the confusion results because of the inherent complexities of the subject, some extraneous sources of confusion may be identified and perhaps corrected.

Much of the difficulty arises simply because of the failure to specify precisely the conditions of the reactor for which the components of the reactivity inventory are stated. The importance of precise specification can be illustrated by an example.

Consider the simplified case of a reactor containing control rods, which is just critical at room temperature with all rods inserted, which has a relatively large temperature defect, and which at room temperature has just enough excess reactivity (assumed to be 0.06) to achieve criticality at 500°F. The curve of excess reactivity as a function of temperature might look something like the curve (a) in Fig. 4-1. The significance of this curve is well defined: if at any particular temperature all control rods were withdrawn, the power would rise exponentially with a period consistent with the reactivity $\Delta\rho_{ex}$, see curve (a′), but this is evidently Now assume that the reactor is brought to some intermediate temperature, say 300°F, and is held critical there by an appropriate adjustment of the control rods. One would now like to assign a "reactivity worth" to the amount of control rod withdrawal required to maintain criticality over the temperature increase from room temperature to 300°F (or alternatively to assign a reactivity worth to the temperature increase itself). The first temptation is to equate it to the change in excess reactivity $\Delta\rho_{ex}$, see curve (a′), but this is evidently incorrect, since the curve of excess reactivity applies only to a core without control rods, whereas the reactivity changes in question take place in a core containing control rods.

There are two possible operational definitions for the reactivity worth in question, either of which might be adopted. Conceptually, one might insert all control rods into the 300°F reactor, measure the (negative) reactivity, and assign this value (with appropriate algebraic sign) as the reactivity worth in question. Alternatively, one might reduce the reactor temperature to room temperature, withdraw rods to the degree which previously gave criticality at 300°F, evaluate the reactivity through a period measurement, and identify this reactivity

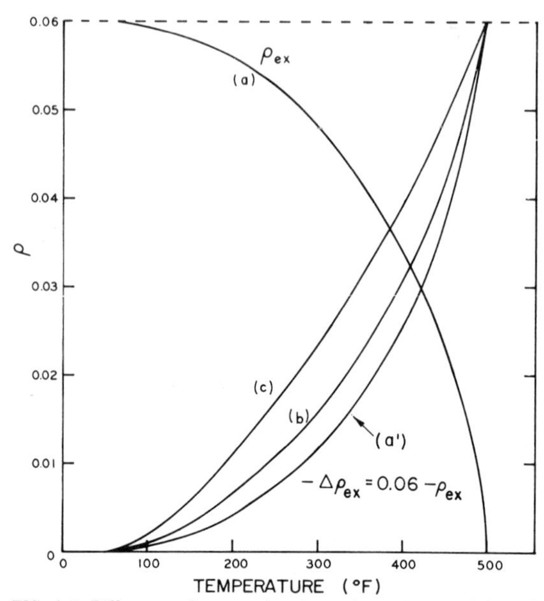

FIG. 4-1 Differences between the change in excess reactivity and the reactivity "worth" of a change in reactor temperature. A curve like (c) might result if the reactivity "worth" of the temperature change were evaluated as follows. The reactor is assumed to be just critical at room temperature with all control rods fully inserted. To determine the worth of the temperature change from temperature T to room temperature the reactor is first heated to T and all control rods are withdrawn as a "bank" to make the reactor just critical. With the rods held in this position the reactor is then suddenly cooled to room temperature, and the period of exponential power rise is measured. The reactivity corresponding to this period is designated as the reactivity worth of the temperature change from T to room temperature. A curve like (b) might result if the same method were used except that the control rods were not withdrawn as a bank, but individually, as required.

with the reactivity worth in question. The two definitions would yield different values for the reactivity worth, and neither would, in general, be equal to the value of $\Delta\rho_{ex}$ at 300°F. Either of the two might be the more appropriate and useful definition, depending upon what the result is to be used for.

Furthermore, regardless of which of the above definitions is adopted, the reactivity worth in question will vary with the type and programming of control elements. If, for example, many small control rods were provided, uniformly distributed in the core, and they were withdrawn, individually, in such a way as to preserve a reasonably uniform pattern, the (negative) reactivity worth of the temperature increase might follow a curve similar to (b) in Fig. 4-1. If all the control rods were withdrawn together, as a bank, the reactivity worth of the temperature increase might follow curve (c). Yet the curve of excess reactivity (a) would remain unchanged in either case.

If the second of the two definitions above were adopted for a particular case, one might (conceptually) construct a reactivity curve such as (c) of Fig. 4-1 by the following procedure. First the reactor temperature would be raised, in steps, from room temperature to 500°F, and the critical control rod position determined at each step. The reactor would then be cooled down to room temperature. In the cold reactor, the control rods would be withdrawn, in successive experiments, to the various positions previously found to yield criticality at the observed temperatures, and the reactivity would be determined at each position by a period measurement. The reactivity measured at each position could then be assigned as the (negative) reactivity worth of the temperature increase which would make the reactor just critical with that control rod configuration. In this procedure, reactivities would be found to be rigorously additive only in the trivial sense that the curve would be continuous and single-valued so long as the specified calibration procedure was followed. If, for example, one attempted to construct the curve by carrying the room-temperature-based calibration only to 300°F, and adding to it a curve developed by calibrations taken from 300°F as a base point, one would produce a slightly different curve. This latter consideration is of great practical importance, for in practice, if one were determining the temperature defect experimentally, one would not be able to develop a curve like (c) of Fig. 4-1 by withdrawing the control rods in progressively larger increments, since the very short periods corresponding to the larger increments would be intolerable. In practice, one might proceed as follows. With the reactor at room temperature, the control rods would be withdrawn some small safe increment and the reactivity worth of this increment would be determined by period measurement. The reactor temperature would then be raised by some increment, the reactor would be made critical by appropriate rod withdrawal, the rods would be withdrawn another additional increment, and the reactivity worth of the rod increment again would be determined by period measurement. This procedure would be repeated until the reactor temperature reached 500°F. The control rods would have been calibrated over their entire range, and this calibration could be used to evaluate the "worth" of the temperature change at the final temperature or at any intermediate temperature. The careful analysis of Henry [52] of the significance of such a calibration has shown that it does not yield a curve identical to (c), but in many cases gives one which is a close approximation. Reference [52] has an excellent discussion of the significance of reactivity specifications, and it should be familiar to all who compute, specify, or measure reactivity changes.

The point of the example above is that, although the temperature defect to 500°F is a reasonably unambiguous quantity in this particular case—simply because in this particular case the control rods are completely withdrawn at 500°F—the temperature defect at some intermediate temperature is not a definitely defined quantity unless one specifies the control rod configuration at that temperature.

In the more complex cases of practical power reactors, still further information may be necessary for giving a complete specification of the reactivity change due to a change in some reactor variable. Thus, for example, the temperature defect will be different when xenon has built up in the reactor than when it has not, and it may be quite different at the end of core life than at the beginning. It seems an obvious conclusion that the designer, in addition to specifying carefully the conditions under which the components of the reactivity inventory have been calculated, should be careful to choose those conditions as the ones which will be important in actual reactor operation. In spite of this, the calculated cases are sometimes based on highly idealized situations which will never occur in practice. To some extent this deficiency in analysis has been fostered by a preoccupation with the need for producing a neatly balanced breakdown of the reactivity inventory components—a breakdown which adds up to the excess reactivity. Actually, such a balance may have little or no practical significance, and often can be rigorously achieved only by the analysis of idealized situations.

It is much more to the point for the designer to analyze the situations which can actually occur, taking into account the control element configurations and power distributions to be expected in practice. Table 4-1 gives an illustrative set of reactivity specifications which perhaps represents the minimum acceptable for characterizing the beginning-of-life reactivity inventory of a reactor with large excess reactivity. Similar characterizations will usually be necessary at other points in the core lifetime, depending upon the individual reactor. Other methods of specifying the reactivity inventory may be equally good if they relate the reactivity components to real operating conditions. The make-up of Table 4-1 is directed toward giving the information needed for a general assessment of reactor characterisitics, and should be applicable to most reactors having substantial excess reactivity. Additional situations should be calculated to predict any reactivity components which can be compared directly with experiment in the particular reactor under consideration and to give any values needed for the analyses of specific hazards visualized in that reactor.

The following two rules should be followed in specifying reactivity changes to ensure that the specification is meaningful and unambiguous:

1. Always specify reactivity changes in terms of the reactivity ρ, as defined in Sec. 2, or by specifying the values of k_{eff} before and after the change in reactor conditions which causes the reactivity change. Both of these specifications are dimensionless, requiring no units; attempts to assign units to them (e.g., %$\Delta k/k$ are only confusing.*

*The ratio ρ/β is a useful quantity which may also be used unambiguously for the specification of reactivity inventory, provided ρ is given its accepted definition (Sec. 2). If β (the effective delayed neutron fraction) can also be given, the net result is obviously the same as that recommended above. The ratio is of course useful even when β cannot be specified, but it does not serve the same purpose in relation to the statics of reactor analysis as does the reactivity ρ alone.

It is confusing that the ratio ρ/β is also referred to as the reactivity. It is given the unit of the dollar: thus, reactivity in dollars = ρ/β and reactivity in cents = 100 ρ/β. It is perhaps regrettable that the necessity for finding a convenient name for the ratio ρ/β has been

TABLE 4-1

Sample Specification Sheet for Important Components of the Reactivity Inventory*

1. Excess reactivity, cold

2. Shutdown margin, cold reactor, at time of maximum reactivity

3. Shutdown margin, hot reactor, at time of maximum reactivity

4. Critical control rod positions; i.e., state the degree of withdrawal of each control rod, when withdrawn according to the program actually to be used in reactor operation, to give criticality under the following conditions of the reactor:

   position a: cold, clean

   position b: hot, clean, zero power

   position c: hot, clean, full power

   position d: hot, full power, equilibrium Xe and Sm

   position e: hot, full power, max. Xe and Sm (if significantly different from d)

5. Reactivity of cold, clean, reactor with rods at position b (see item 4 above) = temperature defect**

6. Reactivity of hot, clean, zero-power reactor, with rods at position c (see item 4 above) = power defect**

7. Reactivity of hot, clean, full-power reactor, with rods at position d (see item 4 above) = Xe + Sm defect**

8. Reactivity of hot, clean, full-power reactor, with rods at position e (see item 4 above) = transient Xe + Sm defect**

9. Reactivity of hot, full-power reactor, with equilibrium Xe + Sm, and all rods out = reactivity available for fuel burnup**

*This table is intended only for specification of the details of the reactivity inventory; it is not intended to cover the information ordinarily needed on control-rod worths.

**Note that there is a degree of arbitrariness in these names which have been applied to the reactivities specified in items 5 through 9; but whether or not the names are judged appropriate, the reactivities specified are the significant ones.

One might also interpret items 5 through 9 in terms of control rod worths, thus:

Item 5 = worth of control rods from configuration a to configuration b
Item 6 = worth of control rods from configuration b to configuration c
Item 7 = worth of control rods from configuration c to configuration d
Item 8 = worth of control rods from configuration d to configuration e
Item 9 = worth of control rods from configuration e to fully withdrawn

However, it must be remembered that these worths would apply for the particular reactor conditions specified, and additional specifications of rod worth would ordinarily be needed.

2. Specify completely the reactor condition before and after the reactivity change in question.

4.1.2 Relation of Excess Reactivity to Safety

As implied by Table 4-1, the cold clean reactor usually must be provided with excess reactivity to counter the reactivity losses accompanying operation: the losses experienced in reaching the operating levels of temperature and pressure; the losses, such as additional losses due to Doppler broadening of neutron resonance absorption lines, which may be associated with raising the power to the operating level; the losses due to the quickly saturable fission product poisons, xenon and samarium; and the losses expected as a result of fuel burnup, due to the buildup of slowly saturable (long-lived) fission products and the change in isotopic composition of the fuel material.

The excess reactivity provided may vary over a rather wide range according to the reactor type and the specific design and purpose of the reactor. The natural-uranium graphite-moderated reactors fall at the lower end of the scale, with typical excess reactivities of about 0.04, while the enriched-uranium, H_2O-moderated reactors are among those having the highest excess reactivities, in the range from perhaps 0.15 to 0.25.

Inasmuch as the built-in excess reactivity is

---

circumvented by defining a variable unit (the dollar) for the reactivity; this peculiarity, however, is not particularly troublesome once it is understood. Another usage defines the "reactivity" in dollars as $(k_{eff} - 1)/\beta$, compounding the confusion.

the source of reactivity for potential accidents, it is probably fair to say that, other things being equal, reactors with low excess reactivity are inherently safer than those with high excess reactivity. Yet in almost all useful reactors, the excess reactivity is quite high enough to support a serious accident, and the relationship between safety and excess reactivity per se is not a strong one. Once it is evident that the excess reactivity of the reactor in question must be well above the effective delayed neutron fraction (β), there is no great incentive, from safety considerations alone, for the reactor designer to minimize the excess reactivity requirement; but it is certainly true that, as excess reactivity increases, the related safety problems of providing adequate shutdown margin and of limiting the range of possible reactivity addition rates to a safe value become increasingly difficult. Thus, in assessing the safety of a reactor, a high value of excess reactivity should not be regarded as inherently unsafe, but should serve to direct particular attention to the question of how well the designer has solved the related safety problems.

4.1.3 Relation of Shutdown Margin to Safety

Two basic safety requirements must be satisfied by the shutdown margin provided in the reactor: the control agencies must be capable of rendering and holding the reactor subcritical at any time despite the effects of any reasonable failures or malfunctions of the control elements or of the reactor proper, and the shutdown capability must be large enough to reduce reactor power at a rate necessary to avoid fuel overheating in the event of certain transient accidents, such as the failure of coolant pumps.

The first of these requirements should be stated in terms of the required number of excess control elements rather than as a required degree of subcriticality when all control elements are operating normally. For rod-controlled reactors, it is usually stated in terms of the stuck-rod criterion: the control system must be capable of rendering the reactor subcritical with some specified number of control rods stuck in the fully-withdrawn position. Just what the stuck-rod criterion should be for acceptable safety will depend upon the particular reactor, the degree of backup provided by auxiliary shutdown systems, and the operating procedures for the reactor. The criterion for a specific reactor should be arrived at by analyses of the conceivable accidents that the reactor might experience. In most reactors a backup emergency shutdown method is provided; in water-moderated reactors this is usually an independent system which can inject soluble neutron absorber (poison) into the reactor water. In some cases (e.g., the Yankee reactor) [53] the control rods are not designed to shut down the cold clean reactor, but the soluble poison system is relied on entirely for this function. In such a case, the safety of the reactor is strongly dependent upon safe operating procedures and careful administrative control in the handling of the soluble poison. Experience with this type of operation is still meager, and generally accepted practices have not yet become apparent.

There are a number of basic differences between a rod control system and a soluble poison control system, which are discussed more fully in a later section, but the important one for the present consideration is that the addition of a soluble poison cannot ordinarily reduce the reactivity nearly as rapidly as a rod system can. A reasonable requirement, when soluble poison is used, would appear to be that a sufficient number of control rods must be in the withdrawn state, at every operating condition, to provide some specified, rapidly appliable, shutdown margin. This requirement would include provision for a stuck-rod margin, and would be established in much the same way as has been discussed above for systems which do not use soluble poisons. Such a criterion, in conjunction with a soluble poison system suitable for use during power operation, would reduce considerably the number of control rods required for a given total excess reactivity. Whether the criterion is adequately safe will depend not so much on the dynamics of reactor shutdown as on the reliability of the soluble poison system.

The problem of satisfying stuck-rod criteria in H_2O-moderated reactors (without soluble poison) is often a difficult one, and to provide shutdown capability with more than one stuck rod often is impracticable unless auxiliary means such as soluble poison are employed. When soluble poisons are available but are not used for normal cold shutdown, a one-stuck-rod criterion is often considered adequate, provided the operating procedures specify immediate correction of a stuck-rod fault if one should occur.

The degree of shutdown margin affects the rate at which reactor power can be reduced in two ways. Since the mechanical speed of control rod scram is a fixed quantity for a given reactor design, the rate of reactivity reduction produced by the scram will ordinarily increase as the total worth of the rods scrammed is increased—i.e., for a given reactor, at given operating conditions, it will increase if the shutdown margin is increased by increasing the strength of the control rods. The benefits which are practically achievable from this effect are often marginal, and this consideration is usually given little weight in the reactor design unless some specific accident can be visualized in which an obvious advantage would result from a high shutdown margin. The more important consideration is usually the rate at which the neutron chain reaction decays as a function of (subcritical) reactivity after the rods have been scrammed. To cope with coolant system failures, the scram system must be able to reduce the subcritical power rapidly enough to prevent fuel element damage by overheating. The required rate may vary widely from reactor to reactor, and will depend upon such considerations as the characteristics of the conceivable coolant failures, the thermal time constant of the fuel elements, the normal specific power of the reactor, the emergency cooling supplied, and the ability of the fuel element to withstand temperatures which are higher than normal.

When the reactor is suddenly made subcritical, as by a scram, the power will drop rapidly to a fraction $1/[1-(\rho/\beta)]$ of the steady state power just prior to scram, and thereafter will decay at a slower rate. In the preceding expression, ρ is the

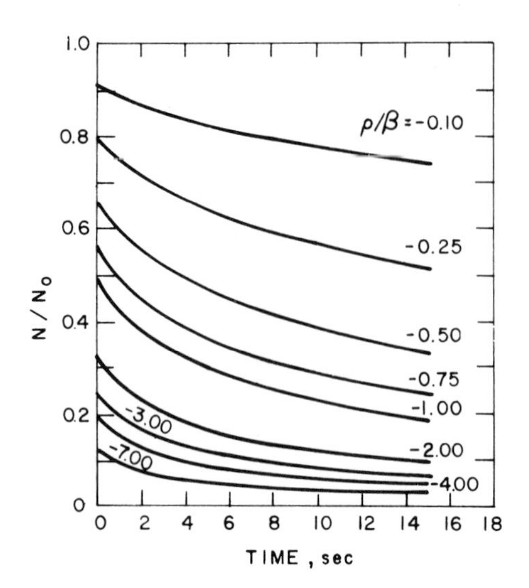

FIG. 4-2 Decay of reactor power from the steady state after an instantaneous reduction to reactivity ρ. N/N_0 is the ratio of the power to the steady-state power before shutdown. The initial power decrease is not instantaneous as implied by the figure, but appears instantaneous because of the rather coarse time scale.

reactivity after scram (inherently negative) and β is the effective delayed neutron fraction. Figure 4-2 [5] shows the decay of reactor power after scram for various values of ρ/β. In many cases, values of ρ/β in the range 4 to 6 are found to produce adequate rates of power decay. These power levels are in addition to the gamma and beta power produced by the decay of fission products, which will amount to approximately 5 percent of the steady state power before scram, and which will decay much more slowly than the neutron chain reaction power and at a rate independent of the reactivity.

The shutdown reactivity required to cope with coolant system failures should not in general be identified with the cold shutdown margin requirement. In most reactors high-power operation of the cold reactor for an extended period of time is not possible. If the reactor can operate for extended periods only at elevated temperatures, then the minimum negative reactivity available for shutdown of the steadily operating reactor will be equal to the minimum cold shutdown margin plus the temperature defect.

4.2 Nuclear Characteristics of Control Rods

The total required reactivity worth of the control rod installation is ordinarily specified by the sum of the absolute values of the cold excess reactivity and the required shutdown margin. This basic requirement focuses design attention immediately upon the question of reactivity worth. Moreover, essentially all safety questions relating to control rods center upon considerations of the reactivity worths of rods or groups of rods. This section is devoted to the treatment of reactivity worth; the other maor consideration, the effect of control rods on power distribution, is covered in Sec. 5.

4.2.1 Black Thermal Absorber Rods; Effect of Rod Shape

The absorber control rod produces a reduction of reactivity by absorbing some of the neutrons which otherwise would cause fission. The reactivity change produced by the absorption of a given fraction of neutrons in a reactor depends upon where in the reactor they are absorbed and upon the energy at which they are absorbed. However, in "thermal" reactors most of the neutron absorption (both in control rods and in other material) occurs in the thermal neutron energy range, and in most power reactors the complement of control rods is sprinkled fairly uniformly over the reactor volume. Consequently some insight can be gained by considering the case of pure thermal-neutron-absorbing rods in a well-thermalized reactor, with the rods distributed evenly over the core column. In such a case, if the control rods absorb a fraction C of all neutrons absorbed in the core, then the reactivity worth of all the rods is very nearly equal to this fraction C*. In this simplified case, the estimation of total control-rod worth reduces to an estimation of the fractional absorption of thermal neutrons by a regular array of absorbers in a multiplying medium which can, to a reasonably good approximation, be considered infinite in extent.

Ordinarily, the designer will use control rods which are nearly "black" to thermal neutrons in order to achieve a high control-rod worth. A rod is black to neutrons if all neutrons that cross its surface boundary are absorbed. Approximate blackness to thermal neutrons requires two characteristics: (1) the thermal-neutron-absorption cross section must be many times larger than the scattering cross section, and (2) the absorber material must have a thickness of several absorption mean free paths.

Figure 4-3 illustrates the changes in neutron absorption when a black, thermally absorbing, infinite slab is inserted in a slot, previously empty, in an infinite thermally absorbing medium. The medium is fed by a thermal-neutron source (e.g., the slowing down of fission neutrons), of strength S neutrons per cm^3, distributed uniformly over the volume of the medium. Prior to the insertion of the slab, the neutron flux has a uniform value ϕ_0 over the medium, the value being determined by the source strength and by the macroscopic absorption cross section (Σ_a) in the medium [see Fig. 4-3 (a)]. When the black slab is present, it absorbs those neutrons which cross its surface. Since the total absorption rate must be equal to the rate of neutron production by the source, the absorption by the slab must cause an equal reduction of the absorption in the medium adjacent to the slab. The neutron flux near the slab drops in such a manner as to produce just the proper reduction of absorption in the medium. The resulting thermal-flux distribution $\phi(x)$ is shown in Fig. 4-3 (b). The fractional difference between ϕ_0 and $\phi(x)$ at any point x (shown as

*The worth is not exactly equal to C because the extra neutron absorption by the rods reduces somewhat the leakage of thermal neutrons from the core.

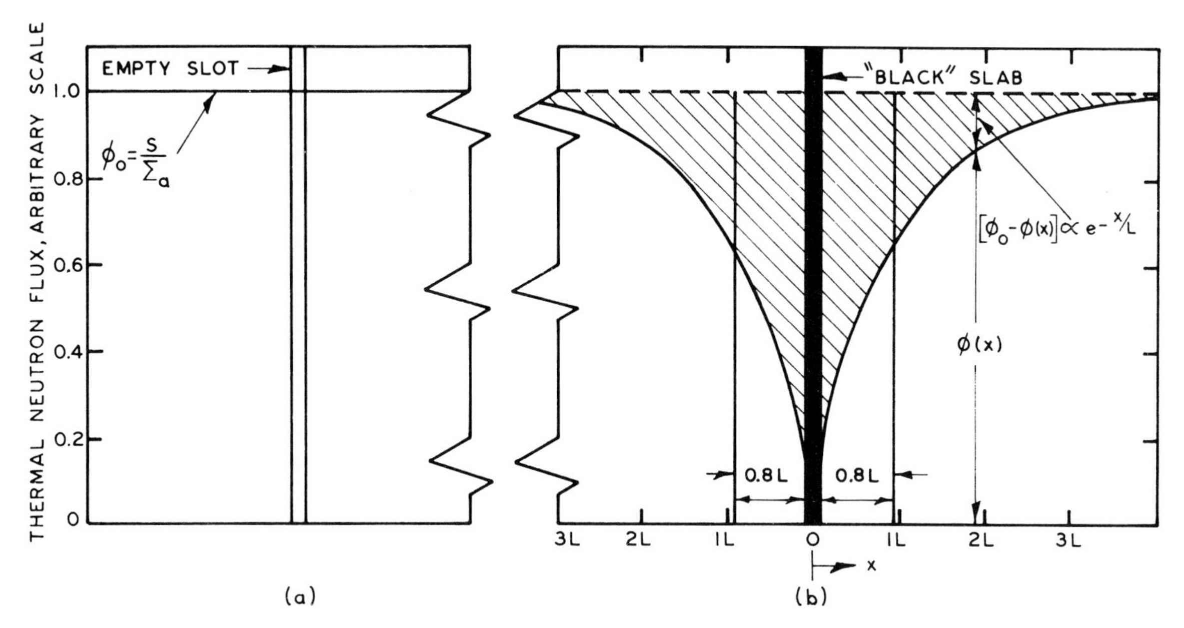

FIG. 4-3 Typical change of thermal-neutron flux distribution when a black thermal neutron absorber (b) is inserted in an empty slot (a) in an infinite medium having a uniform source of thermal neutrons.

a shaded area in the figure) may be regarded as the fractional absorption by the slab of neutrons which would otherwise be absorbed at point x. A short distance from the slab surface the curve $\phi_0 - \phi(x)$ decreases in the form of an exponential, exp (-x/L), where L is the diffusion length of thermal neutrons in the medium surrounding the slab. If this behavior persisted right up to the surface of the slab and if the thermal-neutron flux reached zero at the surface, the area of the shaded region in Fig. 4-3b would be just $\phi_0 L$ on each side of the slab. Thus each side of the slab would absorb a number of neutrons equal to the number produced by the source in that volume of the medium which lies within one diffusion length of the slab surface. Actually the exponential behavior does not persist to the slab surface, and the actual number of neutrons absorbed is somewhat less than this. The actual number can be computed by taking into account the extrapolation length at the control slab surface, which is determined by the transport mean free path in the surrounding medium. If, for example, the surrounding medium had the same thermal-neutron absorption and transport properties as the average core composition of the Experimental Boiling Water Reactor (EBWR), [54] the actual absorption in the slab would be equal to those neutrons produced within a distance of about 0.8 L of the surface.

If other absorber shapes are considered, the results are rather similar in that the black absorber captures a quantity of thermal neutrons equal to those produced within some characteristic fraction of a thermal neutron diffusion length. The exact value of the fraction will depend upon the shape of the absorber and upon the transport mean free path in the surrounding medium, but the variation in size of the fraction is unlikely to be large over the range of practical situations.

In the reactor core the situation is more complex than that considered above, because the thermal neutrons are not supplied by a uniformly distributed source but by the slowing down of fission neutrons which were generated at each point in the fuel at a rate proportional to the thermal-neutron flux at that point. Consequently the production of thermal neutrons near the control rod will be somewhat lower than in regions far from the rod. This is usually not an extremely large effect, however, because the diffusion of the neutrons during the slowing-down process makes the rate of production of thermal neutrons much more uniform than the distribution of fission density.

The net effect of these considerations is that, for most control installations involving a fairly uniform lattice of control rods, a rough idea of the fractional absorption of thermal neutrons can be obtained by assuming that it is equal to the fractional core volume which lies within a distance αL of the control-rod surfaces. The α is a fraction which depends on rod shape and on the transport mean free path in the core; it will seldom be less than 0.5 or greater than 0.8; a value of 0.7 might be representative for most reactors. This approach to the evaluation of control-rod absorption is a crude form of the "absorption area" technique which is widely used for more accurate calculations of control-rod worth. Descriptions of the technique, along with formulas and curves which are useful for evaluation of absorption areas, may be found in references [55] through [57].

The simpler approach may be used to examine, for example, the control rod absorption in the EBWR. [54] In the region of the core occupied by control rods, each rod can be considered to be located at the center of a "cell" having a square cross section with a side length of 12-3/4 in. (32.4 cm), see Fig. 4-4. [54] The rod itself is a cross of 10-in. (25.4 cm) span. If one draws along the rod a line which is everywhere about 0.7L from the rod surface (L = 1.8 cm), the area enclosed between this line and the surface of the rod is 135 cm^2, whereas the cross-sectional area of the cell exclusive of the rod is 1030 cm^2. This rough approximation indicates that the rod absorbs about 13% of the neutrons, and the reactivity change corresponding to this fractional absorption should be 13% of k_{eff}. The quoted value of all the control rods in the EBWR is 12 to 14% of k_{eff}.* Little importance can be attached to the quantitative agreement or lack of it; for, actually, rods are not distributed uniformly over the volume of the EBWR, no allowance has been made for epithermal effects, the core composition is not really uniform up to the surface of the rod, and the rod is not inserted into a nonabsorbing slot but actually displaces water. Nevertheless the estimate does illustrate some of the factors affecting control rod worth. The method of estimating can be used to correlate roughly the rod requirements for specified reactivity worths in thermal reactors of widely varying properties. Table 4-2 compares the estimated worths with reported worths for four quite different reactors. The worth estimates are made simply by examining a cross section of the core with control rods inserted and taking the ratio of the area within 0.7L of the rod surfaces to the area of core external to rod surfaces.

The factors affecting rod reactivity worth can be further illustrated by considering the absorption by cylindrical rods in reactors of widely different compositions. Figure 4-5 illustrates the absorption by rods in a reactor whose core has a low macroscopic absorption cross section for thermal neutrons and a correspondingly long diffusion length. The medium to which the figure applies has average properties similar to those in the core of the original Calder Hall reactor [58]. The moderator is graphite, and the diffusion length in the core is 19.8 cm. If a black rod 2 in. (5.08 cm) in diameter is installed in such a medium, it absorbs a number of thermal neutrons equal to those formed within a distance of about 13.4 cm of the rod surface. If the rod diameter is increased to 3 in. (7.62 cm), the effective distance from which neutrons are drawn increases only slightly, to 14.4 cm. In considering these diagrams, three points are obvious:

1. The volume of the core from which neutrons may be considered to be drawn by the rod is large compared with the volume of the rod. In this respect the rod is quite effective in influencing reactivity, and only a relatively small fraction of the core volume would have to be occupied by control rods in order to achieve a high degree of control rod worth.

2. The number of neutrons absorbed by the rod does not increase in proportion to the size of the rod. The reason for this is obvious. Because of this effect, most reactors which have cores of low average absorption cross section employ rods which are quite small relative to the size of the reactor.

3. Evidently it would not be profitable to install control rods very close together in a low-cross-section reactor, for then adjacent rods would have zones of effective absorption which would overlap and the total number of neutrons absorbed per rod would decrease. Actually the space necessary to avoid important mutual "shadowing" of the rods is somewhat greater than the distance indicated by the circles surrounding the rods in the figure. It would be unusual to find rods closer together than two or three diffusion lengths in any reactor.

*Reference [54]. The worth listed in the reference (calculated) is given as the value of k_{eff} which the rods will hold just critical. The rod worth used here, in terms of that same k_{eff}, is $(k_{eff} - 1)/k_{eff}$.

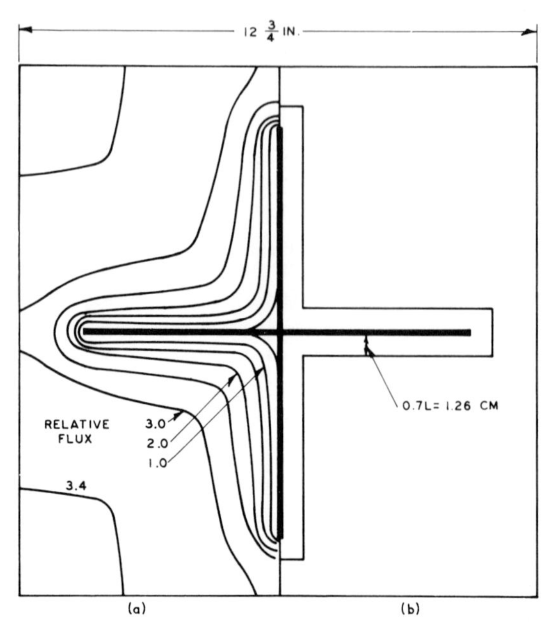

FIG. 4-4 Typical control-rod cell in EBWR. (a) Contours of constant thermal-neutron flux. (b) Approximate area of effective absorption.

Figure 4-6, which is necessarily drawn to a quite different scale, illustrates the absorptions by the same two rods if they were installed in a highly absorbing reactor core. The actual medium considered has the average properties of the cold EBWR core [54]. The main characteristics which are as follows, are quite different than in the previous case:

1. The volume of the core from which neutrons may be considered to be drawn by the rod is not large relative to the rod itself.

2. The increase of absorption with rod size is more nearly proportional to the increase in surface area of the rod.

3. Since the area of effective absorption of the rods is small, rods may be installed relatively close together without important shadowing.

Figure 4-6 illustrates some of the difficulties of devising efficient rod installations for water-moderated reactors. Quite evidently the thermally absorbing cylindrical rod is an inefficient means of compensating excess reactivity, for it occupies a rather large amount of core volume per neutron absorbed. Furthermore, when the rod is withdrawn

TABLE 4-2

Reactivity Worth of Control-Rod Installations in Several Reactors

Core diam., ft.	Control-rod shape	Control-rod size	No. of control rods	Core diffusion length, cm	A (cross-sectional area of core), cm^2	B (cross-sectional area within distance 0.7L of rod surfaces), cm^2	Estimated rod worth, B/A	Reported worth of control rods ($\Delta k_{eff}/k_{eff}$)
EBWR[54] (Experimental Boiling Water Reactor), Boiling H_2O Type								
4	Cross	10-in. span	9	1.8	11,700	1,240	0.11*	0.12 - 0.14**
Adam[59] (Swedish Reactor Experiment), Natural Uranium Oxide - D_2O Type								
11.1	Cylindrical	99-mm diam.	32	14.8	91,000	20,200	0.22	0.15
SRE[60] (Sodium Reactor Experiment), Sodium-Graphite Type								
6	Cylindrical	2.484-in.	8	12.9	26,300	3,480	0.13	0.135
Calder Hall[61], Natural Uranium - Graphite Type								
31	Cylindrical	1.75-in. diam. in 3.25-in. channel	48	19.8***	703,000	36,500	0.052	0.068

*This value differs from the value estimated in the text because there is some core area which is not included in the nine "typical cells" surrounding the control rods.

**In reference [54] the (calculated) worths of the rods are given as the value of k_{eff} which the rods will hold just critical. To get the rod worth, as used here, take $(k_{eff}-1)/k_{eff}$.

***Intermediate core region. See reference [58].

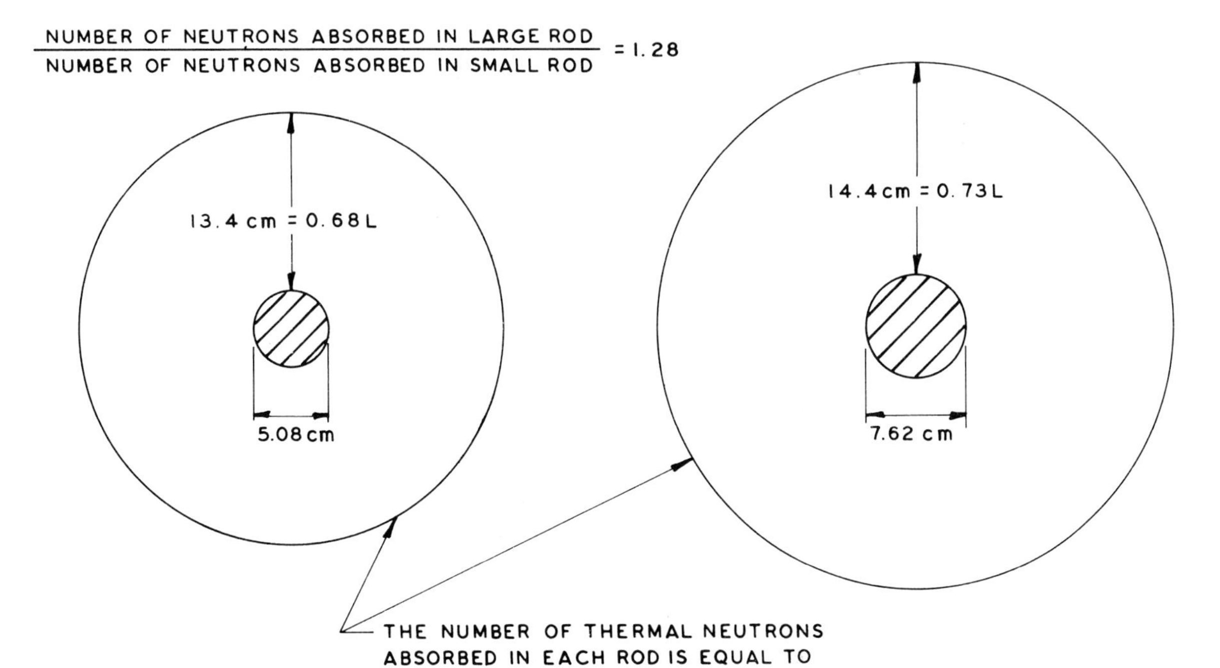

FIG. 4-5 Relative numbers of thermal neutrons absorbed by 2- and 3-in. (5.08 and 7.62 cm) cylindrical black rods when placed in identical infinite media supplied by uniformly distributed sources of equal strength. Neutron transport and absorption properties are the same as in region B of Calder Hall, homogenized.

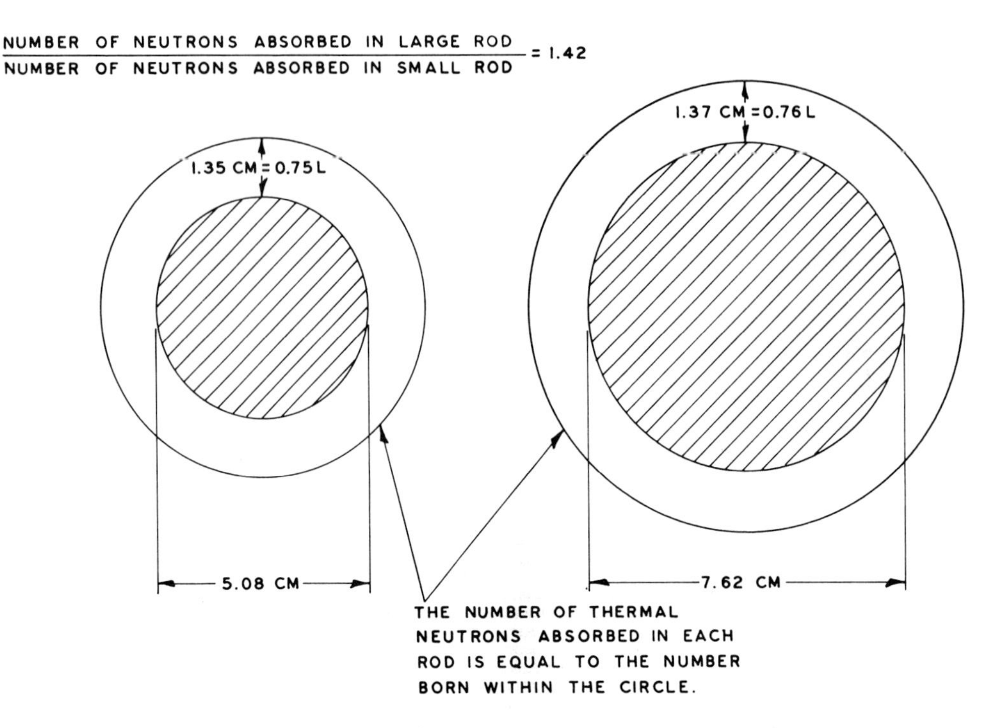

FIG. 4-6 Relative numbers of thermal neutrons absorbed by 2- and 3-in. (5.08 and 7.62 cm) cylindrical black rods when placed in identical infinite media supplied by uniformly distributed sources of equal strength. Neutron transport and absorption properties are the same as in a uniform, cold EBWR core.

from the core, the empty channel which it leaves poses a problem. If the channel is allowed to fill with water, the fast neutrons slowed down in the water will produce a peak in the thermal-neutron flux which will tend to overheat the adjacent fuel elements; and, in addition, the thermal-neutron absorption of the water, enhanced by the peaking of the thermal flux, may reduce the net reactivity gain resulting from withdrawal of the control rod. These are the reasons why rod shapes giving high surface-to-volume ratios, such as crosses, are used in water-moderated reactors. In the water reactor the diffusion length is so small that the rod surface must be brought into proximity to a substantial fraction of the uranium in the core in order to absorb a substantial fraction of the neutrons.

4.2.2 Criteria for Blackness

Usually the designer wishes to achieve the maximum possible neutron absorption per unit area of control rod, and usually it is feasible to incorporate enough highly absorbing isotope in the rod to render it nominally "black" to thermal neutrons. Once the condition of blackness has been achieved for thermal neutrons, further increases in the absorber content of the rod will not increase the absorption of thermal neutrons, but additional absorber may be necessary to allow for burnup of the absorbing isotope during extended operation or may be desirable in order to increase the absorption of epithermal neutrons.

The absorption cross section of a material at any given neutron energy determines how thick a layer of the material must be used in order to be black to neutrons of that energy. For practical purposes a layer of thickness two absorption mean free paths or greater can be considered nearly black. Stevens [62] has shown that an absorbing layer of thickness equal to two absorption mean free paths absorbs something like 97% as many neutrons as a completely black surface. Table 4-3 [62] lists a number of elements and isotopes with high thermal-neutron-absorption cross sections and gives for each the surface density, or number of grams of absorber per square centimeter of surface, required to meet the blackness criterion of two absorption mean free paths.

A further criterion for blackness is that the absorption cross section of the material be much larger than the scattering cross section. Thus large thicknesses of moderately absorbing materials do not make very effective control rods, and the effectiveness of a highly absorbing material may be impaired if it is diluted too strongly with a material which is a good scatterer. Stevens [62] cites the case of a B^{10} slab of thickness equal to two absorption mean free paths. This slab would fail to capture about 6% of the neutrons crossing its surface because of its slight transparency. Such a slab would, of course, be quite thin. If it were diluted by steel sufficiently to form a slab 0.20 in. (5.1 mm) thick, the fraction of neutrons avoiding capture would increase to 8.3%; some of these would pass through the slab, others would be scattered out of the slab before absorption. This degree of dilution increases the scattering cross section to 20% of the macroscopic absorption cross section. It appears that scattering cross sections of 10 to 20% of the absorption cross section are tolerable,

TABLE 4-3

Amounts of Various Absorbers Required for Nominal Blackness to Thermal Neutrons (Two Absorption Mean Free Paths)

Element or nuclide	Average thermal absorption cross section, σ_a, barns	Abundance of important isotopes, %	Surface density for blackness, g/cm^2	Density g/cm^3	Thickness for blackness, cm
Li	62		0.362	0.53	0.68
Li^6	818	7.52	0.024	0.53	0.045
B	651		0.054	2.3	0.023
B^{10}	3,470	18.8	0.0093	2.3	0.004
Rh	130		2.56	12.5	0.20
Ag	54		6.47	10.5	0.62
Ag^{107}	26	51.4			
Ag^{109}	71	48.6			
Cd	2,210		0.165	8.6	0.0192
Cd^{113}	18,000	12.3			
In	165		2.26	7.3	0.310
Sm	4,760		0.100	7.2*	0.0139
Sm^{149}	57,200	13.8			
Eu	3,980		0.128	7.3*	0.0175
Eu^{151}	7,800	47.8			
Gd	39,800		0.0128	7.6*	0.0017
Gd^{155}	60,000	14.7			
Gd^{157}	139,000	15.7			
Dy	950		0.55	8.1*	0.068
Dy^{164}	2,340	28.2			
Er	144		3.78	8.6*	0.44
Lu	96		6.05	?	0.67
Lu^{176}	3,120	2.6			
Hf	91		6.38	13.4	0.471
Hf^{177}	320	18.4			
Re	73		8.28	20.0	0.414
Ir	372		1.68	22.4	0.075
Ir^{191}	820	38.5			
Ir^{193}	104	61.5			
Au	85		7.52	19.3	0.392
Hg	330		1.97	14.22	0.138
Hg^{199}	2,160	16.8			

*X-ray density for oxide.

provided the absorbing and scattering nuclei are intimately mixed so that the competition between scattering and absorption is in proportion to the macroscopic absorption and scattering cross sections. If the scattering and absorbing materials are segregated, the scattering loss of neutrons may be considerably greater. The worst form of segregation is, of course, that in which the scattering material covers the surface of the absorbing material, as when the absorber is encased in a jacket or cladding of low absorption cross section. Stevens shows that an 0.04-in. (1.02 mm) jacket of steel reflects about 8% of the neutrons incident upon the jacket, whereas a 0.10-in. (2.5 mm) thickness of steel reflects about 17%. The actual reduction of rod effectiveness by the scattering effect of the jacket will be less than indicated by these values and will depend on the surrounding medium since some of the reflected neutrons will diffuse back into the rod. Further, any material which lies between the control rod and the fissionable material in the reactor core can produce an effect similar to that of a rod jacket. If the material between the rod surface and the fissionable material is a moderator, then the net effect on rod effectiveness is the difference between the effect of the diffusion barrier which the material represents and the effect of additional neutron slowing-down in the vicinity of the rod surface.

4.2.3 Epithermal Absorption

If the control rod absorbs neutrons of energy above the thermal range, this absorption will contribute also to the reactivity worth of the rod. Naturally, the contribution depends not only on the epithermal behavior of the absorption cross section of the rod material but on the relative number of epithermal and thermal neutrons in the reactor as well. Although the epithermal effect may be nonnegligible in reactors of many types, special efforts to exploit it through the choice of rod materials with high epithermal cross sections are usually justifiable only for reactors with relatively high epithermal/thermal neutron ratios—roughly, reactors in which epithermal neutron fissions constitute 15% or more of the total. Since most H_2O-moderated reactors fall into this class, epithermal considerations frequently play an important part in control rod design for these reactors.

The three most widely used control rod absorbers (boron, cadmium, and hafnium) illustrate the possible range of epithermal absorption behavior. Figure 4-7, reproduced from the data of Stevens, [62] shows the fractional absorptions by slabs containing these materials as functions of neutron energy. The figure gives the fraction of neutrons absorbed, of those striking the slab surface, for four different slabs: a 1/16-in. (1.6mm) thick slab of cadmium, a 0.200-in. (5.08 mm)-thick slab of Hf, a 0.200-in (5.08mm)-thick slab of steel containing 1.0 wt. % B^{10}, and a 0.200-in. (5.08 mm)-thick slab of steel containing 0.15 wt. % B^{10}. The latter case would correspond to about 0.9 wt. % natural boron in the steel. Cadmium is the classic example of the thermal absorber. The 1/16-in. (1.6 mm) thickness is black to neutrons up to about 0.4 ev.

Since Hf is not a very strong thermal absorber, the 0.2-in. (5.08 mm)-thick slab is not really black in the thermal energy region, although most of the thermal neutrons are absorbed. The slab is black in the vicinity of the absorption resonances which occur at 1.10, 2.38, and 7.80 ev. An increase in thickness of the Hf would increase its absorption in the thermal energy range and in the epithermal ranges between the neutron resonances.

Boron-10 is the classic example of the 1/v absorber. An increase in either the thickness of the slab or in the concentration of B^{10} extends the range of blackness to higher and higher neutron energies. The 0.15 wt. % B^{10} slab is about equivalent to the Hf slab for neutrons of energies up to about 0.3 ev; at higher energies, it falls far below the Hf curve. The 1 wt. % B^{10} slab, however, is black up to about 0.7 ev and is about as good as the Hf slab at higher energies.

Most epithermal absorbers are of the resonance type, like Hf. Since these materials characteristically absorb neutrons in discrete energy regions, one of the methods of producing materials with high epithermal absorption efficiency is to mix

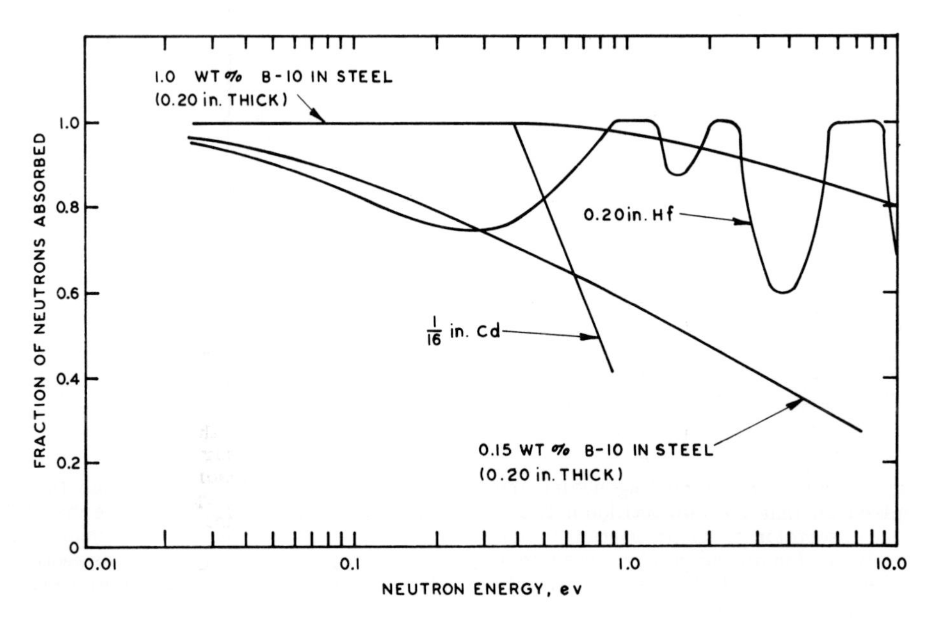

FIG. 4-7 Fraction of neutrons absorbed, of those striking the slab surface, by slabs of various materials.

absorbers in such a way that the resonances of one fall into the energy regions between the resonances of another. The mixture of Cd, Ag, and In is one such combination of this type. Mixtures of the rare-earth elements have also been considered.

The importance of epithermal absorption in H_2O reactors can be inferred from Fig. 4-8, which shows the energy distribution of neutron flux in a typical water-moderated medium, after the calculations of Amster [63]. This flux spectrum, which was computed for water containing 2.9 barns of absorption cross section per hydrogen atom (boron absorption), is much like the spectrum that would be expected in a water-moderated reactor having an H_2O/UO_2 volume ratio of 2.0, with the uranium enriched to 2 wt. % U^{235}. Since the importance of the epithermal energy range is distorted by the logarithmic energy scale, a better idea of the importance of the epithermal neutrons can be had from the integral, $\int \phi(E)\, dE$, which is also plotted in Fig. 4-8. [$\phi(E)$ and $\int \phi(E)\, dE$ are plotted to different arbitrary scales]. It is evident that a substantial portion of the neutron total flux lies above the Cd cutoff at 0.4 ev.

Table 4-4 [64] lists the measured effectiveness for various control rod materials relative to a 0.20-in. (5.08 mm)-thick hafnium sheet, as reported in reference [66]. The measurements were made with small samples, 2 in. by 2 in. by 0.2 in. (5.1 cm x 5.1 cm x 0.51 cm). The spectrum in which the measurements were made is not specified, but it can probably be assumed that it is a typical water-moderated spectrum and that the relative values of effectiveness would be roughly the same for many water-moderated reactors.

Table 4-5 [65] lists the results of relative worth measurements made in the Critical Experiment Facility at Vallecitos Atomic Laboratory. The

TABLE 4-4

Control-Rod Effectiveness* for Various Materials

Material	Absorptiveness,** $\Sigma_a t$	Effectiveness relative to hafnium
3.0 wt. % B^{10} in stainless steel (dispersion of minus 100-mesh particles of 90 per cent boron)	27.2	1.12
Hafnium	2.40	1.00
0.97 wt. % B^{10} in stainless steel (alloy)	9.3	0.98
15 wt. % Eu_2O_3 in stainless steel (dispersion)	8.7	0.96
Indium	3.68	0.93
Silver	1.85	0.88
Cadmium	77.5	0.80
8.7 wt. % gadolinium in titanium	31.1	0.79
Tantalum	0.60	0.71
2.7 wt. % Sm_2O_3 in stainless steel (dispersion)	3.20	0.71
Haynes-25 alloy	0.96	0.68
Titanium	0.16	0.24
Zircaloy-2		0.05
2S Aluminum		0.02

*Determined by R. R. Eggleston and a Critical Experiment Group of Bettis Laboratory on samples 2 by 2 by 0.2 in.

**Macroscopic thermal absorption cross section times thickness.

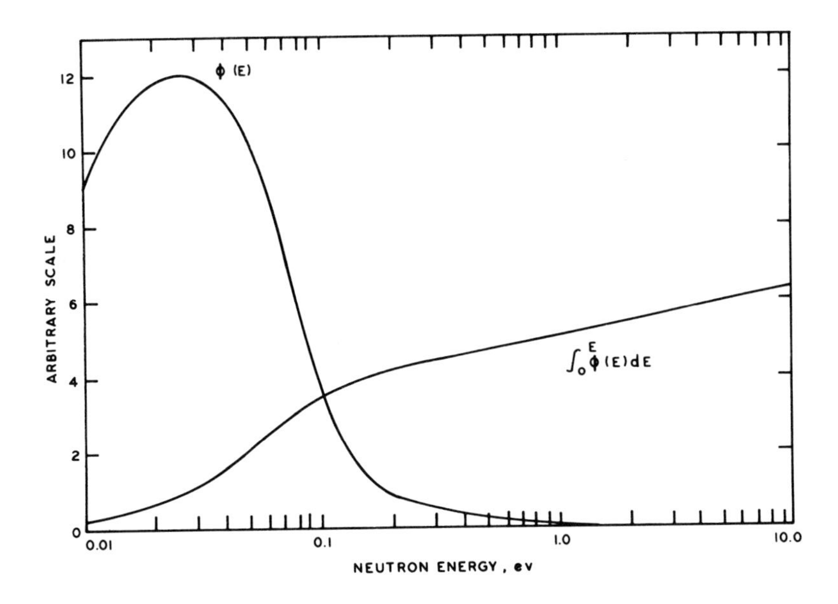

FIG. 4-8 Neutron-flux spectrum in H_2O containing 2.9 barns of 1/v absorber per H atom.

TABLE 4-5

Measured Relative Reactivity Worths of Various Materials in an H_2O-Moderated Assembly

Material	Surface density g/cm^2	Thickness		Relative worth,* bare	Relative worth,* Cd-covered	Equivalent boron surface density, g/cm^2	
		in.	mm			Bare	Cd-covered
Cadmium (Cd)	0.022	0.001	0.025	0.494	0.931	0.0090 ± 0.005	
	0.050	0.002	0.051	0.648	0.931	0.0192 ± 0.0010	
	0.101	0.005	0.127	0.728	0.928	0.0284 ± 0.0015	
	0.555	0.025	0.635	0.811	0.941	0.0440 ± 0.0026	0.0060 ± 0.0051
	0.104	0.051	1.295	0.838	0.948	0.0510 ± 0.0032	0.0100 ± 0.0053
	2.142	0.098	2.489	0.864	0.967	0.0593 ± 0.0040	0.0200 ± 0.0056
	4.389	0.201	5.105		1.000		0.0370 ± 0.0063
	6.585	0.305	7.747	0.928	1.020	0.0890 ± 0.0066	0.0500 ± 0.0071
Cobalt (Co)	5.601	0.252	6.401	0.855	1.046	0.0563 ± 0.0038	0.0810 ± 0.0105
	3.418	0.514	3.912	0.759	1.011	0.0345 ± 0.0018	0.0510 ± 0.0072
	2.187	0.098	2.489	0.663	0.988	0.0206 ± 0.0010	0.0323 ± 0.0061
Dysprosium oxide (Dy_2O_3)	3.047	0.300	7.620	1.039	1.158	0.2050 ± 0.0190	0.2400 ± 0.0250
	1.753	0.174	4.420	0.989	1.119	0.1380 ± 0.0120	0.1660 ± 0.0185
	0.505	0.050	1.270	0.801	1.034	0.0420 ± 0.0024	0.0710 ± 0.0087
Erbium oxide (Er_2O_3)	3.405	0.299	7.595	0.920	1.077	0.0840 ± 0.0061	0.1060 ± 0.0130
	2.009	0.175	4.445	0.839	1.026	0.0520 ± 0.0033	0.0650 ± 0.0083
	0.558	0.049	1.245	0.570	0.979	0.0137 ± 0.0007	0.0290 ± 0.0060
Europium oxide (Eu_2O_3)	3.375	0.299	7.595	1.253	1.332	1.0900 ± 0.0710	1.0000 ± 0.1150
	1.972	0.174	4.420	1.197	1.265	0.6850 ± 0.0570	0.6100 ± 0.0600
	0.561	0.050	1.270	1.032	1.130	0.1940 ± 0.0180	0.1850 ± 0.0200
Gadolinium oxide (Gd_2O_3)	3.388	0.301	7.645	1.042	1.135	0.2100 ± 0.0190	0.1950 ± 0.0210
	1.989	0.177	4.496	0.984	1.082	0.1330 ± 0.0110	0.1180 ± 0.0140
	0.567	0.050	1.270	0.885	1.011	0.0680 ± 0.0047	0.0510 ± 0.0072
Gold (Au)	14.531	0.297	7.544	1.038	1.164	0.2040 ± 0.0190	0.2540 ± 0.0260
	9.606	0.196	4.978	0.987	1.127	0.1360 ± 0.0110	0.1800 ± 0.0200
	4.926	0.101	2.565	0.868	1.076	0.0610 ± 0.0041	0.1100 ± 0.0135
	2.518	0.051	1.295	0.719	1.032	0.0273 ± 0.0014	0.0700 ± 0.0088
	1.318	0.027	0.686	0.564	1.004	0.0127 ± 0.0007	0.0460 ± 0.0068
Hafnium (Hf)	9.175	0.276	7.010	1.106	1.217	0.3530 ± 0.0360	0.4050 ± 0.0390
	6.833	0.206	5.232	1.051	1.189	0.2220 ± 0.0200	0.3200 ± 0.0310
	3.422	0.103	2.616	0.920	1.130	0.0840 ± 0.0062	0.1890 ± 0.0205
	1.476	0.045	1.143	0.726	1.072	0.0284 ± 0.0015	0.1050 ± 0.0130
	0.856	0.025	0.635	0.572	1.039	0.0133 ± 0.0007	0.0760 ± 0.0096
Holmium oxide (Ho_2O_3)	3.365	0.302	7.671	0.745	1.087	0.0314 ± 0.0016	0.1280 ± 0.0150
	1.955	0.176	4.470	0.602	1.038	0.0152 ± 0.0008	0.0740 ± 0.0094
	0.564	0.051	1.295	0.302	0.983	0.0036 ± 0.0003	0.0310 ± 0.0059
Indium (In)	5.554	0.301	7.645	1.036	1.137	0.2000 ± 0.0190	0.2000 ± 0.0215
	3.697	0.201	5.105	0.998	1.105	0.1470 ± 0.0120	0.1470 ± 0.0167
	1.858	0.100	2.540	0.908	1.058	0.0785 ± 0.0057	0.0940 ± 0.0120
	0.928	0.050	1.270	0.785	1.029	0.0386 ± 0.0021	0.0660 ± 0.0086
	0.468	0.025	0.635	0.632	1.003	0.0176 ± 0.0009	0.0450 ± 0.0068
Lutetium oxide (Lu_2O_3)	2.244	0.175	4.445	0.772	1.059	0.0361 ± 0.0019	0.0930 ± 0.0120
	1.605	0.125	3.175	0.693	1.041	0.0241 ± 0.0012	0.0780 ± 0.0099
	0.638	0.050	1.270	0.465	0.992	0.0078 ± 0.0004	0.0360 ± 0.0062
Samarium oxide (Sm_2O_3)	3.357	0.302	7.671	1.063	1.155	0.2490 ± 0.0240	0.2340 ± 0.0240
	1.966	0.175	4.445	1.018	1.113	0.1730 ± 0.0150	0.1580 ± 0.0173
	0.558	0.050	1.270	0.896	1.008	0.0730 ± 0.0052	0.0500 ± 0.0070
Silver (Ag)	7.970	0.299	7.595	1.021	1.172	0.1760 ± 0.0155	0.2710 ± 0.0270
	5.226	0.196	4.978	0.944	1.123	0.0995 ± 0.0076	0.1700 ± 0.0190
	2.746	0.103	2.616	0.798	1.065	0.0414 ± 0.0023	0.0990 ± 0.0125
	1.376	0.052	1.321	0.617	1.022	0.0167 ± 0.0009	0.0600 ± 0.0080

*Edge-corrected

TABLE 4-5 (Continued)

Material	Surface density g/cm²	Thickness in.	Thickness mm	Relative worth,* bare	Relative worth,* Cd-covered	Equivalent boron surface density, g/cm² Bare	Equivalent boron surface density, g/cm² Cd-covered
Tantalum (Ta)	12.733	0.302	7.671	0.855	1.166	0.0680 ± 0.0047	0.2610 ± 0.0265
	8.511	0.202	5.131	0.775	1.124	0.0370 ± 0.0020	0.1740 ± 0.0190
	4.214	0.100	2.540	0.572	1.059	0.0133 ± 0.0007	0.0980 ± 0.0120
Terbium oxide (Tb_4O_7)	2.287	0.175	4.445	0.829	1.067	0.0482 ± 0.0030	0.1010 ± 0.0135
	1.624	0.125	3.175	0.765	1.040	0.0347 ± 0.0019	0.0730 ± 0.0090
Thulium oxide (Tm_2O_3)	2.089	0.175	4.445	0.735	1.056	0.0292 ± 0.0015	0.0890 ± 0.0110
	1.490	0.125	3.175	0.649	1.037	0.0193 ± 0.0010	0.0730 ± 0.0090
	0.580	0.049	1.245	0.418	0.992	0.0064 ± 0.0004	0.0370 ± 0.0063
Tungsten (W)	9.900	0.204	5.185	0.693	1.077	0.0257 ± 0.0013	0.1120 ± 0.0136
	9.581	0.197	5.008	0.675	1.053	0.0237 ± 0.0012	0.0890 ± 0.0110
Ytterbium oxide (Yb_2O_3)	3.584	0.300	7.620	0.579	1.028	0.0138 ± 0.0007	0.0670 ± 0.0084
	2.100	0.175	4.445	0.440	0.994	0.0070 ± 0.0004	0.0345 ± 0.0061
	0.581	0.050	1.270	0.194	0.957	0.0017 ± 0.0002	0.0150 ± 0.0054
Boron glass	0.0478**	0.101	2.565	0.8144	1.0051		
	0.1575	0.104	2.640	1.0062	1.1161		
	0.0335	0.152	3.860	0.7503	0.9903		
	0.0880	0.200	5.080	0.9282	1.0560		
	0.3026	0.206	5.232	1.0867	1.1806		
	0.1206	0.275	6.980	0.9728	1.0882		
	0.4241	0.290	7.365	1.1326	1.2232		
	0.0664	0.302	7.671	0.8756	1.0393		

* Edge-corrected.
** Surface densities of boron.

materials were used in the form of 3- by 6-in. (7.6 cm x 15.2 cm) plates having various thicknesses as recorded in the table. The critical assembly was H_2O-moderated; it was fueled with UO_2 pellets, 0.491 in. (1.247 cm) in diameter, in 0.562-in. (1.427 cm) outer diameter aluminum tubes of 30-mil (0.76 mm) wall thickness. The fuel rods were located on a uniform square lattice, with 0.8-in. (2.03 cm) spacing between centers; the enrichment was 1.6 wt.% U^{235}. Measurements were made of the reactivity worth of each absorbing plate, bare, and of each plate plus a constant cadmium box enclosing it. The relative worth is expressed, arbitrarily, as the ratio of the reactivity worth of the arrangement in question to the reactivity worth of a 0.201-in. (5.10 mm)-thick Cd plate in the Cd box. The results have been edge-corrected to reduce the edge effect to that which would characterize a plate of zero thickness. The equivalent boron surface density is that surface density of natural boron which, if substituted for the sample (bare or in the Cd box as the case might be), would yield the same reactivity worth. The epithermal effectiveness of the materials are illustrated by their Cd-covered relative worths and by the Cd-covered equivalent boron surface densities.

Other measurements on rare-earth oxides and mixtures of rare-earth oxides may be found in reference [66].

4.2.4 Control Rods Containing Moderator

If the control rod itself is capable of slowing down neutrons, it will absorb not only the low-energy neutrons which diffuse into it from the surrounding reactor material, but also those neutrons which enter the rod as fast neutrons and are slowed down inside. As in the case of epithermal absorption, the increase in control rod worth due to moderation in the rod cannot be large unless the number of epithermal and fast neutrons in the reactor is important relative to the number of thermal neutrons. Moreover, only hydrogenous materials are sufficiently effective as moderators to slow down a significant number of neutrons within a rod of reasonable size. Consequently, the use of the moderating control rod has usually been confined to the H_2O-moderated reactors, where it can be very effective. In D_2O-moderated reactors employing clustered fuel elements, D_2O-containing control rods can sometimes be installed between the fuel element clusters. In such a case they do not waste any usable core space, and do not require fueled followers. They may then prove attractive even though the internal moderator does not increase the absorption by a large factor. Such rods are used in the D_2O research reactor at M.I.T.

The usual moderating control rod in a water-moderated reactor consists of a shell of absorbing material enclosing a space within the reactor core which is allowed to fill with water, from the general body of coolant moderator in the core, when the rod is inserted. When the rod is withdrawn a "follower", discussed below, occupies the water-containing space. Usually the rod is a module which fits into a space nearly identical to the space occupied by a normal fuel assembly (Fig. 4-9).

If the smallest diameter of the water-filled space within the rod is not large compared to the slowing-down length($\sqrt{\tau}$ = 5 to 6 cm in cool water) the fraction of neutrons slowed down (and therefore captured) within the rod will be roughly proportional to the ratio of the amount of water within the rod to the total amount of water in the core, and the reactivity worth due to this effect will be proportional to the same fraction. Since part of the space in the remainder of the core is occupied by fuel rather than water, a rod of the type in Fig. 4-9, in an "average" core position, will usually have a worth component due to the absorption of internally moderated neutrons which is somewhat greater than the ratio of the cross-sectional area of the rod to the cross-sectional area of the core; to this worth component will be added a worth component due to thermal neutrons entering the rod from outside which is determined by the principles discussed in Sec. 4.2.1. Usually the former (internal moderation) component exceeds the latter in situations where this rod type is attractive.

A rod of this type must have a "follower" to occupy the water-containing space when the rod is withdrawn; otherwise, when the rod was withdrawn, most of the neutrons thermalized in the remaining water-filled space would be absorbed by the fuel

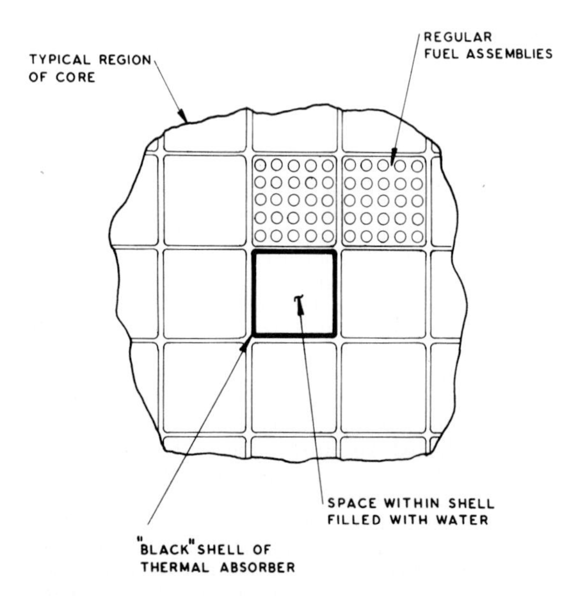

FIG. 4-9 Flux-trap control rod. In a core having a high ratio of epithermal to thermal neutron flux, the reactivity worth of a regular array of such rods distributed uniformly over the core volume is roughly equal to the ratio of the volume of water contained by the rods to the volume of water in the remainder of the core.

elements immediately adjacent to the space, with the result that those fuel elements would overheat. The usual practice is to provide a follower made up of fuel elements similar to the normal elements. These elements (which must be adequately cooled) absorb the fraction of the neutrons thermalized in the remaining water. They not only eliminate, in principle, the power peak but may also provide some additional reactivity to maximize the reactivity change produced by the rod.

Rods of this type have been referred to variously as "MTR-type" (since they were first used in the Materials Testing Reactor) [67], "flux-trap" rods, and "neutron-rectifier" rods. When it is necessary to provide high individual control rod worth in a water-moderated reactor, rods of this type may provide the answer, at the expense of some difficulty in cooling the movable fuel elements in the follower. A reactivity worth of 0.19 to 0.20 was attained in the SM-1 (APPR) [68] with a set of five out of a total of seven such rods. These five rods occupied only about 10% of the total core volume.

From the standpoint of safety, high individual rod worths are not always desirable (see Sec. 4.3). If rods of the moderating type are used the designer must bear in mind that the removal of water from within the rod, as by boiling, could produce a large reactivity increase. Crushing of the absorbing shell of the rod, if it could conceivably occur (for example, in a violent power excursion), would have a similar effect.

4.2.5 Effect of Position in Reactor

When one is concerned with the reactivity worths of single rods or groups of a few rods, the considerations may be considerably more complex than when a complete set of rods may be considered to be uniformly distributed over the core. At a given point in the reactor it is still true that the reactivity worth of a rod is nearly proportional to the ratio of the "absorption area" of rod to the total cross-sectional area of the core. Figure 4-10, for example, shows the calculated reactivity worth of a cylindrical, thermally-black rod installed on the axis of a cylindrical water-moderated (and reflected) core for various core radii R. The worth is very nearly proportional to $1/R^2$. The worth, however, depends also on the position of the rod in the core. A curve like Fig. 4-10 could be constructed for a single rod located at a distance equal to a constant fraction of R from the axis, and the rod worths would be lower than those of Fig. 4-10 by an amount which would become greater as the location was moved farther from the core axis.

Methods of calculating the worths of single rods and groups of rods at arbitrary locations have been published [69, 71]. The following qualitative discussion is intended only to illustrate the physical principles which determine the effects of location.

On the average, when a low-energy neutron is absorbed in the fuel of a reactor, a number, η, of fission neutrons are produced. In some cases this number may be augmented by a multiplier ϵ as a result of neutrons produced by fast fission of fertile material. If this formalism is adopted, the effective

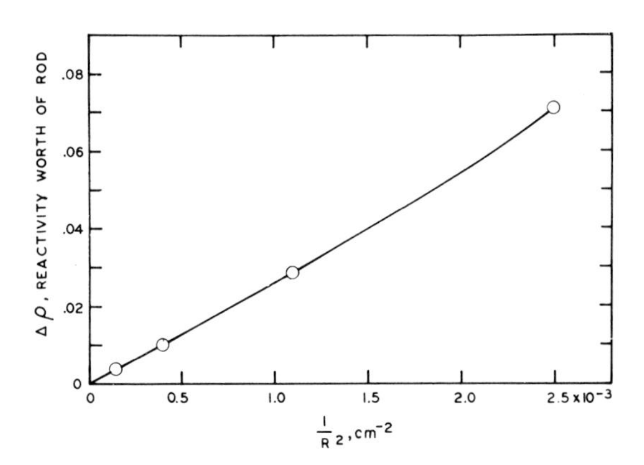

FIG. 4-10 Reactivity worth of thermally black cylindrical control rod of 4 cm (1.57 in.) radius in a water-moderated core of radius R, with infinite water reflector. The reactor is at room temperature, and the volume fractions of materials in the core are as follows: H_2O, 0.604; UO_2, 0.254; Zr, 0.137; void, 0.005. The enrichment is 2.4% U^{235}. The reactivity changes produced by the rod are relative to a core in which the rod channel is filled with material of the general core composition.

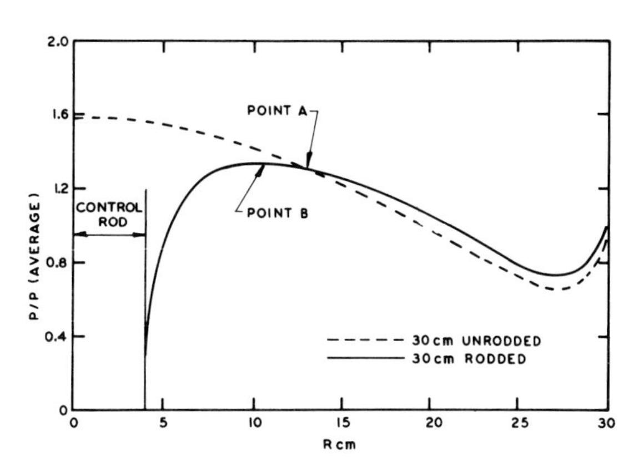

FIG. 4-11 Power distribution in a reflected cylindrical core, with and without a central control rod of 4 cm radius (core same as for Fig. 4-10). Point B is the point of maximum power density when the rod is inserted. At radii greater than that of point A the power density is greater when the rod is in than when it is out. A second control rod installed at a radius greater than that of point A would experience a negative shadowing effect - i.e., would have its worth increased by the presence of the central rod.

multiplication constant k_{eff} may be expressed as follows for a reactor of uniform composition:

$$k_{eff} = \frac{\text{Number of neutrons born in the generation } (n+1)}{\text{Number of neutrons born in generation } (n)}$$

$$= \eta\epsilon \left[\frac{\text{Number of n'th generation neutrons absorbed in fuel}}{\text{Number of n'th generation neutrons born}}\right]$$

$$= \eta\epsilon \left[\frac{\text{Number of n'th generation neutrons absorbed in fuel}}{\text{Number of n'th generation neutrons absorbed}}\right] \times \left[\frac{\text{Number of n'th generation neutrons absorbed}}{\text{Number of n'th generation neutrons born}}\right]$$

$$= \eta\epsilon \left[\frac{\text{Number of n'th generation neutrons absorbed in fuel}}{\text{Number of n'th generation neutrons absorbed}}\right] \times \left[\frac{\text{Number of n'th generation neutrons which do not leak out of reactor}}{\text{Number of n'th generation neutrons born}}\right]$$

Or, in general:

$$k_{eff} = \eta\epsilon F(1 - \mathcal{L}) , \tag{4-1}$$

where F is the fraction of neutron absorptions, per second, which occur in fuel*, and $\mathcal{L}$ is the fraction of neutrons born per second which leak out of the reactor. If any disturbance to the reactor produces a small change in F and/or $\mathcal{L}$, the resulting differential change in the reactivity ρ is given by:

$$d\rho = \frac{d\,k_{eff}}{k_{eff}} = \frac{dF}{F} - \frac{d\mathcal{L}}{1 - \mathcal{L}} . \tag{4-2}$$

Any discrete absorber placed in the reactor as a control rod will usually change both F and $\mathcal{L}$. For strong, localized, control rods, the effect on leakage may be readily apparent. Figure 4-11, for example, shows the power distributions in a uniform reflected core with and without a strong control rod installed on the axis. It is clear that when the rod is in place a larger fraction of the power is generated near the core boundary and proportionately more of the fission neutrons are born near the boundary, so that the average probability of neutron leakage is increased. The worth of the control rod is made up of a component that can be identified directly with the neutron absorption by the rod plus a component due to the increase in leakage.

Even if the control rod of Fig. 4-11 were of infinitesimal strength, however, it would have an effect on leakage which would not be negligible relative to its total effect. The neutrons it absorbs are those at the center of the core, and hence those which would produce fission neutrons with the lowest probability of leakage if they were not absorbed by the rod. Their removal by the rod has a first-order

*For the purpose of this discussion, "fuel" can be considered to be the fissile isotope, the mixture of fissile and fertile isotope, or any typical mixture in the reactor containing fissile isotopes, provided the installation or motion of control rods cannot affect the distribution of absorptions among the components of the mixture, and provided the value of η is chosen consistently.

effect in shifting the average fission density toward the core boundary, and in increasing the average leakage, even when the accompanying change in power distribution is too small to be discernible. Only when a weak control rod happens to be located at that position in the reactor core where the leakage probability has the average value does the leakage component of its worth become zero.

If the reactor does not have a uniform composition, the reactivity change produced by the control rod cannot be considered to involve only the absorption by the control rod and the change in leakage, but it must also take into account the effect on all aspects of the neutron balance of the changed spatial distribution of neutrons. For example, if the reactor contains other control rods, the effect of the additional control rod upon the fractional absorption by the other control rods must be taken into account. If the additional rod is installed far from other rods, its absorption will remove some of the neutrons whose progeny would have a very low probability of being absorbed by the other control rods, with the result that the average neutron population, after the installation of the additional rod, has a somewhat higher probability of absorption in the rods initially installed. Similarly, if the additional rod were installed near one of the original rods, it would absorb some neutrons whose progeny would otherwise have been absorbed by that rod, with the result that the average neutron population, after the addition of the new rod, would have a somewhat lower probability of absorption in the original rods.

Effects of this kind, including the leakage effect considered above, can be taken into account by assigning to the neutrons an importance (to the maintenance of the chain reaction) which is a function of their position in the reactor. Thus a neutron near a core boundary or near a control rod will have an importance which is lower than average, while a neutron far from core boundaries, control rods, or other absorbers, will have an importance which is higher than average; and by an extension of this reasoning, the absorption of a neutron by a control rod at a position of high importance will have a greater effect on reactivity than the absorption of a neutron at a position of low importance. If the fraction of neutrons absorbed by the control rod can be weighted by a suitable importance function, relative to the importance of the average neutron in the reactor, then the worth of the control rod can be considered to be determined simply by this weighted fraction of absorption:

$$\Delta\rho(\text{rod}) = \text{control rod worth}$$

$$= I(r)\left[\frac{\text{Number neutrons absorbed by rod}}{\text{Total number neutrons absorbed in core}}\right]. \tag{4-3}$$

In this equation, I (r) is the importance, which is a function of the position (r) of the control rod in the core.

The fraction of the total neutrons which the control rod absorbs also varies with the position of the rod in the core, since the general level of the neutron flux ϕ (r) varies with position. If the control rod is small or weakly absorbing, it may be a reasonable approximation to assume that the functions I (r) and ϕ (r), except very near the rod, retain the values which they had before the rod was inserted. If so, then the reactivity worth of the rod can be resolved into the three multiplicative factors, I (r), ϕ (r), and W, the last being independent of position and dependent on the size, shape, and material of the control rod and on the characteristics of the core material:*

$$\Delta\rho(\text{rod}) = W\left[I(r)\,\phi(r)\right]. \tag{4-4}$$

In general, the importance function I (r) can be calculated for any case in which a calculation of the flux distribution, ϕ (r) is possible. For thermal neutrons it is given, in the two-group approximation, by the thermal adjoint flux ϕ_s^* (r). The determination of adjoint fluxes and importance functions is discussed in a number of papers on the application of perturbation theory to reactor problems, see references [72] through [74].

When the approximation of Eq. (4-4) can be used, the determination of the reactivity worth of a control rod at any desired position in the reactor may be relatively simple. The worth may be determined for an easily solvable, symmetrical situation, such as the centrally located rod, and the worth at other positions may be assumed proportional to the product I (r) ϕ (r). Or a position may be found in the reactor where the local value of the product I (r) ϕ (r) is equal to the average value over the core. The fractional absorption of the control rod, if it were installed at that position, can be estimated by the "absorption area" technique described in Sec. 4.2.1. The worth of the rod at that position will be given by the expression:

$$\Delta\rho(\text{rod, average position}) = \frac{\text{absorption area of rod}}{\text{cross-sectional area of core}} \tag{4-5}$$

and its worth at other locations may again be estimated by the assumption of proportionality to the product I (r) ϕ (r).

In a uniform bare reactor the function I (r) is identical to ϕ (r), and the weighting factor for control rod worth, I (r) ϕ (r), becomes simply ϕ^2 (r). In reflected reactors with reasonably uniform core composition, I (r) does not usually differ greatly from the fundamental mode distribution in the core, and for estimating or for understanding qualitatively the behavior of control rod installations, I (r) can often be approximated by assuming that it is proportional to the fundamental mode distribution or, as a last resort, proportional to the thermal neutron

---

*For purposes of the present conceptual discussion, Eqs. (4-3) and (4-4) have been written as though the "control rod" were approximated by a point absorber. The symbol r is intended to represent the general three-dimensional position variable. Ordinarily the appropriate approximation to the rod would be a line absorber. In that case, if I and ϕ vary along the length of the line, the total worth of the rod would be determined by integration of WIϕ over the length of the line.

flux. For complex situations which would require detailed calculations of I (r), the use of the perturbation (importance function) approach is often not worth the trouble since it is inherently inadequate for strong absorbers. It is, however, very useful as a guide to understanding the space-dependent worth of control rods.

It is clear from Eq. (4-4) that the worth of a control rod in a uniform reactor will be very much greater near the center than near the edge. Figure 4-12 [71] shows the results of a calculation of the worth of a control rod (3.48 cm diameter) in a bare core (52 cm diameter) as a function of radial position r (cm). The comparison with the fundamental mode distribution J_0 (2.405r/26) indicates that the approximation is rather good for this case. It should be pointed out that in some reflected reactors the maximum worth rods may lie near the core edge, or in the reflector. Care must be taken to determine the situation in each reactor and at all times during core life.

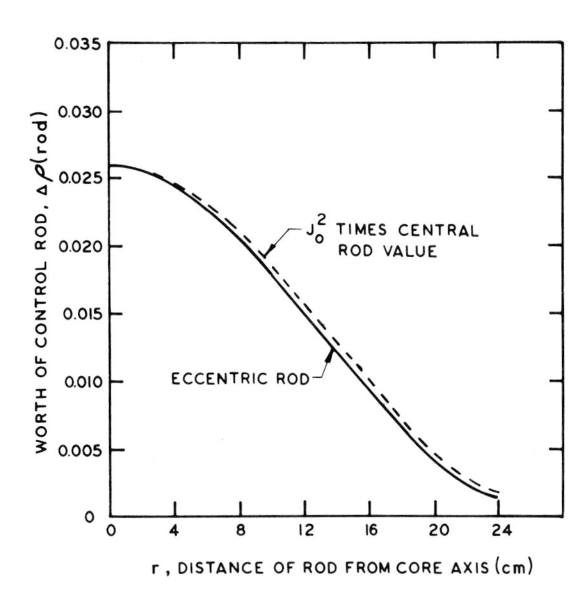

FIG. 4-12 Calculated worth of eccentric control rod vs. distance from core axis. The rod is thermally black, with a radius of 1.74 cm (0.685 in.). The equivalent bare core, of 26 cm (10.24 in.) radius, has the composition of the Bulk Shielding Facility Reactor (fully enriched; aluminum/water volume ratio of 0.65).

4.2.6 Rod Shadowing

It is instructive to consider the addition of a second control rod to the core shown in Fig. 4-11, which already contains one rod on its axis. If the power distributions (which are proportional to the thermal neutron flux distributions) are considered before and after the insertion of the central rod, it is apparent that the relative flux level at radial distances less than A (see figure) are decreased by the insertion of the central rod while those at distances greater than A are increased. If it is assumed that the importance function I (r) varies in approximately the same way, then a second rod installed at a radial distance less than A will have a reactivity worth less than that which it would have if the central rod were not present. In such a case it is said that the second rod experiences a positive shadowing by the central rod. It is more accurate to say that the two rods produce a mutual shadowing of each other, for the total reactivity worth of the two must be the same, regardless of which rod is inserted first. If the central rod is inserted first, the mutual shadowing will manifest itself as an apparent reduction of the worth of the off-center rod when it is inserted. If the off-center rod were inserted first, the mutual shadowing would appear as a reduction in the worth of the central rod when it was inserted. If the off-center rod were installed at a distance greater than A from the core axis, its apparent worth would be greater with the central rod in place than with the central rod removed. In this case, the mutual shadowing of the rods would be said to be negative.

4.2.7 Groups of Rods

Clearly, if one has a single rod to install in a uniform reactor, the highest reactivity worth will result if it is installed at the center of the core (except in some special cases of small, poorly thermalized cores with very effective reflectors). Frequently the designer is faced with the problem of locating a number of rods in the core in such a way as to approximate the maximum total reactivity worth.

If one desired to locate two control rods in the core of Fig. 4-11 and if one arbitrarily decided to locate the first rod on the core axis, the optimum location of the second would be approximately at the peak of the perturbed flux distribution, at point B (Fig. 4-11). The total worth of the two would not be as great, however, as the worth of the two rods when located symmetrically about the axis at a distance apart approximately equal to the distance of point B from the axis.

Figure 4-13 [71] shows the calculated total reactivity worth of two rods, each identical to the single rod considered in Fig. 4-12, installed symmetrically about the axis of the core. The maximum worth occurs when the rods are located about 3.5 cm from the axis; at this location there is some mutual positive shadowing—i.e., the total worth of the two rods is somewhat less than twice the worth of a single rod installed at the same distance from the axis.

For reasonably large groups of rods the optimum (maximum worth) arrangement is somewhat easier to predict than for a few rods. If the rods can be arranged in a simple regular pattern they can always be grouped symmetrically about the core axis, within a volume of such size that the gross* radial power distribution in that volume is nearly constant with r. When this condition has been approximated, the optimum installation has also been approximated.

Figure 4-14 is an example of a practical approximation to such an arrangement. The twenty-one cruciform control rods are assumed to be fixed

---

*i.e. the power distribution which has been "smoothed out" by considering only the average power level within each control rod "cell."

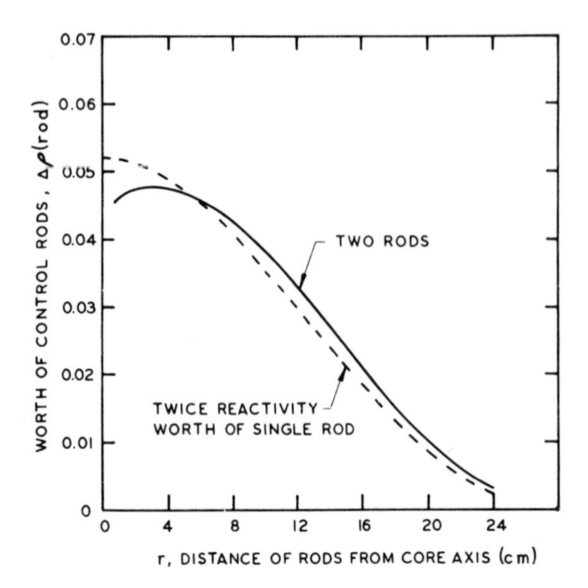

FIG. 4-13 Calculated worth of two symmetrically placed rods. The core is the same as that of Fig. 4-12, and each of the two rods is identical to the single rod of that figure.

in size and material, and the installation pattern to be confined to a square lattice. The problem is to distribute them over the core in such a way as to achieve the maximum total reactivity worth. As the inter-rod distance is adjusted, a spacing will be found for which the gross power distribution over the control rod region is flat, as in curve A. This is the desired arrangement. If the inter-rod spacing is increased, the power distribution will peak at the center (as in curve B) while it will droop at the center if the rod spacing is decreased. In either case the reactivity will increase. That the flat distribution is the optimum may be inferred by considering where one would insert a small additional absorber to produce the largest reactivity reduction in either case B or case C. In case B, one would insert it at the center; in case C, one would insert it at the peak of the power distribution, outside the control-rod zone. In both cases the addition of the absorber would modify the power distribution in the direction of approaching the curve A.*

If the core has an axial buckling B_z^2, the criterion for achieving the flat distribution in the rod zone is, in the one-group picture, simply**

$$k_\infty = 1 + M^2 B_z^2, \qquad (4\text{-}6)$$

where the absorption of the control rods is included in the determination of k_∞ for the zone. This may be done by associating a typical cell with each rod (Fig. 4-14) and using either the absorption area method to determine the rod absorption, as in Sec. 4.2.1, or a diffusion or transport calculation for the unit cell.

In practice, the precise "optimum" arrangement is seldom used. It does not, for one thing, give the maximum shutdown capability in the "stuck-rod" condition. A more important consideration is that the designer seldom has a continuously-variable range of possible control rod spacings. Since the control rod lattice must mesh with the fuel assembly lattice, the designer typically has only two or three possible spacings which lie within the usable range. Nevertheless, the consideration of the ideal situation is useful for the qualitative understanding of the practical installation.

When a reactor contains many control rods and/or when the total worth of the control rods is large, the designer must be on the alert for gross shadowing effects which may have important effects on reactor safety.

A very troublesome effect which is common in H_2O-moderated reactors may be discussed in terms of Fig. 4-14. In this case the absorption area of any one rod is small relative to the total cross-sectional area of the core; hence the reactivity worth of any one rod would be small provided the general level of reactivity of the core were such that it would be approximately critical with only the single rod inserted. If, however, the general level of reactivity is so high that all of the rods must be inserted to hold the reactor subcritical, the worth of a single rod may be much larger. Suppose, for example, it is known that the stuck-rod margin for the core is just slightly less than one complete rod. If one rod is withdrawn, the core will become critical, and the power distribution in the vicinity may look, qualitatively, like that of Fig. 4-15. In effect, the reactor behaves as a small critical core, of size equal to a single control rod cell, surrounded by a "reflector" in which the multiplication is greater than zero but less than unity. In such a case, the reactivity worth of the single rod may be approximated more nearly by the ratio of the absorption area to the area of the cell—i.e., it may be quite large. An alternate view of the situation is that the weighting factor for the particular rod withdrawn is very high because of the flux peaking (and a similar peaking of the importance function) in the cell. The effect of course depends on the size of the cell—more precisely on

*Reference [50] should be consulted for a rigorous discussion of this question. The rigorous condition for obtaining maximum reactivity effect from a given amount of absorber is that the absorber be so distributed as to make the thermal flux flat in the control zone. Only in the one-group approximation, however, can this be achieved by a uniform distribution of absorber in the zone. In actuality, a somewhat higher concentration is needed near the center, especially in small cores.

**Equation (4-6) implies that the central section of the reactor, containing the control rods, would be critical if it were infinite in radial extent. The reactivity of the reactor will depend upon the extent and the k_∞ of the surrounding, unrodded zone. Thus Eq. (4-6) defines only the condition for most efficient installation of rods in the rodded zone—the amount of reactivity controlled depends on the size of the rodded zone relative to the total size of the reactor.

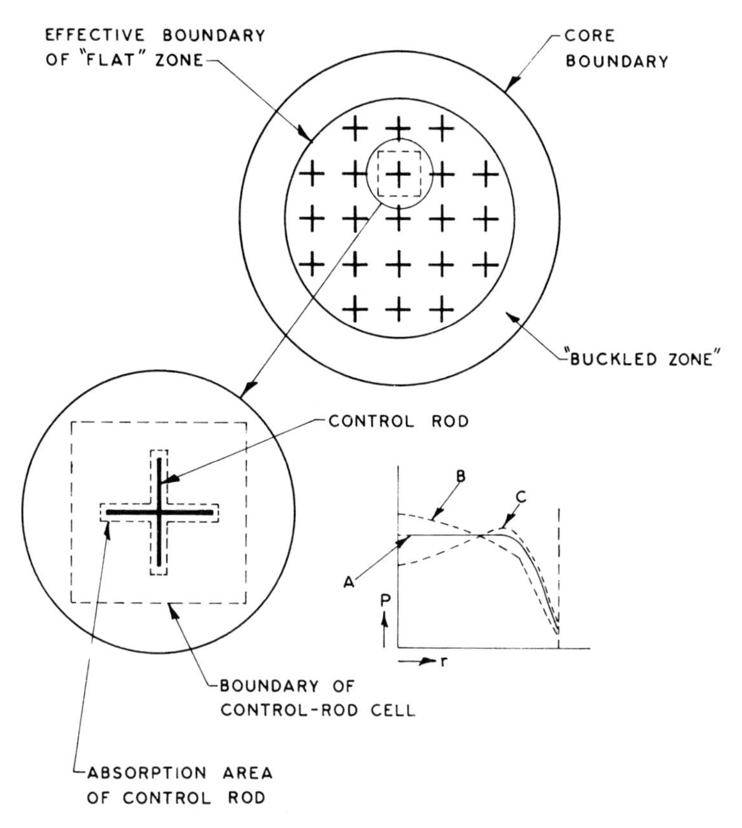

FIG. 4-14 Distribution of control rods for maximum total reactivity worth. The total worth of the rod installation will be greatest when the rod spacing is so chosen that the power distribution is approximately flat in the rod-occupied zone, as in curve A.

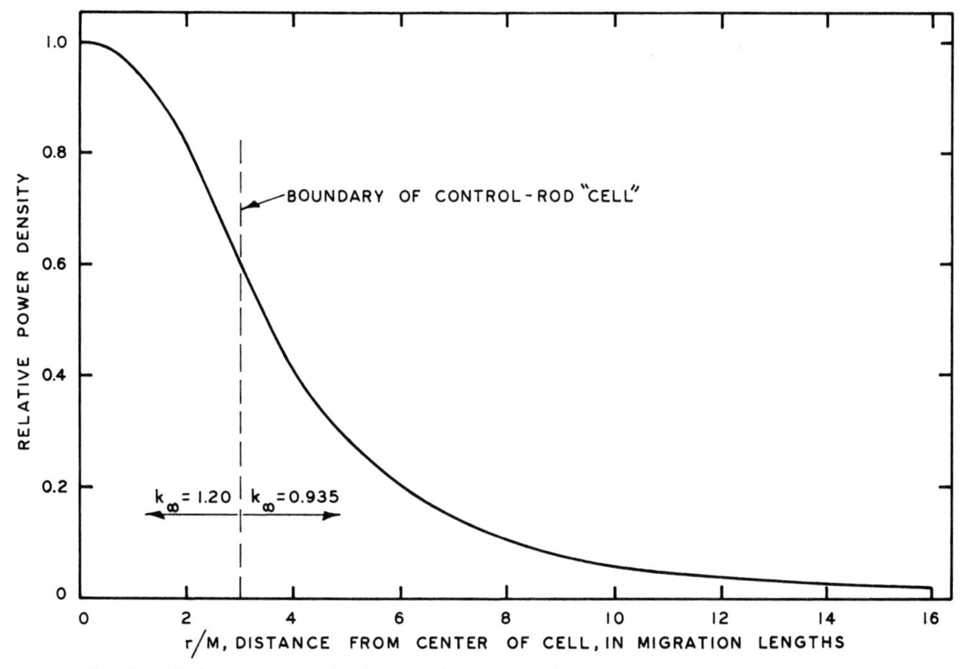

FIG. 4-15 Calculated power distribution in region of local criticality in a reactor core. The core region is assumed to be effectively infinite in extent and to consist of an array of control-rod cells, each having an effective radius of 3M. The value of k_∞ within each cell is assumed to be 0.935 when the control rod is in and 1.20 when the control rod is out: this results in criticality if a single rod is withdrawn. In an H_2O-moderated reactor (cold) M might be about 6 cm (2.04 in.) thus the cell radius considered here would be about 7 in. (17.8 cm).

the dimensions relative to the neutron migration length. The effect is large for large cells containing large rods.

Figure 4-16 illustrates the relationship between the size of the cell from which the rod is withdrawn and the required k_∞ of the remainder of the core to give a just-critical condition. Clearly, there will be little difficulty in achieving a one-stuck-rod shutdown margin if the effective radius of the typical control rod cell is less than 2M, but the chances of achieving this margin are poor if the radius is as large as 3M.

Ordinarily, a reactor would not be designed with a shutdown margin of less than one complete rod. However, a similar, but lesser, effect may occur when criticality is attained upon withdrawal of the second or third rod. If criticality is attained by withdrawing two or three adjacent rods, the effect may be quite large: in fact it is essentially identical to the single-rod effect since so far as the effect is concerned, the two or three adjacent rods may be considered as one large single rod. In most H_2O-moderated power reactors it is extremely difficult to achieve a stuck-rod margin greater than two adjacent rods. In order to avoid the effect of very large rod worth, it is usually necessary to prevent the withdrawal of adjacent rods in sequence, either by suitable design of the control rod drive system or by procedural controls.

4.2.8 Partially Inserted Rods

Ordinarily control rods are inserted axially in a core having flat parallel ends, and the flux and adjoint distributions in the axial direction are approximately sinusoidal. In such a case, if a weak control rod is inserted from one end, progressively farther into the reactor, the differential worth of the rod [dw(z)/dz] will vary with depth of insertion z as the square of a sine function:

$$\frac{dw(z)}{dz} = \frac{2W}{L} \sin^2 \frac{\pi z}{L} , \qquad (4\text{-}7)$$

where L is the equivalent bare length of the core, W is the total worth of the fully-inserted rod, and z is the depth of insertion measured from the extrapolated core boundary. Thus the maximum differential worth is just twice the average differential worth W/L. The reactivity worth w(z) of the inserted portion of the rod is (see Fig. 4-17):

$$w(z) = \frac{2W}{L} \int_0^z \sin^2 \frac{\pi z}{L} \, dz . \qquad (4\text{-}8)$$

These relations will hold rather well for weak rods in most reflected cores, except near the ends, provided z is measured from the end of the equivalent bare core.

If the control rod in question is not weak, that is, if when partially inserted it distorts the general flux distribution in the core, the effect will be to shift the position of maximum differential worth away from the central plane of the core to a position z > L/2 (i.e. to a position farther from the end of the core which the rod enters) and to increase somewhat the ratio of maximum differential worth to average differential worth. The latter effect will not be large, however, unless the flux distortion is rather severe—in other words, unless the rod is quite strong.

When several rods are ganged (operated as a bank), the axial flux distortion produced by rod insertion may be large and the curve of differential rod worth may depart widely from the sine-squared shape. For reactors designed to be operated with banked rods the effect must be taken into account in determining the maximum differential worth of the rods, and in nearly all reactors the effect must be considered in determining scram speeds, since all withdrawn rods will normally enter as a bank when scram is initiated. In some reactors banked control rod withdrawal configurations may lead to extremely large differential worths and constitute a serious safety problem.

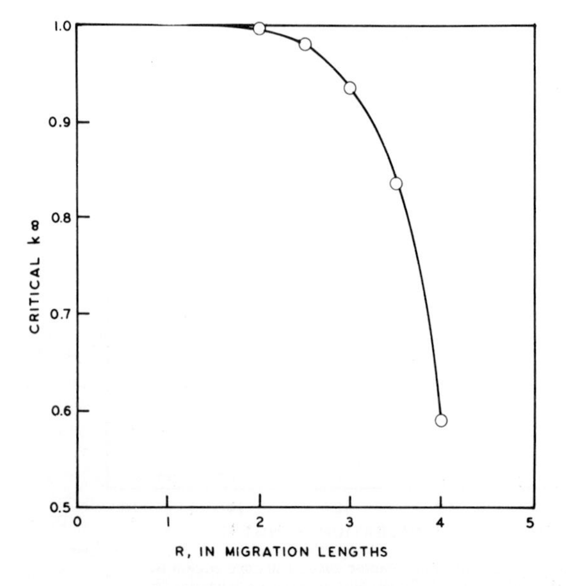

FIG. 4-16 Calculated critical k_∞ of a core region of infinite size, containing a single subregion, of radius R migration lengths, in which $k_\infty = 1.20$.

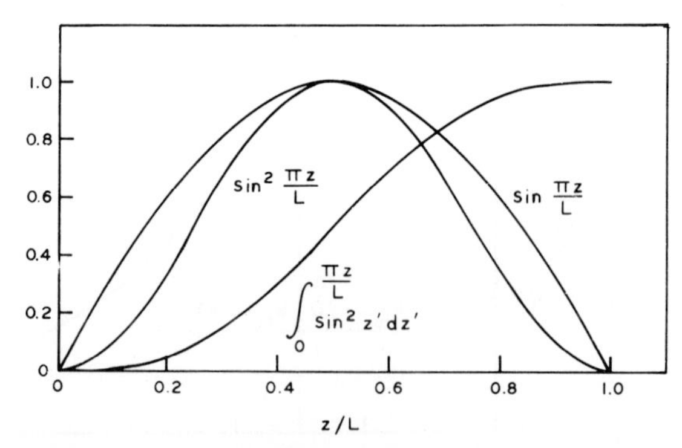

FIG. 4-17 Functions giving approximate worth of a weak control rod as a function of fractional insertion (z/L) in an equivalent bare reactor. L is the equivalent bare length of the core, and z is measured from the extrapolated boundary. The differential worth, given by Eq. (4-7), is proportional to $\sin^2(\pi z/L)$. The reactivity worth of the inserted portion of the rod, see Eq. (4-8), is proportional to the integral (from z = 0 to z = z) of $\sin^2(\pi z/L)$ dz.

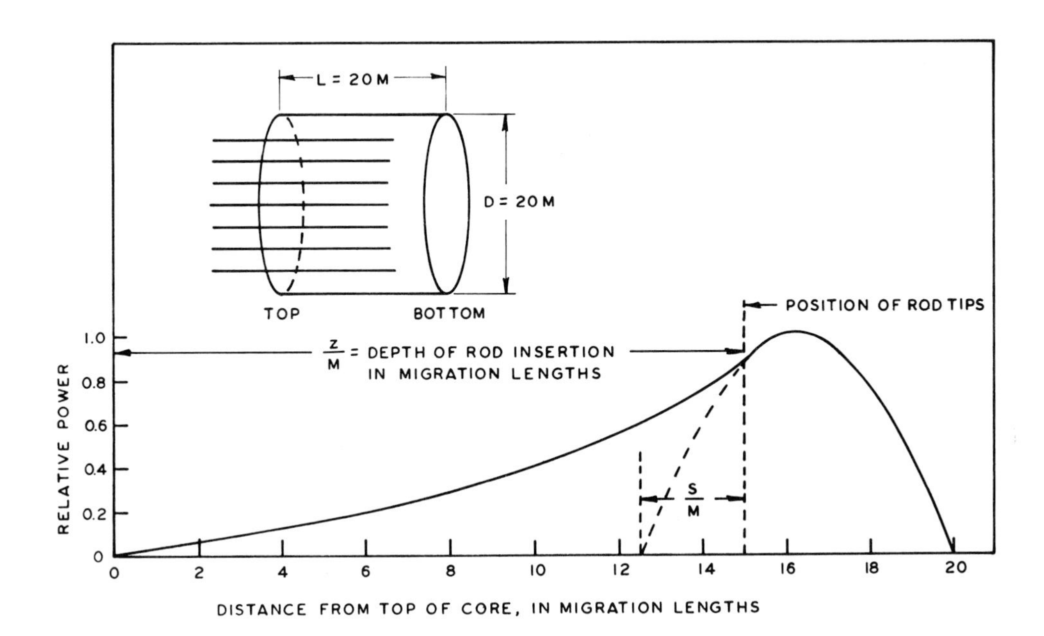

FIG. 4-18 Illustration of axial power distribution in a large critical bare core with a strong bank of control rods partially inserted. k_∞ is assumed to be 1.19, and the extrapolation length S into the rodded region is assumed to be 2.5 M.

For the flux distribution to be very drastically distorted, not only must the strength of the rod or the rod bank be high, but the reactor must also have a length which is large relative to the migration length M. As illustrated in Fig. 4-18, the critical reactor containing a partially-inserted bank of strong rods may be considered to consist of a critical core which occupies the region ahead of the rod tips, and a "reflector" which consists of the portion of the core occupied by rods. The axial geometric buckling of the non-rodded region is determined by the length of the region (L - z) and by the effective "reflector saving" of the rodded region S, which is of course determined by the strength of the rod bank. Criticality can be supported only if the values of k_∞ and M in the unrodded region are such as to satisfy the relation:*

$$\frac{k_\infty - 1}{M^2} = B_r^2 + \frac{\pi^2}{(L - z + S)^2} \qquad (4\text{-}9)$$

or

$$k_\infty - 1 = M^2 B_r^2 + \frac{\pi^2}{\left(\frac{L}{M} - \frac{z}{M} + \frac{S}{M}\right)^2}, \qquad (4\text{-}10)$$

where B_r^2 is the radial component of the buckling [$(2.405/R)^2$ for a cylinder of radius R.]

So long as the situation can be represented as in Fig. 4-18, the differential worth of the rod bank is given by:

$$\frac{d\,k_{eff}}{k_{eff}} = \frac{-dz}{L'} \cdot 2\left(\frac{\pi M}{L'}\right)^2 \frac{1}{1 + M^2 B_r^2 + (\pi M/L')^2} \qquad (4\text{-}11)$$

$$L' = L - z + S$$

The situation cannot be so represented after the rods are inserted so far that the reactor can no longer become critical. At the critical depth of rod insertion the differential worth reaches its maximum value:

$$\left(\frac{d\,k_{eff}}{k_{eff}}\right)_{max} = -2\frac{dz}{L - z + S}\left[\frac{k_\infty - 1 - M^2 B_r^2}{k_\infty}\right] \qquad (4\text{-}12)$$

or

$$\left(\frac{d\,k_{eff}}{k_{eff}}\right)_{max} = -2\frac{dz}{M\pi}\frac{\left(k_\infty - 1 - M^2 B_r^2\right)^{3/2}}{k_\infty}. \qquad (4\text{-}13)$$

The differential rod worth for the case of Fig. 4-18 is given as a function of z/M in Fig. 4-19.

If a single rod is moved independently in a core in which several other strong rods are partially inserted as a bank, the differential worth curve of the single rod may be highly distorted. If the bank of rods is quite strong, the core again will behave as though it were a core of length L - z + S, and the single rod in this short core will have a total worth which is not much different from its total worth in the full-length core when the rod bank is absent. Thus most of the normal total worth may be compressed into a much shorter distance, with the result that the maximum differential worth is increased by a substantial factor. In Fig. 4-19 the

---

*The discussion is of course approximate because the one-group formulation is used as well because of the other simplifications. It is not intended as a presentation of computation methods.

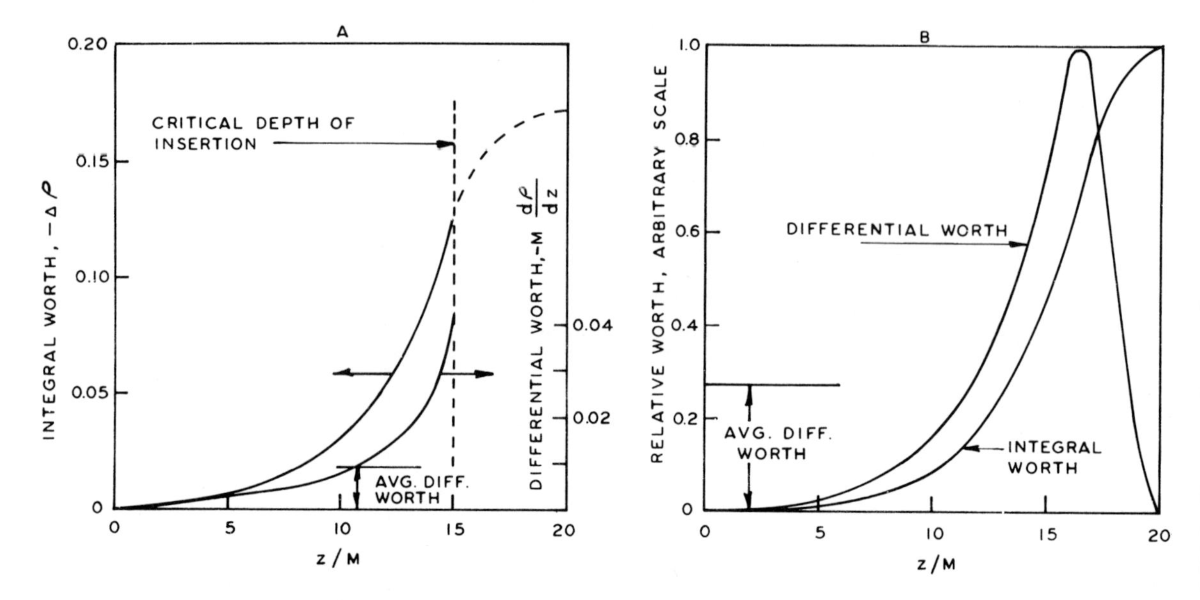

FIG. 4-19 Approximate curves of rod worth for the case shown in Fig. 4-18. Curves A show the differential and integral worths of the rod bank, as functions of depth of insertion, measured in migation lengths. The curves are approximated by assuming that the rodded region of the core acts as a reflector, with reflector saving 2.5 M. Curves B are the relative worth curves for an additional, weak, rod inserted when the rod bank is already in to a depth of 15 M (see Fig. 4-18). Differential worth assumed proportional to square of local power density.

differential worth of a single weak rod in the core illustrated by Fig. 4-18 is also plotted; this has been estimated by assuming that the differential worth at any point is proportional to the square of the power density at that point.

4.3 Design Choices and Specifications Involving Control Rod Worths

The choices of control rod type, shape, size, material, and pattern of installation are strongly involved with many other variables of the core design, and it is not possible to give useful general procedures for arriving at a design. However, there are some aspects of these choices and specifications which involve primarily questions of control rod worth and the variation of worth with the pattern of rod installation and the program of rod motion. These questions are discussed below.

4.3.1 Number of Rods

Although the total worth of the control rod installation is fixed by the excess reactivity and the required shutdown margin, the designer often has the choice of providing the required worth by a large number of weak rods or by a smaller number of stronger rods. There are often many restrictions on the choice because of mechanical and geometrical considerations in the core, as well as because of space limitations in the reactor face or the pressure vessel head. Other things being equal, the designer would usually prefer the smaller number of rods for economic reasons. The nuclear considerations, however, are highly important and usually tend to favor the choice of a larger number of rods. In Fig. 4-20, for example, both arrangements of cruciform rods may be mechanically feasible, and they provide nearly the same arrangement of absorbing surfaces when all rods are inserted, yet arrangement B uses only about half as many rods as A. The nuclear disadvantages of B relative to A are listed below. These disadvantages are typical of few-rod systems

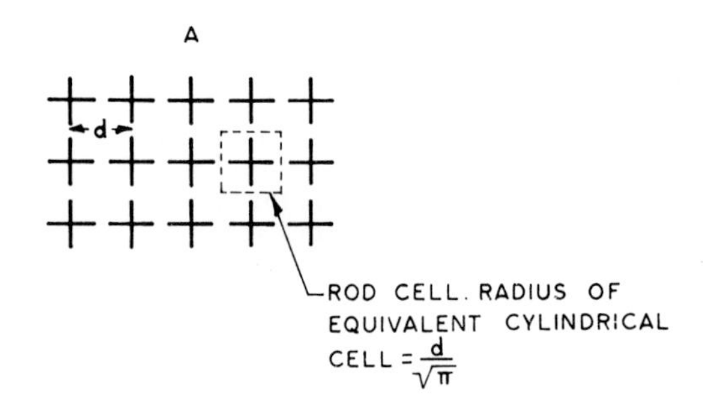

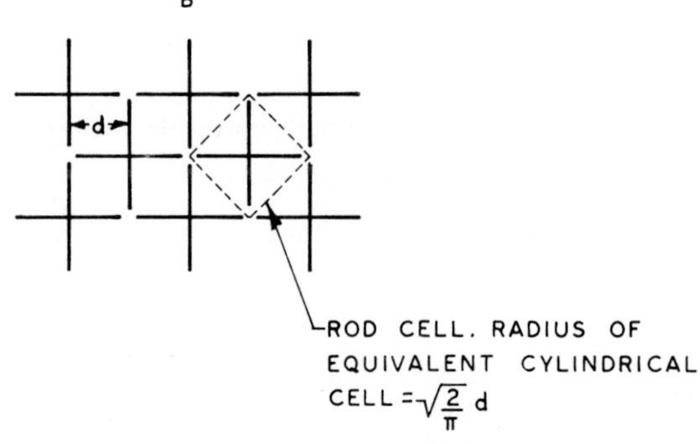

FIG. 4-20 Alternate installations of cruciform control rods.

relative to many-rod systems, although differences in rod type (e.g., simple absorber rods and "flux-trap" rods) must also be considered.

1. Although both systems will provide about the same shutdown margin, the stuck-rod margin with system B is considerably less than with A.

2. When the reactor is in operation, the number of rods withdrawn will typically be larger with system A than with B; hence system A will have more rods available for scram, and failures of individual scram mechanisms will be less serious in system A.

3. The uncontrolled withdrawal of a single rod can add more reactivity with system B than with A.

4. Rod failures which could conceivably add reactivity (e.g., breaking off a rod tip) are more serious with system B.

5. The mutual shadowing of rod pairs and of small groups of rods will be greater in system B than in A, and the control system and operating procedures will have to be designed to accommodate safely a larger range of rod-worth variation.

6. The variation of differential rod worth with depth of rod insertion will be greater for a rod in system B than for one in system A.

7. The opportunities for achieving favorable power distributions through appropriate rod programming are greater with system A than with B.

Of these considerations, only the first can yield an absolute criterion for the acceptability of a particular system; the other disadvantages of system B would be balanced against non-nuclear advantages in arriving at the choice. A first estimate of the acceptability of a given rod pattern can be made on the basis of the relation shown in Fig. 4-16. From the figure it appears that, to meet a one-stuck-rod condition, the radius of the control rod cell cannot be larger than about twice the migration length. In this connection it is to be noted that system B (Fig. 4-20) has a cell radius which is larger than that of A by the factor $\sqrt{2}$, not by a factor of 2.

4.3.2 Rod Program

The rod program refers to the sequence in which control rods are normally withdrawn and re-inserted. In reactors containing many rods, it is impossible to determine what the individual rod worths will be and what shape the differential worth curves will have except in terms of a specific program. Hence the selection of the program enters the design considerations at the same time as the selection of rod types, shape, size, etc. The degree of flexibility allowed in the program may vary with the purpose of the reactor and with other considerations. As the discussions in Sec. 4.2 indicate, the reactivity characteristics of the rods may vary widely with the program; consequently the designer must investigate the possible effects of any variations that are allowed in the program. Similarly, the degree to which any restrictions placed on the program may be circumvented or abrogated during reactor operation should receive serious consideration in the evaluation of safety.

Usually the designer attempts to select a program which will avoid large changes in the reactivity worth of the individual rods, which will avoid highly distorted differential worth curves, which will avoid the occurrence of unusually high or unusually low rod worths through shadowing effects, and which will result in spatial distributions of power approximating the optimum. Often it is found possible and desirable to minimize the spread of reactivity worths among the several rods of the installation by a suitable choice of program. In many cases it may prove desirable to preserve a symmetrical power distribution about the axis of the core. When rods of relatively high individual worth are used this consideration may lead to a degree of banking of the rods which from other considerations is not the most desirable. In some cases, the achievement of some desired reactivity coefficient may affect the choice of rod program. For example, the banking of rods will accentuate the negative reactivity effects due to neutron leakage, and will tend to give the most highly negative temperature and void coefficient for a given operating condition. Strong negative coefficients of this type may be desirable in some cases and undesirable in others.

Special considerations may sometimes require a high degree of flexibility in the rod programming, either in normal operation or for special operations. Research and test reactors of course require a high degree of flexibility to accommodate the varying demands of experiments. Reactors which are susceptible to spatial oscillations of the xenon content require the program flexibility necessary for controlling these oscillations. The use of control rod manipulations for the location of defective fuel elements (Sec. 10.1) may require complete flexibility of the program for this special purpose. It is quite possible for the designer to allow the necessary flexibility, but the range of variation in rod worths must be investigated and allowed for in setting rod operating speeds, procedures, etc.

Ordinarily unusually high rod worths can be avoided if, in starting the reactor, the rods nearest the periphery of the core are always withdrawn first. This is a particularly useful sequence in the cold startup of large water-moderated reactors. Usually, after the reactor has been brought to operating temperature, the possibility of effects such as illustrated in Fig. 4-15 is greatly reduced because of the substantial increase of migration area as the moderator temperature is increased. After operating temperature has been reached, the program can usually be selected to produce some desired effect on the spatial power distribution.

4.3.3 Speeds of Withdrawal and Scram

The core designer ordinarily must specify two mechanical characteristics of the control rod actuation system: the normal speed of rod withdrawal and the required speed (or acceleration) of the rods upon emergency scram. Usually the speed of withdrawal and the normal (non-scram) speed of insertion are the same. In setting this speed the designer must provide for sufficiently rapid reactivity changes to compensate the normal reactivity variations in the core and must avoid an unreasonably slow startup procedure, while avoiding the possibility of rapid reactivity additions through malfunction. Preferably, these requirements would be met with a single value of maximum speed for all rods, but this is not always possible. Obviously, the problem is simplified if the ranges of reactivity worth and differential worth which must be covered are as small as possible; these quantities depend on the control rod design, the pattern of rod installation, and the program. In any case, the crucial item to be predicted is the maximum differential worth which may be achieved within the limits of the possible program.

In arriving at the scram specification, the designer must take into account the possible accidents to be countered by scram and must predict the minimum reactivity worth, as a function of degree of insertion, for the least effective rod arrangement from which scram may be initiated. For this purpose, it is particularly important to take into account the highly skewed worth curve which may apply for a bank of rods in a "long" reactor (Fig. 4-19).

4.3.4 Accuracy of Rod Position and Position Indication

For satisfactory operation of the reactor it is necessary that the control rod mechanism be capable of positioning the rods with sufficient accuracy to preserve steady power, and that the mechanism avoid introducing errors, noise, and hysteresis into the control system through imprecision or backlash in the drive mechanism or in the position indicator. Obviously, the item of importance is the position of the control rod itself, and any imprecision introduced by a linkage between the rod and its drive system or position indicator must be taken into account. Usually the core designer must specify acceptable limits for these quantities. Reactors with strong negative power coefficients of reactivity may impose no very rigid requirements in these respects. In boiling and pressurized water reactors it is often feasible to use drive mechanisms which move the rods, not continuously, but in increments of reasonably large size. In reactors which do not have large negative power coefficients it will be impossible to maintain steady power unless the reactivity can be adjusted quite precisely to the critical value. In arriving at suitable specifications, the designer must take into account the power coefficients and the maximum differential reactivity worths of the control rods.

Aside from questions of maintaining steady power, it is extremely important to recognize that the control rods constitute the most convenient, and often the most reliable, sources of information about conditions in the reactor core. Basically this information consists of control rod positions for the various operational conditions of the core and of reactivity coefficients determined by small changes in the critical positions of the rods. The conditions of importance may be of such a nature that they can affect the reactivity worth of control rods, and consequently it is important to preserve the capability of calibrating the rods by period measurements at any desired time during the life of the reactor. For even rough calibrations, it will be necessary to measure reactivity increments of the order 0.001 with a precision at least as good as 10 percent. Hence the determination of differences in control rod position should be at least as precise as the position change corresponding to a reactivity change of about 0.0001. If one considers for example a rod having a total worth of 0.02 in a core 6 ft. (1.8m) long, the required precision of positioning and position indication would be at least 0.0001 x 6/0.02 = 0.03 ft or 0.36 in. (9.1 mm) if the differential worth of the rod were constant with position. If one allows for a maximum/average worth ratio of the order about 3, the required position precision becomes about 0.1 in. (2.5 mm).

The above requirements apply to relative positions. There is usually no great necessity for high precision in the absolute location of rods, provided the relationship between the actual position and the indicated position remains constant. In any situation, however, where the qualitative behavior of the rod depends strongly on its absolute position, the absolute value should be known accurately. For example, if the absorbing sections of the control rods are considerably longer or considerably shorter than the core, the operator needs to know rather precisely the absolute position of the rod in order to interpret its behavior accurately. In particular, if scrammable rods shorter than the core length must be used, it is important to ensure that the limits of possible motion of the rods are set in such a way that incremental withdrawal of the rod will always increase reactivity (see Fig. 4-21). Recent experience with the Elk River reactor has shown that the characteristics of the control-rod follower must also be considered in determining that no reversal shall occur in the curve of differential rod worth [83]. The Elk River reactor employs Zircaloy followers (less than full core length) on cruciform

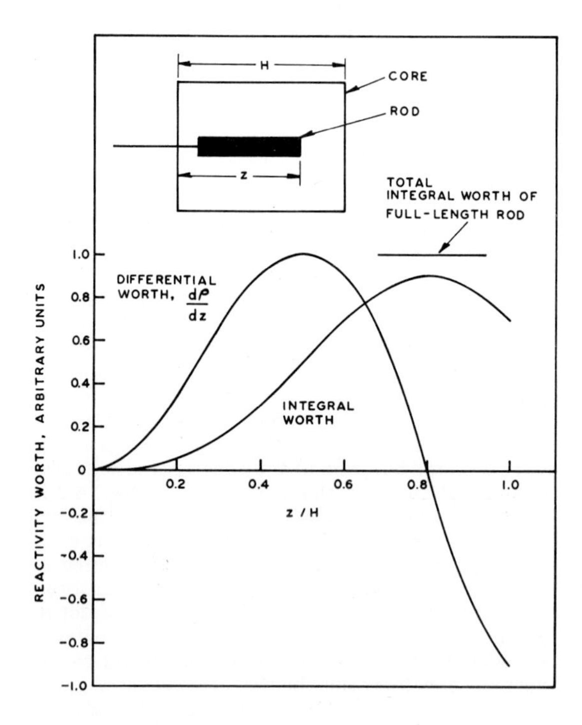

FIG. 4-21 Differential and integral worth curves for a weak control rod of length 60% of core length.

rods. The effects of followers on the worth curve should receive particular attention when fueled followers are used.

4.4 Reactivity Control Methods Other Than Rods

Although absorbing control rods are by far the most widely used method of controlling reactivity, they have certain general disadvantages which have encouraged the use of alternative reactivity control schemes in some cases. Among these disadvantages are the mechanical complexity, and cost of the control rod system, the distortions of spatial power distribution introduced by the control rods, the wasting of neutrons through absorption in control rods, and, in some cases, the difficulty of incorporating a sufficient number of control rods to compensate the excess reactivity. The alternate schemes are sometimes used in combination with some absorbing control rods. In specific applications there may be unique reasons for using alternate methods, as in small fast neutron reactors where absorbing rods are ineffective, or in high-temperature reactors where materials problems make the installation of in-core rods very difficult.

The following paragraphs describe the characteristics of some alternate schemes.

4.4.1 Adjustment of Neutron Leakage

A number of methods have been used for controlling reactivity through the adjustment of the neutron leakage from the core. Obviously a large adjustment of reactivity can be obtained by this method only if a large fraction of the neutrons born in the reactor normally leak out of the core. This is equivalent to saying that the reactor must have a relatively high value of k_∞ if a wide reactivity swing is to be produced by leakage variation, for the fractional leakage of neutrons is given by the expression:

$$\frac{\text{Number of neutrons leaking per sec}}{\text{Number of neutrons born per sec}} = \frac{k_\infty - 1}{k_\infty} \qquad (4\text{-}14)$$

Conceptually, the most obvious way to change neutron leakage is to change the geometry of the reactor core. This is not often a practical possibility in reactors intended for operation at an appreciable power level, but many critical experiments have utilized the principle. These have been experiments in which aqueous solutions of fissile isotope have been added gradually to a core tank until the critical core height is attained.

The effective multiplication constant may be approximated by an expression involving k_∞, the effective dimensions of the core, and the migration length M. The expression for a cylindrical core is typical:

$$k_{eff} \cong \frac{k_\infty}{1 + M^2[(\pi/h)^2 + (2.405/r)^2]} \qquad (4\text{-}15)$$

where h is the equivalent bare height of the core and r is the equivalent bare radius. If the height and the radius of the core are expressed in terms of the neutron migration length M, then

$$k_{eff} = \frac{k_\infty}{1 + \pi^2/H^2 + (2.405)^2/R^2}, \qquad (4\text{-}16)$$

where H = h/M and R = r/M. Evidently one of the terms involving a core dimension, $(\pi/H)^2$ or $(2.045/R)^2$, must have a maximum value which is an appreciable fraction of unity if a large change in k_{eff} is to be obtained through variation of the dimension. That is to say, in any reactor or critical assembly which is to be controlled by varying a core dimension, the minimum value of that dimension must be a small multiple of the migration length M if any large range of reactivity variation is to be obtained. This principle and its relationship to k_∞ are illustrated in Fig. 4-22.

It is to be noted in Fig. 4-22 that the rate of change of reactivity with core length, at the point of criticality, is rather large in terms of the total reactivity swing available. Thus for example in the infinite-diameter case, with k_∞ = 1.5, $H\partial\rho/\partial H$ ($= h\partial\rho/\partial h$) is 0.67; if the core in question were composed of a water solution of uranium, for which M might be about 6 cm and h_{crit} about 27 cm, a change of 1 cm in h would add a reactivity of approximately 0.025. This high rate of change of reactivity with core dimension is typical of cores with short migration lengths and high values of k_∞. In general the rate of change of reactivity at the critical point, for a core of variable height h and effectively infinite in other directions, is given approximately by,* cf Eq. (4-12)

$$\left(\frac{d\rho}{dh}\right) \cong \frac{2}{h}\left(\frac{k_\infty - 1}{k_\infty}\right). \qquad (4\text{-}17)$$

The high rate of change of reactivity with core dimension is the most important consideration in this type of control from the point of view of reactor

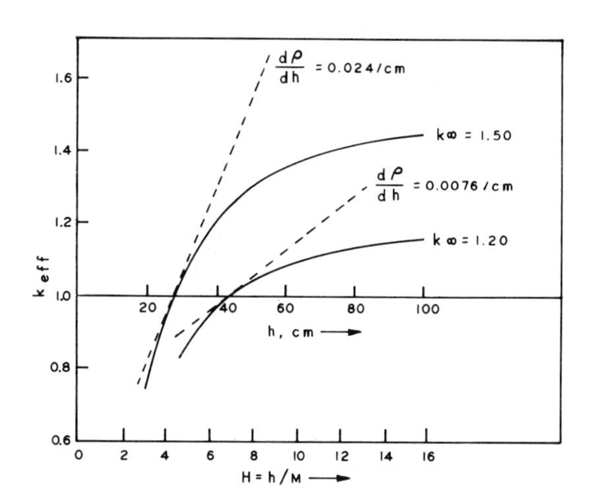

FIG. 4-22 Variation of k_{eff} with thickness h of an infinite bare slab. The curves are general on the h/M scale. The h scale applies for M = 6.33 cm (M^2= 40 cm^2), a representative value for H_2O moderation (cold).

*An alternate expression, which is also useful, is $d\rho/dh = 2\pi^2 M^2/(h^3 k_\infty)$.

safety. Very large rates of change can lead to unexpected reactivity increases. For example, if an absorbing control rod is dropped into a liquid core, the increase in reactivity caused by the increase in liquid level may be greater than the decrease caused by the control rod (Fig. 4-23). A mishap of this type has actually occurred [75].

In many cases the height of a reactor core can be effectively varied by varying the moderator level in the core. When the moderator serves also as coolant, there can be no operating condition at power corresponding to reduced moderator level, but in many instances criticality is approached by raising the water level in water-moderated reactors or critical experiments. In such cases, the typical reactivity variations are essentially the same as those illustrated in Fig. 4-22, and the possibility of rapid reactivity changes with water level must be recognized. In pressure tube reactors, such as the D_2O-moderated types, moderator level may be used to control certain reactivity swings, and moderator dump is often used for emergency shutdown. In large reactors it is difficult to obtain a wide range of reactivity through moderator level adjustment without uncovering a relatively large fraction of the core. Since the power generation will be quite low in those portions of the core which do not contain moderator, this method of control is not usually applicable for large reactivity swings. Moderator dump is of course quite effective as a shutdown mechanism, although it does not reduce reactivity very rapidly in the early stages of dump (see Fig. 4-22). Some studies have been made of H_2O reactors incorporating separate moderator and coolant systems to allow moderator level control in these reactors [76].

The neutron leakage can also be adjusted by changing the properties of the reflector. Needless to say, the reactors must be rather small (in terms of migration lengths) and must have rather effective reflectors if any substantial reactivity change is to be accomplished. Clementine, the earliest fast reactor, used rods of different scattering cross section in the blanket for reactivity control. A high degree of effectiveness was obtained in EBR-1 by moving the entire reflector (usually referred to as

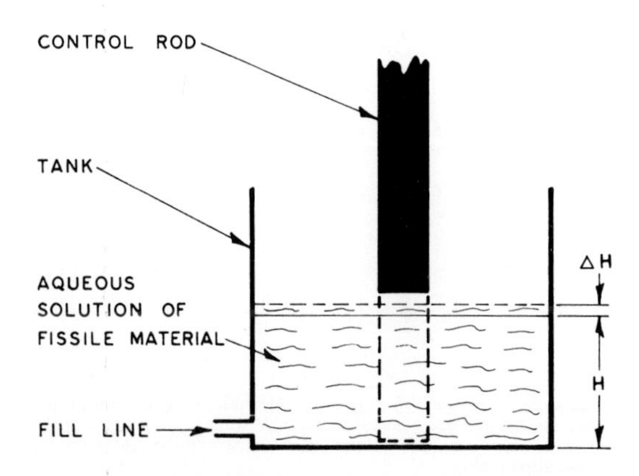

FIG. 4-23 Diagram of reactivity increase caused by control rod insertion. If k_∞ of the solution is high, so that H is rather small at criticality, dropping of the control rod may displace enough solution that ΔH increases reactivity more than the absorption by the rod decreases it.

the blanket) [77]. Absorbing rods were used in the reflectors of early aqueous homogeneous reactors at Los Alamos, and a variable level of heavy water reflector was employed in HRE-I [78]. Other small reactors have used large rotating drums in the reflector, which could place, alternatively, absorber or reflecting material (e.g., beryllium) adjacent to the core.

Leakage is also affected by the density of the moderator, and this is the principal component of the temperature coefficient of reactivity in small highly-enriched water-moderated reactors. This effect is discussed more fully in a later section.

4.4.2 Adjustment of Moderation (Spectral Shift)

Although variation of the effectiveness of moderation changes the neutron leakage, this is only one component of the total reactivity effect in the general case. In reactors of low or moderate enrichment the degree of moderation has a strong effect on the resonance absorption in the fertile material, and in large reactors of low enrichment the resonance effect will ordinarily be larger than the leakage effect. These effects enter into the temperature coefficient of reactivity, which may play an important part in reactor control [84], but the most direct application of moderator variation for control purposes has been in the proposed spectral shift reactor. This reactor is cooled and moderated by a mixture of H_2O and D_2O and reactivity is controlled by varying the H_2O/D_2O ratio. The fuel/water ratio is chosen at such a value that the reactor is highly undermoderated—i.e., a large fraction of the neutrons are absorbed by resonance capture in the fertile material—when the D_2O/H_2O ratio is high. This means that a relative increase in the H_2O content will increase reactivity. The highest D_2O/H_2O ratio is used at the beginning of core life and reactivity is added as needed, as fuel burnup proceeds, by the addition of H_2O. Naturally, many components of the neutron balance are changed as the D_2O/H_2O ratio is changed, but the major effect, in a typical case, is the change in the resonance escape probability. This may be seen in Table 4-6, [82] which gives calculated neutron balances at beginning of life and end of life for a spectral shift reactor fueled with partially-enriched uranium. The reactivity variation with D_2O content is shown in Fig. 4-24, [82].

The spectral shift reactor is an example of a rather highly undermoderated aqueous reactor and requires particular attention to the following safety considerations:

1. The new reactor, if it contained pure H_2O would be highly supercritical (see Table 4-6.) This circumstance must be provided for in the emergency cooling arrangements: i.e., any emergency cooling by H_2O must be done with highly poisoned water.

2. Any accident which could increase the moderator/fuel ratio would increase reactivity substantially.

3. It is possible that reactivity might increase if a fuel element were removed from the core. While this situation is not in itself inherently dangerous, it should be made known to the operator if it exists.

TABLE 4-6

Characteristics of a 330 Mw(e) Spectral Shift Reactor

Neutron Absorptions and Leakage

Isotope	Beginning of life (78 mole % D_2O) Fast and resonance	Beginning of life (78 mole % D_2O) Thermal	End of life (2 mole % D_2O) Fast and resonance	End of life (2 mole % D_2O) Thermal
U^{238}	0.727	0.066	0.479	0.072
U^{235}	0.507	0.493	0.144	0.295
Pu^{239}	—	—	0.094	0.381
Pu^{241}	—	—	0.033	0.053
Pu^{240}	—	—	0.120	0.008
Pu^{242}	—	—	Negligible	Negligible
U^{236}	—	—	0.014	0.001
Xe^{135}	—	—	—	0.036
Sm^{149}	—	—	—	0.012
Fission products of U^{235}	—	—	0.051	0.019
Fission products of Pu^{239}	—	—	0.023	0.010
$H_2O + D_2O$	0.003	0.007	0.004	0.012
Zircaloy	0.005	0.003	0.002	0.004
Stainless steel	0.001	0.001	—	0.001
Leakage	0.022	0.093	0.015	0.096
Total absorptions and leakage	1.928		1.980	

Neutron Production

Isotope	Fast and resonance	Thermal	Fast and resonance	Thermal
U^{235}	0.821	1.016	0.233	0.608
Pu^{239}	—	—	0.168	0.702
Pu^{241}	—	—	0.073	0.104
U^{238} (fast fission)	0.091	—	0.092	—
Total production	1.928		1.980	

Over-all Reactivity Balance

	k_{eff}
Cold, clean, 100% H_2O	1.38
Hot, clean, 100% H_2O	1.28
Hot, clean, 78% D_2O	1.00

Reactivity Control Requirements

	$\Delta k_{eff}/k_{eff}$
Temperature deficit, 130 to 566°F (54 to 297°C)	0.08
Doppler deficit, 566°F (297°C), zero to full power	0.020
Equilibrium xenon	0.03
Samarium	0.01
Fuel burnup and isotope buildup	0.24
Safety shutdown	0.03
Total control requirements	0.41

Reactivity Control Methods

Control in control rods	0.08
Control in soluble poison, 3.4 gm H_3BO_3/l	0.08
Control in D_2O, 78 mole %	0.25
Total control	0.41

4.4.3 Dissolved Absorber

Absorbing materials dissolved in the coolant-moderator have been used frequently in pressurized-water reactors, in conjunction with control rods, as an aid in achieving the required shutdown margin. The injection of highly poisoned solutions as a "backup" emergency shutdown system is common practice in water-moderated power reactors. The use of soluble absorber as a means of reactivity control during operation has been investigated in the Yankee [79] and Saxton reactors, and has been considered for other reactors, including the pressure-tube D_2O-moderated reactors (in the moderator). Approximate quantitative relationships for reactivity worth are given in Sec. 1.2 of the Water Reactor Kinetics Chapter.

The use of soluble absorber during power operation may complicate the coolant-moderator chemistry of the plant. Aside from any hazards that might be connected with such complications—which would depend upon the particular characteristics of the installation—the safety considerations which require attention with soluble absorbers are the following:

1. If the absorber is necessary for shutdown, any emergency cooling water which may be used in the core must be adequately poisoned.

2. The dissolved absorber will make the temperature and coolant- (or moderator-) void co-

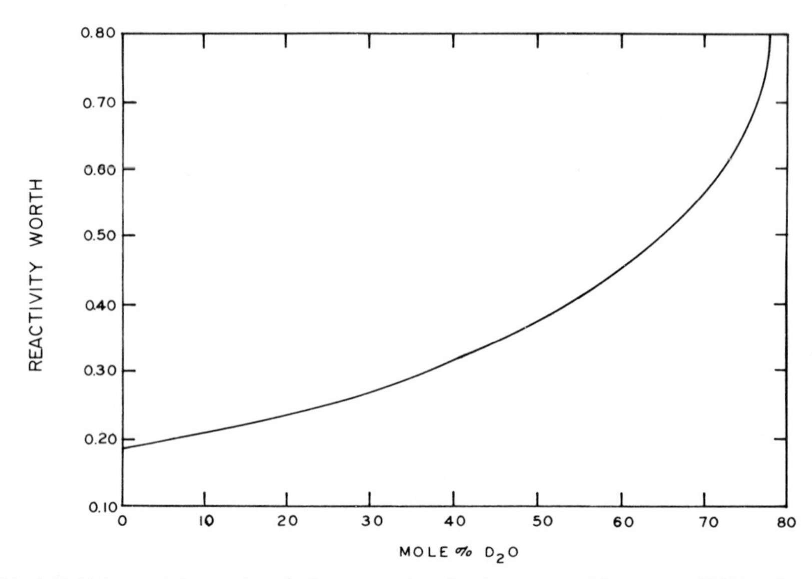

FIG. 4-24 D_2O reactivity worth vs D_2O concentration, for the spectral-shift reactor of Table 4-6. The reactivity worth unit is the reduction of reactivity, in percent k_{eff}, produced by an increase of 1% in the molal concentration.

efficients of reactivity more positive.

3. Hide-out of absorber, fluctuations of reactivity due to variations of concentration, and inadvertent dilution of the absorber (as through the startup of an unpoisoned loop) must be prevented.

4. Possibilities of precipitation of absorber, or removal of plated-out absorber, by changes in temperature or coolant chemistry, must be eliminated. If absorber should plate out on surfaces in the reactor core, the subsequent sudden removal of the plated material, if it could occur, would result in reactivity increase.

5. A dependable means of measuring the ambient concentration of absorber in the coolant or moderator should be provided. The reactor operator should be able to account for all observed reactivity changes to assure safe operation. Reactivity anomalies have been observed in the Yankee reactor, some of which may be attributable to soluble poison effects [79].

4.4.4 Movable Fuel Assemblies

The removal or addition of a fuel assembly to the core of a thermal neutron reactor is not usually a highly effective way of controlling reactivity, although fuel assemblies may be used as followers for absorbing rods. There are no materials, however, which have large absorption cross sections in the high energy neutron spectrum which characterizes highly-enriched fast-neutron reactors. In such reactors movable fuel elements which may be withdrawn from the core to decrease reactivity may prove attractive. This control method is used in EBR-II [80]. In larger fast reactors, which include relatively large fractions of fertile materials or other non-fissile materials in the core, the neutron spectrum becomes relatively "softer," and absorbing rods may be attractive. Boron-containing rods are used in the Enrico Fermi fast breeder reactor [81]. When movable fuel is used, the main effect of removing the fuel assembly from the core is to increase the neutron leakage. This effect may be reasonably important in fast reactors of moderate size, for the total leakage from the core is rather large in such reactors.

In one particular case the use of movable fuel assemblies appears attractive in thermal reactors. This is in the Canadian D_2O-moderated reactors, where highly enriched "booster" rods are used to increase the reactivity of the natural uranium core during xenon transients. This application is particularly attractive, since the booster rods are needed for only short periods of time, and consequently the burnup of enriched uranium is small. At the same time the booster rods provide an excess reactivity which, if it were provided by the natural uranium core, would have to be held down by the wasteful absorption of neutrons during most of the operating period.

The use of movable fuel assemblies for reactivity control does not appear to introduce any special safety problems except for whatever additional difficulty there may be in maintaining adequate cooling for the movable assemblies, and except for any additional probabilities of rod sticking which may be associated with distortions of the movable fuel assemblies.

5 SPATIAL DISTRIBUTIONS OF POWER

The relationship of the (static) spatial distribution of power density in the reactor core to reactor safety is a subtle one. To assess it one must first assess the safety significance of fuel element failures, for the thermal failure of one or more fuel elements is the obvious malfunction which can result from a poor distribution of power density. It has already been said (Sec. 3) that the failure of a small number of fuel elements does not constitute a "reactor accident" provided the reactor containment and waste disposal systems have been designed to accommodate such a number of failures without hazard, and provided there is no danger that failures will be self-propagating from fuel element to fuel element. Indeed, the probability of being able to operate for very long periods without some fuel element failures appears so poor that one might question the wisdom of building any nuclear plant which cannot safely tolerate some fuel element failures.

The core designer always strives for the flattest possible power distribution, for the economic purpose of achieving the highest possible average power density. It is not at all clear that indifferent success in achieving that purpose increases the probability of a reactor accident. To be sure, if the designer overestimates his success in power flattening some fuel elements may fail thermally when the reactor is brought up to full power, or possibly later; but this is not necessarily a serious hazard if proper action is taken to prevent further failures through limiting the permissible operating power or through correction of the core design, and if the initial approaches to full power operation are made reasonably slowly.

On the other hand, if an accident does occur from causes other than fuel element failures, the power distribution may influence the course of the accident; and at the same time thermal failure of fuel elements under these conditions may be more serious than under normal circumstances. Thus, for example, such reactivity-limiting effects as those due to Doppler, coolant-void, and temperature coefficients depend roughly upon the average reactor power, and the maximum/average power density ratio may determine what fraction of the fuel elements sustains thermal damage before the inherent reactivity-limiting effects can terminate a reactivity excursion. Further, in a loss-of-coolant accident the power distribution determines some of the required characteristics of the emergency cooling system; fuel element failures in this case would be more dangerous than in the normal case, since loss of coolant implies that the boundaries of the primary system are no longer effective barriers against fission product dispersion.

It therefore appears that a major safety significance of the power distribution is in its effect on the power coefficients of reactivity, its effect on the actions of those coefficients to determine the courses of postulated accidents, and its effect on the overheating of fuel elements in the course of possible accidents. Another primary safety consideration of flux distribution, especially in large reactors, is its effect on the sensitivity of nuclear instrumentation and on the degree to which local signals will be representative of peak conditions; another somewhat related aspect is the potential shift in effectiveness and calibration of safety controls, both individually and in terms of total system time response and reactivity worth. The flux distribution is also important because it is inherently related to other characteristics of the reactor: without an understanding of the power distribution throughout the core life the operator cannot have the understanding of reactor behavior which is essential to safe operation.

The power distribution in a particular reactor can be determined only by detailed calculations or measurements. The following discussion is intended only to set forth the major effects which determine the distribution. The first two sections treat the distributions in cores which are homogeneous on the local scale. The discussion applies to reactors with truly homogeneous composition and approximately to reactors of "lumped" composition provided one considers only the gross distribution over large regions and ignores local effects such as the power density variations with individual fuel-element assemblies and near control rods. A following section treats these local effects separately. To some degree of approximation the complete power density distribution in a core may be computed as the superposition of local variations upon the gross distribution which is computed for an appropriately "homogenized" composition (see discussion of core cells and homogenization in Sec. 2). The accuracy of this approach varies with reactor type. A final section of the discussion illustrates some of the limitations of the superposition approach in H_2O-moderated reactors, which are among those for which it is least appropriate.

5.1 Distributions in Uniform Cores

In practice not many power reactors employ cores which have uniform composition, even on a gross scale. Almost all incorporate nonuniformities of one or more of the following types: nonuniform distributions of control elements; nonuniform fuel loadings or enrichments; nonuniform distributions of shims or burnable poisons; nonuniform fuel burnup. Nevertheless it is helpful to use the uniform core as a conceptual point of departure for the discussion. The power density distribution in a uniform, bare, cylindrical core of extrapolated radius R and length 2Z is:

$$P(r, z) = P_o J_o\left(2.405 \frac{r}{R}\right) \cos\left(\frac{\pi}{2} \frac{z}{Z}\right), \quad (5\text{-}1)$$

where the origin is taken at the center of the cylinder and P_o is the power density at the origin (the maximum value). If the core is not bare the distribution will usually depart from that of Eq. (5-1) near the core boundaries, but if the reflector saving is not a large fraction of R or Z the ratio of maximum/average power distribution in the core will not be much different from that given by the application of Eq. (5-1) up to the real core boundary. The maximum/average ratios of the J_o and cosine components of Eq. (5-1) are given in Fig. 5-1 as functions of the reflector saving or core extrapolation distance.

5.2 Effects of Variable Composition

Often the designer will attempt to improve upon the power distribution represented by Eq. (5-1) by incorporating a spatial variation of the core composition. This may be a variation in fuel enrichment, in fuel/moderator ratio, or in the density of a burnable poison. It may also be simply a variable absorber content introduced by a suitable arrangement of control rods.

Any flattening of the power density distribution

will increase the neutron leakage from the core, and will therefore "cost" something in reactivity and/or internal conversion ratio. If one is striving for some particular value of the maximum/average ratio which is lower than that for a uniform core, the value can be achieved at the least cost in extra fuel enrichment if the property variation is so arranged that the gross power distribution in a central region of the core is perfectly flat, as indicated in Fig. 5-2. The core is then said to consist of a "flat zone" and a "buckled zone". To a first approximation this may be achieved with uniform, but different, compositions in each of the two zones. The composition of the flat zone is chosen to give zero buckling in the direction of flattening (e.g., $k_\infty = 1.00$ if the core is to be flattened in all directions; $k_\infty = 1.00 + [M^2\pi^2/(2Z)^2]$ if flattening is to be done in the radial direction only). The composition of the buckled zone must be chosen to give criticality with the existing core dimensions and the desired dimension of the flat zone.

Usually the ideally flattened situation cannot be achieved in detail with only two different compositions, for the discontinuity in composition will cause a local power peak at the boundary. If, for example, the enrichment in the flat zone is lower than that

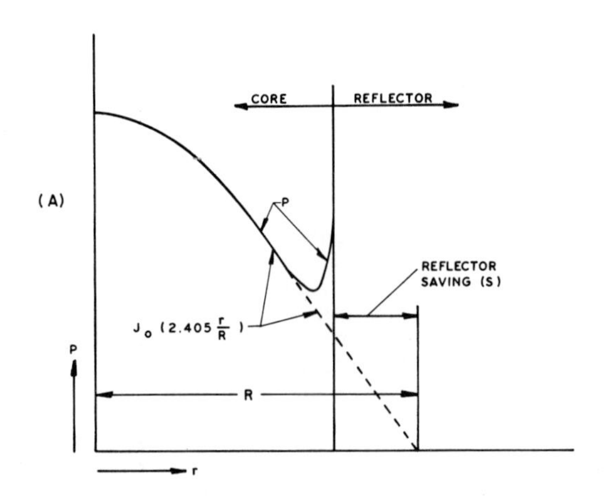

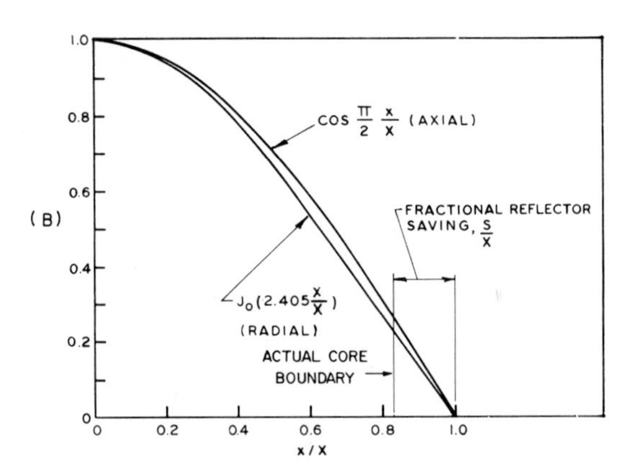

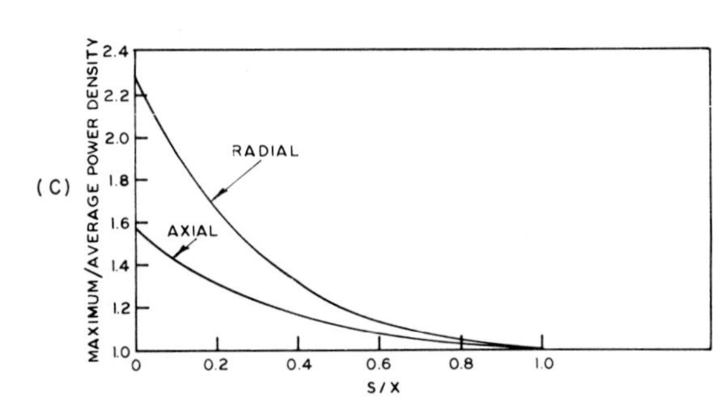

FIG. 5-1 Radial and axial fundamental mode distribution in a uniform cylindrical reactor. The radial component of the fundamental mode distribution is $J_0(2.405r/R)$ and the axial component is $\cos(\pi z/2Z)$, where r and z are measured from the center of the core, and R and Z are the values of r and z at the extrapolated radial and axial boundaries. In a reflected reactor which is otherwise uniform the spatial distribution of power density P may depart drastically from the fundamental mode distribution within about one migration length of the core-reflector interface.

Figure (A) shows this departure for the radial power distribution in a core with strongly moderating reflector. For nonmoderating reflectors the departure may be in the opposite direction.

Figure (B) gives plots of the radial and axial components of the fundamental mode distribution.

Curve (C) shows the effect of the fractional reflector savings S/X on the radial and axial maximum/average power density ratios, with the effects of departures from the fundamental mode distribution neglected. Because of this neglect curve (C) may be a poor approximation when S/X is greater than about 0.2. For large S/X the maximum power density may occur at the core-reflector interface instead of at the center of the core.

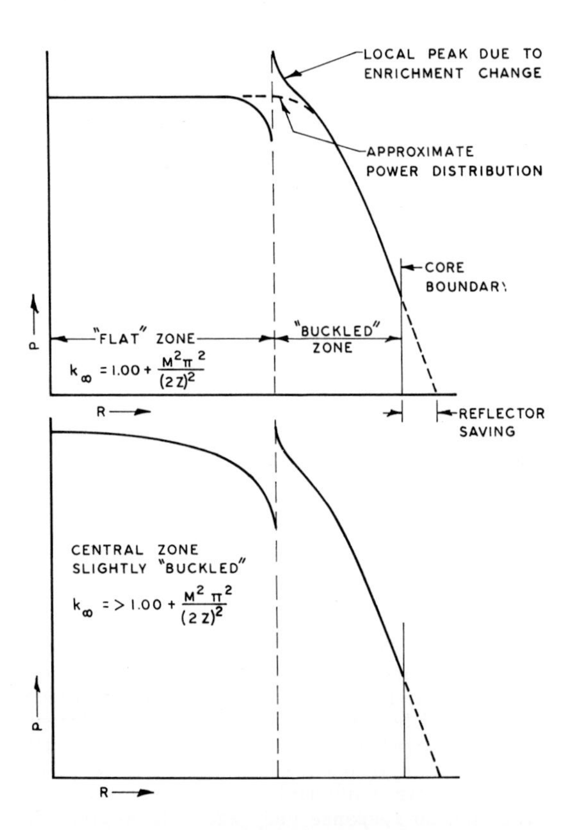

FIG. 5-2 Radial power flattening by the use of two fuel enrichments, (Z = half-length of extrapolated core). Because of the local power peak at the transition between the two zones, due to the higher enrichment in the buckled zone, complete flattening of the central zone does not give the lowest maximum/average ratio. An improvement results from increasing k_∞ (i.e., the enrichment) of the central zone to the point that there is a positive radial buckling, and the power density at the center becomes equal to that at the zone boundary.

in the buckled zone, the ratio (power density)/(thermal neutron flux) must be lower in the flat zone. Since the thermal neutron flux distribution is continuous across the zone boundary, the power density in the buckled-zone fuel elements adjacent to the boundary must be higher than that in the flat-zone elements adjacent to the boundary (Fig. 5-2). The effect of this local peak may be minimized by incorporating a slight buckling in the "flat" zone (for example, by making the enrichment of the inner zone somewhat higher, and the enrichment of the outer zone somewhat lower, than for the idealized "flat" case; see Fig. 5-2), or additional composition changes may be used to produce a more gradual transition.

In certain cases, particularly when both the coolant temperature rise and the film drop in the coolant channel are large, a "roof-topped" axial power distribution may be preferable to a flattened one (Fig. 5-3). Such a distribution lowers the power density near the outlet end of the coolant channel, where the coolant temperature is high. A "roof-topped" distribution can be obtained by suitable variation of the fixed core composition, or by the partial insertion of a bank of control rods, (c.f. Fig. 4-18).

In addition to the variations in composition incorporated by the designer the mere operation of the reactor may introduce spatial variations of reactivity which affect the power distribution. To the extent that the reactor has a power coefficient of reactivity which acts locally, the "zero power" power peaking will be accentuated or reduced at power, accordingly as the coefficient is positive or negative. The Doppler coefficient for example, may introduce a very significant power flattening. Fortunately the power coefficients are, in most reactors, negative. It must be remembered, however, that the effect of any power-coefficient component will be diffused over a distance of some migration lengths from the point of power generation, and a strong negative power coefficient will not reduce appreciably the very localized power peaks such as those due to adjacent water channels.

The boiling water reactor is a special situation in that the steam, which is generated at a rate proportional to power density in the boiling section of the reactor, must traverse the remaining length of the core to escape. Thus the steam-void component of the power coefficient of reactivity is distributed radially according to the power distribution, but axially the component increases continuously with height. The result is a desirable radial flattening, and an axial roof-topping (Fig. 5-4) [85] which is usually not desired in this reactor type. In long cores the axial skewing of the power distribution may become quite severe, and various means have been used or proposed to counteract it, including axial variations in fuel enrichment, fuel-water ratio, and burnable-poison content. An effective means of countering the effect is through the partial insertion of control rod banks from the bottom of the core. This consideration accounts for the use of upward-moving control rods in boiling reactors such as the Dresden reactor [86].

In any reactor which contains control rods during power operation, the pattern of control rod insertion may have a profound effect on the power distribution

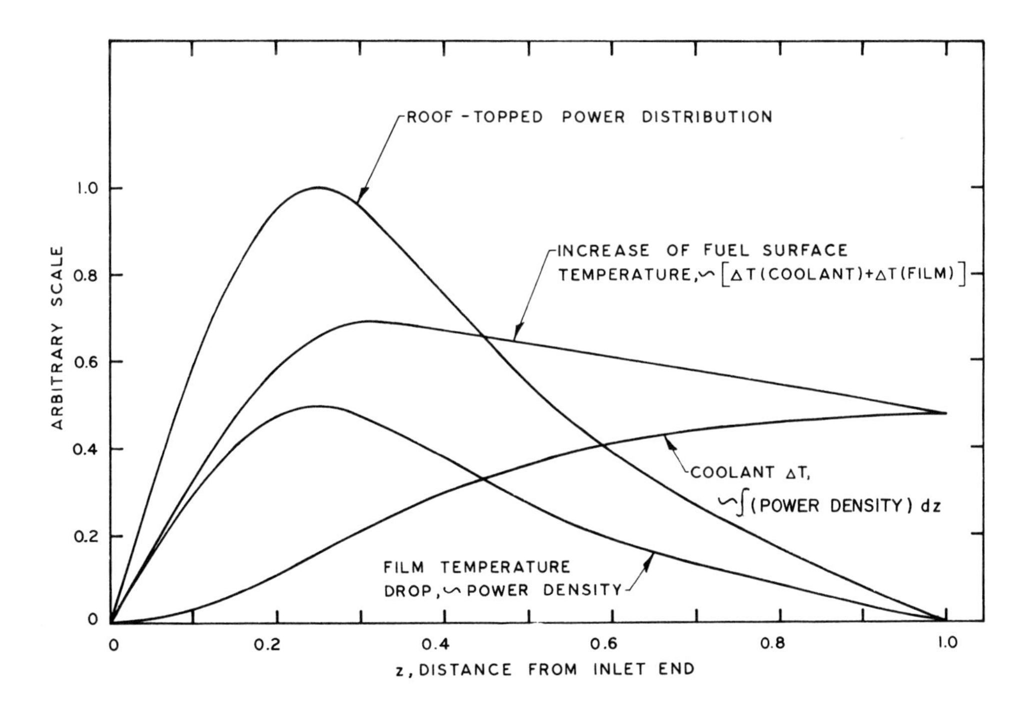

FIG. 5-3 Illustration of effect of "roof-topping" when fuel surface temperature is important. In this hypothetical case the use of roof-topping has shifted the highest fuel surface temperature toward the inlet of the reactor, and has reduced its magnitude considerably below that which would result from a symmetrical power distribution.

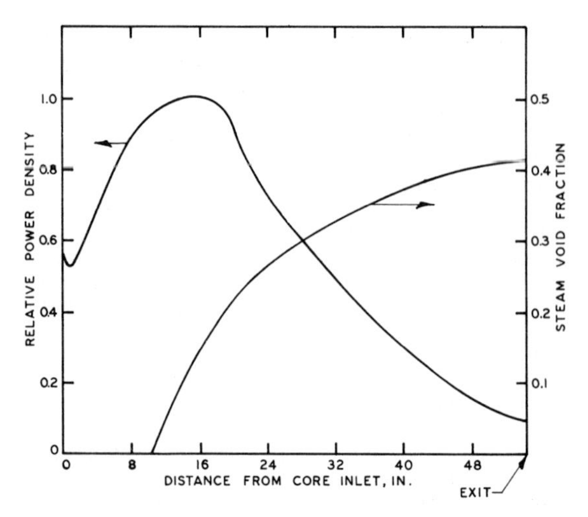

FIG. 5-4 Axial power distribution and void fraction in the Boiling Nuclear Superheater (BONUS) Reactor for a particular operating condition. The pronounced "roof-topping" is due mainly to the steam void distribution, although other effects are also present. The void fraction shown is the volume fraction of steam in the coolant channels, somewhat higher than the fraction of steam averaged over all the water in the core.

(see Figs. 4-11, 4-14, 4-18). Sometimes the rods may be used to improve the distribution, but in any case they will affect it, sometimes quite adversely. These effects must be taken into account by appropriate calculations in any evaluation of power distribution.

Finally, the burnup of fuel and burnable poisons introduces substantial spatial variations in core composition over the core life. Except in reactors with very high conversion ratios the fuel burnup tends to improve the power distribution. This subject is treated in Sec. 8.

At this point it need only be mentioned that if the initial power flattening in the core is achieved by control rods and/or burnable poisons, the improvement afforded by fuel burnup may not be sufficient to replace the flattening effect which is lost as control rods are removed and burnable poisons are depleted. Thus the variation of power distribution with time must be considered in the initial design of the core.

5.3 Local Effects

The power distribution effects considered above do not take into account the fine structure of the reactor core. To do so one must make calculations in detail for the particular arrangement of materials in the core; however, one may consider progressive degrees of inhomogeneity to recognize the more important effects (Fig. 5-5).

Figure 5-5A represents about the closest approach to homogeneity that is realized in current solid fuel reactors. It shows a regular array of single-rod fuel elements in a moderator; the elements may or may not be surrounded by process tubes to direct the flow of coolant. The British gas-cooled, graphite-moderated reactors are representative of this situation. In such a reactor, if no control rods are present, the power distribution will follow closely the ideal cases, such as those given by Eq. (5-1) or those similar to Fig. 5-1 or Fig. 5-2, except that the power density will decrease in the interior of each element, as in Fig. 5-6A. If the element is located in a region where the general power distribution has a gradient (e.g., the buckled zone of Fig. 5-2), this general distribution will be superimposed approximately on the local distribution indicated in Fig. 5-6A, giving an azimuthal variation about the fuel element.

The situation in Fig. 5-5B is much the same as in A, except that the single rod has been subdivided into several smaller rods; there is now a general power depression from the edge to the center of the rod bundle, as well as a local depression within each rod (Fig. 5-6B). If the coolant within the bundle is nearly transparent to thermal neutrons the power depression within the bundle will be nearly the same as the depression in a single rod containing all of the fuel in the bundle. If the coolant is a moderator the depression may be somewhat less, as additional thermal neutrons will be produced by slowing down within the bundle. This type of fuel arrangement is typical of that in the pressure-tube D_2O-moderated reactors, and of that in the graphite-moderated reactors of higher power density.

The arrangement in Fig. 5-5C differs from that of B in that the fuel bundles are large relative to the spaces between bundles. This is typical of the

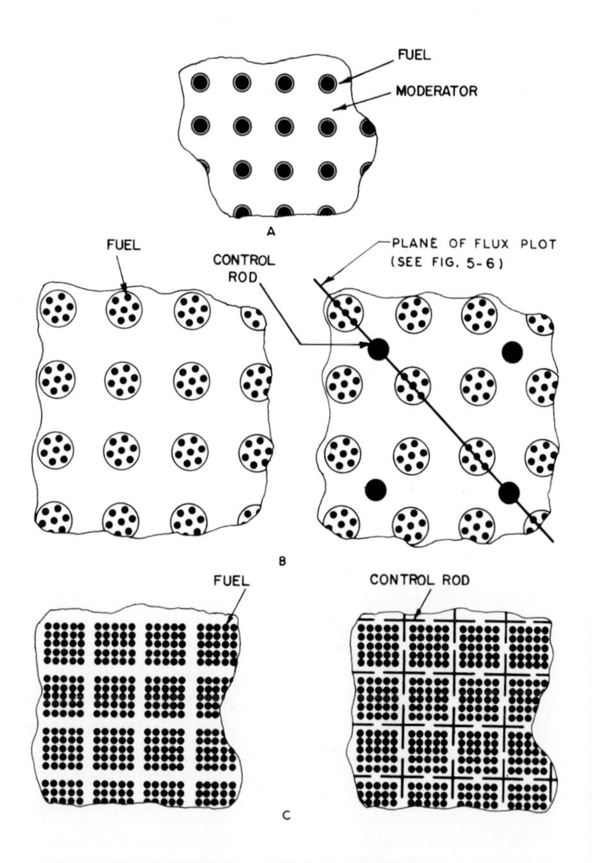

FIG. 5-5 Various degrees of lattice complexity. Top (A) shows regular array of single-rod elements in a moderator. Center (B) shows a regular array of rod-bundle elements and lower (C) is similar to (B) except array is close-spaced.

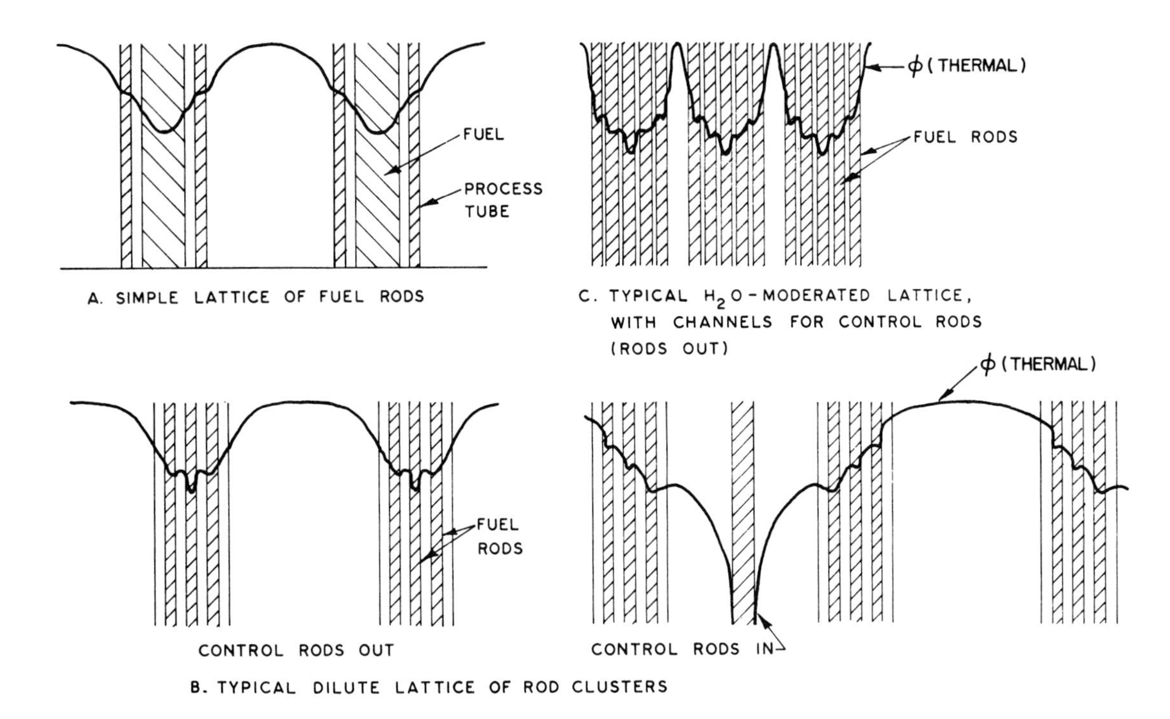

FIG. 5-6 Qualitative local distributions of thermal neutron flux in several reactor types. The power distribution is proportional to the thermal neutron flux in the fueled regions.

H_2O-moderated reactors, where the moderator has a short migration length. The qualitative behavior of the power distribution is the same in (B) and (C), but in (C) the power is said to be peaked at the edge of the fuel bundle, because of the water-channel effect, whereas in (B) the power is said to be depressed in the interior of the bundle. The reasons for the heterogeneity of structure are quite different. In (B) the heterogeneity is used for reactivity reasons and to simplify the mechanical structure, while in (C) it is used simply to provide space for the installation of control rods.

In any of the situations illustrated in Fig. 5-5 one can expect that the absorption of neutrons by the moderator will be a small fraction of the total absorption, otherwise the reactor would be a very poor one. That is to say, most of the neutrons thermalized by the moderator will be absorbed by the fuel. In many reactors, and particularly in those moderated by H_2O, the neutron slowing-down length in the core is longer than the diffusion length, so that many of the neutrons which are thermalized in a given volume of moderator were produced by fissions in fuel elements some distance away, but most of the neutrons thermalized in that particular volume are absorbed by nearby fuel elements. Thus, if one visualizes a situation in which the fuel is removed from one of the positions in Fig. 5-5B (as for example in an on-power reloading procedure) the number of neutrons thermalized in the vicinity of that position will not be greatly reduced, but the fuel will no longer be present to absorb them. Most of them will be absorbed by the fuel in the immediately adjacent positions. Hence the removal of one fuel bundle will increase the power density in the positions adjacent to the empty position. For the situation in Fig. 5-5B the magnitude of the fractional increase would be expected to have an upper limit of 0.25 (1/ [number of adjacent positions]) and might be much less, depending upon the relative values of the diffusion length, the slowing-down length, and the lattice spacing.

The "water-channel peaking" in the H_2O-moderated case may be considered in the same way: the water channels may be considered as portions of a uniform fuel-water lattice from which the fuel has been removed. Since the slowing down length ($\sqrt{\tau}$) may lie in the range 8 to 10 cm in a hot H_2O-moderated power reactor, while the thermal-neutron diffusion length is typically 2 to 3 cm, the effect can be quite large. Thus in Fig. 5-5C, many neutrons from considerable distances away are thermalized in the water channels, but most of the neutrons so thermalized are absorbed in the fuel elements immediately adjacent to the water channels.

For a row of fuel elements immediately adjacent to a water channel, an upper limit to the fractional power peaking* is given by the following expression, which is based on the above considerations:

$$P \leqq P_o \left[1 + \frac{V_H}{V_C}\,(1 - e^{-a/L})\right], \qquad (5\text{-}2)$$

*This analysis considers only the element-to-element peaking, not peaks which may occur at local points within an individual element.

where P is the power generated in an element adjacent to the water channel and P_0 is the power which would be generated in the element if the water channel were not present. The other quantities in the equation are defined in Fig. 5-7. Fuel elements at the intersections of water channels are subject to still larger power peaks, which can be estimated very roughly by the same principles. Figure 5-8 shows the results of a calculation of water-channel peaking which illustrates the principle further. From these considerations one arrives at a simple principle for minimizing power peaks in the design of an H_2O reactor: as far as possible, arrange the design in such a way that each internal fuel element has the same volume of water immediately adjacent to it.

When a thermally-absorbing control rod is installed as in Fig. 5-5, B or C, the power is reduced in the vicinity of the rod. The major portion of the power depression caused by the rod represents the direct absorption of thermal neutrons, and diminishes with distance from the rod with a decrement which is determined by the diffusion length in the core, but there is also a gentler component of the variation resulting from the loss of those neutrons which would have been produced by the chain reaction, as descendents of neutrons absorbed by the rod, if they had not been so absorbed. The change in power distribution produced by a strong rod (cf., Fig. 4-11), or by a number of rods, may be quite large. Usually these effects can be determined only by a calculation of more or less complexity.

Interruptions of the axial regularity of the core structure also introduce local power peaks. The usual interruptions are the ends of fuel elements when elements of part-core-length are used; the effect is usually referred to as end-cap peaking or Wilkins effect. If the elements are in a neutron-transparent coolant the peaking is due simply to the larger solid angle from which neutrons can enter the end of the element. If the coolant is a moderator the effect is another manifestation of the water-channel effect. Sometimes the importance of the effect can be nullified by so choosing the lengths of fuel elements that the ends lie in positions where the general axial flux distribution is well below its maximum value. The effect is of course minimized by minimizing the gap between the ends of fuel-element segments. In some cases it is controlled by using short segments of neutron absorber, or segments of low-enrichment fuel, at the ends of the fuel elements.

5.4 H_2O-Moderated Reactors

In reactors with large migration and diffusion lengths the spatial distribution of power can be visualized as a superposition of local disturbances upon the gross distributions which would characterize the core if its composition were "homogenized" locally; and in many cases the distribution can be calculated by an analagous procedure. In H_2O-moderated reactors, however, the local effects of water channels may be so large, and individual control rods may be so large,

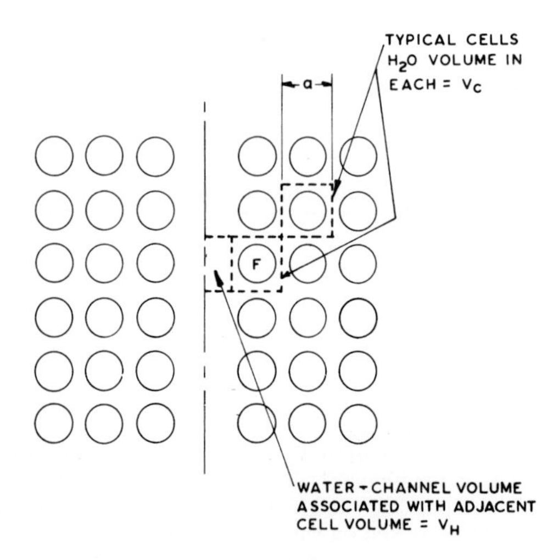

$$P \leq P_0\left[1 + \frac{V_H}{V_C}\left(1 - e^{-a/L}\right)\right]$$

FIG. 5-7 Approximation for upper limit of power peaking by water channel (far from channel intersections). P_0 is the power which would be generated in the element if no water channel were present. L is the effective diffusion length in the lattice, away from the water channel.

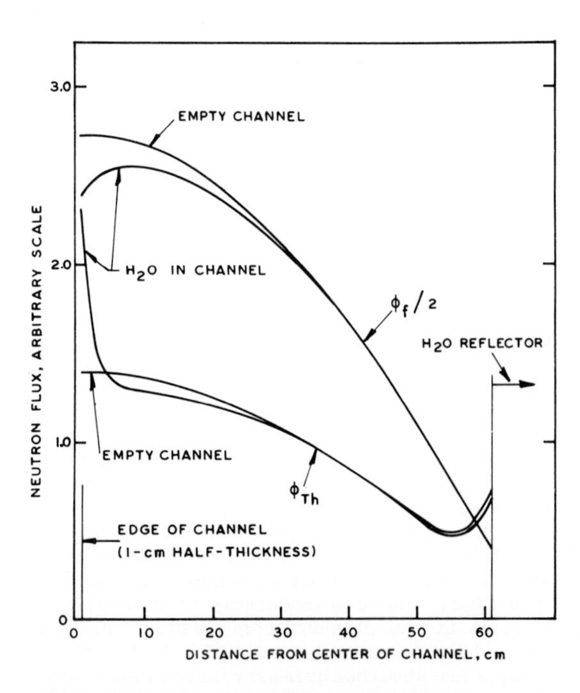

FIG. 5-8a Effect of a water channel at the center of a slab reactor infinite in two dimensions. Figure a shows the results of a two-group calculation. The geometric arrangement is two parallel infinite core slabs, each 60 cm thick, immersed in cold H_2O reflector, and separated by a 2-cm channel which can be either empty or filled with cold H_2O. The composition of the core slabs, in volume fractions, is as follows: H_2O, 0.604; UO_2, 0.254; Zr, 0.137; void 0.005. The enrichment is 2.4% U^{235}. The curves of fast flux (ϕ_f) and thermal flux (ϕ_{th}) have been so normalized that the volume integrals of the two thermal-flux curves are equal. Note that the fast flux has been reduced by a factor of 2 for plotting.

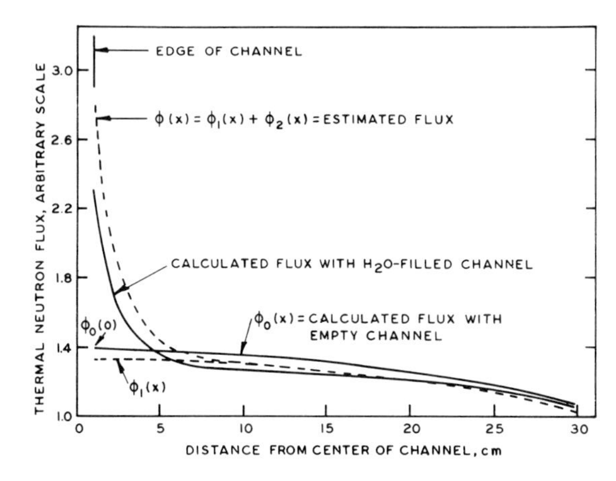

FIG. 5-8b Effect of a water channel at the center of a slab reactor infinite in two dimensions. Figure b is an enlarged plot of the thermal flux near the channel, showing again (the solid lines) the calculated thermal fluxes when the channel is empty and when it contains H_2O. Also shown, as a dashed curve, is an upper-limit estimate of the flux near the water-filled channel, derived by superimposing an approximate water-channel effect on the flux distribution for the empty channel. This estimated curve is given by: $\phi(x) = \phi_1(x) + \phi_2(x)$, where $\phi_1(x) = [1 - (R_o t_h / \ell f_w)] \phi_o(x)$ and $\phi_2(x) = \phi_o(0)[t_h / L f_w] \exp(-x/L)$.
In these equations, $\phi_o(x)$ is the flux distribution when the channel is empty, and $\phi_o(0)$ is its value at the edge of the channel. R_o is the maximum/average flux density ratio for the distribution $\phi_o(x)$; ℓ is the thickness of half-core slab, t_h is the half-thickness of the water channel, f_w is the volume fraction of water in the core slab, and L is the diffusion length in the core slab.

$\phi_1(x)$ is simply a renormalization of $\phi_o(x)$, to make the volume integral of $\phi(x)$ equal to that of $\phi_o(x)$. $\phi_2(x)$ is the flux due to the extra thermalization of neutrons in the water channel. This approximation overestimates the water-channel peaking because it does not take into account the depression of the fast flux by the water channel (cf. Fig. A) or the absorption of thermal neutrons in the water channel itself. The approximation given in the text, Eq. (5-2) results from this same approach, but without the refinement of renormalization represented by $\phi_1(x)$: i.e., it assumes $\phi(x) = \phi_o(x) + \phi_2(x)$.

relative to the core diameter that the "gross" distribution, at least in the radial direction, may be rather effectively obscured by the local effects. In such cases it is difficult to estimate the distribution accurately by superimposition techniques, and detailed two-dimensional calculations covering an entire cross section of the core (or a sector which is typical of the cross section by reason of symmetry) may be required.

Figure 5-9 [87] gives diagrams of a quadrant of the operating Yankee core, with calculated local values of the power density, at the beginning of core life (when some of the control rods are inserted to compensate the operating excess reactivity) and near the end of core life (when all control rods have been withdrawn). The transverse plane illustrated is one which lies near the axial position of maximum power density. The initial enrichment of the core is uniform, and the control rod channels are partially filled by Zircaloy-2 followers when the control rods are withdrawn. The diagrams show the values of relative power density at the center of each fuel assembly, and at the corners which are adjacent to control rod channels. The very large local variations are evident, but they are brought out more clearly by Fig. 5-10, in which selected values of the local power density have been plotted against distance

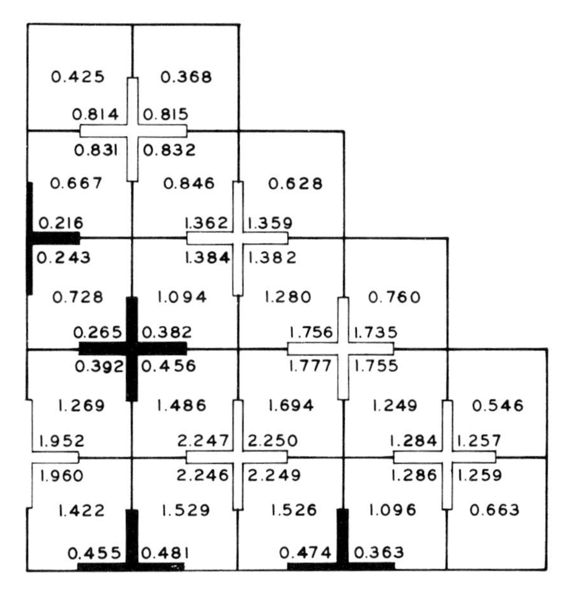

FIG. 5-9a Calculated power densities at the centers and corners of fuel assemblies in one quadrant of the Yankee core: operating condition, beginning of life. The values shown are the power densities relative to the average power density in the core. The darkened slots contain control rods; the other slots contain control rod followers.

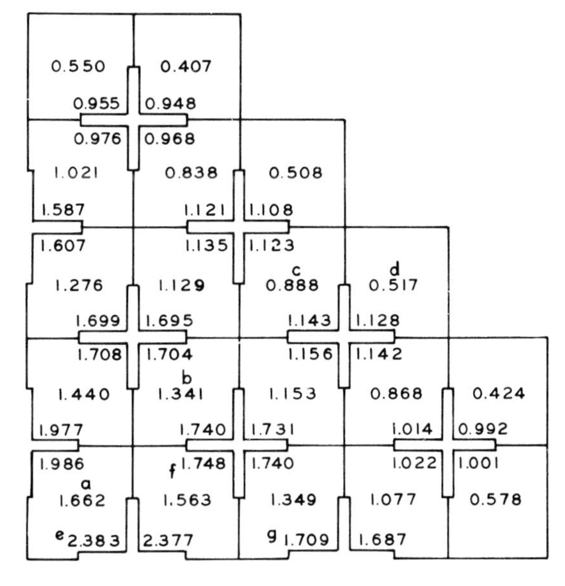

FIG. 5-9b Calculated power densities at the centers and corners of fuel assemblies in one quadrant of the Yankee core: Operating condition, near end of life. The letters by some of the values identify them in the plot of Fig. 5-10.

from the core axis. The distribution defined by the power densities at the centers of the fuel assemblies falls rather close to a cosine distribution (Fig. 5-10) but this is perhaps fortuitous, for at the end of core life the fuel distribution has been considerably modified by preferential burnup in the higher-power assemblies. The significant point is that the local "hot-spots" (points e, f, g) lie far above the relatively smooth curve defined by the points from the fuel assembly centers.

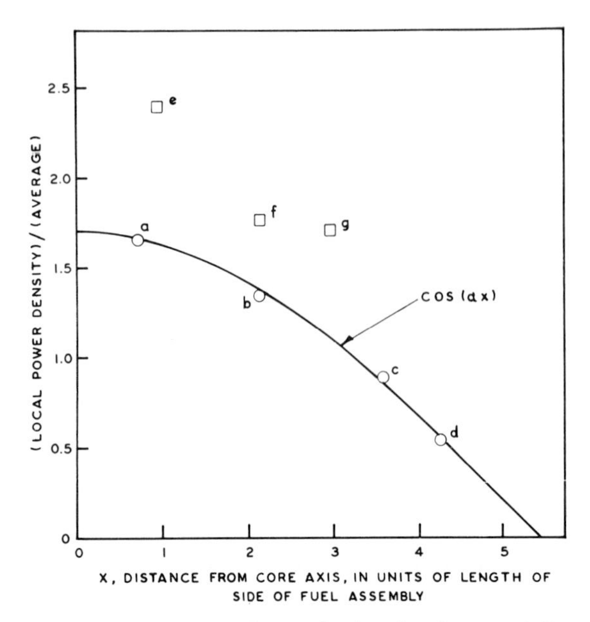

FIG. 5-10 Plot of selected power density values from Fig. 5-9b.

If Fig. 5-9a is compared with Fig. 5-9b it is apparent that the inserted control rods have, in the former case, reduced the levels of power density in the regions very near the center of the core. The maximum/average ratio, however, has been reduced only from 2.38 to 2.25; the main effect has been to shift the maximum power density to a point farther from the core axis. This effect also illustrates the importance of detailed calculations of power distribution, covering the exact operating configurations, for H_2O-moderated reactors.

6 PARTIAL REACTIVITY COEFFICIENTS

Whenever there is a change in operating conditions or in the reactor configuration, a change in reactivity may be induced. Such changes are of great importance in relation to reactor safety: in some circumstances they may initiate reactivity excursions, they determine the course of a reactivity excursion once it has begun, and they determine the degree of stability of reactor operation. The reactivity changes considered here are the "short term" changes; that is, all changes which are not direct results of fuel burnup or fission product accumulations.

These reactivity changes are usually discussed in terms of coefficients, i.e., the reactivity change produced by unit change in the variable of interest. The concept of the reactivity coefficient is very useful in the general understanding of reactor behavior, and it also has a good deal of practical usefulness even though in most cases the reactivity changes are not linear functions of the variables in question, and hence the coefficients do not remain constant over large ranges. The linearization implied by the use of the coefficient is often acceptable because the reactivity variation involved in establishing an unstable regime, or even in initiating a rather severe power excursion, may often be rather small. For the same reason it is often permissible to consider the reactivity changes due to several different variables as independent of each other over a reasonable range, and to evaluate the net reactivity change as the algebraic sum of the partial changes due to the separate variables. That approach is taken in the following rather general discussion, which is intended to call attention to a number of reactivity effects of importance, and to identify their sources in physical processes. Precise theoretical evaluations of reactivity changes and reactivity coefficients for a particular reactor usually require detailed calculations of the reactivity before and after the change in question.

The term reactivity coefficient is used in different senses, sometimes to designate the net reactivity effect produced by a small change in a specific operating variable, and sometimes to designate a particular component of the reactivity change due to a specific physical process. Thus one may speak of the isothermal temperature coefficient of reactivity, meaning the net change in reactivity due to a uniform increase of one degree in the reactor temperature; this change may be the net result of changes in several components of the neutron balance. Or one may speak specifically of the Doppler coefficient of reactivity, meaning the reactivity change produced by a one-degree increase in fuel temperature, as a result of the Doppler broadening of the U^{238} absorption resonances; this change involves only one component of the neutron balance—the resonance escape probability—and gives only one component of the isothermal temperature coefficient. For the purposes of this discussion, coefficients of the latter kind, which may constitute a component of a more complex coefficient, will be called partial coefficients.

In general it is permissible to break the total coefficient down into as many partial coefficients as may be useful, the total coefficient being the algebraic sum of the partial coefficients. But it is important to recognize that each partial coefficient must be evaluated for the same physical situation as the total coefficient. Thus it is permissible to break down the isothermal temperature coefficient for the reactor into a fuel temperature coefficient and a moderator temperature coefficient, but each partial coefficient must be evaluated for the isothermal case: i.e., for a one-degree change in the uniform temperature of the reactor. If the fuel alone were heated to determine the fuel temperature contribution to the coefficient, differential thermal expansion might, for example, produce a bowing of the fuel rods which would affect reactivity, and which would be totally unrelated to the isothermal case.

When total reactivity effects are considered in terms of their partial components, the manner in which the breakdown is made may be quite arbitrary, and one is free to choose that which best suits his purpose. Often one particular breakdown may be most useful for the understanding of the physical processes involved, another may be better suited for the experimental investigation of the effects, and a third (or more often no breakdown at all) will yield the most accurate theoretical prediction.

This section will treat certain important partial reactivity coefficients which involve the response of the neutron chain reaction to arbitrarily assumed changes in temperature, composition, or configuration in the reactor. A following Section, (7), will treat some cases in which these partial effects are related to the practical possibilities of changes within real reactors.

The general treatment of the coefficients is complicated by the circumstance that there are at least three distinct classes of reactors—the "well-thermalized" reactors, the hydrogenous reactors, and the epithermal and fast reactors—which require rather different theoretical approaches for precise reactivity calculations. For conceptual purposes, one may generalize to the extent of writing the reactivity equation for all reactors as has previously been done in (2-1):

$$k_{eff} = \bar{\nu}\,\frac{F}{F+C}\,(1-\mathcal{L}), \tag{6-1}$$

and for many cases the approximation previously given in Eq. (2-17) may go a long way toward illuminating the reactivity effects:

$$k_{eff} = \frac{k}{1+M^2B_g^2}. \tag{6-2}$$

The concept embodied in this equation is useful insofar as k, M^2, and B_g^2 are independent. This is true to first order for many reactors. To about the same degree of approximation it is true that: k depends only upon the ratios of materials in the reactor and upon their placement relative to one another (i.e., k is independent of uniform density changes); M^2 depends on the ratios of materials and is also very sensitive to density changes; and B_g^2 is independent* of reactor composition.

Within the framework of Eq. (6-2) small reactivity variations in the vicinity of criticality can be broken down into the components:**

$$\Delta\rho = \frac{\Delta k_{eff}}{k_{eff}} = \frac{\Delta k}{k} - \frac{M^2B_g^2}{1+M^2B_g^2}\left[\frac{\Delta M^2}{M^2} + \frac{\Delta B_g^2}{B_g^2}\right] \tag{6-3}$$

6.1 Leakage Effects

Reactivity changes due to changes in neutron leakage alone may often be estimated rather easily. An important case is the addition of a fuel assembly to the periphery of a reactor core. Knowledge of this effect is important for safe loading and reloading procedures (note that the equally important case of a fuel assembly added in the core interior is a much more complex situation; it may be understood in terms of the discussion in later sections).

If a fuel assembly is added at the periphery of a cylindrical core of effectively infinite height, its effect may be approximated by assuming that it increases the effective radius of the core in the ratio $1/\sqrt{N}$, where N is the number of assemblies initially in the core. The upper limit of the change in reactivity is given by Eq. (6-3) as:

$$\Delta\rho = \frac{M^2B_g^2}{1+M^2B_g^2}\left(\frac{1}{N}\right) = \frac{1}{N}\cdot\frac{k-1}{k} \quad \text{if core is near criticality.} \tag{6-4}$$

This is the upper limit for a bare cylindrical core of infinite length. The value will be less for a cylinder of finite length, and less for a radially reflected cylinder. In the latter case, if the reflector is effectively infinite, the approximate reactivity change will be given by Eq. (6-4) with N′ substituted for N, where N′ is the number of fuel assemblies which would be contained in the equivalent bare core. For cores which depart drastically from cylindrical symmetry, the worth of a single peripheral fuel element may be larger than that given by the equation. This would be the case for example if the assembly were added in the center of the broad flat face of a long narrow slab reactor.

A second class of leakage effects is that which involves only a change in the average density of the chain-reacting system, without any accompanying change in the dimensions of the system or in the atomic ratios of its constituents. A practical example would be a container, with an overflow pipe, filled with an aqueous solution of fissile isotope. Such a system would resemble closely the Los Alamos water boiler reactor. If the average density of the solution were changed, for example, by bubbling air through it, only M^2 would be changed in Eq. (6-2). Reference to Eqs. (2-17), (2-18), (2-19), and (2-20) shows that both components of M^2 are inversely proportional to products of two macroscopic cross sections; hence M^2 will be inversely proportional to the square of the average density (d) and:

$$\Delta\rho = -\frac{M^2B^2}{1+M^2B^2}\left[\frac{\Delta M^2}{M^2}\right] = -\frac{M^2B^2}{1+M^2B^2}\left[-2\,\frac{\Delta d}{d}\right] = -\frac{k-1}{k}\left[-2\,\frac{\Delta d}{d}\right] \quad \text{near criticality.} \tag{6-5}$$

If the system expands as a unit, so that there is no change in the atomic ratios of constituents, and the volume of the system increases in inverse proportion to the density, the leakage and the reactivity will be affected by both the density change and the volume change; near criticality the reactivity change will be:

---

*This is a reasonable approximation only in reactors of uniform composition. In fact, the concept embodied in Eq. (6-2) is not a useful one unless spatial variations in composition are unimportant enough to be considered as small perturbations of the uniform case.

**Note that if the reactor is not near criticality even the first of the equalities of Eq. (6-3) is invalid, i.e.:

$$\Delta\rho = \rho_1 - \rho_2 = \frac{(k_{eff})_1 - 1}{(k_{eff})_1} - \frac{(k_{eff})_2 - 1}{(k_{eff})_2} = \frac{\Delta k_{eff}}{(k_{eff})_1\,(k_{eff})_2}.$$

$$\Delta\rho = -\frac{k-1}{k}\left[-2\frac{\Delta d}{d} + \frac{\Delta B_g^2}{B_g^2}\right] = -\frac{k-1}{k}\left[2\frac{\Delta V}{V} + \frac{\Delta B_g^2}{B_g^2}\right], \quad (6\text{-}6)$$

where ΔV is the change in volume of the system. B_g^2 will be affected by the expansion in a way that can be determined only when the mode of expansion is known. If the expansion is isotropic, corresponding to an increase Δx in each linear dimension x, $\Delta B_g^2/B_g^2 = -2\Delta x/x$, and near criticality the reactivity change is:

$$\Delta\rho = -\frac{k-1}{k}\left[6\frac{\Delta x}{x} - 2\frac{\Delta x}{x}\right] = -\left[\frac{k-1}{k}\right]\cdot 4\frac{\Delta x}{x}. \quad (6\text{-}7)$$

This situation is approximated in a number of practical cases. It would be the case for the uniform thermal expansion of a "homogeneous" or semi-homogeneous solid fuel reactor if the reactor were assembled without clearances between the units which make up the core. It might be approximated in some cases by other solid moderator reactors and by the uniform thermal expansion of a fast reactor core submerged in a coolant, such as sodium, which is relatively transparent to fast neutrons. Most important, however, it is approximated in any core in which the "expansion" is conceived as the result of the blowing apart of the core by the vaporization of one of its constituents. Thus, for example, if one visualizes a water-cooled core which is blown apart by the rapid vaporization of water in a fast power excursion, the probability of any additional reactivity gain during disassembly, as a result of the increase in core size appears remote, provided only that a decrease in the water/fuel ratio does not cause a large increase in k_∞. That is to say, if the total amount of moderator within the core boundary does not increase as the core expands, the increase in neutron leakage due to the decrease in moderator density will overshadow the decrease in leakage due to the core enlargement.

When the several constituents of the core—e.g., fuel, coolant, moderator—can expand independently, the situation is much more complex. The components of the reactivity change which are due to neutron leakage effects can still be approximated by Eq. (6-3), but there will be additional effects on k due to the change in material ratios, and in many cases these may be the more important ones.

In reflected reactors the neutron leakage may be affected by changes in the reflector, and the magnitude of reactivity changes due to changes in the core may be affected by changes in the reflector savings which may accompany the core changes even if there is no change in the reflector composition.

The opportunities for large accidental increases in reflector effectiveness do not appear very great in power reactors except in those cases where reflector effects are utilized for control. In research and test reactors, however, the reactivity effects of large experiments involving absorbers, voids, or scattering materials which may be installed in the reflector must be given careful attention. The reactivities of small, highly concentrated critical assemblies are very sensitive to reflector effects. This is particularly true for fast-neutron assemblies, which may gain reactivity from the moderating effects of adjacent materials, as well as from their effects on the albedo.

When the change in reflector saving is known the change in geometric buckling of the core (ΔB_g^2) may be evaluated, and the effect on reactivity may be found in Eq. (6-3). The reflector saving increases with increasing reflector thickness, approaching asymptotically a value which would hold for an infinite reflector. In thermal reactors the reflector becomes effectively infinite when its thickness amounts to a few thermal diffusion lengths. Changes in the reflector which occur farther from the core than the effectively infinite thickness can have no appreciable effect on reactivity. An increase in the density of a reflector will increase the reflector saving if the thickness is less than "infinite," but will have little effect if the thickness is "infinite" and if the core diameter is large relative to the reflector saving. An H_2O reflector becomes effectively infinite at a thickness of 6 or 8 inches; for reflectors with low absorption, such as D_2O, beryllium, or graphite, the "infinite" thickness is much greater.

Usually the reflector saving is not very sensitive to core properties except those which change the neutron leakage markedly. When leakage changes because of a change in core composition, the reflector saving will usually change in a direction to partially counteract the effect. Table 6-1 gives calculated reflector savings for a highly-enriched, H_2O-moderated reactor, with an infinite H_2O reflector, at various temperatures and with various void fractions in the core. The saving is plotted in Fig. 6-1 against the slowing-down area (L_f^2) of the core. Note that the points fall near a single line, indicating that the change in density of the reflector, as temperature is raised, has an effect which is small relative to the effect of changes in the core leakage.

6.2 Changes in k

The possible sources of changes in k, the number of neutrons produced per neutron absorbed, are many and varied, but the major ones fall into one or more of three classes: changes of composition, changes of neutron energy spectrum, and changes in the degree of segregation of the constituents of the chain-reacting system.

6.2.1 Composition Changes

The most obvious effects of composition changes are changes in the ratio (absorption rate in fissile isotope)/(absorption rate in other materials). One possible expression for k is:

$$k = \bar{\eta}_f \frac{n_f\bar{\sigma}_f}{n_f\bar{\sigma}_f + \sum_{\text{all } i} n_i\bar{\sigma}_i} \quad (6\text{-}8)$$

TABLE 6-1

Radial Reflector Saving for a Borax Type Reactor* as Calculated by the Two-Group Approximation

Temperature, °F	Void fraction, α	L_f^2, cm^2	Reflector saving, cm
68	0.0	50.6	7.60
68	0.1	60.8	8.41
200	0.0	53.8	7.89
200	0.2	79.4	9.77
200	0.4	127.7	12.81
420	0.0	67.6	9.14
420	0.2	98.8	11.37
420	0.4	158.0	14.95

* The reactor core is a uniform mixture of H_2O, aluminum, and U^{235}
Volume H_2O + void/volume aluminum = 0.42
Σ_a (U^{235}) = 0.0315 at 68°F
Reflector is pure H_2O, infinitely thick, at same temperature as core.

where n_f and n_i, respectively, designate the atomic concentrations of fissile isotope and of the ith nonfissile isotope. σ_f and σ_i refer to the microscopic absorption cross sections of the fissile and nonfissile isotopes, and the bars on σ_f, σ_i, and η_f indicate appropriate averaging over the neutron energy spectrum and spatial variations of neutron flux. If a change in n_f does not change the energy spectrum or the spatial averaging significantly, the effect of the change on k is given by:

$$\frac{\Delta k}{k} = \frac{\Delta n_f}{n_f}\left[\frac{\sum n_i\bar{\sigma}_i}{n_f\bar{\sigma}_f + \sum n_i\bar{\sigma}_i}\right]. \qquad (6\text{-}9)$$

With the same reservations, a change in the atomic concentration n_j, of one of the i nonfissile absorbers, will change k as follows:

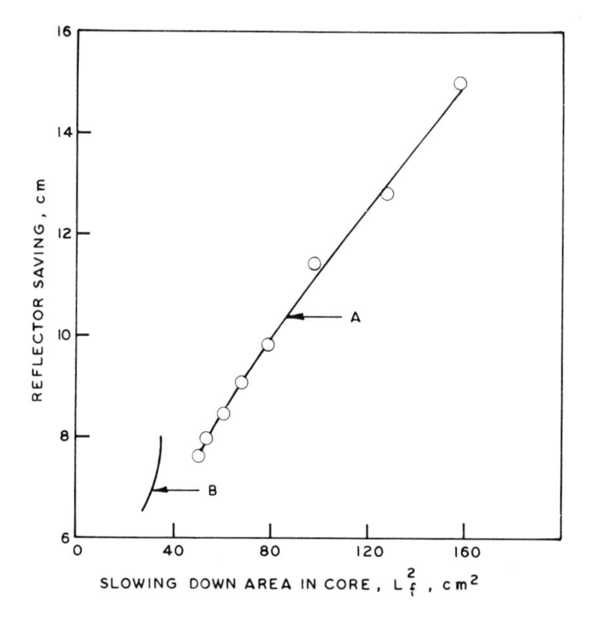

FIG. 6-1 Reflector savings of water-moderated cores, with infinite water reflectors, as functions of slowing-down area in core. Curve A is a plot of the points from Table 6-1. Curve B gives measured values for H_2O-U lattices of low enrichment [88].

$$\frac{\Delta k}{k} = -\frac{\Delta n_j}{n_j}\left[\frac{n_j\sigma_j}{n_f\sigma_f + \sum\limits_{\substack{\text{all } i,\\ \text{including } j}} n_i\sigma_i}\right]. \qquad (6\text{-}10)$$

Thus the fractional change in k is proportional to the fractional change in fissile isotope times the fraction of neutrons absorbed by all other isotopes, and to the fractional change in any nonfissile isotope (j) multiplied by the fraction of neutrons absorbed by that isotope j.

In most cases Eqs. (6-9) and (6-10) will give the first-order effect of a change in composition which involves only materials with unimportant moderating characteristics and unimportant resonance absorption. If these conditions are not met the effects of changes in neutron spectrum and self-shielding* may completely obscure the above effect in the gross behavior of the reactivity.

6.2.2 Spectrum Changes

In moderated reactors where a significant fraction of the fission neutrons reaches thermal energy before being absorbed, there are two possible sources of change in the neutron energy spectrum. A change in the temperature of the moderator will change the mean energy of the thermal spectrum (the thermal base effect) while a change in the absorber/moderator ratio will change the ratio of epithermal to thermal neutron flux, and may also change the shape of the epithermal distribution.

When the reactor may be considered to be "thermal" the spectrum may be approximated, at least for estimating purposes, by a Maxwellian distribution having a characteristic temperature near that of the moderator, plus a slowing-dow[illegible] distribution, generally called the "1/E tail" giv[illegible] by ϕ (E) $\propto$ 1/E [cf. Eq. (2-8)].

A change in the moderator temperature [illegible] produce an approximately equal change i[illegible] characteristic temperature of the Maxwellia[illegible] tribution, while a change in the absorber/mod[illegible]

*See Sec. 6.2.3 for description of self[illegible] effects.

ratio will change the relative magnitudes of the Maxwellian component and the 1/E tail of the spectrum. After the reactor has been built, large changes in the average absorber/moderator ratio are not likely to occur, but changes in moderator temperature may be quite important. (Here the temperature change alone is considered; accompanying changes in density must of course also be taken into account in determining the total effect on reactivity.)

The major reactivity effects of spectrum changes are due to changes in $\bar{\eta}$ of the fertile isotope, and changes in the ratio $\bar{\sigma}_a$ (fissile isotope)/$\bar{\sigma}_a$ (other isotopes). These affect reactivity as indicated by Eq. (6-8). When strong resonance absorbers are present a change in epithermal/thermal spectrum ratio may have a particularly large effect on reactivity. This effect is usually treated separately and specifically, as discussed in Sec. 6.3.

In Fig. 6-2 [89] the values of $\bar{\eta}$ for U^{233}, U^{235}, and Pu^{239} are plotted as functions of temperature for a well-moderated thermal spectrum. In Fig. 6-3 [89] the ratios σ_a (fissile isotope)/σ_a (1/v absorber), normalized to unit at 0°C, are plotted for the same situation. In Figs. 6-4 [90] and 6-5 [91] calculated values of $\bar{\eta}$ are given as functions of fuel/moderator ratio at constant temperature.

It is evident from Figs. 6-2, 6-3, and 6-5 that the effects of neutron spectrum will be particularly large when Pu^{239} is the major fissile isotype present. Since $\bar{\eta}^{49}$ decreases with increasing temperature while $\bar{\sigma}_a^{49}$ increases,* the reactivity effect of an increase in neutron temperature may range from positive to negative over a range of reactors. In low-enrichment reactors, where the absorption by other materials is comparable to that in Pu^{239}, the effect of the increase in $\bar{\sigma}_a^{49}$ will predominate, and the neutron temperature coefficient of reactivity

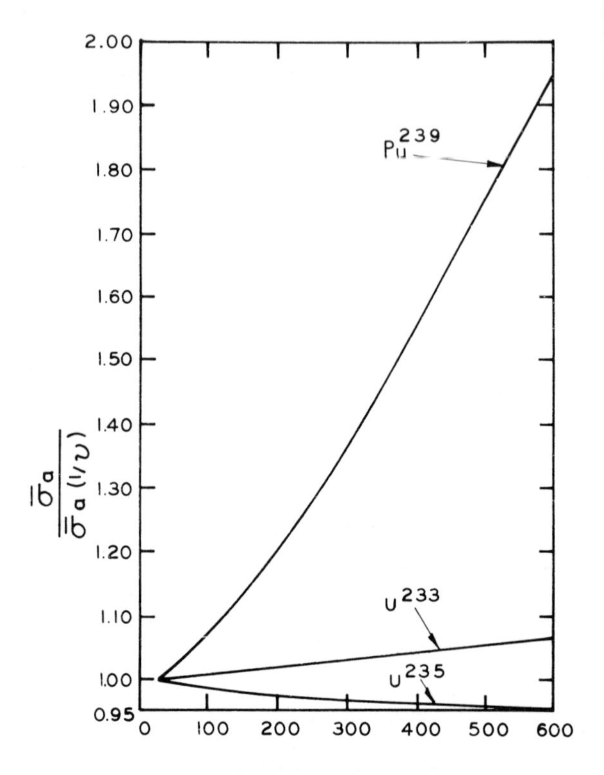

FIG. 6-3 Ratios of absorption cross sections of fissile isotopes to 1/v cross section, as a function of temperature in a well-thermalized neutron spectrum. From Wescott ($\hat{\sigma}_4$ values, r = 0.07).
The ratios are normalized to unity at 20°C. Most absorbers other than the fissile isotopes will have absorption cross sections which vary nearly as 1/v in the thermal and near-thermal energy ranges. Hence the above cross-section ratios indicate how the competition for neutron absorption, between the fissile isotopes and other materials, will vary with neutron temperature.

will tend to be positive; in highly-enriched, high-leakage reactors, the effect of $\bar{\eta}^{49}$ may predominate to give a negative coefficient.

Unusual effects may occur when the fuel is bathed in moderator which is at a temperature far different from that of the main body of moderator. In such a case the spectrum of the neutrons entering the fuel will lie somewhere between that in the

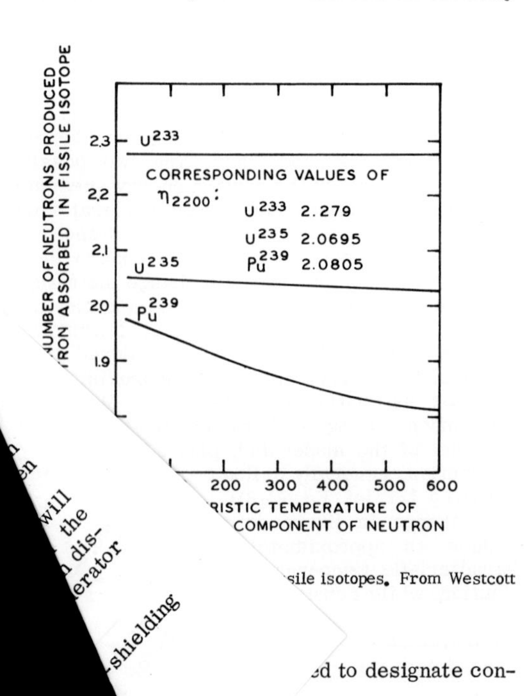

...sile isotopes. From Westcott

...will ...the ...dis- ...erator ...-shielding

...ed to designate con-

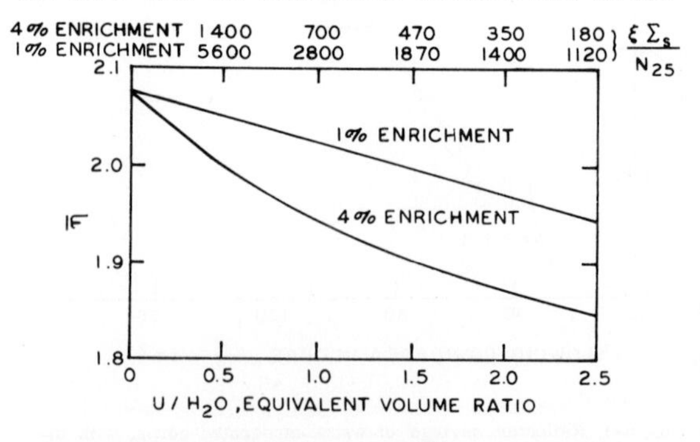

FIG. 6-4 Calculated values of $\bar{\eta}$ for U^{235} in uranium dioxide - water lattices, at room temperature. (Note: the abbreviation "25" in the subscript of N means "U^{235}".)

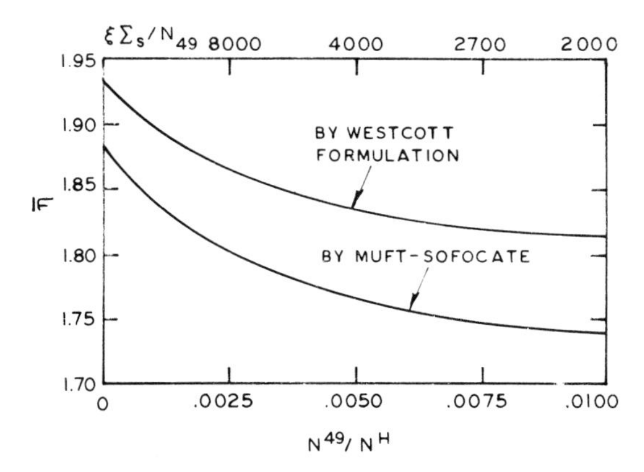

FIG. 6-5 Calculated values of $\bar{\eta}$ for Pu^{239} as a function of plutonium/hydrogen atomic ratio, at 500°K. (Note: the abbreviation "49" in the subscript of N means "Pu^{239}".)

general body of moderator and that which would be characteristic of the moderator adjacent to the fuel. A practical case is that of a water-cooled graphite-moderated reactor, in which the graphite is allowed to operate at a high temperature. Figure 6-6 [92] shows the results of Russian measurements of k_∞ as a function of uranium/carbon ratio in such a lattice. Although an increase of temperature lowers k_∞ in the dry lattice, and the addition of "coolant" water lowers k_∞ in the cold lattice, an increase of graphite temperature increases k_∞ markedly when the "coolant" water is present. This is because the cold water adjacent to the fuel raises the effective value of the absorption cross section in the fuel relative to that of the hot graphite.

In the unmoderated reactors there is no "thermal base" effect, as only a negligible fraction of the neutrons slow down to thermal energy, but the effects of composition on the neutron spectrum may be quite important (see Fig. 2-2). Indeed, in a fast breeder, the spectrum may be a characteristic of the highest importance in determining both breeding performance and safety.

In fast reactors, spectrum effects and leakage effects are often more closely interrelated than in moderated reactors, not so much because the spectrum *per se* is affected by the energy-dependent leakage as because the critical ratio of fissile isotope to other isotopes is often restricted to a narrow range by the magnitude of the leakage. Thus the neutron energy spectra in large fast reactors tend to embrace significantly lower values than those in the small ones because the ratio of inelastic and elastic scatterers to fissile isotope is usually larger also. Naturally, the materials used also have an effect. The calculated spectra, by neutron energy group, are given in Table 6-2 [93] for two large, plutonium-fueled, fast reactors, one using metallic fuel and the other oxide.

The introduction of moderator into a fast reactor will of course lower the spectrum, and the introduction of strong moderation would usually cause a large increase of reactivity, primarily because it would reduce the leakage. A slight increase in moderation, however, may have the opposite effect because its effect on leakage may not be as important as the decrease in U^{238} fission, the decrease in $\bar{\eta}$ of the fissile isotope, and the increase in parasitic absorption cross sections relative to the fission cross section, which result from the softening of the spectrum. Thus in certain postulated large sodium-cooled fast reactors, partial loss of sodium causes an important reactivity increase, mainly because of the spectrum effect [94]. Table 6-3 [95] illustrates some of the effects of the neutron energy spectrum in fast reactors on the reactivity response to changes in composition.

Another significant aspect of the spectrum in fast reactors is its relation to the Doppler effect. At the very high neutron energies the self-shielding of the resonances is small, because of relatively large widths, low maxima, and relatively close spacing. If a significant portion of the neutron spectrum extends down below about 10 kev, however, the Doppler effect in U^{238} may be appreciable.

6.2.3 Changes in Segregation of Core Materials

The third class of effects on k, the segregation of reactor materials, can operate in three different ways. The most common manifestation is self-shielding. Whenever a material absorbs neutrons strongly the average neutron flux within it will be decreased relative to the flux in nearby weak absorbers, and its absorption rate will be lower than might be expected from consideration of its cross section alone. An extreme example would be the absorption of a "black" control rod (Sec. 4.2.1). For relatively strong absorbers which do not approach "blackness" the effect is usually taken

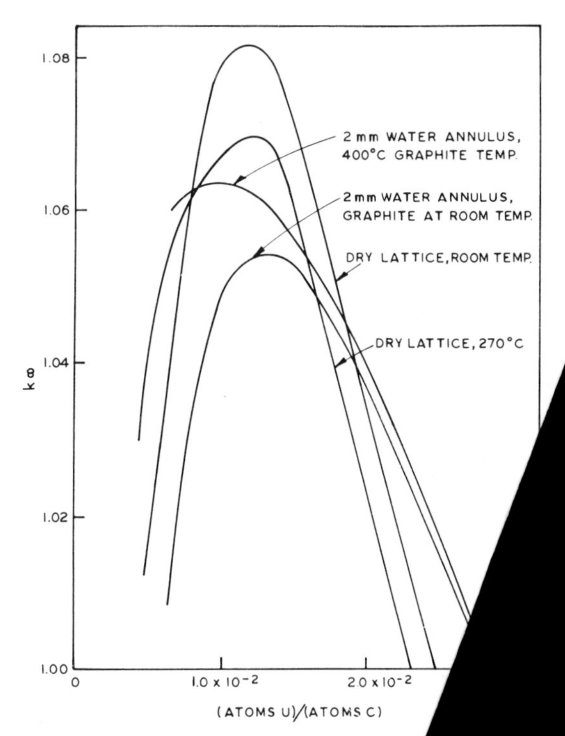

FIG. 6-6 Measured k_∞ for uranium-graphite latti[illegible] cold and hot. These are determined in expone[illegible] employing uranium slugs of about 1.3 in. (3.[illegible] square lattices in graphite.

TABLE 6-2

Calculated Neutron Energy Spectra and Neutron Balances
for Large (2500 Liter) Spherical, Fast-Reactor Cores with U^{238} Blankets

(The fissile isotope is pure Pu^{239}.)

Fuel		Metal		Oxide	
Ratio fuel atoms to diluent atoms, Pu/U^{238}		0.0945		0.210	
Critical mass of fuel, kg		1026		806	
Group	Flux integrals Energy interval, Mev	Core	Blanket	Core	Blanket
1	3.668 - 10.00	1.72	0.12	2.31	0.17
2	2.225 - 3.668	3.36	0.22	4.79	0.33
3	1.35 - 2.225	6.15	0.46	9.02	0.70
4	0.825 - 1.35	10.71	1.31	12.78	1.74
5	0.5 - 0.825	18.98	3.45	19.53	4.22
6	0.3 - 0.5	24.77	5.33	23.58	6.16
7	0.18 - 0.3	23.41	5.69	26.05	6.82
8	0.11 - 0.18	19.81	5.37	23.78	6.67
9	0.067 - 0.11	11.18	3.01	17.03	4.03
10	0.0407 - 0.067	10.67	3.89	15.65	5.24
11	0.025 - 0.0407	4.85	1.60	8.90	2.29
12	0.015 - 0.025	5.98	1.99	11.92	3.12
13	0.0091 - 0.015	2.39	0.76	6.61	1.36
14	0.0055 - 0.0091	1.07	0.30	3.87	0.61
15	0.0021 - 0.0055	0.48	0.13	2.37	0.31
16	0.0005 - 0.0021	0.41	0.054	4.03	0.21
Neutron balance					
Pu^{239} Fissions		0.276	—	0.308	—
Pu^{239} Captures		0.0568	—	0.0823	—
U^{238} Fissions		0.0646	0.0122	0.0323	0.0184
U^{238} Captures		0.347	0.242	0.212	0.342
Coolant (Na) captures		0.00242	0.000259	0.0102	0.000558
Structures captures		0.0200	0.00419	0.0298	0.00563
Internal breeding ratio		1.04		0.54	
Total breeding ratio		1.76		1.42	
Ratio captures to fissions in $Pu^{239} + Pu^{241}$		0.205		0.267	
Escape from blanket		0.00620		0.00811	

into account by disadvantage factors, as discussed generally in Sec. 2.5, and specifically, for resonance and thermal absorption, in Secs. 6.3 and 6.4.

The second effect, which might be called the flux-trap effect, might be considered to be a special aspect of self-shielding, and can also usually be treated by a "disadvantage factor" approach. This ... the peaking of thermal neutron flux in strong ...erators when they are segregated into volumes ... are large relative to the neutron thermal ... length. The strong peaking of thermal flux ...ay enhance its absorption by a factor of ...re. It may sometimes be found, for ...n H_2O-moderated reactor, that the ... of reactivity is positive when the ...vater which is in empty control ...as it is quite negative when the ... "coolant" water, near fuel

...fect which, for want of a ...ed the "fast clumping" ...sion by fast neutrons ...nhanced by placing ...se as possible to ... in thermal re... ...s greater as ...rge, dense, ...vity increases ...ed and clumped ...s segregated fertile isotope, the fission of the fertile isotope is favored, and the reactivity is higher, when the fertile isotope is placed close to the fissile isotope rather than separated from it by some other material. Table 6-4 [96] gives the measured reactivity effects of changing the fuel-plate thickness (i.e., clumping the U^{235}) in three experimental fast-neutron critical assemblies, as reported in reference [96]. The same reference mentions the effect of changing the location of the U^{238} relative to U^{235} in Assembly 34 (Table 6-4). Ordinarily the carbon in this assembly was installed in the form of graphite sheets, 3.2 mm thick, between the U^{235} and the U^{238}. When the positions of the graphite and U^{238} plates were interchanged, to place the U^{238} adjacent to the U^{235}, the result was a reactivity increase of 0.0035.

6.3 Resonance Absorption, Doppler Effect*

Some nuclides, such as U^{238}, absorb strongly and selectively incident neutrons moving with certain speeds relative to the absorbing nuclei. Such nuclides are called resonance absorbers. Similar effects occur in the fissile isotopes, giving resonances in both the absorption and fission cross sections. Often the resonances are very narrow and

*This subsection written with T. J. Thompson.

TABLE 6-3

Characteristics of Four Fast-Neutron Critical Assemblies

Assembly No.	22	23	24	25
Core volume, liters	138.2	148.5	323	435.6
Volume U^{238}/volume U^{235}	7.44	0.0723	9.6	10.4
Composition, volume % U^{235}	9.43	9.27	7.6	7.12
U^{238}	70.12	0.67	72.9	74.05
SS	9.13	9.16	9.3	9.28
Al	0	42.82	0	0
Critical mass, kg U^{235}	243.70	258.09	460.69	581.59
Fission ratios				
U^{238}/U^{235}	0.360	0.0679	0.0284	0.0292[a]
U^{234}/U^{235}	0.294	0.403	0.246	0.253[a]
U^{233}/U^{235}	1.529	1.48	1.44	—
Pu^{239}/U^{235}	1.17	1.18	1.14	1.17[a]
$\Delta k/k$/inhour, $(\times 10^{-5})$	2.23	2.26	2.23	2.23
Danger coefficients, inhours/kg				
U^{235}	261	154.2	168	145[a]
U^{233}	484	282.5	304.2	270.9[a]
Pu	466	275.1	273.4	239.4[a]
Depleted uranium	- 11.6	+ 3.65	- 7.74	- 7.20[a]
Graphite	- 31.0	+ 65.6	- 35.6[a]	- 37.2[a]
Boron[b]	2684	1873	1641[a]	1411[a]
Lifetime, 10^{-8} sec	7	4	8	8

[a] Measured in "bunched core" configuration.
[b] Enriched to 90.7% in B^{10}.

The main feature to be noted is that Assembly 23 contains very little U^{238}, but considerable aluminum and void. Its spectrum is considerably "harder" than those of the other assemblies because of the small amount of inelastic scattering by U^{238}. The "harder" spectrum is indicated by the higher U^{238}/U^{235} fission ratio (the fission ratio is the relative probability of fission per atom present), and gives characteristic differences in reactivity coefficients. Thus the addition of a small amount of either moderator (graphite) or U^{238} will increase reactivity in Assembly 23, whereas the same additions will decrease reactivity in the other assemblies.

TABLE 6-4

Effect of Thickness of Fuel (U^{235}) Plates in Three Cylindrical Critical Assemblies

	Assembly No. 32*	Assembly No. 30*	Assembly No. 34*
Core dimensions (length x dia., cm)	61.1x52.2	62.2x82.8	86.5x91.9
Core volume (liters)	130.7	356	574.4
U^{235} content (kg)	227.5	395	503.0
Core composition, vol. %			
U^{235}, 18.75 g/cm^3	9.26	5.92	4.67
U^{238}, 19.0 g/cm^3	0.7	9.06	10.3
Al, 2.7 g/cm^3	0.0	23.4	25.5
C, 1.43 g/cm^3	0.0	0.0	10.6
SS 304, 7.85 g/cm^3	81.0	24.6	24.6
O, 2.55 g/cm^3	0.0	7.2	0.0
Void		Remainder	
Reactivity with 1/16-in. fuel plates	-0.005	-0.005	-0.005
Reactivity with 1/8-in. fuel plates	0.000	0.000	0.000
Reactivity with 1/4-in. fuel plates	+0.015	+0.015	+0.011

*All cores were surrounded by blankets of 83.3 vol. % U^{238}, 7.3 vol. % stainless steel, at least 30-cm thick.

embrace a very small neutron energy band (see, for example: reference [97], Chapters III, IV, reference [98], Secs. 2.38, 3.63, etc.). For instance, the width at half maximum of the neutron energy band for the 6.7 ev resonance in U^{238} is only 0.0015 ev and the maximum cross section is $\sim 10^3$ times that of nearby values outside the resonance. The very sharp energy dependence of the cross section can lead to reactivity effects which are strongly dependent upon local absorber/moderator ratios. Further, the dependence of the cross section on the relative neutron-nucleus velocity may be so sharp that the velocity of thermal agitation of the absorber atoms becomes significant in determining the absorption probability. In the case of fission resonances, or in the case of a resonance absorber such as U^{238} which is intimately mixed with the fissile isotope, this Doppler effect may give rise to a reactivity coefficient which follows the fuel temperature effectively instantaneously. Needless to say, such a "prompt" coefficient has great significance for reactor safety. For this reason, and because the reactivity effects arising from resonance absorption may be quite subtle, the subjects of resonance absorption and Doppler effect are treated here at some length. Reviews of these subjects may be found in references [97a] through [102].

The phenomenon of self-shielding is a characteristic aspect of both resonance absorption in general and the increase of absorption by the Doppler effect. If a continuous energy spectrum of neutrons is incident upon a macroscopic sample of absorber having strong resonances, those neutrons having energies within the resonance band will be absorbed before they penetrate very far past the surface of the sample, and if the sample does not have moderating properties, the neutron spectrum at all points beneath a thin surface layer will be depleted of neutrons within the resonance energy bands. Under these circumstances a thick sample of the absorber will not absorb many more neutrons than a thin sample. This is an example of self-shielding. If the sample has moderating properties, either because of the inherent properties of the absorbing atoms or because of an admixture of moderating atoms, some neutrons which pass the surface of the sample at energies outside the resonances will be scattered into the resonances while they are inside the sample, and will be absorbed. Thus the addition of moderator decreases the self-shielding. For strong absorbers, however, the ratio of moderator to absorber must be very large to reduce the self-shielding to an insignificant level.

Indeed, it is not necessary to employ a segregated sample of resonance absorber to observe the effects of self-shielding. If the entire medium consists of a uniform mixture of moderator and resonance absorber the neutron flux density will be reduced by the strong absorption in those energy regions corresponding to the resonances, and the total absorption rate will be less than that computed on the assumption of a smooth (e.g., 1/E) slowing down spectrum. The resonances are said to be self shielded, and to cause flux depletion in their energy ranges.

The strong self-shielding of resonance absorbers occurs because the resonances are very "high" (i.e., the value of the cross section at the peak of the resonance is very large) but very narrow. The self-shielding would be reduced if the resonance were broadened, and the total absorption would be increased even if, at the same time, the "height" were reduced so that, for example, the integral of the cross section

$$\int_{\text{resonance}} \sigma(E)\, dE,$$

remained constant. The Doppler effect produces just such a broadening.

As the temperature within a reactor is raised, the atoms of the materials, such as U^{238}, are heated, and their thermal agitation increases. Since it is the relative velocity between the incident neutron and the nuclide that determines the absorption probability, the thermal motion of the nuclide in effect permits neutrons formerly outside of the resonance to be absorbed as though they were in the resonance. Thus a neutron moving a little too slowly to be absorbed by a "colder" U^{238} nucleus may find itself approaching a U^{238} nucleus itself thermally agitated and moving towards the neutron. The sum of the two velocity components towards each other may make the relative neutron velocity such that the incident neutron has an energy within the resonance; this will increase the probability of the neutron being absorbed. In effect, increased thermal agitation broadens the resonance and increases the total absorption that can occur in such a medium. This phenomenon is a manifestation of the familiar Doppler effect which applies to wave motion in general.

It is important to recognize that the Doppler effect simply broadens the resonance but does not change the area under the resonance; that is, the integral of the energy dependent absorption cross section over the energy region remains unchanged. For a 1/E incident neutron energy spectrum and using the Breit-Wigner single resonance formulation this gives;

$$I_o = \int_{\text{resonance}} \sigma_a(E')\,\frac{dE'}{E'} = \frac{\pi}{2}\,\sigma_a\,(\text{max.})\,\Gamma = \text{constant}, \tag{6-11}$$

where Γ is the total width of the resonance level. The integral defined in Eq. (6-11) is known as the "resonance integral" (I_o). It assumes that the energy spectrum of neutrons incident on the resonance absorber varies as 1/E and that there is no depletion of neutrons at the energy of the absorption resonance. This will obviously hold only for the case of an infinitely thin (or infinitely dilute) resonance absorber. In practical cases it is necessary to make use of a quantity called the "effective resonance integral" which is intended to take into account the fact that the actual neutron spectrum is depleted at the resonance energies. If the resonance absorber is mixed intimately with any scattering material that may be present, a general equation can be written for the effective resonance integral. This equation is derived in many textbooks on reactor physics, e.g. references [97] and [98].

$$I_{eff} = \int_{E}^{E_o} \frac{\Sigma_s}{\Sigma_s + \Sigma_a} \sigma_a (E') \frac{dE'}{E'}$$

$$= \int_{E}^{E_o} \frac{1}{1 + \frac{\sigma_a}{\Sigma_s/N_a}} \sigma_a (E') \frac{dE'}{E'} . \quad (6\text{-}12)$$

This equation holds rigorously for hydrogen as a moderator and a very heavy absorber, but otherwise is restricted to absorbers with widely spaced resonances or those with weak resonance absorption. In many practical cases, the actual behavior follows the relationship closely enough that the effective resonance integral is found to be a single value function of Σ_s/N_a within experimental error. Thus the effective resonance integrals for homogeneous mixtures of the fertile materials U^{238} and Th^{232} with moderating scatterers have been determined over wide ranges of mixtures by means of a relatively few experimental measurements, as shown in Fig. 6-7. [116]

In Eq. (6-12) above, the quantity Σ_s/N_a is the macroscopic scattering cross section associated with each atom of resonance absorbing material such as U^{238}. The quantity σ_a, the microscopic absorption cross section of the absorbing material, is a function of energy (as are all the other parameters of the equation) and, of course, is large at a resonance. However, once the absorbing material is picked, σ_a is fixed and the value of I_{eff} is then dependent only on the value of Σ_s/N_a. As the number of absorbing atoms relative to number of scattering atoms increases, Σ_s/N_a decreases and the value of the integral decreases. As the value of N_a is decreased (i.e., more dilute fuel), the value of I_{eff} increases and, in fact, at very low dilution reduces to I_o as given in Eq. (6-11).

From this discussion it is evident that the effective Doppler broadening of a resonance by thermal agitation of the atoms of the absorbing nuclides does not affect the total absorption and hence reactivity, if the concentration of the absorbing atoms is very dilute. As the concentration of the absorbing atoms increases, the value of the effective resonance integral I_{eff} decreases more and more below the I_o value. It is this difference between I_o and I_{eff} which can be exploited in part by the effective Doppler broadening of the resonance. It is also evident why the Doppler effect is more important in a homogeneous reactor system with a high U^{238} content than in one with a low U^{238} content, in a concentrated system more than in a dilute system, and why the effect is still more important in solid-fueled reactors with a high percentage of U^{238} in the fuel. In other words, the effective Doppler broadening of a resonance affects reactivity because of changes in the self-shielding of the resonance-absorbing medium. The greater the local concentration of resonance absorber, the greater the self-shielding effect, and hence the greater the potential reactivity effect.

It is clear that σ_a must vary in Eq. (6-12) as a function of the energy E of the neutrons and also at the resonance as a function of Doppler broadening. This is discussed in Weinberg and Wigner [97] and outlined briefly by Pearce [99] For theoretical reasons outlined in these references, the low-lying resonance levels tend to retain their

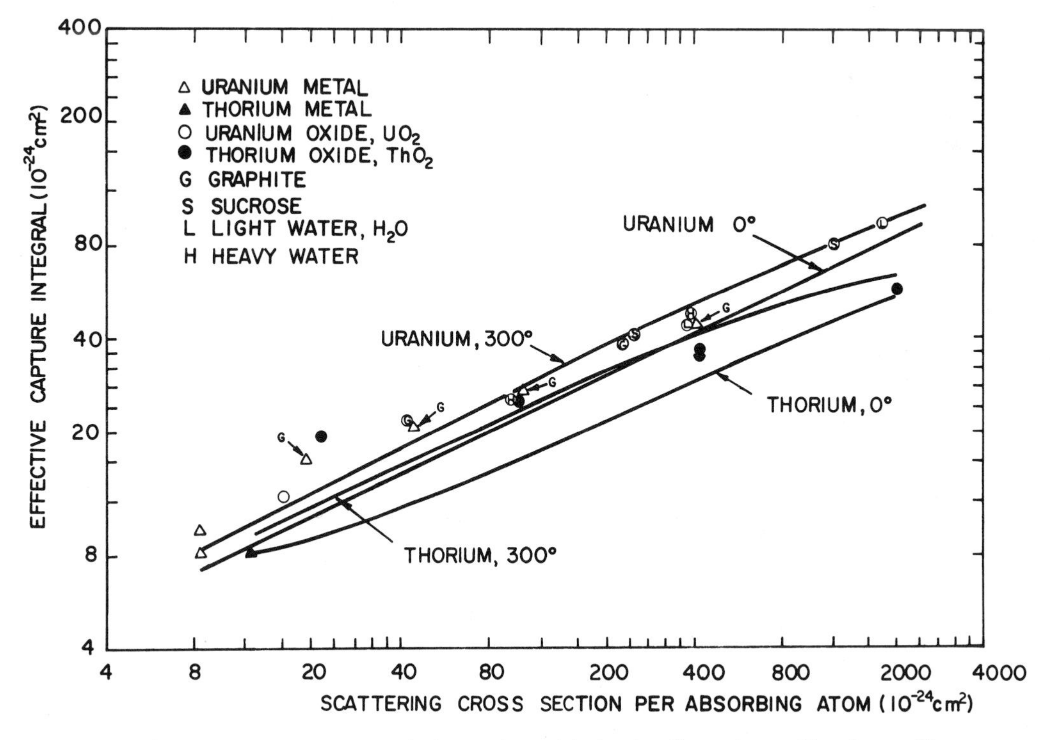

FIG. 6-7 Effective resonance integrals for uranium and thorium in uniform mixture with moderator. The curves are derived theoretically; the plotted points are experimental. Temperatures indicated are °C.

unbroadened shape, even at higher temperatures, and exhibit only a small Doppler effect. Therefore, most of the Doppler effect is due to resonances at energies appreciably higher than those energies where more resonance capture occurs. On the other hand, according to Pearce, the resonances below 100 ev in U^{238} are responsible for 90% of the peak shown in Fig. 6-8a [99a].

For those cases in which other important neutron interactions such as fission do come into play in the energy region covered by resonance absorption, it is useful to specify the resonance absorption, over the energy range E_1 to E_2, in terms of the resonance escape probability (p) for that energy range. The resonance escape probability is the probability that a neutron entering the absorbing-moderating region at any energy E, or lower (down to E_2), slows down to energy E_2 (or below E_2) without being captured by the resonance absorber. In thermal reactors, the concept is a very useful one, and the energy range covered by the resonance escape probability is often taken as the entire energy range from fission energy to thermal. In such reactors, k_∞ is proportional to p, and the partial change in k_∞ due to change in p is simply

$$\frac{\Delta k_\infty}{k_\infty} = \frac{\Delta p}{p} \,. \tag{6-13}$$

The resonance escape probability can be evaluated if the effective resonance integral is known by the following expression:

$$p = \exp\left[-\frac{\bar{\phi}_{R,A}}{\bar{\phi}_{R,M}} \cdot \frac{N_A}{\xi N_M \bar{\sigma}_M} \cdot I_e\right], \tag{6-14}$$

where N_A/N_M is the ratio of absorber atoms to moderator atoms in a representative volume (such as a cell), $\xi \bar{\sigma}_M$ is the average moderating power of the moderator, $\bar{\phi}_{R,A}/\bar{\phi}_{R,M}$ is the ratio of average resonance neutron flux in the absorber to that in the moderator (the disadvantage factor), and I_e is the effective resonance integral of the absorber.

The ratio $\bar{\phi}_{R,A}/\bar{\phi}_{R,M}$ takes into account the possibility that the resonance absorber may be segregated spatially from the moderator. It is the ratio of the general level of the resonance flux in the absorber to the general level in the moderator; the depletion of the flux spectrum in the local energy ranges corresponding to the absorption resonances is included in the effective resonance integral as discussed above. For a homoegeneous mixture of absorber and moderator, $\bar{\phi}_{R,F}/\bar{\phi}_{R,M} = 1$, and I_e for the fertile absorbers, may be taken from Fig. 6-7, or from similar experimental curves.

In those cases where the moderator and the resonance absorber are segregated completely,* theory and experiment show that the effective resonance integral is a function of S/M (cm^2/gram),

*Lumped cases in which some moderator is mixed with the absorber are discussed briefly in Sec. 6.4.

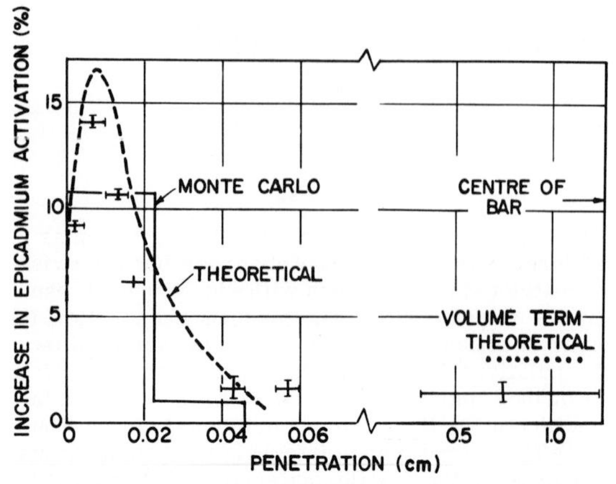

FIG. 6-8a Percentage increase in the epicadmium activation of uranium when the bar is raised from 20°C to 290°C vs. penetration into the bar. Experimental points and prediction in dashed line from Pearce. Dotted line from Dresner's prediction. Solid line from Monte Carlo results of Morton.

the surface-to-mass ratio of the discrete absorber lumps. The various theoretical models of the situation yield either a linear or a square root relationship:

$$I_e = A + B\,S/M, \quad \text{or} \tag{6-15}$$

$$I_e = A' + B'\sqrt{S/M}\,. \tag{6-16}$$

Usually the experimental evaluations of I_e can be fit over the range of practical interest by either relation. Recent experimental values for A′ and B′ are given in Sec. 6-4.

The change in p due to a change in the effective resonance integral is given by:

$$\frac{\Delta p}{p} = (\log_e p)\,\frac{\Delta I_e}{I_e}\,. \tag{6-17}$$

The Doppler effect may be treated as a variation in the effective resonance integral with the temperature of the absorber. Frequently this variation is specified in terms of a Doppler coefficient, ΔD_o. It can be defined as the fractional change in the effective resonance integral per unit temperature change or

$$\text{Doppler coefficient} = \Delta D_o = \frac{1}{I_e}\frac{dI_e}{dT} \tag{6-18a}$$

$$\text{or} \quad \frac{1}{p}\frac{dp}{dT} = \Delta D_o\,(\log_e p)\,. \tag{6-18b}$$

This relationship imposes a linear relationship between the effective resonance integral and the temperature, thus::

$$I_e(t) = I_e(0)\left[1 + \Delta D_o\,(T - T_o)\right], \tag{6-19}$$

where I_e (0) is the resonance integral at some base temperature, T_o °C, and I_e (T) is the resonance integral at another temperature, T °C. This formulation is useful over relatively small temperature ranges, and is also sometimes used to describe the Doppler effect over large, specified temperature ranges when only the end points of the range are significant.

Theoretical calculations indicate that the resonance integral does not vary linearly with temperature, and that, if the formulation indicated by Eq. (6-20) is to be used over large temperature ranges, ΔD_0 itself should be a function of temperature. Studies have indicated that the approximate variation of ΔD_0 might be taken as proportional to T^{-n}, where T is the absolute temperature and values of n have been variously given [104 - 109] as 1/2, 1, or 3/2.

A simpler approach is to abandon the concept of a Doppler coefficient which relates the resonance integral linearly to the temperature, and to search for a temperature dependence which will be compatible with a constant coefficient $\Delta D_0'$. Theoretical investigations [106] indicate that the true variation may not be expressible by a simple relation of this type over an extremely large temperature range, but a considerable segment of the available experimental data appears to be expressible in the following form:

$$I_e(t) = I_e(0)\left[1 + \Delta D_o'(\sqrt{T} - \sqrt{T_o})\right] \qquad T \text{ in } ^\circ K . \tag{6-20}$$

The magnitude of the Doppler change in the resonance integral depends upon the surface/mass ratio of the fuel. Table 6-5, reproduced from reference [99], is a summary of measured Doppler effects, stated in terms of the linear coefficient, ΔD_0, which illustrates the variation with S/M. The coefficient increases with S/M in the range covered by the table.

Hellstrand, et al., [104] have reported measurements of the variation of the resonance integral with temperature, expressed in the form of Eq. (6-20). They give the following variation of $\Delta D_0'$ with surface/mass ratio.

$$\Delta D_o' = 10^{-2}(0.58 + 0.5\ S/M) \text{ for uranium oxide,} \tag{6-21a}$$

$$\Delta D_o' = 10^{-2}(0.51 + 0.5\ S/M) \text{ for uranium metal.} \tag{6-21b}$$

The results of Arnold and Dannels's Monte Carlo calculations for the temperature dependence of the $U^{238}O_2$ resonance integrals have been fitted by Spano to an exponential form which can be converted approximately to the form of Eq. (6-20) and gives:

$$\Delta D_o' = 10^{-2}(0.696 - 0.0262\ M/S). \tag{6-22}$$

Spano has demonstrated that the results of the SPERT-I oxide core tests (see Sec. 7.3.3) show good agreement with calculations based on the Arnold-Dannels value for $\Delta D_0'$ and poorer agreement with the Hellstrand value. He has also shown that these tests agree best with the temperature dependence as shown in Eq. (6-20).

The increase in the Doppler effect on the resonance integral with increasing S/M is illustrated by Fig. 6-8b, [99a], which has been inferred from the work of Egiazarov [110] by R. M. Pearce. [99a] The reason for this variation is clarified by Fig. 6-8a from the work of Pearce. In these measurements, the increase in epicadmium activation of a one-inch diameter bar of uranium was measured, when the temperature was increased from 20°C to 290°C, as a function of penetration into the bar. It may be seen that the increase in activation at the higher temperature is largely confined to a thin layer near the surface. Within a depth of 0.04 cm, the increase has dropped to about 1% and remains almost the same throughout the interior. One would therefore expect the Doppler coefficient to increase as the surface/mass ratio is increased, until the absorber begins to approach the "infinitely thin" condition in which self-shielding of the resonances is unimportant. Thus the Doppler coefficient will pass through a maximum value as S/M is increased: calculations indicate that this value would correspond to a diameter of about 0.2 millimeter for cylindrical rods, [99, 99a] a value considerably lower than the fuel-element dimensions in most practical reactor designs.

On the basis of this brief review of resonance absorption, it is possible to point out some of the characteristic features of the Doppler coefficient of reactivity. First, it is evident that the effective Doppler broadening of a resonance by thermal agitation of the atoms of the absorbing nuclides does not affect the total absorption, and, hence, reactivity, if the concentration of the absorbing atoms is very dilute.

Obviously, an appreciable fraction of the neutron population must lie in the resonance energy range in order for Doppler broadening to have an important effect. For thermal or near-thermal reactors, this means simply that the resonance escape probability (p) must be appreciably lower than unity [see Eq. (6-14)]. In small fast-neutron reactors the concentration of U^{235} must be quite high in order to achieve criticality. In such reactors the mean neutron energy may be quite high—up to 200 kev—and the Doppler effect on U^{238} absorption is small, both because there is not much U^{238} present (relative to U^{235}) and because most of the neutrons exist at energies above the large U^{238} resonances. In such reactors there is a possibility of a positive

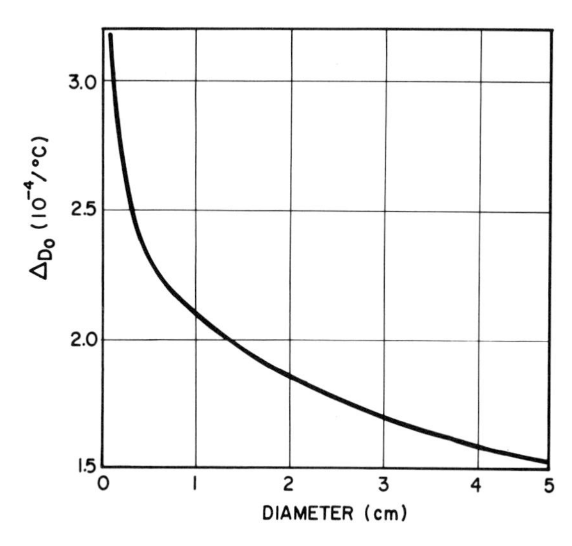

FIG. 6-8b The Doppler coefficient ΔD_0 as a function of the diameter of uranium cylinders, inferred from the work of Egiazarov et al.

TABLE 6-5

Experimental Doppler Coefficients

Material	Method	Reference	S/M (cm^2/gm)	Δ_{D_0} ($10^{-4}/°C$)	$\Delta_{D_0}(1/E)$* ($10^{-4}/°C$)
U	Activation	Mitchell	0	1.5	
		Egiazarov	2.14 0.214 0.053	3.1 2.1 1.6	 1.9 1.4
		Rodeback	0.117	0.9 ± 0.2	
		Pearce	0.084	1.0 ± 0.2	1.0
		Hellstrand et al	0.08 0.27	1.35 ± 0.10 1.56 ± 0.10	1.35 1.56
	Reactivity, small sample	Pearce and Walker	0.066	1.1 ± 0.2	1.1
		Small	0.100 0.147 0.169 0.376 0.399	2.24 ± 0.14 2.17 ± 0.22 2.30 ± 0.18 2.93 ± 0.58 3.59 ± 0.36	1.89 1.80 1.93 2.48 3.14
		Davis	0.062	1.74 ± 0.2	1.54
		Chernick and Kunstadter	0.076	1.1	
	Reactivity, full lattice	Borst		1	
		Leslie	0.076	1.9	
		Adyasevich	0.063	1.95	
		Panasyuk		1.4 ± 0.25	
		BNL-113	0.076	1.5	
		Chernick and Kunstadter	0.076	1.5	
UO_2		Creutz	0 0.177 0.297	2.7 1.6 1.8	
		Mitchell	0.18	0.88	
		Hellstrand et al	0.14 0.23 0.48	1.59 ± 0.10 1.68 ± 0.10 2.00 ± 0.12	
			Cluster	2.0 ± 0.2	
Th		Rodeback	0.189	2.8 ± 0.3	
		Pearce	0.140	3.8 ± 0.2	
		Pettus	0.284 0.148	3.1 ± 0.4	
		Small	0.253 0.276 0.653	3.9 ± 0.3 3.9 ± 0.3 2.6 ± 0.5	
ThO_2		Pettus	0.796 0.767 0.528 0.292	3.8 ± 0.2	

* The experimental coefficient corrected to a 1/E spectrum.

Doppler coefficient of reactivity due to the broadening of fission resonances in the fissile isotopes present (U^{235}, Pu^{239}, Pu^{241}). Calculations and measurements have indicated that its magnitude is small, but any possibility of a prompt positive coefficient of reactivity is a matter of some concern in a fast reactor.

Since the effects of Doppler broadening may become significant in isotopes other than the fertile isotopes initially incorporated in the reactor, it is necessary to consider the effects over the core lifetime and to take into account any other isotopes, such as Pu^{239} and Pu^{240}, which have strong resonances. Usually the net Doppler coefficient will be negative in any reactors in which the content of fertile isotopes, U^{238} or Th^{232}, is much larger than the content of fissile isotope, for the fission resonances in the fissile isotopes are accompanied by capture resonances which tend to counteract their effects; in addition, the shapes, spacings, and energy ranges of the fission resonances are of such a nature as to give lower Doppler coefficients than the capture resonances in the fertile isotopes, at least in thermal or near-thermal reactors. Nevertheless, the effect of fuel exposure should be considered whenever a positive Doppler coefficient is conceivable, or whenever a strong negative coefficient is considered important to the safety of the reactor.

In crystalline solids it may be necessary to take into account the quantum mechanical heat energy distribution within the material. The specific heat of such materials as ThO_2 or ThC, for instance, has not been measured. Work by Brown and Sampson [111] appears to indicate that calculated values of the Doppler effect tend to overestimate in the region of 300° because crystalline binding is not taken into account. Greebler and Goldman [112] also attribute much of their calculational uncertainty for low temperature Doppler effects to lack of knowledge of the Debye temperature of UO_2 and related uncertainties.

In reactor fuels containing two or more resonance absorbers, it is necessary to consider the interaction of resonances of the two absorbers. A resonance of one absorber such as U^{235} may obscure a nearby resonance in another absorber, thus reducing its Doppler effect. Codd and Collins [113] (and also Bhide and Hummel [107] have pointed out that it is improper simply to add the Doppler effect due to Pu^{239} to that of U^{238} in dilute fast reactors. Although the addition may be a fair approximation for neutron energies near 100 kev, it is certainly incorrect for intermediate energies from 100 ev to 5 kev, the most important energy region for many cores. Preliminary calculations have shown that the effect of U^{238} resonances on Pu^{239} resonances makes the Doppler contributions of the Pu^{239} resonances to reactivity negative instead of positive. The effect of Pu^{239} resonances on U^{238} resonances does not appear to be very strong. [113]

6.4 The Four-Factor Breakdown

It has been impractical here to consider the reactivity coefficients of the many possible reactor types in terms of the theoretical models most appropriate to each. For the thermal reactors, however, it is possible to give a somewhat more detailed discussion in terms of the "four-factor" approximation presented in Sec. 2.4. Simplification by approximation is particularly helpful in the thermal case because of the wide range of heterogeneity which must be taken into account. This reaches from compositions which are effectively homogeneous, through lattices which consist of a simple repeating structure of fuel element and moderator (Fig. 5-5A), to those which consist of a regular array of fuel element clusters embedded in moderator (Fig. 5-5B, 5-5C). Most practical situations can be discussed in terms of the three arrangements shown in Fig. 5-5, the homogeneous case being considered a limiting case of 5-5A.

The partial reactivity coefficients which are usually most important in the thermal reactors are those which affect k_∞:

$$k_\infty = \eta f p \epsilon . \qquad (6\text{-}23)$$

Conceptually, this formulation of k_∞ is convenient even for those thermal reactors, such as the hydrogenous reactors, which require different theoretical models for precise calculation of the absolute reactivity.

Once the materials have been selected for a particular reactor, the only changes that can be made in k_∞ are those which result either from changes in temperature or from changes in the ratios of materials (either generally or locally) within the reactor. Of the latter, changes in fuel/moderator ratio are particularly important. Before proceeding to more complex situations it may be helpful to examine the changes in the components of k_∞ as the fuel/moderator ratio is changed in a simple lattice. Figure 6-9 [6b, 115] has been included for this purpose. It shows the measured variations for a simple lattice of uranium rods in H_2O.

It should be observed that the four factors define a method of "keeping books" on the neutron balance rather than a set of well defined physical processes, and there is room for considerable variation in the definitions of the factors. The more subtle variations that have been used are discussed

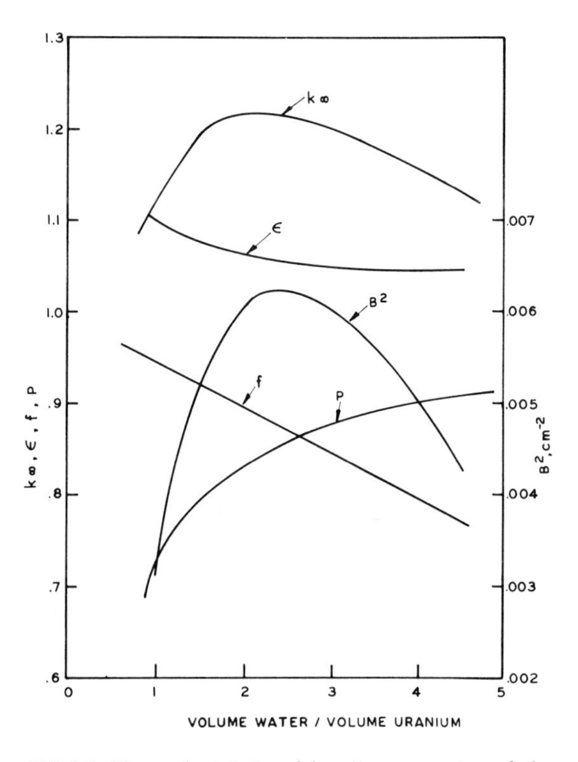

FIG. 6-9 Measured variation of k_∞, its components, and the material buckling, with moderator/fuel ratio in a simple uranium-H_2O lattice. The lattice is composed of uranium rods of 0.600 in. (1.524 cm) diameter, 1.3% enriched, clad with 0.028 in. (0.71 mm) of aluminum. The variation of k_∞ can be understood in terms of the plotted variations of ϵ, p, and f; the variation of η is small. The material buckling (B^2) may be defined as $(k_\infty - 1)/M^2$; it specifies the size of a lattice which would be just critical. The minimum critical size corresponds to the maximum value of B^2; for this case it would occur at a water/fuel ratio of about 2.4.

in reference [114], but here it is sufficient to point out that the definitions used for η and f depend upon what one defines as the fuel. The fuel may be defined as the fissile isotope (in which case η depends only on the identity of the isotope and on the neutron spectrum), as the mixture of fissile and fertile isotopes (in which case η depends on the fertile/fissile ratio as well), as the mixture of fertile and fissile compounds or alloys (e.g., $U^{235}O_2 + U^{238}O_2$), and in other ways. In this section the third of these definitions will be adopted.

The regeneration factor η is defined as the average number of fission neutrons produced per neutron absorbed in the fuel. It depends on only the fuel composition and the neutron spectrum:

$$\eta = \frac{\nu\bar{\Sigma}_F}{\bar{\Sigma}_a\,(\text{fuel})} = \eta_o \frac{\bar{\Sigma}_a\,(\text{fissile isotope})}{\bar{\Sigma}_a\,(\text{fuel})}, \qquad (6\text{-}24)$$

where η_o denotes the value of η for the fissile isotope, and the bars indicate averages over the neutron energy spectrum. The bar is omitted over η because, when used in the four-factor formula, it is by definition an average over both the neutron energy and the materials comprising the fuel. After a reactor has been put into service, one would expect η to change only because of changes in the neutron spectrum and because of changes in composition caused by fuel burnup. The former of these has already been discussed generally, and the latter is discussed later in a separate section (8).

The thermal utilization f is the fraction of total neutron absorption which occurs in the fuel:

$$f = \frac{A_F}{A_F + \sum\limits_{\substack{\text{all non-fuel}\\ \text{absorbers}}} A_i} = \frac{V_F\Sigma_{aF}}{V_F\Sigma_{aF} + \sum\limits_{\substack{\text{all non-fuel}\\ \text{absorbers}}} \frac{\bar{\phi}_i}{\bar{\phi}_F} V_i\Sigma_{ai}}, \qquad (6\text{-}25)$$

where A_F, V_F, Σ_{aF}, and $\bar{\phi}_F$ are the absorption rate in the fuel, the volume of fuel, the effective macroscopic absorption cross section of the fuel, and the average thermal neutron flux density in the fuel, respectively. The corresponding symbols with the subscript i refer to the same quantities for the ith (non-fuel) material in the core. The ratio $\bar{\phi}_i/\bar{\phi}_F$, the disadvantage factor, is 1.0 for all materials in a homogeneous core. In heterogeneous cases, f is usually evaluated for a typical cell in the core (Figs. 5-5, 5-6, Sec. 2-5). Ordinarily the changes that can occur in f, once the reactor has been built, result from changes in the average density of the moderator or coolant (as from boiling or changes in temperature) and changes in the absorption cross section of the coolant or moderator (as from the addition or removal of soluble poison).

If the absorption rate of material j changes from A_j to $A_j + \Delta A_j$, the change in f is [note that this is simply another form of Eq. (6-10)]:

$$\frac{\Delta f}{f} = \frac{\Delta A_j}{A_j} \cdot \frac{A_j}{A_F + \sum\limits_{\substack{\text{all}\\ \text{absorbers}}} A_i} = \frac{\Delta A_j}{A_j}\left[\begin{array}{c}\text{fraction of neutrons}\\ \text{absorbed by material j}\end{array}\right]. \qquad (6\text{-}26)$$

Thus, if soluble absorber is added to a liquid moderator to change its macroscopic cross section from Σ_{mod} to $\Sigma_{mod} + \Delta\Sigma_{mod}$, and if the change is not large enough to cause a significant change in $\bar{\phi}_{mod}/\bar{\phi}_F$, the change in f is:

$$\frac{\Delta f}{f} = \frac{\Delta\Sigma_{mod}}{\Sigma_{mod}}\left[\begin{array}{c}\text{fraction of neutrons}\\ \text{absorbed in moderator}\end{array}\right]. \qquad (6\text{-}27)$$

In general, an increase in the fractional absorption of a given non-fuel material, j, will decrease the flux ratio $\bar{\phi}_j/\bar{\phi}_F$ (disadvantage factor) and hence the effect on f (a decrease) will be less than that indicated by the change in cross section. A special case occurs when a scattering (and possibly moderating) material is added within a segregated fuel "bundle." Aside from the absorption of the added material, its scattering will decrease the diffusion coefficient, and hence increase the flux depression on the fuel bundle and tend to lower f. On the other hand, if the material is a strong moderator it may thermalize enough neutrons to increase the average thermal flux within the bundle, to the extent that f increases. A material with low thermal-neutron diffusion coefficient which is added near, but outside, the fuel bundle impedes the diffusion of neutrons into the fuel and tends to lower $\bar{\phi}_F$; if the material is also a moderator the effect may be counteracted by the additional production of thermal neutrons near the fuel bundle. Figure 6-10 illustrates some of these effects. In this particular case the effect of the added material (D_2O) on the diffusion coefficient is more important than its moderating effect.

The resonance capture of neutrons (usually by U^{238} or Th^{232}) is accounted for in the 4-factor formulation by p, the resonance escape probability. The appropriate equation for p has already been given (6-16):

$$p = \exp\left[-\frac{\bar{\phi}_{R,A}}{\bar{\phi}_{R,M}} \cdot \frac{N_A}{\xi N_M \sigma_M} I_e\right] \qquad (6\text{-}28)$$

For homogeneous mixtures the ratio $\bar{\phi}_{R,A}/\bar{\phi}_{R,M}$ is unity, and I_e, the effective resonance integral is usually evaluated from experimental curves such as those of Fig. 6-7. For cases in which fuel and moderator are segregated, I_e is a function of the surface/mass ratio for any given species of absorber, as indicated in Eqs. (6-15) and (6-16). The results of some of the more recent experimental

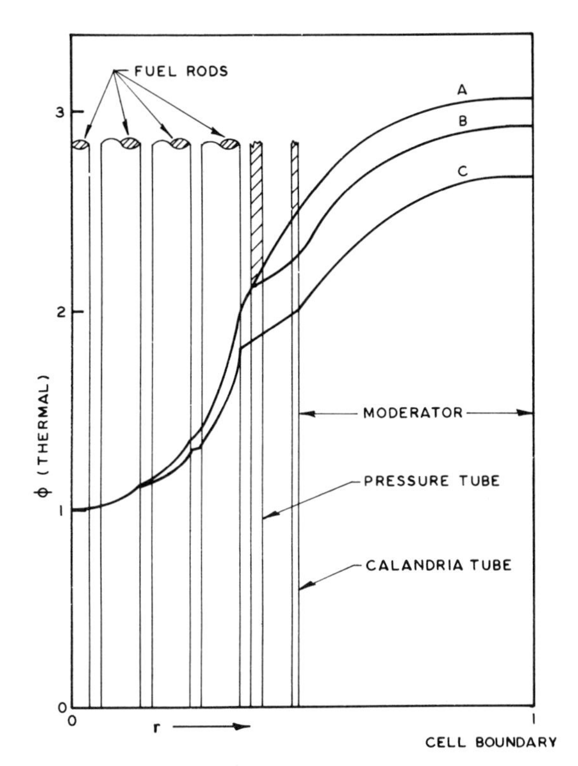

FIG. 6-10 Possible effect of moderator in fuel bundle on thermal utilization. The figure shows possible thermal neutron flux distributions in a 37-rod fuel bundle within a pressure tube which in turn is surrounded by a calandria tube. The flux distributions are all normalized to unity at r = 0.

Curve A is a possible distribution when all open spaces are filled with moderator. Curve B applies when the space between the pressure tube and the calandria tube is empty (the normal situation). The flux in the main body of moderator is lower (and f is higher) than for A because the diffusion of neutrons into the fuel is not hindered by the moderator in the gap. Curve C applies when the moderator is removed also from within the pressure tube. The flux depression in the fuel is now less because removal of the moderator has increased the average diffusion coefficient in the bundle, and f increases again.

This illustrates only one possible behavior — it is similar to that in a well-thermalized D_2O-moderated reactor. If the ratio of fast-to-thermal neutron flux is high, and/or if the moderator within the fuel bundle is hydrogenous, removal of the moderator from the bundle may decrease f because it removes the source of thermal neutrons from within the fuel bundle.

determinations give:*

for U metal: [117] $I_e = 2.95 + 25.8\sqrt{S/M}$

for UO_2: [117] $I_e = 4.15 + 26.6\sqrt{S/M}$

for Th metal: [118] $I_e = (-0.2 \pm 1.3) + (17.2 \pm 2.1)\sqrt{S/M}$

where S is the surface area in cm^2, and M is the mass in grams.

The processes which can change p after the reactor is built are changes in average moderator density, changes in fuel temperature, changes in position of fuel and/or moderator and changes in fuel isotopic composition with burnup.

In a simple lattice (Fig. 5-5A) the effect of a change in moderator density is to change N_M in Eq. (6-28). This ordinarily produces a second-order effect on $\bar{\phi}_{R,F}/\bar{\phi}_{R,M}$ which is in the direction to enhance the effect of the change in N_M. If the latter effect is neglected, the change in p due to a change in moderator density, in a simple lattice, is:

$$\frac{\Delta p}{p} = \left[-\log_e p\right]\frac{\Delta d}{d} = (1 - p)\frac{\Delta d}{d}, \qquad (6\text{-}29)$$

where $\Delta d/d$ is the fractional change in moderator density.

In a complex lattice, containing clusters of fuel elements, the "fuel" is usually defined, for purposes of calculating p, as the material within the envelope of the cluster, i.e., within the boundary which would be established by wrapping a string or a rubber band around the fuel cluster. An approximate effective surface/mass ratio for the fuel cluster may then be determined by: [97]

$$S/M = (S_o + \gamma S_i)/M, \qquad (6\text{-}30)$$

where S_o is the surface of the envelope, S_i is the fuel surface within the envelope, and γ is a weighting factor which depends upon the macroscopic scattering cross section (Σ_s) of the material which fills the spaces between elements in the cluster:

$$\gamma = 2\Sigma_s r\left[1 - P_2(\Sigma_s r)\right], \qquad (6\text{-}31)$$

where r is proportional to the distance between fuel lumps in the cluster (if the fuel cluster can be approximated as a single lump of fuel penetrated by equal circular holes, r is just the radius of a hole) and $P_2(\Sigma_s r)$ is the probability that a neutron which suffered a collision in the "hole" will suffer its next collision again in the "hole," without returning to the fuel.

In a complex lattice, then, an increase in the density of the moderator outside the envelope of the fuel bundle will increase p, by an amount given by Eq. (6-29), while an increase in the density of moderator** within the bundle will decrease p by increasing the effective resonance integral [Eqs. (6-30) and (6-31)]. The above treatment is obviously a reasonable approximation only when the clustering of fuel is very pronounced. For the sort of clustering that is typical of H_2O-moderated reactors, which is brought about only by the provision of channels for control rods, other approximations must be used. The effect of this clustering is significant, and substantial errors may result from neglecting it.

In highly clustered lattices employing moderating coolants, the bowing of the fuel elements as a result of uneven thermal expansions can cause the elements to move closer together or farther apart in the cluster (see Sec. 7.4) giving an increase or a decrease in p, respectively.

*These are the resonance integrals in excess of the resonance integral which would result from extrapolating the 2200 m/sec cross section to higher energies in proportion to 1/v.

**The effect of the material within the fuel cluster is determined by its scattering cross section rather than by its moderating power ($\xi\Sigma_s$) provided ξ is reasonably large. Very heavy scatterers such as lead, however, may have only a small effect on the resonance integral.

The fast fission factor, ε, takes account of the production of neutrons by the fast-neutron fission of U^{238} or Th^{232}. The quantity $\varepsilon - 1$ is the net number of neutrons produced by fast fission (i.e., by neutrons having energies above the fission threshold of the fertile isotope) per neutron produced by thermal fission; it usually does not exceed 0.05 except in very "dry" water-moderated lattices. In reactors employing rather massive, widely separated, fuel elements,* most of the fast fission in a given element is produced by fission neutrons which originated in that element, since the neutrons quickly slow down below the fast-fission threshold once they leave the element and enter the moderator. To the extent that this is true, ε cannot be affected appreciably by the changes in composition which accompany reactor operation. For the case of small, closely spaced fuel elements many fission neutrons which originate in one element may cause fast fission in another. In such a case ε may depend strongly on the density of the moderator between fuel elements, increasing with decreasing moderator density. The H_2O-moderated reactors are the most important example. For the limiting case in H_2O, where fuel and water are mixed homogeneously, ε can be calculated approximately by:

$$\epsilon = 1 + \frac{\left(\nu - 1 - \frac{\Sigma_c^{28}}{\Sigma_f^{28}}\right)\Sigma_f^{28}}{\Sigma_{in} + \Sigma_c - (\nu - 1)\,\Sigma_f^{28}}\,, \quad (6\text{-}32)$$

where the cross sections are all "effective" values, given in Table 6-6 [6a], $\nu = 2.48$ for U^{238}, and the superscript 28 designates U^{238}. Σ_{in} is a macroscopic cross section for scattering of fission neutrons below the fast fission threshold (by inelastic scattering in the case of heavy elements).

6.5 Effect of Strong Absorbers

It has been indicated (Sec. 4.2.1) that the absorption area of a "thermally black" absorbing control rod is a function of the diffusion length in the surrounding medium, increasing in approximately direct proportion to the diffusion length. Consequently, if the diffusion coefficient in the core increases because of thermal expansion of the moderator or the introduction of voids, or if the average absorption cross section of the core decreases because of an increase of effective neutron temperature, the reactivity worth of any control rods which may be present in the core will increase. If the control rods are held in fixed positions, this increase in control rod worth will appear as an additional negative component in the temperature and void coefficient of reactivity for the core. A similar effect will result from any other strong localized absorber in the core, such as an absorbing shim or a sheet of burnable poison. The effect is most pronounced when the absorber is black to thermal neutrons, but it will occur

*Fuel element here means the largest unit of fuel which does not contain moderator—it does not mean a fuel assembly in which there may be also a coolant with moderating properties.

TABLE 6-6

Cross Sections to be Used with Equation (6-32) for Fast Effect in "Homogeneous" H_2O-U Lattices (in Barns)

Material	σ_f	σ_c	σ_{in}
Hydrogen	0.00	0.00	1.38
Oxygen	0.00	0.024	0.171
Uranium	0.30	0.042	2.32

whenever the absorber is strong enough to produce an important local depression of the thermal neutron flux. It may be regarded, alternatively, as a change in the disadvantage factor for the absorber as core temperature or density is changed.

If a number of control rods are inserted partially, as a bank, their primary effect may be effectively to shorten the core, so that the axial leakage is increased, as in Fig. 4-18. Qualitatively, this mode of operation of control rods has effects on the reactivity coefficients which are comparable to those discussed above for individually fully-inserted rods. The details of the effect are however rather different. The major portion of the effect will depend upon changes in the migration length rather than upon changes in diffusion length. Fully-inserted rods which produce appreciable changes in the gross radial power distribution in the core may involve both types of variations, and may affect the density and temperature coefficients of reactivity through both the migration length and the diffusion length. In any event, the presence of control rods or other strong localized absorbers in the core may affect the temperature and density coefficients of reactivity very significantly if they control a large amount of reactivity, and the coefficients calculated for idealized cases, without control rods, may bear little relation to the coefficients which will characterize the actual core containing control rods.

7 COMPOSITE REACTIVITY COEFFICIENTS**

In the preceding section, the short term reactivity changes which may occur in a reactor were broken down into partial coefficients, each of which was associated with a particular component of the neutron balance in the chain reaction. In the practical case, the changes in conditions which cause reactivity changes may involve several components of the neutron balance simultaneously, and for small changes of this kind the net change in reactivity will be the algebraic sum of the changes denoted by the several partial coefficients. Thus, such a simple change as a uniform increase in the temperature of the reactor may affect a number of components of the neutron balance. In a thermal reactor it will change the effective energy of the thermal neutron group, and consequently may change the relative values of the important fission and absorption cross sections. It will increase

**This section is by T. J. Thompson and J. R. Dietrich.

slightly the Doppler broadening of resonances in any resonance absorbers which may be present. Usually it will also be accompanied by a significant change in the density of the moderator (and perhaps the density of the fuel) which in turn will change the fractional leakage of neutrons, the resonance escape probability, the fast fission effect, and the thermal utilization. Thus, all or most of the partial coefficients discussed above may be involved, but still it is useful to characterize the net reactivity change by a coefficient—the isothermal temperature coefficient of reactivity—which specifies the net reactivity change resulting from a one degree increase in the uniform temperature of the reactor. Coefficients of this kind are referred to here as composite reactivity coefficients.

In all discussions of composite reactivity coefficients it is important to specify clearly just what the coefficient under discussion is intended to represent. In the past there has been a regrettable laxity of terminology which at best wastes time in communication and at worst might lead to hazardous misunderstandings. Thus, for example, the isothermal temperature coefficient of reactivity has often been confused with the moderator temperature coefficient of reactivity, and often no distinction is made between the Doppler coefficient of reactivity and the complete fuel temperature coefficient of reactivity. Often it is difficult to find an appropriate name for the coefficient which is under discussion, but it is always possible to give a precise operational definition.

The particular coefficients which are useful in safety considerations perform one or both of two functions. They may be simply coefficients which are easily observed and easily measured in the actual reactor, and their main significance may be as verifications of the calculated characteristics of the reactor. Or they may be important because they specify the reactivity response to changes in operating conditions which can actually occur in practice. Of the latter type the coefficient which is most universally important is the power coefficient of reactivity: that is, the change in reactivity resulting from unit change in reactor power. This characteristic determines how the reactor will behave when it is subjected to a reactivity change (or to a change in operating conditions which induces a reactivity change) from outside. In considering the operation of the power coefficient, it is useful to have in mind the conceptual relationship shown in Fig. 7-1. Here an externally applied change of reactivity is pictured as operating directly upon the kinetics of the neutron chain reaction, to give rise to an incremental change in reactor power. This in turn is pictured as causing a change in reactivity, via the power coefficient of reactivity, which feeds back and combines with the externally applied reactivity, reinforcing or diminishing it accordingly as the power coefficient is positive or negative. This concept is normally used in analyses of reactor stability and reactor transfer functions, which often employ simplifications such as the linearization of non-linear responses; but

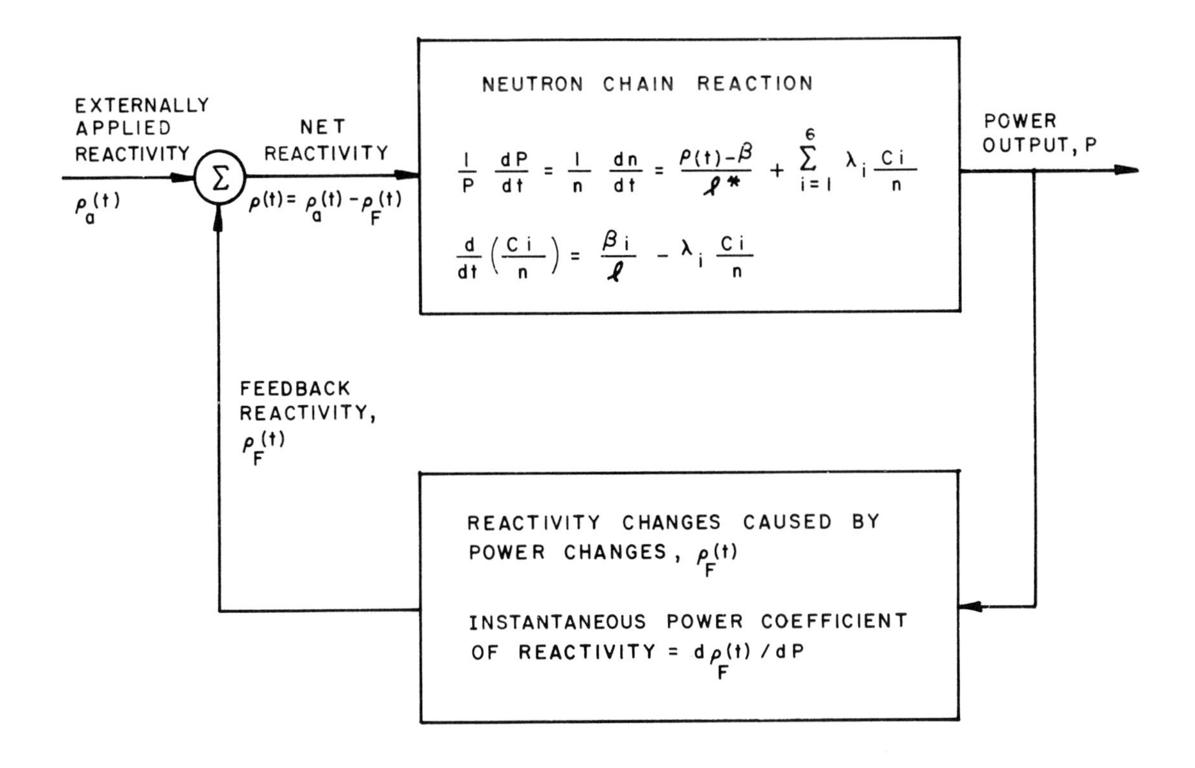

FIG. 7-1 Concept of reactivity feedback through power coefficient of reactivity. The equations in the upper box are the usual neutron kinetics equations, given in standard texts on reactor physics.

the feedback concept is independent of these simplifications and is a valid and generally useful one, at least for the great majority of practical cases in which the space-independent neutron kinetics equations are valid.

The reactivity changes due to changes in reactor power are all caused by power-induced changes in the temperatures, densities, or relative positions of the parts of the reactor. In general, these changes do not occur instantaneously when the power is changed, but are coupled to the power changes by time-dependent dynamic relationships. The net change may be the algebraic sum of a number of component changes such as changes in fuel temperature, changes in coolant temperature, changes in moderator temperature, and changes due to thermal expansion of the reactor structure, each of which may have its own dynamic relationships to the reactor power. Thus, the feedback illustrated in Fig. 7-1 may actually represent a number of feedbacks with different dynamic characteristics; and the power coefficient of reactivity must be considered either to be a composite of several partial coefficients having different dynamic characteristics, or to be a dynamic coefficient characterized by a number of different time constants. Whichever view is taken, the important point is that the power coefficient of reactivity cannot, in general, be specified simply in terms of a reactivity change, but must be characterized also by its dynamic behavior.

In the following sections a number of composite reactivity coefficients are discussed. These are by no means all of the coefficients which may be important, but they are coefficients which in the past have been found to have considerable significance with respect to safety, and which illustrate some of the subtleties which must be taken into account in evaluating reactivity behavior. Some of the coefficients are themselves components of the power coefficient of reactivity.

7.1 Zero-Power Temperature, Density, and Pressure Coefficients

It is useful to distinguish between those changes in operating conditions which are imposed upon the reactor from outside and those which arise internally. Of the latter, all which occur in normal operation arise from changes in reactor power, and their reactivity effects are accounted for by the power coefficient of reactivity. The reactivity effects of the externally applied changes will, in general, vary according to the operating power level. They can be observed as independent reactivity effects only when they are decoupled from the power effects—that is, when they are applied to a reactor which is critical at effectively zero power. Hence they are treated here as zero-power reactivity coefficients.

7.1.1 The Isothermal Temperature Coefficient

The reactivity change caused by a one-degree increase in the uniform temperature of the zero power reactor is called the isothermal temperature coefficient of reactivity. Although it may involve complex changes in the neutron chain reaction it is itself a simple concept and it is a quantity which is usually easy to observe and to measure on an existing reactor. Since it is possible in theory and generally in practice to maintain a reactor system at any desired temperature from room temperature to operating temperature, it is possible to make measurements of this type throughout the entire range of temperatures. In a number of reactor types, this can often be done by heating the moderator system and, hence, the whole system by means of the work put in by the coolant pumps. In small research reactors measurements can be made at powers of a few watts or less. In power reactors it is often necessary to use powers in the range of kilowatts to megawatts. In any case, the heat generated in the fuel should make a negligible contribution to the temperature change in both fuel and moderator if the measurement is to be valid.

To make the measurement by the static method the entire core and moderator system should be at equilibrium at one temperature. The reactor is then made critical or slightly supercritical on a convenient long asymptotic period (see Chapter on Criticality). From the control-rod position information, the inhour equation for the reactor in question, and the reactivity calibration of one control rod, the critical position is ascertained. The same procedure is repeated at another temperature a few degrees different with all control rods in the same configuration except the single control rod which has been calibrated. This rod is used to make the reactor critical or to put it on a long asymptotic period—as before. From the temperature change and the two criticality measurements with the calibrated control rod the isothermal coefficient of reactivity may be calculated. If the two temperatures chosen are close enough together, no net movement of the control rods may be necessary, and the coefficient may be evaluated from the change in reactor period produced by the temperature change. Sets of two measurements may be repeated throughout the desired temperature range to determine the coefficient as a function of temperature.

It is also possible to measure the isothermal temperature coefficient by a dynamic or "drift" method in which the temperature of the system is allowed to change monotonically at a slow rate. While this is done, period measurements are made at constant rod positions to determine the coefficient. This method is often useful in power reactors, especially where time can be saved through its use during the heat-up of the system. [119]

To calculate the isothermal temperature coefficient of reactivity, the reactivity of the core may be suitably evaluated at two different temperatures. The coefficient is equal to the difference in reactivity divided by the temperature difference. Since most reactor calculations are now done by computer codes, this may be a relatively simple matter once the basic method of reactivity calculation has been set up.

Alternatively, one can use perturbation theory to develop a set of equations for the various perturbations caused by the temperature change. These may be superimposed upon each other to give the isothermal temperature coefficient of reactivity. Often it is desirable to calculate local coefficients as functions of position within the core and to sum these effects to obtain the isothermal temperature coefficient. This procedure may yield simultaneously the isothermal coefficient and a set of space-dependent coefficients for use in transient analyses. The calculations for the Enrico Fermi Reactor were done in this way. [120] For example, the reactivity change due to local thermal expansion of sodium in a volume element dV is:

$$\frac{d^2\rho}{dT\, dV} = -D_{Na}V_{Na}\alpha_{Na}C_{Na}(r,z) \quad , \qquad (7\text{-}1)$$

where

D_{Na} = sodium density,

V_{Na} = local volume fraction of sodium,

α_{Na} = volumetric expansion coefficient of sodium, and

$C_{Na}(r,z)$ = sodium material coefficient of reactivity in units of reactivity per unit mass.

Note that $C_{Na}(r,z)$ is the reactivity change per unit mass of sodium weighted by the reactivity importance factor (cf. Sec. 4.2.5). This can be calculated and normalized to give an average worth agreeing with the experimental value observed in a critical experiment, or it could be measured by inserting an appropriate small sample in the mockup core in various positions and obtaining relative local worth values experimentally in this way. The local reactivity worth of U^{238} was surveyed in the Enrico Fermi case in this way.

After all components of the local temperature coefficient of reactivity have been summed, integration of the local coefficient over the core volume then gives the core isothermal temperature coefficient. In non-isothermal cases the appropriate use of this system allows the designer to state a core temperature coefficient by weighting the local reactivity coefficients by the temperature distribution and integrating over the core. It is important to recognize, however, that cases occur in which the local coefficient is not independent of the spatial distribution of temperature.

Reported values of the isothermal temperature coefficients of several reactors are given in Table 7-1. These have been compiled mainly from the hazard reports for the individual reactors.

7.1.2 Coolant and Moderator Coefficients

In some reactors, notably the H_2O reactors, the functions of coolant and moderator are performed by the same unsegregated liquid, and one can make no distinction between a moderator temperature coefficient of reactivity and a coolant temperature coefficient of reactivity. Indeed, so far as direct measurement is concerned, it is usually impossible, in the zero-power condition, to measure the reactivity effect of the water temperature independently of that of the fuel temperature. Conceptually, however, it may be useful to recognize a coolant-moderator temperature coefficient of reactivity, and it may be possible to approximate an experimental value from the isothermal coefficient, since one needs only to subtract out the portion of the isothermal coefficient which is due to the fuel temperature change. This adjustment is usually a small one which can be made with adequate precision by calculation.

In those reactors having moderators and coolants which are thermally segregated-either because they are different materials, as in the sodium-graphite reactor, or because they are separated by a thermal barrier, as in the pressure-tube reactors cooled and moderated by D_2O-one can distinguish and measure coolant-temperature and moderator-temperature reactivity coefficients.

The moderator temperature usually determines the thermal neutron temperature, and gives rise to a component of the reactivity change which depends upon the relative variations of the fission and absorption cross sections in the reactor with neutron temperature. Usually it will go in the direction of a reactivity decrease with increasing temperature unless plutonium makes a significant contribution to the fission rate. However, important neutron rethermalization effects, such as that cited in Sec. 6 may come into play if the coolant is also a strong moderator. Insofar as the temperature change of the moderator proper causes a change in its density, a change in neutron leakage will occur which tends to cause the reactivity coefficient to be negative. If the density change also causes a change in the atomic fuel/moderator ratio this will give rise to a positive component due to the changing absorption of the moderator, a negative component due to the changing degree of moderation and hence changing resonance escape probability, and possibly a negative component due to changes in effectiveness of any control rods which may be in the core.

Certain gas-cooled reactors of the Dragon or pebble bed types have the moderator, or part of the moderator, intimately mixed with the fuel. In these cases, the component of the moderator temperature coefficient of reactivity which is due to the fuel-mixed moderator will be "prompt" relative to the fuel temperature. In such a case the moderator coefficient has a much more important effect on the dynamic behavior of the reactor. Table 7-2 lists Doppler coefficients of reactivity (the important fraction of the fuel temperature coefficient) and moderator temperature coefficients of reactivity for several gas-cooled reactors of the heterogeneous and semi-homogeneous types. In the latter case, the "prompt" portion of the moderator coefficient is also listed where the values are available. For these cases, the total "prompt" coefficient is the sum of Doppler and "prompt" moderator components. It is also worthy of note that the moderator temperature coefficient of reactivity becomes positive with fuel exposure in those reactors which generate plutonium (EGCR, AGR, Calder-Hall).

In those reactors which do not have moderator mixed with fuel, and in which coolant and moderator are segregated, the coolant temperature coefficient of reactivity is usually more important than the moderator temperature coefficient, because the thermal time constant of the coolant is much

TABLE 7-1

Isothermal Temperature Coefficients of Reactivity

(Units: $d\rho/dT$)

Type	Conditions	Coefficient ($°C^{-1}$)
Pressurized Water		
Indian point (internal thorium converter)	500°F (260°C)	-5.58×10^{-4}
Saxton (5.7% uniform enriched core)	80°F (36.6°C)	$+1.80 \times 10^{-4}$
	530°F (277°C)	-9.54×10^{-4}
Yankee (3.4% uniform enriched core)	514°F (320°C) 0 ppm B	-5.58×10^{-4}
	514°F (320°C) 1200 ppm B	-2.34×10^{-4}
Boiling Water		
Dresden (1.5% enriched UO_2 core)	Room temperature	$+5.4 \times 10^{-5}$
	Operating temperature	-2.88×10^{-4}
Elk River (4.3% U^{235} in UO_2-ThO_2)	Room temperature	$+0.6 \times 10^{-4}$
	Operating temperature	-1.5×10^{-4}
Pressurized Heavy Water		
HWCTR, Savannah River (pressurized D_2O, natural U)	428°F (220°C)	-5×10^{-4}
Pressure Tube Heavy Water		
CVTR (two region core, 1.1 and 2.0% U^{235})	Moderator 155°F (68.4°C)	-2.43×10^{-4}
	Coolant 530°F (277°C)	-4.50×10^{-5}
PRTR, Hanford (plutonium recycle test reactor)		-1.06×10^{-4}
Superheat Reactor		
ESADA Vallecitos Experimental Superheat Reactor (H_2O-moderated, cooled by steam and moderator boiling, 5.4% enriched UO_2)	68°F (20°C) (flooded	$+1.80 \times 10^{-4}$
	68°F (20°C) (unflooded	$+1.62 \times 10^{-4}$
	545°F (284°C) (flooded	-2.70×10^{-4}
	545°F (284°C) (unflooded	-3.60×10^{-4}
Sodium-Cooled, Graphite-Moderated		
Hallam	Full power (moderator	$+3.96 \times 10^{-5}$
	Full power (coolant	$+0.90 \times 10^{-5}$
Gas Cooled		
EGCR (2.46% enriched UO_2, He-cooled, graphite-moderated)		-2.4×10^{-5}
Organic-Moderated and Cooled		
Piqua Nuclear Power Facility	Moderator	-9.54×10^{-5}
	Coolant	-1.80×10^{-5}
Fast Reactors		
Enrico Fermi (25.6% enriched U core, depleted U blanket)	Na in core	-4.44×10^{-6}
	Na in blanket	-5.54×10^{-6}
EBR-II (49% enriched U core)	Na density change	-9.1×10^{-6}

shorter. Further, a positive coolant temperature coefficient implies the probability of a positive coolant void coefficient of reactivity and the possibility that loss of coolant, depressurization of the coolant tube, rupture of the tube, or vapor formation will lead to a reactivity increase. On at least two occasions this phenomenon has played a role in accidents at the NRX Reactor in Canada. In this case, the reactor coolant is H_2O and the moderator is D_2O. While it may well be possible to instrument or design around this difficulty, nonetheless its presence considerably increases the risk of an accident which could have at least serious economic effects. It may be hard to avoid entirely in certain types of reactors. The large D_2O-moderated pressure-tube reactors, for example, tend to have positive coolant void coefficients even when cooled by D_2O. Erickson [121] has done some work in trying to get around this problem.

It is usually possible to measure the coolant-void coefficient at zero power, as well as the coolant-temperature coefficient, at least for the limiting case of complete voiding.

The components of the coolant coefficients of reactivity are apt to be varied. As the temperature of the coolant is increased (and its density is decreased) or as voiding occurs, the major positive reactivity effects are likely to be due to the decrease in absorption by the coolant and to the decrease in effective resonance integral of the fertile material as neutron scattering decreases in a fuel cluster. Changes in the fast-fission factor and

TABLE 7-2

Reactivity Coefficients at Operating Conditions

(Coefficients stated as $\frac{d\rho}{dT} \times 10^5$; T in °C)

Reactor	Doppler	Moderator coefficient temperature	Prompt moderator temperature coefficient	Over-all prompt temperature coefficient (column 1 + column 3)	Estimated prompt component of power coefficient*
Heterogeneous					
EGCR					
Initial (uniform enrichment = 1.8%)	-3.38	- 1.97		-3.38	-0.95
Equilibrium (feed enrichment = 2.2%, ave. burnup 5000 Mwd/tonne)	-3.20	+ 3.00		-3.20	-0.83
AGR					
Initial	-2.33	- 3.00		-2.33	-0.56
Arc burnup = 3000 Mwd/tonne	-2.33	+15.0		-2.33	-0.44
Calder Hall					
Beginning of life	-2.4				
End of life	-2.4	+15			
Semihomogeneous					
HTGR					
Beginning of life	-1.8		-0.6	-2.4	Negative
End of life	-1.8		+0.8	-1.0	
Dragon					
Beginning of life	-0.7	- 4.4	-1.1	-1.8	-3.5
MGCR					
Beginning of life	-4.5	Negative			
End of life		Slightly positive			
ML-1	+0.1				
UHTREX	-0.6	-8.0			

* Calculated by assuming a step increase in power with no additional heat transfer from fuel to moderator. Units are 10^{-5}/°C-Mw(t).

in the disadvantage factor of the fuel may, however, also be significant. Figure 7-2 [164, 165] shows the calculated components of the change in k_∞ due to voiding of the D_2O coolant from a fuel bundle in a pressure-tube reactor. The positive effect would,

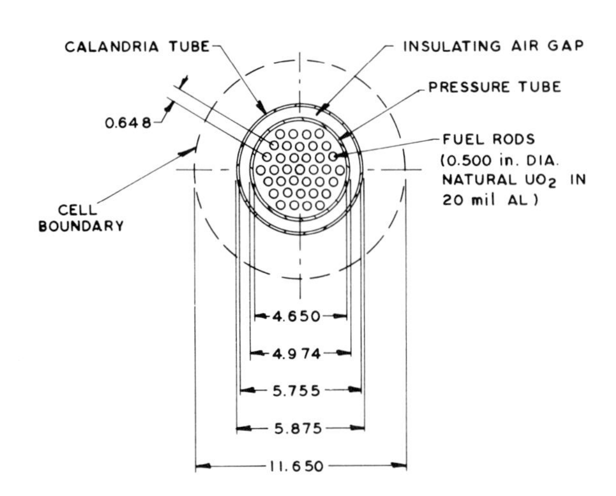

FIG. 7-2 Change in components of k_∞ upon removal of D_2O from a D_2O-cooled fuel bundle in a D_2O-moderated reactor. The calculated components of k_∞ for the bundle shown above are as follows:

Condition	η	ϵ	p	f	k_∞
D_2O in	1.3253	1.0145	0.8644	0.9071	1.0543
D_2O out	1.3253	1.0231	0.8760	0.9131	1.0851

The calculated change in k_∞ of 0.0308 compares with a measured change of 0.0261.

of course, be much larger if the coolant were H_2O in a similar case.

In some cases, coolant-pressure coefficients of reactivity may be measured at zero power and may be significant. For reactors employing coolants which are entirely in the liquid phase this coefficient is likely to be small, but it may be measurable and may give useful theoretical information. Some fast reactors which may entrain gas bubbles in the coolant stream have been observed to have pressure-dependent reactivity effects. Reactors which employ gaseous coolants with appreciable absorption cross sections may exhibit significant coolant-pressure coefficients and these may have direct practical significance, as for example in assessing the effects of loss of coolant pressure. Quite obviously, zero-power pressure coefficients will have no direct relation to the pressure coefficient at power for reactors which normally employ a change of phase in the coolant under power operation.

Needless to say, zero-power measurements can be quite useful in checking the effects of dissolved absorbers in the coolant and/or moderator for reactors which employ soluble poison for control or shutdown.

7.1.3 Significance of Zero-Power Coefficients

Although the zero-power reactivity coefficients are significant insofar as they represent measurable quantities which serve to check the results of calculations, and insofar as they indicate approximate values of coefficients which will apply under conditions of power operation, they may also have some practical significance in their own right. In particular, the integral of the zero-power temperature coefficient, from room temperature to the nominal temperature corresponding to normal power operation, specifies the excess reactivity which must be provided to get the reactor up to operating temperature. This integral is usually called the temperature defect, and that terminology will be used here, although there is some variation in practice. The sum of the temperature defect and the cold shutdown margin gives approximately the minimum shutdown margin available at operating temperature. For reactors using a common moderator-coolant the temperature defect is a straightforward concept, corresponding simply to the reactivity change when the moderator-coolant (and the fuel, since it is in contact with the moderator-coolant) is heated from room temperature to the normal operating temperature of the moderator-coolant. Thus it is the integral of the isothermal temperature coefficient over this temperature range. For reactors in which the moderator and coolant are segregated, a standard definition for the temperature defect has not evolved, and when such a number is quoted, the fuel, moderator, and coolant temperatures to which it is intended to apply should be stated specifically.

In a number of cases, the zero-power reactivity coefficients may give reasonably good approximations to the reactivity effects which may be observed in power operation. Thus for example, the reactivity change due to voiding of the coolant tubes in a pressurized-D_2O-cooled, pressure-tube, reactor may not be greatly different at zero power and operating power. Or, in a pressurized-water reactor, the initial reactivity change experienced in a "cold-water" accident at power may not be much different from that which would be estimated from the isothermal temperature coefficient. But it is important to recognize that the zero-power coefficients are not identical to those under power operation, and to make suitable allowances where they are appropriate.

7.2 The Power Coefficient of Reactivity

The power coefficient of reactivity is defined as the change in reactivity resulting from unit change in reactor power level. The immediate causes of the reactivity changes which accompany a power change are essentially the same as those entering the other reactivity coefficients: changes in temperature and density (or phase) of the coolant and/or moderator, changes in temperature of the fuel and fuel elements, and changes in reactor core configuration caused by the thermal expansion of the parts. These effects can be considered in terms of partial coefficients the same as, or analagous to, those discussed in Sec. 6. However, the effects are not imposed upon the reactor from outside, but originate within the reactor, as a result of the power change; to evaluate them they must be related back to the power change, and to evaluate their effect upon the reactor they must be "fed back" into the reactivity balance as indicated in Fig. 7-1.

Under the influence of reactor power generation a number of physical situations may occur within the reactor which could not be imposed by outside agencies. If, for example, low-conductivity fuel elements (e.g., UO_2) are used, their average temperature may be far above that of the moderator or coolant during power operation, and a Doppler coefficient which makes a relatively minor contribution to the isothermal temperature coefficient of reactivity may make a very important contribution to the power coefficient. Similarly, in a boiling reactor, only power generation is likely to produce the characteristic distribution of steam void within the reactor core. In general it may be said that space-dependencies are much more likely to be important in the power coefficient than in the coefficients related to externally-applied variations of operating conditions. Thus steep thermal gradients may be generated and maintained during power operation, and the power coefficient may have important components related to the mechanical distortion of the reactor structure (fuel rod bowing). Similarly, in H_2O-moderated reactors, it may be necessary to recognize a "coolant temperature" component of the power coefficient which is associated with the water immediately adjacent to fuel elements, and which does not apply to "moderator" water which may flow in control rod channels of other non-fueled spaces.

Two of the possible components of the power coefficient, the Doppler coefficient and the effects of fuel rod bowing, deserve special consideration and are treated at some length in a later portion of this section.

7.2.1 The Static Coefficient

There is of course a reciprocal relationship between the reactivity change produced by a given change in reactor power and the reactivity which must be added to the reactor to change its power by that given amount. Thus, if reactivity is added to a reactor which is initially operating in a steady equilibrium condition, to increase its power, and if the new power level is measured after the elapse of sufficient time for temperature equilibrium to be established throughout the entire reactor system (but after a time which is too short for a significant change in xenon concentration), a power coefficient of reactivity can be defined as the (negative) ratio of the applied reactivity change to the change in power level. This coefficient should be called the static power coefficient of reactivity, but it is often referred to loosely as the power coefficient. Quite obviously it is not simply a characteristic of the reactor proper, but is affected by the heat removal system, which may include certain control elements.

This situation can be illustrated by considering the pressurized-water reactor which is shown schematically in Fig. 7-3. Consider the case in which the reactor is initially in steady equilibrium operation and let the steam flow to the turbine be suddenly reduced by partially closing the throttle valve. If the reactor control rods are not moved the reactor will continue, momentarily, to operate at its initial steady power, but since only a fraction of this power can be disposed of by the throttled secondary system, the excess power will go into increasing the temperature of the water in the primary and secondary coolant systems, and the steam pressure at the turbine throttle valve will

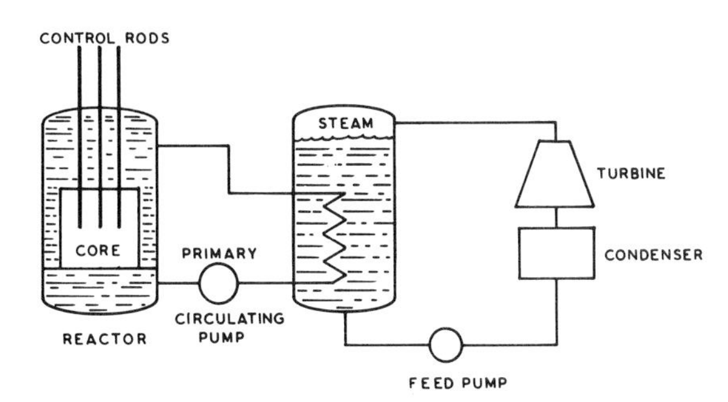

FIG. 7-3 Diagram of pressurized water reactor with elementary power conversion circuit.

rise. As the reactor coolant temperature increases, reactivity will be lost because of the negative coolant temperature coefficient of reactivity and the reactor power will decrease even though the control rods have not been moved. Eventually, the reactor will reach a new steady-state power level, lower than the original, and of such a value as to be just equal to the power carried off by the secondary system at some new, higher, steam pressure. For this case, which is a practically achievable one in many pressurized water systems, a substantial change in power level has been brought about without any change in the externally applied reactivity, and the static power coefficient of reactivity might be said to be zero. Nevertheless, substantial changes may have occurred in the components of the reactivity. In particular, the fuel temperature may have decreased substantially because of the power reduction, and a component of reactivity may have been gained from the decrease in the Doppler effect. This will have been compensated by the loss of reactivity due to the higher moderator-coolant temperature.

If one considers power level changes which are initiated in converse fashion, by movement of the control rods, it is clear that the ultimate effect on reactor power level must depend upon what is done with the turbine throttle valve and with the coolant circulating pumps: thus, the static power coefficient of reactivity can be defined only after the operating characteristics of the external circuits have been specified. In many cases the normal operating characteristics of these circuits have been fixed, by a suitable combination of component characteristics and operating procedures, to regulate the important reactor variable according to a rather precise pattern as power level is changed. In such cases a "normal" power coefficient of reactivity may be defined quite precisely, but it is not necessarily the same as the coefficient which would apply in the case of a system malfunction.

Although some reactors may respond automatically to load changes, without requireing movement of the control rods, over a limited range of power levels, there will usually be a rather definite amount of reactivity which must be added by the control rods to increase the power level of a reactor from zero to the maximum design power when the system is operated in its normal manner. This quantity may be called the power defect of reactivity, and may often be a more significant number than the static power coefficient. It defines the amount of excess reactivity which must be built into the reactor to provide for raising the power to the design level. It is equal to the amount of reactivity which must be subtracted to reduce the power of the steady-state, full-power reactor slowly to zero (hot) when there is no change in xenon concentration during the shutdown. It is not necessarily equal to the reactivity which must be subtracted to reduce the hot reactivity rapidly to zero, as in the case of a scram.

The static power coefficients of reactivity which have been quoted for a number of reactors are listed in Table 7-3.

7.2.2 Dynamic Coefficients

In general the dynamic power coefficients of reactivity are more significant than the static coefficient for safety considerations. Usually, also, the response of reactivity to a power change is a complex dynamic process which cannot be represented by a simple coefficient. Various simplified pictures of the response are used to approximate the reactivity behavior in certain limited areas which may be of particular significance.

Two types of dynamic situations are of particular significance. One of these is the response of the reactor to some large externally-applied perturbation of reactivity or reactivity-related operating variable (e.g., rod-withdrawal accident; cold-water accident). This type of situation is often called a nuclear excursion. The other area of particular interest is the response of the reactor to small fluctuations of reactivity or operating variables of the types that might be encountered in normal operation. This area of interest is concerned with the stability of the reactor—or the stability of the reactor system, which may or may not include the control system. In either type of situation the major simplification which can usually be made is the use of the space-independent equation for the kinetics of the neutron chain reaction. This means that the various physical changes which may occur in the reactor as a result of a power change can be fed back into the dynamic power equation simply by determining their effects on the over-all reactivity, ρ.

In excursion analyses further major simplifications can often be made by identifying certain specific physical situations which are of particular importance, because they represent either conceivable accidents or limiting cases of such accidents, and which embrace only a limited range of dynamic variables. In this way some of the power-dependent reactivity feedbacks may be identified as unimportant—because, for example, their response may be too slow—and the complexity of the situation may be reduced to the point that the dynamic problem may be solved directly.

A particularly useful simplification can be made for the understanding of power excursions which begin from very low power levels, for which the power may be assumed to be increasing on a steady exponential period by the time it reaches a significant level. For this case simple asymptotic solutions may be found for many of the important dynamic processes, such as heat conduction. Although these solutions will not apply once the reactor begins to "shut itself down," and the power

TABLE 7-3

Power Coefficients of Reactivity

(Units: $d\rho/dP$; P in Mw(t))

Type	Conditions	Coefficient, $[\text{Mw (t)}]^{-1}$
Boiling Water		
Dresden (1.5% enriched UO_2 core)	Full power	-7.0×10^{-5}
Pressurized Water		
Yankee (3.4% uniform enriched core)	Measured 80-392 Mw(t) Calculated	-0.33×10^{-4} -0.41×10^{-4}
Pressurized Heavy Water		
HWCTR, Savannah River (pressurized D_2O, natural U)	Moderator Fuel	-2.4×10^{-4} -0.21×10^{-4}
Sodium-Cooled, Graphite-Moderated		
Hallam	Full power	-0.77×10^{-5}
Fast Reactors		
EBR-II (49% enriched U core)	$0 \longrightarrow 22.5$ Mw $22.5 \longrightarrow 62.5$ Mw	-3.2×10^{-5} -6.0×10^{-5}

variation deviates from the exponential, simple calculations based on the asymptotic exponential model can often give considerable insight into the dynamic characteristics of the reactor. Reference [122] gives equations for the temperature distributions in plate-type fuel elements under conditions of exponential power variation.

In stability analyses the whole spectrum of dynamic responses must be taken into account, but simplifications can usually be made by assuming that, normally, large changes in operating conditions will be imposed only slowly upon the system. This allows one to investigate stability by studying the effects of small perturbations at a number of specific operating conditions which sample adequately the range of operating variables. The restriction to small perturbations usually makes it possible to express all the dynamic processes by linear differential equations, and to characterize each process by one or more specific "time constants." The responses represented by the linear equations can be appropriately combined, in the complex plane, by the methods familiar in electric circuit theory—most elegantly by use of the Laplace transform. Stability may be investigated by the Nyquist criterion or variations of it.

Alternatively, the "reactivity feedback" system may be set up in simulated form on an analog computer, along with the kinetic equations of the neutron chain reaction, and the system may be investigated "experimentally." In this case, linearization of the equations may not be necessary, but complete representation of the real physical situation may still be seriously limited by the essentially one-dimensional nature of the usual analog computer. So far as nonlinearities are concerned, those of the neutron kinetics equations are often the most serious: they are usually treated exactly in analog studies.

Figure 2-2 in the chapter on Water Reactor Kinetics, is a simplified diagram of the more important power-dependent reactivity feedbacks in a pressurized-water reactor: it takes into account the effect of the fuel temperature (T) and the average coolant temperature (θ) which is assumed to be the average of the inlet (θ_1) and outlet (θ_2) temperatures. Note that the fuel temperature is determined by two time constants, τ and τ', which are determined by thermal conduction in the fuel and the cladding, respectively (the effect of coolant temperature on fuel temperature is neglected). The exit coolant temperature is determined by the fuel temperature, the inlet temperature, and the inlet velocity, and hence involves time constant τ_c and τ_{01} as well as the fuel-temperature time constants. The time constant τ_{01} accounts for the transit time in the inlet pipe, while τ_c accounts for the core transit time. Note that the diagram is incomplete, in that the inlet and exit pipes are open-ended. To study the response of the complete system the outlet and inlet pipes would have to be connected through the primary pumps and heat exchanger, and the feedback of the complete secondary system through the heat exchanger would have to be included: obviously the complexity of the diagram and the number of time constants would be greatly increased.

The dynamic characteristics of specific reactor types are treated in much more detail in the chapters on Fast Reactor Kinetics, Water Reactor Kinetics, and Solid-Moderator Reactor Kinetics. Here, only a few further remarks are in order.

It must be evident from the preceding discussion that a dynamic power coefficient of reactivity can be stated only for the linearized system, and if so stated it will usually be made up of a number of components, each being defined by an amplitude and one or more time constants. The static power coefficient of reactivity may be contained in the dynamic coefficient, but the dynamic coefficient will not usually be derivable, even approximately, from observation of the static coefficient. The reactivity feedback which is important in large reactivity excursions may have little direct relation to the static coefficient and is not identical with the linearized dynamic coefficient. In some cases it may be approximated from the latter, but in other cases it may be quite different.

In considering reactor stability it is important to recognize that, in the neutron kinetics equation, the reactivity (ρ) determines the fractional rate of change of power. Hence the power coefficient of reactivity which determines the "gain" of the feedback system is the change in reactivity per unit fractional change in reactor power, the fraction being measured relative to the operating power at the moment in question. Thus, a power coefficient of reactivity which is constant with reactor power when measured in $\Delta\rho$ per megawatt will represent a feedback gain which increases linearly with the operating power level. This is why there is a power-level threshold for instability in reactors which do exhibit oscillatory instabilities.

Finally, it is worth noting that there are differences in the vocabularies used to describe different aspects of dynamic behavior. The most obvious of these has to do with the designation of "prompt" reactivity coefficients. The Doppler coefficient, for example, is often termed a "prompt" coefficient, meaning there is no perceptible delay in its action relative to the fuel temperature. This is an important consideration in excursion analysis, where melting or failure of the fuel may be the measure of severity of the excursion. In stability analysis the important consideration is the time constant relative to reactor power. The Doppler coefficient is by no means "prompt" in this sense. Relative to reactor power, the fuel temperature change (and hence the reactivity change due to the Doppler effect) is characterized by a time constant, which is a function of the heat capacity of the fuel, its thermal conductivity, and the heat transfer and transport characteristics of the coolant. For oxide fuel elements the time constant is typically as long as several seconds—not much shorter than the time constant of the "once through" coolant temperature. Indeed, in oxide-fueled reactors the time constant (relative to reactor power) of the "once through"* coolant temperature coefficient of reactivity is usually determined primarily by the thermal time constant of the fuel elements.

7.2.3 Measurements of Power Coefficients

Several methods have been used for measuring the static power coefficient of reactivity, or components related to the static coefficient. Measurements are usually made by utilizing some calibrated means of changing reactivity to compensate for the reactivity change which occurs as the power is changed from one equilibrium level to another. The calibrated compensation may be a calibrated control rod, a calibrated coolant temperature, or the like. The configuration of the reactor including over-all control rod positions, etc. is otherwise unaltered.

As has been pointed out before, the measured static coefficient will depend upon how the other operating variables are controlled as the power is increased.

Measurements at KAPL, [123] on the SIR, have employed a calibrated control rod, with the flow of sodium so adjusted as to maintain the same ΔT across the core. This increases the exit temperature with power level and results in changes in reactivity due to the increased reactor temperature and to the increased temperature difference between fuel and sodium. This over-all temperature increase effect has been found to be expressible as:

$$\delta k_T = \delta T_{Na}(C_T + C_{Xe}\,\$_{Xe}) \quad , \qquad (7\text{-}2)$$

where

δT_{Na} = change in average sodium temperature

C_T = isothermal temperature coefficient

C_{Xe} = the xenon thermal base effect per dollar xenon

$\$_{Xe}$ = reactivity value of xenon present in reactor

The remainder of the reactivity change, (δk_f), may be ascribed to the change in fuel temperature:

$$\delta k_f = \delta k_{measured} - \delta k_T \ . \qquad (7\text{-}3)$$

The fuel power coefficient is then $\delta k_f / \delta P$. The limitations on accuracy are due to: (a) the reactivity changes are small, (b) short-term xenon transients can cause serious errors, (c) system fluctuations, and (d) various reading errors. In addition, the fuel effect may be less than half of the total effect.

Measurements on the Yankee Reactor [124, 125] have utilized two techniques. One of these relies upon calibration of a single control rod to compensate for the change in reactivity as the power level is raised from one equilibrium value to another while a constant average primary coolant temperature is maintained. In the other technique, the average primary coolant temperature is varied to compensate for the change in reactivity as the power is raised from one equilibrium value to another. The first technique requires the calibration of one rod, while the second technique makes use of the isothermal temperature coefficient measurements for interpretation of the results.

Since both techniques involve increases in the ΔT across the core, certain complexities of interpretation arise. An increase in ΔT will affect the neutron migration area and the leakage differently in the lower and upper parts of the core. Since the core is axially asymmetric due to the presence of control rods and to coolant heating, this will affect the reactivity in a manner hard to estimate precisely. Measurements [125] indicate that a change from a four-pump to a three-pump case, with a correspondingly greater ΔT, increases the (negative) power coefficient by 15%. Measurements in which the flow is adjusted to maintain a constant ΔT across the core should yield results which might be used with the isothermal temperature coefficient to provide an insight into the fuel temperature co-

*i.e., the coolant time constant if the coolant inlet temperature is arbitrarily held constant.

efficient of reactivity. In some reactors it may be difficult to adjust the flow to meet this requirement.

In any of these techniques, the reactivity change caused by the power change may be masked partially by xenon poison effects. For ramp power increases, the best measurements are made from the xenon-free core condition, obtainable with a fresh core or after a long shutdown. Then the xenon poisoning will rise for a time at a relatively slow rate and correction for its effect can be easily applied. For decreases in power, the measurement should be started from a steady-state xenon condition. In general, practical considerations limit load increases to ramp increases, while decreases may be either of the step or ramp variety.

In order to measure the reactor thermal power, it is necessary to establish a baseline. This is often done by measuring the steady-state calorimetric data of the output heat, calculating losses in the primary system, and measuring pump input work. This total heat output may be calibrated against the electrical generator output. After repetition of these measurements at various powers, a curve can be plotted of electrical vs. thermal output and the electrical gross output can be used for subsequent measurements.

All measurements of the static power coefficient of reactivity which are made by purely static means depend upon the availability of some means of varying reactivity whose reactivity worth may be assumed to be independent of power level—that is, some means which can be calibrated at "zero" power and can be assumed to hold that calibration when the power is raised to the operating level. This is reasonable in some reactors, but not in all. In boiling water reactors, for example, there is no reactivity control agency whose worth would be expected to be independent of power level, and consequently no reliable measurement of the static power coefficient is possible by conventional means.

A means does exist for these difficult cases, at least for those reactors for which the space-independent neutron kinetics approximation is valid, but it has not been widely used. If a control rod is provided which can be oscillated with a period short relative to the shortest time constant in the dynamic power coefficient, then the rod can be calibrated, at any desired power level, by observing the response of the power to the rapid rod oscillation. Once the rod has been so calibrated, it can be used to produce the known (slow) reactivity changes needed for a measurement of the static power coefficient. If the frequency of oscillation of the rod can be varied, the rod can, of course, be used to obtain the frequency response, or transfer function, of the reactor. This transfer function implicitly contains the complete (low amplitude) dynamic power coefficient of reactivity. Analysis of the transfer function can yield empirical components of the dynamic coefficient, with their time constants, and in some cases these components may be identified with the physical processes occurring in the reactor. The general theory of reactor transfer functions is discussed in reference [135], and some of the many applications to boiling-water reactors are described in references [126] through [132], which have been reviewed in references [133] and [134].

For the experimental investigation of the dynamic behavior in reactivity excursions a number of experimental reactors have been subjected to large, intentional, reactivity increases. These experiments are discussed in the Chapters on Fast Transient Dynamics, and Water Reactor Kinetics.

7.3 The Doppler Coefficient*

The Doppler coefficient of reactivity, which has been discussed generally in Sec. 6.3, is not itself a composite coefficient, but it often is an important component of the power coefficient and one which deserves some discussion in that specific connection, since its evaluation involves a good deal more than straightforward neutron physics. There are difficulties in evaluating the magnitude of the effect even in idealized cases, and the difficulties are multiplied when one attempts to make a precise evaluation in a practical dynamic situation within the reactor.

7.3.1 Measurements of Doppler Coefficient

Experimentally, three different techniques have been used to measure ΔD_0, the Doppler coefficient in a 1/E neutron spectrum. These techniques are discussed in some detail in a review by R. M. Pearce [99]. The first involves activation experiments using cadmium-covered samples or threshold resonance foils in a 1/E spectrum. The second utilizes reactivity measurements made with small samples. The third technique involves reactivity measurements with entire reactor lattices.

The Doppler coefficient obtained in activation measurements using cadmium-covered foils necessarily includes the effects of activations due to the 1/v epicadmium neutron captures not in resonances. The coefficients measured by this method are, in general, slightly smaller than the true Doppler coefficient, since the Doppler coefficient is stated as a fractional change in that resonance integral [Eq. (6-19)] which does not include the temperature-independent 1/v component. A theoretical correction can be made for this effect. [136-139] Cadmium-covered foil measurements may be in error due to the shielding of resonances in the measuring foil by resonances in the cadmium cover. Generally, this effect is believed to be small.

Reactivity measurements made with heated samples in a reactor lattice are subject to a number of uncertainties. If the heated sample raises the temperature of the surrounding moderator it will distort the incident neutron spectrum, and hence change the effective value of η. The neutron energy spectrum within the lattice may not, in any case, approximate very well the 1/E distribution prescribed in the theoretical calculations, although this effect should be considered a deficiency of the

*This section treats mainly the Doppler coefficient of reactivity, as distinguished from the Doppler coefficient of the resonance integral from which it is derived and which is the main object of discussion in Sec. 6.3. In common usage, both are referred to as the Doppler coefficient.

theory rather than of the experiment if the experimental lattice used approximates the practical lattice under consideration. In some lattices the 1/E behavior has been shown to hold quite well. [140] Brown and Sampson have shown that at least in one case they obtained excellent agreement between the foil activation method and the heated sample reactivity method. In their technique using the reactivity method, they alternately oscillated a hot and cold tube of the desired material into a position in the reactor core. Replacement of the resonance absorber by an equivalent amount of boron in precisely the same geometry affords a method of making a 1/v correction. Baker and Jacques [141] have described a similar oscillator technique for measuring the Doppler effect in fast reactors. Utilizing the same general technique, Springer and Carpenter [142] have measured the Doppler effect in thorium, U^{238} and U^{235} in a fast neutron energy spectrum with median fission energy of 190 Mev. They have made measurements which they interpret as presumably due to Doppler broadening. They estimate Doppler coefficients of the following magnitude:

Material	(1/p) (dp/dT)
Thorium	$-5.55 \times 10^{-5}/°C$
Uranium-238	-1.10×10^{-4}
Uranium-235 (93.19%)	-5.58×10^{-6}

Doppler measurements on complete lattices or large segments of lattices also are subject to uncertainties because the method may perturb the neutron spectrum or the flux density distribution. The Doppler effect in full lattice measurements is usually partially masked by moderator temperature effects and by core expansion effects. Chernick reviewed the early complete lattice measurements in 1953. [143]

7.3.2 Theoretical Application to Practical Cases

In predicting the effect of the Doppler coefficient for practical reactor cases, a number of complications arise. For example, if one postulates a reactivity excursion in a specific reactor and desires to evaluate the effect of the Doppler coefficient in compensating the applied reactivity, one would ideally proceed through the following five steps: (1) calculate the radial power distribution in the rod at the time of the power transient; (2) calculate the normal radial temperature distribution in the rod; (3) calculate the change in temperature in the rod from the normal power operation to the transient condition as a function of radius and of time; (4) use this change in temperature to calculate, either by averaging or as a function of depth, the change in I_{eff}, and from this the value of the Doppler shutdown reactivity effect in the rod as a function of time; and (5) repeat the first four steps for rods in the various core regions and compute an appropriate weighted averaging of the reactivity effect to be applied to the entire core.

The first step requires that the neutron flux distribution within the rod be known as a function of neutron energy over the range in which the fission cross section is important. The power distribution within the rod may then be calculated. The value of the fission cross section, and the distribution of fissile isotope within the rod, will vary with the life of the core, the initial fissile isotope (e.g., U^{235}) being depleted preferentially near the surface of the rod and bred isotope (e.g., Pu^{239}) being formed preferentially near the surface. Usually it is assumed that the power distribution is uniform across the rod, but the adequacy of this assumption depends upon both the distribution of fissile isotope and the neutron flux depression in the rod.

Insofar as the power distribution and the thermal conductivity in the fuel rod are uniform, the temperature distribution in the rod will be parabolic under steady state conditions (step 2). One method of correcting the Doppler coefficient for a parabolic temperature distribution is outlined by Greebler and Goldman. [144] In many cases there may be rather large uncertainties in both the magnitude and the distribution of the normal fuel temperature, not only because of uncertainties in the distributions of neutron flux and fissile isotope but also because of uncertainties in the thermal characteristics. This is true especially for unbonded clad fuels with poorly measured heat transfer coefficients that may be changed by radiation and other effects. It might be, for example, that the center temperature of the hottest fuel rod in an oxide-fueled reactor, which could be 4000°F, would have a probable error of 500°F or more.

The third step requires an estimate of the change in temperature in the rod as a function of radius and time during the transient (Fig. 7-4). If the transient is of the step type, it may be reasonable to assume that no heat is conducted outward through the clad, and the temperature change as a function of radius may be obtained simply by calculating the total heat generated in the step transient as a function of radius and converting this by an appropriate specific heat value into temperature changes. Obviously, a rough estimate can be made by averaging the heat generated over the rod. If the transient is a ramp function, significant heat transfer may occur during the transient and the problem of the temperature change as a function of both radius and time becomes quite complex. Reference [146] discusses the problem of calculating the transient temperature distribution in a fuel element of the UO_2 rod type.

The fourth step requires a detailed knowledge of the Doppler broadening and resonance absorption if any use is to be made of the temperature distribution. Unless the uniform temperature assumption can be made, the gross Doppler coefficient as defined in Eq. (6-19) is not definitive. As the power rises during the transient, the temperature change in the fuel rod is not uniform and therefore the measuring of dI_e/dT in its usual sense is no longer unambiguous. One solution would be to define a Doppler coefficient in terms of a fuel increment at radius r and with temperature T by

$$\Delta D_o(r,T) = \frac{1}{I_e} \frac{dI_e(r,T)}{dT}. \tag{7-4}$$

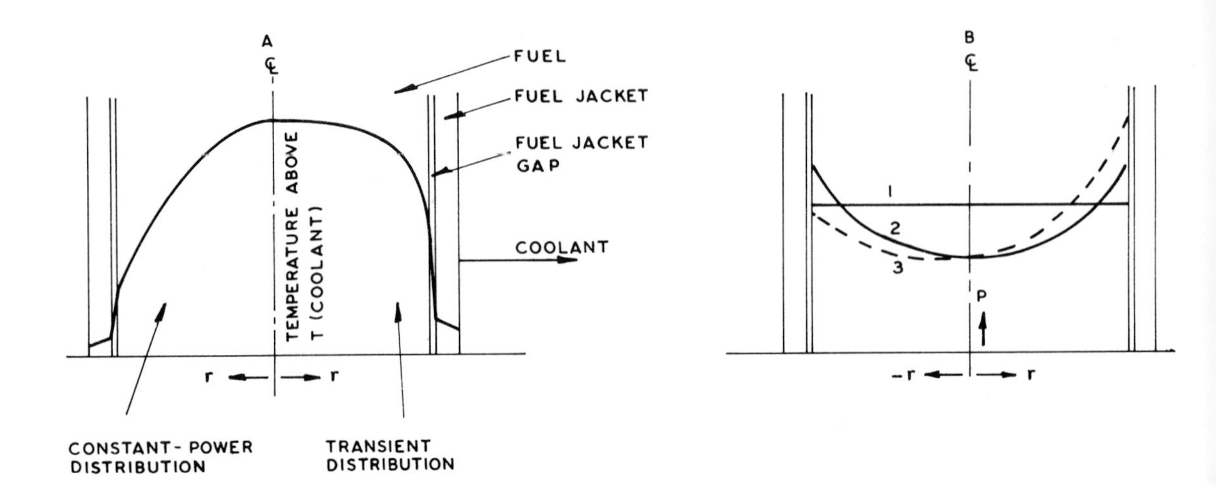

FIG. 7-4 Factors affecting temperature distribution in a fuel rod. In A the approximate temperature distributions shown are plotted to arbitrary scales. The shape of the transient distribution will depend on the rate of change of power, becoming flatter as the rate becomes faster. These curves are based on the assumptions that the power distribution and the thermal conductivity are uniform. Variations in these quantities can be taken into account if they are known. See curves B.

In B, the actual power density (P) will seldom be uniform (1), but will usually be highest near the surface (2) and may be skewed (3).

This would not be expected to be a simple function since the neutron spectrum incident in each successively deeper radial increment deviates more and more from the 1/E spectrum assumed in Eq. (6-19); in fact I_e (r,T) would no doubt be a unique function of the temperature distribution, T(r). Nevertheless, if the function I_e (r,T) could be derived for a particular fuel element in a particular transient, an effective Doppler coefficient, $\Delta D_o(T_x)$ [or ΔD_o (E)] could be specified which would relate the effective resonance integral to either some crucial temperature T_x (e.g., the fuel centerline temperature) or to the energy production, E, in the usual way. It should be emphasized that the entire procedure would have to be redone for various rates of heat loss from the fuel (hence, for various radial temperature or energy profiles in the rods). The effects of nonuniform temperature on resonance absorption are discussed in references [166] and [167].

Step five extends the calculation from the single fuel rod to the entire reactor core. Other rods may develop lower or higher temperatures during the same reactivity transient because of the variations of neutron flux and fuel composition throughout the reactor. An appropriate average value must be found. The temptation is to analyze a rod which accepts the average energy increment during the transient and assume that it is representative. However, this result is likely to be far from correct. The largest flux rise and, hence, temperature rise, normally occurs at or near the center of the core. Thus the largest Doppler shutdown effect occurs there, where the reactivity "importance" is often highest. Another possible perturbing effect is the deviation of the flux distribution in the transient from that observed in normal operation. Only if the transient is such that the physical system is not unduly disturbed by the transient does the flux distribution approximate that of the system in normal operation. A large perturbation of the control rod positions (a possible cause of the transient) or the creation of voids, etc., may cause rather large differences in the flux distribution.

Often it is necessary to divide the entire core into a number of regions, both radially and axially, and to calculate by steps one through four the Doppler coefficient and total Doppler reactivity effect for a representative section of each region. These regional reactivity effects may then be given appropriate relative weights and a weighted average may be taken to give the over-all Doppler reactivity effect.

Care must be taken to insure that this weighting is properly carried out. For instance, the regional values of ΔD_o obtained by experimental reactivity effect measurements within an actual core are already weighted by virtue of the location of the experiment. Such measurements should not be weighted again. Other calculated values based on the measured value should be given weights relative to the measured value in direct relation to their relative core positions. Lacking better information, weighting is often done in proportion to the square of the relative flux or power density. [145]

Alternatively, the over-all effect may be calculated directly by making a multi-region "criticality" calculation in which the various regions have the appropriate reactivities as determined by their individual Doppler behaviors.

Computer codes have been developed which are helpful in practical Doppler calculations. Greebler and Goldman [144] have developed a code, SPARTA, which can be used to calculate spatially-averaged Doppler coefficients and spatially-averaged reactivity changes. The code considers i radial, j axial regions and m annular subregions, and an average temperature $T_{i,j,m}$ is computed for each subregion. The spatially corrected Doppler coefficient is calculated from the subregion temperatures which result from a transient starting from the initial core temperature distribution.

There also exist a number of Monte Carlo codes which calculate Doppler effects or resonance integral effects in varying degrees of approximation. One such code (REP) is that of Richtmyer, Van Norton and Wolfe [147] This code has been modified and used at Westinghouse in calculations of Doppler effects [148]. It gives a radial distribution of capture events within a fuel rod, for a uniform temperature distribution within the rod. Another Monte Carlo code has been devised by Morton [149] for the same general purpose.

7.3.3 Experimental Doppler-Limited Power Excursions

Up to the present time there has been only one set of tests which has actually demonstrated the Doppler effect as a shutdown mechanism. This set was carried out under the AEC-SPERT Program in Idaho Falls, Idaho, during the summer of 1961 [150]. The experimental core, moderated and reflected by H_2O, consisted of a lattice of rods of the following characteristics:

Total length of fuel rod	71.5 in. (1.816 m)
Active length of rod	69.9 in. (1.775 m)
OD of fuel rod	0.500 in. (1.27 cm)
Clad thickness (304-S.S)	0.028 in. (0.071 cm)
Fuel	Compressed UO_2 powder
Fuel enrichment	4.02 wt.% U^{235}
Effective density (87% theoretical)	9.45 g/cm^3
Mass UO_2 per fuel rod	1600 g
Mass U^{235} per fuel rod	56.7 g
Mass U^{238} per fuel rod	1353 g
Number of rods	592
Rod spacing (square)	0.663 in. (1.684 cm)

The uranium enrichment of these rods is comparable to that currently being employed in pressurized and boiling water reactors. The rod diameter, length, clad material and thickness, and number of rods, are in the region of interest. The only real variation from current power reactor practice is the use of powdered UO_2 rather than sintered oxide pellets of somewhat higher density. This has an unknown effect on heat conduction within the powder and to the clad. Since the rod transients started from a relatively low power, the initial temperature within the rod was almost uniform. The flux depression within the elements was probably resonably small [154], and the assumption of a flat initial and final temperature within all fuel elements is probably a reasonable approximation.

The general form of the observed transient for an applied reactivity corresponding to a 9.7 msec period* is shown in Fig. 7-5. [153] The shutdown mechanism effective here appears to be almost totally due to the Doppler effect. The maximum peak power reached in this case was 669 Mw and the energy generation to peak power was 13.51 Mw-sec. It may be assumed that practically all of this energy, E_m, is retained in the UO_2 up to the time of peak power (a good assumption as is indicated by other tests—see Sec. 7.4). At peak power the reactivity $R_c(t_m)$ compensated by the Doppler effect is very nearly equal to the applied prompt reactivity for these short periods. This provides a simple relationship between the average energy developed in the fuel up to the time of peak power and the total shutdown worth of the Doppler effect (plus any other prompt effects) up to that same time. An analysis of the data suggests the empirical relation:

$$R_c(t_m) = -6.2\, E_m^{0.74} \qquad (7\text{-}5)$$

for reciprocal periods α_0 from $150 \lesssim \alpha_0 \lesssim 320$ sec^{-1}, where $R_c(t_m)$ is expressed in cents and E_m is in megawatt-seconds.

By differentiation, the average Doppler energy coefficient of reactivity is

$$\frac{dR_c(t_m)}{dE_m} = -4.6\, E_m^{-0.26} \quad ¢/\text{Mw-sec}\,. \qquad (7\text{-}6)$$

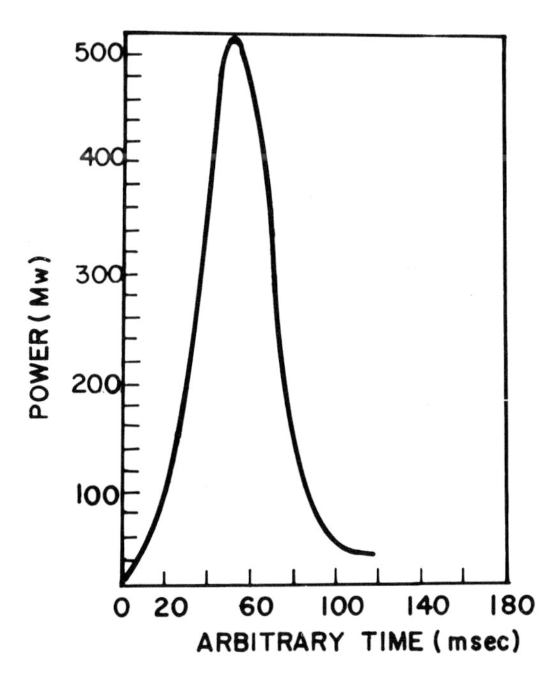

FIG. 7-5 Power record of a reactivity excursion terminated by the Doppler effect in a UO_2-fueled, water-moderated reactor. The initial exponential period was 10.2 msec and the core was constrained against fuel-rod bowing (compare Fig. 7-7).

*This transient is for the restrained core, in which rod bowing had been reduced to an unimportant level. See Sec. 7.4 for other results of the same set of experiments, in which bowing occurred.

7.3.4 Significance of Doppler Effect

In summary, a negative Doppler effect has unusual importance in safety considerations because it is an inherent reactivity limiting effect, prompt with respect to fuel temperature, which, once established for a particular reactor, apparently cannot be vitiated by any conceivable set of circumstances. Further comfort can also be obtained from the evidence of the SPERT I oxide core results [109] which seem to show that the Doppler coefficient varies as $T^{-1/2}$, and therefore indicate that the Doppler effect will remain quite important even at elevated temperatures. The preceding discussion has pointed up some of the difficulties in assessing the practical magnitude of the effect, primarily to point out the need for further investigation. In current practice the evaluation of the Doppler effect often leaves much to be desired. In particular, the first three steps of the evaluation procedure discussed earlier are largely glossed over. The calculations often consider the initial core temperature distribution to be constant, as it would be only in the case of a shut down reactor (which might still be hot). In the same way, the transient energy is considered to be deposited uniformly in the rod region in question. Doppler coefficients calculated in this way are called "isothermal." The temperature dependence of the Doppler effect is often considered only in a very gross way. These gross approximations are often rationalized by citing the rather large uncertainties which would remain even if a careful calculation were carried through. It appears that fundamental work to reduce these uncertainties would be well worth the effort.

7.4 Fuel Rod Bowing

The distortion of fuel elements as a result of thermal-expansion gradients represents a possible source of reactivity changes which may contribute to the power coefficient of reactivity. Such contributions have been observed in several cases for rod-type fuel elements, and in these cases they have had important effects upon the dynamic behavior of the reactor. The bowing of fuel rods adjacent to control rod channels may also interfere with control rod motion, and any bowing which occurs is apt to cause a deviation from the optimum coolant-flow geometry originally designed into the reactor. For these reasons, and because the effect is one which requires close attention from both the reactor physicist and the mechanical designer, it is discussed here in some detail, even though other contributors to the power coefficient are important in a broader range of cases.

If a fuel rod within a reactor core is located in a gradient of the neutron flux, the side of the rod in the higher flux will heat up and expand more than the other side. Consequently, the rod will bow with the convex curvature being in the higher neutron flux. This phenomenon is called rod bowing. If the bowing is general throughout the core it may produce either a shrinking of the effective core diameter, with a concurrent increase in the average fuel density in the core (Fig. 7-6a) or an expansion of the effective core diameter with a concurrent decrease in the average fuel density, depending upon the mode of support of the fuel rods. If the radial power distribution in the core is complex, the bowing may occur in different directions in different core regions (Fig. 7-6b). If the fuel rods are clustered in bundles which themselves are restrained from bowing (e.g., pressure tube reactor) the individual rods may still be subject to a bowing which may either increase or decrease their average spacing within the bundle.

These bowing effects may either increase or decrease reactivity, depending upon the particular reactor design, but often the following considerations will apply:

1. In unmoderated reactors an increase of average fuel density within the core, even at the expense of a decrease in core radius, will increase reactivity.

2. In unmoderated reactors the local clustering of fuel, due to bowing which causes certain groups of elements to come closer together, will usually increase reactivity.

3. In moderated reactors an increase in average fuel density, at the expense of core diameter, may either increase or decrease reactivity, depending upon the initial moderator/fuel ratio and the core size (cf. Fig. 6-9).

If the fuel rods in the reactor core are freely supported at both top and bottom, the general tendency is for the fuel to move toward the region of higher flux, as is shown in Fig. 7-6a. The bowed position is indicated by the dotted line and the flux shape is sketched in. Note that the flux is not always highest in the reactor center. Figure 7-6b shows a case in which the flux gradient is highest at the outer edge. This situation may occur, for instance, in a water-moderated reactor reflected by a pure water reflector and, in particular, represents an actual system which was subjected to an experimental reactivity excursion at the SPERT Facility. [150] In this case, the large number of rods near the edge, their relatively high reactivity worth, the steep outer gradient, and a long unsupported length combined to lead to a very large reactivity increase and a larger transient, in fact, than that caused by the original transient control rod ejections.

If the fuel element is supported rigidly at one end of the core and is free at the other, the same bowing which in Fig. 7-6a led to an increased fuel-to-moderator ratio at the core center can result instead in a decreased fuel-to-moderator ratio concentrated near the bottom, as shown in Fig. 7-6c. Since the control rods usually uncover selectively either the top or bottom of the core, it is clear that the reactivity effect due to bowing of this type may be purposely made to have a large or small effect depending on the relationship between the rigid support plate location, the control rod entry location, and the degree of control rod insertion at the time in question. Thus, in Fig. 7-6c, if control rods enter from the top the bowing reactivity effect is maximized.

One other example is worth mentioning. Figure 7-6d shows rods supported at the center. Such rods will bow as shown. Note that Figs. 7-6a, c, and d show the same flux distribution. Thus, in principle, bowing can be made to increase or decrease the fuel-to-moderator ratio (or the average fuel density in the core) at the discretion of the reactor de-

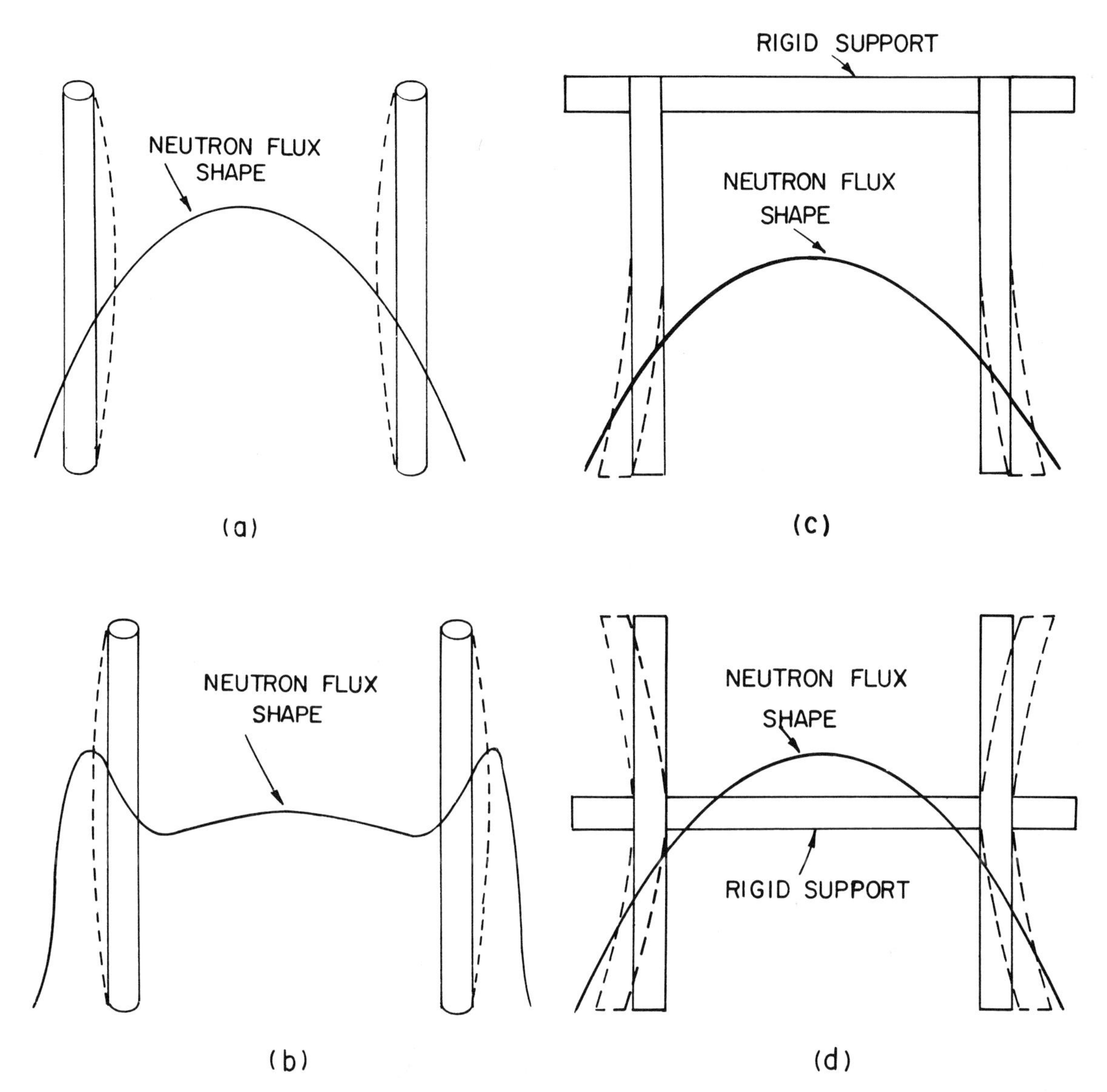

FIG. 7-6 Fuel-rod bowing by neutron flux gradients.

signer. The British [151] have considered using the bowing effect as a means of providing a relatively fast negative power coefficient of reactivity for some of their new fast reactor designs where the Doppler coefficient is small and may even be positive.

In general, the bowing effect is most pronounced in liquid-moderated (or unmoderated) reactors which are small radially and large axially, which have steep flux gradients, and which have separately supported rods. Bowing would be expected to have little effect on solid moderated gas-cooled reactors since fuel bowing will not affect the fuel-to-moderator ratio. Bowing could conceivably have considerable effect in pressure-tube reactors in which coolant and moderator are segregated, and is known to have occurred in a sodium-graphite reactor (see Sec. 7.4.1).

7.4.1 Experimental Observations

The effects of rod bowing are usually observable experimentally only in dynamic tests: these may involve either oscillation techniques or transients. In changes of reactor conditions such as flow, power, temperature, etc. from one steady state to another, bowing effects will tend to be masked or lumped with other effects. This masking was clearly demonstrated in the SRE during tests on Core II [152]. In this reactor the isothermal temperature coefficient was predicted to be slightly positive. No anomalous effects were found during low-power steady-state tests. When the power was increased the reactor proved hard to control manually. Variations in reactivity with coolant flow at low powers began to give indications that mechanical changes in the rod clusters were occurring. In several cases the first clues to the existence of bowing instabilities have come through reactivity effects caused by flow variations. Thus, a change in average core coolant temperature may raise fuel temperatures and cause bowing effects, which will be lumped with the ΔT effect if the measurement is limited to steady-state conditions.

During the initial set of transients carried out on the core composed of UO_2 fuel rods in SPERT I, [150] two power peaks were observed in each transient, as shown in Fig. 7-7. [153] During these tests the five-foot-long fuel rods were supported only at the ends. The first peak is the original power rise caused by the sudden withdrawal of the transient control rod. The shutoff mechanism which is effective in terminating the transient is the Doppler effect. However, about 130 milliseconds after the first peak, a second power rise occurred. This power rise was due to rod bowing. That this is true has been dramatically illustrated by placing an egg-crate structure within the core in such a way as to eliminate most of the bowing effects. One of the resulting power traces is the one shown in Fig. 7-5. Both of these transients occurred in the same core and under essentially identical conditions. The only difference was the degree of constraint of the fuel rods. In the case of the unrestrained rods, the second shutdown is finally achieved through a combination of moderator heating, boiling within the core, and additional Doppler effects. Note that both the bowing transient rise and the final shutoff mechanisms are relatively slow compared to the initial shutdown.

Perhaps the best known example of bowing is that afforded by the EBR-I core. Early tests showed that there existed a prompt positive power coefficient of reactivity obserable whenever coolant flow was varied. These observations led in May of 1955 and again in November of 1955 to a series of transfer function measurements which showed that the reactor could be brought into a resonant condition at certain reactivity oscillation frequencies. A typical curve of reactor power response is shown in Fig. 7-8 [155] Both a damped oscillatory behavior is shown and also a divergent oscillation which was terminated by scram. It should be noted that the steady-state power coefficient was negative, and reactivity had to be added to cause a power rise at constant flow. The time-dependent behavior indicated a prompt positive effect and a much larger and more slowly acting negative component. Analyses have indicated that the positive coefficient was due to fuel rod bowing of the type discussed above, complicated by bowing of the grid plates which positioned the rods at their upper ends. The slower negative coefficient is attributed to radial expansion of the grid plates, possibly enhanced by a further bowing effect which reverses the initial positive effect because of changes in the points of restraint [156-159] Experiments with the EBR-I Mark-III core have clearly demonstrated the prompt positive effect can be eliminated by eliminating fuel-rod bowing. [155, 157]

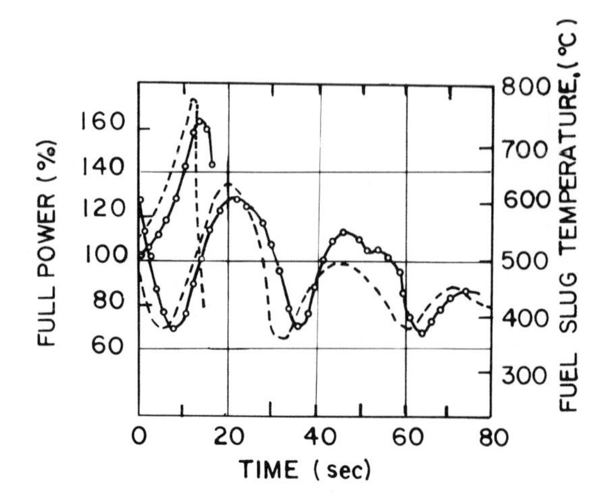

FIG. 7-8 Power (broken line) and fuel temperature records (solid line, encircled points) from EBR-I (Mark-II core) in oscillatory operation. The oscillatory behavior could be initiated by sudden changes in coolant flow under certain operating conditions. They were controllable by manual adjustment of the control elements.

The actual tests on EBR-I terminated with the partial meltdown of the core during an experiment in which the reactor was placed on a short period. In retrospect, it would appear that the meltdown was due more to the unusual procedures required to obtain experimental information about the instability rather than to any insidious behavior of the instability which could have been encountered under normal operating circumstances.

The bowing of the SRE [152] Core II was attributed to fuel cluster bowing, with individual rods bowing either toward or away from the cluster center. In a sense, this is bowing due to microscopic variations in the flux distribution (cf. Fig. 5-6) rather than to the gross flux distribution over the core. This reactor also exhibited sustained oscillations. The fact that these effects were due to bowing was clearly demonstrated by changing the configuration so that no rod movement within the cluster could occur.

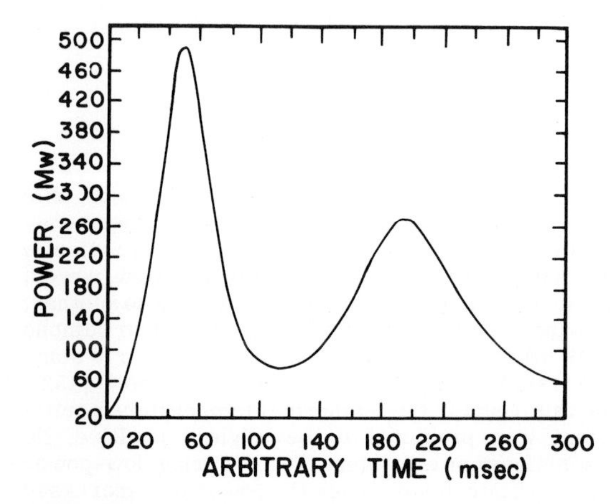

FIG. 7-7 Power record of a reactivity excursion initially checked by the Doppler effect and re-initiated by fuel-rod bowing. The initial exponential period was 10.1 msec in the "unconstrained" core (compare Fig. 7-5).

7.4.2 Calculational Methods

The bowing phenomenon can be treated analytically, taking into account inertia effects as well as a slight variation of the neutron and temperature field in the direction of the axis of the rod. A derivation outlined by W. K. Ergen shows how this may be done. In order to reduce the complexity of the analysis, it will be assumed that the rod axis, though not necessarily perfectly straight to begin with in its initial, isothermal condition, is all contained within one plane, and that the temperature field is symmetrical with respect to this plane. The ends of the rod are assumed to be restrained so that they cannot move perpendicularly to the rod axis, but at least one end is free to move in the axial direction (allowing for rod expansion), and the ends can turn with the bowing of the rod (see Fig. 7-9). This assumption regarding the rod ends could readily be replaced by other assumptions. The deviation of the rod axis from a straight line is considered to be small. The rod cross section does not vary in size, shape, or composition in the axial direction. Further assumptions will be found in the footnotes, as they are introduced.

Two coordinate systems (Fig. 7-9) are used in these considerations. Both have their origin at the lower end of the rod axis. The z-axis of the first coordinate system follows the equilibrium shape of the rod axis. This is a Cartesian system, except that the z-axis is not necessarily perfectly straight.* The xz-plane, i.e., the plane of the paper, is the plane of symmetry mentioned above. In view of the assumed symmetry of the temperature field, the rod axis remains at all times in the xz-plane and can be described by a function $s(z,t)$ the determination of which is the purpose of the analysis.

The z-axis of the second coordinate system follows the time-variable shape of the rod axis. This coordinate system is (except for the slight curvature of the z-axis) a cylindrical system. The azimuth ϕ is taken to be zero in the plane of symmetry. This coordinate system will be used to describe the temperature field $T(r, \phi, z, t)$ as a function of the radial distance r, and of ϕ, z, and t.** (More specifically, T is the deviation from the equilibrium temperature.)

The analysis follows largely Ref. [160] although Ref. [161] gives more accurate but more involved methods. The method is carried out by cutting the rod by a plane $z = z_0$ (z_0 is an arbitrary constant) and by determining the shape $x(z, t)$ of the rod from the dynamic equilibrium of the following three moments:

1. The resisting moment results from the curvature of the rod and the compression stresses on the inside of the bend together with the tension stresses

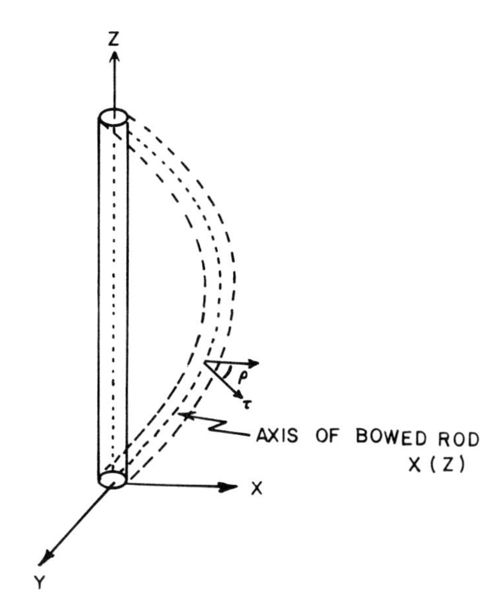

FIG. 7-9 Arrangement of coordinates for analysis of fuel-rod bowing. Solid lines indicate original shape of rod. Broken lines indicate bowed shape.

on the outside of the bend. This moment is (reference [160], pp. 256-261):

$$M_r = EI(d^2x/dz^2) , \tag{7-7}$$

where E is the modulus of elasticity, and I the moment of inertia of the rod cross section.

2. The thermal-stress moment results from the differential thermal expansion of the longitudinal "fibers" of the rod. Each fiber tends to undergo the fractional expansion αT (α is the coefficient of thermal expansion), and the fiber exerts the longitudinal stress $E\alpha T$.*** The distance of a given fiber from the yz-plane is $r\cos\phi$. The product of the stress and the distance, integrated over the cross section of the rod, gives the thermal-stress moment:

$$M_t = E\alpha \int dr \cdot r^2 \int d\phi \cos\phi\, T(r,\phi) . \tag{7-8}$$

3. The moment of the inertia forces results from the inertial resistance of the rod to the transverse accelerations to which its elements are subjected when moving from one shape to the other. The magnitude of these inertial resistance forces, per unit length of the rod, is $-m\ddot{x}$, where m is the mass of the rod per unit length. The forces act like a distributed load on the "beam," represented by the rod. Following the conventional procedure for computing the moment caused by this distributed load, one first computes the reaction force R_1 at the end $z = 0$ (reference [160] p. 154). R_1 is determined by the condition that its moment with respect to the other end, $z = L$ (L = length of the rod), shall be equal and

*In order to simplify the following presentation, the equilibrium shape of the rod axis will be taken to be straight from here on. However, with the above definition of the coordinate system, the results apply to slightly curved equilibrium shapes as well.

**Because the rod axis never deviates much from a straight line, its deflection must be small at all times, and it is reasonable to assume that this deflection does not influence the temperature field T.

***The assumption is made that E and α, which are temperature-dependent, can be replaced by temperature-independent average values.

opposite to the moment of the distributed load with respect to the end z = L:

$$R_1 L = \int_0^L dz\, m\ddot{x}(z) \cdot (L - z). \qquad (7\text{-}9)$$

According to reference [160], p. 158 ff, one must compute the moment, with respect to z, of all forces on one side of z:

$$M_i = -\left[R_1 z - \int_0^z dz'\, m\ddot{x}(z') \cdot (z - z') \right]$$

$$= -m \left[(z/L) \int_0^L dz'\, \ddot{x}(z') \cdot (L - z') - \int_0^z dz' \ddot{x}(z')(z - z') \right]. \qquad (7\text{-}10)$$

M_r, M_t, and M_i are functions of z and the equation of dynamic equilibrium is:

$$M_r + M_t + M_i = 0 \qquad (7\text{-}11)$$

for all z between 0 and L.*

For $M_t = 0$, that is without the forcing term due to the temperature gradient, the equations describe the free oscillation of a rod supported at the ends in the manner discussed above:

$$EI \frac{d^2x}{dz^2} - m \left[\frac{z}{L} \int_0^L dz'\, \ddot{x}(z') \cdot (L - z') - \int_0^z dz'\, \ddot{x}(z') \cdot (z - z') \right] = 0. \qquad (7\text{-}12)$$

This homogeneous equation admits solutions of the form:

$$x = \sin \alpha_m z \sin \omega_n (t - t'). \qquad (7\text{-}13)$$

*That the signs of the moments are given correctly in Eqs. (7-7), (7-8), and (7-10) may be verified by taking specific examples: (a) Let the condition depicted in Fig. 7-9 be a static one, $\ddot{x} = 0$, $M_i = 0$. M_r is negative, because d^2x/dz^2 is negative. Fig. 7-9 is a static condition if T is greater on the right side ($\cos \phi > 0$) than on the left side ($\cos \phi < 0$). Hence, M_t is positive, as required for $M_r + M_t = 0$. (b) Let the temperature field be as in example (a), so that $M_t > 0$. Let the rod start out from its equilibrium position, $M_r = 0$. Then M_i has to be negative to fulfill the equilibrium condition, Eq. (7-11). Since the rod accelerates in the positive $\ddot{x}$ direction, $\ddot{x} > 0$ and the contribution of the reaction force to M_i is negative. This contribution is the dominant one in Eq. (7-10), as can be shown by dividing the $\int_0^L$ into $\int_0^z$ and $\int_z^L$. The $\int_0^z$ is, by itself, greater than the contribution of the distributed forces, and the $\int_z^L$ is positive.

The terms with $\cos \alpha_n z$ are excluded by the boundary conditions at z = 0 and z = L, and these boundary conditions also demand that:

$$\alpha_n = n\pi/L \quad (n = 1,2,3 \ldots). \qquad (7\text{-}14)$$

Terms with $\cos \omega_n(t - t')$ are eliminated by suitable choice of t′, and substitution of Eq. (7-13) into Eq. (7-12) and use of Eq. (7-14) yields through straightforward algebra,

$$\omega_n = \sqrt{\frac{EI}{m}} \frac{n^2\pi^2}{L^2}. \qquad (7\text{-}15)$$

For the stationary case, $\ddot{x} = 0$, $M_i = 0$, we obtain from Eqs. (7-7) and (7-8):

$$x(z) = -\frac{\alpha}{I} \left[\int_0^z dz' \int_0^{z'} dz'' \int dr \cdot r^2 \int d\phi \cos\phi\, T(z'',r,\phi) - \frac{z}{L} \int_0^L dz' \int_0^{z'} dz'' \int dr \cdot r^2 \int d\phi \cos\phi\, T(z'',r,\phi) \right]. \qquad (7\text{-}16)$$

Here, the boundary condition at z = 0 determines that there is no constant term, and the boundary condition at z = L determines the coefficient of z in the second term in the brackets. This case has been treated by C. G. Johnson [162] and H. Chelemer. [163]

In the general case, in which there is a temperature gradient as well as appreciable inertia, it is advantageous to perform** a Fourier analysis of the forcing term with respect to z:

$$\int dr \cdot r^2 \int d\phi \cos\phi\, T(z,r,\phi,t) = \sum_{n=1}^{\infty} a_n \sin(n\pi z/L), \qquad (7\text{-}17)$$

where the Fourier coefficients are given by:

$$a_n = \frac{2}{L} \int_0^L dz \sin\left(\frac{n\pi z}{L}\right) \int dr \cdot r^2 \int d\phi \cos\phi\, T(z,r,\phi,t) \qquad (7\text{-}18)$$

The a_n are, in general, functions of the time t. Equation (7-11) then reads:

**It is assumed that the neutron flux, and hence the heat generation, is zero at the ends of the fuel rod and that there are consequently no terms $\cos(n\pi z/L)$ in the expansion.

$$EI \frac{d^2x}{dz^2} - m\left[\frac{z}{L}\int_0^L dz'\,\ddot{x}(z')(L-z') - \int_0^z dz'\,\ddot{x}(z')(z-z')\right] = -E\alpha \sum_{n=0}^{\infty} a_n \sin\left(\frac{n\pi z}{L}\right) \qquad (7\text{-}19)$$

a particular solution of which is:

$$x(z,t) = \alpha\sqrt{\frac{E}{mI}} \sum_{n=1}^{\infty} \sin\left(\frac{n\pi z}{L}\right)\int_{-\infty}^{t} dt'\,a_n(t')\sin\left[\omega_n(t-t')\right]. \qquad (7\text{-}20)$$

In order to fulfill specific initial conditions (boundary conditions in time), solutions of the form of Eq. (7-13) may have to be added.

It is necessary to know, or to estimate analytically, the temperature distribution within the rod as a function of the time and space variables. Generally, this will be a complicated problem. It is usually acceptable to assume that only the macroscopic flux gradient effects are acting. Actually, the microscopic effects may also enter—depending on fuel spacing and arrangement. Note that control rod positions and control rod channels can play an important role. For more complicated geometries and arrays of rods fastened together, a more complex analysis than that given above may be required and further approximations may be necessary.

8 EFFECTS OF CORE EXPOSURE

When a reactor is operated at a significant power level the composition of the core will change gradually as a result of the neutron-induced nuclear reactions. The important components of the change will be the buildup of fission products, the change in concentration of the various heavy isotopes (Z in the range 90 or greater) and the change in concentration of any burnable poison which may be incorporated in the core. In terms of reactivity, the net result will be a change in the excess reactivity which will cause the critical control rod pattern to change. Accompanying this change there will usually be changes in power distribution in the core and changes in reactivity coefficients, which will result partly from the change in control rod pattern and partly from the composition change. Effects of the former type have been discussed in preceding sections; the effects of the latter type are to be discussed here.

Aside from the significance of the radioactive fission products and higher isotopes themselves as the primary sources of the reactor hazard—a subject treated elsewhere in this volume—the important safety aspects of core exposure are its effect on the following:

a) Shutdown margin,
b) Reactivity coefficients,
c) Spatial distribution of power,
d) Effective delayed neutron fraction.

8.1 Fission Products

The most obvious effect of the buildup of fission products is the resulting decrease in reactivity. This effect is significant in itself and its magnitude determines how important the fission products may be in influencing other characteristics of the reactor. Often it is convenient to express the fission-product content of the reactor in terms of the poisoning, ψ:

$$\psi = \frac{\bar{\phi}_P \bar{\Sigma}_P}{\bar{\phi}_F \bar{\Sigma}_F}, \qquad (8\text{-}1)$$

where $\bar{\phi}_P$ is the average neutron flux to which the fission products are exposed, $\bar{\Sigma}_P$ is the effective absorption cross section of the fission products in that flux spectrum, and $\bar{\phi}_F$ and $\bar{\Sigma}_F$ are the corresponding quantities for the fuel. In almost all practical cases the fission products will occupy the same locations as the fuel, and the ratio $\bar{\phi}_P/\bar{\phi}_F$ will be unity.

If the thermal utilization, f, is defined in the usual way:

$$f = \frac{\bar{\phi}_F \bar{\Sigma}_F}{\bar{\phi}_F \bar{\Sigma}_F + \sum_{\text{all } i} \bar{\phi}_i \Sigma_i}, \qquad (8\text{-}2)$$

where i covers all non-fissile materials, then the fractional change in the infinite multiplication factor, k_∞, caused by the introduction of a poisoning ψ, is given for thermal reactors by:

$$\frac{\Delta k_\infty}{k_\infty} = \frac{\Delta f}{f} \doteq -f\psi, \qquad (8\text{-}3)$$

where the definition of "fuel" must be the same for computing f as for computing ψ. Frequently it is most convenient to calculate ψ relative to the fissile isotope (e.g., $\psi_{25} = \bar{\Sigma}_P/\bar{\Sigma}_{25}$ when U^{235} is the only fissile isotope present). In such a case the thermal utilization relative to the fissile isotope (i.e., f_{25}) must be used in Eq. (8-3) to compute the effect on k_∞. It is then useful to remember that, however the "fuel" is defined,

$$\eta f = \eta_{25} f_{25}; \qquad f_{25} = \frac{\eta}{\eta_{25}} f \qquad (8\text{-}4)$$

for the case where U^{235} is the only fissile isotope present, and analogous expressions hold when other fissile isotopes are present.

Although treatment of the fission product absorption in terms of the thermal utilization implies that the major absorption is in the thermal energy range, it is true that many of the "long-lived" fission products have important epithermal cross sections. The epithermal absorption can be treated in terms of a resonance escape probability, or may be placed in an appropriate epithermal energy group when multigroup calculations are made.

However, for many purposes the thermal utilization treatment is sufficiently accurate, provided an effective absorption cross section is used to include the epithermal absorption as an "effective" thermal absorption.

The two high cross section fission products, Xe^{135} and Sm^{149}, usually must be treated separately from the other fission products. The effects of these two isotopes are discussed in standard reactor engineering texts, [168] and in reference [169]; they will be covered only briefly here. The nuclear reactions involved are:

$$\text{Fission}\,(U^{235}) \xrightarrow{6.1\%} Te^{135} \xrightarrow[<1\,m]{\beta^-} I^{135} \xrightarrow[6.7\,h]{\beta^-} Xe^{135} \quad (\text{direct yield } 0.2\% \text{ to } Xe^{135})$$

$$\xrightarrow[9.2\,h]{\beta^-} Cs^{135} \xrightarrow[2\times10^6 y]{\beta^-} Ba^{135}\ (\text{stable})$$

$$\text{Fission}\,(U^{235}) \xrightarrow{1.1\%} Nd^{149} \xrightarrow[2.0\,h]{\beta^-} Pm^{149} \xrightarrow[53\,h]{\beta^-} Sm^{149}\ (\text{stable})$$

The thermal-neutron absorption cross section of Xe^{135} is about 3×10^6 barns and that of Sm^{149} about 5×10^4 barns. The cross sections of the other isotopes in the chains are moderately low.

Because it is unstable, with a half life of 9.2 hours, and because it has a very high absorption cross section, Xe^{135} reaches equilibrium quickly in any reactor which is operated at steady power. In a few hours the concentration rises to the point that the rate of destruction of Xe^{135}, by radioactive decay and by neutron absorption, is equal to the rate of production. The stable Sm^{149} reaches equilibrium also, after the concentration has become high enough that the rate of destruction by neutron absorption is equal to the rate of production by fission, but the time required to approach equilibrium is strongly dependent on the neutron flux level, and is quite long for low-flux reactors.

The equilibrium value (ψ_0) of the poisoning by Xe^{135} is given by:

$$\psi_0\,(Xe) = \frac{(Y_I + Y_X)\,\Sigma_F}{1 + \dfrac{\lambda_X}{\phi\sigma_X}\,\Sigma_{Af}} = \frac{(Y_I + Y_X)}{1 + \dfrac{\lambda_X}{\phi\sigma_X}} \cdot \frac{\eta}{\nu}, \quad (8\text{-}5a)$$

where Σ_F is the macroscopic fission cross section and Σ_{Af} is the macroscopic absorption cross section of the "fuel"; similarly, ν is the average number of neutrons produced per fission and η is the average number of neutrons produced per neutron absorbed in the "fuel." Y_I is the fission yield of Te^{135} (0.061), Y_X is the direct yield of Xe^{135} (0.002), λ_X is the decay constant of Xe^{135} (2.1×10^{-5} sec^{-1}), σ_X is the effective microscopic absorption cross section of Xe^{135} in the neutron flux ϕ in the fuel.

Since the thermal neutron flux in a thermal reactor is determined by the specific power relative to the fissile isotope, the equilibrium value of the xenon poisoning can be expressed in terms of the specific power. The approximate value, when U^{235} is the fissile isotope, is given by:

$$\psi_0\,(Xe)_{25} = \frac{0.18}{1/P_s + 3.5} = \frac{\bar{\Sigma}_{Xe}}{\bar{\Sigma}_{25}}, \quad (8\text{-}5b)$$

where P_s is the specific power in megawatts per kilogram of U^{235}. Here the energy per fission has been taken as 200 Mev, and the fission and capture cross sections of U^{235} as 577 barns and 106 barns.

The equilibrium poisoning by Sm^{149} [ψ_0 (Sm)] is independent of the flux level, and is just equal to the fission yield, about 0.01.

Because both Xe^{135} and Sm^{149} have precursors with appreciable half lives, the formation of both of these isotopes continues after the reactor has been shut down, but the destruction by neutron absorption does not; hence after the shutdown of a high flux reactor the poisoning by xenon and samarium may build up to a value far above the equilibrium level. For Xe^{135} this transient effect is small for thermal neutron fluxes up to about 10^{13}, but for a flux of 10^{14} the peak transient poisoning reaches a value of about 0.23, and for a flux of 2×10^{14} is about twice that great. After shutdown from high flux operation the xenon poisoning reaches its peak in 10 to 11 hours and thereafter decays to zero, according to the 9.2-hour-half life of Xe^{135}. After shutdown from high flux operation the samarium poisoning builds up to an asymptotic value, and does not decrease (until the reactor is again brought up to power). After shutdown from a thermal neutron flux of 2×10^{14} the samarium poisoning builds up to a value of about 0.036.

The absorption cross section of Xe^{135} has a resonance at 0.082 ev; this results in a cross section which is nearly flat over the thermal range, up to about 0.1 ev, and which drops off rapidly at higher energies. For this reason the xenon effect is small in fast and epithermal reactors. The peculiar shape of the cross section curve will affect the temperature coefficient of reactivity of those high-flux thermal reactors which can operate with large values of xenon poisoning. It is conceivable that it may contribute a positive component to the coefficient for high-temperature thermal reactors, but this has not proved to be a common difficulty. The absorption cross section of Sm^{149}, in the thermal and near-thermal regions, behaves much like that of Xe^{135}; its effect on temperature coefficients of reactivity would be expected to be small because of the relatively low values of samarium poisoning, in terms of the total neutron balance.

The "transient" xenon effect—the tendency for the poisoning to increase when power is decreased (after a period of continued operation)—is in effect a positive power coefficient of reactivity. Insofar as the effect acts upon the reactor as a whole it does not usually cause serious dynamic problems after reactor startup in reactors operating at thermal-neutron flux levels below the 10^{14} range, for the rates of reactivity change are usually slow in terms of the response time for considered action by the reactor operator. For reactors with very high thermal-neutron fluxes the reactivity changes due to xenon transients may be rapid enough to set a significant lower limit on the maximum design speed of control rod motion, and may set other dynamic requirements for the control system and/or the reactor operator.

Any nonuniformity of distribution of the Xe^{135} and Sm^{149} contents over the reactor volume will modify the distribution of power density in the reactor. The equilibrium poisoning effects of these isotopes tend to be uniform over any regions of the reactor which are otherwise uniform in composition, since the equilibrium concentration of Sm^{149} is independent of the flux level and the equilibrium concentration of Xe^{135} is rather insensitive to flux level. The transient effects do, however, depend strongly on flux level, and the power distribution in a reactor which is started up with a high transient concentration of Xe^{135} may be quite different from that of the equilibrium case; the effect is to flatten the power distribution, since the highest poisoning occurs in the regions where the thermal neutron flux is normally highest.

If the transient xenon effect is large enough and the normal fractional leakage of neutrons from the reactor is small enough (roughly, if the reactivity change due to the Xe^{135} transient is comparable to, or larger than, $M^2 B_g^2$ the local positive power coefficient of reactivity due to transient xenon effects may cause spatial oscillations of the local xenon concentration, accompanied by spatial oscillations of the power density. The remedy for such oscillations is the same as the remedy for oscillation of the general power level of the reactor under the influence of the positive xenon power coefficient of reactivity: the oscillating tendency must be countered by appropriate movements of control rods. Again the time scale of the oscillations is long, and there are usually no dynamic problems in exercising control; however, it is essential that information on local power levels be available to guide the control measures (in-core instrumentation) and that a sufficiently fine-grained distribution of control elements be provided to make possible the control of reactivity in the important local regions of the reactor. Spatial oscillations of Xe^{135} are discussed more quantitatively in Sec. 4 of the chapter on Water Reactor Kinetics; recent work on the subject is reported in references [170-179].

Aside from any problems which may arise in connection with spatial oscillations, the more usual ways in which xexon and samarium poisoning interact with safety considerations are the following:

1. The poisoning increases the excess reactivity requirement and the total reactivity-worth requirement for the control elements. This increases the amount of reactivity available for the initiation of accidents, increases the difficulty of maintaining favorable power distributions over all ranges of control element motion, and increases the range of variation of reactivity coefficients over the range of available control-element programs.

2. In reactors which do not incorporate enough excess reactivity to override the effects of transient xenon, there may be an incentive to restart quickly after a scram or other temporary shutdown. When this is the case, special precautions are necessary to ensure that the incentive for fast restart does not override safety considerations. The sense of urgency arising from attempts to restart quickly enough to "beat the xenon peak" has been a factor in several accidents or near-accident situations.

3. The change in reactivity after shutdown, due to the xenon transient, complicates the evaluation of the true shutdown margin. In any cases where the shutdown margin could conceivably decrease after initial startup, it is important to provide means for monitoring the shutdown capacity despite transients in the xenon concentration.

4. Particular precautions should be taken against the loss of shutdown capability after the decay of the xenon concentration below the equilibrium level.

5. In low-power reactors the effect of the gradual buildup of Sm^{149} should be kept in mind in the continuing evaluation of the shutdown capability. The buildup of Sm^{149} can produce a significant reactivity loss over time periods which, while long in terms of reactor history, may be short in terms of fuel burnup.

The fission products other than Xe^{135} and Sm^{149}, for approximate purposes, are often lumped together as the "long-lived fission products" and are assigned some effective cross section, usually expressed in barns per fission. Actually, the half-lives as well as the absorption cross sections of the individual fission product species vary over wide ranges, and the composition of the fission product burden will be different after long irradiation than it is in the early stages of irradiation, both because of the radioactive decay relationships and because of the preferential destruction of the higher cross section isotopes by neutron absorption. Thus the average cross section per atom of fission product will tend to decrease somewhat with exposure, and will be a function of the average flux level at which the exposure occurs. The average epithermal cross section of the fission products is significant relative to the thermal cross section, and therefore the effective absorption will depend upon the neutron flux spectrum in the reactor. For rough purposes, the effective cross section of the fission products in thermal reactors has been taken to be about 50 barns per fission, but the true value may vary from this by as much as a factor of two because of epithermal effects and because of exposure effects. Investigations of effective fission product cross sections up to 1958 are summarized in reference [6] some of the later work is reported in references [180] through [183]

The major effects of the long-lived fission products are their effects upon reactivity, both general and local. The local effects lead to variations in the power density distribution with core exposure, but these effects are usually lumped in with the effects of changing concentrations of the heavy isotopes. The main effect of the long-lived fission products on reactivity coefficients is usually the indirect effect due to the change in pattern of control rod insertion.

8.2 Changes in Heavy Isotope Content

In those reactors which contain fertile materials, the change in isotopic content of the fuel with exposure is a complex one. The nuclear reactions involved for the cases of U^{238} and Th^{232} as fertile materials are shown in Figs. 8-1 and 8-2, respectively, and Figs. 8-3 [185] and 8-4 [6, 187] show the buildup of heavy isotopes in these two materials, as functions of irradiation, for some illustrative cases.

When U^{238} is the fertile isotope, Pu^{239}, because of its high absorption cross section, approaches its equilibrium value after a relatively short irradiation, and reactor cores or fuel elements which

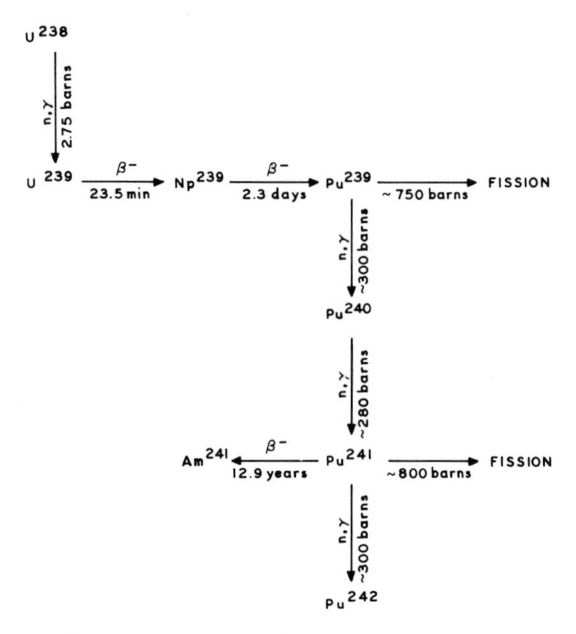

FIG. 8-1 Chain of most important isotopes produced in the irradiation of U^{238}. The cross sections are 2200 m/sec values.

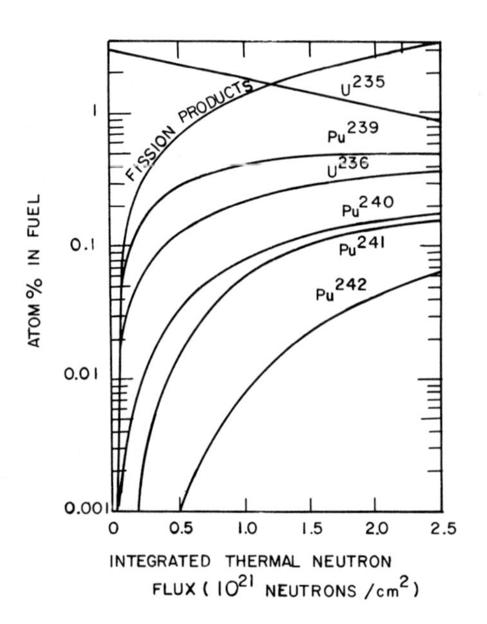

FIG. 8-3 Isotopic composition changes in U^{235}-U^{238} fuel in a pressurized-water reactor. In this particular case the initial fuel enrichment is 3.44% U^{235}, and the initial conversion ratio is approximately 0.67. These characteristics determine the ratios between plutonium isotopes and U^{235} and between the plutonium isotopes and fission products. The absolute values of the plutonium isotope concentration are nearly independent of these quantities in themselves, although they do depend upon the energy spectrum of the neutron flux.

employ U^{238} with relatively high conversion ratios may demonstrate the characteristic effects of Pu^{239} over much of their useful lifetime. One of the important characteristics of Pu^{239} is the strong dependence of its absorption and fission cross sections upon the effective neutron temperature. As has been mentioned previously, this may lead to a positive moderator temperature coefficient of reactivity if the conversion ratio of the reactor is high enough that the fission of plutonium eventually makes an important contribution to the neutron balance.

If the initial conversion ratio is reasonably high, in the early stages of operation each U^{235} atom which is destroyed will be replaced by some large fraction of a Pu^{239} atom having a fission cross section considerably higher than that of U^{235}. Hence the average macroscopic fission cross section of the fuel will increase, and this may cause a reactivity increase with exposure (despite the lower value of η for Pu^{239}) provided the initial fuel enrichment is low enough that the early formation of plutonium can make a significant contribution of the total fissile isotope content. As irradiation proceeds, the Pu^{239} content will approach its equilibrium value, and the reactivity will begin to decrease because the extra fission cross section

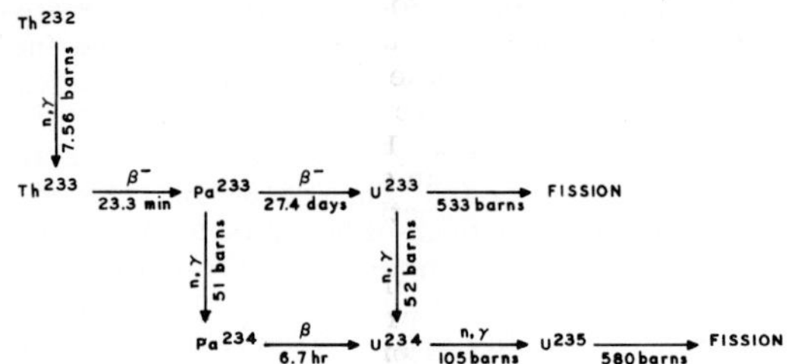

FIG. 8-2 Chain of most important isotopes produced in the irradiation of Th^{232}. The cross sections are 2200 m/sec values.

provided by the buildup of Pu^{241} is not sufficient to compensate for the continuing burnup of U^{235} and the buildup of fission products. Hence, in a reactor utilizing the U^{235}-U^{238} combination, the reactivity may either pass through a maximum with exposure or many decrease monotonically. An initial reactivity increase will be favored by high conversion ratio, low enrichment, and high moderator temperature. The natural uranium graphite-moderated reactors are good examples of these characteristics. Naturally, the effect of plutonium on the moderator temperature coefficient of reactivity will be most pronounced in those reactors which show a reactivity increase with plutonium buildup. Figure 8-5 [186] shows the effect of initial fuel enrichment on the reactivity behavior with exposure at a constant initial conversion ratio.

Both the changes in the local multiplication factor and the changes in fission cross sections which occur with fuel burnup will affect the spatial distribution of power density in the reactor; the spatial variation of the multiplication factor will affect primarily the gross power distribution, while the variation in fission cross section will affect mainly the more localized distribution. Figure 8-6 [185] shows the calculated change in gross power distribution in a pressurized-water reactor after a fuel exposure of 23,000 Mwd/metric ton. The calculation applies to the idealized case in which the initial excess reactivity is compensated by a uniformly distributed absorber: the absorber is removed uniformly to maintain criticality as burnup proceeds. In practice this would correspond to the use of a soluble poison control system. In the more usual case of control by absorbing rods the changes

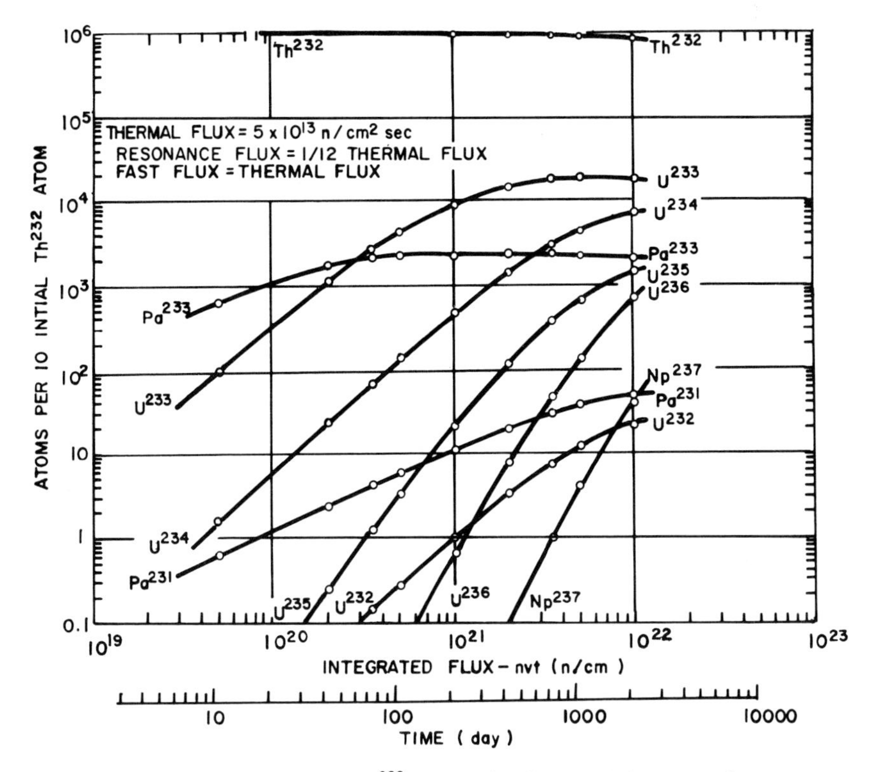

FIG. 8-4 Isotopic composition changes in Th^{232} when irradiated in a particular neutron flux spectrum.

in power distribution due to fuel exposure might be completely masked by the changes caused by the changing pattern of control-rod insertion. Nevertheless, Fig. 8-6 does serve to show that large changes in gross power distribution can result from long fuel exposure, and that these changes may be large enough to shift the point of maximum power density in the reactor core. The latter effect may, of course, be accentuated or otherwise modified by fuel shuffling schemes, as illustrated in Fig. 8-6C.

In general it may be said that, in converter reactors, the eventual effect of fuel exposure is to flatten the gross power distribution. Initially this tendency may be reversed if conditions are such that the initial effect of exposure is to increase reactivity.

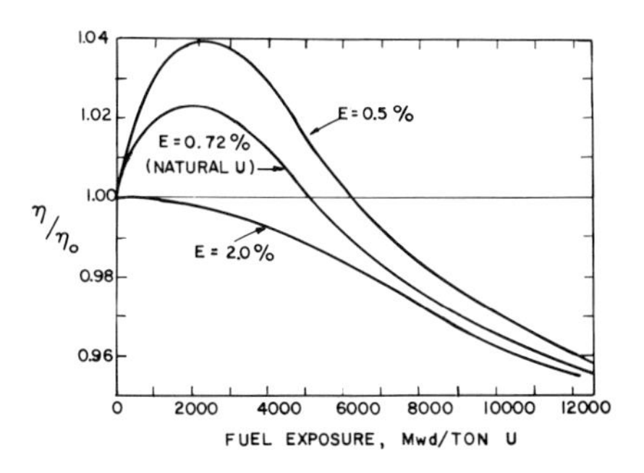

FIG. 8-5 Effect of initial fuel enrichment on long-term reactivity behavior in a U^{235}-U^{238} fueled reactor. The initial conversion ratio is 1.0 for all cases. The change in reactivity is given, very nearly, by the ratio η/η_0, where η and η_0 are the values for the U^{235}-U^{238} mixture.

If the power density is initially flattened by a nonuniform distribution of absorber—as in control rods or burnable poisons—then the net effect on power distribution, as fuel exposure proceeds, cannot be specified by a general statement, but must be computed for the particular case in question. It can only be said that if the initial flattening by absorber distribution is too great, the net effect of exposure will be to give a more peaked power distribution.

Any local power peaking effects which exist, such as the peaks in fuel elements adjacent to water channels, may behave differently than the gross power density distribution as fuel exposure proceeds. As pointed out above, the local peaking is affected mainly by the local value of the macroscopic fission cross section (Σ_f), whereas the gross distribution is affected mainly by the local value of k_∞ ($\sim\eta f$). Since σ_f is considerably larger for Pu^{239} than for U^{235}, while η_{49} is less than η_{25}, and since the buildup of fission products affects k_∞ but not Σ_f, it is quite possible for the local value of Σ_f to be increasing while the local value of k_∞ is decreasing. Thus in some cases fuel exposure may accentuate local power peaks while flattening the gross power distribution. The local peaking adjacent to water-filled control rod channels may of course also be accentuated if the fuel elements adjacent to the channels are protected from burnup

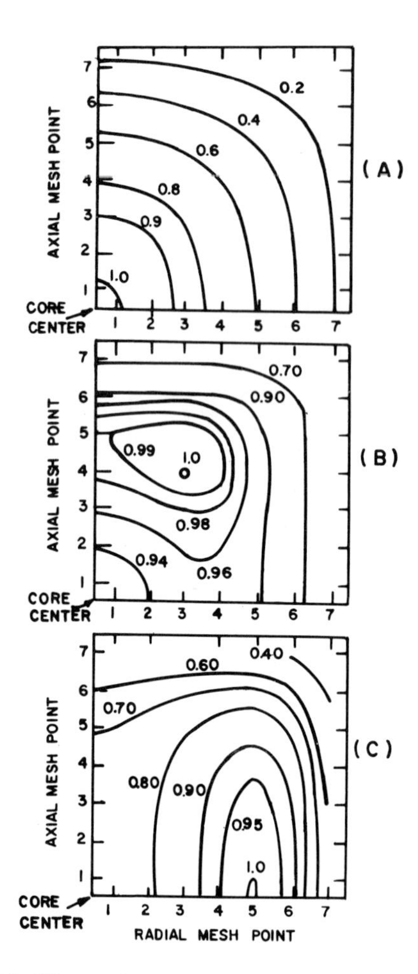

FIG. 8-6 Effect of fuel exposure on power density distribution. The curves, which apply only to a specific reactor, show contours of constant density for a cylindrical pressurized-water core in the r-z plane. (A) gives the distribution at the beginning of fuel irradiation, with the excess reactivity compensated by uniformly distributed absorber, after xenon and samarium have built up to equilibrium concentrations. (B) shows the distribution after fuel exposure to an average value of 23,000 Mwd/metric ton, with fixed fuel positions. (C) shows the distribution for an average exposure of 23,000 Mwd/metric ton when new fuel is added at the core periphery and exposed fuel is shuffled toward the center, on an equilibrium cycle.

during the early stages of reactor operation by the presence of control rods which are later withdrawn. The prediction of the effects of fuel exposure on local power peaks adjacent to control rod channels is a complex computation, not only because of the effects mentioned above, but also because the conversion ratio in these local regions differs from the average ratio for the core: in general it will be lower than the average ratio when the control rod is withdrawn and may be considerably higher than the average when the control rod is inserted.

The fissile isotopes which build up in converter reactors as exposure proceeds are characterized by average delayed-neutron fractions which differ from that of U^{235}–as it happens, they all have substantially lower fractions. Consequently, in a converter reactor initially fueled with U^{235}, the effective value of β (the total delayed neutron fraction) will decrease with exposure, and, where significant, this change must be considered in stability and accident analyses, and in the interpretation of dynamic reactivity measurements. The effect is particularly important in fast reactors, where large fractions of plutonium may build up (or where plutonium may constitute the initial fuel) and where prompt-critical power increases may be very rapid. Fortunately these reactors usually produce a substantial fraction of U^{238} fissions and the high β for U^{238} fission partially counteracts the low value for Pu^{239} fission.

The delayed-neutron fractions for the several fissile isotopes are given in Table 3-1 of the Criticality Chapter and the effective value of β can be computed for any combination of fissile isotopes if the relative fission rates in the members of the combination can be computed. To arrive at an accurate value of β_{eff} it is necessary to recognize that the reactivity worth of a delayed neutron is not the same as that of a prompt neutron, because of the lower average energy of emission of the delayed neutrons. Hence, even if only a single fissile isotope is present the effective value of β is not, in general, equal to the delayed neutron fraction for that isotope: it may be either larger or smaller, depending upon whether leakage effects or fast-fission effects are more important. Reference [184] discusses the reactivity importance of delayed neutrons and computation of β_{eff}. Finally, if the fissile isotope composition varies over the core volume—as for example because of a non-uniform exposure distribution—it is necessary to weight the distribution appropriately, according to its reactivity importance, if an accurate value of β_{eff} is to be obtained. It should be noted in passing that highly accurate values of β_{eff} are not usually required for adequate prediction of reactor safety and stability. An accurate value of β_{eff}, either computed or measured, is, however, necessary for relating experimental dynamic observations on the reactor to reactivity changes.

All of the preceding discussion of fuel exposure effects has centered around those fuel cycles involving plutonium. Those cycles which utilize thorium as the fertile material, and therefore involve U^{233} as the fissile isotope in addition to U^{235}, are in most respects more straightforward; for the variations of the important U^{233} cross sections with neutron energy are not qualitatively different from the behaviors of the corresponding U^{235} cross sections. Thus, as U^{233} builds up in a reactor fueled initially with U^{235}, one does not expect large changes in reactivity coefficients because of this change per se (although there may be substantial changes due to related causes, such as the withdrawal of control rods), nor does one expect an initial reactivity increase unless the initial conversion ratio is greater than unity.

The one characteristic of the cycles involving U^{233} which must be borne in mind is the relatively long half-life (27.4 days) of the U^{233} precursor, Pa^{333} (see Fig. 8-2). Because of the resulting "hold-up" of bred U^{233}, high-flux reactors with substantial conversion ratios for U^{233} will initially lose reactivity much faster than they will after a few weeks of operation. Similarly, during long periods of shutdown, these reactors will gain reactivity as the inventory of Pa^{233} gradually decays to U^{233}. Care must be taken to provide sufficient shutdown capability for this reactivity gain, and to take it into account in any periodic evaluations of shutdown margin that may be made. In these respects the reactivity gain is analogous to that which results from xenon decay, although the time scale is much longer. In particular, the effect must be taken into account if a short-term burnable poison is incorporated to compensate for the initial reactivity loss during the buildup of Pa^{233} to equilibrium.

8.3 Burnable Poisons

Burnable poisons are neutron-absorbing materials of relatively high microscopic cross section, which are converted to low-absorption isotopes by neutron absorption, and which are incorporated in reactors to compensate for part or all of the reactivity decrease that would otherwise result from fuel exposure. Insofar as their function is concerned, they may be mixed with the fuel or the structural materials, or may be installed separately in the form of plates or other suitable shapes. Their use reduces the number of control rods required for a given exposure lifetime of the fuel. If suitably installed, they may decrease the variation of power density distribution with fuel exposure, and make possible more uniform power density distributions over the life of the core.

Needless to say, if the burnable poison is to "burn out" during a core lifetime and thus provide a component of reactivity gain with exposure, it must not be incorporated in the form of "black" absorbers, as the control rods are, but must be in a relatively dilute form. If the poison is quite dilute, so that it produces no local depression of the neutron flux in its vicinity, even when it is initially installed, it will "burn out" exponentially in flux-time:

$$N_p(t) = N_{p,o}\, e^{-\bar{\sigma}_p \phi t}, \qquad (8\text{-}6)$$

where N_p is the number of atoms of the "poison" present, ϕ is the neutron flux, and $\bar{\sigma}_p$ is the effective microscopic absorption cross section of the "poison," averaged over the flux spectrum.

The variation in poison concentration given by Eq. (8-6) is not well suited to matching the reactivity change required to compensate the reactivity loss due to fuel burnup and fission-product buildup—it tends to be too rapid initially and too slow at a later time. It comes closest to matching the reactivity loss in a "non-converter" reactor. In such a reactor, assumed to be a thermal reactor fueled with U^{235}, the reactivity is nearly proportional to k_∞, and for the simple case of uniform composition, where all disadvantage factors are unity:

$$k_\infty = \eta_{25} \cdot \frac{1}{1 + \dfrac{\sum \bar{\Sigma}_i}{\bar{\Sigma}_{25}} + \dfrac{\bar{\Sigma}_p}{\bar{\Sigma}_{25}}}, \qquad (8\text{-}7)$$

where $\bar{\Sigma}_{25}$ is the macroscopic absorption cross section for U^{235}, $\bar{\Sigma}_p$ is the corresponding cross section for the burnable poison, and $\bar{\Sigma}_i$ is the corresponding cross section for the ith non-fissile material; the summation is taken over all materials present in the reactor, except the burnable poison. If the effects of fission products are neglected $\bar{\Sigma}_i$ is constant with fuel exposure, and the criterion that k_∞ stay constant with exposure is:

$$\frac{\sum \bar{\Sigma}_i + \bar{\Sigma}_p}{\bar{\Sigma}_{25}} = \frac{\sum \bar{\Sigma}_i + \bar{\Sigma}_{p,o}\, e^{-\bar{\sigma}_p \phi t}}{\bar{\Sigma}_{25,o}\, e^{-\bar{\sigma}_{25} \phi t}} = \text{constant}, \qquad (8\text{-}8)$$

where $\Sigma_{p,o}$ is the macroscopic cross section of burnable poison at $t = 0$, etc. Evidently Eq. (8-8) can be satisfied for arbitrary t only if $\bar{\Sigma}_i = 0$ and $\bar{\sigma}_p = \bar{\sigma}_{25}$. Even to the extent that such a case is approximated, the behavior of the burnable poison is not particularly desirable, for its fractional burnup over the life of the core is only equal to the fractional decrease of fissile isotope, which cannot be large for criticality reasons: thus the reactor would always contain a rather large amount of burnable poison, and the neutron loss through parasitic absorption in the poison would be large.

To modify the burnup law from that of Eq. (8-6) to a more suitable one, the usual practice is to "lump" the burnable poison sufficiently to produce a significant amount of self shielding. By this means the variation of absorption with exposure can be drastically modified; and in principle almost any (decreasing) variation can be approximated by the use of burnable poisons with several different degrees of self-shielding. In the extreme case, for example, an absorber of high $\bar{\sigma}_p$ might be used, in a local concentration sufficiently high to make the absorbing material initially quite "black." The absorption would then change only slightly with exposure until enough absorber had been "burned" that the material was no longer "black"; thereafter the absorption would decrease rapidly with exposure, approaching closer to an exp $(-\bar{\sigma}_p \phi t)$ decrease as less and less of the absorber remained.

Figure 8-7 shows typical effects of burnable poisons on the reactivity variation with exposure in a converter reactor. Note that the use of a burnable poison which is not self shielded reduces the reactivity "swing" due to a burnup of 10,000 Mwd/ton from about 0.08 to 0.06, while the use of a "50% shielded" poison reduces the swing to about 0.03. The initial increase in reactivity is typical when burnable poisons are used to compensate reactivity changes of important magnitude: it occurs because the burnable poisons typically "burn out" too rapidly at first, and too slowly later, to match the components of reactivity loss.

Burnable poisons have their main safety significance because they are usually designed to yield an initial reactivity increase and because their design usually falls into a difficult area of reactor physics: the calculation of absorption in "gray" absorbers. The theory of burnable poisons has been treated fairly extensively in the literature. References [6] and [188] through [195] are useful papers on the subject. However, the precise calculation of burnable poison effects still presents many difficulties in individual cases, and reactors which incorporate strong burnable poisons usually require particularly careful monitoring of the shutdown margin during their operating programs.

The burnup of burnable poison will ordinarily

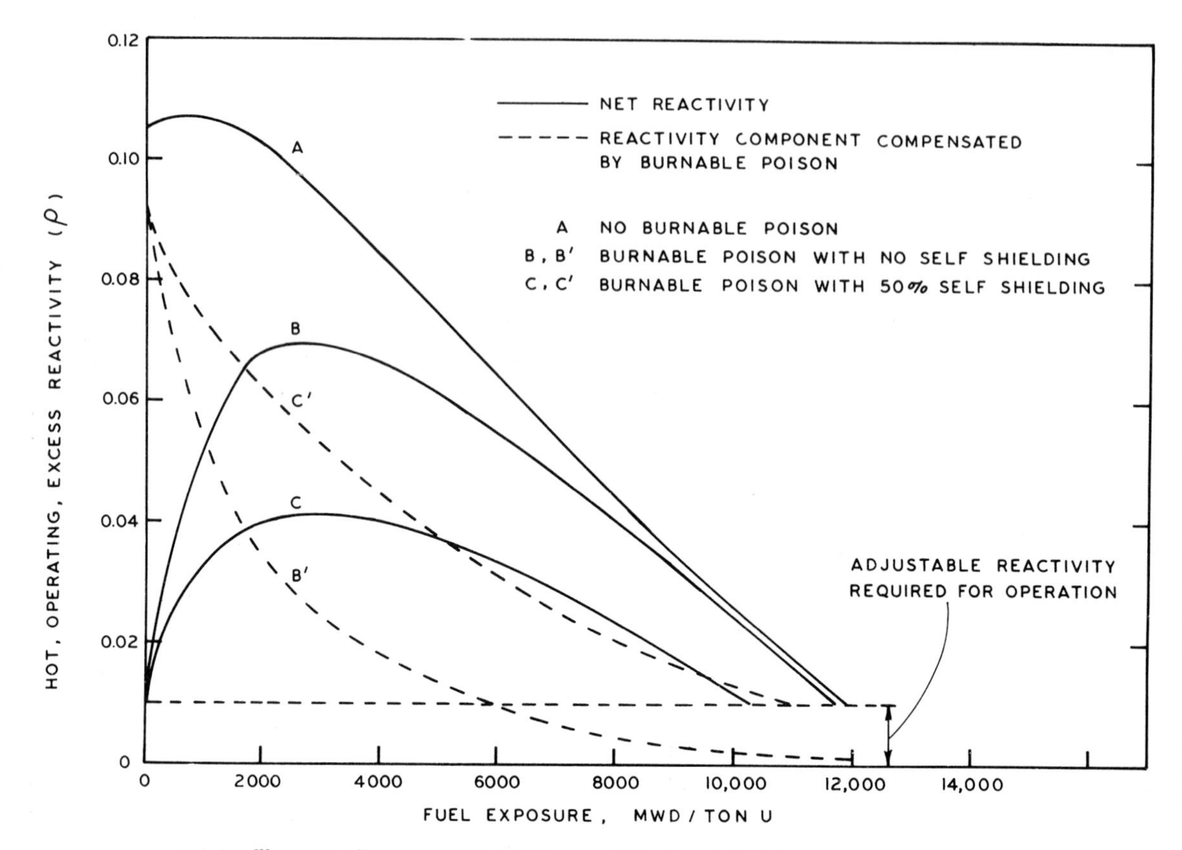

FIG. 8-7 Illustrative effects of burnable poisons on reactivity change with fuel exposure. The curves, which are approximate, apply to a boiling water reactor utilizing UO_2 fuel of 2.3% enrichment with an initial conversion ratio of 0.70. The reactivity shown applies to the hot operating condition, with equilibrium xenon and samarium concentrations. It is assumed that an adjustable excess reactivity of at least 0.01 is needed for operational adjustments, and the loadings of burnable poison are chosen to give this value for the initial excess reactivity. The burnable poison is B^{10}.

When the poison is unshielded it burns out exponentially with flux-time (ϕt). Although ϕt is not proportional to the exposure in megawatt days per ton, it is nearly so over the range considered, and the reactivity compensated by the unshielded B^{10} (Curve B') is approximated by: ρ comp = 0.095 exp(-E/2500), where E is the exposure in Mwd/ton. This rate of burnup is initially much too fast, and the net excess reactivity reaches a peak of about 0.07.

The self-shielded case considered is for 50% self-shielding: i.e., the B^{10} is installed in discrete "lumps," of such concentration that the average thermal neutron flux in the B^{10} is initially only 50% of that in the fuel. This means that twice as much B^{10} is installed as in the non-self-shielded case, to produce the same initial reactivity compensation. The fractional rate of burnup of the B^{10} is considerably less (Curve C'), and a significant quantity of B^{10} remains after the net excess reactivity has passed through its peak and has declined again to the value 0.01: the result is that the useful exposure lifetime of the fuel has been reduced by about 1500 Mwd/ton. However, the use of the self-shielded B^{10} has held the excess reactivity peak to about 0.04.

yield a component of change in reactivity coefficients with fuel exposure. In highly self-shielded cases, in particular, the burnable poison may add a strong negative component to temperature and void coefficients of reactivity because of the dependence of the absorption rate on the diffusion length and the diffusion coefficient in the surrounding medium. Shaping of the power density distribution by the distribution of burnable poison may also have important effects on temperature, void, and power coefficients of reactivity. A corollary is that the reactivity worth of burnable poisons for cold shutdown should not be inferred from the worth in the hot or operating reactor: in some cases there may be very large differences among these quantities.

Finally, if burnable poisons are used to flatten the power distribution in the reactor, the effects of poison burnout on the distribution must be taken into account. If the initial flattening is moderate, the preferential burnup of fuel may maintain, or improve, the distribution as exposure proceeds; but if the power is over-flattened by the initial distribution of burnable poison the maximum/average ratios may increase with exposure.

9 IN-CORE INSTRUMENTATION

In-core instrumentation is defined here as any instrumentation which yields information from local regions of the core, regardless of whether the sensing element is actually located within the core boundaries. In-core instrumentation is closely related to core design because the design must provide for its installation, and because the approach taken in the design to matching local heat removal capabilities with local power densities is determined by the nature of the information provided by the instrumentation. From the point of view of safety, the main significance of in-core instrumentation lies in the improved knowledge and understanding of the core condition and behavior which it affords.

9.1 Detection and Location of Failed Fuel Elements

The detection of fuel element failures is a necessity in most reactors. Failures must be detected at least by the time they have caused significant contamination of the coolant system and preferably earlier so that corrective action can be taken at a convenient time. All detection systems operate on the principle of detecting some fission product (or products) which either has leaked from the defective element or has been formed by the fission of some fuel that has leaked from the element. Usually the problem is to distinguish the activity due to the selected fission product from the general background activity. In some cases this is done by searching for the characteristic lines of the fission product in the gamma-ray spectrum; in others a fission product is selected which emits delayed neutrons, and the neutrons are detected. [205]

The use of a separation process for concentrating the fission product of interest may be effective in circumventing background effects. In the British gas-cooled reactors, for example, the solid decay products from short-lived gaseous fission products are precipitated electrostatically upon a wire which is subsequently counted [224]. At Savannah River a system has been used experimentally which separates fission-product xenon and krypton from the reactor cover gas by gas chromatography [239]. Activity of the main turbine condenser off-gas stream is monitored in boiling water reactor, direct cycle plants for gross gamma and/or selected fission products such as xenon or krypton.

Any background of fission-product activity due to a fissile isotope which has entered the reactor system as surface contamination obviously will interfere seriously with the detection of failed elements. The surface contamination of Zircaloy by U^{235} has plagued failed-element detection systems in pressurized-water reactors employing Zircaloy-jacketed fuel. A study of the sources of contamination has been reported in reference [240] Surface contamination on aluminum-jacketed elements also has been a problem.

The location of a failed fuel element, among the many elements which make up the core, may be a much more difficult task. In general, the principle is to examine the coolant for a fission product, as is done in the failure detection system, and to deduce the location of the failed element by taking the coolant samples from local core regions. The facility and accuracy with which this can be done depends upon the ease with which coolant from specific core regions can be segregated and identified. Thus in a tube-type reactor it is straightforward to identify the tube containing the failed element if all tubes are sampled individually and if the fission-product detected has a half-life which is not long compared to the circulation time of the reactor coolant. The latter requirement may not be easy to meet, however, if the fission products diffuse out of the failed element slowly.

In H_2O reactors the fuel elements are usually not well segregated; and at best, an on-power failed-element location system requires the installation of a complex system of sampling tubes within the reactor vessel. To date the on-power location of failed elements has been only partly successful in H_2O-moderated power reactors. An effective method of identifying leaking fuel elements after reactor shutdown is to withdraw a suspect element into a test container and check for buildup of fission products (radio-iodines) in the water surrounding the element.

An effect has been observed in water-moderated reactors employing UO_2 fuels which may be used to advantage in connection with the location of failed elements. Defective elements have been observed to emit "puffs" of fission-product gases when the power in the elements is increased rather rapidly after a period of shutdown or low-power operation. It has been postulated that water enters the defective element during shutdown and leaches some fission products from the oxide fuel; these products are expelled with the water when power is increased and the oxide becomes hot. Figure 9-1 [205] shows this "puff" as detected by the failed element detection and location system in the Shippingport reactor. This effect may be used to increase the resolution of an in-core failed-element location system. The effect has also been used, in conjunction with control-rod manipulation, for rough indication of the failed-element location by means of a general detection system only. The reactor is operated for a time with one control rod completely inserted, to depress the power in the elements near it. The control rod is then withdrawn, while the general power level is maintained by appropriate adjustments of other rods. If a puff of fission gas is detected by the general system, the indication is that a defective element lies near the control-rod channel. By such "probing" with all of the control rods in sequence, much of the core volume can be investigated.

9.2 Local Power, Temperature, and Heat Removal

The observation of local values of power density, temperature, and coolant flow is desirable in all reactors, but its importance varies from one reactor type to another, as does it practicability. In some reactors it is a virtual necessity, in others it represents an attractive method of maximizing the power capability, and in still others the amount of local information which may be obtainable may not be worth the difficulty of the in-core installation.

In reactors which are subject to spatial oscillation of the Xe^{135} content, information on local power densities or neutron flux densities is needed continuously in order to detect incipient oscillations and to guide the control-rod manipulations for controlling the oscillation. To this extent in-core instrumentation or its equivalent is essential for

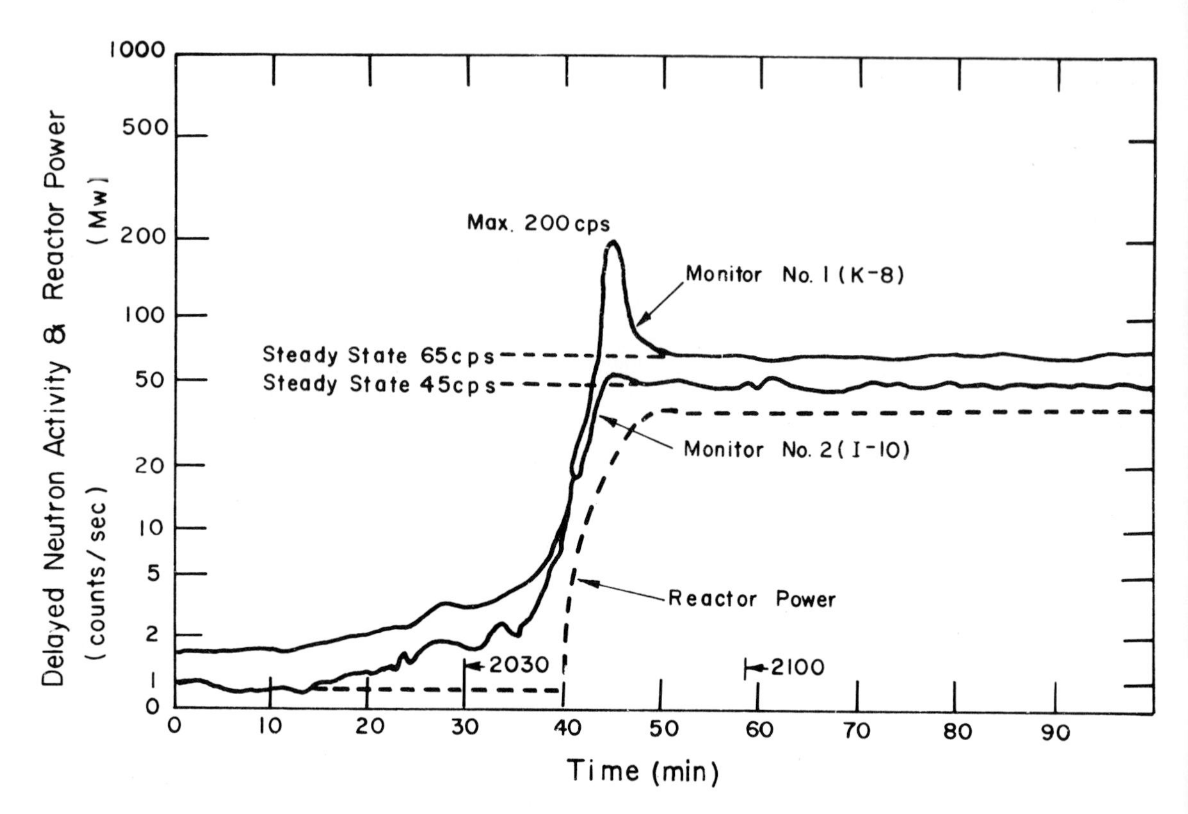

FIG. 9-1 The delayed neutron activity records of blanket assemblies K-8 and I-10 in the Shippingport Reactor during a reactor startup transient performed on July 16, 1960. [Ed. Note: Apparently, the scale of ordinates in the interval 0 to 1 cps or Mw is linear; the scale within each cycle (1-10, 10-100, 100-1000) is logarithmic, although the scale of each cycle differs from that of the other two cycles.]

such reactors. The criteria which determine the possibility of xenon oscillations are discussed in Sec. 4.1 of the chapter on Water Reactor Kinetics. In general, oscillations occur in reactors which are large in terms of the neutron migration length and which operate at high thermal neutron flux levels (i.e., at high specific power relative to fissile isotope). The D_2O-moderated natural uranium reactors tend to fall into this category. Spatial oscillations were also observed in the early operation of the Shippingport reactor [242].

The in-core instrumentation of pressure-tube reactors is usually rather easy to accomplish, and the characteristics of the reactor are such as to allow maximum use of the in-core information. Consequently, the effective use of in-core instrumentation is usually a basic point of departure in the design of such reactors. The more specific reasons for the attractiveness of this approach in the pressure-tube reactors are the following:

1. Coolant flow and temperature rise can usually be measured for individual pressure tubes by instruments that may be located at some distance from the core proper.

2. There are no strong local peaking factors in the power density, consequently, measurements of the thermal neutron flux in the moderator between pressure tubes is a good indication of the power density in nearby pressure tubes. The installation of flux measuring devices in the moderator is generally considerably easier than their installation in a fuel-bearing region.

3. The opportunities for adjustment of local power levels are usually more numerous in this type of reactor than in other types. These adjustments may be made either by the use of appropriate patterns of control rod insertion or by the appropriate shuffling of fuel when an on-power refueling scheme is used.

Some of the solid-moderator reactor types also present favorable situations for in-core instrumentation. In the graphite-moderated gas-cooled reactor, although the measurement of local flux values may be difficult, the measurement of exit coolant temperature from individual fuel channels along with determinations of coolant flow rates in individual channels may yield good measurements of the power produced in each channel; the opportunities for tailoring the power distribution are much the same as in the tube-type reactor.

In liquid-moderated reactors of the pressure-vessel type the core consists of a fuel element matrix bathed in moderator-coolant and the difficulty of in-core measurement is greater and the opportunities for exploiting the information are fewer. This is particularly true in the H_2O-moderated reactors. Aside from the difficulties of operating the instruments in the core environment, it is essentially impossible in many cases to examine local conditions in sufficient detail to monitor those regions which are the crucial ones. Thus, for example, discrete cooling channels are not defined for each fuel element as the coolant is free to move in cross flow from one element to another within an assembly and sometimes within the entire core. Hence a locally measured coolant exit temperature cannot be attributed to a single fuel element.

Further, the thermal neutron flux gradients in H_2O-moderated reactors can be so steep that an individual fuel element may operate at an average power density which is quite different from the power density in its next neighboring elements. Consequently, a flux measurement which will indicate directly the power density in the hottest fuel element must, in effect, be made within that element; this is usually an impossibility. At the same time, the opportunities for exploiting a detailed knowledge of the local thermal situation may be limited. The degree of fuel shuffling which is practical in a water-moderated power reactor is usually rudimentary, and the degree of flexibility available in programming the control rods may be small.

Despite these difficulties, in-core instrumentation has been used to a considerable extent in the large H_2O-moderated reactors. In these applications, the instrumentation system does not evaluate the hot spots directly but does give information useful for reducing the errors which would attend their evaluation by purely computational methods. The advantage of the instrumentation is primarily to evaluate changes in the general power distribution as the result of fuel burnup, control-rod programming, and changes in xenon concentration. The local hot spots must still be found and evaluated by computation. Because of the complexity of the computation which may be necessary to derive hot-spot and hot-channel factors from the instrument indications, the degree to which in-core information may be used to direct the moment-to-moment operation of an H_2O-moderated reactor is often limited. Nevertheless, in-core power instrumentation may make possible a greater flexibility of control-rod programming and, over a period of time, may allow the operator to discover more favorable programs, simply because it allows the investigation of a greater number of situations than would be practicable by calculation alone. The use of extensive in-core power instrumentation tends to become more attractive as reactor size is increased, for the power distributions in larger cores are more sensitive to local variations of reactivity, and the range of possible control rod programs increases as the number of rods increases. Other things being equal, in-core power instrumentation is probably more attractive in boiling than in nonboiling reactors, for the interdependence of steam-void distribution, power distribution, and reactivity increases the labor of determining power distributions by calculation alone.

A current example of in-core instrumentation in an H_2O-moderated reactor is that in the Yankee reactor [241]. The instrumentation consists of 27 chromel-alumel thermocouples for monitoring local coolant outlet temperatures and 8 flux-wire drive mechanisms which can install (and remove) manganese-containing flux wires in the centers of 22 different fuel assemblies. The mechanical features of the installation are described in Sec. 9-4; the locations of the sensing elements are shown in Fig. 9-2. [241]

It has been estimated [241]. that the thermocouple installation, in conjunction with the thermometers which measure average inlet and outlet temperatures, can determine the relative integrated axial power in each individual fuel assembly to within ± 5% provided the flow distribution in the core is known. There may be some uncertainty in the latter distribution, since the fuel assemblies are not shrouded in the Yankee core.

The flux wires, of course, can yield only intermittent data, for the irradiation, cooling, and counting of a set of 8 wires requires over an hour, and the wires can be irradiated only after some time has been allowed for the decay of activity from previous irradiations. Each wire gives a measurement of the axial thermal neutron flux distribution in a single fuel assembly. Calculated corrections must be applied to obtain either the average power density in the assembly or the maximum power density, which occurs at a corner of the assembly adjacent to the intersection of two control-rod channels. These corrections are applied from a set of center/average and center/corner ratios which have been calculated for each fuel assembly for all configurations in the control rod program. The basic calculations for the establishment of these ratios were made by two-dimensional calculations at the beginning of life, using the PDQ code; they were extended through the core lifetime by the TURBO code. Reference to Fig. 5-9 will show that the "corner" corrections are by no means small;

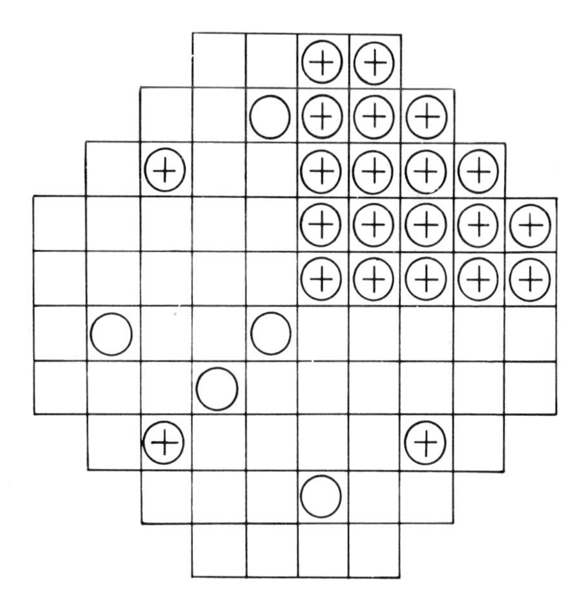

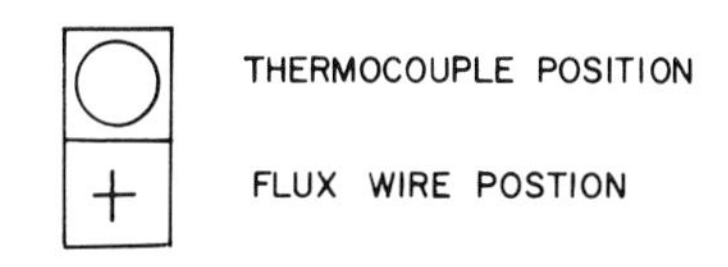

Fig. 9-2 Diagram of in-core detector positions in the Yankee Reactor (See Fig. 5-9 for locations of control-rod channels).

thus the evaluation of the actual "hot-spot" still depends strongly on the results of computation.

The data reduction for a complete set of thermocouple and flux-wire measurements is handled by an IBM-7090 code. It applies the corrections noted above, as well as other corrections such as residual wire activity, and yields the following quantities: [241]

1. an integrated power density distribution in the x-y plane (obtained by an axial integration of the measured distribution,
2. the axial peak-to-average power density in each fuel assembly,
3. the power density and coolant enthalpy rise distribution at each corner point around the control rods and followers (from which the hot channel factors* F_Q and $F_{\Delta T}$ are obtained),
4. the maximum heat flux at the corner points described in 3,
5. the coolant temperature rise in each assembly and a "flow distribution" (this "flow distribution" is the ratio of F_{xy}** obtained from flux wire traces to that obtained from thermocouple readings and is influenced by flow redistribution within the core and mixing near the outlet nozzles,
6. the accumulated burnup, in megawatt hours, along the length and averaged over given segments of each assembly, and
7. the DNB ratio.***

9.3 Control Rods as In-Core Instruments

The usefulness of control rods as indicators of conditions in the core is not often emphasized, but this means of gathering useful information should not be neglected. The critical configuration of control rods, in the operating condition and in various conditions of "zero power" criticality, are continuing sources of information as to the reactivity inventory. The control rods can be used to measure reactivity coefficients throughout the lifetime of a core, and thereby yield not only operating information but information which may have fundamental significance. Finally, in many cases the power distribution within the reactor can be calculated with reasonable precision providing the operating configuration of control rods is known. This is particularly true in the smaller reactors in which changes of core composition due to fuel burnup are not highly effective in changing the general power distribution. The information obtainable from the control rods can be maximized by appropriate attention to the design of the control rod installation, to the selection of the rod program, and to computation methods for following and interpreting the history of control rod motion.

9.4 Practice in Existing Reactors

At this stage of reactor development, neither the design nor the application of in-core instrumentation systems is standardized, although some in-core instruments are available commercially [243]. The best approach to a survey of the current status is to examine the systems that have been provided in reactors which are operating or will soon begin operation. A brief summary for several such reactors is given in the following paragraphs.

9.4.1 First Atomic Power Station, Obninsk, USSR [198,199]

Each coolant channel in the reactor is supplied with an inlet pressure transducer, an inlet flowmeter, and an outlet temperature sensor, which is said to be a thermometer. The arrangement of the instrumentation with respect to a coolant channel is shown in Fig. 9-3 [199]. The flowmeters give continuous indications of individual channel flows and are connected to warning lights which are located on a central panel in the plant control room. The outlet water temperatures from the individual channels are scanned by an automatic system which displays a light signal for any channel whose temperature exceeds a preset value, and a plug panel in the control room permits manual selection of specific channels for continuous temperature monitoring and recording. Automatic shutdown of the reactor can be initiated by abnormal pressure, flow, or temperature signals.

9.4.2 Pressurized Water Reactor (PWR), Shippingport, Pennsylvania [196,197,200-205]

Provision has been made for measuring the coolant flow in each fuel assembly by either a venturi meter or a flow nozzle. The flow in 16 seed assemblies (alternate assemblies in the 32-element seed region) and 20 blanket assemblies is monitored routinely. Experience has indicated that the venturi measurements are, in general, reliable, while the flow-nozzle measurements exhibit some anomalies when compared either to theoretical predictions or to venturi measurements for similar conditions.

Thermocouples are located in the coolant streams at the entrance and exits of individual core subassemblies, and despite some erratic behavior of this system, the temperature readings obtained have made it possible to determine the power distribution in the core. The operational status of the coolant-water thermocouples after six months of operation was as follows: of 51 exit thermocouples, 46 were functioning; of 11 inlet thermocouples, 5 were functioning. The exit-water thermocouples in the seed region, of which there were 24, all performed satisfactorily during the period, while the inlet-water thermocouples, of which there were 6, all failed. Since the inlet thermocouples indicated temperatures generally corresponding to reactor outlet water temperature at all power levels, it is hypothesized that the sheath material of these thermocouples failed, therby permitting coolant water to saturate the mineral insulation and negate

---

*F_Q is the ratio of the maximum power per unit length of fuel element to the average power per unit length.

**F_{xy} is the ratio of the enthalpy rise in a given coolant channel to the average coolant enthalpy rise. The maximum value of F_{xy} is denoted as $F_{\Delta T}$.

***DNB denotes departure from nucleate boiling. The DNB ratio is a ratio of the heat flux which would yield this condition to the maximum operating heat flux. This subject is discussed in the chapter on Heat Transfer.

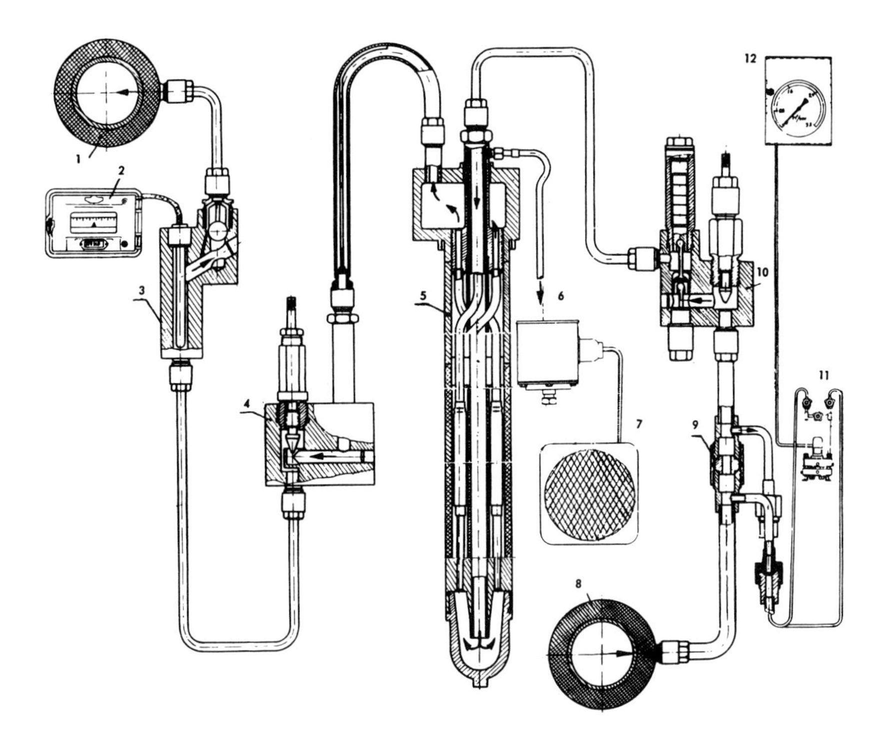

FIG. 9-3 Fuel-channel instrumentation, First Atomic Power Station, Obninsk (USSR). 1. Header, 2. device indicating outlet water temperature, 3. outlet valve with a thermometer, 4. valve, 5. fuel channel, 6. channel cell pressure indicator, 7. signal panel, 8. distribution header, 9. flowmeter, 10. disconnecting device, 11. flowmeter pickup unit, 12. flowmeter (secondary device).

its effectiveness.

The reactor is equipped with a failed-fuel-element detecting and locating system which analyzes independent samples of coolant taken from the outlet and of every blanket assembly. Satisfactory performance was predicated on the ability to detect delayed neutrons from fission products which had escaped through clad defects, but an unexpectedly high background count in the coolant has made the system ineffective during steady-state operation.

9.4.3 Savannah River Reactors, Aiken, South Carolina [206]

Neutron- and gamma-sensitive monitors have been developed for use inside the Savannah River reactors. The neutron monitor, which is shown in Fig. 9-4, [223], uses thermocouples to measure temperature differentials between a uranium-aluminum button and its case, and the gamma monitor uses an ionization chamber as its sensing element. The life of the neutron monitor is ultimately limited by burnup of the uranium, but performance is unaffected by neutron and gamma exposures of up to 10^{21} n/cm^2 and 5×10^{11} r, respectively; theoretical predictions indicate that the uranium burnup at an exposure of 5×10^{22}n/cm^2 will result in a sensitivity reduction of about 10%. The dynamic range of the neutron monitor is about one decade, and its time constant is about 30 sec. The useful life of the gamma monitor may be limited by radiation damage to the insulating material, but the instrument has performed satisfactorily to a total exposure of 3×10^{12}r with synthetic sapphire insulation.

In addition to these monitors a variety of other in-core instrumentation is provided. This instrumentation includes flow monitors and temperature monitors to determine the coolant flow and temperature rise through the individual fuel assem-

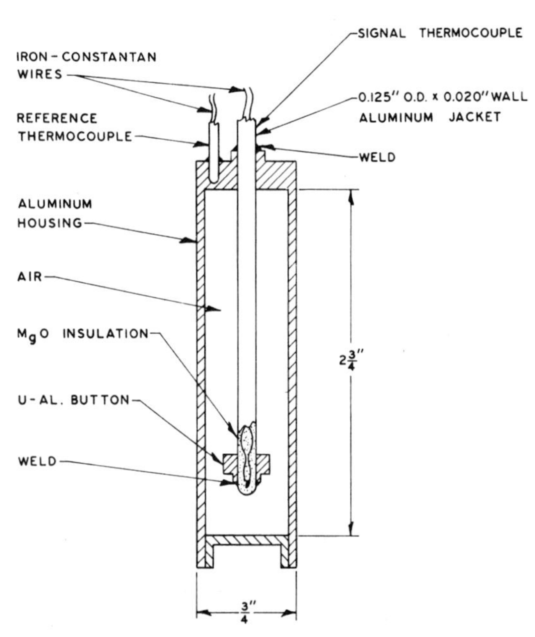

FIG. 9-4 Savannah River in-core neutron monitor.

blies, wire monitors to determine the axial flux distributions in the moderator, and a wide variety of activity monitors to detect and locate fuel element failures.

9.4.4 NRU Reactor, Chalk River, Ontario [245]

A sample line from the exit end of each process tube allows the coolant exit temperature, flow rate and radioactivity to be monitored. The flow rate is monitored by sensing the pressure. A signal is given for any process tube whose pressure drop is appreciably lower or higher than the average of several other rods. Either excessive temperature or unusual flow conditions will cause automatic shutdown of the reactor. For detection and location of fuel element failures, the exit coolant samples are directed to 21 monitoring stations, each sample being directed to 2 different stations. This system allows the 21 stations to identify the source of radioactivity from any single one of the approximately 200 process tubes, since the number of combinations of 21 stations taken two at a time is 210. The monitoring stations employ delayed-neutron monitors whose indications are backed up by gaseous fission product monitors.

9.4.5 Yankee, Rowe, Massachusetts [196, 197, 207-209]

The Yankee reactor is provided with a system for inserting wires into the core, where they can be activated to determine thermal-neutron flux distributions, and with a network of thermocouples for measuring water temperatures at the core outlet. The locations of the sensors are indicated in Fig. 9-2. Since the control-rod program for reactor operation is a symmetrical one, the locations are chosen to cover one quadrant of the core completely, and a few locations are sampled in other quadrants to monitor any asymmetry which may appear. The data gathered with these devices are used to determine core operating characteristics but are not used directly in the control or safety systems. The wire-activation system, which measures neutron fluxes axially and radially, consists of 8 Teleflex drive units which move eight flux wires from their storage containers into 8 of 22 core thimbles, retrieves the wires after irradiation, and passes them through a counting system. A ninth drive unit moves a Co^{60} calibration source past the counting assembly as needed. The 22 thimble tubes, which are at atmospheric pressure inside, extend from above the reactor vessel to the bottom of the core. The thimbles enter the vessel through 2 spare control-rod ports; 11 thimbles penetrate each port and are sealed by Conoseal fittings. Inside the vessel the thimbles are installed as indicated in Fig. 9-5 [208]. Outside the reactor vessel, each thimble is connected to an extension or conduit which leads outside the reactor cavity to the region in which the 8 Teleflex drive units are located. Solenoid-actuated transfer devices permit 6 of the 8 drive units to insert their wires into any of 3 thimbles each, while the other 2 drive units serve 2 thimbles each. Transfer devices are also used to direct wires to the conduits of the counter assembly. Each activation wire can be positioned in the core with 0.5 in. (12mm) accuracy, is approximately 120 ft (36m) long, and bends through angles totaling about 1400 degrees (~ 4 revolutions) in moving through the conduit.

The 27 outlet-water thermocouples enter the reactor vessel through 2 other spare control rod ports, as indicated in Fig. 9-5. The thermocouples are concentrated in one quadrant of the core, as are the activation-wire thimbles, while only a few check points are measured in the other three quadrants. The chromel-alumel thermocouple wires are insulated with metal oxide, and each stainless-steel sheath serves as a pressure barrier so that the individual wires do not require high-pressure insulating seals.

The information furnished by the in-core instrumentation in Yankee Core I was considered a valuable aid in the design of Core II and in establishing the feasibility of the increase in power rating of the reactor. The experience has been quite favorable through the first two core loadings.

9.4.6 Saxton, Pennsylvania [196, 197, 210]

The Saxton reactor is largely an experimental installation and, as such, is provided with extensive in-core instrumentation. This includes flux wires which are inserted and withdrawn by 8 remotely-operated cables and electrical drives, and may include miniature continuously-reading detectors which could be positioned similarly. Flow-distribution information is furnished by pitot tubes positioned at the inlet and outlet nozzles of 5 removable fuel subassemblies and at 9 other positions across the core. Thermocouples are used to measure the average water temperature at the inlet to the core and at the outlets of 14 fuel assemblies, as well as the local temperature distribution in the water at the outlet of one assembly. Locations of the sensing devices are indicated in Fig. 9-6, [210] which shows cross sections through the reactor core at inlet and outlet elevations.

9.4.7 Indian Point, Buchanan, New York [196, 197, 211]

The operating procedures for the Consolidated Edison Thorium Reactor, at Indian Point, include provisions for the installation of special in-core instrumentation during fuel element or control element removal or replacement. The special instruments are 2 neutron count-rate detectors which will be installed in the core and which will operate in the reactor safety system, replacing 2 of the 4 startup-range channels of normal instrumentation.

9.4.8 Carolinas-Virginia Tube Reactor (CVTR), Parr, South Carolina [196,197,212-214]

The Carolinas-Virginia Tube Reactor is equipped with instrumentation capable of determining primary coolant flow and temperature rise in each individual U-tube* and of detecting and locating

*The reactor is a pressure-tube reactor cooled and moderated by D_2O. The pressure tubes are

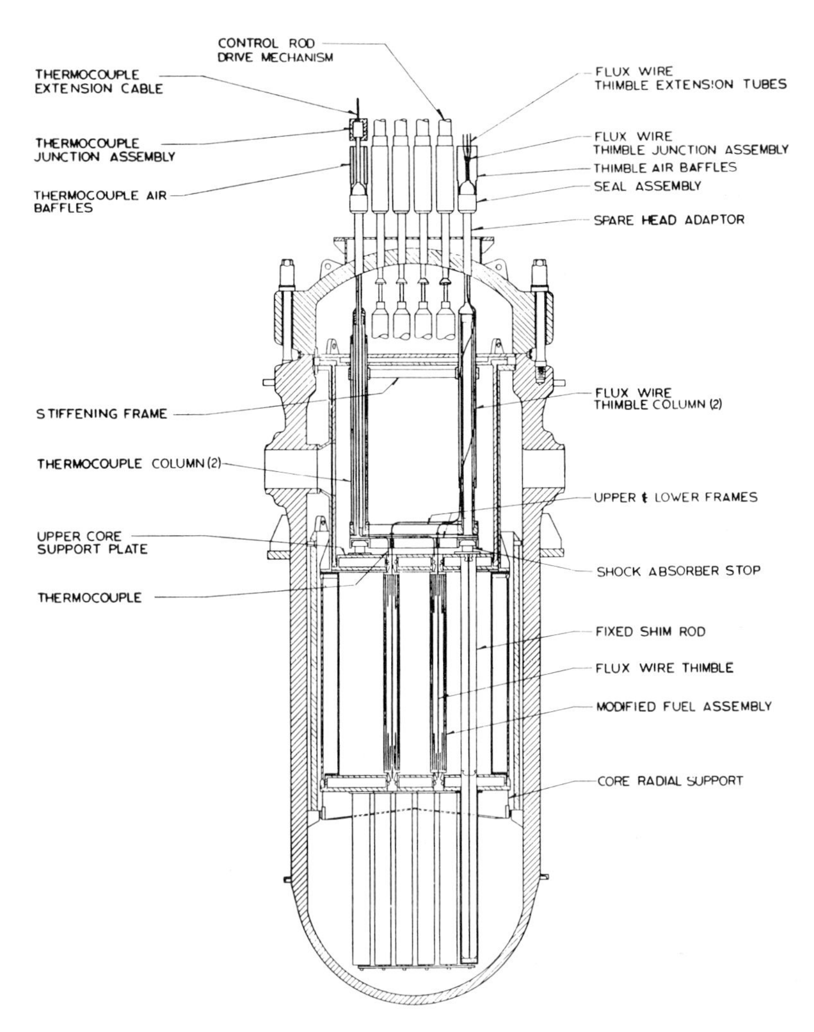

FIG. 9-5 Conceptual arrangement, Yankee core instrumentation.

failed fuel elements. At its outlet, each U-tube is equipped with a venturi tube and pressure tap as well as with a resistance thermometer, while resistance thermometers are installed in the two inlet headers of the primary coolant loop. The flow and temperature information furnished by these instruments is used in a program aimed at determining core characteristics during operation but is not used as input to the reactor safety or alarm systems.

The failed-fuel-element detection and location system consists of two high-pressure, high-temperature sampling loops and one low-pressure, low-temperature sampling loop. The latter loop, an iodine monitor, operates continuously and performs the detection function of the system. The high pressure loops, which are put into operation only after the continuous monitor gives a high-activity-level indication, are used to analyze coolant samples drawn from individual U-tubes through the venturi pressure-tap lines, thereby performing the locating function of the failed-element detection and location system. Samples are returned to the primary coolant circuit.

9.4.9 Dresden, Morris, Illinois [196,197,215-219]

The Dresden reactor is equipped with a system of 64 in-core neutron-sensitive devices which provide 64 meter readings indicative of local thermal neutron fluxes. The system, which is used to actuate alarms and scrams, can be used for relative measurements or can be calibrated to give absolute indications of local power levels at the points reached by the detector. The neutron-sensitive devices are miniature ion chambers with responses which begin at about 1% of rated power and cover a range of 5×10^{11} to 10^{14} n/cm^2 sec. Detector assemblies, each fitting within a 0.5 in. (12.7mm) tube and containing 4 axially-displaced ion chambers, are located in 16 modified fuel bundles dispersed radially through the core. The

---

grouped in pairs and connected at their lower ends to form U-tubes. Coolant flows into the top of one leg of a U-tube and out from the top of the other leg.

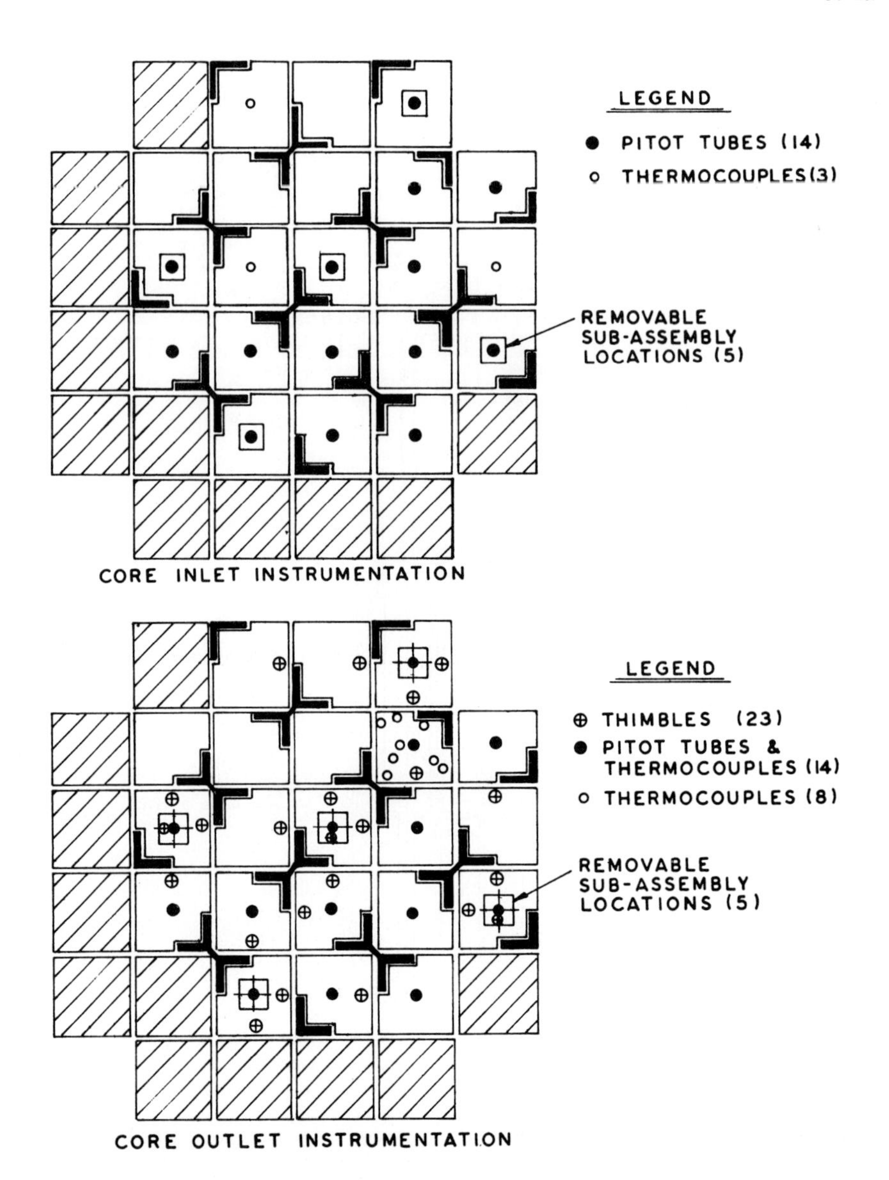

FIG. 9-6 Core inlet and outlet instrument locations, Saxton Reactor.

tubes containing the detector assemblies are installed through the reactor head. The arrangement of a 3-detector assembly, similar to the 4-detector assemblies used in Dresden, is given in Fig. 9-7. The instruments are reported to perform well. Design modifications have been incorporated in the systems as experience has been gained. As of October 1961, modified assemblies covering 32 points in the core had operated without loss of capability since June 1960 and had seen about 7×10^{20} n/cm^2.

9.4.10 Humboldt Bay, Big Rock Point, VBWR, SENN, and JAERI [196, 197, 218-220]

In-core ion chambers have been or will be installed in the Humboldt Bay, Big Rock Point, SENN, and JAERI reactors. The Humboldt Bay and Big Rock Point chambers (24 for each reactor) are installed through the tops of the vessels to positions between fuel bundles, with the electrical leads penetrating the vessels at their bottoms. The Humboldt Bay ion chambers are not connected to the safety system but are provided with individual indicators and warning lights. In addition to the in-core ion chambers described above, the Big Rock Point plant will be equipped temporarily with instrumented fuel bundles containing temperature, flow, and flux sensors housed in a package which replaces four central fuel rods of the bundle. These special assemblies are for the research and development program of this reactor, and are not anticipated for use during regular plant operation thereafter. Prototype instrumented fuel bundles

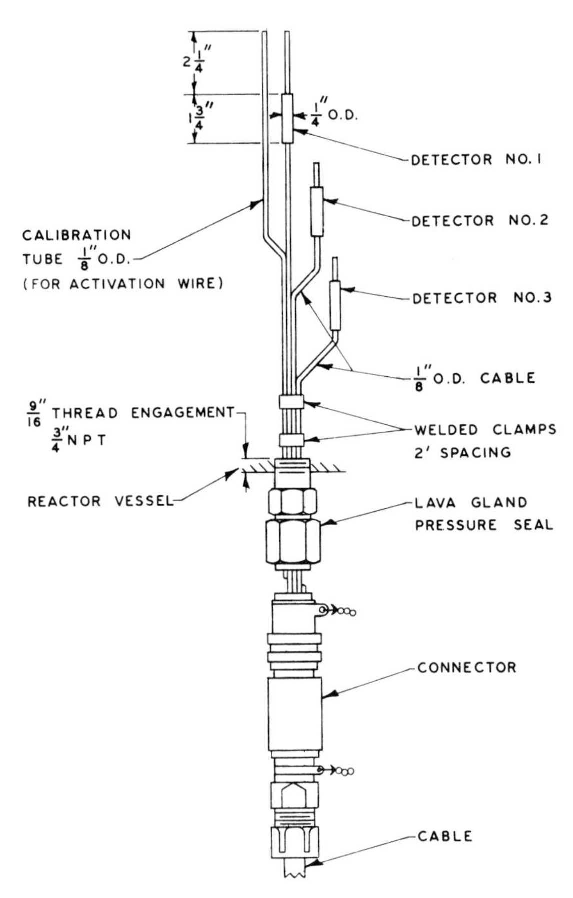

FIG. 9-7 Installation details, in-core ionization chambers. ("In-Core Flux Monitor" data sheet, General Electric Company, Nuclear Electronics Products Section, Atomic Power Equipment Department, San Jose, California).
Note: In the figure the notation 1/8″ means 1/8 inch (3.175 cm) and 2′ means 2 ft (61.0 cm), etc.

have been exposed in the Vallecitos Boiling Water Reactor. The in-core ion chambers in SENN (80 in number) and JAERI (16 in number) are installed through the vessel bottoms into tubes positioned between fuel bundles. A traversing in-core probe, which would replace or supplement the fixed detectors, is also being considered for use in these later reactors. The probe is a fission or boron chamber which moves axially in its own guide tube within the core and which can be withdrawn and routed to other radial positions for continuous axial-flux-profile measurements. In performance, the probe is similar to the fixed detectors.

A recent study of methods to be used for reactor scheduling computation on the Big Rock Point reactor indicates that the monitoring of 24 in-core points is only marginally helpful in following the power distribution during core lifetime. The following quotation is from reference [244].

> "There are two basic procedures that might have been followed for power distribution determination in the Big Rock Point core. The first is full use of every in-core monitoring device available with heavy reliance on analytical and experimental assistance, and the second method is the strictly analytical approach. The method chosen in this report leans toward the analytical method with program options to include to some extent available experimental data and directly observable data. From power calculations considerations, an unfortunate situation exists in the Consumers core—there is a sufficient number of monitors present so that they should be employed in the power determination, and there are too few present so that they can be employed to any great extent. Essentially the problem is to generate a 504 node power distribution given 24 points of information, some of which may be redundant, erroneous, or both. There are many problems of relating an in-core monitor reading to the power level of surrounding fuel bundles. Consequently, an attempt is made to use the in-core monitor readings and experimental data to the greatest extent possible at this time, but this attempt is limited."

9.4.11 BORAX-5, National Reactor Testing Station, Idaho [197,221]

In-core instruments in this fifth reactor of the BORAX series are located in special fuel assemblies. Two such assemblies are supplied for the superheating region of the core and two for the boiling region. The superheater assemblies are provided with thermocouples at the entrance and exit transitions and at the edges of the fuel plates at 20 different points; the signals from these thermocouples can initiate alarms and scrams. A flow-measuring venturi is installed in the riser. The instrumented boiler assemblies have turbine-type entrance and exit flowmeters for measuring liquid as well as two-phase flow. It is expected that average steam-void fractions may be deduced from the entrance and exit measurements. To measure coolant temperatures, a dummy fuel rod houses 12 thermocouples whose junctions project into the coolant stream. Additionally, rather special thermocouples have been developed for measuring central temperature in a fuel rod in the boiling region. These thermocouples, which are Ta-sheathed and BeO-insulated, employ junctions of W vs W-25 wt. % Re.

9.4.12 ESADA Vallecitos Experimental Superheat Reactor (EVESR) [197,222]

The Vallecitos Superheat Reactor is equipped with axial tubes, located at five radial positions in the core, for the insertion and removal of irradiation wires during operation. Instrumentation is provided to give outlet steam temperature from each fuel bundle; the temperatures are read in the control room, where there are 32 indicators. Each of the indicators has high and low adjustable trips for alarm.

9.4.13 Calder Hall, Cumberland, England [196, 197, 224, 225]

The burst slug detection (BSD) system, or

failed-fuel-element detection and location system, in Calder Hall has served as the model for systems in most subsequent gas-cooled reactors. Its operation is based on the principle of collecting electrostatically the ionized atoms of solid isotopes which result from the decay of certain gaseous fission products. The system consists chiefly of a sample tube from the outlet of each reactor channel, a multiport selector valve, a "precipitator" chamber in which the ions of solid isotopes are collected on a wire by electrostatic precipitation, and means for "counting" the wire. For BSD purposes, the 1696-channel core is divided into 8 zones of 212 fuel channels each. The stainless-steel sample tubes (of 1/4-inch nominal bore) from the channels in each zone are collected into a bundle within the reactor vessel and are welded into a stainless-steel plug which in turn is welded to a branch in the pressure vessel wall. The 8 bundles of 212 sample tubes are then passed through the thermal and biological shields to the area where the selector valves, wire precipitators, and ancillary equipment are located. The complexity of the system is illustrated by its use of approximately 40 miles (64 km) of sample tubing, and its importance is attested by the special control room, as large as the reactor control room, which is provided for the precipitator units.

Temperatures of the coolant gas at the exits of about 120 of the 1696 channels in the core are obtained from the thermocouples attached to the sampling tubes of the BSD system. It is occasionally necessary to anneal the graphite core structure of the reactor by allowing its temperature to rise above that which occurs during normal operation. Thermocouples are installed at a sufficient number of points to give a complete picture of the temperature distribution in the graphite. The accuracy of measurement is desired to be within 2°C, but the time constant is not important. Fuel element temperatures were originally measured by thermocouples attached to 48 elements situated in various parts of the reactor, but additional thermocouples have been installed along with a fuel element excess temperature tripping facility. The facility is provided with 3 initiating elements; it gives warning in the event of a single high temperature and causes automatic shutdown if 2 high temperatures are detected. The method of attaching a thermocouple to a fuel element is illustrated in Fig. 9-8 [234]. Wire-activation techniques are used to determine the neutron flux distribution in the core.

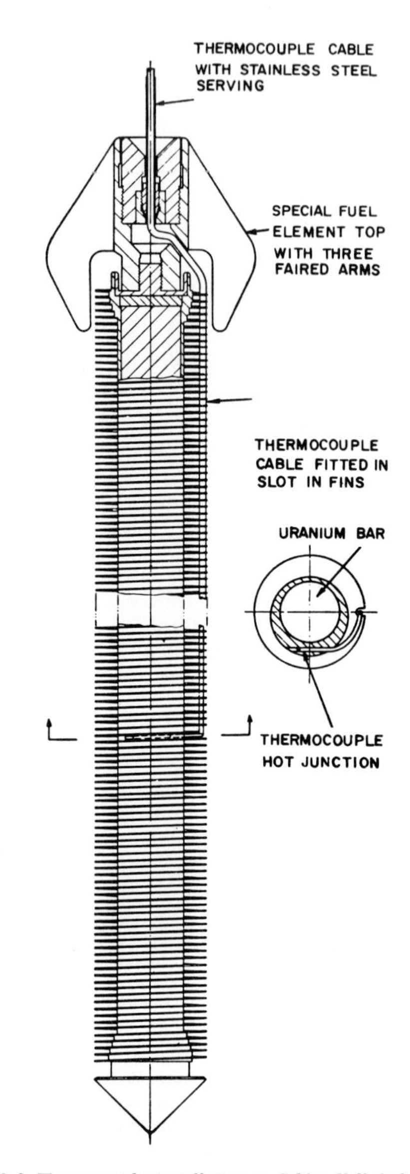

FIG. 9-8 Thermocouple installation in Calder Hall fuel element.

An indication of the degree to which the later British reactors are instrumented is the fact that, in the Berkeley Power Station, a total of 181 measurements may be selected for continuous monitoring and control purposes from 202 fuel-element can temperatures, 230 channel outlet-gas temperatures, and 310 graphite temperatures.

9.4.14 Later British Reactors [196,197,226,227]

The experience gained in operating the Calder Hall plant indicated that later designs should incorporate refinements in in-core instrumentation. In the later reactors (as typified by Hunterson, Hinkley Point, and Bradwell) this has amounted primarily to increases in the quantity of instrumentation. In the earliest reactors, the normal control and alarm instrumentation was concentrated in the zones of maximum fuel temperature and greatest Wigner energy storage; in the later plants, this instrumentation has been increased considerably and many readings are taken continuously throughout the volume of the reactor core.

9.4.15 G1, G2, G3; Marcoule, France [196,197, 228, 229]

Since its first high-power runs, the plutonium-producing, graphite-moderated, air-cooled reactor G1 has been equipped with about 50 thermocouple-containing cartridges which are used to measure temperatures along the longitudinal axis and along a radius of the reactor. Additionally, the temperature of the air leaving each channel is measured every 100 min by a cyclic sampling system. The exit air temperatures are used to indicate where individual adjustments of coolant flow should be made in order to obtain a properly orificed distribution; but, because of the influence on adjacent channels of a flow change in a specific channel, the adjustment process is a long and imperfect one. The G2 and G3 reactors, which are graphite-moderated, natural-uranium fueled, and CO_2-cooled, contain 1200 horizontal channels each and are equipped with BSD systems similar to those of Calder Hall. Gas temperatures at the outlets of the 1200 channels are measured by thermocouples which are connected to a distributor, and graphite temperatures are measured at 48 points.

9.4.16 Experimental Gas-Cooled Reactor (EGCR), Oak Ridge, Tennessee [196, 197, 230, 231]

Coolant temperatures will be measured at about 6 to 12 points at the inlet (bottom) face of the core of the Experimental Gas-Cooled Reactor. The coolant gas temperature will be measured at the outlet of each of 232 fuel channels, and there are 90 thermocouples in the graphite moderator and reflector columns, principally in the hottest region (upper half) of the core. There are 4 instrumented fuel columns, each containing flowmeters as well as about 25 thermocouples which measure cladding temperature, gas temperature, and graphite sleeve temperature. The instrumented columns can be placed in any of 20 fuel channels which are located directly beneath nozzles in the top head of the reactor vessel. The reactor is provided with a burst slug detection system, the sample lines for which are also used by the pneumatic temperature monitor, a system which determines the outlet gas temperature at each of the 232 fuel channels.

9.4.17 Dragon, Winfrith, Dorset, England [197, 232]

The high-temperature, gas-cooled, Dragon reactor will be provided with in-core thermocouples for measuring fuel-element and graphite temperatures. The fuel-element thermocouples will be connected to extension leads, which are permanently installed in the reactor vessel, by means of the device shown in Fig. 9-9. [232] The device consists of a two-part connector: the top part is attached to the fuel element and thermocouples and the bottom part is attached to the fuel element bottom support and the extension leads. Since the bottom part of the connector is permanently installed, it contains rugged fixed contacts only, while the top part, which can be removed and replaced along with the fuel assembly, contains the more delicate spring-loaded contacts. Since the thermocouples will be located in an extremely hostile environment, a rather extensive thermocouple development program is under way.

9.4.18 Sodium Reactor Experiment (SRE), Santa Susana, California [196, 197, 233]

Thermocouples are used to measure temperatures of the coolant within the core tank of the Sodium Reactor Experiment and coolant tempera-

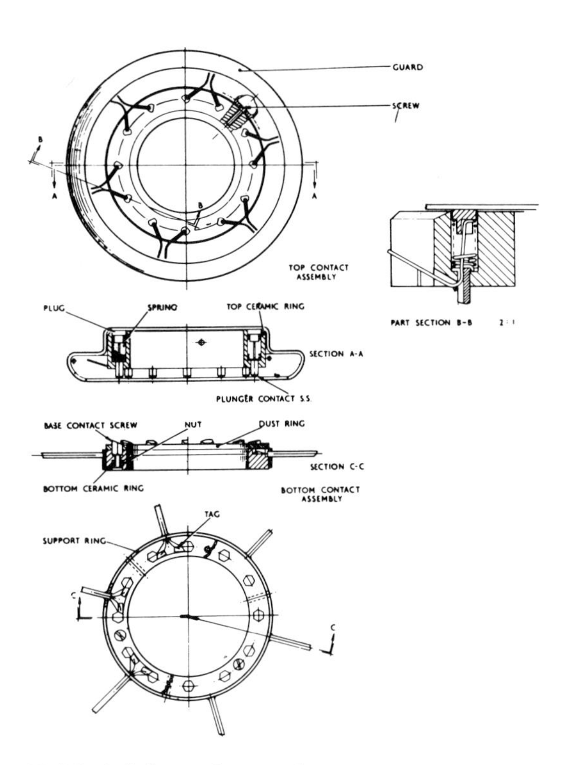

FIG. 9-9 Fuel element thermocouple connectors, Dragon Reactor.

tures within the reactor fuel channels. The reactor uses multi-element fuel bundles, and the fuel channel thermocouples are mounted in the central structural members of these fuel bundles. In order to determine temperatures in the sodium stream which is cooling the moderator, two of the corner channels are fitted with special elements containing thermocouples whose junctions are located at various elevations within the core. Each special element consists of a column of beryllium cylinders suspended on a stainless-steel tube which extends downward from a plug in the top shield; the thermocouples in the assembly are contained within the supporting tube and thus do not contact the liquid sodium.

9.4.19 Enrico Fermi Atomic Power Plant, Monroe, Michigan [196, 197, 234]

In-core temperature measurements in the Enrico Fermi Atomic Power Plant are obtained from thermocouples which are installed in the flow paths from individual core subassemblies. The temperature measurements are used as one set of inputs to the reactor safety system.

Fuel element failures are detected by a wire precipitation system, which monitors the argon cover gas above the sodium coolant in the reactor and one of the primary pumps, and which is similar in detection principle to the Calder Hall burst slug detection system. In this reactor, fuel elements are transported between the core and the loading-unloading ports by a transfer rotor, which is contained within the reactor vessel, submerged in sodium, beside the reactor core; fuel element transfer between the rotor and the core is by a handling mechanism installed above the rotor-core complex. Once a fuel element failure has been detected, the failed element is to be identified by applying the detection system to the cover gas above the rotor, as fuel elements are transferred, one at a time, from core to rotor.

For monitoring, the argon cover gas is conveyed from the sampling points to a shielded cubicle near the reactor building through stainless-steel tubes. There the gas is passed through a sodium vapor trap and cooler, and is pumped to the electrostatic precipitator. The precipitator contains a grounded axial steel wire, while the walls of the precipitator are maintained at a positive potential. Positively changed ions of cesium and rubidium, produced by the beta decay of the fission products xenon and krypton, are attracted to, and deposited on, the wire. The wire is then fed into a counting chamber, where the further decay of the attached cesium and rubidium atoms is detected by a beta-sensitive phosphor and photomultiplier, with an efficiency of 30% or better. A count-rate meter, with ranges from 0 to 10^7 cpm, feeds a panel meter, a strip chart recorder, and a high-count-rate alarm relay. This system differentiates between the fission products and other radioactive species which may be present. Of the latter, the most important by far is Ar^{41}, produced by neutron activation of the cover gas; the solid decay product of Ar^{41} is K^{41}, which is stable.

9.4.20 The Hallam Nuclear Power Facility, Hallam, Nebraska [196, 197, 235]

The Hallam Nuclear Power Facility is equipped with in-core thermocouples which are located as follows: 2 in each fuel element sodium-flow outlet, 16 in the source element, 16 in a specially-instrumented moderator can, 16 in each of 3 temperature elements, 12 in each of 2 "thermocoupled" fuel elements, and 16 on each specially-instrumented control rod. The fuel-element sodium-outlet thermocouples are placed in contact with the liquid sodium to give fast response and are connected to an instrument system designated the reactor fuel channel outlet temperature scanner. This system is composed of two 192-point multi-bank recorders, each scanning the same set of thermocouples. The timing of the scanners is set 96 points apart, and the speed of the over-all system is such that each channel is monitored by one or the other of the scanners every 6.4 minutes. An automatic plant protection system is actuated by coincident signals indicating high outlet temperature or by coincident indications of high rate of change of outlet temperature. For a normal, hot-leg, outlet temperature of 945°F (507°C), the automatic system is set to alarm at channel outlet temperature of 975°F (524°C) and to initiate a scram at 1025°F (552°C). The rate-of-change alarm is given for a step change of ± 15°F (± 8.3°C) or a ramp change of ± 45°F/hr (25°C/hr) while scram is initiated for a step change of ± 25°F (± 14°C) or a ramp change of ± 75°F/hr (± 42°C/hr).

9.4.21 Organic Moderated Reactor Experiment, National Reactor Testing Station, Idaho [197, 236, 237]

The temperatures of 7 fuel plates in the Organic Moderated Reactor Experiment are continuously monitored. Each of the 7 instrumented fuel plates is equipped with 5 thermocouples, any one of which may be selected as the sensor for a temperature indication on the main control panel.

9.4.22 Piqua Nuclear Power Facility, Piqua, Ohio [196, 197, 238]

The organic-moderated reactor of the Piqua Nuclear Power Facility is provided with a failed-fuel-element-location system which detects the presence of a failed element in the core by continuously monitoring the delayed neutron activity of the bulk outlet coolant. It may then be used to determine the position of the failed element by manually controlled sampling of the outlet.

9.4.23 Douglas Point Nuclear Generating Station (CANDU), Kincardine, Ontario [246]

The instrumentation of the pressure-tube-type CANDU reactor includes systems which measure the coolant flow rate and coolant temperature at the inlet and outlet, respectively, of each fuel channel. An automatic scanning system continuously determines the hottest channel temperature and sends a corresponding signal to the plant power controller. An alarm is sounded for either high temperature or low flow in any fuel channel, and the simultaneous occurrence of high-temperature, low-flow conditions in a single channel initiates a scram signal. Temperature and flow data are combined to estimate channel power and to indicate radial flux distribution.

Two separate failed fuel element monitors are planned for the CANDU reactor, [247] one to detect the occurrence of a defect and the other to identify its location.

The detection system employs a wire precipitator monitor in a gas sample which is stripped continuously from the main coolant system. The precipitated activity on the wire is counted by a gamma-sensitive scintillation counter with energy discrimination set for the 1.83 Mev gamma of Rb^{88}. The latter is the daughter of the fission product K^{88}, which has a 2.8-hour half life.

The failure location system consists of a set of delayed neutron detectors which monitor continuously-flowing samples of the coolant from the outlet of each of the 306 fuel channels of the reactor. About 40 detectors are used. Each normally monitors the combined samples from eight fuel channels; however, any one of the eight samples can be directed individually through the detector by means of valves in order to identify the single source of the delayed-neutron activity, once it has been localized within one of the groups of eight fuel channels. A detector consists of a BF_3 counter or fission chamber installed in a tank of ordinary water which serves as moderator; a tube carrying the flowing D_2O coolant sample passes through the tank. The neutron signal comes from the delayed-neutron emission of 22-second I^{137} and 56-second Br^{87}. The background neutrons consisting of N^{17} decay neutrons (4.2 second) and photoneutrons produced by the N^{16} gammas (7.4 second) must be taken into account, but their effects can be reduced to a satisfactory level by suitable choices of delay time and size of sample tube. It is expected that substantial transient increases in the delayed-neutron signal from a defective element can be produced by cycling the reactor power between 75% of full power and full power, if such an increase should prove necessary. This effect is expected to be usable even for those fuel elements which operate at relatively low power (down to about $\int k d\theta = 10$ w/cm). [247]

10 CORE DESIGN: SPECIFIC EXAMPLES

The objective of this section is to discuss, in as general a way as possible, how the mechanical considerations in core design interact with the safety considerations and the nuclear considerations which have been discussed in prior sections of the chapter. A detailed consideration of mechanical systems is given in the chapter on Mechanical Design.

The diverse properties of the materials in the various reactor types, interacting through considerations of compatibility, nuclear performance, and heat removal, result in such a diversity of mechanical arrangements that a truly generalized discussion is apt to be too academic to be useful. The range of core size alone, which is summarized in Fig. 10-1, is sufficient to illustrate this diversity. The approach taken in this discussion is to consider a number of reactors which have been built and to hope to achieve some generality through the variety of situations examined. The intent is not to cover all reactors or all reactor types, but simply to examine enough cases to illustrate the types of safety considerations which enter the design of a reactor core. The reactors are grouped into a few broad classes which are deemed appropriate for this particular discussion. The order of discussion proceeds by degrees from the reactors which are largest and which have the greatest proportion of fixed core structure to those which are smallest and have the least fixed structure.

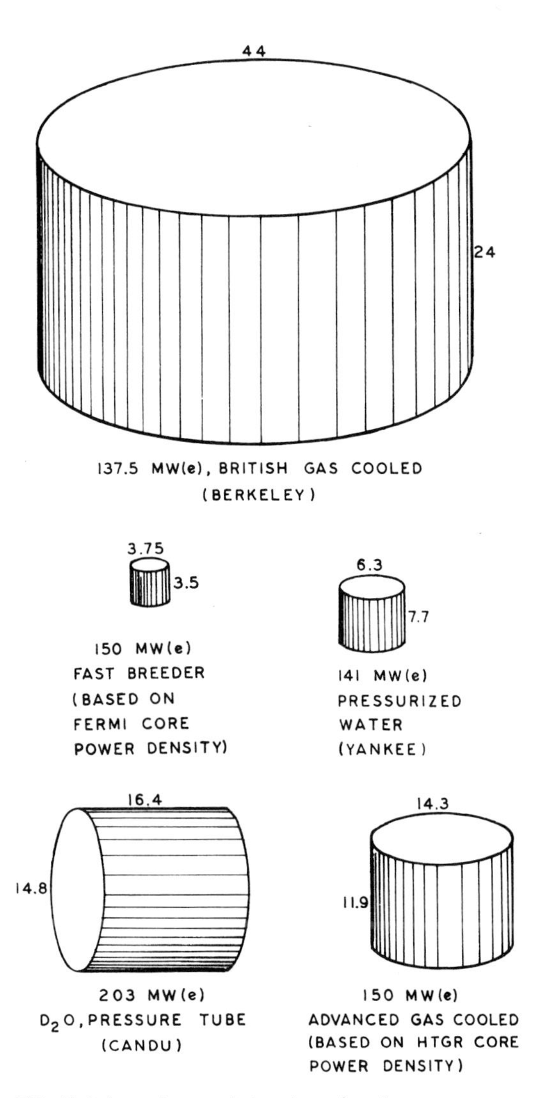

FIG. 10-1 Approximate relative sizes of various reactor cores. Numbers indicate approximate core lengths in feet.

10.1 Solid-Moderator Reactors

This class of reactors includes most of those which are classified as "dry" reactors in the Solid-Moderator Reactor Kinetics Chapter. The discussion here centers around two representatives of the class: the gas-cooled, graphite-moderated reactor and the sodium-cooled, graphite-moderated reactor. Both of these employ a basic core structure which is built up of stacked graphite blocks, as indicated conceptually in Fig. 10-2. In the gas-cooled case, however, the graphite is exposed to the coolant, while in the sodium-cooled case it must be "canned."

10.1.1 Natural Uranium, Gas-Cooled

The principal mechanical design consideration for the graphite structures of the large, natural-uranium, gas-cooled reactors is to assure that any distortions which take place during the life of the plant do not interfere with the removal and charging of fuel or with the operation of the control rods. Since the graphite assembly may be over 50 ft (15.2 m) in diameter by 30 ft (9.1 m) high, may weigh thousands of tons, and may be penetrated by several thousand fuel and control channels, the problem of designing the structure to remain stable despite the effects of irradiation, temperature, and pressure is a considerable one. Dimensional changes and induced stresses caused principally by temperature effects are complicated by the differences in the coefficient of thermal expansion of graphite and the steel support members of the core structure. Irradiation damage produces two effects: individual blocks are subjected to stress concentrations due to differences in fast neutron flux across the block; and the graphite changes dimensions with irradiation, the rate of change with exposure being dependent upon the operating temperature and the method of graphite manufacture.

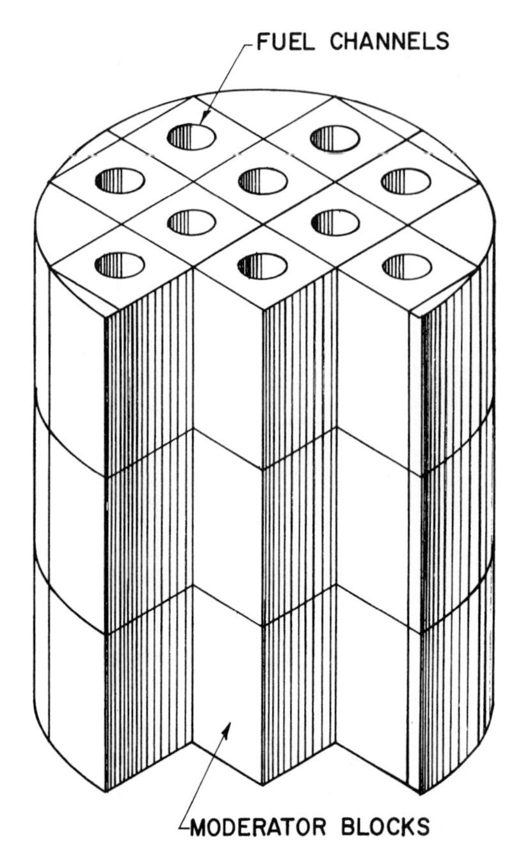

FIG. 10-2 Solid-moderator core concept.

The Calder Hall (196,224) reactors represent a design whose basic features have been used in all of the later British plants. These common features include the reactor vessel, a core support grid, a vertical core, and top access ports. Coolant flow is upward through the core. The core is not attached directly to the reactor vessel but merely sits upon a support plate atop the support grid, which in turn is attached to the vessel. The grid, being an open structure, permits free passage of the coolant, and the support plate is provided with holes at the proper locations to allow entry of the coolant to the channels in the graphite above.

If the graphite columns were placed directly upon the support plate, the differential thermal expansion that occurs because the coefficient of graphite is smaller than that of steel would tend to separate the graphite columns as the system temperature increased. To prevent the separation of the columns, the entire graphite stack is banded circumferentially by a number of restraining members which apply hoop restraint. The members are designed to absorb the strain imposed by differential expansion, but still act as continuous garters around the graphite structure. To reduce the stresses in the lower bands, the designers provide for rolling friction between the graphite stack and the support plate so that the steel support plate can expand at a greater rate than the graphite without exerting high graphite-separating forces. The rolling-friction system employs eighteen 3/4-in. (19.05 mm) -diameter balls under each column of graphite, the balls operating between races at the bottom of the column and above the support plate, as shown in Fig. 10-3. Each column system has a capacity of 12 tons under operating conditions, while the actual load is about 1/2 ton. The balls and races are stainless steel with a hardness greater than 47 Rockwell C scale, to insure that they will retain adequate hardness when the pile temperature is allowed to increase (when annealing out the Wigner energy). The bottom system of bands

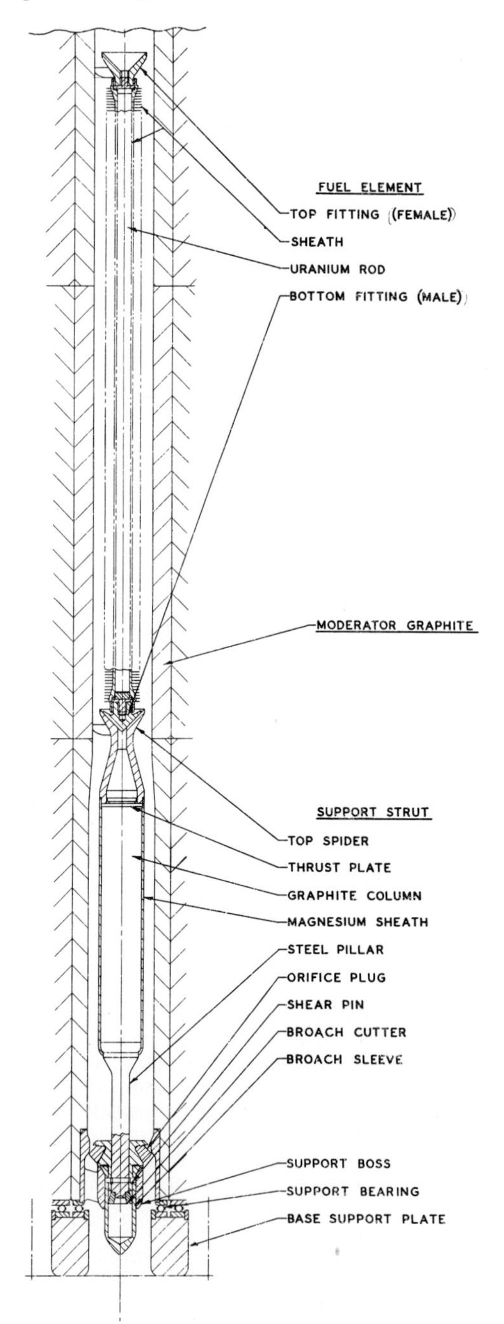

FIG. 10-3A Vertical section, bottom of Calder Hall fuel channel.

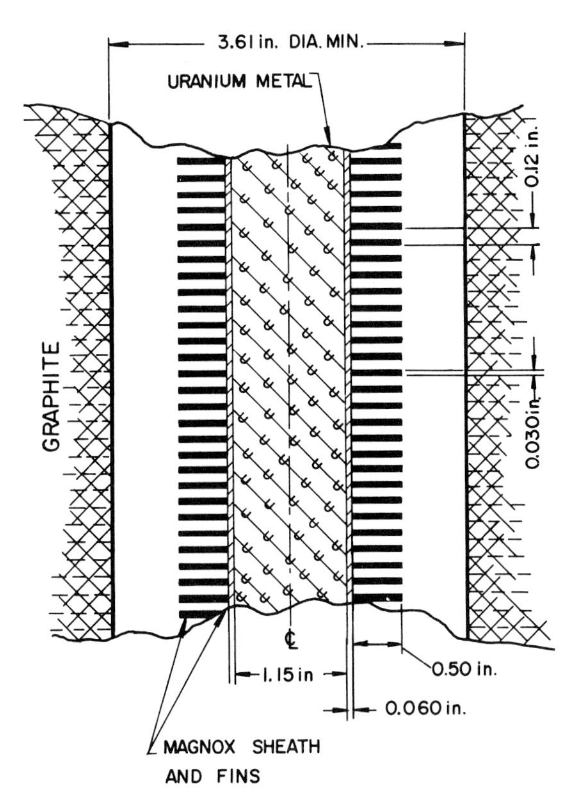

FIG. 10-3B Section of Calder Hall fuel element in coolant channel.

around the graphite stack is provided with several brackets which function as key-ways to engage keys welded to the core support grid structure. These key-ways are radially oriented and serve to maintain the graphite structure in its correct position relative to the access ports in the reactor vessel top.

The top of the graphite stack is covered with heavy castings called "charge pans". These castings, one for each 16 channels, serve to hold down the top layer of graphite blocks, to support the bottom of the fuel-charging chute, and to provide terminal points for the sample tubes of the burst-slug-detection system.

An annular gap, varying in width from 6 to 8 in. (15.2 to 20.3 cm), exists between the graphite structure and the reactor vessel side wall. Although it is necessary for a small amount of coolant to flow through this annulus to cool the structure, far too great an amount would bypass the fuel channels if the gap were left entirely open. Therefore, a seal is installed around the bottom of the graphite and the leakage to the annulus is metered. The seal consists of two horizontal rings of 0.2-in. (5.08 mm) -thick metal which are attached to a flange welded to the vessel and which slide in peripheral slots in the graphite. The maximum differential movement between the seal metal and the graphite is about 0.030 in. (0.762 mm) in the vertical direction and about 1 in. (25.4 mm) horizontally. The metal rings are chromium-plated to reduce corrosion and are provided with bearing strips that contact the graphite. Enough 0.5-in. (12.7 mm)-diameter holes are drilled in the seal to permit a leakage flow equal to 0.5% of the total coolant flow.

Each fuel channel contains a stack of 6 fuel elements and a bottom support strut, the details of which are shown in Fig. 10-3. The fuel elements consist of massive, solid rods of natural uranium metal, 40 in. (101.6 cm) long and 1.15 in. (2.92 cm) in diameter, clad in magnesium-alloy sheaths having transverse fins for extended heat-transfer surface. Each element has a male fitting attached to its bottom end and a female fitting, with radial spacers, attached to its top end. The support strut at the bottom of the column is equipped with a female fitting at its upper end, and proper mating of the fittings results in transferring the load of the entire stack of fuel elements, about 170 lb (77.2 kg), to the support strut. Thus, the components which are subjected to the greatest loads are those located at the relatively cool end of the channel, where the effects of thermal creep are minimized.

The support strut, which also serves as a bottom spacer for the fuel columns, is constructed of materials with low neutron absorption properties as indicated on Fig. 10-3. The strut incorporates a shear pin-broach shock-absorbing mechanism capable of damping the loads that would occur if, during fuel element removal or replacement, a fuel element were accidentally dropped from the top of the reactor. Since the strut can perform its shock-absorbing function only once, it is designed to be removed and replaced in a fashion similar to the fuel elements. The lower end of the strut fits into a support boss, a permanent fitting which contacts the upper race of the graphite column support bearing. Thus, the load of the fuel-channel column is transferred directly to the thrust bearing and not to the graphite structure. The lower end of the support strut is equipped with a horizontal disc which, in combination with an extension of the support boss, forms a flow restriction. The disc can be adjusted before the strut is inserted into the reactor, and in this manner the coolant flow can be orificed to match the radial power distribution.

No top holddown is provided for the fuel column, but the weight of the top element is sufficient to prevent its "floating" upward under the influence of coolant flow. The fuel column is free to expand upward as the system temperature increases, but the reactivity effects accruing from this small increase in the effective core height are unimportant in a system as large as the Calder Hall core.

The cylindrical control rods in the Calder Hall reactors are cable-connected to drive mechanisms which operate through ports in the reactor tops. Gravity provides the motive force for rod insertion, and each rod is equipped with a spring shock absorber that supports the rod in its normal lowered (inserted) position, with the rod top resting on the charge pan atop the graphite core structure. The rods move in unlined holes in the graphite moderator, the holes having a diameter of 3.25 in. (8.25 cm) to accommodate the 1.75 in. (4.44 cm) -diameter rods with adequate clearance. The normal rate of insertion of a control rod is less than 6 in./sec (15.24 cm/sec), even during scram, and the spring shock absorber in the rod provides the normal cushion at the end of stroke. Should a drive cable break, the capacity of the spring shock absorber could be exceeded and an unacceptable force delivered to the charge pan. To prevent this, each rod is provided with shear pins which limit the impact force on the charge pan to 300 lb (136 kg). If the pins are sheared, the rod continues to fall and its kinetic energy is absorbed by a broach-type shock absorber at its bottom end; the absorber contacts a structural member located at the bottom of the control rod channel. This structural member, called a "restrictor", also serves to limit the coolant flow through the control-rod channel. The restrictor is designed so that the differential pressure which would exist after a reactor vessel or coolant loop rupture cannot cause it to rise in the channel and block the entry of the control rod. The control rod and channel dimensions are such that no possible pressure differential can cause the rod to be blown out of the core.

Operating procedures include periodic surveys of the condition of the core structure. Provision is made for the use of gauges and optical instruments which can be inserted through charge ports and which utilize permanent "bench marks" both inside and outside the vessel. Condition of the channels through the graphite is checked by inserting a special television camera in the channels. Techniques have also been developed for flash photography of the inner surface of the pressure vessel dome, giving excellent definition.

In summary, the main safety characteristics of reactors of the Calder type are related to the following salient features of the basic reactor concept and its mechanical design:

1. The use of the moderator as the basic structural material provides a massive core structure which apparently will maintain the fuel/moderator ratio and the core configuration in any conceivable situation except a fuel meltdown. Important reactivity effects due to internal distortions of the core would not be expected and have not been found in practice. Changes in fuel/moderator ratio and in self-shielding factors could result if the fuel elements melt or slump from overheating, but the natural uranium reactors are ordinarily designed with a fuel/moderator ratio which gives very nearly the maximum reactivity, and it would be unlikely that melting or slumping could produce a net increase of reactivity. The possibility might exist if enriched fuel were used. Boron powder is held in a compressed air device for injection against any such contingency.
2. The effects of expansion and growth have been allowed for by a system which applies enough restraint to make the core configuration a determinate one: the core structure — and the reactivity — cannot "wander." This is an important consideration in reactor safety.
3. The problem of maintaining free control rod motion, despite local and general displacements of the core structure, is solved by the provision of generous clearances around the control rods and by the use of a very flexible actuating system (cables). The feasibility of the large clearances results from the nuclear characteristics of the graphite reactor: the total rod worth required is small and the diffusion length is large enough that rods which are very small relative to the core size can be used; thus the clearance does not amount to any significant loss of core volume. Because the clearance space is essentially empty so far as neutron diffusion is concerned, it does not present any barrier to the flow of neutrons into the control rods, and uncontrolled lateral displacements of the rods in their channels do not result in appreciable reactivity fluctuations. The degree of indeterminancy in the positions of individual rods which goes along with the cable drive system is acceptable because the worth of a single rod is small. In the Calder reactors rod positions are indicated to a precision of ± 1 in. (±2.54 cm). Any sticking of a rod which might conceivably occur would be detected by a zero-tension indicator on the cable, but apparently sticking has not been a problem in practice. The net effect of thermal expansion of the cables, their support system, and the core and its support (see Fig. 10-4) no doubt makes a contribution to the temperature coefficient of reactivity when many rods are held in partially inserted positions. The effect, although not a serious one in this case, would normally be investigated as part of the core design.
4. The effects of abnormal pressure differentials, as from a break in the coolant circuit, have been taken into account in the mechanical design, so that they cannot cause dislocations of control rods or other important core components.
5. Any distortions which may occur in the massive single-rod fuel elements are unlikely to have any significant effects on reactivity because they do not change the fuel/moderator ratio, either locally or generally. The coolant passages are large, and the problems of avoiding any possiblity of blocking the passages are straightforward so long as the elements may be assumed to maintain their integrity. The jackets, however, do operate at 430° C (806° F) and the fuel itself is subject to swelling under irradiation. These considerations have imposed the fundamental limitations in the thermal design of the reactors.
6. All the reactivity coefficients of the reactor are relatively small. The only major fast-acting coefficient is the Doppler coefficient. It does not give rise to a large normal power coefficient of reactivity because of the relatively small rate of change of average fuel temperature with reactor power. The coolant density coefficient of reactivity is negligible because of the low neutron absorption cross section and the low normal density of the CO_2 coolant. The moderator temperature coefficient of reactivity is negative if the reactor is loaded with fresh fuel, but becomes positive as plutonium builds up, and finally causes the over-all temperature

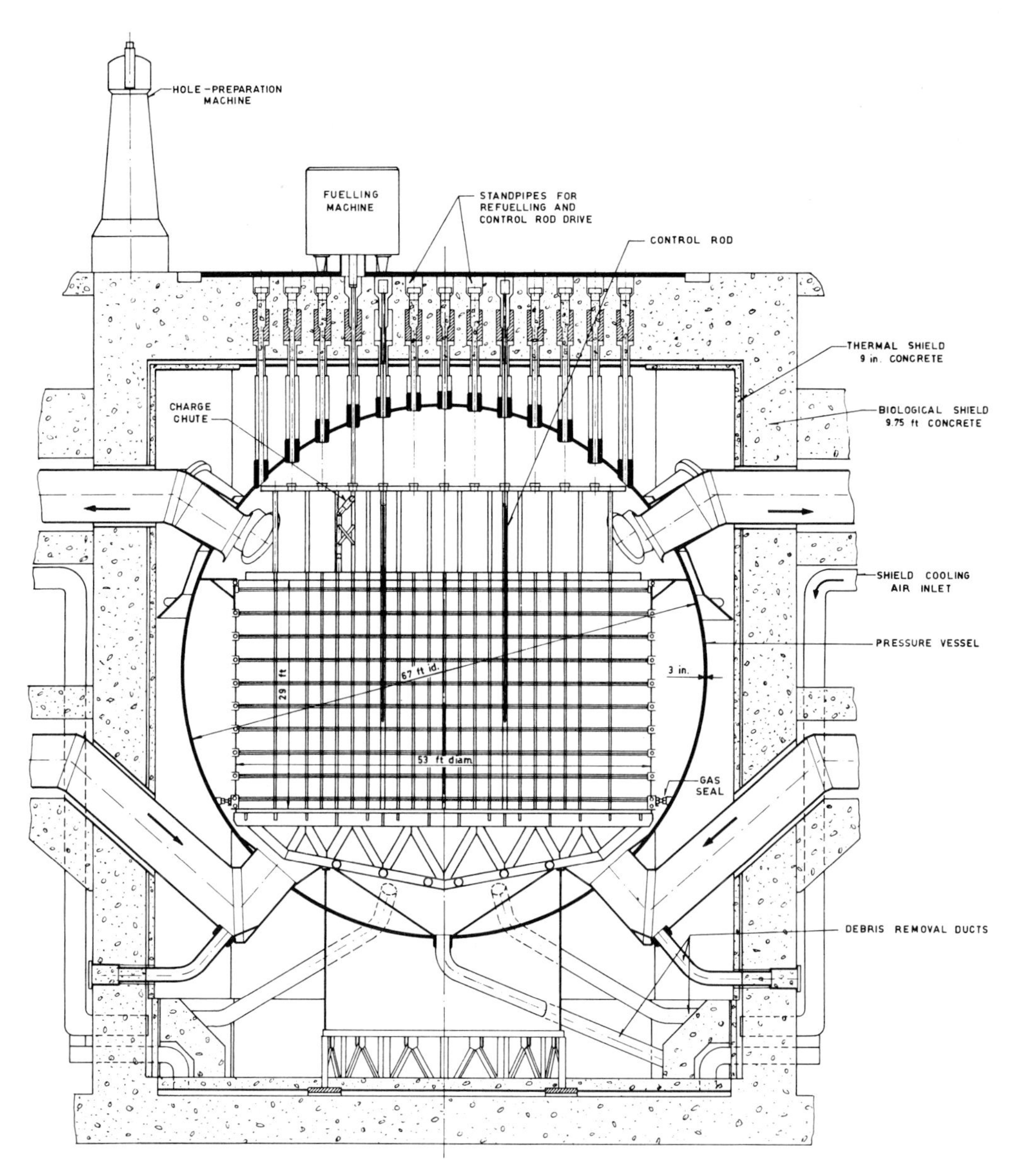

FIG. 10-4 Vertical section, Hinkley Point Reactor.

coefficient of reactivity to become positive. This positive coefficient is tolerable because, when considered as a power coefficient, its time constant is very long.

7. The problem of emergency cooling in the event of a coolant system rupture and accompanying loss of CO_2 coolant is a serious one. Each of the core materials (Magnox, uranium, and graphite) can react rapidly with air at some temperature. This problem is one which must be considered very carefully in the mechanical and thermal design of such a reactor.
8. The generally large scale of the reactor facilitates in-core inspections during the reactor life, and such inspections have been provided for in the design and in the operating plans.

The later British natural uranium reactors retain the basic design concepts of Calder Hall. A vertical section of one of the Hinkley Point [196, 248] reactors, scheduled for operation in 1963, is shown in Fig. 10-4. Its thermal power level is 954 Mw, over four times that of a Calder Hall unit. Other than this increase of power level, the major changes involve the use of helically finned, rather than transverse-finned, fuel jackets, an increase in coolant pressure, and the provision of on-power refueling. The higher pressure and somewhat higher specific power increase the difficulty of some of the safety provisions, but do not change them qualitatively. The on-power refueling is carried out one element at a time, with warning devices to detect the inadvertent lifting of more than one element, and consequently the reactivity changes associated with individual refueling operations are not large enough to constitute appreciable hazards.

10.1.2 Sodium-Cooled

The reactor region of the graphite-moderated, sodium-cooled Hallam Nuclear Power Facility [60, 196, 235, 249] is shown in Fig. 10-5. The reactor, which utilizes slightly enriched uranium fuel, has a gross heat output of 240 Mw and a net electric output of 76 Mw. The moderator assembly approximates a vertical right circular cylinder and is constructed of full-height, vertical graphite elements encased in stainless steel. A horizontal section through a moderator element is shown in Fig. 10-6, which illustrates the scalloped edges employed in the elements to form fuel and control rod channels. Each moderator element consists of six hexagonal graphite blocks stacked one above the other and clad in a single, overall steel sheath, 0.016 in (0.41 mm) thick. The blocks are not connected but alignment is maintained by graphite dowels. As temperature increases, the steel sheath expands relative to the graphite, and the frictional forces between the sheath and the graphite segments lift each seg-

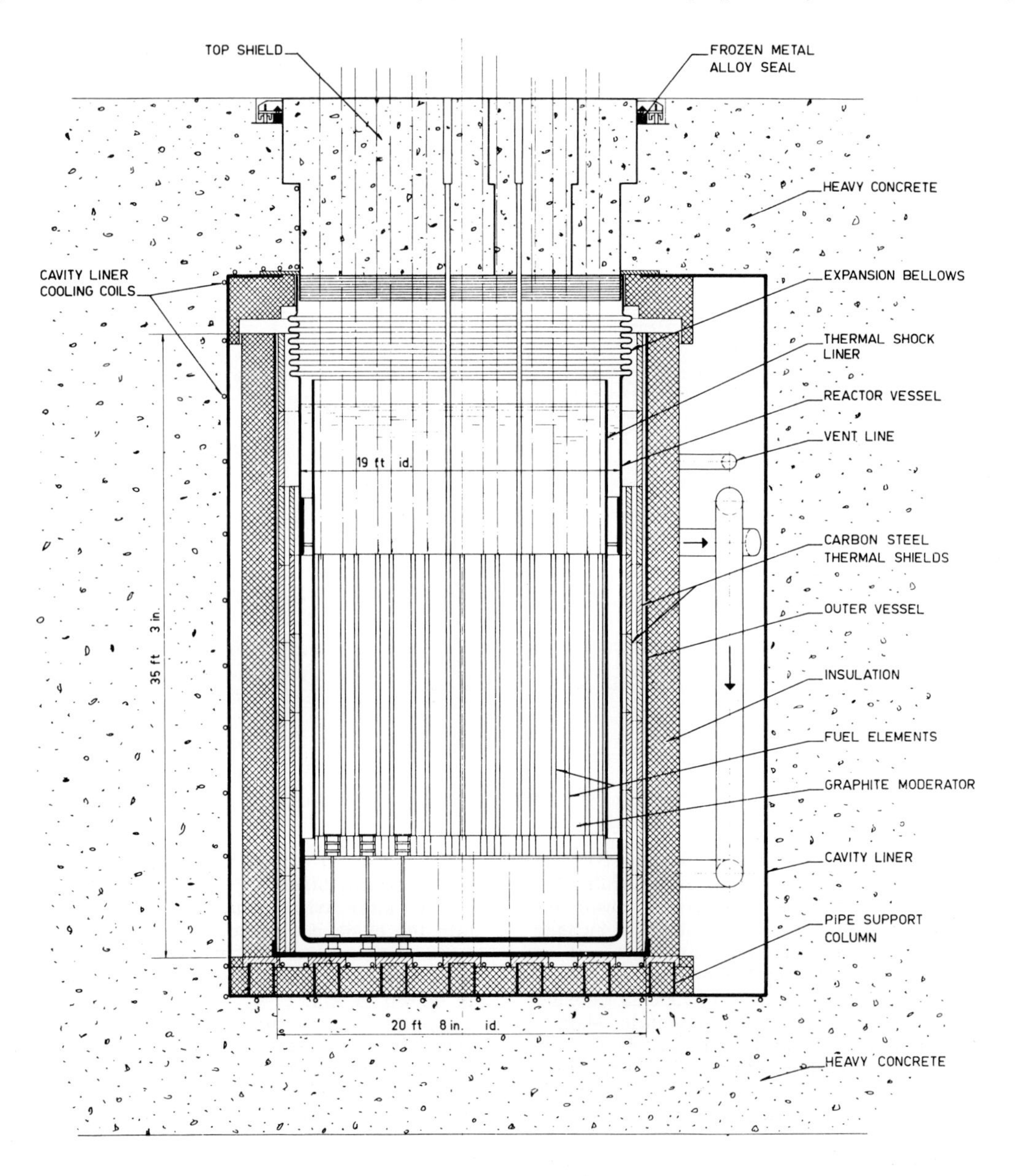

FIG. 10-5 Vertical section, Hallam Reactor

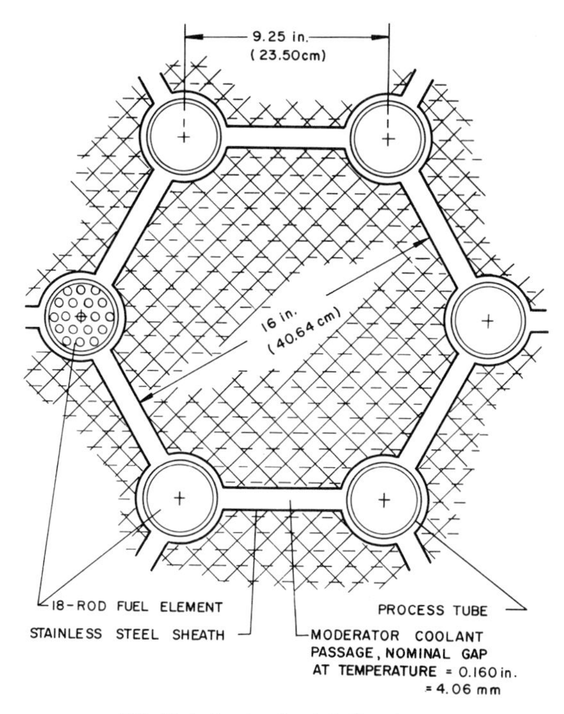

FIG. 10-6 Lattice detail, Hallam Reactor

ment off the one beneath it. Since the segments are free to separate, the frictional stresses in the sheath are limited to an acceptable value. The moderator elements rest on pedestals on a bottom grid plate and are positioned at the bottom by the grid plate. At the top the assemblage of elements is held together by core clamps which are supported by an internal ring welded to the reactor vessel, but a nominal clearance of 0.160 in. (4.06 mm) at operating temperature is maintained between elements by spacers attached to the tops of the sheaths. The moderator is cooled by sodium which flows upward through these clearance passages (Fig. 10-6).

Each fuel assembly consists of 18 stainless-steel-clad fuel rods of U-10 wt.% Mo which surround a central, hollow spacer tube and which are themselves housed in a cylindrical, Zircaloy-2 process tube. The fuel-bearing region of the assembly is provided with end fittings as shown in Fig. 10-7. The element is supported by hanging from the top, and a piston-ring fitting at its bottom plugs into a hole in the core support grid. The piston-ring fitting accommodates the axial growth of the element by permitting downward expansion, and at the same time assures that the upward-flowing coolant is properly directed to the fuel elements and does not leak around the process tube. The top region of the fuel assembly consists principally of a hanging-support device and a coolant flow orifice. The orifice is adjustable during operation, but the action necessary to accomplish a flow change has been made relatively complex in order to prevent too rapid adjustment and a consequent rapid coolant temperature change. The plug-type orifice is designed so that the minimum attainable flow is 26% of the maximum flow, and any adjustment beyond the minumum point results in a flow increase.

Experience with the Sodium Reactor Experiment (SRE) reactor [60] has shown [152] that it is important, in a sodium-graphite reactor, to restrain the elements of a fuel assembly against bowing. The SRE is an experimental reactor which contributed to the design of the Hallam reactor and which is much like it in its general features. The second fuel loading of the SRE consisted of five-rod assemblies in the configuration indicated by Fig. 10-8A. Each fuel rod consisted of a sealed stainless-steel tube 96 in. (2.44 m) long, containing a stack of 12 thorium-7.6 wt.% uranium alloy slugs. The slugs were 0.750 in. (1.905 cm) in diameter by 6 in. (15.2 cm) long. The tube was filled with a sodium-potassium (NaK) alloy, to a level 9 in. (22.9 cm) above the fuel slugs, to serve as a thermal bond. Each fuel rod was spirally wrapped with a 0.091-in. (2.31 mm) -diameter stainless-steel wire on a 10.5-in.(26.7 cm) pitch, as a spacer.

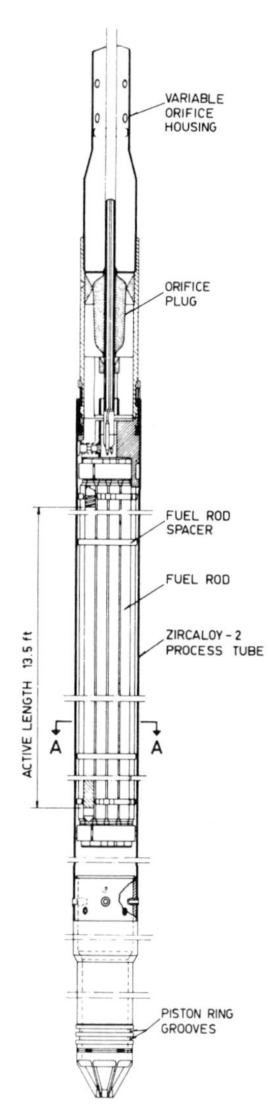

FIG. 10-7 Fuel assembly, Hallam Reactor

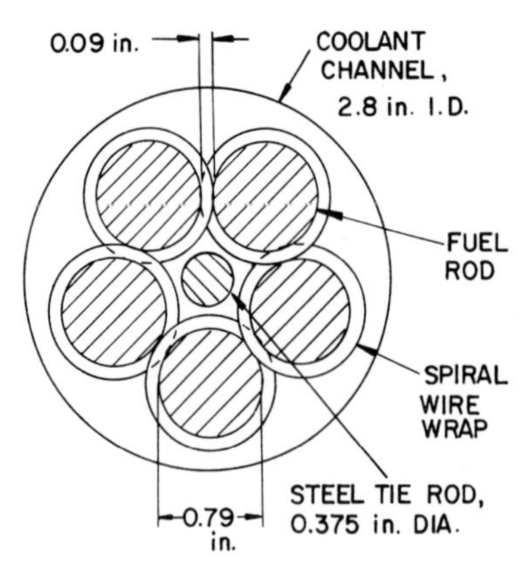

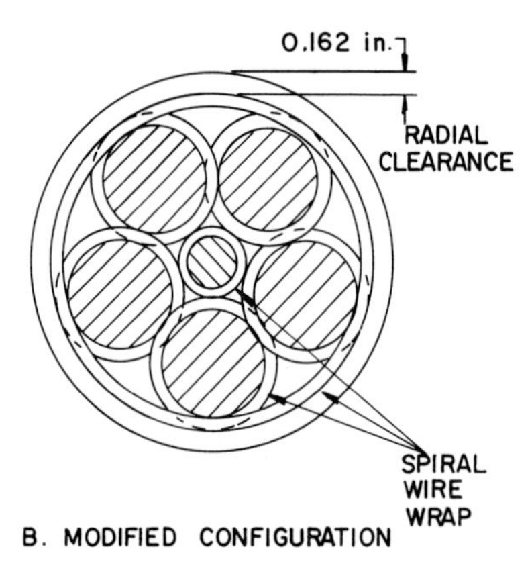

FIG. 10-8 Sections through SRE fuel bundles (second core loading).

The five fuel rods were assembled around a central stainless-steel hanger rod, with supports at the ends only.

When the reactor was operated at power, it was found that the fuel rods bowed outward because the thermal neutron flux depression in the bundle resulted in higher power densities in the fuel nearest the graphite moderator. This deflection decreased somewhat the flux depression in the fuel, and therefore increased the thermal utilization and the reactivity. It also increased the fraction of sodium within the bundle, and therefore decreased the average thermal neutron flux in the sodium; this effect also increased the thermal utilization. At the same time the increased separation of the rods increased slightly the resonance absorption by the thorium. The latter effect decreases reactivity, but the net effect was a reactivity increase which contributed a fast positive component of 18¢/Mw to the power coefficient of reactivity and caused the reactor to be unstable (although controllable). The condition was corrected by providing a helical wire wrapping on the central rod, to restrain inward bowing, and a helical wrapping around the entire bundle to restrain outward bowing (Fig. 10-8B). This effect in the sodium-graphite reactor is important not only because there is a substantial reactivity change associated with the bowing of fuel rods but also because there is no strong fast-acting negative component of the power coefficient to counteract the effect. Bowing is prevented in the Hallam reactor by spiders which are attached at 1-ft intervals along the length of the central spacer tube in the 18-rod fuel assembly; the spiders act as spacers for the individual fuel rods.

The 19 cylindrical control rods enter the Hallam reactor from the top and move within helium-filled, Zircaloy-2 thimbles which extend down from the face of the top shield to the core support grid, and occupy regular lattice positions. The bottom ends of the thimbles are equipped with closed, piston-ring-type fittings similar to those on the bottoms of the fuel elements. These fittings enter holes in the bottom core support structure, thereby fixing the transverse position of the bottom of the thimble while permitting longitudinal expansion. Each control element is connected to a ball-nut-and-screw actuator by a set of three pull rods. Each is provided with three shock absorbers, located in the loading face shield, which operate on a portion of the control rod assembly greater in diameter than the poison portion which enters the core thimbles. Thus, failure of these shock absorbers would not permit the lower end of the control rod to strike the bottom of the thimble but would only exert high decelerating forces on the control rod assembly. The assembly is designed to withstand at least three falls without shock absorbers. If the poison section of the control rod should separate from the upper, shock-absorbing sections, it would fall to the bottom of the thimble; the thimble is designed to absorb the impact of such a fall without failing.

The reactor vessel is protected from thermal shock by shock liners, cylindrical steel shells which surround the core and provide a region of stagnant sodium between the core and the vessel wall to act as a temperature buffer. These are necessary because of the very high heat transfer coefficients which characterize sodium, even at low flow velocities, and because of the relatively high temperature rise of the sodium in traversing the core at full power (335°F or 186°C). In the event of a scram, the vessel wall could be cooled very rapidly if the thermal shock liners were not provided.

The stainless-steel bellows assembly, which is attached to the top of the reactor vessel side wall and seals the vessel to the upper cavity liner, accommodates thermal expansions and contractions.

Since the primary coolant system operates only slightly above atmospheric pressure, large differential pressures would not result from a coolant system rupture. The design of the reactor vessel and reactor cavity is based on the principle of maintaining sodium over the fuel elements at all times, and the vessel nozzle locations and cavity size guarantee this even in the event of primary loop failures or reactor vessel leaks. The sodium piping layout is designed to prevent siphoning from the vessel or cavity region in the event of a leak.

The more important relationships between the mechanical features of the Hallam reactor and its safety performance may be summarized as follows:

1. Although the graphite moderator does not perform the same structural purpose in the

Hallam reactor as in the gas-cooled reactors, it still serves as a permanent solid barrier against any substantial changes in fuel/moderator ratio which might occur for any reason other than melting of the fuel. As in the gas-cooled reactors, it seems unlikely that melting would lead to reactivity increases, but this possibility would normally be investigated in the reactor design, particularly since relatively high enrichments are used.

2. The basic core dimensions are determined by metallic members — the bottom grid plate and metallic spacers and core clamps bearing laterally upon the relatively massive end closures of the moderator sheath cans. Consequently the important core dimensions are not affected by dimensional changes of the graphite.
3. The relative thermal expansions which must be taken into account in assuring free motion of the control rods are much smaller than in the gas-cooled reactors, primarily because of the smaller reactor size. The problems of lateral displacement which do exist are solved by the use of a long and rather flexible connection (a system of three slender parallel hanger rods) between the control rod proper and the drive mechanism, and by the use of a flexible structure for the control rod itself (short annular cans of gadolinium and samarium oxide strung upon a central support tube). The problems of rod actuation in sodium are bypassed by installing the rods in helium-filled thimbles; this is feasible because of the low pressure of the coolant system.
4. There are no problems arising from high pressures or high pressure differentials. On the other hand, thermal shock effects, which are possible because of the very effective heat transfer capabilities of sodium, must be and are guarded against by the provision of thermal shock shields and by the appropriate control of sodium flow.
5. The fuel assemblies are subdivided into relatively small elements (compared to the gas-cooled case) in order to take advantage of the high cooling effectiveness of the sodium. Bowing of these elements, either because of temperature gradients or because of pressure differentials, must be avoided because of associated reactivity changes. Part of any such reactivity change is related to the neutron absorption of the sodium. In the Hallam reactor bowing is prevented by suitable fuel rod spacers. For the maintenance of free coolant passages in the fuel assemblies, the well-known problems of sodium oxidation must be kept under control by suitable handling and cleanup measures. The metallic uranium fuel is expected to grow as a result of irradiation. It is expected that the growth will occur in the longitudinal direction, and space is left, at the upper ends of the fuel element jackets, to accommodate a growth of 6 vol.%
6. When the fuel elements are restrained to prevent bowing, the normally occurring reactivity coefficients in this reactor type are small. It is the absence of strong, rapidly acting negative power coefficients of reactivity which makes reactivity effects due to rod bowing be particularly troublesome. The loss of coolant from the reactor would cause a substantial reactivity increase, but the permanent loss of coolant is considered highly improbable. The possibility of boiling and its reactivity consequences would normally be considered in any safety evaluation of this reactor type.
7. Because the coolant operates at a temperature well below its atmospheric boiling point, it is feasible to design a system in which the possiblility of loss of coolant, to the extent that fuel elements would be uncovered, is essentially eliminated. The problem of emergency cooling then reduces simply to one of removing heat from a large body of sodium. Although the problem of flashing of the coolant is absent, careful attention must be given to the problem of maintaining the sodium temperature above the melting point at all parts in the coolant system.

10.2 Pressure Tube Reactors

A pressure tube reactor consists basically of a large number of pressure tubes which pass through a body of low-pressure moderator; the pressure tubes contain fuel elements and constitute the ducts through which coolant flows, at a pressure substantially higher than that in the moderator space, over the fuel elements. In the field of power reactors, the pressure tube concept has been used mainly for heavy-water-moderated systems, but it is also applicable to solid moderators. The Russian First Atomic Power Station [199], for example, is a pressure tube reactor in which the fuel elements themselves serve as pressure tubes. The discussion here will be limited to heavy water moderation and to heavy water cooling, although various other coolants are possible, and serious development is under way on heavy water moderator concepts employing gas [250] and organic liquids [251, 252] as coolants.*

The basic mechanical design problem in this reactor type is to provide a leak-tight moderator container penetrated by pressure tubes, to insulate the hot coolant thermally from the cool moderator, and to provide for the differential expansion of the pressure tubes relative to the moderator tank. The difficulty of the problem is greatly increased because of the necessity to minimize the neutron absorption by the pressure

---

*The steam generating heavy water reactor (SGHW) now being constructed by the U.K.A.E.A. at Winfrith Heath is a 100 Mw (electrical) power demonstration reactor which uses boiling water on a direct cycle to the turbine. A number of superheat steam channels are also incorporated and the reactor is arranged for on-load refuelling. For fuller description see reference [264].

tubes and by any other structural or insulating material required in the core region. Once the design problems have been solved, the permanent structure of moderator tank and pressure tubes provides a well-defined structural matrix which maintains the gross geometrical fuel-moderator relationship, much as the moderator structure does in reactors of the solid moderator type.

Since each pressure tube constitutes a separate, isolated, coolant flow path, the concept is ideal for the measurement of thermal conditions in each pressure tube and for the control of coolant flow in the individual tubes. Because the moderating properties of heavy water are such as to require relatively large ratios of moderator volume to fuel volume, there is usually ample room for the installation of control rods in the low-pressure moderator between pressure tubes, although the ends of the reactor are usually cluttered with the headers and connecting pipes for the individual pressure tubes, and this may complicate the installation of axially oriented control rods. Since the cooling of the fuel elements does not depend on the presence of the moderator, dumping of the moderator from the reactor tank is a practical and very effective means of shutdown, although it is slower in action than the usual control rod scram. The individual pressure tubes are quite adaptable to the use of an on-power refueling system which can virtually eliminate the reactivity change due to fuel exposure. At the same time, the temperature defect of reactivity tends to be small because the moderator is kept cool, and hence it is feasible in some pressure tube designs to keep the total excess reactivity quite small and to operate without absorbing control rods. This approach has been used in the Canadian NPD-2 reactor.

There are two different basic approaches to the design of a pressure tube reactor, one of which involves a "hot" pressure tube with the thermal insulation outside it, and the other a "cold" pressure tube with internal insulation. The hot-tube concept can of course be used only when the coolant temperature is compatible with the temperature capabilities of the tube material. When the hot-tube approach is used, it is usually convenient to employ it in connection with a calandria design for the moderator tank. That is to say, tubes of low neutron cross section material (usually aluminum of zirconium alloy) having inside diameters slightly larger than the outside diameter of the pressure tube, are welded between the end plates of the moderator tank, to form a water-tight vessel (calandria) penetrated by empty channels into which the pressure tubes can be installed. The gas spaces between the calandria tubes and the pressure tubes then serve as the thermal insulation. In the cold-tube designs, it is not possible to separate so simply the moderator and coolant containers, and the solution of the design problem cannot be stated in general terms. In the following discussion specific reactors embodying the two approaches will be discussed.

10.2.1 Calandria Type — NPD-2

The Canadian NPD-2 reactor, of 82.5 Mw(t) output (20 Mw(e), gross), is a prototype of the 200 Mw(e) CANDU reactor, and hence most of the description applies qualitatively to CANDU also. The basic arrangement of the reactor is illustrated in Fig. 10-9. The attitude of the reactor is horizontal. A double-walled moderator tank of the calandria type is used, the outer region of the tank providing space for a light water reflector. The calandria tubes are thin-walled aluminum tubes, 4-in. (10.16 cm) inner diameter and 0.052 in. (1.32 mm) wall, which are welded into the double-walled tank structure as indicated in Fig. 10-9. The Zircaloy-2 pressure tubes, 0.163 in. (4.14 mm) thick, are installed within the calandria tubes, and are then connected to stainless-steel end fittings, of larger diameter, by roll forming. The pressure tube thus lies in the calandria tube, with a radial clearance of 0.20 in. (5.08 mm), which is maintained by a spiral spring spacer, and is free to expand axially independently of the moderator tank. The type 403 stainless steel end fittings provide connections to individual inlet and outlet coolant lines, as well as end closures which can be operated by an on-power refueling machine. A flow control orifice, which may also be changed by the refueling machine, is installed at the inlet end of each pressure tube.

Normally, a refueling operation on an individual pressure tube involves the replacement of only one of the 19.5-in. (49.5 cm) -long fuel assemblies. For this operation a refueling machine is connected to each end of the pressure tube. After the closures are removed, a new fuel assembly is pushed in by the refueling machine attached at the (coolant) outlet end of the pressure tube, advancing the entire column of assemblies in the pressure tube, and the exposed assembly at the inlet end of the tube is removed by the refueling machine which is connected at that end. When refueling is not in progress, the column of fuel assemblies in the pressure tube is held against the fuel latch assembly by the force of the coolant flow (see Fig. 10-9).

Long-term reactivity changes are held to a very small value by effectively continuously refueling the reactor. The small variations that do occur are controlled by slight variation of the moderator level in the tank and by variation of the moderator temperature over the range 120° to 180°F (49° to 82°C). Upon shutdown or power reduction the transient buildup of xenon causes a sizable reactivity loss. This is compensated, up to a point, by a "booster" control rod, which enters horizontally (on an axis perpendicular to that of the pressure tubes) and which consists of fuel elements of highly enriched uranium with suitable cooling circuits. The booster rod can increase the reactivity by 0.0025. Since the NPD-2 reactor is not large enough to be subject to spatial oscillations of power density, rods are not needed for the adjustment of local reactivity. A number of strategically located booster rods could perform such a function in a larger reactor. For emergency shutdown the D_2O moderator is dumped through large open ports in the bottom of the calandria vessel upon release of a helium balancing pressure by quick-acting valves (see Fig. 10-9). After the moderator is dumped, D_2O is pumped from the dump tank and sprayed on the calandria tubes to

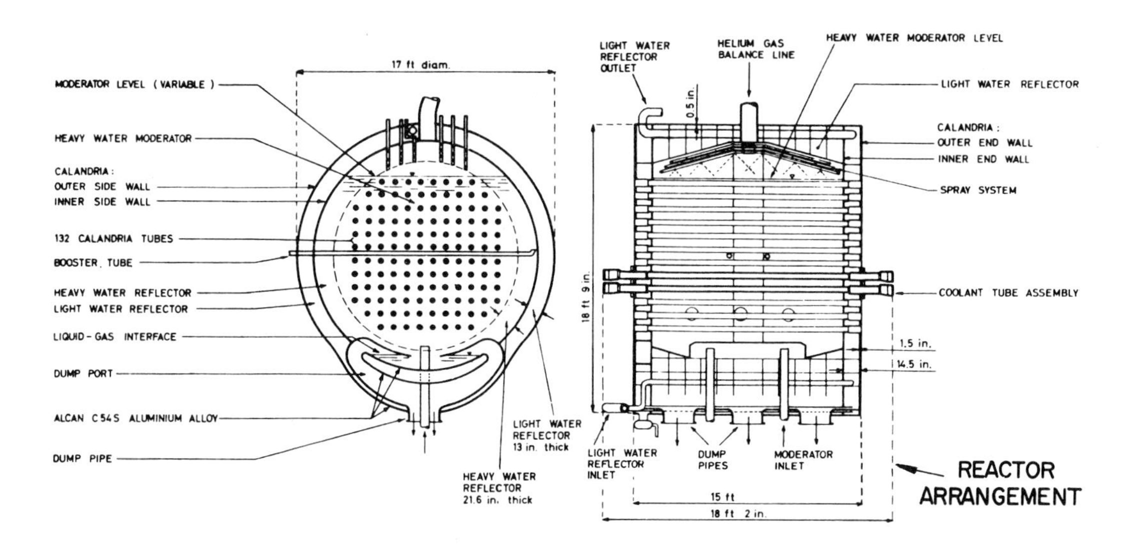

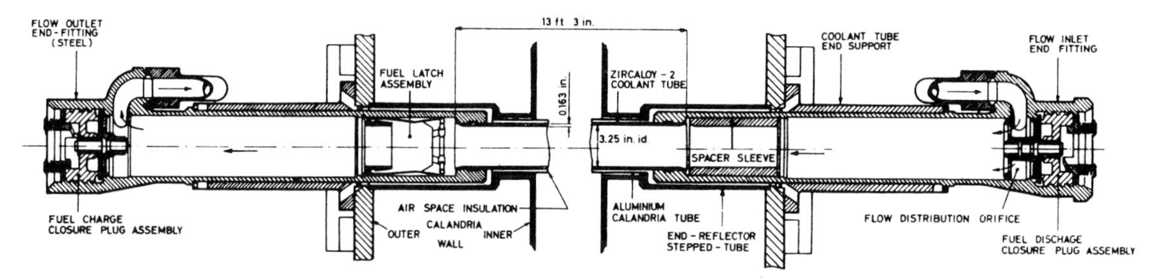

COOLANT TUBE ASSEMBLY

FIG. 10-9 Reactor arrangement and coolant tube assembly, NPD reactor.

keep them cool. All free surfaces of D_2O in the moderator system are blanketed with helium gas.

Two types of fuel assemblies, identical in length and over-all diameter, are used. An assembly of 7 rods of 1 in. (2.54 cm) diameter is used in the peripheral pressure tubes, while an assembly of 19 rods of 0.6-in. (1.524 cm) diameter is used in the higher power tubes near the center of the reactor. The fuel elements consist of sintered, high-density UO_2 pellets (of natural enrichment) in Zircaloy-2 jackets. The 7-rod assembly is illustrated in Fig. 10-10.

10.2.2 Cold-Tube Design - CVTR

The only operating example of the cold-tube design is the Carolinas-Virginia Tube Reactor (CVTR) [253], designed for a thermal output of 56 Mw to the coolant (17 Mw(e), net). It is quite different in concept from the NPD-2; in conjunction with that reactor the CVTR illustrates the wide range of characteristics which may be covered by pressure tube reactors, even when they employ the same moderator and coolant. Basically the reactor consists of a tank of cool heavy water in which are suspended 36 internally insulated, U-shaped pressure tubes. Each U-tube contains 2 fuel rod bundles, one in each leg of the U. The heavy water coolant enters at the top of one leg of the U and leaves from the top of the other leg. The tank hangs in a cavity in the concrete biological shield and supports on its top a heavy base plate on which are mounted the supports for the pressure tubes and the vertically acting control rods. The reactor concept is illustrated in Fig. 10-11.

The use of the U-tube arrangement simplifies the design of the reactor vessel, for it locates all the pressure tube penetrations in the top plate, in the region occupied by the helium blanket gas, where they do not constitute potential sources of moderator leakage. As is the usual practice in pressure tube designs, the low thermal neutron absorption pressure tubes (made of Zircaloy-4 in the CVTR) are connected to stainless-steel extension tubes which provide all the end fittings. The connection is made by a mechanical joint of the Marman Conoseal type. The U-connection at the bottom, although it is also of Zircaloy-4, is joined mechanically to the tubes by Conoseal fittings in order to avoid the problem of hydride embrittlement caused by weld zones. Each U-tube is provided with instrumentation for the measurement of coolant flow velocity and temperature rise. The 28 control rods, 12 of which are "black" and 16 "gray", enter through the top plate, as shown in the diagram, Fig. 10-12.

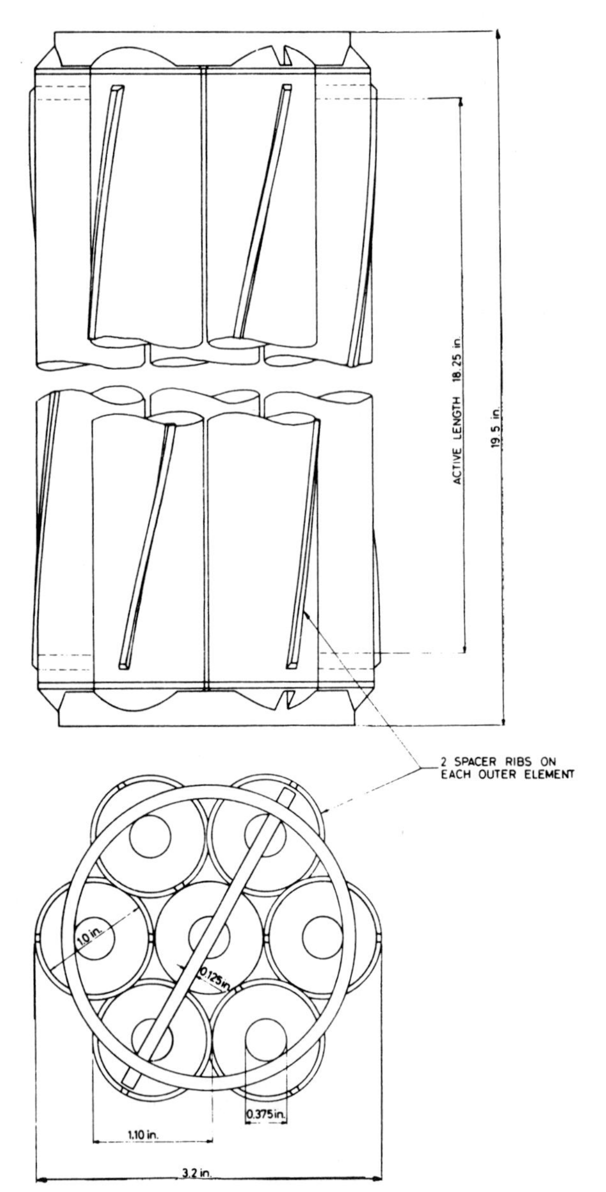

FIG. 10-10 Fuel element, NPD reactor.

The fuel assembly and pressure tube arrangement is shown in cross section in Fig. 10-13. The internal thermal insulation is provided by a series of baffles which define layers of essentially stagnant D_2O. The fuel rods, of full-core length, are spaced in a 19-rod hexagonal bundle by spiral wrappings of Zircaloy wire. The venting of the space outside the hexagonal flow baffle (Fig. 10-13) is so arranged that inlet coolant pressure is exerted on the outside of the baffle, deforming it inward to press the fuel bundle tightly together whenever coolant is flowing. The fuel rods are 0.49 in. (1.245 cm) in diameter, and are made up of sintered UO_2 pellets in free-standing cladding of Zircaloy-4. The fuel bundles near the periphery of the core employ uranium enriched to 2.0% U^{235}, while those near the center use an enrichment of 1.5%. The initial core is loaded for an average exposure of 8400 Mwd/mtU; partial reloading will extend the equilibrium fuel life to 13,400 Mwd/mtU. The initial excess reactivity provided for fuel burnup, and compensated by control rods in the operating reactor, amounts to 0.189.

The individual U-tubes are removable through the core support plate, and this is in fact one of the possible ways of refueling the reactor.

Although the CVTR utilizes the U-tube approach to circumvent the problem of installing cold-wall pressure tubes through the moderator tank, through-tube designs are possible within the cold-tube concept. In reference [164], for example, a design study for a 500 Mw(e) plant is described. Another example is the design of the French reactor [250] EL-4, which, although it employs CO_2 as the coolant, would presumably be applicable in principle to other coolants.

10.2.3 Safety-Related Characteristics

As mentioned above, the characteristics of pressure tube reactors as a class cover quite a large range, even if attention is restricted to a single coolant and moderator. However, a few remarks on the safety-related characteristics can be made which have some generality for the D_2O-moderated, D_2O-cooled type, and which have rather obvious implications for other coolant and moderator combinations.

1. The permanent pressure tube structure may be expected to maintain the gross geometrical relationship between fuel and moderator, without dependence on fuel-element integrity; thus the meltdown or extensive failure of fuel elements would not be expected to lead to large reactivity increases.
2. If pressure tube reactors are compared to pressure vessel reactors, the failure of a pressure tube is much less serious than the failure of a pressure vessel, but it may be also a more likely possibility.* In general, the approach to safe design for the vessel-type reactor is to guarantee that no vessel failure can occur, while for the pressure tube type the approach is to guarantee that, if a pressure tube should fail, no dangerous secondary effects could ensue. For this purpose it is necessary, in addition to providing for adequate cooling after a pressure tube failure, to ensure that the failure cannot lead to secondary failures of other pressure tubes, through the effects of shock waves, or through the "whipping" of a broken tube under the force of escaping steam. In cases where the moderator level

*The latter supposition is regarded by some as debatable, and it may remain so until resolved by experience. In any case, for both pressure vessel and pressure tube reactors, containment design is currently based on the arbitrary assumption of failure of the largest external pipes. The total energy in the system does not differ greatly as between pressure vessel and pressure tube reactors, while the rate of discharge of this energy is most likely to be governed by the number of pumps used, since it is this which tends to fix the size of the largest pipe.

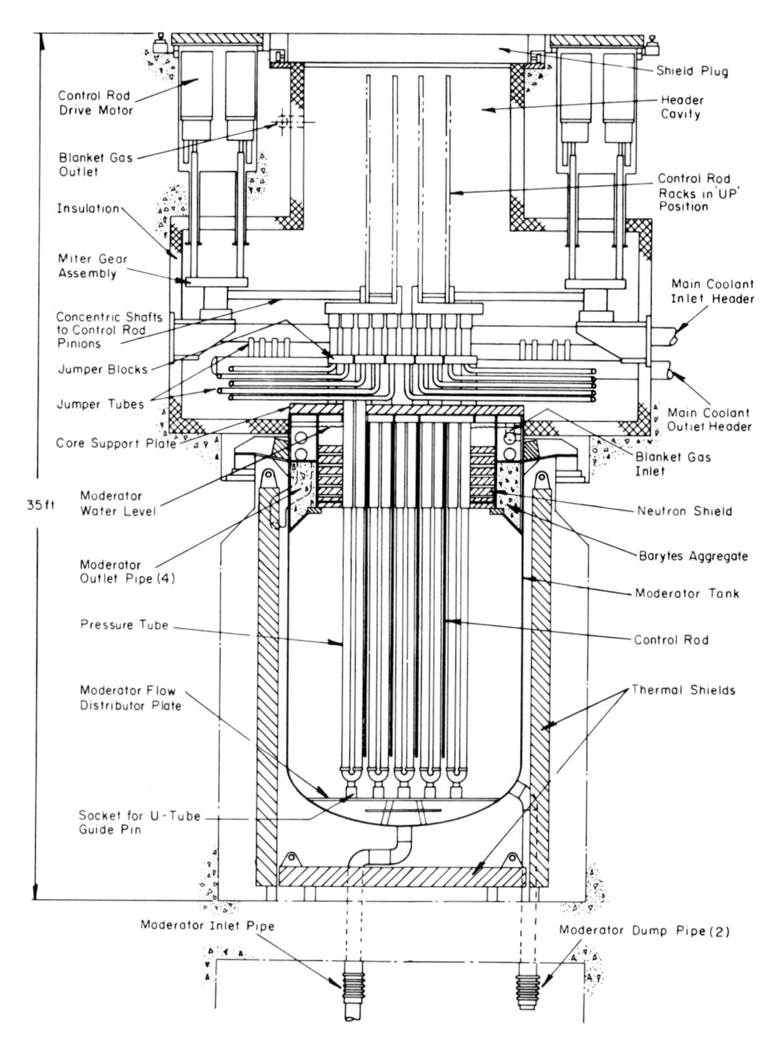

FIG. 10-11 Vertical section, Carolinas-Virginia Tube Reactor (CVTR).

is used to control reactivity, it is necessary to ensure that a steam bubble from a failed pressure tube cannot cause the moderator level to rise sufficiently to result in a dangerous reactivity increase.

3. Usually, pressure tube reactors moderated and cooled by D_2O will have positive coolant temperature and coolant void coefficients of reactivity. The temperature effect is not usually large enough to pose a stability problem, particularly since it usually makes a much smaller contribution to the power coefficient of reactivity than does the (negative) fuel temperature coefficient due to the Doppler effect. The positive void effect must, however, be taken into account in assessing the effects of loss of coolant and loss of pumping power.

   The positive effects tend to be much larger if H_2O is used as a coolant with D_2O moderation. A reactivity increase caused by the vaporization of coolant in some of the pressure tubes contributed to an accident in the NRX reactor [254, 255]; that reactor is a low pressure (research) pressure tube reactor using H_2O as coolant and D_2O as moderator. D_2O-moderated reactors employing hydrogenous organic coolants will also tend to have substantial positive coolant-void coefficients of reactivity, but the probability of trouble from this characteristic is much reduced if the vapor pressure of the coolant is low at its operating temperature, as is usually the case for designs utilizing the organic coolants.

4. The nature of the pressure tube arrangement is usually such as to preclude the possibility of power-connected bowing of the fuel channels themselves. However, the bowing of individual fuel rods in the bundle is a possibility if mechanical restraints are not provided, and such bowing would, in general, cause reactivity changes. Currently, the usual practice is to space fuel rods by spiral wire wraps. When this system is used, it is necessary in any case to elim-

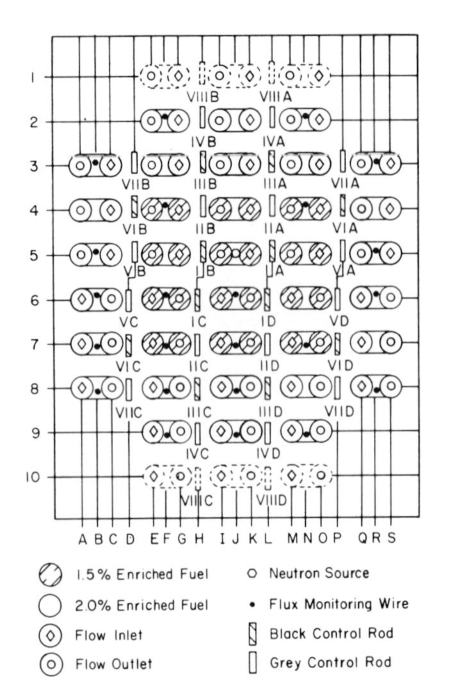

FIG. 10-12 Core cross section, Carolinas-Virginia Tube Reactor (CVTR).

inate the possibility of rod bowing in order to avoid fretting corrosion at the points of contact between the wires and the fuel jackets, and hence no significant reactivity effects due to bowing are to be expected in a properly executed design of this kind.

5. The problems of emergency cooling tend to be more difficult (but by no means insurmountable) in aqueous pressure tube reactors than in aqueous pressure vessel reactors, for a variety of reasons. Conditions are apt to be poor for natural circulation, especially in the horizontal and U-tube designs. In some designs (e.g., NPD-2), the fuel rod spacing may be very small, either to maximize the reactivity for natural uranium fueling, or to minimize the reactivity gain upon loss of coolant. This also may restrict the effectiveness of natural circulation or other emergency cooling measures. The provision of a general spray system for the core, as is done in some pressure vessel designs to provide shutdown cooling in the event of loss of primary coolant, is not effective in the pressure tube case, where coolant must be provided in each individual pressure tube. On the other hand, the design opportunities for limiting the rate of coolant loss, in the event of a major break in the coolant system, are often greater in the pressure tube reactors than in those utilizing pressure vessels.
6. The environment of control rods in pressure tube reactors is usually less severe than in pressure vessel reactors, since the rods need not operate in the reactor coolant. Space for control rods in the core is usually plentiful, and if the problem of penetrating the maze of headers and connector tubes at the ends of the reactor can be solved or circumvented, conditions are usually favorable to designing a reliable control rod system of adequate capacity.
7. The dumping of moderator may be a reliable back-up system for shutdown when it is used in conjunction with shutdown control rods. When moderator dump alone is used for shutdown, careful attention must be given to the speed limitations of the system. In large reactors these limitations are particularly important, since the moderator level may have to drop by a rather large amount before there is any large reduction of reactivity (see Sec. 4.4.1). In "cold-tube" designs cooling may be necessary for the pressure tubes in the event of a moderator dump, and the possible results of failure of the auxiliary cooling system provided for this purpose may be serious enough to override the advantages of moderator dump as a shutdown agency.
8. In the particular case of heavy-water-moderated reactors the prompt-neutron lifetime is usually relatively long, and the γ-n reaction in deuterium provides extra delayed neutron components. Both of these characteristics tend to make the response of the reactor to reactivity variations relatively slow, and confer definite safety advantage in those areas where the dynamic behavior of the reactor is important.

10.3 H_2O-Moderated Reactors

The core of the typical light-water-moderated power reactor consists essentially of a lattice of round fuel rods composed of uranium oxide in metal jackets, interrupted by cruciform control rods spaced in a regular array in such a way that the fuel rods are grouped into square cells having sides of the order of 10 in. (25 cm) long, and more or less surrounded by the arms of the control rods. The core must be provided with a structure, some of which may be permanent and some associated with removable assemblies of fuel rods, to perform the following functions:

1. The long thin fuel rods, typically 1/2 in. (1.27 cm) or less in diameter, and having lengths up to 8 ft (2.44 m) or more, must be held in the prescribed lattice arrangement, but in such a way that they may be removed for replacement.

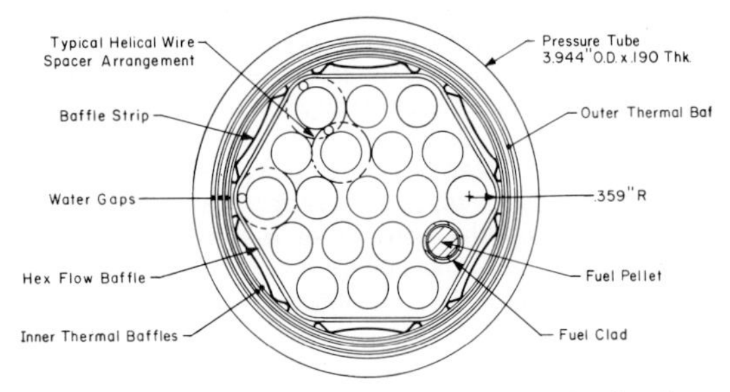

FIG. 10-13 Horizontal section through pressure tube, Carolinas-Virginia Tube Reactor (CVTR).

2. Channels among the fuel rods must be maintained for the passage of the control rods with adequate clearance, and the rods must be guided so that they move freely without deleterious rubbing.
3. The flow of coolant must be properly confined about the fuel rods, and serious bypassing of coolant through the control rod channels or around the outside of the core must be prevented.
4. In some cases a multipass coolant flow arrangement may be used; in such cases the core structure must define and separate the passes.

The basic approach to the structural design is to group the fuel rods into assemblies, which may typically contain from 36 rods to more than 100 rods.

The remaining problem is to provide the structure for holding these assemblies in place to form the core, with proper consideration for the requirements listed above. Several different basic approaches are possible, of which four are discussed below.

1. The fuel rods may be assembled into open bundles, suitably shaped to provide control rod passages, and may be held in place between upper and lower grid plates. In such an arrangement the control rod must be provided with a nonabsorbing follower of nearly the same dimensions as the rod to prevent severe bypassing of coolant flow through the empty control rod channel when the rod is withdrawn. When such a follower is provided, rod guidance is provided by the continually present bearing points in both the upper and lower grids. For the same purpose of preventing serious flow bypassing, the fuel rods must be positioned quite close to the control rod channel. Usually this can be accomplished only through the use of a removable (rather than fixed) upper grid. The use of a removable grid provides excellent positioning and holddown for the fuel assemblies, but usually complicates the procedure of fuel reloading. When such an arrangement is used, care must be exercised in the design to provide adequate support and positioning of the assemblies during the reloading operation, or at other times when the upper grid may be removed.
2. The fuel rods may be assembled into open bundles, but the control rod passages may be defined by shrouds or scabbards which are part of the permanent core structure.
3. A fixed "egg-crate" structure may be provided in the core to define control rod passages and cells for the fuel bundles. The fuel assemblies may be open square bundles of rods which are placed in the cells of the egg-crate structure, usually four bundles to a cell. This arrangement allows control of the coolant flow (e.g., orificing) from one cell to another.
4. The fuel rods may be assembled into bundles which are surrounded by a coolant flow shroud as part of the assembly. These assemblies, held between upper and lower grid plates, define the control rod passages and allow the coolant flow to be varied, by means of orificing, from assembly to assembly.

Usually any of the last three arrangements described above can employ a fixed upper grid through which the assemblies are inserted for core loading. The assemblies must however be held firmly in place. This may be accomplished by holddown fixtures on the upper grid, but a common practice is to provide short leaf-spring protrusions on the sides of the upper fittings of the assemblies; these springs bear against each other or against the webs of the grid to provide a tightly clamped structure once the core is fully loaded. From the point of view of coolant flow confinement, any of the arrangements except No. 1 could be used without long control rod followers. Followers are usually provided however to minimize the peaking of the thermal neutron flux in the control rod channel when the rod is withdrawn. For the same reason, the designer usually tries to make the control rod channel as narrow as possible, even when restriction of the channel is not necessary for control of coolant flow.

The ways in which the above principles are embodied in actual core designs are illustrated, in following paragraphs, by descriptions of a pressurized water (Yankee) and a boiling water (Big Rock Point) design.

10.3.1 Yankee

The pressurized-water reactor of the Yankee Nuclear Power Station [196, 208, 256, 257] employs the "open core" concept described under item (1) above. The use of this approach has allowed the designers to utilize the core volume very efficiently for packing in fuel elements with a minimum loss of space to control rod channels and core structure, and to achieve benefits in core power density.

The Yankee reactor is shown in vertical section in Fig. 10-14. A horizontal section of the reactor is given as Fig. 10-15, and Fig. 10-16 is a large-scale horizontal section of a portion of the core which shows the fuel element and control rod arrangement in some detail. Coolant water enters the reactor through nozzles located above the core region (Fig. 10-14) flows downward through the regions outside the core baffle, then back upward through the core, after which it leaves the reactor vessel through outlet nozzles located at the same elevation as the inlet nozzles. Control rods are top-driven, and the pressure vessel internals are supported from the top.

The basic fixed element of the core is the fuel rod, composed of a closed stainless-steel tube which contains uranium dioxide fuel in the form of cylindrical ceramic pellets. The pellets of 3.4% enriched UO_2 are placed in the fuel tubes in groups of 25. Each group is separated from the next by a perforated stainless-steel disk (0.1205 in. or 7.747 mm thick) coated with braze material. Each disk has a circumferential groove into which the tube is crimped during the loading process to fix its position until the entire subassembly is brazed. The disks are so spaced that each group of 25 pellets is allowed a free expansion space (approximately 0.2 in. or 5 mm). The disks are

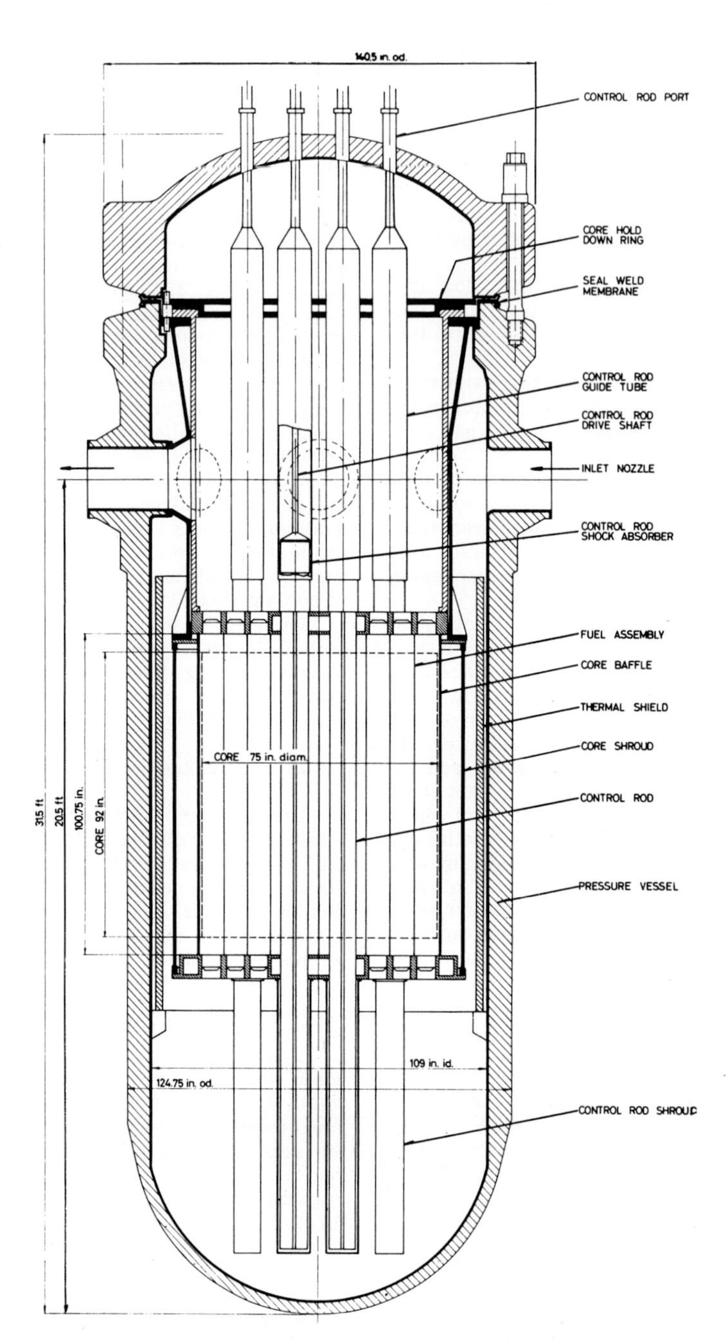

FIG. 10-14 Vertical section, Yankee Reactor.

provided to avoid the possibility of distortion of the fuel by repeated differential expansion between a long uninterrupted fuel column and its surrounding stainless-steel jacket.

The fuel rods are 93.6 in. (2.377 m) long and 0.340 in. (8.63 mm) in outer diameter; the jacket tube, of type-348 stainless steel, is thick enough (0.021 in. or 0.533 mm) to be free-standing under reactor pressure, and is closed at the ends by welded steel plugs. The fuel is in the form of pressed and sintered pellets of UO_2, assembled into the tubes with a clearance of 0.002 in. (0.051 mm) on the radius.

The core is formed by vertical assemblies of fuel rods, arranged to approximate a right circular cylinder; each fuel assembly is basically an 18-by-18-rod assembly, nominally square in cross section, 7.61 in. (19.33 cm) on a side. Peripheral rods are omitted as required from the 18-by-18 matrix to provide slots for the passage of the cruciform control rods, so the actual number of rods in a fuel assembly is either 304 or 305.

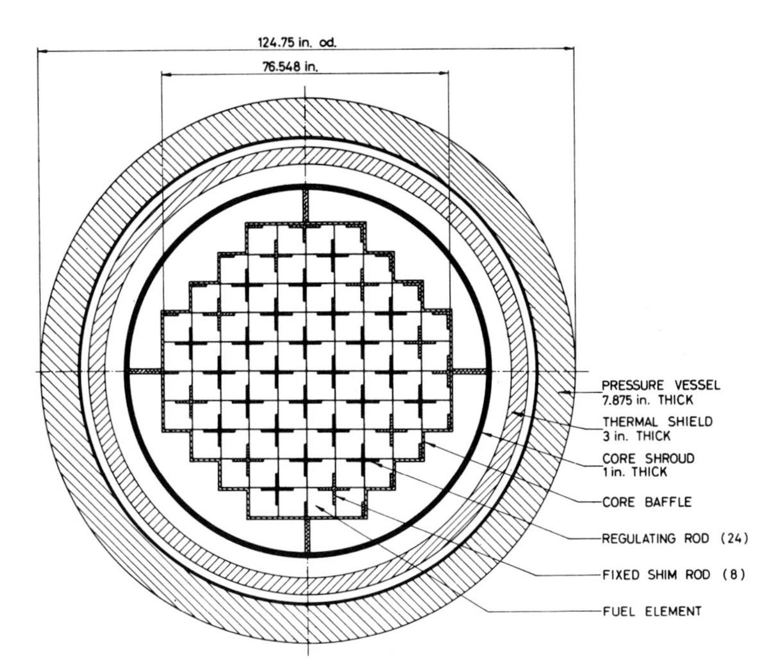

FIG. 10-15 Horizontal section, Yankee Reactor.

Fig. 10-16 shows the 2 typical fuel assemblies and gives pertinent dimensions and clearances for the assemblies as well as a control rod. Either type assembly contains 9 subassemblies, each of which is formed by a nominally 6-by-6 square arrangement of fuel rods; the rods are joined to form the subassembly by 1/2-in. (1.27 cm) -long tubular spacers, or ferrules, brazed between the fuel rods at 8-in. (20.32 cm) axial intervals. Thus, the fuel subassemblies are rigid structures designed to prevent significant thermal bowing of one fuel rod relative to another in the same group. No provision is made for differential expansion between the fuel rod jackets of a single subassembly.

Nine subassemblies are fitted between perforated end plates to form a complete fuel assembly. The center one of the 9 subassemblies is fixed to the end plates, while the surrounding 8 subassemblies are free to float axially, guided at each end by slip-fit plugs which enter holes in the end plates, thereby maintaining proper orientation and transverse positioning. Tie straps, which extend through the assembly from one side to the other but which are fastened only to the outermost rods on each side, are located at several axial positions. These tie straps prevent relative radial movement of the assembly components and cause the assembly to act as an integral unit. The fuel assemblies are designed to prevent excessive thermal bowing due to the radial power gradient across the core and the local flux gradients caused by control rods. With the worst combination of thermal gradients and mechanical tolerances, adequate clearance exists between the fuel assembly and adjacent control rod. Short, transverse rubbing bars or straps are brazed to the outer row of tubes in each of the outer 8 subassemblies, as indicated in Fig. 10-16. Sets of these rubbing straps are located at 4 axial positions along the assembly to prevent contact between control blades and fuel rods or between fuel rods of adjacent assemblies, which could cause hot spots or fretting problems. Nominal clearance between control blades and rubbing straps is about 1/8 in. (3.2 mm). Beyond each end plate, the fuel assembly is provided with a short spacer which accomplishes the geometric transition from the square cross section of the assembly to the hollow, cylindrical shape of the assembly end nozzles. These nozzles are used as handling sockets for the assembly and serve to position it in the upper and lower core support plates.

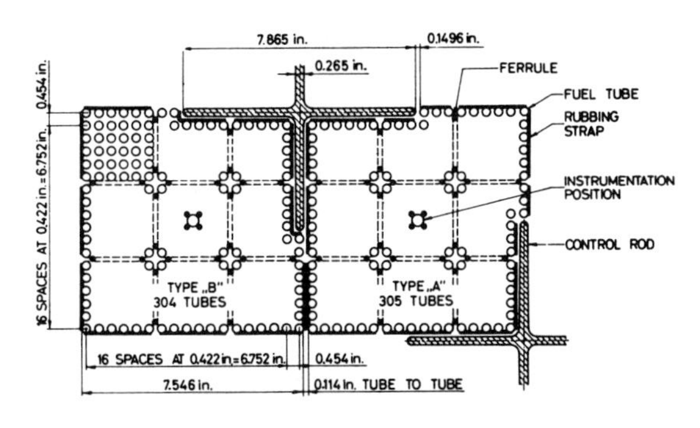

FIG. 10-16 Lattice details, Yankee Reactor.

The cruciform control rods are provided with Zircaloy followers of identical shape, which extend into closed shrouds below the core (see Fig. 10-14). The shrouds, which hang from the lower core support plate, allow the coolant flow into the control rod channels to be restricted to the desired extent.

Each core support plate is actually a rigid sandwich type structure consisting of upper and lower perforated plates joined by a peripheral spacer band and by sleeves which accept the fuel assembly nozzles and accomplish the radial spacing of the assemblies. The plates also position the control rods by means of guide blocks, position and serve as a mount for the control rod shrouds (lower plate), and support the fixed portion of the control rod shock-absorbing mechanism (upper plate). The design of the plate assemblies, which are 8 in. (20.32 cm) high, minimizes thermal stresses by employing relatively light structural members and by providing coolant access to the interiors of the assemblies.

The core baffle, seen in Figs. 10-14 and 10-15, provides the boundary between the incoming coolant which flows downward outside the baffle, and the coolant in the actual core, which flows upward. The baffle extends from the upper to the lower core plate and is stiffened to prevent inward bowing due to hydraulic pressure differentials by webs which connect it to the surrounding, cylindrical, core shroud. The thermal shield is a separate, thick-walled, hollow cylinder resting upon support lugs welded to the inner surface of the pressure vessel. The lower core support plate is doweled and bolted to the lower rim of the core shroud, and the upper rim of the shroud, as well as the upper flange of the core baffle, is doweled and bolted to the bottom flange of the lower core support barrel. The top flange of the lower core support barrel is supported on a circumferential ledge in the pressure vessel wall at the level of the vessel flange.

The upper core support plate is doweled and bolted to the bottom rim of the upper core support barrel, the top flange of which rests upon the top flange of the lower core support barrel. The upper flange of the upper core support barrel supports and positions a guide tube support plate which, in turn, supports the core holddown ring. Thus, the entire core support structure is hung as an integral unit from the support ledge near the top of the vessel. When the vessel head is in place, a compression load is applied to the core holddown ring which causes it to act as a large Belleville spring with an induced load of greater force than results from the pressure differential across the core. Four guide pins provide rotational alignment between the lower core support barrel and the reactor vessel, and four other pins orient the components which rest upon the upper flange of the lower barrel.

In addition to the components described above, a system of thimble tubes for the core instrumentation is located within the reactor vessel. A description of the core instrumentation is given in Sec. 9.4.5, and a vertical section which shows the thimble arrangement within the Yankee pressure vessel is given as Fig. 9-5.

10.3.2 Big Rock Point

The Big Rock Point reactor [196, 258, 259] differs from the Yankee reactor in many important details, yet it too illustrates the basic characteristics of H_2O-moderated reactors which have been summarized earlier. The reactor, which is shown in perspective in Fig. 10-17 is of the direct-cycle, forced-circulation, boiling-type. Water enters the reactor vessel through nozzles located near its bottom, as shown in Fig. 10-17 and in Fig. 10-18. The water flows upward through the core, where steam is produced, and the steam-water mixture leaves through outlet nozzles located near the top of the cylindrical portion of the vessel. Risers carry the steam-water mixture to an external steam drum where the steam is separated. Recirculating water flows through downcomers to the recirculating pumps, which return it to the inlet nozzles of the reactor vessel. Control rods are bottom-driven, and the core structure is supported from the bottom.

The core approximates a cylinder 5 ft (1.52 m) in diameter and 6 ft (1.83 m) high; it is made up of vertical fuel bundles spaced by a fixed egg-crate structure to provide vertical slots for the passage of cruciform control rods. Typical lattice dimensions are given in Fig. 10-19, a horizontal section, and the fuel bundle is illustrated by Fig. 10-20. Each bundle contains 144 fuel rods arranged in a square 12-by-12 array. The rods consist of stainless-steel tubes containing stacks of UO_2 pellets. Of the 144 fuel rods in the fuel bundle, 132 are standard and 12 are special: the special rods, 3 in each corner of the bundle, are smaller in diameter and have thicker jackets than the standard rods (Fig. 10-19) to lower the local power peak which occurs at the corners because of the high water concentration represented by the intersecting control rod slots. The 4 corner rods of the bundle are segmented and also serve to support 3 transverse wire spacer grids.

The stack of UO_2 fuel pellets in all but the 4 corner rods is 70 in. (1.778 m) long, and a 2-3/4-in. (6.985 cm) plenum is provided at the top of the stack to accommodate expansion and released fission gases. A coil spring and wafer in each plenum keep the stack of UO_2 pellets in the proper position, and welded plugs are used to seal the fuel rods at each end. Each of the 4 segmented rods in a bundle consists of 4 sections which are similar to the full-length rods but which have correspondingly shorter active fuel and plenum lengths.

The rods are positioned, at the top and bottom of the bundle, by 1/8-in. (3.175 mm) -thick drilled end plates which are shown in Fig. 10-20. The plates are supported in turn by a cast handle and base at the top and bottom of the bundle, respectively. The designs of the fuel rod end plugs and the holes in the end support members are such that improper assembly is impossible. The fuel bundle is held together axially by 32 of the 144 fuel rods. These 32 rods have threaded end plugs which screw into the base and which are attached to the handle with safety-wired nuts. The other 112 fuel rods have unthreaded end plugs which slide into holes in the end plates. All 144 rods have external coiled springs at their upper end plugs to permit relative expansion. The wire spacer grids provide radial spacing at 3 axial locations in the bundle. The spacers, each consisting of a double layer of crimped wire grids and a surrounding band dimpled at the points of contact with the fuel rods, are positioned axially by the 4 segmented corner rods which fit through sleeves in the corners of the spacers.

As shown in Fig. 10-17, the fuel bundles fit within channels attached to the tops of support tubes. The bundles are not held down; presumably their weight is sufficient to prevent lifting under the forces of coolant flow, even

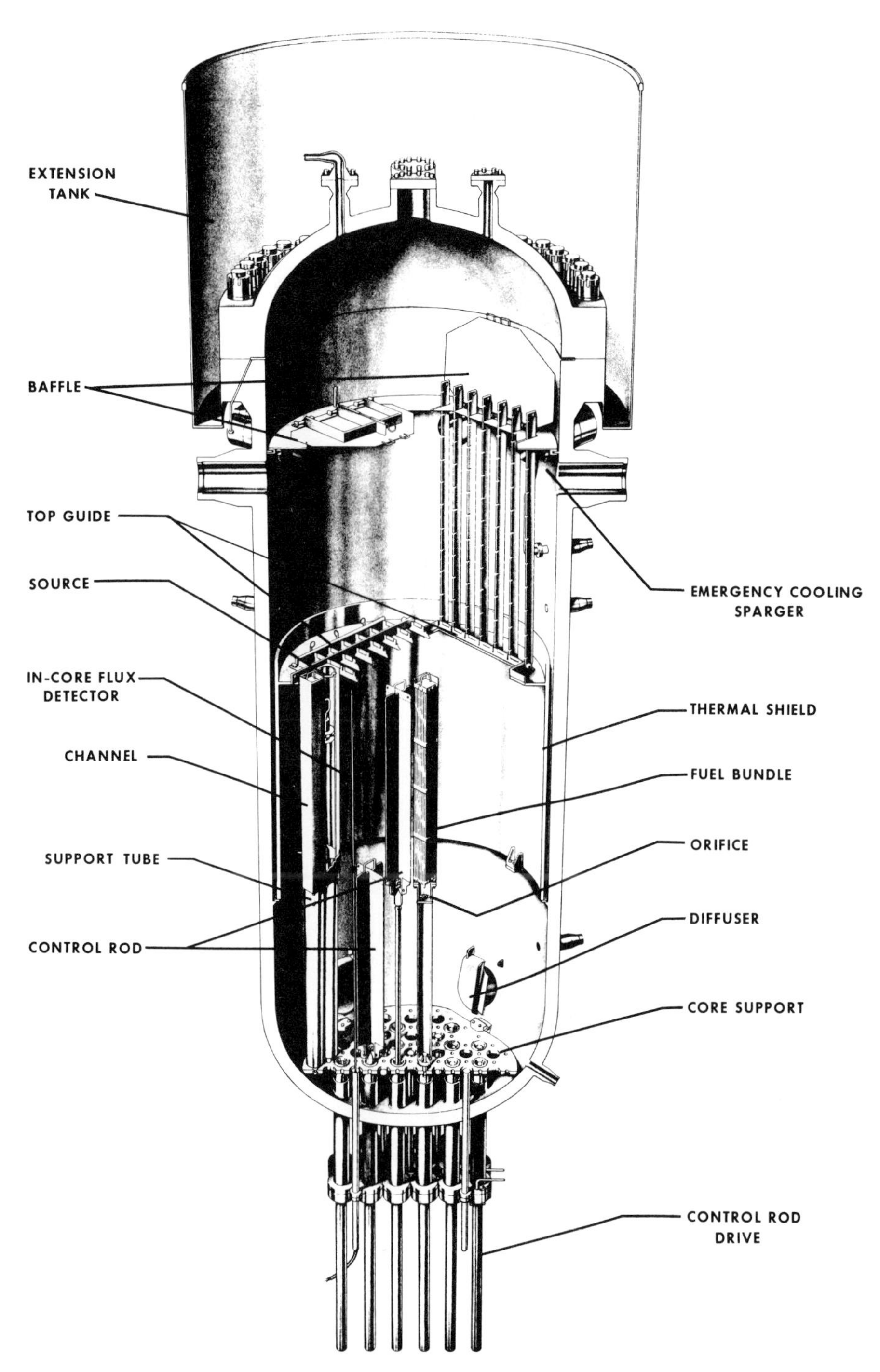

FIG. 10-17 Perspective, Big Rock Point Reactor.

in the event of a break in the outlet coolant line. The channel-and-support-tube assemblies, which also provide guidance for the control rods, are positioned by the core support at the bottom and a guide assembly at the top. The core support is a 1-1/2-in. (3.81 cm) -thick circular plate resting on 4 support pads welded to the reactor vessel near its bottom. The support plate has 4 alignment brackets which engage the support pads and position the plate relative to the control rod drive penetrations in the bottom head of the vessel A minimum radial clearance of 1/8 in. (3.175 mm) is provided where each control rod drive nozzle penetrates the plate. Support-tube adapters, which

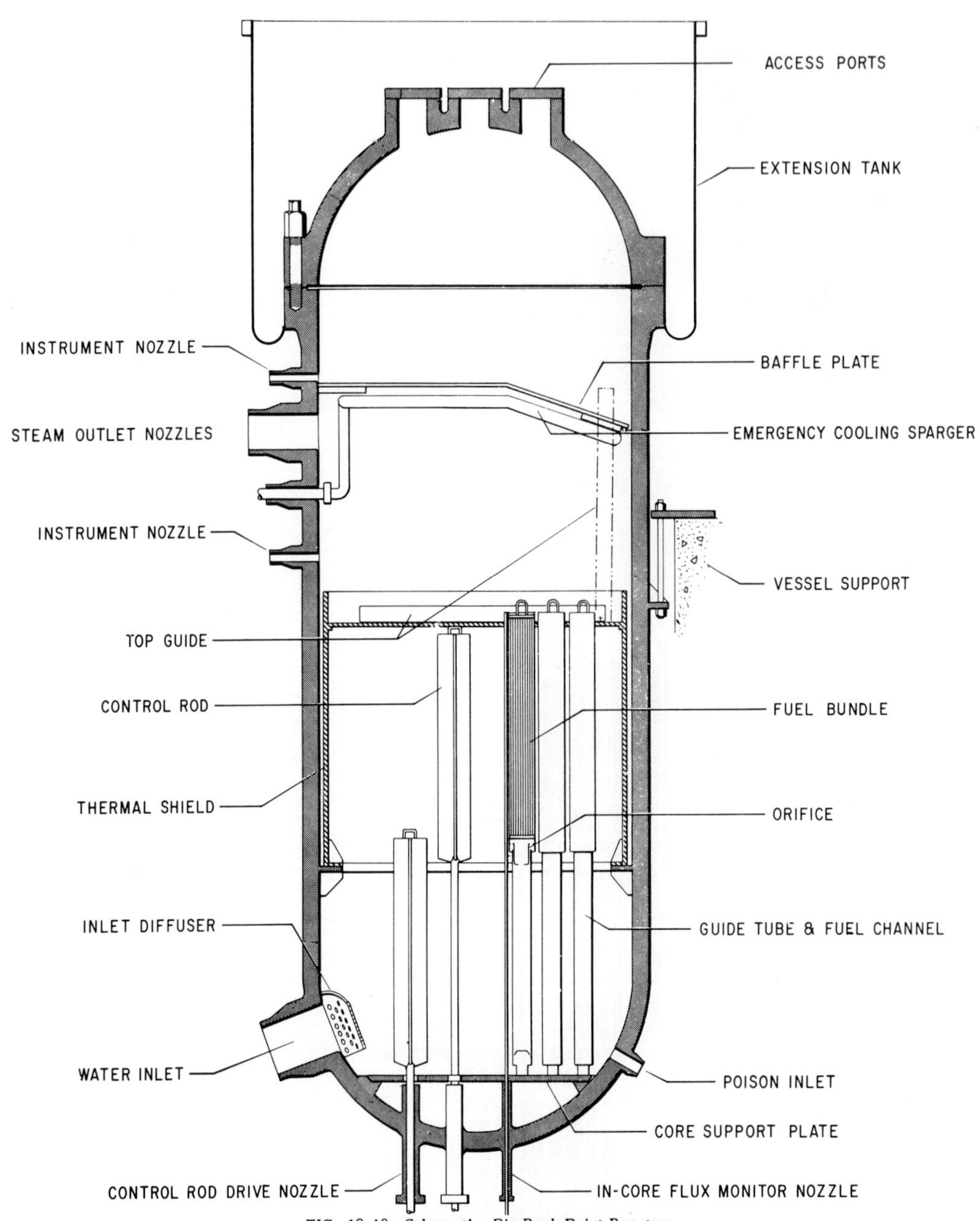

FIG. 10-18 Schematic, Big Rock Point Reactor.

align the bases of the channel-and-support-tube assemblies with the control rod penetrations, are bolted to the support plate. Holes in the support plate permit coolant flow from the region between the plate and the vessel bottom head.

Coolant for each fuel channel enters the bottom of its support tube, which is held clear of the core support plate by the support tube adapters. The support tube is round in cross section and is joined to the square fuel channel by a bolted flange. The fuel channel may be either stainless steel or Zircaloy, depending on the core loading and the shutdown control requirements, and its thickness, given on Fig. 10-19, is sufficient to withstand possible pressure differentials between its inside and outside. Each assembly is provided with a removable flow orifice at the connection between the support tube and the channel. The orifice is locked in place by turning, and spring latches prevent inadvertent unlocking prior to fuel loading. When a fuel bundle is in place in the channel, the bundle base interferes with the orifice handle to provide positive protection against unlocking. Although the orifices can be used to match the flow distribution to the radial power distribution, they are installed primarily to increase the pressure drop experienced by the flowing coolant while in the single (liquid) phase. The effect is to make individual channel flow less dependent on power, thus improving the burnout ratio and parallel-

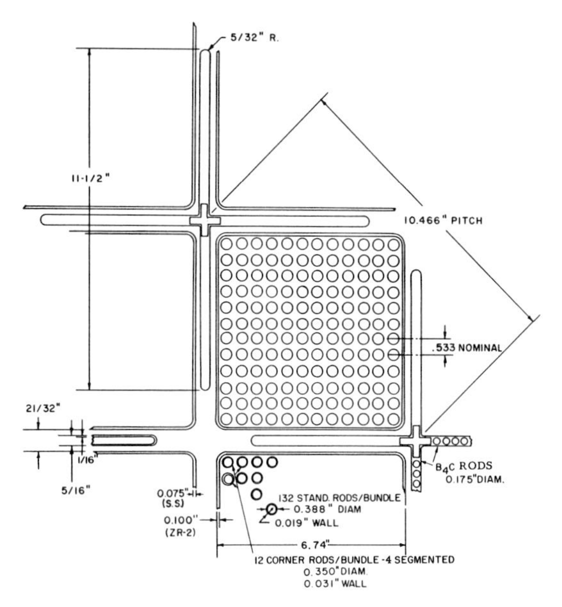

FIG. 10-19 Horizontal section, portion of Big Rock Point Reactor Core.

channel flow stability.

The thermal shield is supported in the reactor vessel by 6 equally spaced pads. Each of the pads is provided with 2 threaded studs of 1 in. (2.54 cm) diameter which secure the shield, and three of the pads have 1-1/4-in. (3.175 cm) -diameter alignment pins. The annulus between the thermal shield and the reactor vessel is closed at its bottom by a seal ring, but holes near the bottom of the shield allow a controlled amount of coolant to flow through the annulus.

A continuous ring and intermittent brackets near the top of the thermal shield provide a mounting surface for the top guide assembly, which is aligned with the shield by 2 pins and is bolted in place. The top guide assembly provides radial support and alignment for the upper ends of the channel-and-support-tube assemblies; the clearance between the channels and the top guide beams controls the coolant leakage flow which bypasses the fuel bundles. When the top guide is in the assembled (horizontal) condition, its beams serve as holddowns for the channel-and-support-tube assemblies. The beams are locked in the assembled position by clamps which can be unlatched to allow the beams to be lifted individually to a vertical position for the removal of channel-and-support-tube assemblies. In Figs. 10-17 and 10-18 beams are indicated in both horizontal and vertical positions. The design of the beam clamps is such that no force is transmitted to the latches.

As indicated by the above description, the top and bottom alignment of the core are accomplished by two independent assemblies. These assemblies are fixed relative to each other and to the control rod drive nozzles by their common attachment to the reactor pressure vessel.

The baffle shown above the steam outlet nozzles in Figs. 10-17 and 10-18 is composed of 4 segments which lock in position but which can be unlocked and swung upward to allow access to the core from the vessel top nozzles.

The control rods are formed by poison tubes contained in a structure composed of a central core and four sheaths which form a cruciform shape, as shown in Fig. 10-21. Heat generated in the poison rods is removed by coolant which flows through the holes shown in the sheaths and by contact transfer from the rods to the sheaths. The control rods are guided in the channels by rollers at the top and bottom of each blade, with nominal clearances as shown in Fig. 10-19.

In addition to the components mentioned above, the Big Rock Point reactor contains in-core flux detectors. The detectors are housed in tubes which penetrate the vessel bottom head through nozzles and which extend upward to a level near the top of the core region. The detector tubes are located between fuel channels, in positions not occupied by control rods, and are equipped with spacers to prevent transverse movement.

10.3.3 Safety-Related Characteristics

The foregoing discussions illustrate typical methods of solving the core design problems of H_2O-moderated reactors. These typical design approaches call attention, by implication, to most of the inherent safety-related characteristics of the water-moderated reactor core. These characteristics can, however, be stated more explicitly as follows:

1. The change in water density with temperature leads to rather large changes in fuel/moderator ratio in these reactors over the range of operating conditions. This variation is mainly responsible for the strong temperature and power coefficient of reactivity which are typical of the water-moderated reactors and which often are so strong as to obscure any coefficients due to other sources in the normal operation of the reactor. Both the sign and magnitude of the reactivity coefficient due to water temperature or water density depend upon the existing value of the fuel/water ratio, becoming more strongly negative as the fuel/water ratio is increased. Thus, the coefficient varies with the operating condition, but for any particular operating condition can be fixed by the reactor designer. If the water is allowed to boil, the range of fuel/moderator ratios covered by the operating conditions is particularly large.
2. In addition to the desire for negative temperature and power coefficients of reactivity, other considerations also tend to encourage the use of high fuel/water ratios. Among these are the desire to pack as much fuel as possible into a pressure vessel of given size in order to maximize the power density, and the desire to maximize the initial conversion ratio in order to minimize the rate of loss of reactivity with fuel exposure. Opposing these considerations is the desire to avoid large temperature and power defects of reactivity. The balance between these considerations varies from design to design.

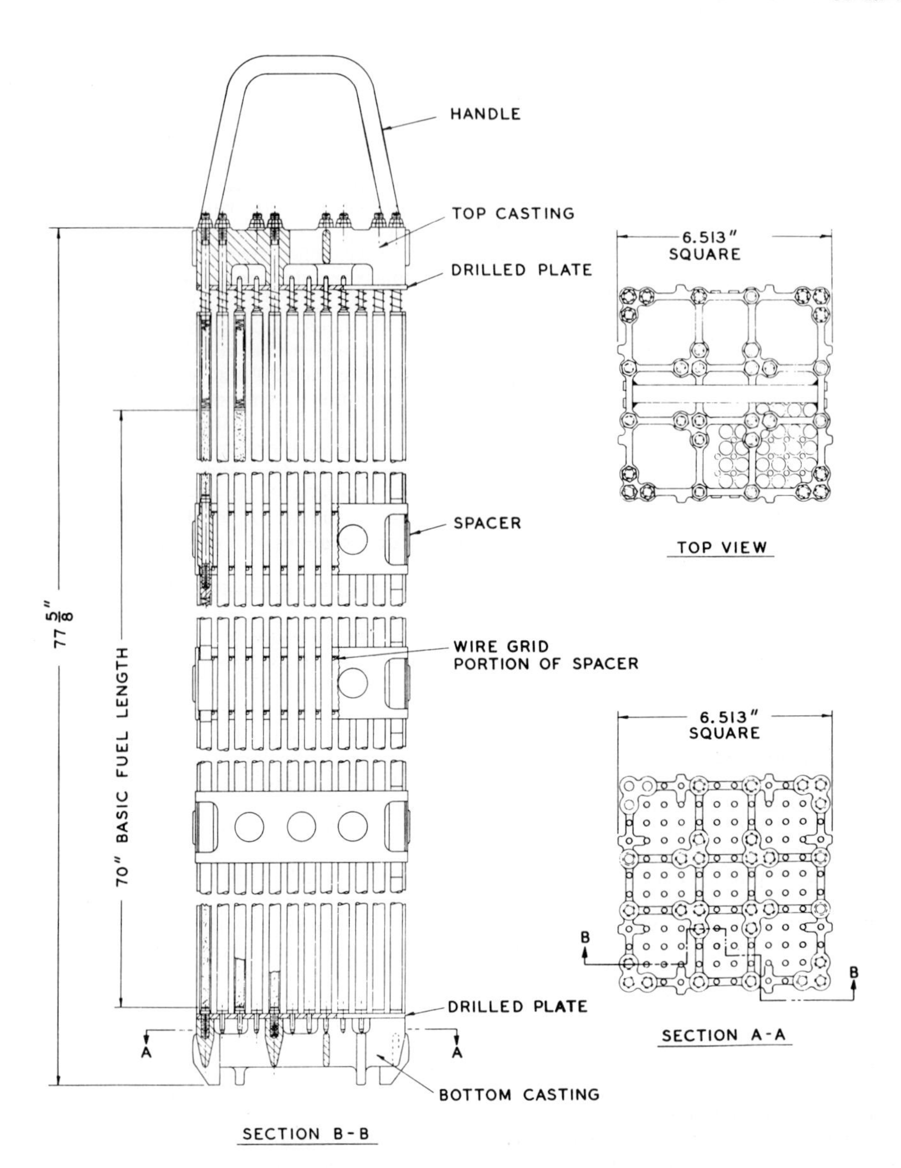

FIG. 10-20 Fuel bundle assembly, Big Rock Point Reactor.

It is not unusual to find designs — particularly in the boiling cases — where the temperature and void coefficients of reactivity are positive when the reactor is cold. Nor is it unusual to find the opposite case — where the coefficients are strongly negative under all conditions, and the temperature defect is quite large.

3. Whatever the situation at room temperature, the power coefficient of reactivity due to water temperature is ordinarily quite strongly negative at operating temperature, and is usually so large as to obscure the components of the temperature or power coefficients due to thermal expansions or distortions of the core structure, and to render them unimportant.
4. The absence of a solid moderator makes the core dependent, for its integrity and the constancy of its composition, on the structure which is devised to hold the fuel elements in their appointed places. Experience shows that reasonably precise positioning of core components can be maintained under the normally operating forces; the primary consideration in maintaining the prescribed positioning is to ensure freedom of movement of control rods rather than for the control of reactivity coefficients. If no attention is given to the restraint of core motion, however, important reactivity effects can result, as in the case of the SPERT-I Oxide Core Experiments [109].
5. It is not easy to predict the response of metal-jacketed oxide fuel elements, typically used in water-moderated reactors, to the expanding and distorting influence of temperature and temperature gradients. In most

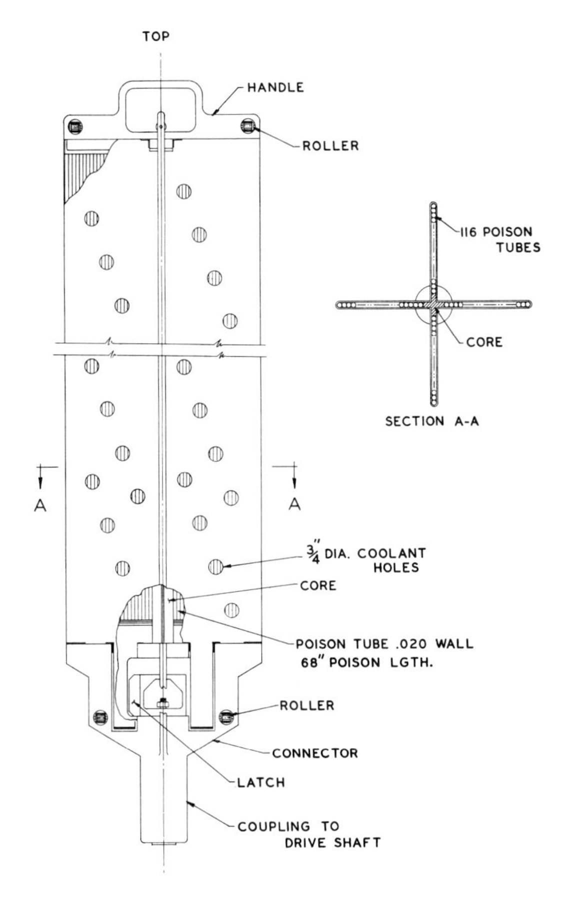

FIG. 10-21 Control rod blade assembly, Big Rock Point Reactor.

cases one would presume that this behavior would be determined by the temperatures and temperature distributions in the jacket rather than those in the fuel itself. When this is true, the time constants which characterize the expansions and distortions will often be not greatly different from the coolant temperature time constant. In these cases, the reactivity changes due to coolant temperature change or coolant voiding are apt to be not only larger in magnitude than those due to mechanical distortions, but may also occur rapidly enough to mask the reactivity effect of mechanical distortions.

6. Because the water moderator has a rather important neutron absorption cross section, economic and performance considerations will usually influence the design in the direction of the rather sparing use of water in the core, and most designs utilizing slightly enriched fuel will have appreciable resonance absorption (p is usually well below 0.9). Consequently fuel temperature changes in these reactors will cause substantial reactivity changes due to the Doppler effect, and the low conductivity and high operating temperature of the oxide fuel assure that this effect will manifest itself in terms of a relatively strong negative Doppler component of the power coefficient of reactivity. The strong Doppler coefficient will also be effective in limiting the results of accidental reactivity excursions. It must be remembered however that the hottest portions of the fuel normally operate at temperatures near the melting point of the oxide, and hence the capability for Doppler limitation of reactivity excursions is much smaller at operating power than at zero power.
7. The large temperature defect of reactivity of water-moderated reactors, along with substantial power defects of reactivity which result partly from the Doppler coefficient and, in the case of boiling reactors, from the reactivity reduction due to steam voids, usually result in a relatively high excess reactivity requirement for these reactors. In most cases, a further large requirement is introduced because long reactivity lifetime is desired. A control rod installation of large reactivity worth is usually required, and the worth of individual rods is often high. Strong mutual shadowing effects can exist which can cause unusually high reactivity worths of individual rods under certain programs of rod withdrawal. It is often impractical to provide a stuck rod shutdown margin of two adjacent control rods. For these reasons, it is particularly important to build in safe limitations on control rod programming, speed of control-rod withdrawal, and manual actuation of control rods during periods of shutdown, reloading, etc.
8. Because the coolant serves also as the moderator, the reactivity is particularly sensitive to changes in coolant characteristics which may occur in the course of coolant circulation. Thus, the cold water accident is a widely recognized source of important reactivity increases. When the neutron absorption characteristics of the coolant may be changed, as when soluble poisons are used for shutdown or operating control, the possibilities of introducing reactivity changes via the coolant flow must be carefully watched.
9. The use of the water coolant at high temperature and pressure has important relations to reactor safety:
   a) It introduces the possibility that any break in the primary coolant system may result in uncovering the reactor core, through the loss of water by vaporization and entrainment with outflowing steam.
   b) It introduces a source of stored energy which imposes the need for designing the containment building to withstand appreciable interior pressure.
   c) It introduces the possibility of large unbalanced forces on the core and control rod structures in the event of breaks in the primary coolant system.
   d) The high pressures and unbalanced forces in the core following a break in the primary coolant system may make it very difficult to guarantee provision of ade-

quate emergency cooling water to those areas of the core which would be steam-blanketed. In addition, the stored heat in the reactor structure, which will be considerably greater than the decay heat to be removed from the fuel, will tend to evaporate any cooling water supply when the reactor pressure is low.

All of these possibilities must be taken into account in the safety aspects of the design of water-moderated power reactors.

10.4 Fast Reactors

The core of the fast reactor, like that of the typical H_2O-moderated reactor, consists, essentially, of only fuel and coolant. Conceptually, the basic problem of mechanical design is the same as that of the H_2O-moderated core, but the differences in detail are so great that they are often more important than the basic similarity. Many of these differences result from the following characteristics of the fast reactor and the coolant, liquid sodium (or sodium-potassium alloy), which has been used in all fast reactors of significant power level to date.

1. Since the coolant does not perform an essential function of neutron moderation it is not necessary to the chain reaction, and hence an increase in fuel/sodium ratio is apt to increase reactivity: this is always the case if the increase in fuel/sodium ratio results from a general compaction of the fuel in the core; it is usually the case if the increase is a local one, corresponding to an increased "lumping" of the fuel, even if the average ratio over the core remains constant; in large cores it may be the case even if the increase in fuel/sodium ratio results from loss of sodium without change in the fuel geometry.
2. The desire for high power density in fast reactor cores tends to result in fuel elements of small cross section which pose difficult support problems. These problems are accentuated by the high coolant temperature which is usually used, and by the high heat fluxes, which are often sufficient to maintain important temperature gradients across materials of relatively high thermal conductivity.
3. The peculiar characteristics of the sodium coolant introduce characteristic problems into the core design: its affinity for oxygen promotes the sticking, or even welding, of metallic surfaces in contact with one another; its optical opacity, its need for protection from oxidizing gases, and its radioactivity after neutron exposure complicate greatly the problem of refueling; its high melting point requires that the reactor system be kept always hot; its very high heat transfer capability, although desirable in a coolant, introduces thermal shock possibilities.
4. There are no strong absorbers of fast neutrons, therefore thin absorbing control rods are not feasible in fast reactors.

Other peculiarities of the fast reactor are discussed in Sec. 10.4.2, but those listed above are sufficient as an introduction to the details of fast reactor core design, which will be illustrated primarily by reference to the design of the Enrico Fermi Fast Breeder Reactor.

10.4.1 The Enrico Fermi Fast Breeder Reactor (EFFBR)

The Fermi reactor, [81, 196, 260, 261] shown in perspective in Fig. 10-22, is a sodium-cooled fast breeder which has an enriched core and a depleted-uranium blanket. Its gross heat generation is 200 Mw, of which 174 Mw are produced in the core: a net electric output of 60 Mw gives the plant a net efficiency of 30%. The general arrangement of the reactor is further illustrated by Fig. 10-23, a vertical section which shows the interconnected arrangement of the reactor vessel and transfer rotor vessel within the primary shield tank. An offset fuel handling mechanism is located in the upper, enlarged section of the reactor vessel and is supported by the reactor vessel shield plug. The plug also locates the core hold-down mechanism, which is hollow to accommodate the control and safety rod drives. Liquid sodium enters the bottom of the reactor at a temperature of 550°F (288°C) and leaves the top at 800°F (427°C); outlet pressure is slightly above atmospheric, and the reactor pressure drop is about 100 psi (~ 7 kg/cm^2).

The Fermi reactor is divided into fissile and fertile regions, as shown in Fig. 10-23 and in Fig. 10-24, a horizontal section. There are a total of 871 lattice positions on a square pitch of 2.693 in. (6.840 cm), with 105 positions occupied by core elements, 32 by inner radial blanket elements, 500 by outer radial blanket elements, and the remainder by control rods, thermal shield elements, etc. Each of the 105 core elements is divided into 3 vertical zones, with the central zone consitituting the core region and the end zones forming axial blankets. The central, or fuel, section of each core element consists of 140 zirconium-clad pins arranged in the pattern indicated in Fig. 10-25, an assembly drawing. The 12 x 12 matrix of pins contains only 140 fuel-bearing members, since stainless-steel tie rods occupy the corner positions of the square. The fuel pins, 0.148-in. (3.76 mm) in diameter and made of an enriched uranium—10 wt.% molybdenum alloy, are clad with metallurgically-bonded, 0.005-in. (0.127 mm) -thick zirconium and are provided with cold-swaged zirconium caps at the top and with cold-swaged stainless-steel caps at the bottom. The lower end of each pin is fixed in a stainless-steel spacer assembly, while the upper end is not axially restrained and can accommodate changes in length resulting from temperature or irradiation effects. Relative radial movement of the pins within an assembly is prevented by a series of 17 transverse egg-crate spacers. The spacers, or stainless-steel grids, are located at 1.875-in. (4.762) intervals along the fuel assembly by the stainless-steel corner tie rods mentioned above.

The end, or axial blanket, zones of the core element each contain 16 rods arranged on the edge of a square; this arrangement can be seen in Section A-A of Fig. 10-25. The axial blanket rods are 0.443 in. (11.25 mm) in diameter, including a

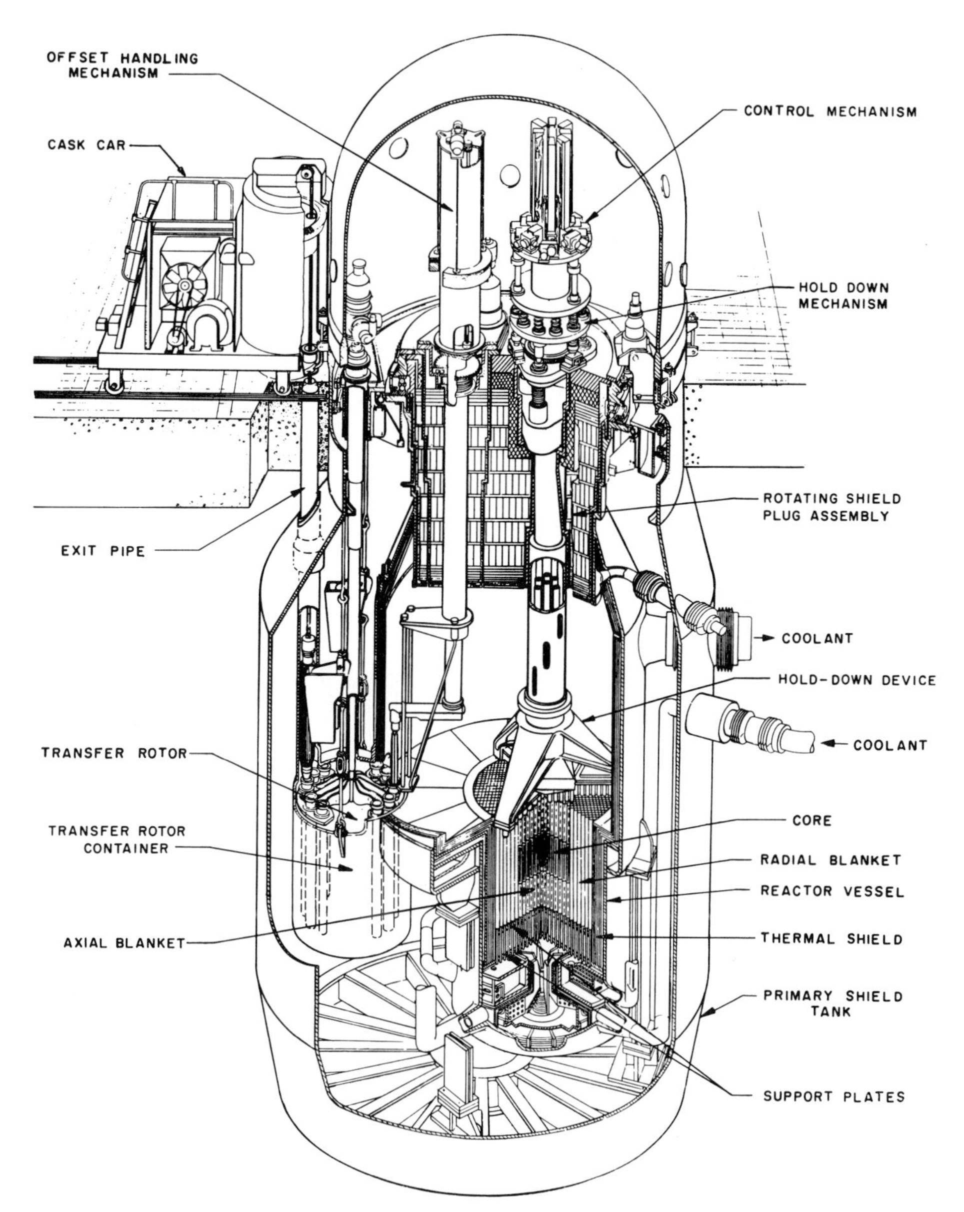

FIG. 10-22 Perspective view, Enrico Fermi Reactor.

10-mil (0.025 mm) -thick stainless-steel sheath; the rod material is a depleted uranium-3wt.% molybdenum alloy, and the radial gap between the rod and the sheath is filled with sodium to provide a low-resistance thermal bond. The rods are positioned around the edge of the square by top and bottom end supports. The central regions of the core-element end zones do not contain blanket rods for several reasons: the elimination of the rods reduces the pressure drop across the assembly, reduces the amount of blanket material that must be reprocessed on a core schedule (frequent element replacement relative to the radial blanket schedule), and, in the event of a core meltdown, provides a fairly unobstructed passage through the lower blanket section for molten fissile material.

The core-pin and blanket-rod bundles of a core element are housed within a square stainless-steel can, as shown in Fig. 10-26, a cutaway drawing of a complete subassembly. The square tube, with rounded corners, is about 6 ft (1.83 m) long, and is provided with a handling lug at the top and with a spring-loaded, telescoping nozzle at the bottom. The axial and transverse positions of the handling lug and the bottom nozzle are fixed when the element is in the core structure, but the spring-loaded, telescoping feature of the bottom nozzle permits axial expansion of the element. The radial power gradient across the core results in a temperature difference between opposite walls of a core subassembly: the wall farthest from the core center

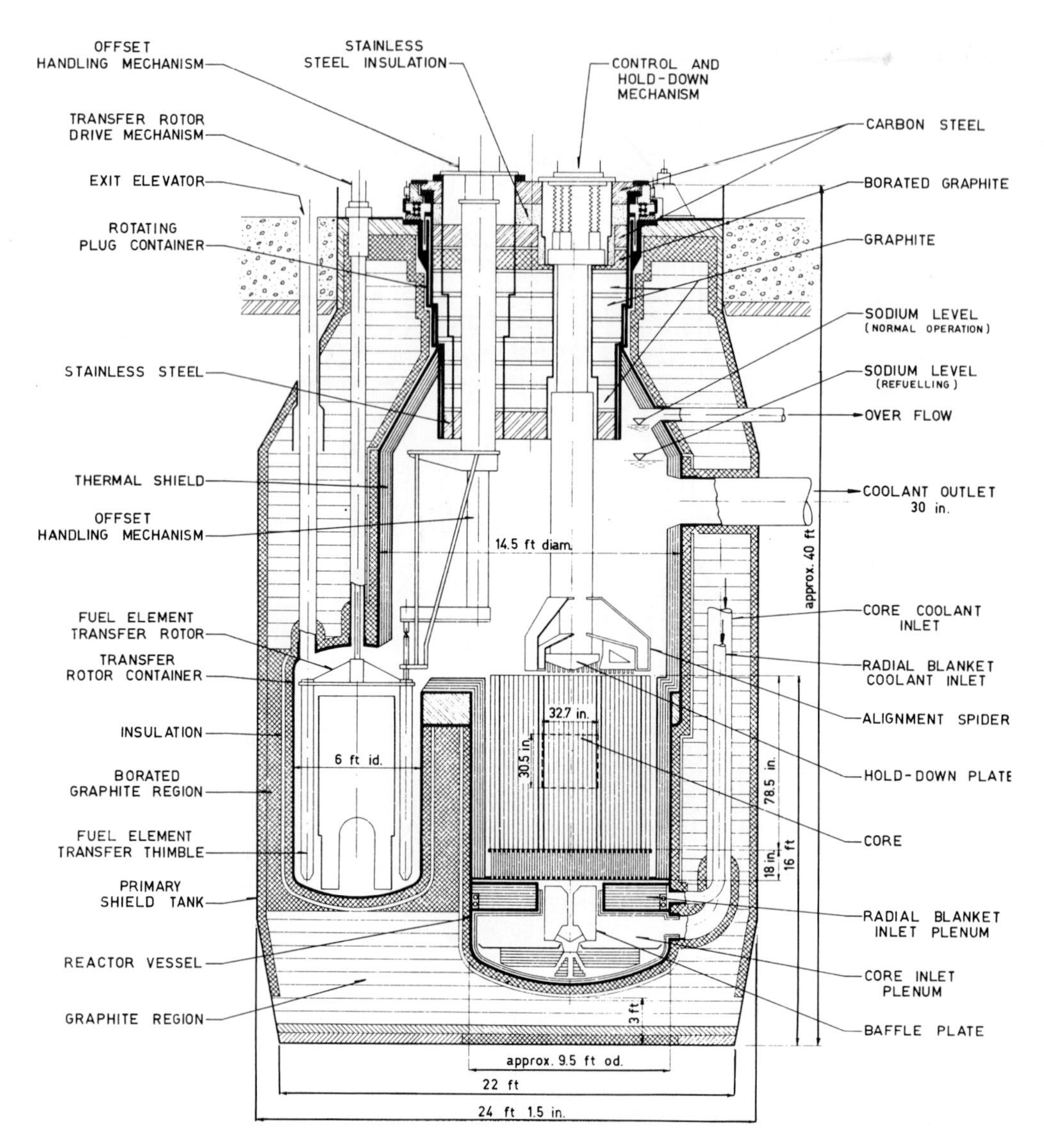

FIG. 10-23 Vertical section, Enrico Fermi Reactor.

is cooler than the wall nearest the center. Under the effects of this temperature difference, the subassembly would tend to bow inward toward the center of the core, thereby effecting a fuel compaction and causing a reactivity increase. To prevent any such bowing in the core, each square subassembly tube is provided with spacer pads which are welded on the corners of the tube at an axial location corresponding to the point of maximum free deflection of the subassembly. The pads of neighboring subassemblies contact one another and prohibit bowing. Figure 10-26 shows these spacer pads and also indicates a cam, located at the bottom of the square portion of the subassembly, which fits in a key-way in the core support plate and prevents rotational movement or misalignment of the subassembly.

Each radial blanket assembly consists of 25 rods, identical to the axial blanket rods except longer, arranged in a 5 x 5 cluster and enclosed in a square, stainless-steel wrapper tube. The rods run the full length of the assembly, and the wrapper tubes are provided with upper handling lugs and lower nozzles, as are the core subassembly tubes. The innermost radial blanket elements, which immediately surround the core elements, may be replaced with core elements at a later date and therefore are in a region of high coolant flow. Since the flow is great enough to lift the elements in this region, the inner radial blanket subassemblies are held down at the top, as are the core subassemblies. These elements all have the spring-loaded, telescoping bottom nozzles to accommodate axial expansion. The outer radial blanket elements operate with a much lower coolant velocity; they are held down by

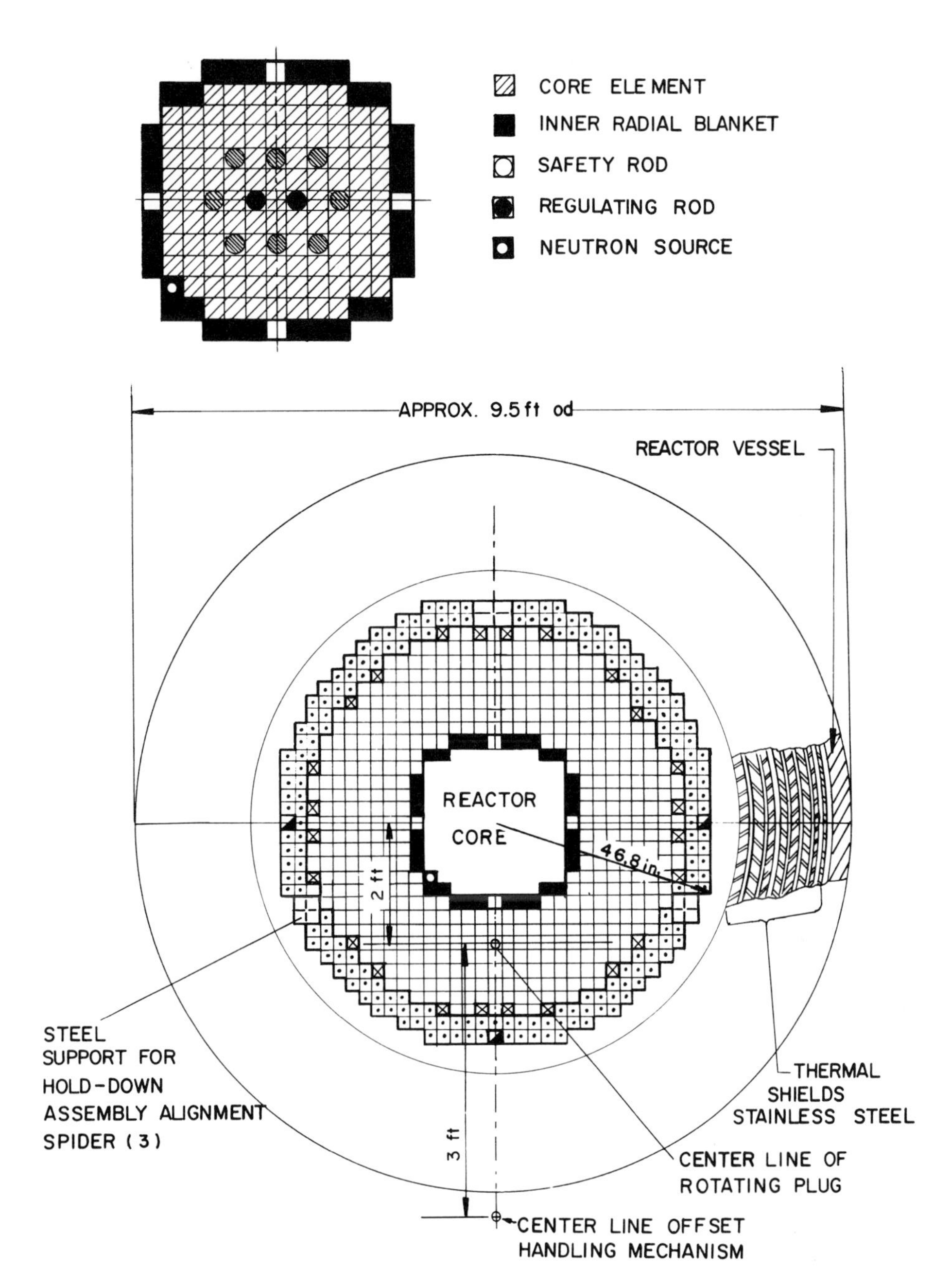

FIG. 10-24 Horizontal section, Enrico Fermi Reactor.

gravity only, are free to expand upward, and do not require the spring-loaded, telescoping nozzles.

As can be seen in Fig. 10-23, the core and blanket are located in the lower portion of the reactor vessel, which is an open-top cylinder with a domed bottom and a diameter of about 9.5 ft (~2.9m). An annular box structure, called the "radial blanket inlet plenum," is welded to the vessel sides at an elevation about 4 ft (1.22 m) above the domed bottom. This structure divides the coolant into two zones, a high-pressure central zone for the core and inner radial blanket, and a low-pressure zone for the outer radial blanket. The pressure zones are provided with individual sodium inlets, as shown in Fig. 10-23, the outer blanket coolant entering the annular box plenum and the main coolant entering the lower reactor vessel at a point below the box structure.

The core is supported by a sandwich-type support plate, which actually consists of upper and lower plates, each 2-in. (5.08 cm) thick, spaced 14 in. (35.56 cm) apart by welded ribs. The upper and lower plates are drilled to receive the lower end fittings (nozzles) of the core and blanket ele-

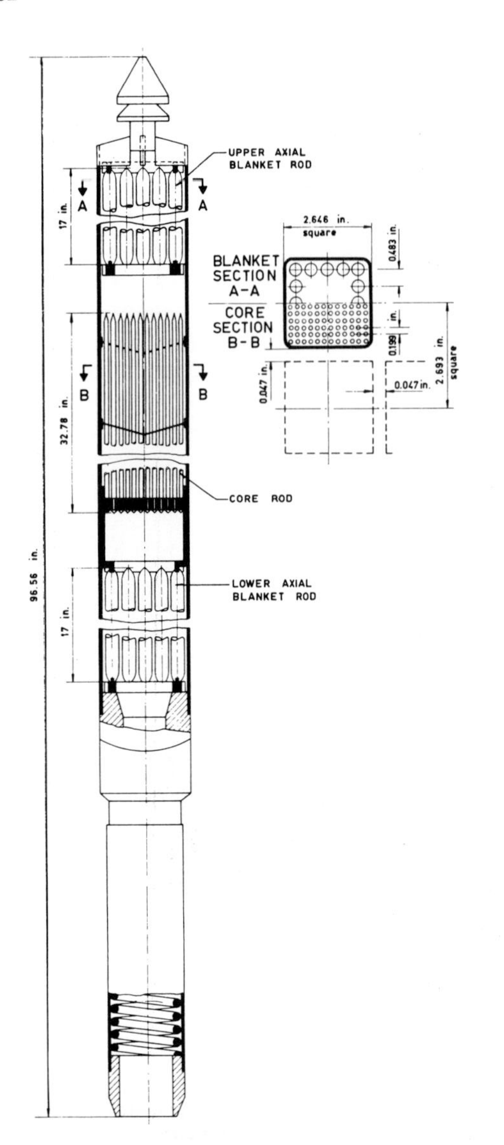

FIG. 10-25 Core element, Enrico Fermi Reactor.

ments. The support-plate structure rests on and is carried by members attached to the annular box structure which forms the radial blanket inlet plenum. Since later cores for the Fermi reactor may require support arrangements different from those for the first core, the central section of the support plate, under the core region, is removable. Sodium that bypasses the core and blanket elements cools the support-plate structure and holes are provided in the ribs of the structure to permit some transverse flow and mixing.

The core and blanket are assembled by plugging subassemblies into the support plate. The nominal diametral clearance between the subassembly nozzles and the support structure holes is 5 mils (0.127 mm). The flow rate through the outer blanket subassemblies is low enough, as mentioned above, that lifting forces will not raise the elements, and they are not provided with a top holddown. The core and inner blanket subassemblies, however, experience large lifting forces and require positive holddown. Since the coolant leaving the reactor is at approximately atmospheric pressure, a disturbance such as the rupture of an outlet coolant pipe cannot cause severe pressure transients which could increase significantly the lifting forces on either core or blanket subassemblies.

The core and inner blanket subassemblies are fixed in position by a holddown head, attached to a holddown column which passes through the upper reactor vessel shield plug. The holddown head, seen in Fig. 10-27, has a three-arm alignment spider, the arms of which are fitted with sockets to engage fittings located atop the three support columns indicated in Fig. 10-24. These support columns are supported and aligned at the bottom by the reactor support plates and are keyed at the top to the reactor vessel wall. The holddown head is equipped with a cluster of holddown extension rods, each of which fits the handling head at the top of a core or inner blanket subassembly. The pitch of the holddown rods is slightly less than the pitch of the holes in the bottom core support plate and slightly less than the element spacing that would occur with the spacer pads on the subassembly tubes just touching. Thus, when the holddown head is lowered into position (it can be moved 9 in. or 22.9 cm vertically to allow the upper shield plug to rotate and position the offset handling mechanism over the core), the core and inner blanket elements are cinched up tightly and no further compaction of the core is possible. The holddown column, which must transmit a force of some 50 tons to the core and blanket elements, is hollow and accommodates thermocouple leads and control rod guide tubes. The thermocouples are used to measure sodium temperatures at selected subassembly outlets and are installed through the fittings shown in Fig. 10-27.

Control of the reactor is achieved through the use of absorber rods: eight such rods, each having a reactivity worth of 0.0075, are used in a safety role and are normally fully withdrawn from the core, while two regulating rods, each with a reactivity worth of 0.0031, perform the actual control function. All rod locations are near the center of the core lattice, the regulating rods being allotted positions on opposite sides of the central fuel subassembly and the safety rods forming, roughly, an ellipse around the regulating rods, as shown in Fig. 10-24. Drive mechanisms are located above the upper shield plug, and drive extensions project downward inside the hollow holddown column. All rods are fitted with handling heads similar to those of the fuel elements, and the rods may be delatched from their drive extensions. A safety interlock is provided by the design of the rods, drive extensions, etc., since the rods cannot be entirely withdrawn from the upper axial blanket region and must be unlatched before the holddown device can be raised and the shield plug rotated for refueling. This required sequence of operations insures that, at the beginning of any fuel-handling procedure, all rods are inserted into the core and a condition of maximum shutdown prevails. The safety rods can be removed by the offset handling mechansim, but an additional interlock feature

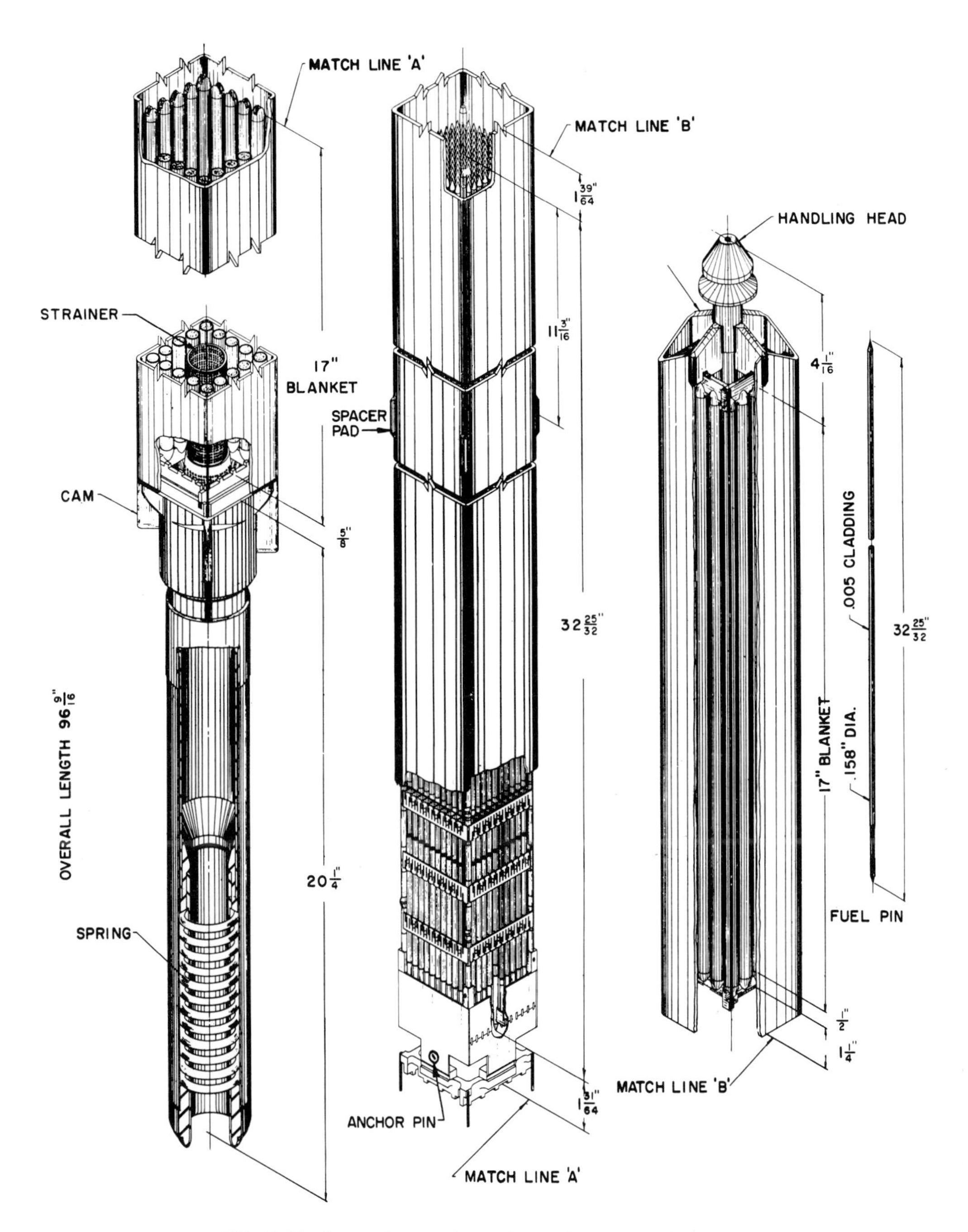

FIG. 10-26 Cutaway drawing of core element, Enrico Fermi Reactor.

prevents pick-up of the rod by the mechanism until the four adjacent core elements have been removed With the four adjacent core positions vacant, removal of a rod cannot cause accidental criticality. Within the core, all control rods move within guide tubes which are contained in square cans the same size as the fuel subassembly cans. The guide tubes are circular in cross section, as are the control rods, and the guides for the safety rods are equipped with hydraulic shock absorbers at their lower ends to cushion the rods at the end of a scram stroke. The regulating rods are not used for scram and do not have a rapid-insertion capability; hence, their guide tubes are not provided with shock absorbers. All guide-tube assemblies have orifices at their lower ends to restrict the coolant flow to a value which is the minimum consistent with the cooling requirements of the rods. This flow does not impart significant lifting forces to the rods, nor does it seriously impede their downward movement.

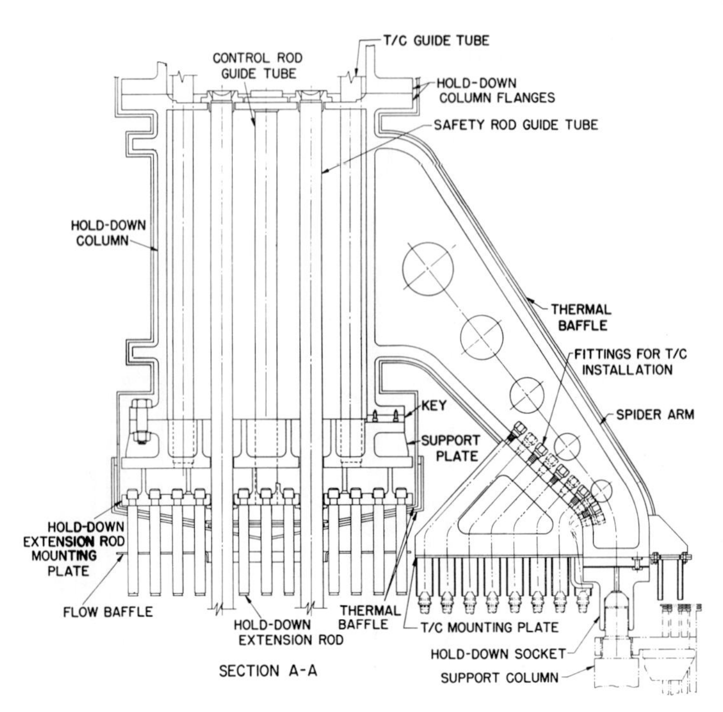

FIG. 10-27 Elevation of hold-down head, Enrico Fermi Reactor.

Although the safety rods and regulating rods differ in design, they both use boron carbide as poison material. In each type of rod, the poison section consists of a bundle of small, thick-walled, stainless-steel tubes containing the boron carbide and fitting within a single, relatively large, tube, which forms the outer surface of the rod assembly. The individual tubes containing the boron carbide are sealed at the ends and act as miniature pressure vessels to contain the quite large amount of helium generated by neutron absorption in the boron.

Actuating force for the scram action of the safety rods is provided by a spring contained in each rod. When it is desired to couple a safety rod to its drive extension for withdrawal from the core, the scram-actuating spring must be compressed, thereby "cocking" the rod and giving it the stored energy for scram movement. After the spring is compressed, a latch couples the rod and drive extension. This latch is held in place by a cam attached to a long extension arm which passes up the center of the drive extension tube. The extension arm is held in a position which locks the latch by an electromagnet located above the shield plug. Upon receipt of a scram signal, or in the event of a power supply failure, the electromagnet is de-energized, thereby releasing the cam shaft. With the cam released, the latch between the safety rod and its drive extension is opened and the compressed spring in the rod is free to extend, thereby scramming the rod.

The reactor vessel is lined with multi-layer internal thermal shields, as shown in Figs. 10-22, 10-23, and 10-24, and the peripheral positions in the lattice structure are occupied by thermal shield rods which are indicated in Fig. 10-24. The radial blanket inlet plenum contains multiple horizontal layers of steel, shown in Fig. 10-23. Generally, in the Fermi reactor, steady-state thermal stresses are minimized by the use of thin sections, and thermal shocks, which are most severe at the reactor inlet, are moderated by the mixing of the coolant in the plenums. In the lower reactor vessel, thin plates separated by layers of stagnant sodium serve as thermal buffers and are located adjacent to the inner surface of the vessel wall. The heat capacity of the large amount of steel in the radial blanket inlet plenum reduces thermal shocks in the box structure, while the large pool of sodium above the core and blanket regions effectively moderates temperature transients in the vessel outlet region by mixing.

Figure 10-28 shows the details of a structure which, in the eventuality of a core meltdown, is intended to disperse the molten fuel and prevent its agglomerating into a supercritical mass. Molten material from the fuel region of core subassemblies should flow down the open, center portions of the bottom axial blanket clusters and onto the conical flow guide seen in Fig. 10-28. The flow gude, which directs the coolant in normal operation, deflects the downward-flowing material to the zirconium liner, or tray, which holds the molten fuel until it spreads out. Should the hot fuel eat its way through all of the several layers of metal (stainless steel) beneath the zirconium liner, it will then be contained by the secondary meltdown section, which amounts to a borated graphite crucible.

The lower reactor vessel, the larger upper vessel, and the transfer rotor container are con-

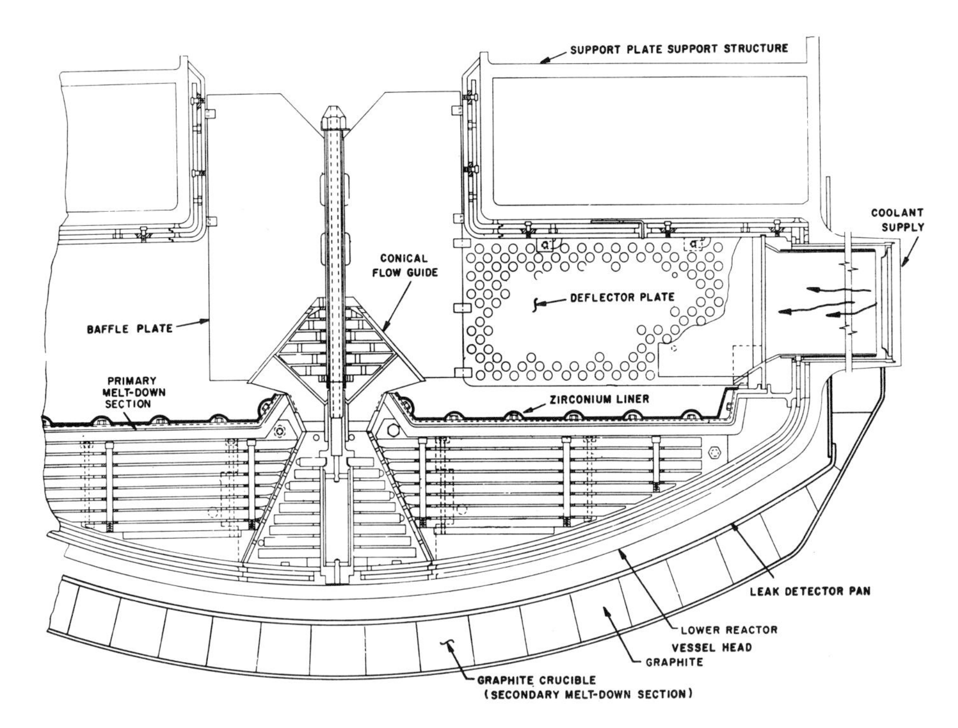

FIG. 10-28 Meltdown section of lower reactor vessel, Enrico Fermi Reactor.

nected by a transition section which consists of a flat plate with welded stiffening ribs. Brackets welded under the transition section, in line with the ribs, are connected to relatively flexible support columns formed from plates 2 in. (5.08 cm) thick and 7 ft (2.13 m) high. These columns, which can be seen in Fig. 10-22, transfer the load of the vessel to the web structure located in the bottom of the primary shield tank, and are so arranged that the centerline of the upper reactor vessel remains fixed despite thermal expansion.

10.4.2 Other Reactors

The other sizable fast reactors for which detailed descriptions are available are the Experimental Breeder Reactor Number II (EBR-II) and the British Dounreay Fast Reactor. Both of these differ in many respects from the Fermi plant. Perhaps the most notable difference of the EBR-II, a 20 Mw(e) experimental plant, [80, 261] is the use of a large primary sodium tank which contains the reactor, fuel-handling equipment, primary heat exchanger, and in fact the entire primary coolant system, within one large pool of liquid sodium. Within the area of core design, the Dounreay reactor [262] shows the most pronounced differences from the Fermi reactor, and is described in some detail below to illustrate the wide range of possible approaches to the design of a fast-reactor core.

A vertical section of the Dounreay reactor is given in Fig. 10-29. The enriched-uranium core, rated at 60 Mw(t), is located in the center of a natural uranium radial blanket, and a natural uranium upper axial blanket is provided. The initial coolant is a 70/30% sodium-potassium alloy (NaK) which, because its melting point is lower than that of sodium, offers operational convenience. Ultimately, the reactor will be cooled with sodium, the better heat transfer agent. The reactor vessel is supported by a top flange which rests on rollers, thereby permitting free radial and axial expansion. The vessel is closed at the top by two eccentric rotating shield plugs which support and position a charge machine.

Coolant flows downward through the core and blanket. The designers enumerate the following advantages for this flow direction:

- There is no tendency to lift the fuel elements, so a hold-down system is not required;
- in the event of a fuel meltdown, coolant flow will help remove molten material from the core; and
- the mechanisms at the top of the reactor are subjected to only the relatively cool inlet temperature.

The principal disadvantages of downward coolant flow are:

- The coolant is flowing against the direction of a natural thermal siphon, so a reversal of coolant flow must occur before natural circulation can furnish cooling in the event of complete pump failure; and
- the rotating shields at the vessel top are subjected to the relatively high inlet pressure, and seal problems are greater.

The primary coolant is circulated by 24 electromagnetic pumps in 24 loops, so the possibility of complete pump failure is rather more remote in this reactor than in most; consequently, the designers attach somewhat less significance to the

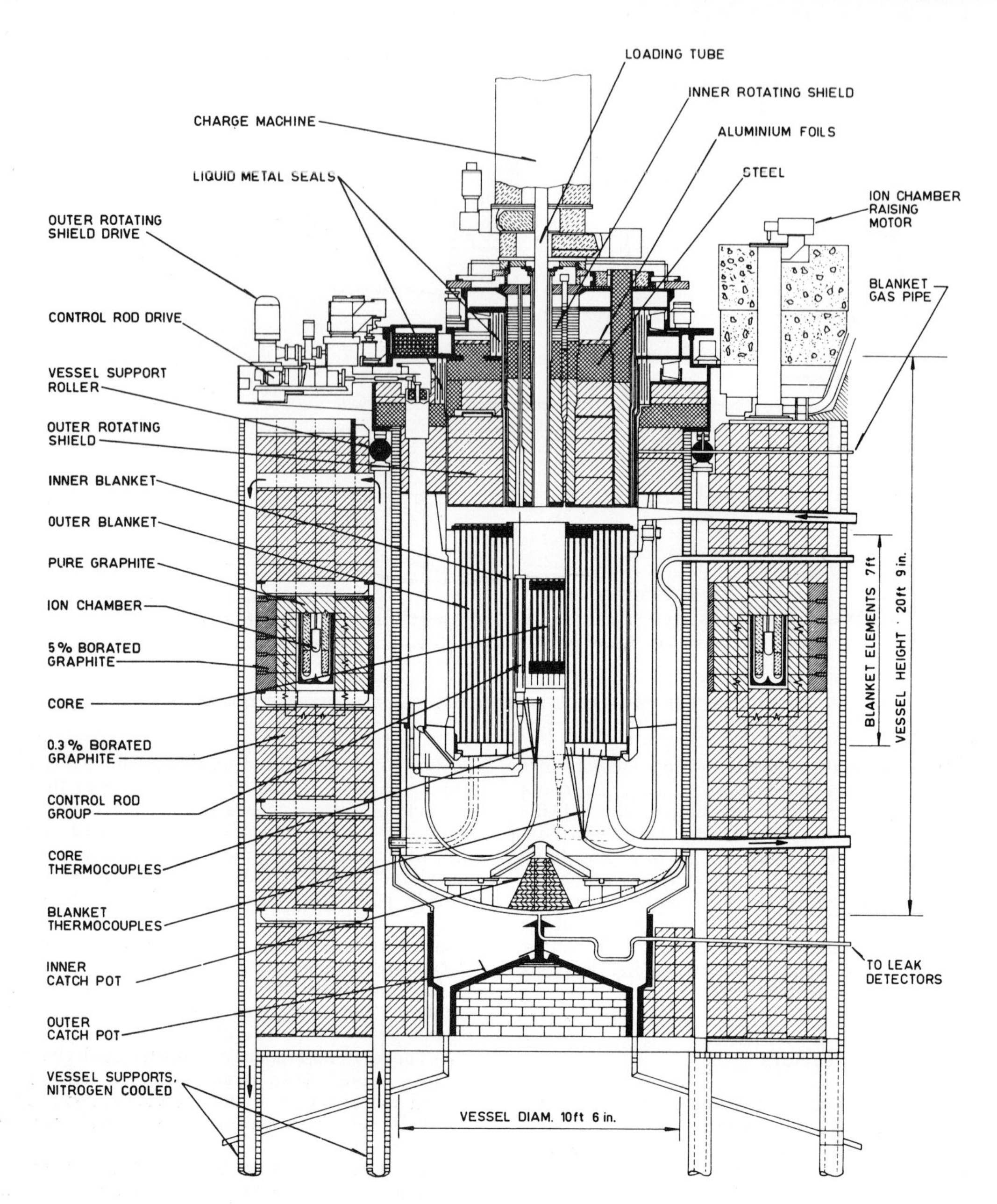

FIG. 10-29 Vertical section, Dounreay Fast Reactor.

disadvantage of flow reversal than it commonly receives.

The Dounreay core element, Fig. 10-30, differs markedly from its counterpart in the Fermi reactor. While the Fermi core subassembly uses multiple, small pin-type elements, the Dounreay core subassembly consists of a single, much larger, annular element; the power densities are comparable in the two reactors, but the maximum heat flux is larger in the Dounreay core by about a factor of 5, i.e. 3.3×10^6 Btu/ft^2-hr vs 0.64×10^6 Btu/ft^2-hr (or 1.04 vs 0.202 kw/cm^2). Enriched uranium, alloyed with molybdenum, is the fuel material, and the annulus has outside and inside diameters of about 3/4 in. (19.0 mm) and 1/4 in. (6.35 mm), respectively. As in the Fermi reactor, provision is made for molten fuel to flow to the bottom of the reactor vessel in the event of a meltdown. In Dounreay, there is no bottom axial blanket. Originally, the fuel element cladding materials were chosen so preferential melting of the inner clad would occur in the event of severe core overheating, leaving the outer clad intact to act as a channel for guiding the molten material downward out of the core region. This technique has now been discarded and both

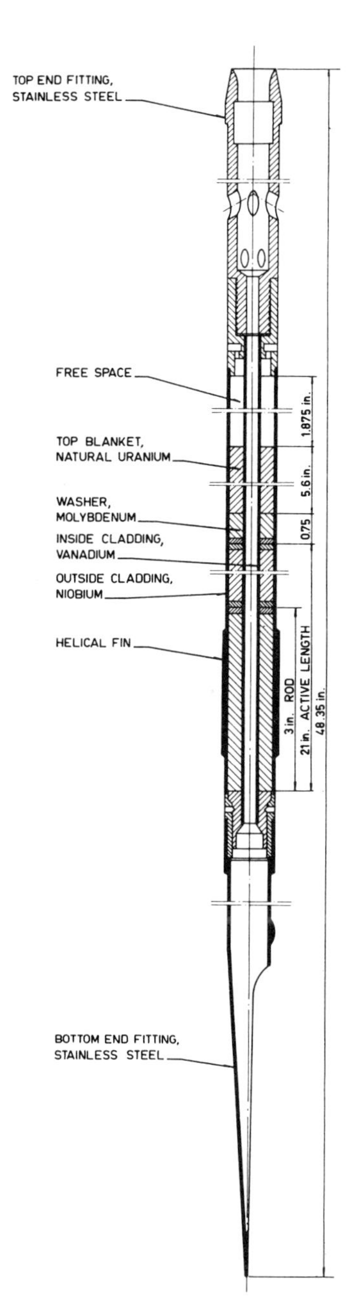

FIG. 10-30 Fuel element, Dounreay Fast Reactor.

inner and outer tubes are made of niobium. Inner and outer catch pots, indicated in Fig. 10-29, are provided to receive and disperse the molten material.

The basic reactivity control is accomplished by moving groups of fuel elements in and out of the core, although three B^{10} absorber rods are available as alternate emergency shutdown devices. The absorber rods are normally held above open lattice positions, but they can also be positioned to fall into the blanket region immediately adjacent to the core. Twelve control assemblies, each containing ten fuel elements, can be moved vertically in and out of the core region by the mechanism shown in Figs. 10-29 and 10-31. Each group of fuel elements is held in a carrier which is actuated from below the core by a horizontal arm located beneath the radial blanket region. The horizontal arm is connected to a vertical drive shaft which extends upward in the annular space between the outer boundary of the radial blanket and the reactor vessel. The drive shaft is connected to the drive mechanism at the reactor top. Scram is accomplished by deenergizing a magnetic coupling, thereby permitting the vertical shaft, horizontal arm, and fuel element carrier to fall under the effect of gravity. A hydraulic shock absorber cushions the falling assembly at the end of its movement.

In small fast reactors where neutron leakage is high, the removal of fuel from the core is quite effective in reducing reactivity. As reactor size is increased leakage becomes less important, and fuel movement becomes a less effective means of reactivity control. At the same time, large reactors are apt to contain larger quantities of scattering material in proportion to fissile material, and hence to have lower neutron energy spectra in which absorbers, such as boron, are more effective. Thus the relative attractiveness of absorbing control rods tends to increase with reactor core size. Operating experience with the components of the Dounreay control system is intended to provide useful information on the performance of the mechanical devices associated with each mode of control.

10.4.3 Safety-Related Characteristics

Opinions may vary as to the most important general characteristics of the fast reactor in relation to reactor safety, but it is certainly true that all considerations of accidents in fast reactors are strongly influenced by the very short prompt neutron lifetime which characterizes the reactor type. This characteristic means that a reactivity only slightly above the prompt critical limit is sufficient to cause power increases which are too rapid to be terminated safely by externally actuated shutdown systems. The reactor design must therefore be such as to preclude the possibility of increasing the reactivity above the point of prompt criticality if the possibilities of meltdown or extensive thermal damage of the core are to be eliminated. In achieving such a design, any inherent rapidly acting power coefficients of reactivity must of course be taken into account: beneficially if they are negative or adversely if they are positive.

Any reactor which is to be a breeder must utilize, as fissile isotope, either Pu^{239} or U^{233}. Both of these isotopes have delayed neutron fractions which are significantly lower than that of U^{235} (see Criticality chapter). Consequently it is apt to be more difficult to stay within the prompt criticality limit in the fast breeder than in nonbreeding reactors which use U^{235} as the major fissile isotope. If a substantial quantity of U^{238} is used as a fertile isotope in the reactor core, a significant benefit to the average delayed neutron fraction may result from the fast fission of U^{238}, which has a large delayed fraction.

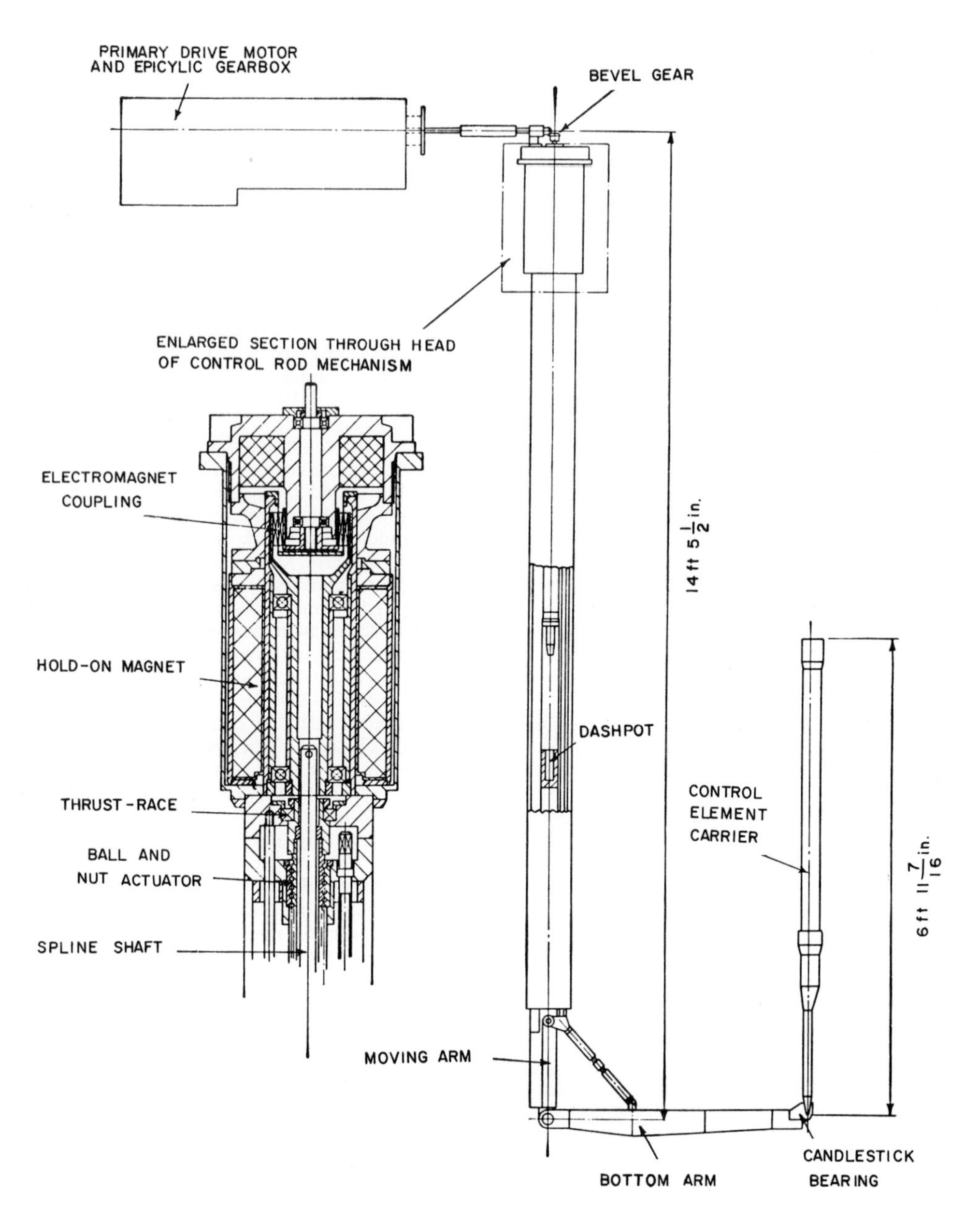

FIG. 10-31 Control element mechanism, Dounreay Fast Reactor.

Bearing in mind these dynamic considerations one can appreciate more fully the relation to reactor safety of other fast reactor characteristics which result more directly from the mechanical and thermal design of the reactor core. The main safety-related characteristics are the following:

1. As mentioned before, any increase in the average fuel density in the core will cause a reactivity increase. Any accident which can cause a compaction of the fuel may produce a very serious increase in reactivity, and even a slight general inward bowing of fuel elements or fuel assemblies may cause reactivity increases sufficiently large to affect reactor stability.

If one visualizes an accident, such as a fuel meltdown, which destroys the integrity of the structure intended to maintain the normal average fuel density, the question of the subsequent reactivity behavior becomes a most difficult and important one. In all but the smallest and most compact fast reactors, the agglomeration of even a fraction of the total fuel into a compact mass will usually result in a highly supercritical assembly. Consequently, in the fast reactor, the dynamic portion of a reactor accident cannot be considered to end with the general melting or thermal failure of fuel elements: on the contrary, it is conceivable that the serious portion of the accident may only begin at that point. The foregoing discussion of reactor designs has indicated the provisions that have been made in the Fermi and Dounreay reactors to assure that the melting of fuel will result in dispersal rather than agglomeration of fuel.

2. Quite evidently, the presence of a strong, fast-acting, negative power coefficient of reactivity would be highly desirable in a fast reactor. The physical mechanisms available for achieving such a coefficient are however few in number, and are not necessarily inherently present in the fast reactor.

   The axial thermal expansion of fuel elements may give a relatively important negative coefficient, and this coefficient may act quite quickly if thermal expansion of the fuel material itself can result in the axial extension of the fuel, and hence in a decrease of fuel density in the core. This should be true for metallic fuel elements of full core length, at least until they have experienced a high degree of radiation damage. For elements which consist of discrete particles or fragments of fuel material confined by metallic jackets—e.g., the oxide fuel element — it is not clear that expansion of the fuel material will actually elongate the fuel element; such elongation may occur only from the thermal expansion of the jacket, which may be of limited magnitude, and in any case would be slow in action.

   Reactors which contain large fractions of fertile material in the core may have negative Doppler coefficients of reactivity which are quite significant and helpful. As the fraction of fertile material is increased (relative to the fissile material), the negative Doppler effect increases both because of the increasing fraction of resonance absorber present and because of the general lowering of the neutron energy spectrum. The effect may be enhanced still further by incorporating a small amount of moderating material to degrade the spectrum still further; this usually decreases the breeding ratio quite significantly, however. Reactors which contain very little fertile material in the core may have slight positive Doppler coefficients of reactivity due to the fission resonances in the fissile material. The magnitude of the effect is usually quite small.

3. The coolant void coefficient of reactivity in sodium-cooled reactors may be either positive or negative depending on the size and composition of the reactor. The coolant temperature coefficient, which is simply the reflection of the void coefficient via the thermal coefficient of expansion, will also vary accordingly.

   The two most important effects of sodium voiding are the increase in neutron leakage and the increase in the general level of the neutron energy spectrum. The former always causes a reactivity loss, but the effect of the latter is complex and depends upon the initial spectrum and the composition of the reactor. It is quite strongly positive in large reactors which contain a substantial fraction of U^{238}, for in these reactors the initial spectrum is relatively "soft", and spectrum "hardening" increases the fraction of fissions in U^{238} as well as the effective η of the main fissionable isotope. Generalizations about the coolant coefficients of reactivity must be made with caution; but it can be said that in small, leaky, reactors, which contain high concentrations of fissile isotope and have rather "hard" spectra, the loss of sodium usually decreases reactivity, while in large reactors, particularly in those with relatively low concentrations of fissile isotope, the loss of sodium may cause an important reactivity gain. There appear to be great difficulties in designing fast reactors of high total power capacity which avoid the problem of a positive coolant void coefficient of reactivity. This subject, and the general subject of reactivity coefficients in fast reactors, are discussed at some length in reference [263]. In its effect, the coolant void coefficient of reactivity is often too slow in action to constitute a major help in alleviating the effects of fast reactivity excursions; but when the coefficient is positive, it is quite fast enough to lead to situations which are conceivably quite dangerous.

4. Because the negative power coefficients of reactivity which result directly from temperature or thermal expansion are at best only weakly negative, and because neutron leakage tends to be high even in fast reactors of relatively large output, the reactivity effects due to thermal distortions, such as fuel rod bowing, may be quite important even though the mechanical deflections producing them may be quite small. In general it may be said that the thermal distortion of fuel elements and fuel assemblies in the fast reactor must be held to a very low level. The difficulty in preventing such distortions is often increased by the small transverse dimensions of the fuel elements, by the high power density which is sought in fast reactor cores, and by the relatively high coolant temperatures that are used. The high power density not only promotes strong temperature gradients but also may limit the degree to which spacers and other mechanical restraints which interfere with coolant flow may be used.

5. Because of the high specific power which is usually sought in fast reactors, a high degree of shutdown cooling is usually required. Failure-proof emergency cooling is essential in the fast reactor, its importance being enhanced by the conceivably dangerous reactivity effects from a core meltdown. The difficulty in providing shutdown cooling also complicates the operation of spent-fuel

handling, and, although one does not visualize catastrophic accidents as the direct result of fuel handling, particular attention must be given to the prevention of fission product release from individual elements during the refueling processes.

6. The introduction of moderating material into or near the core of a fast reactor will usually produce a large reactivity effect. The effect may be positive or negative, depending upon the amount of material added and upon the initial neutron energy spectrum. Because of the high temperature of the sodium coolant it is difficult to imagine highly effective liquid moderators that might be carried into the core of a power reactor by coolant flow. Accidents involving moderators are much more likely to occur in fast critical assemblies or in the dry loading of a fast reactor core.
7. The characteristics of the sodium coolant present typical problems which must be taken care of.
   (a) Means must be provided for keeping the coolant above its freezing point at all times in all parts of the coolant circuit.
   (b) Impurities which may be insoluble and clog coolant passages must be removed from, and kept out of, the sodium. The most obvious of these is oxygen. The design of cold and hot traps for the removal of sodium oxide and other impurities, and of instruments for monitoring the impurity content, are part of the technology of sodium handling. Liquid organic materials may also cause clogging deposits in sodium systems and must be guarded against.
   (c) The chemical reactivity of sodium must be kept under control. The obvious possibilities are reactions with water, which may give explosive quantities of hydrogen, and the burning of sodium in air.
   (d) Thermal shock due to sudden changes in sodium temperature must be guarded against in the core structure and the coolant circuit. It is possible, for example, for the sudden shutdown of the reactor to result in thermal shock to structures near the core outlet, as they are suddenly bathed in cool sodium after having been previously in contact with the hot sodium from the power-generating core.
   (e) The sodium becomes highly radioactive from neutron bombardment (Na^{24}, of 15-hour half-life, emitting gammas of 2.75 and 3.7 Mev), and suitable shielding and control of this radioactivity and of any activiated impurities must be provided.

REFERENCES

1. H. Etherington (Ed.), Nuclear Engineering Handbook, Section 6-2, Reactor Calculations, McGraw-Hill Book Co. Inc. N.Y., 1958.
2. A. M. Weinberg and E. P. Wigner, The Physical Theory of Neutron Chain Reactors, University of Chicago Press, 1958.
3. S. Glasstone and M. C. Edlund, The Elements of Nuclear Reactor Theory, D. Van Nostrand Co. Inc., 1952.
4. R. L. Murray, Nuclear Reactor Theory, Prentice-Hall, N. J., 1957.
5. H. Soodak (Ed.), Reactor Handbook, Vol. III Part A: Physics, Interscience Publishers, N. Y. and London, 1962.
6. "Reactor Physics Constants", USAEC Report ANL-5800, Argonne National Laboratory, 1958. a) p. 133; b) Sec. 4.5.1
7. A. Radkowsky, Chairman of Editorial Board, Naval Reactors Physics Handbook, Vol. I, The Physics of Naval Reactors; Vol. II, The Physics of Pressurized Water Reactors.
8. M. Goldsmith, R. T. Jones, T. M. Ryan, S. Kaplan, and A. D. Voorhis, "Theoretical Analysis of Highly Enriched Light Water Moderated Critical Assemblies", Proceedings of the Second U. N. International Conference on Peaceful Uses of Atomic Energy, Geneva, 1958, Vol. 12, p. 435.
9. M. Goldsmith and T. M. Ryan, "Thermal activation shapes in a water-moderated critical assembly," Nucl. Sci. Eng., 5, (1959)299.
10. H. Kouts, et al, "Physics of Slightly Enriched, Normal Water Lattices (Theory and Experiment)", Proceedings of the Second International Conference on Peaceful Uses of Atomic Energy, Geneva, 1958, Vol. 12, p. 446.
11. R. W. Deutsch, "Method for Analyzing Low-Enrichment, Light-Water Cores," Report GNEC-133, General Nuclear Engineering, Inc., 1960.
12. R. W. Deutsch, "An engineering physics method of calculation applied to light-water critical experiments investigating nuclear superheat," Nucl. Sci. Eng., 13,(1962)110.
13. M. F. Valerino and R. S. Harding, "State of the Art-Physics of Nuclear Superheat Reactors", Report GNEC-282, General Nuclear Engineering, Inc., 1963.
14. G. T. Petersen and F. G. Warzek, "AEC-Superheat Criticals—A Comparison of Experiment and Theory on Uniform Lattices", Report GEAP-3882, General Electric Co., Atomic Power Equipment Dept., 1962.
15. P. W. Davison, et al, "Microscopic Lattice Parameters in Single and Multiregion Cores: A Comparison of Theory and Experiment", Report WCAP-1434, Westinghouse Electric Corp., 1961.
16. D. Hicks, "Nuclear Calculation Methods for Light Water Moderated Reactors", British Report AEEW-R64, 1961.
17. E. R. Cohen, "Exponential Experiments on D_2O-Uranium Lattices", Proceedings of the First U. N. International Conference on Peaceful Uses of Atomic Energy, Geneva, 1956, Vol. 5, p. 268.
18. Y. Girard, J. C. Koechlin, J. Moreau, and R. Naudet, Natural Uranium-Heavy Water Lattices, Proceedings of the Second U. N. International Conference on Peaceful Uses of Atomic Energy, Geneva, 1958, Vol. 12, p. 281.
19. G. Dessauer, "Physics of Natural Uranium Lattices in Heavy Water", Proceedings of the Second U. N. International Conference on the Peaceful Uses of Atomic Energy, Geneva, 1958, Vol. 12, p. 320.
20. P. Pershagen, G. Andersson, and I. Carlvik, "Calculation of Lattice Parameters for Rod Clusters in Heavy Water and Correlation with Experiments", Proceedings of the Second U. N. International Conference on the Peaceful Uses of Atomic Energy, Geneva, 1958, Vol. 12, p. 341.
21. D. W. Hone, et al, "Natural-Uranium Heavy-Water Lattices, Experiment and Theory", Proceedings of the Second U. N. International Conference on Peaceful Uses of Atomic Energy, Geneva, 1958, Vol. 12, p. 351.
22. H. Soodak and R. Sullivan, "A Method for Calculating the Reactivity of D_2O-Moderated Natural Uranium Lattices with Clustered Rod Fuel Elements", Report NDA-2131-38, Nuclear Development Corp., 1961.
23. M. F. Duret and R. Marriot, "A Computer Program for Reactor Studies", Canadian Report AECL-911, 1959.
24. J. L. Crandall, "Status of the United States Effort in D_2O Reactor Physics", Report DP-787, E. I. du Pont de Nemours and Co., 1962.
25. Heavy Water Lattices, International Atomic Energy Agency, Vienna, 1960.
26. P. F. Gast, "Normal Uranium, Graphite-Moderated Reactors: a Comparison of Theory and Experiment—Water Cooled Lattices", Proceedings of the First U. N. International Conference on the Peaceful Uses of Atomic Energy, Geneva, 1956, Vol. 5, p. 288.
27. D. E. Davenport, "Exponential Experiments in Graphite System", Proceedings of the First U. N. International Conference on Peaceful Uses of Atomic Energy, Geneva, 1956, Vol. 5, p. 309.
28. B. Cutts, et al, Graphite-Moderated Thermal Reactor Calculations for the Calder Hall Reactors", Proceedings of the Second U. N. International Conference on Peaceful Uses of Atomic Energy, Geneva, 1958, Vol. 12, p. 612.

29. H. L. Reynolds, "Critical Measurements and Calculations for Enriched-Uranium Graphite-Moderated Systems", Proceedings of the Second U. N. International Conference on Peaceful Uses of Atomic Energy, Geneva, 1958, Vol. 12, p. 632.
30. P. Bacher, et al, "Natural Uranium-Graphite Lattices", Proceedings of the Second U. N. International Conference on Peaceful Uses of Atomic Energy, Geneva, 1958, Vol. 12, p. 666.
31. R. A. Laubenstein, "Exponential Experiments on Graphite Lattices which Contain Multi-rod Fuel Elements", Proceedings of the Second U. N. International Conference on Peaceful Uses of Atomic Energy, Geneva, 1958, Vol. 12, p. 689.
32. D. Okrent, R. Avery, and H. H. Hummel, "A Survey of the Theoretical and Experimental Aspects of Fast Reactor Physics", Proceedings of the First U.N. International Conference on Peaceful Uses of Atomic Energy, Geneva, 1955, Vol. 5, p. 347.
33. W. B. Loewenstein and D. Okrent, "The Physics of Fast Power Reactors; A Status Report", Proceedings of the Second U. N. International Conference on the Peaceful Uses of Atomic Energy, Geneva, 1958, Vol. 12, p. 16.
34. G. E. Hansen, "Properties of Elementary Fast-Neutron Critical Assemblies", Proceedings of the Second U. N. International Conference on Peaceful Uses of Atomic Energy, Geneva, 1958, Vol. 12, p. 84.
35. D. D. Smith and J. E. Sanders, "Experimental Work with Zero Energy Fast Reactors", Proceedings of the Second U. N. International Conference on Peaceful Uses of Atomic Energy, Geneva, 1958, Vol. 12, p. 89.
36. J. K. Long, et al, "Fast Neutron Power Reactor Studies with ZPR-III", Proceedings of the Second U. N. International Conference on Peaceful Uses of Atomic Energy, Geneva, 1958, Vol. 12, p. 119.
37. S. Yiftah, D. Okrent, and P. A. Moldauer, Fast Reactor Cross Sections, Pergamon Press Inc., 1960.
38. A. Radkowsky, Chairman of Editorial Board, Naval Reactors Physics Handbook, Vol. III, The Physics of Intermediate Spectrum Reactors.
39. H. B. Stewart, P. L. Hofmann, and M. L. Storm, "Physics Considerations in the Design of Intermediate Energy Spectrum Reactors", Proceedings of the Second U. N. International Conference on Peaceful Uses of Atomic Energy, Geneva, 1958, Vol. 12, p. 142.
40. E. P. Wigner and J. E. Wilkins, Jr., "Effect of the Temperature of the Moderator on the Velocity Distribution of Neutrons with Numerical Calculations for H as Moderator", USAEC Report AECD-2275, 1944.
41. H. J. Amster and R. Suarez, "The Calculation of Thermal Constants Averaged Over A Wigner-Wilkins Flux Spectrum. Description of the SOFOCATE Code", Report WAPD-TM-39, Bettis, Atomic Power Laboratory, 1957.
42. H. J. Amster, "A Compendium of Thermal Neutron Cross Sections Averaged Over the Spectra of Wigner and Wilkins", Report WAPD-185, Bettis, Atomic Power Laboratory, 1958.
43. M. S. Nelkin and E. R. Cohen, "Recent Work in Neutron Thermalization", Progress in Nuclear Energy, Series I V3, Pergamon Press Inc., 1959.
44. M. J. Poole, M. S. Nelkin, and R. S. Stone, "The Measurement and Theory of Reactor Spectra", Progress in Nuclear Energy, Series 1 V2, Pergamon Press Inc., 1958.
45. C. H. Westcott, "Effective Cross Section Values for Well-Moderated Thermal Reactor Spectra", (3rd ed. corr.), Canadian Report AECL-1101, 1960.
46. J. F. Hogerton and R. C. Grass (Eds.), Reactor Handbook, "Engineering", p. 91, McGraw-Hill Book Co. Inc., 1955.
47. S. Yiftah and D. Okrent, "Some Physics Calculations on the Performance of Large Fast Breeder Power Reactors", USAEC Report ANL-6212, Argonne National Laboratory, 1960.
48. W. K. Anderson and J. S. Theilacker (Eds.), Neutron Absorber Materials for Reactor Control, U. S. Government Printing Office, 1962.
49. R. L. Murray and J. W. Niestlie, "Reactor control-rod theories", Nucleonics, 13, 2(1955)18.
50. A. M. Weinberg and E. P. Wigner, "Reactor Control Statics", Chap. XXII in The Physical Theory of Neutron Chain Reactors, p. 753, University of Chicago Press, 1958.
51. H. Soodak (Ed.), "Reactor Statics", Chap. 4 in Reactor Handbook, Vol. III, Part A, "Physics", p. 200, Interscience Publishers, N.Y., 1962.
52. A. F. Henry, "The Application of Reactor Kinetics to the Analysis of Experiments", Nucl. Sci. Eng., 3(1958)52.
53. Yankee Atomic Electric Co., "Yankee Nuclear Power Station, Technical Information and Final Hazards Summary Report, Report YAEC-167, 1960.
54. "The Experimental Boiling Water Reactor (EBWR)", USAEC Report ANL-5607, Argonne National Laboratory, 1957.
55. H. Hurwitz, Jr. and G. M. Roe, "Absorption of neutrons by black control rods", J. Nucl. Energy, 2,2(1955)85.
56. H. E. Stevens, "Half-Group Theories for Control Rods", Reactor Control Meeting Held in Los Angeles, March 6-8, 1957, USAEC Report TID-7532 (Pt. 1), p. 16.
57. S. Pearlstein, "Application of the absorption area method to three-group diffusion theory problems", Nucl. Sci. Eng., 4(1958)322.
58. G. Packman and B. Cutts, "Basic design of reactor", J. Brit. Nuclear Energy Conf., 2,2(1957)102.
59. I. Wivstad and C. Mileikowsky, "Adam—A 75-Mw Nuclear Energy Plant for House Heating Purposes", Proceedings of the Second U. N. International Conference on Peaceful Uses of Atomic Energy, Geneva, 1958, Vol. 8, p. 228.
60. C. Starr and R. W. Dickinson, Sodium Graphite Reactors, Addison-Wesley Publishing Co. Inc., Reading, Mass., 1958.
61. S. A. Ghalib and J. H. Bowen, "Equipment for Control of the Reactor", J. Brit. Nuclear Energy Conf., 2,2(1957)187.
62. H. E. Stevens, "Nuclear requirements for control materials", Nucl. Sci. Eng., 4(1958)373.
63. H. J. Amster, "The Wigner-Wilkins calculated thermal neutron spectra compared with measurements in a water moderator", Nucl. Sci. Eng., 2(1957)394.
64. "Survey of the Physics Metallurgy and Engineering Aspects of Reactor Control Materials", Report GEAP-3183, General Electric Co., Atomic Power Equipment Dept., 1959.
65. R. A. Becker and J. L. Russell, Jr., "Relative Effectiveness of Reactor Control Materials", Report GEAP-3201, General Electric Co., Vallecitos Atomic Laboratory, 1959.
66. H. F. Johnston, J. L. Russell, Jr. and W. L. Silvernail, "Relative control rod worths of some rare earth oxides", Nucl. Sci. Eng., 6(1959)93.
67. U. S. Atomic Energy Commission, "Light-Water-Moderated Reactor", Chap. 3 in Research Reactors, p. 153, McGraw-Hill Book Co. Inc., N.Y., 1955.
68. Alco Products, Inc., "APPR-1: Design, Construction and Operation," Report APAE-23, Alco Products Inc., 1957.
69. "Theory of Asymmetric Arrays of Control Rods in Nuclear Reactors": Report APAE-48, Alco Products, Inc., 1959.
70. H. L. Garabedian, "Control Rod Theory for a Cylindrical Reactor", Report AECD-3666, Westinghouse Atomic Power Division, 1950.
71. R. L. Murray, "Reactivity Values of Multiple Reactor Control Rods", Preprint 83, for presentation at Nuclear Engineering and Science Congress, Sponsored by Engineers Joint Council and American Institute of Chemical Engineers, December 12-16, 1955.
72. A. M. Weinberg and E. P. Wigner, "Perturbation Theory", Chap. XVI in The Physical Theory of Neutron Chain Reactors, p. 537, University of Chicago Press, 1958.
73. L. N. Ussachoff, "Equation for the Importance of Neutrons, Reactor Kinetics and The Theory of Perturbation", Proceedings of the First U. N. International Conference on Peaceful Uses of Atomic Energy, Geneva, 1955, Vol. 5, p. 503.
74. R. L. Murray, "Perturbation Theory and Application to Reactors", APAE Memo No. 208, Alco Products, Inc., 1959.
75. J. T. Thomas and D. Callihan, "Radiation Excursions at the ORNL Critical Experiments Laboratory, I, May 26, 1954; II, Feb. 1, 1956", USAEC Report ORNL-2452, Oak Ridge National Laboratory, 1958.
76. "Technical Feasibility and Economic Potential of the Variable Moderator Reactor, Final Report", Report ATL-A-109 (Rev. 1), Advanced Technology Laboratories, 1960.
77. H. V. Lichtenberger, F. W. Thalgott, W. Y. Kato, and M. Novick, "Operating Experience and Experimental Results Obtained from a NaK-Cooled Fast Reactor", Proceedings of the First U. N. International Conference on Peaceful Uses of Atomic Energy, Geneva, 1956, Vol. 3, p. 345.
78. J. A. Lane, H. G. MacPherson, and Frank Maslan (Eds.), "Design and Construction of Experimental Homogeneous Reactors", Chap. 7 in Fluid Fuel Reactors, p. 340, Addison-Wesley Publishing Co. Inc., Reading, Mass., 1958.
79. C. G. Poncelot, "Analysis of the Reactivity Characteristics of Yankee Core I", Report WCAP-6050, Westinghouse Electric Corp., Atomic Power Div., 1963.
80. L. J. Koch, H. O. Monson, W. R. Simmons, M. Levenson, F. Verber, E. Hutter, R. A. Jaross, T. R. Spalding, J. R. Simanton, and A. Lovoff, "Construction Design of EBR-II: an Integrated Unmoderated Nuclear Power Plant", Proceedings of the Second U. N. International Conference on Peaceful Uses of Atomic Energy, Geneva, 1958, Vol. 9, p. 323.
81. "Enrico Fermi", Nucl. Eng., 6,64(1961)377.
82. D. Mars and D. Gans, Jr., "Spectral Shift Control Reactor Design and Economic Study", Report BAW-1241, Babcock and Wilcox Co. and Stone and Webster Engineering Corp., 1961.
83. J. R. Fisher and A. Diaz, "Experimental Evaluation of the Fully Loaded Elk River Reactor", Trans. Am. Nucl. Soc., 6,1(1963)65.
84. P. C. Zmola, W. S. Flinn, R. J. Rickert, G. V. Notari, and R. H. Young, "The Unified Modular Plant, A Compact Nuclear Steam Generator for Marine Application", Report CEND-169, Combustion Engineering, Inc., 1962.
85. "Boiling Nuclear Superheater (BONUS) Power Station Final Hazards Summary Report", Report PRWRA-GNEC 5, General

Nuclear Engineering Corp. and The Puerto Rico Water Resources Authority, 1962.
86. G. Sege, "Preliminary Hazards Summary Report for the Dresden Nuclear Power Station", Report GEAP-1044, Commonwealth Edison Co., and General Electric Co., 1957.
87. J. D. McGaugh and R. H. Chastain, "Power Density and Burnup Distributions in Yankee Core I", Report WCAP-6051, Westinghouse Electric Corp., Atomic Power Division, 1963.
88. "Reactor Physics Constants", USAEC Report ANL-5800, p. 369, Argonne National Laboratory, 1958.
89. C. H. Westcott, "Effective Cross Section Values for Well-Moderated Thermal Reactor Spectra", Canadian Report AECL-1101, 1960.
90. J. Bengston and R. L. Hellens, "A Survey of the Beginning of Life Characteristics of Uranium-Fueled, Water-Moderated Lattices", Report CEND-137, Combustion Engineering Inc., 1961.
91. E. Fein and L. C. Noderer, "An Evaluation of Effective Plutonium Cross Sections in Reactivity Calculations", Report CEND-146, Combustion Engineering, Inc., 1961.
92. V. P. Adyasevich, I. M. Frank, O. I. Kozinets, F. L. Shapiro, I. V. Shtranikh, and K. D. Tolstov, "Measurement of Temperature Effects in Uranium-Graphite Subcritical Systems" in Conference of the Academy of Sciences of the USSR on the Peaceful Uses of Atomic Energy, July 1-5, 1955, USAEC, 1956, p. 109.
93. S. Yiftah and D. Okrent, "Some Physics Calculations on the Performance of Large Fast Breeder Power Reactors", USAEC Report ANL-6212, p. 27, Argonne National Laboratory, 1960.
94. D. Okrent, "Breeding, Safety and Fuel Cycles", Paper presented at the VIII Nuclear Congress sponsored by Comitato Nazionale Energia Nucleare, Rome, Italy, June 17-23, 1963. Argonne National Laboratory.
95. W. P. Keeney and J. K. Long, "Dilute Fast Critical Experiments in Zero Power Reactor No. III", Trans. Am. Nucl. Soc., 3,2(1960)327.
96. J. K. Long, A. R. Baker, W. Gemmell, W. P. Keeney, R. L. McVean, and F. W. Thalgott, "Experimental Results on Large Dilute Fast Critical Systems with Metallic and Ceramic Fuels", in Physics of Fast and Intermediate Reactors, Vol. I, Proceedings of a Seminar, Vienna, August 3-11, 1961, IAEA, Vienna, 1962.
97. A. M. Weinberg and E. P. Wigner, The Physical Theory of Neutron Chain Reactors, University of Chicago Press, 1958. a) Chaps. III, VII and XIX.
98. S. Glasstone and M. C. Edlund, The Elements of Nuclear Reactor Theory, (see: Secs. 4.57-4.61, 7.23-7.33), D. Van Nostrand Co., Princeton, N.J., 1952.
99. R. M. Pearce, "The Doppler effect in thermal reactors", J. Nucl. Energy, Pt. A., Reactor Science, 13,3/4(1961)150. 99a) Fig. 6 and Fig. 3; See also: 1) R. M. Pearce, AERE R/R 2806 and 2) R. M. Pearce, "Radial dependence of the Doppler effect in bars of uranium and thorium", J. Nucl. Energy, Pt. A, Reactor Science, 11(1960)136; 3) L. Dresner, ORNL 55-12-147 (1956a); and 4) K. W. Morton, AERE/R 2929 (1959).
100. R. M. Pearce and D. H. Walker, CRRP-581; see also, Nucl. Sci. Eng., 2(1957)24.
101. J. Chernick and R. Vernon, "Some refinements in the calculation of resonance integrals", Nucl. Sci. Eng., 4(1958)649.
102. L. Dresner, "Resonance Absorption of Neutrons in Nuclear Reactions" USAEC Report ORNL-2659, Oak Ridge National Laboratory, 1959. Also: Pergamon Press, 1960.
103. J. B. Sampson and J. Chernick, Progress in Nuclear Energy, Ser. 1, Vol. 2, Pergamon Press, N.Y., 1958.
104. E. Hellstrand, P. Blomberg, and S. Horner, "The temperature coefficient of the resonance integral for uranium metal and oxide", Nucl. Sci. Eng., 8(1960)497.
105. W. H. Arnold, Jr. and R. A. Dannels, "The Doppler coefficient of $U^{238}O_2$, Trans. Am. Nucl. Soc., 3,1(1960)229.
106. A. R. Vernon, "Calculation of the effective resonance integral of U^{238}", Nucl. Sci. Eng., 7(1960)252.
107. M. G. Bhide and H. H. Hummel, "Calculations of the Doppler Coefficient of Large Ceramic-Fueled Fast Reactors", USAEC Report ANL-6601, Argonne National Laboratory, 1962.
108. R. B. Nicholson, "The Doppler Effect in Fast Neutron Reactors", Report APDA-139, Atomic Power Development Associates, 1960.
109. A. H. Spano, "Self-limiting power excursion tests of a water-moderated, low-enrichment UO_2 core", Nucl. Sci. Eng., 15(1963)37.
110. V. B. Egiazarov, V. S. Dikarev, and V. G. Madeev, Conference of the Academy of Sciences USSR on the Peaceful Uses of Atomic Energy, 1955.
111. J. R. Brown and J. B. Sampson, "Measured Doppler Coefficient of Thorium Dispersion in Graphite", Report GA-3422, General Atomic Division of General Dynamics, 1963.
112. P. Greebler and E. Goldman, "Doppler Calculations for Large Fast Ceramic Reactors—Effects of Improved Methods and Recent Cross Section Information", Report GEAP-4092, General Electric Co., 1962.
113. J. Codd and P. J. Collins, "Pu^{239} and U^{238} Resonance Interaction Effects in a Dilute Fast Reactor", EAES Symposium on Advances in Reactor Theory, Karlsruhe, April 23-24, 1963.
114. Heavy Water Lattices, International Atomic Energy Agency, Vienna, 1960. See: D. S. St. John, "Definition of Lattice Parameters"; E. C. Critoph, "Neutron Balance", and B. Pershagen, "Formulae and Definitions Used in Swedish Lattice Calculations".
115. H. Kouts and R. Sher, "Experimental Studies of Slightly Enriched Uranium Water-Moderated Lattices", USAEC Report BNL-486, Brookhaven National Laboratory, 1957.
116. L. Dresner, "The effective resonance integrals of U^{238} and Th^{232}", Nucl. Sci. Eng., 1(1956)68.
117. E. Hellstrand, "Measurements of the effective resonance integral in uranium metal and oxide", Letters to the Editors, Nucl. Sci. Eng., 12(1962)435.
118. A. B. Rothman and C. E. W. Ward, "Effective resonance of thorium metal", Nucl. Sci. Eng., 12(1962)293.
119. J. E. Gray, W. H. Hamilton, and W. E. Wynne (Eds.), "Shippingport Operations from Start-up to First Refueling—Dec. 1957 to October 1959", Report DLCS-364, Duquesne Light Company and Westinghouse Electric Corporation, 1959.
120. PRDC Technical Information and Hazards Summary Report, Enrico Fermi Reactor. Section 1: Reactor and Plant Design, 1961.
121. V. O. Ericksen, author, "Void Effects in Boiling Heavy Water Reactors". Second U. N. International Conference on Peaceful Uses of Atomic Energy, 1958, Vol. 11, p. 466. Private communication to T. J. Thompson.
122. J. F. Silvers and J. R. Dietrich, "Transient Generation and Removal of Heat", in Reactor Handbook, Engineering, p. 141 McGraw-Hill Book Co. Inc., N. Y., 1955.
123. J. R. Stehn (Ed.), "The Physics of Intermediate Spectrum Reactors", Chap. 28 in Vol. III, Naval Reactors Physics Handbook, USAEC, 1958.
124. C. G. Poncelet, "Analysis of the Reactivity Characteristics of Yankee Core I", Report WCAP-6050, Westinghouse Electric Corp., 1963.
125. J. M. Gallagher, Jr., H. W. Graves, Jr., D. Hunter, and J. E. Howard, "Startup Experiment Program for the Yankee Reactor", Report YAEC-184, Yankee Atomic Electric Co. and Westinghouse Electric Corp., 1961.
126. J. A. DeShong, Jr., "Power Transfer Functions of the EBWR Obtained Using a Sinusoidal Reactivity Driving Function", USAEC Report ANL-5798, Argonne National Laboratory, 1958.
127. E. S. Beckjord, "Dynamic Analysis of Natural Circulation Boiling Water Power Reactors", USAEC Report ANL-5799, Argonne National Laboratory, 1958.
128. J. A. DeShong, Jr., and W. C. Lipinski, "Analysis of Experimental Power-Reactivity Feedback Transfer Functions for a Natural Circulation Boiling Water Reactor", USAEC Report ANL-5850, Argonne National Laboratory, 1958.
129. J. A. Thie, "Dynamic Behavior of Boiling Reactors", USAEC Report ANL-5849, Argonne National Laboratory, 1959.
130. J. A. Thie, "Theoretical Reactor Statics and Kinetics of Boiling Reactors", Proceedings of the Second U. N. International Conference on Peaceful Uses of Atomic Energy, Geneva, 1958, Vol. 11, p. 441.
131. A. Kirchenmayer, "On the kinetics of boiling water reactors", J. Nucl. Energy, Pt. A., Reactor Sci., 12,4(1960)155.
132. A. Z. Akcasu, "Theoretical Feedback Analysis in Boiling Water Reactors", USAEC Report ANL-6221, Argonne National Laboratory, 1960.
133. "Boiling water reactors; Stability", Power Reactor Technology, 3,1(1959)42.
134. "Reactor dynamics", Power Reactor Technology, 4,3(1961)29.
135. J. M. Harrer, Nuclear Reactor Control Engineering, D. Van Nostrand Co., N.J., 1963, Chaps. 7 and 8, p. 228-316.
136. J. B. Sampson, "Analysis of Activation Measurements of Th^{232} Resonance in the Peachbottom (40-MW(E)) Prototype HTGR Critical Assembly", Report GA-3069, General Atomic, Division of General Dynamics, 1962.
137. M. K. Drake and J. R. Brown, "Activation Measurement of Resonance Integrals", Report GA-3125, General Atomic, Division of General Dynamics, 1963.
138. A. Weitzberg, I. Kaplan, T. J. Thompson, "Measurement of Neutron-Capture in U238 in Lattices of Uranium Rods in Heavy Water", USAEC Report NYO-9659, 1962.
139. W. Dardenne, M.I.T. Nuclear Engineering PhD. Thesis, AEC-NYO Report, to be published in 1964.
140. C. A. Anderson, Jr., "Measurement of Neutron Energy Spectra with the M.I.T.R. Fast Chopper", M.I.T. PhD. Thesis, August, 1961.
141. A. R. Batten and J. A. J. Jaques, "A Measurement of the Contribution of the Doppler Effect to the Temperature Coefficient of Reactivity in a Fast Reactor", British Report AERE R/M 168, 1958.
142. T. H. Springer and S. G. Carpenter, "Fast Spectrum Doppler Measurements", Proceedings of the Conference on Breeding;

Economics and Safety in Large Fast Power Reactor, Argonne National Laboratory, 1963.
143. J. Chernick, "Temperature Coefficients of Reactivity of Reactors", USAEC Report BNL-1344, Brookhaven National Laboratory, 1955.
144. P. Greebler and E. Goldman, "Doppler Calculations for Large Fast Ceramic Reactors—Effects of Improved Methods and Recent Cross Section Information", Report GEAP-4092, General Electric Co., 1962.
145. P. Greebler and B. A. Hutchins, "The Doppler Effect in a Large Fast Oxide Reactor, Its Calculation and Significance for Reactor Safety", Proceedings of the Seminar on the Physics of Fast and Intermediate Reactors, Vienna, Austria, August 3-11, 1961, IAEA, Vienna, 1962, Vol. III, p. 121.
146. "Simplified calculation on thermal transient of a UO_2 fuel rod", Letters to the Editors, Nucl. Sci. Eng., 11(1961)340.
147. R. D. Richtmyer, R. Van Norton, and A. Wofe, "The Monte Carlo Calculation of Resonance Capture in Reactor Lattices", Proceedings of the Second U.N. International Conference on Peaceful Uses of Atomic Energy, Geneva, 1958, Vol. 16, p. 180.
148. W. H. Arnold, Jr. and R. A. Dannels, "A Monte Carlo Study of the Doppler Effect in UO_2 Fuel", Report WCAP-1572, Westinghouse Electric Corp., 1960.
149. K. W. Morton, "A Monte Carlo Study of the Resonance Absorption and Its Temperature Variation in a Square Uranium-Graphite Lattice", British Report AERE-R-2929, 1959.
150. A. H. Spano, "Self-limiting power excursion tests of a Water-Moderated Low-Enrichment UO_2 Core", Nucl. Sci. Eng., 15(1963)37.
151. F. R. Farmer, United Kingdom Atomic Energy Authority, personal communication.
152. R. W. Woodruff, C. W. Griffin, and R. W. Keaten, "Dectecting and Eliminating Fuel Rod Bowing in the SRE", Report NAA-SR-7705, North American Aviation, 1963.
153. A. H. Spano, J. E. Barr, L. A. Stephan, and J. C. Young, "Self-Limiting Power Excursion Tests of a Water-Moderated Low-Enrichment UO_2 Core in SPERT I", Report IDO-16751, Phillips Petroleum Co., 1962.
154. H. V. Lichtenberger, M. Novick, B. C. Cerutti, R. A. Cameron, D. F. McGinnis, E. N. Pettit, G. K. Whitham, R. A. Haroldson, and L. J. Koch, "Experimental Breeder Reactor Progress Report, April 1, 1951—January 31, 1953", USAEC Report ANL-5023, Argonne National Laboratory, 1953.
155. R. R. Smith, "Recent EBR-I Stability Studies", Proceedings of the Conference on Transfer Function Measurement and Reactor Stability Analysis, USAEC Report ANL-6205, p. 232, Argonne National Laboratory, 1960.
156. R. R. Smith, et al., "A Mechanism Explaining the Instability of EBR-I, Mark II", USAEC Report ANL-6354, Argonne National Laboratory, 1961.
157. R. R. Smith, et al., "Instability Studies with EBR-I, Mark III", USAEC Report ANL-6266, Argonne National Laboratory, 1960.
158. R. R. Smith, R. G. Matlock, F. D. McGinnis, M. Novick, and F. W. Thalgott, "An Analysis of the Stability of EBR-I, Marks I to III, and Conclusions Pertinent to the Design of Fast Reactors", in Physics of Fast and Intermediate Reactors, Vol. III, p. 43, Proceedings of a Symposium, Vienna, August 3-11, 1961, IAEA, Vienna, 1962.
159. "Reactor dynamics and safety", Power Reactor Technology, 5,2(1962)22.
160. S. G. George and E. W. Rettger, Mechanics of Materials, McGraw-Hill Book Co. Inc., N.Y., 1935.
161. S. Timoshenko and J. N. Goodier, Theory of Elasticity (2nd ed.) McGraw-Hill Book Co. Inc., N.Y., 1951.
162. C. G. Johnson, "Thermal Deflection of the Yankee Fuel Assembly from Linear and Non-Linear Temperature Gradients", Report UAEC-80, Westinghouse Electric Corp., 1959.
163. H. Chelemer, "Thermal Deflection of the Yankee First Core Fuel Assembly", Report YAEC-139, Westinghouse Electric Corp., 1961.
164. "Heavy Water Moderated Reactors Evaluation Study, Vol. III, Plant Designs", Report ECNG-1062, American Electric Power Service Corp. and General Nuclear Engineering Corp., 1962.
165. J. H. Hutton, et al., "Pawling Lattice Test Rig—Test Report No. 2, 37-Rod Cluster, 11.1 In. Lattice Pitch", Report NDA-2-31-43, Nuclear Development Corp., 1961.
166. L. Dresner, "Some remarks on the effect of a nonuniform temperature distribution on the temperature dependence of resonance absorption", Nucl. Sci. Eng., 11(1961)39.
167. J. H. Ferziger, "Resonance absorption of neutrons in lumps with nonuniform temperature distributions", Nucl. Sci. Eng., 14(1962)244.
168. S. Glasstone and A. Sesonske, Nuclear Reactor Engineering, D. Van Nostrand Co., Inc., Princeton, N.J., 1955, 1963.
169. H. Soodak (Ed.), Reactor Handbook, Vol. III, Pt. A, "Physics", (2nd ed) Interscience Publishers, N.Y. 1962.
170. K. Mochizuki and A. Takeda, "An analysis of neutron flux spatial oscillation due to xenon build-up in a large power reactor core", Nucl. Sci. Eng., 7(1960)336.
171. J. Chernick, G. Lellouche, and W. Wollman, "The effect of temperature on xenon instability", Nucl. Sci. Eng., 10(1961)120.
172. R. M. Pearce, "Method of studying xenon spatial instability with an analog computer", Nucl. Sci. Eng., 11(1961)328.
173. G. S. Lellouche, "Space dependent xenon oscillations", Nucl. Sci. Eng., 12(1962)482.
174. G. S. Lellouche, "Reactor size sufficient for stability against spatial xenon oscillations", Letters to the Editors, Nucl. Sci. Eng., 13(1962)60.
175. D. Randall and D. S. St. John, "Xenon spatial oscillations", Letters to the Editors, Nucl. Sci. Eng., 14(1962)204.
176. G. L. Gyorey, "The effect of modal interaction in the xenon instability problem", Nucl. Sci. Eng., 13(1962)338.
177. L. M. Shotkin and F. H. Abernathy, "Linear stability of the thermal flux in a reflected core containing xenon and temperature reactivity feedback", Nucl. Sci. Eng., 15(1963)197.
178. S. Lundquist and P. Weissglas, "A method for determination of spatial stability", Nucl. Sci. Eng., 15(1963)474.
179. R. M. Pearce, "Xenon oscillations in finite reactors", Letters to the Editors, Nucl. Sci. Eng., 16(1963)336.
180. R. C. Bolles and N. E. Ballou, "Calculated activities and abundances of U^{235} fission products", Nucl. Sci. Eng., 5(1959)156.
181. S. B. Gunst, E. D. McGarry, and J. J. Scoville, "The reactivity of natural UO_2 irradiated to 6×10^{21} n/cm^2", Nucl. Sci. Eng., 7(1960)407.
182. J. D. Garrison and B. W. Roos, "Fission-product capture cross sections", Nucl. Sci. Eng., 12(1962)115.
183. W. H. Walker and S. A. Kushneriuk, "Yields and Effective Cross Sections of Fission Products and Pseudo-Fission Products", Canadian Report AECL-1054, 1960.
184. L. N. Usachoff, "Equations for the Importance of Neutrons, Reactor Kinetics and the Theory of Perturbation", Proceedings of the First U. N. International Conference on Peaceful Uses of Atomic Energy, Geneva, 1955, Vol. 5, p. 503.
185. M. Benedict, R. T. Shanstrom, S. L. Amberg, N. B. McLeod, and P. T. Steranka, "FUELCYC, a new computer code for fuel cycle analysis, Pt. II, Examples of applications", Nucl. Sci. Eng., 11(1961)386.
186. B. I. Spinrad, J. C. Carter, and C. Eggler, "Reactivity Changes and Reactivity Lifetimes of Fixed-Fuel Elements in Thermal Reactors", Proceedings of the First U. N. International Conference on Peaceful Uses of Atomic Energy", Geneva, 1955, Vol. 5, p. 125.
187. R. P. Schuman and R. L. Tromp, "Calculation of the Composition of Reactor Irradiated Heavy Nuclides", Report IDO-16571, Phillips Petroleum Co., 1959.
188. A. Radkowsky, "Theory and Application of Burnable Poisons", Proceedings of the Second U. N. International Conference on Peaceful Uses of Atomic Energy", Geneva, 1958, Vol. 13, p. 426.
189. D. A. Gavin and D. J. Anthony, "Particle Self-Shielding of Boron Carbide", Report KAPL-2000-8, Knolls Atomic Power Laboratory, 1960.
190. H. Hurwitz, Jr. and P. F. Zweifel, "Self shielding of lumped-poison mixtures", Nucl. Sci. Eng., 1(1956)438.
191. R. C. Dahlberg and F. D. Judge, "The Depletion of Burnable Poison in Endurance Calculations", Report KAPL-2058, Knolls Atomic Power Laboratory, 1960.
192. J. Dwork, P. J. Hofmann, H. Hurwitz, Jr. and E. F. Clancey, "Self-Shielding Factors for Infinitely Long, Hollow Cylinders", Report KAPL-1262, Knolls Atomic Power Laboratory, 1955.
193. J. C. Corelli and F. Feiner, "Burnup of Europium and Dysprosium under Neutron Irradiation", Report KAPL-2000-8, Knolls Atomic Power Laboratory, 1959.
194. A. M. Smith and J. Jeffrey, "Burnable Poisons in Small Power Reactors", Proceedings of the Second U. N. International Conference on Peaceful Uses of Atomic Energy", Geneva, 1958, Vol. 13, p. 447.
195. G. P. Rutledge and R. R. Eggleston, "Self-shielding experiments with gaseous and solid B-10", Nucl. Sci. Eng., 4(1958)530.
196. Directory of Nuclear Reactors, Vol. IV Power Reactors, International Atomic Energy Agency, Vienna, 1962.
197. "Control and instrumentation for nuclear power stations", Nucleonics, 19, 11(1961)133.
198. D. I. Blokhintsev and N. A. Nikolaev (USSR), "The First Atomic Power Station of the USSR and the Prospects of Atomic Power Development", Proceedings of the First U. N. International Conference on Peaceful Uses of Atomic Energy, Geneva, 1955, Vol. 3, p. 39.
199. N. A. Dollezhal, et al., "Operating Experience with the First Atomic Power Station in the USSR and Its Use Under Boiling Conditions", Proceedings of the First U. N. International Conference on Peaceful Uses of Atomic Energy, Geneva, 1958, Vol. 8, p. 88.
200. J. W. Simpson, et al., "Description of the Pressurized Water Reactor (PWR) Power Plant at Shippingport, Pa.", Proceedings of the First U. N. International Conference on Peaceful Uses of Atomic Energy, Geneva, 1955, Vol. 3, p. 235.
201. J. W. Simpson and H. G. Rickover, "Shippingport Atomic

Power Station (PWR)", Proceedings of the Second U. N. International Conference on Peaceful Uses of Atomic Energy, Geneva, 1958, Vol. 8, p. 44.
202. P. G. DeHuff, et al., "Experiences in the Design, Construction and Operation of the PWR at Shippingport", Proceedings of the Second U. N. International Conference on Peaceful Uses of Atomic Energy, Geneva, 1958, Vol. 8, p. 75.
203. J. E. Gray, W. H. Hamilton, W. E. Wynne (Eds.), "Shippingport Operations from Start-up to First Refueling, December 19, 1957 to October 7, 1959", Report DLCS-364, Duquesne Light Co. and Westinghouse Electric Corp., Bettis Atomic Power Laboratory, 1959.
204. C. N. Dunn and R. F. Stratton (Eds.), "Shippingport Operations from Power Operation After First Refueling to Second Refueling, May 6, 1960 to August 16, 1961", Report DLCS-36402, Duquesne Light Co. and Westinghouse Electric Corp., Bettis Atomic Power Laboratory, 1961.
205. C. S. Abrams, P. W. Frank, O. D. Parr, and F. R. Vaughn, "Evaluation of the PWR Core 1 Failed Element Detection and Location System", Report WAPD-TM-330, Westinghouse Electric Corp., Bettis Atomic Power Laboratory, 1962.
206. A. C. Lapsley, "Neutron, gamma measurements for in-pile power monitoring", Nucleonics, 16, 2 (1958) 106.
207. A. J. Campanek and E. C. Tornuzzer, "Neutron activated wires plot fluxes in Yankee cores", Nucleonics, 20,2(1962)44.
208. Yankee Atomic Electric Company, "Yankee Nuclear Power Station, Technical Information and Final Hazards Summary Report", Report YAEC-167, 1959.
209. AEC Analysis of the Yankee Plant Hazards, USAEC Docket No. 50-29, Aug. 31, 1962.
210. Power Reactor Experiments, Vol. II, pp. 237-239, International Atomic Energy Agency, Vienna, 1962.
211. Consolidated Edison Co. of New York, Inc., "Technical Specifications to Accompany Facility Operating License", Part A, USAEC Docket No. 50-3, Sept. 1, 1961.
212. "Carolinas-Virginia Tube Reactor, Preliminary Hazards Summary Report", Vol. II, Report CVNA-27, Westinghouse Electric Corp., 1959.
213. AEC Analysis of the CVTR Plant Hazards, USAEC Docket No. 50-144, p. 38, Aug. 9, 1962.
214. A. Andonedis, "CVTR Failed Fuel Element Detection and Location System Analysis and Description", Report CVNA-80, Westinghouse Electric Corp., 1961.
215. R. K. Baird and J. Forster, "Design and Performance of the Dresden Station In-Core Monitor System", AIEE Conference Paper CP-61-929, submitted May 24, 1961.
216. M. Joslin, "Operating Experience at Dresden", Presented at the Joint Japan-US Atomic Industrial Forum Meeting, Tokyo, Japan, December 5-8, 1961.
217. General Electric Co., Atomic Power Equipment Dept., Nuclear Electronics Application Newsletter, Vol. II, No. 1, June, 1962.
218. J. Forster, "Ion chamber systems", Nucleonics, 20,2(1962)50.
219. A. H. McQueen, "Control and Instrumentation for Nuclear Power Stations," Atomics, 16,2(1963)9.
220. AEC Analysis of the Humboldt Bay Hazards, USAEC Docket No. 50-133, p. 18.
221. Power Reactor Experiments, Vol. II, p. 23, International Atomic Energy Agency, Vienna, 1962.
222. Power Reactor Experiments, Vol. II, p. 63, International Atomic Energy Agency, Vienna, 1962.
223. Proceedings of the Power Reactor In-Core Instrumentation Meeting, Washington, D.C., April 28-29, 1960, USAEC Report TID-7598, pp. 5, 6, 1961.
224. The Journal of the British Nuclear Energy Conference, II(2), 153, 197, 207, 211 (April, 1957).
225. H. G. Davey, et al., "Operating Experience at Calder Hall", Proceedings of the Second U. N. International Conference on Peaceful Uses of Atomic Energy, Geneva, 1958, Vol. 8, p. 16.
226. J. B. W. Cunningham, "Current Re-designs of Calder Hall", Proceedings of the Second U. N. International Conference on Peaceful Uses of Atomic Energy, Geneva, 1958, Vol. 8, p. 420.
227. S. A. Ghalib and J. R. M. Southwood, "The Berkeley Power Station", Proceedings of the Second U. N. International Conference on Peaceful Uses of Atomic Energy, Geneva, 1958, Vol. 8, p. 476.
228. M. de Rouville, et al., "Experience Obtained During Two Years Operation of the Reactor G1", Proceedings of the Second U. N. International Conference on Peaceful Uses of Atomic Energy, Geneva, 1958, Vol. 8, p. 18.
229. France Atome, "Description of Reactors G2 and G3", Proceedings of the Second U. N. International Conference on Peaceful Uses of Atomic Energy, Geneva, 1958, Vol. 8, p. 352.
230. Proceedings of the Power Reactor In-Core Instrumentation Meeting, Washington, D.C., April 28-29, 1960, USAEC Report TID-7598, pp. 84, 84, 1961.
231. Marquardt Corp., Test Report, "Feasibility Demonstration of a Pneumatic Temperature Sensor", Report No. 512-534-123, Aug. 18, 1960.
232. Power Reactor Experiments, Vol. I, pp. 34-36, International Atomic Energy Agency, Vienna, 1962.
233. W. E. Parkins, "The Sodium Reactor Experiment", Proceedings of the First U. N. International Conference on Peaceful Uses of Atomic Energy, Geneva, 1955, Vol. 3, p. 306.
234. "Enrico Fermi Power Plant", Report APDA-124, Atomic Power Development Associates, Inc., 1959.
235. Atomics International, "Final Summary Safeguards Report for the Hallam Nuclear Power Facility", Report NAA-SR-5700, (and suppl.), North American Aviation, Inc., 1961.
236. Directory of Nuclear Reactors, Vol. II, Research, Test and Experimental Reactors, International Atomic Energy Agency, Vienna, 1959.
237. Atomics International, "Organic Moderated Reactor Experiment, Safeguards Summary", Report NAA-SR-2323, North American Aviation, Inc., 1958.
238. Atomics International, "Final Safeguards Summary Report for the Piqua Nuclear Power Facility", Report NAA-SR-5608, North American Aviation, Inc., 1961.
239. W. R. Kritz, "An Automatic Gas Chromatograph for Monitoring of Reactor Fuel Failures, Part IV., Model 2 Design", Report DP-668, E. I. du Pont de Nemours and Co., Inc., Savannah River Laboratory, 1962.
240. W. M. Cashin (Ed.), "Uranium Surface Contamination on Nuclear Reactor Fuel Elements", Report KAPL-2061, Knolls Atomic Power Laboratory, 1959.
241. J. D. McGaugh and R. H. Chastain, "Power Density and Burnup Distributions in Yankee Core I", Report WCAP-6051, Westinghouse Electric Corp., Atomic Power Department, 1963.
242. P. A. Fleger, et al., "Shippingport Atomic Power Station Operating Experience, Developments, and Future Plans", Report WAPD-T-1429, Westinghouse Electric Corp., Bettis Atomic Power Laboratory, 1961.
243. J. M. Harrer, Nuclear Reactor Control Engineering (Chapter 5: Neutron Detectors), D. Van Nostrand Co. Inc., Princeton, N. J., 1963.
244. N. H. Barth, "Physics Calculations for the Consumers Big Rock Point Process Computer", Report GEAP-3932, General Electric Co., Atomic Power Equipment Dept., 1963.
245. J. W. Greenwood, "Contamination of the NRU Reactor in May, 1958", Canadian Report AECL-850, 1959.
246. Atomic Energy of Canada Limited, "Douglas Point Nuclear Generating Station", Canadian Report AECL-1596.
247. J. F. Palmer, "Finding failed fuel in CANDU", Nucleonics, 22, 6 (1964) 39.
248. H. S. Arms, C. Bottrell, and P. H. W. Wolff, "The Hinkley Point Power Station", Proceedings of the Second U. N. International Conference on Peaceful Uses of Atomic Energy, Geneva, 1958, Vol. 8, p. 434.
249. "Design practice: Hallam", Power Reactor Technology, 5,3(1962)39.
250. J. Horowitz, B. Bailly-Du Bois, and R. Naudet, "Le Projet EL-4", D'Energie Nucleaire, 3,6(1961)369.
251. F. W. Gilbert, "WR-1 fits organic coolant to Canadian program", Canadian Nucl. Tech., 2,2(1963)37
252. J. C. Leny, "ORGEL - A European concept", Nucl. Eng., 6,67(1961)508.
253. Carolinas-Virgina Nuclear Power Associates, Inc., "Part B. License Application, Final Hazards Summary Report", Report CVNA-90 (USAEC Docket No. 50-144), Dec. 14, 1961, issued in four volumes, as follows: I. Description of Site and Surrounding Area; II. Description of Reactor and Nuclear Power Plant; III. Accidents and Hazards; IV. Plant Operations.
254. W. B. Lewis, "The Accident to the NRX Reactor on December 12, 1952", Canadian Report AECL-232 (DR-32), 1953
255. D. G. Hurst, "The Accident to the NRX Reactor Part II", Canadian Report AECL-233 (GPI-14), 1953.
256. "Design practice: Yankee Nuclear Power Station", Power Reactor Technology, 4,3(1961)47.
257. "Reactors on the line No. 9 - Yankee", Nucleonics, 19, 3(1961)53.
258. "Final Hazards Summary Report for Big Rock Point Plant", Consumers Power Co., Docket No. 50-155, Nov. 14, 1961.
259. Power Reactor Technology, 7,1(1963)
260. A. Amorosi and J. G. Yevick, "An Appraisal of the Enrico Fermi Reactor", Proceedings of the Second U.N. International Conference on Peaceful Uses of Atomic Energy, Geneva, 1958, Vol. 9, p. 358.
261. J. R. Dietrich and W. H. Zinn, Solid Fuel Reactors, Addison-Wesley Publishing Co., Inc., Reading, Mass., 1958.
262. J. British Nuclear Energy Conference, 6,3(1961).
263. D. Okrent, "Neutron physics considerations in large fast reactors", Power Reactor Technol., 7,2(1964). p. 107.
264. A. Firth and J. E. R. Holmes, "The SGHWR prototype reactor", Nucl. Eng., 9,93(1964)46.

CHAPTER 3

General Reactor Dynamics

E. P. GYFTOPOULOS
Massachusetts Institute of Technology
Cambridge, Massachusetts

CHAPTER CONTENTS*

*Except for a few changes and additions made in proof, this chapter is based on information in the literature or known to the author prior to November 1963.

INTRODUCTION

Designing a safe nuclear plant or maintaining the safety margin of an operating plant which is statically sound requires an effective understanding of the dynamics of the reactor and its associated components and equipment. Such an understanding is necessary in order to ensure that the constraints, imposed either by the materials of the plant components or by the environment in which the plant operates, will not be exceeded at any time.

The constraints that are inherent in various materials in conjunction with the environmental conditions are discussed in other chapters.

The purpose of this chapter is to introduce some general mathematical techniques and experimental procedures that are of import to the problem of understanding the dynamics of nuclear reactors. How these techniques and procedures have been implemented and put to practical use for specific reactors is discussed in the chapters on the kinetics of specific reactor types.

The attempt to understand nuclear reactor dynamics, or for that matter the dynamics of any physical system, involves several closely interrelated steps. The first step is to select a set of variables, the state variables, that are adequate to characterize the physical processes taking place during the operation of the system. Typical nuclear plant state variables are the neutron density, the coolant temperature, the control rod position, etc. The state variables must be directly or indirectly measurable and must be directly or indirectly related to the constraints imposed by the materials and the properties of the environment of the plant. In addition, the state variables must also be chosen so as to preserve a reasonable balance between the desired simplicity of characterization of the system and the required detail of characterization.

It must be emphasized that the problem of selection of state variables is extremely important. From the experimental point of view, the choice of the state variables is influenced by the available instrumentation for the plant since, in principle, all design specifications should be achieved by monitoring and controlling these variables. From the analytical point of view, the selected state variables specify the level of mathematical sophistication necessary for the theoretical visualization of the plant.

The second step in the process of understanding reactor dynamics is to find the time-dependent equations that interrelate the different variables. If all physical processes that partake in the operation of the system are known and if the appropriate state variables have been chosen, this step is relatively easy. The relations between variables are usually based on well-founded conversation laws such as the conservation of energy, mass, elementary particles, momentum, etc. at any instant of time and at every position in the nuclear plant and, therefore, time-dependent balance equations can be written fairly readily.

The third step is to solve the dynamic equations either analytically or with a computer. The objective of this step is either to establish results that can be implemented experimentally, thereby verifying the practicality of the equations, or to predict the range of values of the influential design parameters so that limitations imposed by the constraints can be satisfied. This is a purely mathematical and not a physical problem and it is one of the most difficult. Even for relatively simple systems, the establishment of general solutions is practically impossible. Thus one is forced to consider approximations either in the detail of the characterization of the system or in the range and form of variation of the parameters and variables involved. Approximations lead to practical results provided the results are used in their range of validity. Approximations and the evaluation of their range of validity require a thorough understanding of the physics of the system.

Finally, the practical value of any and all of the preceding steps must be justified by experiment, as already indicated. Any physical concepts or mathematical models or any analytical or computer results, regardless of their degree of elegance and sophistication, are of no consequence, particularly in questions of nuclear safety, if they have not been or cannot be justified and repeatedly verified by experiment.

The subsequent sections indicate briefly how these four general steps are specifically implemented when nuclear reactor safety is the primary concern.

1 NUCLEAR REACTOR DYNAMICS

1.1 General Remarks

From the standpoint of reactor safety, the most important variable is the energy stored in the reactor and its spatial distribution because it can be directly related to all other variables and constraints. In the final analysis, it is the stored energy, its distribution between fuel, moderator, and coolant and its time rate of change that determine the consequences of a serious accident.

Even though the stored energy is such a key variable in reactor safety, it has not been possible to measure it and control it directly. Instead, the energy behavior is inferred from measurements of the neutron population in the reactor, the temperatures of different components, the pressure, etc. The neutron population is indicative of the time derivative of the energy released in the reactor while temperatures and other thermodynamic variables are measures of the energy stored in the various regions of the reactor.

The analysis of the time behavior of the neutron population in a reactor can be treated by a number of alternative models. The common basis of all these models is that they are analytical statements of the fundamental equality:

$$\begin{bmatrix}\text{Rate of change}\\ \text{of neutron}\\ \text{population}\end{bmatrix} = \begin{bmatrix}\text{Rate}\\ \text{of neutron}\\ \text{production}\end{bmatrix} - \begin{bmatrix}\text{Rate}\\ \text{of neutron}\\ \text{destruction}\end{bmatrix}. \tag{1-1}$$

The difference between models lies in the theoretical concepts that are used to express quantitatively the various physical processes that affect the rates of production and destruction. Similar rate equalities can also be used for the other variables which are useful in the study of the behavior of the energy stored in the reactor.

The most general method of analytically implementing Eq. (1-1) is by means of transport theory. This theory allows a precise and detailed representation of all conceivable interactions that neutrons undergo in a reactor with a minimum number of assumptions. The equations that result are very difficult to handle analytically or to implement experimentally because of their generality. Under certain conditions, however, they can be reduced to simpler forms and thus be related in an approximate way to the energy stored in the reactor. One of these simpler forms is the space-independent conventional system of equations of reactor kinetics, which proved a productive tool in studies of reactor dynamics.

The purpose of this section is to derive the conventional form of reactor kinetics equations starting from the basic notions of transport theory. The value of this approach is that all the factors that influence the dynamics of the reactor are put in evidence, the exact meaning of such convenient parameters as reactivity, neutron lifetime, coefficients of reactivity, etc., can be readily indicated and the assumptions and limitations inherent in the conventional form of reactor kinetics are established. The derivation proceeds along the lines indicated by A. F. Henry [1] with some modifications.

1.2 Neutron Kinetics—Transport Theory

Consider a reactor in which the fuel is stationary. For mathematical simplicity and without any loss of generality, assume that there is only one species of fissionable material and that there is a very large number of neutrons in the reactor. The last assumption implies that statistical variations are ignored.

The time-dependent balance equations for the neutron population and the delayed neutron precursors are:

$$\frac{\partial}{\partial t} N(\vec{r}, E, \vec{\Omega}, t) = \int_{\Omega'} d\vec{\Omega}' \times \left\{ \frac{1}{4\pi} f_0(E) \int_0^\infty dE' \nu(E')(1-\beta) v' \Sigma_f(\vec{r}, E', t) N(\vec{r}, E', \vec{\Omega}', t) + \int_0^\infty dE' v' \Sigma_s(\vec{r}, E' \to E, \vec{\Omega}' \to \vec{\Omega}, t) N(\vec{r}, E', \vec{\Omega}', t) \right\}$$
$$+ \frac{1}{4\pi} \sum_{i=1}^{m} \lambda_i f_i(E) C_i(\vec{r}, t) + S(\vec{r}, E, \vec{\Omega}, t) - \vec{\Omega} v \cdot \operatorname{grad} N(\vec{r}, E, \vec{\Omega}, t) - v \Sigma_t(\vec{r}, E, t) N(\vec{r}, E, \vec{\Omega}, t) \tag{1-2}$$

$$\frac{\partial}{\partial t} C_i(\vec{r}, t) = \int_{\Omega'} d\vec{\Omega}' \int_0^\infty dE' \nu(E') \beta_i v' \Sigma_f(\vec{r}, E', t) N(\vec{r}, E', \vec{\Omega}', t) - \lambda_i C_i(\vec{r}, t), \quad i = 1, 2, \ldots m, \tag{1-3}$$

where
$C_i(\vec{r}, t)$ — delayed neutron precursor concentrations
E — relative neutron energy
$f_i(E)$, $(i=0, 1, \ldots m)$ — prompt and delayed neutron energy spectrum
$N(\vec{r}, E, \vec{\Omega}, t)$ — neutron density in the $(\vec{r}, E, \vec{\Omega}, t)$—space
$S(\vec{r}, E, \vec{\Omega}, t)$ — source density in the $(\vec{r}, E, \vec{\Omega}, t)$—space
v — neutron speed
$v\vec{\Omega}$ — neutron velocity
β_i, β — delayed neutron fractions
λ_i — delayed neutron decay constants
$\nu(E)$ — number of neutrons per fission, of energy E
$\Sigma_f(\vec{r}, E, t)$ — fission cross section
$\Sigma_s(\vec{r}, E' \to E, \vec{\Omega}' \to \vec{\Omega}, t)$ — scattering cross section
$\Sigma_t(\vec{r}, E, t)$ — total cross section

The meaning of Eqs. (1-2) and (1-3) may be understood as follows. Neutron interactions depend on the relative energy and direction of motion of neutrons with respect to the target nuclei. The neutron density, $N(\vec{r}, E, \vec{\Omega}, t)$, is the number of neutrons per unit volume, per unit energy and per unit solid angle that are at the position $\vec{r}$ of the reactor, have a relative energy E and a relative direction of motion $\vec{\Omega}$ at time t. In a small volume ΔV, centered around the point $\vec{r}$, the time rate of change of neutrons that have energies between E and $E + \Delta E$ and directions of motion between $\vec{\Omega}$ and $\vec{\Omega} + d\vec{\Omega}$ is

$$\Delta V \Delta E \Delta \vec{\Omega} \frac{\partial}{\partial t} N(\vec{r}, E, \vec{\Omega}, t). \tag{1-4}$$

Neutrons are born in the elementary volume $\Delta V \Delta E \Delta \vec{\Omega}$ of the phase space $(\vec{r}, E, \vec{\Omega})$ through the production mechanisms:

prompt fissions,
elastic and inelastic scattering collisions,
decay of delayed neutron precursors, and
external neutron sources.

Thus, the rate of production in $\Delta V \Delta E \Delta \vec{\Omega}$ can be written as:

$$\left[\int_{\Omega'} d\vec{\Omega}' \left\{ \underbrace{\frac{f_0(E)}{4\pi} \int_0^\infty dE' \nu(E')(1-\beta) v' \Sigma_f(\vec{r}, E', t) N(\vec{r}, E', \vec{\Omega}', t)}_{\text{(prompt fissions)}} + \underbrace{\int_0^\infty dE' \Sigma_s(\vec{r}, E' \to E, \vec{\Omega}' \to \vec{\Omega}, t) N(\vec{r}, E', \vec{\Omega}', t)}_{\text{(scattering collisions)}} \right\} + \underbrace{\frac{1}{4\pi} \sum_{i=1}^{m} \lambda_i f_i(E) C_i(\vec{r}, t)}_{\text{(delayed neutrons)}} + \underbrace{S(\vec{r}, E, \vec{\Omega}, t)}_{\text{(source)}} \right] \Delta V \Delta E \Delta \vec{\Omega}, \tag{1-5}$$

provided that both fission neutrons and delayed neutrons are emitted isotropically and all fission fragments remain in the immediate vicinity of the fission event that produces them. The cross sections are taken explicitly dependent on time in order to account for externally introduced changes in the reactor materials.

On the other hand, neutrons are lost from the elementary volume $\Delta V \Delta E \Delta \vec{\Omega}$ through the destruction mechanisms:

leakage, and

absorption, including scattering-out collisions.

Thus the rate of destruction can be written as

$$[\vec{\Omega} v \cdot \text{grad}\, N(\vec{r}, E, \vec{\Omega}, t) + \quad \text{(leakage)}$$

$$+ v\Sigma_t(\vec{r}, E, t) N(\vec{r}, E, \vec{\Omega}, t)]\, \Delta V \Delta E \Delta \vec{\Omega} \,. \quad \text{(total absorption)} \tag{1-6}$$

It is evident that the combination of expressions (1-4) to (1-6), as indicated by the principle of conservation of particles, Eq. (1-1), yields the balance Eq. (1-2).

A similar procedure can be used to verify the delayed neutron precursor Eqs. (1-3).

Equations (1-2) and (1-3) are basic and quite general. Their generality, however, renders them extremely difficult to use in this form for practical studies. The difficulties stem both from the dependence of the equations on seven independent variables, $(\vec{r}, E, \vec{\Omega}, t)$, and from the fact that the macroscopic cross sections are complicated implicit functions of the behavior of the neutron population throughout the entire reactor. Indeed, the macroscopic cross sections depend on the densities of the reactor materials. These densities are determined by the energy stored in the reactor which in turn is related to the integral of the neutron density, $N(\vec{r}, E, \vec{\Omega}, t)$, and to the fission cross section, $\Sigma_f(\vec{r}, E, t)$, as well as to the mechanisms that are used to extract energy from the reactor. The exact functional relationship between $N(\vec{r}, E, \vec{\Omega}, t)$ and the cross sections is therefore very involved and it is further complicated by the fact that approximately 10% of the fission energy is associated with long range radiations which may deposit their energy at positions far away from where the fissions occur.

In addition to all these difficulties, it must be recognized that the density $N(\vec{r}, E, \vec{\Omega}, t)$ can not be measured readily by experiment and consequently, even though it is the most general and appropriate attribute of the neutron population in the reactor, its practicality is questionable.

In view of these remarks, it is clear that additional simplifying assumptions and approximations are necessary to reduce Eqs. (1-2) and (1-3) to a form that is more amenable to analysis and experimental interpretation. The reduction can be achieved by a variety of mathematical techniques. Which techniques are appropriate is a question of expediency and the answer depends on the reactor type and the particular aspect of reactor dynamics that is under investigation. For example, series expansions in terms of a complete set of eigenfunctions, perturbation or variational methods, multi-group calculations, etc., are different techniques that have been used and are more or less suitable for extracting specific information from the transport theory equations [1-5].

Another approach to the problem of simplification of Eqs. (1-2) and (1-3) is to proceed as in the following section.

1.3 Reactor Kinetics—Conventional Form

As already emphasized, the mathematical complexity of the kinetics equations, derived from transport theory, is due to their explicit or implicit dependence on the neutron energy, direction of motion and position. The kinetics equations would be, conceptually at least, easier to handle if they were dependent only on time. Consequently, it is practical to integrate (average) the energy, direction and position variables out of the transport theory equations. The result of the integration is a set of ordinary differential equations with respect to time.

The question is how to perform the averaging. If the dependence of the functions in Eqs. (1-2) and (1-3) on energy, direction and position were completely known (viz. the neutron and delayed neutron precursor densities were known!), then these dependences could be replaced in Eqs. (1-2) and (1-3) and the equations reduced to a set of ordinary integrodifferential equations with respect to time. Of course, this is not the case and one has to approximate the dependences and then substitute them into the equations. The result of this replacement is that the ordinary integrodifferential equations that ensue are also approximate and the degree of approximation is of the same order of magnitude as the order of approximation of the energy, direction and position dependences.

The error in the ordinary equations that are sought, with respect to the error in the approximation of the various undesirable dependences, can be reduced if the unknown functions in Eqs. (1-2) and (1-3) are first multiplied by some weighting factors and then the equations are integrated with respect to energy, direction and position over the entire reactor. The weighting factors are usually taken from the exact solutions of equations that are adjoint to Eqs. (1-2) and (1-3). In many cases the adjoint solutions are not known either and it is necessary to use approximate weighting factors as well as energy, direction and position dependences and then integrate. This integration procedure is called bilinear averaging and it implies that the error in the resulting time-dependent equations is of second order with respect to the errors made in the approximations of the various dependences and weighting factors.

Consider next how the preceding concepts are purposefully used to derive the conventional form of reactor kinetics from the basic equations of transport theory.

To this end, consider the source-free adjoint equation, corresponding to some arbitrary critical state of the reactor:

$$\int_{\Omega'} d\vec{\Omega}' \int_0^\infty dE' \left[v'\Sigma_{so}(\vec{r}, E' \to E, \vec{\Omega}' \to \vec{\Omega}) \right.$$

$$+ \frac{1}{4\pi} f_t(E)\nu(E')v'\Sigma_{fo}(\vec{r}, E')\Big] N_0^*(\vec{r}, E', \vec{\Omega}')$$

$$+ \vec{\Omega}v \cdot \text{grad } N_0^*(\vec{r}, E, \vec{\Omega}) - v\Sigma_t(\vec{r}, E)N_0^*(\vec{r}, E, \vec{\Omega}) = 0 , \qquad (1\text{-}7)$$

where

$$f_t(E) = f_0(E)(1 - \beta) + \sum_{i=1}^{m} f_i(E)\beta_i .$$

Suppose that Eq. (1-7) can be solved and that the adjoint, steady-state density $N_0^*(\vec{r}, E, \vec{\Omega})$ is a known function of $(\vec{r}, E, \vec{\Omega})$. Express the time-dependent neutron density $N(\vec{r}, E, \vec{\Omega}, t)$ as the product of a time-dependent function and a function whose integral over all $(\vec{r}, E, \vec{\Omega})$ is bounded for all values of time t. Specifically:

$$N(\vec{r}, E, \vec{\Omega}, t) = P(t)N_0(\vec{r}, E, \vec{\Omega}, t) , \qquad (1\text{-}8)$$

$$I = \int_u d\vec{r} \int_0^\infty dE \int_\Omega d\vec{\Omega} N_0(\vec{r}, E, \vec{\Omega}, t) < M = \text{constant}. \qquad (1\text{-}9)$$

Note that Eqs. (1-8) and (1-9) do not involve any assumptions. They are merely definitions of the functions P(t) and $N_0(\vec{r}, E, \vec{\Omega}, t)$ that are, conceptually at least, easy to implement. The essence of these definitions is that any growth tendencies of $N(\vec{r}, E, \vec{\Omega}, t)$ can be expressed as a function of time, P(t), only and that the shape of $N(\vec{r}, E, \vec{\Omega}, t)$ within the reactor and in the energy and solid-angle spaces can be described by the shape function $N_0(\vec{r}, E, \vec{\Omega}, t)$ which may vary with time but in such a manner that both the function and its integral over $(\vec{r}, E, \vec{\Omega})$ remain finite at all times.

Next substitute Eq. (1-8) into Eqs. (1-2) and (1-3) and multiply both sides of the equations by $N_0^*(\vec{r}, E, \vec{\Omega})$ and both sides of Eq. (1-7) by $P(t)N_0(\vec{r}, E, \vec{\Omega}, t)$. Then integrate the resulting equations over the entire volume of the reactor, over all energies and over all solid angles to find:

$$\frac{d}{dt} P(t) = \frac{\rho - \bar{\beta}}{\Lambda} P(t) + \sum_i^m \lambda_i C_i(t) + Q(t) - \frac{\partial \ln(\Lambda F)}{\partial t} P(t) \qquad (1\text{-}10)$$

$$\frac{d}{dt} C_i(t) = (\bar{\beta}_i/\Lambda)P(t) - \lambda_i C_i(t) + [\partial \ln(\Lambda F)/\partial t]C_i(t) ,$$

$$i = 1, 2, \ldots m , \qquad (1\text{-}11)$$

where if

$$\int_u du \ldots = \int_V d\vec{r} \int_0^\infty dE \int_\Omega d\vec{\Omega} \ldots ,$$

$$\rho = \rho(t)$$

$$= \frac{1}{F} \int_u du \Bigg\{ N_0^*(\vec{r}, E, \vec{\Omega})$$

$$\times \Bigg[\frac{f_t(E)}{4\pi} \int_{\Omega'} d\vec{\Omega}' \int_0^\infty dE'\nu(E')v'\delta\Sigma_f N_0(\vec{r}, E', \vec{\Omega}', t)$$

$$+ \int_{\Omega'} d\vec{\Omega}' \int_0^\infty dE'v'\delta\Sigma_s N_0(\vec{r}, E', \vec{\Omega}', t)$$

$$- v\delta\Sigma_t N_0(\vec{r}, E', \vec{\Omega}', t) \Bigg]$$

$$- N_0^*(\vec{r}, E, \vec{\Omega})\vec{\Omega}v \cdot \text{grad } N_0(\vec{r}, E, \vec{\Omega}, t)$$

$$- N_0(\vec{r}, E, \vec{\Omega}, t)\vec{\Omega}v \cdot \text{grad } N_0^*(\vec{r}, E, \vec{\Omega}) \Bigg\} . \qquad (1\text{-}12)$$

$\delta\Sigma_i$ represents the difference between the time-dependent and time-independent cross sections.

$$F = F(t) =$$

$$= \int_u du N_0^*(\vec{r}, E, \vec{\Omega}) \sum_{i=1}^{m} \frac{f_i(E)}{4\pi} \times$$

$$\int_{\Omega'} d\vec{\Omega}' \int_0^\infty dE'\nu(E')v'\Sigma_f(\vec{r}, E', t)N_0(\vec{r}, E', \vec{\Omega}', t) \qquad (1\text{-}13)$$

$$\Lambda = \Lambda(t) = \frac{1}{F} \int_u du N_0^*(\vec{r}, E, \vec{\Omega})N_0(\vec{r}, E, \vec{\Omega}, t) \qquad (1\text{-}14)$$

$$\bar{\beta} = \sum_i^m \bar{\beta}_i . \qquad (1\text{-}15)$$

$$\bar{\beta}_i = \bar{\beta}_i(t)$$

$$= \frac{\beta_i}{F} \int_u du N_0^*(\vec{r}, E, \vec{\Omega}) \frac{f_i(E)}{4\pi} \int_{\Omega'} d\vec{\Omega}'$$

$$\times \int_0^\infty dE'\nu(E')v'\Sigma_f(\vec{r}, E', t)N_0(\vec{r}, E', \vec{\Omega}', t) \qquad (1\text{-}16)$$

$$F\Lambda C_i(t) = \frac{1}{4\pi}\int_u du N_0^*(\vec{r}, E, \vec{\Omega}) f_i(E) C_i(\vec{r}, t) \quad (1\text{-}17)$$

$$F\Lambda Q(t) = \int_u du N_0^*(\vec{r}, E, \Omega) S(\vec{r}, E, \vec{\Omega}, t) \,. \quad (1\text{-}18)$$

Equations (1-10) and (1-11) can be reduced to a set of ordinary differential equations with respect to time by neglecting the term $[\partial \ln(\Lambda F)/\partial t]$. Thus*

$$\frac{dP(t)}{dt} = \frac{\rho - \bar{\beta}}{\Lambda} P(t) + \sum_i^m \lambda_i C_i(t) + Q(t) \quad (1\text{-}19)$$

$$\frac{dC_i(t)}{dt} = \frac{\bar{\beta}_i}{\Lambda} P(t) - \lambda_i C_i(t) \,, \quad i = 1, 2, \ldots m \,. \quad (1\text{-}20)$$

These equations are the conventional reactor kinetics equations that were sought. The next question is what physical meaning, if any, can be attributed to the various quantities appearing in these equations?

To answer the question note first that, apart from the omission of the logarithmic term, these equations would be exact if the calculations implied by the formal definitions of ρ, $\bar{\beta}_i$ and Λ could be carried out. As already emphasized, in general this is not possible and $N_0^*(\vec{r}, E, \vec{\Omega})$ and $N_0(\vec{r}, E, \vec{\Omega}, t)$ must be approximated. Thus, ρ, $\bar{\beta}_i$ and Λ are known only approximately and therefore Eqs. (1-19) and (1-20) are approximate.

One approximation that is often used is to assume that the neutrons in the reactor are describable by a one-group, age-diffusion theory model. Thus, the neutron densities $N_0^*(\vec{r}, E, \vec{\Omega})$ and $N_0(\vec{r}, E, \vec{\Omega}, t)$ become independent of energy and direction. In the calculations of ρ, Λ and $\bar{\beta}_i$ the density $N_0(\vec{r}, t)$ is replaced by the steady-state shape function $N_0(\vec{r})$. In the context of these assumptions, the growth function P(t) is interpreted as proportional to the average reactor power and is experimentally associated with the output of a neutron counter in any position around or inside the reactor. The quantity ρ is interpreted as the reactivity and related to the multiplication factor; Λ is taken as the mean prompt or asymptotic neutron generation lifetime and $\bar{\beta}_i$ as the effective delayed neutron fractions. The prompt neutron lifetime Λ and the effective fractions $\bar{\beta}_i$ become time invariants of the particular reactor under consideration, when the fission cross section is independent of time.

The physical interpretations, however, of P, ρ, Λ and $\bar{\beta}_i$ and the invariance of Λ and $\bar{\beta}_i$ cannot be carried outside the context of the one-group age-diffusion theory model which is representative of only a few practical situations.

To see this clearly note that the normalization factor F is arbitrary. Consequently, even though the kinetics equations are approximately independent of F, the quantities ρ, Λ, and $\bar{\beta}_i$ which are dependent on F cannot be measured independently of each other and therefore they cannot be considered, in general, as physically meaningful or invariant parameters. The arbitrary character of F and its consequences basically stem from the fact that what constitutes a neutron production or a neutron destruction process is entirely a matter of personal choice. Specifically, the quantity that determines the behavior of the neutron population in the reactor is the difference between the production and destruction processes. This difference can be separated into two parts in an infinite number of ways. To each way of separation corresponds a normalization factor F and different definitions of ρ, Λ and $\bar{\beta}_i$. Therefore, ρ, Λ and $\bar{\beta}_i$ cannot be physically meaningful by themselves.

Even if a choice of F is made once and for all, the coefficients $\bar{\beta}_i$ are not identical with the corresponding physical quantities that are established from experimental observations of the neutron decay of fission products. Their values depend not only on the particular reactor under consideration but also on the prevailing operating conditions. They may differ from the physical quantities by 20 to 30%.

Given a choice of F, the conventional equations could be exact if

$$\frac{\partial}{\partial t}(\Lambda F) = 0$$

or

$$\frac{\partial}{\partial t}\int_u du N_0^*(\vec{r}, E, \vec{\Omega}) N_0(\vec{r}, E, \vec{\Omega}, t) = 0 \quad (1\text{-}21)$$

and ρ, Λ, and $\bar{\beta}_i$ can be calculated exactly. Equation (1-21) is true only when the time dependence of the neutron density $N(\vec{r}, E, \vec{\Omega}, t)$ is truly separable from the other variables. Separability of variables is possible only when the reactivity is constant, the source term is negligible and the reactor is on an asymptotic period. Under these conditions $\bar{\beta}_i$ and Λ are also invariant and exactly calculable if the neutron spectrum with respect to the reactor materials and the neutron dependence on direction are known. Under these conditions the growth function P(t) is representative of the average reactor power in the sense that the output of any counter at any position of the reactor is proportional to the average reactor power.

Under all other operating conditions, the conventional equations are approximate because the shape function $N_0(\vec{r}, E, \vec{\Omega}, t)$ and the exact variation of the cross sections are not known and they must be approximated for the calculation of the bilinear averages ρ, Λ, and $\bar{\beta}_i$, and because the term $\partial \ln(\Lambda F)/\partial t$ is omitted.

In view of the preceding remarks, it is important to examine the value of the conventional equations for analytical studies and for comparisons of theory with experiment. By analytical studies it is meant studies of the solutions, or the properties of the solutions, of equations of the form (1-19) and (1-20)

*Note that the omission of the term $\partial \ln(\Lambda F/dt)$ can be avoided by suitable redefinition of ρ and λ_i. The redefinitions are $\rho_i = \rho - \Lambda \partial \ln(\Lambda F)/\partial t$ and $\lambda_i = \lambda_i - \partial \ln(\Lambda F)/\partial t$.

without specific reference to any particular reactor or to any physical requirements and limitations for experimentally achieving these solutions or properties.

For general analytical studies, the conventional equations are extremely useful. To see this clearly, consider the coefficients of P(t) in Eqs. (1-19) and (1-20) as arbitrary functionals of time and the reactor state variables. If the reactor is not increasingly forced externally, these functionals must remain finite or bounded at all times since the shape function is selected so that all growth tendencies of the neutron density are represented by P(t). Physically, this assumption of boundedness is justified because the growth of the neutron density of a chain reacting system is governed by the number of excess neutrons per neutron causing fission. This number can never be, on the average, greater than $\sim$1.5 and, therefore, no matter what the functional dependence of the coefficients of P(t) in Eqs. (1-19) and (1-20) or (1-10) and (1-11) are, these coefficients must be ultimately bounded functions of time. Thus, by assigning to the coefficients a broad variety of reasonable bounded functional representations it is possible to explore the general properties of P(t) that result as solutions of the differential Eqs. (1-19) and (1-20) and thus establish many general trends of reactor dynamics, such as boundedness, stability, and overall transient response. These general trends are extremely useful because they provide the reactor designer or operator with a better understanding of the time behavior of a large variety of reactor operations.

The value of the conventional kinetics equations for comparisons between theory and experiment, however, cannot be qualified in general but must be assessed in the context of the particular experimental conditions under consideration.

When the variations of the state variables of the reactor are small, then $\bar{\beta}_i$ and Λ can be approximated by constant values derived from asymptotic calculations (for example, two-group theory) and ρ can be approximated by first-order approximations of the four-factor formula. The growth factor is interpreted as average power and associated with the output of a counter around or in the reactor. Under these conditions good agreement between theory and experiment is established. For many reactor types, reactivity changes of a few tenths of $\bar{\beta}$ or less fall into this category. The linearized version of the kinetics equations is an adequate representation of the dynamics of the reactor, and the behavior and distribution of the energy stored in the reactor can be predicted to a satisfactory approximation and correlated with experiment. This is the reason why oscillation, autocorrelation or crosscorrelation analysis and tests, which are discussed in Sec. 3, have proved such productive concepts and tools in the study of reactor dynamics.

When the variation of the state variables of the reactor is large, then the physical meaning of the conventional equations is beclouded and comparison of theory and experiment consists more of semi-empirical fitting of experimental results than of good theoretical predictions of experimental data, particularly if the experimental data are in the form of time traces of the outputs of one or more counters around or inside the reactor. This is due to at least two facts.

First, the error introduced through the omission of the term $\partial \ln(\Lambda F)/\partial t$, if this term is omitted, is in general of the same order of magnitude as the error involved in approximating the nonlinear Eqs. (1-19) and (1-20) by a set of linear equations with constant coefficients [5]. This error becomes particularly important during fast transients such as the self-limiting, short period SPERT excursions.

Second, all the quantities involved in Eqs. (1-19) and (1-20) cannot be assigned exact physical meanings. For example, suppose that P(t) is interpreted as indicative of the average power of the reactor and it is associated with the record of the output of a counter. Any correlation of experimental and theoretical values during severe transients, where large variations of state variables are involved would be at best fortuitous. Experimentally, two identical counter records may arise from entirely different neutron density distributions as schematically shown in Fig. 1-1. Theoretically, different neutron density distributions lead to different values of ρ, Λ and $\bar{\beta}_i$ and different solutions P(t). Hence the difficulty of assuming that P(t) is the average reactor power.

For large perturbations of the state variables, it is more appropriate to abandon the concepts of the conventional kinetics equations and work with other types of approximations of Eqs. (1-2) and (1-3) which usually require elaborate computer codes. Not much work has been done in this area. However, the space-time flux synthesis procedures developed by Kaplan [6] and the space-time solution of a two-group, diffusion approximation of Eqs. (1-2) and (1-3) for boiling water reactors discussed by Wolfe and Greebler [7] are essential steps in the right direction. (See also the discussion of the STAB codes in Sec. 5.3 of the chapter on the Kinetics of Solid-Moderator Reactors.)

In spite of the aforementioned difficulties, the conventional kinetics equations are often used to correlate experimental data pertinent to large excursions. In these correlations, Λ and $\bar{\beta}_i$ are taken as invariants, ρ is expressed in terms of some adjustable constants and as a function of power or stored energy, P(t) is interpreted as the power and associated with the output of a counter. The correlations are more or less successful because the many uncertainties involved in the calculations permit the necessary adjustment of the constants to make theory fit the experimental data. Of course,

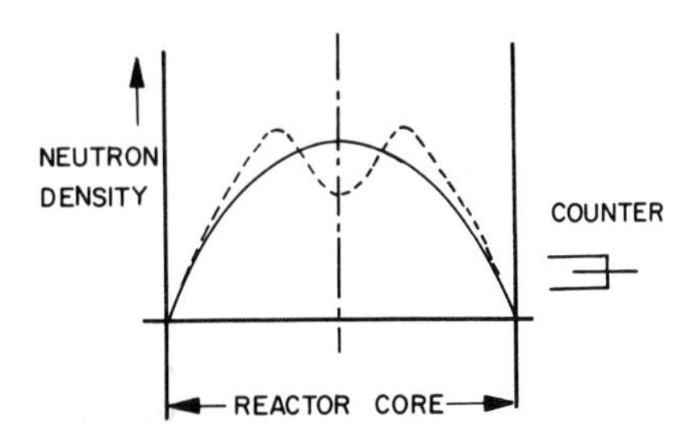

FIG. 1-1 Two different neutron density distributions result in identical counter records. The counter senses only neutron density in peripheral region where distributions coincide.

the subject is by no means well understood yet and more experimentation is needed to permit the establishment of correlations that might be valid for different classes of reactors. There is no doubt that, from a practical standpoint, good semi-empirical correlations are as useful as pure theoretical results.

For a detailed discussion of the use of the kinetics equations in the interpretation of experiments the reader is referred to reference [1].

1.4 Prompt Neutron Lifetime

As already indicated, the prompt neutron lifetime is a concept that is physically meaningful only when the reactor is on an asymptotic period. (For a detailed discussion see also reference [8]). It can then be measured by a variety of small perturbation techniques such as 1/v-poisons, oscillation tests, statistical correlations of neutron fluctuations or pulsed neutron experiments. The values of Λ that are thus established range from milliseconds for heavy water reactors to fractions of a microsecond for fast assemblies.

Even though the prompt neutron lifetime is variable during severe transients and it affects the kinetics equations only in terms of the ratios ρ/Λ and $\bar{\beta}_i/\Lambda$ and not by itself, it is interesting to examine what significance may be assigned to the value of the prompt lifetime derived by means of small perturbation techniques and how this value influences the problems of reactor control and safety.

From the standpoint of controllability during normal reactor operations, the magnitude of the prompt neutron lifetime is unimportant. The important reactor time constant, in this regime of normal reactor operation, is determined by the effective delayed neutron precursor constants $\bar{\beta}_i$ and λ_i and it is several orders of magnitude longer than Λ. Typical values of this time constant are of the order of tenths of a second. Consequently, in this regard, fast, intermediate and thermal reactors present the same operational control problems.

From the standpoint of the power level beyond which stability with respect to small perturbations is lost, the prompt neutron lifetime may be important. For example, for reactors with relatively long heat transfer time constants, such as oxide fuel reactors, the smaller the value of Λ the higher the value of the power level at which the reactor dynamic behavior becomes unstable with respect to small perturbations [9]. Thus, everything else being equal, a reactor with a short prompt neutron lifetime has a higher upper limit of the power level for linear stability than a reactor with a long prompt neutron lifetime. It should be pointed out, however, that instability levels derived from considerations of small perturbations do not necessarily imply that the reactor variables will grow without limit (see also Sec. 2).

From the standpoint of ultimate reactor safety, it is not clear whether the value of the prompt neutron lifetime can be assigned any particular significance or not. Whether a short or long prompt neutron lifetime is the most desirable depends on what is considered to be the most important design criterion. For example, if a reactor is to be controlled externally, say by control rods, then there is no doubt that, for large changes of ρ compared to $\bar{\beta}$ ($\rho/\bar{\beta} > 1$), a smaller value of the prompt neutron lifetime leads to a more difficult problem of controlling or shutting down the reactor. In this regard, thermal reactors are better than fast reactors. On the other hand, if the reactor has an inherent physical safeguard mechanism, then the value of the prompt neutron lifetime may or may not be relevant to the ultimate safety of the reactor. Specifically, given an inherent shutdown mechanism, the significant safety criterion is the total energy, at the end of a self-limited power burst, stored in the region or regions that introduce the shutdown reactivity effect. If the shutdown reactivity effect is of the form $[-\alpha E^n(t)]$ or $[\alpha E(t) - bE^2(t)]$, where E(t) the stored energy as a function of time and α, b, n constants, then the total energy stored at the end of the burst is independent of the value of the prompt neutron lifetime. In other words, reactors with the same constants (α, n) and the same admissible total energy storage can safely withstand the same amount of reactivity, regardless of whether they are thermal or fast. If, however, the shutdown reactivity effect is of the form $[-\alpha E^n(t-\tau)]$, where τ is a pure time delay between the release of energy in the reactor and the appearance of energy storage in the region that introduces the shutdown reactivity effect, then the total energy stored at the end of a power burst is larger the shorter the prompt neutron lifetime. These general results are discussed in more detail in the chapter on Mathematical Models of Fast Transients.

Finally, from the standpoint of possible assembling rates the prompt neutron lifetime may be used as a measure of the maximum assembling rate that can be tolerated. It turns out that, given an energy-dependent shutdown mechanism, the shorter the asymptotic prompt neutron lifetime the higher the maximum tolerable assembling rate. A qualitative discussion of this result is also given in the chapter on Fast Transients. In addition, a discussion of the role of prompt neutron lifetime in fast reactor accidents is given in the chapter on Fast Reactor Kinetics.

In closing this discussion it must be emphasized that the value of Λ derived by small perturbation techniques is not unique. The lack of uniqueness stems from the fact that each reactor can be critical for a large variety of combinations of its state variables and the reactor constituents. To each critical state corresponds a different prompt neutron lifetime. Which value is more appropriate, or a better approximation for a given problem, depends on the particular type of experiment that one tries to interpret theoretically. This question is discussed in reference [1].

1.5 Reactivity

Regardless of whether $\bar{\beta}_i$ and Λ are approximated by some constant values or are considered as variables, the other quantity that determines the time behavior of P(t), the growth factor of the neutron density, is the reactivity ρ that is defined by Eq. (1-12). When ρ is equal to zero* then there

*Strictly speaking it must be required that $\rho_1 = \rho - \Lambda \partial \ln(\Lambda F)/\partial t = 0$.

are no time-dependent changes in the reactor, or in other words the reactor is critical because

$$\frac{d}{dt}\left[P(t) + \sum_{i}^{m} C_i(t)\right] = 0 . \qquad (1\text{-}22)$$

When ρ is different from zero, then P(t) also varies as a function of time.

It is of interest to examine the physical mechanisms that are responsible for the changes of the reactivity ρ and how these changes can be approximated or exactly calculated in terms of measurable reactor state variables.

Inspection of Eq. (1-12) which is the formal definition of reactivity, indicates that the changes in reactivity with respect to an arbitrary critical reference level can be stimulated by the following factors:

variations in the neutron relative energy spectrum,

variations of the macroscopic fission, absorption and scattering cross sections, and

variations of the spatial neutron distribution.

These factors are in turn dependent on variations of the material composition of the reactor, such as variations in fuel, moderator, coolant and structure, and variations of the energy stored in these materials. The dependence of reactivity on the composition of the materials of the reactor stems from the fact that these materials determine the microscopic cross sections that must be used in Eq. (1-12). The dependence of reactivity on the energy stored in these materials is introduced by the fact that this energy determines the thermal agitation and the thermodynamic state of the various target nuclei. Changes in the thermal agitation alter the relative energy spectrum of the reacting neutrons and changes of the thermodynamic state result in variations of the relative target nuclei volume densities which in turn result in variations of the macroscopic cross sections.

Attempts to compute reactivity from Eq. (1-12) lead immediately to the same difficulties that earlier forced the restatement of the transport theory equations in the form of the conventional kinetics equations. In fact, the quantity ρ lumps together in one parameter a major fraction of the computational weaknesses. There are, however, several steps that can be taken which yield very useful practical results without excessive errors.

As a first step, it is found expedient not to compute the reactivity over the entire reactor at once and not to consider simultaneously all the factors that stimulate reactivity changes. Instead, attempts are made to calculate reactivity changes either over distinct reactor regions such as the fuel, the coolant or the moderator or over distinct physical processes such as spectral shifts, void changes, geometric changes, etc., and then to superimpose the different reactivity changes that are so computed. The tacit assumption behind such computational schemes is that the various reactor regions or physical processes that contribute to reactivity are weakly coupled to each other as far as reactivity changes are concerned. This assumption is justified for small changes of reactivity but may lead to intolerable errors for large changes of reactivity.

In the context of this approach to the problem of calculation of reactivity, the various types of reactivity changes are classified in terms of the particular region or physical process under consideration. Thus, terms are used like the fuel, the moderator, or the void reactivity effect, etc.

Even with this expedient isolation of the various contributions to reactivity, however, the computational difficulties are not over. The reason is that the energy stored in the various reactor regions is not readily measurable experimentally and analytically its time-dependent behavior is governed not only by the nuclear interactions but also by the various heat transmission processes that partake in the operation of the reactor. As a second step in the calculation and in order to avoid these difficulties, it is found practical not to relate the previously mentioned reactivity changes or effects directly to the energy stored in the various reactor regions but to other equivalent and measurable thermodynamic variables. Such thermodynamic variables are the fuel or coolant temperature, the void concentration, the coolant pressure or flow, etc.

This approach to the problem of calculation of reactivity introduces another possible classification of reactivity changes in terms of thermodynamic variables. For example, terms are used like fuel temperature reactivity, moderator temperature reactivity, void reactivity, etc.

The relation between measurable thermodynamic variables and corresponding reactivity changes is nonlinear. For small relative changes of the thermodynamic variables, however, the nonlinear relation between a particular variable and reactivity can be approximated by a linear relation such that the reactivity change in question $\rho_i(t)$ takes the form

$$\rho_i(t) = \alpha_i \theta_i(t) , \qquad (1\text{-}23)$$

where $\theta_i(t)$ is a measure of the change of the thermodynamic variable characteristic of the particular region or process with respect to the critical reference reactor and α_i is a constant of proportionality. The constants of proportionality α_i are classified as coefficients of reactivity and thus terms are used like the fuel temperature coefficient of reactivity, the void coefficient of reactivity, the Doppler coefficient of reactivity, etc.

For large changes of the thermodynamic variables, the coefficients of reactivity cannot be taken as constant. They are functions of all the thermodynamic variables. This is another way of saying that the various reactivity changes are coupled to each other. In practice, however, the coefficients of reactivity are often approximated by functions that depend on only one variable. Thus, for large reactivity changes Eq. (1-23) becomes:

$$\rho_i(t) = \alpha_i(\theta_i(t))\theta_i(t) . \qquad (1\text{-}23a)$$

Finally, the third step in the calculation of reactivity is to establish a relationship between the measures of the change of the thermodynamic variables $\theta_i(t)$ and the energy stored in the reactor or, what is equivalent, the difference between the fission energy and the energy carried away by the coolant. This is also a difficult task involving complicated computations. In practice, the $\theta_i(t)$ are usually expressed in terms of P(t). To this end, the energy balance equations are written for each position of the reactor. These equations are, in general, partial integrodifferential equations similar to Eqs. (1-2) and (1-3) and the thermodynamic variables appear as functions of both space and time. The bilinear averaging procedure, described in Sec. 1.3, can be used to reduce the energy-balance equations into a set of ordinary, time-dependent differential equations of the form:

$$\frac{d\theta_i(t)}{dt} = f_i(P(t), \theta_j(t)) , \quad i, j = 1, 2, \ldots n , \qquad (1\text{-}24)$$

where n is the number of the thermodynamic variables appearing in the energy-balance equations and $\theta_i(t)$ are the averaged measures of the change of these variables. In other words, in principle, by solving the set of Eqs. (1-24), the $\theta_i(t)$ can be expressed as functionals of P(t) which in turn is related to the energy stored in the reactor.

The functions $f_i(P(t), \theta_j(t))$ are, in general, nonlinear. For practical purposes they are often approximated by first order or linear equations and thus:

$$\frac{d\theta_i}{dt} = a_i(P(t) - P_0) + \sum_j^n b_{ij}\theta_j(t) , \quad i = 1, 2, \ldots n , \qquad (1\text{-}25)$$

where a_i and b_{ij} are constants. Through an appropriate transformation of variables, the set of Eqs. (1-25) can always be transformed into a canonical form such that all the thermodynamic variables are decoupled. Specifically, if the new variables are denoted by $\theta'_j(t)$, then Eqs. (1-25) can always be transformed into the form:*

$$\frac{d\theta_i'(t)}{dt} = a_i'(P(t) - P_0) - g_i'\theta_i'(t) , \quad i = 1, 2, \ldots n , \qquad (1\text{-}26)$$

where a'_i and g'_i are new constants resulting from a_i and b_{ij} through the performed transformation of variables. The reason for casting Eqs. (1-25) into the form of Eqs. (1-26) is that the latter provide another very important classification of reactivity changes in terms of the constants g'_i. If g'_i is large ($1/g'_i$ small), then the reactivity change associated with θ'_i is classified as "prompt." If g'_i is small ($1/g'_i$ is large), then the reactivity change associated with θ'_i is classified as "delayed." As it will become apparent in Sec. 2, the promptness or delayedness of a reactivity change plays a very important role in the dynamic behavior of the reactor.

The justification for the classification of reactivity as prompt or delayed depending on whether g'_i is large or small, respectively, can be understood by considering a step change $P_1 - P_0$, of $P(t)-P_0$. Indeed then, the solution of Eqs. (1-26) is

$$\theta_i'(t) = a_i'(P_1 - P_0)t \cdot \frac{1 - e^{-g_i' t}}{g_i' t} ; \quad i = 1, 2, \ldots n . \qquad (1\text{-}27)$$

If g'_i is large, then the step change contributes very rapidly to the change of $\theta'_i(t)$ and therefore to the reactivity change $\alpha'_i\theta'_i(t)$ and hence the classification of this reactivity as prompt is appropriate. On the other hand, if g'_i is small, the step change requires a relatively long time to be felt by $\theta'_i(t)$ and the corresponding reactivity contribution and hence the classification of this reactivity as delayed. In an asymptotic experiment where P(t) represents the true average power, the term $a'_i(P_1 - P_0)t$ in Eq. (1-27) is the average energy stored in the ith region and the promptness or delayedness of a reactivity change can be thought of as a measure of the effectiveness of the stored energy to produce a sizable reactivity change in a relatively short or long time.

As a reference measure of the magnitude of the inverse time constants g'_i, it is customary to use some value of the prompt neutron lifetime. When $(1/g'_i)$ is smaller than or comparable to the prompt neutron lifetime, the corresponding reactivity change is classified as prompt. When $(1/g'_i)$ is much larger than the prompt neutron lifetime then the corresponding reactivity change is classified as delayed.

It is clear that the values of the inverse time constants g'_i are not unique but depend on the operating conditions prevailing in the reactor. Of particular importance, in this respect, are the heat transfer processes that carry the heat produced by fission away from the reactor.

For example, if there is good heat transfer between the fuel and the coolant, then reactivity changes due to the coolant are relatively prompt. On the other hand, if the heat transfer between the fuel and the coolant deteriorates as the power increases, then the coolant reactivity changes become delayed. Similar remarks can be made about the fuel. For example, if there is heat transfer from the fuel to the coolant the fuel has a certain time constant. If, however, the heat transfer is interrupted either by failures in the fuel-cladding interface or by the blanketing of the fuel-coolant interface with a poor heat conduction layer, then all the fission energy remains in the fuel and the fuel reactivity changes are felt immediately; the fuel time constant becomes much smaller than the prompt neutron lifetime.

In summary, the difficulties involved in the calculation of the reactivity given by Eq. (1-12) make it necessary to impose various simplifications and approximations. These simplifications lead to the concepts of various types of reactivity changes or reactivity effects and coefficients of reactivity. It is clear that the separation of reactivity into the different components is completely arbitrary. This is the reason why different reactor designers characterize their reactors by a variety of reactivity coefficients and effects.

A detailed description of the various reactivity effects and coefficients and specific procedures for their computation in terms of the thermodynamic variables is given in the chapters on the Reactor

*Note that the same transformation of variables must also be used in the expressions for reactivity changes. Thus, in the linear approximation, new coefficients of reactivity α'_i, with respect to the new variables $\theta'_i(t)$, will result.

Core and the Doppler Coefficient and the chapters on the kinetics of different reactor types.

Even though the computation of reactivity is facilitated by the introduction of the different reactivity effects and reactivity coefficients, it is important to recognize that, apart from the limitations of the procedure that have already been outlined, there are other significant implications that should not be overlooked.

For example, suppose that the only reactivity effect of interest is due to density changes of the moderator. In addition assume that this density varies at the nth power of the temperature of the moderator at each position of the reactor. If the neutron distribution in the reactor is independent of the moderator temperature, then the over-all reactivity effect calculated by using Eq. (1-12) will depend on the nth power of some appropriate average temperature of the moderator. If, however, the neutron distribution is affected by the moderator temperature, which might be the case during severe transients, then the over-all reactivity will depend on some average moderator temperature raised to a power different from n. This example is important because it implies that what is calculated or measured microscopically may not necessarily apply directly to a particular severe reactor transient or to a particular reactor type.

Another important implication of the computationally expedient separation of reactivity into different effects is that sometimes some of these effects and the corresponding coefficients cannot be measured individually. For example, consider a heterogeneous reactor which experiences reactivity changes due to the Doppler effect and to fuel expansion. Both these effects depend on the temperature of the fuel. Any reactor experiment that affects the fuel temperature will disclose both effects at once. Only special experimental arrangements might allow the measurement of each individual effect but then these arrangements do not correspond to the reactor in question. Similar comments can be made about other reactivity effects and reactivity coefficients.

Finally, a few remarks are necessary regarding the omission or inclusion of the term $\Lambda \partial \ln(\Lambda F)/\partial t$ in the definition of reactivity. It is clear that this term depends primarily on the rate of change of the shape function $N_0(\vec{r}, E, \vec{\Omega}, t)$, see Eq. (1-14). If the shape changes are rapid as a function of time, then this term is important. If shape changes are slow then this term may be neglected.

In attempting to approximate the effects of the rate of change of the shape function on reactivity by means of thermodynamic variables, some caution must be exercised with regard to what constitutes a proper approximation. To be specific, suppose that the macroscopic cross sections vary proportionately to some power k of a representative reactor temperature $\theta_m(t)$, and that the shape function varies in proportion to some other power ℓ of the same temperature. Then the reactivity effect experienced by the reactor is

$$\rho_1 = \rho - \Lambda d\ell n(\Lambda F)/\partial t \simeq a\theta_m^{k+\ell}(t) + b\theta_m^{\ell-1}(t)[d\theta_m(t)/dt], \qquad (1\text{-}28)$$

where a, b = constants. The meaning of Eq. (1-28) is that the dependence of the rate effects on the thermodynamic variables is different than the dependence of cross section effects. This result is different than the suggestion made in the literature [10] according to which, if ρ is approximated by, say, $a\theta_m(t)$, then the rate effect should be taken as $bd\theta_m(t)/dt$.

2 ANALYTICAL TECHNIQUES USED IN REACTOR SAFETY STUDIES

2.1 General Remarks

The gross dynamic behavior of a nuclear power plant may be represented by a set of ordinary differential equations, such as Eqs. (1-19) and (1-20), in conjunction with the equations that relate the variations of the quantities ρ, Λ and $\bar{\beta}$, to the changes of the state of the reactor or, ultimately, to the energy stored in the reactor.

As already pointed out, in attempting to use these equations for analytical studies, serious questions of interpretation and computability of various quantities arise. For example, one question of interpretation is whether P(t), ρ, Λ and $\bar{\beta}_i$ can be assigned specific physical meanings. Regardless of the interpretation that is adopted, these quantities will be referred to in this section by their accepted names (power, reactivity, etc.) keeping in mind the comments of Sec. 1.

Concerning the question of computability, it is evident that the computation of the integrals defining ρ, Λ and $\bar{\beta}_i$ during a severe transient encounters major difficulties because, in general, it is not possible to know the exact form of the integrands.

Disregarding these questions, the analytical problem of reactor safety may be approached from a purely mathematical point of view, that is, as the solution of a set of differential or integrodifferential equations of the form given by Eqs. (1-19) and (1-20). If in so doing general requirements can be derived which:

- when satisfied, guarantee the safe operation of the plant,
- can be tested by means of simple nonhazardous experiments, and
- are insensitive to small perturbations of the physical constants of the system,

then the questions of interpretation and computability lose their importance. It is the purpose of this section to show that this is indeed sometimes possible and to present a number of specifications that a large variety of safe reactors should satisfy.

In order to present some general mathematical techniques that are, or can be, used in the conceptual solution of the kinetics equations, in the following discussion it is assumed that the ratios $\bar{\beta}_i/\Lambda$ are physical invariants of each reactor and that reactivity can be expressed as a function or functional of the power or the stored energy.* The discussion is also limited to dynamic problems

*In order to improve the approximation implied by the assumed constancy of $\bar{\beta}_i/\Lambda$ the term

occurring over periods of time of the order of minutes or less.

The assumption about the constancy of $\bar{\beta}_i/\Lambda$ does not affect the qualitative aspects of the results that will be derived.** Also, the assumption about the dependence of reactivity on power is theoretically justifiable and it is consistent with the majority of the existing experimental evidence.

2.2 Linear Version of Reactor Kinetics

In the context of the previous comments, one simple form that the kinetics equations may take is

$$\frac{dP}{dt} = \frac{\rho - \bar{\beta}}{\Lambda} P + \sum_i^m \lambda_i C_i + Q \tag{2-1}$$

$$\frac{dC_i}{dt} = \frac{\bar{\beta}_i}{\Lambda} P - \lambda_i C_i , \quad i = 1, 2, \ldots m \tag{2-2}$$

$$\rho = \rho_e + \int_0^t f(t - \tau)[P(\tau) - P_0]d\tau , \tag{2-3}$$

where P_0 is a reference steady-state power level, ρ_e is the externally introduced reactivity, applied during $t \geq 0$, and $f(t)$ is a kernel determined, in general, from the various coefficients of reactivity and the time constants associated with each coefficient.*** The kernel $f(t) = 0$ for $t < 0$.

The integral in Eq. (2-3) accounts for reactivity introduced by the various reactivity effects which are experienced whenever the reactor changes its state. The implication of this integral is that all causes of reactivity change, such as temperatures for example, are related to power by linear differential equations with constant coefficients and that they are related to reactivity effects by means of constant coefficients of reactivity. This linear interdependence between all processes contributing to reactivity is an approximation and it is appropriate for small changes of power.

Equation (2-1) is nonlinear. The explicit solution of the system of Eqs. (2-1) through (2-3) is a formidable, if not impossible task. Specific solutions can be found, however, by considering limited regimes of reactivity variation.****

2.2.1 Slow Startup

As an example, consider the process of reactor startup. During startup the variation of the energy stored in the reactor system is very small and feedback effects are negligible, provided that the reactor is brought to power over a time interval of a few minutes. Thus, if the externally introduced reactivity is applied stepwise at $t = 0$, the reactivity experienced by the reactor may be approximated by

$$\rho = \rho_0 + \rho_- , \quad \rho_0 = 0 \ (t < 0) , \tag{2-4}$$

where ρ_0 is the step reactivity and ρ_- is the reactivity when the reactor is shutdown. The reactor equations under these conditions reduce to the approximate form:

$$\frac{dP}{dt} = \frac{\rho_0 + \rho_- - \bar{\beta}}{\Lambda} P + \sum_i^m \lambda_i C_i + Q_0 \tag{2-5}$$

$$\frac{dC_i}{dt} = \frac{\bar{\beta}_i}{\Lambda} P - \lambda_i C_i , \quad i = 1, 2, \ldots m , \tag{2-6}$$

where Q_0 is a constant source. If before startup the reactor is at equilibrium, the balance equations are:

$$0 = \frac{\rho_- - \bar{\beta}}{\Lambda} P_0 + \sum_i^m \lambda_i C_{io} + Q_0 \tag{2-7}$$

$$0 = \frac{\bar{\beta}_i}{\Lambda} P_0 - \lambda_i C_{io} , \quad i = 1, 2, \ldots m , \tag{2-8}$$

where P_0 and C_{io} are the steady state values. The solution of the set of Eqs. (2-5) and (2-6) can be easily found in terms of Laplace transforms [11]:

$$s\bar{P} - P_0 = \frac{\rho_0 + \rho_- - \bar{\beta}}{\Lambda} \bar{P} + \sum_i^m \lambda_i \bar{C}_i + \bar{Q}_0 \tag{2-9}$$

$$s\bar{C}_i - C_{io} = \frac{\bar{\beta}_i}{\Lambda} \bar{P} - \lambda_i \bar{C}_i , \quad i = 1, 2, \ldots m . \tag{2-10}$$

If $\bar{C}_i$ is eliminated from Eqs. (2-9) and (2-10) and the steady state conditions (2-7) and (2-8) are used, then it is found that:

$$\bar{P} = \frac{P_0 \left(\Lambda + \sum_i^m \frac{\bar{\beta}_i}{s + \lambda_i} \right) + \Lambda \bar{Q}_0}{\Lambda s + \sum_i^m \frac{\bar{\beta}_i s}{s + \lambda_i} - (\rho_0 + \rho_-)} . \tag{2-11}$$

$\partial \ln(\Lambda F)/\partial t$ may be retained in the equation for $P(t)$, so that reactivity is $\rho_1 = \rho - \Lambda \partial \ln(\Lambda F)/\partial t$, but omitted from the delayed neutron precursor equations.

**The reason for this is that, when the solutions of the kinetics equations are bounded, the reactor system belongs to the class of systems that are classified mathematically as "structurally stable" with regard to the delayed neutron precursors. In other words, under this condition, variations of the values of $\bar{\beta}_i/\Lambda$ do not affect the qualitative properties of the solutions of the reactor equations.

***If the reactor is controlled by an external control system, the kernel f(t) may also include the characteristics of the controller.

****Some representative digital computer codes for numerical solution of the space-independent kinetics equations, Eqs. (2-1) and (2-2), are described in Appendix I of this chapter. Appendix I was prepared by Harold Greenspan of Argonne National Laboratory.

The inverse transform of $\overline{P}$ gives the time-dependent function P(t). The general features of P(t) depend on the roots of the denominator of $\overline{P}$. This denominator is related to the well-known inhour equation. Indeed, when the source level is zero ($Q_0 = \rho_- = 0$), the reactor is initially critical and the inhour equation for step reactivity changes determines the roots of the denominator of $\overline{P}$:

$$\rho_0 = \Lambda s + \sum_i^m \frac{\overline{\beta}_i s}{s + \lambda_i}. \tag{2-12}$$

For $\rho_0 > 0$, Eq. (2-12) admits one positive root and m negative roots. The negative roots correspond to exponentially decaying terms in P(t). The positive root corresponds to an exponentially rising term in P(t). The inverse of the positive root is usually referred to as the asymptotic period. When $\rho_0 < 0$, all roots are negative and the critical reactor is shutdown. For details about the inverse transform of $\overline{P}$ and the experimental interpretation of the inhour equation, the reader is referred to the literature [12, 13].

It is important to realize that the assumption that the feedback effects are negligible is meaningful only when the step changes in reactivity are of the order of cents and the approximation is used only over a limited period of time. Also it must be realized that, in the presence of a noisy source term different from zero, the problem of calculating the power behavior is one of probabilistic estimation rather than of a deterministic computation [14] (see also Criticality chapter).

The zero feedback approximation is useful because it provides the operator of a reactor with a good insight into the temporal behavior of the reactor during a slow startup and it constitutes a simple technique for the calibration of the reactivity worth of different reactor components.

2.2.2 Small Perturbations of Reactivity

Another example of limited reactivity variation is the case when the externally introduced reactivity is perturbed in such a way that the changes of all state variables of the reactor are sufficiently small that nonlinear effects can be neglected. Thus if before the reactivity perturbation is introduced, the reactor is at a critical or operating steady state and if, after the perturbation, each of the state variables is written as the sum of the steady-state value plus an increment, then Eqs. (2-1) through (2-3) reduce to the form:

$$\frac{dp}{dt} = \frac{P_0}{\Lambda}\rho_e - \frac{\overline{\beta}}{\Lambda}p + \frac{P_0}{\Lambda}\int_0^t f(t-\tau)p(\tau)d\tau + \sum_i^m \lambda_i c_i, \quad Q = 0. \tag{2-13}$$

$$\frac{dc_i}{dt} = \frac{\overline{\beta}_i}{\Lambda}p - \lambda_i c_i, \quad i = 1, 2, \ldots m, \tag{2-14}$$

where the lower case letters denote the increments with respect to the steady-state values and where second or higher order terms have been neglected. Note that Eqs. (2-13) and (2-14) constitute the linear version or the first approximation of reactor kinetics equations of any form. In other words, if the variations of all the state variables are small enough, then the dynamics of any reactor can be represented by the set of Eqs. (2-13) and (2-14). Of course, to each reactor and each particular operating point there will correspond a different kernel f(t).

Considering ρ_e as the input and p as the output of the reactor system, the solution of Eqs. (2-13) and (2-14) in terms of Laplace transforms is:

$$\frac{\overline{p}}{\overline{\rho}_e} = \frac{P_0 R(s)}{1 - P_0 R(s) F(s)} = H(s), \tag{2-15}$$

where

$$R(s) = \frac{1}{\Lambda s + \sum_i^m \frac{\overline{\beta}_i s}{s + \lambda_i}} \tag{2-16}$$

is the normalized zero power reactor transfer function and F(s) is the Laplace transform of f(t) [11, 13]. Equation (2-15) is shown in Fig. 2-1 by means of block diagrams.

The inverse transform of $\overline{p}$ can be written as:

$$p(t) = \int_0^t h(t-\tau)\rho_e(\tau)d\tau, \tag{2-17}$$

where h(t) is the inverse transform of H(s) or the reactor system impulse response. The transform H(s) is the reactor transfer function at power. The meaning of Eqs. (2-15) or (2-17) is that, for the types of reactivity variations specified above, the time-dependent power increment can be explicitly calculated provided that the reactor transfer function H(s) or the equivalent impulse response h(t) is known.

The general features of p(t) for a given well-behaved $\rho_e(t)$ depend on the poles of the transfer function H(s), namely, on the location of the roots of the characteristic equation:

$$1 - P_0 R(s) F(s) = 0. \tag{2-18}$$

These poles are often also referred to as the eigenvalues of the first approximation. Note that, in the

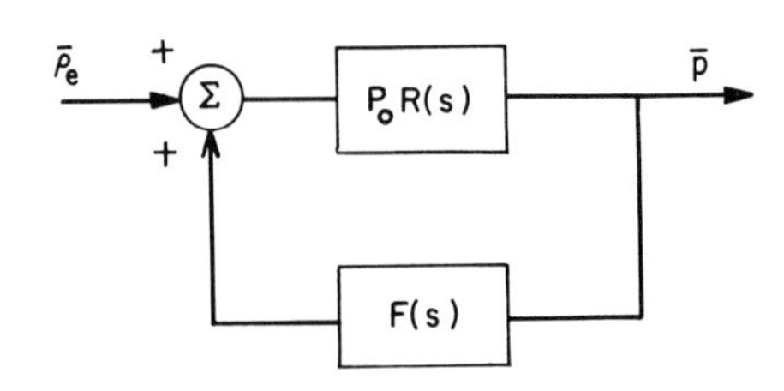

FIG. 2-1 Block diagram of Eq. (2-15).

present case, they depend on the value of the steady-state power, P_0.

If the poles are in the left half of the complex plane, the solution p(t) is asymptotically stable and after a long time reaches the value

$$p(t)_{t\to\infty} = \left[s\bar{\rho}_e \frac{P_0 R(s)}{1 - P_0 R(s) F(s)} \right]_{s\to 0} . \quad (2\text{-}19)$$

A necessary but not sufficient condition for the existence of a physically meaningful limit is that:

$$F(0) = \int_0^\infty f(t)dt < 0 . \quad (2\text{-}20)$$

In terms of reactivity coefficients and associated time constants, the meaning of inequality (2-20) is that the sum of the weighted coefficients of reactivity must be negative. The weighting factors are the time constants of, and the fractions of the energy stored in, the various regions of the reactor that are responsible for contributions to reactivity.

To see this argument clearly, consider the linearized version of the kinetics equations written explicitly in terms of all the state variables that are pertinent to the calculation of the feedback reactivity. These equations can always be written in the canonical form (see also Sec. 1.5):

$$\frac{dp}{dt} = \frac{P_0}{\Lambda}\rho_e - \frac{\bar{\beta}}{\Lambda}p + \frac{P_0}{\Lambda}\sum_j^n r_j\theta_j + \sum_i^m \lambda_i c_i \quad (2\text{-}21)$$

$$\frac{dc_i}{dt} = \frac{\bar{\beta}_i}{\Lambda}p - \lambda_i c_i , \quad i = 1, 2, \ldots m \quad (2\text{-}22)$$

$$\frac{d\theta_j}{dt} = a_j p - g_j\theta_j , \quad j = 1, 2, \ldots n , \quad (2\text{-}23)$$

where θ_j are the increments of the canonical thermodynamic variables and r_j the coefficients of reactivity corresponding to these variables. The system of Eqs. (2-21) through (2-23) is completely equivalent to Eqs. (2-13) and (2-14). In particular:

$$f(t) = \sum_j^n r_j a_j e^{-g_j t} \quad (2\text{-}24)$$

$$F(0) = \int_0^\infty f(t)dt = \sum_j^n r_j(a_j/g_j) \quad (2\text{-}25)$$

Equation (2-25) proves the statement that inequality (2-20) requires the sum of the weighted coefficents of reactivity to be negative with the weighting factors being the time constants, $1/g_j$, and the measures, a_j, of the fractions of the total energy stored in the various regions, j.

This important requirement is also necessary even when the feedback reactivity is not a linear functional of power, provided that to each critical state of the reactor there corresponds one and only one combination of values of the state variables. In other words, suppose that a reactor is critical and then it is excited by an external reactivity, ρ_e. If there is one and only one set of changes of the reactor variables which compensate for ρ_e and lead to another critical state, then it is necessary that inequality (2-20) be satisfied.

If the poles of the transfer function H(s) are in the right half of the complex plane, the solution p(t) of the linearized equations is unstable. This instability, however, is true both mathematically and physically only for small changes of the reactor variables. It must be recognized that the ultimate behavior of the power is governed by the nonlinearities that have been neglected; when the power is growing the assumed conditions for linearization are no longer valid. Nonlinear effects will be discussed later.

The problem of finding the roots of Eq. (2-18), namely, the problem of investigating the stability of linear systems, is discussed in many textbooks [15, 16]. The techniques that have been developed for this purpose are Bode diagrams, Nyquist plots, the root locus, etc. All these techniques lead to identical results cast in different forms, and the choice among them is a matter of convenience or personal preference. These techniques will not be discussed in detail here. It suffices only to note that in the case of a nuclear reactor the poles are a function of the steady state power P_0 and therefore, even though the reactor may be linearly stable for a range of values of P_0, it may become linearly unstable for values of P_0 outside this range.

In view of the approximate nature and limited range of applicability of Eqs. (2-13) and (2-14), the justifiable question is often raised about the real importance of the linearized or transfer function approach to the problem of reactor dynamics analysis, or, stated differently, about the connection, if any, between the exact solution and the one derived from the linearized model.

Several remarks are appropriate with regard to this question. First, the solution of the linearized equations, either in the frequency domain [Eq. (2-15)] or in the time domain [Eq. (2-17)], suggests definite and easily implementable experimental procedures which permit the determination of H(s) or the equivalent h(t), as well as R(s) and F(s). These procedures are discussed later.

Second, from a purely mathematical standpoint (as discussed in many textbooks), much information can be gained from the linear approximation about the nonlinear solution. For example, if one requires asymptotic stability, then the solution of the first approximation must necessarily be asymptotically stable. Also, if one is interested in the solutions for small perturbations, the linear solutions in many cases yield an adequate approximation. Of course, the magnitude of perturbations for which the linear equations are valid, even though mathematically well defined, [17, 18] is often difficult to evaluate quantitatively in practice because of lack of information about the exact form of the nonlinearities or the complexity of the problem. As a rule of thumb, for nuclear reactors, variations of the order of tenths of the steady-state values of the state variables may generally be considered as adequately represented by linear approximations. Beyond that level the nonlinearities become important.

2.3 Nonlinear Reactor Kinetics—Welton's Sufficient Criterion of Stability

Consider again the set of Eqs. (2-1) through (2-3) in which feedback reactivity is a linear functional of power. As already stated, no general analytical solution for these equations has been derived. Specific results have been found only by means of computer codes or analog simulators. Even though there is no general analytical solution, techniques are available for the investigation of the stability of the solutions.

For example, Welton [19, 20] has derived a sufficient criterion which guarantees the asymptotic stability of the reactor power to all bounded variations of external reactivity for all steady state power levels P_0. This criterion states that, if

$$\text{Re } F(jw) = \int_0^{\infty} f(t)\cos\omega t dt \leq 0 , \qquad (2\text{-}26)$$

then the reactor is asymptotically stable. In other words, if the real part of the Laplace transform of the feedback kernel along the $j\omega$-axis is nonpositive, then this is sufficient to assure asymptotic stability of the reactor power with respect to all bounded variations of external reactivity. The meaning of the criterion is that the phase of the negative feedback transfer function [-F(s)] along the $j\omega$-axis is between -90° and +90°. In simple practical terms this implies that the linearized version of the dynamic equations of the reactor without delayed neutrons should admit stable solutions at all power levels P_0 and that the negative feedback transfer function belongs to the restricted class of positive real functions or input impedance type of functions of passive electric networks.

Welton's criterion is very appealing because it relates nonlinear stability to properties that characterize the linear behavior of the reactor. Unfortunately, though, this sufficient criterion is over-restrictive for several reasons.

First, the reactor can be linearly stable without delayed neutrons even if the negative feedback transfer function is not a positive real function.

Second, the sufficient criterion is non-constructive even for simple reactor systems since both theoretical considerations and a variety of experimental results indicate that most reactors do become linearly unstable after the steady state power reaches a particular level. In view of the fact that the criterion is only sufficient, there is no a priori reason to believe that reactors which are not linearly stable at all power levels are necessarily nonlinearly unstable. As a matter of fact, there is a lot of experimental evidence to the contrary.

Third, it is not realistic to require that the reactor be stable at all possible power levels since this can never be achieved in practice due to constraints imposed by the reactor materials.

It is, therefore, important to examine whether, from the analytical standpoint, it is possible to establish less restrictive requirements which are applicable to a larger variety of reactors than Welton's criterion.

To this end many authors [21-23] have considered Liapunov's direct method as an alternate approach to the problem of reactor stability. It turns out that in most of the specific reactor problems that have been treated in the literature by Liapunov's direct method, the derived requirements for asymptotic stability are at best equivalent to, if not more restrictive than, Welton's criterion. This fact has not been generally recognized. It is not the purpose of the present chapter to go into a detailed proof of the preceding statement. Briefly, it may be said that Liapunov's method has been applied to reactor dynamics by neglecting delayed neutrons and transforming the remaining equations into a canonical form similar to that proposed by Lur'e and Letov [24]. Thus, sufficient conditions for stability are derived by means of a variety of Liapunov functions of the same type used by Lur'e and discussed by Popov [25] and others. It can be shown [26] that the sufficient requirements thus derived are at best equivalent to Welton's.

It is felt that the restrictive results derived so far by Liapunov's method do not necessarily represent an inherent limitation of the method but rather a shortcoming of the procedures that were used for its implementation. Liapunov's method is both unique and powerful and should be further pursued for the analysis of questions of nonlinear reactor dynamics.

Whether Liapunov's method or any other technique is used for analysis of reactor stability, there are a number of important practical aspects that must be incorporated in the analysis in order to arrive at useful results. These are discussed in the following sections.

2.4 Some Practical Considerations of Nonlinear Reactor Kinetics

2.4.1 General Remarks

From the discussion in the preceding section, it follows that analytical results on nonlinear reactor stability are not satisfactory. It is the purpose of this section to discuss a number of changes which might be introduced in the approach to the problem of analysis of reactor stability in order to arrive at some practical results.

2.4.2 The Practical Importance of Delayed Neutrons

In most analyses of reactor dynamics, the delayed neutrons are neglected from the kinetics equations. In other words, for analysis purposes, it is assumed that $\overline{\beta}_i = \lambda_i = 0$. The motivation for this omission is the fact that the complexity of the equations is greatly reduced. The justification for the omission is that if a reactor is linearly stable without delayed neutrons, it is even more stable when the delayed neutron precursors are taken into account. This fact has been proved several times in the literature. In addition, it can be also rigorously shown that in the presence of feedback effects, if a reactor can be proven to be nonlinearly stable without delayed neutrons, then it is more stable when delayed neutrons are included in the analysis [27]. Mechanically speaking, the delayed neutrons always add extra dumping into the system.

From the practical standpoint, however, the omission of delayed neutrons is a gross oversimplification. First, it turns out that analytical results derived without delayed neutrons may be so conservative that they are impractical. To appreciate this fact, suppose that a reactor becomes linearly unstable after a certain critical steady-state power level P_{cr}. Assume that this critical level is established through analysis of the reactor model without delayed neutrons and that its value is A. Next, suppose that the same analysis is carried out with the delayed neutron precursor equations included in the linearized kinetics equations. It can be readily shown that, in general, the value of the critical power level is B > A. In fact, in reactors with relatively long heat transmission time constants (of the order of a fraction of a second and above) and prompt lifetimes smaller than 10^{-4} sec, the difference between B and A may be one order of magnitude or more [9]. Needless to emphasize, this is such a substantial difference that the conservative estimate becomes impractical.

Another reason that renders the omission of delayed neutrons impractical is that any experimental verification of theoretical results will necessarily include the effects of delayed neutrons.

2.4.3 The Admissible Operating Power Levels

Many analyses of nonlinear reactor dynamics attempt to establish requirements for stability at all possible operating power levels $P_o (0 < P_o < \infty)$. In other words, these analyses deal with the problem of stability to unrestricted initial perturbations or the problem of global stability.

Designing, however, a reactor to be globally stable or globally asymptotically stable at all operating power levels is an unrealistic and unnecessary target to aim for. It is unnecessary because environmental constraints require that the operating power level P_o does not exceed some specified and limited value, compatible with the materials and heat transmission processes partaking in the operation of the reactor. It is unrealistic because reactors, already built and having an excellent safety record, do become even linearly unstable and yet not destructively so, because of inherent safeguarding nonlinearities.

Similar comments can be made about the energy stored in the various regions of a reactor. Without repeating the arguments, it is obvious that it is not practical to require that a reactor be capable of storing unlimited amounts of energy.

2.4.4 The Feedback Reactivity

Apart from the preceding considerations with regard to delayed neutrons, operating power level and energy storage, there is another aspect of the model for feedback reactivity that has been used so far, which renders results on nonlinear stability questionable. Specifically, it is the practice of assuming that feedback reactivity is linearly related to power, see Eq. (2-3). The implication of this assumption is that the various reactivity effects remain decoupled no matter how large the deviations from equilibrium are and, furthermore, that these effects are always linearly related to the stimuli that initiate them. In addition, Eq. (2-3) implies that every reactor is capable of providing an unlimited feedback reactivity of either sign.

It is clear that all these interrelated implications of the linear model for feedback reactivity are not consistent with the concept of reactivity, as a review of the assumptions behind the formal definition of reactivity immediately reveals (see Sec. 1). In particular, the feedback reactivity cannot grow indefinitely with power because of the limited number of neutrons per fission.

2.4.5 Asymptotic Versus Lagrangian Stability

Quite often, analysis of nonlinear reactor dynamics attempts to establish requirements that are sufficient to guarantee asymptotic stability of the reactor power and of all the other variables with respect to all, or a large variety of, perturbations from equilibrium. These requirements may often be very restrictive, if not impossible for the reactor designer to meet. In addition, even if the reactor is designed to be asymptotically stable, in practice the inherent reactor noise leads to a mode of operation characterized by state variables that are bounded but not asymptotically stable. It is, therefore, of interest to examine other types of stability.

One possibility is to attempt to establish sufficient requirements which guarantee Lagrangian stability rather than asymptotic stability [18]. Specifically, it may be expedient to design a safe reactor so that all state variables remain bounded within certain predetermined upper and lower bounds without concern about their exact temporal behavior within these bounds. Geometrically speaking, designing for Lagrangian stability implies that the designer is satisfied with the assurance that all the reactor variables are confined within a closed region of the multi-dimensional state variable space instead of requiring that they tend to coalesce at one point.

It turns out that the conditions for Lagrangian stability may be more relaxed with respect to the conditions for asymptotic stability [28]. This is to be expected since, so to speak, it is intuitively understandable that it is easier to reach one or all of many points than to reach a single point.

Lagrangian stability is neither better nor worse than asymptotic stability. In certain cases asymptotic stability may be essential to the operation of a system as, say, when it is desired to keep the temperature of a room at exactly 70°F. In other cases, however, a given margin may be just as tolerable. It is felt that this is the case of reactor systems and it is suggested that this fact be incorporated in the design procedures pertaining to stability.

2.5 A Practical Model for Nonlinear Reactor Stability and Some of Its Properties

The purpose of this section is to indicate how consideration of some of the comments of Sec. 2.4

might improve the requirements for nonlinear reactor stability and lead to practical results.

As indicated in Sec. 2.4.4 the physical nature of the fission process is such that the feedback reactivity is an ultimately bounded function with respect to any and all the state variables of the reactor.

Given an equilibrium state of the reactor, feedback reactivity changes around this state might be expressed as:

$$\rho = \rho_e + \rho_f , \quad \rho_f = z + \phi(\theta_1, \ldots \theta_n) ,$$

$$z = \sum_j^n r_j \theta_j , \quad (2\text{-}27)$$

where ρ_e is the externally introduced reactivity, ρ_f is the feedback reactivity, θ_j are the different perturbations of state variables which may introduce a reactivity change, r_j are the coefficients of reactivity when the latter is represented by a linear approximation and $\phi(\theta_1,..\theta_n)$ is a nonlinear function of $(\theta_1,..\theta_n)$ such that the feedback reactivity ρ_f is bounded both for positive and negative values of z. The exact values of the coefficients r_j and the exact analytical expression of $\phi(\theta_1,..\theta_n)$ depend on the reactor type and the prevailing operating conditions. Regardless of the exact form of $\phi(\theta_1,..\theta_n)$, however, it is evident that it must satisfy the inequalities:

$$0 < \phi(\theta_1, \ldots \theta_n) < |z| \text{ for } z < -z_1 ;$$

$$\phi(\theta_1, \ldots \theta_n) < 0 , \quad |\phi(\theta_1, \ldots \theta_n)| < z \text{ for } z > z_2 ,$$

$$(2\text{-}28)$$

where z_1, z_2 are given positive quantities and $\phi(0) = 0$.

The complete set of reactor dynamics equations that must be considered in the light of the comments of the preceding sections is:

$$\frac{dP}{dt} = \frac{\rho - \bar{\beta}}{\Lambda} P + \sum_i^m \lambda_i C_i , \quad (2\text{-}29)$$

$$\frac{dC_i}{dt} = \frac{\bar{\beta}_i}{\Lambda} P - \lambda_i C_i , \quad i = 1, 2, \ldots m , \quad (2\text{-}30)$$

$$\frac{d\theta_j}{dt} = a_j(P - P_0) - g_j\theta_j , \quad j = 1, 2, \ldots n , \quad (2\text{-}31)$$

$$\rho = \rho_e + z + \phi(\theta_1, \ldots \theta_n) , \quad z = \sum_j^n r_j\theta_j . \quad (2\text{-}32)$$

The coefficients a_j, g_j may be taken as constant or variable but bounded. The basic difference between Eqs. (2-1) through (2-3) and Eqs. (2-29) through (2-32) is the introduction of the nonlinear bounded variation of feedback reactivity and the admission of coefficients a_j, g_j that may be variable but bounded.

The set of Eqs. (2-29) through (2-32) has some useful properties that are independent of the exact analytical form of $\phi(\theta_1,.\theta_n)$ and/or a_j, g_j. Before proceeding with the enumeration of these properties, it is helpful to reduce the equations to a more convenient form.

To this end, consider the cases where the externally introduced reactivity is a step, ρ_e = constant, applied at time t = 0.* In addition, for algebraic simplicity assume that $P_0 = 0$. Under these conditions suppose that the reactor admits only two equilibrium states. One state is the shutdown state, $(P = C_i = \theta_j = 0)$, and the other is the state that the reactor will attempt to reach at the end of the transient initiated by the reactivity ρ_e, $(P_\infty, C_{i\infty}, \theta_{j\infty}, \rho = 0)$. The latter is also called the operating equilibrium state and can be determined from the solution of the algebraic system of equations:

$$\rho_e + z_\infty + \phi(\theta_{1\infty}, \ldots \theta_{n\infty}) , \quad z_\infty = \sum_j^n r_{j\infty}\theta_{j\infty} ;$$

$$(2\text{-}33)$$

$$\frac{\bar{\beta}_i}{\Lambda} P_\infty - \lambda_i C_{i\infty} = 0 , \quad i = 1, 2, \ldots m ; \quad (2\text{-}34)$$

$$a_{j\infty}P_\infty - g_{j\infty}\theta_{j\infty} = 0 , \quad j = 1, 2, \ldots n , \quad (2\text{-}35)$$

where the coefficients $a_{j\infty}$, $g_{j\infty}$, if variable, are evaluated at $(P_\infty, C_{i\infty}, \theta_{j\infty})$. Mathematically, the system of algebraic Eqs. (2-33) through (2-35) may admit more than one solution. This possibility, however, is excluded from the present discussion. In fact, when the feedback reactivity is a monotonic function of z, the necessary condition for the existence of one and only one operating equilibrium state is

$$\sum_j^n \frac{r_j a_{j\infty}}{g_{i\infty}} < 0 \quad (2\text{-}36)$$

for all practical operating conditions, as simple inspection of Eqs. (2-33) through (2-35) immediately reveals. This condition is identical with the condition for the existence of the solution of the linear approximation that was derived in Sec. 2.2.2.

If all the variables are measured with respect to the operating equilibrium state and each variable is normalized with respect to its own equilibrium value ($\neq 0$), then the system of Eqs. (2-29) through (2-32) reduces to the form:

$$\frac{dp}{dt} = -\frac{\bar{\beta}}{\Lambda} p + \sum_i^m \frac{\bar{\beta}_i}{\Lambda} c_i + \sum_j^n \frac{\gamma_j}{\Lambda} \theta_j + R(p, \theta_i) ;$$

$$(2\text{-}37)$$

---

*The results that will be derived can readily be generalized to cases where ρ_e is variable but uniformly bounded with respect to time [29].

$$\frac{dc_i}{dt} = \lambda_i(p - c_i)\ , \quad i = 1, 2, \ldots m\ ; \qquad (2\text{-}38)$$

$$\frac{d\theta_j}{dt} = g_{j\infty}(p - \theta_j) + R_j(p, \theta_i)\ , \quad i, j = 1, 2, \ldots, n\ , \qquad (2\text{-}39)$$

where the lower case letters p, c_i denote the dimensionless normalized increments with respect to the operating equilibrium values (P_∞, $C_{i\infty}$), θ_j denotes again the dimensionless normalized increment of θ_j, $\gamma_j = r_{j\infty}a_{j\infty}P_\infty/g_{j\infty}$ and

$$R(p, \theta_i) = \frac{1}{\Lambda}\sum_j^n \gamma_j\theta_j p + \frac{\phi((1+\theta_1), \ldots (1+\theta_n)) - \phi(1, \ldots 1)}{\Lambda}(1 + p)$$

$$R_j(p, \theta_i) = 0 \text{ for } a_j, g_j = \text{constant or}$$

$$= g_{j\infty}\left(\frac{a_j}{a_{j\infty}} - 1\right)(p + 1) \text{ for } a_j, g_j = \text{variable}\ . \qquad (2\text{-}40)$$

In this normalized formulation, the operating state is ($p = c_i = \theta_j = 0$) while the shutdown state is ($p = c_i = \theta_j = -1$). In addition, the stability of the reactor may now be viewed as the stability of the equilibrium states with respect to arbitrary initial conditions of the state variables rather then step perturbations of the external reactivity. Also note that the assumed boundedness of the feedback reactivity and the coefficients a_j, g_j implies that the functions $R(p, \theta_i)$ and $R_j(p, \theta_i)$ vary at most linearly with p and θ_i ($i = 1, 2, \ldots n$) for large deviations from equilibrium.

It is a simple matter to prove that, for the cases of practical interest, the shutdown state is unstable, as it should be, otherwise it would not be possible to start up the reactor.

Regarding the operating state, the following properties are inherent in the form of Eqs. (2-37) through (2-39) and independent of the exact functional dependence of ρ_f, a_j, g_j.

a. Either all state variables are constant or all are unbounded. The assumed uniqueness of the operating equilibrium state implies that regardless of what the initial conditions are for the set of Eqs. (2-37) through (2-39), no solutions exist which consist of some constant and some diverging state variables as $t \to \infty$.

Indeed, suppose that only $\theta_j = M$ as $t \to \infty$. Thus $d\theta_j/dt\big|_{t \to \infty} = 0$ and consequently p would also tend to a constant as $t \to \infty$ [Eq. (2-31)]. But if p tends to a constant then all c_i [Eq. (2-30)] and θ_j [Eq. (2-31)] would also tend to a constant and therefore the limiting values of all the variables would be given by the set of algebraic Eqs. (2-33) through (2-35) which is assumed to admit only one solution.

b. No state variable admits a finite escape time. The form of Eqs. (2-37) through (2-39) excludes the possibility of any of the state variables increasing indefinitely over a finite period of time. In other words, there are no finite escape times.

Indeed, since for large values of p and θ_i the nonlinear functions $R(p, \theta_i)$ and $R_j(p, \theta_i)$ vary at most linearly with p and all θ_i, it is always possible to write that, for

$$K = p^2 + \sum_i^m c_i^2 + \sum_j^n \theta_j^2 \geq H_0\ ,$$

$$H_0 = \text{positive constant}\ , \qquad (2\text{-}41)$$

all the rates of change of the state variables are such that

$$\frac{dp}{dt} \text{ or } \frac{dc_i}{dt} \text{ or } \frac{d\theta_j}{dt} \leq F(p, c_i, \theta_j)K^{1/2}\ , \qquad (2\text{-}42)$$

where $F(p, c_i, \theta_j)$ is some continuous positive and bounded function with respect to all its arguments. Consequently,

$$\frac{d}{dt}K = 2p\frac{dp}{dt} + 2\sum_i^m c_i\frac{dc_i}{dt} + 2\sum_j^n \theta_j\frac{d\theta_j}{dt}$$

$$\leq 2F(p, c_i, \theta_j)K\ , \quad K(t > t_0) \geq H_0\ , \qquad (2\text{-}43)$$

or

$$\int_{t_0}^T \frac{dK}{K} \leq 2\int_{t_0}^T F(p, c_i, \theta_j)dt\ . \qquad (2\text{-}44)$$

Note that if some state variable admits a finite escape time T, the left hand side of inequality (2-44) will tend to infinity while the right hand side will be finite since the integrand $F(p, c_i, \theta_j)$ is a bounded function. Clearly this is absurd and no finite escape times exist.

The practical importance of this result is that growing instabilities, if any, take a relatively "long time" to reach high levels and therefore they might be controllable. In order to be able to qualify quantitatively the meaning of "long time" it is necessary to have additional information about the function ρ_f and the coefficients a_j, g_j. In this connection, the following property is informative.

c. Existence of continuously growing instabilities. Having established that there are no finite escape times, it is permissible to approximate the feedback reactivity by some suitable linear functional of power, or equivalently by a linear combination of the state variables θ_j, provided that the approximation is used over a finite time only. Note that there is no inconsistency here with regard to the discussion of Sec. 2.4.4 because the linear approximation is taken only over a finite time.

Thus, the dynamic behavior of the reactor is

described now by Eqs. (2-37) through (2-39) with $\phi(\theta_1,..\theta_n) = 0$ and suitable average definitions of $\bar{a}_j$, $\bar{g}_j$, $\bar{r}_j$ = constant, or by Eqs. (2-1) through (2-3), where

$$\bar{f}(t) = \sum_j^n \bar{r}_j \bar{a}_j e^{\bar{g}_j t} . \qquad (2\text{-}45)$$

In order for this representation to be meaningful, Eq. (2-36) must be satisfied. In other words:

$$\int_0^\infty \bar{f}(t)dt = \sum_j^n \bar{r}_j \bar{a}_j / \bar{g}_j < 0 , \qquad (2\text{-}46)$$

so that the approximate system does admit one and only one operating equilibrium state as the original system does.

In addition, suppose that

$$\bar{f}(t) < 0 \text{ for all } t > 0 , \qquad (2\text{-}47)$$

i.e., the kernel of the average linear approximation is negative.

Under these conditions no solution of the reactor equations exists which contains a continuously increasing power level. Indeed, if the opposite were true and the power were always increasing and positive, there would exist some time T such that the reactivity seen by the reactor,

$$\int_0^T \bar{f}(t-\tau)p(\tau)d\tau , \qquad (2\text{-}48)$$

would become negative and increasing in absolute value. Then, the power equation indicates that p(t) should be decreasing, a result that contradicts the assumed continuous growth of p(t). Consequently, no solution exists with a continuously increasing power level.

The practical meaning of this result is that, if a reactor is designed so that the linear approximation around all possible operating power levels is such that

$$f(t) < 0 \text{ and } \int_0^\infty f(t)dt < 0 , \qquad (2\text{-}49)$$

then growing instabilities, if any, will be experienced in the form of diverging oscillations. Since the reactor power is physically bounded for negative values ($p(t) > -1$), the positive half-cycle of any oscillation lasts a shorter time than the corresponding negative half-cycle* and, therefore, again, diverging oscillations, if any, take a relatively long time to the extent that corrective action may be taken.

*If this were not true, under the assumed conditions, the feedback reactivity would have a constant sign and thus no oscillation could exist.

Of course, there is an important question here which has not been answered. Specifically, "Do diverging oscillations exist?" or, to put it differently, "What are the conditions that prevent the existence of diverging oscillations?" This question cannot be answered without further knowledge of the functional dependence of feedback reactivity on the state variables. Some special examples have been treated in the literature (see also chapter on Fast Transients).

Even though the existence or nonexistence of diverging oscillations cannot be discussed in general terms, it is felt that the proposed practical model for reactor kinetics and its properties have yielded some useful results. Specifically, the nonexistence of finite escape times, the nonexistence of continuously growing instabilities and the difference in time length of the positive and negative half-cycles of oscillations all tend to provide the designer with the confidence that, if the premises on which these conclusions are based, namely inequalities (2-49), are satisfied, then it may be possible to keep likely accidents under control and to avoid undesirable consequences. The results are also very useful for space-time dynamic studies. They provide the analyst with some general properties which must be satisfied by any approximate space-time representation that may be proposed. For example, suppose that the transport theory equations are approximated and solved in space and time by means of a computer. If the computer solutions indicate a finite escape time, then the analyst knows that he has introduced the wrong approximation.

2.6 A Desirable Model for Feedback Reactivity that Guarantees Practical Safety

The purpose of this section is to derive a general requirement on the behavior of feedback reactivity which guarantees the practical safety of the reactor. Practical safety means a self-limiting transient performance that entails a tolerable amount of energy storage in the reactor with regard to a given maximum external reactivity input.

To this end, given an operating power level P_0, reduce the kinetics equations into one equation of the form:

$$\frac{dp(t)}{dt} = f_0(p(\tau))(p(\tau) + P_0) - \int_0^t d(t-\tau)p(\tau)d\tau , \qquad (2\text{-}50)$$

$$d(t) = \sum_i^m \frac{\bar{\beta}_{io}}{\Lambda_0}\left(\delta(t) - \lambda_i e^{-\lambda_i t}\right) \text{ for } t > 0 ,$$

$$= 0 \text{ for } t < 0 , \qquad (2\text{-}51)$$

where $\bar{\beta}_{io}/\Lambda_0$ are the asymptotic values of $\bar{\beta}_i/\Lambda$ around the operating level P_0, $\delta(t)$ is the delta function and $f_0(p(\tau))$ is a functional of the incremental power p(t) ($f_0(0) = 0$) that accounts for the feedback reactivity and the variations of Λ, $\bar{\beta}_i$ and the normalization factor F (Sec. 1). Eq. (2-50) is quite general and represents the dynamics of any reactor with respect to arbitrary initial perturbations.

Next multiply both sides of Eq. (2-50) by p(t) and integrate the result with respect to time to find:

$$\frac{1}{2}p^2(t) + \int_0^t \int_0^\tau d(\tau - \lambda)p(\tau)p(\lambda)d\tau d\lambda$$

$$- \int_0^t p(\tau)f_0(p(\tau))[p(\tau) + P_0]d\tau = \frac{1}{2}p^2(0) . \qquad (2\text{-}52)$$

The second term in the left hand side of Eq. (2-52) is positive definite, regardless of the values of p(t). This can be proved either formally by using the procedure proposed by Bochner [30] or heuristically by observing that d(t) may be interpreted as the input impedance of a RC passive electric network and the double integral as the total energy supplied to this network.

It is evident from Eq. (2-52) that, if there exists a time $t = T_1$ beyond which the third integral in the left hand side of this equation becomes and remains negative, then p(t) must be decreasing. In addition, if there exists a time $t = T_2 > T_1$ such that the energy stored in the reactor during that time is tolerable, if p(t) has reached a level comparable to the rate of withdrawal of energy from the reactor and/or possibly, but not necessarily, if the time interval T_2 is adequate for external controls to be initiated, then the reactor is for all practical purposes safe.

It is recognized that the above general specifications on the behavior of reactivity do not indicate the explicit dependence of feedback reactivity on the stored energy in the reactor and its distribution throughout the reactor. It is felt, however, that these specifications in conjunction with Eq. (2-52) provide a simple and general way for verifying whether different feedback reactivity models are capable of guaranteeing practical safety. The following example illustrates the point.

A possible model for the functional $f_0(p(\tau))$ is

$$f_0(p(\tau)) = \int_0^t f(t - \tau)p(\tau)d\tau + f_1(p(\tau)) , \qquad (2\text{-}53)$$

where the functional $f_1(p(\tau))$ is such that $f_0(p(\tau))$ is bounded both for positive and negative values. After some elementary algebra, Eq. (2-51) can be written as:

$$\frac{d}{dt}\left[\int_0^t \int_0^\tau k(\tau - \lambda)p(\tau)p(\lambda)d\tau d\lambda\right]$$

$$= p^2(t) \int_0^t f(t - \tau)p(\tau)d\tau + p(t)[p(t) + P_0]f_1(p(\tau)) , \qquad (2\text{-}54)$$

where k(t) is the inverse Laplace transform of the function:

$$K(s) = s + D(s) - P_0F(s) \qquad (2\text{-}55)$$

and D(s) and F(s) the Laplace transforms of d(t) and f(t), respectively. Note that K(s) is the inverse of the reactor transfer function at power $P_0(K(s) = 1/H(s))$.

If it is assumed that

$$\int_0^\infty f(t)dt < 0 , \quad f(t) < 0 ,$$

$$K(s) \text{ a positive real function} , \qquad (2\text{-}56)$$

then the reactor power is ultimately bounded. Indeed, when K(s) is a positive real function, the double integral in the left hand side of Eq. (2-54) is a positive definite function. Its time derivative becomes negative for large positive values of p(t). If $f(t) < 0$ and $\int_0^\infty f(t)dt < 0$ and the power is positive and increasing when $p(t) > p_1$, then $\int_0^t f(t-\tau)p(\tau)d\tau < 0$ and increasing in absolute value. At this level $f_1(p(\tau)) > 0$ and is varying slower than linearly with $\int_0^t f(t-\tau)p(\tau)d\tau$ since $f_0(p(\tau))$ must be a bounded function. Thus, the sign of the right hand side of Eq. (2-54) is determined by the first term and it is negative. Since p(t) is physically bounded for $p(t) < -1$, it is concluded that p(t) is ultimately bounded outside some range $(-1, p_1)$ or, in other words, that there are no diverging oscillations.

In addition, the positive real character of K(s) (or H(s)) guarantees that p(t) is asymptotically stable with respect to small perturbations of the operating power level P_0.

The above conclusions are based on satisfying conditions (2-56) which can be easily implemented either theoretically or experimentally. It must be noted, however, that asymptotic stability with regard to small perturbations and ultimate boundedness do not exclude the possibility of bounded oscillations.

2.7 Space-Dependent Reactor Kinetics

The analysis of reactor stability through investigation of the properties of the space-independent kinetics equations is necessary in order to establish whether the reactor variables have a tendency to grow beyond tolerable levels or not.

As already pointed out, however, conclusions derived from the space-independent kinetics equations pertain only to the time-dependent growth factors of the state variables. These growth factors may be viewed as average attributes of power, temperatures, etc., prevailing in a reactor. Undoubtedly when these average attributes tend to grow beyond tolerable limits the reactor is unsafe. On the other hand, if they remain bounded or if they have an asymptotic value, this does not necessarily mean that the reactor is safe. The reason is that even though, on the average, the reactor variables may be well behaved, local extrema may occur which lead to undesirable consequences. For example, suppose that the power in a bundle of fuel elements behaves like a self-limiting burst which is a desirable state of affairs. Assume, however, that at the same time the energy stored in these elements is so high that before it has a chance to be transmitted to the coolant it melts the fuel. Obviously this may have serious consequences.

The implications of these comments are that guarantees for stability derivable from the space-independent reactor kinetics equations are not the only requirements that must be met by a reactor design. What is mostly needed, after stability in the above sense is guaranteed, is an analysis of space-dependent kinetics or in other words derivation of complete solutions of space- and time-dependent equations. Not much analytical work has been done in this area for the simple reason that the problem is very difficult. Some computer codes have been developed for boiling water reactors [7], and a flux synthesis technique has been proposed [6]. (See also Sec. 5.3 of Chapter 9.)

It is interesting to note here that in attempting to study space-dependent kinetics the concept of reactivity must be abandoned or generalized. The problem must be approached by devising new space- and time-dependent approximations to the basic transport theory balance equations. There is no a priori reason why these new approximations must be cast in terms of the concept of reactivity.

3 MEASUREMENT OF LINEAR DYNAMIC CHARACTERISTICS OF NUCLEAR REACTOR SYSTEMS

3.1 General Remarks

The discussion of Sec. 2 indicates that when the variations of the output power of a reactor system are relatively small, then the time behavior of these variations can be calculated from a knowledge of the input reactivity and the impulse response, h(t), or its Laplace transform, the transfer function H(s). Similar relations can be derived for any other pair of variables viewed as input and output.

When the variations of the reactor state variables are large, then the computation of the exact time behavior of these variables is a much more difficult problem. However, under certain conditions, knowledge of the properties of the linear approximation which corresponds to small variations may yield useful information about large variations.

In view of these remarks it is useful to have experimental techniques which allow the study of the various impulse responses or transfer functions of reactor systems so that analytical results can be verified experimentally.

The purpose of this section is to discuss some of the techniques that can be used for the measurement of linear dynamic characteristics either in the time or in the frequency domain. Such techniques are oscillation, crosscorrelation and autocorrelation tests. In addition, some consideration is given to the problem of measurement of nonlinear characteristics or experimental indentification of nonlinear systems. As already mentioned, this is a much more involved problem and by no means fully understood yet (see also chapter on Mathematical Models of Fast Transients).

3.2 Oscillation Tests

In order to establish the theoretical foundation of oscillation tests, consider the relationship between power and reactivity when both undergo relatively small variations. In the frequency domain this relationship is [Eq. (2-15)]:

$$\bar{p}(s) = H(s)\bar{\rho}_e(s) . \tag{3-1}$$

The partial fraction expansion [11] of $\bar{p}(s)$ contains two types of terms. The first depend on the poles of the input $\bar{\rho}_e(s)$ and the second depend on the poles of the transfer function H(s). Specifically:

$$\bar{p}(s) = \sum_{i}^{q} \frac{k_i}{s - s_i} + \sum_{j}^{r} \frac{k_j}{s - w_j}, \tag{3-2}$$

where k_i, k_j are the residues, s_i are the poles of the input (i = 1, 2, ... q) and w_j are the poles of the transfer function (j = 1, 2, ... r). For simplicity and without loss of generality, all poles have been assumed single so that the residues are derived from the simple relations:

$$k_i = (s - s_i)\bar{p}(s)|_{s=s_i} \quad \text{or} \quad k_j = (s - w_j)\bar{p}(s)|_{s=w_j} . \tag{3-3}$$

The inverse Laplace transform of $\bar{p}(s)$ is:

$$p(t) = \underbrace{\sum_{i}^{q} k_i e^{s_i t}}_{\text{steady-state response}} + \underbrace{\sum_{j}^{r} k_j e^{w_j t}}_{\text{transient response}} . \tag{3-4}$$

If all the poles, w_j, of the transfer function are in the left half complex plane, then after a sufficiently long time all the terms that correspond to these poles, namely the transient response, will become practically zero. After that time, p(t) reaches its steady-state response which is characterized only by the poles of the input and the corresponding residues.

The steady-state response, when the input reactivity varies sinusoidally, is:

$$\rho_e = A \sin \omega t , \quad \bar{\rho}_e(s) = A \frac{\omega}{s^2 + \omega^2} \tag{3-5}$$

$$\text{steady-state } p(t) = k_1 e^{j\omega t} + k_2 e^{-j\omega t} , \tag{3-6}$$

where $k_2 = \bar{k}_1$ = conjugate of k_1,

$$k_1 = \left[(s - j\omega) A \frac{\omega}{s^2 + \omega^2} H(s)\right]_{s=j\omega} = \frac{A}{2j} H(j\omega) . \tag{3-7}$$

Therefore:

$$\text{steady-state } p(t) = A|H(j\omega)|\sin(\omega t + \underline{/H(j\omega)}) . \tag{3-8}$$

The meaning of Eq. (3-8) is that the steady-state response of a reactor, or any other linear system, excited by a sinusoidal input is also a sinusoid of the same frequency. Its relative amplitude and phase with regard to the input are given by the magnitude and phase of the transfer function evaluated at the same frequency as that of the input, respectively.

Equation (3-8) suggests a simple steady-state experiment for the measurement of the transfer function along the frequency axis $s = j\omega$. Indeed, if the reactor is excited by means of sinusoidally varying reactivities of different frequencies, the relative amplitudes and phases of the corresponding steady-state power oscillations yield the magnitude and phase of the transfer function at these frequencies. The experimental results may be plotted either in polar form or as Bode diagrams or in any other convenient diagrams [15].

Knowledge of the values of the transfer function $H(s)$ for $s = j\omega$ is sufficient to determine $H(s)$ for all values of s because this function is analytic everywhere except at its poles and $H(j\omega)$ can be readily analytically continued for all s [31]. In addition, by taking the inverse transform of $H(j\omega)$ it is possible to determine the impulse response, $h(t)$, if it is so desired.

This is the essence of oscillation tests which are steady state experiments resulting in the measurement of linear dynamic characteristics. It is evident that the technique can be implemented to measure the transfer function between any pair of reactor variables one of which is considered as the input and varied sinusoidally and the other as the output.

For details of the experimental setup for the measurement of transfer functions by means of oscillation tests, the reader is referred to the abundant literature on the subject [32, 33].

Transfer function measurements through oscillation tests have been used for the measurement of prompt neutron lifetimes, coefficients of reactivity and associated time constants, cross sections, the prediction of power level at which linear stability is lost, shutdown reactivity, etc.

Before closing this brief discussion on oscillation tests, it is worth indicating some of the limitations that arise when the method is applied to a "real reactor system." The power of a presumably exactly critical reactor is not constant and the steady-state response to a sinusoidal input reactivity is not a true sinusoid. The reason for the deviations from the ideal performance is that practically all physical phenomena are to some degree statistical in nature or subject to small perturbations due to unpredictable environmental changes. For example, the fission process and/or the boiling process vary statistically in a reactor or the criticality may be influenced by atmospheric temperature changes, etc.

The net effect of these inherent or extenally stimulated statistical fluctuations on oscillation tests performed on reactors is that either the input reactivity amplitude may be required to be large enough so that the power oscillations have an amplitude much larger than the statistical fluctuations, or the power oscillations must be Fourier-analyzed in order to extract the fundamental frequency component and reduce the contribution from the inherent fluctuations.

Increasing the amplitude of the input reactivity is undesirable on two counts. First, the output is distorted because the reactor does not behave like a linear system. Of course, it is possible to extract the fundamental component of the output by Fourier analysis but then the end result is a measurement of the describing function rather than the transfer function [33]. Second, it may be unsafe or mechanically difficult to use large amplitude reactivity inputs.

Fourier analysis of the power oscillations is in a sense a crosscorrelation of the output with a sinusoid. Consequently, if crosscorrelation is to be used in order to reduce the errors due to statistical fluctuations, it is not necessary to excite the reactor by a pure sinusoidal input reactivity. Any small amplitude periodic reactivity waveform will accomplish the same purpose. The reason is that any periodic waveform (square wave, saw-tooth, etc.) may be visualized as a sum of pure sinusoids. Each sinusoid results in its own contribution to the steady-state response of the output power. Thus by comparing the fundamental component of the output with the fundamental component of the input, it is possible to measure the transfer function.

Two other implications of the presence of statistical fluctuations in reactor systems are: (a) the need for Fourier analysis of the power oscillations suggests that it may be more appropriate to excite the reactor with an input reactivity that contains a broad band of frequencies with equal amplitude and to crosscorrelate this input with the corresponding output. The result of this operation is again the measurement of the dynamic characteristic between the input and the output (the impulse response) as discussed in Sec. 3.3; (b) the statistical fluctuations contain informatin concerning the dynamics of the reactor. Under certain conditions, this information may be extracted by autocorrelating the power or any other observable variable fluctuations, as discussed in Sec. 3.4.

3.3 Crosscorrelation Tests

The basis of crosscorrelation tests is the convolution integral relationship between input and output [Eq. (2-14)]:

$$p(t) = \int_0^t h(t-\tau)\rho_e(\tau)d\tau = \int_0^t h(\tau)\rho_e(t-\tau)d\tau . \quad (3\text{-}9)$$

If the autocorrelation of a function, $x(t)$, is defined as

$$\phi_{xx}(\tau) = \frac{1}{T}\int_{-T/2}^{T/2} x(t)x(t+\tau)dt , \quad \phi_{xx}(\tau) = \phi_{xx}(-\tau) \quad (3\text{-}10)$$

and the crosscorrelation between two functions, $x(t)$ and $y(t)$, is defined as

$$\phi_{xy}(\tau) = \frac{1}{T}\int_{-T/2}^{T/2} x(t)y(t+\tau)dt , \quad \phi_{xy}(\tau) = \phi_{yx}(-\tau) , \quad (3\text{-}11)$$

where T is the period of the functions x(t) and y(t), when these functions are periodic, or $T \to \infty$ when the functions are stochastic or aperiodic, then the crosscorrelation of $\rho_e(t)$ and p(t) is

$$\phi_{\rho_e p}(\tau) = \frac{1}{T}\int_{-T/2}^{T/2} \rho_e(t-\tau)p(t)dt$$

$$= \frac{1}{T}\int_{-T/2}^{T/2} dt\rho_e(t-\tau)\int_0^t d\lambda h(\lambda)\rho_e(t-\lambda)$$

$$= \int_0^t d\lambda h(\lambda)\frac{1}{T}\int_{-T/2}^{T/2} dt\rho_e(t-\tau)\rho_e(t-\lambda)$$

$$= \int_0^t h(\lambda)\phi_{\rho_e\rho_e}(\tau-\lambda)d\lambda\ . \qquad (3\text{-}12)$$

The meaning of Eq. (3-12) is that the autocorrelation function of the input $\rho_e(t)$ is related to the crosscorrelation of $\rho_e(t)$ and p(t) by the same convolution integral as $\rho_e(t)$ is related to p(t).

If the input reactivity $\rho_e(t)$ is a broad band signal,* its autocorrelation function is approximately equal to a delta function:

$$\phi_{\rho_e\rho_e}(t) = A^2\delta(t)\ ,\quad A^2 = \text{constant}\ , \qquad (3\text{-}13)$$

and Eq. (3-12) yields

$$\phi_{\rho_e p}(\tau) = A^2 h(\tau)\ . \qquad (3\text{-}14)$$

In other words, the crosscorrelation of the input and the output is proportional to the impulse response when the input is a broad band signal. Eq. (3-12) can also be written in the frequency domain. To this end, use must be made of two-sided Laplace transforms [15] because correlation functions are defined both for positive and negative time shifts τ. Thus,

$$\bar{\phi}_{\rho_e p}(s) = H(s)\bar{\phi}_{\rho_e\rho_e}(s) \qquad (3\text{-}15)$$

or

$$\bar{\phi}_{\rho_e p}(j\omega) = H(j\omega)\bar{\phi}_{\rho_e\rho_e}(j\omega)\ . \qquad (3\text{-}16)$$

In other words, the transfer function for $s = j\omega$ is equal to the ratio of the power spectrum of the crosscorrelation of the input and the output over the power spectrum of the autocorrelation of the input.

Crosscorrelation tests are essentially an experimental implementation of either Eq. (3-14) or (3-16). For the details of the experimental procedure, the reader is referred to the literature [34, 35]. For the purposes of this discussion, it suffices to note that the presence of inherent statistical fluctuations is not as restrictive as in oscillation tests [36]. The reason is that the properties of the broad band input are in general completely independent of the statistics of the fluctuations. In fact, through proper choice of the input waveform, the results of crosscorrelation tests may also be unaffected by the nonlinear characteristics even for large input amplitudes. This topic is discussed in Sec. 3.6.

3.4 Autocorrelation Tests

The use of autocorrelation tests, to extract information about the linear dynamic characteristics of a reactor, entails a series of assumptions about the cause of the fluctuations and the statistical attributes of this cause.

To make ideas specific, assume that the power fluctuations p(t) or those of any other variable, around a constant mean value, are due to inherent and statistically predictable variations of reactivity, $\rho_e(t)$. Thus, the relationship between p(t) and $\rho_e(t)$ is given by Eq. (3-9). The autocorrelation function of p(t) is

$$\phi_{pp}(\tau) = \frac{1}{T}\int_{-T/2}^{T/2} p(t)p(t+\tau)dt$$

$$= \frac{1}{T}\int_{-T/2}^{T/2} dt\int_0^t d\lambda h(\lambda)\rho_e(t-\lambda)\int_0^{t+\tau} d\mu h(\mu)\rho_e(t+\tau-\mu)$$

$$= \int_0^t d\lambda h(\lambda)\int_0^{t+\tau} d\mu h(\mu)\frac{1}{T}\int_{-T/2}^{T/2} dt\rho_e(t-\lambda)\rho_e(t+\tau-\mu)$$

$$= \int_0^t d\lambda\int_0^{t+\tau} d\mu h(\lambda)h(\mu)\phi_{\rho_e\rho_e}(\tau+\lambda-\mu)\ . \qquad (3\text{-}17)$$

Eq. (3-17) can best be understood in the frequency domain. Its two-sided Laplace transform is

$$\bar{\phi}_{pp}(s) = H(s)H(-s)\bar{\phi}_{\rho_e\rho_e}(s)\ , \qquad (3\text{-}18)$$

and for $s = j\omega$

$$|H(j\omega)|^2 = \frac{\bar{\phi}_{pp}(j\omega)}{\bar{\phi}_{\rho_e\rho_e}(j\omega)}\ . \qquad (3\text{-}19)$$

The meaning of Eq. (3-19) is that the square of the amplitude of the transfer function H(s) for $s = j\omega$ is equal to the ratio of the power spectrum of the autocorrelation of the output over the power spectrum of the autocorrelation of the input. Under the assumption that $\rho_e(t)$ is an inherent statistical perturbation, the power spectrum of $\phi_{\rho_e\rho_e}(\tau)$ cannot be readily measured. If, however, this spectrum is assumed as flat and equal to a constant over the frequency range of interest, then the measurable spectrum of the fluctuations of p(t) is proportional to the square of the magnitude of the transfer function. In other words, the magnitude of the transfer function can be measured by simply autocorrelating the power fluctuations and finding the power spec-

*The bandwidth of the signal must be equal or broader than the bandwidth of the transfer function.

trum.

Note that autocorrelation tests do not disclose any information about the phase of $H(j\omega)$. The phase can be computed from the magnitude only when the transfer function belongs to the class of phase minimum functions such as the impedances or admittances of linear, passive, lumped parameter electrical networks [15, 37].

The great appeal of autocorrelation tests is that information about the linear dynamic characteristics of the reactor can be derived without externally perturbing the reactor but from the fluctuations that are superimposed on the steady state record of any reactor variable. A major drawback of the method is that it is not always possible to verify whether the power spectrum of the autocorrelation of the presumed statistical excitation is indeed flat white. Thus, resonance tendencies of $\overline{\Phi}_{\rho_e\rho_e}(j\omega)$ might be attributed to the transfer function and erroneous conclusions be derived.

3.5 A Stability Monitor

When the linear or nonlinear dynamic behavior of a reactor system is adequately characterized by the linear characteristics, as in some of the examples presented in Sec. 2, then the crosscorrelation method can be used to continuously monitor the quality of the transient and stability properties of the reactor [38].

To this end, the reactor is continuously perturbed by a small amplitude, broad band reactivity input which is then crosscorrelated on line with the resulting output. The crosscorrelation is performed simultaneously for a large number of time shifts τ. The outputs of the crosscorrelators are depicted on a fluorescent screen and provide a graph of the impulse response, h(t). This graph is readily interpretable to predict any poor transient response or instability tendencies that may be developing in the reactor.

3.6 Representation and Identification of Nonlinear Systems

The purpose of this section is to present some of the techniques that are currently pursued for the analytical representation and experimental identification of nonlinear, lumped parameter, stationary and physically realizable systems which from an engineering standpoint can be visualized in terms of input-output data. These techniques have not yet been fully utilized in the nuclear field but they are suggestive of possible developments and they provide a theoretical basis for studies of nonlinearities.

The discussion is first presented without reference to any particular physical system, type of input or output. Only one input and one output are considered for mathematical expediency. The formalism can be easily generalized to any number of inputs and/or outputs.

Some of the results are used to indicate the effects of nonlinearities in measurements performed in reactor systems.

3.6.1 The Functional Representation of Nonlinear Systems

Suppose that a nonlinear system is excited by an input x(t) which results in an output y(t). A possible interpretation of this visualization is that the input x(t) enters the system, it is processed by the system and then appears as an output y(t). Therefore, measurements of the input and the output contain all the information about the dynamics of the system. For the types of systems under consideration, this information can be stated analytically in terms of a functional,

$$y(t) = F(x(t-\tau)) \; ; \quad \tau < t \, . \qquad (3\text{-}20)$$

The physical meaning of the analytical statement (3-20) is that the present value of the output is uniquely determined only by the past history of the input. The exact form of the functional, of course, depends on the specific system on which the input-output measurements are taken.

The functional F can always be expanded into an infinite series of functionals of the input, a form that is more suggestive of possible approaches to the problems of representation and identification of nonlinear systems. Specifically, it can easily be shown that it is always possible to write [39]:

$$\begin{aligned} y(t) &= F(x(t-\tau)) \\ &= h_0 + \int_0^\infty h(t-\tau)x(\tau)d\tau \\ &+ \int_0^\infty \int_0^\infty h(\tau_1, \tau_2)x(t-\tau_1)x(t-\tau_2)d\tau_1 d\tau_2 + \\ &\ldots + \int_0^\infty \ldots \int_0^\infty h_k(\tau_1, \ldots \tau_k)x(t-\tau_1) \\ &\ldots x(t-\tau_k)d\tau_1 \ldots d\tau_k + \ldots \end{aligned} \qquad (3\text{-}21)$$

In principle, the nonlinear system is now characterized by an infinite set of kernels $h_k(\tau_1, \ldots \tau_k)$. The contribution from each kernel to the output is derived from a generalized convolution operation. Equation (3-21) is called a Volterra expansion. Note that, if the system is linear, ($h_k(\tau_1, \ldots \tau_k) = 0$, $k \geq 2$), the functional expansion reduces to the well-known linear convolution plus the constant h_0.

Functionals of the type appearing in Eq. (3-21) have been studied by Volterra [40]. Wiener has used the functional expansion to investigate nonlinear electrical network problems [41]. Other authors have investigated different properties of functionals and developed a systematic algebra and multi-Laplace transformation theory for a system or combination of systems represented by functional expansions [42-44]. The findings of these authors will not be discussed here. The reader is referred to reference [45] for an excellent summary.

For the purposes of this presentation, it suffices to emphasize that if the kernels $h_k(\tau_1, \ldots \tau_k)$ were known, then the nonlinear system would be completely represented; in other words, if the kernels could be measured, then the nonlinear system would be completely identified.

Equation (3-21), as it stands now, is not always convenient for experimental interpretation. However, it is suggestive of a similar expansion which is more suitable both for theoretical and experimental studies. Specifically, if the expansion were in terms of an orthogonal set of functionals, a number of advantages are evident:

If the expansion is truncated at a finite number of functionals, the input is approximated in the least mean square error sense.

Each member of the expansion is linearly independent of the others and the number of functionals considered.

Each member of the expansion can be determined by use of the orthogonality relationship which is, in essence, a generalized crosscorrelation procedure.

The advantages of an expansion in terms of orthogonal functionals can be achieved when the input is a gaussian white noise, as rigorously proved by Wiener [41] and Chesler [42]. Wiener's theory is briefly summarized in the next section.

3.6.2 Wiener's Canonical Representation of Nonlinear Systems

Wiener's rigorous theory of representation of nonlinear systems is described in his monograph "Nonlinear Problems in Random Theory" [41]. Only some important results are repeated here for convenience.

Consider a gaussian white noise signal $x(t)$ defined over all times from $-\infty$ to ∞ and having a power spectrum equal to unity. Given an arbitrary, symmetrical kernel $K_n(\tau_1, \tau_2 \ldots \tau_n)$ and the functional

$$f_n(t) = \int_0^\infty \cdots \int_0^\infty K_n(\tau_1, \ldots \tau_n)x(t - \tau_1) \ldots x(t - \tau_n)d\tau_1 \ldots d\tau_n \,, \quad (3\text{-}22)$$

it can readily be shown that the average value of this functional over all times is

$$\frac{1}{2T}\int_{-\infty}^{\infty} f_n(t)dt = 0 \quad \text{for} \quad n = \text{odd}$$
$$T\to\infty$$

$$= (2m-1)(2m-3)\ldots(1)\int_0^\infty \cdots \int_0^\infty K_n(\tau_1, \tau_1, \ldots \tau_m, \tau_m)d\tau_1 \ldots d\tau_m$$
$$\text{for } n = 2m = \text{even} \,. \quad (3\text{-}23)$$

On the basis of Eq. (3-23) it is easy to prove that the following G_n functionals are orthogonal regardless of the values of the symmetric kernels $K_n(\tau_1, \ldots \tau_n)$.

$$G_0 = \text{constant}$$

$$G_1(K_1, x, t) = \int_0^\infty K_1(\tau)x(t - \tau)d\tau$$

$$G_2(K_2, x, t) = \int_0^\infty \int_0^\infty K_2(\tau_1, \tau_2)x(t - \tau_1)x(t - \tau_2)d\tau_1 d\tau_2 - \int_0^\infty K_2(\tau, \tau)d\tau$$

$$G_3(K_3, x, t) = \int_0^\infty \int_0^\infty \int_0^\infty K_3(\tau_1, \tau_2, \tau_3) \times x(t - \tau_1)x(t - \tau_2)x(t - \tau_3)d\tau_1 d\tau_2 d\tau_3 - 3\int_0^\infty \int_0^\infty K_3(\tau_1, \tau_1, \tau)x(t - \tau)d\tau_1 d\tau$$

. . . .

$$G_n(K_n, x, t) = \sum_{\nu=0}^{[n/2]} a_{n-2\nu}^{(n)} \int_0^\infty \cdots \int_0^\infty K_n(\tau_1, \ldots \tau_n)x(t - \tau_1) \ldots x(t - \tau_{n-2\nu}) \cdot \delta(\tau_{n-2\nu+1} - \tau_{n-2\nu+2}) \ldots \delta(\tau_{n-1} - \tau_n)d\tau_1 \ldots d\tau_n$$

where $[n/2] = n/2$ for n = even and

$[n/2] = (n-1)/2$ for n = odd and

$$a_{n-2\nu}^{(n)} = (-1)^\nu \frac{n!}{2^\nu (n - 2\nu)!\nu!} \,.$$

In other words, the functional G_1 is orthogonal to all constants, the functional G_2 is orthogonal to all constants and all functionals G_1, etc. The orthogonality is defined over the time variable t:

$$\frac{1}{2T}\int_{-\infty}^{\infty} G_n(K_n, x, t)G_m(K_m, x, t)dt = \delta_{nm} \quad \text{for} \quad n \neq m$$
$$T\to\infty$$

(3-24a)

$$\lim_{T\to\infty}\frac{1}{2T}\int_{-\infty}^{\infty} G_n(K_n, x, t)G_n(L_n, x, t)$$

$$= n!\int_0^{\infty}\cdots\int_0^{\infty} K_n(\tau_1, \ldots \tau_n)L_n(\tau_1, \ldots \tau_n)d\tau_1, \ldots d\tau_n, \tag{3-24b}$$

where $L_n(\tau_1, \ldots \tau_n)$ is also a symmetric kernel.

The orthogonal functionals G_n form a complete set. Therefore, any function y(t) which is square integrable can be expanded in a series of G_n functionals in a unique way:

$$y(t) = \lim_{N\to\infty} [G_0 + G_1(K_1, x, t) + \ldots + G_N(K_N, x, t)] . \tag{3-25}$$

If the gaussian, white noise signal x(t) is the input of a nonlinear system and y(t) is the corresponding output, then Eq. (3-25) constitutes Wiener's canonical representation of the system and the kernels $K_n(\tau_1, \ldots \tau_n)$ are the characteristic kernels of the system. These kernels are related to the kernels $h_n(\tau_1, \ldots \tau_n)$ of the Volterra expansion, Eq. (3-21). To illustrate the point, suppose that the Volterra expansion has only two terms,

$$y(t) = \int_0^{\infty} h_1(\tau)x(t-\tau)d\tau + \int_0^{\infty}\int_0^{\infty}\int_0^{\infty} h_3(\tau_1, \tau_2, \tau_3)$$

$$x(t-\tau_1)x(t-\tau_2)x(t-\tau_3)d\tau_1 d\tau_2 d\tau_3 \tag{3-26}$$

then the Wiener representation has also two terms and the kernels are related by

$$K_1(t) = h_1(t) + 3\int_0^{\infty} h_3(\tau_1, \tau_1, t)d\tau_1 ,$$

$$K_3(t_1, t_2, t_3) = h_3(t_1, t_2, t_3) . \tag{3-27}$$

Therefore, knowledge of either set of kernels is adequate to characterize the nonlinear system.

The important difference between the two representations is that Wiener's orthogonal expansion suggests a simple experimental procedure for the measurement of the kernels $K_n(\tau_1, \ldots \tau_n)$ while the same is not true for the Volterra kernels $h_n(\tau_1, \ldots \tau_n)$. This procedure is discussed in the next section.

3.6.3 Measurement of the Wiener Kernels

The measurement of the kernels $K_n(\tau_1, \ldots \tau_n)$ is effectively an experimental implementation of the orthogonality relationships [Eq. (3-24)]. Indeed, suppose that the gaussian white noise signal x(t) is fed simultaneously into the physical system under investigation and another known system consisting of n pure delays in parallel. All the outputs from the pure delays are multiplied together and the output of the multiplier is also multiplied by the output y(t) of the physical system and then integrated over a long time. The net result of this integration is proportional to the nth Wiener kernel $K_n(\tau'_1, \ldots \tau'_n)$, where τ'_i are the delay times ($\tau'_1 \neq \tau'_2 \neq \ldots \tau'_n$). Indeed, the n delays and the first multiplier may be thought of as a known system with a Wiener representation:

$$x(t-\tau'_1)\ldots x(t-\tau'_n) = \int_0^{\infty}\cdots\int_0^{\infty} \delta(\tau_1-\tau'_1)\ldots$$

$$\ldots\delta(\tau_n-\tau'_n)x(t-\tau_1)\ldots x(t-\tau_n)d\tau_1, \ldots d\tau_n$$

$$= \int_0^{\infty}\cdots\int_0^{\infty} L_n(\tau_1, \ldots \tau_n)x(t-\tau_1)\ldots$$

$$\ldots x(t-\tau_n)d\tau_1, \ldots d\tau_n = G_n(L_n, x, t) , \tag{3-28}$$

where

$$L_n(\tau_1, \ldots \tau_n) = \delta(\tau_1-\tau'_1)\delta(\tau_2-\tau'_2)\ldots\delta(\tau_n-\tau'_n) . \tag{3-29}$$

In other words, the set of n delays and their multiplier can be represented by a special Wiener functional of nth order. If the output of the physical system is visualized as a sum of Wiener functionals, $G_n(K_n, x, t)$, then $G_n(L_n, x, t)$ is orthogonal to all these functionals except $G_n(K_n, x, t)$. Therefore, the output of the integrator is [Eq. (3-24b)]:

$$\int_{-\infty}^{\infty} y(t)G_n(L_n, x, t)dt = n!\, K_n(\tau'_1, \tau'_2, \ldots \tau'_n) . \tag{3-30}$$

This proves the statement that the output of the integrator is proportional to the nth Wiener kernel. It is evident that by changing the values of the delays τ'_i, the entire range of values of $K_n(\tau'_1, \ldots \tau'_n)$ can be measured. Since the number of delays, n, can be chosen at will, all kernels can be measured.

The experimental procedure for the measurement of the Wiener kernels is a generalized crosscorrelation procedure. The necessary crosscorrelations can be performed on line or by means of digital and/or analog computers.

Wiener's representation and identification procedure for nonlinear systems is conceptually very simple. It also shows that, just as gaussian white noise is adequate for the complete characterization of linear systems, it is also adequate for the characterization of nonlinear systems.

The use of gaussian white noise, however, is a

major drawback because of the necessity of extremely long integration times for crosscorrelation. The long times are necessary because of the serious experimental errors that are otherwise introduced. This is the reason why the method has not yet found wide application.

In spite of its shortcomings Wiener's canonical representation is very useful for a variety of theoretical and practical investigations.

In particular, in any practical system the functional representation will necessarily be truncated after a small number of functionals. In addition the useful bandwidth of any physical system is limited. These two observations suggest that it may be possible to design periodic signals defined over finite intervals of time and such that their pseudostatistical properties are similar to those of gaussian white noise up to a certain degree. Thus it will be possible to use the Wiener representation and perform practical measurements. Work along these lines is currently under way with promising preliminary results [46].

It is worth noting that the proposed procedure for the design of test signals for nonlinear systems is identical to the procedure used for linear systems. For example, in linear systems instead of using a gaussian white noise signal, it is adequate to use a periodic signal which over a time interval of one period has an autocorrelation function approximately equal to a delta function. In nonlinear systems the approximate comparison between the periodic signal and gaussian white noise must be carried not only up to the first correlation but also to higher order correlations.

3.6.4 Comparison of Oscillation and Autocorrelation Tests Performed on Reactors in the Presence of Nonlinearities

When the variations of the input to a reactor system are small, the reactor behaves like a linear system and oscillation or autocorrelation tests may yield almost the same information, namely, the transfer function of the reactor. When the variations are large, however, the reactor behaves like a nonlinear system and the question arises as to whether these two types of tests yield the same information.

To answer this question, suppose that the reactor power variations are represented by means of a Wiener canonical expansion:

$$p(t) = \sum_{0}^{\infty} G_n(K_n, \rho_e, t) . \qquad (3\text{-}31)$$

When the reactivity is varied sinusoidally, contributions to the fundamental frequency component of p(t) arise from all the odd-order kernels $G_{2m+1}(K_{2m+1}, \rho_e, t)$ only. Thus, Fourier analysis of the output yields the describing function which depends only on the kernels $K_1(\tau)$, $K_3(\tau_1, \tau_2, \tau_3)$, etc. [33]. On the other hand, if the reactivity is gaussian, white noise and the output fluctuations are autocorrelated, the result depends on all the kernels. Indeed, since the functionals G_n are orthogonal to each other, only functionals of the same order correlate with each other and

$$\bar{\phi}_{pp}(\tau) = \sum_{n}^{\infty} \frac{1}{2T} \int_{-\infty}^{\infty} G_n(K_n, \rho_e, t) G_n(K_n, \rho_e, t+\tau) dt. \quad (T \to \infty) \qquad (3\text{-}32)$$

Consequently, the spectrum of this autocorrelation function is different from the square of the amplitude of the describing function derived by means of oscillation tests.

3.6.5 Use of the Describing Function for Stability Studies

The describing function has been used to derive stability criteria for nonlinear reactor dynamics [47, 48]. The representation of a nonlinear system by means of functional expansions is quite informative regarding the question of the practical value of such criteria.

The describing function depends only on the odd-order kernels. Consequently, stability criteria derived from the describing function are bound to be incorrect when the even-order kernels contribute appreciably to the output of the system.

In fact, stability criteria derived from functional expansions are in general more restrictive than necessary. The reason is that such criteria usually reflect the mathematical conditions for the existence of the particular functional expansion chosen rather than the conditions for the stability of the sum of the functionals.

To see this point clearly, consider the simple function

$$y(t) = \frac{1}{1+t}, \quad t > 0 . \qquad (3\text{-}33)$$

This function is a well behaved function for all values of $t > 0$. Suppose now that y(t) is written in the form of a power series:

$$y(t) = 1 - t + t^2 - t^3 . \qquad (3\text{-}34)$$

The series exists and converges to y(t) only when $|t| < 1$. The implication of this simple example is that if the convergence of y(t) were derived from its power series expansion, the range of t would be unnecessarily limited, even though the actual function y(t) is bounded and convergent for all values of $t > 0$. Since a functional expansion may be thought of as a generalized power series expansion, this example shows exactly the difficulty that renders functional expansions impractical for stability studies of reactor systems.

APPENDIX I REPRESENTATIVE DIGITAL COMPUTER CODES FOR SPACE-INDEPENDENT REACTOR KINETICS

Harold Greenspan, Argonne National Laboratory

A number of digital computer codes have been written, and a variety of techniques [49] have been applied to obtain accurate and efficient numerical solutions for the space-independent, one-energy group, reactor kinetics equation systems.

These equations have been derived in Sec. 2 of this chapter, see Eqs. (2-1) and (2-2). For convenience in discussing the computer codes two forms of the kinetic equations are repeated below:

$$\frac{dn}{dt} = (\rho' - 1)\beta \frac{n}{\Lambda} + \sum_{i=1}^{I} \lambda_i C_i + Q$$

$$\frac{dC_i}{dt} = \beta_i \frac{n}{\Lambda} - \lambda_i C_i , \qquad (A-1)$$

and

$$\frac{dn}{dt} = [k_{eff}(1 - \beta) - 1] \frac{n}{\Lambda} + \sum_{i=1}^{I} \lambda_i C_i + Q$$

$$\frac{dC_i}{dt} = k_{eff}\beta_i \frac{n}{\lambda} - \lambda_i C_i . \qquad (A-2)$$

The nomenclature applicable to this Appendix are:

n = neutron density.
k_{eff} = effective multiplication constant.
$k_{ex} = k_{eff} - 1$.
β = delayed neutron fraction.
$\rho = k_{ex}/k_{eff}$ = reactivity;
$\rho' = (k_{ex}/k_{eff})/\beta$ = reactivity in dollars
C_i = density of precursors for the ith delayed neutron group.
β_i = delayed neutron fraction for the ith delayed neutron group.
λ_i = decay constant of the ith delayed neutron group.
Λ = prompt neutron lifetime.
I = total number of delayed neutron groups.
Q = external neutron source.

A zero subscript to the above quantities (where applicable) indicates an initial value.

The codes cited here (listed alphabetically) have been selected on the basis that they represent a method used, or they have some particularly useful feature like the reactivity-neutron density relation, etc., not available in other codes.

More extensive kinetic code listings and descriptions may be found in references [50] and [51].

A.1 AIREK II, AIREK III codes

The AIREK codes [52, 53] have been written in FORTRAN for the IBM-704, 709, and 7090. These programs solve the kinetic systems Eqs. (A-1) coupled with equations describing the reactivity ρ' either as a function of time or in terms of a feedback function F. The following relation is assumed for F:

$$\frac{dF}{dt} = a_1 F + a_2 n(t) + a_3 n(0) , \qquad (A-3)$$

where a_1, a_2, and a_3 are input parameters. The F equation with $a_3 = 0$ has the same form as the precursor equation; this fact is utilized in the subsequent solution.

AIREK II solves the coupled system using a fourth-order Runge-Kutta method. AIREK III obtains a substantially improved numerical solution by using a modified system of equations and a fifth-order Runge-Kutta procedure. The modified system is derived by assuming $n(t) = \exp[J(t)]$ and dividing Eqs. (A-1) by $n(t)$.

In both programs subroutines are available for the following forms for ρ':

$\rho' = \rho_0'$, ρ_0' being a fixed step input reactivity;

$\rho' = \rho_0' + \sum_m D_m F_m$, D_m being constant and F_m as described in Eq. (A-3);

$\rho' = \rho_0' + f(t)$, with $f(t)$ defined as straight line segments or a trigonometric function;

$\rho' = \rho'(t)$, values of $\rho'(t)$ being listed in tabular form.

The outputs from the code are n, inverse period, C_i, and F_m as functions of time.

A.2 RE 29, RE 129 codes

Both of these codes [54, 55] solve the system Eqs. (A-2) for n(t) given k_{ex}. RE 29 is for the IBM-650 and RE 129 is the IBM-704 version.

Using the precursor equation C_i is eliminated from the neutron equation of Eqs. (A-2). The resulting equations are then:

$$\frac{dn}{dt} = \frac{k_{ex} n}{\Lambda} - \sum_{i=1}^{I} \frac{dC_i}{dt} + Q ,$$

$$\frac{dC_i}{dt} = (1 + k_{ex})\beta_i \frac{n}{\Lambda} - \lambda_i C_i . \qquad (A-4)$$

Both sides of Eqs. (A-4) are then integrated with respect to t, and the quantities under the integral signs are replaced by linear approximations valid in the time interval Δt under consideration. From the resulting expression, $n(t_j)$ is determined successively for a set of time values t_j.

Forms of k_{ex} available with the code are:

(a) $k_{ex} = At + B \int_0^t n(y)dy$;

(b) $k_{ex} = At + B\,[n(t) - 1]$ for $t < t_0$,
$\qquad = B[n(t) - 1]$ for $t \geq t_0$;

(c) k_{ex} specified at discrete time points;

$$\text{(d)}\quad k_{ex} = \sum_m \left\{ A_m(t - t_0)^m + \left[\int_0^t n(y)dy\right]^m \right\} \quad \text{and}$$

$$\Lambda = \sum_m L_m \left[\int_0^t n(y)dy\right]^m \quad \text{for } m = 0, 1, \ldots 4.$$

Forms (a), (b), and (c) are available with RE 29; forms (a), (c), and (d) are available with RE 129. For problems involving an initial step or ramp change in k_{ex}, an optional, artificial start routine is provided to compute the initial values of dC_i/dt and Δn (the variation of n over a time interval Δt).

The output of the code is n, $[n/(dn/dt)]$, k_{ex} and $\int_0^t n(y)dy$ for specified values of time t. In addition, for form (d) the computed neutron lifetime is printed.

Attempts to improve the above codes resulted in developing a new technique, "the method of collocation," and applying it [56] to the solution of the kinetics equations.

In this method Eqs. (A-4) are converted into an integral equation and a solution found by the method of collocation [57]. Included in the numerical scheme as applied to the kinetics equations is an automatic interval-size control.

The above method is the basis of an experimental code called COLLOREK [49].

A.3 RE 126, RE 135 codes

These codes [58] are in FORTRAN for the IBM-704. They solve Eqs. (A-2) for a specified reactivity step. The solution is accomplished by obtaining the roots s_j of the inhour equation and using these in the relation for the neutron density:

$$\frac{n(t)}{n_0} = \sum_{j=1}^{I+1} (A_j + B_j) \exp s_j t + C \qquad \text{(A-5)}$$

A_j, B_j, and C are determined from λ_i, β_i, Λ, s_j, and $k_{ex,0}$, the form depending on whether the equilibrium or nonequilibrium precursor problem is treated.

RE 126 performs this calculation for the case of a reactivity step with equilibrium precursors. RE 135 computes the case of nonequilibrium precursors. In the latter computation the reactor is assumed to have been operating at a steady power n_0 for a buildup time of T_B seconds and this determines C_{i0}.

Both codes use an iterative procedure to determine the roots s_j. The problem is terminated when the difference between the reactivity as determined by the s_j and a specified reactivity is a minimum.

The output is: s_j, A_j, B_j, and also $\frac{1}{n_0}\frac{dn}{dt}$, $\frac{n}{n_0}$, $\frac{n}{dn/dt}$, and deviation from asymptotic period for each time value.

A.4 RE 138 code

This IBM-704 FORTRAN code [59] calculates k_{ex} and time derivatives of k_{ex}, using Eqs. (A-2), given the neutron density as a function of time. Defining:

$$\alpha(t) = \frac{1}{n}\frac{dn}{dt}$$

$$d_i(t) = \frac{\lambda_i \Lambda}{\beta_i}\frac{C_i}{n}, \qquad \text{(A-6)}$$

evaluating the neutron density equation at t_j and t_{j+1}, and subtracting the result yields:

$$\Lambda\Delta\alpha_j = (1 - \beta)\Delta k_j + \sum_{i=1}^{I} \beta_i \Delta d_{ij}. \qquad \text{(A-7)}$$

Assuming that the neutron density can be represented by

$$n = n_j \exp \bar{\alpha}_j(t - t_j) \quad \text{for} \quad t_j \leq t \leq t_j, \qquad \text{(A-8)}$$

where

$$\bar{\alpha}_j = \frac{1}{\Delta t_j} \ln\left(\frac{n_j + 1}{n_j}\right)$$

and setting

$$\alpha_j = \frac{\bar{\alpha}_{j-1} + \bar{\alpha}_j}{2}$$

the precursor equations can be integrated for Δd_{ij}. This is then substituted in the neutron formula to obtain Δk_j.

For each time value the code lists k_{ex}, dk_{ex}/dt, d^2k_{ex}/dt^2, reactor period and $\int_0^t n(y)dy$.

A.5 RTS code

The RTS code [60] is for the IBM-704. In Eqs. (A-2) the C_i are eliminated and the Laplace Transform applied to the result. By means of convolution theorems and roots of auxiliary functions in the transform space the neutron density may be written as:

$$n(t) = n(0) + \sum_{j=0}^{I} A_j \times \int_0^t \exp\,[s_j(t - t')]\; k_{ex}(t')n(t')dt' + \Omega_0(t), \qquad \text{(A-9)}$$

where A_j and s_j are universal constants for the fissionable species to be used for various values of Λ, and $\Omega_0(t)$ is a known function of time.

The RTS code solves Eq. (A-9) recursively at each time point for k_{ex} in the form

$$k_{ex} = P_1(t) + P_2(t) \int_0^t \exp R(t - t')n(t')dt' ,$$

where $P_1(t)$ is either a polynomial in t of specified degree, or may be given as a table; $P_2(t)$ is a polynomial in t of specified degree.

The code has an automatic interval size change procedure.

The quantities $n(t)$, $k_{ex}(t)$, $\int n\,dt$, $\Omega_0(t)$, $C_i(t)$ and other quantities are printed as output for each value of t.

REFERENCES

1. A. F. Henry, "The application of reactor kinetics to the analysis of experiments", Nucl. Sci. Eng., 3(1958)52.
2. E. R. Cohen, "Some Topics in Reactor Kinetics", Proceedings of the Second U.N. International Conference on Peaceful Uses of Atomic Energy, Geneva, 1958, Vol. 11, p. 302.
3. A. Foderaro and H. L. Garabedian, "Two-group reactor kinetics", Nucl. Sci. Eng., 14(1962)22.
4. D. E. Dougherty and C. N. Shen, "The space-time neutron kinetic equations obtained by the semidirect variational method", Nucl. Sci. Eng., 13(1963)141.
5. E. P. Gyftopoulos and J. Devooght, "On the range of validity of nonlinear reactor dynamics", Nucl. Sci. Eng., 4(1961)372.
6. J. A. Bewick, A. F. Henry, and S. Kaplan, "Synthesis Approximations in the Time Direction", Trans. Am. Nucl. Soc., 5(1962)177.
7. W. H. Wolf and R. L. Crowther, "Nonlinear Axial Dynamics of Large Boiling-Water Reactors", Trans. Am. Nucl. Soc., 6(1963)209.
8. J. Lewins, "The Basis and Application of Perturbation Theory of Neutron Diffusion", ScD Thesis, Nuclear Engineering, Massachusetts Institute of Technology, May 1959.
9. E. P. Gyftopoulos, "Transfer Function Representation of Nuclear Power Plants", Proceedings of the Conference on Transfer Function Measurements and Reactor Stability Analysis, USAEC Report ANL-6205, Argonne National Laboratory, 1960.
10. R. L. Moore, "The transfer function of a water-boiler reactor", Nucl. Sci. Eng., 6(1959)157.
11. M. F. Gardner and J. L. Barnes, Transients in Linear Systems, J. R. Wiley and Sons, Inc., N.Y., 1953.
12. S. Glasstone and M. C. Edlund, The Elements of Nuclear Reactor Theory, D. Van Nostrand Co., Inc., Princeton, N.J., 1952.
13. M. Schultz, Control of Nuclear Reactors and Power Plants, (2nd Ed.), McGraw-Hill Book Co., Inc., N.Y., 1961.
14. H. Hurwitz, Jr., "Kinetics of low source reactor startups, Part I", Nucl. Sci. Eng., 15(1963)166; "Part II", Nucl. Sci. Eng., 15(1963)187.
15. J. G. Truxal, Control System Synthesis, McGraw-Hill Book Co., Inc., N.Y., 1955.
16. J. E. Gibson, Nonlinear Automatic Control, McGraw-Hill Book Co., Inc., N.Y., 1963.
17. S. Lefschetz, Differential Equations—Geometric Theory, Interscience Publishers Inc., N.Y., 1957.
18. J. LaSalle and S. Lefschetz, Stability by Liapunov's Direct Method with Applications, Academic Press, N.Y., 1963.
19. T. A. Welton, "Kinetics of Stationary Reactor Systems", Proceedings of the First U.N. International Conference on Peaceful Uses of Atomic Energy, Geneva, 1955, Vol. 5, p. 377.
20. E. P. Gyftopoulos and J. Devooght, "Boundedness and stability in nonlinear reactor dynamics", Nucl. Sci. Eng., 7(1960)533.
21. W. K. Ergen and A. M. Weinberg, "Some aspects of nonlinear reactor dynamics", Physica, 20(1954)413.
22. H. B. Smets, "Low and High Power Nuclear Reactor Kinetics", ScD Thesis, Nuclear Engineering, Massachusetts Institute of Technology, May 1958; and "Problems in Nuclear Power Reactor Stability", Presses Universitaires, Brussels, Belgium, 1962.
23. E. P. Gyftopoulos, "Applications of geometric theory to nonlinear reactor dynamics", Nucl. Sci. Eng., 10(1961)254.
24. A. M. Letov, Stability in Nonlinear Control Systems, Princeton University Press, N.J., 1961.
25. V. M. Popov, "Notes on the Inherent Stability of Nuclear Reactors", Proceedings of the Second U.N. International Conference on Peaceful Uses of Atomic Energy, Geneva, 1958, Vol. 11, p. 245.
26. E. P. Gyftopoulos, "Stability criteria for a class of nonlinear systems", Information and Control, 6(1963)276.
27. E. P. Gyftopoulos and J. Devooght, "Effect of delayed neutrons on nonlinear reactor stability", Nucl. Sci. Eng., 8(1960)244.
28. E. P. Gyftopoulos, "Lagrange Stability by Liapunov's Direct Method", Proceedings of the Symposium on Reactor Kinetics and Control, University of Arizona, March 1963.
29. I. G. Malkin, "Certain Questions on the Theory of the Stability of Motion in the Sense of Liapunov", Am. Math. Soc. Translation No. 20, 1950.
30. M. KyFan, "Les fonctions definies-positives et les fonctions completement monotones", Memorial des Sciences Mathematiques, Fascicule CXIV, 1950.
31. K. Knopp, Theory of Functions, Dover Publications, N. Y., 1945.
32. J. A. DeShong and W. C. Lipinski, "Analysis of Experimental Power Reactivity Feedback Transfer Functions of a Natural Circulation Boiling Water Reactor", USAEC Report ANL-5850, Argonne National Laboratory, 1958.
33. A. A. Wasserman, "Contributions to Two Problems in Space-Independent Nuclear Reactor Dynamics", ScD Thesis, Chemical Engineering, Massachusetts Institute of Technology, January 1962; and IDO-16755, 1962.
34. J. D. Balcomb, H. B. Demuth, and E. P. Gyftopoulos, "A cross-correlation method for measuring the impulse response of reactor systems", Nucl. Sci. Eng., 11(1961)159.
35. V. Rajagopal, "Determination of reactor transfer functions by statistical correlation methods", Nucl. Sci. Eng., 12(1962)218.
36. J. A. Thie, Reactor Noise, Rowman and Littlefield Inc., N.Y., 1963.
37. H. W. Bode, Network Analysis and Feedback Amplifier Design, D. Van Nostrand Co., Inc., Princeton, N.J., 1945.
38. E. P. Gyftopoulos, "A Statistical Stability Monitor" Proceedings of the Idaho Conference on Reactor Kinetics, Report IDO-16791, Phillips Petroleum Co., 1960.
39. E. P. Gyftopoulos, "On the measurement of dynamic characteristics of nuclear reactor systems", Trans. Am. Nucl. Soc., 4(1961)206.
40. V. Volterra et J. Peres, Théorie Générale des Fonctionelles, Gauthiers-Villars, Paris, 1936.
41. N. Wiener, Nonlinear Problems in Random Theory, M.I.T. Press, Cambridge, Mass., 1958.
42. D. A. Chesler, "Nonlinear Systems with Gaussian Inputs", Massachusetts Institute of Technology Research Laboratory of Electronics, Technical Report 366, February 1960.
43. G. Zames, "Nonlinear Operators for System Analysis", ScD Thesis, Electrical Engineering, Massachusetts Institute of Technology, September 1960.
44. D. A. George, "Continuous Nonlinear Systems", Massachusetts Institute of Technology Research Laboratory of Electronics, Technical Report 355, July 1959.
45. J. F. Barrett, "The Use of Functionals in the Analysis of Nonlinear Physical Systems", Statistical Advisory Unit Report 1/57, Ministry of Supply, Great Britain, 1957.
46. E. P. Gyftopoulos and R. J. Hooper, "Signals for Transfer Function Measurements in Nonlinear Systems", Trans. Am. Nucl. Soc., 6(1963)211.
47. H. A. Sandmeier, "The Kinetics and Stability of Fast Reactors with Special Consideration of Nonlinearities", USAEC Report ANL-6014, Argonne National Laboratory, 1959.
48. H. B. Smets, "The Describing Function of Nuclear Reactors", Institute of Radio Engineers, Trans. Nucl. Sci., NS-6(1959)8.
49. H. P. Flatt, "The Numerical Solution of the Reactor Kinetics Equations by the Method of Collocation", IBM Nuclear Computing Technical Bulletin, No. 5, International Business Machines Corp., 590 Madison Ave., N.Y.C. 22, 1962.
50. "Digital Computer Codes", Section 10 in "Reactor Physics Constants", USAEC Report ANL-5800 (2nd Ed.), Argonne National Laboratory, 1963.
51. V. Nather and W. Sangren, "Abstracts—Nuclear reactor codes", Comm. Assoc. Comp. Mach., 2,1(1959): 3,1(1960).
52. A. Schwartz, "Generalized Reactor Kinetics Code—AIREK II", Report NAA-SR-Memo 4980, North American Aviation Inc., 1960.
53. A. Schwartz, "Fifth-Order Runge-Kutta Numerical Integration Formulas—Generalized Reactor Kinetics Code, AIREK III", North American Aviation Inc., Atomics International Memorandum.
54. R. O. Brittan, "Some Problems in the Safety of Fast Reactors", USAEC Report ANL-5577, Argonne National Laboratory, 1956.
55. Report RE 129, AMD Program Library, Argonne National Laboratory.
56. J. J. Kaganove, "Numerical Solution of the One-Group Space-Independent Reactor Kinetics Equations for Neutron Density Given the Excess Reactivity", USAEC Report ANL-6132, Argonne National Laboratory, 1960.
57. Z. Kopal, Numerical Analysis, p. 471, John Wiley and Sons, N.Y., 1955.
58. C. E. Cohn and B. J. Toppel, "IBM-704 Codes for Reactivity Step Calculations", (RE 126 and RE 135), USAEC Report ANL-6134, Argonne National Laboratory, 1960.
59. Report RE 138, AMD Program Library, Argonne National Laboratory.
60. G. R. Keepin and C. W. Cox, "General solution of the reactor kinetics equations", Nucl. Sci. Eng., 8(1960)670.

CHAPTER 4

The Doppler Coefficient

L. W. NORDHEIM
General Atomic, Division of General Dynamics,
San Diego, California

CHAPTER CONTENTS*

1 RESONANCE ABSORPTION IN THERMAL REACTORS

1.1 Introduction

Neutron absorption by heavy nuclei takes place at sharp resonance energies whose width is generally small compared to the neutron energy loss in elastic collisions with moderator material. The resulting "resonance absorption" constitutes one of the fundamental phenomena in reactor physics [1, 2].** Firstly, it affects the reactivity directly; secondly, it is temperature-dependent. The presence of resonances produces sharp neutron flux dips in energy at the resonances and flux dips in space where there are absorbers. The Doppler effect, due to the thermal motion of the absorber nuclei, broadens and flattens the resonances thus reducing the neutron flux dips and with that the self-shielding of the resonances. The resonance absorption, therefore, always increases with increasing temperature. If the material is a pure absorber, such as U^{238} and Th^{232}, the reactivity will be decreased at higher temperatures; in fissionable materials the effect can be either positive or negative depending on the balance between fission and absorption, which both will increase. If the absorber is intimately mixed with the fuel, as it is usually the case for fertile materials, then the Doppler coefficient will be prompt. It thus constitutes one of the most important automatic shutdown mechanisms in reactor excursions as described at several places in these volumes.

In the treatment of resonance absorption one has to distinguish two main cases. First, the case when the resonances are well separated, as is the situation for the fertile materials at neutron energies below 30 kv. This will be treated in Sec. 1; the treatment is suitable for most thermal reactor problems. Since the methods used here are well known and well documented, mathematical details will be omitted. The second case occurs when the resonances are not well separated and when there are overlapping resonances due to different materials. This is of importance mainly for fast and

*Except for a few changes and additions made in proof, this chapter is based on information in the literature or known to the author prior to January, 1964.

**Credit for the first study of this effect and elucidation of the underlying principles is due to E. P. Wigner. His work, performed under the wartime Manhattan Project, was first published in reference [1]. A good account of the general theory for well-separated resonances and of the history of the subject is given in reference [2].

intermediate reactors. Though the treatment at this time is less complete, this chapter will go somewhat more into detail because no other general review of the subject is available.

In practical cases complications may arise from nonuniform temperature distributions. Such effects will not be considered here. It may be mentioned that the case of temperature profiles in absorber lumps has been treated by Dresner [3], who shows that the use of a suitable average temperature is in many cases adequate, and by Ferziger [4], who modifies the escape probabilities (see Sec. 1.3) for nonuniform source and cross sections. The temperature distribution in the large over a reactor is probably best handled by a suitable reactor code.

1.2 Resonance Cross Sections

In this section the principal formulas for the energy dependence of resonance cross sections are collected. They are required for all quantitative calculations, but may be skipped over by readers only interested in the qualitative aspects of the subject.

For the formal theory of nuclear reactions, reference is made to the representations of Lane and Thomas [5] and Vogt [6].

Neutrons with spin $S = 1/2$ and an angular momentum ℓ impinging on a nucleus with spin I can produce reactions in any channel with total spin J compatible with the vector addition of angular momenta. The one-level Breit-Wigner formula for the reaction cross section of type x in the neighborhood of a resonance at energy E_o, for a target at rest, is then given by

$$\sigma_x = \pi \lambda\!\!\!^{-2} g \frac{\Gamma_n \Gamma_x}{(E - E_o)^2 + \Gamma^2/4}, \tag{1-1}$$

where, as usual, Γ is the total width, Γ_n the neutron width, Γ_x the width for reaction x (radiative capture Γ_γ, fission Γ_f), $\lambda\!\!\!^{-}$ the De Broglie wavelength, and the statistical factor g is

$$g = \frac{2J + 1}{2(2I + 1)}. \tag{1-2}$$

The elastic scattering cross section exhibits interference between elastic and resonance scattering, and the cross section is given by

$$\sigma_s = \pi \lambda\!\!\!^{-2} g \left| \frac{i\Gamma_n}{E - E_o + i\Gamma/2} + e^{2i\delta} - 1 \right|^2 + \pi \lambda\!\!\!^{-2} (2\ell + 1 - g) \left| e^{2i\delta} - 1 \right|^2. \tag{1-3}$$

Here δ is the phase shift responsible for the potential scattering. It is given in good approximation by the "hard sphere" model

$$\tan \delta_\ell = \frac{j_\ell(R/\lambda\!\!\!^{-})}{n_\ell(R/\lambda\!\!\!^{-})},$$
$$\delta_o = R/\lambda\!\!\!^{-}, \tag{1-4}$$
$$\delta_1 = R/\lambda\!\!\!^{-} - \tan^{-1} R/\lambda\!\!\!^{-},$$

where j_ℓ and n_ℓ are the spherical Bessel and Neumann functions respectively, and R is the nuclear radius. For extensions beyond the hard sphere model see Vogt [6]. Carrying out the square in Eq. (1-3) yields

$$\sigma_s = \frac{\pi \lambda\!\!\!^{-2} g}{(E - E_o)^2 + \Gamma^2/4} \left[\Gamma_n^2 - 2\Gamma_n \Gamma \sin^2 \delta + 2\Gamma_n (E - E_o) \sin 2\delta\right] + 4\pi \lambda\!\!\!^{-2} (2\ell + 1) \sin^2 \delta. \tag{1-5}$$

The first term is the resonance scattering, the terms containing δ the interference scattering, and the last term the potential scattering. The total cross section, the sum of all reaction cross sections and scattering, is

$$\sigma_t = \frac{\pi \lambda\!\!\!^{-2} g \Gamma_n}{(E - E_o)^2 + \Gamma^2/4} \left[\Gamma \cos 2\delta + 2(E - E_o) \sin 2\delta\right] + 4\pi \lambda\!\!\!^{-2} (2\ell + 1) \sin^2 \delta. \tag{1-6}$$

(The identity $1 - 2 \sin^2 \delta = \cos 2\delta$ has been used.) Note that the resonance term in σ_t itself may become negative. However, the total quantity is, of course, always positive. For small $R/\lambda\!\!\!^{-}$, that is low energies, the $\cos 2\delta$ can be replaced by unity and $\sin 2\delta$ by 2δ, and one obtains the more familiar form for s-waves ($\ell = 0$):

$$\begin{aligned} \sigma_x &= \sigma_o \frac{\Gamma_x}{\Gamma} \sqrt{\frac{E_o}{E}} \frac{1}{(4/\Gamma^2)(E - E_o)^2 + 1} \\ \sigma_s &= \sigma_o \frac{\Gamma_n}{\Gamma} \sqrt{\frac{E_o}{E}} \frac{1}{(4/\Gamma^2)(E - E_o)^2 + 1} + \left(\sigma_o \sigma_p g \frac{\Gamma_n}{\Gamma} \sqrt{\frac{E_o}{E}}\right)^{1/2} \frac{4(E - E_o)/\Gamma}{4(E - E_o)^2/\Gamma^2 + 1} + \sigma_p, \end{aligned} \tag{1-7}$$

where

$$\sigma_0 = \sigma_{00} \Gamma_n^0/\Gamma,$$
$$\sigma_{00} = 4\pi \lambda\!\!\!^{-2} g = 2.6 \times 10^6 g/E_0,$$
$$\Gamma_n = \Gamma_n^0 \sqrt{E/E_0}. \tag{1-8}$$

In presence of thermal motion, a Maxwell-Boltzmann distribution of the target nuclei has to be folded into the relative energy of the incoming neutrons. This procedure leads to the well known result for the partial cross sections*

$$\sigma_x = \sigma_{00} (\Gamma_n \Gamma_x/\Gamma^2) \psi = \sigma_0 (\Gamma_x/\Gamma) \psi \tag{1-9}$$

*A discussion of the accuracy of these expressions has been given by Hinman et al. [7].

and for the total cross section

$$\sigma_t = \sigma_{00}(\Gamma_n/\Gamma)(\psi\cos 2\delta + \chi\sin 2\delta) + \sigma_p$$

$$\sim \sigma_0\psi + (\sigma_0\sigma_{p_0} g\Gamma_n/\Gamma)^{1/2}\chi + \sigma_{p_0}, \qquad (1\text{-}10)$$

where σ_p is the potential cross section and where the shape functions ψ and χ are

$$\psi = \frac{\xi}{\sqrt{4\pi}} \int_{-\infty}^{+\infty} \frac{\exp[-(q-p)^2\xi^2/4]}{1+p^2}\,dp, \qquad (1\text{-}11)$$

$$\chi = \frac{\xi}{\sqrt{4\pi}} \int_{-\infty}^{+\infty} \frac{2p\exp[-(q-p)^2\xi^2/4]}{1+p^2}\,dp. \qquad (1\text{-}12)$$

In these expressions

$$q = 2(E - E_0)/\Gamma, \quad \xi = \Gamma/\Delta, \qquad (1\text{-}13)$$

where

$$\Delta = \sqrt{4EkT/A} \qquad (1\text{-}13a)$$

is the Doppler width, with E the energy, T the absolute temperature, and A the reduced mass of the target nucleus.

Actually the absorber atoms do not form a free gas state, but will be generally bound in a crystalline state. This problem has been studied by Lamb [8] who found that it is a very good approximation to take a Maxwell distribution, however not at the temperature of the crystal, but at a higher temperature which corresponds to the average energy per vibration degree of freedom of the lattice, including the zero point energy.

The shape functions, ψ (symmetric) and χ (asymmetric), are typical for dispersion phenomena as modified by Doppler broadening [9]. There exists an extensive literature about them. We note the following properties (which will be referred to later):

(a) $\displaystyle\int_{-\infty}^{+\infty} \psi(q,\xi)\,dq = \pi; \quad \int_{-\infty}^{+\infty} \chi(q,\xi)\,dq = 0;$

(b) $\psi(0,\xi) = (\sqrt{\pi}/2)\,\xi e^{-\xi^2/4}\,\mathrm{erfc}(\xi/2);$

(c) $\displaystyle\chi(q,\xi) = 2q\,\psi(q,\xi) + \frac{4}{\xi^2}\frac{\partial\psi}{\partial q}. \qquad (1\text{-}14)$

For extreme Doppler broadening, $\Gamma/\Delta \ll 1$, one has the approximation, except in the "wings",

$$\psi \sim (\sqrt{\pi}/2)\,\xi\exp(-q^2\xi^2/4). \qquad (1\text{-}15)$$

For large q, that is in the wings, there exists an asymptotic expansion

$$\psi \sim \frac{1}{1+q^2}\left[1 + \frac{2}{\xi^2}\frac{3q^2-1}{(1+q^2)^2} + \cdots\right]$$

$$\chi \sim \frac{2q}{1+q^2}[1 + \cdots] \qquad (1\text{-}16)$$

Some tables for the ψ functions have been published [10]. For machine calculations several rapid methods are available [11, 12, 13].

A few remarks may be made on the case where the resonances are not too well separated, as encountered in the fissile isotopes even at low energies. In this case the cross section at a definite energy would be influenced by many levels, and there are also several competing final states (channels), corresponding to the different possible reactions (scattering, capture, fission). The formal multilevel-multichannel theory for such cases was introduced by Wigner and Eisenbud [14] and applied to the fissile isotopes by Vogt [15] and others. The procedure, however, is rather complicated and restricted to very small groups of resonances.

An important advance has recently been made by D. B. and F. T. Adler [16]. They have introduced a different representation in which the cross sections appear simply as superposition of independent but shifted quasi-resonances with asymmetric shape factors. In place of Eqs. (1-6) and (1-7) one obtains

$$\sigma_t = \frac{C}{\sqrt{E}}\sum_i \frac{\Gamma_{it}G_{it} + (E_i - E)H_{it}}{(E-E_i)^2 + (\Gamma_i^2/4)} + \sigma_p \qquad (1\text{-}17)$$

$$\sigma_x = \frac{C}{\sqrt{E}}\sum_i \frac{\Gamma_{it}G_{ix} + (E_i - E)H_{ix}}{(E-E_i)^2 + (\Gamma_i^2/4)} \qquad (1\text{-}18)$$

$$\sigma_s = \sigma_t - \sum_x \sigma_x. \qquad (1\text{-}19)$$

In these formulas

$$C = \pi\lambda\!\!\!^{-2}E = 6.52 \times 10^5 \text{ barns ev.}$$

The E_i and Γ_{it} are energies and total widths of the quasi-resonances and the G_{it}, G_{ix}, H_{it}, H_{ix} energy-independent free parameters which can be adjusted to fit the experimental cross sections. This formulation has further the great advantage that the Doppler broadening can be very easily applied leading, for example, to the reaction cross section

$$\sigma_x = \frac{C}{\sqrt{E}}\sum_i \frac{2}{\Gamma}\left\{G_i\psi(q_i,T) + H_{ix}\chi(q_i,T)\right\}, \qquad (1\text{-}20)$$

where $q_i = (E - E_i)2/\Gamma_i$ and ψ and χ are the shape functions, Eqs. (1-11) and (1-12). It is to be expected that in the near future analyses of the cross sections of the fissile isotopes in terms of the Adler formulas will become available.

For use later the integral

$$K = \int_{-\infty}^{+\infty} \psi_k \psi_{k'} dE \tag{1-21}$$

over two resonances k and k′ with different individual total widths Γ_k and $\Gamma_{k'}$ will be required. From the definition of the ψ function it can be shown by a somewhat tedious calculation as pointed out by Hwang [17] that

$$K = \frac{\pi}{2} \frac{\Gamma_k \Gamma_{k'}}{\Gamma_k + \Gamma_{k'}} \psi(Q, \Xi), \tag{1-22}$$

where

$$Q = \frac{2(E_k - E_{k'})}{\Gamma_k + \Gamma_{k'}}, \quad \Xi = \frac{\Gamma_k + \Gamma_{k'}}{\sqrt{2}\,\Delta}. \tag{1-23}$$

In particular, if k and k′ are the same resonance (i.e., Q = 0), one has

$$\int \psi^2 dE = (\pi/2)^{3/2} (\Gamma^2/2\Delta) \exp(-\Gamma^2/2\Delta^2), \tag{1-24}$$

and for different resonances, when the extreme Doppler broadened form Eq. (1-15) can be used for ψ

$$\int \psi_k \psi_{k'} dE \cong (\pi/2)^{3/2} (\Gamma_k \Gamma_{k'}/2\Delta) \exp\left[-\frac{(E_k - E_{k'})^2}{2\Delta^2}\right]. \tag{1-25}$$

It follows further from Eqs. (1-22) and (1-14a) that

$$\int K(E_k - E_{k'})\, d(E_k - E_{k'}) = (\pi^2/4)\, \Gamma_k \Gamma_{k'} \tag{1-26}$$

1.3 The Calculation of Resonance Integrals

An extensive literature is available on this topic [1, 2, 18, 19]. Therefore, only a short review will be given. Heterogeneous geometries will be included from the beginning. The reactor is assumed to be built up of cells of moderator in which elements are embedded containing absorber, fuel, and perhaps some added scatterer. For the moment, it is assumed that the absorber elements are several mean free paths apart so that they do not influence each other. The effective cross sections, as needed in multigroup calculations, are the flux-weighted averages. Thus, for an interval E_1 to E_2 containing one (or several) resonances one will have as effective cross section

$$\tilde{\sigma}_a = \frac{\int_{E_1}^{E_2} \phi \sigma_a dE}{\int_{E_1}^{E_2} \phi dE}, \tag{1-27}$$

where σ_a is the true absorption cross section, and $\phi(E)$ the flux already averaged over the volume of the absorber. Since in absence of the absorber the flux would be 1/E and since for widely separated resonances the flux has almost everywhere its undisturbed value*, one can write

$$\tilde{\sigma}_a = I_i/\ln(E_2/E_1), \tag{1-28}$$

where

$$I_i = \int \phi \sigma_a dE; \tag{1-29}$$
$$\phi = 1/E, \text{ above the resonance}$$

is called the effective resonance integral. In this expression the flux ϕ has to be normalized so that it is 1/E above the resonance. The capture rate is determined by the competition between absorption and slowing down. For a single resonance at energy E_i the capture probability is

$$q_i = I_i/\overline{\xi\sigma_s} \tag{1-30}$$

where $\overline{\xi\sigma_s}$ is the average slowing down power per absorber atom. The escape probability is then

$$p_i = 1 - q_i. \tag{1-31}$$

For a group of resonances then

$$p = \prod_i (1 - q_i) \sim \exp\left[-\sum_i I_i/\overline{\xi\sigma_s}\right], \tag{1-32}$$

provided the individual q_i are small enough**, and the neutron spectrum in absence of the resonance absorbers would be 1/E. In this case $I = \sum I_i$ is called the effective total resonance integral.[i]

To calculate the resonance integral, and with it the effective absorption cross sections, one has to find the average flux. It is determined by the neutron balance equation

$$\phi_o(\sigma_{t_o} + \sigma_m) = (1 - P_o)\left\{\frac{1}{\alpha_o}\int_E^{E/1-\alpha_o} \phi_o(E')\sigma_{s_o}\frac{dE'}{E'} + \frac{1}{\alpha_m}\int_E^{E/1-\alpha_m} \phi_o\sigma_m \frac{dE'}{E'}\right\} + P_1\phi_1\sigma_{s_1} \tag{1-33}$$

*For a discussion of the errors involved see Sec. 2.3.

**The product formation corresponds to multigroup theory, while the exponential form is correct in case of hydrogen as moderator. For a more thorough discussion of escape theory see Weinberg and Wigner, The Physical Theory of Neutron Chain Reactors [20].

Here the index 0 refers to the absorber, m to admixed scatterer (inside moderator) t means total, s means scattering, and $\alpha = 4A/(A+1)^2$ is the fractional maximum energy loss in an elastic collision. (It should be remembered that $\sigma_s, \sigma_m \cdots$ are cross sections per atom of neutron absorber.) The term on the left side of Eq. (1-33) is the number of neutrons removed from the energy interval dE by collisions. The two integrals give the neutrons scattered into dE by the absorber and admixed scatterer respectively. The factor $(1 - P_0)$ is the nonescape probability, i.e. that the next collision is still in the absorber element. The last term represents the neutrons entering from the outside moderator where the flux is ϕ_1. The case of homogeneous mixtures without cell structure is contained in Eq. (1-33) by letting $P_0 = 0$, no escape from the medium, and $P_1 = 0$, no neutrons entering from the outside.

The Eq. (1-33) represents an empty form unless the escape probabilities P_0, P_1, and ϕ_1 are specified. It is the basic assumption in almost all current treatments of resonance absorption, that 1) the outside flux ϕ_1 can be approximated by 1/E, that is, it is not appreciably disturbed; and 2) that the escape probabilities can be calculated in sufficient approximation by calculating these functions for a flat source in the respective media. The results of actual calculations indicate that these simplifications are sufficiently accurate for practical purposes.

With the flat source approximation several results from transport theory are available. The most important relation is the reciprocity theorem

$$P_o(\sigma_{s_o} + \sigma_m) = P_1 \sigma_{s_1} \tag{1-34}$$

as proved by Rothenstein [21] under very general assumptions. This permits at once the verification of the asymptotic solution $\phi_0 = \phi_1 = 1/E$ of Eq. (1-33) valid in regions where there is no absorption and $\sigma_{t_0} = \sigma_{s_0}$, $\sigma_{t_1} = \sigma_{s_1}$ are constant. The neutron balance equation becomes then (we omit the subscript 0 on the flux in the absorber)

$$\phi(\sigma_{t_o} + \sigma_m) = (1 - P_o)\left\{\frac{1}{\alpha_o}\int_E^{E/1-\alpha_o} \phi(E')\,\sigma_{s_o}\frac{dE'}{E'} + \frac{1}{\alpha_m}\int_E^{E/1-\alpha_m} \phi(E')\,\sigma_m\frac{dE'}{E'}\right\} + P_o\,\frac{\sigma_{t_o} + \sigma_m}{E}. \tag{1-35}$$

The escape probabilities P_0 for widely spaced lumps are functions of $\bar{r}/\lambda$, where λ is the mean free path and

$$\bar{r} = 4V_o/S_o, \tag{1-36}$$

is the mean chord length in the absorber; V_0/S_0 is the volume-to-surface ratio. The escape probabilities have been determined for simple geometries (slab, cylinder, sphere) by Case et al. [22]. They depend only weakly on geometry.

For many cases Eq. (1-35) can be somewhat simplified. Since the average energy loss in collisions with a good moderator nucleus generally greatly exceeds the width of a resonance line, one can use in the moderator integral the so-called "narrow resonance approximation" (see next section). This consists in replacing the flux in the integral by its asymptotic value above the resonance, that is, by 1/E. The integral is then simply σ_m/E and one obtains

$$\phi_{t_o}(\sigma_{t_o} + \sigma_m) = (1 - P_o)\,\frac{1}{\alpha_o}\int_E^{E/1-\alpha_o} \phi\,\sigma_{s_o}\frac{dE'}{E'} + \frac{P_o\sigma_{t_o} + \sigma_m}{E}. \tag{1-37}$$

This equation forms the starting point of most calculations of resonance integrals for thermal reactors [2, 18, 21, 23, 24]. For analytic treatments further simplifications have to be introduced. It has been found, however, that the full Eqs. (1-35) or (1-37) can be integrated directly on a fast computer without any further approximations and with reasonable speed [25]. The program calculates Doppler-broadened cross sections from resonance parameters, with inclusion of interference scattering. It then calculates the flux distribution in energy, taking the functions P_0 from provided tables*, and calculates the resonance integrals. The unresolved s-wave resonances are treated statistically and their contribution is evaluated by a separate program.

1.4 Possible Refinements

While the practical results of the procedure of the preceding section have been gratifying [19], it should not be minimized that the several approximations retained may introduce errors which are difficult to estimate. They are

1. Flux recovery between resonances. Resonances above a particular one will depress the flux below. This interference seems to be of little importance for thermal heterogeneous reactors, where most of the slowing down occurs in the moderator. It can in principle be calculated and will be discussed to some extent in Sec. 2, since it is of greater importance for fast reactors.
2. The flat flux assumption for calculating escape probabilities. This is the most serious shortcoming of the present methods and the most difficult to improve upon. For the absorber it is not important when there is little scattering in a resonance and as long as the dimensions of the absorber lumps are relatively small, but may introduce appreciable errors for large lumps.

*The original program has such tables for the standard simple geometries. For more complicated shapes the P_0 values can be separately calculated by a transport code (S_n or Monte Carlo) and the corresponding tables added.

For the outside moderator Wigner and Weinberg [20] introduce a disadvantage factor from a diffusion treatment of the whole resonance region. This rather doubtful procedure leads only to small corrections and it seems to be preferable to omit them.

Rothenstein [21] introduces a correction for wide resonances only, that is, resonances in which energy loss in elastic scattering is small compared to the width of the resonances (see next section). This results after considerable simplifications in a correction

$$\Delta I = -\frac{I^2}{2\sigma_m a_m}, \quad (1\text{-}38)$$

where I is the resonance integral itself. This correction is normally quite small. It is, however, questionable when it should be applied. Iijima [26] has, under many simplifying assumptions, calculated the influence of the space and energy distributions of neutrons for single resonances and for slab geometry and finds a fairly sizeable effect for the lowest resonances of U^{238}. The effect will be smaller for thinner elements and if moderator is admixed with the absorber, and also for cylindrical geometry.

This effect is the only one in which the outside moderator or material plays a role. A really satisfactory treatment has not yet been developed. The differences due to different moderator materials at present are masked by experimental uncertainties.

Better understanding of these effects could be achieved by detailed Monte Carlo calculations for individual resonances and geometries. Extensive calculations of the full resonance integrals by a Monte Carlo method have recently been presented by Levine [27], who found excellent agreement with the results of the analytic calculations, again demonstrating that the approximations in the latter do not have important effects except perhaps for very thick absorber lumps.

1.5 The Equivalence Relations

While it is clear that problems of such complexity as resonance absorption particularly with inclusion of the Doppler effect are best handled by suitable machine programs, there are some approximations of a semiquantitative nature, which provide insight into the mechanism of the effect and which have had historical significance.

The escape probability P_0 plays a basic role for heterogeneous systems. It is, as already mentioned, a function of $\bar{r}/\lambda$, i.e., the ratio of the average chord length to the mean free path in the absorber lumps. It depends also on the shape of the lumps. However, this dependence on geometry is not strong, and this is the reason why experimentally determined resonance integrals generally can be quite well represented as a universal function of the surface-to-mass ratio of the absorber lumps.

An approximation for P_0 can be found by studying its asymptotic properties. For very small bodies, that is small $\bar{r}/\lambda$, $P_0 \to 1$, that is, all neutrons will escape. For large bodies, on the other hand, $P_0 \to \lambda/\bar{r} = \lambda S_0/4V_0$, that is, effectively all neutrons within a quarter mean free path from the surface will escape. (This is the fraction of neutrons escaping from a surface element of a body with uniform neutron distribution within.) These properties suggest the simple interpolation formula, first proposed by Wigner [1]:

$$P_{oR} = \frac{1}{1 + (\bar{r}/\lambda)} = \frac{\Sigma_e}{\Sigma_e + \Sigma_t} = \frac{\sigma_e}{\sigma_e + \sigma_t}, \quad (1\text{-}39)$$

where

$$\Sigma_e = 1/\bar{r}, \quad \sigma_e = 1/(\bar{r}N_o) \quad (1\text{-}40)$$

can be considered as effective macroscopic and microscopic escape cross sections. N_0 is the number density of the absorber atoms. This rational approximation gives systematically too low values for P_0, and with this for the resonance absorption, with errors from 5 to 10% [2]. It leads, however, to interesting approximate correspondence relations in case of admixtures of moderator. With $\sigma_t = \sigma_{t_0} + \sigma_m$ insertion of (1-39) into (1-37) yields

$$\phi(\sigma_{t_0} + \sigma_m) = \frac{\sigma_{t_0} + \sigma_m}{\sigma_{t_0} + \sigma_m + \sigma_e}\frac{1}{a_0}\int_E^{E/1-a_0} \sigma_{s_0}\phi\frac{dE'}{E'} + \frac{(\sigma_{t_0} + \sigma_m)(\sigma_m + \sigma_e)}{(\sigma_{t_0} + \sigma_m + \sigma_e)E}$$

and by multiplying with $(\sigma_{t_0} + \sigma_m + \sigma_e)/(\sigma_{t_0} + \sigma_m)$

$$\phi(\sigma_{t_0} + \sigma_m + \sigma_e) = \frac{1}{a_0}\int_E^{E/1-a_0} \sigma_{s_0}\phi\frac{dE'}{E'} + \frac{\sigma_m + \sigma_e}{E}. \quad (1\text{-}41)$$

This is exactly the equation one would obtain for a homogeneous mixture of the absorber with moderator of scattering cross section $\sigma_m + \sigma_e$. We obtain thus the equivalence relation:

> Heterogeneous assemblies with the same $\sigma_m + \sigma_e$ have the same resonance integral, and a heterogeneous assembly has the same resonance integral as a homogeneous one with $\sigma_m + \sigma_e$ as total scattering cross section.

This equivalence is quite well verified experimentally for the connection between U metal and UO_2. For comparison with the homogeneous case, it is to be noted that the heterogeneous equivalent will have a somewhat higher value of the resonance integral due to the errors in the rational approximation.

It is to be noted, of course, that the same resonance integral does not necessarily mean the same resonance escape probability. The relation between p and I is given by (1-32). Thus, for heterogeneous assemblies the slowing down power is produced not only by the admixed moderator, but also by the external moderator. Therefore, by lumping of the absorber it is possible to decrease the resonance integral (due to less moderator admixture) without decreasing the slowing down power, and thus to achieve a great reduction in resonance capture, as is necessary for natural uranium reactors.

1.6 The Narrow Resonance (NR) and Infinite Mass (IM) Approximations

There are two approximations to the solution of Eq. (1-37) which have been widely used. The first (the "narrow resonance" or NR approximation) is applicable when the energy loss in scattering with an absorber atom $\alpha_0 E$, is large compared to the energy region in which absorption takes place. In this case the quantities in the integral can be replaced by their asymptotic values, i.e., $\phi = 1/E$, and $\sigma_{s_0} = \sigma_{p_0}$, the potential scattering between resonances. One obtains then

$$\phi(\sigma_{t_0} + \sigma_m) = [(1 - P_o)\sigma_{p_0}/E] + [(P_o\sigma_{t_0} + \sigma_m)/E]$$

$$\phi = \frac{1}{(\sigma_{t_0} + \sigma_m)E}\left[\sigma_{p_0} + \sigma_m + P_o(\sigma_{t_0} - \sigma_{p_0})\right], \quad (1\text{-}42)$$

and for the resonance integral

$$I = \int \frac{\sigma_{a_0}}{\sigma_{t_0} + \sigma_m}\left[\sigma_{p_0} + \sigma_m + P_o(\sigma_{t_0} - \sigma_{p_0})\right]\frac{dE}{E}. \quad (NR) \quad (1\text{-}43)$$

Transport theory [22] gives the result

$$P_o = \frac{\lambda}{\bar{r}} G = \frac{S_o}{4V_o N_o \sigma_t} G, \quad (1\text{-}44)$$

where

$$\sigma_t = \sigma_{t_0} + \sigma_m$$

is the total cross section per absorber including admixed scattering, and G the probability that a neutron entering from a uniform outside bath of neutrons makes a collision in the absorber lump. The resonance integral divides into a volume part I_v and a surface part I_s, as introduced originally by Wigner [1]:

$$I_v = \int \frac{\sigma_{a_0}(\sigma_{p_0} + \sigma_m)}{\sigma_t}\frac{dE}{E}, \quad (NR)$$

$$I_s = \frac{S_o}{4V_o N_o}\int \frac{\sigma_{a_0}(\sigma_{t_0} - \sigma_p)}{\sigma_t^2} G \frac{dE}{E}. \quad (NR) \quad (1\text{-}45)$$

The opposite extreme is reached when the resonance is so wide that the absorption is large over an energy region much greater than $\alpha_0 E$. Then the cross sections can be considered as slowly varying functions of energy and be taken out of the integral in (1-37). This would be strictly correct if $\alpha_0 \to 0$, that is, for infinite mass of the absorber. This approximation is, therefore, called the "infinite mass' or IM approximation. This limit gives for the flux

$$\phi\sigma_t = (1 - P_o)\sigma_{s_0}\phi + \frac{P_o\sigma_{t_0} + \sigma_m}{E},$$

or

$$\phi = \frac{P_o\sigma_{t_0} + \sigma_m}{\sigma_t - (1 - P_o)\sigma_{s_0}}\frac{1}{E}$$

and

$$I' = \int \frac{\sigma_{a_0}(P_o\sigma_{t_0} + \sigma_m)}{\sigma_t - (1 - P_o)\sigma_{s_0}}\frac{dE}{E}. \quad (IM) \quad (1\text{-}46)$$

This can again be split up into a volume and a surface term:

$$I'_v = \int \frac{\sigma_{a_0}\sigma_m}{\sigma_t - (1 - P_o)\sigma_{s_0}}\frac{dE}{E}$$

$$I'_s = \frac{S_o}{4V_o N_o}\int \frac{\sigma_{a_0} G}{\sigma_t - (1 - P_o)\sigma_{s_0}}\frac{dE}{E}, \quad (1\text{-}47)$$

where the volume term comes only from admixed moderator, and where the absorber scattering is largely cancelled out from the removal cross section (the denominator).

Introduction of the rational approximation for P_o Eq. (1-39) leads to further simplifications. One obtains

$$I = \int \frac{\sigma_{a_0}(\sigma_{p_0} + \sigma_m + \sigma_e)}{\sigma_{t_0} + \sigma_m + \sigma_e}\frac{dE}{E}, \quad (NR) \quad (1\text{-}48)$$

$$I' = \int \frac{\sigma_{a_0}(\sigma_m + \sigma_e)}{\sigma_{a_0} + \sigma_m + \sigma_e}\frac{dE}{E}. \quad (IM) \quad (1\text{-}49)$$

For the homogeneous case $\sigma_e = 0$, and the above formulas do not contain approximations other than NR or IM.

In earlier calculations [2, 18, 24] of resonance absorption, either the NR or IM approximation has been used, depending on whether the "practical width" Γ_p is smaller or larger than the maximum loss $\alpha_0 E$ in a collision, with the usual definition of Γ_p as the width of the energy interval in which the resonance cross section exceeds the potential scattering. Since the equality occurs in the wings of the resonance, where the line shape approaches the natural form, Γ_p is given with sufficient accuracy by

$$\Gamma_p = \Gamma\sqrt{(\sigma_o/\sigma_{p_0}) - 1} \approx \Gamma\sqrt{\sigma_o/\sigma_{p_0}}, \quad (1\text{-}50)$$

where σ_0 is the cross section at resonance. This criterion gives 6 or 5 resonances in U^{238} and Th^{232} to be treated in the IM approximation. While with many resonances errors introduced this way tend to cancel out [18], it became clear that the errors in individual resonances may be quite large. Chernick and Vernon [24] introduced, therefore, an iteration procedure, which can be carried through only by neglecting Doppler broadening and using the rational approximation for P_0. Goldstein and Cohen [28] introduced a variational procedure to interpolate between the two limiting cases. It seems, however, that the straight numerical procedure of reference [25] offers the best solution, since for quantitative calculations it will in any case be necessary to use computing machines. The general result is that the NR and IM methods generally bracket the true values, with NR giving lower ones, while consideration of interference scattering tends to diminish the difference.

1.7 Approximate Dependence on Geometry

The preceding formulas make it possible to obtain roughly the dependence of the resonance integrals on geometry and admixed scattering material. Inserting the expressions (1-9) and (1-10) into (1-48) and (1-49) and recalling (1-13) one obtains

$$I = \frac{\Gamma_\gamma}{2E_i}(\sigma_{p_0} + \sigma_m + \sigma_e)\int_{-\infty}^{+\infty}\frac{\psi\, dq}{\psi + \gamma\chi + \beta} \quad \text{(NR)}$$

$$I' = \frac{\Gamma}{2E_i}(\sigma_m + \sigma_e)\int_{-\infty}^{+\infty}\frac{\psi\, dq}{\psi + \beta'}, \quad \text{(IM)} \qquad (1\text{-}51)$$

where E_i is the energy at resonance and

$$\beta = \frac{\sigma_{p_0} + \sigma_m + \sigma_e}{\sigma_0}; \quad \beta' = \frac{(\sigma_m + \sigma_e)\Gamma}{\sigma_0\Gamma_\gamma};$$

$$\gamma = \left(g\frac{\Gamma_n}{\Gamma}\frac{\sigma_{p_0}}{\sigma_0}\right)^{1/2}. \qquad (1\text{-}52)$$

The function

$$J(E, \xi) = \frac{1}{2}\int_{-\infty}^{+\infty}\frac{\psi\, dq}{\psi + \beta} \qquad (1\text{-}53)$$

has been extensively discussed by Dresner [2] and Reichel [29] and it has been widely tabulated [2, 18, 30, 31]. Corrections for the interference term $\gamma\chi$ in (1-51) which are generally not too important have been computed by Rothenstein [21]. For $\sigma_m + \sigma_e \gg \sigma_0$, that is high dilution, $J \to \pi/2\,\beta$ (compare (1-14)) and both I and I′ tend to the infinite dilution limit

$$I_\infty = \frac{\pi}{2}\frac{\sigma_0\Gamma_\gamma}{E_0}. \qquad (1\text{-}54)$$

It can also be shown that J, and with it the resonance integrals, always increase with temperature [2]. The Doppler coefficient of reactivity is thus always negative. At finite dilution, the resonance integrals decrease with decreasing dilution and are always smaller than the limit (1-54). These characteristics should carry over to any more accurate treatment of resonance absorption.

Otherwise the function J is a rather complicated one. A simple result can only be obtained for the natural line form, that is T = 0, when $\psi = 1/(1 + x)^2$ and $\chi = 2x/(1 + x^2)$. One obtains then

$$I = I_\infty\sqrt{\frac{\beta}{1 + \beta - \gamma^2/\beta}}, \quad \text{(NR)}$$

$$I' = I_\infty\sqrt{\frac{\beta'}{1 + \beta'}}, \quad \text{(IM)} \qquad (1\text{-}55)$$

where all geometry dependence is contained in the factors under the square root sign. Now β or β' are generally quite small, since the cross sections at resonance, σ_0, are large. We thus have approximately

$$I \sim \sqrt{\beta} \sim (\sigma_{p_0} + \sigma_m + \sigma_e)^{1/2}$$

$$I' \sim \sqrt{\beta'} \sim (\sigma_m + \sigma_e)^{1/2}.$$

This suggests a dependence like $\sigma_m^{1/2}$ for homogeneous mixtures and, since $\sigma_e \sim S/VN \sim S/M$ the surface-to-mass ratio Eq. (1-40), it suggests the functional dependence on geometry

$$I \sim \sqrt{a + b(S/M)} \sim a' + b'\sqrt{S/M}, \qquad (1\text{-}56)$$

where a, b or a′, b′ are constant. This behavior has been verified well by experiment (see Sec. 1.9); the difference between the two forms being unobservable over the range of measurements. The square root dependence on S/M was first derived by Gourevitch and Pomeranchouk [32]. In American studies of the resonance integral, firstly a linear form was preferred [1] following the separation of volume and surface contributions, compare (1-45) and (1-47). However, this is not quite correct since the penetration factor G does introduce an explicit dependence on geometry. In view of the simplifications made it is clear that expressions like (1-56) will hold only approximately and far from the infinite dilution limit, and that the constants may be slightly different for different shapes.

No general statement can be made on the temperature dependence of the total resonance integrals, since many different resonances with widely different characteristics are involved. As will be seen later, both calculations [19, 33] and experiment lead to an approximate behavior like

$$I \cong I_0(1 + \beta\sqrt{T}), \qquad (1\text{-}57)$$

again, of course, only for values sufficiently away from the infinite dilution limit.

The spatial dependence of the resonance absorption is far from uniform. This is due to the neutrons which enter from the outside moderator and which are absorbed in a thin layer because of the large cross section at resonance. Wagner [34] has given a simple treatment, based on neglecting scattering in the absorber lump and Doppler broadening; he obtains quite good agreement with experimental data [35]. His main result is that, under his assumptions, the distribution in depth is essentially independent of the resonance parameters. For a plane surface this dependence is essentially $\sim d^{-1/2}$ (d = depth below surface) except for a very thin surface layer, For curved surfaces (cylinders, spheres) the behavior is similar, though described by somewhat more complicated functions. A study of the spatial distribution, both experimental and theoretical, with inclusion of the Doppler effect has been given by Pearce [36].

1.8 Closely Packed Assemblies

The previous developments were based on the assumption that the absorber elements are so widely spaced that they do not interact with each other. In particular, this means that they are at least about two moderator mean free paths apart.

The case when this condition is not fulfilled was first treated by Dancoff and Ginsburg [37]. When there are no neighboring absorbers then the flux through a surface element of one absorber comes from the whole surrounding moderator. If other absorbers are nearby (considered as black), they will shield part of the flux from the moderator. This effect can be calculated as follows. If the source strength, from slowing down in the moderator, is Q per unit volume and the macroscopic scattering cross section Σ (note that every collision changes the energy and is therefore counted as an "absorption"), then the infinite medium flux will be

$$\phi = Q/\Sigma . \qquad (1\text{-}58)$$

The current through a surface element dS (see Fig. 1-1) is then

$$dI = dS \cdot Q \int d\Omega \int_0^R r^2 dr \exp(-\Sigma r) \frac{\vec{n}\cdot\vec{\Omega}}{4\pi r^2} . \qquad (1\text{-}59)$$

Here $\vec{\Omega}$ is the solid angle, $\vec{n}$ the normal to dS. In this formula $d\Omega r^2\, dr$ is a volume element in the moderator, considered as source element, $dS\,(n\cdot\Omega)/4\pi r^2$ is the solid angle under which dS appears from the source element, and $\exp(-\Sigma r)$ the survival probability from source to surface. R, finally, is the distance to the next absorber element. The integration over r can be carried out:

$$dI = \frac{dS\cdot Q}{4\pi}\int d\Omega(\vec{n}\cdot\vec{\Omega})\int_0^R \exp(-\Sigma r)\, dr$$

$$= \frac{dS\cdot Q}{4\pi\Sigma}\int d\Omega(\vec{n}\cdot\vec{\Omega})[1-\exp(-\Sigma R(\Omega))] .$$

If the upper limit is $R = \infty$ (no adjacent absorber) one has with

$$\int d\Omega\,(\vec{n}\cdot\vec{\Omega}) = 2\pi\int \cos\theta \sin\theta\, d\theta = \pi$$

the well-known result

$$dI/dS = \phi/4 . \qquad (1\text{-}60)$$

In the general case the average flux into the absorber is

$$\frac{I}{S} = \frac{\phi}{4\pi S}\int dS\int d\ell(\vec{n}\cdot\vec{\Omega})\left(1-e^{-\Sigma R(\Omega)}\right) = \frac{\phi}{4}(1-C) , \qquad (1\text{-}61)$$

where the Dancoff correction

$$C = \frac{1}{\pi S}\iint e^{-\Sigma R}(\vec{n}\cdot\vec{\Omega})\, d\Omega dS$$

$$= \iint e^{-\Sigma R}(\vec{n}\cdot\vec{\Omega})\, d\Omega dS \Big/ \iint (\vec{n}\cdot\vec{\Omega})\, d\Omega dS$$

$$= \left\langle \exp(-\Sigma P)\right\rangle \qquad (1\text{-}62)$$

is the fractional reduction in flux through the surface of the absorber due to other absorbers, and equal to the average survival probability, $\langle \exp(-\Sigma R)\rangle$, over all moderator chords. In case of an array of absorbers, the total correction will be a sum over all other absorber parts directly visible from the absorber under consideration

$$C = \sum C_K . \qquad (1\text{-}63)$$

The above derivation assumed that the absorbers are entirely black to all neutrons, which will not be the case for thin elements, particularly in the wings of the resonances. The general case of gray absorbers has been treated by Rothenstein

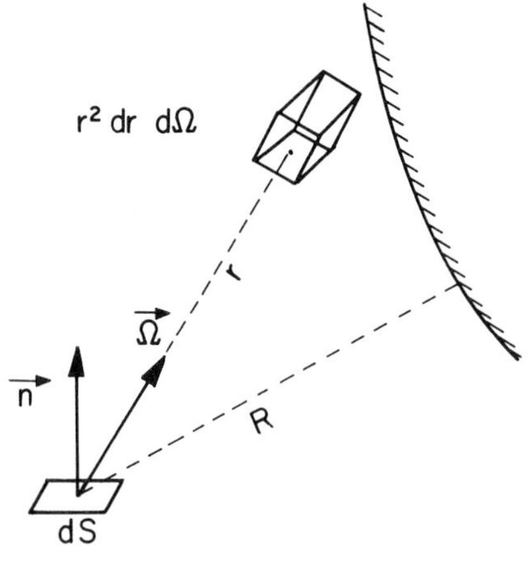

FIG. 1-1 Flux reduction through surface element dS by an adjacent absorber.

[21] and Nordheim [38], who observed that the problem can be solved in principle by redefining the escape probability P_0. This quantity was defined as the probability that a neutron, coming from a uniform source density in the absorber, escapes from it, which means that for large separation of the absorber lumps, it will make the next collision in the moderator. One can then define an effective escape probability P_0^*, for close assemblies, which means just the same, i.e., that a neutron born with a flat distribution in an absorber makes its next collision in the moderator, excluding the cases in which, after traversing one or more moderator sections, it collides inside another absorber element. It is clear that under the flat flux assumptions the general equations of Sec. 1.3 still hold when P_0 is replaced by P_0^*, and it remains only to find an appropriate expression for the latter.

The Dancoff correction depends only on geometry and the properties of the moderator, and not on any resonance parameters. It can, therefore, be calculated for a given reactor as a definite number. To do this, however, is generally a quite complicated procedure. The case of parallel cylinders was first treated in the original paper by Dancoff and Ginsburg [37]. Extensive tables for this case are given in the Argonne National Laboratory Reactor Constants Handbook [39]. General methods for computation have been described by Carlvik and Pershagen [40], who treat also parallel infinite or finite plates and holes of square and circular cross section filled with moderators. For the latter case it can be shown by transport theory [22] similar to above, that

$$P_o = \frac{1}{\bar{r}\Sigma}\left[1 - \langle \exp(-\Sigma R) \rangle\right] = \frac{1}{\bar{r}\Sigma}(1 - C)$$

or

$$C = 1 - \bar{r}\,\Sigma\,P_o\,, \qquad (1\text{-}64)$$

where now P_0 is the escape probability from the moderator-filled hole. This expression for C shows, of course, the correct limiting behavior $C \to 1$ for small holes (small $\bar{r}\Sigma$) and $C \leftarrow 0$ for large ones.

For practical applications of the Dancoff correction it is to be remarked that it is equivalent to a reduction in effective surface, that is, one has to make a replacement in empirical formulas like Eq. (1-56) of S by S (1 - C) or in calculations by introduction of

$$P_o^*(\bar{r}\Sigma) = P_o\left(\frac{\bar{r}\Sigma}{1 - C}\right) \qquad (1\text{-}65)$$

that is, a reduction of the mean chord length in the argument. This is satisfactory if the absorber dimensions are not too small. More generally, Nordheim [38] has derived an approximation for gray absorbers,

$$P_o^* = \frac{(1 - C)\,P_o}{1 - (1 - \bar{r}\,\Sigma\,P_o)\,C}\,, \qquad (1\text{-}66)$$

which gives the correct limit for highly transparent lumps, i.e., $\bar{r}\Sigma P_0 \to 0$, where the correction vanishes.

Relatively simple expressions for C and P_0^* can also be derived for slab geometry [21]. Recently, Fukai [41] has worked out the case of arbitrary arrangements of cylindrical fuel elements in arbitrary geometry and developed a machine program for this calculation. Among his results is that Eq. (1-66) approximates the true value for these geometries within about 1% for a wide range of parameters; an example is shown in Table 1-1. Fukai also has shown that differences between hexagonal and rectangular arrangements are generally quite small, as is to be expected. A relatively simple and quite accurate approximation for the generalized escape probabilities and the Dancoff correction has been given by Sauer [41a].

Experimental verifications of the Dancoff correction are rather difficult to obtain. Recently a good experimental confirmation has been obtained by Hardy, et al. [42].

A case where Eq. (1-66) is strictly correct is the one where the absorber is disposed in grains randomly embedded in a matrix of moderator [43] as it occurs in some modern reactor materials. Here the random arrangement effects the averaging used in the derivation of the above formula. Since the distance between grains may be very small, the Dancoff correction is very important. It is found to be

$$C = \frac{1}{1 + \bar{r}\,N_o\,\sigma_m}\,,$$

where $\bar{r}$ is the mean chord length of the grains, N_0 the density of the absorber nuclei in the grains and σ_m the moderator cross section per absorber nucleus. Insertion of this expression into Eq. (1-66) gives

$$P_o^* = P_o\,\frac{\sigma_m}{\sigma_m + P_o\,\sigma_{t_o}}\,, \qquad (1\text{-}67)$$

This formula has an immediate physical interpretation. $P_0\sigma_{t_0}$ is the absorber cross section corrected for self-shielding in the grains. In Eq. (1-67), therefore, the first factor is the escape probability from one definite grain, and the second factor gives the probability that the next collision is with a moderator nucleus. As will be seen later, the self-shielding in the grains decreases with increasing temperature, and the grain effect thus leads either to an increased Doppler coefficient for the same resonance absorption or to a decreased resonance absorption for the same Doppler coefficient.

A last remark may be made for more complicated geometries, and annular absorber elements, with moderator both inside and outside the elements with a different degree of transparency. It follows then from Eq. (1-62) that one has to take the surface weighted average, that is, to define an average Dancoff correction as

TABLE 1-1

Modified Escape Probability P_0^* According to Fukai and to Approximation Eq. (1-66). Rectangular Lattice, Diameter of Fuel Rods $\bar{r}$ = 0.93 cm. Equal Moderator and Fuel Volume, Moderator Cross Section $\Sigma_m = 1.4916$ cm^{-1}.

$\bar{r}\Sigma_0$	0.1	0.3	0.5	0.8	1	2	4	10
Fukai	0.807	0.580	0.447	0.330	0.279	0.1590	0.0804	0.0326
Eq. (1-66)	0.809	0.584	0.453	0.336	0.285	0.1576	0.0811	0.0327

$$\bar{C} = \frac{\sum_i C_i S_i}{\sum_i S_i}, \quad (1\text{-}68)$$

where the S_i are the different surface parts and the C_i their respective Dancoff corrections.

1.9 Results and Comparison with Experiment

The best available values of the resonance parameters are those for U^{238}. The ones used here are taken from BNL-325, Supplement I [44]. Other necessary parameters employed were:

Radiation width
$\Gamma_\gamma = 2.46 \times 10^{-2}$ ev,
Average reduced neutron width
$\langle \Gamma_n^0 \rangle = 1.74 \times 10^{-3}$ ev,
Average level spacing
D = 18.5 ev.

The infinite dilution value of the resonance integral with these parameters is 270 barns as compared to the best experimental value [45] of 280 ±10 barns.

The results are shown in Table 1-2 for both U metal and for UO_2. The table lists for selected rod sizes, i.e., S/M values, the contributions of the resolved resonances below 1000 ev and the unresolved s-wave resonances below 30 kv. The last line shows the values from empirical formulas which have been given by Hellstrand [46], derived from his measurements, for which an accuracy of ±2% is claimed. The line above gives the calculated values plus an additional contribution of 1.6 barns to account for the p-wave resonances above 1 kv and absorption above 30 kv. This value is entirely reasonable [24]. It will depend on the high energy spectrum of the reactor in which the measurements are made, and it is the only adjustable parameter in the calculation. Since there should be no appreciable self-shielding for these contributions, the proof of agreement lies in the fact, that for a consistent series of measurements a single value should be sufficient to give agreement for all rod sizes and both materials. As can be seen the agreement is excellent and in fact rather better than one would have expected from the uncertainities both in the resonance parameter and resonance integral data.

For UO_2 a line "oxygen correction" has been added. This shows the correction introduced by using the more accurate formula (1-35) in place of the approximation (1-37). The effect is small for thin rods and becomes barely noticeable for the larger thicknesses. It is due entirely to the low lying wide resonances. In fact, the 6.68 ev resonance contributes ~68% and the next two (21 and 36.9 ev) another 28% to the correction.

The results for U^{238} thus indicate that the method of calculation used [25] is entirely adequate and that the various remaining approximations do not invalidate the results.

With respect to the comparison between measured and calculated values of resonance integrals, it has to be remembered that the theoretical values refer to a 1/E spectrum, and also that the 1/v absorption in the epithermal region is not included. Actually the calculations for U^{238} were terminated at low energies, at 3/7 of the energy of the lowest resonance, i.e., at 2.86 ev.

The temperature dependence of the resonance integrals is rather difficult to measure. Older investigations obtained fairly widely divergent results as discussed by Pearce [47]. Calculations by

TABLE 1-2

Comparison of Calculated Values for the Resonance Integrals of U^{238} at 300°K with Hellstrand's Empirical Formulas

Metal, Density 18.7 g/cm^3					
Rod Radius (cm) →	0.1055	0.211	0.422	0.844	1.69
S/M	1.013	0.507	0.254	0.127	0.0634
Resolved	25.29	18.04	12.91	9.31	6.75
Unresolved	1.96	1.67	1.40	1.18	1.03
Total + 1.6 barn	28.85	21.31	15.91	12.09	9.38
$2.95 + 25.8\sqrt{S/M}$	28.91	21.28	15.95	12.18	9.45

UO_2, Density 10.2 g/cm^3					
Rod Radius (cm) →	0.125	0.25	0.50	1.0	2.0
S/M	1.570	0.785	0.393	0.196	0.098
Resolved	34.40	24.62	18.01	13.59	10.74
Unresolved	2.21	1.94	1.68	1.46	1.30
Oxygen Correction	-0.08	-0.14	-0.25	-0.43	-0.68
Total + 1.6 barn	38.13	28.02	21.04	16.22	12.96
$4.45 + 26.6\sqrt{S/M}$	37.76	27.95	21.10	16.23	12.76

Vernon [33] indicated that the resonance integrals are approximately linear functions of $\sqrt{T}$. Hellstrand et al. [48] analyzed their data in terms of best fit to the form

$$I(T) = I(300)\left[1 + \beta(\sqrt{T} - \sqrt{300})\right]. \quad (1\text{-}69)$$

Therefore theoretical values of

$$\beta = \frac{I(T) - I(300)}{I(300)(\sqrt{T} - \sqrt{300})}$$

have been calculated from the data of references [25, 38, 49]. They are shown in Table 1-3 for the various indicated temperature ranges. If the form (1-69) were exact, the values for one definite rod size should be the same for all temperature ranges. As a matter of fact they straggle slightly and in a rather unsystematic fashion. It must be said that there is no a priori reason that the temperature dependence should be like Eq. (1-69). From the data presented here, it appears, however, that this equation gives a good practical interpolation formula. It can, of course, be expected to hold only as long as one is far away from the infinite dilution case, where there is no temperature dependence left.

The values of the temperature coefficient β show further a definite dependence on S/M, which is again slightly irregular and definitely not linear.

Comparison with the measured values of Hellstrand [48] is shown in Table 1-4. The calculated values are graphically interpolated from Table 1-3 for the temperature ranges of the existing measurements, that is 300-600° K for the metal, and 300-1200° K for UO_2. There is agreement well within the uncertainities of the measurements, quoted as ±5%, with the calculated values possibly a shade lower than the experimental ones. It certainly can now be said that there is no discrepancy between theory and experiment. It also seems that the Doppler coefficient of reactivity (1/I) (dI/dT) can now be more easily calculated than measured. The tables presented here will make it possible to obtain a good estimate for all lattices of U metal and UO_2, when the proper value of the effective S/M ratio with inclusion of the Dancoff correction is used.

A further interesting question is the one as to the relative contributions of the individual resonances to the resonance integral and the Doppler coefficient. It is illustrated in Tables 1-5 (a) and (b) based on the data of reference [49]. These tables give for two rod sizes of UO_2 with radius 0.25 cm (rather thin) and 2 cm (very thick) the resonance integrals at 300° K and 1500° K, their absolute values and percentage difference for all resolved resonances, characterized by their energy and neutron width (Γ_γ = 0.0246 ev for all resonances). At the end the totals for all the resolved resonances, the unresolved s-wave resonances, and the grand total are also shown.

It is clear from the figures given that the greatest contribution to the absolute values of both resonance integrals and Doppler coefficient come from the low-lying resonances. The percentage changes vary all over the place and, contrary to statements sometimes made, there is no definite "Doppler region" which contributes preferentially to this effect. It is notable, however, that for the thick rods the contributions to the Doppler effect are shifted somewhat to higher energies as compared to the thin rod (i.e., higher dilution) case. It is also worth mentioning that the Doppler change for the resonances with small Γ_n is individually quite small. This is due to the fact that they are already strongly Doppler-broadened at 300° K and thus show very little self-shielding. This is the reason why p-wave resonances will have a negligible effect on the Doppler coefficient of thermal reactors.

The erratic behavior of the individual resonances makes it clear that simple emprical for-

TABLE 1-3

Calculated Values for the Temperature Coefficient $\beta \times 10^2$ for Uranium, See Eq. (1-69)

U Metal					
Radius (cm)	0.1055	0.211	0.422	0.844	1.69
300-600°K	0.930	0.731	0.613	0.544	0.561
300-900°K	0.961	0.743	0.609	0.528	0.513
Average	0.95	0.74	0.61	0.53	0.53

UO_2					
Radius (cm)	0.125	0.25	0.50	1.0	2.0
300-600°K	1.16	0.913	0.720	0.648	0.570
300-900°K	1.20	0.942	0.760	0.647	0.585
300-1200°K	1.22	0.956	0.762	0.636	0.565
300-1500°K	1.24	0.976	0.771	0.645	0.571
Average	1.21	0.94	0.75	0.64	0.57

TABLE 1-4

Comparison of Calculated Temperature Coefficient $\beta \times 10^2$ with Hellstrand's Measurements

Uranium Metal		
Rod radius (cm)→	0.4	1.4
Calculated	0.62	0.55
Observed	0.64	0.55

UO_2			
Rod radius (cm)→	0.4	0.85	1.4
Calculated	0.81	0.67	0.61
Observed	0.84	0.69	0.63

TABLE 1-5

Contributions of the Individual Resonances to the Resonance Integral and Doppler Effect in UO_2 Rods

(a) Radius 0.25 cm

E_o	Γ_n	I(300K)	I(1500K)	I(1500) - I(300)	$\frac{I(1500) - I(300)}{I(300)} \times 100$
6.7	1.52×10^{-3}	9.8385 + 00	1.1269 + 01	1.4305 + 00	14.54
21.0	8.90×10^{-3}	4.4471 + 00	5.3968 + 00	9.497 - 01	21.36
36.9	3.25×10^{-2}	3.3844 + 00	3.9424 + 00	5.580 - 01	16.49
66.3	2.50×10^{-2}	1.2879 + 00	1.8181 + 00	5.302 - 01	41.16
81.3	2.10×10^{-3}	5.3523 - 01	7.4206 - 01	2.068 - 01	38.64
90.0	9.00×10^{-5}	4.3099 - 02	4.4120 - 02	1.030 - 03	23.89
102.5	6.50×10^{-2}	8.4939 - 01	1.1072 + 00	2.578 - 01	30.35
116.5	1.50×10^{-2}	5.6863 - 01	8.8819 - 01	3.196 - 01	56.21
145.6	8.00×10^{-4}	1.2326 - 01	1.3547 - 01	1.221 - 02	9.90
165.2	3.50×10^{-3}	2.6370 - 01	3.3890 - 01	7.520 - 02	28.52
189.6	1.35×10^{-1}	3.6165 - 01	4.5597 - 01	9.432 - 02	26.08
208.5	5.50×10^{-2}	2.8934 - 01	4.4585 - 01	1.565 - 01	54.09
237.5	3.20×10^{-2}	2.5150 - 01	3.9055 - 01	1.391 - 01	55.31
264.5	2.30×10^{-4}	1.2951 - 02	1.3133 - 02	1.820 - 04	1.41
274.0	2.70×10^{-2}	2.1444 - 01	3.2909 - 01	1.1465 - 01	53.48
291.0	1.90×10^{-2}	2.0022 - 01	2.9420 - 01	9.3980 - 02	46.99
311.5	10.00×10^{-4}	3.7150 - 02	3.8830 - 02	1.680 - 03	4.52
348.0	4.50×10^{-2}	1.5441 - 01	2.3659 - 01	8.218 - 02	53.22
377.0	1.50×10^{-3}	3.6930 - 02	3.8820 - 02	1.890 - 03	5.12
398.5	10.00×10^{-3}	1.1571 - 01	1.4450 - 01	2.879 - 02	24.88
411.0	1.70×10^{-2}	1.2697 - 01	1.7153 - 01	4.456 - 02	35.09
435.0	1.40×10^{-2}	1.1303 - 01	1.4581 - 01	3.278 - 02	29.00
455.0	7.00×10^{-4}	1.2954 - 02	1.3203 - 02	2.490 - 04	1.92
464.0	7.00×10^{-3}	7.7703 - 02	8.9771 - 02	1.2068 - 02	15.53
479.0	4.50×10^{-3}	5.6153 - 02	6.1855 - 02	5.702 - 03	10.15
490.0	10.00×10^{-4}	1.5622 - 02	1.6003 - 02	3.81 - 04	2.44
519.0	3.70×10^{-2}	9.7786 - 02	1.3888 - 01	4.109 - 02	42.02
536.0	5.40×10^{-2}	8.9564 - 02	1.3164 - 01	4.208 - 02	46.99
557.0	10.00×10^{-4}	1.2181 - 02	1.2430 - 02	2.49 - 04	2.04
581.0	4.20×10^{-2}	8.3941 - 02	1.1827 - 01	3.433 - 02	40.90
596.0	6.60×10^{-2}	7.5844 - 02	1.1126 - 01	3.542 - 02	46.70
605.0	6.00×10^{-4}	6.3999 - 03	6.4722 - 03	7.20 - 05	1.13
621.0	3.90×10^{-2}	7.7541 - 02	1.0657 - 01	2.903 - 02	37.44
629.0	9.00×10^{-3}	5.3524 - 02	6.0533 - 02	7.009 - 03	13.09
662.0	1.25×10^{-1}	5.7881 - 02	8.7063 - 02	2.918 - 02	50.41
680.0	1.30×10^{-3}	1.0515 - 02	1.0723 - 02	2.080 - 04	1.98
695.0	5.30×10^{-2}	6.5087 - 02	9.0827 - 02	2.574 - 02	39.55
710.0	1.70×10^{-2}	5.7453 - 02	6.8466 - 02	1.101 - 02	19.16
723.0	1.47×10^{-2}	5.3097 - 02	6.1925 - 02	8.828 - 03	16.63

TABLE 1-5 (Continued)

(a) Radius 0.25 cm (Continued)

E_o	Γ_n	I(300K)	I(1500K)	I(1500) - I(300)	$\frac{I(1500) - I(300)}{I(300)} \times 100$
732.0	4.25×10^{-3}	2.4990 - 02	2.6354 - 02	1.364 - 03	5.46
766.0	9.00×10^{-3}	3.8110 - 02	4.1972 - 02	3.862 - 03	10.13
782.0	3.00×10^{-3}	1.6719 - 02	1.7317 - 02	5.980 - 04	3.58
792.0	1.10×10^{-2}	4.0069 - 02	4.4707 - 02	4.638 - 03	11.57
825.0	6.00×10^{-2}	5.1249 - 02	6.9964 - 02	1.872 - 02	36.53
855.0	1.30×10^{-1}	4.1414 - 02	6.1081 - 02	1.967 - 02	47.49
859.0	6.00×10^{-2}	4.8604 - 02	6.5740 - 02	1.714 - 02	35.27
867.0	2.20×10^{-3}	1.0514 - 02	1.0758 - 02	2.44 - 04	2.32
896.0	1.30×10^{-3}	6.1270 - 03	6.2150 - 03	8.80 - 05	1.44
909.0	9.00×10^{-2}	4.1954 - 02	5.8701 - 02	1.675 - 02	39.92
928.0	3.70×10^{-2}	4.4108 - 02	5.5265 - 02	1.1157 - 02	25.29
940.0	1.95×10^{-1}	3.3265 - 02	4.8304 - 02	1.5039 - 02	45.21
960.0	1.90×10^{-1}	3.2546 - 02	4.7286 - 02	1.474 - 02	45.28
983.0	10.00×10^{-4}	3.9930 - 03	4.0300 - 03	3.70 - 05	0.927
995.0	4.00×10^{-1}	2.7706 - 02	3.6362 - 02	8.656 - 03	31.24
	Resolved Total	2.4620 + 01	3.0167 + 01	5.547 + 00	22.53
	Unresolved	1.9384 + 00	2.2729 + 00	3.345 - 01	17.26
	Total	2.6558 + 01	3.2439 + 01	5.882 + 00	22.15

(b) Radius 2.0 cm

E_o	Γ_n	I(300K)	I(1500K)	I(1500) - I(300)	$\frac{I(1500) - I(300)}{I(300)} \times 100$
6.7	1.52×10^{-3}	4.0102 + 00	4.1601 + 00	1.50 - 01	3.74
21.0	8.90×10^{-3}	1.8598 + 00	1.9665 + 00	1.07 - 01	5.74
36.9	3.25×10^{-2}	1.5540 + 00	1.6180 + 00	6.40 - 02	4.12
66.3	2.50×10^{-2}	6.0299 - 01	6.8953 - 01	8.65 - 02	14.35
81.3	2.10×10^{-3}	2.0839 - 01	3.0601 - 01	9.76 - 02	46.84
90.0	9.00×10^{-5}	3.6554 - 02	4.0298 - 02	3.74 - 03	10.24
102.5	6.50×10^{-2}	4.5584 - 01	4.9687 - 01	4.10 - 02	9.00
116.5	1.50×10^{-2}	2.3161 - 01	3.0366 - 01	7.21 - 02	31.11
145.6	8.00×10^{-4}	7.5387 - 02	9.9145 - 02	2.38 - 02	31.57
165.2	3.50×10^{-3}	1.1210 - 01	1.7079 - 01	5.87 - 02	52.36
189.6	1.35×10^{-1}	2.1267 - 01	2.2862 - 01	1.60 - 02	7.50
208.5	5.50×10^{-2}	1.3954 - 01	1.8130 - 01	4.18 - 02	29.93
237.5	3.20×10^{-2}	1.0442 - 01	1.4385 - 01	3.94 - 02	37.76
264.5	2.30×10^{-4}	1.1822 - 02	1.2545 - 02	7.23 - 04	6.12
274.0	2.70×10^{-2}	8.5793 - 02	1.2639 - 01	4.06 - 02	47.32
291.0	1.90×10^{-2}	7.9276 - 02	1.2180 - 01	4.25 - 02	53.63
311.5	10.00×10^{-4}	2.8552 - 02	3.3689 - 02	5.14 - 03	17.99
348.0	4.50×10^{-2}	6.4017 - 02	9.0899 - 02	2.69 - 02	41.99
377.0	1.50×10^{-3}	2.7666 - 02	3.3189 - 02	5.52 - 03	19.96
398.5	10.00×10^{-3}	5.2681 - 02	8.1461 - 02	2.88 - 02	54.63

TABLE 1-5 (Continued)

(b) Radius 2.0 cm (Continued)

E_o	Γ_n	I(300K)	I(1500K)	I(1500) - I(300)	$\frac{I(1500) - I(300)}{I(300)} \times 100$
411.0	1.70×10^{-2}	5.2273 - 02	8.4119 - 02	3.18 - 02	60.93
435.0	1.40×10^{-2}	4.9150 - 02	7.7585 - 02	2.84 - 02	57.85
455.0	7.00×10^{-4}	1.1485 - 02	1.2433 - 02	9.48 - 04	8.25
464.0	7.00×10^{-3}	4.1663 - 02	5.9986 - 02	1.18 - 02	43.98
479.0	4.50×10^{-3}	3.4779 - 02	4.6540 - 02	1.83 - 02	33.82
490.0	10.00×10^{-4}	1.3451 - 02	1.4839 - 02	1.39 - 03	10.32
519.0	3.70×10^{-2}	4.0089 - 02	6.1676 - 02	2.16 - 02	53.85
536.0	5.40×10^{-2}	3.6887 - 02	5.4872 - 02	1.80 - 02	48.75
557.0	10.00×10^{-4}	1.0736 - 02	1.1672 - 02	9.36 - 04	8.72
581.0	4.20×10^{-2}	3.4691 - 02	5.3263 - 02	1.86 - 02	53.54
596.0	6.60×10^{-2}	3.1613 - 02	4.6304 - 02	1.47 - 02	46.48
605.0	6.00×10^{-4}	5.9643 - 03	6.2550 - 03	2.91 - 04	4.88
621.0	3.90×10^{-2}	3.2478 - 02	5.0149 - 02	1.77 - 02	54.41
629.0	9.00×10^{-3}	3.0557 - 02	4.2817 - 02	1.23 - 02	40.12
662.0	1.25×10^{-1}	2.6243 - 02	3.7425 - 02	1.12 - 02	42.61
680.0	1.30×10^{-3}	9.3051 - 03	1.0094 - 02	7.89 - 04	8.47
695.0	5.30×10^{-2}	2.4610 - 02	4.1513 - 02	1.69 - 02	68.68
710.0	1.70×10^{-2}	2.8821 - 02	4.3100 - 02	1.43 - 02	49.55
723.0	1.47×10^{-2}	2.7970 - 02	4.0864 - 02	1.29 - 02	46.10
732.0	4.25×10^{-3}	1.8540 - 02	2.2464 - 02	3.92 - 03	21.17
766.0	9.00×10^{-3}	2.3736 - 02	3.1817 - 02	8.08 - 03	34.04
782.0	3.00×10^{-3}	1.3553 - 02	1.5552 - 02	2.00 - 03	14.75
792.0	1.10×10^{-2}	2.3907 - 02	3.2812 - 02	8.91 - 03	37.24
825.0	6.00×10^{-2}	2.1825 - 02	3.3190 - 02	1.14 - 02	52.06
855.0	1.30×10^{-1}	1.8206 - 02	2.6998 - 02	8.79 - 03	48.28
859.0	6.00×10^{-2}	2.0848 - 02	3.1701 - 02	1.09 - 02	52.05
867.0	2.20×10^{-3}	9.1295 - 03	1.0032 - 02	9.02 - 04	9.88
896.0	1.30×10^{-3}	5.6344 - 03	5.9643 - 03	3.30 - 04	5.86
909.0	9.00×10^{-2}	1.7898 - 02	2.6354 - 02	8.46 - 03	47.24
928.0	3.70×10^{-2}	2.0251 - 02	3.1454 - 02	1.120 - 02	55.32
940.0	1.95×10^{-1}	1.5588 - 02	2.1392 - 02	5.80 - 03	37.23
960.0	1.90×10^{-1}	1.5067 - 02	2.0967 - 02	5.90 - 03	39.15
983.0	10.00×10^{-4}	3.7697 - 03	3.9212 - 03	1.52 - 04	4.02
995.0	4.00×10^{-1}	1.5385 - 02	1.7434 - 02	2.05 - 03	13.31
	Resolved Total	1.0739 + 01	1.2032 + 01	1.293 + 00	12.04
	Unresolved	1.2979 + 00	1.6714 + 00	3.735 - 01	28.78
	Total	1.2037 + 01	1.3703 + 01	1.666 + 00	13.84

TABLE 1-6

Resonance Parameters for Th^{232}

E_o(ev)	Γ_n(mv) Harwell	Γ_n(mv) Columbia	E_o(ev)	Γ_n(mv) Harwell	Γ_n(mv) Columbia
21.7	1.68 ± 0.15	2.2 ± 0.3	665	26 ± 5	17 ± 4
23.4	3.1 ± 0.2	4.0 ± 0.4	675	195 ± 23	125 ± 20
59.3	3.1 ± 0.3	4.6 ± 0.7	687	55 ± 10	38 ± 8
69	39 ± 5	42 ± 8	701	18	18 ± 4
113	14.5 ± 3	11 ± 1	713	28	25 ± 5
121	24 ± 3	18.5 ± 2	741	196 ± 20	160 ± 20
129	3.5 ± 0.4	3.1 ± 0.3	778	10	10 ± 3
146		0.1 ± 0.03	804	205 ± 25	125 ± 20
154		0.22 ± 0.04	866	13	10 ± 3
170	60 ± 6	58 ± 4	890	32	30 ± 8
193	13.5 ± 2	14 ± 2	942	42	40 ± 10
199	7.2 ± 2	9 ± 2	962	7	7.5 ± 3
221	21.5 ± 4	23 ± 3	982	25	24 ± 8
251	24.5 ± 5	24 ± 3	998	90	70 ± 20
263	23 ± 5	15 ± 3	1011	100 ± 15	
306	not given	20 ± 4	1039	15 ± 3	
330	56 ± 8	55 ± 8	1065	10	
342	33 ± 5	30 ± 5	1076	13	
365	27 ± 5	35 ± 5	1091		
369	29 ± 4	24 ± 5	1109	31 ± 10	
401	10.5 ± 2	9 ± 2	1114		
420	0.5 ± 0.3		1119		
454	3.0 ± 1.5		1138	20	
463	59 ± 10	45 ± 9	1150	23	
489	50 ± 7	50 ± 10	1193	10 ± 4	
510	5 ± 1	3.5 ± 0.7	1203		
528	13 ± 2	15 ± 3	1227	42 ± 10	
540	0.8 ± 0.4	1 ± 0.4	1241	18 ± 5	
570	38	25 ± 4	1248	145 ± 25	
578	2.7 ± 0.7	1.6 ± 0.5	1269	23 ± 4	
598	11 ± 2	11 ± 2	1292	70 ± 20	
618	4 ± 1	5 ± 1	1300	47 ± 10	
656	37 ± 8	32 ± 6			

mulas like (1-69) are the result of a complicated statistics and can have only limited accuracy and validity. It is also not surprising that the functional dependence of the resonance integrals on T and on S/M will not be quite smooth, since the number of contributing resonances is far from infinite.

The second important resonance absorber is Th^{232}. Here the situation is somewhat less clear due to uncertainties in the values of the resonance parameters themselves. In 1960 the Columbia University cyclotron group [50] issued a list of parameters which gave resonance integrals which were about 10% too high [19]. While this is not a large discrepancy, it is outside the expected limits of errors. Meanwhile, the resonances are being remeasured by groups at Harwell, Columbia University, and General Atomic. Preliminary results are not yet in agreement and no final results can, therefore, be reported.

In order to give a review of the present status, some results of calculations by Nordheim [51] are reproduced here. Table 1-6 gives the values of the resonance energies, on which all observers are in good agreement and two values of neutron widths, one by Columbia [50] and one by Harwell group [52]. It is to be noted that the Harwell neutron widths for the lowest resonances are definitely smaller than the Columbia values. With respect to the radiation width, Columbia gives an average of 40 millivolt. The Harwell groups give smaller values with quotations between 18 and 30 mv. In this connection it is to be noted that the theoretical expectation is for very little variation of Γ_γ from level to level, so that in view of the experimental uncertainties in the determination of this quantity it is better to use an average from many levels than to try to assign separate values to the individual resonances. A first check is furnished by the infinite dilution resonance integral. Recent measurements [53] give values of 82± 3 barns. The calculations gave values shown in Table 1-7 under various assumptions for Γ_γ as indicated. It is seen that the Columbia set gives a definitely too high value, the Harwell 18 a too low one, while Harwell 30 and 40 are compatible with the experimental results. A value of Γ_γ = 34 mv would give about the best fit. The difference of 7 barns between Columbia and Harwell for 40 mv = Γ_γ is entirely due to the neutron widths in the lowest resonances; this represents a fairly strong argument for smaller Γ_n there.

The resonance integrals for finite rod sizes of Th metal and ThO_2 are shown in Table 1-8. The formulas on the next to the last line are those given by the latest measurements by Hellstrand and by Weitman [54, 55]. These authors list possible errors as ±5%. A value of 1.9 barns has been subtracted for the p-waves and high energy absorption to give the best over-all fit. It is seen that the Columbia set gives definitely too high values. Harwell Γ_γ = 18 mv gives much too low values and can definitely be ruled out. Harwell 30 and 40 bracket the measured values, while the interpolation Harwell 34 shows an excellent fit. This parameter set reproduces thus all measured resonance integral values and is thus satisfactory from a practical point of view. It is, of course, somewhat dangerous to make deductions on cross sections from resonance integral calculations. However, two features seem to stand out. As compared to the original Columbia set, the neutron width for the lowest resonances have to be reduced somewhat to account for the infinite dilution value while the Γ_γ values have to be reduced somewhat to account for the finite rod sizes. Of course, Γ_n and Γ_γ values can be traded for each other to some extent to give the same resonance integrals. These calculations should, therefore, be repeated when consensus on

TABLE 1-7

Calculated Values for the Infinite Dilution Resonance Integral for Th^{232} for Various Parameter Sets

Parameters	Columbia	Harwell			
Γ_γ(mv)	40	18	30	40	34
I_∞^* (barn)	93.6	67.8	79.7	86.4	82.6

the values of the resonance parameters has been achieved.

The temperature coefficient β for the metal and oxide is shown in Table 1-9. It has been calculated on basis of the Columbia parameter set, but it should not depend strongly on the choice of parameters. Again the form of Eq. (1-69) holds quite well over the calculated temperature range. The coefficient is slightly larger than for uranium. This is due to the absence of a low lying resonance, such as the 0.68 ev resonance in U^{238}, which contributed a large fraction to the resonance integral but somewhat less to the Doppler effect.

Measurements of the thorium resonance integrals and their Doppler coefficients have been made recently in the HTGR critical assembly [56]. The thorium was mixed intimately with graphite and disposed in annuli with inner diameter of 1.5 in. (3.81 cm) and outer diameter of 2.5 in. (6.35 cm) with various C/Th atom ratios. The comparison for the resonance integral itself is

TABLE 1-8

Calculated Values for the Resonance Integrals for Th^{232} at 300°K for Various Parameter Sets

Thorium Metal, Density 11.7 g/cm^3

Rod Radius (cm)	0.138	0.275	0.55	1.1	2.2
S/M	1.24	0.623	0.311	0.155	0.078
Columbia	21.2	15.2	11.1	8.3	6.5
Harwell 40	21.0	15.2	11.1	8.3	6.4
Harwell 30	18.5	13.3	9.7	7.3	5.6
Harwell 18	14.7	10.5	7.5	5.6	4.4
$1.4 + 16.1\sqrt{S/M}$	19.3	14.2	10.4	7.8	5.9
Harwell 34	19.5	14.1	10.3	7.8	5.9

Thorium Oxide, Density 7.1 g/cm^3

Rod Radius (cm)	0.25	0.50	1.0	2.0
S/M	1.125	0.564	0.282	0.141
Columbia	22.1	16.2	12.2	9.7
Harwell 40	21.7	16.0	12.2	9.6
Harwell 30	19.2	14.1	10.6	8.4
Harwell 18	15.3	11.1	8.3	6.6
$3.1 + 15.6\sqrt{S/M}$	20.6	14.8	11.6	9.0
Harwell 34	20.3	15.0	11.3	9.0

TABLE 1-9

Calculated Values for the Temperature Coefficient $\beta \times 10^2$ for Thorium

Th Metal

Radius (cm)→	0.130	0.275	0.55	1.1	2.2
300 - 600°K	1.679	1.509	1.303	1.080	0.923
300 - 900°K	1.637	1.475	1.252	0.996	0.856

ThO_2

Radius (cm)→	0.25	0.5	1.0	2.0
300 - 600°K	1.684	1.536	1.344	1.182
300 - 900°K	1.632	1.499	1.311	1.143

shown in Table 1-10. The cross sections used were the Columbia parameters, slightly modified for the lowest resonances to reduce the infinite dilution value. The agreement is fairly good considering the uncertainities in the resonance parameters and the complicated geometry involved. The measured and calculated Doppler coefficients are shown in Fig. 1-2. Again the agreement can be considered as satisfactory, since the rather difficult measurements themselves have an uncertainty of ~± 10%. An interesting feature is in the temperature dependence of the Doppler effect. While above 500° K the effect goes down with T as expected, it is less T dependent below. This behavior is quite likely real and due to the crystalline binding effect [8] which, at temperatures below the Debye temperature, will increase the effective kinetic energy above the Maxwellian kT due to the zero energy motion.

A final result worth mentioning is the change in resonance integral and Doppler coefficient when the absorber is contained in grains of finite size embedded in a matrix of moderator, as discussed in Sec. 1.8 and reference [43]. Calculations have been made for grains of ThO_2 (density = 8 g/cm^3) in graphite. The value of the effective moderator scattering cross section was σ_m = 1011 barns. Figure 1-3 shows the ratio of the resonance integral for finite grain size to the one with homogeneous distribution, i.e., the self-shielding factor as function of grain diameter D for two different temperatures. The effect is nearly linear in the grain diameter. The shielding decreases with temperature, as was to be expected. It amounts to about 15% for grains of D = 0.06 cm at 300° K.

TABLE 1-10

Resonance Integral for Th^{232} in HTGR Annular Fuel Elements

System	I (Measured) barns	I (Calculated) barns
Infinite dilution	79	79.6
C/Th = 55.7	35.2	34.4
C/Th = 48.7	33.4	31.7

TABLE 1-11

Temperature Coefficients dI/dT and (1/I)dI/dT as Function of Temperature and Grain Diameter D

Temperature	D = 0 cm		D = 0.0237 cm		D = 0.0600 cm	
	dI/dT	(1/I)dI/dT	dI/dT	(1/I)dI/dT	dI/dT	(1/I)dI/dT
(°K)	(barn/°K)	(10^{-5}/°K)	(barn/°K)	(10^{-5}/°K)	(barn/°K)	(10^{-5}/°K)
450	0.0141	31.8	0.0141	33.9	0.0140	36.7
750	0.00882	18.5	0.00892	19.8	0.00908	21.8
1050	0.00641	12.8	0.00661	13.9	0.00671	15.3
1350	0.00501	9.70	0.00516	10.5	0.00533	11.7
1650	0.00409	7.73	0.00424	8.39	0.00440	9.35

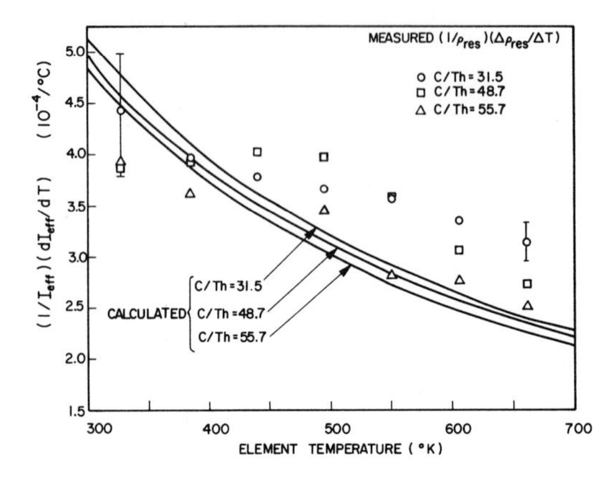

FIG. 1-2 Doppler coefficient for small high temperature gas-cooled reactor with annular fuel elements having several different C/Th ratios.

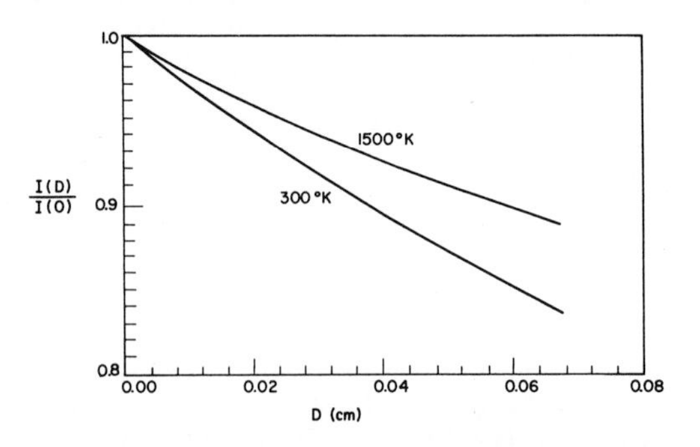

FIG. 1-3 Self-shielding factors as function of grain diameter.

Table 1-11 [43] shows the temperature coefficient of the resonance integral dI/dT and of the reactivity (1/I) dI/dT. It is seen that dI/dT itself does not change much. It increases slightly with grain size at higher T. However, it is to be noticed that the values of I decrease with increasing grain size. One can, therefore, either have the same Doppler coefficient with reduced resonance absorption or with the same resonance absorption (i.e., heavier loading) a larger Doppler coefficient, as measured by (1/I) dI/dT. This is shown in Fig. 1-4 which gives the ratio of the Doppler coefficient in the grain-structured material to the homogenized one for two grain sizes. It is seen that the effect is from 16 to 21% for the larger grain size, by no means a negligible effect.

2 FAST REACTORS

2.1 General Considerations

While the state of our knowledge on resonance absorption and Doppler effect for thermal reactors is satisfactory and existing methods for calculation have been quite successful, the situation with respect to fast and intermediate reactors is much more open and subject to uncertainties. The reasons for this are easy to see.

For well-moderated reactors the neutron spectrum does not deviate much from 1/E. While it is necessary, of course, to take deviations from this behavior into account, as it is done in modern multigroup methods, the general features, nevertheless, remain much the same. In particular, it was seen that the contributions to Doppler effect

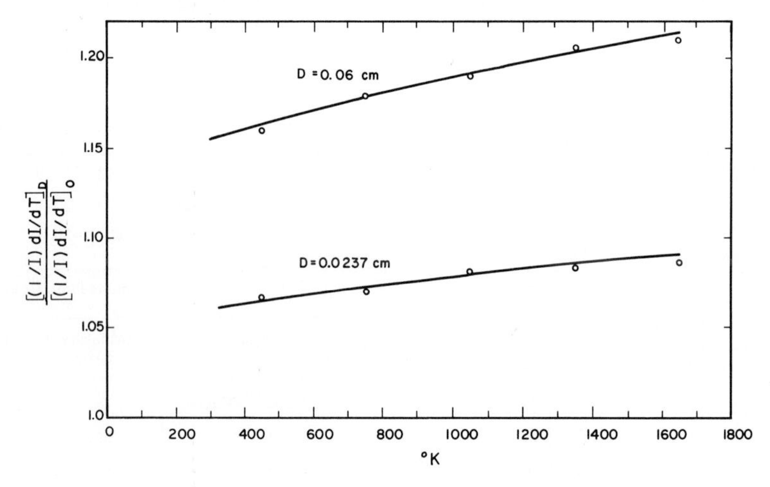

FIG. 1-4 Increase in temperature coefficient at constant resonance absorption for two grain diameters.

and resonance absorption decrease rapidly with increasing energy, and it is the low-lying and well-measured resonances which contribute most. In fast reactors the spectrum is entirely different. Again, as will be seen, relatively low energy resonances contribute most, but they cannot do so if the neutron spectrum there is much depleted. There is, therefore, a critical balance between the spectrum and contributions to the Doppler effect. Because of this any calculation of the latter demands a very accurate description of the neutron energy distribution.

In thermal reactors the contributions of the fissile isotopes to the Doppler coefficient are generally very small and are masked by other larger temperature coefficients. This is due to the neutron width being small compared to the total width (including fission and radiation) of the resonances. At higher energies, however, this is not guaranteed and, since all other temperature effects also become smaller, it is necessary to take the fissile isotopes into account, the more so since the ratio of fissile to fertile isotopes will be generally much larger for fast than for thermal assemblies.

The resonance parameters themselves, both for fissile and fertile nuclides, are generally well-known only for low neutron energies. It is necessary, therefore, to use statistical treatments to a large extent and to extrapolate the statistics of resonances to higher energies. This introduces a considerable amount of guess work, in particular with respect to p-waves, about which not much solid information is available.

The level spacing between resonances decreases generally with energy. It is smaller for p-waves than for s-waves, and much smaller for the fissile nuclides than for the fertile ones. Therefore, interference effect of two kinds may play a role. First, there may be interference between neighboring resonances in the same species, and second, if there is a mixture of isotopes, there may be interference between resonances of different species.

On the other hand, there are also some simplifying features. Heterogeneities will play a much smaller role. This is due to the fact that the cross sections themselves are smaller, and thus less self-shielded, and also to the fact that the composition of fast reactors is generally more homogeneous. Therefore, the treatment here will be restricted to the homogeneous case, and it will be assumed that heterogeneities can be handled adequately by the equivalence relations as described in Sec. 1.5. Also in general, it will be sufficient to use the narrow resonance approximation, since even for uranium the maximum energy loss in elastic collisions is large compared to the width of the resonances above 200 ev. This means that the flux distribution over energy intervals comprising several resonances, but small compared to large changes will be simply

$$\phi(E) \sim 1/\sigma_t , \tag{2-1}$$

where σ_t is the total cross section.

The effective partial cross section for a process x (absorption or fission) is thus

$$\tilde{\sigma}_x = \frac{\int_{\Delta E} (\sigma_x/\sigma_t)\, dE}{\int_{\Delta E} (1/\sigma_t)\, dE} , \tag{2-2}$$

where the interval ΔE is wide compared to the width of individual resonances and comprises many of them, but is small compared to regions where the average flux varies considerably. In the spirit of this approximation all slowly varying functions of E such as 1/E or $\sqrt{E}$ will be replaced by their average values.

For the cross sections, superpositions of generalized resonance cross section, such as Eqs. (1-17) and (1-18), should be taken. In addition to the resonance parts, there will be a nonresonant part σ_n contributed by the cross sections of all isotopes present, including nonresonant admixtures. One can then write as a general formula for the effective cross section

$$\tilde{\sigma}_x = \left\langle \sum_k \int_{\Delta E} \frac{\sigma_{kx}\, dE}{\sigma_n + \sigma_{kr} + \sum_{k' \neq k} \sigma_{k'r}} \right\rangle \Big/ \left\langle \int_{\Delta E} \frac{dE}{\sigma_n + \sum_k \sigma_{kr}} \right\rangle , \tag{2-3}$$

where the sums are over all resonances of all isotopes in ΔE and where the microscopic cross sections of different isotopes have to be weighted with their respective number ratios. The averaging symbols mean that an average has to be taken over both the distribution of resonance parameters and the distribution of level spacings. Of course, if in a given interval ΔE the resonances are known, the sums should be taken over these resonances in place of averaging.

The statistics of these parameters will be discussed in the next section. But even if this is known, the evaluation of an expression as complicated as Eq. (2-3) demands further simplifications and approximations. Here two limiting cases appear as natural starting points.

On the one side, there may be regions where the resonances can still be considered as well separated. Since the cross sections drop off sharply away from the resonances, this means that the cross sections due to other resonances are negligible in the neighborhood of a given one. The expression Eq. (2-3) reduces then to

$$\tilde{\sigma} = \left\langle \sum_k \left(\int_{\Delta_{res}} \frac{\sigma_{kx}\, dE}{\sigma_n + \sigma_{kr}} \Big/ \int_{\Delta_{res}} \frac{dE}{\sigma_n + \sigma_{kr}} \right) \right\rangle , \tag{2-4}$$

where the symbol Δ_{res} means that the integral is to be taken over an interval just containing one single resonance. This approximation, generally denoted as method B, will be treated in Sec. 2.3. It is the same as the volume term of the narrow resonance approximation, discussed in Sec. 1.6. As will be seen in the next section it is applicable to

U^{238} up to energies of 30-50 kv and fissile materials below 1 kv.

The other limiting case occurs when the resonances are largely washed out by the Doppler effect and by overlapping. One can then develop the denominators around an average cross section. This has been accomplished in a remarkable paper by Feshbach, Goertzel, and Yamauchi [57], which will be discussed in Sec. 2.4. Their result is generally denoted as method A. It is sufficient for all isotopes above ~ 30 kv, and thus for very fast assemblies containing not much diluent of the fissile materials. It will remain applicable for fissile materials to energies as low as ~ 1 kv.

Nicholson [58] has observed that application of method A below and of method B above their range of validities gives much too large values for the cross sections. He suggested, therefore, that a suitable point of distinction would be the energy at which the two methods give the same values. This is clearly not too satisfactory. Many fast reactors now under consideration contain diluents, such as oxides, and generous amounts of sodium. Their spectra tail down to fairly low energies where the resonances of the fissile isotopes are neither widely separated, nor entirely washed out, and where well-separated and large resonances of U^{238} are superposed. There will, thus, be interference effects as already mentioned, which require new methods.

The natural way to bridge the gap between the two limiting methods A and B consists in the "nearest neighbor" method in which only the resonance k' nearest to k is taken into account in the general formula Eq. (2-3). This method and another alternate development will be discussed shortly in Sec. 2.6. The whole subject of the Doppler effect in fast reactors has been discussed recently at a conference at the Argonne National Laboratory [17]. An attempt will be made to incorporate some of the results presented there, though only preliminary reports are available.

2.1.1 Doppler Width and Level Spacings

The key to the discussion of the various approximations is furnished by the relative magnitudes of Doppler width, level spacings, and natural widths of the resonance lines. The Doppler width has already been given in Eq. (1-13a) as

$$\Delta = \sqrt{4EkT/A}\,, \qquad (2\text{-}5)$$

which for the heavy isotopes is $\sim 1.2 \times 10^{-3}\sqrt{ET}$, where E is the energy in ev and T the temperature in°K. Then for E = 1000 ev and T = 1000°K, Δ = 1.2 ev.

The average level spacings D_s for all s-waves, as observed at low energies are given in Table 2-1 (from a recent survey by Garrison [59]). The average spacing for different channels and for higher ℓ values can be obtained from D_s from the assumption that the level density for a given compound nucleus is proportional* to (2J + 1). With

TABLE 2-1

Average Level Spacings for Various Isotopes

Isotope	Average D_s (ev)
U^{238}	18.5 ± 2
Pu^{240}	11 ± 2
Pu^{239}	2.7 ± 0.4
U^{233}	0.8 ± 0.2
U^{235}	0.7 ± 0.15
Pu^{241}	1.2 ± 0.3

the help of the statistical factor g, Eq. (1-2) and the relations

$$\sum_J g_J = 2\ell + 1 \quad \text{for given } \ell, \qquad (2\text{-}6)$$

the spacing for any channel J can then be expressed as

$$D_J = D_s/g_J\,. \qquad (2\text{-}7)$$

For instance, for U^{235} with I = 7/2 the spacings for s-waves and channels J = 3 and 4 are

$$D_3 = \frac{16}{7} D_s\,, \quad D_4 = \frac{16}{9} D_s\,,$$

and for all p-waves for any nucleus

$$D_p = D_s/3\,. \qquad (2\text{-}8)$$

It is thus seen that for U^{235} the Doppler width becomes comparable to level spacing already at ~1000 ev, while for U^{238} this washing-out of the cross sections occurs only at much higher energies. On the other hand, it is to be emphasized that the large Doppler width will not invalidate the resonance structure of the cross sections, since this broadening is produced by superposition of the effects of different nuclei with different velocities. A neutron sees an individual nucleus with a well-defined relative energy and thus with the same sharp resonance structure as a nucleus at rest.**

Of course, the assumption that the total cross sections can be considered as the superposition of independent resonances is justified only as long as

$$\Gamma/D \ll 1\,,$$

*This follows, for instance, from the model of the nucleus as an uncoupled, strongly degenerate gas of individual independent nucleons, as discussed most completely by Newton [60]. Some finer effects treated in detail in this reference are not important for the present purpose.

**The collision time, wavelength over velocity, is of order 2×10^{-15} /E (sec/ev) which is very short compared to crystalline vibration periods.

that is, the natural width is small compared to the level spacing of resonances with the same symmetry (i.e., J). Even for U^{235} this ratio is of order 1/10, and the errors due to the effect of level interference should not change orders of magnitude, even though it is known from low energy data that interferences do exist and influence the cross sections.

2.2 Statistics of Resonance Parameters and Level Spacings

The cross section formulas for individual resonances have been discussed in Sec. 1.2. Since, however, for fast reactors the most important part of the spectrum lies in the region where the resonances are not resolved any more, statistical considerations have to be applied and it is necessary to obtain information on the statistics of resonance parameters and level spacing. A very thorough study of this subject has recently been made by Garrison [59]; only a short summary can be given here.

With respect to level widths, Porter and Thomas [61] have made a detailed analysis and given strong arguments that their distribution should be of the form of "chi-square for n degrees of freedom", that is

$$P_n(z)\,dz = \frac{n}{2\,\Gamma(n/2)}\left[\frac{nz}{2\langle z\rangle}\right]^{\frac{n}{2}-1}\exp\left(-\frac{nz}{2\langle z\rangle}\right)\frac{dz}{\langle z\rangle}, \quad (2\text{-}9)$$

where $\langle z\rangle$ is the average of the variable z, n the number of channels in the corresponding emission process, and $\Gamma(n/2)$ is the gamma function. We note that with the above distribution

$$\langle z^2\rangle/\langle z\rangle^2 = 1 + (2/n) \quad (2\text{-}10)$$

The distributions P_n are wider the lower the value of n. For large n they approach the δ-function $\delta(z - \langle z\rangle)$.

For elastic scattering of low energy neutrons there is only one entrance channel, and the function

$$P_1\,dz = \frac{1}{2\sqrt{\pi}}\left[\frac{2\langle z\rangle}{z}\right]^{\frac{1}{2}}\exp\left(-\frac{z}{2\langle z\rangle}\right)\frac{dz}{\langle z\rangle} \quad (2\text{-}11)$$

is in excellent agreement with observations for the reduced neutron widths

$$\Gamma_n^0 = \Gamma_n/\sqrt{E}\;. \quad (2\text{-}12)$$

However, for target nuclei with spins ≠ 0 and $\ell > 0$ waves, some channels can be reached in two ways. For instance, for U^{235} with I = 7/2 and p-waves, the channels J = 3 and 4 can be obtained both by vector addition of I and $j = 1/2(p_{1/2})$ and $j = 3/2(p_{3/2})$, so that the n = 2 distribution would be applicable, while J = 2 and 5 can be reached only in a single way.

Radiative capture is associated with possible transitions to many exited levels; n, therefore, is large, and the radiation width Γ_γ is practically constant, and has been taken as such for all calculations by all authors.

The fission width Γ_f is an example of a few channel distributions. Porter and Thomas [61], on the basis of the then available data, estimated n = 2.3 with a rather wide margin of error. With the data for 60 resolved resonances [62] in U^{235}, one finds

$$\frac{\langle\Gamma_f^2\rangle}{\langle\Gamma_f\rangle^2} = 1.9\,,$$

so that, according to Eq. (2-10), n ~2 is an acceptable value. This has also been verified by Garrison [59], who gives n = 2 ± 0.2 for all fissionable nuclei. There are some indications, also discussed by Garrison, that there are actually two groups of fission resonances, a broad one and a narrow one, with individually much narrower spreads. It has been conjectured that these two groups may be associated with different spins of the compound nucleus. In this case, of course, no single chi-square distribution is likely to hold. It seems also that there is a relative deficiency in levels with very small fission widths. However, the evidence is too far from complete to warrant going into details, and the n = 2 distribution seems to fit moderately well with the available data.

It is further to be remarked, that there is no evidence for any correlation between Γ_f and Γ_n, so that they can be considered as statistically independent.

As to the distribution of level spacings, it was first pointed out by Wigner [63], that it should be much narrower than random due to the repulsion of levels with the same symmetry. He proposed a distribution for the distances η between neighboring levels:

$$P_0(\eta)\,d\eta = (\pi/2)\exp(-\pi\eta^2/4D^2)\frac{\eta\,d\eta}{D^2}\,. \quad (2\text{-}13)$$

Functions of this type are somewhat awkward to handle. Nicholson [58] analyzed, therefore, the 55 resolved levels in U^{238} in terms of the functions of type Eq. (2-9) and found that n = 10 gave a very good representation, which can hardly be distinguished from Eq. (2-13), certainly not within the observational limits of error. This representation is

$$P_{10}(\eta)\,d\eta = \frac{5}{4!}\left(\frac{5\eta}{D}\right)^4\exp(-5\eta/D)\,\frac{d\eta}{D}\,. \quad (2\text{-}14)$$

Besides the distribution of spacings between adjacent levels $P(\eta)\,d\eta$, another function of interest is the probability $\Omega(\eta)\,d\eta$ that there is a level in an interval $d\eta$ from a given resonance, regardless how many levels are in between. For convenience we introduce the variable $y = \eta/D$, that is, we use the average spacing between levels with the same J as the unit of level spacing. The probability $\Omega(y)dy$, according to Wilkins [64] and Hwang [17a], satisfies the integral equation

$$\Omega(y) = P(y) + \int_0^y P(y - y')\,\Omega(y')\,dy'\,. \quad (2\text{-}15)$$

This equation says simply that the probability $\Omega(y)dy$ is the elementary probability $P(y)dy$ that a level appears in dy plus the sum of the probabilities that there were levels in dy' with $y' \leq y$, and the next level falls into dy. Equations of the above type can be solved by means of a Laplace transformation

$$\mathcal{L}(F(y)) = f_s = \int_0^{\infty} e^{-sy} F(y)\, dy \,. \quad (2\text{-}16)$$

The convolution theorem for Laplace transforms, as easily verified by proper changes of integration variables, leads then to

$$\omega_s = p_s + p_s \omega_s \,,$$

or

$$\omega_s = p_s/(1 - p_s) \,. \quad (2\text{-}17)$$

This transformation can be evaluated explicitly for the elementary distributions of type (2-9)

$$p_s(n) = \int_0^{\infty} e^{-sy} P_n(y)\, dy$$

$$= \frac{(n/2)^{n/2}}{\Gamma(n/2)} \int_0^{\infty} e^{-sy-(n/2)y}\, y^{(n/2)-1}\, dy$$

$$= \left[\frac{n/2}{s + (n/2)}\right]^{\frac{n}{2}} , \quad (2\text{-}18)$$

and thus to

$$\omega_s = \frac{1}{\left(\frac{s + n/2}{n/2}\right)^{n/2} - 1} \,. \quad (2\text{-}19)$$

The Laplace transform, Eq. (2-19), can easily be inverted for even n = 2m. In this case

$$\omega_s = \frac{1}{\left(\frac{s + m}{m}\right)^{m} - 1} \,. \quad (2\text{-}20)$$

This transform has simple poles at

$$s_k = m\,(e^{2\pi i k/m} - 1) \,, \qquad k = 1, \cdots m \,. \quad (2\text{-}21)$$

The well-known inversion formula of the transform (2-16) yields then

$$\Omega(y) = \frac{1}{2\pi i} \int_{a - i\infty}^{a + i\infty} e^{sy} \omega_s ds = \sum \text{res} \left\{ \frac{m^m e^{sy}}{\prod_k (s - s_k)} \right\}$$

$$= m^m \sum_{\ell} \frac{\exp(s_\ell y)}{\prod_k{}' (s_\ell - s_k)} \,, \quad (2\text{-}22)$$

where the sum of the residues is over all the poles of ω_s, and Π' means the product over all k with exception of the term $\ell = k$.

From the above formulas one obtains by elementary calculations the results

n = 2, i.e. m = 1, (Poisson Distribution)

$$\Omega(y) = 1 \quad (2\text{-}23a)$$

that is, the distribution is uniform. For

n = 4

$$\Omega(y) = 1 - \exp(-4y) \quad (2\text{-}23b)$$

n = 8

$$\Omega(y) = 1 - 2(\sin 4y)\exp(-4y) - \exp(-8y) \quad (2\text{-}23c)$$

which is close enough to the distribution n = 10 (which seems to fit best the experimental distribution) to be of practical use. Finally, in the limit $n \to \infty$ and $P = \delta(y - 1)$, that is, uniformly spaced resonances, Eq. (2-15) gives directly

$$\Omega(y) = \sum_{k=1}^{\infty} \delta(y - k) \,, \quad (2\text{-}23d)$$

where the sum goes over all integer values except zero. It is to be noted, of course, that by its nature $\Omega(y)$ must be an even function, since the probabilities of finding levels on either side of a given one must be the same. For negative values of y or η/D, therefore, its absolute value has to be used in the above formulas.

2.3 Discrete Resonances

We will now take up the case where the resonances of a given nucleus are well separated, as is the case for U^{238} up to quite high energies. The effective cross section is defined in Eq. (2-2). For the cross section we assume

$$\sigma_x = \sigma_{0x}\psi = \sigma_{00}(\Gamma_x \Gamma_n/\Gamma^2)\psi \quad (2\text{-}24)$$

and

$$\sigma_t = \sigma_n + \sigma_0\psi = \sigma_n + \sigma_{00}(\Gamma_n/\Gamma)\psi \,, \quad (2\text{-}25)$$

where σ_n denotes the nonresonance part inclusive of the contributions of all other materials. We neglect thus the interference scattering which, for U^{238}, is quite unimportant. In case there are heterogeneities they can be included by adding an escape cross section σ_e, see Eq. (1-40), into σ_n. It is customary, though not always justified as explained below, to replace the denominator in Eq. (2-4) by D/σ_n. Then with $q = 2(E - E_0)\Gamma$, see Eq. (1-13), one obtains

$$\tilde{\sigma}_x = \sigma_n \Gamma_x J(\beta,\xi)/D \quad (2\text{-}26a)$$

where

$$J = \frac{1}{2}\int_{-\infty}^{+\infty} \frac{\psi}{\beta + \psi}\, dq \tag{2-27}$$

and

$$\beta = \sigma_n/\sigma_0 \,. \tag{2-28}$$

Thus, the average effective cross section will be effective cross section will be

$$\langle\tilde{\sigma}_x\rangle = \frac{\sigma_n}{D}\langle\Gamma_x J\rangle \,, \tag{2-29}$$

where the average is to be performed over the statistics of the resonance parameters. The explicit form of this average will then be

$$\langle\Gamma_x J(\xi,\beta)\rangle = \int_0^\infty\int_0^\infty P(y_n)\, P(y_f)\, \Gamma_x J(\xi,\beta)\, dy_n\, dy_f \,, \tag{2-30}$$

where $P(y_n)$ and $P(y_f)$ are the distribution functions for the neutron and fission widths. $\langle\tilde{\sigma}_x\rangle$ does depend on temperature in contrast to the average cross section

$$\bar{\sigma}_x = \frac{1}{D}\int \sigma_x\, dE = \frac{1}{D}\frac{\Gamma}{2}\int \sigma_{0x}\psi\, dq = \frac{\pi}{2}\frac{\Gamma_x\sigma_0}{D}$$

with the help of Eq. (1-14), or

$$\langle\bar{\sigma}_x\rangle = (\pi/2D)\,\langle\Gamma_x\sigma_0\rangle = (\pi/2)\,(\sigma_{00}/D)\,\langle\Gamma_n\Gamma_x/\Gamma\rangle \,. \tag{2-31}$$

(Compare Eq. (1-8).) In place of the cross section itself, frequently a temperature-dependent self-shielding factor is introduced through the definition

$$f = \tilde{\sigma}_x/\bar{\sigma}_x = (2/\pi)\,\beta J(\beta,\xi) \,. \tag{2-32}$$

The function J, already introduced in Sec. 1.7, has been widely discussed [2] and tabulated [2, 18], and a method to carry out the averaging over a Porter-Thomas distribution for the neutron width has been developed, see references [18, 25, 65]. The temperature coefficient of the cross sections is calculated as follows. One has with Eqs. (1-13) and (1-13a)

$$\frac{\partial J}{\partial T} = \frac{\partial J}{\partial\xi}\frac{\partial\xi}{\partial T} = -\frac{\Gamma}{2T\Delta}\frac{\partial J}{\partial\xi} \,,$$

and thus from Eq. (2-30)

$$\frac{\partial\tilde{\sigma}_x}{\partial T} = -\frac{\sigma_n}{2T\Delta}\left\langle\Gamma\Gamma_x\frac{\partial J}{\partial\xi}\right\rangle . \tag{2-26b}$$

Useful approximations for $\partial J/\partial\xi$ have been given by Nicholson [58].

It remains to discuss the approximation used for the denominator in Eq. (2-4), which consists in replacing $1/\sigma_t$ by $1/\sigma_n$. Accurately one has

$$\overline{\frac{1}{\sigma_t}} = \frac{1}{\Delta E}\int_{\Delta E}\frac{dE}{\sigma_t} = \frac{1}{\Delta E}\int_{\Delta E}\frac{dE}{\sigma_n + \sigma_0\psi}$$

$$= \frac{\Gamma}{2\Delta E\sigma_n}\int_{-\Delta E/\Gamma}^{+\Delta E/\Gamma}\frac{dq}{1 + (\psi/\beta)} \,, \tag{2-33}$$

where the limits in the integral in the spirit of the approximation should be well outside the region where ψ is of appreciable magnitude. The integral in Eq. (2-33) can be evaluated as follows:

$$\frac{\Gamma}{2\Delta E}\int_{-\Delta E/\Gamma}^{+\Delta E/\Gamma}\frac{dq}{1 + (\psi/\beta)} =$$

$$\frac{\Gamma}{2\Delta E}\int_{-\Delta E/\Gamma}^{+\Delta E/\Gamma}\left(1 - \frac{\psi}{\beta + \psi}\right)dq \cong 1 - \frac{\Gamma}{2\Delta E}J \,.$$

In this expression one can set for ΔE the energy space per resonance, that is the average level spacing D. One obtains thus

$$\overline{\frac{1}{\sigma_t}} = \frac{1}{\sigma_n}\left(1 - \frac{\Gamma}{2D}J\right); \tag{2-34}$$

and in place of Eq. (2-29)

$$\langle\tilde{\sigma}_x\rangle = \frac{\pi}{2D}\left\langle\frac{\Gamma_x J}{1 - (\Gamma/2D)\,J}\right\rangle . \tag{2-35}$$

The averaging would be over the product and not the individual factors, since the fluctuations in the numerator and denominator in Eq. (2-4) occur at the same resonances, i.e., they are correlated.

For U^{238}, $\Gamma \gtrsim 10^{-1}$ ev, $D \sim 20$ ev, and the correction is very small, though J can be larger than one for small values of β. The correction is still less important for strongly heterogeneous thermal reactors, where it would be applicable only to the "volume term", since in the moderator the flux will be largely undisturbed. For the fissile isotopes, however, the smallness of the correction is not guaranteed, and it should be more closely investigated. It always tends toward an increase of the cross sections.

2.4 The Fluctuation Region

The opposite limiting case is reached in regions where the individual resonances are largely washed out due to Doppler effect and overlap and where the cross sections show only small fluctuations around average values, as treated by Feshbach, Goertzel and Yamauchi [57]. It will be appropriate for the very hard part of the spectra, in particular for assemblies containing mostly fissionable nuclei. This method is here treated first because it is relatively straightforward, and some of the ideas form also the basis of later work, where attempts are made to bridge the gap between the two extreme cases.

The fundamental assumption made by Feshbach, Goertzel and Yamauchi is that the cross sections do not fluctuate too strongly, that is everywhere

$$\frac{\sigma_t - \overline{\sigma}_t}{\overline{\sigma}_t} \ll 1 . \qquad (2\text{-}36)$$

In this case one can develop

$$\frac{\sigma_x}{\sigma_t} = \frac{\sigma_x}{\overline{\sigma}_t + (\sigma_t - \overline{\sigma}_t)} = \frac{\sigma_x}{\overline{\sigma}_t} - \frac{\sigma_x(\sigma_t - \overline{\sigma}_t)}{\overline{\sigma}_t^2} \qquad (2\text{-}37)$$

and obtain

$$\overline{\frac{\sigma_x}{\sigma_t}} = \frac{\overline{\sigma}_x}{\overline{\sigma}_t} - \frac{\overline{\sigma_x\sigma_t} - \overline{\sigma}_x\overline{\sigma}_t}{\overline{\sigma}_t^2} . \qquad (2\text{-}38)$$

Similarly one has

$$\overline{\frac{1}{\sigma_t}} = \frac{1}{\overline{\sigma}_t} + O(0) \qquad (2\text{-}39)$$

and thus from Eq. (2-38)

$$\tilde{\sigma}_x = \overline{\sigma}_x - \frac{\overline{\sigma_x\sigma_t} - \overline{\sigma}_x\overline{\sigma}_t}{\overline{\sigma}_t} \qquad (2\text{-}40)$$

Finally, it is clear that any nonresonant contribution to σ_t will drop out in the numerator in Eq. (2-40). The temperature-dependent part of the effective cross section becomes thus

$$\Delta\tilde{\sigma}_x = \tilde{\sigma}_x - \overline{\sigma}_x = \frac{\overline{\sigma_x\sigma_r} - \overline{\sigma}_x\overline{\sigma}_r}{\overline{\sigma}_t} , \qquad (2\text{-}41)$$

where σ_r is the resonant part of the cross section. Actually, only the term $\overline{\sigma_x\sigma_r}$ will depend on temperature. It is, however, of advantage to calculate the whole expression since there is considerable cancellation between the two contributions.

The average cross section $\overline{\sigma_x}$ was already given in Eq. (2-31). Similarly one obtains for the total resonance part of the cross section from Eq. (1-10):

$$\overline{\sigma}_r = \frac{\pi}{2}\frac{\sigma_{00}}{D}\left\langle\frac{\Gamma_n}{\Gamma}\right\rangle \cos 2\delta . \qquad (2\text{-}42)$$

since the integrals over the χ part vanish. The total average cross section for an isotope is, of course,

$$\overline{\sigma}_t = \overline{\sigma}_n + \overline{\sigma}_r . \qquad (2\text{-}43)$$

These averages are, as already mentioned, independent of temperature.

The only quantity not yet evaluated in (2-41) is

$$\overline{\sigma_x\sigma_r} = \frac{1}{\Delta E}\int_{\Delta E}\sum_{k,k' \text{ in} \Delta E} \sigma_{00}^2\left(\frac{\Gamma_n\Gamma_x}{\Gamma^2}\right)\psi\frac{\Gamma_n'}{\Gamma'} \times (\psi'\cos 2\delta + \chi'\sin 2\delta)\, dE , \qquad (2\text{-}44)$$

where the prime denotes that the quantity has to be taken for the resonance k'. It will be shown that the χ' term gives no contribution, and it will, therefore, be omitted in the following formulas. The double sum in Eq. (2-44) is over all pairs of resonances in ΔE. It can be decomposed into a diagonal part (the two resonances are the same) and a nondiagonal one

$$\sum_{kk'} = \sum_{k} + \sum_{k' \neq k}$$

With Eq. (1-24) and the assumption $\Gamma \ll \Delta$, one obtains for the diagonal part

$$\overline{\sigma_x\sigma_r}\Big|_d = \left(\frac{\pi}{2}\right)^{3/2}\frac{\sigma_{00}^2}{2D\Delta}\left\langle\frac{\Gamma_n^2\Gamma_x}{\Gamma}\right\rangle\cos 2\delta\,\frac{\langle\Gamma_n\Gamma_x/\Gamma\rangle\langle\Gamma_n\rangle}{\langle\Gamma_n\Gamma_x/\Gamma\rangle\langle\Gamma_n\rangle} , \qquad (2\text{-}45)$$

where the last identity factor has been added to permit further simplification. With the help of Eqs. (2-31) and (2-42) one can rewrite Eq. (2-45) as

$$\overline{\sigma_x\sigma_r}\Big|_d = \frac{D\,\overline{\sigma}_x\overline{\sigma}_r}{\sqrt{2\pi}\,\Delta}E_x , \qquad (2\text{-}46)$$

where

$$E_x = \frac{\langle\Gamma_n^2\Gamma_x/\Gamma\rangle}{\langle\Gamma_n\Gamma_x/\Gamma\rangle\langle\Gamma_n\rangle} \qquad (2\text{-}47)$$

is a dimensionless quantity. The term with χ in (2-44) drops out since ψ is an even and χ an odd function of $E - E_k$.

The nondiagonal part of the summation over the resonances gives

$$\overline{\sigma_x\sigma_r}\Big|_{nd} = \frac{1}{\Delta E}\int_{\Delta E} dE\sum_k \sigma_{00}(\Gamma_n\Gamma_x/\Gamma^2)\,\psi_k \times \sum_{k'\neq k}\sigma_{00}(\Gamma_n'/\Gamma')\,\psi_{k'}\cos 2\delta . \qquad (2\text{-}48)$$

The statistical treatment of resonance spacing leads then to the replacement of the sum over k′ by an integral over $\eta = E_{k'} - E_k$ using the probability distribution $\Omega(\eta)$, introduced in Sec. 2.2, for a level to lie in the interval of $d\eta$ at an arbitrary distance η from the level E_k. This gives with Eqs. (2-31), (2-42) and (1-25):*

$$\overline{\sigma_x \sigma_r}\Big|_{nd} =$$

$$= \frac{1}{\Delta E}\int dE \sum_k \sigma_{00}\frac{\Gamma_n\Gamma_x}{\Gamma^2}\psi(E-E_k) \times \int d\eta\, \sigma_{00}\frac{\Gamma_n'}{\Gamma'}\psi(E_k+\eta)\cos 2\delta\, \Omega(\eta)\, d\eta$$

$$= \frac{1}{D}\left(\frac{\pi}{2}\right)^{3/2}\frac{1}{2\Delta}\sigma_{00}^2\left\langle\frac{\Gamma_n\Gamma_x}{\Gamma}\right\rangle\langle\Gamma_r\rangle\cos 2\delta\int e^{-\eta^2/2\Delta^2}\,\Omega(\eta)\,d\eta$$

$$= \frac{D}{\sqrt{2\pi}\,\Delta}\,\overline{\sigma_x}\,\overline{\sigma_r}\int e^{-\eta^2/2\Delta^2}\,\Omega(\eta)\,d\eta \qquad (2\text{-}49)$$

It can now be verified that the asymmetric term, containing the shape function χ, Eq. (1-12), see Eqs. (1-10) and (1-20), does not contribute. The integral

$$\int \psi(E - E_k)\,\chi(E - E_k')\, dE$$

is positive when $E_k' < E_k$ and negative when $E_k' > E_k$ and, therefore, an odd function in $\eta = E_k - E_k'$, while $\Omega(\eta)$ is an even function. Consequently, the integral over their product vanishes. Thus, even if such an asymmetric part is introduced phenomenologically, as discussed in Sec. 1.2, it will have no effect in the pure fluctuation region discussed here.

In order to complete the discussion of the temperature-dependent part of the effective cross section, Eq. (2-41), one can use the identity

$$\overline{\sigma_x}\,\overline{\sigma_r} = \frac{\overline{\sigma_x}\,\overline{\sigma_r}}{\sqrt{2\pi}\,\Delta}\int_{-\infty}^{+\infty} e^{-\eta^2/2\Delta^2}\, d\eta \qquad (2\text{-}50)$$

*Note that the following relatively simple results come from using the approximation Eq. (1-25) in place of the accurate formula Eq. (1-24). The two expressions differ markedly only in the wings of the ψ function, that is at relatively large separations of the two resonances. However, it is the underlying assumption in the present treatment that the distances between resonances are not large compared to their Doppler width. Also the cross sections in the wings probably are not properly described by the ψ function due to interference. It is believed, therefore, that a more accurate evaluation, as for instance attempted by Nicholson [58], is not very meaningful.

and obtain from Eqs. (2-41), (2-45), (2-49) and (2-50) the final expression

$$\Delta\tilde{\sigma}_x = \frac{\overline{\sigma_x}\,\overline{\sigma_r}}{\overline{\sigma_t}}\frac{D}{\sqrt{2\pi\Delta}}(E_x - e)\,, \qquad (2\text{-}51)$$

where E_x is given by Eq. (2-47) and

$$e = (1/D)\int_{-\infty}^{+\infty} e^{-\eta^2/2\Delta^2}[1 - D\,\Omega(\eta)]\,d\eta \qquad (2\text{-}52)$$

is also a dimensionless quantity and a function of the variable D/Δ, since $D\,\Omega(\eta)$ is actually a function $\Omega(y)$ of $y = \eta/D$. e is called the spacing correction, while E may be called the fluctuation term.

In case there are more than one entrance channels, that is, several isotopes are present and/or several possible J values for the total angular momentum, one has to sum over all channels, that is

$$\Delta\tilde{\sigma}_x = \frac{1}{\sqrt{2\pi}\,\Delta}\frac{1}{\overline{\sigma_t}}\sum_J \overline{\sigma_x}\,\overline{\sigma_r}\,D_J(E_J - e_J)\,. \qquad (2\text{-}53)$$

The total cross section includes, of course, all contributions of nonresonant processes and of admixtures, and all cross sections are to be evaluated per absorber atom.

It will be shown in the next section that $0 \geq (E-e) \geq 3$. Δ is therefore negative, E is independent of T, while e depends only weakly on it. All of the temperature dependence comes from the Doppler width Δ, and therefore

$$\frac{d\tilde{\sigma}}{dT} = \frac{d\tilde{\sigma}}{d\Delta}\frac{d\Delta}{dT} = \frac{\Delta}{2T}\frac{d\tilde{\sigma}}{d\Delta} \qquad (2\text{-}54)$$

Now from Eq. (2-51) one has

$$\frac{d\tilde{\sigma}}{dT} = \frac{D}{2\sqrt{2\pi}\,T\Delta}\frac{\overline{\sigma_x}\,\overline{\sigma_r}}{\overline{\sigma_t}}[E - e^*]\,, \qquad (2\text{-}55)$$

where

$$e^* = e - \Delta\,(de/d\Delta)\,, \qquad (2\text{-}56)$$

and where the difference between e* and e is in most cases negligible. Equation (2-55) indicates that the contribution to the Doppler coefficient from the fluctuation region is proportional to $T^{-3/2}$, and that the cross sections always increase with temperature. In case of many channels, one has, of course, to perform a summation as in Eq. (2-53).

The contributions to the Doppler effect decrease rapidly with increasing neutron energies for the following reasons: (1) the level spacing D decreases with energy while Δ increases; (2) the phase shift correction cos 2δ in σ_r, Eq. (2-42), becomes of random sign, see Eq. (1-4); (3) higher J values enter which have smaller D and generally smaller reaction cross sections as compared to scattering. The calculations by Feshbach, Goertzel and Yamauchi [57] and Nicholson [58] show that the effect becomes quite negligible above ~200 kev. This fact has also the consequence that d-waves and inelastic scattering are of much smaller importance than the uncertainties in s- and p-wave data, so that there is no point in trying to incorporate their effect. Nicholson [58] gives some estimates on the influence of inelastic scattering which bear this out.

2.5 Evaluation of the Statistical Functions

The final part of the expression for the temperature-dependent part of the cross section (2-51) contains the spacing correction e, Eq. (2-52) and the fluctuation term E, Eq. (2-47).

The quantity e will depend only weakly on temperature as long as $D/\Delta \ll 1$, that is as long as the level spacing is small compared to the Doppler width, and this has to be fulfilled anyhow for the fluctuation method to be applicable. In this case one can obtain an upper bound for e as follows, at least for the functions Ω originating from the elementary level spacing functions of type (2-9) for $n \geq 2$. One has

$$e = \int_{-\infty}^{+\infty} [1-\Omega(y)]\exp(-y^2D^2/2\Delta^2)dy \leq 2\int_0^{\infty} [1-\Omega(y)]\,dy$$

provided $\Omega(y) \leq 1$, as is fulfilled in the cases considered here. One has further, compare Eqs. (2-16) and (2-19)

$$\int_0^{\infty}(1-\Omega)dy = \lim_{s=0}\int_0^{\infty}(1-\Omega)e^{-sy}\,dy$$

$$= \lim_{s=0}\left[\frac{1}{s} - \frac{1}{\{[s+(n/2)]/(n/2)\}^{n/2}-1}\right]$$

$$= \frac{n-2}{2n}\,.$$

This limit is found easily by combining the terms in the large square brackets into one fraction and then differentiating twice both the numerator and denominator. Thus one finds

$$e \leq \frac{n-2}{n}\,. \qquad (2\text{-}57)$$

This result is due to Wilkins [64], who also gave the asymptotic expansion for n = 10

$$e = \frac{4}{5}\left[1 - \frac{1}{50}\left(\frac{D}{\Delta}\right)^2 - \frac{6}{3125}\left(\frac{D}{\Delta}\right)^4 + \cdots\right], \qquad (2\text{-}58)$$

which shows that the deviations from the asymptotic value are quite small as long as D/Δ is not too large. The upper limit of e = 1 would be reached for $n \to \infty$, that is when the spacing of the resonances is uniform, while for n = 2 one would have e = 0. For practical purposes a value of 0.75 to 0.8 would seem to be sufficient without too much error except if the level spacing functions were rather "pathological." If the Doppler width should not be large compared to the level spacing, one will have to resort to numerical calculations, using such expressions as Eq. (2-23c), and also the accurate formula Eq. (1-22) in place of the approximation Eq. (1-25). But in this case, the whole method becomes of doubtful validity.

For the fluctuation term E_x, Eq. (2-47), limits can be obtained as follows. If the partial widths had sharp values, then all averages would simply be the indicated functions of the sharp values and E_x would be equal to unity. On the other hand, at low energies generally one has $\Gamma_n \ll \Gamma$. If the Γ_n are distributed according to a P_n distribution, Eq. (2-9) and the distributions of different widths are uncorrelated, then for small Γ_n

$$\langle \Gamma_n^2 \Gamma_x/\Gamma\rangle \cong \langle\Gamma_n^2\rangle\langle\Gamma_x/\Gamma\rangle$$

and

$$E_x \cong \frac{\langle\Gamma_n^2\rangle\,\langle\Gamma_x/\Gamma\rangle}{\langle\Gamma_n\rangle^2\,\langle\Gamma_x/\Gamma\rangle} = \frac{\langle\Gamma_n^2\rangle}{\langle\Gamma_n\rangle^2} = 1 + \frac{2}{n}\,. \qquad (2\text{-}59)$$

Compare Eq. (2-10). For the most probable and smallest value n = 1, $E_x \to 3$. This would seem to be its maximum possible value, since if finite values of Γ_n are taken into account, it will increase the denominator Γ in (2-47) and thus reduce the value of E_x. The absolute limits seem thus to be

$$0 \leq (E - e) \leq 3\,. \qquad (2\text{-}60)$$

The value 0 would be reached if both the level spacing and the resonance parameters had sharp values with no fluctuations. The other limit +3 would be reached when $\langle\Gamma_n\rangle$ is small compared to Γ; and the level spacings would be distributed according to the unlikely P_2 (Poisson) distribution. In practice, the uncertainty is much less, since e has in practical cases a nearly constant value; and E-e can be expected to vary only between +1 and +2.

An actual evaluation of E, suitable for numerical calculations, can be obtained as follows. If it is assumed, as it is done by all investigators, that Γ_γ can be taken as constant, one can introduce the variables:

$$u = \Gamma_n/\Gamma_\gamma\,,\quad v = \Gamma_f/\Gamma_\gamma$$

$$\langle u\rangle = \langle\Gamma_n\rangle/\Gamma_\gamma\,,\quad \langle v\rangle = \langle\Gamma_f\rangle/\Gamma_\gamma$$

$$\Gamma = \Gamma_n + \Gamma_\gamma + \Gamma_f\,. \qquad (2\text{-}61)$$

One obtains then, for instance

$$\left\langle\frac{\Gamma_n^2\Gamma_f}{\Gamma}\right\rangle = \Gamma_\gamma^2\left\langle\frac{u^2v}{1+u+v}\right\rangle$$

$$= \Gamma_\gamma^2\int_0^\infty du\int_0^\infty dv\,\frac{u^2v}{1+u+v}\,P_n\left(\frac{u}{\langle u\rangle}\right)P_m\left(\frac{v}{\langle v\rangle}\right), \tag{2-62}$$

where P_n and P_m are the distribution functions for Γ_n and Γ_f. The subsitution

$$\frac{1}{1+u+v} = \int_0^\infty e^{-s(1+u+v)}ds$$

gives then

$$\left\langle\frac{\Gamma_n^2\Gamma_f}{\Gamma}\right\rangle = \Gamma_\gamma^2\int_0^\infty e^{-s}ds\int_0^\infty du\left[u^2P_n\left(\frac{u}{\langle u\rangle}\right)e^{-su}\right]$$

$$\times\int dv\left[v\,P_m\left(\frac{v}{\langle v\rangle}\right)e^{-sv}\right]. \tag{2-63}$$

The integrals over du and dv can be carried out

$$\int_0^\infty du\left[\cdots\right] = \langle u\rangle^2\,\frac{1+(2/n)}{[(2s\langle u\rangle/n)+1]^{(n/2)+2}} \tag{2-64a}$$

$$\int_0^\infty dv\left[\cdots\right] = \langle v\rangle\,\frac{1}{[(2s\langle v\rangle/m)+1]^{(m/2)+1}} \tag{2-64b}$$

and thus

$$\left\langle\frac{\Gamma_n^2\Gamma_f}{\Gamma}\right\rangle =$$

$$\frac{\langle\Gamma_n\rangle^2\langle\Gamma_f\rangle}{\Gamma_\gamma}\int_0^\infty\frac{e^{-s}[1+(2/n)]\,ds}{\left[\dfrac{2s\langle\Gamma_n\rangle}{n\Gamma_\gamma}+1\right]^{\frac{n}{2}+2}\left[\dfrac{2s\langle\Gamma_f\rangle}{m\Gamma_\gamma}+1\right]^{\frac{m}{2}+1}} \tag{2-65}$$

and similar expressions for other averages. They involve only single quadratures, and are thus suitable for numerical integration. It is to be noted, that such an averaging process is already required for the computation of the average cross section, which involves $\langle\Gamma_n/\Gamma\rangle$, compare Eq. (2-31).

Some tables for such integrals have been given by Nicholson [58] and Greebler [66]. A frequently used alternate procedure is to replace the continuous distribution P_n by a discrete set of values, and carrying out the averaging process numerically. One can also carry out this decomposition only for the narrower fission distribution and then carry out the integration over the neutron distribution in Eq. (2-62) in terms of tabulated functions [67].

2.6 Intermediate Cases

The various procedures to reduce the general formula Eq. (2-3) to manageable proportions all consist in approximating or developing the bothersome denominator in various ways. In the "discrete method" all resonances except the one in the numerator are neglected. In the "fluctuation method" the denominator is developed around the average cross section $\bar{\sigma}_t$. To bridge the gap between the two extremes, the following two methods are natural.

2.6.1 The "Nearest Neighbor" Approximation

It consists in taking in the denominator of the integral in Eq. (2-3) only the next resonance k' into account. This should be a fair approximation for a rather wide range of conditions since the cross sections fall off so rapidly away from a resonance peak. In practice this means that the integral J, Eq. (2-27), has to be replaced by

$$J^* = \frac{1}{2}\int_{-\infty}^{+\infty}\frac{\psi dq}{\beta+\psi+A\psi'}, \tag{2-66}$$

where

$$A = \frac{\sigma'_{00}}{\sigma_{00}}\frac{\Gamma'_n}{\Gamma'}\frac{\Gamma}{\Gamma_n}\frac{N'}{N} = \frac{g'}{g}\frac{\Gamma'_n}{\Gamma'}\frac{\Gamma}{\Gamma_n}\frac{N'}{N}. \tag{2-67}$$

N'/N is the ratio of the number densities of the respective nuclear species, if they are different for the adjacent resonances, and the arguments of the function ψ' are

$$q' = (q+\Delta q)\,\Gamma/\Gamma';$$

$$\Delta q = 2(E_k - E'_k)/\Gamma\ ;\ \xi' = \xi\Gamma'/\Gamma. \tag{2-68}$$

It would not seem to be difficult to develop a computer program to evaluate the integral Eq. (2-66). In fact, Hwang [17a] states that such a program is available at the Argonne National Laboratory. The complications stem from the multitude of parameters involved. In addition to β and ξ in the original J function, there are three more, Δq, the distance between the levels, Γ'/Γ, and A.

The integral J* is, of course, always smaller or equal to the corresponding J, and the minimum value of J* occurs when the separation of the resonances Δq becomes zero. The difference between J* and J vanishes rapidly with increasing separation. The temperature coefficient, however, behaves differently. The behavior of dH/dT with

$$H = J^* - J \tag{2-69}$$

is shown schematically in Fig. 2-1 following Codd and Collins [17b] and Hwang [17a].

At small separation the Doppler effect is increased for the following reason. ψ' will then largely overlap ψ, and since ψ' will be depressed at higher T, its influence on J* will be reduced. At larger distances, however, the broadening of ψ' will increase the overlap with ψ and thus reduce J* more at higher temperatures than at low temperatures. Thus dH/dT varies from positive to negative before it approaches zero at large separations. The area under the negative part of such a "v curve" is larger than under the positive part, thus presaging a total negative effect when averaged over separations. From here on the procedure will be different for the two cases, firstly, the interfering resonances belong to the same series (same isotope, same compound nuclear spin J) and secondly, they belong to a different series (different J and/or isotope).

In the first case the distribution of the separation between adjacent levels is described by the function P_0 introduced in Sec. 2.2 of forms like Eqs. (2-11) or (2-14). The expression for A, Eq. (2-67) simplifies to $A = \Gamma'_n/\Gamma_n$. Some actual calculations have been made by Hwang [17a], who uses the approximation $\beta \gg \psi + A\psi'$. This gives with development of the denominators in J* and J

$$H = J^* - J \cong -\frac{1}{2\beta^2}\int_{-\infty}^{+\infty} \psi \sum_{k'} A_{k'} \psi' dx + \cdots, \tag{2-70}$$

where the sum over all neighboring resonances has been reintroduced, since it does not lead to further complications. The evaluation then proceeds very similarly to Sec. 2.4, with the exception that the correct value for the integral of the product of two ψ functions, Eq. (1-22), in place of the approximation Eq. (1-25) is used; and further the expression (2-23c) is used for the level distribution Ω. Some of the results will be quoted in Sec. 2.9.

In the second case the distribution of adjacent levels will be random, that is equal probabilities for all separations. The most important case is the disturbance of a fission resonance by a U^{238} resonance. In this case the constant A of Eq. (2-67) may have a rather large value, and the approximation Eq. (2-70) becomes inapplicable. For this case Hwang approximates, therefore, Eq. (2-66) by

$$J^* \cong \frac{1}{2}\int_{-\infty}^{+\infty} \frac{\psi\, dx}{\beta + A\psi'} \tag{2-71}$$

which he evaluates by different approximations for small and large separations. Again some results will be quoted in Sec. 2.9.

2.6.2 The "Adjusted Scattering" Approximation

This method consists of expansion of the denominators in Eq. (2-3) around a modified non-

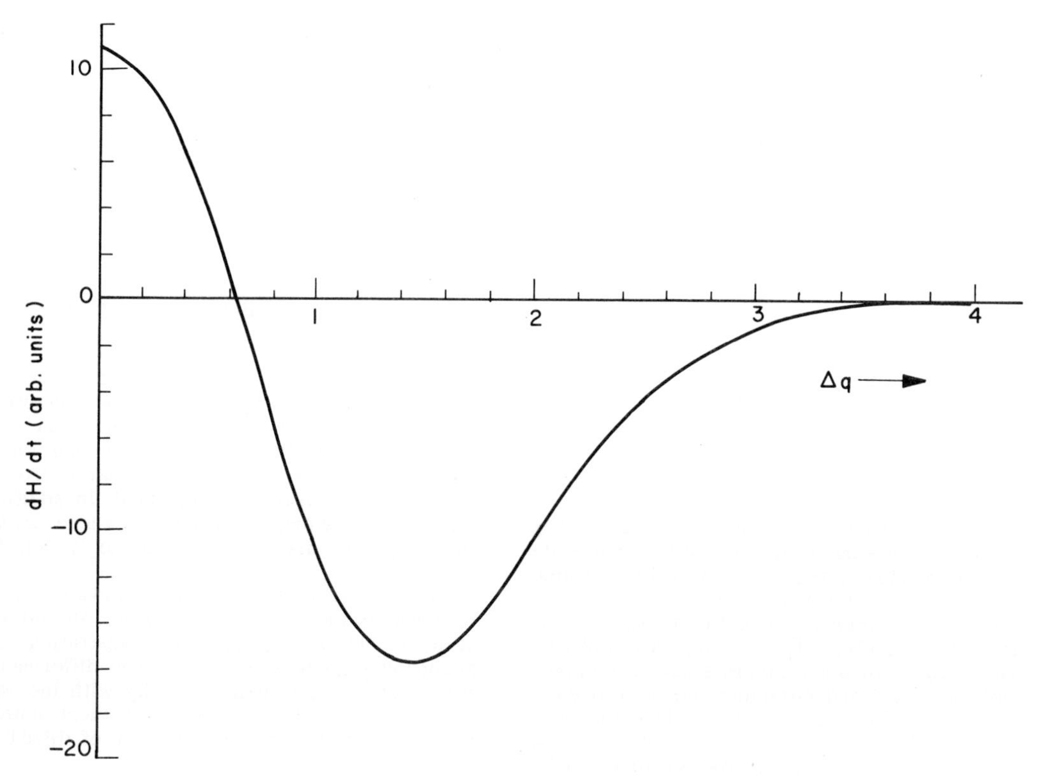

FIG. 2-1 Change of Doppler coefficient due to interference as a function of separation of resonances. Arbitrary units are used.

resonant part σ_n^* to be determined so as to minimize the higher order terms. It has been introduced by Froelich, Ott, and Schmidt [17c] to treat the cases where the U^{238} (or Th^{232}) resonances with $\ell = 0$ are still well separated, but where all other resonances, fissile isotopes and U^{238} with $\ell > 0$, have peaks small compared to the nonresonant part of the cross sections. The interference between different U^{238} resonances is neglected.

The total cross section will be, therefore,

$$\sigma_x = \sigma_r + \sum_{k_o} \sigma_{k_o r} + \sum_{k_i} \sigma_{k_i r} = \sigma_n + \sigma_{rt} , \quad (2\text{-}72)$$

where the index k_o refers to the U^{238} ($\ell = 0$) resonances, and k_i to all other series of resonances which may be present. The individual cross sections have, of course, again to be weighted with the ratios of the number densities of the isotopes present. In view of the preponderant influence of the U^{238} resonances, it is most convenient to refer all cross sections to this isotope. The effective cross sections, Eq. (2-3), become then,

$$\tilde{\sigma}_x = \frac{\sigma_n}{\Delta E} \left\langle \sum_{k_o} \int_{\Delta E} \frac{\sigma_{k_o x} dE}{\sigma_n + \sigma_{k_o r} + \sum_{k_i} \sigma_{k_i r}} + \sum_{k_i} \int_{\Delta E} \frac{\sigma_{k_i x} dE}{\sigma_n + \sigma_{k_o r} + \sum_{k\ell} \sigma_{k\ell r}} \right\rangle , \quad (2\text{-}73)$$

where k_o refers in the first term to the same resonance in numerator and denominator, and in the second term to the nearest neighbor to k_i. The denominators are now written with a modified scattering cross section as

$$\sigma_n + \sigma_{n_o}^* + \sigma_{k_i r} + \left(\sum_{k_i} \sigma_{k_i r} - \sigma_{n_o}^* \right)$$

$$\sigma_n + \sigma_{n_i}^* + \sigma_{k_o r} + \left(\sum_{k\ell} \sigma_{k\ell r} - \sigma_{n_i}^* \right) .$$

The first denominator is then expanded:

$$\int \frac{\sigma_{k_o x} dE}{\sigma_n + \sigma_{k_o r} + \sum_{k_i} \sigma_{k_i r}} = \int \frac{\sigma_{k_o x} dE}{\sigma_n + \sigma_{n_o}^* + \sigma_{k_o r}} \left(1 - \frac{\sum_{k_i} \sigma_{k_i r} - \sigma_n^*}{\sigma_n + \sigma_n^* + \sigma_{k_o r}} + \cdots \right) . \quad (2\text{-}74)$$

Similarly, the second denominator is expanded to obtain:

$$\int \frac{\sigma_{k_i r} dE}{\sigma_n + \sigma_{k_o r} + \sum_{k\ell} \sigma_{k\ell r}} = \int \frac{\sigma_{k_i x} dE}{\sigma_n + \sigma_{n_i}^* + \sigma_{k_o r}} \left(1 - \frac{\sum_{k\ell} \sigma_{k\ell r} - \sigma_{n_i}^*}{\sigma_n + \sigma_{n_i}^* + \sigma_{k_o r}} + \cdots \right) . \quad (2\text{-}75)$$

The quantities $\sigma_{n_o}^*$ and $\sigma_{n_i}^*$ are then so determined that the first terms of the expansion vanish, that is

$$\sigma_{n_o}^* = \left\langle \sum_{k_i} \sigma_{k_i r} \right\rangle$$

$$\sigma_{n_i}^* = \frac{\left\langle \sigma_{k_i x} \sum_{k_{\ell r}} \sigma_{k\ell r} \right\rangle}{\langle \sigma_{k_i x} \rangle} , \quad (2\text{-}76)$$

where the average is, as usual, over all spacing and level width distributions. It is to be noted that $\sigma_{n_o}^*$ will be independent of temperature, while $\sigma_{n_i}^*$ will still depend on it. The full details of this method have not yet been published, but it is clear that the evaluation of the integrals and the necessary averages can proceed along the lines of the development described in Secs. 2.3 to 2.5.

2.7 Reactivity Changes and Adjustment of Parameters

The preceding developments were aimed at the calculation of cross sections as a function of temperature as required for multigroup calculations. In view of the sensitivity of the effect on the spectral distribution, this seems to be the best way to handle it. In place of this, several authors introduce a perturbation theory treatment, which will be sketched briefly.

The temperature coefficient of reactivity is given by the perturbation theory formula

$$\frac{\partial k}{\partial T} = \int \phi(E) dE \left[W_f \nu \frac{\partial \Sigma_f}{\partial T} - W(E) \frac{\partial (\Sigma_f + \Sigma_\gamma)}{\partial T} \right] . \quad (2\text{-}77)$$

Here Σ_f and Σ_γ are the macroscopic fission and absorption cross section. $\phi(E)$ is the flux distribution averaged over the core and normalized to a total fission rate of unity

$$\int \phi(E) \Sigma_f dE = 1 . \quad (2\text{-}78)$$

W(E) is the "value" of a neutron of energy E, that is, the number of fissions it produces, and W_f the value of a fission neutron. Since each fission produces ν neutrons, one of which will produce another fission, the value of a fission neutron is

$$W_f = 1/\nu . \quad (2\text{-}79)$$

Some authors introduce the further approximation $W(E) = W_f$, which is not too bad for many fast reactors. One has then

$$\frac{\partial k}{\partial T} - \int dE\, \phi(E) \frac{1}{\nu} \left[\frac{\partial \Sigma_f}{\partial T} (\nu - 1) - \frac{\partial \Sigma_\gamma}{\partial T} \right] . \quad (2\text{-}80)$$

For the contribution of a fissionable isotope one will have approximately*

$$\frac{\partial \Sigma_\gamma}{\partial T} = \alpha \frac{\partial \Sigma_f}{\partial T} ,$$

and obtain

$$\frac{\partial k}{\partial T} \sim \int dE\, \phi(E) \frac{\partial \Sigma_f}{\partial T} \frac{1}{\nu} (\nu - 1 - \alpha) . \quad (2\text{-}81)$$

With these assumptions the Doppler coefficient for fissionable isotopes is generally positive, since $\partial \Sigma_f / \partial T > 0$. A negative coefficient can result if the value W(E) for lower energies is appreciably larger than W_f. This will not happen for fast reactors, but may occur for intermediate assemblies. For the cross sections such formulas as (2-26b) or (2-55) as applicable will have to be used.

There are many more parameters going into the calculation of the Doppler effect than are actually known. They have thus to be obtained from theoretical arguments and educated guesses. An exhaustive discussion is beyond the scope of this chapter. However, the relevant items will be shortly enumerated. A recent survey of some of them has been given by Froelich, Ott, and Schmidt [17c].

The average level spacings for various isotopes and their dependence on angular momentum have already been discussed in Sec. 2.1. It should be added, that the level spacing decreases with increasing energy. Theory [60] predicts a behavior like

$$D(E) \cong D(0) \exp(-E/\theta) , \quad (2\text{-}82)$$

where D(0) is the spacing near thermal, and θ a kind of effective temperature of order 0.7 Mev. The effect is thus not yet very large at 200 kv, but increases rapidly for higher energies.

All experimental data on resonances come from the low energy region, below 1 - 2 kev for the fertile and below 100 ev for the fissile isotopes. Thus they pertain practically exclusively to s-wave resonances. Even for these there remains quite a bit of uncertainty due to insufficient statistics and uncertainties in the measurements. Data for p-waves have, therefore, to be adjusted from other information such as average cross sections or theoretical models. For the various widths, generally the following remarks can be made.

The energy dependence for the average neutron width is given by

$$\langle \Gamma_{n\ell} \rangle = \langle \Gamma^0_{n\ell} \rangle \sqrt{E}\, v_\ell , \quad (2\text{-}83)$$

where Γ^o_n is the reduced width at 1 ev. Γ_n depends on the orbital angular momentum ℓ through the penetration factor v_ℓ. It is

$$v_0 = 1 ,$$

$$v_1 = \frac{R^2/\lambda\!\!\!^{-2}}{1 + R^2/\lambda\!\!\!^{-2}} , \quad (2\text{-}84)$$

where R is the nuclear radius and $\lambda\!\!\!^-$ the neutron wavelength. In place of adjusting $\langle \Gamma^o_n \rangle$ it is preferable to adjust the "strength function" which is defined for arbitrary ℓ as

$$S_\ell = \frac{1}{2\ell + 1} \sum_J (g_J/D_J) \sum_j \Gamma^0_{n,\ell,j,J} ; \quad (2\text{-}85)$$

for s-waves this is

$$S_o = \langle \Gamma^0_n \rangle / D_s \quad (2\text{-}85a)$$

and for p-waves

$$S_1 = (1/3) \sum_J (g_J/D_J) \sum_j \langle \Gamma^0_{n,\ell,j,J} \rangle . \quad (2\text{-}85b)$$

If, as it will be assumed in absence of any better information, all $\Gamma_{n,j}$ are the same, then with Eq. (2-7) this reduces to

$$S_1 = \langle \Gamma^0_n \rangle / D_s . \quad (2\text{-}85c)$$

The advantage of the strength function is, that it is less sensitive to errors due to missed resonances and that it should be a fairly smooth varying function of the atomic weight A.

The values of Γ_γ are generally assumed to be the same for all energies and all ℓ values.

The fission width Γ_f is only known for s-waves at low energies. To arrive at practical results it is generally assumed that it does not depend strongly on J and ℓ. With this provision one can determine if from the capture-to-fission ratio α as

$$\langle \Gamma_f \rangle / \Gamma_\gamma = 1/\alpha , \quad (2\text{-}86)$$

or more accurately from

$$\bar{\sigma}_f / \bar{\sigma}_\gamma = 1/\alpha . \quad (2\text{-}87)$$

Since α decreases with increasing energy, the Γ_f will increase with energy. As Hummel and Rago [17d] have shown, this adjustment depends quite markedly on the assumed distribution of fission widths [that is, the choice of number n of degrees of freedom in Eq. (2-9)]. The effect on the Doppler

---

*This will not be strictly true when resonances of different species overlap, see the discussion in Secs. 2.6 and 2.8.

coefficient, however, seems to be quite small, when the average fission width is also adjusted so as to give all cross sections in the appropriate energy range.

With all due cautions as to its provisional nature, Table 2-2 gives a list of recommended values with possible errors. It has been taken partially from the compilation by Froelich et al., [17c]. Some comments are given in footnotes.

The fission width has to be adjusted to give the correct averages of $\alpha = \sigma_\gamma/\sigma_f$. Table 2-3 shows the values from the compilation of Schmidt [68]. Their uncertainties are of the order ±20%. The table shows also some tentative values for the average fission width. They were obtained by J. Garrison by using the Γ_γ values from Table 2-2 and fitting σ_γ and σ_f with a fission width distribution with two degrees of freedom in the region where only $\ell = 0$ resonances are important. The $\langle \Gamma_f \rangle$ are rather sensitive with respect to this distribution, and a choice of $n_f = 3$ would give somewhat smaller values, as used, for instance, by Greebler [66] and Hummel and Rago [17d]; it is clear that more experimental and theoretical work is necessary in this area.

Since the Doppler coefficient is so sensitive with respect to the neutron spectrum, other quantities which affect the spectrum, such as α values at high energies and inelastic scattering, will also affect it, even if these data do not enter directly into the calculations.

2.8 Some Representative Results

It has been repeatedly emphasized that the Doppler coefficient in a fast reactor depends sensitively on the neutron spectrum and thus on its detailed composition, inclusive of all scattering and coolant materials. It is, therefore, not possible to give general results, as has been done in Sec. 1.9 for thermal reactors. Thus, one is necessarily restricted to the discussion of some typical examples. This will be done here in decreasing order of hardness of the neutron spectrum.

2.8.1 Hard Spectrum Assemblies

For pure U^{235} a rough estimate can be obtained from Eqs. (2-81) and (2-55) in which, with the conventions made, the macroscopic cross sections will enter in the same way as the microscopic ones. The average effective neutron energy will be somewhat lower than the medium fission energy, that is, it will be of order 100 kev. There are 6 entrance channels (J = 3, 4 for s-waves and J = 2,3,4, 5 for p-waves). One can then decompose the product of the partial cross section as follows:

TABLE 2-2

Provisional Values for Resonance Parameters

Isotope	Orbital Angular Momentum	Strength Function	Radiative Capture width, Γ_γ (ev)	Level Spacing (ev)	Footnote Citation
I. FERTILE ISOTOPES					
Th^{232}	$\ell = 0$	$S_0 = 0.7 \times 10^{-4} \pm 15\%$	0.030 ± 20%	19 ± 10%	a
U^{238}	$\ell = 0$	$S_0 = 0.94 \times 10^{-4} \pm 10\%$	0.0246 ± 10%	18.5 ± 10%	b
	$\ell = 1$	$S_1 = 1.0 \times 10^{-4}$ + 200% - 30%	0.246 + 200% - 10%	$D_{1/2} = 18.5 \pm 10\%$ $D_{3/2} = 9.25 \pm 10\%$	c, d
Pu^{240}	$\ell = 0$	$S_0 = 2 \times 10^{-4} \pm 50\%$	0.034 ± 20%	11 ± 20%	e
II. FISSILE ISOTOPES					
U^{235} (I = 7/2)	$\ell = 0$	$S_0 = 1 \times 10^{-4} \pm 20\%$	0.033 ± 20%	0.7 ± 30%	f
Pu^{239} (I = 1/2)	$\ell = 0$	$S_0 = 1 \times 10^{-4} \pm 40\%$	0.039 ± 15%	2.7 ± 20%	g
	$\ell = 1$	$S_1 = 1 \times 10^{-4}$ + 200% - 30%	0.039 ± 15%		
Pu^{241} (I = 5/2)	$\ell = 0$	$S_0 = 1.2 \times 10^{-4} \pm 20\%$	0.043 ± 25%	1.2 ± 20%	e

[a] The situation with respect to the Th^{232} resonance parameters has been discussed in Sec. 1.9.

[b] The main uncertainty in the S_0 values is due to insufficient statistics.

[c] There is considerable uncertainty in the S_1 values. Bilpuch, Weston and Newson [69] find from fitting the average cross section in the kilovolt region that $S_1 = (0.7 \pm 0.3) \times 10^{-4}$, while optical model calculations by Krueger and Margolis give a larger value, 3.0×10^{-4}.

[d] Theoretically, one would not expect a strong J dependence of Γ_γ. However, Bilpuch, Weston and Newson [69] claim that agreement with experimental capture cross section data in the kilovolt region requires $\Gamma_\gamma (J = 1) \sim 2.5\ \Gamma_\gamma (J = 0)$.

[e] Data from reference [44].

[f] From discussion in reference [58].

[g] From reference [17c].

$$\sum_J (\Sigma_f \Sigma_r) = \left(\sum_J \Sigma_f\right) (\Sigma_r D)_{av} = (\Sigma_f)_{total} (D \Sigma_r)_{av} ,$$

where the first factor is the total fission cross section, and the second the average resonance cross section per entrance channel. One obtains then from Eq. (2-81)

$$\frac{\partial k}{\partial T} = \frac{1}{\sqrt{8\pi}\, T} \frac{(D\Sigma_r)_{av}}{\Delta \Sigma_t} (E - e^*) \frac{\nu - 1 - \alpha}{\nu} \int \phi(E) (\Sigma_f)_{total}\, dE . \tag{2-88}$$

The normalization, Eq. (2-78), would give unity for the integral. The actual value should be somewhat smaller since high energies do not contribute to the Doppler effect, and it will be taken as 0.5. The total resonance cross section (fission plus absorption) is of order 2.4 barns, or per channel 0.4, the total cross section including scattering is of order 12 barns; therefore, $(\Sigma_n)_{av}/\Sigma_{tr} \sim 1/30$. E - e* will be of order 1.5 to 2. D, with inclusion of p-waves, will be ~ 0.2 ev [compare Eq. (2-8)] and Δ from Eq. (2-5) will be $\sim 0.4\sqrt{T}$. With $\nu = 2.5$ and $\alpha = 0.25$ one obtains with all these values

$$\frac{\partial k}{\partial T} \sim \frac{1.7 \times 10^{-3}}{T^{3/2}} \sim 0.3 \times 10^{-6} \left(\frac{300}{T}\right)^{3/2} .$$

While this is an admittedly rough estimate, it shows that the effect is very small indeed, and it will be masked generally by other reactivity effects due to thermal expansion. For Pu^{239}, although the effect will be about 6 times larger due to the wider level spacing (Table 2-1) and due to the existence of only 5 in place of 6 entrance channels (J = 0, 1 for s-waves and 0, 1, 2 for p-waves), it still is small enough to escape easy detection. The temperature dependence of $\partial k/\partial T$ for such assemblies is typical as $T^{-3/2}$, that is the effect decreases rapidly with increasing T.

Admixture of U^{238} will give a negative contribution to the Doppler coefficient. Bethe [67] estimates that the total coefficient will become zero for a 1 to 1 mixture of U^{235} and U^{238}. This estimate is perhaps somewhat more reliable than the absolute value. Since, according to Eq. (2-55) the effect is quadratic in the relative concentrations of different materials, the Doppler coefficient should cancel for a ~ 2.5 to 1 ratio in mixtures of U^{238} with Pu^{239}. These estimates suppose, of course, an energy spectrum so hard that the resonances of U^{238} below 30 kev do not contribute appreciably.

Experiments so far have confirmed the smallness of the effect in highly enriched assemblies. Kato and Butler [72] attempted to measure the Doppler coefficient of U^{235} and Pu^{239} in a mock-up of EBR-1, which contained only highly enriched U^{235} and some structural material and sodium. They oscillated heated samples of 506 g of U^{235} and of 235 g of Pu^{239} in these assemblies. Because of the uncertainties introduced by the thermal expansion of the samples (which were in form of a helix), they were only able to set an upper limit of 0.5×10^{-8} per °K ($\Delta k/\Delta T$) for the Doppler effect of the samples near room temperature, which when extrapolated to the entire critical mass, would result in a coefficient no greater than 1.6×10^{-6} per °K ($\Delta k/\Delta T$). Baker and Jaques [71] made measurements, also by an oscillator technique in the Zeus - 3B core, which has an average U^{238}/U^{235} ratio of 1.72. They estimated that the Doppler coefficient for the entire reactor was zero with an upper limit $\pm 1 \times 10^{-6}$ per °K ($\Delta k/\Delta T$).

TABLE 2-3

Values of α and $\langle \Gamma_f \rangle$ for U^{235} and Pu^{239} in the Unresolved Region

E (ev)	U^{235} α	U^{235} $\langle \Gamma_f \rangle$ (mv)	Pu^{239} α	Pu^{239} $\langle \Gamma_f \rangle$ (mv)
150	0.51	130	0.74	91
375	0.49	134	0.68	97
750	0.47	140	0.63	105
1500	0.45	146	0.59	101
2700	0.44	150	0.55	107
4450	0.43	154	0.52	122
7300	0.41		0.48	
1.2×10^4	0.40		0.44	
2×10^6	0.39		0.39	
5×10^4	0.34		0.21	
10^5	0.28		0.14	
2×10^5	0.22		0.13	

Going now to a slightly softer spectrum, Nicholson [58] has made extensive calculations for a metal version of the Enrico-Fermi reactor. The composition of this reactor was as shown in Table 2-4. The results of his ten-group calculation are shown in Table 2-5. The total Doppler coefficient at the temperature of 550°K was thus -2×10^{-6}/°K In this calculation the discrete method was used for U^{238} in groups 8 to 10. No correction for overlap was made, which should make the effect slightly more negative. It is seen clearly how relatively low energies give most of the effect in spite of the depletion of the spectrum.

A coefficient of order 10^{-6}/°K can be considered as small and of very little influence on

TABLE 2-4

Composition of a Model for the Fermi Reactor

Material	Volume Fraction
U^{235}	0.0612
U^{238}	0.1657
Molybdenum	0.0472
Zirconium	0.0382
Stainless Steel	0.2591
Sodium	0.4286

TABLE 2-5

Calculation of the Temperature Coefficient $\partial k/\partial T$ of the Fermi Reactor at 500°K

Energy Group	E average (kev)	ϕ (E) $(cm)^{-1}$	$\frac{\partial k_E}{\partial T}$ (per °K)
1	6000	10	0
2	1800	13	0
3	900	50	0
4	400	37	0
5	240	32	-2.3×10^{-8}
6	120	50	-19×10^{-8}
7	50	18	-39×10^{-8}
8	17	8	-65×10^{-8}
9	6	1.5	-35×10^{-8}
10	2	0.5	-40×10^{-8}

$\partial k/\partial T$ = Total = -2.0×10^{-6}

the operation of a fast reactor. In hard spectra assemblies changes due to thermal expansion and shape distortion will, therefore, be more important.

2.8.2 Large Reactors with Soft Spectra

In recent times the trend in the development of fast reactors has been towards larger sizes and increased total power output. This means a higher dilution, that is, larger ratios of fertile to fissile materials. Also as to materials, the trend is towards oxides or carbides, which permit higher temperatures. In most designs there is also a large volume fraction of sodium as coolant. All these features tend to soften the spectrum. This leads, of course, to a lower breeding ratio, but it also has the beneficial effect of introducing larger negative Doppler coefficients, of order 10^{-5}/°K, which then become of importance in the operation of the reactors.

Several cases of large reactors have been given by Greebler, Hutchins, and Sueoka [73]. The breakdown of the Doppler coefficent contributions* for a typical example is shown in Table 2-6. This reactor is fueled with Pu and both Pu and U^{238} are in form of their oxides. The atom ratio U^{238} to Pu^{239} is 7.2 and Pu^{239} to Pu^{240} is 2. The volume ratio of fuel to steel is 2 and the sodium volume fraction is 50%. The calculations from group 11 down were done by the discrete method, with no correction for overlapping. The higher energy

*In a more recent report Greebler and Goodman [66] have updated both the method of calculation and the cross section data used. The results, though different in detail, are qualitatively much the same. Table 2-6 has ben retained here for illustrative purpose, since the newer work did not give such a complete breakdown in group structure.

TABLE 2-6

Contributions to the Doppler Coefficient by Various Isotopes in a Large Oxide Breeder at 900°C

Energy Group	Lower Energy E_L (ev)	$\partial k/\partial T \times 10^6$			Total
		U^{238}	Pu^{239}	P^{240}	
1	2.25×10^6	0	0	0	0
2	1.35×10^6	0	0	0	0
3	8.25×10^5	0	0	0	0
4	5.0×10^5	0	0	0	0
5	3.0×10^5	0	0	0	0
6	1.8×10^5	- 0.014	+0.003	0	-0.011
7	1.1×10^5	- 0.031	+0.006	0	-0.025
8	6.7×10^4	- 0.081	+0.017	0	-0.064
9	2.5×10^4	- 0.261	+0.054	0	-0.207
10	9100	- 0.559	+0.116	0	-0.443
11	4000	- 0.866	+0.065	-0.019	-0.820
12	1000	- 3.257	+0.316	-0.138	-3.079
13	300	- 3.900	+1.342	-0.562	-3.120
14	100	- 1.516	+1.664	-0.415	-0.267
15	30	- 0.332	+0.357	-0.247	-0.222
16	5	- 0.083	+0.260	-0.019	+0.158
17	0.4	0	0	0	0
18	0	0	0	0	0
TOTAL	---	-10.9	+4.2	-1.4	-8.1

groups, which were only approximately treated, contribute only ~10% to the total. The overall Doppler coefficient, -8.1×10^{-6}/°K, is much larger than in the previous examples despite the high temperature (1173°K). The breakdown of the spectrum has, unfortunately, not been given. A noteworthy fact is that the Pu^{239} effect is large (larger than the U^{238} effect in group 14) at low energies, but declines more rapidly at higher energies than the U^{238} contributions. It probably would be reduced somewhat by the overlapping effect.

Similar results have been obtained by Froelich, Ott, and Schmidt [17c] for reactors fueled with a mixture of UO_2 and PuO_2 with a volume fraction as follows: 30 vol.% fuel, 30 vol.% steel and 40 vol.% sodium. They find that the Doppler coefficient can be well represented by a formula

$$\frac{\partial k}{\partial T} = C\left(\frac{300}{T}\right)^{\gamma} . \qquad (2\text{-}89)$$

The values for C and γ are shown in Table 2-7. It is seen that the effect has a somewhat lower dependence on T than $T^{-3/2}$, which would come from the high energy region above, compare Sec. 2.4. These calculations were made with the adjusted scattering method and include overlap and interference effects.

TABLE 2-7

Temperature Dependence of the Doppler Coefficient for Oxide Reactors

Volume ratio UO_2 to PuO_2	Sodium density	C	γ
7	100%	-1.34×10^{-5}/°K	0.89
	50%	-1.28×10^{-5}	0.96
	0%	-1.26×10^{-5}	1.06
5	100%	-0.547×10^{-5}	0.88
	50%	-0.529×10^{-5}	0.98
	0%	-0.528×10^{-5}	1.11

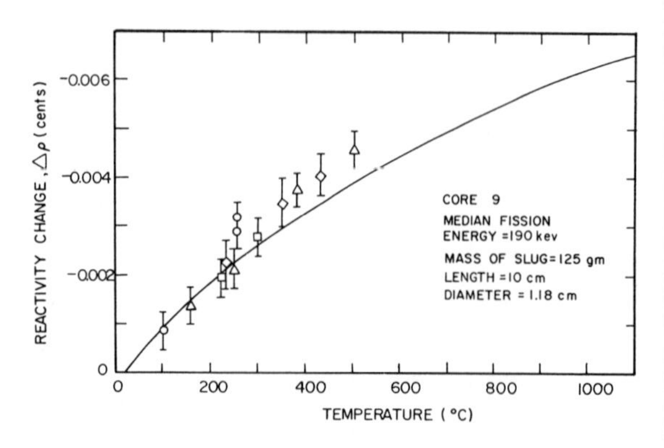

FIG. 2-2 The Doppler effect in thorium metal.

There have been two recent measurements of the Doppler coefficients in spectra corresponding to large fast reactors. Springer and Carpenter [74] have measured Th^{232} and Fisher, et al. [75], U^{238}. Both experiments used zoned critical assemblies in which the central part has the desired spectrum and is surrounded by appropriate buffer and driver regions. Two samples as identical as possible, one hot and one cold, are then oscillated into and out of the assembly and the reactivity change measured. Care has to be taken to ascertain that these changes are due to heating only and not to such effects as thermal expansion or errors in positioning. These effects were proven to be very small in both cases. In measurements of this type the size of the sample will have some influence on the result as pointed out by Greebler [76]. In a small sample the flux spectrum will be that of the surrounding, unmodified by the Doppler broadening, while in a large sample the flux in the resonances will be affected. As shown by Greebler the effect is always in the direction that the smaller sample shows a larger Doppler coefficient than the true one with increases up to 25%.

The composition of the test region in the Th experiment [74] was as follows: U^{233} (in U_3O_8) 24.44 kg, O(in U_3O_8) 4.61 kg, Al 53.96 kg, stainless steel 88.29 kg, and Th 219.59 kg. The Th sample was rather small, a cylinder of 125 g, length 10 cm, diameter 1.18 cm. Results are shown in Fig. 2-2, which is taken from reference [74]. The diverse symbols represent the various runs performed at different days and with different slugs of identical shape and mass. The solid line represents the result of a theoretical calculation. The changes in cross sections with temperature were calculated by the discrete method and then plugged in a multigroup perturbation diffusion theory code. The measured values are probably somewhat too high due to the smallness of the sample and the calculated curve due to the use of the Columbia parameters (see Sec. 1.9) but the agreement is reasonable. A check on the analytical technique was made by calculating the absolute reactivity worth of the sample relative to a void; this gave -1.48 cents, compared to a measured value of -0.132 cents. The contributions of the various parts of the neutron spectrum confirm the trends observed previously. Though the change of cross sections is largest in the energy groups containing the resolved resonances, the flux there is small and more than 50% of the Doppler effect comes from the unresolved regions.

In the U^{238} experiment by Fischer, et al. [75], the composition of the central zone (in atoms/cm^3 $\times 10^{-24}$) was U^{235} = 0.00114, U^{238} = 0.00732, C = 0.00833, Na = 0.01061, and stainless steel = 0.01657. In a second experiment, 40% of the sodium was replaced by graphite; this resulted in a significantly softer spectrum. The Doppler elements were rather large, containing 1.96 kg of U^{238} in the form of UO_2 of 70% theoretical density. The samples had a diameter of 3.5 cm and a length of 28 cm, and the temperatures were 800°K hot and 300°K cold. The measured reactivity change hot versus cold was 0.400 ± 0.005 inhours. Conversion to reactivity gave -0.90×10^{-5} $\Delta k/k$ = 0.462×10^{-5} $(\Delta k/k)$/kg of U^{238}. A 1-D diffusion theory calculation with a recently developed set of ANL cross sections [17d] gave -1.0×10^{-5} $\Delta k/k$. For the softer spectrum assembly with some of the sodium replaced by graphite, the measured effect was -0.68 inhours versus a calculated -0.98.

While the details of the calculations for both experiments, Th^{232} and U^{238}, have not yet been given they show that the effects of these two absorbers, coming practically all from the relatively low-lying discrete resonances, can be measured and calculated with existing methods to perhaps ± 20%.

Recently Springer, Carpenter, and Tuttle [77] have reported measurements on U^{235}. The method was similar to the Th measurements by the same investigators [74], except that the U^{233} in the central region of their assembly was replaced by highly enriched U^{235} (93%). The composition of the test region was as follows: U^{235} (metal) 84.37, U^{238} 6.23, Al 107.53, stainless steel 185.03, Th 299.63, all in kg. The spectrum was thus quite hard with a median fission energy of 195 kev. The test sample (183.81 gm) was highly enriched U^{235} and should thus give results indicative of this material. A large negative temperature coefficient, -5.65×10^{-6} /°K $(\partial k/\partial T)$, was found over the range from 300 to 800°K. This result is rather surprising in view of previous U^{235} measurements [71, 72] and the predictions of theory. A possible explanation would be that the predominatly fission resonances are much broader than the predominantly absorption resonances, since the Doppler effect has much less influence on a resonance which is already quite broad.

This would imply, of course, that the distribution of the fission widths would be quite different from the normal P_n distributions (see Sec. 2.2) and/or that the Γ_γ are strongly negatively correlated with the Γ_f. If the above result is confirmed, it will be of considerable importance for fast reactors. Of course, it does not follow necessarily that U^{233} and Pu^{239} should show a similar behavior.

2.9 Effect of the Uncertainties of the Resonance Parameters

It appears from the preceding examples that the Doppler effect is very small for very fast assemblies, and that the most important contributions for cases where it is appreciable, come predominantly from the low energy part of the spectrum, particularly below 10 kev. This is fortunate in so far as the uncertainties in the resonance parameters increase strongly with increasing energy. It emphasizes, however, the necessity of making calculations as accurate as possible in the low energy range.

The usual discrete method has the shortcoming that it neglects the interference and overlap effects discussed in Secs. 2.1 and 2.6. The importance of these effects was first pointed out by Codd and Collins [17b], who, however, gave only a few sample calculations for energies around 1 kev. Hwang [17a] gave some more detailed results. Table 2-8 gives the change of the cross sections of U^{235} and Pu^{239} in the temperature range of 300 - 1500°K due to Doppler effect calculated with the discrete and the next-neighbor method, and the percentage reduction due to overlapping for selected energies. The assumptions made were a background nonresonant cross section $\sigma_n = 200$ barns, one degree of freedom for the neutron width and two for the fission width, and a strength function $\langle \Gamma_n^0 \rangle / D = 10^{-4}$. Only s-waves were taken into account. Other data used are indicated in the table. It is again seen that the contribution to the Doppler effect decreases rapidly with increasing energy. The reduction due to overlapping is markedly larger for U^{235} than for Pu^{239}, due to the smaller level spacing. This reduction is of order 30% for the upper energy range considered, but much smaller in Pu^{239} at lower energies.

The change in the Doppler coefficient of Pu^{239} due to presence of U^{238} is shown in Table 2-9, also due to Hwang [17a]. The Pu^{239} to U^{238} ratio is 1 to 5 and the assumed background scattering was 300 barns per Pu atom. Only the contribution of the channel J = 1 was calculated. The result is a very drastic reduction of the Doppler effect, which at the energy of 6.5 kev even amounts to a reversal from positive to negative.

The simultaneous action of the two effects (overlap in the same series of resonances and superposition of U^{238}) has not yet been studied.

TABLE 2-8

Change of Cross Sections ($\delta\sigma$) in barns in the Temperature Range of 300 - 1500°K Due to Doppler Effect According to the Discrete Method and with Overlap Correction

U^{235}

$\langle \Gamma_f \rangle = 0.12$ ev, $\langle \Gamma_\gamma \rangle = 0.033$ ev

$D_3 = 1.72$ ev $D_4 = 1.34$ ev

Energy (kev)	$\delta\sigma_f$ Discrete	$\delta\sigma_f$ With overlap	$\delta\sigma_f$ % reduction	$\delta\sigma_\gamma$ Discrete	$\delta\sigma_\gamma$ With overlap	$\delta\sigma_\gamma$ % reduction
12.05	0.0053	0.0035	33.5	0.0024	0.0016	34.3
6.50	0.0140	0.0094	32.7	0.0065	0.0043	33.2
2.25	0.0609	0.0452	32.5	0.034	0.023	32.0
0.61	0.3718	0.2610	29.8	0.211	0.152	28.0
0.20	1.820	0.8550	27.7	0.745	0.557	25.1

Pu^{239}

$\langle \Gamma_f \rangle = 0.099$ ev, $\langle \Gamma_\gamma \rangle = 0.039$ ev

$D_0 = 11.58$ ev $D_1 = 3.86$ ev

Energy (kev)	$\delta\sigma_f$ Discrete	$\delta\sigma_f$ With overlap	$\delta\sigma_f$ % reduction	$\delta\sigma_\gamma$ Discrete	$\delta\sigma_\gamma$ With overlap	$\delta\sigma_\gamma$ % reduction
9.10	0.0172	0.0123	28.3	0.0094	0.0066	29.8
6.50	0.0290	0.0215	26.0	0.0170	0.0126	26.0
2.25	0.1400	0.1135	18.9	0.0800	0.0648	20.2
0.61	0.6500	0.5767	11.3	0.4400	0.3928	10.7

TABLE 2-9

Change of Cross Section ($\delta\sigma$) of Pu^{239} in the Temperature Range 300 - 600°K Due to Doppler Effect and Simultaneous Presence of U^{238}

U^{238}

$\Gamma_f = 0.099$ ev $\Gamma_\gamma = 0.039$ ev

$D_1 = 3.86$ ev

Energy (kev)	$\delta\sigma_f$ Discrete	$\delta\sigma_f$ With U^{238}	% reduction	$\delta\sigma_\gamma$ Discrete	$\delta\sigma_\gamma$ With U^{238}	% reduction
6.50	0.0087	-0.0027	130.4	0.0046	-0.00154	131.2
2.25	0.0413	0.00097	97.6	0.0259	0.00058	97.8
0.61	0.193	0.1081	44.0	0.1372	0.0675	50.8

From the figures given it appears that the Doppler coefficient due to the fissile materials is generally overestimated. There will, of course, also be a reduction in the coefficient for U^{238}, but it can be expected to be quite small. It would seem to be highly desirable to develop a fairly rapid method to estimate these effects and to combine it with a regular reactor code. This is also necessary since sodium has quite marked effects. They come from the general change in the energy spectrum produced by it, the change in sodium density by thermal expansion, and also from the resonance structure of the sodium cross section, in particular the 2.85 kev resonance, which produces corresponding flux depressions. It is thus always necessary to treat Doppler and sodium effects concurrently, see for instance the discussion by Greebler [17e].

A last item of importance is the uncertainty in the calculation of the Doppler effect due to lack of prior knowledge of the nuclear data. This has been studied by Hummel and Rago [17d], and by Froelich, Ott, and Schmidt [17c]. The results of the latter group for a definite reference reactor are reproduced in Table 2-11. Their treatment includes overlap and interference effects. The composition of the reference reactor is shown in Table 2-10. The average fission width was determined by the average capture to fission rate $\bar{\alpha}(E)$. The number of degrees of freedom for the fission width ν_f was taken as one.

Table 2-11 shows the contributions to the Doppler coefficient for the two materials in various energy ranges at a reference temperature of 800°K. This table shows again the predominance of low energies for this type of reactor. In particular, the Pu contribution comes practically entirely from the lowest group. The p-wave contributions for Pu were entirely negligible.

Table 2-12 shows the changes produced in the Doppler coefficient due to changes in the various parameters as indicated in the first column. These changes are about the maximum compatible with the present knowledge.

The most important uncertainties come from the s-wave strength functions, Eq. (2-85a), of both isotopes. In particular, S_0^{49} (the s-wave strength function for Pu^{239}) is the least well known due to the insufficiency of the experimental data. Inaccuracies in the fission data (ν_f, $\bar{\alpha}$) produce total errors of only ~10% since Pu is the smaller contributor to the Doppler coefficient. The influence of the p-wave contributions for U^{238} (last two lines of Table 2-12) is rather weak as an increase in S_1^{28}, the p-wave strength function for U^{238}, by 300% produces only a 20% effect. The tentative conclusion is thus, that the Doppler coefficient for this type of reactor can be calculated with an accuracy of about ±20%. The uncertainties for reactors with a higher energy spectrum will be obviously larger, but the effect itself will become less important.

TABLE 2-10

Composition of Reference Reactors and Average Nuclear Data

Composition:	5 vol. % $Pu^{239}O_2$
	25 vol. % $U^{238}O_2$
	40 vol. % Na
	30 vol. % steel
Nuclear data:	
U^{238}	$S_0 = \langle\Gamma_n^\circ\rangle/D = 1.0 \times 10^{-4}$
	$\Gamma_\gamma = 0.0246$ ev
	$D_S = 18.5$ ev
	$S_1 = \langle\Gamma_n^1\rangle/D_1 = 1 \times 10^{-4}$
Pu^{239}	$S_0 = 1 \times 10^{-4}$
	$\Gamma_\gamma = 0.0387$ ev
	$D_S = 2.64$ ev
	$S_1 = 1 \times 10^{-4}$
	$\langle\Gamma_f\rangle/\Gamma_\gamma = 1/\bar{\alpha}$ for J = 0, 1, 2
	$\nu_f = 1$

TABLE 2-11

Contributions to Doppler Coefficient of Reference Reactor at 800°K

		Pu^{239}		U^{238}		Total	
Group	$E_L - E_H$ (kev)	$\left(\frac{dk}{dT}\right)$ [10^{-7}/°K]	Percent of total	$\left(\frac{dk}{dT}\right)$ [10^{-7}/°K]	Percent of total	$\left(\frac{dk}{dT}\right)$ [10^{-7}/°K]	Percent of total
9-10	40.7-110	0.111	+ 2.1	- 0.931	3.3	- 0.820	3.5
11-12	15-40.7	- 0.052	- 1.0	- 2.296	8.2	- 2.348	10.1
13-14	5.5-15	- 0.094	- 1.8	- 5.274	18.7	- 5.368	23.1
15	2.1-5.5	0.182	+ 3.0	- 6.779	24.1	- 6.597	28.4
16	0.1-2.1	4.719	+97.0	-12.846	45.7	- 8.127	34.9
	Total	$+4.866 \times 10^{-7}$/°K		-28.126×10^{-7}/°K		-2.326×10^{-6}/°K	

TABLE 2-12

Change in Doppler Coefficient for Reference Case Due to Changes in Resonance Parameters

	Pu^{239}		U^{238}		Total	
Case	$\left(\frac{dk}{dT}\right)$ [10^{-6}/ K]	Relative error %	$\left(\frac{dk}{dT}\right)$ [10^{-6}/ K]	Relative error %	$\left(\frac{dk}{dT}\right)$ [10^{-6}/ K]	Relative error %
Average nuclear data	0.487	0	- 2.813	0	- 2.326	0
$\bar{S}_0^{28}$ + 20%	0.455	- 6.6	- 3.234	+15.0	- 2.779	+19.5
$\bar{S}_0^{28}$ - 20%	0.477	- 2.1	- 2.342	- 16.7	- 1.865	- 19.8
$\bar{S}_0^{49}$ + 40%	1.316	+170.2	- 2.813	0	- 1.497	- 35.6
$\bar{S}_0^{49}$ - 30%	0.041	- 91.6	- 2.813	0	- 2.772	+ 19.2
$\nu_f = 1; \bar{\alpha}$ + 20%	0.273	- 43.9	- 2.813	0	- 2.540	+19.2
$\nu_f = 1; \bar{\alpha}$ - 29%	0.757	+ 55.4	- 2.813	0	- 2.056	- 11.7
$\nu_f = 3; \bar{\alpha}$ + 20%	0.421	- 13.6	- 2.813	0	- 2.392	+ 2.8
$\nu_f = 3; \bar{\alpha}$ - 20%	0.286	- 41.3	- 2.813	0	- 2.527	+ 8.6
$\bar{\Gamma}_\gamma^{28}$ ($\ell = 1$) = 60 mv	0.487	0	- 2.974	+ 5.7	- 2.487	+ 7.0
$S_1^{28} = 3 \times 10^{-4}$	0.487	0	- 3.333	+18.5	- 2.846	+22.5

REFERENCES

1. E. Creutz, H. Jupnik, T. Snyder, and E. P. Wigner, "Review of the measurements of the resonance absorption of neutrons by uranium in bulk", J. Appl. Phys., 26(1955)257.
2. L. Dresner, Resonance Absorption in Nuclear Reactors, Pergamon Press, N.Y., 1960.
3. L. Dresner, "Some remarks on the effect of a nonuniform temperature distribution on the temperature dependence of resonance absorption", Nucl. Sci. Eng., 11(1961)39.
4. J. H. Ferziger, "Resonance absorptions in lumps with non-uniform temperature distributions", Nucl. Sci. Eng., 14(1962) 244.
5. A. M. Lane and R. G. Thomas, "R-matrix theory of nuclear reactions", Rev. Mod. Phys., 30(1958)257.
6. E. Vogt, "Theory of low energy nuclear reactions", Rev. Mod. Phys., 34(1962)723.
7. G. W. Hinman, G. F. Kuncir, J. B. Sampson, and G. B. West, "Accurate Doppler-broadened absorption", Nucl. Sci. Eng., 16(1963)202.
8. W. E. Lamb, Jr., "Capture of neutrons by crystals", Phys. Rev., 55(1941)190.
9. M. Born, Optik, Julius Springer, Berlin, 1933.
10. M. E. Rose et al., "A Table of the ψ Function", Report WAPD-SR-506, Bettis, Atomic Power Laboratory, 1954.
11. J. R. Triplett et al., "The RBU Reactor Burnup Code", Report HW-700049, General Electric Co., Hanford Atomic Products Operation, 1961.
12. H. J. Amster, "A differential equation for calculating Doppler broadened resonances", Nucl. Sci. Eng., 11(1961)343. See also E. M. Gelbard, "The Computation of Doppler-broadened Cross Sections", Trans. Am. Nucl. Soc., 6(1963)257.
13. D. M. O'Shea and H. C. Thacher, "Computation of resonance line shape functions", Trans. Am. Nucl. Soc., 6,(1963)36.
14. E. P. Wigner and L. Eisenbud, "Higher angular momenta and long range interaction in resonance reactions", Phys. Rev., 72(1947)29.
15. E. Vogt, "Low-energy cross sections of fissionable nuclei", Phys. Rev., 118(1960)724.
16. D. B. Adler and F. T. Adler, "Cross sections for fissile elements: A simple approach to the multilevel formalism", Trans. Am. Nucl. Soc., 5(1962)53; 6(1963)37; and preprints (unpublished).
17. Proceedings of a Conference on Breeding, Economics, and Safety in Large Fast Power Reactors, October 7-10, 1963, Argonne National Laboratory, Report ANL-6792.
    a. R. N. Hwang, "An Improved Method for Doppler Effect Calculations for Fissile Materials in the Intermediate Energy Region".
    b. J. Codd and P. J. Collins, "Some Calculations Concerning the Overlapping on the Doppler Effect in a Dilute Fast Reactor".

c. R. Froelich, K. Ott, and J. J. Schmidt, "Calculation of Doppler Coefficients of Dilute Fast Reactors" and "Dependence of Fast Reactor Doppler Coefficients on Nuclear Data Uncertainties".
d. H. H. Hummel and A. L. Rago, "Effect of Parameter Variation in Doppler Effect Calculations".
e. P. Greebler, "Recent Improvements in Calculations of Doppler and Sodium Reactivity Effects for Large Fast Reactors".
18. F. T. Adler, G. W. Hinman, and L. W. Nordheim, "The Quantitative Evaluation of Resonance Integrals", Proceedings of the Second U.N. International Conference on Peaceful Uses of Atomic Energy, Geneva, 1958, Vol. 16, p. 142.
19. L. W. Nordheim, "A new calculation of resonance integrals", Nucl. Sci. Eng., 12(1962)457.
20. A. M. Weinberg and E. P. Wigner, The Physical Theory of Neutron Chain Reactors, University of Chicago Press, 1958.
21. W. Rothenstein, "Collision probabilities and resonance integrals for lattices", Nucl. Sci. Eng., 7(1960)162.
22. K. M. Case, F. de Hoffman, and G. Placzek, Introduction to the Theory of Neutron Diffusion, U. S. Government Printing Office, Washington, D.C., 1953.
23. J. Chernick, "The Theory of Uranium Water Lattices", Proceedings of the First U.N. International Conference on Peaceful Uses of Atomic Energy, Geneva, 1956, Vol. 5, p. 215.
24. J. Chernick and A. R. Vernon, "Some refinements in the calculation of resonance integrals", Nucl. Sci. Eng., 4(1958)649.
25. L. W. Nordheim and G. Kuncir, "A Program of Research and Calculations of Resonance Absorption", Report GA-2527, General Atomic, Division of General Dynamics Corp., 1961; TID-4500; see also, G. Kuncir, "A Program for the Calculation of Resonance Integrals", Report GA-2525, General Atomic, Division of General Dynamics Corp., 1961.
26. S. Iijima, "Resonance absorption and the resonance disadvantage factor", Nucl. Sci. Eng., 17(1963)42.
27. M. M. Levine, "Resonance integral calculations for U^{238} lattices", Nucl. Sci. Eng., 16(1963)271.
28. R. Goldstein and E. R. Cohen, "Theory of resonance absorption of neutrons", Nucl. Sci. Eng., 13(1962)132.
29. A. Reichel, "The temperature dependence of effective resonance integrals", Reactor Sci. Technol., 17(1963)281.
30. V. W. Nather and L. W. Nordheim, "Extended Tables for the Computation of the Volume Term of the Resonance Integral", Report GA-2460 General Atomic, Division of General Dynamics Corp., 1961.
31. V. J. Bell, P. A. C. Buckler, and I. C. Pull, "The Calculation of Generalized Doppler Functions", British Report AEEW-R-266, 1963.
32. I. I. Gourevich and I. Y. Pomeranchouk, "Resonance Absorption in Heterogeneous Reactors", Proceedings of the First U. N. International Conference on Peaceful Uses of Atomic Energy, Geneva, 1956, Vol. 5, p. 649.
33. A. R. Vernon, "Calculation of the effective resonance integral of U^{238}", Nucl. Sci. Eng., 7(1960)252.
34. M. Wagner, "Spatial distribution of resonance absorption in fuel elements", Nucl. Sci. Eng., 8(1960)278.
35. E. Hellstrand, "Measurements of the effective resonance integral in uranium metal and oxide in different geometries", J. Appl. Phys., 28(1957)1493.
36. R. M. Pearce, "Radial dependence of the Doppler effect in bars of uranium and thorium", J. Nucl. Energy, Part A, 11(1960)136.
37. S. M. Dancoff and M. Ginsburg, "Surface Resonance Absorption in a Close Packed Lattice", USAEC Report CP-2157, 1944.
38. L. W. Nordheim, "The Theory of Resonance Absorption", Proceedings of Symposia in Applied Mathematics, Vol. XI, 58, American Mathematical Society, Providence, R.I., 1961.
39. "Reactor Physics Constants", USAEC Report ANL-5800 (2nd Ed.) Argonne National Laboratory, 1963.
40. I. Carlvik and B. Pershagen, "The Dancoff Correction for Various Geometries, (USAEC Report AE-16), Aktiebolaget, Atomenergi, Stockholm, Sweden, 1959.
41. Y. Fukai, "First flight collision probability in moderator-cylindrical fuel system", Reactor Sci. Technol., 17(1963)115.
a. A. Sauer, "Approximate Escape Probabilities", Nucl. Sci. Eng., 16(1963)329.
42. J. Hardy, Jr., G. G. Smith, J. A. Mitchell, and D. Klein, "Measurement of the Dancoff corrections", Nucl. Sci. Eng., 12(1962)301.
43. R. K. Lane, L. W. Nordheim, and J. B. Sampson, "Resonance absorption in materials with grain structure", Nucl. Sci. Eng., 14(1962)390, and Addendum, 16(1963)337.
44. D. J. Hughes, B. A. Magruno, and M. K. Brussel, "Neutron Cross Sections", Report BNL-325 (2nd Ed., Suppl. No. 1), Brookhaven National Laboratory, 1960.
45. J. Halperin and R. W. Stoughton, "Some Cross Sections of Heavy Nuclides Important to Reactor Operation", Proceedings of the Second U.N. International Conference on Peaceful Uses of Atomic Energy, Geneva, 1958, Vol. 16, p. 64.
46. E. Hellstrand and G. Lundgren, "The resonance integral for uranium metal and oxide", Nucl. Sci. Eng., 12(1962)435.
47. R. M. Pearce, "The Doppler effect in thermal reactors", J. Nucl. Energy, Part A, 13(1961)150.
48. E. Hellstrand, P. Blomberg, and A. S. Hörner, "The temperature coefficient of the resonance integral for uranium metal and oxide", Nucl. Sci. Eng., 8(1960)497.
49. L. W. Nordheim, "The Contribution of the Individual Resonances to the Resonance Integrals in Uranium and Thorium", Report GA-2563, General Atomic, Division of General Dynamics Corp., 1961.
50. S. Desjardins, J. Rainwater, J. L. Rosen, and W. W. Havens, Jr. "Tentative Results of Br, I, and Th Resonances", Report CU (PNPL)-206, Columbia University, 1960.
51. L. W. Nordheim, "Resonance Absorption", paper presented at a Symposium at the Meeting of the American Physical Society in Houston, Texas, February, 1963, Report GA-3973, General Atomic, Division of General Dynamics Corp., 1963.
52. C. A. Uttley and R. H. Jones, "Total cross sections", from "Nuclear Physics Division Progress Report for the period 1 July to 31 December, 1961", D. L. Allan, Ed., British Report AERE-PR/NP2. See also M. C. Moxon and E. R. Rae in the same report.
53. M. Brose, "Zur Messung and Berschnung der Resonanzabsorption in Gold, Uran and Thoriumfolien", Thesis, Karlsruhe, 1962.
54. E. Hellstrand and J. Weitman, "The resonance integral of thorium metal rods", Nucl. Sci. Eng., 9(1961)507.
55. J. Weitman, "The effective resonance integral of thorium oxide rods", Nucl. Sci. Eng., 18(1964)246.
56. J. R. Brown and J. B. Sampson, "Measured Doppler Coefficient of Thorium Dispersed in Graphite", Report GA-3422, General Atomic, Division of General Dynamics Corp., 1963; also, R. A. Bardes et al., "High Temperature Gas-Cooled Reactor Critical Experiment and Its Application", paper presented at the International Atomic Energy Agency Symposium on Exponential and Critical Assemblies, Amsterdam, September 1963, Report GA-4496, General Atomic, Division of General Dynamics Corp., 1963.
57. G. Goertzel, "An Estimation of Doppler Effect in Intermediate and Fast Neutron Reactors", Proceedings of the First U.N. International Conference on Peaceful Uses of Atomic Energy, Geneva, 1956, Vol. 5, p. 472; see also: H. Feshbach, G. Goertzel, and H. Yamauchi, "Estimation of Doppler effect in fast reactors", Nucl. Sci. Eng., 1(1956)4.
58. R. B. Nicholson, "The Doppler Effect in Fast Neutron Reactor", Report APDA 139, Atomic Power Development Associates, 1960.
59. J. D. Garrison, "A Statistical Analysis of Resonance Parameters", Symposium on Statistical Properties of Atomic and Nuclear Spectra, State University of N.Y., Stony Brook, N.Y., May 3, 1963, in publication.
60. T. D. Newton, "Nuclear level spacings from the statistical model", Can. J. Phys., 34(1956)804; 35(1957)1400.
61. C. E. Porter and R. G. Thomas, "Fluctuations of nuclear reaction widths", Phys. Rev., 104(1956)483.
62. W. W. Havens et al., "U235 Resonance Measurements", Report CU-178, Columbia University, 1958.
63. E. P. Wigner, "Conference on Neutron Physics", USAEC Report ORNL-2309, Oak Ridge National Laboratory, 1952.
64. J. E. Wilkins, "Upper Bound for the Spacing Correction e", Memo, November 1962, private communication.
65. J. H. Ferziger, P. Greebler, M. D. Kelly, and J. W. Walton, "Resonance Integral Calculations for Evaluation of Doppler Coefficients, the Rapture Code", Report GEAP-3923, General Electric Co., Atomic Products Division, 1962.
66. P. Greebler and E. Goldman, "Doppler Calculations for Large Fast Ceramic Reactors and Recent Cross Section Information", Report GEAP-4092, General Electric Co., Atomic Products Division, 1962.
67. H. A. Bethe, "On the Doppler Effect in Fast Reactors", Report APDA-119, Atomic Power Development Associates, 1957.
68. J. J. Schmidt, "Neutron Cross Sections for Fast Reactor Materials", Report KFK-120 (EANDC-E-35U), Part II, Tables, European-American Nuclear Data Committee, 1962.
69. E. G. Bilpuch, L. W. Weston, and H. W. Newson, "Neutron capture cross sections in the kev region", Ann. Phys., 10(1960)455.
70. T. K. Krueger and B. Margolis, "The s-wave and p-wave strength functions", Nucl. Phys., 28(1961)578.
71. A. R. Baker and T. A. J. Jaques, "A Measurement of the Doppler Effect to the Temperature Coefficient of Reactivity in a Fast Reactor", British Report AERE R/M 168, 1963.
72. W. Y. Kato and D. K. Butler, "Measurement of the Doppler temperature effect in an EBR-I type assembly", Nucl. Sci. Eng., 5(1959)320.
73. P. Greebler, B. A. Hutchins, and J. R. Sueoka, "Calculations of Doppler Coefficient and Other Safety Parameters for a

Large Fast Oxide Reactor", Report GEAP-3646, General Electric Co., Atomic Products Division, 1961.
74. T. H. Springer and S. G. Carpenter, "A Measurement of the Doppler effect in thorium in a fast neutron energy spectrum with a median fission energy of 190 kev", Nucl. Sci. Eng., 17(1963)194.
75. G. J. Fisher, H. H. Hummel, J. R. Folkrod, and D. A. Meneley, "Doppler coefficient measurements for U^{238} in fast reactor spectra," Nucl. Sci. Eng., 18(1964)290.
76. P. Greebler, "Measurement of Doppler coefficient by heating a small region of a fast reactor critical assembly", Nucl. Sci. Eng., 18(1964)287.
77. T. H. Springer, S. G. Carpenter, and R. J. Tuttle, "A measurement of a Negative Doppler Effect in U^{235}", private communication to the Advisory Committe on Reactor Physics, April 1964.

CHAPTER 5

Criticality

H. C. PAXTON, G. R. KEEPIN
University of California, Los Alamos Scientific Laboratory,
Los Alamos, New Mexico

CHAPTER CONTENTS*

*Except for a few changes and additions made in proof, this chapter is based on information in the literature or known to the authors prior to July 1963.

The object of this chapter is to relate critical assembly experience, both experimental and analytic, to reactor safety. The emphasis is not on the safety of critical assembly operations per se, because the requirements for experimental flexibility and for a negligible fission-product inventory result in conditions generally inconsistent with those basic to reactor safety.

Attention is directed toward the collection of practical safety information that has existed primarily as scattered lore and toward systematizing the information available and filling in certain gaps in the information required for interpreting safety measurements. Neutron-multiplication measurements as used for safely approaching criticality, are in the category of "scattered lore". New or newly organized information includes analysis of the perturbing effects of photoneutron and delayed-neutron transients on multiplication measurements, delayed-criticality determinations, and reactivity measurements. Finally, there are some remarks about fuel storage at the reactor site.

1 NEUTRON MULTIPLICATION AS A NUCLEAR SAFETY INDEX

Reactivity monitoring is required to confirm the safety of certain operations with reactor fuel, such as reactor loading, approach to criticality, and storage under unevaluated conditions. The most simply measured reactivity index for such monitoring is neutron multiplication or a quantity roughly proportional to it. The basic requirements for this type of measurement are a neutron source, either inherent in the fuel assembly or added to it, and a neutron detector with an adequate and reliably scaled response to neutrons from fission. If these requirements are satisfied, the detector response rate under equilibrium conditions is proportional to neutron multiplication, and its reciprocal approaches zero as the fuel assembly is made to approach the delayed-critical state. The criticality approach curve (e.g., reciprocal response rate vs quantity of fuel, spacing between fuel elements, moderator level, or control-rod position) is then orderly and in principle its form is sufficiently predictable for extrapolation to criticality. Where such extrapolation is possible, neutron response measurements during reactivity buildup can serve as the required safety guide.

In certain special systems, where the ratio between response rate and neutron multiplication is known, a single measurement may be sufficient to indicate the reactivity. Unfortunately, the measurement of absolute neutron multiplication is straightforward only for small fast-neutron assemblies. In the following sections this particularly simple case is considered, primarily to provide a basis for discussing complications which become extreme for large hydrogen-moderated systems. First, however, it is appropriate to consider the meaning of idealized neutron multiplication.

In the discussion that follows, influences of photoneutrons and of nonequilibrium counting conditions are ignored until these perturbing effects are introduced specifically.

1.1 Definitions of Neutron Multiplication

Let us imagine the introduction of source neutrons into a subcritical assembly, such that their energy spectrum and spatial distribution are characteristic of the assembly, i.e., equilibrium spectrum and so-called "normal-mode distribution" are matched. Total multiplication with normal-mode source T_n then is the total number of neutrons appearing in the fissionable material per source neutron.* Assuming S source neutrons, there will be Sk neutrons after the first fission generation, Sk^2 neutrons after the second generation, and so on, so that the total number of neutrons per source neutron after many generations approaches:

$$T_n = \frac{(1 + k + k^2 + \ldots)S}{S} = \frac{1}{1 - k}. \quad (1\text{-}1)$$

*Concepts and terminology were developed by Robert Serber during the early years of the Los Alamos Scientific Laboratory.

In terms of F, the total number of fissions produced per source neutron, and ν, the number of neutrons emitted per fission,** the total multiplication is

$$T_n = 1 + F\nu . \quad (1\text{-}2)$$

The net multiplication with normal-mode source, N_n, is the number of neutrons escaping permanently from the fissionable material per primary source neutron, or***

$$N_n = 1 + F(\nu - 1 - a) , \quad (1\text{-}3)$$

or

$$N_n = 1 + \frac{(\nu - 1 - a)}{\nu} \cdot \frac{k}{1 - k}. \quad (1\text{-}4)$$

Though conceptually simple, both T_n and N_n are difficult (if not impossible) to measure in any assembly.

A more nearly measurable multiplication, particularly in fast-neutron systems, is net multiplication with the source concentrated at a point within the fissionable material, e.g., at the center of a geometrically simple assembly. This, or as close an approach to it as attainable, is the type of neutron multiplication that is used as a safety index. Net neutron multiplication with a central source may be expressed as

$$M = 1 + \frac{\gamma_0(\nu - 1 - a)}{\nu} \cdot \frac{k}{1 - k}. \quad (1\text{-}5)$$

Here γ_0 is the average ratio of effectiveness of central neutrons to normal-mode neutrons, which is nearly constant if k is near unity.**** Thus, in most cases, reactivity or departure from criticality (k=1) may be represented adequately for safety purposes by

$$\frac{1 - k}{k} \sim \frac{1}{M - 1}, \quad \text{or} \quad 1 - k \sim 1/M . \quad (1\text{-}6)$$

Although this approximation is reasonably general, it is of limited usefulness because there is not always a practical means of measuring the true central-source multiplication.

1.2 Multiplication in Fast-Neutron Assemblies

Operationally, the logical definition of multiplication is the ratio of the external counting rate with a neutron source centered in the assembly (multiplied count) to the counting rate with

**That Eqs. (1-1) and (1-2) are equivalent may be seen from the relation $F = k/[\nu(1 - k)]$ which arises from the summation of k/ν, the probability of first generation fission per source neutron, k^2/ν the probability of second generation fission, and so on.

***In the development that follows, $\alpha = \sigma_c/\sigma_f$, the ratio of (suitably averaged) capture to fission cross sections.

****As $k \to 1$, the coefficient $\gamma_0 (\nu - 1 - \alpha)/\nu$ approaches $1 \pm .25$ for a wide variety of fast-neutron assemblies.

fissile material removed* but with the source and other materials undisturbed (unmultiplied count). The multiplication determined in this way is the same as the M of Eq. (1-6) for small fast-neutron assemblies, in which effects of neutron energy change are unimportant. In water-moderated cores, by contrast, it is almost impossible to match neutron spectra for multiplied and unmultiplied counts; consequently, departure from the M of Eq. (1-6) is to be expected.

Where absolute value of multiplication is important, there is some advantage to a neutron detector with a relatively flat response in the appropriate neutron energy range (see Sec. 1.4, Fig. 1-7). When a critical assembly is made with highly γ-active fuel, special precautions must be taken with the instrumentation to assure that the signal is primarily neutron-dependent. This applies to fast and thermal systems and all types of moderators. Successfully used techniques include:

a) Gamma-compensated ion chambers**
b) Fission counters
c) Lead-shielded neutron-sensitive detectors of any kind
d) Li^6 (n-α) detectors.

Equation (1-6) suggests how multiplication measured in the manner just described can be used for fast-neutron systems. In these cases, fraction of critical mass can be judged roughly from a single appropriate multiplication value. Figures

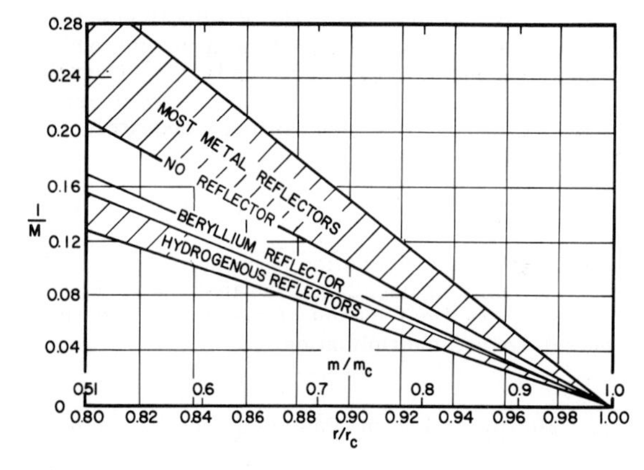

FIG. 1-1 Reciprocal neutron multiplication vs fractional critical radius or critical mass for spheres of U^{235} in various reflectors. External "long" BF_3 counters and a central mock-fission source were used.

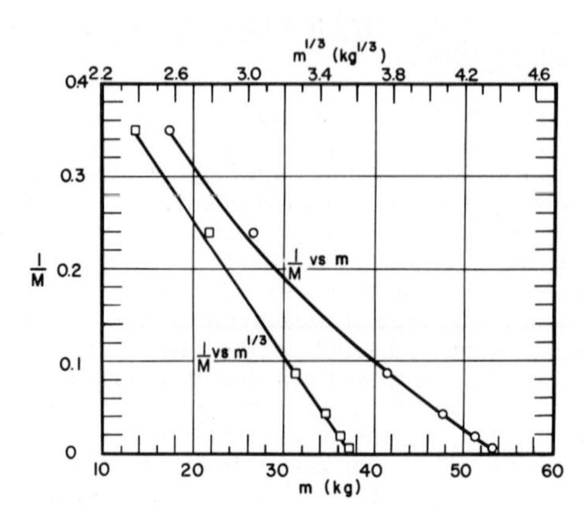

FIG. 1-2 One of the critical approach curves from which Fig. 1-1 was obtained. The unreflected U^{235} assembly was built up of nesting rings in order to maintain an approximately spherical form.

1-1 and 1-2 give observed relations between 1/M and fraction of critical radius or fraction of critical mass for U^{235} metal spheres in various reflectors. These curves also apply fairly well to Pu^{239} and U^{233} cores. For such cores in non-moderating reflectors M = 10 corresponds to $\sim(3/4)m_c$, where m_c = the critical mass, and represents a reasonable limit for hand operations. For a metal core with a thick hydrogenous reflector, the equivalent multiplication at this fraction of a critical mass may appear to be as great as 20.

Some empirical values of S_r and S in the relations $1/M = S_r(r_c - r)/r_c$ and $1/M = S(m_c - m)/m_c$, where r_c and m_c are critical radius and mass, are given in Table 1-1 [2, 2a, 56].

TABLE 1-1

Approximate Empirical Relations Between Reciprocal Multiplication and Other Parameters Related to Reactivity

Assembly	m_c (kg core material)	$S_r = \frac{1/M}{(r_c-r)/r_c}$	$S = \frac{1/M}{(m_c-m)/m_c}$	$\frac{\Delta(1/M)}{100¢}$
bare U^{235} (94%)	52	1.1	0.36	0.009
bare Pu (δ phase)	16.3	0.8	0.3	0.002
bare U^{233} (98%)	16.1	~0.8	~0.3	~0.003
U^{235} (94%) in thick U	17.4	1.3	0.43	0.010
Pu (δ phase) in thick U	5.7	0.9	0.3	0.003
U^{233} (98%) in thick U	5.5	~0.9	~0.3	~0.004
bare U (53-1/2% U^{235})	162	1.2	0.4	0.007
$U^{235}H_3C$ in thick Ni	14.7	1.3	0.44	0.013
$U^{235}H_3C$ in thick U	14.7	1.5	0.50	0.018

Increments of 1/M equivalent to the reactivity interval between the delayed critical and prompt critical states (100¢) are also included for guidance in cases where delayed criticality is to be attained but, as usual, prompt criticality avoided. Data that are included for some assemblies with slightly-moderated cores show that central-source multiplication as we measure it is not too badly distorted for use as a safety index. For these systems, unmultiplied counts were obtained with moderator in place but fissile material removed.

*For convenience, the fissile material sometimes is replaced by natural uranium for the unmultiplied count. With a mock fission source, such use of natural uranium in fast-neutron systems affects the multiplication value by less than 2% but with the higher energy Po-Be source, there may be 20% distortion because of fast fission multiplication in the uranium [1].

**The typical compensated ion chamber is 95-99% compensated; some may be over-compensated. It is necessary to prove neutron dependence of the signal by experiment in the existing gamma field to ensure adequate response.

1.3 Multiplication Problems with Moderated Systems

For D_2O-, Be-, and C-moderated and reflected cores of intermediate neutron energy, and especially H_2O-moderated cores, the distortion of multiplication may be serious as the result of of the difficulty in establishing an appropriate unmultiplied reference system. Multiplication with a normal-mode source would be appropriate, but the more practical central-source multiplication no longer bears a simple relationship to it. In such a situation, the observed multiplication is very sensitive to source spectrum, detector response and geometric details. Even though distorted, however, a plot of the reciprocal of the neutron response is useful for tentative extrapolations to criticality during the buildup of such an assembly.

1.4 Neutron Response Ratios for Water-Moderated Systems

1.4.1 Measurement of Neutron Multiplication

It is not generally possible to determine k from multiplication measurements during the approach to criticality. It is generally desirable to monitor the approach, and analyze the data, to ensure that the approach is orderly and that one is "far from" or "close to" critical. In a laboratory experiment, one may conduct an empirical search for source and detector locations that give a 1/M curve that follows Eq. (1-6). In the typical power reactor, this technique is severely constrained by inflexible hardware, and there is relatively little need for knowledge of k (i.e., the information sought generally is the number of fuel assemblies required to become critical). A proven technique is to use several detectors (3-6) distributed around the fuel and to combine the data into a composite approach curve. In general, many detector moves are necessary, and each detector will indicate a different neutron level due to different source-fuel-detector geometries. Corrections can be made for each by relative flux measurements and empirical weighting factors (see Sec. 1.4.2).

1.4.2 Shape of Reciprocal Count-Rate Curves

As we have observed, the curve of reciprocal neutron-count rate vs the parameter that controls reactivity may be a useful guide for safely approaching criticality, even if the scaling of the rate should be arbitrary. Tentative extrapolations of the curve to criticality (zero reciprocal rate) give conservative estimates provided a "nose-diving" shape (negative curvature), such as shown by curve A of Fig. 1-3, can be avoided. Where reactivity is increased by decreasing the spacing between fissile components of an assembly, this "nonconservative" shape (i.e., extrapolation leads to estimates which are not conservative) may be inherent (Fig. 1-4). In such cases, the shape of the reciprocal count curve must be anticipated in order to choose safe steps of reactivity addition during approach to criticality. There is no serious problem provided the conditions that can lead to the nonconservative shape are recognized in advance.

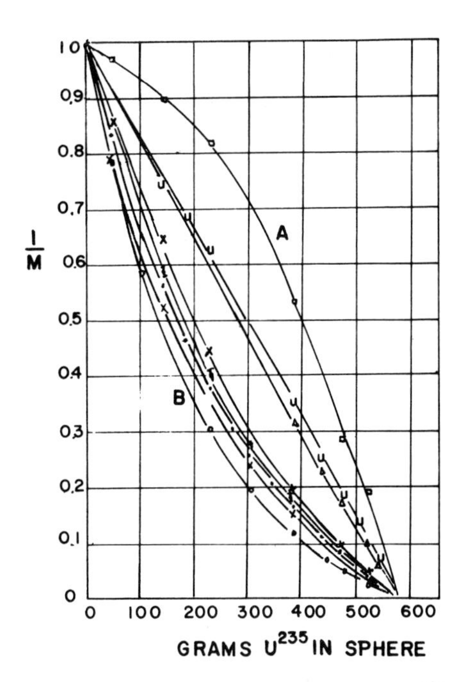

FIG. 1-3 Critical approach curves obtained while U^{235} concentration was increased in the original (low power) Los Alamos water boiler.

It may be noted that the desirable linear approach curve is frequently difficult to attain. Sometimes, a practical substitute is to position detectors so that the curve from one tails out (positive curvature) while the other dives. The former curve, then, guides the conservative choice of reactivity increments, and the pincers-like combination leads to a more realistic estimate of the critical condition.

1.4.3 Uniform Lattices

In the very practical case of building up a uniform lattice of enriched-uranium fuel elements in water, it is relatively easy to avoid a "nose-diving" relationship between reciprocal-count rate

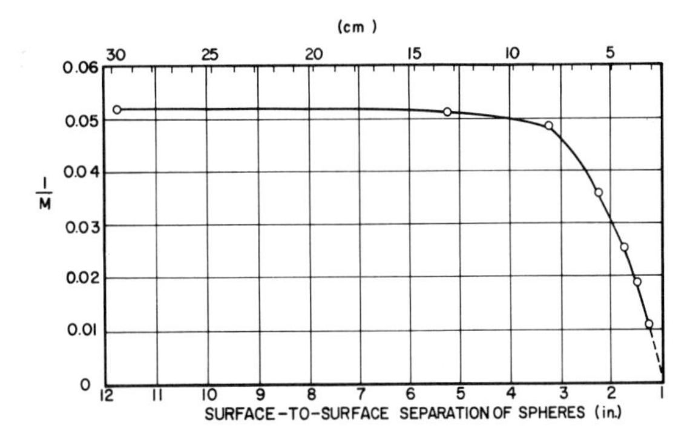

FIG. 1-4 Reciprocal neutron multiplication vs separation between two ~20 kg (~44 lb) enriched-uranium spheres in water.

and number of elements. This is because the reciprocal rate tends to be more nearly proportional to an overall dimension of the lattice than to the mass of fissile material (as in Fig. 1-2). In other words, curve B of Fig. 1-3 is the natural shape and is approximated if the lattice is built outward from the source in a continuous fashion and if the initial neutron attenuation between source and detector is not much greater than an order of magnitude, yet not significantly less than the attenuation between detector and the early fuel additions. The form of curve A is observed if the source and detector are too close together or if fuel loading starts far from the detector and continues toward it. A nose-diving curve also may result if there is departure from the final lattice spacing, particularly if vacant sites are left for later filling at which time the reactivity change per element would tend to be abnormally large.

An example of fuel loading curves in which there is an informative combination of shapes has been provided by C. A. Anderson, Jr. of the Sandia Corporation, Fig. 1-5. The figure applies to the Sandia Engineering Reactor, in which MTR-type fuel elements were loaded from the center outward. Four Pu-Be neutron sources were in the core support plate, somewhat below the fuel. The three detectors that gave similar tailing curves (E, S, W) were within the pressure vessel above fuel locations. As intervening water was 19 in. (48.3 cm) to the nearest source but only 7 in. (17.8 cm) to the four central elements, these counters were particularly sensitive to initial fuel additions. The detector (NW) that gave the nearly linear curve was outside the pressure vessel at a lower level than the others, such that it was separated from source and fuel elements by about the same water thickness (12 in. or 30.5 cm and 10 in. or 25.4 cm, respectively). A combined "pincers-like" criticality approach, as mentioned earlier, is well illustrated.

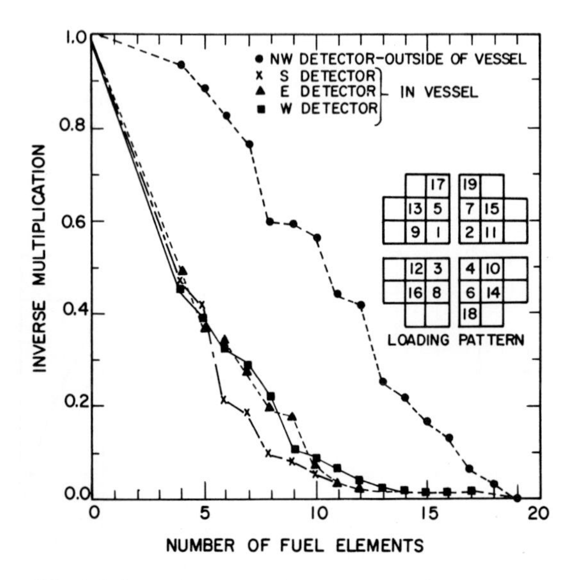

FIG. 1-5 Approach to critical. Inverse multiplication vs fuel loading for the Sandia Engineering Reactor.

If the fuel to be arranged in a lattice contains plutonium, the natural neutron source may be adequate for monitoring. In this case, the neutron source strength increases with quantity of fuel; consequently, the appropriate (reciprocal) index is the neutron counting rate divided by the mass of plutonium in the lattice. With this built-in source a nose-diving response curve is avoided if the detector "sees" initial fuel additions at least as well as those that follow. Figure 1-6 shows a response curve for a lattice of LAMPRE fuel pins (3/8 in. diameter x 6 in. long Pu alloy in 0.027 in. thick Ta, 170 g Pu each)* in polyethylene. Two detectors below the array viewed the fuel pins through 2 in. thick (5.08 cm) polyethylene.

Another common means of approaching criticality with a lattice is to follow neutron count rate as the water level is raised in a complete array of fuel that is initially air-spaced. Here, in addition to the precautions that have been mentioned, it is highly desirable to maintain a constant thickness of water between source and detector throughout somewhat more than the range of water level for which safety monitoring is required. The alternative is a series of unmultiplied counts at various water thicknesses, which is usually tedious because of extreme sensitivity to geometric detail. Although applying to solution systems that are brought critical by raising the level, the multiplication curves reported by Beck, Callihan, Morfitt, and Murray [3] illustrate

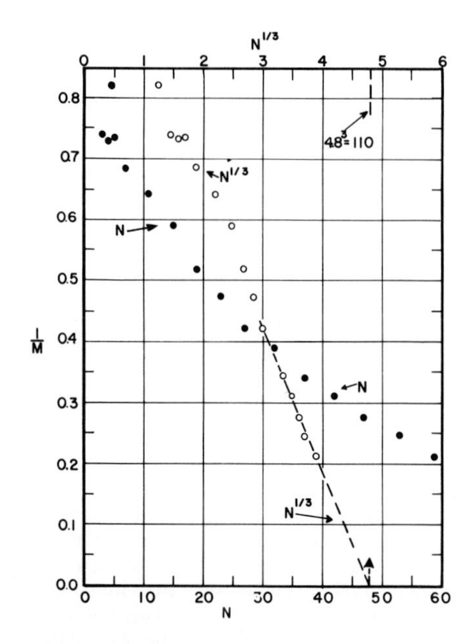

FIG. 1-6 Neutron multiplication of Ta-clad Pu LAMPRE elements latticed in a polyethylene block -- 0.5 in (1.27 cm) minimum spacing. Detectors were "long" BF_3 counters beneath the block; there was no auxiliary source.

---

*0.95 cm diameter x 15.2 cm length Pu alloy in 0.69 mm Ta (1/2 in. or 1.27 cm minimum polyethylene between pins).

distortions that are encountered until the neutron source is covered.

A striking example of the influence on apparent neutron multiplication of detectors with differing energy responses is provided by Dixon Callihan of ORNL. The multiplication curves of Fig. 1-7 were obtained with one BF_3 counter imbedded in the water reflector of the indicated assembly, and another counter external. In one normally-shaped curve of Fig. 1-7 [curve C-2′ (B)] the external counter is surrounded by a paraffin sleeve, but, for the curves that show a minimum in apparent neutron multiplication [e.g., C-2(A)], the counter remains bare. An explanation of the anomalous curves depends upon the fact that the bare detector responds primarily to thermal neutrons, as contrasted with the response of a detector surrounded by hydrogenous material. Until the assembly is covered by considerable water, most thermal neutrons outside the reflector originate as fast neutrons that have leaked out the top and have been thermalized by scattering off the concrete walls, floor, and ceiling surrounding the assembly. Although both internal and bare external detectors give well-behaved reciprocal multiplication curves when the system can be made critical (C), the superposed decrease in fast neutron leakage with water depth upsets the form of the bare counter response curve when the assembly remains subcritical (A and B).

1.4.4 Homogeneous Systems

As suggested by the illustrations of reference [3] neutron counting problems with homogeneous aqueous assemblies are similar to those encountered with water-moderated lattices if quantity of fuel-plus-moderator is the controlling parameter. But

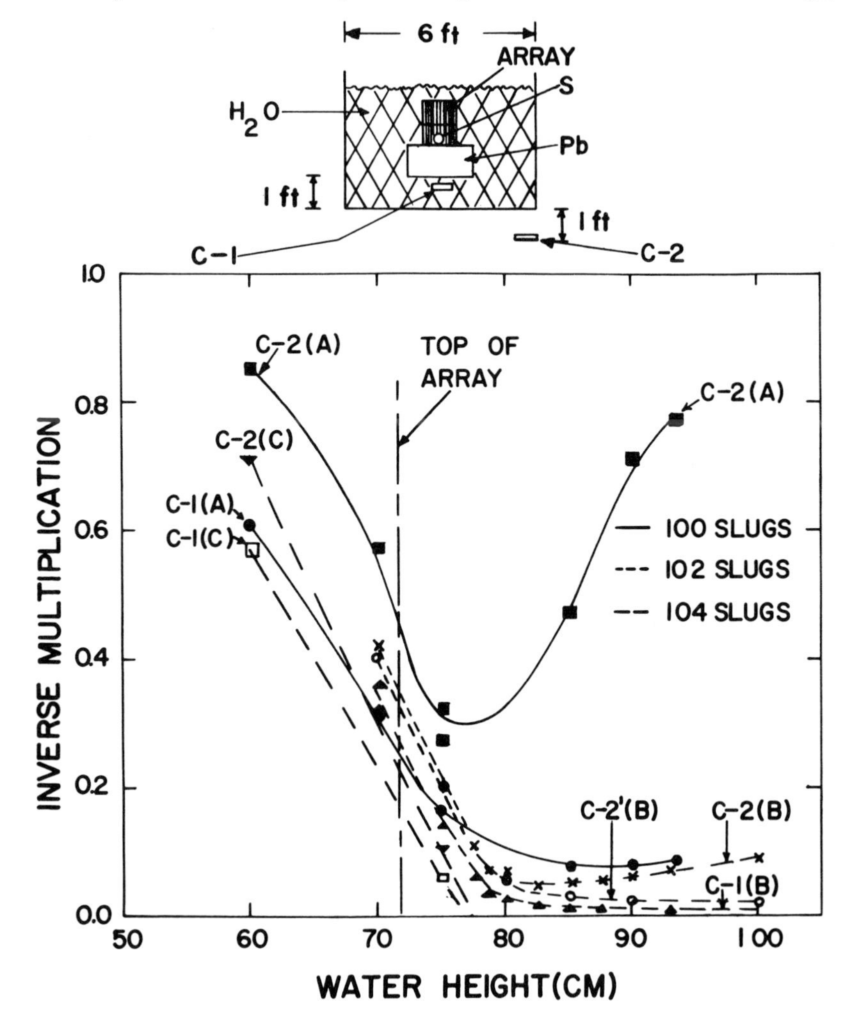

FIG. 1-7 Multiplication curves for lattices of U (enriched to 93% U^{235})-Aℓ slugs. The U-Aℓ has 7.6 wt.% of U. Basic array is a 7 x 8 x 2-high lattice of 1.35 in. diam. x 8 in. long (3.43 cm diam. x 20.3 cm long) cylindrical slugs, axes vertical, 0.75 in. (1.91 cm) surface-to-surface spacing. Experimental arrangement is shown in sketch above the curves. C-1 (A) and C-2 (A) designate the responses of interior counter and exterior counter, respectively, for a 100-slug lattice; C-1 (B), C-2 (B) are the responses of the same counters for a 102-slug lattice and C-1 (C), C-2 (C) for a 104-slug lattice. C-1 is always in water and C-2 is in air. The curve C-2′ (B) designates exterior counter response when counter is in moderator.

a somewhat different situation exists if the approach to criticality is by stepwise increases in concentration of fissile material (at constant volume). In this case Δk may be nearly proportional to concentration so that Eq. (1-6) implies a linear reciprocal multiplication curve, which, in practice, may be distorted to the form of either curve A or B of Fig. 1-3. C. P. Baker and co-workers [4] report that such distortions, observed during startup of the first Los Alamos water boiler, depended on the separation of the detector and the centered neutron source. A nose-diving form resulted for very small separations and tailing-out form for large separations. With the detector within the core at the computed node of the third harmonic of the neutron flux distribution (the most important higher harmonic, assuming a centered source), the linear form of Fig. 1-3 resulted*.

1.4.5 Large, Heavily Poisoned Lattices

The large, heavily poisoned cores of pressurized-water and boiling-water reactors present special problems in tracking reactivity by neutron-multiplication measurement. In the typical cold core, neutron absorbers tend to isolate fuel bundles that individually have high reactivity. The result is that a single neutron detector responds primarily to additions of fuel between it and the source. Until the core is nearly of critical size, the effect may be so extreme that, with a central source, a detector on one side of the core does not "see" additions of a number of fuel bundles toward the far side. In spite of this difficulty, the composite picture from several detectors spaced about the reactor gives reasonable nuclear safety guidance during buildup of most of the core. As the system approaches final size, important additional guidance is provided by curves of 1/M vs position of one or more control rods.

Frequently, loading response patterns established in advance by critical assembly measurements minimize uncertainties during reactor-core buildup. Farrar [5] reports such measurements appropriate to core I of the Consolidated Edison reactor, and Ball and McKinney [6] give similar information for core I of the reactor of the N. S. Savannah. Both sets of data illustrate the peculiarities of response curves that are typical of most water-moderated power reactors.

Multiregion lattices, such as mockups of the hollow square "seed" arrangement of the original Shippingport reactor core [7], also call for special precautions in interpreting response curves. As the center of reactivity can shift violently with loading and as the shielding of distant fuel regions is large, no single detector can be depended on for reliable diagnosis. It is customary to monitor independently the four sides of such an assembly.

*This result was predicted by Enrico Fermi. The fundamental mode, or "first harmonic", increases in relative importance as the system approaches critical.

1.5 Other Observations about Neutron Response Curves

Usually, the problems associated with proper response curves for graphite-, D_2O-, or Be-moderated assemblies are not severe. With a central source and a detector outside the core region, the curve of reciprocal count rate vs core mass extrapolates conservatively (i.e., underestimates the critical mass) when a uniform core is built outward from the axis. The latter condition, however, is not essential for yielding a curve which extrapolates conservatively; this is illustrated by Schemel's report [8] of the stepwise buildup of a nonhydrogenous assembly in which fuel additions were distributed almost randomly over the cross section of the cylindrical core. For central source and external detectors, he observed that curves of reciprocal count against number of fuel elements were nearly linear beyond 25% of the total loading. Initial distortions were smaller for detectors alongside the core than for others at the core ends.

Where an internal source is not practicable, the shape of the critical-approach curve will depend markedly on the counting geometry. Extreme nose-diving is to be expected if source and detector are on the same side of the core, whereas tailing out is customary if source and detector are on opposite sides.

Particularly for large cores, neutron-count relationships may be distorted by shifts in center of reactivity unless loading is symmetric about the source.

A practical means of combining data from multiple detectors to balance such distortions has been communicated to the authors by R. O. Brugge and T. M. Snyder of General Electric Company, APD. Based on their experience with Boiling Water Reactors, the following preferred arrangements, practices and conclusions are presented.

"1. An internal source is preferred, and will produce the best data; this cannot always be done, and external-source data can be handled adequately by subtracting the source-only readings from readings taken with fuel. (This is not necessary in an H_2O-moderated boiling water reactor with an internal source because of the rapid attenuation.)

"2. Data from several detectors are normalized to give equal weight to each detector at the beginning and renormalized at selected fuel configurations (2×2, 2×3, 3×3, etc.)

"3. Normalized values are combined as the sum of the reciprocals because the value changes less with a large change from a single detector than the reciprocal of the sum. (A large change from one detector is usually due to a radical geometric change.)

"4. Ion-chamber-picoammeter systems are considered superior to B^{10} counters because these instruments have a much wider range (8.5 decades) and better long time stability. The wider range permits data to be taken from the source-only level to critical, and beyond, without moving the chamber and thus causing an additional geometric effect.

"5. k_{eff} vs N is calculated for loading steps through critical size. This permits a plot equivalent to $1/M$ vs B_g^2 for any geometric shape.

"6. The sum of normalized reciprocals multiplied by N is plotted vs calculated k_{eff} (1/MN vs k) or vs $1-k_{eff}$.

"7. If the source is internal, a 'source injection efficiency' correction is made to the raw data until the source is completely surrounded by fuel (4 elements). That is, for loadings of 1, 2, and 3 elements, the data are multiplied by 4, 2, and 4/3 respectively.

"Items 2, 3, 6 and 7 are tedious and time consuming; however, loading steps and rod withdrawals to obtain data take 1 to 2 hours each and data processing is not a controlling item. The result is a reasonably consistent forecast of the loading size that will become critical".

A multiplication curve for the minimum critical Dresden reactor core (Fig. 1-8), which was obtained by following the above techniques, appears in Fig. 1-9.

Finally, we should recognize the perturbing influence of irradiated fuel on multiplication during partial reloading of a reactor. The effect, attributable to the extraneous photoneutron source, is most pronounced in reactors containing Be or D_2O. Even the D_2O in ordinary water, however, may be sufficient to result in significant additional distortion of a response curve [8a]. The extreme example occurs when monitoring (by means of an external counter) the reloading of a central fuel zone after removing all old elements in the zone. In this case, photoneutrons from the periphery mask interior changes until the reactor is nearly critical. The reciprocal-response curve remains almost flat through most of the reloading, then suddenly dives toward critical in the final stages. Where practicable, a better procedure for tallying reactivity changes is to replace fuel bundles individually and follow each replacement by a delayed-criticality check.

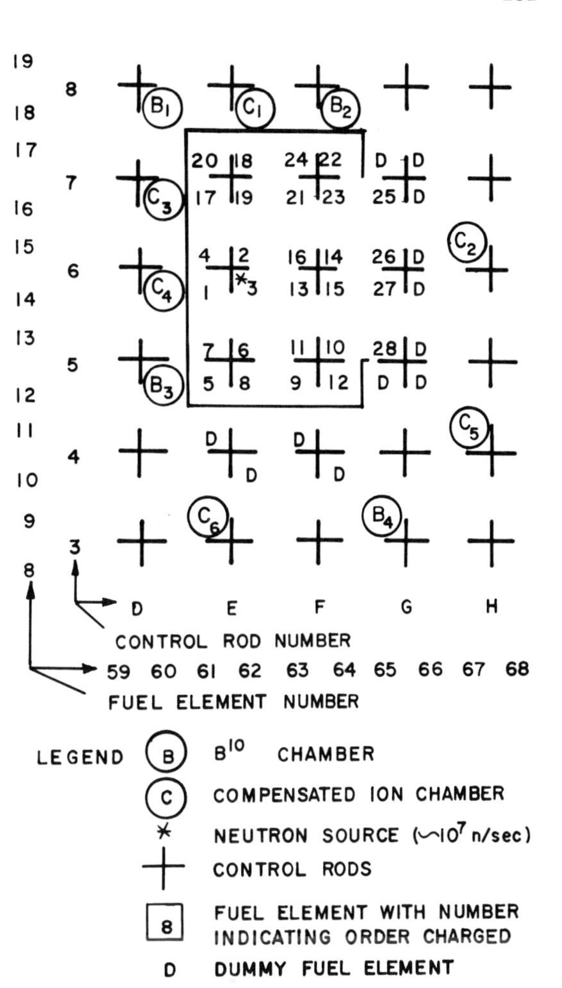

FIG. 1-8 Loading pattern and final chamber locations of the minimum critical Dresden core. Fuel was in bundles of 36 fuel rods in a 6 x 6 square lattice: 1.5% U^{235}, Zr-clad, Zr channels. A control rod cell (4 fuel elements) is ~ 10 in. (25.4 cm) square and the fuel ~ 9 ft (2.75 m) long. Source was an Sb-Be pin, 16 in. (40.6 cm) long and centered axially in fuel element No. 3 and near the xy center of this element.

1.6 General Remarks about the Approach to Criticality

The above emphasis on multiplication values and the form of reciprocal-count curves is not intended to imply that these factors are of exclusive importance in taking a reactor to critical. Usually other guiding information, such as the tentative critical configuration, control rod calibrations, or values of the reactivity worth of fuel, is available from prior experiments or from measurements during the startup of similar reactors. Information of this sort provides independent safety guidance and eliminates complete dependence upon multiplication curves. Nevertheless, because independent safety checks are desirable, it is still important to forsee the nature of approach-curve distortions.

Perhaps more important than the details of instrumentation, data processing and interpretation, is a thoughtfully conceived procedure for the test carried out under close supervision of personnel that are critical of every deviation from plan and expectation. One of the most vital elements for an orderly critical loading is communication and its interplay with procedure. The typical power reactor is isolated from the control room by distance and enclosures. There must be only one man who is responsible and he must have assured communication with the other location. From experience, supervisory control is best located in the control room; this provides first-hand knowledge of all meter readings and access to technical advisors, processed data, etc. The loading supervisor is then "slaved" to the control room supervisor and performs loading steps only when and how directed. He must be constantly alert for a change in orders and continuously inform the control room of the loading status. Confusion is generally minimized with a direct headset connection so that supervisor-to-supervisor communications do not interfere with supervisor-to-workman instructions.

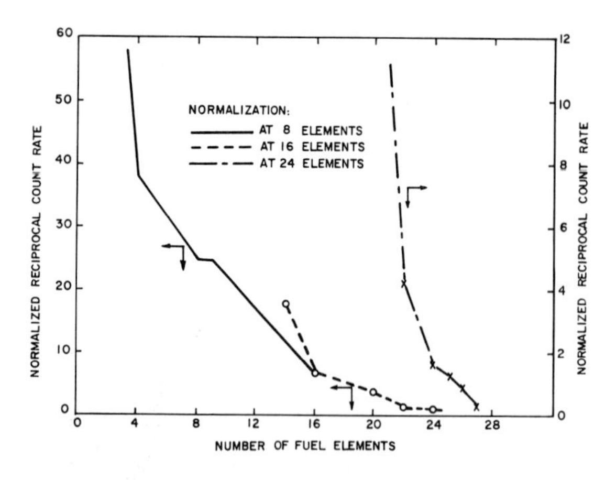

FIG. 1-9 Composite reciprocal multiplication curve in the loading of the minimum critical Dresden reactor core. (See caption of Fig. 1-8 for details.)

Background noise at the loading area is generally high; any generally-audible signals, alarms, speakers, etc., only increase the difficulty of close supervisory control.*

1.7 Requirements on Measuring System

1.7.1 Normal Startup

To reiterate, for quantitative interpretation of multiplication, distortions that are due to space-dependent "higher-mode" effects must be minimized by appropriate location of neutron source and detectors. Characteristically, modal effects, sensitivity to detector location, etc., are difficult to interpret, particularly for systems subcritical by more than a few dollars. In general, all such distorting effects become relatively small near delayed criticality and extrapolated critical points become progressively more reliable as the critical condition is approached (cf. Sec. 1.2).

Although measurement of source multiplication is basically a static technique, there are time-dependent effects which can be important in the interpretation of results. The asymptotic, "true" multiplication following a reactivity change in a subcritical system is observed only after all delayed neutrons (and photoneutrons, if present) have attained equilibrium concentration. Representative results of calculations ** of the transient approach to various equilibrium multiplication values are shown in Figs. 1-10 and 1-11. To illustrate the importance of photoneutrons, these calculations apply to U^{235} reactors that contain D_2O or Be with various proportions of photoneutrons in the delayed fraction. Each system is brought suddenly from an initial equilibrium shutdown condition (assumed 10$ subcritical) to the final multiplication values shown. The parameter varied in the curves is the relative photoneutron effectiveness γ_p/γ, defined as the ratio of average effectiveness (in producing fission) of photoneutrons to the average effectiveness of delayed neutrons. The case $\gamma_p/\gamma = 0$, of course, corresponds to a reactor without photoneutrons.

Photoneutrons [10] with longer lifetime increase greatly the time required to reach equilibrium. (The abundant 3.1 hr group in beryllium [10] is quite apparent in the $\gamma_p/\gamma = 1$ curve of Fig. 1-11 as a straight-line segment in the neighborhood of 10^4 sec.) The data of Figs. 1-10 and 1-11 apply to systems initially in equilibrium; under nonequilibrium initial conditions even longer time intervals are required to reach equilibrium multiplication at the final level.

During a sequence of stepwise additions of reactivity, as in following 1/M during approach to criticality, total asymptotic multiplication will clearly be observed at each step if one waits sufficiently long after each reactivity addition. But if reactivity is added too rapidly—either stepwise or continuously—the observed multiplication will lag the asymptotic multiplication thus giving an underestimate of the instantaneous reactivity of the system. The maximum possible such error in observed multiplication occurs for very large ramp rates when the system is in the near-critical condition. This maximum error is determined solely by the effective total delay fraction, $\gamma\beta$, such that the observed (1/M) exceeds the asymptotic (1/M) by an amount $\leq \gamma\beta$. The expected observed multiplication vs reactivity at various ramp rates (10^{-2} \$/sec to 10 \$/sec) in U^{235} systems with and without photoneutrons have been claculated in detail by the RTS kinetics code; representative results are plotted in Fig. 1-12. It may be noted that the envelope at high values of the ramp rate is directed toward the prompt critical limit instead of the delayed critical limit. Such calculations can provide useful guidance for estimating the accuracy (and safety) of various proposed methods for approaching delayed criticality by the reciprocal-multiplication method.

Light water moderated systems have no significant photoneutrons.*** Power reactors of this type have startup requirements somewhat peculiar to the type. In general, they contain several critical masses and may be made critical with a near-infinite number of flux distributions. This requires rigid control of the rod withdrawal pattern(s), multiple detectors to ensure knowledge of the approximate reactivity condition when less than critical and a slow, controlled approach to criticality. Precise knowledge of k below critical is not practical and is of little interest. The knowledge of flux level and power is affected by the flux distribution and the changing neutron attenuation between the core and the detectors until powers

*The authors are indebted to R.O. Brugge of General Electric for the information in this paragraph.

**All reactor transient calculations of this chapter were carried out by the Los Alamos RTS reactor kinetics code as described in reference [9].

***The authors acknowledge the suggestion of R. O. Brugge of General Electric for the text of this and the following five paragraphs.

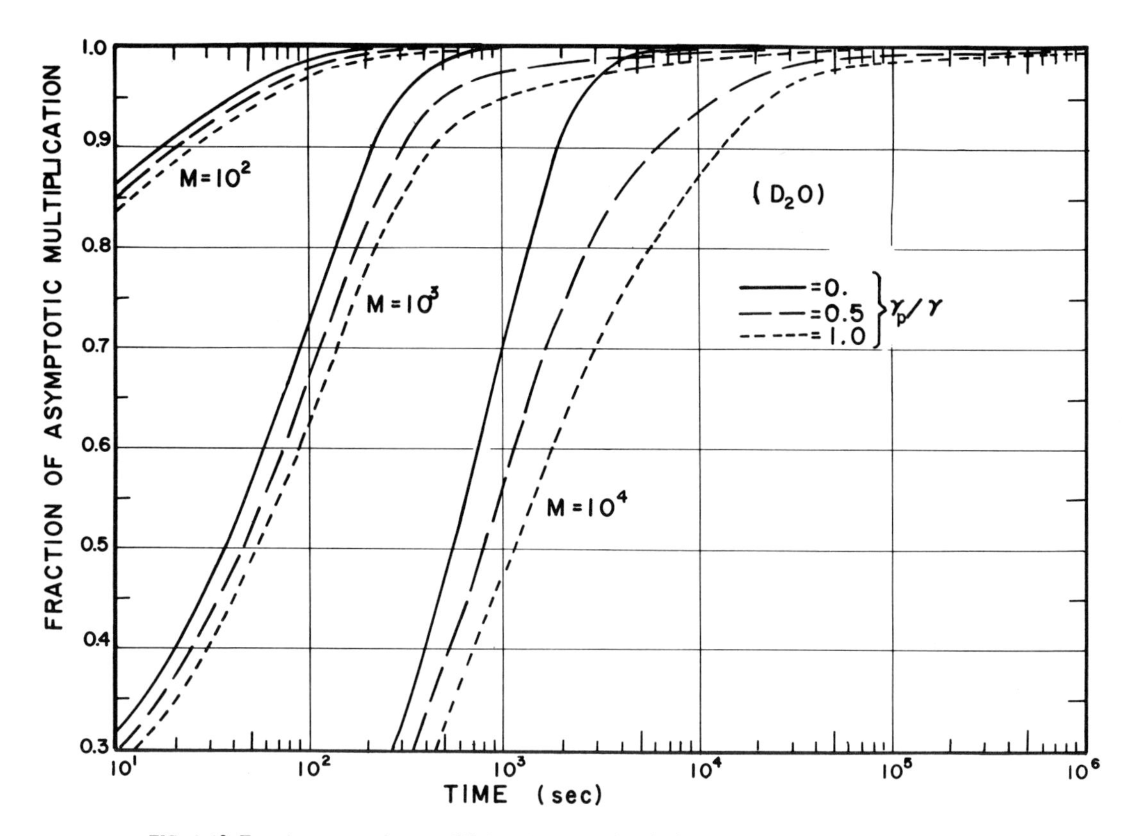

FIG. 1-10 Transient approach to equilibrium (asymptotic) multiplication values for D_2O-moderated systems. Curves are parametric in relative photoneutron effectiveness, γ_p/γ. Time is measured from instant of step change in reactivity.

above a few percent are achieved and rated temperature is reached.

The typical light-water-moderated power reactor has a large uncontrolled k_{eff} and small M^2, especially in the cold condition. Criticality can be achieved by withdrawal of as few as two control rods if they are immediately adjacent; the same core may require the withdrawal of twenty scattered control rods to achieve criticality. Acceptable control rod withdrawal sequences (patterns) must satisfy the following criteria:

1. The flux distribution during the approach to and at critical must be relatively uniform, i.e., the reactor must be brought critical as a whole, not in a localized area. This will achieve a relatively uniform rate of reactivity addition.
2. The set of withdrawn rods at critical must be contained within the set of withdrawn rods at full power. This avoids "swapping" rods after criticality is obtained which is considered prone to operator error.
3. The pattern must provide a relatively uniform flux increase at all neutron detectors—both those used for information only (usually low level counters not in the safety circuit), and primary safety system detectors which, in general, are not sufficiently sensitive to detect flux changes in the early phases of startup. (This criterion is automatically met if number one is met.)
4. The chosen pattern must not create rods of excessive reactivity worth among those remain "in" near critical.
5. The sequence of withdrawal to the critical pattern must be such that it is easy for an operator to follow and thus reduces the probability of a rod withdrawal error. A left-to-right and top-to-bottom sequence is one good choice.

The startup instrumentation and source-core arrangement should be such that the operator can assure himself that the output signal is neutron dependent before he begins rod withdrawal. Many techniques are possible, but the preferred arrangements allow the console operator to perform an unambigous check unassisted. A period-indicating (or doubling-time) meter is of considerable value as an operator aid; a period scram-trip, if used, must be set for a rather short period to avoid spurious scrams from the localized prompt increase when a single detector is used to feed the period system.

Flux level, and flux level scram protection, may be provided by compensated ion chambers or counting systems. Stability, reproducibility, and dependability are of over-riding importance. A fail-safe feature is frequently provided and may, for example, prevent rod withdrawal unless there is an upscale signal or the instrument is on a very sensitive range in the absence of a sufficient signal for response. The flexible flux shape nature

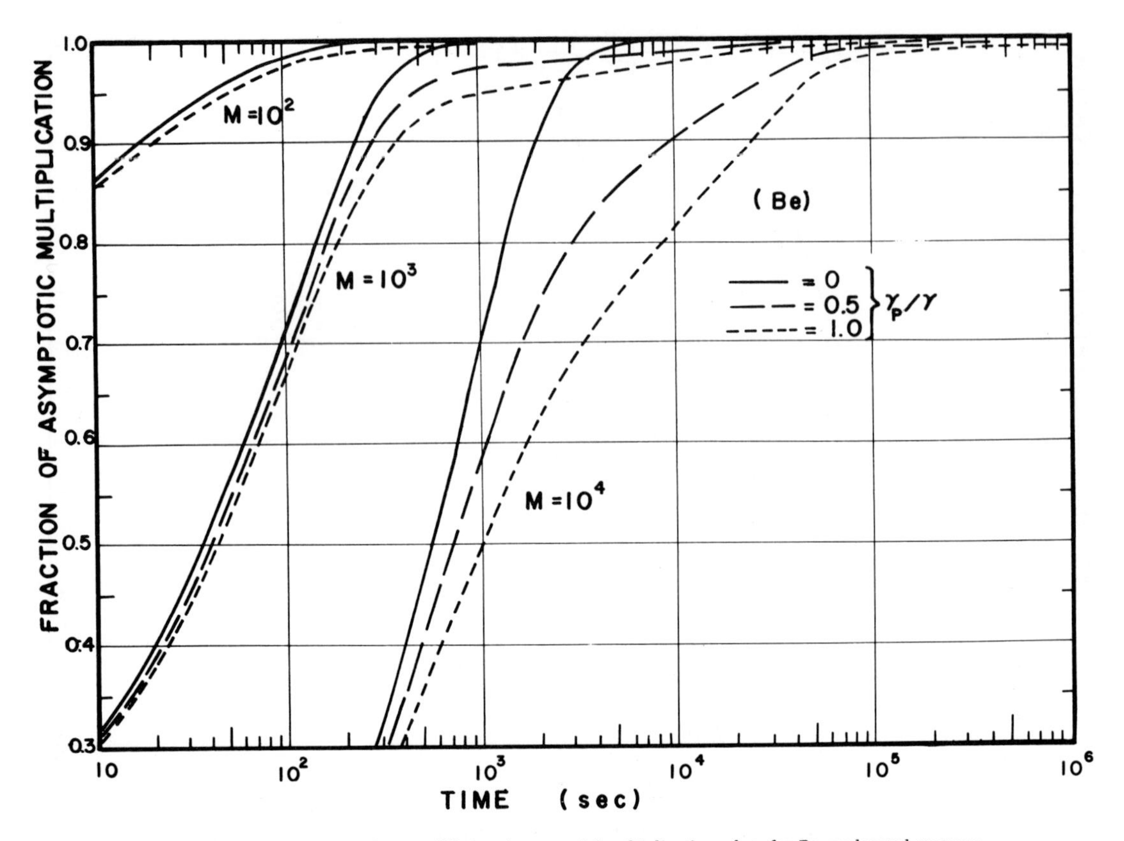

FIG. 1-11. Transient approach to equilibrium (asymptotic) multiplication values for Be-moderated systems. Curves are parametric in relative photoneutron effectiveness, γ_p/γ. Time is measured from instant of step change in reactivity.

of light water reactors requires multiple detectors, and power plant operating requirements frequently lead to complex coincidence logic circuits to prevent a scram caused by a single spurious signal. An infinity of adequate systems is possible, but all must possess the capability of a thorough checkout of their safety functions by plant operating personnel.

The relationship between reactor power and nuclear instrumentation readout is inherently variable in a light water system. Correlation must be determined, instruments adjusted, and scram-trip points verified frequently. Linearity of response is a function of both chamber location and reactor flux-shape. The response must be made reasonably linear in the 50-100^{+}% range to ensure scram trip setting validity.

Two conflicting desires frequently present themselves with respect to startup instruments. One may desire a linear addition of k to the core or a given startup-instrument response to k-additions as critical is approached. As a generality, one is obtained at the expense of the other. Assuming an adequate signal when fully shut down, one can always get a large instrument response by withdrawing control rods near the detectors; however, this inevitably leads to higher rod worths, higher addition rates of k, and nonuniform rod patterns. Uniform-pattern rod withdrawals (procedural or automatic) will provide a low and uniform rate of k-addition which reduces the probability and consequences of operator error or equipment malfunction but yields a lower magnitude instrument response. The weight of experience on routine startups of power reactors favors the latter approach.

1.7.2 Startup under Weak Source Conditions

In nearly all discussions of reactor kinetic behavior, it is tacitly assumed that the system exhibits the "average" kinetic behavior predicted by the reactor kinetic equations. This assumption holds during operation at power and in the case of normal startup with an adequate neutron source level.* On the other hand, with a very weak source (e.g. spontaneous fission of enriched uranium) the power level during the early stages of startup is so low that statistical fluctuations predominate and the kinetic equations are not applicable. At later stages the power level becomes sufficiently high that statistical fluctuations

*The usual criterion for adequate source strength, $(R\tau)^{-1/2} \ll 1$, implies a detector response R from fission sufficiently large that the counting time interval τ is convenient. One can compensate for small R by increasing τ if the counter background is low.

are no longer important. Nevertheless, the effects of fluctuations at the early (low) level persist through the high level stage in the sense that the early fluctuations determine the initial conditions for the high level stage.

The treatment of statistical fluctuations in

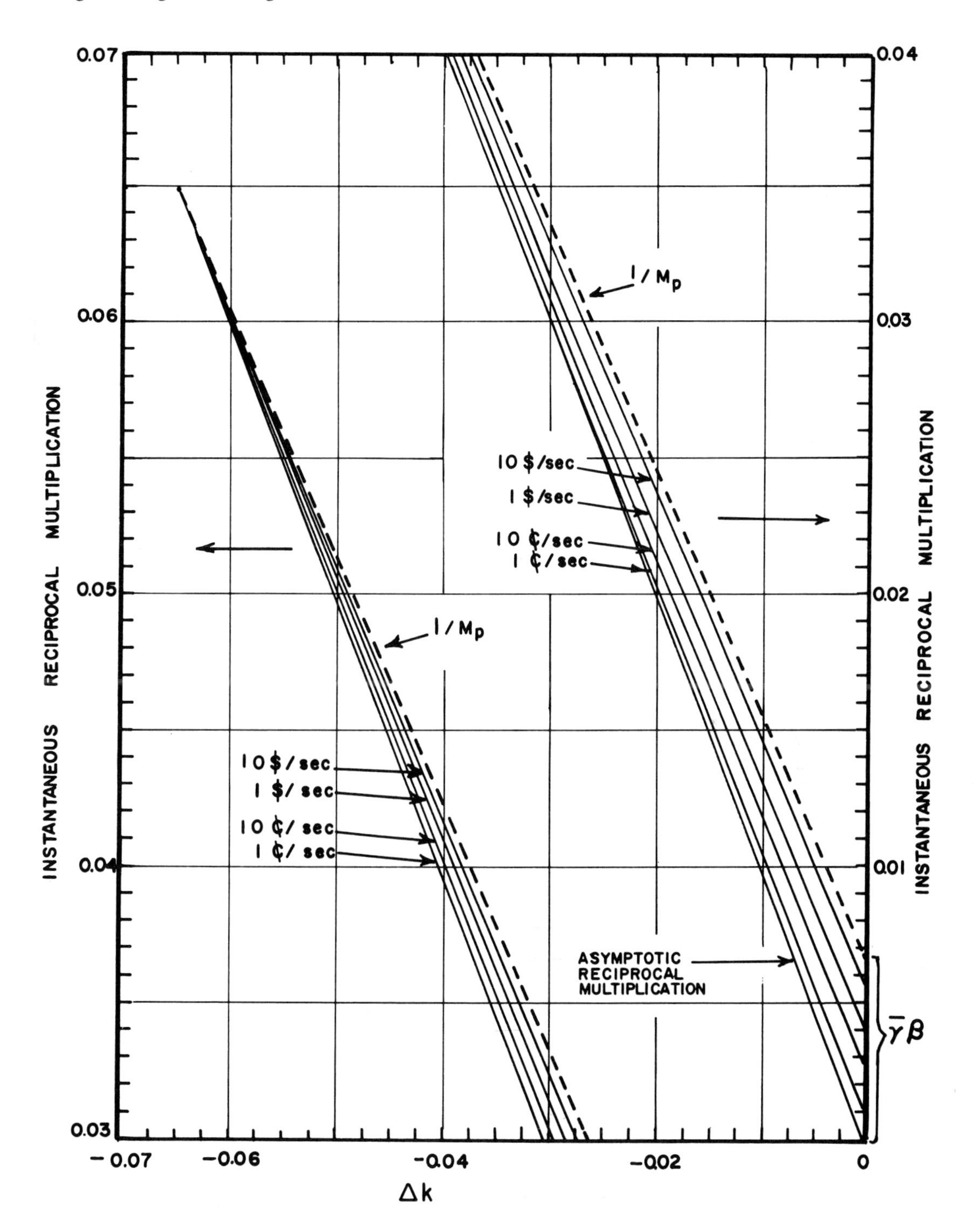

FIG. 1-12 Instantaneous reciprocal multiplication as a function of Δk for various ramp rates between zero (asymptotic reciprocal multiplication) and infinity (reciprocal prompt multiplication). Calculation carried out for system initially 10$ subcritical, $\Delta k \simeq 0.065$, and $\ell = 10^{-5}$ sec. The broken line is the reciprocal prompt multiplication ($1/M_p$). (For $1/M \geq 0.03$ the asymptotic reciprocal multiplication curve and the curve for a ramp rate = 1¢/sec overlap in the figure.)

neutron population is formulated in terms of a general probability distribution $P(n,m_i,t)$, the probability that at time t there are exactly n neutrons and m_i precursors of the type i in the reactor. This probability distribution is developed in terms of probability generating functions [11, 12, 13].

Two practical applications of calculations based upon this model are noteworthy here as they are relevant to nuclear safety. First, for safe reactor startup—or assembly of fissile material—under weak-source conditions, it must be shown that the fluctuating power level does, with a prescribed high probability, exceed a prescribed safe fiducial level (such that fluctuations have negligible effect) before reactivity becomes dangerously high. Hurwitz [12] has recently presented extensive numerical results of probability distributions of power during startup for a wide range of source strengths and ramp reactivity insertion rates. The results of these and similar calculations by Bell [13] show good qualitative agreement with experimental weak-source transient data* on the Godiva (bare U^{235}) critical assembly. Practical interpretation is a schedule of reactivity additions and corresponding (lengthy) periods of observation which allow for uncertainties in reactivity and are consistent with source strength [14].

The second application of the statistical model of reactors relevant to safety pertains to the behavior of pulsed reactors and yields predictions in supercritical excursions. Hansen [15] has treated this subject in some detail, using the method of moments to develop a probability distribution in time of occurrence of fission bursts following step and ramp increases of reactivity. For the case of ramp reactivity insertions under weak-source conditions, Hansen shows that the total energy release in an excursion can greatly exceed (by as much as two orders of magnitude in fast metal systems) the energy release predicted by the reactor kinetic equations.

The problem of statistical chain fluctuations, and related startup accidents is certainly not confined to fast systems (as Godiva). Transient tests of the BSR II core in the Spert 1 facility at NRTS, Idaho (see references [16], [17]) have shown that under weak-source conditions significant delays can exist between the injection of reactivity and the attainment of a prescribed period or power level.

Finally, a word of caution regarding the application of calculations of this type to reactor safety criteria: although "good qualitative agreement" between the predictions of statistical theory and direct experiment has been obtained for the closely coupled Godiva assembly, this does not guarantee similar agreement for loosely coupled reactors or heterogeneous reactors in general. As already noted, normal, safe startup procedure calls for adequate neutron source strength and detector sensitivity to ensure accurate indication of instantaneous multiplication at all subcritical levels. However, in special cases where the weak-source condition may be specifically required, detailed statistical calculations in conjunction with appropriate experimental normalization and conservative interpretation have provided valuable new data for nuclear safety guidance.

*See data of G. E. Hansen and T. F. Wimett in references [12] and [13].

2 SUBCRITICAL REACTIVITY MEASUREMENTS

The methods of reactivity measurement to be discussed in this section are emphasized because of their value in establishing, or confirming margins of safety during such operations as criticality approach or the loading of a fuel shipping cask. Results can be particularly reassuring when the interpretation of a neutron response curve is doubtful. As an example, detector response during the loading of a cask or storage basin may level off (approach a constant value) either because further fuel addition does not change reactivity significantly or because of a progressive failure of the detector to "see" fuel additions properly. An independent means of reactivity measurement may distinguish between these alternatives.

Discussions of extensions of these or similar measurements into the critical or supercritical region, included here for logical continuity, are not intended to detract from these safety applications.

2.1 Response to Reactivity Perturbation

The so-called "rod drop" technique provides one means of measuring increments of reactivity [18]. This technique is based on the fact that a reactivity step decrease (Δk) rapidly reduces the neutron flux in the ratio of the prompt neutron multiplication [$M_p = (1-k_p)^{-1}$] after the step to the prompt neutron multiplication before the step:

$$\frac{n_1}{n_0} = \frac{1-k_{p0}}{1-k_{p1}} = \frac{1}{1+\Delta k(k_1\gamma\beta)}. \quad (2\text{-}1)$$

Here $\gamma\beta$ is the effective delayed neutron fraction, and the subscripts 0 and 1 refer to values before and after the step. Hence the value of the step change of reactivity (rod drop) in units of dollars can be obtained directly from the observed power ratio:

$$\text{Rod value in \$} \equiv \frac{\Delta k}{k\gamma\beta} = \frac{n_0}{n_1} - 1. \quad (2\text{-}2)$$

This provides a reasonably accurate and reproducible technique for control rod calibration. In practice the postdrop level n_1 is determined by extrapolation of the subsequent neutron decay curve back to the time of the drop.

A rod drop experiment may involve large local changes in neutron flux in the vicinity of the rod. To avoid effects of distorted flux in rod drop measurements the neutron detector must be located so that its response remains proportional to the total neutron flux in the assembly throughout the reactivity change. The variation of neutron density $n(t)/n(0)$, as a function of negative reactivity step in units of dollars is plotted in Fig. 2-1 for a U^{235} thermal reactor. The parameter varied in the curves is the time following the negative step from initial equilibrium. Such calculations (appropriate to a given reactor) can be very useful in determining control rod worth. In addition

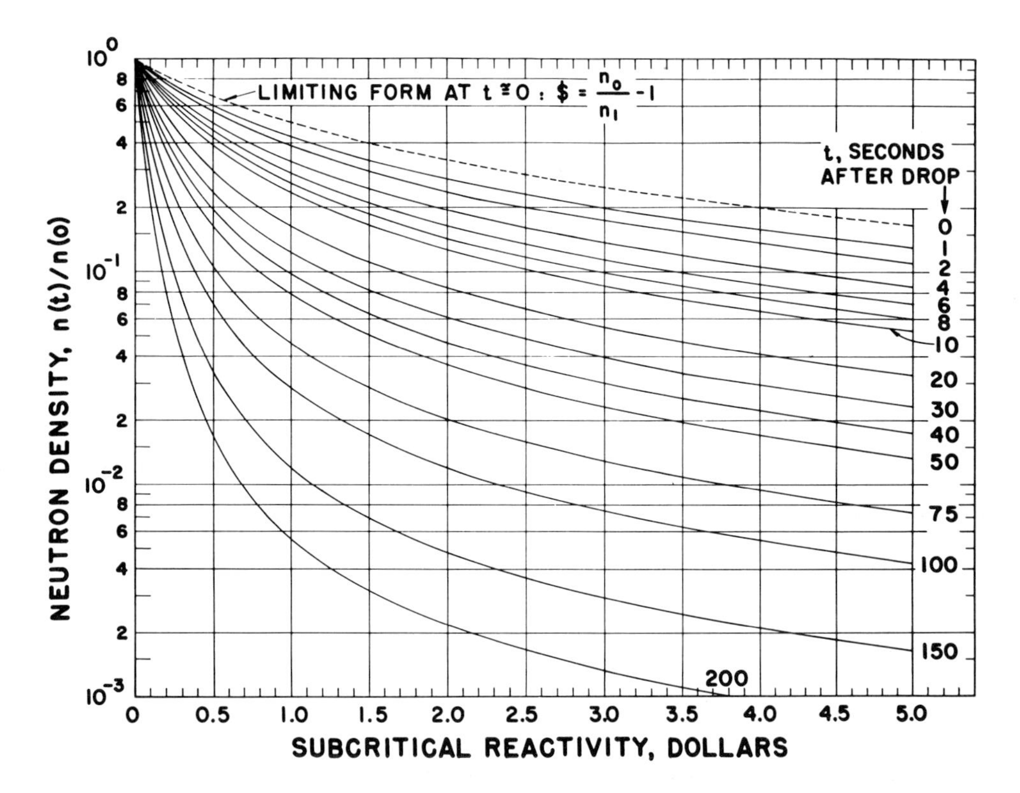

FIG. 2-1 Time decay of relative neutron density n(t)/n(o) following negative reactivity steps. Assumptions: $n(0)$ = constant, $dn/dt = 0$ for $t < 0$, $\ell = 10^{-4}$sec, $\gamma\beta = 0.0065$.

they permit a convenient direct check on the suppression of the influence of higher flux modes by judicious location the the detector. Thus if the detector "sees" a nearly constant flux shape, the same reactivity value should be indicated (via measured $n(t)/n(0)$ in conjunction with Fig. 2-1) for all times following the step.

Modifications of the standard rod drop technique include the integral count method [19] in which the total neutron count following the rod drop is measured rather than the post-drop level. In this case:

$$\text{Rod worth in \$} \cong \frac{n_0 \Sigma_i a_i/\lambda_i}{\int_0^\infty n(t)dt} \cong \frac{12.75\, n_0}{\int_0^\infty n(t)dt}, \quad (2\text{-}3)$$

where the indicated numerical value of $\Sigma a_i/\lambda_i$ is for U^{235}. The integral count method offers the advantages of (1) not requiring a measurement of the neutron decay-versus-time curve and (2) avoiding uncertainties in extrapolating the detector response back to the time of the drop. As with any rod drop experiment, good results are predicated on equal sampling by the detector of pre-drop and post-drop fluxes.

Although the reactor period method (of Sec. 3.2) is by far the most common technique for reactivity measurement, relatively long wait-times are required to ensure that asymptotic period is being observed. In fact, measuring a detailed multipoint control rod calibration curve by the period method may require several hours. A much faster technique is the rod oscillator method in which the oscillating component of the power level, resulting from periodic change of reactivity, is measured and related to the amplitude of the reactivity oscillation. If we denote the oscillating component and the steady-state neutron density by Δn and n respectively, we have the relation $\Delta n/n = W(j\omega)\Delta k/\beta$ where $\Delta k/\beta$ is the amplitude of the reactivity oscillation (in dollars) and $W(j\omega)$ is the reactor transfer function. The complex function $W(j\omega)$ can be calculated from the reactor kinetic equations (assuming appropriate reactor constants are known) or it can be determined experimentally by measuring the frequency response to reactivity oscillations of known small amplitude.

Comparisons of the rod oscillator method with the period and rod drop methods of reactivity measurement have demonstrated that good agreement is possible [20]. When a large number of reactivity calibration points must be measured this can be done most quickly by the rod oscillator technique since (1) the steady-state power remains nearly constant so that no time is lost in waiting for the delayed neutron precursors to return to equilibrium and (2) the average reactor power need not be held precisely constant because the oscillating component is measured rather than the power itself.

2.2 Response to Source Perturbation

The prompt neutron decay constant α is an extremely useful parameter in reactor kinetics. It represents an eigenvalue of a given system which can be both calculated and measured directly and provides a convenient and precise index of reactivity. The basic expression for α is given by the kinetic equations (neglecting delayed neutrons) as

$$\alpha \equiv \frac{1}{n}\frac{dn}{dt} \frac{(k_p - 1)}{\ell}, \quad (2\text{-}4)$$

where $k_p = k(1 - \gamma\beta)$ and ℓ is the prompt neutron lifetime (including leakage). Thus a burst of neutrons injected into a multiplying assembly (which is below prompt critical, $k_p < 1$) decays as:

$$n(t) = n_0 e^{\alpha t} = n_0 \exp[(k_p - 1)t/\ell]. \quad (2\text{-}5)$$

As in any application of the reactor kinetic equations, the above expression represents the composite or average behavior of a large number of individual neutron fission chains. The characteristic prompt decay constant or eigenvalue α can thus be measured by two methods: (1) by observing the simultaneous decay of a large number of neutrons introduced into an assembly as a burst or pulse or (2) by observing the decay of individual chains in succession and continuing this process until a sufficiently large number of chains have been observed to provide a statistically reliable measure of α. Method 1 characterizes the pulsed-neutron technique (to be discussed later in this section) and method 2 characterizes the "Rossi-α" and "variance-to-mean" techniques. The Rossi-α technique requires only a fast-response, sensitive neutron detector and some type of time analyzer with short time channels (e.g. microseconds to milliseconds) to display the decay of prompt neutron population versus time.

Bruno Rossi first suggested what is now called the "Rossi-α" method; its theory (the statistical theory of chain reactor neutron populations) was developed by Feynman, de Hoffmann and Serber [21, 22], and later extended and applied by Orndoff [23]. This theory develops an expression for the time dependence of detector counts observed in a chain-reacting assembly at low power. In particular, one requires the probability of a neutron being detected in the time interval (t, $t+\Delta t$) following the occurrence of a neutron detection ("count") at $t = 0$. An assembly at low power contains only a few prompt neutron chains at any given instant so that a sensitive detector can frequently detect two or more neutrons from the same chain. After a neutron count from a given chain at, say $t = 0$, there is a certain probability that the counter will detect at a later time either a random neutron from some other chain or a "chain-related" neutron from the same chain as that which produced the original count at $t = 0$. The probability of detecting a random neutron is simply $C\Delta t$ where C is the average counting rate and Δt is the time interval of observation (i.e. width of a single channel in the time analyzer). Equation (2-5) states that the neutron population in a fission chain reactor decreases on the average as $e^{\alpha t}$, so we expect the probability of detecting a chain-related neutron to decrease as $e^{\alpha t}$. The total probability of detecting a neutron in the interval t, $t + \Delta t$ after previously detecting a neutron at $t = 0$ is then:

$$P(t) = C\Delta t + Qe^{\alpha t}\Delta t. \quad (2\text{-}6)$$

The term $Q\exp(\alpha t)$ in Eq. (2-6) is thus the probability of detecting chain-related pairs of counts. In a Rossi-α experiment, the function $P(t)$ is measured by a time analyzer; subtraction of the random counting rate $C\Delta t$ then leaves the single exponential term $Q\exp(\alpha t)$ from which the desired chain-related prompt neutron decay constant α is readily determined, either graphically or by computer least-squares fitting procedures. The Rossi-α method is best applied to fast systems near delayed criticality. For slower systems, fission chains are longer, and lower fission rates (with consequent higher detection efficiency) are required to prevent overlapping of chains. The time required for accumulation of sufficient data is roughly proportional to prompt neutron lifetime, making a Rossi-α experiment excessively lengthy for $\ell \gtrsim 100$ μsec.

Rossi-α experiments have been carried out on a number of fast metal assemblies at Los Alamos by Orndoff [23], at Argonne by Brunson, et al. [24], and recently at Oak Ridge by Mihalczo [25]. An extensive series of Rossi-α measurments on a wide range of critical assemblies at the National Reactor Testing Station, Idaho Falls, has been completed recently by Brunson and coworkers [26]. The Rossi-α method is generally capable of very good accuracy (errors of a few percent) in determining the prompt neutron decay constant by means of Eq. (2-6). Representative α data for various types of systems can be found in references [23] to [25]. In addition there have been numerous measurements [27, 28, 29] of α (actually of $\ell/\gamma\beta = \alpha^{-1}$) made at the SPERT facility (Idaho Falls) by examining statistical chain fluctuations around steady-state low-power operation.

From the definition of the prompt neutron decay constant, $\alpha = (k_p - 1)/\ell$, one obtains a simple direct relation between measured α and reactivity in dollar units:

$$\alpha = \alpha_{DC}(1 - \$),$$

where $\alpha_{DC} \simeq -\gamma\beta/\ell$ is the value of α at delayed criticality. Thus in principle the reactivity scale in dollar units is established by a single measurement of α at delayed criticality—or at any known reactivity. In precise reactivity calibrations it is desirable, if possible, to extend Rossi-α measurements not only to delayed critical, but well into the supercritical range. This has been done on the Godiva assembly at Los Alamos by means of a cycling technique [23] in which α-data are accumulated during a short interval of positive period operation, following which the system is disassembled and allowed to decay before resuming another cycle of data taking. As is apparent, the extrapolation of α to zero establishes prompt criticality on any scale that follows reactivity change.

Another method for reactivity determination which has become popular in recent years is the pulsed neutron technique. Although this technique is now used for a wide range of measurements and studies in reactor physics [30, 31, 32] the concern here is with its specific application to the measurement of reactivity. A burst of neutrons is injected into a subcritical multiplying system and the decay of neutron population is observed as a function of time. After neutron thermalization and decay of higher modes of neutron flux, the decay constant of the fundamental mode (α_0) can be measured. From the relation between α_0 and reactivity:

$$\alpha_0 = \alpha_{DC}(1 - \$), \quad \alpha_{DC} = -\gamma\beta/\ell \qquad (2\text{-}7)$$

one obtains the reactivity of the system in dollar units.* The proportionality constant α_{DC} can be determined directly by a pulse decay measurement at delayed criticality. (If the reactivity is precisely known, as at delayed criticality, the pulse method may be viewed alternatively as a precise measurement of the ratio of effective delay fraction to prompt neutron lifetime, a quantity of obvious importance in reactor safety considerations.)

As with all techniques for reactivity determination it is important to ensure that only fundamental-mode decay is being measured. Near criticality, of course, the fundamental mode is enhanced by the fundamental-mode fission source. But in highly subcritical assemblies, modal and thermalization-time effects impose an effective lower limit (something like 15$ subcritical) on the pulsed neutron method of reactivity measurement. The use of small detectors permits discrimination against odd harmonics of the flux distribution by judicious location of the source and "point" detector [33]. When in doubt, the degree of modal purity may be ascertained by (1) checking agreement of measured decay constants at various space points and (2) analyzing by least squares the decay curve at a single space point into its modal components [34, 35].

Thus far the possible variations of prompt neutron lifetime and of effective delay fraction with reactivity have been ignored. In the measurement of very large reactivity changes (several dollars) by the pulse technique it is necessary to correct for variations in neutron lifetime. Such corrections have been applied to accurate measurements of the reactivity worth (> 6$) of central plate-type fuel elements in the Oak Ridge Bulk Shielding Reactor [36]. Additional reactivity measurements on partial fuel element loadings were also reported in the careful work at Oak Ridge [36], which should encourage further application of pulsed neutron techniques to the measurement of large reactivity changes. When reactivity changes are not large, say $\lesssim 0.1\$$, the dependence of lifetime and of delay fraction on reactivity can be ignored, as has already been demonstrated experimentally [36-38].

Detailed comparisons [38-41] of the pulse technique with asymptotic period and rod drop measurements have shown good (within a few percent) agreement for reactivities down to a dollar below delayed criticality. For larger values of subcriticality the pulse method has been found to be definitely superior and in nearly all cases it appears to be "more convenient, rapid and straightforward [38, 39]." The α method furthermore provides its own reactivity calibration (via a delayed-critical measurement) and does not depend upon supplementary measurements to establish the delayed-to-prompt interval of reactivity. Work at Knolls Atomic Power Laboratory [41] suggests extension of the pulsed neutron method to reactivity measurements even in highly heterogeneous systems. Viz: In experiments with distributed poison in highly enriched hydrogen-moderated critical assemblies, measured values of α appear to be insensitive to detector position for reactivities above -5$ whether the detector is in a highly poisoned region, in the nonmultiplying moderator, or in a region of high local multiplication.

Where delayed-neutron parameters are known, reactivity changes may be obtained from corresponding relative fission rates by direct solution of the reactor kinetic equations. This is the operation performed by the usual "reactivity meter", an analog computer that solves for reactivity as it receives signals from a reactor [42]. Because of difficulties with an adequate signal in the far-subcritical region, and uncertainties of results until there is a delayed-critical check, this type of analyzer is seldom used for critical-approach measurements.

Finally, two recent proposals of somewhat different "reactivity meters" are cited. First, the feasibility of using a pulsed neutron source in conjunction with detectors and wave analyzers as a continuous, analog reactivity monitor has been pointed out [43]. The second proposal concerns a new method of continuously determining "shutdown margin" reactivity via statistical analysis of reactor noise [44]. In contrast to conventional methods of monitoring reactivity, both of these schemes would appear to be useful for monitoring reactor startup under poorly established source conditions.

3 CRITICAL AND ZERO POWER OPERATION

On attaining the delayed critical condition in a reactor, a new safety emphasis is required, namely, reactivity measurements must be more precise than were necessary in protecting against gross reactivity errors during critical approach. The concern now is to prepare the groundwork for tracking excess reactivity and shutdown margins through the life of the reactor. Controls must be calibrated accurately and reactivity coefficients established. Fortunately the delayed critical condition is an excellent reactivity fiducial and, in this neighborhood, the difficulties with far-subcritical reactivity determinations tend to disappear.

*In D_2O- or Be-moderated systems, the effective delay fraction may be increased due to photoneutrons from delayed fission gamma activity. Likewise, the fraction would be decreased if there were appreciable prompt neutron contribution from processes such as (n, 2n), (prompt-γ, n), etc. All such effects, properly weighted for spatial and energy dependence, will, of course, be an intrinsic part of measured α_0 values and hence of measured reactivity values in dollar units ($\Delta k/k\gamma\beta$).

The following discussion of effects that must be considered in precision reactivity measurement is presented in this context.

3.1 Precursor Transient Effects in Delayed-Critical Determination

During a typical stepwise approach to delayed criticality by the reciprocal multiplication method, it has been seen that the neutron level following each addition of reactivity should be allowed to stabilize or "settle out" in order to obtain an accurate indication of asymptotic multiplication before proceeding with the next reactivity addition. As shown in Figs. 1-10 and 1-11 this settling-out time (or the time for a period meter to return to infinity) becomes progressively larger as criticality is approached. This phenomenon is all too familiar to anyone who has brought a reactor or critical assembly to delayed criticality from its shutdown state; the sluggishness is especially apparent (and bothersome!) when photoneutrons are present, as in Be- or D_2O-moderated systems. The importance of the delayed critical condition as a precise reactivity reference point calls for a thorough understanding of this effect.

The physical reason for this transient effect is clear. The delayed critical condition is identified by a self-sustained constant power level only when all delayed neutron precursors (including delayed photoneutrons) are in equilibrium, meaning that the formation rate of precursors, $n_0 k_0 \beta/\ell$, equals the decay rate, $\Sigma \lambda_i C_i(0)$. Here β represents the total delayed fraction (including photoneutrons) and the summation index i extends over all delayed neutron and photoneutron groups. At constant power the precursor concentration exhibits an exponential buildup:

$$C_i(t) = C_i(0)\,[1 - \exp(-\lambda_i t)]\,. \qquad (3\text{-}1)$$

This corresponds to an instantaneous precursor decay rate $\lambda_i D_i(t)$ (in effect a "source" of neutrons), which is less than the instantaneous precursor formation rate $n_0 k \beta_i/\ell$ (in effect a "sink" for neutrons). Thus there results a time-dependent deficit in the neutron balance. In this nonequilibrium situation it is desired to obtain the corresponding reactivity variation which is required to keep the system at constant power, as during a precise delayed criticality determination. This variation is easily derived from the kinetic equations which at constant power give

$$\frac{\Delta k - k\beta}{\ell}\, n_0 = -\Sigma_i \lambda_i C_i\,. \qquad (3\text{-}2)$$

Substituting the exponential buildup of $C_i(t)$, and introducing the precursor equilibrium condition gives

$$\frac{\Delta k}{k\beta} \simeq \frac{\Sigma_i \lambda_i C_i(0) \exp(-\lambda_i t)}{\Sigma_i \lambda_i C_i(0)}\,, \qquad (3\text{-}3)$$

which is the desired reactivity (in dollars) as a function of time required to maintain constant power for a delayed critical determination. Note that Eq. (3-3) is just the normalized precursor decay function $(\Sigma a_i \exp(-\lambda_i t)$ with $\Sigma a_i = 1)$ which can be calculated from known delayed neutron and photoneutron data. Since photoneutron group lifetimes extend to hours and even days, it is a practical impossibility to achieve true equlibrium for all photoneutrons and hence practically impossible to attain a true delayed critical condition in Be- and D_2O-moderated systems. Thus in experiments requiring precise determination of delayed critical conditions it is often essential to evaluate reactivity errors due to precursor nonequilibrium (transients) in order (1) to ensure that they are acceptably small under specified experimental conditions or (2) to apply precise quantitative corrections where necessary.

Figure 3-1 shows the calculated reactivity variation in systems which were brought rapidly to constant power (in a few prompt neutron lifetimes) and then held at this constant power level. This calculation was carried out [by the RTS inverse code which solves for $\Delta k(t)$, given $n(t)$] for systems with different effective photoneutron contributions as indicated by the γ_p/γ parameter. For the case of no photoneutrons, $\gamma_p/\gamma = 0$, the error in reactivity becomes negligible within a few hundred seconds. But in Be- and D_2O-moderated reactors, appreciable errors may persist for an hour or more after first reaching the delayed-criticality power level. In precise reactivity-coefficient and perturbation experiments in such systems, calculations (as in Fig. 3-1) can be used to correct observed reactivity data, and thus excessively long delays in delayed-criticality determinations can be avoided.

It should be noted that the calculations in Fig. 3-1 apply to an initially shutdown system with no precursors present. Under these conditions one obtains a maximum error due to precursor transient effects. The presence of precursors (as from previous irradiations) decreases precursor transient effects. Similarly, for systems undergoing successive changes in power level, transient effects in general are smaller the smaller the power level change. When required, the RTS inverse code or its equivalent can be used to calculate detailed errors due to precursor transient effects in any given experimental situation.

3.2 Reactor Period

Small reactivity departures from delayed criticality can best be measured in terms of reactor period. Considerations that enter the interpretation of period in terms of reactivity will now be discussed.

3.2.1 Inhour Relations

The solution of the space-average reactor kinetic equations for constant k (e.g. following a step change in reactivity) is given* by a superposition of exponentials:

$$\frac{n(t)}{n(0)} = \sum_{\nu=0}^{6} N_\nu \exp(\omega_\nu t)\,, \qquad (3\text{-}4)$$

*See Sec. 2.1 of the chapter on "The Reactor Core."

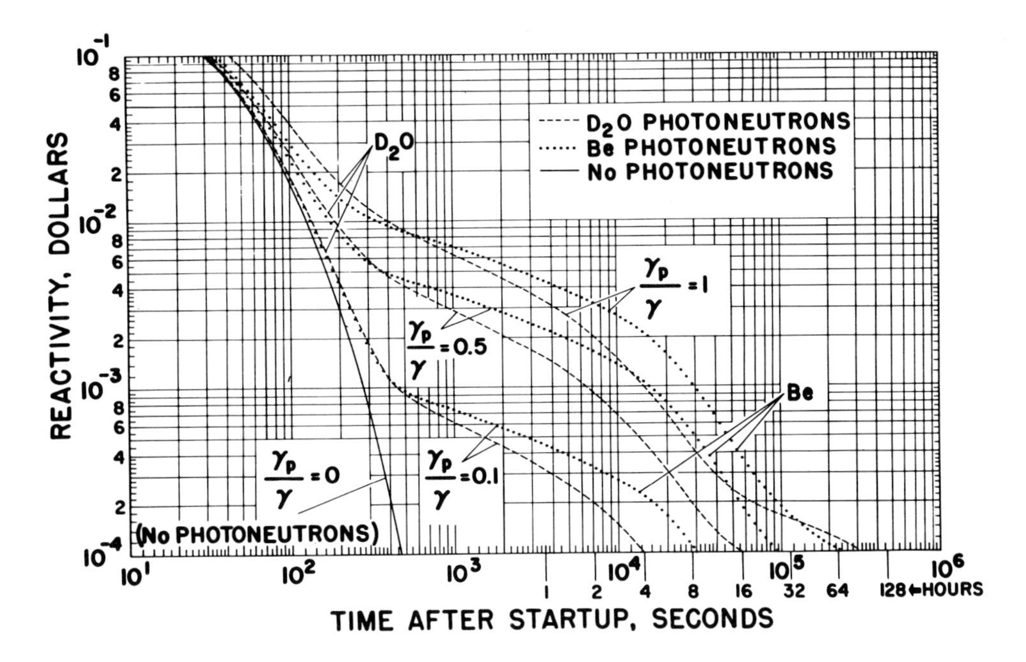

FIG. 3-1 Calculated reactivity variation with time after startup in systems brought rapidly (in a few prompt neutron lifetimes) to constant power and then held at this power level. Systems were initially shut down (no precursors). Effect of D_2O and Be photoneutron contribution is clear from broken-line and dotted curves.

where ω_ν are the roots of the characteristic equation:

$$\omega_\nu = \frac{\Delta k}{\ell} - \frac{\omega_\nu k}{\ell} \sum_{i=1}^{6} \frac{\gamma_i \beta_i}{\omega_\nu + \lambda_i}. \qquad (3\text{-}5)$$

Here β_i and λ_i are the delay fraction and decay constant for the i-group, and γ_i represents the effectiveness (in producing fission) of delayed neutrons of the i-group compared to prompt neutrons. If the ω_ν in Eq. (3-5) is replaced by the corresponding periods $T_\nu = \omega_\nu^{-1}$ and reactivity expressed in units of $\Delta k/k\gamma\beta$ (the dollar), Eq. (3-5) becomes a seventh-order polynominal in T:

$$\frac{\Delta k}{k\gamma\beta} = \frac{\ell}{k\gamma\beta T_\nu} + \sum_{i=1}^{6} \frac{(\gamma_i \beta_i/\gamma\beta)}{1 + \lambda_i T_\nu}. \qquad (3\text{-}6)$$

This is readily solved by digital computers. The computed behavior of Eq. (3-6) for U^{235} is shown in Fig. 3-2 where the "steady state" period ($T = \omega_0^{-1}$) and the six "transient" periods ($T_\nu = \omega_\nu^{-1}$) appear as functions of reactivity in dollars. It is seen that for a subcritical assembly all seven roots of Eq. (3-6) are negative, six of these approaching the delayed neutron periods (τ_i) in the limit of large negative reactivity. For a supercritical assembly the "steady state" period T becomes positive while the remaining negative T_ν again approach the six delayed neutron periods in the limit of large positive reactivity. The time behavior of neutron density (excluding photoneutron effects) is represented by a superposition of these seven periods, as indicated by Eq. (3-4).

The first quadrant of Fig. 3-2 (positive values of both period and reactivity) is the familiar relation, stable period versus reactivity. Because of its practical importance in reactivity determination this "inhour relation" has been evaluated by digital computer for six fission species, using the delayed neutron data of reference [45]. The factors $\gamma_i\beta_i/\gamma\beta$ in Eq. (3-6) were replaced* by the measured relative group abundance values, a_i. The prompt neutron term $\ell/T\,k\gamma\beta$ in Eq. (3-6) is negligible under most operating conditions for fast and intermediate reactors and for $T \gtrsim 30$ sec in thermal reactors.

In general, then, reactor kinetic behavior is determined solely by the delayed neutron periods and relative abundances through the term

$$\sum_{i=1}^{6} \frac{(\gamma_i \beta_i/\gamma\beta)}{1 + \lambda_i T}$$

in Eq. (3-6). This quantity is plotted in Figs. 3-3, 3-4 and 3-5 as a function of stable period (both positive and negative values of T) for six fission species.** The prompt term $\ell/Tk\gamma\beta$ has been

*Calculation of individual γ_i for a particular system implies neutron effectiveness calculations (see Sec. 3.2.2) for which complete multigroup input data are not normally available. For slow and intermediate systems the approximation $\gamma_i \simeq \gamma$ is usually valid, so that $\gamma_i\beta_i/\gamma\beta$ can be replaced by the corresponding relative abundances a_i with neglibible error in period-reactivity relations.

**The period-reactivity calculations of Figs. 3.3, 3-4, 3-5 have been closely corroborated, within ~ 0.1% mean deviation, by an independent method of evaluation, namely, by direct numerical integration (Laplace transform) of prompt burst neutron decay data [46].

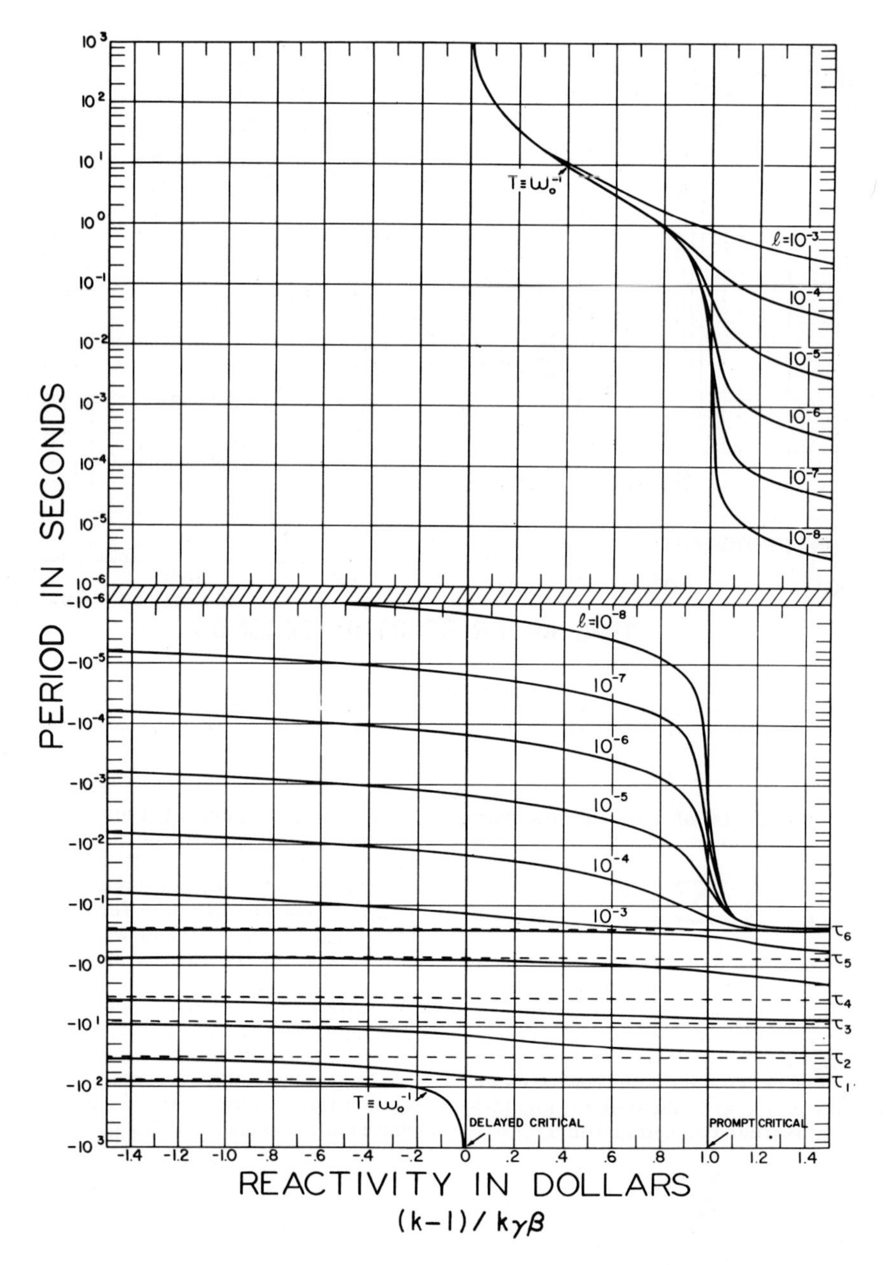

FIG. 3-2 Reactor "stable" - and "transient" - periods vs reactivity for U^{235} delayed neutron data. Dotted lines, $\tau_1 \ldots \tau_6$ are the six delayed neutron mean-lives for U^{235}. Parameter ℓ is prompt neutron lifetime.

added for several representative prompt neutron lifetimes, 10^{-8} sec $\leq \ell \leq 10^{-3}$ sec. Values of $\ell \geq 10^{-5}$ sec are not represented for the threshold fissioning species (U^{238}, Th^{232} and Pu^{240}) because long neutron lifetimes are not characteristic of fast fission systems. Numerical computations were not extended to decay periods shorter than T = −100 sec since here T is determined largely by the longest delayed neutron period and becomes insensitive to reactivity.

3.2.1.1 Influence of Photoneutrons In Be- or D_2O-moderated systems the reactor kinetic equations must be extended to include contributions from the Be(γ,n) and D(γ,n) photoneutrons. In kinetics calculations, photoneutrons can be considered simply as additional groups of delayed neutrons having decay constants λ_j and group fractions β_j. There must be due allowance for buildup factors for each group, $(1 - \exp(-\lambda_j \tau))$, where τ is the effective irradiation-time, and for relative photoneutron effectiveness (γ_p/γ) which represents average photoneutron effectiveness relative to average delayed neutron effectiveness in a given system.

Period-reactivity relations calculated for D_2O-

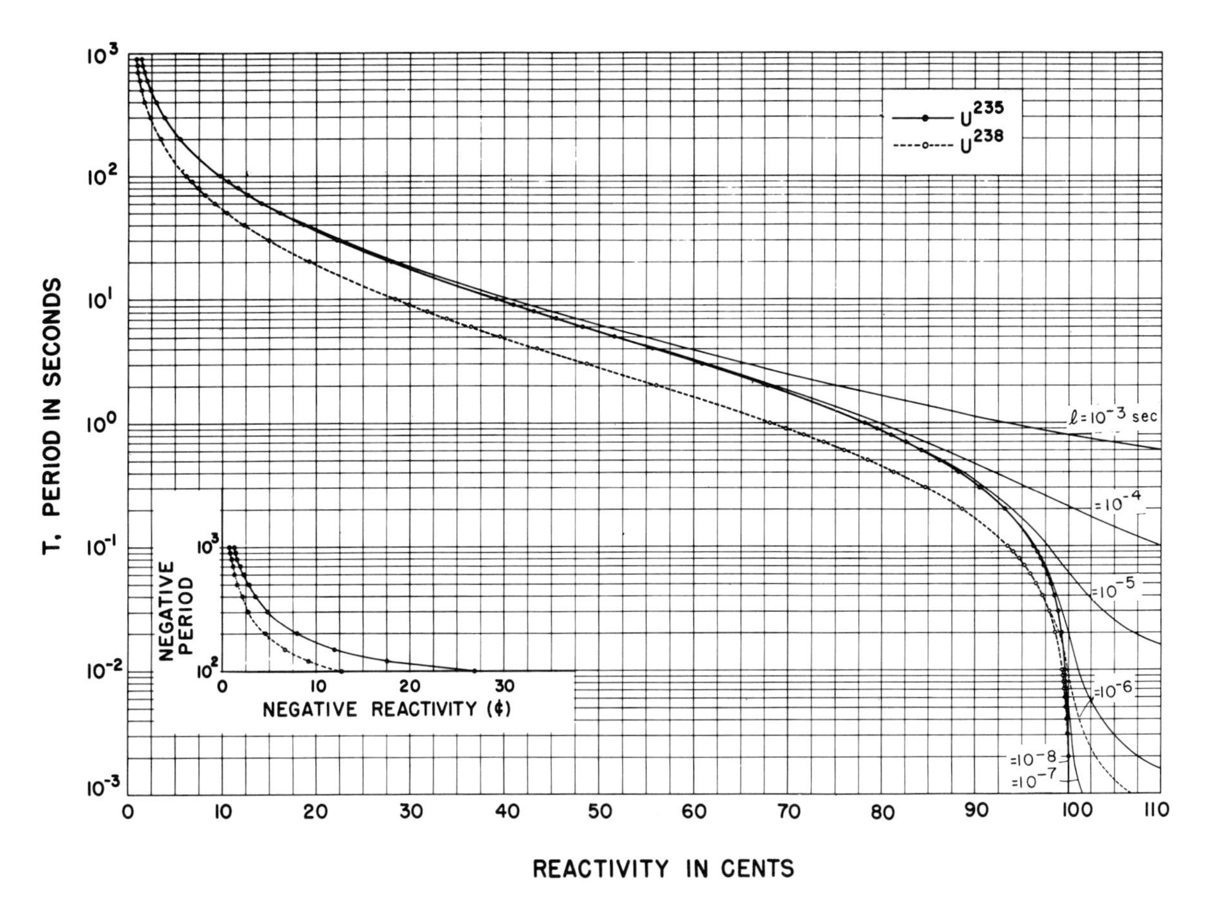

FIG. 3-3 Asymptotic period vs reactivity for U^{235} and U^{238}. Parameter ℓ is prompt neutron lifetime. (Heavy curves were calculated directly from Laplace-transformed prompt burst decay data; corresponding points calculated from delayed neutron periods and abundances.)

and Be-moderated systems are presented in Figs. 3-6 and 3-7. Recent Be photoneutron and D_2O photoneutron data [10] were used in the machine calculations. Four representative values of γ_p/γ were chosen, together with three effective irradiation times, 10^2, 10^3 and 10^{∞}sec (the last representing saturation irradiation for all photoneutron groups).

Effects of undersaturation of the photoneutron groups are apparent from comparison of the saturation-irradiation curves with the curves for irradiation times of 10^3 and 10^2 sec. Note the large decrease in photoneutron contribution to kinetic behavior as the effective irradiation time is decreased from saturation to 100 sec. The delayed neutron group yields are, of course, saturated for irradiation times of 10^3 sec or longer. For the 100-sec irradiation, however, the effect of undersaturation of the delayed neutron groups themselves is perceptible: the $\gamma_p/\gamma = 0$ curve is shifted toward shorter periods from the ordinary period-reactivity curve for delayed neutrons alone, i.e. the $\gamma_p/\gamma = 0$ curve for 10^3- sec or saturation irradiation.

Comparison of Fig. 3-6 with Fig. 3-7 shows that Be-moderated systems tend to be somewhat more sluggish than D_2O systems, following saturation irradiations in both cases. On the other hand, the total D_2O photoneutron fraction $\Sigma\beta_j$ is nearly seven times that for Be, cf. Tables 6 and 9 of reference [10]. This is explained, of course, by the greater mean life of Be photoneutrons compared to D_2O photoneutrons. (The ratio of mean lifetimes is $\tau(\text{Be})/\tau(D_2O) = 200\ \text{min}/24.1\ \text{min} \simeq 8.3$ at saturation.) This ratio decreases somewhat for shorter irradiation times - which are more representative of actual operating conditions - and the period-reactivity relations for D_2O and Be systems then become more nearly alike.

3.2.1.2 Influence of Fission Product Removal
The rapid removal of fission products, notably the delayed neutron precursors, clearly alters the neutron economy and dynamics of any reactor. If a large fraction of the precursors is removed, the effective delayed neutron yield is correspondingly reduced, which can in turn pose serious control problems. Precursor loss can occur at high temperature by diffusion escape of fission-product halogen gases [47, 48], mainly Br and I, which account for some 90% of delayed neutron precursors [10]. Loss can also occur by physical transport of precursors out of the core region, as in circulating fuel reactors [49, 50], particularly where there is direct chemical extraction (e.g. in a continuous fuel-reprocessing loop).

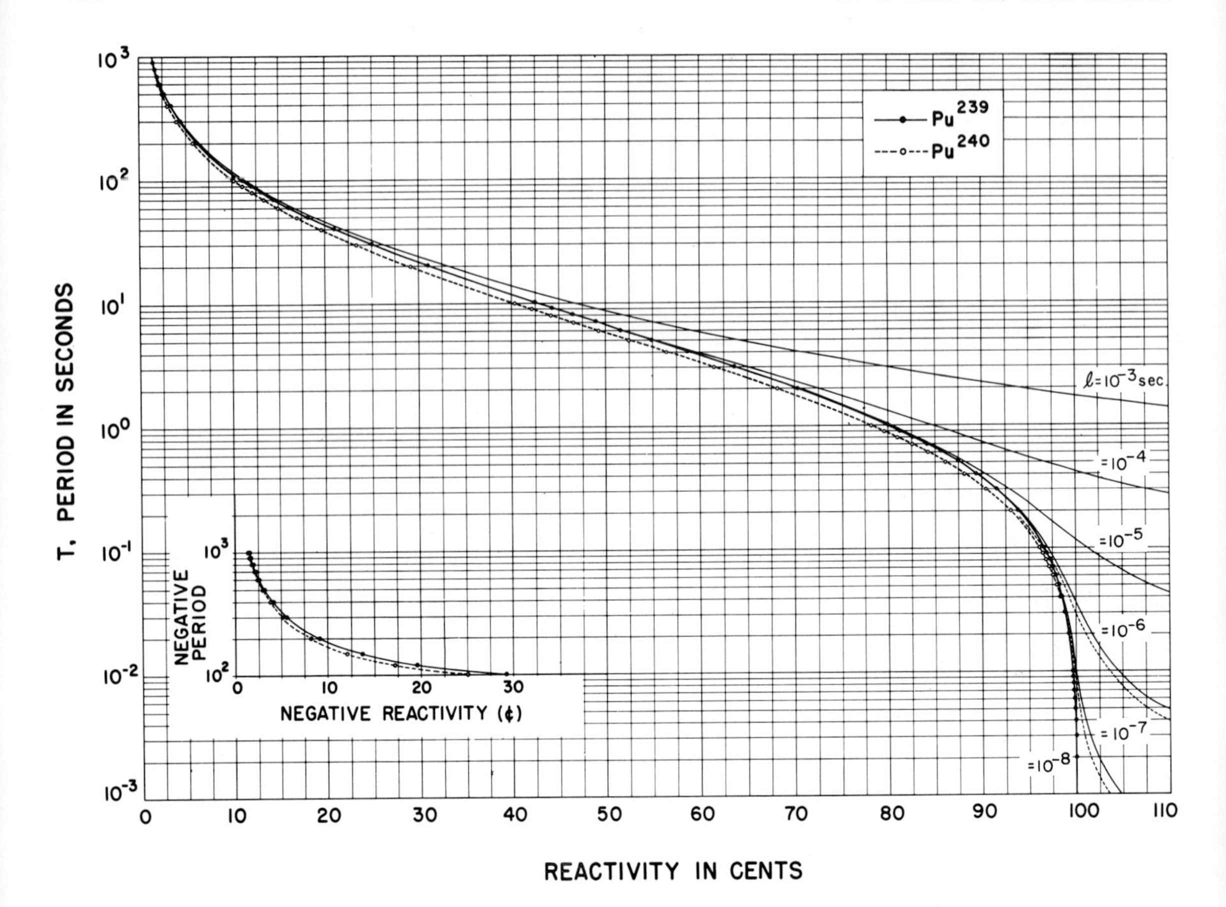

FIG. 3-4 Asymptotic period vs reactivity relation for Pu^{239} and Pu^{240}. Parameter ℓ is prompt neutron lifetime. (Heavy curves were calculated directly from Laplace-transformed prompt burst decay data; corresponding points were calculated from delayed neutron periods and abundances.)

Here we consider the effects of fission product removal by fuel circulation on the period-reactivity relation. We assume, for mathematical simplicity, that the neutron density is constant throughout the core and that the fuel moves with constant velocity (one-dimensional slug flow) through the core and external loop. Let T_c be the transit time in the core and T_t be the total transit time through the core plus the external loop. Solution of the space-average kinetics equations for a system exhibiting exponential time behavior ($n \sim \exp \alpha t$) then predicts [51] the fraction of group-i precursors in the core as:

$$f_i(\alpha) = 1 - \frac{[1 - \exp(-(\alpha + \lambda_i)T_c)][1 - \exp(-(\alpha + \lambda_i)(T_t - T_c))]}{T_c(\alpha + \lambda_i)[1 - \exp(-(\alpha + \lambda_i)T_t)]}. \quad (3\text{-}7)$$

Under steady state conditions ($\alpha = 0$) the fraction of precursors in the core is then

$$f_i(0) = 1 - \frac{[1 - \exp(-\lambda_i T_c)][1 - \exp(-\lambda_i(T_t - T_c))]}{T_c\lambda_i[1 - \exp(-\lambda_i T_t)]}. \quad (3\text{-}8)$$

This quantity has been calculated as a function of $\lambda_i T_c$ for various T_c/T_t, the fraction of the time spent by the fuel in the core. Figure 3-8 shows the results of these calculations, giving the fraction of precursors in the core of a circulating fuel reactor under steady state conditions.

In Eq. (3-7) and in Fig. 3-8, for no circulation ($T_c/T_t = 1.0$), $f_i(\alpha)$ is identically unity, as expected. But for high fuel velocities, or total transit time small compared to delayed neutron mean lifetime, i.e. $T_c \ll \tau_i$, the fraction of precursors in the core is given approximately by the fraction of the total time spent by the fuel in the core:

$$f_i \simeq T_c/T_t \quad \text{for} \quad T_t \ll \tau_i .$$

Note in Fig. 3-8, for example, that the f_i ordinate-intercept values are essentially the same as the indicated values of the parameter T_c/T_t.

The effect of fuel circulation on the period-reactivity relation, Eq. (3-6), can be accounted for by introducing a modified delayed neutron fraction

$$\beta_i^{mod} = \beta_i[1 + \lambda_i T(1 - f_i(\alpha))] ,$$

where T is the stable reactor period and $f_i(\alpha)$ is given by Eq. (3-7). Introducing β_i^{mod} into Eq. (3-6) (with $\gamma_i \simeq \gamma$) gives the rescaled relationship between the stable reactor period T and the reactivity in dollars:

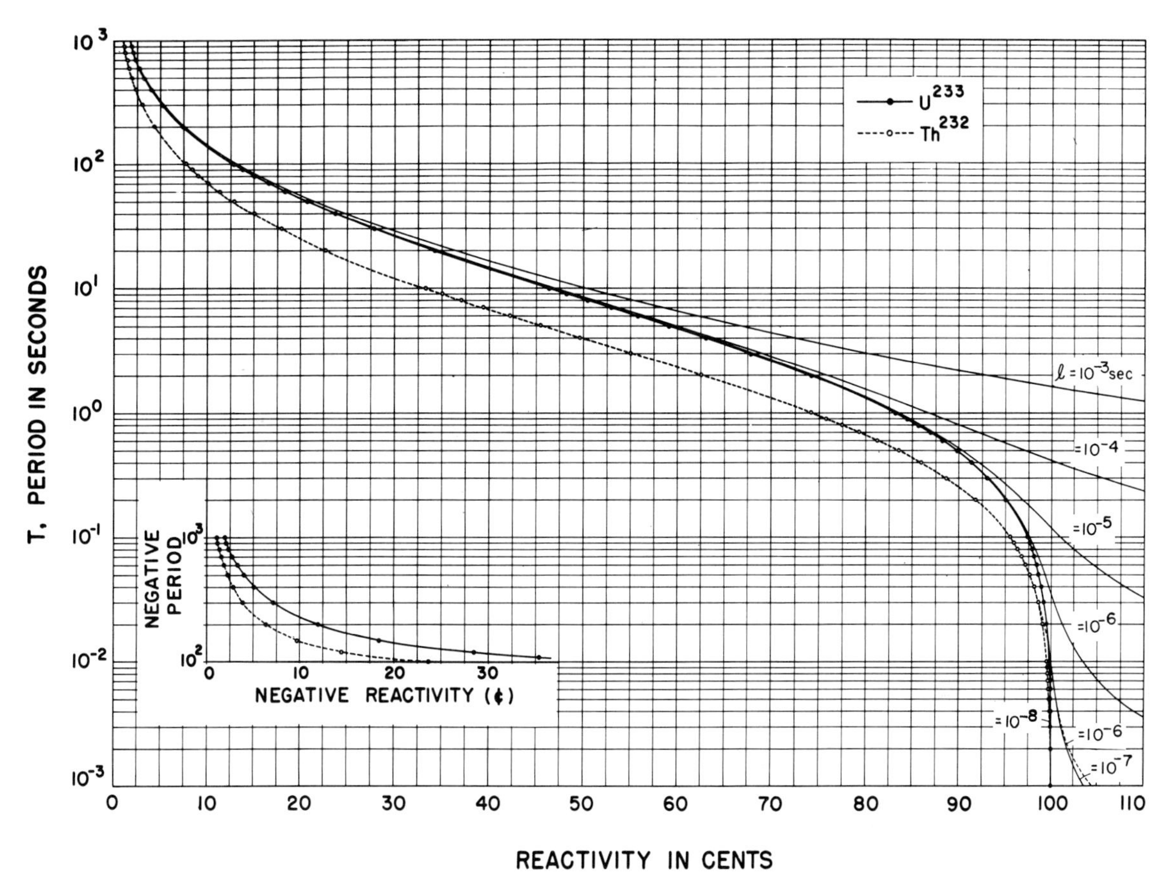

FIG. 3-5 Asymptotic period vs reactivity relation for U^{233} and Th^{232}. Parameter ℓ is prompt neutron lifetime. (Heavy curves were calculated directly from Laplace-transformed prompt burst decay data; corresponding points were calculated from delayed neutron periods and abundances.)

$$\$_{circ} \simeq \frac{\ell}{Tk\gamma\beta} + \sum_i \frac{a_i[1 + \lambda_i T(1 - f_i(\alpha))]}{1 + \lambda_i T}. \qquad (3\text{-}9)$$

This rescaled equation reduces to the usual period-reactivity relation for stationary fuel systems when $f_i(\alpha) = 1$ (no fuel circulation) as it should. However, in circulating fuel reactors, the steady state (delayed critical) condition does not correspond to zero reactivity as defined for a stationary fuel system. Thus at steady state ($\alpha = 0$, $T = \infty$) Eq. (3-9) takes the form

$$\$^{DC}_{circ} \simeq \sum_i a_i[1 - f_i(0)], \qquad (3\text{-}10)$$

which gives zero reactivity for zero fuel velocity $[f_i(0) = 1]$ as it should. On the other hand, for arbitrarily large circulation velocity we have seen that $f_i \to T_c/T_t$ so that Eq. (3-10) reduces to

$$\$^{DC}_{circ} = 1 - T_c/T_t .$$

Hence, to maintain a circulating-fuel system critical requires additional positive reactivity, the amount ranging from zero for a stationary fuel system to a maximum of $1 - T_c/T_t$ for an infinite fuel velocity. Clearly a circulating fuel reactor initially at steady state acquires a positive excess reactivity if the circulation velocity is suddenly decreased. It may be noted that the final reactivity after this restoration of partial delayed-neutron influence cannot exceed prompt criticality.

For circulating fuel systems it is necessary to redefine the term "excess reactivity." Thus let $(\$_{circ} - \$^{DC}_{circ})$ represent the excess reactivity over and above that required for criticality at any specified circulation velocity. This quantity is just the difference between Eqs. (3-9) and (3-10):

$$\$_{circ} - \$^{DC}_{circ} = \frac{\ell}{Tk\gamma\beta} + \sum_i \frac{a_i\{f_i(0) + \lambda_i T[f_i(0) - f_i(\alpha)]\}}{1 + \lambda_i T}. \qquad (3\text{-}11)$$

For long periods ($T^{-1} = \alpha \ll \lambda_i$) from Eqs. (3-7) and (3-8) it follows that $f_i(0) \cong f_i(\alpha)$; hence Eq. (3-11) reduces, approximately, to

$$\$_{circ} - \$^{DC}_{circ} \simeq \frac{\ell}{Tk\gamma\beta} + \sum_i \frac{a_i f_i(0)}{1 + \lambda_i T}. \qquad (3\text{-}12)$$

Thus a circulating fuel system behaves (for long periods) just like a stationary fuel system except that the delayed neutron fractions β_i are reduced to $\beta_i f_i(0)$, with $f_i(0)$ the steady state fraction of delayed neutron precursors in the core (see Fig. 3-8).

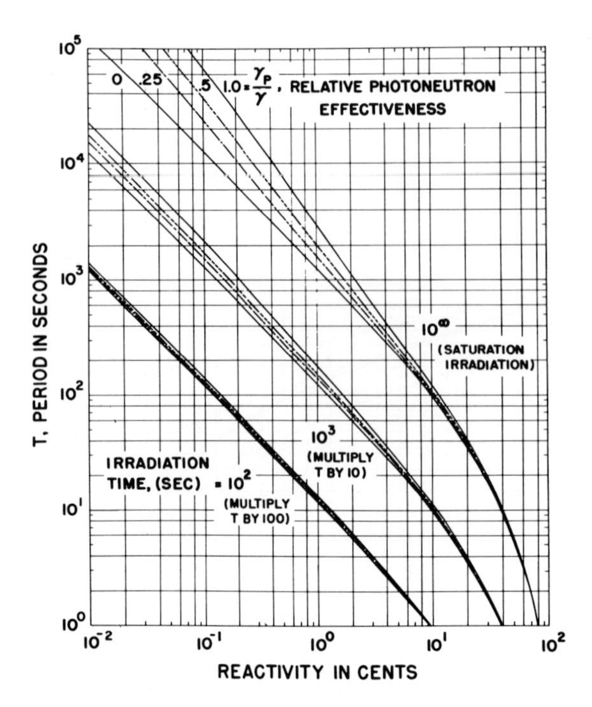

FIG. 3-6 Period-reactivity relations for D_2O-moderated systems with representative values of γ_p/γ(relative photoneutron effectiveness) and effective irradiation times 10^2, 10^3 and 10^∞ seconds. The four values of γ_p/γ indicated for saturation irradiation are plotted similarly for 10^3- and 10^2-sec irradiations. The period scale is to be multiplied by 10 for the 10^3-sec irradiation and by 100 for the 100-sec irradiation as indicated.

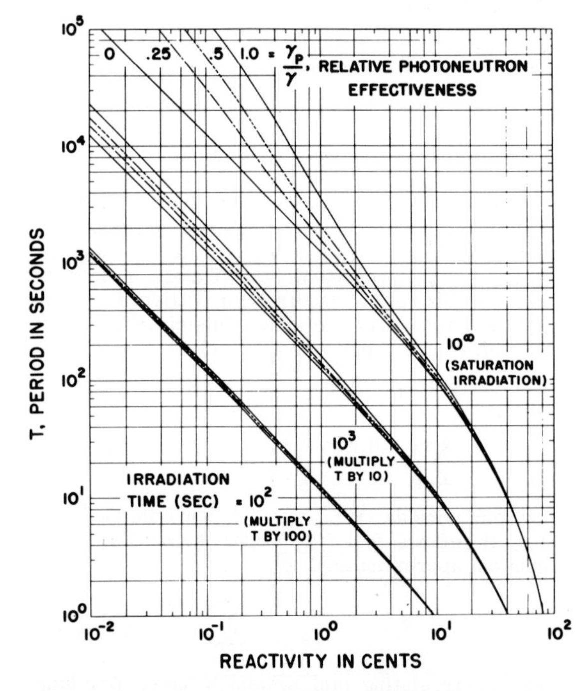

FIG. 3-7 Period-reactivity relations for Be-moderated systems with representative values of γ_p/γ (relative photoneutron effectiveness) and effective irradiation times 10^2, 10^3 and 10^∞ seconds. The four values of γ_p/γ indicated for saturation irradiation are plotted similarly for 10^3- and 10^2-sec irradiations. The period scale is to be multiplied by 10 for the 10^3-sec irradiation and by 100 for the 100-sec irradiation, as indicated.

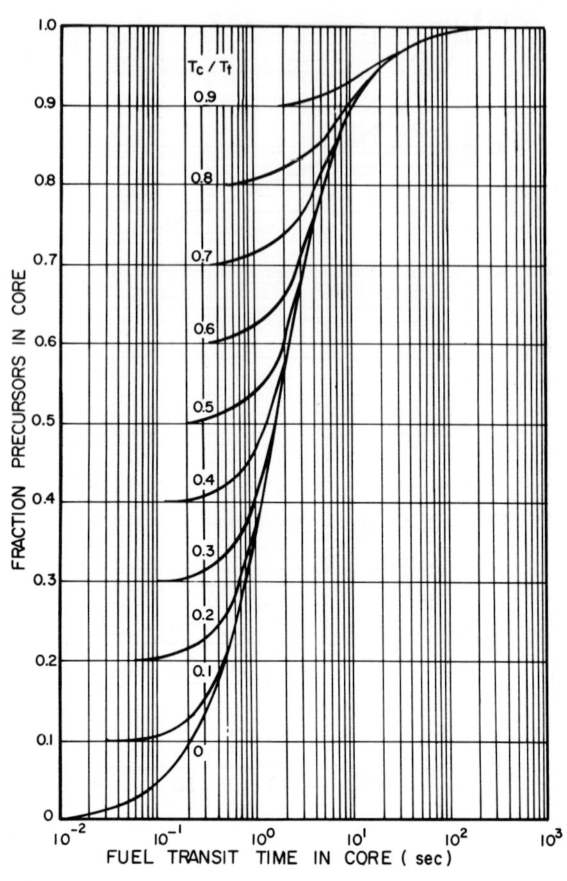

FIG. 3-8 Fraction of precursors in the core of a circulating-fuel reactor under steady state conditions. $T_c \equiv$ transit time of fuel in core; $T_t \equiv$ total transit time of complete fuel circuit (core plus external loop).

The use of reduced delay fractions $\beta_i f_i(0)$ in the conventional inhour equation is permissible either for long periods or for very high flow velocities where the f_i are insensitive to delayed neutron lifetimes (in which case $f_i(0) \rightarrow T_c/T_t$, as we have seen). One consequence of a reduced delay fraction is a corresponding reduction in the value of the dollar unit of reactivity. In general, reduced delay fractions mean a smaller reactivity interval between the delayed and prompt critical conditions and hence a smaller margin of control.

3.2.2 Neutron Effectiveness: Calculation of β_{eff}; Inhour Relations for Composite Systems

For detailed reactor kinetics calculations one requires accurate values of the delayed neutron fraction $\beta = (n/F)/\nu$, where n/F is the absolute yield of delayed neutrons per fission and ν is the average total number of neutrons emitted per fission. Table 3-1 summarizes the "best values" of ν (based largely on the measurements of Diven and Hopkins [52] and Moat et al. [53]) together with the best values of delayed neutron yield n/F [45]. The resulting delayed neutron fractions β (with propagated errors) are presented in columns 4 and 7 of Table 3-1. These

TABLE 3-1

Delayed Neutron Fractions (β)

(Including Tabulated Delayed Neutron Yields n/F and Average Total Neutron Yields per Fission ν)

Fission nuclide	Fast fission (E_{eff} ~ fission spectrum)			Thermal neutron induced fission		
	n/F	ν	β	n/F	ν	β
Pu^{239}	.0063 ± .0003	3.08 ± 0.04	$.00204 \pm .0001_1$	.0061 ± .0003	$2.82_6 \pm .02_1$	$.0021_6 \pm .0001_1$
U^{233}	.0070 ± .0004	2.61 ± .03	$.0026_8 \pm .0001_6$	.0066 ± .0003	$2.46_9 \pm .02_0$	$.0026_7 \pm .0001_2$
Pu^{240}	.0088 ± .0006	3.3 ± .2	$.0026_6 \pm .0002_4$	—	—	—
Pu^{241}	—	—	—	.0154 ± .0015	3.14 ± .06	.0049 ± .0005
U^{235}	.0165 ± .0005	2.59 ± .03	$.0063_7 \pm .0002_2$	.0158 ± .0005	$2.43_0 \pm .001$	$.0065_0 \pm .0002_1$
U^{238}	.0412 ± .0017	2.80 ± .13	.0147 ± .0009	—	—	—
Th^{232}	.0496 ± .0020	2.42 ± .20	.0205 ± .0019	—	—	—

delay fractions are appreciably lower than earlier values,* indicating generally lower margins of reactor control, with consequent important implications for reactor kinetics, control and safety. The lower β values of Table 3-1 have been corroborated by a number of independent kinetics experiments in recent years [2, 10, 56-59].

Note that delayed neutron fractions and yields vary by an order of magnitude among the different fissioning species; moreover the values of "effective" delayed neutron yields can exhibit even greater variation, depending on details of the individual chain-reacting system. It is thus essential to know the relative effectiveness (in producing fission) of the delayed and prompt neutrons from each fissioning species present in a given system. This is particularly important in small, fast systems (with high neutron leakage) or in mixed isotope systems, notable examples being breeder reactors which combine fissionable materials of different delayed neutron (and hence kinetic) characteristics.

Computation of neutron effectiveness is based on established perturbation methods [60-62] and is usually carried out by multigroup diffusion or transport theory codes. Effective prompt and delayed neutron yields, and effective delay fractions have been calculated in detail [10] for representative bare and reflected metal critical assemblies at Los Alamos (cf Table 3-2). Such idealized metal assemblies can provide useful experimental checks on multigroup transport methods for calculating the neutron kinetics of fast breeder systems [63], to give one example. The basic simplicity of these metal critical assemblies permits attention to be focused on certain aspects of the relevant neutron physics without introducing a host of associated engineering problems present in actual power breeders. Thus, the purpose of the calculations to be discussed is to highlight the factors that influence neutron effectiveness and to illustrate the general magnitude of effects.

*For example, in the case of U^{235} thermal fission, $\beta = 0.0065$ in Table 3-1 is some 15% lower than the previously widely used value [54] $\beta(U^{235}) = 0.00755$. Earlier values for other fission species are summarized in Table 9 of reference [55].

Input data for the present calculations are described in some detail in reference [10]. Using a six-energy-group split and multiregion spherical geometry, both actual and effective neutron yields (prompt and delayed) were calculated for all fissioning isotopes present in each of the assemblies listed in Table 3-2.

TABLE 3-2

Six Bare and Reflected Metal Assemblies

Assembly	Composition (of idealized assembly)
Bare U^{235}, "Godiva"	Bare U (93.8% U^{235}) sphere, 8.7 cm radius
Bare Pu^{239}, "Jezebel"	Bare δ-phase Pu sphere, 6.3 cm radius
Bare U^{233}, "Skidoo"	Bare ~98.1% U^{233} sphere, 5.8 cm radius
U-reflected U^{235}, "Topsy"	~8 in. (20 cm) thick natural uranium reflected U (94% U^{235}) sphere, 6.045 cm radius
U-reflected Pu^{239}, "Popsy"	~8 in. (20 cm) natural uranium reflected δ-Pu sphere, 4.4 cm radius
U-reflected U^{233}, "23 Flattop"	~8 in. (20 cm) natural uranium reflected U^{233} sphere, 4.1 cm radius

Table 3-3 summarizes the results of these calculations, (by EDF code written for IBM 704) giving the relative importance of prompt and delayed neutrons from all fissioning species in each of the six metal systems. Fission ratios and total neutron yield ratios are also tabulated. Relative effective yields are obtained from the product of importance and neutron fraction for each isotope q; that is, effective delayed neutron yield $E_d^q = I_d^q \beta^q$ and effective prompt yield $E_p^q = I_p^q (1-\beta^q)$.

The effective delayed neutron fraction for a mixture of fissionable isotopes is given by the ratio of effective delayed neutron yield to effective total neutron yield (prompt plus delayed):

TABLE 3-3

Relative Importance of Individual Fission Events in Six Metal Systems

Assembly	Fission isotope	Fission ratios	Total neutron yield ratios	Relative importance	
				Prompt neutrons	Delayed neutrons
Bare U^{235}	U^{235}:	1.0	1.0	1.0000	1.0342
(Godiva)	U^{238}:	0.0089_4	0.0091_0	0.0090_5	0.0093_6
Bare Pu^{239}	Pu^{239}:	1.0	1.0	1.0000	0.9459
(Jezebel)	Pu^{240}:	0.0316	0.0354	0.0354	0.0334
Bare U^{233}	U^{233}:	1.0	1.0	1.0000	1.0598
(Skidoo)					
U-reflected U^{235}	U^{235}:	1.0	1.0	1.0000	1.0097
(Topsy)	U^{238}:	0.2479	0.2540	0.0852	0.0762
U-reflected Pu^{239}	Pu^{239}:	1.0	1.0	1.0000	0.9507
	Pu^{240}:	0.0077_4	0.0087_4	0.0088_0	0.0083_7
(Popsy)	U^{235}:	0.1032	0.0845	0.0115	0.0095_8
	U^{238}:	0.3795	0.3270	0.0803	0.0695
U-reflected U^{233}	U^{233}:	1.0	1.0	1.0000	1.0395
(23 Flattop)	U^{238}:	0.3000	0.2962	0.0765	0.0702
	U^{235}:	0.0680	0.0639	0.0098_6	0.0086_2

$$\gamma\beta \equiv \frac{\sum_q E_d^q}{\sum_q \left(E_p^q + E_d^q\right)}, \qquad (3\text{-}13)$$

where the summation over q includes all fissioning isotopes in the system. Table 3-4 presents calculated effective delayed neutron fractions $\gamma\beta$ together with "experimental" $\gamma\beta$ fractions (from surface mass increment measurements [56]). Also included in Table 3-4 are measured Rossi-α [23] values (prompt neutron decay constants) at delayed critical and prompt neutron lifetime $\ell = (\gamma\beta)_{calc}/\alpha_{DC}$ for each assembly.

To examine the dependence of delayed neutron effectiveness on delayed neutron spectra, the calculations of Tables 3-3 and 3-4 were carried out using the two available composite delayed

TABLE 3-4

Effective Delayed Neutron Fractions, Rossi α Values and Prompt Neutron Lifetimes for Six Metal Assemblies

Assembly	Fission isotopes present	Fraction of total effective delayed neutron yield	Effective delayed neutron fraction, $\gamma\beta$	Experimental $\gamma\beta$ [56]	Rossi α [23, 56] at delayed critical (per μ sec)	Prompt neutron lifetime = $\gamma\beta/\alpha_{DC}$ (sec)
Bare U-235	U-235	0.9795	0.0066_4	0.0065_9	-1.10	6.04×10^{-9}
(Godiva)	U-238	0.0205	—	—	—	—
Bare Pu-239	Pu-239	0.9560	0.0019_5	0.0019_4	-0.65	3.00×10^{-9}
(Jezebel)	Pu-240	0.0440	—	—	—	—
Bare U-233	U-233	1.0000	0.0028_2	0.0029_0	-1.00	2.82×10^{-9}
(Skidoo)						
U Reflected U-235	U-235	0.8512	0.0069_4	(0.0072)[a]	-0.38_2[b]	1.82×10^{-8}
(Topsy)	U-238	0.1488	—	—	—	—
U-Reflected Pu-239	Pu-239	0.6369	0.0027_7	—	-0.22_9[b]	1.21×10^{-8}
(Popsy)	Pu-240	0.0073	—	—	—	—
	U-235	0.0200	—	—	—	—
	U-238	0.3358	—	—	—	—
U Reflected U-233	U-233	0.7178	0.0035_5	—	-0.27_1[b]	1.31×10^{-8}
(23 Flattop)	U-238	0.2680	—	—	—	—
	U-235	0.0142	—	—	—	—

[a] Estimated by Hansen [2] from measured σ_f (U-235), σ_f (U-238) and delayed neutron yields.

[b] Data of Barton, Balestrini, Chezem and Orndoff.

neutron spectra [10]. This comparison has indicated a difference of less than ~4% in the relative importance of individual isotopes and a slightly smaller percentage difference in computed effective delay fractions $\gamma\beta$ (see reference [10]). All values reported in Tables 3-3 and 3-4 were calculated using the composite data of Batchelor [64] and Bonner [65], which represent the best available measurements of delayed neutron energy spectra.

In the absence of individual γ_i values it is customary in reactor calculations to assume all $\gamma_i = \gamma$. Using the individual delay-group spectra χ_i of reference [10], this approximation can be checked rather closely for the first four delay groups (which constitute some 85% of total delayed neutron yield). For composite systems, γ is computed as the ratio* of effective delay fraction $\gamma\beta$ to "actual" delay fraction β, where

$$\beta \equiv \frac{\sum_q \beta^q \int \sum_j \nu_j^q \sigma_{f_j}^q N^q(r)\phi_j(r)\, dV}{\sum_q \int \sum_j \nu_j^q \sigma_{f_j}^q N^q(r)\phi_j(r)\, dV} . \qquad (3\text{-}14)$$

Similarly $\gamma_i = \overline{\gamma_i\beta_i}/\beta_i$ where $\overline{\gamma_i\beta_i}$ is computed by weighting group-i delayed neutron yield, with the group-i spectrum χ_i rather than the composite delayed spectrum χ_d [10]. A comparison of γ_i and γ values calculated for six metal assemblies is summarized in Table 3-5. Individual γ_i (excepting γ_1) are seen to differ from γ by less than 2% for the bare assemblies. (The very low energy spectrum of delay group 1, relative abundance < 4%, results in larger values for γ_1.) In view of the appreciable uncertainties in U^{235} delayed neutron spectra, the results in Table 3-5 appear to justify the assumption $\gamma_i \simeq \gamma$ for bare metal systems (and probably for "slow" systems which are less sensitive to fast spectrum shape). The three reflected assemblies, on the other hand, exhibit disturbingly large differences among individual γ_i values. For such systems the common assumption $\gamma_i \simeq \gamma$ may be adequate for general reactor calculations, but must be questioned for precise evaluations (e.g. detailed analysis of neutron kinetics in a fast breeder reactor).

Note in Table 3-5 that delayed neutron effectiveness - both the γ_i and γ - exceeds unity for the U^{235} and U^{233} bare systems. In these thermal-neutron-fissioning species the lower energy of delayed neutrons means higher fission probability (greater Σ_f) and higher nonleakage probability, $\exp(-B^2\tau_d)$, while slowing down. A somewhat anomalous situation occurs in the Pu^{239} bare assembly where the γ_i and γ are actually less than unity due to the dip in the Pu^{239} fission cross section in the vicinity of 0.5 Mev. One therefore expects γ_2 to be the smallest effectiveness value for Jezebel, since the second delay-group spectrum χ_2 peaks in the vicinity of 0.5 Mev.

In reflected systems a considerable fraction of fissions occur in the U^{238} of the reflector, giving large relative neutron yields for U^{238}, as shown in column 4 of Table 3-3. On the other hand, the importance of neutrons (both prompt and delayed) born in the reflector is much lower than the importance of neutrons born in the core. For example, in Popsy (U-reflected Pu^{239}) some 25% of all fissions occur in the U^{238} of the reflector, so that over two-thirds of all delayed neutrons in Popsy are from U^{238}. But the relative importance of these delayed neutrons is sufficiently small that U^{238} actually contributes only 33% to total effective delayed neutron yield. Similarly in Topsy (U-reflected U^{235}) over 38% of all delayed neutrons are from U^{238}, yet these account for less than 15% of the total effective delayed neutron yield, as seen in column 3 of Table 3-4. The major contribution to $\gamma\beta$ clearly comes from the core material in three reflected systems. Thus, despite the large relative neutron yields from U^{238} in the reflector, overall delay fractions $\gamma\beta$ are not radically different from $\gamma\beta$ for the corresponding bare assembly.

The accurate determination of effective delay fractions is particularly important in kinetics and reactivity measurements, since $\gamma\beta$ provides the bridge between experimental measurement (of ratios $\gamma\beta/\ell$, $\Delta k/k\gamma\beta$) and theory (in terms of k alone). To illustrate one practical application of effectiveness calculations in reactor kinetics, we compute the detailed period-reactivity relation in the Popsy (U-reflected Pu^{239}) and 23 Flattop (U-reflected U^{233}) assemblies. For a composite system, period-vs-reactivity is closely approximated by linear interpolation between "pure isotope" curves in direct proportion to the effective delayed neutron yield from each fission isotope present. Thus for a mixture of isotopes $q = 1 \cdots n$ (having relative delayed neutron abundances $a_i^1 \cdots a_i^n$ and decay constants, $\lambda_i^1 \cdots \lambda_i^n$) the reactivity is

$$\$(T) \equiv \frac{\Delta k}{k\gamma\beta} \simeq \sum_{q=1}^{n} \sum_{i=1}^{6} \frac{\left(E_d^q \Big/ \sum_q E_d^q\right) a_i^q}{1 + \lambda_i^q T} , \qquad (3\text{-}15)$$

where $\$(T)$ is the reactivity in dollar units, $\gamma\beta$ is the effective total delay fraction (cf Eq. (3-13)) and E_d^q is the effective delayed neutron yield from isotope q. The resulting period-versus-reactivity relations for Popsy and 23 Flattop are shown in Fig. 3-9. Corresponding relations for the pure isotopes (U^{233}, Pu^{239}, U^{238}) are included for comparison.

3.2.3 Requirements on Period Measurements; Time Dependence

When making period measurements for reactivity determination, it is important to avoid a premature observation which may include transient contributions (cf Fig. 3-2) thereby resulting in an observed period smaller than the asymptotic value [66]. The resulting errors in deduced reactivity are often significant, as is shown below.

---

*For reflected systems, delayed neutron effectiveness may also be defined as the ratio of effective delay fraction $\gamma\beta$ to actual delay fraction in the core only. This effectiveness value, labeled γ_{core}, is given in Table 3-5 for Topsy, Popsy and 23 Flattop.

TABLE 3-5

Comparison of Delayed Neutron Effectiveness Values for Six Metal Assemblies

Assembly	γ_1	γ_2	γ_3	γ_4	γ
Bare U^{235} (Godiva)	1.096	1.028	1.050	1.033	1.034
Bare Pu^{239} (Jezebel)	0.963_5	0.942_4	0.947_5	0.944_4	0.945_9
Bare U^{233}	1.123	1.055	1.078	1.059	1.060
U-reflected U^{235} (Topsy)	1.089	0.936_6	0.902_2	0.872_8	0.863_2 (γ_{core} = 1.08)
U-reflected Pu^{239} (Popsy)	0.739_3	0.668_6	0.585_8	0.498_7	0.530_6 (γ_{core} = 1.35)
U-reflected U^{233} (23 Flattop)	1.119	0.803_7	0.737_8	0.603_8	0.650_0 (γ_{core} = 1.33)

The conditions necessary for observing asymptotic period with a specified accuracy have been investigated [67] by solving the kinetic equations for the instantaneous period, $n(t)/\dot{n}(t)$, as a function of time following various step changes in reactivity. Representative results, presented in Figs. 3-10 and 3-11, show the waiting time (following a step change) required to observe a period within 1%, 5% or 10% of the asymptotic ("true") value. These calculations apply to a reactor initially at delayed critical without an external source. Note the long waiting times required following negative reactivity steps. For large negative steps ($|\Delta k| \geqslant \beta$) the instantaneous period approaches asymptotically the longest delayed neutron period. Thus in general for subcritical systems, reactor period does not provide a satisfactory reactivity index and other methods such as reciprocal multiplication or the measurement of prompt neutron decay rate α (see Sec. 2.2) are preferable.

It is useful to know how an error in measured period is propagated through the inhour equation as an error in reactivity. This has been computed for various positive and negative asymptotic periods. The results plotted in Fig. 3-12 again show the large errors associated with negative periods.

For the reactor with constant neutron source (e.g. from photoneutrons, extraneous sources, spontaneous fission, etc.) the influence of transient terms is very nearly the same as for a reactor with no source, and again transients must be avoided.

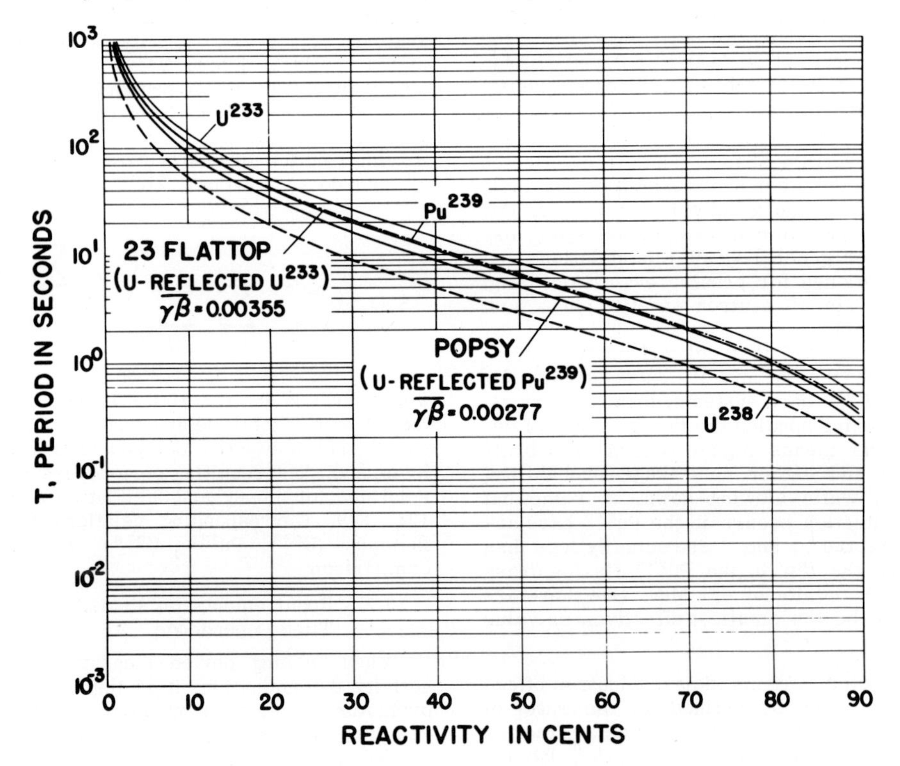

FIG. 3-9 Computed period vs reactivity relations for U-reflected U^{233} and Pu^{239} systems. (See Table 3-2 for isotopic composition of systems.)

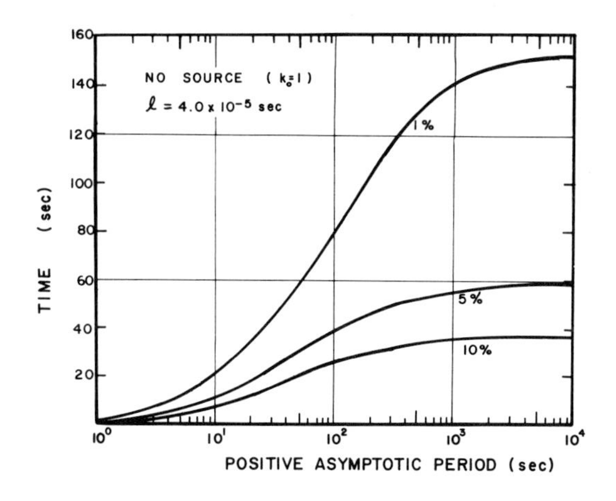

FIG. 3-10 Time after a positive step change in reactivity to observe a period within 1%, 5%, and 10% of the asymptotic value as a function of positive asymptotic period.

However in this case the constant neutron source term itself becomes the major cause of deviations from asymptotic period. Representative calculated waiting times for 5% accuracy in measurements of positive periods are shown in Fig. 3-13. The parameter is the initial reactivity k_0 rather than source strength S or normalized source strength S/n_0. These quantities are related through the initial equilibrium condition:

$$\frac{n_0}{\ell} = \frac{S}{1 - k_0},$$

where $S/(1 - k_0)$ is the multiplied source strength and n_0/ℓ, the ratio of initial neutron density to prompt neutron lifetime, is the equilibrium rate of loss of neutrons from the assembly. The case $k_0 = 1$ corresponds to negligible source strength

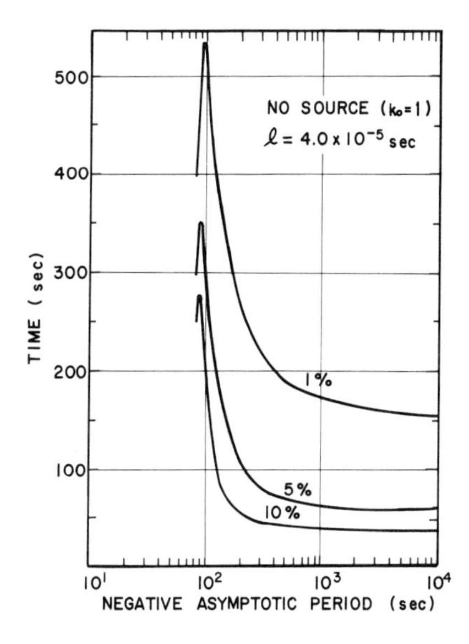

FIG. 3-11 Time after a negative step change in reactivity to observe a period within 1%, 5%, and 10% of the symptotic value as a function of negative asymptotic period.

and represents the influence of transient terms only. Comparison of Fig. 3-13 with the corresponding curve for 5% accuracy in the source-free case (Fig. 3-10) indicates the greatly increased waiting times required when a source is present. As one would expect, longer periods are seen to be most sensitive to source effects.

In general it is not advisable to use calculated error functions, as in Figs. 3-10 to 3-13, to provide accurate correction factors for premature period measurements. Rather, the calculations are intended to serve as a guide for avoiding premature period observations altogether.

3.3 Calibration and Diagnostics

In principle, approximate reactivity calibrations of fuel additions and control devices may be obtained in the course of loading a reactor. Particularly for fast reactors, source and detector locations may be such that changes of reciprocal count rate correlate with reactivity increments, so that predicted calibrations can be checked before the system is made critical. For most reactor types, however, the detector response depends not only on the value of a reactivity increment, but upon the location where the increment is introduced. This sort of distortion of response is almost the rule for heterogeneous water-moderated reactors. In this situation, the opportunity for confirmation of predictions while the reactor is subcritical is restricted to a qualitative check of the anticipated loading response curve [68]. It may be noted that in the subcritical region essentially all techniques are subject to spatial distortions similar to those of neutron multiplication.

In general, as we have seen, precise reactivity calibrations (such as required for confirming the shutdown and hold-down capability of controls) and determination of reactivity coefficients are performed in the neighborhood of criticality. The commonest and usually most satisfactory method is by period measurement in the slightly supercritical region. Normally, safety considerations limit positive-period measurements to the reactivity range from delayed criticality to about 50 cents. Larger intervals can be subdivided, and the reactivity of each step compensated between successive measurements. Interpretation of the summation of smaller increments, however, depends on the way in which the reactivity of each step is compensated - for example, whether localized or distributed. Coupling between a reactivity control being calibrated and that providing compensation is generally complex [69].

The direct determination of a large reactivity change implies a measurement at subcriticality of some quantity that can be correlated with reactivity. Such a reactivity index may be α, the prompt neutron decay constant as determined by pulsed-source or Rossi-α techniques. As noted in Sec. 2.2, whenever large reactivity changes (several dollars) are being determined from the observed α, it is necessary to check for possible changes in neutron lifetime and effective delay fraction between the initial and final configurations.

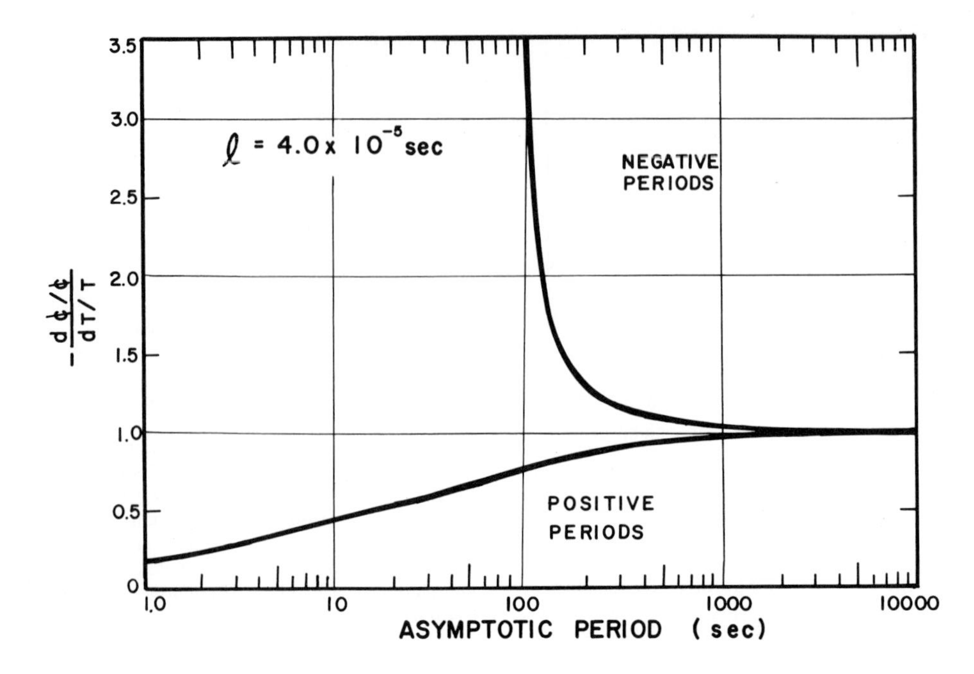

FIG. 3-12 Relative fractional error in reactivity d¢/¢ and asymptotic period dT/T as a function of asymptotic period for positive and for negative periods.

Other useful reactivity indices are neutron multiplication (if measurable) and the rod-drop technique.

In the course of measurements of approach to criticality and in reactivity calibration, there is ample opportunity for successively improved checking of the proper functioning of instrumentation, the mechanical stability of the reactor, and the nuclear characteristics of components. Guides to these characteristics are the consistency of data

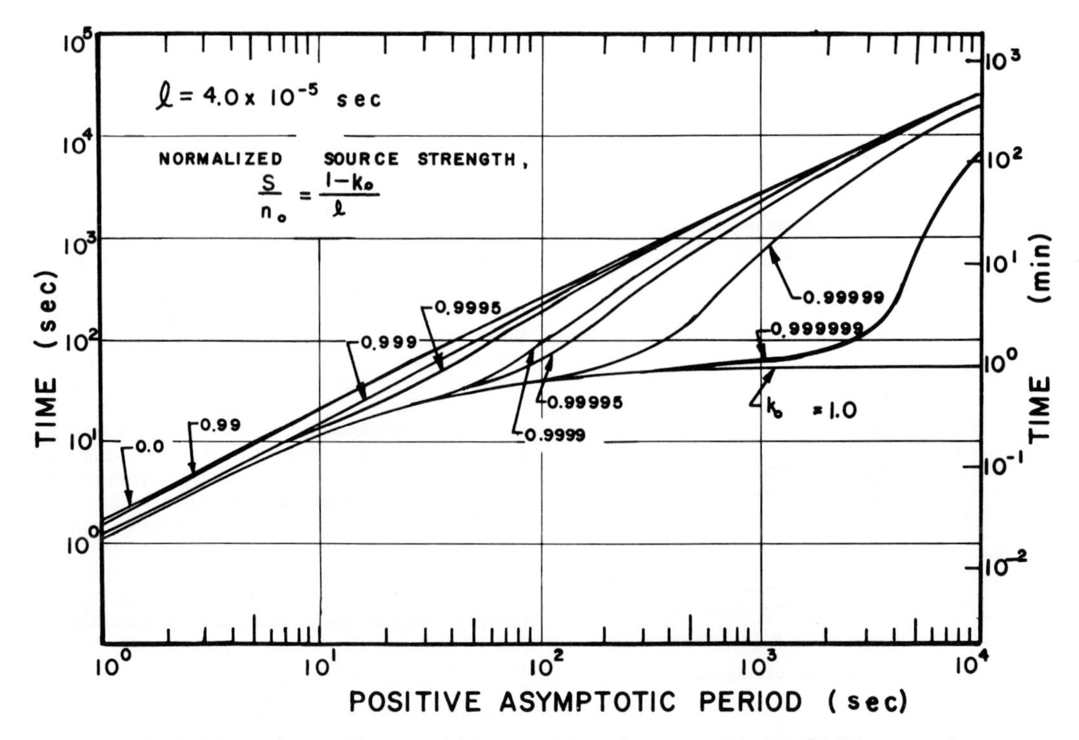

FIG. 3-13 Time after a positive step change in reactivity to observe a period within 5% of the asymptotic value as a function of positive asymptotic period for various values of initial reactivity k_o.

and comparisons between observations and predictions, particularly near delayed criticality. A series of consistency checks during buildup of the reactor core is good nuclear safety practice. Anomalies that appear at any given step should be understood, discrepancies should be resolved, and instrumental or mechanical faults that are indicated should be corrected before proceeding with further reactivity additions.

3.4 Reactivity Addition Rate Considerations

In the analysis of reactivity addition rates, as during the assembly of fissionable material or initial reactor startup, it is useful to examine the neutron level and period as a function of reactivity for various average ramp-rates. Thus Fig. 3-14 presents calculated relative neutron level and instantaneous period as a function of reactivity for various ramp-rates (starting from 10 $ initially subcritical). The extremes of photoneutron effects were calculated for Be and D_2O photoneutrons with $\gamma_p/\gamma = 1$, and compared with results for no photoneutrons ($\gamma_p/\gamma = 0$). As would be expected, ramp response is quite insensitive to the sluggish photoneutrons except for very slow ramp-rates in the vicinity of delayed criticality. (The effect of photoneutrons, from Be or D_2O, is not perceptible on the scale of the plot in Fig. 3-14.) Limiting n(t) response for zero and infinite ramp-rates are shown; corresponding limiting period curves cannot be shown on a logarithmic plot since T becomes negative for subcritical systems as $a \to 0$, and $T \to 0$ as $a \to \infty$. Note that neutron level is quite insensitive to insertion rate for highly subcritical systems, but is strongly dependent upon a as the system approaches delayed criticality. With progressively larger reactivity addition rates, the system passes through delayed criticality at progressively shorter periods. Figure 3-15 is an auxiliary plot of the data in Fig. 3-14, showing instantaneous period vs ramp-rate at various degrees of subcriticality in systems with and without Be and D_2O photoneutrons. This presentation permits easy period interpolation between the decade ramp-rate values of Fig. 3-14. Also this expanded plot shows the period-lengthening effect of photoneutrons for low ramp-rates when the system is near criticality.

From an operational and a safety point of view Figs. 3-14 and 3-15 give the relationship between the four reactor dynamic parameters: period, level, reactivity and ramp rate. Assuming a given initial shutdown condition, specification of any two of these parameters in principle determines the other two, thus providing a description of the overall kinetic behavior of the subcritical system during approach to criticality. In the event of

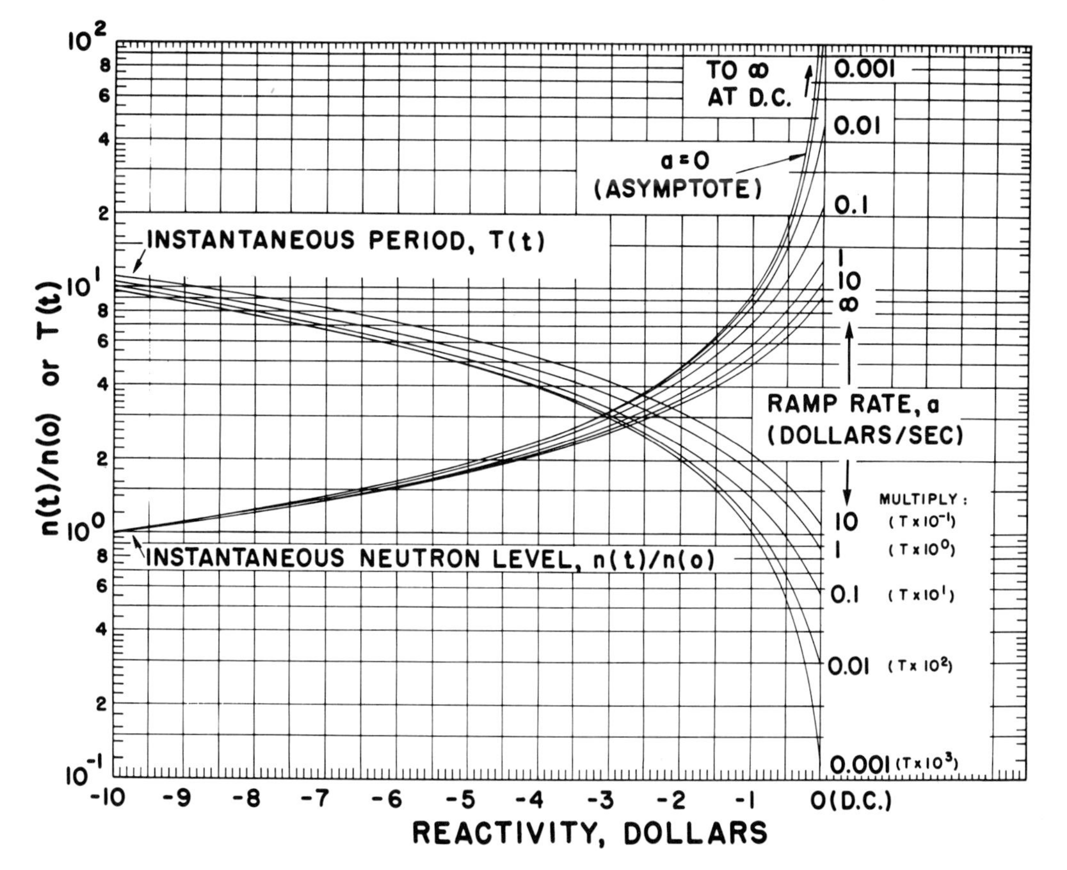

FIG. 3-14 Reactor transient response characteristics (instantaneous period and instantaneous neutron level) vs subcritical reactivity for various average ramp rates. Assumptions: $\ell = 10^{-5}$ sec, reactor initially 10$ subcritical. Note indicated scaling factors for period curves.

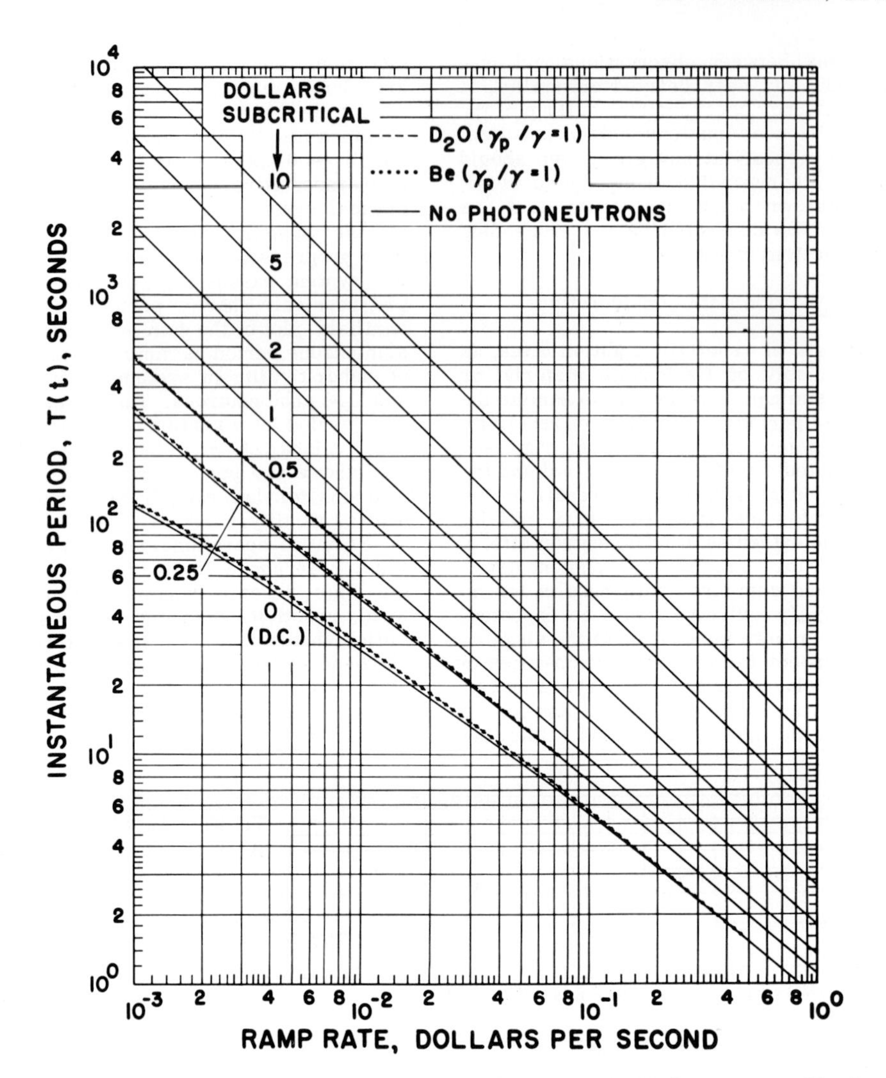

FIG. 3-15 Instantaneous period vs ramp rate for various degrees of subcriticality in systems with and without Be and D_2O photoneutrons; $\ell = 10^{-5}$ sec. (Note: 0$ subcritical corresponds to "D.C." or delayed criticality.)

too rapid insertion of reactivity, and a consequent period monitor trip, for example, calculations as in Figs. 3-14 and 3-15 can be useful in determining the time between period trip and attainment of delayed (or prompt) criticality for a given ramp-rate (see Sensing and Control Instrumentation chapter); in other words, the calculations can be used to establish requirements for safe shutdown.

Use of such parametric relations as presented in Figs. 3-14 and 3-15 for establishing reactivity addition-rate criteria may be illustrated as follows: In this example, suppose that it is desired to start up a reactor by first bringing it up to a preassigned period at a preassigned low power level which is just sufficiently high to permit an accurate period reading. Upon reaching the desired period the reactor will then be held on this period (either manually or by servo-control) until the desired final power level is attained. In this case we want to determine what ramp insertion rate will bring the reactor from initial shutdown (here assumed 10$ subcritical) to the preassigned period at the desired power level, and to determine what will be the reactivity of the subcritical system at this time. Suppose the period is 30 sec and the required low power level corresponds to the level $n(t)/n(0) \simeq 5$. We see from Fig. 3-14 that for this power the system must be between 1$ and 2$ subcritical, regardless of ramp rate. Next, Fig. 3-15 indicates that for a period of 30 sec and a system between 1$ and 2$ subcritical, the ramp rate must lie between 4 and 8 ¢/sec. Having thus bracketed the desired operating range, we quickly find (after one or two iterations) that a ramp rate of ~7¢/sec and the system at ~1.7$ subcritical correspond uniquely to the desired period of 30 sec at the desired level $n(t)/n(0) = 5$.

In the ramp-parametric calculations presented here reactivity feedback effects have been neglected, as these are normally unimportant in zero-power critical assemblies and during the initial, subcritical, stages of reactor startup. Of course ramp-rate calculations are similarly useful for investigating the kinetic response of supercritical systems and of reactors at power. Such calculations have aided analysis of the characteristics (power, energy release, time duration) of hypothetical startup excursions, the consequences of delays or failures in period or level scram systems, etc. [70]. However, the kinetic behavior of supercritical systems at power, including shutdown effects, falls outside the scope of the present chapter; these problems are treated in other chapters.

Figures 3-14 and 3-15 are presented merely to illustrate the utility of parametric kinetics calculations in analyzing reactor transient behavior. In practice there are many departures from idealized linear ramp-rates, for example, due to the variation of control-rod effectiveness with position, rod shadowing and cross multiplication effects. Further, there are several possible perturbations on reactor startup conditions. Fission product poisoning, fuel depletion and reactor temperature all can affect startup as well as operating characteristics. Neutron source contributions from delayed neutrons and photoneutrons of course depend on the recent operating history of the reactor. Although all such perturbing effects can be included in parametric kinetic calculations the number and range of the variables involved precludes further generalization here. Calculations can of course be carried out in any desired detail, assuming adequate neutron source levels so that the reactor kinetic equations are applicable. For the case of a "weak" neutron source, injecting well under one neutron per prompt neutron lifetime, the resulting statistical fluctuations in neutron density must be treated by purely statistical methods, see Sec. 1.7.2.

4 NUCLEAR SAFETY OF FUEL OUTSIDE REACTORS

Criticality experiments provide guidance for safely building up a reactor to its critical state and for maintaining safe control margins during reactor operation. These functions are important in protecting the public as well as in safeguarding the large investment represented by a nuclear power plant.

Criticality experiments also provide guidance for the safe handling of fissile material outside reactors. Operations at the reactor site that fall within this category include local transportation and storage of new fuel, transfer and interim storage of spent fuel, and, ultimately, loading of spent fuel into shipping casks. For certain future power reactor installations, there is also the possibility that irradiated fuel will be processed within the generating complex. All of these activities are potentially hazardous to plant personnel.

The prevention of any direct hazard of these operations to the public and to the power generation system is much more straightforward than similarly motivated preventive measures for the reactor. A major difference is that the designer and operator are not confronted with the inherent stored energy of the functioning reactor system, which is presumed to be capable of driving rapid reactivity additions. Only reasonable care is required to avoid a similar potential in fuel handling arrangements. It is true that any widely felt emergency at the site may have indirect effects on reactor safety, for example, by upsetting reactor operation, and it is primarily for this reason that general nuclear safety practices deserve attention here.

4.1 General Nuclear Safety Criteria

Broad statements of nuclear safety principles appear in the ASA-ANS "Standard for Operations with Fissionable Materials Outside Reactors", which is reproduced as the Appendix to this chapter. Attention is called particularly to the section Nuclear Safety Controls and to the seven items that appear under General Criteria. Principles are also discussed in Part I of the AEC "Nuclear Safety Guide" [71].

In general, the Standard encourages "built-in" nuclear safety control, as by geometry of equipment or by permanent neutron poison, instead of "administrative" control, such as required for batch limitation in large unpoisoned vessels. Experimental bases for nuclear safety limits are preferred, either by direct use of critical data to establish limits or indirectly through verification of specific calculative methods to be used for certain "classes" of fuel.

Conditions to be assumed in establishing limits (such as degree of reflection) should take into account the effects of emergencies and accidental conditions. (There is no point, however, to assuming consequences of disasters that would be expected to mask the effects of an excursion.) For example, the influence of incidental water is usually assumed unless flooding is not a practical consideration.

The "double contingency rule" is generally used as a somewhat subjective guide to the appropriate degree of protection against multiple abnormalities. In the "Standard for Operations with Fissile Materials Outside Reactors" the statement of this criterion is:

> "Process designs should, in general, incorporate sufficient factors of safety to require at least two unlikely, independent, and concurrent changes in one or more of the conditions originally specified as essential to nuclear safety before a nuclear accident is possible."

There is a growing opinion that this rule may be relaxed for adequately shielded operations, as with irradiated fuel, provided there is economic justification. As an example, the use of cadmium-nitrate solution for criticality control of a wet-chemistry process is difficult to justify on the double-contingency basis, for improper reagent identification or improper addition would nullify the control [72]. Yet the consequences of an error of this type in a shielded facility could be minor provided there is an inherent limitation to reactivity addition rates (as through restricted flow rates). Whether or not such relaxation would be appropriate at a reactor site is a matter for specific evaluation.

4.1.1 Nuclear Safety Guides

The most accurate nuclear safety guidance is obtained from criticality experiments specifically directed toward the operation in question. As an illustration, storage conditions for fuel from the Experimental Gas-Cooled Reactor were established by special criticality measurements at ORNL [73]. An alternative to preliminary experiments would be verification of the safety of the actual setup by means of neutron multiplication measurements during introduction of the fissile material (see Sec. 1). An example is the proof test of a compact, poisoned storage rack for spent fuel of the PM-2A reactor (Portable Medium Power Reactor - 2A). Multiplication measurements for the purpose were conducted at the reactor site, Camp Century, Greenland [74].

Where specific information is not obtainable, criticality data from surveys that link experimental points by computation may often be applied conservatively to the actual plant conditions. Nuclear safety limits with this sort of generality, which appear, for example, in the "Nuclear Safety Guide", cover certain situations with an unknown but presumably large degree of conservatism and do not apply at all to others. Frequently, "factors of ignorance" may be reduced by converting the most nearly appropriate criticality data to plant conditions by means of special analytical and empirical techniques. For the needed compilations of data and discussions of computational methods, the most complete reference at present is "Criticality Control in Chemical and Metallurgical Plants" [75], the proceedings of a 1961 symposium at Karlsruhe. Another compilation of strictly experimental criticality data, "Critical Dimensions of Systems Containing U^{235}, Pu^{239}, and U^{233}" being prepared as a companion piece to the "Nuclear Safety Guide", will contain references to most original papers [76].

4.2 Nuclear Safety of Specific Operations

4.2.1 Fuel Processing

The future of local processing of irradiated fuel may be judged by reactor experiments and proposals in which fuel treatment is part of the system. Continuous fission product removal is considered for such widely varying fuels as the aqueous solution of the Homogeneous Reactor Experiment [77] the bismuth alloy of the Liquid Metal Fuel Reactor [78] and the plutonium rich alloys of the Los Alamos Molten Plutonium Reactor Experiment [79]. Further contrast appears between the pyrometallurgical fuel recovery process of the Experimental Breeder Reactor (EBR-II) system [80] and the more conventional solution chemistry methods of Purex [81]. As the entire gamut of nuclear safety problems is represented, we cannot afford detailed coverage. To the earlier general remarks about criticality control may be added the rather obvious comment that nuclear safety restrictions are less severe for nonaqueous processes than for solutions methods. McLendon and Mee [82], for example, refer to a gas-phase process for converting UF_6 to UF_4, which avoids most of the criticality problems associated with more conventional aqueous techniques.

4.2.2 Storage Under Dry Conditions

The following discussion applies to storage arrays of air-spaced fuel where water flooding is not of practical concern. Let us assume that there are no large masses of beryllium, heavy water, or graphite within or bordering the storage regions, and that there is no hydrogenous material within the array. Under these specific conditions there is no criticality restriction on the storage of unirradiated U^{235}-U^{238} fuel in which the U^{235} enrichment does not exceed 5 wt.%. For higher U^{235} enrichments, the total quantity that can be permitted in a storage array depends on the spacing between elements, in other words, on the average density of U^{235} throughout the lattice.

4.2.2.1 Three-Dimensional Dry Storage Arrays It is particularly simple to establish conservative dry storage limits for most fuel elements designed for water-moderated reactors, where the neutron multiplication of an unmoderated element is small. In this situation, an enclosed storage array can be viewed as a homogeneous, low density unit in which the critical mass m_c is approximated by

$$m_c = m_{co}(\rho_0/\rho)^{\sim 1.4}, \quad (4\text{-}1)$$

where ρ is the average lattice density of fuel element material, ρ_0 is the average density of material within the fuel element, and m_{co} is the critical mass of fuel material at full density when reflected like the array [83]. In other terms, ρ/ρ_0 is the fraction of the array volume that is actually occupied by fuel elements. To illustrate conservative application, m_{co} is chosen as the critical mass of a water-reflected sphere of full density uranium metal [84], and Eq. (4-1) is modified:

$$m_c \simeq m_{co}(\rho_0/\rho) = 2 \times (\text{allowable mass}) \quad (4\text{-}2)$$

Resulting "allowable masses" of Table 4-1 can be increased for certain fuel elements if cladding is considered in the composition. Still larger limits might be anticipated for experimentally verified arrays, primarily as the result of more reasonable exponents for the density ratio than the value unity of Eq. (4-2). Of course, the storage limits of Table 4-1 are not necessarily applicable to areas in which flooding is a concern.

The extension of relation (4-2) to Pu or U^{233} is direct. Minimum values of m_{co} are 5.6 kg for a water-reflected sphere of α-phase Pu^{239} (density 19.6 g/cm^3), and ~7.0 kg for a water-reflected sphere of U^{233} [84].

4.2.2.2 Plane Dry Storage Arrays If unmoderated fuel is to be stored in a plane, e.g. a single layer of elements against a concrete wall, there exists a spacing of units beyond which an infinite array is subcritical [83]. (This is unlike a cubic air-spaced array for which there is no limiting critical spacing.) More suitable for generalization

TABLE 4-1

Examples of Conservative Mass Values in Dry Storage Arrays of U^{235} Fuel Elements

Fuel composition	m_{co} (kg U^{235})	Allowable kg of U^{235} in array			
		$\rho/\rho_o = 0.3$	$\rho/\rho_o = 0.1$	$\rho/\rho_o = 0.03$	$\rho/\rho_o = 0.01$
U (93) metal	22.8	38	114	380	1140
U (40) metal	~42	70	210	700	2100
U (20) metal	~66	145	330	1450	3300
U (10) metal	~190	415	950	4150	9500
U (<5) metal	∞	No limit	—	—	—
U (93.5)O_2**	45*	75	225	750	2500
U (40)O_2**	62	103	310	1030	3100
U (20)O_2**	111	185	555	1850	5550
U (10)O_2**	520	865	2600	8650	26000
U (≤5)O_2**	∞	No limit	—	—	—
U (93.5)O_2F_2-H_2O, H/U^{235} = 50	3.25	5.4	16.2	54	162
U (93.5)O_2F_2-H_2O, H/U^{235} = 500	0.82	1.36	4.1	13.6	41
U (93.5)-C, C/U^{235} = 77.5***	89.3*	148	445	1480	4450
U (93.5)-C, C/U^{235} = 316***	45.8*	76	229	760	2290
U (93.5)-C, C/U^{235} = 1270***	14.2*	23.6	71	236	710

*DSN or DTK computation with Hansen-Roach cross sections.

**Density of U (nat)O_2 = 10.96 g/cm^3.

***Density of undiluted graphite = 1.90 g/cm^3.

is the limiting critical surface density of fissile material as measured on the storage plane. Below this value of the surface density, the plane array remains subcritical regardless of extent.

Several series of measurements on plane arrays of highly enriched uranium or plutonium units have been sufficiently complete for extrapolation to the spacing at which the infinite array would be just critical [85, 86]. Corresponding critical surface densities appear in Fig. 4-1 as fractions of the surface density of the infinite critical metal slab (uniform thickness) that is completely reflected by water (30.4 g U^{235}/cm^2, or 16.4 g Pu^{239}/cm^2). (The water-reflected slab is chosen for reference because of the availability of criticality data.) The abscissa, "fraction critical" of unit, is the ratio of the mass of a sphere with the same reactivity as the unit to the mass of the corresponding critical sphere. Commonly, the "fraction critical" of a dry fuel cluster that is designed for a water-moderated reactor is very small. It may be noted that the surface densities of ORNL critical solutions arrays of various shapes fall above the curve of Fig. 4-1 when expressed as fractions of the critical solution slab density, and that the curve appears to be a reasonable limit for infinite plane arrays of this class [87].

Confining our attention to plane storage arrays which are closely reflected on one side only, the ordinate 0.5 of Fig. 4-1 is suggested as a maximum surface density ratio for the storage of reactor fuel of interest. This limit is not valid if an individual item to be stored has large reactivity, that is, if its "fraction critical" exceeds something like 0.3. It is not necessarily adequate for arrays that can be flooded.

Examples of surface-density limits that correspond to the above criterion appear in Table 4-2. They are intended more for orientation than to replace specific studies.

4.2.3 Underwater Fuel Storage

There is insufficient experimental information to serve as a basis for practical general rules on the underwater storage of reactor fuel. The only measurements directed specifically toward this problem were conducted at ORNL on flooded arrays of MTR-type fuel elements [90], Rover reactor fuel [91], and elements for the Experimental Gas-Cooled Reactor [73]. The following summary by Callihan (page 160 of reference [75]) of results with the MTR-type elements indicates the nature of the information that can be obtained and demonstrates the value of such experiments where storage space is at a premium.

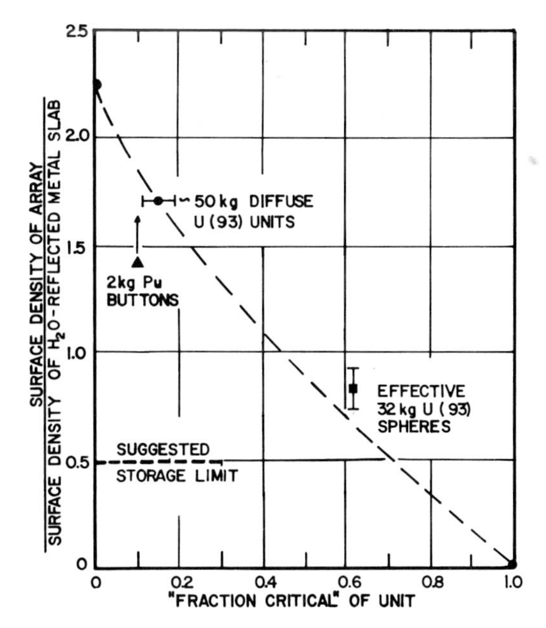

FIG. 4-1 Criticality of infinite plane arrays of metal (U^{235} or Pu), semi-reflected by concrete. The critical surface density of a fully reflected U^{235} (93.5) slab is 30.4 g U^{235}/cm^2 or 28.2 kg/ft^2. The corresponding value for Pu^{239} is 16.4 g Pu^{239}/cm^2 or 15.2 kg/ft^2. [The "fraction critical" of a unit is the ratio: (mass of a sphere with the same reactivity as the unit)/(mass of the corresponding critical sphere).]

TABLE 4-2

Examples of Surface Densities Corresponding to the Limit of Figure 4-1 for Dry Storage of Fuel in Plane Array

Fuel composition	Crit. thickness infinite refl. slab - cm	U^{235} or Pu^{239} density-g/cm^3	Surface density limit-g/cm^2 of U^{235} or Pu^{239}
Pu^{239} metal	0.84	19.6	8.2
U (93.5) metal	1.72	17.6	15.5
U (40) metal	3.82*	7.5	14.3
U (20) metal	7.64*	3.8	14.5
U (10) metal	19.8*	1.9	18.8
U (≤5) metal	∞	—	No limit
U $(93.5)O_2$**	3.8*	8.8	16.7
U $(40)O_2$**	7.3*	3.82	13.9
U $(20)O_2$**	13.9*	1.91	13.2
U $(10)O_2$**	37.5*	0.95	17.9
U $(5)O_2$	∞*	—	No limit
U $(93.5)O_2F_2$-H_2O, $H/U^{235} = 50$	4.45	0.48	1.06
U $(93.5)O_2F_2$-H_2O, $H/U^{235} = 500$	8.4	0.052	0.22
U (93.5)-C, $C/U^{235} = 77.5$***	24.0*	0.468	5.6
U (93.5)-C, $C/U^{235} = 316$***	32.7*	0.117	1.9
U (93.5)-C, $C/U^{235} = 1270$***	36.3*	0.029	0.53

*DSN or DTK computation with Hansen-Roach cross sections [88, 89].

**Density of U $(nat)O_2 = 10.96$ g/cm^3.

***Density of undiluted graphite = 1.90 g/cm^3.

"In general, these elements were constructed of plates containing a uranium-aluminum alloy supported in a frame or box about 7.5 cm^2 and 50-60 cm long. The uranium was enriched to 93.2% in U^{235}. Most of the elements available for these experiments contained 168 g U^{235}, a few had 200 g while still others were loaded with 140 g. These elements were of the type used in the Oak Ridge Research Reactor and in the Bulk Shielding Reactor. The following observations were made in the experiments.

"The number of elements in a critical array submerged in water is a strong function of the spacing between the elements. With this spacing optimum, about 1.2 cm between adjacent fuel sections, the number of elements required for criticality contained approximately 2.6 kg U^{235}, a quantity essentially independent of the mass per element within the above range. Increasing this separation to 3.2 cm increased the critical number to 31, about 5.2 kg U^{235}, while an array of an infinite number at a 5.7 cm separation would probably not be critical. Each array was a single element high.

"No appreciable source neutron multiplication was produced by a submerged, close-packed, 11- x 12-element array in which adjacent rows of elements, in one direction, were separated by a 0.05 cm thick cadmium sheet. The average loading of the 132 elements was 160 g U^{235} each. There was no appreciable multiplication when both the cadmium and the water were removed and the array was surrounded by a 30 cm thick paraffin reflector.

"An infinite number of elements arranged in a plane in water will be subcritical. Two rows of 26 elements each (a total of twenty-four 200 g elements and 28 of 168 g) submerged at optimum spacing were subcritical."

Storage arrays of spent power reactor fuel have been established by computations that are normalized to reactor experiments [92, 93] instead of to storage conditions. For "unpoisoned" storage there is a crucial difference, consistently small water spacings within the reactor versus some large spacings within the storage array. The resulting uncertainties imply that the calculated results must be interpreted conservatively. One type of water-immersed array consists of uniformly latticed fuel bundles, and another consists of a lattice spaced more closely in one lateral dimension than in the other, resulting in parallel planes of fuel that are more or less isolated by water. If the reactivity of an individual fuel cluster is not too large (like one-half the controlling critical dimension) extra inches of water for conservatism cost more with the first arrangement than with the second. This is a principal reason for interest in flooded plane arrays of fuel.

4.2.3.1 Plane Arrays of Fuel Clusters in Water Here, attention is confined to fuel elements for water-moderated reactors. It should be emphasized that there are no direct data on critical slabs of water-moderated fuel clusters for low values of U^{235} enrichment. In the more favorable cases, critical slab thicknesses are obtained by conversion from near-equilateral or elongated shapes. Though extrapolation distances are available from flux traverses, their validity for precise shape transformations has not been established. With these limitations in mind, the minimum critical slab thickness estimates of Fig. 4-2 are used to illustrate a means of establishing criteria for storage in water.

The upper curve of Fig. 4-2, estimated from Westinghouse APD data on stainless steel clad UO_2 fuel elements [92, 94, 95], represents near-optimum rod diameter and pitch. The curve below it is a minimum composite of a survey of unclad-metal lattices by Brown of Hanford [96] and an unpublished analysis of unclad-UO_2 lattices by Barry and French of Westinghouse APD. The suggested limits for plane storage arrays in water, one-half the indicated critical thicknesses, are intended only as an example.* From the limit curve for 0.3 in. (0.76 cm) diameter UO_2 clad in 0.02 in. (0.051 cm) thick stainless steel, it is apparent

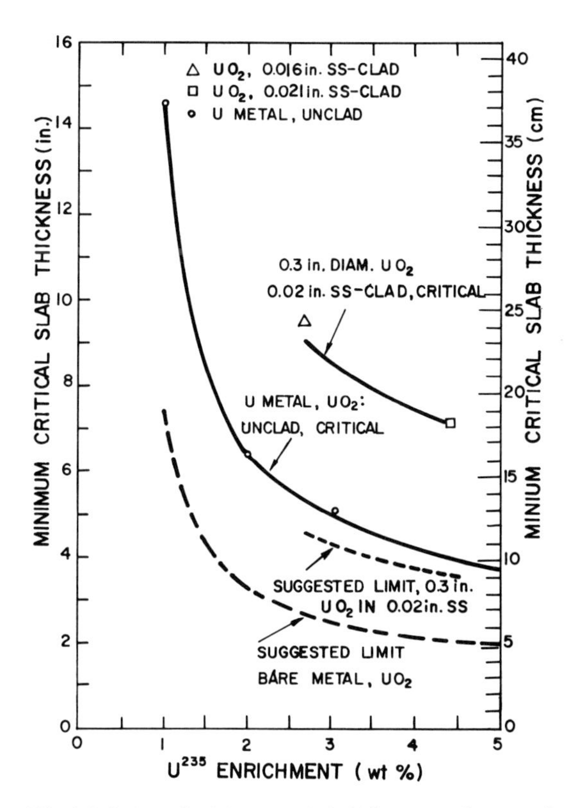

FIG. 4-2 Estimated minimum critical thicknesses and suggested plane storage limits for latticed slabs of U and UO_2 (1 to 5 wt.% enrichment) in H_2O.

*A smaller margin is appropriate provided the guiding critical data are reliable and interaction with other fissile material may be ignored.

that assemblies of such fuel with uranium enrichment of 2.7% can be stored in contact in a plane provided the cluster cross section does not exceed a square with 4 1/2 in. (11.4 cm) sides. According to this criterion, larger fuel bundles would be spaced in the plane. At 4.5% enrichment the limiting cross section for close-packing in a plane would be a 3 1/2 in. (8.9 cm) square. For Zr- or Al-clad elements, of course, the smaller average thicknesses of the lower limiting curve would be appropriate.

Where long fuel assemblies have high individual reactivity (for example, k_{eff} approaching 0.9), a near-isolating thickness of water is required between assemblies, so that the concept of storage in planes breaks down. An array spaced equally in both lateral dimensions is then indicated.

4.2.3.2 Parallel Plane Storage Arrays in Water

Efficient storage of irradiated fuel calls for the paralleling of individually safe planes with near-isolating thicknesses of water between planes, or a smaller thickness of poison and water. The required thickness of unpoisoned water depends on the reactivity of the individual slab and possibly on its composition. In the absence of data that can be generalized, large arbitrary spacings have sometimes been used, like 12 in. (30.5 cm) generally [97] and 8 in. (20.3 cm) for units such as those usually encountered in a processing plant [71]. The latter value has some basis from measurements of interaction between pairs of various reasonably reactive units containing highly enriched uranium (see Fig. 39 of reference [84]).

Results of computations provided by D. R. Smith of Los Alamos, see Table 4-3, suggest that much can be gained in storage efficiency by examining individual cases instead of adopting generally conservative water spacings. Although Smith's method of calculation, the DTK (DSN-type) code and Hansen-Roach cross sections, has been checked against a variety of critical systems (ref. [75] pp. 173-204), appropriate experimental data in the low U^{235} enrichment range are scanty. Consequently, the values of Table 4-3 are considered to be illustrative instead of gospel.

Particularly where water separation distances in storage arrays are to be near minimum, it is apparent that there must be a reliable arrangement for positioning fuel clusters. As interaction can increase rapidly with decreasing water separation, there must be allowance for uncertainties or possible errors in location.

4.3 Poisoned Storage Arrays and Shipping Casks

Because water-moderated power reactors start life as large and heavily poisoned arrays of fuel clusters, computational techniques that reproduce the characteristics of such a reactor should be appropriate for designing other close-packed poisoned arrays of the fuel elements in water. In this case, it is possible to devise efficient heavily poisoned storage tanks and shipping casks for irradiated fuel.

Where confirmed methods of calculation are not available, there is a choice between special criticality or subcriticality measurements, design by

TABLE 4-3

Computed values of Critical Separation between Parallel Slabs in an Infinite Flooded Array

Slab composition	Ratio of individual slab thickness to critical slab thickness (t_c)	Water thickness between slabs just critical, infinite array ($k_\infty = 1$)
U (93) metal	0.5 (t_c = 1.76 cm)	6.0 in. (15.2 cm)
U (93)-H_2O, H/U^{235} = 400	0.5 (t_c = 8.4 cm)	3.4 in. (8.6 cm)
U (5)-H_2O, H/U^{235} = 400	0.5 (t_c = 12.8 cm)	3.3 in. (8.4 cm)
U (5)-H_2O, H/U^{235} = 400	0.75 (t_c = 12.8 cm)	4.7 in. (11.9 cm)
U (5)-H_2O, H/U^{235} = 400	0.9 (t_c = 12.8 cm)	6.5 in. (16.6 cm)
U (2)-H_2O, H/U^{235} = 400	0.5 (t_c = 24.0 cm)	3.0 in. (7.5 cm)

approximate calculation followed by experimental confirmation (as by neutron multiplication measurements during loading), or conservative adaptation of the sparse data that are suitable for generalization. Several series of measurements on uranium lattices in boron solution are sufficiently complete to provide estimates of the maximum boron concentration required to reduce k_∞ to unity (i.e., to reduce buckling to zero) [95, 98, 99]. As this concentration must be optimized with respect to the diameter of fuel rod and lattice spacing, a very extensive study is required for a reliable value at each U^{235} enrichment. Interpretations of existing lattice data appear in Fig. 4-3, along with similar information for homogeneous mixtures obtained with the Hanford Physical Constants Test Reactor [100]. The principal lattice series is for unclad or Al-clad rods, and, as expected, the single point for stainless steel clad rods shows a reduction in boron requirement. In view of uncertain optimization and extrapolation of the data in this figure, it is suggested that the indicated boron requirements be

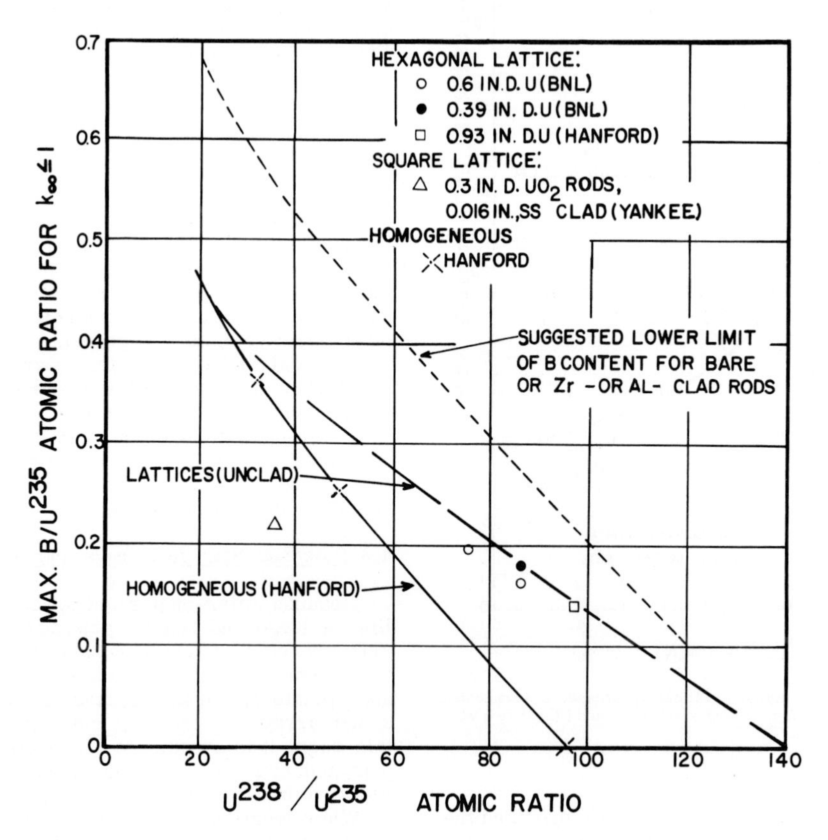

FIG. 4-3 Boron concentration (in H_2O) required to reduce k_∞ to unity. Data are for enrichments of 5% or less. ("0.6 IN. D.U." = "0.6 in. diameter uranium".)

increased by about 50% for general applications in which boron is distributed favorably. The dotted curve represents such lower limits for fuel which is unclad or clad with Zr or Al.

It is apparent that the suggested limiting curve of Fig. 4-3 cannot apply to grossly lumped boron, but is reasonable for boron solution or for uniform layers surrounding fuel bundles or separating slabs of fuel in which the reactivity is not excessive. Practical application, of course, requires assurance that the poison retains its proper disposition. Questions of the deterioration of Boral in water, of retaining proper concentrations in solution, or of stable forms of cadmium, for example, enter the design and surveillance of a storage or shipping facility that depends upon poison.

In the opinion of the authors of this chapter, the status of general information about efficient fuel storage and shipment is unsatisfactory, as is the present rate of improvement. It is clear from this brief survey that increased experimental and theoretical effort can profitably be directed toward these nuclear safety problems. In the absence of more precise knowledge one has to assume factors of safety which are undoubtedly much too large.

APPENDIX

Standard for Operations with Fissionable Materials Outside Reactors (N6.1-1964)

1. INTRODUCTION

All operations with fissionable materials outside reactors must be performed in a manner precluding the establishment of a nuclear chain reaction. The prevention of the chain reaction is important because the concomitant release of energy, principally as heat and radiation, might be lethal to nearby personnel, and it could cause property damage and interruption of process schedules. The extent of the dispersal of fission products in the event of a nuclear chain reaction would depend on the severity of the damage to containers and, in the case of reactor fuel elements, on the fission-product content established by their prior history. Contamination in addition to that resulting from the chain reaction could result from the dispersal of fissionable materials.

The rapid evolution of nuclear technology has made impossible any complete and permanent recording of recommendations for assuring safety under the nearly illimitable number of practical combinations of process conditions. It is appropriate therefore to set forth only generalized criteria on which the solution of problems in this field of specialized safety may be based. In the criteria that follow, industrial nuclear safety means the prevention or termination of inadvertent nuclear chain reactions, that is, conditions of nuclear criticality, in nonreactor environments.

2. PURPOSE

It is the intent of this Standard to provide guidance in the prevention of nuclear chain reactions in all procedures for handling, storing, processing, transporting, and treating fissionable nuclides. Generalized basic criteria are presented to achieve this purpose. Solutions to specific problems may be derived from the literature referenced (see 6, Procedural Guides).

3. SCOPE

The recommendations of this Standard shall be applicable to all operations with fissionable nuclides outside nuclear reactors. The materials to which the Standard and the referenced literature primarily refer are the isotopes of uranium of masses 233 and 235 and the isotope of plutonium of mass 239. Although there are other nuclides that fission, these three do so most readily and constitute the greatest potential hazard.

This Standard does not cover the design of process equipment, shipping containers, etc., beyond the dimensions and capacities necessary for nuclear safety as specified in the references (see 6, Procedural Guides). The integrity of all equipment in the use for which it is designed is assumed. The Standard does not refer per se to the details of any of the procedures for the use of fissionable materials or to processing methods. It does concern limitations which must be imposed on such processes and equipment because of the unique nuclear properties of these materials by which they can support nuclear chain reactions.

This Standard does not refer explicitly to the effects of radiation on man or materials or to the sources of such radiation either natural or the result of nuclear chain reactions.

4. NUCLEAR SAFETY CONTROLS

Procedures of the usual engineering type which govern any operation with fissionable materials must be supplemented by appropriate regulations to ensure nuclear safety in the process. The factors which govern the establishment of nuclear chain reactions and which, singly or in combination, must be controlled to effect safety include the mass and distribution of the fissionable nuclide in the process, the dimension and volumetric limitations imposed by the equipment, the proximity of neutron reflectors, and the physical and chemical properties of the process materials, such as isotopic and chemical concentrations, density, neutron cross sections, and other nuclear properties.

Since not only the fissionable materials, but also their containers and immediate environment contribute to the conditions for criticality, all these factors shall be considered in specifying nuclear safety limits. Maintenance of the specified limits at all times shall be the assigned responsibility of qualified operating personnel. Administrative procedures, with emphasis on safety, shall be established to clearly define this responsibility.

The technical literature contains the results of many experiments performed expressly for the establishment of regulations of the type considered here. It also contains interpretations and applications of data that in some cases are of general interest and in other cases have particular reference to specific processes and operations. These presentations are continually revised and superseded as newly derived intelligence replaces extrapolations and estimates.

5. GENERAL CRITERIA

The following general criteria form a basis for all nuclear safety specifications:

(1) Process designs should, in general, incorporate sufficient factors of safety to require the occurrence of at least two unlikely, independent, and concurrent changes in one or more of the conditions originally specified as essential to nuclear safety before a nuclear accident is possible.

(2) In a general consideration of nuclear safety problems it is assumed that only those substances commonly encountered in nature or usually associated with the process will be mixed with or located near fissionable nuclides. Examples of these materials are concrete, natural water, and wood. Special consideration shall be given to cases in which reactor materials, such as heavy water and beryllium, are present.

(3) Unless the absence of neutron-reflecting materials, including personnel, from areas adjacent to the process is assured, specifications should be based on the assumption that the fissionable material may become surrounded by water to the extent that will yield the maximum nuclear reactivity. This requirement is particularly necessary in the design of shipping containers which might become submerged as a consequence of an accident while in transit.

(4) The importance of incorporating nuclear safety in the design and constitution of equipment by limiting the dimensions to preclude chain reactions is strongly emphasized. In the design of such safe equipment, full advantage can be taken of the nuclear characteristics of the process materials. Careful administrative control shall be exercised to ensure that the specified material properties are maintained throughout the process.

(5) The use of neutron-absorbing materials, such as cadmium and boron, can render equipment and processes safe within the requirement of nuclear safety provided adequate experimental data confirm their suitability and their installation has assurance of permanency. Solid absorbers may be included in the construction and assembly of equipment; the use of solutions of neutron absorbers as components of process streams is less acceptable because of the administrative control required to assure their presence.

(6) Limits of nuclear safety should be established on bases derived from experiments. In the absence of directly applicable experimental measurements, the results of calculations based on reactor theory and made by a method shown to be valid by comparison with experimental data are acceptable provided limits of error can be assigned to them.

(7) Appropriate safety factors shall be included in all specifications derived from data and calculations. The magnitude of the safety factor shall be commensurate with the uncertainty in the basis of the specification, the probability of its violation, and the seriousness of the consequences of a conceivable nuclear accident.

6. PROCEDURAL GUIDES

Specified limits on the factors controlling the nuclear safety of processes involving fissile isotopes are given in the "Nuclear Safety Guide."[1] It is recommended that the limits appearing in the latest revision of the Guide be applied to problems of nuclear safety. The document also contains an extensive bibliography of reports of basic nuclear data and their application to various operations. Additional reference is made to the "Guide to Shipment of U^{235} Enriched Uranium Materials."[2]

---

[1] "Nuclear Safety Guide," Document TID-7016, Office of Technical Services, Department of Commerce, Washington 25, D.C.

[2] "Guide to Shipment of U^{235} Enriched Uranium Materials," Document TID-7019, Office of Technical Services, Department of Commerce, Washington 25, D.C.

REFERENCES

1. F. F. Hart and E. C. Mallary, "Practical Aspects of Pajarito Neutron Multiplication Measurements", USAEC Report LA-1604, Los Alamos Scientific Laboratory, 1953.
2. G. E. Hansen, "Properties of Elementary Fast-Neutron Critical Assemblies", Proceedings of the Second U.N. International Conference on Peaceful Uses of Atomic Energy, Geneva, 1958, Vol. 12, p. 84.
2a. W. U. Geer, E. G. Koontz, J. D. Orndoff, and H. C. Paxton, "Hazards Evaluation for Los Alamos Critical Assembly Facility", USAEC Report LAMS-2698 (Rev.), Los Alamos Scientific Laboratory, 1963.
3. C. K. Beck, A. D. Callihan, J. W. Morfitt, and R. L. Murray, "Critical Mass Studies, Part III", Report K-343, Union Carbide Nuclear Co., 1949.
4. C. P. Baker, F. L. Bentzen, J. Bridge, R. E. Carter, H. Daghlian, G. Friedlander, H. Hammel, J. Hinton, F. de Hoffmann, M. G. Holloway, D. W. Kerst, L. D. P. King, H. M. Lehr, J. H. Midney, R. E. Schreiber, J. S. Starner, and P. H. Watkins, "An enriched homogeneous nuclear reactor", Rev. Sci. Instr., 22(1951)489.
5. M. C. Edlund, R. M. Ball, J. C. Deddens, R. B. Flickinger, J. P. Farrar, and H. S. Barringer, "Consolidated Edison Thorium Reactor, Core I, Zero Power Experiments", Report BAW-173, Babcock & Wilcox Co., 1962.
6. R. M. Ball and A. L. McKinney, "Nuclear Merchant Ship Reactor Zero Power Test, Core I", Report BAW-1202, Babcock & Wilcox Co., 1960.
7. The Shippingport Pressurized Reactor, by personnel of the Naval Reactors Branch, USAEC, Westinghouse Electric Corp., and Duquesne Light Co., Addison-Wesley Publishing Co., Inc., Reading, Mass., 1958.
8. "Report of the Project Physics Section for June, July, August, 1954", Report KAPL-1207, p. 71, Knolls Atomic Power Laboratory, (Classified).
8a. A. L. Colomb, "Photoneutrons and the control of a pool-type reactor", Nucl. Sci. Eng., 9(1960)289.
9. G. R. Keepin and C. W. Cox, "General solution of the reactor kinetics equations", Nucl. Sci. Eng., 8(1960)670.
10. G. R. Keepin, "Neutron data for reactor kinetics. 1. Delayed neutrons from fission. 2. Photoneutrons from D_2O and beryllium", Nucleonics, 20,8(1962)150.
11. E. D. Courant and P. R. Wallace, "Fluctuations of the number of neutrons in a pile", Phys. Rev., 72(1947)1038.
12. H. Hurwitz, D. B. MacMillan, J. H. Smith, and M. L. Storm, "Kinetics of low source reactor startups. Part I", Nucl. Sci. Eng., 15(1963)166.
13. G. I. Bell, "Probability distribution of neutrons and precursors in a multiplying assembly", Ann. Phys., 21,2(1963)243.
14. S. H. Hanauer, "Role of neutron source in nuclear safety", Nucl. Safety, 4,3(1963)52.
15. G. E. Hansen, "Assembly of fissionable material in the presence of a weak neutron source", Nucl. Sci. Eng., 8(1960)709 and references cited therein.

16. F. Schroeder, "Power Excursion Studies", Report IDO-16791, p. 53, Phillips Petroleum Co., 1963.
17. L. A. Stephan, "Transient Tests of the BSR-II Core in the SPERT I Facility", Report IDO-16768, p. 21, Phillips Petroleum Co., 1963.
18. J. M. Harrer, Nuclear Reactor Control Engineering, D. Van Nostrand Co., Inc., Princeton, N. J., 1963.
19. W. S. Hogan, "Negative-reactivity measurements", Nucl. Sci. Eng., 8(1960)518.
20. F. J. Jankowski, D. Klein, and T. M. Miller, "Calibration of control rods", Nucl. Sci. Eng., 2(1957)288.
21. F. de Hoffmann, "Statistical Aspects of Pile Theory", p. 103, The Science and Engineering of Nuclear Power, Vol. II, Clark Goodman (Ed.), Addison-Wesley, Publishing Co., Inc., Reading, Mass., 1949.
22. R. P. Feynman, F. de Hoffmann, and R. Serber, "Dispersion of the neutron emission in U^{235} fission", J. Nucl. Energy, 3(1956)64.
23. J. D. Orndoff, Prompt neutron periods of metal critical assemblies", Nucl. Sci. Eng., 2(1957)450.
24. G. S. Brunson, R. Curran, S. G. Kaufmann, J. McMahon, and L. Pahis, "Measuring the prompt period of a reactor", Nucleonics, 15,11(1957)132.
25. J. T. Mihalczo, "Prompt Neutron Decay in a Two-Component Enriched Uranium Metal Critical Assembly", USAEC Report ORNL TM 470, Oak Ridge National Laboratory, 1963.
26. G. S. Brunson, personal communication, 1963; see also: G. S. Brunson, R. N. Curran, J. M. Gasidlo, and R. J. Huber, "A survey of prompt neutron lifetimes in fast critical assemblies", Trans. Am. Nucl. Soc., 6(1936)56.
27. F. Schroeder (Ed.), "Quarterly Technical Report, SPERT Project, October-December, 1960", Report IDO-16687, p. 5, Phillips Petroleum Co., 1961; "Quarterly Technical Report, SPERT Project, October-December, 1961", Report IDO-16750, p. 9, Phillips Petroleum Co., 1962; "Quarterly Technical Report, SPERT Project, January-March, 1963", p. 5, Phillips Petroleum Co., 1963.
28. R. Johnson, "A Statistical Determination of the Reduced Prompt Neutron Generation Time, Λ/β in the SPERT IV Reactor", Thesis, University of Idaho, 1963.
29. R. Johnson, J. Grund, and H. Larson, "A Summary of Neutron Statistics Measurements on SPERT Reactors", to be published 1964.
30. K. H. Beckurts, "Reactor physics research with pulsed neutron sources", Nucl. Instr., 11(1961)144.
31. "Proceedings of Symposium on Pulsed Neutron Techniques, Berkeley, California, December 1958", Report UCRL-5665, University of California Lawrence Radiation Laboratory, 1960.
32. D. E. Parks, J. R. Beyster, and N. F. Wikner, "Thermal neutron spectra in graphite", Nucl. Sci. Eng., 13(1962)306 and references cited therein.
33. G. de Saussure and E. G. Silver, "Determination of the Neutron Diffusion Parameters in Room-Temperature Beryllium", USAEC Report ORNL-2641, Oak Ridge National Laboratory, 1959.
34. P. B. Daitch, "Theoretical Interpretation of Pulsed Neutrons" in Neutron Physics, M. L. Yeater (Ed.), p. 187, Academic Press Inc., N. Y. 1962.
35. J. R. Beyster, "Neutron Thermalization Measurements Using a Electron-Linear Accelerator" in Neutron Physics, M. L. Yeater (Ed.), p. 215, Academic Press Inc., N.Y. 1962.
36. G. de Saussure, K. Henry, and P. Perez-Belles, "Reactivity worth of the central fuel element in the Bulk Shielding Reactor-1", Nucl. Sci. Eng., 9(1961)291.
37. See reference to work of Simmons and Bohl in J. Bengston, Proceedings of Second U.N. International Conference on the Peaceful Uses of Atomic Energy, Geneva, 1958, Vol. 12, p. 63; see also: B. E. Simmons, "The dynamic reactivity interpretation of pulsed neutron measurements", Nucl. Sci. Eng., 5(1959)254.
38. O. C. Kolar and F. A. Kloverstrom, "Pulsed neutron measurement of control rod worths", Nucl. Sci. Eng., 10(1961)45.
39. E. Garelis and J. L. Russell, Jr., "Theory of pulsed neutron source measurements", Nucl. Sci. Eng., 16(1963)263.
40. N. G. Sjöstrand, "Measurements on a subcritical reactor using a pulsed neutron source", Arkiv för Fysik, 11(1956)233.
41. B. E. Simmons and J. S. King, "A pulsed neutron technique for reactivity determination", Nucl. Sci. Eng., 3(1958)595.
42. C. A. Sastre, "The measurement of reactivity", Nucl. Sci. Eng., 8(1960)443.
43. J. L. Shapiro, "A Reactivity Monitor for Reactor Startup", Trans. Am. Nucl. Soc. 5(1962)180.
44. R. E. Uhrig, "Measurement of Reactor-shutdown Margin by Noise Analysis", in Reactor Kinetics and Control, Proceedings of a Symposium held at the University of Arizona, March 25-27, 1963. AEC Symposium Series No. 2, April 1964.
45. G. R. Keepin, T. F. Wimett and R. K. Ziegler, "Delayed neutrons from fissionable isotopes of uranium, plutonium and thorium", Phys. Rev., 107(1957)1044. A more complete account is given in J. Nucl. Energy, 6(1957)1.
46. G. R. Keepin, "Period-reactivity relations determined directly from prompt-burst neutron decay data", Nucl. Sci. Eng., 5(1959)132.
47. G. A. Cosan and C. J. Orth, "Diffusion of Fission Products at High Temperatures from Refractory Matrices", Proceedings of the Second U.N. International Conference on Peaceful Uses of Atomic Energy, Geneva, 1958, Vol. 7, p. 328.
48. E. A. Bryant, G. A. Cowan, J. E. Sattizahn, and K. Wolfsberg, "Rates and mechanisms of the loss of fission products from uranium-graphite fuel materials", Nucl. Sci. Eng., 15(1963) 288.
49. W. D. Burch, C. H. Cater, D. L. Hathaway, B. S. Maxon, N. R. Williamsen, and O. O. Yarbro, "Immiscible-Liquid-Cooled, Fluid-Fuel Reactor", USAEC Report CF-55-8-188, Oak Ridge National Laboratory (X-10), 1955.
50. R. P. Hammond, R. E. Stanford, and J. R. Humphreys, "Mobile Fuel Plutonium Breeders", USAEC Report LA-2644, Los Alamos Scientific Laboratory, 1962.
51. J. A. Fleck, Jr., "Theory of Low Power Kinetics of Circulating Fuel Reactors with Several Groups of Delayed Neutrons", USAEC Report BNL-334-T57, Brookhaven National Laboratory, 1955.
52. B. C. Diven and J. C. Hopkins, "Numbers of Prompt Neutrons per Fission for U^{233}, U^{235}, Pu^{239}, and Cf^{252}, Physics of Fast and Intermediate Reactors, Proceedings of an International Atomic Energy Agency Symposium held at Vienna, 1962, Vol. I, p. 149.
53. A. Moat, D. S. Mather, and P. Fieldhouse, "The Number of Prompt Neutrons from U^{235} Fission over the Range 0.04 to 3MeV", Physics of Fast and Intermediate Reactors, Proceedings of an International Atomic Energy Agency Symposium held at Vienna, 1962, Vol. I, p. 139.
54. D. J. Hughes, J. Dabbs, A. Cahn, and D. Hall, "Delayed neutrons from fission of U^{235}", Phys. Rev., 73(1948)111.
55. G. R. Keepin, "Delayed Neutrons", Progr. Nucl. Energy, Series I, Vol. 1, p. 191, Pergamon Press, N.Y., 1956.
56. G. E. Hansen, "Status of Computational and Experimental Correlations for Los Alamos Fast-Neutron Critical Assemblies", Physics of Fast and Intermediate Reactors, Proceedings of an International Atomic Energy Agency Symposium held at Vienna, 1962, Vol. 1, p. 445.
57. T. F. Ruane, D. J. Anthony, P. Buck, D. A. Gavin, and H. B. Stewart, "A Measurement of the Effective Delayed Neutron Fraction by the Substitution Method", Trans. Am. Nucl. Soc., 1(1958)142.
58. S. Kaplan and A. F. Henry, "An Experiment to Measure Effective Delayed Neutron Fractions", Report WAPD-TM-209, Bettis, Atomic Power Laboratory, 1960.
59. R. Perez-Belles, J. D. Kington, and G. de Saussure, "A measurement of the effective delayed neutron fraction for the Bulk Shielding Reactor-1", Nucl. Sci. Eng., 12(1962)505; see also: USAEC Report ORNL-3016, Oak Ridge National Laboratory, 1960.
60. A. F. Henry, "The application of reactor kinetics to the analysis of experiments", Nucl. Sci. Eng., 3(1958)52.
61. G. E. Hansen and C. Maier, "Material Replacement Experiments: Theory and Measurements for the Lady Godiva Assembly", USAEC Report LA-1525, Los Alamos Scientific Laboratory, 1953.
62. L. N. Usachoff, "Equations for the Importance of Neutrons, Reactor Kinetics and the Theory of Perturbation", Proceedings of the First U.N. International Conference on Peaceful Uses of Atomic Energy, Geneva, 1946, Vol. 5, p. 503.
63. W. H. Roach, "Computational survey of idealized fast breeder reactors", Nucl. Sci. Eng., 8(1960)621; see also: L. J. Koch and H. C. Paxton, "Fast reactors", Ann. Rev. Nucl. Sci., 9(1959)437.
64. R. Batchelor and H. R. McK. Hyder, "The energy of delayed neutrons from fission", J. Nucl. Energy, 3(1956)7.
65. T. W. Bonner, S. J. Bame, Jr., and J. E. Evans, "Energy of the delayed neutrons from the fission of U^{235}", Phys. Rev., 101(1956)1514.
66. R. O. Brittan, "Some Problems in the Safety of Fast Reactors", USAEC Report ANL-5577, Argonne National Laboratory, 1956.
67. B. J. Toppel, "Sources of error in reactivity determinations by means of asymptotic period measurements", Nucl. Sci. Eng., 5(1959)88.
68. Savannah Nuclear Technology Department, "Savannah Nuclear Power, Summary Test Report, Reactor Operations at Camden", Report NYS-106, Fig. 2, New York Shipbuilding Corp., 1962.
69. G. E. Hansen and C. Maier, "Perturbation theory of reactivity coefficients for fast-neutron critical systems", Nucl. Sci. Eng., 8(1960)532.
70. J. L. Anderson, "Role of LogN period meter in reactor protection", Nucl. Safety, 4, 3(1963)48.
71. Nuclear Safety Guide, USAEC Report TID-7016, Rev. 1, 1961.
72. J. P. Nichols, "Soluble Neutron Poison as a Primary Criticality Control in Shielded and Contained Radio-Chemical

Facilities", USAEC Report ORNL-3309, Oak Ridge National Laboratory, 1962.
73. E. B. Johnson and R. K. Reedy, Jr., "Neutron Multiplication by Experimental Gas-Cooled Reactor Fuel Assemblies", USAEC Report ORNL-TM-433, Oak Ridge National Laboratory, 1962.
74. W. J. McCool and E. W. Schrader, "Startup nuclear testing of a portable nuclear power station at a remote arctic site", Trans. Am. Nucl. Soc. 5(1962)137.
75. "Criticality Control in Chemical and Metallurgical Plants", Proceedings of the Organization for Economic Cooperation and Development, European Nuclear Energy Agency, held at Karlsruhe, November, 1961, OECD Mission Publication Office, 1346 Connecticut Ave., N. W., Washington 6, D. C., and OECD 33 Rue de Franqueville, Paris XVI.
76. "Critical Dimensions of Systems Containing U^{235}, Pu^{239}, and U^{233}", USAEC Report, to be published as companion to "Nuclear Safety Guide" (see ref. 71).
77. Homogeneous Reactor Project, Oak Ridge National Laboratory, "Aqueous Homogeneous Reactor Fuel Technology", Proceeding of the Second U.N. International Conference on Peaceful Uses of Atomic Energy, Geneva, 1958, Vol. 7, p. 3.
78. C. J. Klamut, D. G. Schweitzer, J. G. Y. Chow, R. A. Meyer, O. F. Kammerer, J. R. Weeks, and D. H. Gurinsky, "Material and Fuel Technology for an LMFR", Proceedings of the Second U.N. International Conference on Peaceful Uses of Atomic Energy, Geneva, 1958, Vol. 7, p. 173.
79. M. E. Battat, "Critical Experiments and Nuclear Calculations - LAMPRE-1", Physics of Fast and Intermediate Reactors, Proceedings of an International Atomic Energy Agency Symposium held at Vienna, 1962, Vol. 1, p. 263.
80. A. B. Shuck and J. E. Ayer, "Engineering considerations for remote refabrication of EBR-II fuel elements", Nucl. Sci. Eng., 12(1962)398.
81. E. R. Irish, "Description of the Purex Process", Report HW-60116, Hanford Atomic Products Operation, General Electric Co., 1959.
82. J. D. McLendon and W. T. Mee, "Nuclear Safety Progress in Uranium Hexafluoride Conversion", Report Y-KB-6 (informal), Union Carbide Nuclear Co., 1962.
83. E. R. Woodcock and H. C. Paxton, "Criticality Aspects of Transporation of Fissile Materials", Progr. Nucl. Energy, Series IV, 4, 401, Pergamon Press, N.Y., 1961.
84. H. C. Paxton, "Critical Data for Nuclear Safety Guidance", USAEC Report LAMS-2415, Los Alamos Scientific Laboratory, 1960.
85. C. L. Schuske, "Neutron Multiplication Measurements of Oralloy Units in Arrays", Report RFP-51, Dow Chemical Co., Rocky Flats Plant, Denver, 1955.
86. C. L. Schuske, C. L. Bell, G. H. Bidinger, and D. F. Smith, "Industrial Criticality Measurements of Enriched Uranium and Plutonium, Part II", Report RFP-248, Dow Chemical Co., Rocky Flats Plant, Denver, 1962.
87. H. C. Paxton, "Correlations of Experimental and Theoretical Critical Data", Fig. 6, USAEC Report LAMS-2537, Los Alamos Scientific Laboratory, 1961.
88. W. R. Stratton, "Critical Dimensions of U(93.5)-Graphite-Water Spheres, Cylinders, and Slabs", USAEC Report LAMS-2955, Los Alamos Scientific Laboratory, 1963.
89. D. R. Smith, Los Alamos Scientific Laboratory, personal communication.
90. J. K. Fox and L. W. Gilley, "Critical Experiments with Arrays of ORR and BSR Fuel Elements", Neutron Physics Semiannual Report, USAEC Report ORNL-2609, p. 34, Oak Ridge National Laboratory, 1958.
91. D. Callihan, Oak Ridge National Laboratory, personal communication, 1962.
92. W. H. Arnold, Jr., "Critical Masses and Lattice Parameters of H_2O-UO_2 Critical Experiments: A Comparison of Theory and Experiment", Report YAEC-152, Yankee Atomic Electric Co., 1959.
93. D. B. Wehmeyer and K. E. Roach, "Nuclear safety of UO_2-ThO_2-H_2O systems", J. Nucl. Energy (A & B), 14(1961)189.
94. P. W. Davison, S. S. Berg, W. H. Bergmann, D. F. Hanlen, B. Jennings, R. D. Leamer, and J. E. Howard, "Yankee Critical Experiments - Measurements of Lattices of Stainless Steel Clad Slightly Enriched Uranium Dioxide Rods in Light Water", Report YAEC-94, Yankee Atomic Electric Co., 1959.
95. P. W. Davison, V. E. Grob, D. F. Hanlen, R. D. Leamer, H. Ritz, and E. Santandrea, "Two-Region Critical Experiments with Water-Moderated Slightly Enriched UO_2 Lattices", Report YAEC-142, Yankee Atomic Electric Co., 1959.
96. C. L. Brown, "Calculated Critical Parameters for Slightly Enriched Uranium Rods in Light Water", Report HW-69273, Hanford Atomic Products Operation, General Electric Co., 1961.
97. H. F. Henry (Ed.). "Guide to Shipment of U^{235} Enriched Uranium Materials", USAEC Report TID-7019, 1959.
98. R. C. Lloyd, "Summary Listing of Subcritical Measurements of Water-Uranium Lattices Made at Hanford", Report HW-65552, Hanford Atomic Products Operations, General Electric Co., 1960.
99. H. Kouts, R. Sher, J. R. Brown, D. Klein, S. Stein, R. L. Hellins, H. Arnold, R. M. Ball, and P. W. Davison, "Physics of Slightly Enriched Normal Water Lattices (Theory and Experiment)", Proceedings of the Second U.N. International Conference on Peaceful Uses of Atomic Energy, Geneva, 1958, Vol. 12, p. 446.
100. V. I. Neeley, J. A. Berberet, and R. H. Masterson, "k_∞ of Three Weight Percent U^{235} Enriched UO_3 and $UO_2(NO_3)_2$ Hydrogenous Systems", Report HW-66882, Hanford Atomic Products Operation, General Electric Co., 1961; and: V. I. Neeley, "A Comparison of PCTR and Critical Experiment Determinations of k_∞ for Two Weight Percent U^{235} Enriched UF_4 in Paraffin", Report HW-62031, Hanford Atomic Products Operation, General Electric Co., 1959.

CHAPTER 6

Sensing and Control Instrumentation

A. PEARSON and C. G. LENNOX
Atomic Energy of Canada Limited
Chalk River, Ontario, Canada

CHAPTER CONTENTS*

1 INTRODUCTION

A reactor plant contains many systems that require control, but there are two in particular that are of most concern from the safety viewpoint: the reactor system and the containment system.

The reactor control system maintains the neutron flux and its rate of change at a level that meets the requirements imposed upon the reactor plant. To carry out the control two distinct aspects are usually evident, both of which are included in the term "reactor control": a regulating or operational requirement and a safety requirement.

The regulating system maintains the neutron flux against disturbances in reactivity resulting from poison changes, fuel burnup, temperature and pressure effects, etc. Further, the regulating system maintains the flux at a value demanded by the power production requirements.

The safety system returns the neutron flux and its rate of change to a safe state whenever the regulating function is not properly carried out, or whenever lack of integrity in some part of the reactor system requires a speed of response beyond the capabilities of the regulating system. The safety system must be able to exercise its high speed control properties only in a direction that forces the system to a safer state.

The containment control system is in some respects a backup to the reactor safety system. Control must be exercised over all gaseous and liquid effluents, and over the action of plant personnel, so that radioactivity above preset levels is not permitted to reach the public domain.

This chapter deals with the sensing instru-

*Except for a few changes and additions made in proof, this chapter is based on information in the literature or known to the authors prior to December 1963.

mentation that obtains signals for these control systems and with the instrumentation needed to perform the control functions.

1.1 Control System Philosophies

Because of the emphasis placed on public safety in nuclear power plant design, control philosophies have been formulated [1] that set the broad requirements for a reactor control system. These philosophies have been molded by the experience of different groups of people with diverse reactor types, and hence the philosophies diverge markedly. There are two extremes. One philosophy requires that characteristics inherent in the nuclear design, such as the Doppler effect, assure that unsafe neutron densities cannot be generated, even when external forces attempt to increase the reactivity. A philosophy at the other extreme relies solely on externally generated reactivity effects, such as control rods to maintain control, and freely admits that, if these external controls were all to fail coincident with a reactivity increase, the neutron density could reach dangerous levels.

In general, of course, a position is reached between these extremes. No reactor is without instrumentation and external aids, but their function in maintaining reactor integrity depends largely on the specific reactor design.

The question whether safety is "built-in" also extends to the regulating system itself. If a sensor that measures reactor power for a regulating system fails in a manner that falsely calls for higher power, one might attempt to anticipate the failure and actuate the safety mechanisms directly. On the other hand, one could accept the power rise to the level of an independent trip circuit that shuts down the reactor. In both cases the result of the failure is to return the reactor to a safer condition, but the method of calling the protective circuit into action is fundamentally different.

Philosophies also diverge when the role of the operating staff is considered, especially when considering the operator's role in regulating reactor parameters. Some insist that during the startup procedure, when the reactor is being brought from a low power level into its power-producing range, an operator must manually adjust the regulating controls. Others require the procedure to be entirely automatic, with the operator overseeing the general situation as it advances. When the reactor is operated at power, it is almost universal practice now to rely on an automatic regulating system, but a divergence of opinion exists: some automatic systems are made to require constant operator attention, others to require only periodic attention. There is, however, general agreement that safety systems should be completely automatic and require no human on-the-spot decisions, no doubt because all safety systems require a speed of response and a dependability beyond human capabilities.

In the long term the operating staff is the key to reactor plant safety, for without a continuing effort to maintain standards as designed, deterioration and ultimate failure are very likely.

No matter what philosphy is adopted as the guiding theme for system design, the designer must be convinced himself that he has achieved a safe design. To this end he must invoke appropriate experience, together with statistical evidence, to convince first himself and then a safety authority that not only will the control system meet the requirements posed by any credible situation, but that it is built in a sound manner and is capable of being kept sound for the life of the reactor.

1.2 Control Requirements

The reactor control system must bring the reactor and its auxiliaries safely from a shutdown state to a power-producing state taking into account maximum allowable stresses in all plant components and the demands imposed by the external load.

The control system must also prevent excursions in any parameter that would permit a reactor component to go beyond its rating in the face of any credible disturbances. Included in such disturbances would be loss of load due to turbogenerator or transmission line faults or any failure in the regulating system that would cause reactivity to be inserted, or an absorber to be withdrawn, at its maximum rate (the maximum rate in general being determined by some fundamental means, such as the synchronous speed of control rod drive motors).

These requirements produce a conflict between regulating and safety system design. For example, the maximum rate of withdrawal of reactivity that can be effected by the regulating system, together with the total amount of reactivity available for withdrawal, can be used to set minimum requirements for the safety system. In general, it is required that the speed of response of any controller be limited by some fundamental means in order to provide sufficient time for the safety system to detect any controller failure and to override the fault.

There are also credible failures that cannot be forestalled by the control systems. Loss of integrity of a containment or a coolant system because of undetected flaws can allow rupture to occur at normal operating levels. In such cases the control system is required to actuate dousing systems and emergency cooling systems, to close containment vessels after initial surges have passed, to divert cooling water to delay storage tanks, and to reroute ventilating and cooling air.

In the particular case of fuel cladding failures, the control system must provide evidence of the failure soon enough to permit removal of the element or elements before a more serious failure occurs and the removal becomes potentially a difficult operation.

In loss-of-integrity incidents the main function of the control system is to prevent the spread of radioactive contamination. It must also warn of excessive radiation levels in a manner that provides personnel with the maximum opportunity to evacuate the affected area.

Finally, in considering control requirements, it is important to be precise in stating in detail what is expected of the control system under various fault conditions.

1.3 Specific Considerations in Control System Design

1.3.1 Control Parameters

The specific parameters upon which the control of a nuclear power plant is based depend largely on the reactor type. Measurements associated with four reactor types will be considered as typical examples: the gas-cooled reactor, the dual cycle boiling water reactor, the pressurized water reactor, and the liquid-metal-cooled fast reactor. Diagrams of each type of plant, in Figs. 1-1, 1-2, 1-3, and 1-4, show where the important parameters are measured along with simple regulating loops (indicated by broken lines). Appendix 1 of this book gives values for the various parameters of plants now operating.

Neutron Flux. As a reactor control parameter the neutron flux, although it has its limitations, is unique in two respects. Its response to reactivity disturbances is, for most purposes, instantaneous and it permits the status of the chain reaction to be followed many decades below full power, thereby providing a means to detect reactivity changes even in a subcritical reactor.

Neutron flux is invariably used to initiate safety action in the event of excess reactor power, and it is the only parameter that permits automatic control and startup from the lowest power levels.

In general, neutron flux sensors are placed in small numbers about the periphery of the reactor vessel and measure only a sample of the mean core flux. This is not as a rule a true representation of the mean flux, because of flux distortions caused by unbalanced loading or by insertion of regulating rods. Also, the leakage flux measured by the neutron sensor can be affected by the moderator temperature.

As a consequence of these errors, the ratio of measured flux to mean core flux varies with reactor conditions and direct control of the reactor by means of the measured flux can lead to variations in the mean flux and to excess heating in some sectors of the reactor core. A typical situation is indicated in Figs. 1-5 and 1-6.

The position of flux sensors with respect to several sequentially operated regulating rods [2] and the variation in the measured-to-mean flux with regulating rod position is shown. If in a safety system shutdown action is based unconditionally on neutron flux, the varying trip margin must be taken into account. A further discrepancy appears between measured neutron flux and a thermal power measurement during the transient condition after shutdown when the fission product heating can amount to several percent of full power. In covering the complete operating span from full power to a completely shutdown subcritical state, the neutron flux may vary by a factor 10^6 in heavy-water-moderated or beryllium-moderated reactors or by a factor 10^{12} in graphite-moderated and light-water-moderated reactors. After the initial startup power reactors with beryllium or heavy water moderators always retain a high residual neutron flux due to the (γ, n) reaction (Sec. 2.2.).

Primary Coolant Outlet Temperature (T_0). The

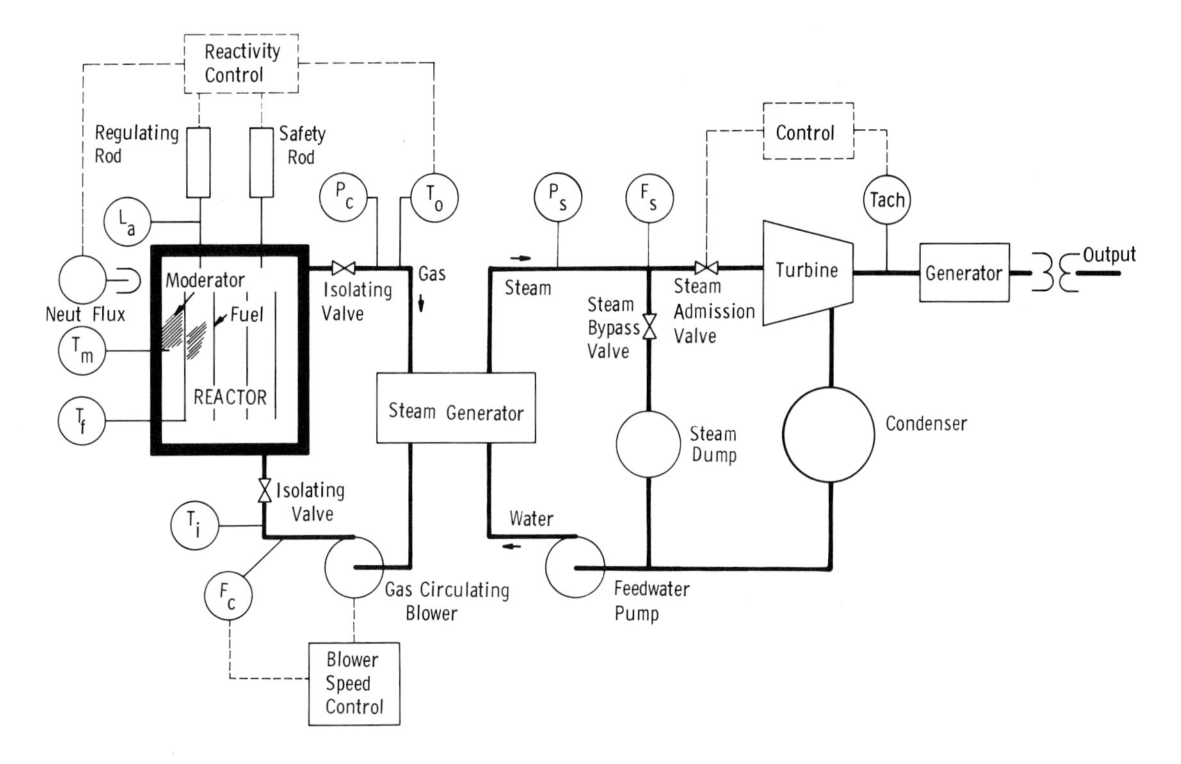

FIG. 1-1 Gas-cooled reactor.

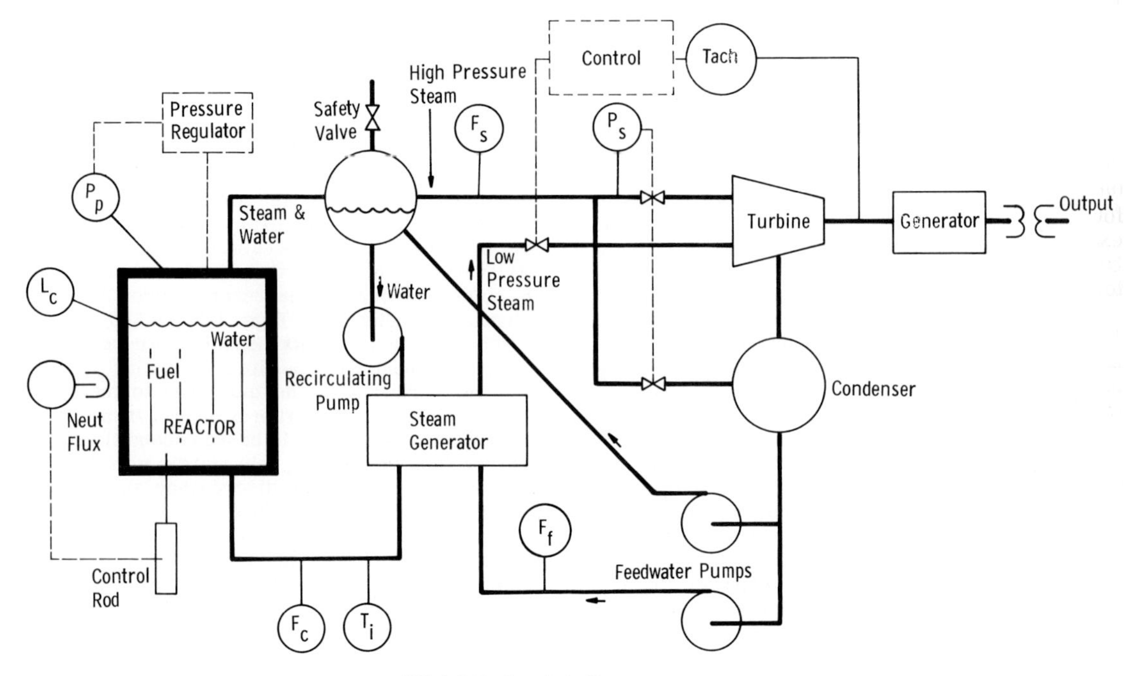

FIG. 1-2 Dual-cycle boiling water reactor.

primary coolant outlet temperature in gas-cooled and pressurized water reactors is directly related to the mean core flux and avoids to some extent the difficulties posed by a varying ratio of measured to mean flux. It is used as a regulating system parameter to calibrate the neutron flux measurement in terms of required average thermal conditions [2, 3]. In slowly acting regulating systems the outlet temperature could be the directly controlled variable [4a].

Both the outlet temperature and its rate of change are important safety system parameters, as a backup for neutron flux measurements to sense power excursions, and as a means to sense a reduction in coolant flow rate.

The response of the outlet temperature to a change in flux or flow is delayed by the thermal capacity of the sensor and by transport lags. This delay may range from seconds to tens of seconds. Because of this the temperature measurement is

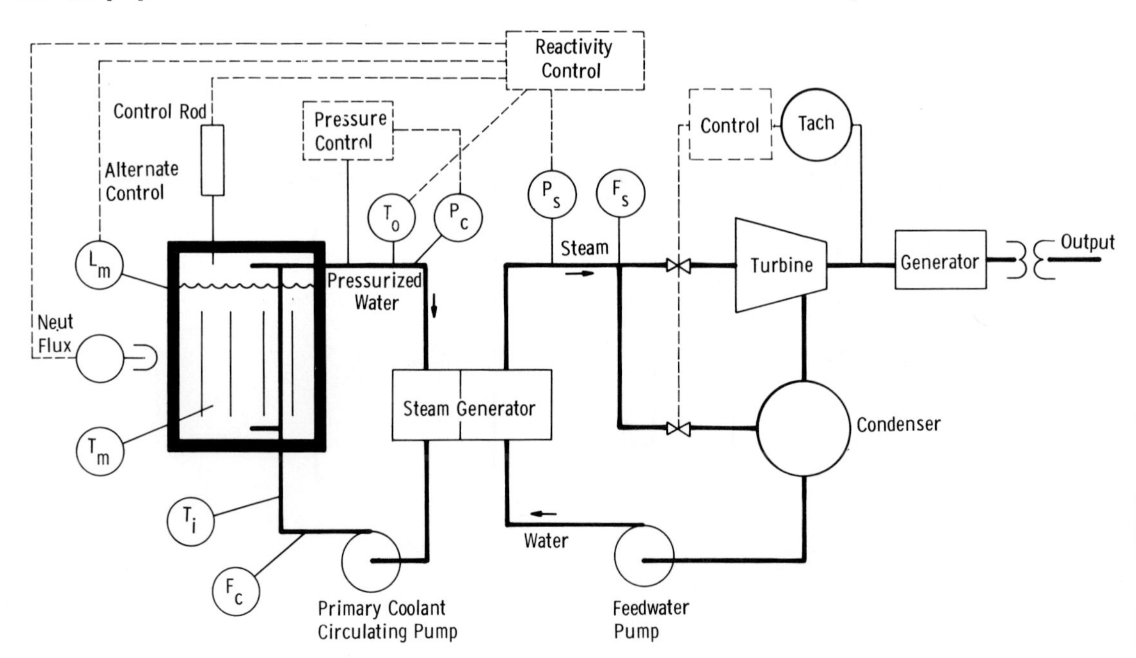

FIG. 1-3 Pressurized water reactor.

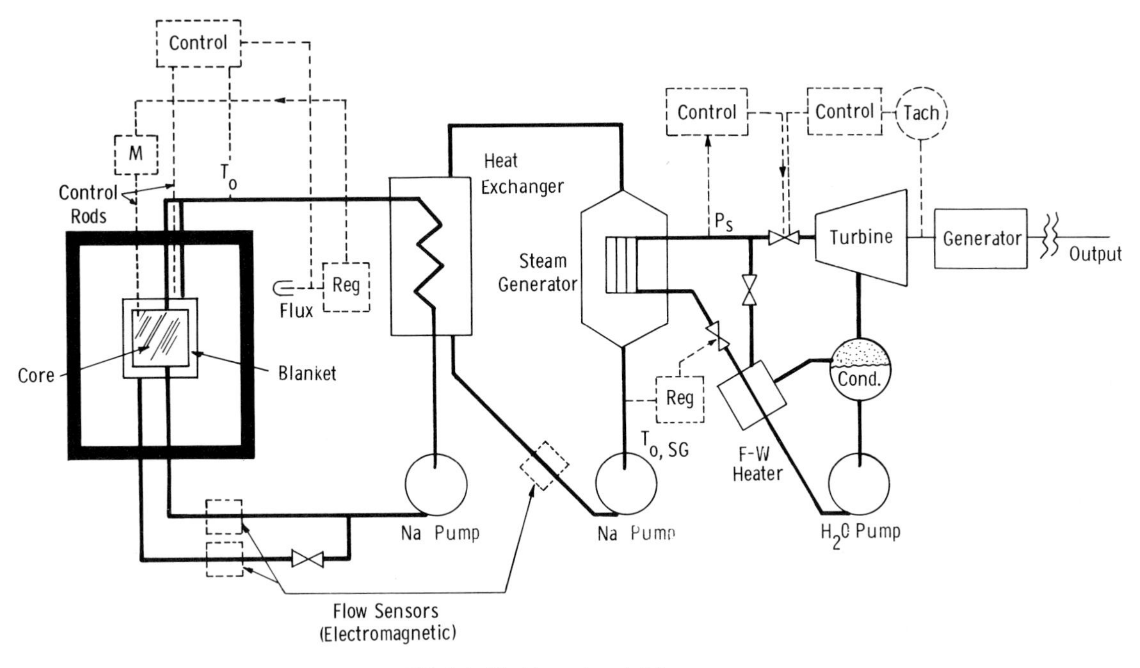

FIG. 1-4. Liquid-metal-cooled fast reactor.

not generally considered a prime source of protection against the fastest credible transients in neutron flux and coolant flow.

Primary Coolant Inlet Temperature (T_i). The coolant inlet temperature is used along with the outlet temperature to determine the coolant temperature rise ($T_o - T_i$) through the reactor or the mean coolant temperature ($T_o + T_i)/2$. Regulating schemes based on these parameters have been discussed [5a, 6a]. These parameters are a function of coolant flow and of reactor power, and in reference [3] these relationships are discussed along with coolant flow control systems for adjusting power production.

Primary Coolant Flow Rate (F_c). In any reactor system the primary coolant flow-rate is a basic safety parameter. In water-cooled reactors the flow rate is measured by sensing pressure drop across an orifice plate, or other means (see chapter on Fluid Flow), or it may be inferred from coolant-pump operating conditions. If the pump shaft is turning at its proper speed, the valves are open, and if the pump motor is absorbing a predetermined fraction of its maximum drive power, then flow is assumed normal.

Loss of coolant flow invariably invokes safety system action.

In water-cooled reactors the primary coolant flow is generally kept constant but in gas-cooled reactors, as mentioned earlier, the coolant flow may be used as a regulating system parameter to

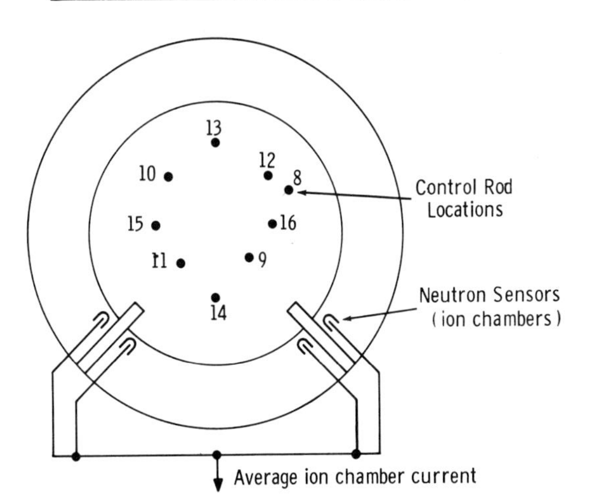

FIG. 1-5 Position of neutron sensors relative to control rods in NRU. (Control rods numbers 1 to 7, normally withdrawn, are not shown.)

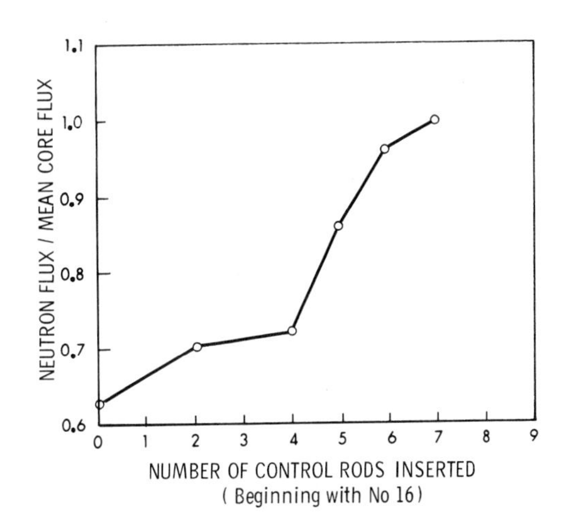

FIG. 1-6 Variation of measured to mean flux with regulating rod insertion. Mean core flux is measured by thermal signals.

regulate power output. By setting the coolant flow as determined by power output requirements, the reactor temperature regulating system will bring the reactor to the correct state of power production. In boiling water reactors the primary coolant flow may be varied as a means of load following.

Primary System Pressure (P_p). In gas-cooled reactors the coolant pressure is monitored to detect leakage from the pressure vessel. The rate of change of pressure is normally used to obtain adequate sensitivity. Rapid loss of pressure implies loss of coolant and safety action must ensue.

The primary pressure in pressurized water reactors is sensed to warn of overpressure or of system rupture. Gross loss of pressure usually requires that emergency coolant systems be invoked as well as normal reactor shutdown action. Rapid reduction in coolant pressure can depress the saturation temperature below the existing temperature and coolant boiling can start. In some reactor designs the boiling mode is not permitted.

Steam Pressure (P_s). In any thermal power plant the pressure of the steam fed to the turbine is a regulating parameter that responds very rapidly to changes in load through movement of the turbine steam admission valve. Pressure deviations are used to regulate either the reactor power [7] or the position of the bypass valve which controls the flow of unwanted steam to a dump condenser [8, 9].

In boiling water systems the steam pressure is essentially the primary system pressure. It is both a safety and a regulating parameter, since any increase in the steam pressure causes void collapse and, when the void coefficient is negative, an increase in power. Pressure regulating systems are normally included to alleviate this difficulty.

Steam Flow (F_s). The quantity of steam flowing from the reactor or steam generator is a direct measure of thermal power output. It is used in some designs [7] to determine correct power demands on the reactor itself (Sec 4.3).

Whether used directly in a regulating loop or not, steam flow can be measured to determine plant efficiencies; although measuring feedwater flow while holding the water level of the steam-drum constant may be easier.

Feedwater Flow (F_f) and Reactor Water Level (L_c). These two parameters have special significance in direct cycle boiling-water reactors. In these reactors the feedwater returns directly from the condenser to the reactor. Temperature changes in feedwater brought about by new load requirements on the turbine are felt directly by the reactors. This, in turn, affects reactivity and power.

The head of water over the reactor core is affected by the void volume, which changes when sudden alterations in load occur. In EBWR [10] the water level is regulated by using the derived parameter $K_1 (L_{ce} - L_c) + K_2 (F_s - F_f)$ where $(L_{ce} - L_c)$ is the error in level and $(F_s - F_f)$ is the difference between steam flow out and water flow in. At equilibrium both these quantities are forced to zero by changing the feedwater flow.

Moderator Temperature (T_m). In some heavy-water-moderated reactor designs the moderator is kept at a low temperature to increase the system reactivity [7]. This entails a separate cooling system and measurement of the moderator temperature for regulation. In graphite-moderated reactors the moderator temperature is used to initiate the process by which the Wigner energy is released.

Moderator Level (L_m). In reactors using moderator level as a reactivity control [7] the power output may have to be reduced if the level is less than full and if the moderator is removed from the vicinity of some fuel channels.

A level measurement is used in the regulating systems of the NPD reactor [7] to compensate for the nonlinear relationship between level and flux at the ion-chamber location. (See also paragraph below on absorber position).

Tachometer (Tach). The speed of the generator is a measure of the load requirement and its deviation from a reference speed is used to control steam flow to the generator. This control is exercised by a "governor".

If the station is operating as a base load plant where the generator speed is essentially constant, the governor is useful to initiate quick control action in the event of the load being disconnected.

Control Absorber Position (L_a). The position of any control absorber is important, both for regulation and safety. In reactors controlled primarily by a negative power coefficient the position of the regulating absorber determines the output power. In other reactors either the absorber position or absorber velocity is directly controlled by deviations from the required power level (see Sec. 4.3).

A reliable indication of whether an absorber is in or out of the core is required. And in the case of a regulating absorber a fine position indication is needed to determined whether the expected reactivity balance has been achieved.

Neutron Flux to Coolant Flow Ratio. In reactors that may normally operate with less than maximum coolant flow, the maximum permissible power must be limited to suit the available cooling. The ratio (neutron flux/coolant flow) [5b] is a convenient parameter upon which to base such control.

Measured Neutron Flux to Thermal Power Ratio. The effect of variations in (measured flux/mean core flux) on trip margin settings may be reduced by calculating neutron flux trip settings from the measured ratio (neutron flux/thermal power) [11]. In the event a disturbance causes a change in this ratio there will be a delay of some seconds, due to lags in the thermal measurement, before the new trip setting is determined.

Individual Fuel Site Parameters. The parameters so far discussed are bulk parameters that sense reactor plant conditions as a whole. To sense spatial distortions in the power distribution in the reactor and to monitor the integrity of individual fuel sites, the individual measurements described in the following paragraphs are necessary.

Fuel Surface Temperature (T_f). In any reactor the surface temperature of the fuel is an important safety parameter. It provides an early warning of possible loss of fuel sheath integrity due to overheating. This parameter is usually measured in gas-cooled reactors.

Coolant Flow Rate in Fuel Channels. Blockage of a single or a small number of fuel channels is not detectable by measurements on the bulk coolant flow. Individual measurements must be made if protection to individual elements is required.

Coolant Outlet Temperature of Fuel Elements on Channels. The temperature at the outlet of each of the fuel channels can give a backup warning that coolant flow is decreasing. However, if coolant flow is suddenly and completely lost, the temperature measurement will be false.

The temperature pattern produced by these measurements allows the power distribution to be monitored. In reactors where zonal instability is a problem these temperatures may be used as a regulating system parameter to control the movement of absorbers positioned to counteract the spatial disturbances.

Radioactivity in Fuel Element or Channel Coolant. With proper processing of the information, the measurement of radioactivity in the fuel channel coolant is a safety parameter and may, in some instances, be classed as a regulating system parameter. If large amounts of activity are detected along with low coolant flow or high coolant temperature, rapid safety action may be required to prevent more serious blockage and failure. If, however, only a small amount of activity is detected, other action such as the routine removal of the fuel element may be indicated.

1.3.2 System Disturbances

Disturbances to the operation of a nuclear power plant fall into four categories. The first is:

1. Disturbances one expects to control by means of the plant regulating system and against which the station output should remain substantially constant.

These are disturbances that show up as variations in the reactivity load of the reactor and require compensating action. The reactor design determines whether this action is brought about by internal reactivity coefficients or by external regulating aids.

The Buildup and Burnout of Xenon Poison. The equilibrium xenon poison in a high flux reactor results in a reactivity load typically about 30×10^{-3}. If such a reactor starts up with no xenon, the approach to the equilibrium xenon condition begins at a rate of about 1×10^{-3}/hr. The largest transient occurs when the reactor is first brought to power after a poison shutdown when the poison burns out initially at a rate of about 44×10^{-3}/hr. Other transients are indicated in Fig. 1-7.

Fuel Burnup. The change in reactivity due to fuel burnup occurs so slowly that no transient problem exists. The magnitude of the change varies widely with reactor type. The dynamics of the reactor itself may also vary as burnup proceeds and require changes in the regulating system.

On-Line Fueling and Discharging. In reactors that are fueled while at power, the total reactivity

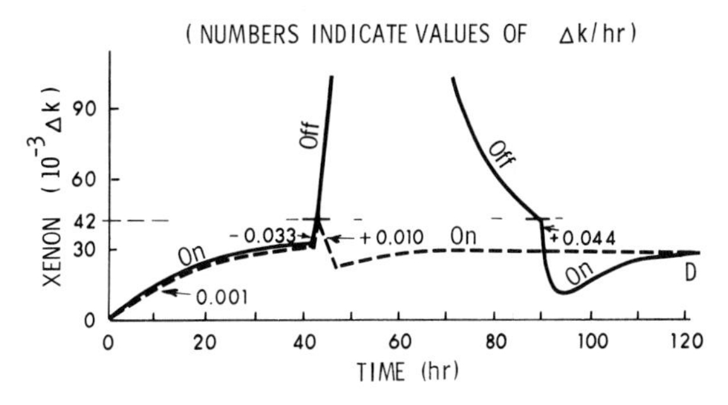

FIG. 1-7 Typical xenon poison transients. In solid curve the reactor is turned off between 40 hr. and 90 hr. In broken curve the reactor is turned off between 40 hr. and 40 hr. 20 min.

of the fuel is kept at a nearly constant level. Insertion of a fresh fuel element produces a transient in reactivity.

If the mechanism required to grasp the fuel is inserted into the core, its loading effect must also be considered.

Booster Rod Removal. In order to overcome, to some degree, xenon poison and permit a startup when normal reactivity controls would be inadequate, positive reactivity can be inserted in the form of slowly moving control elements containing enriched fuel. When the reactor has burned out sufficient poison, the booster elements are removed.

Fuel Temperature Coefficient of Reactivity. The reactivity of the fuel varies with temperature and during changes in reactor power a transient occurs that must be counteracted. Typical values are given in Appendix 1 of this volume.

Moderator Temperature Coefficient of Reactivity. The reactivity of the moderator also varies with temperature. Typical values are given in Appendix 1 of this volume and in The Reactor Core chapter.

Void Coefficient of Reactivity. In boiling water reactors, reactivity is dependent on the void volume. Since the void fraction is sensitive to temperature and pressure, with little time delay, the effect of both must be seriously considered. Load requirements change the feedback of both temperature and pressure to the reactor and these must be kept within the limitation of the regulating system if reactor shutdowns are to be prevented.

The second category of system disturbances is:

2. Disturbances that one expects to keep under control by means of the plant regulating system; a reduction or even loss in station output is tolerated.

Loss of Turbine or Load. If a fault occurs in the turbine or if the load is removed because of a grid fault, the turbine steam admission valve will close with an accompanying increase in steam pressure.

Generally, automatic controls direct unwanted steam to a dump condenser, but the transient felt by the reactor must be considered. In direct-cycle boiling-water reactors the effect is to cause void collapse and increased reactivity.

In systems using a heat exchanger the effect on the reactor is less pronounced. Usually direct control from steam pressure and reactor outlet temperature to the reactivity controls is capable of counteracting the disturbance, thus permitting a controlled shutdown. It is desirable to prevent the reactor from tripping out so that the turbine can be reloaded as soon as the fault is corrected. It is also desirable that the generator continue to supply the station requirements.

Emergency Shutdown. The consequences of an emergency shutdown, due to operation of the safety system, must be considered. If the reactor were at full power prior to the shutdown, the low power level to which the reactor falls may, in some reactors, be sufficient to hold the turbo-generator synchronized to the grid for several minutes. This is desirable since it may avoid having to re-synchronize the generator when the reactor is restarted. If the dump condenser is accepting any part of the load at the time of the reactor trip, the condenser will normally be unloaded in advance of the turbine to avoid having to disconnect the generator from the grid.

The third category of disturbance is:

3. Disturbances that can only be controlled by the safety system; loss of station output is accepted but no loss of integrity in any reactor component is accepted.

Failure of the Regulating System. It is generally required that any regulating system failure forcing its controlled variable to an unsafe level, will be counteracted by safety system action.

Loss of Coolant Due to Pump or Circulator Failure. The stoppage of a primary coolant pump or circulator, due in particular to failure of an electrical power supply or regulating system, is considered a credible operational fault. Within the rundown time of the coolant system, safety action must be complete to prevent fuel failure.

The final category of disturbance is:

4. Disturbance caused by loss of reactor component integrity, during which safety action is expected to prevent dispersal of radioactivity.

Loss of Primary System Coolant Due to a Rupture. If the primary system coolant is lost through rupture, it is often necessary to introduce special safety measures in addition to the normal reactor trip. For example, emergency cooling may be brought into action, along with dousing systems, to reduce pressures in the vicinity of the rupture.

The safety system must also ensure that normal means for penetration into the containment vessel are properly sealed.

1.3.3 Response Time Consideration

There are two distinct aspects to response time considerations in control systems: the speed of the safety system and the speed of the regulating system. In a sense, the requirements conflict because better regulation is usually achieved by a rapidly acting system, but the safety system may be unable to override it.

One must also consider the possibility that a regulating system could control a disturbance when, in fact, safety action should be taken. For example, a blockage might occur in a fuel channel causing boiling and void formation. If the voids introduce a loss in reactivity, the reactor power falls but the flux regulating system returns the reactor to power, and reliance for safety is on the flow and temperature sensors. On the other hand, if the effect of the void is to add reactivity, the power rises and reaches the excess power safety level before regulating action can occur. In either case a regulating system with a too rapid response can defeat (at least temporarily) the safety action.

With the present state of the art, regulating system faults are inevitable, and the speed with which a fault can occur must either be limited by fundamental means or the maximum amount of reactivity must be limited to match the capabilities of the safety system. (See Sec. 2.5).

The speed of response of safety systems is determined by the response associated with the sensors and amplifiers and by the safety mechanism driving force; this latter is often gravity or, in many systems, gravity aided by accelerating springs, downward coolant flow and other forces. From the initiation of the shutdown signal, the flux is generally beginning to fall in from 0.1 to 1.0 sec. Within these limits regulating systems have been designed that permit reactor startup at rates as high as 4%/sec i.e. a period of ≈ 25 sec (see Sec. 2.5.4), and power maneuvering capabilities as high as 2 to 4 Mw/sec for limited excursions. Turbine loading rates are generally the limiting factor, with 20% of full power per minute typical. Reactivity control systems with response times to fractions of a second and capable of moving reactivity at rates as high as 1 to 2×10^{-3}/sec can be shown to be safe, given adequate protective instrumentation, (see also the chapter on Mathematical Models of Fast Transients).

A specific point to be considered is whether the failure of a regulating unit must be detected upon failure or whether the failure can be permitted to cause the controlled parameter to go out of bounds and to bring in safety action indirectly. While, in general, an attempt would be made to anticipate failure and obtain the advantage of time, the question to consider is whether or not the anticipation is fundamental to the safety argument.

Most of the disturbances discussed in Sec. 1.3.2 place minimum speed-of-response requirements on the regulating system if the disturbances are to be adequately controlled.

1.3.4 Accuracy Requirements

Accuracy requirements vary from one part of the system to another. Since any requirement for greater accuracy generally means instrumentation of greater complexity and reduced reliability, it must be given serious attention.

The accuracy required of any parameter should be commensurate with the subsequent use of the information. The aim of the nuclear power plant is to produce electrical energy as cheaply as possible. The various parameters associated with the determination of the efficiency of power production must, therefore, be known with an accuracy in accord with the art of calculating these efficiencies. The thermal power generated by the reactor, the steam pressure, flow, temperature, and quality will enter into the conventional side of this calculation. And, since a fraction of a percent change in efficiency is equivalent to many thousands of dollars per year in terms of power production costs, the accuracy required of these measurements approaches the limit of current technology. Accuracy in itself may not improve efficiency but it will permit significant changes to be detected.

On the nuclear side the main factor governing generating costs is the fuel burnup. This depends not only on the reliability of the fuel but also on the amount of parasitic neutron absorption in the reactor. Reactor core physics is precise enough to establish the reactivity of a cold clean reactor to within 1.0×10^{-3} and so absorber position measurements, at least during the first approach to criticality, should be made with similar precision. (Typically 1 cm of absorber motion represents a reactivity change from 5×10^{-6} to 5×10^{-5} or 1 cm of moderator height represents a change of from 2×10^{-4} to 2×10^{-3}). A negative reactivity of 1.0×10^{-3} has been estimated to decrease the fuel burnup in some cores [12] by approximately 100 Mwd/tonne.

From the point of view of reactor safety any appreciable difference between calculated and measured reactivity values should be accounted for. The discrepancy could be due to some foreign material lodged in the core and capable of falling out, or to insufficient absorbing material in the reactivity control elements.

Similar economic arguments may also apply to measurement of the irradiation experienced by the fuel.

So far the discussion has been concerned mainly with the economic arguments that may be put forward for high accuracy. There are, however, other aspects of the accuracy requirement that can affect the reliability of power-plant operation. Figure 1-8 shows the information flow in a typical control system. Demands are given to a regulating system that force the regulated process to the required state as determined by an appropriate sensor. The sensor output is also presented on some form of visual display. Two sensors are shown. To achieve short-term accuracy (i.e. dynamic accuracy) a fast sensor may be employed; if high accuracy is required, the measurement will generally take longer.

Independently of the regulating system, the regulated parameter is sensed by the safety system and compared with a demand.

If the indicated value of the regulated parameter is not equal to the demand, the operator usually adjusts the demand until the required value is obtained. An error of this sort is not serious providing the regulating system maintains the required value. If, however, the sensor produces an output that varies independently of the parameter it is measuring, then the regulating system will act to maintain the measured output constant and as a result the actual value of the regulated parameter will be varying. This variation may be noticed on an independent sensor such as the safety sensor. The design should ensure that sensor output uncorrelated with real parameter variation is smaller than the dead-band of the regulation.

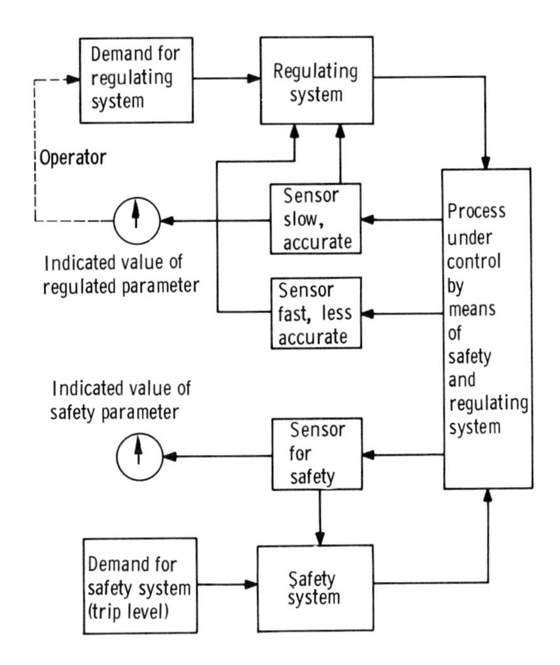

FIG. 1-8 Information flow in typical control system.

For the reason just given and because of normal component tolerances, the value of the controlled parameter as measured by the regulating system sensor will not agree with the value as measured by the safety system sensor. In many cases the measurement for the safety system need not be highly accurate because the basic knowledge upon which the safety level is chosen is not precise; in turn, advantage can be taken of simpler and more reliable instrumentation. Nevertheless the difference between the two measurements, even though within design limits, can be disturbing, and often adjustments are provided to produce agreement. Further, it may defeat the reliability of the safety system if precision were introduced simply to achieve agreement. The requirement is for a design that takes legitimate errors into account by proper presentation of the information.

The difference between the indicated value of the safety sensor output and the trip level is the margin-to-trip. The absolute trip level must be known and displayed but the margin-to-trip is also important. If the trip level setting does not have associated with it the same errors that are present in the indicated value, incorrect action can result when the margin-to-trip is small.

In control systems using redundancy to improve reliability (refer to Sec. 1.4.3) the question of errors between sensors measuring the same parameter arises. The intercomparison of measurements is fundamental to the redundancy principle but it must be possible to apply it without requiring extreme accuracy; for example, to obtain agreement beyond what would be required if the same equipment were used in a single-channel system.

1.3.5 Organization for Control System Design

Before any control and instrumentation design

can be undertaken and successfully applied, an organization capable of ensuring contact among many disciplines is necessary. Figure 1-9 shows the stages and groups involved in the design, construction and operation of a control system along with the interchange of information that is necessary to ensure success.

The talents of many specialists must be used both during the conceptual stage and during the specification of the system. The coordinating of these people is important to reactor safety because the work of one must so often be compatible with that of another.

Further, it is vital that at an early stage the design concepts are documented, for one of the greatest difficulties besetting over-all design is the unavailability of the orginators of design concepts when the design is being put into practice. The engineering of the system will always require modifications to the original specification and a link back to the specialists must be available to reconcile the desirable with the possible. Again, any changes should be included in the documentation.

At this stage in design the importance of some parts of the system can be overlooked because they appear too conventional. The provision of adequate power supplies commonly falls into this category. The placement of interconnecting cabling and the proper layout and grounding of the normal electrical distribution system are often not given the attention they require in light of the sensitivity of nuclear instrumentation. A well-conceived safety system is easily rendered unreliable because sound practices of instrument isolation are not followed from beginning to end.

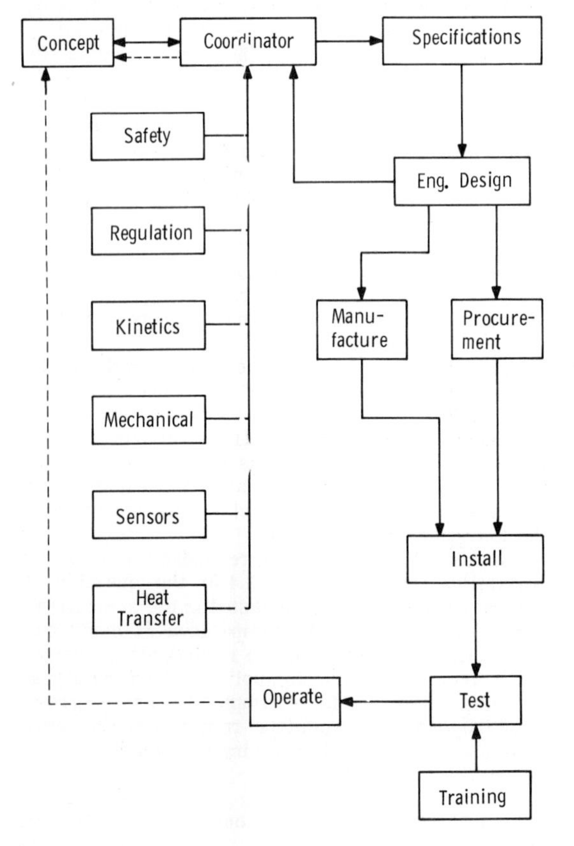

FIG. 1-9 Information flow required to produce successful system.

Prior to or during the installation phase, operations and maintenance people must be trained. A useful part of this training may be production of operating manuals. The timing of installation relative to other construction must be taken into account to ensure that the equipment is not subjected to environmental conditions far in excess of its rating (such as temperature, moisture or dust). Alternatively, proper packaging must be specified.

During the testing period the specialists may again be required and the organization should make this possible. In addition, the normal operating staff should play a significant role during testing and commissioning so they will, at the earliest stage, become acquainted with the symptoms of trouble.

If the organization is to be worth while, then reliability and safety should improve with successive reactor plants. This will only take place if, when the plant is operating, the working system is compared with the original concept and the differences brought to the attention of the specialists. Especially important is the period just before steady operation begins when the work load is abnormally large and as a consequence many difficulties are circumvented without record.

It may take many months [13] before the fault rate falls to a level used in the reliability estimates and even then a continuing effort is required to make certain that any increase in the fault rate is noted and that cause and effect are correlated.

1.4 Objectives of Instrumentation Systems

1.4.1 Philosophy

The philosophy behind the design of an instrumentation system will depend largely on the type of reactor being controlled. Again, there are two extreme situations. If the reactor is being controlled by internal reactivity coefficients and is deemed safe on this account, the objective of the instrumentation will be to provide more flexibility in maneuvering the power level and to add to the safety. If, on the other hand, the reactor requires instrumentation for control because the required speed of response is greater than can be provided by the various reactivity coefficients, the safety of the reactor may depend entirely on the reliability and suitability of the instrumentation; in fact, there have been suggestions that proper instrumentation can relieve the containment requirement [14].

The basic instrumentation scheme will also depend on the ultimate requirements of the plant and on the consequences of unnecessary shutdowns.

In view of such diverse philosophies that might govern the design of an instrumentation system, it is reasonable that no universal approach has been accepted.

A few basic rules for achieving maximum safety have been suggested by various groups. Some of these conflict and so not all can be applied in any one case.

1. Safety instruments should not be used for any other purpose.
2. Each safety parameter should at all times be sensed by at least two independent safety channels.
3. Some designers enlarge on rule (2) and require that two diverse parameters (each with at least two safety channels in operation) be used to sense each fault situation.
4. Instruments should remain operative, even when the reactor is not.
5. Maintenance of instrumentation should only be carried out when the reactor is operating so that the results of any change are immediately known. (This requires special system design consideration).*
6. Instruments must be tested on a regular routine whether the reactor is operating or not.
7. Introduction of test signals should not open the normal signal path unless a return to continuity can be definitely established.
8. The status of the nuclear chain reaction must be known at all times, including the shutdown state.
9. It must not be possible to remove safety instrumentation without causing an alarm or shutdown.
10. No credible instrument fault should be incapable of being exposed by the prescribed test routine.
11. As far as possible all faults should expose themselves.
12. Instruments should fail-safe, i.e., they not only should expose their own faults but should, when faulty, also initiate the safety function for which they are provided.
13. Instrumentation must be tested immediately before it is required in the operating sequence. This usually involves special design features that release the next stage in the operation to the operator or regulating system, only when a successful test has been completed.
14. Instrumentation must not be deliberately incapacitated (jumpered) to permit operation if the safety is reduced below the minimum specified in the reactor hazards evaluation report.
15. The range of any instrument should extend sufficiently beyond any action level to permit transients above this level to be observed.
16. Since any instrument will reach a saturation limit upon the application of a sufficiently large input signal, it is important that this saturation does not casue the putput to fall.

This is by no means an exhaustive list, but it is representative of the type of thinking that prevails.

Any group of experts examining and judging a reactor instrumentation system also exerts considerable influence on the philosophy adopted, simply by passing favorably on a particular scheme.

A safety analysis of the instrumentation should, in general, probe three distinct aspects of the design:

1. Is the system able to achieve its stated purpose?

   Under this heading the instrumentation should be examined by inquiring as to the response to certain situations. Response time, accuracy, and appropriateness of the measurement to a particular situation are questioned. Does the scheme behave as it should in meeting all operational and credible accident situations?

   In this context it is often argued that a safety probe should deal only with accident situations and not be concerned with the number of times the reactor plant is shut down in an attempt to meet safety requirements. However, if the control system does not provide uninterrupted operation, the pressure placed on the operating staff may itself constitute a hazard.
2. To what extent can the system be relied upon to achieve its aim, bearing in mind the possibility of single or multiple instrument failure?
3. What situations may arise to invalidate assumptions about the behavior of an instrument? For example, is it possible for an accident situation, against which neutron flux instrumentation provides a guard, to prevent neutrons from reaching the sensors?

One of the main objectives of the instrumentation system is to provide satisfactory answers to these questions.

1.4.2 Functional Requirements

In order to satisfy the criteria posed in the last section, several requirements must be met. The most common disturbance that the instrumentation will be expected to sense is reactivity addition, attributable to fuel entering the reactor or an absorber being removed or due to reactivity being changed through temperature or pressure effects. The magnitude of a credible reactivity transient will depend largely on the reactor type, but invariably its detection (since the transient could begin at any power level) will require the neutron flux to be monitored down to the most subcritical state. This may typically involve a range of 10^{12} to 1 in neutron flux. The instrumentation response time should be short enough that safety action can be initiated in time to prevent the generation of excessive energy.

If the instrumentation ... ates a fault condition in a time short ... ess) compared with the time it takes ... tem to regain control, the response ... tory and any improvement will ad... rall safety. This is generally a... x (corresponding to $> 0.1\%$ ... at low flux the instrumentation ... ining element and must be ... to ensure adequacy. This is ... ux values ($\sim 10^2$ neutrons/cn ... tion being sensed is arriving ... l amounts and there is an inhe... a trend is known with any cer... a real re-

*Most reactor operating groups permit maintenance during reactor shutdown by using careful checking procedures to ensure proper reconnection.

quirement to be able to say for certain that information being received is from the fission process and not from electrical interference or other unrelated radiation.

The response time required in instruments sensing other parameters must also be considered with respect to the disturbance being postulated.

Most generally there are a number of instrument features that experience has shown to be useful in ensuring adequate control.

1. It should not be possible to set the safety action level beyond the instrument range.
2. It should be possible to check the alarm or trip (level) visually, without interfering with the instrument.
3. If there is an operational requirement to adjust action levels other than for reactor safety, this should be done on an independent instrument.
4. The action level should be tested by bringing the measured parameter to the action level and not by bringing the action level to the value of the measured parameter.
5. If possible, the action level should be tested without disconnecting the instrument from its associated sensor. If this cannot be done, then monitoring should be provided to ensure that the sensor is reconnected.

1.4.3 Reliability: Safety Plus Serviceability [15, 16, 17, 18]

Once there is confidence that an instrumentation system will meet operational requirements imposed by safety and regulation criteria, the question "How much reliance can be placed on it?" must be answered. It must be reliable in two senses, providing both safety and serviceability.

Instrumentation provides safety if, when an unsafe deviation in a reactor plant process occurs, it is infallible in sensing the situation and in advising the safety system.

Instrumentation provides serviceability if it carries out its part in plant regulation without introducing disturbances into the process being controlled and if it never presents false information to the safety system.

In order to produce reliable instrumentation systems many schemes and philosophies have been put forward. Several approaches to the problem are in use, but combinations of diversity, redundancy, coincidence, testability and fail-safe can be found in all of them.

Diversity. The diversity principle requires that as many as possible of the situations against which one is guarding can be sensed by at least two different means.

The parameters, neutron flux and reactor outlet temperature, are diverse means to sense reactor power. An increase in outlet temperature together with a change in power drawn by the coolant pumps would indicate a loss of coolant flow.

It is difficult to meet the diversity requirements in all cases, especially in reactor systems where regulating absorbers can be moved fairly rapidly. In this case neutron flux (or its rate of rise) may be the only parameter that can sense a startup accident, although rate of rise of primary pressure has been proposed as a second parameter in some cases. Less diverse would be the use of rate of change of the logarithm of neutron flux and rate of change of neutron flux to sense a rapid power increase.

The diversity principle in fact attempts to guard against gross oversights or faults associated with a particular parameter. For example, there is a possibility of reduction of the neutron flux to all the neutron sensors together by flooding the space between the core and ion-chambers.

Redundancy. The simplest way to increase the reliability from the safety point of view is to duplicate or triplicate the instrumentation on each measurement. To discuss reliability quantitatively we define unreliability [19] as the fraction of time an instrument is not capable of performing the function required of it. Further, only those faults will be considered that need a test to be exposed (i.e. unsafe faults).

Thus, if an instrument fails f_u times a year in this unsafe manner and on the average remains in this condition for t_u years, the unreliability is $f_u t_u$ (e.g. if $f_u = 1/\text{year}$ and $t_u = 1/50$ year, the unreliability would be 2×10^{-2} or, on the average this instrument will afford no protection for 1/50 of a year per year). This definition is useful for if one multiplies the unreliability by the number of times the unsafe process situation is expected to occur, the product is the number of times per year that one can expect the safety of the system to be defeated.

Since in order to expose an unsafe failure a test must be performed, the average duration t_u of the unsafe fault is one-half the time between tests.

Consider now two similar instruments arranged to perform independently the same function. Figure 1-10 shows diagramatically the situation where the shaded regions indicate that during this interval the instrument was incapacitated, the dots show when the test was performed. The fault could, of course, occur any time between the tests, but only the mean situation is shown.

The unreliability of the system comprising these two instruments is the fraction of time that neither instrument can respond. If the failure rates of each are again f_u/year, the unreliability is $(f_u t_u)^2$.

Figure 1-11 shows the unreliability of systems using up to four instruments for various fault rates and test intervals, where one instrument in the group is required to be operable. As shown in Fig. 1-10 the test is performed in each instrument at the

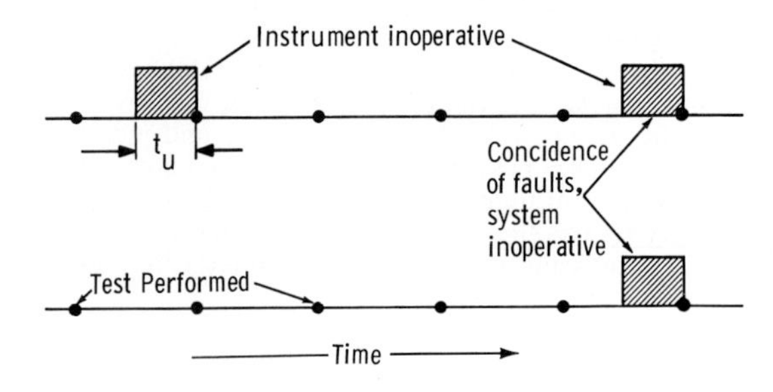

FIG. 1-10 Coincidence of faults in a pair of systems subjected to periodic testing.

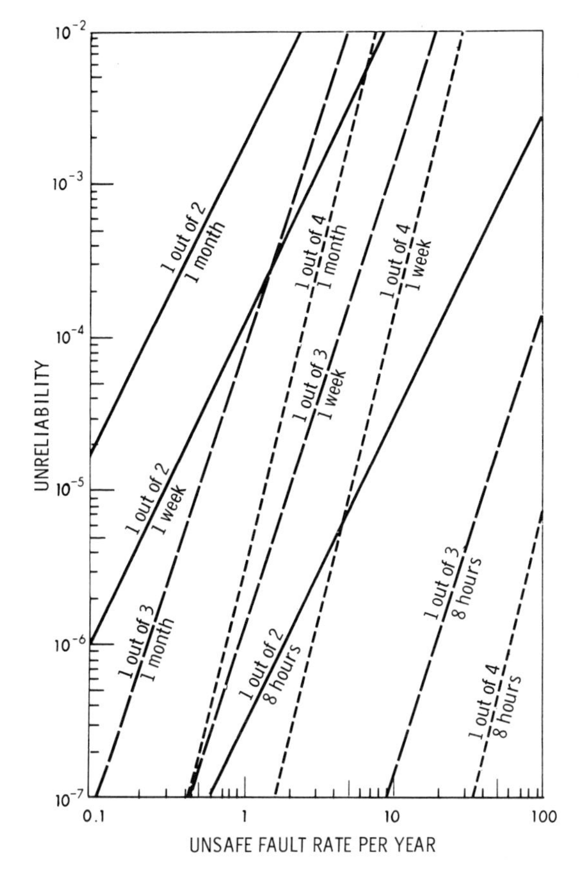

FIG. 1-11 Unreliability of systems as a function of fault rate for various test intervals.

same time. This need not be so. In fact, experience indicates that staggered testing is useful in that it prevents an operator carrying out an incorrect procedure simultaneously on all instruments.

It is evident that, for example, with three instruments, each having an easily attainable fault rate (one per year), an immeasurably small unreliability can be achieved. However, caution should be used when applying unreliability numbers that assume complete independence of the instruments involved. It has been found [13] that unreliability can be greatly increased by causally connected faults brought about by design oversight, environmental conditions, or incorrect maintenance. Modifications necessary to overcome design errors can be a source of faults appearing in all instruments, since the modification is usually a single operation involving like instruments at the same time. An application of the diversity principle can be used to alleviate this situation by sensing the same parameter by instruments of different design but similar function.

Coincidence. Most faults appearing in instrumentation are not unsafe faults, since by design they generally produce a response mimicking precisely the process fault being guarded against. The safety system then takes action to counteract nonexisting process excursions.

As the redundancy principle is used to improve safety, the probability of false safety-action increases in proportion to the number of redundant instruments, and serviceability decreases. A compromise between these competing factors is possible by using coincidence or majority logic in which at least two or more redundant instruments must agree before action is taken.

The unreliability of a system increases with coincidence logic. If two instruments are used so that both must be operable the unreliability of the pair is $2\ f_u t_u$ and for a two-out-of-three connection the unreliability is $3\ f_u^2 t_u^2$. Figure 1-12 shows the unreliability of several coincidence combinations.

The probability of getting a false response from a system of coincidence-connected instruments depends on the safe-failure rate and the duration of the fault. Since by definition safe-faults expose themselves, action can be taken immediately to correct the situation; hence, the duration of the fault is simply the time taken for replacement. Our concern now is with the frequency of false responses. In a one-out-of-two system this is just $2\ f_s$ where f_s is the safe-fault rate. In a two-out-of-two system the false response rate is $2\ f_s^2 t_s$ where t_s is the fault duration. The situation is depicted in Fig. 1-13.

Figure 1-14 shows the spurious response rate of various combinations. Two times are indicated for t_s. One hour is taken as reasonable if replacement is immediate, and twelve hours infers that

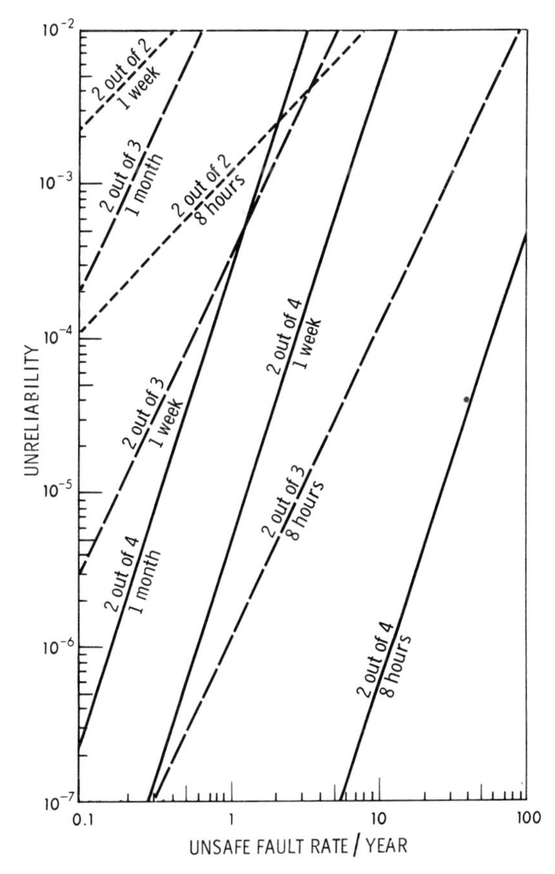

FIG. 1-12 Unreliability of coincidence systems as a function of fault rate for various test intervals.

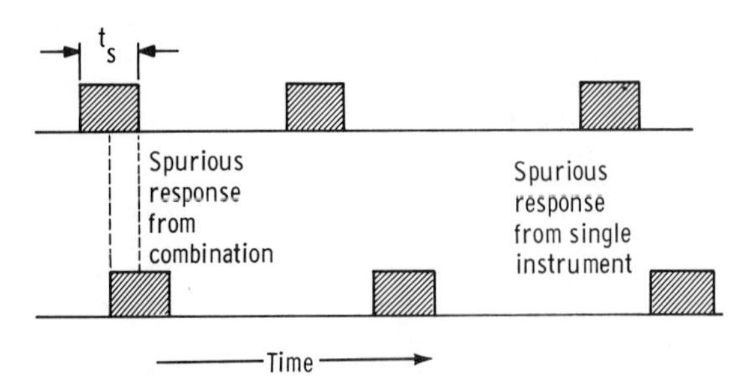

FIG. 1-13 Occurrence of spurious response in a coincidence system.

replacement of all faulty instruments is carried out once per day.

If the conditions on which the redundant coincidence approach is based (i.e. complete independence of the instruments) are fulfilled, the two-out-of-three and two-out-of-four systems offer acceptable safety and reliability with present-day technology.

As an example, consider a typical safety arrangement with two two-out-of-three systems being used to sense a reactor power excursion. Assuming an unsafe fault rate of one per year per channel in each two-out-of-three system and a weekly testing schedule, the unreliability of each

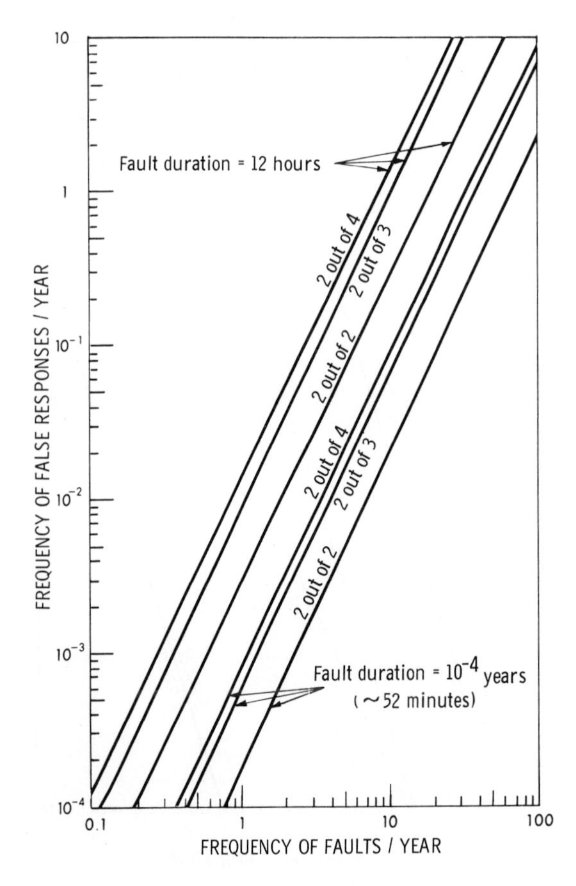

FIG. 1-14 Frequency of false responses of systems for various coincidence arrangements.

system would be 3×10^{-4} (Fig. 1-12) and the rate at which both systems would be unsafe simultaneously is twice the unreliability times the fault rate (Fig.1-13), i.e. 6×10^{-4} per year or once per 1400 years.

Even though such small fault rates can, in principle, be achieved, in practice accidents have occurred due to instrument failure, but failure invariably due to a source of trouble common to several instruments.

In order to give the cross-linked fault the attention it deserves we feel it is wise not to assume that the occurrence of simultaneous faults is a coincidence but to look intensively for the possibility of a cross-linkage through direct connection, maintenance procedures, or environmental conditions. In dealing with several faults in coincident safety and regulating systems we have found that the adoption of this approach has led to the location of several points of inter-channel coupling that were unsuspected or were thought to be of too high impedance to be of concern.

Thus far discussion has implied the use of redundance and coincidence only in the safety portion of the reactor control system. These features have been applied [20] to a regulating system instrumentation (see Sec. 4) to achieve a high degree of serviceability. Redundancy by itself has no value when directly applied to regulating systems because these can produce both safe and unsafe movements of control elements. However, regulating systems have been devised that respond when any single sensor demands an ostensibly safe control maneuver but only when a majority of sensors demands control action in what could be an unsafe direction.

The coincidence system ensures that only majority decisions cause regulating action in either direction.

The serviceability of the regulating system has been brought directly into the safety argument by some reactor safety specialists. It is noted that the product of the regulating system failure rate (in the unsafe direction) and the unreliability of the safety system yields the accident frequency due to regulating-loss incidents. There are several opinions on how the responsibility for preventing this type of accident should be apportioned between the safety system and the regulating system:

G. C. Laurence [19] suggests a reasonable goal for unreliability in the safety system to be 3×10^{-3} and a regulating system failure rate of 0.3 per year, provided that the unreliability of the containment features is no greater than 3×10^{-3}.

E. P. Epler [21] "The primary role of the safety system must be to contain the inevitable startup accident with absolute reliability" ... "The existence of a fully monitored, fast, and reliability safety system ... should be exploited to permit the control system [regulating system by our definition] to be designed for the maximum convenience consistent with good practice".

E. Siddall [14] suggests a goal for the regulating system of no more than 7×10^{-3} failures per year and a safety system that fails no oftener than one in 500 attempts.

Our experience to date essentially supports Epler's point of view. In nine reactor years during which redundancy and coincidence have been in use, a safety system has never been found incapacitated and only once has a safety parameter been incapable of being sensed because of simultaneous unsafe faults.

However, during the same period even with the use of coincidence in regulating systems, simultaneous faults (both from equipment failures and operating errors) have initiated safety system action [22].

The incentive to prevent regulating system failure is tied much more to the direct financial loss caused by an outage rather than to the risk of accident because the safety system may not be operative at the same time as a regulating system fails.

Testability. The most important feature that ensures reliability in an instrumentation system is the ability to apply a test that will expose any credible fault. The test must encompass the chain of instrumentation from sensor to the final control element. Coincidence systems permit this requirement to be achieved conveniently; systems using redundancy only can be designed with suitable facilities for deliberately incapacitating an instrument while it is being tested or repaired.

Testing increases the probability of false safety action, since a test generally energizes the safety action. The number of false responses due to a safe-fault coinciding with a test can be determined from Table 1-1.

TABLE 1–1

Number of False Responses Attributable to a Safe-fault Coinciding with a Test

False response rate per year	Type of system
$2 f_s f_t t_t$	2 out of 2
$6 f_s f_t t_t$	2 out of 3
$8 f_s f_t t_t$	2 out of 4

f_s = safe-fault rate per year

f_t = frequency of testing per year

t_t = duration of test in years

The test duration t_t should be kept short; this imposes requirements both on instrument design and on test methods.

Some reactor systems [23] permit a jumper to short the safety contacts of an instrument under test.

Fail-safe. Fail-safe design, or the art of making all failures appear at the instrument output as a real process fault, is practiced in varying degrees. The techniques used include the application of a continuous test signal that can be readily identified, special attention to specific modes of component failure, and intercomparison of several redundant instruments whose outputs should agree.

However, by their nature, safe-faults as they occur receive design attention and the safe-fault rate tends to decrease as more reliable components replace those producing the faults. This leaves the small unsafe-fault rate and the methods devised to expose them the most important aspects of instrument design.

The fail-safe principle must be applied as intensely as possible to those parts of redundant systems that are necessarily common.

Failure of instrument power or gross instrument disconnection must always revert the reactor to a safe condition.

2 NUCLEAR INSTRUMENTATION

2.1 General Considerations

Nuclear instrumentation is concerned with monitoring the fission process by direct detection of the nuclear radiation flux that is produced. The flux is an almost instantaneous indication of the fission rate and any delay that appears in the measurement is due, generally, either to inadequacies in the means used to sense the flux, or in the production of a useful signal.

It is not always possible, however, to rely solely on the speed of response of a flux measurement to provide safety because of the time lag inherent in the operation of the safety system itself. It is often necessary to anticipate the occurrence of excessive flux levels in time to permit the safety system to counteract them. There are several relations between neutron density (or flux) and the reactivity of the reactor that indicate the usefulness of a wide range flux measurement to aid in this anticipation. When the reactor is shut down the neutron density n is given by,

$$n = S/(1-k) \qquad k < 1 . \tag{2-1}$$

S is the neutron density due to some neutron-emitting source in the reactor, and k is the multiplication factor. (See discussion in Sec. 1.1 of the chapter on Criticality.)

During startup, reactivity is increased at a controlled rate. To obtain an expression containing its rate of change Eq. (2-1) can be differentiated and arranged to yield.

$$\frac{1}{n}\frac{dn}{dt} = \frac{dk/dt}{1-k}$$

= the rate of rise of the logarithm of the flux,

$$= \text{the inverse period} = \alpha . \tag{2-2}$$

This relation is useful because it relates the rate of change of reactivity and the departure from criticality with a readily obtained parameter. Thus, if reactivity is being added too rapidly, the rate of rise of flux can be used to warn that criticality is being approached too rapidly.

This expression, Eq. (2-2), also indicates the usefulness of a measurement of rate of rise of neutron flux for the control of reactor startup. A reactor being brought to criticality by control of $(dn/n)/dt$ has its rate of change of reactivity automatically adjusted to suit the prevailing reactivity and needs no previous assessment of it.

Equation (2-2) also states that the time still to elapse before criticality is reached, i.e. $(1-k)/(dk/dt)$, is equal to the instantaneous period. This is a fairly accurate approximation when the reactor is well below critical but overestimates the time remaining when the reactor is within two or three dollars of criticality. (See Sec. 3.4 of the chapter on Criticality.)

When the reactor is supercritical the inverse stable period is given by:

$$\frac{1}{\tau} = \alpha = \frac{1}{n}\frac{dn}{dt}, \tag{2-3}$$

since $n(t) = n_0 \exp(t/\tau)$. The stable period τ is a function of the neutron mean lifetime, the lifetime of the delayed neutron emitters, and the amount by which the reactor is supercritical. (See also the chapters on Water Reactor Kinetics and General Reactor Dynamics.)

Figure 2-1 [25] gives the relation between inverse reactor period and excess reactivity for various reactor fuel types. If the reactivity is greater than β, the delayed neutron fraction, very rapid rates of rise result, making safety action difficult to carry out. Rate-of-rise measurements monitor the closeness to this condition.

2.2 Neutron Sources and Range of Neutron Flux Levels

The limits of the neutron flux range are determined at the high end by the maximum reactor power and at the low end by the factors in Eq. (2-1). The source term may be amplified by as little as a factor of ten in a typical shutdown situation (i.e. $k \approx 0.9$) so its magnitude is important. Source fluxes orginate from spontaneous fissions, delayed neutrons, photoneutrons and artificial neutron sources.

Spontaneous Fission. In reactors fuelled with natural uranium the spontaneous fission rate of U^{238} provides a useful neutron source, one gram giving about 6.9×10^{-3} fissions/sec [24]. Thus in a reactor containing 40 tonnes of natural uranium fuel and having $k = 0.9$ the source would provide about 6×10^{5} neutrons/sec, yielding a flux of the order of 10^{2} neutron/cm^2-sec, which is about 12 decades below a typical full-power flux.

Delayed Neutrons. Many of the fission products emit delayed neutrons and immediately following a shutdown provide a source term. Figure 2-2 [25] shows the flux decay following step and ramp decreases in reactivity ignoring any poison or temperature effects.

Following an initial rapid drop in flux, the negative reactor period in a natural uranium reactor is limited by the decay of the longest-lived delayed neutron emitter to about 80 sec.

Photoneutrons. In D_2O- or Be-moderated reactors photodisintegration (by gamma rays) provides a large source of neutrons. The decay time depends on the lifetime of the various gamma-emitting fission products that excite the disintegration, and on the power level and the length of time the reacor has been at power. The long-lived gamma emitter La^{140} with a 40-hr half-life, maintains the flux above 10^{-6} of full power for several weeks.

Some types of reactors are shut down by removal of their D_2O moderator [7] and the thermal neutron flux to which most sensors respond is removed. In these cases a moderator level is defined at which a flux indication is first required and interlocks ensure that a minimum flux is

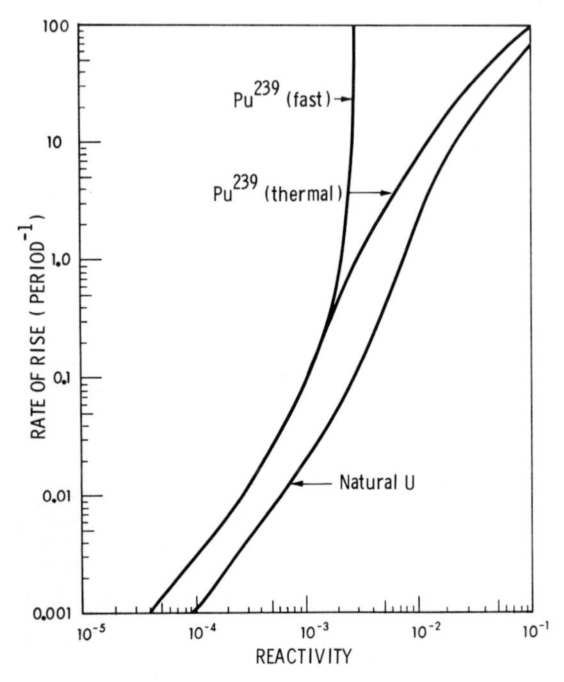

FIG. 2-1 Inverse period vs. excess reactivity.

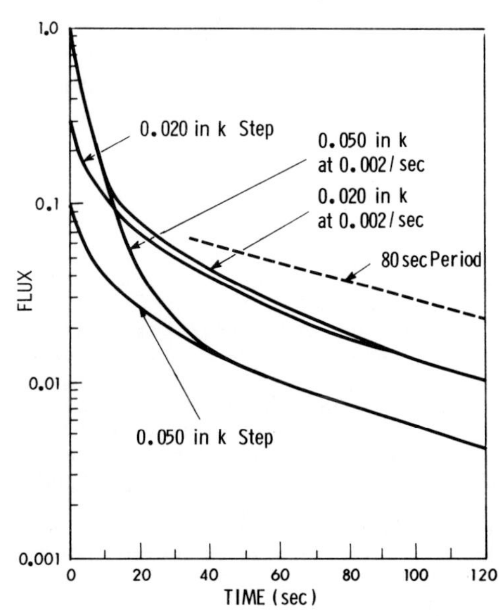

FIG. 2-2 Flux decay following reactivity insertion.

measured before moderator level can be further raised.

Artificial Sources. In reactors with few or no natural neutron sources, artificial neutron sources are used. They depend on either the (α,n), or the (γ,n) reaction. The reactions Ra(α,n)Be, Po(α,n)Be and Pu(α,n)Be provide from 10^6 to 10^7 neutron/sec-curie. The radium source is difficult to handle because it is also a gamma-emitter, the polonium source emits no γ-rays but has an inconveniently short (138 days) half-life, while the plutonium source has a low gamma activity and a long half-life and is also useful for the initial approach to criticality because of low γ-activity and ease of handling.

Several (γ,n) reactions are useful. The gamma-emitting material can have its activity induced by placing it in the neutron flux of the reactor and so be self-sustaining. This is particularly advantageous in reactors that require an artificial source for routine operation, although a special source might be required during initial commissioning. The reactions Na^{24}(γ,n)Be and Sb^{124}(γ,n)Be are most commonly used. Sodium (introduced as the carbonate) yields the higher neutron emission for a given loss in reactivity but is short-lived (half-life = 14.8 hr). If long shutdowns are encountered, the antimony source (half-life = 60 days) is better, although its neutron emission rate is only one-sixth that of the sodium source [25]. The emission rate of the antimony source is about 10^7 neutrons/sec-curie, but the size of the source depends on the reactivity loss that can be tolerated. For a reactivity loss of about 0.2×10^{-3} an increase in the neutron flux range above the spontaneous fission level of about 10^3 has been achieved in graphite-moderated, natural uranium reactors. The Sb-Be source becomes reactivated at high powers.

2.3 Types of Detectors

To cover the wide range of neutron flux discussed in Sec. 2.2 several types of sensors are used. They fall into two categories: (1) Detectors in which the movement of charged particles produces a signal, and (2) detectors in which the thermal effects of radiation are used. Detectors responding to neutrons of thermal energy will be emphasized although gamma radiation may provide a useful measurement [26]. Many texts and reviews [27, 28] cover the fundamentals of neutron detection and so only the pertinent facts will be mentioned briefly here.

2.3.1 The Fission Counter

The simplest form of detector for sensing single ionizing events is the parallel-plate gas-filled ionization chamber of which the fission counter shown in Fig. 2-3 [29] is an example. This type of chamber is widely available. The chamber is made neutron sensitive by coating the cathode with a fissile material (in this case U^{235}). When the U^{235} undergoes fission two fragments are formed. Normally only one of the two fragments formed in fission enters the gas-filled chamber, the other being absorbed by the electrode. A charge of up to 5×10^{-13} cou is available for collection per neutron captured. The sensitivity of the chamber shown, coated with U^{235}(96%) to a thickness of 1.0 mg/cm^2 over the cathode area (120 cm^2), is about 0.2 counts/sec per unit flux.

In addition to the ionized fission fragments which result in the chamber from neutron interactions, the fissile material coating emits alpha particles in the 4.0 Mev range and specific design measures must be taken to ensure optimum discrimination between these two types of ionizing events.

First, the use of a fissile material having a high ratio of fission cross section to alpha activity is desirable to reduce the number of alpha particle signals. In this respect U^{235} is better than U^{233} or Pu^{239} by orders of magnitude.

Second, the spacing between the electrodes should be just less than the mean range of the heavy fission fragments since the energy lost in the first few millimeters by the fission fragment is much greater than that lost by the alpha particle.

Third, the use of a gas filling (90% Ar + 10% CH_4) that aids in the reduction of charge collection time will permit higher time resolution. Collection times of 0.060 μsec have been achieved, making it possible to reduce the counting rate due to alpha particle activity to less than one pulse per minute.

Fourth, the U^{235} coating should be thin, preferably less than 1.0 mg/cm^2 and to obtain optimum pulse height (i.e. the charge collected per pulse) distribution, it should be placed on the cathode only.

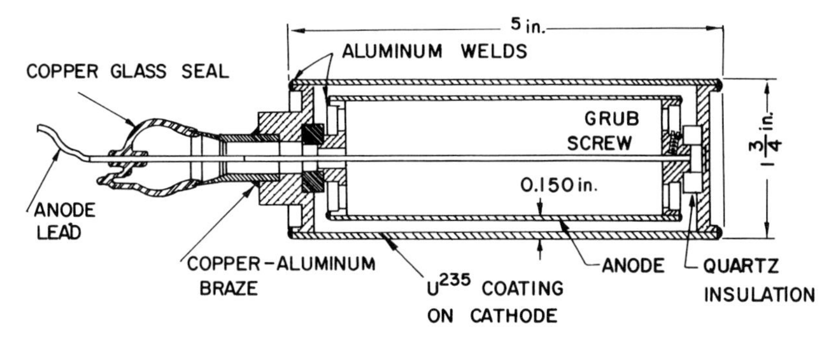

FIG. 2-3 Fission counter.

Figure 2-4 shows the pulse height distribution of the counter described*.

High temperature fission counters have been described [30a, 31] capable of operating at 500°C (932°F) having a coated cathode area of 500 cm^2 and a sensitivity of 0.5 counts/sec per unit neutron flux. Small high temperature fission counters 0.4 mm in diameter with a sensitivity of 8×10^{-3} counts/sec per unit neutron flux have also been produced.

It is also possible to use a fission counter in a d-c mode in which the ionization produced simultaneously by many fragments is swept out as a steady electrical current by the polarizing voltage. For this application [30b] both electrodes were coated with U^{235} to a thickness of 0.3 mg/cm^2. The electrode spacing was 0.3 mm and the sensitivity 5×10^{-16} amp/unit neutron flux. The current due to alpha activity was 5×10^{-10} amp.

The fission counter has several distinct advantages over other forms of sensors to be discussed.

1. It operates with a low polarizing voltage (≈ 200 to 300v). Fission counters have a saturation characteristic similar to d-c ion chambers and the count rate is relatively insensitive to small changes in the polarizing voltage if the voltage is sufficiently high.
2. It can be operated in gamma fields up to 10^5 r/hr (and will tolerate up to 10^7 r without deterioration). However, care must be taken when a fission counter is operated in high gamma fields that the d-c ionization current does not cause the polarizing voltage to fall below the saturation level.
3. The alpha activity, while presenting a design problem, can provide a useful test signal if the associated instrument design permits lowering the discrimination level.
4. Counting rates up to 10^6 counts/sec are possible.

*The pulse height distributions shown in the figures of this section were taken with a low-noise amplifier. The pulse-shaping time constants used are indicated on each figure.

2.3.2 The Proportional Counter

The proportional counter makes use of the high electric field near a fine-wire electrode to provide amplification (by collision processes) of the initial ionization.

A proportional counter is shown in Fig. 2-5. The chamber is filled with BF_3 gas (enriched in B^{10}) and the reaction

$$n + B^{10} \rightarrow Li^7 + He^4 + 2.3 \text{ Mev}$$

is used to detect the neutron. The energy lost by the Li^7 and He^4 ions (i.e. a total of 2.3 Mev) in ionizing the BF_3 gas produces an initial charge of about 1×10^{-14} cou.

With the dimensions shown and a polarizing voltage of 2400 v, the field near the central wire produces a gain in ionization of about 20 so that a maximum of 2×10^{-13} cou of charge are available for collection. A typical set of pulse height distributions as a function of voltage across the counter is shown in Fig. 2-6. The change in gas gain with polarizing voltage is clearly evident.

With the discrimination levels shown, the effect on counting rate as the polarizing voltage is increased is given in Fig. 2-7.

The spread in pulse height is due primarily to the various possible positions of the track of initial ionization in relation to the center wire.

Using boron enriched in the B^{10} isotope (to 90%) the counter shown has a sensitivity of about 30 counts/sec per unit neutron flux.

The main advantage of the proportional counter is its high sensitivity, but it has several disadvantages.

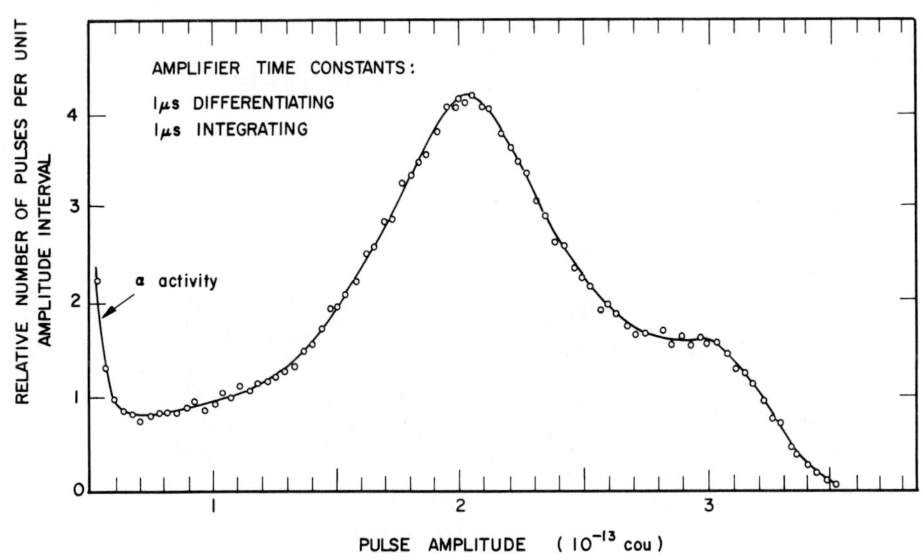

FIG. 2-4 Fission counter: pulse amplitude distribution.

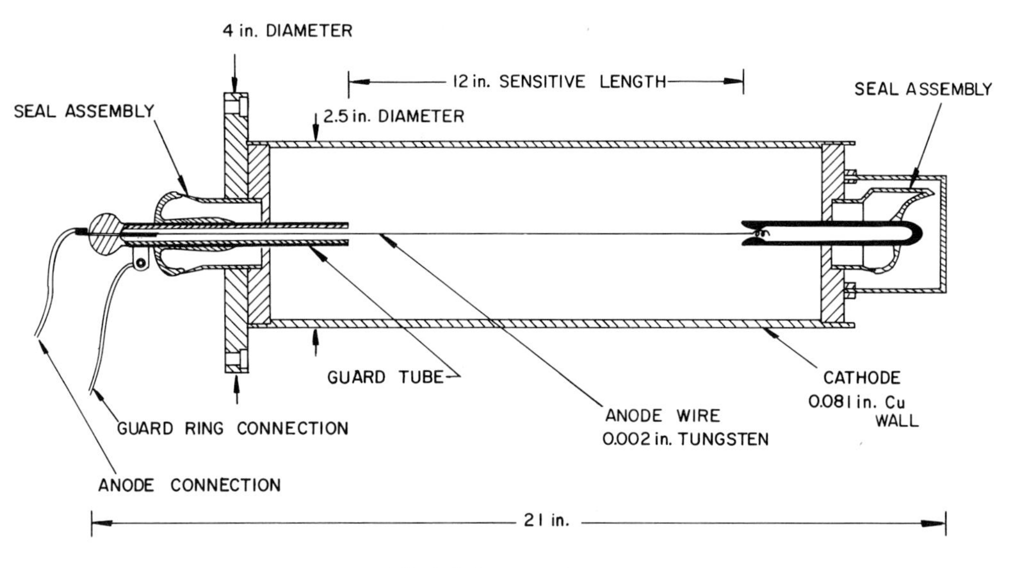

FIG. 2-5 Proportional counter.

1. It requires a high polarizing voltage (1000 - 3000 v).
2. The count rate is limited because of the slow charge collection time (1 - 3 μsec) limiting the counter's effectiveness at higher flux levels. The counter may become inoperative or count with impaired efficiency in such cases, thus giving false readings.
3. The properties of the BF_3 as a counter gas deteriorate rapidly in gamma fields of 10^3 r/hr or more.
4. In gamma fields of the order of 100 r/hr or more the gammas cause spurious counts and discrimination is difficult.

Proportional counters using a boron-coated cathode and filled with argon show less deterioration in gamma fields, but only one of the ionizing particles is available to produce ionization; further, because of partial absorption in the boron coating, an ill-defined pulse height distribution is obtained. Consequently, a proportional counter with boron-coated cathode has a poor counting rate versus polarizing voltage characteristic. Boron-coated chambers have been produced [32] for which little change in characteristics is found after an exposure of 3×10^{16} neutrons/cm^2. (In contrast, BF_3-filled chambers show more deterioration in the plateau and in the resulting pulse-height spectra.) This feature of the solid boron deposit may be useful in cases where the chamber cannot be withdrawn from the reactor core.

If BF_3-filled counters are to be left in the core, deterioration can be reduced by short-circuiting the cathode to the anode [33].

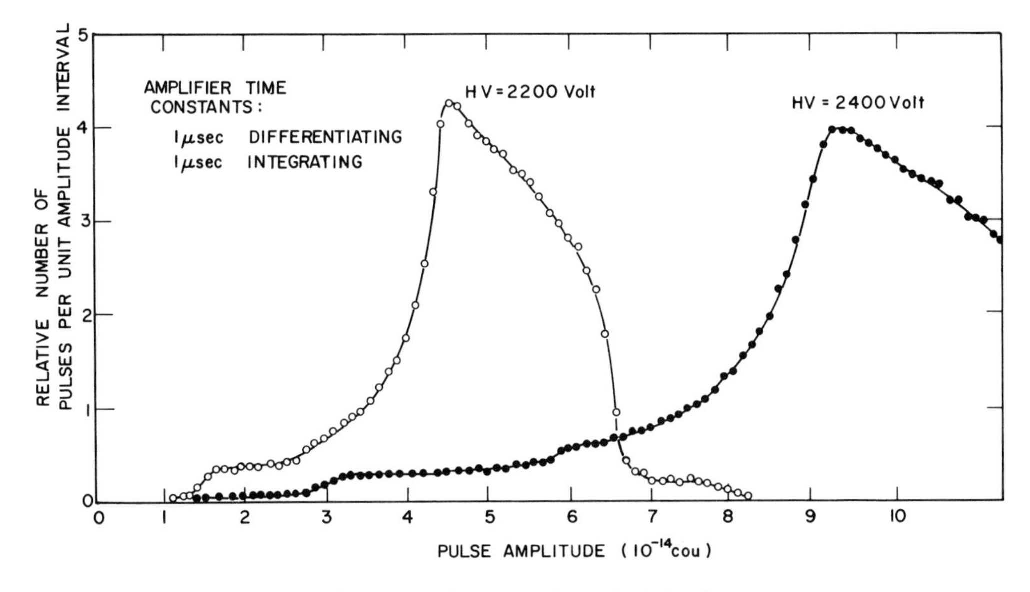

FIG. 2-6 Proportional counter: pulse amplitude distribution.

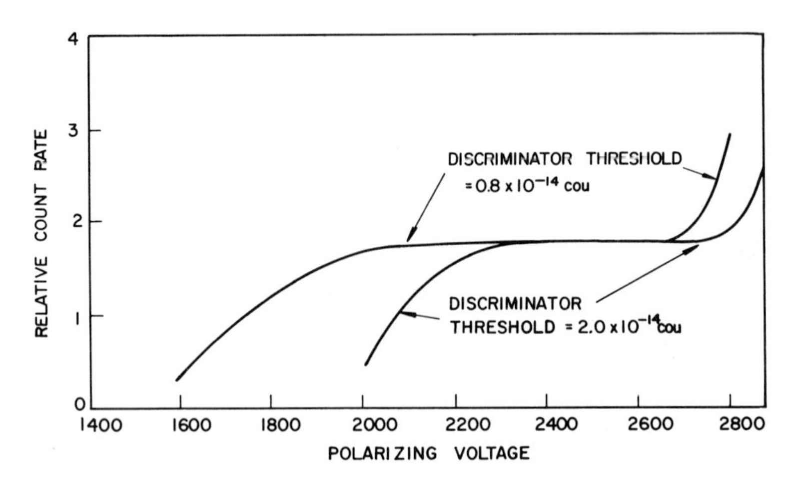

FIG. 2-7 Typical plateau curves for proportional counters.

2.3.3 The Ionization Chamber

At high flux levels the parallel-plate ionization chamber [34, 35, 36] with boron-coated electrodes is the most commonly used sensing element. A chamber of this type is pictured in Fig. 2-8 [30c].

Only one of the Li^7, He^4 pair enters the chamber volume and, hence, only 1.2 Mev of energy is available per neutron captured. In fact, because of absorption in the boron coating and walls, the average energy absorbed by the gas filling per ionizing event is about 0.2 Mev. No gas amplification occurs so that single events cannot be detected in a chamber of this type, but the ionization produced by many events is swept to the collecting electrode and measured as a direct current.

Figure 2-9 shows how the current for a given flux varies with polarizing voltage. Saturation occurs when the electron field is high enough to remove the ionization before recombination can occur.

These saturation curves are important from the safety viewpoint. Unless the chamber is operated in the saturated condition at the highest expected flux (taking into account credible power excursions), the current will not be proportional to flux at the higher flux levels. Figure 2-10 shows the actual deterioration that can occur in the saturation characteristic after some years of operation, due to the increase in recombination rate probably caused by impurity buildup in the chamber gas.

The operating characteristics of several types of ionization chamber are shown in Table 2-1. The higher gas pressure increases sensitivity somewhat but the voltage required to produce saturation is also increased.

Ionization chambers are also sensitive to gamma radiation. The TQU chamber gamma sensitivity is about 1.3×10^{-11} amp/r-hr, so that a gamma field of 1.0 r/hr produces the same current as a neutron flux of 300 neutrons/cm^2-sec.

When the reactor is operating at power, the neutron-to-gamma current ratio is about 10^3 but, when the reactor is shut down, the neutron flux decays much more rapidly than the gamma radiation. To ensure that gamma radiation does not produce a false indication of power, two measures should be considered.

First, it is important to eliminate from ionization chambers materials with high neutron absorption cross sections. If the half-lives of the beta and gamma radiation resulting from the neutron capture are very long, the decaying background current will not necessarily fall with reactor flux. This can result in a misleading signal indication, especially at low neutron flux levels. Also, of course, such chambers will be more difficult to handle and maintain. Manganese and iron are examples of impurities to be avoided.

Second, precautions must be taken in locating the ion chambers to shield them from gamma background fields. (See Sec. 2.4).

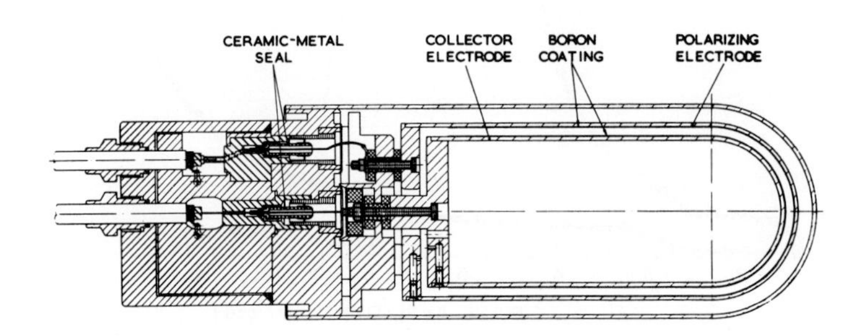

FIG. 2-8 Neutron-sensitive ionization chamber.

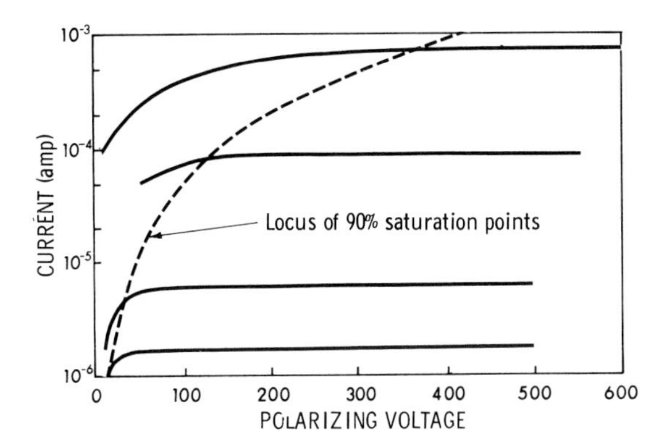

FIG. 2-9 Ionization chamber saturation characteristics. [Hydrogen-filled (225 cm Hg) Plessey PN 1026.]

If the neutron flux is suddenly removed from an ionization chamber (which uses high purity magnesium for its electrodes) the current from activation products decays as shown in Fig. 2-11 [37] where the ratio of current from the chamber to the current at full power is given as a function of time. The various components contributing to activation current are listed in Table 2-2.

The residual background current should at all times be less than the current due to neutrons from the fission process in the reactor. In D_2O-moderated reactors the neutron flux as maintained by photoneutron sources (see Sec. 2.2) permits shutdowns of about 2 months before the current due to the neutron flux approaches the activation currents.

2.3.4 The Gamma-Compensated Chamber

In order to improve the ratio of neutron to

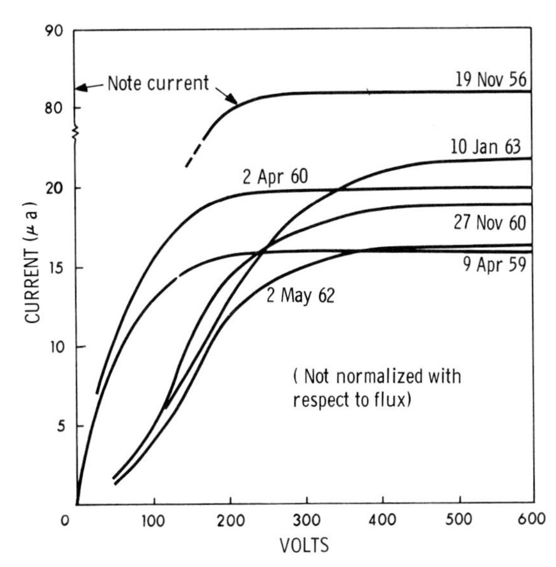

FIG. 2-10 Change in shape of saturation characteristics of ionization chamber with time.

gamma current, it is possible to design a chamber where to some degree the current produced by gamma radiation and by self-activation are cancelled. Such chambers are called compensated ionization chambers.

A gamma-compensated chamber is shown in Fig. 2-12 [30c]. In this chamber a neutron-plus-gamma sensitive volume and a gamma-only sensitive volume are arranged so that only the neutron current is passed to the signal cable. The gamma-sensitive volume in one chamber is adjusted during manufacture to provide a 30:1 compensation in a uniform gamma field. This ensures that in practice,

TABLE 2–1

Neutron Sensitive Ionization Chambers

Manufacturer and Number	Gas Filling	Pressure	Coating	Neutron Sensitivity amp/nv	Gamma Sensitivity amp/(r/hr)
Plessey Nucleonics PNI-1026	H_2	225 cm Hg 75 cm Hg	B^{10}	$2.6 \cdot 10^{-14}$ $1.7 \cdot 10^{-14}$	$6.5 \cdot 10^{-12}$ $2.2 \cdot 10^{-12}$
20th Century Electronics RC7EB	H_2	180 cm Hg	B^{10}	$2.7 \cdot 10^{-14}$	$5.5 \cdot 10^{-12}$
RC6EB (Compensated)	H_2	180 cm Hg	B^{10}	$2.0 \cdot 10^{-14}$	–
Reuter-Stokes RSN-76A	N_2	1 atmos	B^{10}	$3.5 \cdot 10^{-14}$	$4.5 \cdot 10^{-11}$
General Electric 5467870G4	N_2	1 atmos	B^{10}	$2.2 \cdot 10^{-14}$	$1.7 \cdot 10^{-11}$
5467870G11	N_2	1 atmos	B^{10}	$2.2 \cdot 10^{-14}$	$<8.5 \cdot 10^{-13}$

Typical Properties

Size, approx. 14 in. long 4 in. OD
Polarizing volts 100 to 1000 dc (depending upon maximum current, see Fig. 2–10)
Ambient temp. up to 200°C
Total neutron exposure 10^{19} neutrons/cm^2
Maximum neutron flux 10^{10} to 10^{11} neutrons/cm^2-sec.

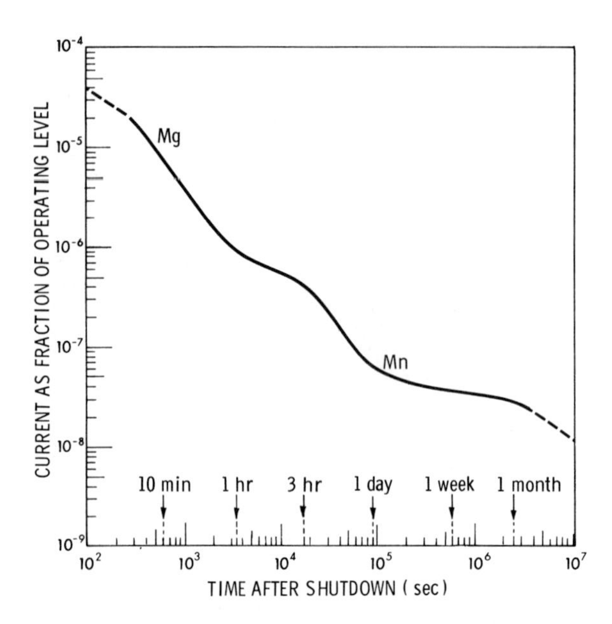

FIG. 2-11 Ionization chamber current (due to activiation) following shutdown.

TABLE 2-2

Components of TQT Activation Current

Half-Life	Initial Activity As Fraction of Operating Current	Element	Source
2.5 min	$5 \cdot 10^{-5}$	Al^{28}	Case
8.75 min	$9 \cdot 10^{-6}$	Mg^{27}	Electrodes
155 min	$9.6 \cdot 10^{-7}$	Mn^{56}	Boron Coating
15 hr	$8.8 \cdot 10^{-7}$	Na^{24}	Boron Coating

with typical nonuniformities produced by location and induced activity, the compensation is at least 10:1.

The polarizing voltage from the gamma-sensitive volume need not be large (i.e. 20 to 150v) since saturation is required only at the low currents where compensation is necessary. Another method of adjusting the compensation is by changing the polarizing voltage to the gamma-compensated electrode of a specially designed chamber [38]. In this case the compensating volume is determined by the electrode voltage.

Care must be taken to ensure that the instrumentation used with compensated chambers is not impaired if for some reason the gamma current exceeds the gamma-plus-neutron current and current flows in a direction opposite to normal. If this occurred, the effective value of the smoothing time constant used in some designs could be made very long (i.e. thousands of seconds). (See Sec. 2.5 for further details). In particular the instrumentation should not find itself unable to operate normally under such circumstances if the neutron current were suddenly to increase.

2.3.5 General Comments Concerning Ion Chambers

Good insulation is necessary between the polarizing electrode and the collecting electrode if currents in the 10^{-12} amp range are to be measured. Either quartz or alumina has proven very satisfactory as insulating material. However, insulators can become brittle after long irradiation and are susceptible to damage.

Cable using polythene insulation is not seriously affected by slow neutrons but becomes brittle after accumulating a gamma dose of about 5×10^9 r [34]. Special cable construction is required to prevent the generation of spurious currents by mechanical shock; type RG 149/u, which has a graphite-coated conductor is suitable. For high temperature in-core installations, cables sheathed in stainless steel and insulated with powdered alumina have been used. (See Sec. 3.1). It may still be necessary to watch that the cable capacity of some of these types does not slow down the response of the amplifier (Sec. 2.5).

2.3.6 The Semiconductor Fission Counter [39]

If a semiconductor junction detector is coated with fissile material, it can be an effective neutron monitor in low flux regions and may be especially

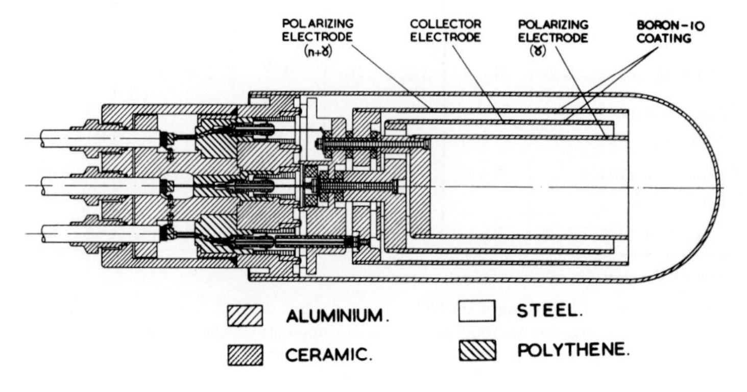

FIG. 2-12 Compensated ionization chamber.

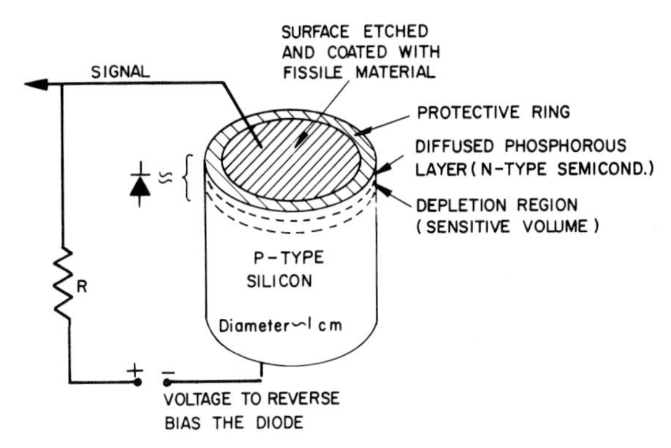

FIG. 2-13 Solid-state neutron detector.

appropriate for flux mapping. A detector with a U^{235} coating is shown in Fig. 2-13.

This type of detector is essentially a solid parallel-plate ionization chamber but releases about 10 times as much charge per electron-volt absorbed as a gas-filled chamber.

The effective plate spacing (i.e. the thickness of the depletion layer formed between the p and n regions by the reverse bias) and hence the detector capacitance is a function of the polarizing voltage. Special circuit precautions are taken (see Sec. 2.5) to ensure that changes in polarizing voltage do not affect the magnitude of the signal obtained.

The pulse-height distribution is shown in Fig. 2-14 where the alpha background from the uranium and the two fission fragment groups can be clearly seen.

The effect of neutron (fast) radiation damage is to decrease the lifetime of ions (i.e. the electron-hole pairs) formed in the silicon and hence to increase leakage currents.

2.3.7 The Neutron Thermometer

One version of the neutron thermometer uses the thermal energy of the fission process. A small cylinder of uranium-aluminum alloy surrounds a thermocouple that detects the temperature rise produced by the fission heat. Figure 2-15 [30d] shows the construction. The time constant is large (~ 30 sec) and the dynamic range is about one decade.

A second type of neutron thermometer [40] is shown in Fig. 2-16 [40]. A boron pellet surrounds each of the chromel-alumel thermocouple junctions and the energy to heat the junctions comes from the B^{10} + n reaction. The sensitivity is about 0.6

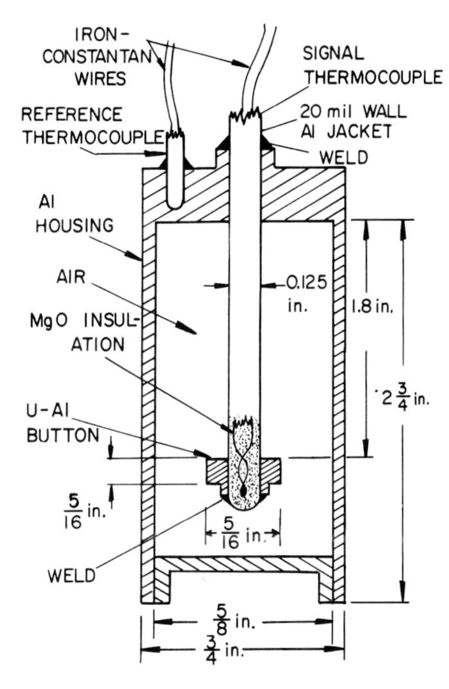

FIG. 2-15 Neutron thermometer using fission heat.

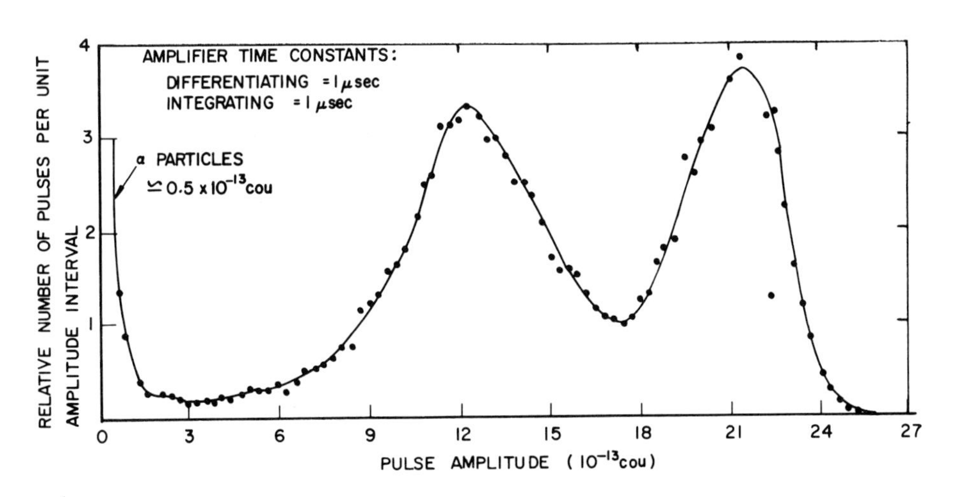

FIG. 2-14 Pulse-height distribution for solid-state neutron detector.

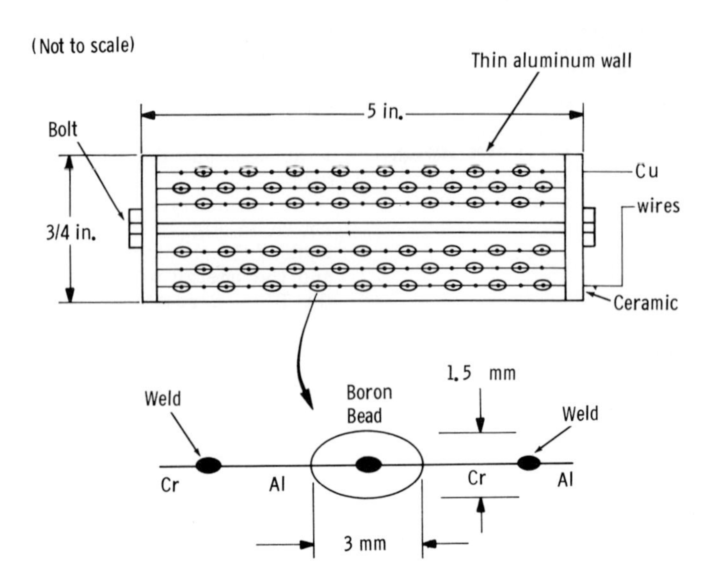

FIG. 2-16 Neutron thermometer using boron pellets.

mv/10^{12} neutrons/cm^2-sec per junction giving a total of 40 mv in the unit shown. The time constant of this assembly is about 13 sec. Other types have been described in the literature [41].

2.3.8 Beta-Emitter Neutron Detector

This device utilizes a material emitting a beta particle following capture of a neutron [42]. Wires of rhodium and vanadium are used as center conductors in a thin cable insulated with metal oxide and sheathed by Inconel. For detectors about 10 cm long a sensitivity of 0.6 μa per flux of 10^{14} neutrons/cm^2-sec has been measured [43]. The long half-lives (i.e. minutes) of these emitters does not permit a very rapid measure of the flux to be obtained. Their advantage is for in-core use where small size and low neutron absorption are required.

2.3.9 The Gamma Thermometer

The gamma thermometer utilizes the heat produced in the gamma-ray capture process. A mass of stainless steel is heated by the gamma flux and the temperature is determined by a thermocouple. The time constant is about 60 sec and the dynamic range one decade. The construction is shown in Fig. 2-17 [30d].

2.3.10 Detector Range

The chart in Fig. 2-18 shows where the various detectors described in this section fit into the normal neutron flux range. The range of fluxes which may be in a reactor varies from less than 10^{-2} to 10^{14} neutrons/cm^2-sec.

2.4 Location of Detectors

The neutron flux detector should be located with three considerations in mind: (1) that it monitors only radiation emanating directly from the fission process, (2) that it produces a signal representative of the average core flux in the required range, or of that neutron flux which is desired, and (3) that it be capable of performing reliably in the environment in which it is located.

It is not necessary to locate the detector in the highest flux. Only when the fluxes are very low and there is a detector sensitivity problem is this desirable. During core loading the most sensitive detectors available are often located at, or close to, the point of predicted maximum flux. The problem is generally to obtain sufficient sensitivity along with adequate discrimination against spurious signals.

At other times geometric attenuation, shielding, and choice of detector type can be used to cover satisfactorily the required range.

It is evident from Fig. 2-18 that neither a single detector nor a group of detectors, in one fixed location, can satisfactorily cover the range of over 10^{12} in flux values that can be encountered in a reactor flux monitoring system.

Figure 2-18 shows that the useful detector ranges are of the order of 10^6 -10^7. Thus combinations of detectors must be positioned or moved in such a way that a suitable combination is always available to supply on-scale signals for control of the system. It has become customary to divide the range of 10^{12} into three regions. One, the startup range, has already been discussed. A

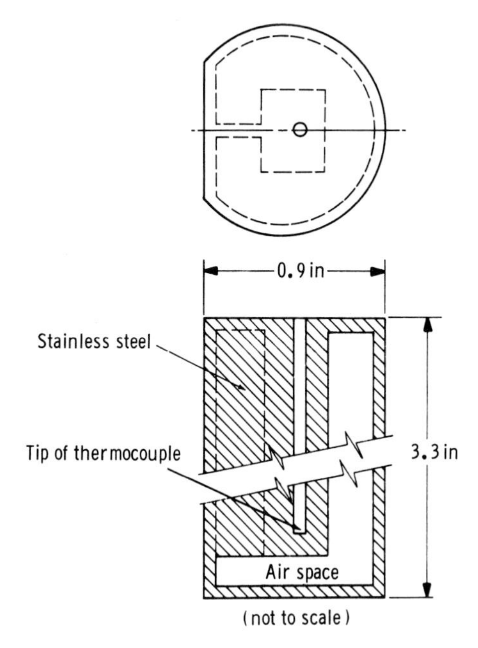

FIG. 2-17 Gamma thermometer.

second, called the "power range", is easy to describe. It is that range where sensible heat is being generated by the system and usually embraces those neutron fluxes corresponding to the region from one percent of full power to full power. It is in this region that thermal instrumentation (thermocouples, flow meters, etc.) provides the necessary (but with slower response time) backup and calibration for the neutron detection system. The third region is between the startup and the power range. It usually extends from about 10^5 -10^6 n/cm^2-sec, to 10^{10} -10^{12} neutron/cm^2-sec. In this region where the rate of rise is most important, compensated ion-chambers (Sec. 2.3.4) find their greatest use.

2.4.1 Detector Location for Initial Reactor Startup

Special consideration must be given to the location and number of detectors required to monitor the first approach to critical (See also Sec. 1 of the chapter on Criticality.) In general the flux levels will be low and it will be necessary to place sensitive detectors at the position of maximum flux within the reactor core to obtain a measurement truly proportional to the fission rate. The fission counter is suitable and most generally used. Often BF_3 proportional counters are also used to provide diversity.

It must be appreciated that, while the initial startup operation may require temporary instrumentation, it should be treated with at least the same concern with regard to safety as the permanently installed sensing equipment. Temporary connections should be avoided by considering the need for startup instrumentation during the design of the safety system. Special provision must also

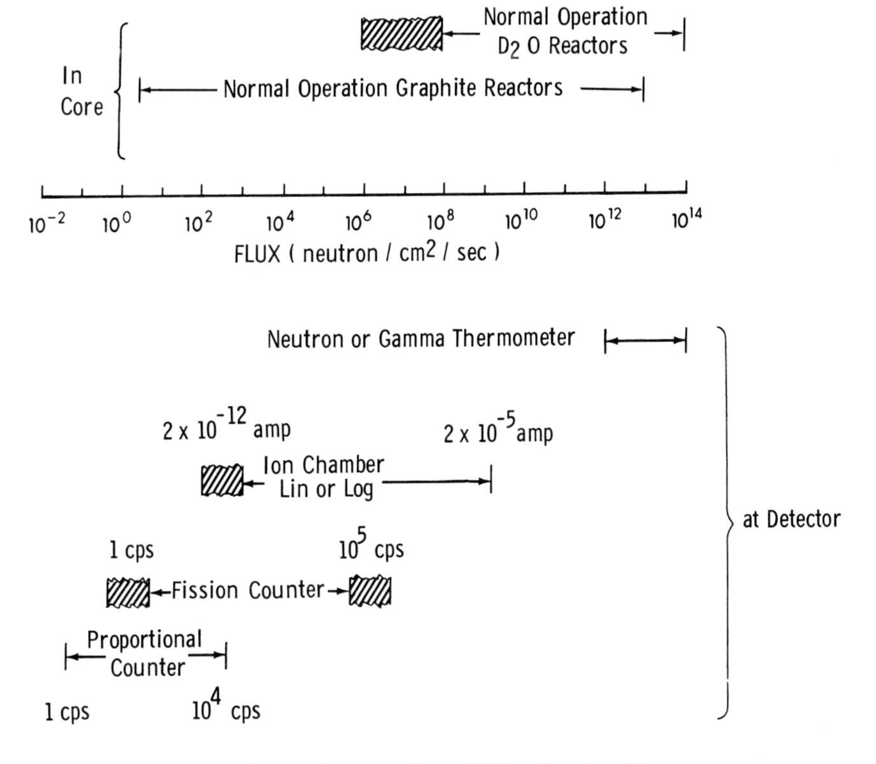

FIG. 2-18 Typical ranges and sensitivities of various detectors.

be made for its connection to the safety system.

Startup can proceed until the flux at the detector location is about 10^5 neutrons/cm^2-sec when, allowing for reasonable overlap, another sensor should be producing a reliable signal. The region up to about 10^5 neutrons/cm^2-sec is normally designated as the "startup" range.

For initial startup an ionization chamber placed in a region where the flux is about a factor of 100 less than the core maximum will generally provide an adequate signal. And, since gamma radiation from fission products and activation will be negligibly small, a reliable signal is available for use until the permanently installed sensors are reliably on-scale. Alternatively, additional range can be obtained by moving the more sensitive counters to a position of lower flux. This, however, would appear to be less desirable from the safety viewpoint since the safety provisions associated with the counters may have to be bypassed to avoid an accidental shutdown due to sudden movement or due to interference caused by cable stressing. This is especially true if period protection has also been associated with the sensors. Deliberate chamber movement also requires careful logging of their position with re-insertion if a shutdown occurs.

If an artificial source is used to enhance the initial flux, its position relative to the detector must be considered to ensure that neutrons (or gamma radiation) coming directly from the source are not detected. Further, reference [6b] indicates that the shape of the reciprocal-of-the-flux measurement vs reactivity or core loading curve used during startup to predict criticality, can be affected by the position of the source, as discussed in Sec. 1.4 of the Criticality chapter.

2.4.2 The Location of Sensors for Wide Range Operation

In reactors that require a wide range monitoring system, (a range of the order of 10^{12}), for every startup the sensors must form part of the permanently installed instrumentation. Efforts have been made to design chambers that can be left in the reactor core. Semiautomatic sensor positioning is often provided and, in fact, may form the basis of a control system [44]. However, it is more desirable to have an installation in which wide range is achieved without requiring movement of the sensor. Since the measurement should include in its range the highest reactor power level, ionization chambers are used to take advantage of the higher flux and consequently of the faster response time that can be achieved. Unlike pulse-operated detectors that can be made to discriminate against a large gamma radiation background by instrumentation techniques, the ionization chamber must be located in a region where the ratio of neutron flux to gamma field is as high as possible to obtain maximum range. Several arrangements for producing the necessary conditions are in use. These methods permit neutron flux to be monitored over about seven decades.

The Shielded Chamber. Surrounding a chamber by several inches of lead attenuates the gamma radiation but has little effect on the slow-neutron flux. With 4 in. (10.2 cm) of lead the neutron-to-gamma ratio measured as a ratio of the ionization chamber currents produced by each is increased by about 200. Typically, then, if 10^{-11} amp is taken as the lowest current that can be measured reliably in a reactor instrumentation system, the gamma field at the chamber should be no greater than 100 mr/hr (20 r/hr outside a 4 in. thickness of lead) at a neutron flux corresponding to 10^{-11} amp.

Whether or not a simple shielded assembly is sufficient depends on the reactor type. In a D_2O-moderated reactor in which a principal neutron source is produced by core gamma radiation, a high neutron-to-gamma ratio results naturally and the addition of the shielded housing permits a seven-decade range to be covered.

The Shielded Chamber with a Local Thermalizing Region. Good moderating nuclei include hydrogen, deuterium, beryllium or beryllia, and graphite. The choice made is usually between hydrogen, in the form of a plastic or water, or graphite on the basis of expense and ease of handling. Fast neutrons will be attenuated much less in passing through a thick pressure vessel wall than thermal neutrons. Ideally then, the system would provide the highest useful neutron flux (i.e. thermal flux) at the detector if all neutrons moving in the general direction of the detector remained fast neutrons until they struck the local moderating region at the detector. Several such arrangements can be visualized and one is shown in Fig. 2-19.

In D_2O-moderated reactors, because of the photoneutron source, a thermalizing region may be used with no shielding immediately around the detector.

The Shielded Chamber Using Scattered Neutrons. A scheme used in the NRU [45] reactor is shown in Fig. 2-20. The chambers are shielded from direct gamma radiation from the core but detect slow neutrons scattered into them. The sensitivity is maximum when the movable lead plug is opposite the center of the sensitive volume of the ion chamber. A similar arrangement is used in NPD but with a hydrogenous scatterer.

Compensated Ionization Chambers. Any of the above arrangements can be improved in varying degree by using compensated ionization chambers. Good compensation may approach the improvement possible with lead shielding but generally the simplicity of shielding is more appealing than the more complex compensated chamber and its instrumentation. Space requirements may, however, dictate otherwise.

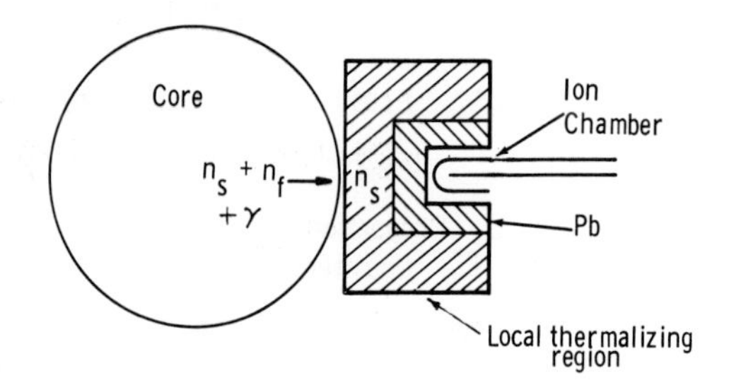

FIG. 2-19 Shielded ionization chamber in local thermalizing region.

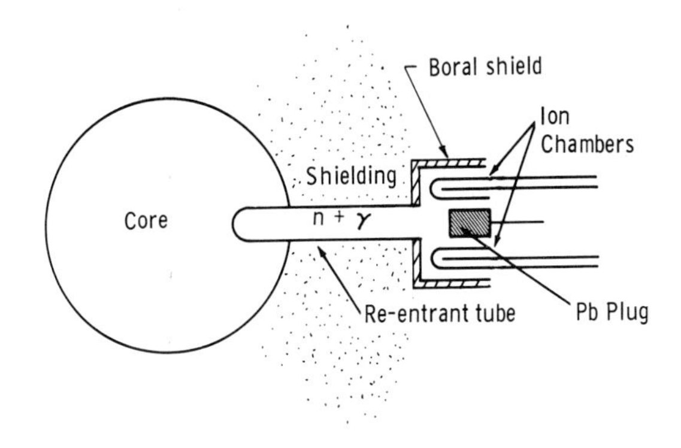

FIG. 2-20 Shielded ionization chamber using scattered neutrons.

2.4.3 The Use of Redundant Sensors

To obtain the required reliability several sensors are normally used for each range. For the various housings described in Sec. 2.4.2 the question arises whether to place extra sensors in the same housing or to have separate housings. From this viewpoint, the possibility of a shielding accident, that could reduce the flow of neutrons to a housing, must be examined. Other features may also affect the independence of the sensors. The housing shields may require cooling. If air from a common source is used, the possibility of moisture or contamination affecting all sensors simultaneously must be considered.

Particularly in large reactors it may be desirable, or even preferable, to locate the various detectors in a number of separate and widely spaced housings. This provides the operator with more information about local neutron flux in various parts of the reactor. Such information is vital in large cores where second-order zonal flux changes can occur. As the designer separates the detectors physically from one another he must rely more heavily on individual testing (or local pairs of detectors) rather than agreement between sensors to ensure reliability.

Shadowing by control absorbers is a major problem and is illustrated in Fig. 2-21. In particular during startup when a large number of absorbers are in the core, only a few sensor locations may be capable of detecting a flux change when the initial absorbers are withdrawn. Large differences between sensors and especially in period signals derived from them can result. (See also Sec. 1.4 of the Criticality chapter.)

The cable-runs from the sensors to associated instrumentation must also be treated independently to guard against common failure from fire or mechanical damage.

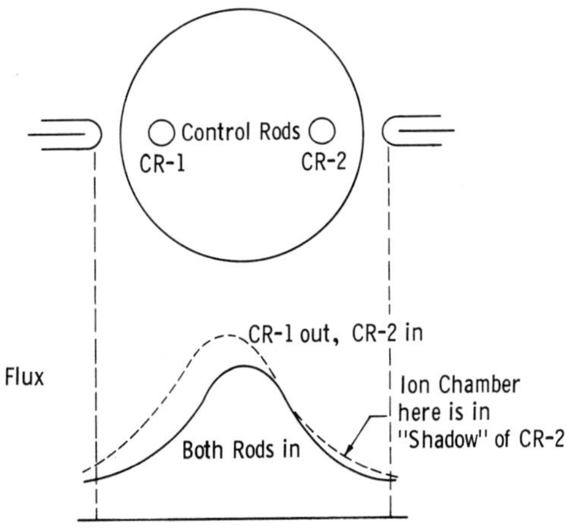

FIG. 2-21 Flux distortion by control rods.

2.4.4 Special Considerations in Reactors Using Changes in the General Geometric Configuration for Control

A number of reactor types utilize moderator dump, reflector change (fast reactors) or some other major geometric or chemical change to aid in reactor control. Placement of ionization chambers must take these effects into account. For example, in reactors using moderator level for reactivity control, the ionization chambers should be placed below the highest moderator level at which criticality is possible. The response of the sensors as the moderator surface passes the chambers during an emergency shutdown and during the ensuing startup must also be considered. Qualitatively, the following situation could occur. As the moderator level falls the neutron flux will drop but while the chambers are covered (see Fig. 2-22) the neutron-to-gamma ratio will be satisfactory. As the chambers are uncovered the slow neutron flux falls to zero and the gamma flux increases because of the removal of the moderator shielding effect. What occurs on startup depends on the length of time the shutdown lasts, but if it is a day or so the signal from the chamber may decrease or may even remain almost constant as the moderator level passes the chambers, if the shielding effect just compensates for the increase due to the neutrons.

A more pronounced effect might be noticed if the chambers were surrounded by a local moderator. When they become uncovered, an increase in fast flux might occur and an increase in signal could take place to beyond the level obtaining just before the chambers were uncovered. This would subsequently decay with the characteristic decay times of the delayed neutron groups. One observed example of this general type is discussed in Sec. 1.4.3 of the Criticality chapter.

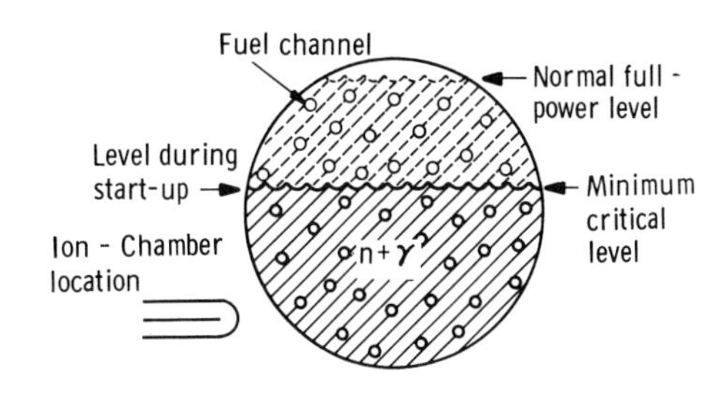

FIG. 2-22 Location of ionization chamber for D_2O level regulation systems.

In reactors which utilize a variable concentration soluble neutron absorber such as boric acid solution errors can also occur. The addition of more absorber will tend to alter the flux shape at the edge of the reactor and to diminish the flux of neutrons reaching the detector. Thus, the amount of shutdown margin could be overestimated unless care is taken in calibrating the chamber against reactor power.

Ambiguous signals of any type are obviously undesirable from a safety viewpoint and ion chamber locations must be chosen to minimize such difficulties.

2.4.5 Flux Propagation Time

Because of the propagation velocity of the thermal neutron flux wave, the dynamic response of the detector to changes in absorber movement is influenced by the location of the detector with respect to the absorber. Figure 2-23 [46] shows the phase of the reactor transfer function. For low frequencies of absorber modulation the change in gain and phase-lag to any position is the normal critical transfer function. At higher frequencies and as the sensor and oscillating absorber are spaced farther apart the phase shift increases. For very close spacing some phase advance has been noted. Similar effects are reported by Wasserman based on observations of the SPERT-I core [47].

2.4.6 Spatial Instability

In large reactors that are susceptible to spatial flux instabilities it may be necessary to measure flux at several locations not only to obtain a better average but also to permit zonal control to be exercised. Figure 2-24 shows how a multi-zone

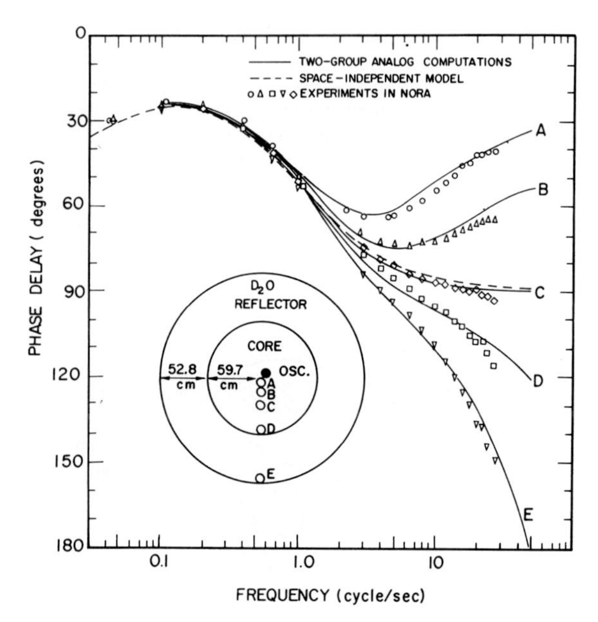

FIG. 2-23 Phase lag in Nora (Savannah) core.

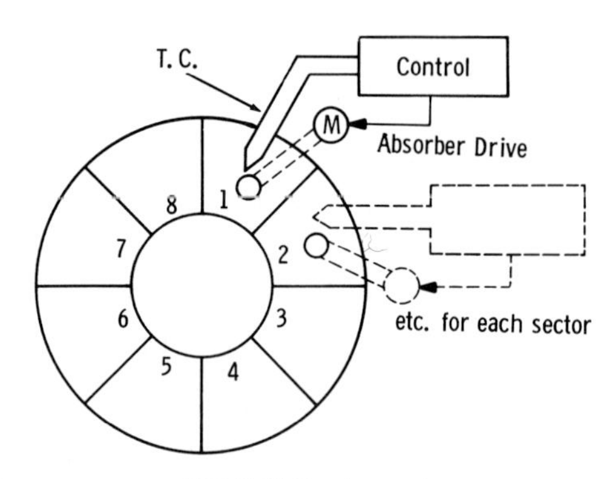

FIG. 2-24 Zonal control.

control system appears using temperature measurements [48]. Each sensor is coupled to a control mechanism regulating essentially only the zone seen by its sensor. In arrangements of this type reliability considerations may require redundancy to be applied at each detector site.

2.5 Power and Period Measurements

2.5.1 The Pulse-Counting Channel

A vital part of nuclear instrumentation is the circuitry used to couple the various pulse-producing neutron sensors to appropriate amplifiers. The charge produced in the various detectors described in Sec. 2.3 is in general large compared with the noise inherent in amplifiers, so that some loss in signal-to-noise ratio can in some cases be tolerated to provide better time resolution and more convenient mechanical arrangements.

2.5.1.1 Counter Pulse Shapes. In order to appreciate the various coupling schemes that have shown most merit, it is necessary to consider a detector as a source of current as well as a source of voltage. A brief discussion of pulse formation is thus worth while.

Figure 2-25a shows a detector with its associated elements. Both a voltage and a current signal are available. A convenient circuit that is

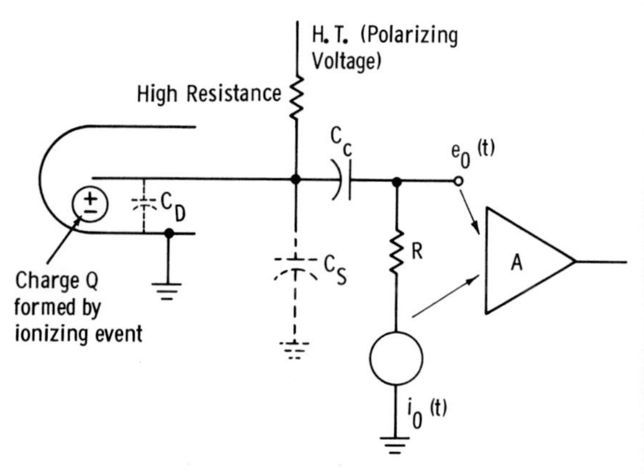

FIG. 2-25a A detector and its elements. C_C = coupling capacitor (large enough to be ignored); C_S = stray capacitance; C_D = detector capacitance; and R = circuit plus amplifier resistance.

equivalent to this for the purpose of studying the voltage output is shown in Fig. 2-25b. The voltage source E(t), representing the detector, generates the waveform shown each time an ionizing event occurs in the counter. The voltage increases linearly with time as a charge Q moves from its point of formation to the collecting electrode in the time T_R. Only this simple case need be considered, although in fact pulse shapes can be complex and variable. The form of the available voltage e(t) will depend on the capacitance C_S and the resistance R. Various situations are shown in Fig. 2-25c. As R becomes smaller the maximum pulse amplitude decreases as does the length of the pulse. In the extreme case amplification becomes difficult.

If, however, the current pulse is examined, the opposite effect is taking place and as the resistance becomes smaller the current output increases. An equivalent circuit showing this situation is given in Fig. 2-25d. In the limit the current pulse is square with an amplitude equal to Q/T_R amperes and a width T_R as seen in Fig. 2-25e.

2.5.1.2 Coupling Methods. In amplifiers using vacuum tubes the resistance R can be very large and the voltage mode is generally used. If the counter is separated from the amplifier, the capacity added by the connecting cable will cause a loss in pulse amplitude. Often 10 to 20 ft (3 to 6 m) is used between the counter and a preamplifier in order to get the circuits outside the reactor shielding. The simplest approach is to have a single vacuum tube closely coupled to the counter in either of the connections shown in Fig. 2-26. The split cascode circuit [49] has excellent signal-to-noise ratio properties, but is subject to gain variation due to changes in the mutual transconductance of the input tube. A cathode follower stage also is shown with a constant-current feed and with means to monitor tube conditions.

Once long voltage pulses have been obtained from such circuits, they can be shaped [49] to suit resolution and signal-to-noise requirements. An advantage in having the pulse length maintained until the pulse is in the amplifier is the flexibility available in the shaping networks and the capability of reducing low-frequency noise that may appear at the amplifier input.

With the adoption of transistors as input elements a new approach was needed. The input impedance of a simple transistor amplifier is low and consequently the voltage signal is small. However, the current flowing into the transistor is large but in the form of a short pulse as shown in Fig. 2-25e with $R(C_S + C_D) < T_R$. The amplification of a pulse of this duration (<0.1 μsec in fission counters and solid-state radiation detectors) is difficult, but if the input circuit is connected as a current integrator its output will rise to a voltage Q/C_F (see Fig. 2-27) at a rate governed by the frequency response of the amplifier. This arrangement is known as a charge amplifier [50, 51].

Two circuits [52, 53] that are commonly used to couple all types of pulse-forming detectors to amplifiers are shown in Fig. 2-28. The coupling amplifier can be considered as having a transfer resistance R_0 as indicated in Fig. 2-27, measured as the incremental ratio, output-voltage/input-current. It is readily shown that the pulse shape obtained is the same as those considered in Fig. 2-25c with R_0 and C_F replacing R and $(C_D + C_S)$. The detector can be separated from the coupling amplifier with an increase in circuit noise as the penalty. Unlike the voltage amplifier, however, the pulse amplitude at the output is essentially unaffected by the cable length (up to 20 or 30 ft or 6 to 9 m).

When semiconductor detectors are used (see

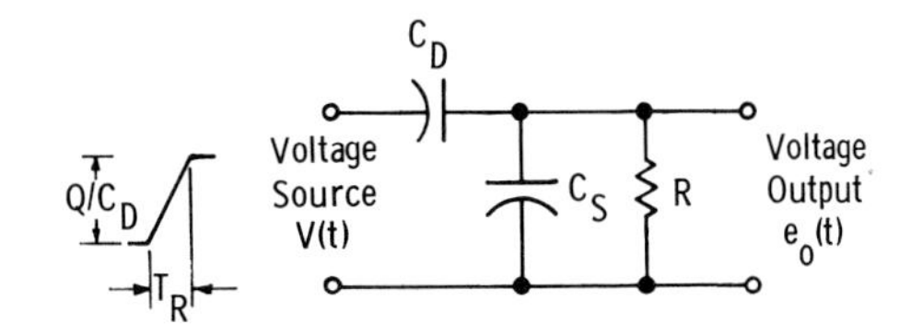

FIG. 2-25b Equivalent circuit for voltage output.

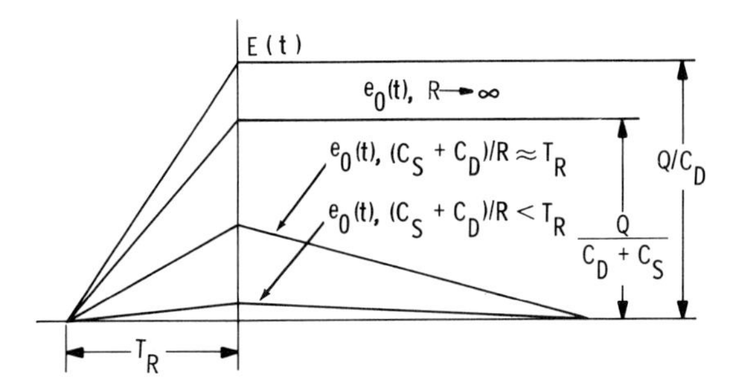

FIG. 2-25c Pulse amplitude variation with R.

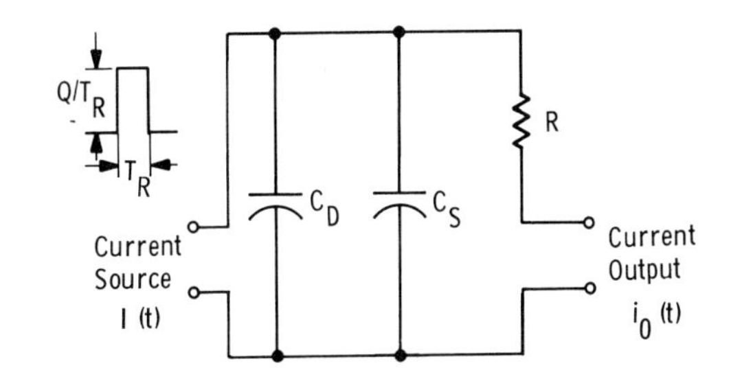

FIG. 2-25d Equivalent circuit for current output.

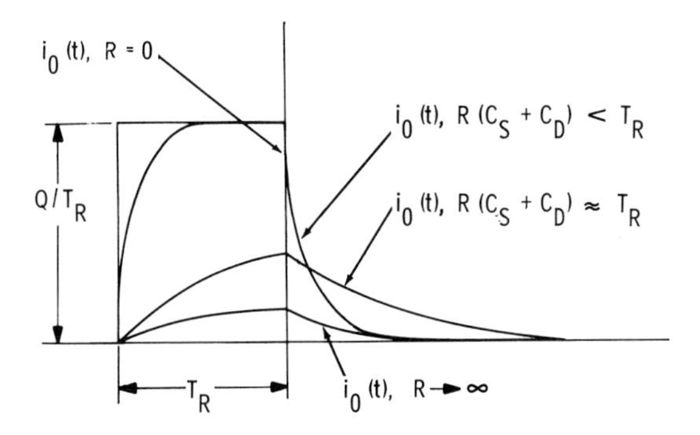

FIG. 2-25e Current pulse shape variation with R.

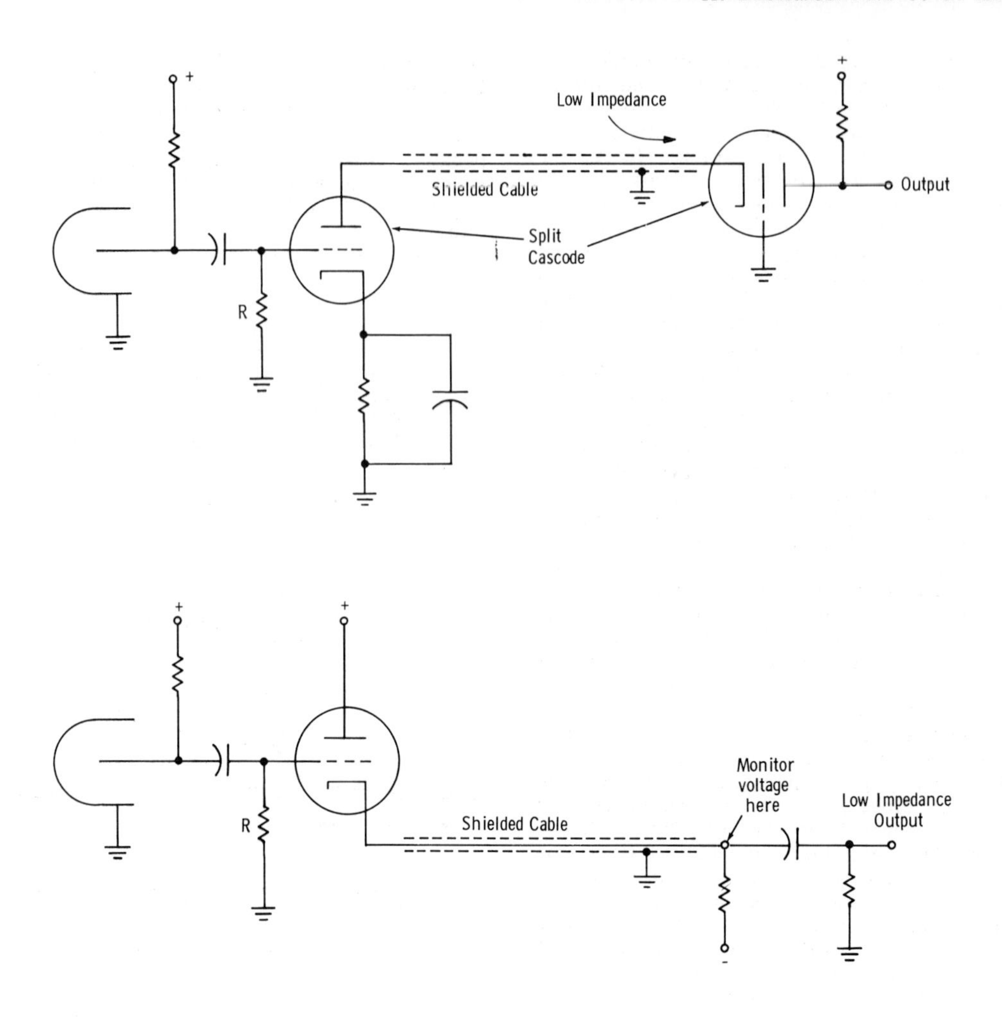

FIG. 2-26 Coupling circuits.

Sec. 2.3) the charge amplifier connection is necessary to avoid changes in pulse amplitude due to variation in the detector capacity, the capacity being a function of polarizing voltage.

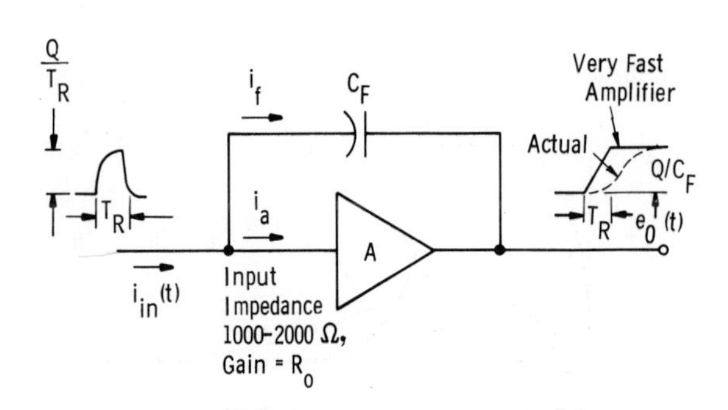

FIG. 2-27 Charge amplifier. (In the circuit, i_f = feedback current, i_a = amplifier input current, $R_o = e_o/i_o$ = gain.)

Since it is desirable to have no active components with the counter because of environmental difficulties (temperature and radiation), coupling schemes have been devised that permit several hundreds of feet ($\sim$ 100 m) of separation between the detector and the amplifier. The containment around reactors is usually opened infrequently and at great expense. To shut down and depressurize during core life because of the failure of a single detector or piece of electronic equipment constitutes a serious economic penalty.

Transformers have been used as coupling elements to adjust impedances so that proper cable termination is provided. Figure 2-29 shows a simple transformer connection and its equivalent circuit. The limitations are due to stray capacitance, which increases with turns ratio, and to the primary inductance which should be high to avoid an oscillating response. This inductance is limited because extra turns produce additional stray capacitance. With these design considerations, the circuit of Fig. 2-30 is one way to provide coupling

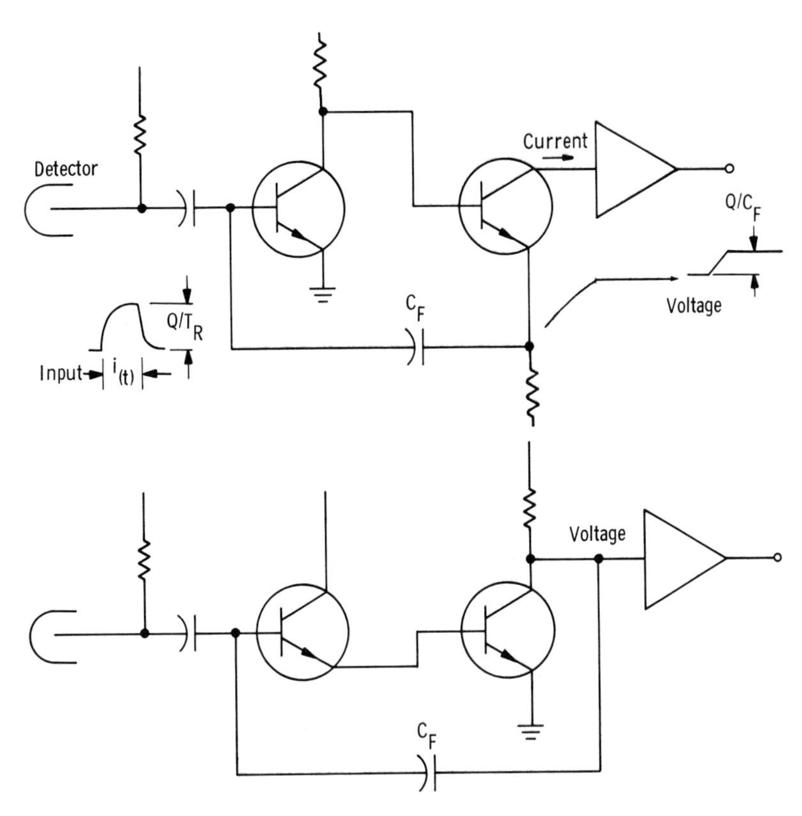

FIG. 2-28 Counter-amplifier circuits. (Corresponding waveforms are same in both circuits.)

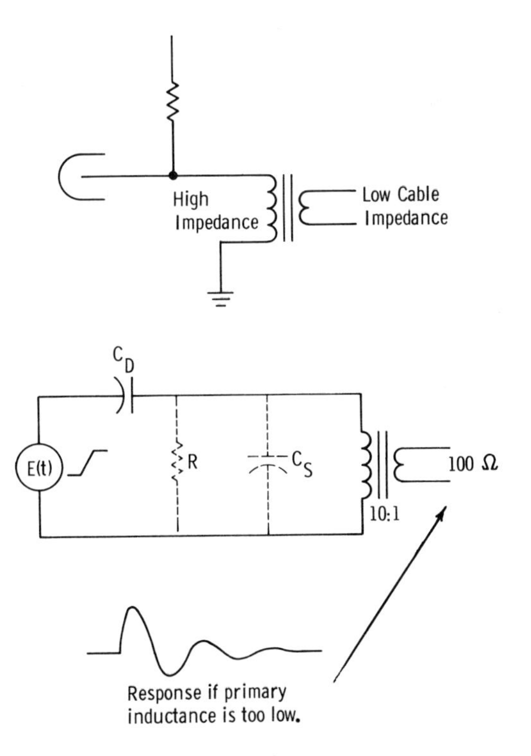

FIG. 2-29 Transformer coupling.

for a fission counter to a cable up to 400 ft (~120 m) in length. This circuit is not satisfactory with a proportional counter because of the longer pulse rise-time although suitable transformer circuits have been described [53].

The pulse height at the end of the cable is a factor of 20 less than a close-coupled voltage amplifier would provide, but is still of sufficient amplitude to exceed amplifier noise. The pulse height spectrum at the output of the transformer-coupled system is shown in Fig. 2-31. The distribution has been broadened by amplifier noise as can be seen by comparison with the spectrum in Fig. 2-5.

Another coupling approach is illustrated in Fig. 2-32; the detector current pulse is used and fed into a low impedance circuit consisting of a grounded-base transistor. This circuit does not provide any amplification but the pulse does become available for integration as shown, although the current pulse itself may be used [54]. The resistance in series with the detector allows proper termination of the cable, or alternatively, the bias current through the transistor can be adjusted until the input impedance is the correct value for termination. The inherent noise in this circuit is high and signal-to-noise ratios similar to those in the transformer-coupled case are obtained. Although the current integrator itself is a low impedance device, impedances low enough to terminate transmission lines have not been obtained.

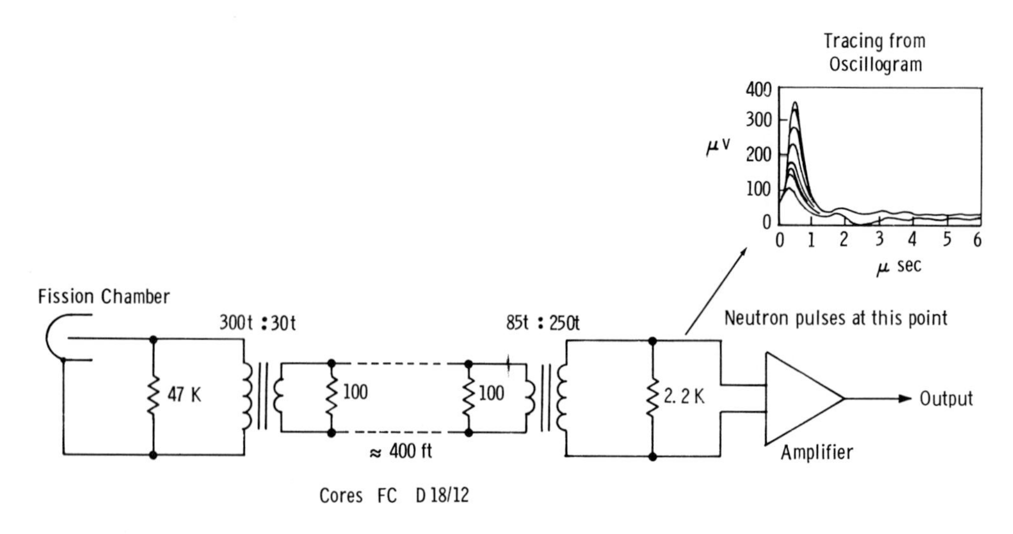

FIG. 2-30 Transformer coupling circuit.

Transistors are prone to failure from voltages or power dissipation in excess of their maximum ratings. The most common of these arise from connection or disconnection of the polarizing voltage. Damage can be prevented by protecting the input transistor with voltage-limiting diodes.

Electrical interference however still remains the most troublesome aspect of nuclear instrumentation and is the least documented of nuclear instrumentation problems. Noise currents flowing in the cable sheath are of most concern and the magnitude of the signal produced on the central conductor by a specific sheath current can be used to specify the noise rejection properties of various cables. Solid sheathed cables provide good interference rejection as does spiral-wrapped solid sheathing and double shields of braid. When installed away from direct sources of interference (cables carrying large currents to contactors, etc.) the shielding arrangement shown in Fig. 2-33 has been successful. The circuit is balanced to earth and uses a twin-core double shielded cable carrying the detector polarizing voltage along with the signal.

It is difficult to specify the amount of interference rejection that has been achieved in a system because standards of types and amplitudes of interference do not exist. In almost every installation, pulse-amplifying systems are made to work after they have been installed by rearranging

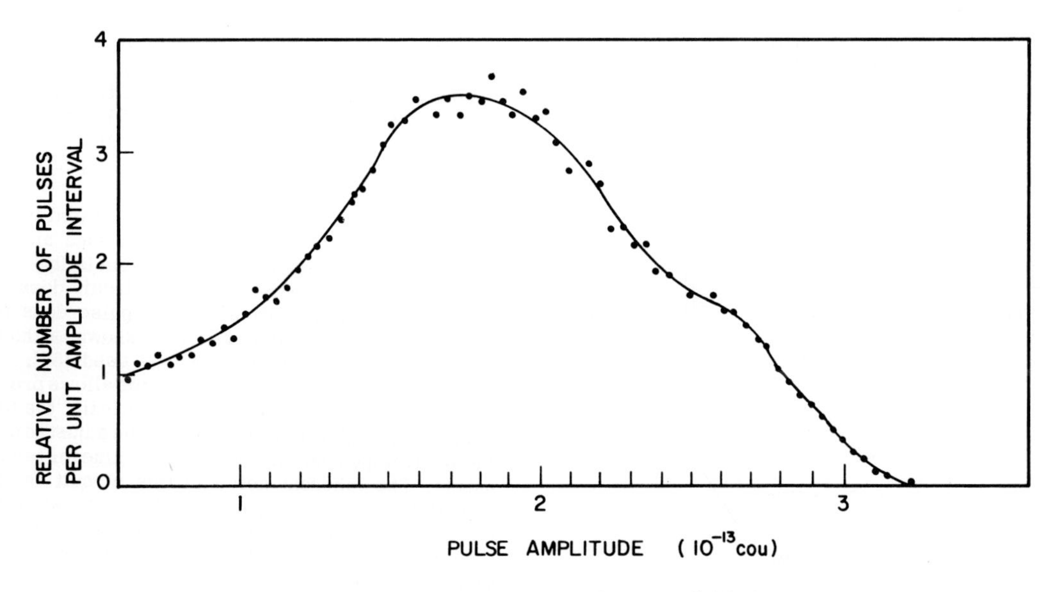

FIG. 2-31 Pulse height distribution from a transformer-coupled fission counter.

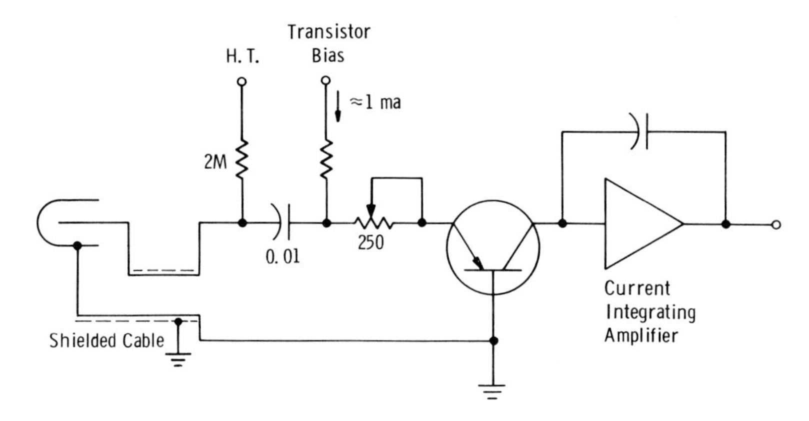

FIG. 2-32 Grounded base input circuit.

grounding points, by adding more shielding, or by adding more amplification at the detector.

A basic study of the problem has not been forthcoming.

2.5.1.3 Pulse Amplifiers. There are many amplifier designs that are entirely satisfactory for increasing the output of the coupling amplifier. The requirements are not particularly stringent, especially with regard to noise performance and linearity, so that thermionic tubes are finding less and less application.

Figure 2-34 shows a well-tried current amplifier circuit [55]. In general, circuits for amplifiers follow trends found in nuclear research electronics and do not incorporate specific fail-safe techniques.

2.5.1.4 Pulse-Height Discriminators. The output of an amplifier connected to a neutron detector contains not only pulses due to neutrons but also smaller pulses due to noise and other types of radiation. These small pulses must be eliminated and a discriminator is used to produce a pulse of standard size whenever the input pulse exceeds a preset threshold. Discriminators have been well-documented by McCollom [56].

The main requirements of a discriminator for reactor instrumentation use are stability, the ability to vary the discrimination level, and to

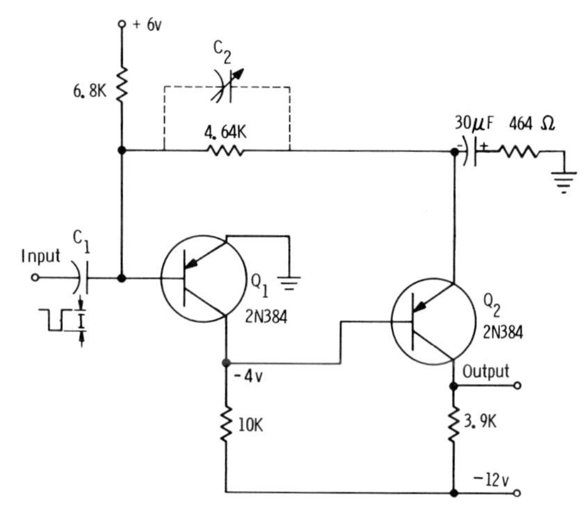

FIG. 2-34 Pulse amplifier.

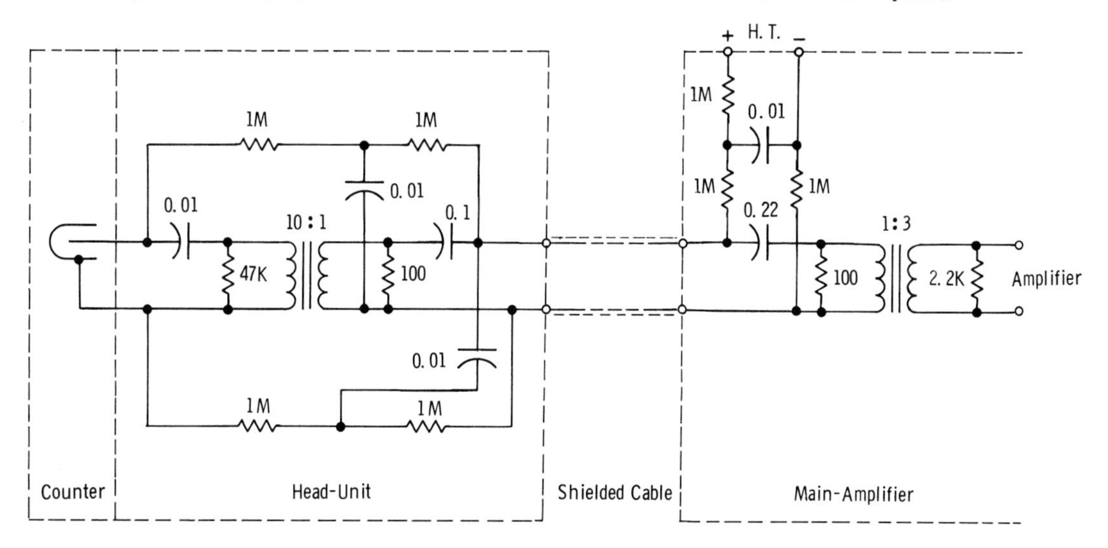

FIG. 2-33 Two-wire shielded connection.

know precisely where it is set. This latter adjustment is vital to correct operation and is usually determined by plotting count-rate from the discriminator output as a function of the discrimination level. A visual inspection of the resulting characteristic (see Fig. 2-35) is made to determine the optimum setting. This procedure requires either a calibrated potentiometer or a direct measurement of the discriminator bias.

A safer procedure for determining the optimum discriminator setting uses a differential pulse-height analyzer connected as shown in Fig. 2-36. Only those pulses are analyzed that operate the discriminator so that a pulse-height distribution is obtained beginning at the threshold level. This method does not require any change in the bias to give assurance that the threshold is correctly set.

When a proportional counter is used, it is customary to use a fixed discriminator setting and to vary the detector polarizing voltage and the same arguments apply. On occasion both the voltage and the discriminator setting are varied to provide the best signal-to-noise ratio, especially when the ratio is relatively low (for instance, because of use of long cables between the detector and amplifier system).

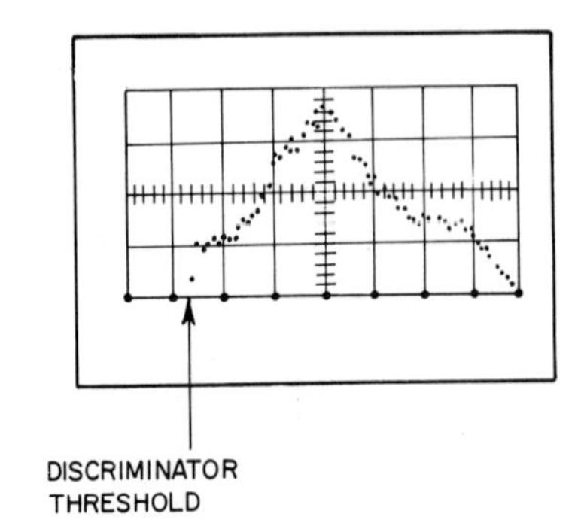

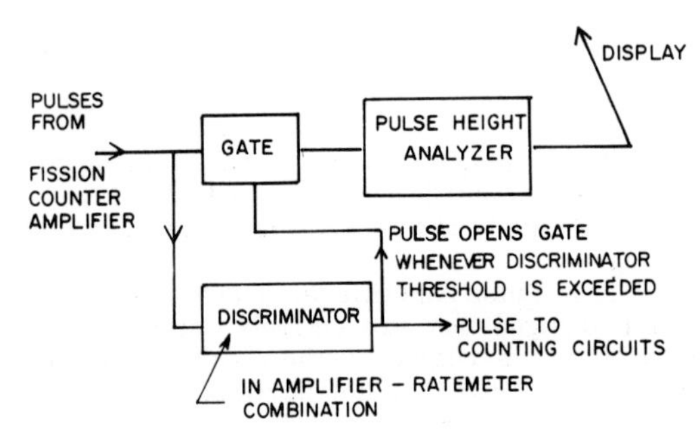

FIG. 2-36 Pulse height analyzer used to determine discriminator threshold. (Upper part of figure is tracing from oscilloscope screen.)

2.5.2 Linear Direct-Current Amplifiers

The current from an ionization chamber is, in general, determined by measuring the voltage it produces across a known resistance. Figure 2-37 is the basic circuit. The value R of the measuring resistance is determined by the noise or stability inherent in the amplifier. If the amplifier input is connected to ground, it is found that even under this condition, with no input signal, the output fluctuates. This fluctuation might be equivalent to an actual input signal of, say, 10 mv and, if we require that this be 0.1% of full scale, the maximum input voltage must be 10 v. And, to measure currents as low as 10^{-12} amp the resistance R would be 10^{13} ohms. In this arrangement the amplifier is simply an impedance converter permitting the voltage developed across a high resistance to be measured across the low output impedance of the amplifier.

The dynamic response of the input circuit depends on the time constant $R(C + C_c)$ which might be as high as 10^4 sec with reasonable lengths of connecting cable. This is intolerably long and, as a solution to the problem, the principle shown in Fig. 2-38 is used almost universally. The amplifier detects the difference between the voltage developed across the resistance R and ground and V is changed to maintain this difference as small as possible. Often V is obtained from a potentiometer driven by a recorder on the amplifier output, but more generally the circuit shown in Fig. 2-39 is used. The feedback forces the input voltage e_i to remain close to ground potential. The advantage of these circuits is that the capacity C is charged at most to a very low voltage and the input time constant is greatly reduced. (It becomes RC/amplifier-gain).

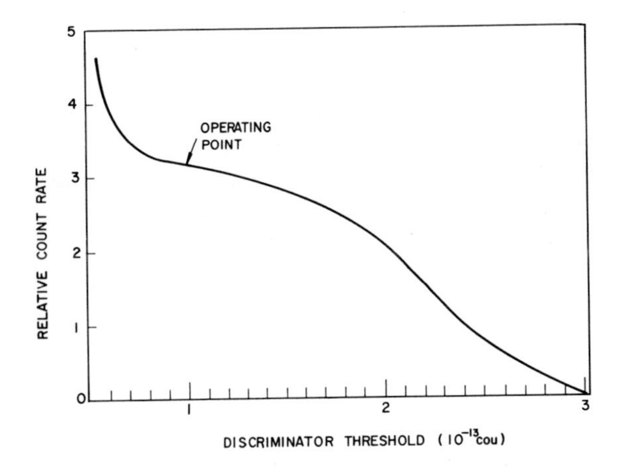

FIG. 2-35 Fission counter: integral bias characteristic.

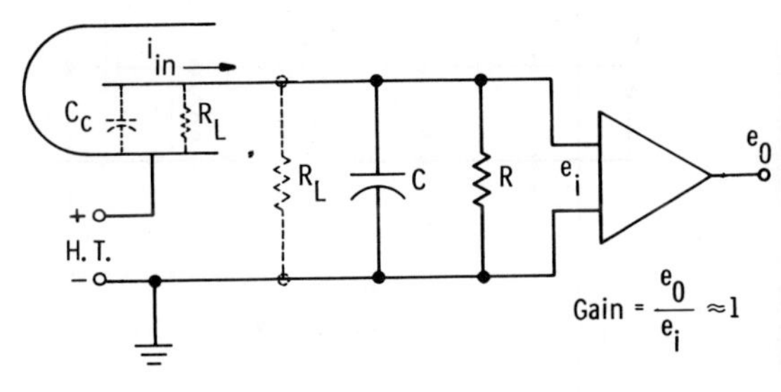

FIG. 2-37 Ionization chamber current measurement. C_c is the chamber capacitance, C the circuit and cable capacitance, R_L the leakage path resistances, R the measuring resistance, i_{in} the ion-chamber current, and e_i the voltage developed by i_{in} across parallel resistance.

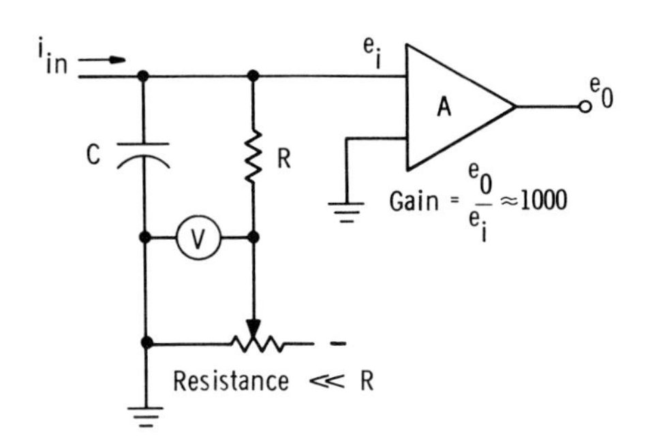

FIG. 2-38 Voltage difference measurement.

The feedback arrangement does not change the argument that requires a high resistance when small currents are to be measured. If an amplifier has a drift equivalent to 10 mv at its input, it will, upon the application of feedback, have an output drift of at least 10 mv. The drift or noise will in fact be greater than this because of the shunting effect of the capacitance C on the feedback signal. This attenuation occurs at frequencies higher than $f = 1/2\pi RC$ and, when R is large, this frequency may be as low as 10^{-4} cycles/sec. As a result the output noise is increased and, to reduce it to reasonable proportions, a capacitance is connected across R. This in turn increases the amplifier response time and the inevitable compromise between noise reduction and dynamic response must be reached. To keep the capacitance as low as possible even with the feedback connection, the first stage in the amplifier is often placed as closely as possible to the ionization chamber.

There are two distinct applications for direct-current linear amplifiers. One is for measuring flux over a wide range and the other specifically for measurements in the range 1.0 to 100% of maximum reactor power.

To measure flux over a wide range the amplifier must, first, be able to detect currents in the 10^{-12}- amp region, which means that the input stage of the amplifier in Fig. 2-39 must be sensitive to currents much less than this, and second, either the noise and drift must be small so that an output of 10^{-5} of full scale is meaningful or the measuring resistance must be increased to maintain the output significantly high as the input signal decreases. Up to nine decades may be covered in this manner.

When a multi-range amplifier is used during a reactor startup, the range switch is usually advanced by the operator as the output approaches either full scale or an alarm or trip level near full scale, and it is an important design requirement that no transients be caused by the switching process. It would be counter to basic safety practice if the amplifier alarm circuit had to be incapacitated each time the range was changed.

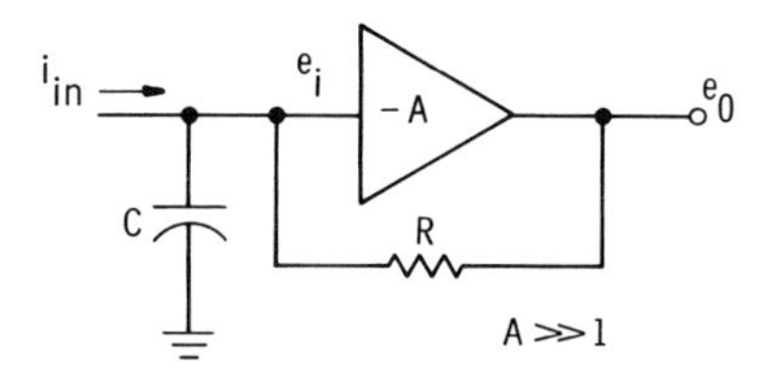

FIG. 2-39 Feedback amplifier.

The requirements imposed on an amplifier which measures ionization chamber currents, corresponding to the thermal power range of the reactor, are less severe than those just discussed, and advantage of this is taken to produce simpler and more reliable circuits.

Several amplifier circuits that meet the various requirements are described below.

2.5.2.1 The Electrometer Tube. The electrometer tube is specially designed to permit the measurement of currents in the 10^{-12}-amp region. An amplifier using an electrometer tube at its input will exhibit a drift equivalent to about 10 mv at the input and does not meet the wide range requirement without range switching. Figure 2-40 shows the electrometer tube in a housing that can be located remote from the remainder of the amplifier. Range switching is achieved by magnetically operated reed switches sealed in glass.

The thermally operated shorting switch that connects the grid to ground is used to measure the offset, due to drift, that is present at any time. The zero control permits the offset to be removed. This test should only be made when an input signal is present so that assurance can be given that the short has been removed.

The low-current measuring ability of the electrometer tube circuit cannot be achieved unless precautions are taken to minimize leakage currents in all parts of the input circuit (the leakage paths in Fig. 2-37, especially from the polarizing electrode to the signal electrode). Spurious voltage produced by strain must also be avoided and light must be excluded from the housing. Moisture or humidity changes may affect the reliability of the installation by changing the leakage currents.

A design difficulty arises when the electrometer tube is remotely located. The capacitance from its anode to ground affects the closed-loop stability of the amplifier and so the distance between the remote housing and the main unit must appear as part of the amplifier specification.

Because part of the drift arises from temperature and voltage supply fluctuation, circuits have been devised using an additional electrometer tube as a balancing element [57].

2.5.2.2 The Chopper Amplifier. Figure 2-41 shows the basic technique for converting dc to ac before amplification. The switch SW_1 (the chopper) periodically allows the current to flow as a pulse into the amplifier and SW_2 operating in synchronism converts the amplified pulse back to dc. Because of the switch characteristics this amplifier can have a drift referred to the input of 50 μv or so but the current sensitivity will be determined by the magnitude of the amplifier current i_A. It must be small compared with the current through the measuring resistance. Typically, with a reasonable a-c amplifier gain and input impedance, currents down to 10^{-9} or 10^{-10} amp are measurable.

The chopper amplifier is used more often as a stabilizing amplifier than as a d-c amplifier. It can be used as in Fig. 2-42 to detect that the input of an amplifier is not zero and to make a correction

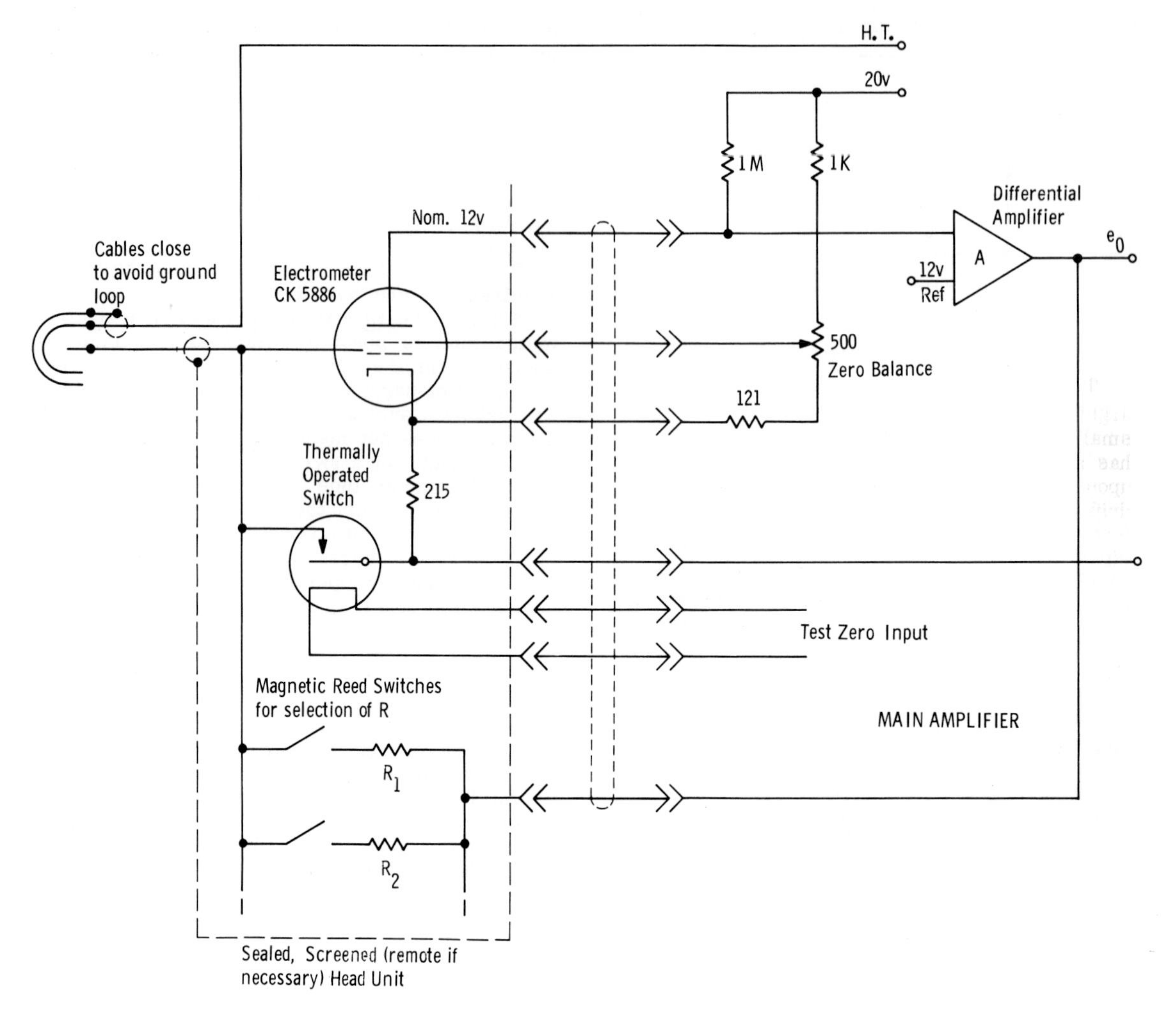

FIG. 2-40 Electrometer-tube amplifier.

independently of the normal feedback in a manner analogous to the manual operation of shorting the input to ground and bringing the output to zero.

Chopper amplifiers have also been used as error-sensing devices in other ways. Figure 2-43 shows a chopper amplifier used to measure the voltage across a current-measuring resistance. A high impedance a-c amplifier is used to drive a standard recording potentiometer motor. The motor drives until the chopper output is zero, i.e.

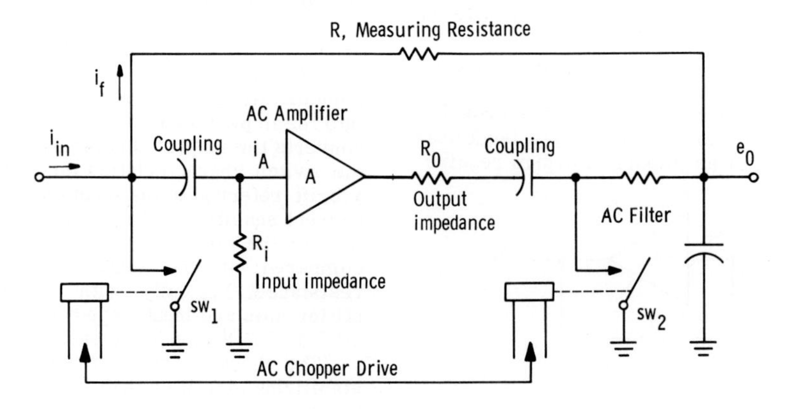

FIG. 2-41 Chopper amplifier.

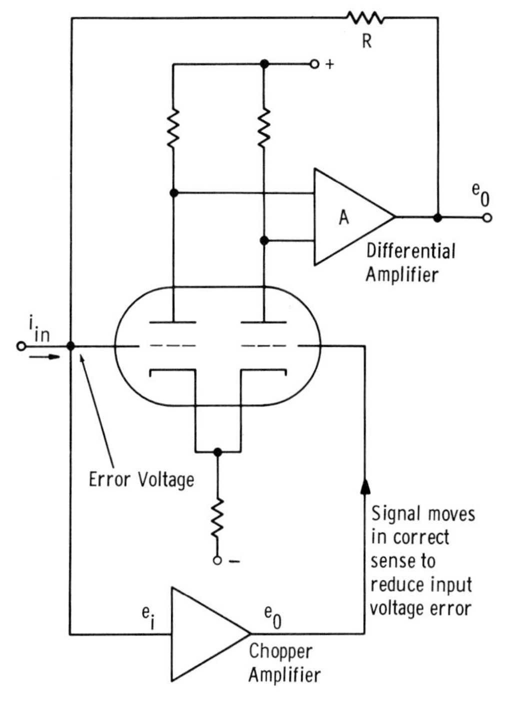

FIG. 2-42 Chopper amplifier stabilization.

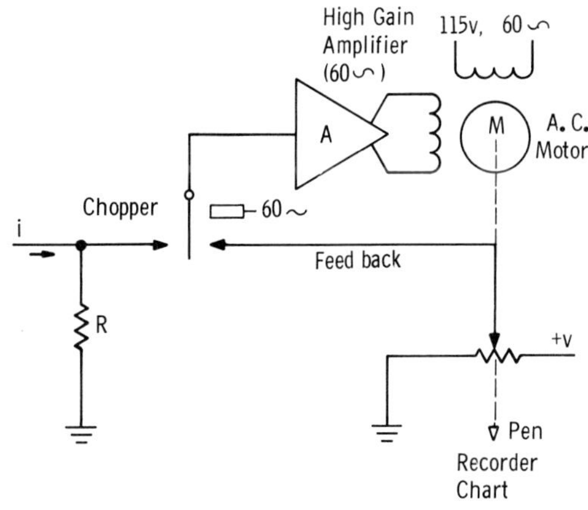

FIG. 2-43 Chopper amplifier with motor output.

when both its fixed contacts are at the same potential. Currents down to 10^{-10} amp are measurable.

The circuit of Fig. 2-44 [59c] shows a similar arrangement for a safety amplifier. A reference voltage that sets the trip level replaces the motor-driven potentiometer voltage. The output from the phase-sensitive detector changes sign when the input signal potential across R exceeds the reference level and forces the polarized relay open. Chopper-type amplifiers do not fail safe without the use of monitoring circuits.

The use of mechanical choppers in safety circuits has not been too common because of their relatively short life (~a few thousand hours)*.

2.5.2.3. Solid-State Capacity Modulator [60]. The trend to use semiconductor elements because of their higher reliability has led to amplifiers

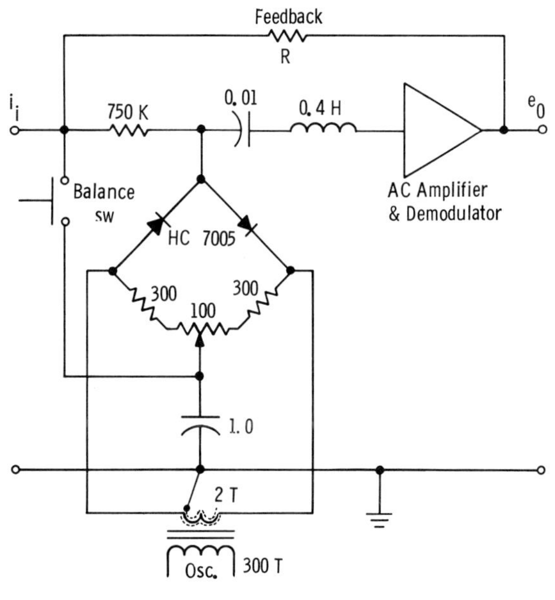

FIG. 2-45 Solid-state capacity modulator.

*The vibrating reed electrometer does not, however, have this limitation.

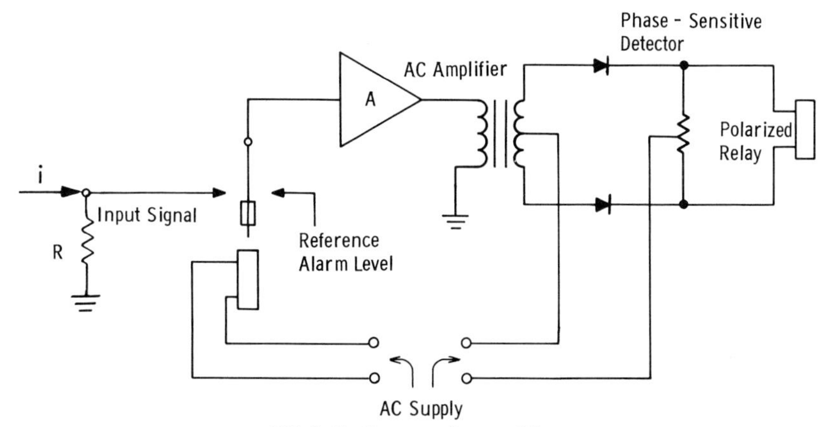

FIG. 2-44 Chopper safety amplifier.

using semiconductor diodes as switches. Figure 2-45 [60] is a form of modulator using the fact that the capacity of a semiconductor diode is a function of the voltage across it. The bridge is unbalanced by the d-c error changing the capacity of the two diodes. Various arrangements for coupling into and out of the bridge have been developed, some circuits drive the bridge through a transformer and other through capacitors. The carrier frequency varies from 10 to 500 kc/sec. There is some evidence that parametric amplification takes place because of the nonlinear voltage-capacity diode characteristic. Signal currents as low as 10^{-12} amp are measurable with noise levels of about 10%.

2.5.2.4 Diode Ring Modulator [61, 62]. A second type of modulator that depends on switching the signal current through semiconductor diodes is shown in Fig. 2-46. Two diodes conduct in either half-cycle of the reference signal; the current in these diodes is defined by the drive transformer and must always exceed the signal current. Leakage currents in the diodes limit the sensitivity to about 10^{-8} amp and although temperature regulation may lower this figure, switching transients and harmonic distortion at null are overriding factors.

2.5.2.5 DC-Coupled Transistor Amplifier [63, 64]. When transistors are used as d-c amplifiers, balanced circuitry is required to reduce the effects of temperature on junction voltage and on leakage currents. A typical circuit is shown in Fig. 2-47. Currents as low as 10^{-5} amp (and voltages down to 10 mv) can be measured before noise and drift (with temperature) are troublesome.

2.5.2.6 Magnetic Amplifiers [65, 66, 67a]. A magnetic amplifier circuit having sufficient stability for measuring ionization chamber currents is shown in Fig. 2-48. The output drift is equivalent to about 1.0 μa at the input. This drift is caused by variations in the voltage and frequency of the

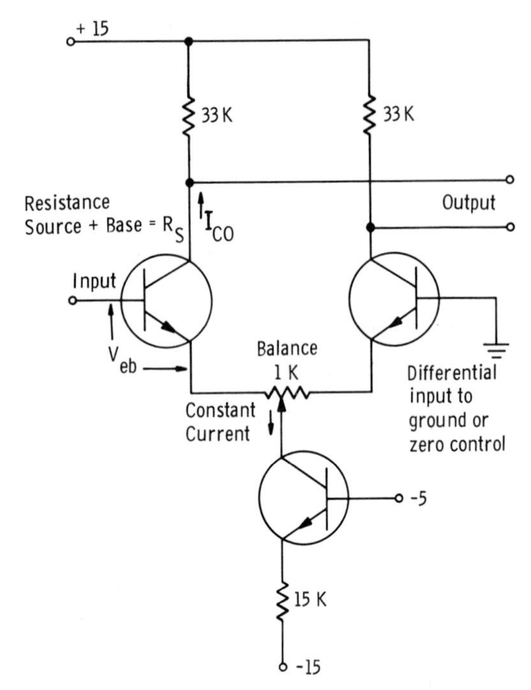

FIG. 2-47 DC-coupled transistor amplifier. In the above circuit drift is mainly due to temperature: $\Delta V_{eb} \cong -2.5$ mv/°C, ΔG (change in base to collector current gain) $\sim 1\%$/°C at $I_c = 1$ ma, and I_{co} increases by a factor of 2 per 8°C temperature change. (Typical values of I_{co} at 25°C are 1 μa for Ge and 10^{-3} μa for Si.)

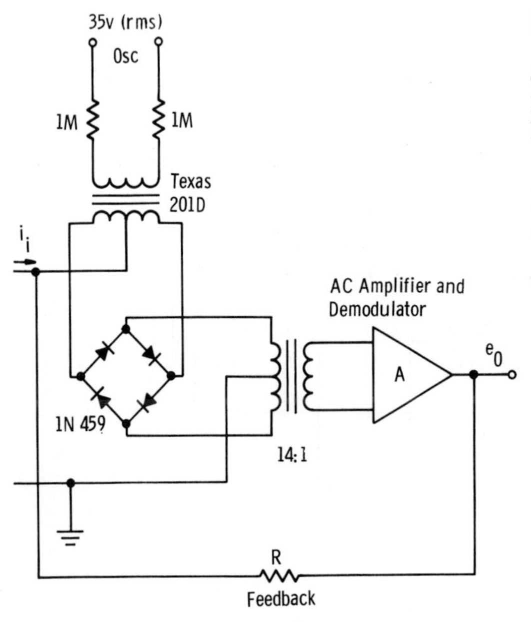

FIG. 2-46 Diode ring modulator.

power source, and by changes in the rectifier and core characteristics (with temperature). Push-pull arrangements are generally used, the output containing odd harmonics only. The second harmonic current flowing in the input winding can be sensed as an output signal (it changes in phase as the input changes polarity). Even harmonics are only generated when the coils are asymmetrically excited, the drift being reduced to about 0.1 μa.

The advantages of a magnetic amplifier are its simple circuitry and ruggedness but its output contains switching transients that cause difficulty if used to feed other circuits.

The response speed is related to the operating frequency (up to 5 kc/sec) and to the driving circuit resistance.

Magnetic amplifiers can be used to amplify directly the current from ionization chambers. They can be used as "operational" amplifiers in reactor regulating systems. More often they are used as two-state alarm circuits to operate relays or to otherwise initiate safety action.

2.5.3 Circuits with Logarithmic Response

A logarithmic presentation of power level is advantageous in that about six decades of power can be measured without range switching. It also avoids the safety worry of not having the amplifier on the range most appropriate to the existing power level.

The usefulness of the quantity,

$$\frac{1}{n}\frac{dn}{dt} = \frac{d}{dt}(\log_e n) , \qquad (2\text{-}4)$$

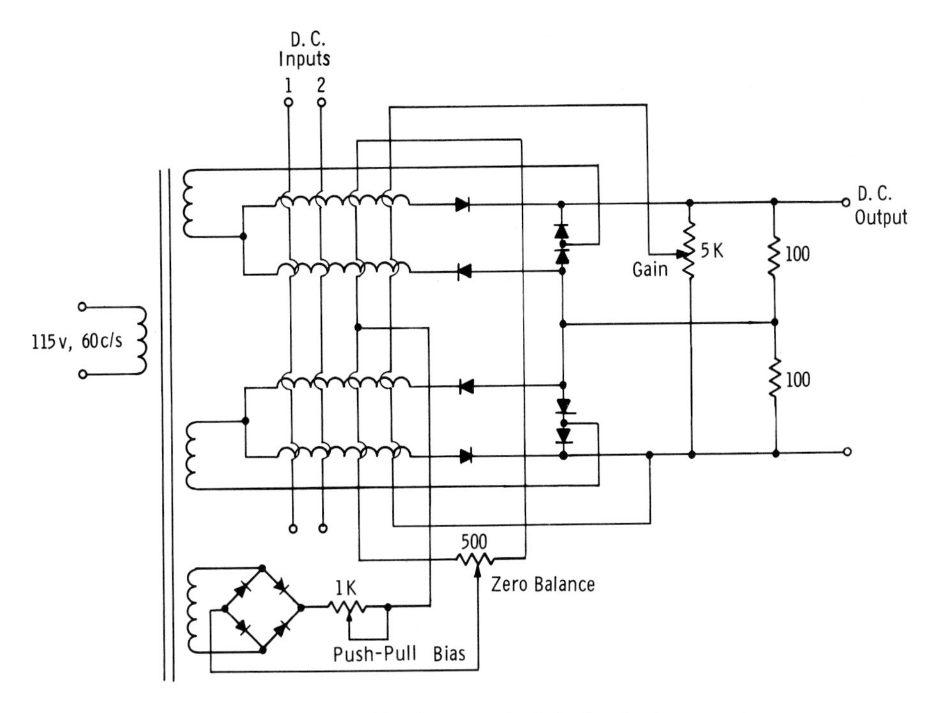

FIG. 2-48 Magnetic amplifier for measuring ionization chamber currents.

has been discussed in Sec. 2.1. This can be readily obtained from the logarithm of the flux signal.

2.5.3.1 Logarithmic Count-Rate Circuits. Two circuits for producing the logarithm of a count rate have found widespread use in reactor instrumentation: the diode pump circuit and the logarithmic element circuit.

The Diode Pump. The diode pump circuit is well described in the literature [68]. Pulses of fixed amplitude are formed by a bistable (B/S) as shown in Fig. 2-49. Considering only the first pump circuit, the action of the diodes is to "pump" a charge equal to $(V-v)C_1$ coulombs into the tank-circuit capacitor for each square pulse (i.e. for every two input pulses) so that the voltage across the tank circuit is

$$v = \frac{V(n/2)T_1}{1 + (n/2)T_1}, \tag{2-5}$$

where V is the input voltage, v the output voltage, n the pulses/sec and $T_1 = R_1 C_1$ (in sec). The summing of the outputs of seven circuits, with time constants as indicated, results in the response of Fig. 2-50.

It should be noted that the components contributing to the upper decades play no part in the response at the lower end of the range. Hence a failure of a component could go unnoticed unless the test routine specifically checks the calibration to full scale.

The Logarithmic Element Circuit [69]. A single diode pump circuit may be used to "pump" a constant charge per pulse into a logarithmic element as shown in Fig. 2-51. Circuit arrangements [68] ensure that the voltage v stays essentially zero.

The current flow into the logarithmic element is

$$i = (n/2)VC. \tag{2-6}$$

Details of the logarithmic circuit are described in the sections to follow.

2.5.3.2 Direct-Current Logarithmic Circuits. Almost without exception logarithmic amplifiers for reactor control derive their logarithmic characteristic from the properties of either a thermionic diode or a semiconductor diode.

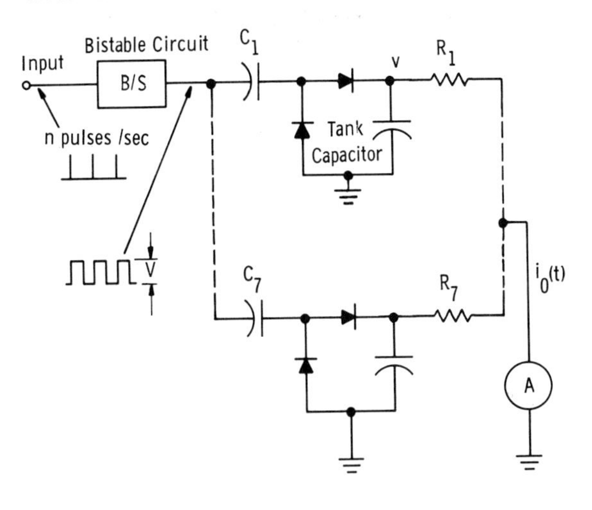

FIG. 2-49 Multiple diode-pump ratemeter.

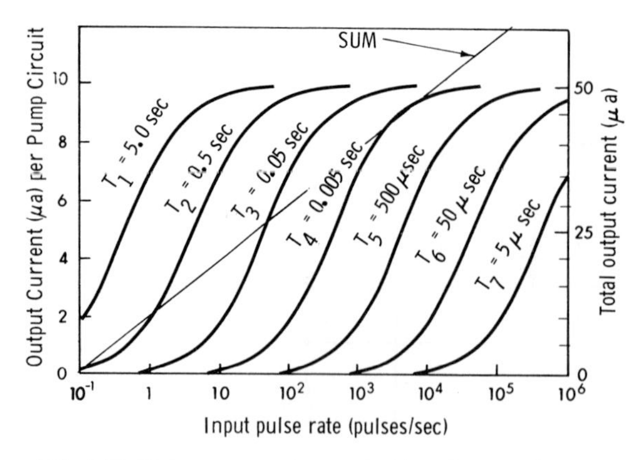

FIG. 2-50 Diode-pump output as a function of input-pulse frequency.

The Thermionic Diode. If a thermionic diode with plane parallel electrodes is operated in the retarding field region, the anode voltage V is related to the diode current i by the relation

$$i = i_0 \exp\left[\frac{e}{KT}(V - v_0)\right], \qquad (2\text{-}7)$$

where $(e/K) = 8.62 \times 10^{-5}$ volts/degree Kelvin, T is the absolute temperature of the cathode, i_0 the saturation current, and v_0 the anode-to-cathode contact potential.

Figure 2-52 shows typical diode characteristics with the currents shown as a function of anode voltage at different temperatures. The constants i_0 and v_0 are indicated [69a].

In all reactor control applications the independent variable is the diode current and the anode voltage is the required signal. From Eq. (2-7) we have,

$$V = v_0 - (KT/e)\log_e(i/i_0), \qquad (2\text{-}8)$$

or, $\Delta V = 86$ mv/neper change in i or 200 mv/decade change in i when T = 1000°K.

The incremental impedance of the diode is of interest because of the part it plays in determining response time. It is

$$\frac{dV}{di} = R_{incr} = \frac{0.086}{i} \sim \frac{1}{11i} \text{ ohms}. \qquad (2\text{-}9)$$

These relations apply to many diodes (and triodes connected as diodes) with cylindrical geometry. The type 9004 diode characteristic at different heater voltage (cathode temperature) is shown in Fig. 2-53. The displacement of the characteristic is due essentially to the temperature sensitivity of i_0. Only a small change in the slope of the characteristic takes place.

The Semiconductor Diode. The forward conduction of a semiconductor junction diode follows the relation

$$i = i_s\left[\exp\frac{eV}{nKT} - 1\right], \qquad (2\text{-}10)$$

where i is the diode current, V the diode voltage, i_s the reverse saturation current, $K/e = 8.62 \times 10^{-5}$ volts/degree Kelvin, and $n \simeq 1$ to 1.4, depending on the diode [30e]. From Eq. (2-10) it follows that

$$V = \frac{nKT}{e}\log_e\left(\frac{i + i_s}{i_s}\right), \qquad (2\text{-}11)$$

and $\Delta V \simeq 30$ mv/neper change in i or 70 mv/neper change in i when T = 300°K.

The incremental impedance of the semiconductor diode is less than the thermionic diode and at 300°K is

$$\frac{dV}{di} = R_{incr} = \frac{0.026}{i} = \frac{1}{38i} \text{ ohms}, \qquad (2\text{-}12)$$

where it is assumed that $i_s << i$. Both the voltage drop across the diode and the reverse saturation current are temperature-dependent. The voltage drop across a silicon junction near room tempera-

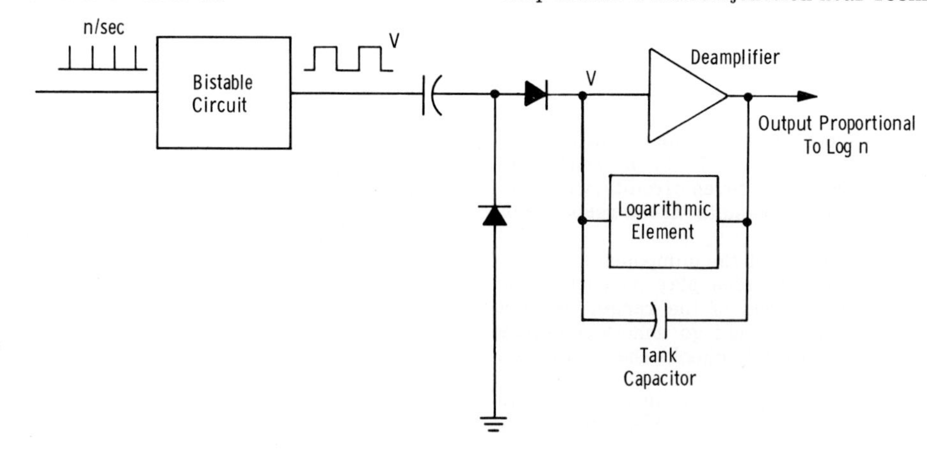

FIG. 2-51 Single diode-pump and logarithmic circuit.

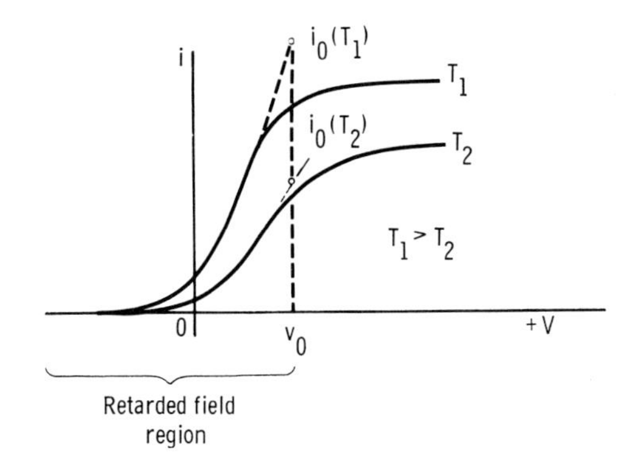

FIG. 2-52 Characteristics of thermionic diode in retarded field region.

ture changes about -2.5 mv/°C, and the reverse saturation current (i_S) doubles for every 7°C (12.6°F) increase in temperature.

These effects are illustrated in Fig. 2-54 which shows the characteristics of a Type FD 300 silicon diode at different temperatures. The characteristic becomes nonlogarithmic at higher currents due to the ohmic resistance of the diode material.

Diode Logarithmic Amplifiers. [57, 70, 71] The diode element may be used in the same manner as the measuring resistance in a linear amplifier. As shown in Fig. 2-55 the element may be either in the feedback path or directly across the amplifier input, the same reasoning applying as in Sec. 2.5.2. The drift problem is severe since the total signal change in covering six decades of input current is only of the order of a volt. Hence the temperature effects discussed in Sec. 5.3.2 represent a large change in terms of the measured neutron flux.

To avoid this difficulty some means of temperature compensation is required and it is generally based on using a similar element in a compensating fashion. Figure 2-56 shows a basic circuit for compensating the drift due to heater voltage in a thermionic diode and Fig. 2-57 shows a compensation circuit for both voltage and current drifts in a silicon junction diode.

Compensation schemes must be examined cri-

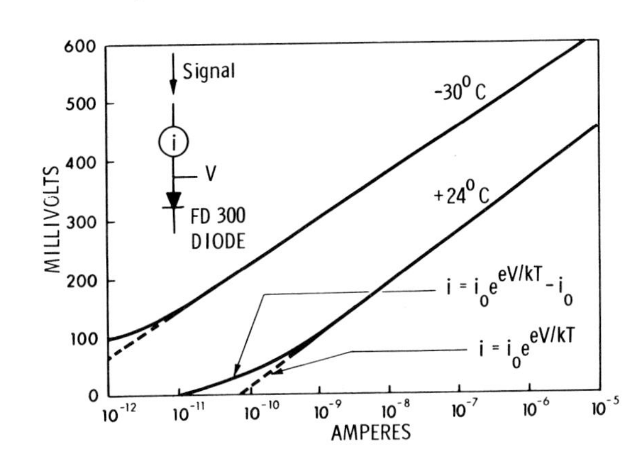

FIG. 2-54 Temperature behavior of semiconductor diode. The logarithmic characteristic extends to $\sim 10^{-2}$ amp.

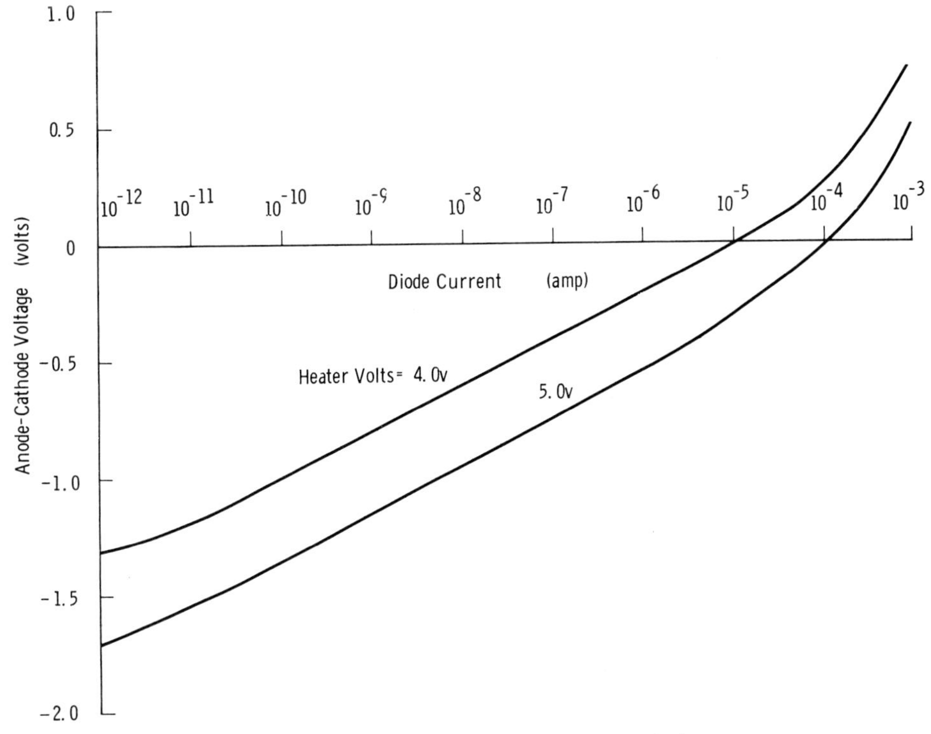

FIG. 2-53 Temperature dependence of thermionic diode.

tically since failure or misadjustment can cause a substantial error in a flux measurement.

The Pentode Logarithmic Circuit [72]. A circuit that appreciably reduces the drift of a diode is shown in Fig. 2-58. The grid-cathode space of a pentode is used as a diode and the signal current flows as positive grid current. The grid voltage (which is proportional to the logarithm of the signal current) is amplified by the pentode. To maintain linearity the plate current is held constant by a feedback arrangement that adjusts the screen voltage. The change in screen voltage is about 3.0 v/decade change in input current. The incremental change in grid voltage is the same as for the thermionic diode, i.e., 0.2 v/decade. Quantitatively, drift reduction can be seen to occur by considering the effect of an increase in cathode temperature. Referring to Figs. 2-56 and 2-53 and assuming that the contact potential remains constant, it is evident that the grid-cathode voltage must become more negative to maintain the constant grid current condition imposed by the ionization chamber. This increasing negative bias tends to stabilize the anode current against an increase in cathode temperature.

In practice the fractional drift (i.e., decades/degree) observed at the screen grid is about 1/15 of that at the anode of a diode. It is, however, general practice to apply some stabilization to the cathode heater current.

A disadvantage of this circuit is that feedback to reduce the input time constant cannot be applied.

Reference [73] discusses several tubes that have been found suitable for this application.

Logarithmic Circuits Using Servo System Techniques. Logarithmic amplifiers have been described

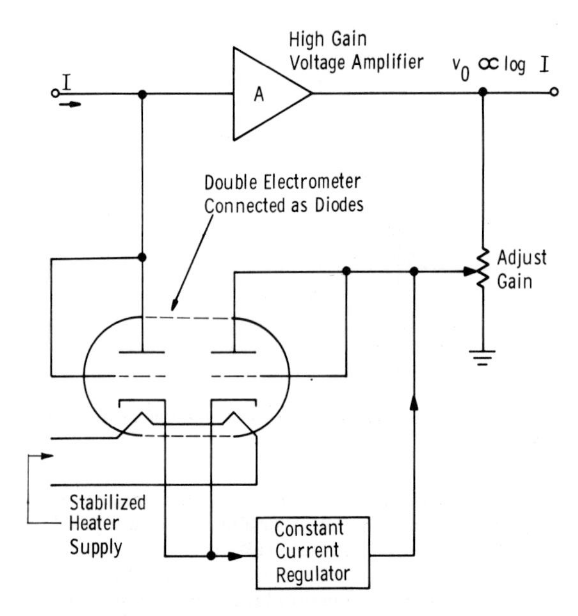

FIG. 2-56 Drift-compensated thermionic diode.

that use servo system techniques either to provide the logarithmic response or for feedback purposes.

A four-decade exponential potentiometer is driven by a servo motor, in one design [74], to a null position, the position being proportional to the logarithm of the input signal. A signal proportional to the rate of change of the logarithm of the input is obtained from a tachometer on the motor shaft.

Another circuit [75] uses a pentode as described

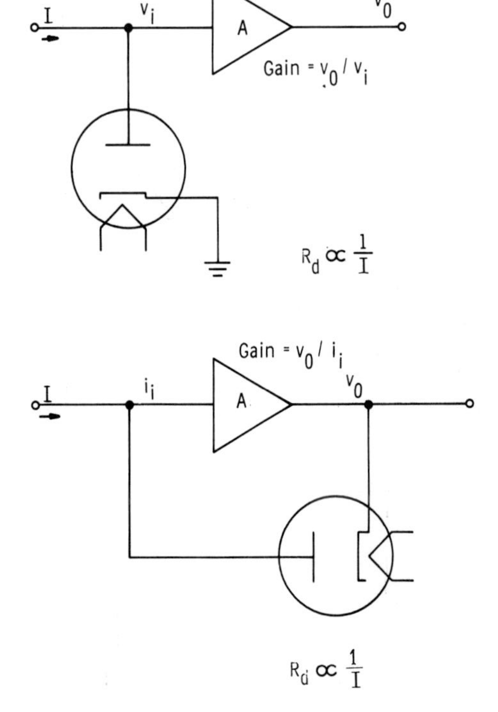

FIG. 2-55 Diode logarithmic amplifiers.

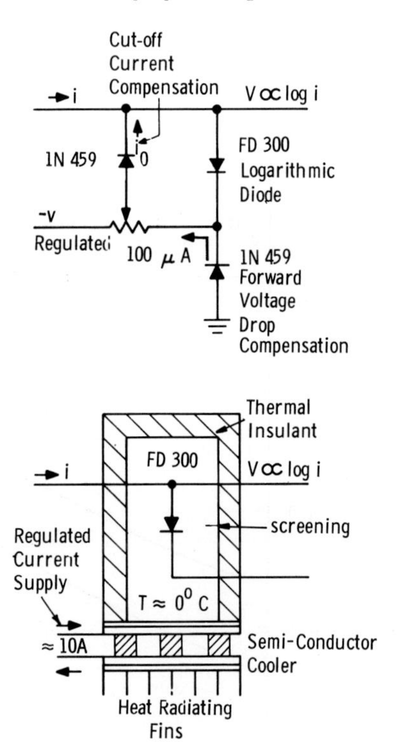

FIG. 2-57 Drift-compensated semiconductor diode.

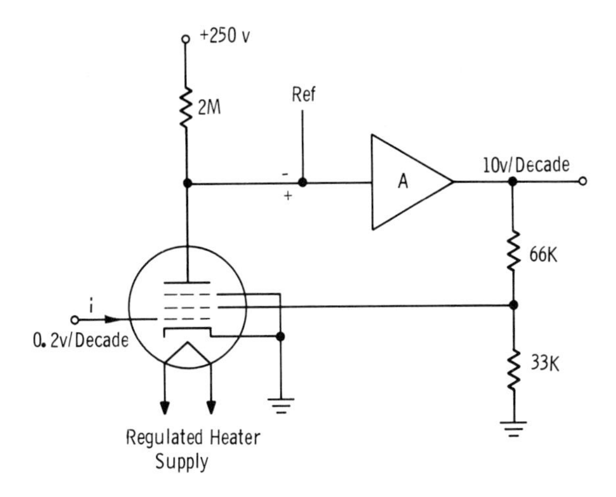

FIG. 2-58 Logarithmic pentode circuit.

previously but with the feedback to the screen grid by means of a servo-driven potentiometer. A tachometer on the motor shaft provides rate-of-change signals.

Neither circuit is used extensively in reactor safety systems.

Mixed Mode Amplifiers. To reduce the number of ion chambers required for flux monitoring it is possible to generate both a logarithmic signal and a linear signal from the same input current.

A basic circuit is shown in Fig. 2-59. The magnetic amplifier has a low impedance and does not affect the logarithmic response of the circuit. It provides linear amplification of the ion chamber current.

There are safety aspects associated with this type of circuit. The presence of the logarithmic signal provides assurance that the ion chamber is functioning before the linear range is reached; however, the two signals cannot be treated as independent safety parameters.

2.5.4 Rate-of-Change Circuits

The rate of change of neutron flux is an important safety and regulation parameter but limits are set on its usefulness by response time and noise con-

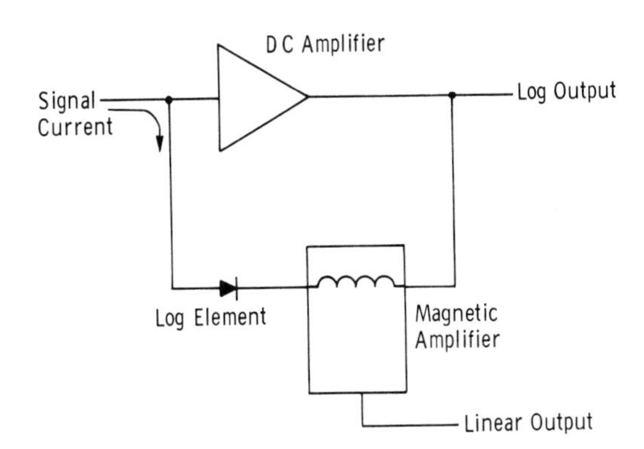

FIG. 2-59 Mixed mode amplifier.

siderations. This section discusses these aspects.

Rate-of-change signals are obtained directly by taking the time derivative of the output of either a logarithmic or a linear flux-sensing amplifier.

2.5.4.1 Differentiating Circuits The differentiating circuit most commonly used to obtain rate-of-change signals evolves as shown in Figs. 2-60(a) to (e) from a simple capacitance-resistance network. When a linearly rising voltage is applied to the input the voltage at the output approaches an equilibrium value dependent only on the rate-of-rise of the input signal and on the differentiating time constant RC.

The time for circuit (a) to settle down is improved by reducing R and by increasing the amplifier gain as indicated in circuit (b) but drift in the amplifier output increases directly with the response time reduction.

The arrangement in (c) not only reduces response time but also achieves low drift (compare with feedback in linear amplifiers, Sec. 2.5.3).

In Fig. 2-60 (d) the capacitor C_1 is added to make the response time independent of the amplifier gain. Typically RC_1 is about ten times RC/G.

Finally, the addition of the resistor R_1 in Fig. 2-60 (e) improves the noise performance of the period channel as explained in Sec. 2.5.4.3.

2.5.4.2 Logarithmic Rate or Period Channel The rate of change of the logarithmic amplifier is useful because it readily yields information about the reactor condition over a wide range of power levels and because it can be related to some simple reactor kinetics notions.

Differentiation of the logarithmic signal yields

$$\frac{d}{dt}(\log_e i) = \frac{di/i}{dt} = \frac{dn/n}{dt}, \qquad (2\text{-}13)$$

where i is the neutron sensor current and is proportional to the flux n.

This quantity is the rate of rise of the incremental flux and is the slope of the logarithmic trajectory at any instant of time as indicated in Fig. 2-61. Since this slope can be defined in many ways a variety of calibrations for rate-of-change meters has become popular.

Reactor period is perhaps the most common designation for the slope since the equation defining a straight line on Fig. 2-61* is identical with the flux excursion of a reactor having a constant excess reactivity.

However, because

$$\frac{d}{dt}(\log_e n) = \frac{1}{\tau}, \qquad (2\text{-}14)$$

an inverse scale results if meters are calibrated in reactor period. If more obvious units are con-

*A straight line on Fig. 2-61 is $n = n_0 \exp(t/\tau)$, where τ is the time constant or period associated with the exponential.

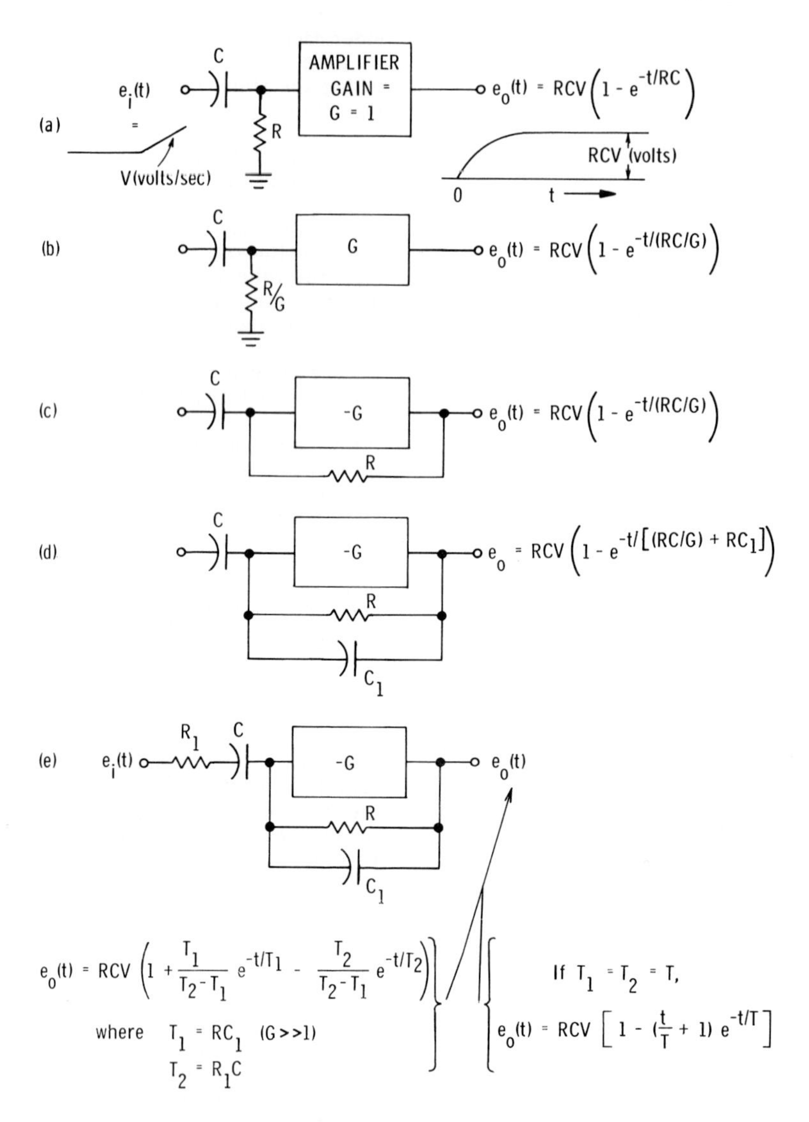

FIG. 2-60 Evolution of differentiator circuit.

sidered for the slope measure, i.e., nepers per second, or bels (decades) per second, or per cent per second, a linear scale results. Table 2-3 relates these quantities to each other.

The quantities "decading time" and "octaves per second" (inverse doubling time) do not appear to be used.

There seems to be no fundamental argument why one method of calibration should be preferred over another, although the linear scales do show increasing values as the situation becomes more unsafe. Further, it is felt that such phrases as "time to double" should be avoided since they tend to obscure the essentially dynamic meaning of the rate-of-change measurement.

To discuss the properties of the rate-of-change channel that are associated with reactor safety, it is convenient to use the functional schematic shown in Fig. 2-62. The specific function carried out by each of the components in Fig. 2-60 (e) is emphasized. Also shown are the three basic logarithmic circuits which have already been discussed (Sec. 2.5.3); any one of these may be connected to the rate-of-change circuitry.

2.5.4.3 Noise in the Logarithmic Rate Channel
Noise in logarithmic rate channels has been the subject of extensive study [76, 77, 78, 79]. This is inherent noise and is due to the randomness of occurrence of the ionizing events in the sensor. Other forms of noise also appear: noise due to reactivity fluctuations caused by control rod vibration, by moderator turbulence, or by coolant boiling. Still another source of noise is in the randomness of the fission process itself which under special circumstances is measurable and useful.

To outline the effects of inherent neutronic noise a few concepts taken from general noise theory [80] will be used.

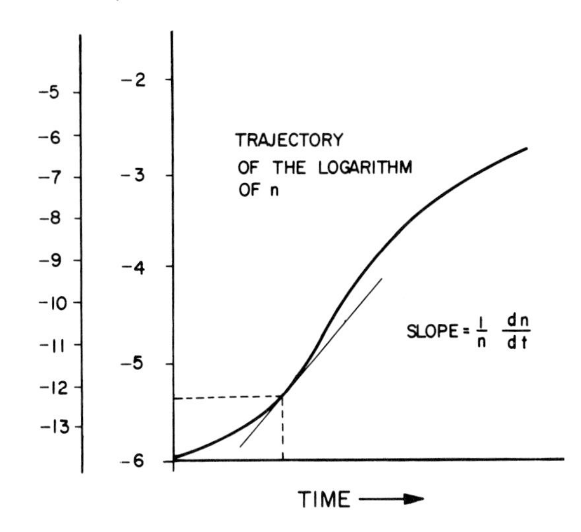

FIG. 2-61 Definition of rate of change of the logarithm of flux.

The mean square noise $\overline{i^2}$ in a bandwidth Δf, associated with a current I made up of randomly arriving increments of charge Q, is

$$\frac{\overline{i^2}}{\Delta f} = 2QI \text{ (amp}^2\text{/cycle of bandwidth)} . \quad (2\text{-}15)$$

The quantity 2QI is called the "spectral density" of the noise and may in some cases be frequency-dependent.

If a noise current with a spectral density ϕ is passed through a network having a transfer function TF, the spectral density at the network output is

$$\phi_0 = \phi\, |TF|^2 , \quad (2\text{-}16)$$

and the mean square noise at the network output is

$$\overline{i_0^2} = \int_0^\infty \phi\, |TF|^2 df . \quad (2\text{-}17)$$

The differentiator in Fig. 2-60 (e) has the transfer function

$$\frac{e_0}{e_i} = j\omega T_{Diff} \cdot \frac{1}{1 + j\omega T_1} \cdot \frac{1}{1 + j\omega T_2} , \quad (2\text{-}18)$$

or

$$\left|\frac{e_0}{e_i}\right| = \omega T_{Diff} \left\{\frac{1}{[1 + (\omega T_1)^2]\,[1 + (\omega T_2)^2]}\right\}^{\frac{1}{2}} , \quad (2\text{-}19)$$

where T_1, T_2 are the time constants

Equation (2-19) is shown graphically in Fig. 2-63 plotted with ωT_{Diff} as the independent variable. Notice that two integrating time constants (T_1 and T_2) are required to produce a finite bandwidth. If T_2 is not added deliberately as in Fig. 2-60 (e), a similar effect will appear, usually at a higher frequency, due to fundamental limitations in the bandwidth of the amplifier. The time constants T_1 and T_2 control bandwidth and hence the amount of noise appearing at the differentiator output in different ways: one by limiting the gain and the other by limiting the high frequency response.

The noise actually applied to the differentiator depends on the type of logarithmic circuit used.

The Multiple Diode Pump Circuit. Consider the first section of the diode pump circuit in Fig. 2-62. Randomly arriving input pulses are applied to the bistable circuit at the rate of n/sec. Hence charge is fed into the tank circuit ($r_1 C_{i-1}$) at a rate

$$\frac{Q_i}{\text{sec}} = \frac{n}{2}(V - v_1) C_1 = i_1 . \quad (2\text{-}20)$$

Assuming that the fluctuations in this current are small enough to permit a linearized approach, the spectral density of the current flowing into the tank circuit is, from Eq. (2-20),

$$2Q_1 i_1 = 2\left\{\frac{nC_1^2V^2/2}{[1 + (n\tau_1/2)]^2}\right\} , \quad (2\text{-}21)$$

and the spectral density of the current flowing into the output amplifier is

$$\phi_1 = \frac{nC_1^2V^2}{[1 + (n\tau_1/2)]^2} \cdot \frac{1}{1 + (\omega T_{i-1})^2} . \quad (2\text{-}22)$$

The mean square noise at the logarithmic rate meter output is from Eqs. (2-17) and (2-19)

$$\overline{e_0} = \int_0^\infty \phi R_2^3 \frac{(\omega T_{Diff})^2}{[1 + (\omega T_1)^2]\,[1 + (\omega m T_1)^2]} df , \quad (2\text{-}23)$$

TABLE 2-3

Relationships Among Period Meter Calibrations

Period	τ(sec)
Doubling time	$0.69\,\tau$
Nepers/sec	$\frac{1}{\tau} = a$
%/sec	$\frac{100}{\tau}$
Decades/min	$\frac{26}{\tau}$
Decades/sec	$\frac{0.43}{\tau}$

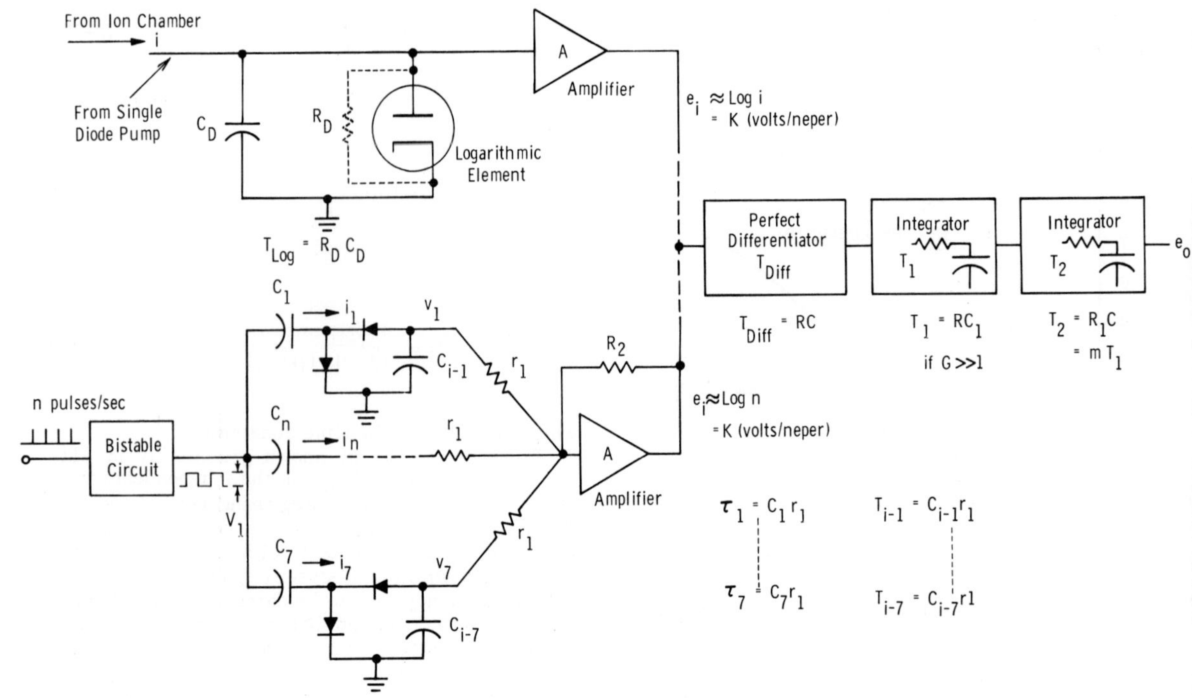

FIG. 2-62 Complete logarithmic rate channels.

where $m = T_2/T_1$.

If the rate meter circuit comprises diode pump circuits having their respective τ_i separated by a decade, the logarithmic constant at the summing amplifier output will be

$$K = \frac{VR_2}{2.30 r_1} \quad \text{(volts/neper)} , \qquad (2\text{-}24)$$

and the contribution of the first section to the fluctuations in the rate-of-change amplifier output,

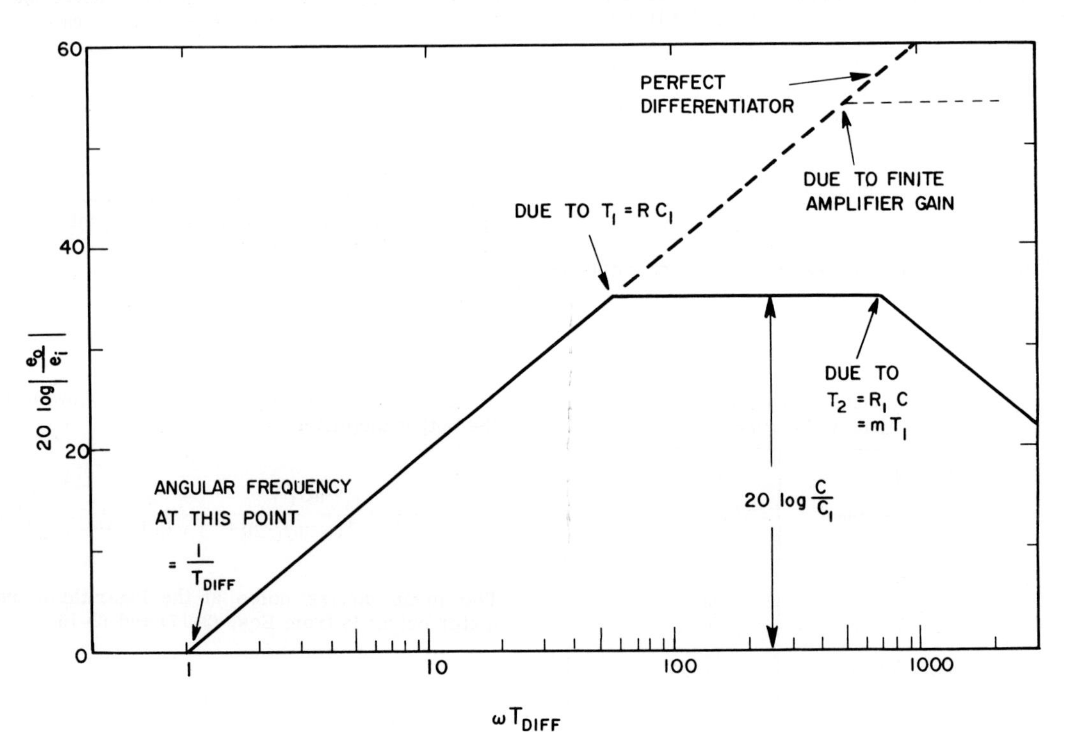

FIG. 2-63 Differentiator transfer function.

expressed in (nepers/sec)^2, is

$$\overline{a^2} = \frac{(n\tau_1/2)}{[1 + (n\tau_1/2)]^2} (2.30)^2 \frac{\tau_1}{2T_1^3} \left[\frac{A^2}{(1+m)(Am+1)(A+1)}\right], \tag{2-25}$$

where $A = T_1/T_{i-1}$.

The noise output due to all diode pumps is obtained by adding together the rms contribution of each since the noise from one is coherent with the noise from any other.

The mean square noise from a single pump section can be obtained from Fig. 2-64.

The gaussian character of the output noise from a single pump section is illustrated in Fig. 2-65 where an amplitude distribution is shown. It was obtained by means of a differential pulse height analyzer. The rms value of the noise can be obtained by taking 0.42 times the full width at half-maximum. The corresponding strip chart record of the same noise signal is also shown but with a reduced rms value because of the decrease in bandwidth caused by the recorder. An approximate measure of the rms value of the noise can be obtained from a chart record by taking one-sixth of the peak to peak amplitude of the envelope.

If fluctuations occasionally appear that are significantly greater than three times the rms value of the normal background they almost certainly originate elsewhere. A sticking control rod can cause such "spikes."

The Logarithmic Element Count-Rate Meter. If the current from a diode pump circuit is passed through a logarithmic element (see Fig. 2-62) the

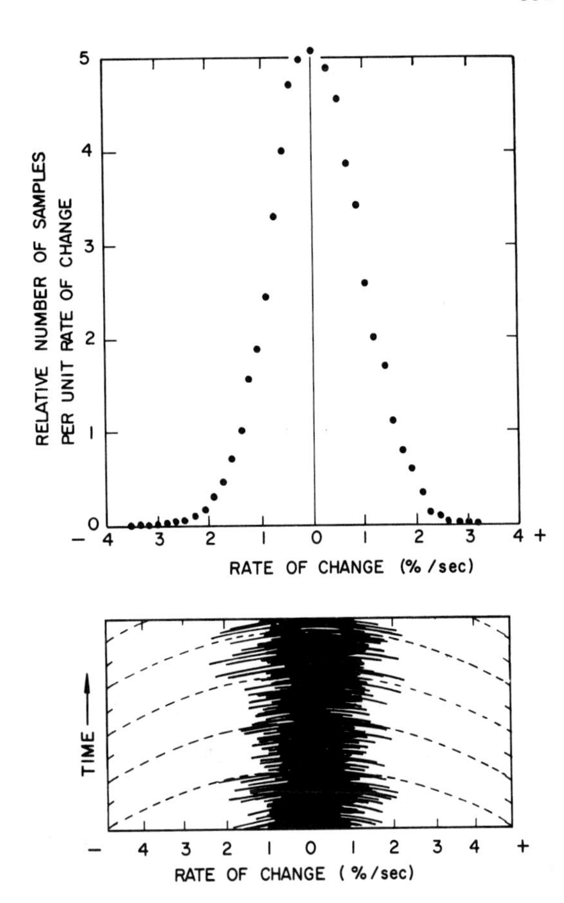

FIG. 2-65 Gaussian character of noise from logarithmic rate channel.

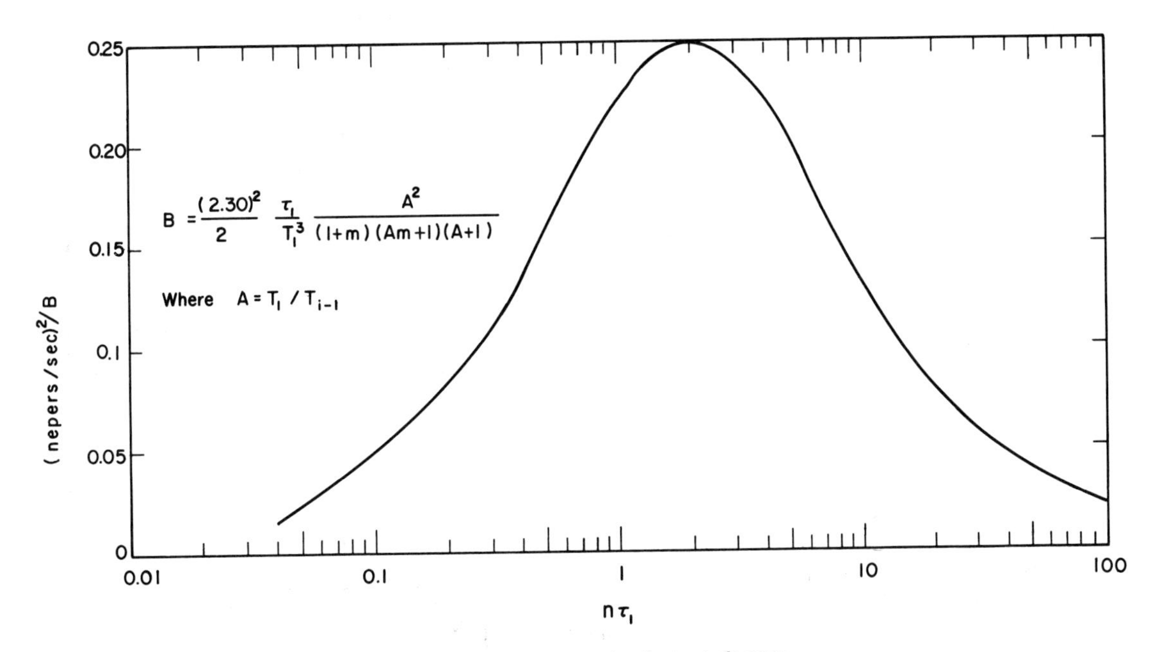

FIG. 2-64 Logarithmic rate noise due to single pump.

spectral density of the noise voltage across the logarithmic element is

$$\phi_D = \frac{2iQT_{Log}^2}{C_D[1 + (\omega T_{Log})^2]}, \tag{2-26}$$

where Q is the charged pumped into the tank capacitor C_D per n/2 input pulses [$i = (n/2)Q$] and $T_{log} = C_D R_D$ [see Eq. (2-9) where $R_{incr} = R_D$]. The noise at the rate-of-change meter output is therefore

$$\overline{a^2} = \frac{Q}{2C_D K T_1^2}\left[\frac{A}{(1+m)(Am+1)(A+1)}\right], \tag{2-27}$$

where

$$A = T_1/T_{Log} = i/(C_D K/T_1) = (nQ/2)/(C_D K/T_1) \tag{2-28}$$

The d-c Logarithmic Element. The expression given by Eq. (2-27) applies directly to the case of a d-c logarithmic element that is obtaining its current from an ionization chamber. The effective value of Q is, however, a function of the charge released in the chamber per neutron absorbed and of the chamber design. Some Q values have been reported [77].

Equation (2-27) is plotted in Fig. 2-66 where the mean-square noise is shown as a function of A. The A scale can be changed to current or count rate by the use of Eq. (2-28). The appropriate Q value should be used to determine the noise.

Table 2-4 contains some typical noise calculations using Q values from reference [77].

2.5.4.4 Reactivity Noise. While the type of noise discussed in the last section is important to an understanding of system limitations, in practice the noise produced by reactivity disturbances. is almost always of overriding importance. Trip levels and smoothing time constants are often determined by the magnitude of such effects.

Figure 2-67 [30f] shows how large these disturbances can be. The tracing is from a neutron flux rate-of-rise channel. The fluctuations are due to the swaying of control elements in the NRU reactor. As the coolant circulating pumps were started one after the other the turbulence caused greater motion. Before high-power operation could be achieved it was necessary not only to put constraints on the lateral motion of the absorbers but also to increase the rate-of-rise trip setting.

Figure 2-68 [30f] is an example of a noise "spike" produced by the sudden movement of an absorber. Obviously this disturbance does not belong to the noise ensemble producing the background fluctuations and any attempt to excuse it on that account could result in overlooking a system fault.

Boiling water reactors are notably "noisy" and period protection at high power is usually dispensed with. Figure 2-69 [81] shows a power trace from a boiling water reactor flux sensor and clearly indicates the magnitude of the effect.

In pressurized water reactors the sudden appearance of increased noise may in fact be an indication of boiling.

Noise originating in the randomness of the fission process itself is discussed in the chapter on Water Reactor Kinetics.

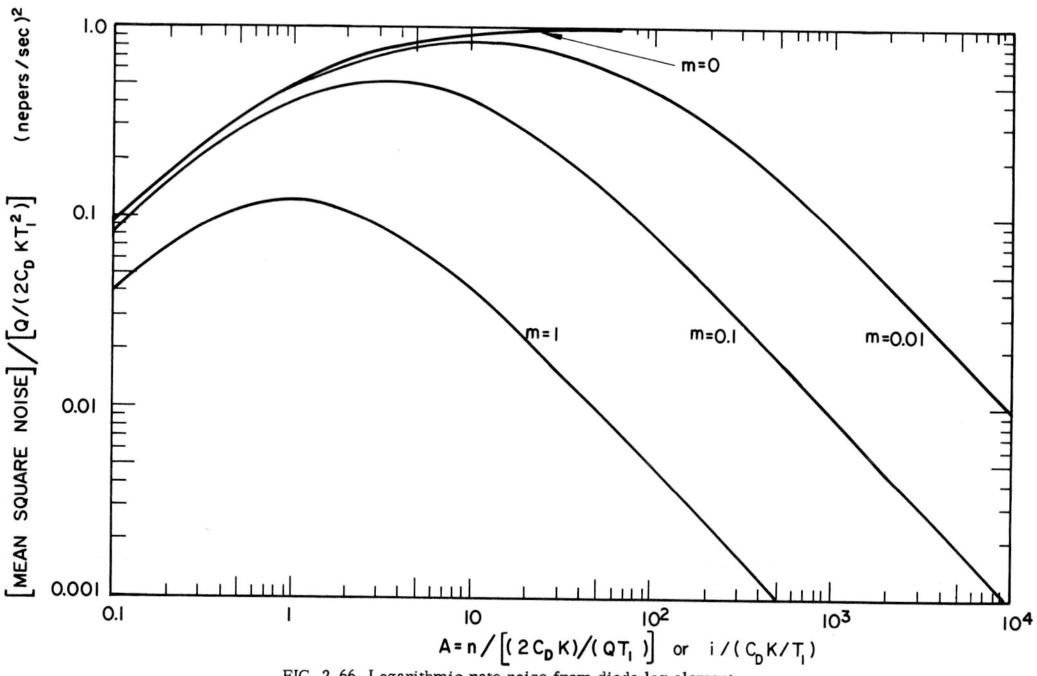

FIG. 2-66 Logarithmic rate noise from diode log element.

TABLE 2–4

Typical Values of Noise in Logarithmic Rate Circuits

C_D (pf)	RMS Noise, α (sec^{-1})	
	$Q = 10^{-14}$ cou	$Q = 3.5 \times 10^{-15}$ cou
100	4.8×10^{-3}	9.0×10^{-4}
1000	1.5×10^{-3}	2.8×10^{-4}
10000	4.8×10^{-4}	9.0×10^{-5}

T_{Log} = 10 sec
T_1 = 1 sec
K = 8.6×10^{-2} v/neper
m = 1

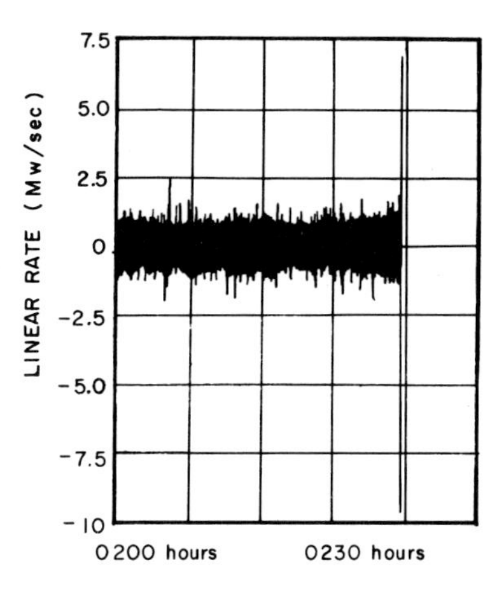

FIG. 2-68 Spike on rate noise. (NRU, Jan. 23, 1959).

2.5.5 Response Time

A basic parameter of any safety system sensing instrument is the time required for the output to reach a specified level after a disturbance appears at the input; or alternatively the level reached by the input when the output passes a specified threshold. This response characteristic is often described loosely by a single quantity, but any such description should be used in a hazards evaluation only when it is certain that deductions made from it are valid.

While the examples given in this section are typical, the aim is not to catalogue response characteristics but rather to aid the reader in probing for an accurate description of the pertinent dynamics of a sensing instrument.

2.5.5.1 The Response of a Linear Amplifier. Consider an amplifier as shown, for example, in Fig. 2-39 whose time-response characteristic is determined by a single integrating time constant. If the input begins to rise, the effect of integration is to produce a delay in the appearance of the output. Figure 2-70 shows the amount by which the input overshoots a desired action level because of this delay. The overshoot is simply BT where B is the linear rate of rise of the input and T is the effective time constant of the amplifier. The delay experienced by the output in reaching an action level is seen to approach a constant value equal to T seconds. Hence the response of a linear amplifier to a ramp input can be described as a simple delay equal to the integrating time constant. If two equal integrating time constants were involved, the delay would be twice as great, and so on.

At high flux levels the delay caused by a sensing amplifier may be in the millisecond region and at low flux levels in the tens of seconds region.

2.5.5.2 The Response of Differentiating Circuits. The response time of a differentiating circuit may be taken as the time required for its output to reach any arbitrarily selected fraction of its equilibrium value. However, to specify completely the response characteristic to interest, the circuit configuration must be known.

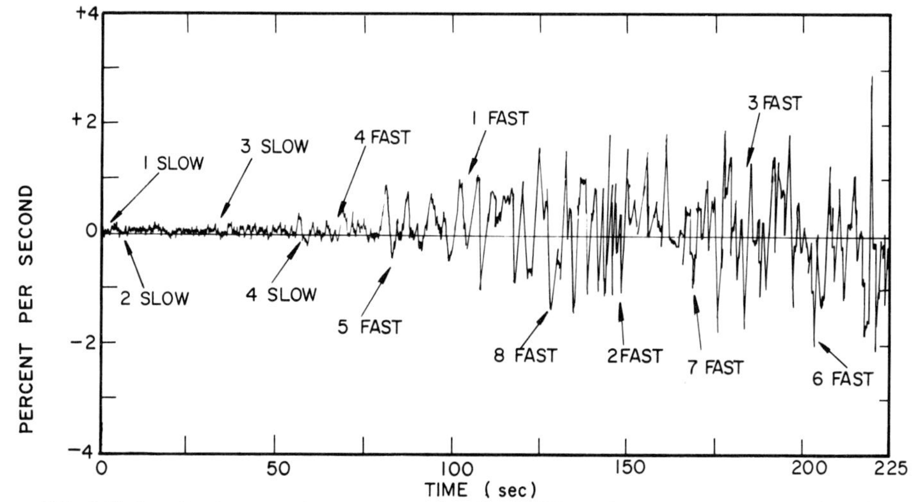

FIG. 2-67 Logarithmic rate with increasing coolant flow (NRU); number of pumps running as indicated.

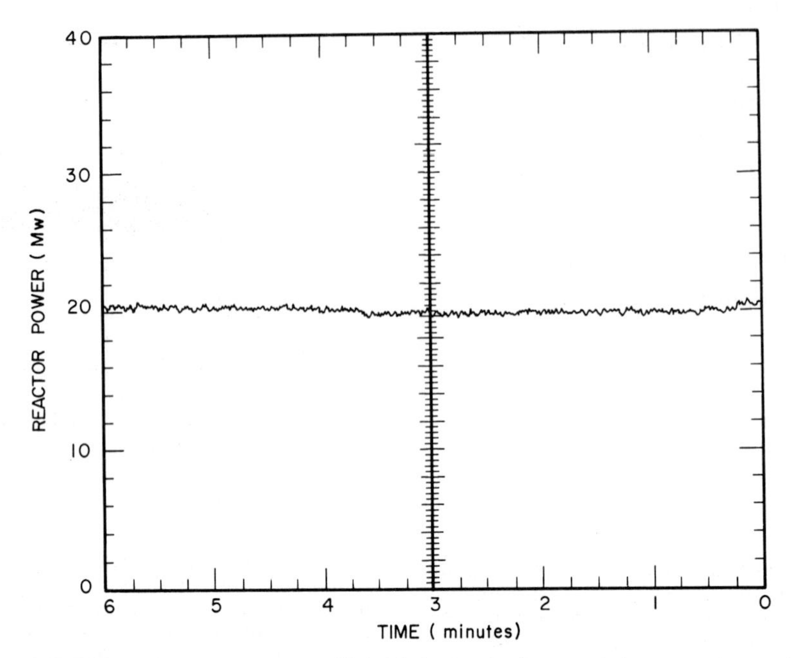

FIG. 2-69 BWR, reactivity noise at power. Typical ionization chamber record when reactor is at equilibrium.

To be specific, the response time T_R will be taken as the time for the differentiator output to reach 63% of its final value. Hence, a differentiating circuit containing a single integrating time constant T_1 (Fig. 2-62) has a response time that is T_1. Two equal integrating time constants would produce a response time of approximately $2T_1$. If the latter circuit were to have the same response time as the single integrator, the time constant of each integrator would have to be decreased to $0.48T_1$. (Interest in double integration stems from the reduction in noise that is possible when used in period circuits [82] (See Sec. 2.5.4)).

Figure 2-71 shows that, even though two circuits may have equal response times, their response characteristics can be different. The time (normalized by the response time) to reach a specified trip level is shown as a function of the ratio (rate-of-rise/trip level) for each circuit.

When the input rate is 1.59 times the trip level, it is reached (by definition of response time) in a time T_R with either circuit. However, for an input

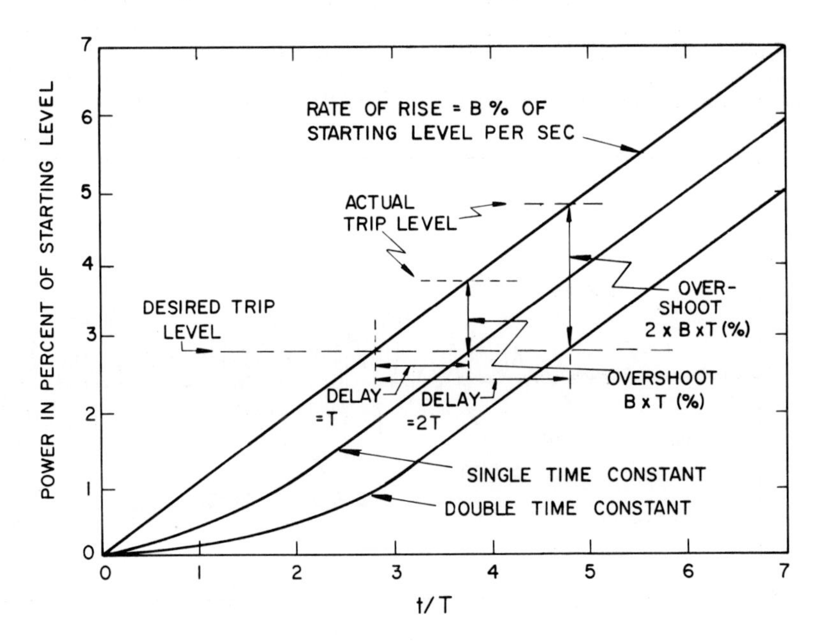

FIG. 2-70 Delay in a linear sensing instrument.

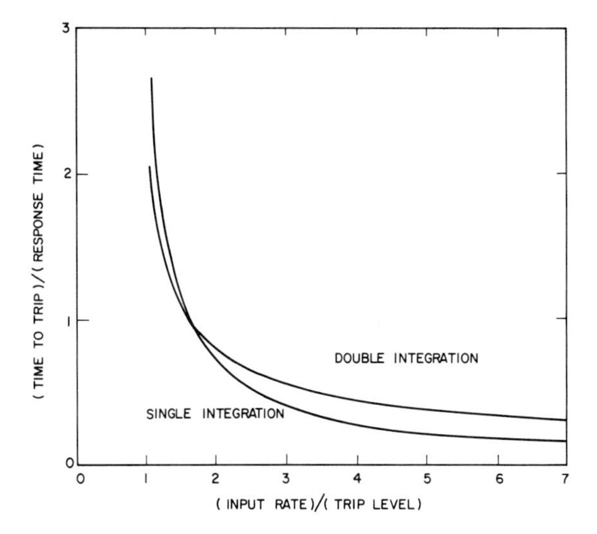

FIG. 2-71 Response characteristics of different circuits having equal response times.

six times the action level, the single integrator circuit responds twice as fast.

To achieve equal performance at the higher rates would require a further reduction in the time constants of the double integrator, and less advantage would be gained in noise reduction.

A differentiator is often used in conjunction with a linear flux amplifier (or temperature measurement) to improve the over-all response. Of interest is the level reached by the safety parameter when its rate of change reaches a trip level.

Figure 2-72 assumes a reactor initially at a power level of 100 Mw. It begins to increase in power at the various rates shown. The rate-of-rise trip level is taken as 0.5 Mw/sec and the dotted curve shows the power excursions (normalized by T_R) that occur at each rate before the rate tripping-threshold is reached.

Notice that as the rate of rise becomes greater the excursion becomes smaller. This action complements the safety performance of a linear amplifier where the overshoot becomes greater as the rate of rise increases.

2.5.5.3 Response of Logarithmic Circuits.

The Logarithmic Count-Rate Meter. When a logarithmic count-rate meter is fed with an exponentially rising counting rate it exhibits a dynamic behavior similar to the linear amplifier. However, at low counting rates the time constants required for smoothing are appreciable and further because the integrating time constants (T_{i-1} etc. in Fig. 2-60) are not all equal, a peculiar response characteristic occurs.

Figure 2-73 shows the indicated value of count rate as a function of time after the application of an exponential input [78, 83]. The time scale is normalized by α, the inverse period of the exponential. As in the linear case a delay occurs but the delay varies with the rate of rise of the input.

If a trip level is set at 10 cycles/sec, the input overshoots this level by the time the trip occurs by a factor of ten (delay 4.2 sec) when the inverse period is 0.5 sec and by a factor of two (delay = 12 sec) when the inverse period is 0.05 sec.

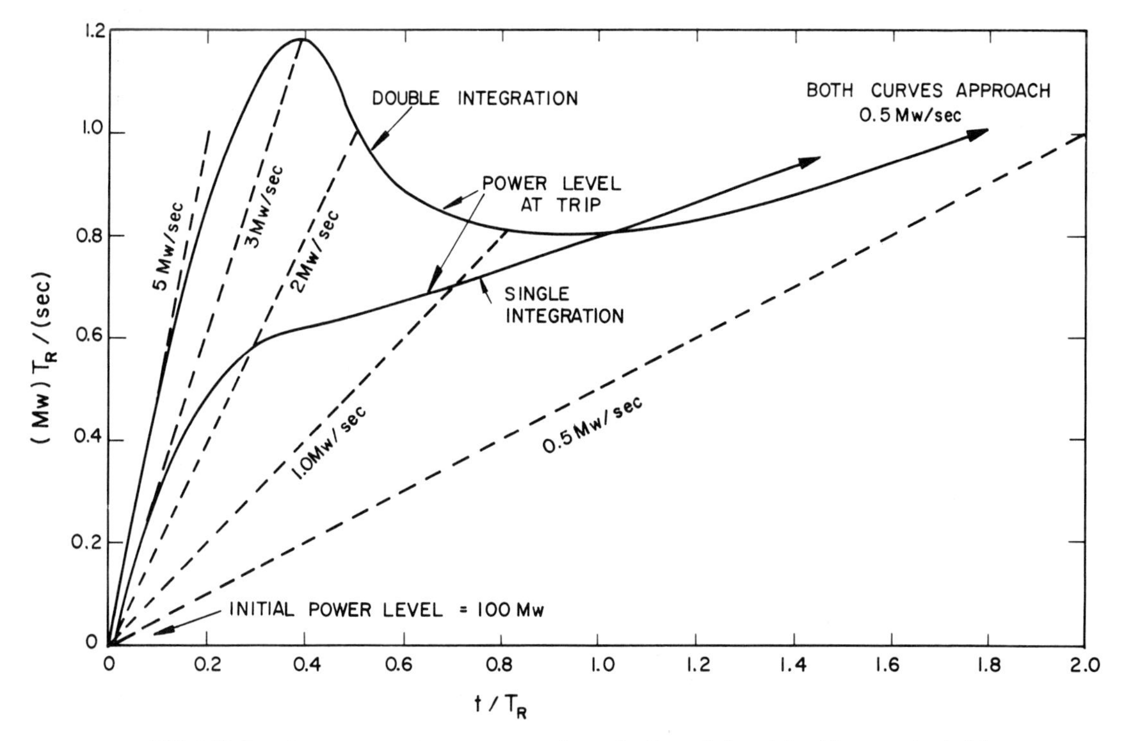

FIG. 2-72 Use of power-rate trip as power-overshoot trip. Rate-of-rise trip level is assumed to be 0.5 Mw/sec.

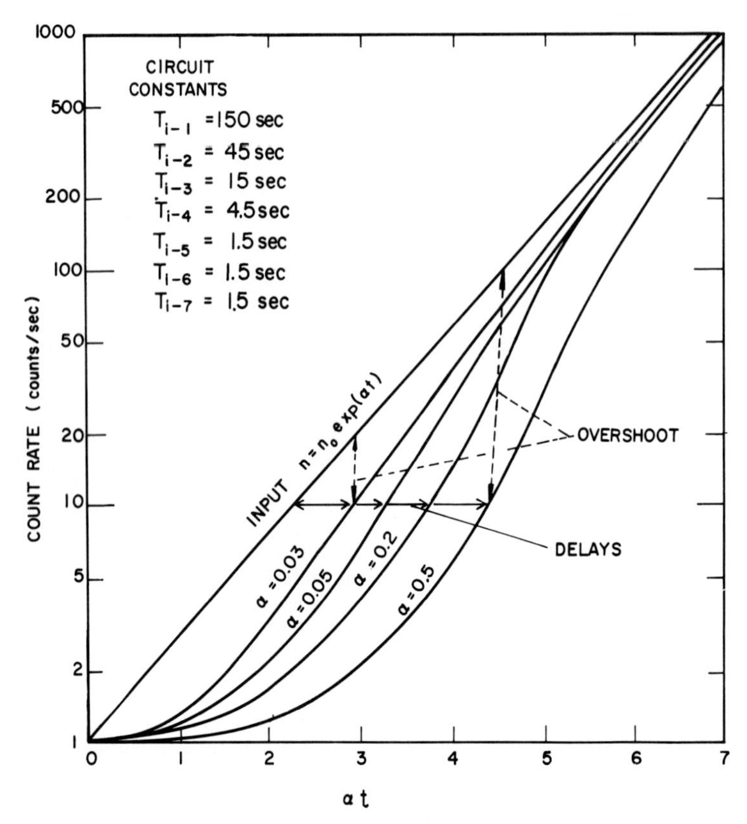

FIG. 2-73 Response of log count-rate-meter to exponential input. (Compare with Fig. 2-70.)

The long delays that occur at low levels do not appreciably affect the response at higher levels even when the transient starts at a low level [68]. This is an advantage gained by reducing the integrating time constants in the pump circuits associated with the higher counting rates.

Period signals obtained from the logarithmic count rate meter are also affected by its transient behavior but they are similar to the effects produced by a d-c logarithmic amplifier described below.

The D-C Logarithmic Amplifier. Only the complete period meter performance will be discussed since the logarithmic amplifier is not commonly used for over power protection.

At low current levels ($\sim 10^{-11}$ to 10^{-9} amp) the response of the period channel is determined by the logarithmic element time constant (T_{Log}) formed by the incremental impedance of the element and the capacity across it (see Sec. 5.3.2) [84, 85]. The response of the period channel to an exponential input with α = 0.1/sec is shown in Fig. 2-74. The only difference amongst the curves is the initial starting condition, and the time constant T_{Log} is sufficient to specify this condition uniquely. However, the usual variable in the time constant expression is I, the initial starting current, and it should be emphasized that this current is due to fission neutrons and not due to background radiation or a bias source. (Ditto [86] has discussed the situation when a bias is present and indicates that the effect on response time is not serious.) When the response time is ~ 1 sec, interaction of the differentiating circuits with other units in a regulating loop may be observed and the discussion in Sec. 4.3.5 becomes pertinent.

In Fig. 2-75 several rates of rise have been shown all starting from the same initial state. These curves are normalized and, if points of equal inverse period are joined, it is seen that any specified level is reached by all curves at the same αt as α increases. This means that the input always rises to the same level for a given inverse period trip setting no matter how high the rate of rise [87].

This result (marked "no delay") is indicated in Fig. 2-76 where the power excursion that occurs before the rate of rise signal reaches 10%/sec (α = 0.1/sec) is shown as a function of the input rate of rise. The figure also includes the effect of adding a fixed delay to the time taken for the rate of rise signal to reach its preset threshold. To extend the region over which the protection is maximum, these delays should be as small as possible.

2.5.5.4 Response of a Period Meter to a Reactivity Step. It is useful to know the effect on a period circuit when a small positive reactivity transient occurs. Such transients are often caused by the sudden movement of a regulating absorber, in whose mechanism excessive friction has developed.

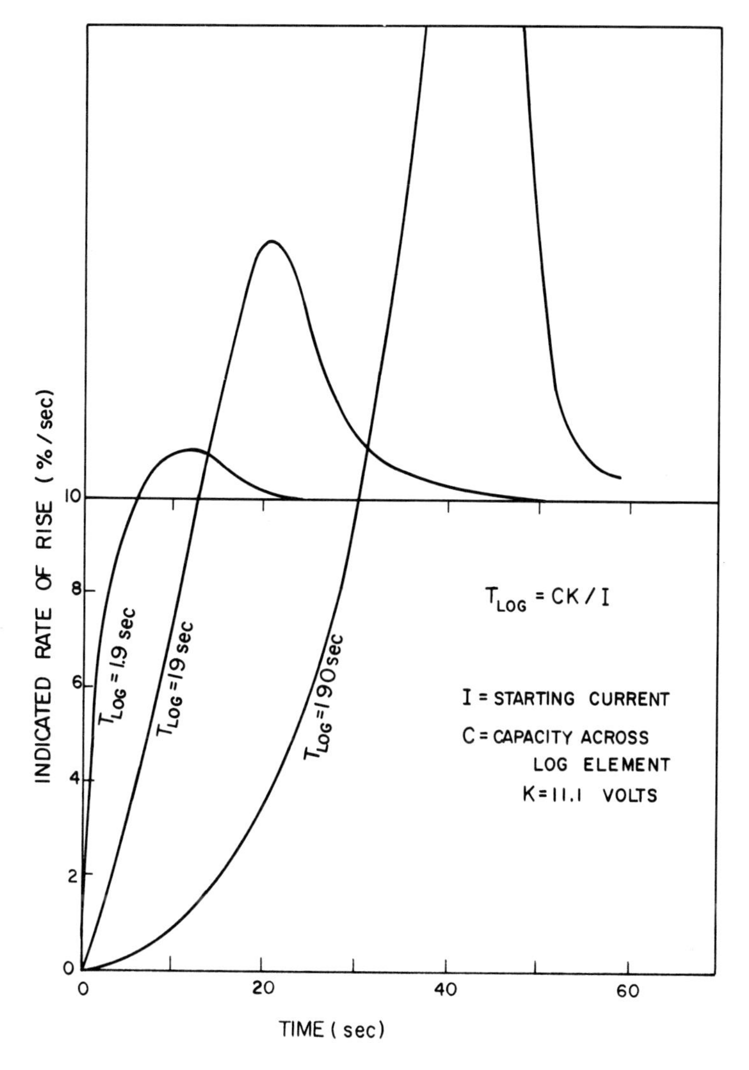

FIG. 2-74 Period channel response to an exponential input.

An estimate of the period meter transient caused by a small reactivity disturbance on a critical reactor can be made by using the approximation

$$\frac{1}{n}\frac{dn}{dt} = \alpha = \frac{\Delta k}{\ell}\frac{\exp(-t/\ell)}{\beta - \Delta k}, \tag{2-29}$$

where Δk is the stepwise reactivity increase, ℓ is the neutron lifetime, and β is the delayed neutron fraction.

The transient presented to the period circuit is given by Eq. (2-29), but the transient as it appears at the output will depend on the value of the integrating time constants in the logarithmic element and on the differentiator.

Assuming that two equal time constants are associated with integration, the shape of the output transient is given by

$$\alpha_{\text{indicated}} = \frac{\Delta k}{\ell}\frac{T_p^2}{(T_1 - T_p)^2}\left\{e^{-t/T_p} - e^{-t/T_1}\left(1 - \frac{(T_1 - T_p)\,t}{T_p T_1}\right)\right\}, \tag{2-30}$$

where $T_p = \ell/(\beta - \Delta k)$ and T_1 is the integrating time constant in each integrator. This assumes operation at a high enough level that the time constant at the input of the logarithmic element can be ignored.

The curves in Fig. 2-77 show the extent of the transient in typical situations.

At low input signal levels where the logarithmic time constant (T_{Log}) predominates,

$$\alpha_{\text{indicated}} = \frac{\Delta k}{\ell}\frac{T_p}{T_{Log} - T_p}\left[e^{-t/T_{Log}} - e^{-t/T_p}\right]$$

$$\simeq \frac{\Delta k}{\ell}\frac{T_p}{T_{Log}}\,e^{-t/T_{Log}}\,. \tag{2-31}$$

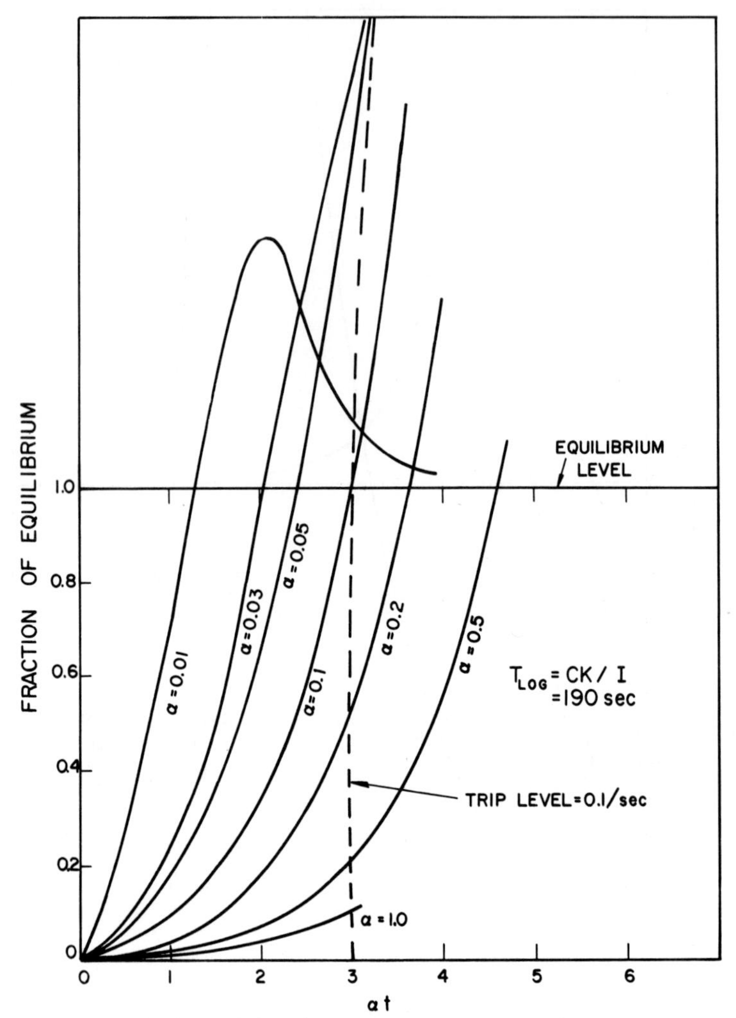

FIG. 2-75 Period channel response to various exponential inputs.

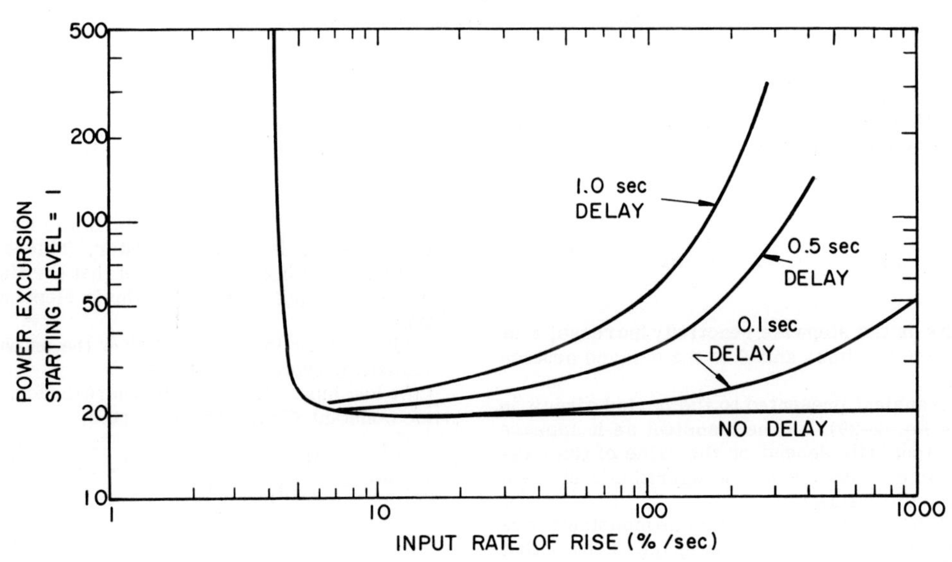

FIG. 2-76 Use of log rate of rise signal as power-overshoot trip.

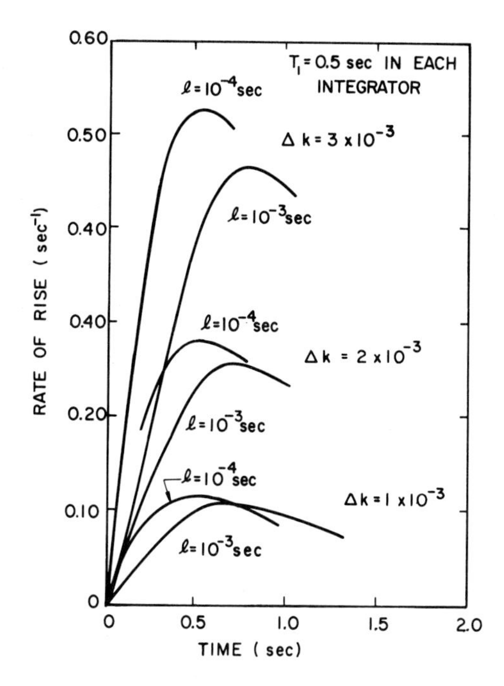

FIG. 2-77 Period circuit response to reactivity transients.

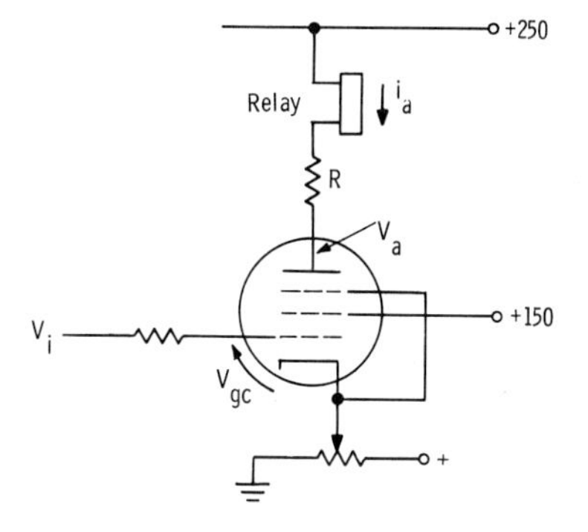

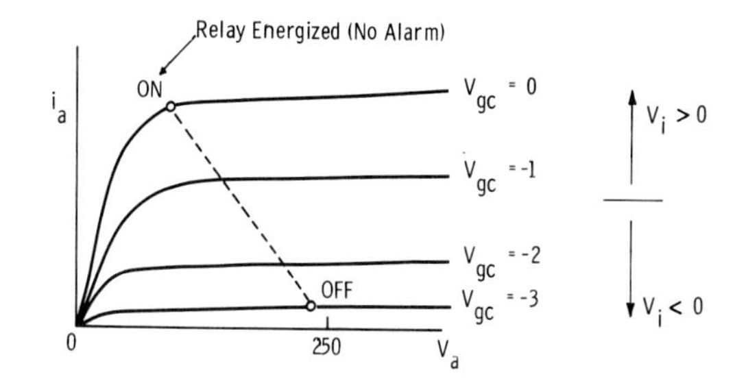

FIG. 2-78 Vacuum tube trip circuit.

If, for example, T_{Log} = 10 sec, $\Delta k = 3 \times 10^{-3}$ and $\ell = 10^{-3}$ sec, then α indicated will reach a maximum of approximately 0.08 sec^{-1}.

2.6 Trip and Alarm Circuits

The action required of a flux or period sensing instrument in a safety channel is to provide a signal whenever the measured parameter exceeds a preset level. This signal is usually either the opening of a pair of relay contacts or a substantial change in a voltage or current.

Many circuits have been developed to perform the trip and alarm function and in some instances fail-safe techniques have been employed to ensure a low unsafe-fault rate. For the most part, however, attempts to build in failure-to-safety features are giving way to circuits where simplicity is the main key to reliability.

2.6.1 Saturated Trip and Alarm Circuits

Saturated trip and alarm circuits are characterized by the operation of a linear amplifying device over a range ten to one hundred times its linear range. An output change with input signal only occurs when the input passes through the linear region.

One of the earliest trip circuits used a pentode tube as shown in Fig. 2-78. With the signal voltage more positive than the cathode, the grid current maintains the grid potential essentially equal to the cathode potential and a constant current flows through the relay coil and the resistance R. The grid voltage remains constant (as the signal decreases) until the signal reaches the cathode voltage. Beyond this point the current in the tube quickly (within 3 or 4 volts) reduces to nearly zero and the relay de-energizes.

This was a fail-safe circuit for cathode emission failure but not from interelectrode short circuits (such as grid to cathode), emission failure being the most likely fault.

Figure 2-79 shows a transistor operating in a fashion similar to the one just described [57]. However, when the signal voltage is below the trip level reference the transistor is not conducting because of the reverse bias on the emitter to base junction. When the emitter voltage exceeds the base voltage by about 400 mv the transistor conducts and the collector voltage changes abruptly from about - 20 v to a positive potential. Diode D_1 prevents the trip level reference potentiometer from loading the sensing amplifier output when the alarm condition exists. This circuit produces a trip signal if the transistor develops a short circuit but is unsafe for an open circuit. The mode of failure in a transistor is not so preferentially oriented toward open-circuit as it is in a tube. However, thermal run-away conditions can occur in transistor circuits which are subjected to over-temperature or excess power. A shorting-type failure usually results.

A differential amplifier [88, 89] may also be used as a comparator for producing a trip signal. If in Fig. 2-80 the signal is below the reference Q_2 will be nonconducting and Q_1 will conduct a current fixed by the constant current source. The collector of Q_2 is at - 25 v. When the signal voltage is above the reference, Q_2 is conducting and its collector is held at zero by D_3. The transistion occurs in an interval of 20 mv about the reference level. The

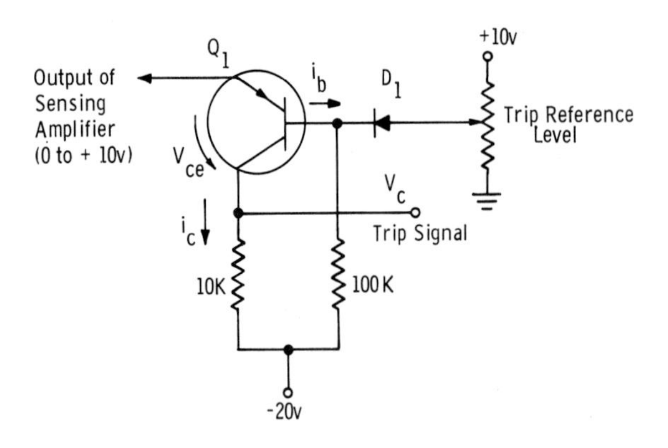

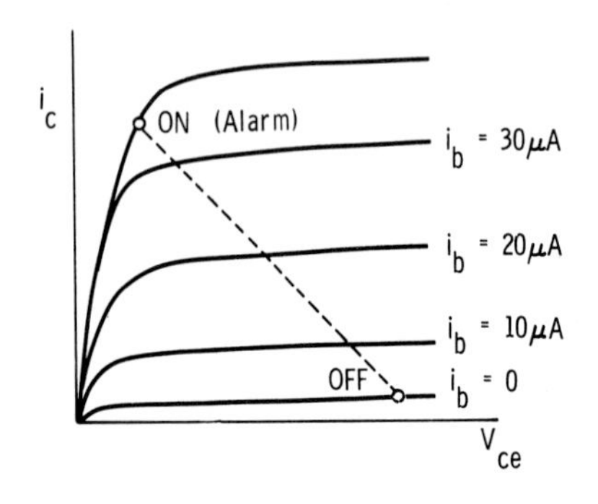

FIG. 2-79 Single-transistor trip circuit.

diodes D_1 and D_2 protect the transistors from overbias conditions.

A circuit that compares two currents is shown in Fig. 2-81 [90]. When the signal current is lower than the reference current, the voltage drop across D_1 ensures Q_1 is not conducting. When the signal current increases and exceeds the reference by about 2 μa, the circuit reverses and the relay is de-energized.

Edge movement meters are used to permit the differences between the trip level and the signal level to be easily seen (see Sec. 1.4.2).

A high gain bi-directional magnetic amplifier (see Sec. 2.5.26) can be used as a reliable current comparator [91, 59a]. The reference current and the signal current are fed into separate control windings. Typically a difference between the two of 1 to 2 μa will cause the amplifier output to be saturated, the direction of saturation depending on which of the two input currents is greater. A polarized relay is used at the output or a normal relay in series with a diode.

The magnetic amplifier can withstand larger electrical overloads than the corresponding transistor circuits but has a slower response if operated at 60 cycle/sec (≈ 50 to 100 msec compared with 100-200 μsec). Furthermore, care must be taken to prevent currents that normally circulate in the control windings from interfering with preceding circuits.

2.6.2 Bistable Trip Circuits

The most common bistable trip circuit is the Schmitt trigger [92, 93, 94]. A typical one is shown in Fig. 2-82. The reference voltage maintains the emitter of Q_2 at a fixed level against which the input signal is compared. As the input passes this level, Q_1 is brought into conduction and the emitter coupling forces the circuit into a new state with Q_2 turned off.

The circuit in Fig. 2-83 is similar to the one above but locks itself on so that resetting is necessary after a trip [94]. The current flow through R is a fail-safe feature. The trip circuit normally operates from a low impedance voltage source (e.g., a feedback d-c amplifier) in which case the current through R has no effect. If, however, the input is open-circuited or if the preceding amplifier fails and exhibits a high output impedance, the current will cause the circuit to trip.

The tunnel diode suggests itself for trip circuitry because of its inherent bistable properties [95, 96]. A proposed circuit [97] is shown in Fig. 2-84 along with characteristics of the diode itself. As the signal current increases (in the negative

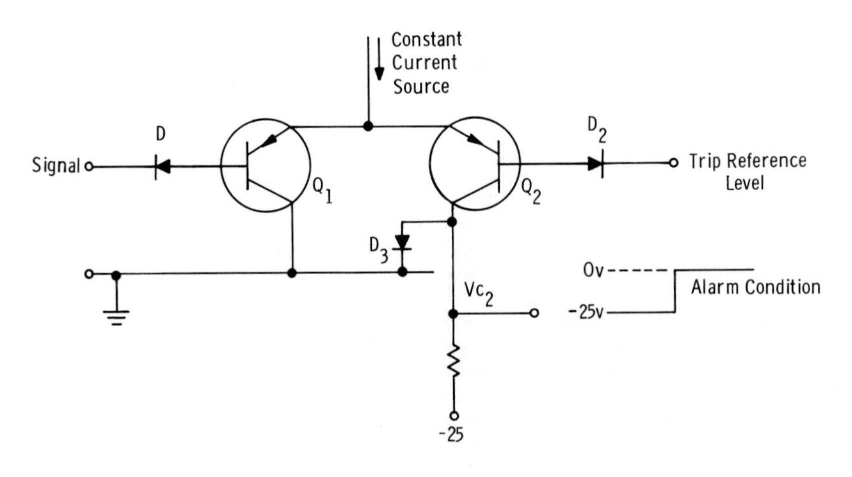

FIG. 2-80 Double-transistor trip circuit.

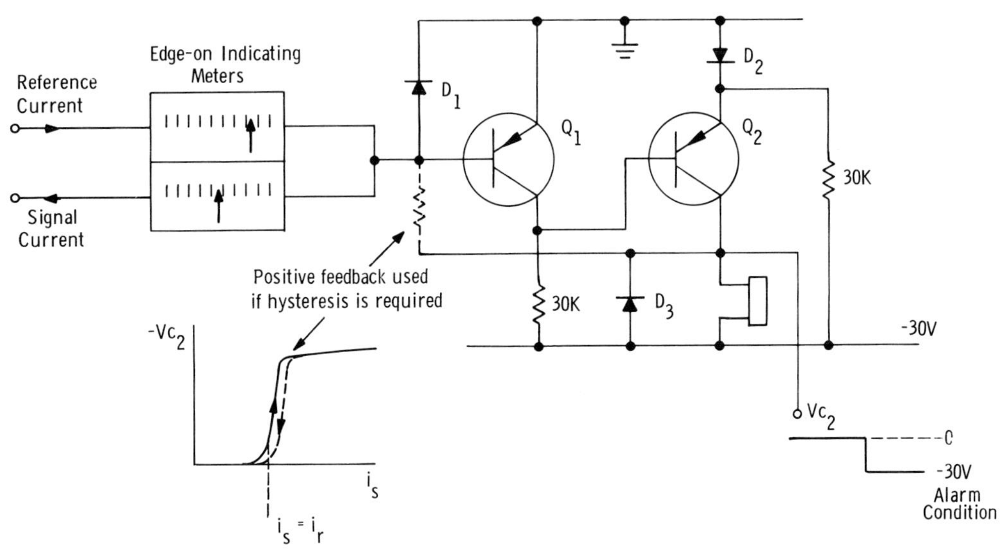

FIG. 2-81 Two-stage transistor trip circuit. (Reference current = i_r, signal current = i_s.)

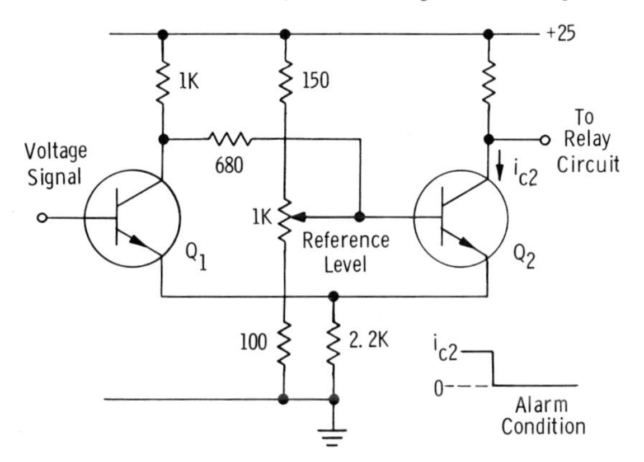

FIG. 2-82 Bistable transistor trip circuit.

direction) a point is reached at which the diode voltage suddenly increases to - 0.4 v from about 30 mv. This is sufficient to turn off the transistor and to present a sudden voltage increase to the trip device. The stability of the trip point is about 1.0 μa.

2.6.3 Meter-Movement Trip Devices

Because of its simplicity and low cost, one of the most appealing trip devices is the meter movement with contacts operated by the pointer. Operationally it is also convenient because the trip level is clearly displayed relative to the signal level. However, by the time this simple concept became a useful instrument considerable complication was necessary.

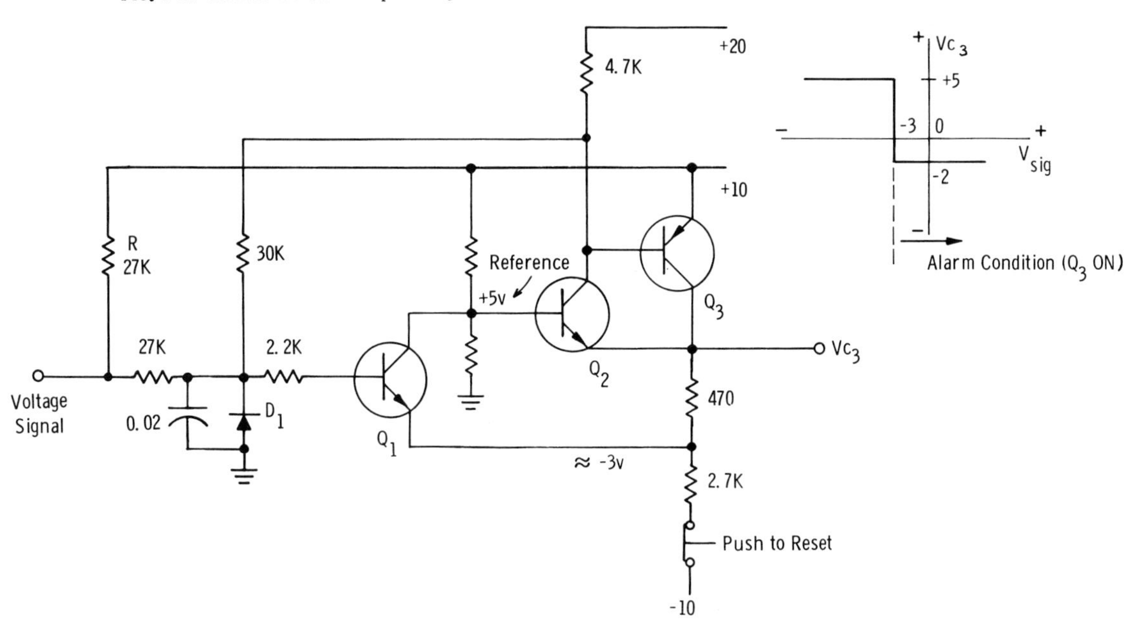

FIG. 2-83 Self-locking bistable transistor trip circuit.

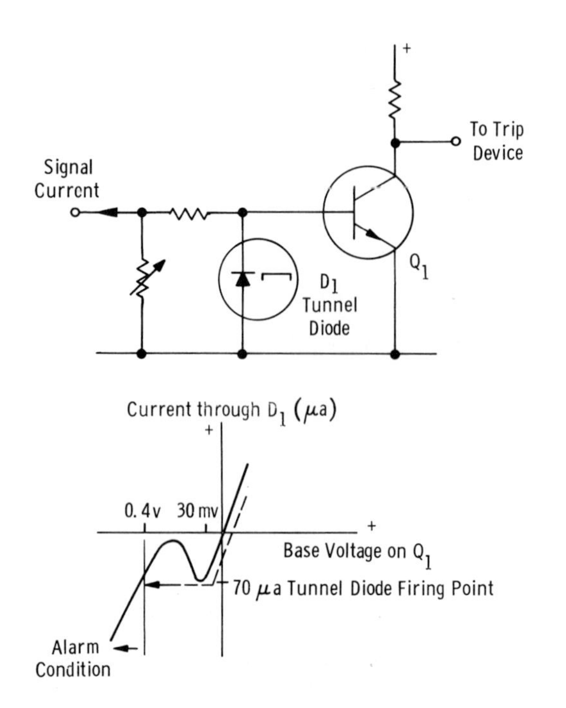

FIG. 2-84 Tunnel diode trip circuit.

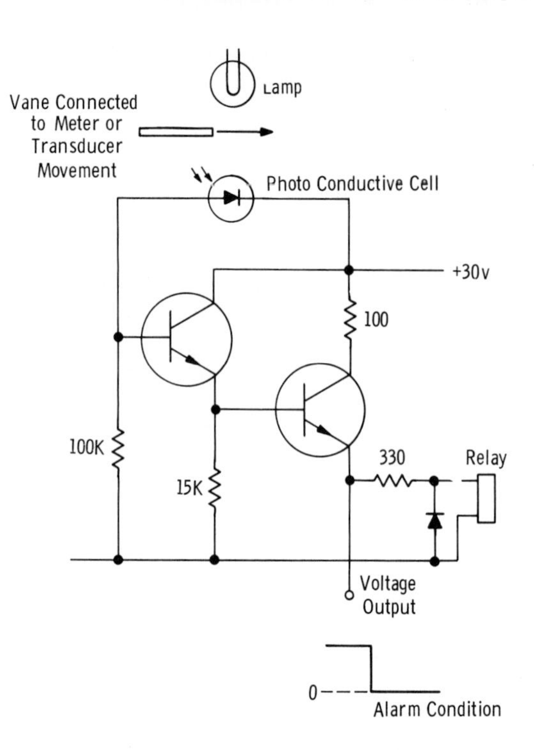

FIG. 2-86 Optically actuated trip circuit.

Simple pressure contacts using only the torque supplied by the signal current were found to be unreliable and self-locking arrangements were developed.

The two most common latching mechanisms are pull-in of the pointer to a contact by means of a magnet [98] and pull-in by means of a positive feedback arrangement that causes a large current to flow through an auxiliary coil as soon as contact is established [99].

Many installations use meter-relays directly in series with an ionization chamber for overpower protection. A circuit that is fail-safe to loss of power is shown in Fig. 2-85.

To avoid contacts other meter movements use vanes carried on the pointer arm either to block the passage of light to a photocell [100] or to interfere with a high frequency electromagnetic circuit [101]. As with other alarm systems, it should not be possible to set the alarm reference above the full-scale signal value.

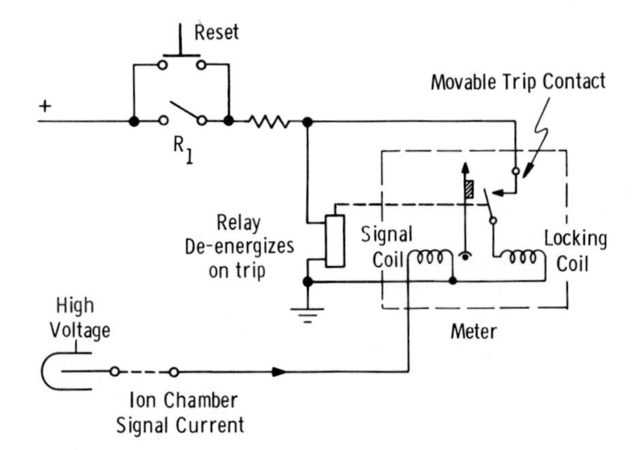

FIG. 2-85 Contacting meter trip circuit. (The contact R_1 is a contact of the relay in the circuit and is open when the relay is de-energized.)

The application of a photodiode type circuit is shown in Fig. 2-86. When the light beam is cut off, the impedance of the photodiode is high, the current in Q_1 falls to a low value, and the relay is de-energized.

3 NON-NUCLEAR INSTRUMENTATION

Many of the measurements required to control a reactor plant are not unique to the operation of a nuclear reactor. However, the radioactive environment in which the measurements must be made has brought about the need for new instrument types [67b] and design philosophy. Once the reactor has been operating some parts (the core region, for example) are accessible only with the greatest difficulty while other areas are accessible only during a reactor shutdown. Under these conditions instrumentation that was once considered reliable, given a certain amount of routine maintenance and adjustment, is no longer adequate.

Two groups of instrumentation are needed, one to measure the bulk conditions in the primary and secondary coolant, and another to measure local conditions throughout the reactor (see the Reactor Core chapter, Sec. 9). The latter requirement calls for a large number of sensors, especially in pressure-tube reactors where in-core conditions are inferred to a large extent from per channel measurements on the coolant.

Much is still to be done in the development of highly reliable and inexpensive sensors for the measurement of such conventional parameters as temperature and flow to meet these requirements.

The radiation levels in the areas where these measurements are made range from 10^6 to 10^{14} neutrons/cm^2-sec and 10^5 to 10^{12} r/hr within the core to 10^5 to 10^7 neutrons/cm^2-sec and 10^3 to 10^6 r/hr in the region of coolant pumps and heat exchangers.

Radiation damage to instruments usually results from the change of physical properties of construction materials. Organic materials [102] and semiconductors [103] are poorest, metals are best. Table 3-1 shows the radiation sensitivity of various electrical components [104, 105]. In addition to radiation effects, high temperature can both affect the stability of mechanical seals and also impair the electrical insulation properties of signal cables.

Data transmission techniques must be considered specifically when signals originate inside pressurized active areas. Figure 3-1 [9a] shows the electrical cable penetration fitting used on the Enrico Fermi Reactor containment vessel. Figure 3-2 [46] shows another approach used at Halden. Both pneumatic and electrical instrumentation systems are used. When pneumatic lines are used to penetrate the vessel, radioacitivity leaks through such lines can develop in case of a maximum credible accident. Pneumatic instruments are generally cheaper and are less sensitive to radiation than electrical systems. However, it is more difficult to change range or zero settings remotely on pneumatic devices [67b]. Many sensors are basically electrical and so are quite compatible (perhaps even without amplification at the sensor) with electrical transmission. Also the availability and increasing requirement for data-handling and analysis on a large scale (Sec. 3.9) encourages the trend to an all-electric system.

TABLE 3-1

Radiation Damage to Electrical Components

Component	Approximate Radiation Limit for Acceptability
Vacuum Tubes	6.10^9 rads
Transistors	10^6 to 10^7 rads
Tunnel Diodes	$> 10^9$ rads
Resistors	5.10^6 rads
Capacitors	5.10^6 rads
Rubbers	2.10^6 rads
Plastics	10^6 to 10^8 rads
Motors	Limited by Insulation Materials
Transformers	
Plugs	
Cables, etc.	

[1 rad = 100 ergs/gm of air = 10^{-5} watt-sec/gm.]

3.1 Temperature

3.1.1 Thermocouples

Thermocouples are used where small size, fast response (less than one-second time constant), radiation resistance and stability against shock and vibration are required. They are generally chosen for in-core temperature measurements, because they can be made substantially resistant to radiation effects [67b, 9a, 5c]. The sensitivity of thermocouples ranges from 10 to 80 μv/°C [107]. Iron constantan (60 μv/°C) and chromel-alumel (40 μv/°C) are commonly used in reactor systems. Iron-constantan couples can be operated satisfactorily up to 750°C (1400°F) and chromel-alumel to 1090°C (2000°F) [108]. Tungsten-(tungsten, 26% rhenium) couples shown promise for use up to 2760°C (5000°F) [109] and can therefore be used to measure the temperature inside fuel rods [110a]. Chromel-alumel couples appear to stand up to high radiation fields better than other types [110b, 110c], satisfactory performance being measured to a total exposure of 4×10^{20} neutrons/cm^2 [111]. Under a similar irradiation a platinum-(platinum, 10% rhodium) couple gives a 3% error due to the transmutation of rhodium to palladium.

Radiation effects on cable insulation are also important. Rubbers suffer brittleness at a dose of 2×10^8 ergs/g, plastics at 10^8 to 10^{11} ergs/g [102, 112]. Ceramic insulants are used for in-core thermocouples [67e]; alumina, and magnesia exhibiting no change up to 10^{20} neutrons/cm^2 [113, 114]. Alumina is preferred in water reactors because magnesia swells in contact with water and causes sheath rupture. Stainless steel or Inconel metal are used for sheathing.

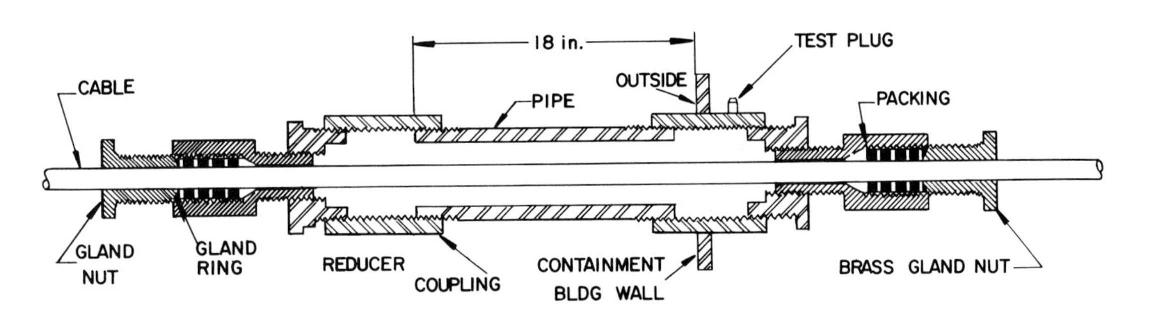

FIG. 3-1 Cable penetration of containment building.

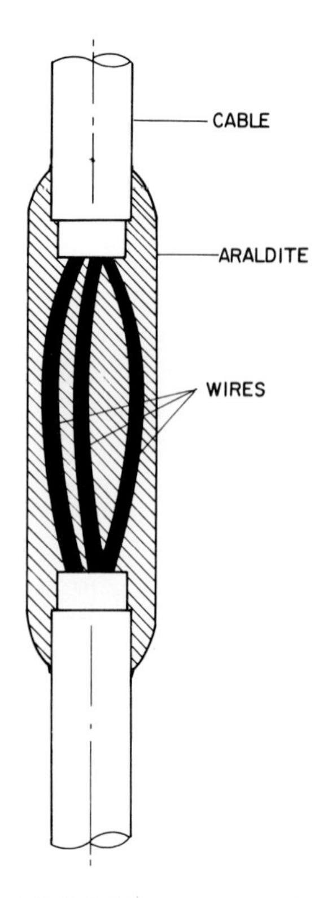

FIG. 3-2 Cable penetration seal.

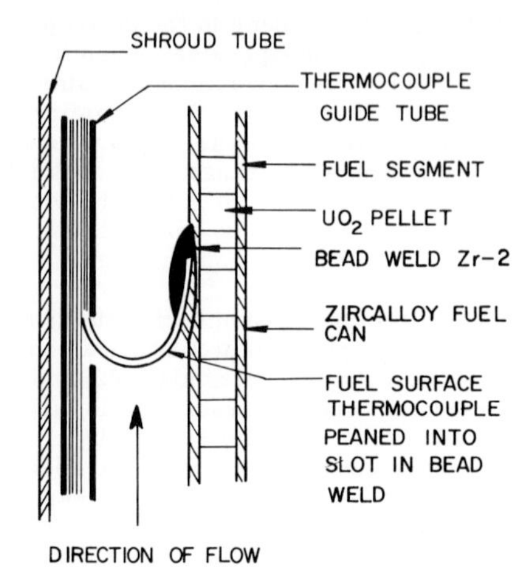

FIG. 3-3 Attachment of thermocouple for fuel surface temperature measurement.

Where minimum response time is required the smaller wire gauges (30 gauge or so, corresponding to 10 mil or 0.25 mm diam. wire) are used. The resulting couple has an outer diameter of about 1/16 in. or 0.16 mm and is fragile [110a, 115]. The response time can vary from less than one second to minutes, depending on the type of metal connection made (a butt-weld being faster than twisted wires). An example of one type of attachment used is shown in Fig. 3-3 [46]. If the couple is in a liquid, then its response is faster than in a gas. The mounting arrangement is also critical in this respect, good thermal conduction from the fluid to the couple, and good thermal insulation from the surrounding metal being necessary for fast response.

The voltage output of the thermocouple is proportional to the difference between the measuring junction and the reference junction; the latter has either to be temperature-controlled or the changes in temperature must be compensated for in the measuring unit. Very often, especially where several measurements are taken with one amplifier unit, copper wires are run to the instrument from a constant temperature reference junction block as shown in Fig. 3-4a. In millivolt meter type instruments, and in some recorders, reference junction compensation is obtained by a local temperature-sensitive resistor.

Measuring devices may be millivolt meters, magnetic amplifiers, vacuum-tube or transistor amplifiers. If the latter are used, chopper stabilization (Sec. 2) is generally required to achieve 1°F (0.56°C) accuracy. Often electronic amplifiers are used in association with a recorder in a balanced configuration (see below).

When low impedance circuits such as meters and magnetic amplifiers are used, the total lead resistance determines the sensitivity.

A high-gain differential amplifier may be used to amplify a thermocouple signal. In recording type instruments where a mechanical pen-movement is obtained, a null balance is achieved at the input by mechanically adjusting the bridge balance. A capacity balance system is shown in Fig. 3-4b. The input switch is carefully constructed to minimize contact bounce and contact potentials.

If a mechanical output is not required, a feedback voltage-amplifier, shown in Fig. 3-4c, is used. The output range is determined by a feedback resistor. An additional reference-junction and zero-offset current enters at the summation point. In the circuit shown above a fail-safe circuit protection is used in case the thermocouple circuit opens. Two resistors R pass current through the thermocouple (negligible heating error occurs because of the low junction impedance, although calibration errors can occur due to lead resistance if the current is not added close to the junction). If the couple open-circuits, enough signal is placed on the input to drive the output up-scale to the alarm condition. It may not be convenient to apply this test to very small, high resistance, thermocouple wires.

One of the chief causes of difficulty in thermocouple measurements is the low signal level, and the effect of grounding the couple at the measuring point (for good thermal contact). Ground currents produce 60, 120 cycles/sec, or higher-frequency signals on both leads (common-mode), that can give rise to differential signals as described below. Careful shielding [116] (solid wrap being most effective) and care in tying down shields to reduce ground loop currents are required [117, 118] since these signals can saturate the sensitive input stages of the amplifier resulting in a false indication. Differential amplifiers with high input impedance (greater than 10^8 ohms) are used where long cable runs (100 ft or 30 m or so) occur, and where multiple grounds cannot be avoided. Ungrounded thermocouples reduce the effect of electrical inter-

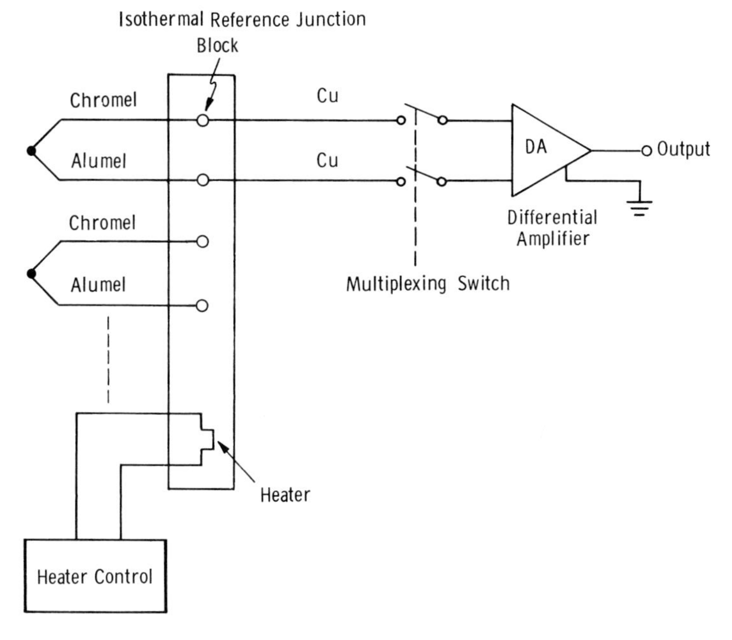

FIG. 3-4a Thermocouple compensation block.

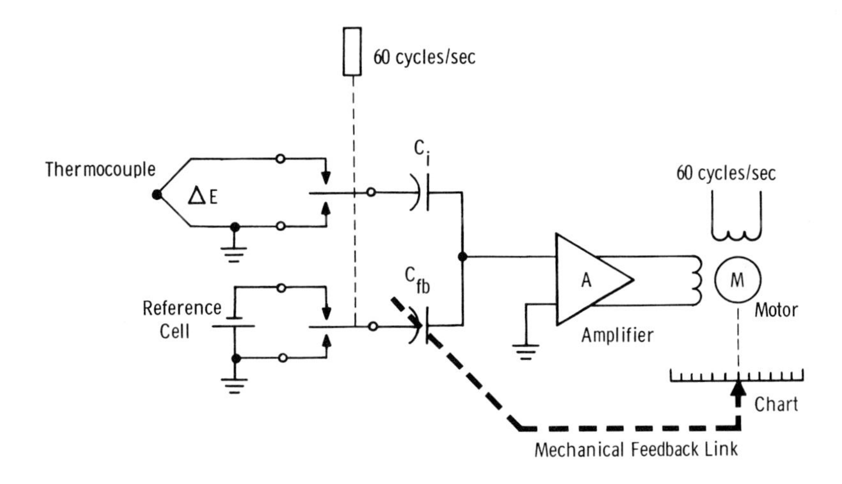

FIG. 3-4b Recording system with mechanical feedback.

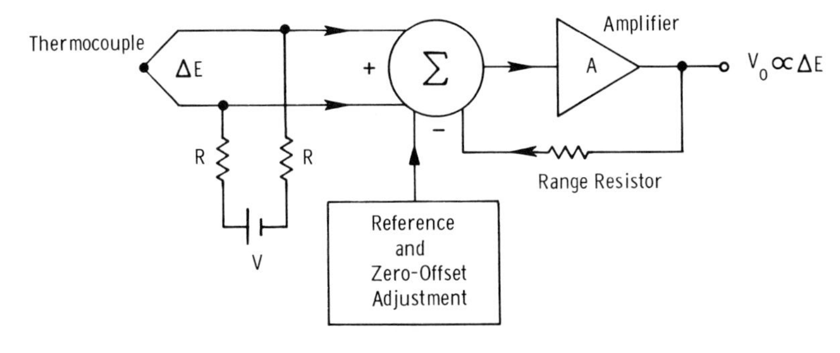

FIG. 3-4c Feedback amplifier.

ference and also permit identification of insulation breakdown that could in itself lead to false signals being generated by induced circulating currents. Signal leads are often twisted to reduce interference [119].

Careful attention to lead impedance and capacity unbalance is required to minimize the effect of common-mode signals. A simple example (Fig. 3-5) shows the magnitude of this effect. It can be seen that amplifier input impedances greater than 10^7 ohms leave a very small error signal, and that to reduce a-c interference the capacity from either side of the input lead to the amplifier ground must be small. Screened, floating input stages are therefore used. One such possible arrangement is shown in Fig. 3-5c. The use of input filtering to reduce high-frequency signals is also used, but only at the expense of system response time. When several inputs are switched into the amplifier in turn by a multiplexer different common-mode voltages can appear. This produces transient errors that can overload the amplifier, and a finite settling time is required before a measurement is valid. In such systems the capacity across the switch contacts should be minimized to prevent unbalance in the common-mode interference. Further remarks concerning multi-input systems are made in Sec. 3.9.

3.1.2 Resistance Temperature Detector

Variation in the ohmic impedance of a metal with temperature gives a very sensitive and stable thermometer. Below 500°C (932°F) resistance temperature detectors are often more accurate than thermocouples [9a, 67c]. Resistance bulbs also give a much larger output signal than thermocouples, simplifying the amplifier circuits. But the need to protect the thin temperature sensitive wires and to insulate them increases the detector response time [108]. The slow neutron flux in a reactor core will cause a transmutation in platinum to gold, this change being less than 3% after a dose of 2×10^{22}/cm^2 has been accumulated [110b]. Gamma heating can also introduce an error. The most common metals used up to 400°C (752°F) are platinum (sensitivity = 0.4%/°C or 0.22%/°F) and nickel (1%/°C or 0.56%/°F). The temperature coefficient of nickel is not linear, but when nickel is used in an unbalanced bridge network the bridge nonlinearity is cancelled. A typical resistance value at 0°C is 200 ohms, giving a sensitivity for platinum of 0.8 ohms/°C or 0.5 ohms/°F. With a bias current of 5 ma through the bulb, an output of 4.0 mv/°C or 2.5 mv/°F is obtained which is substantially greater than that from a thermocouple.

Because the detector resistance is measured in a bridge network good electrical insulation is usually required between the wire and its sheath, glass, mica and various cements have been used [107, 108]. A typical mounting arrangement is shown in reference [5d]. Because of the use of small wire diameters required to achieve high resistance, one of the most common forms of failure is an open circuit being caused by thermal stresses imposed

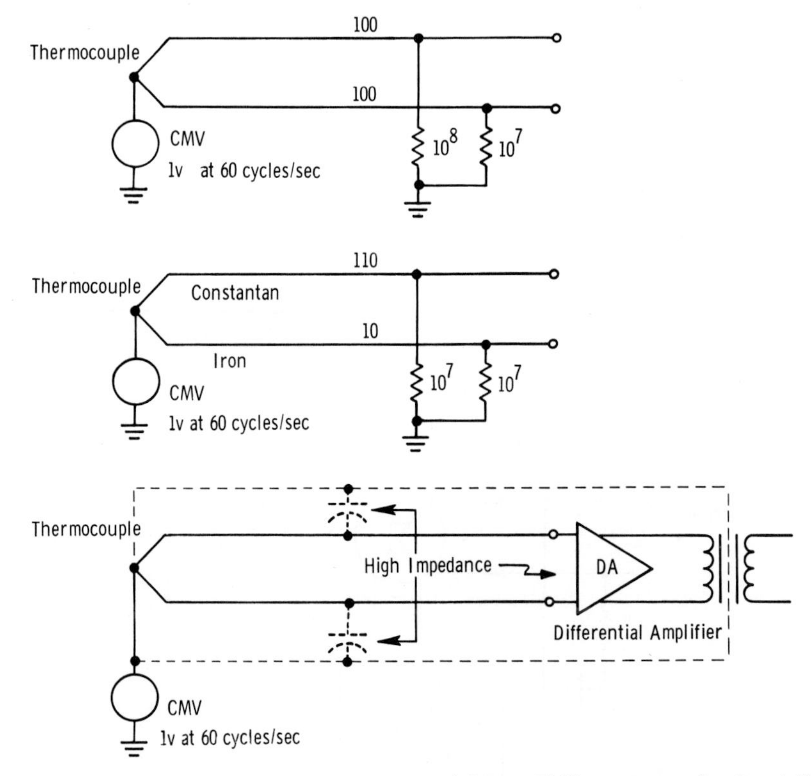

FIG. 3-5 Common-mode rejection due to impedance unbalance. (CMV = common mode voltage.) Top diagram shows case where cable impedances are equal, amplifier impedances unequal. Middle diagram shows case where cable impedances are unequal, amplifier impedances equal. Bottom diagram shows use of screening to reduce capacity unbalance.

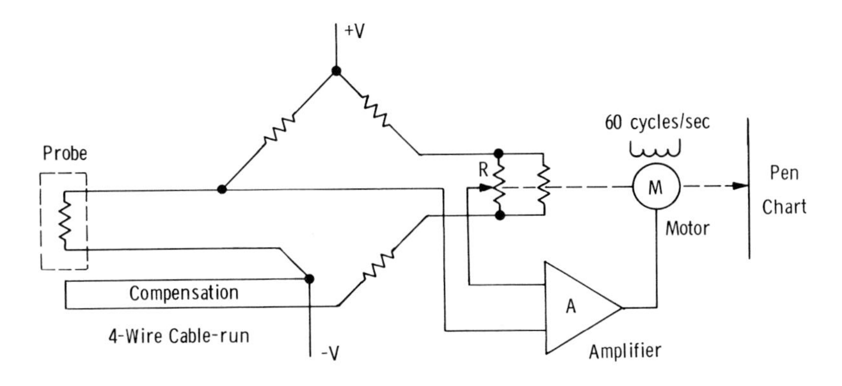

FIG. 3-6 Recording resistance thermometer using balanced bridge circuit.

on the encapsulated wire. Fast-response (0.1 sec) bare-wire elements, coated with a thin layer of ceramic having a thermal expansion coefficient closely matched to that of the wire, have been developed.

Bridge measurements include both resistive and capacity balance (the latter being similar to the thermocouple servo-balance system). A typical arrangement is shown in Fig. 3-6. In this system bridge balance is achieved by positioning the balance arm, R. A second pair of leads are used to compensate for temperature changes in the cable-run. This is often not necessary in low temperature environments, or where the resistance of the detector is high compared to the cabling.

Both a-c and d-c bridge excitation have been used. Direct-current systems use conventional d-c type amplifiers. Filtering of unwanted high frequency or 60 cycles/sec interference is simple. Alternating-current drive can simplify circuit details and selectively avoid certain interference frequencies, one or two kilocycle/sec operation being typical.

Pulse circuits have also been adopted for application in systems where many detectors are scanned in sequence. Samples of the bridge unbalance can be made with and without the bridge voltage applied to eliminate standing errors. If an unbalanced bridge configuration is used, then the output is nonlinear. With suitable drive impedances this may only be 1% over 100°F (56°C) and with the accompanying circuit simplicity this may be quite acceptable. Data handling and computing equipment can at a later stage compensate for such errors. One advantage of the pulse drive [120] (pulse lengths of 100 μsec to 1 msec) is that greater excitation can be applied to the detector without self-heating effects becoming objectionable.

In the circuit shown in Fig. 3-7 a high-level switch applying the excitation voltage is shown together with a low-level or output switch, S_2. Either or both can be employed.

All the pickup problems and common-mode effects, previously mentioned with regard to thermocouples, apply to resistance bulbs. The signal levels are higher, but the high source impedance aggravates ground-unbalance problems. When the output is taken from an unbalanced bridge (i.e., no direct feedback) common-mode voltage signals generate differential errors proportional to the out-of-balance. Common-mode voltages are reduced by insulating the detector and shielding the cable similar to the thermocouple case shown in Fig. 3-5.

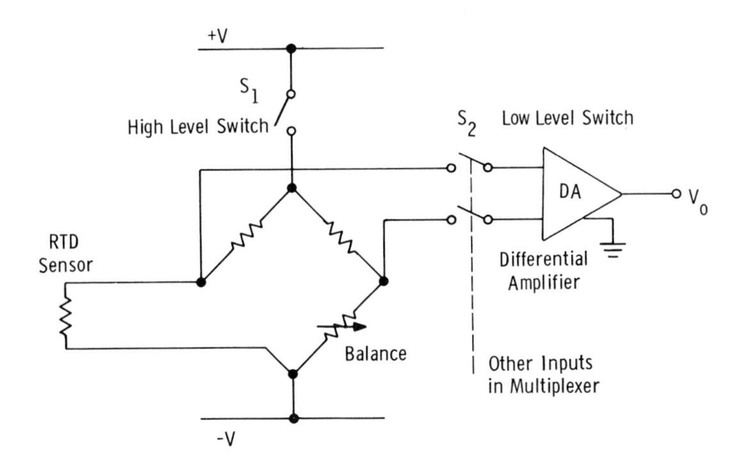

FIG. 3-7 Resistance thermometer: use of switched or pulsed bridge.

3.1.3 Thermistors

Thermistors are semiconductors and have a negative temperature coefficient of resistivity. This coefficient varies as B/T^2 where B is a constant of the material used. The value of this coefficient is from 1 to 5%/°C (0.6 to 2.8%/°F) at room temperature [107]. Their absolute calibration is not as well defined as resistance temperature detectors, and drift with time can occur, but the large output signal gives them special prominence where high sensitivity is required. The smaller sizes can have short time constants (1 sec or so), but also have a higher impedance (10^4 ohms) than the large types (10 ohms). The semiconducting material makes them radiation-sensitive so they cannot be used in very high-field areas. Their chief use is in coarse temperature compensation networks and in control applications where accuracy can be sacrificed for simplicity and sensitivity.

3.1.4 Filled System Thermometers

These are used in applications where simplicity and low cost is a requirement. They have the disadvantage that the sensing and indicating portions must be close together (less than 100 ft or about 30 m). One of the more usual faults in such instruments is a leak in the transmission pipe. Such a fault is unsafe and its possibility may nullify the apparent advantage of a cheap and simple instrument. Both volumetric and pressure changes are utilized [107, 108] (Sec. 3.2.2).

3.1.5 Bi-Metal Thermometers

Direct reading of temperature can be achieved by use of a bimetallic strip connected to an indicator pointer [107]. Such devices provide simple temperature indication but are not so suitable for information transmission. They are often used for alarm circuits in instruments subjected to overheating if ventilation fails.

3.1.6 Strain-Gauge Instruments

Since most reactor vessels and associated piping are fabricated of metals, expansion and contraction occur as a result of temperature changes. Strains set up as a result can be monitored and used as an indication of temperature. Resistance strain gauge units can be bonded onto structures and operated in bridge circuits similar to those described for the resistance bulb. Radiation resistant properties are chiefly limited by insulation material. Gamma heating should also be considered.

3.1.7 Temperature Indication by Volumetric Displacement

The primary fluid is often displaced by thermal expansion, giving a very fast indication of rates of change of temperature if such expansion is monitored. Moderator temperature changes are often important and can be measured by use of level probes (Sec. 3.4) in the moderator tank, or by level or pressure measurement in an overflow tank.

3.2 Pressure

Pressure measurements are required to indicate the operating state of a process, or as a means for indirect measurement of flow or level. Both mechanical and electromechanical instruments are available. Pressure instruments have moving parts with high radiation tolerance but with attendant maintenance and replacement problems. Operating conditions (such as high temperature) often preclude the use of transducers with direct electrical outputs. For temperatures above 200°F (~ 100°C) it may only be possible to place a mechanical diaphragm at the source and to place the pneumatic or electrical amplifiers remotely.

3.2.1 Pressure Sensors

The basic principles involved in the measurement of pressure are adequately covered in the literature [107, 108]. These include measuring instruments utilizing the displacement of bellows, diaphragms, Bourdon tubes and strain gauges. Figure 3-8 [67b] shows an absolute pressure transmitter capable of standing an overpressure of 2000 psi (~140 atm). It has a moving-coil transmitter (Sec. 3.2.2). Figure 3-9 [67b] shows a twisted Bourdon-type pressure instrument with two seals [9a, 5c]. Figure 3-10 [67b] shows a pressure sensor using a strain gauge pickup. These instruments have been evaluated for reactor use by Oak Ridge National Laboratory.

3.2.2 Transmitting Systems

Strain gauge pickups are used in bridge circuits with d-c or a-c excitation similar to the resistance bulbs already described. If a recorder type instrument is used, then the bridge is balanced by mechanical feedback to one or other of the arms.

One advantage of using a mechanical diaphragm and strain gauge combination is that a wide dynamic response is obtained with frequencies up to

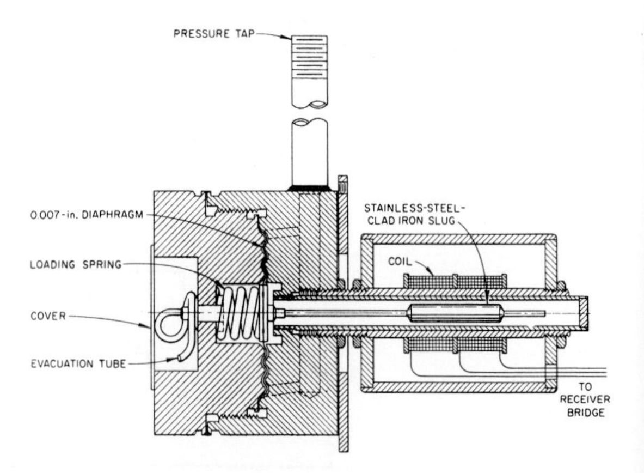

FIG. 3-8 Weld-sealed, electric, absolute-pressure transmitter. (Foxboro Instrument Co.) Material: type-347 stainless steel; range: 0-300 in. H_2O (0 to 560 mm Hg); over-range: 2000 psi (140 kg/cm^2); design temperature: 70°F (21°C); maximum allowable temperature: 140°F (60°C); calibration accuracy: 3%; maximum hysteresis: 1% body rating: design pressure = 4000 psi (280 kg/cm^2), test pressure = 6000 psi (420 kg/cm^2); output: 1 mv/volt input over full range.

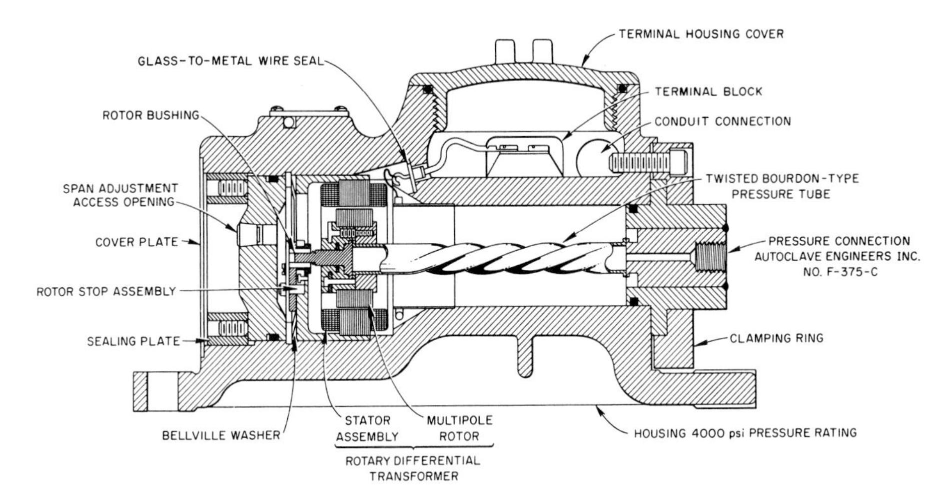

FIG. 3-9 Twisted-Bourdon pressure transmitter. (Norwood Controls Div.) Process-containing parts: type-347 stainless steel; range: 0–2500 psig (0–176 kg/cm^2); output signal: 0 to 8.5 volts, 60 cycle/sec open circuit; error: ± 1%; test pressure: 5000 psig (350 kg/cm^2); pressure rating without rupture: 12,500 psig (880 kg/cm^2); maximum environmental temperature, 200°F (93°C).

several kc/sec being recorded; the fragile wire and its temperature sensitivity are a disadvantage. Miniature unbonded types are available with 1% accuracy, giving 20 mv output with 10 v excitation. The unbonded types are more subject to failure if vibration (due to pumps, etc.) occurs.

One of the most rugged and useful indicators that can be associated with mechanical diaphragm pressure sensors is the magnetic coil or differential transformer. Both linear displacement (Fig. 3-8) and rotary motion (Fig. 3-9) are used. In this device small linear motions of an electrically excited core are differentially picked up by two sensing coils. The output signal changes in phase as the coil passes the balance point. This signal can operate a position-indicating motor with mechanical feedback forcing electrical balance. Alternatively the variation in coil coupling can change the output amplitude of an oscillator, which in turn provides the output signal and feedback force. In force-feedback systems the displacement of the moving coil is held to a minimum, reducing nonlinearity, corrosion and frictional effects. These circuits are largely insensitive to small changes in excitation frequency or voltage and give accuracies of 0.25%.

Variable reluctance pickups and eddy-current pickups both can be utilized in a similar way. Stress can also directly change the permeability of a magnetic core and be used to measure pressure. Such devices can be made small in size.

Small linear displacements can be accurately measured by the change of capacity between two close surfaces. One plate can be the outer wall of the unit, containing the pressurized fluid. Very small displacements (1 mil or 0.025 mm) can be measured. The capacity probe is connected into a tuned circuit where changes in amplitude, frequency or phase are taken as indication of the movement. Methods for minimizing changes in transmission

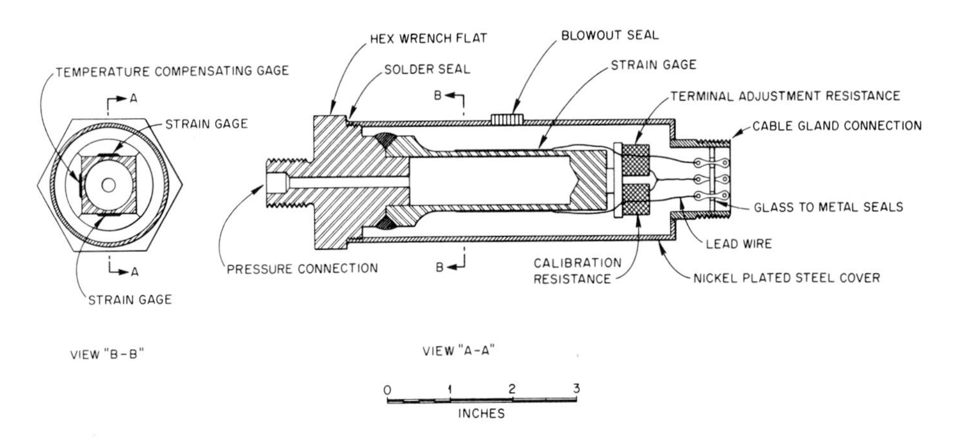

FIG. 3-10 Strain gage pressure transmitter. "SR-4" resistive strain gage used. (Baldwin-Lima-Hamilton) Material in contact with process fluid: type-410 stainless steel; range: 0–5000 psi (0–350 kg/cm^2); accuracy: 1/4% of range; output signal: 1.000 mv/volt; maximum operating temperature: 150°F (68°C); bridge resistance: 120Ω; overpressure test: 200% of range.

cable capacity have been developed [107]. Displacement due to changes in temperature must be carefully considered.

The piezo-electric device operates by distortion of a crystal where the output voltage is proportional to the distortion. These devices (and lately semiconductor devices) are very useful for transient pressure measurements from a few cycles/sec up to several kc/sec. They are well capable of standing shock but not high temperature.

3.3 Flow

In most process systems, and in the reactor system especially, bulk movement of large quantities of fluids are involved. Accurate knowledge of the amount of fluid mass moved is important to establish maximum operating power and peak efficiency.

Some of the problems in flow measurement are concerned with the operating conditions—temperature, pressure, and the homogeneity of the fluid (in steam-water mixtures).

3.3.1 Head Flowmeter

In its simplest form this consists of some method of temporarily increasing fluid velocity so that a local decrease of static pressure is obtained. Flow rate is determined by pressure drop. An orifice plate is the most direct form. Flow nozzles and Venturi tubes are more sophisticated types that are most useful where wet steam is used or where low pressure-losses are required [107, 108]. Pitot tubes with small openings facing the incoming fluid find applications where flow profile measurements are required, but are less useful for wide-range precision measurements. Venturi nozzles with differential pressure cells are used on the Shippingport Reactor [5c].

In measurements involving steam flow condensate, drainage should be controlled so that the contribution to pressure head because of unequal head levels does not cause erroneous ΔP indication.

In multi-channel arrangements common wet-legs may, in testing, become throttled and cause a common fault (see Sec. 2).

Figure 3-11 shows a nozzle arrangement for flow measurement. Since a square-root relationship exists between flow and ΔP, any signal damping that is required must be applied after the extraction of the square-root. Also because of the square-root relationship the dynamic range of a single instrument is restricted to about 4:1. (One method for extracting the square-root from the electrical signal on a reactor feed-water flow system is given in reference [9a].) If a mass-flow measurement is required, then corrections for pressure and temperature should be made, especially for gases whose density is more sensitive to these parameters than liquids. Corrections are also required for coolants where the pressure drop is a function of fluid temperature. This is especially so for organic coolants.

A differential pressure cell that can operate at a high absolute pressure is shown in Fig. 3-12 [106]. This is used on HRE-2 [106].

Two methods for connecting the differential pressure cell are illustrated in Fig. 3-13. A steam flow measurement is shown in Fig. 3-13a. The pipes to the ΔP measuring instrument are filled with equal heights of condensate. A gas flow measurement is shown in 3-13b. Liquid seals are used if the gas contains corrosive elements.

Isolation and equalizer valves (usually manual) are positioned as shown for testing purposes. It should be noted that faulty operation of the equalizer valve disables the instrument. The valves are often used so that the instrument can be removed for calibration. Water is often used for calibration purposes. Drainage of condensate is important, especially if corrosion can result [67d].

Tables showing typical measurement ranges are given in references [107, 108].

3.3.2 Area Flowmeter

This device is basically a float supported by the vertically moving fluid in a tapered vertical pipe, the pipe diameter becoming wider at the top. The float is then positioned to a place which de-

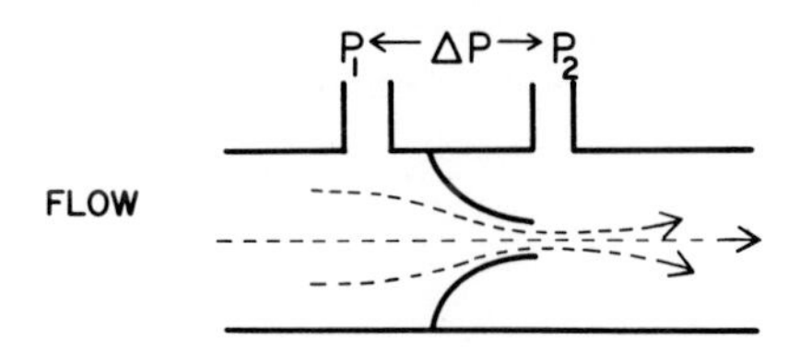

FIG. 3-11 Pressure-drop orifice for volumetric flow measurement.

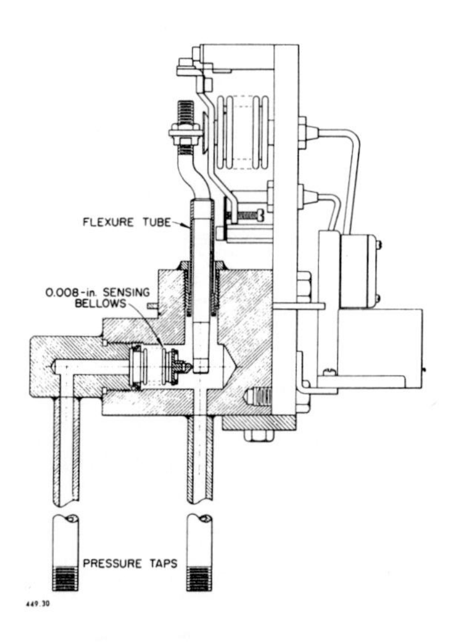

FIG. 3-12 Pneumatic cell for measurement of differential-pressure at high pressure levels. Material: type-347 stainless steel; range: ±50 psi (±3.5 kg/cm^2); overrange: ±100 psi (±7.0 kg/cm^2); design temperature: 70°F (21°C); maximum allowable temperature: 140°F (60°C); calibration accuracy: 2% maximum hysteresis; 1%; body rating: design pressure = 4000 psi (280 kg/cm^2), test pressure = 6000 psi (420 kg/cm^2); output: 3-15 psi (0.21-1.05 kg/cm^2) air.

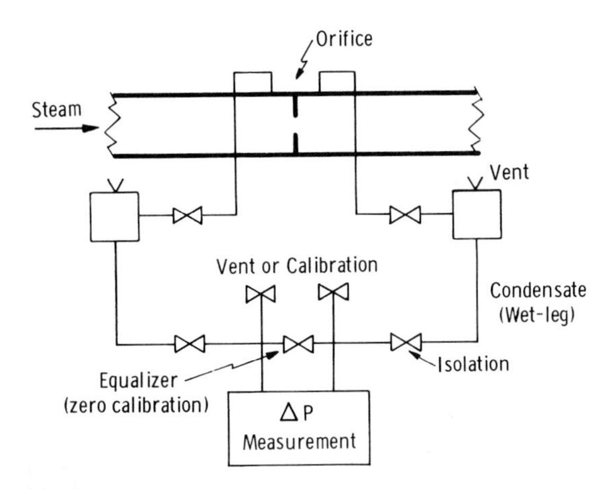

FIG. 3-13a Measurement of steam flow.

pends on the flow as shown in Fig. 3-14 [67b]. In older units the float rotated (rotameters). Various relationships between float height and flow rate can be made, good linearity being possible by controlling the taper and float shape.

In most cases the float position is read visually, however, simple extension rods have been used in combination with penumatic and electronic pick-offs for remote indication or alarm. In the case of magnetic coil indicators care must be taken since foreign particles (e.g., metals) may be attracted by the magnetic field and foul up the passage.

These indicators are useful for low flow-rates, such as may occur in activity monitoring systems, and can cover the range 1 cm^3/min to 400 gpm (25 liter/sec) of water and 10 cm^3/min to 500 cfm (240 liter/sec) of gas.

3.3.3 Displacement Meters

These units are most useful for metering purposes where precise monitoring and direct control (by number of rotations) is required. Such application may be necessary for control of liquid poisons passed into reactors, or control of pH changing liquids. Various types of piston displacement are possible [107, 108].

3.3.4 Heated Bulb Flowmeter

A heated resistance bulb, or a temperature sensitive device surrounded by or preceded by a heater can give an indication of coolant flow [67b]. The sensitivity is much greater in gases than for liquids; calibration is dependent on location, skin effect, flow pattern and thermal conductivities. This method is most useful for coarse alarm following a large change in flow conditions.

3.3.5 Turbine Flowmeters

A rotating turbine and densitometer can be used together to indicate mass flow-rate. The turbine

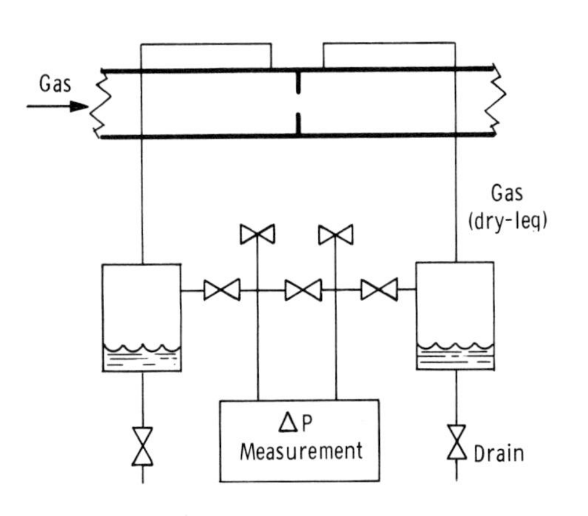

FIG. 3-13b Measurement of gas flow.

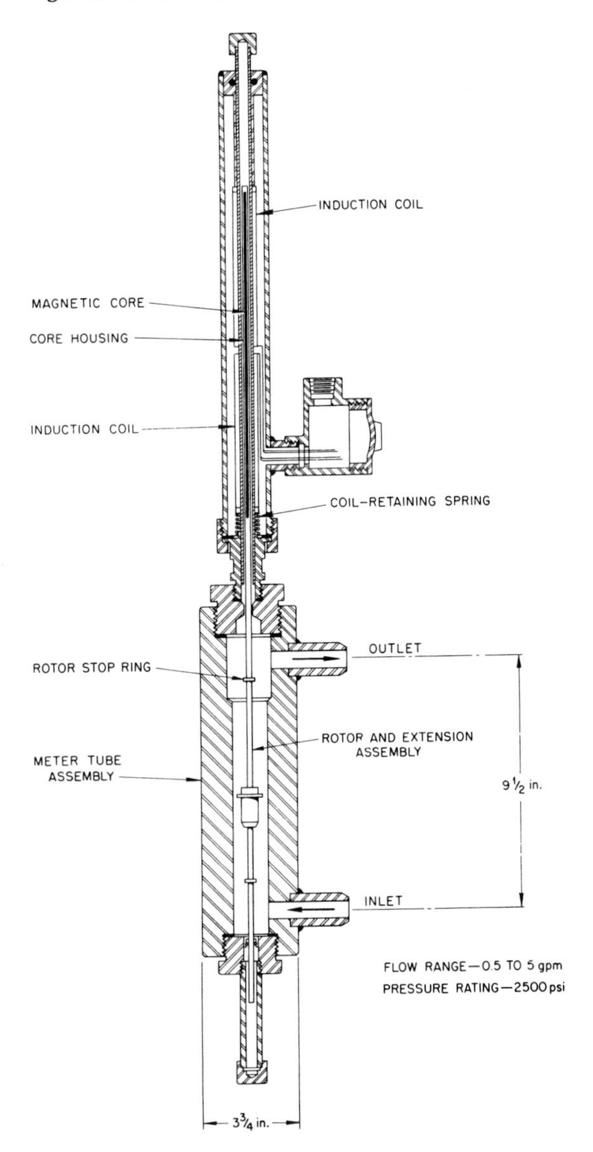

FIG. 3-14 Variable-area flow transmitter.

output is a train of pulses proportional to flow velocity which is converted to an analog signal and then multiplied by a density signal to give mass-rate. A turbine arrangement used in the Halden reactor is shown in Fig. 3-15 [46]. Bearing corrosion or burn-out due to pressure transients restrict the use of these instruments to noncorrosive and smooth flow conditions, and restrict their use to accessible areas.

Methods using impellers and differential transformers together with densitometer sensors have also been described [107].

3.3.6 Acoustic Flowmeters

Acoustic (ultrasonic) measurements have the advantage of not disturbing the existing flow pattern, and of using electronic components throughout.

Crystal transmitter generators are placed down and upstream along the fluid pipe and trains of high frequency signals are passed around each loop, as shown in Fig. 3-16. The difference frequency depends on the fluid velocity [107]. The high cost of the associated electronic equipment has restricted the use of ultrasonic flowmeters. A radiation tolerance of 10^9 r has been noted [104].

3.3.7 Magnetic Flowmeters

This is basically of use only when the fluid is conducting, liquid metal falling into this category [9a, 67e, 121]. Conductivities in excess of 10 μmho are required. The output is linear and low flows (down to 1 ft/sec or 30 cm/sec full-scale) can be measured. Figure 3-17 shows the arrangement. The magnetic field can be direct or alternating (for example, at 60 cycle/sec) giving an output voltage at this frequency proportional to the field and fluid velocity. Because of the small signals involved it is important to establish good electrical contact between the liquid and the signal pickup plates.

3.4 Level Measurement

This is a requirement primarily concerned with liquids. The height of the moderator in D_2O- and H_2O-moderated reactors is probably the most important of these. Others are coolant and feedwater liquid levels, steam boiler and emergency cooling tank levels.

Direct level measurement includes contact with the surface by a probe and ultrasonic (or distance) techniques; indirect methods are pressure-head measurements and electrical capacity based on a change in dielectric thickness.

3.4.1 Visible Level Gauges

These are most useful for direct determination of liquid level and can be applied to moderator level indication requiring precise measurement (such as at startup) or for the calibration of other (indirect) instruments. Levels of a few tenths of an inch (~ millimeters) can be read. If the vessel is pressurized it may not be possible to use thin glass tubing and special constructional precautions should be taken [107]. The temperature of the liquid in the measurement tube should equal that in the measured volume. Remote viewing devices such as television cameras may be employed when direct viewing is difficult. Another possible technique is the use of a servo-driven photocell.

3.4.2 Float Gauges

Various types of float gauges are in use. These comprise some type of ball or light float connected

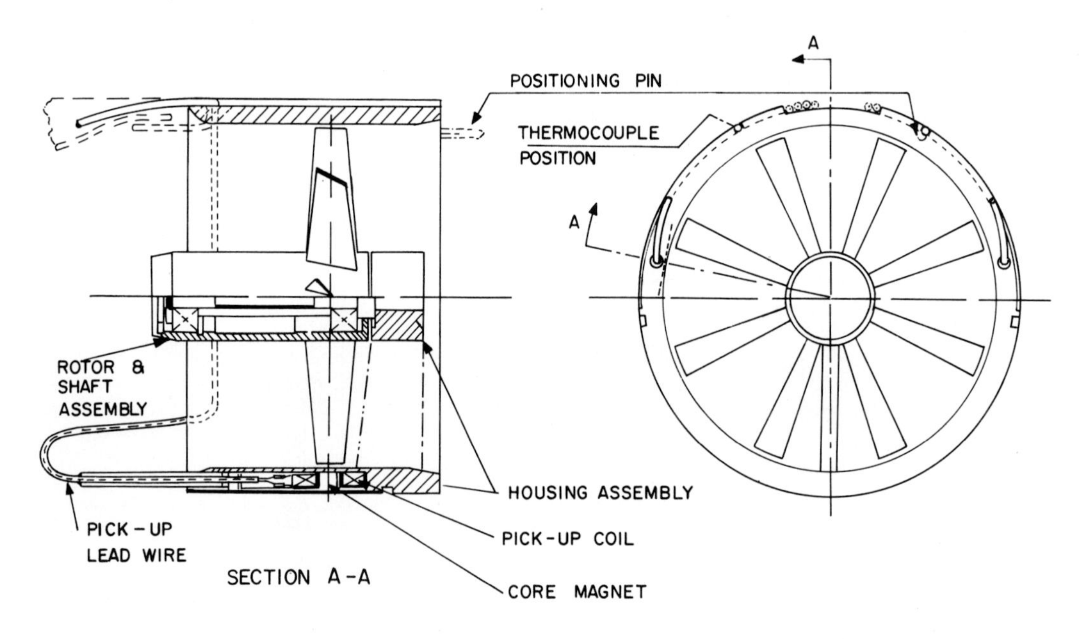

FIG. 3-15 Turbine flowmeter.

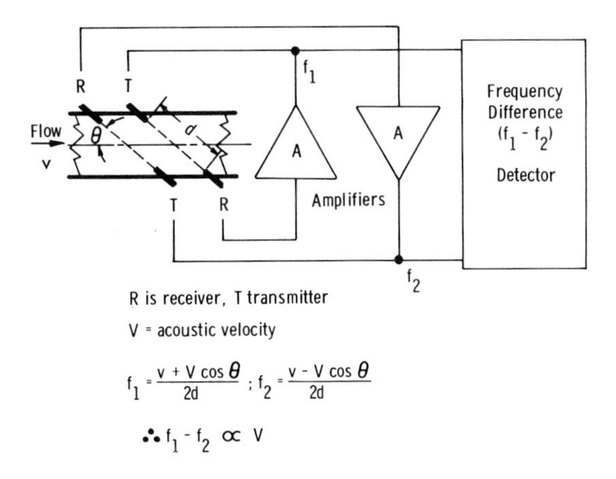

FIG. 3-16 Acoustic flowmeter.

by linkages to pneumatic valves [107] (for direct level control) or to pneumatic transmitters for remote indication (as described under pressure sensors). Some solutions (organics especially) may leave deposits on the float, changing the indicated level. Use has been made of magnets floating around a tube with a magnetic field sensor inside for level measurement. Problems due to sticking can occur. A hydraulically damped gauge with a differential-transformer transmitter developed at ORNL is shown in Fig. 3-18 [67b].

Other floats produce torques in restraining arms and force-balance instruments (pneumatic usually) may also be used.

3.4.3 Differential Pressure

Such units measure the pressure head developed by the liquid; this is one of the more common methods for liquid level measurement. Standard pressure measurements are taken as described in Sec. 3.2 [107, 108], and in such cases other forms of pressure disturbance must be kept out of the system. These can be due to turbulent motion from coolant circulation or from movement of mechanical devices. When measuring moderator level, errors may arise during rapid changes (such as reactor dump) due to shock-wave phenomena. In these systems a return is made from the vessel top, and if the space above the liquid is filled with saturated vapor, then a filled line (wet-leg) return pipe is used to avoid errors due to condensation, shown in Fig. 3-19. The volume of the condensate trap is determined by the volumetric displacement of the pressure-sensitive diaphragm. Entrained gas must be excluded from these systems (especially important in setting-up) since a zero-offset due to a different static head can result. Pressure errors due to changes in temperature can occur in wet-line systems; this is especially significant for organic fluids where the density may change by 20% for a 500°F (or 280°C) change in temperature. Automatic or manual correction by temperature is then required. One advantage of this technique is that the ΔP instrument can be remotely located if radioactivity or other severe operating conditions exist.

Several ΔP cells and transmitters used at ORNL are shown in reference [67b].

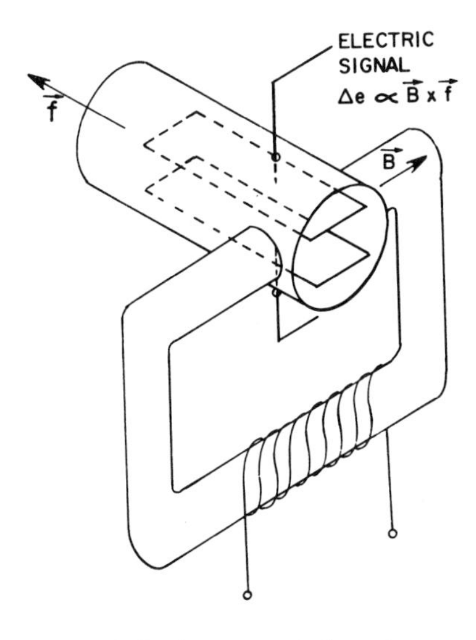

FIG. 3-17 Electromagnetic flowmeter.

3.4.4 Acoustic Methods

Ultrasonic measurement of liquid height by timing a high-frequency pulse of energy can very accurately (to 0.01 in. or 0.2 mm) determine level. Reflection from within the liquid-gas interface or from without can be used. Compensation of propagation velocity for changes in density or temperature may be necessary. Difficulties exist when the surface is disturbed (i.e., in boilers and in some cases the surface level of liquid moderators) since the returning wavefront can be widely scattered.

3.4.5 Radiation Measurement

Two principles can be used: the variation in absorption of radiation by the thickness of material between a source and a counter or the variation of the distance of a floating source from a counter. These relationships are nonlinear.

Radiation counting techniques have been described in Sec. 2.

Care should be taken in a reactor plant to ensure that level techniques based on radiation measurement are not affected by background radiation or by the presence of activity following a process fault.

3.4.6 Contact Probe

Resistance or conductivity probes can be used either directly within the tank or in a tube connected to it. Conductivity probes are usually driven up or down until surface contact is measured. Even low-conductivity liquids can be measured. The technique involves a point measurement where small surface disturbances may be misleading or difficult to measure. A contact probe used on the Halden reactor is described in reference [46]. Others suitable for use on liquid-metal-cooled reactors have also been described [67e].

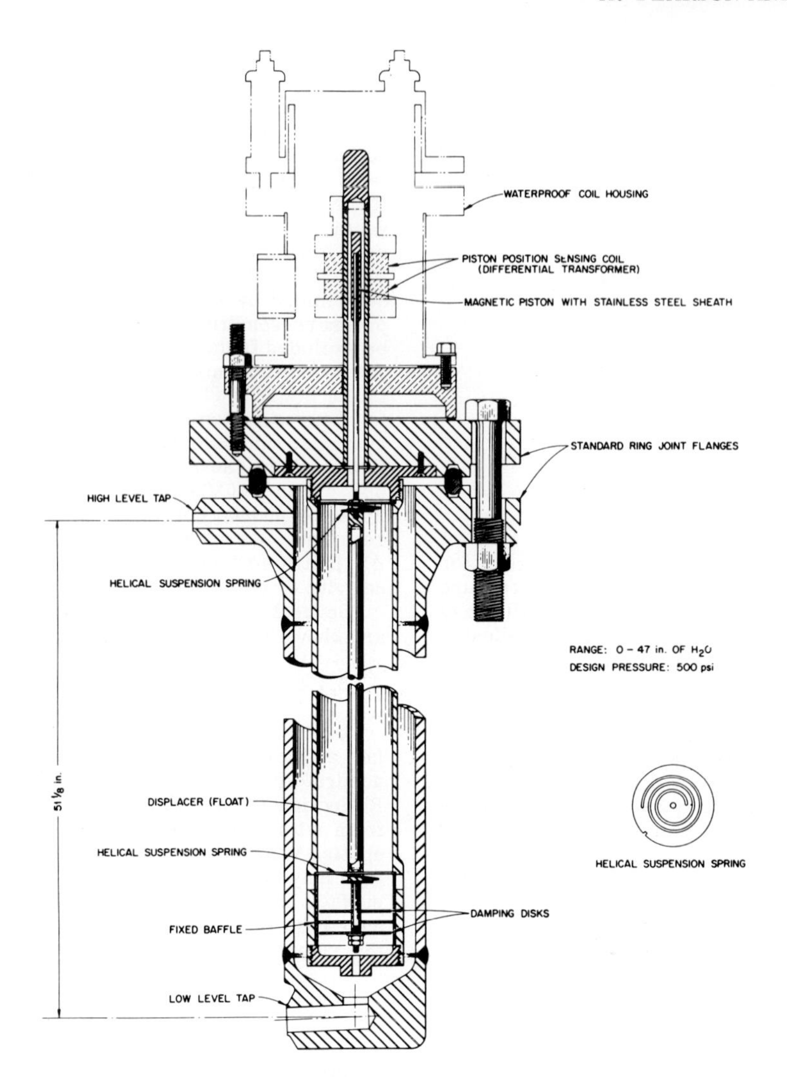

FIG. 3-18 Hydraulically damped displacement-type level transmitter.

"Continuous" probes, made up of a large number of two-element contacts can be used in a fixed position and give discrete indication of small changes in level. Connecting cables must be run to all contacts. Figure 3-20 [67b] shows a level probe used to measure the level of a NaK mixture. The mixture should have a lower conductivity than

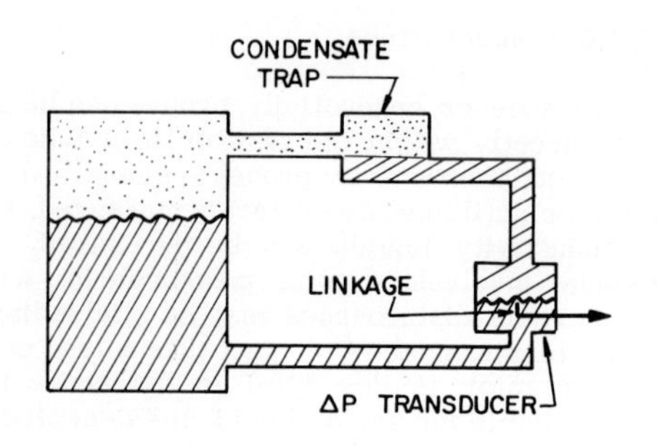

FIG. 3-19 Level measurement using differential pressure.

the pipe; it then forms a short-circuit across the pipe resistance [67f].

Heated temperature sensors are extremely sensitive to removal of liquid coolant and can be used to operate an alarm when the level drops below a pre-set value [67b]. These devices are useful where the liquid-gas interface is not static (such as in a boiler) and where the "ability to remove heat" is a sensitive indication of liquid presence. They are not useful when the liquid exceeds a temperature of about 500°F (~300°C).

3.4.7 Inductive Probe

Inductive probes have been used for level measurement and they are especially useful for liquid-metal fluids having high electrical conductivity. Figure 3-20b shows one arrangement [121], the level being sensed by differentially wound coils, the balance is upset by the amount of liquid effectively short-circuiting the lower coil. Temperature effects must be compensated.

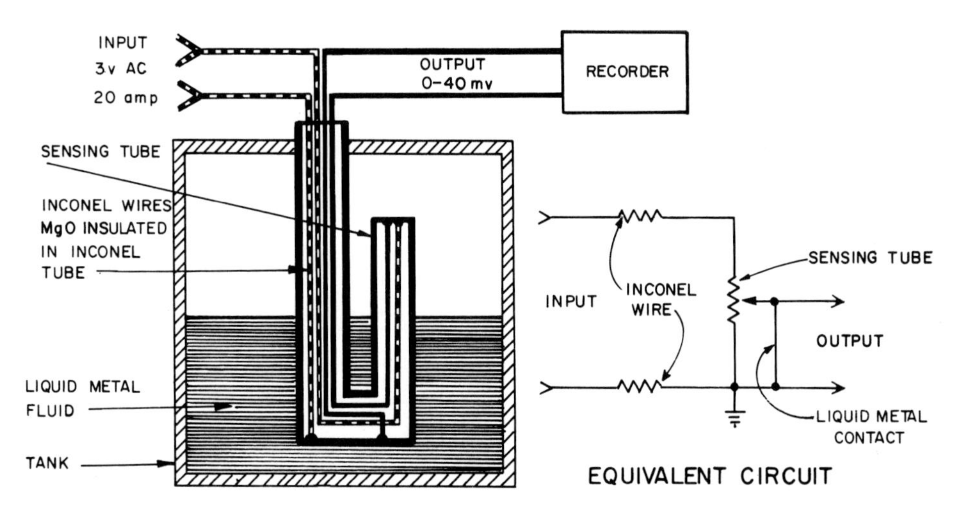

FIG. 3-20a Diagram of J level probe.

3.4.8 Dielectric Measurement

Electrodes inserted into the liquid can be used as capacity plates; variation of liquid height changes the capacity because of its dielectric properties (the dielectric constant of air = 1, that of water = 80). The tank itself can be used as one electrode, generally at ground potential [107]. This should be checked to ensure it is not a potential source of electrical interference with other sensitive electronic units, especially under fault conditions.

3.5 Power Measurement

Reactor output power, turbine inlet power and generator output power are three important measurements because they lead to a determination of plant efficiency.

Output is determined by multiplying temperature rise across the power source by coolant mass-flow. A simple analog computation provides a ready technique, multiplication often being done by a slide wire on a differential temperature recorder.

The power generated by mechanical heating from pumps should be taken into account. Transient differences between the various parameters in the calculation should also be watched and appropriate smoothing used (flow measurements as a rule being obtained faster than temperature).

In these calculations scaling and correction of nonlinearities (flow measured by differential pressure) can be significant sources of error; if such effects can change, consideration should be given to digital calculations where flexibility is more easily obtained.

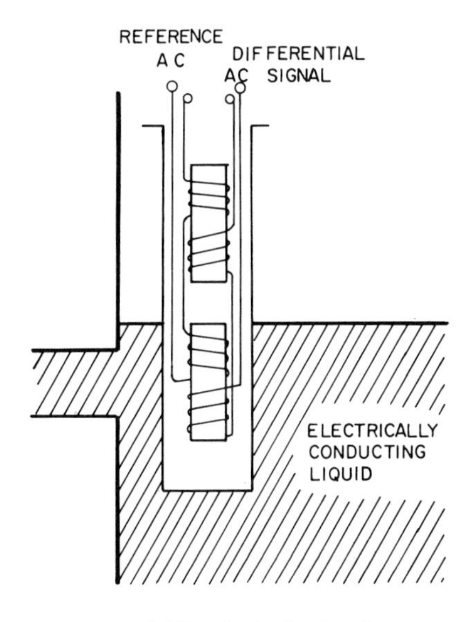

FIG. 3-20b Inductive level probe.

3.6 Steam Quality

This is an important measurement since the steam quality affects turbine operation (wet steam being most common in reactor plants); in the case of boiling (or water-steam) reactors steam quality is proportional to generated power.

Quality is the fractional weight of the vapor. For steam-water mixtures measurement of flow of the vapor and liquid is sufficient provided the mixture is homogeneous (i.e. the slip factor is unity).

Three basic methods are available: A direct measurement as defined by the definition given above; a thermodynamic measurement; and the measurement of some parameter that is directly related to steam quality. The first two methods require the removal of a sample.

3.6.1 Direct

In one method the sample of steam and water is separated by letting the mixture negotiate a sharp bend. Time is required for the accumulation and weighing of the samples. Other methods of separation (by swirling as in a centrifuge) have been suggested. A quality range from 50 to 70% can be treated by this approach.

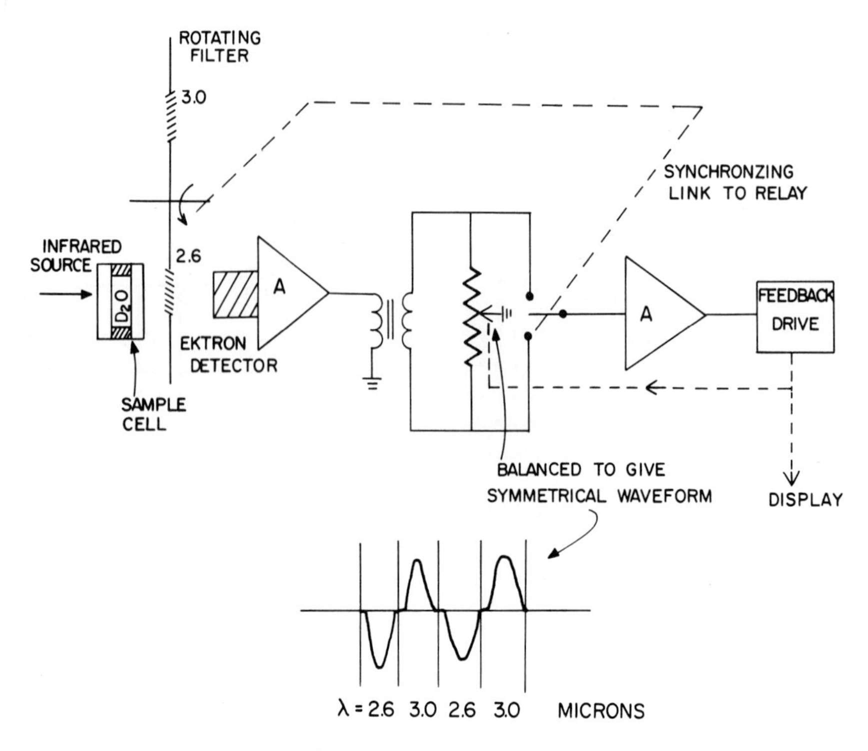

FIG. 3-21 Infrared heavy-water monitor.

3.6.2 Thermodynamic

A throttling calorimeter can be used to dry the mixture. Since the inlet and outlet enthalpies are equal and the inlet pressure and outlet pressure and temperature can be measured, the initial quality can be calculated. This method is most suitable if the quality is high, i.e., 93 to 97%.

Addition of heat by electrical means or cooling with another fluid are convenient ways of changing the quality to a known end state. A small representative sample of the mixture must be taken, otherwise the power transferred becomes inconveniently high. The temperature of the sample indicates the change of state and a measurement of pressure permits the quality to be calculated.

3.6.3 Indirect

Several approaches are being studied in this area [123]. These involve search for an arrangement by which some property of the mixture can be calibrated in terms of quality. Examples are measurement of the pressure head [124, 125] due to a steam-water mixture flowing past a restriction, by gamma ray absorption across a pipe [126, 127], by the change of electrical impedance across a pipe section [128], by the drag of an immersed object [129], and by acoustic velocity measurements. Such techniques involve very different types of analyzing equipment and it is too early to know which will be most useful and reliable.

3.7 Moisture in Coolant Gas

The moisture content of the gas in gas-cooled reactors must be kept to a minimum in order to prevent corrosion. Corrosion could result if moisture in the gas condensed out following a shutdown. Some moisture can appear from graphite if this is used as the moderator or as a thermalizing region for ionization chambers. It has been proposed [130], in filling CO_2-cooled graphite moderated reactors, to pass all the gas through silica-gel, and during operation bleed off 0.5% through a recirculation drying loop. Moisture could then be maintained below 0.01% (below the dew-point at room temperature).

Another reason for monitoring moisture is that any water leak in the high pressure heat exchanger tubes may be detected. Two methods of measurement have been developed [131]. One of these uses an infrared beam and detector with the beam passing through a sample of the gas. A sensitivity of 500 ppm (by volume) was reported. Another technique uses the change in electrical conductivity of P_2O_5 exposed to the moisture in the gas.

3.8 Detection of Leaks into and out of the Heavy Water Moderator

In D_2O-moderated reactors the presence of H_2O in the moderator must be avoided. A few hundred parts per million can incur an economic penalty because of reduced fuel burnup. Further, the cost of D_2O is sufficiently high that any loss from the system must also be avoided. Leakage that might result from defective heat exchanger tubing can be detected by measuring any change in the D_2O content of the H_2O secondary coolant.

While the escape of D_2O can produce a tritium health hazard, the economic incentives to keep system integrity high favor greater safety.

Infrared analyzing equipment has been developed for these measurements. The method is based on the different molecular vibration frequencies of the H_2O, HDO and D_2O molecules. A very strong absorption occurs in light water at 3 microns and at a lower frequency, corresponding to 4 microns, in heavy water. Two measuring techniques can be used: one measures the absorption in the water sample and compares it to a reference at a fixed wavelength, the other method uses several wavelengths to obtain a spectral distribution. The second method is less sensitive to small particulate material than the first. Reference [132] gives a good bibliography on this subject.

A system using rotating infrared filters transmitting wavelengths of 3 and 2.6 microns is shown in Fig. 3-21 [132]. This system balances the output signals by a feedback link to the stepping switch and can detect changes of 10 ppm of light water in heavy water. By suitable choice of wavelength the detection of small amounts of heavy water in light water becomes possible.

The detection of small quantities of D_2O can also be measured by monitoring for tritium activity (see Sec. 5.2.4), this activity is dependent on the radiation to which the D_2O has been exposed.

Other methods for finding leaks have been by the extensive use of moisture (humidity) cells, where electrical conductivity measures the presence or absence of water. A triggering sensitivity of 0.5 mg is possible.

Leak detection methods used in a liquid-fueled reactor have been reported [106].

3.9 Data Logging and Computer Techniques Applied to Core Instrumentation

In many reactor systems a large number of sensor signals require logging. This function is sometimes followed by checks for alarm conditions and the initiation of safety and regulation action. Recent advances in solid-state circuit design make possible systems other than those employing individual amplifiers and alarm circuits for every sensor.

The systems to be described here are being applied to power reactors and generating plants where a centralized computation center can be economically attractive. A centralized facility becomes attractive when one hundred or more sensors of a similar type are being considered [133, 134], i.e., especially when in-core instrumentation is concerned, or when complicated or frequent calculations are required.

If the signals (mostly analog) are converted into binary form, digital computer systems can be readily applied for further data processing. Fairly rapid changes can be made to the operating mode by program control, provided such program changes have been anticipated to some degree beforehand. The advantage of flexibility over the more conventional multi-system analog equipment may bring the greatest dividends by shortening the plant commissioning period [135] when new calculations and display forms not normally used may be demanded. The sharing of several problems with one processor is also an essential development (multiprogramming or time-sharing).

Problems still to be resolved are mostly concerned with input and output equipment. This includes low-level multiplexers, program changing by special control consoles, and information display using methods other than typewritten data-sheets [136]. Use of hardware priority interrupt features and other generally optional computer features, including index registers, and real-time clocks, varies with the application.

There still remains some concern about a large plant having all its operating conditions processed by one computer, and various approaches to reliability [137, 138] are given in this section.

The main emphasis in the systems described here has been placed on gathering plant and especially in-core data, and carrying out computations to improve the operators' knowledge of plant conditions. Very few of these examples have been or are proposed to be connected directly into the reactor safety system. However, considered as a group, they are seen to fit into almost every aspect of reactor plant operation, i.e., into general data-logging, core optimization, fuel-rod activity monitoring, regulation and safety.

3.9.1 Analog Per-Channel Approach

When a large number of input signals are being examined, individual sensors can be connected to amplifiers, indicating units and, if necessary, alarm circuits. Such a system is shown in Fig. 3-22. A shared display unit may be used. Channel separation is maintained up to the alarm circuit outputs.

Two major objections arise to these systems. One is the high cost of individual amplifiers and alarm circuits per point. The other is the adjustment, testing and display problems associated with a typical several hundred-point system. It has also become evident that, in the majority of systems, computation based on the amplified outputs is required, and that normalization techniques (due to nonlinearities and background signals) have to be used.

3.9.2 The Multiplexer Approach

The widespread use of transistorized equipment has made other more attractive approaches possible for logging or recording a large number of similar signals. The sensor signals can be sampled in turn by a "multiplexer" at high rate, amplified in sequence by a common high quality amplifier, and each signal level checked by a common alarm circuit and the relevant information stored for display. The output signals may be digitized either before or after the alarm function. This approach is shown in Fig. 3-23.

If the switches are relays then sampling rates of 1 to 100 cycle/sec are usual; if they are solid-

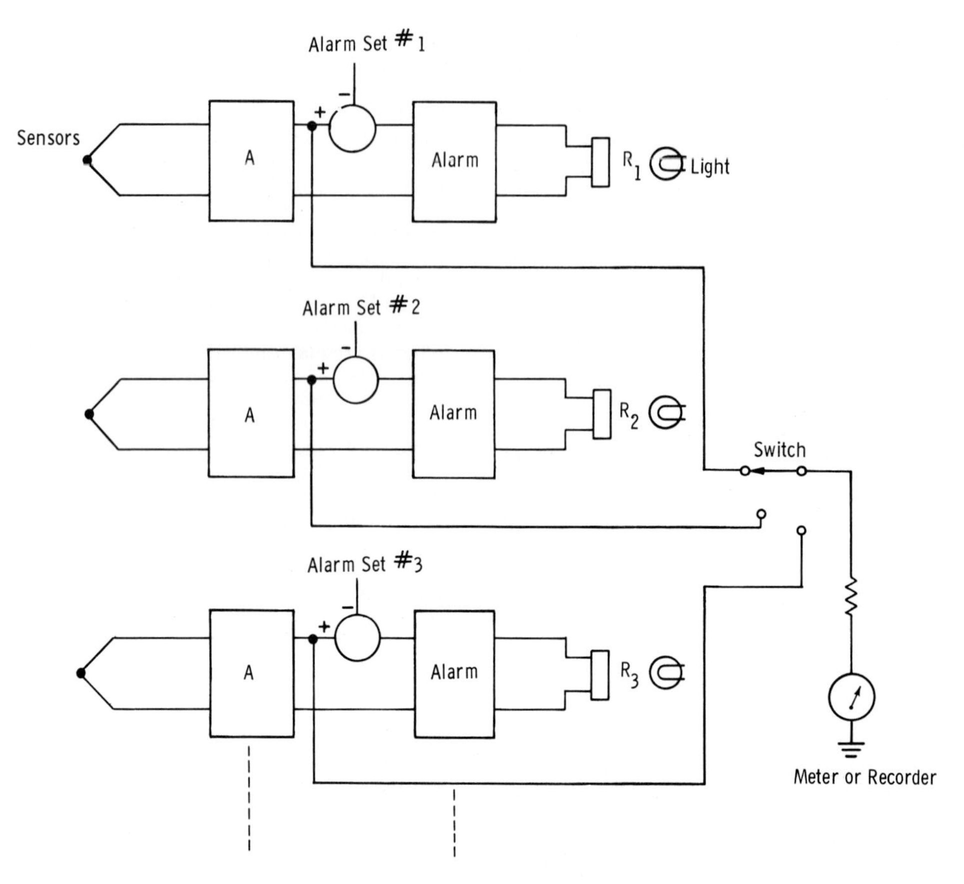

FIG. 3-22 Analog per-channel instruments.

state rates up to 10 or even 50 kc/sec are possible [139]. The higher rates are an advantage if a large number of points are being scanned and short-time resolution is desired for safety or regulation purposes. Also high rates permit a fixed scan sequence to be adopted with some circuit simplification.

The chief disadvantage of the solid-state high speed multiplexer is due to drift and errors in the switches. This is discussed further in the next subsection.

Outputs are in the form of illuminated numerals and/or typewritten sheets. Alarm levels are usually

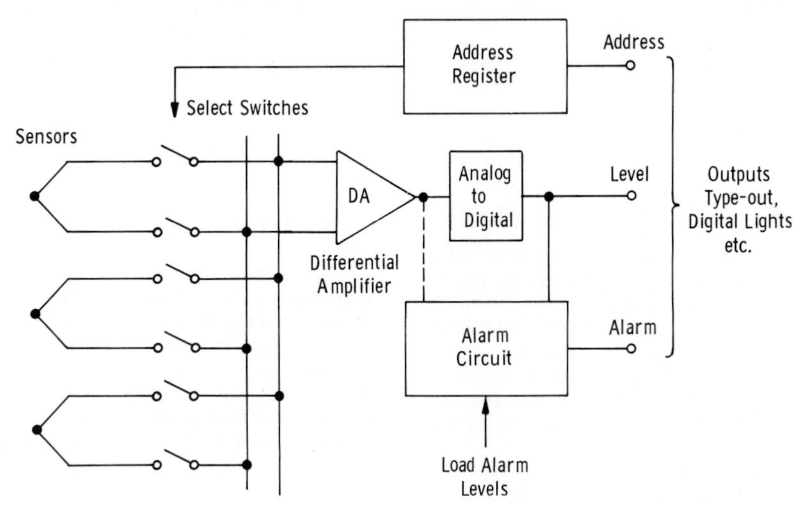

FIG. 3-23 Multiplexer instrument.

fed in from decimal-to-binary conversion switches.

Simpler systems have used a contactor multiplexer and one or more standard multi-point recorders each with a mechanically actuated common microswitch alarm circuit.

3.9.3 Multiplexer Problems

The chief problems associated with logging and computer systems have been with mechanical in-out devices. These are generally inexpensive and can be readily duplicated. However, difficulties associated with matching the input signals to the logging device are not so readily solved, and problems with multiplexer switches are not unusual.

Figure 3-24 shows a multiplexer arrangement usually applicable to relay systems where only one input point is connected at a time to the amplifier. This is known as the "flying capacitor" technique. Isolation of ground loop currents is obtained giving a minimum of interference. Shield lines can also be switched by relay contacts.

A similar approach using solid-state switches would be expensive and so single-pole switching is often adopted. This approach is shown in Fig. 3-25. The advantage of high-speed switching is offset somewhat by the "on" voltage across the switch (typically 100 μv). These switches are not suitable for resolving better than 1 or 2°F (0.6 to 1.1°C) from thermocouples. Also the "off" leakage current (typically 10^{-8} μa) causes errors. Another problem is the greater capacity across the switch (10 pF or so) compared to the relay; this loosely couples each input to the amplifier. High frequency interference must be adequately filtered or screened-out ahead of the multiplexer switches.

The effective input capacity of the amplifier, as shown in Fig. 3-26, must be charged every sampling time. This of course applies to every type of multiplexer with a common output but since the amplifier input capacity is formed by the "off" switches and their signal cables the value is much higher for solid-state multiplexers (1000 pf) than for relay systems (50 pf or so). In order to charge this capacity a source with a sufficiently high capacity or low impedance must be presented by each input line, or the scan rate must be low. This source capacity can conveniently form part of the input analog filter. Filtering is, in any case, usually necessary to prevent signal frequencies higher than one-half the sampling rate from giving false readings [140, 141].

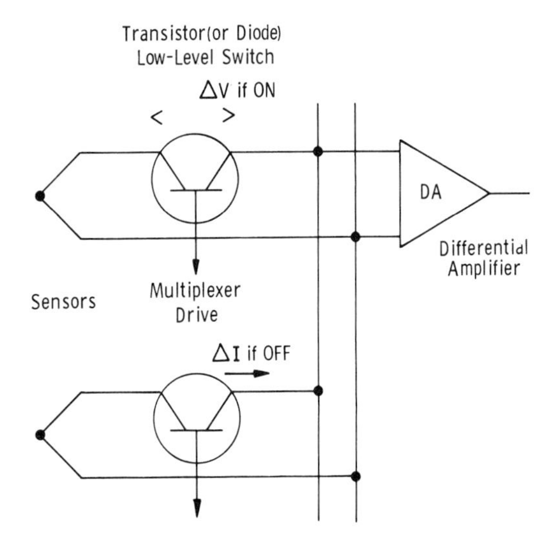

FIG. 3-25 Solid-state multiplexer.

If input signal lines are faulted (for example to the mains), or if a transducer open-circuits, or if a cable line short-circuits, a large (even damaging) error signal may appear. This can charge the amplifier input capacity when that point is sampled, not only giving an incorrect output but also affecting subsequent readings. Limiting circuits should be incorporated into the input filters and amplifier to prevent this.

Other forms of isolation include electrically floating the input switches, as well as using an a-c coupled amplifier. The signal passing through an a-c coupled amplifier would have to be bi-polar

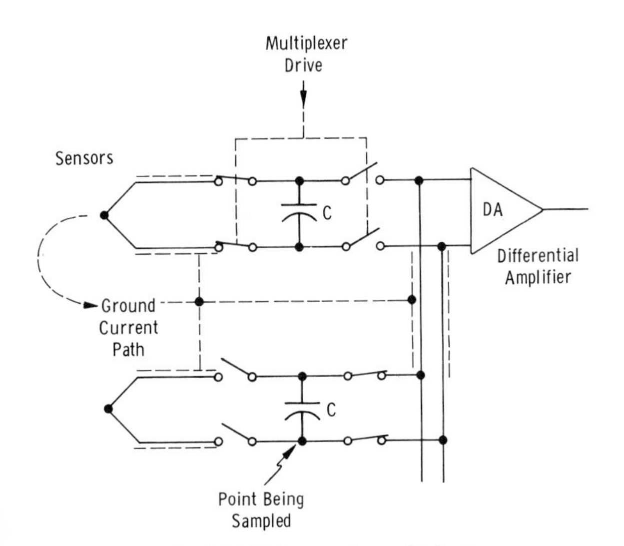

FIG. 3-24 Flying capacitor multiplexer.

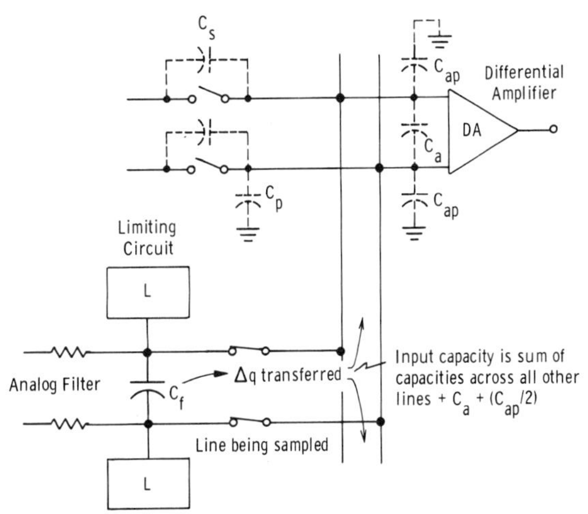

FIG. 3-26 Capacity loading of differential amplifier.

or the amplifier would have to have d-c restoring circuits on its output. As a rule the d-c amplifier connection, although requiring greater care to avoid drift, has had greater success in overcoming gross overloads due to faults.

An important difference between the differential amplifier used for a multiplexer and one used for a single-point measurement is the large bandwidth required for use with a multi-point system. The rise time is set by the scanning rate, the low frequency response by the reduction in pulse amplitude permitted immediately following sampling. This greater bandwidth increases the sensitivity to interference and amplifier circuit noise. Only in single-point systems is it possible to use an amplifier with narrow bandwidth.

Input circuit capacity also tends to convert common-mode interference to differential mode signals, as described in Sec. 3.1.1. Methods to reduce the capacity charge-current include floating the amplifier shield and even driving the shield by the common-mode voltage. The amplifier used must be able to supply sufficient current to charge the shield-to-ground capacity, often difficult if the interference is at a high frequency. The differential amplifier used should have an input impedance of several hundred thousand ohms or more and usually is made of three or more small amplifier units.

Double-banked switches have been used to reduce the input capacity effect. Such an arrangement is shown in Fig. 3-27. The second switch bank may be relay contacts even in a high-speed scanner, as successive banks need only be connected at infrequent intervals.

3.9.4 Computer Systems

The more recently designed systems use a small general purpose computer in place of the special purpose analog-to-digital converter, alarm and display devices described in the previous section. The multiplexer and amplifiers are unchanged but the scan sequence can be controlled by the computer program. The program is executed in sequence and is held in the core memory or other electronic digital store.

The analog-to-digital conversion can be controlled by successive approximation with the computers' arithmetic unit and a sensitive analog comparator. This can be followed by alarm checks with the logical operation "is-the-signal-greater-than-the-alarm?", followed by print-out if it is. The scheme is shown in Fig. 3-28.

An example will help identify the speed of computation required. Consider a continuous scan frequency of 1 kc/sec. One millisecond is available for the calculation associated with each point. Table 3-2 shows the approximate instruction steps for multiplexing, digitizing, checking against three-alarm levels, loading an output buffer and initiating a type-out. Such a program may involve about 37 instructions taking about 555 μsec per point for an average instruction time of 15 μsec. There are several small process computers available today with this computational speed.

3.9.5 Application to Reactor Plants

3.9.5.1 General Data Acquisition System—Shippingport Plant. This system logs and types-out on demand and at preassigned times reactor operating data in engineering units [142]. Flow and temperature signals occupy most of the 300 inputs. The input multiplexer is a slow-speed (5 points/sec) relay unit with program controlled point selection. The system is shown in Fig. 3-29 [146].

3.9.5.2 Core Power Density Computer—Big Rock Point Plant. This computer system was developed and installed to reduce fueling costs by optimization of core operating conditions. Two main calculations are made [143]. The first one is to calculate optimum control-rod positions taking into account fuel exposure and the power distribution. The second one is to calculate the limiting power density by calculating rod center-temperature, steam quality and surface heat flux by using core flow and power distribution data.

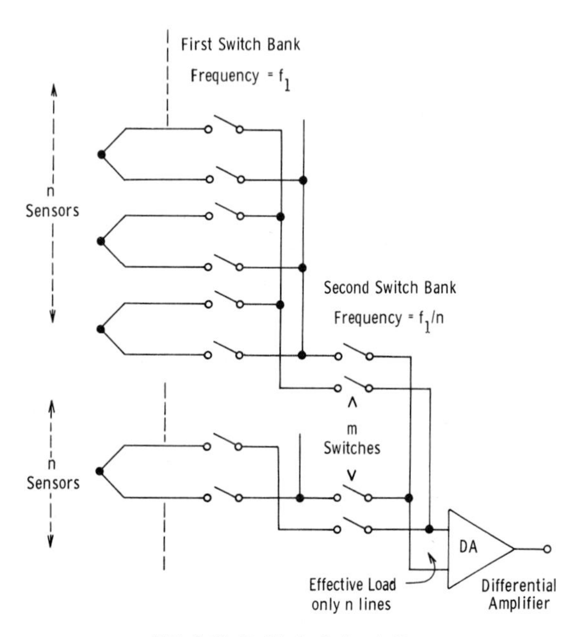

FIG. 3-27 Double banked switching.

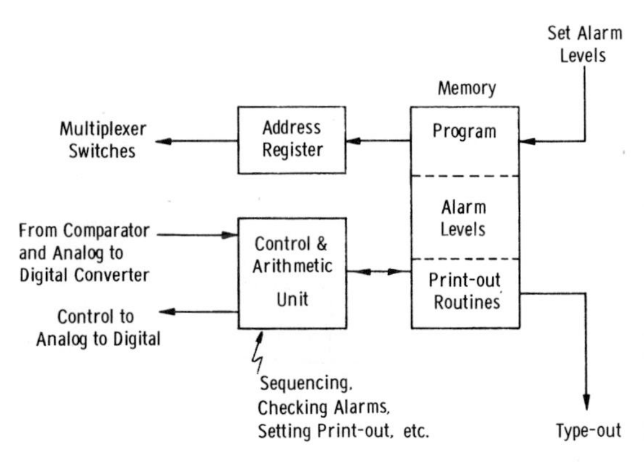

FIG. 3-28 Computer logging instrument.

TABLE 3-2

Computer Program for Logging

Program	Instruction Steps
Sequence	3
A/D Conversion	11
Check 3 Alarms	12
Storing Alarms	3
Initiating Type	3
Restoring Interrupt	5
Total Steps	37
If average time	= 15 μsec
Time per sensor	= 555 μsec

The layout and program function is shown in Fig. 3-30 [143].

The system uses a slow speed (30 points/sec) relay scanner with 115 input signals. Logs on type-out sheets and on punched paper-tape are provided. The computer has a 20-bit word and used 40,000 words of drum memory.

3.9.5.3 Fuel-Rod Activity Monitoring Computers—Chinon Plant. This system was put on the EDF-1 reactor plant which has 1148 gas-cooled fuel channels. The coolant gas from each channel is sampled and the samples are selected by valve systems. All the channels are examined over a 24-min scan cycle time using 12 detectors and coarse scaler channels [144, 145, 146] averaging 4 fuel-rod channels. When high activity is detected, each fuel-rod channel of that group is sampled in turn. The scalers feed pulses directly into the computers. The computer calculates the rate of increase of activity taking into account the changing background activity averaged over a previous reference cycle.

Two computers are used for reliability, connected in parallel and programmed identically. Each machine is an 18-bit word unit with 8,000 words of drum store, as shown in Fig. 3-31 [146].

3.9.5.4 Fuel-Rod Temperature and Plant Regulating Computer—Douglas Point Plant. This computer system [147] has been designed to scan individual fuel-rod temperatures, to alarm on high temperature in coincidence with low flow (from conventional equipment), and through this function the system is connected into the reactor shutdown system. The temperature signals taken as a group give a slow regulation trim signal to the triplicated analog neutron flux regulation channels. In detail these signals are used also to correct flux tilt by controlling the position of absorber rods. This system is shown in Fig. 3-32.

Turbine startup control is also carried out by the computer using conventional turbine analog transducer signals giving load, speed, and vibration data.

Other analog inputs come from 40 count-rate amplifiers fed from neutron sensitive counters. Each counter normally monitors the activity from 8 fuel-rod sample lines, but these lines can be valved so that any combination of sample lines can be checked.

The computer has a 15-bit word length with a core store of 8,000 words and drum store of 48,000 words.

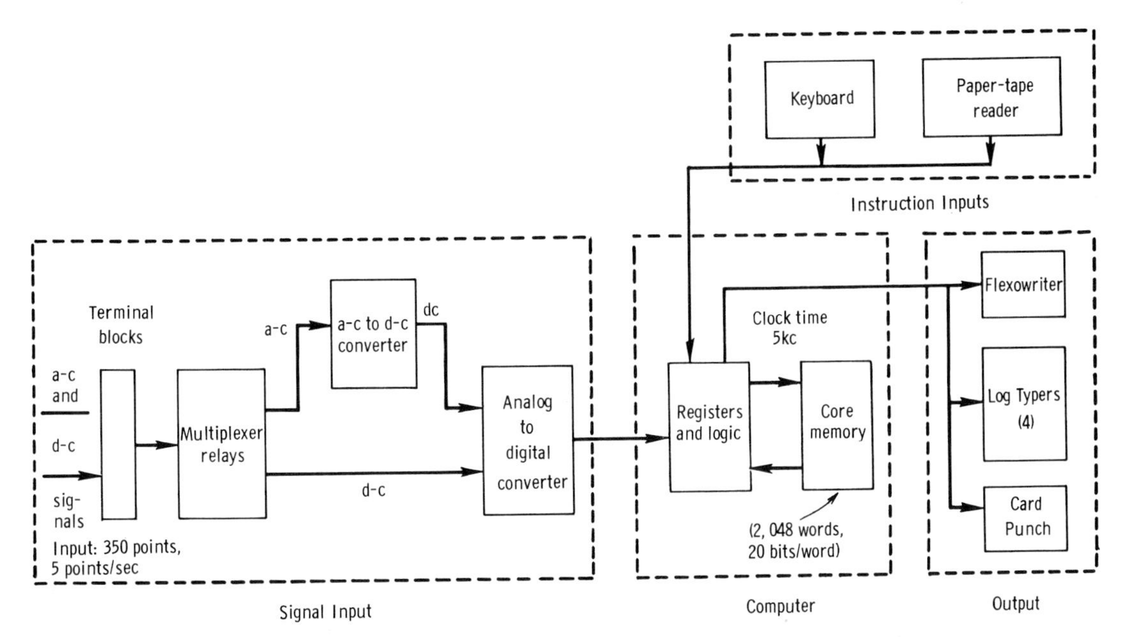

FIG. 3-29 Shippingport data-acquisition system.

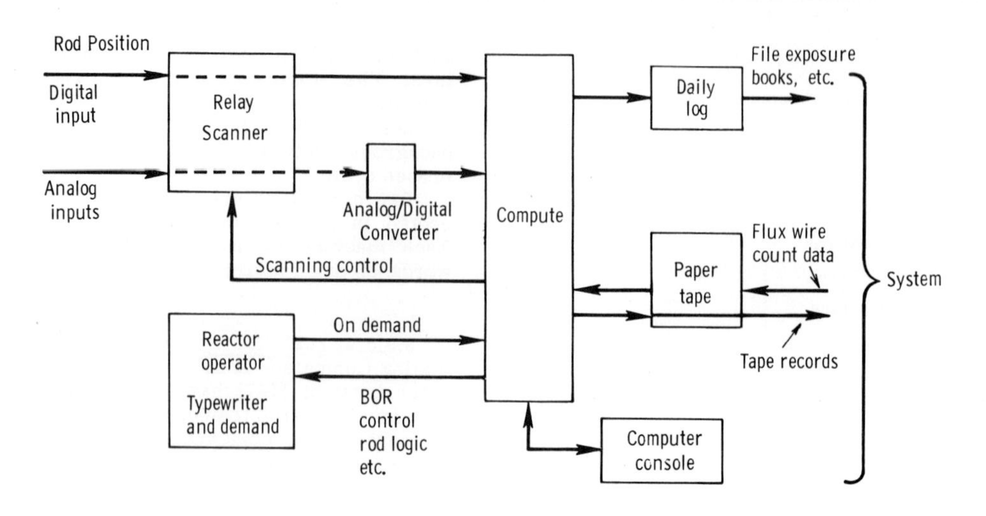

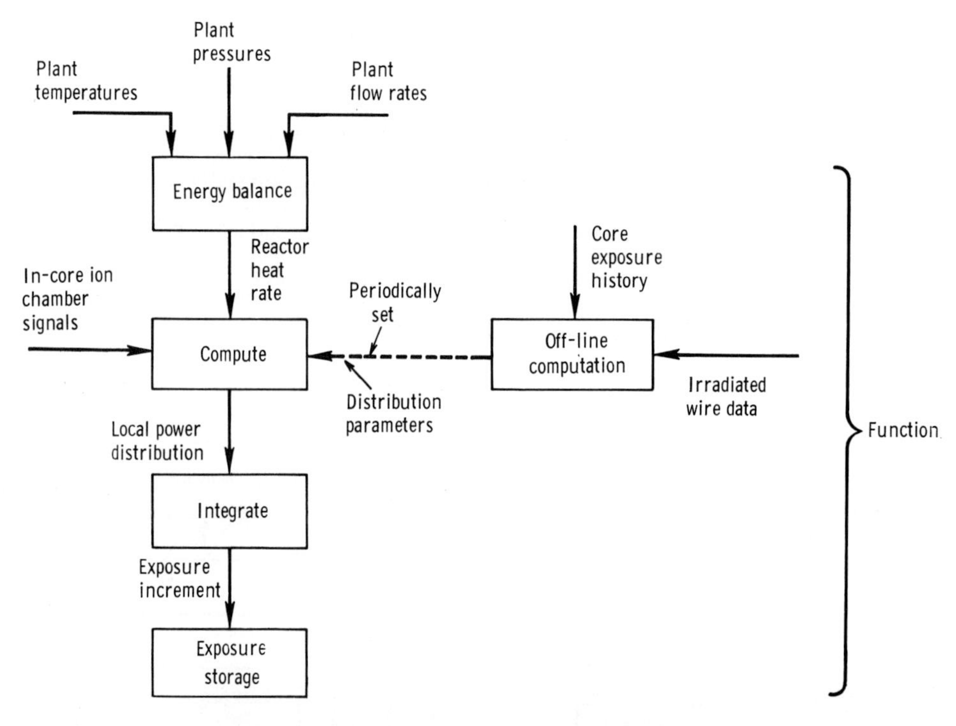

FIG. 3-30 Big Rock Point core-power distribution computing system.

3.9.5.5 Fuel-Rod Temperature Alarm System—Hanford. This system has been under extensive laboratory testing for some time at Hanford [148]. The equipment is all solid-state with analog comparator circuits for alarms and with digital numeric display logic. A cathode-ray-tube display showing isotherm patterns has also been developed where all fuel-rod locations with temperatures above a predialed set-point are displayed in their correct spatial position.

The program is fixed in pin-board core store, but otherwise construction techniques and design are similar to a small special purpose computer.

Inputs are 112 RTD transducers scanned at approximately 5 kc/sec, as shown in Fig. 3-33 [146].

3.9.5.6 Experiment in Monitoring and Computer Control—Chalk River Nuclear Laboratories. The

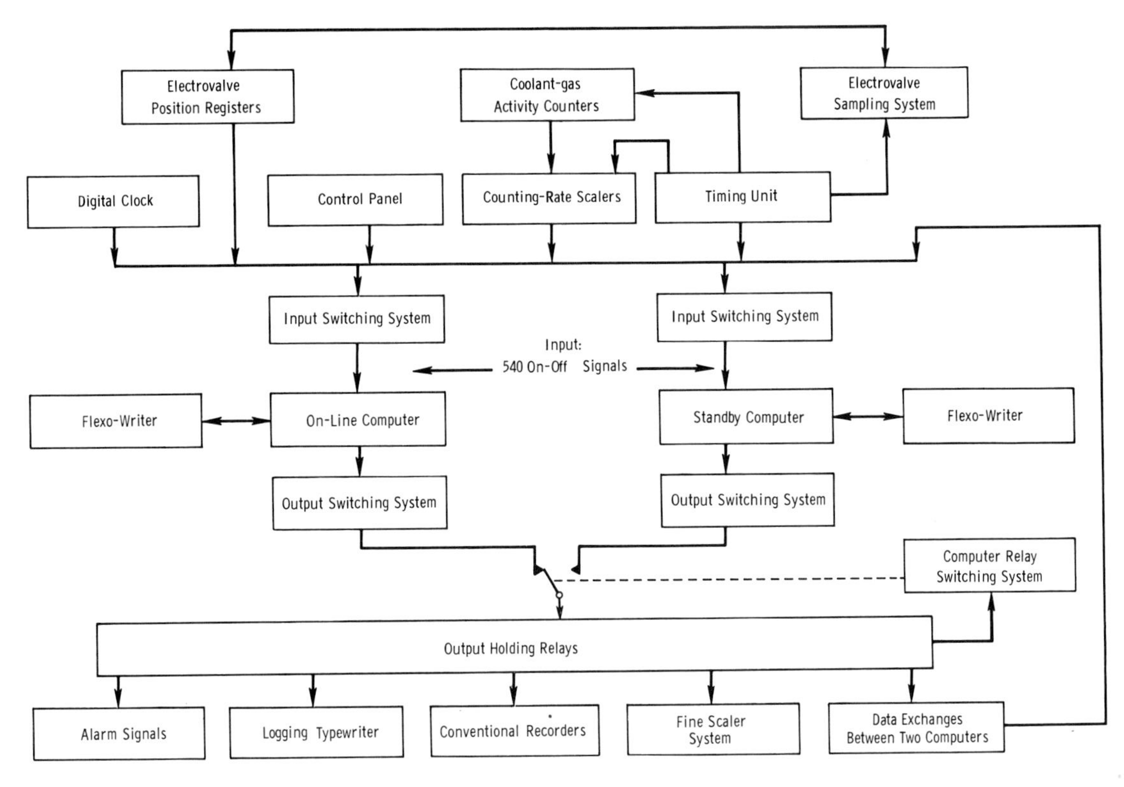

FIG. 3-31 Chinon activity monitoring system.

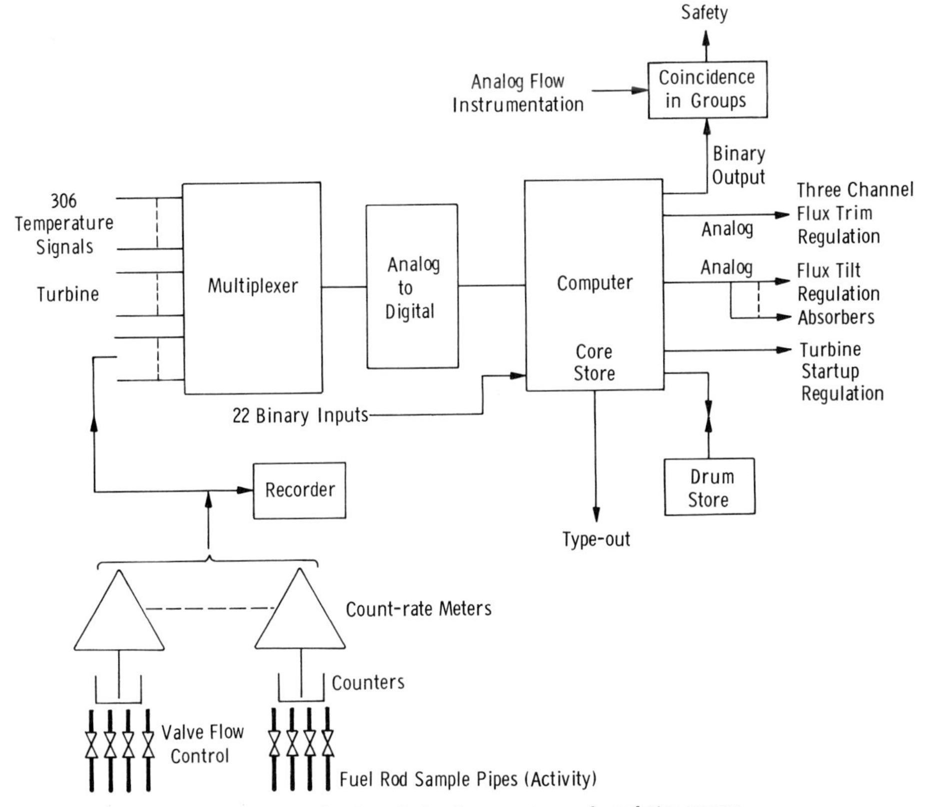

FIG. 3-32 Douglas Point fuel rod temperature and regulation system.

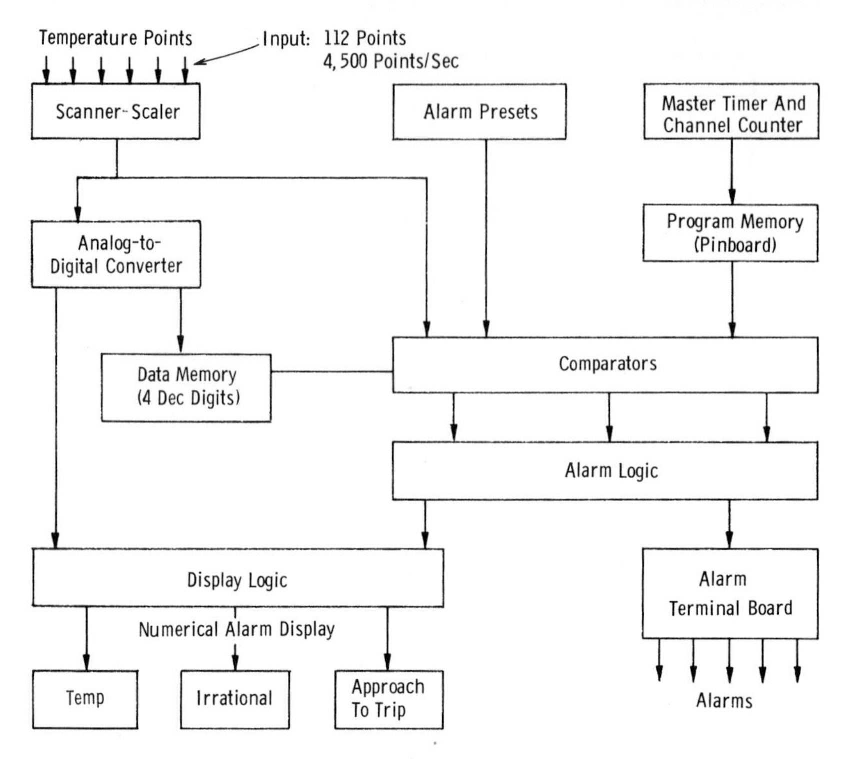

FIG. 3-33 Hanford fuel rod temperature alarm system.

approach adopted in this experiment emphasizes reliability and uninterrupted plant operation by separating the safety and regulation functions. Emphasis is also given to the organization and sorting of data prior to display to facilitate rapid operator interpretation of plant status when the need arises. The computer system (Fig. 3-34) forms a basis for closed-loop regulation experiments based on information obtained from measuring the coolant water temperatures at eight heat exchanger locations [149].

A small digital computer with a solid-state multiplexer is used as a signal monitor to perform the primary safety function. Fifty analog inputs are each sampled at 40 cycle/sec. These signals are from 24 RTD's, 6 neutron flux chambers and a miscellaneous selection of control-rod and other reactor signals. Each input is compared with three independently set alarm levels stored in the computer, and in addition the input is examined before it is converted into digital form to determine whether it is abnormally low or high.

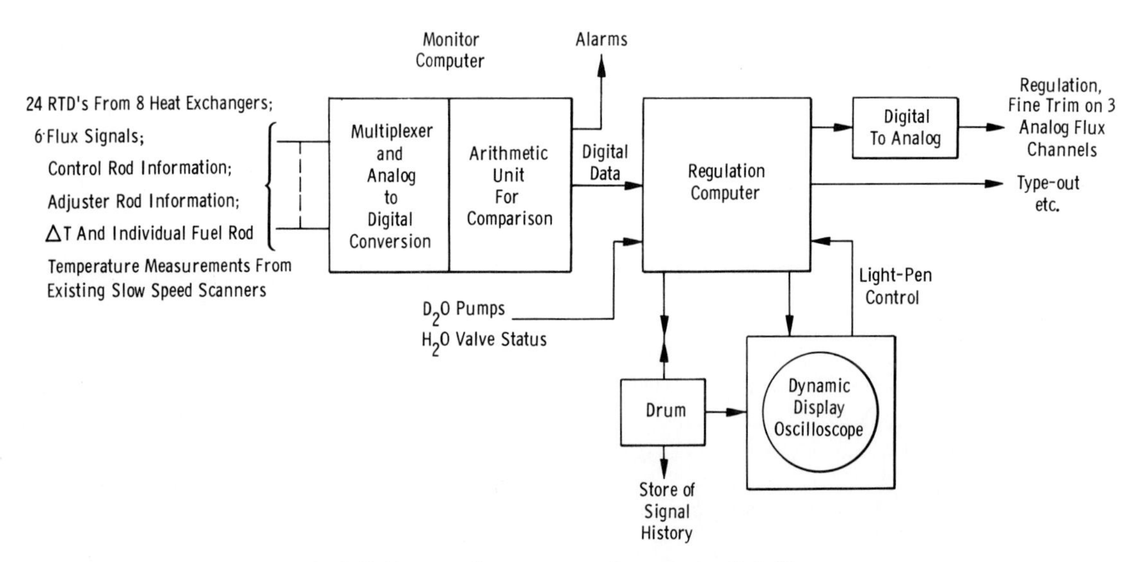

FIG. 3-34 Monitor and computer control experiment, Chalk River.

The monitor computer uses a 12-bit word and has a capacity of 1,000 words in its core store.

An asynchronous one-way digital link is made to a larger regulation and display computer that uses the digital data obtained by the monitor. This computer has an 18-bit word and used 8,000 words of core-store for program storage and calculations. Recent plant history is stored on a 16,000 word magnetic drum store on several time scales [150, 151]. The drum data is displayed on a cathode-ray tube screen either directly or under program control. The object is to create dynamic displays of reactor variables so that the operator can quickly find out the plant condition and observe the recent history following a fault or change of state.

3.9.6 Reliability

Computer system reliability is dominated by three factors, the ability of the program to meet all expected conditions, the technical quality of the hardware, and the competance of the system designers.

Most control applications require the computer to be time-shared [152] among several tasks, and cascaded interrupts have to be carefully taken into account. As with wired-logic systems it often takes time to uncover incorrect behavior under the whole spectrum of conditions. Certain functions can be executed either by programs [153, 154, 155] or by use of special optional hardware [156, 157]. Some examples are program interrupts and temporary data storage, indexing (counting-down), and program protection.

The reliability [137,138] of logic modules used in computer systems is very high. However, in such complicated instruments faults occur due to the high number of individual components present. In a plant relying on a single computer the economic penalty resulting from lack of control may be high. Not only is the fault-rate (the inverse of the mean time between failures) important, but also the average fault duration. Availability (up-time divided by total elapsed time) of 99.5% is now considered possible although control engineers would like 99.95%. For example, 30 errors per year of 10-min mean duration gives an availability of 99.95%, but that many interruptions may be very annoying. On the other hand, a single interruption that can only be corrected by a highly skilled technician or design engineer may involve at least a 24-hr delay, resulting in an availability of less than 99.75%. It is still to be seen whether advantage can be taken of permitted down-time intervals to do maintenance that will ensure 100% availability when the plant is operating. These examples illustrate the advances required in self-diagnostic displays and logic to aid fault interpretation. Generally speaking, single module faults in computers

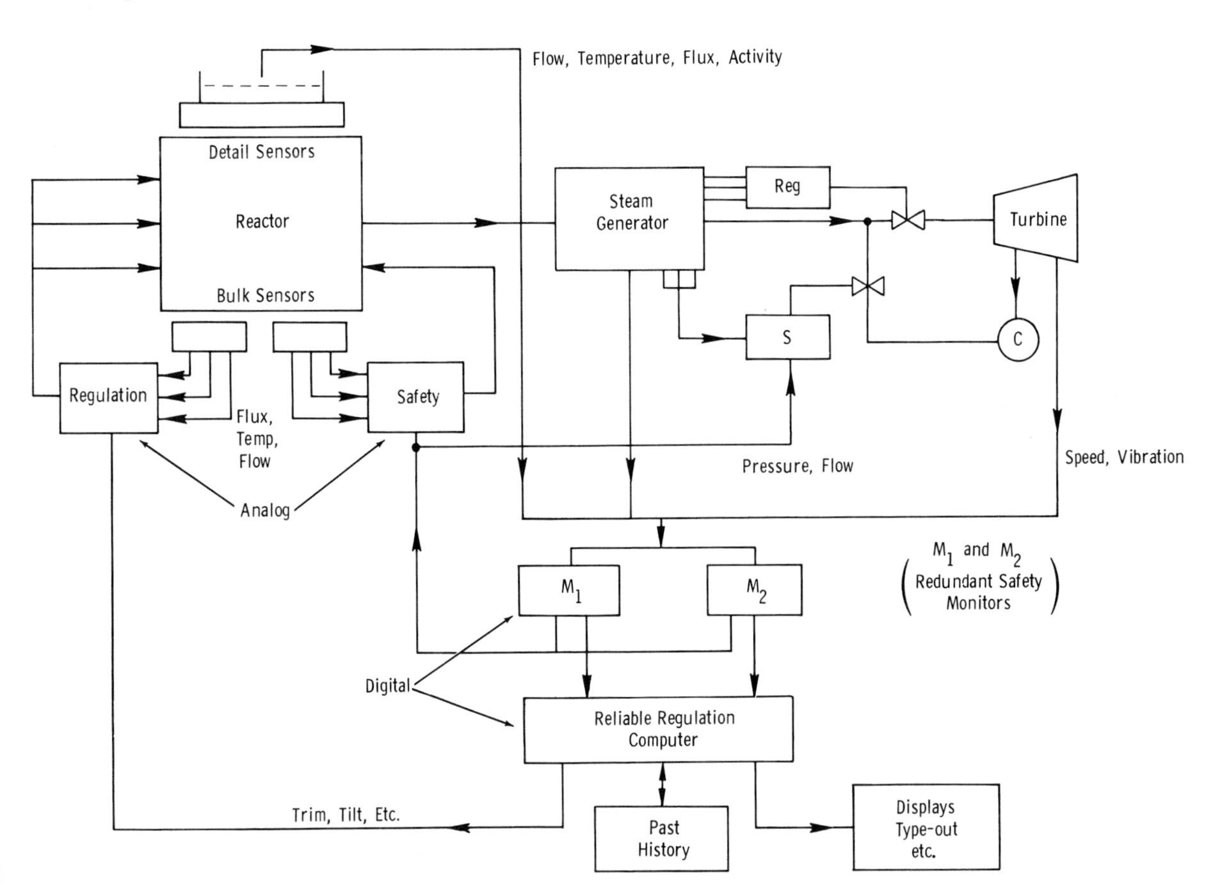

FIG. 3-35 Digital computer instrumentation system for a nuclear power plant.

result in completely erroneous results. Computer logic and operational programs should be designed so that only part of the system, (such as part of memory) becomes inoperative following a fault.

At the present time redundancy techniques as described in Sec. 3 appear to be ruled out for computers on economic grounds. Other approaches for achieving high reliability must be found [149, 158]. However, it does not seem necessary to duplicate all the input sensors for in-core instrumentation. A single fault should not necessitate a shutdown.

An alternative approach using current technology is to improve the reliability of a computer system by using redundancy in the memory blocks, the analog-to-digital converters, and in other important decision units such as the arithmetic unit. The whole system would normally be in use but the less necessary calculations would be dropped when a failure occurred.

A general approach to plant instrumentation using some of these ideas is suggested in Fig. 3-35.

4 REACTIVITY CONTROL INSTRUMENTATION

The output of most of the sensing instruments described in the previous sections is used in some manner to control reactivity.

The instrumentation used in this control system bears a large part of the reactor safety burden since it is here that many redundant and diverse channels of information come together.

Great care must be taken to avoid the possibility of incapacitation of the entire system by a single fault and thus obviate all the advantages of redundancy in the sensing instrumentation.

4.1 Safety Circuits

Almost invariably instruments used to sense unsafe conditions do so by opening an electrical circuit. In general, final safety action is initiated by breaking the circuit to an electromagnet or relay allowing absorbers to fall under gravity. In other cases valves may be opened or closed by means of coiled springs or suitable pneumatic or hydraulic systems used to drive the safety mechanism.

To introduce some basic ideas an elementary actuating circuit is shown in Fig. 4-1. Assume that relay contacts A_1, B_1, etc., are associated with safety parameter sensing instruments and are open when an unsafe condition prevails.* Safety action is initiated when the current through the reactivity control element coil falls below a specified value.

This safety circuit could be rendered inoperative by the faults shown as conduction paths to earth from the contact line or by a single fault that short circuited the contact chain. A single path to earth would not affect the operation but on the other hand would never be discovered by the routine of opening the contacts to test the safety circuit. If no monitoring circuits are provided to determine if such a fault exists a safety analysis must assume that it does exist and hence entertain the possibility that a single fault could incapacitate the system.

Short circuits to earth from both lines to the reactivity control system would cause the circuit breaker to open and release the reactivity control elements so that a deliberate ground connection as shown might be considered to provide a fail-safe feature. However, the integrity of this earth connection must undergo periodic verification as must any link in the safety system chain. Another type of accident can occur, a loose or broken wire can cause another source of voltage to be connected to the circuit and thus negate the safety action.

*$A_1, A_2, \ldots A_n$ indicate the contacts of the relay A; $B_1, B_2, \ldots B_n$ indicate those for the relay B; etc. The contact condition is drawn for the relay de-energized. These conventions are observed in all circuit drawings presented here.

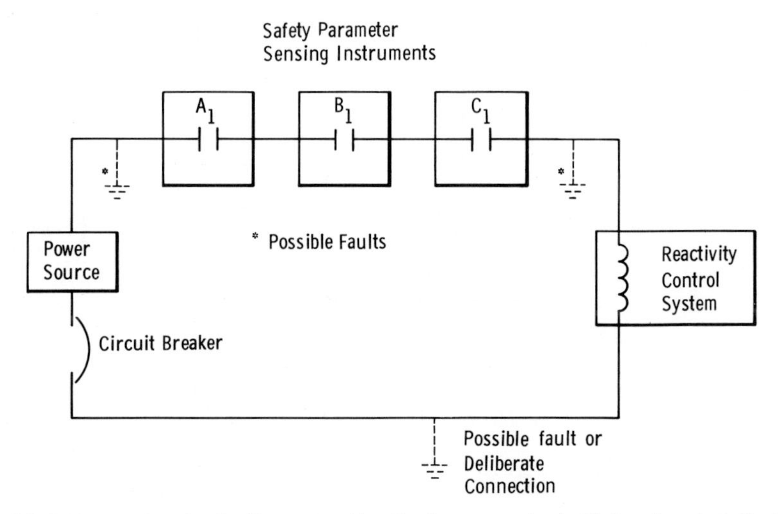

FIG. 4-1 Basic actuating circuit. The contacts A_1, B_1, C_1 are associated with the relays A, B, C. Contacts are shown (in all figures) in the condition assumed when the relay is de-energized.

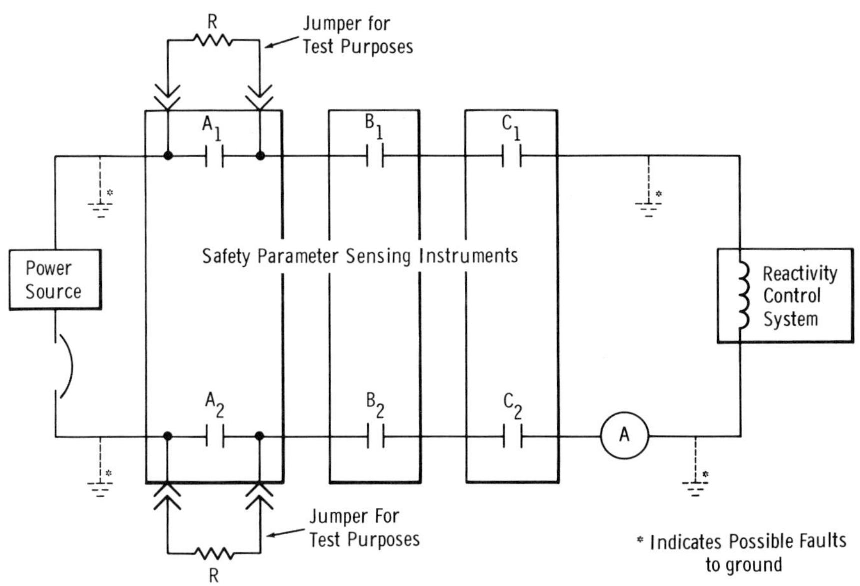

FIG. 4-2 Fault and testing problems.

To answer some of these points the circuit of Fig. 4-2 may be used. Dual contacts are provided so that both sides of the safety circuit are broken.

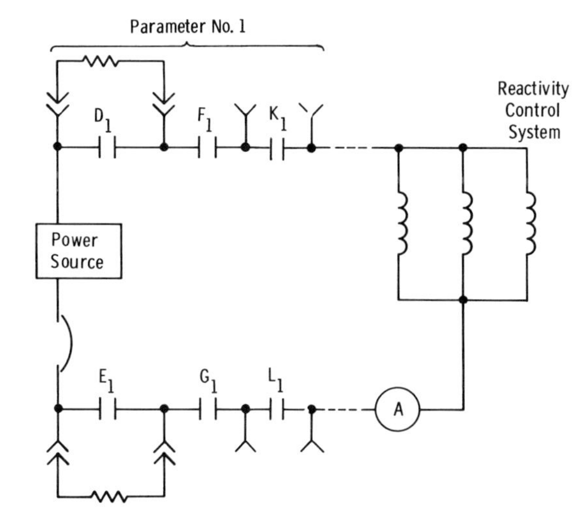

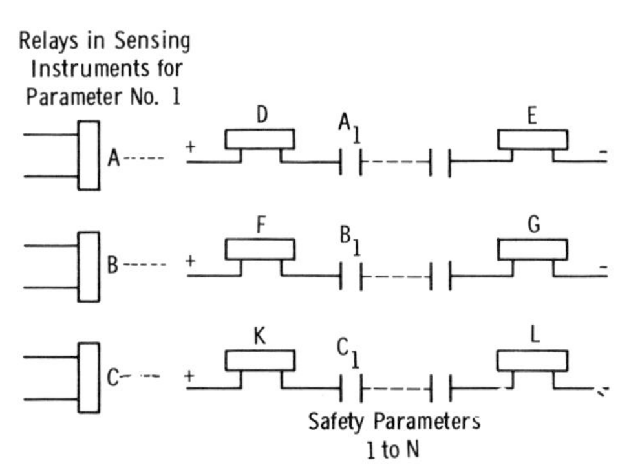

FIG. 4-3 Typical safety system circuit.

However, since initiation of safety action upon release of relay A (or B or C) does not verify that both sets of its contacts have opened, some specific measures must be taken to provide the necessary verification; an obvious way being to test the circuit twice with a deliberate connection across one set of contacts each time.

The resistors and the ammeter are shown in Fig. 4-2 to indicate a method that permits testing without releasing the final control system. If, for example, relay A is de-energized while the resistors are in place as circuit jumpers, the current can be made to fall to a level still above the tripping threshold but sufficiently changed for the difference to be detected by the meter. The magnitude of the current decrease indicates whether or not both sets of contacts open.

Contacts in sensing instruments are often not connected directly into the final safety system but are used to release other relays which may activate warnings and carry out the final action. Figure 4-3 shows the features of a typical safety system. Care must be taken in the physical arrangement of the relays so that the path between the power source and the reactivity control elements cannot conceivably be bridged. The use of redundant relays to provide the dual contact arrangement permits physical separation of the two paths (see Fig. 4-4) so that faults arising from a common cause, such as mechanical shock, maintenance error, or fire, are unlikely to bring about complete failure. With this system there still remains the remote possibility that a voltage supply could somehow be inserted in the line between the last relay and the magnets. Therefore, some systems split the magnets into subgroups activated by independent systems of this same sort.

4.1.1 Coincidence Safety Circuitry

The current trend in safety circuitry is to use coincidence arrangements [16]. These require that

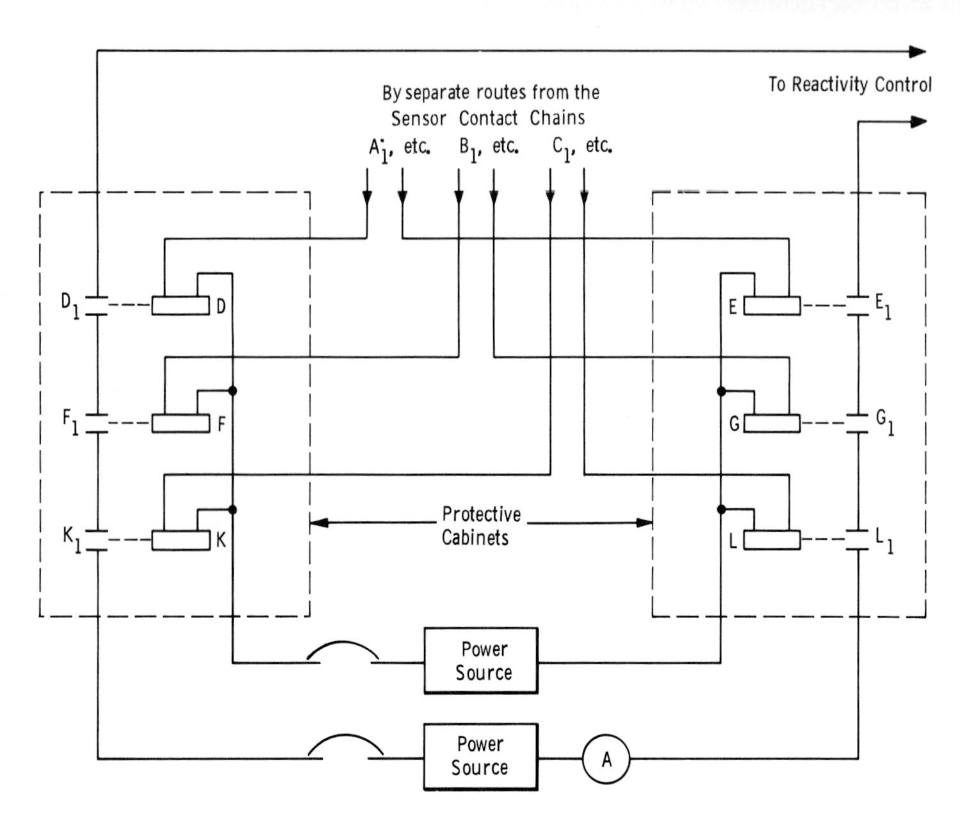

FIG. 4-4 Circuit separation features.

two or more sensing instruments signal an unsafe condition before safety action is taken. This trend reflects the difficulties and economic losses associated with reactor shutdowns (or scrams) due to false signals. Thus, the purpose in using coincident systems is to reduce the number of false shutdowns. Care must be taken to ensure that such systems do not reduce the overall safety afforded by the control system (see Sec. 1.4.3).

While the main reason for adopting coincidence circuitry is to permit the safety system to withstand a safe failure, the testing procedures can be simplified. Deliberate jumpering is not necessary; the coincidence system itself performs the appropriate action.

Figure 4-5 is a coincidence system using 2-out-of-3 logic to operate a single bank of safety elements. Each of the three sensing instruments associated with a safety parameter operates in an independent channel. Two redundant relays D and E belonging to the first channel are used to provide greater safety by operating in both sides of the final control circuit.

The coincidence arrangement is shown in Fig. 4-6. Testing is done by simulating an unsafe condition at each sensing instrument in a channel. The operation of all the redundant contacts in the final circuit can be verified by measuring the current change due to the contacts and resistor combinations.

Another arrangement for the 2-out-of-3 system is shown in Fig. 4-7. Opening one phase of the three phase circuit causes a reduction in current through the safety element electromagnet. The

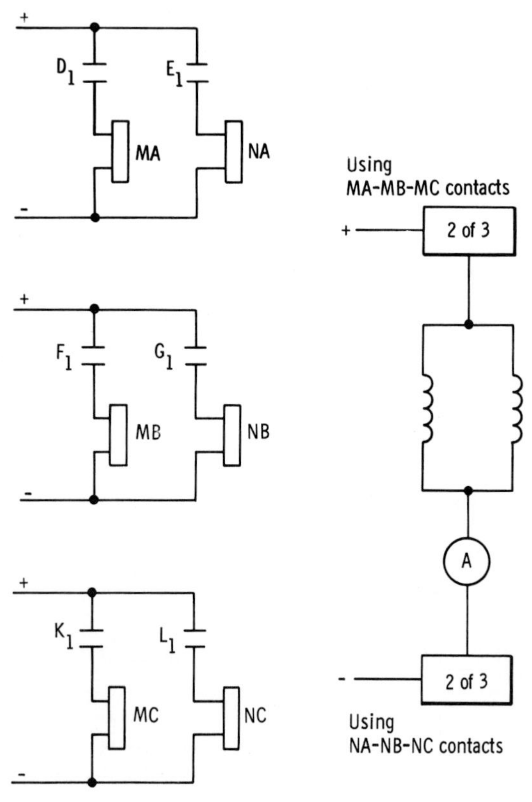

FIG. 4-5 Coincidence system.

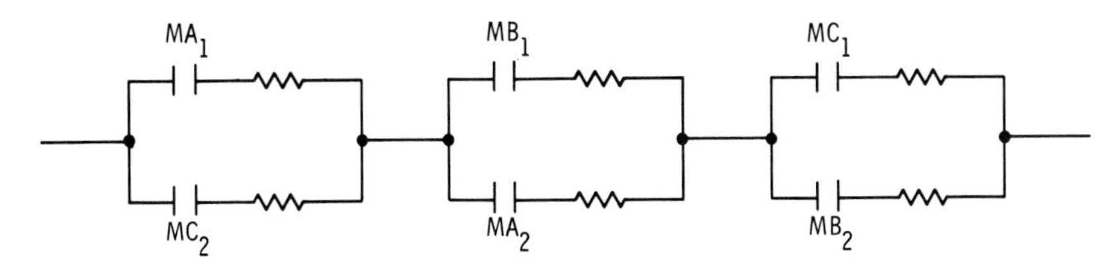

FIG. 4-6 Two-out-of-three coincidence relays.

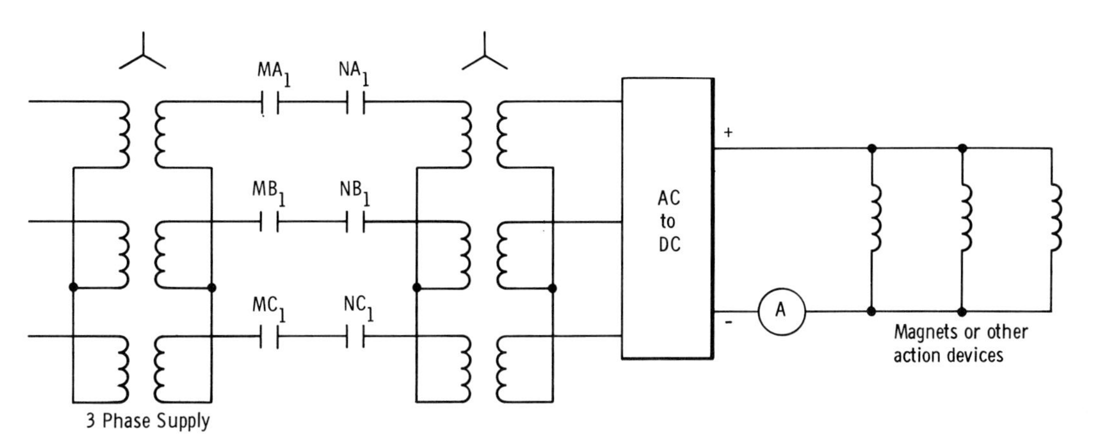

FIG. 4-7 Two-out-of-three coincidence transformers.

reduction is measurable but not sufficient to release the safety system. If any two phases are opened safety action is initiated.

The coincidence circuit of Fig. 4-5 suffers from the disadvantage that safe faults in any two parameters in different channels will cause a reactor shutdown. This is especially significant when a deliberate fault has been placed on one channel for the performance of a test or maintenance (see Table 1-1). To reduce the probability of a false trip from this cause the "local" 2-out-of-3 circuit in Fig. 4-8 can be used.

Each detector channel has a contact in each of the three safety channels. These contacts are arranged so that no line is open when only one instrument is signalling an unsafe situation. Correct operation of all the contacts can be verified by using the three ammeters. Relays D, E and F are connected in a coincidence circuit and to test them a switch is provided to permit each channel to be opened.

An alternative "local 2-out-of-3" scheme as shown in Fig. 4-9 is normally a series arrangement because of the positions of switches SW_1, SW_2 and SW_3 [159]. If a safe fault (say in A) occurs, or maintenance is being carried out, the appropriate switch (SW_1) is thrown shorting out the open contact. The system becomes 1-out-of-2 for the associated parameter and with the appropriate switch on, a single fault in any other parameter will now not cause a shutdown.

Another feature of this circuit is the use of the differential relays G and H. As contacts D_{1-2}, E_{1-2} and F_{1-2} open successively in pairs during a test, current flows in G and H provided all contacts operate properly. Signal lights associated with G and H enable the test to be verified visually.

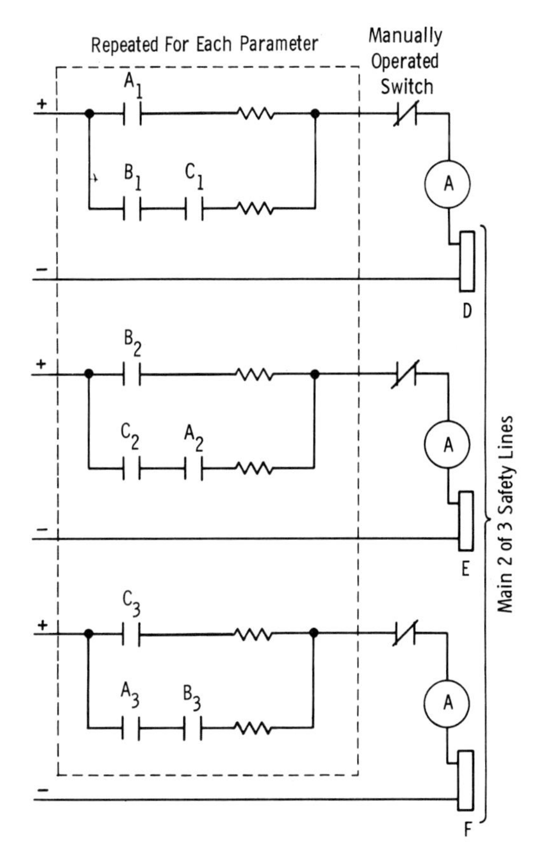

FIG. 4-8 Local 2-out-of-3.

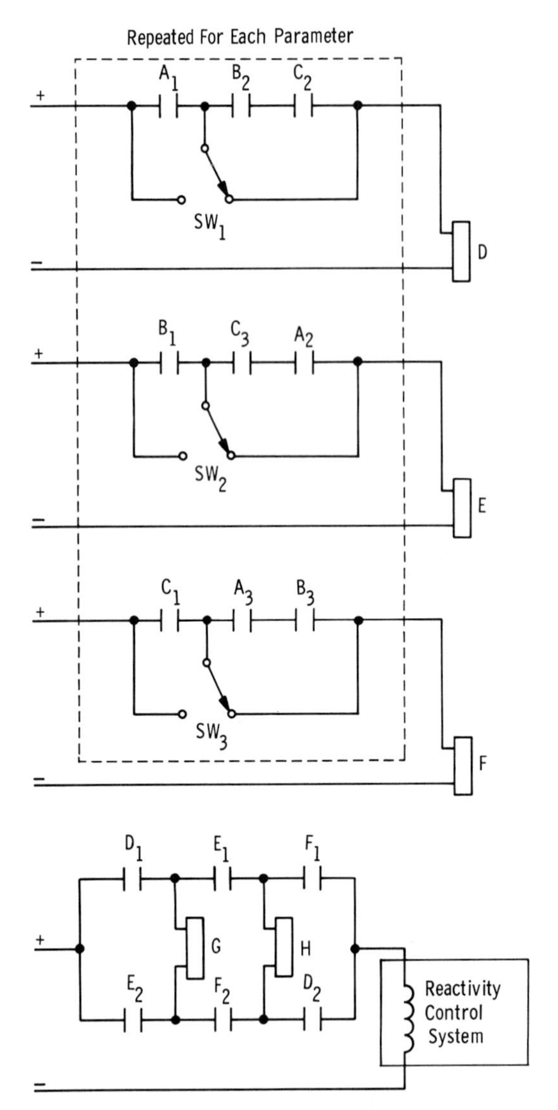

FIG. 4-9 Local 2-out-of-3 with switching arrangement. Contacts A_1, A_2, A_3, B_1, B_2, B_3, C_1, C_2, C_3 are associated with relays A, B, C in sensing instruments.

The "local 2-out-of-3" system, has, however, associated with it one disadvantage. Because of the contact arrangement there is a degree of interconnection among the safety channels and the possibility of a cross-linked malfunction must be weighed against the advantages.

The use of redundancy in a coincidence circuit has been carried to greater lengths [160] as shown in Fig. 4-10. Two independent sets of 3 safety channels are used in order to reduce the possibility of common failure to negligible proportions. Separate release circuits are used for each safety element (i.e. control rod) and more than one type of shutdown mechanism may be desired.

A two-channel arrangement using two sets of sensors in each channel has been described for Dresden [8]. Figure 4-11 shows that connections in both channels A and B must open before safety action takes place (i.e., normally 2-out-of-2) but each line opens on a 1-out-of-2 basis from each pair of sensing instruments, i.e., this is a coincident 1-out-of-2 arrangement. A shutdown indication in sensors 1 and 3 would cause safety action but not 1 and 2. In this system only two shutdown pilot valves are activated, so that the two channel approach is a good deal more simple than a three channel one ending in a two channel output. Simplicity should always be aimed at to reduce components, to permit easier maintenance and testing, and to aid understanding of the system operation.

In reactors where safety action can be obtained by dumping the moderator [20] a coincidence scheme involving the dump control valves is possible. It has the advantage that a test verifies the final element in the safety system chain. The arrangement is shown in Fig. 4-12. A test signal applied to any channel opens two valves without causing the moderator to be dumped. The relay circuitry opens all valves whenever any 2-out-of-3 lines open.

Automatic intercomparison among the outputs of similar sensing instruments is an approach used to obtain fail-safe operation [20]. Any sensor differing from the others by a preset amount is deemed faulty and by means of the circuit of Fig. 4-13 the appropriate safety line is opened or an alarm is operated. If the fault is safe (i.e., it causes the relay in the sensing instrument to de-energize) both contacts in the secondary relay line will open. The interconnection between channels is again a disadvantage because of the possibility of common faults.

4.1.2 Solid-State Safety Circuitry

Since safety system circuitry consists almost entirely of on-off devices the use of solid-state switches is attractive.

However, to replace a relay by a solid state switch in what is essentially a static role seems to offer little advantage [161]. A periodic test would still be required to prove their ability to operate as their mode of failure is as likely to be in the unsafe as the safe direction. Corresponding to welded contacts and jammed armatures in relays a solid-state switch can develop a short circuit as well as an open circuit between its input and output (see Sec. 2.6.1).

There is a little evidence to indicate that a solid-state safety circuit would be more reliable than one using relays because the static application is not a common one.

However, safety circuits have been developed that make use of the dynamic properties of a solid-state switch in order to come closer to the fail-safe ideal.

One approach [162] uses solid-state elements that are automatically tested many times per second by pulses long enough to be identified positively at the reactivity control element but not long enough to release it. The appearance of the pulses at the correct time is assurance that the switches in the safety circuit are opening and closing properly.

Another approach [161] to solid-state logic makes use of the dynamic properties of a transistor in a novel manner. Figure 4-14 illustrates the use of a transistor as an element in a safety chain. Al-

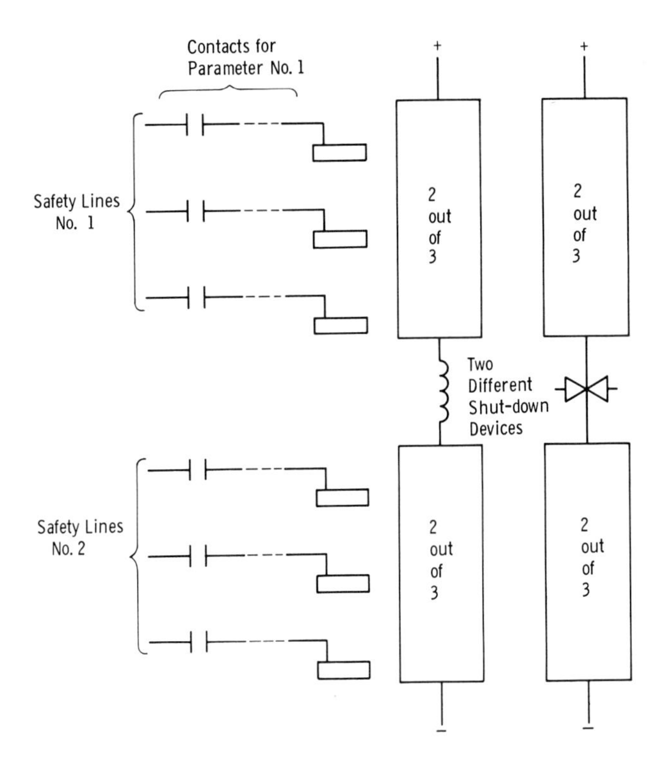

FIG. 4-10 Redundant safety systems.

though the transistor is acting as a switch its ability to function is being continuously tested. Removal of the collector supply voltage, which occurs when the safety amplifier contacts open, causes the loss of the a-c output signal. This in turn removes the only source of power to the reactivity control magnet.

A short circuit between the transistor base and collector, while permitting some signal to pass, would introduce sufficient attentuation to cause a safe failure.

The isolating transistor in series with the magnet provides a high impedance path and reduces release time. A double Zener diode protects this transistor from transient surges.

It is too early to assess the real value of these approaches. The solid-state device is appealing basically because of its lack of moving parts but in addition advantage may be taken of the fact that its operation does not appreciably hasten its deterioration whereas with relays there is a direct correlation.

Nevertheless, properly designed relay circuits are providing entirely satisfactory service and the immediate future of solid-state circuits will depend largely on the behavior of the few systems now being installed and used.

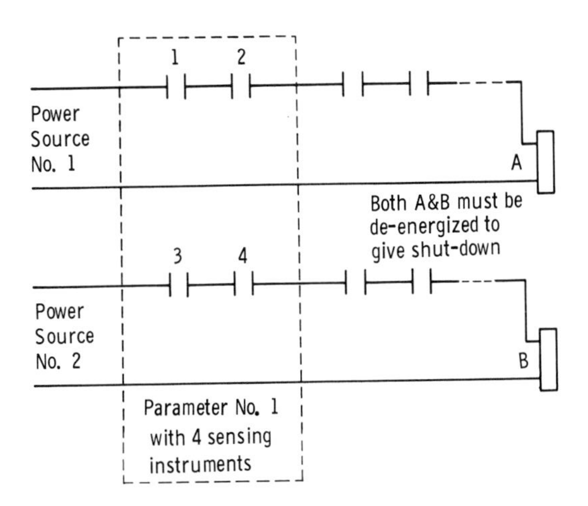

FIG. 4-11 Dresden safety lines.

4.2 Redundant Regulating Circuitry

The use of redundancy was introduced into reactor regulating systems primarily to provide increased serviceability by ensuring that regulation would be unaffected by any single fault either safe or unsafe. In this respect multi-channel regulating systems might be compared with the coincidence circuitry associated with redundant safety channels.

Redundancy is not nearly so prevalent in regulating systems as in safety systems although the redundancy concept was introduced into the design of the regulating system for the NRU reactor at Chalk River some twelve years ago and in fact before the coincidence concept was introduced into its safety circuits.

Pure redundancy can be applied of course to regulating systems, in which case the movement of reactivity to counteract a downward power excur-

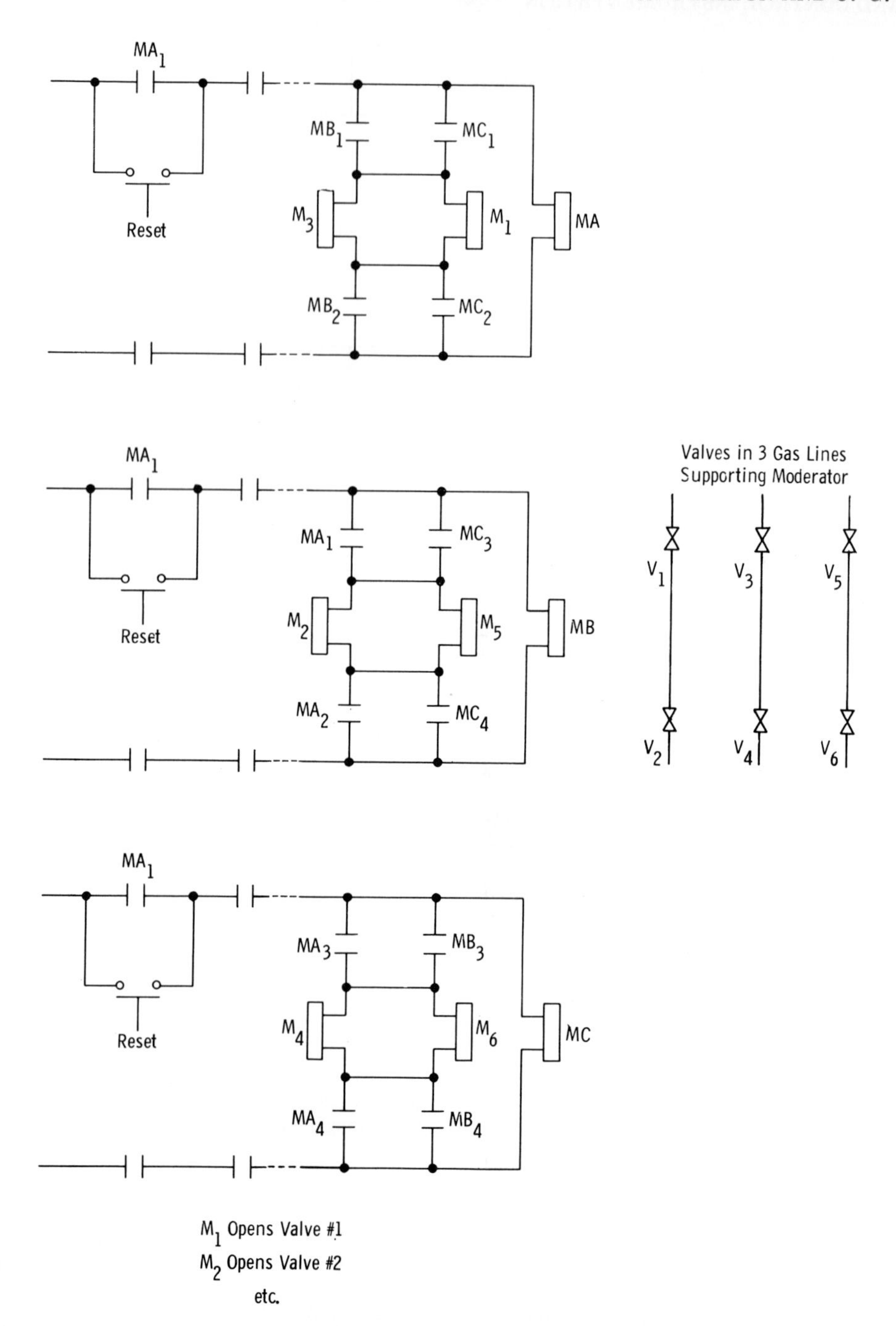

FIG. 4-12 NPD safety system.

sion is permitted only when all channels concur; however, if any channel senses a power increase correcting action is allowed. As with redundant safety systems, the serviceability decreases in proportion to the number of regulating channels used.

Two basic differences between safety and regulating systems should be recognized.

1. Unsafe failures in a safety system do not alter reactivity whereas unsafe failures in the regulating system cause a reactivity increase.

2. There is no deliberate feedback in a safety system. If one redundant channel releases the reactivity elements the other channels are unaffected. In a multi-channel regulating system if a fault occurs in one channel the other channels attempt to counteract it.

These two characteristic differences between the systems affect the manner in which the signals from redundant regulating channels are mixed to act

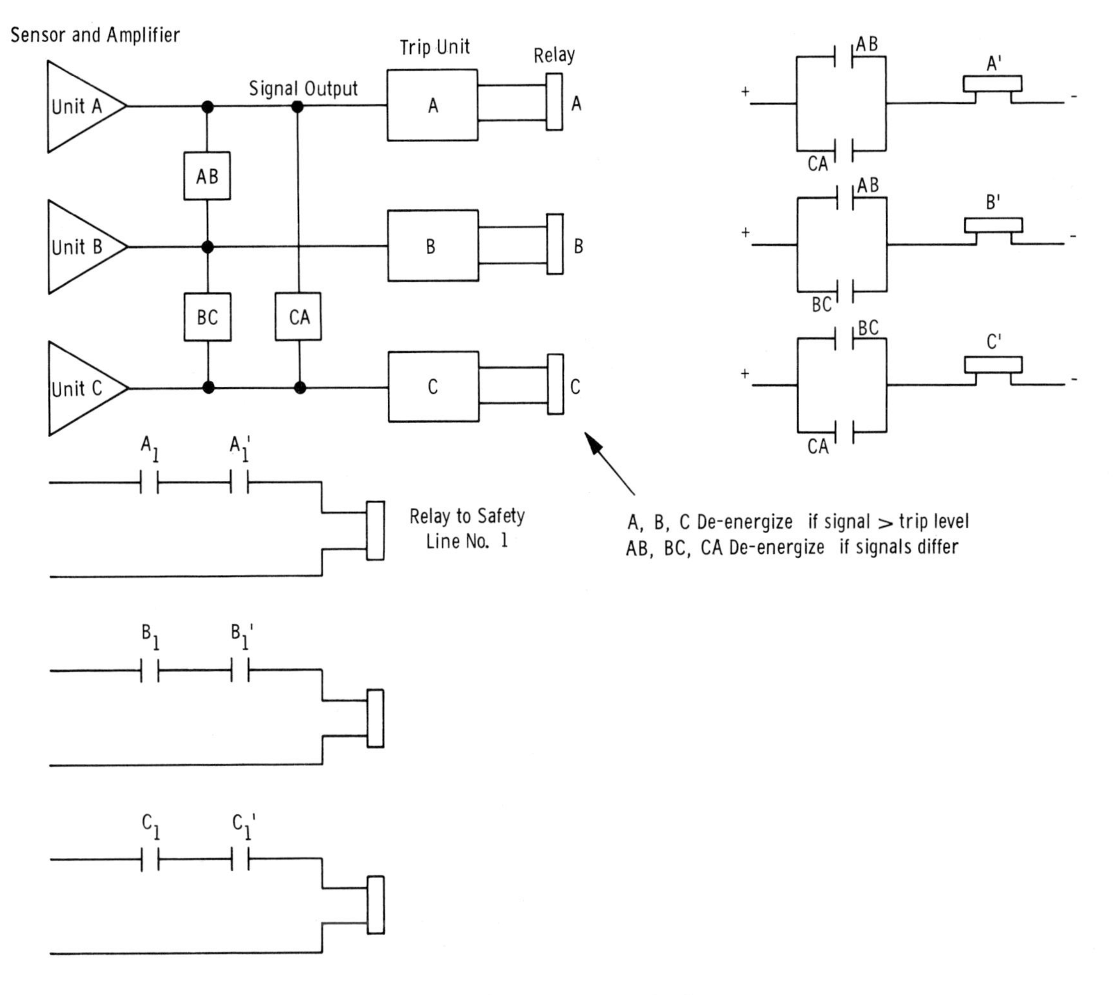

FIG. 4-13 Intercomparison circuits.

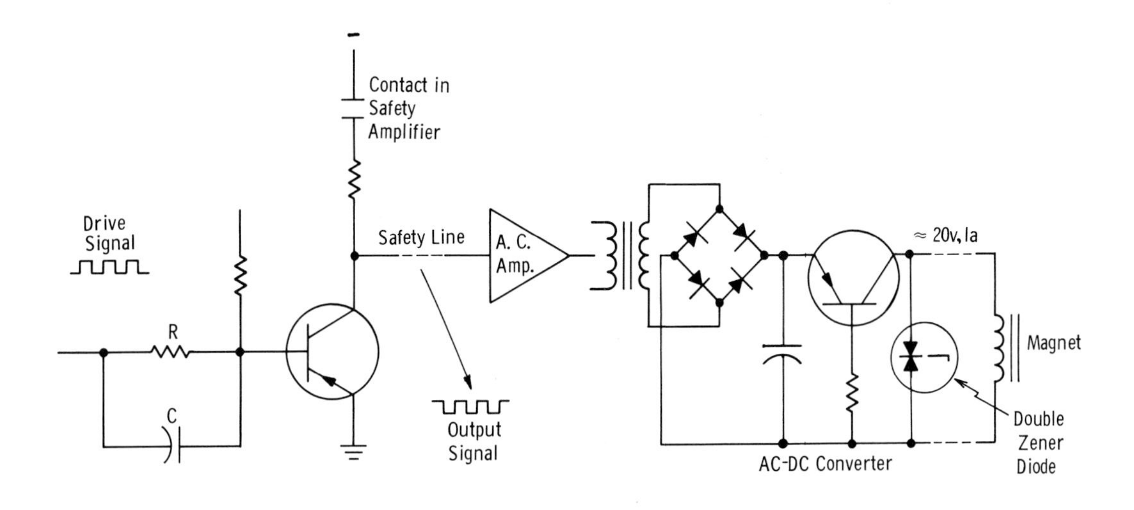

FIG. 4-14 A dynamic solid-state safety circuit.

upon the reactivity of the reactor. Several methods have been used or suggested.

4.2.1 Averaging

Some of the problems arising from the simple averaging of several regulating channel outputs can be appreciated by considering Fig. 4-15. Each channel contains the essential elements for control. The output of the sensing instrument (a neutron flux sensor, for example) is compared with a reference; the difference is amplified and applied to the reactivity control system. In a single channel system the fundamental limit on the rate of change of reactivity is incorporated into the reactivity controller but this is not sufficient if any advantage is to be gained from the use of redundancy. If a fault occurs on the error amplifier of the channel marked A (neglecting for a moment the limiting circuits) and causes its output to exceed twice its normal full scale value, the other two channels will not be able to counteract it even by going to full scale on the opposite direction. As a result, reactivity will be withdrawn (or inserted as the case may be) in an uncontrolled manner. To avoid this difficulty the limiting circuits are used; they form a basic part of the system. The limit circuits prevent any signal greater than that corresponding to maximum rate of change of reactivity (in either direction) from reaching the averaging circuitry. Thus, in the worst case the two sound channels will have to produce an output corresponding to one-half the maximum reactivity rate in order to counteract a faulty channel.

Whether or not this situation can be tolerated depends on the gain of the error amplifier and on the permissible error in the controlled parameter. Typically, the error amplifier gain is ten so that a 10% deviation of the parameter from the reference produces maximum correcting action. Hence, under the fault condition postulated above, the controlled parameter would offset 5% from the reference level.

An important consideration is the system response time. If a sudden unsafe failure occurs in one of the channels the reactivity control element may undergo a transient movement before the other channels respond to counteract the fault. The resulting period transient may reach the trip level. It may be impracticable to decrease the overall system response time and other means may be required to ensure that the fault is not seen by the reactivity controller.

A method used to disconnect a fault is shown in Fig. 4-16. Three error signals are summed in a multi-input magnetic amplifier and the differential relays AB, BC and AC detect any output differing from the others. By arranging the circuit as shown previously in Fig. 4-13 the faulty channel can be disconnected from the sum. Because summation is used to form the average, the loop gain of the regulating system is reduced to two-thirds of its normal value when a channel is removed. Summation as a form of averaging is possible in this case because the normal signal level is zero.

If averaging is done ahead of the error generation point the circuit of Fig. 4-17 can be used. Disconnection of a faulty signal is required to avoid serious distortion of the average. For example, if one sensor amplifier fails and produces zero output instead of 100% of the required value, the other two must increase to 150% to maintain zero error.

The need to average sensor instrument outputs often arises because of real differences among them caused by spatial flux distortion. In the example cited of an error amplifier with a gain of ten, a difference of 10% between sensors would cause a 100% difference between error amplifiers fed directly from the unaveraged sensor signals.

The sensitivity of the differential relays must be adjusted to permit normal differences to be accepted but the larger these differences the less is the system's ability to detect real faults.

To continue with redundant circuitry beyond the averaging point several averages are required as shown in Fig. 4-18.

The internal impedance of the sensor amplifiers now becomes important. Certain types of fault in the error amplifier circuits can cause current feed back into all the amplifiers and the voltage

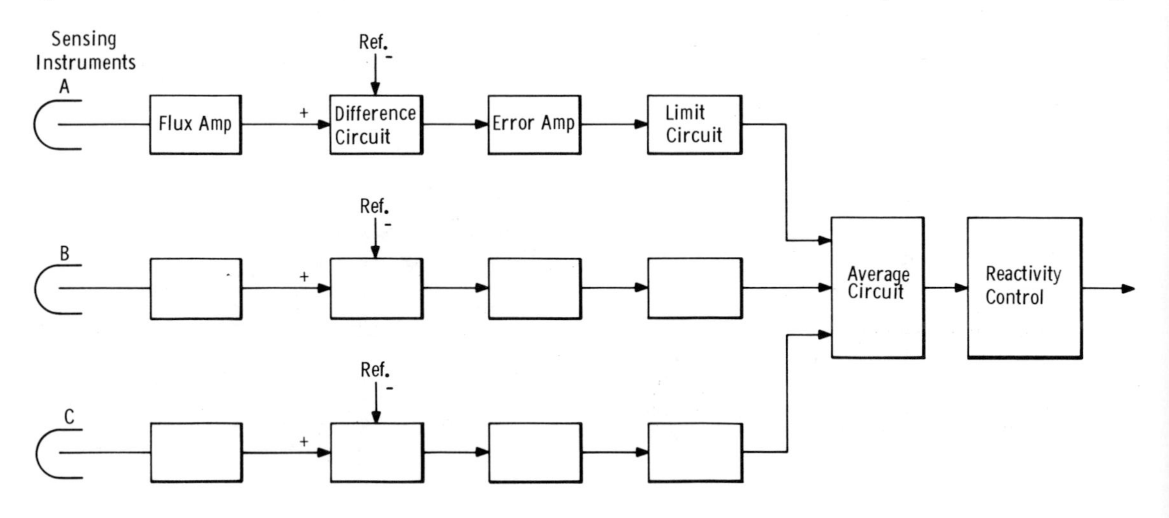

FIG. 4-15 Three-channel regulating system.

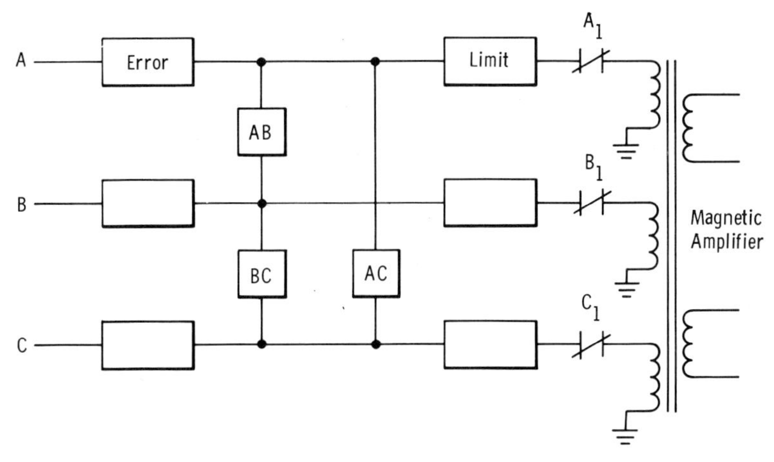

FIG. 4-16 Fault comparator.

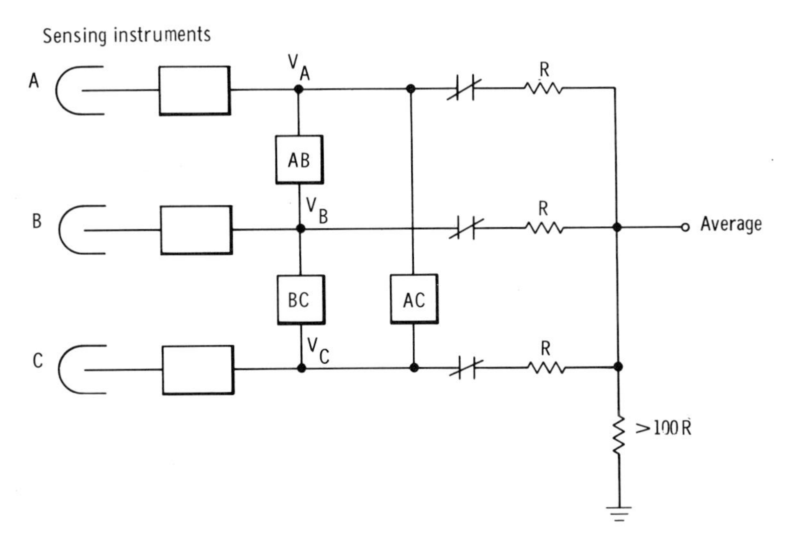

FIG. 4-17 Averaging and fault disconnection at sensing instruments.

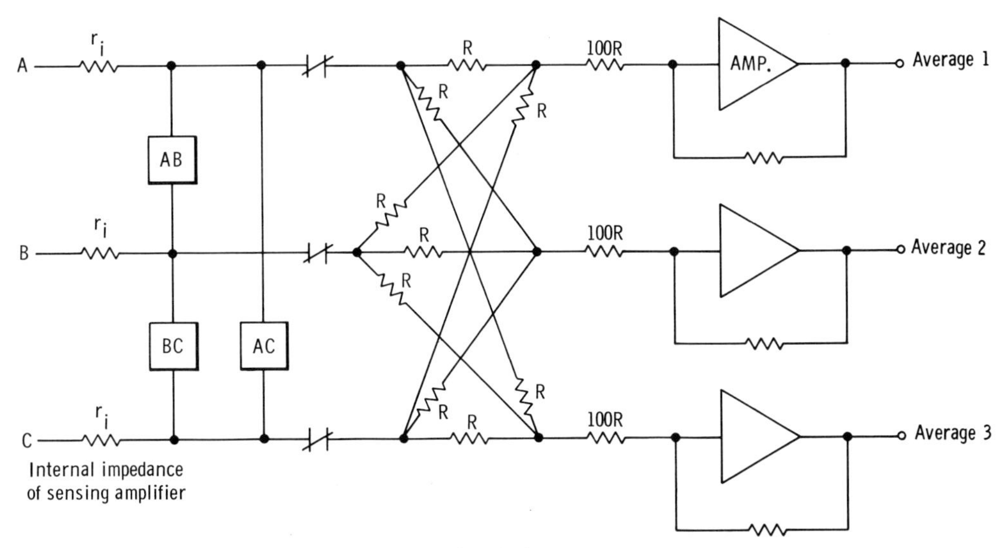

FIG. 4-18 Multiple averages.

drop across each internal impedance is seen by the channels associated with it.

Experience has shown [13] that interconnections between redundant channels should be avoided if a significant improvement in reliability is to be achieved. Attention should be paid to the generation of control signals that do not differ significantly from each other and to the design of regulating systems that do not require large gains between the error production point and the reactivity controllers (see Sec. 4.3).

In reactors using moderator level as the reactivity control good isolation among the redundant channels is possible at the output end. In Fig. 4-19 each error amplifier positions its own drain valve. A fundamental limit on the single channel error is inherent in this system since a valve can do no more than fully open or close. The valve characteristics must be chosen so that such a failure can be overridden by an offset of the two valves still in operation. The result of a valve failure in such an arrangement is shown in Fig. 4-20. The result of the correcting action taken by the remaining two valves can be seen. When valve B fails to the closed position the power increases sufficiently to open the other two valves, when valve B fails to the open position the power must drop; in this case the power continued to fall because the return water-pump was not being operated at maximum rate.

The characteristics required of valves for this type of service is discussed in Sec. 6.4 of the chapter on Mechanical Design of Components.

4.2.2 Redundancy With On-Off Controllers

If an on-off controller is employed for reactivity adjustment the function of the differential relays and disconnect circuits can be performed by the controller switching circuits. Figure 4-21 shows several combinations. In circuit (a) power is only fed to the controller if two channels agree upon the action. Hence a single channel fault is never connected to the controller whereas in the averaging methods described previously, a fault had to be sensed and disconnected.

Circuit (b) requires a majority decision for up-drive (i.e., an increase in reactivity) and a minority decision for down-drive; circuit (c) requires an unanimous decision for up-drive.

Circuit (a) corresponds to the analog system in which two channels have the control capability to override the worst fault the third channel can introduce.

4.2.3 Auctioneering Circuits

The circuits in Fig. 4-22 are used when control is required on the highest (or lowest) output from a set of sensors. If the sensors are all monitoring the same parameter the arrangement is equivalent to the application of pure redundancy (as, for example, in Fig. 4-21 (c)). It requires either all channels to indicate a low value of the parameter or one channel to indicate a high value before regulating action occurs.

Auctioneering circuits are also commonly used in a way that is not strictly an application of re-

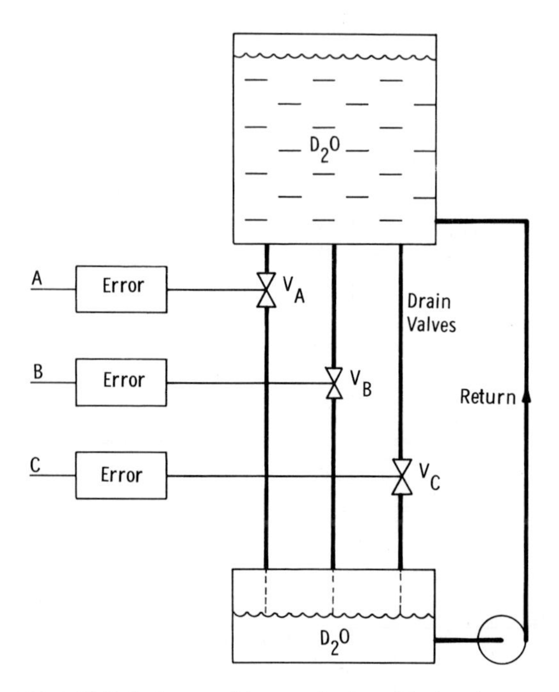

FIG. 4-19 Redundant reactivity control valves. The dots shown at bottom of three drain lines are intended to show the steady flow out of drain valves under equilibrium conditions.

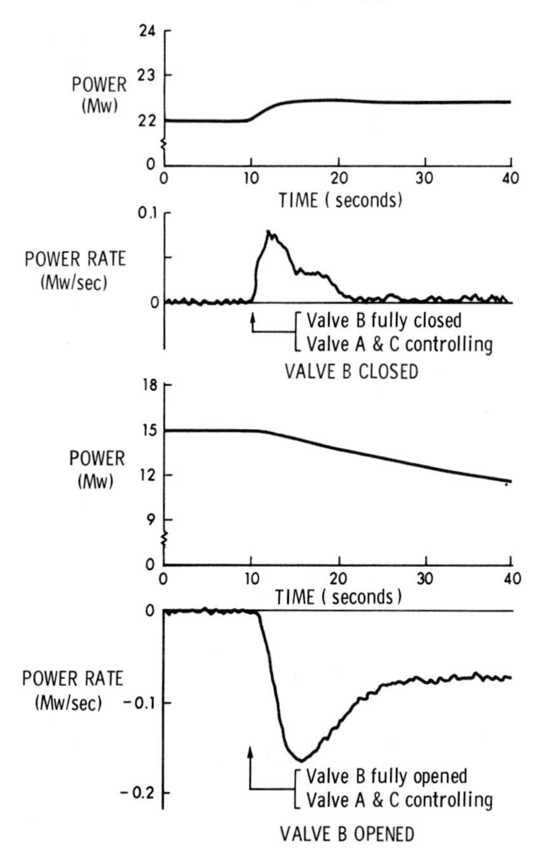

FIG. 4-20 Failure of valve in B channel.

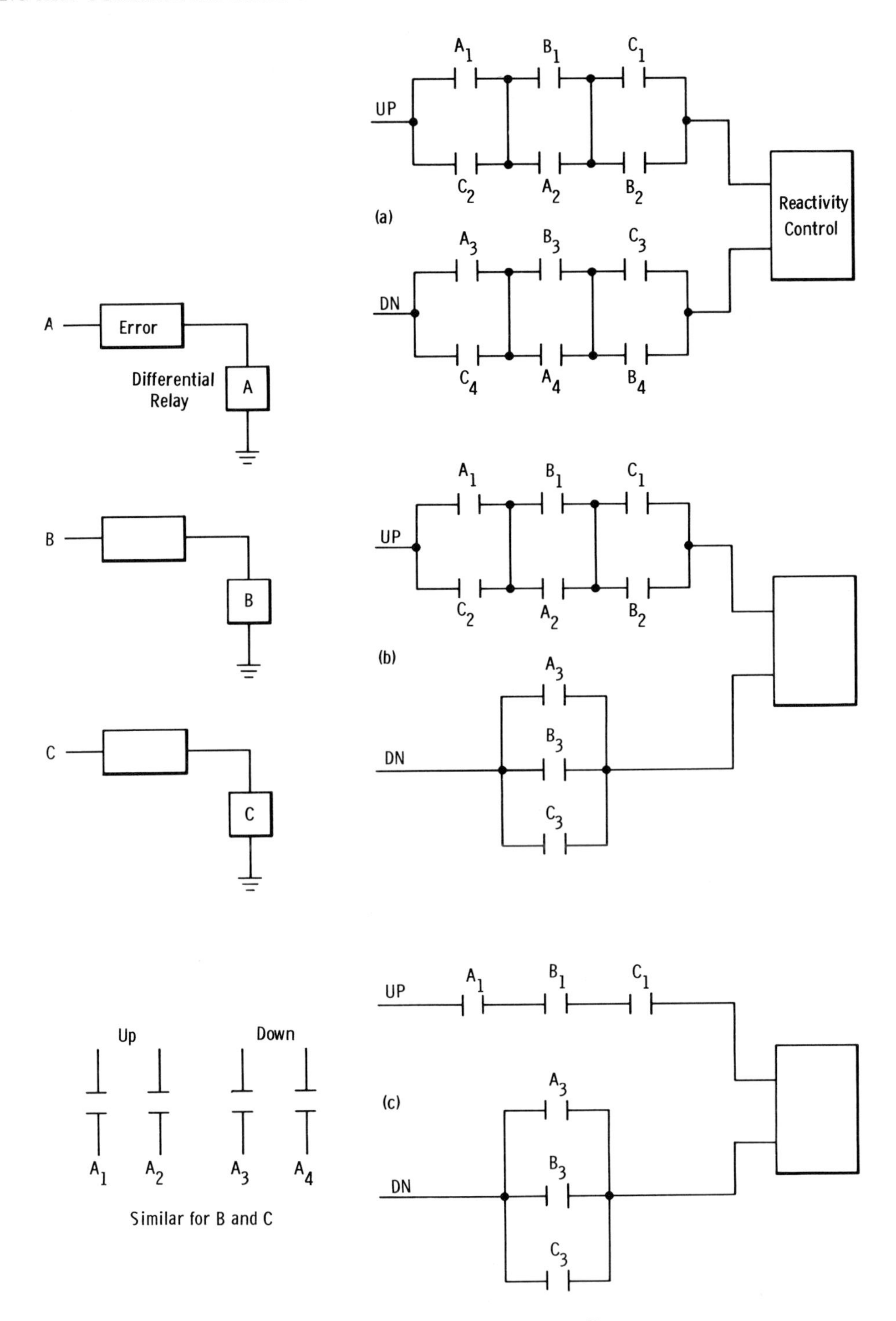

FIG. 4-21 Redundancy with on-off controller.

dundancy. Several sectors of a reactor may be monitored by auctioneering circuits so that the highest level is defined, but not the sector attaining it. To apply redundancy in these cases would mean using the methods of this section for each auctioneered measurement.

4.2.4 Median Circuits

The error that is introduced into an average by a faulty signal can be eliminated by using the median of three signals for control.

The basic circuit is shown in Fig. 4-23 where

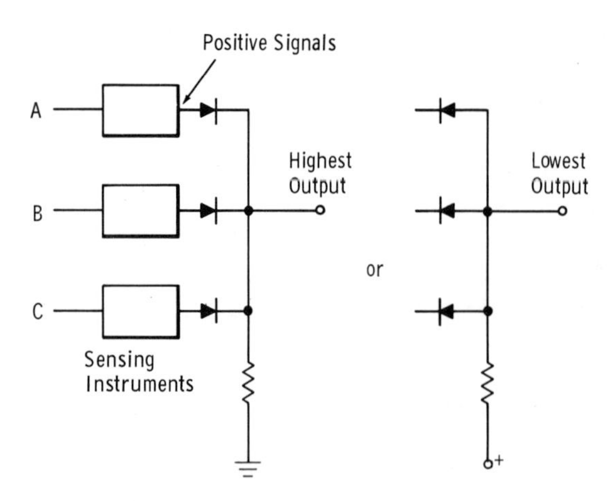

FIG. 4-22 Auctioneering circuits.

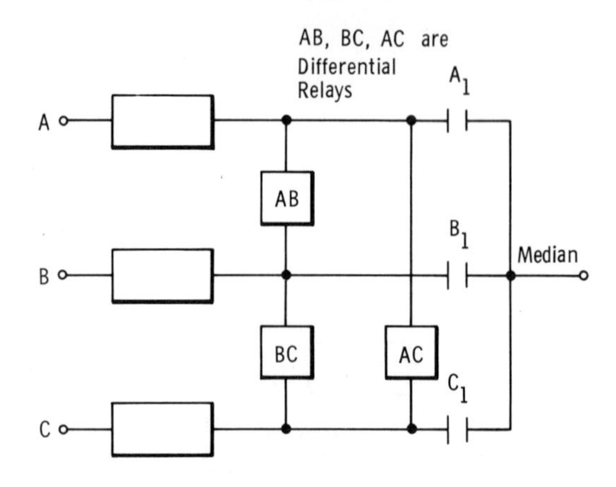

FIG. 4-23 Median circuit.

the differential bistable relays decide which signal is the median and cause it to be connected to the controller. The relay connections required to perform the logic are shown in Fig. 4-24.

Obviously, if the three signals are identical any one may be connected. However, the operation is more easily visualized if one channel is assumed to be the median with the others acting as standby circuits, one on either side, guarding against a

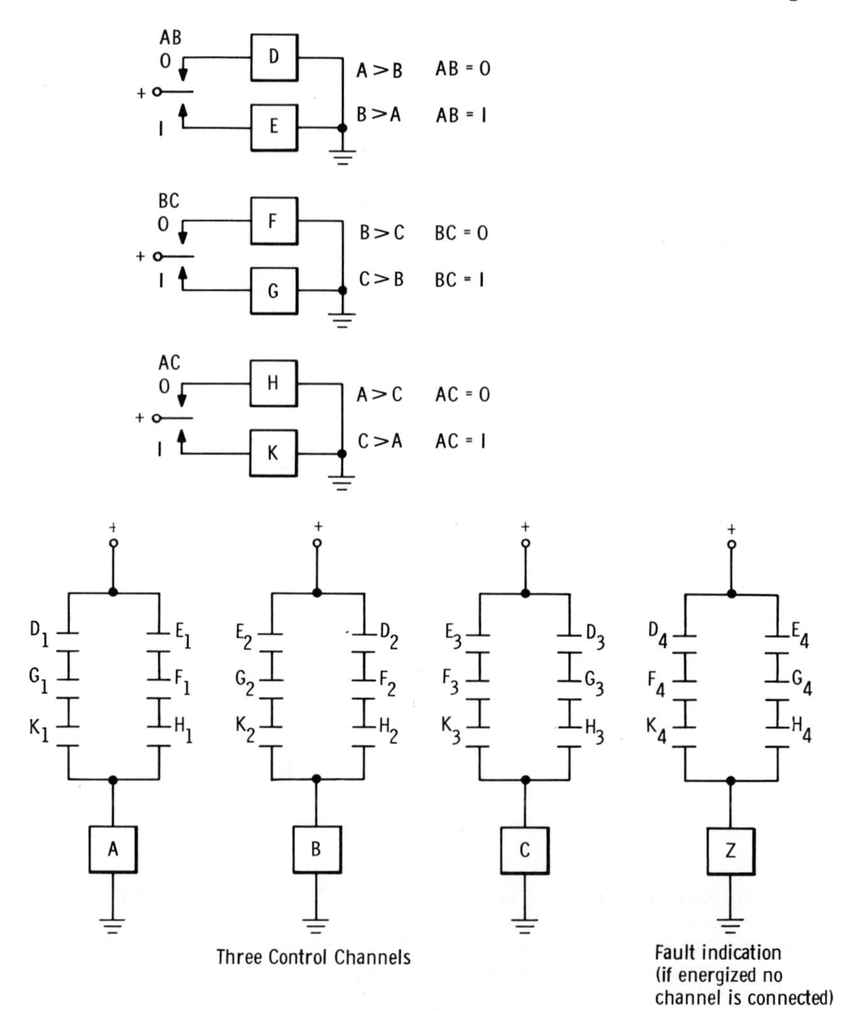

FIG. 4-24 Median circuit logic.

failure in either direction of the controlling signal.

A deliberate sharing of the regulating function amongst the channels is readily performed by a slight adjustment of their respective control points. Such a maneuver on a routine basis would give assurance that each channel was in operating condition.

In Fig. 4-25 a circuit suggested by F. S. Goulding performs median selection using six diodes.

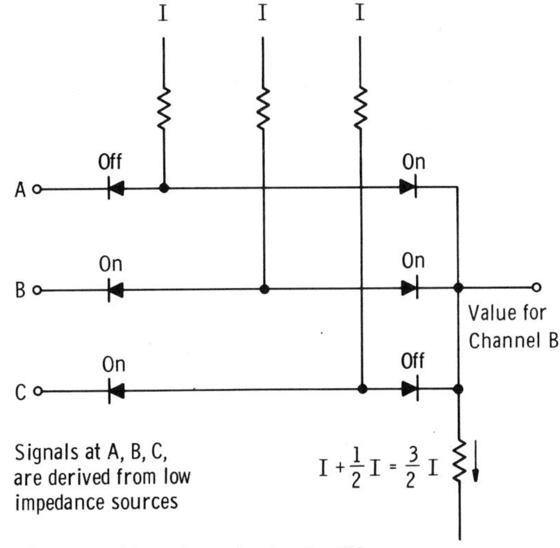

FIG. 4-25 Median circuit using diode logic.

4.2.5 Problems with Integrating Controllers

Basic difficulties arise on multi-channel regulating systems when integrating circuits are used. A typical situation is shown in Fig. 4-26 where a neutron flux controller is reset by thermal signals to provide constant thermal output in the face of spatial flux distortions.

In operation the reference signal applied to the flux channel is varied at a rate proportional to the error between the thermal signal and its reference until this error is zero.

In a single channel system any drift in the integrator will cause an error to appear in the

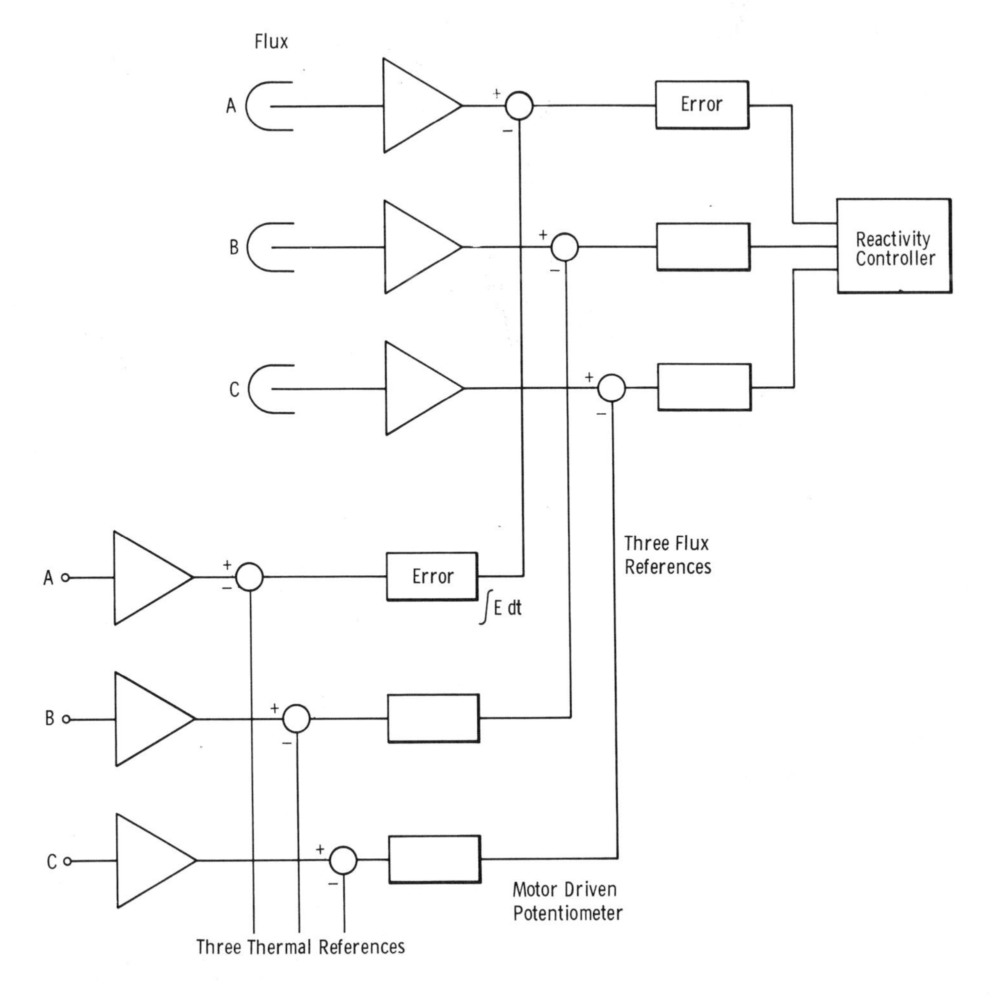

FIG. 4-26 Integrating controllers in reference lines.

thermal signal just sufficient to counteract the spurious signal causing the drift. In a multi-channel system, however, the other two channels will act to counteract the drift and eventually two integrators will be limited at opposite ends of their travel and the third channel will be controlling.

This is a fundamental problem and the result is always the same no matter how small the drift.

Specific design measures can be taken to overcome the difficulties and are described in Sec. 4.3.

4.3 Instrumentation for Regulating Systems

Previous sections have described instruments required to measure reactor parameters. The action taken with these signals in order to regulate reactor operation is now examined.

4.3.1 The Function of Regulation

The first function of regulation is to take the reactor, steam generator and turbine, etc., correctly through the range of power from shutdown to full-load.

The second function is to maintain chosen operating conditions in face of disturbances due to changes in reactivity, output load, or process faults.

Figure 4-27 shows the three reactor states about which most of the discussion in this section is concerned. These are, startup from low power using log-rate (or inverse-period) control, control over the last decade of power by the derivative of a power signal, and regulation at steady power based on a power signal.

If the reactor gets thermally hot during the final approach to power temperature dependent reactivity coefficients can affect the startup. If the coefficients are negative and of sufficient speed

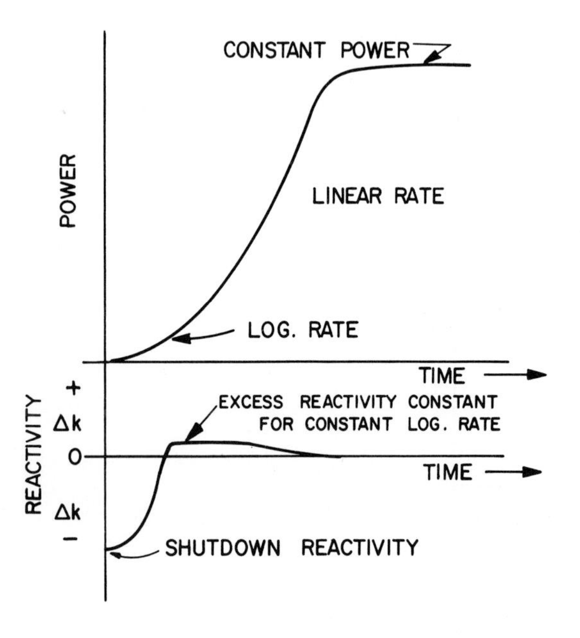

FIG. 4-27 Reactor startup.

and magnitude, self-regulation occurs. The power then rises in proportion to the external addition of reactivity. Such systems may be stable enough to enable simplification of the external reactivity control system. One should know details about the magnitude and time constants of the coefficients and how they may change due to fuel irradiation and changes in coolant state. For further treatment of these subjects see the chapters on General Reactor Dynamics and Water Reactor Kinetics.

In boiling water reactors, coolant voids change the reactivity. These can cause a positive (in the center) or negative (towards the outside) reactivity coefficient.

Both negative and positive reactivity coefficients may arise in graphite-moderated reactors [4b]. Following a rapid rise in power, the initial feedback is by the negative fuel temperature (Doppler) rise (see chapter on General Reactor Dynamics); this is followed several minutes later by a positive coefficient effect associated with an increase in graphite temperature.

As power reactors become larger than about 15 ft ($\sim$5m), spatial instabilities caused by xenon poison changes may become a problem. In such cases zonal regulation can become an important feature. A system need not necessarily be divergent but even a damped oscillation can result in hot spots and require additional damping by a controller.

The startup requirements and system disturbances must be evaluated and the inherent reactivity coefficients and their response times calculated to decide whether, and with what performance specification, an external reactivity regulation system should be used.

4.3.2 Reactor Dynamic Response

The dynamic response of a system is most easily appreciated when shown as the response in time of one or more parameters to an applied step or ramp disturbance. In the case of a reactor the disturbance may be either a small change in reactivity or a small change in demanded power. For the purpose of analysis, however, and especially for ease in determination and correction of instability in closed-loop feedback systems, the response of the unit to sinusoidal disturbances is more convenient. The amplitude and phase response of a linear unit to sinusoidal disturbances over a wide range of frequencies is known as the unit transfer function [163a]. In practice measurements are taken at a number of discrete intervals of frequency. The transfer function and impulsive time response are related by the transform shown in Fig. 4-28 [164a].

The reactor transfer function is very dependent on mean neutron lifetime and on the delayed neutron groups. The mean lifetime limits the rapidity of the flux change following an instantaneous step change in reactivity; the delayed neutron groups have a significant effect on easing the stability problem associated with regulation. An extensive treatment of this subject is given in the chapter on General Reactor Dynamics.

The reactor transfer function is usually expressed as the fractional change in flux per unit of reactivity since the absolute change in flux is

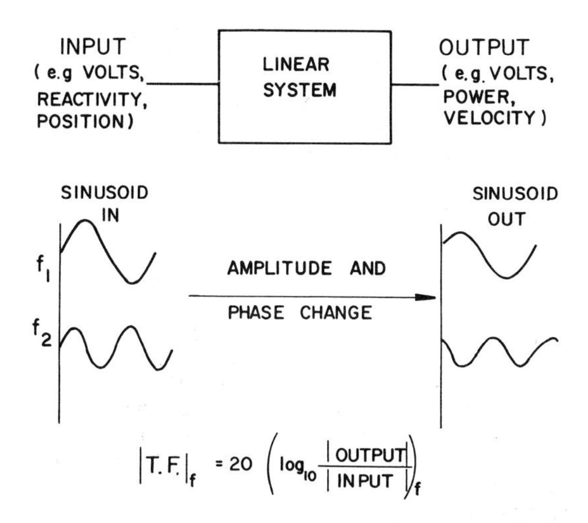

FIG. 4-28 Unit transfer function. The impulse response is L(t) in time domain. The transfer function is H(s) in frequency domain. The two functions are related by the familiar Laplace transform:

$$H(s) = \int_0^\infty L(t) \exp(-st)\, dt.$$

proportional to power. This has considerable influence on the design of automatic regulating systems (Sec. 4.3.3).

When the reactor transfer function is expressed this way, changes in gain and phase (see chapter on General Reactor Dynamics) result as the reactor is brought from the subcritical state (where it is easier to stabilize the loop) to the overcritical case (where the regulation loop is most difficult to stabilize) [165, 166].

Two other reactor effects strongly influence the dynamic response and subsequently the regulation problem. Firstly, changes in fuel, coolant or moderator temperature can affect reactivity. If the change in reactivity is proportional to the change in power or temperature, then the change for a constant percentage change in power is proportional to power and can drastically change the 'total'' reactor transfer function as shown in the chapter on Water Reactor Kinetics. The response is also influenced by the time constants associated with the coefficients, transport delay terms being particularly difficult to stabilize. Such terms can cause negative coefficients to give rise to resonant type transfer functions that may, because of their rapid phase changes, be more difficult to control than positive coefficients. Secondly, changes in flux generate changes in xenon poison concentration that are large but not rapid, and because of the 13-hr Xe lifetime the concentration is influenced for several days. Figure 4-29 shows the rise in xenon poison expressed in percent reactivity versus time from a step reduction in power for a 200 Mw D_2O-moderated natural uranium reactor. Even modest power reductions result in large changes of reactivity, therefore, especially in power generating reactors there is a reluctance to drop power unless absolutely necessary. This places greater importance on accurately assessing the true reactor state by instrumentation.

As reactors become large, xenon-induced spatial instabilities become possible [167, 168, 169,

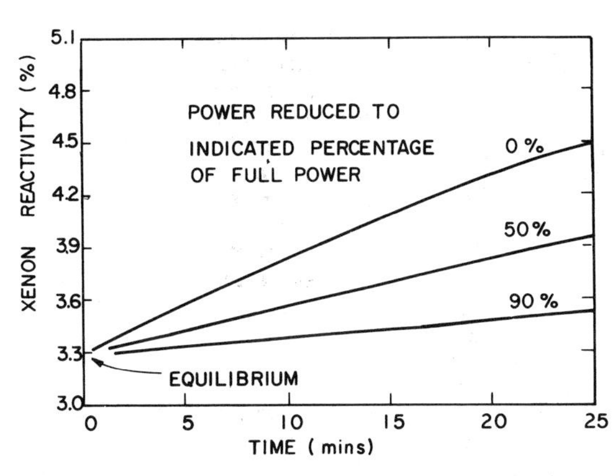

FIG. 4-29 Rise in xenon reactivity following power reduction in a 200 Mw, natural uranium, D_2O-moderated reactor.

170]. Large size and a high flux aggravates this effect. If the core design requires both axial and radial regulation zones, the control problem could become formidable. In reactors with a short core it is only necessary to determine radial power distribution and in-core detectors may not be required. Short core-length is also favored by the thermodynamicist for reducing coolant pressure drops and thermal stress.

4.3.3 Dynamic Response of the Reactivity Control System

The accuracy of regulation is dependent on several factors, among the more important being the speed at which disturbances occur, the finite accuracy with which the regulating device can be positioned, and its maximum acceleration and velocity during motion. The drive method adopted, i.e., a continuous or discontinuous velocity or position drive, also affects the regulation accuracy.

Consider the regulation loop closed by manual or automatic action as shown in Fig. 4-30. In both cases comparison is made between the reactor output and the demand, any error resulting in drive being applied to the reactivity regulating mechanism in such a way as to minimize the error.

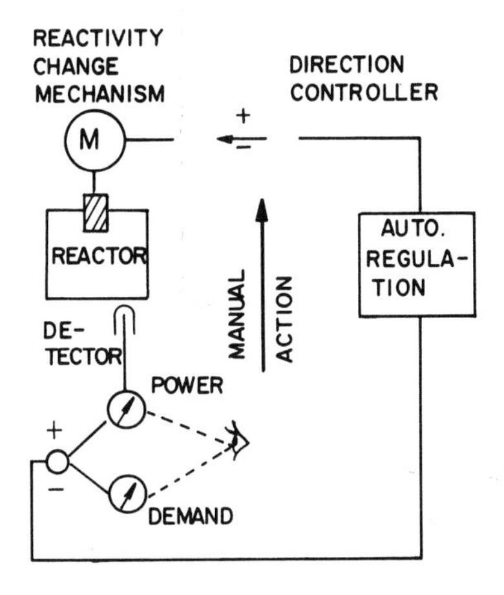

FIG. 4-30 Power regulation.

If the effect of the control system on the reactor transfer function is examined then the response of the controlled reactor can be calculated [171] for step, ramp or other input disturbances. Also, since the reactor gain in Mw per unit of reactivity increases in proportion to power, the forward loop gain of the regulating system should fall as the power is raised. Putting this another way, if the reactor transfer function is plotted as fractional power change, then the regulating system should give a certain correcting rate for a given constant fractional change in power.

Because any power level well above the source power can be obtained with the reactor very close to critical, it is usual to regulate the velocity of the reactivity change mechanism. There are various exceptions to this requirement described in Sec. 4.3.4.

Denoting the reactor transfer function gain by R at some frequency, and the regulating loop gain by C at the same frequency, then the closed-loop gain response is R/(1 + RC)[172a]. This is shown in Fig. 4-31. If the frequency considered is such that RC >> 1, then a sinusoidal reactivity disturbance is reduced by a factor RC when the reactor is under control. The modified reactor transfer function is plotted for various controller gains, expressed as reactivity change/sec per 10% power deviation. For the reactor shown, little improvement is obtained for gains in excess of 10^{-2}/sec for 10% error, as the reactor itself is insensitive to frequencies above 1 cycle/sec. To obtain gains as high as 10^{-2}/sec the regulating system gain at high frequencies must be increased above that used at low frequencies. Whether this can be done depends upon the noise or high frequency flux components (Sec. 2) that might cause circuit limiting. The inertia of the reactivity mechanism generally limits the high frequency response.

If position control is used then the curves would not continue to fall at 6 db/octave at reduced frequency, but would become flat at a frequency equal to the bandwidth of the position controller.

Regulatory or stabilizing effects occur if negative power coefficients with suitable time constants are present. A given change in reactivity then results in a new equilibrium power. Such an effect is obtained in the boiling-water reactor at Dresden where increased subcooling results in a collapse of voids followed by a rise in power, until the void volume returns to the original value. The change in the reactor transfer function due to an inherent negative temperature coefficient as well as to an applied reactivity regulating system is shown in Fig. 4-32.

Table 4-1 shows a list of typical regulatory approaches, their approximate maximum reactivity rates, and bandwidth, in decreasing order of performance. Whether or not this is equal to decreasing order of merit depends upon the imposed conditions such as core access and penetration problems, speed and type of disturbances, and the type of flux distortions allowed. The drive circuits for the externally applied mechanisms also vary considerably in complexity. In many cases combinations of these systems are chosen.

4.3.4 Manual Control

The regulation loop can be closed manually. Observation of the relevant sensor data and knowledge of the required 'program steps' will permit intelligent movement of the reactivity control mechanism.

Just how well the system can be controlled depends both on the response of the human operator and on the reactivity mechanism. Broadly speaking, a man's response is governed by three terms:

read-time computation time	typically 0.5 to 0.8 sec
output response time	about 0.2 to 0.3 sec.

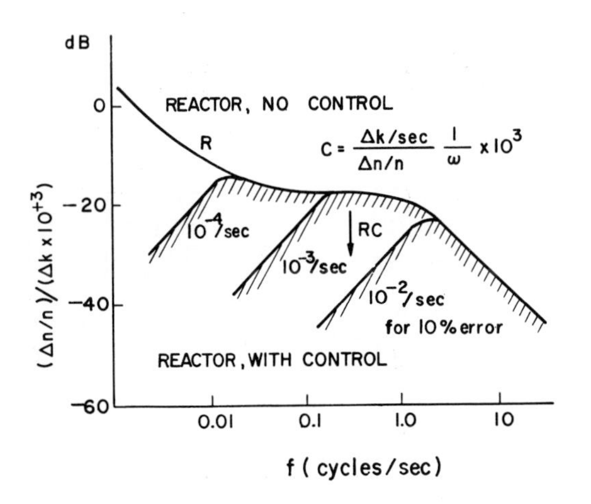

FIG. 4-31 Controlled reactor, transfer function.

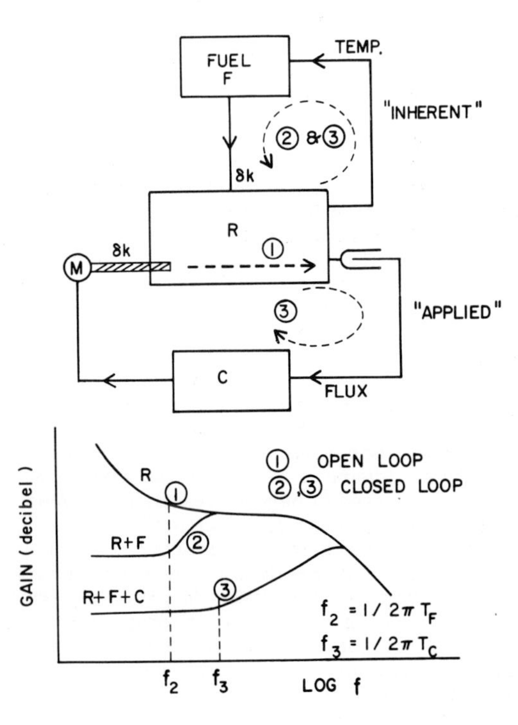

FIG. 4-32 Inherent and applied feedback.

TABLE 4–1

Regulating System Performance

System Approach	Max. Reactivity Rate	Bandwidth
Prompt fuel temperature	Set by power rate. See Chapters 2 and 8	0.1 to 10 cycle/sec depending on fuel time-constant.
Coolant void effect	Set by power rate. See Chapter 8	1 cycle/sec set by flow rate.
Continuous rod drive	Up to 10^{-3}/sec	1 to 10 cycle/sec
Discrete-step rod drive	Up to 10^{-4}/sec	Step rate up to 10/sec.
D_2O moderator level change	5.10^{-5} to 5.10^{-3}/sec	0.1 cycle/sec
Cd or B liquid poison	Slow, set by ion exchange columns.	0.01 cycle/sec or slower depending on circulation rate.

This results, under favorable conditions, in a bandwidth of 0.5 cycle/sec. An equivalent transfer function [173] could be:

$$\frac{1}{1 + pT_1}, \qquad e^{-pT_2}, \qquad 1 + pT_3,$$

where T_1 is data assimilation and reaction time constant, T_2 is the muscular load lag time, and T_3 is the mental compensation or anticipatory function.

It may often occur that reactivity rates are slow, and disturbances of such a nature that seconds or minutes are sufficient for a control decision; under these conditions the human link may operate very well. Unfortunately, it is under rather inactive circumstances that human fatigue and boredom show up and are detrimental to performance. There is also the question of data interpretation, calculation and display. Several practices have been accepted to aid the operator. Lights and bells warn him of sudden changes in state; the controlled variable display and controller are in close proximity; the more important parameters are grouped for display and rapid surveillance. If several units occur in a regulatory loop each having delays or phase lags, then intermediate displays of some other variable (such as control rod velocity or position or rate of change of power) often help the operator anticipate the result of his action. In manual control systems the operator usually finds it easier to control absorber position rather than velocity.

If properly trained, the operator can react intelligently and promptly to an unusual occurrence. In low power research reactors it may therefore be necessary to have an operator in close touch with the controls, and he often exercises manual control especially during startup. In favor of this is that the operator is closely watching his controlled parameter; against it is that other parameters, normally of not much significance, may be indicating an unusual occurrence and go unnoticed.

There is also the problem of smoothing statistical noise (Sec. 2.5) and the effect of this on the response of the instrument. The response characteristic should be known to the operator. This may be especially important where two instruments have overlapping ranges (Sec. 2.3) and differing dynamic response. It is common practice to smooth signals to meters and recording circuits to reduce signal fluctuations, electrical interference, and mechanical wear. Thus, rapid transients may go unnoticed to the operator.

One argument used in favor of the human operator is his sense of anticipation and his ability to run close to alarm levels under special circumstances. It may often be too difficult and expensive to build such ability into an automatic system. However, subtle changes in dynamic response may be smothered by the differences between operating personnel; one of the virtues of the automatic system is its long-term consistency.

4.3.5 Automatic Control

In the simplest case automatic control may be used to aid in regulation at power. The design aim may then be to overcome disturbances due to refuelling, xenon transients, or changes in coolant properties so that the power or power-rates do not deviate outside certain limits. A more advanced task would be the automatic startup of the reactor from the shutdown state.

Often the designer has to use a reactivity controller chosen for certain safety properties, in such a case the problem would be to optimize the system using this chosen device. Economic restrictions may limit the complexity, although best reliability and serviceability tend to be achieved with the simpler systems. In specifying performance there is a tendency to add unjustified safety factors; this is especially true in specifying accuracy and stability. On the other hand, recognition should be given to any cases where high accuracy may be required for an accurate analysis of plant performance.

Automatic control will be covered in two sections, the first concerned with reactor startup, and the second with full-power regulation.

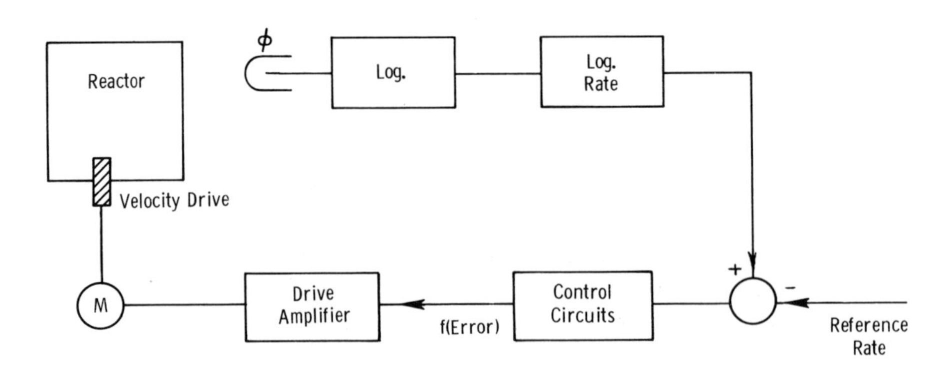

FIG. 4-33 Automatic startup with logarithmic-rate control.

1. Startup

During startup reactor power indication is almost entirely obtained from neutron-flux instrumentation, i.e., the log and period amplifiers of Sec. 2. Multi-range linear amplifiers may be used for indication but are more awkward for automatic control because of the frequent range switching required.

It is usually desired to regulate the inverse period or log-rate since this quantity is closely related to reactivity (see chapter on General Reactor Dynamics). The period signal is compared to a reference quantity and the error used to move the reactivity controller in such a way as to minimize the error. Figure 4-33 shows such a loop. The error is nearly zero if the gain of the control loop is high. The maximum reactivity rate is usually set with safety considerations in mind. Instability may arise because phase changes occur in the reactor transfer function (see chapter on General Reactor Dynamics), and in the log and period amplifiers (Sec. 2) as the power is raised. The conditions for closed-loop stability have been adequately covered in the literature, [163b, 172b, 174].

The most difficult effect to stabilize is that due to the changing time constant used to smooth statistical signal noise in the logarithmic channel (Sec. 2). As the power rises the smoothing time constant becomes smaller, and at some power level will equal a time constant elsewhere in the loop (generally due to the reactivity controller mechanism). At power levels below this level the system response will be slow, above it, fast; at precisely this level the loop will exhibit least stability. The increased reactor gain during the over-critical run-up should also be taken into account (see chapter on General Reactor Dynamics).

The graph shown in Fig. 4-34 [175] is a plot

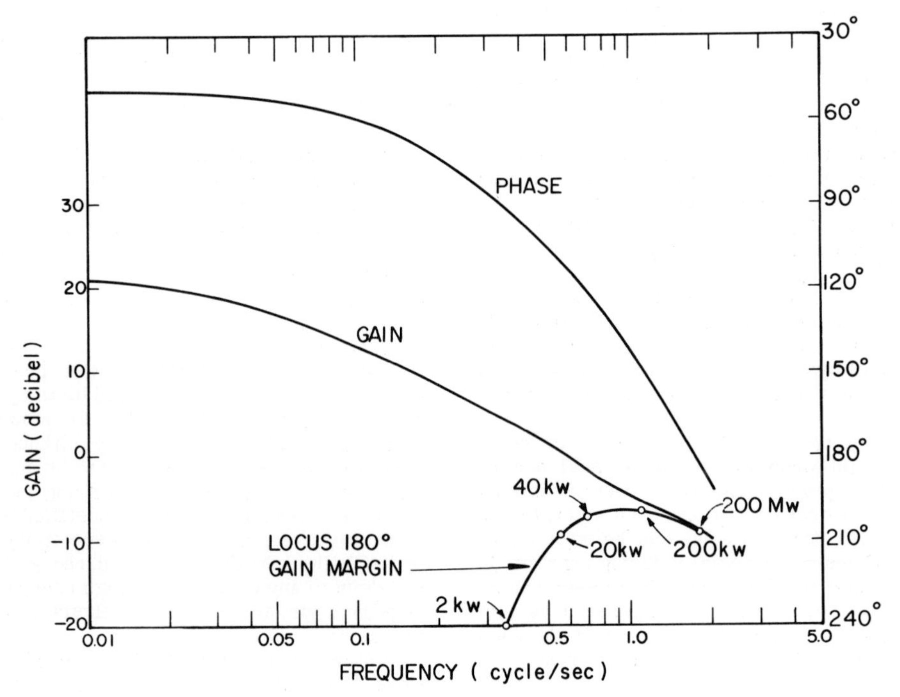

FIG. 4-34 Open-loop transfer function of the logarithmic-rate channel.

of open-loop gain and phase of the NRU reactor logarithmic rate startup channel, showing the gain margin (i.e. the gain at 180° phase-shift) at various power levels. Minimum margin occurs about 3 decades below full-power.

In attaining stability some high-frequency attenuation networks may be required in the control circuits. If such are used note should be taken of any overshoot resulting at the output of the log-rate amplifier from a change in demanded power [166].

By use of these techniques automatic startup from about 10^{-5} of full power to full power is possible. The power range is limited by the scale range of the logarithmic amplifier and the smoothing time constant used at low current levels.

2. Power Regulation

A linear-flux measurement is taken in order to level off the power as the demanded level is reached. It also provides a rapid measurement of power when operating at a fixed level.

In general the control will consist of a comparison between the flux signal and a reference level, where a defined error causes full-speed correcting action.

If regulation is required at various power levels then the circuit gain should vary inversely with power if the total loop gain is to stay constant (Sec. 4.3.3). Figure 4-35 shows two approaches to the comparison of the flux signal to a reference with inverse gain compensation. The references or demand power potentiometers may be adjusted by signals obtained from measurements on moderator level in D_2O reactors, on temperature, on thermal power, or on activity measurements such as N^{16}, in order to take into account changes in neutron flux not proportional to average reactor power.

In order to execute power changes without exceeding thermal stresses the linear rate of power change may also be controlled. Differentiating circuits measuring the rate of change have been discussed in Sec. 2.5, and this is shown incorporated into a regulating system in Fig. 4-36A. Four "states" are shown here to illustrate how the regulation passes smoothly from rate to power control. Figure 4-36B shows a motor drive to the demand-power potentiometer of the previous figure. The advantage of this technique is that no rate measurement is required, useful if the desired rate-of-change is very slow. An additional advantage is that the demand level can be "frozen" if the motor is de-energized.

Figure 4-37 shows the automatic startup of the NRU reactor using inverse period, linear rate and linear power regulation. The effect of the sequential rod withdrawals on these parameters can be seen, also the reactivity noise due to mechanical vibrations of the control absorbers.

Figure 4-38 is an enlarged trace of the regulation at power for the same reactor with and without the automatic regulation loop closed. The attenuation of low frequency changes is clearly seen (see Fig. 4-31).

Figure 4-39 shows an enlarged trace from the NRX regulation system records; this system uses D_2O-moderator-level control by use of triplicated water-valves. A disturbance was introduced by

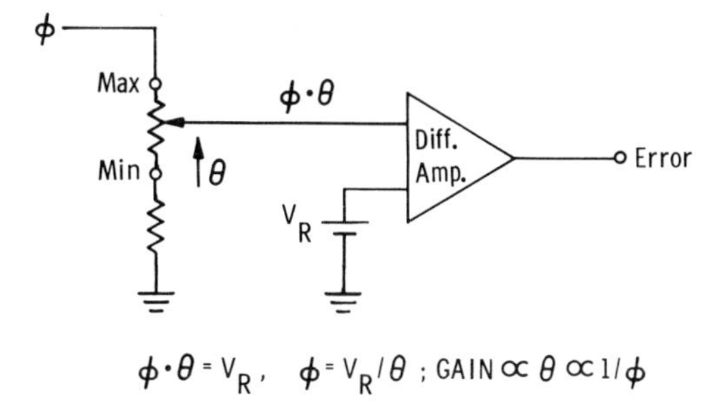

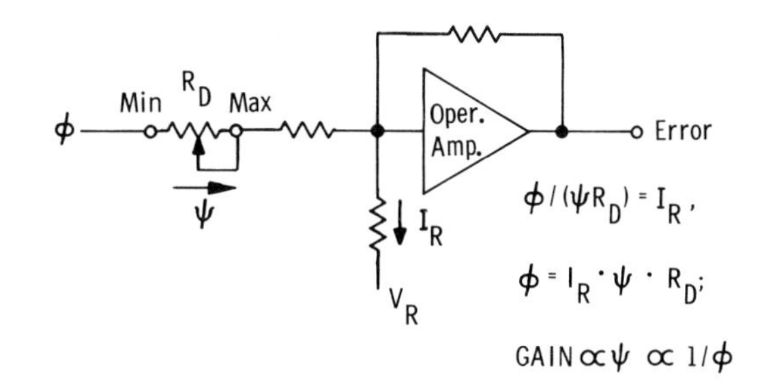

FIG. 4-35 Setting reference levels.

moving an adjuster rod at a fixed velocity. The resultant power off-set is an indication of the regulation loop-gain.

Once full power has been achieved it is often necessary to take into account changes between the flux indicating instruments and true thermal power (or whatever signal is to be finally regulated). In general the flux signal is used because of its rapid dynamic response to reactivity changes (Sec. 2) but final control usually is based on thermal measurements.

Two methods for modifying the flux demand are shown in Fig. 4-40. In A, direct control is achieved by moving the demand potentiometer, or by adjustment of the reference either through a motor driven potentiometer or by an integrating amplifier. In some cases an integrating amplifier is not needed (Fig. 4-41), one of its disadvantages being the drift in output due to finite gain; and due to the error in output accumulated when the reactor is starting up, a clamp circuit may be used to overcome this latter effect.

In B, a system is shown that uses the same demand dial for regulating signal 1 or 2. The advantage of this approach is that demanded changes in signal 2 can be achieved at a speed primarily set by the response of signal 1 [11]. If signal 1 is a flux measurement and signal 2 is temperature, this may enable maneuvers in temperature to take place 10 times faster than possible in A of Fig. 4-40. Dynamic balance of the signals to the divider is important.

Signals from the steam-generating system are required to regulate a plant. A load following approach [176] is shown in Fig. 4-41. Steam-flow to the turbine represents the heat sink, and neutron power (corrected by thermal signals) the heat source. Any difference cools or heats the plant. If the steam pressure equals that demanded then steam flow determines the required neutron power

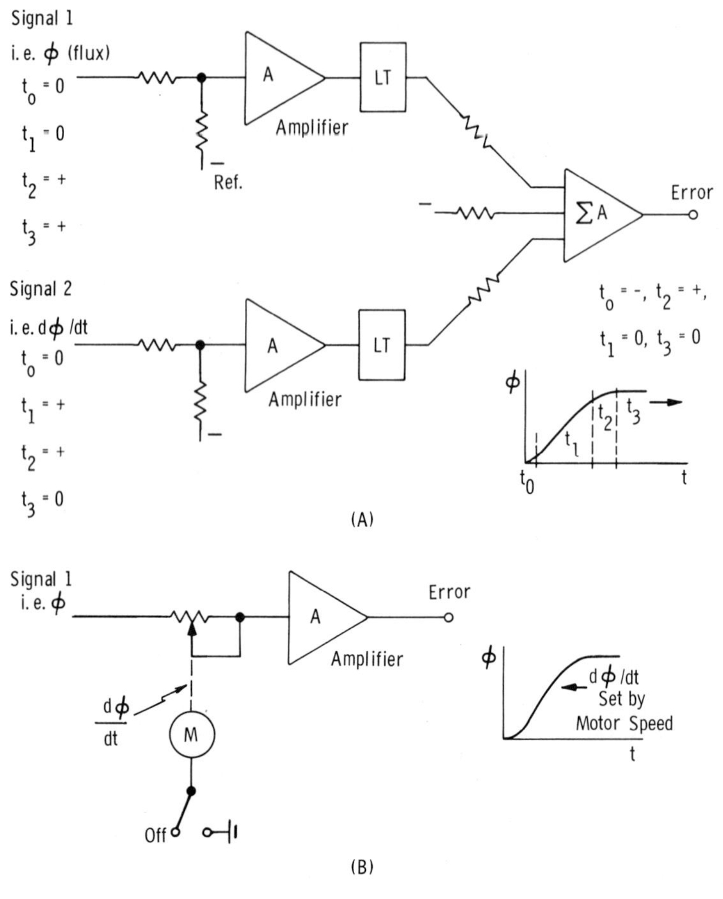

FIG. 4-36 Startup mode changes.

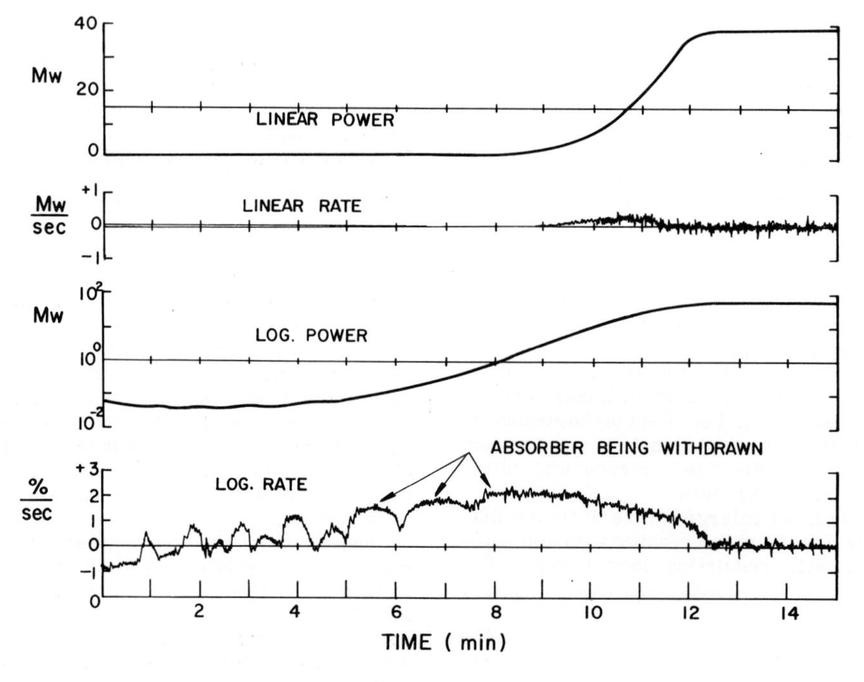

FIG. 4-37 Automatic startup.

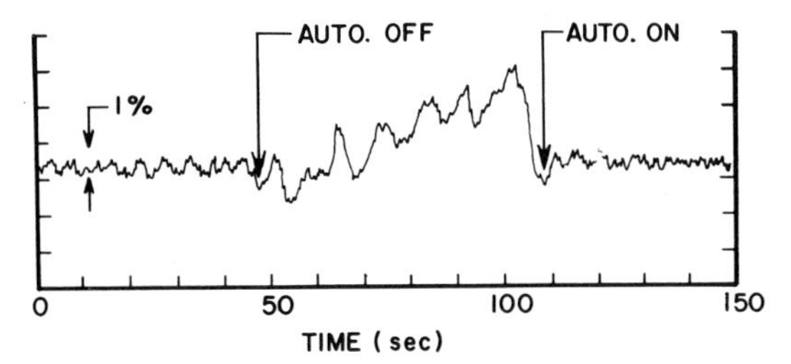

FIG. 4-38 Effect of automatic regulation.

reference level. During startup the steam pressure is less than that demanded so the error determines the plant warming-rate.

In this system other signals controlling moderator level and a nominal low-power point are also set into the summing amplifier. A limit is set on the maximum thermal power fed to the neutron reference, and a reactor coolant outlet temperature signal can override the flux signal if an undesired high temperature is reached. The above analog computing system is triplicated in the NPD reactor with each channel operating two gas valves regulating D_2O moderator level.

The three set-pressure and three plant-warming controls are separately ganged to common shafts at the control desk.

The turbine-generator regulation system for the same plant is shown in Fig. 4-42. After

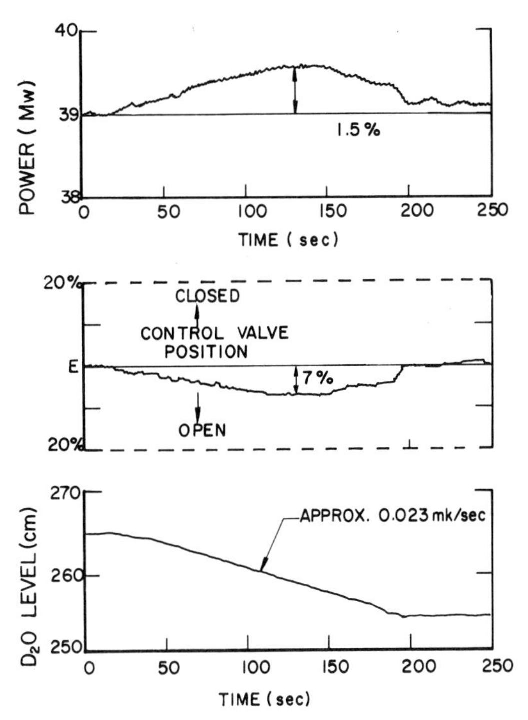

FIG. 4-39 Reactivity disturbance applied to reactor with D_2O level regulation.

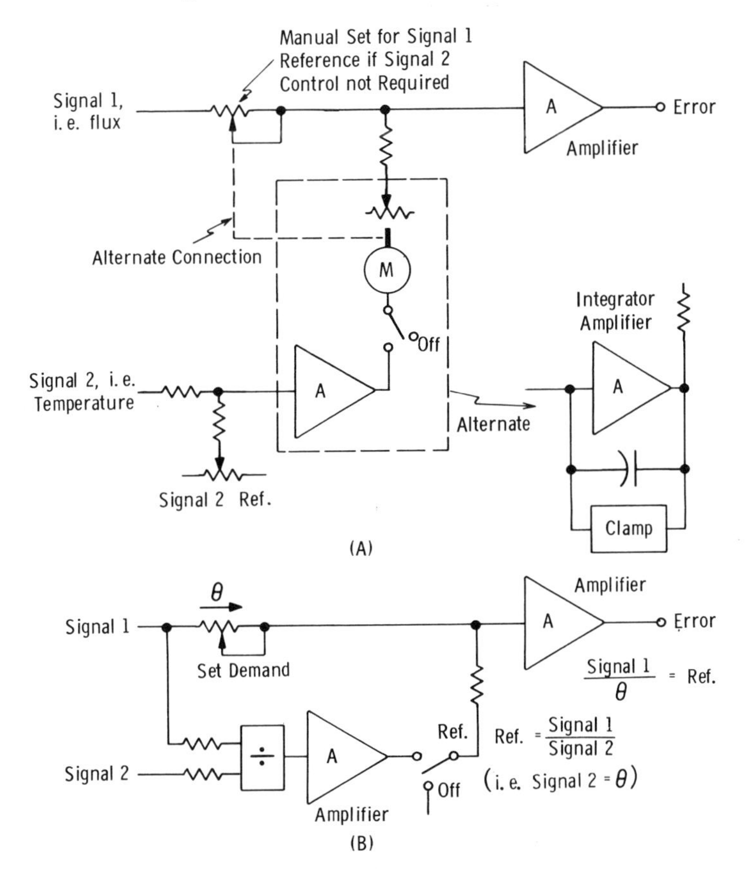

FIG. 4-40 Changes in controlled parameter.

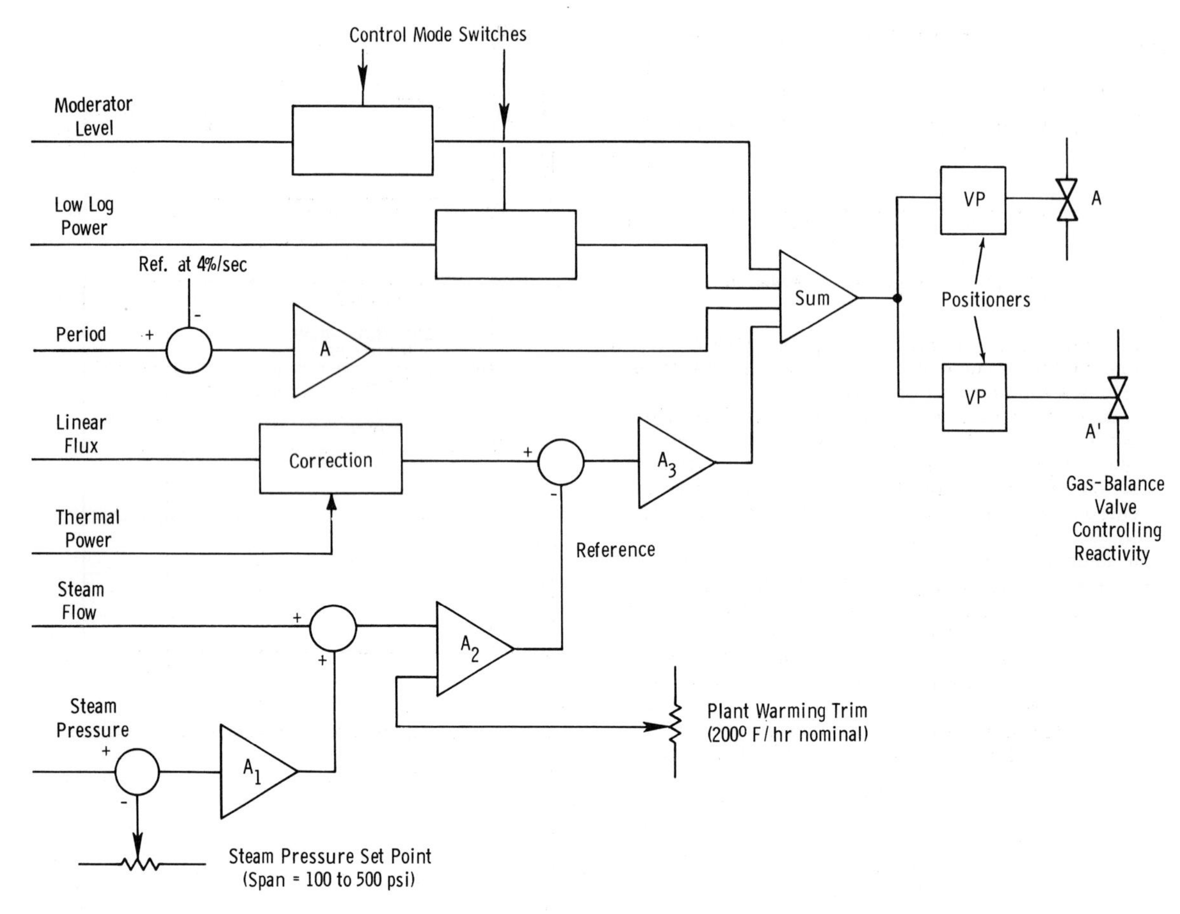

FIG. 4-41 Power station reactor regulating system.

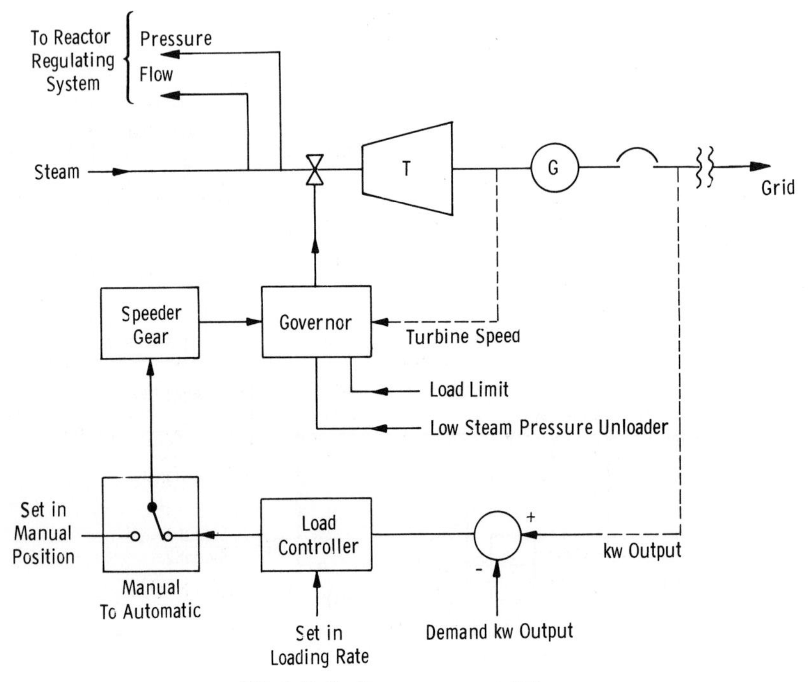

FIG. 4-42 Turbine-generator regulation.

sychronization and closure of the generator breaker-switch, loading can proceed manually or automatically by opening the turbine throttle valve. Loading at rates up to 20% of nominal full-power per minute are allowed. This corresponds to a reactor outlet temperature rate of 1000°F/hr (556°C/hr).

5 RADIATION MONITORING

5.1 General Requirements

In addition to the measurement of radiation for reactor control purposes there are two general radiation monitoring requirements that pertain specifically to the safety of the operating personnel.

1. Operating personnel must be warned when exposure to excessive radiation is occurring. Monitors used for this requirement usually measure radiation in terms of its biological effects.
2. The spread of radioactive contamination into the public domain must be prevented. The monitors used to provide this protection are designed to detect the presence and magnitude of radioactivity, and in some cases to provide identification of the radioactive nuclides.

Since radiation may be of various types and may occur in a variety of areas around a reactor and at various levels of intensity, a variety of both fixed and portable instruments is used. The following are typical of the more detailed requirements.

1. In normal working areas monitors are required to give continuous assurance that radiation levels are within safe limits.
2. In areas where the radiation level could become high due to a local accident, such as a criticality incident in a fuel handling area, or a spill of radioactive material, special high range monitors are required to provide evacuation alarms.
3. The spread of contamination by gaseous and liquid effluents either inside or outside of the plant area must be guarded against by monitors coupled with control measures to reduce the quantity of effluent whenever radiation levels are abnormal. This problem is made difficult because even very small amounts of radioactivity released over a long period of time can concentrate in some animal or plant form. As a further precaution on-line monitors should be backed up by periodic radiochemical analysis both of the effluent and of biological specimens taken from the surrounding area.
4. To prevent the spread of contamination by personnel they must be monitored when leaving those areas of the reactor site where contamination could occur.
5. Portable radiation monitors are necessary to survey regions not protected by permanent instruments and to locate concentrated areas of contamination. Various working surfaces must also be checked to ensure that good housekeeping is practiced where radioactive materials are handled.
6. In areas where the radiation level is known to be high during reactor operation, fixed monitors may be installed to indicate when entry is possible. Portable survey instruments are also used to confirm that entry is safe.
7. When personnel are required to work in high radiation fields, monitors are worn by the individual to measure the accumulated dose.
8. As an over-all check on the utility of the various safeguards outlined above for the protection of personnel, individuals wear monitors that indicate the total amount of radiation received over a week or some longer period. Such checks may also be supplemented by periodic blood and urine analysis.
9. To ensure that the precautions taken to avoid spread of contamination are adequate, monitors should be dispersed about the reactor site and should be in operation before the reactor is brought to power. It may be necessary to supplement these monitors with surveys by portable instruments and by examination of biological specimens taken from the environs.

While many radiation monitoring instruments use the same principles as described in Sec. 2 for reactor control, there is a general attempt to produce simple circuits because of the large number of monitors needed.

Several instruments are described below that embody novel ideas and principles associated specifically with radiation monitoring.

5.2 Area Monitors

5.2.1 Gamma Monitors

Area gamma monitors are designed typically to indicate radiation levels from background ($\approx$ 0.1 mr/hr) to about 50 mr/h with the indication on a logarithmic scale.

While ionization chambers and d-c amplifiers of the type already described have been used, the trend is toward simpler instruments using Geiger tubes and pulse counting circuitry. The charge available from a Geiger tube per pulse is in the 10^{-9}-cou region which is sufficient to operate a simple transistor switch directly. On the other hand, the current from a reasonably sized ionization chamber ($\sim 10^{-13}$ amp/cm^3 per r/hr) at the radiation levels encountered in this application is in the 10^{-13} amp region and requires sophisticated electrometer tube techniques for reliable measurements.

A circuit [177] that will indicate the rate at which a simple switch opens and closes is shown in Fig. 5-1. When the switch closes the capacitors discharge through the diodes. When the switch opens the capacitors charge at a rate determined by the value of the time constants RC and R_1C_1. If the time constants differ by a factor ten the smoothed charging current will be a quasi-logarithmic func-

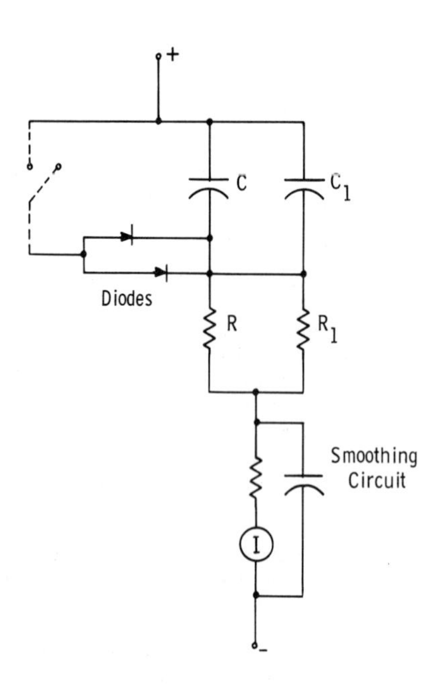

FIG. 5-1 Ratemeter circuit.

tion of the number of times per second the switch closes.

The arrangement of transistors [178] in Fig. 5-2 forms a switch capable of discharging the capacitors each time a Geiger tube pulse appears. The transistors are normally nonconducting but a pulse from the Geiger tube is followed by regenerative action that forces both transistors into the conducting state. When the capacitors are discharged the circuit returns to the nonconducting state.

The circuit [179] shown in Fig. 5-3 also acts as a switch. The current from the Geiger pulse when amplified by two transistors is sufficient to discharge the capacitors rapidly.

The halogen-quenched Geiger tube [180] has a long life especially when a current limiting resistor is used and is ideally suited to many radiation monitoring applications. The sensitivity of the Geiger tube varies with gamma-ray energy and tubes are usually shielded to produce a more uniform dose-rate response. Figure 5-4 shows a shielded Geiger tube and its response characteristic. This tube produces about 20,000 counts/min in a radium gamma field of 50 mr/h. Directional effects can be important especially at lower energies and must be kept in mind when placing instruments. At very high count-rates the dead time of the Geiger tube causes saturation and the mean ionization current has been used to extend the range. Care must be taken that all personnel utilizing such instruments know of these saturation characteristics. A gradual drop in reading from such a device does not necessarily signify a drop in radiation level. For monitors measuring low ambient radiation levels a totalizing scaler, rather than a ratemeter, is more suitable for determining the dose accumulated over a long period.

Some area monitoring applications [30g] require an instrument that will detect a small change in ambient conditions. If a scintillation counter is used with a discriminator set in the low kev region all the gamma-ray quanta from this energy up will be given equal weight. The dose rate calibration is incorrect but a monitor results with a high signal to background ratio. A change of 5% in the ambient radiation level is detectable.

A special application [181] of the scintillation counter area monitor has extended the use to include directional information by fitting monitors with rotating collimators. This permits source location by means of a triangulation procedure.

5.2.2 Airborne Activity Monitors

A specific hazard to be guarded against is the appearance of uranium or plutonium dust particles in the atmosphere. The simplest technique draws air, from the region being monitored, through a filter paper which is subsequently monitored by an alpha sensitive detector. The sensitivity is limited by natural activity (radon and thoron daughters) on

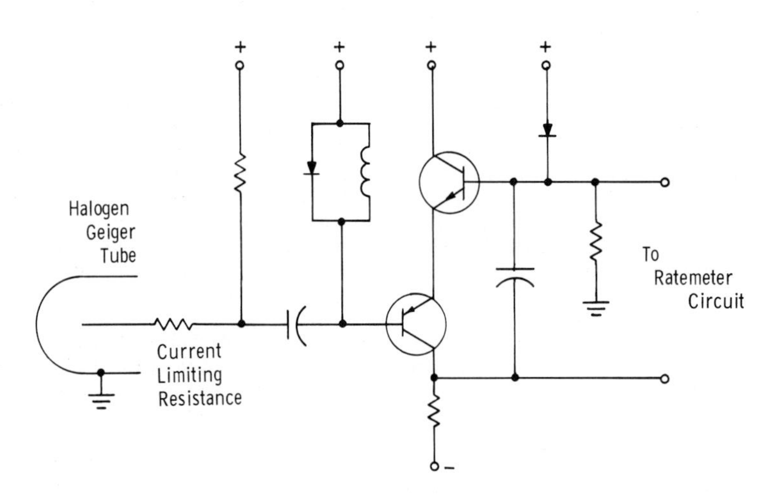

FIG. 5-2 Regenerative trigger circuit.

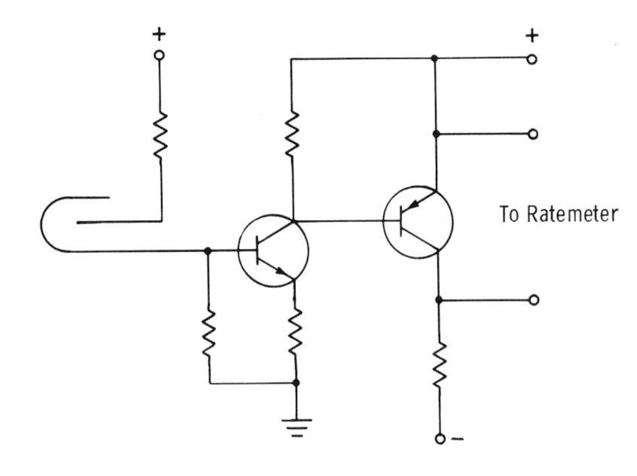

FIG. 5-3 Saturating switch.

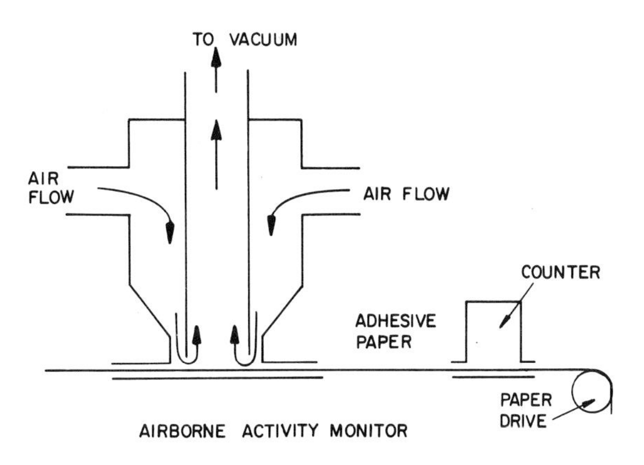

FIG. 5-5 Airborne activity monitor.

dust that is also trapped by the filter. The sensitivity is improved if the radon is allowed to decay for a period before the sample is monitored.

Mechanical and electronic means can also be used to improve the signal to background ratio.

Figure 5-5 shows a scheme [59b, 182] that selectively deposits the heavier particles of uranium or plutonium on an adhesive surface and permits lighter dust particles to be carried away. It has been estimated that background can be reduced by 95% over a simple filter mechanism. The alpha detector uses a zinc-sulfide scintillator coupled to a photomultiplier tube.

It is not possible to distinguish among various alpha particle energies with a zinc-sulfide detector but with the advent of the semiconductor radiation detector [183] it is possible to reduce the background by energy discrimination. A detector [184a] with a sensitive surface area of 3.8 cm^2 shows a resolution of about 300 kev for alpha particles in the 5 Mev region. The detection of 2×10^{-12} $\mu c/cm^2$ may require as much as eight hours of counting after sampling.

Another electronic discrimination technique

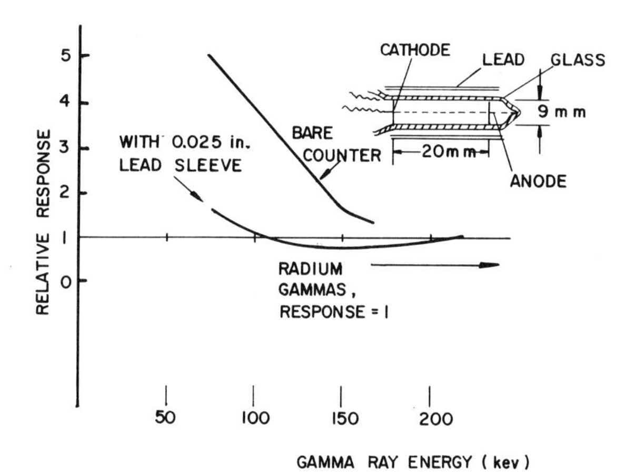

FIG. 5-4 Energy dependence of Geiger counter response.

makes use of the particular mode of decay of radon and thoron. Delayed coincidence counting techniques permit radon and thoron disintegrations to be identified and these counts are subtracted from the total to give the Pu^{239} count. The detection of 2×10^{-11} $\mu c/cm^3$ of Pu^{239} in 90 min is reported [184b].

Particulate matter trapped on filters may be monitored for beta activity. Thin window Geiger tubes or thin plastic scintillators can be used for detection [185a]. Monitoring can be done during the collection period to warn of any unduly large increase in activity concentration.

5.2.3 Neutron Monitors

For the detection of slow neutrons a boron trifluoride proportional counter and charge sensitive amplifier is a satisfactory combination. The permissible tolerance level permits a small counter to be used and still provides adequate speed of response. A typical instrument [186] is shown in Fig. 5-6. It incorporates a counter voltage supply using a high-frequency transistor oscillator and a Cockcroft-Walton type voltage multiplier.

For fast neutron monitoring (>.01 Mev) thermal neutron detectors have been enclosed in moderating material. Such monitors indicate only that neutrons are present and do not provide any dose-rate information.

The most common dose-rate instruments make use of the proton-recoil reaction in a hydrogenous material. The proportional counter shown in Fig. 5-7 [30m] has a polythene liner from which protons are ejected into the sensitive volume of the counter. Since greater biological damage is caused by the higher energy protons the associated circuitry (Fig. 5-8) gives greater weight to the larger pulses. This results in an output counting rate more nearly proportional to the dose rate.

At fast neutron tolerance levels the count rate from a proportional counter of this type is low (a few counts per second) so the time to obtain the radiation level with sufficient accuracy is long. An instrument with a higher counting rate uses a scintillator-photomultiplier combination. The

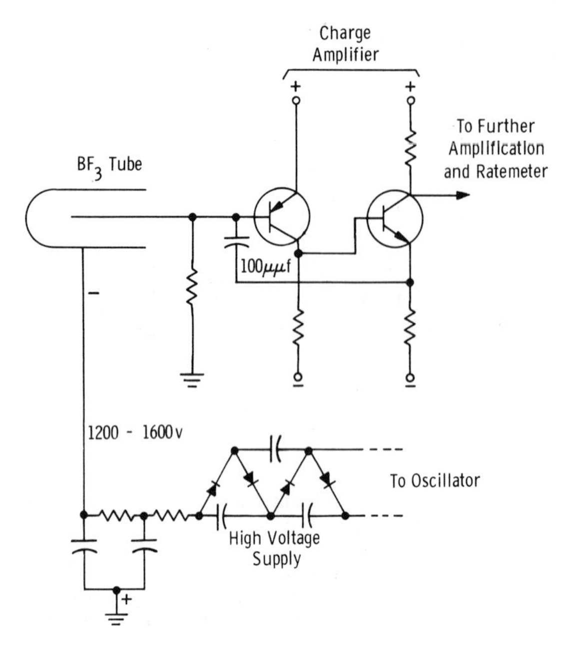

FIG. 5-6 Slow neutron health monitor.

scintillator [187] consists of a Lucite cylinder in which zinc sulfide is dissolved. Fast neutrons produce recoil protons that in turn interact with the zinc sulfide to produce scintillations.

5.2.4 Tritium Monitors [184c,188,189,190]

In D_2O-moderated reactors large amounts of tritium are produced and the escape of any heavy water permits tritium to become airborne in heavy water vapor. Tritium emits low energy beta-rays (up to 18 kev) and is detected by bringing the tritium into the sensitive volume of a radiation detector.

A tritium monitor is shown in Fig. 5-9. A compensated ion chamber arrangement is used to reduce the effect of gamma-ray background and an ion trap ensures that an ionization-free sample of atmosphere enters the chamber. In a four-liter chamber a tritium concentration of 2.5×10^{-5} $\mu c/cm^3$ produces an ionization current of about 10^{-13} amp.

5.3 Radiation-Incident Monitors

The gamma radiation associated with criticality incidents, or spills of radioactive solutions or powders, is used to detect their occurrence. The monitors usually have an extended range (up to several or even hundreds of r/hr) and are designed to be highly reliable at the expense of accuracy if necessary.

Emphasis is placed on not giving false alarms, which reduce confidence in the instrument, since ignoring a warning from such installations could have serious consequences. To this end faults should be indicated by some means other than by sounding the radiation alarm. Even too frequent testing can cause too much familiarity with the alarm signal and some hesitation when an incident has occurred.

Figure 5-10 shows a battery operated criticality incident alarm using the simplest possible circuit [30h]. The ionization chamber current flows as grid current into the electrometer tube and a logarithmic response results at the anode (see Sec. 2). A high and low contact on the meter provides alarms.

Figure 5-11 indicates another approach [191] aimed at high reliability. A plastic phosphor is coupled to a photomultiplier tube. As the radiation level increases, the mean current in the photomultiplier attempts to increase but is maintained nearly constant by the feedback connection to the dynode voltage supply. The dynode voltage is a quasi-logarithmic function of the radiation level and is used as the output. A magnetic amplifier operates the alarm relay and a distinctive audible alarm.

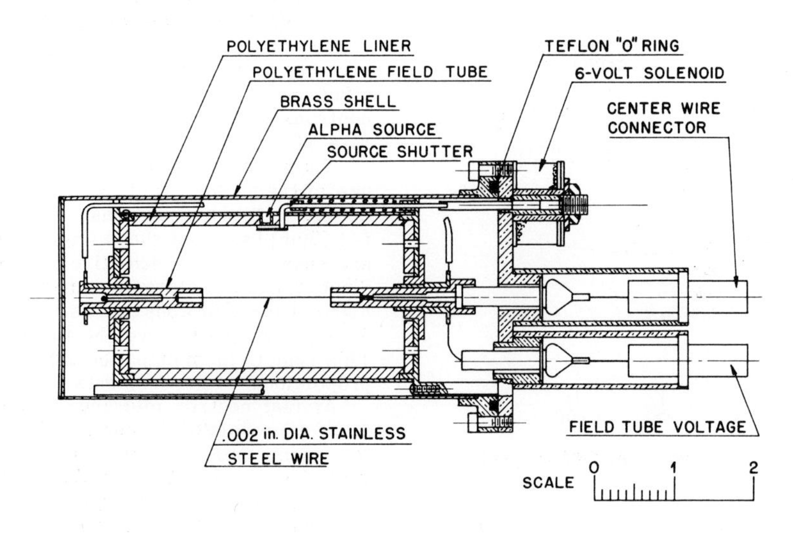

FIG. 5-7 Fast neutron detector.

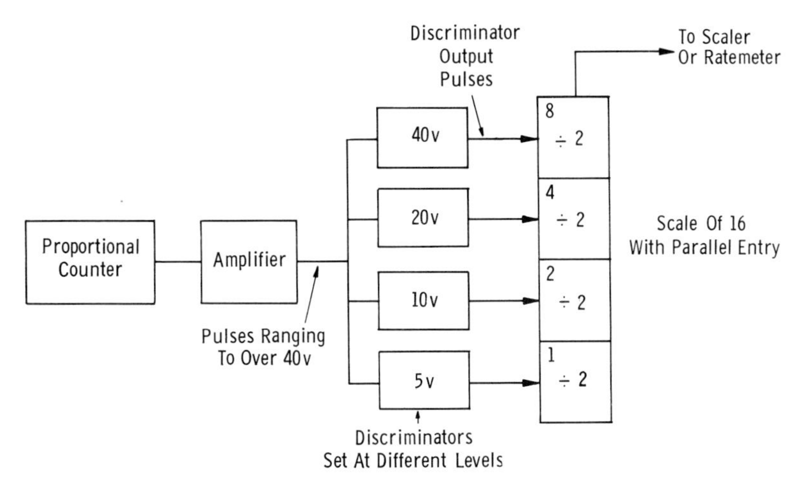

FIG. 5-8 Fast neutron dosimeter.

An interesting method for specifying the sensitivity of incident-monitoring installations is noted by G. Cowper [192]. It states that an alarm should sound whenever a radiation field is detected that corresponds to the appearance anywhere in the protected area of a source equivalent to 10 curies or more of radium. This type of specification emphasizes the need to consider the placement of radiation monitors along with their detailed design.

5.4 Effluent Monitors

Since every reactor comes into contact with the public domain through gaseous and liquid effluents the control of their radioactive content assumes an important role. Two major sources of radioactive contamination exist, fission products within the fuel and activated impurities in cooling or moderating fluids. The chapters on Fuel Elements, Mechanical Design, and Fission Product Release deal with these in detail and with the mechanical lines of defense set up to control their escape. This section deals with methods used to detect their presence.

There are two approaches to effluent monitoring.

1. Relatively high level monitors are located in effluent streams to detect when a containment breach has occurred. Such monitors may perform one of several control actions including fast shutdown (scram) of the reactor, cutback of power or diversion of radioactive streams into storage-delay systems.
2. The effluent is analyzed by sensitive and accurately calibrated monitors, either directly or in conjunction with sampling systems, that are capable of measuring activity at much less than maximum permissible concentration levels. These monitors exercise long term control over what is permitted to leave the reactor site.

5.4.1 Gaseous Effluent Monitors

To monitor effluent gas for particulate radioactive material, samples may be drawn (see the chapter on Radioactive Waste Management) from the effluent stream (usually a vertical stack) and taken to monitors of the type described in Sec. 5.5.2 [185b]. The detection of the gross gamma-ray intensity level in the effluent gas (Sec. 5.2.1

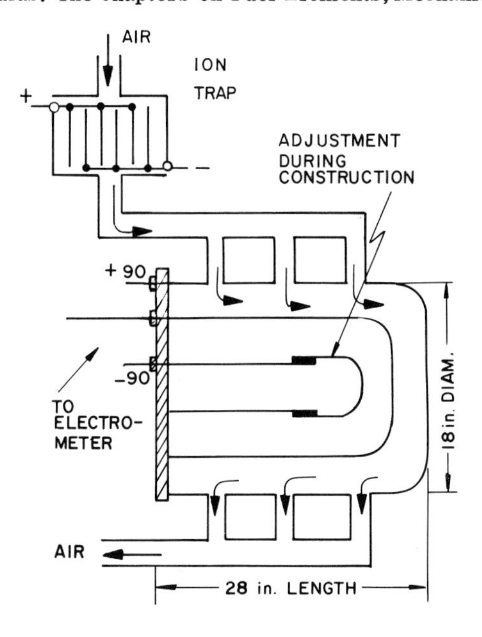

FIG. 5-9 Tritium monitor.

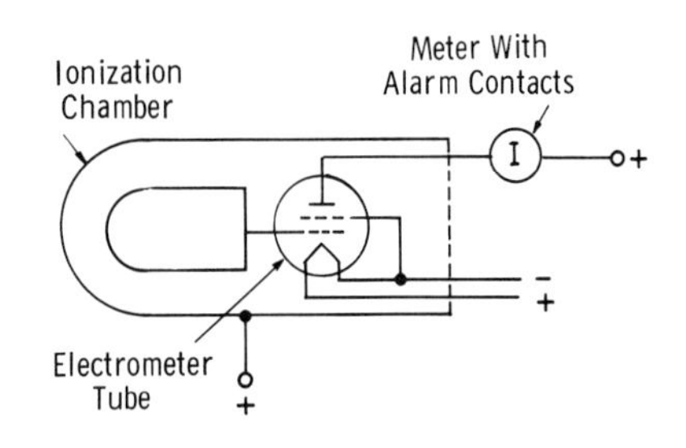

FIG. 5-10 Essentials of a simple criticality monitor.

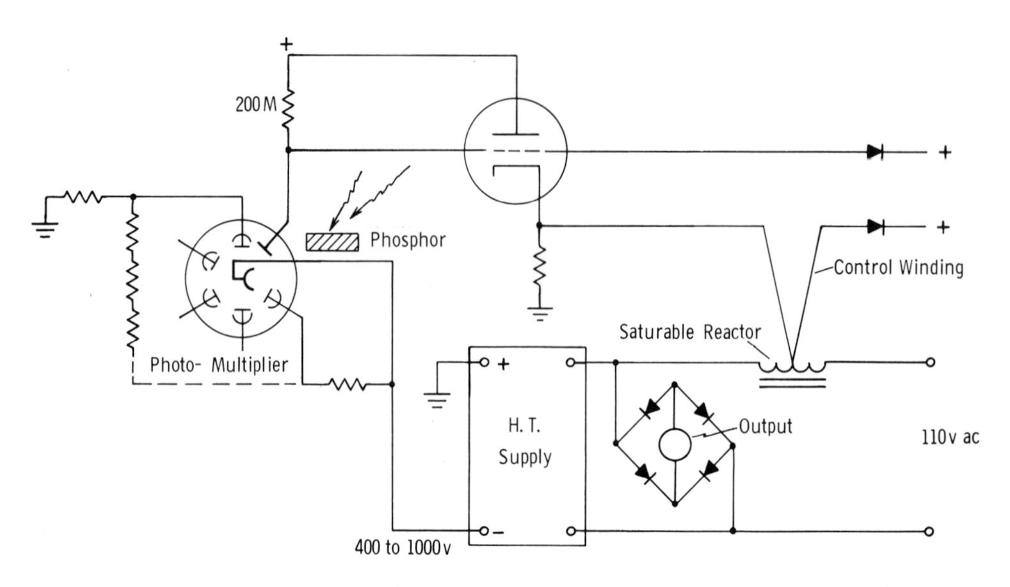

FIG. 5-11 High-range gamma monitor.

again) can be done with a sodium iodide crystal and photomultiplier placed near the top of the stack but out of the main plume to avoid contamination buildup on the detector. Banks of Geiger-Müller tubes are often used for this type of service also. These monitors detect abnormal changes in background. Operating experience usually dictates what is normal.

When gross-gamma monitors are placed inside effluent ducts the housing design must permit easy decontamination.

For the specific detection of I^{131}, gamma-ray spectroscopy can be used [193]. An on-line I^{131} detection system has been described that uses the 81 kev gamma-ray from Xe^{133} upon which to base its action. It is assumed that a fuel rupture permitting I^{131} to be released will also release Xe^{133}. The detectors look directly at the effluent and the normal I^{131} spectrum is obscured by other background gamma-rays except in the 80 kev region. The iodine gamma-ray at 80 kev is relatively weak [194] so the more abundant 81 kev Xe^{133} line was used.

The control system associated with the monitor is shown in Fig. 5-12 and illustrates the application of redundancy and coincidence techniques. When a predetermined level of activity is reached the stack is closed at the top and the integrating chan-

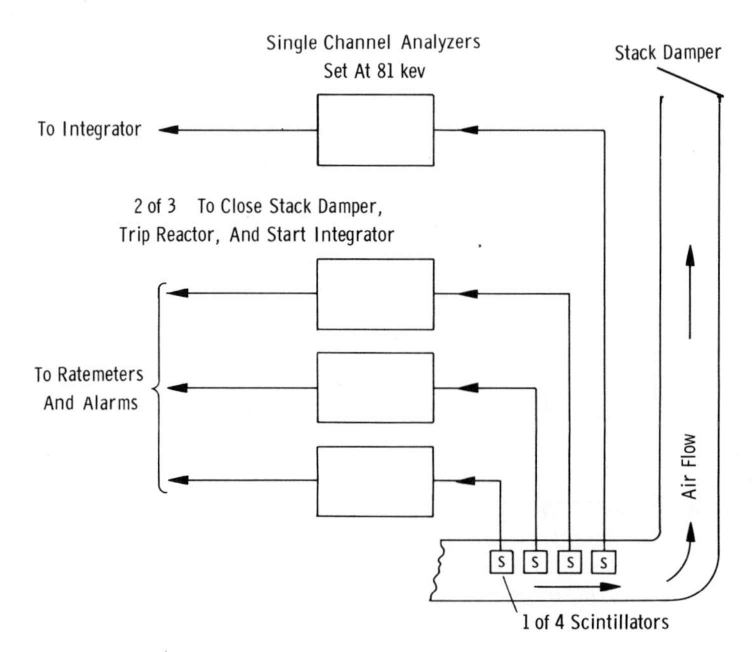

FIG. 5-12 Iodine monitoring and effluent control system.

nel begins to show the activity being accumulated in the system. Radiation monitors control the closure of containment penetration in several reactor systems [195].

Mechanical separation of iodine from other products by means of charcoal filters [185c] enables the detection and identification of small amounts of iodine possible.

A more sophisticated system for radioisotope determination in gaseous effluents [196] draws a gas sample through two filters, one designed to absorb iodine and the other to remove particulate matter (ruthenium). Both filters are examined by both a beta-particle detector and a gamma-ray spectrometer. The outputs are fed into an analog computer where the effect of interference (for example of ruthenium appearing with the iodine in the iodine filter) is removed and the output due only to the iodine and to the ruthenium is presented.

5.4.2 Effluent Water Monitors [197, 198]

The simplest approach to effluent water monitoring is a gross-gamma detector immersed in a well type mount in the effluent pipe. With a sodium iodide crystal and photomultiplier an activity level of about 10^{-9} curie/liter will double a typical background (~ 0.1 mr/hr) counting rate. A Geiger tube would be an order of magnitude less sensitive.

Vibration may be a problem in a well type housing and components associated with sensing head should be mounted accordingly.

A continuous-flow monitoring system sensitive to beta-emission has been described [199,200]. A methane proportional counter is mounted close to the water surface while a second is mounted directly above it. The electronic circuit rejects pulses that are coincident from both thus eliminating cosmic-ray background. Detection limits in the 2×10^{-7} μc/mℓ are claimed for the higher energy beta-emitters such as Sr^{90}.

A "swirl" monitor that takes a sample continuously from the main flow and uses a plastic scintillator for detection [201] is shown in Fig. 5-13. The sample enters the housing tangentially and spirals around the scintillator without touching it. Detection sensitivities in the 10^{-11} μc/mℓ region are possible for Sr^{90}.

Automatic sampling, concentrating, and counting systems are also in use [202] and are being further developed.

5.5 Personnel Monitors

5.5.1 Hand and Foot Monitors

Hand and foot monitors are intended to detect small quantities of beta or alpha emitting contamination on hands and footwear. There are several design problems.

1. The levels of contamination that are considered biologically significant are low (10^{-6} μc/cm^2 alpha and 10^{-4} μc/cm^2 beta).
2. The monitoring process should be rapid otherwise the temptation to by-pass the monitor is great, especially at peak load periods when a large number of personnel are leaving the site.
3. The monitor should be as insensitive to background radiation as possible. False alarms add to the problem in (2).

A monitor designed to reduce background variation [30i] difficulties is shown in Fig. 5-14. This monitor uses Geiger tubes to detect beta activities through accompanying gamma radiation, although beta or alpha sensitive detectors could have been used.

The pulses from the detector are brought to a uniform amplitude and when no one is being monitored the pulses are fed to a ratemeter circuit whose output voltage is proportional to the background count rate. When monitoring begins the pulses are re-routed to the integrating circuit whose output is proportional to the number of counts received from the beginning of the monitoring period. In addition, however, a current proportional to the background is fed from the ratemeter output (now frozen by the action of the switches) into the integrator and subtracts the

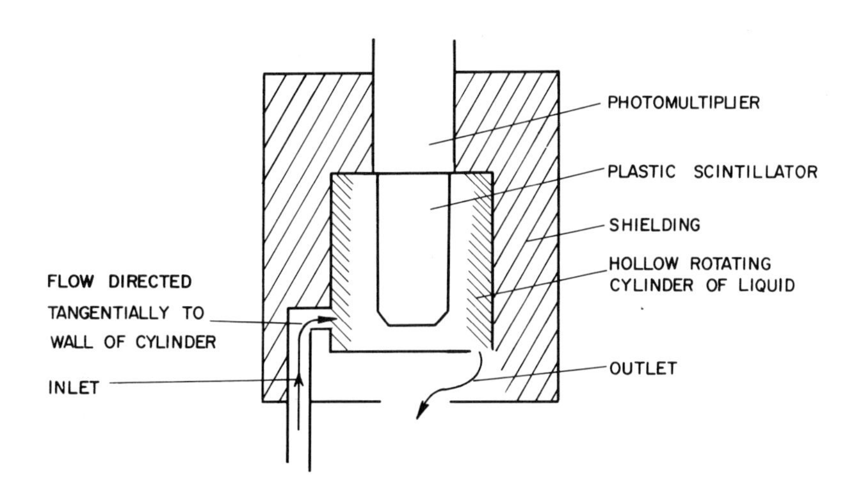

FIG. 5-13 Swirl cell monitor.

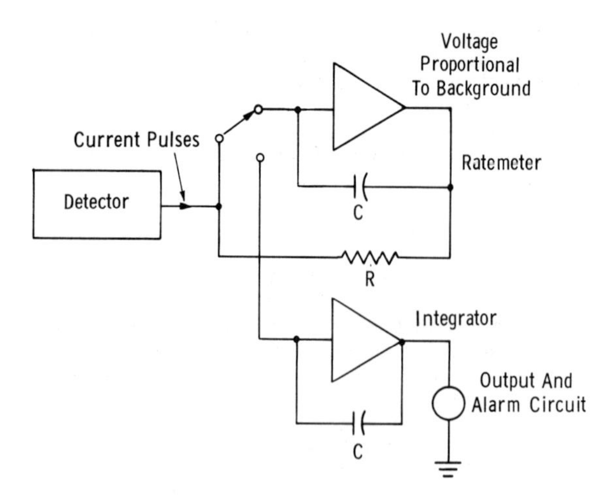

FIG. 5-14 Hand and foot monitor with background compensation.

contribution being made by the background to the integral count.

Monitors usually require a fixed time interval to be allotted to each person and this is only reduced if a high level of contamination is found. A scheme to reduce the time allotted to uncontaminated persons (who are by far in the majority) is shown in Fig. 5-15 [203]. Pulses from the monitoring counter are integrated on the capacitor C. At the instant monitoring begins a current I is also fed into the capacitor. If no counts are coming from the detector the current I charges C to a predetermined negative voltage at which a "clear" alarm is given. If the input counting rate yields a current larger than I the capacitor charges to a preset positive voltage and a "contaminated" alarm sounds. For borderline cases the time could become very long so an arbitrary decision must be made at the end of a specified interval. A reduction in time of about five is expected in cases where a large proportion of the personnel is uncontaminated.

This alarm method has been incorporated with an instrument [204] that uses a thin plastic phosphor coupled to a photomultiplier for beta particle detection. Sensitivity to gamma radiation is reduced considerably by the use of thin scintillators and therefore no other means is used to avoid gamma background fluctuation problems. For alpha monitoring a zinc-sulfide scintillator replaces the plastic.

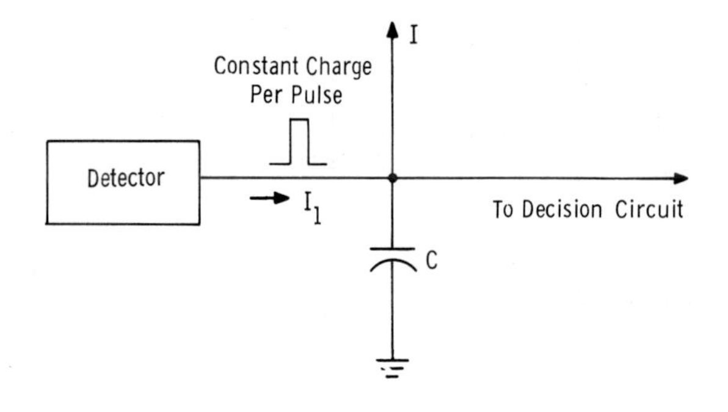

FIG. 5-15 Quick-acting alarm system for personnel monitoring.

5.5.2 Doorway Monitors

Doorway monitors are usually sensitive gamma detectors to prevent personnel with contaminated clothing from leaving the monitored area. A sodium iodide crystal-photomultiplier combination can provide the necessary sensitivity (e.g. the ability to detect the equivalent of 5 μc of radium) and the main design aim after achieving the desired sensitivity is to eliminate spurious responses.

The monitoring scheme shown in Fig. 5-16 has such features. The detectors are mounted in lead shields with windows to view only the passageway. The count rate of pairs of detectors is connected to a rate of change circuit that alarms only if a contaminated person approaches the detectors at a normal walking speed. Slow changes in background are not a problem and the sensitivity is increased about 25 over a simple count-rate sensitive arrangement with a trip level set to accommodate normal background variations. In addition to this precaution the alarm circuit is made active by the light-photocell arrangement only when someone is about to enter the sensitive area of the detector.

5.5.3 Dosimeters

Active dosimeters, or alarming dosimeters, should be worn by personnel working in above normal radiation levels. They provide a warning when either the dose rate or the dose exceeds a preset level. The requirement is for a small, light instrument with an alarm loud enough to be heard over typical factory noise levels.

In very high radiation fields the dose measuring instrument is preferable since primarily total dose is the quantity being regulated. In lower fields (a few mr/hr) the dose-rate instrument may be suitable since overexposure would be unlikely even at levels just below the alarm point. A personal radiation monitor has been described by Borkowski [30j] that provides a visual indication of dose rate by means of a light whose flashing rate is proportional to dose rate. An aural alarm whose pitch is rate-sensitive is also given when the dose rate is excessive. The detector is a halogen Geiger tube operating in the mean current mode. Figure 5-17 indicates the detector current available in two alternate modes of operation, either externally quenched for use with lower fields or self-quenched for use in higher fields.

The simplified circuit in Fig. 5-18 shows the visual indication feature. The Geiger current charges the capacitor C until its voltage reaches the striking potential of the neon bulb which then lights. The capacitor discharges and the process repeats.

Another circuit uses a Geiger tube in a dose measuring instrument (Fig. 5-19). The mean ionization current charges a capacitor. When its voltage reaches a reference level an alarm sounds and the instrument must be reset. The individual Geiger pulses are also available to operate a hearing-aid type earphone.

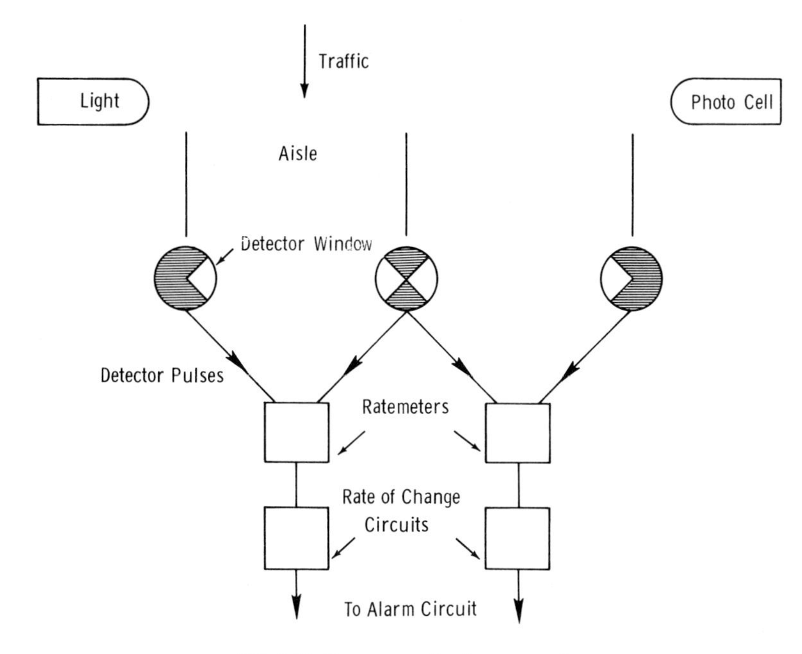

FIG. 5-16 Doorway monitor for personnel.

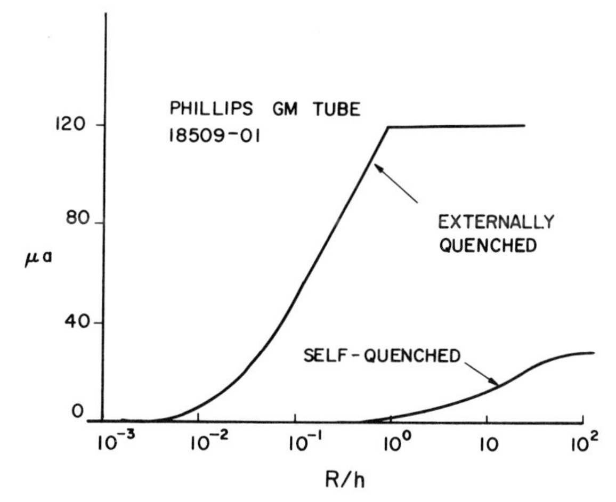

FIG. 5-17 Mean current characteristics of GM-tube.

Another approach [205] uses an ionization chamber in a dose measuring instrument. The electrometer tube (Fig. 5-20) has its cathode several volts positive with respect to the grid at the beginning of the measuring interval and is not conducting. The ionization chamber current flows into the capacitor. When the voltage at the grid is within a volt or so of the cathode, the tube begins to conduct and a regenerative circuit enchances the action. An audible alarm sounds at the same time. The alarm level is about 200 mr. Because of leakage across the capacitor, inaccuracies occur if the dose rate is less than 80 mr/hr.

There are also many passive dosimeters in use, capacitor discharge types and film badges being most common. Their purpose is to permit accurate long term exposure records to be kept. This is an

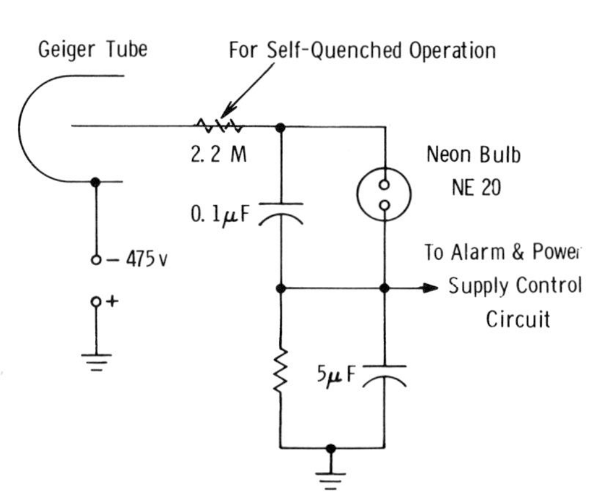

FIG. 5-18 Personal dose-rate meter.

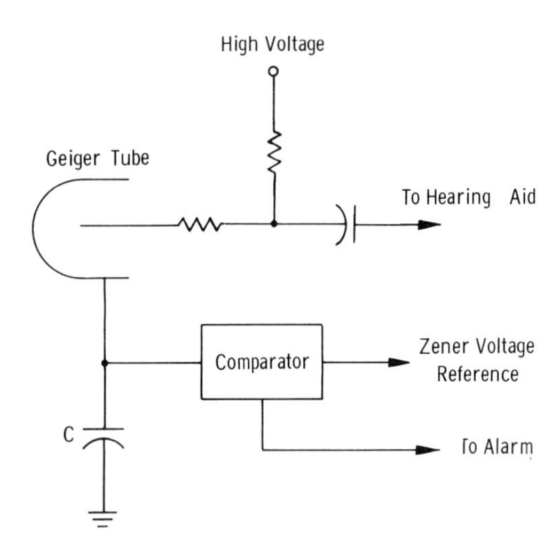

FIG. 5-19 Personal dose meter using Geiger tube.

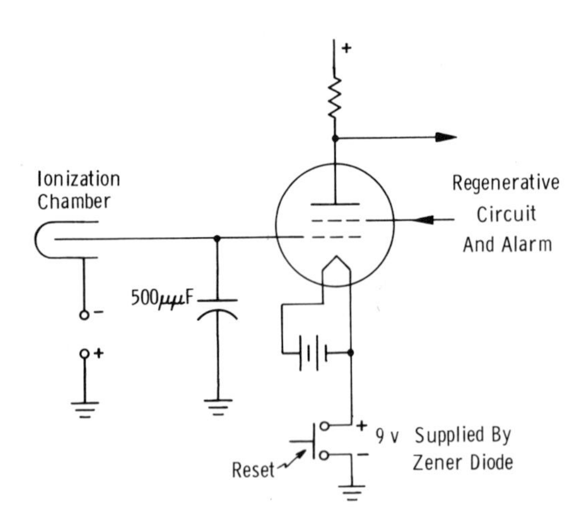

FIG. 5-20 Personal dose meter using ionization chamber.

important safety aspect if an emergency occurs and personnel (with low accumulated doses) are required to do cleanup work.

Indicating quartz fiber dosimeters are also common and while they can be read any time it is not likely they would prevent an overexposure in the event of an unexpected increase in radiation.

5.6 Portable Monitors

5.6.1 Survey Meters

It would be impracticable to cover all radiation situations by means of fixed monitors and portable meters must be used to survey unprotected areas whenever excessive radiation levels are suspected. Portable instruments for gamma radiation, neutrons, and tritium are used and are generally battery operated versions of the fixed monitors already described. The transistor circuitry used on fixed installations chiefly for reliability is also ideally suited to portable applications where small size and low power consumption is a distinct advantage.

Two instruments [206] that are specifically portable and contain some novel features not usually found in fixed monitors are described below.

It is sometimes necessary to measure large radiation intensities up to 500 r/hr or more. Such a monitor has a Geiger tube mounted at the end of a telescopic tube and operates by indicating the mean ionization current from the Geiger tube.

Accurate estimates of beta exposure hazards are difficult with most survey meters. Figure 5-21 shows the basic portions of an instrument designed to measure beta dose rate. The detector is a thin organic scintillator coupled to a photomultiplier. The mean current from the photomultiplier is measured by an electrometer tube circuit and is proportional to dose rate over a wide range of particle energies. The dynode voltages are supplied by successive stages of a Cockcroft-Walton type voltage multiplier.

5.6.2 Contamination Meters

Contamination meters are portable instruments specifically designed to detect surface contamination. Again the circuitry differs little from that already discussed. Thin-window Geiger tubes are common detectors for beta-emitting material and zinc sulfide phosphor screens coupled to phototubes for alpha detection. However, with the greater flexibility brought to instrument design by solid-state components, gamma and alpha spectrometers

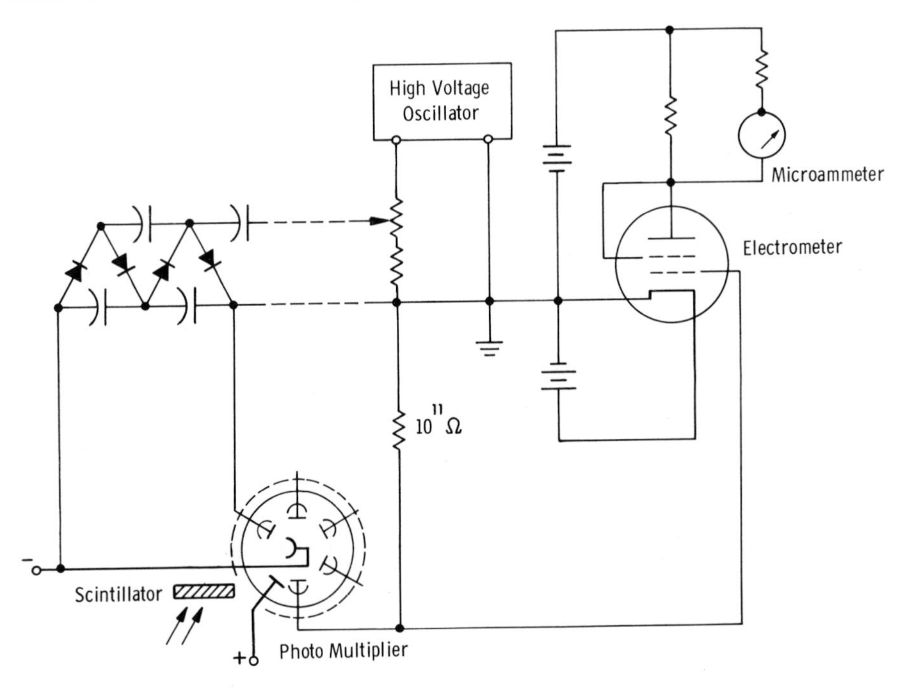

FIG. 5-21 Beta dose-rate meter.

will become feasible tools for the more precise identification of radioactive contamination. A dual purpose contamination meter has been described [207, 208] that uses a thin plastic scintillator for beta detection with a zinc sulfide coating for alpha detection. The discrimination is done by pulse height or by pulse shape.

6 COMPUTATIONAL AIDS

Instruments have been designed to study the neutron kinetic behavior of the reactor, the dynamic performance of the heat exchangers and turbine, or to excite the plant system to measure a particular response. These devices are used for teaching and training, for plant disturbance analysis, system design, to keep track of reactivity and for operational time-planning.

6.1 Reactor Kinetics Simulation

This has generally been done by use of analog computers [6c, 209, 210, 211] with their advantages of flexibility and ease of changing important parameters.

The reactor dynamic equations (see the chapter on General Reactor Dynamics) are taken as follows:

$$\frac{dn}{dt} = \frac{\delta kn}{\ell} - \frac{\beta n}{\ell} + \sum \lambda_i r_i + s \,, \qquad (6\text{-}1)$$

$$\frac{dr_i}{dt} = \frac{\beta_i n}{\ell} - r_i \lambda_i \,, \qquad (6\text{-}2)$$

where $\delta k = k-1$. A set of operational amplifiers is used to satisfy the reactor kinetic equations, as shown in Fig. 6-1.

If the coefficient in the equations governing this analog circuit are equated to those in Eqs. (6-1) and (6-2) the following results:

$$V = n \,, \qquad V_i = r_i \,,$$

$$\theta/R_5C = \delta k/\ell \,, \qquad 1/R_1C_i = \beta_i/\ell \,,$$

$$1/R_4C = \beta/\ell \,, \qquad 1/R_2C_i = 1/R_3C = \lambda_i$$

$$V_s/R_sC = S \,.$$

The amplifiers used to isolate each delayed neutron group can be eliminated if the delayed neutron equation is solved as follows,

$$r_i \lambda_i = \beta_i \frac{n}{\ell} - \frac{dr_i}{dt} \,, \qquad (6\text{-}3)$$

and immediately following a sudden change of flux,

$$r_i \lambda_i = \beta_i \frac{n}{\ell} \,, \qquad (6\text{-}4)$$

so the amplifier circuit shown in Fig. 6-2 can be used.

In Eqs. (6-1) and (6-2) $\delta k = k - 1$; if the equations are to be solved in terms of k, then a useful alternative circuit is shown in Fig. 6-3 which solves the equation

$$n = \int (1 - \beta) \frac{kn}{\ell}\, dt - \int \frac{n}{\ell}\, dt + \sum \lambda_i t_i + \int S\, dt \,. \qquad (6\text{-}5)$$

A practical disadvantage occurs in this circuit because of the inverse relationship between the coupling resistor R and k.

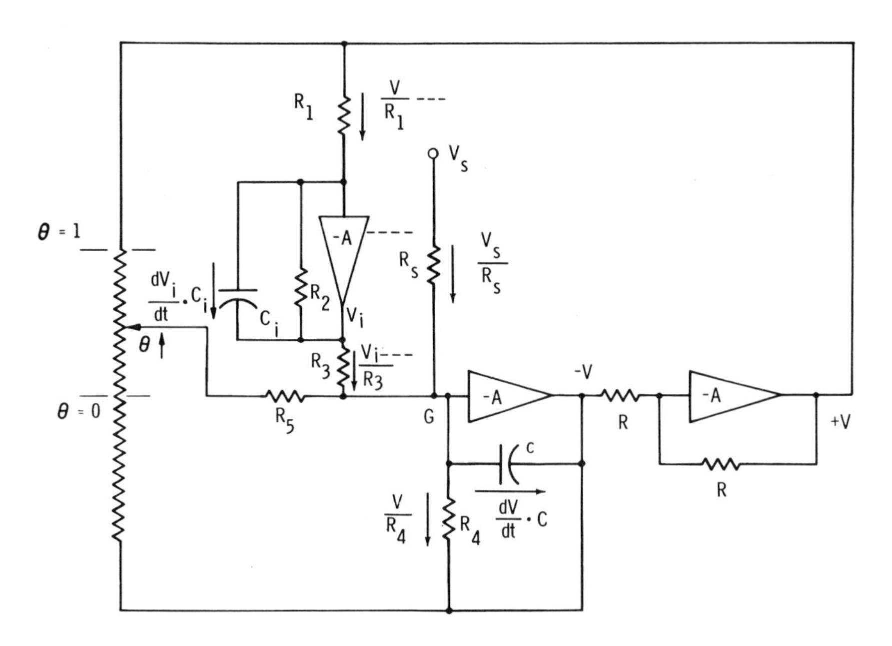

FIG. 6-1 Reactor kinetics simulator.

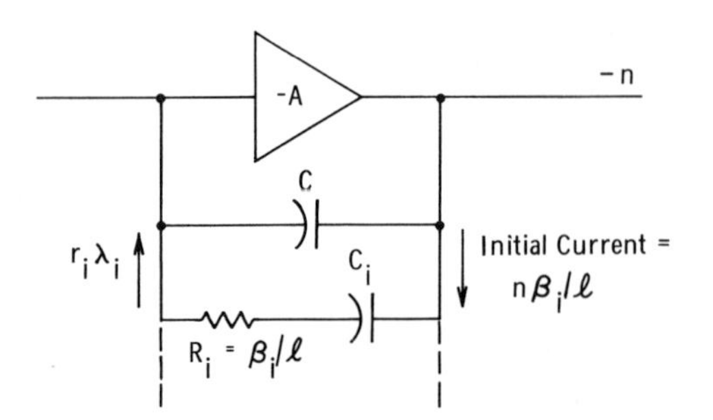

FIG. 6-2 Simulation of delayed neutron groups.

It will be remembered that neutron flux can vary over many decades of power, especially if the source term is small. Electronic amplifiers have a limited range in output voltage. This is because of voltage and current drifts associated with the input circuit components and because of output voltage limitations. By using chopper-stabilized amplifiers and power supplies a useful minimum output of 1 mv is possible. A maximum voltage of 100 v is reasonable. Errors are accumulative in the integrating amplifier and microvolt stability is required if the output power level is to stay constant over several minutes of operation, once balance has been achieved.

It is also important to consider whether logarithmic amplifiers similar to those measuring the ionization chamber currents are to be used with the simulator. Normally such amplifiers operate from a current source, and voltage drift at the input circuit is of little importance. If such an amplifier is used as a wide-range indication of simulator output (a voltage source) the input circuit errors may easily swamp those of the simulator and give erroneous readings of power. If the transient response of a logarithmic amplifier is being measured a large feed resistor ($\sim 10^{10}$ /ohm) is required to define the input current. Small capacities across this resistor markedly affect the dynamic performance of the amplifier and may give erroneous results when rapid transients are being analyzed.

An alternative approach is to produce a logarithmic instead of a linear signal directly from the analog simulator. If inverse period is calculated, this can be integrated to give log n. To achieve this the reactor kinetic equations are rearranged [212].

$$\log n = \int \frac{1}{n}\frac{dn}{dt}\,dt = \int \left[\frac{s}{n} + \frac{(\delta k - \beta)}{\ell} + \sum \frac{r_i}{n}\lambda_i\right] dt \tag{6-6}$$

where,

$$\frac{d}{dt}\left(\frac{r_i}{n}\right) = \frac{\beta_i}{\ell} - \lambda_i\left(\frac{r_i}{n}\right) - \left(\frac{r_i}{n}\right)\cdot\frac{1}{n}\frac{dn}{dt} \tag{6-7}$$

and where r_i/n is a new variable. Equation (6-7) may be solved for $(1/n)(dn/dt)$ using a multiplier for the last term in each delayed-neutron loop. A simple integration yields log n. A multi-decade range is therefore achieved directly.

6.2 Subcritical Reactivity Measurement

The amount by which a reactor is subcritical is an important safety parameter and various approaches have been taken to find a satisfactory method of directly indicating reactivity, especially when the reactor is subcritical.

This discussion does not include reactivity calibration procedures by, for example, measurement of the stable reactor period [115b, 213].

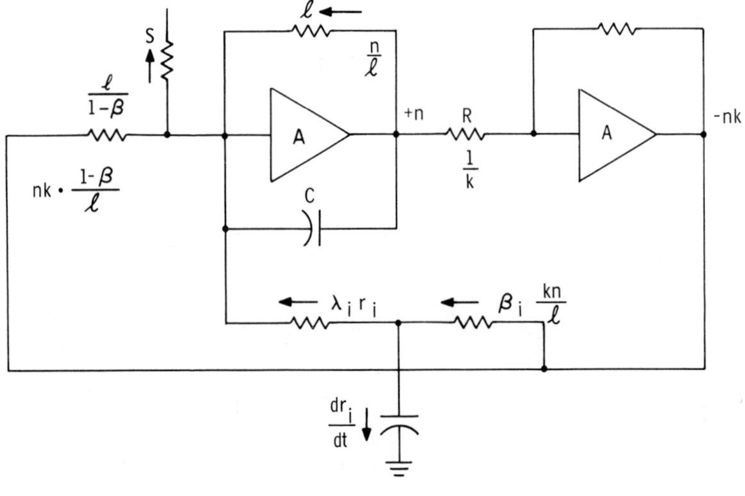

FIG. 6-3 Alternative reactor kinetics simulator.

6.2.1 Using the Kinetic Simulator

The normal kinetic simulator may be placed in a feedback system where the reactivity is calculated from a measurement of the reactor flux [214]. This is shown in Fig. 6-4. The source term must be known, and is assumed constant for short periods of time.

6.2.2 Using Transient Response

The relationship $n = n_0/(1-k)$ can be used to calculate departure from critical. A small step-change in reactivity is introduced and the ratio n/n_0 measured after a time equal to several neutron life-times. The ratio measurement can conveniently be taken by measuring log n_0 just before the transient, and log n after the transient, and subtracting as shown in Fig. 6-5. The sample switches must be programmed from the reactivity drive mechanism. Considerable time is required for this measurement and it is not very accurate when $k < 0.95$.

A method using source excitation and cross-correlation analysis has also been proposed [215].

6.2.3 Transfer Functions

The reactor transfer function can be used as an indication of reactivity [216]. Care must be taken to see that flux tilts and direct shadowing do not spoil the measurement (for example, if the reactivity is modulated by a control rod). Phase rather than amplitude measurements may be more independent of these effects. These measurements are affected as full power is approached by temperature coefficients, poison changes and reactivity noise.

6.3 Heat Exchanger and Plant Simulation

This is most useful in control system design and in plant dynamic studies.

Usually the various units comprising the plant can be described in terms of transfer functions and the whole plant becomes simulated out of a large number of fairly simple circuits. Alternatively the behavior may be described by a set of differential equations, generally with nonlinearities

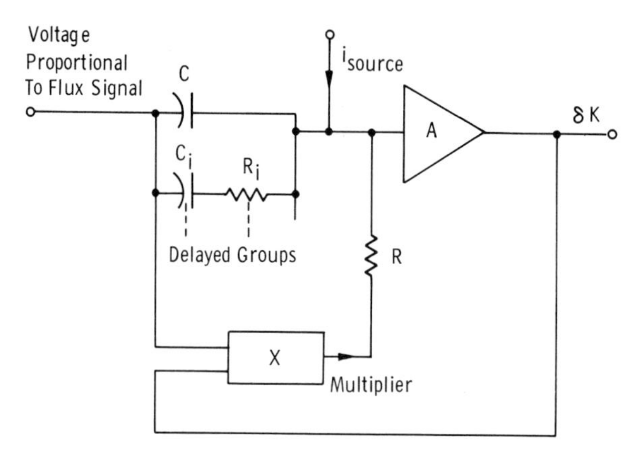

FIG. 6-4 Analog reactivity computer.

and multivariable parameters. The latter technique is more appropriate for digital computer simulation and the former for analog simulation. In practice it has been found hard to simulate the complex arithmetic and intricate couplings of a whole plant on a digital computer. The approach has been to get a feel for the problem on an analog computer and to define areas for more accurate analysis on a digital computer.

A typical plant broken into sections suitable for analysis is shown in Fig. 6-6. In this system 40 to 50 computing amplifiers may be used. One of the difficult areas is in simulation of transport delays $\exp(-p\tau)$, these introduce phase shift with no attenuation and may easily cause hunting or instability. A set [217, 218] of three to eight amplifiers can simulate a delay reasonably well, but a tape transport mechanism is more accurate (where head positioning sets the delay). Another method is by rotating capacitors on a commutator wheel [30k], a fairly economical approach. Transfer functions representing fuel temperature rise may have several terms of the form $(1+pT_1)/(1+pT_2)$, each of which requires an amplifier; often such functions are simulated by only one term, $1/(1+pT_2)$.

An example of the techniques used in deriving an analog simulator circuit for the thermodynamic side of the plant [219, 220] is shown in Fig. 6-7. A very simple approach is shown, where transport delays are not simulated. In this case P_0 may be a function of outlet steam temperature, throttle position and demanded load. The simulation becomes much more complicated if the mean reactor temperature is calculated from both the inlet and the outlet temperature. Reactor moderator temperature may also be required with a reactivity feedback connection.

In the steam generation side nonlinearities occur, especially if a large range in operating conditions requires examination. For this reason simulation is often carried out for a set of discrete conditions using small signal analysis.

If the simulation is required for demonstration or training purposes, it is usual to operate on a real-time scale. For control system analysis a faster time-scale is often used although serious consideration should also be given to the use of the real components to be used in the reactivity control circuits. Whenever possible use should be made of "real" systems. But systems using magnetic amplifiers, for example, do not necessarily permit an approach suitable for use in a simulator using vacuum tubes. In some cases [175] the simulator can be used to check out each part of the plant in turn, using more of the real components as these become commissioned.

In other cases [46] the simulation study has been carried out in parallel with reactor testing to improve the understanding and prediction of dynamic effects. The boiling water reactor program fits into this picture. In this case hydraulic and reactivity feedback loops affect reactor stability [221]; the loops are shown in Fig. 6-8. Care to estimate under what conditions, and in what regions of the core, the void coefficient is negative or positive is required. The significance of pressure control is dependent on the type of feedback control used. Computer analysis was used on EBWR and

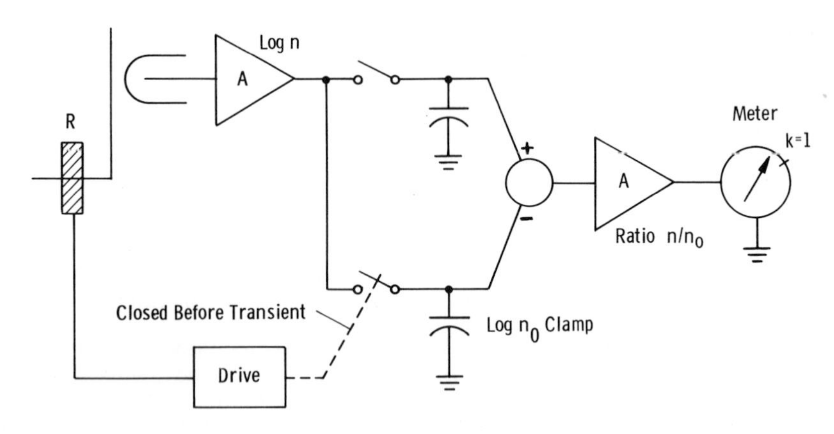

FIG. 6-5 Transient response and reactivity meter.

peaks in the transfer function at high powers were successfully predicted [222].

6.4 Xenon and Samarium Poison Calculations

The calculation of xenon or samarium concentration can be performed in two ways.

1. A signal proportional to reactor flux is fed into an analog computer [223] that continuously solves the poison concentration equations and continuously presents the results.
2. A reactor flux history is kept in a form that permits a signal proportional to it to be fed into a digital or analog computer [224] when a computation is required.

A knowledge of the poison concentration is valuable in several ways. It permits one to determine whether the control absorbers are in approximately the correct position at all times. The values obtained for the xenon and iodine concentration can be used as initial conditions in a calculation to predict the course of poison following a power change [225, 226].

On shutdown the growth of xenon poison can be watched to predict when poison-out will occur, and also after being poisoned out to predict when startup is possible. The equations being solved are:

$$\frac{dI}{dt} = a\Phi - \frac{I}{ti} \qquad (6\text{-}10)$$

$$\frac{dX}{dt} = b\Phi + \frac{I}{ti} - \frac{X}{tx} - c\Phi X \qquad (6\text{-}11)$$

where a, b, c are constants depending on the isotope cross-sections and fission yield [6d].

ti = mean life of iodine = 9.6 hr.

tx = mean life of xenon = 13 hrs.

I, X, represent the concentrations of these

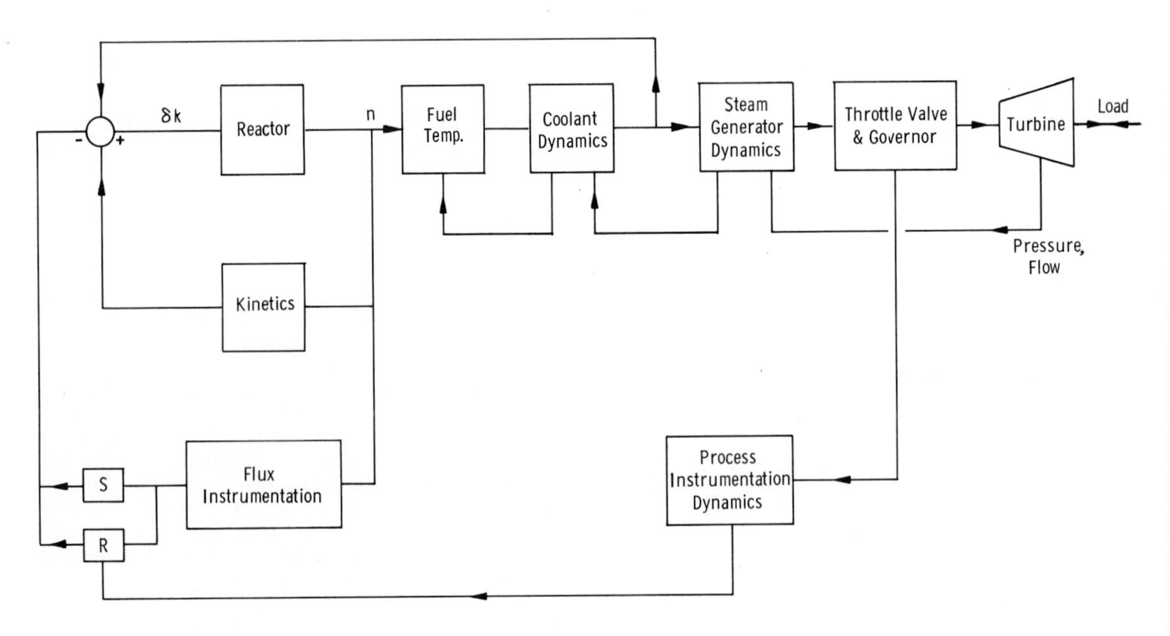

FIG. 6-6 Plant simulation.

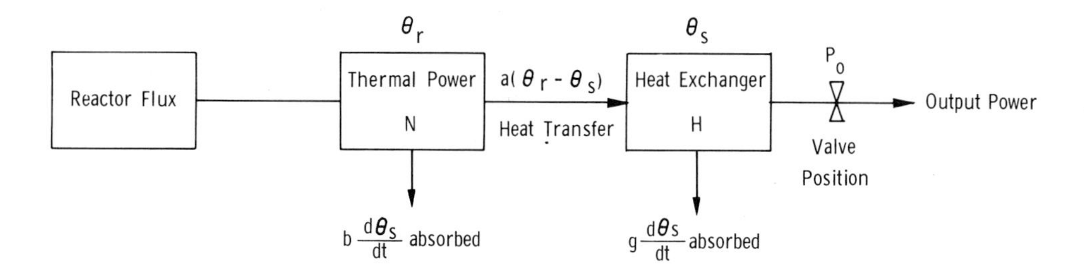

$$N = b\frac{d\theta r}{dt} + a(\theta_r - \theta_s)$$

$$P_o = a(\theta_r - \theta_s) - g\frac{d\theta_s}{dt}$$

$$P_o = T.\theta_s$$

Simulation:

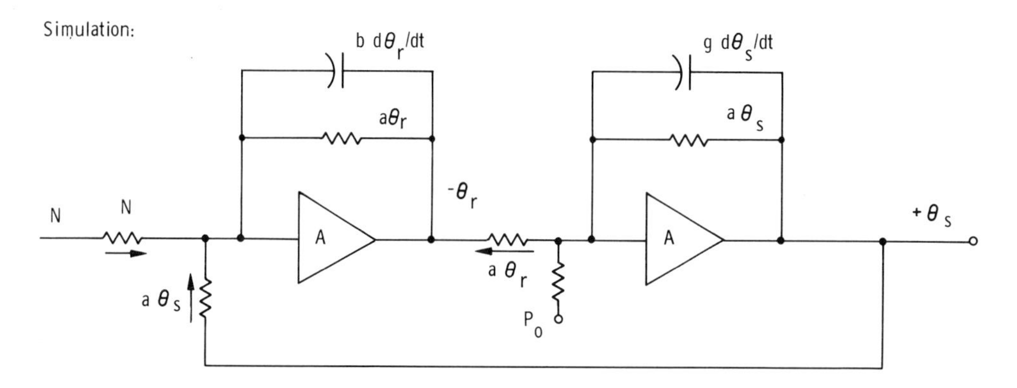

FIG. 6-7 Thermal power simulation. In the figure, θ_r is the mean reactor temperature, θ_s the mean steam temperature, b the reactor heat capacity, a the heat exchanger conversion factor, g the heat capacity of the heat exchanger, and P_o the power from the boiler.

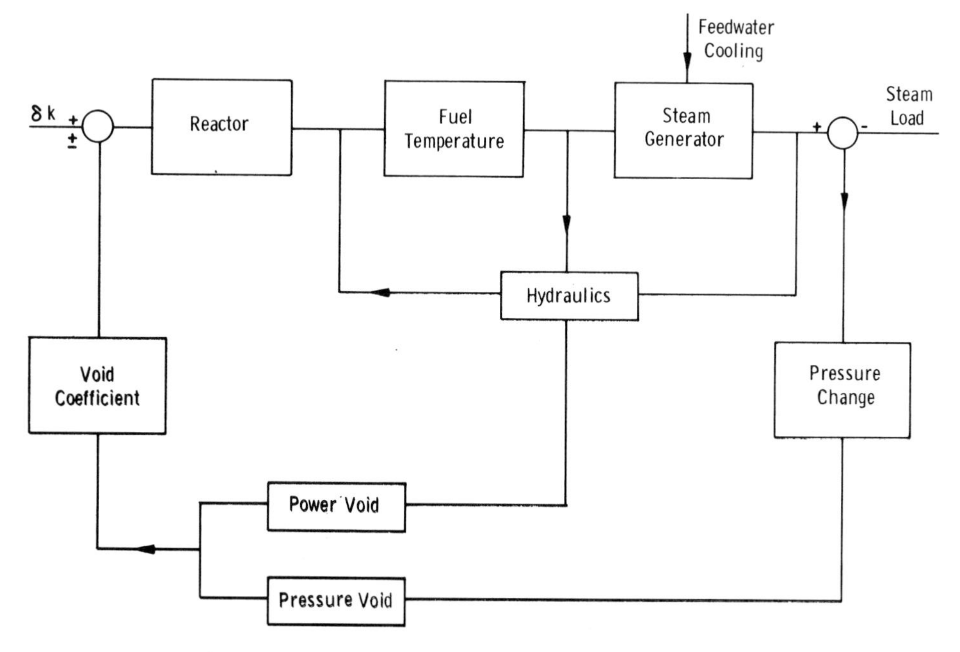

FIG. 6-8 Boiling water reactor dynamics.

isotopes. An operational amplifier simulation of these equations is shown in Fig. 6-9. In the case shown one second on the output represents 30 minutes of real time. It must be remembered that the equilibrium value of xenon may be only about 15% of the peak value following a shutdown and so the peak value generally sets the system accuracy. Methods for discharging the capacitors to re-set the circuit are also incorporated.

For a continuous real-time computation the time constants must be 13 and 9.6 hours. With these large time constants difficulty is experienced in obtaining high enough quality amplifiers (low drift) and capacitors. An alternative approach is to use servo-motor integrators [227] with tachometer feedback. In this case the accuracy is set by the "crawl" speed of the motor. For a typical motor in a feedback loop, speeds of 1% of full speed can be maintained. A better approach is to use two motors running at constant speed but in opposite directions, coupled through a differential gear; an infinite speed range is then possible.

The samarium poison is usually not so significant as xenon. It may be simulated in a similar way. The differential equations are

$$\frac{d(Pm)}{dt} = f\phi - \frac{(Pm)}{tp}, \quad (6\text{-}12)$$

$$\frac{d(Sm)}{dt} = \frac{(Pm)}{tp} - g(Sm)\phi, \quad (6\text{-}13)$$

where f and g are constants [6d]. Samarium is a stable element so no natural decay term exists, the mean life of Pm^{149} being 67 hours.

Both xenon and samarium conditions can be calculated, perhaps most conveniently of all, on a digital computer. All that is necessary is to update the computer store periodically so that an accurate power history is available. Whenever a calculation power is required a prepared program would tabulate the poison concentrations. In large reactors spatial distortions may dictate that for the xenon calculation the flux in several zones must be known, necessitating several calculations. Analog techniques would require several complete computers; the same calculation could readily be made in one digital computer.

The difference equations for the calculation of xenon and iodine concentrations by a discontinuous calculation based on updated values is shown below (for the equations defined at the start of this subsection).

$$\Delta I = (a\phi - I_n/ti)\,\Delta t, \quad (6\text{-}14)$$

$$\Delta X = \left[(I_{n+1}/ti) - X_n(1/tx + c\phi) + b\phi\right]\Delta t, \quad (6\text{-}15)$$

$$I_{n+1} = I_n + \Delta I, \quad (6\text{-}16)$$

$$X_{n+1} = X_n + \Delta X, \quad (6\text{-}17)$$

if

$$\Delta t \ll tx,\ ti.$$

X_n is kept up to date by recalculating I and X at each interval Δt. These equations are in a suitable form for a digital computer.

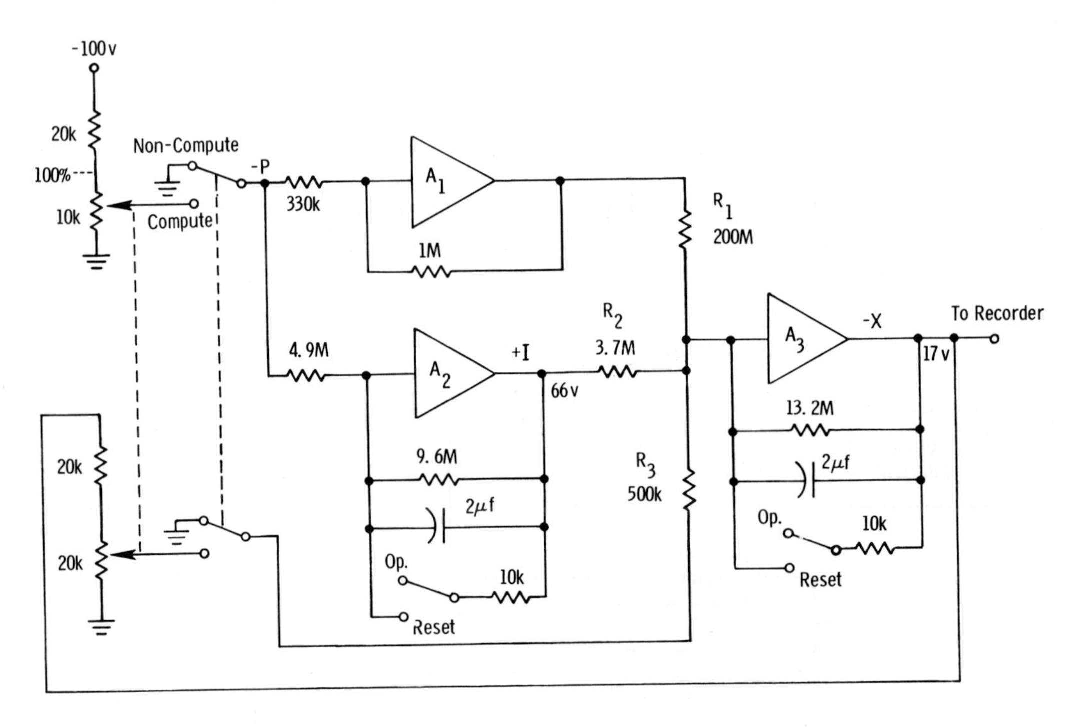

FIG. 6-9 Xenon poison computer.

6.5 Transfer Function Analysis

The reactor transfer function and the changes in it associated with passing from the subcritical to the critical, and to the full power modes has been described in the chapter on General Reactor Dynamics.

Three approaches have been used for transfer function analysis, these are: sinusoidal analysis, multi-frequency analysis, and noise analysis, the titles referring to the type of disturbance applied as the driving function. A fourth approach, using step disturbances is restrictive because of the likelihood of activating safety devices, however, the introduction of small step changes is usually simple and a quick feel for system response can be obtained.

6.5.1 Sinusoidal Analysis

This is often used to investigate the reactor transfer function [228], although at low frequencies an embarrassingly high power movement may result, and at high frequencies danger of a period trip arises. The advantage of this technique is in the middle frequencies where high accuracy is obtained.

An example of the equipment used to perform sinusoidal analyses is shown in Fig. 6-10. In this system reactivity oscillations are obtained by rotating shaped absorbers [301] at various frequencies. The range usually covered is from 10^{-3} to 10 cycle/sec with a rms reactivity value of up to 10^{-4}. The advantage of this approach lies in the rejection of all unwanted frequencies in the analysis equipment. The modulation is detected by a flux amplifier. The resulting signal is then detected by a phase sensitive device and provides in-phase and out-of-phase components of the measured flux. Accuracies of $\pm$ 2% in amplitude and $\pm$ 1° in phase are reported. Most of the frequency adjustment is carried out by changing the gear ratio.

6.5.2 Multi-Frequency Analysis

Because of the finite-time required in the calculation of amplitude and phase at each frequency in the sinusoidal analysis, especially long at low frequencies (0.01 to 0.001 cycle/sec), an approach using several frequencies at one time has been used [229]. In this case the input is obtained by moving the generator in a series of binary steps according to a definite pattern illustrated in Fig. 6-11. The power is concentrated at several discrete frequencies. The resultant output is analyzed to extract the response due to each frequency component, and then the basic time interval is changed to cover another frequency range. Although considerable time is saved one drawback is that information is not available until the calculation analysis has been made. However, seven frequencies can be covered at once. Frequencies down to 10^{-3} cycle/sec can be measured. Accuracies of $\pm$ 5% in amplitude and $\pm$ 5° in phase are reported. Errors due to nonlinearities (i.e. in the reactor transfer function) occur, but are kept small by limiting the amplitude of the fundamental The step motion of the reactivity generator is also simple to form although the bandwidth for mechanical generators is limited to about 0.1 cycle/sec.

6.5.3 Correlation Analysis

This is usually based on analysis of noise before and after being passed through the unit to be analyzed. The noise can be band-limited although it is advantageous for it to be "white".

If we examine the autocorrelation function [230] of a signal.

$$\phi_{xx}(\tau) = \int f_x(t) \cdot f_x(t+\tau)\, d\tau , \qquad (6\text{-}18)$$

and the crosscorrelation function of two signals,

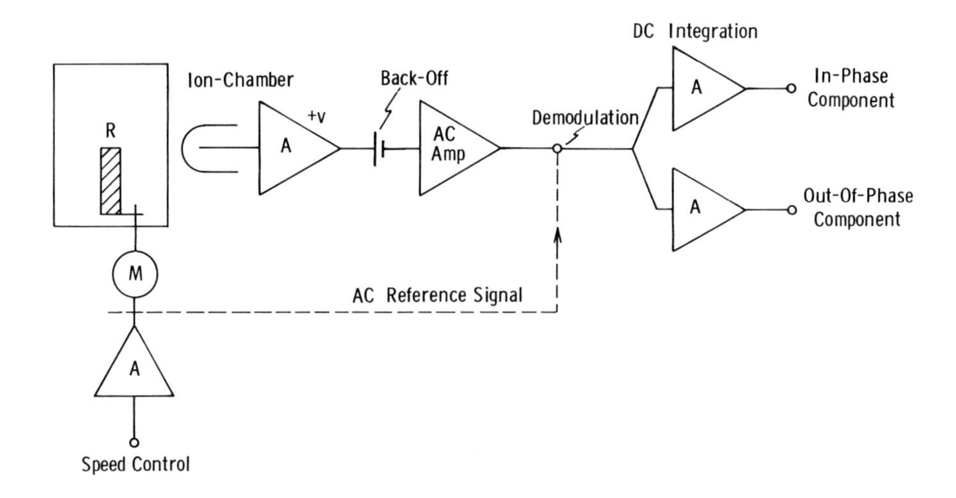

FIG. 6-10 Transfer function analyzer.

$$\phi_{xy}(\tau) = \int f_x(t) \cdot f_y(t+\tau)\, d\tau , \qquad (6\text{-}19)$$

and if

$$\phi_{xx}(\tau) = 1 \quad \text{(i.e. the input is white noise)} , \qquad (6\text{-}20)$$

then,

$$\phi_{xy}(\tau) = \int h(t)\, dt, \quad \text{or the impulse response} . \qquad (6\text{-}21)$$

The response of a system to noise and the impulse response of the same system are shown in Fig. 6-12.

Now consider the input power-spectral density,

$$\Phi_{xx}(f) = \int \phi_{xx} e^{-j\omega} dt \qquad (6\text{-}22)$$

If we take the cross-power spectral density $\Phi_{xy}(f)$ and divide by the input spectral density, the result is the transfer function, i.e.,

$$H(f) = \frac{\Phi_{xy}(f)}{\Phi_{xx}(f)} = \frac{\Phi_{yy}(f)}{\Phi_{yx}(f)} . \qquad (6\text{-}23)$$

Hence, the transfer function can be calculated from the autocorrelation and crosscorrelation functions, or from the power spectral densities directly.

Difficulty in measuring power-spectral densities occur at low frequencies (below 100 cycle/sec) where narrow band-filters are difficult to construct. Digital storage and computational techniques are most attractive in this field; there is no basic low-frequency problem since the signals can be sampled at any desired interval and stored indefinitely. The usual problem is to reduce the statistical errors to tolerable limits by taking the record over a sufficiently long period of time.

Although pure white noise has been generated for tests in a reactor (one case [231] used the signal from a radioactive source to drive an absorber), it has been found more convenient to generate a "quasi-random" noise using a special binary series [232] where the correlation function approximates the unit impulse. An example of such a series is shown in Fig. 6-13 [233]. In this system the input binary signal is crosscorrelated to the output and the reactor transfer function calculated from the measured crosscorrelation function. Using parallel analog integrating channels a reactor impulse response was measured in 140 sec, without disturbing the power more than 1%. Accuracies in gain of 0.5 db, and phase of 2° at 1 cycle/sec are reported. As shown in Fig. 6-13 a binary series on punched paper tape directs the positioning of a small absorber. The tape also operates diode switches controlling the plus or minus input to analog integrators. Five points on the correlation curve are therefore calculated at a time. Using the crosscorrelation technique it is possible to take measurements while other perturbations are occurring [46], and with signals equal to the background noise. Information may be deduced directly from the crosscorrelation response before transformation of the data into the frequency domain.

The transfer function H(s) can be calculated from [164b]

$$H(s) = \Phi_{xy}(f) = 2\int_0^{\infty} \left[\phi_{xy}(\tau) + \phi_{xy}(-\tau)\right] \cos 2\pi f\tau\, d\tau$$

$$- 2j \int_0^{\infty} \left[\phi_{xy}(\tau) - \phi_{xy}(-\tau)\right] \sin 2\pi f\tau\, d\tau , \qquad (6\text{-}24)$$

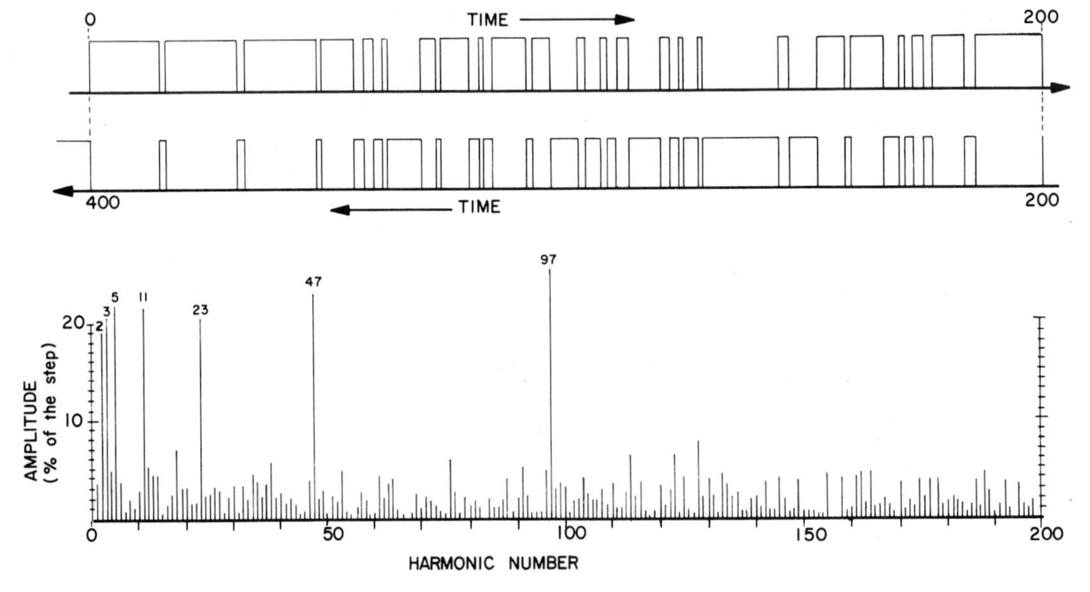

FIG. 6-11 Multi-frequency analysis.

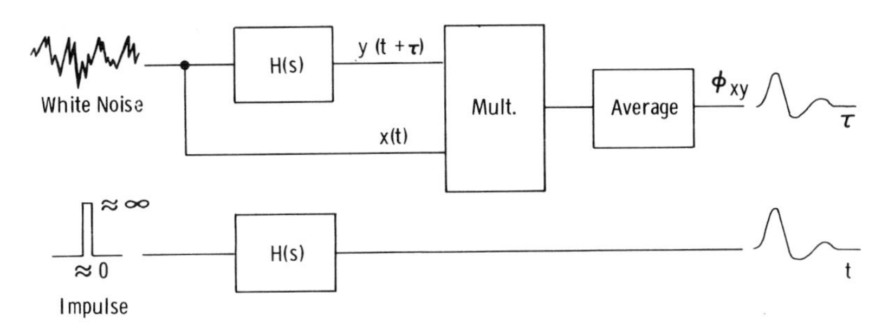

FIG. 6-12 Autocorrelation and impulse response.

and by dividing by $\Phi_{xx}(f)$ if the input is not white noise. Another useful form is [233]

$$H(s) = \Phi_{xy}(f) \cdot$$

$$= \frac{1}{s^2} \sum_i \begin{bmatrix} \text{change in slope} \\ \text{at } i \text{ th point} \end{bmatrix} e^{-p\tau_i} + \frac{1}{s} \begin{bmatrix} \text{value at} \\ \tau = 0 \end{bmatrix} . \tag{6-25}$$

These equations are derived from those listed at the beginning of this subsection. The confidence limits on spectral densities and transfer functions calculated by these techniques should be carefully estimated.

6.5.4 Transient Response

If impulse, step, or ramp functions are used to disturb a system, information regarding the transfer function is available. In practice it becomes difficult to move reactivity elements or coolant valves quickly enough to gather accurate information. However, such techniques are most useful as a simple routine test and can be used to check regions of suspected instability. The equipment is simple. A closed loop-test of an automatic control system can be performed by inserting a step-disturbance. Such an experiment [166] carried out on the NRX reactor is shown in Fig. 6-14. In this case the reactivity is adjusted by moderator level control. Changes in level are made by changing the position of three valves draining the moderator from the calandria to a dump-tank. A constant-speed recirculating pump returns the water. Other tests have been made to check the region of least stability in the startup log-rate mode (Sec. 2), and to examine the response following a system fault [4b].

Although such tests do not necessarily lead to an accurate derivation of a transfer function, the method can indicate whether or not the transfer function used in the analysis was correct.

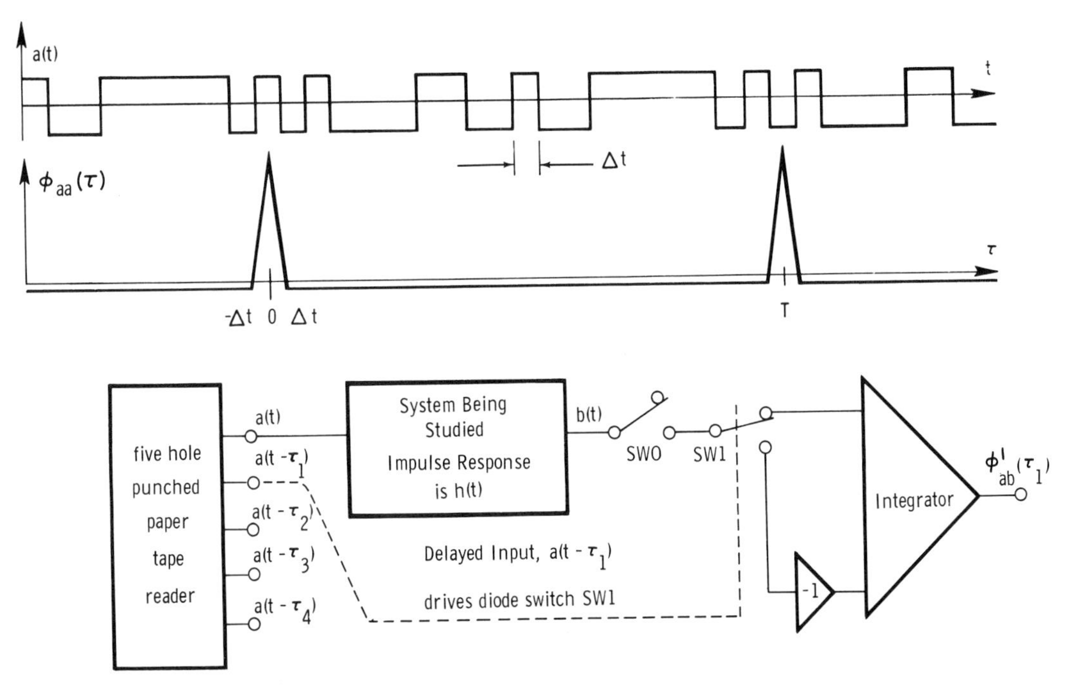

FIG. 6-13 Correlation function analysis.

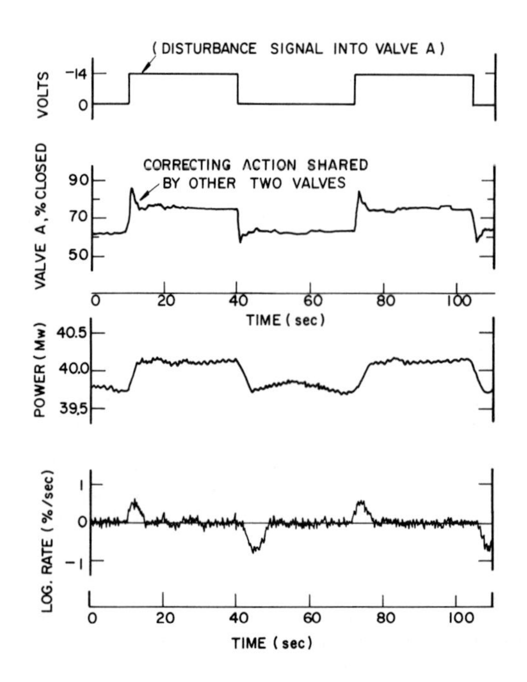

FIG. 6-14 Step disturbance of valve controlling level of moderator.

7 DISTRIBUTION OF ELECTRICAL POWER

7.1 General Considerations

The requirements for plant serviceability, public protection and plant equipment protection have already been established. The design of the electrical distribution system is affected by all three factors and considerable attention has been applied to the sequence of events following a fault; i.e., circuit breakers and power re-routing, starting of alternate power sources, and selection methods whereby only the most important items have access to all possible power sources.

7.2 Electrical Distribution System

The electrical distribution system includes both the outgoing power, alternate incoming power and other methods of power generation for "essential service" operation.

The following areas are recognized:

1. Generator(s) to high voltage outgoing power lines and to station auxiliaries.
2. Incoming power to station auxiliaries and/or essential services.
3. Locally generated power to essential services.
4. "Last-ditch" power sources.

All reactor types require coolant-pump power when operating and for some time immediately following shutdown. This power requirement comprises a large fraction of the local station load. The power consumption varying from 10 to 18% of that generated.

Electrical supplies are also essential to provide power for safety instruments and reactor shutdown circuits. These systems should be highly reliable (see Sec. 4.1).

A representative approach to nuclear station electrical distribution systems is shown in Fig. 7-1. Each of the categories numbered above is shown and marked. Normally the station distribu-

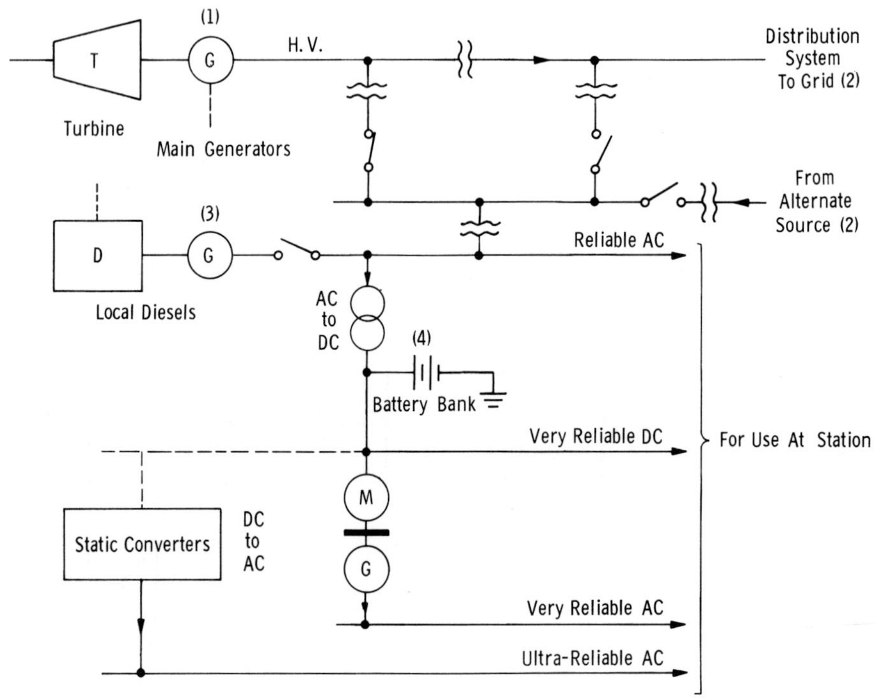

FIG. 7-1 Electrical distribution in a nuclear power station.

tion system is energized by the main-turbine generator, but one or more alternative sources are usually connected through appropriate switchgear. In the case of long transmission systems alternative sources of a-c power may necessitate resynchronizing equipment with attendant delays in power change-over.

Most stations have local diesel-generators (usually more than one) in case all the above mentioned a-c sources fail.

In the final instance reliance is placed on the d-c battery supply. If proper charge currents and maintenance of these supplies is carried out then their reliability is very high. The battery bank can be charged by the a-c system or by the diesel through rectifier units. Loads are sometimes connected directly to the d-c supply. For example, certain safety relays may be energized from this source, and it may also be desirable to drive some of the coolant pumps from the d-c line in emergencies.

A more reliable a-c source can be obtained from the d-c by local motor-generator sets. During fault conditions, loading or load-sheding on these local MG sets may give rise to poor regulation. It may be necessary to take special care that safety instrumentation is not affected by such disturbances.

Three plant installations will be described. These are Dresden, Shippingport and Hunterston. Features of a number of other plants are described in reference [234].

7.2.1 Dresden

Electrical system redundancy is employed to provide station services in face of equipment faults and operating errors. Five sources of auxiliary power are provided [235]. A system schematic is shown in Fig. 7-2. The 4 kv auxiliary power can be obtained directly from the generator or from the 138 kv transmission lines. A third source of power is from a 34.5 kv line used during the plant construction. If these three sources fail, the reactor is scrammed and the diesel generator supplies 480 v (500 kw) service for plant safety. In addition, 125 v d-c battery power is available for equipment control and emergency lighting.

Instrument a-c power (for the safety circuits) is normally supplied through two ac-ac motor-generator sets, with heavy flywheel damping to prevent normal switch-gear delays affecting the output level.

7.2.2 Shippingport

In this system [236] power is available from both the plant generator and from the distribution system. Four 2.4 kv bus-bars are excited to feed the plant auxiliaries, as shown in Fig. 7-3. Interlocking allows a 23 kv 1000 kva transformer to be connected instead of the local diesel generator. Each 2.4 kv bus supplies 1250 kw to one of the four coolant pumps. Half the load is normally taken from the plant generator, and half from the distribution system.

The generator is permanently connected to the main system transformer without the usual associated switchgear. During shutdown the excitation on the generator remains on until the speed is quite low.

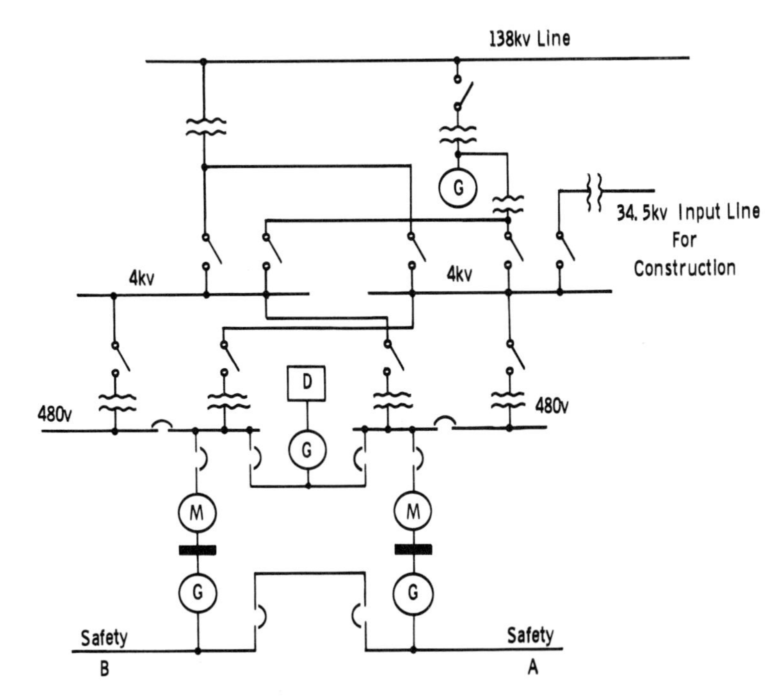

FIG. 7-2 Dresden electrical system.

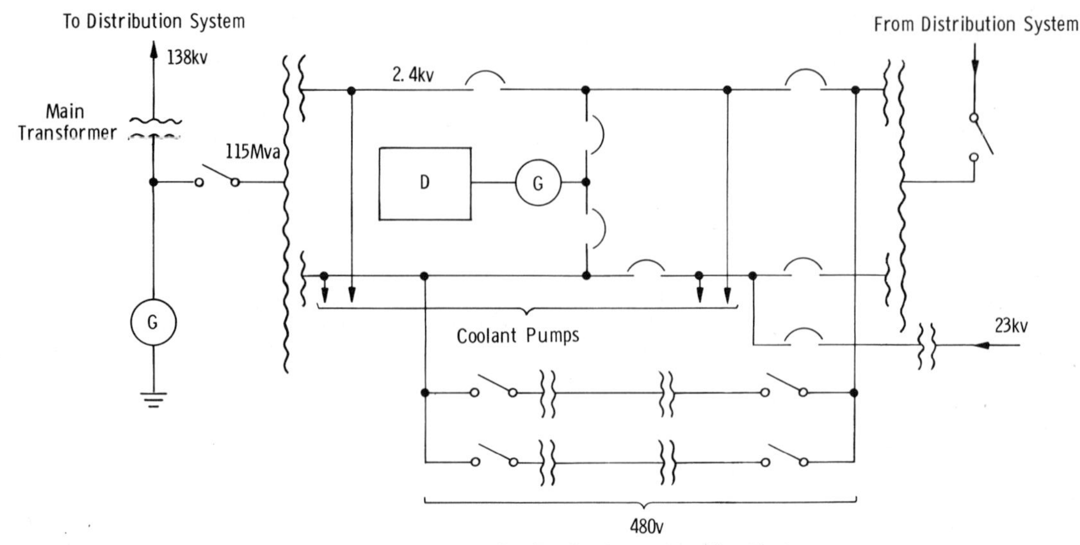

FIG. 7-3 Shippingport electrical system.

Plant service continuity is ensured by three separate d-c battery-fed sources. Each source has its own motor-generator charging and control system, and is floating with respect to ground. These d-c systems also supply current to the 138 kv circuit breakers to re-establish a-c power to the station. The No. 1 control battery can supply the turbine-generator control circuits for 8 to 12 hours; the emergency control battery can supply emergency lighting for just over two hours.

7.2.3 Hunterston

This system is normally run in two halves, each associated with one reactor and three generators [237]. A total of six unit boards supplied directly from the generators and two station boards from the 132 kv distribution lines are used to supply all services as shown in Fig. 7-4. Each unit board can be supplied, through circuit breakers, by power from a station board. The star-points of the unit and generator transformer are resistance-connected to ground to limit faulty earth currents.

The details of the "essential services" supply system are shown in Fig. 7-5. Three diesel-generators are available for use, in case one is shut down for repair and one fails to start. The battery-bank provides essential service to a pair of gas circulators on each reactor, to their oil pumps and to motor-generator sets supplying instruments and safety circuits for a period up to 20 min after failure of all sources of ac. Two

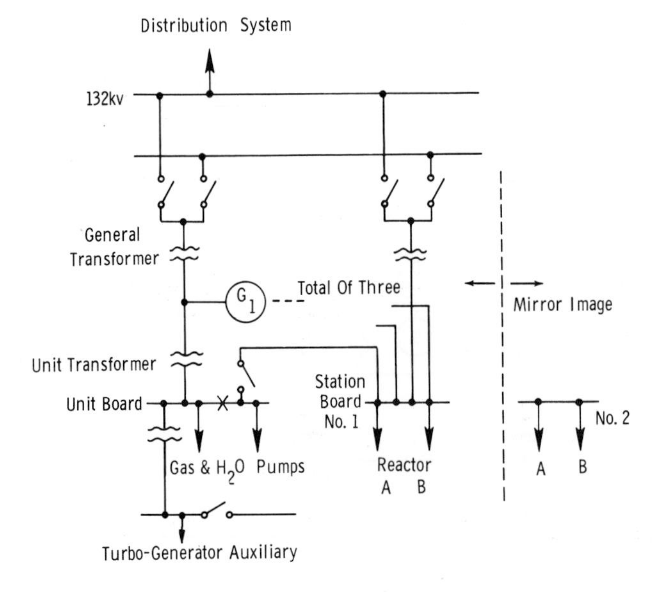

FIG. 7-4 Hunterston electrical system.

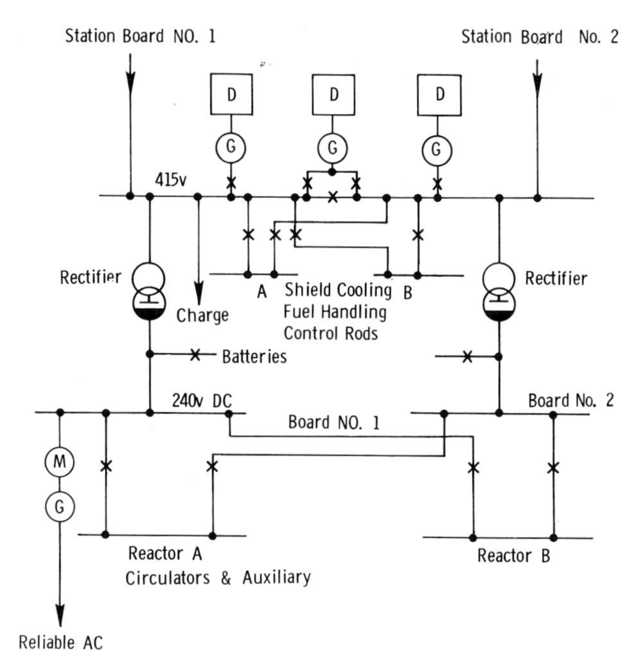

FIG. 7-5 Hunterston: essential services.

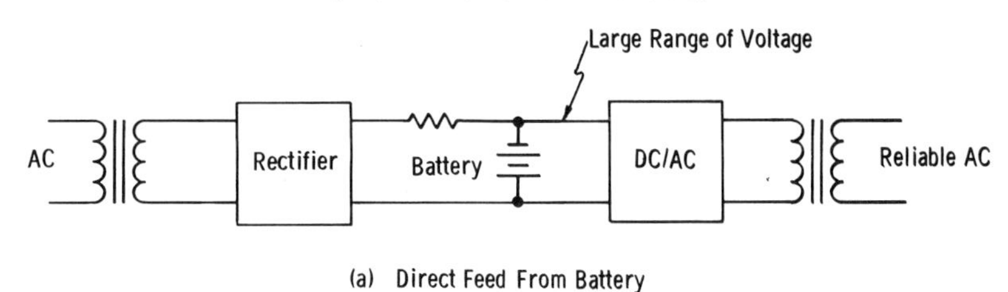

(a) Direct Feed From Battery

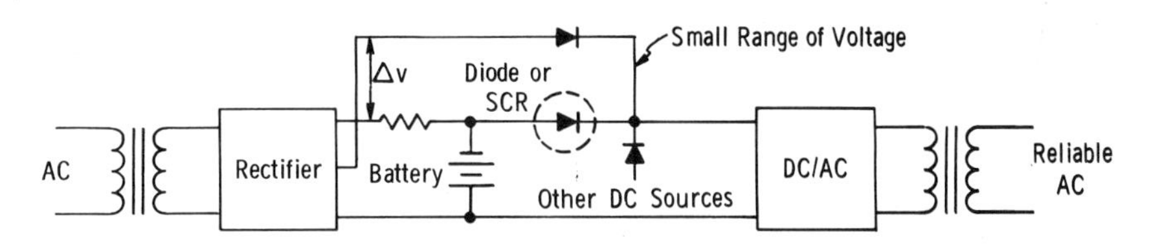

(b) Feed From Battery If Mains Fails

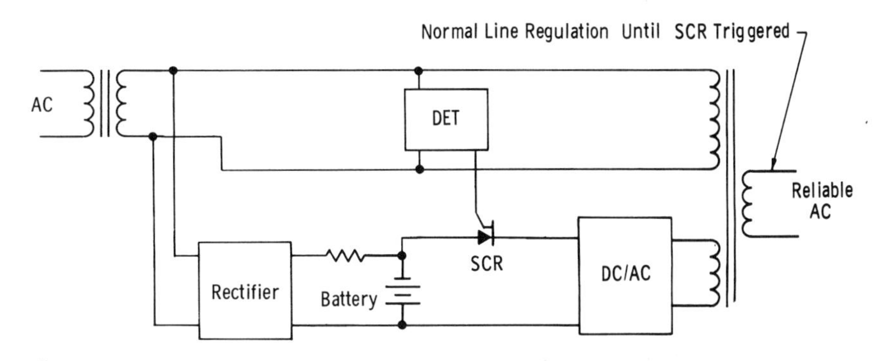

(c) Feed From AC Line Until It Fails

FIG. 7-6 Reliable alternating current using solid-state converters. (SCR = silicon-controlled rectifier.)

physically separated battery systems are used. Currents arising from fault conditions in the dc circuit are too large for economic use of air circuit-breakers. Use is therefore made of protective resistors and time-delayed over-current relays.

When use is made of local motor-generator sets for light electrical loads, care must be taken to evaluate all effects of loads suddenly dropped or added (such as pumps or valves). These may cause unexpectedly large voltage changes (20% or so for a few cycles), and frequency changes. Safety instrumentation operated from such supplies should be designed to operate satisfactorily under these conditions.

7.3 Light Instrumentation Loads

Indication of plant status is especially important following loss of electrical or nuclear power.

Plant safety instrumentation loads may require up to 100 kw of power. Transistorized equipment can often reduce this requirement to less than 10 kw. If regulating circuits, including motors and/or valves are also included the power consumed may be higher.

Two main approaches to ultra-reliable instrumentation power have been adopted.

1. Utilization of reliable battery power (dc) to drive instruments directly.

2. Conversion of reliable battery power to ac and then connecting this to the normal a-c feeder lines.

Battery supplies are usually rated at 100 v or less. Vacuum-tube equipment generally requires from 200 to 500 v dc so that instruments using these devices may have to be excluded from the first approach. Magnetic amplifiers relying on 60 or 400 cycle/sec power may also have to be excluded. Many transistor circuits and small motor circuits could be supplied directly, and in such cases built-in rechargeable cells providing exactly the correct d-c voltages could, as an alternative be used with each instrument. The major objection to the later course has to date been in knowing the correct small recharge currents required for each battery.

In the second case dc to ac converters can supply 50, 60 or 400 cycle/sec at a variety of line voltages. Rotating dc/ac electromechanical converters has the advantage of stored rotational energy, thereby filtering out input supply transients. However, slow response of the regulating controls may reduce response time to load changes, often resulting in poor line-voltage and frequency regulation.

Solid-state converter units are now available to supply power up to 10 kw. These use high efficiency silicon switches, have good regulation characteristics and can, in normal use, be synchronized to the line. Solid-state switches which switch rapidly can generate transients to which some instruments may be particularly sensitive. Maintenance problems are less than with the equivalent electromechanical converters. A variety of system connections are possible [238, 239]; these place different emphasis on the normal source of power and on system redundancy. Some of these are shown in Fig. 7-6.

REFERENCES

1. J. E. Binns, "Design of safety systems for nuclear reactors", Nucl. Safety, 4, 2(1962)63.
2. C. G. Lennox and A. Pearson, "Thermal control of the NRU reactor", Inst. Radio Engrs., Trans. Nucl. Sci., NS-5, 2(1958)68.
3. R. J. Smith, "The Overall Control of Nuclear Power Stations of the Gas-Cooled Thermal Reactor Type", Proceedings of the Second U.N. International Conference on Peaceful Uses of Atomic Energy, Geneva, 1958, Vol. 11, p. 360.
4. J. M. Yellowlees and R. S. French, "Kinetics and Control of the Berkeley Reactors", Proceedings of the Second U.N. International Conference on Peaceful Uses of Atomic Energy, Geneva, 1958, Vol. 11, a) p. 391; b) p. 402.
5. U.S. Naval Reactors Branch, Division of Reactor Development, Westinghouse Electric Corp., and Duquesne Light Co. for the USAEC, The Shippingport Pressurized Water Reactor, Addison-Wesley Publishing Co., Inc., Reading, Mass., 1958. a) p. 285; b) p. 301; c) p. 91 and p. 385; d) p. 386.
6. M. A. Schultz, Control of Nuclear Reactor and Power Plants, McGraw-Hill Book Co., Inc., (2nd Ed.), 1961. a) p. 256; b) p. 371; c) 430; d) p. 160.
7. "The Engineering Design and Operation of the NPD Reactor", A Reprint of Papers Presented at the Winter Meeting of the American Nuclear Society, Washington, D.C., November 1962. Canadian Report AECL-1682, Atomic Energy of Canada, Ltd., 1963.
8. V. A. Elliot et al., The Dresden Nuclear Power Station, Proceedings of the Second U.N. International Conference on Peaceful Uses of Atomic Energy, Geneva, 1958, Vol. 8, p. 508.
9. "Enrico Fermi Atomic Power Plant Revised License Application. Part B, Vol. 7, Technical Information and Hazards Summary Report, Section 1, Vol. 2", Report NP-10458, Power Reactor Development Co., 1961. a) p. 109.13 and b) 109.14 in the above.
10. A. W. Kramer, Boiling Water Reactors, p. 296, Addison-Wesley Publishing Col, Inc., 1958.
11. A. Pearson and C. G. Lennox, "NRU Thermal Power Control", Canadian Report AECL-506, 1957.
12. W. J. Henderson and D. G. Hurst, "The Effect of Flux Flattening on the Economics of Heavy Water Moderated Reactors", Canadian Report AECL-949, 1959.
13. C. G. Lennox, A. Pearson, and P. R. Tunnicliffe, "Regulation and Protective System Design for Nuclear Reactors", Reactor Safety and Hazards Evaluation Techniques, Proceedings of a Symposium held at Vienna, May 14-18, 1962, IAEA Vienna, 1962, Vol. 1, p. 449.
14. E. Siddall, "Statistical analysis of reactor safety standards", Nucleonics, 17, 2(1959)64.
15. M. J. Cowper and D. Wray, "Estimation of Accident Probability Due to Protective System Failure in Power Reactors", British Report AERE-AHSB(S)R35, 1962.
16. E. Siddall, "A Study of Serviceability and Safety in the Control System of the NRU Reactor", Canadian Report AECL-399, 1954.
17. I. M. Jacobs, "Safety Systems for Nuclear Power Reactors", Am. Inst. Elec. Engrs., Trans Paper 57-9-6.
18. M. A. Schultz, "Reactor safety instrumentation", Nucl. Safety, 4, 2(1962)1.
19. G. C. Laurence, "Operating Nuclear Reactors Safety", Reactor Safety and Hazards Evaluation Techniques, Proceedings of a Symposium held at Vienna, May 14-18, 1962, IAEA Vienna, 1962, Vol. 1, p. 135.
20. A. Pearson, E. Siddall, and P. R. Tunnicliffe, "The Control of Canadian Nuclear Reactors", Proceedings of the Second U.N. International Conference on Peaceful Uses of Atomic Energy, Geneva, 1958, Vol. 11, p. 372.
21. E. P. Epler, "A Philosophy of Control System Design", USAEC Report ORNL 2090, Oak Ridge National Laboratory, 1956.
22. D. G. Breckon and J. H. Collins, "Control and Safety in Operation of the NRX and NRU Reactors", Reactor Safety and Hazards Evaluation Techniques, Proceedings of a Symposium held at Vienna, 1962, IAEA Vienna, 1962, Vol. 1, p. 501.
23. M. W. Jervis, "Nuclear reactor safety systems", AEI Engineering, 2, 5(1962)222. (Associated Electrical Industries, Ltd.) (Brit.)
24. O. R. Frisch (Ed.), Progress in Nuclear Physics, Vol. 2, p. 154, Pergamon Press, N.Y. and London, 1961.
25. R. J. Cox and J. Walker, "Control of Nuclear Reactors", Proc. Inst. Elec. Engrs., 103B(1956)577. (Brit.)

26. D. P. Rouix and S. H. Hanauer, "On the Use of Gamma Radiation as a Reactor Control and Safety Parameter", Trans. Am. Nucl. Soc., 6(1963)74.
27. W. J. Price, Nuclear Radiation Detection, McGraw-Hill Book Co., Inc., N.Y., 1958.
28. W. D. Allen, Neutron Detection, Philosophical Library, N.Y., 1960.
29. J. M. McKenzie, "Making fission counters for neutron monitoring", Nucleonics, 17, 1(1959)60.
30. Proceedings of the Sixth Tripartite Instrumentation Conference Held at Chalk River, Ontario, Canada, April 20-24, 1959.
    a) P. G. Salmon, "High-Temperature Fission Chambers", Canadian Report AECL-805, p.37.
    b) M. L. Awcock, "U^{235}-Coated Ionization Chamber, Type IZ/400", Canadian Report AECL-805, p.44.
    c) R. P. Henderson, "RC6 and RC7 Ionization Chambers", Canadian Report AECL-805, p. 20.
    d) A. C. Lapsley and R. R. Ingham, "Internal Flux Measurement", Canadian Report AECL-801, p.55.
    e) F. S. Goulding, C. G. Lennox, and L. B. Robinson, "Analysis of the Non-Linear Behavior of Semiconductor Junctions and Its Applications in Nucleonics", Canadian Report AECL-801, p. 77.
    f) A. Pearson and C. G. Lennox, "Experience with the NRU and NRX Control Systems", Canadian Report AECL-801, p. 111.
    g) L. Cathey, "Area Radiation Monitor", Canadian Report AECL-802, p. 52.
    h) H. S. Sanders, L. H. Cook, Jr., A. L. Burke, and H. V. Hardison, "Criticality-Incident Alarm", Canadian Report AECL-802, p. 106.
    i) W. D. Howell, "Beta-Gamma Hand and Foot Monitor", Canadian Report AECL-802, p. 125.
    j) R. H. Dilworth and C. J. Borkowski, "Personal Radiation Monitor", Canadian Report AECL-802, p. 74.
    k) R. A. Dandl and R. S. Stone, "A Variable Function Delay for Analog Computers", Canadian Report AECL-801, p. 166.
    l) D. Harrison, "A Nuclear-Reactor Transfer-Function Analyzer", Canadian Report AECL-801, p. 42.
    m) G. S. Hurst, "Some Examples of Radiation Dosimetry Activities at the Health Physics Division of Oak Ridge National Laboratory", Canadian Report AECL-802, p. 28.
    n) R. V. Studley "Reactor Flux Monitors", Canadian Report AECL-801, p. 71.
31. J. Duchene, "High temperature fission chambers", Inst. Radio Engrs., Trans. Nucl. Sci., NS-9, 1(1962)123.
32. W. M. Trenholme, "Effects of reactor exposure on boron lined and BF proportional counters", Inst. Radio Engrs., Trans. Nucl. Sci., NS-6, 4(1959).
33. J. M. Harrer, W. C. Lipinski, and R. L. Ramp, "Control Instrumentation", Chapter 8 in Reactor Handbook, Vol. IV, (USAEC in publication).
34. H. Carmichael, "Design of Chalk River Ion Chambers, Types TPA, TPJ and TPD", Canadian Report AECL-293, 1946.
35. W. Abson and F. Wade, "Nuclear Reactor Control Ionisation Chambers", Proc. Inst. Elec. Engrs., 103B(1956)590. (Brit.)
36. W. Abson, R. J. Cox, and A. L. Gray, "Neutron Flux Instrumentation for Research and Power Reactors. A Review of Present Methods and New Development", Proceedings of the Second U.N. International Conference on Peaceful Uses of Atomic Energy, Geneva, 1958, Vol. 11, p. 498.
37. P. R. Tunnicliffe and H. Carmichael, "Interim Report on Development of Ion Chambers for the NRU Reactor", unpublished Canadian Report RPD-64, 1954.
38. H. S. Creary and R. T. Bayard, "An Electrically Adjusted Compensated Ionization Chamber", Report WAPD-60, Bettis, Atomic Power Laboratory, 1962.
39. D. Strominger, "Solid-state fission counters with AETR critical assemblies", Trans. Am. Nucl. Soc., 5(1962)186.
40. T. A. J. Jaques, H. A. Ballinger, and F. Wade, "Neutron Detectors for Reactor Instrumentation", Proc. Inst. Elec. Eng., 100, 1(1953)110. (Brit.)
41. R. A. DuBridge, "In-Core Power Monitoring of Nuclear Reactors", Report GEAP-3914, General Electric Co., Atomic Products Division, 1963.
42. M. G. Mitel-man, R. S. Erofeev, and N. D. Rozenblyum, "Transformation of the energy of short lives isotopes", Atomnaya Energiya, 10, 1(1961)72.
43. J. W. Hilborn, "Self-powered neutron detectors for reactor flux measurements", Nucleonics, 22, 2(1964)69.
44. P. Kovanic and M. Kulka, "Complete Automation of Nuclear Reactor Control by Means of an Automatically Moved Detector", Proceedings of the Second U.N. International Conference on Peaceful Uses of Atomic Energy, Geneva, 1958, Vol. 11, p. 493.
45. W. J. Henderson, P. R. Tunnicliffe, and A. G. Ward, "The Design and Performance of a Proposed Ion Chamber Arrangement to Monitor the Neutron Flux of the NRU Reactor", unpublished Canadian Report RPD53, Feb. 1953.
46. H. Ager-Hanssen, F. H. Enger, O. R. Kasa, and R. Rose, "Experience from the Operation of the Halden Heavy Water Boiling Reactor on Its First Fuel Charge, June 1959 to April 1961", Report HPR-26, Institutt for Atomenergi, Kjeller, Norway, 1962.
47. A. A. Wasserman, "Contributions to Two Problems in Space-Independent, Nuclear-Reactor Dynamics", Report IDO 16755, p. 93, Phillips Petroleum Co., 1962.
48. B. H. Stonehouse, "Some Aspects of the Control and Instrumentation of Gas-Cooled Reactor Power Stations", Instruments and Measurements, Vol. II, p. 1157. Proceedings of the Fifth International Instruments and Measurements Conference, September 13-16, 1960, Stockholm, Sweden.
49. A. B. Gillespie, Signal, Noise and Resolution in Nuclear Counter Amplifiers, Pergamon Press, N.Y. and London, 1953.
50. C. Cottini, E. Gatti, G. Giannelli, and G. Rozzi, "Minimum noise preamplifier for fast ionization chambers", Nuovo Cimento, 10, 3(1956)473.
51. F. S. Goulding, "Transistorized radiation monitors", Inst. Radio Engrs., Trans. Nucl. Sci., NS-5, 2(1958)38.
52. S. C. Baker, "Transistorized linear pulse amplifiers", Inst. Radio Engrs., Trans. Nucl. Sci., NS-6, 2(1959)57.
53. F. C. Loveless and J. Grossart, "Use of Pulse Transformers with Radiation Detectors", Nuclear Electronics 11, Proceedings of a Conference Held at Belgrade, May 15-20, 1961.
54. R. J. Cox and D. Harrison, "Transistors in reactor control instruments", J. Brit. Nucl. Soc., 8, 1(1963)49.
55. F. S. Goulding, "Transistorization", Nucleonics, 17, 6(1959) 64.
56. K. A. McCollom, "Discriminators", Nucleonics, 17, 6(1959) 72.
57. E. J. Wade and D. S. Davidson, "Transistorized log-period amplifier", Inst. Radio Engrs., Trans. Nucl. Sci., NS-6, 2(1959)53.
58. G. B. B. Chaplin and A. R. Owens, "Some transistor input stages for high gain DC amplifiers", Proc. Inst. Elec. Engrs., 105B(1958)249. (Brit.)
59. Fifth Tripartite Instrumentation Conference Held at Brookhaven National Laboratories, October 1956. USAEC Report TID-7543, 1956.
    a) A. Pearson, "NRU Control System", p. 17.
    b) L. Cathey, "Alpha Monitor for Airborne Plutonium", p. 107.
    c) R. J. Cox, "Recent Advances in Reactor Instrumentation in England", p. 6.
    d) J. Kostalos, Jr., "Pressurized Water Reactor Instrumentation Equipment", p. 26.
60. R. Hoge, "Parametric Modulator", Inst. Radio Engrs. Convention, March 1960, unpublished report.
61. N. F. Moody, "Silicon diode modulator", Electronic Engineering, (Brit.) Vol. 28, March 1956.
62. C. G. Lennox, "The Semiconductor Diode Ring Modulator as an Input Element to DC Amplifier", Canadian Report AECL-977, 1960.
63. G. R. B. Chaplin and A. R. Owens, "Some transistor input stages for high gain DC amplifiers", Proc. Inst. Elec. Engrs., 105B(1958)249. (Brit.)
64. A Kemhadjian, "Transistor Amplifier for DC Signals", Mullard Technical Communication 4, No. 30, December 1958, p. 162.
65. M. G. Say, Magnetic Amplifiers and Saturable Reactors, George Newnes, Ltd., London, 1954.
66. H. F. Storm, Magnetic Amplifiers, John Wiley and Sons, N.Y. and Chapman and Hall, Ltd., London, 1955.
67. Nuclear Process Instrumentation and Control Conference, USAEC Report ORNL-2695, Oak Ridge National Laboratory, 1960.
    a) H. E. Darling, "Magnetic Amplifiers for Reliable Process Conrol".
    b) R. L. Moore, "Instrument Transmitters for High-Pressure Aqueous Nuclear Reactors".
    c) B. E. Woodward, "Primary Elements for Boiling-Water Reactors".
    d) E. A. Goldsmith, "Operational Experience with Primary Elements and Valves in Slurry Service".
    e) P. Bliss, "Liquid Metal Instrumentation Practice".
    f) Conference discussion on "Process Instrumentation in High-Temperature Environment", p. 139.
68. E. H. Cooke-Yarborough and W. W. Pulsford, "An Accurate Logarithmic Counting-Rate Meter Covering a Wide Range", Proc. Inst. Elec. Engrs., Pt. 2, 98(1951)196.
69. R. S. Stone and E. J. Wade, "A General Purpose Log Counting Rate Meter - Model B", Report KAPL-1237, Knolls Atomic Power Laboratory, 1954.
69a. G. F. Wall and M. P. Young, "Naval Research Laboratory Research Reactor. Part III. An Eight-Decade Logarithmic

Amplifier for Nuclear Reactor Instrumentation", Report NRL-5025, Naval Research Laboratory, Washington, D.C., 1957.
70. R. R. Hoge and D. J. Niehaus, "Transistorized reactor instrumentation and protective circuits", Inst. Radio Engrs., Trans. Nucl. Sci., NS-6, 2(1959)42.
71. W. H. Jordon, H. B. Frey, and G. Kelley, "An Instrument for Measuring the Logarithms of Neutron Level and the Period of a Pile", USAEC Report ORNL-110, Oak Ridge National Laboratory, 1948.
72. R. J. Cox, "Automatic start-up of nuclear reactors", Inst. Radio Engrs., Trans. Nucl. Sci., NS-3, 1(1956)15.
73. E. Sikorsky, "An evaluation of vacuum tubes for log N amplifiers", Inst. Radio Engrs., Trans. Nucl. Sci., NS-10, 1(1963)142.
74. W. J. Hartin, "Servologarithmic amplifier for reactor instrumentation", Inst. Radio Engrs., Trans. Nucl. Sci., NS-6, 1(1959)11.
75. A. Pearson and Y. J. Fokkinga, "Logarithmic-amplifier period meter", Nucleonics, 17, 2(1959)2.
76. M. W. Grim, B. B. Barrow, and J. C. Simons, "Random fluctuations in period-meter indications", Nucl. Sci. Eng., 1(1956)80.
77. J. F. Hogg, "Noise and Transient Response in Period meters", British Report RIC 2/7, Dec. 1957.
78. A. G. Klein, "Logarithmic rate meters and period meters", Nucl. Sci. Eng., 11(1961)142.
79. H. Christensen and R. B. Stanfield, "The Log Count Rate Period Meter Used with Safety Circuits", Paper presented at the International Convention of the Institute of Radio Engineers, March 1961.
80. J. R. Ragazzini and S. S. L. Chang, "Noise and random processes", Proc. Inst. Radio Engrs., 50, 5(1962)1146.
81. "Boiling Water Reactor Program - EBWR", Papers presented at the Technical Briefing Session Held at Argonne National Laboratory, May 27-28, 1957, USAEC Report TID-7535, 1957.
82. B. B. Barrow, "The Logarithmic-Diode Counting-Rate Meter and Period Meter", Paper presented at the Second Nuclear Engineering and Science Conference, March 11-14, 1957, Philadelphis, Pa. Paper 57, NESC 60.
83. J. D. Cummins, "Pulse Counting Period Meter Output During Start Up Transients", British Report AEEW-R-245, 1962.
84. W. W. Ramage, "Exponential Circuit Response", Report WAPD-RM-154, Bettis, Atomic Power Laboratory, 1951.
85. J. D. Cummins, "Diode Response and Time Derivative of Response to an Exponentially Increasing Input Current", British Report AEEW-M-10, Winfrith, 1959.
86. J. D. Cummins, "Period safety system response calculated for fast transients", Trans. Am. Nucl. Soc., 3(1960)456.
87. C. R. Nixon and A. Pearson, "Logarithmic Rate Measurements at Low Reactor Power Levels", Canadian Report AECL-1004, 1960.
88. R. R. Hoge and D. J. Niehaus, "Transistorized reactor instrumentation and protective circuits", Inst. Radio Engrs., Trans. Nucl. Sci., NS-6, 2(1959)42.
89. J. Furet, "Securite dans le control des Piles Atomiques", Report CEA No. 2174, p. 201, Centre d'Etudes Nucleaires de Saclay, 1962.
90. C. G. Lennox and A. Pearson, "Our changing views on reactor instrumentation", Inst. Radio Engrs., Trans. Nucl. Sci., NS-8, 4(1961)163.
91. See ref. 30n.
92. W. M. Trenholme, "A reactor safety system using transistors and silicon controlled rectifiers", Inst. Radio Engrs., Trans. Nucl. Sci., NS-7, 4(1960)14.
93. "Development of Temperature Protective Circuits for HNPF Full Channel Exit", Report NAA-SR-5313(Rev), North American Aviation, Inc., 1961.
94. E. J. Wade and D. J. Davidson, "Transistorized Trip Circuit for Critical Assemblies", Report KAPL-1910, Knolls Atomic Power Laboratory, 1958.
95. A. K. Jonscher, "The physics of the tunnel diode", Brit. J. Applied Phys., 12(1961)654.
96. C. J. Nuese, "Tunnel diodes", Inst. Radio Engrs., Student Quarterly, February 1962, p. 36.
97. "High Temperature Materials and Reactor Component Development Programs", Second Annual Report, Vol. III, Reactor Component Development", Report GEMP 177C, p. 24, General Electric Co., Flight Propulsion Laboratory, 1963.
98. Sensitrol Meter Relay, Weston Instrument, Division of Daystrom Inc., 619 Freylinghuysen Ave., Newark, N.J.
99. Contacting Meter Relays, Assembly Products, Inc. Chesterland 91, Ohio.
100. Electronic Meter Controller, Thermovolt Instruments, Ltd., 55 Six Point Rd., Toronto 18, Ontario, Canada.
101. Pyrovane Meter, Minneapolis-Honeywell, Ltd., Brown-Instruments Div., Wayne and Windrin Ave., Philadelphia, Pa.
102. C. J. Lyons and R. I. Leininger, "Radiation effects on insulation, wire and cable", Insulation, May 1959, Pt. I, p. 19; June 1959, Pt. II, p. 38.
103. J. W. Harrity, H. Horiye, V. A. J. Van Lint, and E. G. Wikner, "Research in Radiation Damage in Semiconductors", Report GA-1201, General Atomic, Division of General Dynamics Corp., 1960.
104. D. R. Doman, "Gamma Irradiation Effects on Candidate Electronic, Lighting, Optical, Instrument and Structural Components for In-Reactor Monitoring Equipment", Report HW-76263, General Electric Co., Hanford Atomic Products Operation, 1963.
105. J. R. Burnett, "Radiation-tolerant electronic equipment", Inst. Radio Engrs., Trans. Nucl. Sci., NS-6, 4(1959)12.
106. Oak Ridge National Laboratory Staff, "The Homogeneous Reactor Experiment No. 2", Proceedings of the Second U.N. International Conference on Peaceful Uses of Atomic Energy, Geneva, 1958, Vol. 9, p. 509.
107. D. M. Considine, Process Instruments and Controls Handbook, McGraw-Hill Book Co., Inc., N.Y. 1957.
108. H. P. Kallen, Handbook of Instrumentation and Control, McGraw-Hill Book Co., Inc., N.Y., 1957.
109. F. R. Caldwell, "Thermocouple Materials", p. 122, in Temperature, Vol. 3, Pt. 2, "Applied Methods and Instruments", A. I. Dahl (Ed.) (Papers presented at the Fourth Symposium on Temperature, Columbus, Ohio, March 27-31, 1961), 1962.
110. Proceedings of the Power Reactor In-Core Instrumentation Meeting, April 28-29, 1960, Washington, D.C., USAEC Report TID-7598, 1960.
a) p. 90.
b) W. E. Browning and C. E. Miller, "Calculated Radiations Induced Changes in Thermocouple Composition", p. 273.
c) G. F. Levy, R. R. Fouse, and R. Sherwin, "Operation of Thermocouples under Conditions of High Temperature and Nuclear Radiation", p. 277.
111. M. J. Kelly and W. W. Johnston, "Radiation Effects on Thermocouples", USAEC Report ORNL-2787, Oak Ridge National Laboratory, 1959.
112. E. L. Broncato and J. W. Kallander, "Radiation effects on electrical insulation", Electrical Manufacturing, September 1960, p. 157.
113. A. Goodings, "Some Electrical Phenomena Which Occur in Magnesia Insulated Cables", British Report AEEW-R205, 1962.
114. A. M. Ross, "Radiation Damage to Ceramic Materials by Neutrons", USAEC Report ORNL-1852, Oak Ridge National Laboratory, 1958.
115. J. M. Harrer, Nuclear Reactor Control Engineering, D. Van Nostrand Co., Princeton, N.J., 1963. a) p. 134; b) p. 496.
116. E. S. Ida, "Reducing electrical interference", Control Engineering, 9, 2(1962)107.
117. W. Morton, "Quiet Wiring", Electro-Technology, March 1961.
118. J. Jursik, "Rejecting common mode noise in process data systems", Control Engineering, 10, 8(1963)61.
119. R. C. Masterson, "AC pick-up on signal transmission lines", Control Engineering, 10, 7(1963)123.
120. C. G. Lennox and A. Pearson, "An Electronic Data Scanning and Digitizing System", Canadian Report AECL-1232, 1961.
121. R. Smart, "Instrumentation of the Dounreay Fast Breeder Reactor", p. 1171 in Instruments and Measurements, Academic Press, N.Y., 1961.
122. ASME Power Test Code, Instruments and Apparatus, Pt. II, Pressure Measurement.
123. J. H. Kendron, E. E. Stoner, and G. M. Taylor, "Dynamic Void Fraction Measurement System", Report NAA-SR-7875, North American Aviation, Inc., 1963.
124. R. V. Smith, P. C. Wergin, J. F. Ferguson, and R. B. Jacobs, "The use of a Venturi tube as a quality meter", Trans. ASME, J. Basic Eng., Series D, Sept. 1962, p. 411.
125. J. W. Murdock, "Two-Phase Flow Measurement with Orifices", Paper presented at the ASME Winter Annual Meeting, New York, December 1961.
126. M. Petrick and B. S. Swanson, "Radiation attenuation method of measuring density of a two-phase fluid", Rev. Sci. Instr., 29, 12(1958)1079.
127. J. G. Bayly, "Fog Density Measurement by X-Ray Scattering", Canadian Report AECL-1522, 1962.
128. I. Orbeck, "Impedance Void Meter", Norwegian Report KR-32, Institutt for Atomenergi, Kjeller, Norway, 1962.
129. N. Adorni, L. Cravarolo, A. Hassid, E. Pedrocchi, and M. Silvestric, "Measurement of shear stress on the wall of a conduit and its application to the void fraction determination in two-phase flow", Rev. Sci. Instr. 34, 8(1963)939.
130. R. V. Moore and B. C. Goodlet, "The 1951-53 Harwell design study", J. Brit. Nucl. Energy Conf., 2, 2(1957)55.
131. R. C. Vincent, "Experiences of Instrument Maintenance at Calder Hall", p. 1217 in Instruments and Measurements, Academic Press, N.Y., 1961.

132. J. G. Bayly, R. J. Booth, and W. H. Stevens, "An Improved Infra-Red Method of Monitoring Heavy Water", Canadian Report CRRP-1099, 1962.
133. A. L. Giusti, R. E. Otto, and T. J. Williams, "Direct digital computer control", Control Engineering, 9, 6(1962)104.
134. E. W. Yetter and C. W. Sanders, "A time-shared digital process-control system", Instr. Soc. Am. J., 9, 11(1962)53.
135. R. A. Edwards, "Digital monitors for nuclear plants", Nucleonics, 20, 6(1962)85.
136. Computer Controlled Systems. Panel Discussion. Instr. Control Systems, 36, 5(1963) 85.
137. K. R. Knoblauch, "Some Comments on the Direct Digital Control Concept", Proceedings of the Eighteenth Annual Conference, Instrumentation Society of America, September 1963.
138. G. C. Hendrie and R. W. Sonnenfeldt, "What the User Should Know about Reliability of Control Computers", Proceedings of the Seventeenth Annual Conference, Instrumentation Society of America, N.Y., October 1962.
139. J. P. Knight, L. R. Klinger, and D. C. Yoder, "Low-Level data multiplexing", Instr. Control Systems, 36, 8(1963)86.
140. J. T. Tou, Digital and Sampled Data Control Systems, p. 75, McGraw-Hill Book Co., Inc., N.Y. 1959.
141. C. M. Cundall and V. Latham, "Designing sampled data systems", Part I, Control Engineering, October 1962 and Part II, Control Engineering, 10, 1(1963)109.
142. K. R. Stapleford, "The Design of the Shippingport On-Line Computer Data System Program", Paper presented at the American Nuclear Society (Pittsburgh Section) Symposium on the Role of Computers in Nuclear Power Plant Operation, February 24, 1962.
143. E. Beckjord, "Nuclear performance computer for Big Rock Point", Nucleonics, 20, 6(1962)57.
144. J. Prades and Y. Panis, "Digital computers in nuclear power applications", Inst. Radio Engrs., Trans. Nucl. Sci., NS-8, 4(1961)119. (From the Proceedings of the Joint Nuclear Instrumentation Symposium, North Carolina, September 1961.)
145. J. Auricost, R. Chambolle, Y. Panis, and J. Prades, "Digital computers monitor nuclear steam generator", Control Engineering, 8, 3(1961)127.
146. Editorial, "Data acquisition and performance advice", Nucleonics, 20, 6(1962)55.
147. E. Siddall, "Computer Control of Nuclear Power Plants", Presented at the Nuclear Congress, New York, June 4-7, 1962. Paper 29. See also: Canadian Report AECL-1539.
148. R. Sherrard, "Advances in High Speed Scanning, Monitoring, Systems, Hardware and Logic", Report HW-SA-2423, General Electric Co., Hanford Atomic Products Operation, 1962.
149. C. G. Lennox and N. P. Vakil, "A Digital Computer Control Experiment on a Nuclear Reactor", Paper presented at the Fourth Conference of the Computing and Data Processing Society of Canada, Ottawa, May 1964.
150. C. G. Lennox and A. Pearson, "Versatile data displays for reactors", Nucleonics, 18, 10(1960)82.
151. C. G. Lennox and S. Shinmoto, "Digital Data Storage and Display Using a Magnetic Drum and Cathode-Ray Screen", Canadian Report AECL-1742, 1963.
152. M. J. Marcotty, F. M. Longstaff, and A. P. M. Williams, "Time-Sharing in the Ferranti-Packard FP 6000 Computer System", Proceedings of the Joint Computer Conference, Institute of Radio Engineers, Spring 1963.
153. A. S. Robinson, "Control Programming - Key to the Synthesis of Efficient Digital Computer Control Systems", Joint Automatic Control Conference, Am. Inst. Elec. Engrs., Am. Inst. Chem. Engrs., Inst. Radio Engrs., Instr. Soc. Am., Am. Soc. Mech. Engrs., held at Cambridge, Mass., September 6-9, 1960. Paper 60-969.
154. R. J. Coyle and J. K. Stewart, "Design of a real-time programming system", Computers and Automation, 12, 9(1963)26.
155. J. H. Shannon, "Executive control routines for process computers", Control Engineering, 10, 4(1963)85.
156. M. E. Williams, "Computer priority interrupt", Instruments and Control Systems, 36, 8(1963)91.
157. E. P. Gyftopoulos and P. M. Coble, "A Digital Nuclear Control System", American Institute of Electrical Engineers, Winter General Meeting, N.Y., February 1-6, 1959. Paper CP58-1362.
158. E. Siddall, "Highly Reliable Control Systems for a 200Mw Power Reactor", Canadian Report AECL-685, 1958.
159. K. Becker, "Design and Operation of a Reactor Safety System", Proceedings of a Symposium on Reactor Safety and Hazards Evaluation Techniques held at Vienna, May 14-18, 1962. IAEA, Vienna, 1962, Vol. 1, p. 477.
160. E. Siddall, "Highly Reliable Control Systems for a 200Mw Power Reactor", Canadian Report AECL-685, 1958.
161. I. Wilson, R. E. Whyard, and A. B. Keats, "Solid State Safety Circuits for the Dimple Reactor", J. Brit. Nucl. Energy Soc., 8, 1(1963)59.
162. H. K. Hill and G. H. Pinder, "A Completely Transistorized Self-Tested Safety System for a Power Reactor", Am. Inst. Elec. Engrs. General Meeting, San Diego, 1960. See also: Ref. 116b.
163. J. G. Truxal, Control Engineers Handbook, McGraw-Hill Book Co., Inc., 1958. a) pp. 1-29; b) pp. 2-8.
164. J. S. Bendat, Principles and Applications of Random Noise Theory, John Wiley and Sons, Inc., 1958. a) p. 15; b) p. 70.
165. S. Singer, "The Period Equilibrium Effect in Reactor Dynamics", USAEC Report LA-2654, Los Alamos Scientific Laboratory, 1962.
166. A. Pearson and C. G. Lennox, "The Dynamic Behavior of the NRX Control System", Canadian Report AECL-1050, p. 107, 1960.
167. R. I. Vaughan and R. N. H. McMillan, "Some Aspects of Stability Studies Related to the Control of Power Reactors", Vol. 2, p. 1053 in Instruments and Measurements, Academic Press, N.Y., 1961.
168. A. Takeda, (i) "A Method of Analysis of Reactor Spatial Kinetics of Calder Hall Type Reactors" and (ii) "A Study of the Transfer Functions of Spatial Kinetics in the Nuclear Reactor", British Report AERE Trans. 888, 1962.
169. W. M. Barss, "Digital Simulation of Xenon Instability in Reactors", Canadian Report AECL-1226, 1961.
170. A. Hitchcock, Nuclear Reactor Stability, p. 35, Temple Press, London, 1960.
171. M. S. Corrington, T. Morahami, and R. W. Sonnenfeldt, "The complete specification of a network by a single parameter", RCA Review, 15(1954)389.
172. H. Chestnut and R. W. Mayer, Servomechanisms and Regulating System Design, John Wiley and Sons, Inc., N.Y. 1959. a) p. 7 and p. 139; b) p. 124.
173. R. W. Roig, "A comparison between human operator and optimum controller RMS-error performance", Inst. Radio Engrs., Trans. Human Factors Electron., HFE-3, 1(1962)18.
174. E. Mishkin and L. Braun, Adaptive Control Systems, p. 147. McGraw-Hill Book Co., Inc., 1961.
175. C. G. Lennox and A. Pearson, "NRU reactor neutron level control system", Inst. Radio Engrs., Trans. Nucl. Sci., NS-5, 2(1958)64.
176. W. S. Brown, "Reactor Control of NPD - from the Engineering Design and Operation of NPD Reactor", Papers presented at the Winter General Meeting of the American Nuclear Society, Washington, D.C., November 1962. Canadian Report AECL-1682, 1963.
177. F. S. Goulding, "The Design of a Radioactive Contamination Meter", Canadian Report AECL-878, 1954.
178. F. S. Goulding, "Transistor Circuit Design for a Radioactivity Contamination Meter", Canadian Report AECL-435, 1957.
179. A. R. Jones, "A more reliable gamma survey meter", Canadian Nucl. Technol., 1, 3(1962)40.
180. K. Van Duuren, A. J. M. Jaspers, and J. Hermsen, "G-M Counters", Nucleonics, 17, 6(1959)86.
181. K. E. G. Perry and E. George, "The Sensitive Detection, Identification, and Location of Sources of Gamma Radiation", Paper presented at the Eighth Annual Meeting of the Health Physics Society, N.Y., June 10-13, 1964.
182. G. W. C. Tait, "The Annular Impactor", Canadian Report AECL-130, 1955.
183. G. L. Miller, W. M. Gibson, and P. F. Donovan, "Semiconductor particle detectors", Ann. Rev. Nucl. Sci., 12 (1962)189.
184. Papers presented at the Seventh Annual Meeting of the Health Physics Society, Chicago, June 11-14, 1963.
   a) W. A. Phillips and C. L. Lindeken, "Plutonium and Air Monitor Using a Solid State Detector". (See also: Report UCRL-6692.)
   b) D. P. Brown, M. O. Rankin, W. G. Spear, and C. D. Bayne, "A Coincidence-Count α-Particulate Air Monitor", (See also: Report HW-SR-2254.)
   c) G. A. Little, "Ionisation Chamber Air-Sampling System for Tritium".
185. Seventh AEC Air Cleaning Conference, October 10-12, 1961, USAEC Report TID-7627.
   a) C. H. Distenfield, "A Description of the BNL Research Reactor Environmental Air Control Alarm and Monitoring System".
   b) J. F. Manneschmidt, "Equipment and Procedures for Stack Gas Monitoring at ORNL".
   c) R. E. Adams and W. E. Browning, "Removal of Iodine Gas Streams".
186. V. H. Allen, "Transistorized slow neutron monitor uses proportional counter", Canadian Electronics Engineering, April 1958.
187. W. F. Hornyak, "A fast neutron detector", Rev. Sci. Instr., 23, 6(1952)264.
188. J. D. Anthony, "Portable tritium monitor has gamma compensation", Nucleonics, 17, 4(1959)110.
189. G. Cowper and S. D. Simpson, "A Monitor for Airborne Tritium", Canadian Report AECL-1049, 1960.
190. H. L. Butler, "Tritium hazards in heavy-water-moderated

reactors", Nucl. Safety, 4, 3(1963)7/.
191. A. Pearson, "Reactor Instrumentation at Chalk River", Paper presented at the First Nuclear Engineering and Science Congress, December 1955, Cleveland, Ohio, Pergamon Press Symposium Publications, N.Y. and London.
192. G. Cowper, "High-Level Dosimetry Problems in Reactor Operations", Selected Topics in Radiation Dosimetry, Proceedings of a Symposium held at Vienna, June 7-11, 1960, IAEA Vienna, 1961.
193. A. Blanc and J. Lequois, "Appareil enregistreur de l'activite d'iode 131," Proceedings of a Conference on Nuclear Electronics held at Belgrade, May 15-20, 1961.
194. V. Surya Rao et al., "Safety Provisions for a Research Reactor", Reactor Safety and Hazards Evaluation Techniques, Proceedings of a Symposium held at Vienna, May 14-18, 1962, IAEA Vienna, 1962. Vol. 1, p. 251.
195. S. H. Hanauer, "Instrumentation for containment", Nucl. Safety, 3, 1(1961)41.
196. R. A. Harvey, "Some trends and recent health physics instrumentation developments at Hanford", Health Physics, 7(1961)11.
197. D. J. Knowles, "Sampling considerations in monitoring process waste effluents", Nucl. Safety, 5, 1(1963)59.
198. W. M. Hurst, "Monitoring Liquids for Radioactivity", USAEC Report ORNL-1155, Oak Ridge National Laboratory, 1952.
199. H. Gebauer and S. Muller, "Kontinuerliche Direktmessung radioaktiver Substanzen in Wasser," Die Atomwirtschaft, 7(1962)487.
200. H. Fiefer and R. Maushart, "Kontinuerliche Direktubewashung kleener β-aktivitater in Wasser," Atompraxis, 5(1959)431.
201. E. C. Wingfield, "A β - γ monitor for liquid streams", Report DP-177, E. I. du Pont de Nemours and Co., Inc., 1956.
202. E. C. Wingfield and P. R. Liller, "A Monitor for Low Level Radioactivity in Liquid Streams", Report DP-145, E. I. du Pont de Nemours and Co., Inc., 1956.
203. E. H. Cooke-Yarborough and R. C. M. Barnes, Rapid Methods for Ascertaining whether the Activity of a Weak Radioactive Sample Exceeds a Predetermined Level, Institute of Electrical Engineers, 1960.
204. A. R. Jones, "The measurement of radioactive contamination on hands and feet", Nucl. Instr. Methods, 21(1963)75.
205. A. R. Jones, "A portable γ-dosimeter providing audible warning", Health Physics, 4, (1960) 64.
206. G. Cowper, "Recent Advances in Personnel Dosimetry at Chalk River", Canadian Report AECL-1700, 1963.
207. A. R. Jones, "A Portable Rate Meter for Measuring α and β Activity", Canadian Report AECL-1710, 1963.
208. W. Goldsworthy, "Scintillation Hand and Foot Counter", Report UCRL-2088, University of California Lawrence Radiation Laboratory, 1953.
209. P. R. Bell and H. A. Straus, "Electronic Pile Simulator", USAEC Report AECD-2764, 1950.
210. G. J. R. MacLusky, "An Analog Computer for Nuclear Power Studies", Proc. Inst. Elec. Engrs., Pt. B, 104(1957) 433.
211. J. J. Stone and E. R. Mann, "Reactor Controls Simulator", USAEC Report ORNL-1632, 1954.
212. J. P. Franz and N. F. Simcic, "Nuclear reactor start-up simulation", Inst. Radio Engrs., Trans. Nucl. Sci., NS-4, 1(1957)11.
213. J. Weil, "Measurement of reactivity", Nucleonics, 11, 3(1953)36.
214. G. S. Stubbs, "Design and use of the reactivity computer", Inst. Radio Engrs., Trans. Nucl. Sci., 4, 1(1957)40.
215. T. E. Stern, A. Blaguiere, and J. Valat, "Reactivity measurement using pseudo-random source excitation", Reactor Sci. Technol., 16, 11/12(1962)499.
216. P. Schmid, "Absolute Reactivity Measurement from Transient Behaviour of a Sub-Critical Nuclear Reactor, June 1957, RA6-Bericht Nr 1. Reactor Ltd., Warenlinger
217. S. Stubbs, "Transport Delay Simulator", Report WAPD-7-38, Bettis, Atomic Power Laboratory, 1954.
218. J. Gamp, "Etude Systematique de la Generation de Retard au Moyen d'Unites Analogiques Classiques", European Atomic Energy Command, Report EUR-240f, 1963.
219. D. M. Collier, L. A. Meeks, and J. P. Palmer, "The HRE Simulation", USAEC Report ORNL-1572, Oak Ridge National Laboratory, 1954.
220. E. R. Mann and F. P. Green, "Analog Simulation in the Package Reactor Study", USAEC Report AECD-3719, 1954.
221. H. A. Bethe, "Reactor Safety and Oscillator Tests", Report APDA-117, Atomic Power Development Associates, 1956.
222. J. M. Harrer et al., "Performance Evaluation of Direct Cycle Boiling Water Nuclear Power Plants Based on Recent EBWR and Borax Data", Proceedings of the Second U.N. International Conference on Peaceful Uses of Atomic Energy, Geneva, 1958, Vol. 9, p. 264.
223. G. J. R. MacLusky, "The application of analog methods to compute and predict xenon poisoning in a high flux nuclear reactor", Proc. Inst. Elec. Engrs., 104B, 17(1957)443. (Brit.)
224. M. M. L. Gulati, V. Mahajan, A. W. Pereira, and J. Ranganath, "A Xenon Poison Computer", Proceedings of the Second U.N. International Conference on Peaceful Uses of Atomic Energy, Geneva, 1958, Vol. 11, p. 540.
225. Ash, Bellman, and Kalaba, "Reactor Shut-Down Involving Minimal Xenon Poison", Report P-1500(RAND), Remington-Rand Corp., Santa Monica, Calif., 1958.
226. R. E. Kerr and W. C. Lennox, "Extending Reactor Time-to-Poison by Pre-Shut-Down Power Alterations", Canadian Report AECL-1619, 1962.
227. J. J. Paul and J. R. G. Cox, "Xenon poisoning computer", Nucleonics, 16, 5(1958)97.
228. J. M. Harrer, R. E. Boyer, and K. Darwin, "Transfer function of Argonne CP-2 reactor", Nucleonics, 10, 8(1952)32.
229. H. Ager-Hanssen, "Halden operating and research experience", Nucl. Power, 7, 59(1962)57.
230. Y. W. Lee, Statistical Theory of Communication, p. 209, John Wiley and Sons, Inc., N.Y., 1963.
231. V. Rajagopal, "Determination of reactor transfer functions by statistical correlation methods", Nucl. Sci. Eng., 12, 2(1962)218.
232. E. R. Corran and J. D. Cummins, "Binary Codes with Impulse Autocorrelation Functions for Dynamic Experiments", British Report AEEW-R-210, 1962.
233. J. D. Balcomb, H. B. Demuth, and E. P. Gyftopoulos, "A cross correlation method for measuring the impulse response of reactor systems", Nucl. Sci. Eng., 11, 2(1961)159.
234. "Electrical power systems of nuclear power plants", Nucl. Safety, 2, 1(1960)32.
235. W. J. Shewski, "A Look at the Electrical Features - Dresden Nuclear Power Station", American Institute of Electrical Engineers, Fall General Meeting, Pittsburgh, Pa., October 1958, Paper CP58-1219.
236. H. G. Frus, H. A. Thompson, H. A. Von Wassen, and E. J. Woolever, "Electrical and Control Features of the Shippingport Atomic Power Station", American Institute of Electrical Engineers, Fall General Meeting, Pittsburgh, Pa., October 1958, Paper 58-1199.
237. J. Henderson, G. F. Kennedy, and K. J. Wooton, "Electrical aspects of Hunterston nuclear generating station", Proc. Inst. Elec. Engrs., Pt. A., 109, 46(1962) (Brit.)
238. J. L. Fink, J. F. Johnston, and F. C. Krings, "Static inverters for essential loads", Electrical Engineering, May 1963. (Brit.)
239. J. F. Johnston, "Static Inverters", Plant Engineering, November 1963.

CHAPTER 7

Mathematical Models of Fast Transients

W. E. NYER
Phillips Petroleum Company, Atomic Energy Division,
Idaho Falls, Idaho

CHAPTER CONTENTS

1 INTRODUCTION

In assessing the hazards associated with a chain-reacting assembly, one must consider the excursion behavior of the reacting system. A description of a nuclear excursion can be enormously complicated, involving intimate details of the design, construction and operation of mechanical and instrumentation components of the system as well as the physical properties of the materials of the system and the factors changing the neutronic properties of the system. The great variety of reacting systems adds to the complexity of the problem.

However, despite this complexity, it has been observed that certain regularities are exhibited by the data obtained in reactor excursion experiments [1]. Furthermore, many similarities can be noted between the observed features of these excursions and the predictions of various simple mathematical models for reactor excursions [1]. In this chapter some models of nuclear excursions will be considered with the intention of summarizing and comparing their main features, illustrating some general behavior characteristics of bursts, and providing an introduction to the more detailed calculational efforts discussed in other chapters.

The bases on which these models have been postulated vary from situation to situation. In some cases, it is possible to derive a model from a fundamental representation of a reactor system and thereby obtain a description of reactor behavior during an excursion—a description which approaches the full detail outlined in the first paragraph. In other cases, the model serves as a convenient means of breaking the complete problem into parts which can be attacked individually. For example, a detailed calculation can be made in the part of the problem dealing with either the temperature coefficient or the Doppler coefficient, and this, if reducible to a convenient analytical form, may then be combined with a model. In many cases, the difficult task of computing excursion behavior directly from detailed considerations may be supplemented, or in some instances supplanted, by the use of an appropriate qualitative description.

The meaning, then, of the models may range from being a pedagogical device or a handy means of correlating experimental data, to a theory of reactor excursion behavior.

In their most general forms these models contain no information on specific physical mechanisms. Automatically, then, there is a corresponding lack of detail in the results so that extrapolation to different situations is difficult. Furthermore, the identification of parameters in the models with mechanisms responsible for reactivity changes may not be direct.

At the same time, the models have, in a relative sense, certain virtues. The analytical solutions which may be obtained display clearly the relative importance of the various parameters in a given model and exhibit clearly the effect on burst behavior produced by changes in the model. For example, in a given model the relative importance of the prompt neutron lifetime, with respect to the generalized shutdown property, may be easily seen. The changes in the burst behavior resulting from changes in the model can in many cases be identified with changes in the mechanisms responsible for self-limitation of bursts; this identification is of value to a designer in those instances where he has control over design factors that may emphasize one effect in preference over another. Finally, the prediction of the models may be readily compared with experiment.

The experimental situations to be described can, without any significant loss of generality, be limited to the two situations shown schematically in Fig. 1-1 for very fast bursts. In the first case, it is assumed that the excursion is initiated by effectively adding reactivity to the reactor instantaneously, that is, as a reactivity "step". The initial reactor power is usually taken to be low and the initial reactivity state as delayed critical. In a short time the reactor power begins to rise exponentially with a constant period. Ultimately, the energy release produces effects which begin to cause a decrease in reactivity. As these reactivity losses continue, the power passes through a maximum and in a typical fast transient decreases to a low value. Although this idealized representation presents the essential character of a power excursion, an actual excursion may be rich in detail, with many other distinctive features. See, for example, the features pointed out in a discussion by Horning and Corben [2].

The lower part of the figure shows the behavior, on a linear plot, of the logarithmic derivative of the power (the reciprocal period), which is defined as the quantity α. It is only under special conditions that α, which is an observable of major importance, can be readily identified with the system reactivity. It will be assumed throughout this discussion that these conditions are satisfied so that either the computation of reactivity from α or the identification of α directly with the prompt reactivity [3] may be made.

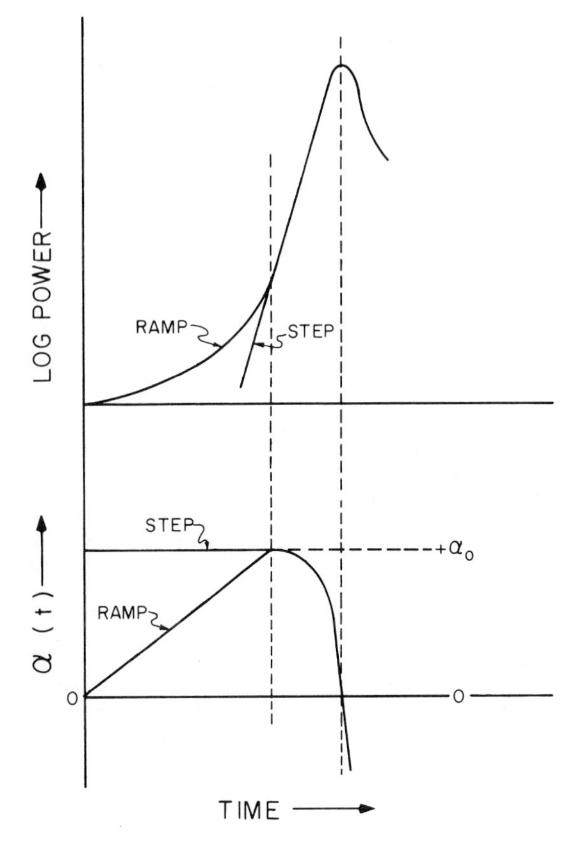

FIG. 1-1 Power and reactivity behavior during reactor excursions.

The second form of power burst, also shown in Fig. 1-1, is that in which the reactivity is introduced linearly with time under initial conditions assumed to be the same as for the step case, i.e., low power operation at delayed criticality. This linear or "ramp" addition of reactivity shown as the curve labeled "ramp" on the lower part of the figure, corresponds to the gradually decreasing period for the power rise indicated on the ramp power curve. In the next stage of the excursion, the energy release brings into effect, to a significant degree, the mechanisms which tend to decrease the reactivity. When the rate of reactivity loss rises to the point that it is equal to the rate of addition of reactivity, the reactor period has its minimum value, that is, α is maximum. From this point on, the development and final quenching of the burst are essentially the same as for a step-transient in which the initial period is equal to the minimum period for the ramp-test.

In both these cases the reactivity functions α are marked by two features. Initially there is negligible departure of the reactivity from the input step or ramp. This is then followed by a very rapid departure which almost abruptly decreases the reactivity to zero. The difference between the step or ramp-inserted reactivity and the instantaneous reactivity at any time will be called the "compensated reactivity".

Although the physical factor of primary importance in such a burst is the time history of the energy release, from both the analytical and experimental point of view it is often more convenient to describe a burst in terms of its power history or other selected burst parameters. The latter can be simply the burst width taken at some fraction of the peak power, the energy release at some particular time such as at the instant of the power peak, the time in periods to the peak, or some combination thereof. These factors are not all of fundamental significance but are merely convenient ways of describing the shape of the burst. In any case, whichever factors are selected to characterize a burst, it then becomes important to describe the way in which the factors depend upon the reactivity insertion and upon the form of the relationship between the power history and the induced reactivity effects. The latter, of course, are determined by the intrinsic properties of the reactor.

One of the most useful models permitting such a description is also the simplest and was the first employed. Because the model may be regarded as a limiting form for many real situations and because it is the starting point for other similar approaches, it will be described in some detail. Various names have been associated with this model in the past but in this discussion it will be referred to as the "Linear Energy Model". This distinguishes it from the variations or generalizations that were subsequently developed and avoids the difficulties of attempting to assign historical credits for what is believed by early workers to have been an obvious step.

2 THE LINEAR ENERGY MODEL

The original development of this model was carried out at Los Alamos where Fuchs [4] used it to estimate the energy release in fast, super-prompt critical assemblies subjected to step- or ramp-insertions of reactivity. He also considered the effect of a threshold. Hansen [5] extended the work to obtain temperatures, pressures, and permissible assembly rates. Hansen's work, which covered a wide range of assembly rates, included thermal, as well as fast, systems. Subsequently, further observations and considerations have been expressed by many others.

2.1 Step Insertions

In this model, attention is focused on the all-prompt neutron approximation for a lumped-parameter system. By definition

$$\frac{\dot{\phi}(t)}{\phi(t)} \equiv \alpha(t) , \qquad (2\text{-}1)$$

where ϕ represents reactor power in some units and the dot represents differentiation with respect to time. The quantity α, as already defined, is the reciprocal of the period τ and is in units of sec^{-1}. In the prompt case and to the accuracy needed, α is related to the effective reduced prompt neutron lifetime and to the reactivity of the system by the equation $\alpha(t) = \$(t)/(\ell/\beta)$, where $\$(t)$ represents the instantaneous degree of super-prompt criticality in dollar units, ℓ and β the effective prompt neutron lifetime and effective delayed neutron fraction, respectively. The reciprocal of ℓ/β, here denoted by α_R, is the prompt alpha at delayed criticality.

In turn it is assumed that the release of energy produces a linear decrease in the reactivity of the system so that $\alpha(t) = \alpha_0 - b\,E(t)$, where α_0 represents the initially injected reactivity, E the energy release and b the reactivity coefficient, that is, the change in reactivity per unit energy release. The reactivity coefficient is expressed in units of α_R and thus contains a factor ℓ in the denominator; i.e., $b = \partial\alpha/\partial E = (1/\ell)(\partial k/\partial E) = \alpha_R (\partial\$/\partial E)$. Then Eq. (2-1) becomes

$$\frac{\dot{\phi}(t)}{\phi(t)} = \alpha_0 - b\,E(t) . \qquad (2\text{-}2)$$

This model has the following properties.

1. The power excursion is a round-topped burst which is symmetrical (in time) about the peak power. E and α also have a symmetry about the peak power. Thus, the changes in the values of the energy and of the reactivity from the beginning to the end of excursion, are twice the respective changes up to the power peak, a result also obtained by Nordheim in his early consideration of the problem of "flashing a pile" [6].
2. The peak power increases as the square of α_0, the reciprocal of the asymptotic period.
3. The energy release at peak power increases linearly with α_0.
4. The peak power and energy for a given α_0 vary linearly with the inverse of the reactivity coefficient b.
5. Since α_0, the added reactivity, and the reactivity shutdown coefficient b both involve α_R in the same way, the energy release at peak power, which is the ratio of these two, is independent of the prompt neutron lifetime. Thus, for a truly prompt quenching mechanism in the linear feedback function, fast reactors will not differ from thermal reactors in the energy release, all other things being the same.
6. The peak power, however, varies inversely as the lifetime.
7. It is of particular importance to notice that the description of the burst requires only two quantities: the reactivity externally added and a reactivity coefficient. Each of these contains the fundamental dynamic parameter of the system, α_R. In effect, all the distingushing properties of the reactor which determine its quenching behavior are represented by the reactivity coefficient, which in this model has a negative effect at all times.
8. If the mechanism by which quenching takes place is a consequence of mechanical displacement of core components, the transient accelerating forces which cause the displacement may be of important magnitude [5]. Calculation of these forces requires specification of a relation between displacement and reactivity, between displacement and energy or specification of an equation of state [7].

If it is assumed that the displacement is linear with the energy release, the second time derivative of the energy may be identified with the transient pressure [8,9]. This is equivalent to specifying a third relation: $\ddot{\alpha} \sim \ddot{E} \sim$ pressure. Then the peak pressure varies inversely as the reciprocal of the reactivity coefficient b and directly as the cube of α_0. The strong dependence of peak pressure on α_0 is plausible on the following grounds. The displacements required for quenching are proportional to α_0 and take place in a time of the order of the burst period, $1/\alpha_0$. Thus, the transient pressure which produces the displacement goes as α_0^3. The peak pressure always occurs before peak power [8]. In this sequence of events, which is distinct from the sequences of events postulated in other models, the burst-governing properties are prescribed in the feedback relation specified by Eq. (2-2), and the pressure-governing condition was added subsequently by the above relation between reactivity and pressure. In this picture, transient pressures do not govern the burst, but are consequences of it.

9. Since the model contains no provisions for the escape of energy, the system remains permanently subcritical at the completion of the burst.

The analytical solutions to Eq. (2-2) are the following [8]:

$$E(t) = \frac{\alpha_0 + R}{b} \frac{(1 - e^{-Rt})}{(Ae^{-Rt} + 1)} , \qquad (2\text{-}3)$$

$$\phi(t) = \left(\frac{2R^2}{b}\right)(A)\,\frac{e^{-Rt}}{(Ae^{-Rt}+1)^2}, \qquad (2\text{-}4)$$

$$\dot{\phi}(t) = \left(\frac{2R^3}{b}\right)(A)\,\frac{e^{-Rt}(Ae^{-Rt}-1)}{(Ae^{-Rt}+1)^3}, \qquad (2\text{-}5)$$

$$[t]_{\phi_{max}} = \frac{1}{R}\ln(A), \qquad (2\text{-}6)$$

$$[t]_{\dot{\phi}_{max}} = \frac{1}{R}\ln\frac{A}{2+\sqrt{3}}, \qquad (2\text{-}7)$$

$$[t]_{\phi_{max}} - [t]_{\dot{\phi}_{max}} = \frac{1}{R}\ln(2+\sqrt{3}). \qquad (2\text{-}8)$$

In the above equations,

$$A \equiv \frac{R+\alpha_0}{R-\alpha_0}, \qquad (2\text{-}9)$$

$$R \equiv \sqrt{\alpha_0^2 + 2b\phi_0}, \qquad (2\text{-}10)$$

$$\dot{\phi} \sim \text{pressure},$$

$$\phi \sim \text{power},$$

$$E \sim \text{energy};$$

the subscript o indicates initital value.

If the approximation is made that the initial power level is very low so that $\alpha_0^2 >> 2b\phi_0$, then

$$R \simeq \alpha_0 + \frac{b\phi_0}{\alpha_0} \approx \alpha_0, \qquad (2\text{-}11)$$

$$A \simeq \frac{2\alpha_0^2}{b\phi_0}, \qquad (2\text{-}12)$$

$$E(t) \simeq \frac{2\alpha_0}{b}\,\frac{1-e^{-\alpha_0 t}}{\left(2\alpha_0^2/b\phi_0\right)e^{-\alpha_0 t}+1}, \qquad (2\text{-}13)$$

$$[t]_{\phi_{max}} - [t]_{\dot{\phi}_{max}} \simeq \frac{1}{\alpha_0}\ln(2+\sqrt{3}) \approx \frac{1.3}{\alpha_0}. \qquad (2\text{-}14)$$

The maximum values of the important parameters are of particular interest and are summarized below. Most of these results may be obtained directly from the differential equations:

$$E_M = \alpha_0/b \quad \text{(independent of } \alpha_R\text{)}, \qquad (2\text{-}15)$$

$$E_f = 2\alpha_0/b \quad \text{(independent of } \alpha_R\text{)}, \qquad (2\text{-}16)$$

$$\phi_M = \alpha_0^2/2b \quad \text{(goes as } \alpha_R\text{)}, \qquad (2\text{-}17)$$

$$p_{max} \sim \alpha_0^3/b \quad \text{(goes as } \alpha_R^2\text{)}. \qquad (2\text{-}18)$$

The subscript M is used to indicate the value of the parameter at the time of maximum power. The subscript f is used to indicate "final" values. (In the above, as well as in many other relations derived here, it is assumed that ϕ_0 is small compared to ϕ_m and ϕ_M.)

Of these parameters, reactor power alone has no importance per se with respect to reactor safety. It is of primary value as an index of behavior which is most readily measured and most reliably compared with theory. However, either excessive energy release or excessive pressure generation are of concern. Thus, according to this model, for a given reactivity insertion and reactivity coefficient, a short lifetime presents no particular disadvantage with respect to energy release, but may do so when transient pressures are to be considered. With respect to the energy release, it is, of course, the total energy that is of greatest importance. However, for experimental reasons as well as convenience, often the focus of attention will be mainly on the energy at the time of peak power.

2.2 Ramp Insertions

The insertion of reactivity as a ramp is often a more realistic approximation to an accident than the situation in which reactivity is inserted as a step. The necessary modifications to the linear shutdown model to incorporate this change are easily made. The reactivity addition will be postulated to be a constant rate a, in units of α/sec. Also, $a = \dot{\$}\,\alpha_R$ where $\dot{\$}$ is the assembly rate in dollars per second. The kinetic equation then becomes the following:

$$\frac{\dot{\phi}(t)}{\phi(t)} = at - b\,E(t). \qquad (2\text{-}19)$$

When the rate of reactivity insertion a just balances the rate of reactivity removal $b\phi$, the function $\alpha(t) = at - b\,E(t)$ has a maximum, as indicated in Fig. 1-1. The maximum of α occurs [10] at a power given by

$$\phi_m = a/b, \qquad (2\text{-}20)$$

which is independent of the starting power and the prompt neutron lifetime and is determined solely by the reactivity addition rate and the reactivity coefficient. The relation provides a gross determination of b from the ramp rate and the experimentally observed power at the time of maximum reciprocal period. The subscript m signifies that the values are to be taken at the time of maximum α. From Eqs. (2-19) and (2-20) it may be seen that ϕ_m is the mean power of the burst up to the time of maximum power. The maximum power is given by

$$\phi_M = \frac{a}{b} \ln \frac{\phi_M}{\phi_0} + \phi_0 , \qquad (2\text{-}21)$$

and

$$\alpha(t)^2 - \alpha_0^2 = 2\left[b(\phi_0 - \phi(t)) + a \ln \frac{\phi(t)}{\phi_0} \right]. \qquad (2\text{-}22)$$

The maximum reciprocal period α_m is then

$$\alpha_m = \left[2a \left(\ln \frac{a}{b\phi_0} - 1 + \frac{b\phi_0}{a} \right) \right]^{1/2} \qquad (2\text{-}23)$$

for $\alpha_0 = 0$. For $\phi_0 << \phi_m$,

$$\alpha_m = \left[2a \left(\ln \frac{a}{b\phi_0} - 1 \right) \right]^{1/2}. \qquad (2\text{-}24)$$

Thus, a weak dependence of α_m on initial power ϕ_0 and reactivity coefficient b is predicted by this model. The strongest dependence is on the reactivity addition rate a. Neglecting shutdown effects, that is, setting $\alpha = at$ in Eq. (2-6) would lead to a reciprocal period,

$$\alpha = [2a \ln (a/b\phi_0)]^{1/2} \qquad (2\text{-}25)$$

at the power $\phi_m = a/b$. This differs very little from that given by Eq. (2-24). Thus, the reactor behaves largely as if there were no shutdown effects up to the time that the power passes through the value given by Eq. (2-7). That is, the reactivity up to this power is essentially the term (at).

By combining Eqs. (2-21) and (2-24), the maximum power in a ramp burst can be shown [10] to also take the form,

$$\phi_M = \phi_0 + \frac{\alpha_m^2}{2b} \frac{\ln (\phi_M/\phi_0)}{\ln (a/b\phi_0) - 1}. \qquad (2\text{-}26)$$

From step theory, for this model, the maximum power is

$$\phi_M = \alpha_0^2/2b . \qquad (2\text{-}27)$$

Consequently, the maximum power resulting from a ramp excursion is equal to that of a step excursion having an initial reciprocal period equal to the maximum reciprocal period during the ramp multiplied by a small correction factor which is the ratio of two logarithmic terms. This is an illustration of one of the ways in which a ramp excursion may be considered to be equivalent to a step-excursion [11]. From Eq. (2-6), the energy at peak power is

$$E_M = at_m/b . \qquad (2\text{-}28)$$

On physical grounds the time to reach maximum power is only slightly longer than the time to reach maximum reactivity, so that $t_M \simeq t_m$, whence

$$E_M \simeq \alpha_m/b \qquad (2\text{-}29)$$

and

$$E_f \sim 2\alpha_m/b . \qquad (2\text{-}30)$$

The principal features of the transient behavior of reacting systems subjected to a ramp addition of reactivity are given below. Many of these results, in particular the important role played by the $\sqrt{a}$, were indicated by Hansen [5], Brittan [12], and McCarthy, et al. [13].

1. The most important parameter in determining the minimum period attained is the reactivity insertion rate. It affects the transient behavior much more strongly than either the initial power or the reactivity coefficient since [Eq. (2-24)] the maximum α varies as the square root of the ramp rate multiplied by a logarithmic factor which involves the ratio of the power at maximum α and the initial power. The power at maximum α [Eq. (2-20)] in turn depends upon the assembly rate and the reactivity coefficient. However, because of the weak dependence of α on the factors involved in the logarithmic term and because of natural limitations on the upper limit to the ratio of powers, the logarithmic term may be treated as a constant over a wide range of conditions. Equation (2-24) can then be rearranged to show the assembly rate in dollars per second ($\dot{\$}$) required to introduce an effective superprompt step-reactivity insertion of magnitude \$:

$$\dot{\$} = \$^2\alpha_R/30 . \qquad (2\text{-}31)$$

The factor of 30 in the denominator is approximate and will vary for particular situations, but for most large bursts this relationship will be correct within a factor of 2. Aside from small variations of the constant this is a general relationship which applies equally well to fast or thermal reactors.

2. A particular feature of a ramp-induced excursion is that, up until the time of occurrence of the maximum in α, the burst shape behaves as if there were no shutdown effects. It has not been demonstrated but it is probably true that this conclusion can be used for a number of forms of the shutdown relationship in addition to the linear model [11]; thus Eq. (2-31) can be used to approximate many situations where the feedback relation is other than the linear one used here.

3. As shown in Eq. (2-21), the difference between peak and initial power is proportional to the ramp rate divided by the shutdown coefficient multiplied by a logarithmic factor involving the initial power. The factor is also slowly varying and the dominant term is the ramp rate.

4. The dependence of the peak power on α_m is the same as that for a step transient except for a correction factor which is the ratio of two slowly-varying logarithmic factors and which has a value of about 1 to 2. This implies that a ramp burst can be treated as a step burst provided that the α_m for the ramp is equal to the α_0 for the step [14].

5. A result of particular importance is that the dependence of the energy release on the prompt neutron lifetime is considerably different from the step case as shown by combining Eq. (2-29) and (2-31). The energy release is proportional to the square root of the assembly rate times the prompt neutron lifetime. Thus, the energy release in the case of a fixed assembly rate is lower in a fast reactor than it would be for the same assembly rate in a thermal reactor.

2.3 Behavior With a Threshold

Many situations can be visualized in which there is an abrupt change in the shutdown properties during the course of a transient. This may arise, for example, through vaporization in some part of the reactor. This may result, as it does in water-moderated plate-type reactors, in a significant change in the amount of energy required to produce a given reactivity change. Thus, there is an abrupt change in the value of the dynamic shutdown coefficient. There may also be such sudden and drastic changes in the internal pressures generated during a transient that the pressure generation may be regarded as a new shutdown mechanism that has been brought into play. In either case, this additional element of behavior may be introduced into the Linear Energy Model by postulating a threshold below which it does not appear.

Fuchs (see also [8,11]) studied the case in which the reactor behaves for a while after the introduction of α_0 as if there were no shutdown mechanisms operating, so that b is zero during this time and the power rise follows a simple exponential path. When a sufficient amount of energy E_0 has been accumulated, b becomes a positive constant and the shutdown effects begin to appear as in the simpler model. The basic Eq. (2-1) becomes:

$$\dot{\phi}/\phi = \alpha_0 - b(E - E_0) \qquad (2\text{-}32)$$

where

$$b = 0 \text{ for } E < E_0 , \quad b > 0 \text{ for } E \geq E_0 .$$

Thus, the threshold acts as an additional reactivity step of magnitude bE_0. For $\alpha_0 \ll bE_0$ the energy release is E_0. For $\alpha_0 \gg bE_0$ the energy release is determined mainly by α_0 and is the same as for the no threshold case.

The solutions to Eq. (2-32) can be shown to have the same form as the solutions for the no threshold case, with the important reservation that the approximation previously made [Eq. (2-11)] regarding the initial power may no longer be valid. It should be noted that the approximations affect both the periods and the magnitudes of excursions.

In particular, the $b\phi_0$ term in the definition of R is now $b(\alpha_0 E_0 + \phi_0)$ and may determine the value of R. If, however, $b(\alpha_0 E_0 + \phi_0) \ll \alpha_0^2$, then the form of the burst after the threshold is reached is unchanged, but the energy release at any time has increased by a constant E_0, which is the energy required to bring the system to the threshold.

If, however, E_0 is large in comparison with the energy release, its effect is more pronounced. Let the energy required to reach threshold be m times as great as the energy required to produce shutdown in the simple model ($=\alpha_0/b$). Substituting $E_0 = m\alpha_0/b = (m/2)\, E_{fs}$, where E_{fs} is the total energy release according to the simple model with no threshold, we obtain for the energy release:

$$E_f = E_{fs}\left[\frac{m + 1 + \sqrt{1 + 2m}}{2}\right]. \qquad (2\text{-}33)$$

Hence, the above statement that, for large threshold energies, the energy release is determined by threshold.

In a system in which the threshold is a phase change from liquid to vapor, the effect of increasing the amount of subcooling while maintaining α_0 constant, is to raise the instantaneous value of the power at the time the threshold is reached. Thus, the time required beyond threshold to produce the necessary energy to effect the shutdown can be made arbitrarily small by sufficient subcooling. Effectively, the time scale for events after passing the threshold is compressed and the slope of the power curve after the peak approaches minus infinity.

Behavior of a system in which shutdown is brought about by physical disassembly caused by high transient pressures is an important problem [5] which is treated in detail by McCarthy and Okrent in Sec. 4 of the chapter on Fast Reactor Kinetics. A qualitative picture of the way in which the burst behavior would be affected under such circumstances is provided by the following discussion.

Assume that the negative reactivity effects are linearly proportional to uniform displacements of some components of the reactor. Then the second time-derivative of the reactivity is proportional to the second time-derivative of the displacement and is thus proportional to the pressure. If it is now assumed that p goes as $(E-E_0)$ for $E > E_0$ and is zero for $E < E_0$, then Eq. (2-1) becomes

$$\frac{\dot{\phi}(t)}{\phi(t)} = \alpha_0 - b\int_0^t\int_0^t (E - E_0)\, dt\, dt' . \qquad (2\text{-}34)$$

Treating the pressure as a threshold follows from the consideration that high transient pressures are usually the result of a vaporization process which requires a fixed energy input prior to the phase change. To indicate the approximate behavior of energy and pressure for cases where $E \gg E_0$, assume further that the burst shape is exponential to the time of peak power. (This is an approximation often made [9,13,15,16].) Then $\phi_M/\alpha_0 = E_M$ and the integral in the above equation is ϕ_M/α_0^3. Thus, at peak power

$$\phi_M \sim \alpha_0^4/b , \qquad (2\text{-}35)$$

$$E_M \sim \alpha_0^3/b , \qquad (2\text{-}36)$$

$$p \sim \alpha_0^3/b . \qquad (2\text{-}37)$$

Noting that $\alpha_0 = \$\alpha_R$ and that $b \sim \alpha_R$; $E_M \sim \$^3\ \alpha_R^2$ which is the parameter $(\Delta k_0^3/\ell^2)$ that is important in the detailed analysis of Okrent. If, further, reactivity is inserted as a ramp, $\alpha_0 \sim \sqrt{a} \sim \sqrt{\$\ \alpha_R}$. Then $E_M \sim \$^{3/2}\ \alpha_R^{1/2}$ and the dependence of yield on lifetime is shown to be weak.

For large values of the threshold and small α_0, the yield is determined by the threshold as in the simpler case described above.

2.4 Summary and Discussion

Comparison of excursion reactor data with the characteristics of the Linear Model shows many points of agreement and disagreement. The early Godiva work as reported by Wimett [17], confirmed for a wide range of experimental conditions the essential features predicted by the Linear Model. However, the conditions under which this model can be expected to apply are limited and only the TRIGA, TREAT, and Godiva-type systems have exhibited Linear Model burst characteristics to a significant degree. Even for these systems, however, the full range of experiments encompasses conditions under which sizeable departures from the simple model occur.

Godiva-type systems show agreement with this model as to burst shape, peak power, and energy release for a considerable range of α_0 [17, 18]. Corrections must be made for time lags in thermal expansion and for reflector effects at high α_0. The reactivity coefficient and energy release can, for these systems, be related to temperature coefficients and temperature rises.

Similarly, the TRIGA reactor excursions can be described as Linear Model bursts, and the reactivity coefficient (see Sec. 4) can be calculated from detailed thermal considerations [19].

The quenching in the TREAT reactor is due to a single temperature-dependent reactivity coefficient which is nearly constant. Thus to the same approximation, the bursts are Linear Model bursts [20].

Three significant features with regard to the transient tests on these systems should be noted. First, for a wide range of α_0, they confirm the predictions of the model. Second, fast and thermal systems are included and different shutdown mechanisms are represented. Third, the reactivity coefficients can be calculated from detailed considerations without requiring use of adjustable constants.

All the excursion reactors display a peak power vs α_0 relationship that clearly differs, but not greatly, from the quadratic law predicted by the model [21].

In contrast with the above situation, there is no reason to expect the Linear Model to apply so well or to be as interpretable in terms of physical quantities when applied to the water-moderated and -cooled reactors because of their more complex quenching properties. For example, the SPERT reactors shut down mainly by changing the amount of water in the core, which can come about in a variety of ways, including steam formation [22]. Thus, there are many shutdown coefficients, not necessarily constant, and a threshold which likely has a time lag. However, the effective dynamic reactivity coefficient resulting from all quenching effects happens to be essentially a constant throughout a burst and over a large range of α_0. Consequently, the departure from Linear Model behavior is not great in several respects, such as the data for peak power vs α_0, which differ from the square law by about the same amount as for Godiva and TRIGA. The more striking differences are in the burst shape and the dependence of energy release on reactivity coefficient.

Early SPERT experiments [14] with ramp insertions of reactivity showed the effective equivalence of ramp and step bursts provided the ramp is characterized by its minimum period. The other important ramp characteristic of weak dependence on initial power is also shown by subsequent ramp studies [23].

Tests of the simple energy threshold model, even on a qualitative basis, have not been performed. This would appear to be difficult to do and there appears to be no case in which this type of behavior is presently important. This is not the case when pressure is assigned a governing role in the burst-quenching.

Thus, in summary, the Linear Model predicts qualitatively the step-burst and ramp-burst behavior for a wide variety of reactors, both fast and thermal, with a wide range of shutdown coefficient values, and with a variety of quenching mechanisms. For certain reactors and for a limited range of α_0, the agreement is quantitative. Experimental investigations have not been made of the important case in which shutdown behavior is governed by transient pressures.

3 NONLINEAR MODELS

The strong similarity between the excursion reactor data and the general predictions of the Linear Energy Model have encouraged studies of other forms of the feedback function. These have been undertaken with the view of characterizing more accurately the processes at work, with the intention of introducing into the mathematical solutions some desired properties or with the intention of determining the relative importance of various effects thought to be of possible importance in determining burst behavior. In a few cases these are refinements of an already accurate picture. In other cases the changes amount to completely new formulations. All have contributed to illustrating important general features of bursts.

3.1 The Zero-Delay Model

While the general characteristics were quite similar, the SPERT data significantly disagreed with the Linear Model on several counts: the burst shape after the power peak was sharper than the predicted symmetrical shape, the dependence of energy release on the variation of the effective shutdown coefficient was weaker than the predicted linear one, and, to a lesser degree, the peak power vs α function was somewhat weaker than the predicted quadratic. Generally, any time delays between release of fission energy and the appearance of that energy as a shutdown effect result in a burst-sharpening. Similarly, weakened dependence on the reactivity coefficient follows from increasing the energy input into a shutdown mechanism faster than the increase of fission energy. Both of these effects exist in SPERT plate-type reactors. The delays result from the time required for heat to be conducted from the point of generation in the fuel to the water, and to superheat water when phase-changes occur. The nonlinear coefficient of expansion of water and the marked density change accompanying steam formation are equivalent to a reactivity coefficient that increases with energy; thus this provides reasons for modifying shutdown effects in the Linear Model so that they are proportional to a higher power of the energy.

Forbes [24] examined a nonlinear model incorporating a delay time and a reactivity reduction proportional to the nth power of the energy developed. The prompt kinetics equation is then written as

$$\frac{\dot{\phi}(t)}{\phi(t)} = \alpha_0 - b'[E(t-\tau)]^n , \qquad (3\text{-}1)$$

where n is a positive adjustable constant and τ is the delay time between the energy release and its manifestation as a reactivity effect. The shutdown coefficient b' used here does not have the same dimensions as that for the Linear Model.

For zero delay time Eq. (3-1) then becomes

$$\frac{\dot{\phi}(t)}{\phi(t)} = \alpha_0 - b'[E(t)]^n . \qquad (3\text{-}2)$$

Note that for $n=1$, the Zero-Delay Model reduces to the Linear Energy Model. The solutions are

$$\phi(T) = \left[(n+1)\frac{\alpha_0}{b'}\right]^{1/n} \frac{n\alpha_0 e^{-n\alpha_0 T}}{\left(1+ne^{-n\alpha_0 T}\right)^{1+(1/n)}} , \qquad (3\text{-}3)$$

$$E(T) = \left[(n+1)\frac{\alpha_0}{b'}\right]^{1/n} \frac{1}{\left(1+ne^{-n\alpha_0 T}\right)^{1/n}} , \qquad (3\text{-}4)$$

where the time scale has been shifted so that T equals zero at the time of maximum power. The power and energy at the time of peak power are given by

$$\phi_M = \frac{n}{n+1}\left(\frac{\alpha_0^{n+1}}{b'}\right)^{1/n} \qquad (3\text{-}5)$$

$$E_M = \left(\frac{\alpha_0}{b'}\right)^{1/n} . \qquad (3\text{-}6)$$

The total burst energy is

$$E_f = (\alpha_0/b')^{1/n}(n+1)^{1/n} ,$$

$$= E_M (n+1)^{1/n} . \qquad (3\text{-}7)$$

3.2 <u>The Long-Delay Model</u>

For this model the delay time τ is assumed to be large compared to the reactor period, $1/\alpha_0$. Equation (3-1) may then be written in the form

$$\frac{\dot{\phi}(t)}{\phi(t)} = \alpha_0 - b'\left[\int_0^{t-\tau} \phi_0 e^{\alpha_0 t}\, dt\right]^n \qquad (3\text{-}8)$$

$$= \alpha_0 - \frac{b'\phi_0^n}{\alpha_0^n}\left(e^{\alpha_0(t-\tau)} - 1\right)^n . \qquad (3\text{-}9)$$

For very small ϕ_0, $e^{\alpha_0 t} \gg 1$ in the region of interest and Eq. (3-9) can be simplified to

$$\frac{\dot{\phi}(t)}{\phi(t)} = \alpha_0 - \frac{b'\phi_0^n}{\alpha_0^n} e^{n\alpha_0(t-\tau)} . \qquad (3\text{-}10)$$

This form is valid only for values of the argument $(t-\tau)$ such that the exponential form for E(t) is appropriate; that is, there shall be no apparent shutdown effect at the time $(t-\tau)$. Then, as shown in reference [24],

$$\phi(T) = \frac{\alpha_0^{(n+1)/n}}{(b')^{1/n}} e^{\alpha_0(T+\tau)} \exp\left[-\frac{1}{n} e^{n\alpha_0 T}\right] . \qquad (3\text{-}11)$$

At the time of the power peak, $T = 0$ and

$$\phi_M = \frac{\alpha_0^{(n+1)/n}}{(b')^{1/n}} \exp\left(\alpha_0\tau - \frac{1}{n}\right) . \qquad (3\text{-}12)$$

The expression for the energy is more complicated [24,25] but by making use of a function q involving the incomplete gamma function, Forbes was able to express the values at time of peak power and the end of the burst as follows:

$$E_M = \left(\frac{\alpha_0}{b'}\right)^{1/n} \exp\left(\alpha_0\tau - \frac{1}{n}\right) q(n, 0) , \qquad (3\text{-}13)$$

$$E_f = \left(\frac{\alpha_0}{b'}\right)^{1/n} \exp\left(\alpha_0\tau - \frac{1}{n}\right) q(n, \infty) , \qquad (3\text{-}14)$$

$$= E_M\, q(n, \infty) . \qquad (3\text{-}15)$$

The values of q (n,0) and q (n, ∞) are functions of n only and independent of α_0. The function can be integrated numerically for some values of n. As illustrated in Table 3-1, the values of q do not vary greatly over the range $1 \le n \le \infty$. Therefore,

TABLE 3-1

q (n, 0) and q (n, ∞) vs n

n	1	2	∞
q (n, 0)	1.72	1.42	1
q (n, ∞)	2.72	2.07	1

Eqs. (3-13) and (3-14) explicitly express the dependence of energy release on α_0, b', n, and τ, but the dependence on n through q (n) is seen to be weak.

Comparison of E^n models with a few SPERT bursts is shown in Figs. 3-1 to 3-4. In Fig. 3-1 it is seen that, up to the time of peak power, the experimental burst is matched rather well by the Threshold Model with n = 0.50, but the post-peak agreement is poor. In Fig. 3-2, the Zero-Delay Model with n = 2 also fits the experimental burst very well up to the time of peak power, but not so well in the post-peak region. For n > 3 the agreement with post-peak behavior is better, but is explicitly worse for the time before the peak. Also, the Zero-Delay case with n = 1, the Linear Energy Model, which predicts many of the general features of burst behavior, does not represent the power burst shape at all well in the short-period region. For the same experimental burst, the best over-all agreement is obtained with the Long-Delay Model with n = 1.5, as shown in Fig. 3-3. Here the agreement is good, not only for the rising part of the curve, but also on the falling side. Figure 3-4 shows exceptional agreement for the Long-Delay Model with n = 2 for a 7.4 msec transient.

Essentially the same information may be obtained by using the reactivity behavior during a burst as the means of comparison between model and experiment. Often this is a more convenient way. Forbes in reference [24] shows such a comparison for the tests of Figs. 3-1 through 3-3.

3.3 Positive Reactivity Coefficients

More general polynominal forms of the E^n Zero-Delay Model can be used [20,24,25,26]. A particular form which arises when initially positive reactivity coefficients are considered will be discussed in this section [27].

The existence of a strong negative coefficient of reactivity in a reactor has been used as a meas-

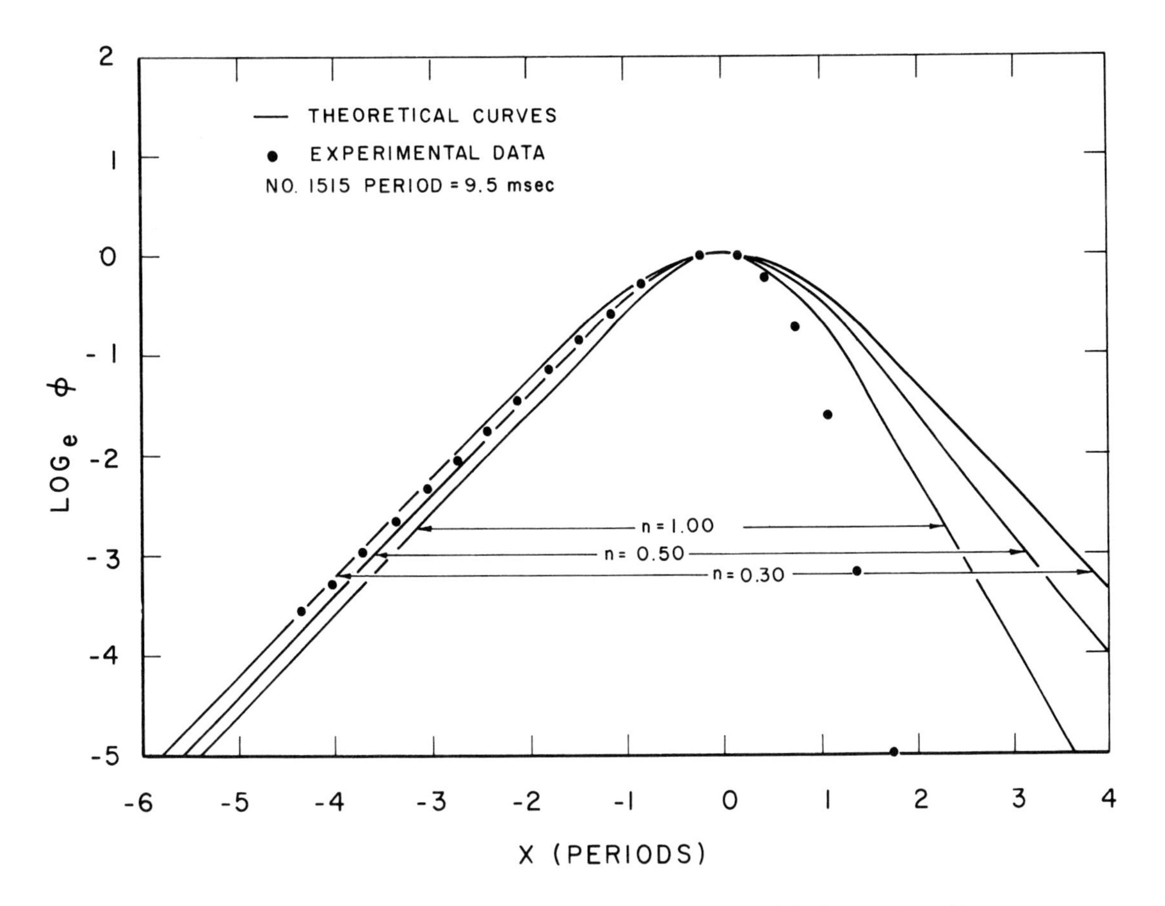

FIG. 3-1 Comparison of dimensionless burst shapes for the threshold model and an experimental burst shape with an initial period of 9.5 msec, as a function of burst parameter n.

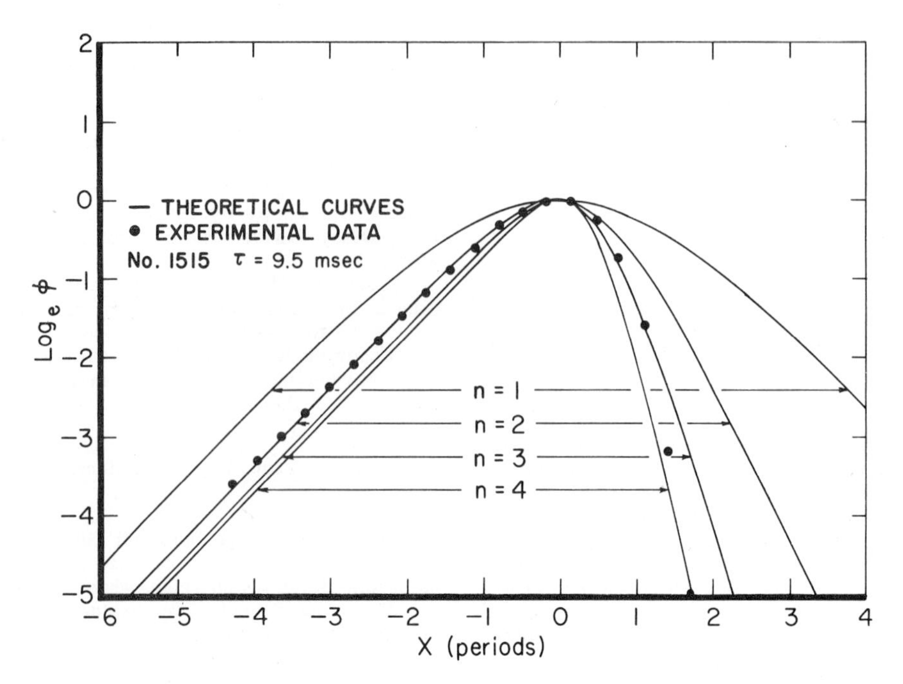

FIG. 3-2 Comparison of dimensionless burst shapes for the zero-delay model and an experimental burst shape with an initial period of 9.5 msec, as a function of burst parameter n.

ure of its safety. However, this may be a somewhat self-defeating requirement for several reasons. For one, in order to override the resultant reactivity defect, a correspondingly large investment in reactivity is required. Clearly, the magnitude of the available reactivity is decreased if the coefficient is permitted to be zero, or even positive, at operating levels below the design operating point. In order that reactor power bursts be self-limiting for the latter case, the coefficient must change from the assumed positive initial value to a negative final value during the burst. That is, the coefficient

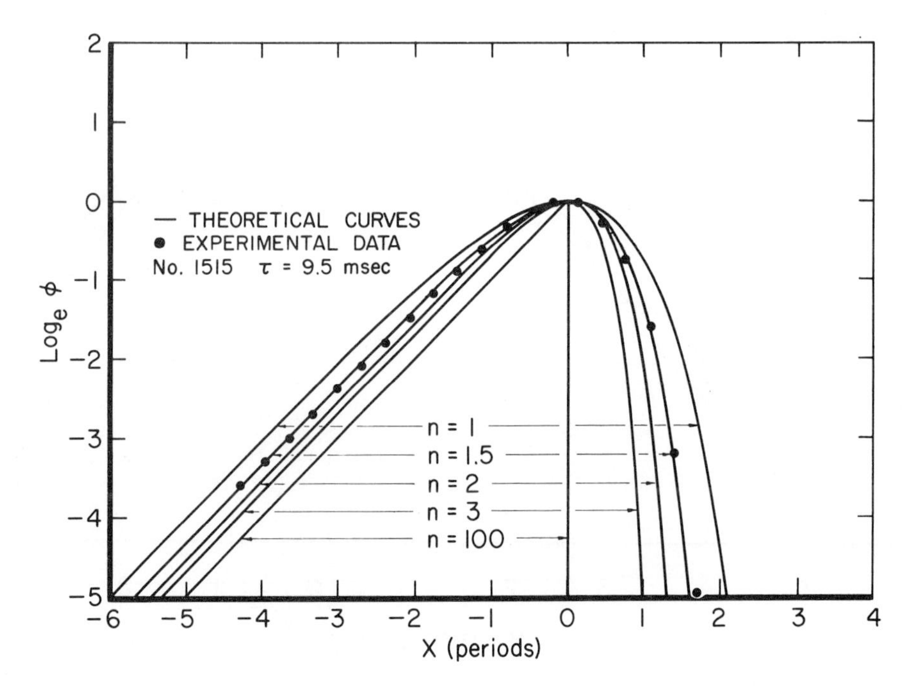

FIG. 3-3 Comparison of dimensionless burst shapes for the long-delay model and an experimental burst shape with an initial period of 9.5 msec, as a function of burst parameter n.

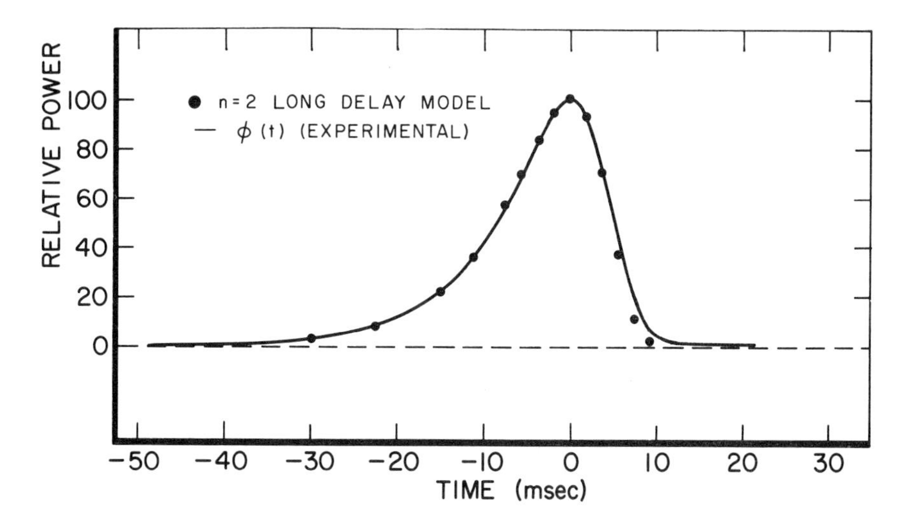

FIG. 3-4 Comparison of power burst shapes for an experimental burst with an initial period of 7.4 msec and the long-delay model with burst parameter n = 2.

is a function of the power history. The effect of this type of reactivity behavior on self-limiting bursts will be illustrated below for a simple case.

For simplicity, it will be assumed that the coefficient varies linearly with the energy release. That is,

$$b(E) = \frac{\partial \alpha}{\partial E} = b_0 - b_1 E \,, \qquad (3\text{-}16)$$

where

$\alpha(t)$ = reciprocal reactor period,

$E(t)$ = total energy release up to time t, with zero defined by $E(0) = 0$,

$b(E)$ = energy coefficient of reactivity, with reactivity expressed in units of reciprocal period,

$b_0 = \left(\frac{\partial \alpha}{\partial E}\right)_{E=0}$ is the initial value of b,

$b_1 = -\frac{\partial b}{\partial E} = -\frac{\partial^2 \alpha}{\partial E^2}$, which is assumed constant, and

$b_0, b_1 > 0.$

With these definitions, it is required implicitly that the energy in the shutdown mechanism be a constant fraction of the total energy release, E. It is convenient to define a quantity

$$E_c \equiv b_0/b_1 \,. \qquad (3\text{-}17)$$

The significance of E_c is that $b(E_c) = 0$, that is, an amount of energy E_c must be released to reduce to positive coefficient to zero. Also,

$$b(E) = b_0 \left(1 - \frac{E}{E_c}\right). \qquad (3\text{-}18)$$

In the present case, $\alpha(t)$ is given by

$$\alpha(t) = \alpha_0 + \int_0^t \frac{\partial \alpha}{\partial E}\frac{dE}{dt}\,dt = \alpha_0 + \int_{E(0)}^{E(t)} \frac{\partial \alpha}{\partial E}\,dE \,. \qquad (3\text{-}19)$$

Substitution of the specific form Eq. (3-16) for $\partial\alpha/\partial E$ and integration results in

$$\alpha(t) = \alpha_0 + b_0 E(t) - (1/2) b_1 E^2(t) \,, \qquad (3\text{-}20)$$

where the initial condition is taken as $E(0) = 0$.

Then the prompt approximation for the kinetic behavior, Eq. (2-1), becomes

$$\frac{\dot{\phi}(t)}{\phi(t)} = \alpha_0 + b_0 E(t) - \frac{b_1 E^2(t)}{2} \,. \qquad (3\text{-}21)$$

Since $\dot{E}(t) = \phi(t)$, this equation may be integrated to give

$$\phi(t) - \phi(0) = \alpha_0 E(t) + (1/2) b_0 E^2(t) \left[1 - \frac{E(t)}{3E_c}\right]. \qquad (3\text{-}22)$$

Two features of this kinetic model should be kept in mind.

1. It does not provide for any escape of energy from the shutdown mechanism; therefore, the only possible steady-state power, in the absence of external manipulation of reactivity, is zero.
2. A power transient will result if $\phi_0 \neq 0$.

The consequence of these two characteristics is that this model best describes those transients resulting from the initial condition of small ϕ_0. Assuming such an initial condition, and using the subscripts m, M, and f to denote the values of functions at the time of maximum reciprocal period ($\dot{\alpha}_m = 0$), the time of maximum power ($\dot{\phi}_M = 0$), and the time after the peak when the power has returned to its initial value ($\phi_f = \phi_0$), respectively, Eq. (3-19) and (3-20) yield the results given below. Since $\phi_f = \phi_0$ is small, values indicated by the subscript f may be regarded as "final" values.

Energy

$$E(0) = 0 ,$$

$$E_m = E_c = b_0/b_1 , \quad (3\text{-}23)$$

$$E_M = E_c\left[1 + \sqrt{1 + \frac{2\alpha_0}{b_0E_c}}\right] > 2E_c , \quad (3\text{-}24)$$

$$E_f = \frac{3}{2} E_c\left[1 + \sqrt{1 + \frac{8\alpha_0}{3b_0E_c}}\right] > 3E_c . \quad (3\text{-}25)$$

Power

$\phi_0 \ll \phi_m$ and ϕ_M (initial condition),

$$\phi_m = E_c\left[\alpha_0 + \frac{b_0E_c}{3}\right] , \quad (3\text{-}26)$$

$$\phi_M = E_c\left[\alpha_0 + \frac{b_0E_c}{3}\right] + \frac{b_1}{3}\left[\frac{2\alpha_0}{b_1} + E_c^2\right]^{3/2} , \quad (3\text{-}27)$$

$$= \phi_m\left\{1 + \left[1 + \frac{2\alpha_0}{b_1E_c^2}\right]^{3/2}\left[1 + 3\frac{\alpha_0}{b_1E_c^2}\right]^{-1}\right\} , \quad (3\text{-}28)$$

$$\phi_M > 2\phi_m ,$$

$$\phi_f = \phi_0 \text{ (by definition)}.$$

Shutdown Coefficient

$$b(0) \equiv b_0 ,$$

$$b_m = 0 ;$$

when $\alpha = \alpha_0$:

$$b = b_0 \text{ for } E = 0 ,$$

$$= -b_0 \text{ for } E = 2E_c ,$$

$$b_M = -b_0\sqrt{1 + \frac{2\alpha_0}{b_0E_c}} < -b_0 , \quad (3\text{-}29)$$

$$b_f = -b_0\left[\frac{1}{2} + \frac{3}{2}\sqrt{1 + \frac{8\alpha_0}{3b_0E_c}}\right] < -2b_0 . \quad (3\text{-}30)$$

Reciprocal Period

$$\alpha = \alpha_0 \text{ for } E = 0 \text{ and } E = 2E_c ,$$

$$\alpha_m = \alpha_0 + \frac{b_0E_c}{2} , \quad (3\text{-}31)$$

$$\alpha_M \equiv 0 \text{ (by definition)} ,$$

$$\alpha_f = -2\alpha_0 - \frac{3}{4} b_0E_c \{1 + [1 + (8\alpha_0/3b_0E_c)]^{1/2}\} . \quad (3\text{-}32)$$

In Fig. 3-5, α, ϕ and b are shown as functions of the energy release E. The two lines b(E) = $-b_0$ and b(E) for $b_0 = 0$ represent, respectively, simple linear and square-law reactivity feedback, as contrasted with the parabolic form used in deriving the preceding expression. Equation (3-25) shows that the minimum energy release, which could result from a power perturbation as well as from a reactivity addition, is greater than three times the energy required to reduce the reactivity coefficient to zero. If the energy required to cause melting in the reactor is taken as a safety limit, clearly E_c must be less than one-third of this limit.

This treatment, while strictly applicable only to transient behavior, may be correlated with steady-state operating conditions by including some means for escape of energy from the shutdown mechanism. That is, operation at finite power results in energy input to the shutdown mechanism. Unless some means for escape of energy is provided, as steady power cannot be maintained.

If it is assumed that the rate of escape of energy from the shutdown mechanism is proportional to energy present in this mechanism, then

$$\dot{E}' = K\phi - \lambda E' \quad (3\text{-}33)$$

where E' is the energy in shutdown mechanism, K the fraction of total power serving as input for E', and λ the relaxation time for escape of E'.

A constant power level requires $\alpha \equiv 0$, and $\dot{\alpha} = b(E') \cdot \dot{E}' \equiv 0$. The only way in which the second condition can be satisfied, unless $b(E')$ vanishes identically, is $\dot{E}' \equiv 0$, which leads to either $\phi \equiv 0$, $E' \equiv 0$, or

$$\phi_s = \frac{\lambda}{K} E'_s , \quad (3\text{-}34)$$

where the subscript s denotes steady-state values. Thus, steady-state power is directly proportional to energy present in the shutdown mechanism, and inversely proportional to the product of escape time and fraction of total energy output entering the shutdown mechanism. Equation (3-34) is of

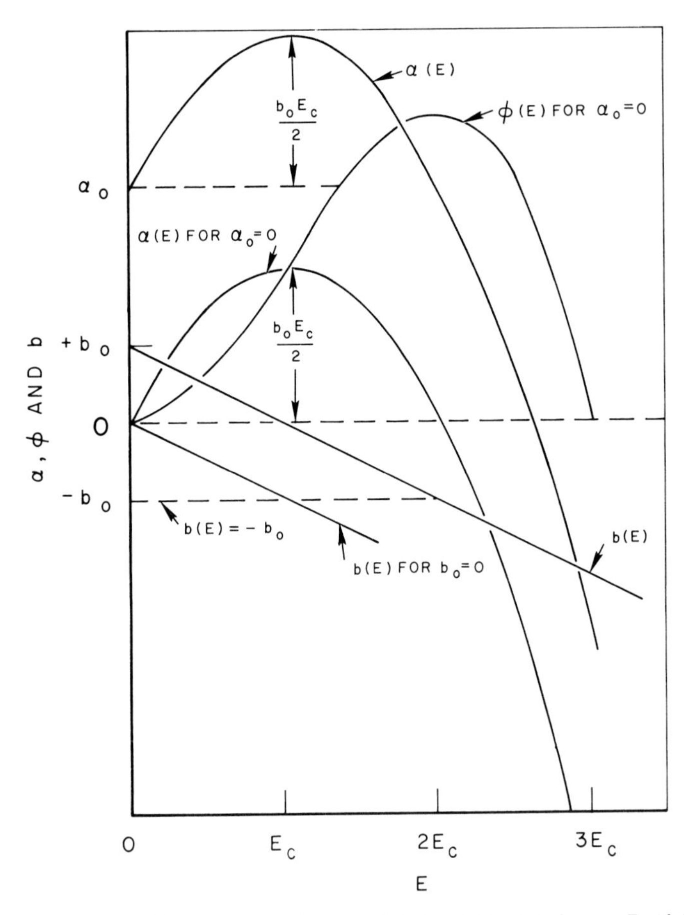

FIG. 3-5 Reciprocal period α, power ϕ, and void coefficient b as functions of energy E with initial void coefficient b_0 a parameter.

significance because E_s', the steady-state energy content of the core, as well as the rate of energy escape, is determined by design parameters of the reactor, such as maximum permissible temperature. If λ, K, and E_s' can be estimated, then ϕ_s may be estimated from Eq. (3-34). Note that Eq. (3-34) is in no way dependent on the form of $b(E')$, but merely says that at only one power level does input to shutdown balance output for a given core energy content.

The form of $b(E')$ is important in that it serves to fix the reactivity which must be available for insertion during startup procedures. Thus, if the design power of a reactor is ϕ_s, the amount of reactivity

$$\alpha_d = \left| \int_0^{E_s'} b(E')\, dE' \right| \qquad (3\text{-}35)$$

is necessary to bring the reactor from $\phi = 0$ to $\phi = \phi_s$. As an example, consider a reactor having temperature coefficient - b and heat capacity H. Then $\phi_s = \lambda H \Delta T / K$ and $\alpha_d = |-bH\Delta T| = bH\Delta T$, where ΔT = core temperature rise.*

The quantity α_d has the further significance that it is a measure of the reactivity which may be available for the initiation of an accident. The worst possible accident in a reactor will usually occur if all the available excess reactivity is inserted suddenly when operating at low power. The resulting power excursion will, in general, have $\alpha_0 >> \lambda$, and therefore escape of energy from the shutdown mechanism during the transient is negligible.

The implication of various forms for $b(E')$ with respect to reactor accidents will now be investigated under the assumptions that a reactor operating at zero power is perturbed by the sudden insertion, at time t = 0, of the amount of reactivity α_d and that all the energy resulting from the excursion remains in the reactor during the burst.

The basis of the comparison of various forms of $b(E')$ used here is that the quantity E_s' will be kept the same for all reactor types; i.e., the question being investigated is "Given a number of reactors, all having the same energy in the core at steady power, what is the effect of changing the form of $b(E')$ upon the consequences of the accident which will take place when all the available excess

*The symbol T is used both to indicate temperature and time measured from peak power, see Eq. (3-3) etc. However, because of the manner of use, there is no ambiguity.

reactivity is inserted at once at lower power?"

For simplicity, K for each system will be taken as unity. The physical basis for this assumption, when considering an essentially adiabatic transient, is that all of the energy generated will go into heating the reactor core, and various phenomena resulting from this heating will bring about shutdown.

The three forms of $b(E)$ considered here and the corresponding forms of α are:

$$b_A(E) = -b_0 \qquad \alpha_A = \alpha_0 - b_0E \tag{3-36}$$

$$b_B(E) = -b_1E \qquad \alpha_B = \alpha_0 - \frac{1}{2}b_1E^2 \tag{3-37}$$

$$b_C(E) = b_0 - b_1E \qquad \alpha_C = \alpha_0 + b_0E - \frac{1}{2}b_1E^2 \tag{3-38}$$

and it will be assumed that the quantities b_0 and b_1 have the same magnitude for each reactor. Initial conditions are $E(0) = 0$, $\phi(0) = 0$. In each case α_0 will be taken as α_d. The comparison is best illustrated in tabular form, see Table 3-2.

The normalization used here has the interesting property that, in each of the three cases considered, energy release up to the time of maximum power has the value E_s. Therefore, the total energy release during the transient is largely determined by the form of $b(E)$ for $E > E_s$. In other words, if energy up to t_M is the same in all cores, then that form of $b(E)$ resulting in the most rapid shutdown thereafter will give the least total energy release.

Note that if $E_s = 2E_c$ in case C, then $\alpha_d = 0$. Such a reactor is unstable at zero power, and thus a power transient can take place with no external manipulation of reactivity. The implication of these results is that if reactor C is perturbed at zero power, it will eventually come to steady operation at the power ϕ_s as given in Eq. (3-34). This behavior is in marked contrast to cases A and B, for which α_d is always greater than zero and which will always return to zero power in the absence of some external reactivity contribution.

If the autocatalytic properties of the reactor considered in case C result in a power excursion, it develops that the total energy release is only $(3/2)E_s$; thus the conclusion anticipated in the first paragraph of this section, that the "worst accident" for a reactor having an initially positive reactivity coefficient can be somewhat less serious than an accident in a reactor with a uniformly negative coefficient.

The point here is that, while it is true that an initially positive reactivity coefficient results in a shorter period during the first part of a power burst than a uniformly negative coefficient under the same initial conditions, the reduction in available reactivity required for the normal operation of a reactor with $b(0) > 0$ can result in a gain in reactor safety for the type of accident postulated here.

E_f represents a lower limit to the total energy release during a transient; therefore, it is desirable that a reactor require considerably more energy than E_f to exceed any imposed safety condition, such as core melting. It is noteworthy that for the three cases cited the smallest value of the final energy E_f is found in the initially positive coefficient case, C. Since the functional form of $b(E)$ determines E_f, this quantity may be further reduced by adopting a suitable form for $b(E)$.

The comparison of these three cases is also striking in terms of the time scale of the burst. In the positive coefficient case the maximum α is $\alpha_0 + (1/2)b_0E_c$. With $\alpha(0) = 0$, the maximum α is 1/4 of the maximum value for the other two cases. This longer period usually represents a gain from a practical safety point of view.

The results of these studies do not, of course, suggest that positive reactivity coefficients are to be taken lightly, but rather that the over-all situation must be examined carefully to arrive at the best adjustment of system requirements.

No experimental program of investigation of systems with initially positive coefficients has been carried out.

3.4 Reactivity Coefficients which Decrease with Energy

In the foregoing discussion of the E^n models it was tacitly assumed that $n \geq 1$. These models all

TABLE 3-2

Comparison of Reactors with Different Shutdown Functions

Parameter	Reactor		
	A	B	C
$b(E)$	$-b_0$	$-b_1E$	$b_0 - b_1E$
$\alpha(t)$	$\alpha_A(0) - b_0E$	$\alpha_B(0) - \frac{1}{2}b_1E^2$	$\alpha_C(0) + b_0E - \frac{1}{2}b_1E^2$
α_d	b_0E_s	$\frac{1}{2}b_1E_s^2$	$\frac{1}{2}b_1E_s^2 - b_0E_s$
E_f	$2E_s$	$\sqrt{3}E_s$	$\frac{3b_0}{2b_1}\left[1 + \frac{2}{3}\left(\frac{b_1}{b_0}E_s - \frac{3}{2}\right)^{\frac{1}{2}}\left(\frac{b_1}{b_0}E_s\right) - \frac{1}{2}\right]$

have the characteristic that as the energy release increases, the shutdown effect will increase to the same or greater extent (depending upon the value of the exponent of energy), that is, will increase without bound. There have been proposed models in which the shutdown effect increases to a lesser extent than the energy release into the system, and again some in which the effect decreases as the energy release increases. These have been developed mainly to handle a specific shutdown mechanism, such as the Doppler effect, or to account for temperature dependence of the effective heat capacity.

Proposed Doppler models have suggested that the Doppler coefficient of reactivity varies with absolute temperature, T, as $T^{-1/2}$ [28], T^{-1} [29], or $T^{-3/2}$[30]. (See the chapter on the Reactor Core). Qualitatively, for these coefficients the corresponding negative reactivity for large energy release varies as $E^{1/2}$, $\ln E$, and $E_0^{-1/2} - E^{-1/2}$ (E_0 is the initial energy content of the system). Thus, the energy release to time of peak power goes as α_0^2, e^{α_0} and $E_0[1-(\alpha_0/b)\sqrt{E_0}]^{-2}$ respectively. Therefore, there is exhibited in these models a strong dependence of the energy release on α_0. Since the shutdown coefficient for the first model goes as $E^{-1/2}$, there is a decreasing amount of reactivity compensation developed as the energy release increases. This effect is accentuated in the second model. In the third model the shutdown effect is such that above a certain energy release, there is no compensation available. That is, there is an upper limit to the compensation possible. (This is also a characteristic of a proposed poison shutdown model [31, 32].)

Of these models, the first one has been used to describe the behavior of an excursion reactor. This is for a case in which the shutdown is due to the Doppler mechanism [33]. This model will also describe the burst characteristics of a system which has a variable heat capacity [34]. For either case the reactor kinetics equation takes the form

$$\frac{\dot{\phi}}{\phi} = \alpha_0 - B[(1 + AE)^{1/2} - 1], \qquad (3\text{-}39)$$

where A and B are constants of the system. The interpretation to be put on these constants depends, of course, upon the particular effect to which the model is to be applied.

The differential equation above can then be solved [33] to yield the following parameters:

$$\text{the shutdown coefficient: } b = \frac{-1/2\, AB}{\sqrt{1 + AE}}; \qquad (3\text{-}40)$$

$$E(T_{max}) = \frac{\alpha_0(\alpha_0 + 2B)}{AB^2}, \qquad (3\text{-}41)$$

$$E_{final} = \frac{1}{A}\left[4\left(\sqrt{\frac{9 + (3\alpha_0/B)}{1 + (3\alpha_0/B)}} - 1\right)^{-2} - 1\right], \qquad (3\text{-}42)$$

$$\phi(T_{max}) = \frac{\alpha_0^2(\alpha_0 + 3B)}{3AB^2}. \qquad (3\text{-}43)$$

When $\alpha_0 << B$, this model will reduce to the linear energy model. The differences between this and the Linear Model are a lowering of the peak power and a broadening of the burst width by a few percent [34,35].

3.5 Burst Shape Properties

As indicated earlier, it is often useful in the consideration of burst behavior to utilize the burst shape, particularly, if it can be represented in a simple, closed form amenable to manipulations in hand calculations. In this section, two such simple approximate forms will be described [25, 36, 37]. They will be used to illustrate some of the systematics of bursts not previously discussed. A parameter to which particular attention will be paid is the negative reactivity generated in the burst. This has frequently been called the "compensated reactivity", k_c.

Values of k_c (peak) calculated for a variety of postulated burst shapes have been compared with the values obtained by machine calculations using experimental power data and the neutron kinetics equations. From these comparisons several conclusions are evident. First, for approximate analyses the values obtained for k_c (peak) using approximations for the burst shape are often sufficiently accurate to justify the use of this more convenient means in preference to laborious machine calculations. Second, in some cases analytical forms for the burst shape can be fit to the experimental burst shapes sufficiently well to be useful in calculations of other quantities of interest during a transient, such as temperature distributions, heat transfer, etc. See for example, the analysis of the SL-1 accident [38]. Thus, this approach may provide clues as to the nature of the feedback function. However, burst shape analysis alone does not provide the required information as to the form of the equation coupling power history with reactivity.

3.6 The Clipped Exponential Burst Shape Approximation

The Clipped Exponential Burst Shape Approximation may be used to approximate a power burst up to the time of peak power. On the diagram shown in Fig. 3-6 is plotted the logarithm of the reactor power as a function of time T, with the time of peak power taken as zero. An actual round-topped power burst is represented by the broken-line with peak power $\phi(0)$. The initial exponential rise of the actual burst is shown extended to the time of peak power where it has the value $\phi_x(0)$. The solid curve made up of the extension of the initial exponential rise and the constant portion of magnitude $\phi(0)$ is the clipped exponential approximation. At the break in the curve, the first derivative of the power changes in a discontinuous fashion to zero. The duration of the clipped portion is seen to be $(\ln r)/\alpha_0$, where r is defined as follows:

$$r \equiv \frac{\phi_x(0)}{\phi(0)} . \tag{3-44}$$

Broad, flat-topped bursts will result in large r and long-duration clipped portions, whereas sharp bursts will have small r and short-duration clipped portions. Thus r can be used as a burst shape parameter. The limiting form for very short bursts is simply an exponential with a vertical back side. This will be called an "exponential sawtooth".

Other mathematical relationships evident from Fig. 3-6 are as follows:

$$\phi(T) = \phi(0) \exp\left[\alpha_0\left(T + \frac{\ln r}{\alpha_0}\right)\right]; \quad T \leq -\frac{\ln r}{\alpha_0} \tag{3-45}$$

$$\phi(T) = \phi(0); \quad -\frac{\ln r}{\alpha_0} \leq T \leq 0 \tag{3-46}$$

$$E(T) = \frac{\phi(0)}{\alpha} \exp\left[\alpha_0\left(T + \frac{\ln r}{\alpha_0}\right)\right]; \quad T \leq -\frac{\ln r}{\alpha_0} \tag{3-47}$$

$$E(T) = \frac{\phi(0)}{\alpha_0} + \phi(0)(T + \ln r/\alpha_0); \quad -\frac{\ln r}{\alpha_0} \leq T \leq 0 \tag{3-48}$$

$$E(0) = \frac{\phi(0)}{\alpha_0}(1 + \ln r) . \tag{3-49}$$

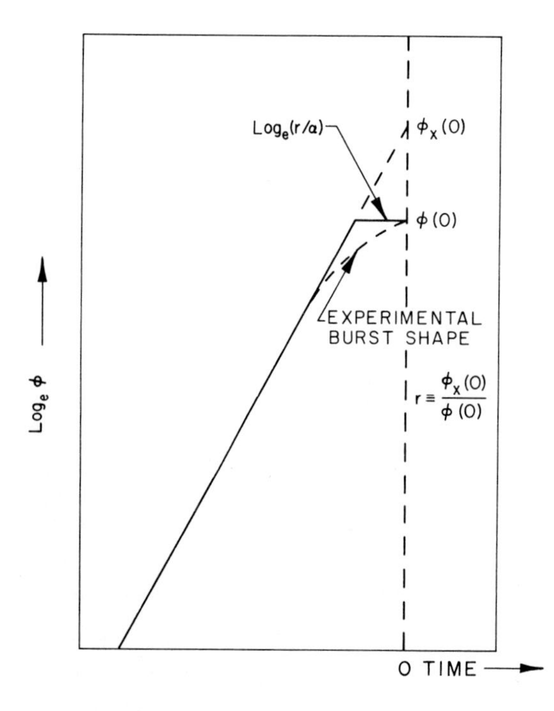

FIG. 3-6 Clipped exponential burst shape approximation.

Note that in the above T = 0 is the time of peak power.

The Clipped Exponential burst shape will, of course, never describe exactly an experimental burst shape. However, if it is matched to an experimental burst as shown in Fig. 3-6, it will always be an overestimate of the energy to the time of peak power, and since the reactivity compensated at the time of peak power is dependent on burst shape, the Clipped Exponential will always provide an upper limit for the value of k_c (peak).

The Clipped Exponential approximation for the burst shape may be combined with the kinetics equation, including delayed neutrons, to yield the following expression for the compensated reactivity:

$$k_c(\text{peak})(\$) = \frac{\alpha_0}{\alpha_R} + \sum_{i=1}^{6} \frac{a_i X_i \left[1 - e^{(-\ln r/\alpha_0)}\right]}{(1 + X_i)} , \tag{3-50}$$

where $X_i \equiv \alpha_0/\lambda_i$ and $a_i \equiv \beta_i/\beta$. This equation is parametric in r and may be used to demonstrate that the burst shape can affect the compensated reactivity function.

A plot of k_c (peak) vs α_0 for various values of r is shown in Fig. 3-7. The upper curve is the $\rho(T_s)$ (i.e., the total excess reactivity introduced at the start of the transient) or inhour curve and represents the upper limit of k_c (peak) which would occur for very broad-topped bursts in which self-shutdown is so gradual that delayed neutrons are essentially in continuous equilibrium with the power level. The lower curve is the short-period asymptote α_0/α_R. While the general shape of the k_c (peak) vs α_0 curves illustrates that there is a maximum and a minimum in the transition region, there is a gradual disappearance of the dip in the regions of very sharp bursts (ln r → o) and very broad burst (ln r ≃ 100). Also, in the limit of ln r = 0 (r = 1), k_c (peak) becomes just the short-period asymptote α/α_R.

3.7 The Two-Term Burst Approximation

A better approximation which represents the entire burst can be made using a series of exponentials as Corben and Forbes have done. The two-term model of Forbes [37] is described below. In this it is assumed that the power burst can be represented by

$$\phi(T) = \phi(0)\left[a \exp(\alpha_0 T) - b \exp(c\alpha_0 T)\right] , \tag{3-51}$$

where a, b, and c are positive constants to be determined such that $\phi(T)$ satisfies three conditions: (1) $\phi(T)$ must coincide with the initial asymptotic power rise; (2) $\phi(T)$ must have the proper value at the peak (T = 0); and (3) $\phi(T)$ must have a zero time derivative at the power peak. As shown in Fig. (3-8), the burst shape in the Two-Term Approximation will be a round-topped burst and r can be defined as in Eq. (3-44).

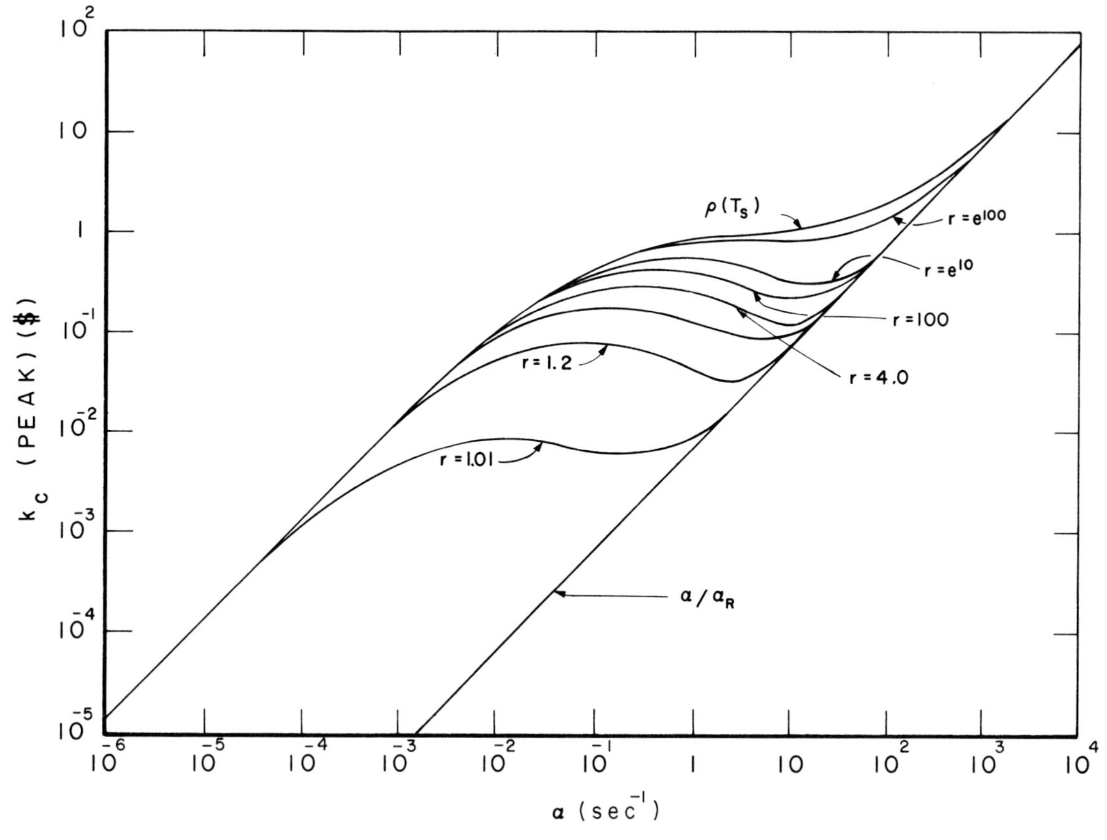

FIG. 3-7 Calculated compensated reactivity at the time of peak power, k_c (peak), vs reciprocal period α_0 as a function of the burst parameter r for the clipped exponential approximation. (Delayed neutron data are taken from LA 2118, July, 1957, by Keepin, Wimett, Zeigler.)

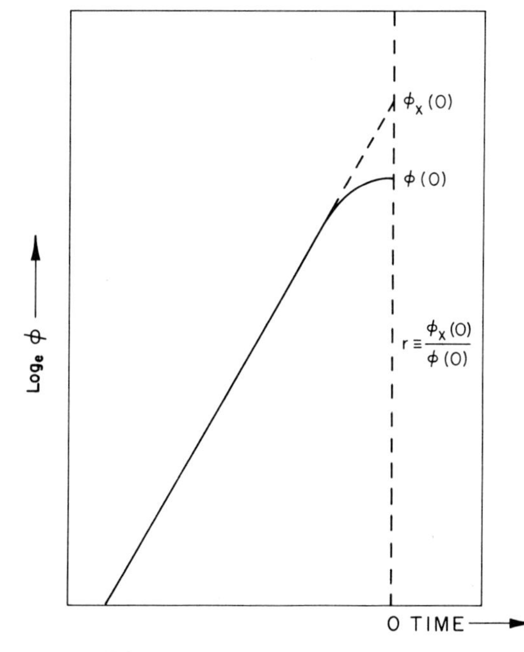

FIG. 3-8 Two-term burst shape approximation.

Applying condition (1) to Eq. (3-51) it follows that for large negative values of T the second term in Eq. (3-51) may be assumed negligible, that is, for $T \ll 0$ the power is rising exponentially as $\phi(0)a \exp(\alpha_0 T)$. Also, $\phi(T) = \phi(0)r \exp(\alpha_0 T)$, from which it follows that a = r.

From the second condition, at the time of peak power T = 0 and $\phi(T) = \phi(0)$, thus $\phi(0) = \phi(0)(r-b)$ from which b = r - 1.

By applying the third condition to Eq. (3-51) and setting $d\phi/dt$ equal to zero:

$$\frac{d\phi}{dt} = 0 = \phi(0)\alpha_0 \left[r e^{\alpha_0 T} - (r-1)c e^{c\alpha_0 T}\right] \qquad (3\text{-}52)$$

from which c = r/(r-1).

Substituting the above relations (a = r, b = r -1, c = r/(r-1)) into Eq. (3-51) yields:

$$\phi(T) = \phi(0)\left[r \exp(\alpha_0 T) - (r-1) \exp\left(\frac{r}{r-1}\alpha_0 T\right)\right]. \qquad (3\text{-}53)$$

The Two-Term burst shape is fairly restricted in its application for matching experimental burst shapes. As an illustration, the limiting burst shapes possible with the Two-Term model are: for r = 1 an exponential sawtooth, and for $r \to \infty$, a somewhat broader burst shape. These two limiting burst shapes are shown in Fig. 3-9 where they are compared with experimental burst shapes. The experimental bursts are from step-transient power excursions in the SPERT-III reactor under the environmental conditions of atmospheric pressure, room temperature, and no forced coolant flow. It is evident from Fig. 3-9 that the shape of the power burst with $r \to \infty$ is still relatively sharp as compared with the experimental bursts which are possible for subprompt critical bursts. As a result, integrating Eq. (3-53) and assuming the initial power to be arbitrarily small gives the result that the maximum energy to the time of peak power for the Two-Term Model is $2\,\phi(0)/\alpha_0$ (for $r \to \infty$) and that the minimum energy is $\phi(0)/\alpha_0$ (for $r = 1$). Thus, the maximum energy is restricted to only a factor of two greater than the minimum energy, also implying a limit in the broadness of the burst shape.

The Two-Term burst shape is quite insensitive to changes in r for $r \geq 10$, and from a practical standpoint, the limiting case of $r \to \infty$ does not mean that the burst shape would be very much broader than the burst shape for $r = 10$.

The compensated reactivity at the time of peak power is given by the following equation:

$$k_c(\text{peak})(\$) = \frac{\alpha_0}{\alpha_R} + \sum_{i=1}^{6} \frac{a_i X_i}{(1 + X_i)\left(1 + \frac{r}{r-1} X_i\right)}, \tag{3-54}$$

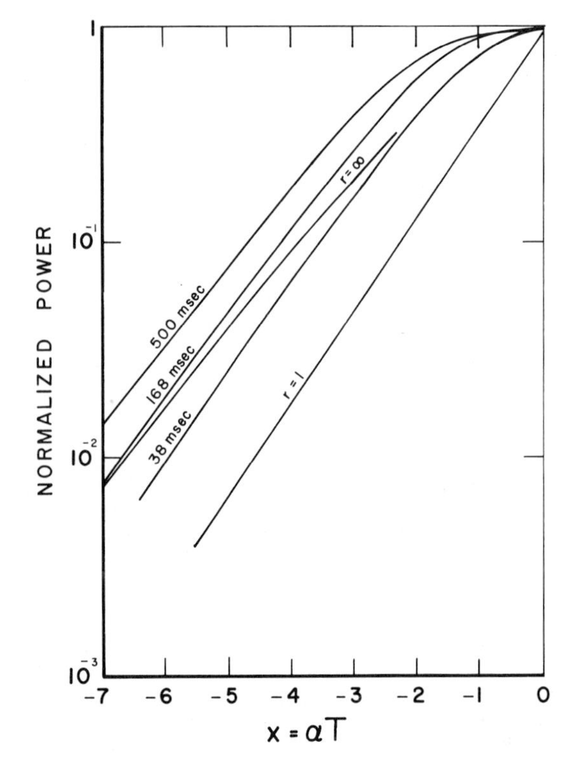

FIG. 3-9 Comparison of the two-term approximation with some experimental burst shapes.

where $X_i \equiv \alpha_0/\lambda_i$ and $a_i \equiv \beta_i/\beta$. A plot of k_c(peak) vs α_0 illustrating the effect of the value of r on k_c (peak) is shown in Fig. 3-10, for values of r from $r = 1.0001$ to $r \to \infty$. This is a better representation of the compensated reactivity than given in Fig. 3-7 because the burst shape is more realistic. However, Eq. (3-49) may overestimate or underestimate k_c whereas Eq. (3-50) always overestimates it. Since for the Two-Term model the burst shape does not change for $r \gtrsim 10$, the curve on Fig. 3-10 for $r \to \infty$ is significant only in that it denotes the limiting values of k_c(peak) which can be obtained by the Two-Term burst shapes.

This approximation was also used by Forbes to investigate the effect of changing the prompt neutron lifetime of the reactor. For $r \to \infty$, the summation term in Eq. (3-54) reduces to the form $a_i X_i/(1 + X_i)^2$. Each term of this series is a function having a maximum value of $a_i/4$ at $X_i = 1$. The asymptotic slope for small X_i is +1 and for large X_i is -1 and the function is symmetric about $X_i = 1$ or $\alpha = \lambda_i$. The summation term of Eq. (3-54) is thus made up of a series of round-topped curves having maxima at $\alpha_0 = \lambda_i$ and amplitudes proportional to a_i. Because of the distributions of λ_i and a_i for delayed neutrons, the sum of these curves is also a round-topped curve having a single maximum. To this is added the first term in Eq. (3-54) to obtain k_c (peak) as a function of α_0. The contributions of the individual terms, the summation term, and the total of all terms for $1/\alpha_R = 7 \times 10^{-3}$ and r approaching infinity are shown in Fig. 3-11.

As r approaches unity (very sharp bursts), the individual terms of the summation become flat-topped curves having half-amplitude points at $\alpha_0 = \lambda_i(1 - 1/r)$ and $\alpha_0 = \lambda_i$. This is illustrated in Fig. 3-12 for different values of r. The peak amplitude of the curve decreases and the width of the flat portion increases as r approaches unity. The net result is a gradual disappearance of the minimum in the k_c (peak) vs α_0 curve for small r, as was seen above in Fig. 3-10.

It is apparent from Fig. 3-11 that the location and prominence of the minimum in the k_c (peak) curve is dependent on the magnitude of α_R. For larger values of α_R than 7×10^{-3} (illustrated in Fig. 3-11) the line will be displaced to the left and the minimum may disappear entirely. Conversely, for smaller values of α_R the minimum will be much more pronounced since the α_0/α_R line will be displaced to the right. The effect of changing the value of $1/\alpha_R$ is illustrated in Fig. 3-13 where the values of $1/\alpha_R$ used are 7×10^{-2}, 7×10^{-3} and 7×10^{-5} sec, which are more or less typical values for heavy water, light water, and fast reactors, respectively. As $1/\alpha_R$ is decreased, the curves show a transition from no dips to rather prominent dips even for the comparatively sharp burst represented by $r = 1.001$.

3.8 Summary and Discussion of Results

The Two-Term burst shape description developed here approximates very closely the SPERT experimental burst shapes only in the short-period (sharp burst) region and thereby provides accurate estimates of the value of k_c (peak) for some cases.

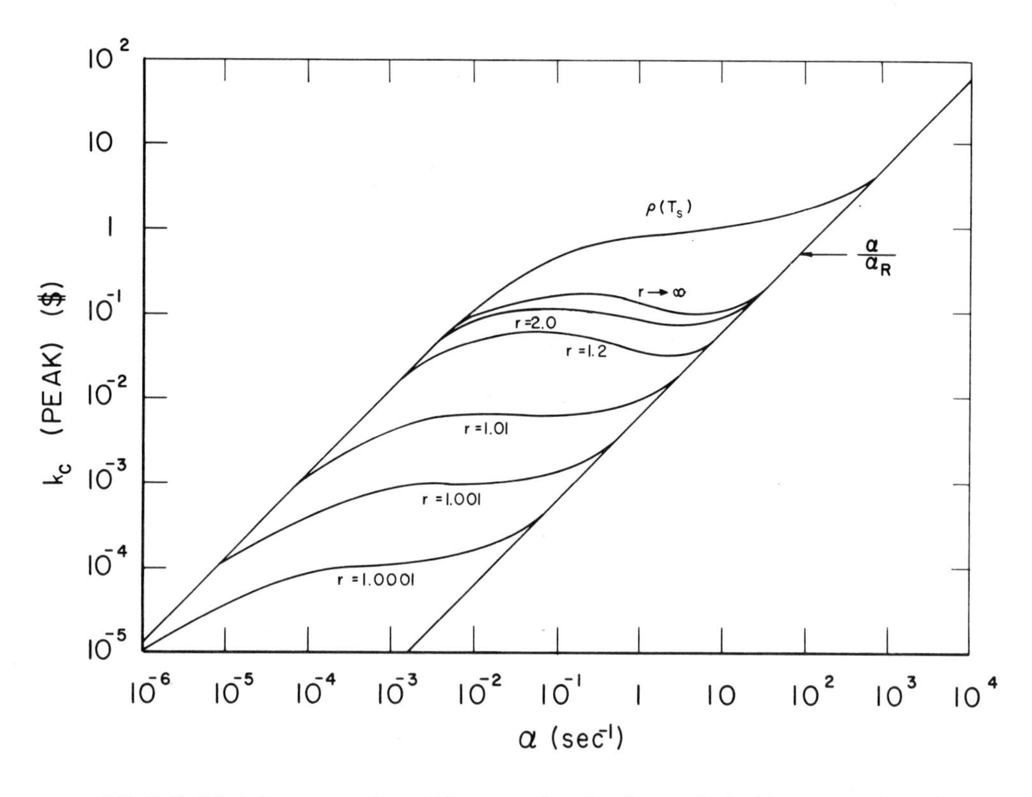

FIG. 3-10 Calculted compensated reactivity at the time of peak power, k_c (peak), vs reciprocal period α_o as a function of the burst parameter r for the two-term approximation. The upper curve ρ (T_S) is the total excess reactivity introduced at the start of the transient (inhour curve).

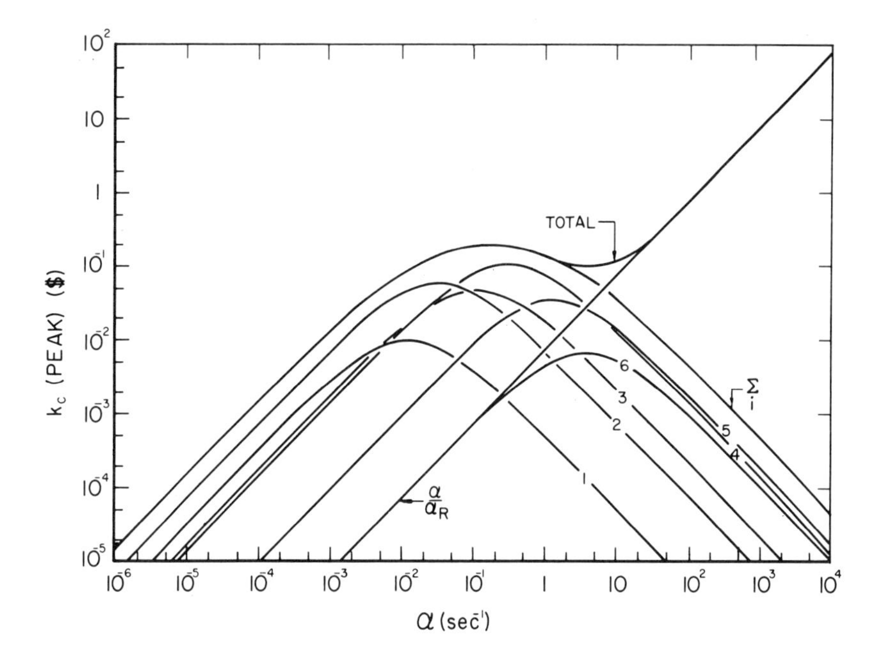

FIG. 3-11 Calculated reactivity compensation at the time of peak power, k_c (peak), vs reciprocal period α_o for the two-term approximation showing the contributions of the individual delayed neutron groups.

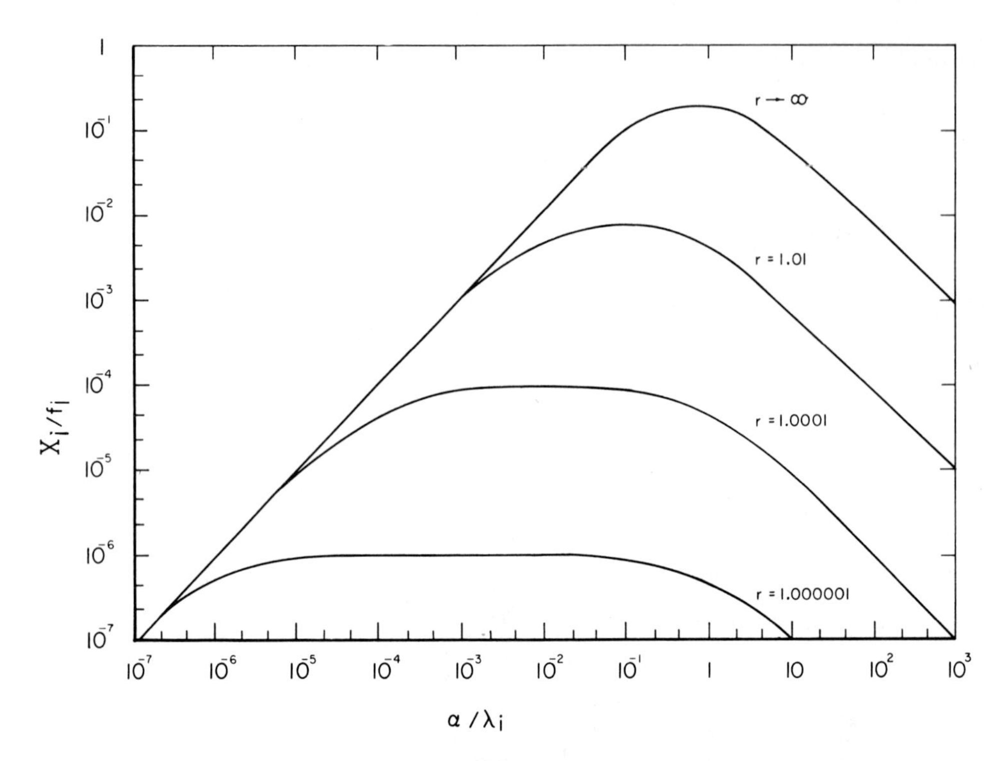

FIG. 3-12 The effect of the burst parameter r on the delayed neutron contribution to the compensated reactivity by the two-term approximation. The ordinate is $X_i/f_i = X_i/(1 + X_i)[1 + (rX_i/r - 1)]$.

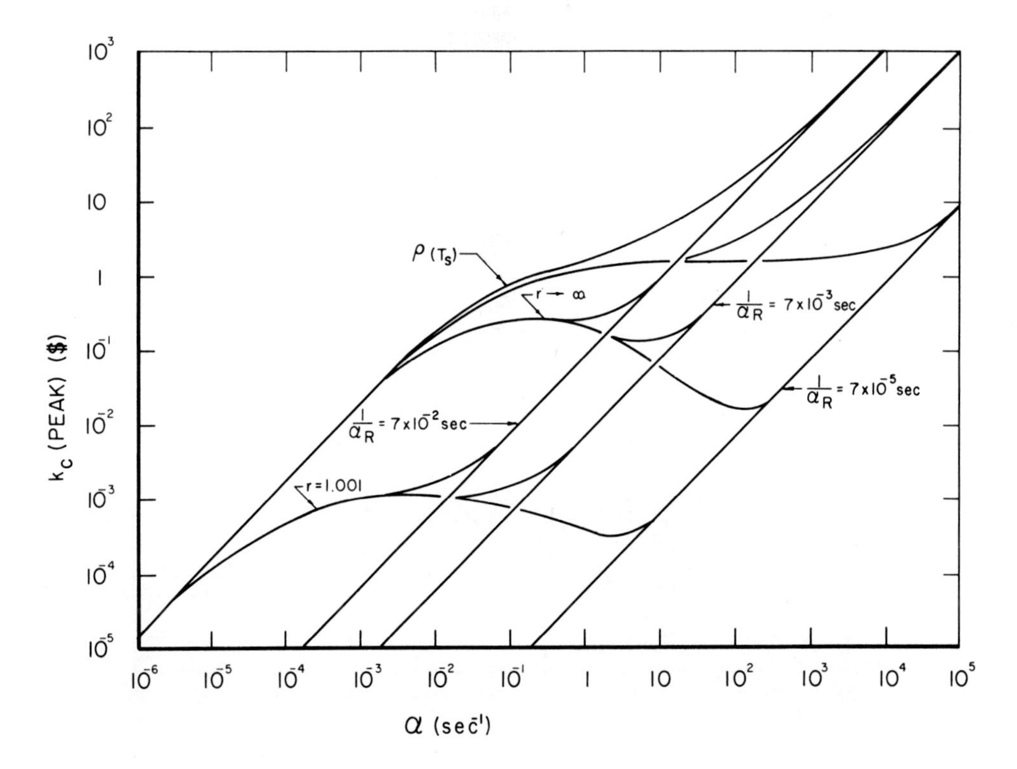

FIG. 3-13 Calculated compensated reactivity at the time of peak power, k_c (peak), vs reciprocal period α_0 as a function of α_R. The upper curve is the total excess reactivity inserted at the start of the transient (inhour curve).

The range of α_0 for which this is true is small and the use of the Two-Term burst shape is rather limited. On the other hand, although the Clipped Exponential burst shapes are never close approximations to experimental burst shapes, they always provide an upper limit for the value of k_c (peak) for any practical α_0, and from the viewpoint of attributing the self-shutdown to physical mechanisms, an upper limit for k_c (peak) is a useful quantity. The behavior of the compensated reactivity at peak power was examined using these burst shape approximations with the following results:

1. The shape of the k_c (peak) vs α_0 curve is dependent on the delayed neutron parameters, the prompt neutron lifetime and the shape of the power burst.
2. The minimum in the k_c (peak) curve will appear for many reactor systems but will not be present in systems which have very long prompt neutron lifetimes ($\simeq 7 \times 10^{-2}$ sec) or which exhibit extremely broad or extremely narrow power burst shapes.
3. This type of reactivity compensation analysis provides little information as to the form of the coupling equation. Physical situations leading to sharp-topped bursts, predicted by r near unity, are those in which the reactivity compensation increases much more rapidly than the energy released by the reactor. Examples are threshold processes such as boiling; nonlinear expansion of moderator such as water or steam; and systems whose reactivity changes are nonlinear functions of density or temperature. A linear energy shutdown model with a threshold leads to a sharp burst which approaches an exponential shape as the threshold is increased. Any reactor having nonlinear shutdown properties which can be represented by having $n >> 1$, will also exhibit k_c (peak) vs α_0 behavior typical of small r results.

4. SOME COMPARISONS OF MODELS AND EXCURSION REACTOR DATA

In the foregoing discussion, the complexities of the reactor behavior have been compressed into a single factor generically called the shutdown coefficient. This mathematical oversimplification has been surprisingly successful in describing the behavior of actual systems. That is, the experimental reactors often behave as if they were governed by a single, or at most, a few quenching mechanisms obeying feedback laws of these simple forms. In this section a few of the main features of these models will be compared with one another and with the excursion reactor data.

Table 4-1 summarizes the previously discussed principal characteristics predicted by the various models. In addition, a burst width parameter W_B has been included. This is defined as the ratio of the product of the reciprocal period and the total energy of the burst to the power at the time of maximum. The symbols are those which have been used previously.

Fig. 4-1 presents for comparison the energy-release functions vs the reactivity for the shutdown

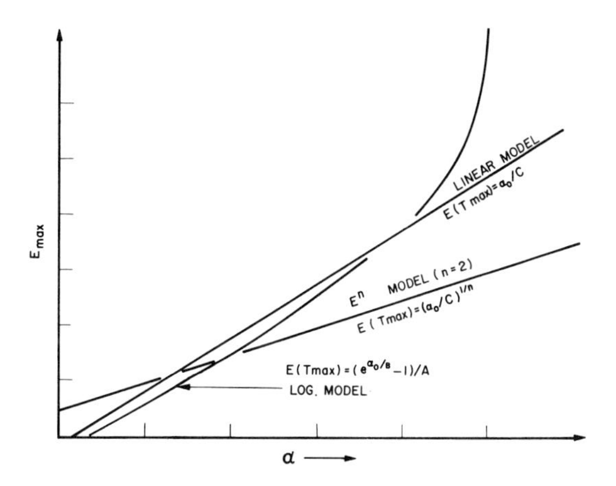

FIG. 4-1 E_{max} vs α_0 for various shutdown models.

models previously discussed. The abscissa is α_0, the initial step-inserted reactivity, and the ordinate is the energy release to time of peak power. The units are chosen for convenience in presentation. Under some conditions the values of the constants in the various expressions for $E(\alpha_0)$ may be such that the choice of form does not seriously alter the results. However, if the range in α_0 is sufficiently great, the differences become important.

The curvature of these lines is determined by the type and complexities of the shutdown mechanism whereas their displacement upward or downward is determined by the magnitude of the reactivity coefficient. One way of judging the relative importance of the two aspects, that is, curvature and displacement, is on the basis of how large α_0 can be made before the energy goes above some selected limit beyond which undesirable or even catastrophic effects occur. This limit could be represented by a horizontal line located arbitrarily in the graph, say at the top of the figure and could represent fuel plate melting, or vaporization or some other effect. If this limit is taken to be the criterion, it is of lesser importance how the energy release function behaves below the limiting value provided it does not cross it, or provided that the point of crossing is sufficiently far to the right that as a practical matter, the step-addition of the required α_0 is not realistic.

Fig. 4-2 presents for comparison the energy-dependent behavior of the previously discussed forms of the shutdown coefficient. Here, also, the values of the constants in the various expressions for b may be such that the choice of form does not seriously alter the results. However, where the shutdown is due to a Doppler coefficient, the variation has been found to be significant [29] and may be sufficient to allow a clear choice between forms of the energy dependence [33].

The excursion reactors exhibit, or come rather close to exhibiting, the behavior characteristics corresponding to the different feedback relations and energy versus α_0 functions represented in these figures. Table 4-2 lists some important types of coefficients along with the reactors which most nearly display the related behavior.

The first coefficient in Table 4-2 is prompt, negative, and remains constant during a burst in-

TABLE 4–1

Comparison of Nonthreshold Models

Parameter \ Model	Linear	E^n, Zero Delay	Quadratic	$E^{1/2}$	ln E	$E^{-1/2}$
$\alpha(E)$	$\alpha_0 - CE$	$\alpha_0 - CE^n$	$\alpha_0 \pm b_1E - b_2E^2$	$\alpha_0 + B - B(1 + AE)^{1/2}$	$\alpha_0 - B \ln(1 + AE)$	$\alpha_0 - B + B(1 + AE)^{-1/2}$
$b(E)$	$-C$	$-nCE^{n-1}$	$\pm b_1 - 2b_2E$	$-\frac{1}{2}AB(1 + AE)^{-1/2}$	$\frac{-AB}{1 + AE}$	$-\frac{1}{2}AB(1 + AE)^{-3/2}$
$E(T_{max})$	α_0/C	$(\alpha_0/C)^{1/n}$	$\frac{b_1}{2b_2}\left[\left(1 + \frac{4b_2\alpha_0}{b_1^2}\right)^{\frac{1}{2}} \pm 1\right]$	$\frac{\alpha_0(\alpha_0 + 2B)}{AB^2}$	$\frac{1}{A}(e^{\alpha_0/B} - 1)$	$\frac{\alpha_0(2B - \alpha_0)}{A(B - \alpha_0)^2}$
E_{final}	$2\alpha_0/C$	$\left[(n + 1)\frac{\alpha_0}{C}\right]^{\frac{1}{n}}$	$\frac{3b_1}{4b_2}\left[\left(1 + \frac{16b_2\alpha_0}{3b_1^2}\right)^{\frac{1}{2}} \pm 1\right]$ $\gtrsim \frac{3}{2}E(T_{max})$	$\frac{4}{A}\left[\left(\frac{9 + 3\alpha_0/B}{1 + 3\alpha_0/B}\right)^{\frac{1}{2}} - 1\right]^{-2} - \frac{1}{A}$	$\approx \frac{1}{A}(e^{\alpha_0/B+1})$ for $AE_{final} \gg 1$	$\frac{4\alpha_0B}{A(B - \alpha_0)^2}$
$\phi(T_{max})$	$\alpha_0^2/2C$	$\frac{n}{n + 1}\left(\frac{\alpha_0^{n+1}}{C}\right)^{\frac{1}{n}}$	$\frac{b_1^3}{12b_2^2}(x^3 \pm 1) \pm \frac{b_1\alpha_0}{2b_2}$ $x = [1 + (4b_2\alpha_0/b_1^2)]^{1/2}$	$\frac{\alpha_0^2(\alpha_0 + 3B)}{3AB^2}$	$\frac{B}{A}\left(e^{\alpha_0/B} - 1 - \frac{\alpha_0}{B}\right)$	$\frac{\alpha_0^2}{A(B - \alpha_0)}$
$W_B = \frac{\alpha_0E_f}{\phi(T_m)}$	4	$\frac{n + 1}{n}(n + 1)^{\frac{1}{n}}$	$\approx \frac{9\alpha_0(x \pm 1)}{2(b_1^2/b_2)(x^3 \pm 1) \pm 18\alpha_0}$	≈ 4 for $\alpha_0 \lesssim B$	$\frac{\alpha_0}{B}\frac{e^{\alpha_0/B+1}}{e^{\alpha_0/B} - 1 - (\alpha_0/B)}$	$\frac{4B}{B - \alpha_0}$

Notes: ϕ = reactor power, $\alpha = \dot{\phi}/\phi$ = reciprocal period; E = energy release during excursion, $E(T_{max})$ = energy released in reaching maximum value of temperature, E_{final} = total energy released during excursion; $\phi(T_{max})$ = maximum power; W_B = burst width parameter; $b(E) = \partial\alpha/\partial E$ = "reactivity coefficient"; α_0 = initially injected value of α or the value of α at the start of the excursion.

dependent of any temperature change, pressure change, or fuel expansion that might be caused by the energy release. To a degree these characteristics are exhibited by the Godiva,TREAT, and TRIGA reactors, but the representation is far from perfect and could be expected to be worse in experiments with larger α_0 because of the increasing importance of time delays and because of possible threshold effects.

The second type of coefficient in Table 4-2 is prompt, negative and increases with energy release. There are no good examples but there is one that at least has a similar component to an important degree. The SPERT metal-plate cores, which are highly enriched, have prompt and delayed components but nevertheless exhibit burst characteristics which can be associated with the type of coefficient that increases with energy release. Some of the reasons for this increase in effectiveness of the coefficient will be discussed.

The last type of coefficient shown is prompt and negative but decreases with energy. The SPERT oxide core, which is a low-enrichment core, exhibits all these characteristics very nicely.

The differences in type of shutdown coefficient manifest themselves as differences in the energy release function, such as shown in Fig. 4-1, and in other important ways such as transient pressure pulses or temperature excursions. In each of these cases reactor design changes which affect the shutdown mechanisms will also affect the relative safety. However, the degree to which safety is affected will depend on the form of the feedback relation and the exact nature of the particular shutdown mechanism, the accident under consideration, and the safety criterion. In some cases the net gain may be appreciable and in others, relatively insignificant.

One interesting type of shutdown mechanism was left off the chart because no reactor exhibits it to a degree worthy of note. That is, one which can be suggested as ideal: its reactivity effect is negative and small (effectively zero), for very long periods and the effect is prompt and strongly negative at short periods. The first requirement means that

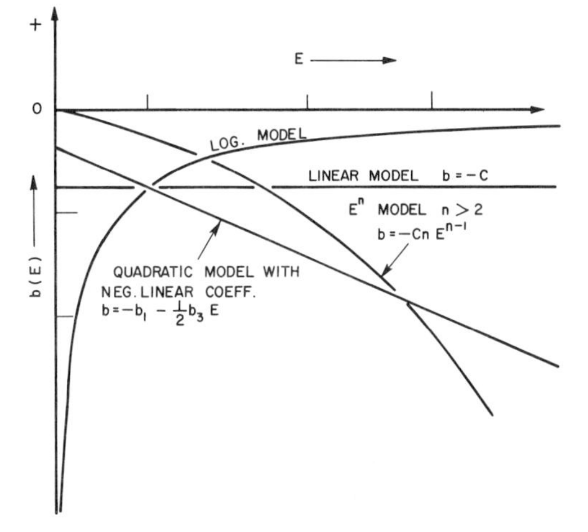

FIG. 4-2 The shutdown coefficient b(E) vs E for various models.

TABLE 4-2

Comparison of Types of Shutdown Coefficients with Various Excursion Reactors

Shutdown Coefficient	Reactors
Prompt, Negative, Constant	Godiva, TRIGA, TREAT
Prompt and Delayed, Negative Increasing with Energy Release	SPERT Metal Plate Cores KEWB
Prompt, Negative, Decreasing with Energy Release	SPERT Oxide Core

the reactivity investment needed to bring this fictitious system to power would be nil and the second requirement is protection against the dangers of large excursions. This is in contrast with the behavior of many mechanisms normally of importance in existing reactors. These exhibit sizeable negative coefficients for slow periods and also have important inherent delays which reduce their effectiveness when reactor periods become short.

As a consequence of both this normal behavior of reactivity coefficients, and also for reasons of plant operating capability, the excess reactivity possessed by the plant at some time in its operating cycle may be many dollars above prompt criticality. Since this investment in reactivity is a potential source for the initiation of a reactivity incident, it becomes important to consider the constraints which exist to limit the magnitude of such occurrences. One such constraint is that the reactivity addition must occur in a finite time rather than instantaneously. This has the important consequence already discussed that the reactivity that can be added is limited by the assembly rate as well as by the total reactivity available. That is, to attain a given period, or what is the same thing, a given instantaneous level of reactivity, there is a minimum assembly rate that must be exceeded.

The importance of this lies in the fact that for every reactivity insertion mechanism there is a natural limitation on insertion rates; this applies equally to insertions by direct human actions such as moving control rods by hand or dropping fuel assemblies into a core, and to insertions by more mechanical actions such as the collapse of voids or accidental control rod withdrawal. In any event, for very large reactivity insertions, the assembly rate requirements and, concomitantly, acceleration requirements become prohibitively high and establish an effective upper limit to the possible net reactivity additions and thus to the possible magnitudes of energy release. It is important to note that in actual cases the reactivity limit set by the attainable insertion rate is often substantially lower than the total available reactivity built into reactors.

These points are illustrated in qualitative fashion in Fig. 4-3 [39]. This shows the reciprocal of the reduced prompt neutron lifetime α_R on the abscissa. The assembly rate in dollars per second is on the ordinate. The lines show the assembly rates required to produce 1, 2 and 5 dollars superprompt criticality (reading from bottom to top) as a function of the characteristic reciprocal

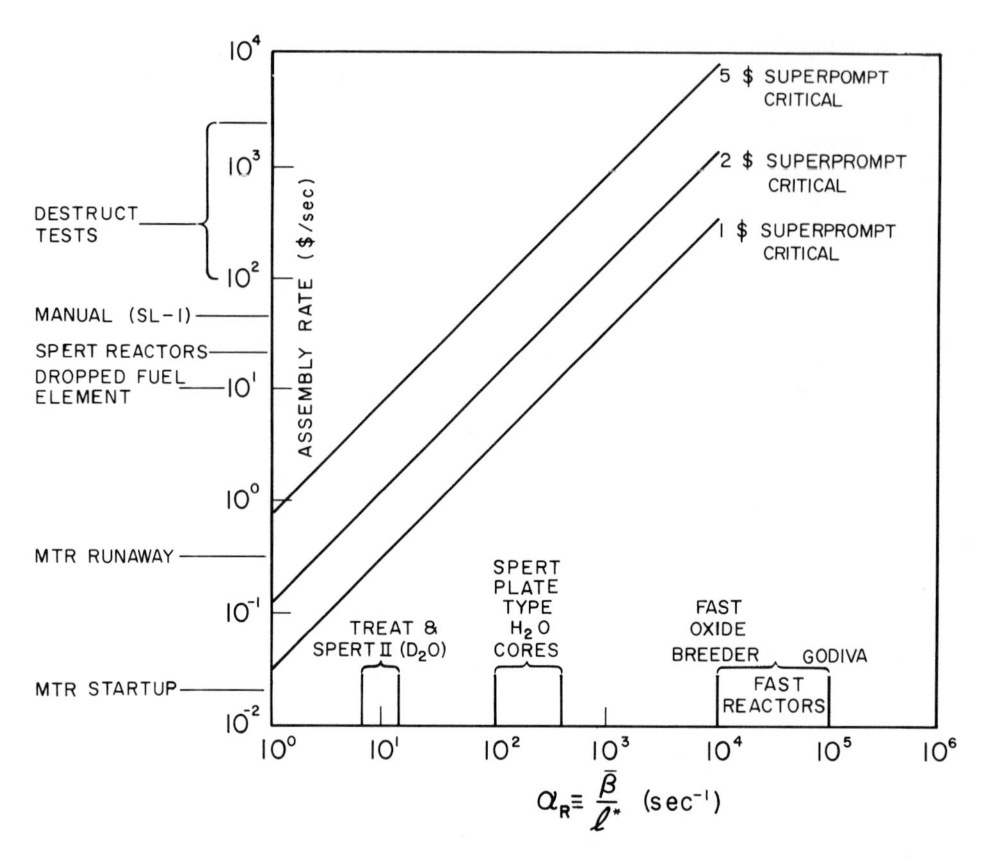

FIG. 4-3 The degree of superprompt criticality attainable as a function of assembly rate and α_R.

period. These lines are computed using the Linear Model and Eq. (2-31) but should apply qualitatively to many other, but perhaps not all, models as well. It should be repeated that two conditions must be met to achieve this degree of supercriticality: the reactivity must be available and it must be inserted at least as fast as indicated on the ordinate. The values of α_R for a number of special cases are shown. These cases can be conveniently grouped according to lifetime and for each group there are characteristic assembly rates required to produce specified levels of superprompt criticality.

Reactors of the long lifetime (small α_R) group would be in the neighborhood of the position indicated for TREAT and SPERT-II with α_R of 8 sec^{-1} and 10 sec^{-1}. These are graphite and D_2O moderated-reflected reactors respectively. The intermediate groups are represented by the SPERT light water, highly-enriched fuel reactors, which range from $\alpha_R = 100$ sec^{-1} to about $\alpha_R = 300$ sec^{-1}. The corresponding prompt neutron lifetimes are in the range about 100 μsec to about 30 μsec. The SL-1, Shippingport, and Yankee reactors are in this range.

At the far right is the fast reactor group ranging from the fast oxide breeder with $\alpha_R \sim 10^4$ sec^{-1} to the very fast solid assemblies such as Godiva with $\alpha_R \sim 10^5$ sec^{-1}

Some special comments can be made on the middle and lower α_R groups; that is, the region of thermal reactor systems. Much of the comment can be applied, in principle, to fast systems, but the actual numbers are so vastly different that great care must be exercised in the application.

Some typical rates are of interest. Normal startup rates for the MTR, for example, are in the range of cents/second. Maximum motor-driven control rod speeds for the MTR amount to only fractional dollar/sec assembly rates. These rates are sufficient to make the reactor superprompt critical but not much more. Over the years higher rates tend to be used as higher performance machines are built and control technology advances. This is likely to be true for power and test reactor systems as well as for the excursion reactors which show very high values. The SPERT reactors require in the order of 25$/sec to produce fast transients and this takes some 20g accelerating force. For the SPERT-I destructive test, for example, the reactor had a capability of 80$/sec which required accelerations of 33g.

The lowest diagonal line shows the assembly rate to raise the system reactivity to 1$ above prompt critical. For the SPERT metal-plate cores this is in the range of several dollars/second, which is of the order of a hundred times the normal startup rate by control rod withdrawal for thermal reactors. Dropped fuel element assembly rates in the neighborhood of about 10$/sec were obtained in SPERT-I tests. The re-entry of moderator to the core region in SPERT-I instability tests was of the order of 15$/sec. To initiate fast step-transients in these reactors the available rates are up to 25$/second. Indications are that a man lifting a control rod similar to the SL-1 could produce rates as high as 20$ to 50$/sec.

A transient rod injection system now under consideration for SPERT will have a maximum assembly rate of about 100$/sec. This is difficult to attain and it requires accelerations of about 100g, whereas the present 25$/sec system requires only about 20g. The 25$/sec insertion for the SPERT-I system is approaching the upper limit to what can be achieved without special measures.

Thus, to produce large excursions, large assembly rates are required. These may be caused by even modest accelerating forces if the agency of adding reactivity is accelerated to a considerable velocity before it becomes effective. The capability of attaining the required velocities must be considered in terms of the time needed for acceleration and the required magnitude of the accelerating force, since either of these can be a natural limitation on any means of inserting reactivity, planned or accidental.

As an example of such limitations consider that the agency for adding reactivity has an effect proportional to its displacement and that it is initially at rest at the point where its effectiveness begins. Then the required accelerating force varies approximately as the cube of the desired excess reactivity [40]; that is, to go from one to two dollars above prompt critical requires that the g forces be increased by a factor of roughly 8 for a given mechanical configuration. This follows from the fact that, for a constant acceleration transient rod drive, the acceleration, and consequently the force, is proportional to the square of the average velocity divided by the distance traveled. If the rod worth is a linear function of position, velocity can be replaced by reactivity rate and distance by total reactivity change. But since the ramp rate must also increase as the square of the reactivity change, Eq. (2-31), this results in a cubed relation between force and reactivity.

This is an over-simplification of any actual situation and the force requirements may be an even higher power of reactivity if the rod effectiveness decreases with distance traveled. Although some force reduction could be gained by "backing-off" to where some initial velocity is acquired before a significant reactivity addition is made, the point is that to achieve larger and larger amounts of superprompt criticality, either in accidents or in pulsed reactors, the attainment of the necessary assembly rates is going to require heroic effort by virtue of the larger accelerations required. The implication with respect to reactor safety is that the chance that any credible accident can occur in which the required assembly forces can exist, gets smaller and smaller with increasing transient severity. However, since it is possible to accidentally produce periods short enough for serious accidents by simple means, it is important to consider such possibilities in detail. Even modest increases in the quenching capability of a system may have significant improvement with regard to safety.

A special design effort was required to yield a test system capable of producing 1 msec periods. This system required insertion rates of about 100$/sec. The present concept has an insertion time of 40 msec and an acceleration of 28g to accomplish the 4$ of reactivity insertion needed for a 1 msec period. Other designs under consideration are in the neighborhood of 600-4000 per second. This region has been explored analytically by G. E. Hansen [5]. The ranges under consideration are labeled "destruct tests" on Fig. 4-3.

In contrast with the design effort required for obtaining these high insertion rates in test reactors, an example of simple means for achieving high assembly rates is afforded by the SL-1 accident. A mockup of the central control rod mechanism demonstrated that it was possible for a man withdrawing the central control rod by hand to achieve somewhere between 20$ and 50$/sec [38,40]. Acceleration considerations would indicate this is very close to the limit attainable by this means. It should also be noted that this rate is decades faster than the normal startup rate designed into many thermal reactors.

One other feature can be noted by rewriting the approximate relations, Eqs. (2-24) or (2-31) in the form $\alpha_m^2 = 30\$\,\alpha_R$ and by making use of the stable period at prompt criticality which is shown in Fig. 4-4 as a function of α_R. (This curve was obtained from an inhour equation whose derivation [41] included allowance for slowing down and thermal lifetimes.) In this figure, the α at prompt criticality is given rather closely by the square root of (1/4) α_R over a wide range of α_R. Using $\sqrt{\alpha_R/4}$ as the value of α_m in Eq. (3-54) yields the assembly rate, $\dot{\$}_{p.c.}$, required in the all-prompt neutron approximation to produce a period equal to the period at prompt criticality:

$$\dot{\$}_{p.c.} \sim \frac{1}{120}. \tag{4-1}$$

Hence, in Fig. 4-3 this would be a horizontal line with a value of about 1¢/sec. This is not shown because of the extremely qualitative nature of the

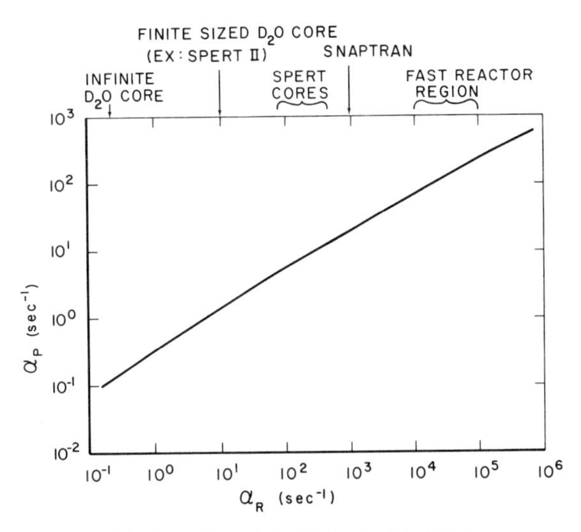

FIG. 4-4 The reciprocal period at prompt critical vs α_R.

argument. However, the result can be used as a guide to distinguish between assembly rates likely to lead to a superprompt critical burst and those likely to lead to superdelayed critical bursts.

On the basis of these considerations it can be argued that credible means of inserting all available reactivity into a reactor as an effective step are difficult to establish. The limitation on the size of possible accidents will more often than not actually arise from natural constraints on achievable assembly rates rather than from the limitation on the total reactivity available. On the other hand, the limitation on the consequences of an excursion depends on the shutdown mechanisms.

This discussion of assembly rates has been based on an extremely simplified picture of the reactivity feedback relationship and many of the properties described may change markedly when time-lags and nonlinearities are taken into account. The role of shutdown mechanisms in governing reactor safety is generally much more complicated than pictured here, but it is unlikely that there is a need to question the validity of certain features which emerge as general properties. First, in cases where stringent requirements are placed on assembly rates that must be achieved in order to increase total reactivity insertion, even marginal gains in the reactivity coefficient may result in substantial increases in safety; thus, there is a strong motivation for improving the effectiveness of the coefficients by increasing their magnitude. Second, it is clear that the delays between the energy release and a reactivity loss may be more important than the magnitude of the coefficient; thus, there is also a strong motivation to understand the details of the mechanism and the feedback relationship determined thereby in order to make a realistic assessment of the effective coefficient under dynamic conditions and to point the way to shortening the delays.

A number of these characteristics are exhibited by the results obtained with excursion reactors and by understanding the physical mechanisms which operate during the bursts.

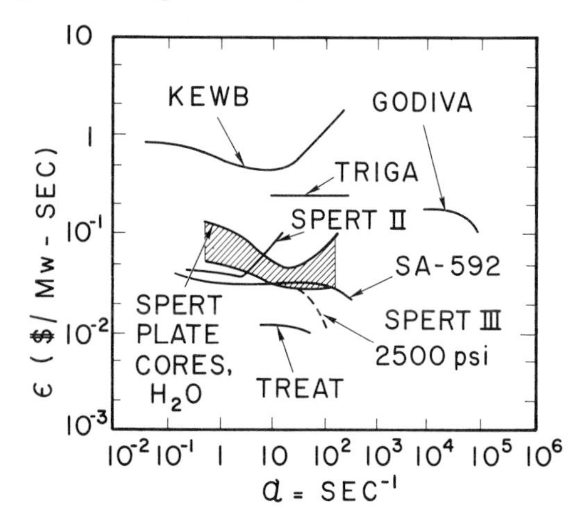

FIG. 4-5 The effective dynamic reactivity coefficient for excursion reactors as a function of α_0.

The step-transient data from the excursion reactors can be compared in the following way. Consider an experimentally determined quantity ϵ, the effective dynamic energy coefficient of reactivity. This is the ratio of the reactivity compensated by the system to the nuclear energy release, evaluated at the time of the power peak. Since a shutdown mechanism will usually have some characteristic time constant it is to be expected that, if ϵ is plotted as a function of reciprocal period, for a system with a single self-shutdown mechanism there would be a region at low alpha where ϵ would be relatively constant and could be determined by nontransient measurements. For a sufficiently short period, the effective value would begin to decrease. If, however, another more effective mechanism begins to act only at short periods, the value of the energy coefficient may level out or even rise, but ultimately for short enough periods, the curve should drop again. A number of the features are exhibited here by Fig. 4-5 which shows this coefficient as a function of reciprocal period for various systems.

The data for the various H_2O-moderated plate-type cores of SPERT and BORAX are contained within the cross-hatched area. In general, for these plate-fueled cores, the effective energy coefficient decreases as the reciprocal period increases and then levels out or slightly rises as periods are reached where boiling is important. That is, as the time scale of the burst is shortened, the effectiveness of the thermal effects, which depend on heat transfer rates, decreases; but in the shorter period region the threshold for the more effective mechansim of steam void formation is passed and the coefficient rises. The various SPERT cores are much alike and there is really very little variation in the coefficient. For this type of system over the range of period available a coefficient of 6 ± 5¢/Mw-sec will describe all the cores for all practicable periods.

The effect on burst behavior of a threshold which brings a new, more powerful shutdown mechanism into operation is more dramatically exhibited by SPERT-II data which show a strong rise in the coefficient as the period decreases. These data lie in the range shown for SPERT H_2O plate-type cores and are shown by the heavy line running through the cross-hatched area of Fig. 4-5. The increase in the coefficient is related to the fact that this system can shut down only by the generation of massive quantities of steam. However, at shorter periods than shown here, it would still be expected that the effective value of the steam coefficient would decrease. The SPERT-III (plate-type cores) high-pressure runs, shown by the broken line at the lower end of the SPERT band, show a marked decrease in ϵ. This is a result of the suppression of boiling as a shutdown mechanism by operating at a system pressure of 2500 psi (170 atm). Actually, the SPERT-III data show two things: the effect of a threshold mechanism like boiling in holding up the value of ϵ as it does in atmospheric pressure tests; and, when the threshold mechanism is absent, it illustrates the generally expected decrease in effectiveness of a shutdown mechanism as the period is shortened.

For this type of reactor, that is, the plate-type SPERT reactors, the principal mechanism for reactivity compensation is the decrease in effective moderator density arising variously from the thermal expansion of the water, thermal expansion of the fuel plates, and from the formation of steam voids. Some of the shutdown mechanisms are prompt, but the primary ones involving heat transfer from the plate to the water are delayed.

For the aqueous homogeneous reactor KEWB, there are no delays due to energy transport as in the plate reactors, but there are delays inherent in void formation.

The more prompt thermal effects present in the homogeneous reactor than present in the plate reactor are partly responsible for the larger coefficient, which is about 50¢/Mw-sec in the prompt-critical range. The onset of the increasingly effective void formation at short periods causes an increase in the coefficient to the order of 2.5$/Mw-sec for 2-msec period tests. The existence of this threshold behavior in both KEWB and SPERT reactors tends to obscure the important fact that those mechanisms which involve a phase change or the transport of energy by conduction or convection to produce a reactivity effect can have, even at relatively long periods, significant delays which are chiefly responsible for the difficulties in predicting reactivity effects. They produce a general tendency already noted for the reactivity coefficient to become less effective at very short reactor periods. This is, of course, opposite to one of the characteristics suggested as ideal for a shutdown mechanism.

In contrast, mechanisms which directly couple a specific neutronic effect with heating of fuel or moderator by absorption of fission fragments, neutron energy, beta-rays or gamma rays, are truly prompt. The TREAT and TRIGA reactors are self-limiting by such a mechanism. In these systems the fuel-moderator mixture is essentially homogeneous. In TREAT, the time-lags in the transfer of heat from the uranium particle fuel to the graphite moderator can be expected to be significant at very short periods. The resultant delay in the quenching would be expected to lead to a decrease in the coefficient and higher energy yields. The experimental TREAT coefficient as shown on Fig. 4-5 is nearly constant at ~1¢/Mw-sec. For TRIGA, the experimental value of the coefficient is about 25¢/Mw-sec over the entire range of tests shown here.

In many practical cases the changes in neutronic properties resulting from expansion, including that of the moderator, are also prompt. However, at short periods inertial effects may prevent the expansion from keeping pace with the temperature increases and, although the temperature rise is prompt, the reactivity effect may be delayed. This type of behavior is exhibited in the well-known case of the Godiva reactor where the delays in the expansion of the metal caused an observed decrease by a factor of two in the value of the coefficient. Since this occurs for periods shorter than about 20 μsec, it is not of practical importance to thermal reactors, but an analogous situation may exist in thermal systems where the shutdown mechanism requires the expulsion of a steam-water mixture. In fast reactors expansion delays of this sort can also be expected when rod-bowing or axial expansion of fuel rods contributes to reactivity effects. The dynamic coefficient for Godiva is several times larger than the coefficient for SPERT cores. However, this quantity is not in itself a reliable index for measuring reactor safety.

There is one more important mechanism for which there are excursion data. This is the Doppler effect which is quite prompt, like the TRIGA and Godiva mechanism, but which exhibits a saturation effect. That is in a manner similar to that shown in Fig. 4-2 for the logarithmic model, the coefficient decreases with energy release. Similar behavior is shown by other mechanisms because of delays, but in this case it is an inherent property. The light solid line at the bottom of Fig. 4-5 labeled SA-592, is for a low-enrichment oxide core in the SPERT-I reactor. In this case the coefficient is approximately constant at about 3¢/Mw-sec for most of the tests. In this region both the Doppler effect and conduction to the water moderator were operating. At higher values of α the loss of conduction heating effects and the decrease in effectiveness of the nuclear mechanism are evidenced by the decrease in the dynamic coefficient.

One important point should be noted here. Although Doppler coefficients in water reactors are generally a factor of ten smaller in magnitude than over-all steady-state temperature coefficients, the dynamic value for this core is only slightly less than that for the other SPERT cores and, in fact, is larger at short periods than for the pressurized SPERT-III case. This shows that the Doppler effect can be the most important shutdown mechanism even in thermal reactors because of its prompt nature [33]. (See also chapter on the Doppler Coefficient.)

The reactors discussed represent a wide variety of types but they are all relatively compact in size. One can expect that larger systems would have characteristically different coefficients on this type of plot. A case in point is the reactor discussed by Greebler [42]. This reactor can be shown to have a dynamic coefficient of about 0.04¢/Mw-sec for an α of about 2000 $\sec^{-1}$. This is about two decades below the value shown for the SPERT-I oxide core. Actually, the coefficients for the two systems are comparable when stated in reactivity changes per degree rise so that the coefficient for the larger system should be smaller in proportion to its higher heat capacity. This emphasizes that broad conclusions about relative safety are difficult to make from this plot. There appears to be no way at present of putting relative safety into a single parameter and displaying it so conveniently.

In the foregoing discussion, the quenching of bursts was treated as taking place solely by the action of inherent physical mechanisms apart from any consideration of the possible contribution from mechanical control systems. Actually, the overall assessment of the performance and safety of a reactor requires consideration of both self-shutdown effects and control system action as modes of limiting excursions. (See the chapter on Sensing and Control Instrumentation.)

Some of the factors involved in evaluating safety when both modes of shutdown are considered may be illustrated by making use of the following simple picture of the course of a fast excursion terminated solely by the action of the control system. Reactivity is inserted in a reactor at a very low initial power level. The power then rises exponentially on a constant period to a trip level where a demand for a scram is initiated by a sensor in the control system loop. The power continues to rise on the same exponential period until a later time when the power burst is terminated abruptly. This abrupt termination approximates the result of rapid movement of the control rods to a point of significant effectiveness. If the elapsed time τ_d between the call for scram and the actual power reduction is a constant, the maximum power level ϕ_M is given by

$$\phi_M = \phi_{trip} \exp(\alpha_0 \tau_d) , \tag{4-2}$$

where ϕ_{trip} is the power level at which the scram is demanded and α_0 is the reciprocal period.

Fig 4-6 shows schematically the calculated behavior of such a scram system with simple time delays of τ_d, $10\tau_d$, and $100\tau_d$, compared with self-shutdown characteristics representative of an enriched plate-type light water-moderated and reflected reactor in the superprompt critical region. Generally, a downward displacement of the curves on the graph can be associated with an improvement in safety. For the case of self-shutdown this is, in principle, accomplished by increasing the magnitude of the (negative) reactivity coefficient. For the case of control system shutdown, this is accomplished by lowering the trip level or decreasing the delay time. It can be seen that for certain combinations of these factors, self-shutdown can be more effective than the mechanical system in limiting the power burst. For other combinations the reverse is true.

The results obtained from experiments with the Oak Ridge National Laboratory Bulk Shielding Reactor II are shown in Fig. 4-7. This was a plate-type reactor which also had a fast-acting safety system [43,44]. In these tests, the subcritical reactor at low initial power was subjected to step insertions of reactivity and the resulting excursions were terminated either by the inherent quenching effects or by the control system. The

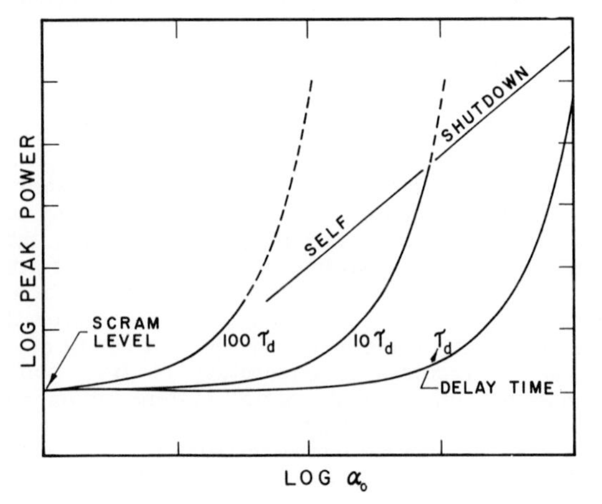

FIG. 4-6 The effect of different delay times on reactor peak power as a function of the reciprocal period α_0.

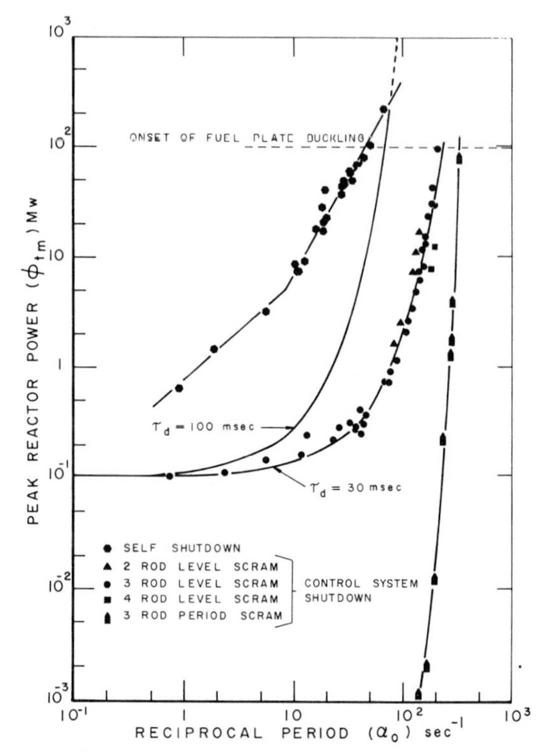

FIG. 4-7 Experimental and calculated values of peak power vs reciprocal period for various shutdown conditions.

results obtained with shutdown by the control system are well described by an expression of the form of Eq. (4-2) with τ_d = 30 msec. The considerable decrease in effectiveness that would result from increasing the delay time to 100 msec is evident. The importance of the delay is further indicated by the behavior of both types of control system scram, which exhibit an effective cut-off of their protective capability for reactivity insertions in which $\alpha_0\tau_d \gg 1$. Clearly, the period scram provided valuable additional protection. Since the tests were conducted without a source in the system, the effectiveness of control system shutdown would be improved under normal initial power conditions.

The foregoing discussions indicate that the experimental data on reactor excursions may be correlated, often very well, by simple models for shutdown by inherent processes or by control systems. However, the applicability of these models to different situations must rest on the degree to which they are understood in terms of the fundamental physical processes at work. Thus, detailed considerations are necessary, either to establish the need for detailed calculations or to be assured that the simple models may, instead, be applied and to what extent.

REFERENCES

1. J. C. Haire and G. O. Bright (Eds.), "Proceedings of the 1960 Idaho Conference on Reactor Kinetics" Report IDO-16791, Phillips Petroleum Co., 1962.

2. W. A. Horning and H. C. Corben, "Theory of Power Transients in the SPERT I Reactor (Final Report)", Report IDO-16446, Phillips Petroleum Co., 1957.
3. A. F. Henry, "Application of reactor kinetics to the analysis of experiments", Nucl. Sci. Eng., 3(1958)52.
4. K. Fuchs, "Efficiency for Very Slow Assembly", USAEC Report LA-596, Los Alamos Scientific Laboratory, 1946.
5. G. E. Hansen, "Burst Characteristics Associated with the Slow Assembly of Fissionable Materials", USAEC Report LA-1441, Los Alamos Scientific Laboratory, 1952.
6. L. W. Nordheim, "Physics Section II", Report CP-2589, pp. 32-36, University of Chicago Metallurgical Laboratories, 1945.
7. W. R. Stratton et al., "Analysis of Prompt Excursion in Simple Systems and Idealized Fast Reactors", Proceedings of the Second U. N. International Conference on Peaceful Uses of Atomic Energy, Geneva, 1958, Vol. 12, p. 196.
8. G. O. Bright, S. G. Forbes, W. E. Nyer, and F. Schroeder, "An Elementary Model for Reactor Burst Behavior", Report IDO-16393 (issued originally as PPC-215, July 1955), Phillips Petroleum Co., 1957.
9. L. Gartner and R. Daane, "The Self-Regulation by Moderator Boiling in Stainless Steel-UO_2-H_2O Reactors", Report NDA-16, Nuclear Development Associates, 1955.
10. W. E. Nyer, "Quarterly Progress Report, January, February, March 1958", Report IDO-16452, Phillips Petroleum Co., 1958.
11. S. G. Forbes, "The Dependence of Reactor Behavior on the Self-Shutdown Mode", Report IDO-16635, Phillips Petroleum Co., 1961.
12. R. O. Brittan, "Some Problems in the Safety of Fast Reactors", USAEC Report ANL-5577, Argonne National Laboratory, 1956.
13. W. J. McCarthy, Jr., R. B. Nicholson, D. Okrent, and V. Z. Jankus, "Studies of Nuclear Accidents in Fast Power Reactors", Proceedings of the Second U. N. International Conference on Peaceful Uses of Atomic Energy, Geneva, 1958, Vol. 12, p. 207.
14. W. E. Nyer, S. G. Forbes, F. L. Bentzen, G. O. Bright, F. Schroeder, and T. R. Wilson, "Experimental Investigations of Reactor Transients", Report IDO-16285, Phillips Petroleum Co., 1956.
15. S. E. Golian et al., "Transient Response of Plane Parallel Fuel Assemblies to Exponential Power Excursions", USAEC Report NRL-4495, Naval Research Laboratory, 1955.
16. V. Z. Jankus, "A Theoretical Study of Destructive Nuclear Bursts in Fast Power Reactors", USAEC Report ANL-6512, Argonne National Laboratory, 1962.
17. T. F. Wimett, "Time Behavior of Godiva Through Prompt Critical", USAEC Report LA-2029, Los Alamos National Laboratory, 1956.
18. J. T. Mihalczo, "Superprompt-critical behavior of an unmoderated, unreflected uranium-molybdenum alloy reactor", Nucl. Sci. Eng., 16(1963)291.
19. R. Scalettar and G. B. West, "Calculations of the Temperature Coefficient and Kinetic Behavior of TRIGA", Report GA-4474, General Atomic, Division of General Dynamics Corp., to be published.
20. D. Okrent, C. E. Dickerman, J. Gasidlo, D. M. O'Shea, and D. F. Schoeberle, "The Reactor Kinetics of the Transient Reactor Test Facility (TREAT)", USAEC Report ANL-6174, Argonne National Laboratory, 1960.
21. "Reactor Physics Constants", USAEC Report ANL-5800 (2nd Ed.), Argonne National Laboratory, 1963.
22. S. G. Forbes, F. L. Bentzen, P. French, J. E. Grund, J. C. Haire, W. E. Nyer, and R. F. Walker, "Analysis of Self-Shutdown Behavior in the SPERT I Reactor", Report IDO-16528, Phillips Petroleum Co., 1959.
23. W. E. Nyer and S. G. Forbes, "SPERT I Reactor Safety Studies", Proceedings of the Second U. N. International Conference on Peaceful Uses of Atomic Energy, Geneva, 1958, Vol. 11, p. 470.
24. S. G. Forbes, "Quarterly Progress Report (January, February, March 1958)", Report IDO-16452, Phillips Petroleum Co., 1958.
25. R. W. Garner, "Review of Several Models for Power Behavior and Reactivity Feedback in Reactor Excursion Studies", Report IDO-16636, Phillips Petroleum Co., 1962.
26. M. S. Dunefeld and R. K. Stitt, "Summary Review of the Kinetics Experiments on Water Boilers", Report NAA-SR-7087, North American Aviation, Inc., 1963.
27. W. E. Nyer, "Quarterly Progress Report (January, February, March 1959)", Report IDO-16539, Phillips Petroleum Co., 1959.
28. E. Hellstrand, P. Blomberg, and S. Horner, "The temperature coefficient of the resonance integral for uranium metal and oxide", Nucl. Sci. Eng., 8(1960)497.
29. R. Greebler et al., "Calculations of Doppler Coefficient and Other Safety Parameters for a Large Fast Oxide Reactor", Report GEAP-3646, General Electric Co., Atomic Products Division, 1961.
30. R. B. Nicholson, "The Doppler Effect in Fast Neutron Reactors", Report APDA-139, Atomic Power Development Associates, 1960.
31. E. T. Clark, "Initial Shutoff in a Reactor Transient", USAEC Report NYO-4726, 1956.
32. R. F. Walker, "Quarterly Progress Report (April, May, June 1959)", Report IDO-16489, Phillips Petroleum Co., 1959.
33. A. H. Spano, "Analysis of Doppler-limited power excursions in a water-moderated oxide core", Nucl. Sci. Eng., 19 (1964) 172.
34. R. Scalettar, "The Fuchs-Nordheim model with variable heat capacity", Nucl. Sci. Eng., 16(1963)459.
35. A. Ghatak and M. Nelkin, "Microscopic prompt-neutron kinetics of the TREAT reactor", Trans. Am. Nucl. Soc., 6(1963)110.
36. R. W. Miller, "Quarterly Progress Report (April, May, June 1958)", Report IDO-16489, Phillips Petroleum Co., 1959.
37. S. G. Forbes, "Quarterly Progress Report (January, February, March 1959)", Report IDO-16539, Phillips Petroleum Co. 1959.
38. "Final Report of SL-1 Recovery Operation (May 1961 through July 1962", Report IDO-19311, General Electric Co., 1962.
39. M. E. Thomas and W. E. Nyer, "Some Special Features of Control Problems of Excursion Reactors", p. 434 of Reactor Kinetics and Control, Proceedings of a symposium held at Univ. of Arizona, March 1963; USAEC Symposium Series 2, April 1964.
40. "SL-1 Accident" U.S. Atomic Energy Commission Investigating Board Report, Annex O, U. S. Government Printing Office, 1961.
41. A. H. Spano, unpublished work.
42. P. Greebler and D. B. Sherer, "Excursion shutdown mechanism for large fast oxide reactors", Trans. Am. Nucl. Soc., 5(1962)158.
43. F. L. Bentzen, "The Merits of Inherent Shutdown vs Mechanical Shutdown of a Plate-Type, Water-Moderated and Reflected Reactor in a Runaway Condition", Report IDO-16722, Phillips Petroleum Co., 1961.
44. J. R. Tallackson, J. B. Ruble, R. T. Santoro, and R. E. Wintenberg, "Performance Tests of the Oak Ridge National Laboratory Fast Safety System", USAEC Report ORNL-3393, Oak Ridge National Laboratory, 1963.

CHAPTER 8

Water Reactor Kinetics

J. A. THIE
Consultant
Minneapolis, Minnesota

CHAPTER CONTENTS*

The reactors treated in this chapter differ from those classified as "solid moderator" reactors and fast reactors because of two characteristics: moderation entirely or partly by a very low atomic weight nuclide and cooling by this water (or organic) moderator. The former characterizes the basic design and physics constants, while the latter is intimately related to the kinetic behavior of these reactors. Fundamentally these reactors in a general way differ from solid moderator and fast reactors in that heating, evaporation, and motion of the coolant strongly affect the reactivity, and hence their kinetic behavior.

1 REACTIVITY CONTROL

1.1 Control Rods

Because of the large values of power, temperature, and burnup coefficients of reactivity typical of water reactors, reactivity control requirements are also large. Even low power and low temperature systems may require substantial control for purposes of experimental flexibility. It is usual to achieve this control with mechanical rods (including blades and crosses), although other means are considered in the ensuing sections.

The worth of a control rod is defined in terms of the change in k_{eff} which the control rod is capable of causing:

for withdrawn position, core has k_{eff} and $\rho = 1-k_{eff}^{-1}$

for inserted position, core has k'_{eff} and $\rho' = 1-k'^{-1}_{eff}$

Customs for specifying worth vary, as the worth might be $k_{eff} - k'_{eff}$, $\rho - \rho'$ or the deviation of k_{eff}/k'_{eff} or k'_{eff}/k_{eff} from unity. These are essentially equal when small (i.e. $\rho \ll 1$, $\rho' \ll 1$) reactivity changes are involved, and in the discussion which follows no distinction will be made. However, when large reactivity changes are treated it is important to specify which definition is used.

Table 1-1 [1] shows total worths of control rods for a number of typical water reactors in their cold-clean condition. From the quotient of the total control rod worth and the total withdrawal time one may obtain an average rate of reactivity addition.

A meaningful measure of safety is the quotient of the effective delayed neutron fraction β_{eff} and this average rate, for this is the time required on the average to go from critical to prompt critical

*Except for a few changes and additions made in proof, this chapter is based on information in the literature or known to the author prior to December 1963.

TABLE 1-1

Control Rod Characteristics in Various Reactor Types

Reactor	Control rods	Withdrawal rate (in./min)	Stroke (in.)	Total rod worth* (%)
APPR	7 square elements	3	22	24.6
BORAX-IV	4 blades	12	25.5	7.5
Dresden	80 crosses	360	106	15.6
EBWR	9 crosses	28	48	12.4
Elk River	13 crosses	30	56	18.3
PWR	32 crosses	11	71.5	25
SL-1	5 crosses and 4 T's	3	32	20
VBWR	7 blades	4 or 12	42	10.5
Yankee	24 crosses	6	90	17

*Note that some variation in the meaning of total rod worth may exist among the various reactors here, as pointed out in the text. It is usually necessary to consult the original calculations or experiments when precise meanings of control rod reactivities are desired.

by continuous rod withdrawal. More significant in hazards analyses, however, is the maximum withdrawal rate when several dollars of available reactivity exists. (If only a dollar or less of available reactivity exists in the control rods, this amount rather than the reactivity insertion rate is the more significant.) This is given by the product of the average rate just defined and the maximum to average factor of the control rods:

$$\frac{\text{maximum reactivity per unit of withdrawal}}{\text{average reactivity per unit of withdrawal}} \cdot$$

The latter factor may be as low as 2 for a core having only one weak rod at the axis of a cylinder. On the other hand, in the Elk River Reactor [2] the factor was measured as 12, the maximum reactivity per unit of withdrawal being achieved when the tip of a slightly off-center rod is approximately halfway between the bottom of the core and the tips of the bank of twelve other rods. This points out that sequences of control rod withdrawal which have not been investigated experimentally or theoretically might create additional hazards if configurations enhancing their worth are encountered.

Many factors can affect the worth of control rods in addition to the flux distribution which is shaped by the rods themselves. Geometrical changes, such as location of fuel elements or of large in-pile experiments, can influence rod worth. One may expect changes in rod worth during core burnup because of factors such as: a) the neutron-absorbing nuclide partially burns up; b) the fissionable and absorbing nuclides in the fuel elements change, thus changing the competition for neutrons; c) the delayed neutron fraction, which depends on the relative distribution of fissions among the different fissile materials, changes. The third effect changes the reactivity worth of rods as expressed in dollars or cents, but not if expressed in percent (or fractional) reactivity.

Common to all water reactors is the influence of neutron temperature and water density. This effect can be seen analytically from the expression for the fully inserted worth of a single black (absorbing) cross of span 2a at the center of a cylinder [3]:

$$\rho - \rho' = \frac{\nu - 1}{\nu\epsilon - 1} \, \frac{1.62a\, B_r^2}{(1 + L^2B^2)\,(\sqrt{3\Sigma_a\Sigma_{tr}} + 2.13\Sigma_a)}, \quad (1\text{-}1)$$

where the neutrons per fission $\nu = 2.48$, the fast effect is ϵ, $L^2 = 1/3\ \Sigma_a\ \Sigma_{tr}$, Σ_a and Σ_{tr} are the thermal neutron absorption and transport macrosopic cross sections respectively, B_r^2 is the radial buckling, and B^2 is the total buckling. The presence of Σ_{tr} in the denominator gives this expression a $(\text{density})^{-1/2}$ dependence on the moderator density. In addition a strictly neutron temperature effect exists because Σ_a varies as $(\text{temperature})^{-1/2}$ and for H_2O moderators Σ_{tr} also varies approximately as $(\text{temperatures})^{-1/2}$. The net result is that the reactivity worth of control rods is larger at elevated temperatures with their associated reduced moderator densities. In reactors in which many control rods are operated as a bank, Eq. (1-6) leads to this same conclusion because of the dependence on the migration area M^2. It must be pointed out that while equations such as these may be useful for giving approximate indications of how several variables influence rod worth, it is necessary to use multigroup neutron diffusion theory with special boundary conditions at the control rods to obtain accurate calculations [4].

Since higher temperature and lower densities lead to stronger rod worths, evidently an additional factor applies at operating conditions on the above cold-clean reactivity insertion rates. Although rod worths 1.3 times the cold values may be typically encountered at operating temperatures, whether or not the maximum reactivity insertion rate occurs at high temperature depends on the control rod configuration encountered there. In boiling reactors there is the additional effect of a nonuniform moderator density distribution which distorts the flux distribution and hence alters the location of rods giving a maximum reactivity insertion rate. Here also detailed multi-region diffusion theory calculations in combination with void distribution calculations are required to specify rod worths under operating conditions.

Other aspects of the behavior of control rods in boiling reactors deserve some attention. Moving control rods in the presence of voids has the effect of replacing a region occupied by control rods by an enhanced void volume in a core of larger volume. Since control rod reactivity is converted to void reactivity and enhanced power, calibration of control rods in the presence of voids is difficult in comparison to zero-power calibration. A feasible method however is to calibrate against known uniform poisons, such as xenon or boric acid.

Since a control rod calibration may be obtained either in the presence of voids or, more simply, at unvoided saturation temperature low power conditions, there are evidently several methods for measuring reactivity in voids by control rods:

$$\Delta\rho = \int_{Z_1}^{Z_2} (d\rho/dZ)_{v_1}\, dZ\ , \quad (1\text{-}1a)$$

$$\Delta\rho = \int_{Z_1}^{Z_2} (d\rho/dZ)_{v(z)}\, dZ\ , \quad (1\text{-}1b)$$

$$\Delta\rho = \int_{Z_1}^{Z_2} (d\rho/dZ)_{v_2}\, dZ\ , \qquad (1\text{-}1c)$$

where Z_1 and Z_2 are control rod positions corresponding to $v_1 = 0$ and v_2 = operating voids respectively. Equation (1-1a) is the recommended experimental definition for three reasons: it is the simplest and most accurate method; it has become traditional to make measurements this way; and the value obtained gives one the knowledge (by use of the inhour curve which gives reactivity period relationships) of the positive period that would result if all the voids v_2 collapsed. In Eq. (1-1b) the integrand includes the effect of the continuous variation of control rod worth as voids increase from 0 to v_2. It would give a result intermediate between Eqs. (1-1a) and (1-1c). In comparing theory and experiment it is important to recognize that Eq. (1-1a) is used in making measurements of reactivities in voids.

1.2 Liquid Poisons

Water reactors can be controlled by one means not practicable in other reactor types: soluble poisons. (See also Mechanical Systems chapter, Sec. 4.7.3.) Typically boric acid H_3BO_3, sodium pentaborate $Na_2B_{10}O_{16} \cdot 10H_2O$, or cadmium sulfate $3CdSO_4 \cdot 8H_2O$ are used, because of solubility and materials compatability. Some power reactors which have had experience with soluble poisons are BORAX-II, BORAX-III, Experimental Boiling Water Reactor (EBWR), and Yankee [5]. In addition to its use as a control device, boric acid is often used as an experimental tool in investigations at zero power. In light water systems the reactivity worth of boric acid is

$$\rho - \rho' = f_w \left(\frac{N_B\sigma_B}{N_H\sigma_H}\right)$$

$$= 3.31 \times 10^{-4} f_w \times [\text{ppm } H_3BO_3]\ , \qquad (1\text{-}2)$$

where f_w is the fraction of absorbed neutrons which are captured by the water (f_w typically may be ~ 0.1). Since 250 ppm (which is about one gram per gallon) is likely to be worth of the order of one dollar, the criterion for safe removal of the poison is that the rate of reducing the boric acid concentration should be small - such as slower than 250 ppm/min.

Concentration reductions of fluid poisons are accomplished either by dilution or by ion-exchange. For both processes the rate of change of concentration is

$$\frac{dC}{dt} = -\frac{\dot{V}C}{V^0}\ , \qquad (1\text{-}3)$$

where C is the concentration and V^0 is the volume of the system containing the poison in a uniform concentration. To assure the uniform concentration required for the validity of Eq. (1-3) water agitation by pumping, stirring, etc., is necessary because the diffusion rates of ions in stagnant water are much too slow. In the case of dilution, $\dot{V}$ is simply the volumetric flow rate of unpoisoned water; for ion-exchanging $\dot{V}$ is the product of the flow rate through the resin and its fractional efficiency in the event the latter is not unity. In Eq. (1-3) if C is replaced by $\Delta\rho$, the negative reactivity held down by the fluid poison, then evidently $\dot{V}\Delta\rho/V^0$ is the rate of addition of reactivity. Since $\Delta\rho$ may be as large as several dollars, the removal time constant or system exchange time, $V^0/\dot{V}$, should be longer than several minutes to have a safe rate of removal. (Actually in typical reactor systems $V^0/\dot{V}$ may be a few hours.)

Injection of fluid poisons is accomplished by pouring, pumping, or forcing under air pressure a concentrated solution of the poison into the reactor water. If a fluid poison system is used as a back-up shutdown device in the event of failure of a mechanical control system, then it is desirable to inject the poison rapidly. However, operator delay-time plus the period for poison injection makes the shutdown time somewhat longer than that experienced in scrams of mechanical systems.

When a reactor is operated with a fluid poison present, a few factors affecting its inherent safety characteristics must be noted. The void coefficient of reactivity is negative in most, but not all, water reactors, and is a reliable shutdown mechanism in severe transients. The presence of a fluid poison causes this coefficient to be less negative and hence less effective. The loss of effectiveness may be computed by comparing the reactivity lost due to void in the reactor with the following expression for the reactivity gained by poison expulsion:

$$\Delta\rho = \frac{\text{void volume in reactor}}{\text{water volume in reactor}} \times C \times \frac{d\rho}{dC}\ , \qquad (1\text{-}4)$$

The possibility of inadvertent dilution resulting in a reactivity increase must be recognized. Lastly, if a leakage neutron flux monitor is in a fluid poison environment, the possibility of inadvertent operation at an undesired power level is increased. This is because the increased poison content of the water desensitizes a detector by diminishing the thermal flux in the environment. Leakage flux detectors, unlike in-core detectors, do not monitor average reactor thermal flux, but rather the product of the leakage fast flux and the thermal-to-fast flux ratio near the detector. It is the latter that is affected by fluid poisons, provided the detector is surrounded by poisoned water.

1.3 Moderator and Leakage Control

This category of control methods is unique with water reactors because of the relative ease with which reactivity effects may be obtained by changing resonance capture, neutron migration area, and geometric buckling. This section discusses the use of these effects as part of the control system. However, the anlysis applies equally whether these effects are deliberately controlled or inadvertently happen.

Control by changes in moderator density is commonly used in pressurized water power reactors [6]. In essence the temperature of the coolant is allowed to rise (or fall) temporarily as the load requirement decreases (or increases). By virtue of the negative temperature coefficient, the reactor power will decrease (or increase) until it readjusts itself to match the load. A quantitative treatment of the theory of such transients appears in Sec. 2.2 of this chapter. A fundamental consideration for core safety in these transients is that with typical temperature coefficients of the order of -2×10^{-4} °F^{-1} (or -3.6×10^{-4} °C^{-1}), only something like 35°F (19°C) of cooling is required to add a dollar of reactivity.

In the design of some boiling reactors special attention has been given to control features which are desirable when load changes occur. Where no special features are present an increase (or decrease) in the power plant demand for steam from the reactor will reduce (or increase) pressure. The immediate effect on reactivity is negative (or postive) because of the change in core void volume, (see Fig. 2-4 and accompanying text). This causes the reactor power to change in opposition to the load change unless control rods are moved to counteract this and to maintain the primary pressure constant.

Therefore some boiling reactors, such as the Dresden reactor [7], use a dual cycle system, i.e., useful power is extracted from both the steam and the water by separate heat exchangers. Where such a system exists the control action just described for pressurized water reactors is derived from the part of the cycle which extracts heat from primary water before returning it to the core. However, in boiling reactors the effect of the temperature of the returning water is to raise or lower the boiling boundary in the core, thus changing the exit void content in the direction demanded. This eliminates, or at least reduces, the need for moving control rods during load changes.

Boiling reactors can also be controlled if desired by varying the void volume in the core. Where forced circulation is used, the void volume can be increased (or decreased) to accommodate the stem generation rates demanded by slowing down (or speeding up) the water recirculation. The Pathfinder reactor is one example of this [8a].

A fundamental safety consideration in these control systems of boiling reactors is the rate at which the core void content can decrease during transients. Some estimate of this may be had by using thermodynamics and hydraulics to compute a void change Δ v for some assumed disturbance to the reactor or its associated equipment. Then the rate of change of reactivity is

$$\frac{d\rho}{dt} = \frac{\Delta v}{\Delta \tau} \frac{d\rho}{dv}, \tag{1-5}$$

where $d\rho/dv$ is the void coefficient and $\Delta\tau$ is the time associated with the void change - usually determined by flow velocities and distances. A more detailed approach to computing these transients is given in Sec. 2.3.

In the "spectral shift" method of control, the balance between neutron moderation and absorption is changed by using different isotopes, thus varying the nuclear macroscopic cross section, rather than the density or volume [9]. Reactivity may be added in an H_2O-D_2O moderated reactor, if the fuel elements are not too widely dispersed, by increasing the H_2O content. This occurs because in a close-packed core the moderating effects of H_2O outweigh its thermal absorption effects. Kinetics considerations here are very similar to those of fluid poisons (Sec. 1.2), except that the reactivity worth of H_2O must be computed from detailed multigroup criticality calculations.

Changes in moderator conditions cause reactivity effects by changing resonance escape, moderator capture, and migration area. Part of the reactivity control obtained by changes in the migration area results from changes in the number of neutrons leaking from the core. Raising and lowering the moderator between the fuel tubes of a pressure tube reactor is an example of this type of control.

For reactors in which the number of neutrons leaking out of the core corresponds to a significant reactivity, control by changing the geometric buckling B_g^2 is suitable. For example, zero power critical assemblies often use lowering and dumping of the water as a means of backing up other shutdown control devices. Where many strong and closely spaced control elements are used as a bank, this may also be regarded as leakage control: a substantially subcritical zone acts as an axial reflector on an unrodded zone.

From the geometry defined in Fig. 1-1 and using the one-group diffusion theory approximation, the effect of height change is

$$\frac{d\rho}{dZ} = \frac{d}{dZ}\left(1 - \frac{1 + M^2B^2}{k_\infty}\right) = 2\pi^2 \frac{M^2}{k_\infty} \frac{1}{(Z + \lambda_1 + \lambda_2)^3}, \tag{1-6}$$

where M^2 is the migration area and k_∞ is the infinite lattice neutron multiplication constant. If the control rod level or the water level is being moved at a rate dZ/dt, then evidently the product of Eq. (1-6) and this rate gives the reactivity rate involved. Where this equation is being used for estimating purposes in H_2O reactors, a typical value for the reflector savings λ_2 of a heavily rodded region is the 17 cm (6.7 in.) observed in the Experimental Boiling Water Reactor (EBWR), and

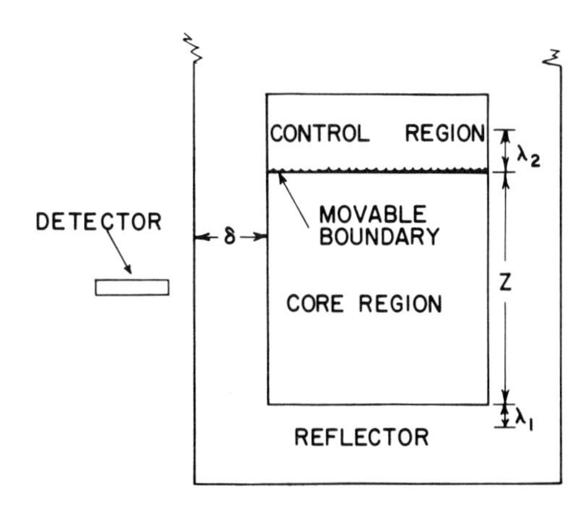

FIG. 1-1. Control of a reactor by changes in its height. The upper boundary of the core region may be defined either by a water level or by the location of the tips of a bank of many upper control rods.

the reflector savings λ_1 at the other end is typically 7 cm (2.76 in.).

The various effects discussed in this section have been analyzed above from a reactivity point of view. One must also consider the practical problems involved in detecting neutron flux changes in reactors using these control methods. If externally-located detectors, positioned as shown in Fig. 1-1, are relied upon for immediate indication of reactor power, it must be expected that their calibration (i.e., the detector current divided by the reactor power) will depend on the product of core leakage and reflector attentuation. If the core neutron age τ and the attenuating cross section Σ of a reflector of thickness δ change, then

$$\frac{(\text{detector reading per Mw})_1}{(\text{detector reading per Mw})_2} = \frac{\tau_1 \exp(-\Sigma_1\delta)}{\tau_2 \exp(-\Sigma_2\delta)}, \tag{1-7}$$

where 1 and 2 denote two reactor conditions of differing temperature or density. Evidently this can be important in water reactors undergoing substantial density changes. It is not unusual to observe that detectors are a factor of 4 more sensitive at operating conditions than at room temperature in some high temperature water reactors. Evidently relying on instruments as power indicators can lead to operation at undesired power levels if nonstandard temperature conditions are encountered and the changing calibration is not taken into account.

1.4 Control in Homogeneous Reactors

Reactors in which the fuel is either dissolved or held in suspension can be controlled by moving fuel into and out of the core region in addition to the preceding methods. In fact, this technique in combination with the control of solution temperature has been used in the Homogeneous Reactor Test (HRT) to provide all the control necessary for operation. To the extent that changing fuel concentration is analogous to changing fluid poison concentration, Eq. (1-3) is applicable.

Normally in homogeneous reactors changes in fuel concentration, and hence in reactivity, are slow. In any event the associated transient can readily be computed approximately. Consider that at some point in the system which precedes the core inlet by a flow time interval τ_1, a change of fuel mass is introduced at the rate dm(t)/dt. The conservation of mass equation for the volume immediately surrounding this fuel injection point is

$$\frac{dm(t)}{dt} = \dot{V}C_{in}(t+\tau_1) - \dot{V}C(t-\tau_2). \tag{1-8}$$

Here $\dot{V}$ is the volumetric flow rate of the loop, $C_{in}(t-\tau_1)$ is the fuel concentration leaving this fuel injection volume expressed in terms of the concentration $C_{in}(t)$ at the core inlet, and $C(t-\tau_2)$ is the fuel concentration entering this fuel injection volume expressed in terms of the concentration C(t) at the core exit. τ_2 is the flow time interval from the core exit to the point at which fuel is injected. This equation is used in conjunction with

$$V^0 \frac{dC}{dt} = \dot{V}C_{in} - \dot{V}C, \tag{1-9}$$

which describes the changes in concentration C in the core volume V^0. The reactivity is a linear function of this concentration. Evidently the significant time constants associated with fuel control transients are the total loop transit time outside of the core $\tau_L = \tau_1 + \tau_2$ and the core transit time $\tau_c = V^0/\dot{V}$.

Abnormally rapid changes in fuel concentration, such as can be associated with criticality incidents in fuel processing plants, are more difficult to compute. Solution channeling occurs and it is necessary to treat the core as having zones of different concentration. The reactivity at any given time may then be computed by multi-region diffusion theory, using the concentration distribution at that time.

In reactors with fuel circulating outside the core, delayed-neutron precursors decay outside the core, resulting in a net loss of delayed neutrons. (See also Criticality chapter, Sec. 3.2.1.2). For example, in the HRT the effective delayed neutron fraction was 0.0057 [10], which is significantly less than that for U^{235} in heterogeneous reactors. The neutron kinetics equations are [11]

$$\frac{d\phi}{dt} = \frac{(1-\beta)k_{eff} - 1}{\ell}\phi + \sum_{i=1}^{6} \lambda_i C_i + S \tag{1-10}$$

$$\frac{dC_i(t)}{dt} = \frac{k_{eff}\beta_i}{\ell}\phi(t) - \lambda_i C_i(t) - \frac{C_i(t)}{\tau_c} + \frac{C_i(t-\tau_2)\,e^{-\lambda_i\tau_L}}{\tau_c}. \tag{1-11}$$

The last two terms in Eq. (1-11) are the removal and return rates of delayed neutron precursors respectively. The effect of these terms on the steady-state solution of these equations is that

$$k_{eff} \cong 1 + (1-a)\beta \tag{1-12}$$

for exact criticality (with $S \cong 0$), where the delayed neutron attenation factor is

$$a = \frac{\bar{\lambda}\tau_c}{\bar{\lambda}\tau_c + 1 - \exp(-\bar{\lambda}\tau_L)}. \tag{1-13}$$

A mean decay constant λ is used for simplicity. More rigorously one may use perturbation theory to obtain a more accurate value for the attenuation factor.

Since $\bar{\lambda}$ is about 0.08 sec^{-1} it is evident from Eq. (1-13) that the loss of delayed neutrons from the core is appreciable when the core transit time τ_c (usually somewhat less than τ_L) is of the order of 10 sec or less. There is an associated loss of safety since less reactivity is required for accidental prompt criticality. Furthermore, changes in τ_c can cause changes in the reactor power as if they were caused by reactivity changes.

2 DYNAMICS

2.1 Reactivity Effects

If the moderator-coolant exerted no effect on the reactor power, then the dynamics analysis would be entirely thermal and hydraulic in nature, resembling analysis of boilers or heat exchangers. However, the heat source is influenced by virtue of various coefficients of reactivity and the neutron kinetics equations, thereby complicating the analysis. Therefore in addition to fundamentally thermodynamic and hydraulic effects, calculation of transient performance must include the neutron kinetics equations with various sources of reactivity. This section treats these reactivities and the following sections couple them into transients.

The reactivity effects listed in Table 2-1 are those which are considered for water reactors in short-term transients, i.e., those lasting minutes or less. Long-term transients (ex. xenon, burnup, etc.) exist, but are not unique with reactors of this chapter and typically do not constitute safety problems as such. It is seen in Table 2-1 that the reactivity effects listed may be classified according to their origin in fuel, moderator-coolant, or elsewhere.

For Doppler effect calculations (see chapter on Doppler Coefficient),

$$\frac{d\rho}{dT} = \frac{d(\ell np)}{dT} = \frac{1}{\sigma_r}\frac{d\sigma_r}{dT}\ell np\ . \qquad (2\text{-}1)$$

This relationship is due to the fact that the resonance escape probability p depends on the resonance capture cross section σ_r as $\exp(-\sigma_r \times$ a constant) and differential reactivity is dp/p. The effect of an increasing fuel temperature T on increasing the microscopic resonance capture cross section depends on the particular nuclide used as fertile material in the fuel. Since $p < 1$, the Doppler effect is always negative for U^{238} and Th^{232}. This reactivity effect, like others to be discussed in this section, must be computed in various regions in the reactor and then summed using importance weighting functions. The center of a reactor is usually the most dominating region for two reasons: its higher importance function and the larger changes in fuel temperature which occur there.

There are a variety of ways for fuel to move and thereby to induce reactivity effects. One way, which represents a possible hazard during fuel loading, is to add a fuel element too rapidly to an almost-critical reactor. The least reactivity would normally be added if the element were at the edge. In the one-group, one-region cylindrical reactor this would be

$$\Delta\rho = \frac{2M^2\,(2.405)^2}{k_\infty\,(R+\lambda)^3}\,\Delta R = \frac{M^2\,B_r^2}{Nk_\infty\,[1+(\lambda/R)]}\ , \qquad (2\text{-}2)$$

where N is the number of fuel elements present within the radius R which has a radial reflector savings λ. However, if the fuel element were added to a water-filled hole near the center of the reactor the reactivity added is likely to be many times that of Eq. (2-2). This is because of the importance of the location and, in the case of H_2O reactors, the elimination of a neutron-absorbing water hole. On the other hand, when resonance escape is low, as in tightly packed D_2O reactors, removal of fuel from the interior may cause a reactivity gain whenever reduction in resonance capture predominates over reduction in fissions.

Fuel motion in homogeneous reactors comes about by virtue of flow processes. Control in this manner has been treated in Sec. 1.5. Anomalous fuel motion has troubled the HRT: precipitation of fuel in certain areas caused power surges [12]. Slurry reactors are also subject to power surges, such as those resulting from sudden increases of the flow velocity when this velocity is of the order of the particles' settling velocity[13].

Thermal expansion of the fuel causes a change in the fuel to coolant-moderator ratio, which may be calculated by purely geometric considerations. This leads to a positive or negative reactivity effect depending on whether the reactor is over-moderated or under-moderated. Distortions, like bowing, may also be thermally induced. (These, along with thermal expansions, are discussed in Sec. 3.5 of the chapter on Fuel Elements, Sec. 7.4 of The Reactor Core and Sec. 1.3 of Fast Reactor Kinetics.) As far as reactivity effects are concerned, phase changes in the fuel, by virtue of associated density and volume changes, are similar to other thermal effects—i.e., are also treated by reactivity calculations as a function of fuel-to-moderator ratio.

Changes in core height ΔZ or in radius ΔR may result from thermal expansion, water level increases, or the addition of fuel solution. Reactivities are given by Eqs. (1-6) and (2-2).

Table 2-1 suggests five ways in which the moderator may be responsible for reactivity effects. In four of these the density or local moderator volume changes and in one the H/D ratio changes. In addition, when the water temperature increases, the neutron temperature increases. This alone usually leads to a net negative reactivity because of the non-1/v nature of the U^{235} cross section and the larger thermal migration area L^2.

TABLE 2-1

Principal Causes of Reactivity Effects in Short-Term Transients in Water Reactors

Fuel effects:	Doppler
	Fuel motion (including bowing)
	Core expansion (including level changes)
	Fuel phase changes
Moderator effects:	Moderator temperature
	Moderator volume (including steam voids)
	Moderator isotopic constitution (including spectral shift)
	Radiolytic gas
	Gas microbubbles
Other effects:	Poison changes
	Neutron shifting

This is true in spite of positive reactivity contributions from the flattening of the intracell flux fine structure and the non-1/v nature of the Pu^{239} cross section if Pu is present.

Steam voids, radiolytic gas, and gas microbubbles all affect reactivity in the same way, although production mechanisms differ. Absorption of the latent heat of vaporization or the energy of molecular dissociation are the mechanisms responsible for the first two. Evidence also exists that gas—either air or previously formed radiolytic gas—which adheres to a fuel element may be released from it by heating [14]. This dilation or growth of microbubbles is distinct from the release of dissolved gases in the water.

A list of the specific effects on reactivity of a change in moderator density is given in Table 2-2. For the most part these also apply to a hydrogen content reduction in the coolant when it is partially replaced by deuterium. Virtually every parameter entering into a criticality calculation is affected. However, the dominating effects are usually moderation at resonance capture energies, thermal capture (for H_2O and organics), and neutron age. Usually the sum of all these component contributions is a negative reactivity effect for density decreases. A typical magnitude for the over-all effect in H_2O boiling reactors at their operating point may be of the order of -0.1% reactivity per percent reduction in moderator density. Because of partial cancellation of positive and negative effects the fractional precision in calculated void coefficients may be poor for coefficients smaller than this unless normalized to experiment.

On the other hand, a positive reactivity effect may result if the (H_2O or organic) moderator thermal capture or fast fission (in D_2O reactors) effects dominate. Thus in the OMRE a positive power coefficient (which caused unstable 0.0016 cycles/sec oscillations) was due to the positive temperature coefficient of reactivity of the moderator between the fuel elements, even though the coolant within the fuel element had a negative temperature coefficient [15]. It is also possible in some reactors to have a positive void coefficient in one region, such as the center, and negative elsewhere.

Boiling reactors with integral superheaters may be best characterized by having separate void coefficients for the boiling and superheating regions. However in the superheating region it is more usual to speak of a total flooding reactivity change: If positive, it is the reactivity gain when the empty superheater tubes are filled with water in the liquid phase. The reason for defining this coefficient is for purposes of having a useful number to characterize the possible hazards associated with the startup or shutdown of the superheater section of this reactor type; it is usual for transition between the liquid and steam phase to take place then.

TABLE 2-2

The Various Components of the Reactivity Effect Due to Decreasing the Moderator Density in a Typical One-Region Reactor with Thick Axial and Radial Reflectors

Quantity	Explanation	Sign of the contribution to reactivity
Fast effect ϵ	Less scattering of fission spectrum neutrons to below the fast fission threshold.	+
Age τ	Less slowing down to thermal and scattering within the reactor boundaries, and thus more fast leakage.	-
Resonance escape p	Increased resonance capture because of less slowing down.	-
	A reduction in the effective resonance capture cross section because of less scattering between surfaces involved in the Dancoff correction.	+
Thermal utilization f	Less thermal neutron capture in the moderator due to less moderator.	+
Thermal migration area L^2		-
f	Additional thermal capture outside of the fuel due to a flattening of the intracell fine structure flux shape.	-
L^2		+
L^2	Less thermal scattering within the reactor boundaries, and thus more thermal leakage.	-
Reproduction constant η	Hardening of the spectrum.	-
Buckling B_g^2	Enhanced reflector effectiveness on a core which is leaking more neutrons. (In the absence of a top reflector, if core size expansion accompanies decreasing moderator density, leakage is also reduced.)	+

The Pathfinder reactor was designed with a calculated flooding reactivity of +0.2% at the beginning of core life and a negative reactivity effect later in core life [8b].

Experimental evidence of the different effect of steam voids in a central integral superheater as compared to their effect in a surrounding boiler zone were found on the BORAX-V [16]. In the process of nuclear heating at low power, 0.8 Mw, larger flux variations were observed, due to density variations in flooded superheater channels, than those variations encountered in operation of the boiler core alone.

The final reactivity effect in Table 2-1 is neutron shifting[17]. When the moderator or moderator-with-fuel solution is forced out both the top and bottom of a core of height Z at the same time, then, because the neutrons or delayed neutron precursors are physically carried out, the reactivity loss is:

for heterogeneous systems,

$$\Delta\rho \cong -w\ell/Z \; ; \tag{2-3}$$

for homogeneous systems,

$$\Delta\rho \cong -w\beta/\bar{\lambda}Z \; . \tag{2-3a}$$

Eq. (2-3) is for heterogeneous stationary fuel reactors (where migrating neutrons are removed) while Eq. (2-3a) is for homogeneous moving fuel reactors (where delayed neutron precursors are removed). Here w is the magnitude of the equal upward and downward velocities of the moderator or moderator with fuel solution. The presence of the prompt neutron lifetime ℓ in the numerator leads to small reactivities for heterogeneous reactors. However, for homogeneous reactors the quotient of the delayed neutron fraction and the mean delayed neutron decay time $\beta/\bar{\lambda} \simeq 0.1$ sec can lead to significant reactivities in Eq. (2-3a). The fuel transport which also takes place in homogeneous reactors has been noted already in Sec. 1.4.

2.2 Dynamics Equations - Heterogeneous Non-Boiling

Power reactor design largely consists of examining the suitability of the reactor and associated equipment under steady-state (usually full power) conditions. However, transients in reactor power which are expected in normal operation as well as under abnormal circumstances must also be investigated. Usually the transients impose more stringent safety criteria on the design. For example, although a steady-state margin of 2:1 may exist between the burnout heat flux and the maximum heat flux, this margin might be reduced during load swings near full power. It is the purpose of these sections to outline the fundamental concepts to be considered in an analysis of transients.

Analysis of the dynamic behavior of a reactor during a given transient process consists of

a) taking into account all the effects which can contribute significantly to reactivity changes;
b) using these reactivity changes in conjunction with the neutron kinetics equations to obtain the time behavior of the power; and
c) using this power along with externally induced changes of the reactor's dynamic variables to establish the effect of the transient on the reactor, especially any possible damages.

Evidently a feedback process exists because the power from b) is one of the factors to be considered in a). In this section both a) and b) will be considered for nonboiling reactors. A further restricting assumption here is that linear differential equations are used to describe the transients. Thus simplified forms of the Laplace transform solutions to these equations may be presented in such a way as to see the fundamental features of the transients.

More exact solutions are possible by making fewer approximations. In fact, a number of codes such as the ART programs [18, 19] are available which numerically solve the conservation of energy, momentum, and mass equations in space and time, (see Fluid Flow chap., Sec. 5.1) the results then being used with reactivity coefficients in the neutron kinetics equations[20]. It is recommended that methods along these lines be used when precision is desired. The dynamic effects described here are the significant ones which these more precise analyses take into account.

There are a number of treatments of the transfer function relating average fuel temperature to reactor power $T(s)/\phi(s)$ [21-25]. The solutions for unclad uniformly heated slab or cylinder (i.e., a single region of homogeneous composition heated by a constant power per unit volume) with constant temperature boundaries are respectively (see also Heat Transfer chapter, Sec. 1.5, for a general treatment of transient heating)

$$\frac{T(s)/T^0}{\phi(s)/\phi^0} = \left(1 - \frac{\tanh\sqrt{\tau s}}{\sqrt{\tau s}}\right)\frac{3}{\tau s} \cong \frac{1}{1+(2/5)\,\tau s} \tag{2-4}$$

or

$$\frac{T(s)/T^0}{\phi(s)/\phi^0} = \left(1 - \frac{2I_1(\sqrt{\tau s})}{\sqrt{\tau s}\,I_0(\sqrt{\tau s})}\right)\frac{8}{\tau s} \cong \frac{1}{1+(1/6)\,\tau s} \, . \tag{2-5}$$

Here the fuel time-constant τ is $c\rho\delta^2/k$ and $c\rho r^2/k$ for slabs of half-thickness δ and cylinders of radius r respectively, c is the heat capacity, ρ the density, k the conductivity, and $s = i(2\pi f)$ is the Laplace variable. These time constants, apart from a constant numerical factor, are the same as those encountered when the surface temperature of an unheated slab or cylinder is suddenly changed[26].

The numerator of these transfer functions, $T(s)/T°$ is the ratio of the spatially averaged fuel temperature change during the transient to the spatially averaged steady-state fuel temperature increase caused by the steady power $\phi°$. The denominator is the ratio of the reactor power change during the transient $\phi(s)$ to the average reactor power $\phi°$. The power referred to in these sections is of course the power generated in the fuel and does not include the small fraction of power generation originating elsewhere in the reactor. Physically, the effect of the time constants, $(2/5)\tau$ or $(1/6)\tau$ is to cause the fuel temperature changes to lag behind slow reactor power changes by approximately these same time intervals. They also cause rapid (i.e. $2\pi f\tau >> 1$) reactor power changes to be attenuated when these are observed by means of the fuel temperature. Generally speaking, large τ values favor safety because of their attenuating these high frequencies, thus reducing the likelihood of rapid oscillations.

The transfer function between changes in the power appearance at the surface of the fuel element ψ, and the reactor power changes ϕ, also contains a delay proportional to τ:

$$\frac{\psi(s)}{\phi(s)} = \frac{\tanh\sqrt{\tau s}}{\sqrt{\tau s}} \cong \frac{1}{1 + (1/3)\tau s} \tag{2-6}$$

or

$$\frac{\psi(s)}{\phi(s)} = \frac{2I_1(\sqrt{\tau s})}{I_0(\sqrt{\tau s})} \cong \frac{1}{1 + (1/8)\tau s} \tag{2-7}$$

for slabs and cylinders respectively.

While the preceding four equations do not take the clad or surface film into account, this is readily done. Thus for slabs, τ in Eqs. (2-4) and (2-6) is replaced by [22]

$$\tau \to \tau \left[\frac{1 + \frac{5}{2}\frac{R' + h^{-1}}{R} - \frac{5\tau'}{4\tau}\left(1 + \frac{R' + 3h^{-1}}{R}\right)}{1 + (3/R)(R' + h^{-1})} + \frac{5}{2}\frac{R' + h^{-1}}{R} + \frac{5\tau'}{4\tau}\left(1 + \frac{2}{R'h}\right)\right] \tag{2-8}$$

or

$$\tau \to \tau\left[1 + 3\frac{R' + h^{-1}}{R} + \frac{3\tau'}{2\tau}\left(1 + \frac{2}{R'h}\right)\right], \tag{2-9}$$

respectively. τ' is $c'\rho'\delta'^2/k'$ for the cladding of heat capacity c', density ρ', thickness δ', and conductivity k'; R and R' are the thermal resistances, δ/k and δ'/k', for the fuel and cladding; and h is the heat transfer coefficient. Evidently the effect of cladding and surface film is to increase the time constant of the fuel.

The effect of a change in power in a uniformly heated slab fuel element of height Z on the outlet coolant temperature θ for the geometry of Fig. 2-1 is computed to be[27]

$$\frac{\theta_2(s)/\Delta\theta^0}{\psi(s)/\phi^0} = \frac{1 - \exp(-\tau_c s)}{\tau_c s}. \tag{2-10}$$

The time constant

$$\tau_c = \frac{Z}{w}\left(1 + \frac{c\rho\delta + c'\rho'\delta'}{c_f\rho_f\delta_f}\right) \tag{2-11}$$

is the effective heat transport time by the coolant whose heat capacity, density, and half-thickness of coolant channel are c_f, ρ_f and δ_f. The steady-state temperature rise $\Delta\theta^0$ of the coolant, having a constant velocity w, for a power ϕ^0 in the half-slab element, is

$$\Delta\theta^0 = \frac{\phi^0 Z}{c_f\rho_f\delta_f A w}, \tag{2-12}$$

where A is the surface area on one side of the slab element.

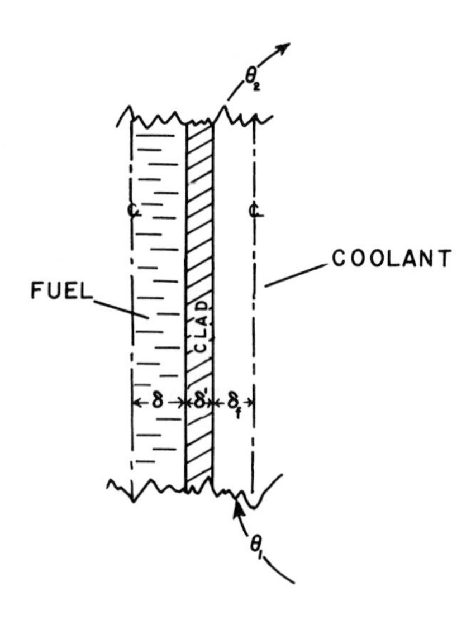

FIG. 2-1. Idealized slab geometry used in obtaining the transfer functions among the dynamic variables of heterogeneous nonboiling power reactor.

The incoming coolant temperature θ_1 influences the outlet temperature θ_2 by a simple time delay τ_c Thus the Laplace transform of the simple physical relationship

$$\theta_2(t) - \Delta\theta^0 = \theta_1(t - \tau_c) \tag{2-13}$$

gives the transfer function

$$\frac{\theta_2(s)}{\theta_1(s)} = e^{-\tau_c s}. \tag{2-14}$$

When the inlet coolant velocity w_1 changes, the transfer function relating this to the outlet temperature has been computed to be[27]

$$\frac{\theta_2(s)/\Delta\theta^0}{w_1(s)/w_1^0} = -\left(\frac{1}{1 + (0.8/Rh)\tau s}\right)\left(\frac{1 - e^{-\tau_c s}}{\tau_c s}\right), \tag{2-15}$$

where the heat transfer coefficient h has been considered to vary as the 0.8 power of the velocity. A comparison of the dependence of Eqs. (2-10) and (2-15) on τ_c with that of Eq. (2-14) shows that power and velocity changes manifest themselves in half the time required for inlet temperature changes to affect the reactor - since

$$[1 - \exp(-\tau_c s)]/\tau_c s = [1 + (\tau_c s/2)]^{-1}$$

and

$$\exp(-\tau_c s) \simeq (1 + \tau_c s)^{-1}.$$

This occurs because the velocity and power changes effectively act in the reactor center rather than at the inlet.

Since the reactor is coupled to the remainder of the plant by pipes, it is evident that the relationships in Eqs. (2-13) and (2-14) apply for temperature transport:

$$\frac{\theta_1(s)}{\theta_0(s)} = \exp(-\tau_{01}s) \tag{2-16}$$

$$\frac{\theta_3(s)}{\theta_2(s)} = \exp(-\tau_{23}s) , \qquad (2\text{-}17)$$

where θ_0 and θ_3 refer to temperatures in a pipe some distances before and after the reactor respectively. Designating the velocities in these pipes as w_{01} and w_{23}, Eq. (2-11) may be used to evaluate τ_{01} and τ_{23} except that the ratio of the fuel element's heat content per degree to the heat content of the coolant is replaced by the ratio of the pipe heat content per degree to that of the coolant. Table 2-3 summarizes these and the other time constants of the reactor system already discussed.

TABLE 2-3

Time Constants Found in Heterogeneous Non-Boiling Power Reactors

Fuel slab or rod	$\tau = c\rho\delta^2/k$ or $c\rho r^2/k$
Cladding	$\tau' = c'\rho'\delta'^2/k'$
Coolant transport in core	$\tau_c = \frac{Z}{w}\left[1 + \frac{c\rho\delta + c'\rho'\delta'}{c_f\rho_f\delta_f}\right]$
Coolant transport in round pipe length Z_{ij}	$\tau_{ij} = \frac{Z_{ij}}{w_{ij}}\left[1 + \frac{c'\rho' 2\pi\,[r + (\delta'/2)]\,\delta'}{c_f\rho_f\pi r^2}\right]$

The physical significance of Eqs. (2-10), (2-14), (2-15), (2-16) and (2-17) is more readily seen in the time domain, although they are written here as a function of s. In all cases they describe a time lag of the numerator of the left hand side behind an excitation caused by the denominator. The time constant or constants on the right hand side determine the length of this time lag. As already noted for the fuel, long time constants favor safety because transient processes in the core then occur slowly, lessening the likelihood of uncontrollable short positive reactor periods.

To apply Eqs. (2-4) through (2-17) to the solution of a reactor dynamics problem, it is necessary to couple the equations to the reactor power by means of the reactivity and the neutron kinetics equations:

$$\frac{\phi(s)}{\phi^0} = G_0(s)\left[\frac{d\rho}{dT}\,T(s) + \frac{d\rho}{d\theta}\,\theta(s) + \rho_{in}(s)\right] , \qquad (2\text{-}18)$$

where the zero power reactor transfer function $G_0(s)$ is the Laplace transform of the neutron kinetics equation. Here $d\rho/dT$ and $d\rho/d\theta$ are average temperature coefficients for the fuel and moderator-coolant, and $\rho_{in}(s)$ is any other perturbing reactivity. In the simplified one-region reactor model here, $\theta(s)$ is taken to be $[\theta_1(s) + \theta_2(s)]/2$. However, it is usual in reactor transient calculations to obtain greater precision by assuming several core regions, each with its own dynamics equations, and all reactivity effects are then summed in Eq. (2-18).

Figure 2-2 couples all these relationships as they would be solved on an analog or digital computer. Each box represents computer equipment having the same transfer function between the incoming and outgoing variables as the reactor itself. Omitted here are less important internal effects, such as the effect of coolant velocity and inlet temperature on the reactivity through fuel temperature. Also all external effects are omitted. Dynamic performance with the particular intent of discovering reactor instabilities of hazardous reactor transient conditions may nevertheless be readily studied: Various assumed reactivity, coolant velocity, or coolant temperature transients may be used to see the corresponding effects on reactor power and fuel temperature.

The coupling between the reactor and its associated power plant may be loose or strong, depending on the transient and the system constants. For example, a pulsating flow caused by an instability originating in the power plant indeed will cause the reactor power to oscillate. However, whether this feeds back into the power plant so as to enhance the flow pulsations depends on whether the coupling in strong or weak. Such a problem as well as others (e.g., reductions in coolant flow, changes in plant load, and resonances generated in pressurizers) should be solved by mathematically coupling the transfer function of the heat exchangers, pumps, turbines, etc. to those of Fig. 2-2.

2.3 Dynamics Equations - Heterogeneous Boiling

Describing the dynamic behavior of boiling reactors is somewhat more complex than describing the dynamic behavior of many other types of reactors because of the presence of an additional (gas) phase in the core and because of the strong influence of system pressure on the reactor. However, as in the preceding section, the use of transfer functions to describe small deviations from steady state makes it possible to obtain relatively simple, though approximate, expressions for the important effects [22, 29, 30, 31]. More exact solutions of boiling reactor transients are possible using digital codes or rather extensive analog computer networks[32]. It is recommended that these approaches be used when precision is essential. The purpose of the discussion in this section, as in the preceding section, is merely to give an insight into the principal effects without encumbering them with undue complexity.

Figure 2-3 illustrates the processes involved here using a simplified model of a boiling reactor in which only two regions in the axial direction are used. Except for the obvious change in the value of the fuel-coolant heat transfer coefficient h between a boiling coolant and a nonboiling coolant, the dynamics of the fuel element for boiling reactors are identical with those of nonboiling reactors. Therefore, the transfer functions between $\phi(s)$ and either T(s) or $\psi(s)$, given in Eqs. (2-4) through (2-7), are valid here. However, starting from the point at which the heat enters the water, the dynamics of boiling reactors is different. As in the previous section, this treatment is confined to the reactor itself. Thus Fig. 2-3 in particular does not show the volumes, pipes, heat exchangers or other equipment receiving the steam or returning the water. Also it is not necessary here to make a distinction between single and dual cycle operation since this manifests itself already in the heat exchangers. However, since changes in returning feedwater are considered here, transients encountered in dual cycle operation may be studied.

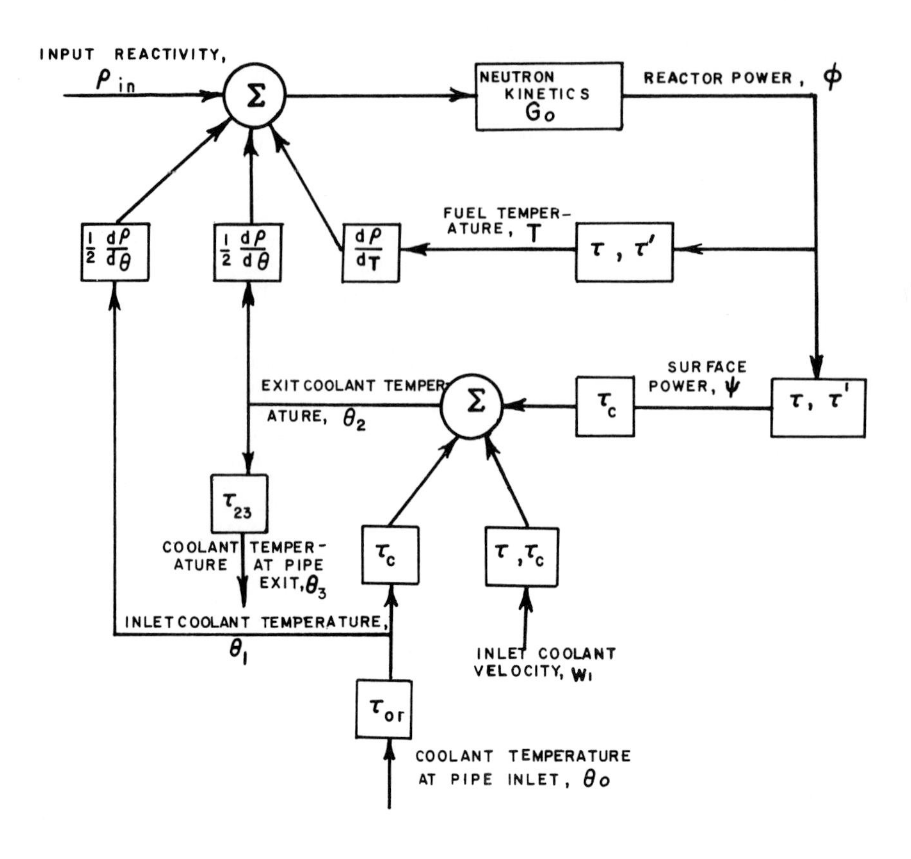

FIG. 2-2. Block diagram showing the interrelationships among all the dynamic variables of a heterogeneous nonboiling power reactor. Table 2-3 identifies the time constants of the transfer functions symbolized by boxes here.

The location of the boiling boundary X^o at steady state for uniform axial heating is determined from

$$\frac{X^0}{Y^0} = \frac{\text{heating of feedwater to saturation}}{\text{evaporation of saturated water}}$$

$$= \frac{H_f - H_1}{H_{fg}}, \qquad (2\text{-}19)$$

where H_f-H_1 and H_{fg} are the enthalpy changes involved in raising the entering liquid to the boiling point and in evaporating it respectively. Transient effects from the power, pressure, or feedwater, listed below, can induce changes in this boundary, which lead to changes in the void volume according to the transfer function

$$\frac{v_1(s)/v_1^0}{X(s)/X^0} = -\frac{1-\exp(-\tau_{c2}s)}{\tau_{c2}s} \simeq -\frac{1}{1+(\tau_{c2}s/2)}, \qquad (2\text{-}20)$$

where τ_{c2} is the transit time of steam in the upper core region i.e., Y° divided by the steam velocity. The quantity v_1° is a fictitious steam volume:

$$v_1^0 = \frac{\phi^0 X^0/Z}{\rho_g H_{fg}} \tau_{c2} \,. \qquad (2\text{-}21)$$

Equation (2-20) shows that the effective void changes in the core are delayed according to the transit time of the steam when these void changes are induced by changes X(s) in the boiling boundary. This is because when this boundary (between the lower single-phase and the upper two-phase mixture) changes, the steam volume associated with this change moves through the upper zone. It arrives at the midpoint of this upper zone a time $\tau_{c2}/2$ after the motion of the boiling boundary.

In the lower reactor zone, the power changes occurring at the surface of the fuel ψ are one of the principal means by which changes may be induced in the boiling boundary. As might be expected, the transfer function contains a lag due to the coolant transport time:

$$\frac{X(s)/X^0}{\psi(s)/\phi^0} = -\frac{1-\exp(-\tau_{c1}s)}{\tau_{c1}s} \simeq \frac{1}{1+(\tau_{c1}s/2)} \qquad (2\text{-}22)$$

If in Eq. (2-11) one uses the distance X^o rather than Z, then the effective coolant transport time τ_{c1} is computed. The effect of this power change on the volume-averaged reactor water temperature θ(s) is

$$\frac{\theta(s)/\Delta\theta^0}{\psi(s)/\phi^0} = \frac{1-(1/\tau_{c1}s)\,[1-\exp(-\tau_{c1}s)]}{\tau_{c1}s/2}$$

$$\cong \frac{1}{1+(\tau_{c1}s/3)}, \qquad (2\text{-}23)$$

where the volume-averaged coolant temperature rise over the entire core is

$$\Delta\theta^0 = \frac{1}{2}\frac{\phi^0 X^0/Z}{c_f W^0(1+\Re)}\frac{X^0}{Z}, \tag{2-24}$$

$\Re$ is the ratio of total flow less feedwater flow to feedwater flow, and W^0 is the steady-state mass flow rate of feedwater.

In the upper zone the power at the surface of the fuel directly causes the steam void volume changes $v_2(s)$. Again the steam transit time is the time lag:

$$\frac{v_2(s)/v_2^0}{\psi(s)/\phi^0} = \frac{1-(1/\tau_{c2}s)\,[1-\exp(-\tau_{c2}s)]}{\tau_{c2}s/2} \cong \frac{1}{1+(\tau_{c2}s/3)}. \tag{2-25}$$

Here the steady-state void volume is

$$v_2^0 = \frac{\phi^0 Y^0/Z}{\rho_g H_{fg}}\frac{\tau_{c2}}{2}. \tag{2-26}$$

As seen in Fig. 2-3, a reactor mass steam flow rate S_2 is due to this upper zone power generation, just as S_1 may be ascribed to the lower zone. Evidently the transfer functions are of unit magnitude:

$$\frac{S_2(s)/S_2^0}{\psi(s)/\phi^0} = 1 \tag{2-27}$$

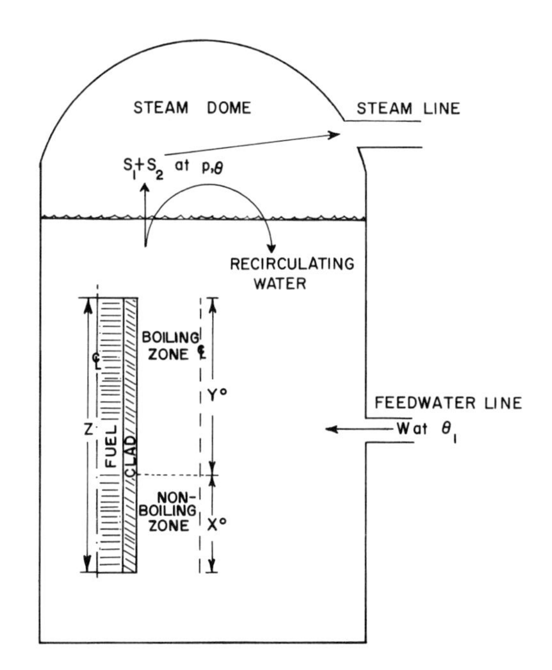

FIG. 2-3. Regions and circulation patterns used in an idealized transfer function analysis of a heterogeneous boiling power reactor.

$$\frac{S_1(s)/S_1^0}{X(s)/X^0} = -1, \tag{2-28}$$

where S_1° and S_2° are the fractions of the total power generated in zones 1 and 2, respectively, divided by the heat of vaporization per unit mass H_{fg}.

Pressure changes must be determined from the steaming rate by using the particular dynamics equations peculiar to each reactor's heat removal system. The effect of a pressure on the boiling boundary is to relocate it immediately according to the new saturation temperature, followed by a pure delay $\exp(-\tau_r s)$ in returning to its original position:

$$\frac{X(s)/X^0}{p(s)} = \frac{(dH_f/dp)W^0(1+\Re)}{\phi^0 X^0/Z}[1-\exp(-\tau_r s)]. \tag{2-29}$$

Here τ_r is the recirculation time, i.e., the average time for all the water in its liquid phase to circulate once around the system. After this time the boiling boundary returns to its original position - as can be shown by again computing X° at the new conditions.

The effect of pressure on coolant temperature is dictated by the differential relation between saturation temperature and pressure $d\theta/dp$:

$$\frac{\theta(s)}{p(s)} = \frac{d\theta}{dp}\left[\frac{Y^0}{Z}+\frac{X^0}{Z}\exp(-\tau_r' s)\right]. \tag{2-30}$$

The effect on the water in the nonboiling zone is not immediate and τ_r' is the time lag associated with the transport of subcooled water from the feedwater distribution point to the core.

Flashing or condensation within the boiling zone is fundamentally due to a rate of change of pressure [whose Laplace transform is $sp(s)$]:

$$\frac{v(s)/v_2^0}{sp(s)} = \frac{\dfrac{m_2(dH_f/dp)}{\phi^0 Y^0/Z}\left[1+\dfrac{2}{3}\left(\dfrac{\rho_g(dH_g/dp)}{\rho_f(dH_f/dp)}-1\right)\gamma_2\right]}{1+\dfrac{\tau_{c2}}{3}\left[1+\dfrac{1}{6}\left(\dfrac{\rho_g(dH_g/dp)}{\rho_f(dH_f/dp)}-1\right)\gamma_2\right]s}, \tag{2-31}$$

where m_2 is the mass of water in the liquid phase in the upper zone, γ_2 is the volume fraction of the steam in the upper zone's coolant passage, and H_g is the steam enthalpy. In the limit of small γ_2 this transfer function differs only by a constant from boiling zone surface power to void transfer function, Eq. (2-25).

The effects of feedwater mass flow rate W and temperature θ_1 on the boiling boundary and average core temperature are essentially contained in Eqs. (2-13) and (2-14) since in boiling reactors the boiling boundary changes inversely as the temperature of water arriving at X^0:

$$\frac{X(s)/X^0}{\theta_1(s)/\theta_1^0} = -e^{-\tau_{c1}s}e^{-\tau_r' s}, \tag{2-32}$$

$$\frac{X(s)/X^0}{W(s)/W^0} = \frac{1}{1+0.8(\tau/Rh)s}\,\frac{1-e^{-\tau_{c1}s}}{\tau_{c1}s}\,e^{-\tau_r' s} \cong \exp\left[-\left(\tau_r'+\frac{1}{2}\tau_{c1}+0.8\frac{\tau}{Rh}\right)s\right], \tag{2-33}$$

$$\frac{\theta(s)}{\theta_1(s)} = \frac{X^0}{Z}\frac{1-e^{-\tau_{c1}s}}{\tau_{c1}s}e^{-\tau_r'}$$

$$\cong \frac{X^0}{Z}\exp\left[-\left(\tau_r' + \frac{1}{2}\tau_{c1}\right)s\right], \quad (2\text{-}34)$$

$$\frac{\theta(s)/\theta^0}{W(s)/W^0} = -\frac{1}{1+0.8(\tau/Rh)s}\frac{1-[(1-e^{-\tau_{c1}s})/\tau_{c1}s]}{\tau_{c1}s/2}e^{-\tau_r's}$$

$$\cong -\exp\left[-\left(\tau_r' + \frac{1}{3}\tau_{c1} + 0.8\frac{\tau}{Rh}\right)s\right], \quad (2\text{-}35)$$

where

$$\theta_1^0 = \frac{H_f - H_1}{[c_f]_{\theta_1}}. \quad (2\text{-}36)$$

To solve a boiling reactor kinetics problem, Eqs. (2-4) through (2-7) and Eqs. (2-19) through (2-36) are combined with reactivity coefficients in the Laplace transform of the neutron kinetics equations:

$$\frac{\phi(s)}{\phi^0} = G_0(s)\left\{\left[\frac{d\rho}{dT}\right]T(s) + \left[\frac{d\rho}{d\theta}\right]\theta(s) + \left[\frac{d\rho}{dv}\right][v_1(s) + v_2(s) + v(s)] + \rho_{in}(s)\right\}. \quad (2\text{-}37)$$

The three reactivity coefficients, $d\rho/dT$, $d\rho/d\theta$ and $d\rho/dv$ are for fuel temperature, water temperature, and steam void volume respectively.

Figure 2-4 couples these equations using the time constants in Table 2-4 as in the manner of their solution on an analog computer [22, 29] As in Fig. 2-2, only the principal effects are shown and dynamic relationships external to the reactor (such as relating pressure to steaming rate) are omitted here. Under certain circumstances, for example when the system is at atmospheric pressure, it becomes important to take into account time delays associated with water acceleration plus the effects of hydrostatic pressure in the core[33]. If steam flows approaching the speed of sound are encountered, effects discussed in Sec. 2.2.2 of the Fluid Flow chapter must be considered.

In addition to using Eq. (2-37) to test for stability, (i.e. whether $\phi(s)/\phi^\circ$ is finite for excitations $\rho_{in}(s)$ at all frequencies), it is commonly used to solve the transients induced by the reactor's heat removal equipment. One of the most common transients studied in hazards analysis is the effect of a sudden reduction or even stoppage of primary steam flow. This would result from a closure of a valve in a primary steam line, (as by a turbine trip in a direct cycle system). The steam flow rate from the core is used to compute the pressure transient in the system according the thermodynamics of the particular heat removal system used by the reactor. The equations

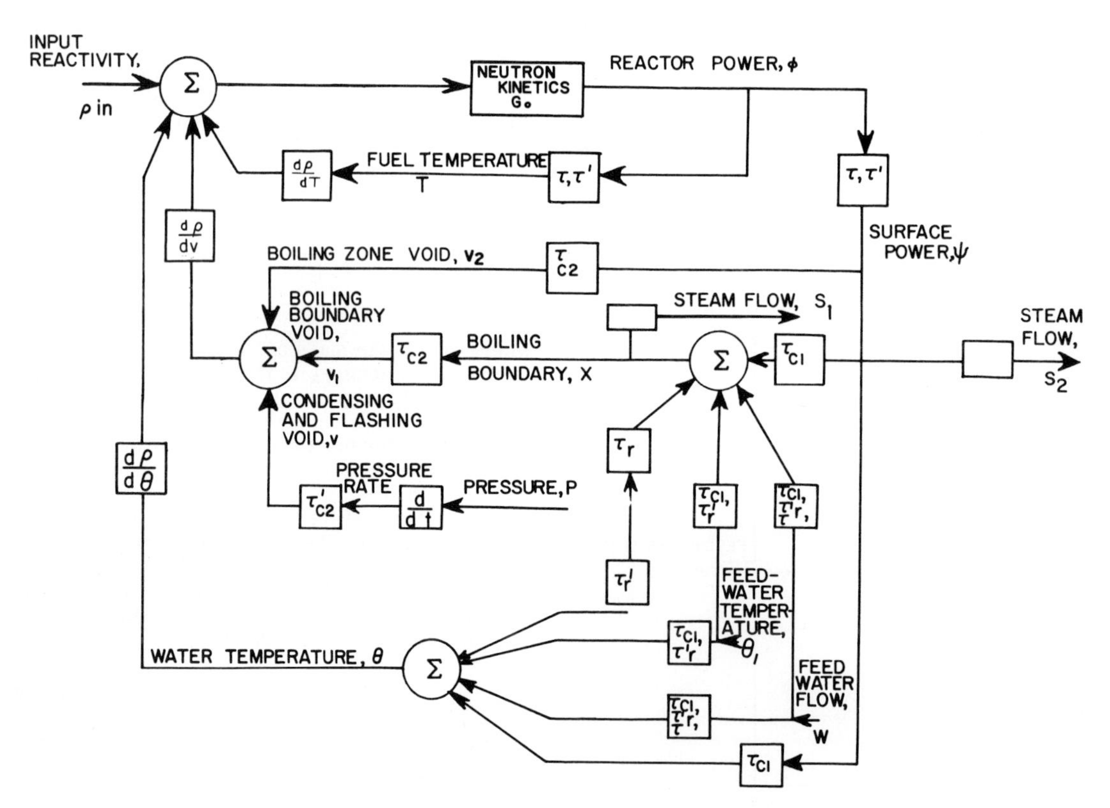

FIG. 2-4. Block diagram showing the interrelationship among all the dynamic variables of a heterogeneous boiling reactor. Tables 2-3 and 2-4 identify the time constants of the transfer functions symbolized by boxes here.

TABLE 2-4

Time Constants in Heterogeneous Boiling Power Reactors

	Formula	Typical value - that for EBWR [22]
Fuel slab or rod	$\tau = c\rho\delta^2/k$ or $c\rho r^2/k$	0.45 sec
Cladding	$\tau' = c'\rho'\delta'^2/k$	0.022 sec
Coolant transport in nonboiling zone	$\tau_{c1} = \frac{X^\circ}{w}\left[1 + \frac{c\rho\delta + c'\rho'\delta'}{c_f\rho_f\delta_f}\right]$	0.64 sec
Steam transport in boiling zone	$\tau_{c2} = Y^\circ/(\text{steam velocity})$	0.35 sec
Recirculation	$\tau_r = \frac{\text{mass of recirculating water}}{W^\circ \Re}$	9 sec
Subcooled water transport	$\tau'_r = \frac{\text{distance from feedwater inlet to core entrance}}{\text{mean recirculating water velocity}}$	4.5 sec

of this section, coupled as shown in Fig. 2-4, may then be solved using this pressure as an excitation function. In general the rising reactor power transient due to void reduction is large if the reactivity in steam voids, the pressure change, and the pressure rate of change are large. Safety in design is thus favored by large steam dome volumes and slow moving steam valves.

The effects of system pressure perturbation on boiling reactors are seen in Fig. 2-5 in which both a pressure decrease and a pressure increase were brought about by opening and closing a steam valve vented to the atmosphere in BORAX-II [34]. This test simulated a safety valve temporarily opening and reseating, or a load increase followed by a decrease. The reactor power decreased as the pressure fell because of flashing, Eq. (2-31), but more so from a lowering of the boiling boundary, Eq. (2-29). However, the power does not drop indefinitely from the negative reactivity effect

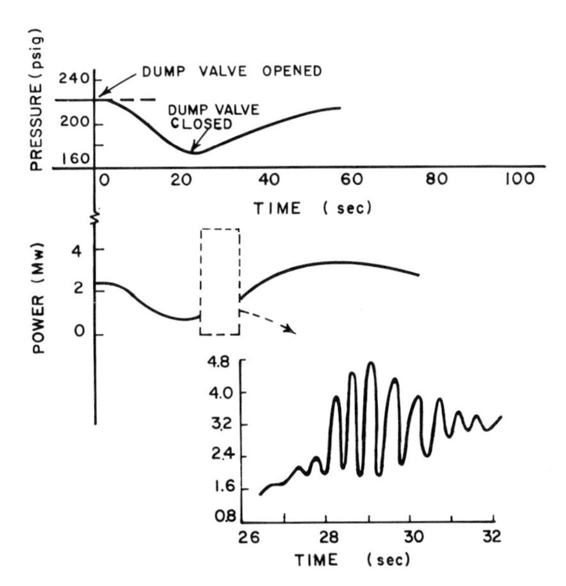

FIG. 2-5. Transient induced in BORAX II by opening and closing a valve dumping reactor steam to the atmosphere.

of these additional voids because the void production by the reactor heat falls off with appropriate delay as the power drops. Upon closing the dump valve as shown in Fig. 2-5, these effects are all reversed. A damped train of rapid power oscillations was induced because the reactor conditions were near the threshold of instability, and the reactivity addition rate shock-excited these oscillations.

2.4 Dynamics Equations - Homogeneous Boiling

Reactors, like the KEWB, differ from the above heterogeneous reactors in their dynamic behavior principally because gas is evolved almost instantly by the slowing down of fission fragments directly in the water coolant. Both steam voids and radiolytic gas may be formed. In the simple model here, no distinction is made:

$$\frac{v(s)/v^0}{\phi(s)/\phi^0} = \frac{1}{1+\tau_d s}\,\frac{1}{1+\tau_c s}. \tag{2-38}$$

Here v° is the void volume produced at steady state by the power ϕ°. The time constants τ_d and τ_c are the gas formation delay time and the mean bubble escape time respectively.

Water heating may be taken into account by

$$\frac{\theta(s)/\theta^0}{\phi(s)/\phi^0} = \frac{1}{1+\tau_c' s}, \tag{2-39}$$

where a temperature rise θ° is produced by ϕ° at steady state. The time constant τ_c' is the quotient of the unit volume heat capacity and unit volume heat removal rate of the coolant.

These two equations together with Eq. (2-37), (omitting its fuel temperature coefficient of reactivity) comprise the description of the dynamic behavior of a KEWB-type reactor. Typical values of 0.7 to 8, 1, and 100 sec for τ_d, τ_c, and τ_c' respectively have been used in KEWB [35], and experiments have been successfully explained.

2.5 Stability by Investigation of Transfer Functions

A stable reactor system, by definition, will have a bounded response to a perturbation of any of its input variables. This means in particular that a series of oscillations of ever increasing amplitude must not be capable of occurring. If the input perturbation is assumed to be a reactivity ρ_{in}, as from a control rod, then a stable system should have a bounded reactor transfer function,

$$G(s) = \frac{\phi(s)/\phi^0}{\rho_{in}(s)}. \qquad (2\text{-}40)$$

This quantity may be evaluated: (a) purely theoretically, using the equations of the form outlined in Secs. 2, 3, or 4 a priori; (b) partially experimentally by using experimental transfer functions between some variables, and relying on the theoretical relationships among others; and (c) entirely experimentally by measuring the response to control rod motion. The frequency range over which the evaluation is to take place must at least encompass resonances at which unstable oscillations might develop. In more extensive studies the frequency range may extend from below $(1/2\pi)$ divided by the system's longest time constant to above $(1/2\pi)$ divided by the system's shortest time constant.

Unstable tendencies in boiling reactors have been investigated in all three of these manners. A priori theoretical calculations [20, 22, 30, 31] indicated a high narrow resonance in G(s) due to the fact that the large negative reactivity effect of steam void formation was delayed 180° in phase at the resonant frequency by the fuel, water, and steam transit times. Rod oscillator measurements [36] confirmed the existence of this resonance and agreed well with the theory of Sec. 3. It was also demonstrated that the instability of SPERT-I could be explained by combining the theoretical zero power transfer function $G_0(s)$ with a measured dk/dv, and a measured transfer function between fuel element heat and water channel voids obtained in an out-of-pile electrically heated loop [37].

The concepts of gain and phase margin in stability theory have been successfully used in the EBWR transfer function measurements to accurately monitor stability in the approach to 100 Mw [39]. Figure 2-6 [22] indicates the equivalent but simpler technique of using measurements of the transfer function's resonance peak to obtain a means of extrapolating to less stable conditions at higher power. Here the reciprocal transfer function and the reactor power are analagous to the reciprocal neutron multiplication and the number of fuel elements in the more familar technique of extrapolating to a critical mass. It is not valid for boiling reactors to extrapolate all the way to the idealized unstable point, $1/G_{max} = 0$, because of the importance of random fluctuations and nonlinear effects, as instability (which then in fact becomes difficult to define uniquely) is approached. Rather than to use as a criterion some arbitrary safety margin away from the condition $1/G_{max} = 0$, it is more meaningful to judge a boiling reactor's stability on a relative basis: The factor by which its G_{max} is less than the G_{max} of a similar reactor which has already operated safely. This is qualitative but at least provides a working guide.

Where rod oscillator experiments are inconvenient, other methods may be used to obtain transfer function information. Such input variables as the feedwater flow may be oscillated. In addition, any variable, especially the reactivity of a control rod, may be changed as suddenly as possible from one value to another. The transfer function obtained from a response to a step in control rod position has been found to agree well with that obtained from rod oscillator tests in the case of the EBWR [31].

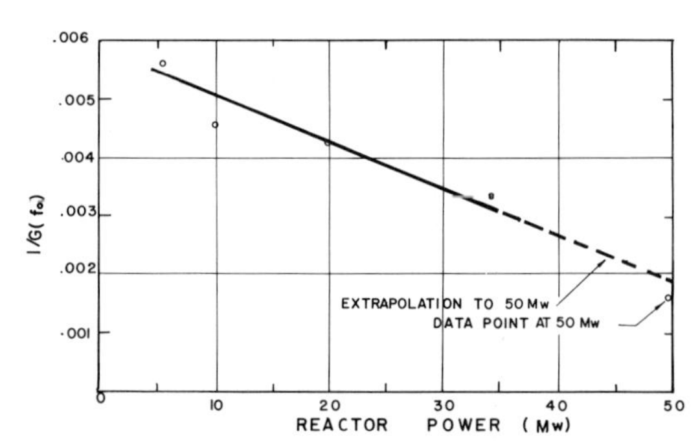

FIG. 2-6. Use of measurements of transfer functions in the EBWR to predict the degree of stability at higher power. The value of the transfer function at its resonant peak predicted by linear extrapolation is indicated by the broken line.

If a reactor has a tendency toward an oscillatory instability, this may be effectively and simply monitored by measuring the damping constant ζ in the damped oscillations which follow these small step excitations. The reactor power takes the form

$$\text{power} = \phi^0 + A\exp(-\zeta 2\pi ft) \sin(2\pi ft + \delta) \qquad (2\text{-}41)$$

in such instances. ϕ° is the final value attained by the reactor power, A is the amplitude and δ the phase angle of the transient oscillation starting at t = 0 on the termination of the excitation. The damping constant so obtained is related to the width Δf of the resonance peak of the transfer function measured with rod oscillator:

$$2f_0\zeta = \Delta f = \text{frequency separation between } f_1 \text{ and } f_2 \text{ on each side of the resonance frequency, } f_0 \text{ (such that } G(f_2) = G(f_1) = G(f_0)/\sqrt{2}). \qquad (2\text{-}42)$$

In boiling reactors tendencies toward instability currently are best judged on a relative basis. Therefore Table 2-5 [36, 38, 39] shows damping constants which have been measured for several reactors. If a reactor similar to one of these has a damping constant (found by calculation of G(s) or by actual measurement) which is not less than those tabulated, then it is at least as stable as the reactors to which it is compared in this table.

The magnitude of the resonant frequency itself deserves attention in stability investigations. It can be predicted theoretically by computing the transfer functions, as indicated above. The reciprocal of the resonant frequency can be estimated from the dominant time constants of the system: usually the sum of the fuel time constant and coolant transit times in boiling reactors. It is advantageous from the standpoint of intrinsic core safety that this frequency be low. Thus the 2 cycle/sec oscillations of Fig. 2-5 may be too fast for a sluggish control system to effect a scram before an un-

TABLE 2-5

Experimental Values of the Damping Constant, Resonance Width, and Resonant Frequency for Two Argonne Boiling Reactors

Reactor	Power Mw(t)	Pressure* (psig)	ζ	Δf (cycles/sec)	f_0 (cycles/sec)
BORAX IV	1.77	0	1.24	.67	.27
	2.54	0	.76	.76	.50
	3.40	0	.17	.27	.80
EBWR Core I	9.7	550	.68	1.51	1.11
	19.9	550	.39	.99	1.27
	50	600	.091	.32	1.76
EBWR Core 2	60	600	.25	.78	1.59
	71	600	.18	.65	1.77

*In c.g.s. units: 0 psig = 1 atm = 1.033 kg/cm^2; 550 psig = 39.70 kg/cm^2; 600 psig = 42.19 kg/cm^2.

desired power level is reached. (This in fact was the case on the EBWR at 100 Mw [39]: excessive capacitance in the high flux scram circuitry prevented scram during large high frequency oscillations.) However resonant frequencies of 0.2 cycle/sec and lower are found in large boiling reactors using oxide fuel, and should an instability develop the control system might more readily cope with it.

2.6 Stability by investigation of noise spectra

Unlike transfer function experiments which use a control rod or other external means of excitation, noise experiments depend on random processes within the reactor system itself to provide the necessary perturbation. If the reactor is being excited by a random reactivity function, whose Fourier spectrum is $|\rho_{in}(f)|^2$, then the measured Fourier spectrum of the reactor power fluctuations is given by

$$|\phi(f)|^2 = |\phi^0 G(f)|^2 \; |\rho_{in}(f)|^2 . \qquad (2\text{-}43)$$

Thus noise analysis gives information about the product of the transfer function and the input reactivity noise spectrum. The advantage of this method lies in the fact that a resonance or unstable tendency in either G(f) or ρ_{in} (f) will be detected in ϕ(f). A slight disadvantage is that most experimental methods merely measure the variation of ϕ(f) with frequency and do not make absolute measurements. This means that Δf and ζ in Sec. 2.5 can be determined, but not G_{max}.

ϕ(f) may be measured by any one of a variety of experimental techniques [40]. In all cases it is required that ρ_{in}(f) be of adequate magnitude so that ϕ(f) will exceed two kinds of noise background: first, random arrival of neutrons at the ion chamber and, second, electronic noise in the measuring equipment. Experiments have always shown these backgrounds to be exceeded for water reactors operating at power, even though the ion chambers are located some distance outside the core [40]. This is because temperature fluctuations in the inlet coolant of research reactors (such as the DMTR in Table 2-6) provides ample reactivity excitation. Also, in the case of boiling reactors, flow or temperature fluctuations in the feedwater mixed with recirculating water is theorized as being an adequately large noise source.

Table 2-6 indicates the variety of results which have been obtained from many noise experiments. Usually the search for an unstable tendency during power increases was a goal of the experiment. Where resonances in the spectrum $|\phi(f)|^2$ were detected, their width Δf, defined as [see also Eq. (2-42)]

TABLE 2-6

Summary of Representative Noise Analysis Experiments Conducted on Operating Water Reactors

Reactor	Power Mw(t)	Resonant frequency cycles/sec	Δf cycles/sec	Standard deviation of fluctuations Mw	Other results
BORAX IV	1.48	0.45		0.047	Resonance grows with increasing power
	3.13	0.8	0.3		
EBWR Core 1	20	1.0			Resonance grows with increasing power
	45.6	1.6	0.2	0.83	
EBWR Core 2	100	1.9	0.35	6.4	Resonance grows with increasing power
N.S. Savannah	55	0.07	0.05		Rolling sea increases noise level and resonance
Dresden	575	0.19	0.6		Some resonance found in flow pulsation in external riser pipes
HRE-2	2.26	0.12	0.05	0.0125	Resonance broadens when uranium is lost from system
DMTR	10			0.02	Fluctuations correlate with those of D_2O coolant temperature

$$2f_0\zeta = \Delta f = \text{frequency separation between } f_1 \text{ and } f_2 \text{ on each side of the resonant frequency } f_0 \text{ (such that } |\phi(f_2)|^2 = |\phi(f_1)|^2 = 0.5\,|\phi(f_0)|^2, \quad (2\text{-}44)$$

is given in Table 2-6. A sign of incipient instability is a decrease of Δf with increasing reactor power. From a knowledge of Δf and f_0, Eq. (2-44) may be used to obtain the same damping constant ζ as would be obtained from rod oscillator tests or step excitation experiments.

It has also been shown [41] possible to detect an approaching instability by measuring the standard deviation of the fluctuation ϕ (t) about ϕ°, the autocorrelation function at zero lag, or the total spectral noise, since these are all equal:

$$\sigma^2 = \frac{1}{T}\int_0^T |\phi(t) - \phi^0|^2\,dt = \int_0^\infty |\phi(f)|^2\,df\,. \quad (2\text{-}45)$$

σ^{-1}, as a function of increasing power, approaches zero as the unstable power is approached.

Though experimental investigations of reactor noise spectra have primarily been undertaken for the purpose of obtaining information about stability of power reactors, another application of importance to reactor safety is in use of noise measurements to determine the shutdown condition of a reactor. It is important to know the reactivity by which the reactor is subcritical, that is, the shutdown margin. (See also Criticality chapter, Sec. 2.) This is usually difficult to measure, but may readily be obtained from the shape of ϕ (f) at high frequencies. This is possible because the spectrum has a frequency dependence which contains k_{eff}:

$$|\phi(f)|^2 \sim \frac{1}{(2\pi f)^2 + [(1 - k_{eff}(1-\beta))/\ell]^2}\,. \quad (2\text{-}46)$$

Spectrum measurements in the University of Florida Training Reactor [42] and the Westinghouse CES reactor [43] have shown that it is possible to measure k_{eff} in subcritical systems using this technique. Although these are water-moderated reactors, this method is applicable to other reactors types.

3 POWER EXCURSIONS

3.1 Possible Causes

Transients involving periods $\lesssim 1$ sec, resulting from the rapid addition of substantial reactivity by one means or another, are termed "power excursions". The periods of power excursions are shorter than desirable in normal operation where period scrams are typically set to trip somewhere in the neighborhood of 1 sec to 20 sec. A characteristic feature of power excursions it the relative inability of the reactor's normal control system to control the event.

In most experimental studies of power excursions the transient is induced by ejecting a control rod from a critical reactor in a fraction of a second. Excursions are also studied by withdrawal of strong control rods at rapid rates. As long as the rate of reactivity addition somewhat exceeds the rate of development of negative reactivity effects due to an increasing power, it is possible to accumulate sufficient excess reactivity to obtain short periods.

Although experimental studies of excursions use control rod excitations, many other means of reactivity addition are possible. In a typical hazards analysis, reactivity addition mechanisms such as the following are postulated:

1) One or more control rods are steadily withdrawn at their maximum speed.
2) Some strong neutron absorber, normally a fixed part of the core, is suddenly withdrawn.
3) Fuel is added (or removed, in the case of some undermoderated cores) at a rapid rate. This could happen when a fuel element falls or a concentrated fuel solution flows into the core.
4) Sudden geometric rearrangements of the core configuration increase reactivity. Mechanical or hydraulic forces may cause a redistribution of fuel, poison, and moderator for a given shape core, resulting in a higher material buckling. Alternatively, the core shape may change, especially in the case of fuel solutions, to one having a lower geometric buckling.
5) Voids in the moderator due to steam, air, or radiolytic gas are removed (or added in the event of a positive void coefficient) at substantial rates. The removal of voids may be due to:
   a) collapse of voids when the system pressure increases;
   b) sweeping out of voids (by the coolant's motion) at a rate faster than the void formation rate;
   c) condensation of voids by cold coolant.
6) Various changes in the materials present in an internal test loop, irradiation hole, or other core vacancy alter the local material buckling sufficiently to add substantial reactivity to the core.
7) An instability develops so that the core spontaneously oscillates between large positive and negative reactivities.

In the sections immediately following, excursions are treated in an idealized manner, i.e. ideal by comparison with the above-listed accidents. Nevertheless, the theory and data will be applicable to the latter more complicated situations, for once reactivity has been added, the mechanics of addition are of little consequence. Then only the neutron kinetics equations, in conjunction with the operation of prompt and delayed reactivity coefficients, determine the course of the transient.

3.2 Significant Variables

The following notation will be used to describe the significant observable variables in power excursions:

ρ_0 = the reactivity (in excess of criticality) injected essentially as a step function into the reactor to initiate the excursion;

α = the reciprocal of the exponential period induced by the injected reactivity;

ϕ = the reactor power, commonly expressed in Mw;

ϕ_m = the peak power reached in the excursion;

$E_m = \int^{t(\phi_m)} \phi\, dt$, the energy liberated by the reactor up to the time of peak power, commonly expressed in Mw-sec;

P_m = the peak pressure attained at a specific location, usually subsequent to the time of peak power;

T_m = the value of the peak temperature of the fuel at a specified location, usually subsequent to the time of peak power.

While no single quantity can be taken as a measure of permanent core damage, it is evident that the degree of bending, melting, and blistering strongly depends on P_m and T_m and temperature gradients.

Figure 3-1 [44] illustrates these quantities as they may pertain to a typical power excursion. It might be noted that until a significant temperature develops, ϕ has a time dependence, exp (αt), with α given by ρ_0 and the inhour relation (since this is a stable period). Concurrent negative reactivity effects then reduce the excess reactivity, thus lengthening the instantaneous period until $\dot{\phi} = 0$ at ϕ_m.

Once ρ_0 has been injected, the course of an excursion depends not only on the dynamic characteristics of the particular reactor but also on the initial conditions at the time ρ_0 is injected. Among the significant initial parameters are the following which have been investigated in SPERT tests: temperature, pressure (atmospheric unless otherwise specified), height of water above the core (e.g. in many SPERT tests this is 2 ft.), and coolant flow rate (no flow unless otherwise specified).

For step reactivity additions and with the initial power level anywhere in the range typical of low power reactor operation (usually below 1 kw), the characteristics of the excursion near peak power are unaffected by the initial power level. However, for ramp reactivity additions, in which reactivity is continually added at a constant rate, the initial power level as well as the ramp rate determines the ultimate severity. This is because the lower the initial power, the more time is available for external reactivity addition prior to the onset, at high power, of inherent shutdown mechanisms such as temperature changes and boiling. The special case of ρ_0 injected (as a step function) at high power levels is closely related to transfer function experiments (see Sec. 2.5).

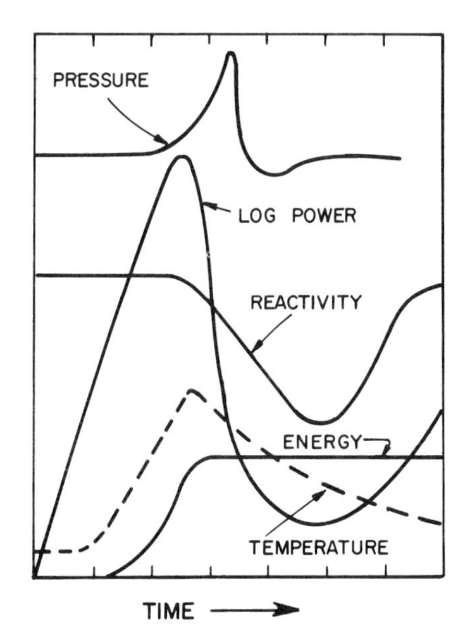

FIG. 3-1. Typical behavior of reactor variables during an excursion.

3.3 Long-Period Experimental Results

It is convenient in discussion power excursions, as in many other phenomena in reactor kinetics, to separate the rapid events in which delayed neutron effects are negligible from slow events in which they are not. This dividing point has been found to be approximately $\alpha \sim 10\ \text{sec}^{-1}$ for light water systems and $\alpha \sim 1\ \text{sec}^{-1}$ for heavy water systems. It is not surprising that as a rule the dominant reactivity shutdown mechanisms also differ in these two period ranges. Although events in the long-period range usually are not important safety problems, these deserve a limited treatment here.

Figure 3-2 [44] illustrates the dependence of ϕ_m and E_m on α in these two ranges. These curves are typical of the thin-plate fuel reactors. The change in slope near $\alpha = 10\ \text{sec}^{-1}$ may be seen in Fig. 3-3 [44]. The experimentally measured "compensated reactivity" shown is the difference between the inserted reactivity step ρ_0 and the instantaneous value of the reactivity at the time of peak power. The latter reactivity was computed from experimental measurements of ϕ (t) and from the neutron kinetics equations. The curves were computed [44] for the various shutdown mechanisms discussed in Secs. 2.1 and 2.2. These curves show

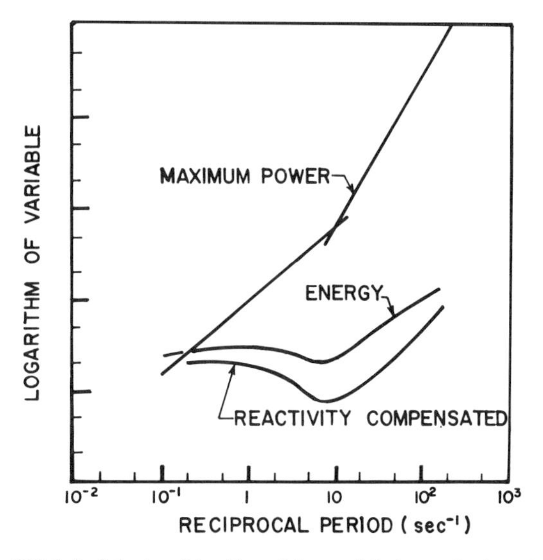

FIG. 3-2. Behavior of ϕ_m, E_m and the reactivity lost at the time of peak power for excursions in a typical thin-plate-fueled H_2O reactor.

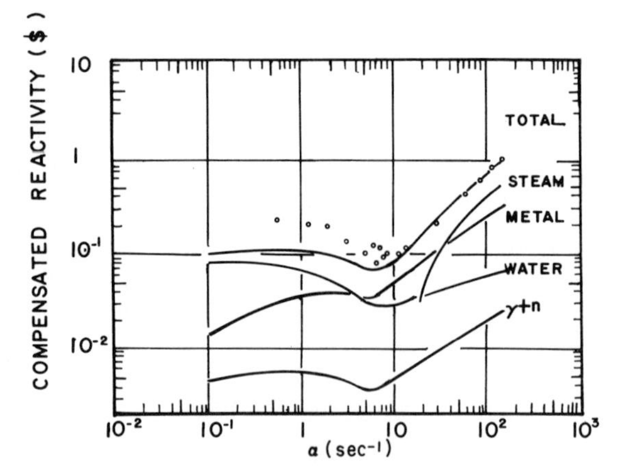

FIG. 3-3 Comparison of the compensated reactivity at the time of peak power for SPERT I-A at 20°C with theoretical calculations. The metal expansion, water heating and radiation heating curves are a priori, while the steam effect was obtained by normalizing theory and experiment at one value of α.

that boiling dominates in short-period excursions, while water and metal temperature expansion effects are dominant in long-period power excursions.

It is seen from Fig. 3-3 that a priori calculations of shutdown mechanisms do not account for all of the measured compensated reactivity in the long period region. This same conclusion was also reached upon a similar analysis of SPERT III data [45]: a suppression of boiling by high pressures showed that at both long and short periods the water heating and fuel expansion effects, while accounting for most of the compensated reactivity, did not account for all of it. It is currently believed that more experimental data or a more refined analysis is needed to resolve this discrepancy. However in the short period region where steam formation is the dominant shutdown mechanism, Fig. 3-3 shows agreement with experiment can be obtained over a wide range of α. It is important to note that the amount of steam formed is not computed a priori but is determined from experiment at one value of α.

It has been verified experimentally [46, 47, 48, 35, 38, 49] for the reactors of Table 3-1 that a single energy coefficient of reactivity may be used with reasonable accuracy to characterize power excursions over a wide range of small values of α. This does not mean that these long period excursions are intrinsically simple. But it does mean that experimental results may be approximately correlated by this one coefficient, which thereby allows one to estimate power excursion effects in reactors similar to those tested. (However higher accuracy may be achieved by assuming a nonlinear reactivity-energy relationship, as discussed in the Fast Transients chapter.)

Use of a linear energy coefficient implies that the equation

$$\rho(t) = \rho_0 + \frac{d\rho}{dE} \int_0^t \phi(t)\, dt \qquad (3\text{-}1)$$

can be used in conjunction with the neutron kinetics equations to predict ϕ_m and E_m at small α, provided the energy coefficient of reactivity ($d\rho/dE$) is known. The quotient of the reactivity and energy curves in Fig. 3-1 gives $d\rho/dE$ for any α. The result typically does not vary appreciably with α. The values of $d\rho/dE$ so computed in Table 3-1 were taken from data at $\alpha = 2$ sec^{-1} for H_2O moderators and $\alpha = 0.4$ sec^{-1} for D_2O moderators, for these α-values are representative of long-period excursions.

TABLE 3-1

Energy Coefficients of Reactivity for Long-Period Excursions

Reactor+	Initial temperature	$d\rho/dE$ ¢/Mw-sec
SPERT I(46) A 17/28*	68°F (20°C)	-7.1
SPERT I(46) B 24/64*	68°F (20°C)	-4.0
SPERT I(46) B 24/32*	68°F (20°C)	-8.5
SPERT II(47)	68°F (20°C)	-4.3
SPERT III(48)	68°F (20°C)	-3.8
KEWB(35)	68°F (20°C)	-50
BORAX IV(38)	68°F (20°C)	-4.3
BORAX IV(38)	203°F (95°C)	-3.8
SPERT I(49) UO_2 Core	68°F (20°C)	-1.2

+Numerical superscripts following reactor name indicate corresponding reference numbers. All are H_2O-moderated except SPERT II which is D_2O-moderated.
*X/Y indicates X plates per assembly in a core having Y assemblies.

Since $d\rho/dE$ is composed of fuel temperature and water temperature coefficients of reactivity together with their heat capacities, it is not surprising that values as low as 1 ¢/Mw-sec and as high as 50 ¢/Mw-sec are found. However, values of $d\rho/dE$ for reactors of a given type (which tend to have similar temperature coefficients and heat capacities) are usually rather similar [50].

3.4 Short-Period Experimental Results

For values of α somewhat in excess of 10 sec^{-1} and 1 sec^{-1} for H_2O- and D_2O-moderated reactors respectively, delayed neutron effects in excursions may be neglected (see also chapter on Dynamics of Fast Transients). The equation describing the transient is therefore

$$\frac{d\phi}{dt} = \frac{1}{\ell}[\rho_0 - \beta - \rho_c(t)]\,\phi\,, \qquad (3\text{-}2)$$

where the compensated reactivity ρ_c (t) is usually a function of t (through ϕ and $\int\phi dt$) and ρ_0 exceeds β. An immediate result of this relation is that the requirement for shutdown, i.e. achieving $\dot{\phi} = 0$ at ϕ_m, is simply

$$\rho_c(t_m) = \rho_0 - \beta = \ell\alpha\,. \qquad (3\text{-}3)$$

Thus, to terminate the initial rapid rise of the excursion only the part of the reactivity above prompt criticality need be compensated by shutdown mechanisms.

By expressing $\rho_c(t_m)$ in terms of ϕ_m and E_m one can evidently predict these quantities for excursions having large values of α by the preceding equation. The three classes of water-moderated reactors which have been investigated (oxide-fueled, thin-plate-fueled, and homogeneous) might be expected to have different shutdown mechanisms at short periods. In the following three subsections experimental data are used to obtain information concerning this fundamental relationship between $\rho_c(t_m)$ and the quantities ϕ_m and E_m for each class of reactor.

3.4.1 Oxide Fuels

Reactors having slightly enriched oxide fuel elements, because of the importance of the Doppler effect (see also chapter on Doppler Coefficient) in conjunction with their long fuel time-constant, may be expected to show excursion behavior which differs somewhat from that of other reactor types. These reactors have a simple shutdown mechanism - one in which the compensated reactivity is related to the energy [51]:

$$\ell\alpha = \rho_c(t_m) = \frac{c}{K}\frac{E_m}{1+\ell\alpha/(4cVT_0)}, \tag{3-4}$$

where c is the Doppler reactivity coefficient per unit volume, K is a proportionality constant, V is the fuel volume, and T_0 is the average absolute temperature in the fuel.

To evaluate c, the temperature dependence of resonance escape p and hence of k_{eff} may be written

$$p = \exp\left[-\frac{I_0(1+\gamma\sqrt{T}-\gamma\sqrt{T_0})}{\xi\sigma_s}\right], \tag{3-4a}$$

where $\xi\sigma_s$ is the average slowing down power per resonance atom, and I_0 is the resonance integral. For the temperature dependence coefficient [51]

$$\gamma = 10^{-2}\left(0.696 - 0.0262\frac{M}{S}\right), \tag{3-4b}$$

where M/S is the mass-to-surface ratio of the fuel in gm/cm^2. Hence

$$c = \frac{1}{V}\frac{d(\ell n\ p)}{dT} = \frac{\gamma(\ell n\ p_0)}{2T_0^{1/2}V} \tag{3-4c}$$

where p_0 is the average initial resonance escape.

K is the energy required for a unit temperature rise of a unit of fuel:

$$K = \frac{\text{average heat capacity of the fuel}}{\text{importance factor} \times \text{fraction of total energy remaining in fuel}} \tag{3-5}$$

where the importance factor is unity for a flat power distribution and was calculated to be 2.1 in the UO_2 core of SPERT I [51]. This means that the average temperature rise must be multiplied by 2.1 to give a reactivity-importance-weighted temperature for use with the coefficient in Eq. (3-4c). The energy remaining in the fuel is much less than 0.5 when $\alpha\tau \ll 1$, where τ is the thermal time constant of the fuel, (see Sec. 2.2 and Table 2-4). Conversely the fraction of energy escaping from the fuel is much less than 0.5 when $\alpha\tau \gg 1$.

According to Eq. (3-4) one may expect experiments in which E_m is measured as a function of α to give results of the form

$$E_m = \frac{K}{c}\ell\alpha\left[1+\frac{\ell\alpha}{4cVT_0}\right]. \tag{3-6}$$

It also has been shown that [51]

$$\phi_m = \frac{K}{c}\frac{\ell\alpha^2}{2}\left[1+\frac{\ell\alpha}{6cVT_0}\right]. \tag{3-7}$$

The validity of these relationships has been established on the SPERT I UO_2 core [49, 51]. It is found for α in excess of 100 sec^{-1} that the Doppler effect is the principal shutdown mechanism. A priori calculations in Table 3-2 [51] of the quantities in Eqs. (3-4a) through (3-5) when used in Eqs. (3-6) and (3-7) give excellent agreement with experiment.

TABLE 3-2

Parameters Calculated for SPERT I UO_2 Core Excursions above $\alpha = 150\ sec^{-1}$

$l/\beta = 3.57 \times 10^{-3}$ sec
$K = 5.34 \times 10^{-7}\,\alpha^{0.16}$ Mw-sec-°K^{-1}-cm^{-3}
$c = 2.86 \times 10^{-10}$ °K^{-1}-cm^{-3}

3.4.2 Thin Plate Fuels

In contrast to oxide fuels with long thermal time-constants and substantial Doppler effects, reactors using thin metallic plates with little U^{238} or Th^{232} have a different shutdown mechanism. The short time-constant of the fuel allows ample energy to be transferred to the water even for α as large as 100 sec^{-1}; boiling becomes the dominant shutdown mechanism. Among the theoretical models in use, an equation similar to Eq. (3-4) is the simplest:

$$\ell\alpha = \rho_c(t_m) = \frac{1}{K^n}cE_m^n. \tag{3-8}$$

Here c is the void coefficient expressed as reactivity per unit volume of uniformly distributed void. The proportionality constant $1/K^n$ is defined so that K is the energy (in Mw-sec) required to produce a unit void volume, with n an empirical exponent. It follows that

$$E_m = K(\ell/c)^{1/n'}\alpha^{1/n}, \quad \text{(in Mw-sec)} \tag{3-9}$$

where n' has replaced n in part of this expression in order to allow additional empiricism in fitting data to this equation. Using the general relationship [53] between E_m and ϕ_m given by Eqs. (3-5),

(3-6) of the Dynamics of Fast Transients chapter, it follows that

$$\phi_m = \frac{n}{n+1} K \left(\frac{\ell}{c}\right)^{1/n'} \alpha^{(n+1)/n}. \quad \text{(in Mw)} \qquad (3\text{-}10)$$

It has been found that for a given set of initial conditions these two equations may be used to describe the ϕ_m and E_m obtained for all large α (i.e. $\alpha > 10$ sec^{-1} for H_2O and $\alpha > 1$ sec^{-1} for D_2O) with an accuracy of the order of ± 30% for the following reactors:

BORAX I, BORAX II,
SPERT I: cores A17/28, B24/64, B16/40, B24/32 and P18/19,
SPERT II, SPERT III, BSR II.

Table 3-3 gives the constants in Eqs. (3-9) and (3-10). Together, these correlate best all the pertinent data from these reactors [34, 44, 47, 48, 54, 55, 56]. The experiments spanned a range of ℓ/c from 0.002 to 0.5 sec/cm³, and involved α-values up to 400 sec^{-1}.

The higher value of K when the initial temperature is well below the boiling point is indicative of the fact that for the same period and reactor characteristics, excursions are much more energetic when the water must first be heated to the boiling point. This is to be expected in this short-period region where boiling is the dominant shutdown mechanism. Fig. 3-4 [47] in its 0 psig (1 atm) curve shows this effect quantitatively for SPERT III.

The effect of varying the pressure independently of the temperature has been studied in SPERT III, (see Fig. 3-4). While not enough data are available at higher pressures and temperatures to justify extending the correlations of Table 3-3, at least the trend in Fig. 3-4 has been verified in other reactors [34, 47]: higher temperature at a given pressure reduces the peak power, while a higher pressure at a fixed temperature, because boiling is suppressed, increases the peak power.

TABLE 3-3

Constants in the Correlation of Power Excursion Data from Thin Plate Cores Using Eqs. (3-9) and (3-10) with l/c in sec/cm³ and α in sec^{-1}

	Initial temperature	K	n'	n
For E_m in Mw-sec	25°C (77°F)	0.18	2.9	1.45
	95°C (203°F)	0.052	3.4	1.16
For ϕ_m in Mw	25°C (77°F)	0.11	2.1	1.27
	95°C (203°F)	0.021	1.7	1.14

[Conversion factor: 1 Mw-sec = 947.8 Btu.]

Experiments on SPERT I with more than the usual two feet of water above the core indicated that ϕ_m increased as the head of water was increased [57]. Nine feet (2.75 m) of water at 95°C (203°F) caused ϕ_m to be a factor of 2 and 1.25 larger for α-values of 11 sec^{-1} and 111 sec^{-1} respectively. On the other hand at 20°C (68°F) both these SPERT I experiments and others in SPERT IV having up to 18 ft (5.5 m) of water above the core showed no measurable increase in ϕ_m even at short periods where boiling is the dominant shutdown mechanism. This difference between the 95 and 20°C tests may be explained on the basis of initial subcoolings (the temperature difference between the water and its boiling point), 9 ft (2.75 m) rather than 2 ft (0.6 m) changed the subcooling from none at all to 7°C (12.6°F) for the former; 18 ft (5.5 m) changed the subcooling from 95-20 = 75°C to 109-20 = 89°C for the latter ~ only 19% more.

Although research and pressurized water reactors operate with high flow rate it might be expected

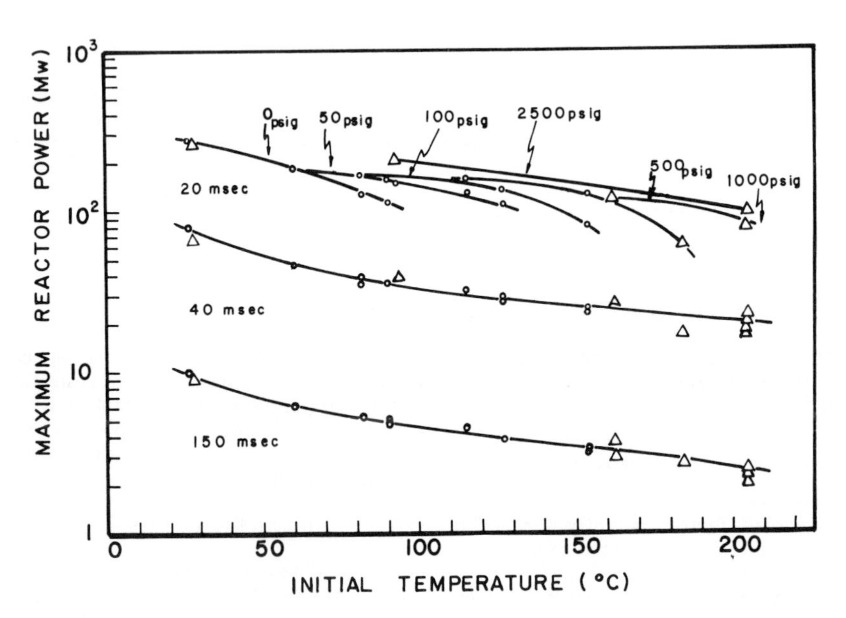

FIG. 3-4. The effect of changing initial conditions of pressure and temperature in SPERT III excursions at an α = 50 sec^{-1}.

that the excursion behavior would not significantly differ from that of the stagnant water experiments whenever

$$w_1/\alpha \ll Z/2 . \tag{3-11}$$

The left-hand side is the distance the coolant moves during one period and the right-hand side is half the core height, that is, the distance over which the coolant density coefficient of reactivity changes appreciably. Otherwise, experiments in SPERT III [59] in which w_1/α was 18 (ft/sec) / 56 (sec^{-1}) = 0.324 ft (9.9 cm) or more (compared to Z/2 = 1.5 ft or 45 cm) showed flow influenced the excursion as indicated in Fig. 3-5 [59]. The lower fuel temperature with flow present is due to improved heat transfer. However, the slightly lower peak power, which is subject to further experimental verification, is not currently understood. After the time of one total coolant transit, 3(ft) / 14.4 (ft/sec) = 0.208 sec, a high equilibrium power level is reached since core temperatures and densities have attained equilibrium values. Similar tests in SPERT IV [58] have confirmed this behavior, although its peak power slightly increased with high flow rates as might be expected when the coolant is removing negative compensating reactivity effects from the core.

The behavior of fuel temperature during excursions is understood quantitatively prior to the onset of boiling and qualitatively thereafter [60]. For low values of α, where the boiling is not extensive, the temperature rise of the fuel surface follows the exponential increase of the reactor power (see Fig. 3-1) while losing heat by conduction to the water. But when boiling starts, the surface temperature then levels out only a few degrees above the boiling point. This is because only a few degrees of superheat are required to achieve the high heat transfer rates of normal (i.e. without vapor blanketing) boiling, which are well above those of conduction. However, at very high values of α, boiling is vigorous enough to cause vapor blanketing of the fuel surface. Then instead of the fuel temperature leveling off as in Fig. 3-1, it climbs rapidly because of poor heat transfer through the steam. In these cases the maximum fuel temperature attained depends primarily on the total energy release, which is approximately $2E_m$.

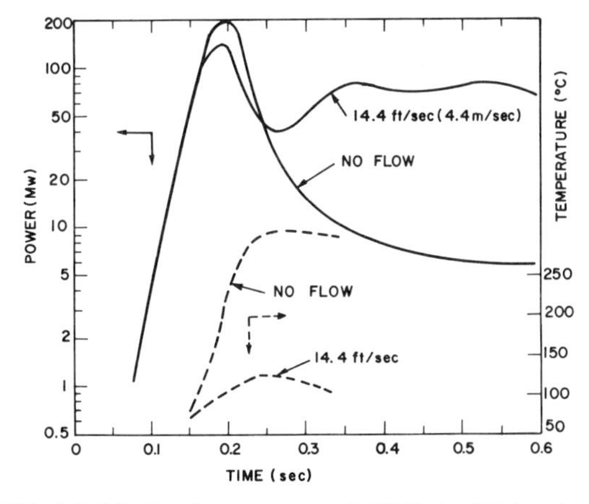

FIG. 3-5 Behavior of reactor power (solid line) and fuel surface temperature (broken line) in 0.020 sec excursions of SPERT III with and without coolant flow. Initial pressure = 2500 psig (175 kg/cm^2), initial temperature = 93°C (200°F).

3.4.3 Fuel Solutions

Reactors in which the fuel is dissolved in the moderator differ from those with fuel elements, in that the energy is more rapidly transformed into a shutdown mechanism. Here boiling and radiolytic gas in combination with a negative void coefficient provide the major reactivity loss at the time of peak power. Some contribution also comes from the negative temperature coefficient. This total reactivity loss also can be empirically represented by Eq. (3-8).

The constants in Eqs. (3-9) and (3-10) have been determined on the KEWB reactor [35] where $\ell = 6.25 \times 10^{-5}$ sec and $c = 4 \times 10^{-5}$ cm^{-3}. The result for excursions at 25°C (77°F) at absolute pressure of 15.6 cm Hg (0.205 atm) is

$$\phi_m = \frac{2}{3} K \left(\frac{\ell}{c}\right)^{1/2} \alpha^{3/2} \quad (K = 0.06,\ \phi_m \text{ in Mw}) \tag{3-12}$$

$$E_m = K \left(\frac{\ell}{c}\right)^{1/2} \alpha^{1/2} \quad (K = 0.074,\ E_m \text{ in Mw-sec}) \tag{3-13}$$

for values of α from 10 to 1000 sec^{-1}. Data taken at absolute pressures of 43.4 and 71.2 cm Hg (0.571 and 0.939 atm) lead to these same expressions but with the K-values 1.5 times larger, this is due to the smaller volume and hence smaller negative reactivity effect of a given quantity of gas.

Excursions initiated at temperatures between 25°C (77°F) and the solution boiling point are less energetic. This is due to less energy being required to raise the solution to the boiling point. Hence boiling plays a larger role in contributing to the total gas which shuts down the reactor.

The damage from excursions in homogeneous reactors can be expected to come primarily from pressure generated as the gases move the water. A pressure referred to as an "expansion pressure" is that measured at the bottom of the vessel containing the solution. It is simultaneous with the reactor power and attains its peak value at the same time as the power. Potentially more damaging however is another pressure termed the "impact pressure", which is that pressure exerted on the top of the reactor vessel when the solution is ejected above its free surface level and hits the vessel. The largest measured value of this pressure was 800 psi (56 kg/cm^2) in a 0.009 sec excursion [61].

3.5 Application of Experimental Results

The preceding sections have summarized the results of many excursions initiated by positive reactivity steps. To apply these to other reactors, the similarity of the postulated accident (including its initial conditions) as well as the similarity of the reactor type itself must, of course, be considered. When these similarities exist, simple methods of interpolating among many experiments are valid, as suggested by Eqs. (3-7), (3-10) and (3-12) and by Table 3-3. If these similarities do not exist, the same methods may at least provide estimates. Estimates of the energy and power of long-period excursions may be made from Eq. (3-1) and Table 3-1.

Often an excursion is excited by a steady rate of reactivity addition. It has been shown [46] (see also Eq. (2-25) in the chapter on Fast Transients) that the resulting ϕ_m and E_m for the first peak of the excursion is the same as that resulting from a step reactivity excitation with

$$\alpha = \sqrt{2a[\log_e(\phi_m/\phi_0)]} \qquad (3\text{-}14)$$

where a is the reactivity added per second divided by the neutron lifetime and ϕ_0 is the starting power. For validity, α must somewhat exceed 10 sec^{-1} and 1 sec^{-1} for H_2O and D_2O systems, respectively.

Where simple interpolation among existing data is not possible or where more precision is desired, the shutdown must be computed from a series of coupled equations specifying the time dependence of the various reactivity mechanisms in conjunction with the neutron kinetics equations. If sufficient detail is used to describe accurately such effects as fuel expansion, fuel bowing, Doppler broadening, water heating, etc., satisfactory predictions of transient behavior may be made. However, if void formation plays a significant role in contributing to the reactivity, this must be incorporated into the theory in some empirical fashion until a better understanding of the basic processes involved is developed into a quantitative theory.

Obtaining ϕ_m or E_m for a postulated accidental excursion is only a means to an end. The non-nuclear problem remains to specify the damage, if any, from resulting pressures and temperatures. Knowing ϕ_m, fuel element temperatures may be computed by using the equations for transient heat transfer. Excursions so energetic as to cause melting and core damage are likely to have the fuel essentially insulated from the water by a vapor blanket. The fuel temperature rise may then be easily computed from the total energy and the heat capacity of the fuel. The computed fuel temperatures may be compared with thresholds for melting or phase changes and these same temperatures may be used to determine if the thermal stresses exceed elastic limits.

Regarding the calculation of possible damage from transient pressures, an uncertainty in the prediction is likely to result from inaccuracies of the pressure estimate as well as from computation of the mechanical effect of the pressure surge on the core or vessel. In regard to predicting pressures, those arising as a result of well-studied reactivity compensating mechanisms are of course easier to predict than those resulting from threshold damage effects.

To make an estimate of the pressure effect accompanying boiling one must know the amount of void causing the water to move with accompanying transient pressures. In applying any particular hydrodynamic model here, there are much pressure data available for checking purposes from the BORAX, SPERT and KEWB experiments discussed above. In the destruction of SPERT I by a 0.0032 sec period pressures in the neighborhood of 400 psi (28 kg/cm^2) were found to accompany the motion of the water out of the top and bottom of the core after ϕ_m was attained [58]. In the destruction of the SL-I it was deduced that pressures of the order of 1000 psi (70 kg/cm^2) near the core accelerated a slug of water upward against the vessel head and moved the vessel several feet upward [62]. Until it is possible to predict these pressures from theoretical considerations, data from these experiments may be used qualitatively to estimate damage in postulated accidents.

Since the experimental data on transients here have been obtained on relatively small cores, the question arises as to the validity of using these data on large cores. Although large core experiments have yet to verify this, it is likely that the application is valid: to the extent that the ℓ/c and α are in the range investigated if voids dominate shutdown; or to the extent that the Doppler effect dominates shutdown. However a simplified analysis is not possible if the flux-tilting effects of Sec. 4 play an important role. Obviously if a small section of a large reactor becomes locally substantially supercritical while the surrounding regions are in themselves subcritical, an appropriate model for analysis is a small reactor reflected by this unimportant subcritical region.

3.6 Effectiveness of Control Systems

In the preceding sections the discussion has been limited to excursions in which the only shutdown mechanisms are those intrinsic with the reactor. However it is usual for reactors to have control rods which scram rapidly in the event the reactor power exceeds some trip level. Scrams actuated when the instantaneous period $\phi/(d\phi/dt)$ falls below a specified value are also a common design feature. Thus the values of the peak power and total energy given above may be higher than those which would be encountered if the control system also assisted in the shutdown process.

The extent to which scramming control rods may be effective in reducing the severity of an excursion depends on the following factors:

1) the conditions of the excursion, particularly the exponential period;
2) the time delay in the action of the scram system;
3) whether the scram is actuated by a certain power level or a certain period, the initial power level playing an important role in the latter case;
4) whether the self-shutdown reactivity mechanisms are rapid and large or slow and small.

An approximate equation for the consequences of a time delay Δt of the scram system is

$$\phi_m = \phi_s \exp(\alpha\Delta t), \qquad (3\text{-}15)$$

where ϕ_s is the power at which the level or period conditions for scram are attained. Here ϕ_m is the peak power that would be attained if other shutdown mechanisms are ignored. In short-period excursions Δt is approximately the sum of electronic delays and the time required for the control rods to enter the core to the point at which the reactor is just prompt critical.

For any postulated excursion, the analysis for finding ϕ_m is straightforward if the scramming action of the control rods is taken to be the only shutdown mechanism. The neutron kinetic equations in combination with a known negative reactivity addition as a function of time are readily solved by analog or digital methods. The ϕ_m obtained may be compared with that resulting from the action of self-shutdown mechanisms alone. In any such comparison it is likely that the added shutdown mechanism by means of control rods gives a lower ϕ_m for small values of α but not necessarily for large values of α.

An experimental comparison of the relative effectiveness of a rather fast scramming system (Δt = 0.030 sec) with self-shutdown was made on the BSR II [63]. For room temperature excursions from low-power levels in this particular reactor a 100 kw power level scram was found more effective than self-shutdown mechanisms for $\alpha < 350$ sec^{-1}. Although this scram can prevent fuel plate buckling for some intermediate values of α, the self-shutdown mechanisms act before the scram for the very large values of α near the fuel melting threshold. It was also found that a scram system with $\Delta t > 0.1$ sec would be ineffective in preventing core damage to the BSR-II since the self-shutdown mechanisms would be more effective even at the threshold of fuel plate buckling. This result is also likely to be true for many other reactors.

The relative effectivenss of a period scram rather than a level scram depends on the particular conditions at the initiation of an excursion. It is readily possible to have a period scram trip at a power level somewhat below that of a level scram particularly if an excursion starts well below the level scram power level. However, ultimately for short period excursions the self-shutdown mechanisms are more effective because they are capable of acting orders of magnitude faster than the inward moving control rods.

Even though a scram system may act slower than the self-shutdown mechanisms in some of the excursions which may damage the reactor, the scram system still has merit. Although an excursion may be terminated at ϕ_m by self-shutdown mechanisms, the termination is only temporary: when the high temperature or void in the core is removed by the coolant after a few seconds, the excursion may repeat itself unless, in the meantime, negative reactivity has been added by some other means, such as by the scram rods. In addition, excursions corresponding to small values of α are much more probable. When these occur in power reactors operating near limits imposed by the center fuel temperature or the surface heat flux, the slow scram system is indeed effective in preventing damage. It might be concluded that while scram system can prevent damage in some transients, it is likely that the self-shutdown mechanisms are more important than control rods in terminating extremely short period excursions.

4 SPATIAL DEPENDENCE IN DYNAMICS

In the preceding sections on dynamics and excursions, it has been assumed that the reactivity, with contributions from all parts of the reactor, affects the reactor power everywhere in the same way. Thus, the reactor power was assumed to be a product of a space function and a time function. In general these assumptions are valid. However, in the special case of large reactors, discussed below, this space-time separability is not valid.

The nonuniform spatial distribution of reactor power, together with the spatial dependence of the reactivity importance functions, already give a spatial aspect to dynamics problems. Although appropriate averages sometimes simplify solutions, it is usually necessary to subdivide the reactor into axial and radial regions to obtain accurate solutions. However, except for very large reactors, the reactivity effects in these transients are somewhat less than the reactivity associated with neutron leakage in the largest dimension. This simplifies the problem for it means that the spatial flux shape remains constant throughout the transient, i.e., the power is the product of the same space function and a time-varying function.

Where the spatial flux shape is not constant during a transient an added hazard to reactor operation must be recognized. The maximum to average power is a function of time and is likely to increase during unstable oscillations. Therefore safety margins for fuel melting can be reduced by a changing maximum to average as well as by a changing power level. This section will point out instances where this type of spatial effect must be considered.

4.1 Xenon Instability

Xenon instability is an example of local reactivities affecting local fluxes, so that the flux distribution is not constant but is time-dependent. If the reactor power increases locally, e.g. during withdrawal of a control rod, xenon burns out enhancing the increase. However, later the larger than normal iodine concentration decays into a larger than normal amount of xenon, thereby locally reducing the power below its normal value. Because of the reduced iodine concentration when it decays a half-cycle later, the power increases again due to the smaller amount of xenon. In order for the xenon to shift from one side of the reactor to the other (or from top to bottom, or from center to edge) in a half cycle the reactivity in xenon must exceed the flux shape stabilizing reactivities due to leakage and negative temperature coefficient.

Fig. 4-1 [64] shows the average flux levels above which xenon instability occurs, assuming a zero temperature coefficient. It is seen that large values for the ratio of core radii squared or core height squared to migration area M^2 lead to instability at lower fluxes, since reactivities associated with neutron leakages are then small. The case of nonzero temperature coefficients has also been treated; in water reactors having negative coefficients the thresholds are larger than shown in Fig. 4-1.

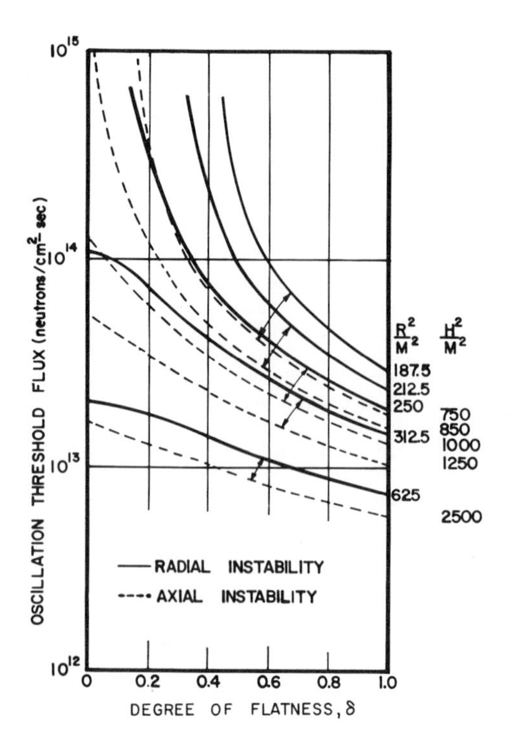

FIG. 4-1. Threshold flux values for xenon instability in cylindrical reactors. Solid and broken lines refer to radial and axial modes of instability respectively. Pairs of curves joined by arrows have 2R = H. The degree of flatness δ is defined in such manner that the power distribution is constant for radii less than $R\delta$.

4.2 Local Coolant Temperature Transients

In large reactors, whether boiling or nonboiling, the problem of a local reactivity perturbation, which causes a transient eventually extending over the entire reactor, requires special consideration. A change in the inlet coolant temperature for fuel elements in a certain region of the reactor is a typical perturbation of this type. In a small closely coupled reactor the flux would change in time everywhere without any spatial shape change - as if the temperature perturbation were uniformly and proportionately distributed over the entire reactor. On the other hand, in an extremely large reactor, whose regions are decoupled from one another, only the region having the temperature change has its flux changed by the perturbation.

The criterion for whether a reactor is large enough to require special treatment comes from comparing the delayed neutron fraction β with the reactivity associated with leakage out the largest dimension. Thus if

$$M^2 (3\pi^2/Z^2) \gtrsim \beta , \qquad (4-1)$$

where M^2 is the migration area and Z is the largest dimension across the reactor, then a significant spatial transient in flux shape is possible. Here $3\pi^2/Z^2$ is the difference between the bucklings of the fundamental and first harmonic one-dimensional flux shapes, which are $\sin \pi z/Z$ and $\sin 2\pi z/Z$ respectively. When equality exists in Eq. (4-1), the initial prompt jump in local flux is only half the change eventually resulting after the delayed neutrons have acted.

A technique for solving these large reactor transients consists in combining flux shapes obtained by multi-region diffusion theory at various times in the transient with the neutron kinetics equations [65]. The reactor power is taken to be $\phi(t)\ S(x,y,z,t)$, where $\phi(t)$ is the time function from the neutron kinetics equations, and $S(x,y,z,t)$ is a shape function which varies as the flux tilts from one region to another during the transient. Normalization is such that $S(x,y,z,t)$ would be unity in small closely coupled reactors. A series of coupled equations, similar to those displayed in Figs. 2-2 and 2-4 may then be solved, with reactivity feedback effects from the various regions summed to give the total reactivity input to the neutron kinetics equations.

A transient in the PWR in which the coolant temperature to one portion of the seed was reduced has been studied in this manner. An analog computer approach similar to Fig. 2-2 was used except that additional relations were introduced to take into account shifting flux ratios among the reactor regions.

4.3 Control Rods in Large Reactors

The effects due to xenon or water temperature described in the preceding sections are fundamentally local reactivity effects which do not immediately propagate over the entire core. Similar methods may be used to investigate the effects of local control rod motion. In fact, in obtaining the true local power peaks during a xenon transient it is necessary to take into account operator action on the control rods.

In the PWR a sustained xenon oscillation can be excited by unbalancing the control rods in the seed: i.e., lifting the rods on one side of the annular seed region, while inserting them on the other [20]. Experimental and theoretical studies of such transients have been made. It is significant that in a reactor potentially unstable with respect to flux tilting, operator action on the control rods can determine the extent to which the instability will develop.

Both theoretical and experimental studies of spatial transients have been carried out for the Dresden boiling reactor [66, 67]. The reactivity associated with leakage in any one direction is comparable with the delayed neutron fraction, and therefore flux shape transients may be excited. Evidence of these spatial effects was obtained by oscillating a control rod at 0.067 cycles/sec while at full power. While nearby in-core ion chambers readily detected the local corresponding flux oscillation, those on the opposite side of the core failed to observe any oscillation. This is in contrast to the much smaller EBWR in which the entire reactor responded essentially "in unison" to a control rod oscillation.

Since M^2 is smaller (typically of the order of 35 to 40 cm^2) for light water reactors at room temperature than for other reactor types, it is apparent that light water reactors at room temperature are most susceptible to spatial effects. The very loose coupling between two virtually

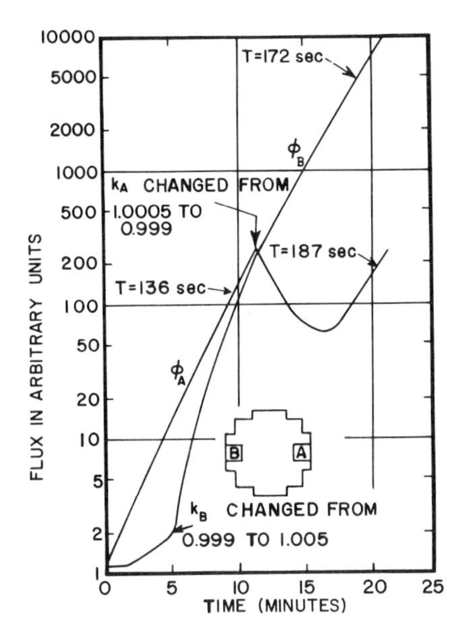

FIG. 4-2. Behavior of flux in each of two loosely coupled regions of the Dresden Reactor (320-element loading).

independently critical regions at opposite ends of a diameter in the Dresden reactor at room temperature is demonstrated by the experiment shown in Fig. 4-2 [68]. In-core ion chambers in two 16-assembly unrodded regions A and B showed each immediately responding to local reactivity changes in control rod positions according to the neutron kinetics equations. But effects in one region do not transmit to the other until much later. Loosely coupled cores have also been studied in the 2-slab loading of the Argonaut, and the theory has been extensively developed [69]. The significance of these effects to reactor safety is twofold:

a) In-core flux monitoring is needed in loosely coupled cores;

b) Operators must recognize the special characteristics of space-time kinetics as differing from the space-independent kinetics of small reactors.

REFERENCES

1. Anon., "Power-reactor control the world around", Nucleonics, 16,5(1958)74.
2. J. Fisher and A. Diaz, "Experimental evaluation of the fully loaded Elk River reactor", Trans. Am. Nucl. Soc., 6, (1963)65.
3. W. H. Arnold, Jr., "Worths of black control rods", Nucl. Sci. Eng., 3(1958)296.
4. "Reactor Physics Constants", USAEC Report ANL-5800 (2nd Ed.), p. 361, Argonne National Laboratory, 1963.
5. J. A. Thie, "Fluid poison control of boiling water reactions", Nucleonics, 16,5(1958)82.
6. Anon., "Reactor control strategies", Nucleonics, 16,5(1958)71.
7. A. W. Kramer, Boiling Water Reactors, pp. 24 and 447, Addison-Wesley Publishing Co., Inc., Reading, Mass., 1958.
8. "Preliminary Safeguards Report, Pathfinder Atomic Power Plant", Report ACNP-5905, Allis Chalmers Manufacturing Co., 1960.
a) 2.56; b) 1.50
9. M. C. Edlund and G. K. Rhode, "Spectral shift control", Nucleonics, 16,5(1958)80.
10. P. R. Kasten, "Dynamics of the Homogeneous Reactor Test", USAEC Report ORNL-2072, Oak Ridge National Laboratory, 1956.
11. J. MacPhee, "The kinetics of circulating fuel reactors", Nucl. Sci. Eng., 5(1958)588.
12. M. W. Rosenthal, S. Jaye, and M. Tobias, "Power Excursions in the HRT", USAEC Report ORNL-2798, Oak Ridge National Laboratory, 1960.
13. K. J. de Jong et al., "A Subcritical Circulating Suspension Reactor", Proceedings of the Second U.N. International Conference on Peaceful Uses of Atomic Energy, Geneva, 1958, Vol. 12, p. 525.
14. S. M. Zivi, "Gas Microbubble Dilation - A reactor shutdown mechanism", Trans. Am. Nucl. Soc., 5,1(1962)161.
15. M. W. Rosenthal, "Operating experience with the OMRE", Nucl. Safety, 2,2(1960)75.
16. "Reactor Development Program Progress Report", USAEC Report ANL-6801, p. 1, Argonne National Laboratory, 1963.
17. B. Wolfe, "Reactivity effects produced by fluid motion in a reactor core", Nucl. Sci. Eng., 13(1962)80.
18. J. E. Meyer et al., "ART - A Program for the Treatment of Reactor Thermal Transients on the IBM-704", Report WAPD-TM-156, Westinghouse Electric Corp., 1959.
19. J. E. Meyer and W. D. Peterson, "ART-04, A Modification Transients on the IBM-704", Report WAPD-TM-202, Westinghouse Electric Corp., 1960.
20. J. N. Grace, "Analysis of a Reactivity Instability Experiment with Boiling and Nonlinear Analysis of Spatial Stability and Flux Tilt Transients", Proceedings of a Conference on Transfer Function Measurement and Reactor Stability Analysis, USAEC Report ANL-6205, p. 174, Argonne National Laboratory, 1960.
21. H. A. Sandmeier and O. M. O'Shea, "Time Constant of cylindrical fuel pin with axial coolant flow", Nucl. Sci. Eng., 5(1958)186.
22. A. Z. Akcasu, "Theoretical Feedback Analysis in Boiling Water Reactors", USAEC Report ANL-6221, Argonne National Laboratory, 1960.
23. M. Iriarte, "An accurate transfer function for dynamic analysis of temperature and heat release in cylindrical fuel elements", Nucl. Sci. Eng., 7(1960)26.
24. L. S. Tong, "Simplified calculation on thermal transient of a UO_2 fuel rod", Nucl. Sci. Eng., 11(1961)340.
25. D. H. Crimmins, "Transient heat transfer from ceramic fuel pins", Nucleonics, 20,8(1962)131.
26. W. H. McAdams, Heat Transmission, (3rd Ed.), p. 35, McGraw-Hill Book Co., Inc., New York, 1954.
27. E. P. Gyftopoulos and H. B. Smets, "Transfer function of distributed parameter nuclear reactor system", Nucl. Sci. Eng., 5(1959)405.
28. E. P. Gyftopoulos, "Transfer Function Representation of Nuclear Power Plants", Proceedings of a Conference on Transfer Function Measurement and Reactor Stability Analysis, USAEC Report ANL-6205, p. 18, Argonne National Laboratory, 1960.
29. E. S. Beckjord, "Dynamic Analysis of Natural Circulation Boiling Water Power Reactors", USAEC Report ANL-5799, Argonne National Laboratory, 1958.
30. J. A. Thie, "Theoretical Reactor Statics and Kinetics of Boiling Reactors", Proceedings of the Second U.N. International Conference on Peaceful Uses of Atomic Energy, Geneva, 1958, Vol. 11, p. 440.
31. J. A. Thie, "Dynamic Behavior of Boiling Reactors", USAEC Report ANL-5849, Argonne National Laboratory, 1959.
32. E. R. Owen, "Combined Reactor and Power System Analysis for a Boiling Water Reactor", Proceedings of a Conference on Transfer Function Measurement and Reactor Stability Analysis, USAEC Report ANL-6205, p. 74, Argonne National Laboratory, 1960.
33. J. A. Fleck, Jr., "The influence of pressure on boiling water reactor dynamic behavior at atmospheric pressure", Nucl. Sci. Eng., 9(1961)271.
34. J. A. Thie, "An Experimental Investigation of Boiling Reactor Safety", USAEC Report ANL-6135, Argonne National Laboratory, 1960.
35. M. E. Remley et al., "Experimental Studies on the Kinetic Behavior of Water Boiler Type Reactors", Proceedings of the Second U.N. International Conference on Peaceful Uses of Atomic Energy, Geneva, 1958, Vol. 11, p. 447.
36. J. A. De Shong, Jr., "Power Transfer Functions of the EBWR Obtained Using a Sinusoidal Reactivity Driving Function", USAEC Report ANL-5798, Argonne National Laboratory, 1958.
37. S. Zivi and R. W. Wright, "Power-Void Transfer Function Measurements in a Simulated SPERT 1A Moderator Coolant Channel", Proceedings of a Conference on Transfer Function Measurement and Reactor Stability Analysis, USAEC Report ANL-6205, Argonne National Laboratory, 1960.
38. B. S. Maxon, O. A. Schulze, and J. A. Thie, "Reactivity Transients and Steady-State Operation of a Thoria-Urania-Fueled Direct-Cycle Light Water-Boiling Reactor (BORAX IV)", USAEC Report ANL-5733, Argonne National Laboratory, 1959.
39. E. Wimunc, "Performance Characteristics of the EBWR", Operating Experience with Power Reactors, Vol. I, International Atomic Energy Agency, Vienna, 1963.
40. J. A. Thie, Reactor Noise, Rowman and Littlefield, New York, 1963.

41. A. Z. Akcasu, "Mean square instability in boiling reactors", Nucl. Sci. Eng., 10(1961)337.
42. R. W. Badgley, "Power Spectral Density of the University of Florida Training Reactor Operating in the Subcritical Region", Thesis, University of Florida, 1962.
43. S. J. Garrett, M. A. Schultz, and C. M. Yang, "Shutdown Reactivity Meter Development Phase I-Feasibility", USAEC Report NYO-10492, 1963.
44. S. G. Forbes, "Analysis of Self-Shutdown Behavior in the SPERT I Reactor", Report IDO-16528, Phillips Petroleum Co., 1959.
45. R. W. Garner, "An Analysis of Nonboiling Reactivity Feedback Mechanisms in Pressurized Power Excursions in the SPERT III Reactor", Report IDO-16819, Phillips Petroleum Co., 1963.
46. S. G. Forbes, "The Dependence of Reactor Behavior on the Self-Shutdown Mode", Report IDO-16528, Phillips Petroleum Co., 1961
47. F. Schroeder (Ed.), "SPERT Quarterly Technical Report, July, August, September, 1961", Report IDO-16726, Phillips Petroleum Co., 1961.
48. F. Schroeder (Ed.), "SPERT Quarterly Technical Report, April, May, June, 1961", Report IDO-16716, Phillips Petroleum Co., 1961.
49. A. H. Spano et al., "Self-Limiting Power Excursion Tests of Water-Moderated Low-Enrichment UO_2 Core in SPERT I", Report IDO-16751, Phillips Petroleum Co., 1962. See also: Nucl. Sci. Eng., 15(1963)37.
50. F. Schroeder, "Power Excursion Studies", Proceedings of the 1960 Idaho Conference on Reactor Kinetics, Report IDO-16791, p. 37, Phillips Petroleum Co., 1962.
51. A. H. Spano, "Analysis of Doppler - Limited Power Excursion in a Water-Moderated Oxide Core", Nucl. Sci. Eng., 1964 (to be published).
52. H. C. Corben and W. Horning, "Theory of Power Transients in the SPERT I Reactor", Report IDO-16434, Phillips Petroleum Co., 1958. See also: Nucl. Sci. Eng., 5(1959)127.
53. R. W. Garner, "Review of Several Models for Power Behavior and Reactivity Feedback in Reactor Excursion Studies", Report IDO-16636, Phillips Petroleum Co., 1962.
54. J. R. Dietrich, "Experimental Investigation of the Self-Limitation of Power During Reactivity Transients in a Sub-cooled, Water-Moderated Reactor", USAEC Report ANL-5323, Argonne National Laboratory, 1954. See also: Proceedings of the First U.N. International Conference on Peaceful Uses of Atomic Energy, Geneva, 1955, Vol. 13, p. 88.
55. J. R. Dietrich and D. C. Layman, "Transient and Steady State Characteristics of a Boiling Reactor", USAEC Report ANL-5211, Argonne National Laboratory, 1954.
56. F. Schroeder (Ed.), "SPERT Quarterly Technical Report April, May, June, 1960", Report IDO-16640, Phillips Petroleum Co., 1960.
57. G. D. Bright and S. G. Forbes, "Miscellaneous Tests with the SPERT I Reactor", Report IDO-16551, Phillips Petroleum Co., 1959.
58. F. Schroeder (Ed.), "SPERT Quarterly Technical Report, January, February, March, 1963", Report IDO-16893, Phillips Petroleum Co., 1963.
59. F. Schroeder (Ed.), "SPERT Quarterly Technical Report, October, November, December, 1961", Report IDO-16750, Phillips Petroleum Co., 1962.
60. F. Schroeder (Ed.), "SPERT Quarterly Technical Report, April, May, June, 1962", Report IDO-16806, Phillips Petroleum Co., 1962.
61. M. E. Remley, "Presentation of KEWB Program", Proceedings of the 1960 Idaho Conference on Kinetics, Report IDO-16791, p. 57, Phillips Petrolum Co., 1962.
62. W. B. Cottrell, "The SL-1 accident", Nucl. Safety, 3,3(1962)64.
63. F. L. Bentzen, "The Merits of Inherent Shutdown vs Mechanical Shutdown of a Plate-Type, Water-Moderated and -Reflected Reactor in Runaway Condition", Report IDO-16722, Phillips Petroleum Co., 1961.
64. D. Randal and D. S. St. John, "Xenon spatial oscillations", Nucl. Sci. Eng., 14(1962)204.
65. N. J. Curlee, Jr., "Nonseparable space-time transients resulting from changes in inlet coolant temperature", Nucl. Sci. Eng., 6(1959)1.
66. E. S. Beckjord and W. H. Harker, "The Dynamics of Large Boiling Reactors", Proceedings of a Conference on Transfer Function Measurement and Reactor Stability Analyses, USAEC Report ANL-6205, p. 189, Argonne National Laboratory, 1960.
67. E. S. Beckjord, "Dresden stability tests", Trans. Am. Nucl. Soc., 3,2(1960)433.
68. R. O. Brugge and A. S. Bartu, "Multiple core behavior of the Dresden reactor", Trans. Am. Nucl. Soc., 3,1(1960)100.
69. R. Avery, "Theory of Coupled Reactors", Proceedings of the Second U.N. International Conference on Peaceful Uses of Atomic Energy, Geneva, 1958, Vol. 12, p. 182.

CHAPTER 9

Kinetics of Solid-Moderator Reactors

H. B. STEWART and M. H. MERRILL
General Atomic, Division of General Dynamics,
San Diego, California

CHAPTER CONTENTS*

*Except for a few changes and additions made in proof, this chapter is based on information in the literature or known to the authors prior to January 1964.

1 CLASSIFICATION OF SOLID-MODERATOR REACTORS

The kinetic characteristics of hydrogen and deuterium-moderated reactors are strongly affected by the relatively large change and rate of change in density with temperature which can occur in most or all of the moderator material. The characteristics of those types of reactors are discussed in the previous chapter. The reactors discussed in this chapter are classified as solid-moderator reactors and are defined by the following characteristics:

a. The neutron spectra are predominantly thermal or intermediate, and
b. The moderator and coolant either contain no hydrogen or deuterium, or the amount, location or physical properties of the hydrogen or deuterium-bearing materials is such that normal density changes in these materials

have a relatively minor effect on the kinetic characteristics of the reactor.

The reactors considered in this class are those moderated predominantly by zirconium hydride, beryllium, beryllium oxide and graphite. These reactors might be cooled by gas, liquid metal, or perhaps water, provided the amount of the water coolant is sufficiently small so that its contribution to neutron moderation is relatively insignificant.

Table 1-1 summarizes the general characteristics of a few reactors included in the classification of solid-moderator reactors. These particular reactors are used in subsequent discussions as illustrations of the general class of solid-moderator reactors. Obviously, a large number of individual reactors need not be discussed in detail since each of them will resemble one of the reactors included in the table. Likewise, other combinations of fuel, coolant and moderator are possible and indeed have been proposed. Because detailed physics and kinetic information is not generally available for these alternate possibilities, they are not used specifically as illustrations in the subsequent discussions.

In addition to the reactors included in Table 1-1, the TRIGA and TREAT reactors are discussed in Sec. 5 since they offer good opportunities to examine the accuracy which can be achieved in calculating the physics and kinetic characteristics of at least some solid moderator reactors.

Since the coolant in the solid-moderator reactor generally does not contribute strongly to the neutron moderation or absorption, the reactivity changes resulting from density changes in the coolant tend to be smaller than those occurring in water-cooled reactors. Consequently, other effects, e.g. moderator temperature, fuel temperature and structural expansion, become relatively more important in describing the reactivity behavior of the solid moderator reactor. In spite of these complications theoretical methods have been developed which allow a relatively accurate description of the nuclear characteristics of these reactors. Some attention is given to these considerations in this chapter and some examples of calculations are presented to illustrate how these methods are generally used to examine the safety characteristics of this class of reactors.

2 REACTIVITY CONTROL OF SOLID-MODERATOR REACTORS

The reactivity control of solid-moderator reactors is divided into discussions of control requirements and control methods. The discussion of control requirements emphasizes the reactor characteristics which lead to the specifications on the total amount and rate of change of reactivity control required. The characteristics affecting the control requirements of solid-moderator reactors are, in general, quite different in relative importance from those of water-moderated and fast-spectrum reactors. The mechanisms commonly used for achieving the required control and the general characteristics of these mechanisms is discussed under control methods.

2.1 Control Requirements

General considerations: The control requirements in a reactor dictate both the total reactivity swing and the limits on rate of reactivity change which must be supplied by the control system. In general, the control requirements for solid-moderator reactors result from reactivity changes imposed by changes in component temperatures, coolant flow and pressure, and changes in material composition due to fuel burnup and fission product formation. Through their effect on control system design, both the probability and potential severity of reactivity accidents in reactor plants can be affected by the control requirements.

The gross control requirements imposed by reactivity changes during reactor startup and subsequent fuel burnup throughout the fuel lifetime are typically between 0.05 and 0.20 $\Delta\rho$. Requirements

TABLE 1-1

Illustrative Solid-Moderator Reactors

Reactor	Fuel/fertile isotopes	Enrichment	Coolant	Moderator	Spectrum
Calder	U^{235}/U^{238}	Natural	CO_2	C	Thermal
EGCR	U^{235}/U^{238}	2.46%	He	C	Thermal
Peach Bottom	U^{235}/Th^{232}	93%	He	C	Thermal
EBOR	U^{235}/U^{238}	62.5%	He	BeO	Intermediate
ML-1*	U^{235}	93%	N_2	H_2O	Thermal
Hallam	U^{235}/U^{238}	3.6%	Na	C	Thermal
SIR	U^{235}	93%	Na	Be	Intermediate
Hanford	U^{235}/U^{238}	Natural	H_2O	C	Thermal

*Although the ML-1 reactor is not a solid-moderator reactor, it has some characteristics typical of solid-moderator reactors and will be used as an illustrative reactor in some subsequent discussions.

on the rate of change in reactivity during startup may arise from considerations of a startup accident or the maximum rate at which the plant can be brought to full temperature. The maximum rate of reactivity change is usually limited to about 10^{-4}/sec, although for some reactors the rate may be as small as 10^{-6}/sec. In contrast to the gross control requirements, the reactivity changes which occur during normal plant power adjustments are usually less than $0.01\,\Delta\rho$. The maximum available rate of change in reactivity during normal operation is dictated by the requirements of plant maneuvering, again within the limitations of safety considerations. Typically, a maximum rate of reactivity change in the range of 1 to 4×10^{-4}/sec is required for manueuvering, although it can be considerably smaller under some circumstances. The control requirements both on gross reactivity and rate of reactivity change are discussed below in some detail, particularly as they relate to solid moderator reactors.

The gross reactivity requirements arise from the following contributions:

1. the reactivity margin required for shutdown;
2. the reactivity change associated with the changes in moderator and fuel temperatures in bringing the reactor from its critical condition at room temperature to its normal operating state;
3. the reactivity changes associated with changes in the coolant conditions between normal shutdown and operating conditions, such as changes in coolant density, pressure and flow;
4. the reactivity changes associated with short and long term buildup or burnout of fission product poisons and heavy element isotopes;
5. the reactivity changes associated with fuel depletion and buildup of new fuel; and
6. the excess reactivity margin required for plant maneuvering in the operating range.

Although the amount of reactivity associated with each of these requirements is of some interest in assessing the performance and safety of solid-moderator reactors, the precise specification of these reactivity swings can be ambiguous. For example, the reactivity defect associated with heating the reactor from room temperature to operating temperature would normally depend on the control rod orientation, xenon poison condition, etc. In spite of these ambiguities, component reactivity swings are frequently indicated for reactors and are considered in this chapter.

The nuclear designer is usually more interested in the following three characteristics:

1. the minimum shutdown reactivity margin that might occur at any time in the reactor history;
2. the excess reactivity available during normal operations; and
3. the amount and disposition of control poisons at all times throughout the reactor history when reactivity is just unity at the normal operating power.

These three characteristics can be calculated without ambiguity.

It is, of course, important to have an adequately large shutdown margin to assure that the reactor can be safely shut down under all possible conditions. The minimum acceptable amount of shutdown margin is discussed in a subsequent section. The minimum shutdown margin usually occurs when the reactor is at room temperature (or its coldest possible temperature if the overall temperature coefficient is negative) and the Xe^{135} has competely decayed. In general, the shutdown margin changes with fuel burnup and may actually decrease with burnup in the early part of the reactor lifetime. This is particularly true for large, graphite-moderated, natural-uranium reactors where the conversion ratio is quite high and the U^{235} destroyed is replaced by Pu^{239} which has a higher fission cross section. It is therefore very important for the nuclear designer to calculate the multiplication constant at successive times in reactor life, with the control rods fully inserted, the reactor at room temperature, or the temperature where the multiplication constant is a maximum, and the Xe^{135} decayed.

The excess multiplication over unity, i.e., $k_{eff}-1$, or k_{ex}, which would be available with all control rods in their most reactive position, is also of great interest in the evaluation of the performance and safety of the reactor. In general, the excess reactivity will be different depending on whether the reactor is at room temperature or at operating temperature and on the amount of xenon present. The nuclear designer is usually most interested in the excess reactivity during normal operating conditions, i.e., at the normal operating temperature and with equilibrium xenon established. The excess reactivity during normal operations must be sufficiently large to assure adequate maneuverability for the plant. However, it usually is desirable to minimize the amount of excess reactivity available within the limitations of operational requirements in order to minimize the control requirements. The requirements for excess reactivity are discussed in greater detail in a subsequent section. Because of the importance of the excess reactivity, the multiplication constant is usually calculated at various time steps throughout the reactor burnup history, with all control rods withdrawn, the reactor at its normal operating temperature, and with equilibrium xenon present. When the excess reactivity becomes zero, or smaller than some minimum value required for plant maneuvering, then the reactor has reached the end of its life for the fuel charge.

The amount and disposition of the control poisons at all times through the reactor history are also important to the nuclear designer because the control rod program usually has a very important effect on the power distribution and fuel burnup pattern.

Because of the importance of the reactivity margins and control poison dispositions, a number of one-dimensional and two-dimensional burnup programs have been prepared by several laboratories, so that part or all of the above information can be obtained from digital computer calculations. Table 2-1 indicates a few of these codes and their characteristics. Figure 2-1 illustrates some typical results of a one-dimensional burnup calculation for the Peach Bottom HTGR [1]. These data were obtained from a four-group diffusion theory calculation using the FEVER code developed at General Atomic [2]. The upper curve represents the calculated multiplication constant at successive time steps. For the case where the control rods were all removed, the reactor was at normal operating temperature and equilibrium xenon was established. The interval between this curve and the k = 1 axis is the excess Δ k available under the normal operating conditions. The lower curve in the same figure represents the multiplication constant at various time steps for the case where the control rods were all completely inserted, the reactor was at room temperature and there was no xenon present. The interval between this curve and the k = 1 axis is a measure of the minimum shutdown margin at each point in the fuel burnup history. At each time-step the program calculates the amount and location of control poisons required to achieve k = 1. The power distribution, fuel inventory and fission product composition of the reactor are also calculated at each time step. A program such as this is very important to the nuclear designer, both as a tool for designing the reactor core and for examining the gross reactivity margins at the various times throughout the reactor fuel life. The isotopic composition data are also useful for more detailed calcuations of the temperature coefficient and kinetic properties of the reactor at various times in its life.

It is emphasized that the solid-moderator reactors may show a completely different behavior in the change of multiplication constant with temperature than that normally found in water reactors. This does not mean that solid moderator reactors as a class are any more or less safe than water-cooled reactors, but it does point to the necessity of examining the characteristics of any reactor under all possible conditions to assure that both shutdown and operational requirements can be met for all of these conditions.

For example, a strong positive moderator temperature coefficient or a positive sodium coolant temperature coefficient could lead to a condition where the excess reactivity might be a maximum and the shutdown margin a minimum at some elevated reactor temperature. This must, of course, be taken into consideration in evaluating both the static and kinetic safety characteristics of the reactor. Hence, both the excess reactivity under operating conditions and the shutdown multiplication under the most pessimistic conditions should be calculated at frequent time steps corresponding to successive periods throughout the fuel burnup history. This information is of paramount importance in evaluating the safety of a reactor design.

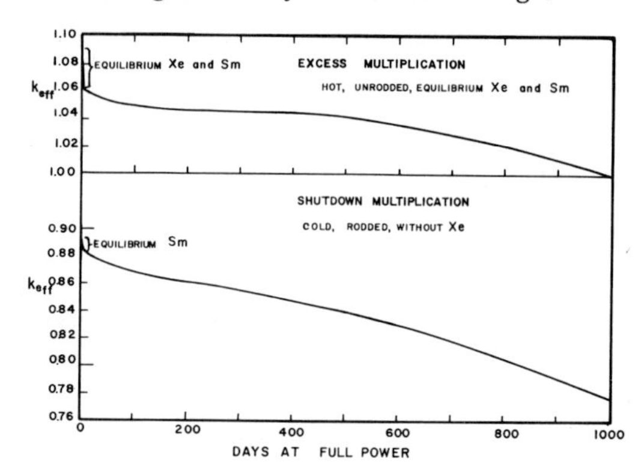

FIG. 2-1. Typical results of one-group burnup calculation for the Peach Bottom HTGR

TABLE 2-1

Digital Computer Programs Used for Diffusion and Burnup Calculations

Type	Max. No. of groups	Machine	Laboratory
One-dimensional diffusion			
AIM-6	18	IBM-7090	Atomics International
GAZE	arbitrary	IBM-7090	General Atomic
WANDA-5	8	S-2000	Westinghouse
ZOOM	18	IBM-7090	Livermore
One-dimensional burnup			
CANDLE	few	S-2000	Westinghouse
FEVER	4	IBM-7090	General Atomic
GAZED	arbitrary	IBM-7090	General Atomic
SIZZLE	18	IBM-7090	Atomics International
Two-dimensional diffusion			
ANGIE	18	IBM-7090	Livermore
CURE	3	IBM-704	General Electric
GAMBLE	10	IBM-7090	General Atomic
KARE	5	S-2000	KAPL
PDQ-5	5	S-2000	Westinghouse
TWENTY GRAND	6	IBM-7090	Oak Ridge
Two-dimensional burnup			
DDB	5	IBM-7090	General Atomic
EQUIPOISE BURNOUT	2	IBM-7090	Oak Ridge
TURBO	4	S-2000	Westinghouse

2.1.1 Excess Reactivity

In this section the implications of excess reactivity and its variation with burnup history are discussed in greater detail. The reactor characteristics which affect the excess reactivity and its time variation are examined and the magnitudes of excess reactivity for some typical solid-moderator reactor concepts are reviewed.

In general, it is desirable to minimize the excess reactivity within the limits of the requirement for adequate reactor maneuverability and endurance. A small excess reactivity is advantageous for the following reasons:

1. The gross control requirements normally depend on the amount of excess reactivity available in the reactor design. It is usually desirable for economic reasons to minimize the number of control rods and mechanisms, hence, the control requirements.
2. Neutron absorption in control poisons wastes neutrons which might otherwise be used for conversion in fertile material. For each $0.01 = \Delta\rho$ in control-poison effectiveness, approximately 0.02 in potential conversion ratio is lost.
3. Control rods frequently cause uneven power distributions and fuel burnup. Not only does the presence of a strong control rod cause a local power depression, but fuel burnup in the neighborhood of the control rod tends to be small relative to the average, so that local hot spots can develop if the control rod is then removed late in the fuel burnup cycle.
4. In general, a large excess reactivity margin increases the possibility of the accidental addition of large amounts of excess reactivity.

Although the excess reactivity margin under normal operating conditions is usually of greatest interest, it is also convenient from the point of view of minimizing control requirements to have a small excess reactivity available at the clean, cold startup condition. Unfortunately, this is usually inconsistent with the requirement for an adequately strong negative temperature coefficient and the necessity to cover the reactivity loss associated with Xe^{135}. The overall temperature coefficient for solid-moderator reactors is typically in the range of -2×10^{-5} to -5×10^{-5} per °C at the beginning of life, with somewhat more positive values usually occurring at the end of the fuel life. The temperature difference between cold shutdown and hot operating conditions may be in the range from 300°C to 1000°C (570°F to 1830°F) depending on the characteristics of a particular reactor, so that the temperature defect for these reactors is typically in the range of 0.005 to 0.05 in $\Delta\rho$. Assuming a xenon defect of $0.03\,\Delta\rho$ and a margin of at least $0.02\,\Delta\rho$ for maneuvering requirements, it can be seen that a cold, clean excess margin of $\Delta\rho = 0.05$ to 0.10 would be typical.

Although it is difficult to effect significant reductions in the temperature defect and xenon defect, the nuclear designer can do much toward reducing the excess reactivity available during normal operations. The methods available include: fertile poisons, semi-continuous fuel loading, removable poison rods, and burnable poisons. The efficient conversion of fertile materials to fissile isotopes is usually the most desirable method of minimizing the reactivity change arising from fuel burnup and fission product accretion. This can usually be done quite effectively in large solid-moderator reactors having a small neutron leakage and in which there is only a small absorption of neutrons in parasitic poisons. The reactivity change due to fuel burnup in this type of reactor can be reduced still more by reloading a fraction of the fuel elements at regular intervals. In addition to partial reloading of the fuel, some reactivity shimming is possible by adding or removing permanent poisons periodically during reactor shutdowns. If removable poisons are used to shim reactivity, it is extremely important that the poison cannot be removed inadvertently, particularly during normal operations or during casualty conditions.

In small reactors with high leakage, or reactors that have large neutron losses to parasitic materials, it is usually impossible to include enough fertile material to realize a substantial conversion ratio. In this type of reactor, it is quite often desirable to use a burnable poison to achieve a relatively uniform excess reactivity throughout the fuel depletion lifetime. The use of burnable poisons has previously been discussed in the Reactor Core chapter. Although a number of elements are possible candidates as burnable poisons, boron is usually used for this purpose. Since the absorption cross section of B^{10} is considerably larger than that of the fissile materials, it has a tendency to burn out too rapidly, thereby causing the reactivity to increase initially. This condition can be corrected, at least partially, by self-shielding the boron. As in the case of the removable poison rods, it is extremely important that the burnable poison cannot be physically or chemically moved from its specified location during normal operations or casualty conditions.

Examples. Figure 2-2 shows the advantages that can be realized by using a boron burnable poison to flatten the reactivity change during fuel burnup. These curves were calculated as an illustrative example using a typical 40 Mw(e) HTGR core composition. It can be seen that the Δk swing in this particular case was reduced from 0.14 to 0.035 using homogeneously distributed boron in the reactor core. By lumping the boron so that the initial self-shielding factor was 0.50, the excess

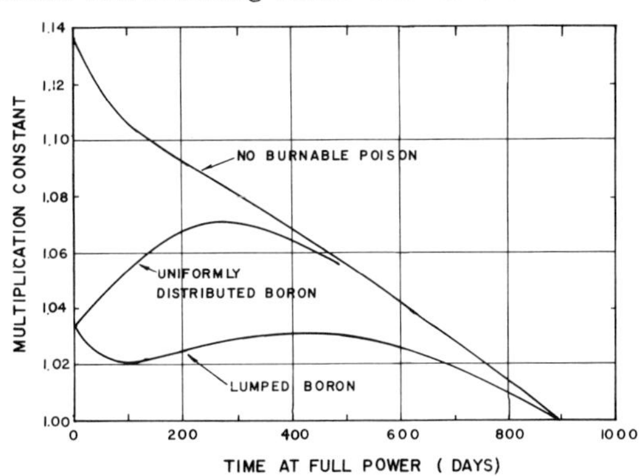

FIG. 2-2. k_{eff} vs lifetime using burnable poisons

reactivity was held nearly flat throughout core life for the same case. This particular case was chosen simply to illustrate how burnable poisons might be used beneficially in a solid-moderator reactor. The principle is, of course, the same as in other types of reactors.

In addition to the improvement in controllability and safety, the use of burnable poisons usually leads to a more uniform power distribution than that available when control rods are used to shim the reactivity.

In high-conversion reactors, the reactivity variation with time is quite different in character for different fuel cycles. Figure 2-3 shows the calculated and measured variation of reactivity with depletion time for the Calder Hall reactor [3]. This reactivity variation is typical of reactors using natural uranium or uranium with very little enrichment. The initial small loss in reactivity results from the buildup of Sm^{149} and other high cross section fission products which saturate very rapidly. Following this initial loss the reactivity increases as the U^{235} depletion is more than compensated by the buildup of Pu^{239} which has a significantly larger fission cross section. Finally the reactivity decreases as the Pu^{239} is depleted and Pu^{240} builds up. The reactivity variation with burnup time is minimized in these reactors by adding or removing fixed poison rods in the reactor [4] to keep the controllable excess reactivity well below $\Delta\rho = 0.01$.

Figure 2-4 shows the calculated variation of reactivity with time for a 500 Mw(e) High-Temperature Gas-Cooled Reactor [5] using the U^{235}/Th^{232} fuel cycle. The reactivity loss in the first 100 days of operation is about $\Delta\rho = 0.015$ compared to less than 0.002 in the Calder Hall reactor. This comes about because the production of the U^{233} is delayed by the 27-day decay half-life of the precursor Pa^{233}. Following this initial reactivity loss, the reactivity decreases more slowly as the fuel load decreases with time. Although the eta for the U^{233} is somewhat better than that for the initial U^{235}, the cross sections are not substantially different, and the overall effect is that the reactivity decreases monotonically for this case.

In the same figure, the reactivity variation is shown for a fuel program where one-fourth of the fuel elements are replaced at each reload period. It can be seen that the total swing in multiplication for this case was reduced from about $\Delta k = 0.06$ to

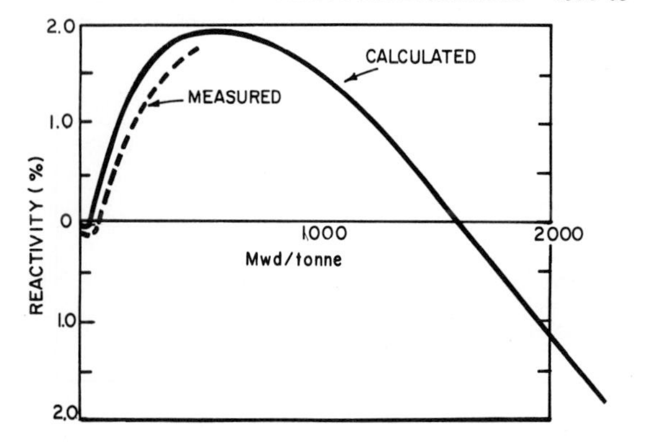

FIG. 2-3. k_{eff} vs lifteime for Calder Hall reactor

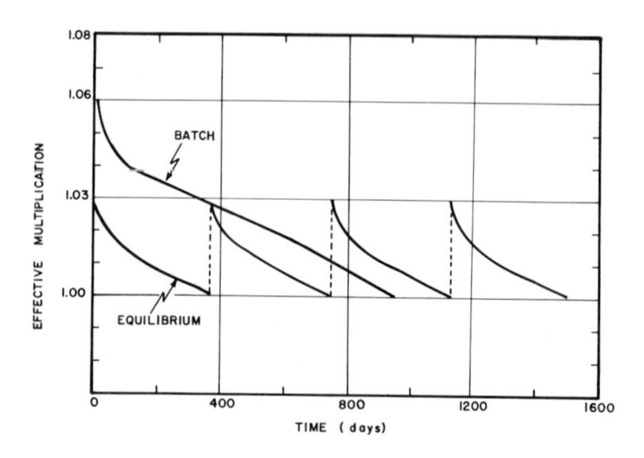

FIG. 2-4. k_{eff} vs lifetime for large HTGR system

0.03 for the partial reloading program and the total fuel burnup lifetime was increased about 50%.

Figure 2-5 shows the reactivity variation with time in a 250 Mw(t), BeO-moderated, gas-cooled reactor [6] using the plutonium fuel cycle. Curves are given for three cases having different mixtures of the plutonium isotopes and two values of specific power, all of which give approximately the same burnup in fissions per initial fissile atom (FIFA). The Pu^{240} acts as both a burnable poison and a fertile material, producing Pu^{241} which has a larger cross section and a somewhat better eta than Pu^{239}. As a result, it is found that the reactivity variation with time tends to be quite small for this fuel cycle. The use of plutonium rich in the isotope Pu^{240} also has some advantages with respect to temperature coefficient, if the reactor is designed appropriately. This particular feature is discussed in greater detail in the section on temperature coefficients.

Calculational Methods. It is important that the nuclear designer have good calculating methods and data so that he can calculate the reactor characteristics at successive intervals throughout the fuel life. In addition, particularly in high conversion reactors where the fuel life approaches or exceeds one fission per initial fissile atom (1 FIFA), an error of 0.01 in reactivity change may result in a significant loss in core life and a substantial penalty in fuel cycle costs.

Some common sources of error in burnup reactivity calculations are as follows:

1. inadequate spatial detail in diffusion theory calculations,
2. inadequate energy detail in describing neutron flux,
3. insufficient detail in treating heavy isotope and fission product isotopic changes; and
4. incomplete treatment of changes in shielding factors.

Zero-dimensional or point geometry calculational programs are commonly used in survey calculations to save both time and money. It is well recognized, however, that fuel burnup tends to occur preferentially in high statistical-weight regions in the reactor, so that the reactivity change with burnup is usually underestimated by ignoring spatial effects. Furthermore, the effect of the control rod programming can have an important effect on the spatial dependence of the fuel burnup and therefore the reactivity change with time. One-dimen-

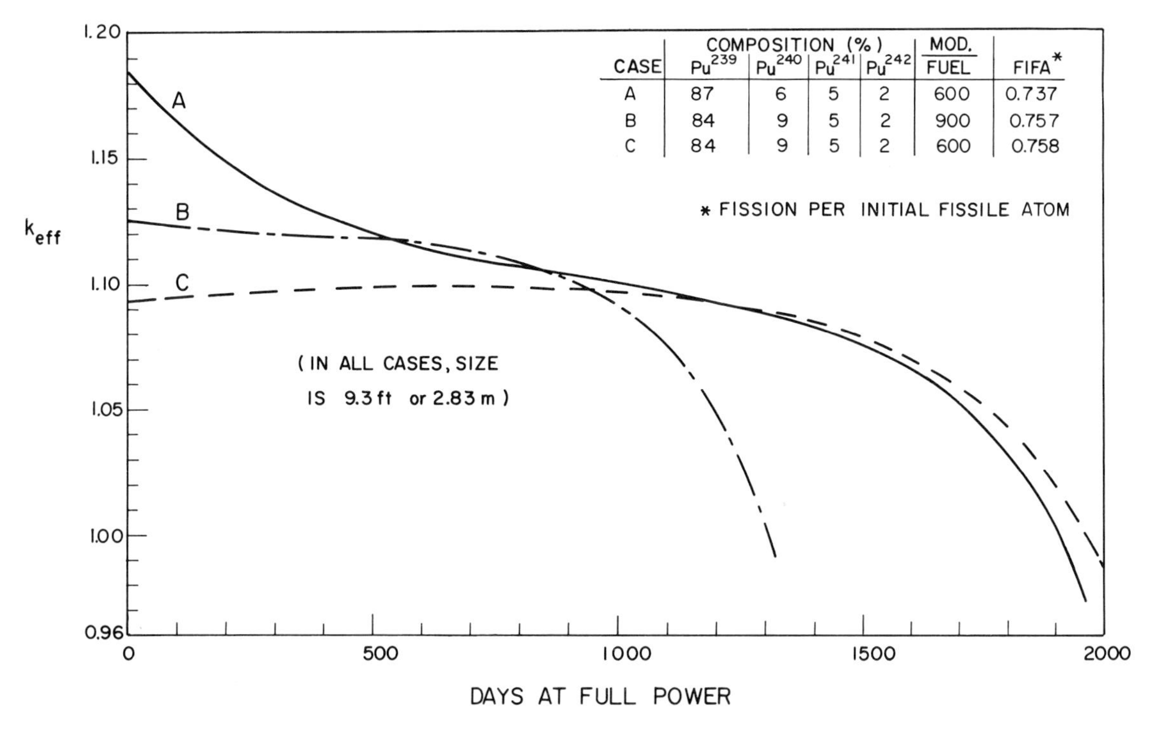

FIG. 2-5. k_{eff} vs days at full power for several plutonium-beryllium oxide reactors

sional burnup programs allow some improvement over the zero-dimensional codes, since they permit the nuclear designer to account for power and burnup variations in the radial or axial direction for cylindrical geometry, or in the radial direction for a spherical approximation. The control rod effect on power distribution can be at least approximated in the radial or axial dimension of a cylinder by allowing poisoned regions to represent rings or banks of control rods. Further improvement is possible in a two-dimensional program which allows a description of control rod poisons, power distribution and fuel burnup distribution in both the radial and axial directions simultaneously.

The magnitudes of the uncertainties associated with various approximations are shown in Fig. 2-6 [7] for a typical 40 Mw(e) HTGR core composition. For this case the reactivity loss is underestimated by about 0.024 for the one-dimensional model and about 0.050 for the zero-dimensional model relative to the two-dimensional calculation. The possible magnitude of the discrepancies in the various approximations can of course be quite different for reactors having different features. The reactor used for the above illustration has spatially distributed poisons and control rod programming to achieve a relatively flat power distribution. Therefore, one might expect the magnitude of the reactivity discrepancies to be smaller than would be the case in some other reactor designs.

The energy distribution of the neutron flux can also be a source of uncertainty in the reactivity calculation, particularly in high temperature, solid-moderator reactors. In gas-cooled and liquid-metal-cooled reactors, the moderator temperature tends to be relatively high and the thermal neutron energy distribution can have profound effects on the absorption rates in most of the fissionable isotopes as well as some of the heavy isotopes and fission product poisons which have non-1/v cross section characteristics.

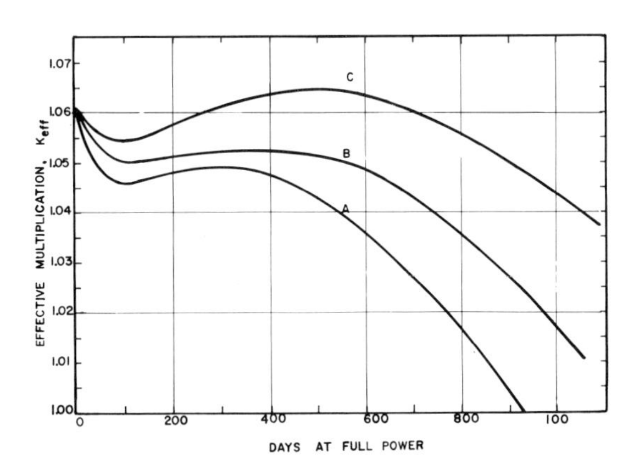

FIG. 2-6. Peach Bottom lifetime for various calculational models (All curves normalized to initial eigenvalue of curve A). A—Two-dimensional (3 fast, 1 thermal group); B—One-dimensional (3 fast, 1 thermal group); C—Zero-dimensional (3 fast, 1 thermal group).

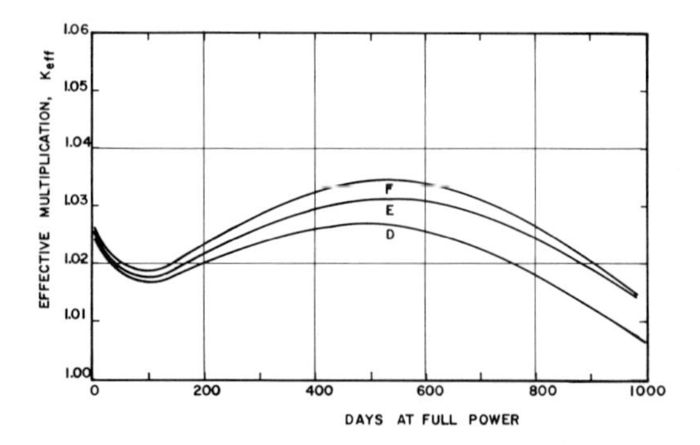

FIG. 2-7. Peach Bottom lifetime for various calculational models D - Zero-dimensional (3 fast groups, 1 thermal group); E - Zero-dimensional (10 fast groups, 1 thermal group); F - Zero-dimensional (10 fast groups, 10 thermal groups).

Although survey calculations are often done with calculational models having a single energy group of neutrons, it should be recognized that this gross simplification can lead to rather large errors in reactivity changes with burnup. Two, three and four energy groups are frequently used in the one- and two-dimensional burnup programs. This increased sophistication is usually introduced to allow a better treatment of the neutron spectrum variation in space in the reactor core. It is usually not adequate, however, to handle the more subtle variation of spectrum with time which arises from the buildup and burnout of isotopes with strong resonances that can change either the thermal or epithermal spectrum or both. Figure 2-7 [7] again shows the reactivity variation for a typical 40 Mw(e) HTGR core composition using different degrees of detail in the representation of the neutron energy spectrum. A part of the effect arises from changes in the thermal spectrum as a result of spectral softening because of the fuel depletion. However, a small effect also can arise from the buildup of U^{234} and U^{236} which have strong resonances at 5.2 and 5.5 ev, respectively. The latter effect is particularly important in high conversion, high burnup reactors [8].

The effect of spectral changes with burnup can be particularly important in reactors using natural uranium, low-enrichment uranium, or plutonium fuel. The Pu^{239}, Pu^{240} and Pu^{241} all have strong resonances in the thermal or slightly epithermal energy region. These resonances, particularly the Pu^{240}, make very substantial changes in the thermal neutron spectrum which affect the relative absorption rates of these as well as other isotopes in a profound way. In order to treat this variation properly it is necessary to use a multigroup thermal spectrum that allows for upscattering as well as downscattering.

The use of multigroup multidimensional burnup programs that allow for upscattering and downscattering is usually prohibitive because of the expense. Therefore, it is convenient to observe the magnitude of the energy effect with a zero-dimensional calculation and make appropriate corrections in the more complex space-energy burnup calculations. An alternative procedure is to calculate new group-averaged cross sections at frequent intervals of a burnup program.

It is, of course, also important to follow isotopic chains explicitly where fission, absorption or decay events can lead to some particular isotope that has a strong effect on reactivity or the reactor temperature coefficients. Although most of the fission products are frequently represented by some aggregate absorption cross section in burnup codes, it is often convenient, both for temperature coefficient calculations and the study of the effect of possible fission product release, to calculate the buildup and depletion of the more important fission product isotopes explicitly with time.

The correct treatment of shielding factors and their changes with fuel burnup can be quite important, particularly in heterogeneous reactors. The disadvantage factor associated with the fuel materials can usually be calculated with some considerable degree of accuracy by using the S_n transport methods, such as the DSN program written by Los Alamos [9]. In burnup calculations it may be necessary to do several of these cell calculations at different time intervals in order to obtain effective homogeneous cross sections which can be used in the diffusion theory calculations at subsequent times in the reactor life. A special case arises for self-shielded burnable poisons where the shielding factor can be calculated from transport theory cell calculations, but can be represented by an empirical formula in the diffusion theory calculations.

Experimental Methods. It is possible, at least in some cases, to study the behavior of solid-moderator reactors at different burnup times by means of critical experiments. As an example, the lifetime of the SIR reactor core was determined from a series of critical experiments done at the Knolls Atomic Power Laboratory [10]. This reactor was a sodium-cooled, beryllium-moderated, enriched-uranium reactor which had predominantly an intermediate energy neutron spectrum. The reactivity defect associated with the temperature swing between room temperature and operating temperature was only about 0.01 ($\Delta\rho$) and was due primarily to the density change in the sodium. As a result the hot operating condition could be represented quite well by simply reducing the density of the materials which mocked the sodium. The burnup characteristics were studied by simply removing fuel, adding poisons to simulate the fission products poisons, and making appropriate adjustments in the control element configurations. The experiment was particularly simple because of the following neutronic characteristics of this reactor:

1. The neutron spectrum was predominantly epithermal so that thermal base effects were unimportant.
2. The fuel was primarily U^{235} which is not critically sensitive to the moderator temperature.
3. The fuel elements had very little self-shielding, and there were no significant heterogeneities in the reactor.

In addition to the above neutronic characteristics, the reactor was sufficiently small to be mocked relatively easily.

In general, solid-moderator reactors are more easily studied by critical experiments than pressurized or boiling water reactors because of the relatively small changes in density of the moderator between room temperature and operating conditions, which thereby minimize the changes in disadvantage factor. However, significant changes in this factor can take place during temperature swings in these reactors because of changes in the thermal neutron spectrum. Furthermore, large reactivity differences can exist between the critical experiment and the operating reactor due to changes in resonance absorption of fertile materials.

Although it is usually not convenient to build full-scale critical experiment mockups of large power reactors, a great deal of information can be obtained from cell experiments following the PCTR procedure developed by Hanford [11]. While this type of experiment is somewhat limited in its capability of mocking a hot, operating reactor condition, particularly with non-1/v absorbers present such as plutonium or Xe^{135}, it is nevertheless a very useful facility for checking the calculational techniques and data used by the nuclear designer.

In summary, good reactivity calculations and excess reactivity information at various times through the reactor burnup life are important to the reactor designer and the operator in order to assure an economic, safe and convenient reactor. These evaluations should lean on good calculational methods, cross section data and critical experiments, where appropriate.

2.1.2 Shutdown Margin

The specification of a safe shutdown margin will normally depend on one or more of the following considerations:

1. The design shutdown margin must be sufficiently large to assure shutdown capability at all times prior to subsequent fuel or poison reloading operations, in spite of reactivity changes which might result from fuel conversion or poison depletion.
2. Particular attention must be given to reactivity increases that can occur in the post-shutdown period as the result of Xe^{135} decay and the buildup of U^{233} from Pa^{233} decay (for reactors using the thorium cycle).
3. The shutdown margin must be sufficiently large to maintain a subcritical condition under postulated credible casualty conditions. These casualty conditions include reactivity increases that might arise from water-flooding, coolant loss, meltdown of core components, loss of core-contained poisons, or relocation of core components, provided any of these events might be expected under accident conditions.
4. The effect of failed or stuck control rods on the shutdown capability of the reactor should receive particular consideration in the specification of the shutdown margin.
5. Under some special conditions, the requirements for post-shutdown heat removal might affect the magnitude of the specified shutdown margin.

In the discussion of excess reactivity requirements it was pointed out that reactivity variations should be minimized throughout the reactor operating history. It is particularly important to minimize reactivity gains which might occur in the shutdown condition. This can be done in the same way as previously noted for controlling excess reactivity, namely, through the use of fertile poisons, semi-continuous fuel loading, removable poison rods, and burnable poisons with appropriate self-shielding.

The importance of assuring a shutdown capability following the decay of Xe^{135} is generally recognized. The effect of post-shutdown Pa^{233} decay on the shutdown reactivity in reactors using thorium as a fertile material should receive equal attention. The reactivity loss at the beginning of a fuel cycle in a thorium converter reactor tends to be large because of the delay in conversion to U^{233} brought about by the 27-day half life of the Pa^{233}. If the reactor is shut down after it has been operating for a long period of time relative to the Pa^{233} half life, then the shutdown margin will decrease somewhat due to both decay of the Pa^{233} and a corresponding gain of U^{233}. As a specific example, the Peach Bottom HTGR has a reactivity gain of approximately 0.02 $\Delta\rho$ when all of the Pa^{233} has decayed to U^{233}. About 0.005 of this is due to Pa^{233} decay and the remainder to increased U^{233} in the reactor.

The shutdown margin must be sufficiently large to cover any reactivity increase that might arise from potential casualty conditions. The failure of a control mechanism to function should be given special consideration. In a well-designed reactor, no single control element should be so strong that its removal could result in a supercritical condition under otherwise normal shutdown conditions. Furthermore, the reactor should be designed so that the probability is extremely small for several control rods to become simultaneously inoperative—particularly when the group of inoperative rods can again lead to a supercritical condition under otherwise normal shutdown conditions. These considerations usually lead to a limitation on a minimum number of independent control elements.

The probability for multiple failures of control rods is discussed in greater detail in the chapter on Mechanical Design. It is emphasized here that the control mechanisms must be completely independent, including the drive signal, the drive mechanism, and the control rod or shutdown element itself. If a failure can be postulated on the basis of interference due to control rod warpage, core warpage or core rearrangement resulting from mechanical shock, temperature effects or radiation effects, then multiple control rod failures are certainly not independent and the probability for multiple failures approaches the probability of an individual failure. These considerations may be particularly important in solid-moderator reactors.

If multiple failures are admissible, then the shutdown margin should be calculated for various combinations of failed control rods. In general, the most serious situation is one where control rods fail in a cluster. Because of core and reflector geometric arrangements the most important region is not always the center of the reactor.

The shutdown reactivity with one or more inoperative control rods is usually done by XY or $R\theta$ diffusion theory calculations using group-dependent boundary conditions obtained either from transport theory cell calculations or perhaps critical experiments. Since the two-dimensional diffusion theory calculations normally require a very large number of mesh points to represent a typical power reactor, it is usually convenient to calculate a fraction (e.g., 1/4 or 1/8) of the core using zero-current boundaries at the edges. If one control rod is removed in the interior of one-quarter of the core, this, of course, corresponds to a full core with four control rods removed. If one-half a control rod is removed from an interface region between two quarter-sections, then the calculation corresponds to two control rods removed from a full core, It is important to recognize, however, that the reactivity swing resulting from the removal of a single control rod in a full core is not necessarily one-fourth the swing, for example, associated with the removal of four control rods symmetrically arranged in the full core. Indeed, in very large cores where the four control rods are distant from each other by several migration lengths, it may be found that the reactivity increase associated with the removal of four control rods in the full core is only slightly larger than that of a single control rod in the full core. This is a very important consideration in calculating the value of a stuck control rod. To do a simplified calculation of a quarter core and to assume that a single rod would be one-fourth the value of four rods, for example, could lead to an error having serious safety implications.

The shutdown margin must be sufficiently large to assure that the reactor will remain subcritical in the event of other plausible casualty or accident conditions. In the sodium-cooled reactors, the coolant usually contributes some net reactivity effect to the reactor multiplication. If the neutron leakage is large (greater than 35%, approximately), then the neutron scattering of the sodium tends to dominate the absorption characteristics and the sodium has a beneficial effect on reactivity. For larger reactors (typical of central power stations) where the leakage is smaller, the absorption characteristic of the sodium dominates the over-all reactivity effectiveness and the coolant is a poison to the reactivity. In the latter case, the removal of sodium from the reactor core increases the reactivity. If it is possible that the sodium might be removed from the reactor, either intentionally or accidentally, while the reactor contains fuel, then the shutdown margin must be sufficiently large to cover the reactivity increase resulting from the sodium removal.

Usually the coolant in gas-cooled reactors is relatively inert, although high-pressure nitrogen and steam are important exceptions. Nitrogen is a neutron poison while steam tends to increase reactivity because of its good scattering and moderating characteristics. Therefore, sufficient shutdown margin must be available to cover a loss of coolant in the nitrogen-cooled reactor. On the other hand, increases in steam pressure can add reactivity to a steam-cooled reactor. Furthermore, all gas-cooled reactors coupled to steam generators must be able to cope with the possible reactivity increase that could arise from a large steam leak into the primary coolant system. Usually the reactivity contributions from these possible accidents are small relative to other control requirements for solid-moderator reactors and, hence, it is not generally difficult to design a control system to accommodate these additional requirements.

Since hydrogen usually increases reactivity, it is very important to prevent water or other hydrogenous liquids from entering the core. Particularly in gas-cooled reactors where the void volume fraction of the core is quite large, the introduction of liquid water could result in very substantial reactivity increases. Usually a gas-cooled reactor and its associated plant can be designed so that the massive introduction of water is practically impossible. Furthermore, the temperature and heat capacity of solid-moderator reactors are usually such that water cannot remain in the liquid phase after the reactor has operated for any appreciable time. Nevertheless, water flooding must be given some consideration in the safety and shutdown analysis of a reactor.

Since solid-moderator reactors tend to operate at relatively high temperatures, the possibility of meltdown and removal of metallic cladding or structural parts from the core must also be considered in establishing the shutdown margin. In particular, the control poison, including burnable poisons, must be contained in a way that meltdown and selective removal of the poisons is not possible. If the central temperature of ceramic fuel elements is sufficiently high so that volatile or gaseous fission products such as xenon or samarium may diffuse to cooler parts of the reactor or system, then the safety margin must be designed to cover any possible resultant reactivity increase.

In addition to the preceding considerations, the uncertainty of the calculational and critical experiment data leading to the design specifications should be factored into the specified shutdown margin. It is sometimes possible to allow for some flexibility in the initial shutdown margin by variations in the initial loading program. However, uncertainties can still arise in the reactivity changes occurring with fuel burnup. Either an adequate shutdown margin must be provided to allow for these uncertainties, or the reactor operator must be prepared for the possibility of reloading the reactor at some intermediate time in the fuel burnup cycle.

In summary, the specification for the shutdown margin of a solid-moderator reactor depends on the potential reactivity increase that can be anticipated for normal and casualty conditions. Table 2-2 summarizes the shutdown characteristics of a number of solid-moderator reactors. It can be seen that the minimum shutdown margin is usually in the range of 0.02 to 0.10 in Δk.

TABLE 2-2

Shutdown Multiplication for Some Typical Solid-Moderator Reactors*

Reactor	Cold rodded multiplication (k_{eff})
EBOR	0.875
EGCR	0.92
Hallam	0.94
Hinkley Point	0.975
Peach Bottom	0.90

* The values shown represent shutdown multiplication factors with all the control rods inserted, the reactor at room temperature, no Xe^{135} poison, and at the beginning of the reactor burnup life.

2.1.3 Control Speed Requirements

During normal operations, the response of the control system should be fast enough to follow reactivity changes brought about by xenon changes and normal temperature changes, yet it must not be so fast that it can create a reactivity hazard. Under emergency conditions, the scram speed must be sufficiently fast to protect the reactor against any damage that might arise from credible accidents.

Usually, solid-moderator reactors used for large central station power-generating plants have specific powers less than 1000 kw/kg and average thermal neutron fluxes less than 5×10^{13}/cm^2-sec. For these reactors, the peak xenon reactivity value is normally less than 0.05 $\Delta\rho$ above the steady state value and the rate of reactivity change during full power operation beginning at the time of the peak xenon is less than 1×10^{-5}/sec, during the Xe^{135} burnout period. Hence, the reactivity rate can normally be controlled without severe requirements on the control system.

For some special-purpose reactors the thermal flux might be considerably larger. As a result, the peak xenon value and the rate of reactivity change during xenon burnout could be much larger; and, indeed, could be so great that the control system would be unable to follow the change without some special provisions. Obviously, the operational requirements of a high-flux thermal reactor within a few hours following a power setback or a reactor shutdown can result in important specifications on the control system. The reactor designer must then demonstrate that the control system not only meets the maneuvering requirements, but also limits reactivity changes to a safe value.

In large central station power plants, the requirements on load changes usually establish the control system rate of reactivity change. The large power stations typically require that the power plant be able to follow changes in load as large as 10% of full power per minute. For a prompt power coefficient $\Delta\rho/(\Delta P/P) = 10^{-2}$, which is typical for normal converter reactors, this requires a control speed $\Delta\rho/\Delta t$ of 10^{-3}/min or approximately 2×10^{-5}/sec. Since the control rod sensitivity usually is quite different for different rod locations and control rod patterns, it is normally necessary to establish the control rod speed on the basis of the minimum sensitivity of reactivity control. Hence, if the minimum control sensitivity is, say, 20% of the maximum sensitivity, then a control system which meets the requirements of $2 \times 10^{-5}/\text{sec} = \Delta\rho/\Delta t$ will have a maximum reactivity rate of 1×10^{-4}/sec. The maximum reactivity addition rate is limited to 2 or 3×10^{-4}/sec for most solid-moderator reactors used in central station power plants.

The maximum rate of reactivity change allowed for a solid-moderator reactor usually depends on the severity of a credible startup accident or a control rod withdrawal accident beginning in the normal power range. A further discussion of these accidents for solid-moderator reactors is included in a later section.

Under emergency conditions where a rapid reactor shutdown is desirable or necessary, the rate of reactivity change required will be much larger. The delay time before scram initiation, the control rod acceleration and velocity and the sensitivity are all important factors. It is desirable to design the reactor with an adequately strong prompt power coefficient so that a maximum credible reactivity addition cannot result in severe damage, even if the shutdown system fails to operate immediately. To achieve this objective the following criteria should be met:

1. The maximum credible reactivity insertion should be limited preferably to less than one dollar.
2. The reactor should have a strong prompt negative temperature coefficient.
3. The temperature margin between normal operating conditions and accident conditions for critical reactor components should be sufficiently large so that damage can not result in time periods short relative to 10 or 100 times the normal scram time.

In addition to the above criteria, the heat capacity of the fuel element can be an important factor in limiting the time rate of power increase. However, the reactor designer usually does not have much flexibility in the choice of this parameter.

If the above objectives are met, then the immediate corrective action depends only on the inherent characteristics of the reactor, and the scram system serves simply to terminate the power generation when the power exceeds the specified level of the reactor plant. The importance of the magnitude of the prompt negative temperature coefficient and the size of the reactivity insertion is illustrated by the results of some simple excursion calculations summarized in Figs. 2-8 and 2-9. In these figures, the curves represent the rise in temperature difference between the average fuel element temperature and the coolant temperature resulting from various step insertions of reactivity. Results are shown for cases where the reactor has a prompt negative temperature coefficient of -2×10^{-5}/°C (-1.1×10^{-5}/°F) and where the prompt coefficient is zero. Figure 2-8 indicates results for a fuel element containing uranium and thorium carbides mixed homogeneously with graphite. Figure 2-9 illustrates the behavior for a typical low-enrichment, uranium oxide fuel element. In the first case, the C:U^{235} atom ratio in the fuel compact was assumed to be 700 for purposes of calculating the heat capacity. This ratio is typical of the fuel compacts in a HTGR-type fuel element. In the second case, it was assumed that the enrichment was 3% U^{235}. In both cases a specific power of 500 kw/kg was assumed.

An extremely simple calculational model was used to generate this data. It was assumed that the heat removal from the fuel elements remained constant and equal to the heat removal prior to the reactivity addition. The reactivity insertion was a step increase. With a prompt temperature coefficient of -2×10^{-5}/°C, which is usually attainable in converter reactors, the fractional fuel element temperature rise is quite modest for reactivity insertions less than 50¢. In the case of the HTGR-type compact, part of the prompt coefficient comes from the heating of the graphite matrix as well as from the Doppler effect in thorium.

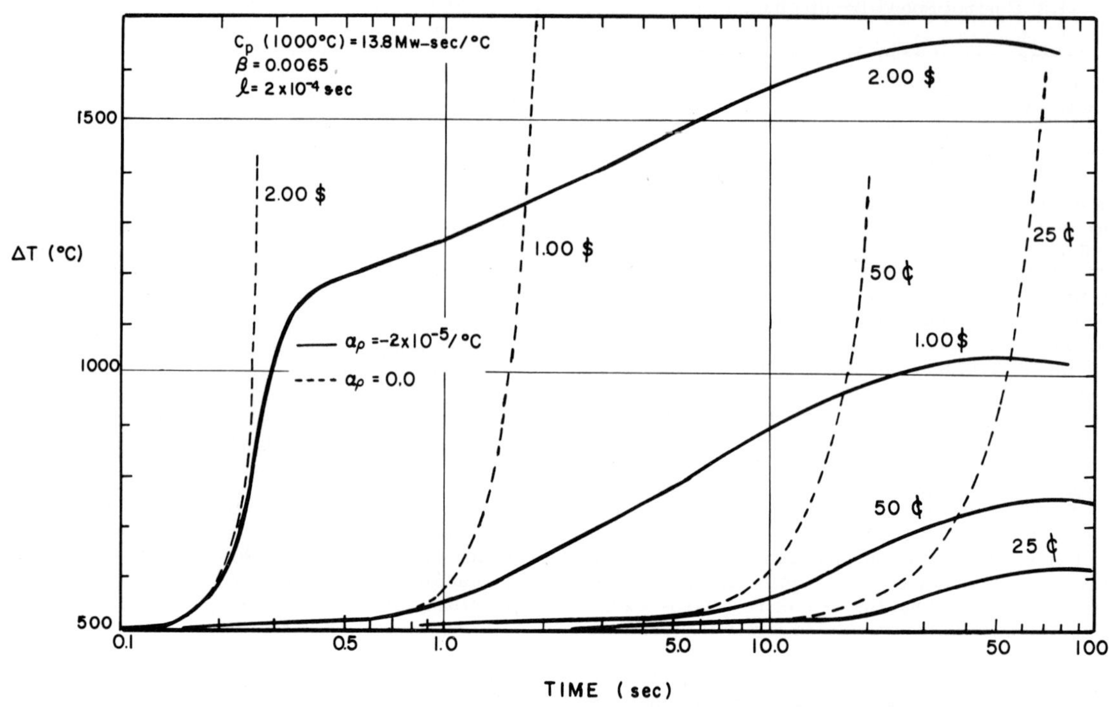

FIG. 2-8. Step insertions for 100 Mw reactor with HTGR-type elements. Constant power removal model.

A comparison of the two figures shows that the larger heat capacity of the HTGR type allows considerably more time before corrective action is required. (The scram delay time in the Peach Bottom HTGR is 0.175 sec and the rods insert in less than 1.0 sec.) However, it must be remembered that the energy storage in the system with the higher heat capacity will also be larger if no corrective action is taken. The maximum temperatures reached are higher for the HTGR type but the melting point of the carbide fuel particles is correspondingly much higher.

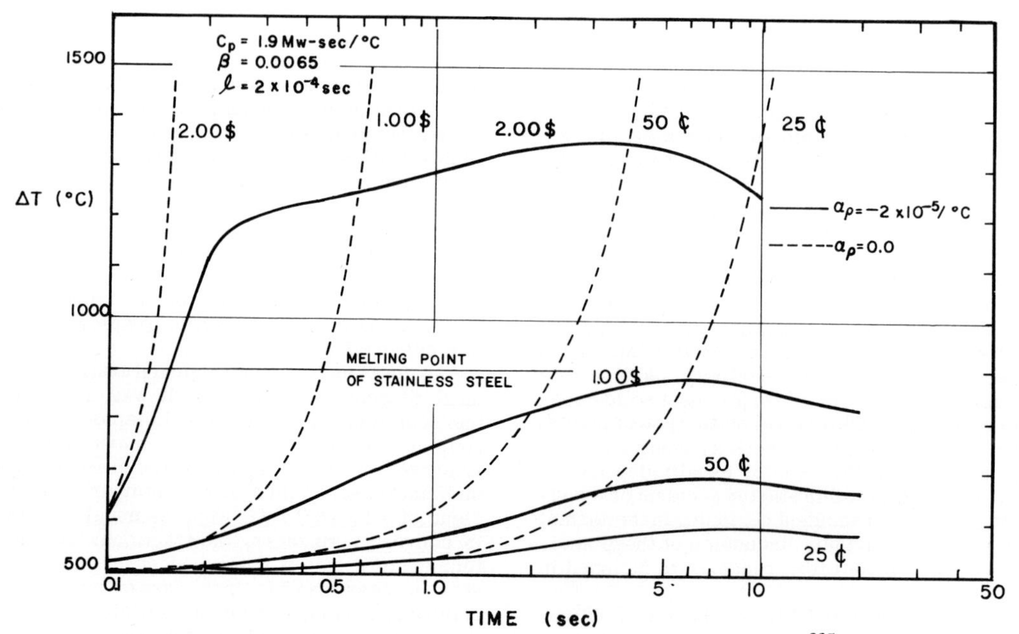

FIG. 2-9. Step insertion for 100 Mw reactor with 3% enriched UO_2 fuel (200 kg U^{235} or 500 kw/kg). Constant power removal model.

High enrichment reactors tend to have a relatively small prompt temperature coefficient and therefore must be designed particularly carefully to avoid the possibility of excessive accidental reactivity insertions.

Hence, the scram reactivity requirements for all reactors require a detailed evaluation of the maximum credible reactivity insertion, the magnitude of the prompt temperature coefficient, the temperature rise of the reactor components and the maximum allowable temperature. The scram specifications involve both the delay time and the rate of reactivity change brought about by the scram. The rate of reactivity change depends both on the control rod or bank speed and its sensitivity. Care must be taken to calculate the shape of the sensitivity curve correctly.

If the total reactivity worth of a single control rod is sufficiently small so that the rod itself does not perturb the flux distribution significantly, e.g. less than 0.01 $\Delta\rho$, then its sensitivity ($\Delta\rho/\Delta x$) will be approximately symmetric about the center plane of the reactor, provided the other control poisons and fuel materials are distributed in a way to produce a symmetric neutron flux curve. If, however, an entire bank of control rods is moved simultaneously, the neutron flux shape changes significantly with the control bank motion and the sensitivity curve is usually strongly asymmetric. Furthermore, the sensitivity is distorted in a direction to decrease the initial reactivity change per unit displacement of the bank as the bank moves into the reactor on scram. Hence, scram reactivity rates calculated on the basis of the simplified symmetrical curve will be overly optimistic. In order to avoid this difficulty, the nuclear designer should always calculate the reactivity sensitivity of the rod bank at various penetrations in the reactor core. This can be done quite well by representing the control rod poisons in the core by a uniformly distributed equivalent poison over the rodded section of the core, and calculating the effect of moving the interface, preferably in a two-dimensional calculation. Some additional sophistication can be introduced by representing rings of control rods by poison rings in an R,Z calculational model.

Figure 2-10 shows a comparison of rod sensitivity curves for a single rod and a control rod bank as reported for the Peach Bottom reactor. The curves clearly show the importance of control bank calculations. The effect is even more spectacular in reactors having a large length-to-diameter ratio. Figure 2-11 shows the rod bank sensitivity in the EBOR reactor where the cross section dimensions are approximately 2 ft × 2 ft (0.6 m × 0.6 m) and the length approximately 6 ft (1.8 m).

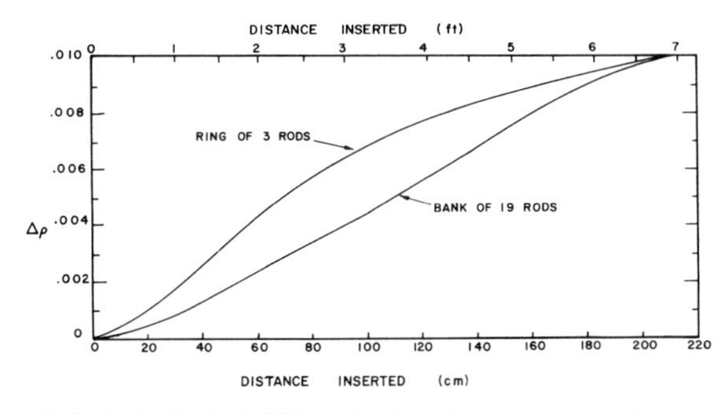

FIG. 2-10. Typical HTGR rod calibration curves normalized to 0.01 $\Delta\rho$

If the reactor has separate control and safety rods and the control rods are partially inserted, then the first part of the safety rod travel will have very little reactivity effect. Again, it is important for the nuclear designer to use appropriate and realistic calculational models to evaluate the scram reactivity curve assuming the most pessimistic operating conditions.

2.2 Control Methods

In the previous section, the control requirements were outlined, the implications on design and operation were indicated, and calculational problems were identified. In this section the mechanics of the control systems for solid moderator reactors are discussed.

In general, the reactor design should ideally have the following design objectives:

1. The reactor design should preclude the possibility of rapid reactivity increases which can lead to severe power excursions. Since the most direct way to introduce reactivity changes is through the control system itself, special consideration must be given to the safe design of this system.
2. The basic shutdown mechanism for the reactor, in the event of a reactivity accident, should be the self-limiting characteristics of the reactor core itself.
3. The fuel elements and other core components should be designed to withstand the maximum credible power excursion limited only by the inherent shutdown characteristics of the reactor.
4. The control system should be capable of terminating the power operation promptly. Although it is desirable to have a control scram system sufficiently fast to limit the size of any possible excursion, it is perhaps worthwhile to emphasize again that the accident potential and reactor response preferably should not be such that a fast scram is essential to avoid serious reactor damage.
5. There must be an adequate shutdown margin under all possible operating and casualty conditions. If this shutdown margin cannot be guaranteed with the normal control system, then a backup emergency control system must be included.

Solid-moderator reactors have been designed with a variety of control mechanisms, including control rods, control cylinders and poison semaphore blades. Since solid moderator reactors tend to operate at high temperatures, the designer must be careful that mechanisms operating in this environment cannot be rendered ineffective or unsafe by a temporary temperature increase from a loss of coolant flow or a sudden pressure change from a loss of coolant pressure. In the case of liquid-metal-cooled reactors, the control system must be designed to avoid the possibility of becoming

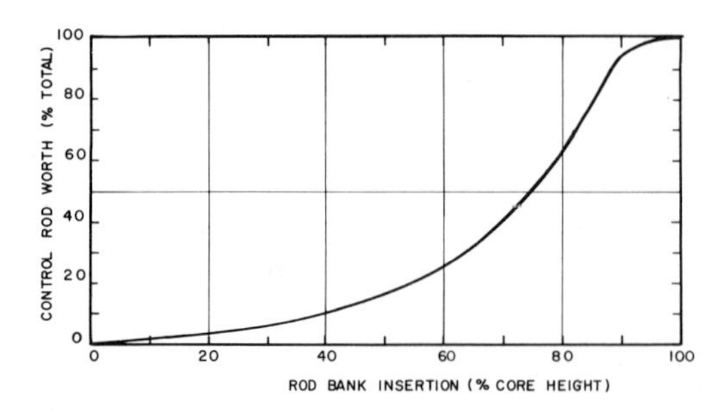

FIG. 2-11. Rod worth vs insertion. Hot clean EBOR 90 kg U^{235}, 70 kg U^{238}.

ineffective through freezing of the coolant or plugging of the channels by contaminants such as oxides or hydroxides.

The number and location of the control mechanisms can have important implications on shutdown capabilities, accident possibilities and power distributions. It is desirable to have a sufficiently large number of control mechanisms so that the failure of any single mechanism does not result in a loss of shutdown capability. Furthermore, the concentration of too much reactivity effectiveness in any single mechanism increases the possibility of a reactivity accident through the malfunctioning of this mechanism. Finally, the control mechanisms can have an important effect on power distribution and therefore component temperatures because of their perturbing effect on the neutron flux. A strong control poison located poorly can result in unacceptable power peaks in another part of the core. A more subtle hazard can come about from differences in fuel depletion rates in the neighborhood of strong poison controls. The power generation rate adjacent to a normally inserted control poison mechanism will be considerably smaller than the power a few diffusion lengths removed from the poison. As a result the depletion of the fuel away from the poison neighborhood will be much greater than that near the poison. Later in the reactor life when the control rod is removed, a local hot spot can develop in the region of high fuel concentration.

The latter problem can be evaluated by cell burnup calculations. If the problem appears to be serious, then it may be necessary to reduce the strength of the control poison or alternate the control program geometry at regular intervals.

Burnable poisons are frequently desirable both to minimize the amount of reactivity swing due to fuel depletion and also to achieve a more uniform power distribution. Since the burnable poison can be distributed over a relatively large volume of the core, its effect on power distribution can be minimized. In fact, it is possible in some cases to use the burnable poison to help shape the power in the core. Under these conditions some caution is necessary to assure a good power distribution after the burnable poison is removed through depletion. Again burnup studies are required to ascertain that power distributions are suitable at all times in the fuel life.

In reactors containing a large number of control rods it is quite often possible to program the rods in a way to improve the power shape. Furthermore, the control rod programming procedure can have some effect on the possibility for reactivity accidents. One common operational procedure for reactor startup is to remove all control rods as a bank. This procedure is useful when the reactor is designed to have very little excess reactivity under normal operating conditions. If bank control is used, it is desirable to limit the rate of reactivity increase so that a continuing bank withdrawal will not produce a rapid power overshoot.

An alternative procedure is to withdraw control rods in groups, maintaining some symmetry during the operation. If the reactor has sufficient excess reactivity, it is often desirable to operate the reactor with most of the control rods either completely withdrawn or completely inserted. In this way the axial power shape has a minimum distortion. Furthermore, it may be possible to choose the radial location of the control rods in a way to improve the radial power shape. It must be remembered, however, that the control rods will have to be removed as fuel burnup proceeds, and the power shape under all lifetime conditions must be explored.

If this latter type of control rod programming is used, then there are some advantages in interlocking the control rod groups in a manner that the maximum reactivity available in any single operation is such that a serious power overshoot is impossible, even if the group is continuously withdrawn to its maximum reactivity condition.

If an accident involving the core or the mechanisms can be postulated where the effectiveness of the control system is lost, then a backup emergency shutdown system is desirable. Systems have been proposed for dry reactors where poison balls or cylinders are dropped into the core, high-strength safety rods are driven in, or reflectors are removed. The safety designer must assess the need for such devices and then insure that the backup system will be available under the conditions existing when the normal system has failed.

3 RELEVANT FACTORS IN THE DYNAMICS OF SOLID-MODERATOR REACTORS

3.1 Introduction

The important nuclear parameters which characterize the dynamic behavior of a reactor are the prompt neutron lifetime, the delayed neutron fraction, and the temperature coefficients. The time responses of the component temperatures leading to reactivity changes through the temperature coefficient are also very important. Time responses are determined by the heat capacities of the various components, thermal conductivities, and heat transfer coefficients, all of which establish the rate of thermal energy storage in the various core components and also determine the degree of damage which may occur during a reactivity transient. The total energy stored as the result of a reactivity insertion will ultimately be limited by control rod corrective measures, by a reactor scram, by the reactivity compensation of the temperature coefficients, or by destruction of the core.

The nuclear parameters have been defined and discussed in some detail in other chapters. The values of these parameters, some of the calculational methods used to obtain them, and their relative importance in the kinetics of solid-moderator reactors will be discussed in this chapter for typical reactors of this type. The major emphasis will be put on temperature coefficients for the following reasons:

1. The temperature coefficients are very important in determining the kinetic behavior of reactors at power.
2. They are particularly important in the ultimate safety of a reactor.
3. In many kinds of solid-moderator reactors the designer has considerable flexibility in specifying the temperature coefficient values without compromising other design objectives.
4. Considerable calculational detail may be required to determine the coefficients accurately.

As illustrated by the tables in the Reactor Core chapter the over-all temperature coefficients of solid-moderator reactors are of the order of $10^{-5}/°C$ compared to $10^{-4}/°C$ for the water-moderated reactors and $10^{-6}/°C$ for the fast reactors. The Doppler coefficients of water-moderated and solid-moderated thermal reactors are generally comparable in magnitude and time response whereas the moderator coefficients of the solid moderator cores are an order of magnitude smaller, owing to the lack of significant thermal expansion and expulsion of moderator from the core. Furthermore, the moderator coefficient may be positive or negative and may have a wide range of time response characteristics. Since the dominant contribution of the moderator coefficient is not from thermal expansion but from changes in the thermal neutron spectrum which influence relative reaction rates and leakage, more sophisticated procedures are usually needed to calculate the moderator coefficient of solid-moderator reactors, and a greater variation in its magnitude through core life may be expected. In the intermediate-spectrum solid-moderator reactors thermal neutron spectrum effects become less important while the effects of thermal expansion of structural components usually become more important. In intermediate energy reactors with highly enriched fuel, the Doppler effect may come predominantly from U^{235} and is smaller and harder to calculate since it involves competition between capture and fission resonances. In some situations it could even be positive.

3.2 Prompt Neutron Lifetime and Delayed Neutron Fractions

Values of the prompt neutron lifetime for solid-moderator reactors range from 10^{-3} sec for the large natural-uranium, graphite-moderated thermal systems to 10^{-5} sec for the intermediate spectrum beryllium-moderated cores, (e.g. EBOR and SIR). The lifetime will vary with fuel loading and will be a function of temperature, although the temperature effect is usually not large.

In the simple case where the shutdown mechanism is a linear function of the total energy yield it can be shown [22] that for an initial reactivity step of Δk dollars the total energy yield is independent of the neutron lifetime and the properties of the delayed neutron fraction. (See chapter on Mathematical Models of Fast Transients.) This is the case, for example, where the only shutdown mechanism is a prompt-responding temperature coefficient. The small effect of the neutron lifetime on the calculated fuel temperature response, for an initial step reactivity insertion of one dollar, is shown in Fig. 3-1. The graphite reactor used in the example of Fig. 2-8 was chosen to illustrate the point. Both the time-dependent power and fuel temperature behavior are illustrated for a range of neutron lifetimes.

The basic data on the delayed neutron fractions (β) of all fissionable nuclides is given in the chapter on Criticality with a discussion of how properly to compute the effectiveness of the delayed neutrons relative to prompt neutrons. The problem of the delayed photoneutrons in beryllium-moderated reactors is also treated there. In many solid moderator reactors the neutron spectrum is predominantly thermal and the fast leakage is small, which makes the effective delayed neutron fraction (β_{eff}) deviate less than 10% from the absolute delayed fraction, and quite often the deviation is within the experimental uncertainty in the β measurements (3 to 5%). In small or intermediate spectrum reactors the relative effectiveness of the delayed neutrons is more important. In the EBOR reactor, for example, a small beryllium-oxide-moderated core, the calculated β_{eff}/β is 1.25.*

Of greater significance in determining reactor safety is the change in β_{eff} due to changes in concentration of fissile nuclides in reactors with large conversion ratios. In a reactor initially fueled with U^{235}-Th^{232}, the buildup of U^{233} reduces the effective β at the end of core life. In a typical high-conversion reactor, the fractional fissions in U^{233} may approach 50%, giving a β_{eff} in the range of 0.004 compared to an initial value of 0.0065. On the other hand, reactors initially fueled with mixtures of the plutonium isotopes (e.g. Fig. 2-5) will experience an increase in β_{eff} as the Pu^{239} is replaced by Pu^{241}, since the delayed neutron fraction for Pu^{241} is 0.0049, while the value for Pu^{239} is 0.0021.

3.3 Temperature Coefficients

In recent years the possibility of accurately calculating temperature coefficients in solid-moderator reactors has been greatly improved by the development of detailed calculational models for the resonance absorption and thermalization processes, by the continuing improvements in cross-section data, and by the availability of high speed computers. In many situations of interest, the reactivity contribution from component temperature changes is easily separable into the two derivatives of k_{eff} with respect to the fuel temperature T_f and the moderator temperature T_m:

$$\frac{dk_{eff}}{k_{eff}} = \frac{1}{k_{eff}}\left(\frac{\partial k_{eff}}{\partial T_f}\right) dT_f + \frac{1}{k_{eff}}\left(\frac{\partial k_{eff}}{\partial T_m}\right) dT_m \qquad (3\text{-}1)$$

---

*Unpublished internal memorandum.

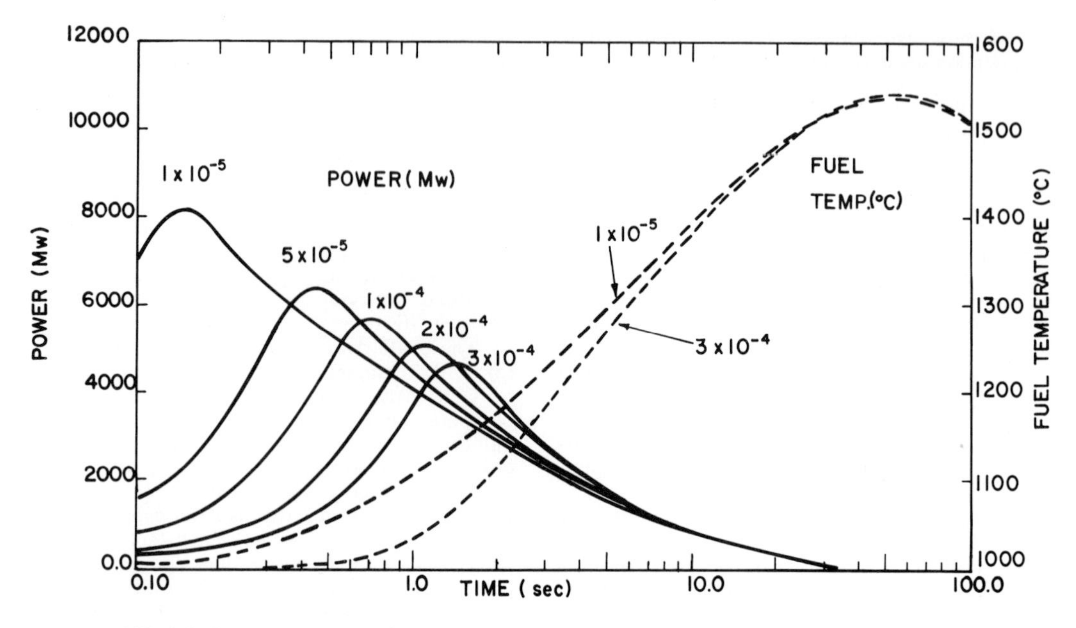

FIG. 3-1. Power vs time and fuel temperature vs time for one dollar step insertions with various prompt neutron lifetimes (indicated in sec adjacent to each curve). Graphite fuel elements from example of Sec. 2.

This separation into fuel and moderator terms also corresponds to the usual computer methods of preparing multigroup cross sections for the reactivity vs. temperature calculations, since these cross sections are averaged over slowing-down spectra, which are only a function of fuel temperature, and thermal spectra, which are only a function of the average moderator temperature. One may then evaluate the isothermal temperature coefficient between two temperatures, for example, by calculating the two values of k_{eff} with fast and thermal multigroup cross sections evaluated at the two temperatures. If the fuel coefficient $(1/k_{eff})(\partial k_{eff}/\partial T_f)$ depends only on changes in the resonance absorption rates of the fuel and fertile materials, then it can be obtained by calculating the k_{eff} when only the fast-group cross sections are allowed to change between the two temperatures. The moderator coefficient $(1/k_{eff})(\partial k_{eff}/\partial T_m)$ can then be obtained for the same temperature increment by subtraction. The entire procedure can be accomplished in a single computer code covering many temperature intervals between room temperature and the highest temperatures conceivable in a transient.

It is important to note that the use of isothermal temperature coefficients in calculating temperature coefficients is simply an expedient. In operating reactors, the temperature of the fuel may be significantly different from the temperature of the moderator, the difference depending on the heterogeneity of the core and the thermal conductivity of various components. Furthermore, the moderator temperature may vary significantly in various parts of a lattice cell, particularly if some of the moderator is intimately mixed with the fuel. In order to describe the kinetic characteristics of a reactor under operating conditions it is necessary to calculate the individual temperature coefficient components over temperature ranges of practical interest for each of the components. The temperature range of interest for the fuel elements is usually somewhat larger than that of the moderator.

The division of the temperature coefficient into a fuel and a moderator coefficient by assigning epithermal effects to the fuel and thermal effects to the moderator has some limitations. For example, if the fuel contains a significant amount of a nuclide with a low-lying resonance, e.g. Pu^{240}, so that the Doppler-broadening of the resonance can have a strong effect on the thermal spectrum, then the fuel coefficient can interact with the moderator coefficient. The method is also limited in validity for liquid-metal-cooled systems in which changes of coolant density with temperature affect both resonance absorption and thermal absorption, although it might be a sufficiently good approximation if the coefficients $\partial k_{eff}/\partial T_f$ and $\partial k_{eff}/\partial T_m$ are not strongly dependent on coolant density over the temperature range of interest. This complication may also occur with the temperature-dependent expansion of other core structural materials in the intermediate spectrum reactors.

Frequently one sees reference to a prompt and a delayed temperature coefficient, or more precisely, a prompt-responding and a delayed-responding temperature coefficient. It is emphasized that any such distinction tends to be arbitrary since the characteristic response times of the delayed effects may vary from seconds to minutes for various reactors. Furthermore, the component of the temperature coefficient identified as prompt-responding is not necessarily the same as the fuel coefficient as defined above. For example, if part of the moderator is intimately mixed with the fuel, then a corresponding part of the moderator coefficient is equally prompt.

Although the development of computer methods has made it no longer necessary or desirable to obtain temperature coefficients by the differentiation of four-factor type prescriptions, a breakdown of the computed derivatives of k_{eff} at appropriate temperatures in terms of reaction rates in important nuclides is useful in understanding the results and in suggesting ways to maintain or increase the negative components of the coefficients by changes in core loadings or fuel element design. Thus, one may write,

$$\frac{1}{k_{eff}}\left(\frac{\partial k_{eff}}{\partial T}\right) = \frac{1}{\eta}\left(\frac{\partial \eta}{\partial T}\right) + \frac{1}{f}\left(\frac{\partial f}{\partial T}\right) + \frac{1}{p}\left(\frac{\partial p}{\partial T}\right) + \frac{1}{\epsilon}\left(\frac{\partial \epsilon}{\partial T}\right) + \frac{1}{P_{NL}}\left(\frac{\partial P_{NL}}{\partial T}\right), \quad (3-2)$$

where T may be T_f or T_m and the five factors are defined in terms of the reaction rates obtained in the multigroup neutron balances. The term P_{NL} refers to the nonleakage probability. The definitions are quite arbitrary and various groupings might be proposed for different reactors. Table 3-1 gives a set of definitions useful for predominantly thermal reactors. In this case P_{NL} has been further divided into fast and slow nonleakage components. Further examples are given later.

It should be noted that temperature coefficient calculations for use in kinetics calculations should be done with the core containing the amount of control poison required to make the reactor critical at the operating temperature or starting point of the transient, although the omission of the control rod contribution to the calculated temperature coefficient is usually conservative.

3.3.1 Doppler Coefficients

The principal contributor to the fuel coefficient $(1/k_{eff})\ (\partial k_{eff}/\partial T_f)$ in most solid-moderator reactors is the Doppler effect in the fertile materials U^{238} or Th^{232}. The reactivity effect of the Doppler broadening of fertile material resonances is expressed predominantly through the $(1/p)\ (\partial p/\partial T)$ term in Eq. (3-2) with second-order contributions in the coefficients of ϵ and P_{NL} attributable to the over-all shifting of the relative fast and thermal fluxes with changes in the resonance absorptions. These second-order contributions will also depend on the definitions used for p, ϵ, and P_{NL}.

The computer methods presently available make it possible to calculate accurately the resonance absorption and Doppler coefficients when reliable data on resonance parameters are available. The GAM-I multigroup slowing-down code, [12] for example, incorporates the numerical procedures of Adler, Hinman, and Nordheim [13] to compute the resonance absorption for a variety of absorber configurations. An improved form of this method which eliminates the need to distinguish between wide and narrow resonances has been developed by Nordheim and Kuncir [14]. Comparisons of calculations using this improved method with the newest experimental data have recently been published by Nordheim [15] and show good agreement for both the resonance integral I and the Doppler coefficient $(1/I)\ (\partial I/\partial T)$ although fewer experimental data are available for the latter quantity. The agreement is particularly good for U^{238} for which the resonance parameters are well known. A more detailed discussion of the procedures and typical results can be found in the chapter on Doppler Coefficients.

The resonance integral in solid-moderator reactors may be primarily a surface absorption effect as in the heterogeneous designs (e.g. Calder Hall) or a volume absorption effect as in the semi-homogeneous designs (e.g. HTGR). In the former case, the Doppler coefficient is most easily varied by varying fuel rod dimensions, while in the latter cases the coefficient is more sensitive to the concentration of the fertile material. In analyzing the effects of such changes the familiar equation

$$\frac{1}{p}\left(\frac{\partial p}{\partial T}\right) = (\ln p)\frac{1}{I}\left(\frac{\partial I}{\partial T}\right), \quad (3-3)$$

which relates the temperature coefficient of p to the Doppler coefficient, is useful although it is only strictly correct when p can be expressed in the form

$$p = \exp\ (-N_a I/\xi\Sigma_s) \quad (3-4)$$

and the p defined from a multigroup neutron balance (e.g. Table 3-1) will in general not be the same.

The division between heterogeneous and homogeneous designs is not always well defined. For example, in the calculation of Doppler coefficients for semihomogeneous types of fuel elements, the size of the absorber particles dispersed in the moderator matrix is an important factor [16]. For particle diameters greater than about 0.01 cm (4 mils) the additional shielding of the fertile material with respect to the moderator becomes significant. The value of I therefore decreases with increasing grain size while the derivative dI/dT is found to increase slightly. As a result, the designer can, by increasing grain size, either keep the same Doppler coefficient with reduced resonance absorption or, alternatively, increase the Doppler coefficient and maintain the same resonance absorption by increasing the fertile loading. The magnitude of the effect is illustrated for ThO_2 particles (density 8 g/cm^3) in a graphite matrix, in Figs. 3-2 and 3-3 [16]. The ratios of I and (1/I) $(\partial I/\partial T)$ for various grain diameters to their values in the homogeneous case are plotted. Since the effect is nearly linear with grain diameter, a single average may be taken over the particle size distribution occurring in an actual fuel element.

Although the most important contributions of the Doppler effect to reactor temperature coefficients are associated with the epithermal resonances of U^{238} and Th^{232}, it is also necessary in some solid-moderator reactors to include the Doppler effect for the fissile nuclides and for the large thermal energy resonances where appropriate. The Doppler broadening of the fissile resonances becomes important in intermediate spectrum reactors and is discussed later in the context of an example. The possibility of significant Doppler broadening of thermal energy resonances is illustrated by Table 3-2 taken from the cross section tabulation of Wikner and Jaye [17]. Values of the parameter ξ, the ratio of the natural width to the Doppler width, are given for the temperature range of interest in the temperature coefficient analysis of solid-moderator reactors. Low values of ξ are an indication of sig-

TABLE 3-1

Definitions of Components of k_{eff} from the Neutron Balance of a Multigroup Code

$$k_{eff} = \eta f p \epsilon P_{NL}^{fast} P_{NL}^{thermal}$$

$$\eta \equiv \frac{\sum_z (\phi_{th})_z (\nu \Sigma_{f,\,th})_z}{\sum_z (\phi_{th})_z (\Sigma_{a,\,th}^{\text{fissile nuc.}})_z} = \frac{\text{Neutron productions from thermal fission}}{\text{Thermal absorptions in fissile nuclides}}$$

where z = core zones

$$f \equiv \frac{\sum_z (\phi_{th})_z (\Sigma_{a,\,th}^{\text{fissile nuc.}})_z}{\sum_z (\phi_{th})_z (\Sigma_{a,\,th}^{\text{all nuc.}})_z} = \frac{\text{Thermal absorptions in fissile nuclides}}{\text{Total thermal absorptions}}$$

$$P_{NL}^{fast} = 1.0 - \sum_{\substack{\text{fast} \\ \text{groups} \\ (i)}} L^{(i)} = \text{Source - fast leakage}$$

$$p \equiv \frac{\sum_z (\Sigma_{out})_z (\phi_{out})_z}{P_{NL}^{fast}} = \frac{\text{Downscatters from group above thermal}}{\text{Fast absorptions + downscatters to thermal}}$$

$$P_{NL}^{thermal} \equiv 1.0 - \frac{L^{(th)}}{\sum_z (\Sigma_{out})_z (\phi_{out})_z} = \frac{\text{Thermal absorptions}}{\text{Thermal source from downscatters}}$$

$$\epsilon \equiv \frac{\sum_z \sum_i (\phi_i)_z (\nu \Sigma_{f,\,i})_z}{\sum_z (\phi_{th})_z (\nu \Sigma_{f,\,th})_z} = \frac{\text{Total productions}}{\text{Thermal productions}}$$

nificant Doppler broadening. It is seen that the following resonances below 3 ev will experience significant broadening:

1. the 2.035 ev resonance of U^{235} at all temperatures,
2. the 1.054 ev resonace of Pu^{240} at all temperatures,
3. the 2.65 ev resonance of Pu^{242} at all temperatures,
4. the 1.257 ev resonance of Rh^{103} at high temperatures,
5. the 1.04 ev resonance of Pm^{147} at high temperatures,
6. the 0.873 ev resonance of Sm^{149} at high temperatures,
7. the 1.1 ev, 1.7 ev, and 2.04 ev resonances of Sm^{151} at all temperatures,
8. the 1.73 ev and 2.456 ev resonances of Eu^{153} at high temperatures,
9. the 2.64 ev resonance of Gd^{155} at high temperatures, and
10. the 2.90 ev resonance of Gd^{157} at high temperatures.

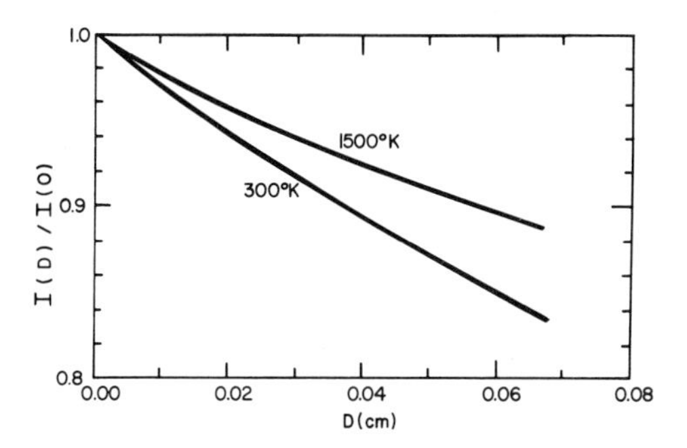

FIG. 3-2. Self-shielding factors as function of grain diameter D.

Of these, only the plutonium isotopes will usually be important in thermal-spectrum reactors, but the capability of including such effects in temperature coefficient calculations should be provided in computer codes for calculating thermal neutron cross sections.

3.3.2 Thermal Spectrum Effects

The moderator coefficient in Eq. (3-1) is primarily determined in solid-moderator reactors by the hardening of the thermal neutron spectrum with increasing temperature and the resultant changes in the effective thermal neutron cross sections. To calculate these changes corectly, an accurate knowledge of the thermal spectrum is required. Such effects as the chemical binding of the moderator atoms, the spatial variation of the spectrum at interfaces or in heterogeneous lattice cells, and the effects of strong non-1/v absorbers may strongly influence the temperature coefficient in particular situations and the calculational methods should include the capability of investigating these effects. Some of the various thermal spectrum models which have been used are reviewed in this section along with examples of the differences in temperature coefficients resulting from their use.

The simplest model for the thermal neutron flux spectrum is the Maxwell-Boltzmann distribution which describes the equilibrium distribution of neutron velocities in the limit of no absorption or leakage and is only a function of the absolute temperature of the moderator. In order to retain the simplicity of this analytic expression in reactor calculations, various prescriptions have been developed which

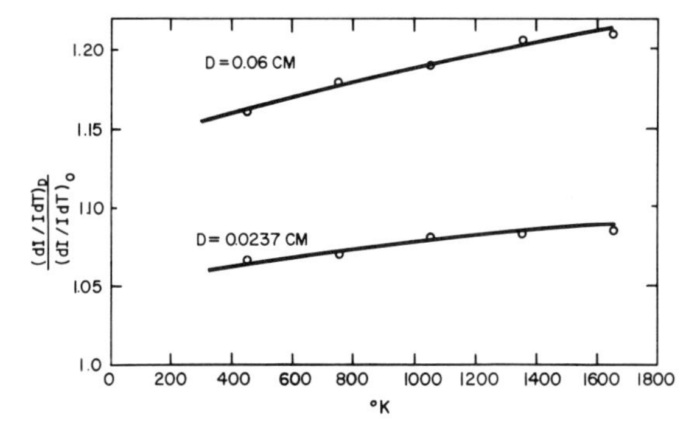

FIG. 3-3. Increase in temperature coefficient at constant resonance absorption for two grain diameters.

attempt to account for the absorption by defining an effective "neutron temperature" as a function of the macroscopic absorption cross section and represent the slowing down source with a 1/E tail joined to the hardened Maxwellian at some energy. Such prescriptions are generally inadequate for temperature coefficient calculations since the changes in total absorption with temperature arising from the non-1/v behavior of the cross sections of important nuclides cannot be accurately fitted with a single hardness parameter. Furthermore, the simplified representation of the spectrum in the joining region, which in many cases is near important cross section resonances, is usually quite inadequate.

An accurate calculation of the thermal neutron spectrum in an infinite medium requires solving the familiar balance equation:

$$[\Sigma_a(E) + \Sigma_s(E)]\ \phi(E) = S(E) + \int_0^{E_c} \Sigma(E' - E)\ \phi(E')\ dE' \qquad (3\text{-}5)$$

in which

$$\Sigma(E' - E) = 2\pi N \int_{-1}^{1} \sigma(E', E, \theta)\ d(\cos\theta) \qquad (3\text{-}6)$$

The integration limit E_c is a suitable cut-off energy above which up-scattering is insignificant. The solution of this equation in various forms is straightforward by computer methods once a model for the scattering kernel $\sigma(E', E, \theta)$ is found. The development of suitable kernels for the various moderators is then one of the central problems of thermalization theory and is also essential for the correct calculation of temperature coefficients in solid-moderator reactors.

The simplest forms of the kernel are obtained when the moderator atoms are assumed to behave as a free gas with a Maxwellian distribution of velocities. The free gas kernel was derived by Wigner and Wilkins [18] who also were able to reduce the spectrum Eq. (3-5) to a second-order differential equation for the limiting case of monatomic hydrogen (A = 1) in which case the kernel becomes particularly simple. A differential equation has also been developed by Wilkins [19] for the case of a heavy moderator atom with chemical binding neglected. This equation (known as either the Wilkins equation or the heavy gas model) has been frequently used in solid-moderator reactor calculations.

Scattering kernels which include the effects of chemical binding have been developed in the last several years for the principal solid moderators (graphite, beryllium oxide and zirconium hydride) and have been found to give excellent agreement with experimentally measured spectra. The general effect of chemical binding is to reduce the slowing-down power of the moderator atoms in the thermal energy region giving a harder thermal neutron spectrum that is less sensitive to temperature. The effects are pronounced at room temperature and

TABLE 3-2

The Ratio (ξ) of the Natural Width (Γ) to the Doppler Width (Δ) for the Thermal Resonances of the Heavy Elements and Fission Products

Material	E_0 (ev)	Γ_γ (ev)	Γ_f (ev)	Γ_n (ev)	Γ (ev)	Doppler width (300°K)	ξ (300°K)	ξ (1200°K)	ξ (2400°K)
U^{233}	0.10	0.056	0.994	0.0000185	1.05	0.00666	157.7	78.8	55.73
	1.45	0.054	0.716	0.000183	0.770	0.00802	96.01	47.98	33.95
	1.76	0.049	0.231	0.000311	0.28	0.00883	31.71	15.85	11.21
	2.30	0.047	0.049	0.000176	0.0962	0.0101	9.52	4.76	3.37
U^{235}	0.273	0.029	0.099	0.0000029	0.128	0.01106	11.57	5.83	4.126
	1.140	0.044	0.1246	0.0000172	0.1686	0.02241	7.52	3.76	2.660
	2.035	0.035	0.012	0.0000077	0.047	0.02994	1.57	0.785	0.555
	3.16	0.311	0.155	0.0000323	0.1861	0.0373	4.99	2.496	1.764
Pu^{239}	0.296	0.0386	0.0554	0.000114	0.094	0.01132	8.3	4.16	2.94
	7.9	0.038	0.042	0.00132	0.080	0.05849	1.37	0.68	0.484
Pu^{240}	1.054	0.034	0.000007	0.00236	0.03636	0.0213	1.707	0.853	0.603
Pu^{241}	0.25	0.043	0.099	0.000055	0.142	0.01034	13.73	6.86	4.85
	4.31	0.043	—	0.000055	—	0.043	—	—	—
Pu^{242}	2.65	0.027	—	0.0018	0.0288	0.0336	0.8571	0.4272	0.3025
Rh^{103}	1.257	0.155	—	0.00078	0.156	0.01248	4.41	2.21	1.56
Cd^{113}	0.178	0.113	—	0.00065 (J = 1)	0.114	0.011	10.36	5.605	3.96
Xe^{135}	0.084	0.0907	—	0.0257	0.1164	0.00802	14.51	7.26	5.13
Nd^{145}	4.37	0.048	—	0.00128	0.0493	0.0558	0.883	0.442	0.3124
Pm^{147}	1.04	0.080	—	0.0000055	0.080	0.0271	2.95	1.48	1.045
	5.43	0.080	—	0.033	0.113	0.0618	1.83	0.917	0.646
Sm^{149}	0.0976	0.053	—	0.00050 (J = 4)	0.0635	0.00822	7.725	3.863	2.72
	0.0873	0.060	—	0.00082	0.0608	0.0246	2.472	1.233	0.872
	4.98	0.067	—	0.00226	0.0693	0.0587	1.181	0.589	0.416
Sm^{151}	1.10	0.062	—	0.00052	0.0625	0.0274	2.281	1.142	0.8054
	1.70	0.062	—	0.00030	0.0623	0.0341	1.827	0.913	0.6449
	2.04	0.062	—	0.00053	0.0625	0.0374	1.671	0.836	0.592
Eu^{153}	0.457	0.097	—	0.00001	0.097	0.0176	5.51	2.76	1.94
	1.730	0.092	—	0.000061	0.092	0.0342	2.69	1.345	0.950
	2.456	0.091	—	0.0013	0.0923	0.0414	2.23	1.12	0.801
Gd^{155}	2.01	0.110	—	0.00028	0.1103	0.0367	3.005	1.502	1.066
	2.64	0.111	—	0.00248	0.1135	0.042	2.702	1.351	0.955
Gd^{157}	0.030	0.100	—	0.00065	0.1006	0.00445	22.61	11.31	8.01
	2.90	0.097	—	0.00036	0.09736	0.0438	2.22	1.11	0.788

become progressively less important at higher temperatures.

The magnitude of the errors introduced in calculations by the neglect of chemical binding effects is indicated in Figs. 3-4 and 3-5 [20] which show infinite medium thermal spectra in a series of U^{235} graphite mixtures computed with the free gas model and with a crystal model developed by Parks [21]. The carbon/uranium ratios cover the range of interest in solid-moderator reactors. Table 3-3 shows the percentage difference in k_{eff} calculated with the two spectral models, and Table 3-4 shows temperature coefficients between 300° K and 600° K. It is seen that large errors in k_{eff} are obtained at low temperatures and large C/U^{235} ratios. In the limit of zero absorption the spectra would be the same for either scattering model and at the other limit of very heavy absorption there are so few thermal neutrons that the scattering model used is unimportant. Hence, there is a maximum error

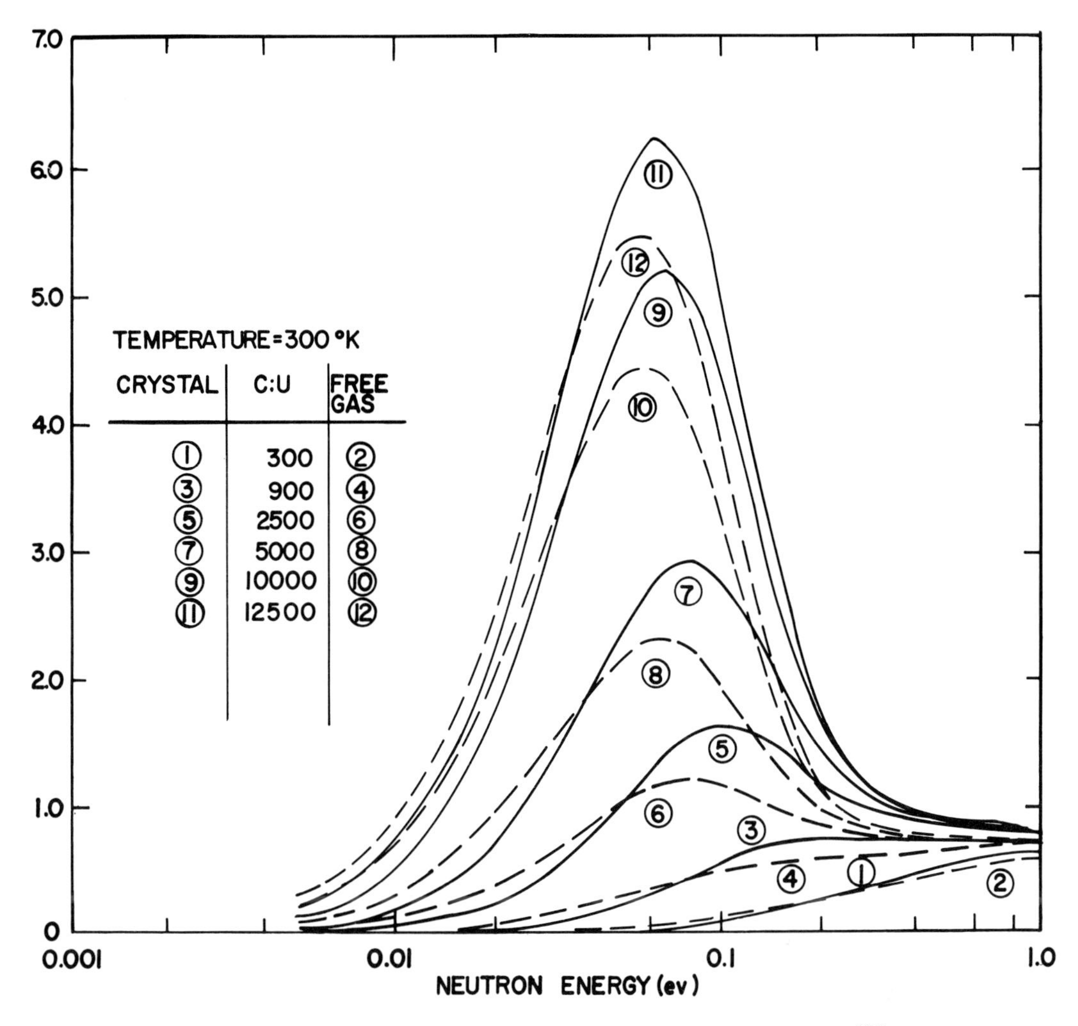

FIG. 3-4. Neutron energy times neutron flux vs neutron energy at 300°K for various $C:U^{235}$ ratios.

somewhere in between which is seen to occur around $C/U^{235} = 10^4$ for a temperature of 300°K and about $C/U^{235} = 2500$ at 1200°K. It is also clear that serious overestimates of the temperature coefficient can result from neglect of chemical binding in the spectral model.

While either the free gas model or a crystal model may be used for temperature coefficient calculations in solid moderator systems at high temperatures (e.g. above 1200°K in graphite) it should not be assumed that further simplification of the free gas model to the heavy gas model will be accurate. In particular, the heavy gas model is not capable of representing the effects of strong resonance absorbers on the thermal spectrum as is illustrated in Fig. 3-6 and Table 3-5 [8]. The plutonium composition in the example is similar to that produced in the Calder Hall type reactors; i.e., 78% Pu^{239}, 17% Pu^{240}, and 5% Pu^{241}.

With adequate kernels available to compute infinite-medium spectra, the remaining major problem of thermalization theory is the inclusion of the spatial dependence of the thermal spectrum which becomes important in highly heterogeneous core lattices such as Calder Hall and in most other reactors at the core-reflector interfaces where sharp discontinuities in temperature and absorption usually exist. These problems require sophisticated diffusion theory or transport theory calculations involving several thermal groups and can be quite expensive even on the fastest computers.

3.3.3 Examples of Temperature Coefficient Calculations for Solid-Moderator Reactors

In this section examples of calculations of temperature coefficients are given for four rather different solid-moderator reactors, each of which illustrates particular calculational problems. Most of the calculations described here were done using a group of IBM-7090 computer codes developed at General Atomic and the DSN transport theory code [9] developed by Carlson at Los Alamos. The features of the less familiar codes are briefly described below to avoid repetition in the subsequent

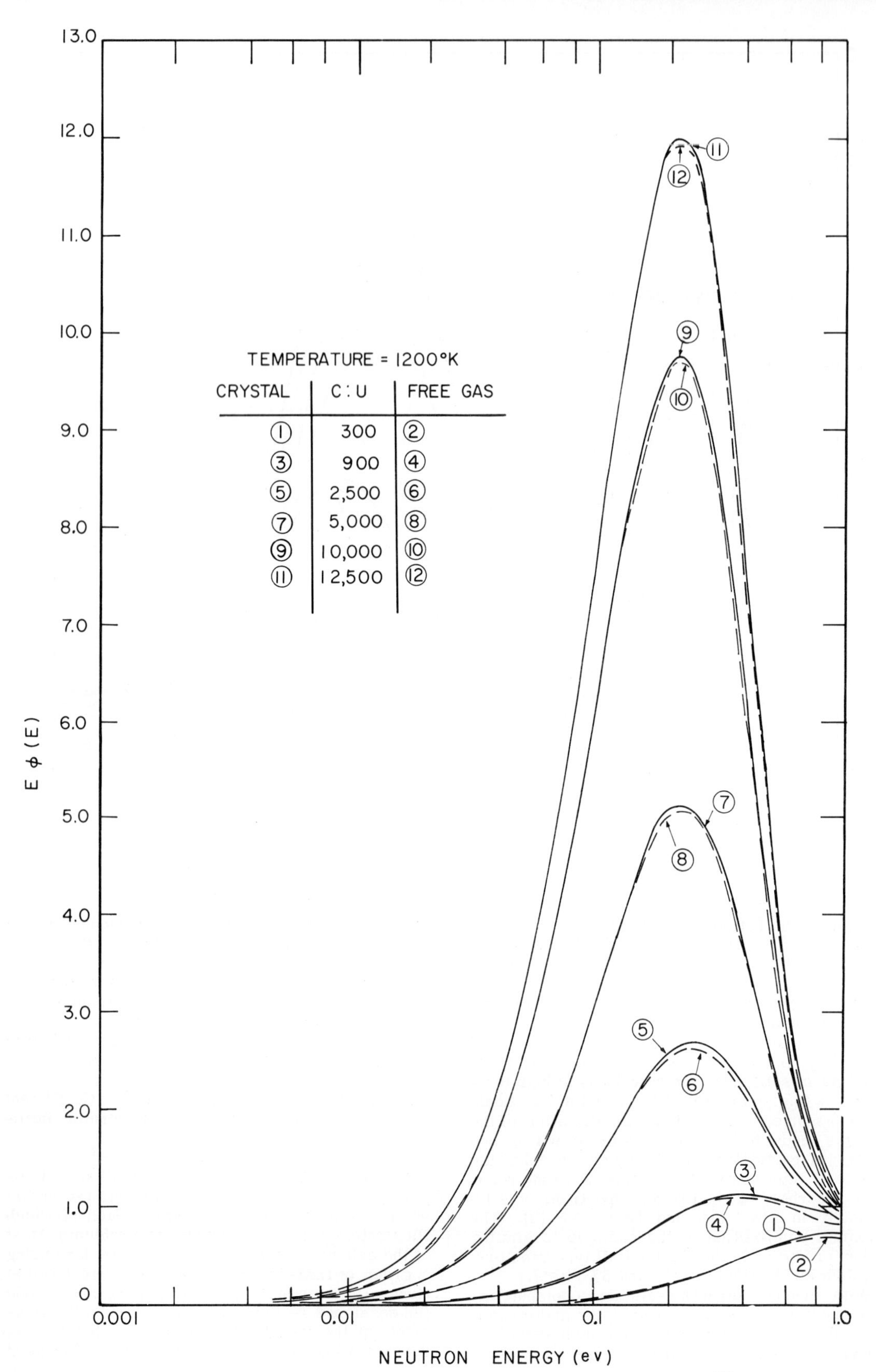

FIG. 3-5. Neutron energy times neutron flux vs neutron energy at 1200°K for various $C:U^{235}$ ratios.

TABLE 3-3

Percentage Difference in Calculated Effective Multiplication Constants Using the Crystal and Free-Gas Scattering Kernels for Several $C:U^{235}$ Atom Ratios

$C:U^{235}$ Atom Ratio	Percentage difference* in k_{eff} 300°K	1200°K
300	-0.162	-0.159
900	-0.787	-0.296
2,500	-1.70	-0.426
5,000	-2.19	-0.405
10,000	-2.45	-0.344
12,500	-2.42	-0.315

*Difference = $100\,[(k_{crystal} - k_{gas})/k_{gas}]$.

TABLE 3-4

Values of $(1/k)\,(\partial k/\partial T)/°C$ Obtained Using the Crystal and Free-Gas Scattering Kernels at Thermal Energies for Several $C:U^{235}$ Atom Ratios

$C:U^{235}$ Atom ratio	$-(1/k)\,(\partial k/\partial T)/°C$ Crystal $(x10^{-5})$	Free-gas $(x10^{-5})$	Difference (%)
300	0.0607	0.242	74.9
900	1.07	2.14	50.2
2500	4.91	7.65	35.8
5000	10.7	14.9	28.2
10000	20.4	25.2	19.0
12500	24.1	28.8	16.3

discussion. Other programs having similar features could have been used, of course, to obtain the results.

GAM-I is a multigroup slowing-down code [12] which computes the slowing-down spectrum in the P_1 or B_1 approximations over the energy range from 10 Mev to 0.414 ev using 68 quarter-lethargy groups. Resonance absorption is treated by the numerical methods of Nordheim. Spectrum-averaged, broad-group cross-section sets are calculated by the code for use in diffusion or transport theory calculations.

GATHER-I is a similar code* for the thermal neutron spectrum which solves the infinite medium spectrum equation, Eq. (3-5), using appropriate scattering kernels for the various moderators and prepares thermal-group cross sections for use in diffusion or transport theory calculations.

GAZE-2 [24] is a one-dimensional multigroup diffusion theory code with full scattering matrix allowed between groups.

GAMBLE [25] is a 10-group two-dimensional diffusion theory code with full scattering matrix between groups.

The examples which will be used to illustrate temperature coefficient calculations of solid-moderator reactors will be the following: (1) the Calder Hall reactor, (2) the Peach Bottom HTGR, (3) the Experimental Beryllium Oxide Reactor (EBOR),

*No published report.

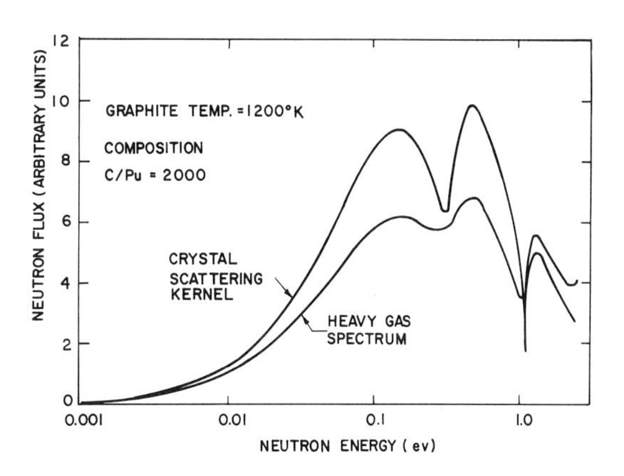

FIG. 3-6. Thermal flux spectra

and (4) the TRIGA reactor. The Calder Hall reactor is illustrative of many large, heterogeneous, solid-moderator reactors using natural or slightly enriched uranium fuel. Other reactors of this type include the Advanced Gas-Cooled Reactor (AGR), the Experimental Gas-Cooled Reactor (EGCR), the Sodium Graphite Reactor (SGR), and the Hanford reactors. The Peach Bottom reactor is typical of the high-temperature gas-cooled reactors (HTGR), which might use either graphite or beryllium oxide as the moderator. The Pebble Bed Reactor is also included in this type of reactor. These reactors are characterized by a homogeneous or semihomogeneous fuel element, usually using the $U^{235}/Th^{232}/U^{233}$ fuel cycle.

The Experimental Beryllium Oxide Reactor is included as an example to illustrate a solid-moderator reactor having predominantly an intermediate-energy neutron spectrum. The TRIGA reactor is an interesting example of a reactor where the temperature coefficient arises primarily from changes in the disadvantage factor with moderator temperature changes. The calculation of the TRIGA temperature coefficient requires some very sophisticated transport theory calculations that must account for thermal neutron spectrum changes between the uranium-zirconium-hydride fuel elements and the surrounding water. This particular example provides a good illustration of the success one is able to achieve in solving difficult neutron thermalization problems.

(1) The Calder Hall Reactor

Although a great deal of analytical work has been done and reported on the Calder Hall reactor, some independent calculations were done as a part of this review in order to illustrate some of the characteristics of the temperature coefficient behavior in this class of reactor. The calculations made use of the computer codes described above and the over-all temperature coefficient has been broken down according to the definitions of Table 3-1. Table 3-6 shows the isothermal temperature coefficient and its components calculated for the beginning-of-life condition in the Calder reactor.

The strong spatial dependence of the thermal spectrum in the heterogeneous lattice cell was treated by 26-thermal-group transport theory calculations from which energy dependent shielding factors were obtained and used to recalculate the

TABLE 3-5

Comparison of Spectrum-Averaged Thermal Neutron Cross Sections Using Various Thermalization Models[a]

Moderator Temperature (°K)	← 300 →			← 1200 →			← 1200 →	
Carbon to U^{235} ratio	← 4000 →			← 4000 →			← ∞ →	
Carbon to Pu ratio	← ∞ →			← ∞ →			← 2000 →	
	Crystal	Free gas	Heavy gas	Crystal	Free gas	Heavy gas	Crystal	Heavy gas
Scattering model	σ (barn)	% diff. relative to crystal	% diff. relative to crystal	σ (barn)	% diff. relative to crystal	% diff. relative to crystal	σ (barn)	% diff. relative to crystal
U^{235} Absorption	321	+14	+17	227	+1.0	+1.5	99.3	-8.7
Pu^{239} Absorption	940	-1.0	+1.7	1,340	-0.22	+0.22	489	-8.8
Pu^{240} Absorption	1,100	+11	+5	982	+0.11	+1.6	1,940	+64
Xe^{135} Absorption	1,540,000	+7.4	+11	800,000	+1.8	+2.0	187,000	-22
Sm^{149} Absorption	41,800	-3.6	-1.1	25,300	+1.1	+0.75	6,680	-22
(1/v) Absorption[b]	0.515	+11	+14	0.374	+0.86	+1.2	0.210	-6.4

[a] The cross sections are averaged over a group from zero to 2.38 ev.

[b] The (1/v) represents a nuclide whose 2200 m/s neutron absorption cross section is unity and varies inversely with velocity.

TABLE 3-6

Beginning-of-Life Isothermal Temperature Coefficient of the Calder Hall Reactor

Temp. °K	k_{eff}	f	η	p	ϵ	P_{NL}^{Fast}	$P_{NL}^{Thermal}$
323	1.03550	0.583935	2.06420	0.819301	1.07997	0.973726	0.997097
530	1.02243	0.581527	2.05685	0.815763	1.08090	0.974185	0.995112
700	1.01462	0.581587	2.04863	0.813486	1.08144	0.974318	0.993502
900	1.00778	0.582242	2.04025	0.811122	1.08192	0.974546	0.991966

Temperature Derivatives ($x10^{-5}$/°C)

Temp. °K	$\frac{1}{k_{eff}}\left(\frac{\partial k_{eff}}{\partial T}\right)$	$\frac{1}{f}\left(\frac{\partial f}{\partial T}\right)$	$\frac{1}{\eta}\left(\frac{\partial \eta}{\partial T}\right)$	$\frac{1}{p}\left(\frac{\partial p}{\partial T}\right)$	$\frac{1}{\epsilon}\left(\frac{\partial \epsilon}{\partial T}\right)$	$\frac{1}{P_{NL}^F}\left(\frac{\partial P_{NL}^F}{\partial T}\right)$	$\frac{1}{P_{NL}^T}\left(\frac{\partial P_{NL}^T}{\partial T}\right)$
426.5	-6.136	-1.996	-1.723	-2.091	+0.4129	+0.2274	-0.9630
615.0	-4.511	+0.06069	-2.355	-1.644	+0.2969	+0.08085	-0.9522
800.0	-3.382	+0.5628	-2.049	-1.455	+0.2184	+0.1169	-0.7736

thermal spectrum and cross sections for the few-group diffusion theory calculations of the temperature coefficient. Thermal spectra were computed with the Parks model for the graphite scattering kernel. The energy-dependent shielding factors in the fuel rod are shown in Fig. 3-7. The importance of including the energy dependence is illustrated in Fig. 3-8 which shows the coefficients of η and f as calculated with a constant shielding factor in the thermal group and with the energy-dependent factors of Fig. 3-7. With an energy-independent shielding factor, the absorption cross section of U^{235} varies in the same manner with energy as the unshielded cross section, falling off more rapidly than 1/v. When the energy dependence of the shielding factor is included, the U^{235} cross section falls off less rapidly than 1/v, especially at higher energies. This gives rise to the positive values of (1/f) $(\partial f/\partial T)$ at high temperatures of the moderator.

The temperature dependence of the effective thermal group cross sections with the energy-dependent shielding factors are shown in Fig. 3-9 for the nuclides of greatest interest. The thermal group extends to 2.1 ev and the spectra at each temperature are given in Fig. 3-10. One can see the following interesting features from these curves: (1) the cross section for the shielded U^{235} falls off less rapidly with moderator temperature than the 1/v cross section of the carbon; (2) the Xe^{135} cross section decreases with temperature somewhat faster than the U^{235} cross section, particularly at higher moderator temperature; (3) the Pu^{239} cross

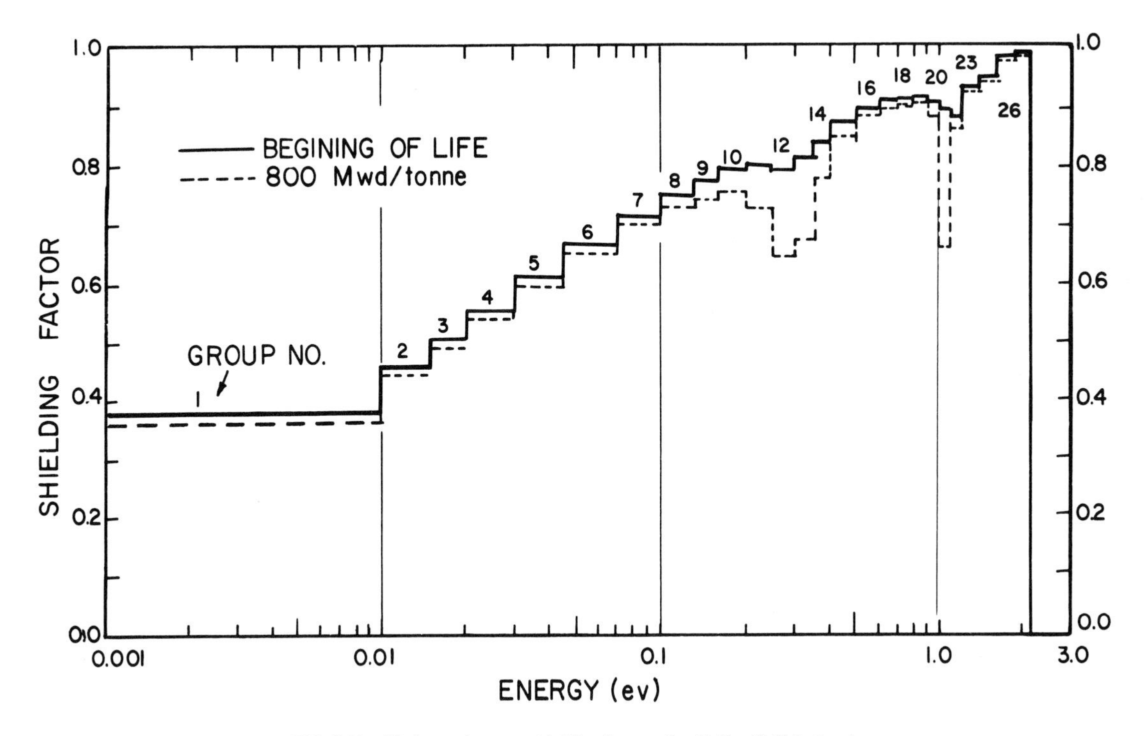

FIG. 3-7. 26-thermal-group shielding factors for Calder Hall fuel rod.

section increases quite rapidly with temperature. The changes to be expected in the coefficient of f as equilibrium Xe and Sm build in and as Pu^{239} is produced are, therefore, evident from these curves. The effects of equilibrium Xe and Sm estimated from these data are shown in Fig. 3-11.

To calculate accurately the plutonium effect would require a reactivity lifetime calculation which accounted for the spatial dependence of the burnup of U^{235} and production of Pu^{239} and included the changes in energy-dependent shielding with increasing concentration of Pu. Rather than attempt this for an illustrative case we have simply calculated the isothermal coefficient for a uniform fuel composition of U^{235} and Pu^{239} corresponding to an

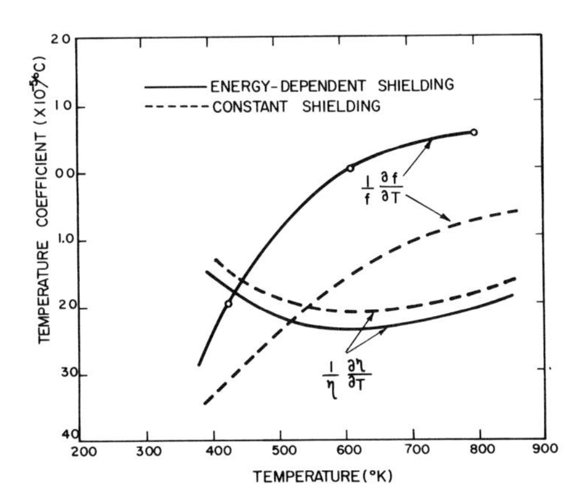

FIG. 3-8. Temperature coefficient of η and f at beginning-of-life with and without energy-dependent shielding in the fuel rod.

exposure of about 800 Mwd/tonne which represents a point halfway through core life for the first Calder Hall core. No equilibrium Xe and Sm or other fission product poisons are present. The fuel shielding factors are shown on Fig. 3-7 with the beginning-of-life values. The effect of the Pu isotopes is obvious. The temperature coefficients are given in Table 3-7. The fuel coefficient reflected in $(1/p)(\partial p/\partial T)$ is of course substantially unchanged while the large positive contribution of $(1/f)(\partial f/\partial T)$ dominates the moderator coefficient.

Comparison of the results of these illustrative calculations with experimental or calculated coefficients published for the Calder Hall reactor are difficult for several reasons, including differences in definition of the factors of k_{eff} and uncertainties as to what average temperature is appropriate for a reported measurement; moreover there are limitations to our calculational model which, for example, does not include control rods or the fixed steel absorbers sometimes used in Calder Hall to reduce excess reactivity. The number of these absorbers present in any experimental measurement is rarely reported. The non-uniform spatial distribution of the Pu produced is also an uncertainty.

In spite of these difficulties, the calculated examples yield results in fair agreement with published experimental values and with some recent work of the British group at Winfrith [26] which has emphasized detailed calculations of the temperature coefficients using transport theory codes and accurate thermalization models. Fig. 3-12 shows a comparison of their calculated moderator coefficients with experiment [27] for the temperature range of 500-600°K (440-620°F). The points labeled "Carlson" refer to transport theory calculations by the

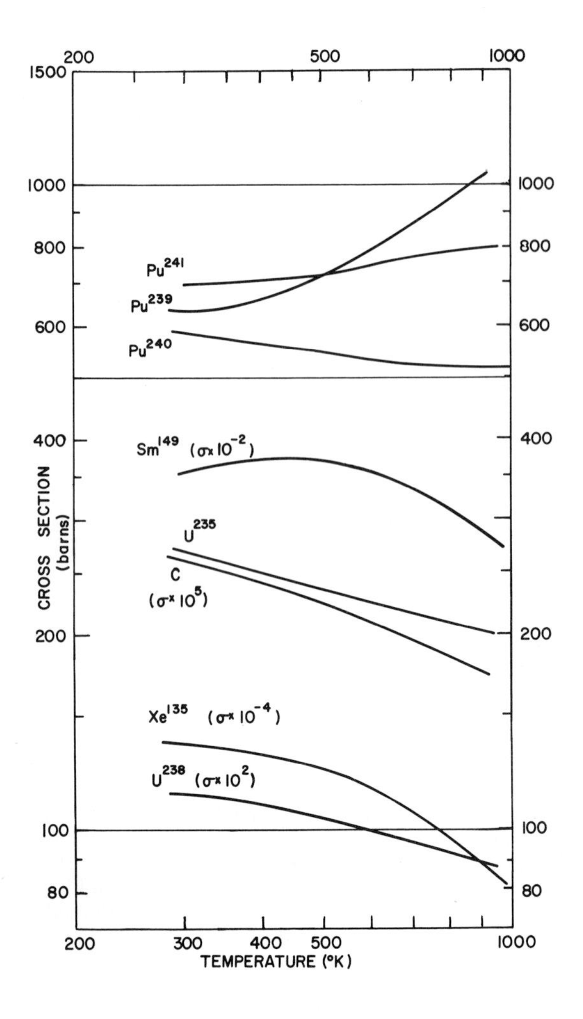

FIG. 3-9. Average thermal group cross sections in Calder Hall spectra. Beginning-of-life; thermal group 0-2.1 ev.

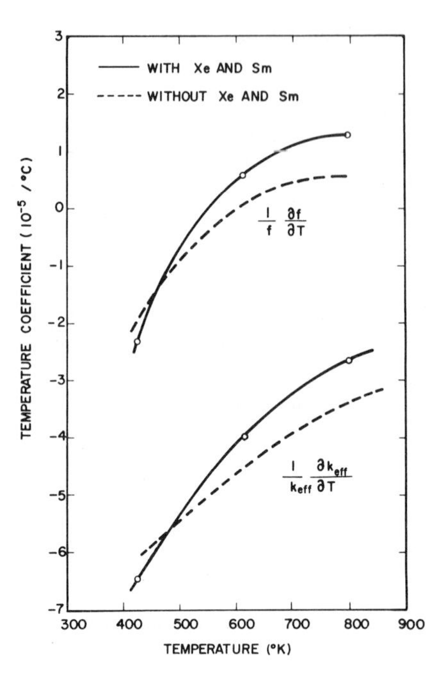

FIG. 3-11. Effect of equilibrium Xe and Sm on the beginning-of-life temperature coefficient.

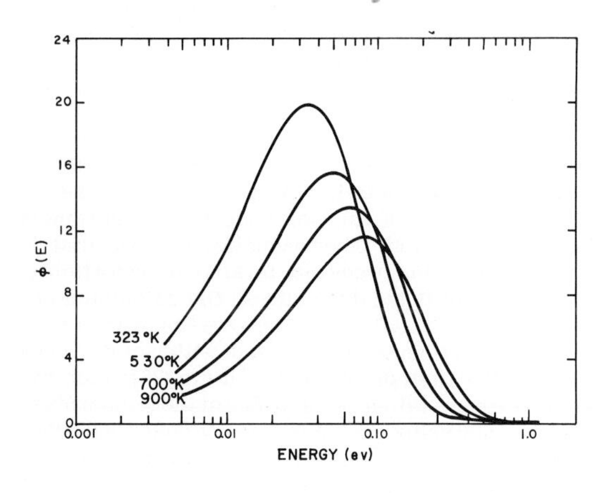

FIG. 3-10. Calder Hall thermal spectra ϕ(E) at beginning of life.

S_n method and the "Spectramet" curve is obtained by an approximation method developed by Leslie which uses diffusion theory in the moderator and collision probability theory in the fuel lump and which is much faster on the computer than the S_n method. The authors feel that the largest possible source of calculational uncertainty is in the cross section data and have concluded that a 1% change in one cross section can lead to an error of 0.3×10^{-5}/°C in the moderator coefficient. They are also attempting more accurate burnup calculations. The uncertainties in the experimental temperature coefficient measurements are stated to be almost as large as the discrepancy indicated between the experiment and the calculations. The moderator coefficient obtained from our illustrative calculations for this temperature range is -3.25×10^{-5}/°C at beginning-of-life. (The moderator coefficient is defined here as $[(1/k)(\partial k/\partial T) - (1/p)(\partial p/\partial T)]$ which is not quite the same definition as that used in the British work.) The moderator coefficient at 800 Mwd/tonne from our estimate is $+3.5 \times 10^{-5}$/°C. Both of these numbers would be increased by about $+1 \times 10^{-5}$/°C if the effect of equilibrium Xe and Sm shown on Fig. 3-11 is included. (It is not known whether Xe was present in the experimental data of Fig. 3-12). A negative correction for the control rods and burnable poisons present would tend to compensate the positive Xe effect.

The behavior of the coefficient illustrated here for the Calder Hall type of reactor is also characteristic of other heterogeneous reactor types (e.g. sodium graphite, EGCR) and similar calculational methods are required for these systems.

TABLE 3-7

Estimated Middle-of-Life Temperature Coefficient of the Calder Hall Reactor
(Not including Xe and Sm)

Temp. °K	k_{eff}	f	η	p	ϵ	P_{NL}^{Fast}	$P_{NL}^{Thermal}$
323	1.06036	.605251	2.04151	.819300	1.07794	.973584	.998056
530	1.05780	.614373	2.01708	.815765	1.07797	.973977	.996619
700	1.06222	.627114	1.99255	.813484	1.07749	.974044	.995673
900	1.06908	.641543	1.96840	.811122	1.07683	.974192	.994934

Temperature Derivatives ($\times 10^{-5}/°C$)

Temp. °K	$\frac{1}{k_{eff}}(\frac{\partial k_{eff}}{\partial T})$	$\frac{1}{f}(\frac{\partial f}{\partial T})$	$\frac{1}{\eta}(\frac{\partial \eta}{\partial T})$	$\frac{1}{p}(\frac{\partial p}{\partial T})$	$\frac{1}{\epsilon}(\frac{\partial \epsilon}{\partial T})$	$\frac{1}{P_{NL}^F}(\frac{\partial P_{NL}^F}{\partial T})$	$\frac{1}{P_{NL}^T}(\frac{\partial P_{NL}^T}{\partial T})$
426.5	-1.168	+7.226	-5.816	-2.089	+0.01108	+0.1948	-0.6962
615.0	+2.453	+12.07	-7.197	-1.647	-0.2602	+0.04038	-0.5582
800.0	+3.219	+11.37	-6.097	-1.454	-0.3058	+0.07615	-0.3716

(2) The Peach Bottom HTGR

The Peach Bottom High Temperature Gas-Cooled Reactor is a graphite-moderated, helium-cooled, thermal reactor with fully enriched uranium and thorium carbides dispersed in graphite compacts inside a graphite sleeve to form semihomogeneous fuel elements. As a reactor concept, the HTGR could include beryllium oxide as moderator, various combinations of uranium, plutonium, and thorium as fuel and fertile material, and, in principle, includes the pebble bed type of fuel elements. From the standpoint of temperature-coefficient analysis the distinguishing features are the semihomogeneous core design, which makes the resonance integral primarily a volume-absorption rather than a surface-absorption effect, and the high operating temperature of the moderator which influences the position of the thermal spectrum with respect to important non-1/v absorbers.

The semihomogeneous fuel design eliminates much of the computational difficulty in temperature coefficient calculations. The adequacy of a few-group diffusion theory calculation has been confirmed for the Peach Bottom Reactor by comparison with more detailed models [28]. It is necessary, however, to represent accurately the spatial dependence of nuclide concentrations, since zoning of fuel and fertile material is used to achieve a flat power distribution. The initial core loadings reported at the time of the Final Hazards Report [29] were 220 kg U^{235} (93.5% enriched), 1450 kg Th, 20500 kg graphite and 1.1 kg of natural boron lumped as a burnable poison.

The beginning-of-life isothermal temperature coefficient of the Peach Bottom HTGR is shown in Table 3-8 in which the components of k_{eff} are separated in a way very similar to the definitions of Table 3-1. The fuel and moderator coefficients, as well as the over-all temperature coefficient, (Eq. 3-1) are plotted in Fig. 3-13. The average operating temperature of the reactor is 1200°K (1701°F).

The average number of neutrons produced η per thermal neutron absorbed in fuel (U^{233} or U^{235}) varies with temperature as $(1/1 + \alpha)$. At the operating temperature of the HTGR the contribution of η to the total temperature coefficient is, in general, small.

In Table 3-8 the temperature derivative of the thermal utilization f is seen to be one of the major negative contributors. It has been enhanced by the addition of 5 kg of Rh^{103} to the fuel compacts. Additional rhodium is formed as a fission product. The rhodium has a large absorption resonance at 1.26 ev which is at a sufficiently high energy to have relatively little effect on the neutron economy at operating temperature but which contributes strongly in the high temperature range where the Doppler effect in thorium falls off. At operating temperature, the negative coefficient of f is primarily due to the fact that the U^{235} cross section decreases with temperature more rapidly than the average cross section of other absorbers. Other important contributors to $(1/f)(\partial f/\partial T)$ are Xe^{135} and the control rods. These contributions are shown in Fig. 3-14 along with that of the Rh^{103}. The behavior of such materials as Rh^{103} and Pu^{240} is suggestive of the great flexibility possible in designing a sufficiently negative temperature coefficient into an HTGR-type reactor without seriously damaging neutron economy. The mixing of the fuel with a substantial portion of the moderator in

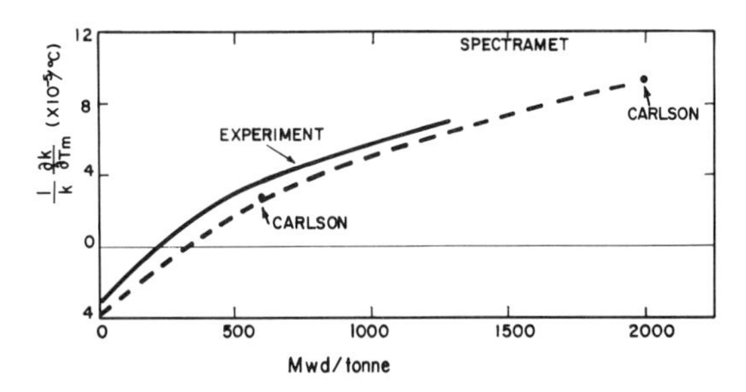

FIG. 3-12. Moderator coefficient of Calder Hall. Point irradiation or 1.42x bulk irradiation.

TABLE 3-8

Components of Beginning-of-Life Temperature Coefficient of the Peach Bottom HTGR (unrodded, no xenon or samarium)

Temp. (°K)	$\frac{1}{k}\left(\frac{\partial k}{\partial T}\right)$	$\frac{1}{\eta}\left(\frac{\partial \eta}{\partial T}\right)$ (x10^-5/°C)	$\frac{1}{f}\left(\frac{\partial f}{\partial T}\right)$ (x10^-5/°C)	$\frac{1}{\epsilon}\left(\frac{\partial \epsilon}{\partial T}\right)$ (x10^-5/°C)	$\frac{1}{P}\left(\frac{\partial P}{\partial T}\right)$ (x10^-5/°C)	$\frac{1}{P_{NL}}\left(\frac{\partial P_{NL}}{\partial T}\right)$ (x10^-5/°C)
450.0	- 7.0	- 1.0	- 1.2	+ 1.3	- 6.1	- 0.0
750.0	- 5.5	- 1.0	- 1.7	+ 1.2	- 3.9	- 0.1
1050.0	- 4.7	- 0.6	- 2.3	+ 1.2	- 2.8	- 0.2
1350.0	- 4.3	- 0.2	- 2.9	+ 1.0	- 2.1	- 0.1
1650.0	- 4.4	- 0.1	- 3.6	+ 1.3	- 1.8	- 0.2

the semihomogeneous fuel elements assures that the time response of these negative effects will be relatively prompt.

The fast fission factor ϵ is defined here as the ratio of the total fission rate to the thermal fission rate. Its variation with temperature is a result of the reduction in the thermal fission cross section along with a gross redistribution in energy of the neutron flux as the temperature is increased. The change in the fast fission factor with temperature yields a positive contribution to the temperature coefficient.

The variation with temperature of the total nonleakage probability P_{NL} results from changes in both the thermal inleakage and the fast outleakage as the core temperature changes. As the core temperature increases, the increased diffusion length tends to flatten the thermal flux gradient and the thermal flux at the core-reflector interface. The inleakage of thermal neutrons is thereby reduced as the core temperature is raised, yielding a negative contribution to the temperature coefficient. The outleakage of fast neutrons is reduced as the core temperature is increased due to a reduced power peak at the core-reflector interface which reduces the source of those neutrons having a high leakage probability. The reduction in fast neutron leakage with increased core temperature is also due to the increased absorptions in thorium at higher temperatures which leaves fewer neutrons available to leak. The fast neutron leakage yields a positive contribution to the temperature coefficient. The contribution of the fast and thermal neutron leakage have alternate signs and tend to compensate each other with the result that the contribution of the total nonleakage probability to the temperature coefficient is small.

The end-of-life temperature coefficient given in Table 3-9 and the fuel and moderator coefficients are plotted in Fig. 3-15.

(3) EBOR

The Experimental Beryllium Oxide Reactor (EBOR) is a heterogeneous, gas-cooled, BeO-moderated intermediate spectrum reactor under construction in Idaho [30]. The fuel elements consist of a central BeO spine of 2 in. (5.08 cm) diameter surrounded by 18 fuel pins of 0.375 in. (0.9525 cm) diameter. The spine and pins are surrounded by a square annular moderator block having outer dimensions 3.5 in. (8.89 cm) and an inner diameter of 2.9 in. (7.37 cm). The coolant gas flows around the pins between the spine and outer block. The fuel material is enriched UO_2 (62.5% U^{235}) diluted with BeO and the pins are clad with Hastelloy. Thirty-six fuel elements are used, forming a core 23 in. (58.42 cm) square and 76 in. (1.930 m) long. Most of the moderation occurs in the reflector, causing a strong spatial dependence in the neutron energy spectrum.

Detailed reports [31, 32] have recently been published on the analysis of the temperature coefficient of this reactor. Because of the competing effects of capture and fission throughout the intermediate spectrum and the relative unimportance of the thermal base effects, as well as the definite

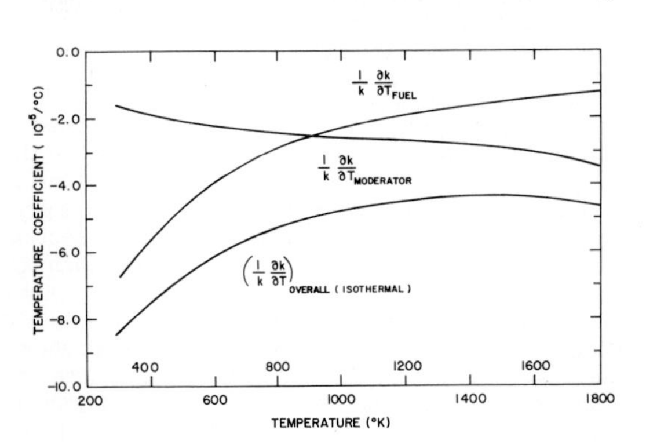

FIG. 3-13. Peach Bottom Reactor: Beginning-of-life temperature coefficient. (Unrodded, no Xe or Sm).

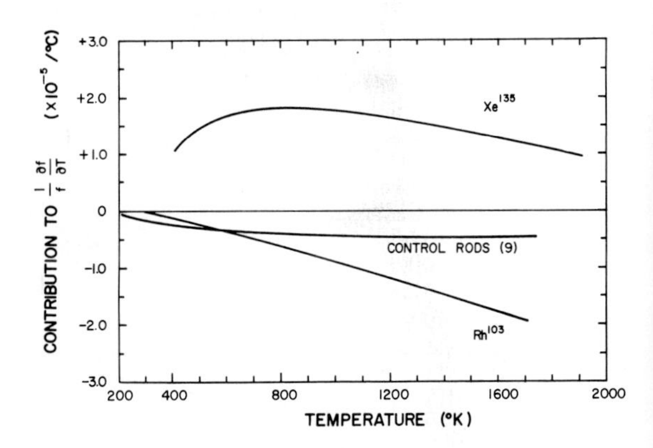

FIG. 3-14. Contributions of Xe, Rh and control rods to the thermal utilization coefficient (1/f) (∂f/∂T). Beginning-of-life core composition.

TABLE 3-9

Components of End-of-Life Temperature Coefficients of the Peach Bottom HTGR After 900 Days of Full Power Burnup

Temp. (°K)	$\frac{1}{k}\left(\frac{\partial k}{\partial T}\right)$ (x10^{-5}/°C)	$\frac{1}{\eta}\left(\frac{\partial \eta}{\partial T}\right)$ (x10^{-5}/°C)	$\frac{1}{f}\left(\frac{\partial f}{\partial T}\right)$ (x10^{-5}/°C)	$\frac{1}{\epsilon}\left(\frac{\partial \epsilon}{\partial T}\right)$ (x10^{-5}/°C)	$\frac{1}{p}\left(\frac{\partial p}{\partial T}\right)$ (x10^{-5}/°C)	$\frac{1}{P_{NL}}\left(\frac{\partial P_{NL}}{\partial T}\right)$ (x10^{-5}/°C)
750	-3.2	-1.0	+1.6	+0.3	-3.8	-0.3
1350	-2.3	-0.1	-0.1	+0.3	-2.1	-0.3
1900	-3.1	+0.2	-1.8	+0.6	-1.6	-0.4

separation in time of the temperature changes in fuel pins, moderator blocks and reflectors, the analysis was quite different from that described above for the Calder Hall and Peach Bottom reactors. The temperature coefficient components are described in terms of the total reactivity change associated with successive temperature increments in various physical components of the core.

(1) Prompt coefficient. This is associated with the fuel pins only. The Doppler coefficient of the fissile and fertile material and the small thermal base effect of the BeO in the pins is included.

(2) Core delayed coefficient. This is the reactivity effect when the remainder of the BeO in the fuel elements changes temperature.

(3) Reflector delayed coefficient. This is the reactivity effect of temperature changes in the reflector.

The algebraic sum of these coefficients is defined as the "total" coefficient. An additional negative effect due to expansion of the grid plate which moves the elements apart is treated separately.

Calculations were done at room temperature (300°K), at 980°K (707°C or 1305°F) to represent operating conditions and at 1366°K (1093°C or 1999°F) to represent accident conditions. Starting with the whole core at one of these base temperatures, a 50°C (90°F) change in fuel pin temperature was made and the reactivity effect computed. Maintaining the fuel pins at the new temperature, the remainder of the core moderator was raised 50°C and the reactivity change computed. Finally, with the whole core at the new temperature, the reflector temperature was raised 50°C and the reactivity change calculated. The procedure was repeated for the other base temperatures. The results are given in Fig. 3-16 for the beginning-of-life case and summarized in Table 3-10 for the beginning and end of life. The changes with burnup are small because of the short reactivity lifetime and the intermediate spectrum.

The reactivity calculations themselves required considerable detail to handle the heterogeneous fuel design and the space-energy problem. The following procedure was used:

(1) Initial fast and thermal spectra were computed for 22 broad groups and used in DSN transport theory cell calculations to determine shielding factors. Shielding for the U^{238} was handled separately in the resonance integral calculation of the GAM-I code.

(2) A second set of spectra were calculated with the shielded cross sections, this time including the U^{238} and allowing it to be shielded by the U^{235} to simulate the interference effects in the resonance energy range. Separate calculations [32] of the U^{235} Doppler broadening, using a modified form of the Nordheim method incorporated in GAM-I, were used at each temperature above 300°K. From these spectrum calculations, cross sections were obtained for all nuclides in twelve broad energy groups.

(3) Reactivity calculations were done in radial geometry with the GAZE code and in X-Y geometry with the 2DXY transport theory code.

The negative temperature coefficient is seen to result primarily from the Doppler effect in U^{235} and U^{238}. The thermal base effect in the rest of the core is very small owing to the intermediate

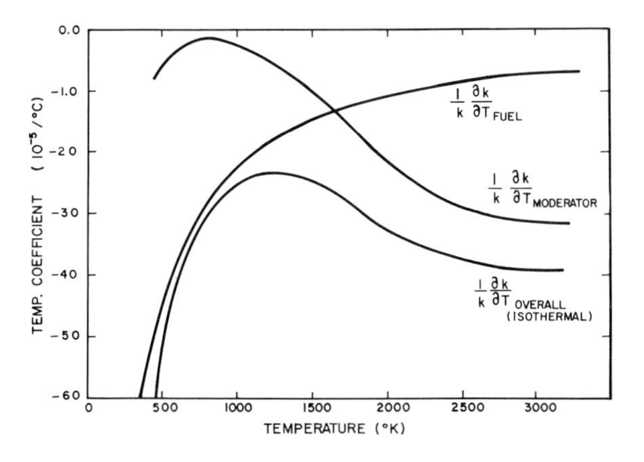

FIG. 3-15. Peach Bottom Reactor: End-of-life temperature coefficients (900 days, Xe + 2 control rods).

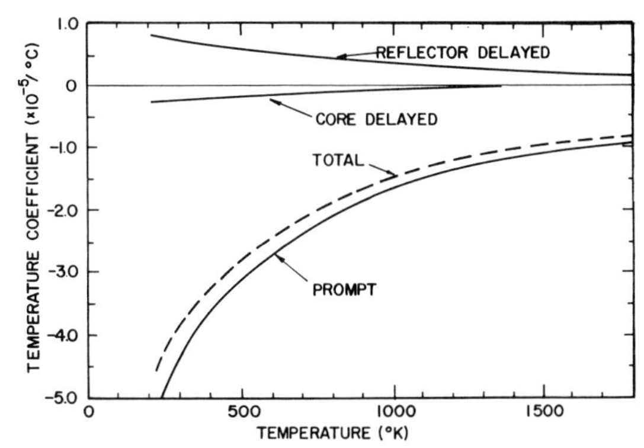

FIG. 3-16. EBOR temperature coefficient.

TABLE 3-10

EBOR 36-Element Core Temperature Coefficients at Operating Temperature with Equilibrium Xenon and at End of Life

	Zero hour, equilibrium Xe ($\times 10^{-5}/°C$)	End of life (10,000 hr) ($\times 10^{-5}/°C$)
Prompt		
U^{235}	-0.5	-0.4
U^{238}	-1.1	-1.0
Total	-1.6	-1.4
Delayed		
Core	-0.2	-0.2
Reflector	+0.3	+0.4
Total	+0.1	+0.2
Total coefficient	-1.5	-1.2

spectrum. Thermalization effects in the reflector are significant, however. It is interesting to note that the positive effect of reflector heating would be overestimated by about a factor of 2 at room temperature if a free gas-scattering kernel had been used in place of the crystal kernel which was actually used.

(4) TRIGA

The TRIGA series of research reactors are not strictly solid-moderator reactors, since they contain a substantial fraction of water, but are included in this chapter because they illustrate the interesting behavior of the temperature coefficient of the solid moderator zirconium hydride and because substantial experimental experience is available with respect to their kinetic behavior. This experience is compared with calculational methods in subsequent sections [33].

The succeeding discussion will be limited primarily to considerations of the TRIGA MARK-III reactor, which is a water-reflected reactor with an active core 15 in. (38.1 cm) long and approximately 20 in. (50.8 cm) in diameter. The core normally contains a maximum of 127 fuel elements, the number depending on the excess reactivity required. The fuel element diameter is 1.47 in. (3.73 cm). The fuel element meat contains 20% enriched uranium mixed with $ZrH_{1.73}$ and is clad with stainless steel. The top and bottom grid plates position the fuel-moderator elements in the core in a circular array allowing a one-third core volume fraction for water. The following discussion of the temperature coefficient evaluation is based on a recent report by Scalettar and West [33].

The transient behavior of TRIGA-type reactors is primarily determined by their large prompt negative temperature coefficient of reactivity. This temperature coefficient results primarily from the hardening of the neutron spectrum associated with the heating of the uranium-zirconium-hydride fuel-moderator elements. The coefficient is prompt because the fuel is intimately mixed with a large portion of the moderator and, thus, fuel and moderator temperatures rise simultaneously. A quantitative calculation of the temperature coefficient requires a knowledge of the energy-dependent distribution of thermal neutron flux in the reactor.

The basic physical processes which occur when the fuel-moderator elements are heated can be described as follows. The rise in temperature of the hydride increases the probability that a thermal neutron in the fuel element will gain energy from an excited state of an oscillating hydrogen atom in the lattice. As the neutrons gain energy from the ZrH their mean free path is increased appreciably. Since the average chord length in the fuel element is comparable with a mean free path, the probability of escape from the fuel element before capture is increased. In the water the neutrons are rapidly rethermalized so that the capture and escape probabilities are relatively insensitive to the energy with which the neutron enters the water. The heating of the moderator mixed with the fuel thus causes the spectrum to harden more in the fuel than in the water. This effect, then, depends on spatial variations of the neutron spectrum over distances of the order of a mean free path with large changes of mean free path occurring due to the energy change in a single collision.

The basis for the calculation of the relevant cross sections for water and zirconium hydride has come from the thermalization work of McReynolds, Nelkin, Rosenbluth and Whittemore, [34, 35] where the atomic motions in the moderator are considered as a superposition of independent simple harmonic motions. For hydrogen in zirconium hydride the dominant mode of motion is the optical mode of the lattice which can be thought of as the motion in an isotropic harmonic potential. The lowest excited state of this motion has been measured to be ~ 0.14 ev. The scattering by water shows less dramatic but quantitatively important effects of chemical binding which can be described with sufficient accuracy for reactor physics by assuming the hydrogen atom motions to be harmonic over the time scale of interest.

TRIGA temperature coefficients have been determined numerically by calculating the change in reactivity associated with a uniform heating of the fuel-moderator elements, the core water and reflector materials assumed to remain at 20°C (68°F).

In calculating the coefficient it is physically convenient and computationally necessary, to separate the contributions due to changes in the core thermal utilization from the contributions due to changes in the over-all neutron balance between core, control rods, and reflector. First an infinite medium of the core composition is considered, and the thermal utilization calculation for an infinite lattice composed two-thirds by volume of TRIGA fuel elements and one-third by volume of water. This is done for a given fuel temperature by a multi-thermal-group DSN cell calculation.

The cell calculations are used to determine group-dependent disadvantage factors which are then used to generate macroscopic cross sections for a homogenized core with the same neutron balance as the original lattice. It is then necessary to homogenize once more to include the effects of the large water slugs (sometimes displaced by aluminum control rod followers) left in the core when the control rods are withdrawn. The calculations for the TRIGA MARK III lattice give a prompt temperature coefficient over the temperature range of 20°C to 400°C (68 to 752°F) of $-8.8 \times 10^{-5}/°C$ ($-4.9 \times 10^{-5}/°F$) associated with the thermal utilization. Somewhat less than 10% of this contribution is associated with the water in the control rod channels.

After determining the macroscopic cross sections for an equivalent homogeneous core, it is then necessary to determine the contribution to the prompt temperature coefficient due to the increased leakage of thermal neutrons into the reflector with increasing hydride temperature. To treat this contribution accurately again requires several thermal groups, but transport effects are no longer of major concern. Thus, reactivity calculations as a function of fuel temperature have been done on the entire reactor using the GAZE and GAMBLE diffusion codes. Essentially, identical results were obtained from the one- and two-dimensional calculations.

In addition to the effects associated with the neutron spectrum, there are two other small effects contributing to the prompt negative temperature coefficient. The first of these is the Doppler broadening of the U^{238} absorption resonances. This is calculated to contribute $-2.3 \times 10^{-5}/°C$ ($-1.26 \times 10^{-5}/°F$) at room temperature and decreases somewhat with increasing temperature to $-1.5 \times 10^{-5}/°C$ ($-0.83 \times 10^{-5}/°F$) at 400°C (752°F). The second effect is the thermal expansion of the fuel element and consequent expulsion of core water. From the measured void coefficient of TRIGA and the known thermal expansion coefficient of zirconium hydride, this effect should contribute somewhat less than $-1 \times 10^{-5}\ \Delta\rho/°C$ to the prompt temperature coefficient, and since the cladding is not bonded to the fuel, will probably not contribute at all until somewhat elevated fuel temperatures are reached. A reasonable approximation is to assume that the sum of the Doppler and fuel element expansion effects contributes a total of $-2 \times 10^{-5}\ \Delta\rho/°C$ to the prompt temperature coefficient, independent of the fuel temperature.

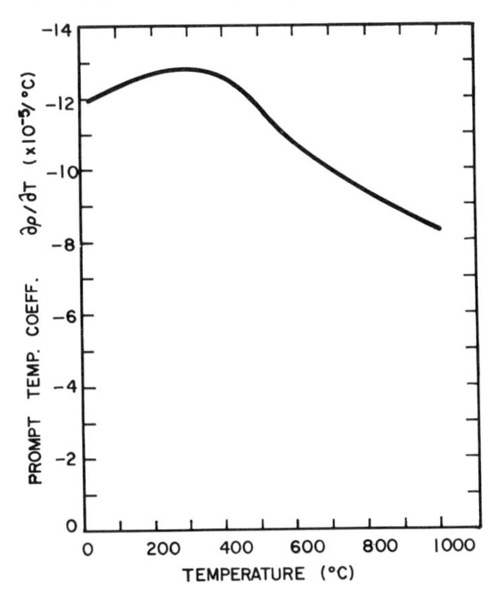

FIG. 3-17. 1 Mw TRIGA: calculated prompt temperature coefficient.

The three contributions considered give a calculated prompt temperature coefficient for the 1 Mw TRIGA of $-12.6 \times 10^{-5}\ \Delta\rho/°C$ averaged between 20 and 400°C (68 and 752°F). A plot of the temperature coefficient vs. temperature is shown in Fig. 3-17. These results are for the rod-free core with the control rod channels filled with water and control rod followers. This calculation, although subject to uncertainties of about 10%, is from first principles. Comparison of calculations with experiment for the earlier TRIGA reactors show agreement within 15%. More detail on TRIGA temperature coefficient calculations is given in an earlier report of Nelkin and West [36].

4 SOURCES OF ACCIDENTS

The sources of accidents which are of concern in the safety analysis of solid-moderator reactors may be divided into those associated with accidental additions of reactivity leading to a power transient and those associated with accidental high temperatures arising apart from power transients. Typical sources of accidents in the latter category are the following:

1. power peaking in a local region following withdrawal of a control rod which had been present for a long period of operation (see Sec. 2 above);
2. flux tilts due to misalignment of control rods which might occur while transferring control from one rod group to another;
3. partial or complete blocking of flow passages;
4. bowing of fuel of moderator sections in regions of high flux and temperature gradients;
5. spatial flux oscillations arising from positive temperature coefficients and/or xenon effects in large loosely-coupled cores;
6. loss-of-coolant accidents; and
7. release of stored energy (Wigner energy release).

Although the accidental operation of parts or all of the reactor core at excessive temperatures has important implications on reactor safety, it is not the purpose of this chapter to dwell on these particular accidents. However, it is seen later in the discussion that some of these accidents can lead to reactivity excursions and in this respect are of importance in the discussion of the kinetics of solid moderator reactors. Hence, with the exception of items 5 and 7 above which are discussed in a subsequent section, the above accidents are discussed only in connection with the subject of reactivity accidents when it is relevant.

In general, the sources and magnitudes of reactivity accidents in solid-moderator reactors tend to be unique since they involve the kinetic parameters, reactivity coefficients, and time constants associated with the types of fuel elements and core designs dictated by the solid moderator concepts. The discussion in this and the next section is directed primarily to the identification and study of reactivity accidents. The present section discusses the causes of these accidents and the typical rates of reactivity addition involved; while Sec. 5 discusses the calculational methods used and typical results obtained.

The principal sources of reactivity changes which could result in an accident in solid-moderator reactors are the following:

- motion of control rods,
- re-arrangement of fuel, moderator, and reflector materials,
- coolant density changes (liquid-metal-cooled reactors and some gas-cooled types),
- moderator density changes,
- coolant temperature changes,
- loss of poison material from core, and
- introduction of reactive material from outside core (e.g. steam).

Each of these possible sources of reactivity accidents will be discussed in some detail. Since the control rods tend to be the only moving parts having any significant reactivity effectiveness in solid-moderator reactors, the first item in the above list represents the most probable mechanism for introducing large reactivity changes in this type of reactor. The coolant in the sodium-cooled reactor might be regarded as an exception to this general statement, but even here the possibility of a change in state or the removal of significant amounts of sodium is quite small.

Large reactivity contributions from the other items listed above may quite often be found as secondary effects in an accident involving the loss of coolant. In such cases the secondary effect may either act to continue or terminate the transient. Each of the items listed above is discussed below with examples of the rates of reactivity addition possible in typical solid-moderator reactors.

4.1 Motion of Control Rods

A discussion of control requirements and control mechanisms characteristic of solid-moderator reactors was contained in Sec. 2 of this chapter. A summary of the considerations of importance in the analysis of possible control accidents is reviewed here.

Accidental withdrawal of control rods may result from operator action or from a malfunction in an automatic control system. In evaluating the rate of reactivity addition, consideration must be given to three factors, i.e., the maximum number of rods that can be moved simultaneously, their reactivity effectiveness, and the rate of reactivity increase that can result.

4.1.1 Number of Rods Which Can Be Moved Simultaneously

The maximum reactivity insertion possible in any single operational step depends on the maximum number of control rods which can be moved simultaneously. In some of the large gas-cooled reactors, e.g. the Calder-type reactors, the excess reactivity under hot operating conditions with equilibrium xenon is limited to about 0.01 $\Delta\rho$, and the control rods are removed as a bank. In order to avoid the possibility of a rod withdrawal accident, the maximum withdrawal speed is limited to a very small value. In reactors where the maximum excess reactivity is larger, e.g. the Peach Bottom HTGR, the rods are usually removed in symmetric groups so that during normal operations most of the control rods are either all in or all out thereby minimizing the axial tilt in power. The detailed control rod program is usually chosen also to minimize radial variations in the power distribution. In this type of program, it is usually advantageous to limit the number of rods that can be removed simultaneously so that the maximum reactivity insertion which could result from any single withdrawal operation cannot cause severe damage to the reactor core or primary system even if a scram should not occur immediately following the power overshoot.

4.1.2 Reactivity Effectiveness of the Rods Withdrawn

In general, the worth of a control rod depends strongly on the number and position of other control rods present. This is most often an important consideration when the rods are withdrawn in small groups at a time. When operated as a bank, the worth is strongly a function of the axial position of the bank. The calibration curve of reactivity vs. distance inserted may change significantly throughout core life, particularly when banked control is used and therefore strongly nonuniform axial burnup of fuel can result. Control sensitivity calculations late in the fuel life must therefore consider any nonuniform burnup of the fuel. Some discussion of the calibration curves for individual rods and rod banks was included in Sec. 2. The worth of individual rods also may increase throughout core life due to an increase in the thermal diffusion length or due to the reduction of shadowing by adjacent rods which are removed during core life. Conversely, the rod worth may decrease due to burnup of the control poison.

4.1.3 Speed of Withdrawal

There is usually a single slow speed at which rods can be withdrawn by the drive motors. The possibilities of faster withdrawal either by malfunction of the drive motor, pressure drop across the rod due to some component failure, or gravity (in the case of bottom-driven rods), must be carefully examined. The control rod characteristics of several solid-moderator reactors are summarized in Table 4-1. It can be seen that the total reactivity swing that is possible in a single operation varies from about 0.005 to 0.20 Δk for different reactors under different modes of operation. Furthermore, the maximum rate of reactivity increase varies between 2×10^{-6} and 5×10^{-4} for these reactors. In the zero-power or startup condition, the rate must be kept slow enough to prevent the reactor from rapidly coming to power on a very short period and damaging fuel elements before some control action can be taken. Using a typical startup reactivity rate given in the table, for example, $\Delta\rho/\Delta t = 3 \times 10^{-4}$/sec, it would require 22 sec to reach prompt critical starting from delayed critical. The maximum power increase over the power level for scram initiation is usually well within acceptable limits under these conditions. In the power range, a suf-

TABLE 4-1

Control Rod Characteristics of Typical Solid-Moderator Reactors

Reactor	Max. No. of rods movable together	Total worth of max. No. (Δk)	Max. rate of withdrawal (cm/sec)	Time for complete withdrawal (sec)	Reactivity rate $\Delta\rho$/sec	
					Maximum (sec^{-1})	Average (sec^{-1})
HTGR (Peach Bottom)	Startup 3*	0.026	1.93	117	3.2×10^{-4}	2.2×10^{-4}
	Power 1	0.009	1.83	117	1.1×10^{-4}	7.7×10^{-5}
EBOR	1	0.005	0.085		1×10^{-4}	
HNPF (Hallam)	19	0.14	0.525	755	3×10^{-4}	1.8×10^{-4}
EGCR	Startup 12	0.198	0.085	5380	9.7×10^{-5}	2.7×10^{-5}
	Power 9	0.124	0.085	5380	8.2×10^{-5}	2.3×10^{-5}
ML-1 (semaphore blades)	Shim-Safety 1	0.017		240	8.2×10^{-5}	6.6×10^{-5}
	Safety					
	Regulating 1	0.006		13	5.1×10^{-4}	3.4×10^{-4}
AGR	All rods 18	0.112	0.0154		1×10^{-5}	
	Automatic 3	0.005			1.6×10^{-5}	
Calder Hall	48	0.067	0.04		1×10^{-5}	

*Below 10% full power.

ficiently prompt negative temperature coefficient can usually compensate reactivity additions to the extent that prompt criticality is never reached. In the HTGR, for example, during a rod withdrawal at full power with an end-of-life temperature coefficient of about -2×10^{-5}/°C (-1.1×10^{-5}/°F) which is typical of most solid-moderator reactors, the maximum net reactivity at any time in the transient is less than $\Delta\rho = 0.001$. In the heterogeneous reactors with similar temperature coefficients and much less heat capacity, the compensation of the inserted reactivity would be even faster.

4.2 Rearrangement of Fuel, Moderator and Reflector Materials

In the case of reactors in which spent fuel is continuously replaced by fresh elements, the possibility of accidental rapid insertion of fresh fuel must be studied. However, the worth of fresh fuel is usually much less than that of any control rod and the loading devices generally operate very slowly. Consequently, this usually is not a critical operation. In the EGCR, for example, [37] the charging machine inserts or removes fuel at speeds of 30 to 48 in./min (76.2 to 122 cm/sec).

It is calculated that the insertion of a single fuel element decreases reactivity in the EGCR due to the fact that the reactor is undermoderated. The maximum change in multiplication due to the complete removal of the fuel assembly in one channel is $+0.001\ \Delta k$. Hence, the fuel assembly withdrawal in this reactor, rather than insertion, tends to be the more significant operation, in terms of safety. However, since the maximum reactivity change is quite small, the operation cannot lead to a large transient in this particular reactor. Although this tends to be true for other solid-moderator reactors also, each reactor should, of course, be examined on the basis of its particular characteristics.

Other possibilities for reactivity changes due to motion of core components can exist in cores which depend on pressure to maintain the configuration. For example, in the HTGR, the side reflector blocks are hinged at the bottom and are held against the fuel elements by the pressure drop between opposite sides of the reflector blocks. A loss-of-coolant pressure could then cause a negative reactivity change of about $0.0014\ \Delta\rho$. Since the reactivity change is small and in the negative direction, this possibility would again be insignificant in terms of safety. In this particular case, it is not possible to increase reactivity by pressure changes once coolant flow has been established. However, possible core component motions arising from coolant pressure and flow changes must, of course, be carefully analyzed in the design of solid-moderator reactors.

The levitation of fuel elements or control rods due to sudden pressure changes in the core must also be examined and can usually be prevented by design. A somewhat more difficult phenomenon to calculate in advance is the possible bowing of fuel elements with changes in power or flow rate. In the SRE, for example, the second core loading with a new design of fuel elements was found to have a prompt positive power coefficient due to the bowing of the fuel elements [38]. The element consisted of a 5-rod cluster suspended from a central stainless steel hanger in each flow channel. Bowing was caused by both radial temperature gradients in the rods due to nonuniform heat generation and by radial pressure differences across the rods at increasing sodium flow rates. The measured prompt power coefficient was about +12¢/Mw as compared to the

-2 ¢/Mw expected from the Doppler coefficient. The effect was much more important in the second SRE core than the first, both because the fuel loading was heavier, thereby allowing larger temperature gradients across the fuel rods, and because the clearance between the fuel elements and the channel was greater, thereby allowing more room for motion. When the fuel elements had been wire-wrapped to eliminate bowing in the second core, a prompt power coefficient of -2¢/Mw was observed.

Effects of this sort are not usually sources of reactivity accidents in themselves but would greatly increase the hazard in a transient initiated by other means. In the case of the SRE, it was not possible to operate the reactor in a stable fashion at power until the bowing had been eliminated. A more dangerous situation might be envisioned in a case where bowing is not detectable in the power range, but could occur suddenly in the course of a transient at higher temperatures. This possibility must be closely examined in reactors in which significant radial fuel motions can occur and particularly in heterogeneous designs using clusters of fuel rods in which small radial motions can greatly change the thermal utilization in a lattice cell. Fairly detailed calculational methods are required for such problems. In the SRE case, for example, Monte Carlo calculations were required to analyze the effect.

A somewhat similar case of fuel element displacement can arise in reactors where a large component of radial flow can occur due to a larger pressure drop along the length of the normal coolant channel than that associated with radial flow plus subsequent leakage upwards through the side reflectors. At increasing power levels with increased coolant flow, the component of flow moving radially between the fuel elements would increase thereby expanding the core radially and decreasing the reactivity unless the fuel elements are sufficiently constrained. A sudden loss of flow at full power would then cause a rapid positive reactivity insertion. It is possible under some circumstances to introduce 25¢ to 50¢ in reactivity in this manner. Other variations of this type can be imagined involving expansion of undermoderated cores, or contractions of overmoderated cores, or changes in coolant or moderator geometries.

4.3 Coolant Density Changes

Changes in coolant density are insignificant for gas-cooled reactors from the standpoint of reactivity changes since, with the exception of nitrogen and hydrogen, the gases commonly used have very small reactivity worth.

In the ML-1 reactor, [39] which is nitrogen-cooled, the pressure coefficient ($\Delta\rho$ per psi) is -3×10^{-6}/psi (or -4.3×10^{-5} cm^2/kg). The total reactivity gain from operating pressure to atmospheric pressure is +0.001 $\Delta\rho$. In the Peach Bottom HTGR the reactivity worth of the helium coolant is less than 0.0002 $\Delta\rho$. To date, hydrogen has only been considered for nuclear rocket propulsion.

Of greater significance for gas-cooled reactors is the change in heat transfer properties associated with large changes in coolant density. For heterogeneous gas-cooled core designs, in which the coolant flows between the fuel and moderator, the heat capacity of the fuel is usually not large and reductions in coolant density or pressure can lead to fairly rapid changes in fuel temperature. In reactors with U^{238} or Th in the fuel, a negative prompt temperature coefficient is very effective in reducing reactivity quickly in the event of a sudden increase in fuel element temperature. The complete loss of coolant is one of the most serious conceivable accidents in gas-cooled reactors even after a scram. Since it is relatively easy to provide a negative prompt temperature coefficient (as discussed above in Sec. 3), this should always be a design objective. Without a negative prompt coefficient, the control system must be relied upon completely to prevent an excursion for even a partial loss of pressure.

In liquid-metal-cooled reactors, the reactivity worth of the coolant is usually significant. In small sodium-cooled reactors, where the neutron leakage tends to be large, the reactivity contribution arising from the scattering characteristics tend to dominate the poison effect of the sodium. In these reactors, a decrease in sodium density or a loss of sodium from the core or parts of the core results in a reactivity loss. However, in the large sodium-graphite power reactors, where neutron leakage is kept to a minimum, the sodium coolant is a poison and a loss of sodium results in a reactivity increase. The measured reactivity worth of all the sodium in the SRE is about +.045 $\Delta\rho$ [40] while the estimated worth for the Hallam reactor was +.05 $\Delta\rho$ [55]. Hence, it is very important to design the reactor in a way to avoid the rapid development of large void volumes in the coolant channels through the core. The loss of sodium by leakage from the core can usually be prevented by locating all the vessel ducts above the top of the core and by designing the reactor cavity to contain the sodium to the required height even in the event of a disaster that could dump the sodium from the vessel.

Sodium voids in the core could be created by boiling in the sodium, which might result either from an excessive power level or an excessively low coolant flow rate. A 10% void distribution over the entire core could introduce a reactivity increase of about 0.005 $\Delta\rho$, and the same void volume at the center of the core could result in an even larger reactivity excursion. Hence, it is very important to prevent boiling in the sodium coolant. In general, the prevention of sodium boiling is not difficult since the margin between normal exit temperatures and the boiling temperature of sodium is quite large. Typically, the average temperature rise through the core is about 350°F (~200°C) and the margin between outlet temperature and the boiling temperature of the sodium is about 700°F (~400°C). Consequently, the power generation in the core could rise to about 300% of the normal power level before boiling would occur. Furthermore, the moderator is generally a very effective heat sink, which tends to absorb some of the extra heat energy in the early part of a power increase.

Of more serious concern is the possibility of sodium boiling resulting from flow impedance in some of the channels. Flow impedance can result from impurities in the sodium thereby causing the deposition of solid sodium compounds, primarily sodium oxide, or conceivably, the swelling of fuel elements blocking the channel. If a sufficiently large fraction of the core flow area is blocked, the resultant boiling and void formation in the core can lead to a significant reactivity increase.

The power excursion in the SRE is a good example of this type of accident [45]. (See also chapter on Accidents and Destructive Tests.) Prior to the power excursion, a small amount of an auxiliary organic coolant, tetralin, had leaked into the sodium loop and decomposed into various other compounds. Steps were taken to remove the organic materials from the system, after the leak was discovered, and the reactor was being brought back to an operating condition when an excursion occurred. On the basis of the analysis of operating data and an examination of the damaged core, it was concluded that plugging occurred in several adjacent coolant channels, leading to local overheating and expulsion of sodium which introduced the amount of reactivity observed.

4.4 Moderator Density Changes

Although this chapter is concerned primarily with solid-moderator reactors, some consideration has been given to gas-cooled reactors with H_2O and D_2O moderators; hence the effect of density changes in this type of reactor will be discussed briefly. Very little data are presently available on the D_2O-moderated, gas-cooled reactor. A considerable amount of information is available however on the ML-1 reactor, a water-moderated, nitrogen-cooled design. In the ML-1 the water moderator has a measured positive temperature coefficient, $+1 \times 10^{-5}/°C$ (or $+1.8 \times 10^{-5}/°F$) from room temperature to the operating range (90°C or 194°F) which precludes a cold water accident [39]. The time constant associated with heat transfer from the coolant to the water moderator is quite large, so that the positive moderator coefficient is rather unimportant in the kinetic stability of the reactor. Voids occurring in water-moderated reactors or gross redistribution of the water could be produced either by boiling or by rupture of a fuel tube which would allow the higher pressure coolant gas to enter the moderator volume. The opposite possibility of moderator water entering the fuel tube following rupture could presumably only occur when the gas pressure is low, i.e., when the reactor is shutdown or duing a loss-of-coolant accident. In the latter case, a steam explosion would be likely, in addition to a positive reactivity insertion, and the exact course of the accident would be almost impossible to predict. It is clear that great care is necessary to insure the integrity of the fuel tubes in these systems.

4.5 Coolant Temperature Changes

Changes in coolant inlet temperature can be visualized in gas-cooled or liquid-metal-cooled reactors due to malfunctions in the coolant circulators or in the automatic control system or by operator error in manual control. The reactivity change brought about by a coolant temperature change would depend on the sign and magnitude of the component temperature coefficients, the relative heat capacities of the coolant fuel and moderator components, and the thermal conductivity between various components. For reactors with multiple coolant loops in which one or more loops can be isolated and subsequently restarted, the possibility exists of producing a decrease in inlet temperature. In the gas-cooled systems, the gas has no significant reactivity worth but, by cooling the fuel, a slow reactivity transient can result if the temperature coefficient is negative. Figure 4-1 (unpublished work) illustrates the behavior of the Peach Bottom HTGR following an assumed 120°F (670°C) step decrease in the inlet temperature starting at full power. This represents closely the transient which would occur if the last feedwater pump were by-passed under certain conditions, assuming no control system response. The maximum rate of reactivity addition ($\Delta \rho$ per sec) during this transient is about 1×10^{-5}/sec, which is much slower than the rate of reactivity change available from normal control rod motions. It was assumed in the calculation of this transient that the lower inlet temperature remained constant, which tends to overestimate the magnitude of the power rise.

In heterogeneous reactors in which the coolant flows between the fuel and moderator and the heat capacity of the fuel is lower, the response of the fuel temperature coefficient would be somewhat more rapid. In the case of the sodium-cooled graphite moderated reactor, a change in the inlet temperature of the sodium affects the reactivity directly because of the sodium density change, as has been previously discussed under Coolant Density Changes. Subsequently, the temperature changes in the fuel element and moderator result in further reactivity changes. Since the sodium in a large sodium-graphite reactor is a poison, the temperature coefficient component arising from the sodium is positive, although usually quite small. The sodium component of the temperature coefficient for the Hallam reactor was calculated to be $0.55 \times 10^{-5}/°F$ (or $0.99 \times 10^{-5}/°C$), for example [55]. The immediate effect of a step decrease in the inlet sodium temperature would be a decrease in reac-

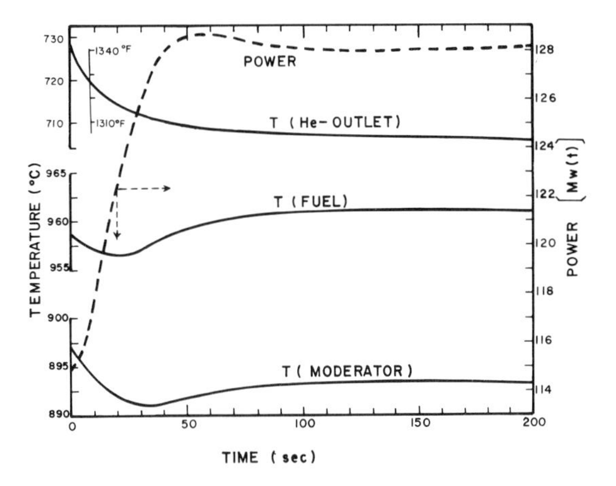

FIG. 4-1. Peach Bottom HTGR - Power level, fuel temperature, moderator temperature, He-outlet temperature as function of time following a step decrease of 123.3°F (68.5°C) in He-inlet temperature.

tivity. Hence, a sudden increase in the inlet sodium temperature might be of greater concern. However, even a sudden increase of 100°C would cause a reactivity change of only 10^{-3} ($\Delta\rho$) due to the sodium itself and therefore would not be serious unless boiling could result, as previously discussed. Since the response time of the fuel is between 2 and 3 sec, the negative temperature coefficient component associated with the fuel would quickly overcome the reactivity effect of a sodium temperature change, but again would not be sufficiently large to impose a serious reactivity hazard. After several minutes the moderator temperature would begin to respond in the opposite direction to the fuel coefficient. Although the moderator coefficient is positive, and therefore tends to lead to some instability, the time response would be so slow that this generally would create no practical problem in reactor control.

In general, a sudden change of coolant inlet temperature, corresponding to the cold water accident in water-cooled reactors, is not of serious concern as a possible source of a reactivity accident in solid-moderator reactors, although some attention must be given to the effect of coolant temperatures variations on the overall control stability of the reactor.

4.6 Loss of Poison Materials from Core

Reactivity changes can be postulated as a result of the loss of poisons such as the loss or migration of fission products, loss of core components by structural failure or meltdown, or loss of a liquid coolant or moderator as discussed above in Secs. 4.3 and 4.4.

It is difficult to conceive of the sudden loss of fission products as the source of a reactivity transient, although such a loss, particularly of Xe^{135}, could conceivably accompany a severe transient initiated by other means. In most of the solid-fuel reactors presently operating, release of fission products involves melting or bursting of cladding which would only occur following loss of coolant or a reactivity insertion not terminated by scram action. The rate of reactivity addition due to loss of the volatile fission products could not be great unless a large fraction of the fuel elements failed simultaneously.

In the semihomogeneous type of fuel elements used in the HTGR and Dragon type of reactor, a fraction of the more volatile fission products are removed by continuous internal purging of the elements. Experimental tests [43] have shown that no rapid release of large amounts of fission products (in a time interval less than several minutes, for example) is possible at temperatures which could be reached in credible reactivity accidents. Figure 4-2 shows the fraction of Xe^{133} retained by Peach Bottom HTGR fuel compacts throughout a severe temperature transient. The fuel particles (uranium-thorium carbide) are contained in a graphite matrix and in this experiment were uncoated. After irradiation in a test reactor, in order to build up some fission products, the fuel body was removed and placed in a furnace. The temperature of the fuel was increased from 1400°C to 2280°C (2550°F to 4130°F) in less than one minute and held there for five minutes. Less than 5% of the Xe^{133} was released during the initial temperature excursion. The temperature was then decreased and some release (about 20%) was observed over about a 10 minute period. The fuel was then held at a temperature of 1850°C (3362°F) for a period of about 16 hours. At the end of this 16-hr period 46% of the original Xe^{133} (after appropriate corrections for decay) was still retained in the fuel body. Following the 16-hr period at 1850°C, the temperature was raised to 2600°C (4712°F), which is above the melting point of the carbide fuel. After two hours at 2600°C essentially all of the remaining Xe^{133} had been released from the fuel body. The conclusion is that rapid release of the fission product xenon does not occur even in very severe temperature transients.

As might be expected, the release from coated fuel particles is even slower. Figure 4-3 shows the results for a similar transient using fuel particles coated with pyrolytic graphite similar to those used in the Peach Bottom and Dragon reactors. Xe^{133}, of course, represents the behavior of Xe^{135} in the reactor. Taking the reactivity worth $\Delta\rho$ of Xe^{135} as +0.03, the maximum reactivity addition rate due to loss of Xe^{135} during the time spent at 3000°C (5430°F) would be about 2.3×10^{-3}/hr, with coated fuel particles, which is clearly insignificant. The situation is similar for other fission products.

On the basis of these studies it appears that fission product release resulting from a temperature excursion is not sufficiently rapid to cause a serious reactivity transient. However, it is important to design the control system such that the reactor can be made subcritical in the event an extended overheating of the core should cause the volatile fission product poisons to ultimately escape from the core.

The loss of core components such as structural members, burnable poisons, etc., which have significant reactivity worth can generally be prevented by careful design.

The loss of fuel element cladding by meltdown is a possible secondary source of reactivity that could occur as the result of a severe power excursion or a loss-of-coolant accident. The maximum reactivity increase could be particularly large in reactors having stainless-steel clad fuel since the total reactivity worth of steel cladding is typically

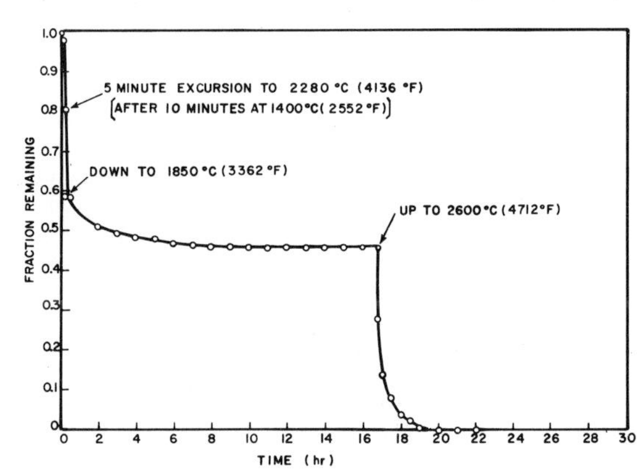

FIG. 4-2. Xe^{133} release from fuel compact prepared from uncoated fuel particles following a simulated excursion to 2280°C (4136°F). $(Th:U)C_2$ = 2.19:1.0.

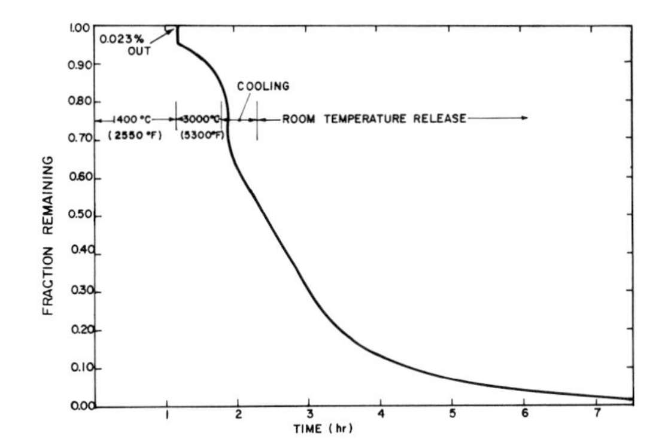

FIG. 4-3. Xe^{133} release from fuel compact prepared with fuel particles coated with pyrolytic graphite following a simulated excursion to 3000°C (5430°F) for 38 min.

between 0.10 and 0.20 ($\Delta\rho$). Generally, the shutdown margin of a reactor is not sufficiently large to cover such a severe accident.

The meltdown of a large fraction of the fuel cladding would, of course, be an extremely serious accident for reasons other than the large reactivity increase. Consequently, solid-moderator reactors, and particularly the gas-cooled reactors, are designed with multiple primary coolant loops, or emergency cooling systems, or both, to assure adequate cooling even in the event of coolant flow failure in a loop or complete depressurization. Furthermore, the control rod rates and protective instrumentation are designed to prevent the possibility of a major fuel element meltdown resulting from a reactivity transient.

Some studies of possible fuel cladding meltdown effects in the EGCR have been reported by the Oak Ridge Laboratory [44]. These studies were admittedly unrealistic in severity but are useful in illustrating the possible magnitude of reactivity increase one could visualize.

The total reactivity worth of the stainless steel cladding in the EGCR is about 0.10 $\Delta\rho$). The complete loss of the stainless steel without a rearrangement of core materials would therefore result in a reactivity increase larger than the initial shutdown margin. However, a complete loss of steel would allow the fuel pellets to slump to the bottom of the core in such a manner that a significant decrease in reactivity would occur as the result of a change in geometry. The net effect is that the resultant reactivity of the arrangement is close to unity for room temperature conditions, and probably subcritical at elevated temperatures. In the more realistic case where a significant part of the steel stays with the fuel, the reactor would be subcritical even at room temperature.

In the event of excessively high core temperatures without any emergency cooling, it is conceivable that the bottom support plate temperature could rise to the point where it would sag, thereby allowing the core to fall away from the control rods. One would normally expect that at such high temperatures the control rod drive cables would also fail, allowing the control rods to drop along with the core. In the same EGCR accident study, the reactivity effects associated with various amounts of support plate sag were examined for various assumptions on control rod dispositions. The studies indicated that even for quite pessimistic assumptions the reactor would remain subcritical. Nevertheless, the importance of adequate emergency cooling is apparent.

While the purpose of this particular discussion is primarily to indicate the sources of reactivity accidents rather than to analyze the course of the accident, it is perhaps pertinent to summarize some results of an accident study to show the extent of fuel element failure one might expect as the result of a severe depressurization accident in the same reactor [44]. For the case of a large breach in the primary coolant circuit followed by a complete loss of coolant flow and a scram after 5 sec, it was calculated that cladding failure would occur in about 3% of the fuel elements within the first 30 sec. The cladding failure would simply be a rupture and no significant amount of cladding would leave the core under these conditions. If emergency cooling is assumed to be initiated within an hour following the depressurization accident, no additional fuel cladding ruptures would occur. The large heat capacity of the graphite moderator provides an effective heat sink for the absorption of heat from the fuel elements. Hence, with reasonable provisions for emergency cooling, the massive loss of fuel element cladding from the core is not a serious concern.

Severe accidents leading to possible fuel meltdown have also been studied for the Hallam Sodium-Graphite Reactor [55]. Again the reactivity worth of the stainless steel is quite large, but in this reactor the fuel itself is also metallic and could melt almost as soon as the cladding, if not sooner. Hence, in this case fuel and poison would probably be removed from the core simultaneously. As discussed previously, local melting could occur as the result of coolant channel plugging followed by sodium boiling, or by boiling due to excessive power generation.

In summary, the meltdown of a large fraction of fuel cladding should be a very improbable accident in a well-designed reactor. However, the possibility of such an accident could be extremely serious, both from the point of view of fission product release and the reactivity excursion that could result. Careful attention should, of course, be given to the identification of design considerations that might lead to cladding meltdown, with primary emphasis on design features to avoid such an accident. In the analysis of cladding meltdown and its possible reactivity effect, the following factors should be taken into consideration:

(1) Melting would take place initially in the highest temperature regions of the core which tend to be the regions of highest reactivity worth. Hence, the location as well as the amount of steel lost should be considered.

(2) If the fuel body is metallic, fuel meltdown and removal may occur practically at the same time as the cladding meltdown. If the fuel body is ceramic, the effect of relocation of the fuel pellets should be evaluated. The effect of melting and rearrangement on ultimate shutdown capability should be examined, although it is usually very difficult to define the exact degree of melting and core rearrangement.

(3) Rearrangement of the core relative to the

control poisons must also be investigated, particularly where the core is supported at the bottom and the control rods at the top of the vessel.

4.7 Introduction of Reactive Materials from Outside the Core

The most likely accident in this category for gas-cooled reactors is the addition of steam or liquid water to the core. This would include steam or water leaks from steam generator failures or water from some other external source.

Steam leaks from a steam generator tube usually add only small amounts of reactivity and represent a greater concern from the standpoint of fuel element damage caused by chemical reactions than from a reactivity standpoint. Introduction of steam into the coolant circuit will also alter the flow and heat transfer properties of the coolant and can lead to overheating and cladding failures apart from the chemical reactions.

In the EGCR, for example, the reactivity worth of steam is $+5 \times 10^{-4} \Delta\rho$ per 100 psi (or $+0.7 \times 10^{-4} \Delta\rho$ per kg/cm^2). In the HTGR the steam coefficient is $+2.3 \times 10^{-4} \Delta\rho$ per lb of steam in the core (or $5.1 \times 10^{-4} \Delta\rho/kg$ steam). The maximum quantity which could exist in the core before relief values opened is less than 50 lb (23 kg), which could add about 0.01 in $\Delta\rho$, but under normal conditions a loop isolation following detection of moisture would limit the amount of steam to a much smaller quantity.

The reactivity rate due to steam leaks is dependent on many factors, e.g. what part of the steam generator the break occurs in, how many tubes rupture, what the flow paths and velocities are in various parts of the core, etc.

The reactivity increase due to complete flooding of the core in a gas-cooled reactor may be very large even for a reactor with an otherwise adequate shutdown margin. Complete flooding would be very difficult, if at all possible, in a reactor at operating temperature. Under these conditions most of the water would vaporize and be expelled from the core which would reduce the reactivity but not prevent fuel damage. It is difficult to make generalizations about the course of such accidents. It is conceivable that water flooding could occur in a gas-cooled reactor during reactor shutdown when the core is relatively cold. In reactors that are undermoderated, either it should be demonstrated that the reactor can be shut down with the core flooded with water, or that insufficient water is available to flood the core to the point where it cannot be controlled, or safety systems should be included to limit the flow of water into the core. In calculating the multiplication constant for water-flooded reactors, it may not be sufficient to examine only the completely flooded case. Cases should also be investigated where the water density is less than one, corresponding to boiling water or high temperature water.

In the case of sodium-graphite reactors, water must obviously be excluded from the core. The possibility of other substances entering the coolant must also be examined. In general, it is found that the amount of hydrogenous material that can be dissolved in sodium is small relative to that necessary to be a reactivity hazard. A more important consideration in sodium-cooled reactors is the coolant channel plugging that can occur as the result of contamination by oxides or organic materials, for example.

The magnitude and potential consequences of an accident originating from any of the reactivity sources discussed above are influenced by other factors apart from the rate of reactivity insertion. The other parameters in the kinetic equations, i.e., delayed neutron fraction, prompt lifetime, and temperature coefficient vary significantly during life in most solid-moderator reactors and the latter two quantities are also temperature dependent. In reactors which use U^{235} as the initial fuel and produce significant quantities of Pu and/or U^{233}, the effective delayed neutron fraction will gradually decrease with burnup. The production of these same isotopes will, in most reactors, also result in a trend toward a less negative or more positive moderator temperature coefficient component. Changes in neutron lifetime with burnup are less significant.

The operating state of the reactor, e.g. shutdown, operating at partial power or full power, is also important since it determines the power and temperature margins allowable before core damage can occur. These margins are also affected by the power distribution (which varies through core life) and by the possible changes in heat transfer properties which may occur after long irradiation. The various setpoints for scrams or other protective actions may also be different at different power levels. The postulated reactivity accidents should be analyzed by starting from the worst conceivable operating conditions for each particular accident and assuming progressive failures of the available scram or shutdown devices to determine the end-point of the accident.

5 ANALYSIS OF EXCURSIONS

5.1 Introduction

In previous sections, control characteristics of solid-moderator reactors have been discussed, the evaluation of kinetics parameters, particularly temperature coefficients, was described, and in the last section the possible sources of accidents in this class of reactors were identified. In this section, the course of reactivity accidents in typical solid-moderator reactors is analyzed, making use of the information developed in the previous sections. In particular, some details of calculational approaches are included.

The principal objectives of reactor transient calculations are to study the changes in power and temperature for various rates of reactivity insertion in order first, to specify the performance of the control system needed for safe and efficient operation and, second, to demonstrate the safety of the plant for the worst conceivable reactivity accidents. With respect to the second objective, it should be noted that, although several types of solid-moderator reactors have been operated for periods of years (e.g. Calder Hall, Hanford, SRE), almost no transient reactor test data exist for large reactivity transients, either in systems of these types or in the newer high temperature concepts. The amount of work done is certainly not comparable to the large body of transient data on water reactors. Hence, more reliance must be placed on calculational models and basic data in demonstrating the safety of these systems. On the other hand, the dynamic response in solid-moderator reactors is generally more sluggish and the reactivity feedback mechanisms easier to deal with in calculations.

The calculational models required for either objective depend on the detail of information desired and to some extent on the reactor type. The early phases of analysis usually focus on ramp and step insertion transients to associate periods and rates of temperature change with the approximate reactivity addition rates which might be realized during normal or abnormal operating conditions. From this information, calculations of required scram insertion rates and scram response times can be determined to a first approximation, depending on the limiting values of temperature (or other parameters) beyond which unacceptable core damage would result. In the case of the heterogeneous reactors with metallic fuel or cladding (e.g., Calder, SGR, EGCR), these limiting conditions usually involve the melting point of the fuel or cladding. In the high temperature ceramic fuel reactors (HTGR, Dragon), the limiting condition might be related to fuel element damage although it is more often associated with the outlet gas temperature which, if too high, could damage other plant components. The rate of temperature change might also be a factor in that it might cause sufficient stresses to crack part of the fuel elements and release fission products. It is possible, particularly in the case of the heterogeneous systems, that the scram speed and response time might be dictated more by the need to protect the reactor against a loss-of-flow condition than by the consequences of conceivable reactivity accidents.

In the preliminary studies, fairly simple calculational models are usually adequate. As the design is developed in more detail and information becomes available on the reactivity-lifetime characteristics, control rod operating program, inventories of the various nuclides, power and temperature distributions throughout life, etc., the conditions for the worst conceivable reactivity transients can be identified. The most severe accident might therefore be found to be associated with some particular operational condition or time in life where one or more of the following characteristics might be important:

The available excess reactivity is particularly large.

The temperature coefficient is least negative.

The inventory of fission products is a maximum.

The delayed neutron fraction is a minimum.

The neutron generation time is a minimum.

Kinetic calculations should then be done in sufficient detail to demonstrate the safety of the reactor against the various possible accidents that can be postulated under the most pessimistic of these conditions. These should include successively the assumptions of (1) normal automatic scram with appropriate delay time, (2) partial scram only, due to multiple failures of rods, (3) manual scram by operator, (4) scram by back-up system if any, (5) no scram, shutdown by temperature coefficient or by meltdown and loss of fuel.

In the following discussion, the dynamic characteristics of several typical solid-moderator reactors are reviewed, with some reference to the calculational models that have been used. Some comparisons of transient measurements and calculations for the TRIGA and TREAT reactors are presented to illustrate the degree of success that can be achieved in calculating the transient behavior of some very simple, solid-moderator reactors using analytic solutions to the kinetic equations. Since these reactors are test reactors, their behavior is uncomplicated by heat removal systems and temperature coefficients for these particular reactors do not have any important delayed components.

As an introduction to the discussion of the transient behavior of larger, power-generating, solid-moderator reactors, a description of some of the digital computer calculational models used for these reactors is presented. The evaluation of the kinetic behavior and dynamic safety characteristics of several of the large power reactors is then discussed including experimental results where available. The heterogeneous, solid moderator reactors are reviewed first since a considerable amount of operational experience is available from at least some of these reactors. Included in this group are the natural-uranium, graphite-moderated, gas-cooled reactors; the enriched-uranium, graphite-moderated, gas-cooled reactors; and the enriched-uranium, graphite-moderated, sodium-cooled reactors. The semi-homogeneous, solid-moderator reactors are then discussed. In this category, we consider only the gas-cooled, graphite-moderated reactor using the $U^{235}/Th^{232}/U^{233}$ fuel cycle.

In each case, the transient behavior is discussed for various postulated accidents including the control rod withdrawal accidents and startup accidents followed by instrument-induced scrams.

Finally, the course of the accidents is reviewed, where information is available, when these accidents are not terminated by a control rod scram. It is generally recognized that the latter type of accident is based on terribly pessimistic assumptions, but is usually given some attention in safeguards analyses.

5.2 Analytic Methods and Their Applications to the TRIGA and TREAT Reactors.

The only solid-moderator reactors for which there are available comparisons of experiment and calculations for large reactivity insertions are the TREAT facility and the series of TRIGA research reactors. In both cases the kinetic response is dominated by a single prompt reactivity feedback mechanism and can be predicted quite well by a form of the Fuchs-Nordheim model, provided sufficient attention is given to careful calculations of the temperature coefficient, the prompt neutron lifetime and the specific heat of the fuel. Although detailed machine calculations are necessary for these input quantities, the maximum temperature and power can be computed from simple analytic methods.

A comparison of theory and experiment for the kinetics of the TRIGA reactors is given in the previously cited report of Scalettar and West [33]. It is first shown analytically and verified by comparison with computer solutions that, during the large power pulses produced by reactivity insertions of several dollars, the delayed neutron source term may be neglected in the kinetic equations and that the heat transfer to the water coolant is negligible. This reduces the kinetics and heat transfer equations to the familar equations:

$$\frac{1}{P}\frac{dP}{dt} = \frac{\delta k_p - \alpha \Delta T}{l}, \tag{5-1}$$

$$\frac{dT}{dt} = \frac{P}{C}. \tag{5-2}$$

where α is the temperature coefficient and C is the heat capacity of the fuel elements.

The maximum power, temperature and stored energy computed from the model are the following:

$$P_{max} = C/2\xi\tau^2, \tag{5-3}$$

$$E_{TOT} = \frac{C}{\xi\tau_o}\left\{1 + \frac{\tau_o}{\tau}\right\} = \frac{2C}{\xi\tau_o}, \tag{5-4}$$

$$T_{MAX} = \frac{E_{TOT}}{C} = \frac{2}{\xi\tau_o}, \tag{5-5}$$

in which $\xi = \alpha/\ell$, $\tau_o = \ell/\delta k_p$ and τ is a period very close to τ_o;

$$\tau = \tau_o \; [1 + (P(t_o)/P_{max})]^{-1/2} \tag{5-6}$$

The time τ_o is a time shortly after the initiation of the pulse when $P(t) >> P_o$ and the neglect of delayed neutrons becomes possible. T_{max} is the temperature reached when the power returns to $P(t_o)$. It has been found that the model can be extended to the case of linearly varying heat capacity, $C = Co + C_1 T$, and temperature coefficient $\alpha = \alpha_o + \alpha_1 T$, and that the solutions are expressed by functions of dimensionless parameters which multiply the constant coefficient solutions Eqs. (5-3) and (5-5). Thus:

$$P_{max} = g(\sigma_o, \rho)\frac{C_o}{2\xi_o\tau_o^2}, \tag{5-7}$$

$$T_{max} = f(\sigma_o, \rho)\frac{2}{\xi\tau_o}. \tag{5-8}$$

The dimensionless parameters σ_o and ρ are defined by

$$\sigma_o = \frac{a_o C_o}{\delta k_p C_1} = \frac{C_o}{C_1}\tau_o\,\xi_o \tag{5-9}$$

$$\rho = \frac{a_1 \delta k_p}{a_o^2} = \frac{\xi_1}{\tau_o \xi_o^2} \tag{5-10}$$

and the functions $g(\sigma_o, \rho)$ and $f(\sigma_o, \rho)$ are shown in Figs. 5-1 and 5-2.

These relations are valid for any reactor for which the Fuchs-Nordheim model is applicable and for which the heat capacity and temperature coefficient can be expressed in the manner indicated. In the case of TRIGA it is found that the variation of heat capacity with temperature can be significant in some situations but that the temperature dependence of the temperature coefficient can be neglected. Figs. 5-3, 4, and 5 show the comparison between experimental results and the calculations and illustrate the effects of variable heat capacity and temperature coefficient for a 3$ reactivity insertion. Fig. 5-6 shows the variation of peak power with initial period for a range of reactivity insertions and again illustrates good agreement between experiment and the simple calculational models.

It has recently been shown by Ghatak and Nelkin [46] that the modified Fuchs-Nordheim model described above provides an adequate description of the kinetics of the TREAT reactors. This is a small homogeneous graphite-moderated reactor in which the fully enriched uranium is uniformly dispersed in the graphite fuel elements. The negative prompt temperature coefficient then comes primarily from increased leakage as the thermal spectrum hardens. Although some unresolved discrepancies remain in the comparison of calculated and experimental temperature coefficients, [47] the relation between peak power and inverse period derived from the modified Fuchs-Nordheim model,

$$\tau^2 P_{max} = \frac{C_o}{ag}\left[1 + \frac{1}{3}\frac{C_1}{C_o}\delta k_p\right], \tag{5-11}$$

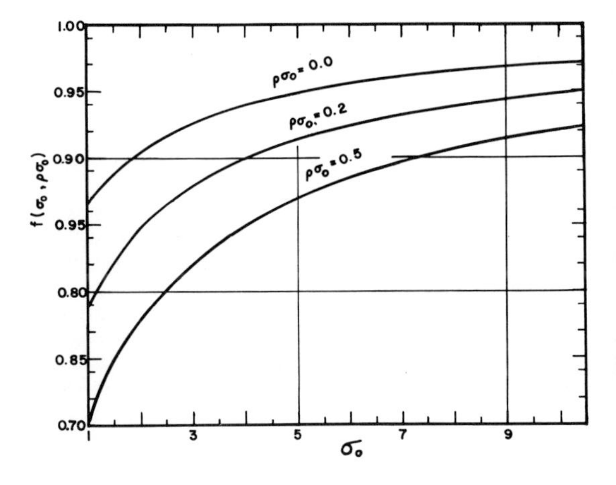

FIG. 5-1. The function $f(\sigma_o, \rho\sigma_o)$ vs σ_o for $\rho\sigma_o = 0.0, 0.2, 0.5$.

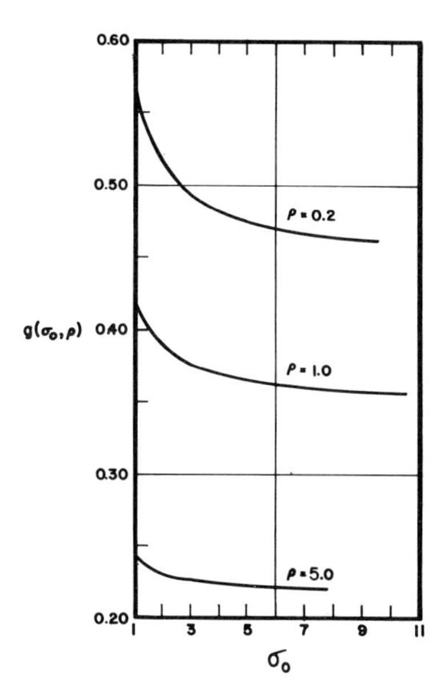

FIG. 5-2. The function $g(\sigma_0,\rho)$ vs σ_0 for $\rho = 0.2, 1.0, 5.0$.

was found to agree with experimental data within 5%. In this equation, g is the peak-to-average-power ratio, and the other quantities are defined above. None of the parameters on the right hand side of Eq. (5-11) are taken from the transient data.

5.3 Computer Methods

The vast majority of kinetics calculations are

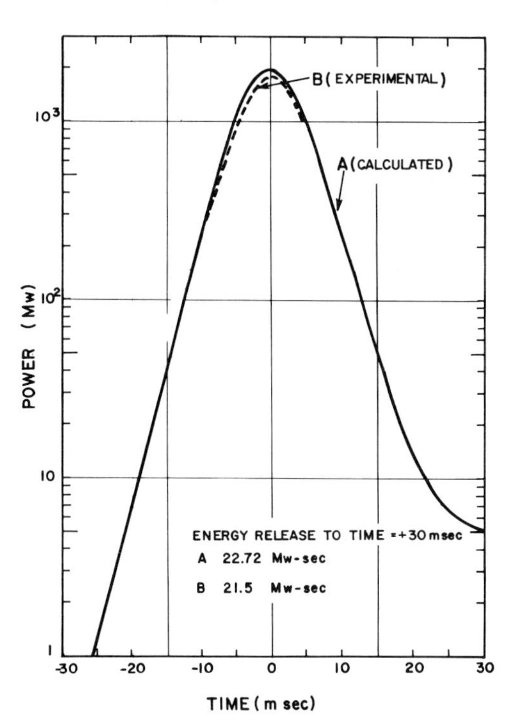

FIG. 5-3. Power vs time for 1 Mw TRIGA, 3$ pulse.

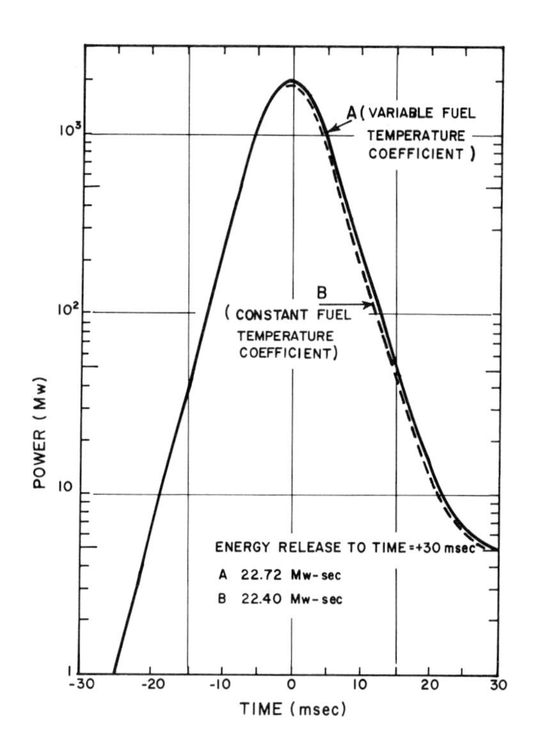

FIG. 5-4. Calculated power vs time for 1 Mw TRIGA, 3$ pulse. Value of maximum average fuel temperature was used in constant fuel temperature coefficient calculation.

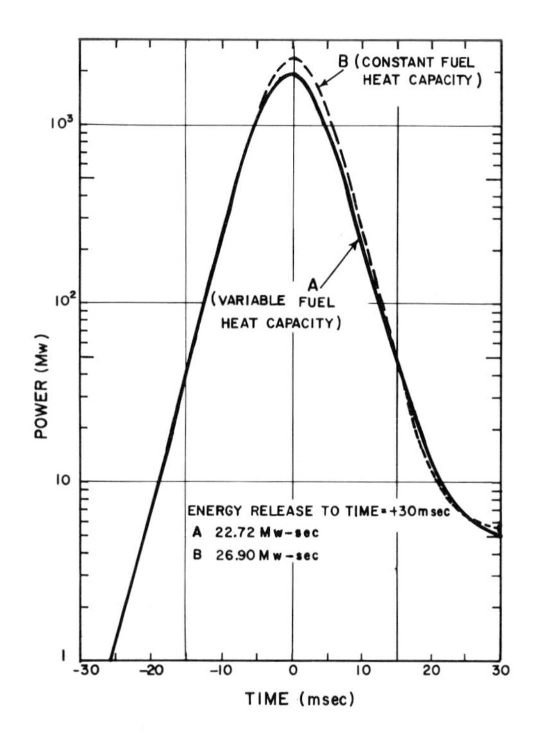

FIG. 5-5. Calculated power vs time for 1 Mw TRIGA, 3$ pulse. Value of maximum average fuel temperature was used in constant fuel heat capacity calculation.

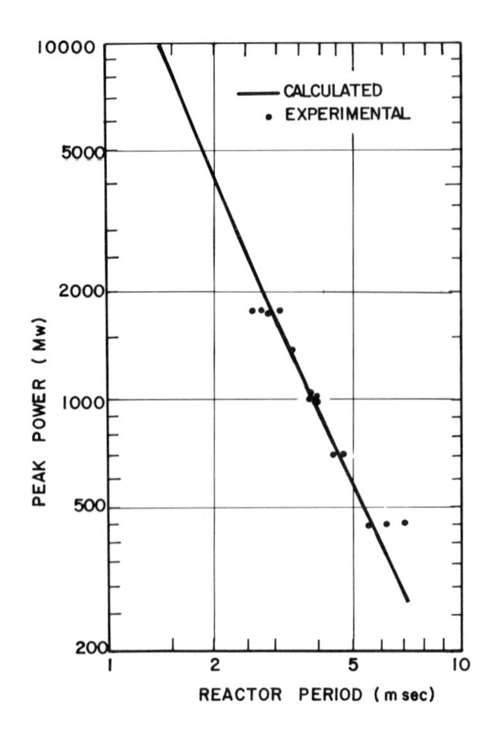

FIG. 5-6. Peak power vs reactor period for 1 Mw TRIGA.

done with the space-independent form of the reactor kinetic equations. However, the calculation of the neutron lifetime, the effective delayed neutron fraction, and the postulated reactivity changes will generally involve multigroup, multidimensional diffusion theory calculations. At any significant power level, some representation of the temperature coefficient is required which in turn requires knowledge of the time-dependent temperatures of the fuel, moderator and possibly the coolant. A heat transfer model of a typical fuel element or lattice cell is then necessary, the degree of detail needed depending on the magnitude and time dependence of the components of the temperature coefficient.

Both analog and digital computer methods have been used in studying reactor transients. Analog simulations are particularly useful in the initial transient studies with ramp and step reactivity insertions and simple lumped parameter models of the fuel element heat transfer. Some difficulties can arise due to drift in analog calculations of very long transients; e.g. in heterogeneous reactors with long moderator time constants. It is also difficult to simulate many decades in flux level in a single problem. Digital computer methods have the advantage that complex variations of reactivity, flow rate, temperature coefficients, and other effects can easily be included as accurately as warranted by the problem and the input information available. The numerical integration of the kinetics equation usually requires very small time steps for accuracy in rapid transients and can be more expensive to solve. Since it is often desired to follow a transient for several minutes, provisions can be made in the program to adjust the time step according to how fast the flux level is changing, to keep the problem from becoming unduly expensive.

It is advantageous in a digital code to program the heat transfer equations or other feedback mechanism calculations in separate subroutines from the neutron kinetics calculations since the latter equations are quite general for all reactors whereas it is very difficult to write a perfectly general heat transfer code. Thus, one kinetics code can be used for a variety of reactor types. Savings in computing time are also possible since it is not usually necessary to integrate the heat transfer equations with time steps as small as those needed for the neutron equations.

Some examples of digital computer codes currently used are given in the chapter on Reactor Dynamics. Two further examples of codes used primarily for calculations of solid-moderator reactors are described in the following paragraphs.

5.3.1 BLOOST* (General Atomic)

The kinetics calculation in the BLOOST code employs the same Runge-Kutta Method used in the AIREK-II code discussed in the chapter on Reactor Kinetics. Much of the input is specified by tables, including the initial reactivity disturbance, the scram (at specified power level and delay time), and the temperature coefficients associated with the fuel and moderator temperatures. The heat transfer subroutine was developed primarily for the HTGR fuel element and computes a two-dimensional temperature map for an entire fuel element. This distribution is averaged to obtain the fuel and moderator temperatures required to enter the temperature coefficient tables.

It is assumed in the heat transfer calculation that axial heat conduction is negligible which tends to be a somewhat conservative assumption. The calculation is really then a set of radial calculations coupled by means of the coolant gas. The gas inlet temperature and flow rate can be varied during the transient, again by means of tabular functions; and the specific heat is a function of temperature. Since the heat transfer calculation is the most time consuming part of the program, several things are done to reduce the running time. An explicit differencing method, which is fast but potentially unstable for long time steps, is used for the short time steps, and a much slower but unconditionally stable inplicit differencing method is used for time steps which exceed a built-in stability criterion. Also, the heat transfer routine is not called for every time step in the Runge-Kutta integration but only when input criteria based on the rate of temperature change and the elapsed time indicate that a new temperature distribution is needed.

The output from the code includes a two-dimensional temperature map printed at given intervals, and the power level, total reactivity, and contributions to reactivity from the initial reactivity insertion, the scram, and the tempera-

*Unpublished

ture coefficients tabulated at each time step.

The neutron kinetics portion of the code provides a general framework into which various heat transfer routines for different types of fuel elements can be incorporated.

5.3.2 STAB (Harwell)

The STAB code [48] is written for the large gas-cooled reactors of the Calder Hall type and, unlike the codes discussed above, includes the spatial variation of the flux shape during the transient. The one-group time-dependent flux equation is solved at each point in a three-dimensional mesh (r, θ, z). Equations for the variation of fuel and moderator temperature, xenon concentration and control poison concentration are included to provide reactivity feedback in the flux equation. Control rod motion in various core regions can be activated by coolant outlet temperature which permits the study of various control rod operating programs to control flux oscillations. The initial steady state fluxes are initially computed in (r,z) geometry assuming azimuthal symmetry, and then expanded to (r,θ,z) geometry for the time-dependent calculation. Figures 5-7 [48] and 5-8 [48] show flow diagrams of the calculational procedure. Some comparisons of calculations with experiments are given in the next section.

5.3.3 Comments on Methods and Models in Kinetics Computations

There is no clear dividing line between analog and digital methods with respect to relative advantages. The degree of conservatism required for kinetics calculations dictate that it must be possible to demonstrate the safety of the system by either method. The added detail of a digital solution is usually justified because of the increased insight it can give into the physical mechanisms, particularly in studies of new reactor types for which experimental data are lacking.

The majority of solid-moderator reactors built to date have heterogeneous core designs in which a cylindrical fuel element is surrounded by an annulus of coolant (gas or liquid) which in turn is surrounded by a block of moderating material. In these reactors there is a significant time lag between a temperature increase in the fuel and a temperature increase in the moderator. In some designs the lag is augmented by a sleeve consisting of a small amount of moderator material located in the flow channel between the coolant and the rest of the moderator. The component of the temperature coefficient associated with the fuel element temperature is thus the only one almost instantaneously available to limit a reactivity accident. The fuel coefficient is usually entirely due to the Doppler effect in the fertile and fissile isotopes. Since most of the reactors of this type operate with natural or slightly enriched uranium, a large negative Doppler effect from U^{238} is obtained. The Doppler effect is smaller in reactors with higher U^{235} enrichment (e.g., EBOR, 63%U^{235}) and could even be positive for some completely enriched systems. Other prompt effects, such as the unshielding of fuel by expansion, may make positive contributions to the fuel coefficient. Since a prompt positive temperature coefficient can usually be avoided in solid moderator reactors by a number of design expedients, an important design objective for this type of reactor should be the assurance of at least a slightly negative prompt temperature coefficient.

The moderator coefficient in low enrichment reactors of this type is often positive after some period of burnup due to the buildup of plutonium and may be sufficiently positive to cancel or exceed the negative fuel coefficient. It can usually be demonstrated in these cases that the long time delay associated with the positive moderator coefficient allows sufficient time to insert control rods or take other action before a dangerous condition is reached.

Most of the published transient calculations for reactors of this type have used lumped parameter heat transfer models which can be solved very rapidly on digital or analog computers. Several detailed presentations of the transient heat transfer equations are available in the literature [49, 50]. Effective average temperatures for the purposes of computing reactivity feedback, are often obtained by flux-squared weighting over the axial and radial directions either for the whole core volume or in regions represented by points in the analog simulation. The procedure is usually adopted without rigorous proof of its applicability and involves assuming a one-group perturbation theory and neglecting changes in leakage with temperature (which would be weighted with the square of the gradient if included). The temperature coefficient is assumed constant to remove it from the perturbation integral. These assumptions are most applicable to the large natural uranium graphite systems, such as the gas-cooled reactors used in the United Kingdom and France.

An excessive concern with the proper averaging procedures for point-model calculations is probably unwarranted in view of the other uncertainties in the calculation. The difficulties of computing reactivity lifetime and temperature coefficients in heterogeneous systems with mixtures of plutonium isotopes, for example, has been commented on in previous sections. Some confidence in this area can be provided by experiments, since in the heterogeneous reactors the fuel and moderator coefficients are well separated in time response and can be measured by pile oscillation tests and other methods. An additional important uncertainty is the difficulty of determining what actual changes in flow rate may be expected throughout the reactor in various transients (e.g. loss-of-coolant accidents).

A recent British paper [51] which discusses the various computational models currently in use and compares some calculated transients with experiments on the Calder Hall reactor observes that, while a lumped parameter or point model adequately describes extremely slow transients (e.g., those governed by xenon poisoning), a spatially dependent model is required to accurately calculate even slow reactivity transients initiated by control rods. This is due to changes in the shape of the

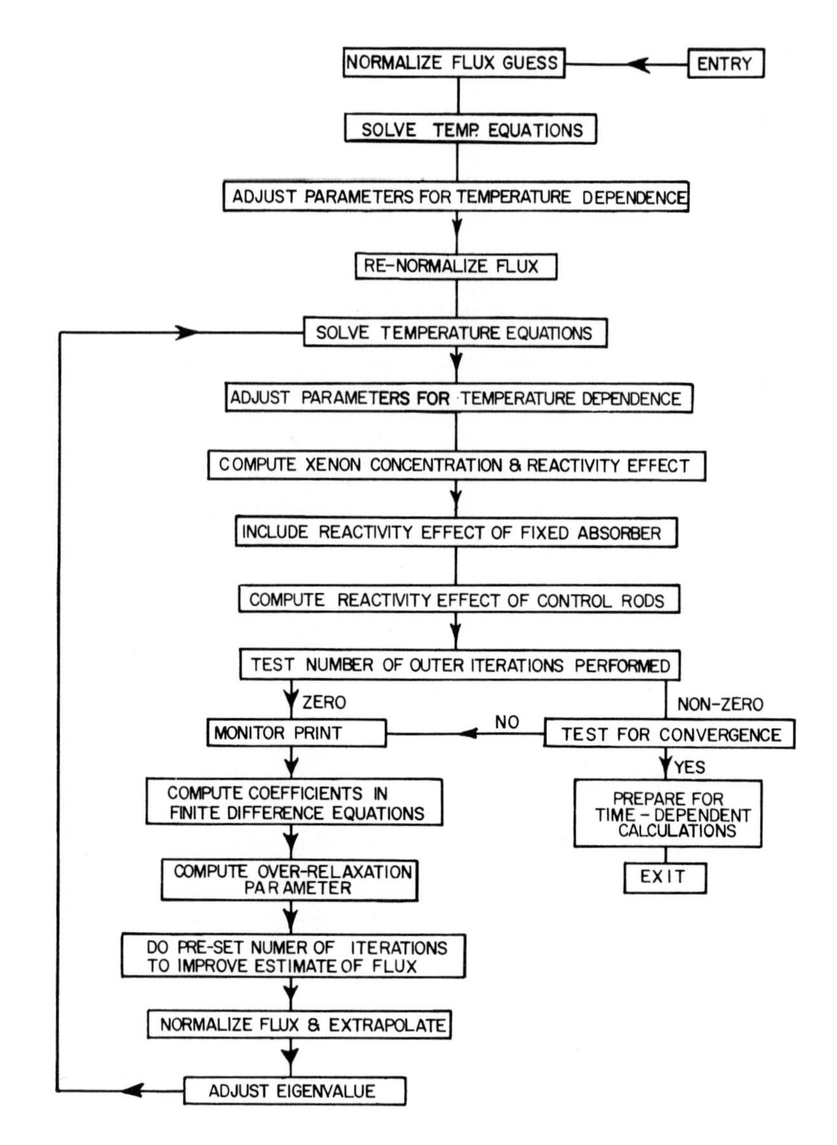

FIG. 5-7. Calculation flow chart for the STAB computer program: steady-state calculation.

power distribution accompanying the change in power level; it would be a particularly severe problem in large reactors with the control rods moved together in a bank. The point model is, however, felt to be adequate for analysis of the control rod transients when one is investigating the effect of varying different parameters.

5.4 Dynamic Behavior of Particular Solid Moderator, Power Reactors.

5.4.1 The Calder Hall Type

A considerable amount of analytical and experimental data have been reported on the dynamic behavior and safety characteristics of the Calder Hall-type reactors. Early results, on the basis of point model calculations, illustrate the general behavior of these reactors under various assumed perturbations in reactivity, coolant temperature and flow conditions. Experiments with the Calder reactor have also demonstrated the dynamic behavior of the reactor. Recent calculations using space-dependent kinetics have been reported showing the excellent agreement that can now be achieved between analysis and experiment.

Figure 5-9 illustrates the results of some point-model calculations showing the power and fuel-element-can temperature responses following a step insertion of 5×10^{-4} in reactivity, assuming no corrective control rod action is taken [23]. Results are shown for two different values of the moderator coefficient, viz., $\alpha_m = -4 \times 10^{-5}/°C$ ($-2.2 \times 10^{-5}/°F$) corresponding to the beginning-of-life condition, and $\alpha_m = +13 \times 10^{-5}/°C$ ($+7.2 \times 10^{-5}/°F$) representing an end-of-life condition. Following the initial prompt increase in power the power increases slowly until the negative fuel

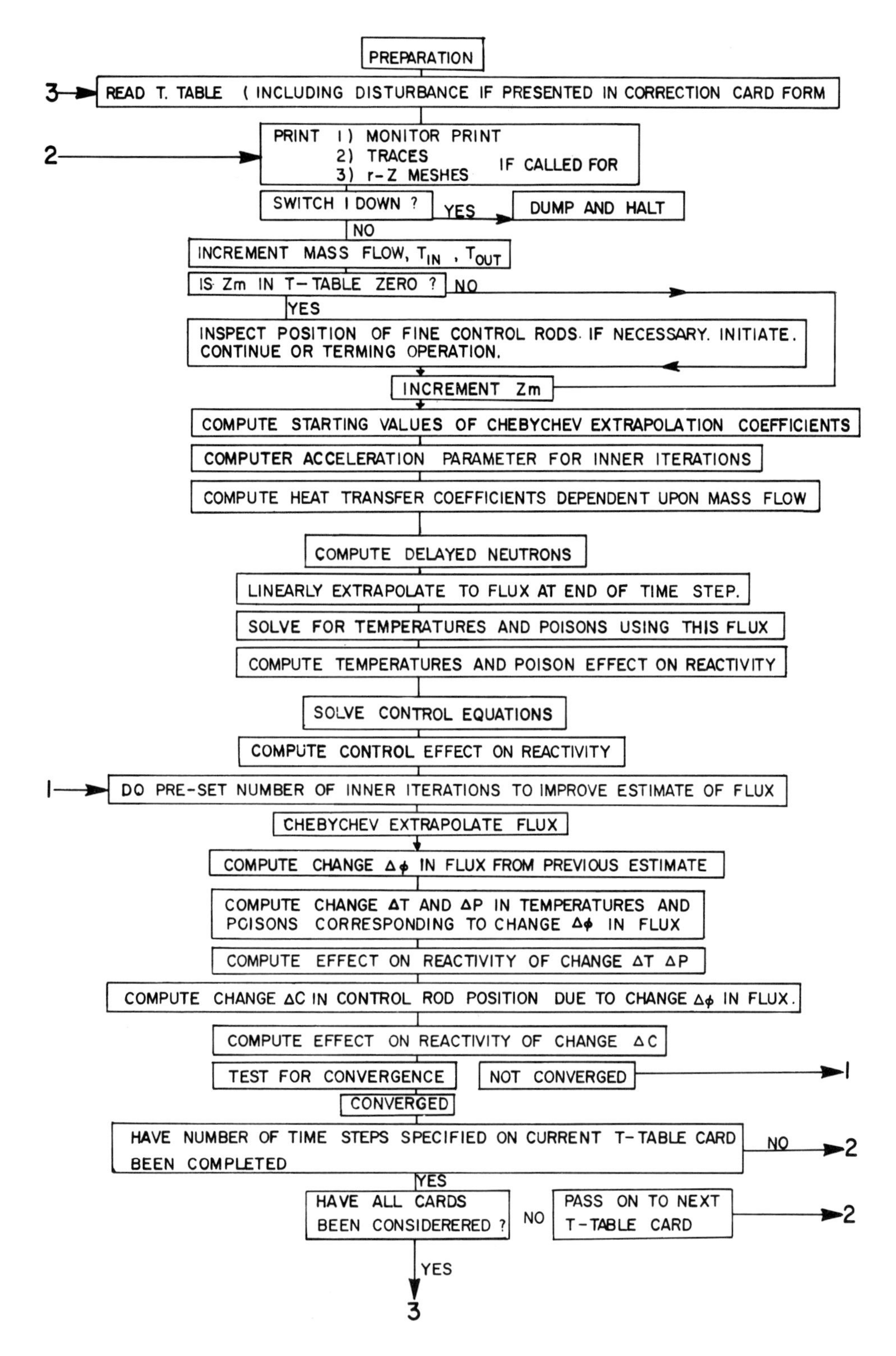

FIG. 5-8. Calculation flow chart for the STAB computer program: kinetics calculation.

coefficient, $\alpha_u = -2 \times 10^{-5}/°C$ ($-1.1 \times 10^{-5}/°F$) compensates the reactivity insertion. After approximately a minute following the step insertion, the temperature change in the moderator becomes significant and the power increases or decreases depending on the sign and magnitude of the mod-

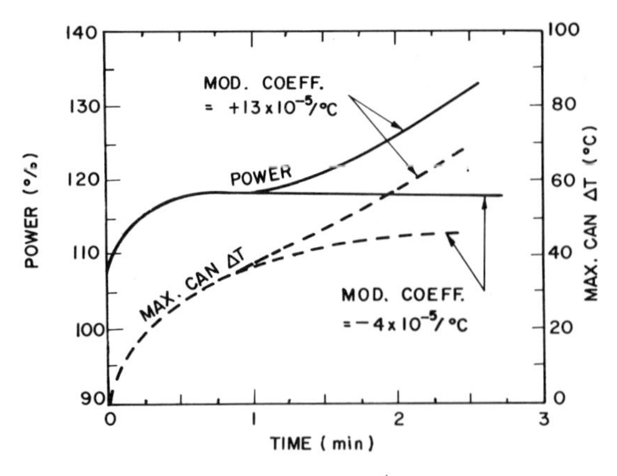

FIG. 5-9. Step insertion $\Delta\rho = 5 \times 10^{-4}$ in Calder Hall Reactor for two moderator temperature coefficients.

erator coefficient. The important point to note is the long delay time associated with the moderator coefficient, and therefore the comfortably long time available for corrective control action.

The temperature limitations in this type of reactor are the melting point of the Magnox can (640°C or 1184°F) and the first metallurgical phase change in the natural uranium metal (660°C or 1220°F). Under normal operating conditions the maximum can surface temperature is about 420°C (or 788°F) and the maximum uranium temperature about 120°C (216°F) higher. In some of the more recent reactors, e.g., the Berkeley reactor, the maximum surface and uranium temperatures have been increased to as high as 473°C (884°F) and 596°C (1104°F), respectively [65]. Even for the higher temperature conditions it appears that this particular accident would not impose serious time limitations on the need for corrective control rod action.

Figure 5-10 illustrates the reactor response to a step reactivity insertion (in this case $\Delta\rho = 1.2 \times 10^{-4}$) for cases where the bulk of the moderator is exposed directly to the coolant gas (no sleeve) and where it is insulated from the coolant by a graphite sleeve [64]. The calculations assume a moderator coefficient of $+15 \times 10^{-5}/°C$ $(+8.3 \times 10^{-5}/°F)$. After a 2-minute interval the power and fuel can temperature increases behave quite differently for the two cases.

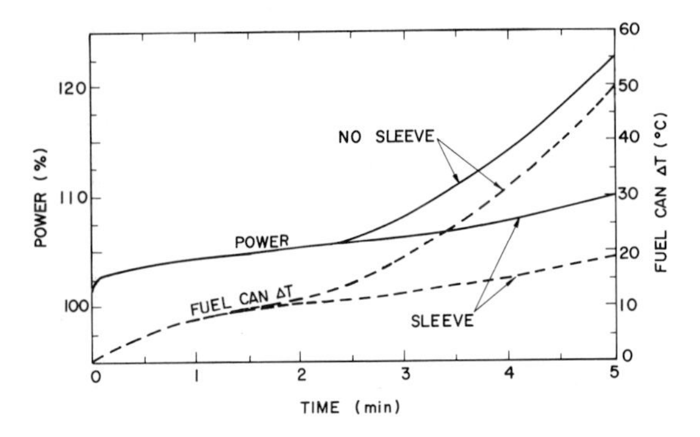

FIG. 5-10. Step insertion $\Delta\rho = 1.2 \times 10^{-4}$ in Calder Hall type reactor with and without moderator sleeve.

Experimental measurements, which have been reported [41] for the Calder Hall reactor, illustrate the transient response of the reactor to changes in reactivity, inlet temperatures and coolant mass flow, as shown in Figs. 5-11, 12 and 13 respectively. The experiments were performed after the fuel burnup had reached a level of 436 Mwd/tonne, and the moderator coefficient had a measured value of $+4.7 \times 10^{-5}/°C$ $(+2.6 \times 10^{-5}/°F)$. The fuel temperature coefficient was determined to be $-2.4 \times 10^{-5}/°C$ $(-1.3 \times 10^{-5}/°F)$ so that the reactor had a net positive over-all temperature coefficient. The initial reactor power level was, in each case, 154 Mw(t).

Figure 5-11 shows the power response to a reactivity change of 1.3×10^{-4} applied during a time interval of approximately one minute. The rate of reactivity change $\Delta\rho/\Delta t$ during this interval was, therefore, about 2×10^{-6}/sec, which corresponds to the maximum withdrawal rate of the control rod bank. The increase in power during the first minute is only 6 Mw(t), or about 4% in power. Following the initial power rise due to the changing reactivity, the power increases slowly as a result of the delayed response of the positive moderator temperature coefficient. It is quite apparent that the control system is very conservatively designed in terms of the control rod withdrawal accident at operating power.

The responses to an increase in inlet temperature and to a reduction in coolant mass flow are shown in Figs. 5-12 and 5-13. The initial effect of the coolant temperature increase is to increase the fuel temperature and decrease the power level. However, this effect is not apparent in the figure because of the slow rate of change in the temperature and the effect of the positive moderator coefficient which dominates the reactor behavior over a time interval of many minutes. The reduction in coolant mass flow leads to a decrease in power level as long as the moderator coefficient is not too large numerically (less than about 2-1/2 times the negative fuel temperature coefficient).

A more recent report [51] compares the results of space-dependent calculations utilizing the STAB

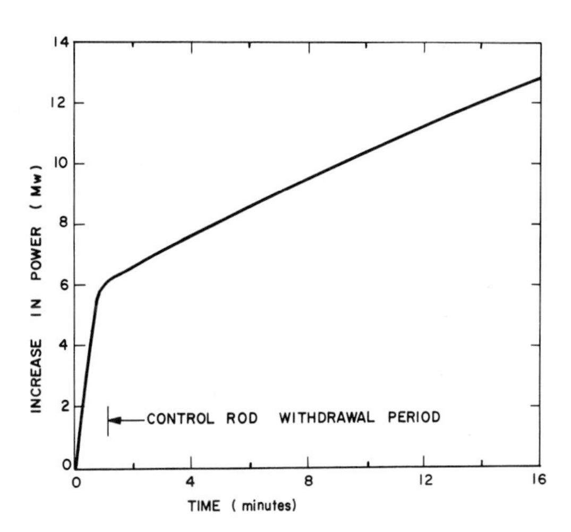

FIG. 5-11. Rod withdrawal transient in Calder Hall Reactor.

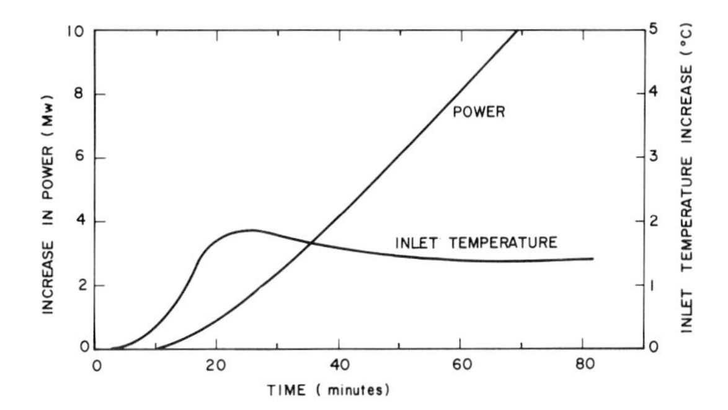

FIG. 5-12. Inlet temperature transient in Calder Hall Reactor.

computer code with experiments on the Calder Hall reactor in which various perturbations including changes in control rod position, inlet temperature and blower speed were made. The experiments and calculations were also done for different fuel exposure times to observe the changes brought about by changes in the moderator temperature coefficient. The general results appear to be similar to those previously reported for point-model studies, but allow the calculation of temperature behavior at various locations throughout the reactor. The excellent results are illustrated by the data shown in Figure 5-14 [51]. This particular figure shows the results of a control rod insertion which reduced the reactivity ($\Delta\rho$) by 1×10^{-4}. The lines indicate the calculated fuel can temperature at a particular point in the reactor, while the crosses indicate measured values.

From a safety point of view, both the calculations and experiments have shown that the control rod withdrawal accident at operating power tends to be a relatively insignificant accident in the Calder reactor. The startup accident could, however, be more severe and has been given considerable attention. Figure 5-15 shows the results of some calculations of reactor response to continuous control rod withdrawal during reactor startup, for two different control withdrawal rates

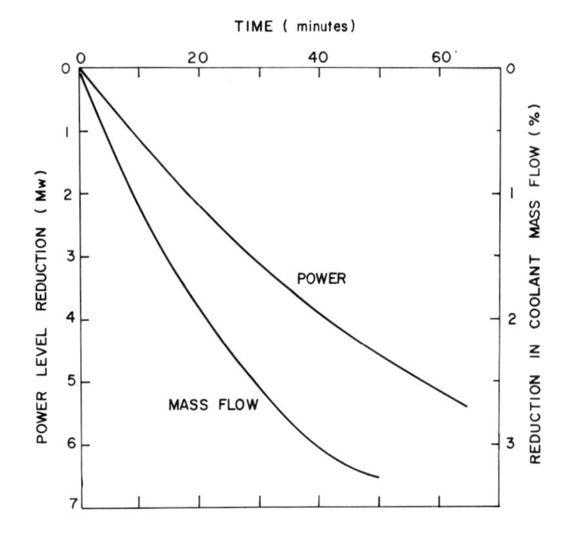

FIG. 5-13. Flow reduction transient in Calder Hall Reactor.

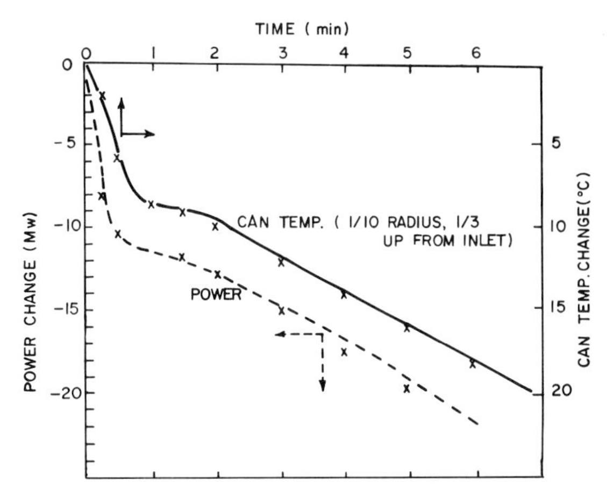

FIG. 5-14. Changes in power and fuel temperature following control rod insertion in Calder Hall Reactor. Comparison of experimental results with a STAB computer code calculation for fuel exposure of 931 Mwd/Tonne.

[62]. An initial power level of 1 watt was assumed for the two cases illustrated, and the coolant flow was assumed to be 100% of the normal operating value. For a withdrawal rate of 2×10^{-6}/sec, corresponding to the reference design, it is seen that the power rise and temperature increase rates are very modest, even for continuous withdrawal. The excess reactivity when the power becomes significant is 0.00156 $\Delta\rho$ leading to a reactor period of about 30 sec and the rate of temperature rise is 2.4°C/min (4.3°F/min). For a withdrawal rate of 1×10^{-5}/sec, the power rise and temperature increase rates are considerably greater, although it is emphasized that the time scale is minutes, thereby giving the impression of a very steep rate curve. In the latter case, the excess reactivity when the power becomes significant is 0.0030 $\Delta\rho$, which corresponds to a reactor period of about 9 sec for U^{235} fuel. The rate of increase in fuel temperature is 24°C/min (43°F/min) in this case.

In order to reduce delays during reactor startup, the rate of control rod withdrawal has been re-

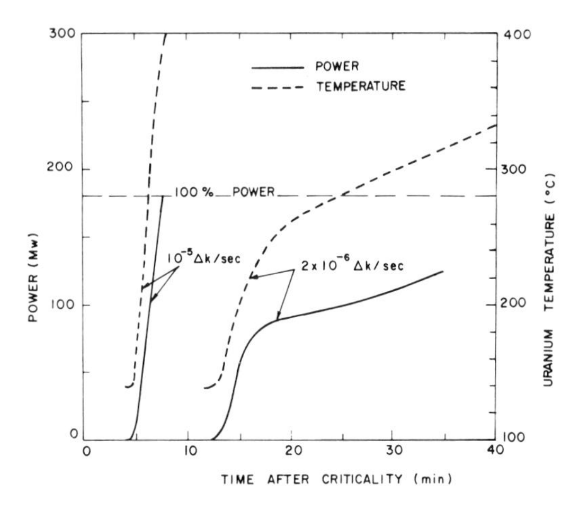

FIG 5-15. Startup transients in Calder Hall Reactor.

examined [52], particularly in view of the operating experience that is now available on the Calder reactors. In addition to multiple flux-monitoring systems, the reactor is protected during startup with fuel element temperature monitors. Hence, the rod withdrawal accident has been examined for different withdrawal rates, assuming that either the flux-monitoring system or the fuel element temperature monitors have failed. Under these conditions the maximum fuel can temperatures for three different withdrawal rates are as shown in Fig. 5-16. In the event of a flux-initiated scram, the fuel can temperature rise is less than 10°C (18°F). Even for the scram induced by the fuel temperatures, the temperature excursion is acceptable in all three cases. Because of other considerations, the control rod speed has only been increased by a factor of three.

A failure of all the coolant circulators, which might result from a power failure, and a large primary coolant system rupture leading to a depressurization accident both tend to be serious accidents in gas-cooled reactors. Figure 5-17 shows the calculated maximum can temperature following the failure of all blowers and a reactor scram [63]. The depressurization accident is reported [62] to be no more severe initially than the loss-of-flow accident in the Calder reactors provided the orifice size of the rupture is no larger than 6 ft (~ 2 m) in diameter. It is, of course, important that the reactor should be scrammed immediately following either of these accident conditions.

5.4.2 EGCR and AGR Reactors

The EGCR and the AGR reactors represent a second generation in gas-cooled reactors. The design differs from the Calder types essentially in that the fuel used is partially-enriched uranium

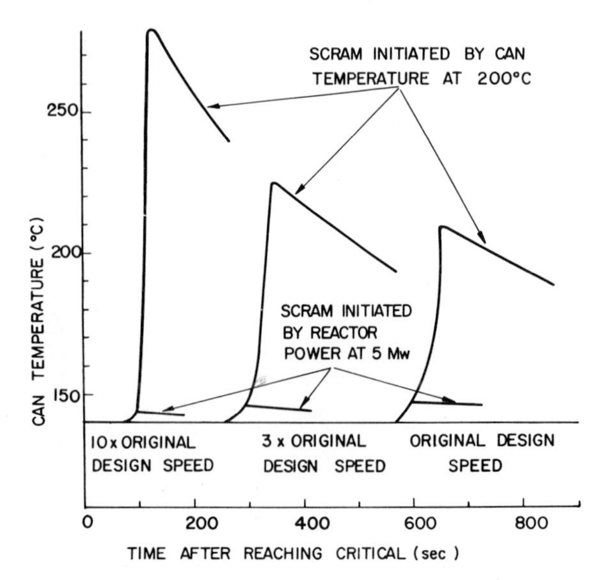

FIG. 5-16. Startup transients in Calder Hall Reactor with various control actions.

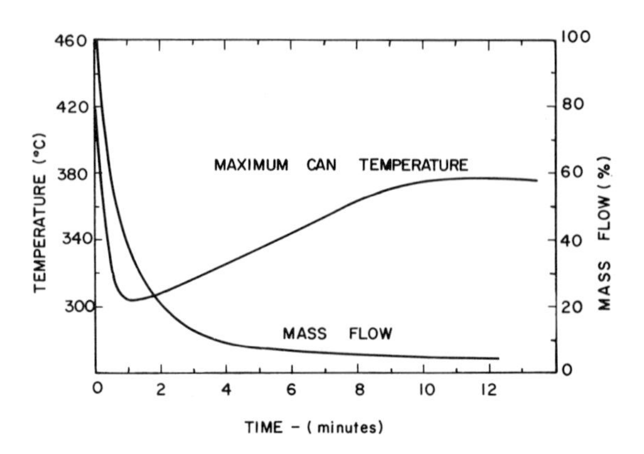

FIG. 5-17. Loss-of-flow transient in Calder Hall Reactor.

oxide in stainless-steel cladding. The EGCR uses a cluster of 7 rods (outer diameter 0.75 in. or 1.91 cm) in a 3 in. (7.62 cm) diameter coolant channel on an 8 in. (20.3 cm) square pitch, while the AGR cluster has 21 rods (outer diameter 0.4 in. or 1.02 cm in a 4 in. (10.2 cm) diameter channel and a 10.75 in. (27.3 cm) triangular pitch. The ultimate temperature limits on the cladding are thus extended from about 650°C (1200°F) to about 1400°C (2550°F) and the fuel melting point is around 2500°C (4530°F). The increase in the margin between operating temperatures and failure temperatures largely compensates the higher density power rating during a power transient.

The kinetic behavior of the AGR-EGCR reactors is similar to the Calder systems except that the use of partially enriched fuel (2.46% in EGCR, 2.5% in AGR) reduces the importance of positive contribution of plutonium to the moderator temperature coefficient. The time response of the moderator coefficient is delayed in both the EGCR and AGR by a separate graphite sleeve around the fuel elements similar to the later Calder type reactors. In the AGR there are actually two sleeves and the time lag is further increased by using a re-entrant coolant flow design. The CO_2 coolant at inlet temperature flows first between the outer sleeve and the moderator blocks and then back through the fuel element channel. An additional objective of this design was the decoupling of the moderator temperature from power changes thereby reducing the problem of controlling flux oscillations. Table 5-1 summarizes the calculated temperature coefficient components and associated time delays for the AGR at an exposure of 10,000 Mwd/tonne.

Some of the EGCR characteristics [37] are reviewed here as an illustration of the safety and dynamic characteristics of this type of reactor. Since the conversion ratio is not as high as that in the Calder Hall type of reactor, the excess reactivity required for fuel depletion is larger and consequently more control poison will be required, particularly at the beginning of the life for a fuel charge. The reactor power distribution, fuel element temperature distribution and control rod sensitivity as a function of position, all depend on the particular control rod program pattern used. The reactor multiplication and power distributions for several control rod programs have been

TABLE 5-1

AGR Temperature Coefficients at 10,000 Mwd/tonne

Material	Contribution to temperature coefficient ($\Delta k/\Delta T$) (x 10^{-5})	Time delay at full power (seconds)
Uranium	-1.58/°C (-0.89/°F)	18
Oxygen (in UO_2)	+0.05/°C (+0.03/°F)	
Steel cans	+0.02/°C (+0.01/°F)	21
Inner sleeve	+0.08/°C (+0.045/°F)	44
Outer sleeve	+0.16/°C (+0.09/°F)	330
Bulk graphite moderator	+1.39/°C (+0.77/°F)	540

calculated. It appears that for operating conditions where there will be a large excess reactivity available, the best power distribution will be achievable with some of the central control rods completely inserted, and the outside control rods operated as a bank.

The control rod sensitivity as a function of insertion has been reported for the case where all 21 control rods are moved as a bank. The sensitivity curve is shown in Fig. 5-18 [37] and illustrates the typical skewing which normally results from bank motion. The maximum sensitivity for the bank motion occurs with the bank inserted more than 80%, a condition which obviously could not occur during normal operations. On the basis of the maximum sensitivity for the bank motion and a maximum bank withdrawal rate of 2 in./min (5.08 cm/min), the maximum reactivity addition rate ($\Delta\rho/\Delta t$) would be 1.9 x 10^{-4}/sec. The maximum rate of reactivity addition that would normally be possible, would be smaller than this value, both because the control rods would not be inserted this far during operations, and some of the control rods would remain completely inserted, at the beginning of life, which would make the sensitivity curve less skewed for the remaining rods.

Transient studies of step reactivity changes and control rod withdrawal accidents have been done using an analog computer. The results of a few cases are summarized in Table 5-2. A step reactivity change of about 10^{-3} $\Delta\rho$ could possibly occur as the result of improper handling of fuel during refueling operations. The effect of a step reactivity change of this magnitude has been examined both at low power and at full power. The accident is self-limiting in the source and intermediate power ranges. At full power, the accident would be terminated by a scram initiated from the flux minus flow system, which activates when the power level exceeds by 10% the level appropriate for the particular flow conditions.

The normal maximum fuel element cladding temperature is calculated to be 1545°F (841°C) with an allowance of 75°F (42°C) for local asymmetries in power. The margin between the normal maximum cladding temperature and 1800°F (982°C), where failure could occur, is therefore 255°F (142°C). In the step accident postulated above, the reactor power level increases from 85Mw to a maximum of 107 Mw during the transient and the temperature increase for the hottest fuel element cladding is only about 20°F (11°C). Although results do not appear to be available for the case of no scram, the average fuel temperature would be expected to increase about 90°F (50°C) with a fuel temperature coefficient of -2.1 x 10^{-5} ($\Delta\rho$/°C), or -1.2 x 10^{-5} ($\Delta\rho$/°F) as reported. Under these conditions the peak cladding temperature rise would be considerably less than the 255°F (142°C) margin available before fuel element damage, even without an immediate scram.

The consequences of the control rod withdrawal at full power have been examined for the case where 10 control rods are withdrawn continuously (one more than the maximum possible in the power range). The reactivity ramp chosen for the evaluation was based on the maximum control rod sensitivity corrected for the fraction of rods being moved. The accident was again terminated by a flux-minus-flow scram with a maximum power of 98 Mw and a fuel cladding temperature rise of about 20°F (11°C). The time from initation of the ramp to the initiation of the scram was 8 sec and the scram delay time was 2 sec.

The control rod withdrawal from the startup source level has also been examined assuming that the period trip is inoperative. For a source level of 0.425 watts and a reactivity ramp corresponding to the withdrawal of 13 control rods, a level scram at 3 Mw would allow the reactor power to reach a maximum of 54 Mw. The maximum cladding temperature would rise from its initial level of 450°F (232°C) to a maximum of 470°F (243°C). The minimum period reached during the transient would be 0.57 sec.

Accidents involving the loss of coolant flow and possible depressurization resulting from a breach in the primary coolant system have received a great deal of attention in the analysis of this class of reactors. Results of such studies for the EGCR were discussed in Sec. 4.6 above.

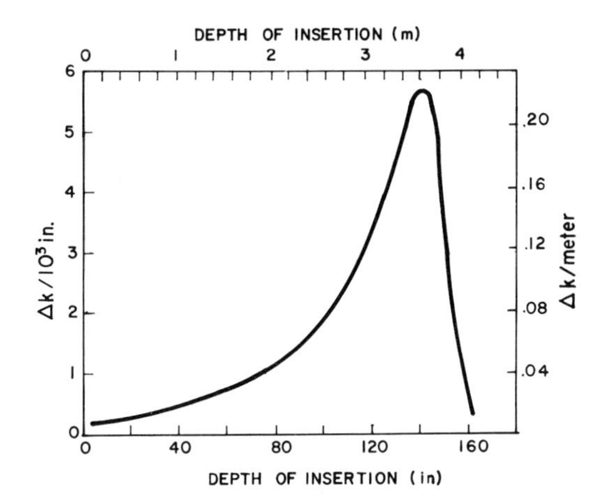

FIG. 5-18. Control rod sensitivity for a bank of twenty-one rods in the Experimental Gas-Cooled Reactor (EGCR). Δk per inch of withdrawal vs depth of insertion.

5.4.3 Sodium-Graphite Reactors (Hallam, SRE)

A third major category of solid-moderator re-

TABLE 5-2

EGCR Reactivity Transients

Reactivity accident	Initial power	Maximum power	Increase in maximum cladding temperature	Time to scram (sec)	Minimum period (sec)
10^{-3} ($\Delta\rho$) step	85 Mw	107 Mw	20°F (11°C)		
Rod withdrawal at power	85 Mw	98 Mw	20°F (11°C)	8	
Rod withdrawal at startup	0.425 w	54 Mw	20°F (11°C)	1096	0.57

actors is the sodium-graphite system. The general kinetic behavior in these system is somewhat similar to the Calder and EGCR types since they are also heterogeneous low-enrichment metal or ceramic-fueled systems with prompt negative Doppler coefficients and delayed moderator coefficients which can be positive under some conditions during core life. The principal point of difference between this reactor and those previously discussed is the liquid metal coolant which has a very high thermal conductivity and can have a significant reactivity contribution. As a result of its high thermal conductivity, the liquid metal coolant temperature tends to follow the fuel temperature quite rapidly. Consequently, the coolant temperature coefficient component can be relatively prompt and is usually positive for reactors where the leakage is small. Furthermore, because of its high conductivity, the coolant can induce rapid changes in temperature in other metal components resulting from fast temperature changes in the coolant. Consequently, it is important to give some attention to matching the coolant flow to the power generation during fast power changes to prevent thermal shock to other reactor components. A rapid decrease in flow rate may be desirable following a scram, and procedures for braking the coolant flow during a scram have therefore been considered under some conditions for sodium-cooled reactors [53].

The characteristics of the sodium-graphite reactor are typified by the Hallam Nuclear Power Facility (HNPF). The calculated temperature coefficient components and their time responses are summarized in Table 5-3 [55]. It is noted that the response time for temperature changes in the moderator is again relatively long, as is characteristic of most heterogeneous, solid-moderator reactors. These temperature coefficient parameters have been used in analog computer studies of the kinetic behavior of the Hallam reactor under various assumed accident condtions.

The protective system setpoints have been established with the following objectives:

to prevent sodium boiling and fuel melting;

to prevent operation at high temperature for long periods of time by limiting the maximum steady-state coolant temperature to 1050°F (567°C);

to limit the peak U-Mo fuel temperature during transients to 200°F (110°C) above the maximum design temperature of 1250°F (677°C); and

to avoid the possibility of a eutectic formation occurring between U-Mo and the stainless-steel cladding by restricting the maximum cladding temperature to 1300°F (704°C) during transients.

Scram signals can be initiated by the following operation abnormalities:

short neutron period,

high computed outlet temperature,

high measured outlet temperature, and

high rate of change of measured outlet temperature.

Other scram signals can be initiated by other plant abnormalities. The second scram signal indicated above is based on simultaneous measurements of the neutron flux level, the sodium flow rate and the sodium temperature measured at the intermediate heat exchanger outlet. It therefore detects simultaneously, abnormal conditions in the power level, sodium flow and sodium inlet temperature. Furthermore, it anticipates the outlet coolant temperature that might ultimately result as the consequence of increasing power, decreasing flow or increasing inlet temperature. Because of the time interval for sodium to reach the core from the heat exchanger, and the heat sink provided by the core, increases in the computed outlet temperature will tend to precede increases in the measured outlet temperature due to any of the three abnormalities previously mentioned.

The transient response to a step reactivity increase ($\Delta\rho$) of 0.8 x 10^{-3} (reactivity equivalent to about 0.5 in. of travel of all 19 control rods at

TABLE 5-3

HNPF Temperature Coefficients

Component	Reactor condition	Temperature coefficient $\Delta\rho/\Delta T$ (x 10^{-5})	Time delay at full power (seconds)
Fuel	BOL*	-2.8/°C (-1.56/°F)	2.5
	EOL**	-3.0/°C (-1.67/°F)	2.5
Coolant	BOL	+1.0/°C (+0.56/°F)	1.5
	EOL	(same)	1.5
Moderator	BOL	+2.4/°C (+1.33/°F)	175
	EOL	+5.2/°C (+2.89/°F)	550

*BOL: Beginning of life.
**EOL: End of life.

their peak worth) without a reactor scram was examined to indicate the general stability of the reactor. The studies indicate a power increase to about 120% of the design base rating within 10 sec following the step insertion. At this point the power reaches a peak temporarily due to the self-limiting prompt temperature coefficient. Following this initial transient, the power increases very slowly due to the positive moderator coefficient. After 2 min, for example, the power would rise to about 135% of the initial design base rating in the absence of a scram. Even in the event of a scram failure, the operator would have several minutes to take corrective action following this accident without endangering the reactor components.

Transients resulting from a postulated control rod withdrawal at power, a startup accident and a loss-of-forced-coolant-flow accident have also been analyzed. In each case these accidents were terminated by reactor scrams. The analog computer results of the control rod withdrawal accident at high power are shown in Fig 5-19 [55]. The control rod withdrawal rate was assumed to be such that a ramp reactivity ($\Delta\rho$/sec) of 3 x 10^{-4} resulted, and the alarm signal and control setback signals were assumed to be ignored or inoperative. The reactor scram occurred as the result of the computed outlet temperature exceeding the trip level of 1029°F (554°C). Because of the thermal inertia of the core, the actual coolant outlet temperature would reach a maximum of only 962°F (517°C) during the transient. The minumum period during the course of the accident is 19 sec, which is not serious. Even if the scram signal from the computed outlet temperature should fail to operate, the reactor would be scrammed on either a rate of change of measured outlet temperature or a measured fuel outlet temperature of 1025°F (551°C) before sodium boiling or fuel meltdown could occur.

Some results of startup accident calculations are shown in Fig. 5-20. Again a continuous reactivity insertion of 3 x 10^{-4} ($\Delta\rho$/sec) was assumed and

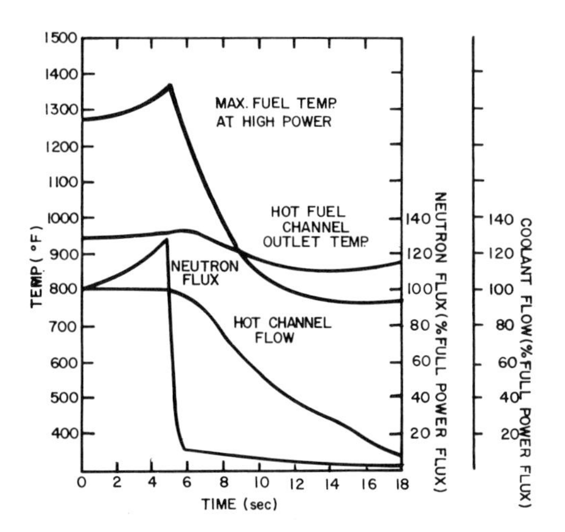

FIG. 5-19. Control rod withdrawal transient from full power in the Hallam Sodium-Graphite Reactor 3 x 10^{-4} Δk/sec ramp.

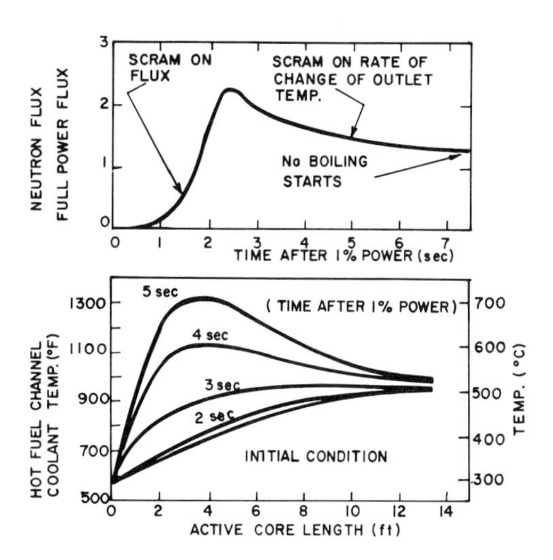

FIG. 5-20. Control rod withdrawal transient from source level in the Hallam Sodium Graphite Reactor. 1% fission power, 2% decay heat power, 3 x 10^{-4} Δk/sec ramp.

the sodium flow rate was assumed to be 2% of full flow for this study. If it is assumed that the control rod setback and scram action resulting from the period signal should both fail, then the transient would be limited by a scram signal from the outlet temperature computer at a flux level of 20% of full power. Under these conditions the flux level would reach a peak that is 58% of full power flux. Because of the relatively fast power rise the integrated power would be rather small and the maximum coolant temperature change would be only 25°F (14°C). Furthermore, the fuel temperature would not reach the normal operating temperature. In the event the outlet temperature computer scram signal should also fail, the reactor would scram on the rate of change of measured outlet temperature about 5 sec after reaching the 1% power level. Under these conditions the peak coolant temperature would be about 1320°F (716°C) and the peak fuel temperature about 1650°F (899°C). A power peak is reached at about 2.5 sec following the 1% power level, after which the fuel Doppler coefficient would reduce the reactivity and power for an interval of several seconds. The hot fuel channel coolant temperature profiles at various times are shown in the figure.

Startup accidents were also studied for slower reactivity ramps, viz., 1 x 10^{-4} ($\Delta\rho$/sec) and 3 x 10^{-5} ($\Delta\rho$/sec.) The energy releases for the slower reactivity rates were somewhat larger than that of the faster rate when all the transients were terminated by a scram at 20% power. It is important to note that in accidents terminated by a scram at a specified power level, the worst accident may not be associated with the largest reactivity rate, since the maximum temperatures are roughly proportional to the integral of the power to the point of scram. A smaller reactivity rate causing a slower rise in power may then allow more heating of the fuel before the scram occurs.

Figure 5-21 [55] indicates the reactor response to a loss of forced coolant flow. The reactor would be scrammed either due to a signal from the relative sodium flow rates in the primary and secondary

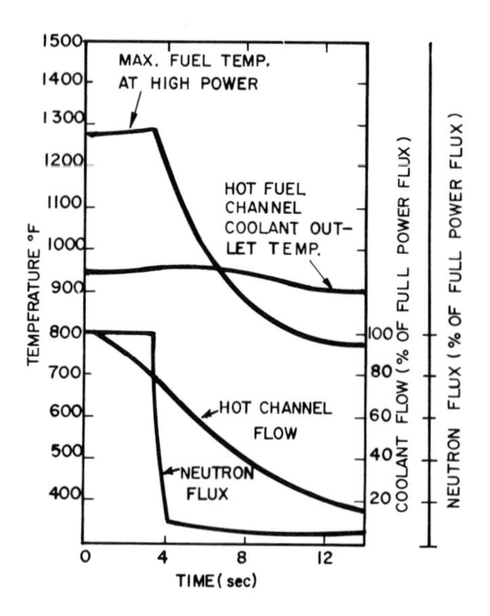

FIG. 5-21. Loss of primary coolant flow accident from full power in Hallam Sodium-Graphite Reactor.

sodium circuits or the computed outlet temperatures, or by signals from the measured outlet temperature (level or rate of change of level). The figure indicates the results following a scram at a computed outlet temperature setpoint of 1029°F (554°C). The temperature transients are well within limiting values.

In the previous section of this chapter it was pointed out that the margin between normal outlet sodium temperatures and the boiling temperature of sodium is usually so large in sodium-graphite reactors that sustained operation at a power greater than 300% normal power would be required to induce boiling in the fuel channels. Although the assumption of such an accidental condition appears to be very unrealistic, some thought has been given to the course of accidents under these conditions [55]. In general, boiling in the hottest channels would tend to increase reactivity because of the positive reactivity coefficient of the sodium. This in turn would cause boiling in a larger fraction of the core and fuel meltdown in the highest temperature regions. The melted fuel metal and cladding would either be swept from the core by the sodium or fall down from the core into the lower plenum if the sodium flow is completely choked in the hot channels. The removal of the melted fuel from the core would ultimately shut the reactor down. However, the rate of propagation of the sodium boiling might be an important factor in determining the severity of the consequences. An extremely rapid propagation of boiling would be undesirable due to the danger of melting the entire core in a less serious accident where some local boiling might occur at a hot spot. On the other hand a very slow propagation at the onset of boiling would allow the accident to persist for some time, thereby resulting in the generation of a large volume of sodium vapor before a significant fraction of the reactor could melt and be rearranged so that it would become subcritical. A rate of propagation between these two extremes would, of course, be most desirable.

The analysis of the boiling sodium problem is just as complicated as boiling water reactor kinetics and much less work appears to have been done on it. Some preliminary studies [54] using an analog computer representation of first three and then seven fuel channels in the SRE investigated the question of whether the choking-off of one channel by boiling would rapidly propagate to other channels as a consequence of the power rise produced by the positive reactivity feedback from the reduced sodium concentration. If this were the case, then meltdown of the entire core would be quite rapid and the amount of sodium vapor produced would be small enough to be condensed in the upper plenum without causing a large pressure increase. The analog study concluded that boiling propagation seems to proceed in a manner so that the shutdown following such a large accident would actually be enhanced by the propagation.

Sodium boiling has also been given some attention in the hazards analysis of the HNPF [55]. It is noted that for boiling at low flow rates the initial production of small amounts of vapor reduces the coolant density and tends to increase flow rate since the vapor-liquid mixture is more buoyant. The somewhat increased flow tends to delay the meltdown of fuel. At higher flow rates (> 30% in this case), frictional effects due to the higher flow velocity required to maintain the flow rate with reduced coolant density tend to predominate and reduce the mass flow rate and choke off the channel. A rod withdrawal accident at full power has been examined for the case of no scram to illustrate the course of a meltdown accident. Again it is emphasized that these assumptions are very unrealistic in view of the multiple scram signals available for manual corrective action. Nevertheless, the example is interesting in tracing the ultimate behavior of the reactor under these very severe conditions.

A control rod withdrawal rate corresponding to a ramp reactivity increase of 3×10^{-4} ($\Delta\rho$/sec) would lead to nucleate boiling in some of the channels after about one minute at which time the power would be 300% of normal. A further increase in power would result in choking of the channels and subsequent meltdown of the fuel in these channels.

5.4.4 Homogeneous or Semihomogeneous, Graphite-Moderated, Gas-Cooled Reactors (Peach Bottom, Dragon, AVR)

Homogeneous or semihomogeneous graphite-moderated, gas-cooled reactors have kinetic characteristics very different from any of the heterogeneous reactors previously discussed. These differences come about primarily because of differences in temperature coefficients and their time responses. Principal examples of the semihomogeneous reactors are the Peach Bottom HTGR, the British Dragon reactor, and the German AVR pebble-bed reactor, all presently under construction. Since these reactors contain only ceramic materials in the core, the problem of fuel cladding meltdown is removed. Furthermore, the margin between operating temperature and melting tem- of the carbide fuel particles is usually quite large, i.e., typically around 1000°C (1800°F), and even for temperatures above the melting point of the carbides it is doubtful that any serious damage would result in a short transient. Cracking of fuel element components could possibly occur due to the large thermal stresses, which could accelerate the release of fission products, but such a release would be slow relative to the release from a canned fuel element in which the fission product gases are kept under pressure. Probably the dominant limiting factor with this type of reactor would be the outlet gas temperature increase as the result of a power excursion, and the effect of this increased gas temperature on other plant components. Because of the thermal inertia of the core components and primary system piping, etc., fast transients do not tend to be sources of difficulty. Instead, sustained operation at a high power and a high outlet gas temperature is a more important factor.

The sort of kinetic behavior to be expected in semihomogeneous gas-cooled reactors can be illustrated by some transient calculations performed for the 40 Mw(e) Peach Bottom HTGR. The core of this reactor contains 804 fuel elements 3.5 in. (8.9 cm) in diameter. An individual element consists of a graphite spine 1.75 in. (4.45 cm) outer diameter, on which are stacked annular fuel compacts 2.75 in. (6.99 cm) outer diameter which contain uranium and thorium carbide particles in a graphite matrix. A 3.5 in. (8.89 cm) outer diameter impermeable graphite sleeve surrounds the compacts. Most fission products are retained in the individual fuel particles which are made of thorium and uranium carbides and coated with pyrolytic graphite. Volatile products which escape the particles diffuse to the outside of the compact and are swept away by a helium purge gas stream flowing between the compact and the sleeve, and carried to an external trapping system. The fuel elements are arranged on a close hexagonal pitch and cooled by helium.

The kinetics of the HTGR have been studied with the BLOOST digital computer code described above. Since the radial dimensions of the fuel element components are of the same order of magnitude as the scattering mean-free-path, it was expected that the average graphite temperature at any time during a transient would determine the thermal neturon spectrum. Multi-thermal-group cell calculations showed this assumption to be true. Figure 5-22 shows several radial temperature profiles in the fuel element following a very large step reactivity insertion which represents the most nonuniform radial temperature distribution conceivable. The spatial variation of the spectrum within a fuel element even for this case was not significant. The radial heat generation profile in the fuel compact is assumed uniform in the code which is not a bad assumption for the nearly homogeneous fuel element.

In addition to the variation of temperature within the fuel element components, there are, of course, important gross temperature variations across the reactor core, both in the radial and axial directions. In general, temperature coefficients were deter-

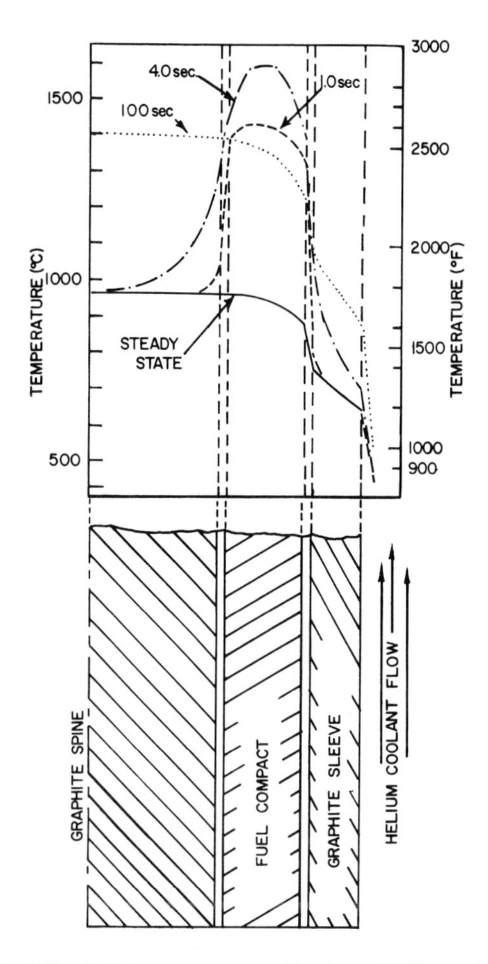

FIG. 5-22. Radial temperature profiles in an HTGR type fuel element at four times following a 0.01 Δk step insertion.

mined by diffusion theory calculations of the multiplication constant at successive temperature points, assuming the core temperature to be uniform and equal to the average temperature at each point. Since the temperature distribution deviates significantly from a uniform distribution, particularly in the axial direction, some attention was given to the adequacy of this assumption.

A two-dimensional power distribution was obtained from the output of a two-dimensional diffusion-burnup code (DDB) for the end-of-life time step and used in a heat transfer code which produced temperature maps for the entire core. By obtaining such maps for several core power levels covering the range of most transients and dividing the core maps into four axial zones, it was possible to associate the average fuel and moderator temperature in each of the axial zones with the corresponding over-all average fuel and moderator temperatures. Spectra and cross-section libraries were prepared for the four axial zones using the radially homogenized end-of-life atom densities from the DDB burn-up code. The whole procedure was quite tedious and could probably only be justified to check the accuracy of simpler models. The resultant temperature coefficient obtained was not very different from the coefficient based on uniform core temperatures. The value of the over-all temperature coefficient, $1/k(\partial k/\partial T)$, was from 8 to 12% more negative, although the contribution from the resonance escape probability, $1/p(\partial p/\partial T)$ was from zero to 17% less negative over the range from 1000 to 1700°K (1341 to 2601°F).

Because of the over-all negative temperature coefficient and the relatively large margin between the operating temperature and the temperature where fuel element damage could occur, the effect of relatively large step increases in reactivity would not be serious, even in the event of no scram action. For a step insertion of $2.5 \times 10^{-3}\ \Delta\rho$ at full power, for example, the power would reach a maximum of about 160% full power and level off at about 116% full power, if a scram should not occur. The temperature rise in the hottest fuel element would be about 133°C (239°F) and the outlet temperature of the coolant would increase by 78°C (141°F). No serious consequences would result.

The results of a control rod withdrawal accident at full power have also been examined [29]. The control rods are interlocked so that only one control rod can be withdrawn at a time. The maximum worth of a control rod at the end of the core life is calculated to be $0.009\ \Delta\rho$ and the withdrawal time is 116 sec. Hence, the average rate of reactivity increase is approximately $8 \times 10^{-5}\ \Delta\rho/\text{sec}$. Actually, the BLOOST program allows for the reactivity insertion rate to vary with time, so that the transient studies on the Peach Bottom reactor were usually done by using the calculated reactivity vs. time for the rod withdrawal. The power and temperature responses to this reactivity insertion are shown in Fig. 5-23 [29]. The accident would normally be terminated by a power scram at 140% full power. Further scrams could be initiated by the gas outlet temperature or by the operator inserting 19 emergency rods. Results are shown for cases where the scram occurs either as the result of excessive power or high outlet temperature. Finally, the calculation was extended to cover the case of no scram action at all. The calculations include the effects of decay heat after shutdown and the reduction of helium blower speed which occurs after a scram.

Limiting temperature conditions in this reactor include the eutectic temperature of the fuel carbides and their coatings (2450 ±50°C or 4530 ±90°F) and a rise in fuel compact temperature of 850°C (1530°F) without any compensating change in the sleeve temperature (which could possibly crack the sleeve due to differential expansion). It is noted that none of these limits are reached even in the case with no scram action assumed.

In computing the maximum temperature in the hottest fuel element it is necessary to note that, since the HTGR is operated with symmetric groups of control rods, the withdrawal of one rod in a postulated accident leaves the core with an unsymmetric configuration which produces a radial flux tilt. For the example given, the initial configuration was three rods 120° apart which causes a radial peak-to-average ratio of 1.19. With only two rods left the peak-to-average is increased to 1.37. The maximum fuel temperature is estimated on the assumption that the tilted distribution holds throughout the transient.

The reactivity feedback contributions from the fuel and moderator components of the temperature coefficients are illustrated in Fig. 5-24 for the no-scram case. It will be noted that in the semihomogeneous fuel elements the time delay associated with the moderator coefficient is much less than that characteristic of heterogeneous designs. In the case of the HTGR fuel element, about 30% of the moderator is in the fuel compact and contributes as a prompt coefficient along with the Doppler effect. Although the calculational model used in the BLOOST code assumes the Doppler coefficient to be associated with temperature changes in the entire fuel compact matrix, it is recognized that it is, in reality, determined by the temperature changes in only the fuel particles. The particles

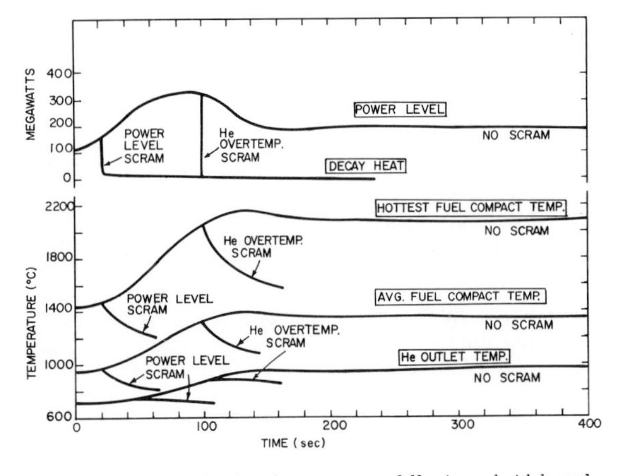

FIG. 5-23. Power levels and temperatures following rod withdrawal transient at full power in the Peach Bottom HTGR. End-of-life core conditions.

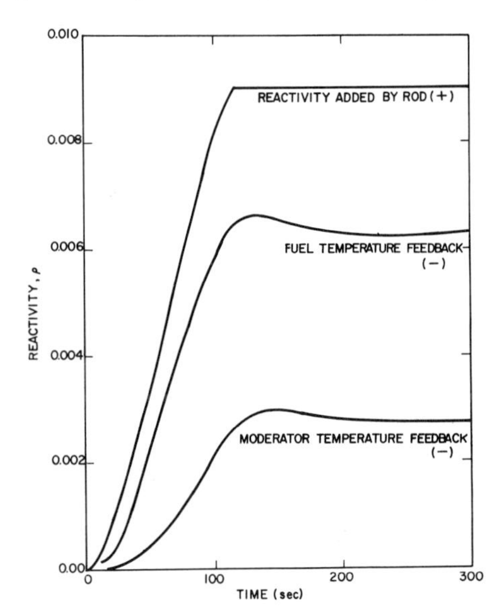

FIG. 5-24. Reactivity feedback from fuel and moderator temperature coefficients for the rod withdrawal transient of Fig. 5-23.

represent a much smaller heat capacity, but can have a significant time constant for heat transfer to the matrix depending on particle size and thermal properties. Thus, if a very rapid reactivity insertion were possible in an HTGR type reactor, one would expect three components of reactivity feedback: from the particles, the compact, and the spine and sleeve respectively. To ignore the particle time constant is always conservative in accident caclulations since doing so delays the negative effect of the Doppler coefficient.

The startup accident also does not appear to be serious in the Peach Bottom HTGR. The fuel elements in the core have a very large heat capacity, viz., 30-40 Mw-sec/°C (17-22 Mw-sec/°F), so that the temperature excursion for a particular energy release is relatively small. This has the advantage that considerable time is available for scram action before excessive temperatures could be reached, but has also the attendant disadvantage that a large amount of energy is stored in the core during a transient, particularly if the period and power scrams should both fail.

The control rods are programmed and interlocked so that a maximum of three rods can be withdrawn simultaneously when the reactor is below 10% full power. The maximum reactivity worth of the control rods in any group is about 0.03 $\Delta\rho$ which is appreciably smaller than the reactivity defect associated with the temperature swing from shutdown temperature to operating temperatures. This reactivity defect is approximately 0.07 $\Delta\rho$. Furthermore, the temperature coefficient is even more strongly negative at the lower temperatures.

The startup accident calculations show that, if the reactor becomes critical at the beginning of a group withdrawal and if the period and power level scrams should fail to function, a continuous group withdrawal would lead initially to a short reactor period which would be limited in about 30 sec when the power level reached the point at which the temperature coefficient feedback is effective. Following this initial transient, the power and temperature would gradually increase as the control withdrawal continued. The final fuel temperature would be well below normal operating temperature and at no time during the transient is the rate of temperature change great enough to cause cracking of the ceramic fuel elements.

6 OPERATIONAL PROBLEMS RELATED TO SAFETY

As was mentioned in Sec. 4, there are several sources of accidents apart from uncontrolled reactivity additions which may result in high temperatures and associated core damage. Among these, the phenomena of (1) spatial flux oscillations due to Xe^{135} and temperature effects and (2) release of stored energy (Wigner release) are of particular interest in some kinds of solid-moderator reactors and are briefly discussed in this section.

6.1 Spatial Flux Oscillations

The general nature of spatial flux oscillations arising from varying Xe^{135} concentrations and or temperature distributions has been discussed in Sec. 4 of the Water Reactor Kinetics chapter. As shown in Fig. 4-1 of that chapter, the occurrence of sustained spatial oscillations is associated with cores having squared dimension (R^2 or H^2) which exceed the migration area M^2 by about a factor of 1000. A rough rule-of-thumb given by Randall and St. John [56] is therefore that if the square of the core dimensions in feet exceeds the migration area in cm^2, then sustained spatial oscillations are possible*. Whether they will occur in a particular case and whether they will be sustained or divergent depends also on the magnitude of the temperature coefficient, the operating flux level and distribution, and the reactivity worth of Xe^{135} in the thermal spectrum.

In a typical graphite core of the Calder Hall type the migration area is about 880 cm^2 which implies that oscillations would become a problem for core dimensions of the order of 30 ft (9.14 m). The first Calder Hall reactor is 31 ft (9.45 m) in diameter by 20 ft 10 in. (6.35 m) high and has been found to have no xenon oscillations [56]. The larger cores of the Calder type have dimensions of the order of 40 ft (12.19 m) diameter by 26 ft (7.925 m) high and are definitely subject to oscillations. The threshold flux level for xenon oscillation is lowered as the moderator temperature coefficient becomes increasingly positive with increasing fuel burnup.

In the Peach Bottom HTGR the graphite core is more heavily loaded and the migration area is about 450 cm^2, which would imply core dimensions of at least 21 ft (6.40 m) before oscillations could be significant. Although the Peach Bottom core is too small for oscillations to occur (9 ft or 2.74 m diameter by 7.5 ft or 2.29 m high), the problem could arise in larger HTGR cores. The threshold flux for oscillations is increased by the over-all negative temperature coefficients typical of HTGR reactors and by the lower reactivity worth of Xe in the harder thermal spectrum.

Most of the other graphite reactors discussed as examples in the preceding section (EGCR, AGR, SGR) are also small prototypes of their respective types and hence too small for oscillation to be a problem.

The periods associated with spatial Xe oscillations are so long (10-30 hr) that they can be easily observed and corrected by control systems or by operator action. Suitable in-core instrumentation and control systems that can independently regulate separate regions of the core are required. With respect to safety, the spatial oscillations in flux are of concern if they result in excessive fuel temperatures, particularly in reactors with metallic fuel elements-which operate near a limiting temperature. Since the oscillations are so slow, it can be argued that no safety problem is involved once the possible modes of oscillation are identified and provided for in the control system. It should be noted that the requirements of such a very

* In metric units this rule of thumb would be: If the square of the core dimensions in meters is more than about 10% of the migration area in cm^2, then sustained spatial oscillations are possible.

flexible control system may in some situations conflict with other safety objectives such as keeping at a minimum the number of control rods movable at one time; it may also increase the cost of the control system. For these reasons, it will be desired in future power reactors to restrict the flexibility of the control system to the minimum requirements for controlling only those modes of oscillation which can actually occur. This objective again emphasizes the importance of an accurate calculation of the temperature coefficient which strongly influences the thresholds for the various modes of oscillation.

6.2 Stored Energy in Graphite

The phenomena of energy storage in irradiated graphite is well known and represents a safety problem in reactors operating with graphite temperature below about 300°C (~ 570°F). A good review of the subject which includes the operational experience obtained in existing reactors is given in Chapters 12 and 17 of "Nuclear Graphite" [57].

For every fast neutron moderated in graphite, about 20,000 atoms are displaced from their lattice positions. If the graphite temperature is subsequently raised above the irradiation temperature, some of the displaced atoms return to their original positions releasing their stored energy. This causes an apparent reduction in the specific heat of the graphite; and if the energy release per unit temperature rise exceeds the actual specific heat, a spontaneous release of stored energy can occur raising the temperature to a value which may be considerably above the starting temperature. Such an occurrence could lead to oxidation of the graphite as in the case of the Windscale fire. In a reactor with a positive moderator temperature coefficient, it would represent a source of rapid positive reactivity insertion, the magnitude of which would depend on what fraction of the core volume reached the temperature at which the stored energy release was initiated and on the previous operating history (integrated flux and operating temperature) in those regions.

In general, the amount of stored energy increases with integrated dose and decreases with higher irradiation temperatures. This behavior is illustrated in Fig. 6-1. It is seen that for irradiation temperatures below 300°C (572°F) the total energy stored saturates at a rather low value. Graphite reactors built to operate below this temperature have been mainly air-cooled research and plutonium production reactors (e.g. X-10, Brookhaven, BEPO, G-1). These have required periodic annealing to release the stored energy in a controlled manner. (The Windscale fire occurred during such an annealing operation in which inadequate temperature monitoring was provided.) In the power reactors of the Calder Hall type, there is significant energy storage in the cooler (~135°C or 275°F) inlet end of the core. It was determined that this will not cause a problem for several years if at all. The three later reactors at Chapel Cross employ graphite sleeves in the core for the purpose of raising the bulk graphite temperatures. The other

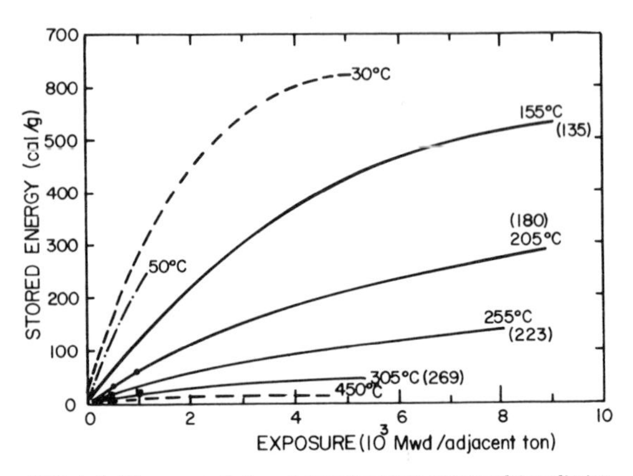

FIG. 6-1. The accumulation of stored energy at several irradiation temperatures. Equivalent Calder irradiation temperatures shown in parentheses for each curve. Curves drawn in solid lines are taken from references [58] and [59] assuming 1 Mwd/adjacent ton (Hanford) = 0.6 Mwd/adjacent ton (Calder). The broken line curves (450°C and 30°C) are from reference [60]. The 50°C curve is from reference [61]. The points indicated by squares are for 159°C and by circles for 200°C: these are taken from reference [61].

graphite-moderated reactors discussed in the chapter (EGCR, AGR, HTGR, SGR) all operate with the moderator temperature sufficiently high that no large quantity of stored energy is expected to accumulate. New core and reflector designs employing graphite should of course always be evaluated with respect to this problem.

REFERENCES

1. K. R. Van Howe, private communication, General Atomic Data.
2. F. Todt, "Fever, a One-Dimensional Few Group Depletion Program for Reactor Analysis," Report GA-2749, General Atomic, Division of General Dynamics Corp., 1962.
3. B. Cutts et al., "Graphite Moderated Thermal Reactor Calculations for the Calder Hall Reactors," Proceedings of Second U.N. International Conference on Peaceful Uses of Atomic Energy, Geneva, 1958, Vol. 12, p. 612.
4. H. G. Davey et al., "Operating Experience at Calder Hall," Proceedings of Second U. N. International Conference on Peaceful Uses of Atomic Energy, Geneva, 1958, Vol. 8, p. 10.
5. S. Jaye, private communication, General Atomic Data.
6. R. K. Lane and A. J. Goodjohn, "Plutonium-Fueled Beryllium-Oxide-Moderated Reactors," Report GA-4599, General Atomic, Division of General Dynamics Corp., 1963.
7. K. R. Van Howe, "Treatment of Leakage and Spatial Effects in Zero-Dimensional Burnup Calculations," Trans. Am. Nucl. Soc., 6,2(1963)279.
8. S. Jaye, "Some Aspects of the Physics of Thorium Fueled High Temperature Gas-Cooled Reactors," Report GA-3704 General Atomic, Division of General Dynamics Corp., 1963.
9. B. Carlson et al., "The DSN and TDC Neutron Transport Codes," USAEC Report LAMS-2346, Los Alamos Scientific Laboratory, 1959.
10. J. R. Stehn (Ed.), Naval Reactors Physics Handbook, Vol. III, Chap. 19, "The Physics of Intermediate Spectrum Reactors," USAEC Government Printing Office, 1958.
11. R. E. Heineman et al., "Experience in the Use of the Physical Constants Testing Reactor (F,S)," Proceedings of the Second U. N. International Conference on Peaceful Uses of Atomic Energy, Geneva, 1958, Vol. 12, p. 650.
12. G. D. Joanou and J. S. Dudek, "GAM-I: A Consistent P_1 Multigroup Code for the Calculation of Fast Neutron Spectra and Multigroup Constants," Report GA-1850, General Atomic, Division of General Dynamics Corp., 1961.
13. F. T. Adler, G. W. Hinman, and L. W. Nordheim, "Effective Resonance Integrals: A Systematic Evaluation for All Temperatures and Geometries of Practical Interest," Proceedings of the Second U.N. International Conference on Peaceful Uses of Atomic Energy, Geneva, 1958, Vol. 16, p. 155.
14. L. W. Nordheim, "A Program of Research and Calculations of Resonance Absorption," Report GA-2527, General Atomic, Division of General Dynamics Corp., 1961.

15. L. W. Nordheim, "A new calculation of resonance integrals," Nucl. Sci. Eng. 12(1962)457; and "Resonance Absorption," (Paper delivered at the American Physical Society Symposium, Houston, Feb. 28, 1963), Report GA-3973, General Atomic, Division of General Dynamics, Corp., 1963.
16. R. K. Lane, L. W. Nordheim, and J. B. Sampson, "Resonance absorption in materials with grain structure," Nucl. Sci. Eng., 14(1962)390.
17. N. F. Wikner and S. Jaye, "Energy Dependent and Spectrum Averaged Thermal Cross Section for the Heavy Elements and Fission Products for Various Temperatures and C:U-235 Atom Ratios," Report GA-2113, General Atomic, Division of General Dynamics Corp., 1961.
18. E. P. Wigner and J. E. Wilkins, Jr., "Effect of the Temperature of the Moderator on the Velocity Distribution of Neutrons with Numerical Calculations for Hydrogen as Moderator," USAEC Report, AECD-2275, Sept. 14, 1944.
19. J. E. Wilkins, Jr., "Effect of the Temperature of the Moderator on the Velocity Distribution of Neutrons for a Heavy Moderator," USAEC Report CP-2481, November 30, 1944.
20. D. E. Parks, J. R. Beyster and N. F. Wikner, "Thermal neutron spectra in graphite," Nucl. Sci. Eng., 13(1962)306.
21. D. E. Parks, "The Calculations of Thermal Neutron Scattering Kernels in Graphite," Report GA-2438, General Atomic, Division of General Dynamics Corp., 1961.
22. H. Hurwitz, Jr., "Approximate analysis of reactor startup incidents," Nucl. Sci. Eng., 6,1(1959)11.
23. R. J. Smith, "The Overall Control of Nuclear Power Stations of the Gas-Cooled Thermal Reactor Type," Proceedings of the Second U. N. International Conference on Peaceful Uses of Atomic Energy, Geneva, 1958, Vol. 11, p. 360.
24. S. Lenihan, "GAZE-2, A One-Dimensional Multigroup Neutron Diffusion Code for the IBM-7090," Report GA-3152, General Atomic, Division of General Dynamics Corp., 1962.
25. J. P. Dorsey, "GAMBLE, A Program for the Solution of the Multigroup Neutron Diffusion Equations in Two Dimensions with Arbitrary Group Scattering, for the IBM-7090 FORTRAN-II System," Report GA-4246, General Atomic, Division of General Dynamics Corp., 1963.
26. D. C. Leslie, "The Spectrox Method for Thermal Spectra in Lattice Cells," British Report AEEW-M-211, Reactor Group, Atomic Energy Establishment, Winfrith, Dorset, England, August 1961.
27. C. F. Griggs and H. M. Sumner, "Notes on the Sensitivity of Calculated Moderator Coefficients to the Model and Cross-Sections Used," British Report AEEW-M-221, Reactor Group, Atomic Energy Establishment, Winfrith, Dorset, England, August 1961.
28. P. U. Fischer and N. F. Wikner, "An Interim Report on the Temperature Coefficient of the 40 Mw(e) HTGR," Report GA-2307, General Atomic, Division of General Dynamics Corp., 1961.
29. Peach Bottom HTGR Final Hazards Summary Report, (to be published).
30. "Preliminary Safeguards Report for the Experimental Beryllium Oxide Reactor (EBOR)," Report GA-2052, General Atomic, Divisions of General Dynamics Corp., 1962.
31. A. J. Goodjohn and A. D. McWhirter, "Temperature Coefficient Calculations for the Experimental Beryllium Oxide Reactor," Report GA-4130, General Atomic, Division of General Dynamics Corp., 1963.
32. A. J. Goodjohn and A. D. McWhirter, "An Approximate Method for Calculating the Doppler Coefficient of Fissionable Materials in Thermal and Intermediate Reactors," Report GA-4016, General Dynamics Corp., 1963.
33. R. Scalettar and G. B. West, "Calculations of the Temperature Coefficient and Kinetic Behavior of TRIGA," Report GA-4474, General Atomic, Division of General Dynamics Corp., (to be published).
34. A. McReynolds et al., "Neutron Thermalization by Chemically Bound Hydrogen and Carbon," Proceedings of the Second U. N. International Conference on Peaceful Uses of Atomic Energy, Geneva, 1958, Vol. 16, p. 297.
35. M. S. Nelkin, "Scattering of slow neutrons by water," Phys. Rev., 119,2(1960)741.
36. M. Nelkin and G. B. West, "Calculations of the Prompt Temperature Coefficient for TRIGA," Report GA-1681, General Atomic, Division of General Dynamics Corp, 1960.
37. "EGCR Final Hazards Summary Report, Volume I-Description and Hazards Evaluation," USAEC Report ORO-586, Oak Ridge Operations Office, 1962.
38. R. W. Woodruff et al., "Detecting and Eliminating Fuel Rod Bowing in the SRE," Atomics International, Report NAA-SR-7705, North American Aviation, Inc., 1963.
39. "Final Hazards Summary Report for the ML-1 Nuclear Power Plant," Report IDO-28560, Phillips Petroleum Co., 1962.
40. R. W. Keaten, "Reactivity Worth of Sodium in Sodium-Cooled Reactors," Report NAA-SR-5904, North American Aviation, Inc., 1961.
41. G. Brown, et al., "Safety Aspects of the Calder Hall Reactor in Theory and Experiments," Proceedings of the Second U.N. International Conference on Peaceful Uses of Atomic Energy, Geneva, 1958. Vol. 11, p. 202.
42. "Additional Safeguards Information for the Hallam Nuclear Power Facility," Report NAA-SR-5700, Suppl. 4, North American Aviation, Inc., 1961.
43. J. R. Brown, D. C. Pound, and J. B. Sampson, "Hazards Summary Report for the HTGR Critical Facility", Report GA-1210, General Atomic, Division of General Dynamics, 1960. See Vol. I Addendum August 1961.
44. "Gas-Cooled Reactor Program Quarterly Progress Report," USAEC Report ORNL-3302, Oak Ridge National Laboratory,
45. F. L. Fillmore, "Analysis of the SRE Excursion of July 13,1959," Report NAA-SR-5898, North American Aviation, Inc., 1961.
46. A. Ghatak and M. Nelkin, "Microscopic prompt-neutron kinetics of the TREAT reactor", Trans. Am. Nucl. Soc., 6,1 (1963) 110.
47. A. Ghatak, "Neutronic Analysis of TREAT", Report GA-3575, General Atomic, Division of General Dynamics Corp., 1962.
48. A. R. Curtis, J. G. Tyror, H. E. Wrigley, "STAB, A Kinetic Three-Dimensional One-Group Digital Computer Code," British Report AEEW-R-77, Winfrith, Dorset, England, 1961.
49. T. J. O'Neill, "Derivation of Reactor Heat Transfer Transient Equations" Proceedings of the Second U. N. International Conference on Peaceful Uses of Atomic Energy, Geneva, 1958, Vol. 2, p. 268.
50. J. H. Bowen and E. F. O. Masters, Nuclear Reactor Control and Instrumentation, Temple Press, London, 1959.
51. P. Jenkinson and D. Smith, "Experimental Investigations of Reactor Kinetics," USAEC Report NP-12440, DTI Extension, AEC, Oak Ridge, 1962.
52. V. R. Goodwin and T. N. Markham, "Increasing Plant Availability", Nucleonics, 20,4(1962)52.
53. J. T. Ream and R. P. Varnes, "Transient Thermal Behavior of Experimental UO_2 Fuel Elements in the Sodium Reactor Experiment (SRE)", Nucl. Sci. Eng., 13,(1962)325.
54. H. H. Cappel, "Multi-Channel Boiling Stability for Sodium Graphite Reactors", Report NAA-SR-6527, North American Aviation, Inc., 1962.
55. "Final Summary Safegurds Report for the Hallam Nuclear Power Facility", Report NAA-SR-5700, North American Aviation, Inc., 1961.
56. D. Randall and D. S. St. John, "Xenon Spatial Oscillations", Nucleonics, 16,3(1958)82.
57. R. F Nightingale (Ed.), Nuclear Graphite, Chapters 12 and 17 Academic Press, 1962.
58. G. B. Greenough, Windscale Laboratories, unpublished data, April 1960.
59. J. C. Bell et al., "Stored Energy in the Graphite of Power-Producting Reactors", Phil. Trans. Royal Soc. London, A254; 361:395 (1962).
60. R. E. Nightingale et al., "Damage to Graphite Irradiated up to 100°C", Proceeding of the Second U.N. International Conference on Peaceful Uses of Atomic Energy, Geneva, 1958, Vol. 7, 295.
61. J. M. Davidson, "Stored Energy in Irradiated Graphite", U.S.-U.K. Graphite Conference held at St. Giles Court, London, Dec. 16-18, 1957, USAEC Report TID-7565 (Pt. 1) p. 11, 1959.
62. E. Anderson and J. H. Bowen, "System Control and Protection", J. Brit. Nucl. Energy Conf. 2,2(1957)212.
63. J. M. Yellowless and P. R. J. French, "Kinetics and Control of the Berkeley Reactors", Proceedings of the Second U.N. International Conference on Peaceful Uses of Atomic Energy, Geneva 1958, Vol. 11, p. 391.
64. J. Miida et al., "Analysis of Dynamic Characteristics of British Type Gas Cooled Reactors", Japan Atomic Energy Research Institute Report JAERI-1006B, June 1959.
65. Directory of Nuclear Reactors, Vol. IV "Power Reactors", International Atomic Energy Agency, Vienna, 1962.

CHAPTER 10

Fast Reactor Kinetics

WALTER J. McCARTHY, Jr.
Power Reactor Development Company, Detroit, Michigan
DAVID OKRENT
Argonne National Laboratory, Argonne, Illinois

CHAPTER CONTENTS*

*Except for a few changes and additions made in proof, this chapter is based on information in the literature or known to the authors prior to December 1963.

1 REACTIVITY CONTROL

1.1 Types of Reactors

The term "fast reactor" is broad and covers a host of reactor types and sizes. Their common attribute is the absence of moderating material in such large quantity as to thermalize a significant fraction of the neutrons. Sodium coolant and oxygen or carbon, which may be present as the oxide or carbide form of the fuel, do affect the neutron energy spectrum, but their moderating power is too small to reduce the energies of any appreciable fraction of the neutrons to thermal values before capture or leakage occurs. In a very small high density fast reactor, inelastic scattering reduces the average fission neutron energy of about 2 Mev only to several hundred kilovolts before fission or leakage occurs. In a very large, ceramic-fueled fast reactor, on the other hand, a large fraction of the fission events occur at energies between 1 and 100 kev, and some neutrons are found in the few electron-volt energy range. In general, inelastic scattering is effective in reducing the energy of neutrons above 100 kev. The number of neutrons found much below this energy depends directly on the ratio of moderating power to total absorption.

All fast systems differ from thermal reactors in that cross sections for fission at these neutron energies are roughly 100 times smaller than at thermal energies. Generally speaking, fission and capture cross sections are slowly varying in this energy range. There are no huge capture resonances, hence no special reactivity effects associated with xenon and samarium. Nor is there any very strong absorber, such as cadmium in a thermal reactor, for use in control rods. Total cross sections are relatively small, neutron mean free paths are relatively long. To a first approximation the reactor can usually be considered homogeneous for purposes of reactor physics analysis. There are no very steep gradients in flux, even near control rods.

There being no thermal neutrons, there is no thermal group whose mean energy will change with changes in moderator temperature. Except for the Doppler effect, the neutron energy spectrum is not directly affected by changes in temperature if the densities remain essentially constant. Small but important changes in spectrum can occur, however, with gross changes in core composition, such as partial or complete loss or expulsion of the coolant (which we shall usually consider to be sodium) from the core. Changes in the relative probability of fission and capture with such a shift in spectrum can be accompanied by important reactivity effects.

Fast reactors are relatively small and, excepting the Doppler effect and the spectral shift on loss of sodium, most changes in reactivity with increase in temperature are associated with changes in leakage probability attributable to sodium and fuel expansion or to small motions of the fuel element. A reduction in the fuel density in the core leads to a loss of reactivity. A reduction in sodium density allows more neutrons to leak, but for large fast reactors this reactivity loss may be counterbalanced by a reactivity gain associated with the hardening of the neutron energy spectrum.

There is a correlation between the size of the reactor, its k_∞, and the mechanisms of control which are available. Very small reactors must be highly enriched, have a high k_∞ (in the vicinity of 2), and hence have a high leakage probability. Large changes in reactivity can then be accomplished by moving reasonable masses of reflecting material next to the core. This method of control has been used in Clementine the Experimental Breeder Reactor I (EBR-I) and in the Soviet BR-5 reactor. As core size increases, k_∞ and the neutron leakage probability decrease; increasingly impractical quantities of reflector material must be moved to achieve the necessary degree of reactor control. Then one ordinarily resorts to fuel movement or to the use of a strong neutron absorber like B^{10}.

The control required can vary considerably with reactor type, design, and manner of operation. Parameters like temperature and power coefficients, internal breeding ratio, maximum allowable burnup and frequency of partial unloading all play a major role.

Reviews of fast reactor physics have been given by Okrent, et al. [1] in 1955, by Codd, et al. [2] in 1956, and by Loewenstein and Okrent [3] in 1958. Several papers on varied topics in fast reactor physics were given at the Second U.N. International Conference on Peaceful Uses of Atomic Energy in 1958 [4, 5, 6, 7] and at the Conference on the Physics of Breeding in 1959 [8].

A major review of work on small fast critical assemblies is given in the work by Paxton in 1959 [9] and in the series of papers by members of Los Alamos Scientific Laboratory [10]. Fast reactor cross sections were examined in detail by Yiftah, et al. [11] in 1960. The most extensive collection of papers in the field of fast reactor physics is to be found in the Proceedings of the Seminar on the Physics of Fast and Intermediate Reactors [12], sponsored by the IAEA in Vienna in 1961.

1.2 Types of Control Needs

The Dounreay [13], EBR-II [14], and Enrico Fermi [15] fast reactors can be considered as a single class of reactor in that (a) they are all fueled with U^{235}, at least in their first loadings; (b) they are all moderate in size and are expected to have small negative temperature coefficients; (c) the allowable fuel burnup is modest so that with frequent partial reloading there will be only a small reactivity requirement for burnup. The first two of these reactors move fuel elements to achieve control, and the third moves boron poison.

The types of control which must be considered in the design of the reactor control system can be classified into the categories of (a) operational, (b) emergency shutdown or scram, and (c) shutdown available during fuel loading. Taking advantage of the relatively large delayed neutron fraction in a U^{235} fueled reactor, of the small temperature and power coefficients, and of the possibility of frequent partial reloading, the designers of the

Fermi reactor adopted the criterion that the excess reactivity available for operational control should be less than one dollar. The anticipated distribution of operational reactivity requirements for the Fermi reactor is given in Table 1-1. This reactor has been designed to have a total of ten control rods, eight of which are classified as safety rods to be used for emergency shutdown, while two are operating control rods. The safety rods are designed to have a total reactivity worth of 8$, or approximately 1$ per rod. The two shim or operational rods have the combined worth of slightly less than 1$. The shutdown existing during loading operations is equal to the worth of about 35 extra subassemblies added at the core periphery. No further means of scramming the reactor during reloading is available in this reactor. All auxiliary methods would have required operation from the bottom of the reactor and afforded severe engineering problems in this particular core design. However, additional protection against insertion of excess fuel during this period is provided by a quick reversal of the subassembly handling mechanism.

Both the EBR-II and the Dounreay reactors maintain some control rods cocked, available for scram, during reloading operations. The EBR-II reactor has a total of twelve combination safety-control rods, worth a total of about 4.5%($\Delta k/k$), [16] plus two additional rods which can be scrammed during reloading. While the excess reactivity available is limited, a rigid upper limit of less than one dollar for operational purposes has not been adopted for this reactor.

Of course, the use of plutonium with its sharply reduced delayed neutron fraction, the desirability (for economic reasons) of longer operating periods between reloading and the recent shift in design interest to reactors having large negative Doppler effects, all tend to make the achievement of a maximum excess reactivity of less than one dollar impractical as a general criterion.

These three moderate-sized reactors have been designed to have negative fuel expansion and sodium density coefficients of reactivity and to be stable. Safeguard analyses [17, 18, 19] show that for the kind of mal-operation and accidents requiring investigation, achievement of completely reliable movement of the safety rods at a reasonable speed is more important than much greater safety rod velocities with a corresponding decrease in dependability. Below prompt critical, the kinetics are controlled by delayed neutrons and the same sort of criteria as used in thermal reactors applies to the time needed for control rod operation. Above prompt critical, mechanical motion of the control rod cannot appreciably affect the course of an accident, at least in its first major energy burst. Hence, greater rates of scram provide only marginal improvement. Of course, future reactors having different reactivity coefficients, different time constants and subject to new kinds of possible accidents, will require re-analysis to see whether this criterion on control rod reliability is still valid.

TABLE 1-1

Operational Reactivity Requirements for Fermi Reactor

	Reactivity in dollars*
Temperature and power coefficients, loading to operating conditions	0.20
Poisoning by fission products	0.02
Net fuel burnup	0.33
Fuel element growth	0.07
Reactor operational control needs	0.30
	0.92

*One dollar, or the effective delayed neutron fraction, is 0.0066 for this reactor.

1.3 Control Requirements as a Function of Reactor Type and Design

1.3.1 Relatively Prompt-Acting Reactivity Effects

1.3.1.1 Doppler Coefficient.* When the temperature of the material of a nuclear reactor is changed, there is a change in the relative velocity distribution of neutrons and nuclei since the velocity distribution of nuclei is changed by the changed thermal motion. With increasing temperature, the change in relative velocity has the effect of widening a cross section resonance and lowering its peak value, keeping the average cross section unchanged. However, since the neutron flux dips at each resonance, the widening of a resonance brings into play the higher flux adjacent to the dip and increases the reaction rate; that is, it increases the effective cross section. The broadening of the fission resonances tends to increase the production of neutrons; broadening of the capture resonances tends to increase capture of neutrons. The resulting effect of the changed reaction rates in the reactivity of a reactor is called the Doppler temperature coefficient of reactivity.

As is known, the EBR-I reactor with the Mark I and Mark II loadings exhibited a prompt positive power coefficient of reactivity [20] which was thought possibly to arise from a positive Doppler coefficient associated with increased fuel temperature. In a small highly enriched fast reactor, the Doppler effect is expected to enhance both fission and capture resonances about equally, resulting in no net change in k_∞ but a rise in reactivity, the latter attributable to the increased probability of absorption relative to leakage.

Although later experiments with the Mark III loading demonstrated that the positive coefficient arose from bowing of the fuel rods, attempts were made to calculate the Doppler coefficient of reactivity, and experiments were undertaken to measure the effect [21,22,23,24,25]. The results of both calculation and experiment indicated that the Doppler coefficient was so small as to be negligible in the EBR-I.

With increasing core size, the fuel is diluted with fertile material, structural material, and coolant. The average neutron energy drops, leading to an increase in the magnitude of the Doppler reactivity effect. Nicholson, [26] Greebler [27], Fossoul [28],

*See also the chapters on The Reactor Core, and The Doppler Coefficient.

and Bhide and Hummel [29] all have calculated these effects for a variety of cores. Nicholson has shown, with approximations, that the contribution to the Doppler reactivity effect for each isotope by a group of neutrons of average energy E can be described by

$$T^{3/2}\frac{\partial k_E}{\partial T} = \frac{A^{1/2}}{4\nu\sqrt{2\pi K}}\frac{\rho^2\phi_E Q_E}{\langle\mu_t\rangle_E E^{1/2}},$$

Eq. 7.4 of reference [26]

where

k_E = portion of multiplication factor contributed by neutrons in group of average energy E,
T = temperature °K,
A = atomic weight,
ν = number of neutrons emitted per fission,
K = Boltzmann's constant,
ρ = atomic density, atoms/cm^3
$\langle\mu_t\rangle_E$ = average total cross section for group of neutrons of average energy of E,
E = average neutron energy of group,
ϕ_E = neutron flux in group of energy E, normalized to give a total fission rate of unity,

$$Q_E = \sum_s \frac{\pi\sqrt{2\pi}\langle\mu_t\rangle\Delta}{\rho\langle S\rangle_s}\left\langle\sigma_{or}\Gamma T\frac{\partial f_E}{\partial T}\right\rangle_s$$

Eq. 3.21 of reference [26]

where

Δ = the Doppler width = 4KT,
$\sigma_{o\gamma}$ = peak resonance cross section for capture,
Γ = resonance width,
S = spacing of resonances,
$\partial f_E/\partial T$ = derivative of self-shielding factor with respect to temperature.

The quantity Q_E has been calculated for U^{235} and U^{238} in reference [26] and for Pu^{239} in reference [28]. Q(E) is plotted as a function of energy in Fig. 1-1. It can be seen that Q(E) increases very rapidly at lower neutron energies. Thus, calculation of the correct energy distribution at the low energy end of the neutron spectrum is of great importance for calculation of the Doppler effect. For a large oxide reactor with 33% fuel, 17% steel, and 50% sodium with U^{238} to Pu^{239} atomic ratio about 7, Greebler [27] calculates about 80% of the fission power to be generated by neutrons above 9 kev, but only 10% of the Doppler coefficient to be derived from this range. About 75% of the Doppler coefficient is contributed by the energy region between 300 and 4000 ev. In this reactor plutonium was calculated to contribute a positive Doppler effect which was only about 40% the magnitude of the negative contribution made by the U^{238}*. The net effect was a coefficient in the

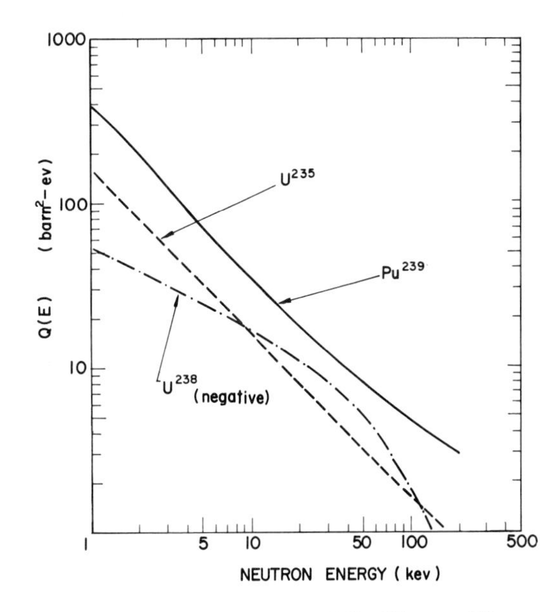

FIG. 1-1 The function Q(E) for U^{235}, U^{238}, and Pu^{239}.

vicinity of -8×10^{-6} Δk/k per°C at the average fuel temperature of 900°C (1650°F).

From Fig. 1-1 it can be seen that Q(E) for Pu^{239} is about twice as large as for U^{235}. The relative size of Q(E) for U^{235} and U^{238} changes as a function of energy. In the vicinity of 100 kev, Q(E) is calculated to be the same for the two uranium isotopes. Hence, one expects that a reactor with all of its neutron energy spectrum at 100 kev and above is likely to have an essentially zero Doppler coefficient if equal amounts of U^{235} and U^{238} are present in the core [23, 26]. For the Fermi reactor with a ratio of U^{238} to U^{235} of approximately 2.7, the coefficient is calculated [26] to be small and negative, namely -1.7×10^{-6} Δk/k per°C at 560°K. In a similar calculation [31] for a large Th^{232}-U^{233} fueled fast reactor, it was found that a minimum ratio of Th^{232} to U^{233} of 1.9 is required to insure a negative coefficient for the fuel mixture. For a 2000 liter reactor of this type using thorium-uranium carbide as fuel and a Th^{232} to U^{233} ratio of approximately 6, the Doppler temperature coefficient at operating temperature was calculated to be -2.6×10^{-6} Δk/k per°C [32].

A summary of some Doppler coefficient calculations made for large plutonium-uranium oxide or carbide fueled reactors is given in Table 1-2 [29]. These reactors all had roughly 52 vol.% sodium, 18 vol.% steel, and 30 vol.% fuel. Calculations were made with and without the sodium present. It was found that for metal fueled systems having the same fuel enrichment and volume fractions, the Doppler coefficient was about one-tenth that calculated for oxide systems. The temperature dependence of dk/dT for these large ceramic systems was found to go roughly as

*Recent theoretical studies indicate that the plutonium Doppler effect should be much smaller, [30 a,b,c,d] due to the interaction between resonances of U^{238} with those of the Pu^{239}. Also the measurement of a negative Doppler effect for the U^{235} in a U^{235}-Th fueled fast critical experiment has been reported [30e]. (See chapter on Doppler Coefficient).

TABLE 1-2

Doppler Reactivity Effects and Coefficients [a]

Computed for Large PuO_2-$U^{238}O_2$ and PuC-U^{238}C Fueled Fast Reactors

Case		300-2500°K		300-750°K		1500-2500°K	
		-%Δk	-Δk/ΔT (× 10^{-5}/°C)	-%Δk	-Δk/ΔT (× 10^{-5}/°C)	-%Δk	-Δk/ΔT (× 10^{-5}/°C)
Oxide							
Ratio Pu to U^{238}	Sodium present?						
1.9	Yes	2.63	1.15	1.09	2.40	0.59	0.59
1.9	No	1.44	0.65	0.63	1.40	0.29	0.29
1.7	Yes	1.81	0.79	0.77	1.73	0.40	0.40
1.7	No	0.92	0.40	0.39	0.87	0.18	0.18
Carbide							
Ratio Pu to U^{238}	Sodium present?						
1.9	Yes	1.95	0.89	0.82	1.82	0.40	0.40
1.9	No	1.11	0.48	0.47	1.05	0.20	0.20
1.7	Yes	1.30	0.59	0.54	1.20	0.29	0.29
1.7	No	0.63	0.29	0.27	0.60	0.12	0.12

[a] Coefficients should be multiplied by 5/9 to give values in $(°F)^{-1}$.

T^{-1} [29]. For the smaller, highly enriched fast reactors the decrease with temperature is expected to be swifter, namely as $T^{-3/2}$ [26]. An example of the way the Doppler effect varies with temperature is given in Fig. 1-2. Here, for each of four large oxide reactors having 30 or 50 vol.% sodium coolant and a core height of 2 or 3 ft, respectively, neutron multiplication factor has been plotted against the average fuel temperature [33]. The slope of these curves at any temperature gives the value of the Doppler coefficient.

It is of interest to note how seemingly unrelated changes in reactor design can produce major changes in the Doppler coefficient. In one large

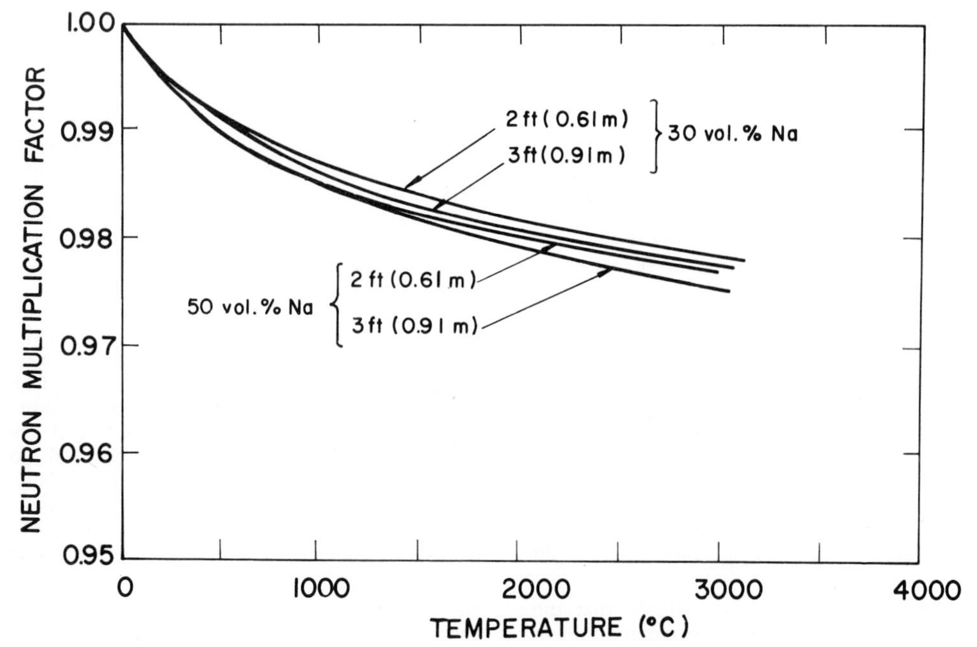

FIG. 1-2 Variation of neutron multiplication factor due to Doppler effect vs average fuel temperature for large oxide-fueled fast reactors. Height of slab-shaped cores and volume fraction of sodium coolant are parameters.

carbide reactor changing the structure material from stainless steel to niobium resulted in a decrease in the Doppler coefficient by a factor of seven [34]. This was a consequence primarily of the sharp reduction in number and importance of neutrons of energy below 10 kev as a result of the introduction of a structural material with a large neutron capture cross section.

The exact amount of reactivity change between loading and operating temperatures which the Doppler effect may cause in a large ceramic fueled reactor will depend on the design. It can go as high as 0.015 ($\Delta k/k$), however [33].

Considerable uncertainties exist in the predictions of Doppler coefficients at the present time. Inaccuracies in the prediction of reactor spectra, inadequate knowledge of the resonance parameters of the fissile and fertile isotopes, and geometric complexities all contribute to the probable error.

1.3.1.2 Reactivity Coefficient of Sodium. In fast reactors like Enrico Fermi reactor, the major contributions to the temperature coefficient of reactivity result from the changes in density of coolant and density of fuel in the core. Both are negative effects in this reactor, that is, the reduction in density upon heating allows more neutron leakage to occur and reactivity is lost.

Actually, the reactivity effect accompanying a reduction in sodium density is the net result of several phenomena.

(1) Neutron capture in sodium: The effect is positive but is usually small in the absence of thermal neutrons. In very large, oxide-fueled reactors, however, there is sufficient capture in the 3 kev resonance to be significant.

(2) Leakage component: This effect is negative. The magnitude of the effect decreases as the core size increases, since leakage is then less likely. For a given core size, the neutron leakage controlled by the sodium is smaller for those cores having higher U^{238} density or having higher density of other materials with large cross sections for scattering or capture.

(3) Spectral shift: Sodium degrades the neutron spectrum at all energies, but is especially effective at low energies, where it has considerable moderating power and where the inelastic effects of other materials are gone. While the spectrum hardening effect which accompanies the loss of sodium depends on reactor design and may result in positive or negative reactivity effects at various neutron energies, the over-all effect of the spectrum shift is positive in those large plutonium-fueled reactors examined thus far [33, 35, 36]. The reason for this trend is rather well illustrated in Fig. 1-3 [35]. Here the neutrons emitted per absorption in all core materials combined is plotted as a function of energy for 1500 liter, plutonium carbide reactors, using steel or niobium in structural parts. This parameter, which is closely related to the reactivity worth of a neutron in a reactor having low leakage, generally increases with increasing neutron energy. This increase is due partly to the high energy

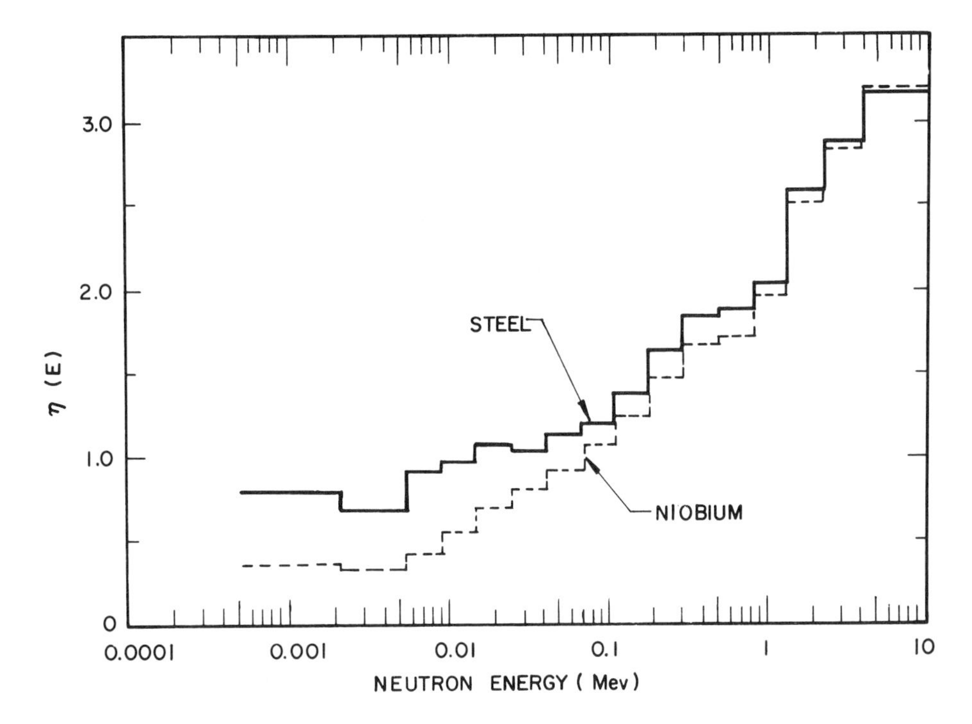

FIG. 1-3 $\eta(E)$, neutrons emitted per neutron absorbed in all materials, vs energy for 1500 liter, PuC-UC-fueled reactors using steel or niobium structure.

fission contribution of U^{238}; partly to the increase with energy of $\nu(E)$, the neutrons emitted per fission in Pu^{239}; partly to the rapid decrease with energy of the capture cross section of U^{238} and niobium; and considerably due to the rapid decrease with energy of the ratio of capture to fission in Pu^{239}.

Using more sophisticated techniques [37a] Hummel et al. [30f] recently reported a similar result: Fig. 1-4 shows the adjoint flux (i.e. the relative reactivity worth of a neutron) plotted versus neutron energy is reduced, or when Pu^{240} and fission products are present.

Since the negative leakage component falls off with increasing reactor size, while the spectral shift may well become more positive, one anticipates that the sign of the sodium reactivity coefficient for a specific reactor design will change from negative to positive as reactor size is increased. This change has been observed in preliminary analyses [35]. An example is given in Fig. 1-5 where the reactivity change effected by removal of 40% of the sodium from plutonium-metal-fuel cores using either steel or niobium as the structural material is plotted as a function of core volume. It is observed that the use of niobium sharply reduces the value of core volume at which the sodium coefficient changes from negative to positive. When the concentration of U^{238} present is reduced or the volume fraction of sodium is increased, the coefficient becomes more negative for a given core volume. Hence, a reduction in the amount of core volume given over to a fuel alloy or the use of a lower density fuel material, leads to a more negative sodium coefficient. This effect is illustrated in Fig. 1-6 where the sodium coefficient is plotted as a function of sodium volume fraction for a series of large plutonium-uranium oxide-fueled reactors [33]. Core height is a parameter for each of the four curves. As sodium volume fraction was increased, the volume available for fuel and the steel structure was correspondingly decreased, maintaining a ratio of fuel to steel volume of 2.0.

Sodium coefficients are difficult to predict accurately by the usual theoretical techniques. The individual scattering resonances of the sodium, iron, carbon, and so forth, must be considered in detail, and new, elaborate computing methods are required [37a, 37b]. In large ceramic-fueled reactors having a considerable Doppler coefficient, the sodium coefficient may be a strong function of fuel temperature [37c, 38]. In general, the coefficient may depend on the fraction of sodium remaining in the core [33]. It is a strong function of the isotopic composition of the plutonium [37d]. The higher isotopes each have a separate effect on the change in reactivity with moderation. The sodium coefficient is made more positive by a buildup of Pu^{240} with its low threshold energy and high cross section for fission contributing greatly to this effect.

The sodium coefficient will also vary considerably as a function of position in the reactor [36, 30g]. The negative reactivity contribution due to the loss of elastic scattering by sodium will primarily take place near the boundary of the core. A change in neutron direction at the center of the reactor has no effect on the leakage probability. The effect of moderation on reactivity, on the

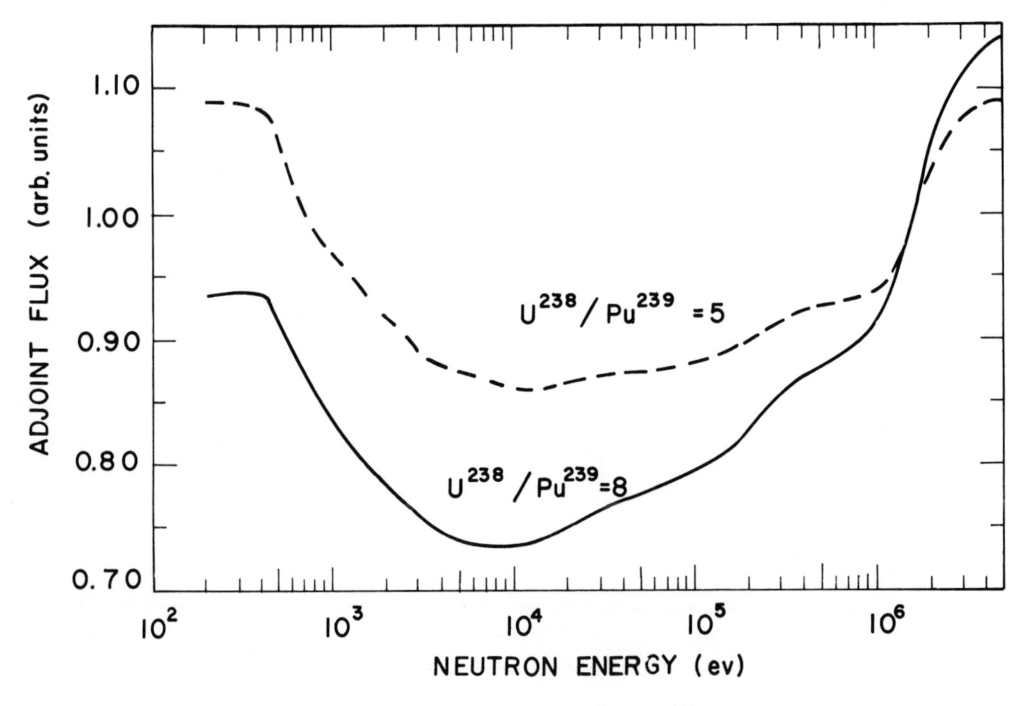

FIG. 1-4 Adjoint flux (neutron reactivity worth) for large $Pu^{239}O_2$-$U^{238}O_2$-fueled reactors. (Reactors have 70 vol.% Na. Effect of Pu^{240} and fission products is not included.)

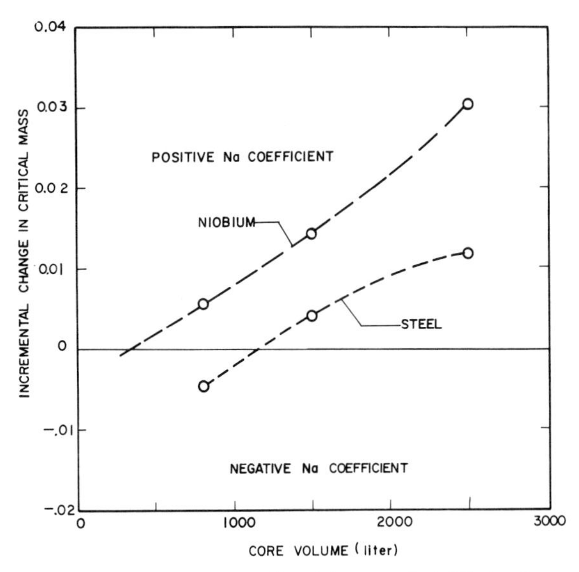

FIG. 1-5 Reactivity change effected by removal of 40% of sodium coolant from plutonium metal-fueled cores using steel or niobium structure. (Reactivity plotted in terms of the fractional increment in critical mass producing the same reactivity effect.)

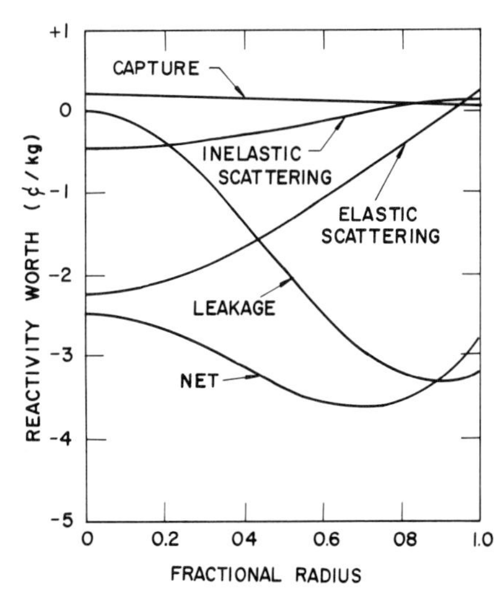

FIG. 1-7 Components of sodium density coefficient of reactivity for Fermi reactor as a function of core radius.

other hand, is likely to be most powerful near the core center. As a consequence, expulsion of sodium from the central regions of a core having an overall zero sodium reactivity coefficient may lead to a considerable gain of reactivity.

The spatial dependence of the sodium void coefficient for the Fermi reactor is illustrated in Fig. 1-7 [36]. Nims and Zweifel calculated that the net sodium coefficient was negative at all core radii. In this relatively small, U^{235} fueled reactor, spectral hardening was calculated not to gain reactivity but to lose it. In a larger reactor, fueled with PuO_2 dispersed in a U-Mo matrix, however, as is shown in Fig. 1-8, only neutron leakage contributed negatively to the sodium coefficient; the net coefficient varied from highly positive at the core center to slightly negative at the core edge.

The most recent results, as of January 1964, on detailed calculations of sodium coefficients in large fast reactors [30a, 30f] will be found in the Proceedings of the Conference on Breeding, Economics and Safety in Large, Fast Power Reactors, USAEC Report, ANL-6792.

It is noted that the sodium reactivity coefficient seems to be negative in U^{233}-thorium fueled reactors in a wide range of types and sizes [35, 39]. This is primarily due to the nearly constant ratio of capture to fission in U^{233}; the variation with energy of neutrons emitted per total absorp-

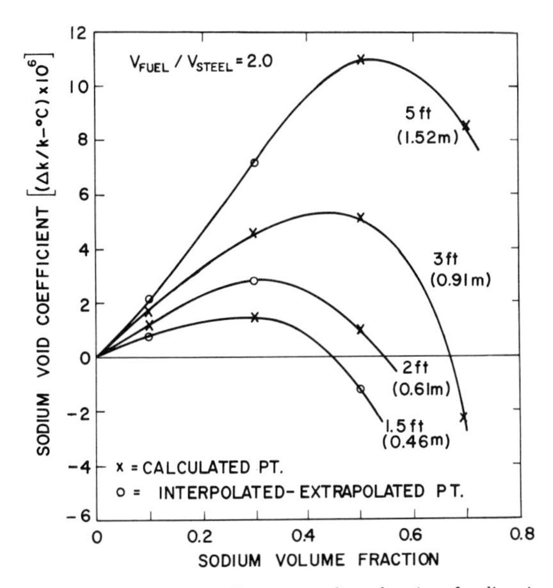

FIG. 1-6 Sodium void coefficient vs volume fraction of sodium in core blanket for several values of height of slab-shaped core.

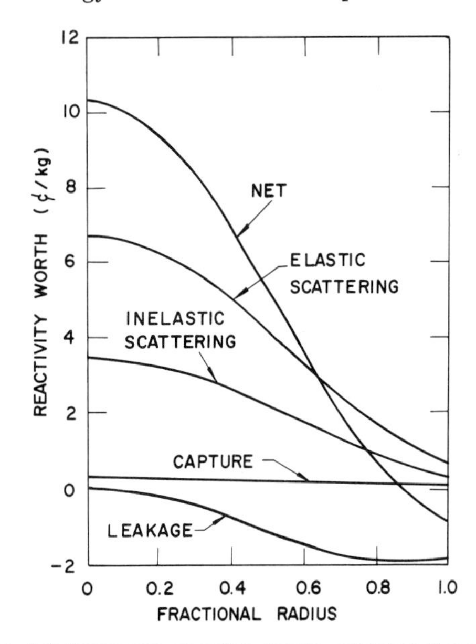

FIG. 1-8 Components of sodium density coefficient of reactivity for an 800-liter cermet-fueled ($Pu^{239}O_2$ in U^{238} + Mo) reactor as a function of core radius.

tion is much flatter than shown in Fig. 1-3. The low fast fission contribution in thorium leads to a similar consequence. Also, thorium metal has a lower density than uranium metal; a metal-fueled reactor having similar volume fractions of fuel has less total scattering and absorption with thorium, leading to a more negative sodium reactivity coefficient.

U^{235} generally provides a sodium reactivity coefficient which is more negative than that provided by Pu^{239}, and is similar to that provided by U^{233} [39]. In the rather small EBR-II and Fermi reactors, the sodium coefficient is highly negative. The calculated values for uniform removal from the reactor are $\Delta k/k = -1.7 \times 10^{-5}$ per°C for EBR-II [18], and -1×10^{-5} per°C for Fermi [19].

1.3.1.3 Temperature and Power Coefficients. Other than the Doppler effect, temperature changes affect the reactivity by thermal expansion, which changes the geometry of the reactor and the density of the material within it. For a fast power reactor like EBR-II or the Fermi reactor, the following is a qualitative and roughly chronological description of events resulting from a small increase in reactivity at some appreciable power level, say 50% of full power. The increase in reactivity causes an increase in the fission rate and hence in the power generated in the fuel and blanket material. The temperature of these materials thus increases, to an extent determined by their thermal capacity and by the ability of the fuel or blanket rod to dissipate heat to the coolant. Eventually the temperature increase causes the fuel to expand radially and axially. The radial expansion of the fuel and blanket material results in compression and then ejection of sodium from the core and blanket, while the axial expansion makes the core and blanket longer and expels some sodium from the hydraulic transition gap at the top of the core. The increased fuel temperature causes more heat to flow to the coolant, causing the average coolant temperature to increase and the density to decrease. The increased coolant temperature causes a flow of heat to the subassembly casings, causing them to expand radially and axially. The radial expansion is cumulative from the core center, resulting in a general outward radial movement of fuel and blanket material. The temperature rise also decreases the density of the envelopes themselves.

The coolant transports the heat from the core and blanket and, since the coolant temperature is greater than before the reactivity increase, the upper blanket temperature is increased more than the amount due to its own increased internal heat generation. The coolant leaving the upper blanket at the new higher temperature may then cause expansion of parts of the structure which orient the core and blanket, thus causing additional radial motion of fuel and blanket material.

If the net effect of the density and geometric effects referred to in the foregoing description is to reduce reactivity, the initial step increase in the multiplication is eliminated and the reactor levels off at a new power level and with a new temperature distribution.

As may be imagined from the complexity of the effects described, quantitative estimation of the relative temperature coefficients is neither simple nor exact. In general, the approach used is to hold all the variables constant except the one under study and then to determine the reactivity effect of temperature on that variable. For example, calculations would be performed using a reactor model containing sodium at a given density and then another calculation would be made holding all else constant but with another sodium density. The reactivity effect would then be expressed as a function of density. For instance, in the Fermi reactor the reactivity effect corresponding to a density change $\Delta\rho$ of the sodium in the core is given as $\Delta k/k \approx 0.03\ \Delta\rho/\rho$. For a large reactor having a positive sodium void coefficient, the numerical coefficient of $\Delta\rho/\rho$ would be negative, of course. Calculations on the effect of axial fuel expansion have given results like $\Delta k/k \approx -0.35\,\Delta L/L$ in EBR-II size reactors, while in very large reactors the coefficient may drop to 0.2 or less.

Multiregion calculations in spherical geometry have been found to provide satisfactory results for most non-spherical systems [3]. If the core has different reflector parameters in the axial direction and the radial direction, it may be advantageous to construct an "axial sphere" and a "radial sphere." Results obtained from these two models are then averaged appropriately. It is often useful to subdivide regions in order to obtain a more detailed picture of the important reactivity effects; the sodium density effect, for example, has been shown above to have a considerable spatial dependence. Where major structural heterogeneities such as a transition gap between core and axial blanket exist, two-dimensional analysis may be necessary to get the proper accuracy. Perturbation techniques can be used but care must be used in their application.

The isothermal reactivity temperature coefficients of EBR-II and the Fermi reactor are given in Table 1-3 [17]. These coefficients have the physical meaning that if the temperature of the entire reactor were changed by 1°C, the reactivity would change due to the effect of each component

TABLE 1-3

Isothermal Temperature Coefficients of Reactivity [a]

Mechanism	$(\Delta k/k)/\Delta T$ for EBR-II ($\times 10^{-6}$/°C)	$(\Delta k/k)/\Delta T$ for Fermi ($\times 10^{-6}$/°C)
Core:		
Axial fuel expansion	-3.9	-2.5
Radial fuel expansion (Na expulsion)	-0.9	-0.6
Density change of coolant and subassembly material	-9.1	-7.1
Structure expansion	-9.7	-6.0
Blanket:		
Density change of coolant and subassembly material	-9.5	-3.3
Growth of uranium	-1.0	-0.5
Structure expansion	-2.0	-0.6
TOTAL	-36.1	-20.6

[a] The coefficients should be multiplied by 5/9 to give values in $(°F)^{-1}$.

and the change would be by the coefficient given.

Although isothermal temperature coefficients of reactivity are of interest, it is important to recognize that power changes do not in general produce uniform temperature changes. Temperature gradients exist, and for a given change in power some parts may undergo temperature changes many times greater than other parts. For instance, the inlet coolant temperature in EBR-II is held constant during operation over the full range of power. Thus, from zero power to full power the lower structure does not change in temperature, whereas the fuel average temperature increases by several hundred degrees. To take these non-isothermal conditions into account, it is necessary to compute the power coefficient for each reactor. This coefficient is a function of a number of particular operating conditions, and great care must be taken in its use. Obviously, at power levels below the level where sensible temperature changes take place, the power coefficient has no real meaning. Likewise, the use of a power coefficient derived for a rate of power change which allows general temperature equilibrium would not be proper during rapid power transients where such equilibrium was not established. For EBR-II and the Fermi reactor, assuming full coolant flow over the entire power range, these coefficients have been calculated to be -3.2×10^{-5}($\Delta k/k$) per megawatt and -0.57×10^{-5}($\Delta k/k$) per megawatt, respectively [17]. The factor of about 5 between the coefficients is due to the smaller isothermal temperature coefficients in the Fermi reactor and to the higher ratio of temperature rise to power which characterizes the EBR-II.

Excluding the contributions from Doppler and sodium density effects, isothermal temperature coefficient and power coefficients are likely to be modest in large sodium-cooled fast power reactors and should not play a major role in setting over-all control rod strengths.

1.3.1.4 Bowing. The previous discussion regarding the geometric and density effects of temperature on reactivity has assumed that thermal expansion of fuel, blanket and structure caused changes in size but not changes in shape. This may not be true because the reactor core structure may bow and/or buckle. This subject is also discussed in Sec. 7.4 of the chapter on the Reactor Core.

Bowing can occur in reactors having a temperature gradient in the radial direction. If the core structure (which may be the fuel itself) is held firmly at the top and bottom of the core the radial temperature gradient causes a differential expansion between the inside and outside of the structure which results in forces tending to bow them toward the center of the reactor. If clearances exist, this bowing can occur and the resulting net fuel movement toward the center increases the reactivity.

Buckling can occur in any reactor which undergoes temperature changes and is simply the radial displacement of fuel or structure caused by axial restraints placed on free thermal expansion. Buckling can be prevented by providing for free axial expansion in the fuel design. If this has not been done, it is possible that the fuel expansion effect on reactivity will reverse as power transients or long term growth consume the available expansion space.

If a uniform radial temperature distribution could be achieved in the core and blanket, bowing would be eliminated. However, as a practical matter, achievement of such a distribution is very difficult and mechanical restraints on bowing have to be made.

In both the Fermi reactor and EBR-II, the individual fuel pins are prevented from bowing by fins or spacers. The bowing forces are transmitted to subassembly envelopes which are strong enough to prevent an appreciable fuel displacement due to the forces developed in the fuel pin. However, the subassembly envelope itself is subjected to the radial temperature gradient of the coolant and differential expansions take place, causing the subassemblies to attempt to bow.

The principal difference between the bowing characteristics of a core consisting of fuel elements which are separate structural entities and a core which is made up of groups of pins assembled in a structurally significant subassembly is in the time behavior of the two cores. Where the fuel provides its own structural strength, bowing forces follow power changes rapidly; when the subassembly is the structure, a relatively long lag behind power may exist. This is due to the fact that the temperature gradient must be established first in the coolant and then by heat transfer in the subassembly envelope. Thus, the envelope heating lags both the fuel temperature and the coolant temperature. The principal concern regarding bowing in the EBR-II and Fermi reactor has therefore not been that a prompt positive coefficient would exist, but that the bowing would cause either instability or a net, slightly positive power coefficient [18].

The exact manner in which bowing occurs depends strongly on the specific design of the reactor. The EBR-II, for example, has no hold-down plate to provide initial radial orientation of the upper ends of the fuel subassemblies. As the radial temperature difference increases, the core subassemblies bow outward freely in varying degree, depending on the radial location, until the clearances are taken up or until the top ends touch the relatively unbowed blanket subassemblies. Then some inward bowing can take place until the projections on the subassembly envelopes are in contact. Further increases in temperature cause the core to expand radially. The clearances which exist between the fuel subassembly and the reactor at the lower support point strongly affect the direction and progression of bowing. The most probable behavior in actual operation appears to be that core bowing first increases, then decreases, and finally increases the core diameter as the power goes from 0 to 100 per cent of full power [18].

The coefficient relating reactivity change and change in core fuel density is a slowly varying function of reactor size. Hence, the same absolute increment in core size resulting from bowing has a much greater reactivity effect in small cores than in large ones. In the EBR-I, fuel element bowing leading to an effective decrease in core radius of 0.04 cm might cause a reactivity increase of approximately 0.3%(Δk/k) [40, 41]. In a reactor having a core volume of 3000 liters, the same increment of 0.04 cm would probably produce less than 1/10 of the reactivity effect. This accident and the bowing problem on EBR-I are discussed further in the Accidents and Destructive Tests chapter.

1.3.1.5 Pressure and Flow Reactivity Effects. Small gas bubbles have been found to accumulate on the surface of the heat generating elements in both the BR-5 and Dounreay fast reactors. As a consequence, considerable change in reactivity has been observed with change in coolant pressure and with coolant flow rate. A series of tests were carried out on the BR-5 reactor in a subcritical condition [42]. On increasing the pressure from the argon cover gas, the reactivity of the system was increased; conversely, with decreasing pressure, the reactivity decreased.

The magnitude of the measured effect was explained by the presence of argon in gaseous form in the core. A series of controlled experiments confirmed this conclusion. Under some circumstances the gas bubbles were compressed with the rise in pressure and the free volume created was filled with sodium, producing an increase in reactivity of up to 0.3%(Δk/k). Under other circumstances, dispersion of the gas bubbles and their partial flotation was produced as a consequence of changes in pressure. On these occasions changes in reactivity were irregular, depending on the degree of dispersion and flotation.

In the process of initially filling the primary loop of the BR-5 reactor with NaK coolant considerable difficulty was experienced with the existence of gas pockets. It was also observed that the design of the pump and the pump reservoir were such as to create favorable circumstances for the formation of a gas-liquid metal emulsion and for the entrapment of the gas upon operation of the pump rotor. The gas pockets in the loop were finally eliminated by the technique of filling the loop with liquid metal while under vacuum. Changes in the design of the pump inlet were made to insure the creation of a quiet metal surface at the pump reservoir. Coolant circulation through the lines of the loop, after incorporating these alterations, showed no noticeable transport of the gas. Nevertheless, the reactivity varied considerably with pressure, as described above.

It was found that the greatest quantity of gas accumulated in the core of the reactor for low coolant flow. For a given flow of coolant, gas would continue to accumulate up to some definite magnitude. As the flow rate increased, the asymptotic quantity of gas accumulation decreased. Also, the time for reaching this accumulation decreased.

Model studies were performed using water flowing through cavities with transparent walls. Small bubbles which were introduced into the water collected on the surface. The quantity of gas accumulated was determined dynamically to be in equilibrium between the rate of adherence of bubbles and the rate of their being washed off by the stream of coolant. The condition of the surface had an appreciable effect on this equilibrium.

The problem of gas bubbles in the coolant has not been completely solved for the BR-5 reactor. However, raising the temperature of the coolant and thus improving the wetability of the metallic surfaces has considerably decreased the magnitude of gas accumulation. The total reactivity held in gas bubbles has been reduced to about 0.03%(Δk/k).

The problem of preventing the existence of gas bubbles in the coolant of future reactors remains one to be solved by proper design. In addition to the possibility of a vortex at the surface with downflow or near the inlet of a mechanical pump, leaky valves and control rod motion have been proposed as possible mechanisms for gas entrainment.

1.3.1.6 Other Coolants. Since the positive character of the sodium void coefficient is primarily attributable to its moderating power, it might be expected that this effect would be reduced by using a heavy coolant. Yiftah and Okrent [35] have calculated coolant void coefficients for several large reactors cooled with a lead-bismuth eutectic. These magic number elements have low neutron capture cross sections and a high threshold for inelastic scattering, making them ideal in this respect. Highly negative coolant void coefficients were obtained with lead-bismuth for reactors in which the use of sodium coolant would lead to a highly positive coefficient.

Lithium has also been examined as a potential fast reactor coolant. [34]. As expected even the weakly absorbing Li^7 isotope gives a more positive coolant void coefficient in large reactors than sodium, as a consequence of its greater moderating power. The use of Li^6 isotope in small amounts in the lithium makes this safety parameter even less attractive. Li^6 is a strong neutron absorber and its expulsion leads to considerable reactivity gains, of course.

Reactors cooled with helium gas are not expected to have sizable reactivity effects associated with coolant pressure changes. With steam or super critical pressure water the picture may be like that with sodium [43].

1.3.2 Long-term Reactivity Effects

1.3.2.1 Burnup and Buildup. The reactivity loss associated with a net uniform burnup of a fraction ΔM/M of the original critical mass is given approximately by the relation

$$\frac{\Delta k/k}{\Delta M_c/M_c} \approx 0.4 \text{ to } 0.5,$$

$$\Delta M_c = M_c \times \frac{\%\text{ burnup of fuel alloy}}{\%\text{ enrichment of fuel alloy}} \times \left(\frac{1}{1+F_c}\right) \times (1+\alpha)(1-\text{IBR}),$$

where F_c = fissions in core fertile isotope per fission in fissile isotope

α = ratio of capture to fission in fissile isotope

IBR = internal breeding ratio
= ratio of fissile atoms produced in core to fissile atoms destroyed (by capture and fission) in core

The change in reactivity is a function primarily of the product of the percent burnup of fuel alloy and the term (1 - IBR). Of course, if the fissile isotope produced is different from that destroyed the effect of the relative reactivity worth of the two isotopes must be included in the equation.

The calculated reactivity change varies widely with specific reactor design. For a large reactor having an internal breeding ratio in the vicinity of unity, the effect would obviously be near zero. For 10% heavy atom burnup in an oxide reactor having an internal breeding ratio in the vicinity of 0.4, on the other hand, a reactivity loss in the vicinity of 0.15($\Delta k/k$) is calculated [37d].

The reactivity gain associated with fissile material bred in a blanket external to the core is much more limited in magnitude. Although considerable quantities of plutonium may be bred here, the reactivity worth of fissile atoms is much reduced. For example, for the same oxide reactor which lost approximately 0.15($\Delta k/k$) due to net burnup in the core, buildup in the blanket added only 0.025($\Delta k/k$) in the same period. Once again, the actual reactivity change is a function of specific reactor design; but since only that fissile material which is produced near the core has appreciable reactivity effects, much less variation occurs for different blanket designs.

1.3.2.2 Fuel Growth under Irradiation. Radial growth of the fuel within its jacket under irradiation should produce only slight reactivity effects, corresponding to a small displacement of sodium bond, for example. Axial growth, on the other hand, corresponds to a net reduction in the fuel density in the over-all core. Its reactivity effect can be estimated in the same way as used for determining that portion of the temperature coefficient which corresponded to heating and axial expansion of the fuel element, see earlier discussion. The relation

$$\Delta k/k \approx -0.3\Delta L/L$$

is appropriate for estimating such reactivity effects for cores having a height to diameter ratio of approximately unity [1, 44]. The numerical coefficient varies appreciably with the ratio of core height to core diameter as shown in Table 1-4 [44].

TABLE 1-4

Fuel Expansion Reactivity Coefficients

Core Size (liters)	H/D	$(\Delta k/k)/\Delta T$ ($\times 10^{-6}$/°C)
50	1	-4.5
	1/4	-0.9
3000	1	-3.4
	1/4	-0.9

Note: All reactors contain 25% PuC-$U^{238}C$, 16% steel and 59% Na in core.

1.3.2.3 Fission Products. Measurements of fast neutron cross sections for most fission products are still very sparse. Theoretical estimates must be made of their average capture cross section as a function of energy, as well as of other cross sections, to enable calculations of their reactivity effects [37d, 45-48]. Estimates of the influence of fission products on reactivity are given in Table 1-5 for three plutonium-fueled reactors having a core volume of 1500 liters [37d]. In each case 0.0008×10^{24} atoms/cm^3 of fission product have been added uniformly to the core of the original reactor. It is seen that in the oxide reactor, which has a small critical mass and a considerably softer neutron energy spectrum, the same number of atoms of fission products produces a considerably greater loss in reactivity. Of course, this amount of fission products does not correspond to the same burnup for each reactor in the example chosen. For the reactor compositions used, 0.0008×10^{24} atoms/cm^3 would accumulate if 3-1/3% of the combined Pu and U^{238} fissioned with the metal fuel, 8.55% for the oxide fuel, and 5.8% for the carbide fuel.

1.3.2.4 Combined Effects. The combined reactivity effects of fission products, blanket buildup and core burnup are listed in Table 1-6 for a metal, oxide, and carbide-fueled reactor, respectively [37d]. The burnups chosen are arbitrary; the numbers can be scaled for other conditions. For the estimates given, the atoms fissioned per cm^3 vary from 0.000467×10^{24} for the oxide to 0.00069×10^{24} for the carbide.

The reactivity losses are very large for the oxide and carbide, primarily as a consequence of their lower internal breeding ratio. Frequent partial reloading will be needed for all these reactors, unless the internal breeding ratio can be raised. Unfortunately, increased internal breeding, which is achieved by introducing more fertile material into the core, is accompanied by a more positive sodium void coefficient, as has been discussed earlier.

1.3.2.5 Protactinium. The buildup of Pa^{233} in a reactor using thorium as the fertile material in the core causes two interesting effects which must be considered in determining reactor control requirements. These are the buildup of reactivity

TABLE 1-5

Influence of Fission Products on Reactivity

Reactor Type [a]	Critical Mass (kg)		Loss in Reactivity $\Delta k/k$
	New Reactor	With Fission Products [b]	
Metal	686	700	0.0115
Oxide	562	592	0.0312
Carbide	613	630	0.0165

[a] 1500 liters, fueled with Pu^{239}, having 25% steel structure, 50% Na and 25% fuel alloy or mixture.
[b] 0.0008×10^{24} atoms/cm^3 of fission product, uniformly mixed in reactor core.

TABLE 1-6

Combined Long Term Reactivity Effects

Reactor type	Atoms/cm^3 Pu plus U (10^{24})	Burnup, % of Pu + U fissioned	$\Delta k/k$			
			Burnup in core	Buildup in blanket	Fission product buildup	Total
800-liter (metal)	0.012	5	-0.054	+0.026	-0.017	-0.045
1500-liter (oxide)	0.00467	10	-0.169	+0.025	-0.037	-0.181
1500-liter (carbide)	0.0069	10	-0.147	+0.031	-0.028	-0.144

after reactor shutdown and change in reactivity with operation.

Protactinium is normally formed in a reactor using thorium as the fertile material in the core by the following chain:

$$Th^{232}(n, \gamma)Th^{233} \xrightarrow[2.33m]{\beta^-} Pa^{233} \xrightarrow[27.4d]{\beta^-} U^{233}.$$

The reactivity of a shutdown core containing Pa^{233} increases steadily because the radioactive decay of Pa^{233} generates the more reactive U^{233}. This process places an additional requirement on the reactor control system.

The magnitude of this effect has been estimated by Goldman [31, 49] for a typical U^{233}-thorium fueled power reactor operating at an equilibrium fuel cycle. The reactor is a right cylinder 3.6 ft (1.10 m) in diameter and length and is surrounded by a 15 in. (0.381 m) thick thorium blanket. It is cooled by sodium and uses U-Th metal alloy fuel elements. One-fourth of the core is replaced by fresh fuel every 15 days, using a four-zone radial shifting scheme. The reactor produces 760 Mw(t), requires a 547 kg loading of U^{233}, and operates with a breeding ratio of 1.33. The power density is 0.74 Mw(t) per liter of core; the specific power is 1.4 Mw(t) per kg of U^{233}; and the average thorium to U^{233} atom ratio is 4.84.

When this reactor is shut down at the end of a normal operating cycle, reactivity is inserted by decaying Pa^{233}:

$\Delta k = \Delta k_0(1-e^{-\lambda t})$, where

Δk = excess reactivity released at time t,

λ = decay constant of Pa^{233},

t = time after shutdown, and

Δk_0 = potential worth of all U^{233} formed from Pa^{233}.

For the particular refueling scheme studied, a decrease in reactivity of 1.76% due to U^{233} burnup occurred between refuelings. It would take 22 days of shutdown at the end of a cycle for decaying Pa^{233} to increase the reactivity by this amount. If the core remains assembled longer, the excess reactivity inserted would require additional shutdown control. The upper limit of reactivity insertion (Δk_0) is 4.1%, the net worth of all the U^{233} formed from the Pa^{233} in the core.

The previous example describes the behavior of a reactor shutdown at the end of a normal operating cycle, but not yet refueled. A reactor which has been refueled and is subjected to a delay in startup or a shutdown shortly after refueling, would also necessitate additional control requirements since there is residual Pa^{233} in the shifted fuel elements. For the reactor considered by Goldman the reactivity inserted by the decaying Pa^{233} in the residual fuel added to the excess reactivity inserted by the fresh fuel can cause a reactivity buildup greater then 1.76% above critical.

An additional effect due to the delay time in formation of bred U^{233} in the core is the increase in the net burnup reactivity change during a cycle. If the Pa^{233} had zero decay time, for example, the reactivity decrease with burnup in the reference reactor would have been 1.1% compared with an actual decrease of 1.76%. This happens because only 40% of the Pa^{233} formed by radiative capture of Th^{232} actually decays to U^{233} while in-pile in the reactor studied.

Even in a reactor with an internal breeding ratio of unity, there would have been a decrease in reactivity with time, since approximately 60% of the thorium absorptions would not actually produce U^{233} until after the fuel was removed in the concept studied. Thus with reactors using thorium as the fertile material in the core, it is always likely that burnup control must be provided.

It is of interest to examine these phenomena in a reactor using U^{238} as the fertile material. The production of fissile material is then described as follows:

$$U^{238}(n, \gamma)U^{239} \xrightarrow[23.5m]{\beta^-} Np^{239} \xrightarrow[2.33d]{\beta^-} Pu^{239}.$$

The half-life of Np^{239} is of the order of 1/12 that of Pa^{233}. Goldman estimates the equilibrium concentration of Np^{239} to be approximately 1/12 that of Pa^{233}; because the half-life of Np^{239} is small compared to the fuel in-pile residence time, the actual concentration would be close to the equilibrium value. This is not the case of the U^{233} breeder, the average Pa^{233} concentration being about 1/2 of the equilibrium level. The average Np^{239} concentration in a Pu^{239} breeder should therefore be about 1/6 that of the Pa^{233} in the U^{233} breeder. Assuming that U^{233} and Pu^{239} have the same reactivity worth, Goldman estimates that the total reactivity associated with Np^{239} is about 0.67% for a reactor of similar design and operating method to his reference thorium-U^{233} breeder. Upon reactor shutdown, the reactivity associated with the Np^{239} would be essentially all released in about 11 days or 5 half-lives. Of course, a similar amount of reactivity would be lost in starting up the reactor and building up this equilibrium quantity of neptunium.

1.3.3 Scram Requirements

The amount of reactivity required for scram purposes is subject to a variety of criteria. These are generally associated with mal-operation of the coolant system, a major error in loading, or some other abnormal operation of the reactor. Clearly, shutdown during reloading must considerably exceed the maximum possible reactivity addition via the introduction of a single subassembly. The amount of reactivity by which the reactor is subcritical should exceed the reactivity available upon cooling down from operating temperature to some ambient temperature. It should exceed the reactivity which would be introduced by total loss of coolant, in the event of a positive void coefficient. For a reactor using thorium as a fertile material in the core, it should clearly exceed the maximum reactivity buildup which might result from the decay of Pa^{233}.

The loss of coolant flow accident provides a more subtle criterion [18, 19]. As coolant flow velocity decays, fuel elements will overheat unless the reactor power is sharply reduced. The extent to which delayed neutrons contribute to the reactor power following scram will depend directly upon the degree to which the reactor has been shut down below delayed critical. Accident analyses on the detailed reactor plant are required to ascertain whether or not the scram is sufficiently rapid and of sufficiently large magnitude.

1.4 Nuclear Aspects of Control and Safety Rod Design

The manner of control which is feasible in fast power reactors depends on a variety of interrelated factors. These include the size and type of reactor, the total reactivity needed, the safety philosophy which is applied, the pertinent mechanical and heat transfer considerations, and the significance of economic factors, including the cost of material and equipment and the relative effects of different control rod designs on breeding ratio.

For the relatively small EBR-II reactor with a core volume of roughly 50 liters, control appeared to be feasible in principle by any of the three usual methods, that is, poison, moving fuel, or reflector control. It was calculated that this reactor would lose about 12%$\Delta k/k$ if sodium coolant were substituted for a 9 cm thick steel or high density uranium reflector [1]; that is for a cylindrical reflector completely surrounding the reactor except at the ends. For an 800 liter reactor of the same design, however, this reactivity effect is cut by a factor of 8 or more for the same thickness reflector. This is due to the reduced importance of neutron leakage to the reactivity of the larger reactor. And, of course, the use of such a thick reflector can be expected to produce significant losses in the breeding ratio, partly from neutron capture and partly from a reduction of the fast fission bonus in U^{238}, if the reflector material is non-fertile.

In the EBR-II reactor, movement of fuel is used to achieve reactor control. Twelve rods at the periphery of the core introduce a total of 4.5%$\Delta k/k$. As U^{235}, U^{238}, and steel move out of the core, sodium coolant rushes in to take its place. The net reactivity effects are a consequence of changes in fission, capture, and scattering cross sections. When control rod worth is calculated in one-dimensional cylindrical geometry, without allowing for the effects of the end blankets, the worth of the rod is generally overestimated by about 25% [16]. Two-dimensional diffusion theory calculations give good agreement with experiment, however. For rods of this type very little interaction or shadowing effects have been found, either theoretically of experimentally.

For the Enrico Fermi reactor, poison in the form of $B_4^{10}C$ is used for both control and safety rods. According to Amorosi and Yevick [50],

> "The principle reason for selecting poison rather than fuel for reactor control is that the mechanical design is less complicated. With poison control, less mass needs to be moved and shutdown involves moving material into the core from above rather than down out of the core, thus reducing the size and complexity of the reactor vessel. A further advantage of poison control over fuel control is that much less heat is generated in the control element for equal reactivity contribution".
>
> "There are two main disadvantages to using poison control. Poison in operating control rods is undesirable because it captures neutrons which could have been used to produce plutonium. To avoid a major breeding loss in the safety rods, they are normally located above the axial blanket so that their effect on neutron absorption is small. Locating them away from the core leads to a second disadvantage, however, namely a delay between the scram signal and the time of appreciable negative reactivity insertion which would not be present if moving fuel were the type of control being used. The safety and reliability afforded by simplified rod design more than compensates for the above disadvantages".

Because of the large mean-free-path for both fission and absorption in a fast reactor, local flux perturbation and consequent local hot spots experienced in some thermal reactors do not exist in the Fermi or EBR-II reactors. The boron in the operating control rods of the Fermi reactor has only a very small influence on the local power distribution. The empty safety rod channels also have little influence [19]. The B^{10} cross section for neutron capture is roughly 2 barns at 100 kev, and large quantities of the isotope are needed to produce appreciable reactivity effects. Li^6, the other potentially strong poison material for fast reactors, has a capture cross section of the same order of magnitude. Metals like tantalum and tungsten have microscopic capture cross sections which are lower than that for B^{10} but their density may be sufficient to make them useful in large fast reactors.

Although the capture cross section of B^{10} for fast neutrons is modest compared to those found for poisons in thermal reactors, enough of the material is used in control rods that some degree of self protection exists. This sort of effect has been observed experimentally, reactivity effects per gram varying as a function of sample size. Rather small shadowing effects have also been observed during mockup experiments of the Fermi reactor safety rods in the ZPR-III critical facility [51].

Generally speaking, one-dimensional multigroup diffusion theory calculations in cylindrical geometry are expected to give rough estimates of the reactivity worth of a particular poison control rod design. Two-dimensional calculations in (r, θ) or (x,y) geometry can be used to provide an estimate of the self-shielding effects. Two-dimensional calculations in (r, z) geometry can then be used to allow for the finite height of the reactor.

Conceptual design studies of large fast power reactors have tended to use B^{10} poison in the form of boron carbide for control rods. In one design of a 500 Mw(e) plutonium oxide-uranium oxide-fueled fast reactor [52] the reactivity requirements were stated to be:

fuel burnup	1.8$
Doppler effect, cold to hot	3.7$
change in core size and sodium density, cold to hot	0.5$
shutdown	8.0$
TOTAL	14.0$

In this reactor, a dollar is worth approximately 0.0038 $(\Delta k/k)$ so that there is about 5.3% $(\Delta k/k)$ total in all rods. To achieve this much control, it was proposed to use eight shim rods, each worth 75¢. These are specified to consist of stainless steel cans two inches in diameter, fueled with boron carbide that is enriched with B^{10} isotope. The composition was to be as follows: 45 vol.% $B_4^{10}C$, 25 vol.% Na, and 30 vol.% steel. In addition to the eight shim rods, eight safety rods, each worth 1$, were to be employed for a shutdown margin and to cover emergency requirements. The safety rods would be constructed using $B_4^{10}C$, just as the shim rods.

In a conceptual design of 315 Mw(e) fast power reactor fueled with Th-U^{233} [32], a considerably different set of control requirements was adopted. As is shown in Table 1-7, the total reactivity required in the control rods was 13.2% $(\Delta k/k)$. Only about 1% $(\Delta k/k)$ was needed for the change in temperature from cold to operating, just as in the plutonium oxide reactor discussed above. Slightly more was allowed for burnup with the U^{233}-fueled reactor. The main difference between the two systems lies in the large allowance made for the effect of Pa^{233} decay. If all the protactinium present in the core and blanket were to decay, a reactivity gain of about 4% $(\Delta k/k)$ could be possible in such a reactor. Hence considerably more shutdown has been provided.

The total amount of reactivity control required for this reactor is much more than for those previously discussed.

TABLE 1-7

Control Requirements 315 Mw(e) Th-U^{233} Reactor

Type of Control	$\Delta k/k$ (%)		
1. Shim Control			
Cold-to-hot reactivity			
Doppler effect	~-0.5		
Coolant density change	-0.1		
Expansion of fuel and structure	-0.4		
Total cold-to-hot change		-1.0	
Burnup, per cycle		-2.3	
Maximum Pa^{233} decay		-1.7	
Total Shim Control			-5.0
2. Safety Control			-8.0
3. Regulating Control			-0.2
TOTAL CONTROL			-13.2

A similar trend toward increasing reactivity requirements in control rods is likely to occur for any of a variety of reasons. Very large negative Doppler coefficients require greater amounts of reactivity control between cold and operating temperatures. Higher burnups between partial unloadings or decreased internal breeding ratios each lead to more reactivity loss to be made up by shim rods. And if reactor designs are such that large reactivity gains accompany loss of coolant, this also considerably increases the reactivity requirements of the control system.

It should be noted that control by the use of movable moderator may prove useful, at least for the operational portions of the control requirements of large fast power reactors. The spectral shift mechanism which contributes to the sodium coefficient will also apply, say, to graphite rods, in the central portions of the core [53].

2 OPERATIONAL BEHAVIOR

2.1 Introduction

The general physical relationships which govern the operational behavior of fast reactors are the same as those applicable to thermal reactors. In somewhat different terms but with the same meaning as has been expressed in the chapter on General Reactor Dynamics, the general equations relating to the time-dependent power behavior are

$$\frac{dP}{dt} = \frac{(\rho - \beta)P}{\ell / k_{eff}} + \sum_{i=1}^{6} \lambda_i D_i \qquad (2\text{-}1)$$

and

$$\frac{dD_i}{dt} = \frac{\beta_i P}{\ell / k_{eff}} - \lambda_i D_i \, . \qquad (2\text{-}2)$$

Here the usual meanings are associated with the variables:

P = power,
ρ = reactivity, $(k_{eff}-1)/k_{eff}$,
β_i = fraction of delayed neutron in i^{th} group,
$\beta = \beta_1 + \beta_2 \ldots . + \beta_6$
ℓ = prompt neutron lifetime,
k_{eff} = effective multiplication constant,
D_i = contribution to power from i^{th} group of delayed neutron precursors, and
λ_i = decay constant for i^{th} group of precursors.

The reactivity ρ has time and temperature dependence which may be expressed as:

$$\rho = \sum_N \rho_N \Delta T_N + F(t) , \qquad (2\text{-}3)$$

where ρ_N = temperature dependent reactivity effects of the nth type,
ΔT_N = average temperature elevation from some base temperature which effects ρ_N, and
$F(t)$ = reactivity added by control system or other external source as a function of time.

The values of ρ_N are determined by essentially static calculations, as is discussed in the previous section, and the product $\rho_N \Delta T_N$ implies an integral over the affected parts of the reactor. The values to be used for the ΔT_N in each case are related to power by such specific characteristics of the system under consideration as thermal conductivity, heat removal rate, coolant velocity, coolant inlet temperature and specific heat.

The value of the prompt neutron lifetime ℓ may range from 5×10^{-7} sec in the case of a large ceramic reactor to 3×10^{-9} sec in the case of a bare plutonium sphere. These values should be compared to the range 10^{-3} to 10^{-5} sec for thermal reactors. However, for values of ρ less than about 0.9$ this short lifetime has only a negligible influence on the operational behavior. This is shown by Figs. 3-2, 3-3, 3-4, and 3-5 in the chapter on Criticality which illustrate values of asymptotic period versus reactivity for various fissile materials and prompt neutron lifetimes.

The values of the total delayed neutron fraction and of the precursor concentrations and decay constants do not differ appreciably for a given isotope whether fission was induced by fast or thermal neutrons. This has been shown in Table 3-1 of the Criticality chapter. However, typical fast power reactors may contain relatively large amounts of U^{238} Th^{232} Pu^{240} and U^{234} in the core and blanket with a neutron spectrum such that a significant number of fissions occur in these isotopes. For instance, in certain reactors 10% to 25% of the total fissions may occur in the U^{238} contained in core and blanket.

The calculation of an effective value for β is complicated in such systems both by the differing spatial distribution of fissions for the various isotopes compared to that of the primary fissile isotope and by the different absolute and relative yields of delayed neutrons from the different sources. Another complication is that, at various times in reactor life, the neutron contributors may be present in different concentrations.

In addition, delayed neutrons have energies below the fission threshold of U^{238} and have been found in many cases to have lower statistical weight than do prompt neutrons.

Effective delayed neutron fractions for various fast assemblies, together with methods for calculation are given in references [40, 54], and are discussed in the chapters on Criticality and The Reactor Core.

The differences between fast and thermal reactors which are important in safety considerations do not occur in comparisons of the delayed neutron kinetics of the two types. The differences lie in two characteristics of fast reactors: the short neutron lifetime as it affects the reactor period for reactivities of β or greater and the fact that core compaction due, say, to meltdown, can lead to very large increases in reactivity. As an example, Fig. 3-2 in the Criticality chapter shows that for a value of ℓ of approximately 10^{-7} sec, the period corresponding to a reactivity of 1.10$ is about 10^{-4} sec. For lifetimes a decade shorter the period would also be a decade shorter. A rate of power increase by a factor of e in 10^{-4} sec is completely beyond the control of any outside agency.

Thus the basic effort in achieving a safe design of a fast reactor has been to assure by all practical means that situations do not arise in which the reactor may exceed prompt cirticality or in which gross meltdown of the core may occur.

This has been done generally by consideration of the probability and the possible consequences of:

1) The emergence of a product $\rho_N \Delta T_N$ which is positive and large enough to cause a severe power transient. This has been termed autocatalytic instability.

2) A time and power-related interaction of two or more of the products $\rho_N \Delta T_N$ resulting in an uncontrollable reactivity oscillation and subsequent power oscillation. This is termed oscillatory instability.

3) The term $F(t)$ in Eq. (2-3) becoming large and positive either by control maloperation or from some discontinuous effect which introduces reactivity suddenly.

4) Local or general cooling failure resulting eventually in extensive core melting.

The first and second of the foregoing relate particularly to the basic stability inherent in individual designs and are considered in this section. The third and fourth points are discussed in Sec. 3.

2.2 Stability Considerations

The stability of fast reactors has been the subject of an amount of analytical effort disproportionate to the number of fast reactors at present in operation [40, 55-57, 37e, 37f]. This has been partly the result of the fact that the EBR-I, the first fast reactor to produce useful power, was afflicted [20, 58] with instabilities of both the autocatalytic and oscillatory types. This fact led not only to a strong effort to try to eliminate the instabilities of EBR-I, but also to understand the basic causes of these instabilities in order to guide the designers of the EBR-II, Fermi, and Dounreay reactors which were concurrently underway.

2.2.1 Autocatalytic Instability

The presence of an autocatalytic instability implies that for a given reactor there is a power level and coolant flow rate for which, if the reactivity is increased above zero, the resulting increase in power level will somehow act to continue to insert reactivity and the power level will continue to rise until the reactor destroys itself by melting or worse. Thus, a reactor having an over-all prompt positive power coefficient will be unstable in this sense.

The EBR-I Mark II loading was observed to have a strong prompt positive power coefficient attributed to inward bowing of the core fuel pins which ordinarily was overridden by negative components of the power coefficient. When the ratio of power to coolant flow or the rate of change of power were abnormally high, the prompt coefficient became dominant and the reactor exhibited autocatalytic instability [59]. The subsequent Mark III loading had a means of preventing such bowing and no autocatalytic instability was observed during an extensive experimental search [40, 37f].

In the group of fast reactors presently operating, steps have been taken to assure that such structural deformations will not result in autocatalytic effects under any circumstances of flow or rate of change of power, both for power levels within the design limits of the reactors and for power levels above the normal limits which might be reached as the result of a mild power overshoot.

At the time of the EBR-I meltdown, there was extensive speculation that the Doppler effect in U^{235} was responsible for the positive power coefficient, but it is now known that this effect was not large enough to have been responsible [24, 25]. However, for many of the large reactor designs currently under study, the Doppler effect is expected to be considerably larger due to the more degraded spectrum, and may be the net result of positive and negative components. Particularly in designs where a fissionable isotope having a positive Doppler effect is thermally separated from an absorber isotope which is counted on for Doppler compensation, care should be taken that at some rate of power rise the temperature of the absorber isotope does not lag so far behind the fissionable isotope that compensation is lost.

The current investigations of large fast reactors using sodium as coolant have disclosed a wide variety of design in which the coolant expansion coefficient of reactivity is positive [36, 30f, 30g, 39]. Such a situation puts a heavy burden on the reactivity effects associated with the fuel itself since if these effects are small or insignificant the over-all effect on reactivity of the coolant expansion may be positive and coupled closely to the power level. The fuel coefficient for a case where the sodium coefficient is quite positive must not only be clearly capable of preventing an over-all positive effect when the core is new but also near the end of life when long term effects have modified reactor characteristics. Situations involving the expulsion of coolant must also be analyzed.

2.2.2 Oscillatory Instability

At present, it is believed that no purely nuclear oscillatory instabilities can occur in the fast reactors now in operation or under serious consideration. No problems analogous to the xenon burn-out instabilities which can occur in large thermal reactors have been identified but the future objectives of very large, high burnup, economic fast reactors may very well lead to loosely coupled cores which are more susceptible to such instabilities than the present generation of small reactors.

In the absence of such nuclear instabilities, any oscillatory instability would have to originate within the reactor through an adverse coupling of the thermal reactivity feedback effects with the neutron kinetics.

Nicholson [60] has investigated a possible source of instability due to induced longitudinal vibration of fuel pins in a reactor in which axial fuel pin expansion reduces reactivity. He found the frequency of possible concern to be far above the natural frequency of any reasonable fuel design. Because of the multitude of reactivity effects in fast reactors, the analysis of the possible coupling of feedback with neutron kinetics is difficult and involves to a very large degree many of the designs details of the specific reactor under consideration. The following analysis, due to Storrer [19] is given as an example of a particular approach to stability analysis taken in the case of the Enrico Fermi Reactor. Analyses with the same object have been made for the Dounreay reactor [57] and for EBR-II [37e].

Temperature feedback can be considered as the sum of two components: the internal feedback, caused directly by a power change assuming a constant reactor inlet coolant temperature and the external feedback, caused by a change of the reactor inlet coolant temperature fed back around the coolant loops from a change of the reactor outlet coolant temperature. The relations involved in the kinetics of the reactor, including the feedback, can be represented schematically by the signal flow diagram of Fig. 2-1. The variables corresponding to each node are listed on the figure. Each branch linking two nodes represents the transmittance of a transfer function linking the corresponding variables.

By using conventional techniques, the reactor transfer function G_p at power P is

$$G_p = \frac{\partial P/P}{\partial R_i} = \frac{G_0}{1 - PG_0\left[X + \dfrac{E_1E_2E_3}{1 - E_2E_4}\right]} \qquad (2\text{-}4)$$

One can define a total power coefficient:

$$X_t = X + \frac{E_1E_2E_3}{1 - E_2E_4}. \qquad (2\text{-}5)$$

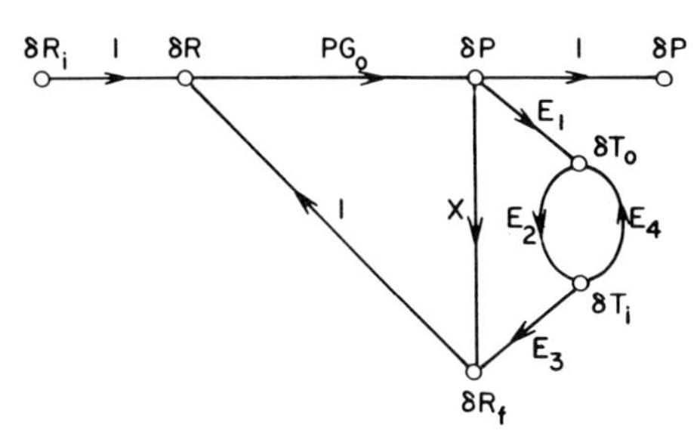

FIG. 2-1 Reactor kinetics, signal flow diagram. The symbols are defined as follows:

R = total reactivity = $R_i + R_f$

R_i = inserted reactivity; R_f = feedback reactivity

P = reactor power

T_o = reactor outlet coolant temperature

T_i = reactor inlet coolant temperature

$G_o = (\delta P/P)/\delta R$ = zero power reactor transfer function

X = power coefficient of reactivity (constant T_i assumed)

$E_1 = \delta T_o/\delta P$, assuming constant T_i

$E_2 = \delta T_i/\delta T_o$ = the transfer function for the transmission of a temperature signal around the coolant loops. Part of this transmission is around the primary loop only, and part is around the primary and secondary loops in series, if there are any secondary loops.

$E_3 = \delta R_f/\delta T_i$, assuming constant P

$E_4 = \delta T_o/\delta T_i$, assuming constant P

$E_3(0)$ = isothermal temperature coefficient of reactivity

$E_4(0) = 1.$

Here X is called the internal power coefficient and $E_1E_2E_3/(1-E_2E_4)$ the external power coefficient. The transfer function can then be written as

$$G_p = \frac{G_0}{1 - PG_0X_t}. \tag{2-6}$$

The feedback appears as a sum of two components acting in parallel. The system will be stable if $P < P_c$ where P_c is the minimum power for which G_p has a pure imaginary pole $i\omega_c$; this can be expressed as

$$1 - P_cG_0(i\omega_c)\,X_t(i\omega_c) = 0. \tag{2-7}$$

At $P = P_c$ the reactor would exhibit an oscillatory instability at angular frequency ω_c. At frequencies higher than a certain value f_1, the external feedback is for practical purposes completely attenuated and can thus be neglected. At frequencies lower than a certain value f_2, all temperatures inside the reactor respond to power or inlet coolant temperature changes in a quasi-steady state fashion, and hence X, E_1, E_3, and E_4 are equal to their steady state values X(0), $E_1(0)$, $E_3(0)$, and $E_4(0)$. From analog simulation studies it was found that f_1 and f_2 are coincident for the Fermi reactor and have the value of about 0.01 cycle/sec. Since $f_2 = f_1$, the entire frequency range can be covered by the ranges $f < f_1$, and $f < f_2$, with the two ranges meeting at $f_1 = f_2$.

According to the definition of f_1 and f_2, the total power coefficient X_t of Eq. (2-5) can be simplified as follows:

$$X_t = \begin{cases} X & \text{for } f > f_1, \quad (2\text{-}8) \\ X(0) + \dfrac{E_1(0)E_2E_3(0)}{1 - E_2E_4(0)} & \text{for } f < f_2. \quad (2\text{-}9) \end{cases}$$

Since the two frequency ranges cover all frequencies, it is sufficient to investigate separately whether instability is possible in either of the two ranges.

2.2.2.1 Stability Analysis in the Higher Frequency Range. In the frequency range $f > f_1$, $X_t = X$. The stability criterion given by Eq. (2-7) can be expressed in terms of amplitude and phase, with the feedback separate from the neutron characteristics:

$$P|X(i\omega_c)| = \frac{1}{|G_0(i\omega_c)|} \tag{2-10}$$

and

$$\phi_X(i\omega_c) = \pi - \phi_{G_0}(i\omega_c), \tag{2-11}$$

where ϕ_X and ϕ_{G_0} are the phase lags of X and G_0 respectively. Instead of Eq. (2-11), the following expression may be used:

$$\tau_b(i\omega_c) = \frac{\phi_X}{\omega_c} = \frac{\pi - \phi_{G_0}(i\omega_c)}{\omega_c} \tag{2-12}$$

At any frequency ω, the phase lags of X and G_0 can be expressed as functions $|X|$ and $|G_0|$ respectively as

$$\frac{\pi - \phi_{G_0}(i\omega)}{\omega} = f\left(\frac{1}{|G_0(i\omega)|}\right) \tag{2-13}$$

and

$$\tau_b(i\omega) = \frac{\phi_X(i\omega)}{\omega} = g(P|X(i\omega)|). \tag{2-14}$$

The functions f and g are single-valued explicit functions; however, in practice they are usually treated as implicit functions of the variable ω.

Equation (2-13) is dependent only on the neutron kinetics characteristics of the system and Eq. (2-14) is dependent only on the feedback reactivity characteristics of the system. At the critical point for instability, $\omega = \omega_c$, $f = g$, and the curves for these two equations intersect. Hence, Eqs. (2-13) and (2-14) are equivalent to Eqs. (2-10) and (2-12) at the critical point. For any reactor, Eq. (2-13) can be calculated as a function of the argument with good accuracy, since it involves only neutron kinetics parameters. This equation is represented on Fig. 2-2 for the Fermi reactor, and also for a reactor with a neutron lifetime of 10^{-4} sec. The two curves diverge only for frequencies beyond the range where instability is most likely.* Since the dollar was chosen as the unit of reactivity, the intermediate part of the curve, where the abscissa is unity, would not be affected by a different value of the delayed neutron fraction. At low frequencies the curve would only be slightly affected by the different delayed neutron characteristics of other fissionable isotopes. Hence for all practical purposes the neutron kinetics curve of Fig. 2-2 is independent of the reactor type of interest.

In order to investigate whether or not a reactor is stable at power P, there is also plotted on Fig. 2-2 the feedback reactivity curve with coordinates $P\,|X(i\omega)|$ and $\tau_b(i\omega)$, where ω is varied from zero to infinity. This curve should be analyzed in relation to the neutron kinetics characteristics curve shown on Fig. 2-2. If at some value of ω the points of both curves have the same ordinate, the relation of Eq. (2-12) is satisfied and the phase

*At high frequencies where $f_2 \gg f_2$ the amplitude X of the feedback is so attenuated and $1/G_0$ is so large that instability is not likely.

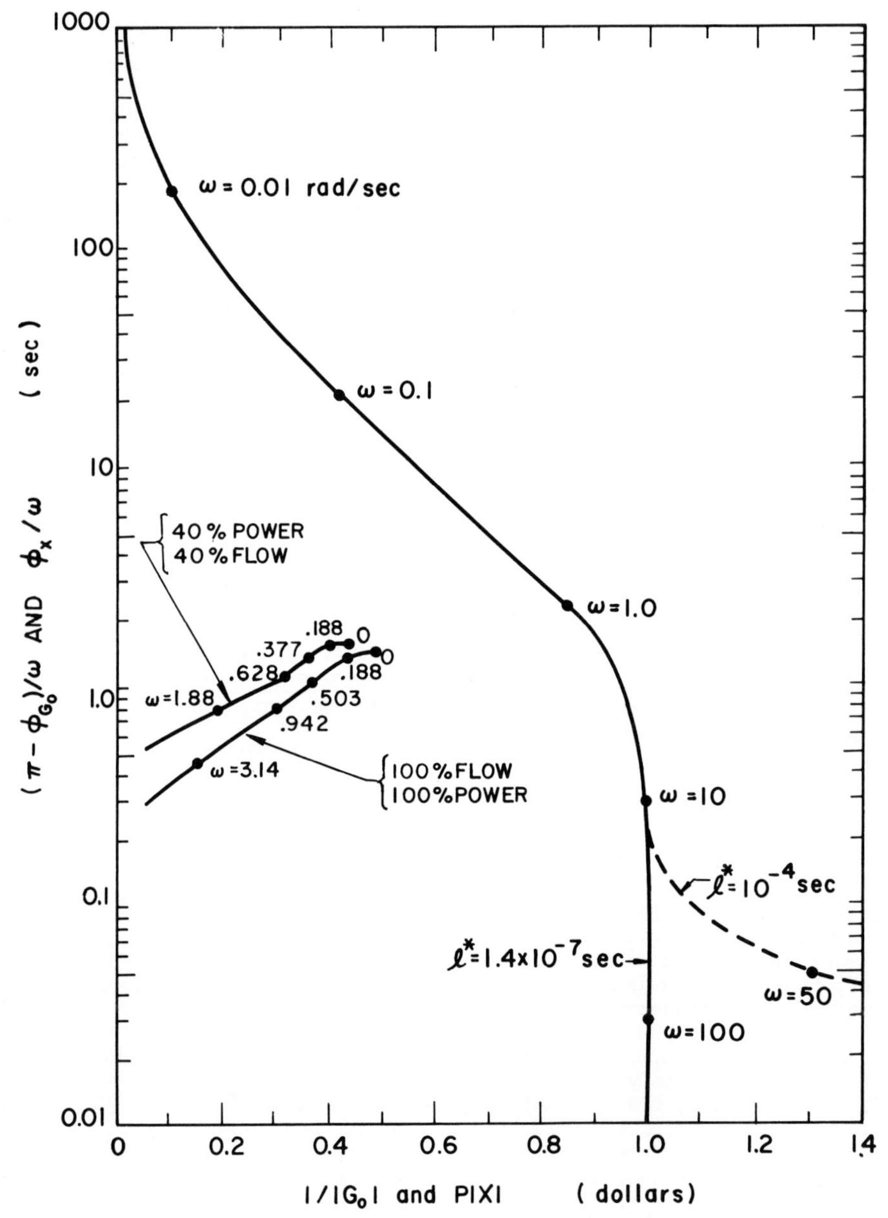

FIG. 2-2 Reactor stability criterion. (The quantity $\tau_b = \phi_X/\omega$ is plotted versus $P|X|$, etc. See text.)

is that required for pure oscillatory instability If at that frequency the point on the feedback curve is at the left of the point on the neutron kinetics curve shown on Fig. 2-2 the power P is smaller than the critical power P_c. This is nothing more than the conventional Nyquist criterion. presented graphically in such a way as to keep separate the more accurately known neutron kinetics characteristics and the less accurately known feedback characteristics.

In most reactors it is sufficient to consider the zero frequency point, with coordinates $P|X(0)|$, $\tau(0)$, to ascertain that the reactor will be stable, if some assumptions are satisfied regarding the frequency dependence of the feedback. These assumptions are:

$$\left.\begin{aligned} |X(i\omega)| &\leq |X(0)| \\ \tau_b(i\omega) &\leq \tau_b(0) \end{aligned}\right\} \qquad (2\text{-}15)$$

for all ω, where $\tau_b(0)$ is defined as the limit of $\tau_b(i\omega)$ as $\omega \to 0$. If these assumptions holds a reactor is clearly stable if the zero frequency point $P|X(0)|$, $\tau_b(0)$ is on the left of the neutron kinetics curve shown on Fig. 2-2.

The assumptions are valid for most reactors. If the feedback followed exactly a transport lag model, one would have

$$X(i\omega) = X(0)\, e^{-i\omega\,\tau_b(0)}, \qquad (2\text{-}16)$$

so that $|X(i\omega)| = |X(0)|$ and $\tau_b(i\omega) = \tau_b(0)$. This is the limiting case of the assumption of Eq. (2-15) and the external feedback curve on Fig. 2-2 would be condensed in a point.

Actually the thermal reactivity feedback in solid fuel reactors results from a combination of heat transport and heat conduction so that the amplitude $|X(i\omega)|$ of the feedback is attenuated with increasing frequency. Also, as a result of heat conduction, the quantity $\tau_b(i\omega)$ decreases with increasing frequency. For example, in the simple first-order heat conduction model, one has

$$X(i\omega) = \frac{X(0)}{1 + i\omega\,\tau_b(0)}, \qquad (2\text{-}17)$$

so that

$$|X(i\omega)| = |X(0)| \cos \phi_X(i\omega) < |X(0)| \qquad (2\text{-}18)$$

and

$$\tau_b(i\omega) = \frac{\phi_X(i\omega)}{\omega} = \frac{\tan^{-1}[\omega\tau_b(0)]}{\omega} < \tau_b(0)\,. \qquad (2\text{-}19)$$

The assumptions of Eq. (2-15) possibly would not be valid if the total negative power coefficient consisted of a large positive prompt component and of a larger negative delayed component, as was the case in EBR-I, Mark II, or if the feedback was not of a purely thermal nature but was amplified by a mechanical or hydrodynamic resonance at some frequency. Note that, if the feedback is linear with power, $P|X(0)|$ is the so-called power override, or the reactivity required to bring the reactor from zero power to power P at constant inlet coolant temperature.

For the Fermi reactor the calculated values of $P|X(0)|$ and $\tau_b(0)$ at full power and full flow are respectively 49¢ and 1.4 sec. As shown on Fig. 2-2 this point is well on the stable side of the curve. The safety margin is seen to be large, from 49 to 92¢ for $P|X(0)|$ and from 1.45 to 14 sec for $\tau_b(0)$. The assumptions of Eq. (2-15) mean that, for increasing values of ω, the point would move toward the left and the bottom of the graph, well away from the curve as shown. Even if the assumptions of Eq. (2-15) were not valid, they would have to be in error by a large amount if the point were to reach the curve. With the feedback mechanism of the Fermi Reactor, i.e., where there is no net positive component in the power coefficient, the assumptions of Eq. (2-15) are valid, and the simple stability criterion is just as good as that which would be obtained by a detailed calculation of $X(i\omega)$ based on the same feedback mechanism.

As Storrer has shown in reference [61], the behavior of $P|X(i\omega)|$ and of $\tau_b(i\omega)$ can present some anomalies at very low frequencies if some component of the power coefficient has a very large time constant so that it comes into play only at very low frequencies. In the Fermi Reactor the hold-down plate expansion coefficient, with a time constant of the order of minutes, is such a component. One can avoid the anomaly by taking for $P|X(0)|$ and $\tau_b(0)$ the fictitious limit which is reached when the frequency is decreased to f_1. This frequency is low enough for all the other components of the power coefficients to have reached their steady state limits, while it is too high for the hold-down plate expansion coefficient to come into play. The values quoted above for the Fermi reactor are those fictitious limits, rather than the zero frequency limits.

$\tau_b(0)$ increases somewhat with decreasing flow, [61] while $P|X(0)|$ decreases somewhat if P is reduced in the same proportion as the flow in order to keep the same coolant temperature rise. For instance, at 40 percent flow, $\tau_b(0) = 1.6$ sec.

For many reactors now in operation, both fast and thermal, the point representing the asymptotic feedback characteristics is on the left of the curve shown on Fig. 2-2, so that these reactors satisfy the stability criterion presented here. In boiling water reactors, the power override, which is a good measure of the magnitude of $P|X(0)|$ and which consists mostly of reactivity compensated by voids, can attain many dollars. The zero frequency point is then well on the right of the curve and a detailed analysis of the frequency dependence of $X(i\omega)$ is required to determine stability.

If plutonium or U^{233} were substituted for U^{235} in the reactor with a resultant decrease in β, and if the power coefficient were about the same in terms of absolute units of reactivity, the numerical value of the power coefficient and the abscissa of the points representing the feedback characteristics of

Fig. 2-2 would be multiplied by about a factor of two if the dollar is used as a unit of reactivity. Since the neutron kinetics curve of Fig. 2-2 would remain approximately unchanged, this substitution would reduce the critical power level for instability by about a factor of two.

2.2.2.2 Stability Analysis in the Lower Frequency Range. In the frequency range where $f < f_2$, $X_t = X(0) + [E_1(0)E_2E_3(0)/(1-E_2E_4(0)]$ and the power P_c, at which the denominator of Eq. (2-4) becomes zero, should be determined:

$$1 - P_c G_0 X_t = 0 . \qquad (2\text{-}20)$$

Noting that $E_4(0) = 1$ and defining α as

$$\alpha = \frac{E_1(0)\, E_3(0)}{2X(0)} , \qquad (2\text{-}21)$$

one obtains

$$X_t = X_o \left[1 + \frac{2\alpha E_2}{1 - E_2} \right] . \qquad (2\text{-}22)$$

As shown below, the relations

$$0 < \alpha < 1 , \qquad (2\text{-}23)$$

$$|E_2| < 1 \qquad (2\text{-}24)$$

are valid for any frequency and, with these two conditions, Eq. (2-20) can never be satisfied for any power or frequency. Hence, system stability is assured at any power level.

The relation $0 < \alpha < 1$ can be obtained in the following way. If the steady state reactivity feedback caused by a unit power change, X(0), is identical to the reactivity feedback caused by an isothermal temperature increase of the whole reactor equal to the increase of the average coolant temperature, one would have:

$$X(0) = \frac{E_1(0)\, E_3(0)}{2} , \qquad (2\text{-}25)$$

since $E_1(0)/2$ represents the increase of the average coolant temperature per unit power increase and $E_3(0)$ is the isothermal temperature coefficient of reactivity. By combining Eqs. (2-21) and (2-25), one can see that $\alpha = 1$. Equation (2-25) would approximately hold for a symmetrical homogeneous reactor. Since, in a symmetrical heterogeneous reactor the fuel temperature at power is higher than that of the coolant, X(0) is certainly larger than in the homogeneous case and α must be less than 1. Note that $\alpha > 0$ means simply that the power coefficient and the isothermal temperature coefficient have the same sign.

For the Fermi reactor the calculated values are [19];

$E_1(0) = 1.25\,°F/Mw$ at full flow
$E_3(0) = 0.291¢/°F$
$X(0) = 0.2458¢/Mw$ at full flow

and thus $\alpha = (1.25 \times 0.291)/(2 \times 0.2458) = 0.74$ at full flow. At reduced flow the value of α is somewhat higher, but still less than unity.

E_2 is the transfer function giving the change in reactor inlet coolant temperature resulting from the transmission around the coolant loops of a change in the reactor outlet coolant temperature. The gain E_2 of this transfer function is smaller than unity, since a passive thermal system cannot act as an amplifier.

From Eqs. (2-20) and (2-22), it is seen that, since the phase of the zero power transfer function G_0 never exceeds 90°, and since X(0) is a negative number, Eq. (2-20) can only be satisfied if the phase of $\{1 + [2\alpha E_2/(1-E_2)]\}$ is at least 90°. That this is impossible, and that therefore instability is impossible when $f < f_2$, is demonstrated in the following paragraph.

The function $1 + [2\alpha E_2/(1-E_2)]$ is equal to $[1 + (2\alpha - 1)E_2]/[1-E_2]$; for the case where α has its maximum value of unity, this expression reduces to $(1 + E_2)/(1-E_2)$. From a simple geometric construction it can be seen that the phase of $(1 + E_2)/(1 - E_2)$ can never exceed 90° when $|E_2| < 1$, whatever the phase of E_2 itself is. In Fig. 2-3, the geometric constructions of $1 + E_2$ and $1 - E_2$ are given for an arbitrary phase angle for E_2. The particular phase angle that was used is not important because the results are independent of the phase angle of E_2. The phase angles of $1 + E_2$ and $1 - E_2$ are indicated as ϕ_1 and ϕ_2 respectively. The former is measured in a counter clockwise direction and is considered to have a positive value; the latter is measured in a clock-

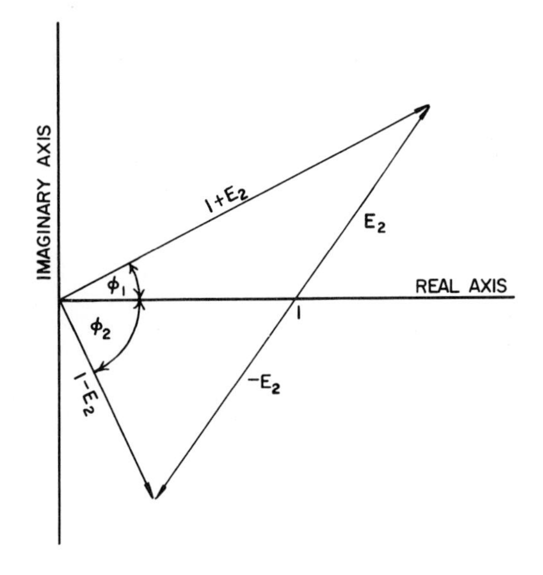

FIG. 2-3 Geometric construction for determination of phase angle.

wise direction and is considered to have a negative value. The phase angle of $(1 + E_2)/(1 - E_2)$ is thus $(\phi_1 - \phi_2)$. Consider now the case when $E_2 = 1$. For this case ϕ_1 and ϕ_2 will be at their respective maxima in an absolute sense, since by physical reasoning, $|E_2|$ can never be greater than unity. Therefore, $(\phi_1 - \phi_2)$ will also have a maximum when $|E_2| = 1$. Now when $|E_2| = 1$, the line formed by the vectors E_2 and $-E_2$ can be considered to be the diameter of a circle and the vector from the origin to unity a radius. Then that diameter and the vectors $1 + E_2$ and $1 - E_2$ become the sides of a triangle inscribed in a semicircle. By geometric reasoning this triangle is a right triangle and the angle $(\phi_1 - \phi_2)$ is a right angle. Thus, for this case, $(\phi_1 - \phi_2) = 90°$. By inspection it is obvious that when α is less than unity and $|E_2|$ is less than unity the included angle at the origin between the adacent sides of the resulting triangle will be less then 90°. This conclusion is valid for any value of the phase angle of E_2. Thus, there is no possibility of instability when $f < f_2$.

2.2.3 Comments on Possible Sources of Nonlinearities.

The foregoing example of the type of analysis which has been applied to a specific fast reactor design assumes that the response of the reactivity-affecting phenomena is linear with power and temperature or can be assumed to be linear over a limited power range. This is the assumption made in the case of the Doppler effect.

For the EBR-I, EBR-II, Dounreay and Fermi reactors, which have been subjected to a detailed stability analysis, this assumption is reasonably well based. Over the range of interest fuel expansion and coolant expansion are linear enough to justify the assumption.

It is in the calculation of the contributions of the structural components to the power feedback that a serious possibility of nonlinearity arises. For example, the effect of clearances should be considered. The core and other components of a reactor require mechanical clearances in order to permit assembly. These clearances may be eliminated to a certain extent before operation begins by either mechanical design, e.g. the tight core concept of the Fermi reactor [19], or by deliberate capitulation to the problem, e.g. the freely bowing core of EBR-II [18]. However, regardless of the steps taken to minimize clearances, it is still likely that during low power operation structural deformations will be influenced by existing clearances and up to some point of power or temperature or temperature gradient it is likely that the structural effects on reactivity will be somewhat different than beyond that point.

Effects of temperature on the elastic properties of the structure, possible local boiling and other phase changes, as well as nonlinear fuel expansion should also be considered for a given design.

2.3 Power Coefficients

The previous sections have illustrated the need for detailed knowledge of the response of a given reactor to change in power, flow, or inlet temperature.

In principle, this requires the calculation for each reactivity-affecting phenomenon of the value

$$X_N(i\omega, \lambda) = \int^V \rho_N(\vec{r})\, T_N(\vec{r}, i\omega, \lambda)\, d\vec{r},$$

where

$X_N(i\omega, \lambda)$ = complex, flow-dependent power coefficient of reactivity,

$\rho_N(r)$ = reactivity contribution per degree temperature rise of phenomenon N at point r, and

$T_N(\vec{r}, i\omega, \lambda)$ = complex, flow-dependent transfer function between power and temperature at point $\vec{r}$ [62].

The integral is to be carried out over the volume V which includes all points $\vec{r}$ relevant to ρ_N.

Except for simple cases the value of $X_N(i\omega,\lambda)$ is difficult to obtain and a detailed knowledge of the structure of the reactor and of the fuel element design is required in order to set up the interrelated heat transfer and mechanical or hydrodynamic relations involved. However, the asymptotic power coefficient, i.e., the value of $X_N(i\omega, \lambda)$ when $\omega \to 0$ is somewhat less difficult to obtain. The calculation of this characteristic will be illustrated for two cases; first where relatively simple relationships exist between ρ_N and T_N, and the second where the physical effects of a temperature change on a reactivity-affecting phenomena are more complex.

2.3.1 Case 1: No Mechanical Effects

The core sodium expansion effect on reactivity at some point dV can be expressed

$$d^2k = -\rho_{Na} V_{Na} \alpha_{Na} D_{Na}(r, z)\, dV\, dT, \qquad (2\text{-}26)$$

where

d^2k = change in reactivity due to sodium expansion,

ρ_{Na} = sodium density at some base temperature,

V_{Na} = volume fraction of sodium at dV,

α_{Na} = volumetric expansion coefficient of sodium per degree temperature rise, and

$D_{Na}(r, z)$ = local sodium reactivity worth in terms of reactivity per unit mass.

$D_{Na}(r, z)$ is the distribution of sodium worth throughout the core, obtained by multigroup calculations, or measurements in a critical experiment. At present considerable uncertainty exists [30f, 37d] in the calculation of this coefficient.

The local sodium expansion temperature coefficient of reactivity is given by:

$$\frac{d^2k}{dT} = -\rho_{Na} V_{Na} \alpha_{Na} D_{Na}(r, z)\, dV. \qquad (2\text{-}27)$$

If the temperature of a fuel pin or plate is raised it will expand both laterally and longitudinally. The lateral expansion forces sodium out of the core resulting in a local reactivity effect

$$d^2k = -2\alpha_f V_f \rho_{Na} D_{Na}(r, z)\, dV\, dT\,, \qquad (2\text{-}28)$$

where

α_f = linear coefficient of thermal expansion for fuel pin including cladding, and
V_f = volume fraction of fuel in core weighted to account for cladding.

The lateral expansion effect on reactivity is then

$$\frac{d^2k}{dT} = -2\,\alpha_f V_f \rho_{Na} D_{Na}(r, z)\, dV\,. \qquad (2\text{-}29)$$

The longitudinal expansion of a fuel pin or plate effectively moves a portion of the fuel to the end of the core. This produces a reactivity effect

$$d^2k = -\alpha_f V_f \rho_f \left[D_f(r, z) - D_f\left(r, \frac{H}{2}\right)\right] dV\, dT\,; \qquad (2\text{-}30)$$

here

α_f = linear coefficient of expansion of fuel,
V_f = volume fraction of fuel, and
$D_f(r, z)$ = local fuel reactivity worth in terms of reactivity per unit mass.

Then the axial expansion of fuel results in a local temperature coefficient

$$\frac{d^2k}{dT} = -\alpha_f V_f \rho_f \left[D_f(r, z) - D_f\left(r, \frac{H}{2}\right)\right] dV\,. \qquad (2\text{-}31)$$

The over-all Doppler temperature coefficient for a reactor is given by the change in reactivity resulting from the change in effective cross sections corresponding to the temperature change. The dependence of the Doppler coefficient on the absolute temperature T is specific to a particular reactor design as was discussed in Sec. 1. The dependence generally varies between $T^{-1/2}$ and $T^{-3/2}$. The effect thus needs to be averaged both over the diameter of a fuel pin and over the core radius.

Greebler [27] gives the following relationship for the effective Doppler temperature coefficient of a large ceramic reactor assuming a 1/T dependence on absolute temperature

$$C_1 = \frac{(dk/d\bar{T})_{eff}}{(dk/d\bar{T})} = \frac{\bar{C}_2\bar{T}\iint [P(r, z)]^2 \dfrac{dT(r, z)}{T(r, z)d\bar{T}}\, r\, dr\, dz}{\iint [P(r, z)]^2\, r\, dr\, dz}\,, \qquad (2\text{-}32)$$

where the integration is over the interval from $r = 0$ to $r = r_{max}$ and from $z = 0$ to $z = z_{max}$. Here C_2 is a correction factor for the radial temperature distribution in the fuel pins. $\bar{T}$ is the average absolute fuel temperature, and T(r,z) and P(r,z) are the fuel temperature and relative power density at core coordinates r and z.

Having once obtained suitable expressions for local temperature effects such as Eqs. (2-27), and (2-31), it is necessary to carry out the integration over the relevant reactor volume in order to obtain values of the over-all isothermal temperature coefficient for each phenomenon. To obtain the over-all power coefficients from the local temperature effects the product of the local temperature effects and the derivative of the local temperature with respect to power must be integrated over the entire core.

From simple heat transfer and heat balance analysis the following functional relationships can be developed for the steady state temperature distributions in the coolant and fuel:

$$T_c = \frac{fP}{F} + T_{c_1}\,, \qquad (2\text{-}33)$$

$$T_f = \frac{fP}{F} + gP + T_{c_1}\,, \qquad (2\text{-}34)$$

where

F = coolant flow rate in units of heat capacity per unit time per degree temperature change,
P = total core power,
T_c = coolant temperature,
T_f = local average fuel temperature, and
T_{c_1} = coolant inlet temperature.

Here f and g are analytical expressions particular to a given reactor describing the axial and radial power distribution. For the Enrico Fermi reactor [19]

$$f = \frac{B_rR}{4\pi}\,\frac{[\sin(B_zz) + \sin(B_zH/2)]}{\sin(B_zH/2)\, J_1(B_rR)}\, J_0(B_rr) \qquad (2\text{-}35)$$

and

$$g = \left[\frac{B_zB_rR}{4\pi Mh}\right]\left[\frac{\cos(B_zz)\, J_0(B_rr)}{\sin(B_zH/2)\, J_1(B_rR)}\right], \qquad (2\text{-}36)$$

where

r, z = radial and axial position coordinates,
B_r, B_z = (radial buckling)$^{-2}$, (axial buckling)$^{-2}$,
H, R = core height, core radius,
M = total number of fuel pins, and
h = average fuel pin heat transfer coefficient.

The average coolant temperature can be expressed as

$$T_c = \frac{P}{2F} + T_{c_1} \qquad (2\text{-}37)$$

and the average fuel temperature as

$$T_f = P\left(\frac{1}{2F} + \frac{1}{MHh}\right) + T_{c_1}\,. \qquad (2\text{-}38)$$

Then the derivative of coolant temperature with respect to power becomes

$$\frac{dT_c}{dP} = \frac{f}{F}\,. \qquad (2\text{-}39)$$

The derivative of the fuel temperature with respect to power is

$$\frac{dT_f}{dP} = \frac{f}{F} + g\,. \qquad (2\text{-}40)$$

2.3.2 Case 2: Consideration of Mechanical Effects

Most structural deformations which affect reactivity in fast reactors are induced by changes in the coolant temperature although it is possible that in a very high power density reactor, gamma heating of the structure could have a noticeable effect. The calculation of such deformations is complicated since they depend on heating of the coolant at points in the core which may be far removed from the actual deformation. An example will be given of the general method of calculation which was used in the estimation of core expansion and bowing effects in the Fermi reactor.

The fuel subassemblies in the Fermi reactor are supported axially and radially at both ends - at the top by holddown fingers and at the bottom by core support plates. With such an arrangement the radial temperature gradient which exists in the core tends to make the subassemblies bow into a shape convex toward the center of the core, moving the fuel inwards and giving rise to a positive power coefficient of reactivity. In order to prevent this, spacer pads were welded to the subassembly walls a little above the core midplane. The position of the holddown fingers is offset toward the core centerline so that permanent contact of the pads is assured when the holddown device is lowered. The radial temperature gradient therefore causes bowing of the subassemblies into an S-shaped curve and the pad location was chosen to equalize the reactivity effect of inward and outward fuel movement.

The core is actually a complex and statically indeterminate structure, but it was assumed to be an elastic structure with elastic supports obeying Hooke's Law and the principal of superposition was applied, allowing the deflection resulting from any distribution of temperature gradients to be evaluated as a sum or integral of the deflections caused by the local temperature gradients. Since the flow is the same in all the coolant channels in parallel, the temperature gradients throughout any horizontal cross section will change by the same fraction with a change in power, independent of the radial location. In treating the bowing coefficient as a local coefficient, it is therefore sufficient to consider only one independent variable, the axial component. The temperature gradients at a given axial coordinate and also the resultant deflections throughout the core and their reactivity effects can be expressed in terms of one characteristic quantity. It is convenient to choose for this quantity the following expression:

$$T_B(z) = \overline{T}_{ca}(z) - T_{cao}(z)\,, \qquad (2\text{-}41)$$

where $\overline{T}_{ca}(z)$ is the average temperature of the core subassembly wall at axial position z, and where $T_{cao}(z)$ is the value which the wall temperature would have at axial location z in a fictitious zero-power channel. If the inlet coolant temperature is constant, $T_{cao}(z)$ is equal to the inlet coolant temperature, a local bowing coefficient of reactivity $\rho V(z)$ can then be defined, so that the total reactivity effect of bowing is given by:

$$R_B = \int \rho_B(z)\, T_B(z)\, dz\,. \qquad (2\text{-}42)$$

Here the integral is carried out over the whole length of the subassemblies. After the shape of the bowing had been estimated, perturbation theory was used to calculate the function $\rho_B(z)$. It was found to be positive in the lower part of the core and negative in the upper part of the core and in the upper axial blanket. In the calculation of $\rho_B(z)$ the elasticity of the holddown fingers and the compressibility of the subassembly wall at the pad location were taken into account.

Since the subassembly walls are in contact with one another at the spacer pads, heating of the walls at that location will expand the whole core structure, thus contributing a negative coefficient of reactivity. The core expansion and resultant reactivity effect is a function of the average value and also of the radial distribution of the wall temperature at the pad location. This reactivity effect can be expressed as follows:

$$R_{cE} = \rho_{cE_1}\overline{T}_{cap} + \rho_{cE_2} T_{capo}\,. \qquad (2\text{-}43)$$

Here, $\overline{T}_{cap}$ is the average wall temperature at the pads, T_{cap} is what the wall temperature would be at the pads in a fictitious zero power channel and R_{cE} is the reactivity effect of core expansion. When the inlet coolant temperature is constant, T_{capo} is equal to the inlet coolant temperature. The two temperatures, $\overline{T}_{cap}$ and T_{capo}, characterize the temperature distribution at the pads. In the calculation of ρ_{cE_1} and ρ_{cE_2} the elasticity of the holddown fingers and the compressibility of the cans at the pad location were taken into account.

Similar reactivity coefficients were also calculated for isothermal heating of the lower support plates and of the holddown mechanism, taking into account the elasticity of the core structure in both cases.

In the calculation of the isothermal temperature coefficient of reactivity for core expansion, the coefficient is equal to the sum of ρ_{cE_1} and ρ_{cE_2} since, in this case, $\overline{T}_{cap}$ and T_{capo} are equal.

In calculating the power coefficient effect of core expansion at the pads, it is seen that the contribution of core expansion at the pads is

$$\frac{dR_{cE}}{dP} = -\rho_{cE_1}\frac{dT_{cap}}{dP}, \qquad (2\text{-}44)$$

where the value of the temperature derivative is the same as that given in Eq. (2-39). At equilibrium, the average wall temperature at the pads is equal to the average coolant temperature at the pads.

The contribution of bowing to the steady-state power coefficient of the reactivity is given by

$$\frac{dR_B}{dP} = \int \rho_B(z)\,\frac{dT_B(z)}{dP}\,dz\,. \qquad (2\text{-}45)$$

Here, $T_B(z) = \overline{T}_{ca}(z) - T_{cao}(z)$ as defined previously and the integration is carried out over the whole length of the subassemblies. At equilibrum, the can temperature is equal to the coolant temperature and $dT_B(z)/dT$ is proportional to the axial temperature profile of the coolant, varying from zero at the inlet to an outlet temperature which is determined by detailed calculation of the variation of temperature along different fuel channels. The detailed calculation of the structural distortions and the methods used are given in reference [63].

2.3.3 Time Dependence of Power Coefficients

The foregoing discussion of the calculation of different types of power coefficients has been based on a steady-state temperature distribution throughout the core. If the over-all power coefficient of any given reactor were negative and the time response of all the components of that power coefficient were identical then the problem of analyzing for fast reactor stability would be much simplified. This is not the case, however, and to enable the analysis of the stability of a given design it is necessary to have a thorough understanding of the relationship of the feedback mechanisms in time. As an example the instability of the EBR-I remained a mystery for some time, even though early in the analysis, Bethe [55] had predicted that the time delay associated with the delayed negative component of the power feedback should be approximately 10 sec. At that time no physical explanation from first principles of such a time constant was available. It remained for Storrer [62] to show that low frequencies of oscillation, a temperature signal transported by the coolant travels slower than the coolant by the factor $(\tau_f + \tau_c)/\tau_c$, where τ_f is the time constant for the fuel and τ_c is the time constant for the coolant. An evaluation of this ratio for the EBR-I, revealed a value of about 10 sec.

Storrer has considered five models of successively increasing complexity describing the temperature distribution throughout a fast reactor as a function of time and position [62]. The first model is a first-order or lumped model which assumes that the shape of the temperature distribution throughout the reactor remains the same during a transient as it is during steady-state operation, so that one characteristic temperature can be chosen to represent the over-all response. A first-order differential equation with constant coefficients was obtained which gives rise to a first-order time constant which is descriptive of the system.

The second model considered assumes that the fuel and the coolant are lumped separately and that the shapes of the temperature distribution in the fuel and the coolant remain the same during a transient as in steady-state operation. Two characteristic temperatures are considered, one for the fuel and one for the coolant, and a set of two first order differential equations is obtained. This second order system is solved exactly and then approximations to the solutions are made so as to be able to associate a first-order time constant to the fuel and coolant separately. It is found that the coolant time constant obtained for this model is the same as that found for the first model.

The third model considered is an axially continuous model in which the fuel elements are assumed to be lumped radially. The model is reasonably exact, except it assumes that the shape of the temperature distribution during a transient in the fuel element cross section is the same as it is in steady-state, or in other words, that heat flow across the fuel-coolant interface is related to the local average fuel and coolant temperature as it is in steady-state operation. This model is quite realistic when the fuel elements are thin. It gives rise to partial differential equations so solutions cannot be described in terms of simple time constants.

The fourth model considered is the axially lumped and radially continuous one which had been analyzed in detail by Sandmeir [64]. No assumptions are made regarding the radial temperature distribution across the fuel elements.

The fifth model considered is the so-called "exact" model, in which no assumptions are made regarding the temperature distributions. However, minor assumptions are made that there is no heat conduction in the axial direction in either fuel or coolant, that there is no axial mixing of the coolant and that all material properties, such as densities, specific heat, thermal conductivity, heat transfer coefficients, and velocity, are independent of temperature.

The four approximate models are based on assumptions which are borrowed from steady-state theory. As expected, the solutions obtained from these models are good approximations to the exact solutions at low frequencies. In comparing the exact solutions to the lumped solutions, the important conclusion is drawn that the solution for the average temperatures which were obtained from the three axially lumped models are asymptotically correct for inlet coolant temperature oscillations but are not correct, even asymptotically, for power oscillations.

The important thought to be left with the reader is that great care must be used in assigning time dependence to power feedback. In particular it is necessary to understand the time scale of the transients which are to be analyzed using the time constants which are calculated.

2.4 Experimental Methods of Investigating Feedback Relationships

The experimental methods which have been used to investigate the feedback relationships of fast reactors are no different in principle than those used for thermal reactors. Basically, the technique is to change the reactivity of the reactor by some amount in one of a number of ways and to observe the time response of the power [55]. The reactivity change may be made by control rod motion, by inlet temperature variation or by a flow change. Depending on the desired information the reactivity change may be slow or fast, large or small.

The most popular technique for obtaining the reactor transfer function,

$$G(i\omega) = \frac{G_0(i\omega)}{1 - PX(i\omega)\,G_0(i\omega)},$$

is the use of periodically varying reactivity which is small in magnitude to avoid nonlinearities and capable of operation over a wide range of frequencies. This is the so-called reactivity oscillator. $G_0(i\omega)$ is measured at zero power where only neutron kinetics is involved and then $G_P(i\omega)$ is measured at various powers, P. The feedback transfer function $X(i\omega)$ can then be obtained directly as

$$X(i\omega) = \frac{1}{P}\left(\frac{1}{G(i\omega)} - \frac{1}{G_0(i\omega)}\right),$$

where the notation is the same as that used previously.

The oscillator may consist of a rotating eccentric poison or fuel section which alters the reactivity more or less sinusoidally during rotation. A fairly small oscillator is desirable so as to fit in the reactor easily and disturb the power distribution as little as possible. This limits the distance which the fuel or poison can move radially in the course of one rotation and requires a fairly large absolute amount of reactivity in the oscillator for a small amplitude of reactivity variation. For instance in the Fermi reactor the total worth of the oscillator rod is about 65¢ and the resulting reactivity amplitude on rotation is only 2¢.

This characteristic of the Fermi oscillator may lead to problems in future, larger reactors. It is desirable to have available a cyclic frequency ω ranging from 0.01 cycles/sec to perhaps 10 cycles/sec, in order to cover the range of possible instabilities and to reach high enough frequencies so that the plateau in the value of the transfer function is reached. With somewhat higher frequencies it may be possible to obtain a measure of the prompt neutron lifetime [54]. The upper limit on frequency imposes severe mechanical constrictions on all devices, especially reciprocating devices when the sodium environment and usual distance of the core from accessible areas is considered.

The experimental convenience of the reactivity oscillator is probably sufficient to assure continued development of mechanical techniques for accomplishing its functions.

A technique which offers many attractive diagnostic features is the variation of inlet temperature as a function of time. This results in heat flow patterns which may result in different effects from those which would result if the reactivity were changed directly. The programming of any but fairly small temperature changes to the inlet presents some difficulty in the more complicated power reactor designs and in fact may be limited by thermal stress considerations in the reactor inlet piping or other components. Abrupt changes in flow may be used under some conditions to change the reactivity by changing the axial temperature distribution in the reactor but the axial temperature distribution in the reactor but here again in large systems it is usually quite difficult to change the flow rapidly enough to cause short term transient temperature distributions.

Several variations have been used for experimenting by use of control rod motion. In one technique the reactor is simply set on a given period and allowed to rise in power to some limit. The conduct of a number of experiments over a wide range of periods can provide much feedback information. Unfortunately a large part of the desired information can only be obtained by using periods short enough so that serious safety questions may be posed.

In addition design limitations on rate of change of temperature may restrict the rate of change of power level to such low values that important kinetics information cannot be obtained. This technique is obviously not the one to use in any case until any possibility of a prompt positive power coefficient has been ruled out by other experimentation.

A second possibility using control rod motion is to impose different ramp rates of reactivity introduction beginning at various power levels. In addition, small step changes in reactivity and exponential introductions have certain analytical advantages.

The basic rule for conducting experiments should always be that a new reactor should not be operated at a new power level until the results of operation at lower power levels have been extrapolated by reliable methods to show that no dangerous instability exists at the new higher power level. In reference [55], Bethe developed the theory of the use of the reactivity oscillator technique and proposed a criterion for establishing the safety of proceeding to a higher power level after conducting transfer function measurements at a lower power. If at the lower power P_1, the maximum ratio of the transfer function over all frequencies explored to that at zero power is

γ, then the power level to which one could go with good assurance of stability is $P_2 = P_1 \gamma/(\gamma - 1)$. Thus if at P_1 it is found that the maximum ratio of the P_1 transfer function and the zero power transfer function is 1.1, this criterion indicates that no instability would occur up to $P_2 = 11\ P_1$. This criterion should not be used blindly of course because the basic assumption of linearity of the power feedback is implicit in it. However, the use of steps of a modest fraction of the ratio γ/γ-1 greatly increases the safety of this method of progressive increase in power.

The technique of noise analysis has been applied to the Dounreay reactor and has many highly attractive features for fast reactor use. Boardman [65] has considered two applications of noise analysis, as a means of determining power-feedback relationships and as a monitoring device. He concludes generally that much development is necessary in applying the techique but that it appears to be useful for detecting resonances of small amplitude, and for monitoring the kinetic characteristics of the reactor for changes due to burnup or other physical effects. It is pointed out that the determination of the reactor transfer function by this technique appears of dubious validity unless it can be assured that the major sources of reactivity noise have a white spectrum, at least over the frequency band examined.

2.5 Discontinuities in Operation

Several phenomena have been conceived which might come into effect during operation in a more or less discontinuous way, perhaps as a result of a local temperature, power, or flow anomaly or as a result of a change in the general environment under which operation is proceeding. If such discontinuities affect reactivity, either directly or indirectly, it is clear that consideration must be given to them in design. Four examples are cited below, but it should be emphasized that careful consideration must be given to each individual design to try and detect any potential sources of discontinuities which may be peculiar to that design.

2.5.1 Sodium Boiling

In the analysis of potential power or flow excursions in a fast reactor, many situations can be envisioned in which the temperature of the sodium reaches or exceeds the boiling point of approximately 1600°F (870°C) at atmospheric pressure.

Nicholson [60] was the first to point out that boiling of sodium in the reactors presently being considered is a severe safety problem because, under many conditions of interest, the boiling cannot be stable above a certain power level which depends upon the details of the reactor design. The onset of boiling in a coolant channel may lead to coolant starvation in that channel with subsequent melting of the fuel. There are many factors that enter the problem, but the single most important feature that distinguishes the behavior of the sodium-cooled fast reactors from other power reactors which can boil in a stable manner is the low pressure of the coolant. The large specific volume of the vapor at these low pressures would require, for stability, velocities too high to be sustained by the available pump pressure. It is possible to calculate the pump pressure required for two-phase flow as a function of the mass flow rate, the heat input, and the exit pressure. The methods are all approximate and there is very little experience in applying them to sodium systems. However, the fluid flow properties of sodium are sufficiently similar to those of water so that the methods of calculation developed for water may apply.

Two-phase pressure drop calculations have been made for the Fermi reactor [17]. In these calculations, mixed flow was assumed with no liquid vapor slip. Frictional losses are accounted for by the use of two-phase friction factors based on the average density and viscosity of the mixture. The basic equations assumed were:

$$\Delta P_M = \frac{W^2}{A^2 g}(V_0 - V_i) \tag{2-46}$$

$$\Delta P_F = \frac{f\rho V^2 \Delta h}{g D_E} \tag{2-47}$$

$$\Delta P = \overline{\Delta P_M} + \Delta P_F \tag{2-48}$$

where, with consistent units, ΔP is the pressure drop, A the flow area, f the friction factor, ρ the fluid density, g the gravitational constant relating mass to forces, D_E the hydraulic diameter, W the mass flow rate, h the length of fuel element for which ΔP is evaluated, $(V_o - V_i)$ the difference between the specific volume at the outlet of Δh and at the inlet of Δh, and V is the fluid velocity.

Figure 2-4 shows the result of these pressure drop calculations. The normal operating pressure drop in the Fermi reactor at full power is about 65 psi (4.43 kg/cm^2) but the flow must be 1/3 normal for boiling to occur at normal full power so that the pressure drop that the pump can supply under these conditions is only about 10 psi ($\sim$0.70 kg/cm^2) This is the condition described by the uppermost curve of the figure. It is seen that no point on the curve corresponds to stable boiling. Indeed, it seems certain that one would have to go to conditions of power and flow below even the lowest curve shown which is 25% normal full power, before the condition of stable boiling could be achieved. But at about 5% or 10% of normal full power the natural convection effects will become strong enough to permit a stable boiling.

Calculations were conducted to determine the heat removal potential of EBR-II for boiling with natural circulation [17]. Assuming no slip between sodium vapor and liquid and a coolant inlet temperature of 700°F (371°C) it was found that 50% of full power can be removed at the onset of boiling and a peak removal rate of 85% of full power occurs when the exit sodium vapor volume fraction is 0.5.

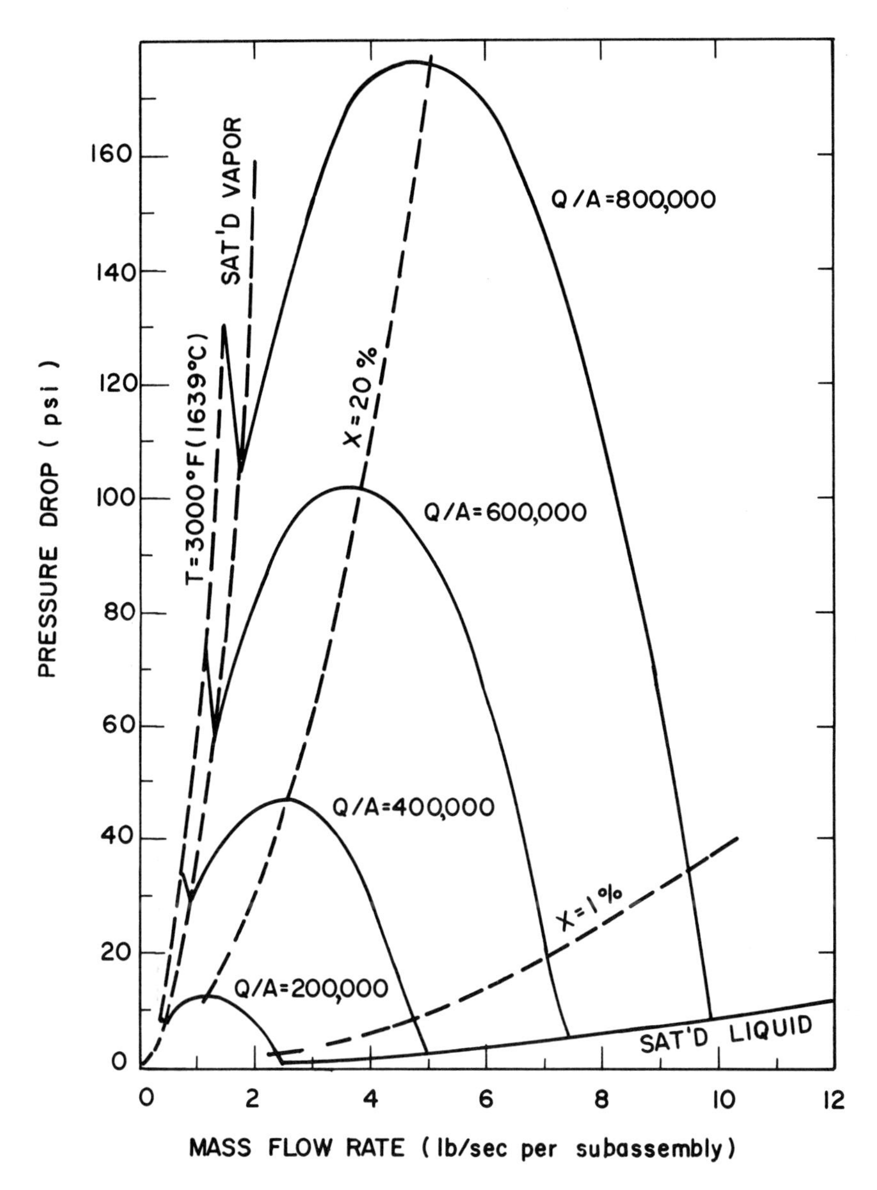

FIG. 2-4 Two-phase flow of sodium and sodium vapor.

While the estimates for the two reactors appear to be widely different, the deviation can largely be resolved by consideration of the design differences between the two reactors. The normal full power per unit flow area is lower for the EBR-II and the chimney effect is greater. Also, the estimation of natural convection flow can only be trusted to within a factor of two or so.

The assumption that there is no liquid vapor slip is somewhat pessimistic. Water systems have been analyzed [66] assuming a constant vapor-liquid slip ratio throughout the coolant channel. (See also the Fluid Flow chapter.) This method can be applied to the sodium boiling system with some modifications as required by the higher ratio of pressure drop to absolute pressure that are considered in such systems.

The momentum term is given by:

$$\Delta P_M = -\frac{W}{Ag}\,\Delta[XV_v] - \frac{W}{Ag}\,\Delta[(1-x)]\,V_\ell \qquad (2\text{-}49)$$

where X is the vapor quality, V_v the vapor velocity, and V_ℓ = liquid velocity.

If a constant slip ratio $S = V_v/V_\ell$ is assumed, then the pressure gradient equation can be integrated along the length of a coolant channel. For a better estimate of the friction term, the Lockhart-Martinelli friction multipliers can be used. For reasonable assumed slip ratios this method gives qualitatively the same results as the more simplified method shown above, in which boiling cannot be stable in this type of reactor except at lower power levels where natural circulation effects become important. In reference [19] a detailed analysis is made of the effects of coolant starvation to a single subassembly in the Enrico Fermi reactor. This starvation is envisioned as occurring due to blocking of the entire subassembly or to a gradual reduction in the flow through the subassembly as a result of either fuel swelling or gradual accumulation of foreign material within the fuel subassembly or its entrance nozzle. If the flow of coolant were suddenly stopped in one subassembly, it is shown that the fuel would melt in a matter of two or three seconds. If then the fuel were to be moved axially from the bottom upwards due to sodium pressure and from the top downwards due to gravity, an increase in reactivity for a central subassembly in the Fermi reactor of approximately 30¢ is predicted.

For reactors having a positive sodium void effect, at least in the central portions of the core, there may be additional important reactivity effects associated with sodium boiling. Analysis of accidents influenced by such a positive sodium void effect will be considered in Sec. 3.

2.5.2 Fuel Phase Changes

In fast reactors fueled with metallic fuel rods or plates, consideration should be given in design to the effect of phase changes of the metal as temperatures vary, or as time and burnup accumulate. An example is given in reference [67] pertaining to the fuel to be used in the first loading of EBR-II.

The fuel in this case is uranium 5 wt.% fissium alloy in which a phase transformation from a predominantly alpha phase to a predominantly gamma phase occurs at approximately 1020°F (550°C). In this case, the transformation if completed within the core of EBR-II would result in a fuel material density decrease of approximately 1.4%. If this density change were isotropic, the long term reactivity drift which would occur during the transformation would require about 0.14% ($\Delta k/k$) compensation. This transformation is slow and the total effect on reactivity even if the material should revert to the alpha phase would be small. However, in each particular reactor design, situations where a similar phase change could occur should be evaluated in the early consideration of the fuel to be used.

2.5.3 Reversible Density Changes - Swelling and Ratcheting

Uranium oxide or mixed oxides of uranium and plutonium appear to present attractive possibilities as fuel for fast reactors. A number of different reactivity discontinuities have been conceived for high burnup of fuels of this type. The one of most concern is the gradual expansion of the uranium oxide within the fuel pin or plate as burnup is accumulated, resulting in a long term reactivity decrease similar to that observed in LAMPRE and described in Sec. 2.6.6.

The occasion for concern with respect to a phenomenon of this sort is that a relatively mild power or temperature transient might cause a reversion of the less dense fuel to its original state. If this reversion were rapid and if the total amount of reactivity regained were large, the effect of the original transient could be magnified greatly.

A similar problem has come to be termed "ratcheting". This problem is conceived as the gradual elongation of a column of fuel pellets inside a cladding tube within the core of a fast reactor. It is imagined that, with each temperature increase the tube undergoes, the fuel elongates and then mechanically is unable to return to its original position when the temperature is decreased. Thus over a long period the effective length of the column of pellets would increase and the reactivity would decrease. If this small change in reactivity were accommodated each time by a change in control rod setting, one could gradually build up a large inventory of reactivity.

2.5.4 Collapse of Voids in Coolant

The Dounreay reactor and the Russian BR-5 reactor both observed relatively large amounts of reactivity tied up in voids due to the entrainment or solution of cover gas within the primary coolant [42,68]. From a safety standpoint such a situation leads to the potential for the sudden insertion of the reactivity which is being displaced by the voids. This reversion to the original reactivity state could occur due to a change in flow, temperature, or pressure which reduced drastically the amount of void within the coolant. In future reactor designs it may be that the primary sodium system will be operated under pressure in order to relieve the problem of unstable sodium boiling in fuel channels. In such a case if a large amount of gas were dissolved in the core and if the sodium expansion coefficient were positive, a reduction in cover gas pressure would allow a large positive void component of reactivity to be introduced to the reactor. Since it is always difficult to absolutely guarantee against a pressure decrease in a large system, it would appear undesirable to permit appreciable cover gas entrainment in a pressurized reactor having a large positive sodium expansion coefficient.

2.6 Fast Reactor Operating Experience

A large number of small, very fast, solid metal assemblies of different types have been operated, primarily to obtain basic neutron physics and neutron kinetics data; e.g. Godiva, Topsy, and Jezebel [3,5]. The pertinent kinetics experience from these reactors is summarized in the Criticality chapter.

In addition, many critical assemblies of various composition have been run, both as engineering mock-ups and as research experiments in programmatic investigations of fast reactor physics [3,7]. Thus, the ZPR-III critical assembly was used to mock-up EBR-II, the Fermi reactor, and Rapsodie, while the Zeus [6] facility was used for a similar purpose in the case of the Dounreay reactor. The operation of these devices has been extremely important and has furnished data which enable much more reliable calculations of the temperature and power coefficients of the respective reactors.

At present, however, operating experience on fast reactors is extremely limited. Only six fast reactors of sufficiently high power density to exhibit typical temperature feedback characteristics have been in operation as of January 1964. Of these, four, the EBR-I, the Dounreay fast reactor, the Russian BR-5 and LAMPRE, have operated at their design power levels and temperatures. The Fermi reactor and EBR-II became critical in August and November 1963, respectively, and operation has been at power levels too low to result in any temperature feedback effects.

2.6.1 The EBR-I

The EBR-I, Argonne National Laboratory's Experimental Breeder Reactor Number 1, was designed in 1948-1951 with two fundamental objectives in mind. The first objective was demonstration of the ability of a fast neutron reactor to breed, i.e., to produce more fissionable material than is destroyed. The second objective was a demonstration of the feasibility of the use of liquid metal cooling systems in power-producing reactors. Both of these objectives were fulfilled during operation of the reactor through the first two core loadings, designated Mark I and Mark II.

However, two dynamic characteristics were observed during experimental operation, both of which warranted further investigation. One of these was the presence of a prompt positive power coefficient of reactivity which was over-balanced by a slower negative coefficient; the second was an oscillatory instability which became evident under abnormal operating conditions when the ratio of power to flow was 50% greater than design conditions.

In each of these two core loadings, the fuel elements consisted of enriched uranium slugs, NaK-bonded to stainless steel tubes 0.448 in (11.4 mm) outer diameter, spaced 0.494 in. (12.55 mm) on centers. The tubes were positioned at the bottom by the tube-plate, and above the core by a lower shield plate, the elements passing through holes 0.460 in (11.7 mm) in diameter. The Mark-I loading had spacer ribs on each steel jacket but considerable clearances still existed from fuel element to fuel element. In the Mark-II loading the spacer ribs were omitted, thus permitting 0.046 in (1.17 mm) separation of the rods.

The isothermal temperature coefficient, as measured over the range from 38°C to 200°C (100°F to 392°F) of the Mark-I reactor was -1.37 inhours/°C (393 inhours = 0.01 Δk); in the Mark-II loading, it was essentially the same. The power coefficient of both loadings was also negative as evidenced by the additional reactivity required to maintain an increased power level with constant coolant flow and constant inlet temperature.

However, early in the operation of Mark I it was noted that, even at low power, a sudden decrease in flow caused an increase in power, followed by a slow decrease to a level below the initial power. The converse was also true, an increase in flow, with the dependent reduction in fuel rod temperature, caused a sudden drop in power followed by a gradual increase in power to a level above the initial level. Asymptotically, the behavior agreed with the observations of the negative steady state power coefficient mentioned previously, but it was obvious that a prompt positive temperature or flow coefficient existed, i.e., a positive coefficient with a short time constant as compared to the constant of the asymptotic negative coefficient.

A second evidence of the same phenomena is shown in Fig. 2-5. In the Mark-I loading, as a preliminary investigation, the reactor was made critical at very low power and with convection cooling only. The reactor was allowed to rise in power on a relatively fast period of 32 sec. When fuel slug temperatures started to increase, a progressive decrease in period was measured [40].

It was during an attempt to run the same experiment in the Mark II core, that the reactor was not shut down in time, and the power rose sufficiently high to melt part of the fuel [59].

A positive Doppler coefficient of reactivity was thought to be a possible source of this prompt positive coefficient. However, calculations and rough preliminary measurements indicated that the Doppler effect in EBR-I should be too small to account for the effects observed. A second possible explanation of the prompt positive effect was that the fuel rods bowed in the temperature gradient created by the fission distribution across the core. Sufficient clearances existed between rods to permit bowing which could account for the effects observed.

The tendency toward oscillatory instability which was observed early in the operation of the Mark II core was investigated by means of transfer function measurements. Some typical results are given in Fig. 2-6, 2-7, and 2-8. A series of theoretical efforts to explain these results culminated in a suggestion by Kinchin [56] that the reactivity feedback consisted of a system having two power coefficients. The positive coefficient, said to be due to bowing, had a short time constant characteristic of core rods. The negative coefficient was attributed to structure and was given an appropriate longer time constant. This model did not fit the data perfectly. The calculated peaks were much less

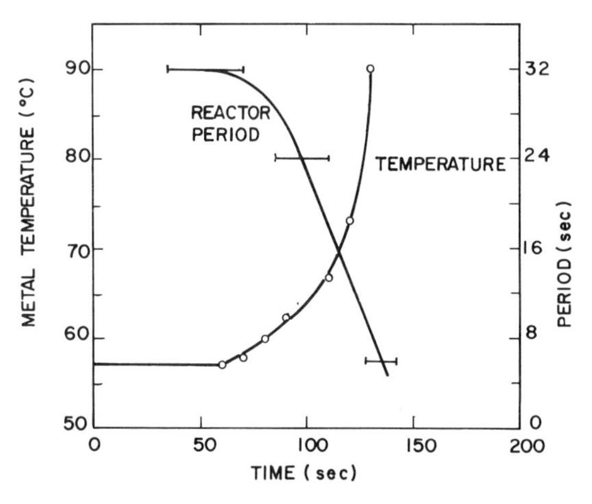

FIG. 2-5 EBR-I Mark I period change with metal temperature.

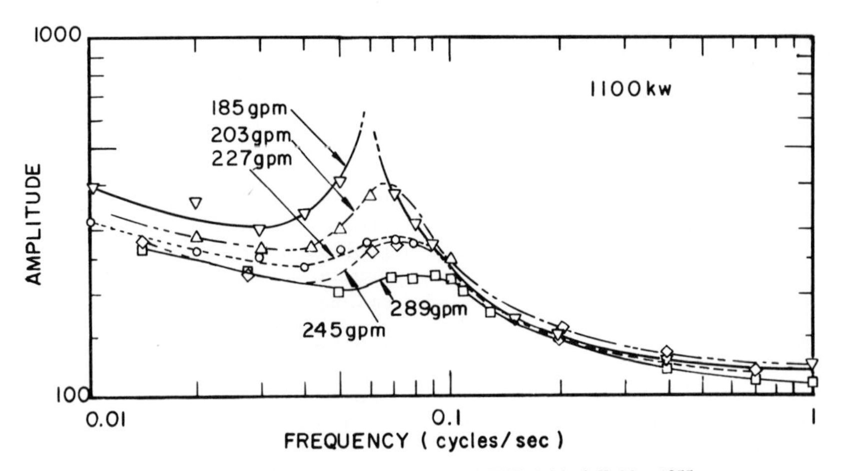

FIG. 2-6 Transfer function measurement in EBR-I, Mark II, May 1955.

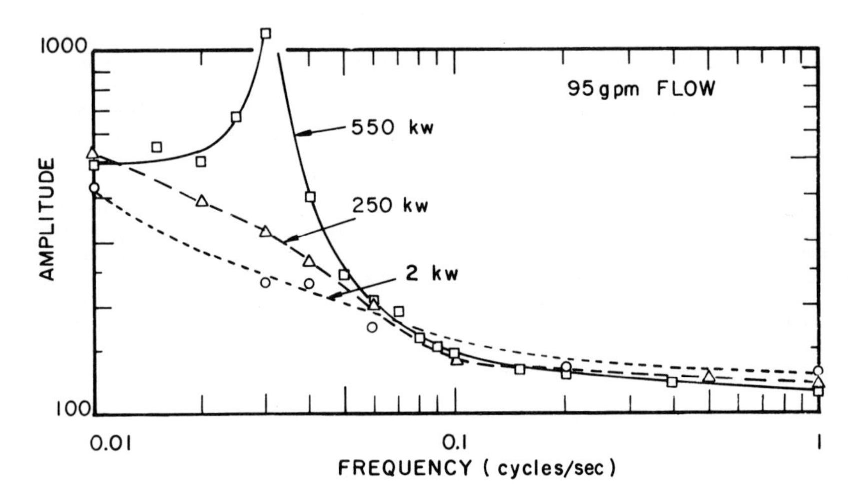

FIG. 2-7 Transfer function measurement in EBR-I, Mark II, May 1955.

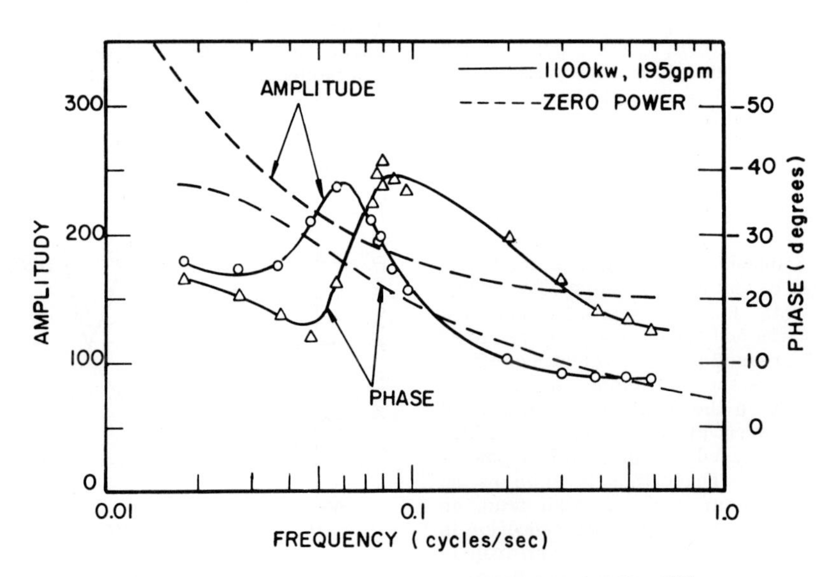

FIG. 2-8 Transfer function measurement in EBR-I, Mark II, May 1955.

sharp than the observed ones. Kinchin suggested that a transport time delay for the passage of coolant between the core and the tube plate above the core would sharpen the calculated peaks and produce better agreement.

Bethe [55] adopted such a model and solved for the parameters which best fit the experimental data. He found a positive power coefficient of 0.8×10^{-3} ($\Delta k/k$) per Mw, a negative coefficient of 3.8×10^{-3} ($\Delta k/k$) per Mw, and a transport time delay of 10 sec. His model for this flow condition became:

$$\frac{\Delta k}{k/\mathrm{Mw}} = \frac{0.8 \times 10^{-3}}{1 + 0.6 i\omega} - \frac{3.8 \times 10^{-3}\, e^{-10 i\omega}}{(1 + 0.6 i\omega)(1 + 1.5 i\omega)}$$

This analysis suffered from a lack of a physical mechanism from which the coefficients could be calculated. Also, it predicted some higher frequency resonances which were not observed. Various hypotheses have been put forward to explain these effects. Eventually theoretical techniques developed by Storrer [62] coupled with careful examination of the behavior of the shield plate under thermal transients seem to have provided a reasonable explanation of the delayed negative coefficient. The positive coefficient fit reasonably well with bowing concepts.

To prove that bowing was the cause of the prompt positive coefficient in the Mark II cores, to demonstrate that stable fast reactors could be designed and operated, to provide extensive experience in stability experiments, and to provide a large body of accurate information for analysis, the "rigid" Mark III core of the EBR-I was constructed and tested. In this reactor the uranium fuel was coextruded with its zirconium jacket. Ribs on each jacket were fitted to provide intimate contact with neighboring fuel elements, and the entire assembly was clamped. A major experimental stability program has been completed on the EBR-I, Mark III core, and the final report issued [69]. The following is a summary of the results obtained.

"From the results of stability studies it may be concluded that the operational characteristics of a fully-ribbed and rigid Mark III core loading were governed by feedback processes which guarantee safe and stable operation under normal operating conditions. The stabilizing ribs increased stability in two ways:

(1) The ribs provided strong radial coupling between fuel rods, and between fuel rods and hexagonal cans, with the consequence that a large radial contribution to the power coefficient of reactivity was realized.

(2) The ribs eliminated, for all practical purposes, positive feedback effects arising from the inward bowing of fuel rods during power increases. Absolutely no evidence of positive reactivity effects was noted for the fully-ribbed and rigid core.

"The extrapolations of full power, full flow test data indicated that the reactor could be brought into a resonant condition only at power levels exceeding 1000 Mw. Nominal full power operation was limited to 1.2 Mw by design. The hypothetical instability condition was only of academic interest since the limiting power level was not controlled by instability but by heat removal capability. In a practical sense, the fully ribbed Mark III loading could never have been made unstable.

"The observed non-linearities in the power coefficient of reactivity posed no serious operational problems over the range of power levels associated with the tests. Except for the region between zero power and 200 kw, where the power coefficient was small, but still negative, the power coefficient was always very strongly negative. Non-linearities however, complicate the interpretation of test data in terms of a single unified model which could be applied over wide ranges of power, flow, and temperature conditions. The origin of the non-linearities appears to have been a consequence of power and temperature sensitive clearances existing between fuel rods and hexagonal cans (as many as 36 fuel rods can be accommodated in a single hexagonal can), and between the hexagonal cans. Attempts to define such systems mathematically were further frustrated by differences between the expansion coefficient of fuel and structural materials. The association of such extremely tenuous clearance systems with power coefficient non-linearities cannot be placed on a firm analytical basis. The descriptive interpretation of the effects caused by such clearance systems is nevertheless important, not only for an understanding of Mark III behavior, but as a guide for non-linear phenomena in other fast reactors.

"The power coefficient was also shown to be sensitive to inlet coolant temperature. Significantly larger static and dynamic power coefficients of reactivity were associated with increased coolant inlet temperatures. Feedback time dependence was insensitive to variations in coolant inlet temperature. This effect was probably due to the strong temperature dependence of the uranium-zirconium fuel alloy expansion coefficient. The effects of increased feedback at higher temperature were relatively minor.

"The influence of structural feedback was also observed in Mark III, fortunately to a much smaller degree than in the earlier Mark II design. At the higher inlet temperatures, approximately 13% of the static power coefficient was associated with a feedback process operating with a 400-second time constant. Because of the extremely long time constant, it is assumed that the physical process was associated with some massive portion of the structure at some downstream location. The existence of this delayed component had little effect on stability because it could not be observed with frequencies greater than 0.003 cps.

"Attempts to enhance and measure possible feedback effects arising in the radial breeding blanket were unsuccessful. It may be concluded that the assumed possible feedback mechanisms due to the pre-heating of inlet coolant by the transfer of heat between core outlet and blanket inlet in series coolant flow were non-existent or extremely small. Pre-heating conditions were similar for both Mark II and Mark III; therefore, it may be concluded that strong blanket feedback mechanisms were not the source of the Mark II delayed negative component.

"As the coolant flow was reduced (at constant power), three feedback phenomena varied:

(1) The power coefficient increased inversely proportional to the change in flow rate.

(2) The time constant for fuel expansion increased.

(3) The effective transport lag increased.

All of these changes tend toward instability since they tend toward increased feedback amplitudes and phase lags at the important frequencies. Extrapolations of one-third coolant flow test data concluded that the reactor would reach resonance instability near 10 Mw; this is more than twenty times the power level consistent with one-third flow. Coolant flow reductions at constant power tend to lower the indicated resonance power level. Extreme flow reductions were inadvisable, not only because of the lowered resonance power level but also because fuel element temperatures might have exceeded design limits.

"The results of removing the stabilizing ribs from fuel elements demonstrated the existence of a strong positive reactivity effect which was associated with the inward bowing of fuel rods. The extrapolation of kinetic data obtained by removing the stabilizing ribs from one-third of the fuel elements gave $+2.06 \times 10^{-6}$ $(\Delta k/k)$/kw for the full core with ribs removed. Since the prompt negative component of reactivity associated with fuel, coolant, and blanket expansion was -2.21×10^{-6} $(\Delta k/k)$/kw, the response of the "non-ribbed" reactor would have still been governed by a very small, but negative, prompt power coefficient.

"Removal of the stabilizing ribs also produced an increase in the magnitude of the delayed structural component of the power coefficient. An empirical model describing the dynamic and static behaviors of the partially-ribbed core gave -2.21×10^{-6}, $+0.543 \times 10^{-6}$, and -0.873×10^{-6} $(\Delta k/k)$/kw for the respective prompt negative, rod-bowing, and delayed structural components of the power coefficient. These values may be correlated with those deduced empirically for the prompt negative and delayed structural power coefficient components for the fully-ribbed core, namely, -2.21×10^{-6} and 0.330×10^{-6} $(\Delta k/k)$/kw, respectively. Rib removal, therefore, introduced two feedback processes: (1) rod-bowing which was prompt and positive; and (2) a negative and extremely delayed structural component. It was clear, however, that the strong increase in the negative structural term did not affect the resonance stability of the partially-ribbed core (and the same is indicated for extrapolations to the non-ribbed core), since reactor oscillations at low frequencies (~ 0.02 cycle/sec) could not sense this delayed effect.

"The Mark III core, whether fully-ribbed and rigid, fully-ribbed and loose, and even partially ribbed, was extremely stable, a major departure from Mark II experience. The most undesirable feature of the Mark II had been the prompt positive, rod-bowing reactivity effect. The inclusion of stabilizing ribs and a system of tightening rods in the Mark III eliminated, or at least minimized beyond detection, similar effects in Mark II.

"The presence of a delayed negative component to the power coefficient of reactivity was another important observation in Mark II. This appears to have been the result of delayed expansions in the structure of the lower shield plate. Experiments with a dummy shield plate have established the existence of radial movements of the shield plate ligaments which, because of their contiguity with the fuel rod extension, effectively modified the core size following power changes. This effect was clearly negative and operated with a time dependence consistent with that analytically deduced from the Mark II test data. Structural features such as the Mark II shield plate were eliminated in Mark III and no significant, long-delayed reactivity coefficient was observed in Mark III.

"In summary, the Mark II instability was enhanced by the prompt positive and delayed negative power coefficient components which were the result of design peculiarities. The elimination of the perforated shield plate system and the addition of the stabilizing ribs to the Mark III fuel rods have resulted in a reactor whose performance was stable under all credible operation conditions."

After the completion of the operations using the Mark III U^{235} core, a plutonium loading was fabricated and installed and operated. A curious reactivity effect was observed during Mark IV operation. The external cooling loop is capable of maintaining inlet NaK coolant temperatures of either 70°C or 200°C (158°F or 392°F). When the reactor has been operated using the 70°C inlet temperature, a given critical control rod position is observed. Then upon switching to the 200°C (392°F) inlet temperature, it is found that the first startup results in criticality at the same control rod position as when the inlet temperature was 70°C (158°F) but that the second startup invariably reaches criticality with substantially less control rod removed; i.e., the core has gained reactivity. The same results are obtained if the process is reversed.

No satisfactory explanation is yet available to explain this effect but investigation is centered on phase transformations and mechanical expansion anomolies. The complicated phase structure of plutonium metal leads one to suspect a density increase due to a phase change at the higher temperature.

2.6.2 The Dounreay Fast Reactor

The United Kingdom Atomic Energy Authority has constructed and is operating the Dounreay Fast Reactor which is located on the Pentland Firth near Thurso, Caithness, at the northern extremity of Scotland. The reactor [37h,71] is fueled by 45% enriched uranium and cooled by downflow of 70–30% NaK alloy. The original core was made up of an assembly of 367 fuel elements in the shape of hollow cylinders clad on the outside by niobium and on the inside by vanadium. The fuel elements are assembled to form a roughly right circular cylinder 21 in. (53.3 cm) in diameter and height. A core tube nest is used which gives lateral restraint to the fuel elements at the center, top, and, bottom of the core. This arrangement allows individual elements to bow under thermal gradients but prevents any cumulative effects.

The reactor was initially brought critical on November 14, 1959, and has been operated since then up to its design power level of 60 Mw(t). Operating experience has been reported and three pertinent items of experience are summarized in the following [68,72].

Gas Entrainment: Nitrogen is used as the cover gas over the NaK coolant in the Dounreay reactor. Shortly after the reactor achieved criticality, and when the coolant flow was increased for the first time to its full design value, the reactor was observed to suddenly lose about 1.0% reactivity. At the same time the coolant volume was observed to increase by 3-4%. Investigation revealed that nitrogen was being drawn into the coolant due to by-passing from the high pressure inlet region to lower pressure regions downstream. These conditions were corrected. An insignificant amount of gas apparently still remained in the NaK as was shown by pressurizing the reactor to 35 psig (2.4 atm gage pressure) with a resulting increase in reactivity of only 0.7×10^{-4} ($\Delta k/k$). This could be accounted for if only 1/4 liter of gas were uniformly dissolved in the NaK in the core.

Reactivity Feedback: Reactivity measurements were made at various power levels and coolant flow rates using a calibrated control rod. A series of measurements was made with core flow at 20% 30%, and 40% of full flow and the power varied in steps. The results are shown in Fig. 2-9. The linear behavior indicated that the fuel elements were fully restrained by the core structure. Another series of measurements was made in which the coolant flow rate was varied while the power was held constant. These results are shown in Fig. 2-10. Below about 20% flow, the curve is highly nonlinear and buoyancy forces acting on the downflowing coolant are suspected of causing flow maldistribution with accompanying deviations from the temperature distribution occurring at higher flows.

In addition to the measurements made using the calibrated control rod a reactivity oscillator was also used for a series of measurements to determine the frequency response of the reactor. The oscillator consisted of a vertical rod inserted in the core which, when rotated, moved an eccentrically placed boron absorber in a sinusoidal manner relative to the core center.

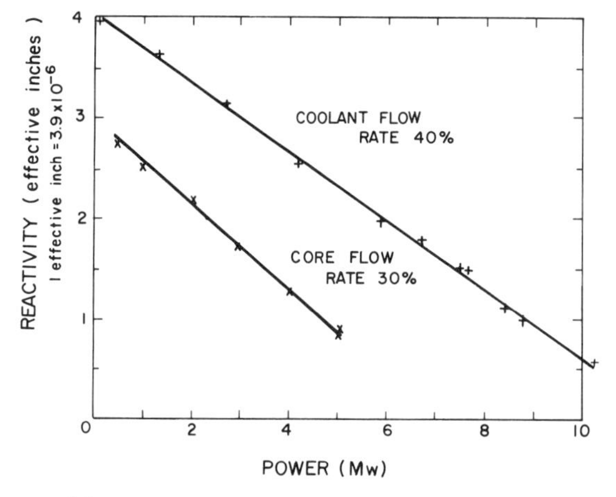

FIG. 2-9 Change of reactivity with power at constant coolant flow rate, Dounreay reactor.

The amplitude of the frequency response of the reactor to the reactivity oscillator is shown in Fig. 2-11 for 7 Mw power and 22% and 30% flow. The response shows no sign of any instability. In Fig. 2-12 the phase response corresponding to Fig. 2-11 is shown. The increase in phase lag at frequencies above about 3 radians/sec is contrary to theory since the lag should be asymptotic to zero at high frequencies. This anomaly was originally attributed to torsional distortion of the oscillator rod drive at high frequencies together with the increased effect of neutron thermalization in the graphite surrounding the neutron detectors. It was later discovered that the phase lag did in fact decrease with frequency and that instrumentation errors were at fault.

Reference [72] makes the following general conclusions regarding the kinetic studies up to 7 Mw.

"The amplitude of the frequency response is always less than that obtained at low power. This, and the absence of any signs of a maximum, shows that the reactor is very stable over the range of conditions tested. There is no evidence which points to a possible instability at higher powers than that so far achieved. The reactivity feedback is reduced linearly with power, showing that no significant bowing of fuel elements has occurred.

"The general features of the separated feedback were as anticipated; some small differences, however, are not yet explained, although suggested causes can be given.

"These differences were obscured by the comparatively low power, torsional distortion in the oscillator rod at high frequencies and some sticking of the rod at low frequencies. The operational necessity for full coolant flow in the blanket did not permit closer examination of the low frequency anomalies."

As of January 1964 no published references were found on stability measurements at higher power.

Noise Analysis: Boardman [65] has reported on the first known fast power reactor noise measurements and analysis which were carried out on the Dounreay reactor in May, 1963.

The measurements on which the analysis is based were made during a 30-min period when the reactor was operating at 60 Mw with a coolant flow of 3×10^{-6} lb/hr (1.31×10^{-6} kg/hr), corresponding to a velocity of about 16 ft/sec (4.9 m/sec). Fluctuations in the coolant inlet temperature were observed using a thermocouple mounted in the inlet plenum. The autocorrelation function for this signal is shown in Fig. 2-13, together with the expression

$$\Psi(\tau) = 0.440 \exp(-|\tau|/4.8) - 0.0067 \exp(-|\tau|/0.6) ,$$

which describes the function. The 4.8-sec time constant is attributed to the thermocouple while the 0.6-sec time constant is unexplained. The inlet temperature noise is then interpreted as noise having an autocorrelation function

$$\Psi(\tau) = 0.4 \exp(-|\tau|/0.6) ,$$

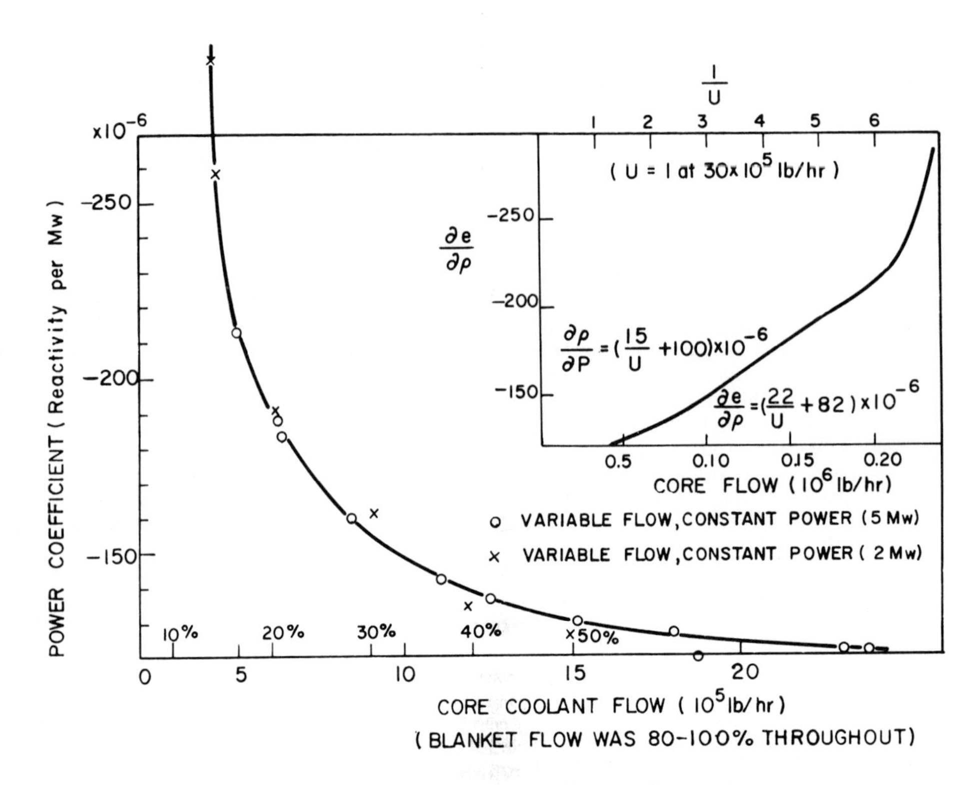

FIG. 2-10 Variation of power coefficient with coolant flow, Dounreay reactor.

as measured by a thermocouple having a 4.8-sec time constant.

The autocorrelation function of fluctuations in reactor power was obtained and is shown in Fig. 2-14 in two parts, from $\tau = 0$ to $\tau = 1.5$ and from $\tau = 1$ to $\tau = 25$. The ordinate is expressed as percentage fluctuation of power. There is a conspicuous resonance at 6 cycles/sec. It is shown in the reference that the method developed by Storrer [62] predicts theoretically a resonance at almost exactly this value.

The autocorrelation function for the core outlet temperature was also determined but its meaning was inconclusive. Cross correlations were obtained between coolant inlet temperature and power and between the inlet and outlet temperatures. The power fluctuations appeared to be only slightly due to the inlet temperature fluctuations but it appeared that there was a small positive coefficient associated with such temperature fluctuation of magnitude about $5 \times 10^{-6}/°C$ ($\sim 3 \times 10^{-6}/°F$).

2.6.3 The Russian Fast Reactor - BR-5

The BR-5 is a plutonium-oxide-fueled sodium-cooled fast reactor designed for a power level of Mw [73]. Power operation began in June, 1959, and has continued intermittently until the present. The fuel in the original core consists of sintered plutonium oxide contained in 0.016 in. (0.41 mm) thick stainless steel tubes about 0.2 in. (5.1 mm) in di-

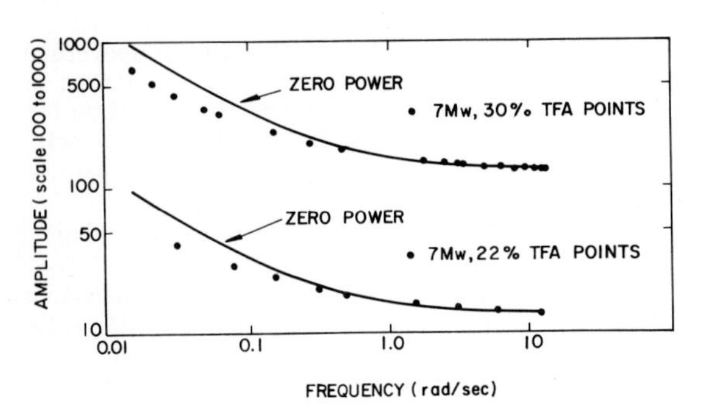

FIG. 2-11 Amplitude of response function |G| at 7 Mw, Dounreay reactor.

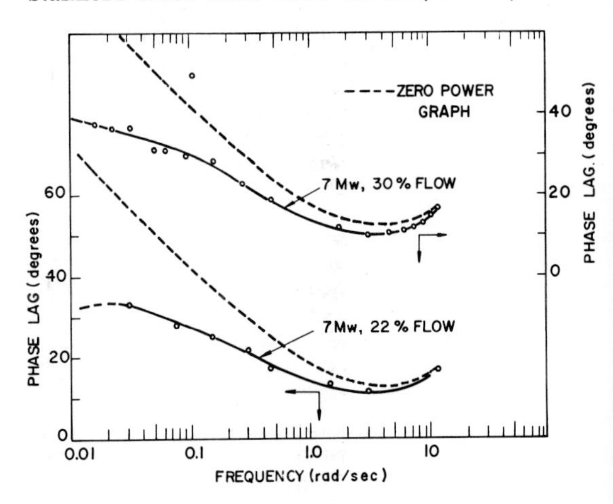

FIG. 2-12 Phase of response function |G| at 7 Mw, Dounreay reactor.

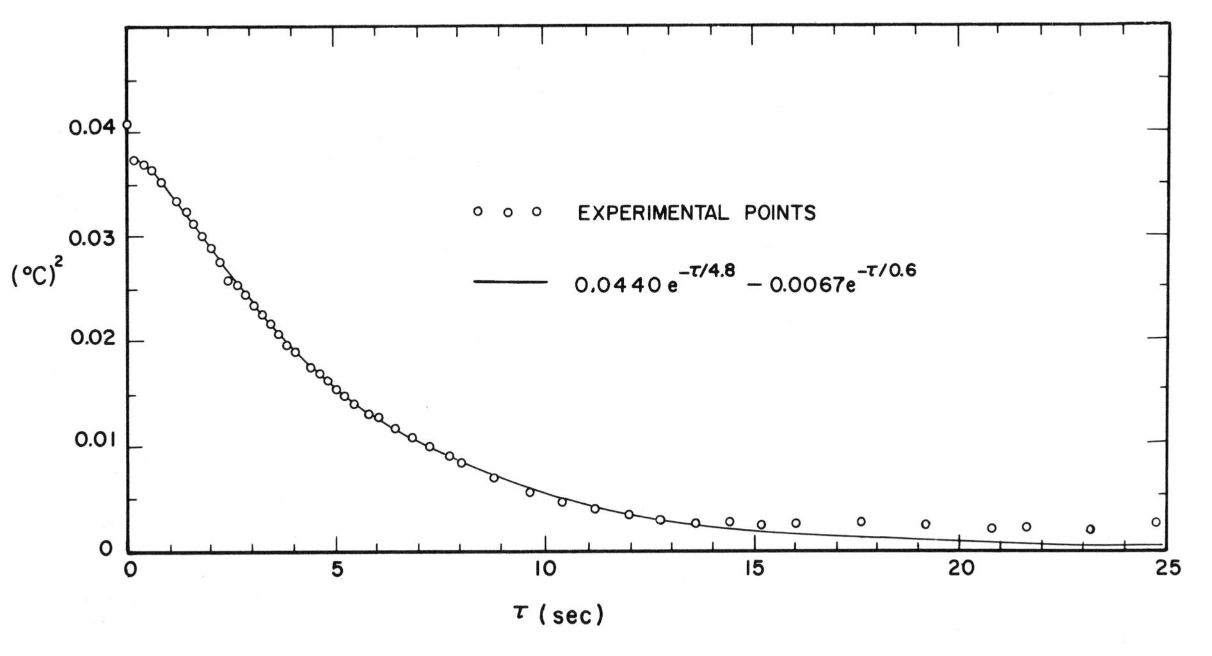

FIG. 2-13 Autocorrelation function of Dounreay reactor inlet plenum thermocouple.

ameter. The plutonium oxide region is about 11 in. (28cm) long. Nineteen such tubes comprise a fuel element, being assembled in a 0.020 in. (0.51 mm) thick hexagonal stainless steel can 1.023 in. (26.0 mm) across flats. A wire is wound helically around alternate tubes to act as a spacer. Eighty-eight fuel elements make up the full core. The fuel elements are located in a bottom tube plate but are free to move laterally at the top. The fuel tubes are welded to the bottom of a subassembly, but are free to move at the top. Sodium flow is upward.

The temperature coefficient of reactivity has been measured relative to the inlet temperature and found to be $-28 \times 10^{-6}/°C$ ($15.6 \times 10^{-6}/°F$), in good

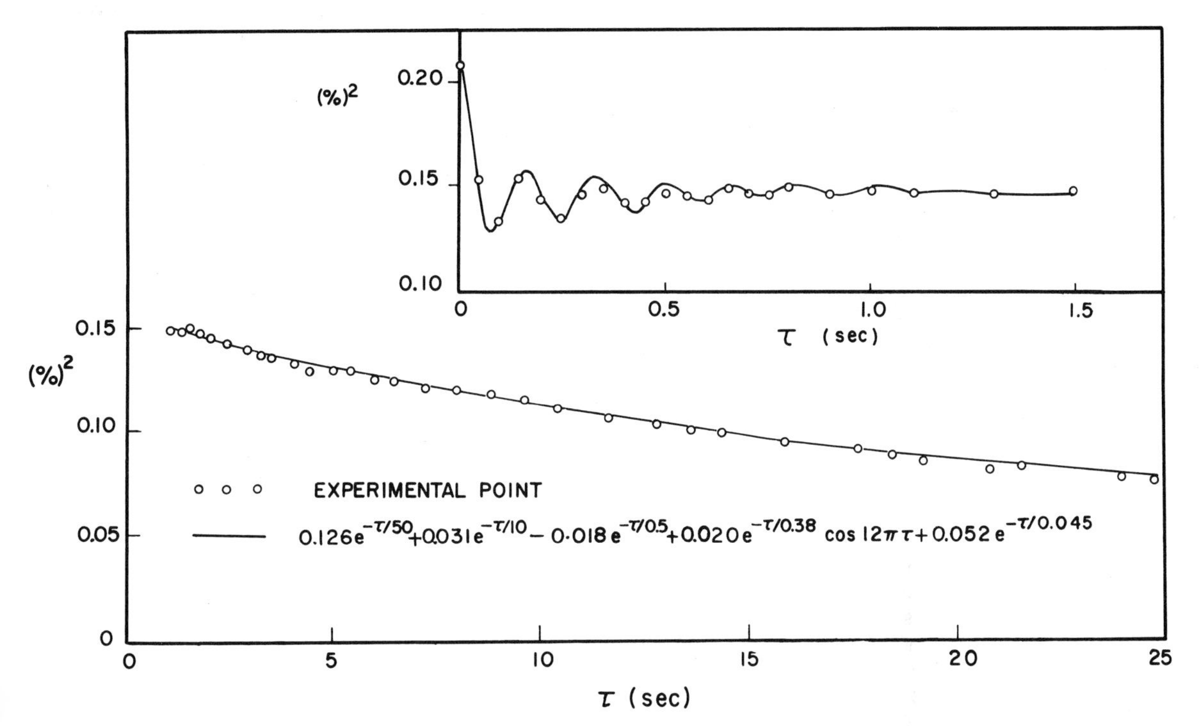

FIG. 2-14 Autocorrelation function of Dounreay reactor power.

agreement with design predictions. The asymptotic power coefficient of reactivity also is said to agree with calculations.

Transient experiments have been reported [37i] which indicate that, although the asymptotic power coefficient for forced flow is negative and has a magnitude of about -2×10^{-4}/Mw, there is a substantial and relatively prompt positive effect. This positive effect shows up quite soon after a step reactivity increase at full flow and gradually is overwhelmed by slower negative effects. However, under natural circulation conditions, the effect persists over times longer than 60 sec. It is not clear from the reference either what the actual test conditions were or what explanation is proferred for the positive effect.

The BR-5 also underwent a period of operation [42] during which over 2$ of positive reactivity was present in the form of cover gas bubbles in the core. No unusual transient effects were observed but some concern for safety was apparent. No reactivity oscillator tests have been run as yet, but it is understood that such tests are planned.

2.6.4 LAMPRE

LAMPRE is the first fast reactor designed to use a fuel which will be liquid under operating conditions. The core of the reactor is an array of about 140 Ta capsules containing 90% Pu - 10% Fe alloy fuel [74]. Each capsule is 0.43 in. (10.9 mm) in diameter and the array forms a core in the shape of a right cylinder about 6 in. (15.2 mm) in height and in diameter. The core is cooled by upward flowing sodium and heat removal capability up to 1 Mw is provided. The prompt neutron lifetime is about 8×10^{-9} sec. The total isothermal temperature coefficient has been measured and is reported at several temperatures. At 160° C (320° F) with the fuel alloy frozen in the β phase, the total coefficient was measured as -1.3¢ /°C (-0.72¢/°F), of which -0.5¢ /°C (-0.28¢/°F) were due to fuel expansion, -0.5¢ /°C due to Na expansion and ~0.4¢/°C (~0.22c /°F) structural expansion. When the system temperature was raised to 480°C or 896°F (the fuel alloy melts at 411°C or 772°F) the total coefficient increased to -2.9¢ /°C (-1.6¢ /°F) of which -2.0¢/°C (-1.1¢ /°F) was due to fuel expansion, the other components remaining unchanged.

The power coefficient has been evaluated by comparing the net change in reactor power following a power demand made by reducing the inlet temperature to the net change in reactivity required to meet the power demand. At full flow of 133 gpm (8.4 liter/sec) the measured power coefficient was found to be about -.50¢/kw increasing to about -.57¢/kw at one-half flow.

In Fig. 2-15, the response of the reactor to a relatively slow decrease in inlet temperature is shown. The reactor was initially operating at 100 watts, essentially zero power, and a 4°C (~7°F) inlet temperature change was imposed more or less linearly over a 15-min period. The power is seen to overshoot its eventual asymptotic value of 10 kw by about 40% but the essential stability of the system is obvious and illustrates

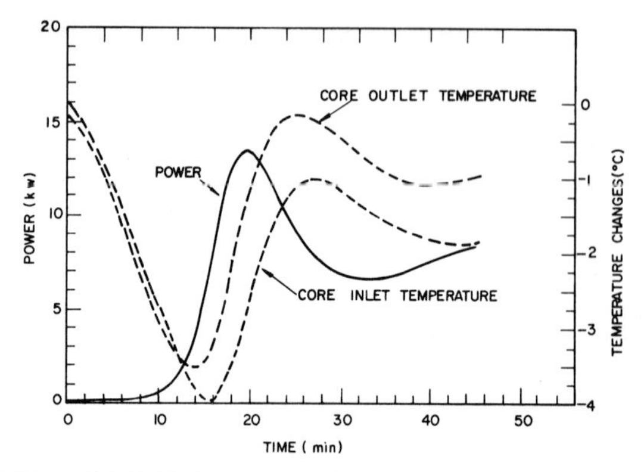

FIG. 2-15 LAMPRE response to 10 kw power demand from 100 w.

the strong damping effect of the strong prompt negative fuel expansion coefficient.

In Fig. 2-16, the effect of a power demand made when the reactor was already operating at power is illustrated. Here the inlet temperature was reduced about 16°C (29°F) over a 4-min. period, resulting in a smooth increase in reactor power from 45 kw to about 145 kw with no overshoot.

Experiments at various power levels from 1 watt to 400 kw have also been carried out using a reactivity oscillator over a frequency range from 10^{-1} to 10^{+2} radians/sec.

The transfer functions obtained are given in Fig. 2-17. The gain curves all show the increase in stability which results as the power level is raised and verifies the greater damping observed in the response of the reactor in Fig. 2-16 as compared to that in Fig. 2-15.

The transfer function data fit fairly well a simple feedback function of the form

$$G(i\omega) = \frac{x(i\omega)}{1 + i\omega\tau}$$

where $G(i\omega)$ is the frequency dependent transfer function, $x(i\omega)$ the reactor power coefficient, ω the input frequency (radians/sec), and τ a characteristic time constant (sec).

In the following tabulation the effect of sodium flow on these parameters is seen.

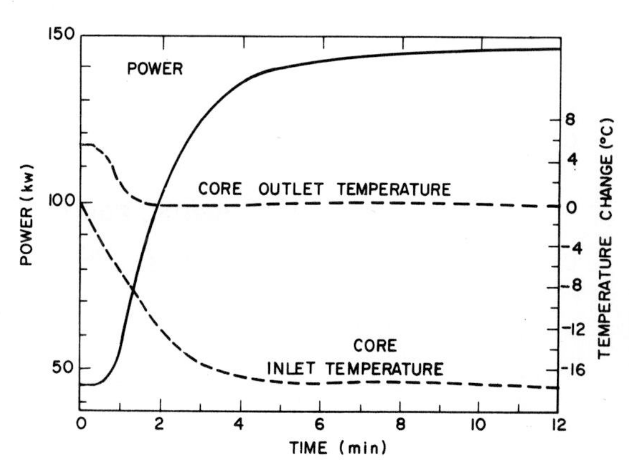

FIG. 2-16 LAMPRE response to 100 kw power demand from 45 kw.

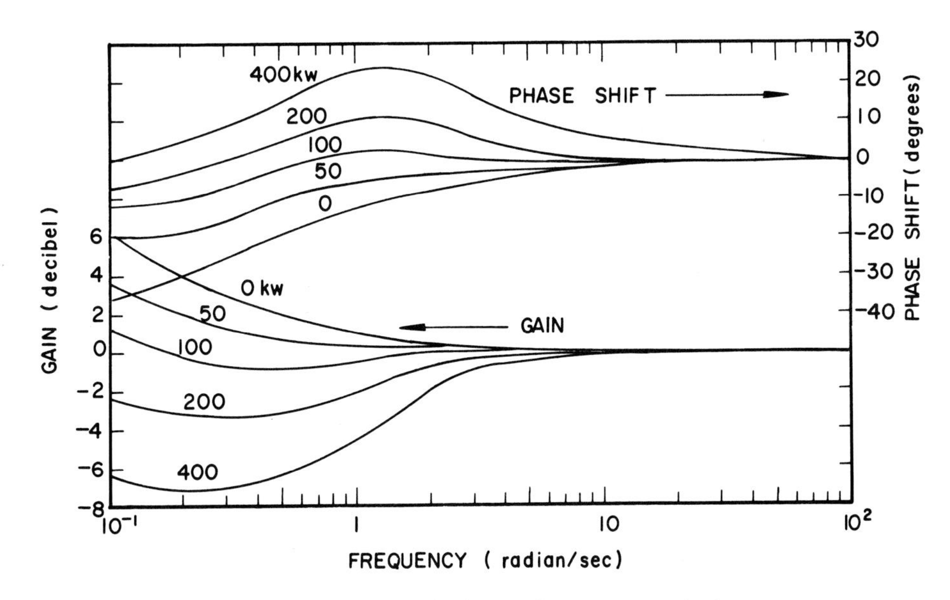

FIG. 2-17 LAMPRE transfer functions for various power levels.

Constant	*Full Flow*	*Half Flow*
x(iω)	– 0.47 ¢/kw	– 0.63 ¢/kw
τ	1.3 sec	1.8 sec

The short time constants found confirms that the principal contributions to the transfer function are due to core effects. The values of the power coefficient are in fair agreement with those obtained by power demand experiments.

An extremely interesting and characteristic aspect of the operating behavior of LAMPRE is the effect on reactivity of fission gas bubbles in the liquid fuel [75, 77]. The loss in reactivity as burnup accumulates is shown in Figure 2-18 for the initial core loading (Core I) and for the second loading (Core II) which began operating in April, 1962. The high rate of loss at ~55 Mwh burnup on Core I was due to high power operation. The Core II loading was characterized by higher purity than Core I and reached a total burnup of over 100

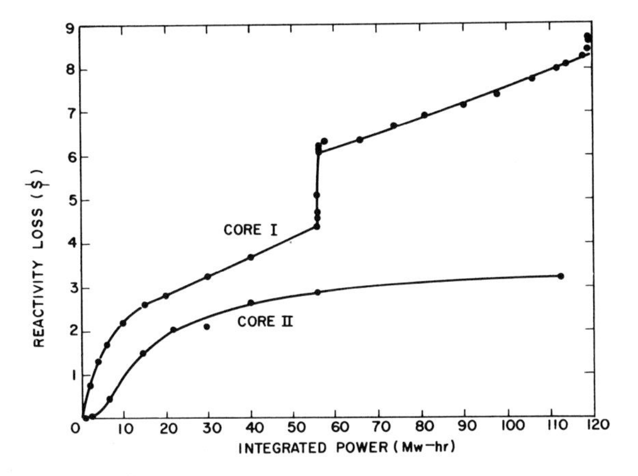

FIG. 2-18 Loss of reactivity in LAMPRE during operation.

Mwd during which a reactivity loss of 12$ accumulated [76]. From this loss it has been estimated that a volume fraction of 0.05 fission gas bubbles was contained in the fuel at the end of life.

2.6.5 The EBR-II

The EBR-II is designed for a thermal power capacity of 62.5 Mw [14]. Sodium flow is upward through the core which initially consists of pins of uranium -5 wt. % fissium alloy. The enrichment is 48.4%. The pins are grouped in subassemblies containing 91 pins and the core consists of an array of 70 subassemblies which are fixed at the bottom but can move to a limited extent at the top.

Wet criticality was achieved on November 11, 1963. As of December 18, 1963, the only experimental measurement pertinent to this chapter was that of the isothermal temperature coefficient of reactivity. This was found to have a value of - 1.78 inhour/°C (- 1.00 inhour/°F) and to be almost linear over a temperature range of 72°C (130°F). This value is only slightly greater than the calculated value of - 1.91 inhour/°C (- 1.01 inhour/°F) [16]. The comparison of the above experimental value, which was determined with sodium in the core, to the value found before sodium was placed in the reactor is of some interest. The dry value was found to be - 1.08 inhour/°C (- 0.60 inhour/°F) which gives some indication of the effect of sodium on the reactivity.

Extensive reactivity oscillator and other kinectics experiments are planned.

2.6.6 The Enrico Fermi Reactor

The Enrico Fermi [19] reactor is the largest fast reactor yet built in terms of size and power capacity. It is located near Monroe, Michigan, about 30 miles south of Detroit.

The reactor is designed for a power level of 200 Mw with its initial core. This core consists of pins of U-10 wt.% Mo alloy, enriched in U^{235} to 25.6%. The pins are clad with zirconium by co-extrusion and are 0.158 in. (4.01 mm) in diameter and 30.5 in. (77.4 cm) long. Subassemblies of pins are comprised of 140 pins assembled inside a square stainless steel tube with an internal grid structure to maintain proper pin spacing. The coolant is sodium which flows upward through the subassemblies. Each subassembly is fixed radially at the bottom by a bottom structure plate and vertically and radially at the top by a hold-down device. Spacer pads are located at about the core midplane and the action of the hold-down forces these pads into contact before reactor operation begins.

The reactor achieved criticality on August 23, 1963, with a loading of 99 subassemblies. Since that date the initial phases of an extensive nuclear test program have been completed.

The isothermal temperature coefficient of reactivity has been measured over the range 400°F-600°F (204°C-316°C) and found to be nearly constant with a value of -0.87 inhour/°F (-1.57 inhour/°C), compared to the calculated value of approximately -1.0 inhours/°F (-1.8 inhour/°C).

Reactivity effects resulting from changes in sodium flow have been found to be below the level of observability over the range from 0 to 9×10^6 lbs/hr (4.1×10^6 kg/hr). The reactor argon cover gas pressure was raised from its normal value of ~0.4 psig to ~10 psig (~0.03 atm to ~0.68 atm gage pressure) to determine whether any cover gas bubbles were dissolved or entrained in the sodium. No effect was observed within the 1 inhour measurement accuracy under the combinations of primary and secondary sodium system flows which were used.

An extensive series of experiments to verify the calculated stability of the reactor is planned. A reactivity oscillator will be installed and operated over a wide range of flow and power combinations. Measurements will be made to enable noise analysis.

3 DYNAMICS OF MODERATE ACCIDENTS

3.1 Introduction

Early considerations of the safety of fast reactors centered about the possibility of an explosive energy release occurring in the core, either as a result of a large and sudden external reactivity insertion or of the action of a positive power coefficient which was both strong and closely coupled to the power as a function of time.

As designs for actual reactors evolved it appeared that it was possible by proper design to preclude beyond all reasonable doubt the ability of the control or fuel handling system by itself to insert reactivity rapidly enough to cause an explosive event. The study of possible causes of positive power coefficients which were strong enough to cause rapid internal reactivity insertions was stimulated considerably by the demonstrated autocatalytic behavior of the EBR-I under certain power and flow conditions. The Doppler effect in U^{235} and mechanical distortions are now believed to be well enough understood so that dangerous positive power coefficients due to these phenomena can be precluded by proper design. In 1959 while carrying out calculations towards design of a large advanced power breeder it was discovered that under certain conditions a reduction in density of the sodium in a large fast reactor could result in an increase in reactivity. To a large degree concern regarding this phenomenon has supplanted earlier concerns regarding the Doppler effect in U^{235} and mechanical distortion of fuel in the consideration of autocatalytic, accident-initiating mechanisms.

As designs progressed, investigations of abnormal characteristics were made and it was found that conditions could exist in which the melting point of the fuel was exceeded. Since the concentration and amount of fissionable material in a fast power reactor is sufficient to produce a number of critical masses if arranged compactly, a number of investigations have been made to consider the redistribution and reassembly problem. This area of meltdown investigation remains one of the central problems in fast reactor safety studies today and will be discussed in detail in Sec. 4.

The nature of the meltdown problem is so complicated that theoretical and experimental work is proceeding to attempt to determine whether the maximum explosive events which have been postulated for presently operating reactors and for reactors now under design are overly pessimistic, or indeed, if they are overly optimistic.

In the following sections a number of accidents which have been analyzed to determine the possible results of various reactor mal-operations will be presented. For the most part these accidents do not lead to gross melting of the cores of the reactors in question. This type of accident will be termed a moderate accident, but a number of situations will be shown in which the possiblility of gross fuel melting is very large, either due to the extreme nature of the initial assumptions or to the dependence of the ultimate outcome of the accident under investigation upon details of behavior which are open to considerable question. In the study of reactor accidents one of the most difficult phases is the selection of initiating events which have some measure of credibility. At the beginning of a design this phase is considerably more simple than when the design is largely complete, since the designer naturally proceeds in such a direction as to eliminate the most credible accident which is apparent at a given time. By the time the design is complete, therefore, the only accident-initiating situations which should remain for study are those which have been overlooked or those which require multiple failure of elements of the design in order to provide the initiating circumstances.

The investigation of even accidents that appear incredible is an important part of the design of a safe reactor. In addition to providing the basic material upon which the safety of the design will be judged by regulatory agencies and responsible management, much can be learned about the behavior of the system by subjecting it in calculation to extremely rigorous circumstances. It has been found possible in many cases to make relatively simple additions to the reactor safety system which will render a particular set of circumstances much more unlikely than before. In addition, phenomena may be discovered that had previously been overlooked. Thus, a number of the accidents which will be discussed require complete control and safety system failure coincident with either coolant failure, contamination of the coolant with moderator, sudden temperature changes, or sudden reactivity insertions, all conditions which the designs are expressly intended to avoid.

3.2 Results of Reactivity Insertions at Rates Higher Than Allowed by Design

The characteristic prompt neutron lifetimes of fast reactors lie in the range between approximately 5×10^{-7} and 3×10^{-9} sec. With such short lifetimes the question of the possibility and consequences of the reactor reaching prompt critical is of obvious importance. For this reason, early in the design of a new reactor or in the consideration of a new type of fuel for an existing design it is customary to investigate the response of the system to reactivity insertion rates which are far higher than any likely to be encountered in practice. By determining the behavior of the system during the resulting transient a measure of the strengths and weaknesses of the fuel and of the degree of conservatism in the design may be obtained. Such investigations, if carried out in sufficient detail, permit an understanding of the coupling between power and feedback reactivity during very rapid rates of change of power.

A number of early studies were made to estimate the effects of such reactivity introductions. Daane [81], in examining a reactor with about the same temperature coefficients due to longitudinal fuel expansion as the EBR-II, found that after taking into account mechanical and thermal inertia, fuel expansion alone was sufficient to stabilize a prompt critical accident if the linear reactivity insertion rate* was less than about 50$/sec. The characteristics of such a rapid insertion are shown in Fig. 3-1. The power rises slowly until the reactor reaches prompt criticality at about 20 msec after the start of the accident. Then there is an abrupt rise in power to about fifty times the full load level. The rapid fuel expansion causes shutdown from this power. Since few delayed neutrons are present from the first burst, power falls rapidly until the external reactivity insertion again causes prompt criticality; then the power increases again but reaches a lower peak because the second burst begins from an initially higher power level. The end results of such an accident would depend strongly on details of the reactor as well as on the time over which the reactivity insertion was allowed to persist. The total amount of reactivity available is only important in the terminal stages of the analysis, after the first burst.

For accidents occurring while the reactor is at power the situation is considerably different.

*In this and following references to accident studies for EBR-II and Fermi the value of $\beta \approx 0.007$.

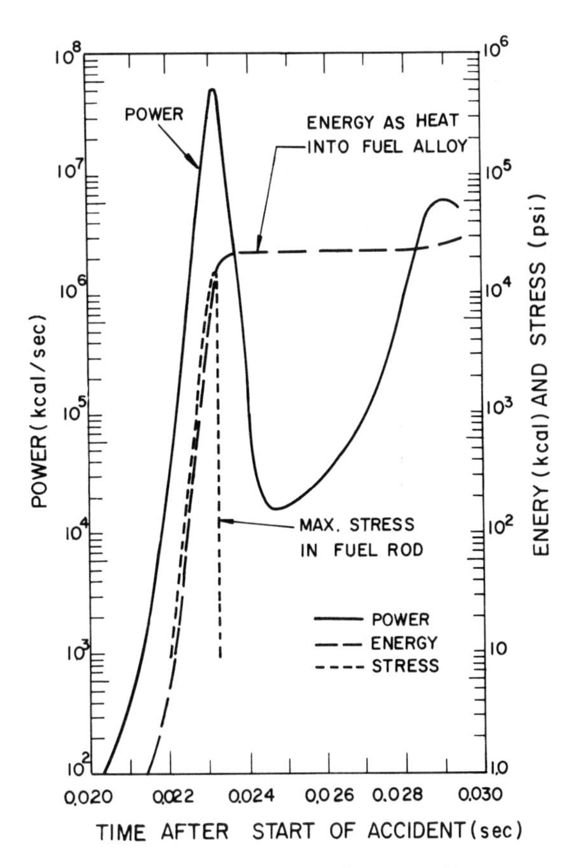

FIG. 3-1 Fifty dollar per second startup accident.

Figure 3-2 shows the results of a calculation made on the same model as was used for Fig. 3-1 except that a 50$/sec insertion rate from full power was assumed. The power and temperature follow the reactivity insertion closely and, although the power rises rapidly, no power peaks are generated.

Taking into account uncertainties in the assumptions, particularly regarding the elastic properties of the fuel, calculations of the type illustrated have given considerable confidence that accidents which involve rates of reactivity insertion of less than about 10$/sec are not likely to cause prompt critical excursions if some prompt negative power coefficient is operable.

A number of current investigations are underway in preliminary attempts to assess the basic safety of some of the large, oxide-fueled fast reactors, in which sodium expansion causes a positive reactivity effect in all or part of the core and in which the fuel Doppler coefficient is an important part of the prompt acting negative effect. Cohen [30g] reported recently on the study of two large fast reactors capable of producing 2500 Mw thermal power. The important characteristics of these two reactors are compared in Table 3-1 [30g].

The results of an addition of 1.50$ reactivity in 10 msec is shown for the two cases in Figs. 3-3 and 3-4. For Case A, which is the hard spectrum core, the energy release in the power pulse is about 2000 Mw-sec. Case B, which has twice the

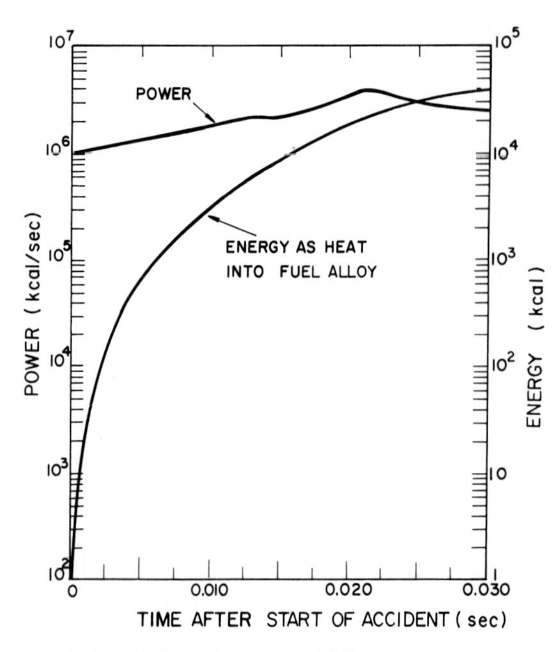

FIG. 3-2 Fifty dollar per second full-power accident.

Doppler coefficient of Case A, releases only 450 Mw-sec. The transient fuel temperatures for both cases are shown in Fig. 3-5.

For Case A there would probably be cladding deformation and perhaps cladding failure. In the reference it is pointed out that if the fuel should be ejected into the sodium, an additional amount of reactivity would be added rapidly, both due to cooling of the fuel with the consequent reduction in the Doppler effect and due to vaporization and expulsion of the sodium. It is pointed out that in this eventuality reactor A would probably be destroyed before the control system would have time to react. If, however, the fuel stayed intact, the sodium temperatures would rise according to the curve shown in Fig. 3-6. Unless a scram terminated the affair for Case A, all channels would reach boiling at the same time, in about 400 msec. The initial boiling would occur at about 480 msec at the top of the core and would produce a negative reactivity effect. A comparison of the two cores shows that core A with the hard spectrum and the smaller Doppler coefficient would almost certainly be destroyed and that Core B would almost certainly survive. The essential difference between the behavior of these two cores is the difference in Doppler coefficient, although the more positive

TABLE 3-1

Comparison of Hard and Soft Spectrum Cores for 1000-Mw(e) Fast Oxide Reactors

	CASE A	CASE B
Volume % BeO	0	8.3
UO_2-PuO_2	33.3	25
Steel	16.7	16.7
Sodium	50	50
Core Height	2.0 ft (0.610 m)	2.0 ft (0.610 m)
Core Diameter	11.8 ft (3.65 m)	13.7 ft (4.18 m)
Axial Blankets	1.25 ft (0.381 m)	1.25 ft (0.381 m)
Radial Blankets	1.0 ft (0.305 m)	1.0 ft (0.305 m)
Core Power, Mw(t)	2150	2150
Total Power, Mw(t)	2500	2500
Mean Energy of Power Spectrum (kev)	230	100
Fraction of Fissions Below 9 kev	0.12	0.28
Total Breeding Ratio	1.34	1.12
At.% (Pu^{239} + Pu^{241})	11.5	14.7
Doppler Coefficient T(dk/dT)	-0.006	-0.012
Δk for Sodium Loss From Core and Blankets [a]	+0.0070	-0.0051
Sodium Power Coefficient Δk/1%ΔP at Full Power	+0.04¢	-0.03¢
Doppler Power Coefficient Δk/1%ΔP at Full Power	-0.8¢	-1.8¢

[a] $\beta \approx 0.004$ for both Case A and Case B.

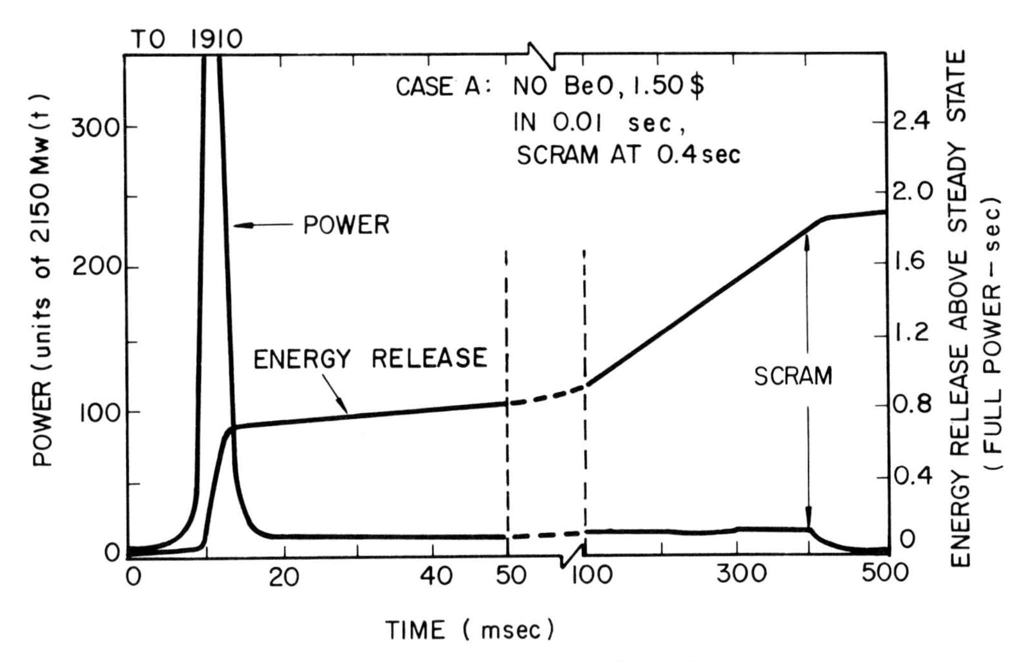

FIG. 3-3 Transient power and energy release - Case A.

sodium void coefficient of Core A also plays some part in the response.

The normal rates of reactivity variation in EBR-II and Fermi reactor are several orders of magnitude less than 10$ per second; nevertheless, a number of hypothetical cases have been examined for each reactor during which reactivity variations are assumed to occur at higher than designed rates. Three representative cases are as follows:

Spectrum or leakage variation. If a material with strongly different scattering properties than the coolant were suddenly introduced into the reactor via the coolant, a reactivity increase could take place. The addition of small amounts of hydrogen into a fast reactor results in two competing reactivity effects, a reactivity gain due mainly to the decreased leakage at lower energies and a reactivity loss due to the decrease in k_∞ at the lower energies [37j].

For the Fermi reactor it has been estimated

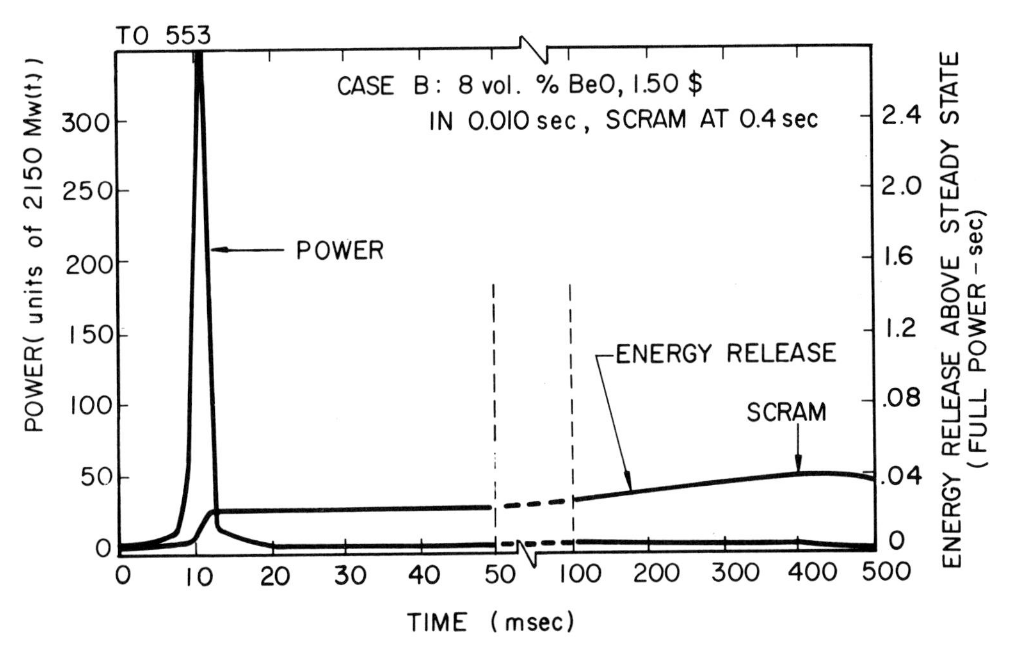

FIG. 3-4 Transient power and energy release - Case B.

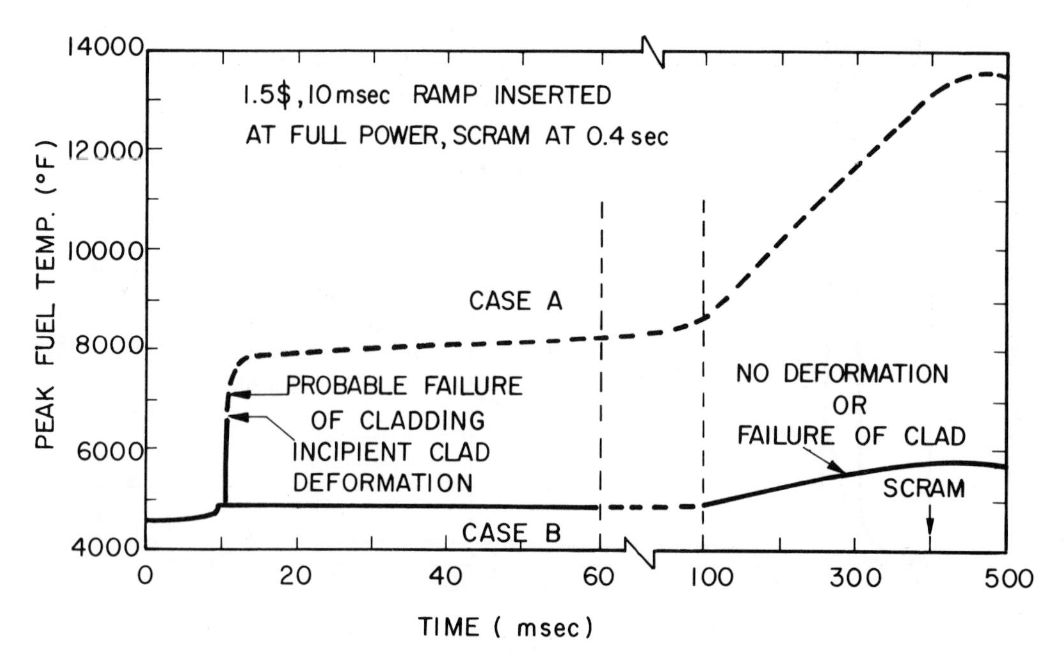

FIG. 3-5 Transient fuel temperatures for Cases A and B.

that a solid wave of material with the same hydrogen density as water would, if it entered the core at the sodium velocity of 20 ft/sec (6.1 m/sec), cause a rate of reactivity increase of 800$/sec. To prevent such an occurrence, as well as the more likely case of a dilute hydrogen insertion via coolant flow, considerable pains have been taken to exclude hydrogenous materials from the primary cooling system. A concentration of approximately 0.01% by volume of hydrogenous material such as oil in the sodium coolant would cause a reactivity increase of only a few cents.

Sudden decrease in coolant inlet temperature. A sudden decrease in the temperature of the coolant at the reactor inlet could cause a reactivity increase due to the negative temperature coefficient. For the Fermi reactor, Fig. 3-7 illustrates the effects of a decrease of 250°F (139°C) in the inlet temperature at a rate of 12.5°F/sec (6.94°C/sec), which is far higher than the cooling system design allows. If this new inlet temperature were maintained constant thereafter, the power would rise to about 2-1/4 times normal but no fuel melting or sodium boiling would occur.

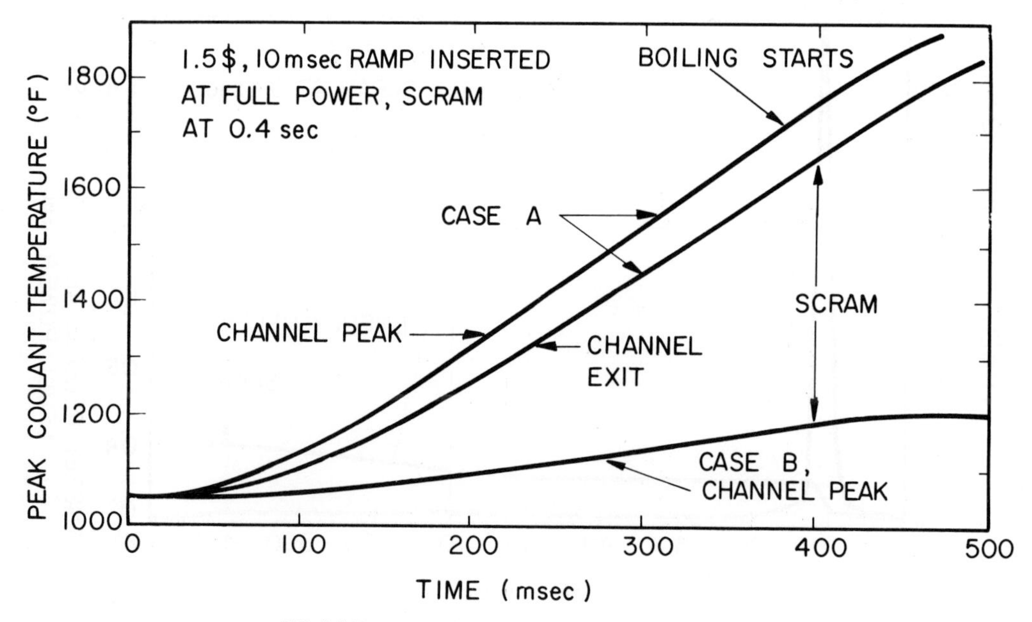

FIG. 3-6 Transient sodium temperatures for Cases A and B.

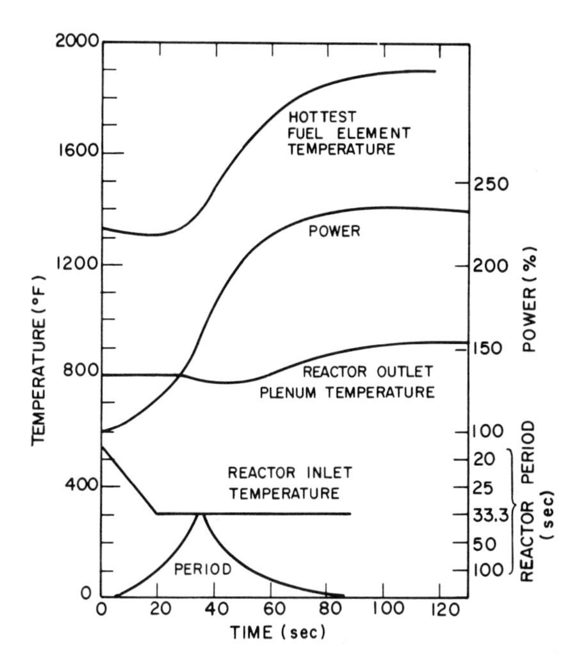

FIG. 3-7 Effect of reactor inlet temperature decrease at full power.

Dropping a fuel subassembly during loading: Despite the fact that the EBR-II is to be approximately 4% subcritical during fuel loading operations, an accident has been studied in which a central fuel subassembly worth 2.65$ is dropped into the core which is assumed to be just critical [18].

In order to determine the characteristics of the reactivity feedback during accidents capable of causing very short periods, an analysis of fuel pin behavior was conducted, the results of which are shown in Fig. 3-8. The energy generated divided by the fractional increase in fuel element length is plotted versus the period of an exponential power rise. For periods greater than 1 sec the curve is flat indicating an essentially equilibrium temperature distribution. For somewhat shorter periods the ratio of length change to energy generated decreases, indicating that more heat is staying in the fuel due to heat transfer time lags, thus causing the fuel to get hotter and expand more. For periods shorter than 10^{-3} sec the curve bends upward sharply as inertial effects begin to impede fuel expansion. Analysis of this accident indicated that reactivity feedback would only occur due to fuel growth because of the extremely short period of the excursion. The reactor was found to become prompt critical about 194 msec after the fuel element dropped. Within about 5.5 msec thereafter, the average fuel temperature had risen to 1200°C (2192°F) and the rate of rise at this time was about 160°C/msec (288°F/msec). Dickerman [30h] has examined this situation and, taking into account the radial variation in subassembly power, the over-all time spread in reaching the failure threshold for the fuel pins was estimated to be about 4 msec. Thus the core would be melting coherently; fuel movement from failed jackets would begin at essentially the same time across the whole core.

3.3 Reactivity Insertions at Maximum Rates Permitted by Design

The designs of EBR-II and the Fermi reactor both limit the rate at which reactivity may be inserted to values of the order of 10^{-2} $/sec. The previous discussion has indicated that this rate is well below that which would cause any violent energy release. However, one accident which has a fairly high probability of occurrence compared to other types previously discussed is a so-called startup accident. In this accident the reactor is presumed to be subcritical at the outset with all the negative reactivity inserted. Reactivity is then inserted at the rate permitted by the control or safety rod actuators until the reactor is as reactive as its loading permits. All automatic period and level trips are presumed inoperative. This accident is often used as the model for determining settings of the source and in intermediate range safety system trips.

The course of such an accident for the Fermi reactor is illustrated by Fig. 3-9. It is assumed for this calculation that the reactor was initially at zero power and at a uniform 525°F(274°C) shutdown temperature [82]. At time zero the two regulating rods began to withdraw at their runaway speeds of 1¢ /sec for the regulating rod and 1¢ /min for the shim rod. At 46 sec the fast rod is fully withdrawn and thus ceases to insert reactivity. At 145.2 sec after the rods begin withdrawing, the power level reaches 3 Mw. At this time 48.42¢ of reactivity are inserted. As can be seen from Fig. 3-9 the rate of rise of power is relatively low in the power range and temperatures are not reached which would damage the fuel.

The results of such an accident are dependent on how much total reactivity is available in the control system. In the example given this value was only about 0.006 ($\Delta k/k$), and the asymptotic power level could only reach about 280 Mw as compared to the design value of 200 Mw.

The ratio of feedback reactivity to externally introduced reactivity at any time is also pertinent since it determines the minimum period reached during the accident. In this example for .01$/sec introduction, the maximum excess reactivity is about 0.70$ and the minimum period ~ 1 sec.

For higher control rod rates or lower values of the pertinent feedback mechanisms the minimum period would be shorter. In addition, the ratio of the total reactivity available in the control rods to the total feedback reactivity that the reactor can produce short of damage, melting or introduction of other reactivity-introducing phenomena determines the ultimate results of the accident if control rod motion continues until mechanically checked.

3.4 Accidents Initiated by Failure of Effective Cooling

The previous examples of accidents have been initiated by reactivity increases of one form or another. In order to illustrate a mechanism by which core damage or melting could result through failure of effective cooling a study has been carried out for the EBR-II reactor.

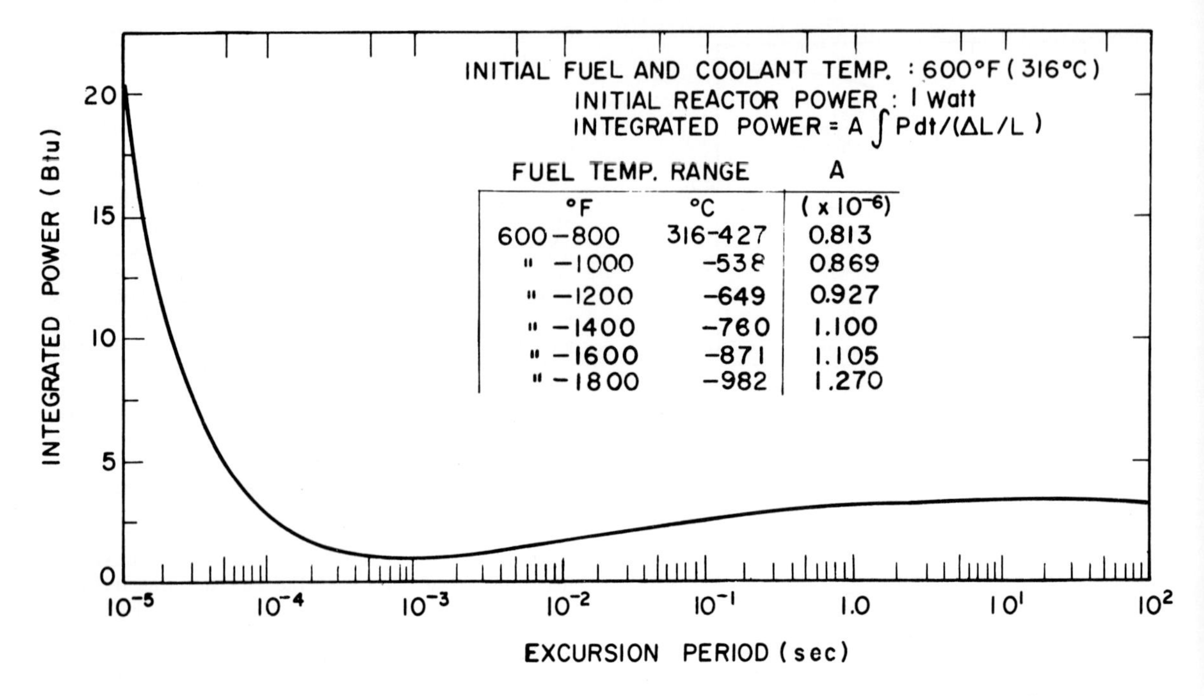

FIG. 3-8 Ratio of integrated power to fractional change in EBR-2 fuel length as a function of reactor period with no coolant flow.

Three pumps are provided in the primary system for EBR-II: two large "main" pumps and one small auxilliary pump. Each main pump is supplied independently with electrical power from a rectifier operating on the building power supply.

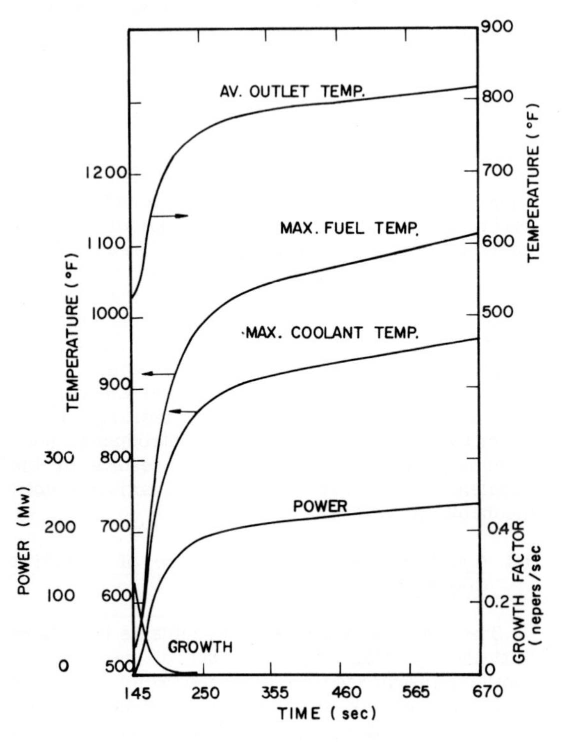

FIG. 3-9 One cent per second startup accident.

The small pump is battery powered. For the calculation of the accident results it was assumed that pumping power of all three pumps is suddenly lost and that, in addition, all twelve control rods are simultaneously rendered completely inoperative so that they remain fixed in their initial positions. The coolant flow rate is presumed to decay as shown in Fig. 3-10 and the reactor power is assumed to remain constant throughout the flow transient at the design power level of 62.5 Mw.

Dickerman [30h] has reported on the distribution and time of the meltdown at different fuel elements within the core. The EBR-II core consists of 61 subassemblies, each of which contains 91 fuel elements. The subassemblies are grouped into rows depending on location with respect to the core centerline. Sodium coolant enters the bottom of each subassembly through an orifice which is preset to vary the flow rate from row to row to minimize differences in coolant outlet temperature. Within a subassembly elements can be classed as a) internal pins which have an effective coolant cross section area of .11 cm^2 (0.017 $in.^2$) and b) peripheral pins which have an effective coolant cross section area of .18 cm^2 (0.028 $in.^2$) Five groups of fuel pins were studied: 1) central row-internal element 2) central row -peripheral element 3) fourth row-internal element 4) fourth row-peripheral element (nearest to core center) 5) fourth row- peripheral element (farthest from core center). In EBR-II the central and fourth rows are those of maximum coolant outlet temperature respectively. Figure 3-11 shows the maximum fuel-cladding interface temperatures as a function of time for the foregoing five cases. On the basis of analysis of meltdown results, a fuel cladding interface temperature of 1015°C

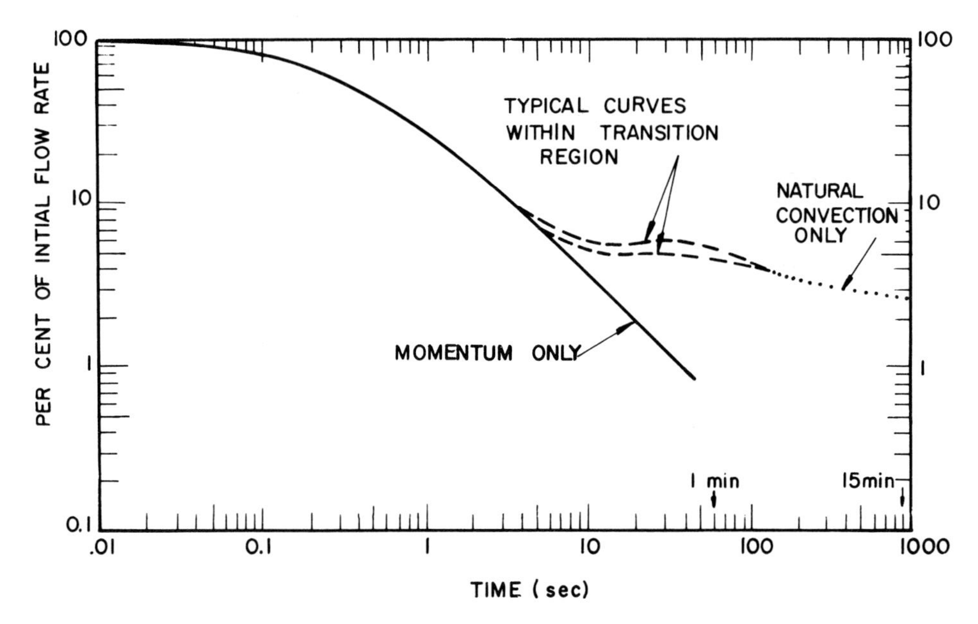

FIG. 3-10 Primary system coolant flow rate vs time after cessation of all EBR-2 pumping power, based on 100% initial reactor power and coolant flow rate.

(1859°F) was taken as the threshold for element failure by cladding dissolution and a temperature of 1200°C (2192°F) was taken to be the threshold for failure due to bursting of the cladding. Penetration time for penetration of the cladding by molten fuel was assumed to be of the order of one second. Figure 3-11 shows that the time spread within subassemblies is larger than that between subassemblies in rows one and four. Failure of one internal element can cause failure of a nearby peripheral element, or vice versa, by formation of molten fuel-cladding eutectic. The time scale for element failures in this hypothetical accident is dominated by the cladding penetration time. The time scale would be of the order of 2 sec for each of the two subassembly locations, and of the order of 3.6 sec for all the elements in both subassembly rows. Thus, in such a drastic accident as this, but where super-prompt-criticality has not been reached, there is considerable time diversity between the melting of fuel in different portions of the reactor. As another example of failure of effective

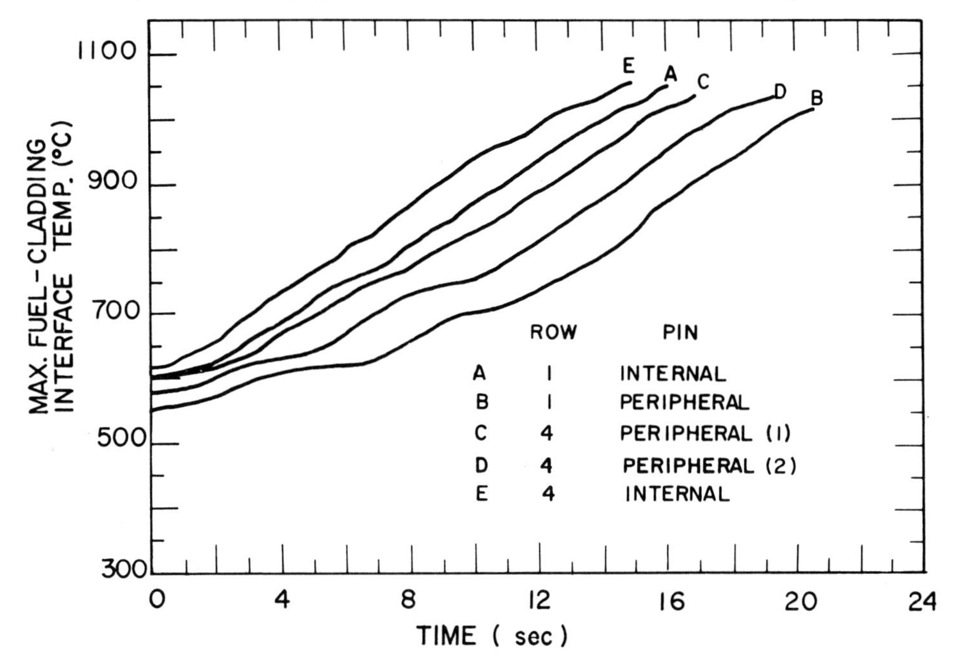

FIG. 3-11 Time required to reach given temperature after flow failure for specified pins.

cooling, the analysis of Cohen [30g] on the two reactors whose basic features are described in Table 3-1 will be quoted. This reactor has a sodium void coefficient which is positive and is perhaps representative of the type of fast reactor design which will be considered in the future.

"It is realized that the calculation of the sodium-void coefficient is a very imperfect art. The numerical values in the figures are in particular subject to uncertainties, perhaps as much as 100%. However, the general trend of the curves is probably real. The conclusions which will be drawn depend more on the general aspects of the curve than on the particular values. Figure 3-12 shows the reactivity changes associated with the progressive axial voiding of the core above the midplane. Voiding the outer radial blanket and outer periphery of the core creates a negative reactivity contribution, so in Fig. 3-12 these are supposed not voided. Since these are the low-power regions of the reactor, this hypothesis is prudent. It will be noticed that in either case the reactivity peaks when the core is partially voided, and then decreases as the blankets are voided. Thus, voiding the upper part of the core and the upper blanket gives a negative reactivity contribution.

"Figure 3-13 shows that reactivity changes associated with progressive radial voiding of the core. The solid curves represent the case where the void extends through the core and both blankets; the dotted curve represents the case where the void only extends through that central part of the core (approximately two-thirds) which gave a peak in the previous figure. From these curves, we can deduce approximately the reactivity traverses for particular patterns for sodium voiding.

"In general, the result of a mild transient or a mild loss of flow would at most cause boiling to occur in the upper blanket, tending initially to shut either reactor down. Let us investigate more drastic occurrences.

"Consider first, the loss of flow or failure to scram on a loss of flow signal. Initially, every channel has (by orificing) essentially the same subcooling at each point along the channel. No coastdown was assumed for the calculation. In this case, the hottest channel gives the most heat to the stagnant sodium. Therefore, boiling will begin at the central channel. The center of the hottest channel will reach zero subcooling first, and the sodium void will appear in the center of the core. (For finite coastdown, the initial void will be displaced towards the top of the core and the time for the initial voiding delayed.) Figure 3-14 shows the ensuing pattern and time scale for sodium voiding for Case A. Figure 3-15 gives the same pattern for Case B. Although the actual void pattern will be sensitive to the details of the coastdown rate and to the hydraulic assumptions, in any event the void pattern after only a small volume of core is voided will closely resemble progressively expanding radial voids with both core and blanket empty. The reactivity traverses will both follow quite closely the lower curve of Fig. 3-13 with an initial reactivity increment in both cases. As shown in Fig. 3-16 reactor B tends to shut itself down. The peak reactivity insertion in reactor A is about 2$ at 500 msec after the first voiding. For reactor A under these assumptions, even assuming a 400 msec scram initiated at the first reactivity increment, a power excursion and fuel temperature of 10,000°F would be added to whatever meltdown problems loss of flow itself creates."

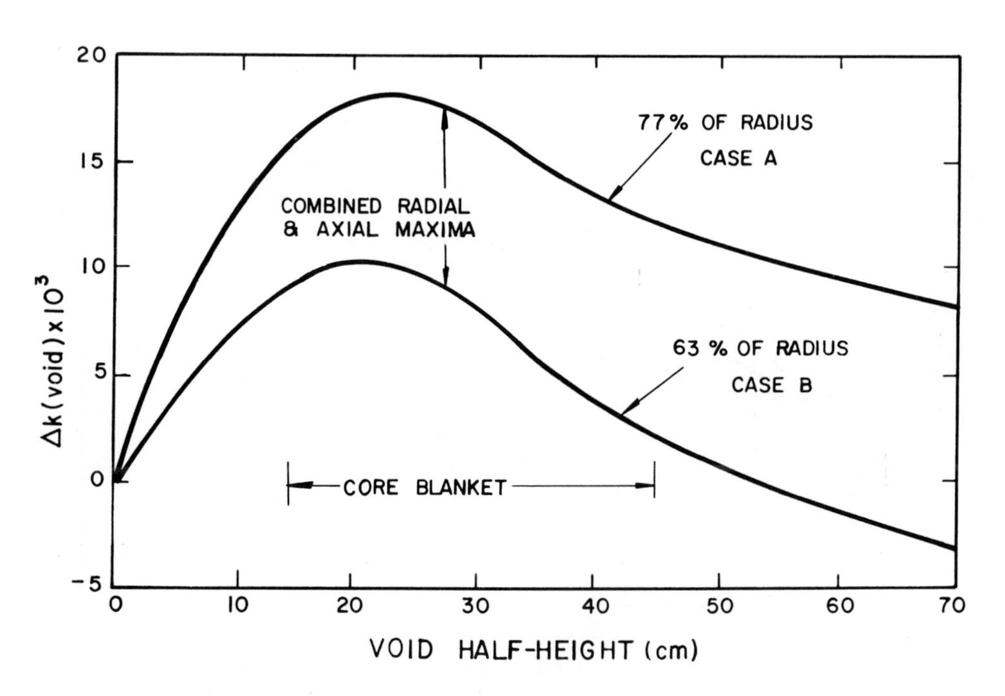

FIG. 3-12 Spatial dependence of sodium loss and reactivity change for axial voiding.

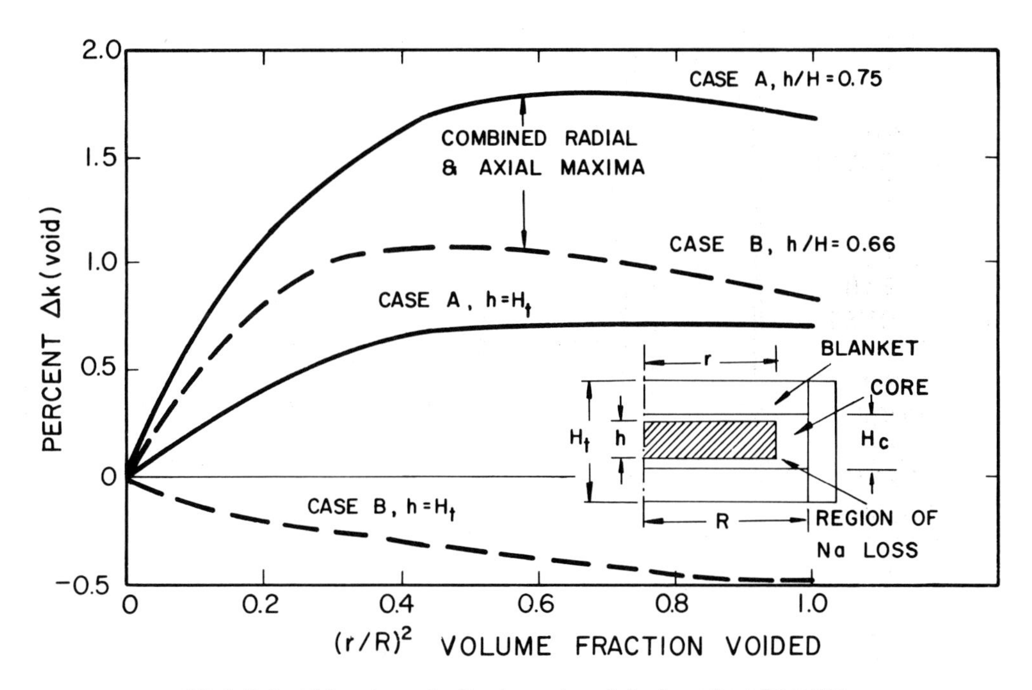

FIG. 3-13 Spatial dependence of sodium loss and reactivity change for radial voiding.

3.5 Calculational Methods

The calculational methods and techniques used in the calculation of moderate accident vary greatly in complexity and range. Both digital and analog techniques have been used extensively [54,83]. Models for excursions are discussed in the chapter on Mathematical Models of Fast Transients.

3.5.1 Digital Techniques. (See also Appendix to the chapter on General Reactor Kinetics)

Argonne National Laboratory has used a digital technique which allows the retention of a large amount of physical meaning for different accident situations. This technique was utilized for a parametric study in which both reactivity insertion rate

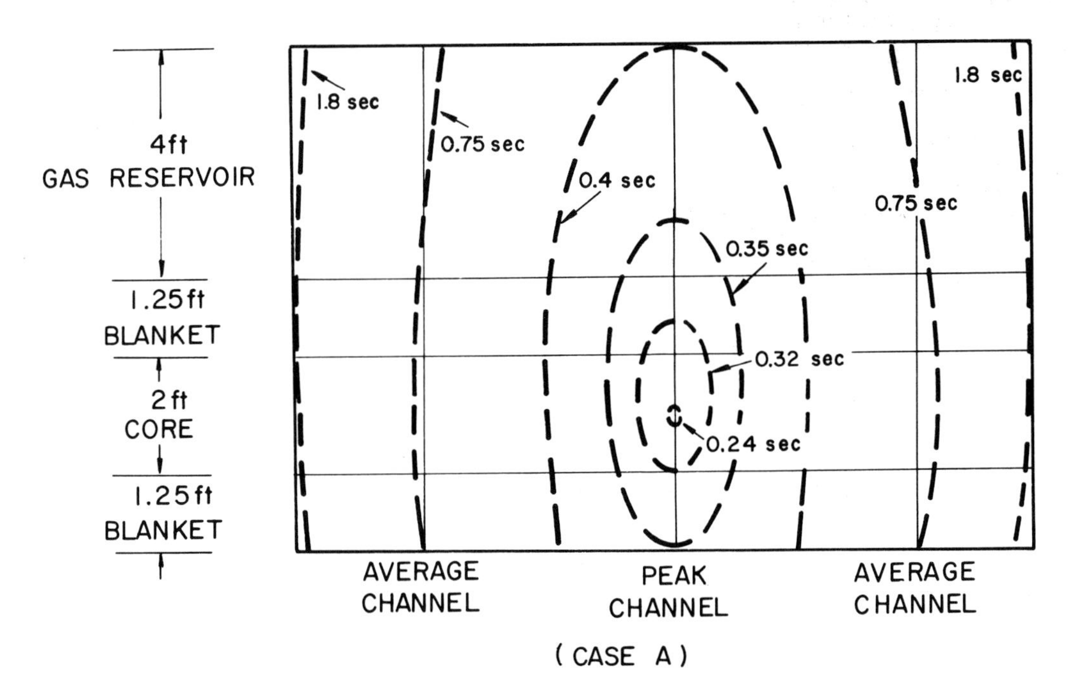

FIG. 3-14 Pattern of sodium voiding for instantaneous loss of flow at full power - Case A.

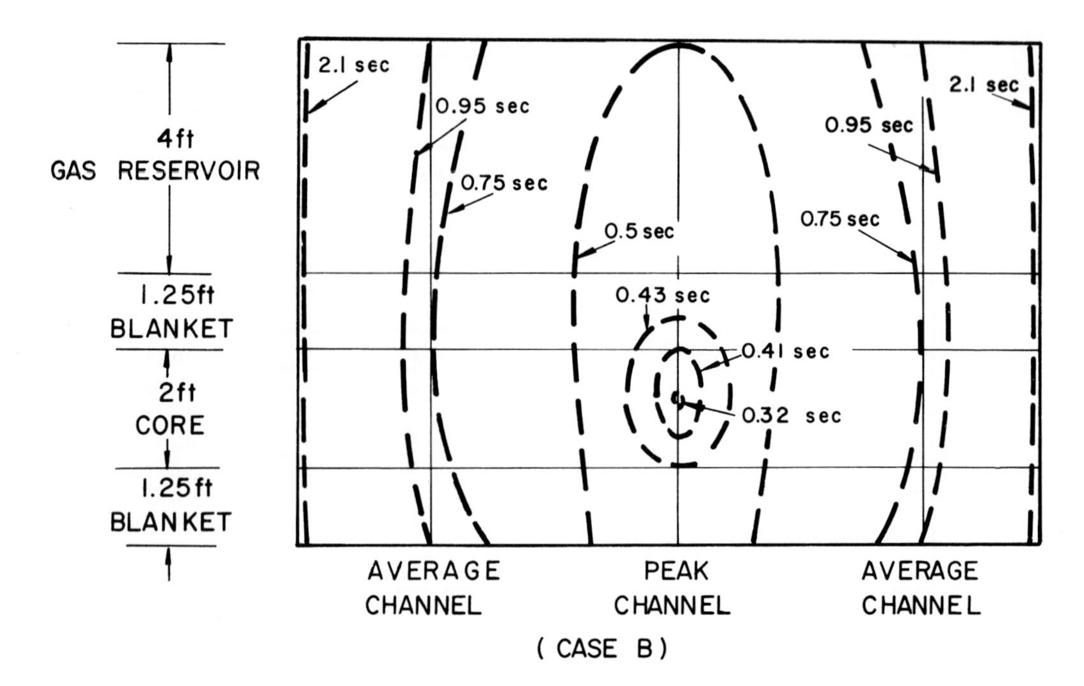

FIG. 3-15 Pattern of sodium voiding for instantaneous loss of flow at full power - Case B.

and reactivity feedback were taken as parameters during safeguards analyses of EBR-II [18]. The study used the calculated average material density effects on reactivity shown in Table 3-2.

Consider first $k_{ex} = k_{eff} - 1 = f(n, t)$. A portion of f(n,t) can be represented by the form At where A is a constant describing the rate of inserting reactivity. If the shutdown mechanism is related to the expansion of the system and hence the temperature change, another part of f(n,t) must be related to the energy stored in the system. This is directly proportional to $\int_0^t n(t)dt$, the constant of proportionality being given as B. The latter must be precisely defined.

The information in Table 3-2 is sufficient to give information concerning the reactivity changes associated with perturbations of the system. It is then possible to relate a given perturbation with a particular temperature change by suitable application of thermal data. Knowing this relationship it is then necessary to obtain the representative temperature change of a given time-dependent situation. In its simplest form, this is simply the ratio of the energy produced per unit volume to the heat capacity per unit volume. Appropriate factors must be introduced to account for heat removal by coolant and for varying temperatures throughout the system.

The energy produced per unit volume is given by $\Sigma_f v \int_0^t n dt$ where Σ_f is the average macroscopic fission cross section and v is the appropriate average neutron velocity. It is possible to obtain the product $\Sigma_f v$ directly from the one-group relationship $\ell = 1/v \Sigma_f \nu$ where ℓ is the prompt neutron lifetime and ν the number of neutrons per fission.

In summary, the excess reactivity as a function of time in its simplest form may be given as:

TABLE 3-2

Reactivity Effects of Changes in Material Density

[in units of $(\Delta k/k)/(\Delta\rho/\rho)$]

	Uranium Density	Sodium Density	Iron Density
Core	0.55	0.03	0.028
Inner blanket	0.047	0.007	0.016
Outer blanket	0.01	0.0006	0.0021
Structural gap	-	0.013	0.026
Upper plus lower blanket	0.017	0.0074	0.015

For axial core expansion into structural gaps: $\Delta k/k \sim 0.27\, \Delta L/L$.

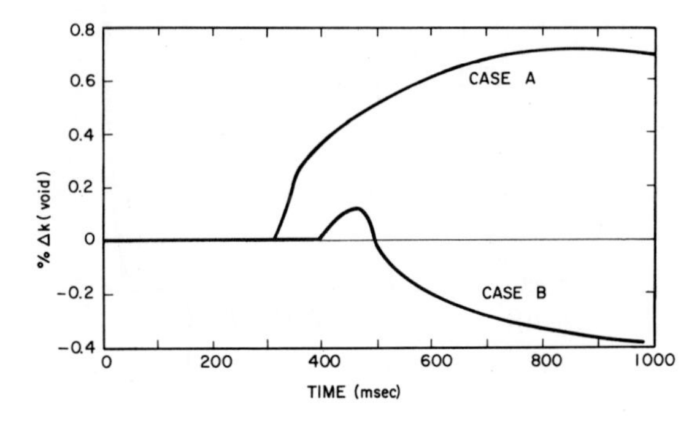

FIG. 3-16 Reactivity changes following instantaneous loss of flow at full power - cases A and B.

$$k_{ex}(t) = At - B\int_0^t n\,dt + Bn_0t\,.$$

B consists of a product of several terms which may be given as:

$$B\int n\,dt = \left(\frac{dk}{dq}\right)\left(\frac{dq}{dt}\right)\Delta T = \left(\frac{dk}{dq}\right)\left(\frac{dq}{dt}\right)\left(\frac{\Sigma_f v\int_0^t n\,dt}{S_v}\right),$$

where

$\frac{dk}{dq}$ = reactivity change corresponding to a perturbation,

$\frac{dq}{dt}$ = perturbation corresponding to a given temperature change,

ΔT = temperature change

S_V = heat capacity per unit volume, and

$\Sigma_f v\int_o^t ndt$ = heat produced per unit volume.

The results of the study are presented in Figs. 3-17 to 3-22 inclusive. The calculations were done assuming the flux n = 1 at t = 0. The power then can be considered as arbitrary until application is made to a specific accident employing a preassigned temperature feedback. In Figs. 3-17

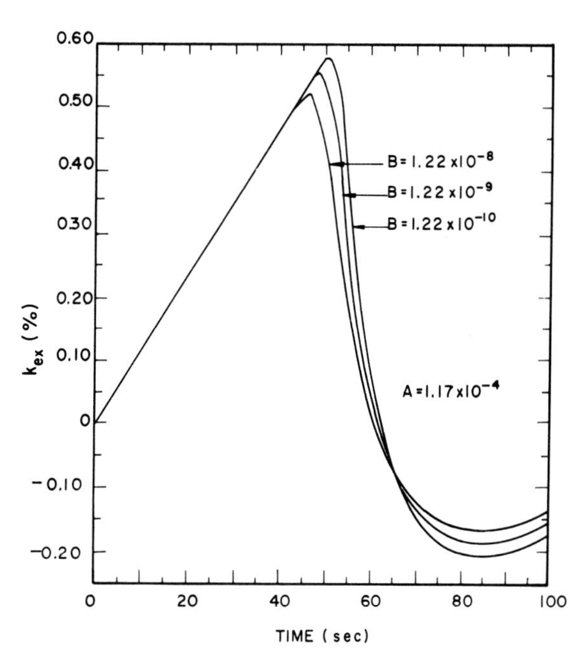

FIG. 3-18 Excess reactivity vs time for several feedback coefficients and fixed insertion rate.

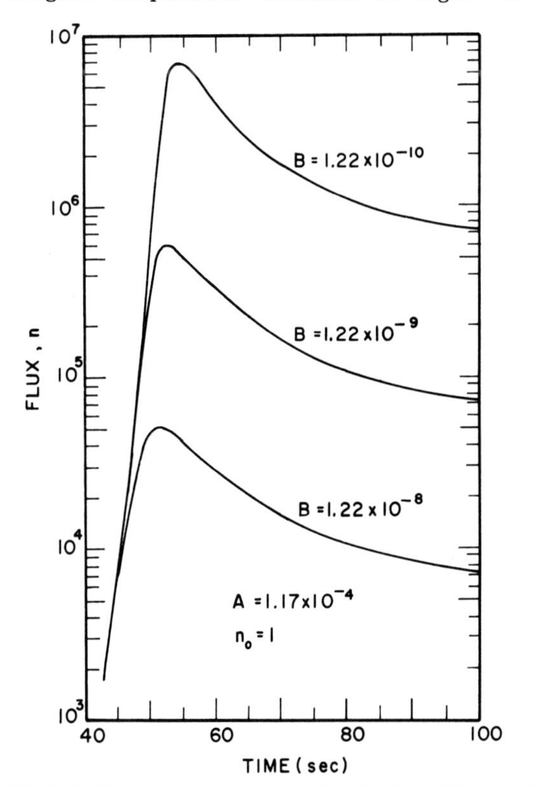

FIG. 3-17 Flux versus time for several feedback coefficients and fixed insertion rate.

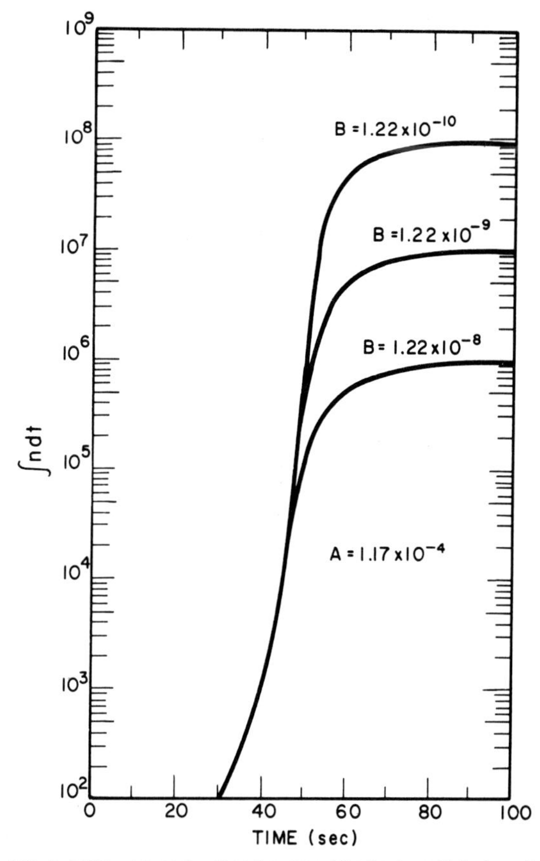

FIG. 3-19 Flux integral vs time for several feedback coefficients and fixed insertion rate.

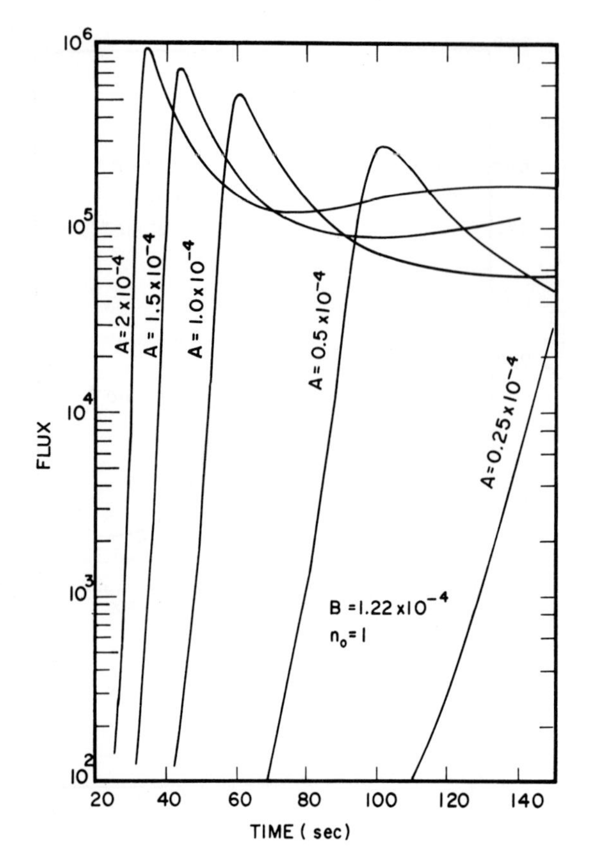

FIG. 3-20 Flux vs time for several insertion rates and fixed feedback.

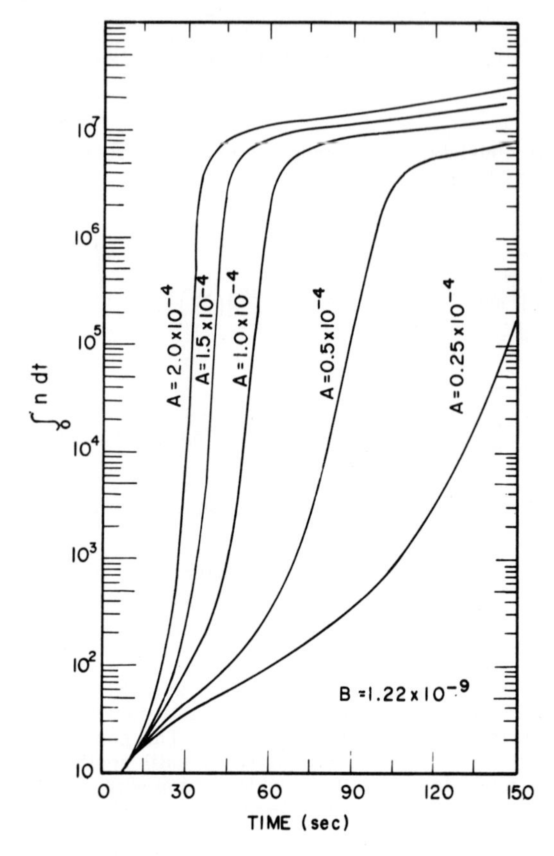

FIG. 3-22 Flux integral vs time for several insertion rates and fixed feedback.

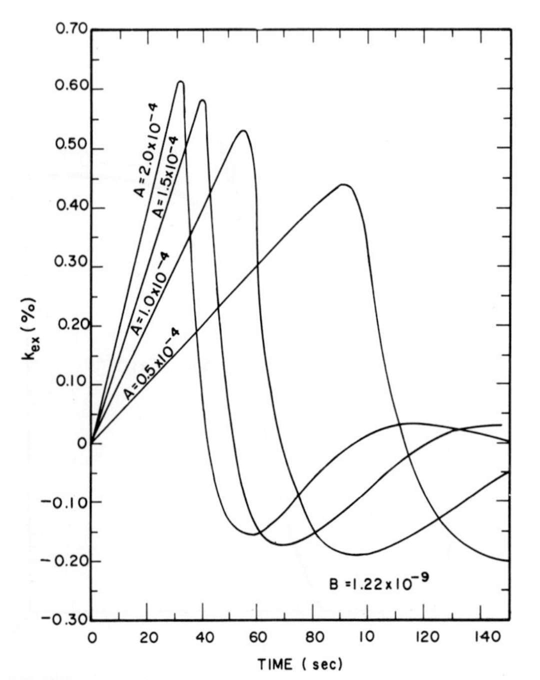

FIG. 3-21 Excess reactivity vs time for several insertion rates and fixed feedbacks.

through 3-19 (for several values of B and fixed A) n, k_{ex}, and $\int_0^t$ ndt are plotted against time. In Figs. 3-20 through 3-22 (for several values of A and fixed B) the same quantities are plotted against time.

From Fig. 3-17 it is seen that reducing the feedback by a factor of ten for a fixed insertion rate, raises the maximum flux by a factor of ten. The effect on the flux integral, Fig. 3-19, is the same.

The results of this study were applied to specific accidents, one example of which will be given here. In this accident the safety rods, which contain fuel and which are withdrawn during a scram, are assumed to be driven into the just critical reactor.

The situation is considered where the reactor is at source power, barely subcritical with the fuel sections of the safety rods located just below the core. Then, somehow the fuel sections of the safety rods move into the core at a speed of 2 in./min (5cm/min). No coolant is flowing. Essentially all of the heat generated is retained by the core and coolant. Some loss of heat occurs due to convection.

Each of the safety rods has been calculated to be worth about 1% in reactivity. Hence a total Δk of 2% is moving into the system. Δk is assumed to be a linear function of position. Hence, for a 14-in. (35.5 cm) stroke the amount being added will be $.02/7 \times 60 = 0.000048$ (Δk/sec) $\approx 5 \times 10^{-5}$ (Δk/sec) = A.

It is necessary now to assign a value to the temperature feedback coefficient B. The assumption that initially the system is at source power with no coolant flow simplifies the analysis. Since the rate of insertion is slow, it may be assumed that the average coolant, fuel and clad temperatures are equal in any region, though varying in time. Thus the heat capacity per unit volume can be expressed as $S_v = \Sigma V_i^F S_{vi}$ where V_i^F represents the volume fraction of material i in the core and S_{vi} represents the heat capacity per unit volume of the pure material i.

To calculate the reactivity change associated with a given temperature change it is necessary to calculate $\Delta k/\Delta t = \Sigma(\Delta k/\Delta q_i)\,(\Delta q_i/\Delta T)$ where q_i represents the perturbations. For core effects only this has been calculated to give $\Delta k/\Delta T \approx 1.28 \times 10^{-5}/°C$ ($0.71 \times 10^{-5}/°F$) using the data of Table 3-2.

In Fig. 2-8, the pertinent result of this analysis have been summarized.

The energy generated, divided by the fractional increase in fuel element length is plotted versus period for the assumption of an exponential power rise. For periods greater than 1 sec, the curve is flat, indicating essentially an equilibrium distribution of temperature. For shorter periods the ratio decreases, indicating more and more heat is staying in the uranium. However, for periods shorter than 10^{-3} sec, the curve starts bending up sharply, as inertial effects begin to take over.

For the core:

$$\frac{\Delta k}{k} = \left(\frac{\Delta k}{k}\right)_{\substack{\text{density}\\\text{changes}}} + \left(\frac{\Delta k}{k}\right)_{\substack{\text{dimension}\\\text{changes}}} + \left(\frac{\Delta k}{k}\right)_{\text{other}}$$

The category "other" includes Doppler effect, to which a coefficient is assigned, and bowing, which is neglected here. Assuming a linear expansion coefficient of expansion equal to 14×10^{-6} in./in.°C ($7.8 \times 10^{-6}/°F$) for steel and uranium, and a density coefficient of $2.9 \times 10^{-4}/°C$ ($1.61 \times 10^{-4}/°F$) for sodium, then for a one centigrade degree change in temperature

$$\begin{aligned}\Delta k/k = {} & [(-0.55 - 0.028)(14 \times 10^{-6}) \\ & - 0.03(2.9 \times 10^{-4})]_{\text{density}} \\ & + [0.27(14 \times 10^{-6})]_{\text{dimension}} + [0.2 \times 10^{-6}]_{\text{Doppler}},\end{aligned}$$

where the Doppler coefficient is an average coefficient between 300°C (572°F) and 1100°C (2012°F) obtained by using the $-T^{2/3}$ relation. Thus

$$\Delta k/k = -1.28 \times 10^{-5}/°C = -0.710 \times 10^{-5}/°F\,.$$

This coefficient is subject to considerable uncertainty. The sodium density effect, which contributes more than half the total coefficient, is particularly doubtful.

One minor effect has been neglected which may possibly give a small positive contribution. The uranium also expands radially within its jacket, forcing some sodium from the core into the upper gap area above the core, where the worth of coolant conceivably may be slightly greater. In addition the effects of coolant convection during an accident have also been neglected.

The core has not been permitted to expand radially, over-all, since the bottom structure, which determines fuel element location, will change dimensions only as the bulk sodium changes temperature.

Of course, some heat will be generated in the blanket regions. For the conditions of no coolant flow, it appears reasonable to take the temperature rise distribution through the reactor proportional to the energy distribution during the accident. The reactivity contribution from the blanket areas per degree temperature rise in the core will be the blanket coefficient per degree multiplied by the relative temperature rise.

The following coefficients have been calculated for the various regions:

$$\left.\frac{\Delta k}{\Delta T}\right|_{\text{inner blanket}} = -0.30 \times 10^{-5}/°C$$

$$\left.\frac{\Delta k}{\Delta T}\right|_{\text{outer blanket}} = -0.03 \times 10^{-5}/°C$$

$$\left.\frac{\Delta k}{\Delta T}\right|_{\substack{\text{upper blanket plus}\\\text{lower blanket}}} = -0.26 \times 10^{-5}/°C$$

$$\left.\frac{\Delta k}{\Delta T}\right|_{\text{gaps}} = -0.42 \times 10^{-5}/°C$$

(The coefficients are -0.167, -0.017, -0.144, -0.233 in units of $10^{-5}/°F$, respectively).

The temperature rise in each region per degree rise in the core is a function of the power distribution and the heat capacities.

$\Delta T_{\text{inner blanket}}$ = 0.045°C per °C rise in the core,

$\Delta T_{\text{outer blanket}}$ = 0.01°C per °C rise in = the core,

$\Delta T_{\text{upper and lower blanket}}$ = 0.02°C per °C rise in the core,

ΔT_{gap} = 0 (neglecting convection and conduction),

resulting in over-all feedback effect of

$$\left(\frac{\Delta k}{\Delta T}\right)_{\text{total}} = -1.3 \times 10^{-5}/°C(-0.72 \times 10^{-5}/°F)$$

for the slow excursion with no coolant flow.

Knowing $(\Delta k/\Delta T)_{total}$, it is then necessary to calculate the core temperature change as a function of time. This will simply be the ratio of the heat absorbed per unit core volume to the specific heat per unit core volume.

The latter has been calculated

$$S_v = 0.534 \text{ cal/cm}^3\text{-}^\circ\text{C (or 0.01931 Btu/in.}^3\text{-}^\circ\text{F)} .$$

The heat absorbed per unit volume can be presented by

$$\gamma\Sigma_f \int_0^t \phi \, dt = \gamma\Sigma_f v \int_0^t ndt .$$

It is possible to obtain $\Sigma_f v$ directly for a particular system from the expression

$$\ell \sim \left\langle \frac{1/v}{\nu\Sigma_f} \right\rangle,$$

where the brackets indicate proper weighting of fluxes and adjoints. The prompt neutron lifetime ℓ is known to be about 0.8×10^{-7} sec, and ν is known to be about 2.5 . Hence

$$\Sigma_f v \approx 5 \times 10^6 \text{ sec}^{-1} .$$

Since each fission liberates approximately 0.76×10^{-11} cal,

$$\gamma\Sigma_f v = (.76)(10^{-11})(5)(10^6) = 0.38 \times 10^{-4} .$$

Thus,

$$\Delta T = \frac{1}{S_v} \gamma\Sigma_f \int_0^t ndt = \frac{0.38 \times 10^{-4}}{.534} \int_0^t ndt$$

$$= 0.71 \times 10^{-4} \int_0^t ndt \, *$$

and the feedback is given by

$$\frac{\Delta k}{k}(t) = \left(\frac{\Delta k}{\Delta T}\right)(\Delta T) = B \int_0^t ndt$$

$$= -(1.3 \times 10^{-5})(0.71 \times 10^{-4}) \int_0^t ndt$$

$$= 0.92 \times 10^{-9} \int_0^t ndt .$$

*Since n is taken as unity at t = 0, the source power is 1.9 cal/sec or 8 watts.

The consequences of this accident can now be examined. The temperature of the system while loading operations are underway should be about 315°C (600°F). If the melting point of the fuel is 1130°C (2066°F), a ΔT of 815°C (1467°F) on the average will melt the fuel pins. Since there is a maximum to average power generation of about 1.5, an average ΔT of 815/1.5 = 545°C (981°F) will see melting begin at the hot point. Hence,

$$\int_0^t ndt = \frac{\Delta T}{7.1 \times 10^{-5}} = \frac{545}{7.1 \times 10^{-5}} = 7.7 \times 10^6$$

when melting begins. From Fig. 3-22, for a B of 1.22×10^{-9} and $A = 5 \times 10^{-5}$, a flux integral of 7.7×10^6 is just reached at the end of the first pulse. However, this feedback coefficient is about 30% too large. From Fig. 3-19 it is seen that the integral is inversely proportional to B for fixed A. Hence, the flux integral for this case can be expected to reach roughly 10×10^6, and melting would began about 120 sec after the beginning of the excursion, should no scrams be operative. The reactivity would be small or possibly negative at this time. The future course of the incident would depend on what new reactivity mechanisms take part.

Should a period scram have been operative, with a scram setting of 5 sec, $(\Delta k/k = 0.0037)$, Fig. 3-21 shows that the signal would have been given at about t = 75 sec, while the power was still very low. The incident would be terminated with no damage. If a power level trip were the shutdown mechanism, its time of action would depend on the scram setting. A level of 1000 watts would be reached at approximately t = 70 sec, as indicated in Fig. 3-20. If the setting were 65 Mw, however, the signal would never have been given before melting began.

While the net bowing effect is expected to be small, an initial decrease in core radius is considered possible as the power rises, if the most

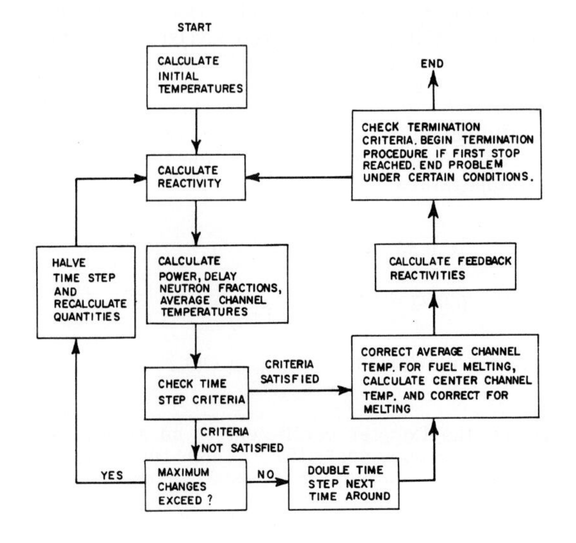

FIG. 3-23 Simplified flow chart - FORE.

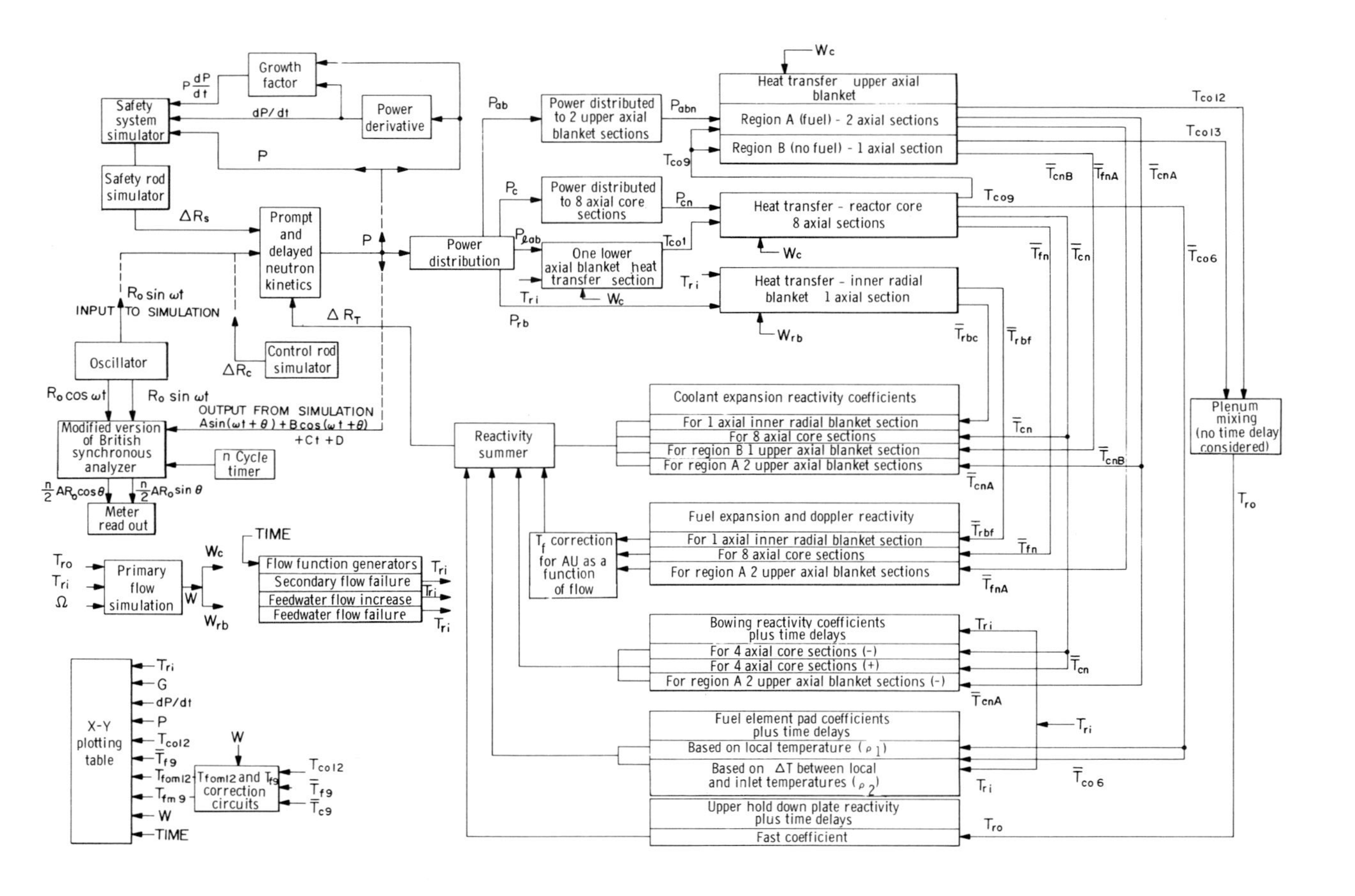

FIG. 3-24 Block diagram of analog computer used for analysis of Enrico Fermi Reactor. In the above, P is the total power (Mw) and W the coolant flow rate (lb/sec) with the various subscripts denoting location (subscript ab = axial blanket, lab = lower axial blanket, c = core, cn = core section n, rb = radial blanket). T denotes temperature (°F) with subscripts denoting position ri = reactor inlet, ro = reactor outlet, cin = average at coolant inlet to n^{th} section, con = average at coolant outlet of nth section, comn = average of coolant outlet of nth section of hottest subassembly, fom 12 = fuel pin centerline at outlet of nth section of hottest subassembly, fon = average of fuel pin cross section at outlet of nth section. $\overline{T}$ denotes average temperatures with $\overline{T}_{cn}$ = average coolant temperature at n^{th} section midplane, $\overline{T}_{fn}$ = average fuel pin cross section temperature at midplane of nth section (not corrected for variation of AU with flow), $\overline{T}_{cnA} = \overline{T}_{cn}$ for region A of upper axial blanket, etc. The ΔR are changes in reactivity with subscripts indicating source (s = safety rod simulator, c = control rod simulator) and with subscript T = total. R_o is the amplitude of sinusoidal reactivity excitation. The quantity AU is heat transfer area (A) times heat transfer coefficient (U) in Btu/sec-°F. The temperature coefficient is $\rho = \Delta k/(k\Delta T)$.

pessimistic assumptions are made. The maximum inward displacement (about 0.014 in. or 0.36 mm) would occur when the burst energy reached 8 Mw-sec, corresponding to a ΔT of roughly 75°C (135°F). The reactivity gain associated with this bowing would be about 0.001 ($\Delta k/k$), the same as would be lost from other temperature effects. Hence, under the worst conditions there would be no feedback during the first 75°C temperature rise. However, during the next 75°C the bowing effect would reverse itself and thereafter continue negative so the net contribution of bowing over the entire excursion would be to raise the feedback coefficient and diminish the burst strength.

More recently a number of very flexible digital programs have been written for the analyses of fast reactor excursions. An example of this type of program is FORE, which was prepared by the General Electric Company for the Transac 2000 [84]. A simplified flow chart for this program is shown in Fig. 3-23. The program is limited to cases in which the reactor is not substantially destroyed and was chiefly designed to study transients lasting a few seconds or fractions of seconds.

3.5.2 Analog Methods

For many system transients and for a limited range of reactor transients, analog techniques may be used. An analog representation used for many of the transient analyses for the Enrico Fermi reactor is shown in Fig. 3-24 [82].

The limited range over which power transients may be analyzed using analog equipment without undue complexity and an excessive number of amplifiers restricts their use considerably. However, for slower system transients in which other components besides the reactor are involved, an analog computer is extremely useful, especially if the computer can be used in the control system designs and for operator training for a given plant.

4 DYNAMICS OF SEVERE ACCIDENTS*

4.1 General Considerations

The term "severe accident" is used herein to cover situations which lead to an explosive release of nuclear energy in the reactor. We shall be dealing primarily with the calculation of those reactivity excursions which are not terminated until some of the fuel has been vaporized and appreciable pressures built up in the core. These pressures will cause the whole core or a portion thereof to expand, leading to a reduction of reactivity and a consequent termination of the excursion.

Assuming grossly pessimistic conditions, calculations made for the EBR-II and Fermi reactors have led to the prediction of destructive effects equivalent to those which would be produced by an explosion of up to a thousand (or even more) pounds of TNT [17]. The extent of an explosion which is considered to be containable within the containment structures provided for these two reactors is in the same range. Hence, both refinement of calculating methods and more careful analysis of the factors leading to a severe accident have been demanded.

Of course, the calculation of a severe accident in a real power reactor means dealing with complex geometry and irregularities in composition. The conditions and cause of the accident are generally not known precisely, and the course of the accident prior to the large nuclear burst is usually very complex. Nevertheless, idealized calculations provide some gauge of the severity of the accident. And the evaluation and comparison of methods of calculation provide an estimate of the errors introduced by the mathematical techniques, once an idealized form of the actual situation has been adopted.

Fast reactors gain reactivity if the fuel is rearranged in a denser configuration. About half the core of a typical fast power reactor is occupied by fuel alloy and solid structural material. The other half is given over to the coolant. Hence very large amounts of reactivity are potentially available if this fuel alloy can be reassembled in a considerably denser configuration. It is in this manner that most hypothetical bad accidents are arranged. The core is assumed to have lost its rigidity, the fuel becomes pasty or molten, and either the forces of gravity or those of a high velocity or high pressure fluid cause much of the fuel to reassemble in a more reactive position.

What turns out to be significant in estimating the explosive yield of a severe nuclear accident is not the total reactivity available but the rate at which it can be added to the assembly. It is assumed that a source of neutrons is always present, and that the reactor is operating at low power at the beginning of the accident. (Otherwise, one could introduce arbitrary amounts of reactivity, effectively as a step-function). The power rise is slight until prompt critical has been reached. Beyond this point the power rises as $\exp[(\Delta k/\ell)t]$, where Δk is the reactivity above prompt critical and ℓ is the prompt neutron lifetime. (Prompt neutron lifetime may range from several $\times 10^{-7}$ sec for large, ceramic-fueled reactors to 10^{-8} sec for small, highly enriched systems). The time scale now becomes so short that heat conduction and delayed neutron effects can be neglected. If the fuel is in a molten or pasty state, the usual reactivity loss mechanism associated with fuel element expansion may not be operative, and considerable energy must be added before the pressure buildup starts fuel movement and loss of reactivity at a rate sufficiently great to override the cause of the accident. (If the fuel element is initially intact, it is assumed that the initiating cause of the accident overrides any reactivity losses due to fuel expansion in the solid state. Doppler reactivity effects associated with fuel heating will be discussed later). The maximum reactivity of the excursion (Δk_{max}) will be given very closely by the product of dk/dt, the rate at which reactivity is being added, and t_1, the time interval between prompt critical and the time when appreciable reactivity loss due to fuel motion begins. Once pressures are generated, the re-

*See also chapters on Mathematical Models of Fast Transients and on Accident and Destructive Tests.

activity excursion is over in about 2 e-folding times, that is, in hundreds of microseconds, and very little additional reactivity can be added mechanically by the initiating cause. The pressures involved are so great that questions of mechanical strength are not important and can be ignored. Inertial effects are paramount in determining the acceleration of the fuel alloy.

The explosive energy yield depends strongly on Δk_{max} and therefore on the rate at which reactivity is being added; on the prompt neutron lifetime, and on the ratio of the energy threshold for appreciable pressure generation to the initial power of the reactor. Other things being equal, the explosive energy increases with reactor size. Of course, not all the energy generated is available for destructive work.

The temperature threshold for significant pressure generation is generally not the boiling point of the fuel at atmospheric pressure but that temperature where all the space freely available to the fuel is occupied by it as a single phase, usually as a liquid under pressure. As a consequence, the void present in the reassembled core becomes a crucial parameter. The fuel must expand until it occupies its original volume plus available void volume before appreciable motion in a macroscopic sense begins.

4.2 Initiating Methods

Most studies of hypothetical severe accidents in the past have dealt with reactivity addition rates in the range 0.05 to 7.5 ($\Delta k/k$) per second. The corresponding maximum reactivities reached above prompt critical lie in the range 4×10^{-4} to 10^{-2} ($\Delta k/k$) [17]. However, any mechanism which can bring the reactor to super-prompt-critical may be of interest. For should there be some autocatalytic effect like a positive Doppler coefficient of appreciable magnitude, the attainment of prompt criticality may automatically lead to explosive disassembly.

A partial list of initiating methods follows:

1. Loading or dropping a fuel element into just-critical reactor [18].
2. Scramming a control rod whose reactivity effect has been inadvertently reversed.
3. Rapid introduction of moderating material in the form of water [81], as a large plug of oil in the coolant or as a loading error in some unusual experiment [85].
4. Large scale inward bowing of fuel elements due to thermal stresses [18,40].
5. Moving reflecting material toward the core of a bare or poorly reflected reactor [86].
6. Removal of all neutron sources, in which case any slow reactivity insertion mechanism may simulate a step-function addition [87].
7. Bunching of fuel, providing a self-multiplication effect.
8. Positive coolant void coefficient [30f,35,36, 37b,39].
9. Slumping of fuel material with its jacket [33].
10. Inward net expansion of fuel upon heating. This might result from a central gap in a two-halve critical assembly [85]. It might result from the net motion of a multi-piece fuel element which had separated, possibly due to ratchet effects. It might result from the manner of support of a multi-piece fuel element. Or it could be the consequence of a void space at the core center, resulting from the previous transport of fuel material from hot to cool areas.
11. Positive Doppler coefficient [21-23,26,27, 30b,37c]. As mentioned previously, a positive coefficient of this sort need not be large enough to initiate the accident and bring the reactor from delayed to super-prompt critical. If it were the sole rapid-acting reactivity coefficient once the reactor attained prompt criticality, even a modest reactivity contribution could appreciably aggravate the accident.
12. Core meltdown and redistribution of fuel. This will be discussed in detail in Sec. 4.3.

4.3 Various Aspects of the Meltdown Problem

The concentration and amount of fissionable material in fast power reactors is usually sufficient to produce a number of critical masses, if the fuel were re-arranged more compactly. Under pessimistic assumptions, such reassembly might result from melting of the fuel at a rate of assembly sufficient to generate a nuclear explosion equivalent in destructive force to many hundreds of pounds of TNT. Thus, the meltdown problem is of considerable interest.

As work progressed on designs for actual metal-fueled fast reactors, the possibility of releasing explosive energy due to either a large and sudden external reactivity insertion or due to a large positive power coefficient was found to be less serious than expected. Moderate sized, fast power reactors have low excess reactivity requirements and negative reactivity coefficients. In addition, there is close coupling between reactivity and temperature in metal-fueled power reactors. As a consequence, it appears possible to design such a reactor system so as to preclude beyond a reasonable doubt that enough reactivity can be inserted into the intact reactor to cause an explosive event.

However, a molten core does not have the usual negative reactivity coefficients associated with expansion of the fuel and the core structure. Behavior of a molten mass of material under changed geometric conditions must be treated. It is informative to consider some meltdown conditions and the hypotheses which lead to reassembly into a super-prompt-critical configuration. Three very pessimistic cases which have received attention in the past [17] are discussed briefly below.

4.3.1 Case A. Rapid Loss of Coolant

A sudden major break occurs in the primary sodium coolant system and all the coolant disappears from the reactor almost immediately. The reactor is shut down, but fission product heating and delayed neutron effects melt the fuel elements. These collapse as a unit under the force of gravity, and as a higher uranium density region forms in the bottom of the core volume the assembly becomes supercritical.

In one early study of the Fermi reactor, detailed calculations of the progress of melting in the core were made [88,89]. It was assumed that the uranium escaped its cladding immediately upon becoming molten, ran down to the bottom of the core, freezing when it impinged on the cooler lower axial blanket section. As the molten material left the core center and collected at the core bottom the reactivity initially fell further below criticality; as the process continued, however, the reactivity rose, with appropriate contributions to the energy input from delayed neutrons. The reactivity was calculated as a function of time and the reactor was found to become delayed critical again 5.4 sec after the accident started. At this time the rate of reactivity insertion was calculated to be 0.2 $\Delta k/k$ per sec. The reactivity insertion rate at prompt critical was not computed.

This hypothetical accident was considered to be completely unrealistic, first, since the sodium was assumed to vanish instantaneously and second, since double or triple-walled containment of the sodium in the reactor vessel is thought to make loss of sodium from the core impossible. Later studies assumed a loss of pumping power in the absence of control rod scram [19]. The course of the accident then depends on the presence or absence of stable boiling heat transfer. Without this capacity for improved heat removal, melting of the fuel element alloy begins 27 sec after the onset of the accident. If it is assumed that the fuel escapes and falls to the bottom of the core immediately upon melting, and if it is further assumed that the density of the settled fuel is that of normal unirradiated material, reactivity insertion rates at the time of going prompt critical are estimated as lying between 0.2 and 0.7 $\Delta k/k$ per second. The effect of sodium coolant flow is ignored in this estimate.

In a third estimate for the Fermi reactor it was assumed that all the sodium left the core almost instantaneously, the core melted in place and then began to settle uniformly under the action of gravity. Assuming the collapsing begins with the reactor slightly subcritical, it was found that a reactivity insertion rate of approximately 0.4 ($\Delta k/k$) per second resulted at prompt critical (Bethe-Tait accident) [90].

4.3.2 Case B. Coolant Boil-Away from Core Center (Jankus-Okrent accident) [18]

The power level somehow is too high and the sodium boils away from the core center. The uranium from the middle of the core trickles down into the lower part of the core and is retained there, producing a region abnormally dense in enriched uranium at the core bottom with a large gap at the core center. (The reactivity falls initially, then starts to rise.) At the worst possible moment, the upper portion of the core falls as a single unit, producing a prompt critical configuration at the highest possible insertion rate.

For the EBR-II reactor the minimum reactivity configuration was calculated to result when 17% of the middle of the core had melted out and was distributed uniformly over the lower part. The reactivity loss associated with this change was 12% in k_{eff}.* If the top section is assumed to fall freely as a single unit at just this point, a relatively long time for free fall is available and the reactivity insertion rate at prompt critical is calculated to be 4.5 ($\Delta k/k$) per second. One can be still more pessimistic. If the reactor is assumed to begin the melting in a highly subcritical state, say with k_{eff} equal to 0.94, thanks to removal of the control rods, it was possible to calculate an even greater reactivity insertion rate at prompt critical; namely, about 7 ($\Delta k/k$) per second [17].

4.3.3 Case C. Coolant Expulsion and Return

The sodium is expelled from the core temporarily by boiling, then rushes back moments later at high speed. During this brief interval the fuel elements undergo some degree of failure. The on-rushing sodium carries the meltdown fragments with it, collecting the fuel rapidly into a more compact configuration [81].

These are extreme assumptions. They could be altered appreciably, yet remain pessimistic in the light of our present-day knowledge. A major difficulty encountered in making such analyses is in specifying a realistic mode of core meltdown. On becoming aware of any realistic mode, the designer adds more safety factors to render this particular accident less possible, even incredible. But even when a cause of meltdown is provided, accurate specification of its course in detail is frightfully difficult. The position of the fissionable material at each and every instant must be determined throughout the course of the incident. Only then can one specify with precision the variation in reactivity with time, which is the essential data needed for fixing the outcome of the event.

The motion of the fuel is subject to the interplay of many complex forces and events. The following list of questions comes to mind when one tries to solve the problem for reactors, such as EBR-II or Fermi, which have one piece, unrestrained, metal fuel pins as fuel elements. With a change in fuel element type some questions would be changed but most would remain similar to those on this list.

How does the meltdown occur?
- What is the reactivity at the time of meltdown, and what external reactivity effects are still being brought into play?
- Exactly what is the reactor power and temperature distribution when fuel element failure begins?

What is the condition of the coolant system?
- Are the pumps off? If so, for how long?
- Is coolant present? If so, is it static, oscillating or flowing under pressure?
- If the coolant is leaking, at what rate?
- What is the temperature of the coolant?

What is the contribution to reactor heating due to fission products?

---

*These estimates were made from criticality calculations assuming a one-dimensionl, slab-geometry. Two dimensional calculations gave a considerably smaller reactivity drop [67].

What is the distribution of this source throughout the core?

What is the mode of failure of the fuel cladding?

Does the cladding give way at a flaw?

Does the eutectic formation produce failure before the uranium becomes molten if the cladding is steel?

Does the sodium bond, if present, generate sufficient pressure to cause the cladding failure? If the cladding holds uranium in the molten form, for how long? Is this a function of uranium temperature?

How does the sodium coolant around the fuel element effect the mode of failure of the pin?

What part does sodium boiling play in pin failure?

If the pin has been irradiated, what role do the fission gases play?

How is pin failure a function of burnup, of fuel alloy and of rate of heating?

Of what import is the effect of partial reloading on the coherence of fuel element failure?

When failure occurs, where and at what velocity does the uranium escape?

Where does the molten uranium go?

If no sodium is present, does the molten uranium pour down through the core and lower blanket, or freeze somewhere on the way?

If sodium is absent from the core but present in the lower blanket, can the uranium penetrate the lower blanket?

If the pumps are still active, what sort of sodium boiling pattern will be present in the core?

Will the molten fuel move upwards, downwards or both?

If sodium is present in the reactor, with the pumps off, what sort of forces will the boiling of the sodium exert on the fuel?

Will there be oscillations and sodium boiling, a "chugging"?

If the fuel melts from the center and freezes at the core bottom, can the remaining upper section fall essentially as a unit to create high rates of reactivity addition?

If the molten fuel runs down through the blanket, can it reassemble below in a critical configuration?

How is all of this a function of reactor geometry and design?

There are few answers available to these questions. A few calculations have been made of the progress of core melting under completely arbitrary assumptions as to the initial cause and with great simplification of the actual heat transfer and of structural details. In addition, some controlled experiments have been performed in which portions of the complex overall problem could be studied.

Experiments on single pins in TREAT have shown reasonable agreement with theoretical predictions of the temperature of cladding failure and its cause [30h,37k]. Sodium-bonded metallic uranium pins are observed to fail primarily due to dissolution of the steel by uranium over broad areas. Uranium metal pins, coextruded with zirconium cladding so as to have a metallurgical bond between pin and clad, may fail due to fine cracks in the cladding, or from local dissolution. Uranium metal pins, sodium bonded in tantalum or niobium jackets, suffer jacket failure because of the buildup of internal pressure, arising from the generation of sodium vapor and from the heating of the inert gas atmosphere placed above the bond during fabrication. Uranium oxide pellets canned in stainless steel, niobium or tantalum fail from buildup of internal pressure. The types of failure observed and the temperature at which failure occurs are in general agreement with theoretical predictions. Of course, these observations were all made in the absence of a flowing sodium coolant. The presence of such a coolant, or a major change in the time scale of heating, could significantly alter these results.

Upon escaping from a Fermi-type zirconium-clad element, in the absence of sodium coolant, the uranium runs down under the influence of gravity with little lateral velocity. The same effect occurs in an unbonded EBR-II type fuel pin. In the sodium-bonded pins, however, once the energy input passes a certain threshold the uranium is pushed out laterally at high velocity by the vaporized bond. It is reasonable to expect failure at the hottest point in the jacket, with all the uranium in the pin escaping the pin in the vicinity of this position.

The observed behavior of the sodium-bonded pins upon failure suggests another mechanism for rapid reactivity insertion upon core meltdown. Namely, this rapid, concentrated expulsion of all the fuel in the pin might occur coherently over a large number of fuel pins. Relatively few pins at the core center will provide sizable reactivity changes if their fuel all escapes at their midpoints. Calculations for the EBR-II reactor, for example, show that 3% ($\Delta k/k$) is available if the 7 central subassemblies fail and send all their fuel to the core center. If all 61 subassemblies did so, 14% ($\Delta k/k$) is available. Even if failure occurs near the top of the core, a sufficient number of neighboring pins failing simultaneously can make the reactor supercritical. Obviously, the time scale and position of pin failure must be known if one is to make a realistic appraisal of the hazard. And the role of all factors in pin failure, including past irradiation history, must be used in analyzing the situation. With relatively slow rates of overheating, fuel element swelling under the action of fission product gases may be sufficient to burst the jacket before the fuel melts. The melting fuel would no longer be subject to high sodium vapor pressures and its escape from the jacket would take a different course. In experiments involving the rapid electrical melting of pre-irradiated uranium metal fuel pins, the escaping fuel has been found to have a foamy, frothy appearance [19]. Low density fuel could not readily reassemble into a more reactive configuration, and the course of an excursion might be appreciably altered. It is also of interest to note that for a pre-irradiated metallic fuel pin not having a sodium bond, the fuel apparently does not seem to escape with a high lateral velocity, as it did under the action of bond sodium vapor.

The transient boiling of sodium or other coolants

and the effect of the boiling on the molten fuel dispersal are problems which remain to be solved. Interesting model studies with electrically heated pins in water have cast some light on the situation [19,30i]. Using system parameters scaled to resemble the Fermi Reactor, an oscillating behavior of the flowing water was observed when the heat input in the steel pins exceeded the energy which could be absorbed with steady flow. Lead pins were then heated electrically to melting temperatures, and failure in the presence of this oscillating water flow was observed. This limited series of experiments provides some empirical information on fuel (as simulated by lead) dispersal and on coolant behavior, but theoretical prediction remains to be accomplished.

Each fuel element type will present new questions of meltdown behavior. The relative times of sodium boiling and fuel melting could be drastically different for a high melting point fuel. The manner of jacket failure would presumably be different for the so-called "strong can, weak-pin" metal fuel concept. And for fuel elements having a low average fuel density within the fuel container, the possibility of fuel melting and slumping within the individual cans may require consideration.

4.4 Methods of Calculation of Severe Accidents

The calculation of the course of a severe accident in a fast reactor involves the solution of coupled thermodynamic, hydrodynamic, and neutron transport phenomena in some idealized formulation of the actual conditions. The power rises, generating heat, which generates pressure which moves material, which changes pressure and reactivity, which affects power. Numerical methods for solving such problems on high speed computers have been developed [91,92]. These methods will be described later in Sec. 4.4.6. In addition, an extremely useful analytic approach has been devised by Bethe and Tait [90] and since modified by Jankus [17,371,93] and by Nicholson [94]. This so-called Bethe-Tait (or modified Bethe-Tait) method is described in some detail in Sec. 4.4.2.

Common to both methods of calculation is the use of an effective step input of reactivity to initiate the accident. This input is determined as described in the following section.

It should be emphasized that both the numerical calculation and the Bethe-Tait approach represent idealized versions of the actual situation. The inherent errors introduced by idealizations of geometry and by assumptions concerning the equation of state, for example, are difficult to assess. In what follows, when evaluations are made of various methods of solution, and when the errors introduced by certain mathematical approximations or by certain additional physical assumptions are treated, such discussion is intended only to guide those interested in comparing these techniques. Computational errors of 5% or 20% may be discussed, knowing full well that the inherent errors involved in idealizing the general problem are much greater.

4.4.1 Reactivity Insertion before Disassembly

A severe accident is terminated by disassembly of the core under the action of high pressures generated in the fissioning isotopes. The high pressures are attained because the energy is generated in a very short time and the inertia of the core materials resists the disassembly process. Since there is usually a certain amount of void space left in the core from loss of coolant, the fuel expansion can at first take place locally. Actually sodium coolant is highly compressible so that even if it is present, there will be some effective void space to fill before high pressures can be generated. After there has been sufficient thermal expansion of the molten fuel to fill the voids, pressures rise very rapidly. Thus, there is an initial phase of the power excursion during which the disassembly process can be ignored and the power increase is determined by solution of the reactor kinetic equation (without delayed neutrons) for a given reactivity insertion rate.

Usually, the amount of reactivity available in a severe accident is quite large. For example, as much as 10% ($\Delta k/k$) might be potentially available in a meltdown. But the rate at which the reactivity can be inserted is limited; and the short neutron lifetime above prompt critical and the correspondingly rapid rate of energy generation limits the maximum excess reactivity that can be reached. The reactivity reduction rate during explosive disassembly in a fast reactor is always very much greater than the reactivity addition rate causing the burst. Thus the mechanical reactivity addition rate during the burst is unimportant. If the reactivity addition rate and the initial power are known, and if one can estimate the power level at which appreciable pressures will be generated, this is sufficient to fix the reactivity inserted mechanically before disassembly terminates the burst.

Let us take the reactivity addition rate after prompt critical is reached to be (dk/dt) and the prompt generation time to be ℓ. Then, if no reactivity reduction mechanism is operative, the time rate of change in power n is

$$\frac{n}{n} = \frac{t}{\ell}\left(\frac{dk}{dt}\right). \tag{4-1}$$

Thus the power at time t after prompt critical is

$$n(t) = n(0)\exp\left[\frac{1}{2}\,\ell^{-1}\,\frac{dk}{dt}\,t^2\right]. \tag{4-2}$$

The amount of heat developed in the interval from $t = 0$ to $t = t_1$, the time available for mechanical insertion of reactivity, is

$$Q_1 = \int_0^{t_1} n\,dt = n(0)\int_0^{t_1} dt\,\exp\left[\frac{1}{2}\,\ell^{-1}\,\frac{dk}{dt}\,t^2\right]. \tag{4-3a}$$

Integrating by parts gives

$$\approx n(0)\ell\left(\frac{dk}{dt}t_1\right)^{-1}\exp\left[\frac{1}{2}\ell^{-1}\left(\frac{dk}{dt}\right)t_1^2\right]$$

$$\left[1+\ell\left(\frac{dk}{dt}\right)^{-1}t_1^{-2}+\cdots\right], \quad (4\text{-}3b)$$

since

$$(dk/dt)\ t_1^2/\ell \gg 1 . \quad (4\text{-}3c)$$

One can rewrite this equation in an alternate form:

$$\ell^{-1}\left(\frac{dk}{dt}\right)t_1^2 = \ln\left[Q_1^2 n^{-2}(0)\ell^{-1}\frac{dk}{dt}\right]$$

$$+\ln\ln\left[Q_1^2 n^{-2}(0)\ell^{-1}\frac{dk}{dt}\right]+\cdots . \quad (4\text{-}4)$$

The reactivity that has been inserted up to time t_1 is

$$\Delta k_1 = t_1\frac{dk}{dt} = \sqrt{\ell^{-1}\left(\frac{dk}{dt}\right)t_1^2}\sqrt{\ell\left(\frac{dk}{dt}\right)}, \quad (4\text{-}5)$$

where ℓ^{-1} (dk/dt) t_1^2 is given by Eq. (4-4). It depends only logarithmically on Q_1 and is quite insensitive to it if the excursion is very large ($\ln [Q_1^2 n^{-2}(0)\ell^{-1}(dk/dt)] \gg 1$).

For example, if one assumes that $Q_1 = 1.1 \times 10^3$ joule/g, that the initial power at prompt critical is $n(0) = 3.3 \times 10^{-4}$ watt/g, that the prompt neutron lifetime is 10^{-7} sec and that the reactivity insertion rate varies from 0.072 to 7.2($\Delta k/k$) per sec, then $\ln [Q_1^2 n^{-2}(0)\ell^{-1}(dk/dt)]$ lies between 43.5 and 48.15. For this range $\sqrt{\ell^{-1}(dk/dt)t_1^2}$ lies between 6.87 and 7.2. A factor of e (= 2.718) change in Q_1 or $n(0)$ changes $\sqrt{\ell^{-1}(dk/dt)t_1^2}$ by only 2%. The same factor of e in either ℓ or (dk/dt) changes $\sqrt{\ell^{-1}(dk/dt)t_1^2}$ by only 1%.

For the example chosen, t_1 varies from 0.0081 to 0.00085 sec, the smaller time corresponding to the largest reactivity addition rate. The reactivities inserted in these times would vary from .00058 to .0061 ($\Delta k/k$).

Since the amount of reactivity which can be added mechanically before disassembly is relatively insensitive to the energy level needed for disassembly, an exact choice of this energy Q_1 is unnecessary. It is convenient to set $Q_1 = Q^*$, the energy threshold for significant pressure generation, although an estimate of $Q_1 = eQ^* = 2.718\ Q^*$ might be better.

Rather than $n(0)$, the power at prompt critical, only the power at the beginning of the excursion may be known initially. For the reactivity addition rates of interest herein, the ratio of power at prompt critical to that at delayed critical is given sufficiently well by the relation [95].

$$\frac{n(0)}{n(\text{delay crit.})} = \beta\sqrt{\frac{\pi}{2\ell(dk/dt)}}, \quad (4\text{-}6)$$

where β = effective delayed neutron fraction.

Should the reactivity excursion start from below delayed critical ($k_s < 1$) at some source power n_s, for large reactivity addition rates [81]

$$\frac{n(\text{delay crit.})}{n_s} = \frac{1-k_s}{\beta} . \quad (4\text{-}7)$$

A rigorous treatment of the transition from source power to prompt critical for a constant reactivity addition rate may be found in reference [96].

4.4.2 Bethe-Tait Method [17,18,90,93]

In fast reactors the reactivity is reduced by the expansion of the reactor. During expansion the density and macroscopic cross sections change and, for the moderately small reactivities considered here, the reactivity can be calculated by first order perturbation theory. If the reactor can be described by one group diffusion theory the flux ϕ is found from

$$\nabla\cdot D\nabla\phi + [(\nu_0-1)\Sigma_f - \Sigma_c]\phi = 0 , \quad (4\text{-}8)$$

where D is the diffusion coefficient and Σ_f and Σ_c are the macrosopic fission and capture cross-sections respectively. From first-order perturbation theory when the change in diffusion coefficient is δD, in macroscopic fission cross section $\delta\Sigma_f$ and in macroscopic capture cross section $\delta\Sigma_c$, the change in reactivity is

$$\Delta k - \Delta k_{max} = \int\left\{-\delta D(\nabla\phi)^2 + \left[(\nu_0-1)\delta\Sigma_f - \delta\Sigma_c\right]\phi^2\right\}dV \div \int \nu_0\Sigma_f\phi^2 dV . \quad (4\text{-}9)$$

Since the macroscopic cross sections during expansion are proportional to the density, the equation of continuity gives

$$\delta\Sigma_f = -\nabla\cdot(\Sigma_f\vec{u})$$

$$\delta\Sigma_c = -\nabla\cdot(\Sigma_c\vec{u})$$

$$\delta D = \nabla\cdot(D\vec{u}) - 2\vec{u}\cdot\nabla D ,$$

where $\vec{u}$ is the displacement vector. After substituting these relations into Eq. (4-9) integration by parts removes $\vec{u}$ from under the divergence sign.

Then, using Eq. (4-8) to replace the fission and capture cross sections by the diffusion coefficient,

$$\Delta k - \Delta k_{max} = \int \vec{u} \cdot \left\{ 2\nabla D(\nabla\phi)^2 + D\nabla(\nabla\phi)^2 - 2\nabla\phi\nabla \cdot (D\nabla\phi) \right\} dV \div \int \nu_0 \Sigma_f \phi^2 dV ,$$

$$= \int 2\vec{u} \cdot [(\nabla\phi \times \nabla) \times D\nabla\phi] \, dV \div \int \nu_0 \Sigma_f \phi^2 dV . \quad (4\text{-}10)$$

For spherical symmetry the preceding expression becomes simpler:

$$\Delta k - \Delta k_{max} = -\int 4uD^{-1} \left(D \frac{d\phi}{dr}\right)^2 r dr \div \int \nu_0 \Sigma_f \phi^2 r^2 dr . \quad (4\text{-}11)$$

Clearly Eqs. (4-10) and (4-11) can also be rewritten in multigroup notation if the subsequent calculations require more precision. Instead of $D(d\phi/dr)^2$ we would have then $\sum_i D^i (d\phi^i/dx)\,(d\phi^{i*}/dx)$ and instead of $\nu_0 \Sigma_f\ \phi^2$ we would have $\sum_i \nu_0^i\ \Sigma_f^i \phi^i \phi^{i*}$, where ϕ^{i*} is the adjoint flux and summation (i) extends over all energy groups. Now F, the fraction of fissions in the core, is

$$F = \int_0^b \nu_0 \Sigma_f \phi^2 r^2 dr \div \int_0^\infty \nu_0 \Sigma_f \phi^2 r^2 dr .$$

Equation (4-11) can be rewritten

$$\Delta k - \Delta k_{max} = -4F \int \rho u \frac{1}{D\rho} \left(D \frac{d\phi}{dr}\right)^2 r dr \div \int_0^b \nu \Sigma_f \phi^2 r^2 dr , \quad (4\text{-}12)$$

where ρ is the material density. If the flux in the core can be approximated by a parabola, $\phi = 1 - qr^2/b^2$, where b is the radius of the core, then $D(d\phi/dr) = -2qr/(3\Sigma_{tr} b^2)$ in the core and the adjacent portion of the blanket. Σ_{tr} is the transport cross section in the core. If, in addition, the microscopic transport cross section (Σ_{tr}/ρ) is the same in the core and in the blanket and $\nu_0 \Sigma_f$ is constant in the core, these constants can be taken out from under the integral signs and, performing the integration in the denominator, the following is obtained:

$$\Delta k - \Delta k_{max} = - \frac{16q^2F}{\Sigma_{tr}\nu_0\Sigma_f b^7 \left(1 - \frac{6}{5} q + \frac{3}{7} q^2\right) \rho_c} \int \rho u r^3 dr . \quad (4\text{-}13)$$

This is an expression for the reactivity reduction, where the integral extends over the whole reactor, but the coefficients in front of it refer to the parameters of the core alone. In the multigroup formulation this expression is still valid, provided one performs the integral in the numerator correctly, that one replaces the constant $4q^2/\Sigma_{tr}$ by an average value of

$$\frac{1}{r^2} \frac{d}{dr} \left[r \sum_i D^i \frac{d\phi^i}{dr} \frac{d\phi^{*i}}{dr} \right]$$

in the core and provided that the expression

$$\sum_i \left(\frac{\Sigma_{tr}^i}{\rho}\right) \left(D^i \frac{d\phi^i}{dr}\right) \left(D^i \frac{d\phi^{i*}}{dr}\right)$$

does not experience an abrupt discontinuity at the core-blanket interface. Differentiating (4-13) twice with respect to time, then substituting the force for the time rate of change of momentum,

$$\rho \ddot{u} = -\frac{\partial p}{\partial r} , \quad (4\text{-}14)$$

and finally integrating by parts, a relation is obtained between the second derivative of the reactivity and the pressure:

$$\ddot{k} = - \frac{48q^2F}{\Sigma_{tr}\nu_0\Sigma_f b^7 \left(1 - \frac{6}{5} q + \frac{3}{7} q^2\right) \rho_c} \int p r^2 dr . \quad (4\text{-}15)$$

Thus $\ddot{k}$ is proportional to the total pressure generated.

Expressions (4-13) and (4-15) are useful for calculation of the reactivity reduction. It is difficult, of course, to use them directly in the general case, since the displacement or the pressure are not known accurately. However, approximations may be made in limiting cases. If an e-folding time were much larger than the time needed for sound waves to traverse the core, $\ell/\Delta k_{max} \gg b/c$ (where c is the wave velocity), it could be assumed that a quasi-static situation has been reached, the pressure remains negligible, and the resulting displacements and the reactivity change can be computed from Eq. (4-13). This is not the case here. On the other hand, if the multiplication ratio is very large and $\Delta k_{max}/\ell \gg c/b$, it can be assumed that the expansion is negligible, the pressure in Eq. (4-15) can be computed as from a constant volume process, and $\ddot{k}$ can be determined from the pressure. This method will over-estimate $\ddot{k}$ somewhat, the more so the smaller the value of $\Delta k_{max}/\ell$.

Assuming that the integral in Eq. (4-15) extends over the core only and that the spatial dependence of power density is maintained, the energy generated at a point of space and time is

$$E(r, t) = Q(t)N(r) . \qquad (4\text{-}16)$$

Now if the power has a parabolic shape in the core and vanishes outside,

$$N = 1 - \left(qr^2/b^2\right) \quad \text{for } r < b$$
$$N = 0 \quad \text{for } r > b , \qquad (4\text{-}17)$$

one can change variables from r to N under the integral in Eq. (4-15) and obtain

$$\ddot{k} = -\frac{15}{4}\frac{A}{b^2\rho_c}\int_{1-q}^{1} p(NQ)\sqrt{1-N}\, dN . \qquad (4\text{-}18)$$

where A is defined as:

$$A = (32/5)(\sqrt{q}\,F)\Big/\left[\Sigma_{tr}\nu_0\Sigma_f\left(1 - \frac{6}{5}q + \frac{3}{7}q^2\right)b^2\right] ,$$

With this expression $\ddot{k}$ can be determined as a function of Q if the pressure is given as a function of internal energy for the material of the core.

For simplicity it will be assumed in the following that the pressure is negligible until the energy density reaches a certain value Q* and that the pressure increases linearly thereafter:

$$p(E, \rho_c) = 0 , \quad \text{for } E < Q^* ,$$
$$p(E, \rho_c) = (\gamma - 1)\rho_c(E - Q^*) , \quad \text{for } E > Q^* . \qquad (4\text{-}19)$$

Here the notation of Bethe and Tait is followed, using $(\gamma - 1)$ as the coefficient in a linear relation. Then performing the integral in Eq. (4-18) it is seen that

$$\ddot{k} = 0 \quad \text{for } Q < Q^* , \qquad (4\text{-}20a)$$

$$\ddot{k} = -A(\gamma - 1)\frac{Q^*}{b^2}\frac{Q}{Q^*}\left(1 - \frac{Q^*}{Q}\right)^{5/2}$$
$$\text{for } Q^* < Q < Q^*(1 - q)^{-1} , \qquad (4\text{-}20b)$$

$$\ddot{k} = -A(\gamma - 1)\frac{Q^*}{b^2}\cdot\frac{5}{2}q^{3/2}\left[(1 - 0.6q)\frac{Q}{Q^*} - 1\right]$$
$$\text{for } Q > Q^*(1 - q)^{-1} . \qquad (4\text{-}20c)$$

The coefficient A is defined after Eq. (4-18). These equations together with the definition of reactivity above prompt critical,

$$\Delta k = \ell \frac{d}{dt}\ln\dot{Q} , \qquad (4\text{-}21)$$

form the kinetic equations. Using proper initial conditions:

$$\Delta k(0) = \Delta k_{max}$$
$$\Delta\dot{k}(0) = 0$$
$$\dot{Q}(0) = (\Delta k_{max}/\ell)\, Q(0) ,$$

one sees that the ratio of the final energy to Q* is a function only of

$$\frac{b^2}{A(\gamma - 1)Q^*}\frac{\Delta k_{max}^3}{\ell^2} = X \qquad (4\text{-}22)$$

and q if $Q/Q^* > (1 - q)^{-1}$. The coefficient A, does not vary much from reactor to reactor.

The direct solution of Eq. (4-20a, b, c) and Eq. (4-21) requires some relatively simple numerical procedures. However, an approximate value for the total energy generated can be obtained easily by assuming that the energy increases exponentially,

$$Q = Q^* \exp(\Delta k_{max}\, t/\ell) , \qquad (4\text{-}23)$$

until the reactivity obtained from Eq. (4-20) vanishes. The value of Q at this time is a fair approximation of the correct total value, since by neglecting the energy generated after the peak in power, one roughly compensates for the overestimate of energy generation during the power rise.

Using these assumptions, one finds that for small values of X (that is, weak explosions) the energy yield (here, loosely defined as (Q-Q*), the total energy generated less the threshold energy) is given by

$$(Q/Q^*) - 1 = (15.75X)^{2/9} + 0.5707(15.75X)^{4/9} \quad \text{if } X \ll 1 . \qquad (4\text{-}24)$$

For very large values of the argument (X), the approximate energy yield is given by

$$(Q/Q^*) - 1 = \frac{2}{5}q^{-3/2}\frac{X}{(1 - 0.6q)} + \frac{1}{2(1 - 0.6q)}\left[\ln\frac{2}{5}q^{-3/2}\frac{X}{(1 - 0.6q)}\right]^2 + \cdots$$
$$\text{if } X \gg 1 . \qquad (4\text{-}25)$$

This approximate analytical method, which is basically that first used by Bethe and Tait [90], gives results considerably in error. The ratio of the excess energies resulting from an accurate solution of the Bethe-Tait formulation to those obtained with these relatively crude assumptions is plotted versus X in Fig. 4-1. Except at very large X, the error is considerable.

A more detailed analysis of the error introduced by this roughly compensating pair of assumptions has been performed with the exact numerical, high speed computer program described in Sec. 4.4.6 [91]. As seen in Fig. 4-2, the rough calculation gives an energy too low by a factor of 1.5 to 2, except for very large explosions, where $\Delta k^3_{max}/\ell^2$, the key variable in the argument X, is large. (For EBR-II, $X \approx 1.1 \times 10^4(\Delta k_0^3/\ell^2)$).

Nevertheless, the asymptotic solutions given in Eqs. (4-24) and (4-25) are of considerable interest, They can be used to obtain simply estimates of the dependence of the energy yield on various factors entering into the explosion.

It is seen that the energy yield increases with an increase of the argument X. The argument increases with increase of core radius b. Thus, larger reactors would develop more energy per unit mass, as well as more total energy, if the maximum reactivity and the lifetime were the same.

The argument X is also proportional to $\Delta k^3_{max}/\ell^2$. If there were no reactivity producing phenomena other than those used in the analysis (like Doppler, for example), then according to Eq. (4-5)

$$X \propto \Delta k^3_{max}\ell^{-2} \propto \left(\frac{dk}{dt}\right)\left[\frac{1}{\ell}\frac{dk}{dt}\right]^{1/2}. \qquad (4\text{-}26)$$

Thus X varies as the square root of the lifetime of the reactor, if the accident is caused by the insertion of reactivity at a constant rate. At the lower insertion rates, in particular, where by Eq. (4-24) the energy yield $[(Q/Q^*) - 1]$ varies with X roughly as $X^{1/4}$, yield will vary with neutron lifetime roughly as $\ell^{-1/8}$, increasing to a $\ell^{-1/2}$ dependence only at very large X. This marked insensitivity to neutron lifetime for ramp type reactivity insertion rates is explained in part as follows: With longer neutron lifetimes there is considerably more time between prompt critical and the point where significant pressure generation begins; thus, Δk_{max} is correspondingly greater for the same dk/dt. Numerical calculations [94] confirm this insensitivity.

The dependence of yield on dk/dt for small X should go like $(dk/dt)^{3/8}$, increasing in sensitivity with X until for very large X, the yield would vary like $(dk/dt)^{3/2}$. This type of dependence again is confirmed at the lower end of the scale numerically [94].

By eliminating the argument X from Eq. (4-22) and (4-24) (dropping the term in $X^{4/9}$ in the latter), one can obtain an expression for the energy yield (or excess energy) dependence on Q*, for small X (or large Q*), namely

$$Q - Q^* = \left[\frac{63}{4}\cdot A\cdot\frac{\Delta k^3_{max}}{(\gamma-1)}\frac{b^2}{\ell^2}\right]^{2/9}(Q^*)^{7/9} + \ldots \qquad (4\text{-}24a)$$

Thus for large Q* the excess energy varies almost linearly with Q*.

Similarly, by using Eq. (4-22) and (4-25) one can obtain an expression for the excess energy valid for large X (small Q*)

$$Q - Q^* = \frac{2}{5}q^{-3/2}\frac{A}{(1-0.6q)}\frac{\Delta k^3_{max}}{(\gamma-1)}\frac{b^2}{\ell^2}. \qquad (4\text{-}25a)$$

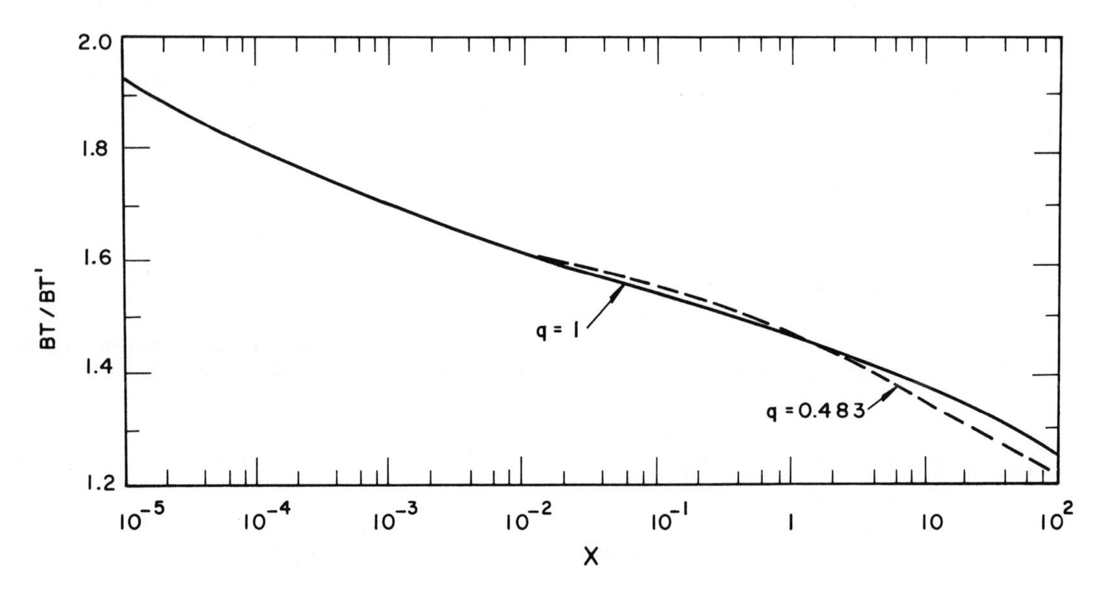

FIG. 4-1 Ratio of excess energy calculated by accurate Bethe-Tait approach to that calculated by approximate Bethe-Tait method.

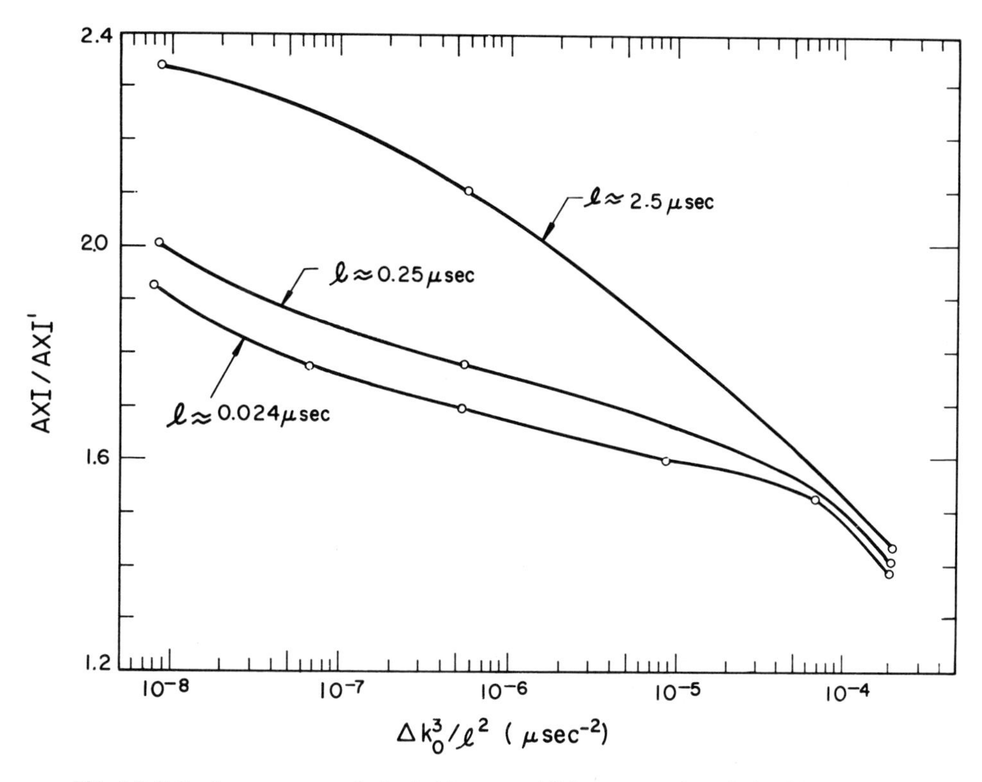

FIG. 4-2 Ratio of excess energy calculated with accurate AX-1 program to that calculated by approximate Bethe-Tait method, as simulated in AX-1 ($X \approx 1.1 \times 10^4 \, (\Delta k_0{}^3/l^2)$).

It is seen that for large X (small Q*), the excess energy becomes nearly independent of Q*. (There is still a logarithmic dependence of Δk_{max} on Q*.)

For very small X, the excess energy varies slowly with the equation of state parameter, i.e., as $(\gamma - 1)^{-2/9}$. For very large X, the excess energy is more sensitive to this parameter.

4.4.3 Generalization of Bethe-Tait Formulation

Basic to the Bethe-Tait formulation is the neglect of expansion and propagation of the pressures in the calculation of pressure. But the previous derivation of an expression for reactivity reduction employed several nonessential assumptions which can be eliminated in a more general formulation [93,94]. The derivation was based on a first-order perturbation of the one-group diffusion equation. It assumed a parabolic shape of the flux and power, and equal microscopic transport cross sections in core and in blanket. The calculation further assumed a sharp threshold in pressure production, a linear relation between pressure and energy above the threshold and the absence of any other temperature-dependent reactivity effects except expansion due to pressure gradients.

We shall here outline a general formulation of the expression for reactivity reduction, then enumerate a number of specific computational elaborations of the original method. Following Jankus [93], we will assume that the core is composed of a single, homogenous material, the blanket of another, and that there may be variation of density of these materials. We will define the worth of reactor material, w(r), as the decrease in reactivity due to removal of unit mass of material at the position r. The worth will be continuous in the reactor except for the interface of the core and the blanket. It is determinable from experiment or from multigroup calculation. The change in the reactivity due to an increase in density, $\delta\rho(r)$, of the core material is

$$\Delta k - \Delta k_0 = \int \delta\rho w_c(r) dV + \cdots .$$

Change in the reactivity caused by changes at the interface of the core and the blanket still has to be evaluated. Thus the ellipsis (...) here stands for these often substantial terms. If the change of density is caused by motion of the material,

$$\delta\rho = \nabla \cdot (\rho u) ,$$

where u is the displacement. Substituting this relation into the preceding one and integrating by parts over the whole core, we see that

$$\Delta k - \Delta k_0 = \int \rho u \cdot \nabla w_c dV + \cdots . \qquad (4\text{-}27)$$

Now, taking the second derivative with time and substituting for $\rho\ddot{u}$ its value from the equation of motion:

$$\rho\ddot{u} = -\nabla p , \qquad (4\text{-}28)$$

where p is the "gauge" pressure, we have

$$\Delta\ddot{k} = \ddot{k} = -\int \nabla p \cdot \nabla w_c dV + \cdots . \qquad (4\text{-}29)$$

Integrating by parts, we obtain

$$\ddot{k} = \int p\,(\nabla^2 w_c)\,dV - \int p\nabla w_c \cdot dS + \cdots , \qquad (4\text{-}30)$$

where dS is an element of surface in the core very close to the blanket. Now we will consider the reduction of reactivity at the interface of the core and the blanket. We will assume that the velocity of wave propagation is relatively small so that reflected and transmitted waves will not have penetrated far from the boundary. Then, if the pressure inside the core is p, the pressure at the interface may be determined approximately using slab geometry. The interface pressure assumes an intermediate value

$$p_i = p\,\frac{\rho_b c_b}{\rho_b c_b + \rho c} , \qquad (4\text{-}31)$$

where ρc and $\rho_b c_b$ are wave impedances of the core and of the blanket. In the region of intermediate pressure, the core and blanket particles are traveling at the speed

$$\dot{u} = \frac{p}{\rho_b c_b + \rho_c} ,$$

and the region itself is increasing with appropriate velocity into the core and into the blanket. Thus the motion of the core part increases the first derivative of reactivity at the rate

$$\int \rho\dot{u} \cdot \nabla w_c c\,dS = \int p\,\frac{\rho c}{\rho_b c_b + \rho c}\,\nabla w_c \cdot dS$$

and motion of the blanket

$$\int \rho_b\dot{u} \cdot \nabla w_b c_b dS = \int p\,\frac{\rho_b c_b}{\rho_b c_b + \rho c}\,\nabla w_b \cdot dS .$$

The sum of these two contributions is

$$\int p\nabla w_c \cdot dS - \int p\,\frac{\rho_b c_b}{\rho_b c_b + \rho c}\,(\nabla w_c - \nabla w_b) \cdot dS .$$

Upon adding this to Eq. (4-30), we obtain the total reactivity reduction as

$$\ddot{k} - \int p(\nabla^2 w_c)dV - \int p\,\frac{\rho_b c_b}{\rho_b c_b + \rho c}\,(\nabla w_c - \nabla w_b) \cdot dS . \qquad (4\text{-}32)$$

If the gradients of worth at the interface are the same for core and blanket materials, the second term vanishes and we have a simple expression for the reactivity reduction.

The original one-group treatment can be placed in similar form. We had expressed the reactivity change by Eq. (4-10).

$$\Delta k - \Delta k_0 = \int \rho u \cdot \left\{ \frac{2}{\rho}(\nabla D)(\nabla\Phi)^2 + \frac{D}{\rho}\nabla(\nabla\Phi)^2 - \frac{2}{\rho}(\nabla\Phi)\nabla \cdot (D\nabla\Phi) \right\} dV \div \int \nu\Sigma_f\Phi^2 dV .$$

This equation, which can be written in a form similar to Eq. (4-27).

$$\Delta k - \Delta k_0 = \int \rho u \cdot f dV , \qquad (4\text{-}33)$$

was valid over the whole reactor, even if composed of different materials. However, instead of ∇w_c of Eq. (4-27), we had

$$f = \frac{1}{(D\rho)} \cdot \left[\nabla(D\nabla\Phi)^2 - 2(D\nabla\Phi)\nabla \cdot (D\nabla\Phi)\right] \div \int \nu\Sigma_f\Phi^2 dV . \qquad (4\text{-}34)$$

Now, applying to Eq. (4-33) the same procedure used in deriving Eq. (4-30), but taking the volume of the whole reactor for the domain of integration, we obtain

$$\ddot{k} = \int p(\nabla \cdot f)dV . \qquad (4\text{-}35)$$

Since pressure is being generated only in the core, we subdivide our domain of integration into three regions: interior of the core, a pillbox-shaped volume enclosing the interface, and the rest of the blanket. The integral over the blanket vanishes if the velocity of wave propagation is small enough. The integral over the pillbox does not vanish if f is discontinuous (and if the pressure wave has reached the edge of the core). Upon evaluating the latter by the divergence theorem, we obtain

$$\ddot{k} = \int_{core} p(\nabla \cdot f)dV - \int p_i(f_c - f_b) \cdot dS , \qquad (4\text{-}36)$$

where p_i is the pressure at the interface, and f_c and f_b are the values of function f on the core and the blanket side at this location. It is obvious that the surface integral may contribute a considerable fraction of the reactivity reduction if $(f_c - f_b)$ is not small. Equation (4-36) is equivalent to Eq. (4-32) if the pressure at the interface is assumed to be given by Eq. (4-31) and the reactor core is composed of one distinct material, the reactor blanket of another. However, Eqs. (4-35) and (4-36) are valid also when compositions as well as the density of core and blanket vary spatially. If the core is composed of several layers of different materials, we still can assume that $f = \nabla w$ within each layer for evaluation of Eq. (4-35). But we have to add an appropriate surface term, as in Eq. (4-36).

Of course, the expression for f, Eq. (4-34), simplifies considerably for a one-dimensional reactor. For a spherical reactor it becomes equal to

$$f = -\frac{4}{r}\frac{1}{(D\rho)}\left(D\frac{d\Phi}{dr}\right)^2 \Big/ \int \nu\Sigma_f\Phi^2 dV . \qquad (4\text{-}37)$$

4.4.4 Extensions of Bethe-Tait Formulation

Some numerical procedures are necessary for solution of the Bethe-Tait formulation, even assuming a parabolic flux and power distribution. Jankus [93] has used a Runge-Kutta routine; Nicholson [94] has used a simple iterative procedure which is said to converge rapidly.

Nicholson has modified the formulation to allow polynomial representation of the power density and reactivity worth distributions, an additional reactivity feedback (Doppler effect) taken to be a function of average core temperature, and a more elaborate equation of state. This all has been programmed for a high speed computer [97]. Nicholson has also examined the propriety and importance of the choice of boundary condition for the core-reflector interface. He has compared (numerically) the assumption of a free surface at the core boundary with the assumption of a reflected core with a continuous pressure gradient at the boundary and found the difference in calculated energy yield to be small for reasonable reactivity insertion rates [94].

Nicholson has examined the effect of the assumption of parabolic power shape on energy yield. It is expected that this assumption will underestimate the energy release in a severe accident. This happens because the parabolic pressure gradient leads to a maximum reactivity gradient near the edge of the core. In reality, the reactivity gradient in a typical fast reactor may flatten near the core, leading to a much smaller gradient here, where pressure gradients and displacements are relatively large in severe accidents. A numerical investigation of this effect for a specific example indicated errors of 15 or 20% are possible from this cause [94].

Delayed neutrons need not be considered in the prompt critical region; however, it is necessary to be certain that the combination of initial power level and rate of reactivity increase are such that the reactor indeed reaches prompt critical.

4.4.5 Two-Dimensional and Nonuniform Effects

The first derivation of the Bethe-Tait formulation assumed spherical symmetry, and most applications are made in this geometry. However, some configurations resulting from core meltdown do not lend themselves to spherical representation. There are two principal new effects which arise in non-spherical, non-uniform situations. First, the equation of state predicts that for a given energy input per unit mass the pressure in a high density region may be much higher than in a low density region. This means that the reactivity effects produced by expansion of high density material can, in some cases, shut the reaction down before there is any significant pressure generation in the low density region.

The second new effect when nonuniform compositions are considered is that it is possible that the maximum in the distribution of reactivity worth and the maximum of the pressure distribution may not occur at the same point in space. Then the slope of the reactivity and pressure distributions could be of opposite sign in some part of the core, so that movement of material in that part would tend to produce an increase rather than a decrease in reactivity.

Nicholson and Stephenson [98] developed an explosion code in two-dimensional cylindrical geometry patterned after the Bethe-Tait formulation. Measurements of reactivity worth and power distribution had been made on two hypothetical meltdown assemblies. This information allowed a direct solution for $\ddot{k}$ by integration of Eq. (4-29) and hence a solution of the problem. Subsequent comparison of two dimensional diffusion theory with the results from critical experiments indicated that such calculations would be reasonably accurate in predicting reactivity worth and power distributions for complicated geometries [99].

4.4.6 Direct Numerical Solution of the Explosion Problem

It has been pointed out that even in its most general form, the Bethe-Tait method does not allow for the effects on pressure of local expansion and wave propagation. To handle these and other aspects accurately, direct numerical solutions of the coupled neutron transport hydrodynamics equations have been developed, using high speed computing machinery. The AX-1 code [91] is written for the IBM-704 and is readily available. Given a spherically symmetric, superprompt critical system, this program computes the variation in time and space of the specific energy, temperature, pressure, density, and velocity. It computes, as a function of time, the reactivity (in the form of alpha, the inverse period), the power, the total energy, and the position of the boundaries of the various shells into which the system has been subdivided. All delayed neutron effects are ignored, and no allowance is made for transfer of heat by conduction or radiation. The input information includes the initial reactivity or geometry, the initial velocities and temperatures of the mass points, the composition and disposition of materials, the appropriate constants of the equation of state, and the microscopic neutron cross sections. For calculational purposes the spherical assembly is divided into a number of hypothetical spherical shells or mass points. The neutron transport of this system is calculated in conventional fashion, using the S_4 method, thereby providing a power density distribution across the radial network, as well as the alpha of the system. From the neutron transport calculation one goes to the thermodynamics and hydrodynamics portion of the code to calculate the variation of power, temperature, pressure, density, and velocity with time.

The calculation proceeds initially like the usual S_n calculation. After computing average cross sections for each of the spherical shells in the mixture code, the program proceeds either to a calculation of alpha ($=\Delta k/\ell$) for the specified configuration or to a scaling of the reactor radii to provide the alpha originally specified. Before proceeding to the hydrodynamics, the code also computes k_{eff} ($=1+\Delta k_{ex}$) for the initially converged configuration, if so requested.

Then, for one or more short time intervals Δt, (short with respect to $alpha^{-1}$), alpha is considered to remain constant while the power varies as exp. $\alpha\Delta t$. From the pressure gradients in the system, the average accelerations of the mass points are computed and, hence, the new velocities at the end of the time interval. These, in turn, lead to the new radial positions of each shell boundary at the end of the time interval. The solution is performed in a Lagrangian co-ordinate system, i.e., the mesh is embedded in the material and follows it along throughout its motion.

During the time interval, energy is added to the system (the average power times Δt) and this is distributed among the shells in accord with the previously calculated fission distribution. After allowance is made for the work done by or on a shell in expansion or compression, the net change in internal energy is computed, and from the internal energy a new pressure and temperature are obtained. The so-called viscous pressure, obtained by a mathematical procedure of von Neumann and Richtmyer, is included to permit thermodynamic and hydrodynamic calculations in the presence of a steep shock front.

When calculation of thermodynamic and hydrodynamic changes during the time interval Δt is complete (a "hydrocycle"), a series of tests is run and the program proceeds with another hydrocycle or goes back to the neutron transport calculation. To control the pace of a problem, the code continually examines the magnitude or rate of change of certain crucial parameters, and varies Δt of a hydrocycle or the number of hydrocyles per neutron cycle accordingly. This latter number begins at unity and is allowed to build up gradually if the forces present are not changing alpha too rapidly or modifying the density of a mass point radically. When the power variation in a hydrocycle or the change in alpha between neutron cycles gets so large as to damage the accuracy of the solution, the pace of the calculation is slowed automatically - or stopped in extreme cases.

In the first formulation of the AX-1 code the hydrostatic pressure p depends linearly on the temperature θ and the density ρ:

$$p = \tau + \beta\theta + \alpha\rho .$$

τ is chosen to have a rather large negative value, depending on the desired threshold for pressure generation. The hydrostatic pressure is set to zero by the code whenever the above expression yields a negative value. Calculations usually begin at low temperature and density, thus requiring the production of an appreciable amount of energy to generate non-vanishing pressures in the core

The computing program has been run with an equation of state similar to that of the Bethe-Tait formulation above [90], also with a Van der Waal's equation of state.

4.4.7 Equation of State

There is no time for heat conduction during the actual burst; only the fissionable isotope and those elements or isotopes intimately mixed with it will get very hot. For other materials, an isentropic equation of state is probably sufficient. Since both the EBR-II and Fermi reactors have uranium metal fuel elements, almost all pertinent work has been on the equation of state for uranium.

The equation of state is a relation between temperature, pressure and density, or between another appropriate set of thermodynamic variables. To simplify analysis in the Bethe-Tait formulation, a simple linear relation between energy density and pressure (with density remaining constant) has generally been used. The AX-1 code uses a linear relation between temperature, pressure and density, plus a linear relation between specific heat and temperature. The dependence of excess energy on the equation of state parameter γ for the Bethe-Tait formulation is given in Sec. 4.4.2.

Actually, there is little experimental information available on the equation of state for uranium, and the region of interest (temperatures up to 20,000°K or more, and pressures up to half a megabar) is extremely difficult to handle theoretically. Among the things measured is the saturated vapor pressure below 2000°K [100]. The melting and boiling points and the heat of vaporization near the melting point are reasonably well known, and measurements on shock waves in solid uranium provide a pressure-temperature relation up to high pressures, but for relatively modest temperatures and a limited density range [101,102].

In much of the pressure-density range of interest, uranium exists as a mixture of liquid and vapor phases. The critical values (V_c, T_c and P_c) are therefore of great interest. Elrod used the law of corresponding states and empirical information on a series of nonmetallic, low-boiling-point liquids to estimate T_c = 14,200°K, P_c = 5,130 atm [103]. Brout [104] accepted his estimate of T_c but used the empirical facts that for most substances measured, $V_c/V_{solid} = 3.11$ and $P_cV_c/RT_c = 0.3$, thereby obtaining $V_c = 40$ cm^3 /mole and P_c = 8650 atm. (Elrod later revised his critical temperature estimate to $T_c = 15{,}150°K$ [105], but this has not been factored into a Brout type analysis). Recently Grosse [106] has proposed an alternate method of correlation (with mercury as the only well-measured metal). This proposal would lead to a critical temperature of 12,500°K for U.

At temperatures above the critical point, two approaches have been used to estimate the equation of state for uranium. Stratton [92] has assumed it obeys a Van der Waal's equation while Brout [104] has assumed that it obeys the law of corresponding states, so that the empirical data on other, nonmetallic materials can be applied. Cowan [107] reviewed these two attempts briefly as follows:

"Stratton kept Brout's values $T_c = 1.206V$ and $V_c = 40\,cm^3/mole$ but used in place of empirical data the Van der Waal's equation

$$P^* = \frac{8T^*}{3V^* - 1} - \frac{3}{V^{*2}} \qquad (4\text{-}38)$$

for which $P_cV_c/RT_c = 0.375$. He therefore had to use a critical pressure 25% greater than Brout's value: $P_c = 10{,}800$ atm = 0.0110 megabars. In the two-phase region the Van der Waal's equation is inadequate, and Stratton used the empirical vapor-pressure relation

$$\log_{10}P = -0.292 - \frac{2.010}{T} \qquad (4\text{-}39)$$

where P is in megabars and T in volts. Although Eq. (4-39) is presumably valid only for low T (the experimental range is $T < 0.15$ volts), by a strange coincidence it gives at T_c precisely Stratton's value $P_c = 0.0110$ megabars; thus it can be used throughout the two-phase region without arithmetical inconsistency. The boundaries of the two-phase region (i.e. the saturated liquid and saturated vapor curves) are thus given by the intersections of Eq. (4-39) with the negative slope (on a PV plot) portions of Eq. (4-38).

"There are evidently appreciable differences between Brout's and Stratton's equations of state - a 25% difference in P_c, for example. At first thought one might be inclined to favor Brout's since it is based on empirical data whereas Stratton's is purely hypothetical. However, Brout was extrapolating to higher temperatures by more than a factor of 10. Moreover, he was trying to obtain an equation of state for a metal from those for nonmetals, and it is by no means certain that these are identical. Hence, the uncertainties in both Brout's and Stratton's results are probably as great as the differences between them, and there is probably not really any impelling reason for choosing one over the other.

"There is an obvious source of error in both equations of state, and that is the omission of any contribution of ionization to the pressure, energy, etc. For the materials (CO_2, NH_3, CH_4, etc.) on which Brout based his calculations, T_c is so low that ionization is negligible even for $T^* = 3$ or 4; but for uranium, $T_c = 1.2$ volts and ionization effects are very likely significant, if not at T_c then certainly at 3 or 4 times higher temperatures. Indeed, the calculations to be made below show that the electron pressure at the critical point may be more than twice Brout's value; thus Stratton's value of P_c may well be more accurate than Brout's."

Nicholson [94] chose to ignore the contribution of ionization and electron pressure, primarily on the assumption that an equation relating pressure with energy, rather than with temperature, should be relatively insensitive to ionization. He has adopted Brout's approach, extended and expanded it to provide an estimate of $(\gamma - 1)$ for the range of uranium density, $\rho = 7.5$ to $18.7\ g/cm^3$. He has also added a saturated vapor pressure term so that the pressure is non-zero below Q^*, the normal threshold energy.

In a recent paper Skidmore and Morris [108] reported some new equation of state measurements involving the passage of shock waves through reduced density uranium. While a precise interpretation of the experimental results was difficult, these authors concluded that "the results suggest that Brout's calculations on the thermodynamics of uranium in the critical region, based on corresponding state considerations, are valid if an electronic contribution is added to his energy values."

Hence, the situation on the equation of state for uranium remains obscure - and practically nothing has been published for other materials in the range of interest. For quick and ready reference, the equation of state for uranium is given here as used in two different studies.

(a) The original Bethe-Tait formulation [90]

$p = (\gamma - 1)(E - Q^*)\rho$, with $\gamma = 2$.

(Nicholson [94] has estimated how γ depends on ρ.)

(b) In the first formulation of AX-1 [91]

$$p = \alpha\rho + \beta\theta + \tau$$

where θ is the temperature in kev

p = pressure in megabars

ρ = density in g/cm^3

$$\frac{\partial E}{\partial \theta} = C_v = A_{cv} + B_{cv}\theta$$

The constants recommended by Stratton were

$$\alpha = .02873\ cm^2/\mu sec^2$$

$$\beta = 278.46\ g/cm\text{-}\mu sec^2$$

$$A_{cv} = 12.163\ cm^2/\mu sec^2\text{-kev}$$

$$B_{cv} = 5.78 \times 10^3\ cm^2/\mu sec^2\text{-kev}^2$$

The constant τ is chosen in each problem to provide the desired threshold energy or temperature. For energies below Q^* the pressure is set equal to zero.

4.5 General Results

4.5.1 Accuracy of Bethe-Tait Formulation

Calculations of energy yield by the Bethe-Tait method are reported in Table 4-1 and Fig. 4-3

TABLE 4-1

Ratio of Excess Energy to Threshold Energy (Q-Q*)/Q by Bethe-Tait Method

X	q = 1	q = 0.7	q = 0.5	q = 0.3
10^{-5}	0.2981	0.2981	0.2981	0.2981
10^{-4}	0.4902	0.4902	0.4902	0.4902
10^{-3}	0.8431	0.8431	0.8431	0.8487
10^{-2}	1.561	1.561	1.562	1.633
10^{-1}	3.267	3.268	3.325	3.774
10^{0}	8.398	8.461	9.090	11.69
10^{1}	29.82	31.07	36.47	54.60
10^{2}	160.5	176.9	226.7	384.6

The dimensionless quantity X is defined in Eq. (4-22)
The parabolic shape factor q is defined in Eq. (4-17)

[93]. Tabulated and plotted are the ratio of the energy density over threshold at the core center to the threshold value (Q-Q*)/Q* as a function of the dimensionless quantity X with q, the parabolic power shape constant, as a parameter. The total excess energy can be obtained by integrating the excess energy density over core volume, keeping in mind the power distribution and the fact that for weak explosions the pressure generation region falls within the core boundary.

Since the Bethe-Tait method ignores the effect of expansion and wave motion on pressure, it can be expected to be most accurate for sharp explosions with small reactivity insertions. A measure of the errors introduced by these assumptions is provided by calculations using the accurate AX-1 method, in which the dimensionless parameter X or ($\Delta k_0^3/\ell^2$) was held constant, while varying the lifetime, ℓ. In Fig. 4-4, the total excess energy yield (defined in this case as the total energy generation less the energy generation up to the incidence of pressure generation at the core center) is plotted against $\Delta k_0^3/\ell^2$ with ℓ as a parameter. In Fig. 4-5, the center peak pressure is plotted against the same abscissa.

A further measure of the errors associated with the Bethe-Tait neglect of expansion effects is provided in Table 4-2 [93]. Here, excess energy

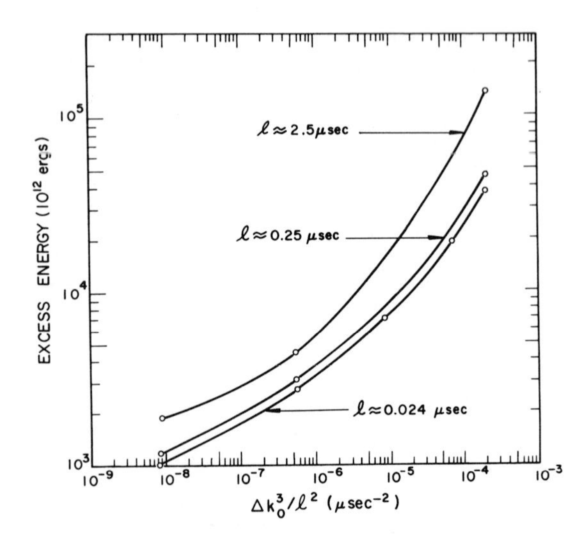

FIG. 4-4 Energy yield in excess of threshold for various neutron lifetimes; E* = 2,100 x 10^{12} ergs.

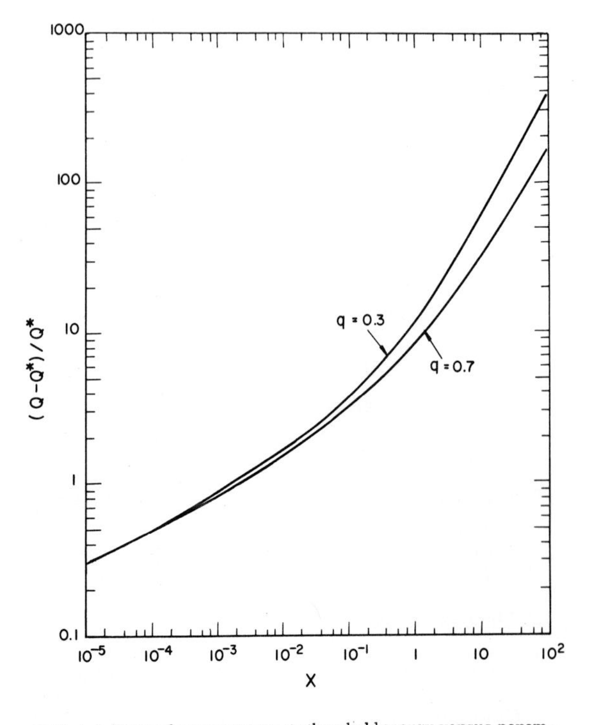

FIG. 4-3 Ratio of excess energy to threshold energy versus parameter X by Bethe-Tait approach.

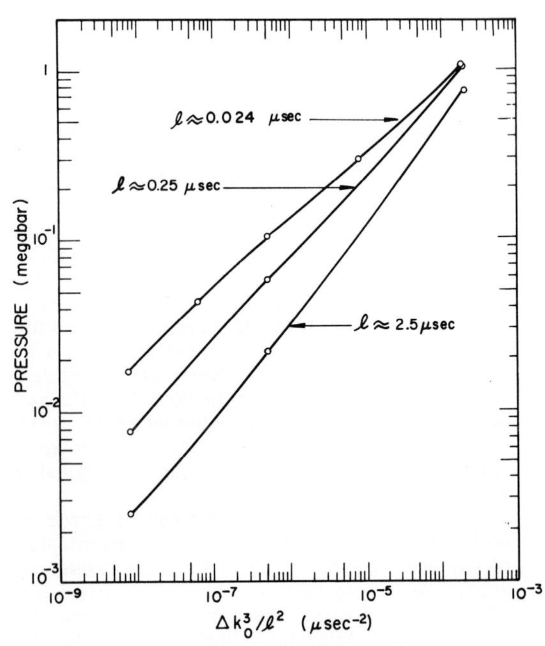

FIG. 4-5 Peak pressure at core center for various neutron lifetimes.

TABLE 4-2

Effect of Variation of Reactivity Input

$\Delta k_0{}^3/\ell^2$ (μsec^{-2})	Δk_0	Excess energy yield (10^{12}ergs) [a]	Center peak pressure (megabars)
7.003×10^{-9}	0.000141	853	0.0208
	0.000560	921	0.00895
	0.004088	1486	0.00479
4.45×10^{-7}	0.000141	2338	0.1423
	0.000560	2378	0.113
	0.004088	2561	0.049
1.665×10^{-4}	0.000141	22892	1.570
	0.000560	23040	1.544
	0.004088	23752	1.391

[a] 10^{13}ergs = 1 megawatt-second.

and pressure are compared for three series of calculations. In each calculation the quantity $\Delta k_0{}^3/\ell^2$ was held constant while Δk_0 was increased. With Δk_0 increased by a factor of ~ 30, the excess energy yield increases by 70% for the weak explosions. The very violent explosions showed much less sensitivity to the relative choice of Δk_0 and ℓ. The effect on peak central pressure is much greater than on energy yield.

Actually, by analyzing the local expansion during an explosion in detail, and translating this into reactivity effects, Nicholson [94] has estimated that this effect alone may introduce errors into the Bethe-Tait calculation of energy yield of about 3%. The error in central core pressure from this effect alone could go up to 15%.

It has been shown that wave propagation at constant velocity in a core having a parabolic flux distribution will not affect the energy yield [18]. In an actual reactor, with a variable speed of sound and non-parabolic flux, some deviation can occur. Using AX-1, Jankus has confirmed numerically the small influence on energy yield getting an appreciable difference only in an example where the wave had time to traverse the entire core before termination of the burst [93]. The effect on peak central pressure was greater, as is shown in Fig. 4-6, where the variation of this quantity with wave velocity was plotted for three different initial reactivities. There is also some evidence of an appreciable effect on the kinetic energy produced by explosions of the same total energy yield; hence, when considerations of damage and containment enter, the accurate AX-1 calculation may be required.

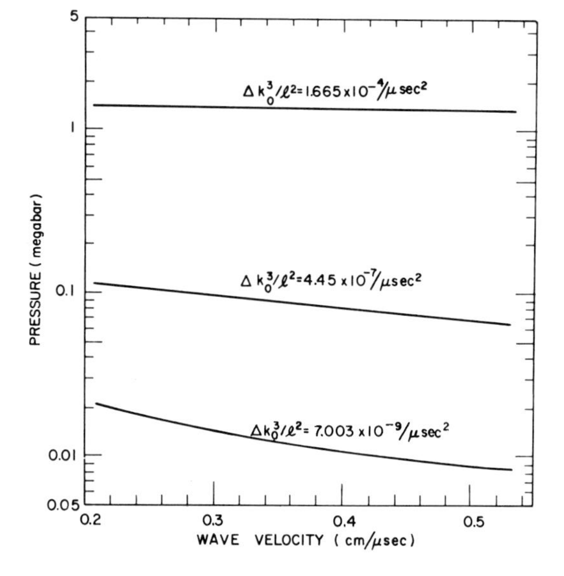

FIG. 4-6 Variation of peak pressure at core center with wave velocity.

4.5.2 Influence of Doppler Effect

The Doppler effect refers to the change in reactivity of a reactor caused by the change in effective neutron resonance cross sections of the reactor materials with temperature [21]. In the fast power reactors that have been built or are now under construction, e.g., EBR-I, EBR-II, Enrico Fermi and Dounreay, the Doppler effect has been predicted [18,19,23,24] to be considerably smaller than the other temperature effects on reactivity which occur during slow temperature transients. However, during severe accidents significant material movement is held back by inertia until high pressures have been generated, whereas the Doppler effect is essentially an instantaneous function of the temperature. The Doppler coefficient falls off with temperature, normally as T^{-1} or $T^{-3/2}$ [26, 29, 30], so that most of the reactivity change is introduced at lower temperatures. Therefore, most of this reactivity change is introduced during the short period of time just preceding the onset of pressure generation, where the power level is high, but the energy Q* has not been reached. Thus one can approximate its influence on the explosion as follows:

Compute Δk_{max} in the manner previously outlined, using Eqs. (4-4) and (4-5). Compute the Doppler reactivity increment between starting and threshold temperatures and add to (or subtract from) the reactivity value inserted mechanically at insertion rate dk/dt. This net reactivity Δk_{net} now is entered into the explosion calculation.

If significant additional reactivity is introduced by the Doppler effect at temperatures above the threshold, a revised estimate may be needed. If the Doppler coefficient is highly negative, the reactivity loss from this effect may balance the reactivity gain introduced by the source of the excursion. In this case, the methods outlined previously break down. The explosion is likely to be much weaker. Vapor pressures introduced prior to the reaching of the threshold energy may become significant. Another computational technique is needed [94,30j].

Some of the fast breeder reactors now being studied [30g,33,37b] for possible future construction have a large fraction of their neutron energy spectrum in the lower energy range where resonance effects are stronger. Hence, the Doppler coefficients are higher. The coefficient calculated [33] for one large, oxide-fueled fast reactor was -1×10^{-5}/°C at normal operating conditions. This is to be compared to the value of -1.7×10^{-6}/°C predicted [19] for the Fermi reactor. The coefficients for EBR-I and EBR-II are thought to be smaller yet.

In moderate sized, metal-fueled fast reactors the fissile isotope has been calculated to have a positive Doppler effect which is just canceled by a negative effect from the fertile isotope when the enrichment is in the range of 30 to 50% [26,31]. Recent calculations [30b] and measurements [30e] suggest that the Doppler effect may be negative even with higher enrichments.

Nicholson [26] computed the effect of Doppler effect on energy yield, adding (or subtracting) the additional reactivity term as a function of average core energy directly to that resulting from pressure buildup. The results for several assumed coefficients as a function of reactivity insertion rate in a Fermi-like reactor are given in Table 4-3. The effect is considerable. At very low rates of reactivity increase, a small Doppler effect ($\Delta k_D = -0.00064$) can reduce the energy release by a factor of two. At higher reactivity insertion rates, the same coefficient has only a 20% influence, since a greater proportion of reactivity enters mechanically. It is particularly interesting that a large negative Doppler coefficient can hold the energy release to a low level. Conversely, a moderate positive Doppler effect might seriously aggravate the normally weak explosion expected for low reactivity insertion rates.

Wolfe, et al. [30j] have performed further calculations on the influence of the Doppler effect in a severe accident. As is to be anticipated, they found that in the presence of a sufficiently negative Doppler effect, the excursion was terminated by the Doppler effect itself. The relatively low pressures which were built up then had time to cause disassembly of the core. These pressures built up before attainment of the threshold energy for rapid pressure buildup, since the equation of state utilized in the calculations allowed for vapor pressure. Such low pressures do not play a strong role in the disassembly process for a high yield accident in the absence of a strong, negative temperature coefficient.

The preceding analysis of this chapter (Sec. 4.4.2) indicates that with no Doppler effect, the energy release for a given reactivity insertion increases as fast or faster than the core mass. Thus, the reactivity insertion rate calculated to produce an energy release of the order of 500 lbs equivalent high explosive in the 400 liter Fermi reactor might lead to something 10 or 20 times as violent in a very large reactor. No calculations have been reported for such large reactors. But, the results obtained by Nicholson [94] and by Wolfe [30j] indicate that the presence of a large negative Doppler coefficient can reduce such yields very greatly.

At this stage of our knowledge of the course of violent disassemblies in large fast reactors of complex geometry, perhaps a word of caution should be added. The possibility that only a portion of such a reactor melts, undergoes a relatively mild explosion which acts to compress other parts of the core extremely rapidly, thus instigating a very much larger energy release, needs further investigation.

TABLE 4-3

The Influence of Doppler Effect on Calculated Energy Release

(Values of energy release are in 10^{16} ergs or 10^3 Mw-sec)

Δk_D	dk/dt					
	0.012	0.056	0.18	0.62	1.9	6.4
+.00064	-	-	-	4.5	-	-
0	1.1	1.7	2.4	3.8	6.4	14.5
-.00064	0.5	1.2	1.8	3.1	5.4	-
-.00192	-	0.4	-	2.0	-	9.2
-.0064	-	-	-	0.75	-	3.6

Δk_D = total amount of Doppler reactivity effect which would result if temperature went from initial temperature to infinity. Not all of this reactivity gets in.

4.5.3 Effect of Saturated Vapor Pressure

Nicholson [94] has studied the effect of neglecting the existence of saturated uranium vapor pressure prior to the attainment of threshold energy Q* for typical explosion calculations in the Fermi reactor. For low reactivity insertion rates (.012 to .056($\Delta k/k$)/sec), he finds the total energy release to be lowered considerably (~40%). For very large dk/dt however, the effect becomes small.

4.5.4 Time Width of Excursions

One may define the "width" of a power burst as the ratio of the integrated power (the energy yield, E_t) to the maximum power, P_{max}, reached. Measured in terms of the initial period ($\ell/\Delta k_o$), this width is given by $\Delta k_o E_t/\ell P_{max}$.* This quantity, as calculated by the Bethe-Tait approach, is plotted against X as the full lines in Fig. 4-7 for two choices of q, the flux shape parameter [93]. The "width" increases from unity at small values of X up to 2.9 for very large values of X.

The open circles in Fig. 4-7 are widths obtained from the more accurate AX-1 calculations. At each value of X, three different combinations of Δk_o and ℓ leading to the same X were used. It is seen that the widths from AX-1 are always larger than those in Bethe-Tait approximation; the deviation increases with increasing values of Δk_o.

For purposes of orientation, it is noted that in the Fermi reactor the very large reactivity insertion rate of 6.4 ($\Delta k/k$)/sec leads to a burst having a time width at half maximum of 30 μsec (and a pressure of $\sim 4.5 \times 10^{-5}$ atm) [94]. This is a pressure similar to that which might be encountered in a TNT explosion.

*Burst shape estimates for a variety of simpler models are given by Nyer in the chapter on Mathematical Models of Fast Transients.

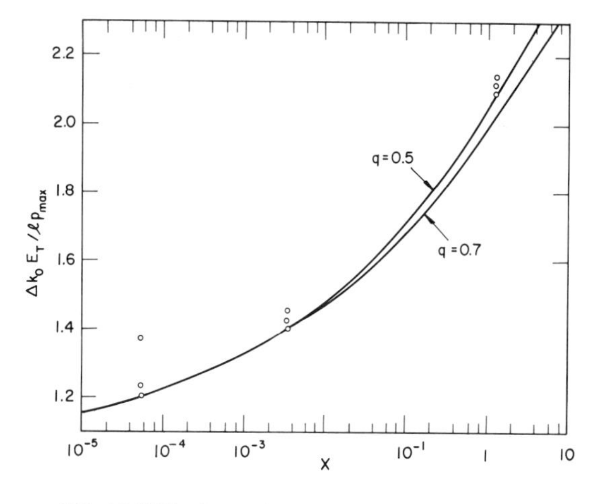

FIG. 4-7 Width of power burst in terms of initial period.

4.5.5 Variation of Blanket Density

During the derivation of the Bethe-Tait formulation, it was found that the assumption of equality of microscopic transport cross sections in core and blanket led to an energy yield depending only on the equation of state in the core and entirely independent of the density and equation of state in the blanket. Under this assumption displacements of the blanket reduce reactivity just like displacements of the core, and the smaller displacement obtained with a heavier blanket is exactly compensated by a correspondingly larger macroscopic transport cross section of the blanket material. Using AX-1, Jankus [93] has confirmed that, for a uniform core and short neutron lifetime, this effect was found to hold reasonably well.

Nicholson [94] has also studied the influence of blanket density on energy yield and found it to be small. Even a free surface was found to give results similar to a well-reflected one, the free surface condition giving an excess energy only 15% lower at high reactivity insertion rates (such that parameter X = 1 for the Fermi reactor).

Of course, to the extent that changing blanket density or composition changes the prompt neutron lifetime or core flux shape or other significant parameters, there is a change in excess energy yield. The conclusion stated at the beginning of this sub-section means that with everything else fixed, an arbitrary change in blanket density has no influence, given equal microscopic transport cross sections.

4.5.6 Effect of Initial Power Level on Yield

We have seen earlier that the reactivity inserted (Δk_{max}) depends only logarithmically on the initial power level. Numerical calculations for the Fermi reactor [94] confirm this insensitivity. For an insertion rate of 0.06 (Δk/k) per second, a variation in initial power from 10 watts to 10^8 watts only reduced the energy release by 40%.

4.6 Specific Results

4.6.1 Parameter Study Results for EBR-II and Fermi

Using the Bethe-Tait approach, the maximum pressure and the ratio of maximum energy density to threshold energy density (Q*) have been calculated for various reactivity insertion rates and three values of $(\gamma-1)Q^*$. The pressure in megabars and the term [(Q/Q*) -1] are given for EBR-II in Table 4-4. The calculation for the Fermi reactor gives energy densities similar to those for EBR-II, since the increase in core radius is compensated in part by the increase of the prompt neutron lifetime. However, the total energy produced in the Fermi reactor is larger since the volume of the core is larger. To compare these two reactors, values of $(\gamma-1) = 1$ and $Q^* = 10^{10}$ erg/g have been chosen and the maximum reactivity, maximum pressure and total energy developed have been calculated for widely varying reactivity insertion rates. The results are given in Table 4-5, using the parameters as shown in Table 4-6.

4.6.2 Re-evaluation of a Calculation for EBR-II

In the Hazards Summary Report for EBR-II [18] a reactivity insertion rate of 600$/sec [4.38 ($\Delta$k/k)/sec] is estimated for gravity collapse following meltdown under the most pessimistic assumptions. The threshold model (with a high threshold energy) leads to a burst yield of 825 lb (375 kg) high explosive equivalent, of which 80% or 660 lb (300 kg) is estimated to be available as explosive energy.

Stratton has scrutinized this hypothetical accident and concluded it is likely to have a distinctly non-threshold character [92]. He suspects it would be more reasonable to assume that the region in the lower part of the core is a high density homogeneous mass, in which all empty space is filled with uranium or a mixture of uranium and cladding

TABLE 4-4

Maximum Pressure (p) Developed and Ratio of Excess Energy Density to Threshold Energy (Q/Q*)-1 in EBR-II for Various Reactivity Insertion Rates (in $/sec = 0.0073/sec) Calculated Assuming Three Different Values for $(\gamma - 1)Q^*$ (in 10^{10}ergs/g)

dk/dt in $/sec	p (in megabars)			(Q/Q*)-1		
10	0.036	0.064	0.109	0.95	0.84	0.72
20	0.048	0.079	0.137	1.27	1.04	0.90
50	0.074	0.119	0.21	1.96	1.56	1.37
100	0.103	0.167	0.27	2.7	2.2	1.76
200	0.163	0.27	0.38	4.3	3.6	2.5
500	0.32	0.45	0.67	8.3	5.9	4.4
1000	0.57	0.78	1.08	15.0	10.3	7.1
$(\gamma - 1)Q^*$ in 10^{10}ergs/gram	0.5	1	2	0.5	1	2

TABLE 4-5

Maximum Pressure (p) and Total Energy (E) Developed in EBR-II and the Fermi Reactor Assuming $(\gamma - 1) = 1, Q^* = 10^{10}$ ergs/gram

dk/dt in $/sec	$\Delta k_{max} \cdot 10^3$	EBR-II		Fermi		
		p in megabars	E in 10^{16} ergs	$\Delta k_{max} \cdot 10^3$	p in megabars	E in 10^{16} ergs
10	0.55	0.064	0.65	0.82	0.073	3.4
20	0.79	0.079	0.72	1.18	0.099	4.0
50	1.26	0.119	0.91	1.89	0.157	5.3
100	1.81	0.167	1.13	2.70	0.22	6.7
200	2.60	0.27	1.63	3.86	0.31	8.9
500	4.16	0.45	2.4	6.22	0.58	15.1
1000	5.96	0.78	4.0	8.90	1.03	25.0

metal. If as much as 2/3 of the critical mass falls within this high density region, the same high reactivity insertion rate is calculated to give only about 1/20 of the previous energy yield. Hence, the energy yield calculated for a specific accident can be very sensitive to the formulation of initial conditions.

4.6.3 Uniform Core Collapse Under Gravity in Fermi

The hypothetical accident in which the core collapses uniformly under the action of gravity, the lower core boundary being held stationary, has been examined in some detail for the Fermi reactor [19]. The Bethe-Tait approach was used, including refinements such as a multigroup calculation of reactivity worth distribution and an equation of state in which $(\gamma - 1)$ as well as Q^* depend on fuel density. The energy density at the core center and the total energy release were calculated for a given power level as a function of $\Delta V/V$, the fractional collapse in core volume required to reach prompt critical. A reduction in reactivity of 0.00036 due to a negative Doppler effect (Δk_D) was assumed to take place, so that the maximum reactivity inserted by core collapse (Δk_{max}) was reduced to $\Delta k_{net} = \Delta k_{max} - \Delta k_D$ at the beginning of pressure generation.

The calculated results, assuming an initial power level of 100 Mw, are given in Table 4-7. Had an initial power of 1) Mw been assumed, energy releases up to 8×10^8 calories would be calculated.

TABLE 4-6

Parameters for EBR-II and the Fermi Reactor

Parameter	EBR-II	Fermi
$3\Sigma_{tr}\nu_0\Sigma_f$	$1.43 \times 10^{-2}/cm^2$	$0.80 \times 10^{-2}/cm^2$
ρ_c	$7.6\ g/cm^3$	$7.5\ g/cm^3$
b	25.0 cm	44.3 cm
q	0.48	0.62
ℓ	0.8×10^{-7} sec	2×10^{-7} sec

It is noted that the reactivity getting into the reactor rises with $\Delta V/V$, since a greater compression required to go critical corresponds to more time for free fall, and a higher reactivity insertion rate. However, the energy yield first rises, then falls off with increased $\Delta V/V$, since the higher density core leads to a lower threshold energy for pressure generation.

4.6.4 Two-Dimensional Explosion Calculations for Fermi

Two hypothetical meltdown configurations of the Fermi reactor which lend themselves to study in cylindrical geometry have been calculated, using a two-dimensional form of the Bethe-Tait approach [19]. The two assemblies were chosen as configurations possibly having autocatalytic effects associated with nonuniform densities and irregular geometries. A diagram of the fuel distributions chosen is given in Fig. 4-7. These configurations are idealizations of fuel rearrangements that might result from certain hypothetical events. In configuration 1 of Fig. 4-7, it is assumed that some of the material in the central part of the core melts and is swept away by the coolant stream. Then for some reason the remainder of the material in the central subassemblies settles into a high density mass at the core bottom. The material in the outer subassemblies has in the meantime remained stationary and uniform.

In configuration 2, it is assumed that all of the material in the central subassemblies melts and is expelled from the core. This is followed by melting of the peripheral subassemblies which then collapse into the annular region depicted.

For three arbitrary reactivity insertion rates, ranging from 0.6 to 6$(\Delta k/k)$ per second, the nonuniform configurations were calculated to give about a factor of two lower energy release than a uniform configuration with the same insertion rate. Configuration 1 gave about 70% the energy release of configuration 2. The primary reasons for the reduced yield were (1) the maxima in power and reactivity worth came at the same place in these configurations, so that the initial motion produced reduced reactivity (2) the high density regions had a lower effective threshold than in the uniform case.

TABLE 4-7

Summary of Energy Release Calculations for 100 Mw Initial Power in Fermi Reactor

$\Delta V/V$	0.1	0.2	0.3	0.4	0.5
Reactivity Insertion Rate $(\Delta k/k)$/sec	0.35	0.56	0.79	1.06	1.42
ρ (g/cm^3)	7.8	8.8	10.0	11.7	14.1
$\gamma - 1$	0.87	0.98	1.10	1.31	1.59
Q^* (ergs)	1.1×10^{10}	0.98×10^{10}	0.87×10^{10}	0.69×10^{10}	0.40×10^{10}
Δk_{max}	0.00104	0.00131	0.00155	0.00180	0.00208
Δk_{net}	0.00068	0.00095	0.0012	0.0014	0.0017
X	0.0018	0.0050	0.0100	0.019	0.045
$\frac{Q - Q^*}{Q^*}$	1.0	1.3	1.5	1.8	2.3
Q	2.2×10^{10}	2.25×10^{10}	2.18×10^{10}	1.93×10^{10}	1.32×10^{10}
Energy Release (g cal)	6.3×10^8	6.5×10^8	6.3×10^8	5.6×10^8	3.9×10^8

4.7 The Destructive Capacity of a Nuclear Energy Burst

4.7.1 Qualitative Considerations

The destructive capacity of a nuclear energy burst in a fast reactor is strongly dependent upon the magnitude of the energy release. A very weak excursion will merely raise the temperature of the fuel slightly and do no damage at all. At somewhat higher energy release there may be some damage to the fuel itself and, of course, at sufficiently high energy, pressures are generated which disrupt the surrounding materials.

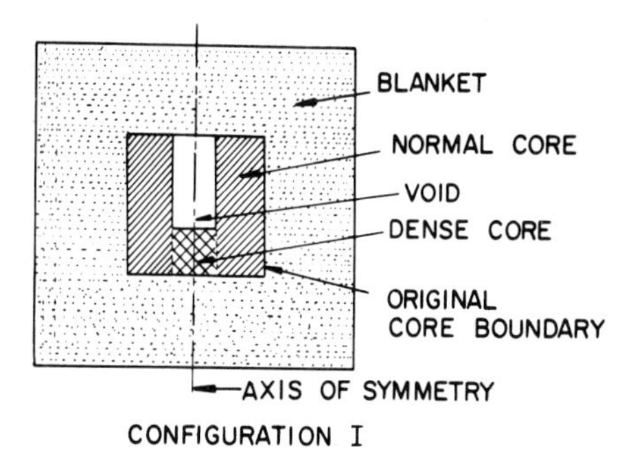

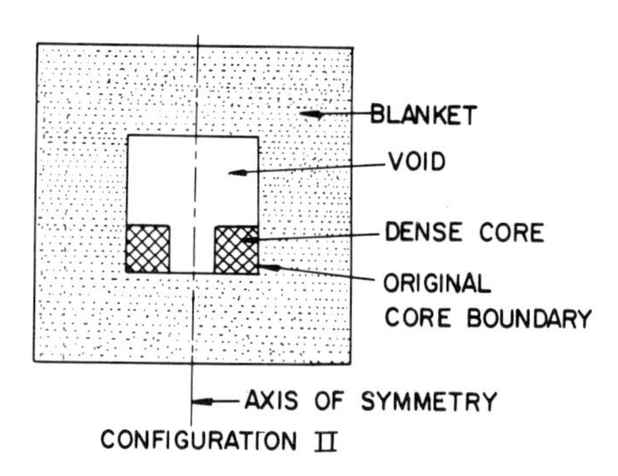

FIG. 4-8 Models of nonuniform configurations used in energy release calculations.

In one of the experiments with Godiva [92], the reactor was accidentally put on a shorter period than was intended and the assembly was permanently distorted, an accident of the second type mentioned in the preceding paragraph. This accident has been analyzed [92] to determine the high explosive equivalent of the energy burst, which was calculated to be 10 grams of TNT. This was determined by an estimate of the total strain energy involved in the distortion of the core. The high explosive equivalent was then defined to be the amount of TNT which could do an amount of work equivalent to that amount of strain energy. The effects of actually detonating that amount of TNT would, of course, be somewhat different.

In a more severe type of accident, perhaps the materials surrounding a reactor core would be accelerated outward at high velocity and strike some containment structure. In this case one might consider the destructiveness to be defined by either the damage to the containment structure or by the total kinetic energy delivered to the fragments before they struck the containment structure. In the latter case, the evaluation of the destructiveness would depend upon the nature of the material surrounding the core and in the former case it would depend both on these materials and on the containment structure itself.

It seems more desirable to define some quantity related to destructiveness that would depend only upon the energy release and the nature of the core materials and not on the surroundings. For this purpose, it is assumed that the energy is released into the core materials, while they are held at constant volume by their own inertia (i.e., it is assumed that the energy release is exceedingly fast) and then a calculation is made of the total amount of work that the materials can do by undergoing an adiabatic expansion down to some selected pressure p_1, such as one atmosphere, or perhaps 100 atmospheres. Thus

$$D = \int_{p_0}^{p_1} p dv$$

is chosen as the parameter of destructiveness. This choice is not without some difficulties. For example, two different accidents having the same value of D, but involving different sized cores, or different core material mixtures will not have precisely the same destructiveness. However, it seems to be the one parameter that most nearly defines the destructiveness for the range of accidents involved in containment studies for fast power reactors.

The evaluation of D requires a knowledge of the equation of state of the core material;

$$p = p(V, T) ,$$

and also a relation between temperature and internal energy, which can be written,

$$dE = C_v dt ,$$

where, however, C_V is not a constant specific heat, but rather for these highly non-ideal-gas systems is a function of both temperature and pressure. The change in internal energy is of course simply

$$dE = pdv$$

and, therefore,

$$dE = dD ,$$

and these coupled equations must be solved to determine the value of D.

The equation of state has been discussed above in connection with the energy release calculations. One must exercise considerable caution here, because in the energy release calculations one often uses a simple linear expression for the equation of state or possibly some other simple analytic expression that is convenient for the energy release calculations. Such simple expressions are usually only valid for small changes in the volume, while the evaluation of D may involve a thousand-fold increase in volume. To determine D one must obtain an equation of state valid over the entire range of volumes in the expansion.

The value of D will increase with increasing energy release for a particular reactor core. Also, in general, the ratio of D to the total energy release will increase. In the Godiva accident this ratio was 0.014. In very severe accidents where the temperature is raised to many tens of thousand degrees Kelvin the ratio can approach a large fraction of unity. Only in this latter case is the destructiveness roughly proportional to the total energy release. It is not possible to make further generalizations about the damage from a nuclear excursion, and each particular case must be evaluted in detail.

Even when the mechanical work fraction is estimated the actual results of this available work are most difficult to predict. For this reason model experiments have been carried out for the Fermi reactor [109,110], and for the Dounreay reactor. Here scale models of the reactors were used in conjunction with fluids representing the coolant in attempts to define upper limits on missile ejection and vessel distortion resulting from energy releases. High explosive charges, appropriately scaled, were used to simulate the energy release in credible nuclear excursions.

4.7.2 Quantitative Estimates

Jankus [93] has provided a method of estimating an upper bound for the energy available for mechanical work within the framework of the Bethe-Tait method. Assuming that the energy density distribution at the end of significant heat generation still has a parabolic shape and that the threshold energy density does not change with specific volume, the energy density above threshold energy density

$$Q[1 - (qr^2/b^2)] - Q*$$

can be converted to work in an isentropic expansion. Spatial integration of the above expression gives an upper bound for mechanical work:

$$W_{max} = \frac{4\pi}{3} b^3 \rho_c Q* \left(\frac{Q - Q*}{Q*}\right)^{5/2} \left(\frac{Q*}{qQ}\right)^{3/2} \cdot \frac{2}{5}$$

$$\text{if } Q* < Q < Q*(1 - q)^{-1}$$

$$W_{max} = \frac{4\pi}{3} b^3 \rho_c Q* \frac{Q}{Q*} \left[(1 - 0.6q) - 1\right]$$

$$\text{if } Q > Q* (1 - q)^{-1} \qquad (4\text{-}40)$$

Jankus [93] has evaluated the fraction of the energy release available as work for a series of cases, and also discusses corrections which can be made to these estimates.

When a numerical solution of the explosion, using the AX-I code, is performed, the kinetic energy released is computed automatically as part of the solution.

NOMENCLATURE (Sec. 4)

A	coefficient, See Eq. (4-18).
b	core radius
c	wave velocity

D	diffusion coefficient
E(r,t)	energy density
f	term similar to ∇w (see Eq. 4-34)
F	fraction of fissions in core
k	effective multiplication factor
k_s	multiplication factor at source power
Δk	excess reactivity
Δk_D	Doppler reactivity increment (or coefficient)
$\Delta k_o = \Delta k_{max}$	maximum excess reactivity (in absence of Doppler effect)
Δk_1	reactivity introduced by time, t_1
Δk_{net}	$\Delta k_{max} \pm \Delta k_D$
ℓ	prompt neutron lifetime
n(t)	power density
n_s	shutdown or subcritical power
N(r)	spatial distribution of power density
p(r,t)	pressure
P_{max}	maximum power
q	parameter in parabolic power shape equation
Q(t)	energy density
Q^*	threshold energy density for pressure generation
Q_1	energy generated up to time t_1
r	radial position in core
t	time
u(r)	displacement due to pressure
V	volume
w	reactivity worth
X	parameter (see Eq. 4-22)
dk/dt	reactivity insertion rate
α	inverse prompt period, $= \Delta k/\ell$
γ	parameter in equation of state
ϕ	flux
ν_o	neutrons per fission
ρ	density (a function of space and time)
ρ_c	initial core density
Σ_c	capture cross section (macroscopic)
Σ_f	fission cross section (macroscopic)
Σ_{tr}	transport cross section (macroscopic)

REFERENCES

1. D. Okrent et al., "Fast Reactor Physics", Proceedings of the U. N. International Conference on Peaceful Uses of Atomic Energy, Geneva, 1955, Vol. 5, p. 347.
2. J. Codd et al., "The Physics of Fast Reactors", Chapter 9 in Prog. Nucl. Energy Series I, Vol. 1, Pergamon Press, Inc., N. Y., 1956.
3. W. B. Loewenstein and D. Okrent, "Physics of Fast Power Reactors", Proceedings of the Second U. N. International Conference on Peaceful Uses of Atomic Energy, Geneva, 1958, Vol. 12, p. 16.
4. A. I. Leipunsky et al., "Fast Neutron Reactor Physics", Proceedings of the Second U. N. International Conference on Peaceful Uses of Atomic Energy, Geneva, 1958, Vol. 1, p.3.
5. G. E. Hansen, "Elementary Fast Neutron Critical Assemblies", Proceedings of the Second U. N. International Conference on Peaceful Uses of Atomic Energy, Geneva, 1958, Vol. 12, p. 84.
6. R. D. Smith and J. E. Sanders, "Zero Energy Fast Reactor Experiments", Proceedings of the Second U. N. International Conference on Peaceful Uses of Atomic Energy, Geneva, 1958, Vol. 12, p. 89.
7. J. K. Long et al., "Fast Neutron Power Reactor Studies with ZPR III", Proceedings of the Second U. N. International Conference on Peaceful Uses of Atomic Energy, Geneva, 1958, Vol. 12, p. 119.
8. "Proceedings of Conference on Physics of Breeding", USAEC Report ANL-6122, Argonne National Laboratory, 1959.
9. L. J. Koch and H. C. Paxton, "Fast reactors", Ann. Rev. Nucl. Sci., 9(1959)437.
10. H. C. Paxton (Ed.), "Fast-neutron critical assemblies and related topics", Nucl. Sci. Eng., 8(1960)523.
11. S. Yiftah, D. Okrent, and P. A. Moldauer, Fast Reactor Cross Sections, Pergamon Press, Inc., N. Y., 1960.
12. Physics of Fast and Intermediate Reactors (3 Vols.), International Atomic Energy Agency, Vienna, 1961.
13. H. Cartwright et al., "Dounreay Fast Reactor", Proceedings of the Second U. N. International Conference on Peaceful Uses of Atomic Energy, Geneva, 1958, Vol. 9, p. 316.
14. L. J. Koch et al., "Construction Design of EBR II", Proceedings of the Second U. N. International Conference on Peaceful Uses of Atomic Energy, Geneva, 1958, Vol. 9, p. 323.
15. J. R. Dietrich and W. H. Zinn, Solid Fuel Reactors, Addison-Wesley Publishing Co., Inc. Reading, Mass., 1958.
16. W. B. Loewenstein, "The Physics Design of EBR II", Proceedings of the Symposium on Physics of Fast and Intermediate Reactors, International Atomic Energy Agency, Vienna, 1961, Vol. III, p. 263.
17. W. J. McCarthy et al., "Studies of Nuclear Accidents in Fast Power Reactors", Proceedings of the Second U. N. International Conference on Peaceful Uses of Atomic Energy, Geneva, 1958, Vol. 12, p. 207.
18. L. J. Koch et al., "Experimental Breeder Reactor II (EBR-II) Hazard Summary Report", USAEC Report ANL-5719, Argonne National Laboratory, 1957.
19. "Enrico Fermic Atomic Power Plant, Technical Information and Hazards Summary Report", Part B, Vol. 7, USAEC Report NP-11526, DTI Extension, Oak Ridge National Laboratory, 1961.
20. H. V. Lichtenberger et al., "NaK-Cooled Fast Reactor (EBR)", Proceedings of the First U. N. International Conference on Peaceful Uses of Atomic Energy, Geneva, 1955, Vol. 3, p. 345.
21. G. Goertzel et al., "Doppler Effect in Intermediate and Fast Reactors", Proceedings of the First U. N. International Conference on Peaceful Uses of Atomic Energy, Geneva, 1955, Vol. 5, p. 472; see also: H. Feshbach et al, "Estimation of Doppler effect in fast reactors", Nucl. Sci. Eng., 1(1956)4.
22. A. M. Lane, J. E. Lynn, and J. S. Story, "An Estimation of the Doppler Effect in Fast Neutron Reactors", British Report AERE T/M 137, 1956.
23. H. A. Bethe, "On the Doppler Effect in Fast Reactors", Report APDA-119, Atomic Power Development Associates, Inc., 1957.
24. W. Y. Kato and D. K. Butler, "Measurement of the Doppler Temperature Effect in an EBR-I Type Assembly", USAEC Report ANL-5809, Argonne National Laboratory, 1958.
25. A. R. Baker and T. A. J. Jaques, "A Measurement of the Contribution of the Doppler Effect to the Temperature Coefficient of Reactivity in a Fast Reactor", British Report AERE R/M 168, 1958.
26. R. B. Nicholson, "The Doppler Effect in Fast Reactors", Report APDA-139, Atomic Power Development Associates, Inc., 1960.
27. P. Greebler and B. A. Hutchins, "The Doppler Effect in a Large Fast Oxide Reactor—Its Calculation and Significance for Reactor Safety", Proceedings of a Symposium on Physics of Fast and Intermediate Reactors, International Atomic Energy Agency, Vienna, 1961, Vol. III, p. 121.
28. E. A. Fossoul, "Effet Doppler du plutonium 239 dans les reacteurs rapides", Proceedings of a Symposium on Physics of Fast and Intermediate Reactors, International Atomic Energy Agency, Vienna, 1961, Vol. III, p. 139.
29. M. G. Bhide and H. H. Hummel, "Doppler Coefficient of Fast Reactors", USAEC Report ANL-6601, Argonne National Laboratory, 1962.
30. "Proceedings of a Conference on Breeding, Economics and Safety in Large Fast Power Reactors", USAEC Report ANL-6792, Argonne National Laboratory, 1963.
    a. P. Greebler, "Recent Improvements in Calculations of Doppler and Sodium Reactivity Effects for Large Fast Reactors".
    b. H. H. Hummel and A. L. Rago, "Effect of Parameter Uncertainties on the Doppler Effect of Large, Ceramic-Fueled Fast Reactors".
    c. R. N. Hwang, "An Improved Method of Doppler Effect Calculation for Fissile Materials in the Intermediate Energy Region".
    d. J. Codd and P. J. Collins, "The Influence of Resonance Overlapping on the Doppler Effect in a Dilute Fast Reactor".
    See also: J. Codd and P. J. Collins, "Pu^{239} and U^{238} Resonance Interaction Effects in a Dilute Fast Reactor". Presented at the EAES Symposium on Advances in Reactor Theory held at Karlsruhe, 1963.
    e. T. H. Springer and S. G. Carpenter, "Fast Spectrum Doppler Measurements".
    f. H. H. Hummel, K. Phillips, and A. Rago, "Studies of Sodium

Void Coefficient of Large Fast Reactors in Spherical and Infinite Slab Geometry".

g. K. P. Cohen et al. "Reactor Safety and Fuel Cycle Economics Considerations for Fast Reactors".
h. C. E. Dickerman et al., "Meltdown Studies".
i. P. R. Huebotter, "Studies of Simulated Sodium Boiling and Fuel Meltdown".
j. B. Wolfe et al., "Influence of the Doppler Effect on the Meltdown Accident".

31. A. J. Goldman, "A Feasibility Study of Fast U^{233} -Th Breeder Reactors", Report NDA-2134-3, Nuclear Development Corporation, 1960.
32. R. Hankel, "Evaluation of U^{233} Thorium Fuels", Report NDA-2-64-3, Nuclear Development Corp., 1962.
33. P. Greebler et al., "Calculation of Doppler Coefficient and Other Safety Parameters for a Large Fast Oxide Reactor", Report GEAP-3646, General Electric Co., 1962.
34. K. M. Horst and B. A. Hutchins, "Comparative Study of Plutonium-Uranium Carbide and Plutonium-Uranium Oxide Fuel in a Sodium-Cooled Fast Reactor", Report GEAP-3880, General Electric Co., 1962.
35. S. Yiftah and D. Okrent, "Some Physics Calculations on the Performance of Large Fast Breeder Power Reactors", USAEC Report ANL-6212, Argonne National Laboratory, 1960.
36. J. B. Nims and P. F. Zweifel, "Sodium Temperature Coefficients in Fast Reactors", Trans. Am. Nucl. Soc., 2, (1959) 172; see also: "Preliminary Report on Sodium Temperature Coefficient in Large Fast Reactors", Report APDA-135, Atomic Power Development Associates, Inc., 1959.
37. Proceedings of an International Atomic Energy Agency Symposium on the Physics of Fast and Intermediate Reactors held at Vienna, August, 1961.
    a. H. H. Hummel and A. L. Rago, "An Accurate Treatment of Resonance Scattering in Light Elements in Fast Reactors", Vol. I, p. 231.
    b. M. G. Bhide and H. H. Hummel, "Reactivity of Coefficients of Sodium in Some Large Fast Reactors", Vol. II, p. 177.
    c. D. Okrent, "A Review of the Nuclear Aspects of Fast-Reactor Safety", Vol. III, p. 155.
    d. D. Okrent, "Performance of Large Fast Power Reactors Including Effects of Higher Isotopes Recycling and Fission Products", Vol. II, p. 271.
    e. H. Hummel and L. T. Bryant, "Stability Analysis of EBR-II", Vol. III, p. 107.
    f. R. R. Smith, et al., "An Analysis of the Stability of EBR-I, Mark I to III, and Conclusions Pertinent to the Designs of Fast Reactors", Vol. III, p. 43.
    g. F. Storrer, "Influence Curves for the Calculation of the Structural Distortion Effect on Reactivity", Vol. III, p. 3.
    h. D. C. G. Smith, "Physics of the Dounreay Fast Reactor", Vol. III, p. 241.
    i. A. I. Leipunski et al., "Physical Characteristics of BR-5", Vol. III, p. 315.
    j. H. Soodak, "The Reactivity Effect of Hydrogen Addition in Fast Reactors", Vol. II, p. 165.
    k. C. E. Dickerman, D. Okrent, and E. Sowa, "The Fast Reactor Safety Program in TREAT", Vol. III, p. 171.
    l. V. Z. Jankus, "A Theoretical Study of Destructive Nuclear Bursts in Fast Power Reactors", Vol. III, p. 209.
38. C. E. Dickerman and D. Okrent, "Letter to editor", Nuclear News, 3(1963)21.
39. D. Okrent, "Neutron physics considerations in large fast reactors", Power Reactor Technology, to be published in 1964.
40. F. W. Thalgott et al., "Stability Studies on EBR-I", Proceedings of the Second U. N. International Conference on Peaceful Uses of Atomic Energy, Geneva, 1958, Vol. 12, p. 242.
41. D. Okrent, "Kinetics of EBR-I and EBR-II", USAEC Report ANL-6205, Argonne National Laboratory, 1960.
42. M. C. Pinchasik et al., "Operating Experience with the BR-5 Fast Reactors", in Power Reactor Experiments, Vol. 1, p. 375, International Atomic Energy Agency, Vienna, 1961.
43. R. E. Peterson and S. L. Stewart, "Moderating Segmentation: A Core Design for Counteracting Positive Coolant Coefficients in Large Fast Reactors", ANS Winter Meeting, 1963; Trans. Am. Nucl. Soc., 6, (1963)260.
44. P. J. Persianni, "FARET Experimental Program", USAEC Report ANL-6708, Argonne National Laboratory, 1963.
45. P. Greebler et al., "Statistical evaluation of fission products absorption cross sections at intermediate and high energies", Nucl. Sci. Eng., 2(1957)334.
46. U. L. Businaro et al., "On the dependence upon energy of the fission product poisoning of U^{235}", J. Nucl. Energy, 4(1957) 319.
47. P. A. Moldauer, "Fast Neutron Cross Sections", USAEC Report ANL-6122, Argonne National Laboratory, 1959.
48. J. D. Garrison and B. W. Roos, "Estimation of Fission Products Capture Cross Sections", Trans. Am Nucl. Soc., 3 (1960)299.
49. A. J. Goldman, "Reactivity effects of protactinium-233 buildup in U^{233} fast breeder reactor", Nucl. Sci. Eng., 10(1961)91.
50. A. Amorosi and J. G. Yevick, "Enrico Fermi Reactor", Proceedings of the Second U. N. International Conference on Peaceful Uses of Atomic Energy, Geneva, 1958, Vol. 9, p. 358.
51. C. E. Branyan, "Critical Studies for Fermi Reactor", USAEC Report ANL-6629, Argonne National Laboratory, 1962.
52. K. M. Horst et al., "Core Design Study for a 500 Mwe Fast Oxide Reactor", Report GEAP-3721, General Electric Co., 1961.
53. D. Okrent, "Breeding, Safety and Fuel Cycles", in the Proceedings of the Eighth Nuclear Congress held at Rome, June, 1963; see also: D. Okrent, "Nuclear Considerations in the Selection of Materials for Fast Power Reactors". Nuclear Metallurgy, Vol. IX, AIME, 1963.
54. "Reactor Physics Constants", USAEC Report ANL-5800 (2nd Ed.), Argonne National Laboratory, 1963.
55. H. A. Bethe, "Reactor Safety and Oscillator Tests", Report APDA-117, Atomic Power Development Associates, Inc., 1956.
56. G. H. Kinchin, "The Instability of Fast Reactors", British Report AERE/RP/M83, Harwell, England, 1956.
57. R. T. Ackroyed et al., "Stability Considerations in the Design of Fast Reactors", Proceedings of the Second International Conference on Peaceful Uses of Atomic Energy, Geneva, 1958, Vol. 12, p. 230.
58. R. O. Brittan, "Some Problems in the Safety of Fast Reactors", USAEC Report ANL-5577, Argonne National Laboratory, 1956.
59. R. O. Brittan, "Analysis of the EBR-I Meltdown", Proceedings of the Second U. N. International Conference on Peaceful Uses of Atomic Energy, Geneva, 1958, Vol. 12, p. 267.
60. R. B. Nicholson, "A high frequency reactor instability mechanism", Nucl. Sci. Eng., 3(1958)620.
61. F. Storrer, "Analysis of the Power Feedback Relations in Fast Reactors" , Proceedings of the Conference on Transfer Function Measurements and Reactor Stability Analysis, USAEC Report ANL-6205, Argonne National Laboratory, 1960.
62. F. Storrer, "Temperature Response to Power, Inlet Coolant Temperature and Flow Transients in Solid Fuel Reactors", Report APDA-132, Atomic Power Development Associates, 1959.
63. F. Storrer and T. A. Doyle, "Calculation Method Used to Determine the Reactivity Effects Due to Thermal Bowing of Subassemblies, Thermal Expansion of the Hold-Down Plate and Radial Expansion of Core", Technical Memorandum TM-31, Atomic Power Development Associates, Detroit, Michigan, 1961.
64. H. A. Sandmeier, "The Kinetics and Stability of Fast Reactors with Special Considerations of Nonlinearities", USAEC Report ANL-6014. Argonne National Laboratory, 1959.
65. F. D. Boardman, "Noise Measurements in the Dounreay Fast Reactor", Transactions of the Symposium on Noise Analysis in Nuclear Systems, University of Florida, November 4 - 6, 1963.
66. P. A. Lottes and W. S. Flynn, "A method of analysis of natural circulation boiling systems", Nucl. Sci. Eng., 1(1956)461.
67. L. J. Koch, W. B. Loewenstein, and H. O. Monson, "Addendum to EBR-II Hazards Summary Report ", USAEC Report ANL-5719 (Addendum), Argonne National Laboratory, 1962.
68. J. L. Phillips, "Operating experience with the Dounreay fast reactor - 1 and 2", Nucl. Power, July-Aug., 1962.
69. R. R. Smith et al., "Stability Studies with EBR-1, Mark III", USAEC Report ANL-6266, Argonne National Laboratory, 1960.
70. R. O. Haroldsen, F. D. McGinnis, and G. K. Whitham, "Operational and Maintenance Experience of the EBR-I Liquid Metal Cooled Reactor", Paper distributed at the American Nuclear Society Conference on Problems of Operating Research and Power Reactors, October 21 - 23, 1963, Ottawa, Ontario, Canada.
71. "Symposium on the Dounreay Fast Reactor", J. British Nuclear Energy Conference, 6(1961)161.
72. K. Henry and D. C. G. Smith, "Operating experience with the Dounreay fast reactor", J. British Nuclear Energy Society, 1, 3(1962)186.
73. A. I. Leipunsky et al., "Experimental Fast Reactors in the U.S.S.R.", Proceedings of the Second U. N. International Conference on Peaceful Uses of Atomic Energy, Geneva, 1958, Vol. 9, p. 348.
74. "LAMPRE-I Final Design Status Report", USAEC Report LAMS-2833, Los Alamos Scientific Laboratory, 1962.
75. "Quarterly Status Report on LAMPRE Program for Period Ending May 20, 1962", USAEC Report LAMS-2720, Los Alamos Scientific Laboratory, 1962.
76. "Quarterly Status Report on LAMPRE Program for Period Ending August 20, 1963", USAEC Report LAMS-2973, 1963.
77. "Quarterly Status Report on LAMPRE Program for Period Ending August 20, 1961", USAEC Report LAMS-2620, Los Alamos Scientific Laboratory, 1961.

78. "Enrico Fermi Atomic Power Plant, Monthly Report for Period August 23-September 22, 1963", Report EF-1, Power Reactor Development Co., 1963.
79. "Enrico Fermi Atomic Power Plant, Monthly Report for Period September 23-October 22, 1963", Report EF-2, Power Reactor Development Co., 1963.
80. "Enrico Fermi Atomic Power Plant, Monthly Report for Period October 23-Noverber 23, 1963", Report EF-3, Power Reactor Development Co., 1963.
81. R. Daane, et al., "Progress Report on the Study of Low Probability, High Hazard Fast Reactor Accidents", Report NDA-14-107, Nuclear Development Corp. of America, 1955.
82. C. D. Taulbee et al., "Transient and Oscillator Analysis for the Enrico Fermi Atomic Power Plant", Bendix Corp., Report 1653, Research Laboratories Division, 1960.
83. "Codes for Reactor Computations", Proceedings of the Seminar on Codes for Reactor Computations, International Atomic Energy Agency, Vienna, 1961.
84. P. Greebler et al., "FORE, A Computational Program for the Analysis of Fast Reactor Excursions", Report GEAP-4090, General Electric Co., San Jose, 1962.
85. R. O. Brittan et al., "Hazard Evaluation Report on ZPR-III", USAEC Report ANL-FWT-105, Argonne National Laboratory, 1955.
86. W. R. Stratton, "A Review of Criticality Accidents", Progress in Nuclear Energy, Series IV, Vol. 3, p. 163, Pergamon Press, 1960.
87. T. F. Wimett and J. D Orndoff, "Applications of GODIVA II Neutron Pulses", Proceedings of the Second U. N. International Conference on Peaceful Uses of Atomic Energy, Geneva, 1958, Vol. 10, p. 449.
88. H. G. Elrod, Jr., "The Distribution of Melting in a Slow Meltdown Accident", Report NDA-14-170, Nuclear Development Corp. of America, 1957.
89. J. H. Ray and A. D. Krumbein, "Calculated Criticality of the Enrico Fermi Reactor During a Hypothetical Meltdown Accident", Report NDA-14-182, Nuclear Development Corp. of America, 1957.
90. H. A. Bethe and J. H. Tait, "An Estimate of the Order of Magnitude of the Explosion when the Core of a Fast Reactor Collapses", British Report UKAEA-RHM (56)/113, 1956.
91. D. Okrent et al., "AX-1, A Computing Program for Coupled Neutronic - Hydrodynamics Calculations", USAEC Report ANL-5977, Argonne National Laboratory, 1959.
92. W. R. Stratton, T. H. Colvin, and R. B. Lazarus, "Analysis of Prompt Excursions in Simple Systems and Idealized Fast Reactors", Proceedings of the Second U. N. International Conference on Peaceful Uses of Atomic Energy, Geneva, 1958, Vol. 12, p. 196.
93. V. Z. Jankus, "A Theoretical Study of Destructive Nuclear Bursts in Fast Power Reactors", USAEC Report ANL-6512, Argonne National Laboratory, 1962. See also: Ref. 37m, pp. 209-38.
94. R. B. Nicholson, "Methods for Determining the Energy Release in Hypothetical Reactor Meltdown Accidents", Report APDA-150, Atomic Power Development Associates, 1962.
95. R. A. Daane and W. H. Jens, "Fast Reactor Accident Scaling Laws", Report NDA-14-84, Nuclear Development Associates, 1955.
96. J. E. Wilkins, Jr., "The behaviour of a reactor at prompt critical when the reactivity is a linear function of time", Nucl. Sci. Eng., 5(1959)207.
97. J. W. Stephenson and R. B. Nicholson, "Weak Explosion Program", Report ASTRA-417-6.0, Advanced Scientific Techniques Research Associates, Milford, Conn., 1961.
98. J. W. Stephenson and R. B. Nicholson, "Reactor Explosion Program", Report ASTRA-417-3.0, Advanced Scientific Techniques Research Associates, Milford, Conn., 1960.
99. D. O'Shea, D. Okrent and J. M. Chaumont, "Some Calculations Pertaining to Fast Reactor Safety", USAEC Report ANL-6501, Argonne National Laboratory, 1962.
100. E. G. Rauh and R. J. Thorn, "The vapor pressure of uranium", J. Chem. Phys., 22(1954)1414.
101. M. H. Rice, R. G. McQueen and J. M. Walsh, "Compression of solids by strong shock waves", Solid State Phys., 6(1958)1.
102. J. Viard, "Adiabatique Dynamique de l'Uranium", Édition du Centre National de la Recherche Scientifique, Paris, 1961.
103. H. Elrod, "A Method for the Estimation of Critical Temperatures and Pressures", Report NDA-14-121; "Estimation of the Critical Temperature and Pressure Uranium", Report NDA-14-122, Nuclear Development Corp., 1956.
104. R. H. Brout, "Equation of State and Heat Content of Uranium", Report APDA-118, Atomic Power Development Associates, 1957.
105. H. Elrod, "Revisions of Estimated Saturation Properties of Uranium", Report NDA-14-139, Nuclear Development Corp., 1956.
106. J. V. Grosse, "The temperature range of liquid metals and an estimate of their critical constants", J. Inorg. Nucl. Chem., 22(1961)23.
107. R. D. Cowan, "On the Equation of State of Uranium at Low Densities and Temperatures", Los Alamos Scientific Laboratory. (Unpublished).
108. I. C. Skidmore and E. Morris, "Experimental Equation of State Data for Uranium and Its Interpretation in the Critical Region", Symposium on Thermodynamics of Nuclear Materials, International Atomic Energy Agency, Vienna, 1962, p. 173. See also E. Morris, "The Equation of State of Uranium in the Critical Region", British Report AWRE 0-61/62, Aldermaston Weapons Research Establishment, 1962. (Depository libraries only.)
109. J. F. Proctor, "Energy Partition of Water-Cased Explosions in an Idealized Model Reactor Vessel", NOLTR-62-155, U. S. Naval Ordnance Laboratory, 1960.
110. W. R. Wise, Jr., "Possible Jump of Rotary Shield Plug for Enrico Fermi Atomic Power Plant", Letter Report, U. S. Naval Ordnance Laboratory, May 15, 1962.

CHAPTER 11

Accidents and Destructive Tests

T. J. THOMPSON
Massachusetts Institute of Technology
Cambridge, Massachusetts

"What Is Past Is Prologue"

CHAPTER CONTENTS*

*Except for a few changes and additions made in proof, this chapter is based on information in the literature or known to the author prior to February 1964.

PREFACE

In reviewing the information available on the various accidents, it has appeared to the author that the accident reports from Windscale and Chalk River were particularly outstanding. They presented, in a logical manner, the facts concerning what happened and drew frank conclusions. The reports were informative enough to supply the reader with an artificial backlog of experience so that he might be forewarned of certain problems. Since the intent of this chapter is the same, the author has tried to continue the tradition of those earlier reports in an abbreviated fashion.

Because a report of an accident loses much of its value if there are no conclusions, some of the more important ones have been drawn using those from the original reports, those of other reviewers, and those of the author. In a number of cases the conclusions of this chapter augment or even differ from those of the original reports. In those cases it should be understood that the author is presenting his personal conclusions. It is hoped that the authors of earlier reports and the groups involved will understand that the sole intent is to emphasize the lessons to be learned.

1 INTRODUCTION

While there have been relatively few accidents of a serious nature involving nuclear reactors, yet each one points out certain lessons worth preserving. It is the intent of this chapter to set down the recorded facts concerning these accidents, emphasizing the causes of the accident and the lessons to be learned.

The author earnestly believes that such a record will do much to prevent similar accidents from happening again, thus make the safety record of reactors even better than it is. To support this view, one can cite the SL-1 accident which almost certainly occurred because an operator or operators manually withdrew a control rod. It is generally believed that the SL-1 accident was the first case in which manual withdrawal occurred. In reality, it is the third case which has led to a recorded accident. Yet, partly because no one has bothered to set down the lesson, a reactor designed several years after the first such accident carried the same flaw, and a serious accident resulted.

In reviewing the significant accidents (including destructive tests) which have occurred to date, it has appeared logical to divide them into three different categories: those in critical facilities where no fission product burden is likely, those in reactors and involving reactivity changes, and fuel failures in reactors.

The first category, accidents in criticality facilities, is discussed in Sec. 2 and is summarized in Table 2-1. For the most part, there are excellent summaries of these accidents [1, 2, 3, 4] up until about 1960. A few of special interest which have happened since then are described briefly in Sec. 2.

The second category, reactor accidents and destructive tests, is discussed in Sec. 3 and summarized in Table 3-1. Only those incidents involving criticality are included. Three incidents, the BORAX-1 experiment, the EBR-I Meltdown and the SPERT Destructive Test, involved planned transient experiments and, therefore were not really "accidents" in the sense of being totally unexpected. They are included, however, for completeness and for the information they have provided. Another, the Windscale accident, involved reactivity only as one aspect of the initiating cause. Yet, it was a reactor accident from which much can be learned, and it is therefore included.

The third category, fuel element failure, is included to provide information on experience with fuels in operating reactors. The information is summarized in Table 4-1 and brief reviews are presented in Sec. 4 outlining the more serious of these accidents. In general, the more minor fuel element failures and experiences of a generally favorable nature are not reported. Thus, those discussed here may be viewed as typical examples, but by no means do they constitute a complete review of all fuel failures. It will be noted that, as experience is gained, the number of fuel failures generally diminishes with each core loading—except, of course, in the cases where the new core loading represents quite a difference in design or material.

Section 5 outlines briefly important examples of failure experience that have arisen. Section 6 summarizes some of the more general findings.

In addition, a section describing some important examples of failure experience is included as well as a section of general conclusions. The radioactivity evolved in a number of these accidents and the measures taken to carry out appropriate cleanup is discussed in Sec. 7 of the chapter on Radioactive Waste Management.

2 CRITICALITY ACCIDENTS OUTSIDE OF REACTORS

2.1 General

This section summarizes the reported criticality accidents which have occurred outside of nuclear reactors through 1963. In general, the incidents listed are limited to those in which the reactivity exceeded prompt critical. It is also limited to those cases where unforeseen events occurred, cases of a truly accidental nature.

Descriptions (of the pertinent facts) of the accidents which occurred before early 1961 are contained in two excellent summaries by W. R. Stratton [3a, 3b]. In his OECD symposium paper [3b] Stratton analyzed in some detail the characteristics of the bursts where such information exists. Table 2-1 summarizes the principal features of all of these accidents. The table is arbitrarily divided into three parts: metal assemblies in air, heterogeneous assemblies, and hydrogenous solutions. The cause and the quenching or shutdown mechanism are noted in abbreviated form. In criticality accidents, it is usually found that there is one (or at most two) simple cause for the accident, so that a table is sufficient to point out the cause. Critical assemblies are relatively simple devices and the accidents which have occurred to date could not be said to be very complex.

Because the basic causes for the accidents in criticality facilities are closely related to the causes of accidents in reactors, both are summarized together in Sec. 6 of this chapter. Since Stratton's review [3b], there have been five additional accidents to September 1964, three in criticality facilities and two in chemical plants. These are described briefly.

The accidents discussed in Secs. 2.2, 2.3 and 2.5 happened in well-designed laboratories for criticality experiments and resulted in no excessive radiation exposures and very little or no damage. The accidents discussed in Secs. 2.2 and 2.3 occurred in general purpose criticality facilities. The accident discussed in Sec. 2.5 occurred in a facility designed to study fast bursts. The public safety was not involved in these accidents, nor is it likely to be in such situations. Much more likely in accidents of this kind is radiation injury to the employees directly involved, particularly in accidents where for some reason personnel are allowed to be present in the area where the nuclear radiation burst takes place.

Of the eight deaths to date from accidents involving the fission chain reaction, three have resulted from exposure to a burst from a critical assembly. In each of these cases there was no compelling reason for personnel to be present in the area at the time of the burst. As shown in Table 2-1 two deaths were at Los Alamos and were due to hand-stacking critical assemblies. These accidents happened near the end of World War II and perhaps can be understood on the basis of military urgency. The third happened in Yugoslavia where six persons were working in a room with a critical assembly in it. The situation was realized after the system had been critical some 3 to 7 minutes and then only when someone smelled ozone. Fortunately, only one of the six persons

Table 2–1

CRITICALITY ACCIDENTS OUTSIDE OF REACTORS

Date	Location – Name	Active Material	Geometry	Total Fissions	Cause	Personnel (P) Radiation Dose (R) Quenching Mechanism (Q)	Damage	References
Metal Assemblies in Air								
21 August '45	LASL	6.2 kg δ-phase Pu	Sphere, tungsten carbide reflected	$\sim 10^{16}$	Hand-stacking reflector, last brick fell from hand	P-2; R-$\cong$800 (fatal), 20 rep; Q-thermal expansion	None	[3,4]
21 May '46	LASL	6.2 kg δ-phase Pu	Same core, Be-reflected	$\sim 3 \times 10^{15}$	Hand-stacking system - screw driver used as wedge to hold up Pu hemisphere - slipped	P-8; R-$\cong$900 (fatal), 185, 116, 93, 41, 26, 18, 18 rep; Q-expansion	None	[3,4]
18 April '52	LASL Jemima	92.4 kg U-metal (93% U^{235})	Unreflected, multidisk cylinder	1.5×10^{16}	Computation error made independently by two people, no graph made of reciprocal multiplication	P-none; Q-expansion	Slight warping of pieces	[3,4]
3 Feb. '54	LASL Godiva	53 kg U-metal (93% U^{235})	Unreflected sphere	5.6×10^{16}	Incorrect operation - reactivity added too fast before chain started	P-none; Q-expansion	Slight warping of pieces	[3,4]
12 Feb. '57 (holiday)	LASL Godiva	54 kg U-metal (93% U^{235})	Sphere, partially unreflected (experiment)	1.2×10^{17}	After reactivity adjusted, graphite-polyethylene mass slipped or slumped closer to sphere	P-none; Q-expansion	Warping, oxidation, near melting at center	[3,4]
26 March '63	UCRL-Livermore Kukla	Over 25 kg U^{235} metal	Nesting cylinders, reflected by Be and polyethylene	$\sim 4 \times 10^{17}$	Shell suddenly slipped down on central cylinder	P-none; Q-expansion, melting	15 kg U burned, 10 kg melted	[12a,12b] Sec. 2.6
Heterogeneous Critical Assemblies								
6 June '45	LASL	35.4 kg U-metal ($\sim$83% U^{235}), 0.5 in. cubes	Pseudosphere in polyethylene box, H_2O-reflected	$\sim 3 \times 10^{16}$	H_2O leaked into box - no scram method provided	P-3; R-66, 66, 7.4 rep; Q-boiling of H_2O	None (cubes used in three days)	[3,4]
1 Feb. '51	LASL	2 cyl. U-metal (93.5% U^{235}) 24.4 kg solid 38.5 kg hollow	Adjacent cylinders, Cd surface, paraffin-filled, H_2O-moderated	10^{17}	Solid cylinder lifted on scram signal pneumatically - passed close to second cylinder, adding reactivity as passed by	P-none; Q-scrams of Cd sheet between two cyl; H_2O drop, cyl. lift cont.	Oxide formed-flaking and blistering	[3,4]
2 June '52	ANL ZPR-I	6.8 kg U^{235}-oxide particles in plastic	Inhomogeneous cylinder H_2O-reflected	1.22×10^{17}	Manual withdrawal of central control rod	P-4; R-136, 127, 60, 9 reps; Q-plastic heated, bubbles	Plastic destroyed	[3,4,14]
3 July '56 (day before holiday)	LASL Honeycomb	58 kg U-metal foils (0.002 and 0.005 in.) in slabs of graphite	Split core of 2 flat slabs 1/2 movable on tracks	3.2×10^{16}	Too rapid assembly of unit	P-none; Q-scram ejected Be-rods, reversed carriage on track. No other good Q available	None	[3,4]
15 Oct. '58	Vinca, Yugoslavia	3,996 kg U-metal, D_2O-moderated	Clad rods, lattice, unreflected	2.5×10^{18}	Subcritical multiplication as function of D_2O level, personnel working in room	P-6; R-400, 700, 850, 850, 850, 1,100 rem (fatal), ozone smell gave first warning	None	[1,3]

15 March '60	Saclay, France Alizé	Partially enriched U H_2O-moderated	Pool reactor, variable fuel normally 4 control rods	3×10^{18}	Mistake in control rod withdrawal. Instrument errors.	P-2; R-slight overdose	None	[6a,6b] Sec. 2.3
10 Nov. '61	ORNL crit. facility	75 kg 93% U^{235} paraffin mod. and reflector	Split core, pseudo-sphere	$10^{15} - 10^{16}$	Split core assembled at fast speed instead of slow; judgment, procedure errors	P-none; Q-expansion	None	[7,8,9] Sec. 2.4
Before 1955	USSR	Unknown	Unknown	Unknown	Unknown	P-2; R-450, 400 rad	Unknown	[1,3,4]
Hydrogenous Solutions								
11 Feb. '45	LASL Dragon	$U^{235}H_3$ pressed in Styrex, solid cubes	Subcritical with central cylindrical hole. U-slug dropped through hole	$\sim 6 \times 10^{15}$ (no accident, simply overheated)	System overheated during burst as slug dropped through central hole	P-none; Q-thermal expansion	Spoiled cubes by swelling, blistering	[3,4]
16 Nov. '51	Hanford Crit. assmb.	1.15 kg $PuO_2(NO_3)_2$ in 63.8 liter H_2O	Sphere 93% full, unreflected	8×10^{16}	Withdrew control rod in fast steps without sufficient wait between steps for power and instru. to catch up. Vessel was not full – if it had been full, expulsion of liquid would have terminated excursion earlier	P-none; Q-expansion, microbubbles, neutron temp., poison rod scram	Building contaminated – cleaned up in a few days	[3,4,14]
26 May '54	ORNL Crit. assmb.	$\sim$18.3 kg $U^{235}O_2F_2$ in 55.4 liter H_2O	Cylindrical annulus, unreflected	1×10^{17}	Tilting of center poison cylinder to position of less worth	P-several; R-max. 0.9 rem; Q-expansion, microbubbles, neutron temp.	3 day cleanup	[3,4,14]
1 Feb. '56	ORNL Crit. assmb.	$\sim$27.7 kg $U^{235}O_2F_2$ in 58.9 liter H_2O	Shallow cylinder, unreflected (flat slab)	1.6×10^{17}	Falling scram rod set up wave and change in H_2O depth > effect of scram rod	P-several; R-max. 0.6 rem; Q-expansion, microbubbles, neutron temp.	Vessel bottom distorted	[3,4,14]
16 June '58	ORNL Y-12 Processing plant	$\sim$2.5 kg $U^{235}O_2(NO_3)_2$ in 56 liter H_2O	55 gallon drum on concrete floor	$\sim 1.3 \times 10^{18}$ (several oscillations)	Wash water added to $UO_2(NO_3)_2$ soln. in non-failsafe vessel (during plant inventory)	P-8; R-estimated doses 461, 428, 413, 431, 298, 86.5, 28.8 rem; Q-expansion, microbubbles, neutron temp., boiling	None	[3,4]
30 Dec. '58 (holiday)	LASL Plutonium processing plant	3.47 kg $PuO_2(NO_3)_2$ in $\sim$191 liter H_2O	Cylinder, H_2O-reflected below, 38 in. diameter 8 in. thick (flat slab being critical)	1.5×10^{17}	Agitator started and created crit. geometry, 8 in. layer barely subcritical. (Plant inventory in progress)	P-3; R≅12,000 ±50% (fatal), 134, 53 rem; Q-mixing	None	[3,4,11]
16 Oct. '59	NRTS Chemical processing plant	34 kg 93% U^{235} as $U^{235}O_2(NO_3)_2$ in $\sim$800 liter H_2O	Cylinder on concrete floor	$\sim 4 \times 10^{19}$ (repeated excursions, boiling	Enriched solution siphoned from safe to unsafe geometry during non-routine plant operation	P-19; R-no gamma or neutron dose; beta doses of 50 r, 32 r, small for 17 others;	None	[3,4b]
25 Jan. '61	NRTS Chemical processing plant	$\sim$8 kg $U^{235}O_2(NO_3)_2$ in 40 liter H_2O	2 ft diameter cylinder, 5 in. deep layer, flat slab geometry	$\sim 6 \times 10^{17}$	Air pressure forced soln. from safe to unsafe geometry during non-routine plant operation	P-none; Q-expansion, microbubbles, neutron temp.	None - system back in operation in one hour	[3,4]
7 April '62	Hanford Chem. proc. plant	$\sim$1.5 kg Pu in H_2O solution	Cylinder, no reflector	$\sim 8.2 \times 10^{17}$ (37 hr. crit.)	Soln. sucked from sump into tank	P-3; R-110, 43, 19 rem; Q-boiling and evaporation	Difficult to shut down	[10] Sec. 2.5
24 July '64	Wood River Jn.	$U^{235}O_2(NO_3)_2$ soln.	cylinder, no reflection	not known	Soln. poured into nonsafe tank	P-1; R > 700 rem (fatal)	slight	Sec. 2.6

died. Such accidents should not happen in well organized and controlled criticality laboratories. The usual type of accident that may occasionally happen during the literally thousands of experiments being carried out is exemplified by those described in Secs. 2.3 and 2.5.

In recent years fast burst facilities have been designed, built, and operated which have devices to deliver, in a controlled fashion, fission pulses having an energy release comparable to many of the accidents listed here. The accident discussed in Sec. 2.5 occurred in such a facility. The problems of control of burst reactors are discussed in the Criticality chapter and also in several reviews [4a, 4b].

The accidents described in Secs. 2.4 and 2.6 took place in chemical processing plants. There have now been six such accidents since mid-1958. The systems involved are exceedingly complex and the type of personnel who work with such systems may be accustomed to working in a somewhat imprecise manner. Even if they are aware of the hazard, the idea of criticality in an innocent-looking tank of liquid may still seem unreal, and the likelihood of accident remote.

In general, there are two types of situations existing in such chemical reprocessing plants. In the first, fissile material free of fission products is being processed. There is usually little or no personnel radiation shielding; consequently, in the event of a criticality accident the working operators are very likely to be exposed to hazardous radiation.

Of the six chemical plant accidents to date, four have been with fissile material free of fission products. All four of these accidents resulted in high radiation doses to personnel. One of these at Los Alamos on December 30, 1958, was an accident in which the operator involved received a fatal dose estimated to be 12,000 r. In each of the accidents it was the actions of personnel present that caused the accident and the personnel were trapped in the accident they caused.

In contrast, the handling of fissile materials containing large quantities of fission products must be done behind heavy radiation shielding. If the shielding is sufficient to protect personnel from the normal fission product burden, it is likely to be sufficient to protect them from the radiation accompanying a criticality burst. Twice to date this has proved true. On the other hand, a criticality burst releases energy which could conceivably drive dangerous quantities of radioactive fission products to the area surrounding the plant and thus create a public hazard. In a plant handling fissile material and fission products the safety problem is mainly one of preventing fission product release to the outside. A "clean" (i.e., no fission products) plutonium processing plant may involve both the hazard of radiation from a burst and the hazard of released material since plutonium itself is highly toxic.

2.2 Alizé I Reactor Accident [5,6]

Alizé I is a light-water-moderated and cooled pool type reactor designed for criticality experiments at power levels from 1 to 100 watts. A variety of fuel types and enrichments may be used. Control is accomplished with a number of identical control rods, normally four, located in positions dependent on the experiment in progress. The primary purpose of the reactor includes use in lattice studies, critical mass studies, buckling measurements, and temperature coefficient measurements between 5 and 95°C (41 and 203°F). It is located at Saclay in France.

The experiment being carried out on March 15, 1960 required that a stable reactor period be established at a very low power level. Accordingly, the reactor was made critical and a definite critical rod configuration and position was established. The reactor was then shut down. A calculation was made, based on rod reactivity worths, to establish the rod position configuration necessary for the desired period. After a suitable decay time to reduce the delayed neutron background, the rods were withdrawn to the predetermined position. However, for reasons not completely clear, a rod previously only partially withdrawn was fully withdrawn in the second instance putting the system on a short period. The error could have been due to a miscalculation, a misunderstanding of the first configuration, or to a number of other possible causes. The experimenter was working alone.

It is reported that "the period trip had been switched out of the circuit by mistake and that the power trip levels did not operate, no doubt because their trip level was exceeded too rapidly."[6]* The power proceeded to rise in a period of approximately 1/2 sec to a level of nearly 10 Mw at which time the Doppler effect due to the heating of the partially enriched fuel was sufficient to offset the reactivity added and the power dropped back to just under 1 Mw. Since the instrumentation available on Alizé itself was completely ineffective, the first indication of the problem was given by the instrumentation of an adjacent reactor which alarmed on a short period and showed a rising power trace. Immediate investigation of this odd phenomenon in the adjacent reactor quickly led the operators to Alizé as the cause of the trouble. It was scrammed manually approximately 70 sec after the peak of the power burst.

Comments, Conclusions, Recommendations

(1) The withdrawal of control rods to predetermined positions is often used as a means of setting up a prescribed pattern, but it is normally carried out under the rules and using the instrumentation usual for reactor startup.

(2) Adequate on-scale safety instrumentation information and scram trips are absolutely essential to the safe operation of a reactor.

2.3 ORNL Criticality Excursion [7]

This excursion occurred on November 10, 1961, in the critical experiment laboratories at ORNL

*It may also have been negated by ion chamber saturation or some other reason.

behind shielded walls and therefore there was no overexposure of personnel. The facility is described in reference [8] and the accident in [7] and [9]. The basic experimental device incorporated a vertically moving divided system in which the lower section moved upward by a hydraulic piston acting through a magnetic coupling. Air pressure applied to the hydraulic system controlled the initial speed of approach. As the moving section reached a given preset point it actuated a switch, reducing the approach speed. It was necessary to adjust the preset point manually. The speed of approach could also be reduced by a manually controlled hydraulic system, but neither of these means of adjustment was at the remote control point.

During this experiment a mass of about 75 kg (~ 165 lb) of 93% U^{235}, which was split into two parts, was being used. It was moderated and reflected by paraffin. The initial speed of approach was 16 in./min (0.41 m/min) and the change to slower speed was set to trip when the parts were 1.94 in. (4.93 cm) apart. In previous runs the system had been unreflected and subcritical under all conditions. In this run a large paraffin reflector was added.

When the test was started, a blue glow was seen while the system was still in the fast part of its assembly motion. This pulse actuated the scram mechanism which withdrew the lower half downward at an initial speed averaging 12 in./sec (0.30 m/sec) for the first inch, 60 in./sec (1.52 m/sec) for the next 9 in. (22.9 cm) and 10 in./sec (0.25 m/sec) for the rest of the travel. This dropped reactivity at a rate of about 100 $/sec in the first inch of travel and shut the system down.

Later measurements indicated that the system was delayed critical at 2.7 in. (6.86 cm) separation and, at the faster speed, was adding reactivity at a rate of 8.6 $/in. (3.38 $/cm) or 2.3 $/sec. Thus, there was a time span of 435 msec between delayed and prompt critical. The instrument response time is believed to be shorter than this and thus it is likely that prompt critical was not reached. In the first 50 msec after the magnet disengaged (~1 msec needed for collapse of the magnetic field and ~ 50 msec to disengage the magnet) the reactivity dropped 5 $.

The fission release was 10^{15}-10^{16} fissions. A 10^{16} fission burst with no heat losses would raise the uranium temperature by only 35°C (63°F) and, indeed, no paraffin was melted.

Comments, Conclusions, Recommendations

(1) Too rapid assembly was the primary accident cause coupled with a poor position indicator in the control area and a sluggish instrument system. These items were improved [9].

(2) Procedures were changed to require measurements of closure speeds prior to the experiment [9].

(3) Split-bed assemblies of this general type present special hazards due to the effect of the rapidly changing geometry on reactivity. The rate of addition of reactivity should decrease as the system approaches criticality. Instead, if the motion of the table is linear, the rate of addition of reactivity increases as the two halves approach.

2.4 Hanford Recuplex Criticality Excursion [10]

The Recuplex System is a multipurpose recovery unit operated by the Chemical Processing Department of the Hanford Plant and used to recover plutonium from various processing solutions. The accident occurred on Saturday, April 7, 1962 in a large hood used for solvent extraction. Plutonium is so toxic that all handling of plutonium bearing solutions and solids is done in large volumes sealed off from inhabited areas. Normally isolation is achieved by means of clear plastic walls mounted on a metal frame. Such a device is called a "hood" or "dry box." It may be of any height and length and is usually narrow enough to permit access to all points inside through appropriate openings for rubber glove access. The intent is to protect personnel from physical contact with plutonium and these hoods are normally not shielded to protect personnel from neutron and gamma radiation.

Within the solvent extraction hood in question were a number of tanks, pipes, valves, and extraction columns. For several weeks prior to the accident the extraction operation in process was the cleanup of the hood in which the accident took place. Callihan's report [10] describes the situation graphically and should be required reading for all reprocessing plant personnel. He writes: "This housekeeping chore was demanded by an accumulation of liquid and solid materials, and some in-between phases, on the floor of the box. The accumulation apparently included both organic and aqueous solutions of plutonium and reagents from leaks in the process plumbing; there were also neoprene gloves and plastic containers that had been temporarily disposed of by dropping them to the floor. All in all, the result was a conglomerate mess that was somewhat rich in plutonium. Much foreign material had collected on the walls of the hood and thereby reduced visibility of the interior, a factor of some significance. The cleanup entailed adding successive volumes of aluminum nitrate solution and nitric acid to the floor." The transfer from the floor to tank K-9 was accomplished by vacuum suction through a temporary plastic tube extending via two shutoff valves from the tank to a floor sump. The cleanup was completed, but the plastic hose was not removed. Tank K-9 was not of the "always critically safe" variety and it was therefore necessary to ensure safety by procedural controls and by adding cadmium nitrate solution.

In the same hood was another tank, J-1, which could receive plutonium solutions. Unknown to the operators, this tank had an overflow which permitted solution to run out directly onto the floor of the hood—a very unusual design since it is normal to contain plutonium-bearing solutions in a piping system. In retrospect, it appears likely that tank J-1 was filled and overflowed onto the floor some 48 liters of solution with a concentration of 45 g/liter (2160 g of Pu). The liquid on the floor was observed before the accident and described as being "clean" and not dark-colored as plutonium solutions are.

Tank K-9 was unquestionably attached to the vacuum line and one of the two valves connected in series to the temporary sump line was clearly open and found open after the accident. (This valve remained open after the transfer of a few "tens of liters" of aqueous phase dilute in plutonium into tank K-9.) The other valve was found closed after the accident, but must have been open for the accident to occur, although "no evidence of the valve being open was presented by operating personnel." The open set of two valves in the plastic sump line permitted concentrated plutonium solution to be sucked up by the vacuum into K-9, unknown to the operator, until the system became critical.

The pattern of the excursion seems to indicate that reactivity was added more slowly than that achievable by adding 48 liters in the time available. It is postulated that the mixing of the concentrated solution sucked up from the sump with the dilute solution occurred gradually after the addition. There was an observed blue flash, radiation alarms sounded and the area was evacuated at once. From the local assembly point, all personnel were transferred to the area first aid building and a "Quick Sort" examination was made. In this test a Geiger-Muller tube is placed against the subject's abdomen and he bends over the tube. The neutron-induced activity in the body is detected. This test identified at once the three employees with high doses and rejected all others, with one possible exception. The doses finally determined by more sophisticated methods for these three were 110, 43, and 19 rem. None of the other 19 persons in the building received more than 2 rem.

The tank with its solution remained critical for a total of some 37 hours and generated about 8×10^{17} fissions (20% in the first half hour). A difficult aspect of the situation was that there was no way to approach the tank safely to carry out any shutdown action. Persisting neutron and gamma fields showed that the system was still critical, and absence of α-activity in the air showed that the plutonium was contained in the hood. A recording neutron device about 350 ft (107 m) away showed full-scale deflection for the first 28 minutes, but displayed the rest of the excursion as an exponential decrease for the second half hour followed by a gradual decrease for the next 34 hours and then a rapid drop to background in the next 24 hours. A one-handed wheeled robot with a closed circuit television eye was invaluable in ascertaining the situation. Using a directional monitor, it determined which tank was critical. It placed lights and instruments, read meters, moved equipment, turned valves, and in general made it possible to permit human entry with reasonably assured safety.

It is hypothesized by Hanford that the system gradually went critical and developed an initial pulse of about 10^{16} fissions. Repeated pulses, each terminated by microbubble formation, then occurred as the solution heated to boiling after about 20 minutes. (Boiling was at 60°C (140°F) under the conditions existing.) Boiling reduced the liquid volume exponentially until 2-3 liters had boiled off (one hour after the start). Then evaporative cooling and heat losses chilled the solution and increased reactivity, thus establishing a quasi-stable state which persisted until evaporation finally made the system subcritical.

Comments, Conclusions, Recommendations

(1) It appears unlikely that accidents in processing plants can constitute a serious hazard to the general public. Only where large amounts of plutonium or of fission products are involved in the process could a public hazard be even remotely conceivable. Much more likely, but still highly improbable, would be an accident involving personnel working at the facility.

(2) There are a number of methods to prevent accidents of this type or to reduce their probability of occurrence. These include:

a) use of radiation shielding to protect personnel;
b) use of vessels with "always critically safe geometries," raschig rings, boron steel grids, or other positive geometric or nuclear poison means of preventing criticality;
c) use of remote control in operations; and
d) use of procedural controls in limiting the quantities of fissile material present at each stage of the process.

The first three methods, when properly used, provide positive protection for personnel. The use of personnel radiation protection usually implies an automatic or remotely controlled process with its accompanying expense and complexity. Because of this expense, plant designers are often forced to choose the fourth method, that is, procedural control of the quantities of fission products. In making this choice designers should be fully aware that the certainty of the other three alternatives is not present and therefore a larger risk is being taken. Conservative design practice would recommend that one of the other three methods should be used for steps in the process where unknown quantities of fissile materials could accumulate or where system cleanup residues are deposited. The problem of the details of the design and operation of such plants is an art in itself and is beyond the scope of this book. See, for example, the symposium in [3b].

(3) If personnel are allowed to work in sections of the plant where criticality might occur, each region where they may be present would be much safer if two exits—preferably in opposite directions—are provided. If there is only one means of egress a man may find himself cut off from escape by a critical container and a high radiation level. In event of an accident, he might then be confronted by the need for a decision to run past the source or to get as far away as he can and remain there. This consideration played a part in the Y-12 accident, the LASL Pu-solution accident, and the Hanford Recuplex accident. Of course, if wise use is to be made of the choice of exits, adequate instrumentation must either be provided to the individuals involved or be located in the region considered so that personnel will know which way to go. Drills should be held to ensure that personnel have practice in taking proper action.

(4) During plant design, as well as during operation, adequate provisions should be made to provide means to shut off a criticality situation. This can be provided, for instance, by a robot of the type described in the Recuplex accident report coupled with appropriate advance thought on the requirements of the mutual interaction required

between system and robot to achieve shutoff. There should be no need to risk exposing personnel to high radiation levels in shutting off a criticality accident.

(5) It is useful, and in fact necessary if personnel are working in a potential criticality area, to have a means of designating which tank is critical; this is an aid to evacuation and helpful in planning re-entry.

(6) Good housekeeping is important to safety. In particular, all operators must be indoctrinated with the need to be precise and careful in handling solutions containing fissile material.

2.5 The LRL Critical Facility (Kukla) Excursion [12a, 12b]

The Livermore Research Laboratory Critical facility (Kukla) consists of a shielded vault with fully controlled and positively interlocked personnel access and with an adjacent control room having a thick shielding view window and a television camera for visual observation of the experimental vault area.

At about midnight on March 26, 1963, an experimental critical assembly underwent an unexpected prompt critical burst of about 4×10^{17} fissions.

The experiment in progress was the measurement of the subcritical multiplication of two concentric cylindrical shells and a central solid cylinder of U^{235} metal, surrounded by concentric reflectors of beryllium and polyethelene. The shells and solid cylinder were all close fitting.

The beryllium and polyethylene reflector completely surrounded the two hollow uranium shells, the ends of the cylinder being closed with circular plates of the same composition. There was a one-inch air space between the reflectors and the uranium shells. The one-inch air space below the cylinders was approximated by a styro-foam spacer resting on the bottom reflector. The bottom section had appropriate holes to admit the solid uranium cylinder on a remotely controlled ram.

The procedure for each multiplication step was:

insert the ram upward a predetermined amount;
measure the multiplication;
extrapolate to determine the next insertion limit;
lower the ram;
enter the vault and reset the mechanical ram stops; and
exit from the vault.

The operator had completed seven multiplication steps and the eighth was in progress. He had inserted the ram most of the way and had stopped to adjust instrument scales when a "sharp pop" was heard through the vault intercommunication system. The television screen used to watch operations in the vault went blank. When the television image returned, it showed material burning and falling from the assembly.

It is estimated that about 10 kg of the uranium in the assembly melted and resolidified. Approximately 15 kg burned. The total fissions which occurred is estimated to have been approximately 4×10^{17}. The assembly had fallen to the floor. Some polyethylene reflector was scorched and melted. Two ion chambers encased in paraffin were knocked from the assembly machine. Part of their paraffin was melted. Other combustible material in the vault was unscorched. The physical damage was limited to the assembly machine, although the entire vault interior was contaminated with fission products and uranium.

There was no serious activity release although the vault was not a sealed containment structure. About one hour later a survey indicated about 1×10^{-9} μc of β-γ activity per cm^3 of air at a point 350 m (1150 ft) north of the building. The peak airborne activity in a section of the building near the vault was 2×10^{-7} $\mu c/cm^3$ shortly after the event. No alpha-emitting material was found outside of the building.

Personnel exposure doses ranged from zero to 120 mrem of gamma rays with no neutron exposure. These were film badge readings and so may have been high since badges are normally worn for several days. Nasal wipes, thyroid scans, and urine analyses were negative.

It is believed that the central solid cylinder was slightly out of line with its mating hollow cylindrical shell. Apparently as the ram and solid cylinder moved upwards it lifted the shell in each of three successive assembly steps, thus giving misleading multiplication measurements. During the last step the shell again lifted but fell back again when the ram was almost fully inserted and while the operator was changing scales. It was not possible to observe this action.

Comments, Conclusions, Recommendations

This experimental run was the first (of literally thousands carried out in this facility) that resulted in a serious accident. It is likely that there will be a few mishaps in such a large number of experiments, almost every one of which is different. Therefore it is necessary to carry out these experiments as though each is a potential serious excursion. The proper procedure was followed in this case and as a result there were no serious personnel exposures and no serious radioactivity release outside the experimental vault. This accidental excursion illustrates very well the relatively minor nature of the consequences of even quite energetic excursions in properly designed critical assembly vaults. The protection of personnel remains the primary concern in such facilities.

2.6 The UNC Wood River Junction Incident*

The scrap recovery facility of the United Nuclear Corporation at Wood River Junction, Rhode Island, is used to recover enriched uranium from unirradiated scrap material resulting from the fabrication of nuclear fuel.

As a part of the recovery operation trichloroethane (TCE), an organic liquid, is used at the plant to remove solvent from the production stream. This solvent contains small traces of uranium. When the used TCE is taken from the stream it may contain as much as 1,000 ppm of enriched uranium.

*Statement filed August 24, 1964, AEC Public Document room, 1717 H St., N.W., Washington, D.C.

Table 3–1

REACTOR ACCIDENTS INVOLVING CRITICALITY OR REACTIVITY CHANGES

Date	Location, name	Active Fuel Coolant, Moderator	Geometry	Total Fissions	Cause (C) Quenching Mechanism (Q)	Person (P) Radiation dose (R)	Damage	Ref. in this chapter
Dec. 1949	LASL (water boiler) crit. 1944, 1950	1 kg U^{235}, $UO_2(NO_3)_2$ in 13.6 liter H_2O	Sphere, graphite reflector	$3\text{-}4 \times 10^{16}$	(C) Manual withdrawal of 2 control rods (Q) Expansion and rise of neutron temperature	P-1: R-2.5r gamma	None	Sec. 3.2
12 Dec. 1952	Chalk River, Canada NRX, criticality 1947, full power May 1948	Natural uranium rods H_2O-cooled, D_2O-moderated	Rod lattice, graphite reflected	0.6×10^{20}	(C) Control rod mal-operation, safety circuit failure-complex (Q) Dump of D_2O moderator	P-none, except in clean-up; many P got small doses, highest 17r, most less than 3.9r	Core badly damaged, removed, replaced	Sec. 3.3
22 July 1954	NRTS Idaho BORAX I (transient tests 1954)	93% enriched U^{235} in U-Al plates (MTR type), H_2O-moderated	H_2O reflector, swimming pool excursion reactor	4.68×10^{18}	(C) Estimate of expected excursion low (Q) Steam void disassembly	P-none R-none	Core destroyed	Sec. 3.4
3 Oct. 1954	Hanford Production Reactor First one critical September 1944	Natural uranium rods H_2O-cooled, graphite-moderated	Process tube type–large graphite reactor	Local over-heating	(C) Water leak changed reactor pattern, short period occurred (Q) Control rod changes then scram	P-none R-none	Some fuel elements failed or were damaged	Sec. 3.5
4 Jan. 1955	Hanford – KW Reactor	. . . . same	. . . same . . .	Fuel failure, local melting	(C) Blockage of cooling water in process tube – initial start-up power decrease noted – rods withdrawn	P-none R-none (Q) Scram on over-pressure	Graphite channel removed by hole cut in shield	Sec. 3.5
Jan. 1955	Hanford Production Reactor	. . . . same	. . . same . . .	No over-power	(C) Misestimate of ρ-instruments (Q) Rod run in by operator	P-none R-none	None	Sec. 3.5
29 Nov. 1955	NRTS Idaho EBR-I Mark-II Operations in 1951	0.5 in. U^{235} rods, NaK-cooled fast reactor	Compact core Nat. U blanket	4.7×10^{17}	(C) Estimate of expected results low – earliest scram attempts not effective (Q) Shutoff by second scram; fuel bowing a factor	P-none R-minor	Core melted, little other contamination	Sec. 3.6
9 Oct. 1957	England-Windscale Operations in July 1950	Natural uranium rods air-cooled, graphite-moderated	8-sided stack of graphite $50 \times 50 \times 25$ ft 25 ft fuel channels	Graphite-uranium fire	(C) Wigner energy release, U-burning triggered by nuclear overheating (Q) Flooding with H_2O	P-none serious R-widespread radioactivity, milk over 200 mi^2 area destroyed	Severe core damage reactor not rebuilt	Sec. 3.7
18 Nov. 1958	NRTS – HTRE-3 crit. October 1958	Enriched uranium gas-cooled, solid moderator	Horizontal cylinder	Not known	(C) System on auto-control with faulty instrumentation (Q) Meltdown slump and/or scram	P-none R-some site contamination	Core melted, basic system undamaged	Sec. 3.8

24 July 1959	Santa Susanna, Cal. SRE crit. 1957 at power May 1958	2.8% U^{238} slugs in SS-clad rods Na-cooled graphite-moderated	Pseudo-cylinder, graphite reflector	2×10^{19} (in last minute)	(C) Coolant channel blockage by impurities overheating, perhaps fuel bowing (Q) Manual scram	P-none R-release was $\sim 0.3\%$ of core activity inventory	12 of 43 elements melted, core removed and replaced	Sec. 3.9
3 April 1960	Waltz Mill, Pa. WTR crit. 1959	93% enriched U^{235} U-Al plates-cylinder H_2O-cooled-moderated	Pseudo-cylinder, water reflected	Overheat of one element	(C) Undercooled, perhaps faulty fuel – negative auto control response (Q) Manual scram	P-none R-minor release to site	1 element melted, $\$10^6$ to clean up damage	Sec. 3.10
3 Jan. 1961	NRTS SL–1 crit. August 1958	93% enriched U^{235} Al-U plates, boiling, H_2O-cooled-moderated	Pseudo-cylinder 5 rods, B-Al strips control	1.5×10^{18}	(C) Manual withdrawal of central control rod (Q) Expansion, boiling, core evaporation	3P-all fatal R > 800r/hr in bldg.; in recovery 14P got R > 5r	Core destroyed, vessel rose 9 ft, reactor dismantled	Sec. 3.11
5 Nov. 1962	NRTS SPERT I (destructive tests)	93% enriched U^{235} Al-U plate type H_2O-moderated	Pseudo-cylinder with transient ρ-rod, open tank		(C) Planned experiment–reactivity transient as planned, energy release effects more destructive than planned (Q) Expansion, steam void, no melting involved	P-none R-minor site contamination	Core destroyed (destruction planned on this test or next)	Sec. 3.12

Up until July 17, 1964, the method of recovering this uranium from TCE was to place 2 or 3 liters of TCE solution in a 5 in. (12.7 cm) diameter, 11 liter, critically-safe polyethylene bottle. About 6 liters of sodium carbonate solution were added. The bottle was then shaken by hand and the liquids allowed to separate. The enriched uranium, now dissolved in the sodium carbonate solution, was then removed by means of a separatory funnel.

The weight and size of the container full of liquid made the mixing difficult for the operators and soon an operator suggested another easier method. This method made use of mechanical agitation by a stirrer in a non-critically-safe sodium carbonate make-up tank. This 30 gal stainless steel tank was 18 in. in diameter with a 26 in. vertical wall and a standard dished bottom. The normal function of this tank was solely to make up sodium carbonate solution for a solvent washing operation. It was not designed for any use involving uranium solutions.

The new procedure was discussed with a supervisor before it was put into effect and was approved informally by the supervisor provided that the concentration of uranium was known. A limit of 800 ppm for use of the procedure was informally established. The new method was used by two shifts beginning on July 17, 1964, and the operator in question utilized it on July 17 as well as on July 24, the night of the accident.

Early on the morning of July 23 an evaporator failed to operate properly. It was discovered that a pipe to the evaporator was plugged with uranium nitrate crystals. Steam was used to unplug the line. When the line was unplugged, material from the evaporator system was drained into several 5 in. diameter, 11 liter bottles.

It is presumed that one of these bottles of highly concentrated U^{235} solution was responsible for the accident. Evidently, at approximately 18:05, July 24, through some mischance the operator in question emptied one of these bottles into the non-critically-safe make-up tank which already contained approximately 15 gal of aqueous sodium carbonate solution.

When nearly all the liquid had been poured into the tank the excursion occurred, solution was expelled from the vessel, and radiation alarms sounded. The operator fell backward to the floor, arose, and ran from the building to an emergency shack about 200 yds from the main building. He died at 19:20, July 26, approximately 49 hr after the incident. It has been estimated that the operator received more than 700 rem of gamma radiation and an additional large fast neutron dose.

3 ACCIDENTS AND DESTRUCTIVE TESTS INVOLVING REACTIVITY CHANGES IN REACTORS

3.1 General

This section summarizes (through 1963) the reported accidents and destructive tests involving reactivity changes in reactors. Table 3-1 outlines the principal features of all of these accidents. The last column of the table indicates the subsection of this chapter in which each accident is discussed.

In this chapter a reactor is defined as a fission

chain reacting system capable of developing substantial amounts of heat on a sustained basis. Some of the reactors listed could have been called critical facilities and vice versa, but even the doubtful ones like BORAX-I and SPERT-I were capable of developing substantial power if desired. In at least two cases, BORAX-I, and the SPERT-I tests, the incidents are better described as destructive tests. They are included here because they present interesting data and complete the nuclear transient record.

In most regards the incidents discussed show no particular uniformity other than some barely discernible trends in their causes. The fragmentary generalizations that may possibly be drawn from these accidents are discussed in Sec. 6. As Table 3-1 shows, a wide diversity in the type and size of reactor is involved. The accidents and destructive tests involve the whole spectrum of reactivity changes from prompt critical excursions (Water Boiler, BORAX-I, SL-1, SPERT-I) down to almost no reactivity involvement at all (Windscale, where a nuclear reheat reached a power level somewhat too high). The quenching mechanisms range from thermal expansion through microbubble formation, void formation, flooding with nuclear poison, core meltdown, to violent fuel vaporization. There have been a number of persons who have received minor overexposures in reactor accidents or in the post-accident cleanups; the only serious exposures and the only deaths to date from nuclear reactor accidents are the three operators who died in the SL-1 accident. Except for this one case, the contrast between exposures to individuals from reactor accidents and those from, say, fresh fissile material criticality accidents is marked. As was mentioned in Sec. 2.1 this is undoubtedly due to the more extensive personnel shielding provided around reactors.

The damage resulting from these accidents and destructive tests also exhibits a wide range of variability from no damage (Water Boiler), through minor core destruction (Hanford, SRE, WTR), to essentially complete core destruction (NRX, EBR-I, HTR-III), and finally to abandonment or destruction of the reactor (BORAX-I, Windscale, SL-1, SPERT-I). The only accident to date which required subsequent action to protect the general public was the Windscale accident where local milk supplies were temporarily contaminated. No accident to date has involved direct radiation overexposure to the general public. It is interesting to note that, with the possible exception of the SRE accident (see Sec. 3.9), not a single one of these accidents happened when the reactor was running normally at power as designed. Of the thirteen accidents considered here, three were planned transient tests (BORAX-I, EBR-I, SPERT-I). Two of the remaining ten incidents occurred in supposedly shutdown and harmless reactors (Water Boiler, SL-1). The Windscale accident was really a graphite radiation damage bake-out problem and largely non-nuclear in character. Five of the accidents occurred during start-up (NRX, Hanford-3, HTRE-3). The WTR was undergoing a boiling limit test. The SRE was engaged in trying to solve a coolant flow obstruction problem in the core, but was at the same time trying to run normally.

3.2 The Water Boiler Criticality Excursion [3a, 3b, 13, 14]

The water boiler is a 12 in. (30.5 cm) diameter stainless steel sphere containing 13.6 liters of a water solution of uranyl nitrate reflected by graphite [13, 14]. At the time of the excursion the reactor was being remodeled for operation at higher power. Two new control rods had been placed in the system in addition to the three existing control rods.

Normally the rods are raised remotely from the control room when the control panel is activated by a key switch. In this case the panel was shut off and the employee who had built the new control rod mechanisms decided to test the drop times of the new rods. He opened the enclosure at the top of the reactor, and manually lifted the rods—neglecting the possibility that this action might cause criticality. After several tests of each rod individually (no criticality resulted since one rod was sufficient to maintain a subcritical margin) the employee decided to check the relative drop times of the two new rods and pulled both rods at once. He held them out for about 5 sec and dropped them simultaneously. A short time later he repeated the test. This action increased the reactivity to about 3 cents [3a, 3b] above prompt criticality, 0.86% in Δk, corresponding to a period of 0.16 sec. The power is estimated to have risen to a very broad peak of $2\text{-}3 \times 10^{16}$ fissions/sec and to have remained there for about 1.5 sec. The excursion was not detected immediately as all the instrumentation was turned off except one direct-reading thermometer which showed a temperature rise of 25°C (45°F) equivalent to a yield of $3\text{-}4 \times 10^{16}$ fissions. The strong negative temperature coefficient of -0.034% $\Delta k/°C$ (-0.019% $\Delta k/°F$) probably limited the transient before the rods were put back. The operator received 2.5 r of gamma radiation. No damage was done to the reactor.

Comments, Conclusions, Recommendations

(1) Reactors should be designed so that it is very difficult or impossible to raise rods by hand, regardless of whether the reactor satisfies the stuck-rod criteria or not. (Note that it was the second rod withdrawn which caused criticality in this case.) Both the Water Boiler and Clementine (the first fast reactor, also located at LASL) had this same design problem. As a result of this accident, the enclosure at the top of the reactor was provided with a lock which was accessible only to two senior members of the group. At the same time the controls of Clementine were locked in a similar manner.

(2) No employee should be allowed to work alone or isolated in any operation that could be dangerous to the employee, the equipment, or the general public.

(3) Whenever operations are to be carried out that could conceivably change reactivity, the instrumentation system should be turned on and someone should be observing its behavior who is in communication at least with those carrying on the operation.

3.3 The NRX Reactor Accident [15, 16a, 16b, 17, 18]

The NRX Reactor is a heavy-water-moderated, light-water-cooled, research and testing reactor, using natural uranium fuel [15]. It is capable of operating at power levels up to 30 Mw. The reactor had 12 shutoff rods which operated on the basis that 7 rods in the down position were sufficient to hold the reactivity below critical for any approved change of fuel and load. The shutoff rods were thin steel tubes filled with boron carbide. The rods were driven into position through their 10-ft (3 m) travel by air pressure derived from a piston at the head of the rod. The air pressures were manipulated by electrical controls. In order to seat the solenoid valves which held the air pressure, an additional control room push button (No. 3) was provided which increased momentarily the solenoid current thereby more firmly seating the valves to prevent leak-off of the air. If the rods were driven in by air pressure, their travel time for half-insertion was 1/3 to 1/2 sec, whereas without air pressure they normally took 3 to 5 sec to drop the full 10 ft. Each rod was instrumented so that a red light showed on the control desk when the rod was fully up.

The rods were grouped as shown in Table 3-2: Group No. 1 was called the "safeguard bank" and the number of rods in that group was at least one greater than the number in any other group. The safeguard bank was brought up normally only from a condition in which all of the shutoff rods of the other banks were down. At some time prior to the accident, this bank had been interlocked in such a way that it was impossible to withdraw other rods before the safety bank was withdrawn.* "Owing, however, to defects in these switches and their being subject to flooding which could make them a hazard, this 'safety' circuit was not in operation at the time of the incident. The added responsibility was accepted by the operating supervisor."

"The design reason distinguishing the safeguard bank is that, for safety, no shutoff rod may be raised unless either (a) more than 7 shutoff rods would be left fully down, or (b) more rods are available for quick release than are being raised at any time. To make startup possible, some rods must satisfy condition (a) and not (b), and, if the total of shutoff rods is only 12, no more than 4 may be set for condition (a). All other rods must satisfy condition (b). To achieve a safe startup in the shortest time, as large a number as possible and the most highly effective rods were in the safeguard bank. The reason for allowing always one more than the minimum safe number is to allow for one undetected failure in the safety system." [16a]

To operate the rod banks, four push buttons were required. Push button 1 raised Bank 1. Push button 2 raised automatically and sequentially the remainder of the rods. Push button 4 was mounted on the wall panel to the left of the desk and charged air to the heads of the shutoff rod assemblies. It was the release of this air which drove the rods down. Push button 3 increased temporarily the current to the solenoid valves, as mentioned previously.

*The chronological history of events is taken for the most part from the report of W. B. Lewis [16a].

Table 3–2

NRX Reactor Shutoff Rod Groupings

Bank No.	No. of Rods	
1 (safeguard)	4	– Push Button 1
2	3	Push Button 2
3	2	Push Button 2
4	1	Push Button 2
5	1	Push Button 2
6	1	Push Button 2

At the time of the accident on December 12, 1952, only one fuel rod was air-cooled and that was a fresh unirradiated element. An experiment was being conducted on the reactivity of the reactor at low power levels. The object of the experiment was to compare the reactivity of long-irradiated fuel rods with that of fresh fuel rods. A number of rods had either a reduced H_2O coolant flow or else temporary cooling provisions.

"The immediate chain of events which led to the accident began with an error by an operator in the basement who opened by mistake three or four bypass valves on the shutoff-rod air system, thereby causing three or more shutoff rods to rise when the reactor was shut down. The supervisor at the control desk noticed this because the red lights came on. He phoned to the operator in the basement to stop and went down himself to investigate and rectify the situation, leaving his assistant at the control desk."

"He recognized the operator's mistake and was horrified at the possible consequences if the operator had continued to open these wrong valves (actually he could not have opened all valves since some handles had been removed for safety). The supervisor rectified all valves and checked air pressures."** "He assumed that all shutoff rods would drop back into position, but, on account of unexplained mechanical defects, it is apparent from subsequent events and inspection that two or three did not drop back, although they slipped down sufficiently to clear all the red lights on the control desk."

"The supervisor then phoned his assistant to press buttons 4 and 1. He had intended to say 4 and 3, but under normal circumstances 4 and 1 should have been safe (all the shutoff rod red lights were out). His assistant therefore did so. Having to leave the phone to reach simultaneously with two hands the two buttons, he could not be recalled to correct the mistake. Button 3 not having been pressed, the air pressure brought up by button 4 leaked away."***

**Note that it might have been possible for an ingenious operator to raise all the rods by transferring handles from one valve to another if these valves were of the usual type.

***In a recent (April 13, 1964) private communication to the author, W. B. Lewis has kindly supplied additional information on the reasons for

"Up in the control room it was soon evident when the first bank of shutoff rods was raised by button 1 that the reactor was above critical, which was of course a complete surprise."

"It takes a few seconds for this to be apparent. There was surprise but no alarm for the next step would be to trip the reactor and thus drop back the shutoff rods. This the assistant did about 20 sec after pushing button 1. But two of the red lights stayed on, and in fact only one of the four rods of the first bank dropped back into the reactor and that over a period of about 1-1/2 min. Even though, as it appeared, the air pressure had leaked from the header, all shutoff rods should have nevertheless dropped back under gravity."

"The galvanometer spot indicated that the power level was still climbing up. The assistant telephoned the supervisor in the basement urging him to do something to the air pressure to get the rods down."

"Others in the control room were worried: the physicists, the assistant superintendent of the reactors branch, and a junior supervisor. At least two thought of the last resort; namely, to 'dump the polymer' (D_2O). All were familiar with the process as it had been done the previous day for experimental purposes. The assistant superintendent gave the word; one of the physicists was already reaching for the dump switch and beat the others to it."

"However by this time the reactor power was up in the tens of megawatts, and the dumping took a few seconds to become effective. Then a fear arose that they might be dumping too fast as the helium pressure had dropped back sharply, and they envisaged danger of collapsing the calandria by vacuum. The assistant superintendent halted the dumping after about 1 min but after a little thought resumed. However, in 10 to 30 sec after starting to dump, the instruments were back on scale, and the power rapidly dropped to zero. The assistant superintendent went to report to the superintendent, but the consequences were only beginning."

"In the basement the door into the chamber under the reactor (the lower header room) was open. Through this an operator saw water gushing down, and immediately he called the supervisor. Their instant reaction was to suspect any water as being heavy water; therefore the supervisor and operator rushed in with a bucket and collected a sample, which was soon found to be light water but radioactive."

"The assistant superintendent, returning to the control room, was met by an operator who reported a rumble and a spurt of water up through the top of the reactor."

"Then the air activity began, and automatic radiation-level alarms sounded in the reactor building. A phone call to the control room from the adjoining chemical extraction plant reported atmospheric activity off-scale and requested the emergency stay-in procedure. The sirens for this were sounded. The radiation hazards control branch got busy reading instruments, making surveys, and collecting reports. Some minutes later the activity inside buildings with forced ventilation was found higher than outside; therefore on the advice of the Biology and Radiation Hazards Control Director the Project Head gave the order for the plant evacuation procedure, and that went into effect."

"Meanwhile in the reactor system not earlier than 30 sec before the dump began, helium began to leak at a rate of 140 cu ft/min. After 3-1/4 min, by which time the reactor power had been down to a negligible level for 2 min, the reserve gasholder was almost empty. Then suddenly in less than 30 sec the 585 cu ft gasholder rose to its fullest extent. The change of direction of motion of the gasholder was so abrupt on the record and its motion so well-timed by pen marks at 15-sec intervals that it can be deduced with certainty that within a period of 15 sec the gasholder became connected presumably to a mass of gas at high enough pressure to give a large acceleration to the massive four-ton gasholder.

"About the same time that the gasholder was forced up, the radiation level in the reactor building became high. Respirators were issued to those in the control room. All not concerned with the reactor operation were evacuated from the building."

"Holding discussions in gas masks is difficult so after a few further minutes those concerned with reactor operation also went to an adjacent building and planned further steps, returning to the reactor building to put them into effect." [16a]

A further post-accident investigation [17] has led to the conclusion that the unusual sequence of events which happened initially left the reactor supercritical by about 0.6$ and that the power rose rapidly. Control rods slowly dropping back into the reactor core made it appear that the power would level off at about 20 Mw(t). However, the reduced cooling rate which was being used for the test with some rods was insufficient at this power level and boiling in the H_2O cooling channels followed at a power level estimated to be about 17 Mw(t). Unfortunately, this reactor has a positive coolant void coefficient and it is surmised that boiling caused the expulsion of light water from the coolant annulus, thus increasing the reactivity by about 0.2$. The power began to rise again on a period estimated as being between 10 and 15 sec. It is estimated that the power was between 60 and 90 Mw(t) when the D_2O was dumped, thus shutting down the reactor. The reactor power was greater than 1 Mw(t) for no more than 70 sec and the total energy release is estimated to have been about 2000 Mw-sec or about 0.6×10^{20} fissions [16b].

Figure 3-1 [16b] shows a map of the core, in-

---

the location of the buttons. He says, "...the difficulty was not that either button was inaccessible to the telephone, but the two buttons were spaced apart and caused the operator to set down the telephone to push the two buttons simultaneously. The placing of these buttons had been deliberate to emphasize the special nature of the double operation. It would not normally be carried out by any one depending on telephone communication. In the event, it is clear that the design choice was wrong. It emphasizes the extreme care necessary in designing interlocks."

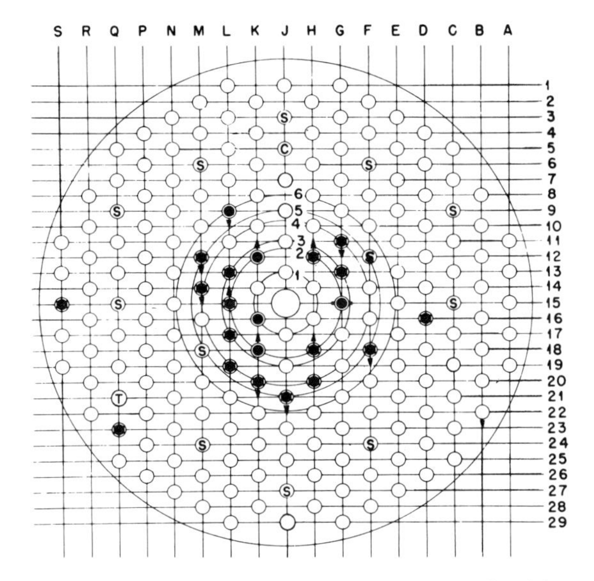

FIG. 3-1 Damaged elements in the NRX core. The plan of the lattice shows rods with abnormal cooling arrangements (indicated by small arrows, ↑ = temporary cooling upflow, ↓ = temporary cooling downflow); the rod at G-15 is air-cooled. The open circles indicate those rods not ruptured. The black circles with a white annulus (L-9, K-12, G-15, K-16, K-18) are the fresh rods whose outer sheaths were ruptured; similar circles with segmented annuli (e.g., G-11, H-12, M-12) are irradiated rods whose outer sheaths were ruptured. An encircled S, C, and T indicates the position of a shutoff rod, control rod, and thorium rod, respectively.

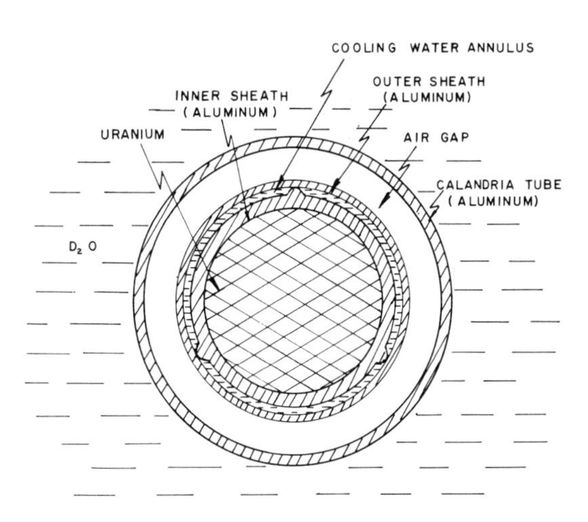

FIG. 3-2 NRX fuel tube cross section. Outer diameter of calandria tube is 2-3/8 in. (6.03 cm).

dicating those elements which ruptured as a result of the accident. At the time of the accident, only one rod (G-15) was air-cooled as mentioned earlier. The only rod which failed in Circle 1 was a fresh rod in K-16. (The parts of a typical fuel-tube system are shown in Fig. 3-2 [16a].) This rod had a small hole in the outer sheath and the calandria tube had a much larger hole. It is surmised that the heat from the very severe rupture of L-15 may have helped to cause the breakdown of K-16. In several instances failure of fuel in one pressure tube caused damage in an adjacent pressure tube. The six rods in Circle 2 were a part of the test being conducted and all were being fed with an upward stream of cooling water with a lower head than usual. As a result, these elements evidently voided first, causing additional local flux increases (in a sense, a local reactivity effect) and causing melting in these elements to be most severe. Other elements with temporary cooling melted shortly thereafter. Damage to some of the normally cooled rods in Circles 3 and 4 appears to indicate that the light water in a number of rods inside Circle 5 was boiling and being expelled by steam. Because of the fuel resistance to the escape of water, the steam pressure built up within the cooling section and at least two rods, M-14 and K-20, broke apart and the upper portions of their shielded sections leaped a foot or so into the air. The air-cooled rod, G-15, which is in Circle 2, the circle of maximum flux, did not damage the calandria tube so far as could be seen. The central parts of the aluminum sheaths melted and ran down the rod, congealing between the calandria tube and the rod. This aluminum formed a barrier on top of which molten uranium formed an ingot contained by the calandria tube. At this point, heat transfer between the calandria tube and heavy water apparently preserved the tube itself. Since there was no damage to the calandria tube in this case, it is evident that steam considerably influenced the course of the accident. Steam under pressure in the rods caused the initial rupturing of the outer sheaths. There appeared to be considerable chemical reaction, and not just melting of the materials present. It cannot be decided whether or not the exothermic aluminum water reaction played a significant role. From the analysis, it seems unlikely.

Even though the metal-water reaction probably was not significant, there was a chemical reaction of some sort as evidenced by the behavior of the helium gasholder. This helium gasholder normally maintained a pressure of 12 in. water (22.5 mm Hg) above the heavy water and had a capacity of 550 ft^3 (15500 liter). Normally, when heavy water is dumped, helium flows from the storage tanks to the calandria and the gasholder merely rides on the system. Loss of helium pressure was noticed during the dumping. A possible interpretation was that the blockage in the return helium pipe caused a partial vacuum as the calandria emptied, and dumping was stopped momentarily to prevent collapse of the calandria. Dumping was resumed when the gasholder was seen to be emptying. Just as the gasholder was almost empty, it suddenly jumped to its full height of 48 in. (122 cm), as shown by the position recorder. It stayed at that position for some time...probably jammed there...and then fell in a series of steps backwards. The gasholder is connected to the rest of the system by 2-in. (5.08 cm) pipes, with an equivalent length of about 75 ft (~23 m). The weight of the dome is 4 tons (3630 kg) and its area is 133 ft^2 (12.4 m^2). In order that the dome be lifted in a normal manner, it is necessary that a supply of over 500 ft^3 (14200 liter) of gas be delivered within 30 sec. Hurst [16b] suggests that the release of uranium from its sheath was accompanied by the evolution of hydrogen. Much of this hydrogen escaped into the calandria and may have been augmented by further reaction of uranium in the calandria. The helium, escaping from the calandria through several holes, may have picked up hydrogen and burst into flame where the gas came in contact with the air. The escape of the helium continued until the gasholder dome reached the lower limit of its travel. This prevented further emission of helium and air then entered the calandria and the hydrogen-oxygen explosion took place.

The core and calandria were damaged beyond repair. Approximately 10^4 curies of long-lived fission products were carried to the basement in about one million gallons of cooling water. The calandria and the contents still remaining in it after initial salvage operations were eventually bagged and dragged away for burial. The auxiliary equipment was decontaminated and the reactor was put back into operation in about 14 months with a new and improved calandria and core [18]. During the cleanup a large number of people participated in order to hold down the dose to any one individual. In general, health physics control was able to adhere to 3.9 r total dose per man. The highest total dose received by anyone was 17 r.

Comments, Conclusions, Recommendations

(1) System design and interlocks should make it impossible for an unauthorized and unassisted operator, either by mistake or on purpose, to withdraw or to influence the performance of control rods.

(2) The instrumentation should record the position of the control and the safety rods at all times, including the position during scrams...not just the limits of travel of the control rods. Since many rods fall free under gravity, this is often a difficult or impossible design requirement, but it should be a design goal.

(3) It should always be possible to maintain two-way communications between groups carrying out related key operations. In essence this means that there must be a satisfactory public address-call box system over which one can talk and listen while making use of his hands. A telephone does not answer these requirements.

(4) Vital controls should be arranged on the control console so as to be easily accessible in an emergency.

(5) The design of reactors with positive reactivity coefficients which can be rapidly brought into play during transients, or by other reasonably ordinary perturbations of the system, should be avoided or is at least open to serious question. The post-accident analysis seems to indicate that the transient would probably have been terminated with little or no damage if it had not been for the positive voiding effect. It is certainly possible to operate such reactors safely as long as their behavior is close to normal. However, if a difficulty or perturbation of normal operations should develop it is very likely to be aggravated by positive reactivity coefficients and a minor incident can easily turn into a major accident.*

(6) From a safety point of view, it is good design and operating practice to ensure that a relatively small perturbation does not create a major effect. In this experiment the cooling was reduced for all of the six rods in circle 2, symmetrically placed, in order to give an adequate signal and, to permit interpretation, they all had to be in identical situations. It is likely that they all voided almost simultaneously, greatly increasing the severity of the transient that followed. What would have been a ramp if they had voided over a longer period became almost a step addition of reactivity. In retrospect, it might have been better, from the safety point of view, to carry out the experiment with fewer rods involved at each stage.

3.4 BORAX-I Destructive Experiment [19, 20, 21, 22]

BORAX-I [19, 20] was a swimming pool type reactor utilizing 0.020 in. (0.508 mm) thick aluminum clad U-Al plates in an MTR type fuel element. It was the first reactor designed for studies of transient behavior and for that reason was located at the National Reactor Testing Station in Idaho.

The reactor consisted of four quadrants separated by gaps to accommodate five cadmium control blades. Each blade was connected to the control mechanism through spring-loaded magnetic couplings. The central blade was cruciform in cross section and so designed that the cocked or raised position was with the cadmium in the core. On appropriate signal the transient was initiated by ejecting the central blade downward out of the core from a predetermined partially down (slightly out of the core) starting position. The other four blades, which filled the gaps between the quadrants, were raised above the core before the experiment started and were injected into the core to terminate the experiment. Each blade traversed the height of the core in about 0.2 sec.

The reactor core was contained in a tank 4 ft (1.22 m) in diameter and 13 ft (3.96 m) high and filled with water. In turn the reactor tank was contained in a larger shield tank sunk part way into the ground with earth piled around it for additional shielding. Adjacent to the shield tank was a concrete-lined pit housing the equipment for filling and emptying the tank and for heating the water.

In a highly successful set of experiments carried out in the summer of 1954 considerable information had been gained in regard to the nature of transients in this type of reactor (see chapters on Mathematical Models of Fast Transients and Water Reactor Kinetics). These tests led to the development of the USAEC's SPERT program. During the course of these experiments in which successively shorter periods were studied, the fuel elements began to show signs of hard use. Some bulged or

*In a private communication (quoted in a letter dated April 13, 1964 from W.B. Lewis to the author) D. G. Hurst states, "The author takes too strong a stand on positive reactivity coefficients. The substandard cooling and delayed dump were much more important. Perhaps the opposite conclusion could be drawn, i.e., even though channels were being emptied of light water by boiling there was no difficulty in shutting down by dumping ..."

were otherwise deformed and had to be forced back into shape. In view of these indications that the end of the reactor core "lifetime" was imminent and realizing that the reactor would become unusable in autumn due to cold weather, it was decided to carry out the most violent transient possible by ejecting the central transient blade entirely out of the core (4% k_{eff}) with the water initially at room temperature.

The record shows that the ejection of the rod was only about 80% complete when the reactor reached its peak power and shutdown mechanisms became effective. The minimum period measured was 0.0026 sec. The total energy release as determined by cobalt foils in the core was 135 Mw-sec. Violent mechanical effects spoiled most of the records after the peak power was reached. Analysis of the damage indicated pressures at least as high as 6000 psig (400 atm) and probably higher than 10000 psig (670 atm).

The motion pictures taken during the excursion and a study of the debris provided information of several types. The reactor tank was burst by the blast and most of the shield tank contents were ejected into the air. The control rod drive mechanism, mounted on a heavy plate at the reactor top and weighing about a ton (~1000 kg) was thrown 30 ft (9.1 m) into the air and fell to one side. Recognizable fuel fragments were thrown as far as 200 ft (61 m) away from the reactor. Essentially all of the fuel could be accounted for within 350 ft (107 m) of the reactor. A wind of 8 mph (3.6 m/sec) at ground level and 20 mph (8.9 m/sec) 250 ft (76 m) above ground was blowing at the time of the experiment. Fifteen minutes after the experiment the total beta-gamma activity 3 ft (0.91 m) above ground 0.8 mile (1.3 km) downwind was 5 mr/hr.

The shutdown mechanisms appeared to be similar to those of earlier excursions in the series, namely, core mechanical expansion, microbubble formation, boiling, and fuel melting and vaporization. High speed motion pictures showed a light flash lasting 0.003 sec as peak power was reached, but the flash was gone before any ejected material left the top of the tank.

The somewhat unexpected destructiveness of the test resulted in losing some information which might have been gained.

Comments, Conclusions, Recommendations

(1) Instrumentation and other means for obtaining information from transient tests should be planned on the basis of overestimates of the possible destructiveness of the tests if maximum information is to be gained. In this sense planning for a test is like making a conservative safeguards estimate.

(2) The fact that the ejection of the rod was only about 80% complete at the time of peak power can well be attributed to the difficulties involved in adding (to this reactor) large amounts of reactivity rapidly enough. It may also be that there is a limiting amount of energy that can be released by a transient in this type of reactor before the transient will terminate itself. (See comments, Sec. 6.3.)

3.5 Hanford Reactor Incidents [23]

A group of large graphite-moderated, water-cooled plutonium production reactors are located at the Hanford Site in Washington. Although most of the information concerning them is classified, a review of three incidents related to reactor safety has been declassified and is given in reference [23]. There are no reports of subsequent incidents involving these reactors, although a criticality incident in the reprocessing plant occurred on April 7, 1962. (See Sec. 2.4.)

October 3, 1954

Previous to October 3, 1954, a water leak had wet a considerable quantity of graphite in the lower part of one of the reactors. On startup, the reactivity effects could only be roughly estimated since both the quantity and the distribution of the water were unknown. Other startups had been made under similar conditions without trouble and so no difficulty was anticipated.

Normally, a reactor physicist prepares a reactor criticality prediction curve in advance, and that was done in this instance. At the end of one hour the reactor was on a two-minute period and proceeding normally. Rods were then withdrawn the calculated amount to place the reactor on a 30 sec period. However, a mis-estimate had been made and the reactor was placed on a 12 sec period instead. The console operator noted the rapid power rise and started inserting control rods. At about the same time the coolant-outlet-tube temperatures on the upper near side of the reactor were reported to be approaching their limit. The cooling water pressures of a number of tubes were dropping. The Chief Operator, being appraised of the situation, ordered more rods inserted to reduce local heating. At about this time low water pressure in one or more of the tubes caused a scram. This sort of scram had occurred before, and so after about 45 minutes the startup was again resumed and the normal operating level was attained without incident. However, a number of fuel element failures occurred in the same region the next day and they continued from then on until the fuel was discharged.

Subsequent investigation showed that two of the limited number of low-level startup chambers were positioned close together and underneath the reactor. This was near the region where the water had leaked. Also, the movement of leakage water away from a warmer zone can create a positive reactivity feedback mechanism. Hence, the signal to both chambers was attenuated by the shielding effect of the water. Also that part of the pile was not as reactive as the upper dry section. Thus, the fundamental cause of this accident must be laid to faulty location (or response) of the low-level neutron detectors.

A secondary cause was the fact that the Chief Operator, while experienced, was new at this particular reactor and perhaps did not have "the feel" of the reactor yet. (See Sec. 3.7—Conclusions.) In addition, he was required to supervise the shifting of scales of certain power level instruments at about the time of the mishap.

January 4, 1955

During the pre-startup program of the new "KW" reactor, it was necessary to determine its reactivity characteristics under a variety of conditions. Some of the tests required that several hundred process coolant tubes be blocked off with solid neoprene disks. At the end of the test it was necessary to see that all disks were removed, and the procedures called for a count of the disks as removed. When it was believed all were out, a test was carried out on the gauge system that monitors water pressure on each tube. This test showed that seven disks had been left in either a tube inlet, or outlet, or both. An eighth disk was missed somehow, or else the gauge was overlooked. There was still another test which might have revealed its location, namely, the process tube pressure. On this occasion the pressure recorded for the tube definitely indicated a blockage, but the Shift Supervisor failed to catch it, and an instrument mechanic adjusted the gauge to give a midscale (and false) reading.

The initial startup of the new reactor on January 4, 1955 proceeded slowly and in stepwise fashion. On January 5, after a stepwise increase, the power decreased sharply because of a reactivity loss. The control rods were withdrawn to regain and hold the desired power level. About 12 minutes later an alarm indicated water in the reactor recirculating-gas duct system. The supervisor, believing this to be an indication of a process tube leak, started a high speed traverse (search) of the outlet water temperatures to locate the leaking tube. Before the traverse was complete, the reactor was automatically scrammed by coolant pressure in the tube row with the blocked tube. When inspected, no tubes showed abnormal readings, but since there were indications of a ruptured fuel element, no restart was made. Monitoring of the reactor rear face water lines confirmed that there had been a rupture.

It became apparent after two days of efforts at removal that this was not an ordinary ruptured slug accident. All methods were tried to no avail until finally on January 14 it was decided to cut a hole in the rear shield wall of the reactor and to remove the entire graphite channel. It took almost six days to drill through the shielding, and final removal and cleanup were not completed until January 28. The cost of one neoprene disk...$550,000.

January 6, 1956

On January 5, 1956 the reactor startup began after a shutdown to test for process-tube water leaks. Although the flux chambers gave some indications of residual water in the lower part of the reactor, there did not appear to be any loss in reactivity due to this. Startup was normal and then the reactor was shut down to remove some poison rods. When startup was then attempted it was discovered that there was not enough reactivity to accomplish it. It was believed that water had condensed out of the reactor gas onto the cool process tubes and the nearby graphite and thus lowered reactivity. After a shutdown of seven hours there was enough xenon decay to carry out a normal startup. A cooling water pressure fluctuation then caused the reactor to scram at 3 a.m. on January 6.

The Operations Supervisor telephoned the assigned physicist, waking him up, to get a prediction of startup reactivity from scram on a cold secondary startup. The cause of the scram was ascertained and corrected. Hurrying to get started (before graphite cooling and xenon buildup lowered reactivity so much that a several hour wait would be necessary for xenon to decay again), a criticality prediction curve was prepared. The safety-rod withdrawal was started about 20 min after the call to the physicist. While the rods were being withdrawn the physicist called back, revising his first prediction to take into account more severe reactivity loss rates. He also indicated that the reactor might go critical in 50 to 60 min.

The Chief Operator continued rod withdrawal on the basis of the new prediction. At 3:35 a.m. it was observed that the proportional neutron counter gave an unexpected indication of criticality and rod withdrawal was stopped. Within 30 sec the counter reached its maximum counting rate and jammed. Suspecting a faulty counting system, he switched to a less sensitive chamber and it too jammed. He then realized the situation and started inserting control rods as rapidly as possible without scramming the reactor. The power rise was halted within 60 sec after rod insertion was started and the maximum power reached was only 5% of the nominal operating level. Normal startup continued and no damage resulted since the Chief Operator had sensed the situation and taken action in time.

In addition to the physicist's error, the scram flux monitor chamber had not been set back to the startup region as it should have been and thus automatic power level scram would not have occurred until full power was reached (about 2 minutes after the abnormal chamber response was observed). Since the period was fairly short, no doubt some damage would have occurred before the automatic scram became effective.

Comments, Conclusions, Recommendations

(1) Detailed consideration should be given to the positioning and sensitivity of startup and power level neutron detectors to ensure that the operator has a full knowledge of core behavior at all times. This is particularly important in large reactors where it is possible that one part of the core may be behaving differently from another. The very large graphite reactors have perhaps the greatest problem in this regard. That is one reason for the criticality prediction done normally at Hanford. This lack of exact knowledge of the heating in all parts of the core played a part in the Windscale accident also. (See Sec. 3.7.) In controlling their large gas-cooled reactors the British tend now to use thermocouples extensively rather than neutron counters to keep track of conditions in various parts of the pile.

(2) Supervisory personnel should not be assigned duties which may distract them from their primary functions during key operations.

(3) Reactors should give an alarm and probably

also scram on fast negative periods as well as on positive periods. An unexplained drop in reactivity is often an excellent indication of a large core perturbation and a forerunner of some change in state and perhaps a meltdown.

(4) Procedures for special tests, and in fact for all operations, should be carefully thought out and written down so as to insure that the reactor is returned to safe operating condition under all circumstances. Check lists and signatures confirming responsibility should be emphasized. Components to be inserted into a region for test and later removed should probably be handled like the surgeon's sponge count.

(5) Difficult operations or procedures requiring alertness or thought should not be undertaken at times when those participating are not at their best.

In the Hanford accident, the physicist who made the most important error was called from his bed at 3:00 a.m. It may not be just chance that two Hanford accidents, the SL-1 accident, and probably others occurred just after long holidays. Perhaps a corollary might be—never start up a reactor in early January.

(6) Reliance on procedural safety and proper action should take second place to well-thought-out interlocks wherever possible. In this case it would be wise to interlock the ion chamber with the rod withdrawal switch so that it would be necessary to have it properly set before rods could be withdrawn, if this were practical. However, it is necessary to move the control rods in either direction with the ion chamber in its least sensitive position. Thus a simple interlock is not practical; a sequential one activated on low chamber level after shutdown was deemed too complex and less reliable than operator procedural control.

(7) When an instrument records a non-safe condition, that instrument should be believed and action taken. Do not assume that the instrument is wrong—you can check later. The Chief Operator, whose behavior otherwise was exemplary, should have taken action to insert rods when the first chamber showed criticality, not the second.

3.6 The EBR-I Meltdown [24-36]

The EBR-I Reactor is a NaK-cooled, unmoderated reactor, the Mark II core of which was fueled with an alloy of 94% enriched U^{235} and 2 wt.% zirconium. It was designed to operate at a nominal thermal power of 1400 kw and to deliver up to 200 kw electrical power. Several detailed descriptions of the reactor exist [24, 25a, 26, 27]. Only those parts pertinent to the accident or its analysis will be described here.

The enriched Mark II core consisted of a hexagonal array of up to 217 fuel rods arranged as shown in Fig. 3-3 [25a]. Each rod was 0.384 in. (0.975 cm) in diameter and the enriched center section of each rod consisted of two slugs each 4-1/4 in. (10.80 cm) long. The rods were NaK-bonded within stainless steel cladding tubes 0.448 in. (1.14 cm) outer diameter and 0.020 in. (0.051 cm) thick. A natural uranium section in each rod extended 4-1/4 in. (10.80 cm) below the enriched section and 8 in (20.32 cm) above it. Thus, the core was 8.5 in. (21.59 cm) high and about 7-1/2 in. (19.05 cm) across the faces of the hexagon. The core was separated from the inner blanket by a plenum wall to direct coolant flow. The inner blanket consisted of an array of natural uranium rods 0.937 in. (2.380 cm) in diameter and 20-1/4 in. (51.43 cm) long.

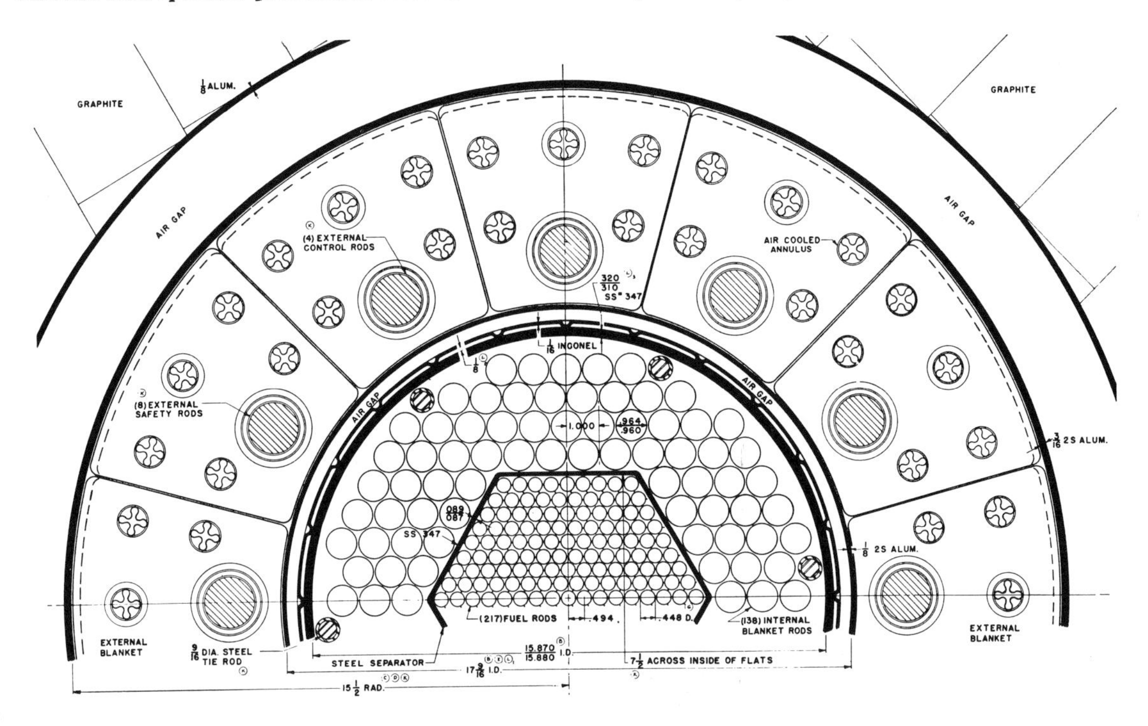

FIG. 3-3. Cross section of the EBR-I core and blanket. Dimensions are in inches.

The rods of the core and inner blanket were supported at the top by a series of close-tolerance plates, the most important of which was the 4 in. (10.16 cm) thick lower shield plate at the bottom of the stack as shown in Fig. 3-4 [27]. Each fuel rod passed through each plate individually. The holes in the lower shield plate were 0.460 in. (1.168 cm) diameter giving a radial clearance of 0.006 in. (0.15 mm). The fuel rods were positioned at the bottom by a grid plate with triangular holes which engaged positioning pins at the bottom of the rods with a nominal clearance of 0.005 in. (0.13 mm). The fuel-bearing section was the lower 21 inches (53.34 cm) of the tubes with the section above, "the handle", consisting of a solid stainless steel rod appropriately fluted to permit flow through the lower shield plate. During shutdown, the NaK in the core and inner blanket could be isolated from the rest of the primary system and a natural convection loop set up to a small by-pass heat exchanger. Flow in the system through the core in normal operation was adjustable from zero to full flow.

An air-cooled outer blanket of natural uranium bricks surrounded the sides of the core and inner blanket. There was an 81 kg (178.6 lb) natural uranium safety plug supported by air pressure below the inner tank (shown in Fig. 3-5 [25a]) worth 0.07% Δk. The outer blanket was so arranged that it also could be dropped downward by means of a hydraulic table either slowly (in normal operation) or rapidly (in scram) providing an additional shutdown worth of 8.9% Δk, by far the most important means of shutdown. In addition there were twelve 2 in. (5.08 cm) O.D. natural uranium shim-safety rods in the outer blanket as shown in Fig. 3-3. The total worth of the twelve shim-safety rods was 0.3% Δk. Four of these outer blanket rods (0.1% Δk) were used for normal operating control and could be accurately positioned, while eight (0.2% Δk) were used for quick-acting safety rods.

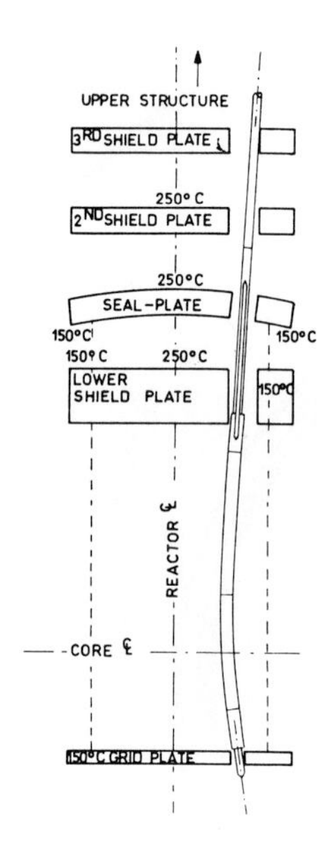

FIG. 3-4 Fuel rod positioning in the EBR-I core.

The EBR-I Reactor was designed principally to study the physics, particularly breeding performance, and behavior of fast reactors of its general type. It was also the first reactor to produce electrical power. A fruitful experimental program had been carried on for several years prior to the accident. During this program it was found [25a] that the power coefficient of reactivity of the reactor had a relatively large positive component which undoubtedly existed under all conditions but which could only be demonstrated when the ratio of power to cooling rate was high and when the rate of change of power was high. A slow negative coefficient of larger magnitude was also observed. The transfer function for the reactor was obtained experimentally by oscillator tests at high power, intermediate power, and very low power and with varying ratios of power level to coolant flow rate.

There was still some doubt as to the origin of these coefficients and in November 1955 it was decided to repeat an experiment done earlier "in which the reactor was started from critical at a very low power (a few watts) on a period of about one minute, and the power allowed to rise [28]. In the previous experiment the ratio of power to the time derivative of the power was found to decrease continuously until it had reached a value of about six seconds, whereupon the experiment was terminated. Because of time lags introduced by the use of the slow power indicator (optical galvanometer) and the thermocouple recorder in this first experiment, the proper relation between reactivity, temperature and power was not determined."

For the new version of the experiment, a two-channel Brush recorder was chosen in order to secure more rapid instrument response. The power level signal was to be taken from a compensated BF_3 ion chamber. A thermocouple attached to one of the central fuel slugs was to record on the second Brush recorder channel. A second fuel slug temperature was recorded on a Leeds and Northrup Speedomax Recorder which had a full scale sweep time of two seconds. A third fuel thermocouple, the one normally used to record temperature during operation, was attached to a Brown 24-hour circular chart recorder. The pile-period meter was disconnected from the scram circuit "since it did not have the flexibility that would permit it to be used for excursion experiments."[29] The power level scram circuits were at their normal settings for 1400 kw operation [25b, 28].* The shutdown natural circulation loop was in operation at a flow of about 1.5 gpm (5.7 liter/min) and a total connected volume of 50 gal (189 liter) of NaK was in use. During operations the current from a BF_3 chamber was

*The report of Brittan [28] states that "the power level scram circuits were used with the trip setting well above the normal operating level of 1150 kw." This value for the normal operating power level appears to be in error since other sources [25a, 25b] set the value at 1400 kw.

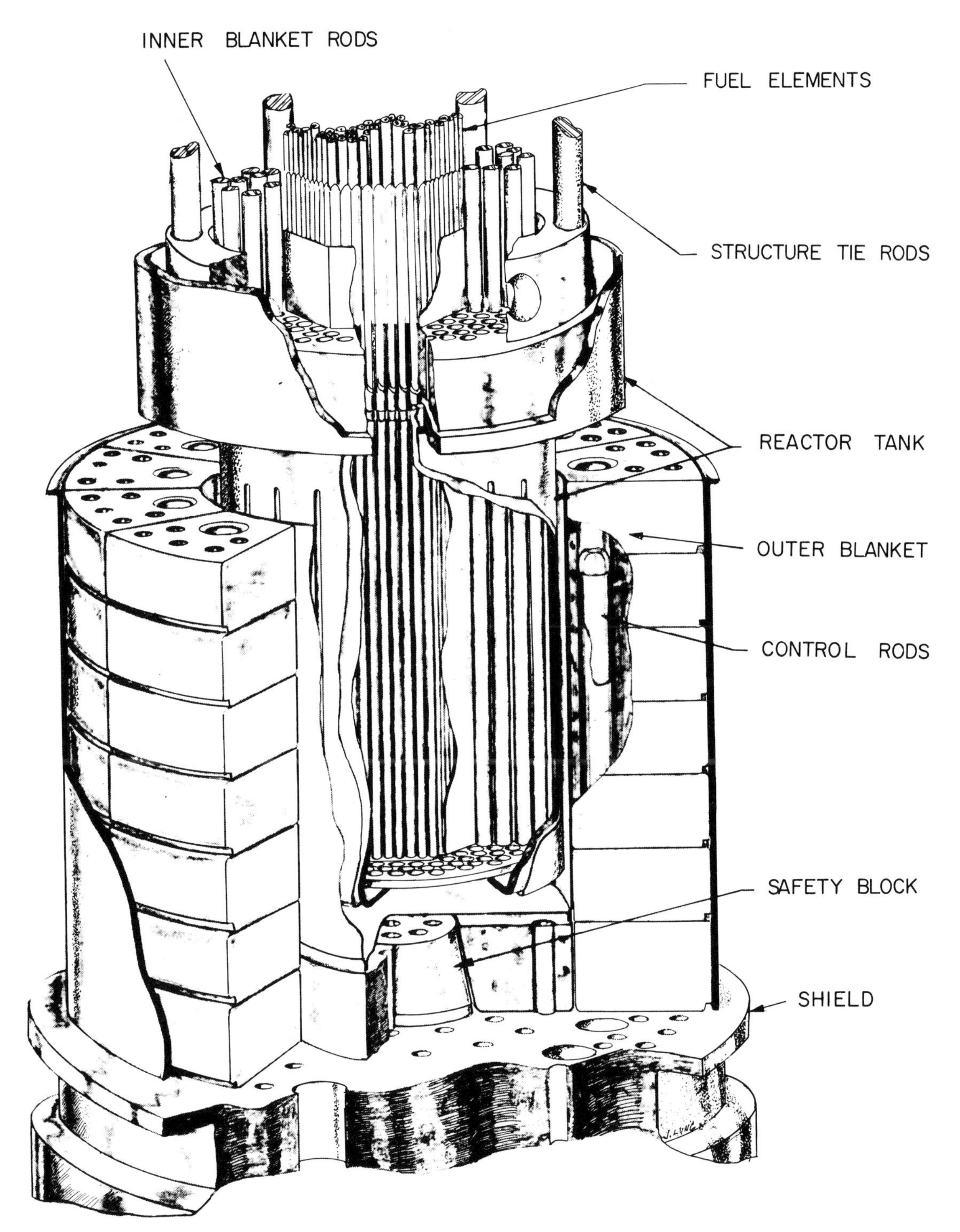

FIG. 3-5 Cutaway drawing of the EBR-I.

used with a vibrating reed type amplifier to drive a strip chart recorder thus giving a valid signal from source level up to about 30 kw. Above that power two optical galvanometers with different sensitivities were used. The galvanometers overlapped the range of the vibrating reed unit, but were, of course, less accurate in the low ranges. A BF_3 counter was also used in the source and very low power range. The core inlet and outlet temperatures and the loop flow rate were recorded on 24-hour circular chart recorders. The temperature in the core at the start of the experiment was 65°C (149°F).

The intent was to place the reactor on a long positive period and to terminate the experiment on a short positive period. The run in question was planned to be the first of a series of several runs [25b] which were to approach gradually a condition permitting the temperature to rise to 500 or 600°C (932 or 1112°F). In this first run it was only intended to get a measurable change in the fuel temperature (~100°C or 180°F) and to check out the equipment in preparation for the more severe transients.

"Because it was desired to obtain the temperature coefficient of the fuel only, it was necessary to shut off the flow of the liquid-metal coolant. Therefore, it can be said that the experiments were done not on an operating reactor but actually on a critical assembly.... Increasing the temperature of the uranium rapidly involved very considerable chance of distorting the fuel element jackets, especially since uranium metal and stainless steel form a eutectic at about 725°C (1337°F). This is very little above the temperature to which it was desired to pulse the fuel elements. ...The temperature of the uranium was to be permitted to rise to approximately 500°C (932°F). This meant that the shutdown of the reactor would have to be rapid and would have to be initiated at the proper instant. Because of the rapid temperature rise required by the experiment, an error of one second in initiating the shutdown could be expected to give trouble." [29]

Following the normal procedure the reactor was made critical at an initial power level of about 11 watts. Then the power level trace on the Brush recorder was calibrated against the normal instrumentation. Using the control rods, the reactivity was increased until the reactor was on about a 50 sec period. The rods were stopped just short of 200 sec after they were first moved to add reactivity (time mark zero). The power level was then about 50 watts. The power level increased with about this constant period until 320 sec after time mark zero. Then the period (as defined by the ratio of the instantaneous power level to the time derivative of the power level) began to decrease at an increasing rate. At 320 sec from mark zero the power level was about 500 watts. At 497 sec after mark zero, the power level was approaching half of the normal full power as observed on the galvanometer [28].* The fuel element temperature, as indicated on the Leeds and Northrup recorder, was beginning to increase.

"In previous experiments of this type, it had been possible to interrupt the excursion and return the reactor to low power by making use of the motor-driven control rods. These subtract reactivity slowly. In this excursion, the technician at the control panel was expected to use the fast-acting shutoff rods upon receipt of a spoken instruction from the scientist in charge, who was observing closely special neutron and temperature recorders." (The Brush Recorder channels).... "Upon receiving the instruction to shut down the reactor, the technician repeated the use of the slower control rods. The staff scientist, as soon as he realized the situation, reached over and pressed the rapid shutoff button and, simultaneously, the automatic power-level trips responded to activate the shutoff rods."[29]

"....The safety rods and plug were scrammed at about 499 sec from mark zero. This caused the optical galvanometer to hesitate only momentarily. When it became evident that the reactor was not shut down, but that the power began increasing above normal operating power, the reactor scram button was pushed and at the same time an instrument scram was indicated aurally and visually. On the reactor scram signal the natural uranium outer blanket cup dropped from around the core and internal blanket."[28]**

"Immediate examination of the Brush trace indicated that although it had gone off scale for about 2 sec the power level was rapidly decreasing and the reactor was subcritical [28]." The thermocouple attached to the Brush Recorder had apparently become shorted out and was recording NaK temperatures well away from the fuel region. The time and nature of the failure can only be conjectured. It reported a maximum reading of about 200°C (392°F). The Leeds and Northrup recorder showed that some time before or during the experiment this thermocouple stopped reading fuel slug temperature and, instead, read a position close to a fuel slug since its maximum reading was about 350°C (660°F). The traces of these two records were typical of those recorded when a steep front heat wave passes through a medium and past a point where the temperature is being recorded. The dual Brush traces indicated that the heat pulse hit the thermocouple about one second after the power maximum was recorded,

*Lichtenberger [25b] doubts that this reading was taken and notes that even if it were taken, it would be meaningless as the time constant of the galvanometer was too long. If the galvanometer read half normal power level, the actual power level would have been much higher. Lichtenberger states "..the power level at this time must have been determined from the Brush recorder."

**The description of these few seconds are quoted from the only two known available information sources in the literature. In order to clarify what appears to be a discrepancy and to ensure that the record is set straight, H. V. Lichtenberger was asked to comment and stated [25b], "Both quoted versions of the means of terminating the excursion are generally correct. The safety plug and the 8 safety rods were scrammed and caused the period to become negative and the power to fall temporarily. However, it was necessary to drop the outer blanket to effect the shutdown." This author has found that personnel working at the reactor called the scram of the safety rods and plug the "slow shutdown" and the full scram of the safety rods and plug plus the blanket cup the "fast shutdown." Viewed in the light of these definitions both the Zinn and the Brittan account agree and complement each other.

that is about when the power trace came back on scale. It is the opinion of Brittan [28] that if the man operating the Brush recorder had changed the attenuator at the proper time as he had done previously in the same experiment, a full record of the power transient would have been obtained. The normal operating thermocouple record was the only one of the three that had operated as planned. However, in this case the slow response of the instrumentation again precluded an accurate recording of the maximum temperature reached and the maximum recorded was about 650°C (1200°F).

The record of the power trace showed that the power surge might have raised the temperature of the fuel high enough to cause damage, although the thermocouple data described above were inconclusive. Visual examination of the reactor vessel and outer blanket revealed no damage. Therefore, in an effort to cool off the elements as soon as possible, and to check the flow rate, the pumps were started and the primary loop put into operation. It was known, of course, that natural convective circulation would provide adequate cooling unless there was extensive damage. The high pressure drop through the core, found when attempting to establish normal coolant flow, immediately indicated extensive core damage. Brittan [28] reports a NaK radioactivity level alarm but Lichtenberger reports there was no alarm [25b]. He points out that the very slight contamination of the NaK system "was then and still is somewhat surprising."

From the off-scale power trace and the last apparent period a quick estimate showed that the power increase could have continued for perhaps one second and reached a value greater than 15 Mw. Zinn [29] states that it was after a period of fifteen minutes when radioactivity was first observed in the cooling system some distance from the reactor and in the ventilation exhaust ducts of the building. From the location of the detector in the cooling system it was inferred that gaseous radioactive fission products had diffused through the gas blanket connections to the detector location. How much of the time, if any, during this period the pumps were on is not clear from the reports. Since the special thermocouples for this experiment were not installed in a gas-tight manner into the core, it is not surprising that some of the radioactivity diffused outward from the core vessel into the room.

Immediate steps were taken to make a complete health physics survey of the reactor area. The building was evacuated until an adequate health physics survey could be made. This survey indicated a low-level contamination by fission product gases emanating from the hole in the shielding through which the thermocouple wires passed. The building was reoccupied without further incident.

Brittan [28] has described the analytical duplication of the experimental power trace given by the Brush recorder. This was done by the use of one-group, space-independent kinetics equations set up in ANL digital computer codes RE 29 and 31. By small reiterative corrections, a power-time relationship was obtained which matched the experimental results within 1% between time zero and 500 sec, when the power trace ran off the chart. The rapid reduction of reactivity due to the shutdown action which occurred between the 499th and 500th second agrees with the rod and plug worth. The data is shown in Table 3-3 [28] and Figs 3-6 [28] and 3-7 [28].

According to this analysis the effect of reduction of reactivity due to the safety rods and plugs was first noticeable at about 499.35 sec. But about 0.3 sec later the increasing positive reactivity effect had overridden this attempt at shutdown and was again rising. The data has been extrapolated onward through the rest of the transient as indicated by the figures and table.

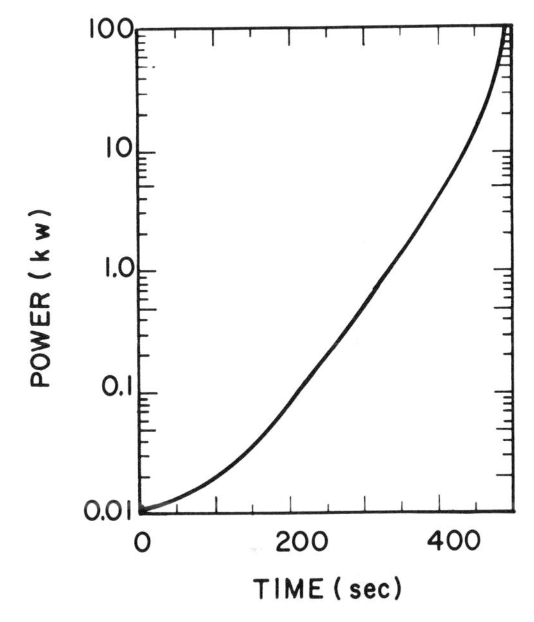

FIG. 3-6 EBR-I meltdown power level.

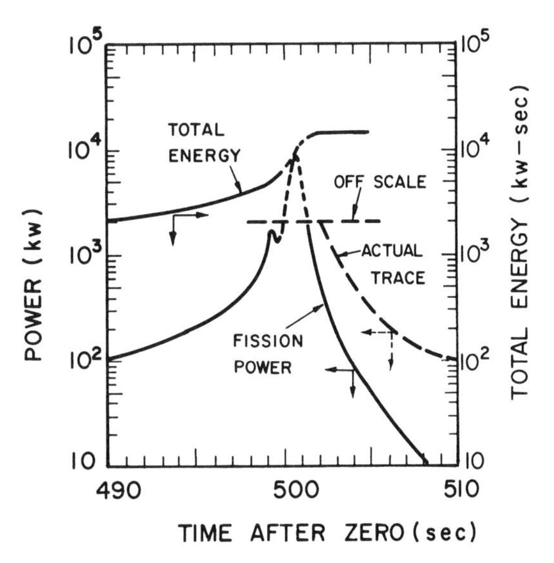

FIG. 3-7 EBR-I meltdown power peak.

Table 3–3

EBR–I Meltdown: Analytical Duplication of Experimental Power Trace

Time (sec)	P power (kw)	Total energy release (kw-sec)	P/(dP/dt) "period" (sec)	% excess reactivity
0	0.011	0	∞	0.0000
50	0.013	0.59	189	0.0275
100	0.019	1.37	105.7	0.0550
150	0.034	2.64	69.1	0.0825
200	0.082	5.30	49.7	0.1100
250	0.209	12.10	53.8	0.1100
300	0.529	29.3	53.9	0.1100
320	0.767	42.1	53.7	0.1100
340	1.123	60.8	51.6	0.1120
360	1.67	88.3	49.3	0.1147
380	2.54	129.7	46.1	0.1190
400	3.97	193.6	43.4	0.1240
420	6.42	295	39.9	0.1300
440	10.97	464	34.7	0.1400
460	20.6	767	28.4	0.1554
480	49.5	1,405	18.31	0.1931
485	68.7	1,696	13.25	0.2174
490	106.3	2,120	9.95	0.2509
495	215	2,870	5.12	0.3180
498	483	3,820	2.15	0.4300
499	1002	4,490	0.882	0.4984
499.2	1200	4,720	0.617	0.5296
499.35	1718	4,950	0.970	0.5575
499.37	1750	4,990	±∞	0.5527
499.4	1728	5,030	−0.832	0.5495
499.47	1470	5,150	−0.425	0.4925
499.5	1384	5,190	−0.727	0.4835
499.68	1250	5,420	±∞	0.4370
499.7	1252	5,450	8.03	0.4385
499.9	1507	5,710	0.730	0.4731
500	1750	5,890	0.613	0.4991
500.1	2090	6,080	0.514	0.5259
Off Scale Extrapolated data for cup starting down at 500.45 seconds				
500.2	2590	6,310	0.426	0.553
500.3	3360	6,600	0.346	0.581
500.4	4650	7,000	0.273	0.610
500.45	5660	7,250	0.273	0.625
500.5	6600	7,500	0.345	0.632
500.55	7500	8,000	0.475	0.634
500.6	8200	8,600	0.740	0.632
500.65	8800	9,200	1.670	0.626
500.695	9000	9,700	±∞	0.617
500.75	8800	10,200	−1.300	0.604
500.8	8300	10,500	−0.860	0.586
501	5600	11,700	−0.560	0.485
501.5	1700	13,300	−0.490	−0.125
502	500	13,900	−0.480	−1.15
503	80	14,000	−0.480	−4.28

i	β_i	λ_i
1	0.00025	14.3
2	0.00085	1.612
3	0.00241	0.456
4	0.00213	0.1535
5	0.00166	0.0315
6	0.00025	0.01246
	$\beta =$ 0.00755	

Prompt neutron lifetime: $l = 0.04\ \mu$sec

Since the exact time at which the outer blanket cup started down was not known, several different times were tried. These gave the values shown in Table 3-4 [28]. It will be noted that a difference of only 0.15 sec in the startdown time more than doubles the peak power reached. Brittan concluded that the scram signal was generated at some time between 500.05 and 500.20 sec. In addition, tests show about a 100 msec circuit time and about 200 ± 100 msec valve relief time for the hydraulic cup actuation. This gives the most probable starting time for the outer blanket cup travel as 500.45 sec.

Table 3–4

EBR–I Peak Power and Total Energy for Various Shutdown Times

Cup starting down-time (sec)	Peak power reached (kw)	Total energy released (kw-sec)
500.35	5,600	10,700
500.40	6,600	12,000
500.45	9,000	14,000
500.50	12,200	16,000

This same study also investigated the possibility that the expansion of core materials observed in post-irradiation examinations of the core might have caused sufficient loss of reactivity to have caused the shutdown prior to the time that the outer blanket cup was dropped. It was concluded that the core damage was less than that to be expected if shutdown was by the mechanism of core material expansion. Thus, the shutdown was almost certainly due to the dropping of the outer blanket cup. The experiment therefore did not give any information either way as to whether core expansion of the type observed was by itself a sufficient mechanism to shut off a fast reactor during a transient of this type.

The core assembly was removed as a unit utilizing design facilities provided in event that integral removal should prove necessary. It was shipped to ANL for examination and dissassembly [28, 29]. This examination showed that about 40 to 50% of the core had melted. Observations indicated that fuel in the central region had melted and formed a porous structure. It has been postulated that this structure could have resulted from the vaporization of entrained NaK. An experiment was performed [30] to see if the metallographic structure observed could be matched by mixing molten uranium with NaK. This was partially successful but some doubt still remains. Brittan [28] states that the "temperatures in the central region of the core certainly exceeded NaK boiling temperatures for several seconds after the burst." He estimated that the temperatures in the central 40–50% of the core reached 1130°C (2066°F). This probably resulted in the rapid penetration of the fuel jackets and the formation of a eutectic at the same time, for the steel cladding was found intimately mixed with the uranium. The NaK coolant boiling forced the central core material outward. Uranium was found both above and below the core center with penetrations several inches into the outer core regions between unmelted fuel elements. This probably blocked coolant channels to some extent.

Contact with coolant eventually chilled and froze the mass, blocking further outward movement. Movement of the core material outward tended to reduce the density of the center and increase that in the outer unmelted zone. The melted and refrozen material which apparently came mostly from the outer melted zone formed a pot of sorts which tended to hold and collect the materials falling from the upper part of the core as they became molten or moved within the core. The eutectic mixture was depleted in the center of the core as it melted and tended to flow selectively to the bottom. Partially melted rods from above apparently dropped into the eutectic mass below.* The material penetrated higher into the upper blanket than it did below. This has been attributed to the inertia of the upwards convective flow of NaK and to the asymmetry of the power distribution, the center (maximum) of which was slightly above the core center.

Another, and perhaps more important, mechanism that may help to explain the porous structure, the low density of the center of the core mass and the upwards movement has been advanced. The fuel in the reactor had accumulated some fission product burden (~0.1% burnup) before this experiment started. During the melting of the fuel, the gaseous and highly volatile fission products would be released and form gas bubbles. This foaming or frothing might satisfactorily explain the observed appearance of the core. Such a phenomenon has been observed to occur in TREAT work with irradiated fuel sections. This mechanism has particularly important safety implications since it can reduce the chance of reassembly of a critical mass; but at the same time it increases the chance of blockage of more coolant channels and more melting within a core, even without a nuclear transient.

The problems which were being investigated at the time of the meltdown were the fast positive and the slow negative coefficients of reactivity observed under certain conditions of coolant flow and rate of change in power. (A brief history of the theoretical work on these effects to 1958 is given by Thalgott [26, 31]. Further discussion appears in the chapter on Fast Reactor Kinetics, Sec. 2.5.5.1.) Work with the EBR-I through the spring of 1961 is outlined briefly in reference [32]. The accident still left the questions as to the origins of these effects in part unanswered. Previous to the accident it was believed that rod bowing was probably not the correct mechanism to explain the positive effect because its magnitude appeared too small. It was believed that coolant flow effects in the blanket must be occurring to amplify the bowing effects.

*The detailed description of what went on is, obviously, to a large extent conjecture. Lichtenberger believes that Brittan's description implies that the core material was molten longer than he believes to be the case. Lichtenberger does not understand how the events described in this sentence could occur.

In early 1956 J. R. Dietrich [33] suggested a mechanism of thermal lags dependent on local heat transfer. Following this, in June 1956 G. H. Kinchin [34] advanced a theory utilizing a fast positive coefficient attributed to bowing and a slow negative coefficient attributed to "structure". H. A. Bethe adopted a similar model and solved for a fit to the largest experimentally obtained resonance [35]. He found a prompt positive power coefficient of $+0.8 \times 10^{-3}$ $(\Delta k/k)$/Mw and a negative coefficient of -3.8×10^{-3} $(\Delta k/k)$/Mw with a transport delay time of 10 sec. (Section 2.5.5.1 of the chapter on Fast Reactor Kinetics discusses these analyses in more detail.)

Measurements carried out with EBR-I, Mark III, whose rods were tightly clamped to avoid bowing, have rather conclusively demonstrated that the fast positive coefficient observed in EBR-I, Mark II was due to rod bowing. (See Sec. 7.4 of the chapter on The Reactor Core and Sec. 2.5.5.1 of the chapter on Fast Reactor Kinetics for further discussion of this effect.) Elimination of the shield plate system of support and tight clamping of the core in Mark III eliminated both coefficients [26, 27]. By successively reducing the number of rods that were tightly clamped, it was possible to demonstrate the reappearance of the fast positive effect. (Fig. 3-4 shows a schematic diagram of two stages of the bowing in the Mark II loading).

However, the slow and larger negative effect was not further demonstrated in the Mark II loading. Smith [27] has presented a model which explains this coefficient by means of a "dishing" of the 4 in. plate which tightly supports the upper ends of the active fuel section. The "dishing" results from higher coolant temperatures on the bottom surface of the plate than on top. As a result the fuel rods were presumably bowed outward resulting in a slow negative reactivity effect. From a review of the discussion of Smith's paper at the IAEA Seminar on the Physics of Fast and Intermediate Reactors and from discussions at other conferences such as those recorded on pages 309-310 of reference [36] it is clear that not everyone accepts this explanation.

Comments, Conclusions, Recommendations

(1) Instrumentation for an experiment, or for any operation for that matter, should be adequate to cover any range of variables that could conceivably be reached during the experiment or operation. This is an extremely difficult but worthwhile goal to strive for, in view of large range of power covered and the short time constants required. In this case useful information was lost due to faulty thermocouple installation, off-scale instruments, and instruments with too long response times.

(2) If an experiment is one which must be carried out on a very fast time scale, adequate provisions should be made to insure that appropriate actions are taken automatically with essentially no chance of failure. This would imply completely automatic control and several months of careful planning and preliminary experiments. Unfortunately, it was during the first of these preliminary manually controlled experiments that a mistake was made and the wrong button pushed, allowing the reactor to enter a region where only fast automatic control could have prevented damage. The accident emphasizes the care which must be taken in such instances to avoid trouble. The same experiment had been started previously and three runs made safely. Even if an experiment has been run before, and hence rehearsed, it is still a fact that both people and equipment are subject to failure and such failures should be anticipated and adequate safety provisions made.

(3) Reference [29] states that "the excursion was one in which the power doubled every 0.27 sec. The temperature of the uranium was to be permitted to rise to 500°C. This meant that the shutdown of the reactor would have to be rapid and would have to be at the proper instant. Because of the rapid temperature rise required by the experiment an error of one second in initiating the shutdown could be expected to cause trouble." This statement is certainly true since, as is pointed out, an eutectic forms between uranium and stainless steel at 725°C (1337°F). Table 3-3 shows that the first shutdown action occurred at ~499.35 sec with the power at 1718 kw, already well above the nominal reactor power level of 1400 kw. In the next second, in spite of the insertion of the shutdown rods, the power level more than doubled again and some melting probably began to occur. In the next one-third second the power level more than doubled again. Certainly in such a transient the likelihood of being able to reach a fuel temperature close to 500°C (932°F) without reaching 725°C (1337°F), where melting starts, by any means other than highly specialized instrumentation and control is very small. Even assuming well-rehearsed signals and practice drills, it is still unlikely that melting could have been avoided by use of voice signals and visual observations unless the fast shutdown had occurred several seconds earlier, and then 500°C would probably not have been reached.

(4) Experiments of this sort, even with very careful planning and fully automatic instrumentation, should be carried out in a series of gradually increasing steps extrapolating only a small fraction of the way each time. This was planned in this instance, but the operator mistake changed the time scale.

(5) The most serious conceivable consequences of such experiments must be fully investigated before the work is started. If the consequences do not involve danger to personnel, either on-site or off-site, the value of the experiment must then be weighed against its economic costs and its usefulness to the technology. This was done no doubt in the present instance, but still the point is worth reemphasis.

(6) The experiment and the subsequent studies have shown that the reactivity effects observed are not due to anything inherent in the basic nature of fast reactors which would preclude being able to design a fast reactor that will be both stable and safe.

(7) The experiment and the resulting work have emphasized the importance of geometric, fluid-flow, and heat-transfer effects. Once these effects are clearly understood, they may be employed to

improve the stability and safety of such reactors.

(8) The possible effects of geometry, fluid flow, and heat transfer on reactivity should be thoroughly investigated in all reactor designs and the designs should be altered to insure that the response to these effects will result in a safe reactor in all cases. After construction and before use at power, experimental investigations by means of oscillator tests or some other reliable method should be carried out to check the response. No dangerous instabilities should exist anywhere within the conceivable range of variables.

(9) It is essential to plan any experiment very carefully, but especially those which are to be carried out on assemblies that are known to have unusual response characteristics. While clearly this experiment was not one which might endanger the public, and probably not even the operators themselves, nonetheless, failures of the type which resulted are very costly to clean up and often have far-reaching implications. Thus, care must be taken to be certain that the nature of the experiments is clear to the experimenters themselves, to their parent organizations, and to the public both before and after they are carried out. Experiments must be carried out to explore the limits of reactor safety, and from time to time these will (and should) involve destructive tests. Such tests can be carried out at appropriate sites without endangering the general public as was the case here. As much as possible such tests should be designed to answer a specific question or questions in a clearly and unambiguous way.

3.7 Accident at Windscale No. 1 Pile [37, 38, 39]

The Windscale No. 1 Pile is an air-cooled, graphite-moderated, plutonium production reactor using natural uranium fuel slugs. Graphite on bombardment with fast neutrons undergoes extensive radiation damage. This subject is discussed in the chapter on Materials and Metallurgy as well as in the chapter on Solid-Moderator Reactors. Suffice it to note here that as a result of neutron bombardment of graphite in the temperature region below approximately 200°C to 300°C (392°F to 572°F) the graphite "grows" in dimensions, its electrical and thermal conductivity diminish, and it tends to store energy. Since this energy storage was first suggested by Dr. Eugene Wigner, it is often called "Wigner energy" or Wigner's "disease." If the graphite is then heated slowly, the stored energy is released as additional heat. Thus, the addition of some heat tends to trigger the release of even more heat.

The situation is an unstable one and can cause or intensify a safety problem in graphite-moderated reactors operating below about 300°C (572°F). It is common practice in this type reactor to "anneal the graphite." In this process, the pile is gradually heated in one manner or another and the stored energy is gradually released and the radiation damage in a large measure is annealed out. The release of this extra heat tends to increase even further the overall temperature of the pile and this release mechanism is very often therefore called "Szilard's fever", in deference to Dr. Leo Szilard who first suggested this effect.

At the time of Windscale accident, the United Kingdom group used what might be termed a static technique for graphite annealing. They normally shut off the air cooling completely and raised the temperature of the pile by nuclear heating for a short time followed by complete nuclear shutdown of the pile. They then admitted small quantities of air only if needed for cooling. The Brookhaven National Laboratory utilizes what might be termed a damped dynamic method of handling this problem. They make use of fission heat from reactor operation at low power to raise the temperature of the pile and at the same time continue to operate the air-cooling system, but at a reduced flow rate. In this way, they can reduce the heating by shutting down the reactor and, at the same time, increase the cooling by opening further the cooling ducts. In either of these methods, it is necessary to insure that the graphite temperature does not rise too high since it will interact with the air coolant and oxidize the graphite to CO and CO_2 which is carried up the stack. In addition, it is necessary to remove the afterheat from the uranium fuel elements, which themselves may catch fire and burn in the air coolant if the temperature is sufficiently high.

In the account which follows, much of the text will be quoted from the Prime Minister's report to Parliament in November of 1957 [37]. Other discussions are available in references [38] and [39].

In September of 1952 a spontaneous release of Wigner energy occurred in the Windscale pile while the pile was shut down. This led to a rise of temperature of the graphite, but the rise was not dangerous and there were no harmful effects. As a result of this early incident, a procedure was instituted for controlling the release of Wigner energy. Eight releases had been carried out by the end of 1956 on this pile. On three of these occasions, it was found necessary to apply a second nuclear heating when it was observed that there was not sufficient heat to cause the release of Wigner energy.

"On the present occasion, the pile was shut down at 01:03 on 7th October, and the main blowers switched off in preparation for the Wigner release. <u>All necessary steps were taken to verify that the pile was completely shut down.</u> (Emphasis added.) The thermocouples were checked and those which were unserviceable were replaced. The pile was first made divergent for the Wigner release at 19:25 on 7th October 1957 and the nuclear heating was stopped early the following morning. Some hours later, it appeared to the operators then in charge of the Wigner release operation that the graphite temperatures were dropping rather than rising, and it was therefore decided to apply further nuclear heating."

"The Committee of Inquiry carefully examined the temperature records on this point. Undoubtedly some of the graphite temperatures were falling and there were some in which no Wigner release was apparent: but a substantial number of graphite thermocouple readings showed steady increases. The Committee would not, therefore, endorse the observation that the general tendency was for the graphite temperatures to be dropping rather than rising."

"However, acting on the observations which he had made, the Physicist in charge decided to boost the release with a second nuclear heating. He had no Pile Operating Manual, with special sections on Wigner release," (emphasis added) "to help him, nor had he had the benefit of sufficiently detailed instructions."

"The pile diverged for the second nuclear heating at 11:05 on Tuesday, 8th October. The uranium thermocouples showed a temperature increase lasting for about 15 minutes after the pile diverged. The Committee of Inquiry studied the temperature-time records and decided that the maximum rate of rise of temperature for a few minutes was several times greater than that which is permitted in normal operations. However, the maximum uranium temperature recorded (emphasis added) did not exceed the maximum normal operating temperature."

"The rapid rate of rise of the uranium temperatures was seen and the control rods were run in to reduce the pile power and thus allow the uranium fuel to cool somewhat. However, in the opinion of the Committee of Inquiry, the damage leading to the accident had probably been done. One or more of the uranium cartridges in the lower front part of the pile had failed."

"There were two separate and distinct faults of instrumentation affecting the second nuclear heating. First, the uranium thermocouples were situated at the region of highest uranium temperatures during normal operations and were not at the position of maximum uranium temperatures during Wigner release. Because of this, the pile operator was unaware that the uranium fuel in the middle front regions of the pile was probably hotter than was indicated by the readings he was taking, and the rate of rise of temperatures was certainly greater than he was observing. A second deficiency of instrumentation was that the pile power meter, whilst correct for normal operations, read low in the situation prevailing. The 'feel' of the pile was different and the second nuclear heating was applied too quickly."*

*In normal operation, the peak fluxes and therefore the peak temperatures, tend to lie near the geometric center of the pile. Since this is the hottest section of the pile, it probably also had the least stored Wigner energy of any section of the pile, except at the very edge. Thus, on a reheat, the hottest section during the Wigner release may not be in the center, but somewhat out from the center, and it will be influenced by control rod configuration, coolant channel flow pattern, and fuel element loading. Note also that there is such a thing as the "feel" of the pile. Each reactor has a characteristic response and "feel" which differentiates it from other reactors in somewhat the same way that each car "handles" differently. Getting this "feel" is a part of operator training and one of the principal functions of an operator is to observe any differences in the "feel" of a reactor which may give indications of changing or even unsafe conditions.

"The picture that the Commitee pieced together revealed that, as a result of the second nuclear heating, graphite temperatures gradually rose through Wednesday, 9th October. This led to the oxidation of the uranium which had been exposed by the overheating. The exposed uranium smoldered throughout the course of the day, 9th October, and gradually led to the failure of other cartridges and their combustion, and to the combustion of graphite. By Thursday evening the fire had spread and was affecting about 150 channels in a region of rectangular cross section."

"During Wednesday, 9th October, the graphite temperatures showed considerable variation but the general tendency was for the temperature to increase following the second nuclear heating. One graphite temperature in particular continued to rise steadily. By 22:00, the temperature had risen to such a value that the pile Physicist in charge was required, by a written instruction which he had, to have certain operational measures taken. These measures were taken as follows. The chimney base and the inspection holes had been shut at 21:00. The effect was to cause the chimney to draw a little air through the pile and thus cool it, rather than to draw air through the chimney base and the inspection holes. At 22:15 the fan dampers were opened for 15 minutes to give positive air flow through the pile. The dampers were opened again for 10 minutes at 00:01 on the 10th October, for 13 minutes at 02:15 and for 30 minutes at 05:10. These damper openings had a cooling effect on all graphite temperatures except the highest one, where the rise was merely arrested."

"At 05:40 on the 10th October, at about the end of the fourth damper opening, the pile stack activity meter near the filter at the top of the stack showed a sharp increase. This was noted by the Physicist who was then on duty but no special action was taken because he regarded it as a normal consequence of first movement of air through the pile and up the stack."

"Between 05:40 and 08:10 the pile stack meter showed falling readings, but at 08:10 the activity began to rise steadily. Soon after noon, high activity was reported on the roof of the Meteorological Station."

"During the morning of 10th October, the graphite temperature started to rise again, after the temporary halt caused by the damper openings, as explained above. At 12:10 therefore, the dampers were opened for 15 minutes, and at 13:40 were opened for 5 minutes. During the openings, the pile stack meter showed a sharp increase."

"These effects suggested to the operating staff the existence of one or more burst cartridges. At 13:45 the shutdown fans were switched on as a necessary preliminary to an attempt either to use the scanning gear to detect any burst cartridges or, alternatively, to blow the pile cool. At 14:30 the turbo-exhauster was switched on in order to scan for the burst slug. It was then found that the scanning gear was jammed and could not be moved. The Pile Manager stated that at the end of previous Wigner releases the scanning gear could not be moved, presumably because of overheating. On the present occasion it was immovable despite the fact that the Maintenance Section had worked on it the previous day, 9th October, and had moved it."*

"Being unable to operate the scanning gear, recourse was had to another activity-measuring instrument which could take samples of air coming from the pile provided that the turbo-exhausters were running. This meter showed a large reading."

"At this stage, the Works General Manager was informed by the Pile Manager that there appeared to be a bad burst and he instructed that the affected channel should be identified and discharged as soon as possible."

"Since the scanning gear could not be moved, it was decided to remove a charge plug in the charge wall in front of the pile, and inspect the channel showing the highest temperatures. Certain steps had to be taken before entry could be effected. An air count had to be taken, to ensure that there were safe working conditions; and the operators themselves had to have a change into protective clothing. Soon after access had been secured to the charge hoist, and the hoist raised to the correct height, the charge plug in the pile wall giving a view of the four uranium channels including the one where the thermocouple reading was highest, was removed. The uranium cartridges were seen to be at red heat."

"The first move was an attempt to discharge the glowing channels but they were found to be wedged owing to the distortion of the cartridges; and prolonged efforts continuing until the following morning were only partially succesful in clearing these channels. The next move was to create a fire break by discharging channels adjacent to those which were red hot. This succeeded in containing the area of the fire for some hours but did nothing to reduce the temperature of the affected area."

"During the night, an attempt was made to bring down the temperatures by use of CO_2 brought over from the Calder Hall site, but this proved ineffective because by then the affected region was too hot."**

"Meanwhile, thought had been given to the possible use of water and by about midnight the decision was reached that if other measures continued to prove ineffective, water should be used. This was the major decision and it was supported by the Director of Production and the Deputy Works General Manager. The Works General Manager, the Director of Production and the Deputy Works General Manager all had in mind that although the fire break appeared to be containing the fire to the 150 affected channels, there was the possibility that further Wigner releases might occur in other regions of the pile, and add to the total heat in the pile."

"Soon after midnight, the Chief Constable of Cumberland was warned of the possibility of an emergency and men in the factory were warned of an emergency with instructions to stay indoors and wear face masks."

"By 03:44 the water hoses were ready to be coupled at 15 minutes' notice. At 07:00 it was decided that water should be used but that before it was turned on, all factory labor should be under cover. Some delay was therefore necessary while the shift changed over at 08:00. Water was first turned on at 08:55 but without immediate effect. About an hour later the shutdown fans, which had been kept in operation in order to maintain tolerable working conditions on the charge hoist (where the unloading operations were being attempted), were closed down. This operation, combined with the continued use of water, caused the fire to subside. Pouring of water continued for over 24 hours until, by Saturday afternoon, the pile was cold."

The pile, as a result of this accident, was a total loss. There was some release of activity to the surrounding countryside, and considerable publicity resulted because some iodine got into the British milk supply [40]. The radiation levels observed are discussed in the chapter on Control of Radioactive Wastes, Sec. 7.7.

Comments, Conclusions, Recommendations

(1) It is essential to know the state of all sections of a reactor at all times. This may be done by a variety of means including neutron flux detectors and temperature measuring devices. This is a difficult and complex problem on a very large reactor such as those at Hanford or at Windscale. It is necessary to know whether a given section of the reactor has the proper local power density; whether it is shut down, just critical, or supercritical; whether it is getting proper coolant flow; and whether the coolant is removing heat properly. As has been pointed out in Sec. 3.5 local temperature distributions often give the best indication of the current local condition. A too high power density, a high rate of rise of power, a low coolant flow, or poor heat transfer will all be reflected in a local temperature increase. (Use of local temperatures as a primary base for control also reflects a primary concern over fuel element melting.) When a local temperature is high it is then necessary to ascertain by means of neutron monitors, flow indicators, and coolant temperature indicators what is the cause of the change. The discussion in reference [37] clearly reflects concern for this problem. It also recognizes as a prime cause of the accident the lack of instrumentation at the proper points for the Wigner release.

(2) The operating staff of a reactor must know the weaknesses of their instrumentation...what it will not detect as well as what it will detect. The failure to understand that the instrumentation was not providing a true picture of the situation played a large part in the accident.

(3) Adequate instructions for all operations must be provided to all operators. These should be kept up to date and all personnel informed of changes at once.

(4) Lacking proper information, a supervisor

---

*Note that the functioning of a particular piece of equipment appears to be dependent upon the immediate past history of the pile itself.

**It is implied that the use of carbon dioxide if applied early in the accident might have prevented the damage which finally resulted.

or operator should not take action until such information is forthcoming. The supervisor in this case proceeded on a course of action without adequate information.

(5) In any reactor design it is possible that thermal expansion, radiation effects, or some other effect will cause a relative shifting of various parts of the reactor. Care should be taken in the design to ensure that such shifting cannot lead to unsafe conditions by immobilizing vital equipment—particularly control rods. In this case apparently the shifting in the pile caused by the Wigner releases twice made it impossible to move the scanning gear.

(6) It is usually better to take decisive safety action early rather than to "wait and see." In this case it is implied in reference [37] that if CO_2 had been used early in the accident, the reactor might have been saved.

(7) The release of the Wigner energy constitutes a potential safety hazard since the mechanism incorporates a positive feedback effect. Graphite systems which operate at low temperatures and at relatively high fluxes must be operated and annealed with due regard to this phenomenon. While authorities differ somewhat in their views as to the relative importance of the Wigner release and the second nuclear heating in the Windscale accident, it seems clear that the release at least augmented the seriousness of the accident. Any phenomenon with a positive feedback effect on the system, which could be the cause or augment seriously the consequences of an accident should be avoided in reactor design.

3.8 The HTRE-3 Excursion [41,42]

This reactor excursion is of special interest since it is apparently the only one which is due solely to instrumentation.

The Heat Transfer Reactor Experimental facility was a core test facility located at the National Reactor Testing Station in Idaho and designed to provide a means for testing high temperature reactor cores. The basic facility included shielding, cooling provisions, instrumentation, and other components necessary to operate a reactor. It was intended that a reactor core be installed, tested, removed, and replaced with another, and so on. The core in which the partial meltdown occurred was the third core tested in the facility. The design of the reactor itself is not important to the discussion of the accident. E. P. Epler has discussed this excursion in reference [41].

The basic control instrumentation, which was a part of the permanent facility, had the usual complement of neutron-sensing chambers including three uncompensated level safety chambers in the scram circuit. The circuit for one of these chambers is shown in Fig. 3-8 [41]. Initially, the circuit did not include the filter circuit consisting of the 1 MΩ resistor in series and the 4 μf condenser to ground. This filter was added when it was found that the high voltage supply or its connecting cable was adding electronic noise

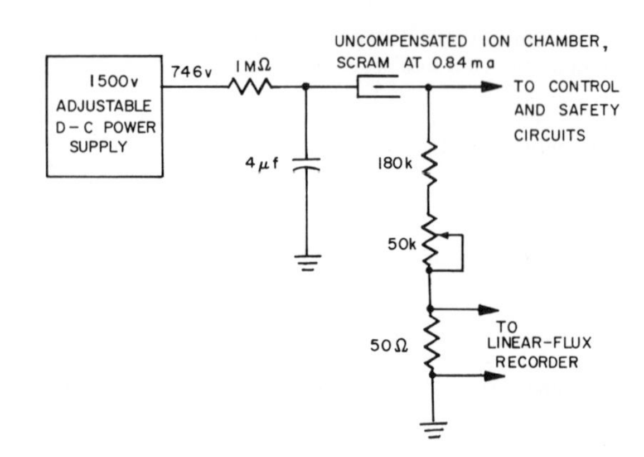

FIG. 3-8 HTRE-3 ion chamber circuit diagram.

to the circuit [42].* Since the filter was evidently quite successful in reducing the noise in the circuit (and hence the likelihood of false scrams and spurious period or power indications) similar filters are reported [41, 42] to have been installed on both the compensated and uncompensated chambers in all the circuits. The units were tested and behaved properly not only during testing but also during use with the first two core installations.

Later, and as a completely separate matter, it was decided to change the startup method from manual startup to automatic startup in the power range. The servo unit designed for this purpose required a larger driving current, and a higher current neutron sensing chamber is reported to have been installed [42]. It is also reported [42] that the other chambers were changed at the same time to this higher current variety.

The planned reactor test called for a greatly reduced coolant flow so that heating rate measurements could be made. Several chambers, including one of the three safety chambers had been replaced with heat-rate sensors. However, either of the other two duplicate power level chambers would produce a scram at a current of 0.84 ma. During the afternoon of November 18, 1958, the reactor was operated at 60 kw and behaved as anticipated. Preparations were made for a 120 kw run, the highest power attempted to date with that arrangement. The ion chambers were to be fully inserted in their most sensitive position and the circuit constants in <u>both</u> the remaining linear flux channels were altered so that the servo-control unit could operate over the range from 15 to 150 kw using one of the two linear flux channels as a detector as well as using it for a safety channel. This would enable full power to be achieved well within the range of the chamber at 80% full-scale or 120 kw. The scram level was set at 180 kw.

The run proceeded routinely on manual control until the power range was reached at what the operator estimated was 10% of the desired full power. Control was then switched to the servo mechanism which was set to hold the power rate

*The problem of noise in such a circuit is a very common one and it is often solved in this way.

of rise to a 20 sec period. The recorder trace shows that, at the time of switching, the reactor was already on a 20 sec period. However, at the moment of switching there must have been a mismatch between the true power as indicated by the chamber and the 10% power demanded by the servo, because the trace shows a stepwise jump from 4 to 10% of full power, as indicated at zero time in Fig. 3-9 [41]. This step jump in power caused the period to become much shorter. The period recorder shows that the change in period from a 20 sec period to something near 5 sec was so fast that the recorder response could not keep up with it. Since the period scram was set for 5 seconds it must not have reached that value. The recorder shows that the system leveled off on an 8.5 sec period. As shown in Fig. 3-9 the flux was held constant by the servomechanism for approximately 20 seconds and then it began to rise again, but this time on a 10 sec period rather than the 20 sec period programmed on the servo.

At 30% full power the log N recorder pen "pegged" (or reached the top limit of its scale) as was evidently part of the plan. The only instruments remaining on scale were the period recorder and the linear flux recorder, both apparently attached to circuits of the type shown in Fig. 3-8. Approximately 40 sec after the 20 sec steady 10% power level plateau had been left, 80% full power was almost reached and the servo should then have held the power steady at 120 kw. Instead, the flux as shown on the power level recorder began to fall off rapidly. The servo, in response to this behavior withdrew shim rods. The operators puzzled as to what was going on, while this condition existed for about 20 sec. At that point the reactor scrammed automatically and three seconds later the operator, having finally decided to take action, scrammed the system manually.

It is generally believed that the scram resulted from a melting of thermocouple connecting wires. The thermocouples for this particular experiment were not located in the region of maximum temperature but the leads to the thermocouples did pass through the hottest region. The circuits were fortunately arranged so that if all ten monitored thermocouple circuits were opened, the reactor would be scrammed. It may also be

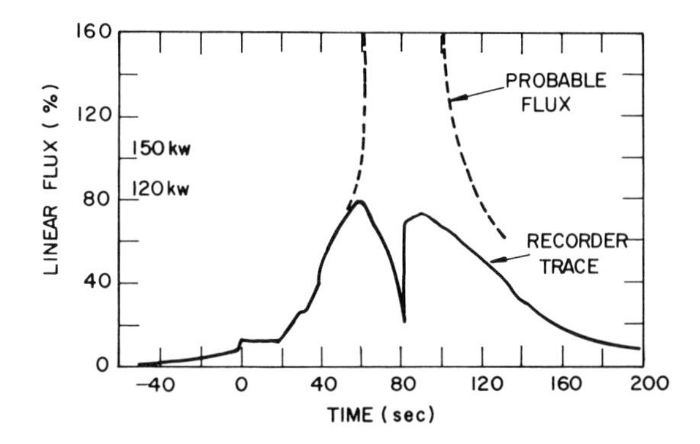

FIG. 3-9 HTRE-3 transient power level plot.

that fuel melting and rearrangement preceeded even the automatic scram since more than 2% $\Delta k/k$ was lost by this means. On scramming the recorder trace first showed an increase in power and then a normal fall-off of power such as occurs after any scram.

The entire accident occupied at total time of 80 seconds from start to automatic scram. In effect, the operator pushed a button and sat back and watched in amazement as the accident happened. The accident was already set up and, once the initial triggering button was pushed, it committed itself and shut itself down without any further human aid.

Investigation showed that several things were wrong with the instrumentation. The primary cause of the accident was a drop in the ion-collecting voltage across the detection chamber of the servo system with increasing neutron flux. This drop in voltage reduced the current observed at the servo unit, and accounts for the strange dip in the middle of the power trace in Fig. 3-9 from 19 to about 95 sec. This strange behavior is well known to ionization chamber specialists, but the average reactor operator is not likely to quickly grasp its significance. In this case the reasons why it happened are complex and the chain of causes had been building up for months.

First, the filters which were added to reduce circuit noise were allowed to remain in the circuit. The 1 megohm resistor effectively isolated the chamber from the voltage supply for very high chamber ionization currents. Under those circumstances most of the dc series voltage drop from the voltage supply through the 1 megohm resistor, the chamber, and the small resistors to ground occurred across the 1 megohm resistor. With increasing flux, and hence ion chamber current, the voltage across the chamber plunged sharply and the current collected fell rapidly. It is likely that there were also space charge shielding effects in the chamber. In effect, the chamber shut itself off for very high fluxes.

Second, it is reported that the original chambers were removed and replaced by higher current ion chambers to drive the servo units. This was a second and separate step taken for a totally different reason than the first and at a completely different time. Third, the voltage of the high voltage supply reported to be common to all chambers had been erroneously set at 746 volts instead of 1500 volts. <u>All</u> of these steps were carried out on what turned out to be the <u>only chambers operating</u> during this particular test. A later calculation [41] indicated that the presence of the filter resistor would have limited the maximum chamber current to 0.746 ma while the 180 kv power trip point was equivalent to 0.84 ma, and thus the reactor could never have scrammed by a signal from these chambers.

The effect of the filters in the gamma compensation ion chambers is also of interest. It so happened that the filter resistor installed in the voltage supply of the gamma compensation chamber was smaller than that installed in the principal chamber voltage supply. Thus, with increasing ion current, the voltage in the principal chamber was reduced more than that in the gamma compensation chamber, and the ion current from the principal chamber fell more rapidly than that from the gamma compensation chamber. Thus, with increasing flux the net current would appear to drop to negative values since the gamma compensation current is subtracted. The result could, and did, appear as a negative period - when actually the period was positive.

Comments, Conclusions, Recommendations

(1) There should always be at least two on-scale neutron signals preferably from different types of chambers with different types of electrical circuits—for example, a fission chamber detector utilizing a battery power supply and operating a galvanometer, and an ion chamber with its own voltage supply and amplifier-recording unit. The reactor instrumentation system should satisfy the Principle of Diversity discussed in the chapter on Sensing and Control Instrumentation, Secs. 1.4.1 and 1.4.2.

(2) Safety system circuitry should not be made a part of an experiment or of a control unit. Use of instruments for both safety and control reduces the number of independent safety circuits and thus violates the Principle of Redundancy discussed in the chapter on Sensing and Control Instrumentation, Sec. 1.4.2.

(3) Records should be kept of all changes made to any part of the system including the circuitry. Before any changes are made, a review (by someone or some group completely knowledgeable of the given reactor and the points at issue) should be carried out of the effect that the proposed change will have on the existing system. This review should also consider any previous alterations which may affect the situation.

(4) Reactor instrumentation should be designed to trip, or at least sound an alarm, on fast negative periods as well as fast positive periods.

(5) The use of automatic controls to carry out a rise to power may be somewhat questionable unless the system has been repeatedly checked and found to be completely reliable in taking appropriate action in the event an unexpected input signal is obtained. In particular, in approaching a new and higher power level slow and careful manual operation by competent operators is likely to be the more conservative approach. Especially this is true when the instrumentation is untested in the desired operating range. Manual operation using slow steps would probably have given the operator a chance to catch the problem in time. It is difficult to build judgment into an automatic startup method which is as sound and broad as that of an alert and knowledgeable human unless the situation is quite routine. (The situation here contrasts somewhat with that discussed in Sec. 3.6 where extremely rapid operations were required. Even there, the system, when automated should have been fully checked out and tested and then operated on a series of small extrapolations in performance.)

(6) Electronic circuitry and chambers should be tested under the operating conditions in which they will be used. If this is not possible, it should be recognized that they are being tested by the operations themselves and extra precautions, care, and alertness should be observed. Epler [41] says "failure to test under planned conditions must result ultimately in an unplanned test."

(7) The safety system should have adequate monitoring and alarms or scrams on such items as circuit continuity, proper high voltages, line voltages, etc.

(8) In general, it is better to utilize a smaller signal from an ion chamber further away from the neutron source with a suitable amplifier, than it is to overdrive an ion chamber and risk operating in a region where the chamber response is far from linear with increasing neutron flux.

3.9 The SRE Fuel Element Damage Accident [43-49]

The Sodium Reactor Experiment (SRE) was built at Santa Susana, Calif. to aid in the development of the sodium-cooled, graphite-moderated reactor concept for civilian power use. This 20 Mw(t) plant went critical in April of 1957 and first generated electrical power on July 12, 1957. As shown in Fig. 3-10 [43], the reactor core region is divided into hexagonal cells of zirconium-clad graphite 11 in. (27.94 cm) across the flats and 10 ft (3.05 m) high. The outer units serve as reflectors and the inner ones contain fuel in a central cylindrical tube 2.80 in (7.11 cm) ID. The fuel elements in the core in question (Fig. 3-11) [43] were made up of 7 rod clusters, each rod a 6 ft (1.83 m) column of 6 in. (15.2 cm) long uranium slugs in a 0.010 in. (0.25 mm) thick stainless steel tube. The 0.010 in. (0.25 mm) annulus was bonded by NaK and there was a helium-filled space above. The outer six rods were wrapped with a stainless steel helical wire to prevent rods from touching each other or the process channel. The flow of the sodium coolant is upward through the core and out to the heat exchanger.

"The design and construction philosophy of the system emphasized the use of conventional, commercially obtainable components wherever possible. To this end, the coolant circulating pumps are simple adaptations of hot oil pumps. The stuffing box (see Fig. 3-12) [43] was replaced by a cooled annulus around the shaft to freeze sodium and thus seal liquid sodium in the pump casing from the supporting bearings on the shaft and the drive motors. The primary and secondary systems each have a 6-in. and 2-in. pump and three of these have had erratic operating experience. . .The 6-in. 1500 gpm* pump in the primary system...was the main offender in the fuel-element damage incident that occurred in July 1959. Difficulties had been experienced from time to time with binding similar to that of the 2-in. pump, and, on two occasions, the auxilary coolant (Tetralin) in the freeze seal has leaked into the main sodium coolant stream. This was detected by identifying hydrocarbon vapor in the atmosphere above the reactor pool. The first of these occurrences,

*1500 gpm = 94.5 liter/sec

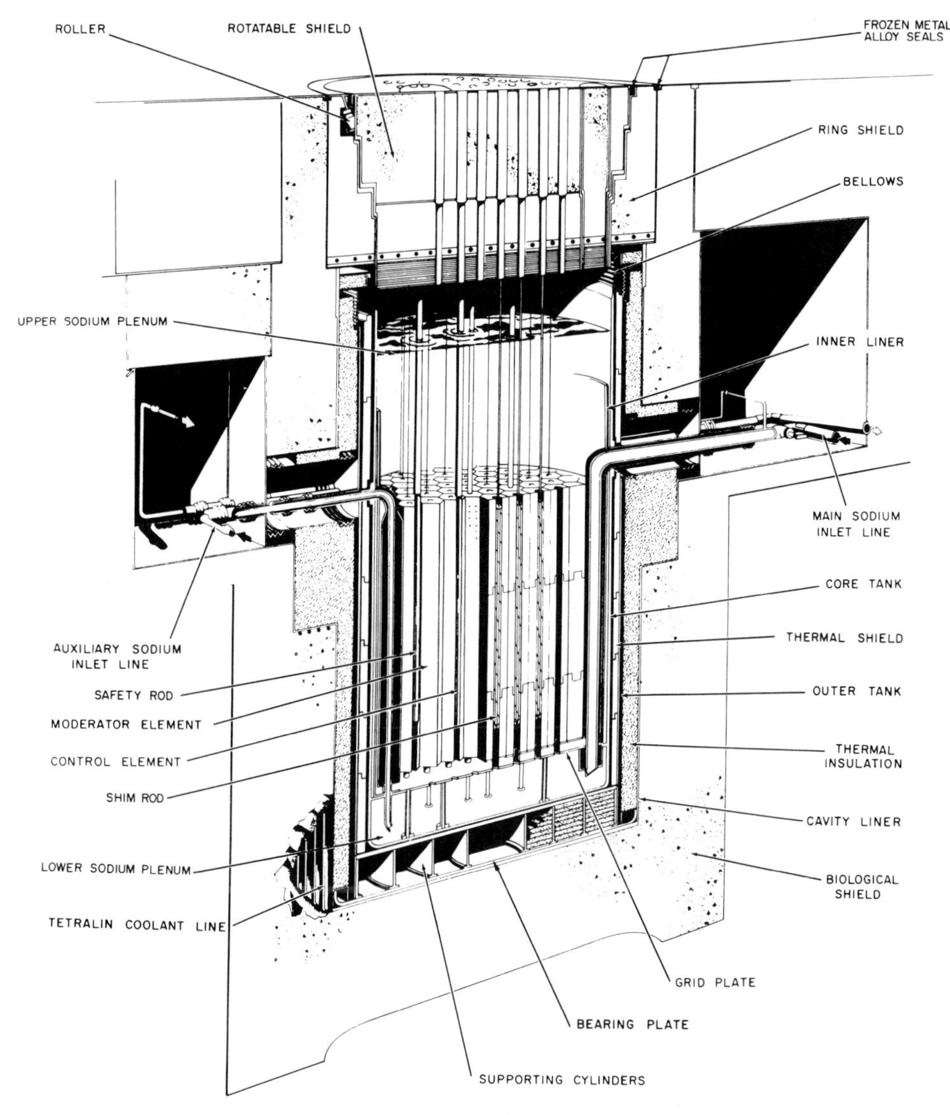

FIG. 3-10 Cross section of the SRE.

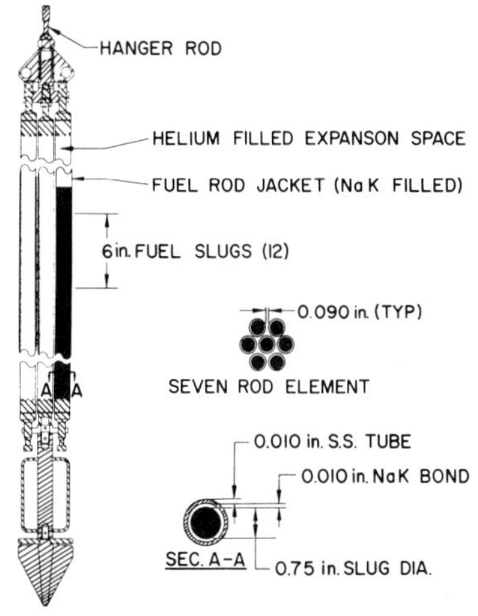

FIG. 3-11 SRE fuel element.

in April 1958, resulted in an insignificant amount of leakage before the freeze seal was repaired. This leak was caused by a pinhole in the freeze-seal casting. The second such leak occurred in May 1959 and resulted in somewhere between 2 and 10 gal of Tetralin being admitted to the primary sodium stream [44]. This leak was traced to failure of a thermocouple well. The Tetralin-cooled shaft seal was then replaced with a NaK-cooled seal arranged in such a way that two independent barriers would have to fail before mixing of NaK and the primary sodium could occur. As will be mentioned later, however, sufficient Tetralin had already been admitted to the system to create the condition that damaged the fuel assemblies [45]."

Some comments on plant performance pertinent to safety or to fission product retention taken from reference [43] are mentioned here although most of them had no direct bearing on the accident itself.

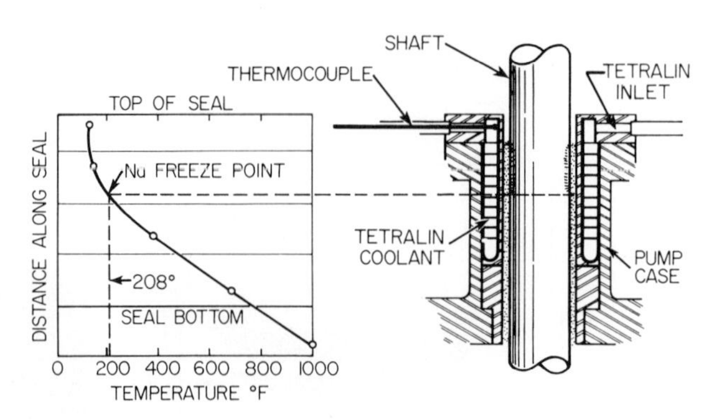

FIG. 3-12 SRE pump shaft freeze seal.

"—Unalloyed uranium metal is an unsatisfactory fuel material for a high-temperature reactor because of a tendency to swell. Fuel rods irradiated to about 1000/Mwd/ton...showed an increase in diameter of 2 to 4 mils. The...NaK annulus... had been occupied by the swelling uranium and the can slightly distended."

"Modification and maintenance of the sodium system have been accomplished with ease and safety. Piping can be cut and welded by first freezing the contained sodium. There have been no sodium fires during any operation involving cutting or welding piping containing frozen sodium. The cold traps and hot traps have performed well in removing sodium oxide; no difficulty was experienced in maintaining the oxide concentration below 10 ppm."

"The major difficulty with the instrumentation has been a series of spurious scrams caused by fluctuations in voltage from the power supply."* Reference [43] presents six pages of graphic operating and scram history for the reactor.

Since the circumstances involved in the accident extended over the entire period of time from November 29, 1958 to July 26, 1959 it seems appropriate to outline briefly the pertinent facts chronologically as is done in more detail in reference [43]. Quotations are from that reference.

Run 8 After a shutdown of about two months for repairs and modifications (involving several transfers of the sodium from the reactor to the fill tank which was known to have considerable sodium oxide), the reactor was taken to 3.6 Mw on November 29, 1958. The fuel outlet temperatures, which usually showed a spread of less than 100°F (62°C) showed much higher values (415-800°F or 259-500°C). This was attributed to high oxide content. On December 12, two elements which had been excessively hot were removed and washed. Both had black material on them. "Jiggling" by moving the element up and down one inch or less (in its position in the reactor) was found to improve heat transfer, but washing helped even more. Power was increased to 12 Mw (December 18) and 14 Mw (December 19) and the run continued to December 23. After shutdown, 15 elements were washed and more cold-trapping** was done. The run continued again from December 27 to January 29 at 20 Mw maximum power with jiggling. On January 7 a sample of the cover gas showed the presence of napthalene (and therefore Tetralin) in the system, something not suspected before—although there had been a prior leak "in June 1958."*** (It was not known if any Tetralin had entered the primary sodium system from the earlier leak caused by a crack in the bearing housing casting on the main primary pump.) The run was terminated on January 29 "because the desired exposure of 600 Mwd was attained."

Run 9 The run started on February 14 and continued until February 26. The reactor was run at 20 Mw with continuing difficulties with fuel exit temperature spread leading to shutdown and more washing and cold-trapping. "Reactor operations were resumed February 20. Examination of the records of the shim-rod positions (which was made after run 14) indicated that an increase in reactivity of 1/2% had occurred. Such an increase is expected because of the xenon decay during a shutdown of this length. However, a preliminary calculation of this effect indicated an expected increase in reactivity of 1%. It is believed that this discrepancy is due to approximations made in calculations of the xenon correction. Similar discrepancies are noted in later runs...."

"There were two reactor scrams caused by

*Frequent spurious scrams and warnings tend to dull the operators' sensitivity and to lead operators to ignore anomalous behavior. At the same time, preventing scrams is also a dangerous practice. Somewhere between these two extremes there is a reasonable level for the rate of scrams per year, but it must be judged on an individual basis, depending on the stage in the reactor operations, the type and use of the reactor, consequences of a scram, complexity of the reactor, etc.

**Cold-trapping to purify Na or NaK is discussed in the chapter on Chemical Reactions.

***The evidence would seem to indicate three Tetralin leaks - one in April, May, or June 1958 (the reports differ in the time), one in December 1958 during run 8 (no mention is made of the source of this one or its repair), and one during run 13 in May-June 1959.

excessive temperature drop across a moderator can and several scrams caused by power line transients. The reactor has a long history of scrams due to the latter effect." Run 9 "was terminated after the desired exposure of 125 Mwd was achieved." After shutdown, the fuel element in reactor channel 56 was examined. The orifice plate had a thin black deposit. The fuel element was washed and replaced in the reactor.

Runs 10, 11, 12 (March 6-7, March 16-April 6, May 14-24) These runs showed some continued improvements in the fuel exit temperature spread. "An examination of the records of shim rod position (made after run 14) shows that at the start of run 11 a loss of reactivity of 1/4% had occurred. The loss may have been due to the replacement of a thimble." In Run 11 a further series of scrams due to flow fluctuations occurred and a reactivity discrepancy of 1/3% attributed to xenon occurred again. During Run 11 the radiation level in the main sodium gallery seemed high although this was not observed until ten days after the run. At the end of this run a filter was installed in the primary system which collected considerable carbon containing material. During Run 12 a planned outlet temperature of 1065°F (574°C) was reached for 1 hour at 6 Mw power and steam was produced at 1000°F (538°C). A check after shutdown showed no measurable change in fuel dimensions.

Run 13 (May 27-June 3) Except for a sodium flow rate scram, the run was considered normal at a power of 20 Mw until 09:00, May 30. Then, several abnormalities were seen including, a slow three-day rise in inlet temperature from 545°F (285°C) to 580°F (304°C), an increase in log mean temperature difference across the intermediate heat exchanger indicating impaired heat transfer, a rise in temperature over a 20-min period from 860-945°F (460-507°C) in one fuel rod, some increases in exit fuel temperatures, a jump of 30°F (17°C) at 22:30, May 30, for moderator delta T compared to earlier 18°F (10°C) fluctuations, and one or two other temperature probe effects. In addition, "although it was not noted at the time because the reactor was on automatic control, an examination of the record of shim rod position (made after Run 14) showed that a shim rod motion corresponding to a reactivity increase of about 0.3% had occurred. This change was gradual and extended over a period of about 6 hours. Following this, the reactivity showed a steady increase of about 0.1% over the next three days of operation."

By June 2, it was obvious that the heat transfer characteristics had been impaired in the primary system and the cause was believed to be a Tetralin leak. The odor of Tetralin was detected in the pump casing of the main primary pump. The run was terminated on June 3, and after a 10 day interval to allow radioactive sodium to decay, the pump was removed. A leak was discovered in the wall of the thermocouple well of the freeze gland seal (see Fig. 3-12) where a dislodged piece of hard plating material wore through the wall as the shaft rotated.

"Seventeen fuel elements were visually examined by means of a television camera and found to be slightly dirty, but in good condition." An attempt was made to wash one element, but during the operation a pressure excursion occurred which severed the hanger (see Fig. 3-11) and lifted the shield plug out of the wash cell. It is believed that hydrocarbons from the breakdown of Tetralin could cause sodium to be trapped in the holddown tube on the hanger rod by blocking the sodium drain holes. This sodium then reacted with the wash water. "As a result of this incident, no further washing was done." It was decided to "strip" the Tetralin and organics which would volatilize by passing nitrogen gas through the sodium system. The process had been used on October 12, 1958 to remove Tetralin from the system after the first leak. The stripping began June 17 and continued until July 5. The sodium system temperature was 350°F (177°C) initially and was raised to 425°F (218°C) by the end to help the removal. In all 400,000 ft^3 (11,300 m^3) of nitrogen were used and 3 pints (1.42 liter) of Tetralin and 1500 cm^3 (91.5 in.^{3}) of napthalene crystals were removed. The system was then purged for ten hours with 4700 ft^3 (133 m^3) of helium and argon. The primary pump was reinstalled with a NaK-cooled freeze seal in place of the Tetralin-cooled seal, and the system was prepared for operation.*

Run 14 (July 12-26, 1959) The run was begun with the anticipation that the situation would be similar to that experienced in Run 8. The reactor was made critical at 06:50, July 12. At 08:35 as the reactor was slowly increasing in power to 0.5 Mw large fluctuations of 10°F (5.6°C) were noted on the moderator delta-T recorder. Normally, even at 20 Mw these were less than 5°F. The fuel exit channel temperatures started to show a spread of about 200°F (111°C). Operation continued at less than 1 Mw until 11:42 when a scram occurred due to loss of auxiliary sodium flow. Criticality was reestablished at 12:15 and operations continued at slowly increasing power levels with fuel exit channel temperatures from 510 to 770°F (265 to 410°C). Fluctuations of 30°F (17°C) in the moderator delta-T at 1.5 Mw were observed. At 15:30 reactor room air monitors showed a sharp increase in activity. The radiation level over the sodium level coil thimble in channel 7 rose to 500 mr/hr. Air filter and stack activities increased. The reactor cover gas pressure was lowered from 2 psig to less than 1 psig in an

---

*After the final accident during Run 14, the use of nitrogen at this point caused considerable concern about the possibility of nitriding of the stainless steel and zirconium and thus promoting fuel and moderator can failures. Tests seemed to indicate that nitriding will occur in preference to carburizing at 1200°F (649°C) in carbon-bearing sodium with a nitrogen cover gas. Apparently it will even occur after a helium purge has supposedly swept all of the nitrogen out. The nitrogen evidently is held by the carbon and calcium impurities. Measurements, coupled with the known solubility of carbon in sodium [46] showed that the system had been saturated in carbon ever since Run 8.

effort to reduce the level. By 17:00 the radiation level over core channel 7 reached 25 r/hr. Accordingly, at 17:30 power reduction began, at 20:57 the reactor was shutdown, and the sodium probe was removed from channel 7 and replaced by a shield plug.

The reactor was brought to criticality at 04:40 July 13 with exit fuel temperature scram set-points lowered to 800°F (427°C). At 13:30 it was observed that the moderator delta-T followed a rise in the sodium outlet temperature and that the moderator temperature did not respond properly to an increase in sodium flow. It was concluded that little sodium was leaking across the grid plate for moderator coolant. At 17:28 a planned increase in reactor power from 1.6 Mw began in order to deliver heat to the electrical substation. "At the start the power level persisted in rising somewhat faster than expected even though control rods were being slowly inserted in an attempt to hold it back." (The solid curve in Fig. 3-13 [43] indicates the course of the power trace during this time.) By 18:07 with the power at 2 Mw, a negative period of about 45 sec was observed and power fell to 2.4 Mw in about 3 min. Control rod withdrawal was started, the reactor was critical at 18:11 and power rose to 3.0 Mw by 18:21. Then, as the power was increasing more rapidly, rod insertion began, but, in spite of this, power continued to rise. At about 18:24 three positive transients were observed with about 50-sec periods and at 18:25 a 7-1/2 sec period was indicated. The reactor should have started an automatic power set-back at a 10-sec period, but did not, and the operator scrammed it manually. The automatic electronic period scram did not act, as it was set for a five second period. The peak power indicated was 24 Mw.* (No particularly high temperatures were recorded during the transient.)

Later examination of the period set-back mechanism (a mechanical actuation by means of a cam in the period recorder) showed that it worked properly only if the period decreased at a slow rate, but would not operate if the period decreased rapidly.

"Recovery from the scram was made cautiously. Criticality was attained at 19:55. Approximately 2-1/2 hours after the scram, reactor power reached 2.0 Mw." Rod positions were now 52 in. (132.1 cm) out rather than 49.5 in. (125.7 cm) as before the scram, but the difference was, at the time, attributed to xenon. The rods returned to 50.5 in. (128.3 cm) by 02:00 at a power of 4.0 Mw. It was decided that the excursion had not affected the reactor adversely. Operations continued until 13:00 when a scram was caused by a short-circuit introduced into the demand circuit for the primary pump being prepared for a flow oscillation test. The reactor was made critical quite rapidly (13:11) and operations continued.

*The final report [43a] corrects this value to about 14 Mw, noting that "The 24 Mw value was obtained by a linear extrapolation of the log N recorder chart from power levels of about 2 Mw, and linear extrapolation is not valid."

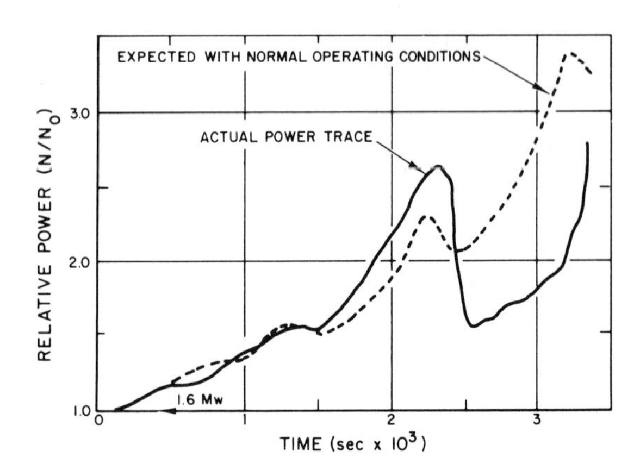

FIG. 3-13 Machine calculation I of the SRE power excursion. Zero time is 17:28 on July 13, 1959.

It was decided to pressurize and vent the reactor atmosphere once to reduce the radioactivity caused by the xenon in the cover gas. At 05:50 July 15 the pressure was reduced from 1.8 psig (1.12 atm) to 0.6 psig (1.04 atm), repressurized to 3.0 psig (1.20 atm), and then reduced to 1 psig (1.067 atm). Lowering the core pressure caused an increase of about 0.01% in reactivity and raising the pressure had the reverse effect. This is not a normal effect on this reactor. "Operation was continued at a power level of approximately 3 Mw."

A review of fuel exit temperature spread on July 15 showed that it would be useless to try "to get the Edison turbine generator 'on the line' since the maximum power level attainable would probably be less than 4 Mw." This would not permit operation at the desired high inlet temperatures while circulating through the steam generator. Alterations were made and on July 16 at 07:04 the reactor achieved criticality at a rod position showing a substantial loss of criticality since the beginning of the run. Intermittent operation continued at less than 2 Mw until July 20. Several tests on pressure effects, plugging temperatures, and sodium level variations were carried out. On July 18, the motor-generator set which supplies the vital bus stabilized power failed and operations were resumed with the unstabilized Edison supply. On July 20, the reactor power was increased to 2.5 Mw to raise loop temperatures gradually to 700°F (371°C). On July 21 at 02:10 a scram was caused by a fast period indication. (Apparently it was attributed to unstabilized power.) The reactor was critical again at 02:25. At 06:45 radioactivity in the reactor began to build up. At 09:45 flow was lost in the main secondary loop causing a scram. The secondary loop was restored to service, the vital bus put back on the repaired motor-generator set, and operations continued at power levels up to 4.5 Mw, sodium flow rates up to 1500 gpm (94.5 liter/sec), and reactor outlet temperatures to 790°F (421°C).

On July 23, it was decided to shut the reactor down and it was scheduled for 17:00 July 24. Until 09:00 July 23, reactor outlet temperatures were kept between 700 and 800°F (371 to 427°C) although a few reached the 900 to 1000°F (482 to 538°C) range. At 09:50 July 23, a reactor scram was caused by a fast period indication. It was attributed to an electrical transient and the reactor was critical again at 10:15.

Between 00:00 and 08:00 on July 24 it was noted (while jiggling elements to dislodge foreign material) that the elements in channels 10, 12, 35, and 76 were stuck while it was known 10 was free on July 22. A scram was caused by a fast period indication at 12:50 on July 24 and was attributed to an instrument transient and the reactor was made critical at 13:14. Accidental loss of auxiliary primary flow caused a reactor scram at 15:40. The reactor was critical again at 15:56. Cold-trapping was put back in service when the outlet temperature reached 510°F (266°C) and the primary plugging temperature gradually dropped from 455 to 350°F (230 to 177°C) within about seven hours.

On July 26 it was noted that the fuel in channels 12 and 35 were no longer stuck and 76 was somewhat free, but 10 was still stuck. The reactor was finally shut down on July 26 at 11:20 after logging 16 Mwd in Run 14. Post-run examination of the core showed that 10 of the 43 assemblies in the core had undergone severe melting of the cladding. The top and bottom halves of these ten elements were separated. The zone of failure was between one-third and two-thirds of the length of the element measured from the top. The accident showed that iodine released from the elements was very effectively retained in the sodium coolant. In fact, no activity except the noble gases was detected in the cover gases [45].

In reference [47] Fillmore has considered the transient which occurred on July 13 in detail. The general features of the transient include a slow but steady rise in power partially compensated for by control rod insertion, a sharp drop in power, followed by another short interval of slow but steady rise, and a final fast rise terminated by scram. These features are shown by the solid line in Figs. 3-13 and 3-14 [43]. On these graphs t = 0 at 17:28 hours, July 13, 1959. The dotted line of Fig. 3-13 shows the course the power trace would have been expected to take if the normal reactivity coefficients and control rod worths were in effect. These calculations were made using the AIRER IBM Code.

By introducing seven ramp and step changes in reactivity, Fillmore was able to get the agreement shown in Fig. 3-14. The general features can then be explained at least semiquantitatively. The slow but steady rise at a rate of +0.04% in a 3 minute ramp in spite of gradual control rod insertion and the negative Doppler effect is attributed to an abnormal rise of the temperature of the moderator which has a reactivity coefficient of $+1.7 \times 10^{-5}/°F$ ($+3.1 \times 10^{-5}/°C$) and perhaps also to some sodium vapor formation in partially plugged channels. (The sodium void coefficient is positive although the fuel temperature coefficient was considered to be $-1.1 \times 10^{-5}/°F$ or $-2.5 \times 10^{-5}/°C$.) The fast negative excursion is attributed to rod insertion reducing the increase in power rate and causing the collapse of sodium bubbles and void regions, thus improving coolant contact with the fuel. This cooling in turn reduced fuel temperature and caused a gain of reactivity because of the Doppler effect, and perhaps sodium void collapse, so that the control rods needed only to be withdrawn a little to again start the reactor up on a slow steady rise. This was followed by a fast transient which added +0.3% in a 5 to 10-sec ramp. It is postulated that this was caused by the more or less simultaneous voiding of about 10 partially plugged fuel elements. Study of the damaged fuel elements [48] seems to show that thermal cycling occurred at temperatures above the α-β phase transition of uranium which would lead to fuel ruptures and also steel-uranium eutectic formation. It is therefore postulated that several channels underwent one or more cycles of heating and sodium vapor formation followed by void collapse. The cycles in the various elements

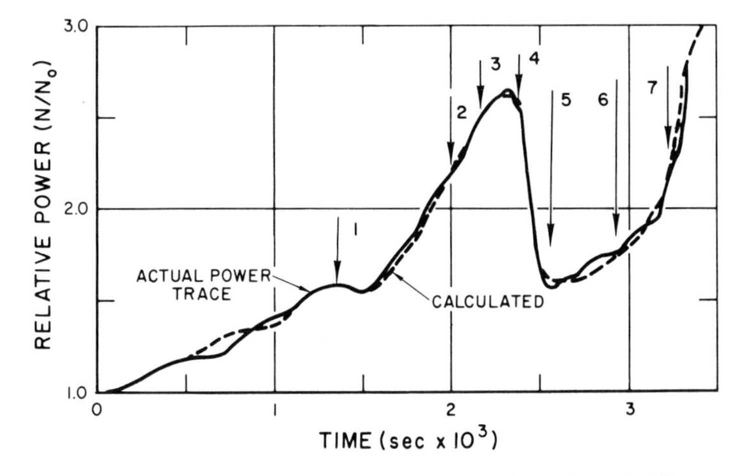

FIG. 3-14 Machine calculation III of the SRE power excursion. Numbers indicate the following sequence: (1) start ramp $\Delta\rho/\Delta t$ of $+0.11 \times 10^{-4}$%/sec, thus introducing total $\Delta\rho$ of +0.0077%; (2) hold first ramp constant at +0.007% and start second ramp $\Delta\rho/\Delta t$ of -0.3×10^{-4}%/sec; (3) hold first two ramps constant, start third ramp of $+0.1 \times 10^{-3}$/sec, thus introducing total $\Delta\rho$ of +0.02%; (4) hold all three ramps constant (at +0.0077%, −0.006% and +0.02%, respectively), introduce step $\Delta\rho = -0.06$%; (5) introduce step $\Delta\rho = +0.02$%; (6) introduce step $\Delta\rho = -0.005$%; (7) introduce step $\Delta\rho = +0.03$%. Zero time 17:28 on July 13, 1959.

are postulated as acting independently at first with periods of the order of 2 minutes. But the rise in power to 5 Mw was sufficient so that "all cycling channels were affected" and voiding in all occurred more or less simultaneously. Further, one can postulate that there was a reactivity interaction between channels and that voiding in one led to heating and voiding in others and so on.*

Comments, Conclusions, Recommendations

(1) Systems should be designed utilizing compatible components, materials, and fluids in all possible cases. In this particular case, the choice of Tetralin for cooling a bearing which could leak into the sodium system was a primary cause for the accident. If this is not possible, then "Where materials are potentially incompatible under normal service conditions, special means must be provided to separate them. Where there are compatibility changes with abnormal operating conditions, e.g. temperature, special attention is required for monitoring and controlling such conditions." [48a]

(2) At the time the choice of fuel had to be made for this reactor the knowledge of fuel performance was not so good as it is now and there were not so many choices. It was known, however, that uranium metal had an α-β phase transition with accompanying swelling at about 1220°F (660°C) and that uranium formed an eutectic with stainless steel at about 1340°F (725°C). Thus, while the use of a stainless steel-uranium metal element may have been the only choice, nevertheless phase transition and eutectic formations greatly lowered the temperature at which fuel element failure and melting occurred. Designers should always try to select materials and combinations of compatible materials that will stand up as well as possible under abnormal conditions. During the planned 1065°F (574°C) run (Run 12), the central fuel element temperatures must have been quite close to the α-β phase transition point in a number of elements.

(3) It is not good practice to use, even as temporary additives, materials or gases which may remain in the system and cause deleterious effects later. Thus, the use of nitrogen as a stripping agent for Tetralin led to a concern over nitriding later. The total cost of the time spent and the tests run was probably greater than if a chemically inert noble gas had been used in the first place.

(4) There were several instrumentation problems. The principal one involving a period setback has been described. The practice of incorporating such an important item of equipment as a period setback as a part of a recorder is questionable. Recorders require easy access for maintenance and ink-refilling, and often have very poor safety characteristics. For instance, if the slidewire breaks, the recorder goes to one extreme of its travel. This can lead to an unsafe situation, (especially if the recorder is being used to drive an automatic control unit for the reactor—as happened in one case). In addition, there was evidently much to be desired in the normal behavior of the system. Reference [43] mentions several times the fact that "the reactor has a history of spurious scrams due to apparent period transients". This problem is often encountered during the initial tests of a reactor and is often due to either improper grounding of electrical components, electrical noise from other equipment on the same power line, or power surges, or all three. Usually it can be eliminated in a short time without sacrificing period trip sensitivity. This instrumentation system had been used since 1957 and by 1959 the problem should have been solved. The fact that the operators are described as achieving criticality after scrams in time intervals of 15 to 25 minutes would seem to indicate that no thorough investigation of possible causes could have been made. In fact, if Fillmore's explanation of the fast transient effect is correct, there may be some reason to believe that some of the several period scrams which occurred in Run 14 were genuine, although "No evidence was uncovered to indicate that there were indeed genuine scrams other than during the transient at 18:25 on July 13" [48a].

McDonald and DeVan [49] give the opinion that, "the reactor instrumentation under the immediate surveillance of the operators was inadequate to indicate excessive fuel element temperatures, the blocking of coolant passages, and fission product leakage. As a result, the operators did not consider such indications (where they existed) serious enough to warrant shutting down the reactor. Since the SRE is a 'developmental facility built to investigate fuel materials' it would appear that additional instrumentation, as well as closer technical management, might have reduced the damage to the SRE core."

(5) From the evidence available in this accident it would appear that fission products other than the noble gases are retained well in sodium. This, in a sense, affords an additional safeguard unless the sodium should then become exposed to air in such a way as to become a fire hazard.

(6) The circumstances which eventually led to this accident began as early as spring, 1958, when the first Tetralin leak occurred. A second leak occurred in Run 8 on November 29, 1958, and problems continued until July 24, 1959. During that time so many difficulties were encountered that, at least in retrospect, it is quite clear that the reactor should have been shut down and the problems solved properly. Continuing to run in the face of a known Tetralin leak, repeated scrams, equipment failures, rising radioactivity releases, and unexplained transient effects is difficult to justify. Such emphasis on continued operation can and often does have serious effects on safety and can create an atmosphere leading to serious accidents. It is dangerous, as well as being false economy, to run a reactor that clearly is not functioning as it was designed to function.

*This description, if correct, illustrates the type of reactivity effect which can be very serious since the mechanism depends upon a positive internal feedback, increasing without limit, until terminated by disassembly - unless terminated first by an external means such as control rods.

In the long run, reactor economics as well as reactor safety will demand adequate continuing maintenance at all times, and this will include early shutdown and proper action whenever there is the least doubt concerning the situation. Management can and must establish this sort of attitude.

3.10 Westinghouse Test Reactor Accident [50-53]

The Westinghouse Test Reactor is a light-water-cooled and -moderated reactor utilizing highly enriched uranium fuel. The fuel element consists of three cylindrical layers surrounding a central thimble. Each layer is made up of 0.052 in. (1.32 mm) of aluminum-uranium alloy, clad with 0.0365 in. (0.927 mm) of aluminum on either side. At the center of each fuel element is a thimble which permits the insertion of a sample. Unless a special experiment is involved, the central sections are loaded with either aluminum or cobalt slugs. The use of these aluminum and cobalt slugs helps with the shimming of the reactor for control purposes.

A diagram of the core is shown in Fig. 3-15 [50]. The nine control rod positions are indicated by the black circles. Each control rod consists of an aluminum-clad cadmium cylinder with a normal fuel element follower. Thus, on the full withdrawal of a control rod from the core a regular fuel element containing 199 g of U^{235} is left in that position. In Fig. 3-15 the number before the dash indicates the top to bottom row starting at the top of the diagram, and the number after the dash indicates the position numbered from left to right in any given row. The number beneath the line indicates the number of grams of U^{235} estimated to be present at the time of the accident. Each of the fuel elements originally contained 199 g of U^{235}. The core has a 36 in. (0.914 m) active height and a main active diameter of approximately 28 in. (0.711 m) with an aluminum-water ratio of 1. At the original licensed power of 20 Mw the average thermal neutron flux in the core was approximately 5.2×10^{13} neutrons/cm^2-sec.

The fuel element that failed (Position 6-5) contained a special experiment designed to monitor fast neutron fluxes. The experiment consisted of a set of 7 hairlike nickel wires, each separately encapsulated in a quartz capsule and held in recesses of a 3/8 in. (9.52 mm) diameter aluminum rod. The rod in turn was encased in a 1/2 in. (12.7 mm) OD aluminum tube with 1/8 in. (3.18 mm) weep holes drilled through the wall at 4 in. (10.16 cm) intervals. This assembly was placed in the central thimble of the fuel element in the normal manner. A flow orifice at the bottom of the thimble section limited the flow to that required for cooling this rod. Other experiments were located in Positions 5-6, 7-3, 7-6, 8-5 and 11-1. Other fuel elements contained aluminum and cobalt, as indicated in Fig. 3-15.

The coolant flow is down through the core and up through the thermal shield. The design specifications required a flow of 15,000 gpm (950 liter/sec) at 20 Mw with a core outlet pressure of 93 psia

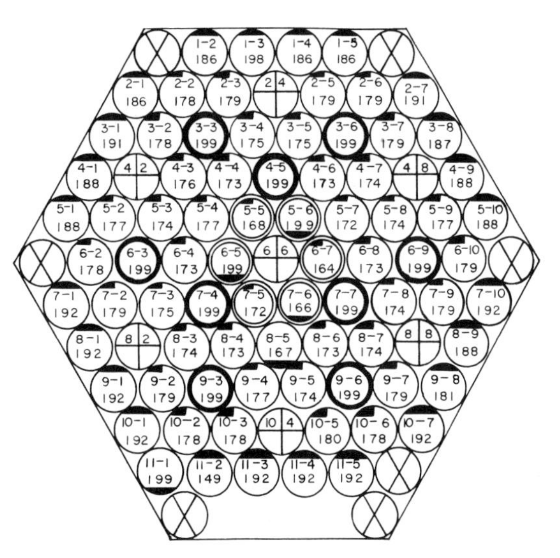

FIG. 3-15 WTR core cross section. Normal circles indicate fuel elements, heavy circles (e.g., 3-3, 3-6, etc.) indicate control rods. Circles enclosing an X are plugged, those enclosing a + are high-pressure thimbles. The contents of the irradiation volumes are denoted by darkened areas at top and bottom of individual fuel elements: black strip across top = standard V-basket with cobalt, black half-strip across top = solid aluminum V-basket, black half-strip at bottom = experiment. Upper numbers indicate position, lower numbers denote number of grams of U^{235} estimated to be present at the time of the accident. (Note: For simplification letter prefixes used in reference [50] to identify positions have been omitted in the above figure.)

(6.4 atm) and a core pressure drop of 15.5 psia (1.05 atm). The core inlet and outlet temperatures were 115 and 125°F (46.1 and 51.7°C) respectively. The normal surface heat flux under these conditions was estimated to be 1.13×10^5 Btu/hr-ft^2 (35.6 watt/cm^2) with a flow velocity of 21.3 ft/sec (6.50 m/sec). These numbers were calculated for 57 fuel assemblies in the core. At the time of the accident there were 78 fuel assemblies in the core including control rods.

The site ventilating systems, the various waste collection tanks, the evaporator-condenser tanks, and several other components all vent their non-condensable gases to a 100 ft (30.5 m) vent stack located between the Process and Reactor Service Buildings.

At the time of the accident the reactor staff was engaged in a series of tests to determine the feasibility of increasing the power gradually to 60 Mw(t). Reference [50] provides a detailed chronology of events. The quotations in the following abbreviated account are taken from there. According to the license amendment of January 8, 1960, to permit operation at a maximum of 60 Mw(t), certain restrictions were imposed. These restrictions were [50]

"1. Westinghouse shall retain the bubble formation apparatus and the special detection channel described in the application in the reactor during the power escalation program until stable operation at 60 Mw(t) power level has been established;

2. The ratio of the maximum heat flux in the reactor to the burnout heat flux shall never exceed one-half;

3. The reactor shall not be operated in such

a way that the ratio of core steam void volume to core coolant volume exceeds one percent; and

4. When the reactor is being operated with the automatic control system, the magnitude of boiling induced neutron level perturbations shall not exceed 5 percent or whatever lesser value is necessary to prevent erratic behavior of or oscillatory interaction between the boiling phenomenon, the reactor power level, and the automatic control system."

The restrictions also required that the bulk water temperature from the element in the first fuel ring of the core would not be permitted to exceed 220°F (104.5°C) during the escalation steps. This latter requirement was satisfied by instrumenting a fuel element in Position 7-6 with 3 aluminum-clad fiber glass insulated chromel-alumel thermocouples projecting into the nozzle space below the element. These thermocouples were connected to a printing data-logger which was used to record the temperatures. Quoting further from reference [50]: "During the early stages of the power escalation program, the readings of these thermocouples agreed with those predicted theoretically... However, within a few days, a drift upward was observed in the thermocouple readings which was not related to the gross power and flow values in the reactor. The thermocouple readings continued to drift upward and eventually two of them rose to a value considerably in excess of 220°F (104.5°C). (Investigation subsequent to the failure of the fuel element indicated that these symptoms were common for water leakage into aluminum clad thermocouples.)

"At this point the thermocouples were assumed to have failed and the reactor was shut down while new thermocouples replaced the old ones. The new thermocouples then read temperatures which agreed with predicted values. With time the second set of thermocouples exhibited a similar drift and eventually indicated in excess of 220°F."[50] Plots of the ratio of the bulk temperature rise in fuel element 7-6 to the bulk temperature rise of the reactor vessel as a whole should remain constant independent of reactor power and coolant flow rates so long as conditions within the reactor remain the same and control rod positions did not affect the power distribution within the core. With the failure of the second set of thermocouples, the reactor was shut down. A third set was installed using stainless steel clad, magnesium oxide insulated thermocouples. The second set as well as the third set (in position 5-6) were in the core at the time of the accident.

Prior to undertaking the power escalation program, a study was made to establish the values of power and flow for the various steps proposed. During the power escalation program the total primary coolant flow for normal operation was presumed to be twice the core flow shown in Fig. 3-16 [50]. Figure 3-16 shows the predicted power at which boiling was expected to start as a function of core flow. Design specifications called for 15000 gpm total flow of which 11200 gpm was core flow, or a total flow of about 1.3 times the core flow.* Presented on this curve are experimental points observed during the escalation test. "During the actual power escalation, using the bubble detector, the flow was reduced to 83 percent of this normal value." [50] The power was raised in 5 Mw steps beginning with 20 Mw and proceeding upwards to 45 Mw during the period from Feb. 21, 1960, to March 7, 1960.

Reference [50] describes the program. "As the program progressed and the power level was increased, boiling noise was indicated at slightly higher flows than were predicted by Fig. 3-16. At this time, however, it was established that boiling was occurring in the access tubes and was not distributed over the core.

*In a private communication, a Westinghouse reviewer writes "The reactor was not operated on flow rate but on pressure drop across the core. Bypass flow was dependent on the type and number of experiments. Proper core flow was determined by pressure drop across the core." Use of such a system might make it advisable to establish minimum core pressure drop limits and to scram on a low signal from this instrumentation rather than on a low flow signal.

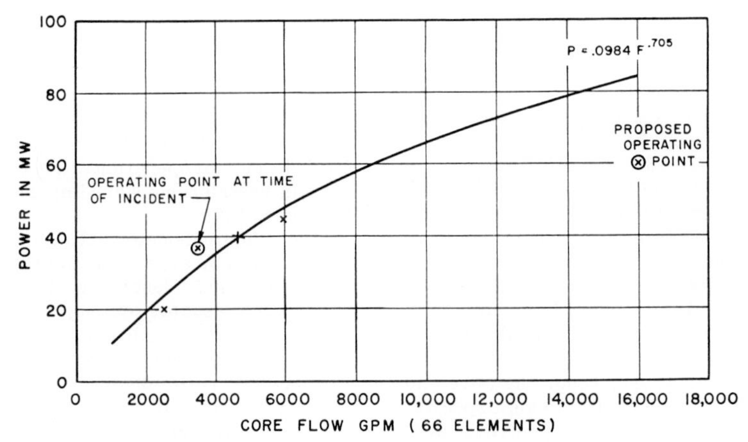

FIG. 3-16 WTR transient. Power to initiate boiling vs core flow. (Note: 1 gpm = 63.09 cm^3/sec.)

"Boiling in the access tube in position 5-8 was substantiated by traversing a thermocouple axially along the tube from the bottom to the top of the core region and observing the temperature...A maximum temperature of 248°F (120°C) was measured in the region of maximum power production in the reactor. Saturation temperature of water under atmospheric pressure and 22 ft (~0.65 atm) of head is 240°F (115.6°C). Consequently, boiling in this tube took place over several inches of length.

"It was believed that boiling in the access tubes was obscuring the detection of the initiation of boiling in the fuel channels, and a decision was made to remove these tubes at the next shutdown. Meanwhile, several runs had been made using the boiling detector, and the system had been calibrated as to amount of void corresponding to noise pattern and amplitude of noise in terms of power output."

It has been estimated that the boiling noise detector which looked at the fluctuations in total reactor power could disclose the presence of boiling only when the amplitude of the power fluctuation exceeded approximately 100 kilowatts.

Initially two points on the curve were checked for low flow value as shown in Fig. 3-16. These two points were 2,500 gpm (158 liter/sec) at 20 Mw and 5,500 gpm (347 liter/sec) at 47 Mw. These two points were obtained with 4 access tubes present containing a 15 ft head (~1 atm) of stagnant water and each within the core. The access tubes extended from outside of the reactor vessel into the fuel assembly location and were provided for easy placement and removal of samples within the core. Thus, boiling within these access tubes occurred first. Since this boiling noise could obscure the behavior of boiling within fuel elements, the access tubes were removed and replaced by regular fuel elements during a shutdown which occurred from March 14 to April 2. During the shutdown some new fuel assemblies were added to the core including the one described earlier that failed in position 6-5 containing a fast neutron flux monitor.

The reactor was started up on April 2 and criticality was achieved at 7:10 a.m. The power level was increased and reached 40 Mw at 1:44 p.m. and was run at this level for one and one-third days.

Reference [50] continues: "With two-thirds of the total flow passing through the core and a radial peak to average power production of 1.33, the expected temperature compared with the measured ones were:

	Computed	Measured
ΔT across reactor vessel	18.5°F	18°F
ΔT across position 5-6	37 °F	36-39°F
ΔT across first channel	25 °F	25-28°F
ΔT across second channel	47 °F	47-55°F

...At approximately 19:00 hours on April 3, the reactor power was reduced to 30 Mw in preparation for the test to be conducted...The alarms, cutback and scram points were reset in accordance with these requirements. At about 20:00 hours primary coolant flow was gradually reduced to 5,250 gpm.

"Figure 3-17 [50] is a copy of the nuclear power recorder chart covering this and the following time interval. As can be observed from this chart, the reduction in flow was accompanied by a slight reduction in power caused by the temperature coefficient. This power dip was compensated for by the automatic control system...

"At 20:20 hours the power level was raised to approximately 35 Mw and allowed to settle to approximately 34 Mw as measured by the nuclear power instrumentation. Thermal power calculations were performed prior to and subsequent to raising the power level using both the reactor flow and core ΔT, and the reactor flow and the instrumented fuel element ΔT. In addition, after the expected delay, the thermal power was displayed by the thermal power recorder. These data are presented below:

Nuclear Power	Core ΔT × Flow	Thermal Power Instrumented Fuel Element ΔT × Flow	Thermal Power Recorder
30 Mw	32.9 Mw	29 Mw	30 Mw
35 Mw	37.8 Mw	36 Mw	34.5 Mw

"Instructions were then given to the reactor operator by the shift supervisor to increase the power level to 40 Mw. To increase the power level, the automatic control system called for additional rod withdrawal. All nine control rods were banked at about 62% at this time. Control rod No. 9 which was on automatic control was, like the other rods, in a low differential worth region of its travel and shortly reached 85% withdrawn. Automatic control was then manually switched to rod No. 8 which also was withdrawn to 85% and the automatic control was manually switched to rod No. 7. About this time (approximately 20:35 on Fig. 3-17) it was observed that

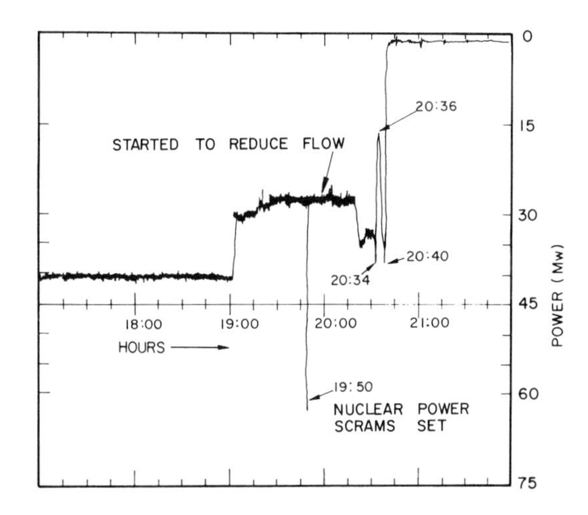

FIG. 3-17 WTR transient. Power level from neutron recorder chart. (Note: the trace has been redrawn and does not show the small fluctuations nor discontinuities associated with rapid movement of the pen.)

the power level was falling and the operator, under instructions of the shift supervisor manually withdrew rods No. 1, No. 2, No. 3 in turn, each 2% in travel. This movement, together with reactivity added by the automatic control system, returned the reactor to approximately 37 Mw. Just before the reactor reached 37 Mw, the demineralized water monitor channel alarmed. This alarm was acknowledged and almost immediately thereafter, several other alarms indicated high radiation levels in the various monitored areas. At approximately 20:40 the power demand set point was reduced, followed immediately by manual reactor cutback, and at 20:44 the reactor was manually scrammed. It was suspected and later confirmed that a fission break had occurred and that the accumulation of fission products in the head tank was producing radiation levels in the plant areas sufficiently high to produce alarms. The plant was evacuated..." The Brush Recorder trace from the bubble detector during the accident is shown in Fig. 3-18 [50].

To explain the cause, reference [50] states, "The power reduction shown in Fig. 3-17 at 20:34 is believed to have occurred as a result of a decrease in reactivity caused by the fuel element failure meltdown and subsequent blockage of the coolant channel. The blockage is presumed to have voided the water channels by the production of steam and bulk boiling in the failed element. A consideration of previously measured void coefficients for an element in this position would indicate a loss of 0.3 to 0.6% reactivity if all channels were voided. This reactivity loss cannot be explained by simple bulk boiling with open channels at the top and bottom because the reactivity would then have been reinserted when boiling ceased at the reduced power level at 20:36. In addition the overall reactor temperature coefficient must have added approximately 0.18% reactivity as soon as the power level was reduced to 17 Mw.*

"Another possible source of permanent reactivity loss is displacement of the fuel contained in the failed section of the element. An upper bound on this effect is that the worth of a fuel element in the failed position is 0.9%. This number obviously is a gross overestimate of what could have happened since the fuel element was not completely displaced from its core position. Later observation of the failed fuel element indicated a considerable amount of burnt up debris was lost from the element. That portion of the element which was severely damaged was worth about 0.6-0.7% in reactivity but only a small portion of this worth was lost. The total reactivity added by means of the control rod withdrawals previously described is approximately 0.6% Δk. Thus, the reactivity changes caused by the voiding theory and possibly by the loss of a small amount of fuel are consistent.

"It is believed that the element failed at approximately the same time as it voided, and later examination confirmed the melting and permanently blocking of the channel. The element remained voided after the reduction in power because of a lack of water about the blocked portion and because of the presence of steam above the blockage."

"...Figure 3-18 is a reproduction of the boiling detector Brush Recorder trace during the time interval under discussion** If time zero is taken at 20:34 corresponding to the first peak in power level of Fig. 3-18, then the reduction in power level caused by the failure and voiding appears to have occurred in about eleven seconds. That a permanent block was established can be seen in that over the next three minutes the noise level remained approximately constant. Then as rod motion forced the power level back up, the boiling noise pattern increased in amplitude, but didn't quite return to the before failure amplitude when the cutback was initiated. An independent observation indicated that an alarm on top of the reactor, actuated by the radiation from the head tank, went off at approximately 240 seconds on Fig. 3-18. This further confirms the other deduction that the element failed at about 20:34 at the first power peak rather than at 20:40, the second peak in power."

As will be noted from Fig. 3-18, the scale of the sensitivity of the boiling indicator was unfortunately reduced shortly after the first power peak. Between the time of the first power peak at 20:34, shown in Fig. 3-18, and six seconds later, power oscillations of 0.8 - 1.0 Mw are shown on the record. The void created at the time of these oscillations is the apparent cause for the reactivity reduction and the sharp power level drop.

Comments, Conclusions, Recommendations

(1) Reference [50] states in conclusion: "A fuel element failure occurred in the WTR on April 3, 1960. The effects of this failure were not measurable off the WTR site. The cause of the failure cannot be established beyond reasonable doubt, but it is reasonable to believe that a normal fuel element operating under the specified test conditions would not have failed. No information has been found that indicates that the stated operating conditions were not being met. A strong possibility exists that the failed element was not normal. Fuel element specifications and inspection in the past have been too lax, and rigid control will be required in the future."

The study presented as a part of reference [50] indicates clearly that on examination new elements showed lack of bonding between the clad and the fuel element meat which could certainly have con-

*The temperature coefficient is not a prompt coefficient and, without a much better calculation than evidently was done it would be difficult to say positively how much of the rise again to power from 17 Mw was due to each of the factors of void collapse, temperature coefficient, and control rod withdrawal. The calculation discussed in Sec. 3.9 on the SRE accident shows this difficulty very well.

**In a private communication a Westinghouse reviewer has stated that the boiling noise instrumentation was located at the reactor shield during this test and was not in the control room.

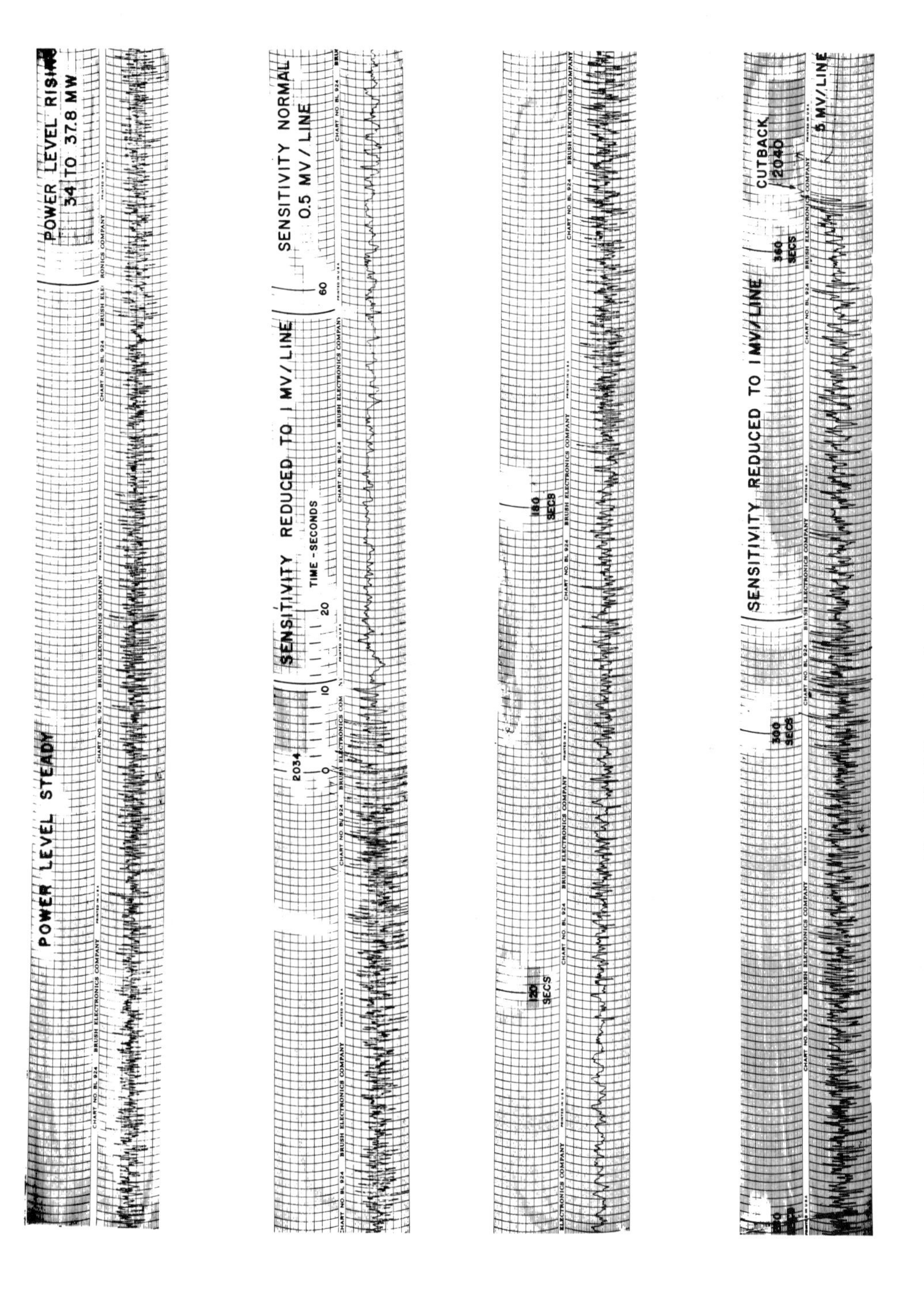

FIG. 3-18 WTR transient. Recorder trace of boiling detector.

tributed to the fuel element meltdown since heat transfer from the fuel element meat through the clad and into the water would be seriously impaired. In some elements sectioned the void was not between the meat and clad but in the U-Al alloy.

(2) The WTR along with many other reactors utilizes downflow in the coolant through the core. The downflow through the core means that steam formation within a channel will increase the pressure drop along the channel and tend to reduce the flow within the channel. In particular, since the pressure drop across the core is low, this may be quite an important effect in this particular core. Flow instability caused by voiding can therefore play an important part in such a situation. Calculations of flow instability in this case are difficult. It is perhaps worthwhile to consider carrying out tests on this sort of arrangement to observe performance of such assemblies. Arguments can be made that such an instability is unlikely since no flow blockage occurred during the previous two boiling tests which were performed at low flow rates and high power densities. Considerations of such instabilities and their consequences should be a routine part of normal reactor design. (Recent work in the reactor laboratories at Grenoble, France, which is scheduled for publication in 1964, is of interest.)

(3) There are several possible lessons which can be drawn from this occurrence in the field of reactor instrumentation and control.

First, the experiment which preceded the accident and, in fact, the accident itself clearly pointed out the difficulties in trying to interpret thermocouple readings within reactor cores. In two successive sets of thermocouples proved unreliable. Figure 3-19 [50] shows the recorded temperatures in access tubes in position 5-8 taken during a 40 Mw run at a flow of 11000 gpm. According to reference [50] the measurement of the boiling in the access tube was demonstrated by the fact that a maximum temperature of 248°F was measured by the thermocouple traverse of the

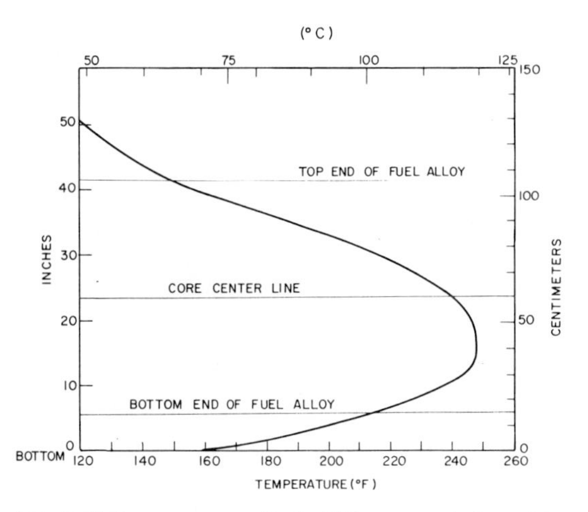

FIG. 3-19 Temperature profile in WTR access tube in position 5-8 during 40 Mw, 11000 gpm run.

tube. It was then calculated that the pressure head at the hot spot was 22 feet of water. Since the saturation temperature of water at 22 feet is 240°F it was then concluded that boiling had occurred in a limited region because the recorded value of the temperature had exceeded the calculated boiling point by 8°F. From the reports it is not clear that due consideration had been given to such effects as radiation heating in the thermocouple, possible errors in the pressure at that point in the access tube, etc. The access tubes were removed just prior to the run in which the accident occurred and it is not clear from the reports whether boiling still occurred at a higher flow rate than expected. Thus, from the evidence in the reports alone, it could be questioned whether or not boiling did occur in the access tubes. In general it would appear to be sound practice not to rely too heavily on in-core thermocouple measurements for vital safety information where absolute values of temperature are required.

Second, the use of the boiling detector based on total reactor power fluctuations is subject to considerable error in interpretation. It is quite conceivable that a small amount of boiling throughout the core will give the same general void behavior as a large amount of boiling or even total voiding in a very small area of the core. Also, the fluctuations in power level resulting from boiling appear to be somewhat large in the actual observed case.

(4) Figure 3-17, which shows the power trace, indicates that the first peak in power level occurred at 20:34. Westinghouse believes that failure occurred at that time rather than later at the second power peak. The use of a radiation monitoring system close to the core would probably have prevented such heavy contamination of the system with fission products by causing an earlier alarm and an earlier shutdown. If the Westinghouse conclusion as to the time of failure is correct, the system ran some 360 seconds at power with a failed fuel element dissipating fission products to the system before the reactor was cut back in power. Earlier shutdown also would have hastened the time when cooling flow could have been reduced and the radioactive core portions more effectively isolated.

(5) Figure 3-18 shows that the person in attendance at the boiling noise indicator apparatus (at the shield face) reduced the amplitude during the most crucial part of the developing situation, thus indicating that he was aware of the increase in boiling noise.

WTR-49 does not mention whether or not the operators in the control room were informed of the increase in boiling noise on the Brush Recorder. In fact it does not mention where the boiling noise detector was located. Certainly information of increased boiling noise should have been passed at once to the control room and, since the indication was not displayed in the control room and since gross boiling was not expected, should have resulted in an immediate scram or power set-back order. While this might or might not have prevented the melting, it would have resulted in a much faster power reduction.

In general it would seem prudent to locate the

information read-out for an apparatus intended to perform an important role in the safety of the power escalation experiments in the control room in order to provide the maximum information directly to the operators for use in evaluating the situation moment by moment. Faulty communications or lack of communication could well negate the usefulness of the information available at the reactor face but not in the control room.

The changing of scales of an important piece of recording apparatus during critical times should only be undertaken if the instrument is clearly pinned, or about to be, and then only if all personnel involved in subsequent decisions know it is being done. There should be no chance that the readings will be misinterpreted in making subsequent decisions. In this accident it might be that the amplitude scale would have been as well left as it was. (In the EBR-I accident, it has been pointed out that a change of scale would have much improved the situation.)

(6) During the short period of minutes from about 20:33 to 20:40 six rod withdrawals or automatic control rod withdrawal selections were made by the operator on command of the supervisor. It is not clear from the Westinghouse report exactly where in the sequence of events these withdrawals occurred. It does appear to be clear that the last three manual withdrawals were carried out on the basis of three successive judgements by the supervisor in spite of the fact that something was obviously wrong since the power was falling rapidly during a rod withdrawal. This accident clearly pointed out, perhaps for the first time, that a sudden loss in reactivity within a reactor may also be dangerous unless proper action is taken. The accident has led to the installation in a number of reactors of negative-period meters which result in a scram or alarm on a fast negative period. The interaction between such a phenomenon as core voiding and automatic control systems must be carefully considered. In particular, the amount of reactivity made available in a given limited time interval (seconds or minutes) to an automatic control system (or for that matter to the operators) should be limited with careful thought given to the consequences of the addition of this amount of reactivity during the time period considered. (This difficulty is often partially answered by the use of a trip which disconnects automatic control if the demand signal differs too much from the actual signal.) Use of manual control during the power escalation program might possibly have caused the situation to be sensed sooner and perhaps might have prevented the accident or reduced its consequences.

(7) A careful study of the reactor fuel configuration at the time of the accident leads to some interesting observations. In Fig. 3-15 essentially undepleted fuel elements of 198 or 199 grams of U^{235} each are indicated in 13 positions. Of these thirteen positions, nine are control rod fuel followers and always occupy the same position relative to the poison sections to which they are attached. In addition these nine were only 62% in the core at the time of the accident. As a result of these two considerations it seems unlikely that these nine could have played a part in an accident of this type. Two of the remaining four undepleted elements are at the edge of the core. Only elements 5-6 and 6-5 are close to the center in an otherwise rather highly depleted region. Element 5-6, according to WTR-49, contained a series of thermocouples in small stainless steel tubes which traversed the entire length of the element. The only element added just prior to the accident was that in position 6-5. According to WTR-49, element 6-5 contained basically only aluminum in the central axial region of the fuel element. The relative reactivity worth of the neutron absorbing materials in the central axial region of these two elements is not given in WTR-49. The relative flux depression is also not presented. However, from the description of these two elements it would appear that 5-6 contained more neutron absorbing material and hence had a lower flux than 6-5. Thus it appears that the logical element to fail from the nuclear heat generation viewpoint is the one which failed.

As stated before, control rods were initially withdrawn 62% of their length, corresponding to approximately 24 in. (61 cm). A study of Fig. 3-20 indicates that the fuel element melt occurred approximately 12 in. (30.5 cm) from the top or 24 in. (61 cm) from the bottom of the fuel element in question. The region of melting is rather sharply delineated. A study of the core locations of the control rods indicates that Rod No. 1 (position 7-4), the third from the last rod withdrawn, is immediately adjacent to this element and that the position of its tip just prior to the accident corresponds roughly to location of the melted zone. The second from the last rod withdrawn was No. 2 in position 4-5 adjacent to fuel element 5-6. By this time element 6-5 had certainly voided and may well have melted.

In summary, a new fuel element without built-up fission products and with little neutron absorption in the axial experiment basket, located in the innermost region of the core, surrounded by depleted elements, and with xenon built up only along the lower part of its length, suddenly had a fresh region exposed near the peak of axial flux as well as radial flux in which no xenon had been built up at all. Even if the element were not faulty originally, it appears likely that this element might have been the first to melt and might have melted where it did. If this element also had a particularly serious defect in fabrication, as appears likely, that made the situation even worse. Thus, while calculations are quite difficult and the exact time sequence will always remain uncertain, it is apparent that a number of contributing factors may have played a part in the fact that this particular element, and no other, melted.

The planning of the location of fuel elements, the planning of sequence of motion of control rods, and all other operations either on long-term or short-term basis must be made with a variety of such considerations as outlined above in mind. In a system such as this with many variable parameters, it is essential that all possible changes be considered carefully at the outset and that these considerations be reviewed by other knowledgeable and competent men not directly involved.

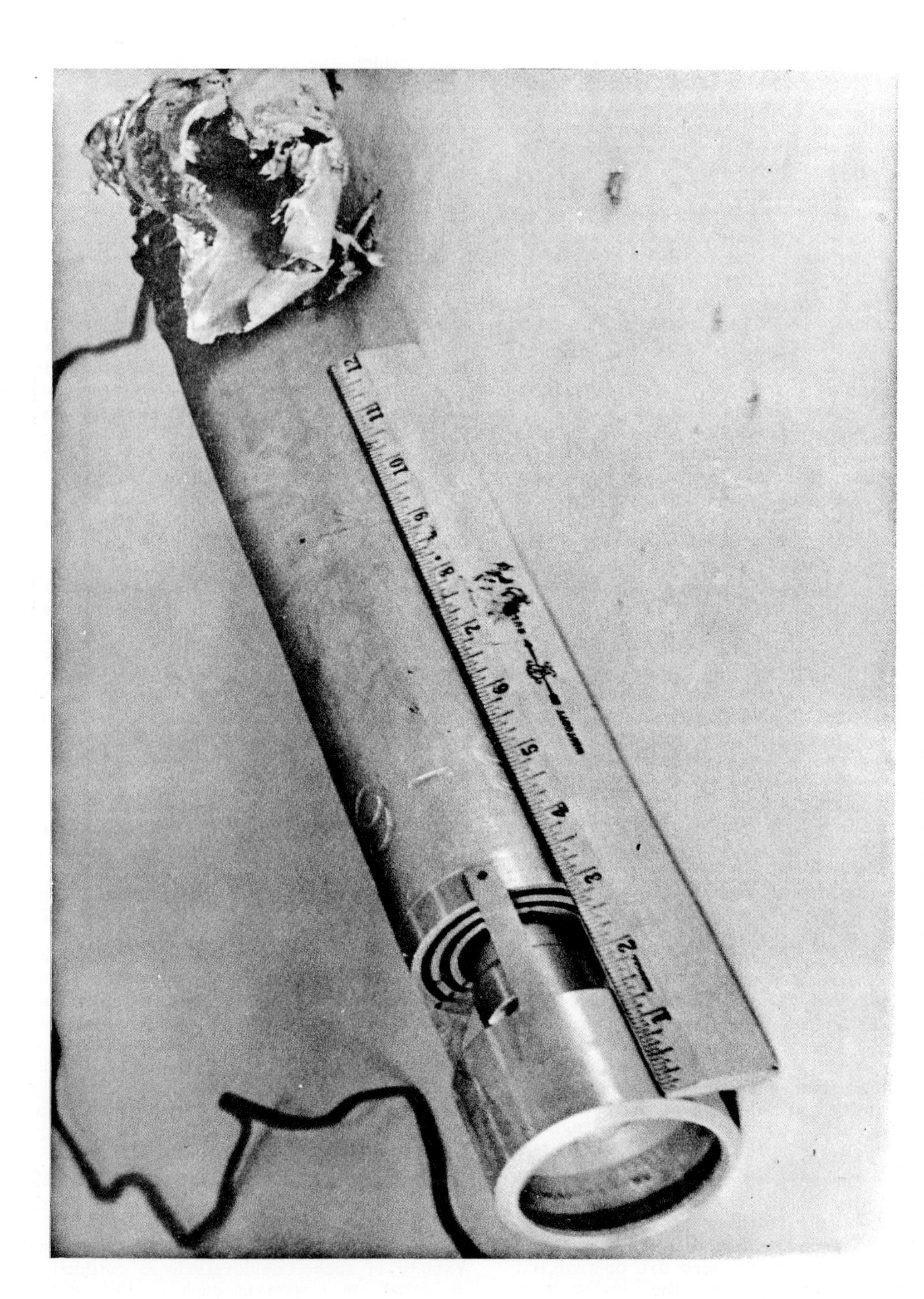

FIG. 3-20. View of upper section of failed fuel element in WTR.

(8) In selecting processes for the fabrication of U-Al alloy fuel elements it is wise to select a method to avoid hydrogen inclusions such as vacuum casting or cycling the melt above and below its melting temperature.

(9) "The existing head-tank vent system, which permitted fission-product gases to be discharged external to the containment vessel, was a serious design deficiency." [51]

3.11 The SL-1 Accident

3.11.1 Introduction

The first power reactor accident resulting in fatalities occurred on January 3, 1961 at the SL-1 Reactor (Stationary Low Power Reactor No. 1) at the AEC National Reactor Testing Station, Idaho Falls, Idaho. Three reactor operators were fatally injured while carrying out a supposedly routine maintenance operation on a shutdown reactor. Since these three men were the only individuals present, and virtually no recording instrumentation was operating at the time, there will always be some uncertainty about the exact sequence of events leading up to the accident. However, the general nature of what occurred was reasonably clear within a few days after the accident and the subsequent dismantling and detailed examination of the reactor has verified and amplified this understanding to the point where there can be almost no doubt of the course of the accident itself or of its primary initiating cause.

In the following sections, the information regarding the accident will be summarized and references will be given to more detailed reports. Section 3.11.2 outlines briefly a chronological history of the reactor and the events leading up to the accident. Section 3.11.3 gives the chronology of the accident and the post-accident operations. The observed fission product release and radiation levels are discussed in Sec. 3.11.4. Section 3.11.5 describes in more detail the features of the reactor and its history and experience pertinent to the accident. In particular, the control rods and their part in the accident are discussed in detail. Section 3.11.5 also discusses additional information gained in dismantling the reactor. Section 3.11.6 discusses an analysis of the features of the transient. Section 3.11.7 summarizes the information and Sec. 3.11.8 presents certain conclusions which have been and can be drawn regarding the accident.

3.11.2 Chronological History of the Reactor and the Accident

The SL-1 Reactor was a natural recirculation boiling water reactor designed to operate at a thermal power of 3000 kw. It was capable of producing 200 kw (net) of electricity and 1.37×10^6 Btu/hr for space heating. It was designed for use at remote military installations as a power and heat source. Design work on this reactor began at the Argonne National Laboratory in 1955. Site work began in the fall of 1956.

In March of 1958, the AEC Advisory Committee on Reactor Safeguards reviewed the design of the Argonne Low Power Reactor (later designated the SL-1 Reactor). The Committee based its review primarily on discussions with ANL personnel and the AEC Hazards Evaluation Branch and on a review of pertinent documents, the principal one being the "Hazards Summary Report on the Argonne Low Power Reactor (ALPR)" [54]. The Committee stated to the Chairman of the AEC the following conclusion:

> "In view of the low power level proposed for this reactor (3 Mw thermal), the remote location at which it is to be operated, and the broad experience of the Argonne National Laboratory with boiling water reactors of this type, the Committee agrees with the conclusion of the Hazards Evaluation Branch report of March 4, 1958, that this reactor can be constructed and operated at power levels up to 3 Mw, at the National Reactor Testing Station without undue risk to the health and safety of the public."

Initial criticality was achieved on August 11, 1958. Argonne National Laboratory, the prime contractor, carried out the reactor design, performed the initial criticality and startup tests and completed a 500-hour full-power test in December 1958.

A report of the initial startup tests is contained in ANL-6078 published in May 1961 [55]. Mr. Shaftman, the author of ANL-6078, also presented a paper which summarized the test results at the ANPP Reactor Analysis Seminar held at the Martin Company, Baltimore, Maryland [56].

On February 5, 1959, the Argonne National Laboratory turned over contractual responsibility for the plant to Combustion Engineering, Inc. (CEI).

According to Annex E of the Report on the SL-1 Incident, January 3, 1961 by the General Managers Board of Investigation [57] "Combustion Engineering, Inc., through its Nuclear Safety Committee, conducted a thorough appraisal of the facility and its proposed operation, including reactor safety, by CEI on March 19, 1959." One military visit to the reactor in January, 1959, resulted in a report which is quoted in Annex E as stating that the reactor plant was "substandard in areas of operation, design and construction, safety and maintenance."[57] This military report was evidently not circulated.

The Army Reactors Branch of the AEC did state in March 1959, at the time the reactor responsibility was transferred to Combustion Engineering that the procedures and manuals which were turned over to CEI by ANL were not satisfactory for use by CEI. CEI was requested to prepare revised material. The material submitted by CEI was accepted as a basis for the start of reactor operations, but CEI was to further develop and modify the operating manuals and procedures after obtaining actual operating experience. "Reactor operating procedures, completely satisfactory to the AEC, have never been completed by Combustion Engineering, Inc., although they have been in the process of preparation and revision since mid 1959."[57]

There appears to be no evidence that any report of any safety problems at the reactor was circulated at any time. The only general reviews of reactor safety were those mentioned above. The only possible exception to this was CEND 1005, the Combustion Engineering report [57] which discusses loss of boron from burnable poison strips in the core. This report was circulated and even mentioned briefly in the November 1960 issue of Nucleonics. Evidently this report was not considered to have raised any safety questions, since no safety reviews were carried out as a result of its issuance.*

From its initial startup, the SL-1 was used to gain operating experience, develop plant performance characteristics, obtain core burnup data, train military personnel in plant maintenance and operation, and test components planned for use in subsequent reactors of this type. Combustion Engineering continued this program. Their initial test operation took place on March 6, 1959, and cold-critical experiments began on March 30, 1959. A 1000-hour sustained power run was concluded in July 1959. Important shutdowns for maintenance and inspection occurred in August 1959, January 1960, November 1960, and on December 23, 1960. Fuel elements were first removed for inspection in September 1959 and later in October 1959, August 1960, and November 1960.

The deterioration and bowing of the boron poison strips (which were tack-welded to the fuel elements) were discovered in 1959. During the August 1960 inspection large sections of the boron strips were observed to be missing. The fuel elements in the center of the core (see Sec. 3.11.3) were extremely difficult to remove by hand. Removal caused boron plate sections to fall off and caused flaking-off of material. A large number of flakes were collected in the bottom of the vessel. "As a result of these circumstances, it was felt that further removal of fuel elements might cause further loss of boron, so that no further inspections were conducted."[57] A short experimental and analytical program was carried out in September of 1960 to evaluate the situation [57].

The loss of approximately 18% of the boron and consequent reduction in reactivity shutdown margin (about 2% $\Delta k/k$) led to the insertion of strips of cadmium in two of the T-rod control shrouds on November 11, 1960. (See Sec. 3.11.5 for more details on core reactivity.) Operations at power continued until December 23, 1960, when the reactor was shut down for the holidays and for maintenance. At that time, the reactor had achieved 932 Mwd of core exposure of which 79 Mwd occurred between November 15 and December 23, 1960. Approximately 40% of the core life had been expended.

In a report to the SL-1 Board of Investigation in August of 1962, Zinn stated that Combustion Engineering had recommended to the AEC, and the AEC had approved, the design and construction of a new core for SL-1 which would have improved the SL-1 Reactor "by calling for adequate shutdown margin with any one rod removed, by providing for replacement of the boron-aluminum strips with boron steel encased in stainless steel, by providing for the installation of newly designed control rod drive mechanisms, and by replacing aluminum with stainless steel as structural material."

During the last shutdown period, the only operation planned which directly involved the core was the insertion of 44 cobalt-aluminum flux-measuring rods into selected positions in coolant channels between fuel element plates throughout the core. The purpose of these flux-measuring rods was to obtain a flux distribution at power to study the effects of the absorber changes which had been made and to plan for operation at higher power levels. Since the sole access to the core was through the control rod nozzles in the reactor vessel head, it was necessary to remove the control rod drive assemblies. The first shift on January 3, 1961 carried out this operation. The day shift carried the installation of the flux monitors to completion under the supervision of Combustion Engineering personnel.

The 4:00 p.m. to midnight shift consisted of three military personnel, namely: the shift supervisor (a qualified chief operator), his operator-mechanic assistant (a qualified operator) and a trainee. This crew was assigned the task of reassembling the control rod drives in position on the reactor top. The reactor was shut down, depressurized, and cold (85°C or 185°F).

3.11.3 Chronology of the Accident**

First indication of trouble at the SL-1 reactor was an automatic alarm received at Atomic Energy Commission Fire Stations and Security Headquarters at 9:01 p.m. January 3, 1961. The alarm was immediately broadcast over all National Reactor Testing Station (NRTS) radio networks. At the same time, the personnel radiation monitor at the Gas Cooled Reactor Experiment gate house, about one mile distant, alarmed and remained erratic for several minutes.

Upon the receipt of the alarm, which could have resulted from excessive temperature, high radiation, a missile impact, or a pressure surge in the region above the reactor floor, the Central Facilities AEC Fire Department and AEC Security Forces responded. A health physicist from the Materials Testing Reactor area was called at this time.

The fire engines and security forces arrived at the SL-1 site at approximately 9:10 p.m.

---

*Indeed, evidence seems to indicate that this problem had little or nothing to do with the accident which ultimately occurred, unless it created anxiety on the part of the operators concerning sticking rods and thus created the attitude leading to the accident.

**This section is taken in a slightly abbreviated form from the first complete press release. The material is presented essentially in its entirety because it illustrates the problems of communication, and the decisions which must be made—often without adequate knowledge—in the event of a serious accident. Even for a well organized group the operation presents many problems.

(See Fig. 3-21 [59].) Security patrolmen opened the gates in the site area fence and later the south door of the SL-1 Administration Building. Firemen equipped with self-contained breathing units and radiation survey meters went through the administration building and the support facilities building in search of the operators and evidence of fire. They detected radiation levels up to 25 r/hr (the highest reading on their meters) and withdrew. No fire or smoke nor any personnel were seen in the support facilities or administration building. The searchers did not enter the reactor building proper.

At 9:17 p.m. the MTR health physicist arrived at the SL-1. He and a fireman, wearing self-contained breathing units made another trip through

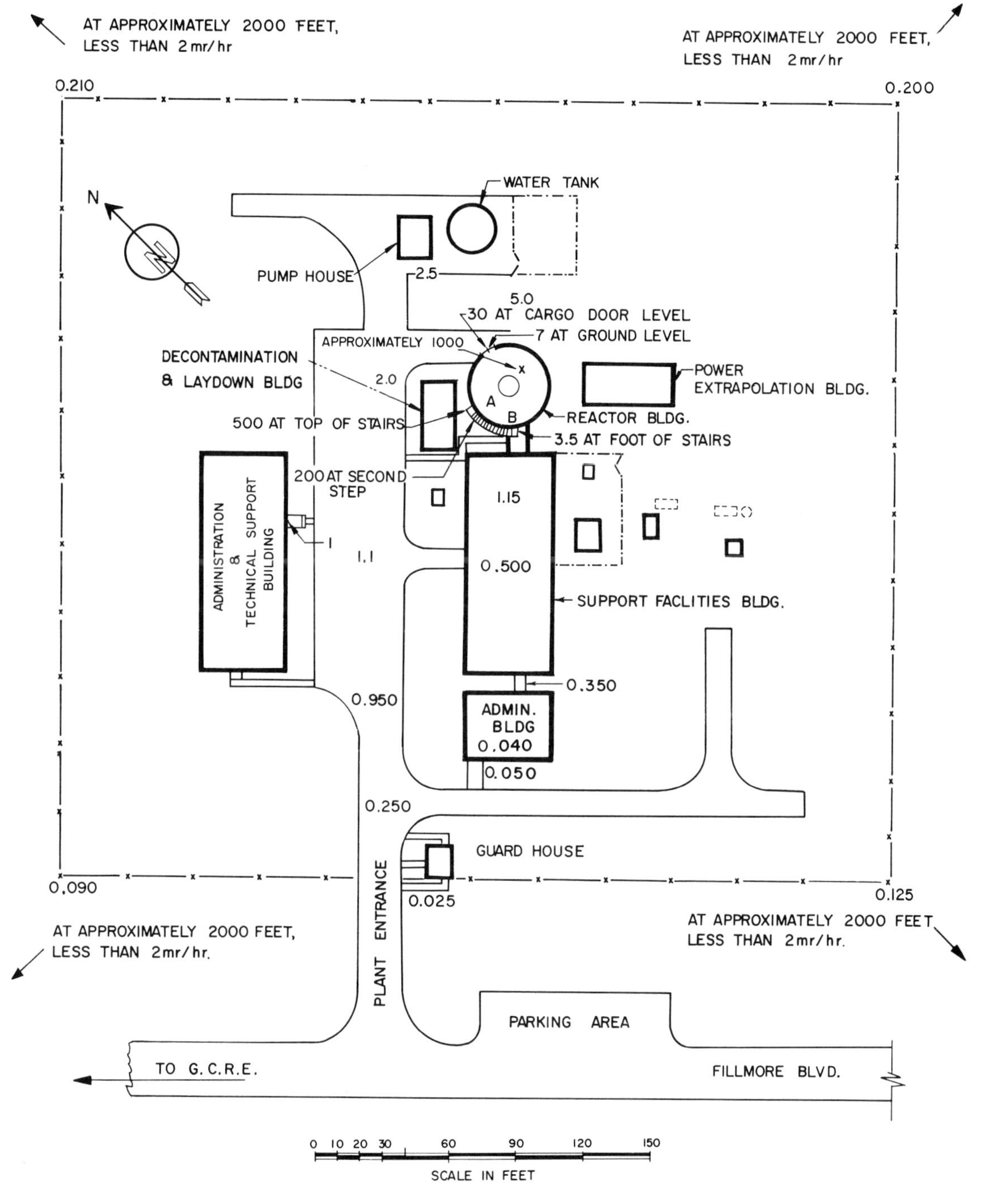

FIG. 3-21 Plan of SL-1 area. Numbers indicate radiation levels in roentgens per hour on January 7, 1961.

the administration and support facilities buildings and as far as the foot of the stairs to the operating floor of the reactor building, where they encountered a radiation level of 25 r/hr, the limit of the survey meter they were using. They retreated from the reactor building and thoroughly searched the administration and support facilities buildings looking for the three men believed to be on duty. They saw no one, nor any smoke or fire. During this search they encountered radiation fields of from 500 mr/hr to 10 r/hr.

By this time a radio check to other NRTS installations confirmed that the three SL-1 operators had not gone to any of them, so it was now presumed they must be in the reactor building.

At 9:35 p.m. two more MTR health physicists arrived, already in protective clothing. One of them, with two firemen and with a 500 r/hr range survey meter, went up the stairs of the reactor building until a 200 r/hr radiation field was encountered. This group withdrew from the building to plan a course of action based on radiation levels noted. Then, with AEC approval, the other MTR health physicist and an AEC fireman went to the top of the stairs and took a brief look at the reactor floor. Observed radiation levels of the order of 500 r/hr forced their quick withdrawal. They saw some evidence of damage but no bodies.

By 9:36 p.m. key personnel had been notified of the SL-1 accident. Following notification, many personnel who played key roles in the rescue efforts at SL-1 had to travel from Idaho Falls to the SL-1 site, a distance of 41 miles. At 10:25 Idaho Operations (AEC) designation of a Class I Disaster was broadcast over the NRTS radio network.

When four Combustion Engineering personnel, including the SL-1 Plant Health Physicist, arrived, they decided to enter the 500 r/hr field. The four Combustion Engineering men, having verified that the three military men on duty had not left the site, prepared to enter onto the reactor operating floor.

At approximately 10:35 p.m. the Combustion Engineering supervisors for plant operations and health physics, wearing self-contained breathing units and carrying two 500-roentgen scale Jordan type radiation survey detectors entered the reactor operating floor for less than two minutes. They saw two men, one moving. They withdrew and returned with two more Combustion Engineering men and an AEC health physicist.

Two of the group picked up the man who was alive and put him on a stretcher at the head of the stairs. The other three of the group observed that the second man was apparently dead. The group got the stretcher down the stairs and out the west door within three minutes of entry, and put the stretcher in a panel truck. The man was taken in the panel truck to meet the ambulance, transferred, and taken to the junction of Highway 20 and Fillmore Blvd. where the AEC doctor was met. When the doctor examined the casualty at 11:14 p.m. he pronounced him dead and the ambulance returned with the body to the SL-1 site pending a decision on the temporary disposition of the body.

At about 10:48 p.m. another group, made up of two military and two MTR personnel, entered onto the reactor floor briefly to locate the third man. They located him lodged in the ceiling above the reactor (in a radiation field later determined to be approximately 1000 r/hr) and determined that he was dead and did not attempt to remove him.

The recovery group went to another facility for preliminary decontamination. Gamma exposures of the five-man group ranged from 23 to 27 roentgens. As the groups were returning from the other facility they stopped long enough to permit one military man and one AEC health physicist to go through the support facilities building and close doors to lessen the chance of a fire starting and spreading in the disaster area; they did not enter the reactor building on this trip.

Having concluded that the remaining two operators were dead, the AEC-IDO health physicist suspended rescue efforts and ordered all personnel back to the roadblock established on Fillmore Blvd. at Highway 20.

Between midnight and 3 a.m. on January 4 approximately 30 people who had been engaged in the emergency at the SL-1 area were admitted to the dispensary for secondary decontamination. These personnel included firemen, security patrolmen, and military personnel. Preliminary badge readings and urine sample analyses for these 30 people were received around 3:30 a.m. and indicated that all personnel could be released. To assist in the above-mentioned decontamination processes, four health physicists came to the dispensary from the MTR and ETR area.

At approximately 6 a.m. on the morning of January 4 a team of five men removed the body from the ambulance located in the SL-1 area. The body was disrobed in order to remove as much contamination as possible at the site. The body was replaced in the ambulance, covered with lead aprons for shielding purposes, and transported to the Chemical Processing Plant where surface decontamination was attempted. Individuals involved in the disrobing and transfer process received a maximum exposure of 770 millirems gamma. Prior to decontamination the reading from the first body was approximately 400 r/hr at the head region, approximately 100 r/hr at the feet, and from 200 to 300 r/hr over the remainder of the body. First efforts to decontaminate the body resulted in no significant decreases in the readings. The other bodies, when recovered, exhibited similar high radiation levels and could not be decontaminated easily. (At the time of preparation for burial the contact dose rates were still all above 1 r/hr and as high as 50 r/hr for the three bodies [59].)

At approximately 4 p.m., January 4, 1961, preparations began to recover the second body from the reactor operating floor. The body was located in an area where radiation levels were estimated to be approximately 750 r/hr.

A recovery team consisting of six military personnel and two AEC health physicists proceeded, after having been extensively briefed, rehearsed, and attired in protective clothing, to the entrance of the SL-1 compound at about 7:30 p.m. Of this group, two military men and the two health physicists entered the Support Facilities Building through the side entrance into the maintenance workshop area. A blanket was placed on the floor of the control room.

Because of the high radiation levels to be encountered, the maximum permissible working time on the reactor operating floor was limited to one minute. One health physicist was assigned to hold a stop watch and time the actual entrance to the reactor operating floor, signaling the two-man recovery team when their time was up. The other health physicist remained in the support facilities building to check the body for radiation after its removal from the reactor building.

Having been briefed as to the location of the body to be recovered, the two-man team entered the reactor operating floor and proceeded directly to the body. One man picked up the victim's legs while the other grasped the shoulders and they moved rapidly out of the high radiation area and down the stairway. Their one-minute limit in the reactor area did not expire until they were part way down the stairway. The two men continued down the stairs and placed the body on the blanket in the control room.

The second two-man team entered the Support Facilities Building and went to the control room where they picked up the body by the four corners of the blanket and carried it out of the SL-1 compound. The work clothing was removed from the body, which was then placed in an ambulance standing by for the purpose at 8:08 p.m. The ambulance proceeded with the body to the Chemical Processing Plant where facilities had been prepared to receive it. The third two-man military team proceeded into the Support Facilities Building and onto the reactor operating floor for the purpose of attempting to gain some more information about the status of the remaining body and the reactor.

The short periods of time that these recovery teams were in the high radiation areas on the reactor operating floor resulted in gamma exposures of from 1 rem to about 13 rem.

On Thursday evening, January 5, an official photographer entered the radioactive reactor compartment to photograph the scene. Radiation fields greater than 500 r/hr were reported by the accompanying health physicists. The photographer, wearing protective clothing and breathing apparatus, was allowed 30 seconds to complete his assignment. By entering the reactor compartment only long enough to trigger his camera and withdrawing to a less radioactive area to change film and make adjustments, the photographer was able to obtain an interior photograph which assisted AEC investigating teams in making plans to recover the third body and in evaluating damage to the reactor operating area. Maximum radiation exposure of these two men was less than two roentgens (gamma) of radiation.

Because of the high radiation fields (above 500 r/hr) personnel could not climb onto a beam to free the third body which itself was highly contaminated with radioactive material.

The plan for removal of this third body was to position a large net (5 ft × 20 ft or 1.5 m × 6 m) under it and attempt to lower the body onto the net. The net itself was fastened to the end of a crane boom. The large doors on the reactor building that are used for moving equipment in and out of the building were opened to permit the crane to position the net just below the body. A closed-circuit TV camera had been placed in the reactor building to help position the net.

When the net was in position, teams of two men each were to move in quickly and try to lower the body onto the net. Because of the radiation fields, each team had less than a minute to make their attempt at freeing the body.

Due to a malfunction of the television equipment, it was necessary to use the first team of men to check that the net was properly positioned; they accomplished their mission in less than their allotted time.

Four additional teams were used to accomplish the mission of freeing the body and lowering it onto the net. A sixth crew, outside the building, was used to move the crane which held the net. The third body was removed from the building at 2:37 a.m. on January 9, 1961. The estimated doses received by the men entering the reactor building to free the body ranged from 2.5 to 7.5 rem.

Recovery operations were completed at 4:42 a.m. January 9, 1961.

During the period 23 persons received radiation exposures above 3 r and extending up to 27 r total body exposure. Of the total, 14 received exposure of 3 to 12 r, 6 were in the 12 to 25 r range and 3 above 25 r. Precautionary medical checkups on these men did not disclose any clinical symptoms [59].

The recovery of the third body terminated the period during which an NRTS disaster plan was in effect (Phase 1). During this period the reactor was determined to be subcritical, but whether or not there was water in the core and the general condition of the core (including control rod positions) were unknown.

Phase 2, establishing the nuclear status of the core, was carried out by Combustion Engineering between January 10 and May 22, 1961. They established that the vessel contained no water, measured radiation levels, and secured photographs and television views of the tank interior.

Phase 3, the dismantling of the reactor, and the study of the evidence was undertaken by General Electric. This phase began May 23, 1961 and extended until November 1962.

3.11.4 SL-1 Postaccident Radiation Levels*

Monitoring at the site and in potentially contaminated areas downwind from the site was begun immediately after the accident [59]. No activity above background except in the immediate vicinity of the SL-1 area was detected by an aerial monitoring survey early on January 4. Some four flights in the first nine days were carried out with some activity (about 125 counts/sec

*See also chapter on Radioactive Waste Management.

above a 220 counts/sec background) detected out to distances of three to four miles. The activity was generally to the south and southwest of the site, as would be expected from the prevailing wind direction.

Sagebrush sampling showed counting rates (per gram) up to 600-700 counts/min for areas about 2 miles (~1.2 km) downwind of the site with backgrounds of about 10-20 counts/min per gram. Release estimates based on crosswind traverses indicate that an upper limit of about 10 curies of I^{131} had become airborne by midmorning of January 4, and 20 curies had become airborne 24 hours later. In addition, it is estimated that another 50 curies at least were released between January 6 and January 30, 1961 [59]. This is an appreciable release of radioactivity but certainly under the conditions at the site not alarming or dangerous. As Dr. Frank Pittman stated [59], "Even though the SL-1 plant is not contained, it appears that practically all of the fission products have been confined to the operating floor and to the immediate area surrounding the power plant building. The airborne activity level at the control point 3/4 miles (~1 km) from the plant, on January 14-16, 1961, showed an iodine level of one-half the maximum permissible concentration for a 168-hour week. On January 19-20, 1961, the levels were 1/30 of the maximum tolerances. There is some indication that airborne activity is still being released (January 24, 1961) since these levels have not been decreasing more rapidly. But they are now at very low levels, and I remind you again that this is from an uncontained reactor where no effort was made to have airlocks or seals or anything of that nature."

To complete the radiation picture, Fig. 3-21 shows superimposed on the site layout the radiation levels observed on January 7, 1961. The decay rate of the gamma activity at the site after January 5 was determined to be exponentially decreasing with a 31-day half-life; the exponential fitted the observed levels to within 10% for several months after the incident. The activities observed are those of the relatively longer-life fission products.

Estimates [60] based on the measured radiation levels on the operating floor and on the debris removed from the reactor building appear to show that 3-5% of the fission product inventory escaped from the pressure vessel. A revised estimate might bring this to as high as 10%. The fission product inventory was estimated to be about 10^6 curies of the long-life isotopes. Of this total only about 0.01% escaped from the reactor building. General Electric [61] estimates that less than 0.5% of I^{131} inventory was found in the desert. It should be pointed out, however, that almost all the energy released in the accident was that in the nuclear transient. The system was not pressurized and the fission product after-heat was not sufficiently high to cause the post-accident core to remain molten or to cause further melting when the coolant was lost.

3.11.5 Technical Description and Information Gained from Recovery Operation

General. The general site plan of the SL-1 reactor area is shown in Fig. 3-21 [59]. The reactor building was cylindrical, and normal access to the reactor operating floor was provided by a stairway from the adjacent Support Facilities Building which housed the reactor control room. The reactor installation was intended as a prototype of an essentially uncontained installation for use in the arctic. The reactor building was a simple steel shell with 0.25 in. (0.65 cm) thick steel walls 48 ft (14.5 m) high and 38.5 ft (11.5 m) in diameter. The doors were not leak-tight and no attempt was made to consider special containment features. The reactor vessel, the fuel storage well, and the demineralizer were located in the lower third of the building and were shielded by gravel as it might be in an arctic installation. The middle third of the building contained the turbine-generator, feedwater pumps and equipment, switchgear, and the shielding blocks which covered the reactor head during operation. The shielding blocks were moved by an overhead crane and were displaced outward radially from the reactor top at the time of the accident in order to provide access to the control rod mechanisms. A recirculating air-cooled condenser was located in the upper third of the building.

The reactor vessel itself was 4.5 ft (1.35 m) outside diameter and 14.5 ft high (less the head). The vessel wall was 0.75 in. (1.91 cm) thick SA-212 steel clad with 0.1875 in. (0.475 cm) stainless steel. It was designed for a pressure of 400 psi (~25 atm). The reactor configuration is shown in Fig. 3-22 [54]. The piping connections consisted of a feedwater spray ring inlet, a separator return line, a purification purge line, a boron spray supply line, and an exit steam line. The reactor core was located well towards the bottom with the chimney region for boiling and steam separation above it. The control rods within the core were connected to extension rods and in turn to rack gears within bell-shaped housings on top of the reactor vessel head. Each rack gear was activated by a pinion gear on a shaft leading through magnetic clutch unit to a drive motor outside of the shielding blocks on the reactor top. Above the reactor vessel head was a metal enclosure filled with thermal insulation and then above that metal punchings, boron oxide, and gravel. Next to the side walls of the pressure vessel was a layer of magnesia brick for thermal insulation.

The reactor core configuration in use at the time of the accident is shown in Fig. 3-23 [53]. The core could accommodate up to 59 fuel assemblies and one source assembly. There were positions for five cruciform-shaped control rods and four T-shaped rods. The control rods were fabricated 0.060 in. (1.52 mm) thick cadmium, mechanically clad with aluminum-2% nickel. (All aluminum-nickel used in the core was Alcoa M-388.) They each had an overall span of 14.25 in. (36.20 cm) and an effective length of about 34 in. (86.4 cm). The cadmium was perforated at intervals by 0.50 in. (1.27 cm) holes. The aluminum-nickel clad was dimpled at those points and spotwelded to the other face. The aluminum-nickel clad was edge welded to complete the

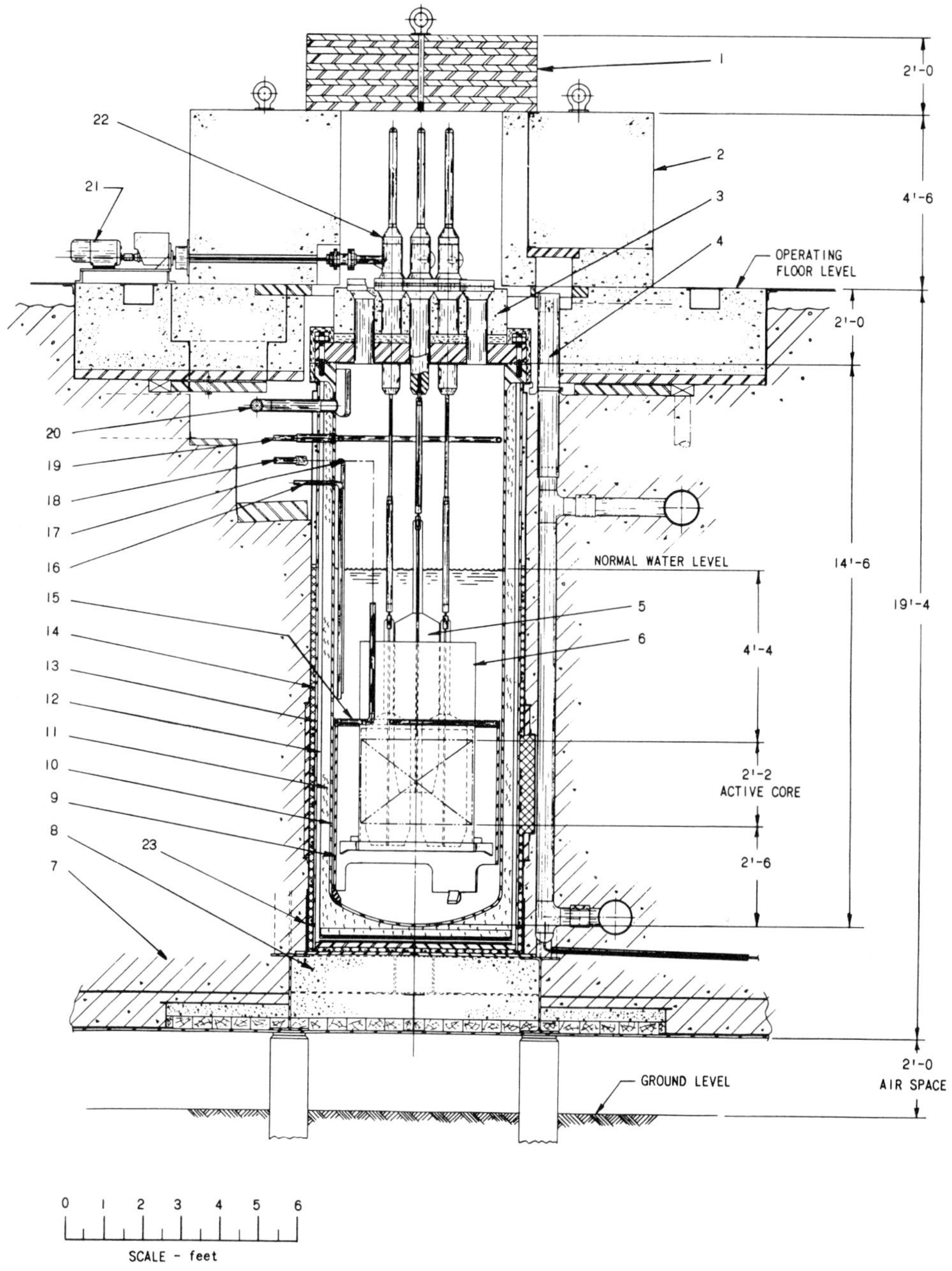

FIG. 3-22 SL-1 reactor installation, vertical section. Numbers indicate principal parts, as follows: 1 - laminated top shield; 2 - concrete shield; 3 - dry shield mixture; 4 - instrument well; 5 - control rod; 6 - core structure; 7 - gravel; 8 - dry shield mixture; 9 - thermal shield; 10 - pressure vessel (4.5 ft outer diam.); 11 - insulation; 12 - air space; 13 - support cylinder; 14 - lead thermal shield and cooling coils; 15 - feedwater spray ring; 16 - separator return line; 17 - purification purge line; 18 - feedwater inlet; 19 - boron spray ring; 20 - steam line; 21 - control rod drive motor; 22 - control rod drive; 23 - Boral shield.

units. Each fuel assembly consisted of nine 0.120 in. thick flanged fuel plates assembled to side plates by spot-welding. A cross section of the element is shown in Fig. 3-24 [53]. Each fuel plate contained a 0.050 in. (1.27 mm) thick fuel portion of aluminum nickel uranium alloy clad by 0.035 in. (0.89 mm) thick aluminum-nickel on both sides.

The goal of the reactor operation was to achieve a three year core life at 3 Mw power. As a means of achieving this end, burnable poisons were chosen. Originally it had been intended to disperse fully enriched B^{10} in the fuel matrix. Because of the developmental problems, it was finally decided to speed up fuel element procurement by eliminating

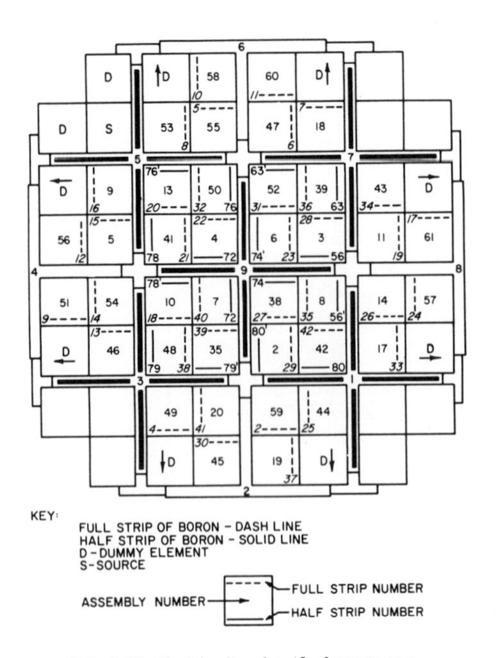

FIG. 3-23 SL-1 loading for 40-element core.

the boron from the fuel itself. Instead, aluminum (X 8001) strips with 0.6 wt.% B^{10} were chosen. These strips were welded to the side plates of the elements in the manner shown in Fig. 3-24 [53]. This method proved to be very flexible and fuel was moved about within the core until a satisfactory loading was obtained. A description of these experiments is contained in reference [55] published in May 1961. Basically they led to selection of two reference cores—a reference 40-element core and a reference 59-element core for use at higher power. The 40-element core with slight modifications, as shown in Fig. 3-23, was the one in the core at the time of the accident. It will be noted that the 16 center elements have an additional half boron strip attached. The strip was spot-welded to the bottom half of the opposite side plate of these elements.

Control Rod Mechanisms. The control rod mechanism is shown in more detail in Fig. 3-25 [61]. The design of this mechanism was such that to accomplish disassembly or assembly it was necessary to move manually the control rod blades within the core. Since assembly was essentially the reverse of disassembly, and since disassembly is easier to describe from Fig. 3-25, it will be done that way here. (Although assembly was the operation being carried out at the time of the accident, the accident could conceivably have happened during disassembly as well and, in fact, this would appear to be even more likely.) In its down position, the rack gear does not extend above the top of the spring housing. The pinion gear and its shaft are disconnected and removed, thus allowing the rack gear to be inserted or withdrawn to any position.

The top of the rack gear terminates in a threaded stud. In normal operation a large washer and a nut slotted for a cotter pin are screwed down tight on this stud. The washer limits the downward motion of the rod when it contacts the top of the spring housing. In order to remove the pinion support housing from the reactor vessel head by sliding it up over the rack gear, it was necessary to remove the nut and washer, shown in Fig. 3-25 at the top of the rack (control rod shown in full out position). To accomplish this, a special tool was attached to the rack by threading it onto the stud above the washer and nut. The design of the mechanism made it necessary to raise the rack and with it the control rod almost 4 in. (10.16 cm) in order to raise the washer and nut high enough above the coil spring housing to insert a C-clamp between the washer and spring housing. With the clamp tightened and resting against the spring housing in order to hold up the rod, the nut and the washer could then be removed. Specifically, the disassembly instructions stated [57]:

" 9. Secure special tool CRT No. 1 on top of rack and raise rod not more than 4 inches. Secure C-clamp to rack at top of spring housing.

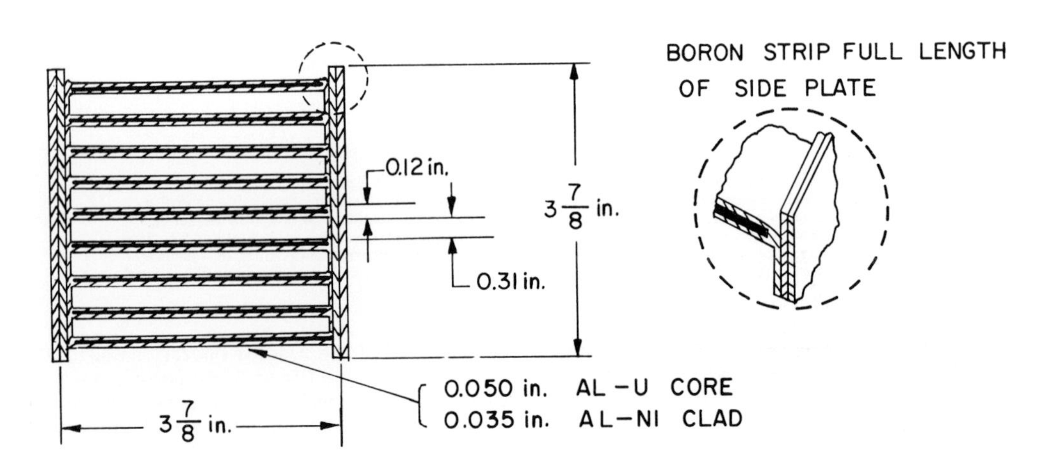

FIG. 3-24 SL-1 fuel element cross section.

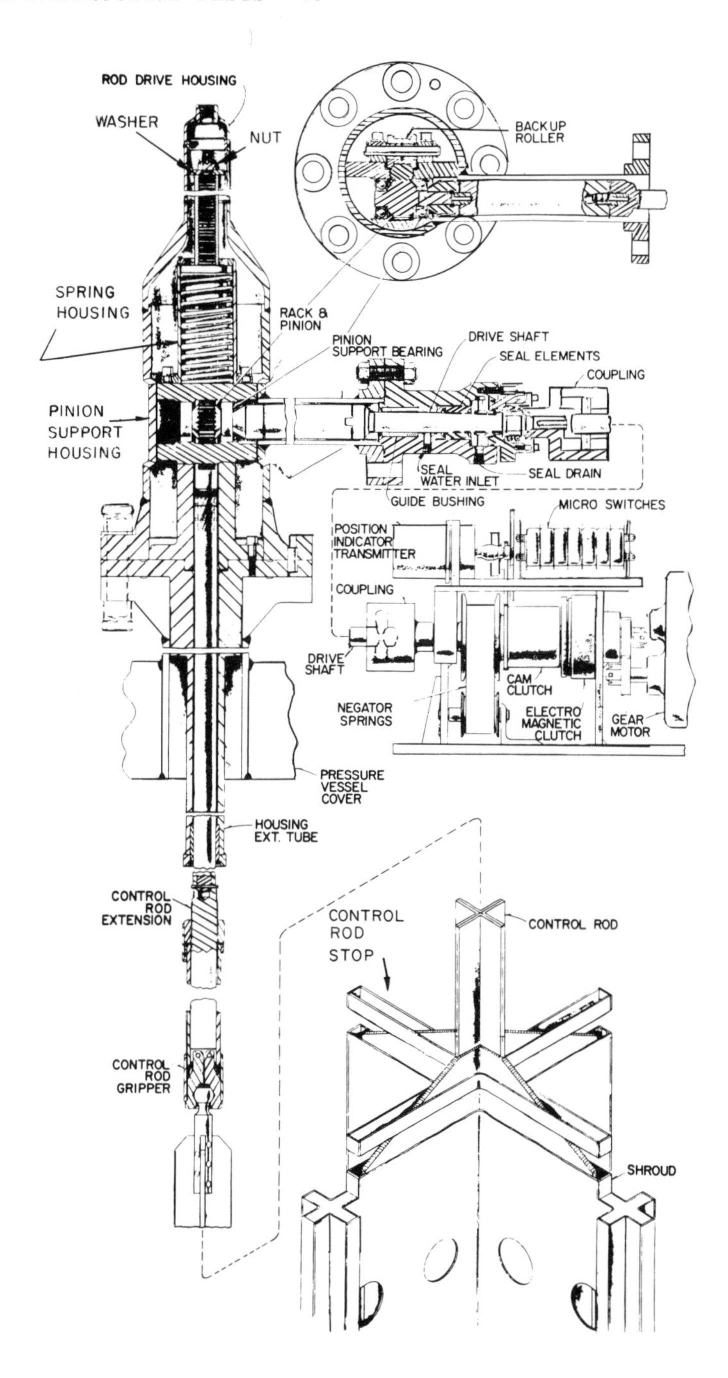

FIG. 3-25 SL-1 control rod mechanism and plug.

10. Remove special tool CRT No. 1 from rack and remove slotted nut and washer.
11. Secure special tool CRT No. 1 to top of rack and remove C-clamp, then lower control rod until the gripper knob located at upper end of fuel element makes contact with the core shroud."

The reassembly instructions stated:

"1. Assembly of the rod drive mechanism, replacement of concrete blocks and installations of motor and clutch assembly are the reverse of disassembly..."

The quotes given above are taken from the reactor operating instructions [57] and are all the written instructions pertinent to this operation that appear to have been available to the operator. There was no position stop and it was possible for an operator to raise the control rod higher and even all of the way out of the core manually. The rise of four inches was to be estimated by the operator.*

Fig. 3-26 [53] shows the three key positions of the control rod during this operation. According to this drawing, the procedure allows 1-3/8 in. (3.49 cm) of rack to extend beyond the top of the spring stop for location of the clamp as shown in Fig. 3-26(b). This is certainly adequate space to use for clamping a C-clamp of the usual work shop variety that was used. The 5/8 in. (1.59 cm) spring deflection below the top of spring housing shown in Fig. 3-26(c) added to the 1-3/8 in. (3.49 cm) for clamping gives a total of about 2.0 in. (5.1 cm) that the rod had to be raised as a bare minimum to allow disassembly as quoted in the operating instructions above. However, in reassembly, Fig. 3-26 clearly shows that the control rod had to be raised considerably more than four inches—in fact 5-45/69 in. (14.49 cm)—to position the rod for clamping. Since the operator probably judged the distance of the rack above the top of the spring in any case, this discrepancy between requirements and operating instructions is probably not important, but it does point out the inaccuracies of directions supplied to the operators.

While the mechanism postulated for the accident from the beginning was the uncontrolled withdrawal of the central control rod, the best evidence that this was indeed the cause of the accident is summarized in Section III of reference [61].

The central control rod shield plug mechanism was found after the accident lying across the top of the pressure vessel. Examination of this unit and other equipment showed [61]:

(a) The upper parts of the mechanism, including the springs, spring housing assembly, pinion gear housing, pinion gear, and roller bearings were in essentially good condition. The shield plug itself, which is shown as integral with the housing extension tube in Fig. 3-25, had incurred severe deformation at two points on its flange. Later matching of marks on this flange with others in the ceiling (an angle bracket in the fan room floor and a condenser corner in the fan room above the reactor) showed that this plug had entered the ceiling and fallen back on top of the reactor.

(b) The housing extension tube shown in Fig. 3-25 was severely collapsed near the top where it joined the main plug body and at the lower end of the tube two or three inches above the bearing housing located there. The collapse near the bearing housing bound the control rod extension firmly and the 4-3/4 in. (12.07 cm) bound section had to be cut out.

(c) It required an initial force of 5020 lb. (2280 kg) and a continuing sliding force of 3600 lb. (1635 kg) to push the control rod extension from the housing extension tube section collapsed on it.

(d) "Visual inspection and stereomicroscopy of the surfaces of the pieces cut from the section

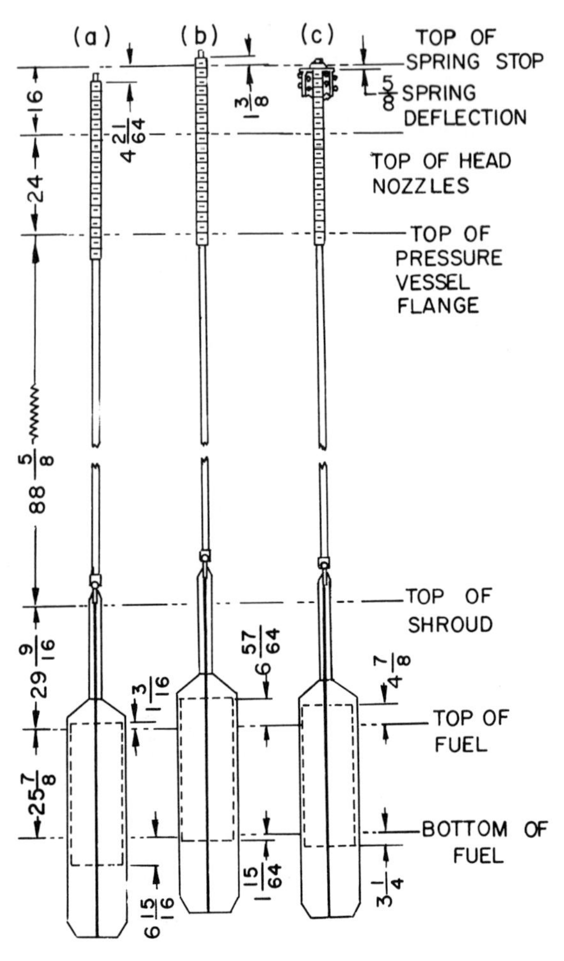

FIG. 3-26 SL-1 control rod cadmium overlap for various positions: (a) with scram stop washer and nut removed, (b) with rack above spring stop and clamped for scram stop washer and nut installation, and (c) zero position for reactor operations. (Broken lines indicate cadmium control poison positions relative to the core.) Dimensions are in inches.

*It is interesting to note that an increase in height of the rack stud by 4 in. (10.16 cm) at most and a corresponding increase in the heights of the bell housings could probably have made it unnecessary mechanically to lift the rods by hand at all. The action of screwing down the slotted nut could have been utilized to transfer the load of the control rod extension and rack from the core shroud to the coil spring shown in Fig. 3-25. No tool would have been needed and it would have been difficult—requiring the fashioning of a special tool—to lift the rod at all. Thus, it is likely that simple mechanical changes could have made it difficult, if not impossible, for this type of accident to happen—even if it were still possible to make the reactor go critical by withdrawing a single rod. (Springs of the type shown in Fig. 3-25 which absorbed scram force and held the rod at its zero position, exert force in such a direction that if the upper retainer of the spring housing were to fail and if the spring exerted forces greater than the weight of the control unit, it would be driven out—tending to suddenly increasing reactivity. Since the dead weight of the rod depresses the spring slightly, the springs in this particular case were weak enough to alleviate this worry.)

disclosed the very important fact that the final travel of the rod was upward through the shield plug after the seizing action started." [61]

"Prior to forcing apart the control rod extension and the housing extension tube from the 4-3/4 in. (12.07 cm) bound section, a portion of the extension tube was cut away and removed without using any force. A portion of the slide marks extending onto this piece showed burring indicating predominantly upward movement of the rod with respect to the plug, although there was some evidence of downward movement as well."..."Inspection of the rod after removal from the extension tube indicated no appreciable downward movement had occurred after seizure." [61]

(e) The lower end of the control rod extension terminated in a threaded stud. This study shows evidence at the thread root diameter of a gradual pre-accident fatigue failure which had progressed over 46% of the total cross sectional area of the stud, as well as a fresh impact-tensile failure during the accident.*

(f) The central control cruciform blade was found lying on top of the core with its shroud collapsed around it. The bottom of the follower was bound in the shroud 4.50 in. (11.43 cm) above the end of the shroud. (In the normal scram position the bottom of this follower extended 15-1/2 in. (39.37 cm) below the shroud. The control rod, therefore, was evidently 20 ± 0.5 in. (50.8 ± 1.3 cm) withdrawn at the time of the collapse. The control rod blade was firmly bound by the collapsed shroud and had not moved after the shroud collapsed, as evidenced by the imprint of the shroud pressure equalization holes on the control rod surface and by spattered molten metal in the same areas only. This also shows that the shroud collapsed before at least some metal was molten.

(g) Slide marks are evident on the control rod extension along its length and extending up into the seized section discussed in (c) and (d) above. When the housing extension tube collapsed on the control rod extension it made a unique impact impression with the tube itself and the rod. The impact impression on the tube was found separated 24 in. (61 cm) from that on the extension rod. The evidence is very convincing that after the seizure the control rod extension was driven downward to its final position relative to the plug by the ensuing events.

(h) Reconstruction of the severed pieces of the housing extension tube show that at the time the mechanism was recovered, the control rod was 2.50 in. (6.35 cm) withdrawn from its normal scram position with respect to the plug and housing extension tube. Therefore, it would appear that at the time of collapse of the housing extension tube, the control rod extension was 26.50 ± 0.25 in. (i.e., 24 in. + 2.5 in.) withdrawn.

(i) A broken C-clamp was recovered from the fan room floor above the reactor. The C-clamp was open to the diameter of the rack. The section of the gear rack still attached to the extension rod was 11 in. (2.79 cm) long. Stereomicroscopy detected many transverse stress cracks on the flat face of the rack gear along its entire length. General Electric [61] concluded that the gear rack sustained an impact fracture in a section weakened by fatigue. The matching section of this gear rack complete with nut and washer attached, was found on the fan room floor above the reactor. The under side of the washer was imprinted with an outline of the top of the rack and the washer was deformed downward at one point by an impact. The threaded upper end stud of the rack was broken off at a cotter pin hole and the broken end of the stud was found still in place in the control rod handling tool.**

From this information several conclusions can be drawn. First, since the washer and nut and handling tool were clearly attached to the rack gear, and since the C-clamp was propelled upwards into the loft and broken with its opening correct for clamping the rack, the stage of reassembly is clear. The rod had already been lifted, clamped with the C-clamp, the tool had been removed, the washer and nut had been replaced on the rack gear stud, and the tool replaced on the stud above the nut. Thus, every indication is in agreement that at the time of the accident

*Eventually this rod would have failed, even without the accident. It would have failed safe, unless the mechanical stops, shown in Fig. 3-25 in the lower right hand corner, were inadequate to withstand the impact of a control rod drop without shock absorbers. (The shroud wall thickness was 5/32 in. or 3.97 mm.) In that case, the rod would have fallen through the core and caused an increase of reactivity, although not a transient if the reactor were already shut down. From the drawings [54] it appears that the rod would have fallen out of the core about 13 in. (33 cm) in such an event. Because of the control rod cadmium overlap this would have been equivalent to about 8 in. (20.3 cm) of the core length unprotected by cadmium. If the reactor had been critical at power with the other rods withdrawn, a rough estimate made from the zero-power experimental results [55] indicates that 1.50 to 2.00$ in reactivity would have been added suddenly. This would have created a nuclear transient, but one that would likely have been retained without damage within the capabilities of the system. Estimates of the reactivity effects are difficult in this case since a section of the core would be affected that had previously achieved little burnup and in which no xenon and samarium had been built up. The designers of the reactor had worried about this possibility because this central control rod is the only one with a long follower—about 17 in. (43.2 cm) compared with about 4.50 in. (11.43 cm) for the rest.

**Metallurgical examination of the gear rack by General Electric disclosed that the rack conditions were of the type usually resulting from overstressing during the finish grinding or some cleaning procedure prior to chrome plating. A fabricator representative supplied information that similar cracks have been observed in 17-4 ph alloy after the use of certain cleaning procedures prior to chrome plating.

there only remained the operation of removing the clamp and lowering the rod into its final fully-in position. It should be emphasized that, from the post-accident evidence, there is no apparent reason why the rod was raised again.*

Tests by General Electric on the time needed to raise a rod in the identical situation showed the results given in Table 3-5 [61]. The control rod withdrawal times from the clamped position to 20 in. (50.8 cm) out, ~50 msec, are short enough to assure that the rod reaches the 20 in. position before steam is formed. The tests showed that even "casual efforts" averaged about 75 msec, still probably sufficiently fast to allow the accident to occur. These tests also showed that even had the operator realized that he had done the wrong thing, he could not have pushed the rod back in again fast enough to have changed the course of the accident significantly.

While other mechanisms have been postulated, it is clear that the accident resulted primarily from withdrawal of the central rod. The other control rods were found after the accident clamped in the full in position by their respective shrouds due to the core expansion during the accident. They could not have participated in any major way to the reactivity addition. The whole of the central control rod blade in its collapsed shroud was found on top of the core. Therefore, at some time in the accident it was fully withdrawn. The mechanism generally accepted has some inconsistences which can be plausibly explained, but not in a foolproof way. For instance, the central control rod blade and its collapsed shroud indicate an initial withdrawal of 20 in. The matching seizure points on the control rod extension and the extension tube indicate a withdrawal of 26.50 in. (67.31 cm), and, even more baffling, the final position of the seized control rod extension corresponds to a 2.50 in. (2.35 cm) withdrawal from the normal scram position—a position completely in keeping with the normal operation. It is, therefore, necessary to postulate that the pressure of the water hammer evolved in the accident ejected the central shield plug mechanism, and at about the same time crushed the housing extension tube into the extension rod. The inertia of the control rod blade and its collapsed shroud, or a collision of the rack gear and the attached tool with an object, caused the extension rod to be driven or pulled down and out of the housing extension tube for a distance of 24 in. (61 cm) so that the extension rod ended up in normal position for the operation—2.50 in. withdrawn. This is almost too close to the normal operational position to be believable, but no other explanation which can be logically defended has been offered.

*It is possible that it was raised because the rod might have shown tendencies to stick although this rod had shown few such tendencies. (Others had shown this tendency before the accident—see Table 3-6.) Perhaps the operator decided to exercise the rod without thought as to the consequences of the action. It is also possible, however, that an operator in anger, in a moment of careless fun, emotional instability, or in an act of deliberate sabotage raised the rod suddenly. But all of these are sheer conjecture.

Table 3–5

Manual Withdrawal of the Central Control Rod (SL-1) With Maximum Effort

(One man at a time pulled rod, except in condition (3). Handling tool gripped near bottom unless otherwise stated.)

Condition	Average Time in Milliseconds from Critical (16.7 in.) to 20 in.
(1) C-clamp in place	49
(2) No C-clamp, rod resting on spring	46[a]
(3) No C-clamp (2-man withdrawal)	45
(4) Stuck rod, quickly released	56
(5) With C-clamp, handling tool grasped in middle	77[b]
(6) With C-clamp, handling tool grasped at top	94[c]

[a] 375 msec from fully inserted to 20 in. position

[b] Could not pull rod past 27 in. position

[c] Could not pull rod past 22 in. position

Post-accident analysis by ORNL [62] showed the burnup of cadmium in the central blade was about 9%. The sampling of one unirradiated rod showed evidence that the cadmium had undergone melting during the welding of the cruciform section. This rod had been stored in a pool and was found to be full of water. Investigation showed that the design had an intentional opening to the interior through the control rod extension. In this rod, corrosion attack was evident on the cadmium and on the aluminum cladding. Only one of four spot welds was holding. During the examination of rods from the reactor after the accident it was also observed that few spot welds were holding. Many of the edge welds of the post-incident blades were either cracked or split open.** Corrosion within the blades was enhanced by water in some cases, but is not believed to have had any effect on the accident. Several welds in the control rods and many in the shrouds failed or showed evidence of failure due to lack of weld penetration.

The 17-4 ph steel used in the rack gears and some other components showed surface stress cracks, and, in fact, the threaded stud on the central control rod was about half cracked through at the time of the accident.

Lustman [57] points out that the riveted connection at the top of the rod extension piece (see Fig. 3-25) is the most probable point of failure in case the rod should drop on the shrouds.

Fuel. Lustman discusses in Annex R of reference [57] the metallurgical difficulties which were encountered in the development, fabrication, and use of the fuel elements and boron strips.

**The presence of water within the blade could lead to steam formation and blade swelling and binding at some critical time. If the spot welds were weak, the blades could expand and the two faces of any blade could shift relative to one another.

The yield strength of irradiated fuel specimens was almost double that of unirradiated specimens (80,000 vs 120-180,000 psi) and the Rockwell hardness had increased. The percentage elongation of irradiated specimens is about one-third that of unirradiated specimens. Corrosion of cladding during operation for about half of the core life appeared to be quite uniformly 0.0025-0.0035 in. (64-89μ) thick. Examination of the fuel carried out in September 1960 seemed to show that the surfaces were clean and reasonably free of crud except in the boron strip areas. Lustman believed that the activity levels of Na^{24} from the Al^{27} (n,α) reaction were higher than they should be (2.11×10^6 dis/min-ml at 3 Mw). He also believed that it was significant that the activity levels went up 120% to 4.72×10^6 dis/min-ml when the power level was only raised 60%.*

In order to weld the boron strips in place without changing fuel dimensions, the initial thickness of the side plates was machined from 0.120 in. (3.05 mm) to 0.055 in. (1.40 mm). This exposed one of the edge bonds of the fuel plate to water at a nominal thickness of only 0.1 in. (2.54 mm) of aluminum from the meat. Lustman believes that this is the principal reason why fission product activity was observed throughout core life. The situation can be seen in Fig. 3-24.

One phenomenon within the fuel that was helpful in studying the temperatures reached was the change in appearance of the fuel meat in the heat-affected zone. In some cases there were three distinct zones observed: a high temperature zone which appeared to be dark gray in photographs of fuel sections, an intermediate temperature zone which appeared to be almost black, and a lower temperature zone light gray in color. In addition, of course, there also sometimes appeared a region where melting and vaporization followed by catastrophic expansion had occurred. The dark region observed has been attributed [61, 62] to the phase change in Al-17%U alloy which occurs at 1194°F (640°C).

Boron-Aluminum Strips. The boron strips were fabricated by mixing X-8001 alloy and boron powders, Annex R of [57]. The mixture was encased in an X-8001 can and sealed. It was then hot-extruded to a rectangular cross section which was subsequently rolled to shape. By this procedure the aluminum-boron meat was roll-bonded to an aluminum-nickel cladding which ended up as 0.002 in. (0.051 mm) thick. Strips were then cut from the finished sheet, leaving the edges exposed and the strips were then fusion spot welded to the fuel assemblies. The same technique was also used in BORAX III and is described in reference [54].

As a result of the bowing of the boron strips observed in September 1960 ANL initiated some tests. Lustman [57] reports that fuel plates of the SL-1 type grew 1 in. in their 27 in. (68.6 cm) length and also bowed on corrosion testing in out of pile 600°F (316°C) water. A similar growth was not noted at 450°F (232°C). Boron-aluminum strips 20 in. (50.8 cm) long grew 0.035 in. (0.89 mm) in testing for 14 days in 600°F water and X-8001 strips grew 0.117 in. (2.97 mm) in length. B-Al strips tack welded to X-8001 plates (presumably in the fashion used on the SL-1 fuel) bowed 0.060 in. (1.52 mm) after 14 days in 500°F (260°C) water and 0.118 in. (3.00 mm) when tested in 600°F (316°C) water. Lustman observes that "it is thus apparent that corrosion of the SL-1 fuel elements, unaccompanied by irradiation, would cause the poison plate bowing observed during the interim examinations." Further, "it is probable that the buckling observed August 27, 1959 at about 200 Mwd of operation is caused by corrosion growth,"—but "corrosion growth would not be expected to greatly embrittle the poison strips."—"On the other hand, irradiation would markedly decrease ductility at boron depletions above 0.1 at .%. (About 1 at.% boron depletion can occur in the SL-1 poison strips.) Corrosion of the strips would tend to become increasingly more rapid, the more the plates became embrittled and cracked, because of the exposure of new corroding surfaces at the crack. The increase of aluminum surface exposed would cause additional corrosion at an accelerated rate; the increase in Na^{24} activity in the coolant from about 6×10^5 dis/min-ml early in 1959 to 2×10^6 late in 1960 may be indicative of such progressive change in the burnable poison strips. It is plausible to postulate that progressively more rapid deterioration of the poison strips during the September 30 to December 23 period directly related to the cause of the incident."

Post-accident examination of these strips [62] revealed that they were covered with a tenacious corrosion film. The hard, brittle black material was identified as α-basic aluminum oxide. The original composition of the poison plate material was 0.423 + 0.016 wt.% boron (enriched to 92.94 + 0.22 % B^{10}) in X-8001 aluminum alloy. Analysis of samples for wt.% boron and isotopic composition correlated with fission product analysis, which gives nvt in the adjacent fuel elements, gives a reasonable picture of the performance of these strips. Analysis seems to show that samples from areas with less than 30% burnup show no external signs of deterioration while samples with more than 56% burnup show very bad deterioration of a laminated character along the rolling axis of the strip. Edge corrosion 0.0025 - 0.0035 in. (64 - 89μ) thick was observed on all strips on microscopic examination. There was a corrosion similar to edge corrosion in laminations along the rolling axis but deep in the specimen on the highly ir-

*Lustman considered it rather surprising that no provisions were incorporated in the plant to analyze reactor water for suspended solids since this was the first plant to use X-8001 aluminum alloy. Lustman concluded that while plant chemistry seemed well controlled, "supplementary chemistry data which would have been of considerable value in development of the SL-1 type reactor, and in assessing the performance capabilities of the new type of cladding employed, the fuel elements, the burnable poison plates, the control rods, and other developmental items were greatly restricted both because of the number of technical personnel assigned and the equipment available."

radiated areas. It appeared that the bulk area of the specimen between laminations was not affected by the radiation corrosion. Yield strength of irradiated specimens appears to have increased about 30%. The density of irradiated strips seems to be slightly decreased. Lustman [57] states that no attempt to find boron in the primary water system from the deterioration of the strips was successful.

Control Rod and Fuel Operational Experience. The design and assembly and disassembly of the control rod drive has been described in the previous section. From initial operations onward, intermittent and increasing difficulties were encountered in the proper operation of these rods. A detailed examination of the SL-1 operating logs [53] indicates that the SL-1 rod drive mechanisms performed a total of 4300 movements. In 84, or 2%, of the movements unsatisfactory performance occurred, of which 46 involved a failure of a control rod to fall freely in a scram. It was then necessary to utilize the mechanical drive to assist or drive the rod in. Thirty-three cases of sluggish or sticking rods were observed in November and December of 1960.

During the first part of core life, and perhaps later, the difficulty was apparently centered at the seals where the drive shaft entered the rack and pinion housing (see Fig. 3-25). The presence of "crud" and also the rate of flow of seal water affected performance. More efficient filtration of seal water helped, and performance specifications were met so long as seal water flow was at the design value.*

The performance of the system during the fall of 1960 became gradually worse. The AEC General Manager's Board of Investigation Report [57] sets forth the situation:

"On the one hand it was postulated by several witnesses that the bowing of the boron strips attached to the fuel elements exerted sufficient lateral force to result in reduction of the clearance within the control rod shrouds, restricting the free motion of the blades. On the other hand, several witnesses felt there was no evidence for such closing of the shrouds, but that there might be some accumulation of crud on the shroud and blade surfaces; and that exercising the drives tended to prevent sticking of the rods in the shrouds. It was also indicated that the higher power operation, which took place only after November 1960, and the addition of the cadmium strips required further withdrawal of the control rods than had been previously required. Consequently, the drives were being used in a new region of the mechanical structure, where closer tolerances, or other differences, caused increased difficulties with rod motion.

"The only known interferences within a shroud were:

1. A crimp or similar bend was observed in the top edge of the No. 1 shroud. A special stainless steel wedge-shaped tool was designed and used to straighten out this defect.
2. A dummy control blade made of aluminum was fabricated for insertion and irradiation in the No. 4 shroud. On initial insertion the blade could not be fully inserted. The wedge-shaped tool was used on this shroud also, but since it could not be inserted within the shroud, the actual remedy for insertion of the blade was to cut a portion off the bottom of the blade.

After the incident a review was made of the operating logs from September 1, 1960, through December 23, 1960, by members of the military cadre. The data set forth in Annex J** give all recorded examples of control rod performance.

"According to testimony presented before the Board, all orders in the night order book for the instruction of reactor operating personnel are given by either the operations supervisor or the plant superintendent with the supervisor's or assistant supervisor's concurrence, and the following orders reflect the efforts of the operations group to maintain the rods in an operable status by frequent exercise:

12/20/60, by the plant superintendent-

'Each shift will perform a complete rod travel exercise at approx. 4 hours after the start of shift. This rod exercising will be required of each shift until further notice.'

12/21/60 by the operation supervisor-

'Perform a complete rod travel exercise on the graveyard and subsequent shifts.'

12/22/60 by the plant superintendent-

'Do not perform control rod exercises during 2.56 Mw power run' (Testimony indicates that a special power run to get equilibrium data was in progress at this time.)

"A review of the operating log #13 reflects

---

*The seal is made at a point still directly connected to the control rod through the rack and pinion gear. Thus, it can directly affect scram performance. While the seal is made on a rotating shaft which is better than a sliding shaft for reliability, still, it is not good design to allow any seal joint to be located directly in the scram unit where it may bind and totally negate scram action. On the other hand, the shaft seal location was better where it was than in a place where it would cause the vessel steam pressure to be exerted across the seal in such a way as to enhance the expulsion of the rod.

In his testimony to the SL-1 Board of Investigation in August of 1962 Zinn discussed new information developed in tests at Combustion Engineering on the operating rod mechanism. These tests, conducted by W. N. Wilhelm, indicate that the motion of the rack gear in its housing can cause reduced steam pressure in the upper part of the housing and indirectly slow the rod drop. The hydraulic effect which causes the action was dependent on the amount and temperature of the water in the shaft channel. It appears likely that the presence of crud in the water might augment the effect. A pressure equalizer tube to the housing eliminated the effect. Thus, many of the control rod troubles experienced on SL-1 might possibly have been eliminated by a relatively simple design change.

**See Table 3-6.

Table 3–6

Operations Log History of SL-1 Control Rods: Summaries of Typical Entries, September to December 1960

Date	Power (Mw)	Rod Coolant (gal/hr)	Performance (Numbers indicate position in inches)
			CONTROL ROD No. 1
Sep 11	0	180	Rod stuck at 20 position and would not drive in or out with 180 gal/hr flow; dropped when coolant was secured.
Oct 17	2.7	120	Dropped from 18 to 16; driven to 18 before freeing.
Nov 20	2.89	110	Raised from 20; stuck momentarily at 21.6 and then continued driving out to 22.
29	0	100	Dropped from 17; stuck at 15; driven to 12 before freeing.
Dec 7	3.0	100	Dropped from 19.4 to 0.4; driven to 0 before freeing.
19	3.06	100	Stuck at 20. Had to drive rod out (to 28) with a pipe wrench before freeing.
19	0	100	Dropped from 16 to 0 in 19 sec.
19	0	20	Dropped from 28 to 0 in 1.13 sec.
Dec 23	0		Stuck at 19.35; driven to 16 before freeing.
			CONTROL ROD No. 3
Dec 19	3.26	100	Stuck at 25; driven to 16 before freeing.
19	0	100	Dropped from 16 to 0 in 1.22 sec.
19	0	20	Dropped from 30 to 0, but hung momentarily at 6. Total drop time 1.185 sec.
23	0		Dropped from 19.35 to 18.85; driven to 9 before freeing.
23	0		Dropped from 9 to 0 in 0.5 sec.
			CONTROL ROD No. 5
Sep 11	0	180	Stuck at 18. Would not drop at 180 gal/hr; dropped clean with 0 flow.
11	0	180	Dropped from 18 to 10; driven to 0 before freeing.
Nov 15	2.71	80	Stuck at 16.5. Would not drive out electrically.
19	2.8	120	Stuck on drive out at 19.4.
20	2.95	120	Dropped part way (from 22.), stuck, then dropped to 0.
Dec 3	0	100	Stuck (at 5.5) while being driven out (from 0). Had to be forced by hand.
17	3.08	115	Stuck at 20. Had to be driven out by hand.
19	3.06	120	Would not drive on drive-out. Had to use pipe wrench
19	0	100	Stuck at 25.
19	0	100	Dropped from 16 to 0 in 0.52 sec.
			CONTROL ROD No. 7
Sep 11	0	180	Dropped from 20 to 10; driven to 0 before freeing.
Nov 18	2.7	100	Dropped from 19 to 2; hung momentarily at 7.
20	3.0	120	Dropped from 22.0 to 3 in 6.965 sec.
27	2.65	100	Dropped from 19.3 to 3; driven to 0 before freeing.
Dec 7	2.87	100	Dropped from 19.1 to 0 in 3.39 sec.
12	2.5	100	Dropped from 19.2 to 3; driven to 0 before freeing.
20			Dropped from 30 to 0 in 1.71 sec.
23	0		Stuck at 19.35; driven to 0 before freeing.
			CONTROL ROD No. 9
Oct 8			Drove out to 18 and pressure diff. went to + 15 psig. Drove in manually.
Nov 4			Drove out of auto. (18 in.)
6			Could not get nut off the top of rod.
7	0	0	Nut cannot be removed unless heat is applied.
17			Jumped out of auto.
27			Hung momentarily before dropping from 18.2.
27	2.65	100	Dropped from 18.5 to 0 in 1.28 sec.

that the aforementioned orders were complied with by the operators. On December 23, 1960, when the reactor was secured, the operating log #13 includes, in part, the following:

'0825 Dropping rods to secure reactor
Rod drop times
#1 no drop
#3 dropped 1/2 in. and stuck
#5 clean drop in 0.82 sec
#7 no drop
#9 clean drop in 0.81 sec
'0827 Driving rods 1, 3, and 7 to zero
'0830 Controlling bypass steam flow to cool down to 2° F/min
'0835 Rod #3 dropped from 9 in. to 0.5 sec
Rod #1 dropped from 16 in. to 9 in. in 1.3 sec.'

"Testimony indicates that this behavior was worse than usual, and that the assistant operations superintendent remembered commenting that this was probably because of the preceding operation (with no rod exercising). The operating procedures called for 'scram-testing' the control rods before and during nuclear startup of the reactor. Rods were dropped individually from a prescribed height before going critical, and also from another height after achieving operating temperature and pressure in the reactor vessel. Prescribed times for full insertion were given. If the prescribed times could not be met, reactor operation was not to proceed. Testimony indicates that if a rod did not meet the drop-time criterion, the test was repeated.

"Review of the experience with control rod performance indicates that this behavior was probably not as bad as had been experienced on some previous occasions, however. A complete record of performance, obtained from the operating logs, is attached as Annex J*. The CEI project manager and the CEI assistant director of the Nuclear Division testified that they were not aware of any significant difficulty with the operation of the control rods and also were not aware of the entries in the log books over the past several months describing these difficulties."

*Table 3-6 summarizes typical entries from the operating logs.

The reasons for the poor response of the control rods during the last part of the reactor life are still not clear according to the reports quoted above. In reference [58] it is stated, "Removal of all elements was difficult. The previously reported expansion of the boron side plates appeared to have increased. As each fuel element was removed from the core a cloud of reddish material, believed to be iron oxide, and a flocculent material formed about the element." Again, in discussing Element 42 in Position 55, "Removal of this element was very difficult. Both the full boron strip and the half boron strip were missing from the fuel element and portions of these plates were recovered as well as the remaining portion of the full boron strip on fuel element No. 8 in Position 56. The No. 1 side plate was buckled between the fuel plates and the top support piece. Further, it states, "The test as originally conceived included inspection of five fuel elements. The test was terminated after three elements were examined because the corrosion of the boron side plates was found to be so severe that further removal of elements with the resultant disruption of boron could cause an unsafe condition in the reactor." Reference [58] recommends finally, "The results of this test indicate that further removal of fuel elements from the core might enhance the redistribution of the boron in the core to such an extent that an unsafe condition in the reactor might exist. It is therefore recommended that this test be limited to a final inspection at the end of core life."*

Reactivity Changes. As stated earlier, the design goal for the reactor was a three-year core life. By use of burnable poison in the form of the boron-aluminum strips spot welded to the fuel, it was hoped that the reactivity changes as a function of core lifetime would not be beyond the capabilities of the control rods to provide both adequate reactivity and a safe shutdown margin at all times. Even the early prediction indicated that the boron would burn out somewhat too rapidly and that the core would gain reactivity until somewhere near mid-life. However, the observed reactivity gains, as indicated by the position of the control rods, see Fig. 3-27 [57], were considerably greater than predicted. Much of the information of this figure was derived from reference [58].

The figure shows the measured banked rod positions for various critical situations as a function of core lifetime. From these data and control rod worth curves estimates of the shutdown margin were made. In addition, as discussed in the last section, physical examination of the fuel showed that the boron strips were bowing badly between spot welds and some pieces were flaking off or had corroded away. Luke and Cahn [58] estimated that 18% of the boron had fallen from the core and that a shutdown margin of about 2% remained. They further stated that calculations indicated that, if all the boron were lost, the reactor would be supercritical with all rods in by 3.3%. They recommended insertion of cadmium strips into the T-rod slots. Accordingly, on November 11, 1960, three cadmium strips were inserted into each of two T control rod slots (Nos. 2 and 6). It is interesting to note that they state "the rate of loss of boron has been constant over the past 300 Mwd of operation. Indications are that this shutdown margin will continue to decrease, thus requiring remedial action."** Figure 3-27 does show an apparently steady drop in rod position with lifetime from 320 Mwd on. No actual cold core banked rod criticality measurements were made after 711 Mwd. Therefore the last part of the cold reactor curve is based solely on the hot equilibrium xenon measurements at 2.56 Mw.

In the September 1960 tests [58], it was found that criticality could be established by withdrawing the central control rod alone to a distance of 14.3 in. (36.3 cm). From the S-shaped integral curve of the central control rod worth (total ~4.8%) and the fact that about 1% of negative reactivity was added by the boron strips, it has been variously estimated that the cold shutdown margin on December 23, 1960, was between 2 and 3% corresponding to a central rod withdrawal critical position estimated to be between about 14.5 to 19 in. (36.8 to 48.3 cm). General Electric [61, 62], who have studied this problem most thoroughly, estimate that a critical position of 16.7 inches (40.6 cm) is consistent with the observations. All of these answers must to some degree involve conjecture since no measurements of criticality were made after 711 Mwd and the reactor had not been critical since Dec. 23, 1960; moreover, operations involving minor control rod movements and tool contacts with the core had been carried out since then.

Pressure Vessel. After the accident all piping to the pressure vessel was found to be sheared off at the vessel outer face. The vessel and the dry core were lifted as an integral unit by a crane through a hole made in the top of the building and placed in a shielded flask on a trailer. It was then taken to the General Electric hot shop where the vessel and its contents were examined in detail. The vessel flange was observed to be severely distorted outward at the bottom of the flange. The tank was found to be bulged from an original circumference of 14.14 ft (4.310 m) to 14.47 ft (4.410 m) in a region corresponding to the original top of the core.

*With these comments and recommendations in mind, there is clearly some reason to believe that a crew of conscientious operators might attempt to exercise the central rod without thought of the nuclear consequences involved. It will be noted from Table 3-6 [57] that the central rod (No. 9) for the first time gave indications of sticking on November 27, 1960, and that the new order to exercise rods went into effect on December 21 only to be temporarily countermanded the next day. Thus, the problem could logically have been much on the minds of the operating crew on the night of January 3, 1961.

**There appears to be little basis for assuming that the corrosion and flaking would continue to be at a uniform rate. In fact, disturbing the core by such operations as putting in the flux wires or even by running might cause major losses to occur, although this can only be conjecture.

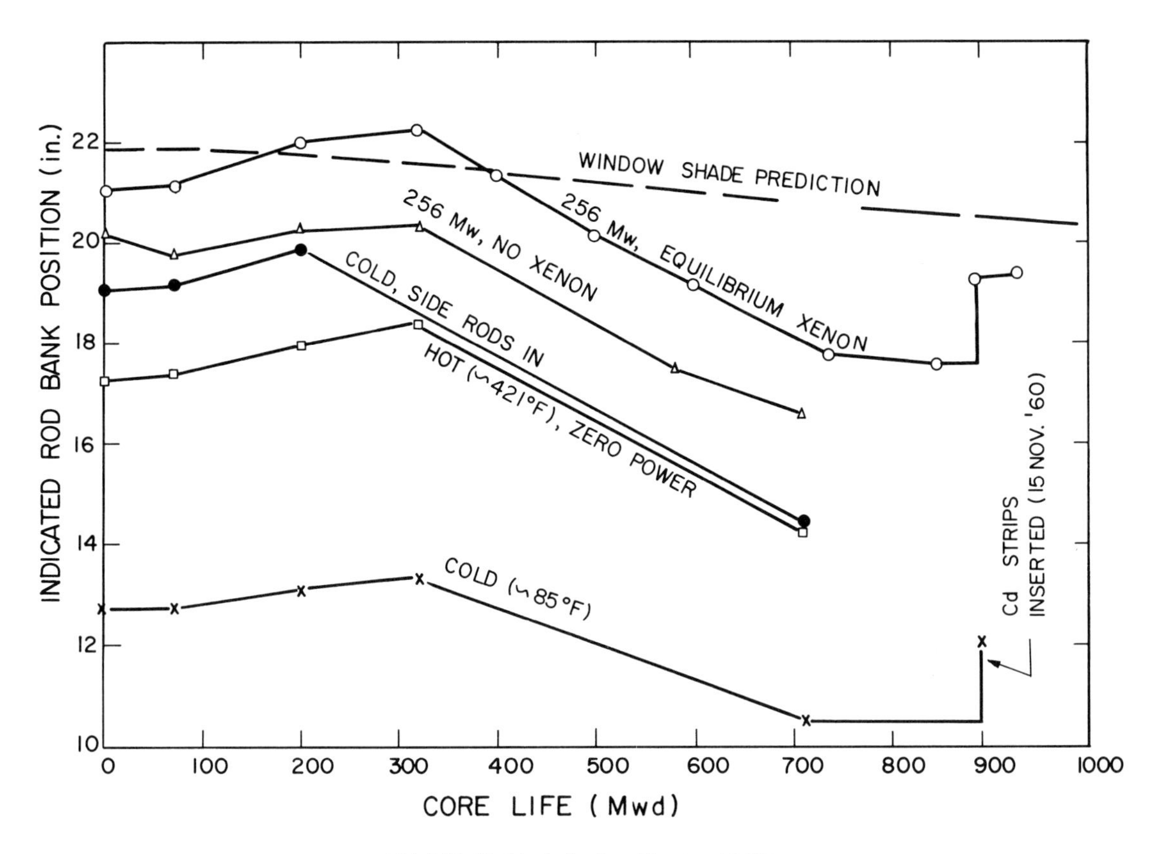

FIG. 3-27 SL-1 banked rod position vs. core life.

There was an even more severe bulge at the top of the vessel where the circumference had increased to 15.18 ft (4.627 m).

Samples cut from the vessel walls at various points were checked for soundness of welding, tensile strength, bending strength, and metallographic changes. Welding samples near the center of the vessel appeared sound. A sample near the top showed evidence of transverse cracks and porosity. The cracks were internal and only became visible as surface cracks after the bend specimen was cut out. The cracks were in the heat-affected zone caused by the welding of stainless steel to mild steel. The evident cause of the crack was the 7% elongation which occurred when the vessel expanded due to transient pressures during the accident. A sample taken from this section failed in bending tests.

Tensile strength and hardness properties of samples taken from the vessel correlate very well with the amount of cold working which occurred due to elongation during the accident. Radioactivity analysis of samples taken from the vessel wall show conclusively that radiation damage was not a factor during the relatively short time the vessel was in use. At the point in the pressure vessel closest to the core, the integrated fast neutron flux as shown by using the $Fe^{54} \rightarrow Mn^{54}$ reaction in samples of the carbon steel vessel was $\sim 8 \times 10^{17}/cm^2$. The integrated thermal neutron flux in the same region as shown by the $Co^{59} \rightarrow Co^{60}$ reaction was $\sim 1.6 \times 10^{18}/cm^2$.

The evidence appears conclusive that during the transient the vessel itself containing the core rose 109.5 ± 1.0 in. (2.781 ± 0.025 m) above its normal position and then fell back into approximately its original position. Insulating magnesia blocks used well down on the vessel sides and an insulation banding strip were found on the reactor room floor. Iron punchings used in shielding were found scattered about. The drive shaft coupling on the No. 5 seal housing, which remained bolted to the lid, collided with the drive shaft of the overhead crane. Other corroborating evidence exists.

It is also necessary to accept this occurrence in order to explain the fact that a number of samples taken from the operators indicate conclusively that they were exposed to an integrated thermal neutron flux of the order of $10^{10}/cm^2$ and an integrated fast neutron flux as high as $10^{13}/cm^2$. The hypothesis of uniformly dispersed uranium outside of the vessel cannot account for this order of magnitude exposure. It could only have happened if somehow an essentially bare core were brought into the vicinity of the victims rapidly enough that delayed neutrons were being emitted in copious quantities. The reactor core itself may have been well above the floor level and much higher than its normal position in the vessel during part of the transient.

Energy Release and Transient Period by Flux Wire Measurements. The best record of the nuclear energy release is probably given by the 45 flux wires that had been inserted into the core just before the accident. The General Electric group spent considerable effort in removing and identifying these wires. Each wire contained either 10 or 13 cobalt-aluminum pellets. Of the total of approximately 500 pellets, 352 were recovered, identified, and analyzed [61]. Some difficulty was encountered because the exposure received was only about 10^{-4} of that intended in the power test and because fission products contaminated the pellets. From the integrated flux plots obtained and a knowledge of the original fuel loadings and burnup (coupled with a fuel disadvantage factor of 0.91 and the fact that only 86% of the fission energy is deposited in the plates), the General Electric group estimated that 133±10 Mw-sec of nuclear energy was released in the burst. Data from the very center of the core was not obtained because damage there was so severe as to preclude identification of the pellets.

These flux wire results prove conclusively that there was a nuclear excursion and, further, that the peak energy release occurred in the core center radially and axially below the center, as would be expected if the central control rod were suddenly withdrawn.

An independent way to check the period of the reactor just prior to the initiation of shutdown is available through the flux wire data coupled with the regions of melting observed. In regions of limited melting and at the edge of regions where complete melting had occurred it is possible to estimate reasonably well the energy input that occurred. The flux wire information and a knowledge of the heat transfer properties of the system provide enough information to predict the maximum temperature that the center of a plate will reach for a given period. Calculations made for several different periods indicate that the most reasonable period is 4-5 msec. Certainly the period cannot have been much outside of that region.

Other Evidence on the Cause of the Transient. Evidence from the post-accident examination [61] clearly shows that all four off-center control rods remained in their normal scram position during the entire transient. All but the central one were found locked in their shrouds at the scram positions.

A careful investigation by the Stanford Research Institute [63] found no evidence of sabotage in the form of explosives or effects attributable to explosives. A hydrogen explosion in the space just below the lid is also considered impossible on the basis that it would have ejected all of the control rods (since all had washers and nuts screwed onto the rack gears) and, in fact, would have ejected the four off-center rods first since they were lightest in weight and since none of the mechanisms were bolted down. On this same basis any other preliminary energy bursts causing rod ejection and followed by the nuclear excursion can be rejected.

3.11.6 Transient Analysis

In a highly enriched water-moderated reactor such as the SL-1, the principal inherent shutdown mechanism which will terminate an otherwise uncontrolled transient is the removal of water from the core region, thus reducing the neutron moderation and increasing leakage. This reduction can be effected in one of three ways: by expansion of the fuel and water itself due to heating, by steam formation due to boiling, and by melting and vaporization of the fuel itself causing a violent expulsion of water from the core.

As a part of the post-accident analysis of the SL-1 accident, the General Electric group undertook an extensive analog computer analysis of the transient. The effort was aimed, first, at trying to pin down more accurately the time scale and sequence of events in the SL-1 accident and, second, at trying to identify those features which might differentiate this reactor and this accident from other transients studied in this general type of reactor. The details of the calculations and methods are presented in Section IV of reference [61] and in Section III and the Appendix of reference [62].

Following SPERT results and methods, they assumed that the burst could be represented by a two-term model (see chapter on Mathematical Models of Fast Transients) of the form:

$$\phi(t) = \phi(0)\left[\exp(\alpha t) - \left(\frac{\gamma-1}{\gamma}\right)\exp\left(\frac{\gamma}{\gamma-1}\,\alpha t\right)\right]$$

where $\phi(0)$ = measured peak power

α = (reactor period)$^{-1}$

$\phi_x(0)$ = extrapolated peak power

$\gamma = \phi_x(0)/\phi(0)$.

Peak power occurs at t = 0. Variations in γ between 1.25 and 10 have little effect on the fraction of energy released before peak power was reached, the change being about 5% at most. In a number of calculations General Electric used $\gamma = 1.5$ which fits the SPERT power burst shapes well during the power increase but drops too sharply for SPERT bursts during shutdown. However, it is believed that this rapid shutdown did occur in the SL-1 transient. This is partially inferred from the fact that almost all of the molten fuel material had a sponge-like appearance. In only three instances was a glazed appearance observed. The sponge-like appearance has been shown to be the result of sudden chilling [62], while the glaze results from slow cooling as might occur if the shutdown were more gradual.

General Electric calculated the amount of reactivity that might be compensated by expansion of the fuel and moderator and obtained the results shown in Table 3-7 [62]. The average plate surface temperature rise is 40°C (72°F) in these calculations corresponding to a plate surface temperature rise of about 115°C (207°F) in the peak flux region—about the temperature overshoot to be expected before boiling begins. This is about 35°C (63°F) above the assumed saturation temperature. Since the delayed neutron fraction in the SL-1 is about 0.70%, the prompt excursions will be limited by thermal expansion for periods longer than about 125 msec. (The SPERT 1-A reactor core has a much larger surface area and negative temperature coefficient, and this per-

Table 3–7

Reactivity Compensation Due to Expansion of SL–1 Fuel and Moderator when Average Plate Surface Temperature is 40°C (72°F) above Ambient

Period (msec)	Inverse Period (sec^{-1})	Inserted Reactivity (%Δk/k)	Compensated %Δk/k in Metal Expansion	Compensated %Δk/k in Conduction to Water	Compensated %Δk/k in Water by Neutron Heating	Total of Listed Compensated Reactivities (%Δk/k)
2	500	3.6	0.080	0.002	0.016	0.098
4	250	2.2	0.042	0.002	0.008	0.052
5	200	1.89	0.038	0.002	0.008	0.048
10	100	1.29	0.028	0.003	0.006	0.037
20	50	1.01	0.024	0.005	0.005	0.034
40	25	0.85	0.021	0.007	0.004	0.032
50	20	0.82	0.021	0.008	0.004	0.033
70	15	0.79	0.020	0.009	0.004	0.033
80	12.5	0.77	0.020	0.009	0.004	0.033
100	10	0.75	0.020	0.011	0.004	0.035
150	6.7	0.73	0.020	0.013	0.004	0.037
200	5	0.72	0.020	0.014	0.004	0.038

mits more compensation of reactivity by thermal expansion in that reactor. It can be controlled by expansion alone for periods as low as 50-100 msec.)

Water heating by conduction drops rapidly at shorter periods as it should. (It should be noted that in slow transients where water heating by conduction is important the water velocity in the core and the water volume and heat exchange characteristics of the entire system should be taken into account.) With shorter periods the importance of metal expansion and water heating by neutrons increases as it should. However, for transients with periods much less than 100 msec the shutdown provided by this mechanism is totally inadequate.

If thermal expansion does not limit the excursion, steam formation occurs. (This problem is discussed much more completely in the chapter on Heat Transfer, Sec. 3.3.2.) Steam does not form at once when the saturation temperature is reached. In analyzing the delays involved, General Electric has used the data of Shrock [63] who has investigated these transient boiling effects to a considerable extent. He attributes them in large part to two causes. First, a thermal resistance at the interface which depends on the nature of the surface and flow and is perhaps period-dependent, and second, the amount of superheating which exists in the water. The first effect should dominate for long-period transients and the second for short-period transients. Plotting Schrock's data and some SPERT data, General Electric find that, for very short periods, boiling begins 3 to 5 msec after the plate surface temperature exceeds the saturation temperature by 20 to 40°C (36 to 72°F).

The transient considered for the SL-1 was initiated from an ambient temperature of about 68°F (20°C). If the system had been near saturation temperature General Electric believes that it is likely that temperature overshoot rather than a fixed time delay would dominate, even for short periods. Once started, boiling evolves rapidly and both boiling and the violent disruption of the core caused by melting and vaporization of fuel should result in much more asymmetric transients than expansion. This is, in fact, what is observed. The out-of-pile data developed at SPERT [64] seem to indicate that almost immediately after the first sign of boiling the void volume reaches approximately 0.1 mm^3/mm^2 of surface area, providing a reasonably prompt mechanism once boiling starts. Since the void coefficient of the SL-1 is 1×10^{-4}% ($\Delta k/cm^3$) and it is necessary to compensate for 1.65% prompt excess reactivity (2.4% less 0.7% delayed neutron effect and less 0.05% due to expansion effects), something like 1.65×10^4 cm^3 of steam will be needed. An allowance should be made for the fact that the void coefficient is averaged while the voids will occur first selectively in the center—their region of highest reactivity worth—and hence somewhat less steam may be needed. The worth of the void coefficient also changes with void fraction present.

If 0.1 mm^3/mm^2 of steam occurs promptly after steam starts to form (3-5 msec after the plate surface temperature reaches the saturation temperature), 1.65×10^6 cm^2 of fuel plate surface area will be needed to terminate the excursion. The SL-1 has only 4.2×10^5 cm^2 of fuel plate surface area and therefore fast-acting steam voids are not sufficient.

On the other hand, if the same amount of reactivity is added to the SPERT 1-A core whose void coefficient is 6×10^{-4}% $\Delta k/cm^3$, then 2,750 cm^3 of steam are needed to compensate for the prompt excess reactivity. On the same basis as above, this would require 26×10^4 cm^2 of plate area. The SPERT 1-A core has approximately 40×10^4 cm^2 and so, conceivably at least, steam voids could limit the SPERT 1-A reactor on this transient.

The SPERT destructive tests described briefly in Sec. 3.12 indicate that a violent reaction occurred which was thermal in nature. Both in the SL-1 accident and in the SPERT destructive tests, fuel temperatures were reached within the fuel element meat which melted, and in the SL-1 accident even vaporized, the meat in a substantial fraction of the core.

In the case of the SPERT destructive tests the data clearly indicate that nuclear compensating reactivity was due to steam voiding as predicted above. In fact, the peak of the nuclear transient was well past before the violent core disassembly occurred. In that case, the shutdown mechanism was clearly due to steam, but when the molten metal of the meat had melted through the aluminum clad, the interaction of the cold water and the molten metal apparently caused a violent reaction—largely nonchemical in nature (see the chapter on Chemical Reactions). There is little evidence that any fuel element vaporization occurred.

In the case of the SL-1 reactor, some metal vaporization certainly occurred. Forty-seven percent of the fuel in the central 16 elements was destroyed, and 20% of the total core was destroyed. Metal vaporization would cause the sudden expansion of the flat fuel plates whose yet unmelted clad sides would, for an instant, resemble a flat box full of molten and vaporized metal. Debris from some fuel elements shows this expansion effect. This expansion would expel water from the core and act as a shutdown mechanism.

This phenomenon was probably the first and most important shutdown mechanism in the Sl-1 accident. Of course, almost immediately, and not really separable from it, the intimate mixing of molten vaporized metal with the water would cause a further and much more violent reaction of steam formation and some metal-water reaction. Of the 34.5 kg (75.8 lb) of fine material found at the bottom of the core, 1.5 kg (3.3 lb) was α-Al_2O_3—an appreciable amount, indicating that some metal-water reaction had occurred. The findings regarding metal-water reactions for this accident parallel those for the SPERT destructive test.

Since the delay time for initial steam formation has been indicated to be of the order of 3-5 msec, it is likely that the shutdown mechanism was almost entirely due to the melting and vaporization of the fuel meat itself, along with the prompt expansion effect. It is likely that orderly boiling, as such, had little to do with it.

Whether steam formation or metal melting and vaporization provide the predominant shutdown mechanism depends on a number of factors, including the clad thickness, its thermal conductivity, the meat diluent and its concentration, any thermal resistances, and perhaps other variables. The General Electric group [62] has calculated the ratio of fuel element center to surface temperature as a function of reactor period for both the SL-1 and for SPERT-BORAX type plates. This information is shown in Fig. 3-28 [61]. This plot is made relative to the ambient temperature base at the start of the transient. Thus, raising the ambient temperature closer to saturation will shorten the delay time before steam forms in appreciable quantities, and higher temperature ratios (shorter periods) must occur before plate metal vaporization will become the dominant shutdown mechanism.

From the graph by interpolation, at a period of 3.0 msec the ratio is ~5.5 and a surface temperature of 420°C (788°F) calculated by General Electric to occur 5 msec after the surface temperature reaches saturation with $\Delta T = 144°F = 80°C$, gives a peak center temperature of ~2300°C (~4200°F). This is the temperature where vaporization begins (2040°C melting point minus 20°C ambient starting temperature plus 310°C temperature equivalent of the heat of fusion.) For SPERT or BORAX plates, a 5-msec delay after saturation is reached gives a surface temperature of 970°C (1778°F) on a 2.0 msec period which is the period at which vaporization of fuel meat first occurs in that reactor. Figure 3-29 [61] shows a General Electric temperature profile plot in an SL-1 fuel plate for a 4 msec transient at the moment of peak power. Two separate sets of curves are shown. The lower set is melting in the meat center but not at the edges of the meat nor in the clad. Before the end of the transient the melting will have spread into the clad but not through it. Such a plate section is said to be at the "threshold of destruction." The regions of limited melting observed in the SL-1 core show clearly as dark areas in photomicrographs [62] of cross sections of heat-affected plates. The peak energy region curve shown has already melted throughout the meat and is on the verge of vaporizing at the center. This vaporizing of metal will cause expansion and destruction of the element and is the most likely source of the principal shutdown mechanism which will go into effect within a fraction of a millisecond after the distribution shown here is valid. The final shutdown will occur with a large fraction of the meat in this plate vaporized and most of the clad melted away.

The analog calculations of the General Electric group [62] were carried out with a number of different sets of assumptions. In all the sets a feedback of compensating reactivity due to moderator and plate expansion was assumed proportional to the nuclear energy released. Since the total compensating reactivity was small for this effect, other shutdown mechanisms were also employed. In the "A series" it was assumed that steam voiding was the sole additional mechanism with a feedback linearly proportional to energy input after an initial delay. The initial reactor period was assumed to be 3.6 msec and the delay before steam void formation began (simulated in the analog calculation by an appropriate trigger) was varied in the various runs. In all cases runs were carried out so that the total energy released was 130 Mw-sec. Two typical runs A-2 and A-4 are shown in Table 3-8 [62]. With zero steam void feedback delay, the transient power curve shape was symmetric, as indicated by E(peak)/E(total) being ~50%. As the delay time between the time when the plate surface reached saturation and the trigger time for the beginning of the linear boiling feedback was increased, there was some increase in asymmetry in the power curve shape as indicated in runs A-2 and A-4. Also the transient continued longer, more steam void shutdown reac-

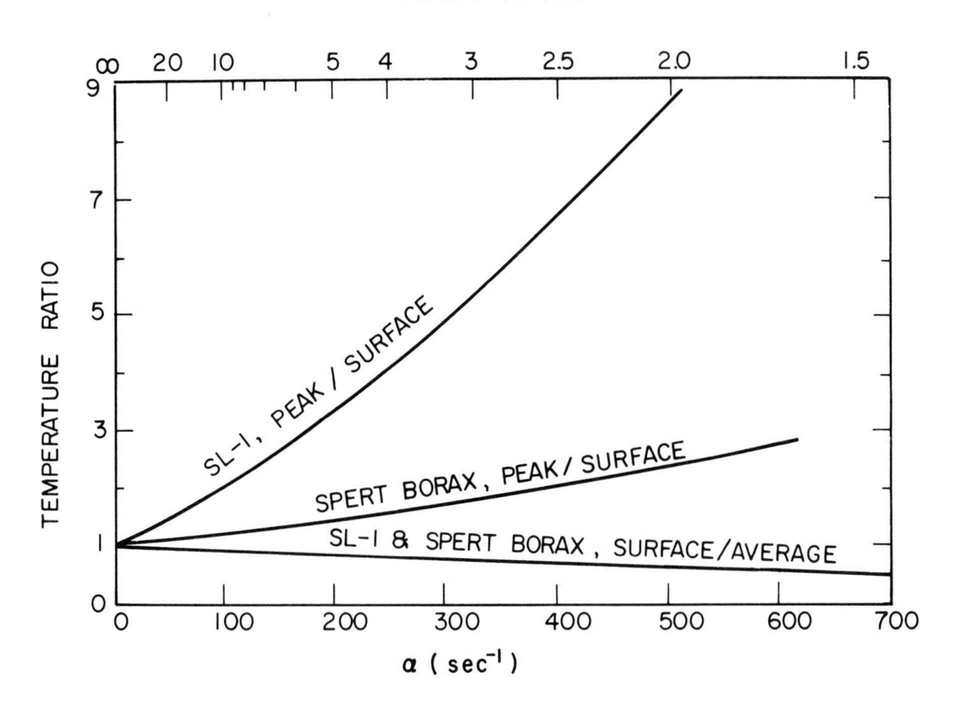

FIG. 3-28 SPERT, BORAX and SL-1 fuel plate temperature ratios vs. transient periods.

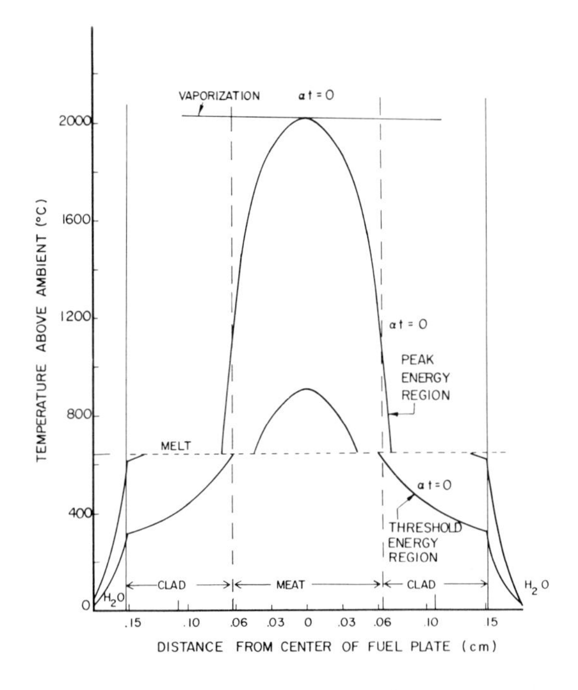

FIG. 3-29 Temperature distribution in SL-1 type plate at time of peak power for 4 msec transient. Both peak energy density region and "threshold of destruction" region are shown.

tivity was required, and the total energy in the transient tended to increase, as would be expected.

Another series ("B") of runs was made with the sole shutdown mechanism (in addition to expansion) being due to fuel plate metal vaporization. Two sample runs, B-1 and B-2, are shown. It was assumed that this mechanism resulted in a step input of compensating reactivity proportional to the energy input in the transient at the time the mechanism was triggered. This would amount to assuming an infinite delay in steam void formation. This resulted in greater asymmetry in the power curve and much higher peak powers. When the triggering temperature was set much higher in another run, corresponding to a 0.5 msec delay in the effectiveness of the mechanism (4892°F or 2700°C center temperature), the peak power was 26.4×10^3 Mw, the E(peak)/E(total) was 78%, and the compensated reactivity was 8.0% which made the reactor subcritical almost at once. It should be noted that the plate surface temperatures in the B type cases shown were over 500°C (932°F) when the fuel melting mechanism was activated. This is unrealistic and clearly points to the fact that both steam voiding and plate melting and vaporization shared in the shutdown of the reactor in the actual case.

A series of runs ("C") were made utilizing both steam voiding and fuel vaporization as the major parts of the final shutdown and three examples, C-2, C-3, and C-6, are shown in Table

Table 3–8

Results of Typical Analog Computer Simulated Excursions
Ambient Temperature 26°C (79°F)

Run number	A-2	A-4	B-1	B-2	C-2	C-3	C-6	D-1	D-3	D-5	E-1	E-2
Initial period (msec)	3.6	3.6	3.6	3.6	3.6	3.6	3.6	5.0	5.0	5.0	2.4	2.4
Type of feedback*	S	S	V	V	S,V	S,V	S,V	S	S,V	S,V	S,V	S,V
Delay of steam void feedback (msec)	2.6	4.4	∞	∞	3.8	5.0	3.4	4.4	3.5	3.8	2.3	1.4
Plate surface T at trigger (°C)	132	210	>500	>500	185	250	160	170	140	150	175	125
Feedback from steam at end (% ΔM)	3.5	4.0	–	–	0.8	2.8	3.7	2.8	1.0	0.9	1.2	2.7
Center T at metal vapor trip (°C)	–		1950	2100	1950	1950	1950	–	2100	2100	3000	2000
Step feedback initially (% ΔM)	–		3.4	3.9	2.8	1.2	0.3	–	0.6	6.0	4.8	1.0
Metal vapor feedback at end (% ΔM)	–		4.2	5.7	4.7	1.8	0.5	–	1.3	6.6	7.6	2.9
E_{peak}/E_{total} (%)	51	56	56	66	56	58	56	57	46	88	64	50
Power at peak (10^3 Mw)	10.0	13.9	18.5	21.3	17.2	17.8	17.2	11.7	11.6	10.7	39.1	39.4
Total energy (Mw-sec)	130	130	130	130	123	123	123	144	185	107	130	245

*S – steam voiding
V – vaporization of fuel plate metal

3-8. Run C-2 is intended to show the effect of a strong feedback due to the almost simultaneous bursting of a large number of plates in the core. It will be seen that the power bust shape and total energy released are not affected much. The 1950°C (3542°F) trip temperature was low due to a calibrating error. The effect is probably not of major importance. Attempts were made to vary the initial period as is shown in the D series and E series runs. These attempts show that shorter or longer periods do not give results agreeing with the evidence. Run D-1 used a 5 msec period and steam void shutdown only. Run D-3 decreased the delay time for steam voiding and the magnitude of the steam void compensating reactivity effect to allow for the center vaporization effect observed. When this was done, the 130 Mw-sec total energy constraint was exceeded unless the compensating reactivity effect of the fuel vaporization was made very strong as in D-5. Then the steam vaporization occurred at fuel plate temperatures that were too high, as in B-1 and B-2. Good fits can be obtained with a 5 msec period but only at the expense of either too high a plate surface temperature at the time of initiation of boiling or else too little core fuel vaporization to account for the observed fuel effects in the core and the large negative reactivity step that would be required. The 2.4 msec periods shown in E-1 and E-2 either do not agree with the 130 Mwsec total energy constraint or show very large amounts of total shutdown reactivity (~10%) and too high plate surface temperatures at boiling.

The General Electric group has used this model to predict the excursion behavior of the three reactor cores shown in Fig. 3-30 [65, 65a, 19]. While the model is not too sophisticated, it would appear that it predicts the energy release within 20-30%. The group believes that this model can be improved considerably by the use of detailed flux distributions within the cores being investigated.

The transient analysis has served to demonstrate that the period of the SL-1 excursion was probably close to 4 msec. It has shown that the shutdown mechanism is likely to have been a combination of thermal expansion, boiling, and vaporization of metal. It has given some insight into the shape of the transient. Last, but not least, it has provided some encouragement along with recent SPERT work that it is possible now to predict the course of such transients in at least one class of reactors.

3.11.7 Summary

The immediate cause of the SL-1 accident can be said to have been the manual withdrawal of the central control rod of the reactor. This action gave rise to a nuclear transient whose general characteristics are given in Table 3-9 [61].

The General Electric group has estimated in Sec. IV of reference [61] that the withdrawal of the central rod would make the reactor critical at 16.7 in. (42.42 cm). They then calculate that the excess reactivity released in continuing the withdrawal to 20 in. or 50.3 cm (the position in which the shroud and the central rod were collapsed together) is $2.4 \pm 0.3\%$ Δk. They believe that power then rose on a period of approximately 4 msec until steam voids and core vaporization terminated the accident at a power of about $1.9 \pm 0.4 \times 10^4$ Mw. At this time, the center meat temperature in the plates in the center of the core had just reached the vaporization temperature 2060°C (3767°F). The thickness of the clad (0.035 in. or 0.889 mm) was so great that none of the outer surfaces had yet reached the melting point. From the start of power decrease to the end of the excursion, 5%

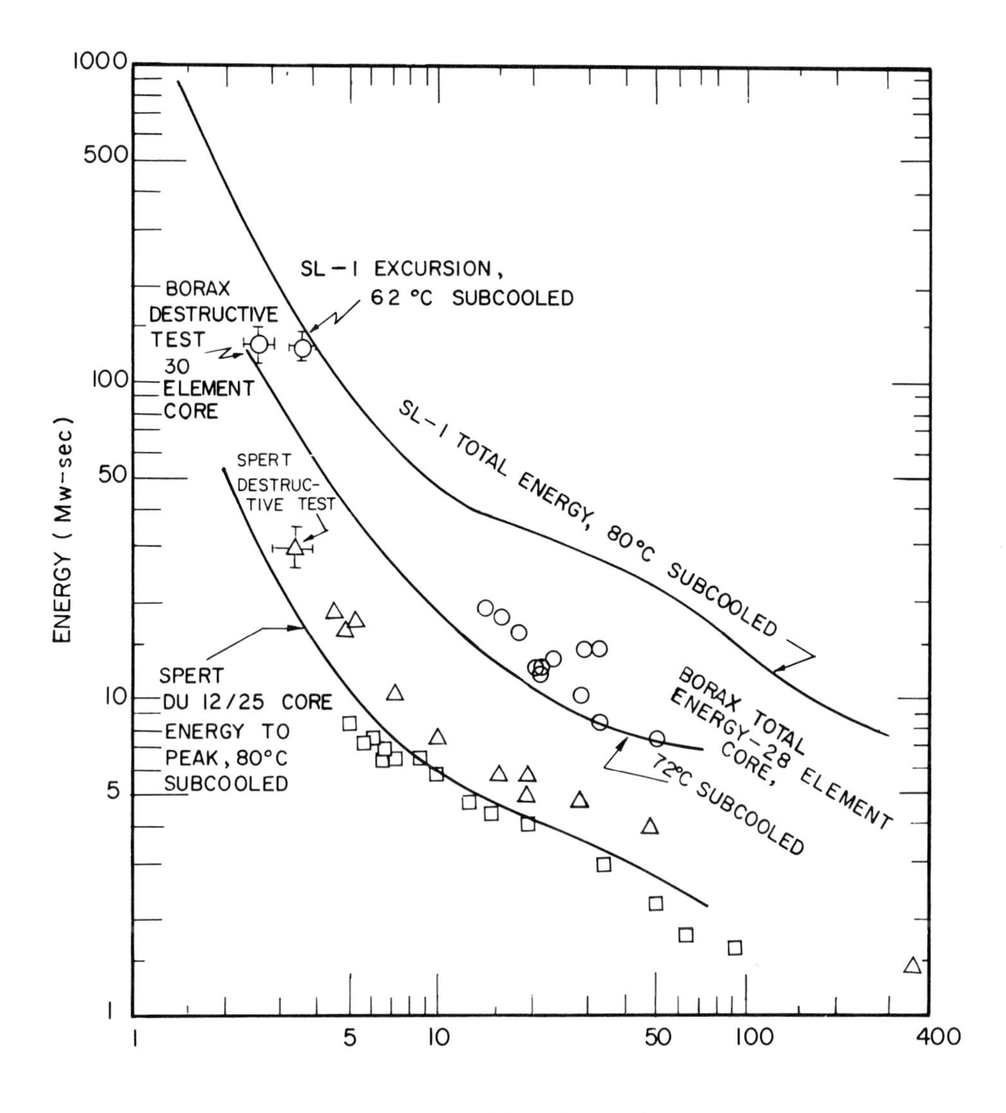

FIG. 3-30 Predicted and measured nuclear energy release vs. period for BORAX-I, SPERT-I, and SL-1. Total energy unless otherwise specified. Circles are BORAX data from reference [19], squares SPERT-I DU 12/25 data from [65a] and triangles SPERT-I destructive test data from [65].

a single slug. The water level in the tank was about 2.5 ft (76.2 cm) below the top of the vessel and the slug, therefore, had this distance to acquire kinetic energy. This slug hit the bottom of the of the plate area in the central 16 elements reached the vaporization temperature and this caused more steam production and violent destruction of this region. About 20% of the entire core shows melting proceeding to the clad surfaces. General Electric estimates that the total nuclear transient energy was 133 ± 10 Mw-sec and that no more than an additional 33 Mw-sec of energy (best estimate 24 ± 10 Mw-sec) was released in chemical reactions between the molten or vaporized metal and water.

The formation of the steam void terminated the nuclear transient, but it also created a high pressure region. The pressure wave front which developed no doubt spread out in all directions, striking the vessel side walls next to the core first and bulging them, then striking the bottom head and giving a net downward force on the vessel,* and finally accelerating upwards the entire mass of water above the core. It appears likely that the water moved upwards more or less as

*Apparently no one has looked into this downward force and one can only conjecture as to whether this downward force was sufficient to sever the pipe connections to the tank. It is difficult to judge the resistance to such a shock provided by the vessel supports.

Table 3–9

Chronological Summary of SL–1 Excursion

Time	Event
− 500 msec	Central control rod withdrawal starts.
− 120 msec	Reactor goes critical with rod at 16.7 in. (40.6 cm).
. . .	Central rod at 20 in. (50.8 cm), period = 3.9 ± 0.5 msec, (2.4 ± 0.3) % Δk.
0	Peak of power burst: (1.9 ± 0.4) × 10^4 Mw.
. . .	Portion of plates reach vaporization temperature, 2060°C (3740°F): ~ 5% of center 16 elements
~ 2 msec	Prompt nuclear energy release ends; total nuclear energy of excursion = (133 ± 10) Mw-sec [+ (24 ± 10) Mw-sec in metal-water reaction].
. . .	20% of plate area destroyed; center 16 elements 50% melted; central shroud and control blade ejected from core.
. . .	Water column above core accelerated by average pressure (500 psi or ~ 35 atm) to velocity ~ 160 ft/sec (~ 49 m/sec).
34 msec	Water slams against lid of vessel. Maximum pressure ≃ 10^4 psi (~ 700 atm).
. . .	Head shielding ejected. Plugs ejected with velocity of 85 ft/sec (26 m/sec) or less. Vessel rises, shearing connecting pipes. Guide tubes collapse. Nozzles and vessel expand.
160 msec	First plug hits ceiling.
. . .	Two-thirds of water expelled. 5-10% fission products expelled.
800 msec	Vessel hits ceiling. Total kinetic energy involved ≃1% of total energy released.
. . .	Insulation ripped from vessel.
2000 to 4000 msec	Vessel comes to rest in support cylinder.

vessel head with such force that it collapsed, or partially collapsed, the stainless steel guide tubes on the control rod followers (estimated to require 10,000 psig or about 700 atm), bulged the pressure vessel badly, distorted the upper flange, and caused the vessel to rise 9 ft, 1.5 in. (278.1 cm) above its normal position. All the loose control rod plugs (1, 3, 4, 7, 8, and 9) were ejected upwards and most penetrated the ceiling above.

Evidence seems to indicate that two operators were standing on the lid at the moment of the accident. One was thrown to one side as the vessel rose. The other was impaled on the No. 7 plug and was carried up to the roof of the reactor room where he remained suspended. The third man was also a casualty because of radiation and flying debris. Two of the casualties died instantly as a direct or indirect result of blast damage and a fatal head wound precluded the third man's survival. There was evidence of flash burns to limited areas of their bodies.

The water was, for the most part, ejected from the vessel, carrying with it released fission fragments. The kinetic energy which produced all of the mechanical damage and was responsible for the death of the three operators was less than 1% of the total energy released.

After the transient, the vessel sank back to near its original position and all was quiet. Two to four seconds had elapsed since the central rod started out.

3.11.8 Comments, Conclusions, Recommendations

General

The subject matter of every chapter contained in these volumes is important to the various phases of the SL-1 history and accident. Materials and metallurgy, fuel elements, mechanical systems, fluid flow, heat transfer, chemical reactions, reactor kinetics, criticality, and sensing and control instrumentation are all involved. In order to bring out these relationships, some conclusions and recommendations will be briefly stated below in subsections labelled to correspond to chapter titles.

Materials and Metallurgy

(1) The burnable poison chosen, a boron-aluminum alloy, was an unfortunate choice. In reference [57] Lustman concludes, "the selection of unclad B-Al strips for the poison application, without prior or concurrent irradiation evaluation, does not appear defensible, certainly not with present knowledge, and probably not with the information available at the time of selection."

The bowing and corrosion of these strips had two bad effects. First, the loss of boron led to a reduction in the shutdown margin. Second, from the evidence it would appear that the bowing of the strips is the most likely cause of the difficulty encountered in moving fuel and, more important, perhaps one cause of control-rod sticking during scrams. It can be conjectured that the operators' concern over this sticking could have led them to try to exercise the central rod and thus to initiate the transient.

(2) The use of 17-4 ph steel with incorrect heat treatment for components in the control rod drives, while apparently not a factor in the accident, could have led to another, although considerably less severe, accident. Coupled with Lustman's comment that the riveted connection at the top of the rod extension piece was the probable point of failure, it seems possible that the final fracture of the 17-4 ph steel stud on the central control rod (already 64% of the cross sectional area had failed) could have caused the central rod to drop through the core and result in a smaller, but nonetheless prompt-critical transient.

(3) Lustman [57] makes the general observation: "The highly developmental nature of the various core components such as the cladding, fuel alloy and fabrication method, which received their first utilization in SL-1, the control rods, whose design and operation conditions were unique to SL-1, and the poison strips, of a type which had never previously been utilized, appears incompatible with the use of the SL-1 facility without an extensive accompanying test, evaluation, and examination program."

(4) The General Electric group who disassembled the core state in their conclusions [62], "Metallographic examination of the core materials also revealed many areas of faulty welds and excessive corrosion. Although none of these problems are presumed to have had any influence on the incident itself, the progressively decreasing shutdown margin which resulted from corrosion and other losses of the poison is not in the best interests of safety. Furthermore, the buckled fuel strips and poor welds are not conducive to satisfactory reactor operation."

"All materials considered for use in the core of reactors should undergo rigid quality control both during their selection and subsequent fabrication. In addition, long term environmental tests prior to reactor operation are frequently appropriate. In cases where such environmental tests are not deemed necessary to the development program, reactor materials should be periodically examined during the operating phase to reconfirm their suitability."

Fuel Elements

(1) In general, the fuel elements themselves appear to have behaved well except for continuing, fairly low-level fission product leakage. This has been attributed to the machining of a nominal 0.065 in. (1.65 mm) from the fuel element flanges. This exposed one of the fuel edge bonds with only a nominal 0.1 in. (2.54 mm) between the water and the active fuel meat. It is normally not good practice to have the meat so close to a bonded edge in the sandwich type of fabrication.

(2) The analysis of the accident showed that the clad thickness of 0.035 in. (0.89 mm) tended to increase the severity of the transient considerably over that obtained in the BORAX and SPERT-type cores with their 0.020 in. (0.51 mm) clad. Thus, the thickness of the clad as well as any thermal barriers or gaps (such as faulty bonding) must be considered in analyzing the effects of fast nuclear transients. As a general rule, the clad should not be any thicker than is necessary for reliable retention of fission products and for any mechanical support requirements. Since the reduction in clad thickness is normally a step towards better core economics, this conclusion will no doubt be closely followed.

Mechanical Systems

(1) Clearly the principal design error in this reactor was the use of a control rod mechanism design which required that the rod itself be raised by hand without any positive stops as a part of assembly and disassembly. This, coupled with the fact that a single rod withdrawal could make the reactor critical, placed the entire responsibility squarely in the hands of a human operator who could make a spur-of-the-moment decision. In this case a positive mechanical limit could have been used, or the suggestion mentioned in the footnote on page 662 seems feasible.

As a general rule, reactor designs should be such that it is impossible, or at least extremely difficult, to take any unpremeditated action which could lead to serious consequences, whether the action be taken through error, ingenuity, or malice.

(2) It would have been possible to design the same reactor with more control rods and thus reduce the reactivity worth per rod. Where very few rods are employed, it is almost always because the mechanical designer and his argument of increased costs for more mechanisms and vessel penetrations has won over the nuclear designer who would always like a multitude of small, slightly absorbing rods if he could get them. Often, in a reactor design requiring separate penetrations, the maximum number of rods that can be used is limited by the nozzle requirements of the Pressure Vessel Code. It is necessary to set a lower limit on the minimum acceptable number of control rods. While this is a complicated matter involving a number of judgments, one aid is the "Stuck-Rod Criterion". It should be impossible for a reactor to be made critical in its most disadvantageous situation on the withdrawal of a single rod. Conversely, it should always be possible to shut down the reactor with one rod stuck in its outermost position. If it is possible that rods or mechanisms might interact so that several could be stuck in the out position, then the number of rods included in the stuck rod criterion should be increased accordingly. The SL-1 did not follow this criterion.

(3) The fact that the pressure vessel rose from

its support some nine feet, severing all of the attached piping, should be a warning to any reactor designer that, if he is relying on emergency shutdown systems such as boron injection and cooling water spray systems, he must make sure that no water hammer or similar force can be developed which is sufficient to move the pressure vessel and sever the piping connections. The supports of the pressure vessel and other components must not only support the weight, but anchor them as well, so that they will resist not only vertical but horizontal transient forces as well. For instance, if a double break of a horizontal main steam pipe is the accident being considered, will the pressure vessel remain fixed in place with all other piping intact in spite of the reaction force from the ejected water and steam? If the answer is "no", then the designer should not rely on the integrity of his spray-cooling and injection systems. It is worthwhile noting that pressurized-gas-cooled systems will have similar problems.

(4) During the accident, six loose shield plugs on the reactor top were ejected as missiles. One impaled an operator, and all six inflicted damage on the ceiling of the room and in the fan room above. Containment barriers present might have been breached by these missiles. Certainly they had lethal potential for any humans present. This reemphasizes the problem of missile protection and the necessity for this protection in order to have a reliable containment in event of a serious accident and in order to protect any personnel who may work around the reactor. This latter comment is particularly pertinent to research and test reactors where personnel are normally present around the reactor.

Fluid Flow

(1) Perhaps the single lesson to be learned concerning fluid flow here is the importance of water hammer. In this accident, the water hammer initiated by a pressure estimated to be 500 psig (~30 atm) of steam developed a water hammer action of such magnitude that it caused hydraulic pressure up to 10,000 psig (about 700 atm).

The importance of the water hammer and mechanical shock effects must be carefully considered in all liquid-moderated reactors. It is clear that not enough is known about this subject and more work should be carried out as soon as possible to put the results on a more quantitative basis. The preliminary work done at the Aberdeen Proving Grounds [61] to try to mock up the SL-1 situation, while not convincing, is encouraging and at least indicates the direction that must be taken.

General Electric [62] postulated that, had the vessel been completely filled with water at the time of the accident, it is improbable that the vessel would have been ejected upward by water hammer; the additional water would also have provided more shielding for the personnel. This practice is worth considering for almost all liquid-cooled or -moderated reactors. However, safety evaluations and operating procedures should take account of necessary maintenance when the reactor vessel is incompletely filled, and of accidents in those reactors operating partly filled (such as boiling water or superheat reactors).

Heat Transfer

(1) The accident itself and the analytical work which has followed [61, 62] again demonstrate and reemphasize that the magnitude of heat transfer barriers between the fuel meat and the water plays an important role in fast transients. Poor heat transfer slows down the onset of the one effective nondestructive shutdown mechanism for all highly enriched water-cooled and water-moderated reactors—steam formation. Thus, poor heat transfer means that the total input energy of the transient may be greatly increased and the consequences much more drastic. The SPERT destructive test showed that extensive melting of fuel can have dramatic effects even after the nuclear transient is over (see Sec. 3.12). The SL-1 transient could not be limited by orderly steam formation in boiling and required both that mechanism and also metal vaporization and fuel destruction for shutdown.

Far too little is known about the facts concerning fluid flow and heat transfer in mixtures of molten and vaporized metal with water. When there is a change in state of any reasonable fraction of a major reactor component—specifically the fuel—the entire situation is drastically changed and a new set of circumstances must be considered. In one way, a fuel element plate of the SL-1 or SPERT type caught in a transient is, for a moment, like a small flat box full of molten metal at a higher temperature than the melting point of the box itself.* The molten metal will melt its way through the box and suddenly confront the cold water over large areas almost at the same instant. Certainly local chaos will result and it is small wonder that there was so much finely powdered aluminum found spread everywhere within the vessel after the SPERT destructive test. (See Sec. 3.12.)

In the event of metal vaporization, the flat cladding box is rapidly bulged by the suddenly vaporized meat and the chaotic situation is created almost instantaneously after the vaporization occurs. The shutdown is more rapid than in the case of melting alone. Much more work needs to be done on both melting and vaporizing of fuels. In either case the principal shutdown mechanisms inherent in a given reactor system depend in a fundamental way on the heat capacity and the heat transfer characteristics of the system.

Chemical Reactions

(1) The evidence to date from BORAX I, the SPERT destructive tests, and the SL-1 is that the metal-water reaction has played a part in each, but not the principal part. In each case it has been very much overshadowed by the energy of

---

*Preparation of such a box and injection of molten metal rapidly into it might be one way to study such a system.

the nuclear excursion itself. However, it has still not been proved that a larger fraction of the potentially available chemical energy cannot be released. If and until larger releases can be disproved on a scientific basis, the metal-water reaction must continue to be regarded as a potential source of excursion energy in those cases where it is thermodynamically feasible.

(2) In this accident the mechanical damage and even the major part of the release of fission products by ejection of the water from the vessel was caused by less than 1% of the total energy released. The nuclear energy release is partially in the form of gamma rays and neutrons which deposit their energy widely and not so effectively in the central core area. Even the fission heating first puts energy into heating the solid fuel elements before causing vapors to be released. Chemical energy release, for low total energy releases, will tend to be much more effective per unit energy in accelerating masses than nuclear energy. (If the nuclear energy release is large, a larger fraction of it may go into kinetic energy and the statement becomes less valid.)

(3) Corrosion effects on the basic core materials, particularly the boron-aluminum strips, had a definite role in this accident.

Reactor Kinetics

A number of important points can be made for the type of reactors utilizing highly enriched uranium fuel in plate form and moderated and cooled by water. Some of these points can, perhaps, be generalized to other water-cooled and water-moderated reactors, and perhaps some even to all reactors.

(1) This accident has pointed out again that when enough reactivity is added to a core so that very short periods ensue, even quite short delay times in the application of shutdown mechanisms become very important. The operator or operators who withdrew the control rod could not have substantially altered the course of the transient even if they had realized that a mistake had been made and if they had let the rod fall back at once under gravity. Since the rods fall on scram under the force of gravity, it is clear that scram action would also have been ineffective. This can be generalized by saying: In reactors where large amounts of reactivity can be added suddenly, causing very short periods, it is likely that the time delays in sensing the rise in flux and in scram action will make the scram action ineffective and that the designer must place reliance on fast inherent shutdown mechanisms. To satisfy this requirement he should concentrate his attention in design on ensuring that these rapid inherent mechanisms such as Doppler effect, steam voids, bowing, etc., exist and are adequate to terminate any conceivable excursion before the core is seriously damaged.

Thus in the SL-1 case, improvements in the speed of the steam void shutdown mechanism response could have been achieved by thinner clad as discussed earlier in the Heat Transfer comments above, and also by having the water less subcooled. It was necessary to heat the water next to the plates from approximately 100°F (38°C) to above the saturation temperature before the mechanism was effective. This means that, from the transient shutdown viewpoint, it is better to operate a water reactor as close to the boiling point as possible. Thus, operating a PWR with nucleate boiling may help to ensure that no serious transients will occur. On the other hand, such operation may be running close to the critical heat flux region where burnout could conceivably occur in the event of a local increase in power density as discussed in the chapter on Heat Transfer.

(2) In the event that fuel melting or vaporization occurs in more than a few percent of the core, it is quite likely that a local chaotic condition will occur and that core destruction will result (at least in uranium-aluminum plate-type elements), whether or not the melting is also the true shutdown method. In this event, it is clear that the reactor kinetics behavior observed before the chaos occurred will be totally inadequate and inappropriate to describe the situation afterwards. Futher, it becomes necessary to investigate in some detail (and very conservatively since the answers will be very approximate) the behavior of the core after the chaotic condition is reached. In particular, this will be important in cores which are greatly undermoderated.

(3) The reasonable success in analyzing the course of transients of the general character of the SL-1, (and the SPERT destructive tests and BORAX also) indicates that it should be possible to estimate quite closely the course of any given transient in this type of reactor, at least up to the chaotic condition caused by fuel melting or vaporization. Since these phenomena lead to rapid shutdown, the entire course of such transients is reasonably well known. The region of destructive transients can then be quite well outlined for this type of reactor. It may be that extension of these methods to other types of reactors can be made on the basis of limited TREAT-type experiments.

Sensing and Control Instrumentation

(1) The principal difficulty with the instrumentation was the fact that it was turned off at the time that an operation involving handling of the control rods within the reactor was being carried out. If movement of the core during the insertion of the flux wires had caused boron to fall from the core, and if the reactor were near critical, the change would have shown on an audible count-rate meter. Further, if readings had changed from those prior to the shutdown period in an unexpected way, this could have been ascertained by taking count rates. While it is doubtful if an alarm alone would have altered the course of the accident, consideration of the implications should lead design and operating personnel to observe the following precautions, all of which were well known and used many places before this accident took place.

a) An audible warning in the form of a count-rate monitor (say a counting register clicking) from a valid neutron signal within the reactor should be provided for any personnel working in a dangerous area or at a potentially hazardous task near a potential critical assembly of any sort. In addition, an alarm may be attached to sound on a set period reading or count rate. Personnel ordered to work in such areas should request that such a count-rate signal be provided and they should be trained to sense immediately increases in count rate. Persons should not be asked to work in potentially dangerous areas on faith alone.

b) During operations involving core changes such as fuel or control rod movements which could conceivably involve large changes in reactivity, the nuclear instrumentation should be activated and an operator should be in attendance at the reactor control console and in communication with the working group.

c) All operations that potentially involve large changes in reactivity such as control rod movements, fuel changes, etc., should be done slowly and in such a way as to preclude sudden large additions of reactivity. Where possible, this should be done by interlocks, limits on speed of movement of such things as control rods and cranes, and assurances that the action of gravity cannot cause such large reactivity additions. Only as a last resort should hand operations relying on the operators' discretion be used.

(2) In design of the instrumentation system consideration should be given to making special instrumentation capable of giving on-scale records of such transients available. What is needed is a logarithmic response flux recorder with a fast response and a normal slow chart speed and a much faster chart speed which is actuated by a trip from an above-normal flux.

(3) Radiation detectors capable of reading high levels should be available—at least in portable form—and in accessible places.

(4) The use of the central control rod on automatic control and the fact that the log shows that the rod drove out of the reactor some six or seven times in the autumn of 1960 is rather disconcerting. In general, it is this author's opinion that automatic control should be limited to quite fine adjustments and should not be so used that its response to a large reactor change can lead to dangerous reactor conditions. The circuits should be so designed that it will always fail by driving in the rod, or at least by not driving it out beyond its position at the time of failure.

Containment, Fission Product Release, Fission Product Leakage

(1) About 20% of the core plate area containing about 40% of the fission products was destroyed. Of this, about 5% resolidified onto the intact portion of the plates, leaving 15% of the core plate material unrecovered in identifiable plate form. The missing 15% of the fuel should have contained about 30% of the total fission product inventory and 1790 grams of U^{235}. Of this, 1104 grams of U^{235} were recovered from the bottom. It appears that 5-10% of the total inventory escaped from the reactor vessel and less than 0.5% of the I^{131} and a negligible fraction of the nonvolatile inventory was found in the desert. Note that, according to reference [62], between 5 and 15% escaped from the reactor vessel and that, even though this large fraction of the inventory had already penetrated two of the usual fission product barriers, little escaped to the outside. While this reactor was not designed with thought to containment, the rudimentary building did provide considerable containment. This fact is very heartening and should encourage all reactor designers—even those with no specific containment requirements—to make a few low-cost provisions to minimize building leakage.

(2) The accident emphasizes again that such things can happen at almost any phase of the operation. Thus, at least some containment provisions should be in effect at all times when it is conceivable that criticality or other occurrences resulting in the loss of fission products could take place.

(3) The low release of fission products, even after a relatively large fraction was released from the core, again points to the need for more definitive knowledge concerning the fate of the various fission products after they are emitted from the molten fuel. This knowledge is hard to gain, especially under anything like the conditions which prevail in a serious accident. There remains much work to be done before it can be certain of what will happen in a given case.

(4) The fact that most of the fuel which melted was observed to be porous and did not have a glazed condition may be due to sudden chilling as General Electric states in reference [61], but it may also be due to the fact that the melted region had a relatively high fission product inventory (16% burnup average). Thus, this appearance may have been due to the so-called "foaming" of highly irradiated fuel when it is suddenly made molten and refrozen quite rapidly before the fission product gases and volatiles have had a chance to boil off. Again, more work is needed to study this foaming process since it may be very important in fast reactors. In reactors of the SL-1 type it is hard to believe that the foaming can have any adverse effect since the expansion will deny the region to water.

Radioactive Waste Management

(1) General Electric [61, 62] states that recovery operations were seriously hampered by the limited access to the reactor floor and the fan room above. They recommended easier accessibility and that all surfaces be smooth, water-tight, painted, and with adequate drains and seals for inaccessible areas. These recommendations tend to conflict with other requirements (especially those of integral containment with a minimum of openings) including cost requirements. In general plants have not been, nor should they be, designed to make disassembly easy after an accident as severe as this one. However, many of the measures suggested do have considerable merit. Concrete in areas where contamination can be a problem should be smooth and painted. Surfaces should be easily cleaned if possible. Equipment should be placed with maintenance in mind. If these precautions for safe and efficient operation

and maintenance are followed, most of the problems of a disassembly after an accident will be minimized.*

(2) General Electric recommends high standards of cleanliness and good housekeeping as they had great difficulty with decontaminating and sorting out the equipment that was needlessly present.

(3) An emergency or alternative access to all areas should be provided.

(4) In areas where a high radiation operation could conceivably be carried out the crane should have all motions electrically powered and should have provisions for quick hookup of a remote or semiremote operating cable.

(5) Noninflammable building insulation should be used. The use of water cooling rather than oil cooling would also reduce the fire hazard.

Emergency Plans

(1) This accident again points out the need for clear emergency plans and adequate supplies. The first access was limited partially because no one other than the three operators knew what the situation was and no records from instrumentation existed and very few notes existed in the log. No radiation detectors with sufficient range were available at the site and no emergency supplies of health physics equipment.

Each reactor should establish one or more emergency depots remote from the reactor and clearly accessible. Each should have high level radiation detectors in operative order, self-contained breathing units, respirators, special clothing, up-to-date drawings as might be required in emergencies, and procedures for operations which might be necessary.

(2) The lack of an adequate written set of operating instructions and procedures seems to have played some part in the entire situation. Evidently, those available during the original startup and operation were deemed inadequate by the Office of Army Reactors, and more adequate ones were never prepared during the ensuing operations. It is clear that an adequate and up-to-date set of operating instructions, a set of standard procedures, and a set of check lists is vital to the safe and reliable operation of any device as complicated as a reactor.

General Concluding Note

As has been pointed out in the introductory chapter, most accidents involve design errors, instrumentation errors, and operator or supervisor errors. The SL-1 accident is an object lesson in all of these. There has been much discussion of this accident, its causes, and its lessons, but little attention has been paid to the human aspects of its causes. There is a tendency to look only at what happened and to point out deficiencies in the system without understanding why they happen, why certain decisions were made as they were. Post-accident reviews should consider the situation and the pressures on personnel which existed before the accident. This section is presented to point out some of the factors which were involved in the decisions which were made. It is hoped that others may take another path when they are faced with the same decisions and the same pressures.

The design and operations attitude which permeated the project from its inception appears to have been more or less a direct result of the objectives set for the reactor, coupled with a limited budget, a tight time schedule, and constantly changing sets of personnel. The reactor was intended as "a prototype for the purposes of testing the operation and serving as a training center." [54] The requirements eventually developed called for 200 kw of electrical power with some expansion capacity and 400 kw of space heat. Because of the remote location and difficulties of refueling, it was desired that the core life be several years, and 3 years' equivalent was finally fixed upon as the goal. It was necessary that the reactor be capable of being operated by military personnel.

The original plan was to use nondevelopmental materials and fuel designs, but the long core life chosen and higher temperatures dictated the choice of X-8001 and a burnable poison. Pressures of schedule and developmental difficulties prevented putting the boron directly in the fuel element plates, and instead, side plates of partially clad boron-aluminum were chosen as being acceptable and providing more flexibility in selecting the final operating loading. (It must be remembered that nuclear calculations are often not too accurate on a new reactor concept. In fact, even rigidly specified duplicate cores on the same reactor may vary as much as 0.01 Δk in reactivity.) The flexibility was indeed used as discussed in reference [55]. No doubt the choice of the number and type of control rods was based on simplicity and ruggedness. The design clearly realized that there were some reactivity problems connected with the central rod. This rod had a much longer follower whose length appears to have been chosen to prevent a serious transient in the event the rod fell through the core and out the bottom. While the point cannot be easily checked and thus it is pure conjecture, it is possible that means were considered to make the assembly and disassembly safer and rejected on the basis of extra cost or, more likely, time schedule. The immediacy of a schedule delay or increased costs often outweighs the threat of a vague and improbable possibility. Other design errors were present and have been mentioned in the discussion. Most of these seem to point to choices dictated by the design objectives or made on the basis of cost and schedule limitations.

Once the design choices had been made, and the reactor built and tested, other factors

*It is interesting to note how strongly the General Electric group feels regarding these points. They were the ones who bore the brunt of the cleanup difficulties and it has clearly influenced their point of view. For instance, they had to remove and bury 15,000 ft^3 (420 m^3) of gravel used as vessel shielding and contaminated during the accident.

no doubt began to play a part. The initial tests seemed quite successful, giving everyone considerable confidence in the reactor. The design and startup groups, having completed their principal tasks, moved on to other work, sometimes not finding time to prepare adequate reports and operating instructions.

On relatively short notice, a new contractor took over responsibility for the plant. The transfer of responsibility for devices as complicated and individual as prototype reactors remains a difficult procedure at any stage, from initial design through to operation. It should only be done in cases of absolute necessity. Then it should be carried out over an extended period and in such a way that it is absolutely certain that all pertinent knowledge has also been transferred. It does not appear to this author that three months is an adequate period. Then too, the size of the staff which now took the responsibility appears to have been very small to carry out the many tasks assigned including becoming familiar with the reactor, preparing adequate operating instructions, training military personnel, and planning and supervising tests. Whether this was a budgetary choice or not is not known.

It should also be pointed out that the lines of responsibility for this particular project were especially confusing. The Department of Defense requested of the AEC that the reactor be built. The design, test, and initial operation was the responsibility of the Argonne National Laboratory. Pioneer Service and Engineering Company was the architect-engineer, and the Fegles Construction Company carried out the construction. The over-all coordination and direction was the responsibility of the Programs Division of the AEC Chicago Operations Office. Later, the operation and in particular the Combustion Engineering part of the operation fell under the jurisdiction of the Idaho Operations Office of the AEC. The Division of Reactor Development of the AEC and its Army Reactors Office exercised over-all program responsibility. The AEC Idaho Operations Office and its Military Reactors Division, as well as the Army Reactors Office in Washington, participated in decisions regarding the amount of supervision to be used at the reactor. In testifying before the Joint Committee on Atomic Energy of Congress, C. A. Nelson, Chairman of the General Manager's Board of Investigation said, "The complexity of the chain of command for the SL-1 may explain, in part, the lack of effectiveness of the existing organization in communicating with higher levels of supervision regarding these substandard conditions." In the same hearing, Commissioner R. E. Wilson said, "The lines of responsibility within the AEC for health and safety, from the General Manager down to the operators of the reactors, were not clear and definite in several respects as they should have been, nor were the levels at which certain safety and operating decisions should be made spelled out."

While there were many people who, at least in principle, had some safety responsibility for varying periods of the reactor's life, no one or no single group exercised any continuous direct responsibility over all phases of its life. In fact, it would appear that there was no one who was totally cognizant of the situation and at the same time had the authority, responsibility, and knowledge necessary to appreciate the problem and take decisive action. It is clear, and many people have later said so, that the reactor should have been shut down pending resolution of the boron difficulties and the general deterioration of the control rod operation. In fact, no one did so or even brought the malfunctions to the attention of any responsible safety group. In the climate that existed before the accident, it is likely that if one man had decided that the reactor should be shut down for safety reasons, he would have been ridiculed and would almost certainly have had an unfriendly response since he would have had to say some rather harsh things to accomplish his purpose.

The type of organization selected and its functioning has a fundamental role to play in reactor safety. The situation can be helped if four basic rules are adhered to as closely as possible:

(1) Insofar as possible, design, construction, and operation should all be the responsibility of one organization in order to ensure continuity and continued responsible judgment of the situation.

(2) The organizational responsibilities in regard to safety and all facets of operation should be clearly and unambiguously laid out. A line organization should be used, not a committee.

(3) Safety reviews should be made by a competent group outside of the operating organization on a regular basis. The safety organization should be such that these reviews are not repeated by competing safety groups so as to unduly harass the operating group and thereby reduce safety.

(4) The ultimate responsibility for reactor safety should rest and must always be allowed to rest on the immediate supervisory organization at the reactor. In the final analysis, the reactor shift supervisor and, in turn, the operator at the reactor console should have the authority to shut down the reactor if either believes it to be unsafe.

3.12 The SPERT-I Destructive Series [64-69]

General. During the summer and fall of 1962 a series of self-limiting power excursion tests were carried out at SPERT-I utilizing an 0.020 in. (0.51 mm) thick U-Al alloy plate-type element clad with 0.020 in. of aluminum. There were 25 elements in a 5 × 5 array in the core. There were one central transient rod and four other safety rods, each located in a separate quadrant. The double plate-type control rods operated in slots in special fuel elements. The water gaps between plates in the core were 0.179 in. (4.55 mm). This facility and core and the SPERT tests as a whole followed a logical development from BORAX-I tests described in Sec. 3.4.

A series of 54 tests were carried out using 5 core loadings, some of the later ones made up in part from undamaged fuel plates salvaged from earlier ones. The results for Cores II to V are summarized in Table 3-10 [65]. The tests showed plate buckling at periods of the order of 6 to 9 msec and a ripple pattern which would have led to heat transfer difficulties if the ele-

Table 3–10

SPERT I: TRANSIENT TEST DATA (RUNS 22 TO 54) ON DU-12/25 CORE

Run No.	τ (msec)	α (sec^{-1})	$\Phi(t_m)$ (Mw)	$E(t_m)$ (Mw-sec)	E_T (Mw-sec)	$\theta(t_m)$[a] (°C)	θ_{max}[a] (°C)	Date (1962)
				Core II				May
22	880	1.14	0.980	–	–	54[b]	94[b]	10
23	19.3	52	120	3.40	4.70	125	132	10
24	9.0	111	395	4.70	7.10	157	300	10
25	7.5	133	555	5.55	8.80	158	355	11
26	6.0	167	890	7.20	13.2	185 210[b]	545 560[b]	11
27	8.1	124	505	5.45	9.20	177	325	16
28	5.0	200	1130	8.35	17.5	320	585	18
				Core III				June
29	49.5	20.2	19.1	1.84	3.20	125[b]	129[b]	6
30	7.6	132	510	5.20	9.40	172 250[b]	400 465[b]	6
31	6.9	145	620	6.40	10.9	190 290[b]	440 490[b]	8
32	4.6	218	1270	8.90	19.0	240 420[b]	570 680[b]	11
				Core IV				July
33	880	1.14	0.760	1.79	–	64[b]	94[b]	23
34	118	8.48	4.50	1.00	–	59 67[b]	102 109[b]	24
35	21.7	46.1	107	3.40	5.05	125 149[b]	128 155[b]	24
36	8.9	112	460	5.05	7.65	151 230[b]	242 350[b]	26
37	880	1.14	0.780	2.21	–	69[b]	87[b]	26
38	112	8.93	5.05	1.18	–	63 68[b]	99 117[b]	26
39	22.4	45.8	105	3.60	5.40	120 146[b]	124 154[b]	27
40	8.7	115	470	5.25	8.20	167 216[b]	234 305[b]	27
41	6.8	147	635	5.75	9.20	232[b]	355[b]	30
42	6.9	145	640	6.00	9.70	156 246[b]	340 430[b]	31
								August
43	6.9	145	660	6.35	10.4	160 243[b]	340 440[b]	1
44	6.9	145	665	6.60	10.4	161 250[b]	325 440[b]	2
				Core V				October
46	1310	0.76	0.390	1.49	–	65[b]	80[b]	18
47	243	4.12	1.43	0.91	–	61[b]	93[b]	19
48	140	7.14	3.80	0.94	–	62[b]	112[b]	22
49	21.1	47.4	110	3.60	5.40	148[b]	153[b] 220[c]	22
50	9.6	104	350	5.00	7.15	195[b]	232[b] 350[c]	22
51	7.0	143	620	6.45	10.2	223[b]	305[b] 455[c]	22
52	9.6	104	380	5.10	7.25	208[b]	229[b] 360[c]	24
53	9.7	103	360	5.05	7.45	204[b]	234[b] 350[c]	24
								November
54	3.2	313	2250	13.8	30.7	430[b]	615[b] 1230[c]	5

[a] Temperature measured at E5–7W–3
[b] Temperature measured when E5–7W–3 was not the highest recorded
[c] Maximum temperature measured using special "fuel cell" thermocouples

ments had been used later in a reactor at power [65]. This pattern was shown to be due to thermal expansion stresses in the plate. During slightly shorter period transients (6.4 to 6.0 msec), the plates appeared to have softened and remained in a plastic state for several days. In a 5.0 msec transient (17.5 Mw-sec) a total of 7 plates showed some melting and fusing together, and 52 plates (~2% of the core area) showed melting and fusion on a 4.6 msec period (19.0 Mw-sec). Corrosion and formation of an oxide coating was found to give decreased heat transfer and was suggested as a cause for concern in regard to melting in operating cores.

The total burst energy E_t, the maximum fuel plate temperature in the central fuel assembly, and the maximum pressure are shown in Fig. 3-31 [69] as functions of the reciprocal period α of the step transient tests. A least squares fit of the peak power data in the range $10<\alpha_0<300$ sec^{-1} gives the peak power (in Mw):

$$P(t_m) = 0.14\,\alpha_0^{\,1.73}\,.$$

The break in the temperature curve at about 140°C (284°F) indicates boiling; the break at 600°C (1110°F) indicates melting and heat of fusion effects as well as a tendency towards failure. The pressure data were obtained from detectors in the reflector water and represent pressure bursts in the region of peak power, and therefore correspond to the self-shutdown mechanism of boiling. They do not show the final pressure of the Run 54 destructive test.

The Destructive Test-Run 54. In the final test on a 3.2 msec period (30.7 Mw-sec) all 270 plates showed melting to some degree with the average melt 35%. The performance of Run 54, from the nuclear viewpoint, was very close to predictions. Evidently, the nuclear characteristics of the shutdown were essentially identical to the earlier transients and consisted of fuel and moderator thermal expansion and boiling. However, after the nuclear transient had already been terminated by these shutdown mechanisms for about 15 msec, there was a violent pressure release resulting in total destruction of the core. The time behavior of the transient is shown in Fig. 3-32 [69]. Note the gradual termination of the transient after peak power is reached. This tends to verify that the shutdown mechanism was the gradually developing boiling rather than a violent disassembly as indicated by the very sharp pressure rise which came later.

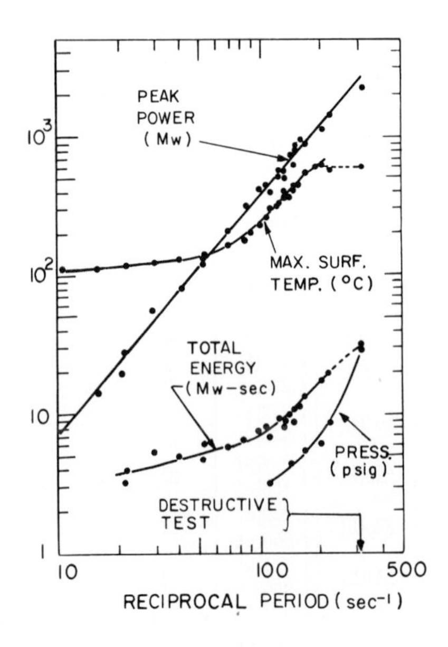

FIG. 3-31 Peak power, energy, maximum temperature, and pressure vs. reciprocal period for SPERT-I.

The data obtained indicated a period of 3.2 msec, a peak power of 2300 Mw, a nuclear energy release to time of peak power of 14 Mw-sec, a total burst energy of 31 Mw-sec, and an initial pressure pulse of about 35 psig (2.4 atm). A special in-pile capsule with fuel material in it indicated a temperature of about 1200°C (2200°F) in the core. The data indicate that the elapsed time between the initiation of the large pressure pulse and its peak was only about 0.5 msec. Pressures of 3000-4000 psig (~200-270 atm) occurred in the reflector water around the core. It is believed "that the blast pressure was produced by a self-propagating steam explosion that resulted from the dispersal of the molten plates into the water throughout the core." [68]

Approximately 20 kg (44 lb) of "spongy" metallic debris were recovered from the tank. About 0.4 kg (0.9 lb) of α-Al_2O_3 was produced corresponding to 0.4% of the core aluminum or about 3.5 Mw-sec of energy in the metal-water reaction (~10% of the nuclear energy). The metal-water reaction in this case is believed to have proceeded slowly with a reaction time of 1 sec [69].

Since the energy of 19 Mw-sec developed in the next shortest period transient caused no such destruction, it is clear that the threshold of this violent destruction is quite sharp and that a factor of two in total energy deposited is very important.

Slightly Enriched UO_2-Stainless Steel Clad Core Tests [70]. In late 1963 a new series of destructive tests was started using 4% enriched UO_2 fuel swaged to 87% theoretical density in 0.5 in. (1.27 cm) diameter, 6 ft (1.83 m) long type-304 stainless steel tubes 0.028 in. (0.71 mm) thick. The core contained 590 pins. Some 14 transients were carried out with periods ranging from 1.31 sec to 2.2 msec. Only the shortest period transient results will be discussed briefly.

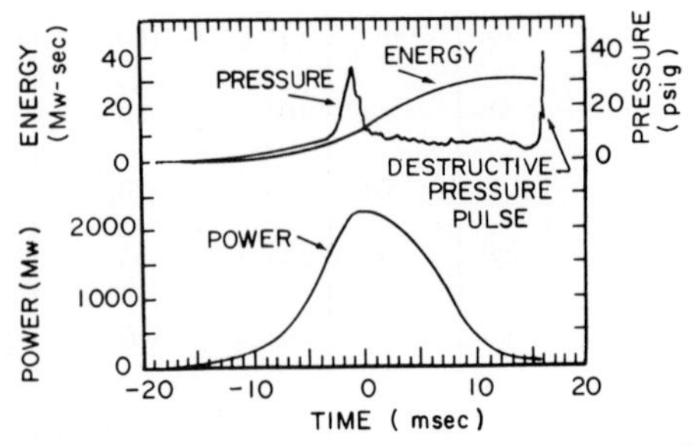

FIG. 3-32 Time behavior of the power, energy, and pressure during the SPERT-I destructive test.

In the November test [70a] the core was initially in room temperature water at a power of 0.2 watts. The withdrawal of the transient rod added 2.70$ in reactivity and put the system on a 2.2 msec period. A peak power of 20,600 Mw was reached with an energy release to peak power of 90 Mw-sec, and there was a total energy release of 160 Mw-sec. The peak water pressure reached in the core was 60-70 psig (4-5 atm). Two of the 590 pins were ruptured and about 1/4 of them were bowed as a result of the test. One of the two pins which ruptured was found later to have holes in it and at least that one was probably waterlogged before the test. The other split as if it broke along the tube weld seam. Runs before the short period test had shown the presence of fission products in the water, indicating also that waterlogging could have been the cause of the burst of the two rods. There is some evidence that the rod failures were due to prior leaks in these pins which allowed water to enter. The hottest fuel surface temperature at the time of peak power was 13 °C (23 °F) above ambient and the maximum fuel surface temperature reached was 556°C (1000°F) above ambient (about 30 msec after peak power).

On April 14, 1964, a second short-period test was run on this same type fuel [70b]. In order to add more reactivity this run was initiated utilizing a ramp followed by a step addition of reactivity. This method added 3.3$ in reactivity and resulted in a 1.55 msec period. It is estimated that a peak power of 35,000 Mw was reached with an energy release to peak power of 95 Mw-sec. The total energy release was estimated at 165 Mw-sec. A sharp pressure peak of 135 psig (~10 atm) was observed to occur about 1 msec before the time that peak power occurred. Two rods were found to have burst. Neither showed evidence of holes this time, but a hot cell examination of one showed a crack in an end plug. It appears likely that waterlogging again occurred. One rod burst on one side above a restraining grid and on the other side below the grid.

The predicted results, based on a calculated Doppler model which was successful for the previous series of tests, were 42,000 Mw maximum power and 155 Mw-sec to peak power. The peak power was predicted to occur approximately 2 msec later than it actually occurred. The evidence seems to indicate that the violent rupture of two waterlogged pins resulted in an early shutdown, terminating the transient even before the Doppler effect did.

The results show the effectiveness of the Doppler coefficient as a shutdown mechanism and also seem to indicate that rupture of a rod and the resultant steam formation caused by the rupture does not spread through the core. They may also indicate a way to improve the rapidity of steam void formation as a shutdown mechanism.

Comments, Conclusions, Recommendations

(1) The sudden onset of total core destruction for only a factor of two increase in total energy deposited was a surprise. It emphasizes the need to carry on such extrapolation tests at remote sites such as NRTS.

(2) These tests indicate evidence for the existence of three triggered shutoff mechanisms. Two of these are essentially nondestructive to the core. The first is the Doppler effect whose effectiveness, while long known, is here demonstrated for the first time. The second is water or some other material in intimate contact with the fuel within the cladding. This intimate contact eliminates the time delay for the heat to go through the cladding and essentially makes local boiling a fast shutdown mechanism, although likely not having such great reactivity worth. The third, fuel meat melting and vaporization, is destructive to the core but appears also to be a useful one to limit the maximum energy release in a transient. (See Sec. 6.3.)

4 FUEL FAILURES

4.1 General

The principal purpose of this section is to provide some information on experience with fuels in operating reactors. In general, accidents involving fuel and not concerned with reactivity changes group themselves into three categories—loss of coolant (L), loss of coolant flow (F), and design or metallurgy problems (M). Table 4-1 summarizes some of the recorded experience.* Since fuel element failure is to be expected to some degree in normal operation, this list by no means records all failures, but rather gives some typical examples and includes those which were relatively serious. Additional examples and further discussion are given in the chapter on Fuel Elements.

In Table 4-1 twenty-two cases can be classified as falling under one or the other of the three categories mentioned above and are identified in the last column. Four, M(H), have occurred in homogeneous reactors and perhaps cannot be classified as fuel failures in the usual sense. The high percentage of vessel failures in these reactors is an indication of the difficulties of the corrosion problems that exist in these highly ionized solutions. Of the 18 remaining, four can be said to be due to loss of coolant (L), three to loss of flow (F), seven involve metallurgical or design inadequacies (M), and four can be said to reflect average operating samples (S) of fuel deemed to be behaving normally.

In general, those accidents involving loss of coolant or loss of coolant flow have much more serious consequences than those due to metallurgical factors. This is to be expected since normally the metallurgical failures, while they may occur frequently if the fuel design and metallurgy is poor, will occur statistically spread out in time and not all at once. Only if many elements are on the verge of failure and the reactor is suddenly perturbed so as to cause a large number to fail

*Information given in this table is taken primarily from references [1] and [2], as well as from [71] to [76] inclusive.

Table 4–1

Fission Product Release Through First Barrier: Fuel Failure or Homogeneous Reactor Tank Failure

	Location, Reactor	Active Material, Coolant, Moderator	Description; Cause (C); Quenching mechanism or shutdown detection method (Q)	Damage	Resulting changes	Category (see text)
1947–48	ORNL, X-10 reactor (2 accidents)	Nat. U slugs, Al-clad, air-cooled, graphite-mod.	(C) Fuel slug "grew", swelled, failed, not detected, plugged channel, others overheated, failed; plugging tended to restrict flow and chance to detect failure.	Channel only abandoned in later operations	Improved f.p. detection system	M–F
24 Dec. 1952	LASL Clementine (fast reactor)	Pu, steel clad, nat. U slug at ends, Hg-cooled	(C) Nat. U slugs swelled with irradiation, burst clad, released Pu to Hg (2nd time); also nat. U safety block swelled, stuck immovable; Pu-Hg formed pyrophoric soln.	Pu dissolved in Hg, contaminated system	Reactor dismantled, replaced with Omega West Reactor	M
1953	Moscow, TR, Academy of Sciences	Nat. U, Al-clad, D_2O-cooled, D_2O-moderated	Failure of clad released activity into system and He blanket. In replacement, element fell and damaged vessel	Low activity in primary system	Not known	(?)
1952–54	Moscow, R.P.T.	Enriched U, Al-clad, H_2O-cooled, graphite-moderated	During 1952–53, 30 elements failed causing shutdowns. During 1954 only 7 elements failed.	Shutdowns, minor contamination	Improved fuel, tightened operating limits	M
June 1954	NRTS, MTR	Enriched U, Al-clad, H_2O-cooled, H_2O-moderated	(C) Buckling outer concave plate contacted next element, reduced cooling, meltdown attributed to hydraulics and thin clad.	Up to 3 r/hr	System cleaned up, elements strengthened	M
May 1955	Raleigh, N.C water boiler	Enriched uranyl sulfate solution in stainless steel vessel	(C) Solution chilled and ppt. formed; in redissolving, chlorides added, corrosion resulted, tank leaked.	Solution flowed into reflector and coolant	Core removed, replaced by H_2O-plate-type elements	M(H)
July-Aug. 1955	Chalk River NRX	Experimental Pu fuel rod, Al-clad, H_2O-cooled, D_2O-moderated	(C) Rod lost coolant and overheated (perhaps local increase in flux); jet of molten Pu penetrated to D_2O (Q) Dumped D_2O	Coolant and tank active, Pu on tank bottom	Down 7 weeks, used flotation on Pu	L
Nov. 1955	Hanford Production Reactor	Al-clad, nat. U slugs, H_2O-cooled, graphite-moderated	Ruptured slug forcibly loosened, flushed out by high pressure H_2O. Flash of flame at face of reactor as element ejected.	Reactor face contaminated, local area	Fog sprays installed to reduce contamination	F
June 1956	Saclay, France EL-2	Nat. U, Al-clad, CO_2-cooled, D_2O-moderated; neutron converter–horizontal U cylinders, Al-clad, H_2O-cooled	Fuel clad rupture. First of several. (C) Pushing core to high burnup (when burst fuel detector warned, reactor normally shut down)	Contamination of CO_2		M–F
6 Oct. 1957			Fuel clad rupture – followed for several days before removal	Same		
26 Nov. 1957			Converter meltdown – U melted caused rupture of H_2O into reflector	Contaminated H_2O and room (during removal)	Safety of converter system improved	
27 Dec. 1958			Usual clad rupture, but reactor ran longer afterwards	Contamination	Pre-alarm horn installed	
16 Feb. 1959			Severe rupture, immediate scram, no time to find ruptured slug, went to 5kw, took 2hrs. to explore and locate. Air entered, activity went up, U burned. (C) Carbonized oil in tube blocked CO_2 flow.	Lost tube and rod, complete disassembly required	Reactor unloaded and new and better fuel installed.	
26 Oct. 1956	Marcoule, France, G-1	Nat. U, Al-clad, air-cooled, graphite-moderated	Clad failure exposed U slug to air, oxidation, fission pdts dispersed in air system. Max. concentration outside below tolerance.	Ruptured fuel contamination		M

Oct. 1956	LASL, LAMPRE	Uranyl phosphate soln. in H_2O (homogeneous)	(C) Excessive corrosion of internal heat exchanger by solution. Leakage of primary solution into secondary system.	Corroded through	System eventually abandoned	M(H)
4 April 1958	ORNL, HRE-2	Uranyl sulfate in H_2O (homogeneous)	(C) Corrosion through core blanket wall, attributed to flow pattern and differential heating, deposit of U on wall.	No escape of activity	Repaired hole, changed flow pattern	M(H)
Feb. 1960			(C) Same basic cause – 2nd hole	No escape of activity	AEC dropped system	M(H)
13 April 1958	Saclay, France EL-3	1.6% U^{235} rods, Al-clad, D_2O-cooled, D_2O-moderated	(Improved fuel over EL-2 due to 2% molybdenum in U, hollow elements clad and cooled outside only, well-bonded; no failures as above to 1960.) Alarm sounded, checked to see if signal valid, scrammed. Activity masked search for damaged element. Pulled all. (C) Loose fuel cell bottom allowed vibration, broke off at top, still in D_2O flux – no coolant flow	Fuel cell broken off, required special recovery; internal contamination	Showed only rare f.p. gases in He cover gas; installed He activity monitor and scram	F
28 May 1958	Chalk River NRU	Nat. U, Al-clad, D_2O-cooled, D_2O-moderated	Three elements failed, one violently, 2nd element distorted and could not be properly withdrawn, 2 pieces fell out and burned.	Severe contamination in building, some nearby	Circuitry, procedure changes	L–M
July 1958	Munich, Germany FRM	Enriched U^{235}, Al-clad, H_2O-cooled and -mod.	Fuel element released f.p., element was identified	Slight contamination	Replaced element	M
1958– July 1959	England Calder Hall	Nat. U, Magnox-clad, CO_2-cooled, graphite-moderated	Over a period 27 out of 90,000 elements failed. Since many reactors anticipate up to 1% failure this is normal.	No damage	No changes	S
10 Nov. 1959	Basel, Switzerland AGN–201	Enriched U^{235} in solid homogeneous plastic, H_2O-cooled	Failure of polyethylene matrix around fuel	Small gas release into H_2O coolant		
21 Dec. 1959	Hanford Prod.	See earlier entry	Fuel rupture – typical example happens once or twice per year.	Some desert contamination	No action	S
1956–60	Marcoule, France, G–1	See earlier entry	From 1956 to 1961, 7 slugs with thermocouples, 25 special, 13 normal leaked.	Negligible	Normal operation	S
1961	Fontenay-aux-Roses, France TRITON	Swimming pool, enriched U, H_2O-cooled and -moderated	Fuel cladding rupture	Slight contamination	Fuel replaced	S
12 Dec. 1961	NRTS, ETR	93% U^{235}, U-Al plates, 1.5 g B in each element, H_2O-cooled, moderated	(C) After shutdown clear plastic "viewing box" left in tank. Sank to top of elements (down flow was blocked). Startup normal until spike on N^{16} detector (partially ignored) cut power from 90 Mw to 70 Mw. When other systems erratic, manually scrammed.	18 plates in 6 elements melted.	All viewing boxes have colored sides – more careful check	F
13 Nov. 1962	NRTS, MTR	93% U^{235}, U-Al plates, H_2O-cooled, moderated	(C) Rubber from seal broke loose in system trapped in fuel element inlet, reduced flow, local boiling and melting occurred. (Q) Scrammed on differential pressure trip and N^{16} detection	One plate partially melted	Made ΔP scram more sensitive, changed procedures	F

at once can there be a serious safety problem. To date this has not happened. It is more likely as core performance is pushed harder and as flux distributions are flattened, since then the variation in core conditions as seen by the various fuel elements is much reduced and all elements are in the same flux which is being pushed as high as possible by economic factors.

Three cases are described in detail in Secs. 4.2, 4.3, and 4.4. Two of these involve loss of coolant flow at high power levels because of flow orifice plugging. The third involves loss of coolant and happened outside a reactor in a fuel transfer operation. It is believed that these accidents represent typical examples of those which are most likely to happen and which would have serious consequences.

4.2 The NRU Loss-of-Coolant Fuel Element Accident [71, 72]

The NRU Reactor at Chalk River, Canada, is a heavy-water-cooled and moderated engineering and research reactor designed to operate at 200 Mw. The core consists of about 200 fuel assemblies, each made up of five flat metal bars of natural uranium, each clad with 1S aluminum contained in a 2.5 in. (6.35 cm) diameter tube. The initial loading was made by slipping aluminum sheaths over the bars and then drawing them through a die to press them into contact. There had been 13 failures of the original rods and considerable contamination of the aluminum tank system by the plate-out of uranium on the aluminum and also by fission product formation. This problem negated much of the usefulness of the monitor system designed to detect faulty fuel elements [71, 72].

After a week of steady operations the reactor was automatically shut down on Friday May 23, 1958, by an excessive rate-of-rise signal. No reason could be found for this occurrence and the shift supervisor prepared to start the reactor up again. The first four attempts at restart were terminated by automatic trips arising as secondary effects from one of the safety features. On the fifth attempt everything was normal for the first five minutes, only to be shut down again by rate-of-rise signals and, this time, by other alarms, including one indicating high radioactivity in the coolant circuit. A momentary flash of light from the full-out signal on three rods was observed although they should not have been all the way out. Some of the other signals were later discovered to be due to a pressure transient in the reactor vessel arising from the violent failure of a fuel rod. It is postulated that the pressure transient in the reactor could have changed the D_2O water level and generated a rod withdrawal signal causing the transient. The reason for the rate of rise transient is unknown. However, it was discovered that although the rate of rise of power was correct just before the last rate-of-rise transient tripped the reactor, the power level had exceeded the demand level. No conclusive reason for the power going beyond the set point was found but a particular selector switch could have caused the fault if certain of its contacts had not been made when the operator switched from 'low power' to the 'normal' mode of operation (3%/sec to 1 Mw/sec).

It is postulated that a cladding defect, perhaps caused by the first transient, allowed water to seep into the element during the shutdown and cooling. Then on restart, this "waterlogged" element burst violently because the rate of rise was 3%/sec, instead of 1 Mw/sec, due to the faulty selector switch. Perhaps if the failed fuel system had not been contaminated, the failed element would have been detected after the first shutdown.

Preparations were made for the removal of the ruptured element [71]. Normally, the fuel removal flask is positioned over the proper element and is sealed to the uncapped hole by a metal snout on the end of an extendable bellows. Reactor water can then flow up to the bottom of the flask. The empty chamber in the flask is filled with heavy water and the main valve at the bottom of the flask chamber is opened, creating a single heavy-water-filled passage from the flask into the reactor vessel. A guide tube is driven down from the upper shield to the fuel assembly. The rod assembly is then raised into the upper shield where it is allowed to decay 15 minutes as it is cooled by reactor D_2O rising through the guide tube. It is then drawn on up into the flask chamber where it is cooled by a recirculating system built into the flask. Another chamber in the flask, previously loaded with a fresh element, is rotated into position, and the new element is lowered through the guide tube into the reactor. The guide tube is then raised to its normal position in the upper shield, the main valve is closed, the bellows seal is broken between the flask and deck plate, the flask shielding skirt is raised, and the flask is removed to the storage block by means of a large crane.

One of the presumably damaged fuel elements was removed easily. However, difficulty arose in trying to get the guide tube to slip down over the second one. Therefore, preparations were made to remove the guide tube mechanism plug with the fuel. The fuel was raised into the flask, but then it was discovered that the heavy water had drained from the flask and there was no cooling for the element. Normally, the guide tube directs sufficient upwards pressure from the core region to keep water in the flask. (Later investigation showed that without the guide tube, heavy water fails to reach the pump inlet by 2-3 ft.) In order to ensure that the element was not blocking the flow, it was raised completely above the main valve but no water came and the element refused to drop back into the lower section. Attempts to refill the flask failed because of gas-locked pumps and lack of time.

The operators then attempted to move the flask to a point just off the reactor deck where a hose was available with emergency H_2O cooling. On the first attempt the drive motor stopped because of an interlock which prevented its going while the snout was in its extended position. The snout was raised and the flask could be moved. Another interlock was so set that it opened the main valve in the flask if there was not enough D_2O there. By this time the sensing unit for this action was uncovered and the main valve opened. This action is also attributed to a faulty hydraulic valve [71]. Attempts to close the valve were futile because of another interlock which prevented valve operation unless the snout was down. Thus, interlocks were so arranged that either position of the snout would cause trouble in this emergency operation.

By this time the fuel was very hot and starting to disintegrate. A small piece of fuel fell on the reactor top and a much larger piece, three feet long, fell into a maintenance pit and both pieces burned. By this time H_2O cooling was connected to the flask and contaminated water leaked through the open valve as the flask was finally moved to the storage block. Sand was brought into the facility and dumped into the pit to quench the burning uranium. The pit fortunately provided shielding for the high level gamma radiation. A 1000 r/hr meter held over the edge pegged at the top of its scale, and estimates ranged up to 10,000 r/hr for the actual level. The highest exposure to personnel was 19 r to one man. Little radioactivity was released from the building, but the building was badly contaminated inside by fission products from the uranium fire. Over the next two months some 600 men took part in the cleanup. Fragments of the fuel were found in the bottom of the reactor tank as well as in the flask, on the reactor top, and in the pit. The area of detectable contamination outside was confined to about 100 acres (4×10^5 m^2) directly adjacent to the building.

Comments, Conclusions, and Recommendations

(1) "A highly active element was transferred to a less safe position without first following the standard policy of ensuring that all equipment was operating properly." [71]

(2) "...The presence of certain 'safety' interlocks actually increased the hazard by preventing actions that, although not normally desirable, needed to be carried out urgently under the emergency conditions at the moment. One might conclude that reactor designers should pay special attention to the function and detailed mode of operation of all safety interlocks, making sure that, as far as possible, the restrictions they create are necessary and desirable under all conditions." [71] Perhaps there should be an interlock override switch with a special key provided for such occasions.

(3) Designers should check not only normal operation of the system, but also other foreseeable types of operation that might be necessary in emergencies. The goal should be to design equipment capable of surmounting all such challenges. (See Sec. 5.2 of this chapter.)

(4) The removal of highly irradiated fuel from a high flux reactor is a very critical safety problem since, in event of loss of coolant, two of the most effective containment barriers, the fuel clad and the reactor primary system containment, are breached at once. In this case, once the fuel was known to be damaged, it might have been wise to let it cool down in the reactor over the weekend, rather than to attempt removal at once and late on a Friday night.

4.3 The ETR Fission Break Incident [73, 74]

This incident is discussed in more detail in references [73] and [74]. During the two days preceding the accident a "sight box" made of clear plastic was floated on top of the core tank pool to aid in viewing the core. This box was triangular and 8 in. (20.3 cm) on a side and 5 in. (12.7 cm) deep. Prior to startup a senior reactor engineer, a reactor engineer, and the shift supervisor visually inspected the core top structure to see that there were no foreign materials there. This was a regular check list procedure. At that time the water surface was also scanned and the sight box was not seen. Apparently before the inspection the sight box had sunk and settled to the top of the core, which is cooled by downward water flow. The reactor top was replaced and startup was begun.

Criticality was achieved at 22:55 (10:55 p.m.) December 11, 1961. The reactor power level was up to 90 Mw by 01:00 December 12. This level was maintained until 01:45 when a "spike" occurred on the N^{16} primary water recorder. (This unit or one similar is used as a secondary power-flow indicator on several H_2O or D_2O-cooled reactors.) However, there was some doubt as to the validity of the pulse as the unit had been behaving erratically for several months just previously, and so power was reduced to 70 Mw(t) but not scrammed. After two minutes at 70 Mw(t) the reactor was returned to 90 Mw(t).* Reactor nuclear instrumentation, including the servo error signal, safety levels, and the countrate meter, then became erratic. A radiation alarm sounded and area monitors showed increased levels, and the reactor was manually scrammed at 01:52.

Post-accident core examination after cleaning by vacuuming disclosed that melting had occurred in small portions of six fuel elements in one quadrant of the reactor where there had been inadequate cooling. After removal of the visible debris and inspection and replacement of the faulty elements the lid was replaced and water was circulated for several hours. The core was reinspected and a few more foreign particles were removed. The lid was replaced and the reactor was again at full power at about 02:00 December 15. However, activity levels remained high with total beta-gamma contamination running 500,000 dis/min-ml compared to 60,000 dis/min-ml before, the gaseous discharge fission product rate was 800 curies/day (20 curies/day before), and the particulate was 55 curies/day (4 curies/day before). A second cleanup was carried out during the period December 18 to 23. At this time it was discovered that lint and bristles from mops and brushes used in the first cleanup were in the primary system, and some were lodged in fuel

*An inspection of the charts on flow, and the power traces later showed that a cross-check and close inspection of the other charts at the time of the first N^{16} pulse would probably have led to a decision to shut down then.

element entrance passages. These were removed and the system was returned to normal.

Comments, Conclusions, Recommendations

(1) Operating procedures should provide a method for retaining a total check (a "sponge count") of items used in reactor maintenance.

(2) Side plates of "sight boxes" should be made of colored material easily visible in water. Preferably, equipment of this type should be unsinkable; perhaps thin plastic which will easily fail should be used [74]. More generally, all procedures being newly instituted and those already in operation should be reviewed initially and occasionally thereafter in order to understand the consequences of possible wrong operator actions, equipment failures, or other unusual situations. In retrospect, the possibility of this happening should have been foreseen and corrected before the accident.

(3) Installation of a flow recorder with full scale coverage of the narrow range around the operating flow would improve the sensitivity to slight flow variations [73].

(4) Cleaning methods should be such that no residue or material is left in the reactor [73].

(5) The iodine monitor should have an alarm to alert the operators to lack of flow through the detector. Installation of a backup monitor should be considered [73].

(6) As a result of this accident a statistical study of the probabilities of simultaneous malfunctions of the reactor instrument systems was undertaken to give the operators guidance [73].

(7) Whenever an instrument indicates a problem, even if it has been malfunctioning occasionally, the safe path should be taken until it is clear that the instrument is indeed malfunctioning. Other instrument indications should be checked closely as the effect may not be very large.

4.4 The MTR Fission Break Incident [75, 76]

On November 13, 1962 the MTR Reactor had been operating at 40 Mw since November 2 except for five short-duration power reductions. At 14:26 (2:26 p.m.) the fuel assembly differential pressure monitor annunciated to indicate that the difference between the pitot tube total pressure and the static pressure had been reduced by 2 psi (0.13 atm). At this time the servo-controlled regulating rod moved erratically between 15 and 17 in. (38.1 and 43.2 cm) withdrawn. Fifteen seconds after the first alarm the N^{16} detector and the low differential pressure trip initiated a scram. One minute after the scram the seal tank activity monitor in the primary cooling-water exit line increased from 100 mr/hr to 9.8 r/hr. Other radiation alarms sounded, including those in the outlying areas. At 14:31 personnel were evacuated from the reactor and its environs. Survey teams determined that all alarms were due to systems containing primary system water. Personnel then reentered the reactor area but were watched by health physicists in areas where piping was present. Flow was continued to reduce deposition of fission products on the walls, a procedure which had been found useful before. Inspection of the pitot tube indicators located one element whose outlet pitot total and static pressure differed by only 8 psi (0.53 atm) instead of 13 psi (0.87 atm).

The primary reactor water was flushed out and the top plug was removed at 20:25. A piece of black material was detected in the suspect fuel element. The material covered an estimated 40% of the area of the top of the plates. Small pieces were also found on two other elements. On November 14, the core loading was changed, but the fuel was not removed, the top plug returned, and the cooling water started at 22,500 gal/min as usual. No more debris was found after two hours circulation, and on November 15 the Reactor Physics and Engineering Staff and the MTR Safeguards Committee approved return to power with the following stipulations (there was still the worry that there might be more debris and this was after the ETR incident described earlier):

(1) Primary cooling water was to be circulated for one hour prior to rod withdrawal and checks made to see that all differential pressure monitors agreed with readings made when the core was known to be clean.

(2) Satisfactory results were to be obtained for shim rod drop tests.

(3) The output of the reserve servo system was to be monitored on a fast recorder to check for the onset of nucleate boiling. Reactor power was to be increased in 5 Mw steps to 30 Mw and then 2.5 Mw steps to full power.

(4) The fuel element differential pressure monitoring system was to be given rigid surveillance during the remainder of the cycle.

The restart went without incident, but at full power it was clear that the earlier suspect element had a fission break and the element was transferred to the MTR hot cell for examination.* The seventh plate in the element had exhibited some melting due to partial blocking on both sides of this plate, but the sixth and eighth plates on either side appeared sound.

During the cycle shutdown, the seal gasket on the seal tank floating roof was found to be in poor condition and replacement of a major portion was required. On analysis the gasket material seemed to be identical with that found in the element.

On November 25, 1962, a scram occurred again on differential pressure and some "red rubber" debris was found on one element. The same thing occurred on December 24 and "red rubber" was found in four elements. During the shutdown January 2-5, 1963, the primary side of the flash evaporators was inspected and numerous pieces of debris were found in the nozzles.

Comments, Conclusions, Recommendations

(1) It is highly desirable to provide filter screens of such a mesh as to pass only those sizes smaller than the narrowest opening in the fuel in which debris may lodge. These filter screens

*It is not clear why this was not done during the first opening of the core since fission products were detected then.

should be inspected periodically to give prewarning of deterioration.

(2) It is desirable to design systems without the use of gaskets made of materials which will deteriorate and then may enter the system.

(3) If screens are not present continually in the system, the design should provide a means of inserting them in event it is necessary to filter out such debris.

(4) Reference [76] points out that a local review committee knowledgeable of the reactor and at the time not directly involved in operations played an important role by developing in less than one day the special procedures outlined above for returning to power.

5 OPERATING EXPERIENCE WITH NON-CORE COMPONENTS

5.1 Introduction

This section will discuss briefly some examples drawn from operating experience gained with reactor primary system components outside the core. In no sense is it intended to present a complete review of all the experience to date, but rather it is intended to present a few selected examples to bring out various important points discussed in other chapters.

Section 5.2 presents, by example, a mechanical design fundamental. Section 5.3 discusses swimming pool experience regarding loss of coolant and control rod problems and presents some results of interest in loss-of-coolant accidents for such reactors. Some major and minor problems in material and mechanical failure are presented in Sec. 5.4. Section 5.5 discusses briefly a hydrogen-oxygen explosion which was not the result of an accident but could have been the cause of a severe one. Section 5.6 presents an example of a material compatibility problem.

5.2 Fuel Rehandling Mechanism at the Shippingport Reactor [77]

No accident was involved in the occurrence which is described here. The occurrence is described to point out a lesson for reactor designers.

During the initial tests at the Shippingport Reactor, a fuel element handling device was being observed in action by the original design group and by others. The fuel element lifting device was so designed that it could be released from the fuel only when the fuel was properly inserted into the grid structure or when the fuel was totally free from the grid structure. As the fuel was being lowered into the grid from above, a senior staff member inquired whether anyone present had a screw driver. No one had a screw driver and a man was dispatched to obtain one. Meanwhile the senior officer had wandered away to look at something else, leaving the design crew to wonder at this strange request.

As the design crew thought about it, one member suddenly had an idea. He saw that if the screw driver were used to wedge the fuel element in a partially inserted position in the grid the design was such that the tool could not be unlatched or moved in any way. Thus any obstruction which interfered with the proper insertion of the fuel element would freeze the entire system and eliminate any possibility of taking further action. They immediately modified the design so as to provide an override which, in event of an emergency, could be used to release the tool with the fuel element in any position.

Comments, Conclusions and Recommendations

While this is a relatively simple example, it illustrates quite clearly that the designer must consider the behavior of the system on which he is working under all possible circumstances, not only under normal operating conditions but also under abnormal operating conditions including failure. If the system is to fail it should fail in a fail-safe manner and in such a way that the system is repairable and replaceable.

5.3 Swimming Pool Reactor Experience

5.3.1 Loss of Water From the Pool [78,79,80]

Very seldom, if ever, has a swimming pool reactor been constructed which did not at some time in its life develop leaks through the concrete. Various means have been utilized to plug these leaks with varying success. Normally, the leaks are small and of no safety consequence—unless the pool water should become quite radioactive, in which case the possibility of contaminating ground water becomes a concern.

However, in three cases, two in the autumn of 1963, leaks of serious proportions have developed. As might be expected, the leaks did not develop because of leakage through the pool walls. In the first instance [78], the mistake made was the insertion of a long shielding plug into a relatively short horizontal air-filled beam thimble which faced the core lattice in the University of Michigan reactor. (The reactor has horizontal beam thimbles of different length reaching in to the edge of the core.) The short thimble was ruptured and water started to rush out from the pool around the edge of the plug and onto the experimental floor. Fortunately, the plug was not fully ejected from the shield and quick work on the part of the personnel present finally got the water flow stopped. The Michigan University Reactor is a two-position pool so the core was placed in the other section and, with the help of a skin diver, repairs and replacement of the damaged thimble were effected.

At the University of Virginia Reactor sometime between midnight and 6 a.m. October 18, 1963, a break occurred in the piping of the demineralizer room [79]. The break was not a fracture but a separation of a screwed fitting in the plastic piping. At 6 a.m. a graduate student discovered the break and promptly reported it. By the time the break was shut off the water level in the pool had dropped about one foot and the low water float switch beneath the reactor bridge had been actuated. The operation of the float shut off the demineralizer pump, but the leakage continued by back flow through the pump into the demineralizer room. Even if back flow had continued as long as possible, the level would have been maintained several feet above the active core.

In November 1963 a supervisor at the Texas Agricultural and Mechanics College Reactor entered the reactor building to find that the pool water level had dropped eight feet [80]. A gasket in the demineralizer tank access hole had failed and water had flowed, ultimately, to the hot sump. From there it was automatically pumped to a holdup tank from whence it overflowed, letting most of the water spill onto the ground. From there it went to a dry gully. The pool water level had an activity of 43 $\mu\mu$c/ml. No serious consequences resulted.

5.3.2 Fuel Temperatures Reached on Loss of Coolant in the LITR [81, 82, 83]

An early set of experiments done at Oak Ridge has provided some idea of the power levels at which melting might be expected in a tank-type or a swimming pool reactor due to loss of coolant [81, 82]. The Low Intensity Training Reactor (LITR) was employed for these tests using 15 MTR-type elements and three control-safety rods in a 3×6 lattice reflected by beryllium on all but one face. The experimental procedure consisted of running the reactor at a constant specified power for the desired length of time, shutting the reactor down by draining out the water (no drop of control rods), and then following the fuel plate temperature in the hottest element until a maximum was passed.

About five minutes before the run was to be terminated, the cooling water pumps were stopped and the inlet and outlet valves closed. A port was opened in the top of the tank as a vacuum vent. Then a 6 in. (15.2 cm) remotely operated valve was opened. It required 2.5 min to lower the level from the tank top to a point 1 ft (30.5 cm) above the fuel plates. At that point the drop in water level began to affect the reactivity and rapidly shut the reactor down. It required only 12 sec more for the water level to drop below the fuel plates and 30 sec more for all water to drain from the tank. The valve and the port in the top were then closed. Temperature readings were continued for two hours, which in every case up to 300 kw was time enough for the maximum temperature to be reached and passed. Nine 2-hr power runs at specified levels from 0.5 kw to 300 kw were carried out. The effect of longer runs was measured by runs of 2, 6.5, and 24 hours at 150 kw.

Later experiments extended these measurements to 1250 kw and 150 hr operation. These produced fuel element temperatures as high as 249°C (480°F) and end-box temperatures above boiling. This later set of experiments used 21 elements instead of 15 and the results had to be normalized by comparing neutron flux measurements in the central elements (1.5 times the maximum temperature observed in the 21-element case).

The observed points extrapolated to infinite running time for a 15-element core are plotted in Fig. 5-1 [82]. It is evident that the 15-element core should melt on loss of water after operating at power levels above about 1500 kw. With a 21-element core the maximum power before melting on loss of coolant would be about 2250 kw.

The added points in Fig. 5-1 show the central fuel plate temperatures reached in tests in which water was sprayed over the fuel at a rate of 2 to 6 gpm (0.13 to 0.38 liter/sec) until a 500 gal (1893 liter) capacity tank was empty. Spraying kept the fuel plate temperatures below 212°F during the 2.5 hr it continued. It was demonstrated in these tests that the LITR could reject heat continuously without melting fuel at a rate of 8 kw. Beall [82] estimates that at least twice this rate is possible before melting occurs.

The equation obtained for the fission product power in a single MTR-type element with 140 g of U^{235} was [81]:

$$q = 14P\,[t^{-0.2} - (t + T)^{-0.2}]\,,$$

where q is in Btu/hr, P is the reactor power (kw) before shutdown, t the time (sec) after shutdown, and T the operating time (sec) before shutdown.

Beall concludes that as much as 75% of the fission product heat is lost by conduction to other parts of the reactor before the fuel reaches a maximum temperature. He also points out that calculations given by Poppendiek and Claiborne [83] estimate melting at much lower power levels. He believes that the difference is primarily due to improvement in thermal conductivity across heat conduction gaps due to water in the gaps. Thus, most of the heat seems to be carried away by aluminum conduction to the metal base plate (weight = 1000 lb or 454 kg).*

5.3.3 Control Rod Experience [78]

In the University of Michigan pool reactor [78], and in other reactors as well, it has been observed that some of the relatively flat blade-type control rods have become waterlogged and have expanded to assume a more cylindrical shape. The expansion appears sometimes to be due to gas formation in boron-containing compounds and sometimes to corrosion gas evolution.

This expansion has led to binding of control rods within their guides in the fuel elements. There then exists the possibility that the fuel element within which the control rod moves will be picked up as the rod is picked up, only to be later dropped back into the core resulting in a nuclear transient. This same type of control rod expansion might also prevent a control rod from

---

*Recently R. Panter, AERE Harwell, has completed work for the "Dido"-class reactors (which use MTR-type fuel plates in various configurations) on fuel element temperatures during loss-of-flow accidents and during fuel transfer operations.

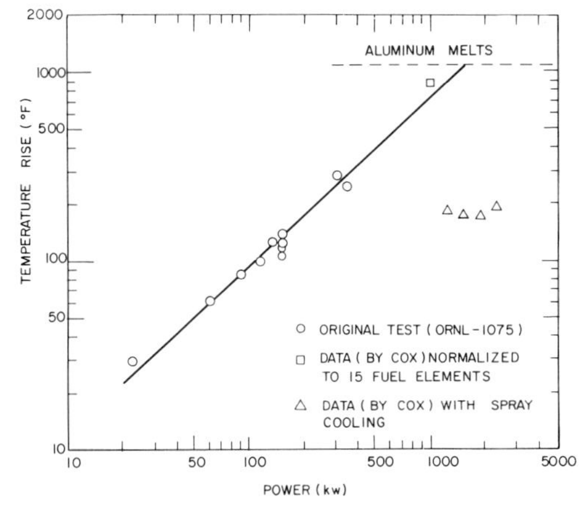

FIG. 5-1 Temperature rise of center fuel element vs. total reactor power for infinite running time. LITR—15-element core, 3 control rods.

dropping back into the core. This problem is discussed further in the Mechanical Design chapter.

Comments, Conclusions, and Recommendations

(1) It is wise in any reactor where loss of water can occur relatively easily (mostly in research and test reactors) to provide alarms to alert personnel to the problem and to take preventive action. If the reactor is to be unattended for an extended period perhaps the alarm should give a switchboard or telephone alert.

(2) It is well to remember that in many reactors—especially research and test reactors—the potential exists for ejection of plugs from reactor access holes. They may be ejected either by water pressure under relatively normal operating conditions or during the course of an accident. Such plugs should be secured in the shield by means of some strong latching device. It is also possible that beam tubes can collapse causing serious reactivity increases.

(3) At least one radiation alarm should remain in its sensitive state in each potential radiation area around a reactor—even when the reactor is off.

(4) The potential to melt fuel exists for any reactor which is light-water-moderated and -cooled and uses plate-type MTR elements and operates at power levels of over 1.5 to 2.0 Mw. Thus, a sudden loss of coolant from such cores during full power operation should be regarded as a potentially serious accident. The beta-gamma after-heat can give core melting. It is likely that single elements and widely spaced elements will not melt until somewhat higher power levels are reached since there is less interaction heating between elements.

(5) Special care should be taken that control rods are well designed, fabricated, and used so as to ensure that no expansion or distortion occurs in such a manner as to reduce reactor safety.

5.4 Material and Mechanical Failures

Evidence continues to show that mechanical and material failures can and do occur. At least four of these failures have had the potential for causing a serious accident, although none actually did occur. Other failures of a more minor nature occur more frequently and are usually not reported. Five well-documented examples of this more minor type are briefly described as typical.

SM-1 (APPR) Closure Bolt Failure [84]. After 10.5 Mw-years of operation of this 10 Mw(t), 1200 psig (82 atm) pressurized water reactor, the vessel head was removed for a core examination. At that time it was discovered that two adjacent vessel head studs were broken. If additional studs had failed, a serious accident could have resulted. The breaks were determined to have been caused by stress corrosion cracking. The Type 410 stainless steel studs were originally heat-treated by heating to the range 1750 to 1800°F (954 to 982°C) and held for 0.5 hr, oil-quenched, tempered at 1000°F (538°C) for 3 hr, and then tempered to Brinell hardness number (BHN) 262 (Rockwell C Hardness (R_C) = 26.5). Examination of the bolts showed that hardness ranged from 220 to 382 BHN (18 to 41 R_C). (During the same inspection it was also found that the cladding of the boron absorbers in four shim rods was severely cracked in the high burnup region. This was attributed to helium gas formation.)

Vallecitos Boiling Water Reactor Main Steam Line Valve Failure [85]. On March 9, 1959, the valve plug separated from the valve stem in the main steam pressure-reducing valve. The failure almost instantaneously stopped the flow of steam while the reactor was operating at 30 Mw(t). The reactor pressure rose, and the core void fraction decreased. This gave a positive reactivity effect and the reactor power rose to 35 Mw(t), the set point for an overpower scram, and the reactor scrammed in 5-10 sec after the valve failure. The reactor pressure had risen by about 50 psig (2.7 atm) above the nominal 1000 psig (68 atm) operating pressure. No serious consequences resulted.

SPERT-III Pressurizer Failure [86, 87]. On October 26, 1961, during a series of test runs on the SPERT-III pressurized water reactor, the system was brought to 221°C (430°F) and 2460 psig (167 atm). About three hours later, smoke was observed coming from the vicinity of the pressurizer. A normal shutdown was started and the fire department was alerted. The plant was cooled down and depressurized without need for the fire department or further incident.

Inspection showed that the smoke came from the fabric covering of the blowdown-line riser, which was in line with a major steam leak in the pressurizer. A 3/8 in. (0.95 cm) wide, 2-3/4 in. (6.98 cm) long hole was found in the central girth seam weldmetal. A 1 in. (2.54 cm) diameter bolt which tightened a stabilizing band around the vessel was broken. Corrosion on the broken face of the bolt and subsequent investigation indicated that the bolt broke in an earlier and separate expansion of the vessel.

The pressurizer is a 33 in. (83.82 cm) inner diameter, 16-2/3 ft (5.08 m) high all-welded vessel of ASTM A-264, grade 3, 0.04% max. carbon steel with 304 L stainless steel fittings and internal cladding. The backing plate was ASTM A-212, grade B, firebox quality carbon steel. The wall thickness was 2.95 in. (7.49 cm) including 1/8 in. (0.318 cm) of cladding. Normally submerged electric heaters in the pressurizer controlled the system pressure. The pressurizer liquid level was measured by a differential pressure instrument with temperature (density) compensation. A signal when the pressurizer liquid level was low actuated a make-up pump, and a high level signal operated a blowdown valve.

It was determined that the water level indicators were capable of functioning incorrectly and that the water level had dropped, thus exposing heaters in the steam atmosphere above. These exposed heaters raised the vessel wall to temperatures approaching 557°C (1000°F) and caused the failure in what appeared to be three different stages.

VBWR Main Coolant Recirculation Pipe Failure [88]. The primary recirculation piping system of the Vallecitos Boiling Water Reactor (VBWR) consists of four pump suction lines and two discharge lines attached to the reactor vessel, all containing water at about 288°C (550°F). The ten-inch (25.4 cm) diameter piping has an 0.64-in. (1.626 cm) wall and was fabricated of ASTM A240, type-304 stainless steel plate-rolled, welded, solution heat-treated, and inspected with liquid dye penetrant and radiography. The pipe was bent hot and the bends subsequently solution heat-treated and water-quenched. The fabrication and erection were carried out in accordance with Section I (Power Boilers) of the ASME Boiler and Pressure Vessel Code.

The four pump suction pipes are welded to pressure vessel nozzles located about halfway up the reactor vessel 20 in. (50.8 cm) above the reactor supports and in opposing N-S pairs with the members of each pair about 36 angular degrees apart. Each suction pipe has a 1-in. diameter feedwater injection pipe welded to it at the 12 o'clock position about 8-in. (20.32 cm) from the reactor vessel. The purpose of these pipes is to introduce 21 to 54°C (70 to 130°F) feedwater from a feedwater header into the system. The northerly directed pair bends around the vessel and all four of these pipes pass south outward through tight fitting holes in a 3-ft (91.44 cm) thick shielding wall. The wall consists of high density concrete bricks caulked with ferrophosphate sand. The southerly directed pair passes through the wall approximately 20 in. (50.8 cm) from the reactor vessel surface.

The two discharge pipe nozzles are located near the top of the vessel east and west on opposite sides. They too pass through the brick wall at a higher level. The discharge nozzles are approximately 85-in. (216 cm) above the reactor supports.

On September 3, 1962, after approximately five years of operation there were indications of steam leakage in the VBWR reactor containment building basement. On six separate occasions between September 3 and September 12 the basement was checked for steam leakage. On two of these occasions the reactor was depressurized and some steam leaks discovered in valve packings were repaired, but still there appeared to be additional steam escaping from an unidentified leak. Accordingly, the reactor was shut down on September 12 and the shielding was removed revealing a 2-1/2 in. (6.35 cm) long pipe crack in one of the southerly directed suction pipes at a position about 2 in. (5.08 cm) from the reactor vessel nozzle weld.

Tests with liquid penetrants revealed no other external cracks in any of the six pipes. However, ultrasonic tests indicated other cracks deeper than 1/32-in. (0.79 mm) on the inside of the same pipe for a distance of about 18 in. (45.7 cm) from the vessel nozzle and indications of other cracks in the other three suction lines but none in the two discharge lines. A study of ultrasonic development of the leaking line as shown in reference [88] shows disconnected short traces in several apparent lines for about 5/6 of the circumference of the pipe.

On the basis of this preliminary examination a length of about 24 in. (61 cm) of the leaking pipe was removed, decontaminated to reduce radioactivity, and pickled to remove corrosion and oxide film. A "crazed" pattern of cracking was observed on the inside of the pipe corresponding roughly to the path the injected (room temperature) feedwater from the one-inch pipe would take. In addition, a series of circumferential cracks existed in the upper half of the pipe in the first three inches from the pipe nozzle. The observed crack, which was about 2.5 in. (6.35 cm) long on the outer surface, "was about 6 in. (15.24 cm) long on the inner surface. Even though the crack was apparent for 2-1/2 in. (6.35 cm) on the outside of the pipe, the crack was not continuous completely through the wall. Sections through the wall revealed that many of the 'crazed' cracks extended into the wall as much as 1/8 to 3/16 in. (0.32 to 0.48 cm). Some of the cracks which are predominately oriented circumferentially extend into the wall as much as 1/4 in." Microscopic examination showed primarily transgranular cracks but with some intergranular cracks also.

The microstructure appeared normal for annealed type-304 stainless steel with little evidence of carbides. The tensile, hardness, and chemical composition tests all revealed typical results for 304 SS. The cracking occurred in base metal and appeared to have no relation to welds or to use of defective material.

It was discovered that the ferrophosphate sand used as caulking had solidified into a solid mass thus causing the pipes to be held even more firmly in the concrete shielding wall. Analyses and tests revealed that the feedwater injection through the one-inch lines had caused a very large amount of stress-strain cycling. Temperature gradients as high as 197°C/cm (900°F/in.) along the inside pipe surface were estimated. The thermal stresses calculated on an elastic basis were as high as 126,000 psi (~8900 kg/cm^2) giving a strain of about 0.5%. Examination of the operating records showed that over 1000 injection cycles occurred during the five years of operation.

It was also determined that the restraint of the brick shielding wall caused varying high mechanical stresses to be exerted on the pipe as a whole and that as many as 500 cycles of this nature might also have occurred. The mechanical stresses calculated on an elastic basis were as high as 50,000 psi (~3500 kg/cm^2) resulting in a strain of about 0.2%. Corrosion is not believed to have played an important part in the failure.

As a result of this discovery, the four pump suction lines (and the two discharge lines as well, although apparently undamaged) were replaced between the reactor vessel and the shield wall. The shield wall was modified to permit unrestricted pipe movement while still providing adequate shielding. The feedwater injection system was redesigned to inject feedwater in a 3-in. (7.62 cm) diameter pipe through a thermal sleeve into a 6-in. (15.24 cm) nozzle near the top of the reactor vessel. Once inside the vessel the feedwater is distributed to a ring header system with dip tubes designed in such a way that thorough mixing occurs and no cold water can reach any important component of the system.

Comments, Conclusions, Recommendations

(1) Appropriate detection means (such as microphones, thermocouples in the building, fire warning detectors, etc.) coupled with alertness and dogged determination to find all the leaks, in this case, led to the early location of a serious flaw in the system. While it is likely that the cracking might not have led to what many reactor operators believe to be almost incredible—the complete severance of a primary pipe—nevertheless, the situation developing was uncomfortably close to that accident.* A pressure surge of modest proportions in such a system might be enough to trigger such an accident. Alertness for unusual behavior symptoms, prompt and full exploration of these symptoms, and fully adequate restoration of the system to normal safe operation are the heart of safe reactor operations.

(2) The thermal and mechanical analysis of the system must be very thorough and detailed to ensure that no excessive stresses can occur and that no excessive thermal or mechanical cycling can occur.

(3) In-place inspection of the installation should be made from time to time if possible to ensure that shifting of equipment or components (in this case ferrophosphate sand) or careless placement of some blocking obstacle has not set up an undesirable restraint on the system. Regular entry into the containment for "in person" inspection of a reactor system at temperature and pressure (the reactor may be shut down for shielding reasons) is a valuable aid to safety.

Savannah River Reactors—Stainless Steel Failures [89, 90]. Four independent failures in stainless steel components which are part of the primary system in the Savannah River heavy-water-cooled and -moderated systems will be discussed briefly. These failures occurred in systems in which special care had been taken to include rigid specifications, fabrication, and testing. They serve to show that even where great care is taken some failures will occur.

The first type of failure occurred in several heat exchangers made of type-304 or 304 LC stainless steel. These exchangers are 33.5 ft (10.21 m) long and 88.5 in. (2.25 m) in diameter. They are shell-and-tube type units with process river water in the shell side and D_2O in the tube side. To minimize the chance for leakage between heavy and light water, the exchangers have double tube sheets at either end with a leakage collection region between. Two exchangers had been removed from service at the time of the report (about four years of use) and examined.

It was found that the type-304 stainless steel tubes were covered with heavy adherent deposits in the leak collection space. Many tubes at the D_2O inlet end were cracked by stress corrosion. All the cracks were initiated from the outside of the tube. No cracked tubes were found at the D_2O exit end or in the main body of the exchanger. Deposits on the tubes contained about 300 ppm chloride content. Moisture, condensed droplets, and dried traces of running water were found in a number of the moisture collection volumes. The D_2O is of high purity with a pH of 5.7 (controlled by nitric acid) and with less than 0.006 ppm of chlorine. The cooling water is Savannah River water which is untreated except for intermittent additions of chlorine for algae control. The chloride content is normally 2 to 9 ppm. Evidently some river water containing chlorides had leaked through the header from time to time and had deposited on the stainless steel tubes, causing the observed stress-corrosion cracking.

(In a number of heat exchangers in other reactors, including sodium heat exchangers, tube failures have resulted from vibration due to faulty hydraulic design or from thermal cycling, or from both causes. Heat exchangers, while normally not a major safety problem, continue to be a troublesome operational problem.)

The second failure was a hole about 1/8 in. (0.318 cm) diameter through a 0.109 in. (0.277 cm) thick AISA 304 stainless steel header wall. The sides of the hole were attacked intergranularly. Nearby, a second 1/8 in. diameter localized plug of intergranular corrosion was discovered. The grains appeared to be completely separated from the rest of the material and only held in place by mechanical interlocking. Examination revealed that the inner and outer surfaces of the wall sections were etched intergranularly to a depth of about 0.0003 in. (0.0076 mm). Scattered areas on both surfaces were intergranularly corroded to a depth of 0.010 in. (0.254 mm). It is believed that this corrosion resulted from pickling during fabrication. The de-ionized water and CO_2 present on either face are believed to be noncorrosive. Many arc strikes were present with crater cracks in the arc strikes, and lack of fusion was found between adjacent weld passes. The hole and the plug region were in areas of severe sensitization, apparently due to welding. Evidently, the area was at sensitization temperatures (800-1600°F or 427-871°C) for a prolonged period of time or for several short periods.

The third case resembles the second in cause.

*It should be pointed out, however, that the cracking in this instance very likely would not have led to a failure of the "guillotine" type, since the crack through the pipe (and others starting to penetrate) were only in part of that half of the circumference subjected to tension. The half that was in compression showed no crack propagation from the crazes.

A crack developed in one outlet nozzle of the main reactor tank. The nozzles are really pipes of special shape, each varying from an almost rectangular cross section at one end to a circular cross section at the other as the pipe makes a 90-degree bend. They were fabricated by welding together formed pieces of 1/2 in. (1.27 cm) thick AISI 304 stainless steel plate. The crack was about 4-1/2 in. (11.45 cm) long extending entirely through the metal, permitting leakage of D_2O. Again the cracking was intergranular and in severely sensitized metal. Apparently pickling solutions used during fabrication caused intergranular attack to a depth of 0.0075 in. (0.019 cm). Repair required excavation of a hole through the shield and installation of temporary shielding because of high radiation levels.

The fourth failure was due to a crack in a 2 in. (5.08 cm) reactor drain line. The location was such that the piping section could not be isolated from the reactor for repair. The crack extended around approximately one-fourth of the circumference of the weld joining the 2 in. (5.08 cm) and a 6 in. (15.24 cm) pipe. It was at the edge of the weld. Metallurgical examination showed that the crack was fatigue failure, probably caused by vibration of the piping.

Comments, Conclusions, Recommendations

(1) While there is apparently no record of an ASME Code Vessel Closure failing in such a way as to release the vessel head as a missile when the vessel is operated according to its design specifications, nonetheless, if it occurred, the resulting accident would clearly be more severe than the usual postulated maximum credible accidents. Therefore, special care should be used in specifying, fabricating, testing, and using closure studs or bolts. In particular, three areas require special attention. The materials must be carefully selected, and it must be verified that they are proper. The heat treatment must be carried out according to specifications and verified. The method of bolt tightening must be chosen with due consideration to material properties, stresses, etc., and bolt tightening must be properly carried out each time. The use of an improper temperature for bolt heaters as a means of tightening bolts is one way to exceed design specifications, for example.

(2) It must be assumed that valves will on occasion fail, or fail to work. To the author's knowledge at least three containment closure valves of a supposedly fail-safe type (including two at M.I.T.*) have failed to give complete closure on occasion. (See also Sec. 5.7 of this chapter.)

(3) There is good reason to believe that failures of major piping and vessels may occur from time to time. In particular, either the SPERT-III pressurizer failure or the VBWR main steam line failure could have resulted without too much change in conditions in something close to the maximum credible accident usually postulated for most reactors—namely, total loss of coolant. It appears to be much less likely that such piping or vessel failures will be brittle failures in which missiles will be formed since the systems are in general well above the nil-ductility transition point when at design temperatures and pressures.

(4) The susceptibility of stainless steel to intergranular and stress corrosion has only become well known in recent years. Overpickling and overheating during fabrication so as to cause sensitization can lead to failures in areas other than weld areas. It is normally assumed that the plate is flawless. Experience does not always bear this out. Even in carefully constructed systems failures will continue to occur.

(5) Since failures will occur, it is well to provide means for the early detection of such failures and for easy repairs or replacement where possible. It is likely that most failures will start as small cracks, as did those at Savannah River and VBWR. Only if there is a sudden pressure surge while such a crack exists is it at all likely that the maximum credible accident will occur. Therefore, early detection is vital.

5.5 Antarctic Pressurized Water Reactor—Hydrogen Explosion [91]

During normal operation of the PM-3A reactor in the Antarctic, an explosion occurred in the water-filled shield tank which surrounds the reactor proper. No damage resulted to the reactor itself or to any major component. It is surmised that radiation from the reactor caused the breakdown of water into hydrogen and oxygen gas which collected above the water surface. Later some of this leaked out and reached the area where an electrical spark set it off.

Comments, Conclusions, Recommendations

(1) In a high level radiation field, water or heavy water may break down into hydrogen and oxygen, thus creating an explosion hazard. (Hydrogen-oxygen reactions are discussed in detail in the chapter on Chemical Reactions.) Such an explosion may be the cause of a more serious accident. In the design of a water-moderated or water-cooled system, care must be taken to deal adequately with this problem. Possible methods include recirculation of the cover gas through a recombiner, purging, and use of a hydrogen or oxygen cover gas to provide recombination within the reactor itself at mixture concentrations outside of the explosive limits.

(2) Potential gas-pocket traps such as upward pointed thimbles, etc., should be avoided. As the PM-3A accident showed, the accidents may not occur in the primary system itself. (Since downward pointed thimbles tend to collect radioactive crud, avoidance of all such thimbles where possible seems worth considering.)

---

*During an annual leak test, two large butterfly-type valves used for containment closure at the M.I.T. reactor were found to be leaking (after approximately five years of use) because of wear and changes in valve seating. The resulting air leakage was large enough to cause the leakage specification (1 vol.%/day) to be exceeded (1.11 vol.%) and to require temporary use of the back-up emergency valves.

(3) Electrical sources such as heaters, electrical probes, etc., may also be the source of hydrogen and oxygen through the electrolysis of water. These units should also be designed, installed, and operated with due regard to possible electrolytic dissociation of water.

(4) Electrical make-break contacts may provide the spark for such explosions.

5.6 N. S. Savannah—Control Rod Hydraulic System Leakage [92]

"During the initial testing of the hydraulic drives provided for the N. S. Savannah control rods, many instances of hydraulic oil leakage were experienced" [92]. Some leakage has persisted throughout the use of this system, the average being 8 to 10 gal/month (30.3 to 37.9 liter/month). This leakage, falling on hot pipes caused small fires during early testing. It was discovered that the use of noninflammable hydraulic fluid was not possible because it would dissolve the insulation on all of the power, control, and instrument wiring in that region. Therefore, "the decision was made during the program conducted by the shipyard to maintain the oxygen concentration of the containment at less than 10%, through the addition of N_2 to the containment, at times when the hydraulic system was pressurized and the primary system was over 400°F" [92]. This addition has prevented any fires, but it has reduced the ease of maintenance and, more important perhaps, the frequency of inspections within the containment.

Comments, Conclusions, Recommendations

(1) The example given here is another and rather subtle example of lack of compatibility of materials. All of the materials used in a given system, including solids, liquids, and gases, must be compatible with the environment in which they are to work. The materials must be compatible with each other if there is any conceivable possibility that they will come in contact with each other.

(2) In general, visual and audible inspection of the containment of a reactor and all of its components is important to reactor safety. Experience has clearly shown that such inspections uncover many problems before they become serious. It is almost essential that inspections be made of all parts of a system where failures could occur while the system is at temperature and pressure. Since most plants can be raised to temperature and pressure without nuclear heat, such inspections can be made without higher radiation levels than are observed in normal reactor shutdown maintenance.

(3) The use of inert gas in any compartment is always dangerous to personnel. A man entering such a region accidentally may become trapped and killed by oxygen starvation before he realizes the danger. Although there is no direct connection with reactor safety, it may be worth while to point out that at least two persons have died from this means at atomic energy installations.

5.7 PWR Containment Valve Malfunction [93]

During the performance of a "Reactor Plant Integrity Test, on November 9, 1963 it was determined that the hydraulic butterfly valves could not be closed by actuating pressure supplied only from the drive accumulators. (Note: The drive accumulators are hydraulic cylinders, charged with approximately 1100 psig, to provide emergency closure of the hydraulic butterfly valves when the makeup accumulators or hydraulic pumps are unable to provide the necessary force.)" [93]

"Investigation revealed that the relay which is energized by the makeup header low pressure switch did not properly operate the four solenoid-operated valves which actuate the drive accumulators to operate the hydraulic butterfly valves on receipt of a 'close' signal" [93]. The relay was repaired and apparently worked properly, but still the butterfly valves failed to close as they would be required to in case of an accident involving loss of AC power in order to give containment integrity.

On November 22, another series of investigations were conducted to no avail and special temporary emergency instructions were issued to cover the situation. On November 26, the "drive accumulator pilot operated check valve" [93] was removed, inspected disassembled, cleaned, assembled, replaced, and tested (November 27) to no avail. Other tests on the rather complex system continued.

From this series of tests it was finally concluded that the pilot connections to the valve in question were reversed. A test quickly showed this to be the case. The installation was not in conformity with the drawings. There is no indication in the report of how long this situation had existed.

Comments, Conclusions, Recommendations

(1) All components important to safety and all modes of operation of those components should be checked regularly to ascertain that they are operating correctly.

(2) Systems shuld be designed as simply as possible in order to facilitate reliable operation and simple testing and repair.

6 CONCLUSIONS

6.1 General Comments

While at first glance it would seem that these accidents bear little resemblance to one another, closer study brings out some rather remarkable similarities, in spite of the fact that there are so few accidents that any statistical analysis is not sensible. Some of these similarities no doubt have significance and all of them are worth mentioning.

As pointed out in Chapter 1 almost all of these accidents involve three principal types of causal ingredients:

a design flaw in the reactor, its interlocks, or its basic procedures,

an instrumentation flaw, and

a "human error," i.e., an error by operators or supervisors at the time of the accident.

These three ingredients can be seen cropping up

again and again. A design flaw, usually long buried and forgotten, lies hidden—lying in wait as it were—ready for the right set of circumstances to emerge. Faulty instrumentation (or often lack of instrumentation) fails to sound the alarm in time. And in those few action-charged moments of decision, usually after prolonged periods of boredom, the operator or supervisor makes the wrong choice or, almost as bad, he makes no choice at all. This is the pattern.*

To some extent the accidents reflect specific causes such as manual withdrawal of control rods, etc. It is often true that these problems have continued to be sources of trouble because designers have not known of their existence. If so, it is the principal purpose of this chapter to eliminate this deficiency by supplying an artificial backlog of experience. On the other hand, some of the problems one might expect to find, such as excessive power levels, are not represented in the accident causes. This is no doubt due to good education, sound design, and cautious operation with respect to those known problems.

The list of accident causes given in Table 6-1 should not be taken as the complete list of the problems of reactor safety. It is simply the list of those causes which apparently have not received enough attention thus far to be eliminated as statistically identifiable causes for accidents. The other causes, such as operation at too high power levels, too fast approach to criticality, etc., exist and must not be forgotten or they too will appear on later lists of this kind.

A study of the accidents cited in this chapter indicates that almost all were caused not by one failure but by several sequential, sometimes dependent and sometimes independent, failures. Thus, these accidents provide small comfort to those who say that more than one or two sequential independent failures or malfunctions in equipment or operations is incredible. Most dangerous of all is the operating philosophy that there are several independent sequential barriers to prevent a given accident and that, therefore, the failure of any given barrier is not serious and repairs to that barrier can be postponed indefinitely. Each true safety barrier to an accident should be treated as if it were the last one—for indeed it may be.

In Chapter 1 accidents were also classified as industrial, economic, and public. It was pointed out that critical assembly accidents are almost invariably of the industrial type involving only personnel working directly on the project and usually involving little or no financial loss. The safety effort in critical facilities should be devoted mostly to protecting the operating personnel from radiation exposure. In all eight fatalities which have occurred to date very high radiation exposures have been an important factor or, in all but the three SL-1 cases, the only factor in the deaths.

The only critical facility accidents where there could be other worries besides radiation exposure to operating personnel are those involving the use of assemblies containing plutonium, or large fission product inventories, or toxic materials. Such assemblies could conceivably involve expensive cleanups (Economic Accidents) or could endanger the general public (Public Accidents). However, plutonium is not a gas and experience to date with its dissemination is not alarming.

As mentioned above the "economic" and "public" accidents will normally involve large inventories of fission products or toxic or radioactive materials such as beryllium or plutonium. As more reactors are built, chemical plants for fuel processing will become a source of concern since they will carry large inventories of such material. Since May 1958 there have been six criticality accidents in "non-fail-safe" fissile material fluid containers. All six of these reported cases involved fissile material salvage operations where inventory uncertainties are more likely and where the operations are less routine. Four of them involved plutonium or fission products. The four reported unshielded cases all resulted in high radiation doses to one or more workers. In no case to date has there been a public accident involving this type of plant and, in general, the cleanup expenses have been low.

Fuel failures resulting from loss of coolant or loss of flow can constitute a serious economic or public accident. While not many accidents of this nature have been reported, the seriousness of their consequences is evident. An accident in which appreciable fuel melting occurs is a serious one, will certainly result in economic loss, and may involve serious radiation hazards to operating personnel or to the general public. The seriousness of this type of accident to the general public will normally not be as great as that of an accident involving criticality changes.

In a boiling or pressurized water reactor, loss of coolant, by itself, will normally result in depressurizing the system, and this will then be followed by the melting of fuel by fission product after-heat if remedial actions are not taken. The accident will be "sequential" in nature and the energy release involved in depressurizing will not aid in dispersing fission products although it may, along with after-heat, play a role in leakage rates from containments. Loss of coolant to hot spent fuel being transferred outside the reactor may permit the escape of larger fractions of fission products than in an in-core loss-of-coolant accident since the fuel is already outside of the primary system. (For even if the primary system is ruptured, say by a pipe break in a loss-of-coolant accident, it still provides some containment for the fission products.) However, since fuel is normally transferred in small lots, the quantities involved are reduced.

Loss of flow will normally result in a far less severe accident than loss of coolant if the reactor is immediately scrammed. It is likely in that case that the situation will result in little or no damage, and in any event, the fission product release from the fuel is still retained within the

---

*In only one instance, the HTRE-3 accident, can it be said that just one of these three factors—instrumentation flaws—played the causal role almost single-handed. Even there the operator could probably have taken action in time to prevent the meltdown if he had decided at once to shut down.

Table 6–1

Causes of Accidents

Cause	Location (Date) Facility or Experiment
1. Personnel working in critical area (all fatalities)	LASL (21 Aug. '45, 21 May '46) Pu sphere; Vinca, Yugoslavia (15 Oct. '58) D_2O critical; LASL (30 Dec. '58) Pu soln.; NRTS (3 Jan. '61) SL-1; UNC (24 July '64) U soln.
2. Personnel working with non-safe fluid geometry	ORNL (16 Nov. '58) 55-gal. drum; LASL (30 Dec. '58) Pu soln.; NRTS (16 Oct. '59, 25 Jan. '61) fuel reprocess; Hanford (7 April '62) Pu soln; UNC (24 July '64) U soln.
3. Loss of coolant*	Canada (12 Dec. '52) NRX; Hanford (4 Jan. '55) KW Reactor; Canada (July–Aug. '55) NRX; Canada (23 May '58) NRU; Santa Susana, California (13 July '59) SRE; Waltz Mill, Pa. (3 April '60) WTR
4. Loss of flow	ORNL (1948) X-10; NRTS (June '54) MTR; Hanford (4 Jan. '55) KW Reactor; Saclay, France (26 Nov. '57) EL-2; Saclay, France (13 April '58) EL-3; Saclay, France (12 Feb. '59) EL-2; NRTS (12 Dec. '61) ETR; NRTS (13 Nov. '62) MTR
5. Scram of control rods or control method causes accident	LASL (1 Feb. '51) critical; ORNL (1 Feb. '56) critical; LASL (3 July '56) Honeycomb; Vinca, Yugoslavia (15 Oct. '58) D_2O critical
6. Reactivity inserted too fast – source and startup	Hanford (16 Nov. '51) critical Pu soln.; LASL (3 Feb. '54) Godiva; Hanford (3 Oct. '54) production; Hanford (6 Jan. '55) production; LASL (3 July '56) Honeycomb
7. Positive feedback effects, an important factor	Canada (12 Dec. '52) NRX; Hanford (4 Oct. '54, 4 Jan. '55) production; Canada (July, Aug. '55) NRX; NRTS (29 Nov. '55) EBR-1; United Kingdom (9 Oct. '57) Windscale No. 1; Santa Susana, Calif. (13 July '59) SRE
8. Instruments caused accident	NRTS (18 Nov. '58) HTRE-3
9. Instruments off	LASL (Dec. '49) Water Boiler; Vinca, Yugoslavia (15 Oct. '58) D_2O critical; Saclay, France (15 March '60) Alizé; NRTS (3 Jan. '61) SL-1
10. Power decrease indicated, control rods withdrawn	Hanford (3 Oct. '54) production; NRTS (18 Nov. '58) HTRE-3; Waltz Mill, Pa. (3 April '60) WTR
11. Flat slab geometries or two units approaching	LASL (21 May '46) Pu hemispheres; LASL (1 Feb. '51) crit. cylinders; LASL (3 Feb. '54) Godiva; ORNL (1 Feb. '56) $U^{235}O_2F_2$ soln.; LASL (3 July '56) Honeycomb; LASL (30 Dec. '58) Pu soln.; ORNL (10 Nov. '61) critical
12. Experiment not well planned, parts performed unexpectedly	LASL (4 June '45) hand-stacked crit.; LASL (18 April '52) Jemima; ORNL (26 May '54) homog. crit.; NRTS (29 Nov. 55) EBR-1; LASL(12 Feb. '57) Godiva; UCRL (26 March '63) Kukla
13. Mis-estimates of effects of reactivity	LASL (11 Feb. '45) Dragon; LASL (18 April '52) Jemima; NRTS (22 July '54) BORAX-1; NRTS (29 Nov. '55) EBR-1; NRTS (5 Nov. '62) SPERT-1
14. Control rods withdrawn manually or by abnormal means	LASL (Dec. '49) Water Boiler; ANL (2 June '52) ZPR-1; Canada (12 Dec. '52) NRX; NRTS (3 Jan '61) SL-1

* In NRX and WTR incidents, boiling caused loss of coolant and fuel melting. The SRE incident could also be categorized as a loss of flow accident.

primary cooling system unless subsequent overpressure ruptures it.

In most reactors the accidents with the greatest potential for serious effects are those involving reactivity changes. This type of accident can occur while fission heat is already being generated in the fuel, or such heat may be generated because of the reactivity accident. This type of accident has the potential of utilizing all the forms of available energy to assist in dispersing the fission product burden of the core through the various containment barriers. Accidents involving reactivity changes do not have the advantage of being "sequential" in character.

To date nine cores have been destroyed or seriously damaged. Of these only two (BORAX-I and SPERT-I Destructive Test) can be said to have been destroyed on purpose as a part of a test. Three reactors have been put out of action by accidents and never revived. Two of these, Windscale No. 1 and the SL-1, were deemed beyond reasonable repair. The third, Clementine, had really reached the end of its useful life and was no longer believed to be a competitive research tool. It was dismantled and replaced by another higher flux research reactor—the OWR. These nine cases of core destruction all represent economic accidents of a serious nature involving radioactivity cleanup, down time, and rebuilding.

To date no accident has seriously involved the health and safety of the general public. The accident which came closest to doing this was the Windscale accident, which contaminated an area of about 200 square miles (52,000 hectares) around the reactor with a temporary low concentration of radioactive fall-out affecting the local milk supply. It cannot be said to have affected seriously the health and safety of the general public.

From the nuclear viewpoint and from the control viewpoint critical assembly and power reactor accidents have many similarities. Therefore, while recognizing that there are important differences, both of these types will be considered together in order to provide a larger number of accident cases. Since the principal concern here is to prevent future accidents, the conclusions which follow deal mostly with causes. There are no fewer than 13 different principal contributing causes which have occurred in at least three different accidents as shown in Table 6-1. The accidents dealing with personnel in potential high radiation areas around critical facilities and those due to loss of coolant and loss of flow have already been discussed.

A surprisingly large fraction of reactor accidents to date have involved control rods. A review of the frequency of these accidents might well lead one to propose eliminating control rods or perhaps using them solely as shutdown rods. On three different occasions, of which the SL-1 is the third, control rods withdrawn by hand resulted in prompt criticality. In the fourth case listed a worker actuated several rods by air pressure in a completely unorthodox manner.

In four cases the control method itself caused the accident. The first three cases are particularly odd in that scram action by pushing the scram button was found to result not in shutdown, but rather in an increase in reactivity before any shutdown at all occurred. In two cases this resulted in prompt critical bursts. This is a particularly treacherous type of accident as operators are trained to rely on "scram actions" as safe.

In a number of cases control rods have been withdrawn too rapidly, particularly in the low startup range where signals sometimes do not differ much from noise and where operators often get the feeling there is no danger. There is a particularly dangerous time during startup where many reactors switch from a period rate of rise or some relatively fast rate at low powers to a slower approach to full power as sensible heat is developed. If the operator neglects to make this switch (assuming procedural control) at the proper time, full power is appraoached too rapidly. This problem has played a part in at least three accidents or near-accidents to large reactors with which the author is familiar. The problem is even more acute if the reactor is being restarted and the operators are hurrying to get it up to power before xenon buildup cathches them and shuts the reactor down. There is a tendency to overlook procedures in such situations.

In all, some seven accidents have apparently involved positive feedback mechanisms which existed either temporarily or permanently in the reactor. One of them (EBR-1), and perhaps another (SRE), involved thermal core structure distortions (fuel rod bowing). It appears likely that a positive sodium void effect had a part in the SRE accident. If this is true, five accidents would then involve positive coolant void effects, one a graphite-moderated sodium-cooled reactor, two in graphite-moderated, water-cooled reactors, and two in a D_2O-moderated, H_2O-cooled reactor. The seventh accident involved a non-nuclear positive feedback mechanism—release of stored energy as a result of the Wigner effect in the graphite of the Windscale pile.

Instrumentation problems have played some part in every accident listed. In four cases there was essentially no instrumentation turned on at all. In at least two of these accidents, properly operating instrumentation would have prevented the accident. In the case of the SL-1, this is not certain. In one accident involving manual withdrawal of a control rod (ZPR-1) the instrumentation responded too late to be of use in terminating the accident. A water dump acted, but only after the transient was terminated by bubble formation in the plastic material that formed part of the fuel matrix.

In the HTRE-3 accident the instrumentation directly caused the accident and, in fact, set up an automatic accident which brought itself about and shut itself down with no other help from the operator than pushing the originating button.

In three cases, the incident involved a situation where the reactor flux dropped, or appeared to drop. In two of these the indication was true; mishaps in the core had caused losses in reactivity. In all three cases, if the reactor had been scrammed at the instant the drop in flux had been observed, damage would have been much less and perhaps there would not have been any damage.

The sensitivity of flat slab geometries and split core assemblies to minor geometric perturbations (where high leakage core faces are exposed to changing conditions) has been mentioned previously. Many people consider this the most dangerous type of critical assembly. Certainly there have been a number of accidents involving them. Note, however, that this type of unit is now only used by design in special remotely controlled critical assembly facilities. Two of the listed accidents of this type occurred outside such facilities and both resulted in fatalities.

A number of experiments have been poorly planned or sloppily executed. No doubt poor planning and sloppy execution—not thinking ahead—will always be a cause of accidents, but the effort must continue to eliminate this type of accident. One of the critical assembly accidents listed here occurred one day after a member of the group involved stated to a group meeting that, "We simply must take more care in carrying out these experiments. We are getting too sloppy and we will have an accident if we are not more careful." A change "to reduce the dependence on procedural controls" was recommended one month before the Los Alamos solution criticality accident happened [11] and the changeover was in progress at the time of the accident. In another case poor arithmetic (and independently checked at that!) is blamed [2].

Some five experimental tests have given results that were in part unexpected. Since two of these were fast transient tests exploring new regions, it is not surprising that the effects were larger than were expected. These tests were all carried out in specially prepared remote facilities and in a manner so as to prevent possible injury to personnel both on and off site. They demonstrate the difficulties in making large extrapolations into new unexplored regions and the reasons for carrying out these tests at remote sites.

A surprisingly large fraction of these accidents happened at night, on weekends, or during startups after a vacation period. A typical example is the SL-1 accident which happened on the evening shift

January 3, 1961, after a 10-day holiday shutdown. Enough others can be cited to indicate that judgment and alertness may be affected adversely on late shifts or on shifts where morale is likely to be low or where attention is wandering because of holidays or other reasons.

To date no serious accident has happened to any reactor which was operating at its normal operating conditions at its rated power. Almost all accidents have happened during startup or under special test conditions. In part, this may account for the large number of accidents just after holidays, since many reactors restart at such times.

6.2 Conclusions and Recommendations (See also those at the end of subsections on specific accidents.)*

(1) The single central goal of reactor safety is to prevent the release of fission products to the environment.

(2) Every reactor designer should strive to create a reactor system which is safe in spite of human errors, malice, or ingenuity. This implies that the system be designed as nearly fail safe as possible and that the core itself contain adequate inherent shutdown mechanisms such as voids, Doppler effect etc.

(3) All materials and components selected for use in reactor systems should have adequate safety margins on such parameters as tensile strength at operating temperatures, corrosion rates, compatibility, etc. No material used in any part important to the safety of the system should be used without a prior or concurrent and adequate testing program. In cases where such environmental tests are not deemed necessary, inspections at intervals frequent enough to ensure the integrity of the part should be carried out as a minimum. Quality control of all key components should be rigidly enforced during all stages of design, selection, and construction. Steps must be taken to be sure that the plant is built as designed and that all components continue to perform in accordance with the design objectives and specifications.

(4) Great care should be taken to utilize materials and fluids which are compatible from the chemical, corrosion, and structural viewpoints. Required materials which are not compatible should be separated by appropriate barriers.

(5) Accidents usually occur because of multiple and often apparently unrelated causes. It is not enough to place reliance on one simple safety barrier or procedure.

(6) Procedural control is at best a poor substitute for design ingenuity in setting up the first line of defense. That is to say, procedural controls should not be relied upon as the only, or even primary, safety barriers. Whenever possible interlocks and positive mechanical barriers should be designed into the system to prevent unsafe actions. When such a system has been designed, procedural controls should then take over to ensure that the system is used and maintained correctly. In this sense, procedural controls also play a vital role. There is, of course, a point beyond which interlocks and barriers become so burdensome as to tempt operator ingenuity to violate them. This also has happened and should be avoided.

(7) Any solid configuration of fuel, absorber, or moderator should be so subdivided that the accidental or planned movement of any single piece will not cause an unacceptably large increase in reactivity.

(8) All possible mechanical movements and configurations in the core in both normal and conceivable abnormal conditions must be considered to see whether they are potential sources of large positive reactivity effects. In particular such movements should not increase the reactivity in an uncontrolled or rapid manner. The planned movement of any components within the core should be carried out so that only small portions of the core materials with safe incremental worths can be moved at one time and then only in a controlled manner.

(9) Those components of the system which control or strongly influence the insertion of reactivity into the system or the removal of reactivity from the system deserve particular attention in design, construction, and operation. The close coordination and mutual understanding of the reactor physicist, the metallurgist, and the mechanical designer are essential. The nuclear consequences of any normal reactor operation or any abnormality should be such as to be easily controllable by the available control systems.

(10) It should be impossible to withdraw by hand or other means in an unpremeditated manner control rods, the withdrawal of which could lead to criticality. This can be prevented by appropriate mechanical interlocks or by other design methods.

(11) In the United States it has become common practice to provide a shutdown margin sufficient to allow for the failure of a single control rod. This "Stuck-Rod Criterion" may be stated that it should be impossible for a reactor to be made critical in its most disadvantageous situation with only a single rod fully withdrawn. Consequently, it should always be possible to shut down a reactor with one rod stuck in its outermost position. If it is possible that rods or mechanisms might interact so that several could be stuck in the out position, the number of rods used in the Criterion should be increased accordingly.

(12) In Canada, Great Britain, and to some extent in the United States a "cocked-rod" rule is used. A control rod or rods should be cocked in the out position during all core changing operations ready to drop into the core in the event that a nuclear transient should result from any operation. In particular, the reactivity worth of the cocked rods should be greater than any conceivable amount of reactivity resulting from an accident during the planned operation. In some reactors such a rule is not possible and special care must be exercised during such operations as loading to ensure that the core is well below critical. (If the reactor

*The conclusions stated are those of the author of this chapter and do not necessarily represent those of the authors of the chapters on the various topics.

design differentiates safety rods from control rods, the cocked rods should be safety rods.)

(13) It should be ascertained that the actuation of the scram system cannot result in an increase in reactivity by any conceivable means—including control rod drop out, momentary increases in reactivity as a function of position, etc.

(14) It is the usual practice in the United States (and in Canada where moderator dump is not available or could fail) to have some sort of emergency shutdown system (of a totally different type than the normal operational control and shutdown system) available as a last ditch safety measure.

(15) The fact that positive relatively prompt feedback effects have been a factor in seven accidents would seem to indicate that it would be well to avoid such positive feedback effects in reactor design where possible, especially if the mechanism can be actuated by more or less normal failures or events likely to occur in everyday operations such as loss of flow in a channel, etc. If such effects cannot be avoided, special precautions and instrumentation are likely to be required in order to ensure adequate and rapid enough response of external controls. It is worth while noting that large negative reactivity effects can have inverse positive effects which are also dangerous (e.g. the cold water accident postulated to large pressurized water reactors with large negative temperature coefficients).

(16) In the event that a new shutdown mechanism is introduced at some stage of an accident or substantial portions of the core enter a new state—say involving a change in state of a fraction of the moderator—the kinetics problem must be considered as a new one and fully investigated on a new basis.

(17) In the design and operation of experimental criticality facilities, reactors, fuel plants, or chemical processing plants, it is necessary to be continually on guard to uncover configurations which can by accident become supercritical. This subject is discussed in more detail in the chapter on Criticality. Some special precautions can be stated:

(a) Any containers in areas where fluid fissile material solutions may exist that are not the "always safe" variety for fluids discussed by the Nuclear Safety Guide (TID 7016) [94] are potentially dangerous. The only plant utilizing solutions of fissile materials which cannot have a criticality accident is one in which every container, pipe, and conceivable volume for solution collection is of the always safe variety for solutions by virtue of its geometry, raschig rings, formed steel grids, or other positive means to prevent such accidents. The likelihood of criticality situations in fissile material salvage operations or other operations involving uncertain inventories and seldom used procedures is increased. Therefore, it would seem wise to ensure that <u>all</u> operations of this nature be conducted carefully and in geometrically safe containers.

(b) Solution handling systems of "clean" fissile material should be carefully designed to protect operating personnel from criticality radiation. The means available include safe containers, remote control operations, shielding, and procedural controls. Of these, procedural controls are the least reliable. Fuel reprocessing plants normally protect personnel by shielding in any event because of the fission products present.

(c) Assemblies involving large leakage "flat slab" geometries in regions of high neutron importance or "split core" units where fissile, moderator, or reflector materials can be added to the large leakage face, are particularly dangerous and should be avoided in reactors intended for research or power use. (This statement includes the successive geometries involved in initial core loadings, see (d) below.) On the other hand a thin flat slab tank turned up on its edge so that solution is added to the thin upper edge can be used as a fail safe geometry if the tank is thin enough. In experimental criticality facilities such assemblies should be treated with great respect. Solutions in the form of flat slabs or split cores have been responsible for several accidents and one fatality. The assembly of split core units requires great caution as the units approach one another and the neutron leakage changes rapidly. Some of these units in criticality facilities do not have inherent shutdown mechanisms other than disassembly.

(d) Reactor fuel loadings should be carried out in such a way as to add small increments of reactivity as the critical configuration is believed to be approached. This must be done by careful prior planning with due regard to the particular reactor. Most often loading is best done by building a cylindrical array outward from a centrally located source. As small increments of fuel as possible should be added to the periphery of the cylinder in such a way as to maintain its cylindrical form at all times. On the other hand, unusual assemblies such as highly undermoderated assemblies may require a different and special treatment.

(e) Fast burst facilities should be caged or otherwise secured to ensure that extraneous material does not accidentally come too close to the assembly.

(f) Criticality loadings should be carried out by remote means with no personnel present who are not adequately shielded. No alterations or loading should be made by hand on any potentially critical assembly. Every effort should be made to prevent personnel from entering an area where criticality is conceivable under the conditions existing at the time.

(g) All safeties designed into potentially critical systems should "fail safe." The fail-safe principle is a goal towards which all designs should strive but which few will achieve. Where possible the design should make use of completely reliable forces such as gravity.

(h) Great care should be taken in experimental critical burst facilities in the correct design and use of neutron sources and final reactivity adjustments to ensure that the burst is properly initiated and of the desired magnitude.

(i) The presence of small monitored neutron sources in certain potential criticality areas might be used to give warning of impending criticality and might also tend to limit the extent of any transient which did result.

(j) Plutonium (and perhaps other fissile materials) has a tendency to form a tenuous deposit on surfaces. This may later be exposed to the proper conditions for dissolution, thus releasing unexpectedly large amounts of plutonium into the system. This is a principal safety lesson from the LASL accident of December 30, 1958 which caused a fatality.* The maximum expected inventory in the room was 125 grams of plutonium—a quantity which could not have gone critical. Techniques have been developed to detect such deposits.

(k) Systems should be designed for simplicity and convenience of operation and maintenance.

(18) The safety effort in criticality facilities utilizing "clean" uranium assemblies should be devoted primarily to protecting on-site personnel from radiation exposures.

(19) A critical facility or plant which utilizes assemblies or handles materials containing appreciable amounts of plutonium, fission products, or toxic materials not only has the same safety problems in regard to plant personnel as in a facility involving "clean" uranium assemblies but also has hazards which could affect the public safety.

(20) Recent work on brittle fracture effects seems to indicate that there are four important variables which affect this phenomenon,** These are temperature, stress, fast neutron radiation dose, and the size of any flaws existing in the metal. For a given stress a piece of metal with a large flaw in it will give brittle fracture at a higher temperature than one with a smaller flaw. A highly stressed piece of metal will brittle fracture at higher temperatures than one with lower stresses, and so on.

Thus, the region around the central zone in a reactor vessel is subject to neutron bombardment, and the brittle fracture problem is mainly one of radiation effects and existing flaw size. Residual stresses probably play little part there. However, in the vessel head region, residual stresses plus bolting stresses may be important and lead to conditions more serious than those in the central belt, even though almost no neutrons reach the vessel lid.

(21) Experiments and tests must be carefully and conservatively planned. Plans should be written out, appropriate calculations made, appropriate instrumentation prepared, and personnel roles reviewed and rehearsed for the tests. Results obtained should be plotted and followed step by step as the experiment progresses. Any deviation from the expected should result in a previously rehearsed shutdown procedure and then a thorough review of why the deviation occurred. Tests exploring new phenomena or extrapolating into potentially dangerous regions of behavior should be carried out at remote sites so as not to endanger the public and with adequate protection for participating personnel. In order to gain the most information from the fewest tests, the tests should be thoroughly instrumented based on the supposition that the effects observed will be greater than expected. Potentially dangerous tests should not be carried out on reactors intended for research, testing, or power use.

(22) The goal of reactor instrumentation should be to supply information to operators and control units which correctly represents the true picture of the core under all conditions and at all times.

(23) The principles of Redundancy and Diversity should be observed (see the chapter on Sensing and Control Instrumentation).

(24) Interlocks and other means should be provided to see that the instrumentation is functioning and capable of giving warning in time whenever any operation is being carried out that could conceivably involve core reactivity changes.

(25) Appropriate communication channels should be provided to cover all possible operations. The system must allow two-way conversations without requiring use of the hands.

(26) Safety circuitry should never be used for any other purpose since it is, by definition, utilized to ensure the safety of the system.

(27) It is good practice to have at least two on-scale signals from the reactor primary information source—usually neutron flux signals.

(28) If possible, control rod positions should be recorded at all times—even during scram.

(29) It is sound practice to scram on sudden drops in indicated power (or short negative periods) as well as sudden increases in power (or short positive periods).

(30) Reactor startup rates should be carefully controlled both by design and by method of operation. A valid neutron flux signal should be displayed on the control panel before withdrawing rods. If the source level is very low so that it cannot be detected, the startup must proceed very cautiously. (See chapter on Criticality.)

(31) In reactors which start up utilizing a relatively fast period rate of rise for low levels followed by a slower rise in percent of power/sec in the sensible heat range as full power is approached, it would be well to interlock the transfer from one mode to the other so that power cannot rise above some fixed low value until the proper transfer is made. Since the operation of transferring from a less safe to a more safe rate of rise requires a positive action, failure to act or failing to act in time results in an unsafe condition. The basic system without interlocks is not fail-safe.

(32) The prime function of properly trained operators is to be alert and ready for emergencies

---

*Private communication from H. Paxton.

**Dr. W. B. Lewis suggests the addition of two other factors. Brittle fractures of large bolts at NRU have been attributed to the action of an indium coating, effectively in the molten state, which seems to promote crack propagation. He also cites the 1963 British Royal Society Bakerian Lecture by A. H. Cottrell which stressed the importance of atmosphere on crack propagation. Lewis suggests another possible factor, namely, the neutron-produced helium in boron-containing stainless steel. (See Proceedings of the Third UN International Conference on Peaceful Uses of Atomic Energy, 1964.)

at all times. They must be able to combat successfully the boredom of normal dull operation so that they respond quickly and with good judgment during the few vital seconds when their skills may be desperately needed.

(33) All difficult and critical operations should be planned for, or postponed to, periods during which the best personnel are available in their most alert and responsive state.

(34) No operation requiring judgment should be undertaken or prolonged when the personnel are tired. During initial startups and difficult operations, frequent breaks should be taken and personnel should get plenty of rest.

(35) Arrangements should be made to have extra personnel and the best personnel present for startup and special tests, as these are the most dangerous periods in a reactor's operation.

(36) No person should be allowed to work around a reactor or in a potential criticality area or in any dangerous place alone.

(37) The knowledge that several safety barriers exist is often used as an excuse for not promptly taking remedial action if one barrier is breached. It is a sound rule to believe that the barrier being risked is the last one, or at least the next-to-last.

(38) The effectiveness of shutdown systems should always be checked as a part of startup procedures.

(39) The design, construction, and operating plans should be aimed at making the operator feel safe, needed, and alert. Thus he should be adequately shielded and know it, he should have adequate exits if needed (preferably at least two), he should have work of importance to do, and it should be frequent enough to prevent boredom.

(40) If there are design weaknesses or instrumentation weaknesses in the system it is vitally important that the operating staff know of them and are alert for trouble.

(41) An up-to-date set of operating instructions and procedures is vital to safe operation.

(42) From the safety viewpoint, fuel elements should be designed to have sufficient mechanical strength and rigidity so that they will not move in an unplanned manner which might cause core reactivity changes. At the same time the arrangement, the cladding material, and the cladding thickness should be chosen so as to provide the optimum heat transfer to the coolant.

6.3 Nuclear Excursion Energy Limits

A study of the three destructive excursions to date (BORAX, SL-1, SPERT) indicates that there is considerable evidence for the existence of a plateau in the magnitude of total energy deposition when plotted against the reciprocal period of the excursion for an accident in the type of reactor involved in all three cases. That limiting plateau may be imposed by the vaporization of an appreciable fraction of the fuel meat. Such a plateau of energy released, if it can be shown to exist for a given type of reactor, would give reactor engineers a design base for a maximum conceivable accident and would put a much firmer foundation under the requirements for the precautions necessary to ensure the safety of the general public for any given reactor.

Of course, it is clear that if sufficiently large amounts of reactivity can be added fast enough, the energy released must rise again for very large reciprocal periods. However, it may well be that the region of practical reactor interest lies below the upper bounds of the plateau, at least for some types of reactors.

It should be understood that the vaporization of fuel in this case acts as a trigger and not just as a more effective quenching mechanism. In a heterogeneous system the transient energy developed is deposited in the fuel. In many such systems an effective shutdown or quenching mechanism is voiding or loss of coolant-moderator. This loss may have a large negative reactivity effect. Therefore, if a means can be found to trigger the rapid and intimate mixing of fuel and moderator so that the heat already deposited in the fuel is transferred to the moderator and results in its rapid expulsion, shutdown will be rapid. Fuel vaporization may supply such a means.

The transient energy needed for shutdown, and much more, already may exist in the core, but due to the heterogeneous nature of the core, no shutdown occurs. All that is needed is a trigger. The amount of energy released in a transient may be controlled by the actuation of a trigger and the energy released may be relatively constant when plotted against increasing reciprocal periods if the trigger works reliably at some fixed point of total energy deposition.

Run 54 of the SPERT destructive series discussed briefly in Sec. 3.12 of this chapter shows that the violent steam formation was totally independent of the reactor kinetics of the nuclear transient and was due solely to the heat energy deposited and the subsequent thermal effects. Further, it shows that the reaction is quite rapid once it is initiated. (Reference [68] estimates that the elapsed time between the initiation of the pressure pulse and its peak was only 0.5 msec.) The SPERT-I core reached the melting point over 35% of the plate area, but no part of the core reached the vaporization temperature. (The highest core temperature was estimated to be about 1200°C (2200°F) [67].) Thus, the violent reaction was probably delayed by the time it took for the meat to melt through the clad. When appreciable vaporization occurs in the core, as in the SL-1 accident, a mechanism is provided to immediately rupture the clad and allow intimate contact between vaporized and molten fuel and water. For the case where appreciable vaporization occurs, it appears likely that the delay time is cut short by an order of magnitude or more.

Section 3.4 of this chapter cites evidence that the BORAX-I transient rod was only about 80% out of the core when the shutdown mechanism became effective in that transient. Thus, it was limited not by the reactivity available, but by either the withdrawal rate or by the shutdown mechanism, which apparently included some fuel vaporization.

Consideration of Fig. 3-30 (page 675) shows that while SPERT-I and BORAX-I tend to give somewhat higher energy values than the General Electric calculations would predict, the SL-1 accident tends to fall lower, perhaps indicating that the SL-1 accident had already reached the limiting value.

As discussed in the chapter on Mathematical Models of Fast Transients—Some Comparisons of Models and Excursion Reactor Data—the nature of the shutdown mechanism in these cores changes from thermal expansion to various types of boiling and finally to core melting and vaporization. A quantity ϵ is defined there as the ratio of the reactivity compensated by the system to the nuclear energy release, evaluated at the time of peak power. It is pointed out that a plot of ϵ as a function of α_0, the reciprocal period, for a single shutdown mechanism involving certain time delays will in general, decrease as α_0 increases. This is because a more or less fixed time delay, such as that required for boiling to commence, allows the power to rise higher before the shutdown occurs as the period is made shorter. When the periods become so short that fuel vaporization occurs, heat transfer is no longer an important factor. At those values of α_0, ϵ may rise for all values of α_0 which are likely to be achievable. The time delays in the shutdown mechanism will change from perhaps 3 to 10 msec (for boiling or melting-through of cladding) to a fraction of a millisecond. The shutdown time should then be dependent primarily on the inertial resistance offerred by the particular system to expansion over a limited volume. The inertial resistance is likely to be a slowly varying quantity in the region of interest.

Since the compensated reactivity resulting from the void created by the expanding vaporized metal plates and the intimate interaction of water and metal vapor is large, this mechanism may be effective in shutting down even large reactivity additions. Unless the water-filled core is very much more undermoderated and there exists the possibility of reassembly in a more favorable geometry, the system is likely to be so disrupted by the fuel vaporization that it will remain subcritical.

Thus, to the two limits placed on transients in the chapter on Mathematical Models of Fast Transients (the amount of reactivity available and the maximum rate of addition of reactivity conceivable), it may be possible to add a third—the amount of energy that can be released during a transient in a given core of this type. This third limit would be set by the amount of heat that must be deposited by a nuclear excursion in order to vaporize an appreciable fraction of the core. It appears likely that the fraction of the core that need be vaporized is fairly small, since the most likely role of the phenomenon is that of a trigger to violently disrupt the core and allow molten fuel to come suddenly in contact with the water. It is evident from the SPERT-I test that a core with only melted fuel in it is quite capable of creating a violent enough disturbance to shut down the reactor. However, in that case the time delay is too long and too uncertain. Extremely rapid fuel vaporization and the accompanying expansion act as a trigger and shorten the delay.

If such a plateau of achievable energy exists, it will certainly simplify the task of safeguards analysis for fast transients and also put a realistic upper limit on the magnitude of this type of accident. The energy evolved in the maximum conceivable nuclear transient would then simply be that required to vaporize, say, 10% of the core in a nuclear transient (while, of course, the rest of the core is heated as well) plus an allowance for metal-water reaction. The calculation would be relatively simple since the geometry will not change at all during the time the power increase is occurring and since no heat transfer is involved. It must be remembered that it is necessary also to demonstrate that the available reactivity for the transient is not so great that the upper bound of the plateau is exceeded.

If the plateau exists it will have a novel effect on reactivity limits. The excess reactivity that designers would have to worry about would then be that which is available between prompt critical and that amount of reactivity which will just provide a transient with enough energy to reach the energy limit. Thus, if the equivalent to the energy limit should be 3% Δk for a given core, then that reactor could just as safely carry 12% Δk as it could carry 3% Δk, as long as 12% Δk did not carry the transient beyond the energy plateau region. Thus, safeguards analysis and discussion would be limited to a discussion of the consequences of this maximum accident which is then quite well defined, and then to the limits on rates of addition of reactivity and the amount available up to the "limiting" amount.

In order to test this idea it would be necessary to destroy several cores with the same characteristics by adding successively greater and greater amounts of reactivity at extremely rapid rates. The addition must be rapid enough so that all of the reactivity is inserted well before the shutdown mechanism can become effective. As is discussed in the chapter on Mathematical Models of Fast Transients, this is difficult and may prove to be impossible, but at least the greatest rate of reactivity addition that could be obtainable in any conceivable reactor under any circumstances should be used. Whether there is such an energy plateau or not will then be very simply answered by whether or not the energy release in the various excursions is anything like a constant quantity.

If there is such a plateau in water-moderated plate-type cores of highly enriched fuel, there may also be such limits for other types of reactors, although they may be at such large energy release values as to be of little value either in providing a useful limit to the energy in the maximum accident or in providing guidance in the management of reactivity. Probably the next type that would be fruitful to consider would be the uranium oxide cores. Studies of the Doppler effect in such cores have already begun at SPERT. These tests have already demonstrated the expected usefulness of the Doppler effect in limiting nuclear transients. However, the violent reaction accompanying fuel melting and vaporization are totally thermal in nature and, no doubt, will still occur if enough energy can be deposited. Since in some operating cores evidence of some center melting has already been observed, the Doppler shutdown mechanism may not satisfactorily limit large excursions in operating reactors of this type before fuel vaporization can occur. Then, as with the plate-type elements, the sudden vaporization and expansion of a small fraction of the fuel to burst its cladding should trigger the interaction of other melted and vaporized fuel elements and create a void in much the same fashion that happened in the three tests to date.

It is of interest to note that apparently the bursting of two pins in the SPERT oxide core test of April 1964 (Sec. 3.12) had a beneficial effect on the rate of shutdown. This seems to indicate that a material with a high vapor pressure like water in direct contact with the fissile material acts promptly to supply negative reactivity by pin bursting and boiling. Perhaps the mechanism could be used to augment or replace the Doppler effect in certain reactors. While water within the cladding would be a difficulty in an oxide-fueled power reactor with its high fuel temperature, nonetheless another material might be found which could be placed in some fuel rods to enhance their bursting and shutdown effect in order to terminate a serious transient more rapidly.

TREAT tests of fuel vaporization and bursting should provide the information to predict the uranium oxide temperatures necessary to give vapor pressures high enough to burst the fuel. (No doubt, the fission products built up during core life will have an effect on the results of these tests.) SPERT transient tests on cold cores and operating cores at temperature and pressure should provide the necessary full core transient information. Plotting the maximum fuel temperatures reached in the transients as a function of α_0 should give a curve that could then be extrapolated to find the α_0 required to give the necessary fuel temperature (really total energy deposition) to achieve the vapor pressures in the fuel to burst the cladding. This should be the energy-plateau limited transient for this type of reactor, if there is such a limit. The transient tests will also give the relation between the reactivity inserted and the total energy deposited or central fuel temperatures reached. The picture should then be complete and on this basis it may be possible to say something also about the limiting accident and about reactivity in this type of reactor. The oxide will, of course, interact with water quite differently from the metal as discussed in the chapter on Chemical Reactions. Such tests as those outlined above, particularly if they eventually lead to destructive tests, should give much insight into the importance of the metal-water reaction in the destructive transients observed to date.

Normally the cores for this type of reactor are very large. It is thus possible to deposit in them relatively large amounts of energy without serious effects. Thus it is not the total Mw-sec deposited in a transient that is important, but the energy density that is important, the Mw-sec/cm^3. Reliable central fuel temperature measurements are then probably the best indication of energy density deposited in a transient.

Other systems such as liquid-metal-cooled fast reactors may undergo similar violent voiding reactions between the coolant liquid metal and vaporized fuel. Whether the effect will result in a positive or negative reactivity change will depend on the reactor in question. Conceivably, if the void effect gives a positive reactivity change, the violent reaction described could for an instant during the interaction be made even worse by it. (These considerations are discussed in the chapter on Fast Reactor Kinetics.) These effects should be investigated in such reactors as well as in water-cooled reactors.

Even if the energy limit considered in this section should be proved not to exist, it is necessary to continue to search for such limiting processes or ways of putting limits on conceivable reactor transients. The present discussion illustrates the type of limit that is sought and the methods by which such limits could be either proved or disproved.

REFERENCES

1. H. B. Smets, "Review of Nuclear Incidents", Progress in Nuclear Energy, Series X, Vol. 3, Law and Administration, Pergamon Press, N.Y. and London, 1962. See also: H. B. Smets, "Review of the Nuclear Reactor Accidents", Vol. 1 p. 89 of Proceedings of a Symposium on Reactor Safety and Hazards Evaluation Techniques, International Atomic Energy Agency, Vienna, 1962. (Identical but limited to reactors.)
2. H. C. Paxton, "Critical assembly booby traps", Nucleonics, 16, 3(1958)80.
3a. W. R. Stratton, "A Review of Criticality Accidents", Progress in Nuclear Energy, Series IV, Vol. 3, Technology, Engineering and Safety, Pergamon Press, N.Y. and London, 1960.
3b. W. R. Stratton, "A Review of Criticality Incidents", Proceedings of a Symposium on Criticality Control in Chemical and Metallurgical Plants, OECD, Paris, 1961.
4a. H. C. Paxton (Ed.), "Fast-neutron critical assemblies and related topics", Nucl. Sci. Eng., 8(1960)523.
4b. P. H. Pitkanen, "Hazards of fast-burst facilities", Nucl. Safety, 4, 1(1962)18.
5. D. Bretin, F. Girard, and J. Charles, "L'Incident Servenu à la Pile Alize, le 15 Mars 1960", Report SEC-90, September 1960.
6. E. P. Epler, "Failure of Alize I reactor safety system", Nucl. Safety, 5, 2(1964)172.
7. D. Callihan, "Criticality Excursion of Nov. 10, 1961", USAEC Report ORNL-TM-139, Oak Ridge National Laboratory, 1962.
8. E. R. Rohrer et al., "New Criticality Excursion Machines", USAEC Report ORNL-3193, p. 168, Oak Ridge National Laboratory, 1961.
9. J. R. Buchanan, "ORNL criticality excursion", Nucl. Safety, 3, 4(1962)95.
10. D. Callihan, "Accidental nuclear excursion in Recuplex operation at Hanford, April 1962", Nucl. Safety, 4, 4(1963)136.
11. H. Paxton et al., "Nuclear Criticality Accident - LASL December 30, 1958", USAEC Report LAMS-2293, Los Alamos Scientific Laboratory, 1959.
12a. Private communication, AEC Division of Operational Safety, October 9, 1963.
12b. R. L. Kathren, W. C. Day, D. H. Denham, and J. L. Brown, "Health physics following a nuclear excursion: The LRL incident of March 26, 1963", Health Physics, 10(1964)183.
13. L. P. D. King, "Design and Description of Water Boiler Reactors", Proceedings of the First U.N. International Conference on Peaceful Uses of Atomic Energy, Geneva, 1955, Vol. 2, p. 372.
14. D. F. Hayes, "A Summary of Accidents and Incidents Involving Radiation in Atomic Energy Activities, June 1945-December 1955", USAEC Report TID-5360, 1956. (Supplements were issued in Aug. 1957, Sept. 1959, Dec. 1961.)
15. D. G. Hurst and A. G. Ward, "Canadian Research Reactors", Progress in Nuclear Energy, Series II, Reactors, Pergamon Press, N.Y. and London, 1956.
16a. W. B. Lewis, "The Accident to the NRX Reactor on December 12, 1952", Canadian Report AECL-232, Atomic Energy of Canada, Ltd., 1953.
16b. D. G. Hurst, "The Accident to the NRX Reactor, Part II", Canadian Report AECL-233, Atomic Energy of Canada, Ltd., 1953.

17. W. J. Henderson, A. C. Johnson, and P. R. Tunnicliffe, "An Investigation of Some of the Circumstances Pertinent to the Accident to the NRX Reactor of December 12, 1952", Canadian Report NEI-26, Atomic Energy of Canada, Ltd, Chalk River, 1953.
18. G. W. Hatfield, "A reactor emergency with resulting improvements", Mech. Eng., 77(1955)124.
19. J. R. Dietrich, "Experimental Investigation of the Self-Limitation of Power Driving Reactivity Transients in a Subcooled, Water-Moderated Reactor", USAEC Report ANL-5323, Argonne National Laboratory, 1954.
20. "Reactors", Nucleonics, 13, 9(1955)40.
21. J. R. Dietrich, "Experimental Determinations of the Self-Regulation and Safety of Operating Water-Moderated Reactors", Proceedings of the First U.N. International Conference on Peaceful Uses of Atomic Energy, Geneva, 1955, Vol. 13, p. 88.
22. J. R. Dietrich and D. C. Hayman, "Transient and Steady State Characteristics of a Boiling Reactor", USAEC Report ANL-5211, Argonne National Laboratory, 1954.
23. J. R. Buchanan, "Hanford reactor incidents", Nucl. Safety, 4, 1(1962)103.
24. W. E. Unbehaun, "History and Status of the EBR", USAEC Report AECD-3714, Chicago Operations Office, 1953.
25a. H. V. Lichtenberger et al., "Operation Experience and Experimental Results Obtained from a NaK-Cooled Fast Reactor", Proceedings of the First U.N. International Conference on Peaceful Uses of Atomic Energy, Geneva, 1955, Vol. 3, p. 345.
25b. H. V. Lichtenberger, W. H. Zinn, private communication, April 1964
26. F. W. Thalgott et al., "Stability Studies of the EBR-I", Proceedings of the Second U.N. International Conference on Peaceful Uses of Atomic Energy, Geneva, 1958, Vol. 12, p. 242.
27. R. R. Smith, R. G. Matlock, F. C. McGinnis, M. Novice, and F. W. Thalgott, "An Analysis of the Stability of EBR-I, Marks I to III", Proceedings of the Seminar on the Physics of Fast Reactors held at Vienna, August 3-11, 1961, International Atomic Energy Agency, 1962, Vol. III, p. 43.
28. R. O. Brittan, "Analysis of the EBR-I Core Meltdown", Proceedings of the Second U.N. International Conference on Peaceful Uses of Atomic Energy, Geneva, 1958, Vol. 12, p. 267.
29. W. H. Zinn, "A letter on EBR-I fuel meltdown", Nucleonics, 14, 6(1956)35.
30. J. H. Kittel et al., "The EBR-I Meltdown - Physical and Metallurgical Changes in the Core", USAEC Report ANL-5731, Argonne National Laboratory, 1957.
31. F. W. Thalgott, "Fast Reactor Safety: EBR-I", Proceedings of a Conference on Fast Reactors, November 20-21, 1957, USAEC Report TID-7548.
32. L. A. Mann, "EBR-I operating experience", Nucl. Safety, 3, 2(1961)68.
33. J. R. Dietrich, "EBR Oscillations", USAEC Report TID-5351, 1956 (Decl. 1957).
34. G. H. Kinchin, "The Instability of Fast Reactors", British Report RP/M83, Harwell, Berks., England, 1956.
35. H. A. Bethe, "Reactor Safety and Oscillator Tests", Report APDA-117, Atomic Power Development Associates, 1956.
36. J. C. Haire and G. O. Bright (Eds.), "Proceedings of a Conference on Reactor Kinetics held at Sun Valley, Idaho, October 12-14, 1960, Report IDO-16791, p. 309, Phillips Petroleum Co., 1962.
37. "Accident at Windscale No. 1 Pile on 10th October, 1957", Presented to Parliament by the Prime Minister by Command of Her Majesty, British Report Cmnd. 302, Her Majesty's Stationery Office, London, November 1957.
38. "Accident at Windscale No. 1 pile on October 10, 1957", Nucleonics, 15, 12(1957)43.
39. J. Metera, "On the Course and Cause of the Reactor Damage at Windscale", Nukleonika, III, 2(1958)34. Trans. from Polish, USAEC Report AEC-tr-4145, 1961.
40. H. J. Dunster, H. Howells, and W. L. Templeton, "District Surveys Following the Windscale Incident, Oct. 1957", Proceedings of the Second U. N. International Conference on Peaceful Uses of Atomic Energy, Geneva, 1958, Vol. 18, p. 296.
41. E. P. Epler, "HTRE-3 Excursion", Nucl. Safety, 3, 2(1959)57.
42. AEC Division of Compliance - private communication.
43. A. A. Jarrett, "SRE Fuel Element Damage", Report NAA-SR-4488, North American Aviation, Inc., 1959.
43a. R. L. Ashley, et al., "SRE Fuel Element Damage, Final Report" Report NAA-SR-4488 (Supplement), North American Aviation, Inc., 1961.
44. R. W. Dickinson, "Coolant block damages SRE fuel", Nucleonics, 18, 1(1960)107.
45. R. W. Dickinson, "SRE operating experience", Nucl. Safety, 2, 3(1960)54.
46. J. G. Gratton, "Solubility of Carbon in Sodium at Elevated Temperatures", Report KAPL-1807, Knolls Atomic Power Laboratory, 1957.
47. F. A. Fillmore, "Analysis of SRE Power Excursion of July 13, 1959", Report NAA-SR-5898, North American Aviation, Inc., 1961.
48. J. L. Baliff, "Examination of Damaged SRE Elements", Report NAA-SR-4515, North American Aviation, Inc., 1961.
48a. S. Siegel, private communication, May 1, 1964.
49. W. B. McDonald and J. H. Devan, "Sodium reactor experiment incident", Nucl. Safety, 1, 3(1960)73.
50. -Staff, WTR, "Report on WTR Fuel Element Failure, April 3, 1960", Report WTR-49, Westinghouse Electric Corp., 1960.
51. R. B. Korsmeyer, "Westinghouse Testing Reactor incident", Nucl. Safety, 2, 2(1960)70.
52. R. J. Catlin, "Health Physics Aspect of the Westinghouse Testing Reactor Fuel Element Meltdown", Report WTR-55, Westinghouse Electric Corp., 1960.
53. A. N. Tardiff, "Some Aspects of the WTR and SL-1 Accidents", Proceedings of the Symposium on Reactor Safety and Hazards Evaluation Techniques Sponsored by the International Atomic Energy Agency, Vienna, May 14-18, 1962, Vol. 1, p. 43. See also: IDO19308.
54. A. Smaardy (Ed.), "Hazard Summary Report on the Argonne Low Power Reactor (SPR)", USAEC Report ANL-5744, Argonne National Laboratory, 1957.
55. D. H. Shaftman, "Zero Power Experiments on the Argonne Low Power Reactor (ALPR)", USAEC Report ANL-6078, Argonne National Laboratory, 1961.
56. D. H. Shaftman, "Proceedings of the ANPP Reactor Analysis Seminar", The Martin Col, Baltimore, Maryland, October 11, 1960.
57. C. A. Nelson, et al., "Report on the SL-1 Accident, January 3, 1961", General Manager's Board of Investigation", USAEC Information Release No. 3-326, September 24, 1962. See also: "SL-1 Recovery Operations, January 3-May 20, 1961", Report IDO-19301, (CEND 1007), Combustion Engineering, Inc., Idaho Falls, Idaho, 1961.
58. C. W. Luke and H. Cahn, "Evaluation of the Loss of Boron in the SL-1 Core I", Report CEND-1005, Combustion Engineering, Inc., 1960.
59. "Remarks by Dr. Frank K. Pittman, Director, Division of Reactor Development, USAEC", for presentation to representatives of industry, AEC Press Release, Germantown, Maryland, January 24, 1961.
60. "IDO Report on the Nuclear Incident at the SL-1 Reactor, January 3, 1961, at the National Reactor Testing Station", USAEC Report IDO-19302, Idaho Operations Office, 1962.
61. "Final Report of SL-1 Recovery Operation, May 1961 through July 1962", Report IDO-19311, General Electric Co., Aircraft Nuclear Propulsion Dept., 1962.
62. "Additional Analysis of the SL-1 Excursion - Final Report of Progress, July through October 1962", Report IDO-19313, General Electric Co., Aircraft Nuclear Propulsion Dept., 1962.
63. V. E. Schrock et al., University of California, Berkeley, Series 163, No. 2, January 1961. See also: S. J. Green, "Estimated Film Boiling Heat Transfer Coefficients at Burnout", Report WAPD-TH-132, Bettis, Atomic Power Laboratory, 1955. (OR: H. A. Johnson, V. E. Schrock, F. B. Selph, J. H. Lienhard, and Z. R. Roxztoczy, "Transient Pool Boiling of Water at Atmospheric Pressure", ASME Paper 29, Part 2, Sec. A.)
64. "SPERT Quarterly Progress Report, July-September 1958", Report IDO-16512, Phillips Petroleum Co., 1958.
65. J. Dugone and D. D. Wieland, "Fuel Plate Experience during SPERT-I Destructive Test Series with an Aluminum Clad, Plate-Type Core", Report IDO-16885, Phillips Petroleum Co., 1963.
65a. A. H. Spano and R. W. Miller, "SPERT-I Destructive Test Program Safety Analysis Report", Report IDO-16790, Phillips Petroleum Co., 1962.
66. R. W. Miller, A. Sola, and R. N. McCardell, "Report on the SPERT-I Destructive Test Program on an Aluminum Plate-Type Water-Moderated Reactor", Report IDO-16883, Phillips Petroleum Co.
67. M. R. Zeissler, "SPERT-ID-Core Data Summary", Report IDO-16886, Phillips Petroleum Co., 1963.
68. A. Sola, R. Scott, Jr., R. K. McCardell, and E. Feinauer, "Reactor Kinetic Behavior and Motion Picture Documentation of the Destructive Test", Trans. Am. Nucl. Soc., 6(1963) 137.
69. R. W. Miller, A. H. Spano, J. Dugone, D. D. Wieland, and J. E. Houghtaling, Experimental Results and Damage Effects of Destructive Test", Trans. Am. Nucl. Soc., 6(1963)138.
70a. News Item, Nucl. News, December 1963, p. 30.
70b. W. K. Ergen, "SPERT-1 'destructive' test with UO_2 fuel", Nuclear Safety, 5, 3 (1964) 231.
70c. F. Schroeder, private communication received May 19, 1964.
71. J. W. Greenwood, "Contamination of the NRU Reactor in May 1958", Canadian Report AECL-850, Atomic Energy of Canada, Ltd., 1959.
72. A. F. Rupp, "NRU reactor incident", Nucl. Safety, 3(1960)70.

73. F. R. Keller, "Fuel Element Flow Blockage in the Engineering Test Reactor", Report IDO-16780, Phillips Petroleum Co., 1962.
74. J. R. Buchanan, "ETR fission-break incident", Nucl. Safety, 3, 4(1962)93.
75. E. H. Smith (Ed.), "MTR Progress Report Cycle No. 182, October 29-November 19, 1962", Report IDO-16831, Phillips Petroleum Co., 1962.
76. R. A. Costner, Jr., "MTR fission-break accident", Nucl. Safety, 4, 4(1963)144.
77. J. Palladino, private communication, June 1963.
78. H. Gomberg, private communication, 1961.
79. University of Virginia, letter to AEC Division of Licensing and and Regulation, dated October 18, 1963.
80. Texas A and M letter to AEC Division of Licensing and Regulation, dated November 20, 1963.
81. S. E. Beall, "An Experimental Determination of Fission Product Heating after Shutdown in the LITR", USAEC Report ORNL-1075, Oak Ridge National Laboratory, 1951.
82. Private communication from S. E. Beall, Oak Ridge National Laboratory, September 17, 1963.
83. H. F. Poppendiek and H. C. Claiborne, "A Transient Heat Transfer Analysis of the MTR Mockup", USAEC Report ORNL-976, Oak Ridge National Laboratory, 1951.
84. J. R. Coombe, "SM-1 Pressurized Water Reactor Operating Experience", Nucl. Safety, 4, 2(1962)100.
85. O. H. Klepper, "Vallecitos Boiling Water Reactor Operating Experience through March 1960", Nucl. Safety, 3, 1(1961)80.
86. J. R. Buchanan, "SPERT-III Pressure-Vessel Failure", Nucl. Safety, 3, 4(1962)91.
87. R. E. Heffner, et al., "SPERT-III Pressurizer Vessel Failure", Report IDO-16743, Phillips Petroleum Co., 1962.
88. W. R. Smith, Sr. and J. B. Violette, "The Cause of Failure and the Repair of the VBWR Recirculation Piping", GE-APED-4116, General Electric Co., 1962.
89. R. H. McKane, H. C. Minton, and J. W. Wade, "Stainless Steel Failures in Savannah River Reactor Areas", Report DP-539, E. I. du Pont de Nemours and Co., Inc., 1960.
90. W. R. Gall, "Heavy Water Leaks at Savannah River", Nucl. Safety, 3, 1(1961)87.
91. J. R. Buchanan, "PM-3A Nuclear Power Plant Fire", Nucl. Safety, 4, 3(1963)86.
92. R. O. Mehann, "N.S. Savannah Operating Experience", Nucl. Safety, 4, 4(1963)126.
93. "Monthly Operating Report - Shippingport Atomic Power Station, November 1963", Report DLCS-5001163, Duquesne Light Co., 1964.
94. "Nuclear Safety Guide", USAEC Report TID-7016 (Rev'd), 1961.

APPENDIX 1

Tabulation of Parameters Relevant to Safety for a Number of Power Reactors

The tables on the following twenty-four pages are based on responses to a questionnaire sent to many reactor installations in mid-1963. Responses were received between August 1963 and February 1964.

Since these data have been collected from a short-answer questionnaire, the reader interested in using the data for quantitative studies of a given reactor is urged to consult original sources or the technical staff at the reactor in question.

Careful study of the tables should provide a reliable "feeling" for the numbers involved in the reactor physics and control aspects of nuclear reactor safety.

The reactors have been grouped somewhat arbitrarily into six categories:

I. Pressurized Water Reactors (Shippingport, Indian Point, Yankee, PM-1, SM-1, HWCTR)
II. Boiling Water Reactors (Dresden, Big Rock Point, Humboldt Bay, BONUS, Pathfinder, BORAX-V, EVESR)
III. Gas-Cooled Reactors (Peach Bottom, EGCR, EBOR)
IV. Sodium-Cooled Reactors (Fermi, EBR-II, Hallam)
V. Pressure-Tube Reactors (CVTR, CANDU, NPD-2)
VI. Other Reactors (Piqua, ETR, GETR, PRTR).

In each of the six tables the information is organized under four headings:

General Description (reactor name, location; identification of fuel, moderator, coolant; operating temperature, pressure; effective core size; thermal and electric power)
Nuclear Parameters (conversion ratio, reactivity coefficients, neutron lifetime, effective delayed neutron fraction, refueling period)
Reactivity Inventory (reactivity tied up in Xe and Sm, temperature defect, fuel burnup, shutdown margin, reactivity tied up in voids and experiments, total reactivity inventory)
Control (control and poisons: worths, mechanical arrangement, rate of withdrawal and insertion, method of use, source of scram force, method of deceleration, shaft seal, rate of rods run-in)

In addition, references cited in the questionnaire responses are listed at the end of each table.

Some responses did not lend themselves to convenient tabulation. This has been met by the use of footnotes. However, in many instances the entries may not be clear, especially concerning the conditions under which the data were measured or calculated. Again, the reader is urged to consult original sources.

To increase the information density many abbreviations have been used, e.g. n.a. = not applicable, op. = operating or operations, enr. = enriched, max. = maximum, EOL = end of life, BOL = beginning of life, SH = superheater, etc. It is hoped that these are intelligible.

The cooperation of the various reactor groups in responding to the questionnaire is gratefully acknowledged.

TABLE I. PRESSURIZED WATER REACTORS

General Description

Abbreviated Name	Full Name; Location	Fuel	Moderator; Coolant	Operating Temp. (°F)	Operating Press. (psia)	Effective Core Size[a]	Power Thermal Mw(t)	Power Elec.[b] Mw(e)
Shippingport	Shippingport Atomic Power Station; Shippingport, Pa.	Seed: 93% U^{235}-Zy Blanket: U	H_2O ; H_2O	500[c]	1800[c]	h = 6 ft d = 6.8 ft	231	68 60
Indian Point	Consolidated Edison, Thorium Reactor; Buchanan, N.Y.	1344 kg 93% enr. UO_2 17207 kg ThO_2	H_2O ; H_2O	500	1510	h = 8.27 ft d = 6.47 ft v = 7705 liter	585[d]	163[d]
Yankee	Yankee Nuclear Power Station; Rowe, Mass.	3.4% enr. UO_2 cyl. pellets in ss tubes	H_2O ; H_2O	514	2015	h = 7.7 ft d = 6.3 ft	540	167 157
PM-1	Portable Medium Power Plant No.1; Sundance, Wyo.	93% enr. UO_2 (ss matrix) fuel tubes ss-clad	H_2O ; H_2O	463	1300	h = 2.5 ft d = 1.90 ft	9.37	1.25(net)
SM-1	Stationary Medium Power Plant No.1; Ft. Belvoir, Va.	93% enr. UO_2, flat ss-clad plates with UO_2-B_4C-ss matrix meat	H_2O ; H_2O	440	1200	h = 1.813 ft d = 1.85 ft v = 139.5 liter	10.74	2.05 1.93
HWCTR	Heavy Water Component Test Reactor; Aiken, S.C.	Driver fuel: a tube 9.3% enr. U^{235} in Zr, with B-ss plate targets	D_2O ; D_2O	392 to 491	1000 to 1200	d = 6.5 ft (incl. reflector)	60	n.a.

[a] Cylindrical unless otherwise noted; h = height, d = diameter, v = volume.

[b] Where two figures are given, the first is gross and the second is net.

[c] The original operating temperature and pressure were 523°F (273°C) and 2000 psia (140 kg/cm^2), respectively; however, after 1693 equivalent full-power hours (EFPH) of operation on Core 1, Seed 1, these values were changed to the ones given here. The values have remained the same since that time, see reference [2] at the end of Table I.

[d] These are reactor power figures. Superheaters contribute 112 Mw(e).

TABLE I (continued)

Nuclear Parameters

Item	Shippingport	Indian Point	Yankee	PM-1	SM-1	HWCTR
k_{eff}		1.005 (full power) 1.026 (zero power) (both at 500^oF)	Use k_{eff} as close to 1.00 as can be reasonably be accomplished	Core reactivity (max.) $\%\rho = 12.5$	Excess reactivity, 0 Mw-yr, 70^oF = 23.8%; k_{eff} = 1.04 (init. loading, 450^oF)	
Conversion ratio	Seed: n.a. Blanket: >1.0 core 1 seed 1, first portion seed 2; <1.0 from latter portion core 1, seed 2 until end of core operation	0.46	0.5	n.a.	n.a.	n.a.[g]
Doppler coefficient (per ^oF)	See footnote a.	Ambient (68^oF): $-4.1 \times 10^{-5} \Delta k$ Op. temp. (500^oF): $-3.3 \times 10^{-5} \Delta k$ at zero power; $-1 \times 10^{-5} \Delta k$ at rated power[d]	Ambient, zero power: $-0.31 \times 10^{-4} \Delta\rho$; operating ($514^oF$), full power: $-0.23 \times 10^{-4} \Delta\rho$			$-4 \times 10^{-6} \Delta k$ (660^oF)[h]
Fuel temperature coefficient (per ^oF)	See footnote a.	Same as Doppler coefficient	Same as Doppler coefficient			
Moderator temperature coefficient (per ^oF)	See footnote a.	Ambient (68^oF): $-3.1 \times 10^{-5} \Delta k$ Op. temp (500^oF) $-1.4 \times 10^{-4} \Delta k$ at beginning of life; $-0.8 \times 10^{-4} \Delta k$ at end of life	Same as over-all temperature coefficient		At ambient temperature, $-0.26 \times 10^{-4} \Delta k$; at operating temperature $-2.2 \times 10^{-4} \Delta k$	At ambient temperature (68^oF), $-0.56 \times 10^{-4} \Delta k$; at operating temperature (482^oF), $-2.8 \times 10^{-4} \Delta k$
Coolant temperature coefficient (per ^oF)	See footnote a.	Same as moderator coefficient	Same as over-all temperature coefficient			Zero
Over-all temperature coefficient (per ^oF)	For seed 2: ambient (150^oF), zero power -0.45 to $-0.15 \times 10^{-4} \Delta\rho$; op. temp., zero power -2.3 to $-1.3 \times 10^{-4} \Delta\rho$; full power ($500^oF$) -2.0 to $-0.8 \times 10^{-4} \Delta\rho$. (See footnote a.)	Ambient (68^oF): $-7.2 \times 10^{-5} \Delta k$ Op. temp. (500^oF) $-1.7 \times 10^{-4} \Delta k$ at beginning of life; $-1.1 \times 10^{-4} \Delta k$ at end of life	Ambient temperature, 1000 ppm boron in coolant: $-0.5 \times 10^{-4} \Delta\rho$; operating temperature, full power, equilibrium Xe: $-2.5 \times 10^{-4} \Delta\rho$	At ambient temperature, $-0.9 \times 10^{-4} \Delta\rho$; at operating temperature, $-1.9 \times 10^{-4} \Delta\rho$	Same as moderator coefficient	Same as moderator coefficient

CONTINUED

TABLE I (continued)

Nuclear Parameters (concluded)

Item	Shippingport	Indian Point	Yankee	PM-1	SM-1	HWCTR
Void coefficient	Not known[b]	$5.7 \times 10^{-6}\,\Delta k/cm^3$ (See footnote e.)	$-2.2 \times 10^{-3}\,\Delta\rho$ per % void at operating temperature	0.37 k_{ex} per % core void fraction		Bulk moderator: $-3.8 \times 10^{-3}\,\Delta k$ per % void
Pressure coefficient (per psi)	At 450°F, pressure range 1250 to 1800 psia: 0.9×10^{-6} to 1.8×10^{-6}	At 500°F: $+2 \times 10^{-6}\,\Delta k$ at beginning of life; $+1.2 \times 10^{-6}\,\Delta k$ at end of life	$+2.5 \times 10^{-6}\,\Delta\rho$ at operating temperature, operating pressure	$\sim 1 \times 10^{-6}$	1¢/100 psi at 115°F	$3 \times 10^{-6}\,\Delta k$ at 1000 psi, 464°F
Power coefficient	See footnote c.	1.8¢/%	Zero power: $-0.6 \times 10^{-4}\,\Delta\rho$/Mw(t); full power: $-0.2 \times 10^{-4}\,\Delta\rho$/Mw(t)			Full power: moderator (428°F) $-2.4 \times 10^{-4}\,\Delta k$/Mw; fuel (662°F) $-0.21 \times 10^{-4}\,\Delta k$/Mw
Other reactivity coefficients			Boron coefficient: at ambient, 1500 ppm B in coolant: $-0.80 \times 10^{-4}\,\Delta\rho$/ppm; at operating temperature, 500 ppm B in coolant: $-0.70 \times 10^{-4}\,\Delta\rho$/ppm			Addition of H_2O: $-0.022\,\Delta k$/% H_2O
Mechanical effects on reactivity	Not known	None	None calculated or measured			None
Neutron lifetime	56 μsec	22 μsec	20 μsec at operating conditions	14 μsec	20 μsec	2 msec
Effective delayed neutron fraction	0.0077	Initial 0.0064, eff. half-life 9 sec; final 0.0054, eff. half-life 9.52 sec	At operating temperature: 0.0071 no burnup; 0.0058 discharge burnup	0.0075	0.0073	0.0075
Refueling period	For core 1, seeds 1 to 4, the equiv. full-power days were, respectively: 241.9, 329.2, 305.4, 279.2* (* estimated).	Core B replaces A after 470 full-power days. B to contain 3 regions; at equilibrium displaced towards center every 285 days	Once per year	After 18.7 Mw-yr per ton[f]	After 16-17 Mw-yr per ton[f]	Drivers: 200 full-power days; individual test: once per three months

a The measured values for the temperature coefficient include the effects of Doppler coefficient, fuel temperature coefficient, moderator-coolant temperature coefficient. Values for each of the above coefficients are not known but all are believed to be negligible except the moderator-coolant temperature coefficient. Measured values of the temperature coefficient of reactivity are given in references [2,3,5,6]. The values given for seed 2 above are typical.

b The effect of steam voids is believed important only during an accident condition. Information concerning the phenomenon of steam voids is given in reference [9].

c The power coefficient of reactivity is considered insignificantly small when compared with variations in the reactivity due to changes in Xe with power level.

d Average fuel temperature = 1870°F(1021°C).

e Conditions: Specific volume at 1500 and 1125 psi, yielding pressure coefficient. Volume of water = 3.385×10^6 cm³.

f Varies with U^{235} and burnable poison loading of core.

g Several types of test fuel are irradiated to long exposures.

h Test fuel temperature range: 350°C(662°F) to 500°C(932°F). Value given is example.

TABLE I (continued)

Reactivity Inventory

Item	Shippingport	Indian Point	Yankee	PM-1	SM-1	HWCTR
Xe, Sm				2.5% ρ		
Max. Xe after shutdown	0.0290 to 0.0500$\Delta\rho$ [a]	-0.0072Δk [g]	-0.009Δk	-0.3% ρ	-4%	-12% k [n]
Steady-state op.	$\Delta\rho$(Xe) = -0.0180 to -0.00031 [a]; $\Delta\rho$(Sm) = -0.0070 [b]	- 0.023Δk(Xe), - 0.007Δk(Sm) [g]	- 0.023Δk(Xe) - 0.007Δk(Sm)	-2% ρ (eq. Xe worth)	-2%	-4.0%
Xe override allowed	All times except near end of seed life [c]	7.5 hours	All times at up to 50% full power	Negligible	Transient Xe easily overridden	Anytime at 212°F, except near end of driver life
Temp. defect (amb. to op. temp.)	Seed 2: -0.5 to -2.1 [d]	-0.034Δk	-0.025Δk (with B) [k]	- 6.46% ρ		-7% k
Fuel burnup	-0.110$\Delta\rho$ (seed 1)	-0.054Δk [h]	-0.061Δk	30.3% of U^{235} at 18.74 Mw-Yr	37% average; 80% maximum	-10% k
Shutdown margin	Seed 2: 0.0650 [e]	0.02Δk (450°F) [j]	0.03Δk at op. temp.; 0.05Δk with B [m]	0.33% ρ one rod fully withdrawn	All 7 control rods = $10.2 (0 Mw-Yr, 70°F)	10% k
Others	See footnote f		Power defect (0 to full power) at op. temp. = 0.023Δk		Some dummy fuel elements decrease ρ	Expts.-vary; mod. purity [p]
Total reactivity		0.141Δk	0.114Δk at op. temp., clean, 0 power, no control	11.92% ρ max. core		10% k at op. power with fresh driver charge

a Minimum and maximum $\Delta\rho$ are measured values, see [2,3,9].

b Value of $\Delta\rho$(Sm) is calculated [1].

c Xe override is allowed at any time after shutdown; however, near the end of each seed life a Xe override is not always possible. During periods of very high or very low Xe concentrations, the maximum allowable power level is reduced

d For seed 1 the temperature defect, ambient (135°F) to operating (525°F), was 1.1 to 2.6 %$\Delta\rho$, see [2]. Seed 2 data are for an op.temp.= 500°F 3. A value of 3.0 %$\Delta\rho$ has been obtained for seed 3 [4].

e Theoretical value for seed 1 = 0.0820. Values for seed 2(above), seed 3(0.0710), seed 4(0.0780) are estimated, based on tests and theor. seed 1 value [1].

f Periodic tests are performed on the reactor to determine temperature and pressure coefficients of reactivity, control rod worth, effective neutron flux, etc. Results are used to predict core performance and to compare with analytical results for verification of calculation methods.

For seed 1, the seed fuel loading was 75 kg. For seeds 2,3,4 the loading was increased to 90 kg and 170 gm of B were added to the seed fuel elements as burnable poison, to compensate for increased fuel loading [3]. The addition of B to seeds 2,3,4 resulted in a reduction in the reactivity loss rate near the beginning of seed life (which for seeds 2 and 3 was 0.1×10^{-4} per equivalent full-power hour). From the middle of seed life to the end, the reactivity loss rate was approximately constant at 0.2×10^{-4}/EFPH.

g Operation time = 10 days.

h Fuel burnup and buildup of fission products includes complete transient Xe override for about 600 full-power days of operation.

j Minimum subcritical margin at 450°F with all control rods inserted.

k 0.025Δk going from ambient temperature, 1000 ppm boron dissolved in coolant, to operating temperatures, 1000 ppm boron dissolved in coolant.

m At operating temp. using control rods; at ambient temp. using control rods and 1000 ppm boron in coolant.

n Near end of driver lifetime.

p Reactivity varies: experiments consist of different complements of test fuel. Moderator purity change: As of Aug.1,1963, purity was 99.1 mole % D_2O, equivalent to a loss of 1 % Δk from initial value.

TABLE I (continued)

Control

Item	Shippingport	Indian Point	Yankee	PM-1	SM-1	HWCTR
Rods	0.2560Δρ	21 rods (95.4% Hf-4.5% Zr): 0.112Δk at $68^{o}F$, 0.140Δk at $501^{o}F$	0.121Δk (amb. temp., 1000 ppm B in coolant); 0.151Δk (op. temp., no B)	6 rod bank, -0.1906Δρ ($68^{o}F$)	7 rod bank, integral worth = \$34.0±2.0 at $70^{o}F$, 0 Mw-yr	Control rods: -0.24Δk at $482^{o}F$; safety rods -0.09Δk at $482^{o}F$
Other control methods	Boron burnable poison. In case of stuck rod(s), $K_2B_4O_7$ is added [a]	B(200-225 ppm) in ss fuel clad 0.022Δk; Sol. poison (68° to $450^{o}F$) 0.028 Δk; [d] Fixed shim rods, [e] B in ss: 0.027Δk	Temporary poison - B dissolved in coolant; normal 1000 ppm B for cold shutdown = 0.089Δk	72 burnable boron poison tubes; 18 ss burnable poison rods [h]	Burnable boron poison	None
Mechanical arrangement, worths	32 control rods, 4 groups: max. ρ insertion rate in 10^{-4}Δρ/sec = 1.28, 1.35, 1.07, 1.33 [b]	5 rings (each has 4 rods), total worths(Δk)(inner to outer) are 0.036, 0.025, 0.026, 0.025, 0.028, respectively [f]	24 cruciform shaped (80% Ag, 15% In, 5% Cd alloy); 6-group arrangement with worths (Δk): 0.017, 0.020, 0.030, .010, 0.028, 0.046	6 y-shaped safety and shim rods, europium titanate in ss; 6 rods=0.1906Δρ, 5 rods=0.1214Δρ, 4 rods=0.0785Δρ; 4 rods=minimum worth [j]	2 safety rods, 5-rod shim bank; integral worth at $70^{o}F$, 0 Mw-yr: \$27.4±1.0 shims, \$34.0∓2.0 shims, and safeties	6 safety rods; 12 outer, 6 inner control rods; worth of safety rod = 0.016Δk ($464^{o}F$), 0.014 ($68^{o}F$); worth of ring rod = 0.020 ($464^{o}F$) 0.017 ($68^{o}F$)
Withdrawal	Varies from 2.75 in./min to 8 in./min, initially continuous, then stepwise; sequential withdrawal: I(16 rods), II(8), III(4), IV(4) [c]	Max. 0.375 in./sec or 0.00014Δk/sec; continuous, variable velocity; withdrawn by concentric groups	3/8 in/step, 16 steps/min, max. worth = 0.0003Δk/step; withdrawn in groups, specified sequence	2 in./min normal; 0.06 in./step; normally withdrawn in groups	3 in./min, 0.0188 Δρ/in. for the 5-rod bank; safety rod bank (2), shim bank (5), individually, safety and shims (7)	Continuous at 30 in. per minute, maximum =$5x10^{-4}$Δk/sec; withdrawn in group of 4, group of 2, group of 4, etc.
Insertion	1.0 sec for rod to fall in; driven in by gravity (150 lb); total transit 69 in. See reference [10]	5.5 ft/sec average; driven in by 3250 lb (of which 600 lb is weight)	Gravity fall: terminal velocity = 6 ft/sec, average vel. = 4.7 ft/sec; weight of drive-in 380 lb plus magnetic force 750 lb [g]	4.8 ft/sec average	0.63 ft/sec initial, 1.75 ft/sec final; available for drive-in - 20 ft-lb at 0 rpm	13 ft/sec initial, 1.0 ft/sec final, 4.5 ft/sec average; available for drive-in - 830 lb at stall
Method of use	See discussion of withdrawal	12 fixed shim rods,+ 21 safety rods, for regulation and safety	All 24 rods used for normal control	No specific designation for each rod	Both shim and safety	Safety rods fall free; control rods drive in
Source of scram force	Gravity	Primary loop pressure and rod weight	Gravity	Gravity + Belleville spring	Gravity and positive down-drive motor	Gravity. Rod decoupled by deenergizing magnetic clutch

CONTINUED

TABLE I (concluded) Control (concluded)

Item	Shippingport	Indian Point	Yankee	PM-1	SM-1	HWCTR
Deceleration method	Damping action of primary water on special damper nut	Water and oil snubber (extremely reliable)	Dashpot, uses coolant for hydraulic fluid	Dashpot	Dashpot (very reliable)	Hydraulic dashpot, spring backup
Shaft seal	No penetrations in press.vessel for control rod shafts	Labyrinth type with a water buffer	n.a. (magnetically operated mechanism)	n.a. (magnetic jack)	Labyrinth seal	Each rod, labyrinth ring [k]
Rate of run-in of rods	Same as withdrawal rate	0.44 in./sec (= 26 in./min)	Same as withdrawal rate	2 in./min or 6 in./min		0.5 ft/sec (= 360 in. per min) control rod drive-in
Fluid chemical control	None during normal operation	Boric acid for control under 425°F (218°C) at shutdown; amount to meet two stuck rods of max. worth when cold	To control op.to amb. temp.swing and for extra shutdown margin at ambient temp.: boric acid dissolved in coolant (~1000 ppm at amb. for 0.089 Δk)			For emergency shutdowns: $K_2B_4O_7 x 4H_2O$ in D_2O; holds down worth of control system 0.24 k.

a Addition requires several hours. Sufficient solution is available to produce a cold subcritical shutdown of at least 200×10^{-4} margin of reactivity [7].

b Maximum safe reactivity insertion rate (seed 3) was 1.38×10^{-4}/sec, see [8]. In general, except for location, all rods are alike.

c For seeds 1,2,3,4 on core 1, initial criticality attained (at op.temp.,equilib. Xe) with group I rods withdrawn; group II served as controlling rods, group I as safety rods. As reactivity depleted, group II withdrawn; when fully withdrawn they become safety rods and group III controlling, etc.

d Soluble poison in coolant removed above 450°F.

e 12 rods of stainless steel containing 1 wt.% of B; the B was enriched to 80 wt.% B^{10}.

f Each group (ring) has 4 rods except outermost which includes the central rod.

g Latter available for non-standard conditions.

h Stainless steel with 0.27 wt.% B(natural).

j Eu_2O_3(equiv. to 30 wt.%) dispersed in ss.

k Individual seal for each rod. Labyrinth ring, pressure breakdown type with high-pressure, cold water seal injection.

REFERENCES

Shippingport

[1] The Shippingport Pressurized Water Reactor, Addison-Wesley, 1958.

[2] DLCS 364, Shippingport Operations, From Start-Up to First Refueling.

[3] DLCS 36402, Shippingport Operations, From Power Operations After First Refueling to Second Refueling.

[4] Westinghouse Atomic Power Division, PWR Project, Test Evaluation, Shippingport PWR-1 Seed 3, Physics Test Results from 150 EFPH

Results at Zero EFPH. Report WAPD-PWR-TE-126, dated Jan.22, 1963.

[5] Westinghouse Atomic Power Division, PWR Project, Test Evaluation, Shippingport PWR-1 Seed 3, Physics Test Results from 150 EFPH to End of Seed 3 Life. Report WAPD-PWR-TE-127, dated April 30, 1963.

[6] Westinghouse Atomic Power Division, PWR Project, Test Evaluation, Shippingport PWR-1 Seed 4, Physics Test Results at Zero EFPH. Report WAPD-PWR-TE-155, dated May 28, 1963.

[7] Shippingport Atomic Power Station Operating Manual, Volume II, Chapter 7 as revised Fe. 16, 1961.

[8] Westinghouse Atomic Power Division, PWR Project, PWR Hazards Summary Supplement. Report WAPD-PWR-RD-(A)120 dated Sept. 1961.

[9] Westinghouse Atomic Power Division, PWR Project, PWR Reactivity Accidents. Report WAPD-SC-542, dated Oct.1957.

[10] Refueling Procedure for Shippingport Atomic Power Station, Disassembly of Control Rod Mechanisms, S-C2-023, dated July 11, 1963.

Yankee: YAEC-109, YAEC-132, YAEC-136, YAEC-154, YAEC-167, YAEC-183, YAEC-184, WCAP-6050, WCAP-6051 were cited.

PM-1: MND-M-1853, MND-M-1858, MND-M-1866, MND-M-2935 were cited.

SM-1: APAE-2(Rev.1), APAE-7, APAE-96 cited.

HWCTR: DP-600, DP-600(suppl.A) were cited.

ACKNOWLEDGMENT Data for PM-1 and SM-1 reactors were compiled by P. M. Krishna, U. S. Army Engineer Reactors Group, Fort Belvoir, Virginia.

TABLE II BOILING WATER REACTORS

General Description

Abbreviated Name	Full Name; Location	Fuel	Moderator; Coolant	Operating Temp. (°F)	Operating Press. (psia)	Effective Core Size[a]	Power Thermal Mw(t)	Power Elec. Mw(e)
Dresden	Dresden Nuclear Power Station; Morris, Illinois	Type I (352 assemblies) 1.50% enr. UO_2 in Zy-2; Type II (107 ass'ys) 2.51% enr. UO_2 in ss[b]	H_2O; H_2O	546	1015	h = 9 ft, d = 9 to 10 ft	686[c]	220[c]
Big Rock	Big Rock Nuclear Power Plant; Big Rock Point, Michigan	UO_2 (3.2 wt.% enr. in U^{235})	H_2O; H_2O	550	1015	h = 5.83 ft, d = 5.21 ft	157[d]	50[d]
Humboldt Bay	Humboldt Bay Power Plant, Unit No.3, Humboldt Bay, California	UO_2 (2.57 wt.% enr. U^{235})	H_2O; H_2O	549[e]	1020[e]	h = 6.58 ft, d = 6.47 ft	165[e]	50[e]
BONUS	Boiling Reactor Nuclear Superheat Project; Punta Higuera, Puerto Rico	Boiler: 4 UO_2 (natural) subass'ys, 60 UO_2 (2.4% enr.) assemblies Superheater: 32 UO_2 (3.25% enr.) assemblies	H_2O; H_2O[f]	Boiler, 540; SH, 900	965	2.95 ft x 2.95 ft x 4.55 ft	50	17.5
Pathfinder	Pathfinder Atomic Power Plant; Sioux Falls, South Dakota	Boiler: UO_2 (2.2 wt. % enr.) Superheater: UO_2 (93 wt.% enr.)[2]	H_2O; H_2O[f]	Boiler, 489; SH, 725[g]	615[h]	6 ft x 6 ft	Boiler, 157.4; SH, 31.5	58.5 (net)
BORAX- V	Boiling Reactor Experiment No. 5; NRTS, Idaho	Boiler: UO_2 (4.9 wt. % enr.), ss clad Superheater: UO_2 (93 wt.% enr.) in ss matrix, ss clad	H_2O; H_2O[f]	Boiler, 489; SH, 850	Boiler, 615[j]; SH, 608[j]	h = 2.0 ft, d = 3.25 ft	20[k]	3.5[k]
EVESR	ESADA Vallecitos Experimental Superheat Reactor; Pleasanton, California	UO_2 (enr.=4.0, 5.4, 7.5 wt.% with average = 5.4 wt.%)	H_2O; H_2O[f]	542	975	h = 5 ft, d = 3.5 ft	12.5	---[m]

a h = height, d = diameter of cylindrical volume

b For year of 1963. Fuel noted are "regular" rods. Type I fuel is in 6 rod x 6 rod assemblies, type II in 7 x 7 assemblies in which 9 are "special corner rods" (2.1 % UO_2, 97.9% ThO_2 with U as 93.5% enr. U^{235}).

c Plant capability limited by turbine, not by reactor.

d Ultimate rating: 75 Mw(e), 240 Mw(t).

e Uprating to 230 Mw(t) is expected, which would produce ~70 Mw(e) corresponding to 1130 psig, 558°F.

f Coolant is steam in the superheater. (Note: "SH" = superheater.)

g Superheater design temp.= 825°F.

h Pressure at reactor dome = 600 psig and at superheater outlet =535 psig.

j Boiler, 600 psig; superheater outlet = 593 psig.

k Plant dissipative capacity = 40 Mw(t). Electric power = turbine capacity.

m Power dumped to condenser.

TABLE II
(continued)

Nuclear Parameters

Item*	Dresden	Big Rock	Humboldt Bay
k_{eff}	k_∞ varies from 1.132 to 1.151 for type I fuel, from 1.131 to 1.139 for type II[a]		See footnote e
Conversion ratio	0.60(0% void)[b], 0.63(20% void), 0.65(40% void)	0.395 net Pu(all isotopes)per atom U^{235} destroyed, average over 10,000 Mwd/ton	Wt.Pu at discharge/Wt.U^{235} destroyed = 0.48 for 0.30 void fraction
Doppler coefficient (per °F)	-0.9×10^{-5}(amb.), -1.2×10^{-5}(op.temp.), -1.4×10^{-5}(full power)	-1.47×10^{-5}(68°F,0 void), -1.03×10^{-5} (550°F,0 void), -1.15×10^{-5}(550°F, 20%V)	Included in fuel temperature coefficient
Fuel temperature coefficient (per °F)	Same as above	Same as above	$-1.5 \times 10^{-5} \Delta k/k$(68°F), $-1.3 \times 10^{-5} \Delta k/k$(549°F)
Moderator temperature coeff.(per °F)	$-3.7 \times 10^{-5} \Delta k$(68°F)[c] $-12.0 \times 10^{-5} \Delta k$(546°F)[c]		$-0.7 \times 10^{-4} \Delta k/k$(68°F) $-3.3 \times 10^{-4} \Delta k/k$(549°F)
Coolant temperature coeff.(per °F)	Same as above		Same as above
Over-all temperature coeff. (per °F)	Same as moderator coeff.[d]	$+2.2 \times 10^{-4}$(87.5°F, clean) -1.4×10^{-4}(550°F, clean)	Moderator coefficient includes nominal Doppler coeff.,assuming fuel temperature = moderator temperature
Void coefficient (per percent void)	Calculated cold values, type I fuel[f]: -0.77×10^{-3}(0% void,clean), -0.80×10^{-3} (8% void,clean), -3.1×10^{-3}(0% void, 4000 Mwd/t), -4.35×10^{-3}(8%v.,4000Mwd/t)	68°F, 0% void: $-1.8 \times 10^{-3} \Delta k/k$(clean), $-0.9 \times 10^{-3} \Delta k/k$(end of life); 550°F, 20-50% void: $-1.7 \times 10^{-3} \Delta k/k$(clean), $-1.4 \times 10^{-3} \Delta k/k$(end of life)	-2.30×10^{-3} $\Delta k/k$ operating conditions, fresh fuel with poison curtains
Pressure coefficient	$+6 \times 10^{-3}$/ % pressure		$+0.48 \times 10^{-4} \Delta k$/psi
Power coefficient	-8.0×10^{-4}/% change in power		Not available
Neutron lifetime	60 μsec	40 μsec	38.1 μsec (fresh fuel)
Effective delayed neutron fraction	0.0071(initial),0.0055(equilibrium)	0.0058 at 7000 Mwd/ton	0.00758(operating conditions with 30% void, fresh fuel)
Refueling period	10 months at 100% C.F., ~14 months at 70%	~9 months	~9 months (after equilibrium conditions)

* Mechanical effects on reactivity are unknown or negligible (for these three reactors).

a Assembly lattice data are given in Appendix A of amendment to DPR-2. E.g., k_∞ (uncontrolled) values are 1.132(68°F, 0% void),1.151(546°F, 0% void),1.145(546°F,20% void)for Type I fuel, including effects of end connectors and vertical leakage. Corresponding k_∞ (controlled) values are 0.946,0.884,0.852 respectively.

b See reference 2 at end of Table II.

c Values given here are for initial operation (1961). At end of cycle 1, prior to refueling, experimental values were $+1.5 \times 10^{-5}$ at 68°F, zero at 300°F, -4.0×10^{-5} at 400°F. Initial operation, cycle 2, mixed system, values were $+1.5 \times 10^{-5}$ at 68°F, zero at 160°F, and -6.5×10^{-5} at 450°F.

d Since the Doppler or fuel temperature coefficient cannot be added to the moderator coefficient in any meaningful way.

e The k_{eff} and Δk_{eff} values used for the reactivity coefficients are representative of the operating core and consequently <u>do</u> include control effects.

f Values for type II fuel: -0.75×10^{-3}(0% void, clean), -0.66×10^{-3}(8% void, clean), -0.21×10^{-3}(0% void,4000 Mwd/ton), -0.21×10^{-5}(8% void,4000Mwd/ton).

CONTINUED

TABLE II (continued) Nuclear Parameters (concluded)

Item	BONUS	Pathfinder	BORAX-V	EVESR
k_{eff}		k_{eff} derived from 2-group diffusion theory equations		
Conversion ratio	Boiler: 0.52; SH: 0.38[a]	~0.50(initial value)	Experimental reactor.	~0.2
Doppler coefficient (per °F)	$-1.0x10^{-5}\Delta\rho(68°F)$; $\Delta\rho$ due to Doppler=$-0.4x10^{-3}$(68 to 540°F) $-0.9x10^{-3}$(zero to full power at 540°F)	$-1.22x10^{-5}\Delta k$(cold, 0 voids), $-1.15x10^{-5}\Delta k$(hot, 0 voids), $-0.92x10^{-5}\Delta k$(full power)	$-0.6x10^{-5}\Delta\rho$(calculated)	-0.70 to $-0.83x10^{-5}\Delta k/k$ at ambient[g]; essentially same at operating temperature[g]
Fuel temp.coeff.		Fuel element expansion included in core temp.coeff.	Not determined separately	Same as Doppler
Moderator temperature coeff.(per °F)		$-0.4x10^{-5}\Delta\rho$(15 psig,70°F), $-4.0x10^{-5}\Delta\rho$(600 psig,489°F)	Not determined separately	Varies from +0.9 to $1.8x10^{-4}\Delta k/k$ at ambient[h]; from -0.7 to $-2.0x10^{-4}$ $\Delta k/k$ at operating temperature [h]
Coolant temp.coeff.	See over-all temp. coeff.	See moderator temp. coeff.	Not determined separately	Steam temp. coeff. negligible
Over-all temperature coefficient(per °F)	-3.5 to $-6.1x10^{-5}\Delta\rho$ at 68°F;[b] $-30x10^{-5}\Delta\rho$ at 540°F [c]	$-3.0x10^{-5}\Delta k$(68°F), $-8.1x10^{-5}$ Δk at 450°F	Average, measured from 80 to 489°F: $-9x10^{-5}\Delta\rho$	Same as moderator temp.coefficient
Void coefficient (per percent void)	$-1.2x10^{-3}\Delta\rho$(540°F, voids in boiler coolant)[c]	$-0.3x10^{-3}\Delta k$(hot, 0 voids), $-1.0x10^{-3}\Delta k$(50% power), $-1.6x10^{-3}\Delta k$(100% power)	$-2.5x10^{-3}\Delta\rho$(calculated) at 600 psig, 489°F, average from 0 to 10% voids	At 68°F: +0.55 to $00.57x10^{-3}\Delta k/k$ per percent void; at 545°F: -0.77 to $-1.50x10^{-3}\Delta k/k$ per % void[j]
Pressure coefficient	$-3.9x10^{-4}\Delta\rho$/psi at 540°F [d]	$+1.5x10^{-5}\Delta k$/psi(600 psi,20% power); $+4.6x10^{-5}\Delta k$/psi at 600 psi, full power[e]	Not determined	Negligible
Power coefficient	$\Delta\rho$ = 0.042 for zero power to full power	$-1.92x10^{-4}\Delta k$/Mw (full power)[f]	$12.7x10^{-4}\Delta\rho$/Mw, calculated average from 0 to 20 Mw(t)	$-1.6x10^{-3}\Delta k/k$ per Mw(t), average zero to full power
Mechanical effects	$-10^{-5}k_{eff}$/°F at 540°F		Not determined separately	Negligible
Others				Maximum positive effect in $\Delta k/k$ for unflooding SH passages at 68°F is +0.007, for flooding SH passages at 545°F is 0.031.
Neutron lifetime	54 μsec(hot); 48μsec(cold)	60 μsec	30 μsec (calculated)	40 μsec(68°F), 30μsec(545°F)
Effective delayed neutron fraction	0.00735	0.0077(initial), 0.0069(EOL)[a]	0.0071(calculated)	0.0068(68°F), 0.0071(545°F)
Refueling period	Partial refueling every 8 months of full-power operation; total is achieved after four partial refuelings	For 80% load factor: boiler - 1/3 core refueling ~6 months; superheater ~ 9 months	Variable (experimental reactor)	~3 years for 100% burnup (depends on SH fuel test program)

[a] SH = superheater. EOL =end of core life.

[b] Calculated temp.coeff. of reactivity at 68°F are $+5.3x10^{-5}k_{\infty}$/°F for boiler, $-9.1x10^{-5}$ for SH. Measured over-all temp.coeff.of core at 68°F is $-3.5x10^{-5}\Delta\rho$/°F when critical by withdrawal of corner rods and $-6.1x10^{-5}\Delta\rho$/°F when critical by withdrawal of blade rods.

[c] Over-all core coefficient, unrodded core.

[d] This is static coefficient, i.e.change in ρ produced by 1 psi change in saturation pressure at 540°F. The dynamic coefficient is $0.77x10^{-6}\Delta\rho$-sec/psi = change in reactivity caused by a pressure rate change of 1 psi/sec.

[e] Other values (in $10^{-5}\Delta k$/psi): +3.3(400 psi, 20% power), +7.5(400 psi,100% power), +0.98(800 psi,20% power), +3.3(800 psi,100% power).

[f] Defined as $(\partial k/\partial v)(\partial v/\partial P) + (\partial k/\partial T)(\partial T/\partial P)$ at constant power, where v=void volume, T= fuel temp., P= power. Other values (in $10^{-4}\Delta k$/Mw) are -2.00(hot,zero voids), -2.84(50% power).

[g] At ambient temp. with control rods in, the Doppler coeff.(in $10^{-5}\Delta k/k$ per °F)are -0.83(flooded),-0.82(unflooded); with control rods out the values are -0.72(fl.),-0.70(unfl.). Corresponding values at 545°F are -0.82,-0.81,-0.72,and -0.69, respectively.

[h] Values (in $10^{-4}\Delta k/k$ per °F) for control rods in: +1.0(flooded,68°F),+0.9(unflooded,68°F),-1.5(fl., 545°F),-2.0(unfl.,545°F). Corresponding values for rods out:+1.8,+1.6,-0.7,-1.2,respectively.

[j] Values (in $10^{-3}\Delta k/k$ per % void) for control rods in: +0.55(flooded,68°F),+0.13(unflooded,68°F), -0.77(fl.,545°F),-1.07(unfl.,545°F). Corresponding values for rods out:-0.06,-0.57,-1.22,-1.50, respectively.

TABLE II
(continued)

Reactivity Inventory

Item	Dresden	Big Rock	Humboldt Bay
Xe, Sm	Maximum after shutdown = 0.06 Δk; steady-state operation 0.04 Δk; override time after shutdown = 7 hours	0.03 Δk	Maximum Xe increase after shutdown=0.007Δk; steady-state operation (Xe,Sm)=0.031Δk. Peak occurs 6-6hr ; no reactivity allowance (accommodated by fuel burnup allowance).
Temperature defect ambient→operating	+ 0.001Δk	0.025Δk	0.015Δk
Fuel burnup	- 0.059Δk	Fuel burnup and maneuvering: 0.127Δk	0.123Δk
Shutdown margin	Increases with exposure. With strongest rod withdrawn: 1%, All rods inserted: 0.045Δk or greater.		0.032Δk
Voids at operating conditions	-0.023Δk average	0.019Δk maximum	0.021Δk average
Others	No reactivity tied up in experiments. Doppler -0.01Δk		Allowance for residual control at end of fuel cycle = 0.010 Δk.
Total	0.161Δk	0.202 Δk	Total excess Δk =0.200. This plus shutdown margin = 0.232 Δk total required control.

Item	BONUS	Pathfinder	BORAX-V	EVESR
Xe, Sm	Maximum effect after shutdown: Δρ(Xe)=0.035, Δρ(Sm) = -0.009. Steady-state operation: Δρ(Xe)=0.027, Δρ(Sm)=-0.009. Xe override not a limitation.	Maximum effect after shutdown: Δρ(equilib.Xe)=1.87; steady-state operation: -(0.0273+0.0103)=-0.376	Maximum Xe effect after shutdown: 0.032Δρ. Steady-state operation:(Xe)0.0249ρ, (Sm) 0.00715ρ[a]. No Xe override allowed.	0.022 Δk/k. Maximum Xe after shutdown= 0.015Δk/k. Steady-state operation = 0.014Δk/k. No Xe override allowed.
Temperature defect ambient→operating	Δρ =-0.066	-0.0251 Δρ(including Doppler)	-0.039ρ(measured, 80--489°F)	0.023Δk/k(545 to 400°F, flooded)
Fuel burnup	10,000 Mwd/ton		Experimental reactor. n.a.	4000 Mwd/ton, 0.058Δk/k
Shutdown margin	Δρ = 0.03(68°F), 0.09(540°F)	~ 5%	~0.081ρ$_f$ at 80°F (superheater flooded)	0.035Δk/k
Voids	Average Δρ =-0.033	-0.0340Δρ	At 10% voids: ~0.025ρ (calc.)	0 to 14.5% voids: 0.016 Δk/k
Others		Superheater draining:+0.0051Δρ	Experiments --variable[b]. SH flooding: 0.0086ρ(80°F), 0.0118ρ(489°F)measured.	Expts.~0.005Δk/k[c]; fuel Doppler(1100°F→545°F)=0.004Δk/k
Total		-0.0916Δρ	SH drained: available excess (measured)=0.093ρ(80°F), 0.054ρ(at 489°F)	0.158Δk/k

a Values are calculated at 20 Mw(t).

b Six flux wire thimbles and five miniature ion chamber thimbles: -0.001ρ(estimated). Fuel rotating oscillating rod in reflector: +0.0012ρ(fuel in, estimated), -0.0003ρ(fuel out, estimated)

c Tensile specimens, thermocouples and leads, flux wire guide tubes, etc.

(NOTE: SH = superheater)

TABLE II (continued)

Control

Item	Dresden	Big Rock	Humboldt Bay
Control rods	Δk available with fuel type I:[a] 0.164(amb.), 0.232(546°F), 0.256(546°F, 20% voids)	32 rods plus 28 ss channels = 0.254 Δk	0.173Δk
Temporary poison	None	Sodium pentaborate solution designed to provide 2000 ppm in primary water, equivalent to ≥ 0.25Δk	0.059Δk in boron steel curtains
Other controls	Sodium pentaborate added in less than 20 sec providing 0.20Δk(type I fuel) and 0.17Δk(type II fuel)	SS channels provided for additional shutdown control of initial core	None
Arrangement	80 control rods, poison blades, cruciform, containing B_4C in ss tubing. Worth of each rod varies from 0.002 to 0.004Δk.[b]	Withdrawal pattern to make complete withdrawal of any rod contribute ≤ 2.5% $\Delta k_{eff}/k_{eff}$.	32 safety rods, each associated with 4 fuel assemblies. Indiv. rod worth during operation 0.009Δk average, 0.022Δk max. With abnormal rod pattern: max. potential is 0.036Δk.
Withdrawal	Continuous or stepwise at operator's option: 8.8 in./step, 0.7 step/sec(av.). Withdrawn in groups of 4 with radial symmetry.	Continuous: 3in./sec(0.0052Δk/k per sec); stepwise, 3 in./step[d]. Withdrawn individually.	Continuous, 3in./sec(0.0004Δk/sec av., with abnormal rod pattern 0.004Δk/sec). Stepwise, 3in./step, 1 step/sec, Δk as for continuous. Rods withdrawn individually.
Insertion(scram)	5.0 to 7.0 ft/sec(initial), 0.5 ft/sec (at end stop) for zero vessel pressure. Higher rates if vessel at pressure.	Maximum time to 90% of fully in position = 2.5 sec.	Average= 2.37 ft/sec(0.042 Δk/sec). 0.105 ft/sec(first 3% of stroke); maximum 4.56 ft/sec(0 to 78% of stroke).[e]
Method of use	Control rods are used both as shim and safety rods, but not designated as such.	All rods used the same --serve as control, safety and shim.	Combined safety and shim.
Source of scram force	At zero vessel pressure, 4200 lb from accumulator; larger at vessel pressure.[c]	~3300 lb derived from high press. H_2O from accumulator or reactor pressure.	4400 lb from bag-type gas-water accumulators (1 for each 2 control rods) pressurized with water to 1400psig in addition to reactor pressure.
Deceleration	Orifice arrangement on main drive piston. Partial backup with spring washers in drive.	Restriction holes at end of stroke. Good reliability.	Reliable hydraulic buffer.
Shaft seal	None. Drive mechanism in a thimble at reactor pressure.	Not used. Control rod drive operates at reactor pressure.	None.
Rod run-in rate	12 to 18 sec (fully out to fully in).	Continuous insertion of each rod by operator at 3 in./sec.	All in in 2.5 sec.
Fluid chemical control	Backup system: sodium pentaborate in water at ambient (15 wt.%). Holds down ~0.15Δk.	For secondary shutdown only. Sodium pentaborate used to hold down ~0.25Δk.	Backup system 40 wt.% sodium pentaborate to hold down 0.21Δk.

a Corresponding values for type II fuel are: 0.131, 0.189, 0.205, respectively.

b Blade width is 6.5 in.; rods are rectangularly spaced 9.962 in. center-to-center.

c An accumulator (a tank containing a free piston) provides scram force to each control rod drive. Bottom side of accumulator is precharged with 500 psig air, then top is filled with 1400 psig water. The compressed air provides energy for a drive scram. On scram, air is vented to an atmospheric tank and full 1400 psig from accumulator applied to drive.

d Step insertion is limited to 1 step per actuation of switch.

e 3% of stroke takes 0.19 sec; 78% takes 1.12 sec.

CONTINUED

TABLE II
(concluded)

Control (concluded)

Item	BONUS	Pathfinder	BORAX-V	EVESR
Control rods	Total available Δρ: control rods, 0.183($540^{\circ}F$), 0.171($68^{\circ}F$) shims, 0.088($540^{\circ}F$, control rods out), 0.038($68^{\circ}F$, rods in)	Δk = 0.15(cold), 0.20(hot). Boiler core shims, available Δk ~ 0.04	0.184ρ(measured) at $80^{\circ}F$	At $400^{\circ}F$, flooded (at point of minimum shutdown): 12 main control rods = 0.158Δk, 4 auxiliary scram rods = 0.007Δk. Total =0.165Δk.
Temporary poison	Emergency soluble poison with Δρ=-0.055, all rods out, $68^{\circ}F$	Liquid boron ~0.30 Δk	Boric acid for control rod calibration and backup shutdown. Poison rods(ss or ss-B) sometimes used for flux flattening or reactivity adjustment.	8 ss channels (worth 0.02Δk) are available to replace Zy channels to shut down under all core conditions and with any one control rod stuck.[a]
Arrangement	17 safety rods. Individual worths(Δρ): 0.012(central), 0.02 (side), 0.014(corner), 0.0074 (blade).	20 control rods in 5 groups of 4 each. Within each group all rods kept at same height ±1 in. Central group controls SH.	9 safety rods: 1 central(cruciform), 4 intermediate(cruciform), 4 outside(T-shaped) with worths(at $80^{\circ}F$, 0 power)=0.022ρ, 0.0303ρ, 0.0101ρ, respectively.	4 ss central rods(max. 0.027Δk, min. 0.011Δk), 8 ss(with B inserts) outer rods(0.033 to 0.013Δk), 4 auxiliary peripheral scram rods(0.005 to 0.001Δk).
Withdrawal	Continuous: 0.108 in./sec(side rod), 0.166 in./sec(all others), max. 0.0001Δk/sec per rod. Stepwise: ~2.5 in./step, 0.0015 Δk/step. Withdrawn by groups: corner rods, side rods, central rod.	Continuous, individually or as group. 4 central rods, 8 outer: ~72 in./min. Other rods: ~72 in. per 3 min. Rates correspond to <10¢/min. Max. rod worth (cold core) = 0.05Δk.	At 0 power, $80^{\circ}F$: central rod 0.1 in./sec(0.017Δk/sec), intermediate rod 0.1 in./sec(0.024Δk/sec), outside 0.1 in./sec(0.008 Δk/sec. Withdrawn indiv. or in groups at 0.1 in/sec except intermed. group(0.025 in./sec).	Continuous: 0.33 in./sec(1.83×10^{-4} Δk/sec average for max. worth rod). Withdrawn individually: 4 auxiliary scram rods first, 8 outer rods second, 4 inner rods last.
Insertion (scram)	Final velocity ~18 in./sec; average velocity ~14 in./sec	Control rods bottomed (including dashpot action) in <2 sec. SH rods not scrammed but run in.	5.8 ft/sec(0.09ρ/sec) for first inch(at 600psig, $489^{\circ}F$). Average 9.6ft/sec(0.576ρ/sec) for full 25-inch stroke.	Main control rods: ≤0.3sec(9% in), ≤0.77sec(50% in), ≤4sec(90% in). Aux. scram rods: ≤0.6sec(50% in), ≤0.8sec(90%). Av. ~1.6ft/sec(0.042Δk/sec)[b]
Method of use	Blade bank regulates power split between boiler and SH. Corner banks - shutdown; side and central - shim and regulating.		Control, safety	Shim and safety
Source of scram force	Gravity. If rod sticks, automatic drive-in.	Gravity and scram assist spring	Spring and gravity and reactor press. (734lb, 600psig; 270lb, 0)	Stored air press., each drive has separate air tank. Total 2460 lb.
Deceleration	Rotary vane hydraul. shock abs.	Dashpot	Hydraulic(water) dashpot.	Orifice in drive cylinder exhaust
Shaft seal	Press. breakdown rotary seal		Linear shaft seal	Reciproc. floating ring seal
Rod run-in rate	Same as withdrawal		6 in./min	20 in./min
Fluid chemical control	114 gal 20 wt.% aq. soln boric acid for emergency shutdown. Cold, clean reactor, all rods out - would be 1.6% subcrit.	19% soln disodium octaborate tetrahydrate for safety shutdown system. Δk to be held = 30% with reactor closed, 7% with reactor open to shield pool.	H_3BO_3 used in zero-power control rod calibrations or for backup shutdown (~10%ρ).	Backup shutdown: ≥250 gal 13.5% sodium pentaborate solution. Under shutdown (minimum) conditions liquid has worth 0.16 Δk.

[a] Plus a Δk shutdown margin >1%. (A Δk shutdown margin ≥0.3% must be demonstrated.)

[b] Average insertion rate for 12 control rods for the core condition of minimum total worth ($68^{\circ}F$, flooded).

REFERENCES

Dresden: Reports GEAP-0878, GEAP-1044, GEAP-2071, GEAP-3009, GEAP-3053, GEAP-3076, GEAP-3082, GEAP-3106, GEAP-4299 and License DPR-2 were cited.

Big Rock Point: Final Hazards Summary Report, Volume 1, dated Nov. 1961; Technical Specifications for the Big Rock Point Nuclear Plant, dated July 5, 1962.

Humboldt Bay: Final Hazards Summary Report (and revisions); Technical Specifications.

BONUS: Final Hazards Summary Report, PRWRA-GNEC-5.

BORAX-V: ANL-6302, ANL-6689, ANL-6707 were cited.

Pathfinder; Safeguards Report, ACNP-5905 cited.

EVESR: Final Hazards Summary Report, APED-3958.

TABLE III GAS-COOLED REACTORS

General Description

Abbreviated Name	Full Name; Location	Fuel	Moderator; Coolant	Operating		Effective Core Size	Power	
				Temp.	Press. (psia)		Thermal Mw(t)	Elec. Mw(e)
Peach Bottom	Peach Bottom Atomic Power Station; Peach Bottom, Pennsylvania	U, Th carbide particles dispersed in annular graphite compact with graphite inner core and sleeve	Graphite; helium	927°C (1700°F)	340	height=7.5 ft diam. =9.15 ft	115	40
EGCR	Experimental Gas-Cooled Reactor, Oak Ridge, Tennessee	2.46 wt.% U^{235} in UO_2	Graphite; helium	562°C (1043°F)	300	height =14.5 ft diameter =11.84 ft	84.3	29.5 (gross)
EBOR	Experimental Beryllium Oxide Reactor, NRTS, Idaho	UO_2(62% enr.)-BeO pellets, Hastelloy clad	BeO; He	704°C (1300°F)	1120	1.92 ft x1.92 ft x6.33 ft	10	0

Nuclear Parameters

Item[a]	Peach Bottom	EGCR[d]	EBOR[d]
Conversion ratio	0.42 (average)	0.5 to 0.65	0.14
Doppler coefficient (per °C)	-8.0×10^{-5} at ambient temperature, -2.4×10^{-5} at operating temperature[b]	-3.7×10^{-5} at ambient temperature, -2.1×10^{-5} at operating temperature	-4.3×10^{-5} at ambient temperature -1.7×10^{-5} at operating temperature
Fuel temperature coefficient (per °C)	Ambient[c] temp.: -6.8×10^{-5}(BOL), -6.5×10^{-5}(EOL); operating temp.[c] (with equilib. Xe): -1.9×10^{-5}(BOL), -1.85×10^{-5}(EOL)	Same as Doppler	Same as Doppler
Moderator temperature coefficient (per °C)	Ambient: -1.62×10^{-5}(BOL), -1.3×10^{-5}(EOL); operating: -2.65×10^{-5}(BOL), -0.5×10^{-5}(EOL)	Ambient and at 7000 Mwd/ton: -4.5×10^{-5}; operating temp.: -2.4×10^{-5} (becomes positive after 5000 Mwd/ton exposure)	Ambient temp.: -0.2×10^{-5}; operating temperature: -0.2×10^{-5}
Over-all temperature coefficient (per °C)	Ambient: -8.4×10^{-5}(BOL, no Xe), -7.8×10^{-5}(EOL, no Xe); Operating: -4.5×10^{-5}(BOL, no Xe), -2.9×10^{-5}(BOL, with Xe), -2.3×10^{-5}(EOL, with Xe)	Ambient temp.: -8.3×10^{-5} Operating temp.: -4.5×10^{-5}	Ambient temp.: -3.8×10^{-5} Operating temp.: -1.5×10^{-5}
Power coefficient (per Mw(t))	10% power: $-2.0\times10^{-4}\Delta\rho$(BOL), $-1.5\times10^{-4}\Delta\rho$(EOL); full power: $-1.07\times10^{-4}\Delta\rho$(BOL), $-0.75\times10^{-4}\Delta\rho$(EOL)		Zero power: $-2.93\times10^{-3}\Delta k$/Mw; full power: $-1.17\times10^{-3}\Delta k$/Mw
Other reactivity coefficients (per °C)			Reflector: $+0.7\times10^{-5}$(amb.), $+0.4\times10^{-5}$(op.) Bulk expansion of core structure: -1.1×10^{-5}
Neutron lifetime	Hot, operating: 0.21 msec(BOL), 0.23 msec(EOL)	0.4 msec	22 μsec
Effective delayed neutron fraction	0.0065(BOL), 0.0050(EOL)	0.0068	0.007 including Be(γ,n)
Refueling period	First core (batch load) ~3 years	First core 2.5 years. One channel/ 14 days at equilibrium	10,000 hours (1.14 years)

a Note temp. coeff. of reactivity are in per °C. Coolant and pressure coefficients are negligible.

b BOL(beginning of life) value=EOL(end of life) value.

c Reactivity change due to temp. of fuel particles alone. Including thermal base contribution of graphite in the compacts (~30% of quoted moderator coeff.) gives prompt coeff.: -2.8×10^{-5}/°C(BOL), -2.0×10^{-5}/°C(EOL).

d BOL, EOL coeff. approx. identical (small burnup).

TABLE III

(continued)

Reactivity Inventory

Item	Peach Bottom	EGCR	EBOR
Xe, Sm	Xe and Sm: 0.03Δk; max.incremental Xe increase after shutdown=0.003Δk(at 5 hr); steady-state operation Xe =0.023Δk; no provision required for Xe override	Maximum Xe after shutdown = 0.031Δk; steady-state operation,0.028Δk; Xe override allowed Δk = 0.003	Maximum Xe after shutdown < 0.01Δk operating; steady-state operation 0.05Δk; can always override Xe
Temperature defect ambient→operating	0.06 Δk	0.046Δk	0.0226Δk
Fuel burnup	0.06Δk	0.087Δk	0.007Δk
Shutdown margin	0.10Δk at 300°K(27°C or 81°F)	0.080Δk	0.125Δk (cold)
Others	None	Experiments: variable from +0.04Δk to -0.02Δk	Other fission products: 0.0247Δk; heavy isotopes: 0.0060Δk; Li^6: 0.0027Δk
Total	0.15Δk	0.161Δk	Total operating excess = 0.0527
Control			
Control rods	36 rods, total worth 0.23Δk	Total worth 0.26Δk	4 cruciform rods: Δk/k is 0.216(cold), 0.213(hot)
Temporary poison	1 kg natural boron lumped burnable poison, 0.06Δk	None	$CdNO_3$ solution flooding =0.48Δk/k --for emergengy cooling only
Others	19 emergency shutdown rods worth 0.12Δk (hot); 55 thermally releasable absorbers worth 0.10Δk(hot)	None	Reflector removal (backup for scram system) = 0.12 Δk/k
Mechanical arrangement, worths	55 rod locations: 36 control rods (CR), 19 emergency shutdown rods(ESR).Central position (1ESR), 4 concentric hexagonal rings(6CR;6CR,6ESR;12CR,6ESR;12CR,6ESR). Max.worth single rod = 0.01Δk; minimum worth ~0.002Δk	21 safety rods in a square array. Worth varies between 0.010 and 0.025 Δk, depending on rod position and core configuration.	4 cruciform rods. Δk/k is 0.040(alone), 0.053(with others), 0.075(withdrawn and others inserted)
Withdrawal	Continuous. 0.72 in./sec($7.7x10^{-5}$Δk/sec average)for central rod. Withdrawn by groups of 3 (120° apart).Above 10% power only one rod in a group is moved at a time.	Continuous. Varies between $0.1x10^{-3}$ and $1.5x10^{-3}$Δk/sec (all rods).Withdrawn both individually and in groups	Continuous: $2.7x10^{-5}$Δk/k per sec. Stepwise: 2 in./step, 1 step/min, 0.0015 Δk/k per step. Withdrawn in 2-in.steps, each rod in turn.
Insertion (scram)	Stroke is 7 ft. 80% of stroke in 0.8 sec. Max.velocity 10 ft/sec, max.deceleration < 32 ft/sec², initial upward acceleration =100ft/sec². Control rods: average 0.23Δk/sec. Emergency rods: $4.8x10^{-3}$Δk per sec (0.133 ft/sec). Drive-in force: 2500 lb (CR), 10000 lb (ESR).	Initially $0.5x10^{-3}$ to $5x10^{-3}$Δk/inch with acceleration = 0.6 g or 19.2 ft/sec². Total rod drop is 14.5 ft.	Average velocity =3.8 ft/sec(0.125 Δk/k per sec). Final velocity = 4 ft/sec(0.04 Δk/k per sec). Peak reactivity insertion = 0.44 Δk/k per sec . Drive-in force =1800 lb.
Method of use	Control rod groups withdrawn from outside to center of core. All withdrawn CR available for scram. Emergency rods always withdrawn during power operation.	Shim and safety	Shim and safety

CONTINUED

TABLE III (concluded)

Control (concluded)

Item	Peach Bottom	EGCR	EBOR
Source of scram force	Control rods: axial piston hydraulic motor with hydraulic accumulator. Emergency rods: electric motor with separate battery supply for each rod.	Gravity	Helium pressure drop 400 psi over pneumatic motor
Deceleration method	Throttling of oil supply to hydraulic motors. Emergency snubbing device at end of stroke.	Regenerative electric braking	Friction disc brake plus compressor action in motor.
Shaft seal	Containment of rod drive unit at higher pressure than core. Sealing only necessary when drive unit is removed for servicing.	Not used	Not used, rods in pressure envelope
Rate of run-in of rods		0.3 in./sec	6 in./min (0.1 in./sec)
Fluid chemical control	None	None	$CdNO_3$ in H_2O for emergency removal of afterheat if decay-heat removal blower and emergency air blower are ineffective

REFERENCES

Peach Bottom: Final Hazards Summary Report

EGCR: Final Hazards Summary Report, Reports ORO-586, ORO-606.

EBOR: Reports GA-2052, GA-4413, GA-4130 .

TABLE IV SODIUM-COOLED REACTORS

General Description

Abbreviated Name	Full Name; Location	Fuel	Moderator; Coolant	Operating Temp.	Operating Press.	Effective Core Size	Power Thermal Mw(t)	Power Elec. Mw(e)
Fermi	Enrico Fermi Atomic Power Plant; Lagoona Beach, Michigan	Zr-clad pins containing U-10wt.%Mo with U enriched to 25.6% U^{235}	None; Na	Outlet: 800 F Inlet: 550 F	Atmospheric	Diameter=30.5 in. Height =30.5 in.	200	65.9 (gross)
EBR-II	Experimental Breeder Reactor No. 2; NRTS, Idaho	U-5% Fissium, with U enriched to ~49% U^{235}	None; Na	Outlet: 900 F Inlet: 700 F	Atmospheric	Diameter=19.06 in. Height =14.22 in.	62.5	20 (gross)
Hallam	Hallam Nuclear Power Facility; Hallam, Nebraska	140 U(3.6% enr.)-10 wt.%Mo; 8 U(3.7% enr) carbide; 2 U(4.8% enr) carbide	Graphite; Na	Outlet: 945 F Inlet: 610F	1 to 6 in. H_2O (gauge)	Diameter 13 ft, Height 13.25 ft	256	86

Nuclear Parameters

Item	Fermi	EBR-II	Hallam
Conversion ratio	Core, 0.29; blanket, 0.87; total =1.16	1.2	~0.4
Doppler coefficient of reactivity	Isothermal range: -0.259 to -0.247×10^{-5} Δk/k per °C Power range: -0.247 to -0.168×10^{-5}	~$+0.04\times10^{-5}$ Δk/k per °C at operating temp.	-2.1×10^{-5}/°F(-3.8×10^{-5}/°C) at ambient; -1.6×10^{-5}/°F(-2.9×10^{-5}/°C) at op.temp.
Fuel temperature coefficient	Isotherm.range: -0.516×10^{-5} Δk/k-°C; Power range: -0.663×10^{-5} Δk/k per °C	-0.39×10^{-5} Δk/k per °C at operating temp.	-1.5×10^{-5}/°F(-2.7×10^{-5}/°C) at ambient, -1.0×10^{-5}/°F(-1.8×10^{-5}/°C) at op.temp. (prompt coeff.--fuel and coolant)
Moderator temperature coefficient	Not applicable	Not applicable	$+2.4\times10^{-5}$/°F($+4.3\times10^{-5}$/°C) at ambient, $+1.3\times10^{-5}$/°F($+2.3\times10^{-5}$/°C) at op.temp.
Coolant temperature coefficient	Isotherm.range: -0.444×10^{-5} Δk/k per °C Power range: -0.416×10^{-5} Δk/k per °C	Core: reduced Na density -0.87×10^{-5} Δk/k per °C [a]	$+0.6\times10^{-5}$/°F($+1.1\times10^{-5}$/°C) at ambient, same value at operating temperature
Mechanical effects on reactivity	Ambient: -2.146×10^{-5} Δk/k per °C Operating: -2.313×10^{-5} Δk/k per °C	Fuel: axial growth -0.39×10^{-5} Δk/k-°C, radial growth -0.09×10^{-5} Δk/k-°C [b]	None
Over-all temperature coefficient	Isotherm.range: -3.365 to -3.353×10^{-5} Δk/k per °C Power range: -3.639 to -3.560×10^{-5}	(Isothermal at operating temperature) -4.2×10^{-5} Δk/k per °C	Isothermal at 400°F: $+0.9\times10^{-5}$/°F($+1.6\times10^{-5}$/°C) At 850°F: $+0.3\times10^{-5}$/°F($+0.5\times10^{-5}$/°C)
Void coefficient	No voids involved in normal operation, but voids would decrease reactivity.	Can be obtained from coolant temperature coefficient	No voids experienced during normal operation. Coefficient not calculated.
Pressure coefficient	No pressure variation involved	Negligible	None
Power coefficient	-1.62×10^{-5} Δk/k per Mw at full power	Without bowing: -3.5×10^{-5} Δk/k per Mw [c]	Prompt: -2.1×10^{-5}/Mw (at 38 Mw), -0.3×10^{-5}/Mw (at 256 Mw); steady-state: -0.8×10^{-5}/Mw (at 38 Mw), $+0.3\times10^{-5}$/Mw (at 256 Mw)
Neutron lifetime	Average neutron generation time=84msec, prompt neutron lifetime = 0.14 μsec	0.08 μsec	U-Mo core, unburned: 370 μsec
Effective delayed neutron fraction	0.0066	0.0072	U-Mo core, unburned: 0.0069
Refueling period	9 to 14 days at full power	Experimental conditions will set period [d]	Steady-state cycle with U carbide: replace 1/9 of core every 8 or 9 full-power months

a Corresponding values for other regions of coolant are -0.38(gaps), -0.21(axial blanket), -0.2(inner blanket), -0.017(outer blanket) --all $\times10^{-5}$ Δk/k per °C.

b Core structure: -0.039(axial growth), -0.97 (radial growth), -0.036(gaps). Other values: -0.17(rad.gr.inner blkt structure), -0.066 (ax.gr.U inner blkt), -0.04(rad.gr.U and clad in inner blkt), -0.034(rad.gr.rad.blkt structure). (Rest are between -0.024 and -0.003.) [All values are $\times10^{-5}$ Δk/k per °C.]

c Including bowing, coeff. is -3.5(0-25 Mw), +1.0(25-34 Mw), -4.0(34-62.5 Mw). [All values are $\times10^{-5}$ Δk/k per Mw.]

d Fuel cycle based on 2% burnup of all atoms in fuel alloy (~< 4% burnup of U^{235} atoms).

e Refuel with U-carbide: 1/3 of core after ~ 1 full-power year, next 2/3 of core within next full-power year.

TABLE IV (continued)

Reactivity Inventory

Item	Fermi	EBR-II	Hallam
Fission product poisons		All fission products - 0.002Δk/k	Max.Xe after shutdown: 0.0185Δk Steady state (256 Mw): equilib.Xe 0.0167 Δk, equilib.Sm 0.0085Δk (total= 0.025Δk) Xe peak 4.8 hr after shutdown (from 256Mw)
Temperature defect ambient operating	0.0040Δk (60.0¢)	~0.002 to 0.003 Δk/k (Ambient is 700°F)	Reactivity increase = 0.0011Δk
Fuel burnup	0.0010 Δk or 15.4¢ (weekly unloading)	0.020 Δk/k	0.038Δk
Shutdown margin	0.0607Δk ($9.2)	Safety rods, 0.013Δk/k; control rods ≥0.03	Actual: 0.06.Required: Na worth+1% = 5%
Others	Control margin 0.0011 (16.6¢)	Irradiation growth (4%) 0.011Δk/k. Build-up of Pu -0.0092Δk/k	Control margin 0.005
Total	0.0668Δk ($9.2+0.92 = $10.12)	Cannot be specified as loading may easily be altered to accomodate reactivity requirements.	Total rod worth = 0.128. Excess =0.068[a] (0.025Xe,Sm +0.038burnup +0.005control)

[a] This is for 140 U-Mo fuel (at 400°F). 10 U carbide elements to add ~0.005Δk/k.

Control

Item	Fermi	EBR-II	Hallam
Control rods	0.0061 Δk total	0.048 Δk/k (All control rods are fueled and are withdrawn to decrease reactivity)	0.128 Δk
Other control	None	Safety rods 0.013 Δk/k	None
Mechanical arrangement, worths	2 control rods (B_4C)near center: worth (operating) 46¢ or 0.0030Δk each. 8 safety rods uniformly spaced around at about 7 in.from vertical centerline: av.0.0097Δk(unshadowed),0.0076Δk(shadowed)	2 safety rods (0.0060Δk/k per rod) on one mechanism, located near core center. 12 control rods (0.0035Δk/k per rod) on 12 separate mechanisms, located near core boundary.	19 control rods on a hexagonal lattice. In 5 groups: 1(center), 3(intermediate) 3(intermed.), 6(outer), 6(outer).(Intermediate and outer groups are alternate-rod arrangements.) Worths of groups: (Δk) 0.0065,0.0086,0.0086,0.0066,0.0050 resp.
Withdrawal	Continuous: safety 1.6 in./min(<6.6x10^{-5} Δk/sec),shim 0.4 in./min(<0.13X10-5Δk/sec),reg.rod 1 to 10 in./min (<6.6 x10^{-5}Δk/sec).Safety rods withdrawn as a group. Two safety rod stops,each 2 min long,at subcrit.levels of 0.0030Δk, 0.0016Δk.	(Insertion, i.e. increasing reactivity) Safety rod: 0.033in./sec(<5x10^{-5}Δk/sec). Control rod:0.083 in./sec(<4x10^{-5}Δk/sec). Movement continuous, no steps.	Continuous motion. 12.4 in./min, maximum allowed Δk/sec =0.0003. In normal operation the central rod and one outer group withdrawn together, other rods either full in or full out. Rods withdrawn in groups.
Insertion (scram)	Average rod velocity 5.83 ft/sec(0.1102 Δk/sec or $16.7/sec). Final velocity = 8 ft/sec ($25/sec).	(Withdrawal = scram) Max.vel.control rods =8.3 ft/sec(0.05 Δk/k per sec per rod). Max.vel.safety rods=5 ft/sec(0.09Δk/k per sec per rod).Over first 10 in. of 14 in. stroke: 4 ft/sec(contr.),3 ft/sec(safety).	Total rod movement = 13 ft. Drop time(to full-in) 0.900 to 1.015 sec. Time to actuation of latch switch: 45-60 msec. Free fall: 0.9g minimum.
Method of use	Safety rods out before critical; control rods to adjust power.	Control rods for shim, regulation and scram, safety rods for scram and for additional protection during fuel handling.	All rods drop on scram. Any may be used for shim. No distinction between shim and safety.
Source of scram force	Holding electromagnet deenergized. Spring-loaded safety rod driven in under gravity	Control rod: 310 lb weight + 210 lb force from compressed air cylinder; Safety rod: 2670 lb weight + 900 lb spring force	Magnet latch deenergized for gravity fall
Deceleration method	Hydraulic dashpot; no moving parts.	Oil dashpot	3 hydraulic cyl. per control rod

TABLE IV (concluded)

Control (concluded)

Item	Fermi	EBR-II	Hallam
Shaft seal	Double bellows	Steel bellows: air on outside, cover gas on inside with low pressure differential	Rod drops inside He-filled thimble. Pressure of He (15psig) prevents Na in-flow.
Rate of rod run-in	Follow-down drive: 10 ft/min(safety rods)		13 ft in approx. 1 sec
Fluid chemical control	None	None	None

REFERENCES

Fermi: Enrico Fermi Atomic Power Plant, Power Reactor Development Company Technical Information and Hazards Summary Report, Vol.1-7, Sec. VI, Evaluation of Hazards.

EBR-II: ANL-5719(May 1957), ANL-5719 Addendum (June 1962), ANL-6299, ANL-6383, ANL-6462, ANL-6614.

Hallam: Summary Safeguards Report, NAA-SR-5700.

TABLE V PRESSURE-TUBE REACTORS

General Description

Abbreviated Name	Full Name; Location	Fuel	Moderator; Coolant	Operating Temp.	Operating Press.	Effective Core Size	Power Thermal Mw(t)	Power Elec. Mw(e)
CVTR	Carolinas-Virginia Tube Reactor; Parr, South Carolina	Inner region(1/3): UO_2 (1.5% enr.); outer region (2/3): UO_2(2.0% enr.)	D_2O; D_2O	530°F[a]	1500 psi[b]	Diameter =5.75 ft Length = 7.95 ft	65[c]	17 net
CANDU	Douglas Point Nuclear Power Station; Douglas Point, Ontario, Canada	Natural U as UO_2 pressed and sintered pellets	D_2O; D_2O	560°F[d]	1310 psi[d]	Diameter =14.80 ft Length = 16.43 ft	655.1	220 gross
NPD-2	Nuclear Power Demonstration Reactor No.2, Des Joachims, Ontario, Canada	Natural U as UO_2	D_2O; D_2O	530°F	1070 psi	Diameter = 10.50ft Length = 21.92 ft	81	22

a This is average temperature of the primary coolant (505°F - 555°F). The moderator temperature = 155°F.

b This is the primary coolant pressure. The moderator is at atmospheric pressure.

c This is the design power level. The initial power level is 44.3 Mw(t).

d The temperature is that of the coolant outlet; the pressure is that at the outlet header.

TABLE V
(continued)

Nuclear Parameters

Item	CVTR	CANDU	NPD-2
k_{eff}	Calculated by PDQ and CANDLE codes	Ratio total neutron leakage to fission neutron source strength (with adjoint weighting) = 0.0337	1.065
Conversion ratio	0.44	At equilibrium irradiation, fissile atom production rate/destruction rate = 0.79	0.76
Doppler coefficient of reactivity	$-(0.59$ to $0.62)\times10^{-5}/^{\circ}F$ at full power	$-0.78\times10^{-5}\Delta k/k$ per $^{\circ}F$ at operating temperature	$-1.9\times10^{-5}k/\%$full power at operating temperature
Fuel temperature coefficient of reactivity	Same as Doppler	$-1.11\times10^{-5}\Delta k/k$ per $^{\circ}F$(zero irradiation) $-0.5\times10^{-5}\Delta k/k$ per $^{\circ}F$(equilib.irrad.)	$-3.1\times10^{-5}k/\%$ full power at operating temperature
Moderator temp.coefficient of reactivity	$-1.3\times10^{-4}/^{\circ}F$ at $155^{\circ}F$	$-7.78\times10^{-5}\Delta k/k$ per $^{\circ}F$(zero irradiation) $-0.39\times10^{-5}\Delta k/k$ per $^{\circ}F$(equilib. irrad.)	$+2.5\times10^{-5}k/^{\circ}F$ at operating temperature
Coolant temp.coefficient of reactivity	$-2.1\times10^{-5}/^{\circ}F$ at $530^{\circ}F$, 1500 psi	$-0.72\times10^{-5}\Delta k/k$ per $^{\circ}F$(zero irradiation) $+1.7\times10^{-5}\Delta k/k$ per $^{\circ}F$(equilib. irrad.)	$-2.0\times10^{-5}k/^{\circ}F$ for fresh fuel, ambient temperature
Over-all temp. coefficient of reactivity		$-9.56\times10^{-5}\Delta k/k$ per $^{\circ}F$(zero irradiation) $+0.78\times10^{-5}\Delta k/k$ per $^{\circ}F$(equilib. irrad.)	
Void coefficient of reactivity	At full power and 99.75% D_2O:[a] $-0.0028\Delta k/k$ per % void(moderator); $-0.0216\Delta k/k$ per % void(coolant)	Calc.effect of total coolant void: $+0.0082\Delta k$(amb.temp.,zero irradiation) $+0.0030\Delta k$(operating temp.,equilib. irrad.)	Total coolant void: +0.0069k(inital loading) +0.0024 equilib.conditions
Pressure coefficient of reactivity	$+2.1\times10^{-7}$/psi (coolant) at $530^{\circ}F$,1500psi	-1×10^{-7}/psi(coolant) at full power	
Power coefficient	-0.88×10^{-4}/Mw at full power	$-0.0047\Delta k$ per full power change	$3.1\times10^{-5}\Delta k/\%$ full power
Other reactivity coefficients	None	None	Moderator level coefficient: $6.4\times10^{-5}\Delta k$ per cm at half tank; 0 at full tank
Neutron lifetime	0.56 msec	0.8 msec at power, equilib. fuel	65 msec at full power, full irradiation
Effective delayed neutron fraction	0.0070(beginning of life), 0.00476(end of life), including n,γ reaction		0.0046 at full power, full irradiation
Refueling	(Design) first core- 18 months, subsequent- 2 months	Average life of fuel bundle = 2.3 years, changes - 5 bundles per day	Continuous, on-power fueling at 1.2 fuel bundles per day

[a] Corresponding values for 97.75% D_2O are -0.0022(moderator), -0.0213(coolant) $\Delta k/k/\%$ void

TABLE V (concluded)

Reactivity Inventory

Item	CVTR	CANDU	NPD-2
Xe, Sm	0.043Δk	Max.Xe effect after shutdown =0.120Δk; steady state-effect 0.0282k(Xe),0.059 (Sm); 30 min. allowed for Xe override	Max.Xe effect after shutdown= 0.047k; steady-state effect 0.024k; 30 min allowed for Xe override
Temperature defect ambient operating	0.047Δk	Zero power,hot, to full power, hot, decrease in k is 0.0048.	0.0045k at equilibrium
Fuel burnup	0.189Δk	k(0) -k(equilib.) = 0.0438 (both k include Xe and Sm	6300 Mwd/tonne U
Shutdown margin	0.043Δk	Shutdown by moderator dump	Shutdown by moderator dump
Other	Moderator isotopic degrading effect, 2% H_2O: 0.010Δk	Movable load: 0.0010 to 0.0015k	
Total	0.332Δk		

Control

Item	CVTR	CANDU	NPD-2
Control rods	Cold Δρ =0.271; hot Δρ =0.313	0.003 k	None
Poison	None	Dissolved boric acid, as required up to about 0.08 k	None
Other control	None	8 booster rods worth a total of 0.010 k average	Moderator can be dropped completely. Booster rod worth 0.0025k can be inserted to override Xe poison
Mechanical arrangement, worths	Safety rods: 12 black, 16 grey. Estimated average worths (Δρ): 0.015 black rod , 0.008 grey rod. Worth very dependent on location and withdrawal sequence.	No safety rods.	No safety rods.
Withdrawal	Continuous only: max.allowable Δρ/sec $=8x10^{-4}$. Groups of 4 are driven by electric motor through shafts and rod-and-pinion drive trains.		
Insertion(scram)	Average velocity 5 ft/sec, maximum velocity 10.8 ft/sec. Scram force -gravity (free fall). Deceleration by hydraulic dashpot backed up by spring package.		
Shaft seal	Not used. Entire system enclosed in helium atmosphere	Not applicable	Not applicable
Rate of rods run-in	100 in./min		Not applicable
Fluid chemical control	None	Shim control (dissolved boric acid in moderator, see above)	None

REFERENCES (CVTR) Final Hazards Summary Report for the Carolinas-Virginia Tube Reactor Volumes II and III. CVNA-90, December 1961.
Development and Evaluation of Thermal and Hydraulic Parameters for the CVTR Fuel. CVNA-95, June 1962.
Thermal and Hydraulic Design and Evaluation of the CVTR Fuel Assembly. CVNA-115, June 1962.
The Nuclear Design of the CVTR Reactor. CVNA-147, July 1962.

ACKNOWLEDGMENT

Data for CVTR were supplied by M. A. Bell of Carolinas Virginia Power Associates,Inc.

TABLE VI OTHER REACTORS

General Description

Abbreviated Name	Full Name; Location	Fuel	Moderator; Coolant	Operating Temp.	Operating Press.	Effective Core Size	Power Mw(t)	Power Mw(e)
Piqua	Piqua Nuclear Power Facility; Piqua, Ohio	U(1.94 wt.% enr.)-Mo alloy	Terphenyl; terphenyl[a]	575°F (outlet)	105 psig	Diameter =4.17 ft Height = 4.49 ft	45.5	11.4
ETR	Engineering Test Reactor; NRTS, Idaho	Element(19 plates): 400 g U^{235}(93.5% enr.)+ 2.2g B, aluminum clad	H_2O; H_2O	120°F (max.inlet)	200 psig (max.inlet)	2.5 ft x2.5 ft x 3.0 ft	175	
GETR	General Electric Testing Reactor; Pleasanton, California	U^{235}(93.5% enr.) in plate-type fuel	H_2O; H_2O	125°F to 145°F(coolant outlet)	135 psi	Diameter = 2 ft Height = 3 ft	30	
PRTR	Plutonium Recycle Test Reactor; Richland, Washington	UO_2 and Pu-Al, or UO_2-PuO_2[b]	D_2O; D_2O	530°F(coolant), 150°F (moderator)	1025 psig (coolant)	Diameter = 7 ft Height = 7.25 ft	70	

a Santowax O.M.P.

b 45 to 75 UO_2, 10 to 42 Pu-Al; or mixed PuO_2-UO_2 up to 1.5% PuO_2. 19-rod cluster, rods 0.504 in. diameter clad with 0.030 in. Zircaloy-2.

TABLE VI (continued)

Nuclear Parameters

Item	Piqua	ETR	GETR	PRTR
k_{eff}			Max. k_{eff} of 1.115 is used.	
Conversion ratio		Nil	Not applicable	0.80 to 1.00[b]
Doppler coefficient of reactivity	$-1.1\times10^{-5}\Delta k/^{\circ}F$ (calc.)[a]	Nil	Not applicable	$-(1.3\pm0.1)\times10^{-5}/^{\circ}F$ (calc., 68°F to 689°F, original 3-zone loading)
Fuel temperature coefficient of reactivity	Same as Doppler coeff.	Nil	Not applicable	$-0.92\times10^{-5}/^{\circ}F$ (meas., 104°F to 500°F) $-1.2\times10^{-5}/^{\circ}F$ (meas., 68°F to 932°F)
Moderator temperature coefficient of reactivity	$-5.3\times10^{-5}\Delta k/^{\circ}F$ (calc.)	$-3\times10^{-5}\Delta k/k$ per °F	$-10\times10^{-5}\Delta k/k$ per °F	$-2.4\times10^{-5}/^{\circ}F$ (meas., 68°F to 140°F, in critical tests) voids not included
Coolant temperature coefficient of reactivity	$-1.0\times10^{-5}\Delta k/^{\circ}F$ (calc.)	$-3\times10^{-5}\Delta k/k$ per °F	$-10\times10^{-5}\Delta k/k$ per °F	$-0.70\times10^{-5}/^{\circ}F$ (meas., 140°F to 347°F in critical tests)
Mechanical effects on reactivity	None	Small gross effect from uniform expansion of core, negligible	Not applicable	None known
Over-all temperature coefficient of reactivity	$-4.5\times10^{-5}\Delta k/^{\circ}F$ (350°F, meas.) $-9.0\times10^{-5}\Delta k/^{\circ}F$ (550°F, meas.)	$-3\times10^{-5}\Delta k/k$ per °F	$-10\times10^{-5}\Delta k/k$ per °F	Not applicable
Void coefficient of reactivity	$-4.0\times10^{-4}\Delta k/\%$ coolant void (calculated)	Varies over core. -0.05 to $-4.0\times10^{-6}\Delta k/k$ per cm^3 void in reactor fuel	$-30\times10^{-4}\Delta k/k$ per % void in water (at 30Mw(t))	$-(10.5\pm2.5)\times10^{-5}\Delta k/\%$ coolant void (clean, cold, original loading) $-4.5\times10^{-5}\Delta k/\%$ coolant void (clean, cold, 0.48% PuO_2 in UO_2)
Pressure coefficient of reactivity	None (measured)	Not applicable	None	Not applicable
Power coefficient of reactivity	0 to full power: $0.004\Delta k$ (calculated)		None	$-0.6\times10^{-5}\Delta k/k$ per Mw, 0 to 70Mw, assumed linear
Neutron lifetime	0.57 msec (575°F, with 61 fuel elements, 2 poison elements)	0.05 msec with experiments, 0.08 msec without experiments	0.05 msec at 30 Mw(t)	0.82 msec
Effective delayed neutron fraction	0.00069 (beginning of life)	~0.00755	0.0073 at 30 Mw(t)	(Including photoneutrons) No Pu: 0.00830; 35% fissions in Pu: 0.00720; 50% fissions in Pu: 0.00541
Refueling period	Zone refueling at irregular intervals	6 weeks	~35 days	Variable. Fuel shuffles about every 20 full-power days.

[a] The measured reactivities given here are those obtained by converting to Δk the reactivities in dollars obtained by solving the inhour equation.

[b] 1.5×10^{20} nvt in 0.48% PuO_2-UO_2 fuel, 16.5% Pu^{240}. Conversion ratio dependent upon fraction of fissions in Pu, which is not precisely known.

TABLE VI (continued)

Reactivity Inventory

Item	Piqua[a]	ETR	GETR	PRTR
Xe, Sm	Max. Xe effect after shutdown: 0.01Δk. Steady-state effect: 0.032Δk.	Max. Xe effect after shutdown too large to measure. Steady state effect ~0.04Δk/k. Xe override varies with op. time.	Max. Xe effect after shutdown 0.16Δk/k. Steady state: 0.033 Δk/k. Xe override allowed varies.[b]	Max. Xe effect after shutdown 0.155Δk. Steady state operation 0.045Δk. Xe override allowed: 2 hr, 0.090Δk.
Temperature defect ambient → operating	0.013Δk (measured) 0.0106Δk (calc., 360F → 550°F)	~0.003Δk/k; varies with experimental and fuel loading	0.005Δk/k	0.037Δk (3-zone loading); 0.030Δk (uniform PuO_2-UO_2)
Fuel burnup	0.0190Δk per 1000 Mwd/tonne U (calculated)	~0.17Δk/k holddown; average burnup of fuel ~25%	Up to 7% burnup	0.010Δk
Shutdown margin	0.0427Δk (measured)	~0.12Δk	≥0.03Δk/k	Complete draining of moderator.
Experiments	None	Probably average to zero since fuel and poisons are included	Maximum reactivity loss of 0.06Δk/k	
Other	Flux flattening and control margin: 0.018Δk	Burnable poison initially ~0.12Δk/k	Sm negative reactivity effect: peak -0.012Δk/k, equilibrium effect -0.007Δk/k	
Total	0.068 (measured)	Initial excess reactivity ~0.22Δk/k	0.175Δk/k	0.137Δk clean, cold, original 3-zone loading; 0.130Δk clean, cold, 1.5% PuO_2 in UO_2 loading

[a] These values are preliminary.

[b] With 8.7%Δk/k excess at startup: start of cycle equilibrium, 1.2 hr; after 10 days operation, 0.7 hr; after 15 days, 0.3 hr; after 20 days, 0.0 hr.

Control

Item	Piqua	ETR	GETR	PRTR
Control rods	0.105Δk available (measured)	~ 0.10Δk available	0.175Δk/k available	Moderator level control: min. level at critical = 72 in.; permitted operating range 101 in. to 109 in.; 8 in. ≃ 0.005Δk.
Poison	2 poison fuel dummys are loaded with worth (relative to having fuel in their location) 0.029Δk	Burnable poison of natural B initially (new core) 0.12Δk/k	None	Movable shim rods: 18 channels, 3 sectors/channel; Inconel, 1.03 in. radius, 36 in. long; 0.065Δk total with uniform mixed oxide loading
Mechanical arrangement, worths	13 safety rods in 3 rings: 1 in center (0.015Δk), 6 intermediate (0.010 to 0.0125Δk), 6 outer (0.0025 to 0.0054Δk). Values of Δk are for 61 element core.	Four Cd safety rods with fuel followers. 12 Ni shim rods, 1 Cd shim rod each with fuel follower. Worths vary from 0.004 to 0.035 Δk/k.	6 scram rods. Worths 0.024, 0.044, 0.016, 0.017, 0.047, 0.027. Total = 0.175Δk/k.	No safety rods. Safety shutdown by moderator dump.
Withdrawal	Continuous: max. 0.00008Δk/sec. Stepwise: 0.050 in./step, 101 steps/minute. Withdrawn individually, 2- or 3-rod gang	Continuous: 0.25 in./sec. Stepwise: 1 step/5 sec. Safety rods withdrawn individually. Grey rods in groups or individually.	Continuous: 0.067 in./sec, max. 0.075Δk/k per sec. No stepwise withdrawal. Can be withdrawn individually or all together.	Moderator level rate of rise varies up to 0.021 ft/sec with max. reactivity input near operating level about 0.00015 Δk/k per sec [a]

CONTINUED

TABLE VI (concluded)

Control (concluded)

Item	Piqua	ETR	GETR	PRTR
Insertion (scram)	Gravity driven. Average velocity about 6 ft/sec. Deceleration by spring shock absorbers.	Magnet release, gravity plus downward water flow forces rod through about 39 in. Deceleration in 9 in. by hydraulic dashpot. Time from release to stop = 0.250 to 0.300 sec (with full flow and pressure).	Driven by gravity and flow of reactor coolant. Average velocity = 10 ft/sec (0.58Δk/k per sec). Deceleration by hydraulic shock absorber.	In 1 sec k_{eff} reduces to 0.980 as moderator is dumped.[b]
Method of use	All rods, rod drives identical. Operational pattern; central rod in, outer ring of rods out, symmetric bank of 3 inner ring rods either fully in or fully out, other inner ring rods banked for critical.	Used as both safety and shim.	Only scram rods are used which serve as shim and safety rods.	Not applicable
Shaft seal	Not needed, electrical connections	Seal is on drive only.	A double-garlock, spring-loaded shaft seal	Not applicable
Rate of rod run-in	About 0.050 in./step and 101 steps/minute.	0.25 in./sec	4 in./minute	Not applicable
Fluid chemical control	None	None	None	None

[a] As the moderator level increases the maximum rate of rise varies and the maximum rate of reactivity addition varies. From 6 ft to 9 ft the data are: 6 ft(0.011ft/sec, 0.0002Δk/sec), 7 ft(0.018, 0.00025), 8 ft(0.021, 0.0002), 9 ft (0.021 ft/sec, 0.00015Δk/sec).

[b] In 2 sec, k_{eff}=0.955; 3 sec, k_{eff}=0.900; 4 sec, k_{eff}=0.75; 5 sec, k_{eff}=0.45; and in 6 sec, k_{eff} is essentially zero. The corresponding Δk values are -0.025, -0.055, -0.15, -0.30, -0.45, respectively.

REFERENCES

(Piqua) Final Safeguard Summary Report for the Piqua Nuclear Power Facility, NAA-SR-5608, August 1961.

(PRTR) Plutonium Recycle Test Reactor: Final Safeguards Analysis, HW-61236, October 1959. Supplements 1 (April 15, 1960), 2(Nov.15, 1960), 6(Feb.1,1963) are relevant.

Plutonium Recycle Test Reactor: Critical Test Results, HW-61900BA, December 1961.

APPENDIX 2

Abbreviations Used In Text

Listed below are brief identifications of abbreviations used in the text of this volume. At the end of this list there is a tabulation of designations for computer codes and programs mentioned in the text. For identification of abbreviations used at United States atomic energy projects, the report "Nuclear Reactors Built, Being Built, or Planned in the United States as of December 31, 1963," U.S.A.E.C. Document TID-8200 (9th revision), is invaluable. Other abbreviations may be found in Volume II, III and IV of the "Directory of Nuclear Reactors" published by the International Atomic Energy Agency, Vienna.

Adam	Swedish reactor experiment, R-3
AECL	Atomic Energy of Canada, Ltd.
AGN-201	Designation of Aerojet-General Nucleonics solid-homogeneous research reactors
AGR	Advanced Gas-Cooled Reactor, Windscale, England
AIME	American Institute of Mining, Metallurgical, and Petroleum Engineers
Alizé	a critical facility at Saclay, France
ALPR	see SL-1
ANL	Argonne National Laboratory, Argonne, Illinois
ANPP	Aircraft Nuclear Propulsion Program (USA)
ANS	American Nuclear Society
APDA	Atomic Power Development Associates, Detroit, Michigan
APPR	see SM-1
Argonaut	Argonne Nuclear Assembly for University Training, Argonne, Illinois
ASA	American Standards Association
ASME	American Society of Mechanical Engineers
ASTM	American Society for Testing and Materials
AVR	AVR (Arbeitsgemeinschaft Versuchs-Reaktor) Atomic Power Station, Jülich, Germany
BEPO	British Experimental Pile Operation Research Reactor, Harwell, Berkshire England
Berkeley	Berkeley Nuclear Power Station, Berkeley, Gloucestershire, England
BGRR	Brookhaven Graphite Research Reactor at BNL
Big Rock Point	Big Rock Nuclear Power Plant, Big Rock Point, Michigan
BNL	Brookhaven National Laboratory, Upton, Long Island, New York
BONUS	Boiling Nuclear Superheat Reactor, Puenta Higuera, Puerto Rico
BORAX	Boiling Reactor Experiment (followed by numeral designating experiment number), NRTS, Idaho
Bradwell	Bradwell Nuclear Power Station, Bradwell, Essex, England
BR-5	Fast Breeder Reactor, Obninsk, USSR (also designated as SBR-5 for "Soviet Breeder Reactor, BR-5")
BSD system	Burst Slug Detection system
BSR-1, BSR-2	Bulk Shielding Reactor (2 cores) at ORNL
Calder Hall	Calder Hall Reactors, Calderbridge, Cumberland, England
CANDU	Douglas Point Nuclear Power Station, Kincardine, Ontario, Canada
CEI	Combustion Engineering, Inc.
CES	Critical Experimental Station (Westinghouse), Waltz Mill, Pennsylvania
Chapelcross	power reactors at Chapelcross, Annan, Dumfriesshire, Scotland
Chinon	see EDF-1
Clementine	fast reactor at LASL
CVTR	Carolinas-Virginia Tube Reactor, Parr, South Carolina
DFR	Dounreay Fast Reactor, Dounreay, Caithness, Scotland
DIDO	research reactor at Harwell, Berkshire, England
DMTR	Dounreay Materials Testing Reactor, Dounreay, Caithness, Scotland
DNB	departure from nucleate boiling
DNPS	see Dresden
Douglas Point	see CANDU
Dounreay	see DFR, DMTR
Dragon	a critical experiment at LASL (not be be confused with Dragon Reactor)
Dragon	Dragon Reactor at Winfreth, Dorset, England
Dresden (or DNPS)	Dresden Nuclear Power Station, Morris, Illinois
EBOR	Experimental Beryllium Oxide Reactor, NRTS, Idaho
EBR-1, EBR-2 (or EBR-I, EBR-II)	Experimental Breeder Reactor No. 1, etc. at NRTS, Idaho
EBWR	Experimental Boiling Water Reactor, Argonne, Illinois
EDF-1	power reactor (Electricite de France) at Avoine, near Chinon, France
EGCR	Experimental Gas-Cooled Reactor at ORNL
EL-2, EL-3, EL-4	research reactors at Saclay, France
Elk River (or ERR)	Elk River Reactor, Elk River, Minnesota
Enrico Fermi	see Fermi
ESADA	Empire State Atomic Development Authority ("Empire State" = New York), see EVESR
ETR	Engineering Test Reactor, NRTS, Idaho
EVESR	ESADA Vallecitos Experimental Superheat Reactor, Pleasanton, California
Fermi	Enrico Fermi Atomic Power Plant, Lagoona Beach, Michigan
FNR	Ford Nuclear Reactor, University of Michigan, Ann Arbor, Michigan
FRM	research reactor at München, Germany
G-1, G-2, G-3	plutonium production reactors at Marcoule (Gard) France
Godiva	a critical assembly at LASL
Hallam (or HNPF)	Hallam Nuclear Power Facility, Hallam, Nebraska
Hanford	plutonium production facility, near Richland, Washington
Hinkley Point	Hinkley Point Nuclear Power Station, Hinkley Point, Somerset, England
Honeycomb	a critical experiment at LASL
HPCTR	Hanford Physical Constants Test Reactor, near Richland, Washington
HRE-1, HRE-2	Homogeneous Reactor Experiment No. 1, etc., at ORNL
HRT	Homogeneous Reactor Test
HTGR	High Temperature Gas-Cooled Reactor, see Peach Bottom
HTRE	Heat Transfer Reactor Experiment (followed by number), NRTS, Idaho
Humboldt Bay	Humboldt Bay Power Plant, Unit No. 3, Humboldt Bay, California
Hunterston	Hunterston Nuclear Generating Station, Hunterston, Ayrshire, Scotland
HWCTR	Heavy Water Components Test Reactor, Aiken, South Carolina
IAEA	International Atomic Energy Agency, Vienna, Austria

ID	inner (or inside) diameter
Indian Point	Consolidated Edison Thorium Reactor, Indian Point, New York
JAERI	Japan Atomic Energy Research Institute (also used in reference to JPDR, q.v.)
Jemima	a critical experiment at LASL
Jezebel	a critical assembly at LASL
JPDR	Japan Power Demonstration Reactor, Tokai-Mura, Japan
KAPL	Knolls Atomic Power Laboratory, Schenectady, New York
KEWB	Kinetic Experiment on Water Boilers at Santa Susana, California
Kukla	a fast critical experimental facility at LRL
LAMPRE	Los Alamos Molten Plutonium Reactor Experiment at LASL
LASL	Los Alamos Scientific Laboratory, Los Alamos, New Mexico
LITR	Low Intensity Test Reactor at ORNL
LMFR	Liquid Metal Fuel Reactor
LOPO	Low-Power Water Boiler at LASL
LRL	Lawrence Radiation Laboratory (University of California), Livermore, California
MGCR	Maritime Gas-Cooled Reactor
MITR	Massachusetts Institute of Technology Reactor, Cambridge, Massachusetts
ML-1	Mobile Low Power Plant No. 1, NRTS, Idaho
MTR	Materials Testing Reactor, NRTS, Idaho
NCSCR-1	North Carolina State College Reactor No. 1, Raleigh, North Carolina
Nine Mile Point	Nine Mile Point Plant (Nuclear Power Plant) Oswego, New York
NPD-1, NPD-2	Nuclear Power Demonstration Station No. 1, etc., Des Joachims, Ontario, Canada
NRTS	National Reactor Testing Station, Arco, Idaho
NRU	NRU Reactor, Chalk River, Ontario, Canada
NRX	NRX Reactor, Chalk River, Ontario, Canada
OD	outside (or outer) diameter
OMRE	Organic Moderated Reactor Experiment, NRTS, Idaho
ORNL	Oak Ridge National Laboratory, Oak Ridge, Tennessee
OWR	Omega West Reactor at LASL
Oyster Creek	Oyster Creek Station (Nuclear Power Plant), Oyster Creek, New Jersey
Pathfinder	Pathfinder Atomic Power Plant, Sioux Falls, South Dakota
PCTR	see HPCTR
Peach Bottom	Peach Bottom Atomic Power Station—HTGR, Peach Bottom, Pennsylvania
Piqua (or PNPF)	Piqua Nuclear Power Facility, Piqua, Ohio
PM-2A	Portable Medium Power Plant No. 2A, Camp Century, Greenland
PM-3A	Portable Medium Power Plant No. 3A, McMurdo Sound, Antarctica
Popsy	a critical assembly at LASL
PRDC	Power Reactor Development Co., Detroit, Michigan
PRTR	Plutonium Recycle Test Reactor, Richland, Washington
PWR	Pressurized Water Reactor, see Shippingport
Rapsodie	experimental fast neutron breeder reactor, Cadarache, France
Rover	space propulsion experiments (U. S. A.)
RPT	Reactor for Physical and Technical Research, Moscow, USSR
RTD	resistance temperature detector
Saxton	Saxton Nuclear Experimental Reactor Project, Saxton, Pennsylvania
SENN	Societa Elettronucleare Nazionale, Rome, Italy ("SENN reactor" = Gorigliano Nuclear Power Station at Punta Fiume on the Garigliano River in Italy)
SER	Sandia Engineering Reactor, Sandia Base, New Mexico
SGHN	Steam Generating Heavy Water Reactor, Winfrith, Heath, Dorset, England
SGR	Sodium Graphite Reactor
Shippingport	Shippingport Pressurized Water Reactor, Shippingport, Pennsylvania
SIFTOR	Safety Information for the Technology of Reactors
SIR	Submarine Intermediate Reactor Mark A, now SIG, West Milton, New York
Skidoo	a critical assembly at LASL
SL-1	Stationary Low Power Plant No. 1 (formerly ALPR), NRTS, Idaho
SM-1	Stationary Medium Power Plant No. 1 (formerly APPR-1), Ft. Belvoir, Virginia
SNAP	Systems for Nuclear Auxiliary Power (followed by number to designate device)
SNAPTRAN	SNAP Transient Test (followed by test number), NRTS, Idaho
SPERT-1, SPERT-2, etc.	Special Power Excursion Reactor Test No. 1, etc., NRTS, Idaho
SRE	Sodium Reactor Experiment, Santa Susana, California
SUPO	Superpower Water Boiler Reactor at LASL
Tokai-Mura	see JPDR
Topsy	a critical assembly at LASL
TQT, TQU	a designation for particular types of ionization chambers (AECL)
TR	heavy water research reactor, Moscow, USSR
TREAT	Transient Reactor Test Facility, NRTS, Idaho
TRIGA	U-Zr hydride, pool reactor, General Atomic
Triton	research reactor at Fontenay-aux-Roses, France
23-Flattop	a critical assembly at LASL
UCRL	University of California Radiation Laboratory, Berkeley, California
UFTR	University of Florida Training Reactor, Gainesville, Florida
UHTREX	Ultrahigh-Temperature Reactor Experiment at LASL
UNC	United Nuclear Corporation, White Plains, New York
UVAR	University of Virginia Reactor, Charlottesville, Virginia
VBWR	Vallecitos Boiling Water Reactor, Pleasanton, California
Windscale	plutonium production reactors at Windscale, Sellafield, England
WTR	Westinghouse Testing Reactor, Waltz Mill, Pennsylvania
X-10	an area at ORNL (original reactor site)
Y-12	an area at ORNL (former site of electromagnetic separation plant)
Yankee	Yankee Nuclear Power Station, Rowe, Massachusetts
Zeus	Zero Energy Uranium System at Harwell, Berkshire, England
ZPR	a critical facility at ANL (followed by number designating particular arrangement)

The following designations are used for computer codes and programs mentioned in the text:

AIM-6	GAZE
AIREK-II	GAZE-2
AIREK-III	GAZED
ANGIE	KARE
ART	PDQ
AX-1	PDQ-5
BLOOST	RE-126
CANDLE	RE-135
CURE	RE-138
DDB	REP
DSN	RTS
DTK	SIZZLE
EDF	SPARTA
EQUIPOISE BURNOUT	STAB
FEVER	TURBO
FORE	TWENTY GRAND
GAM-I	WANDA-5
GAMBLE	ZOOM
GATHER-1	

Page references for above codes are given in Index.

Index*

*Reactors are referred to by their abbreviated names, see Appendix 2. The reader is urged to refer to the initial page of each chapter where the chapter outline is presented.